Volume 3

Coastal Engineering

New York, New York 10017-2398

Abstract:

This proceedings contains over 370 papers presented at the 25th International Conference on Coastal Engineering which was held in Orlando, Florida, 2-6 September, 1996. The book is divided into six parts: 1) Characteristics of coastal waves and currents; 2) long waves and storm surges; 3) coastal structures; 4) coastal processes and sediment transport; 5) coastal, estuarine and environmental problems; and 6) case studies. The individual papers include such topics as the effects of wind, waves, storms and currents as well as the study of sedimentation, erosion and beach nourishment. Special emphasis is given to case studies of completed engineering projects. With the inclusion of both theoretical and practical information, these papers provide the civil engineer and professionals in related fields with a broad range of information on coastal engineering.

Library of Congress Cataloging-in-Publication Data

Coastal engineering 1996 : proceedings of the twenty-fifth international conference, September 2-6, 1996, The Peabody Hotel, Orlando, Florida / edited by Billy L. Edge.
p. cm.
"Held under the auspices of the Coastal Engineering Research Council of the American Society of Civil Engineers."
ISBN 0-7844-0242-6
1. Coastal engineering--Congresses. 2. Ocean waves--Congresses. I. Edge, Billy L. II. Coastal Engineering Research Council (U.S.) III. International Conference on Coastal Engineering (25th : 1996 : Orlando, Fla.)
TC203.5.C6184 1997 97-10664
627'.58--dc21 CIP

Library of Congress Catalog Card No: 97-10664
ISBN 0-7844-0242-6
Manufactured in the United States of America.

Cover: One year after nourishment at Palm Beach, Florida. Photo courtesy of Applied Technology & Management, Inc. and Aerial Photography, Inc.

CONTENTS

Part I Characteristics of Coastal Waves and Currents

Part II Long Period Waves, Storm Surges and Wave Groups

Part III Coastal Structures

Part IV Coastal Processes and Sediment Transport

Part V Coastal, Estuarine and Environmental Problems

Part VI Case Studies

CHAPTER 199

Wave impact beneath a horizontal surface.

D.J.Wood † and D.H.Peregrine ‡

Abstract

Many coastal structures and natural coasts have openings, overhangs and projections which are open to impact by incident water waves. The sudden impact of a wave on a rigid surface leads to a rapid rise of pressure and consequent violent water motions. We consider the wave impact on the underside of a projecting surface. The example discussed is that of a flat deck close to the mean water level. A pressure-impulse approach is used, which has the advantage that given a solution for one problem it is possible to select pressure-impulse contours which give the solution to related problems. The pressure gradient on the underside of the deck is especially strong near the seaward edge of the impact region, so this is a region where any projections on the structure's surface may be subject to strong shearing forces. On the other hand the maximum pressure-impulse is at the landward end of the impact zone, it is here that the deck is most likely to be 'blown' upward.

Introduction

There are a number of circumstances in which the effect of the upward impact of a wave beneath a rigid horizontal surface needs to be estimated. For offshore oilrigs the lack of good estimates of such upward impacts leads to designs where the main platform of rigs is built to be out of reach of 'green water'. This may not be an option for some coastal structures, including piers and jetties, and temporary works in inter-tidal zones. Here we present pressure-impulse calculations for an impact on a horizontal surface near the surface of water of finite depth. For convenience we refer to the rigid surface as a deck.

In studies of wave impact on a wall Bagnold (1939) was the first to note that although pressure measurements show great variability between nominally identical wave impacts the integral of pressure over the duration of the impact,

† Ph.D. student, School of Mathematics, University of Bristol, University Walk, Bristol, BS8 1TW, UK, (*Deb.Wood@bristol.ac.uk*)

‡ Professor of Applied Mathematics, School of Mathematics, University of Bristol, (*D.H.Peregrine@bristol.ac.uk*)

the pressure-impulse, is a more consistent measure of wave impact. This has been exploited theoretically by Cooker and Peregrine (1990 a,b, 1992, 1995) who show that the pressure-impulse and its distribution is not sensitive to the shape of the impacting water except in a region very close to the structure. Chan (1994) and Losada, Martin, and Medina (1995) show good agreement with experiment for wave impact on a wall.

Mathematical model

The geometrical simplifications we make may be seen in figure 1. The water is taken to be of finite depth CD= a, and to impact the horizontal deck BC of length L with an upward velocity V. The free surface outside the deck is taken to be flat, as BA, and to stretch to infinity. However, as indicated below alternative surface shapes are easily found by choosing different contours of pressure-impulse. The boundary conditions on CD given in figure 1 indicate that the problem can be reflected in the vertical plane of CD, corresponding to impact on a horizontal surface of length $2L$ with a central plane of symmetry.

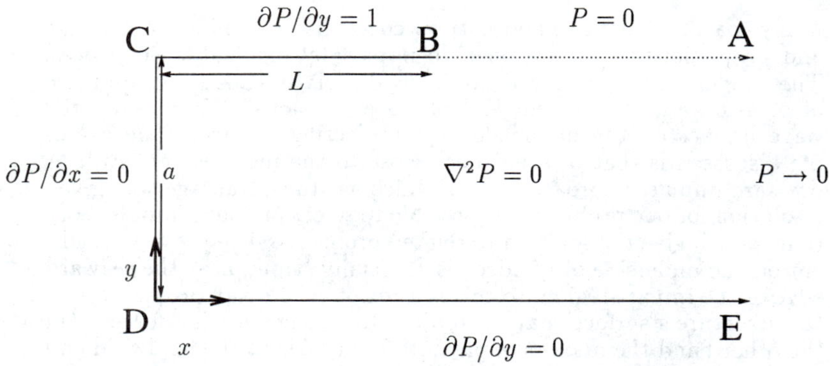

Figure 1: The problem to be solved.

Let the pressure be $p(x, y, t)$, then the pressure-impulse is

$$P(x,y) = \int_{t_b}^{t_a} p(x,y,t)dt \tag{1}$$

where t_a and t_b are the times after and before impact respectively, such that the time interval (t_b, t_a) is short compared with all other time scales in the problem. The main approximation is that during this short time the fluid motion changes so rapidly that the equation of motion may be approximated by

$$\frac{\partial \mathbf{u}}{\partial t} = -\frac{1}{\rho}\nabla p \tag{2}$$

where $\mathbf{u}(x, y, t)$ is the velocity field, and ρ the density which is assumed to be constant and uniform. The neglect of the convective terms $(\mathbf{u}.\nabla)\mathbf{u}$ is consistent except in any small region near the impact where jets may form.

Integration of (2), with respect to time, yields

$$\mathbf{u}_a - \mathbf{u}_b = -\frac{1}{\rho}\nabla P \quad (3)$$

Incompressibility gives $\nabla.\mathbf{u}_a = \nabla.\mathbf{u}_b = 0$. Therefore we find

$$\nabla^2 P = 0 \quad (4)$$

in the region of the water.

The boundary condition at the free surface is that the pressure must be constant and continuous therefore $P = 0$. At the walls and on the bed, the normal velocity must be zero before and after impact, therefore using equation (3), $\partial P/\partial n = 0$, where n is the normal direction. As the water meets the deck BC, the water has vertical velocity V, which could be any function of x, and after impact the water has zero vertical velocity. Therefore, again using equation (3), we have $\partial P/\partial n = V$. For simplicity, we choose V to be constant. We make the problem dimensionless by choosing units for which $V = 1$ and $L = 1$.

Infinite depth solution

The problem of a wave hitting upwards under a deck jutting out from a wall, is mathematically equivalent to a plate dropping onto a body of water and setting the water in motion. Also when considering solving Laplace's equation we can use the direct analogy with the velocity potential of irrotational flow. If we consider the complex potential for a flow past a plate then we just need a change of reference frame to find the complex potential of a moving plate in a stationary fluid. With a complex potential $w = \phi + i\psi$, then $\partial\phi/\partial x = 0$ on $x = 0$, and $\partial\phi/\partial y = 1$ along the plate. These are the conditions that are required by P, and so the lines of constant pressure-impulse are given by lines of constant ϕ. The solution may be found in Lamb (1932, section 71), and in Milne-Thompson(1962, section 6.3), for a fluid flowing past an ellipse. If we allow one of the semi-axes to shrink to zero then we have a plate instead of an ellipse in the flow. Finally choosing the plate to be perpendicular to the flow, the length of the plate to be 2, and the velocity 1, we get an expression for the complex potential of a stream flowing past a plate:

$$w = -\sqrt{1 - z^2} \quad (5)$$

where the origin is taken to be the centre of the plate.

If we subtract the complex potential for a stream from this expression we have the potential for a moving plate. As the velocity of the stream is $(0, -1, 0)$, we must therefore subtract iz to get:

$$w = -iz - \sqrt{1 - z^2} \quad (6)$$

This solution is symmetric about the centre of the plate. This means that we can consider a line drawn perpendicular to the plate from the centre of the plate, to be a wall, bringing us back to the original problem of the water hitting a deck jutting out from a wall. Hence we have an expression for the pressure-impulse:

$$P = \mathrm{Re}(-iz - \sqrt{1 - z^2}) \quad (7)$$

This is the infinite depth solution. Figure 2 shows contours of pressure-impulse. The total impulse on the deck is $\pi/4$, in dimensional terms $\pi\rho VL/4$.

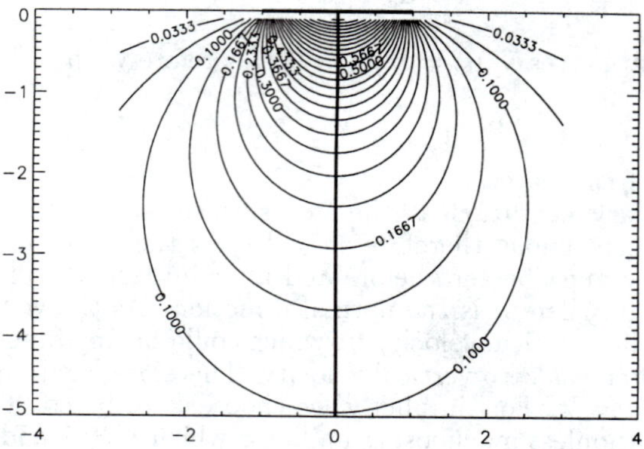

Figure 2: Infinite depth solution. Total impulse on deck (0,1) is $\pi/4$.

Infinitely long deck.

As a becomes small the effect of the free surface on the solution under the plate becomes small. This means it is possible to solve in that region by neglecting the condition at the free surface. Hence we solve Laplace's equation on a strip where $\partial P/\partial y = 1$ along the top, $\partial P/\partial n = 0$, where n is the normal direction, along the left-hand edge and bottom.

The solution is given by:

$$P = \frac{1}{2a}\left[y^2 - x^2\right] + K \tag{8}$$

where K is a constant which depends on the behaviour of P near $x = 0$, where this approximation fails. Figure 3 shows the case when $a = 0.1$, and K is set to zero.

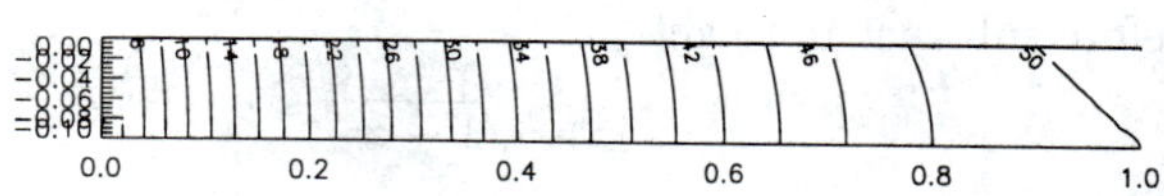

Figure 3: Analytic solution when a is small.($a = 0.1$,$K = 0$)

In practice the 'filling flow' solution of Peregrine and Kalliadasis (1995) may be more relevant to this case.

More general solution

Consideration of the boundary conditions in Figure 1, or the solution (7) shows that at B there is a square root singularity. This singularity causes problems for many numerical solution methods. However, one way to eliminate the problem of the singularity is to map the original problem using conformal maps as follows. First map to a half-space, then use another conformal map to perform a shift and stretch so that by using a final conformal map we can bend the problem back to a semi-infinite strip but with the boundary conditions shifted round to a convenient position, i.e. shift the boundary conditions on the deck round to the vertical wall.

Let the original plane in which the problem is posed be the z plane. The first map we need is $w = u + iv = \cosh(\pi z/a)$. This gives the problem shown in figure 4. As we only use conformal maps P continues to satisfy Laplace's equation throughout.

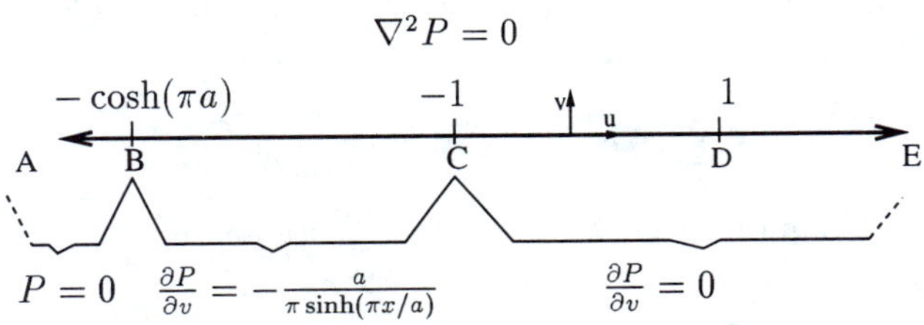

Figure 4: The problem in the w-plane after the first complex map.

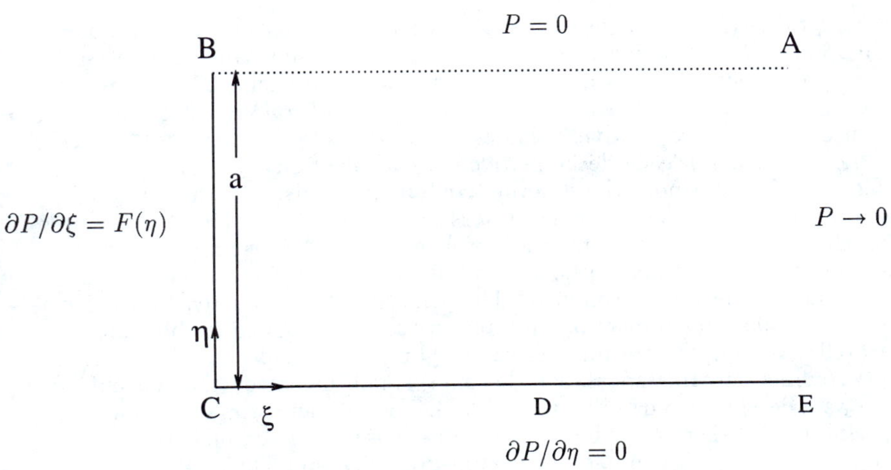

Figure 5. The final problem to be solved in the ζ-plane, where $F(\eta) = -\sin(\pi\eta/a)/(M\sqrt{b^2-1})$ with $b = [\cos(\pi\eta) - N]/M$.

We then use a translation and magnification to shift B to -1, and C to 1. The map required is $h = f + \mathrm{i}g = Mw + N$ where $M = 2/(\cosh(\pi/a) - 1)$ and $N = M + 1$. The last step is to map this problem back to the strip. The final map required is $\zeta = \xi + \mathrm{i}\eta = a\cosh^{-1}(h)/\pi$. This gives the problem as shown in figure 5, in a form with the singularty eliminated by being placed at a corner.

We solve Laplace's equation in this region by separation of variables. Let $P = f(\eta)g(\xi)$, giving $f'' = -\alpha^2 f$ and $g'' = \alpha^2 g$, where α is a constant. Solving for f, using the boundary condition that $f = 0$ at $\eta = a$, and $\partial f/\partial\eta = 0$ at $\eta = 0$, gives $f = A\cos(\alpha_n\eta)$ where $\alpha_n = (n+1/2)\pi/a$. We now solve for g, using the condition that $P \to 0$ as $\xi \to \infty$. This gives $g = R\mathrm{e}^{-\alpha_n\xi}$. Hence we have an expression for the pressure-impulse:

$$P = \sum_n A_n \mathrm{e}^{-\alpha_n\xi}\cos(\alpha_n\eta) \tag{9}$$

Finally we use the condition that $\partial P/\partial\xi = -\sin(\pi\eta/a)/(M\sqrt{b^2-1})$, where $b = [\cos(\pi\eta) - N]/M$ along $\xi = 0$ to get expressions for the A_n. Using this condition we get:

$$-\sum_n A_n\alpha_n\cos(\alpha_n\eta) = -\frac{\sin(\pi\eta/a)}{M\sqrt{b^2-1}} \tag{10}$$

The final step is to multiply both sides by $\cos(\alpha_m\eta)$, and integrate along the line $\xi = 0$ to get:

$$A_m = \frac{2}{\alpha_m a}\int_0^a \frac{1}{M}\frac{\sin(\pi\eta/a)\cos(\alpha_m\eta)}{\sqrt{b^2-1}}d\eta \tag{11}$$

Similar results can be found for any velocity distribution $V = V(x)$.

Results and discussion

The integral in (11) is evaluated by using a numerical routine. For the cases of $a = 0.5$ and $a = 2.0$ taking thirty terms in the sum, gives an accuracy of 4 and 12 decimal places respectively. The distribution of pressure-impulse in the water beneath the deck is shown for deck width to depth ratios of 0.5, 1.0 and 2.0 in figures 6,7 and 8 respectively. The values of the total impulse on the deck and on the wall beneath each deck are given in each caption.

In figures 6,7 and 8 note the differing contour intervals, and the increasing impulse on the deck as the water depth a is decreased. The value of total impulse on the deck is given as a function of a in figure 9. This trend is for the impulse from impact of a given velocity and area to increase as the body of impacting water becomes more confined. The same trend is described by Cooker and Peregrine (1995) for impact on an interior wall of a rectangular box and by Topliss (1994) for impact within a circular cylinder. Consideration of flow in the most confined circumstances, as a becomes small, has given the concept of 'filling flows' (Peregrine and Kalliadasis, 1995). Further, an estimate of how the compressibility of dispersed air bubbles, such as those entrained in waves during breaking, may soften wave impact is given in Peregrine and Thais (1996).

The results are in dimensionless units, for practical use the dimensional pressure-impulse is needed; that is

$$P^*(x^*, y^*) = \rho VLP(Lx, Ly), \tag{12}$$

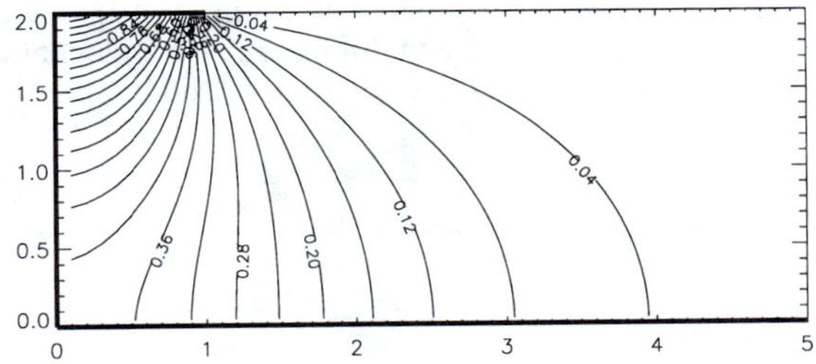

Figure 6: Pressure-impulse contours with $a = 2.0$. Total pressure-impulse on the deck and wall respectively are 0.81 and 1.02

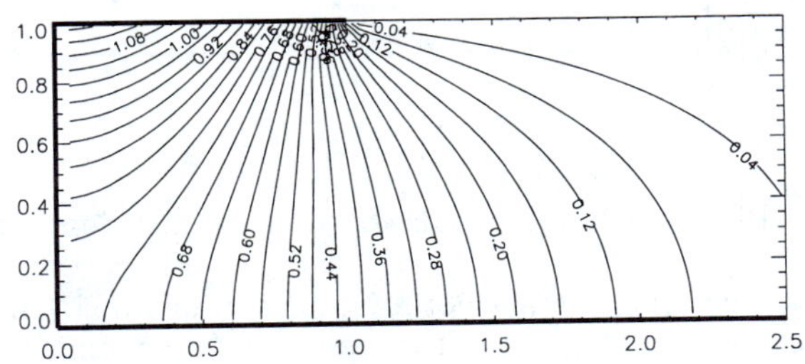

Figure 7: Pressure-impulse contours with $a = 1.0$. Total pressure-impulse on the deck and wall respectively are 0.92 and 0.74.

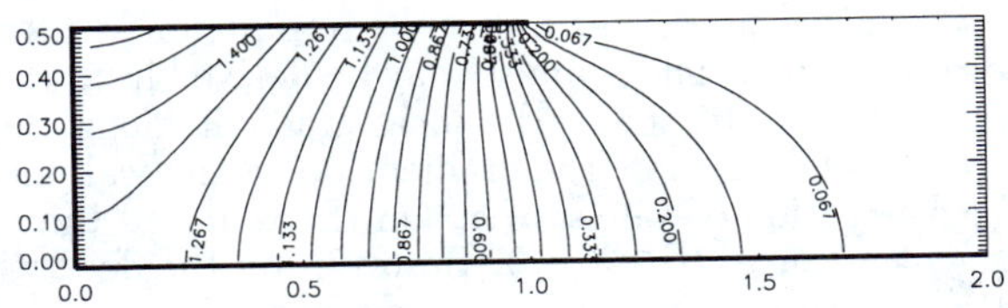

Figure 8: Pressure-impulse contours with $a = 0.5$. Total pressure-impulse on the deck and wall respectively are 1.193 and 0.44.

where * denotes the dimensional quantities. Whilst ρ and L will generally be known, V the vertical velocity of impact will need to be estimated. A simple method of estimating V is to first estimate how high a wave would be in the absence of the deck. Suppose it would have a height ΔH above the deck level. In simple projection of a particle this would require a velocity of $\sqrt{2g\Delta H}$. This is a reasonable, somewhat conservative, estimate for V.

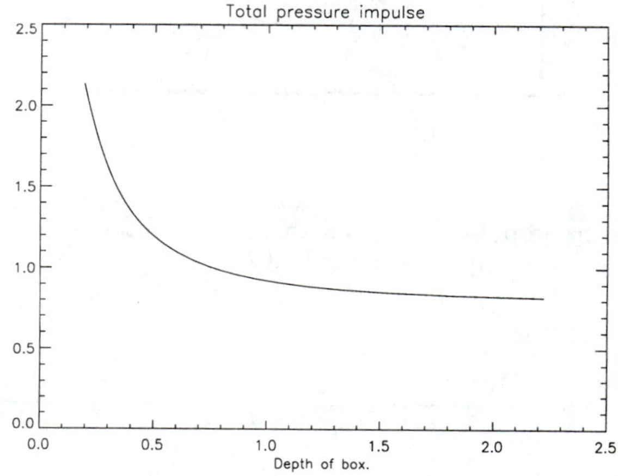

Figure 9: Total impulse on deck against depth a.

Note, the above solutions are not appropriate for impact from jets, e.g. see Cooker and Peregrine (1995), where the semi-infinite rectangular impact on a wall is equivalent to half of a plane jet and section 3.5 gives the solution for a circular jet. However, the solutions can be used for waves which are not nearly level with the deck as the figures indicate. By subtracting the appropriate constant from P, any of the contours of P can be chosen as an alternative free surface. Although such a surface tends downward rather than towards a horizontal level, this is not of great significance as long as the shape within roughly unit distance is appropriate. See Cooker and Peregrine (1995) for some examples.

Clearly the results presented here can be used to estimate the impulse and the spatial distribution of a wave impact. In addition, as illustrated by Cooker and Peregrine (1992) it is possible to estimate the impulse on smaller bodies on and near the wave impact area. The impulse may be derived from the local pressure-impulse gradient. Figure 10 shows the local gradient along the surface of the deck, and figures 11 and 12 show the gradient down the wall and along the bed respectively for a selection of values of a. On the wall and the bed the pressure gradient is tangential since $\partial P/\partial n = 0$. However, on the deck where the impact occurs $\partial P/\partial n \neq 0$ so that there is also a component of impulse perpendicular to the deck and downward.

Consideration of the gradient of pressure-impulse gradient near the edge of the deck shows alarmingly high values because the mathematical solution has a singularity at the edge of the deck. Clearly a better approximation is needed

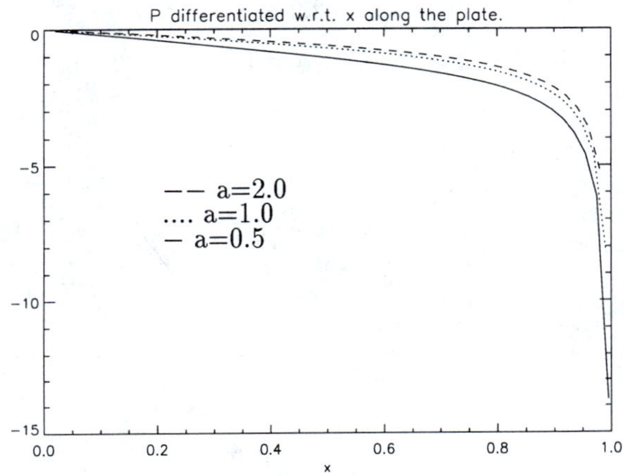

Figure 10: $\frac{\partial P}{\partial x}$ along the deck.

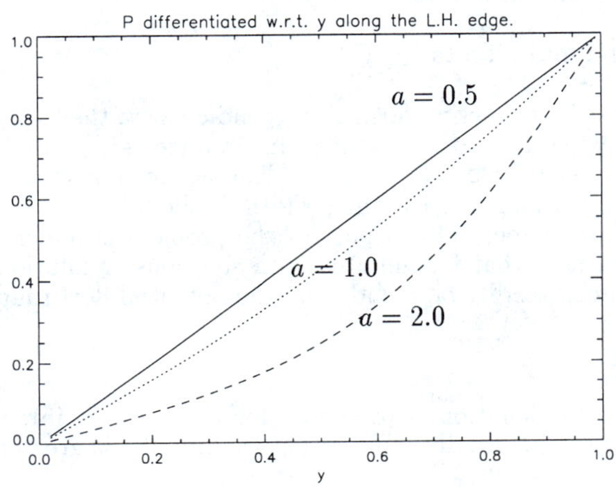

Figure 11: $\frac{\partial P}{\partial y}$ along the left hand wall.

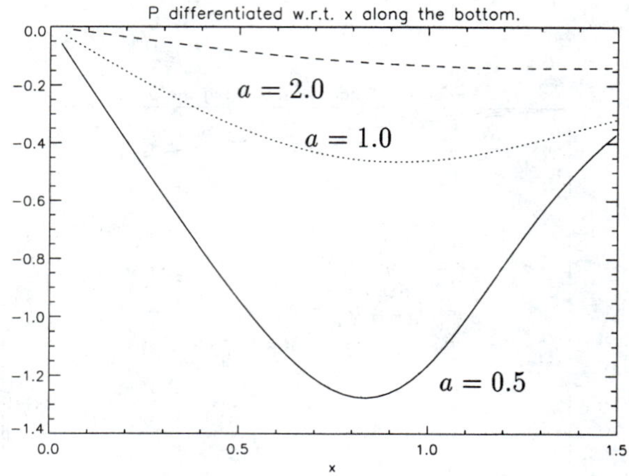

Figure 12: $\frac{\partial P}{\partial x}$ along the bottom.

there. One simple way of obtaining more realistic values is to consider how the solution is obtained for the infinite-depth case, $a = \infty$. There, the solution for the flow past a plate is used. This solution is a limit of flow past an ellipse. Thus a somewhat better solution could be obtained from the flow past a slender ellipse. In any case, it seems reasonable to conclude that attachments beneath a deck are vulnerable to especially large impact forces if they are near the edge of the deck, or the edge of the impact zone.

Three-dimensional effects

All the above work assumes uniformity perpendicular to the (x, y) plane, or some rigid boundaries parallel to that plane. In practice this is unlikely, and three-dimensional effects may be important. That is the impact area on the deck, rather than being a long strip of finite width L, should be taken as a finite area of an appropriate shape. This aspect of the problem is under study. For the present, we just note that for infinite depth solutions, a solution for impact on an elliptic area can readily be found from the potential flow round an ellipsoid.

Conclusions

A readily evaluated solution is presented for the pressure-impulse from waves hitting a deck from below. It is found that the impulse is greater if the water is shallow.

The same results may be useful for estimating the effects of upward impact by liquid confined within a container.

Acknowledgements

Support of the U.K. EPSRC and the European Commission, Directorate XII, Science, Research and Development, contract number MAS3-CT95-0041, is gratefully acknowledged.

References

BAGNOLD, R.A. (1939). Interim report on wave pressure research. *J. Inst. Civil Engrs* **12**, 201-226.

CHAN, E.S. (1994). Mechanics of deep water plunging-wave impacts on vertical structures. *Coastal Engng* **22**, 115-133.

COOKER, M.J. & PEREGRINE, D.H. (1990a). Violent water motion at breaking wave impact. *Proc. 22nd Internat. Conf. Coast. Engng., ASCE* 164-176.

COOKER, M.J. & PEREGRINE, D.H. (1990b). A model for breaking wave impact pressures. *Proc. 22nd Internat. Conf. Coast. Engng., ASCE* 1473-1486.

COOKER, M.J. & PEREGRINE, D.H. (1992). Wave impact pressure and its effect upon bodies lying on the sea bed. *Coastal Engng.* **18** 205-229.

COOKER, M.J. & PEREGRINE, D.H. (1995). Pressure-impulse theory for liquid impact problems. *J. Fluid Mech.* **297** 193-214.

LAMB, H. (1932). *Hydrodynamics.* 6th Edn. Cambridge University Press.

LOSADA, M.A., MARTIN, F.L. & MEDINA, R. (1995). Wave kinematics and dynamics in front of reflective structures. *Wave forces on inclined and vertical wall structures. ASCE* editors N. Kobayashi & Z. Demirbilek 282-310

MILNE-THOMPSON, L.M. (1962). *Theoretical Hydrodynamics.* Macmillan.

PEREGRINE, D.H. & KALLIADASIS, S. (1996). Filling flows, coastal erosion and cleaning flows. *J. Fluid Mech.* **310** 365-374.

PEREGRINE, D.H. & THAIS, L. (1996). The effect of entrained air in violent water wave impacts. *J. Fluid Mech.* **325** 377-397.

TOPLISS, M.E. (1994). Water wave impact on structures. Ph.D. dissertation, University of Bristol.

CHAPTER 200

WAVE IMPULSE PREDICTION FOR CAISSON DESIGN

Michael J.A. Walkden[1], Peter J. Hewson[2] and Geoffrey N. Bullock[3]

Abstract

Problems associated with the measurement of wave impact pressures are described and the poor comparison between the results of previous studies is outlined. It is argued that the force impulse form and magnitude are more relevant parameters for caisson design than maximum impact pressures. Data from small and large scale laboratory measurements are used to develop a simplified design approach for caisson breakwaters based on the limited magnitude of the force impulse. Comparison between large scale data and predictions based on small scale results show reasonable agreement.

Introduction

The form of breakwater selected for a particular site depends upon many physical, economic and political factors. Rubble mound constructions are popular in shallow water because they have proved reliable and easy to repair but they are expensive in deep water. Due to the growth of ocean going vessels and the associated need to provide mooring facilities in deeper water there is increasing interest in vertically sided breakwaters. Historically these have had a masonry construction, although in recent decades cellular caissons have provided a preferred alternative. The most common failure mode of caisson breakwaters has been sliding due to the impact of breaking waves (Oumeraci, 1994, Franco, 1994). Such impacts are highly complex events which can not yet be described theoretically. Small scale laboratory wave impacts are therefore a valuable source of information for both designers and those seeking to improve design methods. Due to the sensitivity of breaker shape to wave and physical boundary conditions and the transient and localised nature of impact pressures, results are frequently test specific. If this problem was successfully addressed then confidence in predictions of prototype loading conditions based on such small scale laboratory tests would be greatly enhanced.

---

[1] Research assistant,

[2] Principal lecturer,

[3] Professor and Head of School, University of Plymouth School of Civil and Structural Engineering, Palace Court, Palace Street, Plymouth, PL1 2DE UK

Scatter of impulse pressures

Even the pressures generated by nominally identical consecutive breakers of a monochromatic wave train are only predictable in a statistical sense. Hayashi and Hattori, (1958) stated that "The initial shock pressure varies so enormously from test to test that it almost seems to have nothing to do with our theoretical considerations". There are many possible reasons for this, some due to random fluctuations in wave form, others due to test conditions and procedures.

Denny (1951) investigated high frequency random fluctuations in the wave form and took measurements of peak pressure from waves of various heights with two different levels of disturbance. This showed that the distribution of peak pressures normalised to wave height was identical for waves of different heights and similar disturbance, but for waves with less disruption, the distribution included much higher relative pressures. This effect may to due to differences in entrapped aeration, variations in the velocity field or some other process. The degree of disturbance and therefore the degree of pressure attenuation will depend on test facilities such as the wave generator and absorption system, and test procedures such as whether a solitary or regular wave train is tested, and how long the water is left to calm between tests.

Variation in pressure records between experiments will be compounded by effects specific to the wave channel and others resulting from the choice of equipment and procedure. The breaker shape depends, for the two dimensional case, mainly on the incident wave conditions, period, height and depth, and the physical boundary conditions mainly foreshore topography and permeability. Any difference in these between tests could be expected to alter impact pressures. The characteristics of the pressure transducers and their position are also significant because peak pressures are generally highly transient and localised. The number of transducers, their size and positions are significant to the spatial resolution, whilst frequency response and sample rate influence temporal resolution.

Impulse repeatability

The wide variation in peak pressure was first addressed by Bagnold who noted that the pressure rise impulse (see Figure 1) was far more repeatable. This observation was used to develop a concept of an 'effective length' of a piston like body of fluid extending horizontally into the wave from the front face. Because it was believed that the generation of the impulse was caused by momentum focusing through the pocket, it was the pocket dynamics and essentially its thickness which determined the impulse duration. Bagnold was unable to prove his model by measuring the air pocket. Also it is now known that an air pocket is not a necessary condition for the occurrence of an impulse load. The work has since been developed by other researchers, including Kirkgoz who measured a similar effective length for flat fronted waves as Bagnold found for air pocket waves, although with wide variation, and Weggel, *et al* (1970) who tried to show, but were unable to establish, that effective length varied with wave steepness.

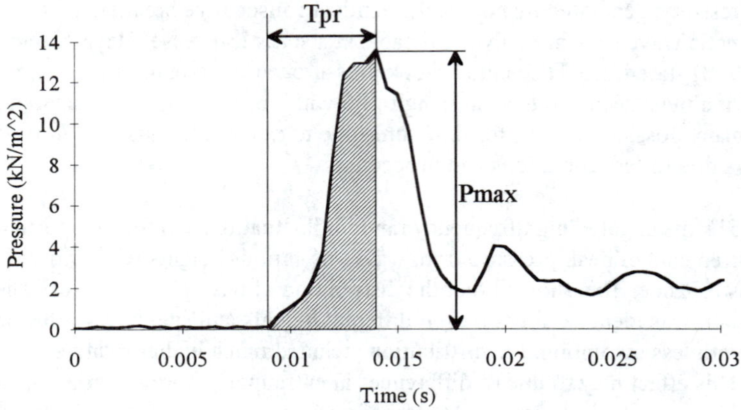

Figure 1. Definition of pressure rise impulse (shaded area)

Pmax and Tpr

The consistency of wave pressure impulse has often been expressed as a relationship between the maximum impact pressure Pmax, and the time for the pressure to rise to it, Tr of the form;

$$Pmax = K.Tpr^{H}$$

where K is a constant. Published examples include;

Pmax = 232Tpr^{-1}, Weggel *et al*, (1970).

The upper envelope of 12 laboratory 2D measurements with a beach slope 1/20 and a discontinuous wave wall. Regular waves were used to form one breaker type, the heights and periods are not specified although the steepness is given as 70 mm/sec. Six pressure transducers were used which were located 19 mm apart. No information is provided on their characteristics or sample rate or the measures taken to ensure a calm water surface.

Pmax = 250Tpr$^{-0.9}$, Kirkgoz (1990).

The best fit line to 70 laboratory 2D measurements with a beach slope of 1/10 and a continuous wave wall. Regular waves were used in still water depth of 0.61 m with an average height of 0.259 m and a 2 second period, these conditions produced plunging breakers. 10 pressure transducers with a 19 mm active surface diameter at 30 mm centres were sampled at 40 kHz. 1 hour was left between each test to allow the water surface to calm.

Pmax = 261Tpr$^{-0.65}$, Witte.

The upper limit function to approximately 100 laboratory 2D events with a beach slope of 1/6. Regular waves of height 0.25 m and period 2 seconds were generated in a still water depth of 0.61 m (depth at wall 0.16 m). 25 waves were generated per train to prevent re-reflection. Steep, vertical and air pocket breaker types were examined. 14 pressure transducers with a minimum gap of 0.015 m and surface diameter of 3.7 mm were sampled at 62.5 kHz.

Pmax = 400Tpr$^{-0.75}$, Hattori *et al,* (1994).
The upper limit function to a large number of laboratory events including a range of breaker shapes. The beach slope was 1:20 and the water depth at the wall was 50 mm. Regular wave trains were generated with an energy absorbing paddle with heights which varied from 40 to 120 mm and periods of 1.5, 1.7 and 2.0 seconds. Six pressure transducers were sampled at 5 kHz and located at 10 mm centres around still water level.
Pmax = 485Tpr^{-1}, Weggel *et al,* (1970).
The test conditions were as described above, except the wave steepness was 17.9 mm/sec.
Pmax = 3100Tpr^{-1}, Blackmore and Hewson (1984).
The upper limit function to approximately 80 full scale marine wave impacts on a sea wall. The wave heights ranged from 0.8 to 1.3 m with periods of from 2.67 to 8.5 seconds. The waves had mostly broken at impact. Seven 25 mm diameter pressure transducers were positioned at around 1 m intervals.

These functions appear because, as Bagnold noted, the pressure impulse is limited, if Pmax is large, then Tpr tends to be small and *vice versa.* There are too many variables between each test to determine the effect of each on K and H although there seems to be some relationship between K and the size of the wave. For example the constant of 3100 calculated from the prototype sea wave impacts of Blackmore and Hewson is significantly higher than those derived from the laboratory data. The fact that this relationship between Pmax and Tpr has been different in each case is probably due, at least in part, to the difficulties of recording consistent impact pressures between tests.

This pressure impulse appears to be a more consistent feature of impact loading that the pressure maxima, it may therefore be a more reliable basis for design if the structure response to the impulse is properly understood.

Structural Dynamics.
Structural response to loading is fundamentally important to design. Goda used a dynamic model to develop his method for breakwater design (Goda, 1974 and 1994) which has since become the standard in many countries. Oumeraci and Kortenhaus (1994) created a similar model which was calibrated with near prototype measurements recorded in the Grosser Wellen Kanal (GWK), see below. This was used to show how the structure responded to different force time histories. It was demonstrated that, for impacts without significant air pocket generated force oscillations, the dynamic response could be accurately predicted by simplifying the force time history to a triangular impulse described by the maximum force, Fmax, the impulse duration, Td and the rise time Tr. The results were presented in the form of dynamic amplification factors, ratios of force maxima to the equivalent static load which would have caused the same displacement (Fstat). The dynamic amplification factors depend upon the ratios of Td and Tr to the natural period (Tn) of the structure

in question. In this way structural dynamic response can be accounted for and the problem for the designer is reduced to ensuring integrity under a known static load.

There are a series of problems to be overcome before adopting this approach. The form and magnitude of the force impulse has to determined from the incident wave conditions and the physical boundary conditions. At present this can only be achieved through physical model tests. The natural variation in the magnitude and form of the impulses measured during such tests must therefore be accounted for and a reliable and appropriate method of scaling must be determined.

Scale effects
A comparison of similar wave impacts at significantly different scales have been conducted by Sakakiyama (1994) for the case of a submerged breakwater in front of a vertical wall, and by Fuhrboter (1986) for sloping revetments. To the authors knowledge no such data has been published for the case of a vertical wall experiencing direct wave breaking. The case for the existence of scale effects appears to be that field measurements provide relatively smaller peak impact pressures than have been measured in the laboratory and scaled with the Froude law. Aeration is often suggested as a probable cause although it seems probable that the above mentioned difficulties associated with measuring consistent impact pressures are at least partly responsible.

Aeration
Entrained aeration has a strong effect on the compressibility characteristics of water. It affects the rate of momentum exchange between the wave and wall (Blackmore and Hewson, 1984) and reduces the speed of sound in the fluid, altering the propagation of pressure waves (Topliss, 1994, Griffiths, 1994). It depends strongly on the chemistry and biology of the water. Bubbles of air in fresh water tend to be formed larger than in sea water due to increased surface tension, they also coalesce more readily and burst more easily at the surface. No direct measurements have yet been made of breaker aeration so its effects can only be assumed, however it has been shown that artificial aeration can reduce impact pressures (Crawford *et al*, 1994, Walkden *et al*, 1995).

If entrained aeration does affect impact pressures this does not necessarily mean that the impulse magnitude is affected. The force impulse represents the momentum of a proportion of the wave. Although aeration reduces water density, the wave mass remains essentially the same due to bulking. Also the wave kinematic pattern is probably not strongly effected, there is therefore reason to believe that the wave impulse might be reasonably independent of entrained aeration and so possibly free of associated scale effects.

Physical experiments
Measurements of wave impact forces were conducted at both small and near prototype scale in order to develop a way of accounting for wave impulse loading, for

impacts without significant air pocket oscillations. The following subjects were investigated;
The magnitude and form of the force impulse,
Linearisation of the impulse,
The significance of K and H,
Scale effects,
Transformation from dynamic loads to equivalent static loads.

University of Plymouth (UoP) small scale wave impact tests
Experiments were conducted in a 20 m long, 1.2 m deep and 0.9 m wide flume fitted with an energy absorbing wedge type wave paddle (Bullock and Murton, 1989). Waves shoaled over a 1:4.5 sloping impermeable foreshore and broke onto a vertical wall made of 25 mm acrylic sheet. Five sealed gauge Kulite type 219 pressure transducers with an active surface diameter of 18 mm were used to record the pressures at the wall at 2 kHz. Three of these had a pressure range of from 0 to 100 kN/m^2, the others were designed for pressures of up to 400 kN/m^2. The transducer centres were 40, 55, 70, 90 and 110 mm above the toe of the wall, although during experiments the transducer at 90 mm was damaged. Data was recorded on a P.C. by means of a Microlink logger. Video records were made of each test.

GWK large scale wave impact tests
Experiments were conducted in Europe's largest wave flume, the GWK. The physical boundary conditions were not geometrically similar to those of the UoP flume but it was felt that adequately similar breaker geometry would be attainable. The GWK is a large concrete 'U' channel, with its base at ground level, internally it is 324 m long, 5 m wide and 7 m deep. Waves broke onto a sand filled concrete caisson mounted on a rubble foundation over a shallow sloping sand bed. Instrumentation was mounted on and built into a spar which was then attached to the caisson face. Four Kulite type 219 pressure transducers were installed which had a pressure range of 0 to 400 kN/m^2 (see above). They were positioned at 200 mm increments, 0.72 to 1.32 m from the top of the rubble mound. An additional pressure record was made with one of the transducers belonging to the Franzius Institut which was already built into the caisson and located 0.54 m above the rubble mound. Cabling was contained within the spar to protect it from the impacting waves. Pressures were recorded at 2 kHz . Video records of each test from two static and one mobile camera provided information on the breaker forms. Thirty eight tests were conducted in total in water depths ranging from 3.1 m to 3.3 m, including regular and pseudo random wave trains and solitary waves. Wave heights ranged from 0.5 to 1.1 m with periods from 3.5 to 5.5 seconds.

Results
Due to space limitations only the results from two tests will be shown, one conduced in the UoP flume, and one carried out in the GWK. In the former case 350 waves of height (H) 0.11 m, period 1.25 s and 79 mm water depth at the wall, broke with a near vertical, slightly overhanging wave front.

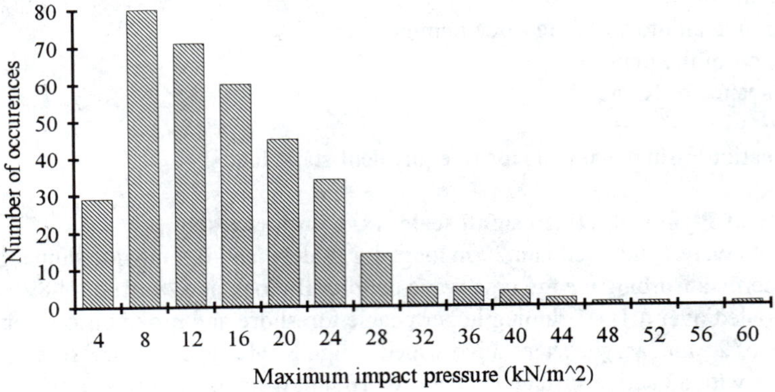

Figure 2. Maximum impact pressures from one transducer (UoP Data)

The data shown in Figure 2 illustrates the typically large spread in Pmax. The mean value is 13.4 kN/m^2, with a range from 2.3 kN/m^2 to 58.6 kN/m^2 . If this test had been conducted in order to model a design wave impact it might be concluded that the design event may produce a pressure maxima of anything from 2.13 ρgH to 54.3 ρgH. It is not clear how such results should be interpreted.

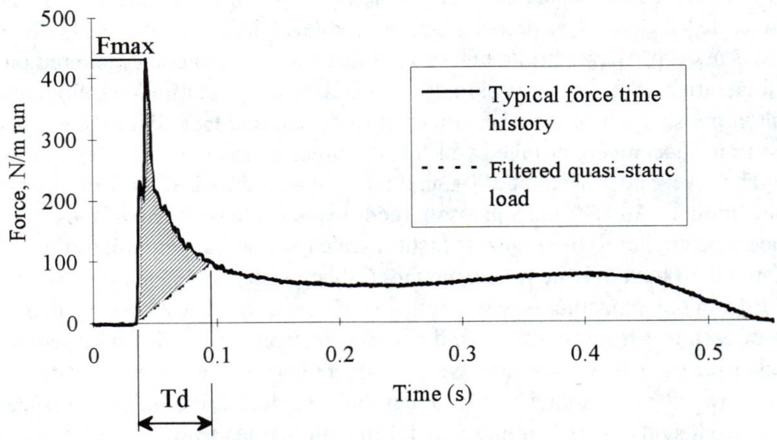

Figure 3. Typical force time history, the shaded area represents the impulse (UoP data)

In order to obtain the impulses, the force time histories were first calculated by spacewise integration of the pressures. The impulse is a dynamic load involving rapid variations in force, therefore a simple low pass filter was used to remove the quasi static force component, see Figure 3. The distribution of the impulse was found to be compact, Figure 4 shows the results from the UoP test, normalised to the total momentum exchange caused by the wave at the wall. This implies that accounting for the impulse magnitude, with for instance a statistical confidence limit, may be fairly straightforward.

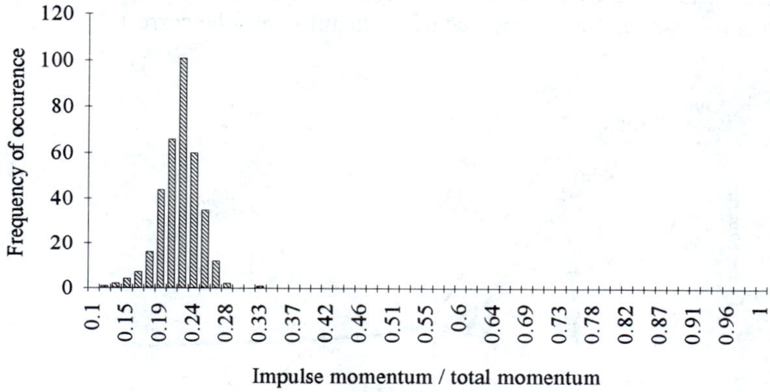

Figure 4. Frequency distribution of force impulse normalised to total momentum (Impulse momentum fraction, UoP data)

Partly because of this consistency, if the data is represented in the Fmax Td domain it clusters around a similar function to those found in the Pmax Tpr domain, as shown in Figure 5.

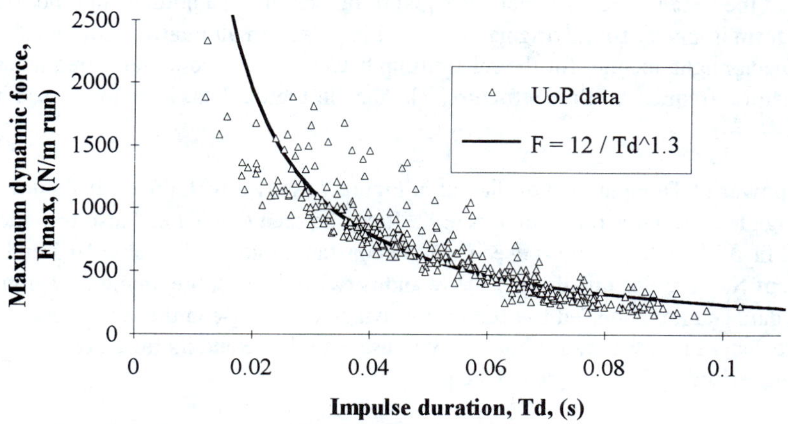

Figure 5. Relationship between maximum dynamic force and impulse duration

The values of Fmax and Td in Figure 5 are not suitable to be used to predict the structure response following the method described by Oumeraci and Kortenhaus. This is because the impulse is linearised to a triangle with base Td and height Fmax. Some of the data shows both a high peak force and a long duration which if used as a basis for inearisation results in an unrealistically high impulse magnitude and an over-prediction of structure response. This is because the form of the impulse tends to be concave as the force drops, as can be seen in Figure 3. In order to realistically predict structure response the magnitude of the impulse must be correct.

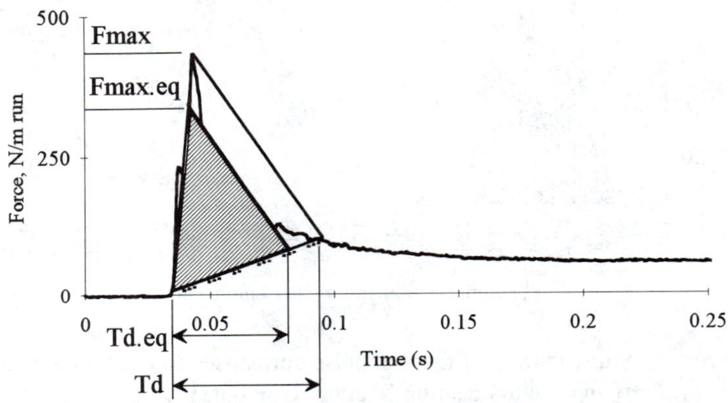

Figure 6. Method of processing the linearised force impulse

It was therefore decided to process the values of Fmax and Td to parameterise both the magnitude and the form of the impulse. Figure 6 demonstrates the method which was used. Td.eq and Fmax.eq are equivalent values which were calculated so that the area of the triangle they described is equal to the impulse magnitude, and the triangle they form is similar to the original linearised impulse. An alternative more ideal approach might account for the relationship between removed impulse frequencies and the natural frequency of the structure. The data in Figure 7 has been processed in this way.

The power of Td.eq in the best line fit in Figure 7 (shown with the dashed line) found through least squares regression, is 0.92. It can be seen that $Td.eq^{-1}$ also provides a good fit. If the latter value is used then the constant assumes the value 16.3 and has units of Ns/m run or impulse per metre and is twice the average impulse magnitude. This data can now be used to account for dynamic response in the way Oumeraci and Kortenhaus suggest because both the impulse magnitude and its form are parameterised by Fmax.eq and Td.eq.

Comparison with large scale test results

Similarity of physical boundary conditions between the large and small scale experiments was not possible. The small scale foreshore was impermeable and had a regular slope, whereas the large scale foreshore was permeable and included a berm. If the incident wave conditions were geometrically similar then the breaker shapes would not be and no comparison between the results would be meaningful. It was therefore decided to base a scale comparison on the breaker shape and the wave momentum. The video records were used to match similar breaker shapes at small and large scale, then the wave momenta were used to obtain the scale ratio. The average momentum measured for the UoP waves was 39.2 Ns/m run and for the GWK data, 4545 Ns/m run, which provides a length scale of 6.7.

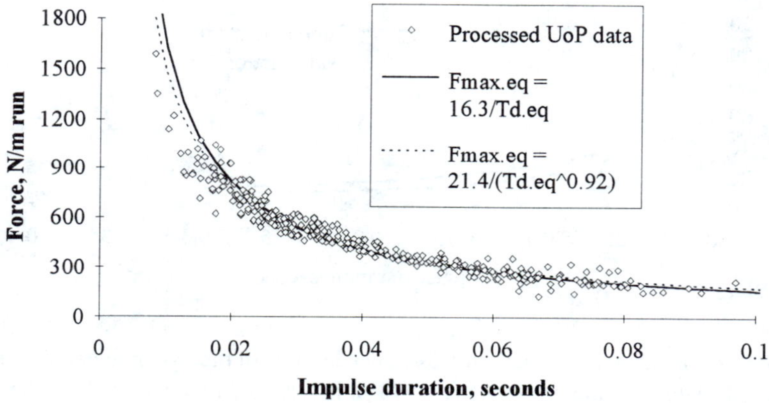

Figure 7. Impulse momentum and shape parameterised by F eq and Teq

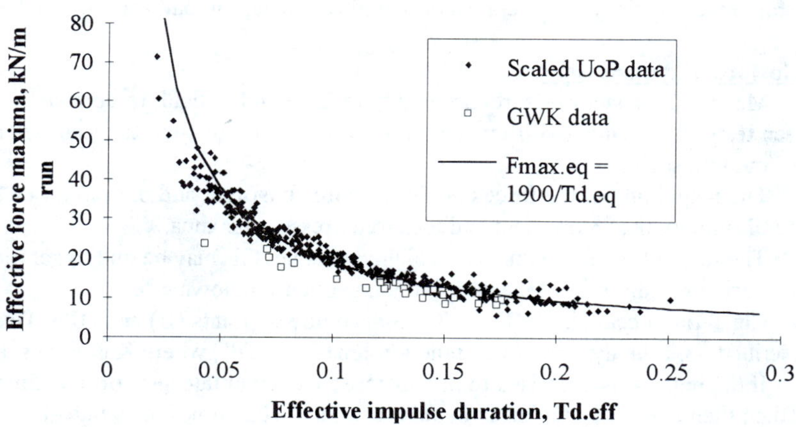

Figure 8. Comparison between scaled UoP data and GWK data

The values of Fmax.eq and Td.eq measured in the UoP flume were scaled according to Froude and compared to values obtained from the GWK test in Figure 8. The match is not perfect but when the lack of boundary condition similarity is considered the results might be considered close.

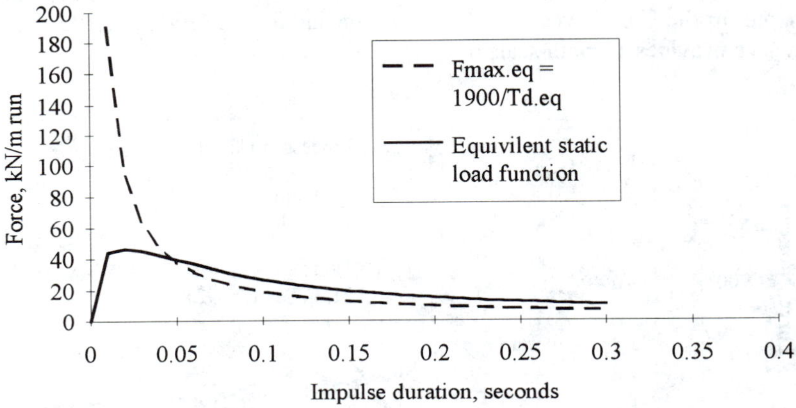

Figure 9. Equivalent static load function calculated from the dynamic amplification factors of Oumeraci and Kortenhaus, 1994. It is assumed that Tn = 0.1 and Tr = 0

In Figure 9 the dynamic force function Fmax.eq = 1900/Td.eq has been multiplied by a dynamic amplification function to obtain the equivalent static load function. The maxima of this (in this case 45 kN/m run) might be considered the largest equivalent static force the impulse can generate and therefore the design load for this breaker.

Conclusions and discussion

(I) Maximum impact pressures are highly variable and difficult to compare between tests. This is due to differences in test procedure and conditions as well as natural random scatter.

(II) The magnitude of the force impulse is more consistent and more relevant to structural stability than a transient and localised pressure maxima.

(III) The form of the force impulse is highly variable, this may be due to aeration effects, variations in breaker kinematics or some other unknown effect.

(IV) Clustering occurs in the Fmax Td domain due to points (II) and (III). This can be described with the dynamic force function $Fmax = K.Td^{H}$, where K is a constant.

(V) If the impulse is linearised to parameterise its magnitude and form by Fmax.eq and Td.eq then K is twice the mean impulse area, and H assumes the value -1.

(VI) Scale effects can be reduced if experiments are conducted in large flumes however unless sea water is employed, similarity in aeration characteristics and therefore compressibility can not be assumed.

(VII) It was not possible to properly assess scale effects on the impulse because of a lack of geometric similarity, however a reasonably close comparison between large scale results and predictions based on small scale tests was obtained suggesting that scale effects on the impulse may be minimal.
(VIII) If Tr is assumed to be zero, and the structure natural frequency is known, then the maxima of the product of the dynamic force function and a dynamic amplification function might be considered the maximum static load for design purposes.

The analysis and interpretation presented is highly simplified and is intended to illustrate the basis of an approach. Areas of necessary further development include;
(I) Allowing non zero rise times.
The rise time can be measured fairly simply and used to select a more appropriate, dynamic amplification function. The ratio Tr/Td may be affected by scale.
(II) Accounting for significant air induced force oscillations.
The entrapment of a long and large oscillating air pocket is recognised as resulting in a significant and dangerous loading condition. It is probably a relatively rare event because it requires a normal direction of approach and a relatively clean wave form. A similar danger may arise due to group dynamics of bubble clouds.
(II) The mean dynamic force function was used to predict equivalent static loads.
For design purposes the loading should be calculated on the basis of the worst case with a chosen probability of occurrence within the design life. If the model test conditions have been chosen to represent events with a known return period then the smooth distribution of measured impulses should allow a confident estimate of a realistic design impulse magnitude.
(III) The method of impulse definition.
The use of a low pass filter to define the impulse seems to work well. The cut off frequency should be decided upon with a consideration of the way the structure will respond to that frequency. Similarly the process of obtaining Td.eq and Fmax.eq could be developed with more consideration of the structure dynamics.
(IV) Directionality.
The waves angle of approach has not been dealt with although its inclusion may be relatively straightforward. This method is based on momentum transfer at the plane of the caisson face which might be expected to vary relatively simply with angle of approach.
(V) Accounting for the structure response.
The conversion of dynamic loads to equivalent static loads has been based on the dynamic amplification factors published by Oumeraci and Kortenhaus (1994). These were derived by reducing the description of the caisson motion to a simplified single degree of freedom model. This was intended to "constitute a useful guide for judgement in developing design load specifications". There are many limitations which have been discussed in their paper regarding their model and modelling of structural dynamics in general. In the light of their comments it would be preferable, though more complex, to use a numerical model rather than a dynamic amplification function to account for structure response.

(VI) Assessment of the effect of breaker shape on the impulse momentum fraction. This varies with breaker type so it will reach a measurable maxima at some breaking condition. This might be multiplied by the design wave momentum in order to calculate a worst case impulse momentum. From this the constant K is obtained, and so the dynamic load function is known. The effect of variations in the incident wave and physical boundary conditions on the impulse momentum fraction should be easier to asses than for impact pressure maxima because of its more compact distribution. It is also far less sensitive to the characteristics of measurement equipment and test procedures. For these reasons scale effects may be also be easier to asses.

The main advantage of this approach is that it provides a means of interpreting the results of physical model tests in a way which is meaningful for structural design. This means that nature can be left to resolve the complexities of the breaker shape, a problem which can not yet be fully solved through numerical modelling.

Future work

Measurements of entrained and entrapped air are in progress at the University of Plymouth in order to examine their effect on the impulse form and magnitude and wave impact pressures. This work is being conducted both in the laboratory and in the field (Crawford, *et al* 1994, Walkden, *et al*, 1995 and Bird *et al*, 1997). It is hoped that the results will help explain the distribution of data along the dynamic load function, provide insights into the relationship between entrapped air and the force impulse and provide data with which to test possible scale effects.

Acknowledgements

The work described in this paper was in part carried out under contract MAS2-CT92-0047 of the European Union's Marine Science and Technology Programme. This, together with co-operation and facilities made available by the Franzius Institut, Universitat Hannover and the Technischen Universitat Braunschweig are gratefully acknowledged.

References

Bagnold, R.A., (1939), "Interim report on wave pressure research". J. Instn. Civ. Engrs, Vol. 12, p. 201.

Bird, P.A.D., Crawford, A.R., Hewson, P.J., and Bullock, G.N. (1997), "Measurement of wave impact pressures and sea water aeration". Paper in print for Coastal Engineering.

Blackmore, P.A. and Hewson, P.J., (1984), "Experiments on full-scale wave impact pressures". Coastal Engineering, 8, pp 331-346.

Bullock, G.N. and Murton, G.J., (1989), "Performance of a wedge type absorbing wave maker". Journal of Waterway, Port, Coastal and Ocean Engineering, Vol.115, No.1, pp.1-20.

Crawford, A.R., Walkden, M.J., Bird, P.A.D., Bullock, G.N., Hewson, P.J. and Griffiths, J., (1994), "Wave impacts on sea walls and breakwaters". Proc.

Coastal Dynamics, pp 656-670.

Denny, D.F., (1951), "Further experiments on wave pressure". J. of the Institution of Civil Engineers, Vol. 35.

Fuhrboter, A., (1986), "Model and prototype tests for wave impact and run-up on a uniform 1:4 slope". Coastal Engineering, 10, pp 49-84.

Franco, L., (1994), "Vertical breakwaters: the Italian experience". Coastal Engineering 22, pp 31 - 55.

Goda, Y. (1974), "New wave pressure formulae for composite breakwaters". Proc. 14th int. conf. Coastal Engineering, pp 1702 - 1720.

Goda, Y., (1994), "Dynamic response of upright breakwaters to impulsive breaking wave forces", Coastal Engineering 22, pp 135-158.

Griffiths, J., (1994), "The effects of aeration on wave impacts". Ph.D. Thesis, University of Plymouth. U.K.

Hattori, M., Arami, A., and Yui, T., (1994), "Wave impact pressure on vertical walls under breaking waves of various types". Coastal Engineering, 22, pp 79-114.

Hayashi, T., and Hattori, M., (1958), "Pressure of the breaker against a vertical wall". Coastal Engineering in Japan, pp 25 - 37.

Kirkgoz, M.S., (1990), "An experimental investigation of a vertical wall response to a breaking wave impact". Ocean Engineering, 17(4): 379-391

Oumeraci, H., (1994), "Review and analysis of vertical breakwater failures - lessons learned". Coastal Engineering 22, pp 3 - 29.

Oumeraci, H. and Kortenhaus, A., 1994, "Analysis of the dynamic response of caisson breakwaters" Coastal Engineering 22, pp 159-183.

Sakakiyama, T. and Ogasawara, M., (1994), "Scale effect on impulsive wave forces and reduction by submerged breakwater". Coastal Dynamics, pp 628 - 642

Topliss, M.E., (1994), "Water wave impact on structures". Ph.D. thesis, University of Bristol, U.K.

Walkden, M.J.A., Crawford, A.R., Hewson, P.J., Bullock, G.N., and Bird, P.A.D., (1995), "Wave Impact Loading on Vertical Structures", Proc. Coastal Structures and Breakwaters '95, London.

Weggel, J.R., Hall, W. and Maxwell, C. (1970), "Numerical model for wave pressure distributions". ASCE. J. of waterways, harbours, and coastal engineering division, Vol. 96, No. WW3, pp 623 - 642.

Witte, H.H., "Wave impact loading on a vertical wall with respect to structure response". Report for the Federal Waterways and Research Institute - Coastal Department - Coastal Department. 30 p.

CHAPTER 201

WAVE OVERTOPPING RATE AND REFLECTION COEFFICIENT FOR OBLIQUELY INCIDENT WAVES

Yoichi MORIYA[1] and Masaru MIZUGUCHI[2]

Abstract

Recently, wave overtopping over a vertical wall for normal incidence were treated analytically on the basis of the wave energy flux concept (Mizuguchi, 1993). In this study, the Mizuguchi's model for normal incidence is extended to oblique incidence. Incident wave angle effect on the wave overtopping rate is investigated. Validity of the extended model is checked experimentally. Satisfactory agreement between the extended model and the experimental results was found for the wave overtopping rate.

1. Introduction

Wave overtopping for oblique incidence has been studied experimentally in several papers. The influence of the incident wave angle on wave overtopping tests over a vertical wall was studied by using regular waves by Inoue and Tuchiya (1971) or using irregular waves by Takayama et al. (1984). Recently, several model tests have been performed, for example, de Waal and van der Meer (1992) ; Juhl and Sloth (1994), and so on. However, the quantitative estimation of wave overtopping rate does not agree among the previous studies. The disagreement partly comes from the varying wave height in front of the wall.

Recently, for normal incidence, a set of simple equation to estimate the wave overtopping rate over a vertical wall and the reflection coefficient were derived

---

[1]Research Engineer; Penta-Ocean Construction Co., Ltd., Institute of Technology, Yonkucyo 1534-1, Nishinasuno-machi, Nasugun, Tochigi-Prefe., 329-27, JAPAN

[2]Professor; Dept. Civil Eng., Chuo Univ., Kasuga 1-13-27, Bunkyo-ku, Tokyo, 112, JAPAN

(Mizuguchi, 1993). The model, which does not include any experimental constants, is based on the energy flux concept with an assumption of partially standing waves. The validity of the model was confirmed by experiment (Yokoyama and Mizuguchi, 1993).

The purpose of this study is to investigate the influence of the incident wave angle on wave overtopping rate and reflection coefficient, by extending the model for normal incidence to oblique incidence. Validity of the extended model is checked experimentally.

2. Theoretical Approach

Wave field in front of vertical wall for oblique incidence

Figure 1 shows the coordinate system. When oblique waves incident a vertical wall, wave field in front of the vertical wall fully apart from the wall end is free from scattering waves forms bi-directional wave field. We supposed that wave field is uniform along the wall.

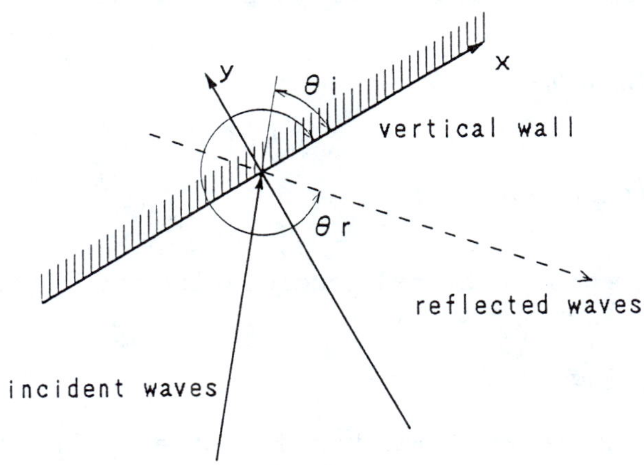

Figure 1 Coordinate system.

A finite amplitude wave theory is needed in evaluating the overtopping rate for waves with large height, especially in estimating the threshold wave height for overtopping. Since coastal structure are normally built in shallow water areas, the wave theory to be employed is the linear superposition of the first order cnoidal waves. The interaction between the incident waves and the reflected waves may be negligible as the interaction time is short. Surface elevation η in front of the vertical wall may be given by following formula:

$$\eta = \eta_i + \eta_r \tag{1}$$

$$= \frac{H_i}{2}\left[cn^2(\varphi_i;\kappa) - \overline{cn}\right] + \frac{H_r}{2}\left[cn^2(\varphi_r;\kappa) - \overline{cn}\right] \tag{2}$$

in which φ is the phase:

$$\varphi_j = \frac{K(\kappa)}{\pi}\left\{kx\cos\theta_j + ky\sin\theta_j - \sigma t\right\} \qquad j=i,r \tag{3}$$

where, H is the wave height, k is the wave number, σ is the angular frequency, θ is the wave angle, K(κ) is the complete elliptic integral of the first kind of modulus κ . The symbols cn denotes Jacobian elliptic function with modulus κ , and an overbar represents the average value over one period. The subscript i and r denote the quantity of the incident waves and the reflected waves, respectively.

Horizontal velocities of x and y direction, u and v, in front of the wall may be given by following formula:

$$u = \sqrt{\frac{g}{h}}\left(\eta_i\cos\theta_i + \eta_r\cos\theta_r\right) \tag{4}$$

$$v = \sqrt{\frac{g}{h}}\left(\eta_i\sin\theta_i + \eta_r\sin\theta_r\right) \tag{5}$$

where, g is the gravitational acceleration and h is the water depth.

Wave overtopping

Figure 2 shows schematic illustration of the wave overtopping over a vertical wall. For normal incidence, wave overtopping discharge is evaluated using an ideal fluid flow model over a sharp-edge weir by Kikkawa et al. (1967). When oblique waves are incident on a vertical wall, surface rise of the wave crest progresses along the wall. Wave overtopping for oblique incidence is supposed to be a kind of overflow of the surface rises over the wall. Then wave overtopping discharge, q_e, for oblique incidence averaged over one wave period may be given by following formula:

$$q_e = \frac{2\sqrt{2g}}{3T}\int_{\eta_e > H_c}\left\{\left(\frac{{v_w}^2}{2g} + \eta_e - H_c\right)^{\frac{3}{2}} - \left(\frac{{v_w}^2}{2g}\right)^{\frac{3}{2}}\right\}dt \tag{6}$$

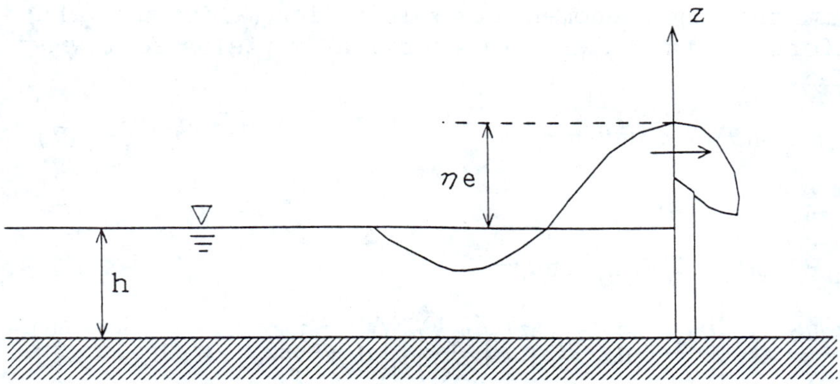

Figure 2 Schematic illustration of wave overtopping over a vertical wall.

where, T is the wave period, H_c is the crown height of the vertical wall above the still water level, η_e is the overtopping waves at the wall (η at y = 0) and v_w is the approaching velocity perpendicular to the wall.

Energy flux conservation

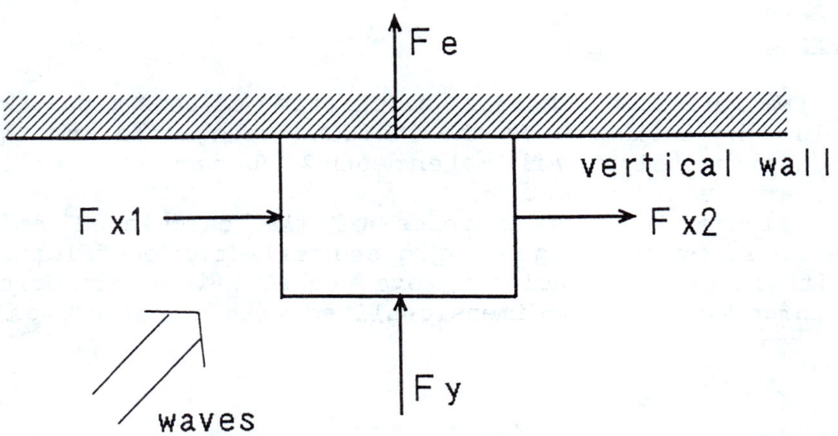

Figure 3 Schematic illustration of the energy flux conservation.

We consider the balance of the energy flux among the incident waves, the reflected waves and the wave overtopping flow. Figure 3 shows schematic illustration of wave energy flux conservation. Here, F_x and F_y are the energy flux of x and y direction and F_e is the energy loss due to overtopping. We assume that the phenomenon of x direction (along the wall) is uniform, so the balance of energy flux yields following formula:

$$E_i c_g \sin\theta_i + E_r c_g \sin\theta_r = F_e \tag{7}$$

or

$$f_{2i} H_i^2 \sin\theta_i - f_{2r} H_i^2 K_r^2 \sin\theta_i =$$
$$\frac{2\sqrt{2g}}{3T\sqrt{h}} \int_{\eta_e > H_c} \left\{ \left(\frac{v_w^2}{2g} + \eta_e - H_c \right)^{\frac{3}{2}} - \left(\frac{v_w^2}{2g} \right)^{\frac{3}{2}} \right\} \left(\eta_e + \frac{1}{2g} \left| u_w^2 + w_w^2 \right|_{z=0} \right) dt \tag{8}$$

where, K_r is the reflection coefficient defined as the wave height ratio and u_w is the velocity along the wall. f_2 is the ratio between the energy and the wave height square on the first order cnoidal wave theory (Isobe, 1985), and is 1/8 for the small amplitude wave theory.

Equation (8) gives the reflection coefficient, when incident wave height, period and angle are prescribed together with geometry or the freeboard of the vertical wall. The wave overtopping discharge can be estimated from eq. (6) by using the reflection coefficient.

Numerical calculation

We performed numerical calculation of this model. We examine the influence of the incident wave angle and the incident wave height while water depth 20.0 cm and wave period 1.2 s are kept constant.

Figure 4 shows examples of the calculated non-dimensional overtopping rate q_e/q_0 and reflection coefficient K_r with respect to incident wave angle. Wave overtopping discharge, q_e, is non-dimensionalized with q_0 defined as

$$q_0 = \frac{H_i \sqrt{gh}}{2}. \tag{9}$$

We performed the calculation using not only the first order cnoidal wave theory but also the small amplitude wave theory. It is clear that the wave overtopping rate and the reflection coefficient slightly decrease with the decrease of the incident wave angle. Significant decreases of more than 10%

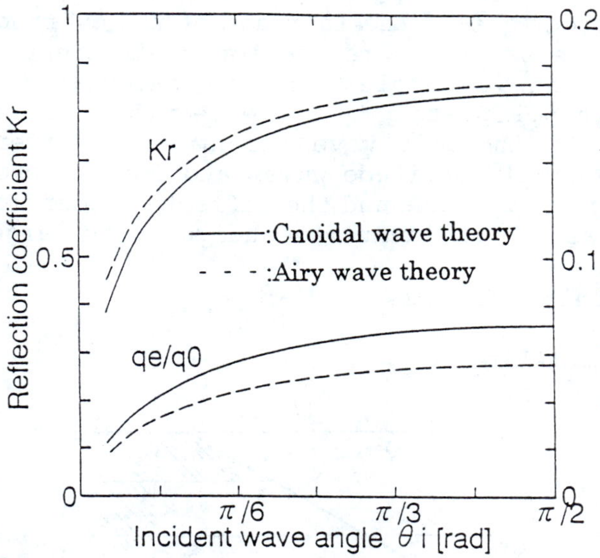

Figure 4 Examples of the calculated non-dimensional overtopping rate q_e/q_0 and reflection coefficient K_r with respect to incident wave angle.(calculation condition h=20cm, T=1.2s, H_i=6cm, H_c=3cm)

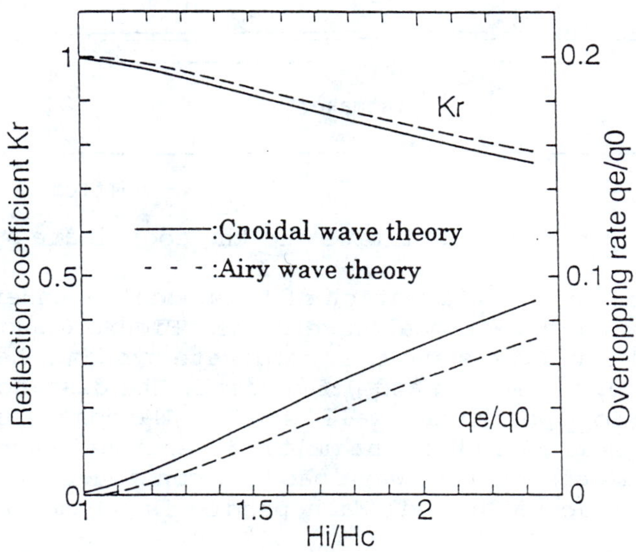

Figure 5 Examples of the calculated non-dimensional overtopping rate q_e/q_0 and reflection coefficient K_r with respect to incident wave height.(calculation condition h=20cm, T=1.2s, $\theta_i=\pi/3$, H_c=3cm)

are expected only for the incident wave angle less than $\pi/4$. Nonlinear effect yields the increase of the overtopping rate and the decrease of the reflection coefficient.

Figure 5 shows examples of the calculated non-dimensional overtopping rate q_e/q_0 and reflection coefficient K_r with respect to incident wave height. The difference between for the small amplitude waves and for the cnoidal waves in the overtopping rate and the reflection coefficient becomes larger as the incident wave height gets larger.

3. Experimental Confirmation

Experimental facilities

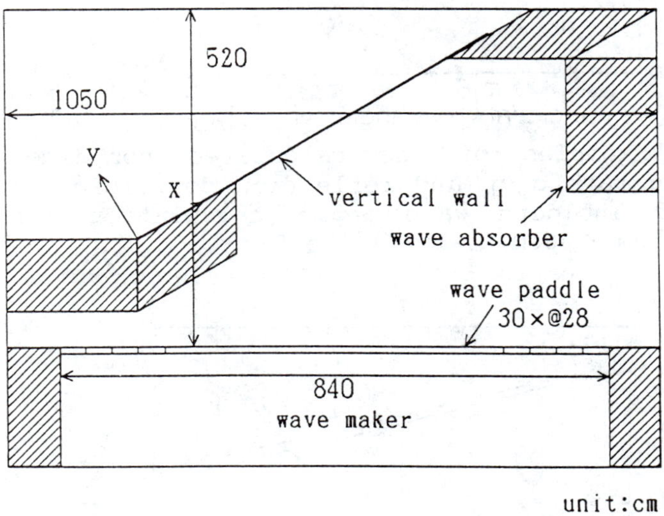

Figure 6 Experimental facilities and the coordinate system.

Experimental confirmation of this model is undertaken by using a multi-directional wave basin. Figure 6 shows the experimental facilities and the coordinate system. The basin is 10.5 m wide, 6.8 m long and 0.6 m depth. The distance from wave paddles to opposite end-wall is 5.2 m. The concrete floor of the basin is almost flat. The multi-directional wave maker consists of 28 piston-type wave paddles continuously linked or 29 sets of actuator rod. Each paddle is 30 cm wide and 50 cm high.

A vertical wall with 12.0 cm high and 540 cm long is installed in the basin. The influence of the scattering waves from the wall ends may be suppressed by placing wave absorbers around the wall ends.

In a multi-directional wave basin, generating a uniform wave field with the target value of wave height and angle is important. However, it is well known that both wave height and angle of generated regular waves show considerable spatial variation. The main cause is the finite length of the wave maker. The effect of the discontinuity at both ends of the wave maker may be suppressed by employing linear decrease of wave paddle amplitude (Mizuguchi, 1994). We applied this ends control method.

Water depth h = 10.0 cm, the crown height of the wall H_c = 2.0 cm, and a wave period of T = 1.0 s are kept constant. Incident wave height and incident wave angle are varied.

Wave field in front of the wall

Non-overtopping experiments (incident wave height H_i =1.0 cm) were carried out in order to check the uniformity of the wave field in front of the vertical wall. Wave gauges are installed 1.0 cm distant perpendicularly to the wall. An array of wave gauge and electro-magnetic current meter measures the surface elevation and the horizontal velocity at the same point 1/12 of wave length perpendicularly apart from the wall.

Figure 7 shows the wave height and the total amplitude of the horizontal velocity distribution in front of the vertical wall for non-overtopping cases. Wave field in front of the wall can be assumed to be uniform at the center of the wall. Where we measure wave overtopping discharge for all the cases.

Separation method between incident and reflected waves

A separation method is developed to evaluate the reflection coefficient and the reflected wave angle from the measured data. The method assumes the linear superposition of the incident waves and the reflected waves, both of them being long waves.

Mathematically, we have eq. (10), eliminating η_i and η_r from eqs. (1), (4) and (5),

$$\frac{\sqrt{h/g}u - \eta\cos\theta_i}{\sqrt{h/g}v - \eta\sin\theta_i} = \frac{\cos\theta_r - \cos\theta_i}{\sin\theta_r - \sin\theta_i} = \alpha \qquad (10)$$

where, α is theoretically a constant, although α vary for a measured data. Symmetrical reflection ($\theta_r = 2\pi - \theta_i$) gives $\alpha = 0$. Incident waves and reflected waves are calculated by following formula,

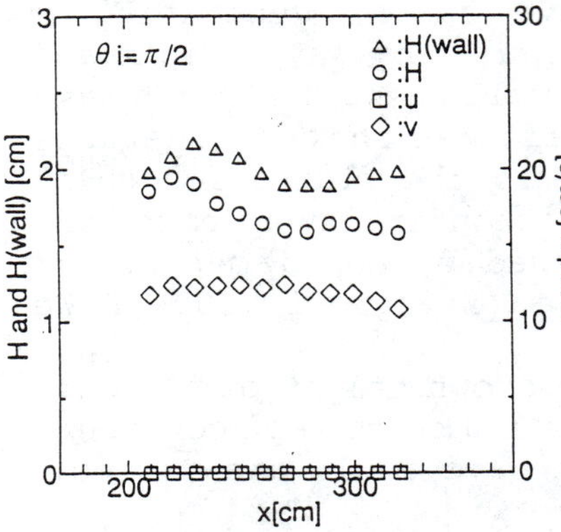

(a) $\theta_i = \pi/2$

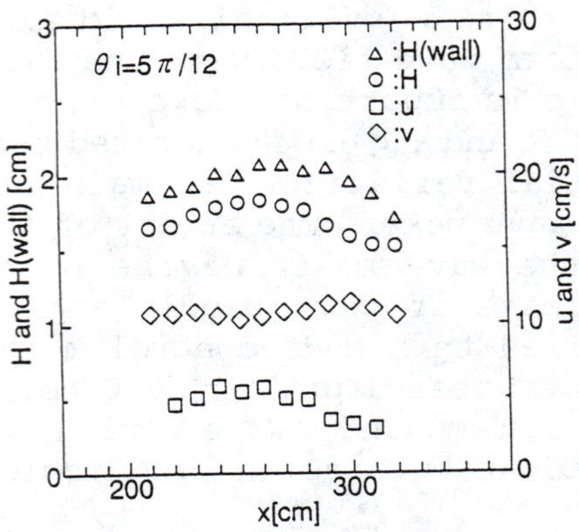

(b) $\theta_i = 5\pi/12$

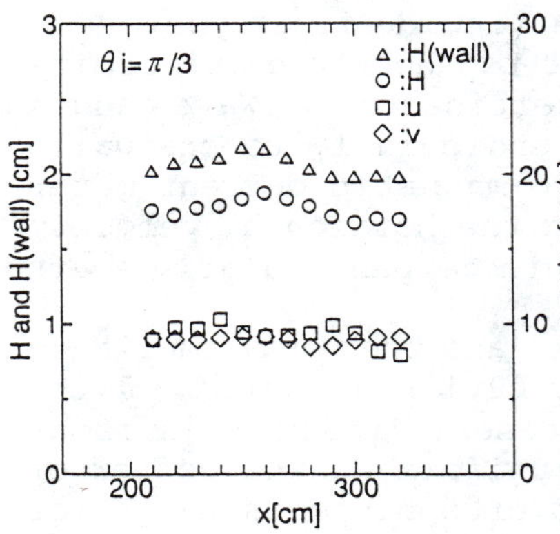

(c) $\theta_i = \pi/3$

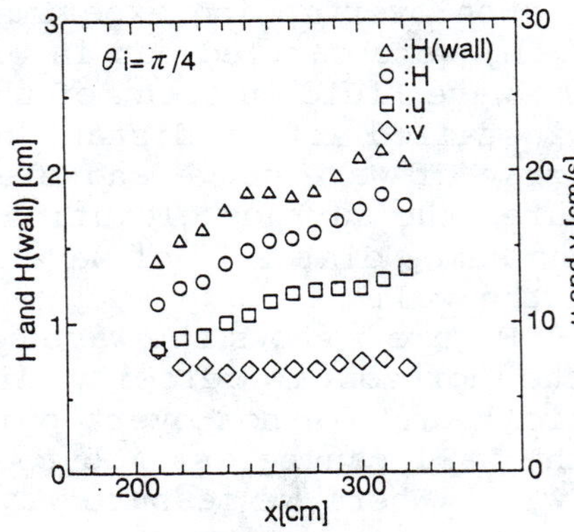

(d) $\theta_i = \pi/4$

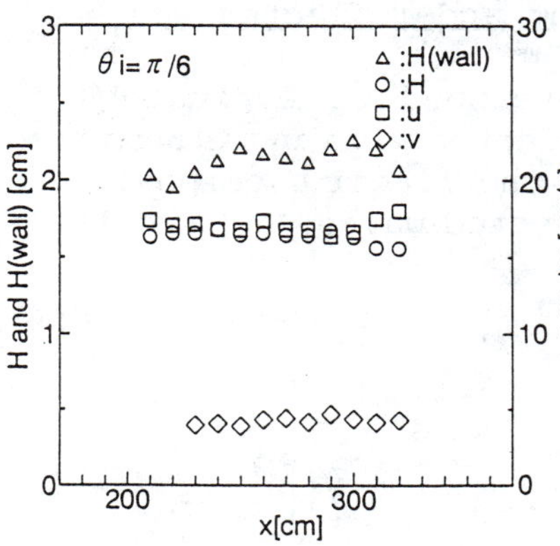

(e) $\theta_i = \pi/6$

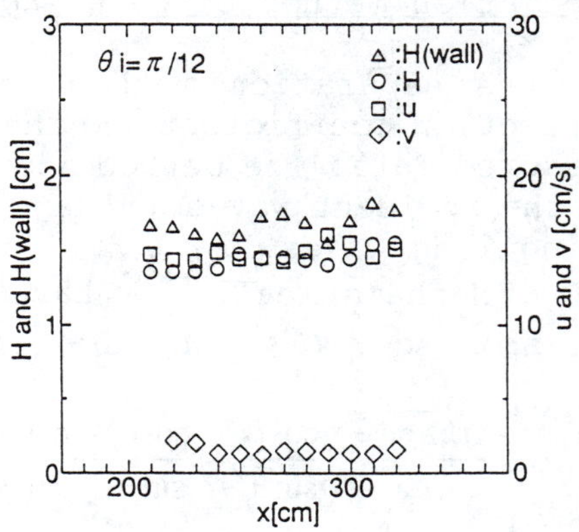

(f) $\theta_i = \pi/12$

Figure 7 Wave height and total amplitude of horizontal velocity distribution in front of the wall.

$$\eta_i = \frac{\sqrt{h/g}v - \eta\sin\theta_r}{\sin\theta_i - \sin\theta_r}\left(= \frac{\sqrt{h/g}u - \eta\cos\theta_r}{\cos\theta_i - \cos\theta_r}\right) \tag{11}$$

and

$$\eta_r = \eta - \eta_i\,. \tag{12}$$

Reflection coefficient is the ratio between the incident wave height and the reflected wave height.

Wave overtopping rate and reflection coefficient

Incident wave heights are H_i = 2.0, 2.5, 3.0 and 3.5 cm in wave overtopping experiments. The wave overtopping discharge are measured by using a wave gauge in the box installed just behind the wall.

Figure 8 shows examples of measured data of surface elevation η and horizontal velocities u and v. Figure 9 shows separated incident waves and reflected waves calculated from equation (10) and (11) together with the time series of α, as using data shown in Fig. 8. Reflected wave angle is calculated by using mean value of α passed from -1 to 1.

Figure 10 shows reflected wave angle obtained from experimental data. When incident wave angle is large, the reflected wave angle shows almost symmetry. However, when incident wave angle is small, the reflected wave angle shows large discrepancy from symmetrical reflected angle. We may consider the reason why the separation method for incident and reflected waves suffers strongly influence from experimental noise. Figure 11 shows incident wave height obtained from experimental data. Incident wave height is almost the same as the target value. When incident wave angle is large, the validity of the separation method was confirmed.

Figure 12 shows measured and calculated wave overtopping rates and reflection coefficients with respect to the incident wave angle. Wave overtopping was not observed, for incident wave height H_i = 2 cm of all incident wave angles and H_i = 2.5 cm of incident wave angle $\theta_i = \pi/4$, $\pi/6$ and $\pi/12$. When wave overtopping occurs, good agreement between the model and the experiment results is found for the wave overtopping rate. The reflection coefficients for large incidence ($\theta_i = \pi/2$, $5\pi/12$ and $\pi/3$) show reasonable values. For small incident wave angle, the reflection coefficient obtained from experimental data did not plot.

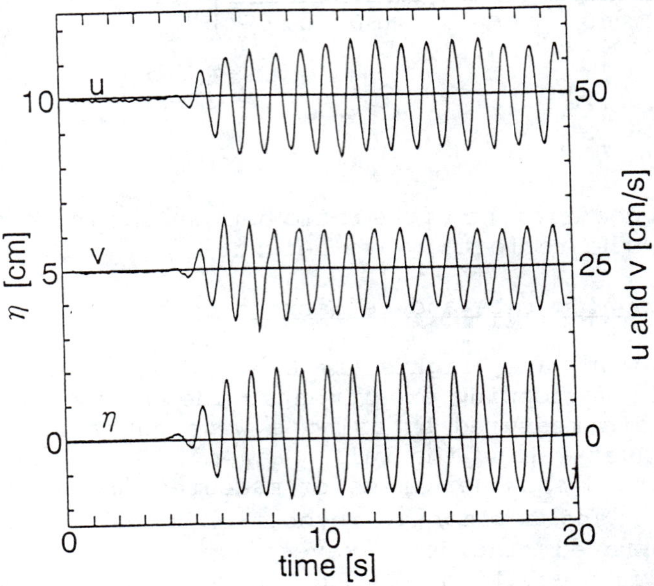

Figure 8 Examples of measured data of surface elevation η and horizontal velocities u and v. ($h=10cm, T=1.0s, H_i=3.0cm, \theta_i=\pi/3$)

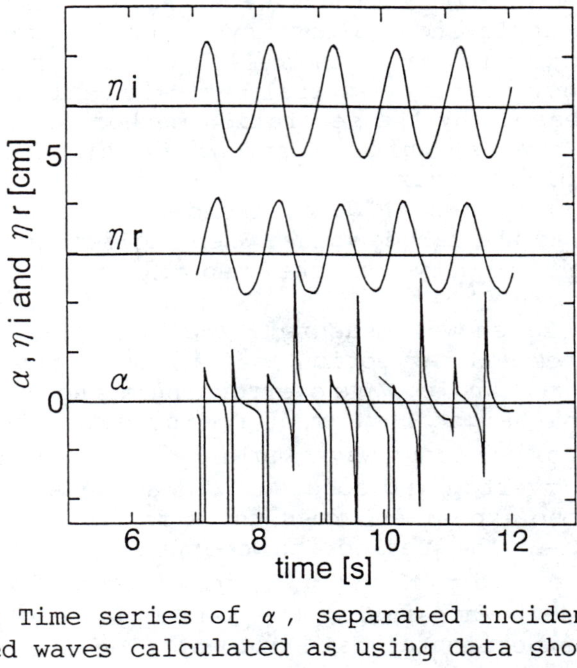

Figure 9 Time series of α, separated incident waves and reflected waves calculated as using data shown in Fig. 6.

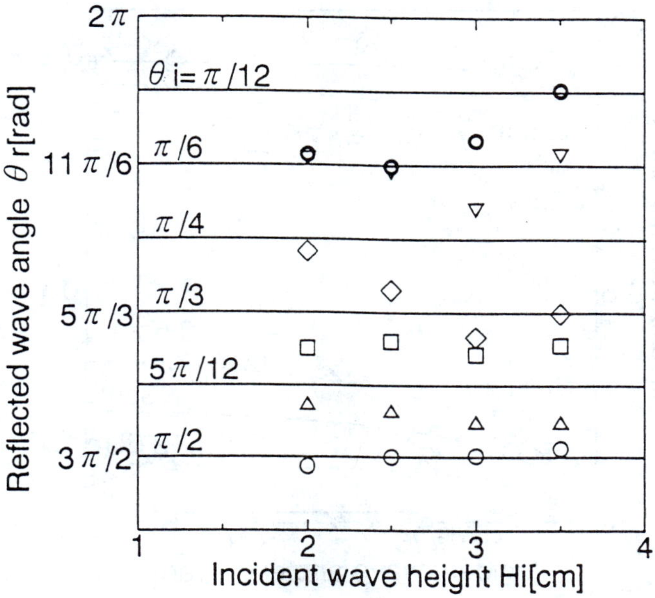

Figure 10 Reflected wave angle obtained from experimental data.(○: $\theta_i = \pi/2$, △: $5\pi/12$, □: $\pi/3$, ◇: $\pi/4$, ▽: $\pi/6$, ◎: $\pi/12$)

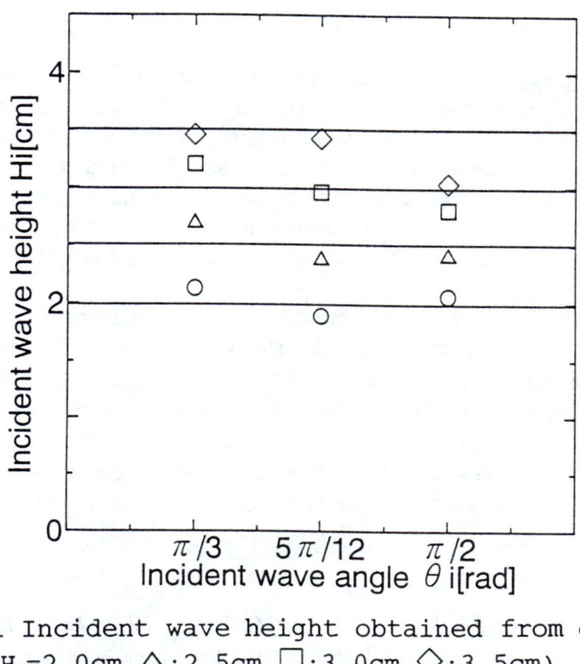

Figure 11 Incident wave height obtained from experimental data.(○: H_i=2.0cm, △:2.5cm, □:3.0cm, ◇:3.5cm)

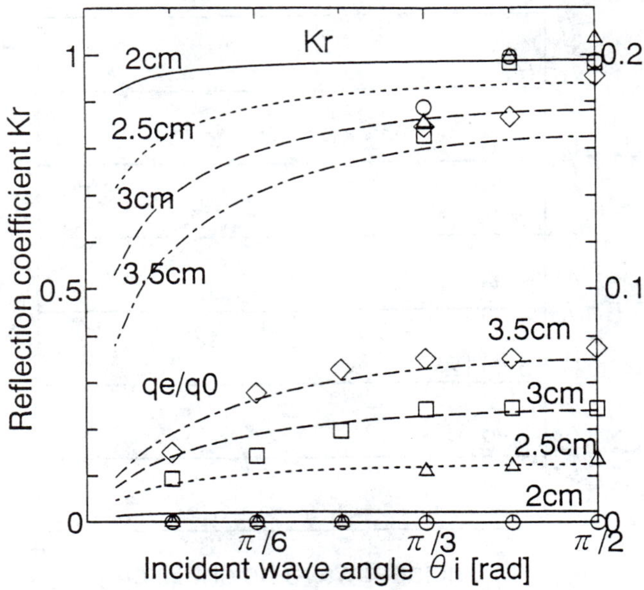

Figure 12 Measured and calculated wave overtopping rates and reflection coefficients with respect to the incident wave angle. (○:H_i=2.0cm, △:2.5cm, □:3.0cm, ◇:3.5cm)

4. Conclusion

Wave overtopping over a vertical wall for the oblique incidence were treated analytically by extending normal incidence (Mizuguchi's) model. Wave overtopping rate and reflection coefficient can be estimated from the extended model which does not include any experimental constants. The results of numerical calculation show the slightly decrease of the wave overtopping rate with the decrease of the incident wave angle.

Experimental confirmation of this model was undertaken by using a multi-directional wave basin. Satisfactory agreement between the model and the experimental results is found for the wave overtopping rate. The reflection coefficient for large incidence ($\theta_i = \pi/2, 5\pi/12$ and $\pi/3$) shows reasonable agreements. The validity of the model was confirmed at least for large incident wave angle.

Reference

de Waal, J. P. and J. W. van der Meer(1992):Wave runup and overtopping on coastal structures, Proc. 23rd, Conf.

Coastal Eng., ASCE, pp.1758-1771.

Inoue, M. and Y. Tuchiya(1971): Experimental study on wave overtopping over a vertical wall for oblique incidence, Proc, 18th, Japanese Conf. on Coastal Eng., JSCE, pp.259-264. (in Japanese)

Isobe, M.(1985): Calculation and application of first order cnoidal wave theory, Coastal Eng., Elsevier, Vol.9, pp.309-325.

Juhl, J. and P. Sloth(1994): Wave overtopping of breakwaters under oblique waves, Proc. 24th, Conf. Coastal Eng., ASCE, pp.1182-1196.

Kikkawa, H., H. Shiigai and T. Kohno(1967): Basic study on wave overtopping for coastal embankment (1), Proc. 14th, Japanese Conf. on Coastal Eng., JSCE, pp.335-338. (in Japanese)

Mizuguchi, M.(1993): Wave overtopping rate over a vertical wall and reflection coefficient, Coastal Eng. in Japan, Vol.36, No.1, pp.37-47.

Mizuguchi, M.(1994): Uniformity of wave field produced by multi-directional wave maker, Abst. 24th, Conf. Coastal Eng., ASCE.

Takayama, T., T. Nagai, K. Nishida and T. Sekiguchi(1984): Experiments on oblique random wave overtopping rates over seawalls, Proc. 31st, Japanese Conf. on Coastal Eng., JSCE, pp.542-546. (in Japanese)

Yokoyama, K. and M. Mizuguchi(1993): Experimental study on wave overtopping over a vertical wall and reflection coefficient, Proc. Coastal Eng., JSCE, Vol.40, pp.676-680. (in Japanese)

CHAPTER 202

Sediment transport and wave reflection near a seawall.

Jonathon R. Miles[1], Paul E. Russell[1], David A. Huntley[1].

Abstract

This paper describes results of a field experiment to examine the effect of wave reflection on suspended sediment transport in front of a seawall. High frequency measurements of wave elevation, velocity and suspended sediment concentrations were made simultaneously on a natural beach and in front of a seawall at Teignmouth in South Devon (U.K.) in June 1995. Wave reflection at the natural beach was found to be dependent on frequency; low frequency waves being preferentially reflected while incident waves were dissipated. At the seawall the incident wave reflection coefficient was 0.9 indicating only a small amount of dissipation. The doubling of energy over the sea bed was found to greatly increase the suspended sediment concentrations in the water column, although the amount of this increase depended on the water depth. A data analysis technique was developed which allowed the incoming and outgoing wave contributions to the sediment transport to be analysed. In these accretionary conditions incoming waves transported sediment onshore in both wall and beach cases, while in the wall case sediment transported offshore by the outgoing waves balanced the onshore transport. Sediment build up which was observed at the top of the natural beach was not observed in front of the wall. Sediment maintained in suspension in front of the wall was available for longshore transport, and this was enhanced by the presence of the wall.

Introduction

Seawalls have been used for many years as a method of coastal protection on eroding shorelines. It has been suggested that the reflection of wave energy over the beach fronting the wall actually helps to erode the beach (Silvester 1977), but the processes controlling this erosion remain poorly understood. There has been considerable debate over the effect of the seawall on the beach in recent years

[1] Institute of Marine Studies, University of Plymouth, Drake Circus, Plymouth, U.K., PL4 8AA.

(Kraus, 1988; Kraus and McDougal, 1996), with experimenters carrying out laboratory experiments with moveable beds (e.g. McDougal et al., 1996) or carrying out surveys of defended and undefended coastlines (e.g. Griggs and Tait, 1988). Researchers into beach processes have identified driving forces behind beach morphology change using high frequency point measurements of suspended sediment concentrations and water velocities (e.g. Jaffe et al., 1984). This approach has been applied to the reflective seawall environment for the first time and the method and initial results are presented in this paper.

Method

To compare similar wave conditions in a surf zone with a reflective wall environment a location was required with a wall adjacent to a natural beach. The site chosen was Sprey point at Teignmouth in South Devon, UK which has a 7m high seawall fronted by a beach of slope 1/15. The sand in front of the wall is medium quartz sand with D_{50}=0.24mm. The beach is macro-tidal and morphodynamically intermediate. A rig designed to minimise interference with the hydro and sediment dynamic conditions whilst providing a stable base for instrumentation was built and bolted to the wall. An extendable rod was positioned on the front of the rig, extending close to the bed, to which the instruments were attached. Two EMCMs, (Electro Magnetic Current Meters) two OBSs (Optical Backscatter Sensors) and a PT (Pressure Transducer) were attached to this rod, 1.5m from the wall in a vertical array. A rig of instruments was also dug in to the adjacent beach on a similar depth contour so that comparisons could be made. This beach site had one EMCM, one OBS and one PT.

All of the *in-situ* monitoring devices were logged at a frequency of 8Hz and were pre-filtered between 3-4Hz. EMCM and PT offsets were determined in the field at the beginning and end of each tide. Surveys were carried out daily to establish beach morphology. Sediment samples were taken for size distribution analysis. During the period of observation and recording, in June 1995, the wind was generally light and the waves were, on average, 30cm in height at the breakpoint with an average period of 4 seconds. The robust design of the rig should enable measurements to be made in waves of up to 2m. The layout of the site is shown in figure 1:

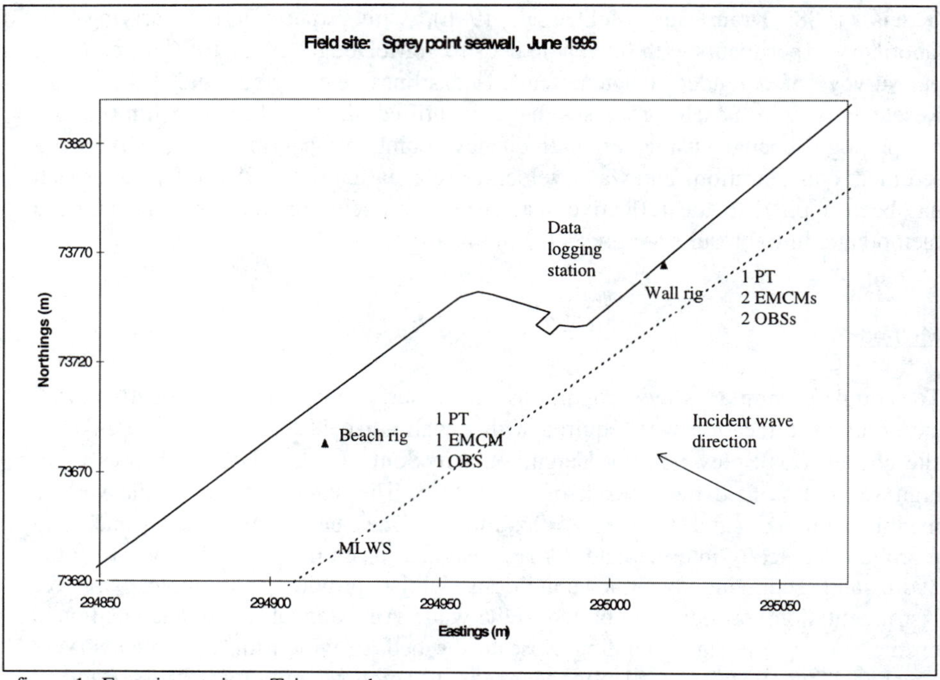

figure 1. Experiment site at Teignmouth.

Instruments were calibrated both before and after the experiment, and the calibrations were applied to the raw data. EMCMs were calibrated in a tow tank at the Royal Naval Engineering College (RNEC) at Manadon, Plymouth (U.K.). Pressure transducers were calibrated for water depth in a 3m deep tank at the RNEC. OBSs were calibrated in the laboratory using sediment samples taken from the experiment site. To calibrate the OBSs a paint stirrer was used to suspend sediment in a tank while samples of the beach sediment were filtered from next to the OBS head. This calibration of the OBSs provided an excellent linear correlation between suspended sediment concentrations and voltage output.

Time series

Time series of surface elevation and suspended sediment concentration are shown in figure 2. Time series have been selected so that water depths are similar in the wall and beach examples, in this case the water depth is approximately 0.8m. Instruments suspended from the wall are 1.2m away from it. This puts them closer to the antinode at the wall than the node 2.7m away from the wall (for a wave period of 4 seconds). The maximum surface elevation deviation from the mean is therefore clearly larger than that observed at the beach rig. Suspended sediment data which is compared in these examples is taken from OBSs at similar heights, both being approximately 18cm from the bed. The scale of the axis of the wall OBS is reduced so that the data fits the plot. In similar incident wave conditions levels of suspended sediment are clearly much larger at the wall, reaching 4 kg/m^3 at this height above the bed, while at the beach OBS sediment concentrations only reach 0.6 kg/m^3.

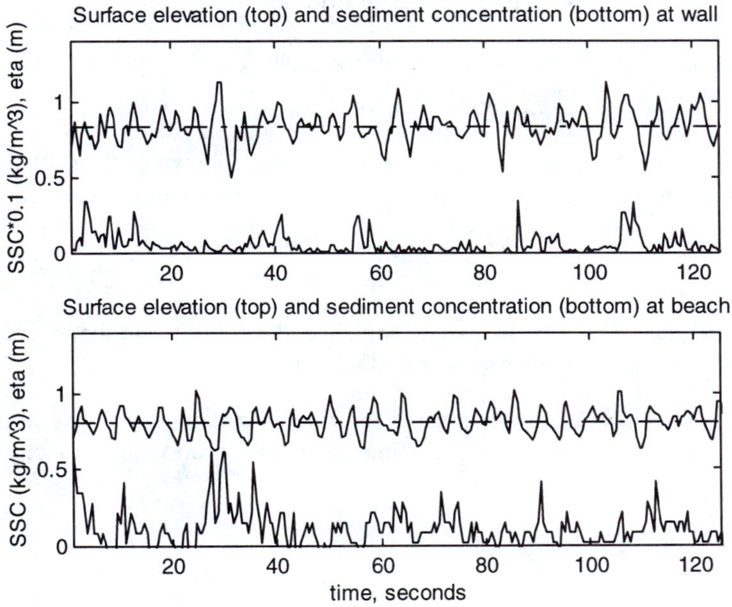

figure 2. Time series of surface elevation and suspended sediment concentrations at the wall and beach rigs. N.B. The sediment concentrations in the wall time series have been scaled down by a factor of 10.

Theory

Incoming and outgoing elevation time series were obtained from elevation and velocity time series using the method of Guza, Thornton and Holman (1984). The separation of incoming and outgoing waves forms the basis for most reflection analysis techniques which involve co-located elevation and velocity sensors. The analysis can also be extended to obtain the incoming and outgoing velocity time series. This allows the frequency dependent oscillatory sediment transport associated with the incident reflected waves to be found.

The derivation of Guza et al.'s (1984) time domain technique arises from linear shallow water theory:

The velocity potential for progressive water waves is as follows:

$$\phi = \frac{-a.g}{\sigma}.\frac{\cosh k(z+h)}{\cosh kh}\sin(kx-\sigma t)$$

The surface elevation can be obtained from this by differentiating with respect to time, and the attenuation term tends to 1 in shallow water.

$$\eta = \frac{1}{g}\left[\frac{\partial \phi}{\partial t}\right]_{z=0} \qquad \text{giving} \qquad \eta = a\cos(kx-\sigma t)$$

A similar equation for velocity is obtained by differentiating with respect to x

$$u = -\frac{\partial \phi}{\partial x} \qquad \text{gives} \qquad u = \frac{gak}{\sigma}\cos(kx-\sigma t)$$

By dividing the elevation and velocity equations the following time domain transformation can be obtained relating velocity and surface elevation (provided the water is shallow).

$$\eta(t) = \frac{c}{g}u(t)$$

It is necessary to transform to the frequency domain and use full linear theory if this relationship is to be applied in intermediate or deep water.

Progressive wave surface elevation traces for waves incoming to a beach and outgoing from it can be obtained from the time series of elevation and velocity as follows:

$$\eta_{in}(t) = \left(\eta(t) + \frac{c}{g}u(t)\right)/2$$

$$\eta_{out}(t) = \left(\eta(t) - \frac{c}{g}u(t)\right)/2$$

In this case, a positive velocity implies an onshore flow, while a negative velocity an offshore flow. In order to obtain a frequency dependent reflection function (FDRF) it is necessary to carry out a spectral analysis of the incoming and outgoing elevation traces separately. After obtaining the power spectra of the incoming and outgoing elevation time series, $S_{ii}(f)$ and $S_{oo}(f)$ respectively, the frequency dependent reflection coefficient is simply:

$$R(f) = \sqrt{\frac{S_{ii}(f)}{S_{oo}(f)}}$$

This technique was examined for signal noise related bias by Huntley et al. (1995) who found that when the coherence between elevation and velocity was high, the bias in the FDRF was low. Two other techniques were discussed by Huntley et al. (1995). These were a frequency domain method by Tatavarti (1989) and a Principal Component Analysis method (Tatavarti et al., 1988). All three techniques were applied to the data in order that the frequency dependent reflection coefficient for the beach and wall cases could be identified. The three different methods were found to offer similar results. This gives confidence in the time domain method which is the most simple to program and extend for incoming and outgoing sediment analysis.

In order to understand the effect of the reflected wave field on the suspended sediment transport, it is possible to carry out a cross-spectral analysis of the incoming and outgoing waves with the sediment.

It is first necessary to apply the transformation from elevation to velocity to obtain incoming and outgoing velocity time series. The appropriate transformations are:

$$u_{in} = \frac{g}{c}\eta_{in}$$

$$u_{out} = -\frac{g}{c}\eta_{out}$$

A wave crest (elevation maximum) correlated with onshore flow therefore represents a wave crest travelling onshore, while a wave peak correlated with an offshore velocity represents a wave crest travelling offshore. The equations for incoming and outgoing velocity are therefore:

$$u_{in} = \frac{1}{2}(u + \frac{g}{c}\eta)$$

$$u_{out} = \frac{1}{2}(u - \frac{g}{c}\eta)$$

It is next necessary to assume that the oscillatory sediment flux can be split into incoming and outgoing components. Jaffe et al. (1984) assumes a similar principle when considering the breakdown of the mean and oscillatory fluxes. They assume that the total velocity can be split into a mean and oscillatory component:

$$U = \bar{u} + u'$$

and that the sediment concentration is also separable into mean and oscillatory parts:

$$C_s = \bar{c}_s + c'_s$$

The time average of the flux therefore reduces to the mean flux and the flux coupling:

$$\overline{UC_s} = \overline{\bar{u}\bar{c}_s} + \overline{u'c'_s}$$

Signals are routinely de-meaned, and the remaining step is to split the oscillatory velocity into incoming and outgoing components before crossing with the suspended sediment to obtain the flux:

$$u' = u'_{in} + u'_{out}$$

The time average of the incoming and outgoing oscillatory flux is therefore:

$$\overline{u'c'_s} = \overline{u'_{in}c'_s} + \overline{u'_{out}c'_s}$$

Huntley and Hanes (1987) identified that the frequency dependent sediment transport could be found by taking the co-spectrum of the oscillatory cross shore velocity with the sediment concentration time series. The frequency dependent sediment transport associated with the incoming and outgoing waves is therefore obtained from the co-spectrum of the incoming or outgoing velocity time series with the sediment time series. The contributions of the incoming and outgoing waves to the sediment transport can therefore be identified.

Frequency dependent wave reflection

Three methods were used to determine levels of energy incoming and outgoing from the natural beach and the wall. Incoming and outgoing wave spectra were obtained from the incident and reflected time series obtained using Guza et al.'s (1984) method. The frequency dependent reflection coefficient was then determined using this method and two frequency domain methods. All three methods require that the current meter and pressure transducers are co-located and log data simultaneously. The analysis was first applied to data from the beach rig (figure 3).

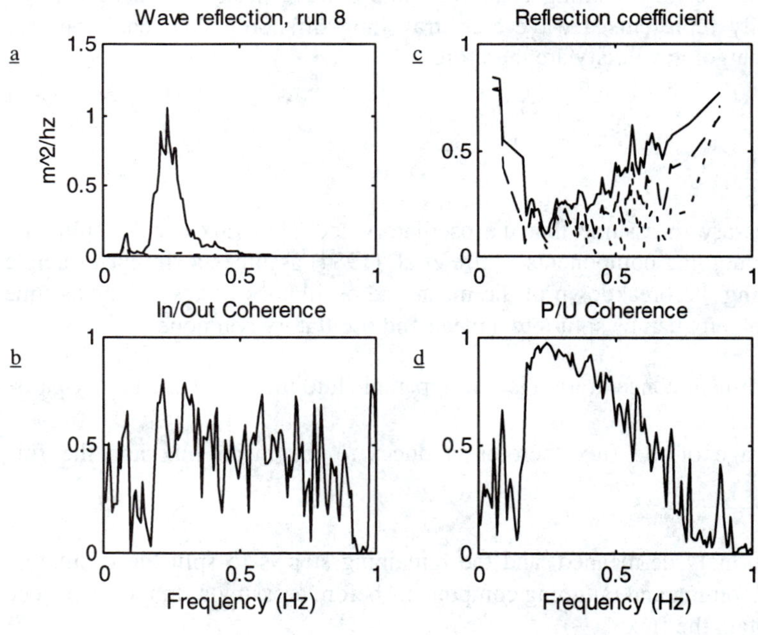

figure 3. Frequency dependent wave reflection at the beach. a: Incoming and outgoing wave spectra calculated using the time domain method. b: Coherence between calculated incoming and outgoing time series. c: Frequency dependent reflection coefficient calculated using time domain (solid), frequency domain (dashed) and principle component analysis (dotted) methods. d: Coherence between elevation and velocity. Water depth is 1.28m.

The energy spectra show clearly that the incident waves (frequency 0.25 Hz) are dissipated by the natural beach, while at a lower frequency (0.1Hz) wave reflection is taking place. By dividing the square root of the spectral estimates, the FDRF for this time domain method was obtained, and this shows a reflection coefficient of 0.1 for the incident waves on the natural beach while for the low frequency waves the reflection coefficient is 0.8. Of the three lines on the reflection coefficient figure above (top right). the lines show the methods of the time domain method (top), the frequency domain method (middle) and the Principal Component Analysis (bottom). The three methods give good agreement where there is good coherence between elevation and velocity.

A similar analysis was carried out to data from the wall rig. The results are shown in figure 4.

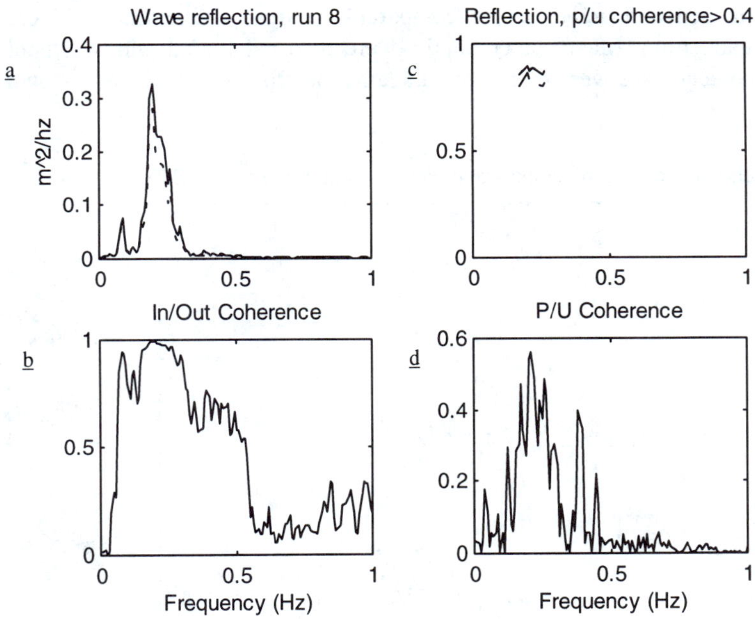

figure 4. Frequency dependent wave reflection at the wall. a: Incoming and outgoing wave spectra calculated using the time domain method. b: Coherence between calculated incoming and outgoing time series. c: Frequency dependent reflection coefficient calculated using time domain (solid), frequency domain (dashed) and principle component analysis (dotted) methods. Reflection estimates are plotted for values of P/U coherence > 0.4 so that there is 95% confidence in the coherence between P and U. This is necessary as cross spectral analysis is carried out in the determination of the FDRF using the frequency domain methods. d: Coherence between elevation and velocity. Water depth is 1.30m.

Similar amounts of energy were found in the spectra of both incoming and outgoing wave time series, indicating that reflection of the incident waves was occurring at the wall. At incident wave frequency all three methods of determining the frequency dependent reflection coefficient gave a value of 0.9 at the wall. The wall is in fact slightly sloping at the position of the rig and this may account for the reflection coefficient being less than unity.

Mean suspended sediment concentrations

The next part of the investigation was to examine the effect of the increase in reflection coefficient at the shoreline on concentrations of suspended sediment. To do this, data from OBSs on the beach and wall rig were compared. During the experiment OBSs were carefully positioned on each rig so that they were at the same height above the bed and are therefore comparable. Mean concentrations were calculated for each run. Mean water depths were also calculated and sediment concentrations in each case were then plotted against depth. The results are shown in figure 5.

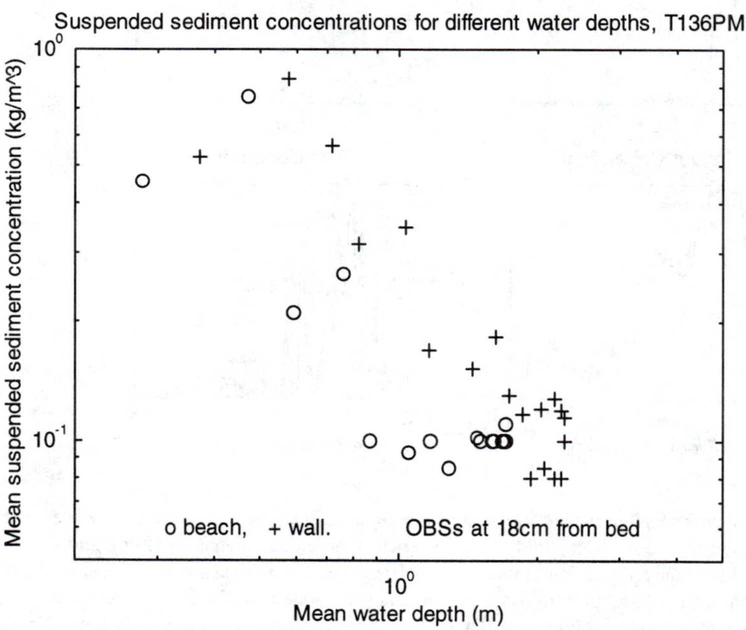

figure 5. Mean suspended sediment concentrations - comparison of data from wall and beach OBSs. Both instruments were mounted approximately 18cm from the bed. Incident waves were approximately 30cm high at the breakpoint.

The results show a greater concentration of sediment was in suspension in front of the wall than was in suspension on the natural beach when water depths were the same. Incoming wave conditions did not vary significantly during this tide, and the results therefore imply that it is the increase in wave reflection at the shoreline which must be responsible for increasing the concentrations of suspended sediment in front of the wall. The distinct linear trend in the data, especially in the wall rig data suggests that it may be possible to parameterise the mean suspended sediment concentrations in this region in terms of the reflection coefficient, water depth, height above bed and incident wave height. Further experimentation with a larger array of instruments would be necessary for this however.

Frequency dependent sediment transport

The incoming and outgoing frequency dependent sediment transport was obtained for both wall and beach rigs for each run of data logged. A clear picture emerged in all runs as to the nature of this transport and this is shown in figure 6.

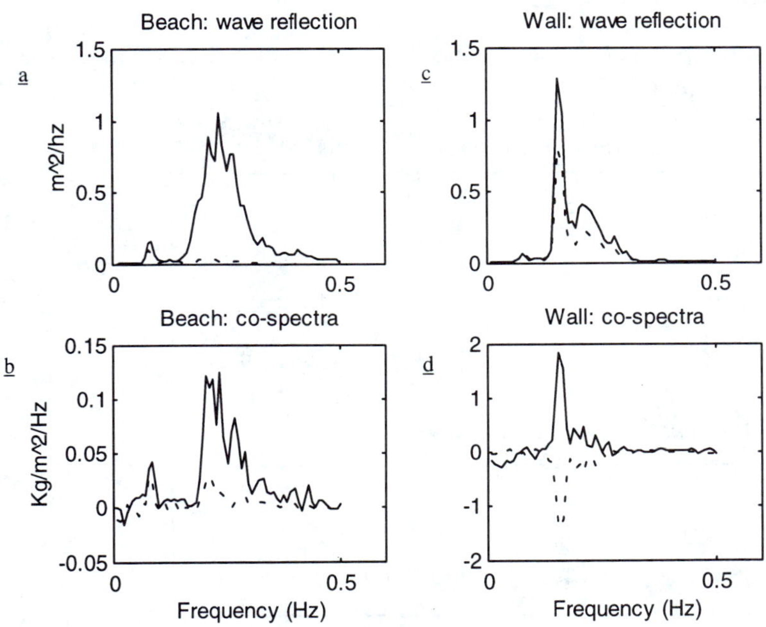

figure 6. Incoming and outgoing spectra and co-spectra in beach (a,b) and wall (c,d) cases. In the energy spectra plotted above, the solid lines represent incoming energy, the dashed lines represent outgoing energy. In the co-spectra, positive is onshore, solid lines are the cospectra between incoming waves and the sediment, dashed lines are cospectra between outgoing waves and sediment. Water depth is 1.3m

In the beach case (figure 6b), incoming waves are found to transport sediment onshore. This result is similar to that observed by Huntley and Hanes (1987). Suspension events occur at the same time as onshore directed flows, and the resultant transport is onshore. There is little energy outgoing from the beach however, and there is therefore little transport associated with the offshore directed waves.

At the wall, the situation is rather different (figure 6d). The incident waves are still responsible for onshore sediment transport, while the outgoing waves give rise to offshore transport. The onshore transport by the incoming waves is effectively halted by the outgoing waves. The net cross-shore oscillatory transport is therefore reduced. The fact that there is more sediment in suspension in front of the wall is also evident in this analysis as the magnitudes of the co-spectra are considerably larger in the wall case than in the beach case.

Longshore transport comparison

Mean longshore transport rates were calculated for both wall and beach rigs. The results are shown in figure 7.

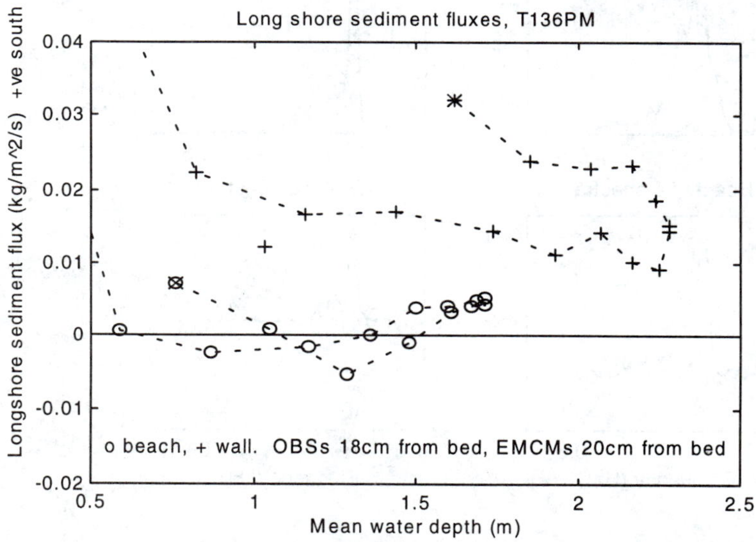

figure 7. Longshore transport rates measured at the beach and wall.

Longshore transport rates in front of the wall were found to be greater than at the beach rig at all times, and this was attributed to both the increase in suspended sediment concentrations and also an increase in the longshore current which passed in front of the wall. The south-westward direction of longshore transport in the wall case is a result of the oblique wave approach from the east (see figure 1).

During the period of observation, morphology changes were monitored using surveying techniques. Increased levels of longshore transport along the Sprey point wall resulted in the development of a small bar to the south west of the wall. (extending SW from point 294990,73740; see figure 1). Shoreward of the beach rig, a net accretion was observed. This is likely to have resulted from the onshore transport by the incident waves. At the wall rig however there was no net accretion or erosion. The balance in transport between the incoming and outgoing waves prevented accretion taking place in this region, while increased levels of sediment in suspension led to an enhanced longshore transport of sediment in front of the wall.

Conclusions

This paper has described initial results of a field experiment to examine the effect of wave reflection on suspended sediment transport in front of a seawall. High frequency measurements of wave elevation, velocity and suspended sediment concentrations were made simultaneously on a natural beach and in front of a seawall. The doubling of energy over the sea bed was found to greatly increase the suspended sediment concentrations in the water column, although the amount of this increase depended on the water depth. A data analysis technique was developed which allowed the incoming and outgoing wave contributions to the sediment transport to be analysed. Incoming waves transported sediment onshore in both wall and beach cases, while in the wall case outgoing waves balanced the onshore transport. Sediment which was maintained in suspension in front of the wall was available for longshore transport, and this was enhanced by the presence of the wall.

Symbols

- a wave amplitude
- c wave celerity
- c_s sediment concentration
- f wave frequency
- g gravitational acceleration
- h water depth
- k wave number $(2\pi/\lambda)$
- t time
- u horizontal cross shore velocity
- x horizontal co-ordinate
- z vertical position. $z = 0$ at the surface, $z = -h$ at the bed
- $R(f)$ frequency dependent reflection coefficient
- S_{ii} power spectrum of incoming wave time series
- S_{oo} power spectrum of outgoing wave time series
- T wave period

σ angular frequency ($2\pi/T$)
η surface elevation
λ wavelength
ϕ velocity potential
$_{in}$ denotes incoming wave
$_{out}$ denotes outgoing wave

References

Griggs, G.B. and Tait, J.F., 1988. The effects of coastal protection structures on beaches along northern Monterey Bay, California. Journal of Coastal Research, Special Issue No. 4, 93-111.

Guza, R.T., Thornton, E.B. and Holman, R.A., 1984. Swash on steep and shallow beaches, Coastal Engineering, 708-723.

Huntley, D.A. and Hanes, D.M., 1987. Direct measurement of suspended sediment transport. Proceedings, Coastal Sediments '87, ASCE, 723-737.

Huntley, D.A., Simmonds, D. and Tatavarti, R.S.V.N., 1995. On the measurement of coastal wave reflection using co-locted elevation and current sensors. Proceedings, Coastal Dynamics, 57-68.

Jaffe, B.E., Sternberg, R.W. and Sallenger, A.H., 1984. The role of suspended sediment in shore normal beach profile changes. Proceedings, Coastal Engineering, 1983-1996.

Kraus, N.C., 1988. The effects of seawalls on the beach: extended literature review. Journal of Coastal Research, Special Issue No. 4, 1-28.

Kraus, N.C. and McDougal, W.G., 1996. The effects of seawalls on the beach: part 1, an updated literature review. Journal of Coastal Research, 12 (3), 691-701.

McDougal, W.G., Kraus, N.C. and Ajiwibowo, H., 1996. The effects of seawalls on the beach part II, numerical modelling of SUPERTANK seawall test. Journal of Coastal Research, 12 (3), 702-713.

Silvester, R., 1977. The role of wave reflection in coastal processes. Proceedings of Coastal Sediments '77, American Society of Civil Engineers, 639-652.

Tatavarti, R.S.V.N., 1989. The reflection of waves on natural beaches. Unpub. PhD thesis. Dalhousie University, Halifax, Nova Scotia, Canada. 175pp.

Tatavarti, R.S.V.N. and Huntley, D.A., 1987. Wave reflections at beaches, Proceedings, Canadian Coastal Conference., Quebec City, Canada, 241-256.

Acknowledgements

The authors would like to thank Dr Mark Davidson at the University of Plymouth for helpful suggestions throughout the study, and Mr Peter Ganderton for his assistance with the field-work phase of the project.

Initial nourishment at Palm Beach, Florida. Photo courtesy of Applied Technology & Management, Inc. and Aerial Photography, Inc.

PART IV

Coastal Processes and Sediment Transport

One year after nourishment at Palm Beach, Florida. Photo courtesy of Applied Technology & Management, Inc. and Aerial Photography, Inc.

CHAPTER 203

WAVE ENERGY DISSIPATION IN KELP VEGETATION

Alfonse Dubi[1] and Alf Tørum[2]

ABSTRACT

A laboratory experiment was carried out to investigate wave energy dissipation in a coastal kelp field of *Laminaria hyperborea* with a purpose of finding the damping rate. Parameters measured were wave heights at 8 locations along the direction of wave propagation, forces on artificial kelp plants, and velocities in between the plants in 4 different water depths. It was found that in 4, 6, 8 and 10 meter water depths, the damping coefficient was 0.0094, 0.0032, 0.0014 and 0.0005m^{-1} respectively. Results show that the damping rate varies with population density, size of plant, water [1]depth and wave period.

1. INTRODUCTION

Through a variety of processes, waves lose energy. Among the commonly known processes are wave breaking, bottom friction, reflection and interaction with porous beds and soft muds. Waves propagating through vegetation also lose energy due the work done on the vegetation.

Following Ippen (1966), the wave amplitude attenuation can be evaluated through the steady-state conservation of wave energy flux equation:

$$\nabla \cdot \left(E\vec{C}_g \right) = -E_D \quad \text{(1.1)}$$

where $E = \frac{1}{2}\rho g a^2$ is the wave energy density, ρ is the density of water, C_g is the group velocity and E_D is the energy dissipation rate. The local wave amplitude a is found to decay exponentially (see for example, Asano et al., 1992) as

$$a(x) = a_o e^{-k_i x} \quad \text{(1.2)}$$

where k_i is the damping rate (also known as the damping coefficient) of the wave height with distance and $a = a_o$ at $x = 0$.

In a study on the interaction between ocean waves and a kelp farm, Dalrymple et al. (1982) arrived at a wave height attenuation formula of the form:

---

[1] Research Fellow, Institute of Marine Sciences, University of Dar Es Salaam, P.O. Box 668 Zanzibar, Tanzania.

[2] Professor/Head of Research, Dept. of Structural Eng., NTNU/SINTEF-NHL, N-7034 Trondheim, Norway.

$$\frac{a(x)}{a_o} = \left(\frac{1}{1+\alpha_d x}\right) \qquad (1.3)$$

where

$$\alpha_d = \frac{2C_D}{3\pi}\left(\frac{d_k}{b}\right)\left(\frac{a_o}{b}\right)\left[\sinh^3 kd + 3\sinh kd\right]\left[\frac{4k}{3\sinh kD(\sinh 2kD + 2kD)}\right] \qquad (1.4)$$

in which only the real part of the wave number k is used. C_D is the drag coefficient assumed constant over depth, d is the height of plant above the bottom, d_k is the plant diameter, and b is the spacing between plants. This approximation, however, takes plants as rigid cylinders; an assumption which results in assigning different values to the drag coefficient to cover the ignorance of the plant motion. Dalrymple et al. (1982) applied Eq.(1.4) for waves propagating through *Macrocystis pyrifera* kelp forest using known values of $C_D = 1.0$, $d_k = 0.3$ meters, $b = 1.5$ meters and $D = 15$ meters. They found that incident wave heights of 6.1 meters with 10-20 seconds periods could be reduced by 50% over a distance of 800 meters. This is equivalent to $\alpha_d = 0.0013m^{-1}$ and applying the exponential law of attenuation, $k_i = 0.0009\ m^{-1}$.

Recently, Kobayashi et al. (1991) developed an analytical model, hereafter known as the Kobayashi model, to describe a 2D problem of small amplitude waves propagating over submerged or subaerial vegetation. They assumed the effect of the vegetation to be expressible in terms of the drag resistance against the fluid motion. The vertical component of the drag resistance and the proximity effects of the surrounding strips on C_D were neglected. Although an analytical solution was obtained, the calibrated drag coefficient C_D varied widely due to the fact that the swaying motion of an individual vegetation stand had been neglected.

Later, Asano et al. (1992) improved the Kobayashi model by including the interaction between the wave and the vegetation motion. Whereas the Kobayashi gave the calibrated values of C_D to the order of 0.1, the model of Asano et al. produced C_D of the order unity and the vertical force was neglected while the horizontal drag force was assumed to be dominant and linearized as

$$F_x = \rho D_k u_2 \qquad (1.5)$$

where D_k is a linear force coefficient determined by the least squares method, u_2 is the horizontal particle velocity in the vegetation zone. Solution of the linear model gives the dispersion equation as

$$\omega^2 = gk\frac{k\tanh kh + \alpha_k \tanh\alpha_k d}{k + \alpha_k \tanh\alpha_k d \tanh kh} \qquad (1.6)$$

in which ω is the angular frequency, d is the vegetation height, h is the water depth above vegetation and

$$\alpha_k^2 = \frac{\omega k^2}{\omega + iD_k} \qquad (1.7)$$

In the case of weak damping, $\omega >> D_k$, separation of the real and imaginary parts of the dispersion equation yields the following

$$\omega^2 = gk_r \tanh k_r(h+d) \quad \text{.......(1.8)}$$

which implies that the presence of vegetation does not change the real wave number k_r as long as $\varepsilon_k \ll 1$. The damping rate is then approximated by

$$k_i = \varepsilon_k k_r \frac{2k_r d + \sinh 2k_r d}{2k_r(h+d) + \sinh 2k_r(h+d)} \quad \text{.......(1.9)}$$

$$\text{where } \varepsilon_k = \frac{D_k}{2\omega} \quad \text{.......(1.10)}$$

On the other hand, applying the energy conservation equation, Eq.(1.1), and invoking Eq.(1.2), the Kobayashi model gives the damping rate as

$$k_i = \frac{D_k}{C_g}\left(\frac{2k_r d + \sinh 2k_r d}{2\sinh 2k_r(h+d)}\right) \quad \text{.......(1.11)}$$

A more recent theoretical model based on limited experimental data is that of Wang and Tørum (1994) in which the dispersion equation is given by

$$\omega^2 = \frac{\tanh kh - \dfrac{i}{\alpha f_x}\tanh k_s h_s}{1 - \dfrac{i}{\alpha f_x}\tanh k_s h_s \tan kh} \quad \text{.......(1.12)}$$

$$\text{where } \alpha = \sqrt{\left|\frac{f_z}{f_x}\right|} \text{ is the force ratio} \quad \text{.......(1.13)}$$

and $k_s = \alpha k$; f_x and f_z are linearized horizontal and vertical force coefficients respectively, h_s = height of plant. The complex wave number is solved by iteration for known parameters and an assumed value of $\alpha = 0.8$.

Results from this model show that the damping coefficient increases with increasing wave period until it reaches an asymptotic values of just below *0.0015*m^{-1}, *0.0020*m^{-1} and *0.0025* m^{-1}, for densities of 8, 12 and 16 *plants/*m^2.

In this paper three different ways of finding the damping rate are used and compared. The first way is to get an analytical expression of the energy dissipation rate E_D using the linearized dissipative force and then apply Eq. (1.1). The second way, is to get a numerical value of the dissipation rate from time series of the horizontal velocities measured in between the plants and the corresponding horizontal force on the plants and apply Eq.(1.1) again. The third way, is to solve the complex dispersion equation for the complex wave number k by iteration as propsed by Wang and Tørum, (1994).

2. THE DAMPING RATE

2.1 Energy dissipation and the damping rate

Energy dissipation considered to be real and positive takes place mainly in the vegetation zone and partly in the boundary layer at the interface and near the solid bottom. For a given volume of the vegetation zone, energy dissipation during a time period T is given by (e.g Madsen, 1974)

$$E_D = \frac{1}{T}\int_t^{t+T} dt \int_V \vec{F}\cdot\vec{U}dV \qquad (2.1)$$

where $\vec{F} = \vec{F}$ (x,y,z,t) = $(F_x,0,F_z)$ is the dissipative force vector in the vegetation zone and $\vec{U} = (u_2,0,w_2)$ is the complex velocity vector for the 2D case. Defining the force and velocity vectors as (**i, j, k**) in the (x, y, z) directions as

$$\left.\begin{aligned}\mathbf{F} &= F_x\,\mathbf{i} + F_z\,\mathbf{k}\\ \mathbf{U} &= u\,\mathbf{i} + w\,\mathbf{k}\end{aligned}\right\} \qquad (2.2)$$

and substituting into Eq.(2.1) we get

$$\left.\begin{aligned}E_D &= \frac{1}{T}\int_t^{t+T} dt \int_{-(h+d)}^{-(h+l_k)} (F_x.u_2 + F_z.w_2)dz\\ &= \frac{1}{T}\int_t^{t+T} dt \int_{-(h+d)}^{-(h+l_k)} F_x.u_2(1+\alpha^2(\frac{w_2}{u_2}))dz\end{aligned}\right\} \qquad (2.3)$$

The second term under the integral in Eq.(2.3) is found to be at least of the order α^2 and if we assume that $w_2/u_2 = O(\alpha^2) << 1$, which is generally true in coastal waters (Mei, 1982), the time averaged energy dissipation then becomes

$$E_D = \frac{1}{T}\int_t^{t+T} dt \int_{-(h+d)}^{-(h+l_k)} F_x.u_2 dz + O(\alpha^4) \qquad (2.4)$$

The dissipative force is linearized and expressed as

$$F_x = \rho f_{dx} u_2 \qquad (2.5)$$

where f_{dx} is a linear drag force coefficient (Dubi, 1995) given as

$$f_{dx} = \frac{4n_k\alpha_r F_\lambda k_s|1-A_m|\cosh^2 k_s h_f}{\rho(2k_s d + \sinh(2k_s d))} \qquad (2.6)$$

where n_k is the number of plants per unit horizontal area, α_r is a force reduction factor accounting for group shielding and is found experimentally as a fitting coefficient equal to 0.08. F_λ is an empirical force coefficient evaluated at the canopy level, i.e. at

$z = -(h + l_k)$ and $h_f = (d-l_k)$ where l_k is half length of the kelp frond. A_m is the velocity amplification factor obtained from the solution for the vegetation motion. The complex horizontal velocity in the vegetation zone is given in Dubi and Tørum (1994) as

$$u_2 = -\frac{iCk}{\omega f_x} \cosh(k_s(h+d+z)) \exp[i(kx - \omega t)] \qquad \text{2.7)}$$

where

$$C = \frac{ga}{\cosh(k_s d) \cosh(kh) \left[1 - \frac{i}{\alpha f_x} \tanh(k_s d) \tanh(kh) \right]} \qquad (2.8)$$

Substituting Eq.(2.7) into Eq.(2.4) we get the linearized form of energy dissipation, after time averaging, as

$$E_D = \frac{\rho f_{dx}}{2} \int_{-(h+d)}^{-(h+l_k)} |u_2|^2 dz = \frac{\rho f_{dx} |C|^2 |k|^2}{2\omega^2 |f_x|^2} \frac{\sinh(2k_s h_f) + 2k_s h_f}{4k_s} \qquad (2.9)$$

in which only the real part is considered. Substituting Eq.(2.9) into Eq.(1.1) for the 2D case we have

$$\frac{\partial}{\partial x}(EC_g) = \frac{1}{2}\rho g C_g \frac{\partial a^2}{\partial x} = -Ba^2 \qquad (2.10)$$

in which C_g is assumed to be constant due to constant water depth and

$$B = \frac{\rho g^2 |\alpha k|^2 f_{dx}}{2\omega^2 F^2 |\alpha fx - iG|^2} \left(\frac{\sinh(2k_s h_f + 2k_s h_f}{4k_s} \right) \qquad (2.11)$$

where $F = \cosh k_s d \cosh kh$ and $G = \tanh k_s d \tanh kh$. The solution of Eq.(2.10) is

$$\frac{a}{a_0} = e^{-\frac{B'}{2}x} \qquad (2.12)$$

where $B' = B/\tfrac{1}{2}\rho g C_g$. Equating Eq.(2.12) and Eq.(1.2) we find the damping rate to be

$$k_i = \frac{g|\alpha k|^2 f_{dx}}{2\omega^2 F^2 |\alpha f_x - iG|^2 C_g} \left(\frac{\sinh(2k_s h_f + 2k_s h_f}{4k_s} \right) \qquad (2.13)$$

in which we shall adopt $|k| \cong k_r$ derived from the solution of the dispersion equation.

2.2 Solution of the complex dispersion equation and the damping rate.

The dispersion equation contains complex unknown variables k and f_x. The angular frequency ω is real and assumed to be known. The solution for the complex dispersion equation is found by iteration when Eq.(1.12) is rewritten as:

$$F(k, f_x) = \omega^2 (1 - \frac{i}{\alpha f_x} \tanh(k_s d) \tanh(kh)) - gk(\tanh(kh) - \frac{i}{\alpha f_x} \tanh(k_s d)) = 0 \;.....(2.14)$$

The iteration scheme is carried out as follows:

(i) As a starting approximation we apply Eq.(2.14) for very small damping force coefficient, that is, for $f_x \rightarrow -i$, leading to an approximate solution for the wave number k_o for given α, ω, h and d. Then $k_1^{(0)}$ = real(k_o)

(ii) With an initial guess $k_2^{(o)}$ for the imaginary part of k, we solve Eq.(2.14) with $k^{(o)} = k_1^{(o)} + ik_2^{(o)}$ as initial value input . The solution gives $k_r = Re(k)$ and $k_i = Im(k)$. The convergence criterion for the case is

$$\left| \frac{k^{(n)} - k^{(n-1)}}{k^{(n)}} \right| \leq e(k) \;.....................(2.15)$$

with $e(k)$ being an arbitrary small number which in our case is set equal to 10^{-5}. Results of the damping coefficient computed from the dispersion equation shall be compared with those obtained from the energy dissipation equation.

2.3 The damping rate from measured force and horizontal particle velocity

The damping rate can also be evaluated by making use of the measured force and horizontal velocity time series. Integrating the product of these two quantities and time averaging gives the energy dissipated per unit time as

$$E_D = \frac{1}{AT} \int_0^T F_m U_m \cos^2 \omega t \;\; dt = \frac{1}{2} \frac{F_m U_m}{A} \;.....................(2.16)$$

where F_m is the total measured force amplitude and U_m is the measured horizontal velocity amplitude. To get energy dissipation per unit horizontal surface area we have divided by $A = 8\ m^2$, which is the horizontal surface area of the shear plate used in the experiment. Inserting Eq.(2.16) into Eq. (1.1) we find that

$$k_i = \frac{1}{A} \frac{F_m U_m}{\rho g C_g a_o^2} \;.....................(2.17)$$

in which ρ, g , C_g and maintain the same definition as in previous formulae, converted to full scale. For irregular waves we shall take significant force and significant velocity.

3. LABORATORY EXPERIMENT

The experiment was carried out in a 33 m long, 1 m wide and 1.6 m high wave tank as shown in Figure 1. The width of the channel was partitioned to give a width of 0.5m. Five thousand models (scale 1:10) of *Laminaria hyperborea* plants were fixed on the bottom over a span of 9.3 meters. This represented a density of about 0.12 plants per horizontal square centimeter in the laboratory or 12 plants/square meter in the field. The setup consisted of eight capacitance wave gauges to measure surface elevations, one shear plate which was fixed flush with the bottom, to measure the horizontal force and a mini-current meter which was inserted in between the plants 4 centimeters above the shear plate to measure horizontal particle velocities. One of the wave gauges was placed above the shear plate. The first wave gauge, taken to be located at $x = 0$ and the last taken to be located at $x = 7$m, were fixed about 1.2 meters inward from the inner and outer boundaries respectively.

We started with 6m water depth and tested with regular waves with periods ranging from 4-14 seconds, full scale. With the same water depth, irregular waves with peak periods ranging from 4-14 seconds were tested. For each wave period, several wave heights were tested.

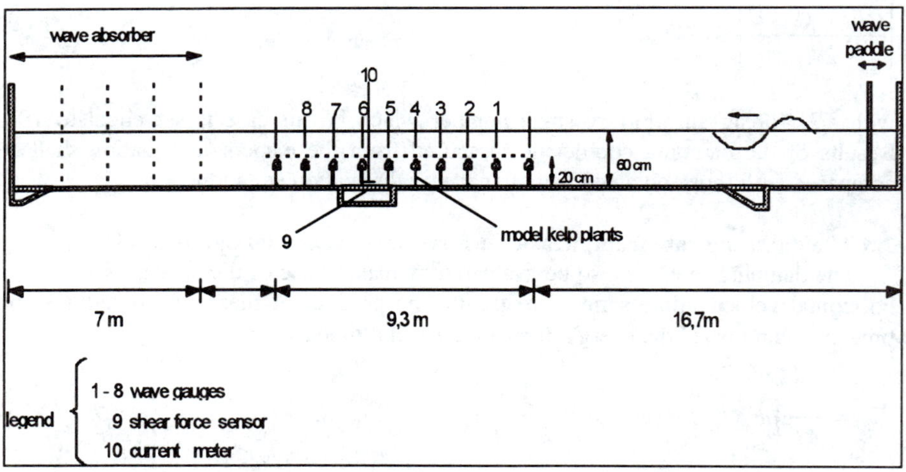

***Figure 1** Laboratory experiment layout*

4. RESULTS

(1) The exponential decay model given by Eq.(1.2) was fitted to measured data by regreesion analysis based on the least squares method. Figure 1 shows an overview of the exponential decay of the wave heights with respect to distance from the origin. Figure 3 shows theoretical variation of the damping rate with wave period T obtained from Eq.(2.14) and Eq.(2.13) for total water depth D = 6m using the following kelp

plant properties: total kelp height d = 2m, frond half length l_k = 0.5m, density of kelp ρ = 1120 kg/m^3, mass of frond and stipe = 0.75 kg/m, linearized spring constant k_o = 20N/m, linearized (empirical) force coefficient F_λ = 30N/(m/s), added mass for both frond and stipe C_a = 1, number of plants per horizontal square meter n_k = 12plants/m^2. For the given parameters the damping rate using Eq.(2.13) increases with the wave period for small values of T to a maximum and then decreases gently for larger values of T to an almost asymptotic value. On the other hand, the damping rate from the dispersion equation decreases rapidly for larger values of T. In the same figure, we also compare the damping rates derived from Eq.(1.11) given by Asano et al. (1992) and the solution of the dispersion equation given by Wang and Tørum (1994).

(2) Figure 4 shows a comparison of theoretical values obtained from Eq.(2.13) with those obtained experimentally for irregular waves in a water depth of 4m. Theoretical models by Asano et al. (1992) and Wang and Tørum (1994) are also compared to the measured data points.

(3) Figures 5 through 8 show a comparison of theoretical values of the energy dissipation equation with measured values from regular and irregular waves in water depths of 6, 8 and 10 meters. From these figures we observe several things:

First, the experimental damping coefficient varies more widely for shorter peak periods and almost converges (narrower scatter) for longer peak periods.

Second, for longer peak periods the theoretical damping coefficient is almost independent of the peak period despite a gradual decrease.

Third, when we compare the damping rates in 4 and 6 meters water depths, we find larger values for the shallower water. We may conclude that the shallower the water, the higher the damping coefficient. This observation is also in agreement with field results by Mork (1995).

Fourth, The theoretical model by Asano et al. (1992) is good only for the prediction of the damping rate for small damping. For large damping, the model grossly overestimates the damping rate. The model by Wang and Tørum (1994) underestimates the damping rate for small waves and overestimates it for large damping. For the whole range and especially for practical wave periods (6-14 seconds), the present model is very good.

Fifth, the damping rate obtained from the experiment for regular waves shows a very wide scatter. However, the theoretically predicted values appear to fit within the range of scatter as shown in Figure 6.

Sixth, as the water depth increases, the energy dissipation equation (Eq.(2.13.)) tends to over-estimate the damping coefficient, while the dispersion equation (Eq.(2.14)) comes closer and closer to the measured values. The reason for this discrepancy is not clear. However, since the empirical force coefficient F_λ was evaluated at 6m water depth, in the absence of force data for other water depths, it may be that this coefficient is variable with water depth.

(4) Figures 9 and 10 show the general functional relationship between the damping coefficient and total water depth and the number of plants per square meter respectively. For all wave periods k_i = 0.0658exp(-0.5*D).

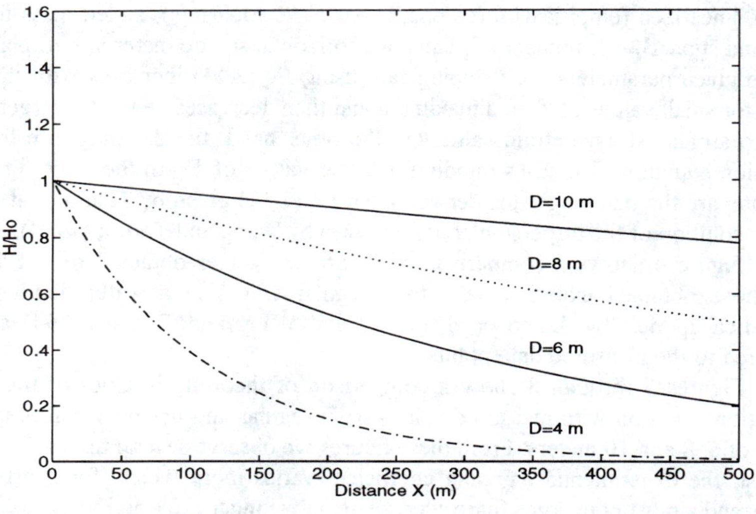

Figure 2 Comparison of exponential decay for different total water depths.

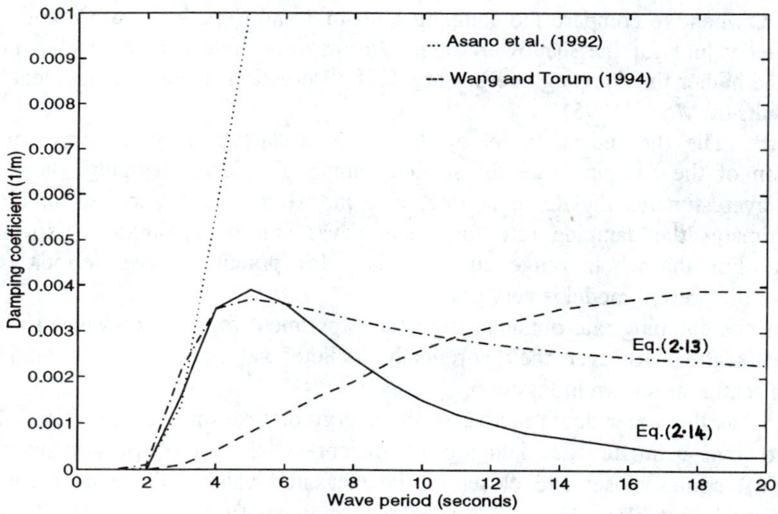

***Figure** 3 Comparison of theoretical models of Asano et al. (1992), Wang and Tørum (1994) and the present model (Eq.(2.13) and Eq(2.14)). Total water depth = 6 m.*

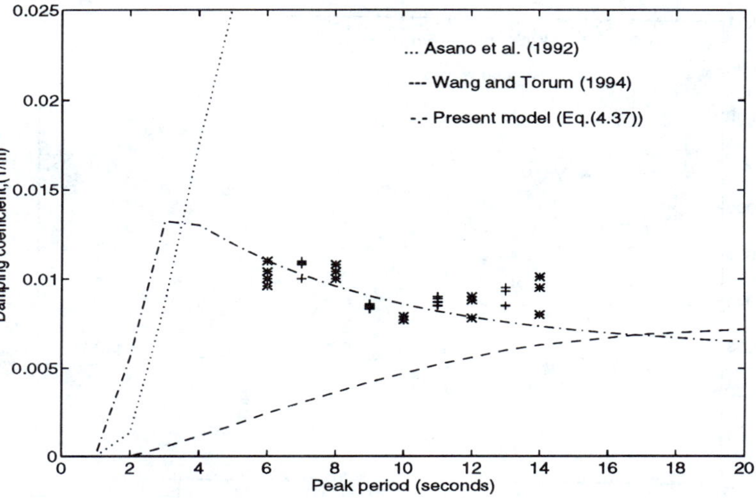

Figure 4 *Comparison of theoretical values of previous models and the present model (Eq. (2.13) with measured values (*+). Total water depth = 4 m*

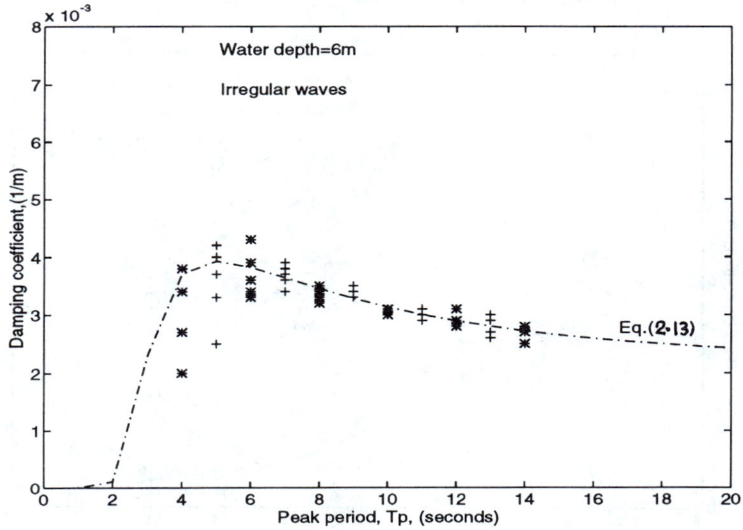

Figure 5 *Comparison of theoretical values of Eq. (2.13) with measured values (*+). with irregular waves. Total water depth = 6 m*

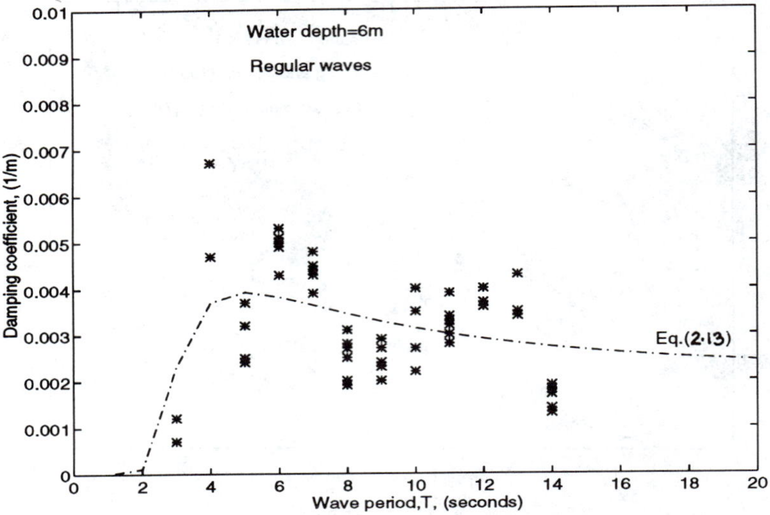

Figure 6 *Comparison of theoretical values of Eq. (2.13) with measured values with regular waves (*+). Total water depth = 6 m*

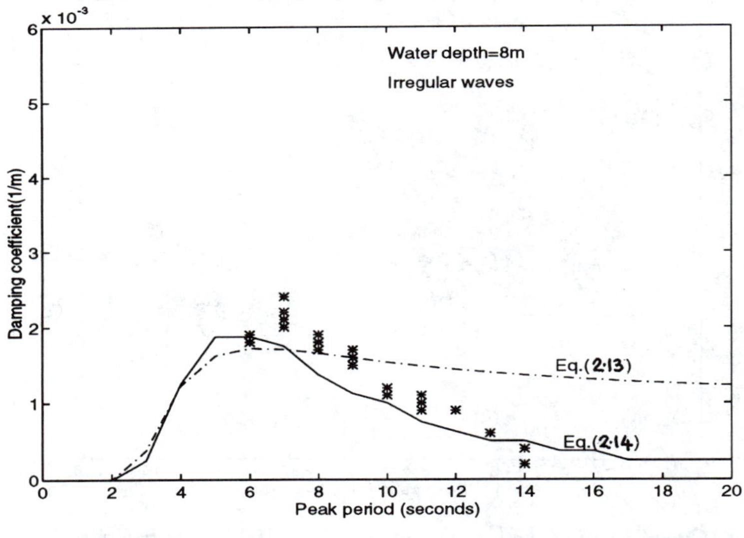

Figure 7 *Comparison of theoretical values of Eq.(2.13) and Eq. (2.14) with measured values (*) with irregular waves. Total water depth = 8 m*

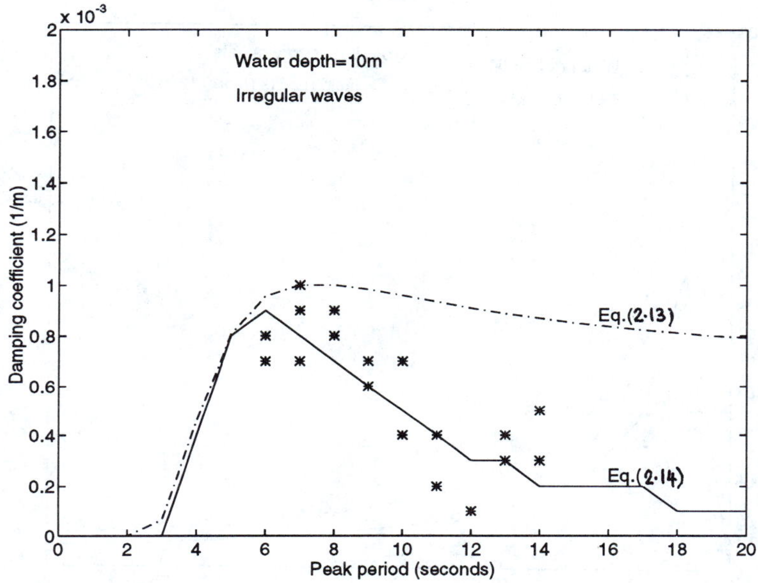

Figure 8 *Comparison of theoretical values of Eq.(2.13) and Eq. (2.14) with measured values (*) with irregular waves. Total water depth = 10 m*

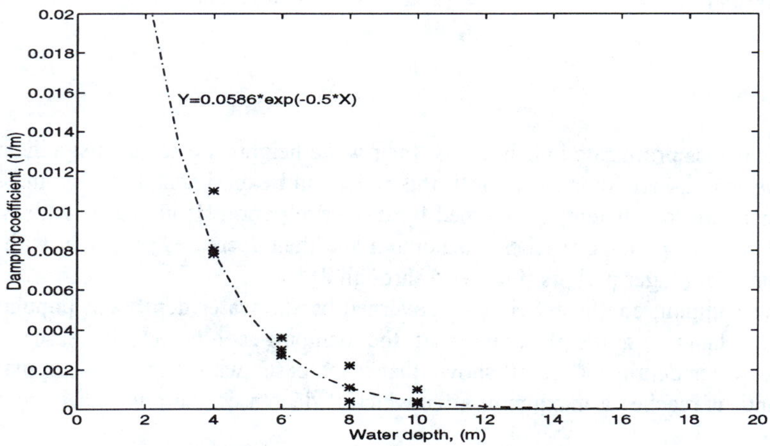

Figure 9 *General functional relationship between damping coefficient and total water depth* $D \geq d$*, T=7, 10 and 14s. For all wave periods* $k_i = 0.0658 \exp(-0.5*D)$

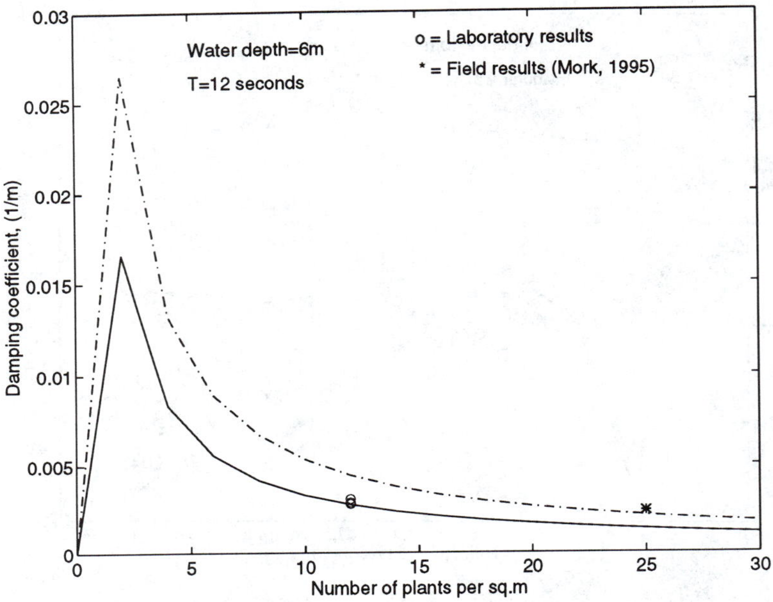

Figure 10 *Variation of damping coefficient with number of plants per square meter. Total water depth = 6 m, ——— d =2m, -·-·-·-· d=3.2m*

CONCLUSION

(1) As waves propagate in kelp fields, their wave heights are reduced significantly and wave lengths are shortened. Both theoretical and experimental studies indicate that the damping coefficient is governed by the wave period. It increases with wave period for shorter periods, reaches a maximum and then decreases gradually with the wave period for longer periods (Figures 5 through 8)

(2)The damping coefficient is also governed by the water depth and population density of plants. Figures 9 shows that the damping coefficient decreases with increasing water depth. Figure 10 shows that it increases with increasing population density until it reaches a maximum, after which it decreases sharply with increasing density.

(3) Basing on the theoretical and experimental results, wave height attenuation is more substantial in shallow water than in deep water. In fact it is almost negligible in water depths greater than 10 meters. For a population density of 12 $plants/m^2$, significant heights of irregular waves will be reduced by say 50% over a distance of 74, 216, 495 and 1390 meters in 4, 6, 8 and 10 meters mean water depth respectively.

(4) The linear assumption of the wave force in relation to the particle velocities leads to an exponential decay (damping) coefficient of the wave height with respect to distance.

ACKNOWLEDGEMENT

We wish to acknowledge with thanks the financial support we received from the Norwegian Directorate for Nature Management and the Norwegian Universities' Committee for Development Research and Education (NUFU). The first author is also grateful to the IOC of UNESCO for the travel grant to the Conference in Orlando, Florida.

REFERENCES

1. Asano, T., Deguchi, H., and Kobayashi, N. (1992). "Interaction between water waves and vegetation". *Proc. 23rd International Conference on Coastal Engineering*, Venice, Italy, pp. 2710-2723.
2. Dalrymple, R.A, Hwang, P.A. and Perlin, M. (1982): "Ocean Engineering aspects of coastal kelp farming". Report no. J-III, Coastal and Offshore Engineering and Research, Inc., Newark, Delaware.
3. Dean, R.G. and Dalrymple, R.A . (1984): *Water wave mechanics for engineers and scientists*. Prentice-Hall Inc., New Jersey.
4. Dubi, A.M. (1995): "Damping of water waves by submerged vegetation. A case studty on Laminaria hyperborea.". Dr. Ing. Thesis. University of Trondheim, The Norwegian Institute of Technology. Trondheim, Norway. ISBN 82-7119-859-9
5. Dubi, A.M. and Tørum, A., (1994): "Wave damping by kelp vegetation". *Proc. 24th International Conference on Coastal Engineering*, Kobe, Japan, pp. 142-156
6. Ippen, A.T. (1966): *"Estuary and Coastline hydrodynamics"*. McGraw-Hill Book Co. Inc., New York.
7. Kobayashi, N., Raichle, A.W., and Asano, T. (1991). "Prediction of wave attenuation by vegetation and seaweed". Research report no. CACR-91-07, Center for Applied Coastal Research, Dept. of Civil Engineering, University of Delaware, Newark, Delaware 19716
8. Madsen, O.S. (1974). "Wave transmission through porous structures". *Journal of Water Ways Harbors and Coastal Engineering Division, Proc. of the ASCE, Vol. 100, No. WW3,* May, 1974 pp. 169-188
9. Mei, C.C. (1989): *"Applied Dynamics of Ocean Surface Waves"*. World Scientific Publishing Co., Singapore, pp. 740.
10. Wang, H. and Tørum, A. (1994). 'A numerical model on beach response behind coastal kelp fields". SINTEF-NHL report no. STF60 A94092

CHAPTER 204

EXPERIMENTAL STUDY ON THE EFFECT OF GRAVITY DRAINAGE SYSTEM ON BEACH STABILIZATION

Hiroshi KANAZAWA [1), Fumihiko MATSUKAWA [1)]
Kazumasa KATOH [2)] and Iwao HASEGAWA [3)]

Abstract

In order to develop a new beach preservation technique which has two functions of protecting the beach in a storm and enabling utilization of beach during calm sea, we consider the system of controlling a groundwater level by burying a permeable layer under the foreshore. Since most of the components are buried underground, people on the beach do not notice the presence of the system, enabling preservation of the beach in its most natural state. The greatest challenge of the system is, however, whether installation of the permeable layer alone will fully prevent erosion. Authors therefore conducted an experiment to evaluate beach stabilization effect of the present system. The result indicates that the permeable layer buried in the sand can prevent rise in the groundwater level and the wave set-up. The critical wave height for the foreshore erosion is 1.66 times higher than that on the beach without the permeable layer.

1. Introduction

Against severe natural conditions in Japan, we have constructed seawalls with wave breaking blocks, jetties, detached breakwaters, and so on. These protection measures have stopped further erosion, and also protected the hinterland and people from large waves in a storm. However, in a calm sea condition, the structures keep the people away from the waterfront. Therefore, we are developing a new beach stabilization technique, which has functions of disaster prevention in a storm and beach utilization in a calm. In order to develop such a new beach stabilization technique, first of all a beach erosion mechanism

---

1)Yokohama Investigation and Design Office, the Second District Port Construction Bureau, Ministry of Transport, Kitanaka-dohri 5-57, Naka-ku, Yokohama, 231, JAPAN
2)Port and Harbor Research Institute, Nagase 3-1-1, Yokosuka, 239, JAPAN
3)ECOH co., Tsuruya-cho 2-19-4, Kanagawa-ku, Yokohama, 221, JAPAN

during stormy weather was investigated in the field (Katoh and Yanagishima: 1992, 1993). These investigations elucidated the relation between rise in the watertable due to wave run-up and the beach erosion. That is, when waves run-up beyond the berm crest in a storm, the sea water stays for a good while on the horizontal area of berm, which accelerates the infiltration of water into the beach. As a result, the watertable becomes higher, and the water exfiltrates through the surface of foreshore. The seepage level of water corresponds to the critical level of berm erosion. Taking this result into consideration, there are two countermeasures for reducing the beach erosion in a storm. One is a construction of a structure offshore to decrease the wave energy and to suppress the wave run-up on the beach. Another is to lower the watertable by some method.

As a latter measure, we consider a permeable layer setting up under the beach. In a storm, waves run-up to the high level and the water infiltrate into the beach. However, the groundwater may be gravitationally drained to the offshore through the layer. Concerning to the effect of permeable layers, two- and three-dimensional experiments were conducted with regular waves to compare the groundwater levels and the profiles on the beach with and without the permeable layer.

2. Details of Experiments

2.1 Aims of Experiments

Aims of two- and three-dimensional experiments are as follows:

(1) Two-dimensional experiments : A purpose is to know the possibility of controlling watertable and reducing the beach erosion for three different kinds of permeable layers.

(2) Three-dimensional experiments : There are two purposes in the three-dimensional experiments. One is to confirm the effect of the convergence type drainage system, in which the water gathered by the permeable layer is drained offshore through one pipe. Another is to evaluate the effect of permeable layer on beach stabilization quantitatively.

2.2 Conditions and Methods in Two-dimensional Experiments

The experiments were carried out in a experimental flume of 38.0 m long, 0.5 m wide and 1.5 m high. The movable profiles were made with sand of 0.135 mm in a median grain size. The permeability coefficient of sand is 6.28×10^{-3} cm/s, which is the result of permeability test.

As the permeable layer, three types shown in Figure 1 were tested. Features of these types are as follows, respectively.

(1) Gravel Permeable Layer (GPL) : Gravels of 13 - 20 mm in size were laid with the thickness of 10 cm from the backshore to the breaker zone (Figure 1, top). Plankton nets were interposed between the sand layers and the gravel layer to prevent sand from inrushing into the gravel layer. The sand layer above the permeable layer was 10 cm thick. A still water level was situated inside the permeable layer of horizontal part.

(2) Gravel Permeable Layer with Drainage Pipe (**GPL-DP**) : The thickness of the permeable layer and the sand layer were decreased to half that of GPL, respectively (Figure 1, middle). The permeable layer in the backshore was therefore situated above the still water level. A drainage pipe (44 mm in inner diameter) extended from the permeable layer near the shoreline toward offshore. The pipe was buried along the wall of experimental flume, with its outlet vent located at the wave breaking point.

(3) Permeable Pipe with Drainage Pipe (**PP-DP**) : Three permeable pipes were used in place of the permeable layer(Figure 1, bottom, Figure 2). The permeable pipe was a spring coil of 15 mm diameter wrapped up with a plankton net as shown in Figure 3. Groundwater percolates into the pipe through the plankton net. Three permeable pipes were buried under the beach with a 15 cm interval in the traverse direction. To protect the permeable pipes, a gravel layer was placed above the pipes. This gravel layer did not act as a permeable layer because sand was filled in the voids of gravel layer. The drainage pipe extending toward the offshore direction from near the shoreline was the same as that used in the GPL-DP.

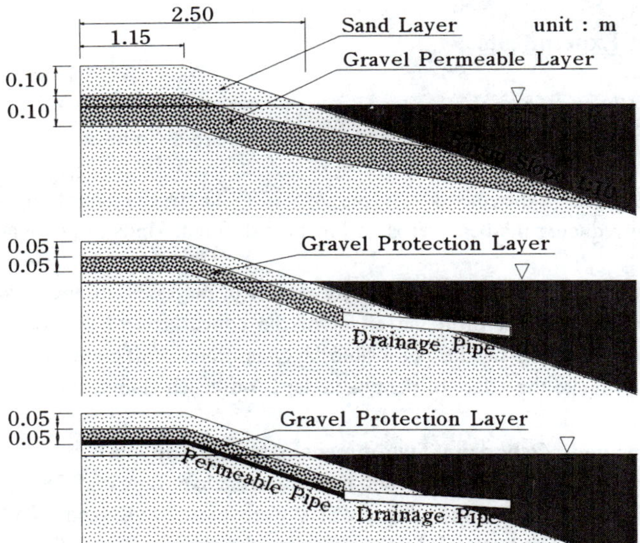

Figure 1: Types of Permeable Layer in Two Dimensional Experiment.

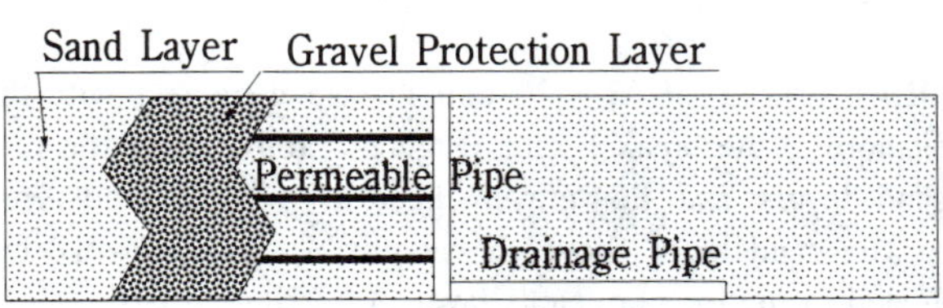

Figure 2 : Plane view of PP-DP.

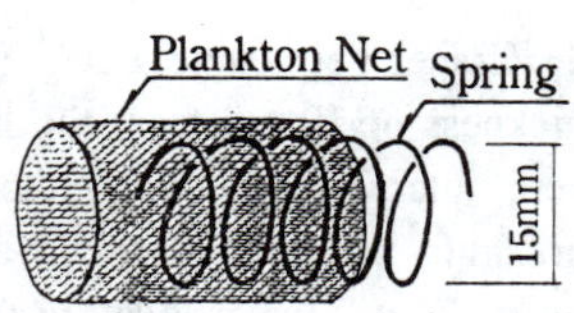

Figure 3 : Structure of Permeable Pipe.

Table 1 : Waves in Two Dimensional Experiments.

H_o (cm)	T (s)	H_o/L_o	GPL	GPL-DP	PP-DP	Without a Permeable Layer
8.4	1.34	0.03	○	○	○	○
10.0	1.79	0.03	○	○	○	○
15.0	1.79	0.03	○			○

The wave conditions in the two dimensional experiments are listed in Table 1, in which the cases with a circle were conducted. In the experiment, the level of groundwater in the section from the backshore to the shoreline, wave set-up in the surf zone, and the profile changes were measured. To measure the groundwater level and the wave set-up, a manometer using a vinyl tube 6 mm in inner diameter was used. Measurements were carried out with an interval of 25 cm along the wall of the flume in the cross-shore direction. Using a sand level meter, the profile changes were measured along three measurement lines in the traverse direction with an interval of 10 cm. The mean value of the profiles along the three measurement lines is calculated and used for further analysis of the profile changes. These result are compared with those obtained in the profile without permeable layer. In the profile without a permeable layer and the profile with GPL, a dye (methylene blue) was injected close to the glass wall of the flume. The seepage velocity inside the permeable layer and the sand layer above the permeable layer respectively were estimated from the travelling distance of the gravitational center of the dye patch and its elapsed time.

2.3 Conditions and Method in Three-Dimensional Experiments

Test conditions were primarily the same as those in the two dimensional experiments except the following.

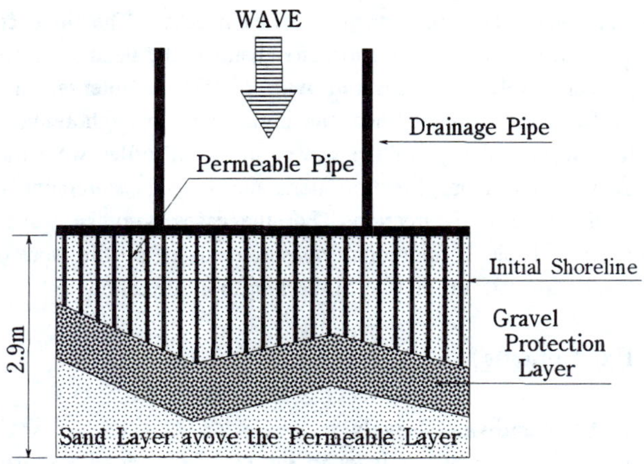

Figure 4 : Layout of Permeable Layer in Three-Dimensional Experiments.

A wave basin (35.0 m L × 1.1 m H × 15.0 m W) was divided into two sections (7.5 m W) to enable simultaneous experiments on the two beaches with PP-DP and without the permeable layer. Figure 4 shows the layout of PP-DP: 100 permeable pipes measuring 2.9 m in length and 15 mm in inner diameter were buried with a 7.5 cm longshore interval and eight drainage pipes measuring 44 mm in inner diameter with a 94 cm longshore interval. The outlet vent of the drainage pipe was positioned at the wave breaking line. The median grain size of the sand used in the experiments was d_{50} = 0.16 mm and the permeability coefficient was k = 1.14 × 10^{-2} cm/s.

Table 2 : Waves and Profile Changes in Three-dimensional Experiments.

T (s)	H_o (cm)	Value of C	Classification of Profile Changes: Without a Permeable Layer	Classification of Profile Changes: PP-DP
1.34	1.5	2.0	Ⅱ, Ⅲ	Ⅲ
	3.0	4.0	Ⅰ, Ⅱ	Ⅱ, Ⅲ
	4.5	6.0	Ⅰ	Ⅱ
	6.0	8.0	Ⅰ	Ⅰ
	7.5	10.0	Ⅰ	Ⅰ
	☆ 8.4	11.2	Ⅰ	Ⅰ
2.01	1.8	1.8	Ⅲ	Ⅲ
	3.7	3.7	Ⅰ, Ⅱ	Ⅱ, Ⅲ
	5.4	5.5	Ⅰ	Ⅱ
	7.2	7.4	Ⅰ	Ⅰ
	9.0	9.2	Ⅰ	Ⅰ
	☆ 12.5	12.8	Ⅰ	Ⅰ

The wave conditions used in the experiments are listed in the first two columns of Table 2. The waves were acted for one hour in each case. The three columns on the right in the table give information on the shoreline changes; the details are given later.

The groundwater level was measured with a 1.25 m interval in the longshore direction along the two measurement lines, the initial shoreline (shoreline at the time of still water) and the line near the tip of the wave run-up. Profiles were measured with a 0.125 m interval in the cross-shore direction along the three measurement lines set with a 2.25 m interval in the longshore direction. For the cases with the wave height values marked with asterisks in Table 1, additional measurements of profile were done along the measurement lines set with 0.25 m interval in the longshore direction.

3. Results of Experiments

3.1 Comparison of Groundwater Levels

Figure 5 shows the groundwater level in the two-dimensional experiment in which waves of 8.4 cm in height and 1.34 s in period were acted on the beach without the permeable layer. The reference level for elevation on the ordinate is the initial still water level. The groundwater level begins to rise immediately after waves are acted, and substantially levels off after 11 minutes. The peak of the watertable near the location of 2.0 m substantially coincides with the tip end of the run-up, and the water level exceeds the still water level by 2 cm. The wave set-up at the shoreline located at 2.5 m is 1 cm above the still water level.

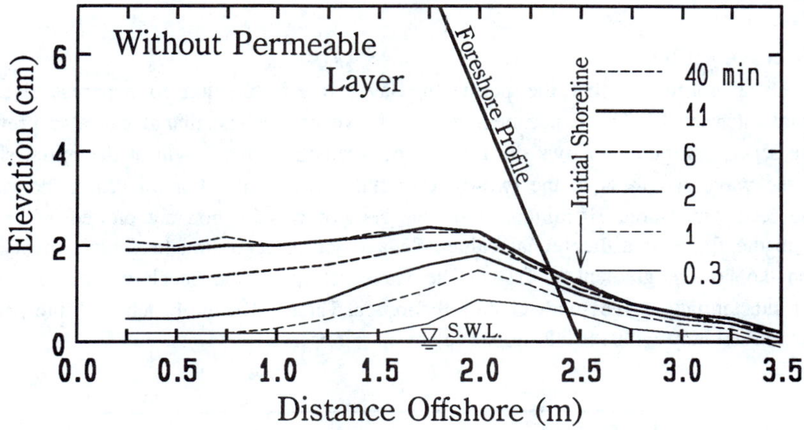

Figure 5 : Changes of Groundwater Level with Time in the Beach without the Permeable Layer (two-dimensional experiment).

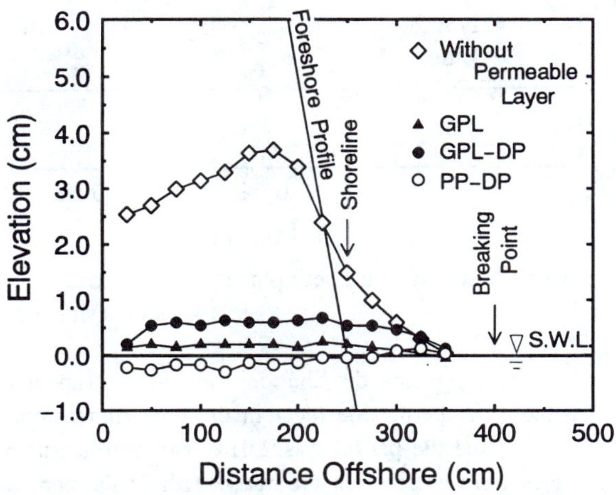

Figure 6 : Comparison of Groundwater Levels between Beaches with and without the Permeable Layer (two-dimensional experiment).

Figure 6 shows the distributions of groundwater level and the wave set-up measured ten minutes after the waves of 10.0 cm in height and 1.79 s in period were acted on the beaches without the permeable layer and with three types of permeable layer. Because of the severe wave conditions, rise in the groundwater level and the wave set-up on the beach without a permeable layer are greater than those shown in Figure 5. In the beaches with the permeable layers, both the rise in the groundwater level and the wave set-up are quite insignificant. In particular, the groundwater level in the beach with PP-DP is lower than the initial still water level. This is considered to be the effect of wave set-down because the terminal end of the drainage pipe on the offshore side is located at the wave breaking point.

3.2 Wave set-up

It is confirmed that the permeable layer has a function to suppress rise in the groundwater level. The experiment results, however, suggest that it can also prevent the wave set-up. Figure 7 shows the relation between the wave set-up at the initial shoreline and the wave steepness in the two-dimensional experiment. For all cases, the values of wave set-up measured 10 minutes after the start of wave action are plotted. The broken line in the figure is a theoretical value of wave set-up according to Goda's theory (1975) when the bottom gradient is 1/10. The wave set-up on the beach without a permeable layer substantially agrees with Goda's theoretical value. On the beach with the permeable layer, particularly with PP-DP, no wave set-up is observed.

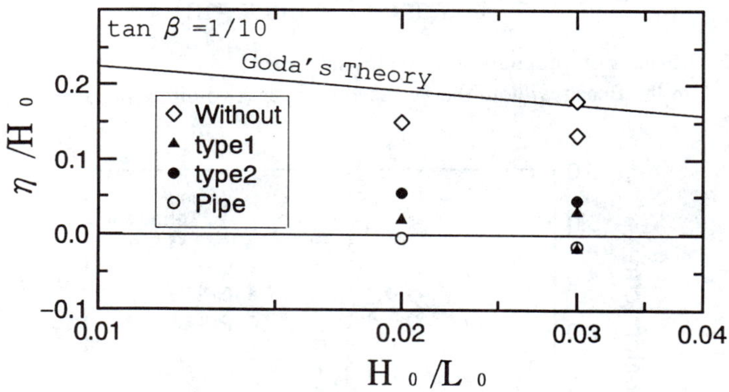

Figure 7 : Relation between Wave set-up and Wave Steepness (two-dimensional experiment).

Figure 8 shows the longshore distributions of wave set-up at the shoreline on the beaches with and without the permeable layer in the three-dimensional experiments. The wave height was 12.5 cm and the period was 2.01 s, and the measurements were taken 30 minutes after the waves were acted. On the beach without the permeable layer, the wave set-up agrees with the theoretical value of Goda (1975) which is shown by a horizontal line in the figure. On the beach with the permeable layer, on the other hand, the average wave set-up is about 0.5 cm. The relative position between the measurement point of groundwater level and the position of the drainage pipe, which is shown by ↑ in the Figure 8, changes from a location to a location. The wave set, however, is considered to be constant in the longshore direction, although a slight fluctuation is observed. This result indicates that groundwater within the range of 0.94 m along the longshore direction could be discharged by one drainage pipe, suggesting that convergence type drainage system is feasible.

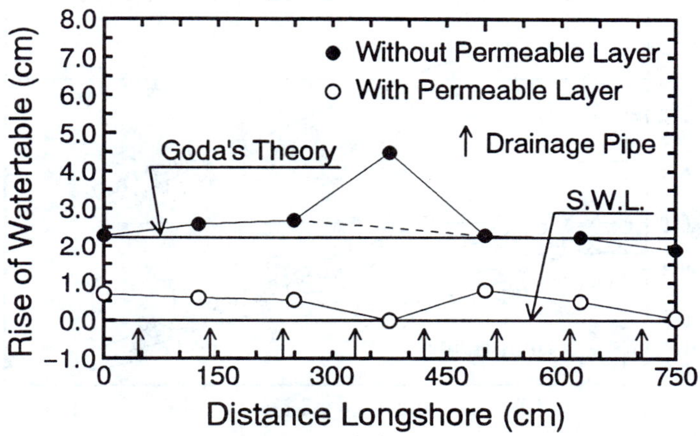

Figure 8 : Comparison of Groundwater Levels between on the Beaches with and without the Permeable Layer (three-dimensional experiment).

3.3 Seepage velocity in the Sand Layer and the Permeable Layer

Figure 9 shows the seepage velocity in the sand layer and the permeable layer. The top portion shows the values for the beach without the permeable layer, the bottom portion those for the GPL. The velocity in the profile without the permeable layer is smaller than 0.15 mm/min. The velocity in the sand layer located above the still water level is directed onshoreward, whereas the flow in the sand layer below the still water level is directed toward the offshore along the sea bottom. In the beach with the permeable layer, the seepage velocity is two to three times that in the beach without the permeable layer, with a offshoreward velocity of about 30 cm/min (about 1000 times the velocity in the sand layer; note that velocity vectors are in logarithmic expression in the figure). It must be noted that the flow in the permeable layer fluctuated with the same period as wave motions, but the time average velocity is shown in Figure 9.

A reason of the fast velocity in the permeable layer is studied. Firstly, a numerical simulation of two dimensional unsteady flow was conducted to evaluate the amount of water infiltration into the beach due to the wave run-up. The simulation was carried out for the entire model beach. The water infiltration into the beach in the portion from the shoreline to the run-up level was given, so that the calculated groundwater level agreed with the measured value. The rate of infiltration into the beach without the permeable layer was estimated to be 1.1 cm^3/min/cm. As for the beach with the permeable layer, if the same amount of water is assumed to infiltrate into the beach due to the wave swashing, the seepage velocity in the permeable layer (10 cm thick) would be 0.37 cm/min, given the void ratio of 0.3. Thus, the fast velocity in the permeable layer cannot be explained solely by this infiltration. In the following consideration, therefore, the experimental fact that no wave set-up is observed on the beach with the permeable layer is taken into account.

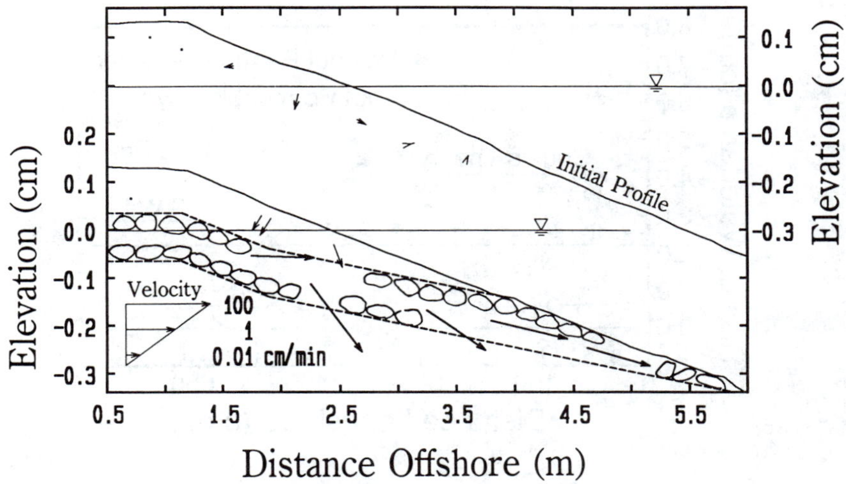

Figure 9 : Seepage Velocity in the Sand Layer and the Permeable Layer (two-dimensional experiment).

Assuming two-dimensional steady-state condition and ignoring the term of horizontal diffusion, the equation of motion in the breaker zone is

$$\frac{d}{dx}\left(\rho U^2 (h+\eta) + Sxx\right) = -\rho g (h+\eta)\frac{d\eta}{dx} - \tau_b \quad \cdots\cdots\cdots\cdots (1)$$

with the offshore direction taken along the positive x-axis, where U is the mean velocity, h water depth, η wave set-up, Sxx radiation stress and τ_b bottom shear stress. Since there is no wave set-up, $\eta=0$. Then, by assuming $\tau_b=0$, the equation (2) can be easily integrated.

From the radiation stress given by the equation below,

$$Sxx = -\frac{3}{16}\rho g \gamma^2 \tan^2\beta \quad \cdots\cdots\cdots\cdots (2)$$

which is approximated by the linear shallow-water theory by assuming a uniform slope ($h = x\tan\beta$) and a constant proportionality between the wave height and depth ($\gamma = H/h$), then it holds:

$$U = -\frac{\gamma}{4}\sqrt{3 g x \tan \beta} \quad \cdots\cdots (3)$$

If it is assumed that a flow which compensates the onshoreward discharge ($U \times h$) is generated in the permeable layer, then the velocity Vi in the permeable layer is

$$Vi = \frac{\gamma}{4 \lambda D}\sqrt{3 g x^3 \tan^3 \beta} \quad \cdots\cdots (4)$$

where λ is the porosity (=0.3) and D is the thickness of the permeable layer (=10 cm). By assuming γ = 0.78, Vi at a distance of 5 cm offshore (h=0.5 cm) from the shoreline is calculated to be 75 cm/min, which is almost 2.5 times larger than the value actually measured in the permeable layer. This is probably because the velocity is overestimated in equation (3). The calculated value would be much closer to the actual measurement if considerations are given to the factors that the onshore velocity in the surf zone is subject to the bottom shear stress, that any small rate of wave set-up would act to decelerate the velocity, and that there is the effect of the permeability coefficient in the sand layer above the permeable layer. It can be said that the fast velocity in the permeable layer is subject to the strong influence of the gradient of the radiation stress.

3.4 Profile changes

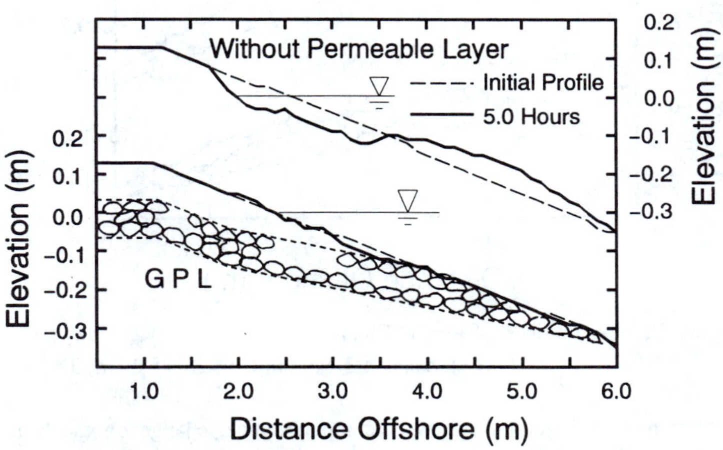

Figure 10 : Comparison of Profile Changes (two-dimensional experiment : H=8.4 cm, T=1.34 s).

Figure 10 shows the profile changes on the beach without the permeable layer and with GPL, respectively acted with waves of 1.34 s in period and 8.4 cm in height for five hours. The profile without the permeable layer eroded in the range including the initial shoreline, and the sand deposited on the offshore forming a bar. The shoreline receded for about 0.5 m. On the beach with GPL, on the other hand, hardly any profile change is observed except for a small berm slightly on the onshore side of the shoreline. No change in the shoreline position is observed. In this case, the permeable layer is effective in preventing erosion near the shoreline.

Figure 11 shows the profile changes acted with waves of 1.79 s in period and 15.0 cm in height for five hours. Because of severe wave conditions, the beach without the permeable layer eroded in the range from 1.3 m to 4.3 m. At a point about 4.1 m, there is formed a trough, of which on the offshore side a bar is formed. Although similar erosion is observed on the beach with GPL, no trough is formed since the erosion do not extend deeper than the permeable layer. Sand from the layer above the permeable layer is transported toward offshore direction and deposits on the outlet vent of the permeable layer. This phenomenon was observed from an early stage of wave action. The drainage capacity of the permeable layer was reduced by the clogging of the outlet vent. The erosion occurring at the initial stage of wave action was attributable to the insufficient drainage capacity of the permeable layer, but the eventual erosion similar to that occurring on the beach without the permeable layer was probably because the outlet vent was clogged.

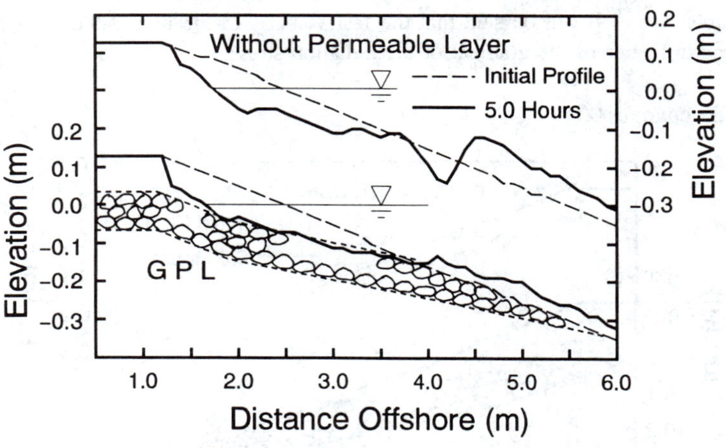

Figure 11 : Comparison of Profile Changes
(two-dimensional experiment : H=15.0 cm, T=1.79 s).

The top part of Figure 12 shows the topographical change on the beach with the permeable layer, PP-DP. Waves of 8.4 cm in height and 1.34 s in period were acted for one hour. The locations of drainage pipes are clearly shown in the figure, but as they were actually buried underground, only the outlet vents were visible. Because of the wave incident angle of 90° to the shoreline, the profile change is two dimensional and there is no three dimensional effect of draining the water offshore through the eight drainage pipes. As for the changes in the cross-shore direction, the foreshore is eroded and a bar is formed offshore at about 4 m. The outlet vents of the drainage pipes was located on the immediately onshore side of the bar. However, the vents were not clogged by the sand because the terminal end of the drainage pipe was set slightly above the sea bottom (see Figure 1). Then, the groundwater controlling effect of the permeable pipes is included even in this foreshore erosion. To demonstrate this point, we conducted a similar experiment using a beach without the permeable layer. The result is shown in the bottom part of Figure 12. On the beach without the permeable layer, erosion on the foreshore is extensive, with a substantial amount of deposition on the offshore bar.

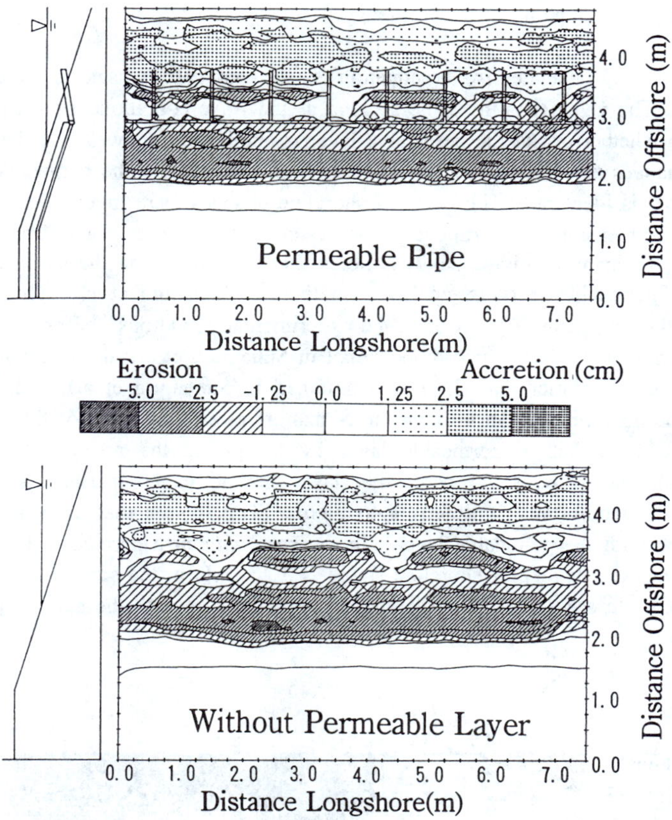

Figure 12 : Comparison of Topographical Changes (three-dimensional experiments : H=8.4 cm, T=1.34 s).

As discussed above, in some cases erosion do occur on the beach with the permeable layer, although its extent is slightly less. To have quantitative understanding on the effect of permeable layer, we conduct an analysis using the constant C of the beach profile type classification (Sunamura and Horikawa, 1974) which is evaluated by the equation (5), in which an averaged profile is considered in each case.

$$C = (\tan \beta)^{0.27} \left(\frac{d}{L_0}\right)^{-0.67} \left(\frac{H_0}{L_0}\right) \qquad \cdots\cdots (5)$$

According to Sunamura et al, profile changes can be classified into three types depending on the value C. Characteristic features of each type with regard to the change of the shoreline position can be described as follows:

$C > 8$ Type I : Recession
$8 > C > 4$ Type II : Stable
$4 > C$ Type III : Advancement

The value C calculated by the equation (5) for each case is listed in the third column of Table 2. The value C is the same so long as the wave conditions are identical,without regarding to whether or not the permeable layer is employed. However, different types of shoreline changes are observed on the beaches with and without the permeable layer even if the value C is identical. The types of shoreline changes were investigated for all cases in the three-dimensional experiment. The results are listed in Table 2 and shown in Figure 13. Referring to Figure 13, it is noted that the value C in the case of experiment without the permeable layer is smaller than that of Sunamura et al. This is probably because of the following differences in the experimental conditions: the present experiment is the three-dimensional one (two-dimensional in Sunamura et al); the foreshore slope was 1/10 (many cases gentler than 1/10 are contained in Sunamura et al); and, waves were acted for one hour (more than 40 hours in Sunamura et al). Because of these differences, we evaluate the effect of permeable layer by comparing the results obtained for the beaches with and without the permeable layer instead of comparing with the results obtained by Sunamura et al. Referring to Figure 13, it is seen that the value C is larger on the beach with the permeable layer. In the case without a permeable layer, the mean value of C in the category of shoreline stable is 3.2, while in the case with the permeable layer it is 5.3. In short, the value of C is larger in the case with the permeable layer by a factor of 1.66.

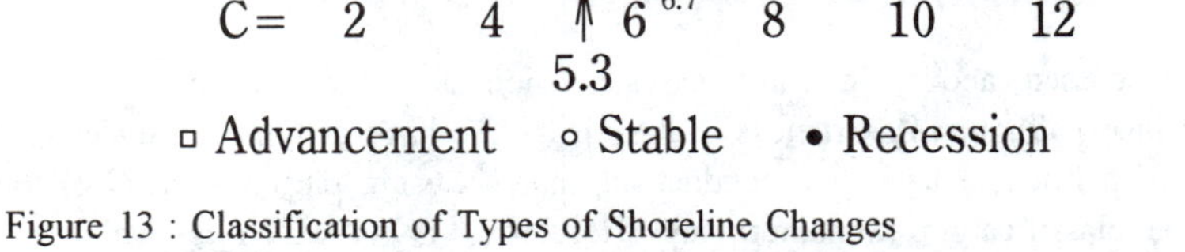

Figure 13 : Classification of Types of Shoreline Changes in Three-Dimensional Experiment.

When the changes in the value C are viewed from a different angle, the following can be said. When a beach with the permeable layer is subjected to waves having a value C_a, shoreline changes of the beach will be the same as those observed in the natural beach (without the permeable layer) subjected to waves of $C_a / 1.66$. In other words, the permeable layer is effective in reducing the value C_a to about 60 %. This effect can be achieved also by a wave controlling structure. Generally, a wave controlling structure can control the wave height and not the wave period. Thus, given a constant bottom gradient, grain size and wave length (period) in equation (5), the value C will be in direct proportion only to the wave height. This in turn means that reduction of the value C to about 60 % attained by the permeable layer is comparable to the effect of a structure which is capable of decreasing the wave height to about 60 %. This wave decreasing effect is achieved by a submerged breakwater with a slightly larger crown depth with slightly narrower crown width. It should be noted, however, that a permeable layer has additional effect in that hardly any wave set-up near the shoreline occurs, which is a function not obtained by a submerged breakwater.

4. Conclusion

Effects of the permeable layer are as follows:

① Substantially no wave set-up occurs at the shoreline.
② Rise in the groundwater level can be suppressed to a greater extent.
③ Erosion near the shoreline can be mitigated.
④ According to our experiment, combination of permeable pipe and drainage pipe is most effective.
⑤ Effect of permeable layer in respect of shoreline changes is comparable to that obtained by reducing the wave height by about 40 %.

It can be concluded that the effect of permeable layer is demonstrated by our laboratory model experiments. Presently, we are conducting filed experiments at a beach in order to complete the present technique. (Katoh and Yanagishima, 1996).

REFERENCES

Yoshimi Goda(1975) : Deformation of Irregular Waves due to Depth-Controlled Wave Breaking, Rep.of PHRI, Vol.14, No.3, pp.59-106.

Katoh, K. and S.Yanagishima(1992) : Berm Formation and Erosion, Proc.of 23rd ICCE. pp.2136-2149.

Katoh, K. and S.Yanagishima(1993) : Beach Erosion in a Storm due to Infragravity Waves, Rep.of PHRI, Vol.31, No.5, pp.73-102.

Katoh, K. and S.Yanagishima(1996) : Field Experiment on The Effect of Gravity Drainage System on Beach Stabilization, Proc of 25th ICCE.

Sunamura, T. and K.Horikawa(1974) : Two-Dimensional Beach Transformation Due To Waves.Proc.of 14th Coastal Eng.Conf., ASCE, pp.920-937.

Zienkiewicz, O.C.(1977) : The Finite Element Method Third Edition, McGraw-Hill, New York, 787p.

CHAPTER 205

Field Experiment on the Effect of Gravity Drainage System on Beach Stabilization

Kazumasa KATOH [1] and Shin-ichi YANAGISHIMA [2]

Abstract

To have a coastal protection work of high quality, we are developing a new beach protection system, that is a beach drainage system, in which the groundwater is naturally drained to the offshore through a permeable layer setting up under the beach. To examine the drainage function of the permeable layer and its effect on the beach stabilization, a field experiments have been conducted during storms. As a result, it is confirmed that the permeable layer drain the groundwater garvitationally into the surf zone and has a function of decreasing a speed of erosion in a storm, and the eroded foreshore is recovered quickly in a calm.

Introduction

As Japan has a long coastline and exposed to severe natural conditions, strong efforts on the coastal protection works such as detached breakwaters, jetties, seawalls are necessary to prevent the coastal disasters. However, these concrete hard structures keep the people away from the waterfront. The view of coastal regions are worse than they were. Over against this the demand for a better quality of protection gets greater, and multipurpose use of valuable coastal zone comes to be required.

In order to make the coasts what people currently desire to be, we started from the basic research on the changes of sandy beach in a storm because a sandy beach is fundamentally essential to create the desired coast. The series of researches have been reported in the ICCE (Katoh and Yanagishima, 1988,

Chief [1] and member [2] of Littoral Drift Laboratory, Port and Harbour Research Institute, Nagase 3-1-1, Yokosuka, 239, JAPAN

1990, 1992; Katoh, 1994). According to the results of basic researches, a shoreline rapidly recesses and a berm on the foreshore erodes in one or two days during a storm, which is due to the development of infragravity waves. When infragravity waves run up beyond the berm crest in the storm, the sea water stays for a good while on the horizontal area of berm, which accelerates infiltration of water into the beach. As a result, the water table becomes higher, and the water exfiltrate through the surface of foreshore. The seepage level of water corresponds to the critical level of berm erosion.

Taking these results into consideration, there are two kinds of measures for maintaining the sandy beach against the severe natural conditions. One is a construction of structure in the offshore such as detached breakwaters to diminish the wave energy and to suppress the wave run–up on the beach. Another is to lower the water table by some method. The system of lowering the level of water table by pumping, which is called the sub–sand filter system, has been tested in the laboratories and in a small scale field experiment in a calm (Chapell et al., 1979; Vesteby, 1991). There is a problem in this method, that is to say, a continuous pumping of the groundwater is required for lowering the water level. Then, an alternative drainage system is considered, in which the ground water is naturally drained to the offshore through a permeable layer setting up under the beach. We carried out an experiment on the function of permeable layer under the beach in a flume (Katayama et al., 1992), of which the final result is reported by Kanazawa et al. (1996). Davis et al. (1992) developed the prototype beach drainage system which incorporated an array of shore normal strip drains made of geotextile fabric enclosing a plastic core in an "egg–carton" configuration, and examined its function for lowering the water table and stabilizing the beach in the field. The results of these researches strongly suggest a possibility of gravity drainage system which enhances the stabilization of beach. If so it will be possible to maintain the beach of beautiful view without any artificial obstacles, because the whole system may be built under the ground.

To examine the drainage function of permeable layer under the foreshore and its effect on the beach stabilization, especially during a storm, a field experiment has been conducted.

Field experiment

1. Site of field experiment

The site of field experiment is an entirely natural sandy beach, being exposed to the full wave energy of the Pacific Ocean (see Figure 1). The foreshore slope is about 1/50, and the median diameter of sand on the beach is 0.18 mm which is nearly constant during the calm and storm conditions (Katoh and Yanagishima, 1995) . The tide range is about 1.5 m. According to the analysis of shoreline locations which were read from eleven aerial photographs taken during a period from 1947 to 1984, the shoreline advanced about 50 m during a period from 1975 to 1979, which was due to the construction of Kashima in 1974. However, it has been stable since 1979. On this beach, Port and Harbour Research Institute, Ministry of Transport, has constructed the Hazaki Oceanographical Research Station (HORS) in 1986 for carrying out field observation in the surf zone even under the sever sea conditions. The research pier is a 427 meters long concrete structure supported by 0.8 meter diameter concrete-filled steel piles in a single line, at 15 meters interval. Although the beach is stable, the field experiment is being carried out on this beach because the basic data obtained in HORS have been accumulated since 1986.

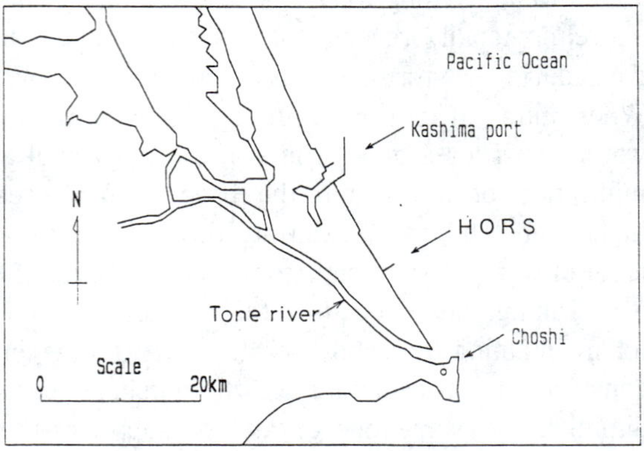

Figure 1 Site of Field Experiment.

2. Permeable layer and its execution

As the permeable material, the unit such as shown in Figure 2 is produced with the two sheets of corrosion-proof expanded metal in a factory, being wrapped with a geotextile fabric. A dimension of the unit is 1 meter wide, 2 meters long, and 20 centimeters thick. There is nothing inside the unit. The units can be connected by using couplers in the longitudinal direction. In August 1994 the permeable layer was set up under the beach in six columns from the foreshore to the backshore, being 30 meters apart in the longshore direction from the pier, as shown in Photo. 1. The elevation of permeable layer is about 3 meters below the beach surface, which is just below an envelope of the beach profiles overlapping for the last decade. The area is 88 meters in the cross-shore direction and 7.8 meters in width, which was enclosed by sheet piles when the permeable layer was being set. The pipe of 0.4 meter in

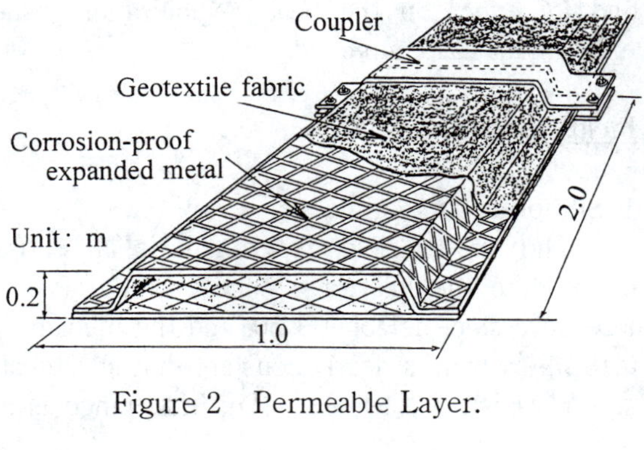

Figure 2 Permeable Layer.

Photo. 1 Execution of Permeable Layer.

diameter has been connected to the sea-side end of permeable layer to drain the ground water into the middle surf zone, 144 meters offshore from the shoreline. The pipe has been buried in the sea bottom with keeping the outlet in the sea. After setting, sand was put on the permeable layer so that we had the same profile as before. The sheet piles at the sea and land sides have been removed. The sheet piles along the longitudinal sides, however, were left without removing in order to make two-dimensional situation in the first experiment. In the second experiment, all the sheet piles were removed.

3.Items of measurements

In the first two-dimensional experiment, two representative measuring lines were taken into account in the cross-shore direction; one was coincided with the center line of the area where the permeable layer was buried, another was the line on the natural beach which was parallel to the first and 29 meters apart in the longshore direction. Along these lines, profiles of groundwater table and beach were surveyed once a day during a calm or every several hours in a storm, respectively. In the second three-dimensional experiment, a topography and plane distribution of water table were measured once a day in an area of 110 meters in the cross-shore direction and 75 meters in the longshore direction, including the area of permeable layer. The drained discharge through the pipe was continuously measured by a supersonic type discharge meter. Waves were measured near the shore line and the offshore.

Result of the first two-dimensional experiment

After about one month of setting up, on 18th September, 1994, a typhoon

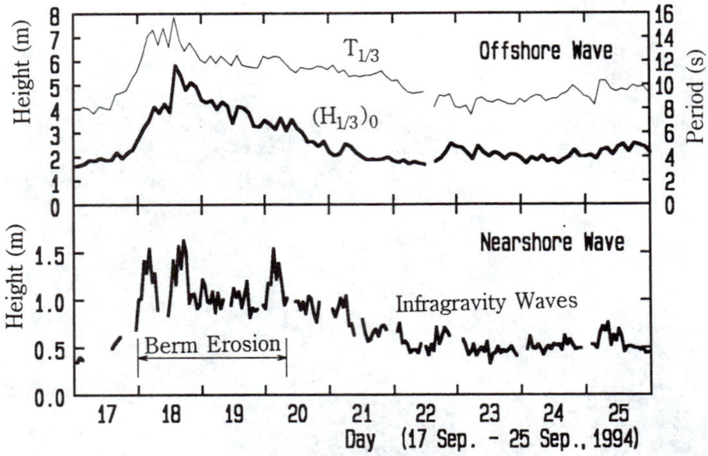

Figure 3 Waves in Typhoon No.9424.

attacked the experimental site. Waves at the offshore and infragravity waves near the shoreline are shown in Figure 3. The extreme value of significant wave height was 5.8 meters. The infragravity waves developed well, being about 1.5 meters in height near the shoreline, which run up to the backshore. The berm on the beach eroded during the days from 18 to 20, which is denoted in the Figure 3.

Photo. 2 was taken from the rooftop of laboratory at HORS on 19th September, on the second day of the storm, when an infragravity wave was running up on the drained beach. Small poles dug into the beach at the cross-shore intervals of 5 meters were arranged along a center line of the area where

Photo. 2 Infragravity Waves Running on the Drained Beach.

Photo. 3 Dried up Area on the Drained Beach.

the permeable layer was set up. Under this situation, sea-water infiltration into the beach was considered to be enhanced in the drained area, because on the drained beach we recognized a very interesting phenomenon that air bubbles came out actively from the ground through the surface of beach, which had never been seen in the laboratory. It is not difficult to infer the situation that the water was infiltrating into the beach because the effusion of air bubbles was a result of being replaced with the infiltrated water into the ground. Photo. 3 is a picture taken about one minute after the situation of Photo. 2, when the infragravity wave was running down to the sea. The beach surface of drained area, of which both sides were cut off by the sheet piles, was already dried up, while the surface of natural beach was still wetted.

Figure 4 is a comparison of profiles of water table between the natural beach and the drained beach, measured on the day of 18th September when the beach began to erode. Infragravity waves had already developed well, and run up to the high elevation denoted by an arrow which was higher than the usual H.W.L. of 1.4 meters. The elevation of groundwater table in the natural beach, which is denoted by triangles, rose up

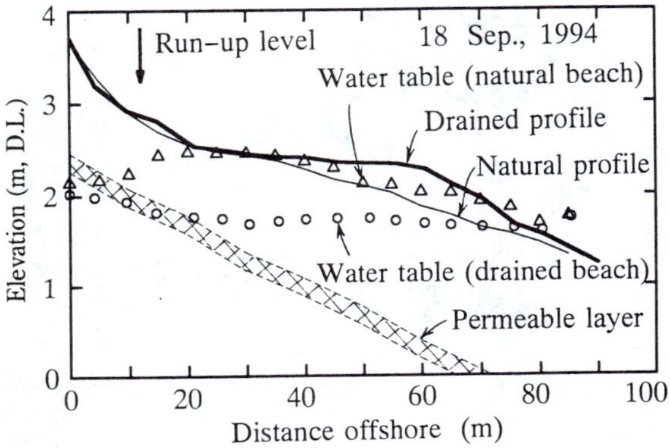

Figure 4 Comparison of Water Tables.

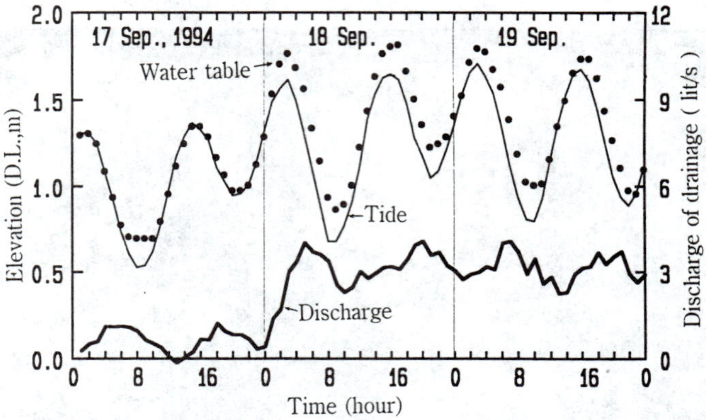

Figure 5 Changes of Tide Level, Water Table, and Drained Discharge.

to the beach surface. In other words, the ground was completely saturated with the water. On the other hand, that in the drained beach which is denoted by circles was 0.6 meter lower in maximum than the beach surface. The area between the beach surface and the water table in the drained beach was unsaturated, which means that the groundwater was gathered to the permeable layer and drained offshore through the drainage pipe.

Figure 5 shows the changes of a tide level at the outlet of drainage pipe, the water table at the point of −25m in the drained area, and a mean discharge in the drainage pipe for 20 minutes. On the day of 17th September, before the storm, the discharge was small. From the day of 18th September on, during the storm, The discharge was large, being about 30 lit/s. The groundwater level changed with a lag of about one hour behind the tidal change. As a result, the level difference between the groundwater table and the tide level became large during the ebb tide, which produced larger discharge.

Figure 6 shows the actual fluctuations of ground water level in the drained

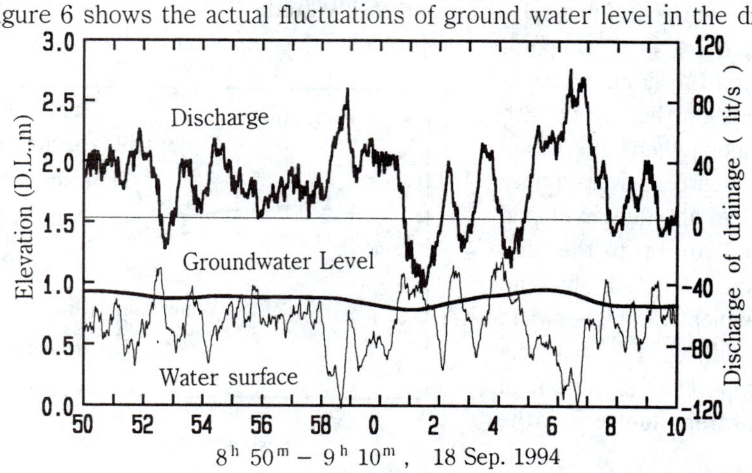

Figure 6 Fluctuations of Groundwater level, Water Surface, and Discharge.

beach and the drained discharge in the surf zone during the segment of twenty minutes in the storm. The level of water surface at the point of outlet in the surf zone, which is the result of running average with 10 seconds, is also presented in the same figure. These elevations were measured on the day of 18th September, the first day of storm. The fluctuation of groundwater level was smooth and small, while that of surface elevation was large, being 70 cm in amplitude and 1 to 2 minutes in period, which was due to the infragravity waves in the surf zone. As the surface level at the outlet was lower than the groundwater level in the better part of time, the water in the pipe usually flowed offshoreward with the discharge of 100 lit/s in maximum, which was equivalent to the mean velocity of 80 cm/s in the pipe. However, the surface level at the outlet was sometimes higher than the groundwater level when the crest of infragravity waves is passing over the point of outlet. When the surface level at the outlet was higher than the groundwater level, the water flows backward, or in the onshore direction.

In Figure 7, the mean discharges are plotted against the level difference between the ground water level and the surface level at the outlet. These values are the averaged ones for every 20 minutes before and during the storm. The averaged discharge was in roughly proportional to the difference of water level, which means that the water was gravitationally drained into the surf zone.

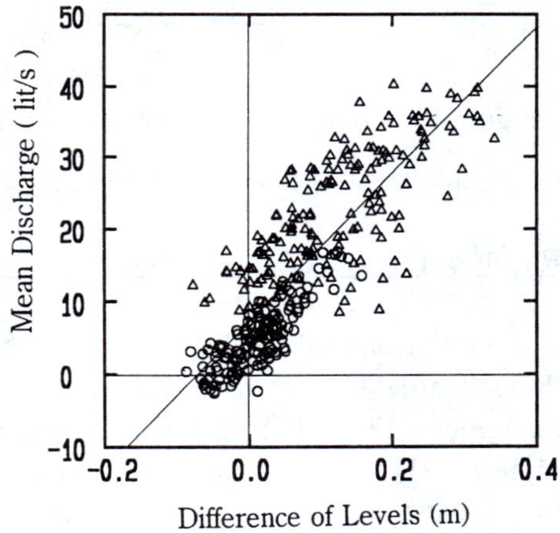

Figure 7 Relation between the Level Difference and the Drained Discharge.

Figure 8 shows the successive comparisons of beach profiles on the natural beach and the drained beach. On the two day before the storm, the day of 16th September, berms were formed on the foreshores of both the natural and the drained beaches (Figure 8, top) due to the relatively calm wave conditions during the three weeks before the storm. The berm on the drained beach was developed better, which is considered to be the effect of drainage. On the first day of storm, the berm on the natural beach eroded to the constant slope as seen in the middle figure, while the berm existed on the drained beach. In the bottom figure, one hour after, the berm still existed on the drained beach, which gradually eroded with time. Around 16 h of 18th, September, which was 13 hours after the time of profiles shown in the bottom figure, the tide level became highest and the height of infragravity waves also became the largest. Under this severe condition, the berm on the drained beach eroded completely to be the same constant slope as that of natural beach. In short, the effect of

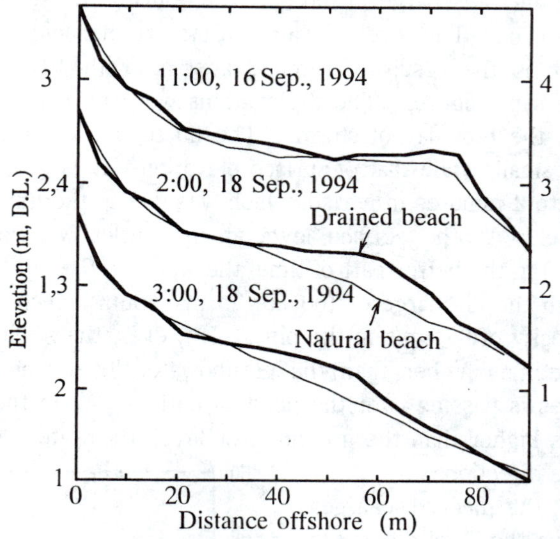

Figure 8 Comparisons of Profiles on the Natural and the Drained Beaches.

permeable layer is to decrease the speed of beach erosion in the storm.

<u>Result of the second three-dimensional experiment</u>

In the second field experiment, the sheet piles along the both sides of drained area had been removed. About one year after, on the day of 17 September 1995, a large typhoon No.9512 passed near the experimental site. Figure 9 shows the changes of the offshore significant waves and the height of infragravity waves near the shoreline. The extreme height of significant wave occurred at 14h on the day of 17th September, being 6.0 m. The height of infragravity waves near the shoreline increased from about 1.0m at 7th on the same day to 1.8m in maximum at 15h, which is the largest in the last 9 years

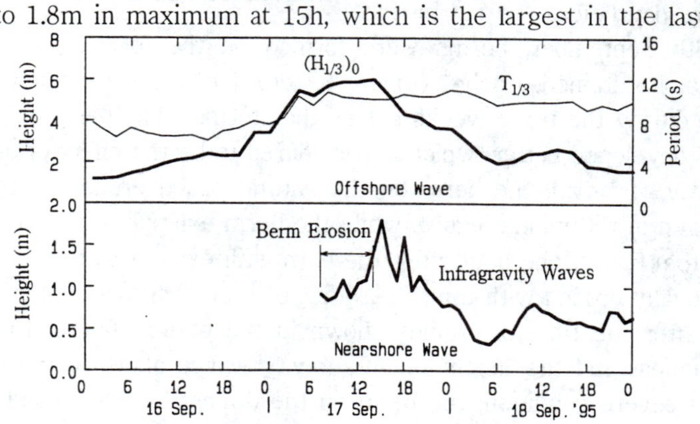

Figure 9 Waves in Typhoon 9512.

since the observation of infragravity waves have started at HORS in 1987.

Figure 10 shows a plane distribution of water table on the day of 16th, September. The water table above the permeable layer was lower than that of the circumference. Waves run up to the location of y=0m, where the water table was higher than 2.0 m in the areas apart from the permeable layer and lower than 1.5 m above the permeable layer.

Figure 11 shows the topography around the drained area, which was surveyed at 7h on the day of 17th, just before the storm. The slope is gentle in the area from y=−50m to 0m, while it is steep in the area offshoreward from y=0m, because a berm is formed on the foreshore. At this time, waves run up to the location of y=−55m, which was observed visually during surveying.

In order to survey the wide area in a short time, nine researchers took part in surveying. During the time from 7h to 14h, on the day of 17th September, however, we could not succeed in obtaining the data of topography because it was very dangerous for us to survey due to the large run−up of infragravity waves and a strong wind velocity of 29 m/s in average.

Figure 12 shows the topography around the drained area which was surveyed after the extreme condition, at 14h on the day of 17th. The foreshore has almost completely eroded to be a plane beach of the two dimensional state of 1/40 in slope. A contour line of 2m has recessed by 30

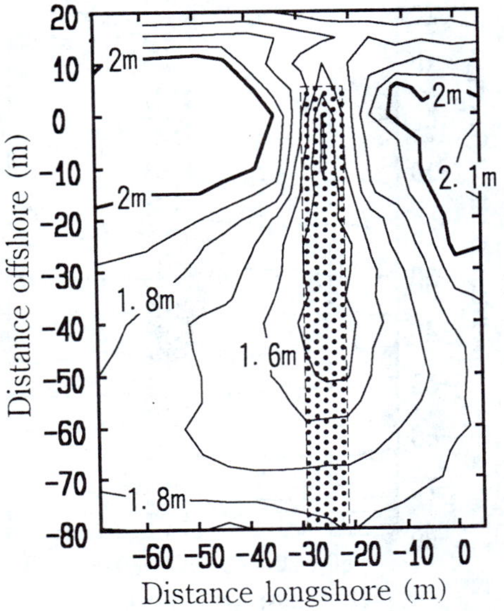

Figure 10 Water Table before the Storm (16th Sep., 1995).

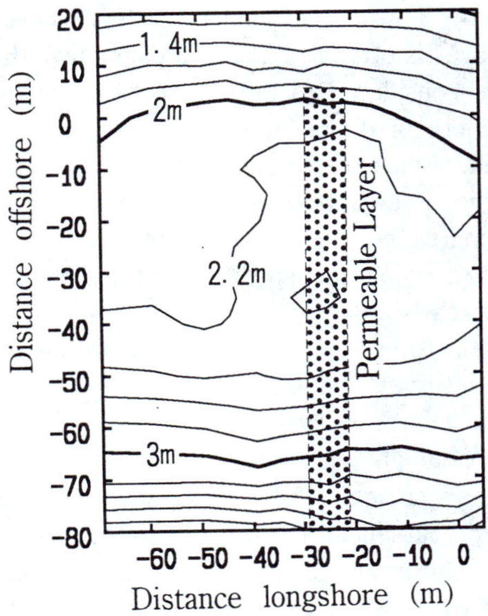

Figure 11 Foreshore Topography, just before the storm

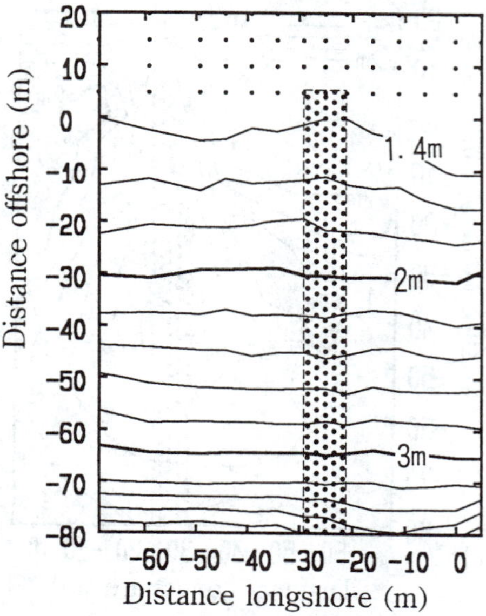

Figure 12 Foreshore Topography, after Erosion.

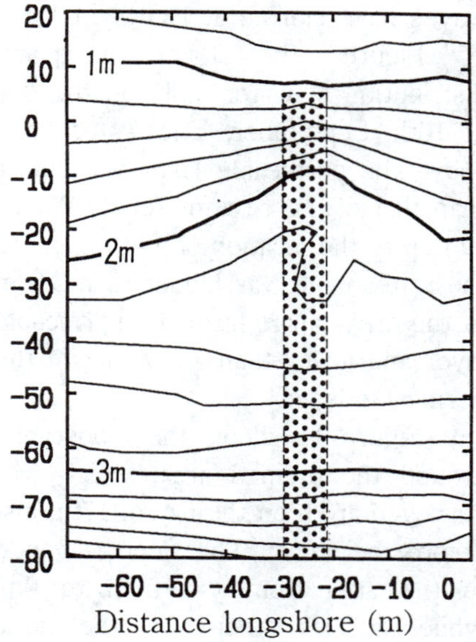

Figure 13 Foreshore Topography, One Week after the Storm.

meters from y=0m (in Figure 11) to y=−30m. No effect of permeable layer is observed in this topography. It is considered to be due to the narrow width of drained area, 7.8 meters, and the attack of extreme large waves. As the interruption of surveying was 7 hours, the effect of permeable layer on decreasing the speed of beach erosion was not confirmed.

After the storm, it was continued to survey the topography in the same area once a day. Figure 13 shows the topography on the day of one weak after the storm, 25th September. The contour lines of 1.4m to 2.2m advanced offshoreward, being convex in the area above the permeable layer. As a result, in the area above the permeable layer a slope was gentle in the area fihigher than 2m in elevation, while it

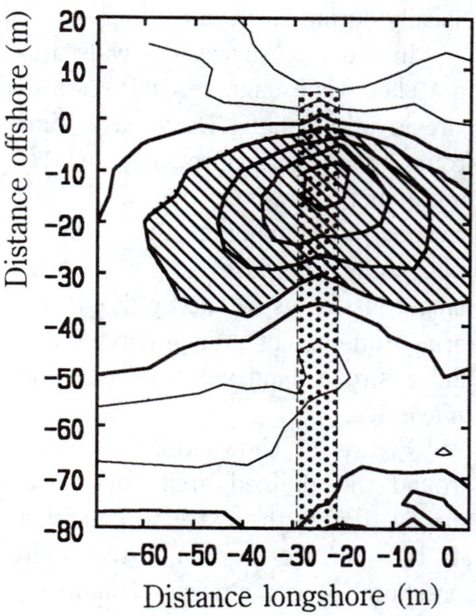

Figure 14 Change of Topography during one week.

was steep in the area lower than 2m. In short, a new berm was formed on the foreshore above the permeable layer.

The rate of sand accumulation during this one week has been calculated, of which result is shown in Figure 14. The area of sand accumulation more than 10 cm is hatched. In the middle of this area, on the drained beach, the thickness of accumulation is more than 40 cm. Then, it can be said that there was a quick recovery on the drained beach. The effect of permeable layer extended to the both lateral areas for about 20 meters, which suggests a possibility of beach protection by setting up the belts of permeable layers in stripes. On the lateral beaches where there was no permeable layer below, sand accumulated gradually and the berms were formed finally, being as the same topography as that before the storm (see Figure 11).

Conclusions

The effects of permeable layer set up under the beach are as follows:

(1)The permeable layer can drain the groundwater gravitationally through the drainage pipe into the surf zone even in a storm.

(2)It has a function of decreasing the speed of foreshore erosion in a storm.

(3)On the drained beach, the eroded foreshore is recovered quickly after a storm.

References

1) Katoh,K. and S.Yanagishima(1988):Predictive model for daily changes of shoreline, Proc. of 21st ICCE, pp.1253–1264.

2) Katoh,K. and S.Yanagishima(1990):Berm erosion due to long period waves, Proc. of 22nd ICCE, pp.2073–2086.

3) Katoh,K. and S.Yanagishima(1992):Berm formation and berm erosion, Proc. of 23rd ICCE, pp.2136–2149.

4) Katoh,K.(1994):Difference between waves acting on steep and gentle beaches, Proc. of 24th ICCE, pp.315–329.

5) Chappell,J., I.G.Eliot, M.p.Bradshow and E.Londsdale(1979):Experimental controle of beach face dynamics by watertable pumping, Eng. Geol., 14, pp.29–41.

6) Vesterby,H.(1991):Coastal drain system, Proc. Inter. Conf. Geotec. Eng. Coastal Development, pp.651–654.

7) Davis,G.A., D.J.Hanslow,K.Hibbert and P.Nielsen(1992):Gravity drainage:A new method of beach stabilization through drainage of the watertable, Proc. of 23rd ICCE, pp.1129–1141.

8) Katayama,T. et al.(1992) :Lowering of watertable with drainage layer under the foreshore, Proc. of Coastal Eng., Vol.39, JSCE, pp.871–875 (in Japanese).

9) Katoh,K. and S.Yanagishima(1995):Changes of sand grain distribution in the surf zone, Coastal Dynamics'95, pp.639–650.

CHAPTER 206

On the change of velocity field in nearshore zone due to coastal drain and the consequent beach transformation

Michio Sato[1], Tadahiro Fukushima[2], Ryuichiro Nishi[3]
and Masahiko Fukunaga[4]

ABSTRACT

It is essential to understand the details of the mechanisms by which a coastal drain works in order to build up the method which enables us to estimate discharge requirement for effective operation in field conditions. Some measurements on the changes of quantities which may be relevant to the function of coastal drain are discussed. Then, an attempt to implicate the current components produced by drainage in a numerical model of beach profile evolution is discussed.

INTRODUCTION

The coastal drain has been considered to be an effective soft approach to coastal erosion control and beach erosion. From the practical point of view, it is indispensable to know how to estimate discharge requirement for effective operation in field conditions. Small scale physical experiments give us qualitative information on some physical aspects related with the coastal drain. There seems, however, to be a difficulty in estimating discharge needed for a coastal drain system based on experimental results due to scale effects. Hence, when we consider the remarkable recent progress of numerical models on beach profile evolution, it will be a practical way to construct a numerical model which enables us to estimate beach profile change under the operation of a coastal drain system. To do this, it is essential to understand the details of the mechanisms by which the coastal drain works.

Authors have shown the following results on forced coastal drain experimentally(1994),

1) seaward area of a shoreline is most appropriate for the installation of a drain pipe
2) as the discharge increases, the drain system becomes effective even for storm wave conditions.

Figure 1 is an example of beach profiles under erosive wave condition. In this case, sediments on the initial shoreline moved mainly toward the upper beachface and the bottom level lowered. But, the erosion in the shoreward part of the drain pipe was

[1] Professor, Department of Ocean Civil Engineering, Kagoshima University, 1-21-40 Korimoto, Kagoshima-shi, 890 JAPAN
[2] Miyazaki prefectural government
[3] Research Associate, Department of Ocean Civil Engineering, Kagoshima University
[4] Kagoshima prefectural government

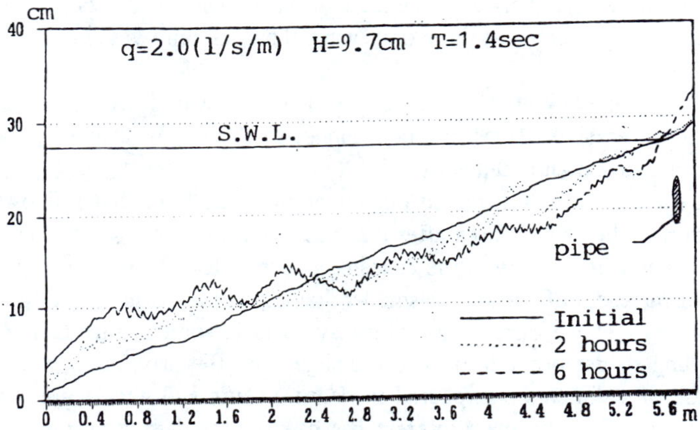

Figure 1 Beach profile change (1)

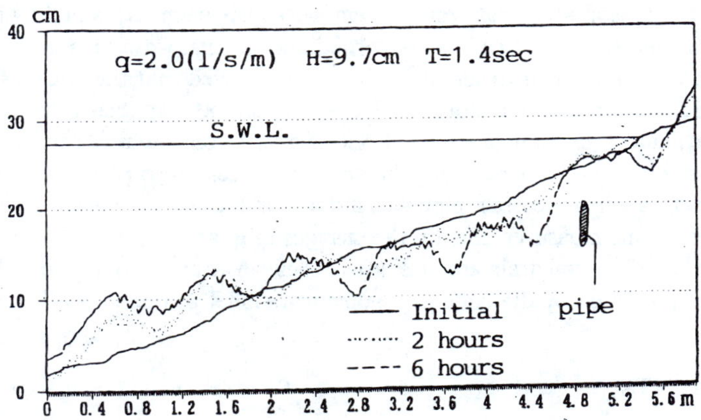

Figure 2 Beach profile change (2)

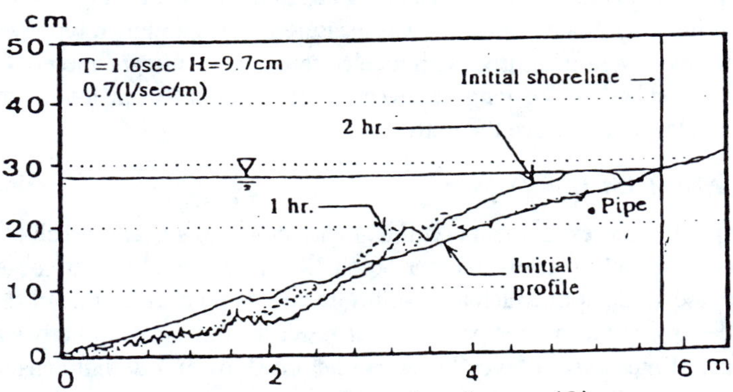

Figure 3 Beach profile change (3)

restrained on the whole. This example made us have expectation the landward area of a drain pipe to be protected at the worst, when the drain pipe was installed in seaward area of a shoreline.

Figure 2 shows an example of the case in which drain was done under the shoreline. Erosion due to erosive waves attained almost to the initial shoreline for the same drainage as the case mentioned above.

Figure 3 is the case which attracted our attention specially in that it showed remarkable accretion in the inshore area after wave action of two hours. The wave condition was less steeper compared with the cases in Figures 1 and 2, but it was not accretive one at all for the cases of without or smaller drainage. This result suggests that, when a storm passes by a shore, similar situation may occur in some stages of the storm passage according to wave condition and drainage. And the process which brought the accumulation of sediments is expected to retard erosion and accelerate restoration of an eroded beach. So, our main interest has been directed to determine under what conditions the accretion takes place.

Generally, the reason a coastal drain system works has been explained by relating effects of drainage to the net sediments carried due to runup-down wash processes of waves on a beachface. Our observation, however, showed that the changes of the hydraulic conditions in the nearshore zone were important. Especially, shoreward flow component induced by drainage in and out of a surf zone, which carried suspended sediments and migrated a bar toward the beachface, seemed to play the dominant role for the accretion when the drain pipe was installed in the seaward area of a shoreline.

In this paper, some measurements on the changes of quantities which may be relevant to the function of coastal drain are discussed. Then, an attempt to implicate the current components produced by drainage in a numerical model of beach profile evolution is discussed.

EXPERIMENTS

The changes of wave runup, wave setup and mean velocity fields due to coastal drain were investigated experimentally by using a wave flume of 13m long, 0.4m wide and 0.4m deep (Figure 4). Water elevation was measured by capacitance type wave gages and velocity fields were measured with an electromagnetic current meter. The measurements were limited in the range between the level about 1cm lower from wave trough and the level 2cm above the bottom.

RESULTS AND DISCUSSIONS

Wave set-up : In their experiments on the gravity drainage system which utilize permeable layer artificially placed under a beach, Kanazawa et.al.(1996) observed no wave set-up, the gradient of which in on-offshore direction provides one of the driving force of undertow. We also anticipated similar result for our cases in early stage. Our measurements of mean water level, however, did not show any definite change in the mean water level as shown in Figure 5. The difference may be considered to reflect the difference of mechanisms which make each system effective.

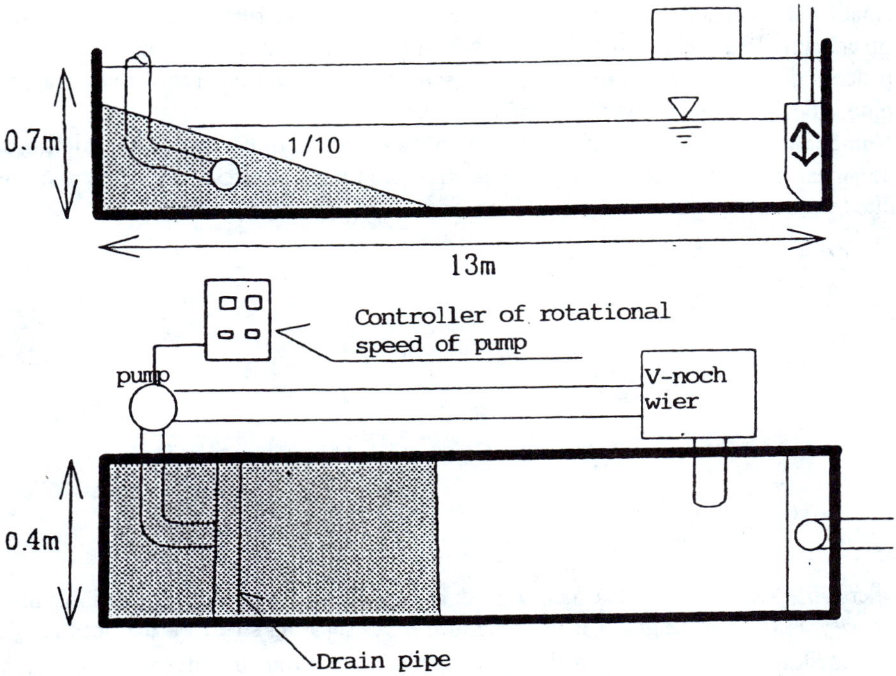

Figure 4 Experimental setup

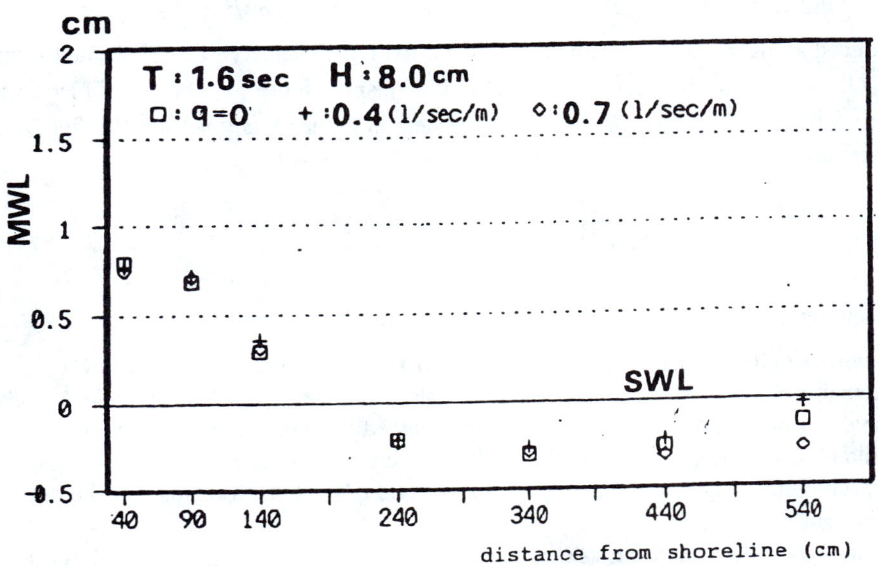

Figure 5 Wave setup

Wave runup : Wave runup was measured visually. The change of wave runup was small. This was supported by the numerical model of wave runup including the effects of coastal drain (Figure 6). These results reveals the change of runup-down wash process due to coastal drain is not necessarily the dominant mechanism when a drain pipe is buried in the seaward part of a shoreline.

Numerical computation was done based on the equations of continuity and motion for nonlinear shallow water waves implicating infiltrating flow from the bottom surface due to drain. The non-dimensional form of the equations are as follows.

$$\frac{\partial h}{\partial t}+\frac{\partial}{\partial x}(hu)=-v^* \qquad (1)$$

$$\frac{\partial}{\partial t}(hu)+\frac{\partial}{\partial x}\left(hu^2+\frac{h^2}{2}\right)+\theta h+f_w|u|u+uv^*=0 \qquad (2)$$

$$x=\frac{x'}{\sqrt{gh'T'}},\quad t=\frac{t'}{T'},\quad h=\frac{h'}{H'},\quad u=\frac{u'}{\sqrt{gh'}},\quad v^*=\frac{v^*}{H'/T'}$$

$$\theta=\frac{\tan\theta'}{H'/\left(\sqrt{gH'}T'\right)},\quad \tau_b=\frac{\tau'_b}{\rho gH'^2/\left(\sqrt{gH'}T'\right)}$$

where h is water depth, t is time, u is fluid velocity in x-direction, v^* is infiltration velocity, H is wave height, T is wave period, τ_b is shearing stress on the bottom, g is the gravitational acceleration and f_w is friction factor. Prime denotes the dimensional variables.

The infiltration velocity v^* was given as the velocity $-(a'q')/\{\pi(x'^2+a'^2)\}$ of the flow induced along x'-axis by a source at $z'=i\,a'$ with the intensity of $-q'/2\pi$ and a sink at $z'=-i\,a'$ with the intensity of $q'/2\pi$ in z'-plane (Figure 7). The non-dimensional form of the velocity for the coordinate system in Figure 8 is given by the following equation.

$$v^*=-\frac{aq}{\pi\left\{(x-x_0)^2+a^2\right\}} \qquad (3)$$

In conducting the numerical computation, Kobayashi et.al.(1987) was referred to.

Mean velocity : Measurements of the mean flow field over the beach in the wave channel were carried out. The beach was plane uniform one of 1/10 slope initially. Wave action of several minutes, however, changed the beach profile to the extent we could ignore. Then, the bed was flattened in every several minutes. Nevertheless, the measurements seems to contain the influence due to the bottom change in the course of measurement.

Figure 9-A shows the mean velocity distribution induced by waves without drain. Figure 9-B and C are the velocity field induced by drain without wave action. And Figure 9-D and E are the velocity field due to waves and drain.

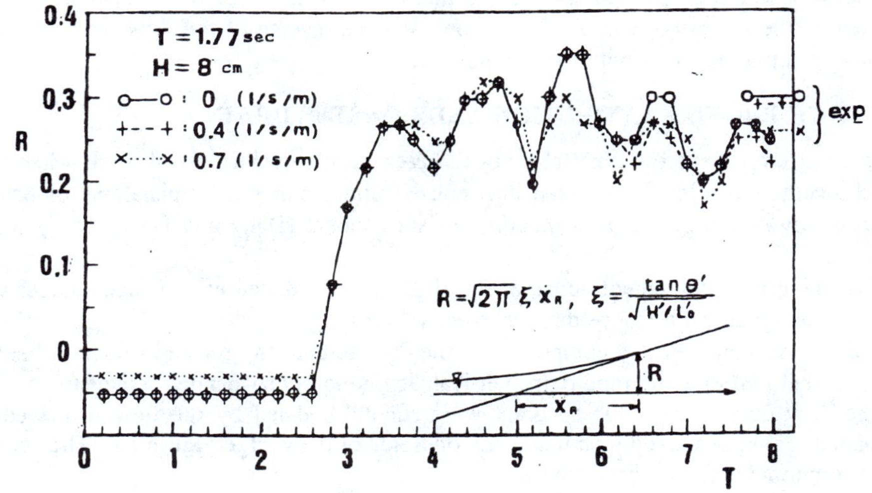

Figure 6 An example of measured and calculated runup

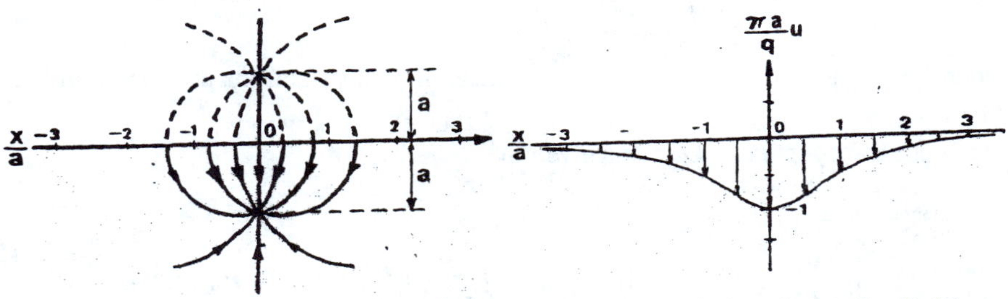

Figure 7 Infiltration velocity model used in the calculation of wave runup

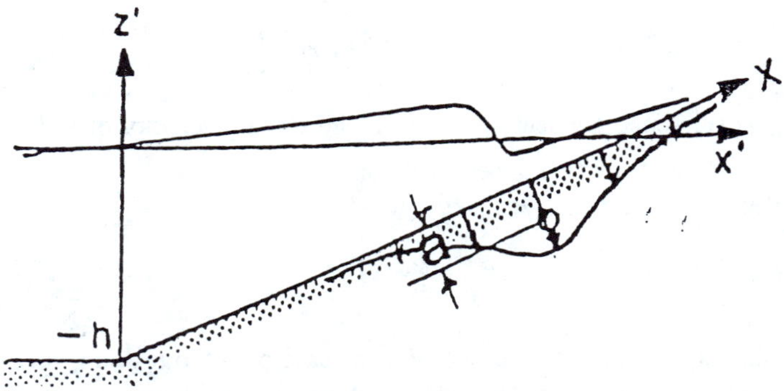

Figure 8 Difinition sketch

From these measurements we can see that the flow in the direction of offshore decreases the intensity near the drain pipe. And the reverse of the flow direction was observed in close proximity to the pipe.

BEACH PROFILE EVOLUTION AND COASTAL DRAIN

In an attempt to know what changes in beach profile evolution will result when the following two effects are taken into consideration, numerical simulations of beach profile evolution based on a modified existing model (Dally and Dean(1984)) were done.

1. the horizontal velocity component of the flow produced by drainage reduces the wave induced offshore-directed mean current,
2. the vertical velocity component of the flow adds to the settling velocity of sediment particles and modify the vertical distribution of suspended sediments.

The flow components due to drain were simply modeled by the flow in a wedge shaped region produced by a point sink on a side of the wedge(Figure 10). The velocity components (u, v) are given by

$$u - iv = \frac{m\pi}{\theta} \cdot \frac{(z/a)^{\pi/\theta}}{z\left\{(z/a)^{\pi/\theta} - a\right\}} \tag{4}$$

where $z = x + iy$ and $i^2 = -1$. The calculated flow distributions are shown in Figure 11.

Sediment transport rate Q_{SS} was given by

$$\begin{aligned} Q_{SS} &= \int_{-h}^{0} \bar{u}(z) \cdot C(z) dz \\ &= \int_{D_p - h}^{0} \{u_1(z) + u_{Dx}(z)\} \cdot C(z) dz + \int_{-h}^{D_p - h} \{u_2(z) + u_3(z) + u_{Dx}(z)\} \cdot C(z) dz \end{aligned} \tag{5}$$

where $\bar{u}$ is mean velocity, $C(z)$ is concentration of suspended sediments, u_{Dx} is x-component of the velocity induced by drain (Figure 12).

The concentration of suspended sediments was given by

$$C(z) = C_A \exp\left\{-15(w + u_{Dz}) \cdot (z - z_A) \big/ \left(h\sqrt{\tau/\rho}\right)\right\} \tag{6}$$

where C_A is reference concentration, w is the settling velocity of sediments, u_{Dz} is z-component of the velocity induced by drain.

Beach profile evolution was calculated by

$$\frac{dh}{dt} = \lambda \frac{dQ_{SS}}{dx} \tag{7}$$

Figure 13 shows the calculated results. For the case of without drain, specious results were obtained. Addition of the flow due to drain makes the sediment transport rates several times larger in magnitude and the sediment transport direction onshore over the whole beach area. And accretion of the shore-side area of the pipe was obtained.

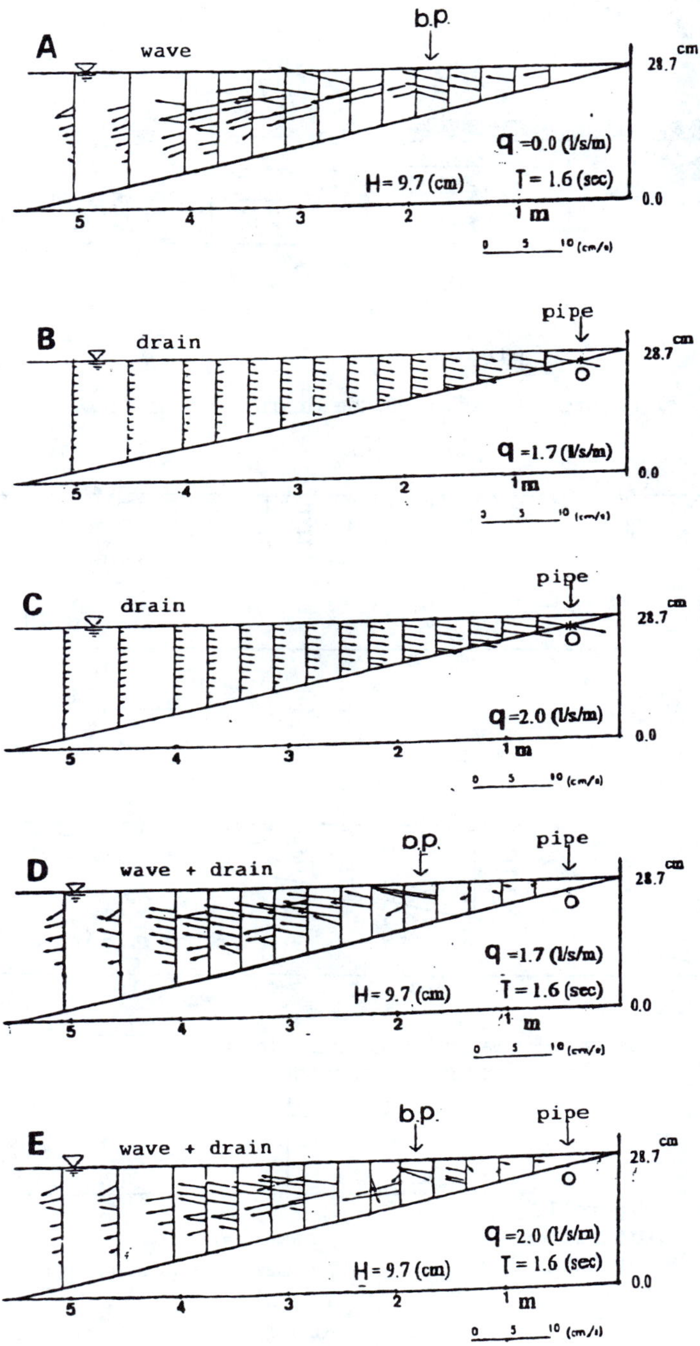

Figure 9 Measured mean velocity

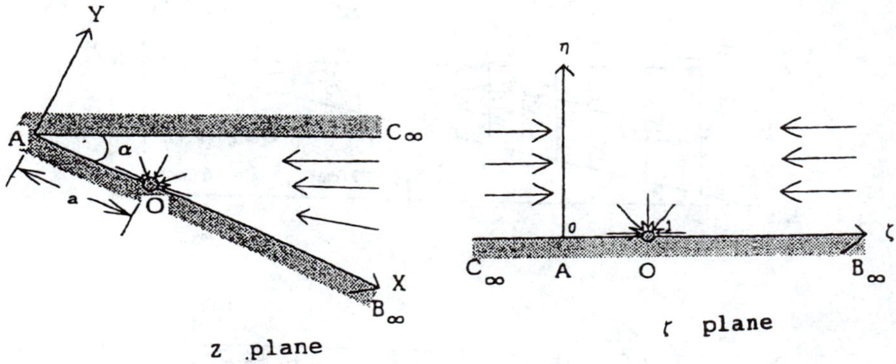

Figure 10 Model of the drain induced flow

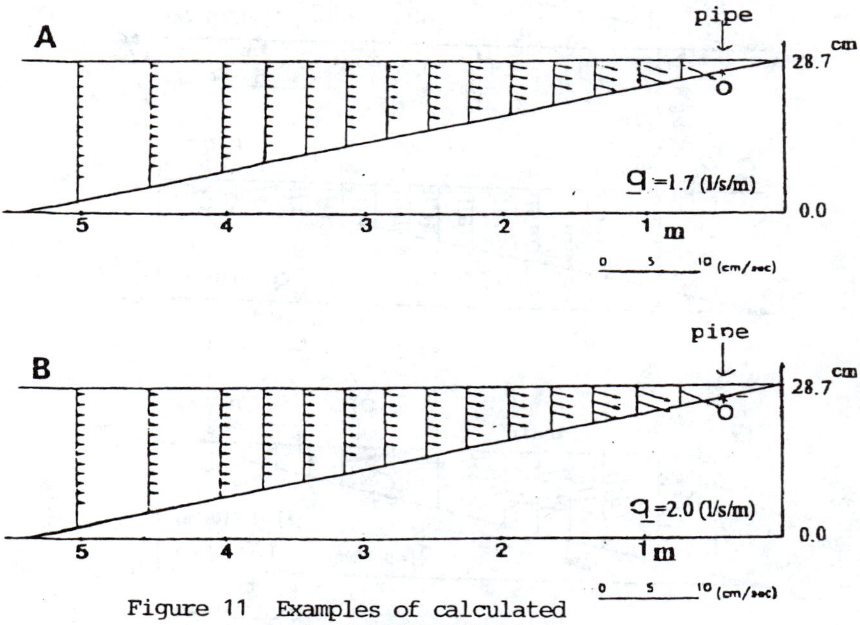

Figure 11 Examples of calculated velocity distribution

Figure 12 Difinition of variables in Eq.(5)

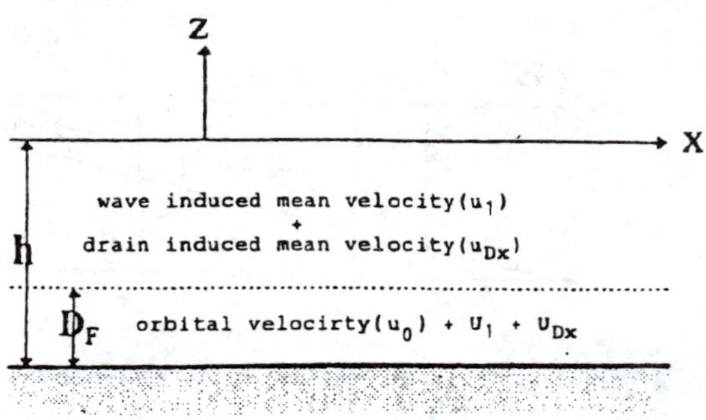

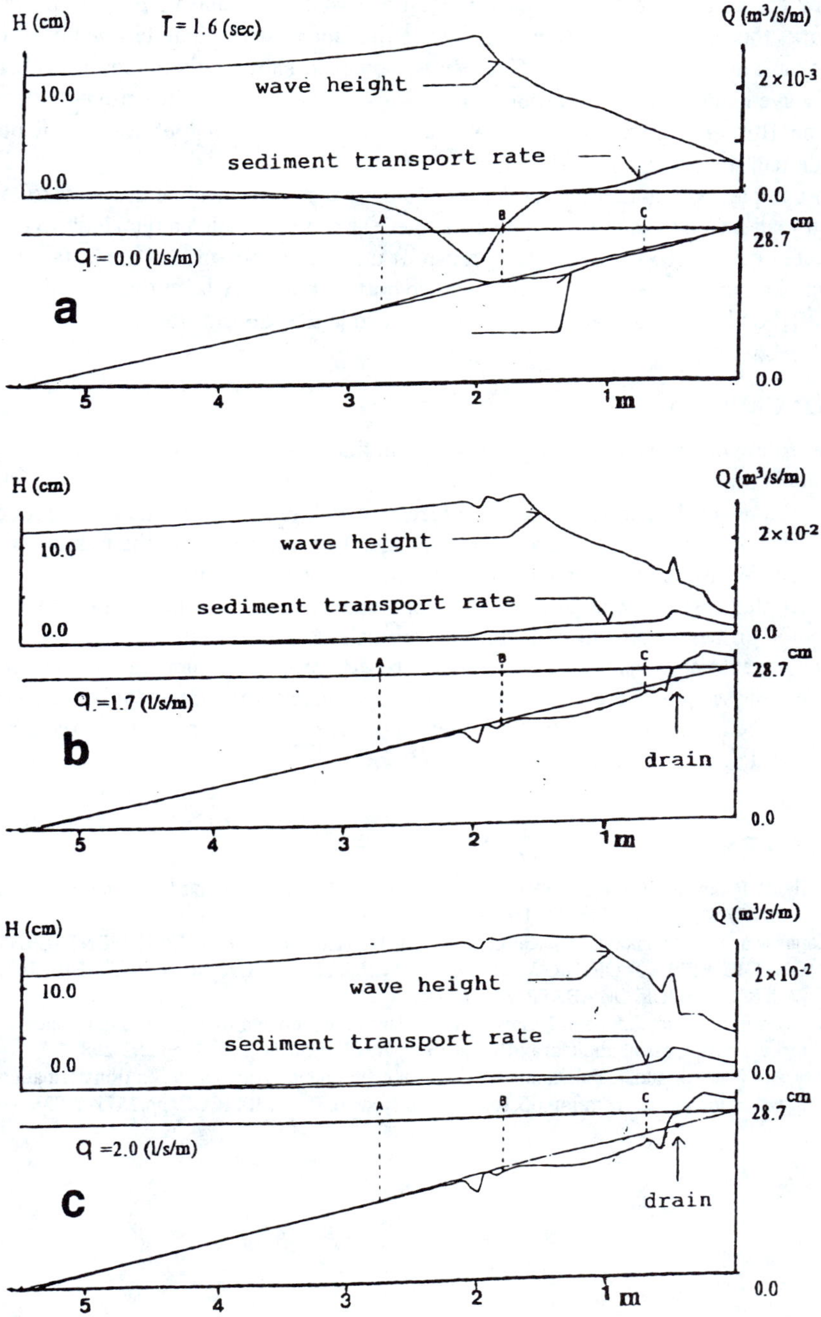

Figure 13 Calculated beach profile evolutions

Therefore, the calculated results show some features of beach profile evolution and they are considered to account for one of possible mechanisms how a coastal drain works though in a qualitative sense, when the pipe is installed under the bottom of the seaward area of a shoreline. The results, however, show fairly bumpy transformation of wave height and bottom topography. Some trials to improve the situation have been done. But, they are not successful so far. Besides, this model does not include bed load sediment transport.

To complete the numerical model for the coastal drain, it will be essential to take the bed load into consideration. However, there are some problems remained to be investigated on the bed load sediment transport under the operation of a coastal drain system like the change in flow within a wave bottom boundary layer over a drain pipe due to suction by forced drain and its effect on bed load sediment transport.

CONCLUSIONS

The following conclusions were drawn from the study:

When a drain pipe is installed in the seaward area of a shoreline,

1. the mean flow in the offshore direction over the whole beach area is reduced by the onshore current induced by drainage. This will results in the reduction of offshore suspended sediments transport in stormy conditions.
2. In this case, the changes in wave runup and wave setup, which are connected with the processes on a uprush zone, were small.

Then, an attempt to implicate the current components produced by drainage in a numerical model of beach profile evolution is discussed. To complete the numerical model, however, there are some problems remained to be investigated, especially on the effects of forced drain on bed load sediment transport.

REFERENCE

Dally, W.R. and R.G. Dean (1984): Suspended sediment transport and beach profile evolution, Jour. of ASCE, WW., Vol. 11. 110, No.1, pp. 15-35.

Kanazawa, H., Matsukawa, F., Katoh, K. and I. Hasegawa (1996) : EXPERIMENTAL STUDY ON THE EFFECT OF GRAVITY DRAINAGE SYSTEM ON BEACH STABILIZATION, 25th ICCE, BOOK OF ABSTRACT, pp.632-633

Kobayashi, N., Otta, A.K. and I. Roy (1987) : Wave reflection and run-up on rough slopes, J. Waterway, Port, Coastal and Ocean Engineering, ASCE, Vol. 113. No.3.,pp.282-298

Sato, M., Hata, S. and T. Fukushima (1994): An experimental study on beach transformation due to waves under the operation of coastal drain system., Proc.24th ICCE.,pp.2571-2582

CHAPTER 207

EVALUATION OF A BEACH DEWATERING SYSTEM: NANTUCKET, USA

William R. Curtis[1], Jack E. Davis[2], Ian L. Turner[3]

Abstract

A commercial multi-segmented beach dewatering system became operational December 1994 on the eastern shoreline of Nantucket Island, Massachusetts, USA. The objective of the installation was to cost-effectively stabilize a critically eroding reach of shoreline with minimal environmental impact. To assess the potential for wider use of beach dewatering in coastal erosion management, a comprehensive and independent monitoring program was developed and implemented to objectively evaluate the beach dewatering project. Observational methods and analyses are discussed relative to the ability to differentiate between the influence of the dewatering segments on beach response, in contrast to 'natural' shoreline response to coastal processes. Economics, measurement needs, measurement accuracy, and temporal and spatial resolution of data acquisition are discussed and evaluated.

Introduction

Nantucket Island is the eastern most member of the Elizabethan Island chain located 48 km southeast of the New England coast, USA (Figure 1). The eastern shoreline of the island is exposed to direct attack from the high-energy wave environment of the North Atlantic. Mean offshore significant wave heights are on the order of 2 m. However, offshore significant wave heights in excess of 5 m are frequently measured, particularly during winter storms. The maximum wave height offshore measured between 1984 and 1993 was 11.6 m (Hubertz, 1995). Historical charts and aerial photography indicate that, for at least the past 150 years, the beach face, dune line and bluff face have experienced episodic accretion and recession (Tiffney et al., 1991). Present estimated erosion rates along the eastern shoreline range from 0.8

---

[1]Research Oceanographer, [2]Research Hydraulic Engineer, USAE Waterways Experiment Station, Coastal and Hydraulics Laboratory, 3909 Halls Ferry Rd., Vicksburg, MS 39180, USA.

[3]Assistant Research Scientist, University of Maryland, Laboratory for Coastal Research, College Park, MD 20742, USA.

m/yr to 4.6 m/yr, depending on location. An analysis of wave refraction patterns suggested that episodic erosion events are due primarily to the focusing of storm wave energy by an offshore shoal complex along the south eastern shoreline of Nantucket (Weishar et al., 1991).

Bluff and dune recession are presently encroaching on private and public facilities of several Nantucket coastal communities. On the eastern shore of the island, where coastal storm damage has been catastrophic, increased coastal development has led to serious consideration of beach management and land use practices.

In 1990, the Siasconset Beach Preservation Fund (SBPF), a private organization of Nantucket homeowners, sponsored an evaluation of shoreline stabilization options. Results of the evaluation revealed that structural shoreline protection alternatives such as large stone revetments and cement seawalls are not economically feasible to construct at Nantucket. In addition, state and local regulatory agencies firmly restrict use of permanent shoreline structures. Beach nourishment was also evaluated, with a projected cost of $20 million (US) with an annual maintenance cost of $1 million (US) to protect approximately 4.27 km of shoreline. In light of these costs, the SBPF opted to privately fund the installation of three commercial beach dewatering-system segments. The dewatering-system segments cost $1 million (US) to construct. Projected annual maintenance costs are approximately $100,000 (US) for the three segments.

Installation

The three beach dewatering segments were developed and installed by Coastal Stabilization, Inc., under license to the Danish Geotechnical Institute (DGI) and became fully operational in December 1994. The project at Nantucket Island is the third commercial installation of beach dewatering technology in the United States, and one of several prototype or full scale facilities installed on North Atlantic coastlines by DGI or under license to DGI (Ovensen and Schuldt, 1992; Lenz, 1994).

Each STABEACH segment consists of buried, perforated drain pipes installed in a shore-parallel orientation below the beach face swash zone. The pipes slope toward wet wells located in the back shore region of the beach. Two dewatering segments drain by gravity, while drainage at the third segment is aided by a vacuum pump. Water collected in the wells is pumped offshore through a 35.5 cm discharge pipe by a 25 hp pump. The stated operational concept of the system is to induce continuous draw down of the watertable at the beach face, enhancing the depositional effect of wave uprush and reducing the erosive effect of wave backrush during natural accretionary periods.

In total, the project spans 9.8 km of shoreline (Figure 2). The northern most segment is referred to as Lighthouse North (*LHN*). The segment immediately to the south of *LHN* is referred to as Lighthouse South (*LHS*). The southern most segment is referred to as Codfish Park (*CFP*). The three systems are separated by stretches of beach outside the influence of groundwater draw down. Figure 3 illustrates the plan-view design of the *LHN* system. Constructed STABEACH segments at *LHS* and *CFP* are similar in concept. STABEACH system parameters are presented in Table 1. Each drain pipe connects to a 1.2 m diameter wet well.

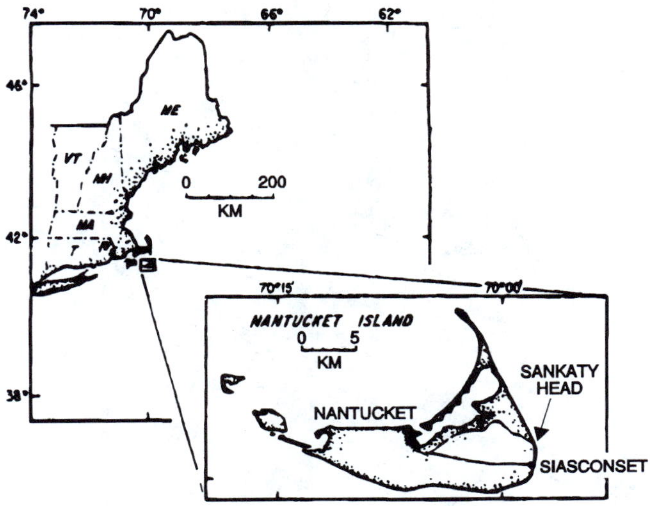

Figure 1. Location of Nantucket Island relative to New England mainland (after Oldale, 1987).

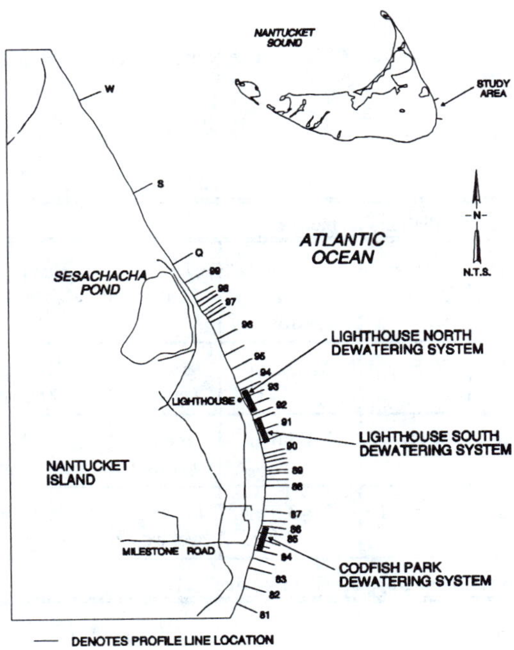

Figure 2. STABEACH segment and profile transect locations (after Coastal Planning and Engineering, Inc., 1996).

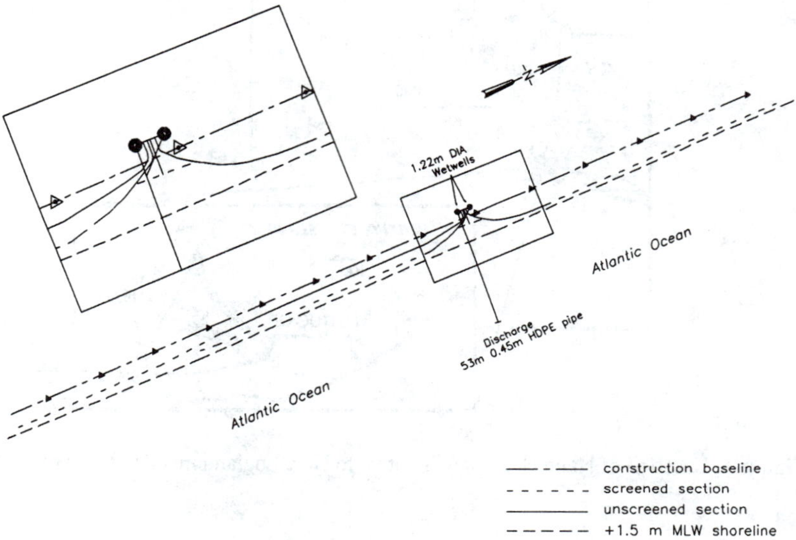

Figure 3. Plan-view schematic of Lighthouse North STABEACH segment.

Table 1. STABEACH Installation Parameters

Segment	Drain Length	Drain Diameter	Drain Elev. (-MSL)	Drain Method	Pump Capacity	Initial Flow Rate
LHN	405 m	30.5 cm	2.4-3.6 m	gravity wet well	22716 lpm 6000 gpm	52 lpm/m 4.2 gpm/ft
LHS	309 m	30.5 cm	2.4-3.6 m	gravity wet well	22716 lpm 6000 gpm	30 lpm/m 2.45 gpm/ft
CFP	357 m	30.5 cm	2.4-3.0 m	low vac. wet well	11350 lpm 3000 gpm	28 lpm/m 2.25 gpm/ft

Note: gpm = gallons (US)/minute, lpm = liters/minute.

Prior to installation of the STABEACH segments, a series of tests were performed by Coastal Stabilization, Inc., to characterize hydrogeological properties of the project site. Subsurface borings identified vertical and horizontal variation of grain-size distributions. Hydraulic conductivity tests were used to optimize vertical and horizontal placement of drain lines and optimize discharge pump design. Table 2 presents average subsurface soil properties for each STABEACH location.

Table 2. Average hydrogeologic properties of project area (Coastal Stabilization, Inc., 1994).

Location	Soil Type	Hydraulic Conductivity (calculated)
CFP	med-coarse sand, no fines clean, uniform	0.233 cm/s
LHS	fine-coarse sand, clean	0.089 cm/s
LHN	fine-med sand, silt/clay	0.2 cm/s

Monitoring Program

Laboratory and prototype field demonstration studies suggest that beach dewatering technology has the potential to stabilize a shoreline. However, the practical application of the technology as a viable means of coastal protection has yet to be adequately documented (Turner and Leatherman, 1996).

To assess the potential for the wider use of beach dewatering in coastal erosion management, a comprehensive and independent monitoring program was developed and implemented by the US Army Engineer Waterways Experiment Station (WES) to objectively evaluate the STABEACH project underway on Nantucket Island. The primary emphasis of the monitoring program is to evaluate whether observed beach changes can be attributed to the dewatering segments. If successful, data from the monitoring program will be used to address quantitative guidelines for the design, construction and operation of future beach dewatering-systems.

Elements of the monitoring program include measurements of beach profiles, bathymetry, hydrogeology, sediment characteristics, coastal processes and system parameters. System parameters such as discharge, operational period, and power requirements have been measured at each dewatering segment for the duration of the monitoring program. The elements of the monitoring program are described in more detail below.

Quarterly Beach Profiles

Beginning in November 1994 (pre-construction), forty-two transects have been established and surveyed on a quarterly basis (Figure 2). The project extends 9.8 km with survey transect 81 at the south end and transect W at the north end. Profile transects are regularly spaced at 305 m intervals, with the exception of 122 m intervals in the vicinity of system drain pipes. For reference, *CFP* system extends from 67 m south of transect 81 to 44 m north of transect 84; *LHS* system extends from 131 m south of transect 91 to 83 m north of transect 91; *LHN* system extends from 46 m north of transect 92 to 122 m north of transect 93.

Each transect extends from the crest of the dune or base of the coastal bank to wading depth (-1.5 m MLW) and was surveyed with a total station and prism rod system. Elevations were surveyed at inflection points in the profile slope and at 6 m intervals from the baseline to wading depth.

To date, quarterly beach profile data have been analyzed for volume and shoreline change. Quarterly measurements document seasonal variability, and comparisons to the pre-construction surveys document the longer-term trends. Figures 4a and 4b present the variation in shoreline position (as defined by the 0.0 m MLW contour) for June 1996 data compared to pre-construction (November 1994) data. When considering profile transects 81-96.5, dewatered beaches have experienced an average shoreline recession of -5.2 m and non-dewatered beaches an average recession of -8.2 m between November 1994 and June 1996. North of transect 96.5, the averaged shoreline has accreted since project installation (0.1 m).

Figures 4a and 4b suggest that rates of recession generally decrease from south to north along the project area. Figure 5 confirms the inference as volumetric changes of the beach profiles between November 1994 and June 1996 range from -115 m^3/m at profile 81 to +12.5 m^3/m at profile W. The trends are consistent with longshore currents which are generally from south to north and dominated by tidal currents (Gutman et al., 1979).

Bi-Monthly Beach Profiles

Spatially and temporally dense data were acquired and are being analyzed to understand small scale variations in beach morphological dynamics as influenced by the system segments. Beach profiles were obtained at 15 day intervals for a six month period (June-November 1996) at eleven transect locations in the vicinity of and outside the influence of the *CFP* segment. Transect locations (82.6, 83.0, 83.5, 84.0, 84.3, 84.6, 85.0, 86.0, 86.4, 87.0, 87.4) were spaced at 122 m intervals in the longshore direction and had a cross-shore spatial resolution of 1.5 m from the dune to -0.9 m MLW. The objective of the high resolution profiling initiative is to correlate beach profile response with concurrent dewatered and non-dewatered groundwater elevation measurements. Through a statistical correlation of profile data with groundwater measurements, an assessment of the system-induced manipulation of beach groundwater on morphological response will be quantified.

Bathymetry

The forty-two transects established for beach profiles were extended offshore using a ship-mounted digital fathometer interfaced to the Coastal Oceanographic HYPAC program. Hydrographic surveys were conducted November 1994, December 1995 and September 1996 using standard fathometer techniques. In September 1995 and June 1996 Scanning Hydrographic Operational Airborne Lidar System (SHOALS) surveys were conducted (Lillycrop et al., 1996). Complete SHOALS surveys of the *LHN* and *LHS* offshore regions were obtained on both occasions, while surveys offshore

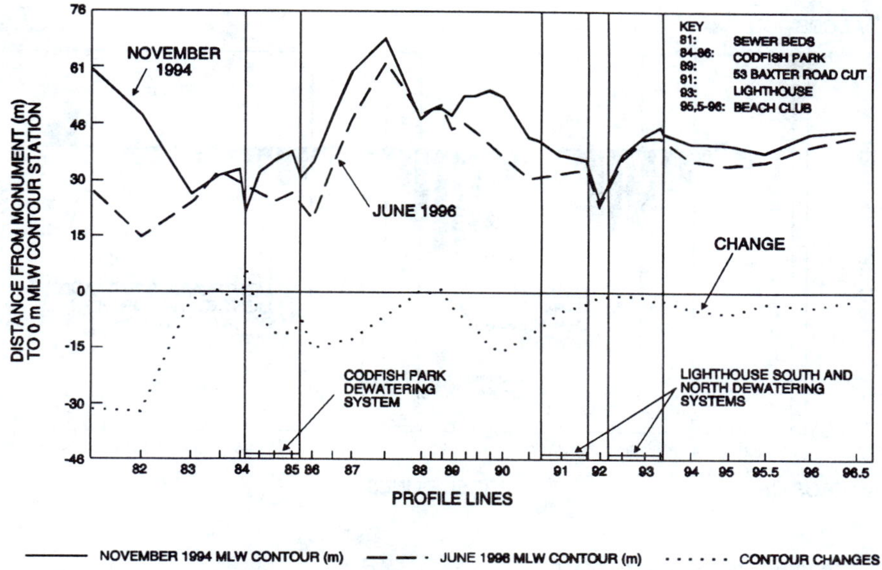

Figure 4a. Contour changes at elevation 0.0 m MLW: November 1994-June 1996 (after Coastal Planning and Engineering, Inc., 1996).

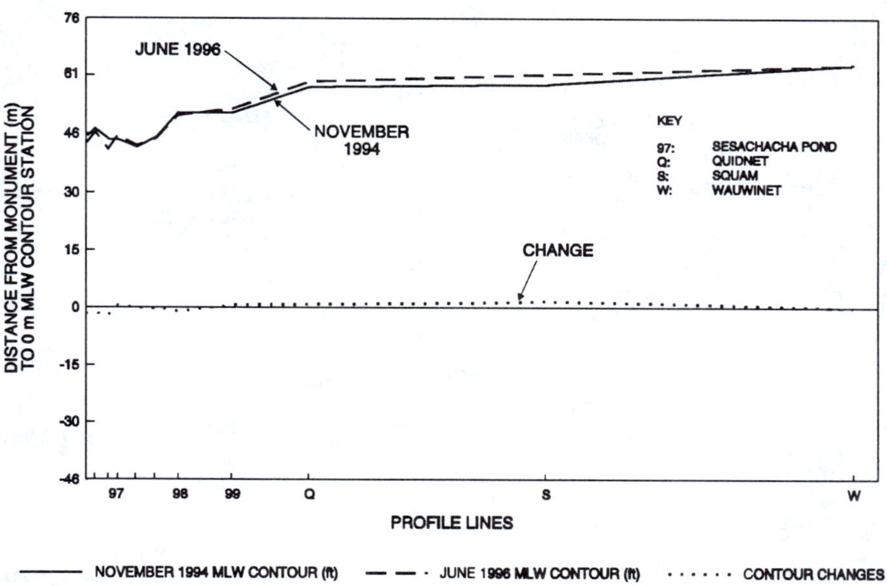

Figure 4b. Contour changes at elevation 0.0 m MLW (continued): November 1994-June 1996 (after Coastal Planning and Engineering, Inc., 1996).

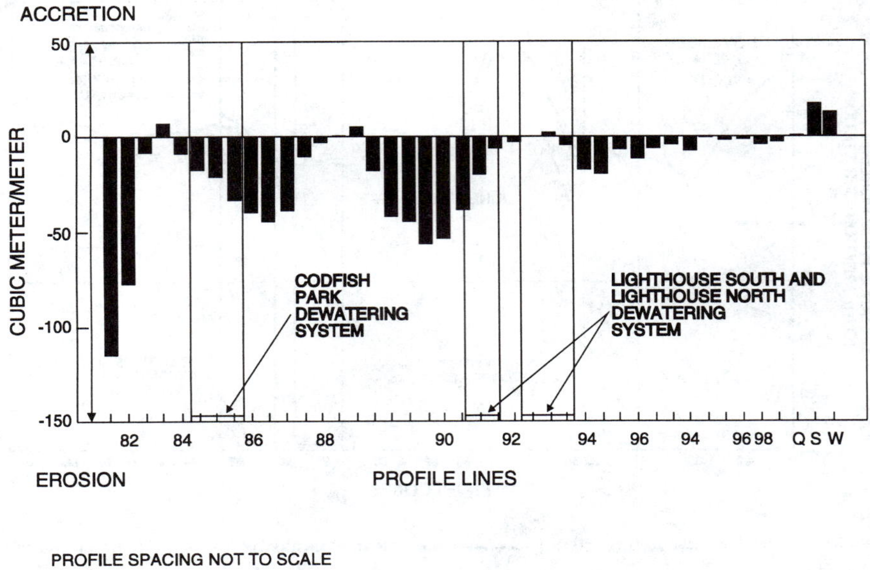

Figure 5 . Volumetric changes at -1.5 m MLW: November 1994-June 1996 (after Coastal Planning and Engineering, Inc., 1996).

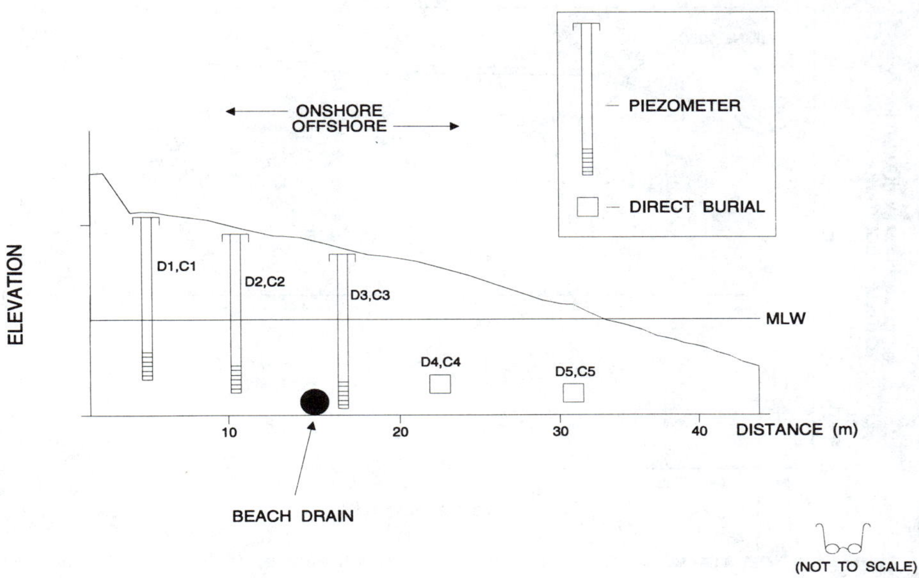

Figure 6. Cross-sectional schematic of the groundwater monitoring installation at Codfish Park STABEACH segment and adjacent control beach.

CFP were unobtainable due to water clarity and weather limitations. Data obtained using standard fathometer and SHOALS methods will be used in combination with the beach profiles by comparative analysis to assess the response of morphological processes to the STABEACH segments.

Groundwater Elevation

Two cross-shore transects of buried pressure sensors were installed in December 1995 and re-installed in May 1996 following storm damage at the project site to monitor the variations of watertable elevation below a dewatered and non-dewatered beach (Figure 6). Groundwater monitoring transects are located at *CFP* and at the non-dewatered beach between *CFP* and *LHS*. Sensors are either directly buried or encased in screened wells and connected to recording instruments. Data loggers sample watertable fluctuations every 30 seconds and average them over 5 minutes at 15 minute intervals. A barometer and internal thermometers are used to correct absolute pressure measurements for low-frequency barometric and temperature fluctuations. The objective of the measurements is to determine the influence of a STABEACH segment on beach groundwater dynamics and address key issues such as the effect of dewatering on bed saturation characteristics at the beach face, the influence of dewatering during storms and during post-storm recovery of the beach profile, and the effect of dewatering on the local fresh groundwater supply. The second transect of pressure sensors outside the influence of the *CFP* system will be used as a control for comparative analysis.

Other Morphological Monitoring Elements

Additional WES-sponsored monitoring include quarterly controlled aerial photographic surveys and use of a video imaging system. Aerial photographic surveys will be used to compliment analyses of shoreline and coastal dune position for the project.

A digital imaging system was installed on a lighthouse located on the bluff above the *LHN* segment. The cameras were located approximately 55 m above the beach. Three cameras were needed to photograph the full extent of the beach within view of the lighthouse. The imaging system is capable of quantifying beach changes and patterns of incident wave breaking within the camera's field of view (Lippman and Holman, 1989). The system obtains instantaneous images of the beach (Figure 7) on the hour during daylight. In addition, the system produces a 10 minute time-averaged image (Figure 8). The time averaged image is produced by sampling at a rate of 3 Hz for 10 minutes and applying an averaging routine to produce a statistically stable image. Images will be used to analyze variations in foreshore morphology at a dewatered and adjacent non-dewatered beach.

Wind and Wave Parameters

Deep water wave parameters are measured on the continental shelf at approximately 500 km east and at 223 km southeast of Nantucket Island. The moored discus buoys are maintained by the National Data Buoy Center and provide directional wind spectra, non-directional wave spectra and climatic data. Unfortunately, no shallow

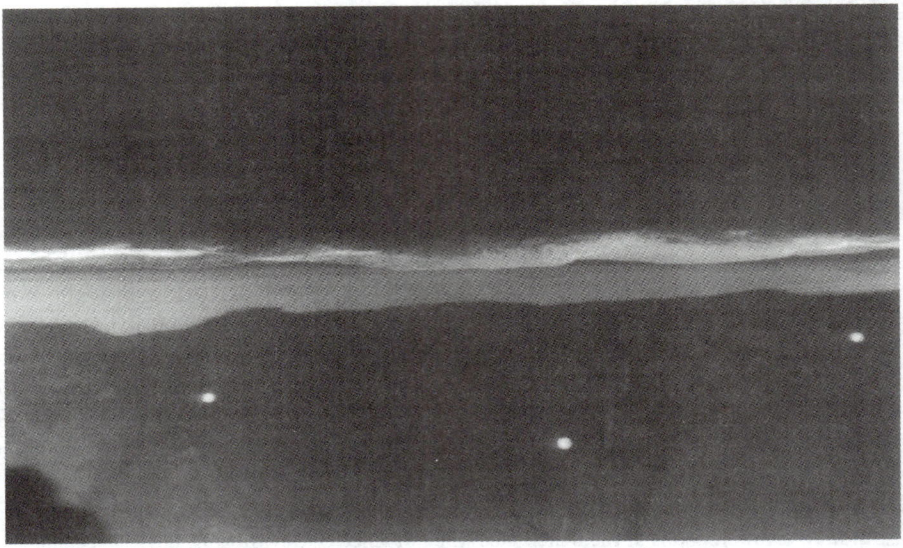

Figure 7. Instantaneous digital image of Lighthouse North dewatered swash zone.

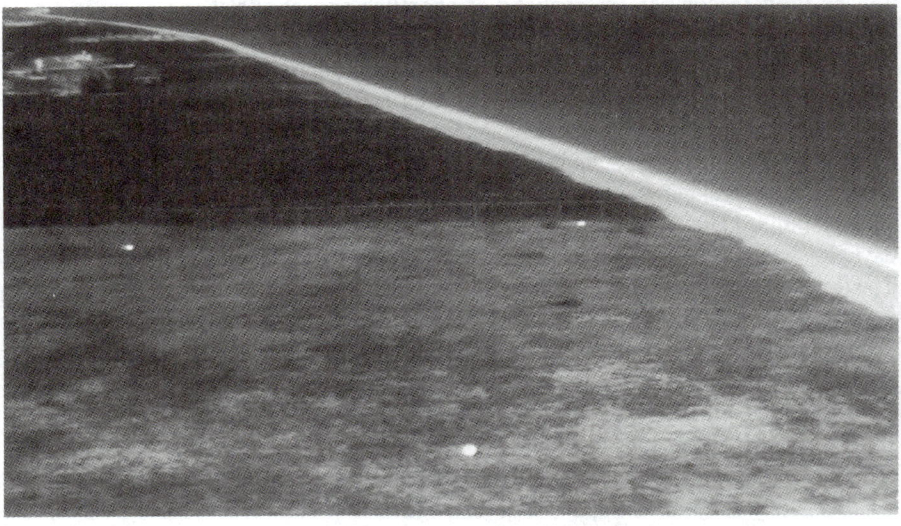

Figure 8. Time-averaged digital image of Lighthouse North dewatered swash zone and adjacent non-dewatered shoreline to the north.

or intermediate depth wave parameters are measured in the nearshore region. However, given the NDBC buoy data, shallow or intermediate water depth equivalents can be estimated. In addition, the SBPF sponsors data collection of local climate variables (e.g., wind speed, direction and barometric pressure).

Monitoring Elements of Public Concern

The SBPF sponsors environmental impact monitoring of the STABEACH segments. Parameters of public concern are measured quarterly and include the impact of beach groundwater draw down on the public fresh water supply, on dune vegetation communities, on beach meiofauna, and on system discharge quality.

Discharge Quality

Salt water is discharged offshore from all three dewatering segments. Salinity of the discharged water ranges from 21-33 ppm and is comparable to nearshore ocean water (30-33 ppm). Bacteria levels of water discharged offshore are an order of magnitude below the standard acceptable for safe swimming. Measured nitrate levels are slightly higher at the *CFP* discharge (0.21-0.54 mg/l) than the *LHS* and *LHN* discharges (0.07-0.29 mg/l). The close proximity of residential septic systems to the dewatering segment is responsible for the higher nitrate measurements at *CFP*. Variations in the elevation of the public fresh water supply aquifer are monitored at 0 m, 70 m, 104 m, 453 m and 640 m landward of the *CFP* wet wells. Monitoring results follow the fluctuation trends in the elevation of the regional aquifer landward of the wet wells, as measured by the United States Geological Survey (USGS) (Figure 9).

Maritime Vegetation and Meiofauna

Monitoring transects were established at *CFP* and *LHS* for evaluation of the impact of groundwater elevation draw down on dune vegetation communities. Variations in the vegetative communities are not influenced by dewatering-system operation. Since project construction, variations in the vegetative communities are primarily due to vegetation lost during wave erosion events. Attempts to quantify dewatering effects on beach meiofauna are not definitive. Spatial and temporal variability in taxonomic richness, diversity, evenness and total number of organisms per sample was significant at dewatered and non-dewatered sample stations (Figure 10). The influence of the STABEACH system on local meiofauna communities is not detectable from the available monitoring data.

Summary

The objective of this paper is not to present definitive conclusions as to the effectiveness of the STABEACH installations at Nantucket Island. Rather, the emphasis has been to convey elements of an independent monitoring program which was established to evaluate beach dewatering technology as a means to stabilize sand on a receding shoreline in a high wave energy environment. Shoreline position of Nantucket Island is highly variable when considering temporal scales ranging from several hours to decades. Monitoring efforts were established to identify trends in shoreline

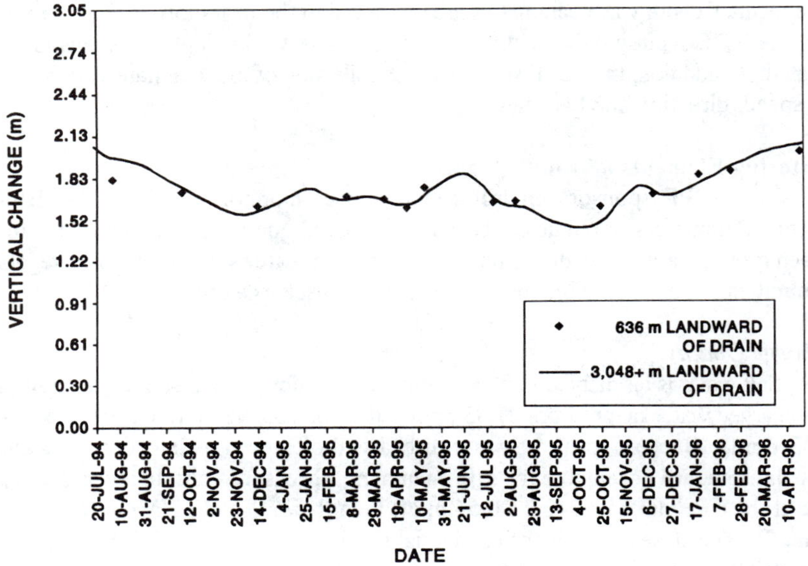

Figure 9. Vertical changes in groundwater elevation landward of Codfish Park STABEACH segment (after Fugro East, Inc., 1996).

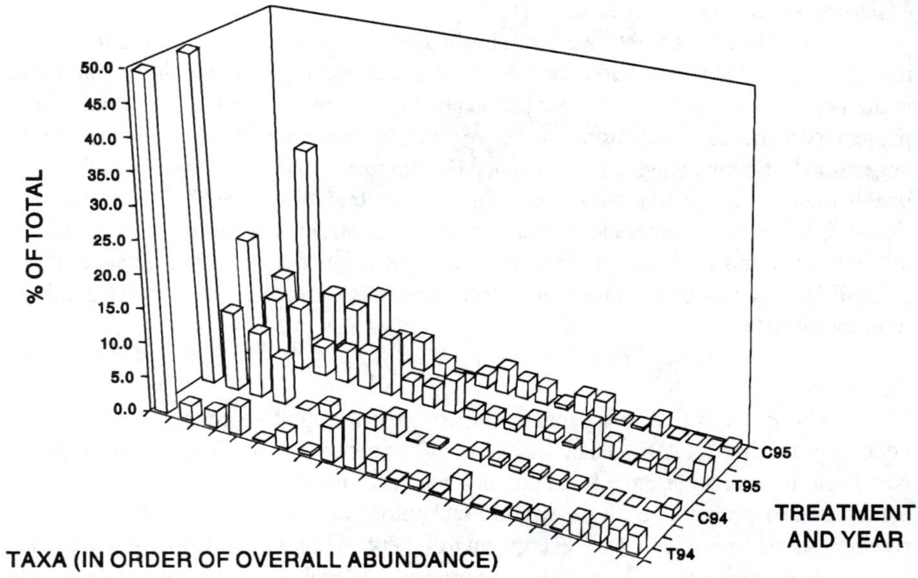

Figure 10. Meiofauna count for dewatered and non-dewatered beaches: 1994 and 1995 (after Fugro East., Inc., 1995).

morphodynamics on varying temporal scales as influenced by operational STABEACH segments.

WES-sponsored monitoring was initiated in June 1995 and is scheduled to terminate in May 1997. After the data collection is terminated, a comprehensive evaluation of use of the STABEACH system as a shoreline stabilization method at Nantucket Island will be presented in the literature.

Three STABEACH segments have been in operation at Nantucket Island for two years. Operation of the *LHN* segment has been nearly continuous since installation. Discharge pumps located at *LHS* were damaged during a storm in January 1996 and returned to operation at 70% capacity in July 1996 and 100% capacity in September 1996. The pump which draws water through the drains at the *CFP* segment was rendered inoperable February 1996. It is speculated that the *CFP* segment operated at 50% capacity under gravity drainage, until the pump returned to operation in August 1996.

Following 18 months of quarterly monitoring, visual and statistical analyses confirm that both dewatered and non-dewatered beaches have highly variable seasonal profile fluctuations. Quantification of the influence of drainage of the beach face on sand stabilization at the project site is presently under investigation.

SBPF-sponsored monitoring reveals that the quality of discharged water has negligible impact on water quality offshore of the dewatering-system segments. Groundwater draw down attributed to the STABEACH segments has had no discernable impact on local vegetative and meiofunal communities. In addition, the drainage of the beach face has no adverse impact on the local public freshwater aquifer.

Acknowledgments

The Authors would like to acknowledge Coastal Planning and Engineering, Inc., Boca Raton, Florida, USA for providing shoreline analysis and beach profile volume change calculations and Fugro East, Inc., Sandwich, Massachusetts, USA for providing environmental analysis information. Results and information presented herein, unless otherwise noted, were obtained from research conducted under the Construction Productivity Research Program of the US Army Corps of Engineers by the US Army Engineer Waterways Experiment Station. Permission to publish this paper was granted by the Chief of Engineers.

References

Coastal Planning and Engineering, Inc. (1996), “Coastal engineering analysis of the Siasconset, Nantucket Island erosion control project”, *Sixth Qtr. Rpt: June 1996*, Coastal Planning and Engineering, Inc., Boca Raton, FL.

Coastal Stabilization, Inc. (1994), “Geotechnical and hydrological investigation and design for the installation of a STABEACH system: Nantucket, Massachusetts”, *Int. Doc.*, Coastal Stabilization, Inc., Rockaway, NJ, 11p.

Fugro East, Inc. (1996), “Winter quarterly report (December 1995-February 1996): Siasconset erosion control project, Nantucket, Massachusetts”, *Int. Doc.*, Fugro

East, Inc., Sandwich, MA.

Fugro East, Inc. (1995), "Summer quarterly report (June-August 1995): Siasconset erosion control project, Nantucket, Massachusetts", *Int. Doc.*, Fugro East, Inc., Sandwich, MA.

Gutman, A.L., M.J. Goetz, F.D. Brown, J.F. Lentowski, and W.N. Tiffney (1979), "Nantucket shoreline survey: An analysis of Nantucket shoreline erosion and accretion trends since 1846", *MITSG 79-7*, Massachusetts Institute of Technology Sea Grant College Program, Cambridge, MA.

Hubertz, J.M. (1995), "Variation of measured meteorologic and oceanic variables off the U.S. Atlantic coast: 1980-1994", *4th Int'l. Workshop on Wave Hindcasting and Forecasting*, Banff, Canada, October 16-20, 1995, 45-57.

Lenz, R.G., (1994), "Beach face dewatering - a tool for coastal stabilization", *Proc. 1994 Nat'l Conf. on Beach Pres. Tech.*, 27-52.

Lippman, T.C. and R.A. Holman (1989), "Quantification of sand bar morphology: A video based technique on wave dissipation", *J. Geophys. Res.*, 94(C1), 995-1011.

Lillycrop, W.J., L.E. Parson and J.L. Irish (1996), "Development and operation of the SHOALS Airborne Lidar Hydrographic Survey System", SPIE: Laser Remote Sensing of Natural Waters: From Theory to Practice (unpublished).

Oldale, R.N. (1987), "Wisconsinian and pre-Wisconsinian drift and Sangamonian marine deposits, Sankaty Head, Nantucket, Massachusetts", *Centennial Field Guide-Northeastern Section*, Geological Society of America, 221-224.

Ovensen, N.K. and J.C. Schuldt (1992), "Beach Management System", *Int. Doc.*, Danish Geotechnical Institute, 20p.

Tiffney, W.N., L.L. Weishar and J.C. Andrews (1991), "Nantucket Island's shifting shoals and moving shores: Nearshore bathymetry controls beach deposition and erosion", *Coastal Zone '91*, American Society Civil Engineers, N.Y., vol. 3, 2421-2433.

Turner, I.L. and S.P Leatherman (1996), "Beach dewatering as a 'soft' engineering solution to coastal erosion - a history and critical review", In press, *J. Coast. Res.*

Weishar, L., W.N. Tiffney, N. Kempe and M. Fields (1991), "Shoreline response to offshore shoals and storms: Low Beach, Nantucket, MA", *Coastal Zone '91*, American Society Civil Engineers, N.Y., vol. 3, 2435-2449.

CHAPTER 208

SEDIMENT TRANSPORT PROCESSES AT OCEAN BEACH, SAN FRANCISCO, CALIFORNIA

Robert T. Battalio, P.E.[1] & Dilip Trivedi, Dr. Eng., P.E.[1]

The San Francisco estuary inlet and tidal bar is a highly complex system that strongly influences littoral processes on adjacent beaches. Ocean Beach is about 4 miles long, located near the entrance to the estuary (Figures 1 and 2). The purpose of the study was to investigate the extent and causes of shoreline variability at Ocean Beach, and identify possible sand transport processes and directions based on available information (Moffatt & Nichol Engineers, 1994; 1995).

Geographic Description of Vicinity

Ocean Beach is a sandy beach on the west side of San Francisco, California, USA (Figures 1,2 and 3). A linear dune behind the beach forms an embankment covering the West Side Sewer Transport Box, supporting the Upper Great Highway, and protecting the urban residential area from the Pacific Ocean. Seawalls have been constructed where the embankment eroded frequently. These structures are the most obvious indications of impacts by man dating back to the late 1800's (O'Shaughnessy,1924; Olmsted and Olmsted,1979; Berrigan,1985a; O'Doherty,1994). Perpendicular to the Great Highway are a series of streets named alphabetically from north to south (eg. Noriega, Ortega, Pacheco, Quintara, etc.) which are often used as location references.

The offshore bathymetry and shoreline to the north and south are very irregular, affecting wave propagation and littoral sand transport (Figure 1). The Golden Gate is located to the north of Ocean Beach. This channel connects San Francisco Bay with the Pacific Ocean. The large tidal prism (2.3×10^9 cubic meters, m^3 for a typical tide range of 1.5 meters) and rocky shoreline results in a relatively constricted channel (0.084 square kilometer) and strong currents (typical peaks about 10 kilometers per hour) (USACE, 1990).

---

1)Civil/Coastal Engineer, Moffatt & Nichol Engineers, 131 Steuart Street, Suite 300, San Francisco, California, 94105

The San Francisco Bar is a horseshoe-shaped, sandy shoal with a radius of about 5 kilometers (3 miles). The Bar crest depths are as shallow as 7.3 meters (24 feet) at low tide. Every winter, larger storm seas and swells induce wave breaking along the crest of the Bar. There are several sand shoals just east of the Golden Gate in San Francisco Bay. Sand beaches exist along the south boundary of the Golden Gate.

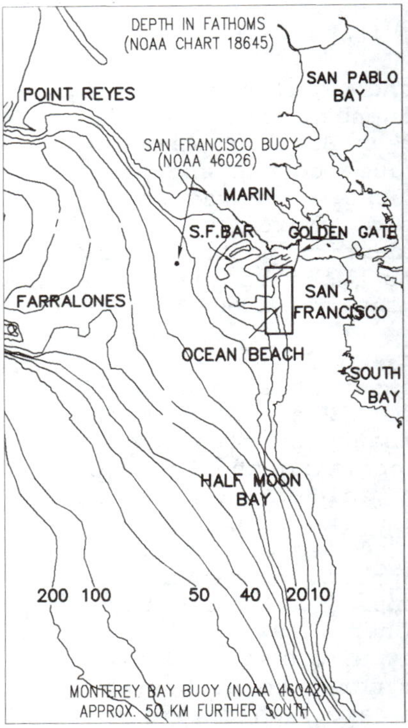

Figure 1: Site Location

Figure 2: Aerial Photo of Site

Sediment Characteristics

Median grain sizes of sand samples from the vicinity were reported by Trask(1958), Moore(1965), Johnson(1971), USACE(1974;1993), Galvin(1979a), Woodward-Clyde Consultants (1979), Ecker(1980), and Noble(1985). At Ocean Beach, the median grain sizes were typically 0.2 to 0.6 millimeters (mm), but occasionally between 0.6 and 1.6 mm. Grain sizes vary significantly along the beach and seasonally, with larger grain sizes in the spring (Trask, 1958; Galvin,1979), especially near the base of the San Francisco Bar. Median grain sizes on the crest of the San Francisco bar are the smallest (0.1 to 0.2 mm), with coarser sizes inside the crest (0.2 to 0.7 mm). Median grain sizes inside the Golden Gate are similar.

Johnson (1971) summarizes mineralogy studies in the area.

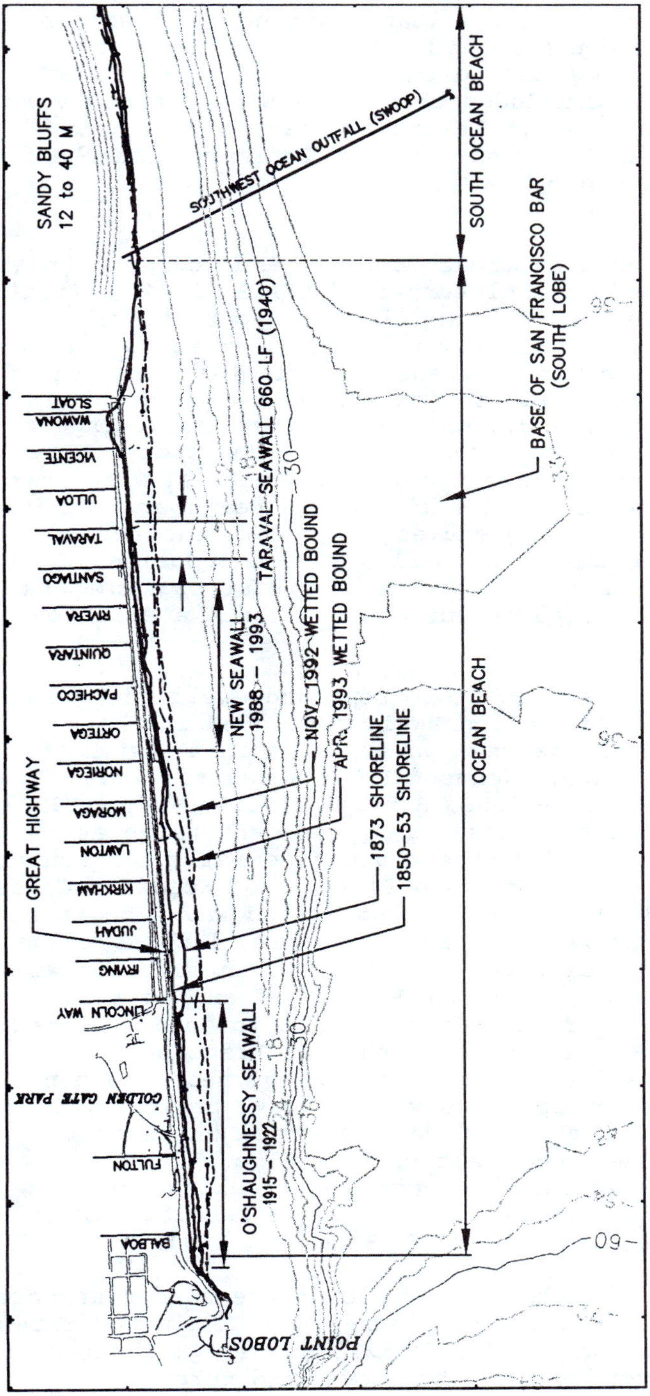

SCALE IN FEET

0 1000 3000 5000

NOTES:

1. DEPTH CONTOURS (FT) SHOWN BASED ON MEAN LOWER LOW WATER, FROM 1954 – 1956 HYDROGRAPHIC SURVEY.

2. PLANIMETRIC MAP OBTAINED FROM HAMMON, JENSEN & WALLEN, NOVEMBER 1992.

Figure 3: Study Area

Similar heavy mineral content in the sands were found at Ocean Beach and nearby areas (San Francisco Bar, Golden Gate, and beaches south to Pedro Point), but not for offshore areas or beaches to the north or farther south. Johnson(1971;1977) concluded that there was probably very little net sand transport at the boundary of the area defined by the San Francisco Bar and Entrance, and the beaches south to Pedro Point.

Currents

Detailed data regarding tidal currents and sediment grain sizes can be found in Enclosure 1 of USACE (1974). Tidal current charts show directions that "fan-out" to cross the San Francisco Bar roughly perpendicular to the crest. At Ocean Beach, ebb tide currents are roughly parallel to shore and southward. Flood tide currents are roughly northward. Due to the diurnal inequality, the strongest currents are associated with the larger ebb tides. Peak annual tide ranges are about 2.5 to 2.8 meters, with peak ebb currents between 2 and 4 kilometers per hour. During these strong flows, large eddies form near headlands, causing nearshore currents to flow in the opposite direction. Currents in the surf zone are strong, complex and variable. Detailed measurements are not available.

Waves

Ocean Beach is exposed to storm seas and swell generated in the Northern Pacific, and swell generated in the Southern Pacific. Large wave power is incident, with the greatest in the winter (Johnson,1977; Woodward-Clyde Consultants,1978; Cross,1980; Berrigan,1985b, Wiegel,1992). Wave refraction causes focusing of incident waves at Ocean Beach for most directions and especially for longer wave periods (Street, Mogel and Perry,1970; Woodward-Clyde Consultants,1978; USACE,1993). Long period waves begin to refract around the 200-fathom (360 meters) line noted on Figure 1. The complex bathymetry results in multiple wave travel paths incident to Ocean Beach, and exposure to waves incident from north of the geometric shadow formed by Point Reyes. Refraction over the San Francisco Bar contours causes a focusing of wave energy near the base of the Bar, even for relatively short wave periods (Figure 4). Figure 2 is a high-altitude photograph which shows wave crossings due to refraction (see also, Vincent et. al., 1994). The result is an amplified breaking wave height at the location of wave crossing, resulting in a very peaked wave form (Figure 5, Battalio, 1994).

Prior estimates of wave-induced alongshore sand transport do not agree in direction nor magnitude. Estimated rates are typically around $5x10^5$ to $1.5x10^6$ m^3/yr ($7x10^5$ to $2x10^6$ cubic yards/yr, yd^3/yr) for the gross, and zero to $6x10^5$

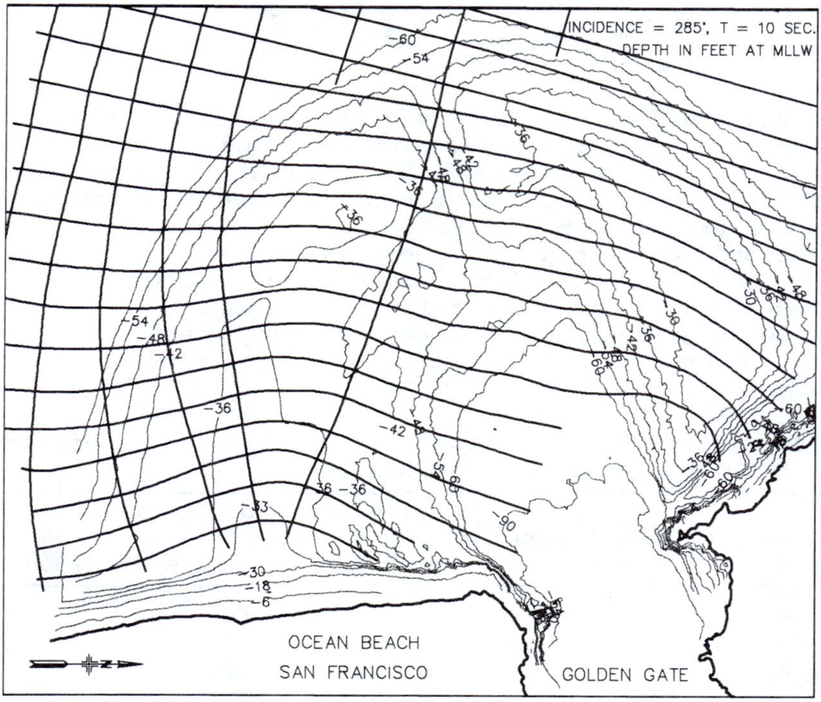

Figure 4: Wave Refraction Over San Francisco Bar (Source Woodward Clyde Consultants, 1978)

Figure 5: Grant Washburn Observes Breaking Wave Heights near Taraval Street, Ocean Beach, San Francisco, (Winter 1992). For scale, the surf board is 9 feet long. Photo: Tim Britton

m^3/yr (8x10^5 yd^3/yr) for the net (Street et. al.,1970; Kamel,1962; Johnson,1979; USACE,1979; Galvin,1979; Domurat, Pirie, and Sustar,1979).

A five-year data set of daily visual observations of breaking wave heights was reviewed (Washburn, 1996). Table 1 summarizes the cumulative distribution of breaking wave heights from these data, which were observed near the base of the San Francisco Bar (Taraval Street). These observations were compared to deep water significant wave heights recorded by the San Francisco Buoy and the Monterey Bay Buoy (see Figure 1 for locations): A time series correlation analysis produced coefficients of 0.79 and 0.76, respectively, indicating a strong correlation (maximum coefficient is 1.0).

Table 1: Cumulative Distribution of Breaking Wave Heights, Taraval Street, Ocean Beach (1991-1996)

Breaking Wave Height Meters	(Feet)	Frequency of Occurrence Larger Waves (% of Daily observations)
1.8	(6)	50%
3.0	(10)	25%
4.6	(15)	10%
7.3	(24)	1%

The method of Hands & Allison (1991) was used to gage the potential for onshore sand transport from the Bar to Ocean Beach. An average wave period of 12 seconds, a depth of 9 meters, and the cumulative distribution given in Table 1 were used. Calculations show a strong potential for onshore sand transport, as concluded by others (Kamel, 1962; Galvin, 1979).

Wind
Strong onshore (mostly northwest) winds occur at Ocean Beach. Prior to major development in the 1900's, the area behind Ocean Beach was an extensive dune field (Olmsted and Olmsted,1979). Sand losses due to inland transport by wind are estimated to be zero to 23x10^3 m^3/yr (30x10^3 yd^3/yr) prior to 1975, and zero to 7.6x10^3 m^3/yr (10x10^3 yd^3/yr) since then. The reduction is based on the City's practice of returning sand to the beach since 1975.

Man's Activities
Significant beach and dune fill occurred during the period 1882 to 1929, when the natural dunes were re-graded to form a straight embankment for the shore-parallel Great Highway (Olmsted and Olmsted,1979). This resulted in a seaward shift of the shoreline of about 60 to 90 meters (200 to 300 feet) (Figures 3 and 6). The estimated net

volume changes to the beach and dunes by man since 1929 were summarized for the sediment budget (Table 3). Comparison of the shoreline data with the timing and location of sand placement indicates a northward littoral drift from the base of the Bar, which agrees with Johnson (1979).

The need for safe navigation to and from San Francisco Bay requires a dredged channel through the San Francisco Bar. The Bar Channel is authorized to a depth of 17 meters (55 feet) below mean lower low water. Since the Bar Channel was last deepened in 1975, annual dredging has totaled about $4.6x10^5$ m^3 ($6x10^5$ cy) of sand. The sand is dumped on the south part of the Bar (see Figure 10). Prior to 1971, the dredged sand was reportedly dumped about one mile southwest of the Bar Channel entrance in about 24 meters (80 feet) of water. The dump location was changed so that the sand would stay in the littoral zone (USACE, 1974).

Prior to 1971, at least 20 million m^3 (26 million yd^3) of sand was dredged. Since 1971, about 11 million m^3 (14 million yd^3) were dredged from the channel area and reportedly placed near the crest of the south side of the Bar: Of this total, about 4.6 million m^3 (6 million cy) was attributed to deepening of the channel.

Shoreline, Dunelines, and Related Volume Changes

Shorelines and toe-of-dune lines from historic maps and aerial photographs were used to investigate changes and as input to the sediment budget analysis. Five shorelines digitized from smooth sheets for the period 1850 to 1944, and eleven shorelines and toe of dune positions from aerial photographs for the period 1938 to 1993 were mapped (Moffatt & Nichol, 1994). The mean high water shoreline was used from the maps. The wetted bound visible in the aerial photos was used to estimate the shoreline position, which was found to approximate the Mean Higher High Water elevation of about +1.8 meters (6 feet) Mean Lower Low Water datum.

Time histories of the shoreline positions and beach widths for Ocean Beach and South Ocean Beach are shown in Figures 6 and 7. The Ocean Beach data show a large seaward offset between 1900 and 1929, while South Ocean Beach data do not. This is attributed to the construction of the embankment supporting the Great Highway, and related beach nourishment. The Ocean Beach data show large fluctuations about an accretion trend, while South Ocean Beach data show smaller fluctuations about a slow erosion trend. Both areas show erosion to the early 1970's, and a recovery in recent years.

Beach and shoreface volume changes were estimated by applying a conversion factor of volume change per beach area change. A range of 8 to 14 m^3/m^2, (1.0 to 1.7 yd^3/ft^2) of beach change was used to calculate the values for Beach and Shoreface in Table 3. The higher value was developed based on a review of profile "closure" elevations and the assumption that the shoreface geometry remains essentially constant as the shoreline migrates. The minimum conversion factor was based on the assumption that shoreline changes were due to erosion of sand placed by man on the upper beach and dune, per Galvin(1979) and Ecker (1980). Volume changes for the dune face were calculated using a conversion factor that ranged between zero and 6 m^3/m^2 (0.74 yd^3/ft^2) of dune line change.

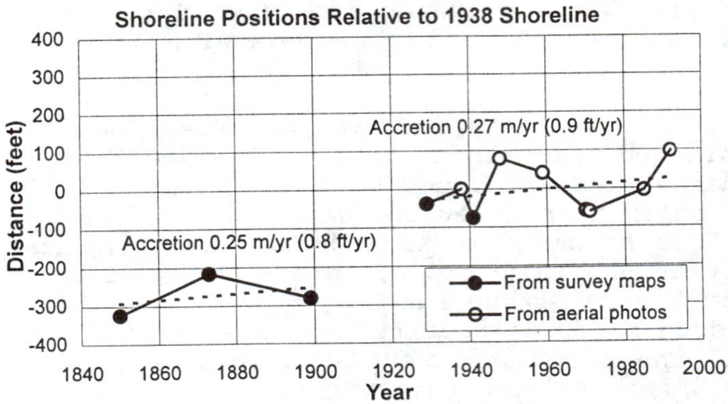

Figure 6: Shoreline Change Data Ocean Beach

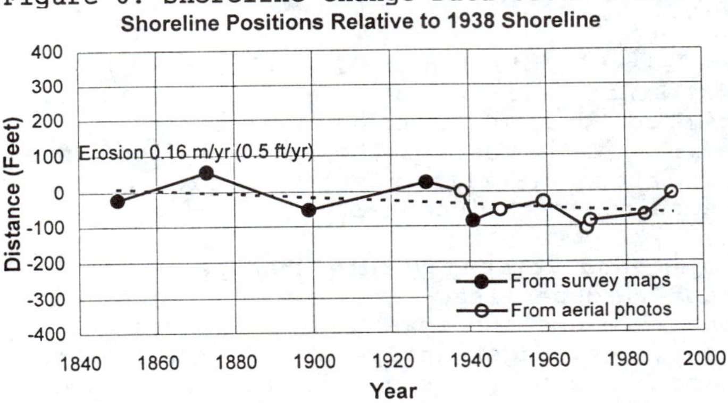

Figure 7: Shoreline Change Data So. Ocean Beach

San Francisco Bar Depth Changes

Prior studies of historic changes to the San Francisco Bar have been identified by Gilbert(1917), Homan and Schultz(1963), and Johnson(1965). Figure 8 shows the depths surveyed in 1900. Figure 9 shows the depths surveyed in 1956, and major depth changes since 1900.

Prior to 1900, the Bar had a relatively consistent "cuspate" shape. By 1956, outer portions eroded and inner portions accreted. Significant changes include the dredged channel, which is deeper, the area closer to Ocean Beach, which is shallower, and the Bar crest, which

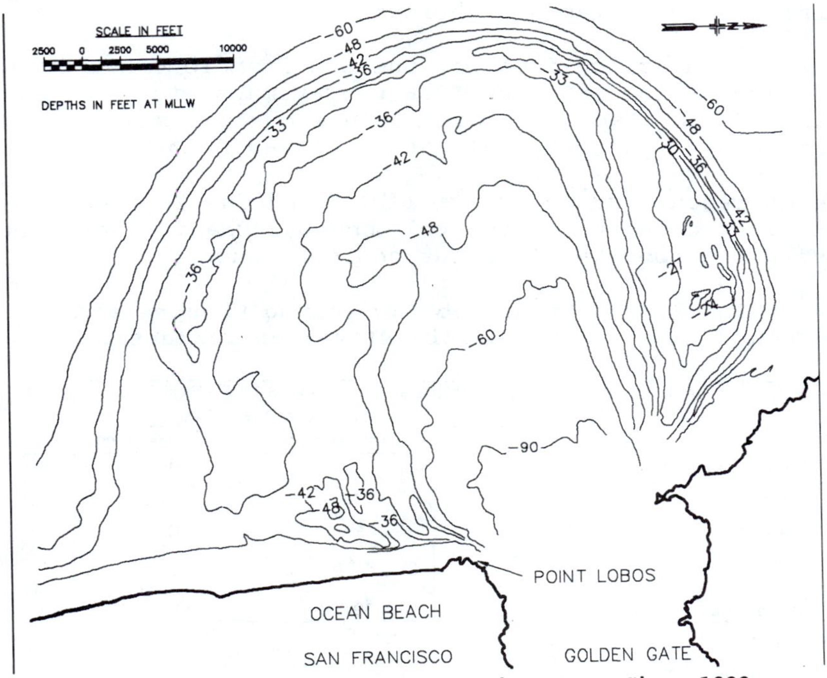

Figure 8: Bathymetry of San Francisco Bar, Circa 1900

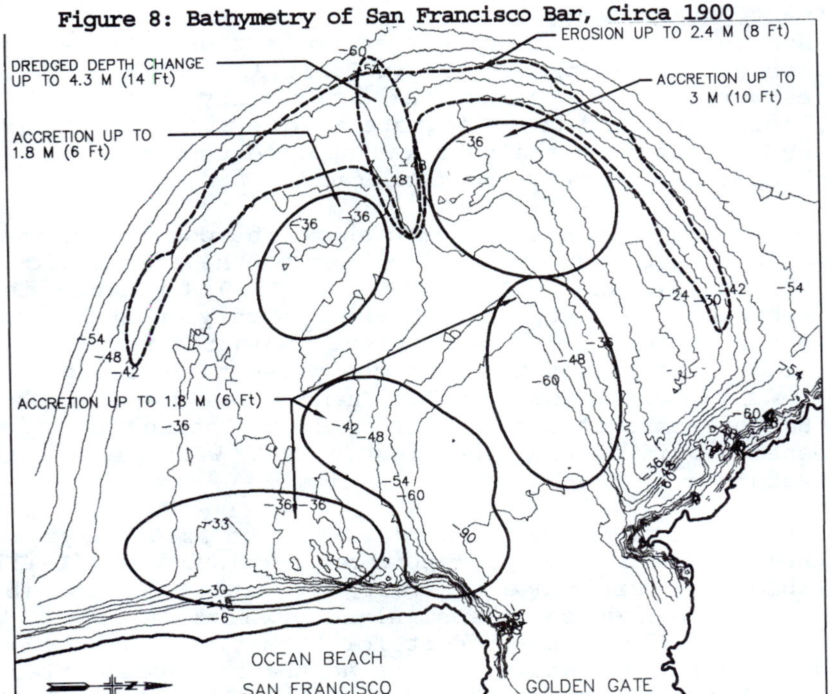

Figure 9: Bathymetry of San Francisco Bar, Circa 1956 and Major Depth Changes 1900 - 1956

generally shows a radial "shrinking."

Volume changes calculated for the San Francisco Bar by a comparison of the 1900 and 1956 survey maps are provided in Table 2. The Table includes volume changes adjusted for two arbitrary systematic elevation corrections.

Since no recent survey is available, and due to the sensitivity to vertical control, the Bar changes were not directly incorporated into the sediment budget.

TABLE 2: San Francisco Bar Volume Changes and Sensitivity to Systematic Elevation Corrections

LOCATION	VOLUME CHANGES FOR 1900-1956 TIME PERIOD Million Cubic Meters		
	Calculated	0.09 m correction	0.12 m correction
Four Fathom Bank	4.5	0.0	-1.5
Channel	-10.2	-12.4	-13.2
South Bar	19.2	14.6	13.1
Net Change	13.5	2.2	-1.5
Average Rate (m3/year)	241,000	39,000	-28,000

Sediment Budget Analysis

The sediment budget was developed to best use the available data, as summarized in Table 3. A range of values were used to reflect the uncertainty of the available data, and to facilitate a sensitivity analysis. The sediment budget analysis was applied to the Ocean Beach shoreface, beach, and dune face.

Based primarily on the historic shoreline positions, the sediment budget analysis was applied to the time period of 1929 to 1971 and 1971 to 1992. By 1929, the shoreline and embankment had been established roughly in its present location. An erosion trend dominated to 1971, becoming an accretion trend to the present. Starting in 1971, sand dredged from the Bar Channel was disposed on the Bar offshore of Ocean Beach, and the Channel was deepened significantly. Starting in 1975, wind blown sand was returned to the beach by the City.

For 1929-1971, a net deficit of $1.3x10^4$ to $1x10^3$ m3/yr ($1.7x10^4$ to $1.4x10^3$ yd^3/yr) is calculated (**Subtotal (I-II)** in Table 3). This range is small, and within the possible range of wind-induced losses inland, so that it is possible to balance the budget for this time frame by assuming zero net sand transport at the area boundaries, or that net input at one boundary (onshore from the Bar), is balanced by losses past other boundaries (north past Point Lobos, and/or south past South Ocean Beach). Local knowledge indicates erosion is occurring in the area south of Ocean Beach, while sand accumulation is a problem in several areas along the north waterfront of San Francisco. It is therefore considered more likely that losses from Ocean Beach are passing Point Lobos and migrating into San Francisco Bay.

TABLE 3: Sediment Budget for Ocean Beach, San Francisco (Annualized for Listed Time Periods)

No.	ITEM	RANGE OF SAND VOLUME CHANGES (Thousands of Cubic Meters per Year)					
		1929-1971		1971-1992		1929-1992	
		Low	High	Low	High	Low	High
I	**VOLUME CHANGES**						
1	Beach & Shoreface	(7.3)	(12.6)	103.4	175.9	29.6	50.2
2	Dune Face (data from 1938)	0.0	(3.1)	0.0	(10.9)	0.0	(5.7)
	Net Volume Changes	(7.3)	(15.7)	103.4	164.9	29.6	44.5
II	**SAND INPUT (LOSSES)**						
3	Net Beach/Dune Nourishment (+) Mining (-)	5.5	9.1	36.4	36.4	15.8	18.2
4	Wind Blown Loss to Inland	0.0	(23.7)	0.0	(7.3)	0.0	(18.2)
	Net Input (Losses)	5.5	(14.6)	36.4	29.1	15.8	0.0
	Subtotal (I - II)	(12.7)	(1.1)	67.0	135.8	13.8	44.5
III	**Estimated to Balance Budget**						
6	Minimum Northward Transport Past Point Lobos	(12.7)	(77.6)	(12.7)	(77.6)	(12.7)	(77.6)
7	Minimum Onshore Transport From San Francisco	0.0	76.5	79.7	213.4	26.6	122.1
	Subtotal	(12.7)	(1.1)	67.0	135.8	13.8	44.5
IIII	**Total (I - II - III)**	0	0	0	0	0	0

For 1971-1992, a net surplus of $6.7x10^4$ to $1.4x10^5$ m3/yr ($8x10^4$ to $1.8x10^5$ yd^3/yr) is calculated, which is a significant change from the previous period. This rate of input can only be explained by alongshore transport from the area south of Ocean beach, or onshore transport from the Bar. However, only the onshore transport potential from the Bar changed markedly around 1971, due to the change in dredging practices.

Postulated Rates to Balance Sediment Budget

The third part **(III)** of Table 3 shows one possible solution to the sediment budget. Assuming a range for onshore transport from the Bar during 1929-71 to be zero to $7.6x10^4$ m^3/yr ($1x10^5$ yd^3/yr), the loss due to alongshore transport past Point Lobos is calculated to be about the same as the input from the Bar. For the period 1971-92, the rate of alongshore transport past Point Lobos is assumed to be the same as for 1929-71. The required onshore sand transport rate from the bar is then calculated to be $8x10^4$ to $2.1x10^5$ m^3/yr ($1x10^5$ to $2.8x10^5$ yd^3/yr).

Figure 10 presents the 1971-92 rates and postulated sand transport for the Ocean Beach, Golden Gate cell. The net alongshore transport rate at Ocean Beach is estimated to be northward. The rate decreases with distance north, with sand accumulating in the beach and dunes. South of the base of the Bar, the net alongshore transport rate is unknown, but expected to be near zero.

The sand that passes Point Lobos is believed to mostly continue alongshore through the Golden Gate. Some is lost to maintenance dredging and dunes. The remainder is speculated to return to Ocean Beach by ebb tide currents, probably diverting from shore at several locations. Some or most of the sand is believed to migrate back to the San Francisco Bar prior to returning to Ocean Beach.

These rates are presented as long-term, average annual rates that could be substantially different in any particular year. Further study is needed, including

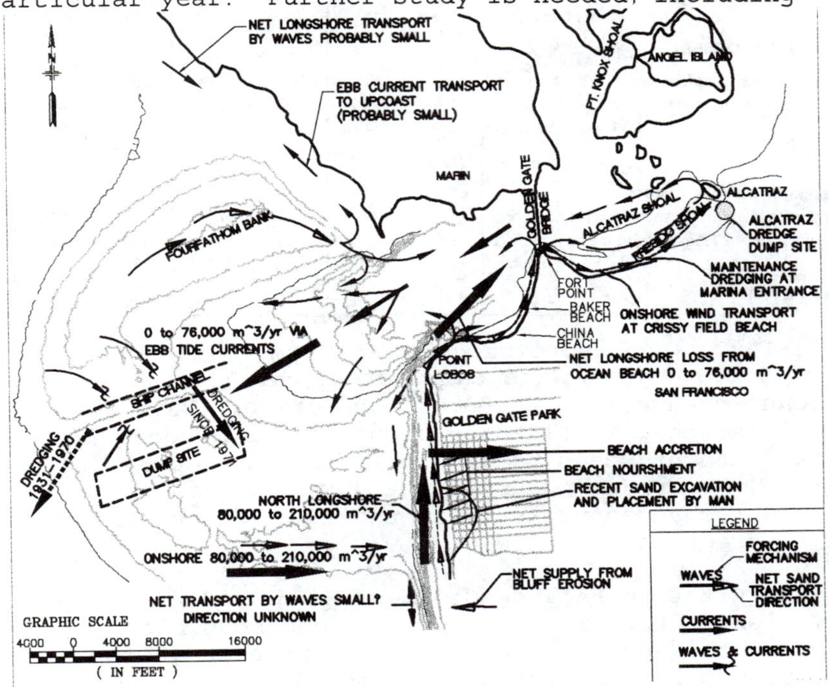

Figure 10: Residual Sediment Transport at Ocean Beach

recent Bar changes, sand transport and sediment budget in the Golden Gate area, and sand transport and sediment budget south of Ocean Beach to Pedro Point, Pacifica.

Acknowledgements
A Technical Review Panel reviewed the methodology and results, and provided guidance. Members included USACE (George Domurat, Tom Kendall and Len Madalon), City and County of San Francisco (Dave Jones), U.S. Geological Survey (John Dingler), Professors Robert Wiegel, Paul Komar and Ray Krone,

and individuals familiar with the area (Dr. James Walker and Dr. Philip Williams).

References

1. Battalio, R.T. (Oct 1994) "Estimating Breaking Wave Height at Ocean Beach, San Francisco," Shore and Beach, Vol. 62, No. 4, pp. 33-36.
2. Berrigan, P.D. (Apr 1985a) "Seasonal Beach Changes at the Taraval Seawall," Shore and Beach, Vol. 53, No. 2, pp. 3-9.
3. Berrigan, P.D. (Jan 1985b) "The Taraval Vertical Sea Wall," Shore and Beach, Vol. 53, No. 1, pp. 3-7.
4. Cross, R.H. (July 1980) "Ocean Wave Statistics for San Francisco," Shore and Beach, Vol. 48, No. 3, pp. 26-29.
5. Domurat, G.W., Pirie, D.M. and Sustar, J.F. (Oct 1979) "Beach Erosion Control Study, Ocean Beach, San Francisco, California," Shore and Beach, Vol. 47, No. 4, pp. 20-22.
6. Ecker, R.M. and Towill, Inc. (1980) Ocean Beach Sand Replenishment Program, prepared for City and County of San Francisco. October 1980
7. Galvin, C.G. (1979) Coastal Processes and Sediment Budget at Ocean Beach, San Francisco, prepared for the San Francisco Wastewater Program.
8. Gilbert, G.K. (1917), Hydraulic-Mining Debis in the Sierra Nevada, United States Geological Survey Professional Paper 105.
9. Hands , E.B. and Ellison, M.C. (1991) "Mound Migration in Deeper Water and Methods of Categorizing Active and Stable Berms," Coastal Sediments '91, Seattle WA, June 25-27, 1991.
10. Homan, W.J. and Schultz, E.A. (1963) Model Tests of Shoaling and of Dredge Spoil Disposal in San Francisco Bay, US Army Corps of Engineers, San Francisco District, California, January 28 - February 1, 1963.
11. Johnson, J.W. (Oct 1965) "Nearshore Sediment Movement- Central California Coast," presented at Coastal Engineering Specialty Conference, ASCE, Santa Barbara, CA, October 11-13, 1965.
12. Johnson, J. W. (1971) Bottom Sediment Characteristics Near Entrance to San Francisco Bay, University of California, Berkeley, Hydraulic Engineering Laboratory Technical Report HEL-24-3, April 1971.
13. Johnson, J.W. (May 1977) Shoreline Characteristics, Ocean Beach-San Francisco, Berkeley, California, May 10, 1977.
14. Johnson, J.W. (1979) Littoral Processes At Ocean Beach, San Francisco, California, November 28, 1978.
15. Kamel, A.M. (Jan 1962) Transportation of Coastal Sediments, University of California, Berkeley, IER, Series 185, Issue 1.
16. Moffatt & Nichol, Engineers (Jan 1994) Shoreline Mapping for Ocean Beach, San Francisco, California, Prepared for USACE San Francisco District.
17. Moffatt & Nichol, Engineers (July 1995) Sediment Transport Processes Study, Ocean Beach, San Francisco, California, Prepared for USACE San Francisco District.
18. Noble Coastal & Harbor Engineering and Dr. Per Bruun (July 1985)

Great Highway - Ocean Beach, Coastal Engineering Report, Seawall Design, prepared for Clean Water Program - City and County of San Fransicsco, July 9, 1985.

19. O'Doherty, John, (October 1994), "Micheal Maurice O'Shaughnessy-San Francisco City Engineer, 1912-1932," Shore and Beach, Vol. 62 , No. 4, pp. 17-23.

20. O'Shaughnessy, M.M. (1924) "Ocean Beach Esplanade, San Francisco, California, Paper No. 1539," Transactions, American Society of Civil Engineers Vol LXXVII 1924

21. Olmsted, R. and Olmsted, N. (Feb. 1979) Ocean Beach Study: A Survey Of Historic Maps And Photographs, City of San Francisco, California, February 23, 1979.

22. Street, R.L, Mogel, T. And Perry, B. (1970) "Computation of Alongshore Energy And Littoral Transport," Proceedings 12th Coastal Engineering Conference, Washington D.C., VII, Ch 57, ASCE, Sept 1970.

23. Trask, P.D. (1958) Beaches Near San Francisco, California, 1956-1957, Technical Memorandum No. 110, Beach Erosion Board, Office of the Chief of Engineers, US Army Corps of Engineers.

24. USACE (June, 1974) Dredge Disposal Study, San Francisco Bay and Estuary, Appendix A: Main Ship Channel (San Francisco Bar), US Army Corps of Engineers, San Francisco District, California.

25. USACE (1979) Ocean Beach, San Francisco, California, Feasibility Report, Beach Erosion Control Study, US Army Corps of Engineers, San Francisco District, California, July 1979.

26. USACE (December, 1990) Long Term Management Strategy for Dredged Material Disposal in the San Francisco Bay Region, Phase I: Evaluation of Existing Management Options. US Army Corps of Engineers, San Francisco District, California.

27. USACE and City and County of San Francisco (Sept 1996) Final Report: Ocean Beach Storm Damage Reduction Study, US Army Corps of Engineers, San Francisco District, California.

28. USACE (September, 1993), Wave Studies at Ocean Beach, San Francisco, US Army Corps of Engineers, San Francisco District, California.

29. USACE (1995), Dredging Records for San Francisco Bar Dredging, Construction Operations Division, US Army Corps of Engineers, San Francisco District, California.

30. Vincent, Charles L., Robert F. Jenson and Richard Goldstein, (Oct. 1994) "Wave-Current Interaction at an Inlet," Shore and Beach, Vol. 62, No. 4, pp. 13-15.

31. Washburn, Grant (September, 1996), Daily Visual Observations (Aug 1991 - 1996) of Breaking Waves, Ocean Beach, San Francisco, Unpublished. Notes provided to Moffatt & Nichol Engineers for analyses.

32. Wiegel, R. L. (January 1992), "Coastal Forum: Some Complexities of Coastal Waves, Currents, Sand and Structures," Shore and Beach, Vol. 60, No. 1, pp. 21-33.

33. Woodward Clyde Consultants (Jan. 1978) Coastal Engineering Evaluation, Sub-Task 3-2, Southwest Ocean Outfall Project, City and County of San Francisco, prepared for PBQ & D, Inc. San Francisco, California, January 20, 1978.

CHAPTER 209

TWO TREATMENTS OF SHORE EROSION IN EXTREME FLOODS ON U.S. GREAT LAKES

Robert J. Hallermeier[1]

Abstract

This investigation addresses empirical evidence on wave action and shore erosion during extreme floods on Great Lakes coasts. Historical information shows that record coastal floods on the Great Lakes usually involve moderately extreme storms during relatively brief intervals with overall lake level much above the long-term average. The 100-year flood on U.S. lakeshores appears most likely to be accompanied by coastal wave heights with recurrence interval of about: three years on the four upper Great Lakes; or one-half year on Lake Ontario. Shore changes during the 100-year flood seem susceptible only to a coarse statistical estimate because available studies reveal Great Lakes erosion to be extremely variable in onset and amount. Two simplified erosion treatments are outlined here, one applying average annual recession rate for a locality, and the second, an empirical relationship originally developed for seacoast storm effects. The latter treatment appears verified by evidence from a Lake Michigan study, and such estimates of episodic erosion cross section are much lower than for extreme events on U.S. seacoasts.

INTRODUCTION

Conclusions developed in the following material are intended for application in the U.S. National Flood Insurance Program (NFIP). The NFIP includes the generation of maps quantifying expected hazards for the local base flood, an event having average annual probability of one percent (or mean recurrence interval of 100 years). Suitable coastal assessments must take into account distinctive characteristics of extreme events along Great Lakes shores, reflecting the particular basin configurations along with the marked variations of mean water levels over long terms.

To demonstrate the distinctive circumstances on the Great Lakes, Figure 1 displays annual mean water levels since 1860 for Lake Huron, with very large oscillations apparent but no clear trend during this century. That behavior strongly differs from the historical sea levels on the Atlantic ocean, where a rising trend is the most notable feature and other variations appear relatively minor. These signals indicate a

---

[1] Senior Coastal Scientist, METS Division, Dewberry & Davis, 8401 Arlington Boulevard, Fairfax, Virginia 22031-4666, U.S.A.

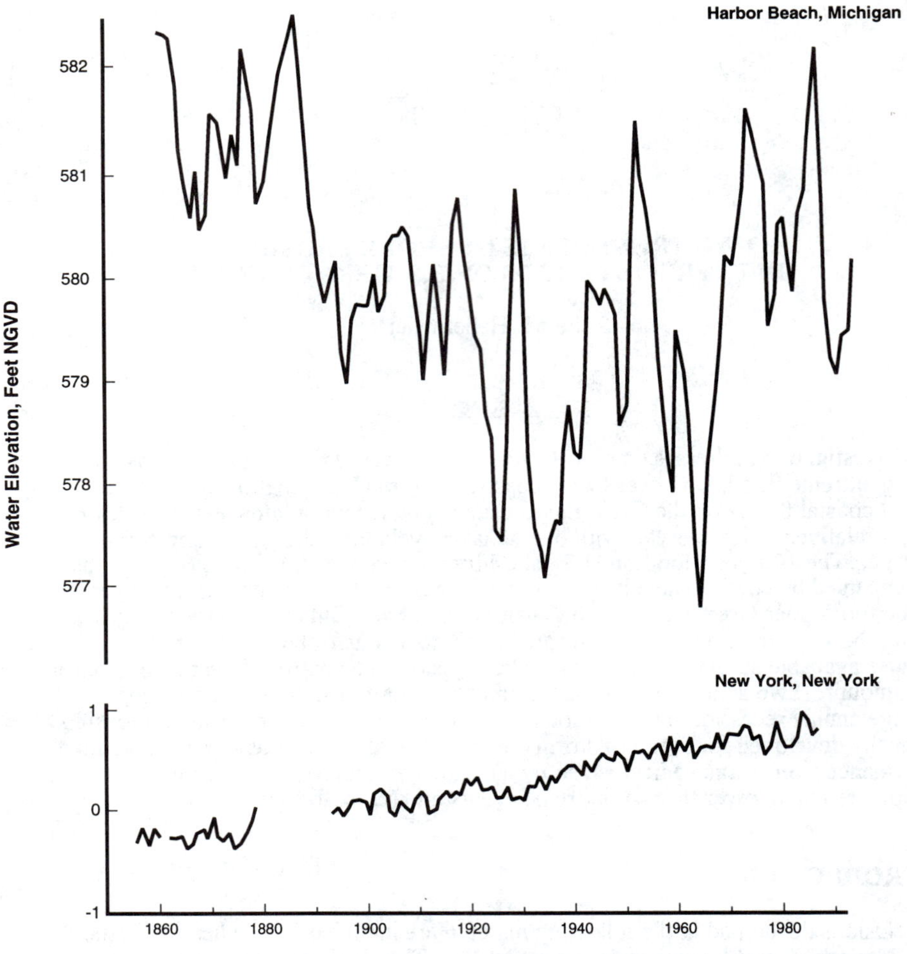

Figure 1. Annual mean water elevations on Lake Huron and Atlantic seacoast

markedly different context exists for coastal processes and extreme floods on the Great Lakes, in comparison to seacoasts.

Until recently, only stillwater inundation was considered in mapping coastal flood hazards for the Great Lakes, but documented flooding during the 1985-1987 highs clearly established the importance of additional wave effects. For NFIP treatments of U.S. Great Lakes shores, a full suite of methodologies for hazard assessment now addresses local stillwater flood elevations, shore erosion expected during the base flood, wave dimensions including growth or decay in coastal areas, and wave runup elevations at barriers to flooding. The U.S. Army Corps of Engineers (USACE) has documented conclusions on extreme local stillwater elevations from long-term records (USACE, 1988) and has provided a simplified treatment for wave runup at the most common types of shore barrier (USACE, 1989). This paper outlines the development of guidance on two other aspects of flood hazards: incident wave action

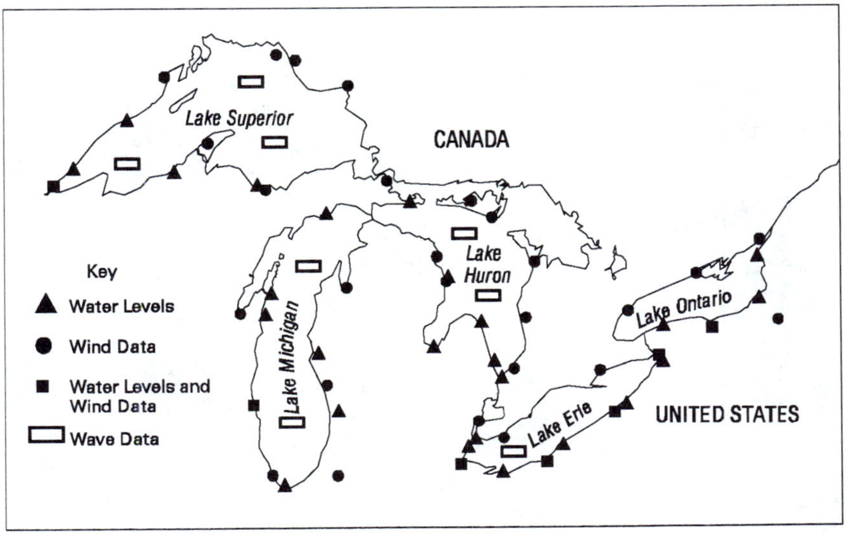

Figure 2. Pertinent data locations for Great Lakes investigations

and resultant shore erosion expected to accompany a base flood on the Great Lakes. These topics are interrelated, since shore erosion from wave action with flood waters alters the crucial boundary condition for storm impacts, thus controlling magnitude and penetration of wave effects over flooded land.

The following considerations are entirely empirical in basis, using best available information on recent extreme events. Quantitative analyses (Dewberry & Davis, 1995) have examined storm wave heights, flood elevations, and erosion magnitudes documented at about 100 varied sites on the U.S. Great Lakes. The necessary order of treatment begins with wave conditions during floods and proceeds to erosion considerations.

WAVE ACTION DURING FLOODS

Wave conditions during record Great Lakes floods have been computed in the recent USACE hindcast covering 1956 to 1987 (e.g., Hubertz and others, 1991). That interval includes 20th-century highs in mean lake levels during 1973-1975 and 1985-1987, along with record instantaneous elevations at many water-level gauges. Before combining this disparate information on waves and water levels for extreme events at particular sites, it is appropriate to examine the reliability of wave hindcast results. Figure 2 provides a map locating important data sites.

The USACE hindcast procedure applies measured winds at scattered land stations (Figure 2) to define wind fields over each lake, and then computes resultant wave action at many deep-water sites. Validation possibilities for hindcast computations are restricted to limited nearshore measurements and to several long-term buoy sites near the lake centers (Figure 2). The reported verifications of hindcast waves tend to

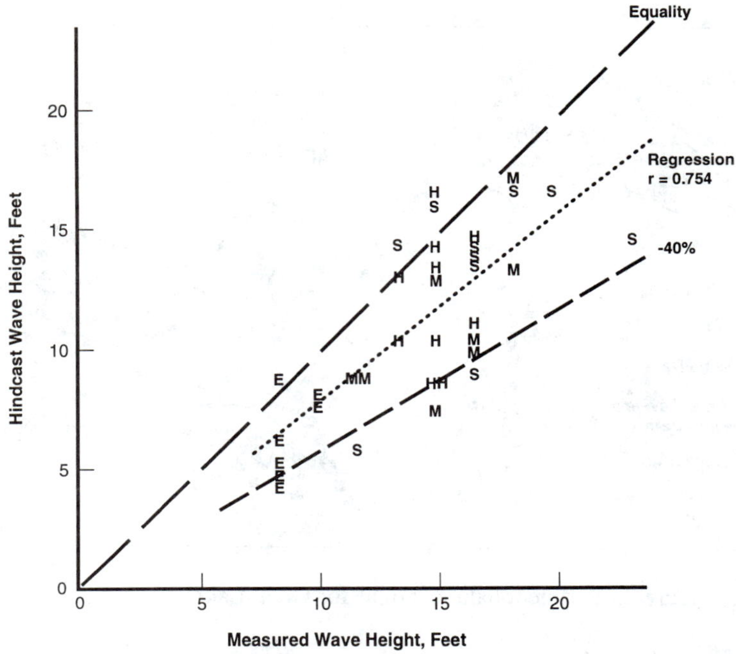

Figure 3. Hindcast versus measured extreme wave heights on four upper Great Lakes; separate lakes identified by initials used in plot.

focus on overall distributions of parameters, consistent with a comprehensive long-term hindcast. However, the present application utilizes brief intervals of storm winds, surges and waves, so the hindcast reliability for such extreme conditions merits evaluation. This is in line with the caveat stated by Hubertz and others (1991): "It is recommended that separate studies be done for specific storms at specific sites for more detailed information."

A convenient evaluation of the hindcast may be based on 35 extreme wave heights tabulated in a published summary of buoy data (Gilhousen and others, 1991). Those 1979-1987 measurements pertain to eight buoy sites on the four upper Great Lakes, and include all tabulated wave heights among the extreme 1% at each site. Hindcast wave height for each case is taken as the highest value during the same wind episode at a nearby computation point. This present selection yields a fairly balanced sample of extreme conditions, with separate lakes identified by the initial used in plotting results on Figure 3. Measured wave heights in this sample are typically on the order of 4.5 m (or 15 feet). Comparison of hindcast and measured wave heights yields a correlation coefficient sufficient to confirm a linear relationship with certainty. However, the hindcast does not accurately reproduce these measured conditions, being 20% low typically and having an error band of ±20%. Part of the hindcast bias may be explained by a lower peak value with the three-hour hindcast interval versus hourly measurements, but the error band remains notable. That sizable random scatter of hindcast wave heights appears related to errors in interpolated winds at buoy sites, as shown in Figure 4 using measured windspeeds also tabulated for a typical 7 of the 35 cases (Gilhousen and others, 1991).

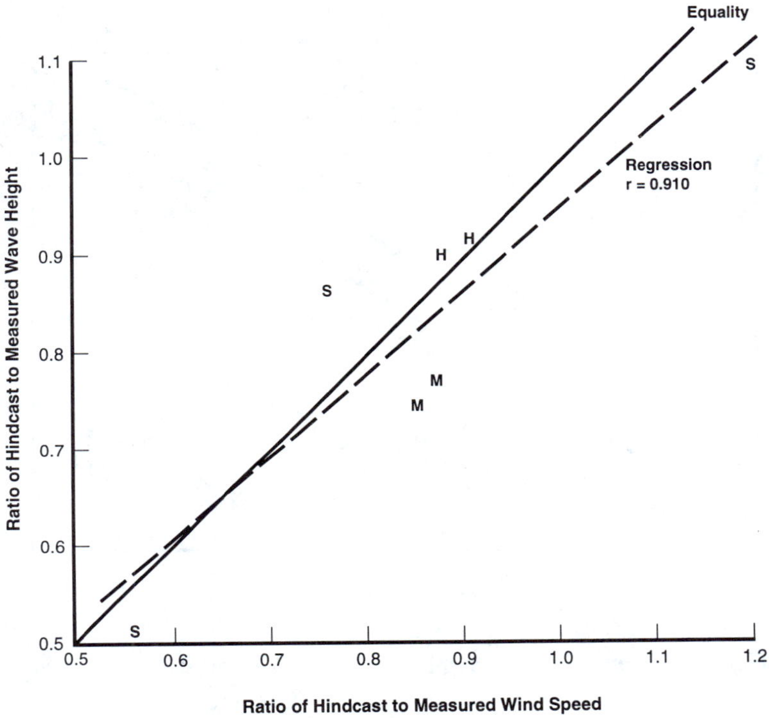

Figure 4. Relation of wave height and windspeed errors in Great Lakes hindcast.

According to this modest but objective selection of storm events, there are considerable errors in the USACE hindcast of extreme wave conditions on the Great Lakes. The data buoy sites may present a notable challenge for wave hindcasts, given relatively large separations from wind data locations. Hindcast results may generally be more reliable at nearshore sites, but appreciable variations in accuracy still seem likely in view of the scattered stations providing wind data. To apply USACE hindcast information in assessing wave action accompanying extreme floods, it seems advisable to maximize the variety of sites and events considered together, aiming at a simplified conclusion with generic pertinence to Great Lakes shores.

That program is carried out by considering in one group the record flood during 1956-1987 at each of 32 water-level gauges (Figure 2), along with wave hindcast results at a nearby computation point for the same episode. To use evident measures of extremity, both peak measured water elevation and largest computed wave height for each event are converted into recurrence intervals from the type of results shown in semi-logarithmic form on Figure 5. There, purely exponential recurrence relations as straight lines are fully defined by the medians of monthly and yearly highs, plotted at recurrence intervals of two months and two years, respectively. Such simplified analysis provides the advantages of a direct basis in annual probability along with a recurrence estimate for any value; note that results for relatively common flood elevations relate smoothly on Figure 5 to reported USACE conclusions on extremes. This example of recurrence relations is remarkable because each line has nearly

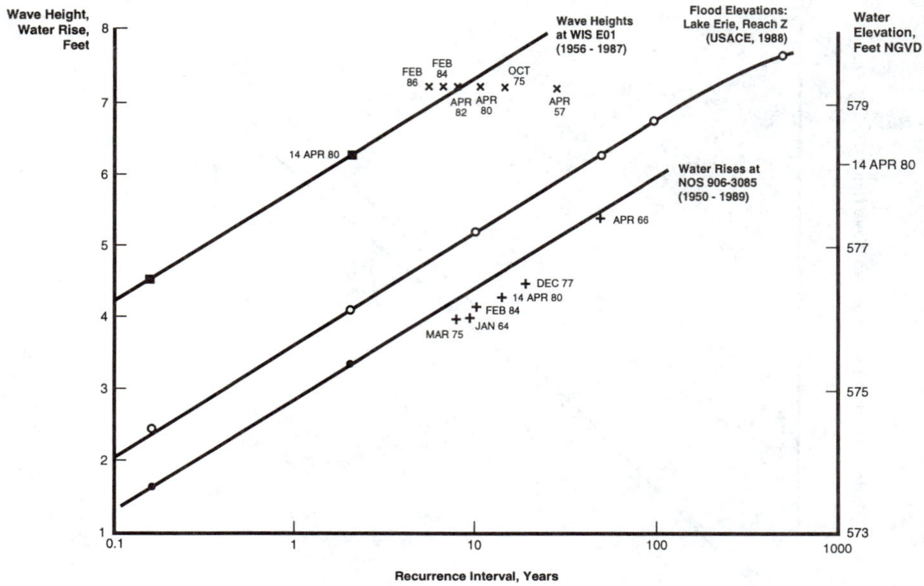

Figure 5. Summaries of extreme effects on Lake Erie at Toledo, Ohio

identical inclination. The inclination is representative of flood elevations on any of the Great Lakes except Lake Superior, where the line would tilt much less; for wave heights, the line has about the minimum inclination to be found on the Great Lakes, while water rises typically define a line with less inclination.

Figure 6 presents a logarithmic cross-plot of recurrence measures for the 32 extreme events, again distinguishing the separate Great Lakes by initial. Measured flood elevations for these cases typically have recurrence intervals on the order of 50 to 100 years. Immense scatter is evident among these results, in part reflecting the notable errors possible in hindcast wave heights and those recurrence intervals. The ± 20% estimate based on Figure 3 converts to various sizable ranges in recurrence intervals for wave heights here, depending strongly on the hindcast result and the local recurrence relation; a usual error bar for representative results extends over a factor of ten. However, a very wide range in wave extremity certainly can be associated with a given flood elevation on the Great Lakes, considering that significant variables include mean lake level and lake seiches, in addition to storm winds.

From close inspection of available evidence, floods on Lake Ontario differ from effects on the four upper lakes in always being associated with very common hindcast wave conditions; in contrast, notably extreme waves occasionally occur during many floods on the other Great Lakes. Deleting the Lake Ontario floods in Figure 6, remaining results yield a sizable positive correlation between recurrence intervals for wave height and flood elevation. The regression result shown in Figure 6 has statistical significance at the 5% level (but not the 1% level; Sachs, 1984), and indicates wave height recurrence of about 1 year for the 100-year or base flood. Since the correlation has limited significance, this evidence may also be examined

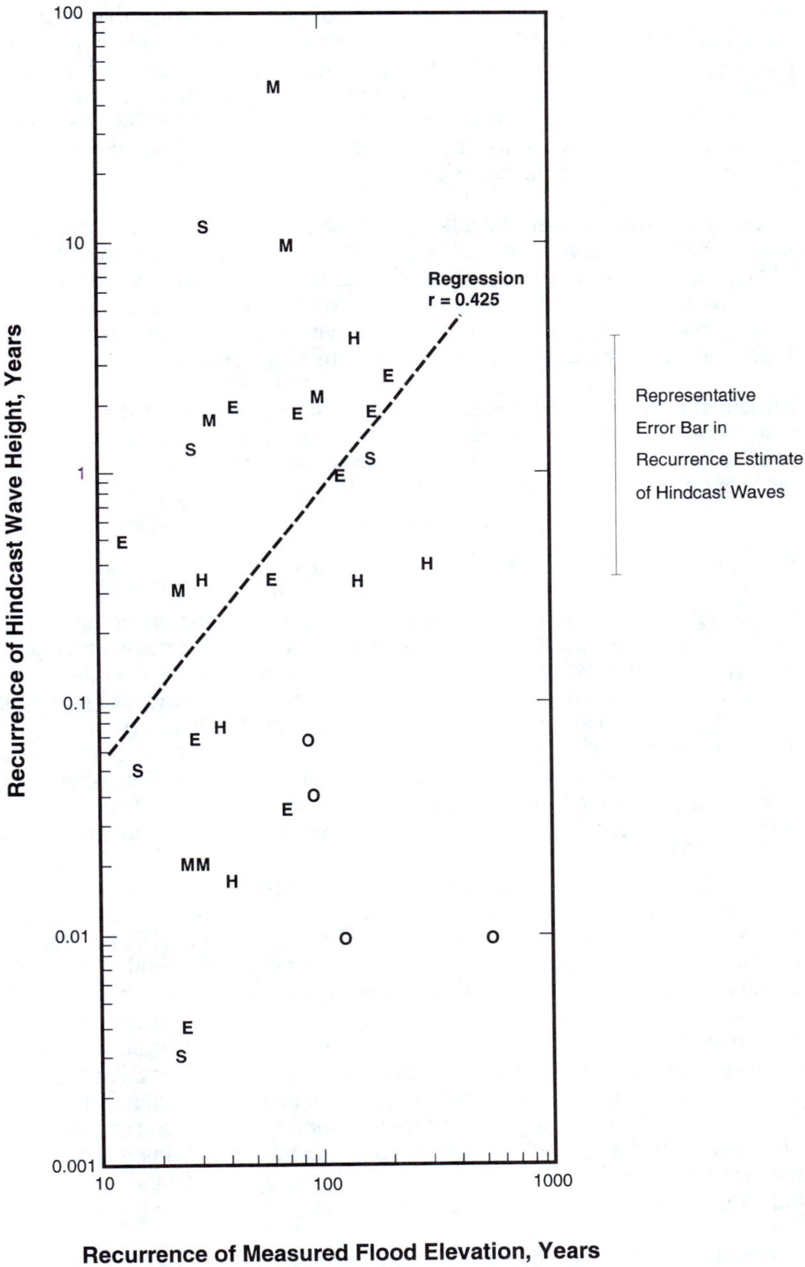

Figure 6. Parameters describing 32 recent extreme floods on Great Lakes; displayed regression result omits events on Lake Ontario (O).

without regard to trend: the most common condition with these record floods is about a 2-year wave height, according to the band of results in that vicinity. Such wave conditions occur during events from mid-November through mid-April, when storm winds can overwhelm the seasonal water levels least conducive to flooding. Taking into account the hindcast bias, these considerations indicate a prudent conclusion to be that a 3-year wave condition (as defined here by an exponential recurrence relation) accompanies the base flood at typical U.S. shore sites on Lake Erie, Huron, Michigan, or Superior.

Hourly water-level measurements demonstrate that lake seiches rather than storm surges are dominant in extreme flood elevations on the U.S. shore of Lake Ontario. There, the large seasonal excursion of mean lake level enhances flooding during the usual May-July high, when onshore storm winds rarely occur. An appropriate conclusion is found to be that wave height with a ½-year recurrence interval accompanies the base flood on U.S. shores of Lake Ontario.

Present conclusions must be recognized as highly generic regarding the likely wave action during extreme floods on the Great Lakes, and as somewhat tailored for routine NFIP application. Additional analysis or historical evidence may provide more appropriate estimates for specific shore sites.

EROSION EFFECTS AND TREATMENTS

Detailed measurements of shore changes pinpoint the apparent complexity of Great Lakes effects over space and time. A striking attribute is erosion variability over all but the longest terms, so that coastal changes during extreme conditions seem coherent only in a statistical sense over appreciable shore reaches. Many factors can be important in Great Lakes erosion processes, but the characteristic result appears to be quite intermittent progress in the fall and removal of dune or bluff sediments, yielding a markedly slowed or interrupted replica of typical seacoast storm effects. This discussion will address just a few erosion studies on sandy coasts of Lake Michigan, considered representative for various erodible lakeshores.

Two intensive studies, Hands (1979) and Birkemeier (1981), documented erosion along the eastern shore of Lake Michigan during the 1964-1973 interval when the increase in mean lake level was about 1.5 meters. Hands (1979) reported shoreline positions at irregular time intervals of a year or more, for a total of 30 transects spanning about 50 km near Pentwater, Michigan. Alongshore variability of retreat rate was extreme as lake level generally rose over 8 years, but the average or median rates over available transects were rather steady in time, regardless of mean lake level or storm climate variations. Those findings do not appear sensitive to the spatial or temporal frequency of measurements, or a later study area extension. As lake level declined over the last study year, shore accretion typically reversed about two years of retreat. The basic results might be summarized as indicating a lagging response to the increasing erosive stress with rising water, but only in an overall sense; changes on individual transects appear too variable in time for simple interpretation.

In the other study, Birkemeier (1981) summarized survey results at four-week intervals during 1970-1974, for 17 profiles roughly 20 km apart. Shore erosion appeared somewhat random in time and location: various numbers of profiles showed bluff or terrace retreat in greatly varying amounts over separate intervals. However, erosion was also clearly seasonal, in phase with the annual storm cycle but not with

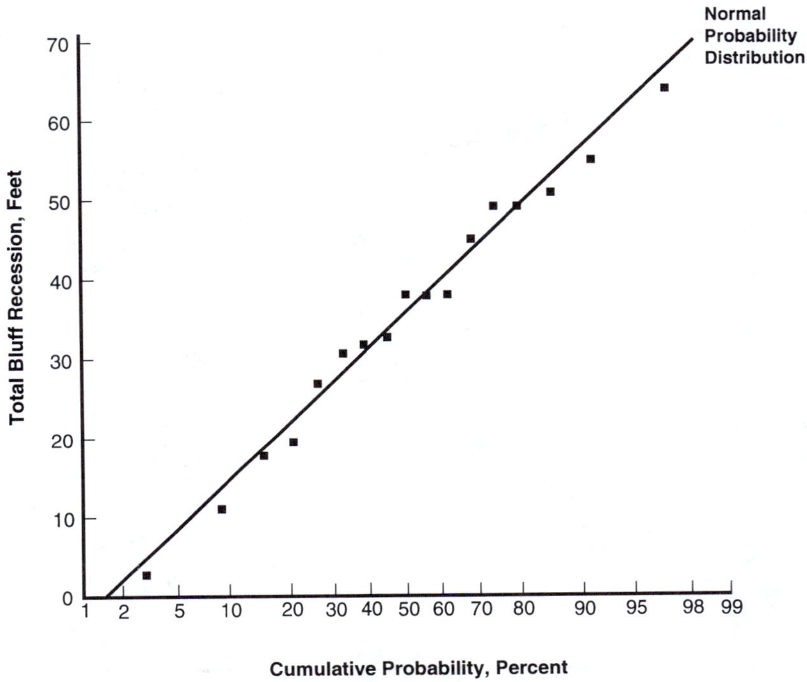

Figure 7. Bluff recessions during 1970-1974 on 17 Lake Michigan profiles.

the lake level cycle. Through the complete study term, changes accruing through sporadic episodes disclose some orderly variation over the entire data set: total recessions on individual profiles conform to a normal probability distribution with sizable dispersion, as shown in Figure 7. With no clear geographical trends, nor much correlation between adjacent profiles, the effectively random scatter of recorded erosion seems to result from the interplay of many independent factors, with insufficient time to develop an equilibrium response.

Even long-term changes on the Great Lakes can require careful interpretation due to the dependence of shore erosion on mean lake level, as revealed by results from two recent studies of Lake County, Illinois. Jibson and others (1994) addressed rates and processes of bluff erosion based on measurements for 1872-1937 and 1937-1987. Average retreat rate over the region showed no significant difference for the two periods, during which mean lake level or precipitation was nearly identical. However, Chrzastowski and others (1993) studied the same shore reach and timespan, but with three intervals isolating sizable elevation differences for Lake Michigan. Documented results reveal significant correlations between the mean lake level and representative movements of sandy shoreline or coastal bluff top, as demonstrated in Figure 8. Since differences from the long-term mean can be larger over briefer intervals, mean lake levels apparently contribute to sizable departures from an average erosion rate.

With this brief outline of factors and variabilities in Great Lakes erosion, it seems clear that statistical viewpoints can most conveniently provide useful summaries or projections. In turn, projections for short-term or storm erosion may be expected to

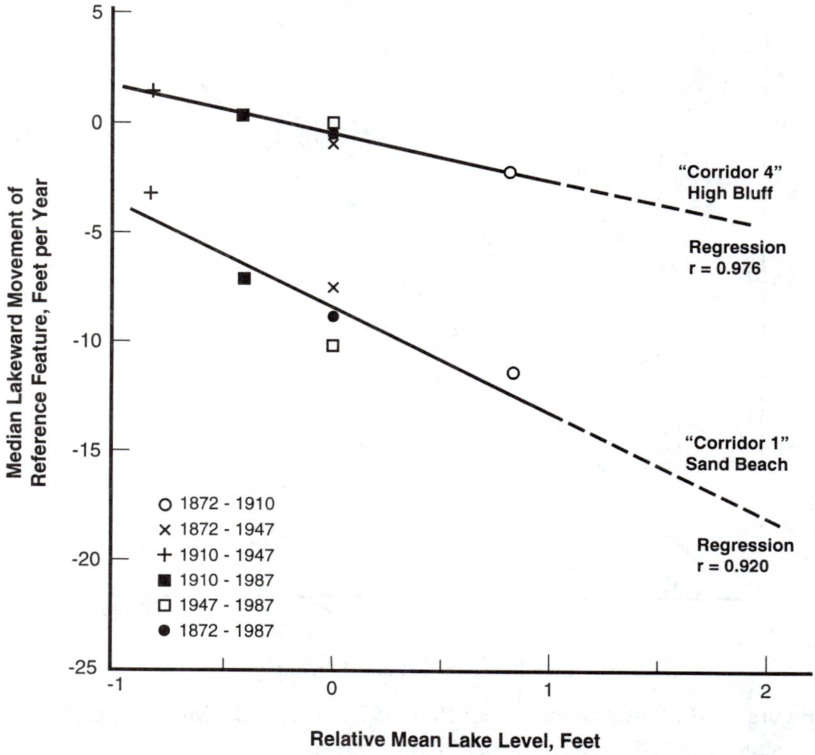

Figure 8. Representative erosion rates versus mean Lake Michigan levels.

attain only order-of-magnitude agreement with actual effects, but suitable procedures for erosion estimates might be rather simplified or generalized in basic character. Two treatments outlined here apply either average local recession rate, or storm erosion quantities from a generic database, the two evident alternatives. The independent treatments take into account distinctive attributes of the Great Lakes.

The first alternative applies the local value of A, average annual rate of shore recession over a long term. This development presumes available topography to be defined during usual lake levels, and addresses cumulative erosion until the base flood is likely to be encountered. As mean lake level rises to highs conducive for the base flood, shore recession is more rapid than the long-term average, and described as a rate of (2.5 A). If that rate occurs over about one-third of a long term, it implies that (0.25 A) is characteristic otherwise, and such a range of values appears appropriate according to available Great Lakes studies. Projected erosion before the base flood equals the accelerated rate times an expected waiting time of 4.8 years with high lake levels. That waiting time is supported by the normal probability distribution defined by data in Figure 9: actual intervals of high lake level before the record flood during 1972-1987, at 31 gauges on the four upper lakes. Given the notable storm waves likely to accompany an extreme flood on the four upper lakes, the additional allowance of (3A) for storm erosion in the base flood appears appropriate. These considerations yield (15A) as projected site erosion on Lake Erie, Huron, Michigan,

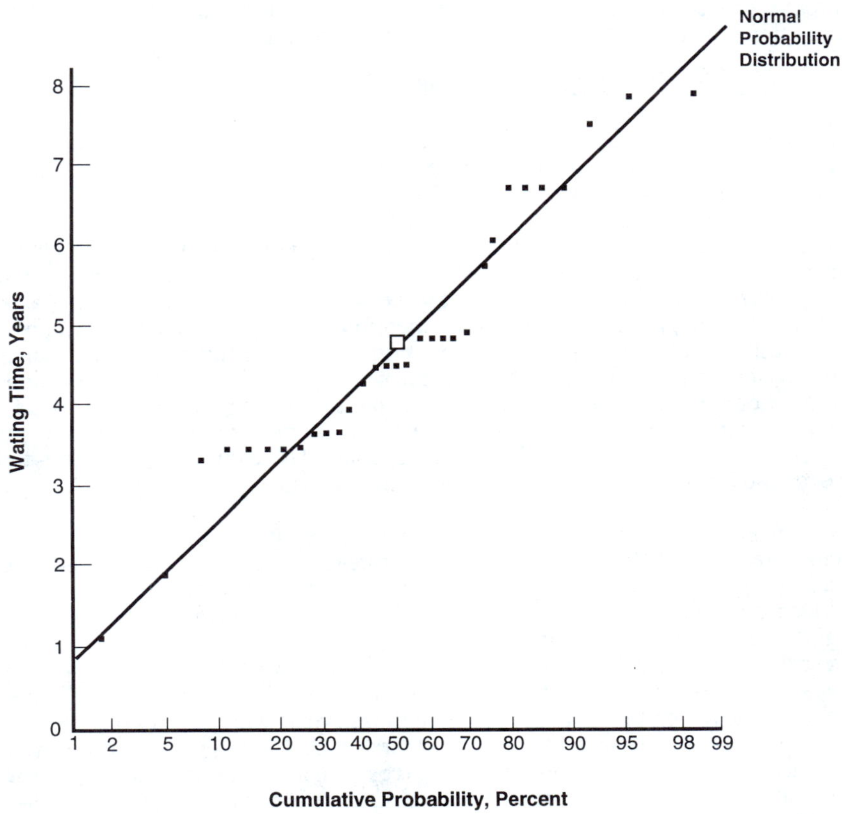

Figure 9. Waiting times for record floods at 31 sites on four upper Great Lakes, after onset of monthly mean levels higher than long-term mean.

or Superior; and (12A) on Lake Ontario. These recession distances describe a parallel retreat of the shore profile from the reference feature to which A pertains.

Such treatment seems relatively straightforward, but the result does not appear validated by measured Lake Michigan erosion reported by Birkemeier (1981). On 17 profiles in nine separate Michigan counties, total recessions over 4.3 years of high lake level show essentially no correlation with average long-term recession rates documented for the individual counties. Although no extreme flood occurred within the study area over this term, cumulative recessions make the present estimation method seem of dubious usefulness. This might be caused by the multiplication of two very uncertain quantities, in regard to short-term applicability at specific sites.

The second alternative considers only the storm effects likely to be associated with the base flood, using an empirical relationship for expected erosion in duneface retreat on U.S. Atlantic and Gulf coasts (Hallermeier and Rhodes, 1988). The relationship expresses eroded cross section above flood elevation in terms of recurrence interval for the flood, amounting to 50 m^2 erosion in a 100-year event:

$$\text{Erosion } [m^2] = 8 \text{ (Recurrence Interval [yr])}^{0.4} \quad (1)$$

This erosion treatment has proved to be suitable for NFIP applications in defining coastal flood hazards, and has been validated by effects recorded in more recent extreme events (e.g., Wolf and others, 1993).

Along U.S. seacoasts, wave condition and flood elevation generally may be expected to have comparable recurrence intervals for most episodes in a sizable sample of extreme events, but that is not the case for the base flood on the Great Lakes, as established by Figure 6. This application therefore employs a blend of individual erosion estimates for the wave and flood recurrences, using the geometric mean of cross sections defined by the basic relationship in Equation (1). Resultant erosion for the base flood, with a ½-year wave condition on Lake Ontario and a 3-year wave condition on the four upper lakes, is estimated as: 17 m^2 for Lake Ontario sites; and 25 m^2 on Lake Erie, Huron, Michigan, or Superior. These quantities are taken to be eroded amounts above local 100-year stillwater elevation.

Results from this erosion treatment appear consistent with storm erosion quantities documented in the Lake Michigan study by Birkemeier (1981). Erosion was recorded on a majority of the 17 profiles only during two intervals, covering notable storms in mid-December 1971 and in mid-March 1973. As defined by USACE hindcast wave heights and measured water elevations: the 1971 storm had ½-year waves and a 1-year flood, giving an erosion estimate of 7 m^2; and the 1973 storm combined 4-year waves and a 1½-year flood, for an erosion estimate of 11 m^2. Average bluff cross sections removed on the eroding profiles in those events were 8 m^2 during December 1971 (for 10 profiles) and 10½ m^2 during March 1973 (11 profiles). The quantitative agreement of estimates with representative erosion magnitudes in these cases provides appreciable validation of the present viewpoint, since base flood estimates for the Great Lakes lie between these amounts and many documented seacoast episodes. However, serious uncertainties in predicting the onset and amount of short-term erosion at Great Lakes sites remain worth emphasizing.

CONCLUSIONS

Empirical evidence provides an extensive basis for simplified summaries of the wave action and erosion expected to accompany the base flood on U.S. shores. Lake Ontario sites may be expected to experience wave heights having recurrence interval of ½ year and eroded cross section of 17 m^2 above flood elevation. Wave heights having recurrence interval of 3 years and erosion of 25 m^2 are projected for sites on Lake Erie, Huron, Michigan, or Superior. These results pertain only to storm-induced flood episodes, and are intended for routine application in flood hazard assessments within the National Flood Insurance Program.

Acknowledgments. The reported investigation was performed by Dewberry & Davis as a technical evaluation contractor for the National Flood Insurance Program, Federal Emergency Management Agency. The conference presentation and preparation of this paper were supported jointly by Dewberry & Davis and the Federal Emergency Management Agency. Present summaries and interpretations represent views of the author. The author wishes to express thanks for extremely responsive assistance provided by Edward Hands, U.S. Army Coastal Engineering Research Center, Vicksburg, Mississippi, and by John Herron, Great Lakes Section, National Ocean Service, Silver Spring, Maryland.

REFERENCES

Birkemeier, W. A., 1981. Coastal Changes, Eastern Lake Michigan, 1970-74, Miscellaneous Report No. 81-2, U.S. Army Corps of Engineers, Coastal Engineering Research Center, Fort Belvoir, Virginia.

Chrzastowski, M. J., Erdman, A. L., Stohr, C. J., and Terpstra, P. D., 1993. Illinois Pilot Erosion-Rate Study. Open File Series 1993-3, Illinois State Geological Survey, Champaign, Illinois.

Dewberry & Davis, 1995. Basic Analyses of Wave Action and Erosion With Extreme Floods on Great Lakes Shores. Fairfax, Virginia.

Gilhousen, D. B., Meindl, E. A., Changery, M. J., Franks, P. L., Burgin, M. G., McKittrick, D. A., 1990. Climatic Summaries for NDBC Buoys and Stations, Update 1. U.S. Department of Commerce, National Oceanic and Atmospheric Administration, National Weather Service, National Data Buoy Center, NSTL, Mississippi.

Hallermeier, R. J., and Rhodes, P. E., 1988. Generic Treatment of Dune Erosion for 100-year Event. Proceedings 21st Coastal Engineering Conference, ASCE, pp. 1197-1211.

Hands, E. B., 1979. Changes in Rates of Shore Retreat, Lake Michigan, 1967-76. Technical Paper No. 79-4, U.S. Army Corps of Engineers, Coastal Engineering Research Center, Fort Belvoir, Virginia.

Hubertz, J. M., Driver, D. B., Reinhard, R. D., 1991. Hindcast Wave Information for the Great Lakes: Lake Michigan. Wave Information Studies Report 24, U.S. Army Corps of Engineers, Coastal Engineering Research Center, Vicksburg, Mississippi.

Jibson, R. W., Odum, J. K., and Stande, J. M., 1994. Rates and Processes of Bluff Recession along the Lake Michigan Shoreline in Illinois. Journal of Great Lakes Research, Vol. 20, pp. 135-152.

Sachs, L., 1984. Applied Statistics, A Handbook of Techniques (2nd Edition). Springer-Verlag, New York.

U.S. Army Corps of Engineers, Detroit District, 1988. Revised Report on Great Lakes Open-Coast Flood Levels. Detroit, Michigan.

U.S. Army Corps of Engineers, Detroit District, 1989. Great Lakes Wave Runup Methodology Study. Detroit, Michigan.

Wolf, Lisa J., Bowen, J. D., and Hickey, K. A., 1993. Dune Profiles Before and After Storm Events in Coastal Massachusetts. Proceedings Hydraulic Engineering '93 Conference, ASCE, pp. 269-274.

CHAPTER 210

LONG-TERM BEACH RESPONSE TO SHORE STABILIZATION STRUCTURES ON THE OREGON COAST

G.E. Hearon[1]
W.G. McDougal[2]
P.D. Komar[3]

ABSTRACT

Increased development along the Oregon coast has led to heightened concern over beach erosion. As a result, more shore stabilization structures have been erected in recent years. A long-term field monitoring program, involving seven typical structures located on the central Oregon coast, was initiated in the spring of 1986 to quantify the effects of shore stabilization structures on the surrounding beach and adjacent properties. All structures are rip-rap revetments or seawalls. Volumetric changes of the subaerial beach at each site were examined at scales ranging from the entire beach surrounding the structure to the beach in the immediate vicinity of the structure. When the data coverage and site geometry permitted, a far field control section was utilized to compare the volumetric response of a portion of beach relatively uninfluenced by the presence of the structure with that of the beach on the two structure flanks. Measurements are presented for the seven structures over the observation period of ten years to document the long-term effects the structures are having on the fronting beach and adjacent unprotected properties. Ten years of monitoring has revealed that the structures at these seven sites are having no adverse impacts on the surrounding beach or adjacent properties. This can be attributed to the limited amount of wave attack that the structures have experienced. The structures are built high on the profile, and experience wave attack only during the most severe storms.

---

[1] Coastal Engineer, Coastal Frontiers Corp., 9420 Topanga Canyon Blvd., Suite. 101, Chatsworth, CA 91311, USA
[2] Professor, Civil Engineering Department, Oregon State University, Corvallis, OR 97331, USA
[3] Professor, College of Oceanic and Atmospheric Sciences, Oregon State University, Corvallis, OR 97311, USA

INTRODUCTION

The structures selected for the monitoring program are located between Neskowin and Florence; spanning a distance of nearly 100 miles of the central Oregon coast. Structures were selected in locations such that the number of factors that could affect the long-term beach change were reduced primarily to the structure itself. As a result, the majority of the structures were chosen such that there were no headlands, jetties, or river mouths in the immediate vicinity, and the adjacent properties were unprotected for at least three structures lengths on either side of the structure. The structures are backed by dunes, sea cliffs, or bluffs; all common coastal environments in Oregon. Table 1 shows the characteristics and site codes (abbreviations) of the seven sites selected for the field monitoring program. The study was initiated to aid the Oregon State Parks Division and DLCD, which regulates the construction of coastal structures in their permit review process.

Table 1 Site and Structure Characteristics

Site	Site Code	Structure Type	D_{50}(mm)	Length (m)
Pacific Sands Motel	PSM	Timber Seawall	0.30	70
Gleneden	GE	Rip-Rap Revetment	0.53	40
Pacific Palisades	PP	Rip-Rap Revetment	0.31	300
Pacific Shores	PS	Rip-Rap Revetment	0.20	20
C&L Ranch	CL	Rip-Rap Revetment	0.20	25
San Marine	SM	Rip-Rap Revetment	0.22	150
Driftwood Shores	DS	Rip-Rap Revetment	0.21	30

DATA COLLECTION

Field data, in the form of multiple beach profiles, were collected intermittently from the fall of 1986 to 1990, and resumed in the winter of 1995. On average, the database contains fifteen surveys per site. An effort was made to obtain at least a winter and summer survey for each year. Numerous transects were measured with an EDM in order to monitor the topography of the beach surrounding each structure. Weather permitting, the surveys provided coverage at least five structure lengths on either side of the structure, and seaward to the swash zone. Transects were spaced on the order of 20 m to 30 m on planar beaches with little three-dimensionality in the longshore direction. For more three-dimensional beaches displaying cusps, transects were spaced accordingly to achieve an accurate representation of the beach.

ANALYSIS METHODS

The beach profile data were analyzed using contour plots, surface plots, contour of difference plots, and results from sand volume calculations. The profile data were interpolated to a uniformly spaced rectangular grid using MatLab codes, from which the plots and sand volume calculations were generated. To obtain the maximum use from the gridded data and explore all aspects of the structures effect on the surrounding beach, several scales were examined. A large scale (Scale I) was examined by considering the largest common area of beach covered in the field monitoring program. To reduce seasonal contamination and noise in the analysis, a intermediate scale (Scale II) was employed, defined as two structure lengths to either side of the structure and one structure length seaward. Finally, to concentrate on effects in the immediate vicinity of the structure, a scale consisting of the two structure flanks and beach immediately fronting the structure (Scale III) was examined. Each scale was divided into three sections; north, center, and south. When possible, the north and south survey sections were defined with the same sizes, thereby allowing a direct comparison between the two sections.

This approach was used primarily for analyzing the changes in sand volume. The volume of sand on the beach was calculated in reference to the 1929 NGVD mean sea level. Volumetric response, or the magnitude to which the sand volume is changing due to erosion and accretion events, was examined with the results of the volumetric calculations. The flanks on either side of the structure are typically considered the most susceptible to erosion; therefore, the sand volume of the north and south flanks, and the volumetric changes of the sand at these flanks over the observation period were of particular interest. Comparison of the sand volumes at these flanks with a far field control section, relatively uninfluenced by the presence of the structure, was an important analysis tool. The far field section provides a baseline to which volumetric responses close to the structure can be compared. The location of the far field section was always taken at the same distance from the bluff and defined with the same plan area as the flank section.

Contour and surface plots provide a means to observe seasonal changes at the sites, such as profile slope or formation of a berm or rip current embayment, as well as an indication of how the beach changes or adjusts in the vicinity of the structure. Figure 1 shows a contour and surface plot from Gleneden, obtained in June 1989. Contour of difference plots were used to show the specific areas of erosion and accretion between two sequential sets of beach profile data. They are useful in determining the degree to which the beach in the vicinity of the structure was responding to erosion and accretion events. Used in conjunction, this array of tools, contour and surface plots, contour of difference plots, and sand volume calculations, provide a detailed account of the long-term changes at the sites.

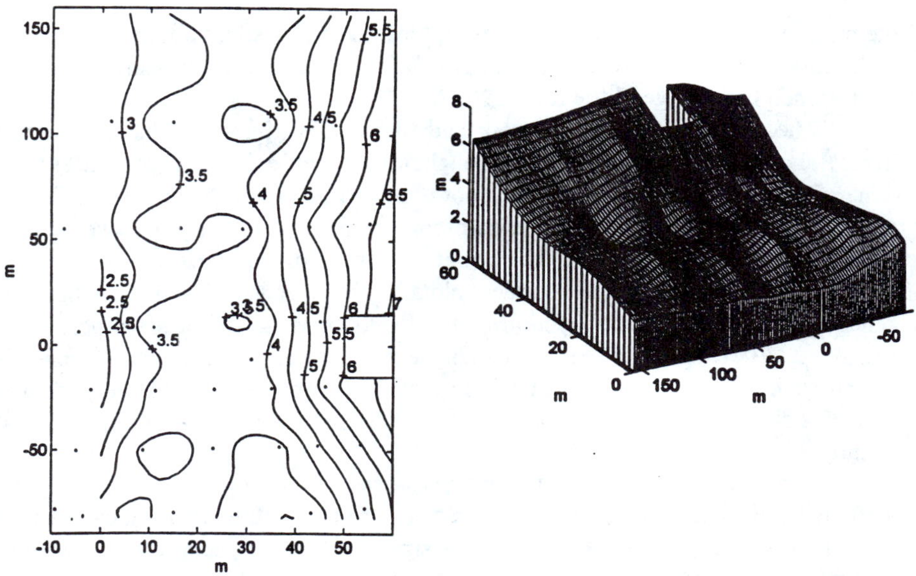

Figure 1 Contour and Surface Plot for Gleneden, June 1989

RESULTS

Analyses were undertaken by examining the contour and surface plots and the sand volume calculations for each site at of the aforementioned scales. One site will be analyzed in greater detail to illustrate the methodology and techniques used throughout the study to derive conclusions for all seven sites. A complete analysis of all sites is given in Hearon (1995) and Sturtevant (1987). Data for the remaining six sites will be presented in two forms, the volume of sand at Scale I over the observation period, and the results of the far field baseline comparison. Research has shown that the structure flanks are susceptible to erosion, therefore the results of the far field comparison are an effective indicator of structure induced impacts. The far field comparison data will be displayed in a normalized form, the volume of sand at the structure flank divided by the volume of sand in the far field control section ($V_{flank}/V_{far\ field}$). It is stressed that conclusions for each of the sites were derived by utilizing all analysis tools discussed in the previous section, and not solely from the data presented herein.

C&L Ranch (CL)

C&L Ranch provides an excellent open coast site. The structure is a 25-m long rubble mound revetment that protects and stabilizes the backing bluff. There are no headlands, jetties or river mouths in the vicinity. The bluff is shore parallel and unprotected for several structure lengths to either side. The land slide potential of the bluff is high. Evidence of several recent localized slides was observed during the 1995 data collection. Aerial photographs indicate that the beach and bluff line at C&L Ranch has changed little from 1965 to 1983.

The complete series of contour and surface plots indicates that the beach at C&L Ranch is very flat and planar throughout the year, exhibiting very little three-dimensionality. A summer berm was not observed in any of the surveys. The flat, dissipative beach at C&L Ranch is common on the central Oregon coast. A slight increase in sand elevation in the vicinity of the structure was observed in the majority of the contour and surface plots. The increased sand elevation was generally at the beach face junction of the flanks or at the seaward corners of the structure. The increased sand elevation at the seaward corners would generally extend several meters seaward of the structure. However, it cannot be concluded from this slight increase in sand elevation that the structure was retaining sand in its vicinity.

Figure 2 shows the volumetric responses of the north, south, and center sections for Scale I. The north and south sides of the structure are of equal area, but the center section is not of comparable size. The three sections all responded similarly to each other and to the total volumetric of Scale I shown in Figure 3. It appears as though the beach on the north side of the structure responded less sensitively to major erosion and accretion events than the beach on the south side of the structure. Similar to the total volumetric response of Scale I, all three sections of beach shown in Figure 2 displayed a balanced volumetric response fluctuating seasonally about a nearly constant mean and experienced a negligible net change in sand volume over the observation period. The volumetric response at Scale II was nearly identical to that of Scale I, showing seasonal fluctuation about a nearly constant mean.

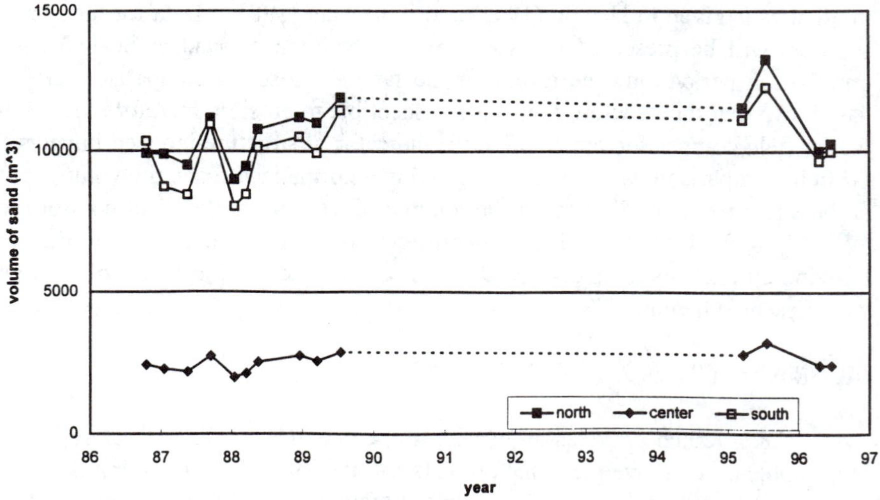

Figure 2 C&L Ranch Section Sand Volumes, Scale I

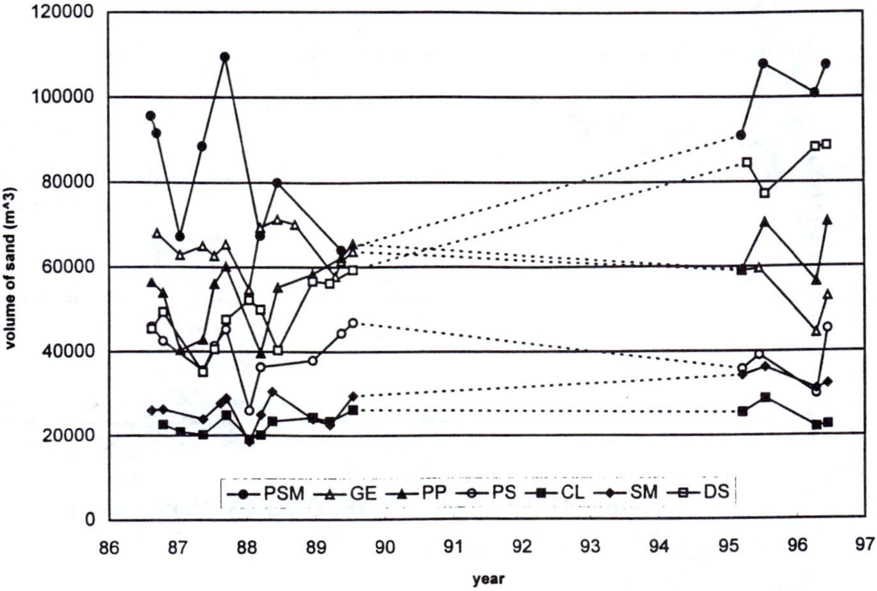

Figure 3 Total Sand Volumes at Each Site, Scale I

Far field sections were used both to the north and south of the structure. Figure 4 shows the comparison of the north far field and the north flank of the structure. The volumetric responses of the two areas were remarkably similar. The sensitivity to erosion and accretion events was approximately the same for both the north flank and the north far field. The north flank experienced a slight net gain in sand volume over the observation period, whereas the north far field section experienced a slight net loss. As Figure 5 indicates, the volumetric response of the south far field and the south flank of the structure were reasonably similar. The south far field section appeared too be slightly more sensitive to erosion and accretion events than the south flank. Both far field and the flank experienced a moderate net loss in sand volume over the observation period. Each of the structure flanks and corresponding far field volumes displayed a balanced volumetric response and negligible net change in sand volume over the ten year period.

The beach at C&L Ranch was very stable. The volumetric response of the beach followed a pattern of fluctuation between summer accretion and winter erosion. Erosion events were followed by periods of comparable recovery. The volumetric response of the beach was very similar for Scale I and Scale II, indicating the absence of seasonal features such as berms. The far field baseline comparison indicated that the structure flanks are responding similarly to the beach uninfluenced by the structures presence. The south flank appeared to be more stable than the south far field, and the north flank experienced a net gain in sand volume over the observation period, whereas both far field sections experienced a slight net loss.

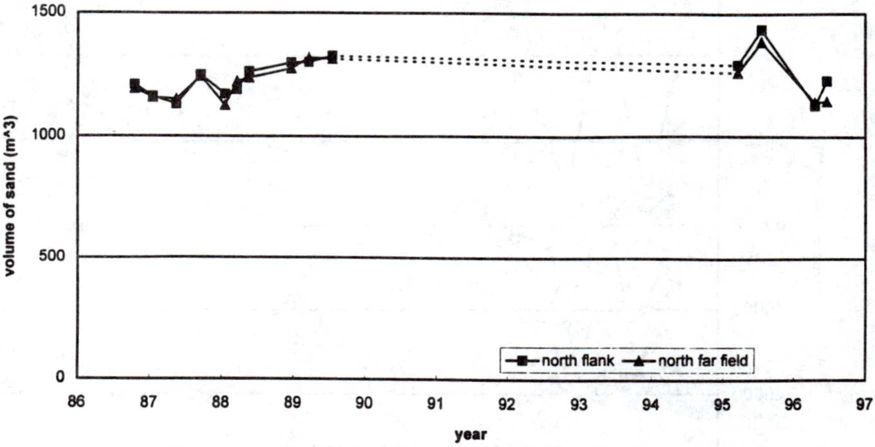

Figure 4 C&L Ranch North Far Field Comparison

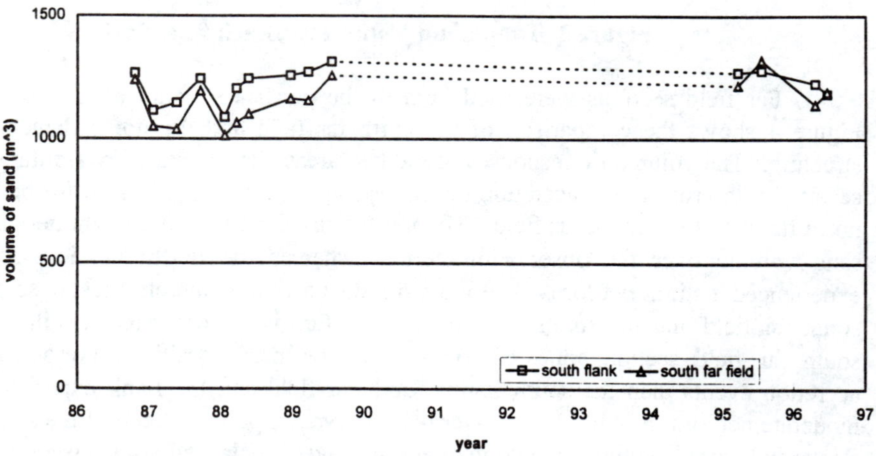

Figure 5 C&L Ranch South Far Field Comparison

Pacific Sands Motel (PSM)

Pacific Sands Motel, located in Neskowin, is the northern most site considered in this study. The structure is a timber seawall fronting the Pacific Sands Motel. The seawall is constructed of railroad ties with limited steel lateral supports and a 4 cm x 25 cm timber cap. The upper portion of the beach is characterized by low dunes with non-continuous low bluffs to the south of the structure. Proposal Rock, located to the south of the structure, is the predominate morphological feature in the area. Evidence suggests that Proposal Rock has a considerable influence on the beach processes at the site, mainly in the formation of a rip current embayment. Concentrated foot traffic at the north end of the seawall results in lower a local dune elevation and sparse vegetation at the foot path, making it difficult to separate structure induced flank erosion from foot traffic induced erosion. Aerial photographs for Pacific Sands Motel show that the coastline has not changed much since the mid-1960's.

Examination of the contour and surface plots reveal that during the summer, generally May through September, when the beaches are subjected to low wave energy, the beach at PSM is characterized by a well defined berm and cusps. The berm is often altered by a rip current embayment formed by the presence of Proposal Rock. During periods of high wave energy, typically November through April, the beach becomes very flat in order to dissipate the increased wave energy. The seawall and dunes are most susceptible to wave attack during these periods.

Figure 3 gives the results of the sand volume calculations for Scale I. Profile data for January 1989 and February 1988 were not used in the Scale I calculations due to insufficient range of coverage. The volume of sand fluctuates between winter lows and summer highs about a slightly increasing mean. The beach experiences a sizable net gain in sand volume over the observation period. The volumetric response of Scale II was very similar to that of Scale I.

Comparison of the volume of sand at the structure flanks with far field control sections, in plots similar to Figures 4 and 5, indicates that the flanks responded less sensitively to seasonal erosion and accretion. This data is presented in a normalized form in Figures 6 and 7. The north flank contained a greater volume of sand than the far field section throughout the observation period. A portion of the north far field control section is located landward of the mean dune line. The dune line tends to migrate seasonally, and at the time of the April 1988 survey the dune line had receded several meters. This results in the +2.40 spike in the V_n/V_{nff} data indicating a much larger volume at the north flank in comparison to the north far field. As shown in Figure 7, the volume of sand at the south flank deviates only slightly from that of the south far field. Sand volumes at the south flank are greater than those of the south far field section over the vast majority of the observation period. The results of the far field baseline comparison indicate that the structure flanks appear to be more stable than far field control sections partially because the seawall tends to stabilize the position of the dune line at its flanks. Overall, the south side of the structure was very stable, and although the north side occasionally underwent large volumetric changes, it was fairly stable over the long-term.

Gleneden (GE)

The Gleneden site is located in the town of Gleneden. The structure is a rubble mound revetment approximately 40 m long. The beach is backed by bluffs approximately 10 m in height. The revetment has two well defined flanks as the structure extends several meters from the bluff. This site provides a good open coast scenario, as the beach extends several miles in either direction without any significant obstructions. A beach access is located approximately 90 m to the south of the structure. A cut, several meters in width, was excavated through a low section of the bluff and partially paved to create this access. The access is significant to the site because a great deal of the storm water runoff from the neighborhood is discharged onto the beach at this point. Depending on the time of year, high flows associated with this runoff scour a creek through the beach face. A considerable volume of sand can be removed through this hydrologic process. Aerial photographs of Gleneden indicate that the coastline has changed little from 1963 to 1983.

The beach at Gleneden is characterized by a well defined berm and cusps in the low wave energy summer months (Komar and McDougal, 1988). A berm of some magnitude usually formed in the early spring and remained until October, however, a slight berm was often present through most of the year. A dominant reflective profile at Gleneden can be partially attributed to the presence of coarser sands on the beach (Shih and Komar, 1994).

The results of the sand volume calculations for Scale I are given in Figure 3. Profile data from January 1989 were not incorporated into the Scale I analysis due to lack of data range in the cross-shore direction. The volumetric response of the beach in Scale I shows the typical pattern of winter erosion followed by summer accretion or recovery. A sizable net loss in sand volume over the observation period had occurred primarily as a result of the severe erosion in May 1996. Recovery from this erosion event indicated from the July 1996 volume was significant, however, this data point probably does not represent full recovery. Oregon experienced a late summer in 1996, and beaches commonly experience accretion well into October.

Due to the drainage area located to the south of the structure, only a north far field control section was utilized for the baseline comparison. Normalized far field comparison data shown in Figures 6 and 7 indicate that the sand volume at the flanks was similar to that of the north far field section throughout the observation period. On average, both the north and south flanks of the structure contained greater volumes than the far field section. During the most severe erosion event recorded over the ten year monitoring period, May 1996, both flanks retained over 10% more sand than the north far field section. This is an indication that the flanks are more stable than the far field control section, and appear to be less sensitive to seasonal fluctuations. Although the beach at Scale I and the far field control section experienced significant net losses in sand volume, the north and south flanks of the structure were not impacted as severely by comparison. The north flank actually recorded a slight net gain in sand volume.

Pacific Palisades (PP)

Pacific Palisades is located just south of Lincoln Beach. The site is on the north side of Fishing Rock, a rocky point of land jutting out approximately 150 m into the ocean. Fishing Rock is not considered a littoral cell boundary, but it does inhibit longshore sediment transport and has a substantial effect on the beach processes at the site. The bluff line north of Fishing Rock forms a crescent shape, indicative of a hooked beach. A rubble mound revetment follows the curve in the bluff for approximately 300 m. The rubble blends into the bluff face and there are no discernible flanks on either end of the structure. The south end of the structure essentially extends to the base of Fishing Rock. The analysis of Pacific Palisades was limited to investigations of Scale I and Scale II. A far field investigation was not feasible considering the crescent shaped backing bluff and the lack of identifiable flanks. Aerial photographs indicate relatively little change in the coastline from 1962 to 1983.

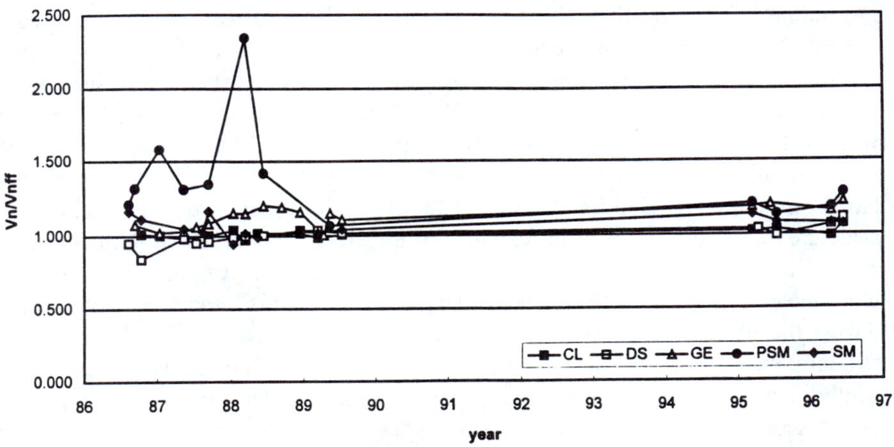

Figure 6 Normalized North Far Field Comparison

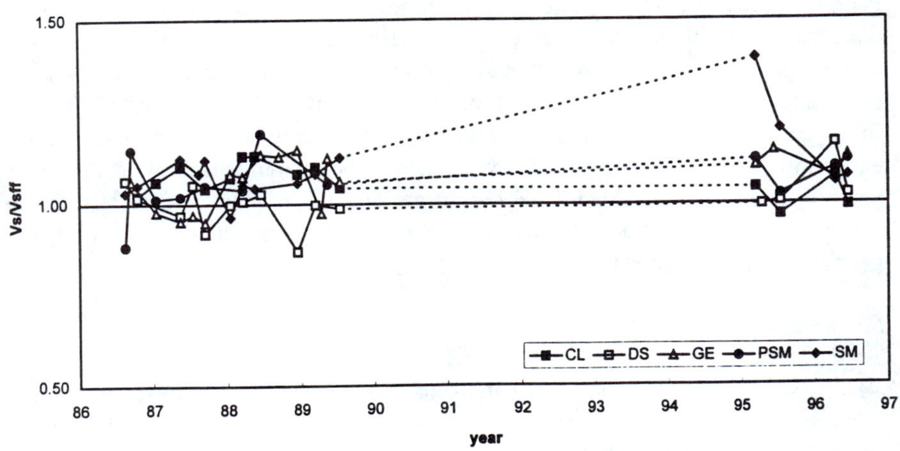

Figure 7 Normalized South Far Field Comparison

When the beach at Pacific Palisades was subjected to low energy waves, usually May through October, the beach was typified by a well defined berm, cusps, and a rip current embayment. The rip current embayment, present throughout most of the year, was most likely caused by Fishing Rock. The sand volume of the beach at Scale I, shown in Figure 3, fluctuates seasonally about a nearly constant mean. The beach experienced a slight net gain in sand volume over the observation period.

Pacific Shores (PS)

Pacific Shores is a sea cliff backed site located approximately three miles south of Newport. The dominant morphological feature at the site is a section of the sea cliff that protrudes further seaward than the adjacent cliffs on either side. This protruding section of cliff, which is probably subjected to wave attach on a regular basis, is composed of an erosion resistant material. The cliff line south of the protrusion, where the structure is located, forms a crescent shape similar to that of Pacific Palisades. The original structure selected for the field monitoring program was a rubble mound revetment measuring approximately 20 m in length. However, since the initiation of the study in 1986, a new 140 m long rubble mound revetment has been constructed immediately north of the original structure. The newer structure was probably built in response to localized land slides which occur at the site. The two structures were not joined together, thus leaving a 15 m gap of unprotected sea cliff between them. The addition of this enormous new structure during the observation period and in such close proximity to the original, smaller, structure made for a complicated analysis. Aerial photographs show that parts of the southern portion of the protruding cliff have been eroded since 1965. It is also possible that parts of the cliff were removed as part of the construction of the newer structure.

The complete chronology of contour and surface plots suggests the beach at Pacific Shores was very flat and planar throughout the year. A summer berm was not present, and there was no evidence of a rip current in any of the surveys. The contour plots did not indicate a trend of sand accumulating in front of any part of the structure. The volumetric response of the beach at Scale I, shown in Figure 3, followed a typical pattern of summer accretion and winter erosion, with the exception of 1989 when the beach experienced a prolonged period of accretion. Ten years of monitoring yielded a negligible net change in sand volume. A far field comparison was not feasible due to the addition of the new structure and the geometric nature of the backing bluff.

San Marine (SM)

San Marine is located between Walport and Yachats. The beach is backed by a low bluff, which is very irregular, containing several indentations and moderately protruding areas along a lengthy front. These are rather small scale features on the bluff, which is essentially shore parallel. The structure is a rubble mound revetment that measures approximately 150 m in length. The bluff line position is several meters further seaward at the structure. The bluff appears to be fairly stable with respect to slides. Aerial photographs for San Marine indicate that the coastline and the bluff were remarkably stable from 1962 to 1983. Several of the unique and irregular features in the bluff, which is composed of an erodable material, are present in both the 1965 and 1983 photos. This suggests that waves seldom reach the bluff line.

The beach at San Marine was very flat and dissipative throughout the year. The majority of the Scale I contour and surface plots indicate a slight rise in elevation near the structure flanks. This slight rise may be attributed to the fact that the bluff is recessed several meters on either side of the structure.

Seasonal fluctuations about a slightly increasing mean are apparent in the volumetric response of the beach at Scale I, Figure 3. The beach at San Marine experienced a considerable net gain in total sand volume over the observation period. This can be attributed to significant recovery following the winter erosion events.

Far field control sections were defined both to the north and south of the structure. Normalized results of the far field comparison, shown in Figures 6 and 7, indicate that the volume of sand at the north flank deviates little from the volume of sand at the far field control section. A trend a increasing volume in the south flank in comparison to the south far field is apparent in Figure 7. This trend can be attributed to a volumetric response about an increasing mean for the south flank, and a volumetric response about a nearly constant mean for the south far field section. Volumes for both the north flank and the north far field section fluctuated about an increasing mean. A gain in net volume over the observation period and the stability of the bluff indicated in the aerial photos suggest that the structure is having no adverse affect on the surrounding beach or adjacent properties.

Driftwood Shores (DS)

Driftwood Shores, the southern most site included in this study, is located at Heceta Beach just north of Florence. The structure is a low rubble mound revetment measuring approximately 30 m in length. There are large dunes to the south of the structure and the area north of the structure is more similar to low hills with heavy vegetation than dunes. The structure is located at a low point in the dune. The backshore area behind the dune is actually lower than the structure and many parts of the upper profile. A house is situated several meters behind the structure. The house is not built in the backshore depression, but is very low relative to the other houses in the vicinity.

The beach exhibits a great deal of local three-dimensionality. The small scale three-dimensionality of the upper portions of the beach appears to be aeolian dominated. The complete chronology of contour and surface plots indicates that the beach was reasonably flat and dissipative during much of the year. The dunes to the south were evident in several of the surveys, but absent in the majority. It was likely that the dunes migrated back and forth in the cross-shore direction during the observation period. There were no typical summer profile features evident in the contour and surface plots, although it was generally in the summer months when the south dunes were present in the contour and surface plots.

The volumetric response of the beach at Scale I, contained in Figure 3, shows seasonal fluctuation about an increasing mean, and a dramatic net increase in sand volume over the observation period. Driftwood shores was the only site which experienced accretion during the May 1996 period. The beach appeared to be gaining sand volume during the active portion of the study, 1986 through 1990, but the most substantial net gain in sand volume occurred between 1990 and 1996. Results from plots similar to Figures 4 and 5 show that the flanks responded nearly identical to the far field section over the ten year period. As indicated in Figures 6 and 7, the flank volumes deviated little from the respective far field volumes throughout the study. Similar to the volumetric response at Scale I, both flanks experienced a considerable net gain in sand volume.

CONCLUSIONS

All seven sites on the Oregon coast behaved in a similar manner. The volumetric response of the beach at each of the sites fluctuated seasonally between low sand volumes in the winter and higher sand volumes in the summer. Considerable recovery following periods of erosion resulted in net gains in sand volume over the observation period at Scale I for three sites (PSM, PP, DS), negligible changes at three site (PS,CL, SM), and a significant net loss in sand volume at only one site (GE).

The structure flanks at each site, including Gleneden, were stable over the observation period.. Sand volumes at the structure flanks deviated little from those of the far field control sections used in the baseline comparison. The average $V_{flank}/V_{far\ field}$ value for all seven sites, regardless of north or south flank, was 1.094 with a coefficient of variance of 0.102. This indicates that the structure flanks were retaining greater volumes of sand over the observation period than the far field control sections. In general, the structure flanks experienced negligible net changes in sand volumes, and often slight net gains in sand volume over the observation period.

There were no significant negative effects caused by the structures at any of the seven sites. This can be attributed to the lack of wave attack that these structures experience (Ruggiero et. al., 1997) The structures were built high on the profile against the base of the bluff or dune. Each site has a wide buffer beach fronting the structure which dissipates the high wave energy found on the Oregon coast and prevents waves from reaching the structure or bluff except during the most severe storms. The structures tend to serve as bluff retention and stabilization structures as much as shore protection structures. Structures built against bluffs appear to be doing an excellent job supporting the bluff and preventing landslides, which can be brought about by undercutting of the bluff or cliff by wave action, but many other non-beach process related factors are also recognized (Komar and Shih, 1994).

When examining the long-term effects of seawalls on beaches, the analysis must be site specific. Susceptibly to wave attack should be considered when permitting or building a shore stabilization structure on the central Oregon coast.

REFERENCES

Hearon, G.E. 1995. *Long-Term Beach Response to Shore Stabilization Structures*. M.S Engineering Report, Oregon State University.

Komar, P.D., and McDougal, W.G. 1988. Coastal Erosion and Engineering Structures: The Oregon Experience. *Journal of Coastal Research*, Special Issue 4: 77-92.

Komar, P.D., and Shih, S.M. 1991. Sea-Cliff Erosion Along The Oregon Coast. *Proceedings Coastal Sediments '91*, American Society of Civil Engineers, 1558-1570.

Shih, S.M., and Komar, P.D. 1994. Sediment, Beach Morphology and Sea Cliff Erosion Within an Oregon Coast Littoral Cell. *Journal of Coastal Research*, 10(1): 144-157.

Sturtevant, M.A., McDougal, W.G., and Komar, P.D. 1987. *The Impact of Shoreline Stabilization Structures on Beaches and Adjacent Properties*. Sea Grant Report, Oregon State University.

Ruggiero, P., Komar, P.D., and McDougal, W.G. 1997. Extreme Water Levels, Wave Runup, and Coastal Erosion. *In these Proceedings*.

CHAPTER 211

Profile Volumes as a Measure of Erosion Vulnerability

Timothy W. Kana, Ph.D.,[1] and Ram Krishna Mohan, Ph.D., P.E.[2]

ABSTRACT

Profile volumes provide an indication of the condition of barrier islands and beaches. While site specific, they can be used along any littoral shoreline as a measure of the relative "health" of one section compared with another. Assuming shorelines develop an equilibrium profile around which the beach cycle and cross shore changes occur, profile volumes integrate small scale perturbations and provide a target sediment volume for the site between reference contours. Profile volumes may be considered for various parts of the profile as follows:

- *Dune volume above storm surge levels* (after FEMA) as a measure of the reservoir of sand required to prevent over topping.

- *Mean sea level (MSL) barrier cross-section* — the section that must be fully scoured to produce a full breach channel.

- *Inner beach profile* — the section that represents the condition of the recreational beach.

- *Inshore profile* — the section that represents the core "base" of the barrier island.

By comparing versions of these sections from site to site, especially where there have been known breaches and variations in beach erosion rates it is possible to develop systematic criteria for the health of the shoreline in question. Profile volumes provide a straight-forward technique for diagnosing barrier erosion problems. A case study from Westhampton Beach and Fire Island (New York) illustrate the application.

---

[1] Senior scientist, CSE-Baird, A Division of W.F. Baird & Associates, Ltd., PO Box 8056 Columbia SC 29202

[2] Senior engineer, Gahagan & Bryant Associates, 9008-O Yellow Brick Road, Baltimore, MD 21237

INTRODUCTION

Profile volumes, that is, the unit quantity of sediment contained between defined contour limits in the littoral zone, can be used as predictors of erosion vulnerability, breach vulnerability, and nourishment requirements. They are especially useful for analyzing sites with few data or sites where surveys are spaced widely in time. The profile volume for a site, ideally, should incorporate the foredune (or the entire backshore seaward of buildings where development is a concern), berm, intertidal beach, and lower foreshore to closure depth.

Multiple transacts surveyed between common boundaries along shorelines having similar orientations and exposures provide a set of profile volumes that can be normalized and compared statistically by simple means (Fig. 1) (Kana et al., 1984; Kana, 1993). The unit volume of any one profile to closure integrates all small-scale perturbations in slope and bar formation, and becomes a reasonably pure measure of the littoral condition. Seasonal or beach-cycle-related variations are accounted for in the volume. Therefore, the phasing of surveys with storms or seasons is less critical. Where the setting is a narrow barrier beach, the profile volume method also provides a measure of vulnerability to breaches.

FIGURE 1.

The reference profile illustrates how average unit-width profile volumes can be developed on a site-specific basis for analyses of the condition of the beach from one section to another, as well as from one survey to the next (after Kana et al., 1984).

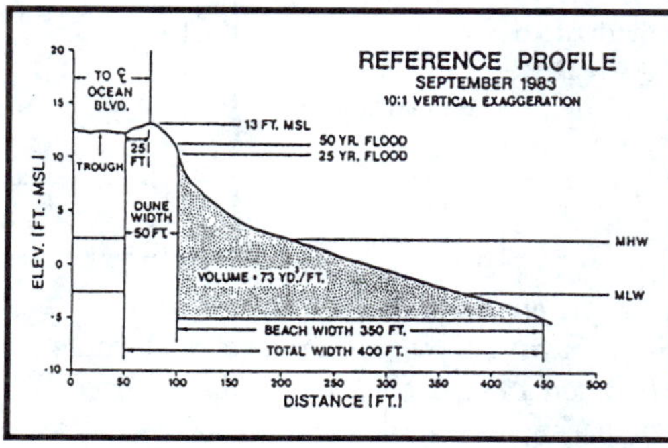

Barrier Sediment Volumes

There are no universally accepted criteria defining the condition of a barrier beach to withstand erosion and storm surges without breaching. In the United States, FEMA (Dewberry & Davis, 1989) has developed rough guidelines relating to dune preservation during storms. FEMA considers a particular volume of sand above the still-water flood level and seaward of the dune crest as being the minimum necessary to prevent overtopping and washovers for a particular return-period storm. Dutch coastal engineers (TAW, 1985; Van de Graaff, 1986) further consider the amount of freeboard above the still-water surge level needed to absorb wave runup without overtopping and breaching. These criteria are based on probabilities and are poorly documented because there are so few sites where all necessary data are available to quantify the relationship. In most cases, dunes either remain standing with considerable excess sediment landward of poststorm scarps, or they are washed out. Thus, establishing threshold minimum quantities to sustain surges without (dune) breaching is difficult.

The present criteria of FEMA are given in a rating curve (Fig. 2). The minimum cross-sectional areas recommended above the still-water surge elevation (for a particular return period) equate to unit volumes of sand (cubic yards per fool) as given on the graph. In simple terms, the 100-year storm would require about 20 cy/ft (50 m^3/m) above its still-water surge level, whereas a 5-year storm would require about 6 cy/ft (15 m^3/m) for protection. FEMA representatives emphasize these criteria are for guidance only and may not apply for certain dune configurations (R. Hallermeier, Dewberry & Davis, pers. comm., March 1994). In other words, the absolute quantity of sand above the surge level must be configured to absorb wave runup effectively; otherwise, breaching may still occur.

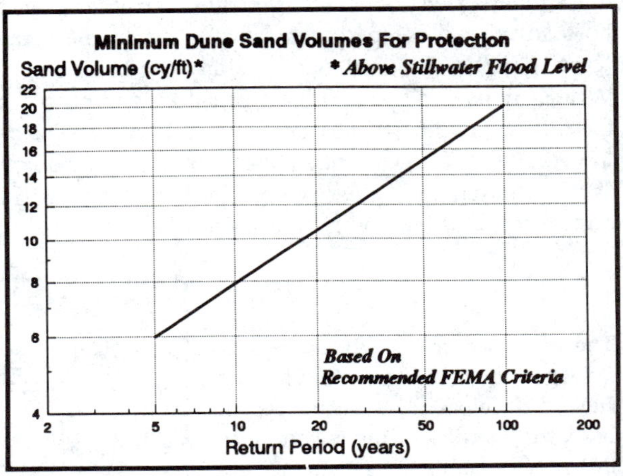

FIGURE 2.

FEMA minimum criteria for dune unit volumes above flood still-water level as a function of storm-return period (based on guidelines in Dewberry & Davis, 1989). Standard deviation for the 100-year storm is ±10 cy/ft, confirming a wide range of "safe" values based on experience.

In the following example application, the authors have computed reference sand volumes above particular datums to relate the condition of a barrier to FEMA criteria. Unit-volume analysis is also useful for comparing the overall condition of the barrier island, including the underwater portion of the profile. By comparing barrier sections that have experienced breaches with unbreached sections, it is possible to develop site-specific criteria relating to the potential vulnerability of future breaches along nearby sites.

EXAMPLE APPLICATION

Winter storms (northeasters) on 11 December 1992 and 13 March 1993 caused widespread damage along the south shore of Long Island, New York (Fig. 3). In terms of beach erosion, structural damage, and mainland flooding, these two storms were the worst for this part of the coast since the well-known 4-6 March 1962 northeaster (USACE, 1963). Over 100 houses were lost along Pikes Beach (Westhampton, immediately downdrift of the groin field) as a result of continued erosion and a breach through the barrier in December. The 1992 breach was left open for over ten months and contributed to locally higher tides in Moriches Bay (J. Tanski, unpubl. data,

SUNY, Stony Brook, March, 1993) before it was closed by the U.S. Army Corps of Engineers by hydraulic dredge (contractor: Great Lakes Dredge & Dock Co.).

The breach at Westhampton demonstrated the problem of unstable inlets near development and the potential for increased flooding around the lagoon shoreline. Concerns over the possibility of a similar breach through Fire Island, the next barrier island downdrift of Westhampton Beach, and a possible increase in the tide range within Great South Bay led to a study funded by the New York Coastal Partnership (Kana and Krishnamohan, 1994). The purpose of the study was to assess the vulnerability of Fire Island to storm breaches and determine the likely physical impacts along Great South Bay (Fig. 3). The study was multidisciplinary and drew on historical shoreline data, developed an updated sediment budget for Westhampton and Fire Island, and used available predictive models of storm surges to evaluate flooding potential. One analysis (the subject of the present paper) evaluated profile volumes in an attempt to develop site-specific criteria for barrier sections that had breached in the past and to identify other potential sites for breaches.

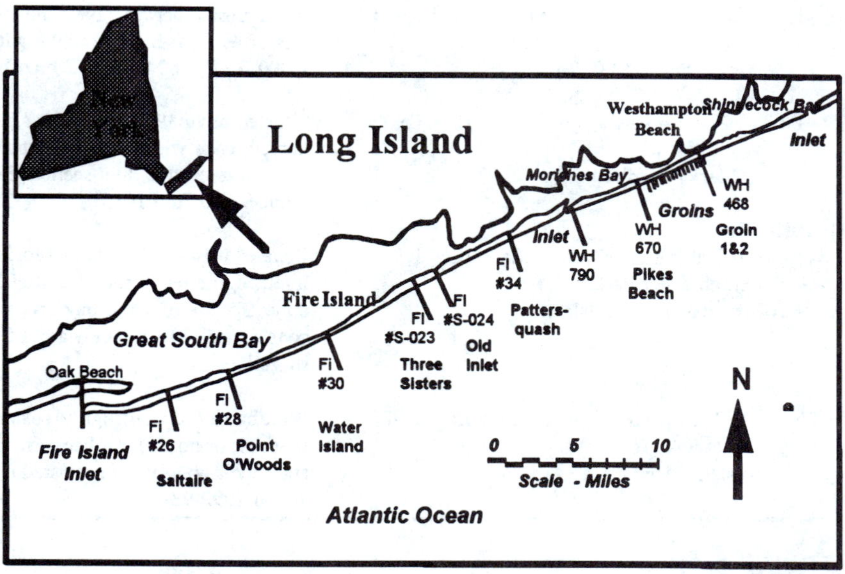

FIGURE 3. General location map of representative barrier-island cross-sections developed for the 1994 study (after Kana and Krishnamohan, 1994).

Methodology

Representative cross-sections through Westhampton Beach and Fire Island were developed from USACE profiles (RPI, 1985), unpublished surveys, and planimetric maps (see Fig. 3). These were selected to represent the range of conditions including physical dimensions of the barriers and historical erosion trends. The profiles extend from the bay shoreline across the barrier and out to approximately -30 ft below MSL. Table 1 describes some of the cross-sections, localities, and significant features [see Kana and Krishnamohan (1994) for details].

TABLE 1. Representative barrier island cross-sections, relevant physical dimensions, and key distinguishing features. [*(1) Width at mean sea level. (2) Highest dune elevation. (3) Distance from MSL to -20 ft contour.]

USACE Transect ID	Locality	1979 Dimensions (ft) (1)*	(2)*	(3)*	Key Features
WESTHAMPTON BEACH					
468+00	Between groins 1 and 2, east end of groin field	825	21	1,400	Rapid accretion and growth of a new foredune to 15 ft MSL between 1962 and 1979; continued accretion through 1993.
670+00	2,700 ft west of groin 15, compartment 143B	700	10	1,600	Rapid erosion between 1962 and 1979, loss of an 18-ft-high foredune within 1,000 ft of 1962 and 1992 breaches.
790+00	2,000 ft east of east jetty, Moriches Inlet	450	15	1,950	Situated btwn 1980 breach and east jetty; breach widening and easterly migration eroded the section btwn January and July 1980.
FIRE ISLAND					
34	6,700 ft east of Smith Point Park Pavilion, eastern Fire Island	1,225	18	1,350	Zone of washovers west of Moriches Inlet; high erosion rates; artificial dune (spoil area from bay dredging) constructed 400 ft landward of the shoreline.
~26	West walk at Saltaire, western Fire Island	2,050	15	1,375	Wide barrier section; highly erosional; first row of houses lost between 1967 and 1979; area of concentrated damage in 1992-93.

Westhampton Profiles

In 1962, Westhampton had sustained its worst storm since the 1938 hurricane (storm of record). The March northeaster breached the barrier along Pikes Beach and caused extensive structural damage. Prior to the storm, erosion had reduced the width of the barrier and chronic problems of overwash forced periodic sand scraping to clear Dune Road. The USACE (1958) had been developing a shore protection plan which proposed construction of 21 groins and concomitant nourishment. By December 1962, emergency repairs had closed the storm breach and had added almost one million cubic yards of additional sand to Westhampton Beach (USACE, 1963). Construction of a portion of the groin field was initiated in 1964 and the final four of 15 groins were completed around 1969.

Two transacts illustrate dramatically different results along Westhampton Beach between 1962 and 1979--USACE station 468+00 (Fig. 4a) situated at the updrift end of the groin field between groins 1 and 2, and USACE station 670+00 (Fig. 4b) situated 2,700 ft downdrift of groin 15. In 1962, both transects represented narrow sections of the barrier island. In fact, the updrift station (468+00) was barely 600 ft wide from the ocean to the bay shoreline and contained a single, well-developed foredune to 20 ft above sea level. The downdrift station (670+00) was slightly wider at about 725 ft with a smaller foredune reaching about 17 ft.

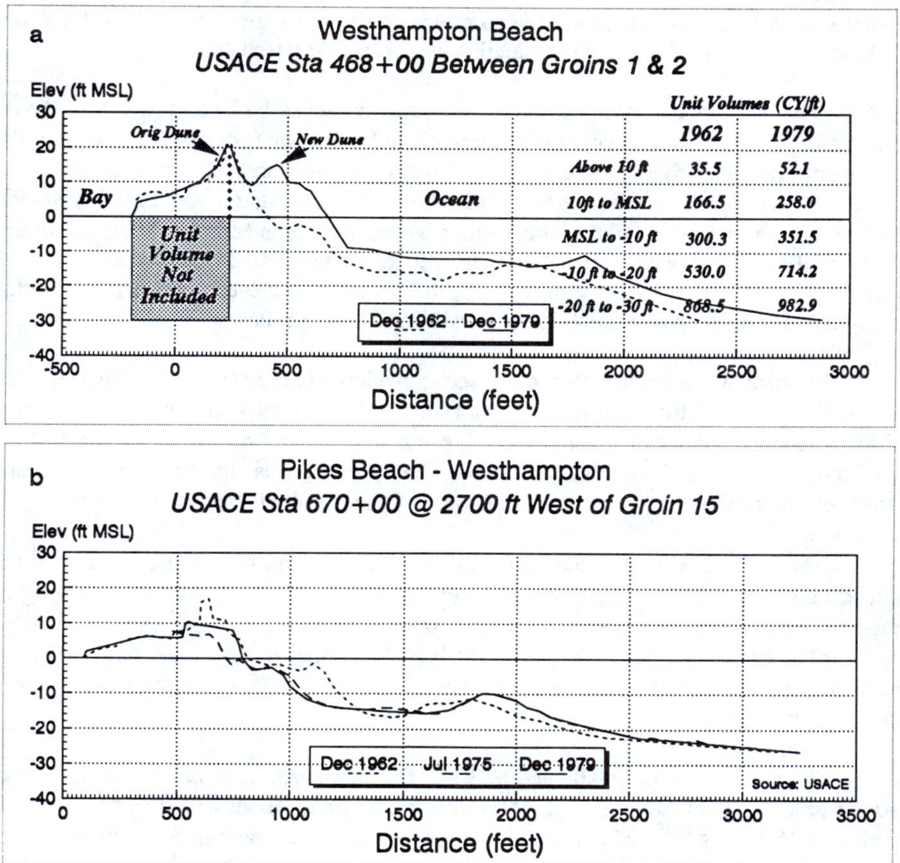

FIGURE 4. Representative, barrier-island cross-sections from the east end of the Westhampton groin field (a) and Pikes Beach (b), illustrating opposite trends in shoreline change between 1962 and 1979. Unit volumes refer to the profile areas between the indicated contours.

By 1979, station 468 + 00 had gained a second foredune and the shoreline built seaward, increasing the barrier width to about 825 ft (Fig. 4a). Associated with the dune and upper beach building was a large increase in sand volume below MSL. This trend of large-scale accretion has continued through 1993 as shown in nearby profile 461 + 25 at groin 1 (Fig. 5). By January 1994, groins 1 and 2 (500-ft-long structures when built) were completely buried.

In contrast, transect 670 + 00 lost dune elevation and width between 1962 and 1979. The trends shown in Figure 4b actually incorporate several beach fills during the period. By July 1975, the barrier had narrowed to about 600 ft and the maximum elevation at the transect was only 7 ft above sea level. Nourishment is believed to have restored part of the beach between 1975 and 1979, but the dune elevation remained no higher than 10 ft MSL. An aerial photo from August 1981 (Fig. 6) in the vicinity of station 670 + 00 shows the recession of the shoreline downdrift of the groin field and a number of houses left standing on the active beach.

The 1992 northeaster caused at least two breaches between transect 670 + 00 and groin 15. The easternmost breach immediately downdrift of groin 15 became the dominant and only channel and expanded from several hundred feet wide after the storm to over 2,000 ft wide by the summer of 1993 (Westhampton Fire Marshall's Office, unpubl. reports). Field observations by the authors in March 1993 and January 1994 confirmed deposition of a flood delta in Moriches Bay in connection with the breach channel. Around November 1993, the breach was closed by hydraulic dredge using a USACE-approved offshore borrow source ~1.5 miles south of the breach.

A third Westhampton Beach transect considered here is from USACE 790 + 00, ~2,000 ft east of the east jetty of Moriches Inlet. This undeveloped section of Cupsogue spit consists of a narrow barrier segment flanked on the landward side by the navigation channel to Moriches Inlet. During the 1960s, the bay channel meandered closer to Cupsogue spit and contributed to scour along the bay shoreline.

By 1979, portions of the spit about 0.5 mile north of the jetties had been reduced in width to 300 ft from ocean to bay. In January 1980, continued narrowing of the spit combined with higher-than-normal wave action breached the barrier (Kana et al., 1981). Between January and July, the breach channel widened and migrated west until it removed the last part of the spit and coalesced with the navigation channel between the jetties.

Transect 790 + 00 is situated between the January breach point and east jetty. At the time of the breach, section 790 + 00 had a width of about 450 ft and a dune elevation of 15 ft (Fig. 7). Much of this section was artificial, having been filled several times in connection with dredging projects in the bay. As the breach widened, it migrated through section 790 + 00, washing out the beach and depositing a large volume in the bay channel.

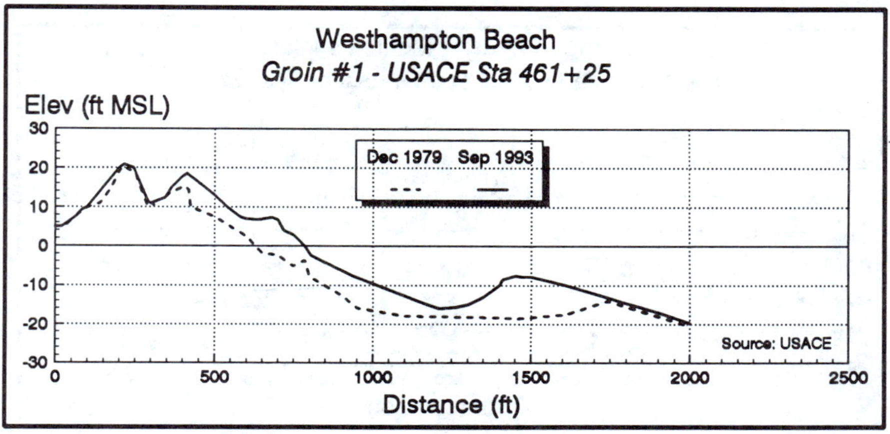

FIGURE 5. Comparative profiles at groin 1 (easternmost) for December 1979 and September 1993. Data courtesy of U.S. Army Corps of Engineers.

FIGURE 6. Pike's Beach and Westhampton Beach in August 1981, looking east. Groin field is at top of photo.

As Figure 7 shows, by July 1980, the elevation across 790+00 was about 4 ft below sea level. The breach was closed in January 1981 by hydraulic dredge, using barrier and littoral sediments that had accumulated in the bay channel, and by trucks using sand from an inland source.

Unit volumes were computed through each barrier cross-section by 10 ft contour intervals (example values given on Fig. 4a). Conveniently, the predicted 100-year still-water surge elevation along this microtidal coast nearly equals +10.0 ft MSL (FEMA, 1987).

Unit volumes above the +10 ft contour generally reflect the condition of the dunes. At Westhampton, the volumes above 10 ft are much greater at section 468+00 (groin field) than the sections near recent breaches. If 468+00 is typical of the groin field, dune unit volumes approached 50 cy/ft by 1979 and are higher today (see Fig. 5). In contrast, stations 670+00 and 790+00 yield an average volume of about 10 cy/ft with zero volume occurring in three of the six surveys used in the analysis.

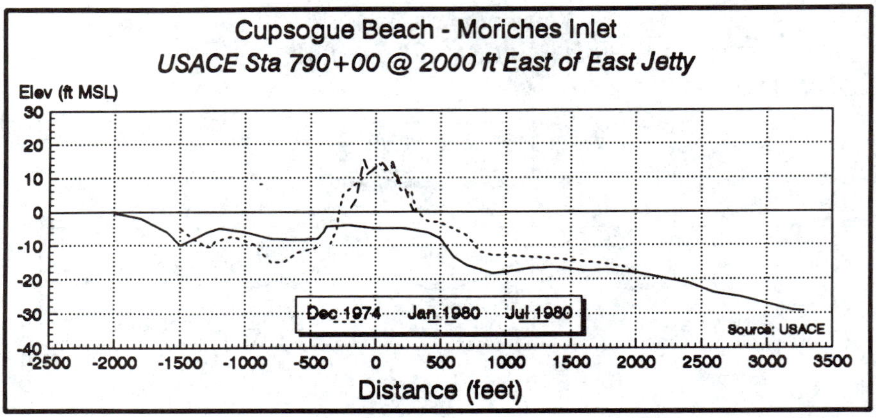

FIGURE 7. USACE station 790+00 situated ~2,000 ft east of Moriches Inlet. Erosion in 1980 occurred after the January 1980 breach to the east migrated through this section.

Between +10 ft and mean sea level, the barrier cross-section volume is largely a function of barrier width. Of the few surveys used for Westhampton, a typical section volume is about 160 cy/ft.

Relative to the total section volumes above MSL, the net changes for the three Westhampton stations generally represent a high percentage of the maximum section volume. For example, station 468+00 gained 108 cy/ft between 1962 and 1979, a volume which is 35 percent of the 1979 section volume (i.e., a gain of over 50 percent of the original unit volume). Station 790+00 lost 100 percent of its 1974 section volume due to the January 1980 breach; station 670+00 lost more than reflected (see Fig. 4) because of remedial nourishment between surveys.

Underwater sand volumes for Westhampton Beach stations similarly reflect the trends above sea level. Typical unit volumes from MSL to -10 ft seaward of the foredune are about 125 cy/ft and from -10 ft to -20 ft are about 425 cy/ft. The net change for 468+00 (groins) between MSL and -20 ft was 235 cy/ft, nearly 50 percent more sand than the 1962 volume. Assuming 461+25 (see Fig. 5) is similar, 468+00 has probably gained an additional 240 cy/ft in these lenses between 1979 and 1993.

Station 670+00 actually contained almost 20 percent more sand than 468+00 between MSL and -20 ft in 1962 before the groins were built. By 1979, this section contained about 75 percent of the unit volume at 468+00 (between MSL and -20 ft); a greater loss occurred between MSL and -10 ft; unit volumes between -10 ft and -20 ft increased by about 5 percent. Most likely, the change would have been much greater if there had been no beach fills.

Station 790+00 had a healthy underwater profile prior to the 1980 breach. As the breach widened and eroded the section, unit-volume losses between MSL and -20 ft were 260 cy/ft, an exceedingly rapid loss over the five-year survey interval. The upper lens from MSL to -10 ft was reduced to a section of about 50 cy/ft by July 1980, a volume representing the underwater sill through the breach. The lower shoreface (-10 ft to -20 ft) eroded severely, reducing the section volume to 330 cy/ft (about 80 percent of the same lens at station 670+00).

Based on the sand volume results in Table 2, typical Westhampton Beach values (rounded) are as follows:

Dune Volume	Mean (cy/ft)	Std. Dev. (cy/ft)	(No. of Surveys)
Above +10 ft MSL	20	±20	(8)
+10 ft to MSL	150	±77	(8)
MSL to -10 ft*	125	±45	(7)
-10 ft to -20 ft*	425	±75	(7)

[*Seaward of dune crest of earliest survey]

The results show high standard deviation because one station built up rapidly and one incorporated breach conditions. Thus, the "healthy" profiles reflect volumes incorporating the additional volume indicated by the plus standard deviation. Unhealthy profiles vulnerable to breaching reflect mean volumes minus the standard deviation. These results provide benchmarks for comparison with Fire Island profiles.

TABLE 2. Barrier-island unit-profile volumes (cy/ft) from the bay to -30 ft NGVD (above MSL encompasses entire barrier width; below MSL extends from foredune seaward). [Numbers in parentheses represent number of surveys.]

	Above 10 ft	10 ft to MSL	MSL to -10 ft	-10 ft to -20 ft	-20 ft to -30 ft
WESTHAMPTON					
468+00 (avg)	43.8 (2)	212.2 (2)	154.4 (2)	450.6 (2)	754.2 (2)
Breach sections (avg)	12.6 (6)	139.6 (6)	115.6 (5)	420.6 (5)	811.2 (4)
FIRE ISLAND					
Minimum	3.1	210.0	90.7	310.9	665.5
Maximum	90.0	475.1	176.3	459.2	786.8
Mean	32.3 (11)	343.3 (11)	134.3 (10)	407.7 (10)	739.6 (9)

Fire Island Profiles

Six transacts across Fire Island (two of which are shown in Figure 8) illustrate the typical range of conditions (see Table 1). Unit volumes have been computed for the same contour intervals and boundaries as Westhampton (Table 2). Some differences are immediately apparent. Above mean sea level, Fire Island profiles typically contain more sediment. The average unit volume above +10 ft is 32 cy/ft (±27.7 cy/ft standard deviation) for 11 surveys. Point O'Woods (USACE-28) and Saltaire (near USACE-26) had the lowest dune volumes, ranging from 3.1 cy/ft for the 1993 Saltaire profile to 7.1 cy/ft for the 1967 Point O'Woods profile. Three Sisters dunes off Bellport represented the maximum with 90 cy/ft above +10 ft (1979). Between +10 ft and MSL, Fire Island sections averaged about 345 cy/ft (±98.6 cy/ft standard deviation), much of which is due to the wider backshore. In fact, volumes in this lens

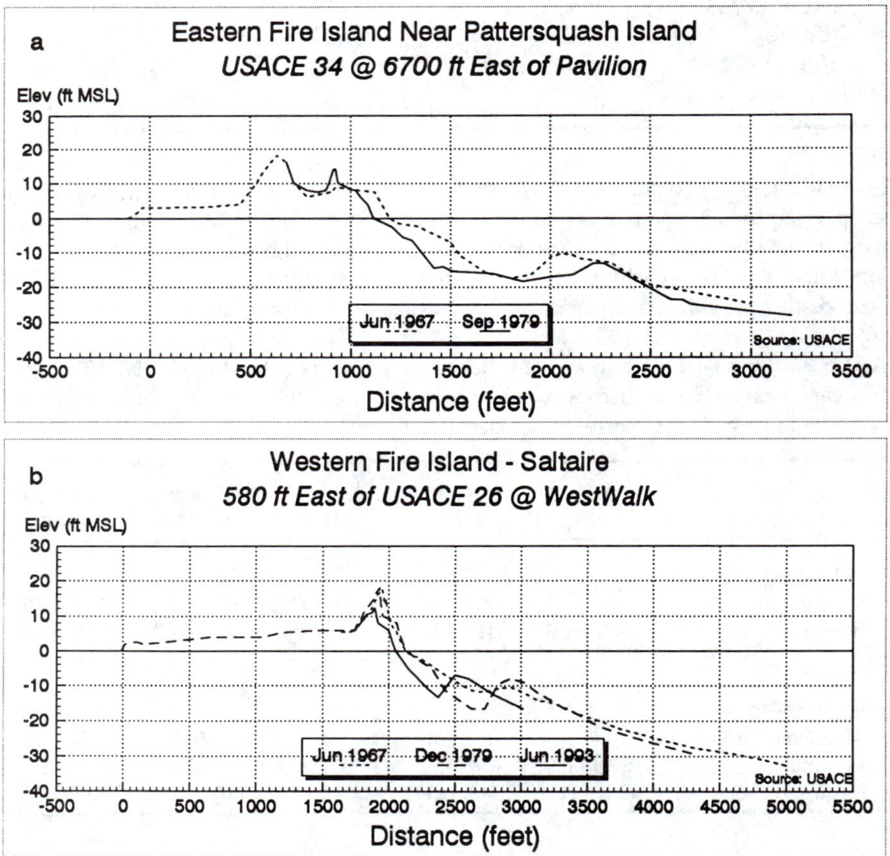

FIGURE 8. Representative, barrier-island cross-sections from (a) Pattersquash Island area east of Smith Point Park (locality for numerous washovers) and (b) west end of Saltaire, where major erosion of the underwater profile has accelerated erosion of the foredune.

are roughly double those of Westhampton (Table 2). Below mean sea level, unit volumes were very similar to Westhampton's, averaging -135 cy/ft between MSL and -10 ft, and ~400 cy/ft between -10 and -20 ft. This is expected because the landward computation boundary in each case is the dune crest.

Highest erosion rates were at the Pattersquash section (near USACE-34) and at Saltaire (near USACE-26) (see Fig. 8). The net change for these stations, respectively, ranged from about 10 cy/ft to 50 cy/ft above MSL and from 200 cy/ft to 150 cy/ft below MSL. In 1979, station USACE-34 was reduced to a volume of 415 cy/ft below MSL (to -20 ft) compared to an average of -550 cy/ft for other stations.

Aerial overflights by the authors and reports from officials of Suffolk County Department of Public Works (SCDPW) indicate the area around the Pattersquash Island transect continued to overwash through January 1994. But at this station, the setback of the artificial dune appears to have prevented a more landward shift of sediment. The artificial dune ends less than 1,000 ft to the east of transect 34, suggesting that more easterly areas are likely to experience deeper penetration of overwash along the back barrier. The authors observed a number of large washovers in that area during a January 1994 overflight. Officials at SCDPW confirmed these trends along eastern Fire Island over the past ten years. The results of the profile volume analysis suggest washovers will be more likely along Fire Island when the profile volumes between MSL and -20 ft go below 500 cy/ft relative to the foredune and when the foredune has negligible volume above +10 ft MSL.

The Saltaire section (near USACE-26) bears some similarities to USACE-34 because of its rapid erosion since 1967. While the barrier is much wider at Saltaire, its backshore elevations are low, averaging less than 5 ft above sea level (Fig. 8b). A single foredune has existed along the oceanfront. In 1967, the dune contained about 24 cy/ft above the +10 ft contour, about 75 percent of the average for the selected Fire Island transacts. By June 1993, dune volume above +10 ft was reduced to only 3 cy/ft. This low volume is offset by over 450 cy/ft between +10 ft and MSL (a total approximately three times that of Westhampton sections). Below mean sea level, station 26 lost about 155 cy/ft between 1967 and 1993, reducing its volume seaward of the 1967 dune crest to 450 cy/ft. As discussed in a regional sediment budget analysis (Kana, 1995), rapid erosion in the western Fire Island area is believed the result of depletion of offshore shoals associated with an earlier position of Fire Island Inlet. This has left the foredune much more vulnerable to erosion. An ~80 ft recession of the foredune since 1967 has destroyed (or caused to be removed) the oceanfront structures at this site, leaving the second row buildings on the oceanfront in 1993.

On the basis of backshore volumes above mean sea level, the Saltaire section is less likely to breach than Pattersquash because it contains almost twice the volume. But the Saltaire foredune is more likely to recede than Pattersquash's artificial dune because of the low unit volumes seaward of it and the lesser setback. Pattersquash's artificial dune profile could be created at Saltaire, but to be stable under present offshore conditions, it would have to be set back at least 150 ft from the 1993 dune crest. Planimetric maps indicate there are at least three rows of houses within 150 ft of the present dune crest at that locality.

SUMMARY — CASE STUDY

The remaining four sections along Fire Island represent intermediate conditions and are not discussed in this paper (see Kana and Krishnamohan, 1994, for details). By the measure of profile volumes through 1979, Fire Island cross-sections generally contain much more sand above MSL than Westhampton Beach sections. The difference in dune volumes between the two islands represents only a fraction of the total lens of sand comprising the barrier. The bulk of the subareal barrier is contained between MSL and +10 ft. Westhampton profiles contain roughly half the volume of Fire Island profiles above mean sea level. Transects that are breached along Westhampton typically contained less than 150 cy/ft above MSL before breaching.

While absolute volumes above mean sea level are not the only parameters to rank the condition of the barrier and establish vulnerability to breaching, they provide one quantitative measure to compare with historical breach areas.

REFERENCES

Dewberry & Davis. 1989. Basis of assessment procedures for dune erosion in coastal flood insurance studies. Prepared for FEMA; Dewberry & Davis, Inc., Fairfax, VA, 57 pp. + appendices.

FEMA. 1987. Flood Insurance Study, Village of Westhampton Beach, Suffolk County, New York. Federal Emergency Management Agency, Washington, DC (unpublished).

Kana, T.W. 1993. The profile volume approach to beach nourishment. In D.K. Stauble and N.C. Kraus (eds.), *Beach Nourishment Engineering and Management Considerations*, ASCE, New York, NY, pp. 176-190.

Kana, T.W. 1995. A mesoscale sediment budget for Long Island, New York. *Marine Geology*, Vol. 126, pp. 87-110.

Kana, T.W., and R. Krishnamohan. 1994. Assessment of the vulnerability of the Great South Bay shoreline to tidal flooding. Final Report, New York Coastal Partnership, Babylon, NY; Coastal Science & Engineering, Columbia, SC, 153 pp. + appendix.

Kana, T.W., M.O. Hayes, and S.A. Covell. 1981. Fire Island to Montauk Point, Long Island, New York, reformulation study: sediment budget — phase I, Moriches Inlet. Report for N.Y. District, USACE, Contr. DACW51-81-C-0030; Research Planning Inst., Inc., Columbia, SC, 82 pp.

Kana, T.W., S.J. Siah, M.L. Williams, and W.J. Sexton. 1984. Analysis of historical erosion rates and prediction of future shoreline positions, Myrtle Beach, South Carolina. Project Report for City of Myrtle Beach; CSE (as subdivision of Research Planning Inst., Inc.), Columbia, SC, 130 pp.

RPI. 1985. Fire Island Inlet to Montauk Point, Long Island, New York: sediment budget analysis. Summary Report for New York District, USACE; Research Planning Inst., Inc., Columbia, SC, 85 pp. + app.

TAW. 1985. Guidelines for the evaluation of safety of dunes as coastal defense. Unauthorized translation of Leidraad voor de beoordeling van de verligheid van duinen als zeewering, Technical Advisory Committee on Sea Defenses (Technishe Adviscommissie noor de water keringen), Dutch Government Printing Office, 16 pp.

USACE. 1958. Atlantic Coast of Long Island, New York (Fire Island Inlet to Montauk Point). Cooperative Beach Erosion Control and Interim Hurricane Study. Appendices, U.S. Army Corps of Engineers, New York.

USACE. 1963. Report on Operation Five High disaster recovery operations from 6-8 March 1962 storm. U.S. Army Corps of Engineers, North Atlantic Division, Civil Works Branch.

Van de Graaff, J. 1986. Probabilistic design of dunes; an example from The Netherlands. Coastal Engineering, Vol. 9, pp. 479-500.

CHAPTER 212

BREACH GROWTH: EXPERIMENTS AND MODELLING

by

Harry de Looff [1], Henk J. Steetzel [2], Arie W. Kraak [3]

ABSTRACT

Since 1989 research has been contucted in the Netherlands on the issue of breach growth. This contribution describes the set-up and results of wave basin experiments which had the aim to obtain insight into the process of growth of the width of a breach. From analysis of the recorded data (flow characteristics in the breach and the development of the breach) an empirical relation for breach width growth has been derived. This relation has been implemented into a (conceptual) numerical model which is briefly described in this document also.

1. INTRODUCTION

The Dutch Technical Advisory Committee on Water Defences (TAW) is aiming for a design method for water defences in which the optimal safety level of these defences is related to the risk of flooding. Risk is defined as the product of the probability of failure and the expected damage costs in case of failure. One of the key factors in the damage costs assessment is the speed and rate of inundation

---

1) Rijkswaterstaat, Department of Road and Hydraulic Engineering, P.O.Box 5044, 2600 GA Delft, The Netherlands

2) ALKYON Hydraulic Consultancy & Research bv, PO Box 248, 8300 AE Emmeloord, The Netherlands (formerly DELFT HYDRAULICS)

3) Rijkswaterstaat, National Institute for Coastal and Marine Management/RIKZ, P.O.Box 20907, 2500 EX The Hague, The Netherlands

of a polder. The latter is predominantly determined by the discharge rate through the breach [Kraak et al., 1994].

Research regarding breach growth in a sandy dike-like structure started in 1989 with an exploratory field experiment. Further, flume experiments have been carried out for the initial phases of the breach erosion process (lowering of the dike crest) [Steetzel & Visser, 1992]. More recently wave basin experiments were done to investigate the breach widening process. Finally a large scale field experiment has been carried out in order to investigate the three-dimensional aspects of breach growth i.e. widening of the breach in combination with the development of a scour hole [Visser et al., 1995].

One of the main instruments to predict inundation and damage costs assessment is a theoretical model for breach growth and inundation. Recently, as a first step towards that model, a conceptual numerical model for breach growth prediction in a sanddike has been developed.

The present contribution is focused on the wave basin experiments regarding the widening of the breach and the conceptual numerical model of the breach growth process.

2. WAVE BASIN EXPERIMENTS

2.1 Set-up of the experiments

The set-up of the experiments is similar to the experiments which have been carried out to investigate the initial phases of breach growth (lowering of the crest) as is described in [Steetzel & Visser, 1992]. The cross-sections of the dike-stucture are identical in both experiments. The only difference is the use of a wave basin instead of a wave flume, with the purpose of investigating the widening of the breach.

All model tests have been carried out at DELFT HYDRAULICS and started with a sandy dike-like structure situated transversely across the wave basin. Because a breach develops symmetrically to the centre of the breach, it was sufficient to investigate only one side of the breach. The breach development, originated from minor initial overflow through a small pilot channel in the crown of the dike, was recorded using three simultaneously operating video camera's and a large number of precise instruments to obtain time series of both the water levels and the velocities in the area next to the breach. By using a recirculating system with an optional pump capacity up to 1.0 m^3/s, the upstream water level was kept as constant as possible. In this way the widening of the breach could be studied in detail until the major part of the dam was eroded.

In order to complete this description of the experiment set-up, in figure 1 a planview of the wave basin lay-out is shown.

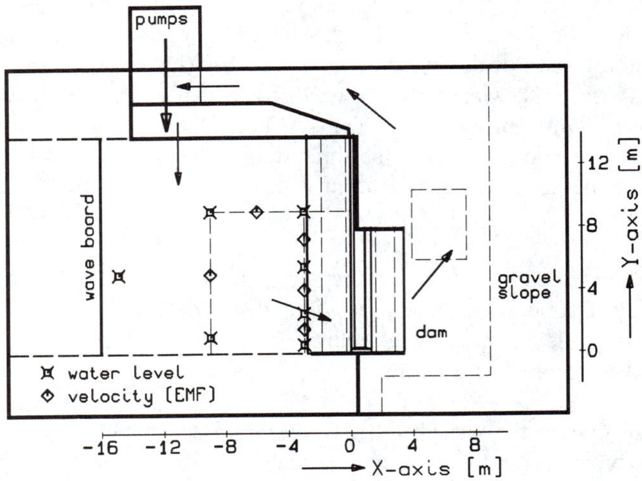

Figure 1: Planview of the wave basin lay-out

2.2. Description of the tests

The influence of several parameters on the breach erosion process has been investigated, viz. the width and height of the dike, the level of the foundation, the seaward bottom slope and the additional effect of (perpendicular) wave attack.

A total of 8 experiments have been performed in which the above mentioned parameters have been varied. Figure 2 gives an overview of the typical cross-sections used for the experiments. The focus lies here on the widening of the breach; in order to avoid effects of depth-growth, a treshold construction has been used for the experiments T1 .. T7 (as can be seen in figure 2). In the test T8 the treshold construction was absent; the only depth limitation is the floor of the basin. In this latter situation, a full three-dimensional experiment was conducted also.

In short the description of the experiments:

T4 = reference case
T1 = as T4, level of treshold is 0.1 m higher;
T2 = as T4, smaller crest width (0.7 m vs. 1.3 m);
T3 = as T4, lower (0.19 m) crest level
T5 = as T4, higher (0.2m) crest level
T6 = as T5, with wave attack (all other tests were without wave attack);
T7 = as T4, with a milder outer slope
T8 = as T7, treshold equals basin floor.

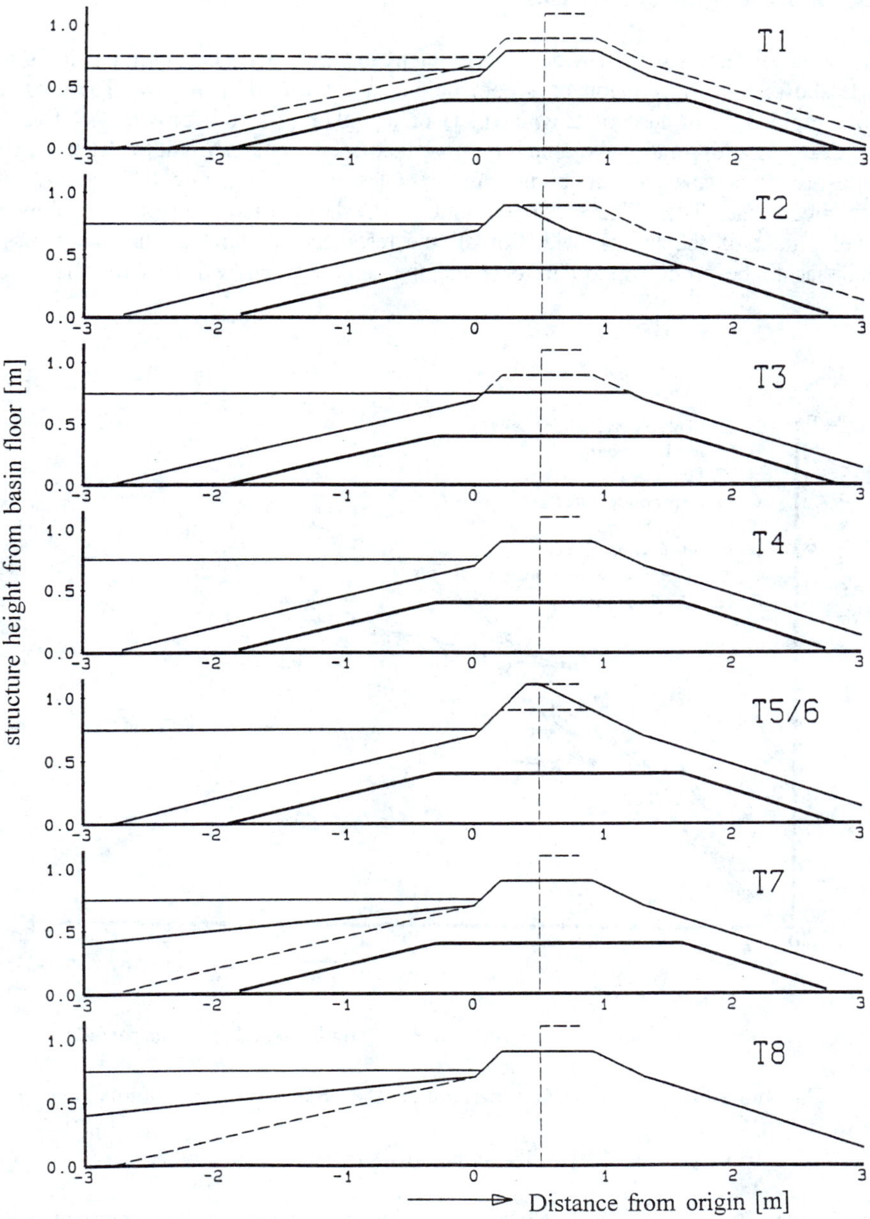

Figure 2: Overview of the cross-sections used for the tests T1 .. T8

2.3 Results of the experiments

The observed breach width development in time (B(t)) is presented in figure 3. It shows that the erosion of a dam having a narrow (T2) or low (T3) crest is more rapid than in case of a wider (T4) or a higher (T4, T5) crown. Relative to the case T5, additional wave attack (T6) yielded a larger breach width, whereas the opposite effect was present in the case of a less steep foreshore (T7, relative to reference case T4). The case without a treshold (T8) shows the slowest development of the breach width of all the tests due to the fact that a lot more sand has to be eroded in order to create the same breach width as for the other tests.

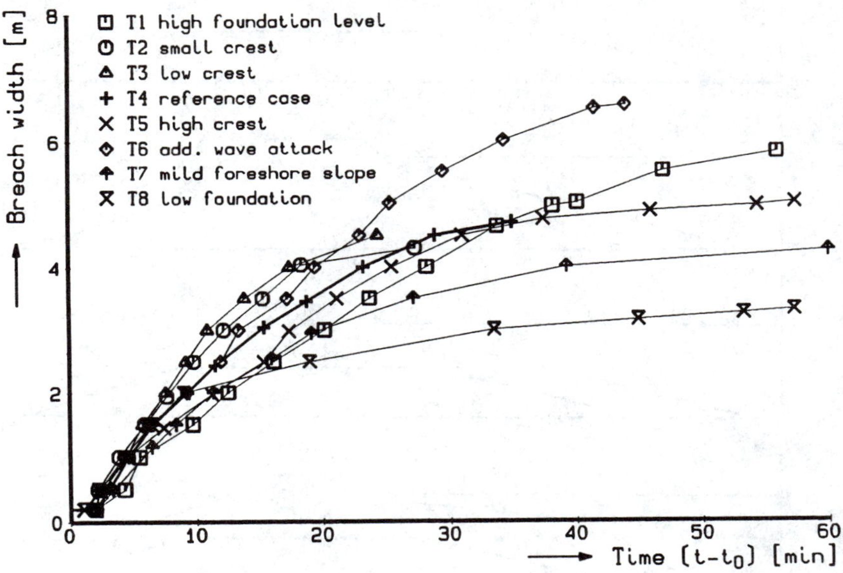

Figure 3: Overview of observed breach growth for all tests performed.

For modelling reasons a new parameter, the breach growth velocity, has been defined as:

$$ß(t) = dB(t)/dt \quad [m/min] \tag{1}$$

To get an impression of this parameter, figure 4 gives its development as a function of time. It shows that a milder outer slope (case T7) yields a smaller breach growth velocity in respect to the reference case (T4).

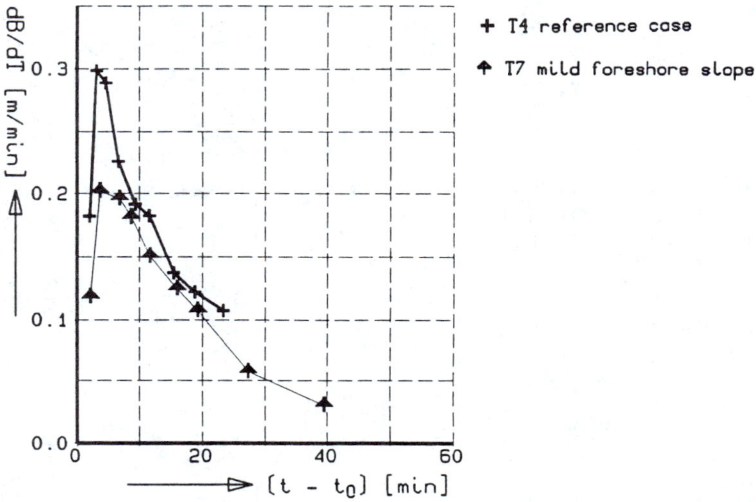

figure 4: development of the breach growth velocity as a function of time; shown is the effect of a mild foreshore slope

To get additional physical insight into the background of the breach growth velocity, this parameter is linked to the recorded flow characteristics in the breach.

Firstly, the relation with a characteristic water velocity in the breach ($v_{breach}(t)$), has been investigated. Figure 5 shows for the reference case (T4) and for the case T7 (milder outer slope) the breach growth velocity as a function of the water velocity in the breach. It can be seen that in order to obtain the same breach erosion velocity (dB/dt) as for the reference case, a higher water velocity in the breach (v_b) is required for case T7 (milder outer slope).

One of the conclusions is that breach width growth stops (and starts) when a certain critical water velocity in the breach is reached. From the experiments (laboratory and field) a critical value of 1.0 m/s can be derived

Secondly, the erodable volume is important. Therefore the volumetric breach growth velocity has been defined as:

$$dV/dt = \beta(t)*A_d \quad [m^3/min] \tag{2}$$

in which A_d = eroded cross-section of the dam as is illustrated in figure 6.

Figure 7 shows the volumetric breach growth velocity as a function of the water discharge through the breach. In this figure all the tests performed are present, exept for T6 (waves) and T8 (3D). A more or less linear relation can be derived.

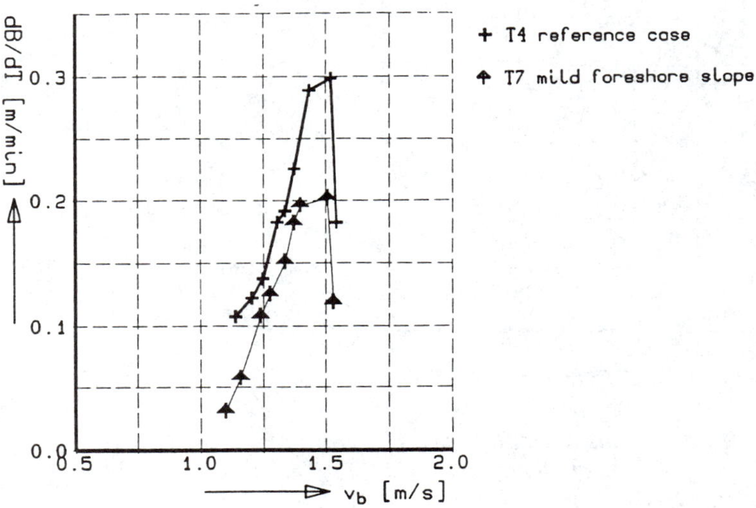

figure 5: development of the breach growth velocity as a function of the water velocity in the breach, shown is the effect of a mild foreshore slope

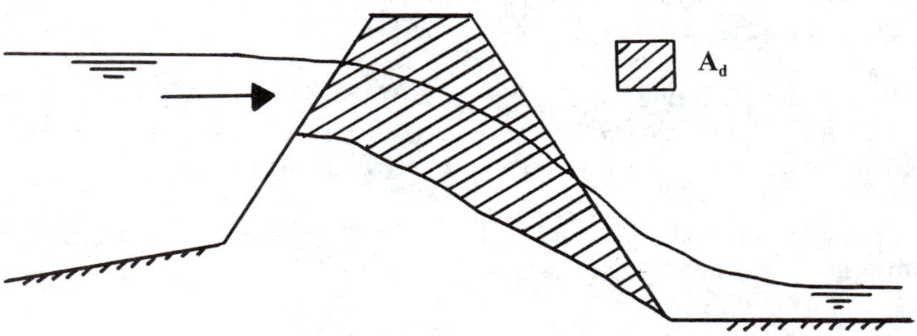

figure 6: definition of the eroded cross-section A_d

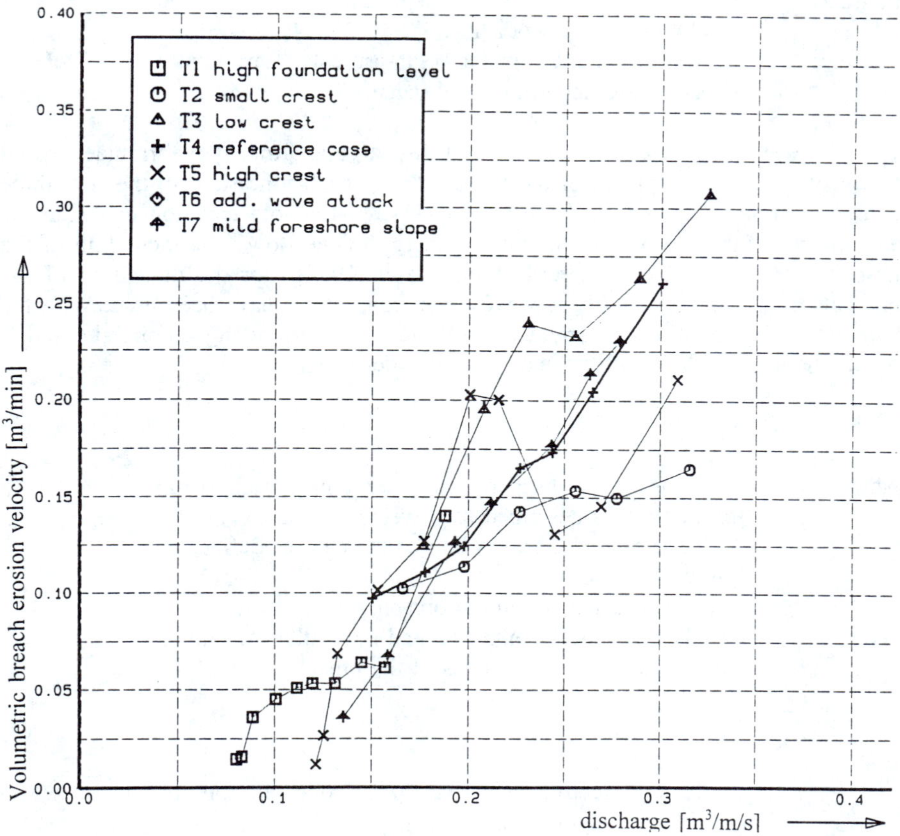

figure 7: volumetric breach erosion velocity as a function of the water discharge in the breach

3. NUMERICAL MODEL FOR BREACH GROWTH

The breach erosion process is characterized by a complex three dimensional behaviour in which low as well as very high velocities and thus erosion rates occur. The Froude number varies typically in the range of 0.03 to 3 and the Shields parameter varies from 0.03 - 50 [Visser et al., 1995]. In order to describe this complex process in a numerical model, five different phases in the breach erosion process have been distinguished [Visser, 1994]:

Phase 1: steepening of the inner slope;
Phase 2: narrowing of the dike crest;
Phase 3: erosion of the dike crest;
Phase 4: breach widening under super-critical flow and development of a scour hole;
Phase 5: breach widening under sub-critical flow until velocities are under a critical value.

In every phase a characteristic set of discharge and transport formulae is used to calculate the development of the breach during each time step. In the first three phases the rate of erosion is described by a Visser-Bagnold equation in which the erosion rate is proportional to the kinetic energy of the flow. The modelling of the first three phases has been described in [Visser, 1994]. For the modelling of the phases 4 and 5, the results of the wave basin experiments have been used. With the parameters and relations mentioned in chapter 3 the following empirical relation for the breach width growth velocity has been derived.

$$\beta = C_b * (d_b * q_b)/A_d \quad [m/s] \tag{3}$$

where: q_b is the water discharge rate per unit breach width $[m^3/m/s]$; q_b is described by a weir formula.
A_d is the eroded cross-section of the dam as defined in figure 6
d_b is the water depth in the breach
C_b is a coefficient, related to the sediment concentration near the edges of the breach

$$C_b = \bar{c} * b_r/d_b \quad [-] \tag{4}$$

where: $\bar{c}$ = depth averaged sediment concentration near the edges of the breach
b_r = width of the zone near the edges for which $\bar{c}$ is defined

At present there is a test-version of the numerical model available. This model covers all the above mentioned phases of the breach growth process. The main input parameters of the model are:

* cross-section of the sand-dike (array of (x,z) values):
  - erodable cross-section
  - cross-section with initial breach
  - ir-erodable cross-section (sill)
  - breach and material characteristics:
    - initial breach width
    - mean grain size diameter D_{50}
    - sediment fall velocity w

- porosity n
- bottom roughness r

* outside boundary:
- waterlevel as function of time (array of (z,t) values)
- optional: in case of e.g. a lake: areas/volume of the outer basin
- optional: in case of river: external input discharge

* inside boundary (in the polder):
- initial waterlevel
- area of the polder that will be inundated

During 1996 and the first half of 1997 this model will be completed. The results of the large-scale field experiments (1989 and 1994) and findings of the 1953 and earlier flood disasters will be used to calibrate and verify the model.

REFERENCES

Kraak, A.W., Bakker, W.T., Van de Graaff, J., Steetzel, H.J. and Visser, P.J., (1994). Breachgrowth and its place in damage assessment for a polder. *Proc. 24th ICCE*, Kobe, Japan.

Steetzel, H.J. and P.J. Visser (1992), Profile development of dunes to overflow, *proc. 23rd ICCE*, Venice, Italy.

Visser, P.J., (1994). A model for breach growth in sand-dikes. *Proc. 24th ICCE*, Kobe, Japan.

Visser, P.J., Kraak, A.W., Bakker, W.T., Smit M.J., Snip W.D., Van de Graaff, J., Steetzel, H.J. (1995). A large-scale experiment on breaching in sand-dikes. *Proc. Coastal Dynamics'95*, Gdansk/Gdynia, Poland.

CHAPTER 213

THE IMPACT OF AN EXTREME EVENT ON THE SEDIMENT BUDGET: HURRICANE ANDREW IN THE LOUISIANA BARRIER ISLANDS

Jeffrey H. List[1], Mark E. Hansen[2], Asbury H. Sallenger, Jr.[2], and Bruce E. Jaffe[3]

Abstract

This paper examines the influence of Hurricane Andrew on the sediment budget of an 80-kilometer section of the Louisiana barrier islands west of the modern Mississippi delta. Because long-term bathymetric change has been extensively studied in this area, excellent baseline data are available for evaluating the impact of Hurricane Andrew. Results show that despite the high intensity of the storm and a storm track optimally positioned to impact the study area, the storm did not have an overwhelming influence on the sediment budget when compared to the changes occurring over the previous 50 years. For the Louisiana barrier islands, a 50-year record appears to be adequate for averaging the long-term contributions of both major and minor storm events to the sediment budget.

Introduction

The coastal sediment budget, i.e., the identification of sediment sources, sinks, and transport pathways, is essential information for predicting rates of shoreline erosion and for planning beach nourishment projects (e.g., Kana, 1995). Unfortunately, most investigators have had to assume that the sediment budget, derived from past measurements, is an adequate representation of future trends as well. This assumption is weakened by the commonly unknown influence of changing and difficult to predict processes, such as the impact of infrequent but high-energy storm events. Few studies have been able to assess their potentially important impact. In one exception, Morton et al. (1995) monitored changes in beach volume associated with a major hurricane along the Texas coast, concluding that large storms can

[1] Oceanographer, U.S. Geological Survey, 384 Woods Hole Rd., Woods Hole, MA 02543, USA.

[2] Oceanographer, U.S. Geological Survey, 600 4th St. S., St. Petersburg, FL 33701, USA.

[3] Oceanographer, USGS, 345 Middlefield Rd. MS-999, Menlo Park, CA 94025, USA

significantly alter the long-term sediment budget and hence, shoreline evolution. They demonstrate that short term sediment budget data (< 10 yr) are inadequate for modeling and predicting future changes because of the variable influence of great storms. It remains unclear whether longer term data, which would potentially include the effects of many major storm events, would be more representative of future conditions.

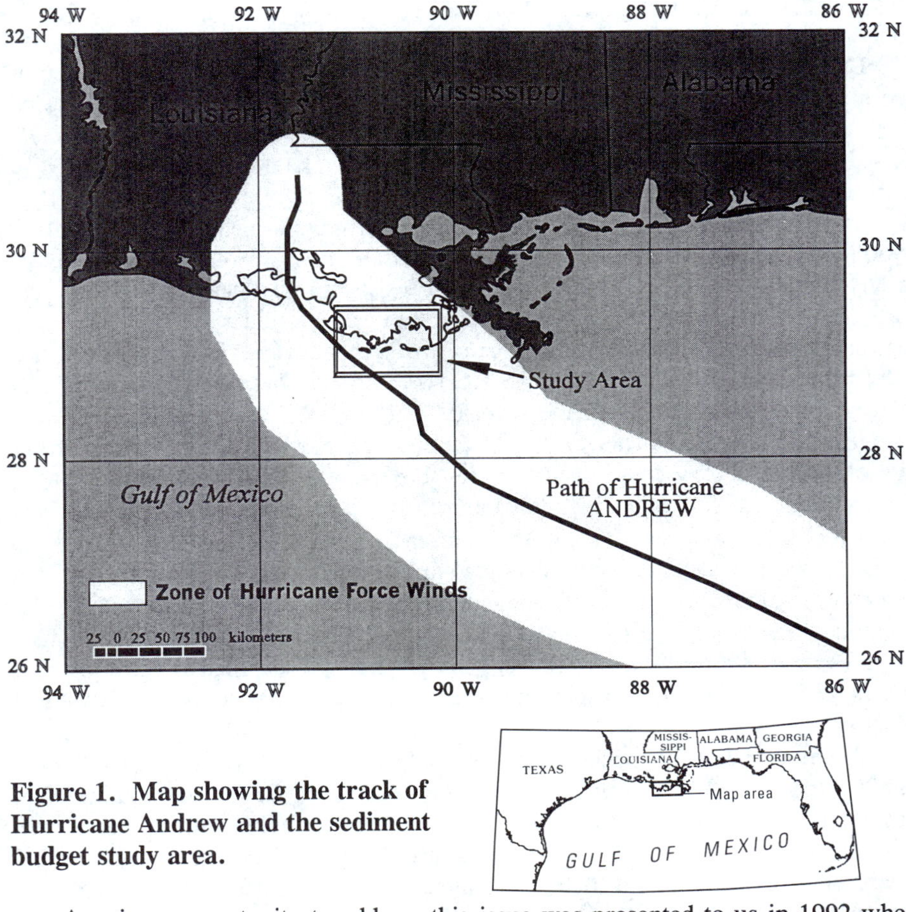

Figure 1. Map showing the track of Hurricane Andrew and the sediment budget study area.

A unique opportunity to address this issue was presented to us in 1992 when Hurricane Andrew impacted the Louisiana coast with wind speeds of more than 50 m/s (Grymes and Stone, 1995), offshore significant wave heights of more than 10 m (DiMarco et al., 1995; Stone et al., 1995), and a maximum storm surge of more than 2.5 m (Halford, 1995). This Class 3 storm on the Saffir-Simpson scale passed just west of a series of barrier islands (Figure 1) which had been the focus of an intensive study on the physical and geological processes responsible for the rapid erosion and disintegration of these islands (the Louisiana Barrier Island Erosion Study, Sallenger et al., 1987, 1991). In response to Hurricane Andrew, several components of the original study were repeated, including measurements of shoreline position, nearshore profiles (Dingler, 1995), and near-island bathymetry. In this paper we compare the bathymetric changes in a six-year period including Hurricane Andrew with the bathymetric changes observed for the 50 years prior to the storm, and evaluate the impact of Hurricane Andrew on the sediment budget.

Data Sources and Processing Techniques

Bathymetric surveys from three time periods were used as the basic source of sediment budget information considered here. Two surveys, from the 1930's and 1980's, were processed or collected as part of the U.S. Geological Survey's Louisiana Barrier Island Erosion Study (1986-1991). The third bathymetric survey was conducted in 1993/1994 following Hurricane Andrew.

The data sources, processing techniques, and error assessment for the 1930's and 1980's surveys are described in detail by List et al. (1991, 1994), and will only be briefly summarized here. The 1930's survey consists of soundings obtained in digital form from the National Geophysical Data Center in Boulder, CO. Data were digitized by the National Ocean Service of the National Oceanic and Atmospheric Administration (NOAA) from U.S. Coast and Geodetic Survey (USCGS) hydrographic smooth sheets for the years 1933, 1934, 1935, and 1936. The 1980's bathymetry was constructed from soundings obtained in digital form from hydrographic surveys conducted in 1986, 1988, and 1989 by companies under contract to the USGS. In addition to bathymetric data, shoreline data were obtained for each of these time periods to provide a mean high water (MHW) line to constrain the bathymetry surrounding the study area's many small islands. Shorelines for the 1930's were digitized from USCGS topographic smooth sheets, while for the 1980's, shorelines were photo-interpreted and digitized from National Aeronautics and Space Administration (NASA) high-altitude photography. A detailed description and analysis of shoreline data for the Louisiana barrier islands are found in McBride et al. (1992).

The post-Andrew bathymetric data were collected by the U.S. Geological Survey in 1993 and 1994, using an Innerspace 448 echosounding fathometer and a Global Positioning System (GPS) for navigation. Although we originally intended to correct soundings for tide effects using post-processed differential GPS vertical measurements, processing difficulties resulted in unreliable depth values. Thus the data were corrected using the same methods as were used in the 1980's survey, i.e., by applying corrections based on local tide observations. Shoreline data assumed to represent the MHW line were also incorporated into the 1993/1994 data set; post-Andrew shorelines were digitized by the Louisiana State University in 1993 (Penland, pers. comm.). The 1993/1994 bathymetric survey was processed exactly as were the 1930's and 1980's surveys, as described by List et al. (1994). Figure 2 shows the 1993/1994 bathymetric coverage area; for comparative purposes, analysis of the 1930's and 1980's data is restricted to the same area.

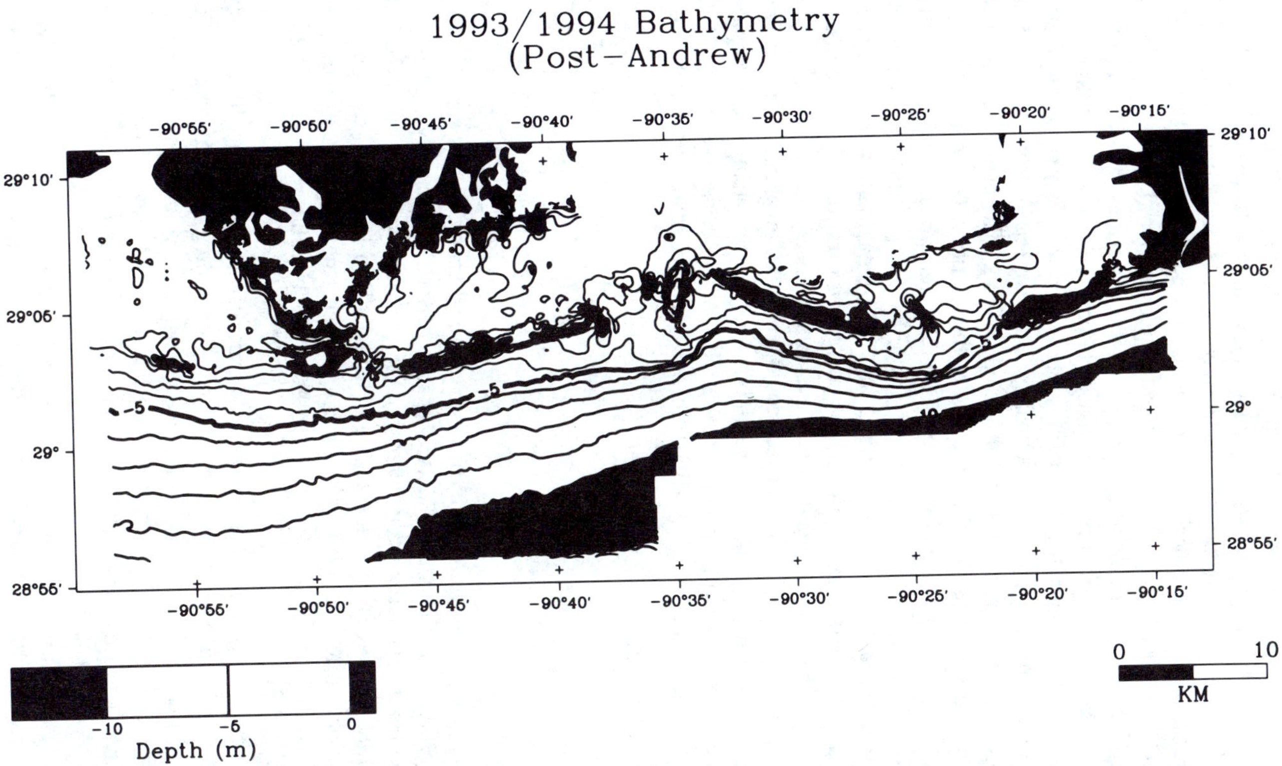

Figure 2. Post-Hurricane Andrew bathymetry collected in 1993 and 1994, showing the coverage area for this study. Depths are relative to Mean Low Water. For comparison purposes, data from the 1930's and 1980's data sets are analyzed within this same region.

After constructing surface models of each survey period's bathymetry using a minimum-tension gridding technique (Briggs, 1974; Dynamic Graphics, Inc.), we calculated bathymetric change grids for the 1930's to 1980's and 1980's to 1993/1994 periods. These change grids, with a spacing of 135 m (see List et al., 1994), form the basis for this study's comparison between long-term changes and changes occurring during a six-year period which included a major storm. Because of the approximate 1 cm/yr relative sea-level rise in the study region (Penland and Ramsey, 1990), a sea-level correction was necessary to prevent a strong bias toward erosion and to permit calculations of erosional and accretionary volumes. As continuous tide records were not available for the entire study period, we employed an alternate means of correcting for relative sea level using an area of sea floor which was judged to have minimal accretion or erosion (Jaffe et al., 1991). The error associated with this datum correction constitutes a large part of the overall data uncertainty.

The total error associated with sea-floor change calculations is a complex combination of many potential sources of error, including sounding measurement error, tidal correction problems, vertical and horizontal datum inconsistencies, computer gridding errors, and non-synoptic sources of data for each survey period. Although some errors can be quantitatively evaluated, others, such as the tidal datum correction problem mentioned above, could only be evaluated qualitatively. Therefore, a combination of quantitative and qualitative approaches were employed to assess the error in bathymetric comparisons (List et al., 1994). As a conservative estimate, a ±0.5 m range of sea-floor change was designated as a zone of no significant change.

Results

Regional

Figures 3 and 4 show the patterns of erosion and accretion associated with the 1930's to 1980's and 1980's to 1993/1994 change periods respectively. Overall, the strikingly coherent and large-scale patterns of erosion and deposition observed in the 1930's to 1980's change period are lacking in the 1980's to 1993/1994 change period. (For a description and analysis of the changes occurring in the 1930's to 1980's change period, see List et al., 1991, 1994, 1997a, 1997b; Sallenger et al., 1992; Jaffe et al. 1989, 1997).

Figure 5 quantifies the total volumes of erosion and accretion within the area enclosed by the data limit polygon in Figure 4, excluding areas of no significant change. In Figure 5A, both erosional and depositional volumes are significantly smaller for the change period encompassing Hurricane Andrew, indicating that the storm did not have an overwhelming influence on the sediment budget of the study area.

However, as the 1980's to 1993/1994 change period represents only about six years compared to the 50 years in the 1930's to 1980's change period, a better comparison between these two periods may be the rates of volume change, as shown

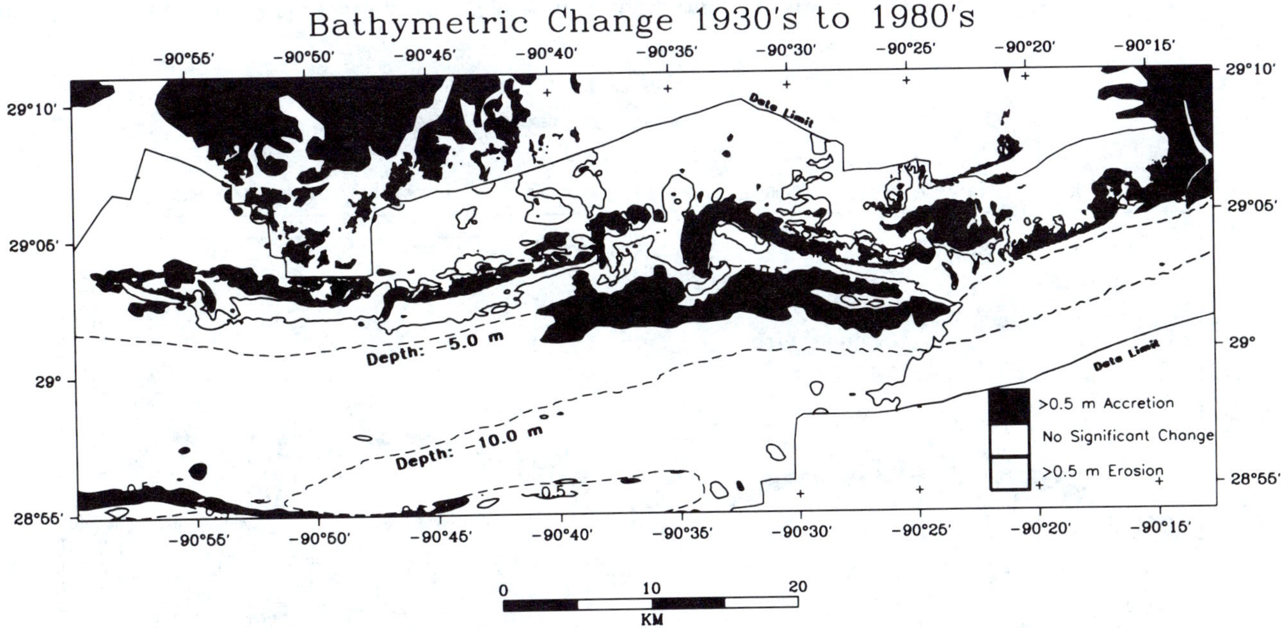

Figure 3. 1930's to 1980's bathymetric change within the complete study region. The -5.0 and -10.0 depth contours are from the 1980's bathymetry. The region encompassed by the line marked "data limit" is the area of overlap between the 1930's and 1980's bathymetries.

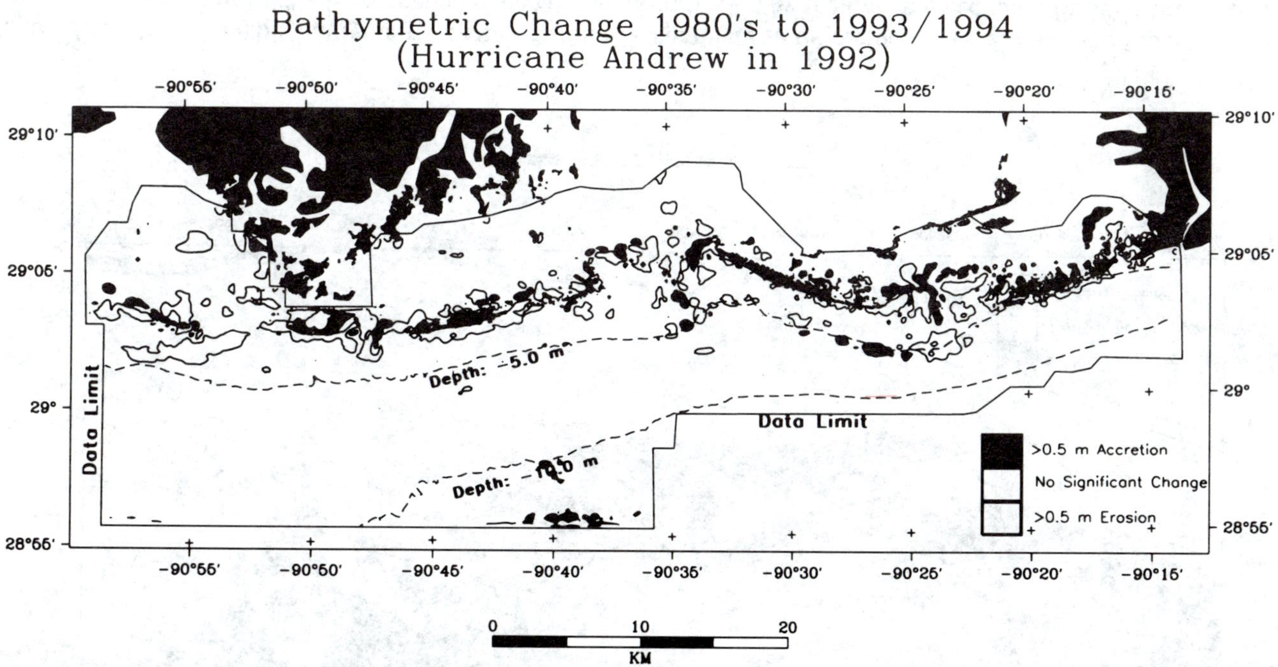

Figure 4. 1980's to 1993/1994 bathymetric change within the complete study region. The -5.0 and -10.0 m depth contours are from the 1993/1994 bathymetry. The region encompassed by the line marked "data limit" is the area of overlap between the 1980's and 1993/1994 bathymetries.

in Figure 5B. This figure shows that two change periods' erosional and deposition rates do not differ significantly, within the error bars resulting from the assumption of a ±0.5 m zone of no-significant change (after List et al., 1994).

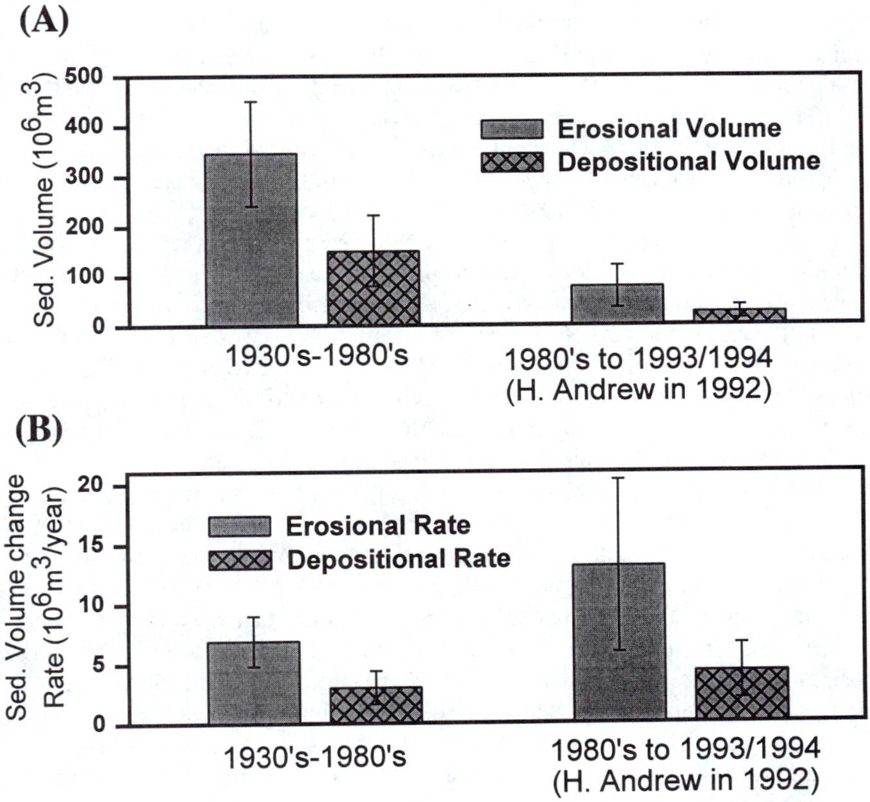

Figure 5. Volumes of sediment change within the area enclosed by the data limit region shown in Figure 4, excluding areas of no significant change. Error bars are based on the assumed ±0.5 m range of no significant change following List (1994). For comparison purposes, the same data limit region is applied for calculating volumes within the 1930's to 1980's change period. (A) Total erosional and depositional volumes for the 1930's to 1980's and 1980's to 1993/1994 change periods. (B) Corresponding yearly rates of erosion and deposition, using intervals of 50 and 6 years for the 1930's-1980's and 1980's-1993/1994 change periods respectively.

Eastern Portion of the Study Area

Figures 6 and 7 focus on changes in the eastern portion of the study area. Again, the 1980's to 1993/1994 change period exhibits few of the large and coherent patterns of erosion and deposition observed in the 1930's to 1980's change period. Still, it is noteworthy that there are *any* coherent patterns of erosion and deposition in the 1980's to 1993/1994 period , given the short six year period of change and the ±0.5 m range designated as no-significant change. While the extensive area of offshore erosion present in the 1930's to 1980's comparison is lacking in the 1980's to 1993/1994 period, many of the smaller patterns, especially in depositional areas, appear to duplicate the later change period, albeit at reduced magnitudes.

Four of these depositional areas in the 1930's to 1980's change period are labeled A through D in Figure 6, with 1980's to 1993/1994 deposits interpreted to be of the same origin similarly labeled in Figure 7. Area A represents a shoreface deposit that appears to extend from erosional areas to the east. In the 1930's to 1980's change period, this depositional area extends from the extensive area of shoreface erosion it adjoins to the east to the next chain of barrier islands to the west, suggesting a mode of sediment bypassing of a wide inlet complex (Figure 3 here, see also Jaffe et al., 1991, 1997). In the 1980's to 1993/1994 change period, deposit A is much less coherent, but still occupies approximately the same location along the 5 m depth contour. Erosion of an ebb-tidal delta to the east suggests a source for the deposit in this change period (compare the bathymetry in Figure 2 with the bathymetric change in Figure 4). Area B represents a bay-side deposit that seems to be associated with landward migration of small barrier islands within a large inlet complex (similarly, they may be, in part, migrating flood-tidal delta deposits). In the 1980's to 1993/1994 period, depositional area B comprises several small areas that are clearly associated with the landward migration (and submergence) of small barrier islands (the depositional area between the two parts of B is associated with inlet migration). Area C represents another bay-side region of accretion, similar in both change periods. This deposit takes the form of a flood-tidal delta, but arguably could also be related to strong landward-directed flows associated with storm surges. Finally, area D in both change periods is a result of inlet infilling at the terminus of a westward migrating barrier island.

Similar to Figure 5, Figure 8 gives the total volume and yearly rates of deposition within areas A-D for the 1930's to 1980's and 1980's to 1993/1994 change periods. Again, while the total volumes associated with the 1980's to 1993/1994 period are much less than in the 1930's to 1980's period (Figure 8A), the *rates* of accretion are not significantly different (Figure 8B).

Discussion

Because the average rates of change in the period bracketing Hurricane Andrew were not significantly different from the average rates of change in the previous 50 years, we cannot conclude with certainty that *any* of the bathymetric changes are

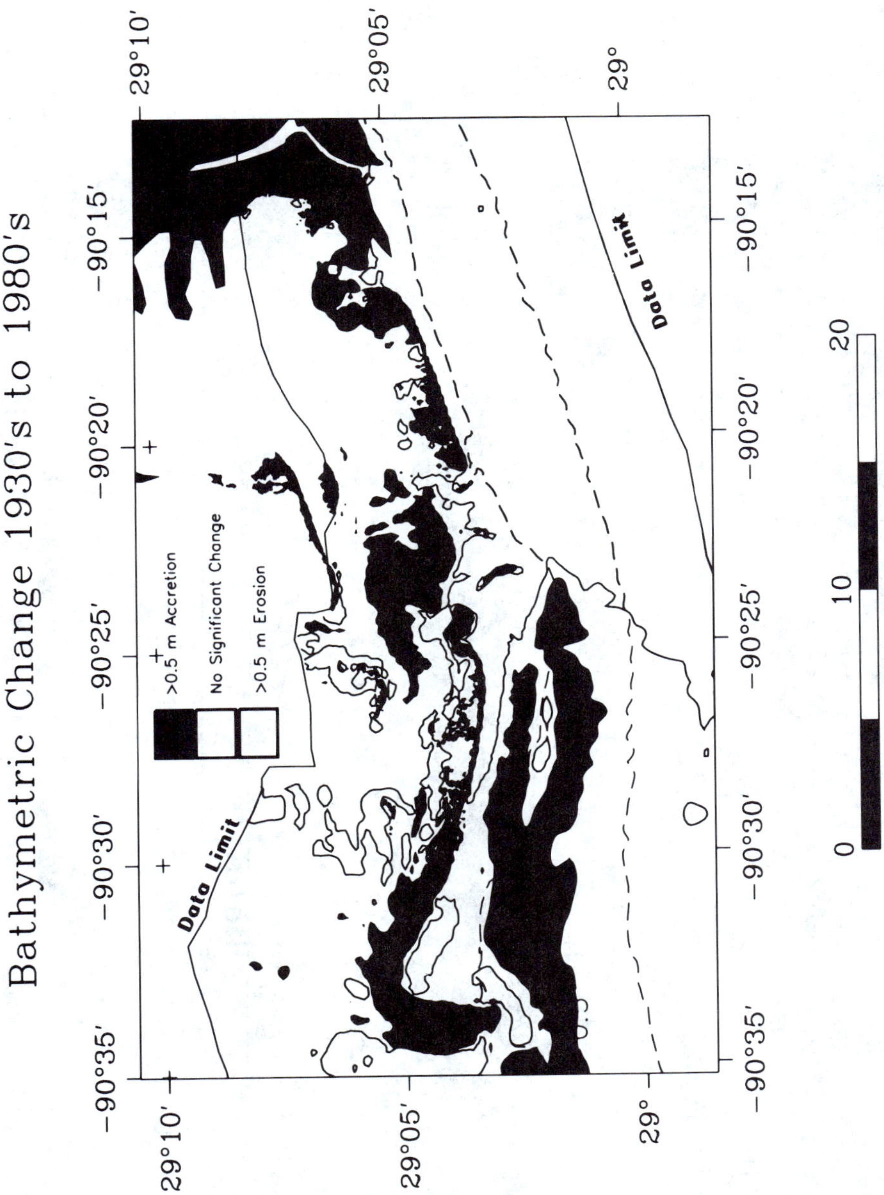

Figure 6. 1930's to 1980's bathymetric change within the eastern portion of the study region. The depth contours (unlabeled) and data limit boundary are the same as in Figure 3. Letters A through D represent areas of accretion for which volumes are shown in Figure 8. In the volume calculations shown in Figure 8, area A is limited to the area east of longitude -90°34', and Area D is limited to the area west of the adjoining barrier island.

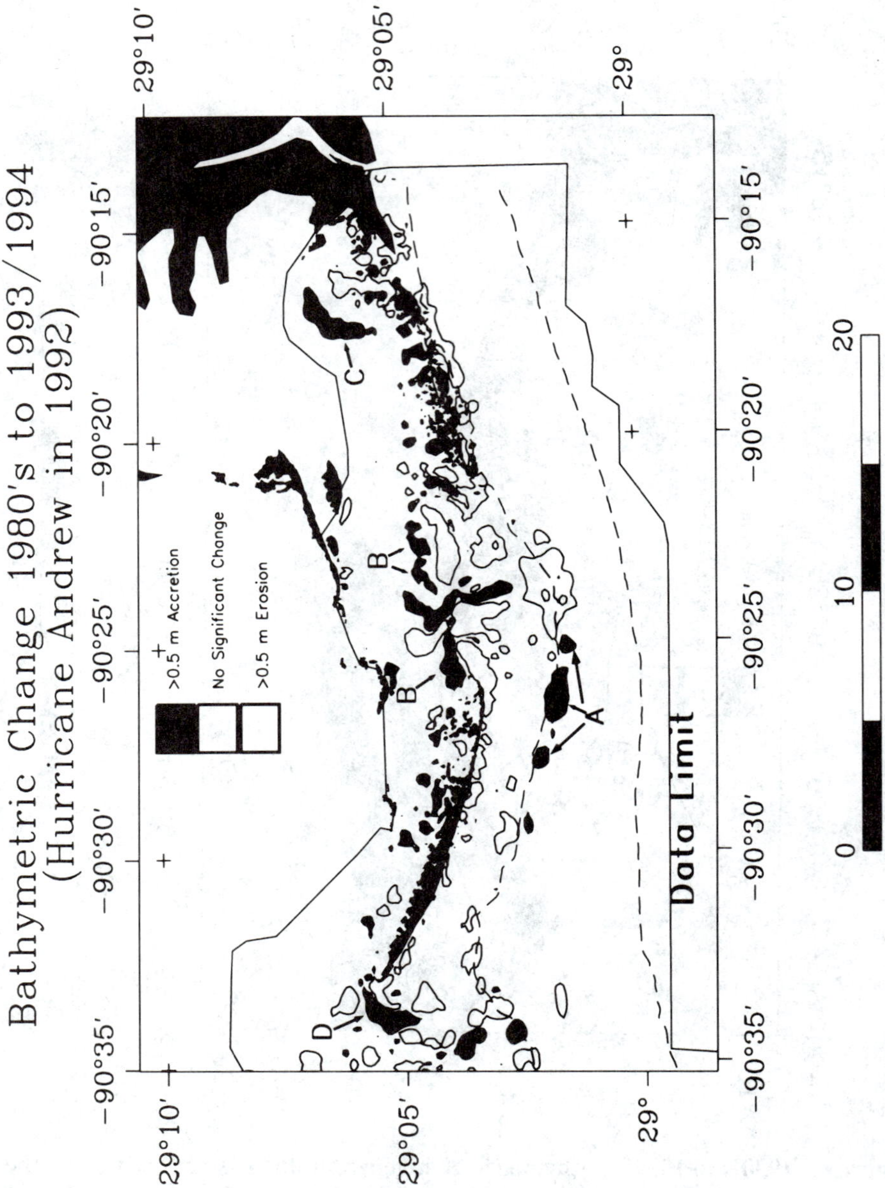

Figure 7. 1980's to 1993/1994 bathymetric change within the eastern portion of the study region. The depth contours (unlabeled) and data limit boundary are the same as in Figure 4. Letters A through D represent areas of accretion for which volumes are shown in Figure 8.

attributable to Hurricane Andrew, rather than typical storm conditions that occur much more frequently. However, as Hurricane Andrew was without doubt the largest storm in the 1980's to 1993/1994 change period, it is reasonable to assume that many of the changes are, in fact, due to this storm. With this assumption, we surmise that Hurricane Andrew had a moderate, through not overwhelming, contribution to the area's sediment budget.

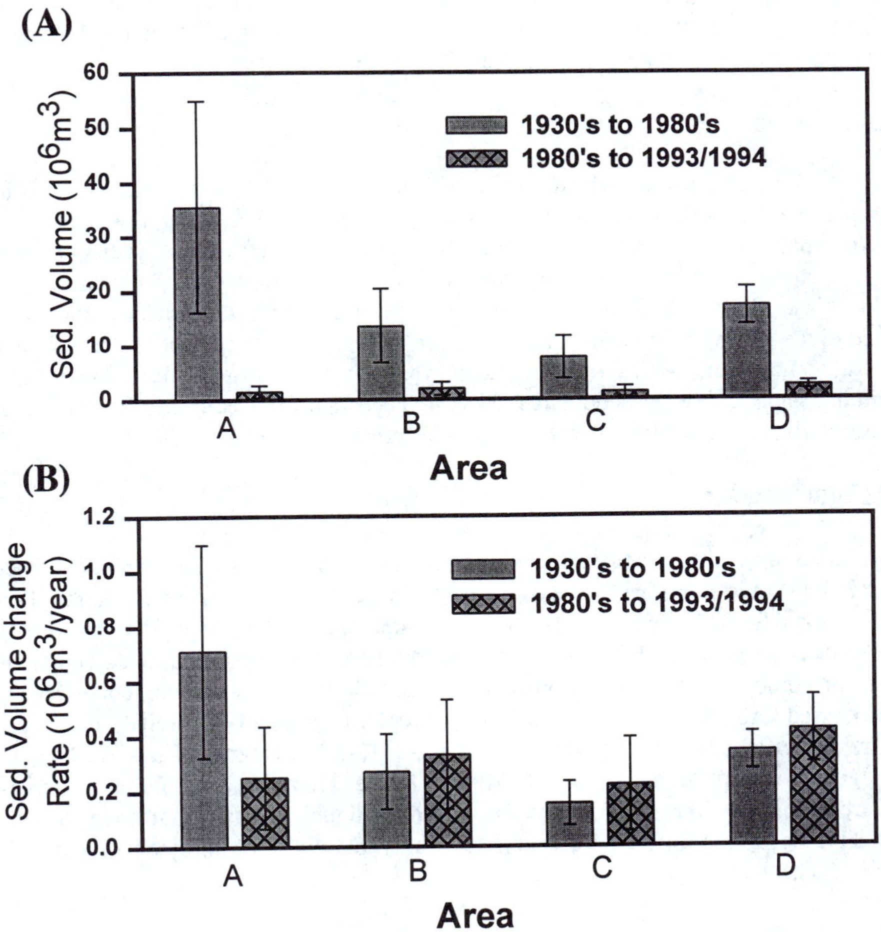

Figure 8. (A) Depositional volumes and (B) rates of deposition within the labeled deposits shown in Figures 6 and 7. Rates are found by dividing the depositional volumes for each region by the 50 and 6 year intervals for the 1930's to 1980's and 1980's to 1993/1994 change periods respectively. Error bars are found as in Figure 5.

Because the rates and patterns of change were similar in both change periods examined here, it seems likely that the processes were similar as well. An examination of the historical database of hurricane tracks (Jarvinen et al., 1984) shows that while Hurricane Andrew might have been the *most* intense storm experienced by the study area in the last 50 years (in terms of both magnitude and track), there were many other hurricanes that could have had a comparable impact. In terms of average rates of change, the 6-year period encompassing Hurricane Andrew (and many minor storms) appears to be representative of the long-term average of processes occurring in the previous 50 years.

Conclusions

Despite Hurricane Andrew's high intensity and a track that was near optimal for impacting the study area, the storm did not have an overwhelming influence on the sediment budget of an 80-km section of the Louisiana barrier islands when compared to the changes occurring over the previous 50 years. However, with the assumption that most of the bathymetric changes observed in the pre/post-Andrew survey were due to the storm, we conclude that the storm had a moderate impact on the sediment budget; however, other storms in the 50-year period prior to Andrew likely had a similar impact. For this study area, 50 years appears to be adequate for averaging the cumulative contributions of both major and minor storm events.

Acknowledgments

This work was conducted as part of the Hurricane Andrew in Louisiana Project and the Louisiana Barrier Island Study, cooperative studies between the U.S. Geological Survey, Coastal and Marine Program, and the Louisiana State University. We thank Shea Penland, Randy McBride and others at the Louisiana State University for providing shoreline-position data. The bathymetric data were collected and processed through the efforts of numerous people, including Greg Gabel, Thang Phi, Graig McKendrie, Rob Wertz, Todd Holland, Carolyn Degnan, Clint Steel, Tracy Logue, Dorothy Hopkins, Keith Dalziel, Lance Thornton, and Jennifer Bexley. Karen Morgan is thanked for contributions to the figures. Any use of trade names in this publication is for descriptive purposes only and does not imply endorsement by the USGS.

References

Briggs, I. C., 1974. Machine contouring using minimum curvature. Geophysics, 39: 39-48.

DiMarco, S. F., Kelly, F. J., Zhang, J., and Guinasso, N. L.Jr., 1995. Directional wave spectra on the Louisiana-Texas shelf during Hurricane Andrew. Jour. Coastal Res., Special Issue No. 21, 217-233.

Dingler, J. R. and Reiss, T. E., 1995. Beach erosion on Trinity Island, Louisiana caused by Hurricane Andrew. Jour. Coastal Res., Special Issue No. 21, 254-264.

Grymes, J. M. III and Stone, G. W., 1995. A review of key meteorological and hydrological aspects of Hurricane Andrew. Jour. Coastal Res., Special Issue No. 21, 6-23.

Halford, K. J., 1995. Estimating the dynamic water-level surfaces associated with Hurricane Andrew crossing the Louisiana coast. Jour. Coastal Res., Special Issue No. 21, 265-279.

Jaffe, B. E., List, J. H., and Sallenger, A. H., Jr., 1997. Massive sediment bypassing on the lower shoreface offshore of a wide tidal inlet; Cat Island Pass, Louisiana. Marine Geology,

Jaffe, B.E., List, J.H., Sallenger, A.H., Jr., and Holland, T., 1991. Louisiana Barrier Island Erosion Study: correction for the effect of relative sea level change on historical bathymetric survey comparisons, Isles Dernieres, Louisiana. U.S. Geological Survey Open File Report 91-276, 33 pp.

Jaffe, B. E., Sallenger, A. H., Jr., and List, J. H., 1989. Massive sediment bypassing of a wide tidal inlet: Cat Island Pass, Louisiana. Transactions, Gulf Coast Assoc. Geol. Soc., 39: 403-411.

Jarvinen, B.R., Neuman, C.J., and Davis, M.A.S., 1984. A tropical cyclone data tape for the North Atlantic basin, 1886-1983: contents, limitations, and uses. NOAA Technical Memorandum NWS NHC 22,

Kana, T. W., 1995. A mesoscale sediment budget for Long Island, New York. Marine Geology, 126: 87-110.

List, J. H., Jaffe, B. E., and Sallenger, A. H., Jr., 1991. Large-scale coastal evolution of Lousiana's barrier islands. Proceed. Coastal Sediments '91, Amer. Soc. Civil Eng., New York, pp. 1532-1546.

List, J. H., Jaffe, B. E., Sallenger, A. H.Jr., and Hansen, M. E., 1997a. Bathymetric Comparisons in the Louisiana barrier islands: processes of large-scale change. Jour. Coastal Res., in press.

List, J.H., Jaffe, B.E., Sallenger, A.H., Jr., Williams, S.J., McBride, R.A., and Penland, S., 1994. Louisiana Barrier Island Erosion Study: atlas of sea-floor changes from 1878 to 1989. U.S. Geological Survey Misc. Invest. Series I-2150-B, 82 pp.

List, J. H., Sallenger, A. H., Jr., Hansen, M. E., and Jaffe, B. E., 1997b. Accelerated relative sea-level rise and rapid coastal erosion: testing a causal relationship for the Louisiana barrier islands. Marine Geology,

McBride, R. A., Penland, S., Hiland, M. W., Williams, S. J., Westphal, K. A., Jaffe, B. E., and Sallenger, A. H., Jr., 1992. Analysis of barrier shoreline changes in Louisiana from 1853 to 1989. In: S. J. Williams, S. Penland, and A. H. Sallenger, Jr. (Editors), Louisiana Barrier Island Erosion Study--Atlas of Barrier Shoreline Changes in Louisiana from 1853 to 1989. U.S. Geological Survey Misc. Invest. Series I-2150-A, pp. 36-97.

Morton, R. A., Gibeaut, and J.C., Paine J. G., 1995. Meso-scale transfer of sand during and after storms: implications for prediction of shoreline movement. Marine Geology, 126: 161-179.

Penland, S. and Ramsey, K. E., 1990. Relative sea-level rise in Louisiana and the Gulf of Mexico: 1908-1988. Jour. Coastal Res., 6: 323-342.

Sallenger, A. H., Jr., List, J. H., Jaffe, B. E., Penland, S., and Williams, S. J., 1992. Regional coastal erosion research and beach preservation. Proceed. 5th Annual Nat. Conf. on Beach Preserv. Tech., Florida Shore and Beach Preserv. Assoc., pp. 115-134.

Sallenger, A. H., Jr., Penland, S., Williams, S. J., Jaffe, B. E., and Suter, J., 1991. Louisiana Barrier Island Erosion Study: further results. Proceed. Coastal Zone '89, Am. Soc. Civ. Eng., New York, pp. 1503-1516.

Sallenger, A. H., Jr., Penland, S., Williams, S. J., and Sutter, J. R., 1987. Louisiana Barrier Island Erosion Study. Proceed. Coastal Sediments '87, Am. Soc. Civ. Eng., New York, pp. 1503-1516.

Stone, G. W., Xu, J. P., and Zhang, X., 1995. Estimation of the wave field during Hurricane Andrew and morphological change along the Louisiana coast. Jour. Coastal Res., Special Issue No. 21, 234-253.

CHAPTER 214

BEACH MAINTENANCE STRATEGIES, A COMPARISON FOR A RECREATIONAL BEACH

M. Pluijm[1], G. Kant[1], R. Nolten[2], A.P. de Looff[3]

ABSTRACT

This paper describes the morphological and economical evaluation of the beach nourishment maintenance strategy of an eroding recreational beach at the Dutch coast. In addition to the results of this evaluation two alternative strategies were analysed. In order to assess the economical feasibility of the alternatives appeared to be essential to estimate the morphological effectiveness of the alternative strategies. Therefore, an extensive numerical modelling scheme was set up. The numerical modelling played a key role in the understanding of the existing morphological system and subsequently in the assessment of the effectiveness of the alternatives on a long term basis. The morphological effectiveness of the alternatives, being offshore breakwaters and a 450 m long groin, was calculated in terms of reduction of required beach nourishments. The objective of the study was not only to find an optimal coastal morphology strategy but also to find the best strategy which combines a sound beach protection with effects valuable for the specific recreational function of the beach, including an effective way to deal with eventual lee-side erosion. Possibilities were studied to shift the erosion to the dune area in the north which would allow a more flexible and dynamic way of maintenance.

---

[1] Frederic R. Harris B.V., Badhuisweg 11, P.O. Box 87875, 2508 DG The Hague, The Netherlands.

[2] Ministry of Transport, Water Management and Public Works (Rijkswaterstaat), Directorate Zuid-Holland, P.O. Box 556, 3000 AN Rotterdam, The Netherlands.

[3] Ministry of Transport, Water Management and Public Works (Rijkswaterstaat), Road and Hydraulic Engineering Division, P.O. Box 5044 Delft, The Netherlands.

SITE DESCRIPTION

Scheveningen Beach (Figure 1) is located on the west coast of The Netherlands. Scheveningen Port is located just south of the beach with its breakwaters reaching 550 and 725 m into the sea. During the entire year but especially in summertime, the sandy beach attracts thousands of locals and holiday making tourists.

The beach consists of sand with a characteristic grainsize (D50) of about 200 μm. The slope of the beach is approximately 1:40. The -10 m MSL contour lies about 1300 m offshore. Scheveningen is also protected by a seawall with a length of 2.5 km. The beach level at the toe of the seawall is about +3 m MSL. North of Scheveningen the land is protected by a massive dune area.

Scheveningen Beach is eroding continuously. In the beginning of the century a system of 150 m long groins (spacing 500 m) was built but the erosion was not stopped. Up to now five beach nourishments were carried out starting in 1969 (see Table 1).

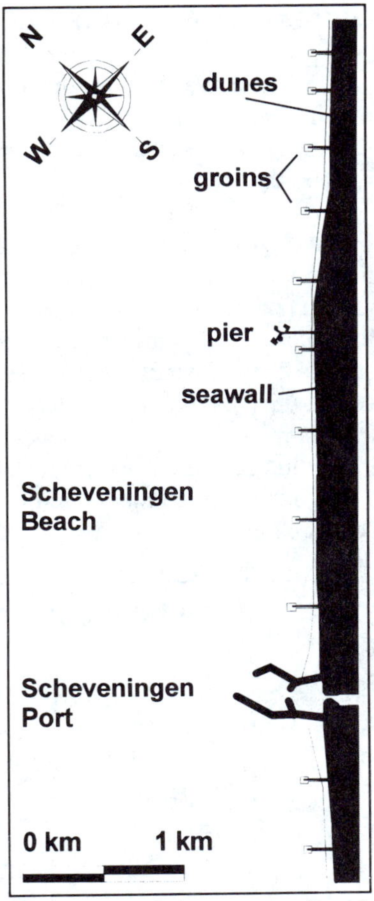

Figure 1 *Scheveningen Beach*

The Dutch west coast is known as a morphological dynamic system. The wave and current climate result in massive longshore sediment transports in northern as well as in southern direction. The average yearly sand transport is directed to the north and is estimated, for this part of the coastline, at 500,000 m^3/year.

year	1969	1975	1985	1991	1996
volume [m^3]	45,000	700,000	330,000	1,000,000	850,000

Table 1 *Beach Nourishment Volumes at Scheveningen Beach*

HYDRAULIC BOUNDARY CONDITIONS

The average wave climate on the southern North Sea is presented as a wave rose diagram in Figure 2. The most persistent conditions develop from south-western directions whereas the highest extreme conditions are to be expected from north-western directions.

The average amplitude of the diurnal tide for this part of the Dutch Coast is 1.7 m. The tidal variation induces a significant tidal flow along the coast. Typical flood currents are in the order of 0.6 m/s, typical ebb currents are 0.4 m/s. Peak flood currents may reach up to 0.9 m/s and ebb currents to 0.75 m/s. However, during extreme south west storms (directed parallel along the shore) the wind-driven current can give a significant contribution to the flow especially in shallow areas (surf zone). Consequently, the longshore sediment transport along this part of the Dutch coast is dominated by the interaction between the waves (disturbance, wave-driven current) and the tidal current. The wind-driven current effect pays a minor but nevertheless a not to be neglected contribution.

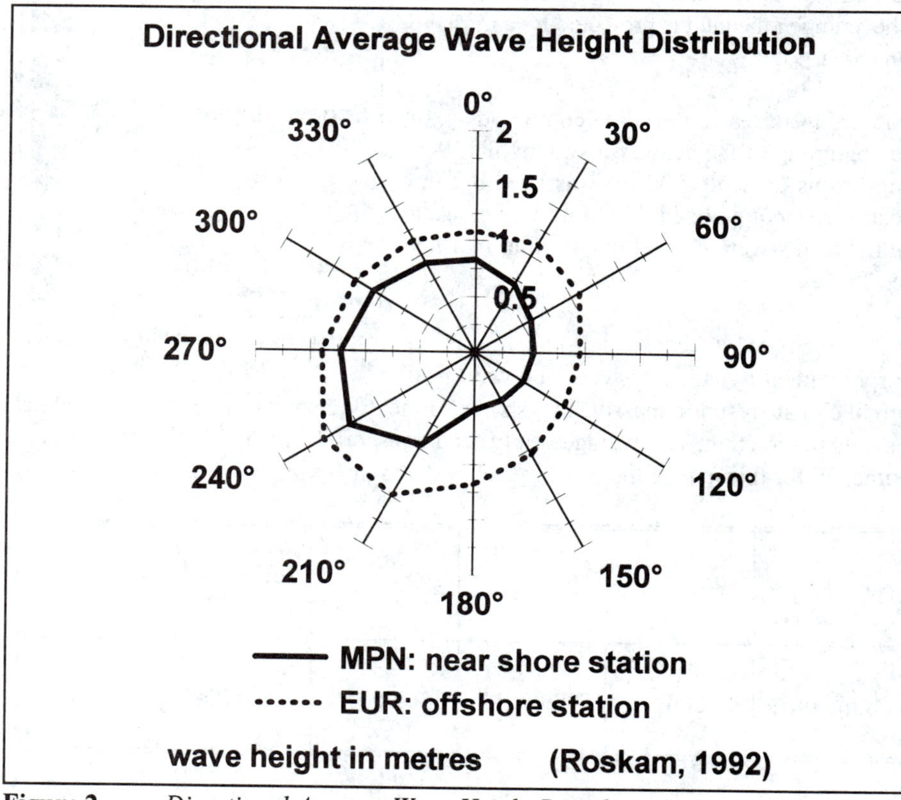

Figure 2 *Directional Average Wave Height Distribution*

EVALUATION PRESENT BEACH NOURISHMENT STRATEGY

The evaluation of the present beach nourishment strategy concentrated on the behaviour of the executed nourishments and the sand transports along the coast. Regular measurements of the Ministry of Transport, Water Management and Public Works were analysed giving insight in the behaviour of the beach nourishments. Based on the measurements and sediment transport calculations it was concluded that (*i*) especially the first nourishments were used as restoration of the steepened cross sectional profile, that (*ii*) the aims of the nourishments were initially met but (*iii*) that the lifetime of the nourishments was almost completely determined by extreme wave conditions, which resulted in a rather random nourishment behaviour. The characteristics of the morphological system do not allow a complete natural restoration of the beach after an erosion causing storm. The unpredictability of the wave conditions results in a capricious development of the beach. This has already led to an unexpected narrow beach in the summer having negative effects on the tourist industry.

NUMERICAL MODELLING ANALYSIS

Based on the conclusions of the beach development evaluation one- and two-dimensional wave, flow, and sediment transport models were applied (DHI, 1995 and DH, 1992). The well calibrated models of the wave- and flow climate were combined in sediment transport simulations using the Bijker sand transport formulae. The model results were used to understand the importance of the various sediment transport mechanisms and to finally formulate a well based hypothesis of the morphological system. After the formulation of the hypothesis simulations were made of alternative coastal protection structures. These were, among others, offshore breakwaters and a 450 m long groin perpendicular to the coastline. The effects of the structures were brought in relation with the hypothesis of the existing system in order to quantify the predicted impacts on the coastal development.

existing situation

The key questions after the evaluation of the nourishment strategy were (*i*) what are the physical reasons for the erosion and (*ii*) where is the sediment going to. It was clear that the position of the beach, the breakwaters of the port, the seawall and the geometry of the beach were essential elements in the morphological system. With the use of the mentioned numerical simulations the following conclusions (schematically presented in Figure 3) could be drawn:

(1) The net north going longshore sediment transport is not blocked by the breakwaters of the port but is passing these obstacles. It was shown that under

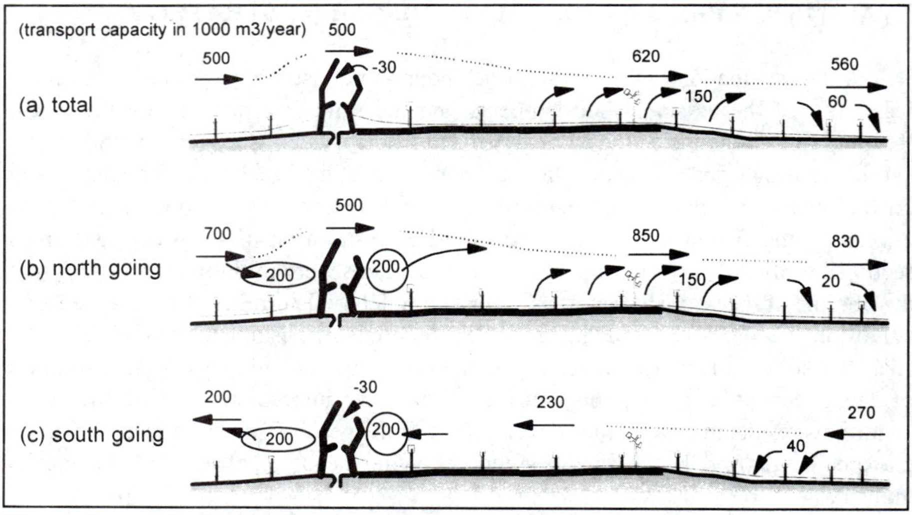

Figure 3 *Sediment Transport Capacities, Existing Situation*

south-western wave conditions the contracted current (tide-, wave- and wind-driven current) in combination with the orbital wave motion result in a substantial sediment transport in front of the breakwaters. This transport is developing in front of the breakwater tips in a water depth of -8 m MSL. Evidence that this process exists is the absence of a sedimentation zone south of the breakwaters. The passing of the net northern transport around the breakwaters is very important for the development of Scheveningen Beach. This transport is the main feeding mechanism of the beach.

(2) It was found the sediment transport north of the port is about 10% higher than south of the port. The reason for this was found in the bathymetry of both areas. The beaches north of the port appear to be more gentle sloping. This results in a wider surf zone north of the port which in this case (transport driven by the combination of waves and tidal current) generates a higher sediment transport. This effect can be observed while standing on the breakwaters.

(3) Because of the forced orientation of Scheveningen Beach by the seawall a strong local longshore transport gradient develops at the northern side of the beach.

(4) The natural beach restoration after a north-western storm is distorted due to the presence of the breakwaters. The timescale of storm events is in the order of hours whereas the timescale of the restoration is weeks to months. In this part of the Netherlands the time averaged flow along the shore is directed to the

north. Sediment deposited on the foreshore during a storm event will therefore contribute to the beach restoration of a beach section somewhat to the north. The presence of the breakwaters of the port distorts this mechanism which results in an incomplete natural restoration of the beach or, in other words, erosion of the beach.

These four main conclusions are the basis of the hypothesis of the morphological system of Scheveningen Beach. The hypothesis is schematically presented in Figure 3. In the figure the annual yearly north going and south going transports are shown as well as the total average yearly transport.

offshore breakwaters

One of the alternative measures which was defined was offshore breakwaters. The impacts were studied on the basis of the developed hypothesis of the existing morphological system. The simulations were repeated for the schematized offshore breakwaters. The main result of the computational effort was that offshore breakwaters do not fit in the morphological system of Scheveningen Beach. Indeed the structure reduces the wave action and thus the sediment transport on the beach. But the structure also redirects the sediment which is transported past the port's breakwaters along the seaward side of the structure. This transport was found to be the most important feeding mechanism of the beach. It is therefore expected that the construction of offshore breakwaters as defined for this study will result in an even stronger erosion on the beach. The predicted sediment transport patterns after the construction of the offshore breakwaters are presented in Figure 4.

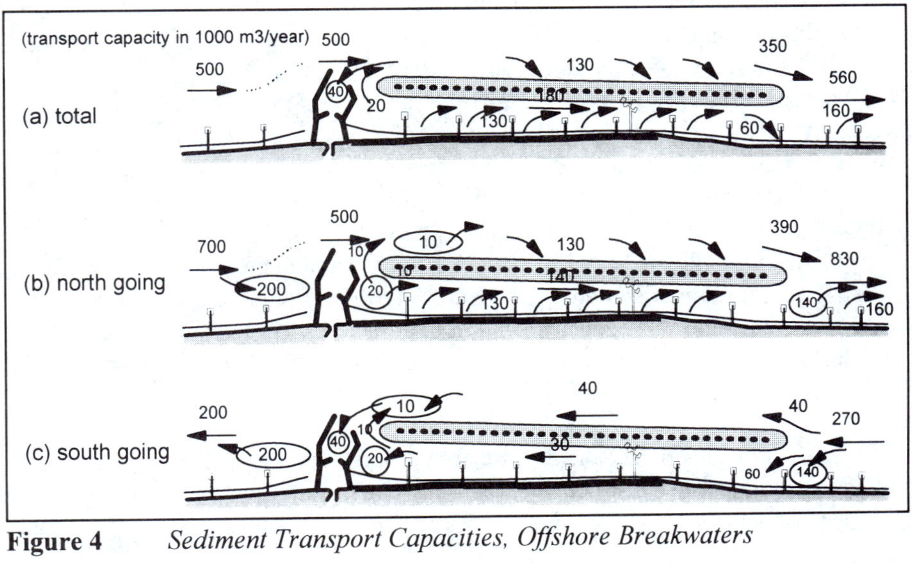

Figure 4 *Sediment Transport Capacities, Offshore Breakwaters*

450 m long groin

Another alternative was the construction of a 450 m long groin at the northern side of Scheveningen Beach. This would prevent the build up of the transport gradient due to the shoreline orientation. The simulations show (Figure 5) that indeed positive results for the beach development are to be expected. The redistribution of sediment due to the local gradient will be prevented and part of the erosion of Scheveningen Beach will be shifted towards the north. This was one of the objectives of the study. A relatively wide dune area is located north of the Beach. Here, the opportunity exists to handle the erosion in a more flexible dynamic way. The construction of a long groin would result in a less frequent nourishment need. This predicted effectiveness of the structure was used to compute the cash flow over a long term period. In this way an economical comparison could be made to the present nourishment strategy including maintenance and re-nourishment needs.

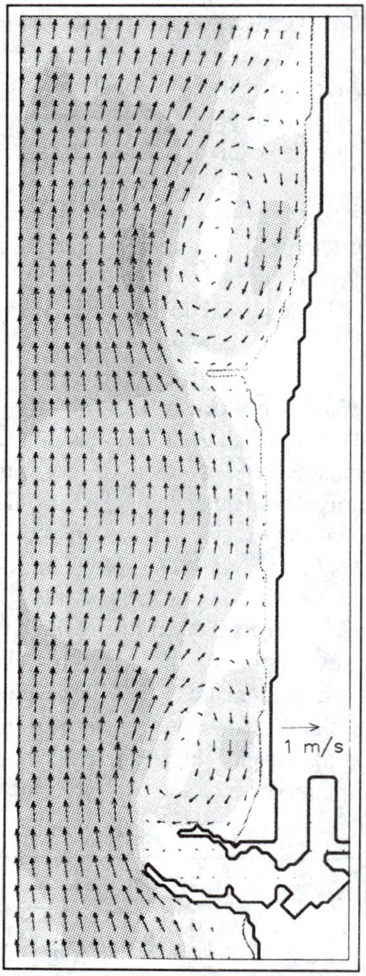

Figure 5 *Flow model result 450 m long groin*

COST COMPARISON

The long term costs of investment of the various beach maintenance schemes were evaluated by means of a Nett Present Value (NPV) computation (Pluijm et al., 1994). In this calculation all the costs of an alternative are capitalized over a period of 50 years. Construction costs, interest, inflation, maintenance but also the nourishment need (the results of the morphological studies) are taken into account in this analysis. The results of the NPV cost comparison are presented in Figure 6. It can be seen that with the present rates for sand it is economically more attractive to proceed with the present nourishment strategy. However, when the prices of sand would increase the construction of a long groin would then become an attractive alternative.

For this cost comparison the extra benefits for the local tourism industry have not been taken into account. The construction of a groin would result into two advantages over the beach nourishment strategy being (*i*) a more predictable development of the beach (coastline stability) and (*ii*) a wider beach. In view of the recreational function of the beach this is an important factor.

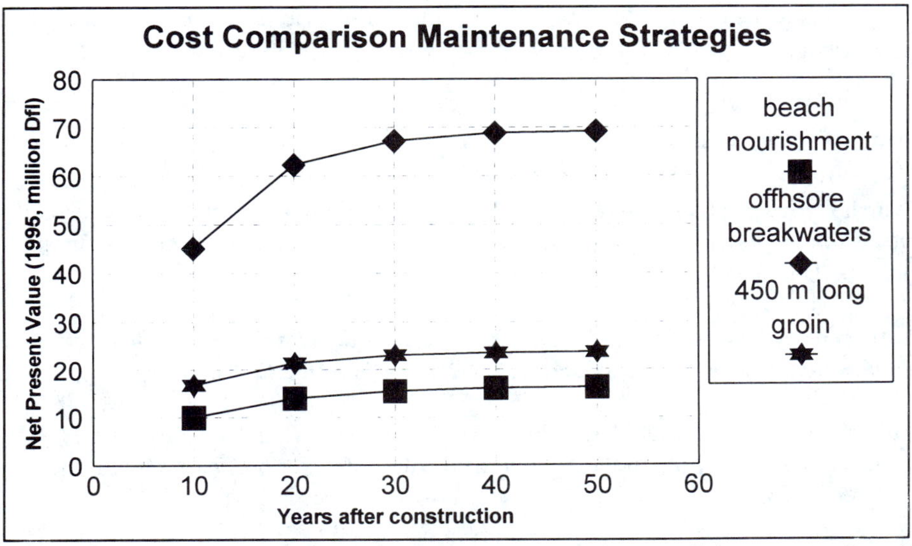

Figure 6 *Cost Comparison Beach Maintenance Strategies Scheveningen Beach*

CONCLUSIONS

Numerical models are of significant importance to develop insight and understanding of the role of each of the morphological processes. The models help to understand the development of a particular beach section, to define the governing processes and to indicate the beach developing sediment transport patterns. Based on this understanding of the morphological system a well-based assessment can be made of impacts of a structure on each of the processes. In many cases this comprehension will allow a reliable prediction of coastline development.

In this paper the methodology has been presented which was adopted to compare alternative beach maintenance strategies. A key element in this evaluation is the nourishment need which will be required when one of the strategies is implemented and which is an important factor in the long term costs. It has been described how numerical simulations models were adopted to assess the morphological effectiveness of the alternatives. These results were then used as input for the cost evaluation resulting in a well based comparison of various beach maintenance strategies.

ACKNOWLEDGEMENT

This study was executed with the full support of the Ministry of Transport, Water Management and Public Works, Directorate Zuid-Holland which is gratefully acknowledged..

REFERENCES

Danish Hydraulic Institute (1995); *Mike 21: A modelling suite for two dimensional simulation of flows, wave propagation and sediment transports*; Danish Hydraulic Institute; Denmark.

Delft Hydraulics (1992); *Unibest-LT, a software package for simulation of longshore sediment transport capacities*; Delft Hydraulics; The Netherlands.

Pluijm, M. (1994); *Offshore Breakwaters Versus Beach Nourishments, A Comparison*; Proceedings of the 24th International Conference on Coastal Engineering, 1994; pp3208-3222; Kobe; Japan.

Roskam, A.P. (1992); *Wave Measurements and Statistical Analyses, MeAO-92-26 and Correspondence (in Dutch)*; Ministry of Transport, Water Management and Public Works (Rijkswaterstaat); Tidal Water Division; The Netherlands.

CHAPTER 215

Littoral Impact of Ocean City Inlet, Maryland, USA

Julie Dean Rosati[1], M. ASCE, and Bruce A. Ebersole[2], M. ASCE

ABSTRACT

Ocean City Inlet, Maryland presents a data-rich site for evaluating the total littoral impact of an inlet system with a significant adjacent beach response. Analysis of adjacent beach and bay shoreline data, adjacent beach profiles, ebb and flood shoal evolution, and dredge/fill history provided a database with which the inlet's littoral impact to adjacent beaches was quantified. The even/odd method, which decomposes shoreline changes into their symmetric (even) and antisymmetric (odd) components about a point of significance, was applied to the volume change dataset. This analysis indicated a present-day (1996) alongshore impact distance of 8 to 13 km from the centerline of the inlet. A second method, which relates the total "inlet sink" volume to adjacent island (ocean and bay) volume change, was also applied. The total inlet sink volume ranged from 10.8 to 15.6 million m^3, and reflects the volume of material captured by the inlet which is assumed to have been derived from adjacent beaches. The inlet sink method indicated that, at most, 10.8 million m^3 can be realized within $\pm$14.2 km (data limit). The paper concludes that the alongshore impact of Ocean City Inlet most likely extends beyond $\pm$14.2 km.

1.0 EVOLUTION OF OCEAN CITY INLET

Ocean City Inlet is located on the Atlantic coast of the United States (Figure 1), and was created by a hurricane on 25 August 1933 which breached the outer barrier island. In 1922, groins were constructed on Fenwick Island (to the north) to stabilize this increasingly-popular beach resort. Assateague Island (to the south) remained in a natural state, and, in February 1920, Sinepuxent Inlet breached the island approximately 4.8 km south of the existing inlet, and had migrated southward about 0.8 km until it was closed by another storm in May 1928. After the hurricane of August 1933, jetty construction commenced in September and the inlet quickly scoured to 3 m depth and 76 m width (Underwood and Hiland 1995).

A study by Dean and Perlin (1977) and Dean, Perlin, and Dally (1978) documented sediment transport moving north along northern Assateague Island, which deposited in the inlet channel. As a result of the study, in 1985 the south jetty was raised and

---

[1]Research Hydraulic Engineer, [2]Supervisory Hydraulic Engineer, US Army Engineer Waterways Experiment Station, Coastal and Hydraulics Laboratory, CEWES-CR-P, 3909 Halls Ferry Road, Vicksburg, MS 39180-6199

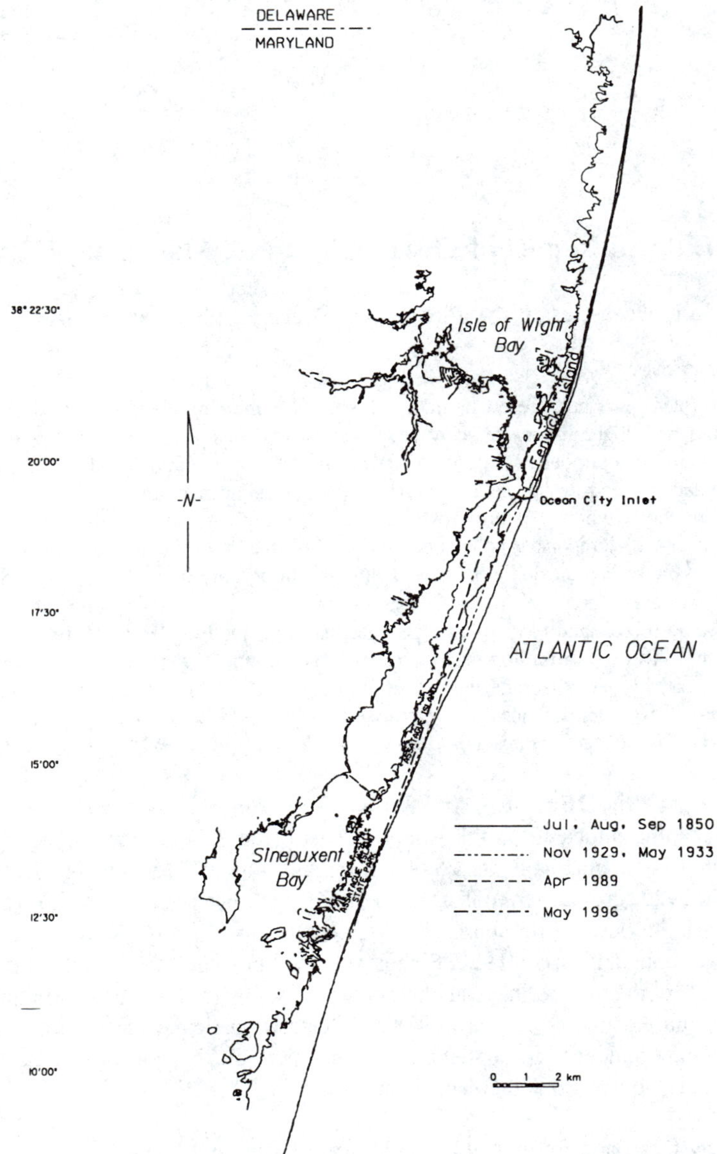

Figure 1. Historical shorelines, Ocean City Inlet, Maryland, USA

tightened, and the northern Assateague Island channel bank was stabilized with three detached breakwaters. The most recent significant engineering event affecting the littoral system has been the placement of approximately 7.4 million m^3 of beach fill on Fenwick Island from June 1988 through April 1995, which extended from the just north of the inlet to the Maryland-Delaware state line.

2.0 DATA ANALYSIS

Shoreline position, adjacent beach profile, ebb shoal bathymetry, and flood shoal data sets as well as dredging events were analyzed and used to create a volume change database. A range of parameters which were applied in sensitivity evaluations were also defined, and are discussed below.

2.1 Variation of Shoreline Position with Time

Both ocean and bay shoreline position were recorded with respect to an alongshore coordinate, *x*, which was set to zero at the centerline of the inlet and used a right-handed coordinate convention (i.e., negative *x* indicates north; positive *x* indicates south). The shoreline position represents the high water line, and was calculated for each time period at a 50-m alongshore cell spacing. Results of the shoreline change analysis are presented in Table 1.

Table 1. Summary of Shoreline Change Analysis: Average Rates and Associated Standard Deviation (m/yr)

Time Period	Alongshore Extent: Ocean (km)	Alongshore Extent: Bay (km)	Fenwick Island		Assateague Island	
			Ocean	Bay	Ocean	Bay
1850-1929/33	-15.1 to 14.2[1]	-13.7 to 3.3	-0.37 ± 0.98	~0[2]	-1.5 ± 1.7	2.8 ± 1.8
1929/33-1995/96	-15.1 to 14.2	Varies; range -13.5 to 16.6	0.44 ± 0.80[3] -0.21 ± 0.91[4]	0.75[5]	-2.9 ± 2.2[3] -2.9 ± 2.7[4]	2.5[6]

[1] Full extent was -19.4 to 23.5 km; the analysis was limited here to equal the alongshore extent for the post-inlet period.
[2] Calculated value was a small negative number. Because it is not likely that pre-inlet bay processes would erode the bay shoreline, a near-zero bayshore change rate was assumed for the pre-inlet time period.
[3] Shoreline change rate reflects advancement due to beach fill.
[4] Adjusted shoreline change rate to remove advancement due to beach fill.
[5] Calculated by prorating 1929-1965 and 1965-1980 bay shoreline change rates. Standard deviation values are unavailable.
[6] Calculated by prorating 1933-1965, 1965-1980, 1980-1989, and 1989-1995 bay shoreline change rates. Standard deviation values are unavailable.

Ocean shoreline change rates for Assateague Island extending 14.2 km south of the inlet nearly doubled from a pre-inlet (1850-1929/33) erosion rate averaging -1.5 ± 1.7 m/yr to an average post-inlet (1929/33-1996) erosion rate averaging -2.9 ± 2.7 m/yr (latter rate excludes shoreline advancement due to documented beach fill of 0.9 million m^3). For both the pre- and post-inlet time periods, overwash processes were significant along Assateague Island, with bay shoreline change indicating accretion. For the Fenwick shoreline extending approximately 15.1 km north of the inlet, the ocean shoreline responded to construction of the jetties by decreasing the pre-inlet erosion trend (-0.37 $\pm$ 0.98 m/yr for the pre-inlet time period to -0.21 $\pm$ 0.91 m/yr in the post-inlet time period (latter rate excludes shoreline advancement due to documented beach fill of 8.1 million m^3)). For the pre-inlet time period, the bay portion of Fenwick Island was comparably stable. The post-inlet time period reflects some bay shoreline accretion, which occurred prior to 1980 (a majority of the Fenwick bay shoreline in the vicinity of the inlet has been developed since 1980).

2.2 Profile Data

Beach profile data were analyzed to provide a relationship to convert shoreline change rates into volumetric change rates. Shoreline change within a given alongshore cell is estimated to represent a horizontal translation of the profile over its active depth. The absolute sum of the berm elevation and the depth at which profile changes are insignificant, the closure depth, is one method of estimating the active profile depth. A second method relates the profile volume change to shoreline movement for a given time period. Sensitivity testing of values from both methods indicated that the difference was not significant (less than 2-percent difference in total volumes). Thus, the first method was applied herein. For Fenwick and Assateague Islands, ocean beach profile data were analyzed to estimate active depth. Fenwick Island pre- and post-fill profile data were also analyzed to estimate the percentage of Fenwick Island beach fill remaining in 1996, which was required to correctly account for alongshore movement of this material from Fenwick Island into other regions of the project.

In evaluating the initial performance of the Fenwick Island beach fill, Stauble et al (1993) estimated the depth of closure for Fenwick Island using profile data measured from the spring of 1988 through the winter of 1992, and concluded that 6.1 m relative to National Geodetic Vertical Datum (NGVD) was a representative value. Because coastal processes at this depth are most likely similar offshore of Fenwick and Assateague Islands, this depth was also assumed to be representative of Assateague Island. However, this active depth was not used for southern regions of Fenwick Island and northern region of Assateague Island which were sheltered by the ebb tidal shoal. For these regions, the depth at the shoreward edge of the ebb shoal was used in active depth calculations. Volume change within the ebb shoal was accounted for separately.

Stauble et al (1993) estimated the active berm elevation for Fenwick Island as 3.0 m NGVD, resulting in an estimate of active depth for regions of Fenwick Island outside the shelter of the ebb shoal equal to 9.1 m. For Assateague Island, the sensitivity of applying pre- versus post- rehabilitation jetty berm crest elevations was insignificant; therefore, post-rehabilitation jetty berm crest elevations were applied herein. The estimate of active depth for the regions of Assateague Island unaffected by the ebb tidal shoal (from 2 to 14 km south of the inlet) ranged from 8.3 to 8.5 m.

Seven profiles from June 1988 (immediately pre-fill) and May 1996 (present-day condition) were analyzed to determine the rate of beach fill “loss” from the profile.

Volume change for each profile was multiplied by an alongshore distance represented by that profile to properly "weight" results at each profile location. Results from this analysis indicated that approximately 17 percent of the placed material was lost over the active depth. An additional 10 percent of material was originally placed in a dune feature (above 3.0 m NGVD), and this entire volume was assumed to remain on the profile. Because the fill material was similar to or coarser than the native sediments, it was assumed that any fill loss moved alongshore rather than offshore. The portion of the 1988-1995 Fenwick Island beach fill that was assumed to move alongshore, totaling approximately 1.7 million m^3, was assumed to deposit and remain within the inlet sink. Shoreline change calculations assumed a beach fill "loss" rate of approximately 2-percent per year.

Profile data for the bay shoreline were only available for the northwest corner of Assateague Island (Dean, Perlin, and Dally 1978), and represented conditions prior to sand-tightening and raising of the south jetty and construction of the detached breakwaters. These data extended into the channel, and reflect stronger current conditions than the rest of the bayshore. Based on estimates of bay active depth of approximately 4 m for this small, higher energy, region of the bayshore, the sensitivity of results to a range of slightly smaller bay shoreline active depths, thought to be more representative of conditions along the majority of the bay (1 m, 2 m, and 3 m) was evaluated. The sensitivity of volume calculations to the choice of bay active depth was found to be significant for certain cases (see Section 3.4).

2.3 Ebb Shoal Evolution

Bathymetric data from 1933, 1976, and 1995 were utilized to define the ebb shoal planform "footprint," depth at the landward edge of the footprint (used in estimating active depth), and volume. Visual inspection of a bathymetric data set or a bathymetric contour plot to delineate an ebb shoal is rather subjective. For this study, the availability of a pre-inlet (1933) bathymetry enabled application of a more objective procedure, as modified from that discussed by Hicks and Hume (1996). First, an idealized "non-inlet" bathymetry was created by using the 1933 bathymetric data set, but replacing the 1933 shoreline and nearshore beach profile data along the northern region of Assateague Island with present-day shoreline and bathymetry which were typical of that found outside the influence of the ebb tidal shoal. Using this idealized non-inlet bathymetry, "residual" topographies were created by subtracting it from the subsequent time periods. Next, the ebb tidal shoal footprint for each residual topography was estimated by four different individuals, and residual ebb shoal volumes were calculated. Four independent estimates of the ebb shoal footprint gave an indication of the error associated with personal subjectivity.

Estimates 1, 2, and 4 applied similar methods of using the 1-m residual contour to delineate the ebb shoal footprint. This contour appeared to provide a planform shape of the ebb tidal shoal that reflected onshore and alongshore evolutionary trends of the shoal as evidenced from bathymetric data. Estimate 3 used the 0-m residual contour to delineate the shoal, and therefore these volume and area estimates are higher. Estimate 3 chose the 0-m residual contour to fully capture bathymetric regions which may have accreted. These regions resulted in a footprint which included more of the areas reflecting growth and lengthening of finger shoals, and deposition in troughs, which are probably only partly represented by Estimates 1, 2, and 4. All residual volumes were calculated by using the selected footprint and determining the volume above a 0-m residual contour within this polygon. Bathymetric data from 1995 did not fully cover the ebb tidal shoal. Therefore, 1933-1976 residual ebb shoal data were used to complete coverage on the outer edges of the ebb tidal shoal for the 1933-1995 residual bathymetry. Results are presented in Table 2.

Table 2. Ebb Tidal Shoal Changes, 1933-1995

Estimate Number	Volume (million m^3)	Area (million m^2)	Volume Change Rate (m^3/yr)
1, 2, 4	9.8, 9.9, 9.9	3.0, 3.0, 3.0	158,000, 159,000, 158,000
3	12.9	4.2	207,000

For estimates of total ebb shoal volume from 1933 to 1996, two ebb shoal volumes were used: a minimum value was obtained by extrapolating the average of Estimates 1, 2, and 4 to 1996 (approximately 10 million m^3); and a maximum value was obtained by extrapolating Estimate 3 to 1996 (approximately 13 million m^3) . For each estimate, depths at the landward edge of the ebb tidal shoal footprint were applied in calculating active depth for regions of Fenwick and Assateague Islands sheltered by each ebb shoal. In sensitivity testing, the differences between active depths for the 10 and 13 million m^3 ebb tidal shoal footprints were found to have an insignificant impact on final calculations. Therefore, results are presented herein for active depths as indicated by the 10 million m^3 ebb shoal footprint.

2.4 Flood Shoal Evolution

Information about deposition in the bay is necessary to account for losses to the littoral system. Unfortunately, detailed bathymetric data provide only partial coverage for Isle of Wight and Sinepuxent Bays. For this study, the flood shoal/bay accretion rate calculated by Dean and Perlin (1977) and Dean, Perlin, and Dally (1978) using 1931 to 1972 bathymetry was used, 19,600 m^3/yr. It is likely that the 1972 through 1996 flood shoal accretion rate has decreased somewhat, perhaps half this value, for a present-day estimate of total flood shoal/bay volume equal to 1 million m^3. This estimate is based on the belief that flood shoal accretion would probably be greatest during the initial years after inlet formation, and would decrease with time. The sensitivity of calculations presented herein was evaluated with two other flood shoal/bay accretion estimates. A maximum estimate assumed that the original rate continued through the present, for a total flood shoal/bay volume of 1.2 million m^3. A minimum volume of 0.76 million m^3 was estimated by assuming that no additional bay/flood shoal accretion has occurred since 1972.

2.5 Dredging History

Various sources were culled to determine the dredging and placement history for Ocean City Inlet. A total of 24 dredging events between 1947 and 1995 indicated a channel shoaling rate of approximately 56,000 m^3/yr, with many dredge placements on northern Assateague Island. For this study, a minimum channel shoaling volume between 1933 and 1996 was calculated by assuming that only the documented dredging events occurred (2.7 million m^3). A maximum channel shoaling volume assumed that the 56,000 m^3/yr rate continued from 1933 through 1996 (3.5 million m^3). Adjacent beach placement of the dredged material was accounted for in the shoreline change database.

3.0 LITTORAL IMPACT OF OCEAN CITY INLET

3.1 Volume

The "inlet sink volume" captured by Ocean City Inlet is defined as the volume of material within the inlet which is assumed to have been derived from adjacent ocean and bay beaches. A range of volumes representing the Ocean City Inlet sink was calculated by summing the minimum and maximum estimates for ebb shoal volume, flood shoal volume, and channel shoaling volume, subtracting maximum and minimum estimates for the original 1933 Ocean City Inlet breach, and subtracting the volume of Fenwick Island beach fill which was assumed to move alongshore and deposit in the inlet. The initial loss of material due to the island breach in 1933 was estimated using typical ocean and bay profile shapes (best estimate: 0.53 million m^3, with a range between 0.43 and 0.97 million m^3).

The minimum inlet sink volume was estimated as 10.8 million m^3 (ebb shoal: 10 million m^3 + flood shoal: 0.76 million m^3 + shoaling volume: 2.7 million m^3 - initial breach: 0.97 million m^3 - alongshore movement of Fenwick Island beach fill: 1.7 million m^3). The maximum inlet sink volume totaled 15.6 million m^3 (ebb shoal: 13 million m^3 + flood shoal: 1.2 million m^3 + shoaling volume: 3.5 million m^3 - initial breach: 0.43 million m^3 - alongshore movement of Fenwick Island beach fill: 1.7 million m^3). Thus, **the total volume of material captured by Ocean City Inlet which can be attributed to adjacent beaches is estimated to be between 10.8 and 15.6 million m^3.**

3.2 Alongshore Impact: Overview. Two methods were used to estimate the distance alongshore for which the inlet has affected adjacent beaches: the even/odd procedure (Section 3.3), and an examination of the inlet's net sink effect and adjacent island response (Section 3.4). Both methods accounted for a background erosion rate, and assumed that pre-inlet volume change trends would have continued through the present if the inlet had not been created. Thus, pre-inlet (1850-1929/33) volume change rates were subtracted from post-inlet (1929/33-1996) volume change rates. The effect of beach fill (which is reflected in the shoreline position) was removed from the signal by subtracting an estimate of the beach fill volume remaining from the volume in each 50-m shoreline calculation cell. An estimate of beach fill reflected in the shoreline position was estimated by assuming a beach fill "loss" (or alongshore movement) rate of approximately 2-percent per year (discussed in Section 2.2).

The assumption of a background erosion rate is reasonable, as evidenced from the Fenwick Island shoreline change rates from 1929/33-1996 (Table 1 with beach fill advancement removed), and the trend for relative sea level rise in this region of the U.S. (1.1 mm/yr for Lewes, Delaware, north of Ocean City Inlet, Marine Board 1987). In addition, the 1850 to 1929/33 reorientation of the Fenwick-Assateague Island barrier island from a convex to a more "linear" shoreline orientation (see Figures 1 and 3a) most likely is a long-term, large-scale response which is continuing through the present.

3.3 Alongshore Impact: **Even/Odd Method.** The even/odd method is a simple analytical procedure discussed and applied by Dean and Work (1993) and Bodge (in preparation) among others, and can be solved from the linearized treatment of shoreline evolution (Pelnard-Considere 1956). The method decomposes shoreline change (or, as applied in this case, volume change) data into their symmetric (even) and anti-symmetric (odd) components about a point of significance. For application at Ocean City Inlet, the center of the inlet was chosen as the point about which data were decomposed into even and odd components. Volumetric change data were used so that the effects of beach fill could be removed, and so that the alongshore variation in active depth could be incorporated. Due to space limitations herein, the reader is directed to Dean and Work (1993) or Bodge (in preparation) for a description of even/odd calculation procedures.

The even and odd functions are indicators of those shoreline or volumetric changes which have been symmetric or anti-symmetric, respectively, about the point of significance. For an inlet with the centerline chosen as the point of significance, the even function reflects changes in shoreline position (or volume change rate) which have occurred symmetrically about the inlet, such as shoreline retreat or advance due to cross-shore transport (due to storms and relative sea level change), and symmetric shoreline retreat due to sediment feeding the ebb and flood tidal shoals. An example of an anti-symmetric change common to stabilized inlets, which is reflected by the odd function, is impoundment on the updrift beach at the expense of the downdrift beach. The alongshore point at which the odd function returns to a zero value is an indicator of the alongshore distance influenced by the anti-symmetric effects, such as impoundment, which is a project-induced impact. The alongshore point at which the even function approaches a constant value is an indication of the alongshore distance influenced by symmetric inlet-induced effects such as shoreline retreat due to feeding the ebb and flood tidal shoals.

Figure 2a shows the post-inlet (1929/33-1996) even/odd analysis, and Figure 2b shows results for the post-inlet rate adjusted by the background erosion rate as defined by the pre-inlet (1850-1929/33) data set. Negative distances indicate shoreline to the north along Fenwick Island. For both Figures 2a and b, the "total" curve shows the updrift

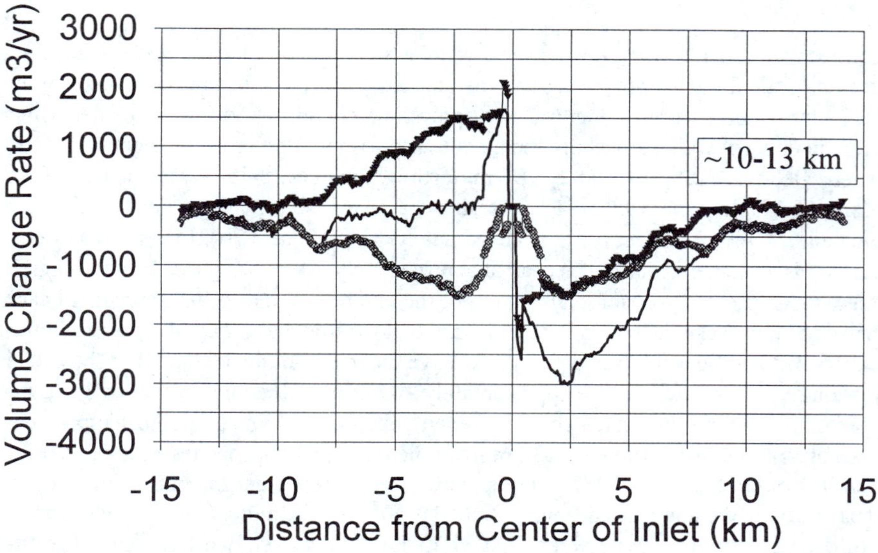

(a)

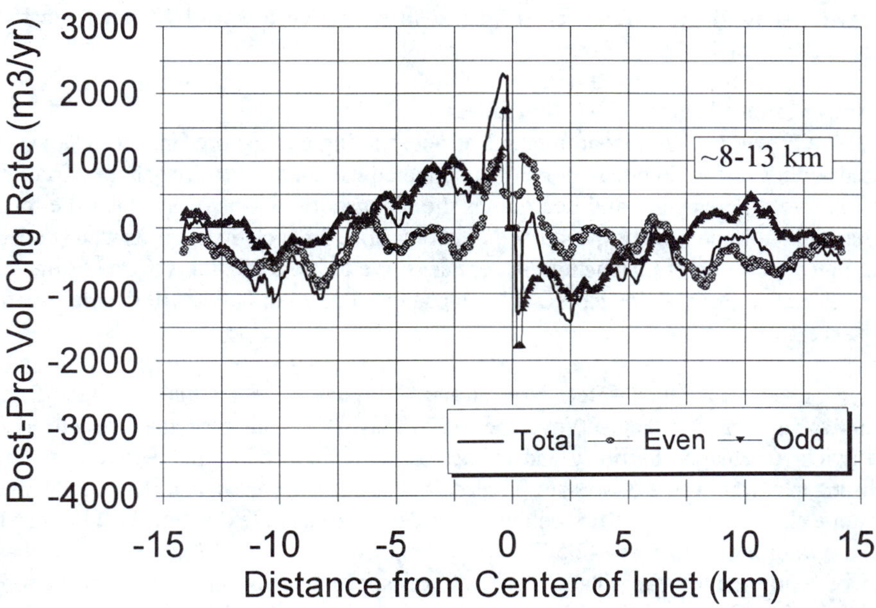

(b)

Figure 2. Total volumetric change, even, and odd functions for (a) post-inlet volume change rates (1929/33-1996); and post-inlet minus pre-inlet (1850-1929/33) volume change rates accretion that has occurred as a result of inlet stabilization and the groin field along Fenwick Island. It is interesting to note the erosional peak in the post-inlet volume change rate ("total" curve in Figure 2a) at approximately 8 km updrift of the inlet, which agrees in concept with a longshore sediment transport nodal point in the vicinity of York Beach, Delaware (approximately 18 km north of the inlet), with a variation of 10 km (Mann and Dalrymple 1986). For Assateague Island, except for a peak in the total adjusted volume change curve at approximately 1 km (Figure 2b), the entire shoreline response has been erosional. The peak at approximately 1 km in Figures 2a and b agrees well with the presently-observed bridging of the ebb tidal shoal to the downdrift beach. The "odd" function indicated in Figures 2a and b illustrates the strong alongshore impact that the inlet and Fenwick Island groin field have incurred on the adjacent beaches. This function returns to a zero value approximately 10 km from the inlet in Figure 2a, and at approximately 8 km in Figure 2b. The "even" function in Figure 2a returns to a near-constant value at approximately 13 km from the inlet, and approaches a constant value in this region for Figure 2b. If Figure 2a were taken to represent the inlet's impact (i.e., no background erosion existed from 1933 to 1996), then the alongshore impact distance would be interpreted as between 10 and 13 km. However, we have evidence that background erosion has occurred during the post-inlet time period. Thus, considering both the even and odd functions, **the alongshore impact distance of the inlet (and Fenwick Island groin field) is estimated to be between 8 and 13 km from the centerline of the inlet.**

3.3 Alongshore Impact: Net Sink Effect

Overview. A second method of determining alongshore impact distance is discussed by Bodge (in preparation). This method calculates the alongshore variation in total volume change, and determines the alongshore distance for which the total inlet sink volume can be accounted. Alongshore impact distance as determined through application of this method is very sensitive to the total sink volume estimate, due to the fact that the alongshore volume changes due to the inlet will decrease with distance from the inlet.

Pre- and Post-Inlet Volumetric Change. Shoreline change from August/September 1850 to November 1929/May 1933 indicated that the pre-inlet Fenwick-Assateague barrier island in the region $\pm$ 7 km from the centerline of the existing inlet was in an erosive state (Figure 3a). The island (ocean and bay) cumulative volume change rate for what would become Fenwick Island (extending 14.2 km north of the inlet location) was -42,500 m^3/yr during this time period (independent of bay active depth; the bay shoreline change rate was approximately zero during this period, see Table 1). The cumulative island pre-inlet volume change for 14.2 km of Assateague Island ranged between -137,700 and -59,300 m^3/yr, depending on bay active depth.

In the post-inlet (November 1929/May 1933 through March 1995/May 1996) time period, Fenwick Island stabilized with cumulative island volumetric change rates ranging from -16,500 to +4,400 m^3/yr, depending on bay active depth. However, cumulative

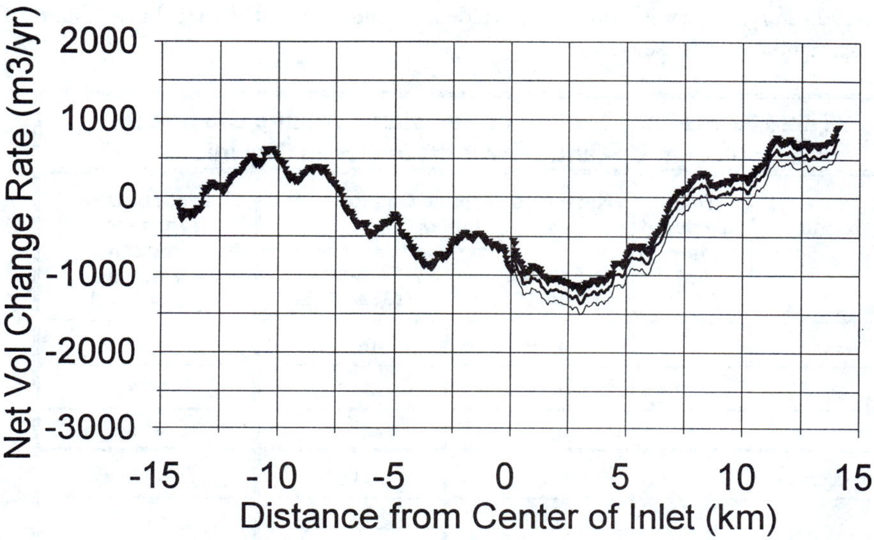

(a)

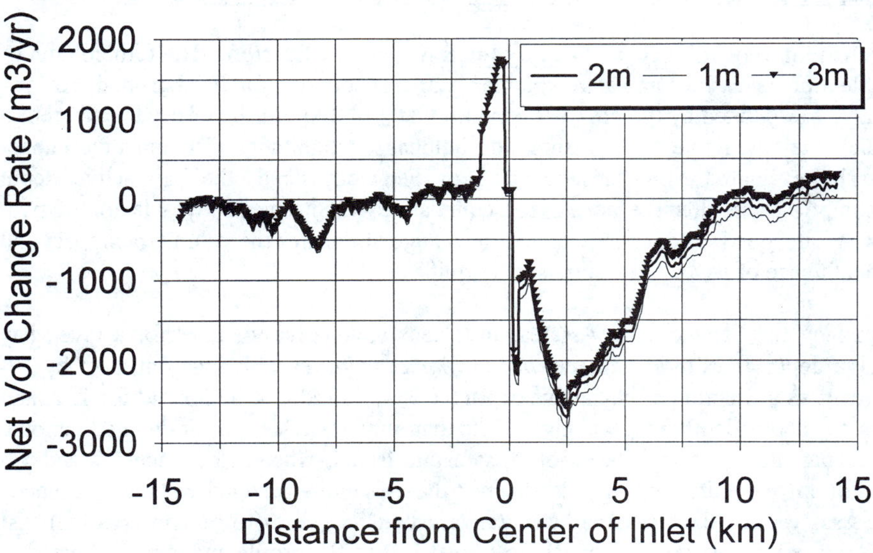

(b)

Figure 3. (a) Pre-inlet (1850-1929/33) and (b) post- inlet (1929/33-1996) island (bay and ocean) volumetric change rates as a function of bay active depth (1, 2, or 3 m) volumetric erosion rates along Assateague Island increased, ranging from -293,000 to -224,000 m³/yr depending on bay active depth (Figure 3b). The first four columns of Table 3 summarize these results.

Table 3. Pre- and Post-Inlet Net Island Volume Change for 14.2 km Up- and Downdrift of Ocean City Inlet

Island	Bay Active Depth (m)	Net Island[1] Volume Change Rate (m^3/yr)		Present-Day Normalized[2] Cumulative Volume (million m^3)
		Pre-Inlet (1850-1933)	Post-Inlet (1933-1996)	
Fenwick	1	-42,500	-16,500	+1.6
	2	-42,500	-6,000	+2.3
	3	-42,500	+4,400	+3.0
Assateague	1	-137,700	-293,000	-9.8
	2	-98,500	-259,000	-10.1
	3	-59,300	-224,000	-10.4

[1] Net ocean and bay volume change.
[2] Post-inlet minus pre-inlet rate multiplied by the number of years since inlet creation (1996-1933=63 years).

Present-Day Impact Accounting for Background Erosion. The fifth column of Table 3 presents the cumulative volume change, accounting for "background erosion" which was defined by the pre-inlet condition, at $\pm$14.2 km of the inlet. The degree to which the bay active depth assumption influences magnitudes of cumulative impact can be evaluated from these data. For Fenwick Island, the bay active depth assumption can alter the normalized cumulative volume by nearly a factor of two. For Assateague Island, ocean volumetric change dominates the volumetric impact and the influence of bay active depth is less significant.

Summing the Fenwick and Assateague Islands cumulative volumes for a given bay active depth gives ***total adjacent beach impact volumes*** which vary at most by 10-percent as a function of bay active depth (-8.2, -7.8, and -7.4 million m^3 for 1, 2, and 3 m bay active depth, respectively). Note that only 0.9 million m^3 of dredged material is documented as being placed on Assateague Island, whereas between 2.7 and 3.5 million m^3 were dredged. If it is assumed that this entire dredged volume was placed on Assateague Island (as has been standard practice), and (to be conservative) that this entire volume moved alongshore or was lost to the profile and is not reflected in the shoreline data, then an additional loss of 1.8 to 2.6 million m^3 can be assigned to the adjacent island losses. Using the upper value of 2.6 million m^3, it can be concluded that the volumetric impact of Ocean City Inlet realized on adjacent ocean and bay shores within 14.2 km from the inlet, is *at most* 10.8 million m^3. This estimate approaches the minimum total inlet sink volume, which was estimated to range from 10.8 to 15.6 million m^3. (However, note that the 10.8 million m^3 adjacent island loss was developed with the *upper* dredge volume estimate, whereas the 10.8 million m^3 sink volume was developed with the *lower* dredge volume estimate.) Thus, with a pre-existing background erosion rate which is defined by the pre-inlet condition, it is likely that the inlet's influence exceeds the available data limit (14.2 km).

4.0 SUMMARY AND CONCLUSIONS

A database representing 146 years of beach response in the vicinity of Ocean City Inlet, Maryland, was created to reflect adjacent beach, bay, and inlet volume change. Analyses were conducted to evaluate the total sink volume of the inlet, and the alongshore distance associated with this volumetric impact. The total inlet sink, defined as the volume captured by the inlet which can be attributed to the adjacent ocean and bay shorelines, was estimated to range between 10.8 and 15.6 million m^3. The even/odd method was applied to ocean shoreline change (adjusted by a background erosion rate) since inlet formation, and indicated an alongshore impact distance (which includes the effects of the Fenwick Island groin field) of approximately 8 km from the centerline of the inlet. The inlet sink analysis, assuming that pre-inlet trends continued through the present, indicated, *at most*, that 10.8 million m^3 can be realized along 14.2-km of up- and downdrift ocean and bay shorelines. This analysis indicated that the alongshore impact distance most likely exceeds the available data limit ($\pm$14.2 km from the centerline of the inlet).

5.0 ACKNOWLEDGEMENTS

The research discussed herein was partially supported by the *Inlet Channels and Shorelines* work unit of the Coastal Inlets Research Program, Coastal and Hydraulics Laboratory, Waterways Experiment Station, US Army Corps of Engineers (USACE), and by the *Ocean City Water Resources Study* funded by the USACE District, Baltimore. Permission to publish this paper was granted by the Chief of Engineers. Thanks to reviewers Mark B. Gravens and Nicholas C. Kraus.

6.0 REFERENCES

Bodge, K. R. 1996 (in preparation). "Sediment Management at Inlets/Harbors," draft Coastal Engineering Manual Chapter V-6, USACE Waterways Experiment Station, Coastal and Hydraulics Laboratory, Vicksburg, MS (draft dated August 21, 1996).

Dean, R. G., and Perlin, M. 1977. "Coastal Engineering Study of Ocean City Inlet, Maryland," *Proceedings*, Coastal Sediments '77, American Society of Civil Engineers, pp. 520-540.

Dean, R. G., Perlin, M., and Dally, W. 1978. (March). "A Coastal Engineering Study of Shoaling in Ocean City Inlet," prepared for USAED, Baltimore, under contract with the University of Delaware, Newark, DE, 135 p.

Dean, R. G., and Work, P. A. 1993. "Interaction of Navigational Entrances with Adjacent Shorelines," ***Journal of Coastal Research***, Vol 18, p. 91-110.

Hicks, D. M., and Hume, T. M. 1995. "Morphology and Size of Ebb Tidal Deltas at Natural Inlets on Open-sea and Pocket-bay Coasts, North Island, New Zealand," *Journal of Coastal Research*, Vol 12(1), pp 47-63.

Mann, D. W., and Dalrymple, R. A. 1986. "A Quantitative Approach to Delaware's Nodal Point," ***Shore and Beach***, Vol 54, No. 2, pp 13-16.

Marine Board. 1987. "Responding to Changes in Sea Level: Engineering Implications," National Academy Press, Washington, D.C.

Pelnard-Considere, R. 1956. "Essai de theorie de l'evolution des formes de rivage en plages de sable et de galets," *4th Journees de l'Hydraulique: Les Energies de la Mer*, Question III, No. 1, pp. 289-300.

Stauble, D. K., Garcia, A. W., Kraus, N. C., Grosskopf, W. G., and Bass, G. P. 1993. "Beach Nourishment Project Response and Design Evaluation: Ocean City, Maryland, Report 1, 1988-1992," Technical Report CERC-93-13, USAE Waterways Experiment Station, Vicksburg, MS 39180-6199.

Underwood, S. G., and Hiland, M. W. 1995. "Historical Development of Ocean City Inlet Ebb Shoal and its Effect on Northern Assateague Island," final draft report prepared for the USACE Waterways Experiment Station under Work Order No. DACA39-92-M-198 and the US Department of the Interior, National Park Service under Inter-agency agreement IA4000-1-0022.

CHAPTER 216

EXTREME WATER LEVELS, WAVE RUNUP AND COASTAL EROSION

P. Ruggiero[1] , P.D. Komar[2] , W.G. McDougal[1] and R.A. Beach[2]

ABSTRACT

A probabilistic model has been developed to analyze the susceptibilities of coastal properties to wave attack. Using an empirical model for wave runup, long term data of measured tides and waves are combined with beach morphology characteristics to determine the frequency of occurrence of sea cliff and dune erosion along the Oregon coast. Extreme runup statistics have been characterized for the high energy dissipative conditions common in Oregon, and have been found to depend simply on the deep-water significant wave height. Utilizing this relationship, an extreme-value probability distribution has been constructed for a 15 year total water elevation time series, and recurrence intervals of potential erosion events are calculated. The model has been applied to several sites along the Oregon coast, and the results compare well with observations of erosional impacts.

INTRODUCTION

Much of the Oregon coast is characterized by wide, dissipative, sandy beaches, which are backed by either large sea cliffs or sand dunes. This dynamic coast typically experiences a very intense winter wave climate, and there have been many documented cases of dramatic, yet episodic, sea cliff and dune erosion (Komar and Shih, 1993). A typical response of property owners following such erosion events is to build large coastal protection structures. Often these structures are built after a single erosion event, which is followed by a long period with no significant wave attack. From a coastal management perspective, it is of interest to be able to predict the expected frequency and intensity of such erosion events to determine if a coastal structure is an appropriate response. Predicting the frequency and magnitude of the most extreme erosion conditions is also important for the establishment of setback lines for coastal developments.

Field studies have been undertaken along the Oregon coast to better understand the processes involved in wave-induced sea cliff and dune erosion. The

---

[1] Ocean Engineering Program, Civil Engineering Dept., Oregon State Univ., Corvallis, OR 97331

2 College of Oceanic and Atmos. Sciences, Oregon State Univ., Corvallis, OR 97331

goal is to develop simple methods for estimating the susceptibilities of coastal properties to wave attack. Wave-induced property erosion generally occurs when wave runup reaches the junction between the beach face and the sea cliff or dune. However, depending on a number of wave, water level and morphological parameters, both the elevation attained by wave runup and the elevation of the beach-face junction are highly variable. This paper presents a simple probabilistic method for predicting the frequency of occurrence of cliff or dune erosion. The predicted frequencies are then compared with the observed shoreline stability for several beaches along the Oregon coast.

Other researchers have attempted to determine the risk of coastal properties to flooding using probabilistic techniques. For example, Gares (1990) demonstrates a method for using runup to determine flooding risk on the New Jersey shoreline. Runup is calculated using the composite slope technique presented in the *Shore Protection Manual* and probability curves are produced assuming that maximum runup occurs at times of maximum storm surge. In the current work, an attempt is made to characterize extreme runup statistics on the high energy, yet dissipative beaches common in Oregon where the storm surge is small. Extreme total water levels, hence erosion risks, are then predicted utilizing a long term data base consisting of both measured tides and measured wave characteristics.

MODEL DEVELOPMENT

There are two main components, in Figure 1, which combine to generate total water levels (Shih *et al.*, 1994). The first is the tide elevation, ζ; all tidal and land elevations presented in this paper will be referenced to the U.S. National Geodetic Vertical Datum (NGVD) of 1929. Superimposed on the measured tide level is the vertical component of wave runup, R, which consists of both wave setup, $\overline{\eta}$ and the swash fluctuations, S, about the setup. Wave induced erosion of a sea cliff or dune will occur only when the total elevation of the water at times of maximum runup exceeds the elevation of the beach-face junction ($\zeta + R > E_j$). Once again, the objective of this study is to predict the frequency with which the total water level, E, reaches or exceeds the elevation E_j of the junction between sea cliff, dune or coastal protection structure and the beach face. To achieve this goal, we construct extreme-value probability distributions (e-v pdf) for water levels measured by tide gauges, water levels due to wave runup, and finally for the combined total water levels.

Extreme Water Levels

Measured tides are often greater than predicted due to the occurrence of a storm surge, or the effects of water temperatures, currents and atmospheric disturbances such as an El Niño. A 24 year time series of hourly measured tides taken from a tide gage in Yaquina Bay, Oregon, has been analyzed to investigate some of these processes. The yearly maxima have been fitted to a Fisher-Tippet Type I extreme-value probability distribution, commonly known as the Gumbel extreme value distribution, and return intervals have been computed for extreme tides. A

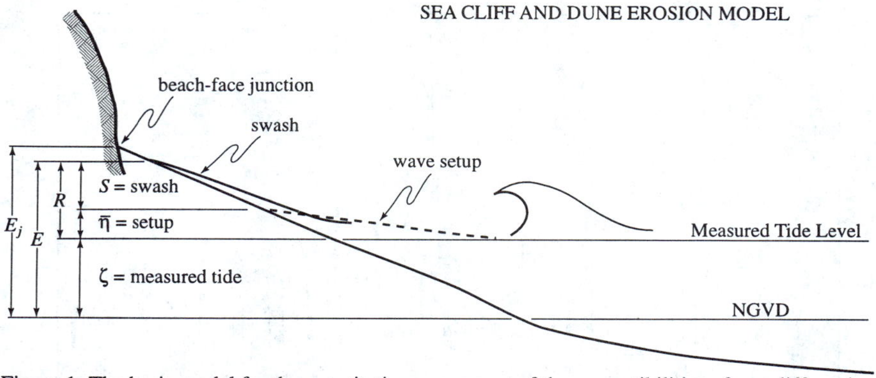

Figure 1: The basic model for the quantitative assessment of the susceptibilities of sea cliffs and dunes to wave-induced erosion.

predicted tide time series has been generated for the same 24 year period, and the predicted yearly maxima have also been fit to a Gumbel distribution. Shih *et al.* (1994) noted that for long return periods there are significant differences, on the order of 0.5 m, between predicted and observed extreme tides.

The difference between measured and predicted tides has been computed for the entire 24 year data set. The auto-correlation of this raw residual time series shows a roll off in correlation at a lag of approximately 48 hours. This lag corresponds well with the typical storm duration on the Oregon coast. The raw residuals were then filtered using a 48 hr low pass filter, eliminating measurement noise from the signal. The standard deviation of the low pass filtered residual time series is approximately 13 cm, giving a typical elevation for storm surge on the Oregon coast. Figure 2 shows the 24 year residual time series after applying a 1 month low pass filter in which single storms as well as spring and neap tides have been averaged out. Surprisingly, there is no distinct seasonal pattern in the figure. The extreme residuals evident in the smoothed time series, those in 1973, 1977, 1982-83, 1992 and 1993, all correspond to times of well-documented El Niño events. Although these events are usually associated with increased wave energy, both the residual and the measured tide are not significantly correlated with wave height throughout the period of overlap between the tide and wave data, a period of approximately 15 years. This observation, along with the relatively low contribution of storm surge to extreme total water levels on the Oregon coast, suggests that models such as that proposed by Gares (1990) may not be applicable to the Oregon coast. Models in which extreme measured water levels and extreme runup occur at the same time, although well suited for coastlines which experience hurricanes and Nor'Easters which generate significant storm surge, are not generally suitable to conditions typical of the Oregon coast.

Wave Runup

The second water-level component in the sea cliff and dune erosion model, superimposed on the measured tide, is wave runup. It is necessary to have the

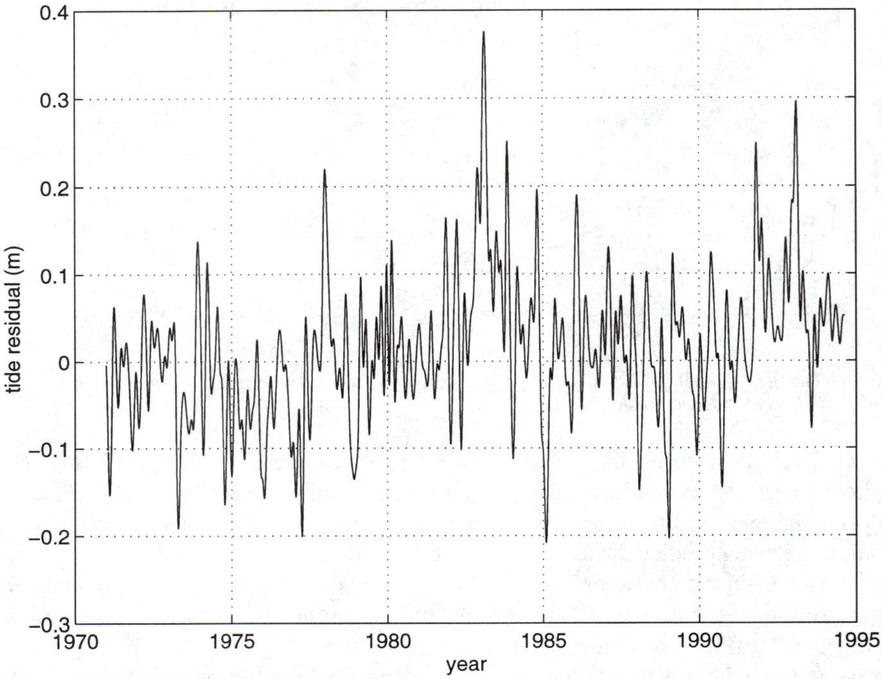

Figure 2: Thirty day low-pass filtered residual time series of measured minus predicted tides in Yaquina Bay, Oregon.

capability to predict extreme runup statistics utilizing measured wave parameters and beach morphology characteristics. Data bases containing a decade or more of measurements of wave heights and periods are now becoming readily available for use by engineers, scientists and planners, and much effort has been spent in finding simple relationships between deep-water wave parameters and runup on beaches. A well known data set of extreme runup statistics is that of Holman and Sallenger (1984) and Holman (1986), who measured wave runup using video techniques on the relatively steep beach of the Field Research Facility, Duck, North Carolina. Among other runup statistics, Holman (1986) examined the 2% exceedence elevation for runup maxima, non-dimensionalized by wave height, and found a dependence on the dimensionless surf similarity parameter, ξ_o, commonly referred to as the Iribarren number, the ratio of the beach slope to the deep-water wave steepness. The resulting relationship is

$$R_{2\%}/H_s = C\,\xi_o = C\,\beta/(H_s/L_o)^{(1/2)} \quad \textbf{(1)}$$

where C is an empirical constant determined by the measurements, β is the beach slope, H_S is the deep-water significant wave height, and L_O is the deep water wave length given by $L_O = (g/2\pi)T^2$ where g is the acceleration of gravity and T is the wave period. This result has been proposed by many researchers, including Battjes

(1971) and van der Meer and Stam (1992). In an attempt at further simplification, Holman's data were re-analyzed by Douglas (1992), who found that removing the beach slope term from Eq. (1) does not cause any reduction in the ability to predict runup on beaches. By analyzing a wide variety of Australian beaches, Neilson and Hanslow (1991) found that the relationship proposed by Holman explained results from field experiments on relatively steep beaches, $\beta > 1{:}10$, while for flatter more dissipative beaches with $\beta < 1{:}10$, runup depended only on the wave steepness. Many Oregon beaches are extremely dissipative with beach slopes much less than 1:10. The resulting Iribarren numbers are usually less than for beaches studied by other researchers, including the most dissipative in the Neilson and Hanslow (1991) study.

To test the dependence of extreme runup statistics on measured wave and beach morphology characteristics, video techniques were used to measure wave runup on several Oregon beaches over a broad range of dynamic conditions. Deep-water significant wave heights ranged from 1.9 to 4.6 m, periods from 7 to 17 seconds, and foreshore slopes from 1:100 to 1:20. This field program culminated in February 1996 at Agate Beach on the central Oregon coast, with a major investigation into the dynamics of high energy dissipative beaches. During this experiment, three video cameras were used to measure wave runup, and the overlap between the cameras allowed for continuous coverage of runup measurements over approximately 2 km of beach. Detailed analysis of the longshore variability of runup on dissipative beaches is underway. Runup elevation time series were extracted from all video data using image-processing techniques developed at the Coastal Imaging Lab of Oregon State University (Holman and Guza, 1984; Holland *et al.*, in press). For each data record, the tide has been removed and extreme runup statistics have been computed after identifying the local maxima of the shoreline elevation time series. Although there is a distinction in the processes which force wave setup and swash fluctuations, for most engineering applications the measure of interest is the extreme statistics associated with the total runup. Therefore, all runup statistics presented include both setup and swash.

The Oregon measurements of $R_{2\%}$ are non-dimensionalized in the form of Eq. (1), and compared with the data of Holman (1986) in Figure 3. The Iribarren number distinguishes between the dynamically different nearshore systems from which the data were taken. The Oregon data falls in the extremely dissipative range of Iribarren numbers, while the FRF data range from dissipative to reflective. Although the Oregon data are of the expected order of magnitude, any linear predictive model such as Eq. (1), derived from Holman's data, would tend to under predict $R_{2\%}$ on the flatter Oregon beaches, an observation noted earlier by Shih *et al.* (1994). The data from Oregon have been re-plotted in Figure 4a, which suggests that extreme runup statistics do not depend on the Iribarren number over this dissipative range. Figure 4b shows that the Oregon runup data has virtually no dependence on the beach slope, and Figure 4c shows a lack of dependence of $R_{2\%}/H_s$ on the wave steepness. Although not shown, the extreme runup also does not depend on the wave period. This is surprising, as runup has a first order dependence on the wave

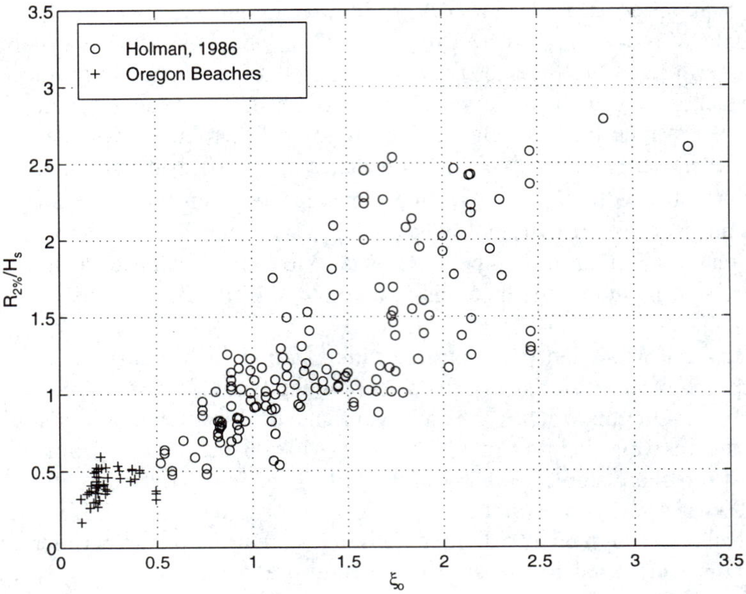

Figure 3: Comparison between non-dimensional extreme runup statistics obtained by Holman from the FRF and from Oregon beaches.

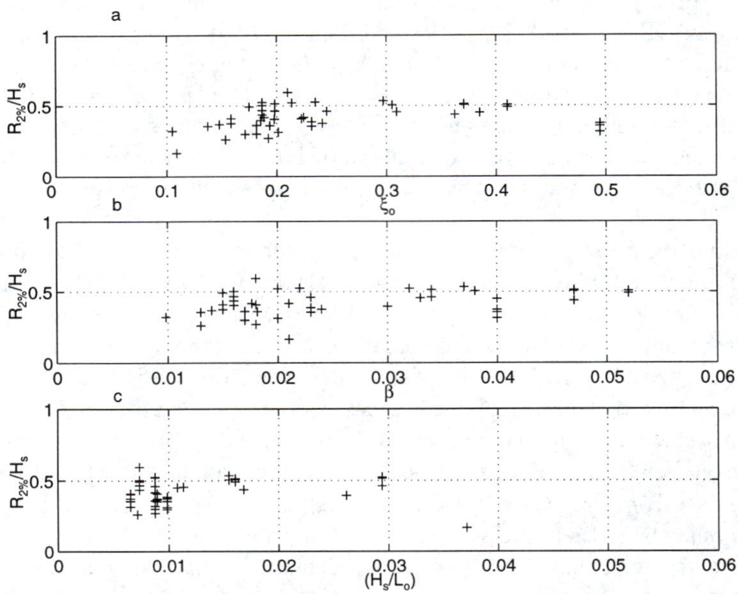

Figure 4: Relationships between non-dimensional extreme runup statistics and a) the Iribarren number, b) beach slope, and c) wave steepness.

period in the dimensional form of Eq. (1). The best parameterization of $R_{2\%}$ for the Oregon data is a simple dependence on the deep-water significant wave height as shown in Figure 5. The best-fit straight line matched to the Oregon data is

$$R_{2\%} = 0.5\, H_s - 0.22 \text{ (m)} \qquad (r^2 = 0.72) \tag{2}$$

A linear relationship between runup elevation and wave height on relatively low energy dissipative beaches was also found by Guza and Thornton (1982) in Southern California, and by Aagaard (1990) in Denmark and Australia. Guza and Thornton (1982) found that the significant vertical swash excursion, defined as 4σ, where σ^2 is the total variance of the runup elevation time series, was approximately 70% of the significant wave height, a slightly stronger dependence on wave height than found for the Oregon data and the results presented by Aagaard (1990). The FRF data of Holman (1986) are shown again in Figure 5 to emphasize the fact that the data are derived from dynamically very different systems as the Holman data clearly falls above the Oregon data. More research is needed to identify regimes where different models for extreme runup statistics are applicable.

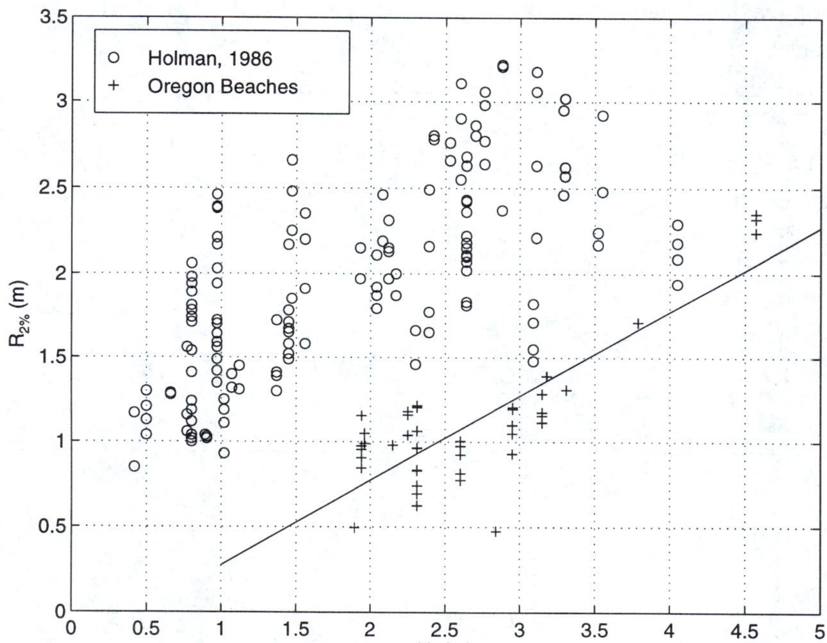

Figure 5: Dependence of extreme runup statistics on the deep-water significant wave height.

The extreme runup statistics presented here, as in the work done by other researchers, are monochromatic representations of spectral phenomena. In fact, all of the Oregon runup data, particularly that on the very low sloping Agate Beach, were dominated by infragravity energy ($f < 0.05$). Spectral peaks typically occurred at periods ranging from approximately 100 seconds to 200 seconds, and usually more than 90% of the runup elevation variance fell in the infragravity band. This infragravity dominance helps explain why runup statistics are independent of the incident band wave period. The incident band energy was saturated for most data runs, and the dependence on wave height shown in Figure 5 can almost entirely be explained by the infragravity band energy and its dependence on the wave height. Again, this is a similar result to that found by Guza and Thornton (1982).

Waves and Total Water Levels

In the previous section we showed that the best predictor of extreme runup on dissipative Oregon beaches is simply the deep-water wave height. A 15-year wave data set from the Coastal Data Information Program buoy offshore from Bandon, Oregon, has been used to estimate the extreme wave climate of the Oregon coast. This irregularly spaced data set has been interpolated to hourly measurements to match the measurement interval of the Yaquina Bay tide measurements. A Gumbel extreme-value distribution has been fitted to this data, and the resulting recurrence intervals for extreme waves are shown in Figure 6. This extreme value pdf has been produced by using all of the wave data above 6.5 meters, rather than just the yearly maxima (Muir and El-Shaarawi, 1986).

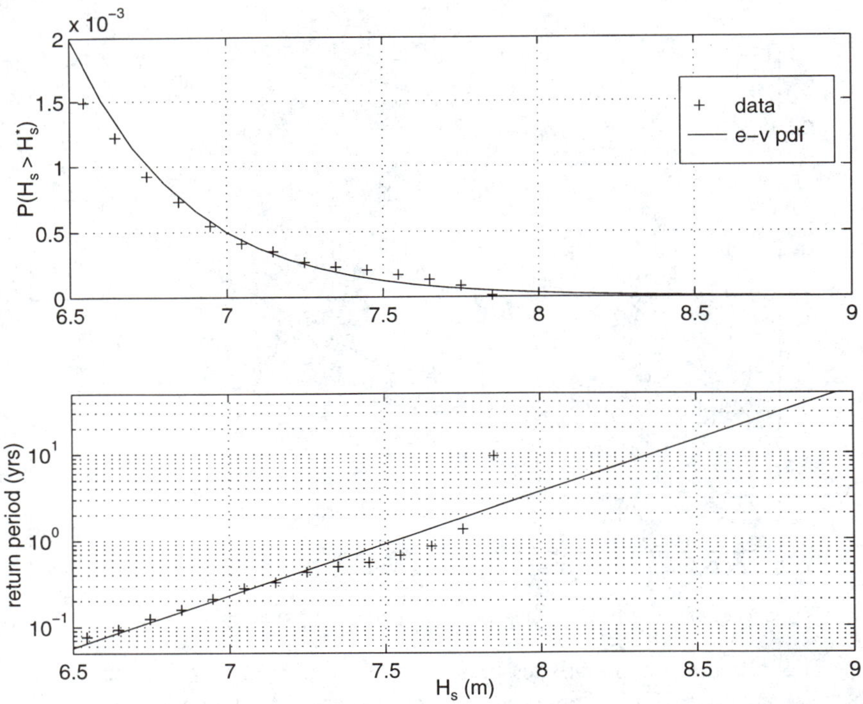

Figure 6: Extreme-value probability distribution and recurrence intervals constructed from 15 years of wave height data obtained by the CDIP buoy offshore from Bandon, Oregon.

Figure 6, as well as results presented by Tillotsen and Komar (in press), clearly demonstrate the intense nature of the wave climate experienced along the Oregon coast. The linear relationship between wave height and extreme runup statistics given in Eq. (2) can now serve as a transfer function to construct an extreme value distribution for $R_{2\%}$ from the distribution of wave heights, and is shown in Figure 7.

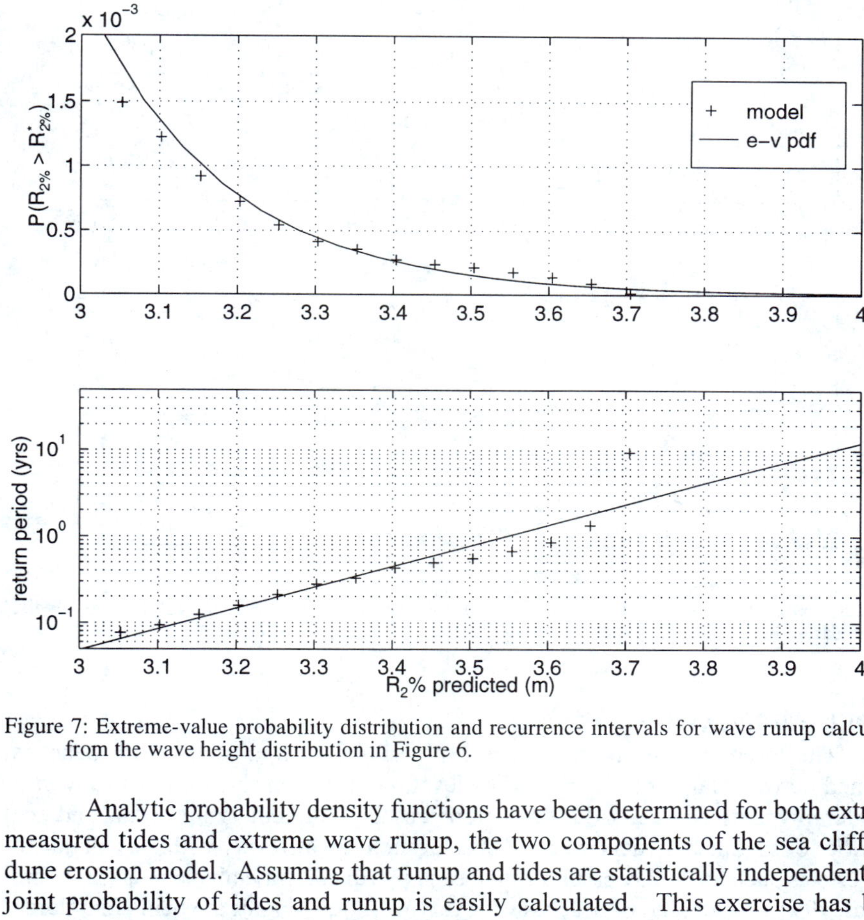

Figure 7: Extreme-value probability distribution and recurrence intervals for wave runup calculated from the wave height distribution in Figure 6.

Analytic probability density functions have been determined for both extreme measured tides and extreme wave runup, the two components of the sea cliff and dune erosion model. Assuming that runup and tides are statistically independent, the joint probability of tides and runup is easily calculated. This exercise has been completed, and contours of equal probability of occurrence have been generated. However, a more useful and direct method for determining the statistics of extreme total water levels is to apply the above model for wave runup to the wave component of the joint time series of waves and water levels. This joint time series is constructed from the time periods in which the wave data and tide data overlap. In doing this, we generate a runup time series which can be added to the measured tide to give a simulated total water level time series. An extreme value distribution is determined from this time series directly, using all total water levels above 2.8 m, and is shown in Figure 8. The recurrence intervals of the total water level reaching or exceeding any elevation can now be determined.

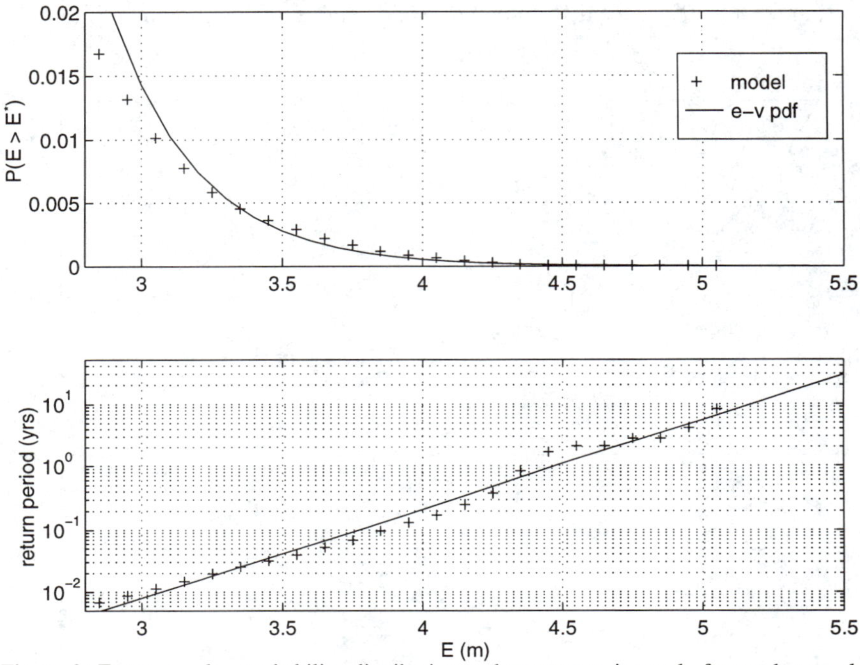

Figure 8: Extreme-value probability distribution and recurrence intervals for total water levels, measured tides plus wave runup.

MODEL APPLICATION

The Oregon coast is divided into a series of littoral cells, with beaches confined between large headlands. The elevation of the beach-face junction varies considerably between cells due to differences in sediment grain sizes, and the quantities of sand in a particular cell. This elevation can also vary within a cell due to local effects such as the lowering of the beach by rip current embayments (Shih and Komar, 1994). A number of beaches with varying morphologies are being monitored in Oregon to determine these elevations relative to NGVD, as well as to quantify typical summer and winter profiles and long term morphology changes. The model has been applied to 13 sites along the central Oregon coast, accounting for 9 of the 13 littoral cells. The beaches are backed by sea cliffs, dunes and shore protection structures and all have beach slopes within the range of applicability of the runup model. The model allows for the determination of the frequency with which the total water level exceeds the elevation of the beach-face junction, thus being a good indicator of potential erosion.

Table 1: Impacts per year as compared to beach stability observations.

Site	Backing Feature	β	E_j *(m)*	Impacts per year (hrs)	Observations
Jump off Joe	Sea Cliff	.034	2.90	173	severe erosion
Nye Beach	Sea Cliff	.034	3.70	13	stable
Beverly Beach	Sea Cliff	.043	4.02	4	erosion
Oceanside	Dune	.023	3.60	18	stable/erosion
South Beach	Dune	.026	4.12	3	accretion
Manzanita North	Dune	.025	4.20	2	stable/accretion
Manzanita South	Dune	.038	6.30	---	heavy accretion
Nestucca Spit	Dune	.046	6.50	---	heavy accretion
C&L Ranch	Sea Wall	.030	3.15	77	severe erosion
Pacific Shores	Sea Wall	.039	3.65	15	erosion
San Marine	Sea Wall	.030	3.75	11	erosion
Pacific Palisades	Sea Wall	.052	5.30	---	accretion
Driftwood Shores	Sea Wall	.033	7.50	---	heavy accretion

Table 1 summarizes the results as applied to the sites on the Oregon coast. The table lists the average winter beach slopes and the average elevations of the junction between the beach and the respective backing feature. The return interval, the fraction of time when the 2% exceedence elevation of runup maxima superimposed on the tide reaches or exceeds E_j has been calculated. However, since we have used all of the data above a threshold value from our simulated total water elevation time series to construct the extreme pdf shown in Figure 8, we can convert return intervals to the more convenient unit of hours of wave runup impact per year. This column in Table 1 gives an estimate of the number of hours per year in which $R_{2\%}$ exceeds E_j for a particular beach, thus being a reasonable proxy for potential erosion at the site. The final column in the table lists basic field observations concerning the morphologic stability of the particular site. The first site, Jump Off Joe, is backed by the remains of a massive landslide which protrudes out onto the active beach profile, hence the very low E_j. The toe of the landslide is actively eroding and is reached by swash during most higher high tides in the winter. Nye Beach, immediately to the south of Jump off Joe, is relatively stable and backed by a vegetated sea cliff. The model predicts that the sea cliff at Jump off Joe will be hit by wave runup much more often than any other site including the cliff at Nye Beach, agreeing well with observations.

SUMMARY AND DISCUSSION

With an appropriate model relating runup elevations to deep-water wave conditions, the frequency of occurrence of sea cliff or dune erosion can be predicted using historical wave and tide records. Coastal regulators, the anticipated users of the model, will have a quantitative method to determine the susceptibilities of properties to erosion and thus a rationale to establish setback distances for coastal developments. The model can also aid in the development of dune-management plans which balance the role of dunes in reducing future property losses versus the development pressure currently being experienced on the Oregon coast. For example, from this data set it appears that shorelines subjected to less than 1 hour of attack per year tend to be stable, while those with more than 10 hours per year tend to experience erosion. In its present form the model may be used to evaluate the need for shore protection structures and may potentially assist in their design. Future model development will attempt to include impact forces in order to predict erosion rates of both sea cliff and dune backed shorelines. The major weakness of the model is the simple relationship between extreme runup statistics and wave height. The model takes no account of the possibility that the functional relationship between runup and wave height may change under more extreme conditions than those measured. The model also ignores large-scale morphology, including offshore bars and bathymetry, as well as large-scale alongshore variability. More runup data are needed on beaches between the dissipative extremes presented here, and the more intermediate to reflective systems to help identify appropriate runup models for a particular beach state.

ACKNOWLEDGMENTS

We would like to thank Rob Holman for the use of his extensive data set. This work is a result of research supported by the NOAA Office of Sea Grant, Department of Commerce. The Agate Beach field experiment and data analysis was supported by ONR Grant # N000-14-94-11196.

REFERENCES

Aagaard, T., 1990, "Swash Oscillations on Dissipative Beaches - Implications for Beach Erosion," *Journal of Coastal Research*, Special Issue No. 9, pp. 738-752.

Battjes, J.A., 1971, "Run-up Distributions of Waves Breaking on Slopes," *ASCE Journal of Waterways Harbor Coastal Engineering Division*, 92, pp. 91-114.

Douglass, S.L., 1992, "Estimating Extreme Values of Run-up on Beaches," *Journal of Waterway, Port, Coastal, and Ocean Engineering*, Vol. 118, No. 2, pp. 220-224.

Gares, P.A., 1990, "Predicting Flooding Probability for Beach/Dune Systems," *Environmental Management*, Vol. 14, No. 1, pp. 115-123.

Guza, R.T. and Thornton, E.B., 1982, "Swash Oscillations on a Natural Beach," *Journal of Geophysical Research*, Vol. 87, No. C1, pp. 483-491.

Holland, K.T., Holman, R.A., Lippman, T.C., Stanley, J. and Plant, N.G. (in press), "Practical use of Video Imagery in Nearshore Oceanographic Field Studies,"*Journal of Oceanic Engineering,* Special Issue on Image Processing for Oceanic Applications.

Holman, R.A. and Guza, R.T., 1984, "Measuring Runup on a Natural Beach," *Coastal Engineering*, Vol. 8, pp. 129-140.

Holman, R.A. and Sallenger, A.H., 1985, "Setup and Swash on a Natural Beach," *Journal of Geophysical Research*, Vol. 90, No. C1, pp. 945-953.

Holman, R.A., 1986, "Extreme Value Statistics for Wave Run-up on a Natural Beach," *Coastal Engineering,* 9, pp. 527-544.

Komar, P.D. and Shih, S.-M., 1993, "Cliff Erosion Along the Oregon Coast: A Tectonic-Sea Level Imprint Plus Local Controls by Beach Processes,"*Journal of Coastal Research*, Vol. 9, No. 3, pp. 747-765.

Muir, L.R. and El-Shaarawi, A.H., 1986, "On the Calculation of Extreme Wave Heights: A Review," *Ocean Engineering*, Vol. 13. No. 1, pp. 93-118.

Nielson, P. and Hanslow, D.J, 1991, "Wave Runup Distributions on Natural Beaches," *Journal of Coastal Research*, Vol. 7, No. 4, pp. 1139-1152.

Shih, S.-M., Komar, P.D., Tillotsen, K.J., McDougal, W.G. and Ruggiero, P., 1994, "Wave Run-up and Sea-Cliff Erosion," *Proceedings 24th Coastal Engineering Conference, American Society of Civil Engineers*, pp. 2170-2184.

Shih, S.-M. and Komar, P.D., 1994, "Sediments, Beach Morphology and Sea Cliff Erosion within an Oregon Coast Littoral Cell," *Journal of Coastal Research*, Vol. 10, No. 1, pp. 144-157.

Tillotsen, K.J. and Komar, P.D., in press, "The Wave Climate of the Pacific Northwest (Oregon and Washington): A Comparison of Data Sources," *Journal of Coastal Research*.

van der Meer, J.W., and C.-J. Stam, 1992, "Wave Runup on Smooth and Rock Slopes of Coastal Structures," *Journal of Waterway, Port, Coastal and Ocean Engineering*, American Society of Civil Engineers, vol. 118, pp. 534-550.

CHAPTER 217

A morphological "mixed-type" model for the Ebro delta coast

Agustín Sánchez-Arcilla and José A. Jiménez[1]

Abstract

A conceptual model to explain and simulate the long-term (decadal) Ebro delta coast evolution is presented as well as the approach to be followed to convert this conceptual model into a numerical one able to quantitatively predict the deltaic coast behaviour. Moreover, a first running sub-model which simulates most of the changes suffered by the outer coast is described in details. This sub-model is a long-term scale version of a one-line coastal model in curvilinear coordinates. The developed model has been verified and predictions of coastline evolution under different scenarios are also presented.

Introduction

The impact assessment of management policies and/or climate change on coastal zones requires comparing the coastal evolution under the present management/climate scenario with respect to the evolution under a different (targetted) scenario. To do this, it is necessary to develop proper tools to predict coastal evolution at adequate temporal and spatial scales.

According to the analysis of scales in coastal morphology done by Sánchez-Arcilla and Jiménez, (1994a, 1996), the representative scale for this kind of tools must be the large spatial scale and the long-term temporal one. In this case, long-term refers to a temporal scale of decades and large scale refers to a length of tens of kilometres up to the entire littoral cell. This implies building models able to adequately reproduce/simulate coastal processes and the corresponding behaviour at this scale, *i.e.* long-term/large scale morphological models.

One of the main questions to solve in large scale coastal modelling is the level of accuracy in the description of the driving terms to be considered. This will determine the

[1] Laboratori d'Enginyeria Marítima (LIM), Catalonia University of Technology (UPC), c/Gran Capitá s/n, 08034 Barcelona, Spain (arcilla@etseccpb.upc.es, jimenez@etseccpb.upc.es).

aggregation level of the processes included in the model. According to deVriend *et al.* (1993) three main approaches exist in large scale modelling:

(i) *input data reduction*, in which it is assumed that large scale processes can be modelled by using process-oriented-models (usually employed to characterise short-term morphological processes) fed by aggregated input data representative of the process to be modelled;

(ii) *model reduction*, in which the model is re-formulated according to the scale of interest (and avoiding shorter scale processes or retaining its integrated effect) and;

(iii) *behaviour-oriented models*, which describe coastal changes without considering the underlying physical processes.

The model here presented is a kind of combination of approaches (i) and (iii) and it is thus called "mixed-type" model. The concept *mixed-type* model refers to a morphological set of equations derived by combining two approaches: (i) the underlying physical processes, particularly at large time scales -*i.e.* subject to an averaging from which only the "residual" effect of small scale processes is retained- and (ii) the actually "observed" morphological evolution, which integrates all "driving" agents into an a priori assumed "predictable" behaviour -from which governing equations could also be derived in an heuristic manner-.

Thus, the aim of this work is to introduce a morphological *mixed-type* model developed for the Ebro delta coast to predict the large scale deltaic evolution. The model is based on the analysis of (i) the existing information on "driving" agents (basically river regime, wind-wave and meteorological climates and mean sea-level fluctuations) and associated physical processes (see *e.g.* Jiménez *et al.* 1996) and (ii) the observed Ebro delta coast evolution at different scales (Jiménez and Sánchez-Arcilla, 1993; Sánchez-Arcilla and Jiménez, 1994; Jiménez, 1996).

After introducing this model, a first sub-model is developed quantitatively. This sub-model deals with one of the main processes involved -the longshore sediment transport- and is used to predict the deltaic coastline behaviour for different scenarios.

Conceptual model

The term conceptual applied in this context refers to a model explaining all the processes governing the Ebro delta coastal behaviour at different temporal and spatial scales as well as their dependence with the acting driving terms and for a defined set of boundary conditions.

This means to develop a model for the Ebro delta coast evolution using all the available relevant information where the coastal response at the different scales is isolated and quantified and, moreover, integrated into a general framework able to explain the coastal evolution as a whole.

This model has been developed for the Ebro delta coast after a detailed analysis of forcing agents, induced processes and associated coastal responses at three main scales: medium-term (Jiménez and Sánchez-Arcilla, 1993), episodic events (Sánchez-Arcilla and Jiménez, 1994) and long-term (Jiménez *et al.*1993). These components have been integrated into a generic model in which the processes are linked through sediment fluxes (across- and alongshore) and the coastal response is aggregated (Jiménez, 1996).

Figure 1 shows a cross-shore section of the model in which the main considered processes are presented. The coast is thus divided alongshore into a series of cells, each one composed by different units linked by sediment fluxes.

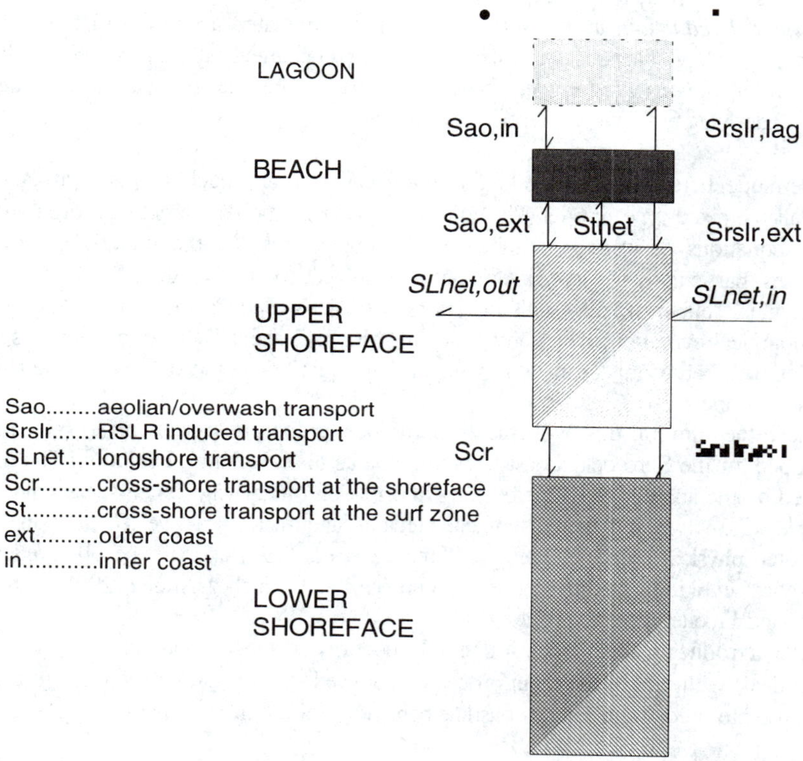

Figure 1. Cross-section of the conceptual model and considered sediment fluxes.

Full details on the conceptual model as well as on the way of quantifying the different sediment fluxes are given in Jiménez (1996).

As a sample description, the most apparent conclusions arising from the developed conceptual model is that, at present times, the Ebro delta coast is subject to a reshaping process in which the net longshore wave-induced sediment transport pattern is the main driving term. This longshore sediment transport pattern (fig. 2), although with a peak frequency located at the medium-term scale (*i.e.* over several years) has a very important residual effect at a longer time scale, since it can be shown that it is also one of the main agents responsible for the sediment redistribution along the coastline on the "long-term" scale (*i.e.* over decades). At this long-term scale, longshore transport will redistribute the sediment along the coast, whereas river sand supplies (severely limited at present), cross-shore sediment exchanges at the shoreface, RSLR induced changes and aeolian and overwash transports (over the dune rows and subaerial coast) will determine the global sediment budget.

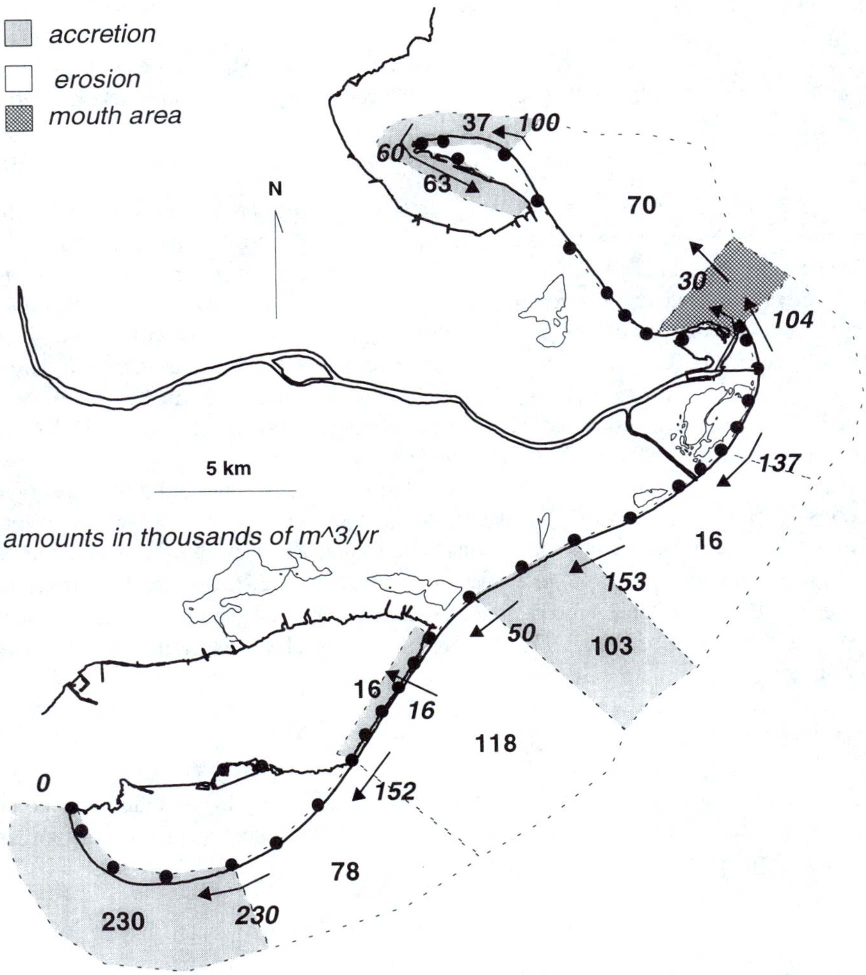

Figure 2. Net longshore sediment transport pattern and associated sediment budget at the medium-term scale (adapted from Jiménez and Sánchez-Arcilla, 1993).

The resulting morphological model has been developed using all the identified "transport mechanisms" relevant for the decadal scale. To estimate the "predictable" coastal evolution, a numerical "box-type model" (*i.e.* based on sediment budget considerations) is being built. This general model is composed by a series of sub-models considering the different processes to be afterwards coupled.

In what follows, the sub-model to predict the coastal "reshaping" due to the existence of a longshore sediment transport gradient (as the only driving term) along the coast is presented and used to simulate/predict the Ebro delta coastline evolution.

Longshore sediment transport parameterization

The net longshore sediment transport rate along any coast, $Sl_{net,i}$, can be expressed as the summatory of contributing transport rates induced by wave conditions acting on that coast,

$$Sl_{net,i} = \sum_{w=1}^{w=n} Sl_{w,ii} f_w \qquad (1)$$

where $Sl_{w,i}$ is the local longshore transport rate induced by the w waves and f_w is the occurrence frequency of those waves, subscript i indicates the position along the coast.

Due to the selected approach - the model to be developed is a combination of *input data reduction* and behaviour oriented models which represent a kind of *result reduction,* equation (1) will be solved at the maximum aggregation level here considered. This approach implies to consider the net longshore sediment transport as the result of a *representative morphological wave, i.e.* integrating the effects of the existing wave climate on the Ebro delta coast from a longshore transport standpoint (see *e.g.* DeVriend *et al.*, 1993 for review of wave data reduction approaches).

The way of parametering sediment transport must be done considering the functional relationship between the type of transport to be modelled and the wave parameters controlling such transport. Since the net longshore sediment transport is the "object" to be parameterized and considering that, in previous studies, the CERC formula has proven to be simple while reasonably reproducing the obtained net longshore transport pattern (Jiménez and Sánchez-Arcilla, 1993; Jiménez, 1996), it has been selected to parameterize this transport.

Transport parameterization

Due to the above mentioned assumptions (maximum aggregation and functional relationship given by the CERC formula), the local net longshore sediment transport in each control point i can be represented by

$$Sl_{net,i} = C K_i H_{b,i}^{2.5} \sin[2(\alpha_{b,i} - \alpha_{s,i})] \qquad (2)$$

where C groupes all the constant terms of the CERC formula, K_i is the CERC calibration factor, $H_{b,i}$ is the "morphological wave height" at breaking in each point along the coast, $\alpha_{b,i}$ is the local wave angle at breaking and $\alpha_{s,i}$ is the local shoreline orientation.

In this approach, the CERC calibration factor, K_i, is permitted to vary alongshore to account for local factors which may influence the transport capacity such as differences in the sediment grain size, sediment availability, beach slope, etc.

Equation (2) is expressed in terms of breaking wave characteristics so that, offshore waves must be propagated towards the coast to evaluate wave parameters at breaking.

This would be equivalent to employing a short term evolution model coupled with a wave propagation model.

In order to obtain a simpler (parametric) formula to be included in a long term evolution model, it will be expressed as a function of deep water wave characteristics as

$$Sl_{net,i} = C K_i Kw_i H_o^{2.5} \sin[2(\alpha_o - \alpha_{s,i})] \quad (3)$$

where Kw_i is a local constant including wave propagation effects. For a "steady" incident wave climate and for coastal changes small enough not to affect wave propagation, the local values will be "steady" as the coast evolves.

Thus, the parameterized net longshore sediment transport rate will be composed by four components: (i) a global constant ($K1 = C H_o^{2.5}$) including all the common terms such as "universal" constants and offshore morphological wave height; (ii) a local constant (K_i) -the calibration factor of the CERC formula- accounting for local effects such as sediment availability, grain size variations, beach slope, etc.; (iii) a local constant (Kw_i) representing the modifications of wave characteristics due to wave propagation and (iv) a sinusoidal function in which the argument varies as the coastline orientation does. In a compressed manner, the net longshore sediment transport rate is given by

$$Sl_{net,i} = Ksl_i \sin[2(\alpha_o - \alpha_{s,i})] \quad (4)$$

where $Ksl_i = K1\ K_i\ Kw_i$.

Transport constants

The application of equation (4) requires to know the net longshore sediment transport rates along the coast of study, Sl_{net}, the actual coastal orientation for which transport rates are estimated, α_s, and, a deep water "representative wave" angle, α_o.

Assuming that the first two variables, transport and orientation are known as it has been already pointed up in the previous section, it remains the problem of the selecting wave angle.

Fig. 3a shows the directional wave climate off the Ebro delta coast in which three main directional components can be seen. From these, the NW component is not relevant for morphological processes due to the coastline orientation since it will represent calm conditions for coastal dynamics (NW waves are generated by local winds blowing offshore). The other two components, E and S, have similar occurrence percentages being Eastern waves slightly more frequent.

In order to see the relative importance of each one of the two components, their energetic content was estimated through their respective offshore wave energy flux. This flux was calculated as the yearly-integrated value of $H_s^2\ T_p$ -being H_s the significant wave height and T_p the peak period- (see Fig. 3b). The obtained pattern of wave energy flux indicates the effectiveness of waves acting on the Ebro delta coast. It is clear that, the "representative" morphological wave, must be close to the East component. This is clearly reflected in the Ebro delta coastal morphology, where the presence of the two spits indicates a net longshore transport directed towards the north northwards of the river mouth, and towards the south southwards of the river mouth.

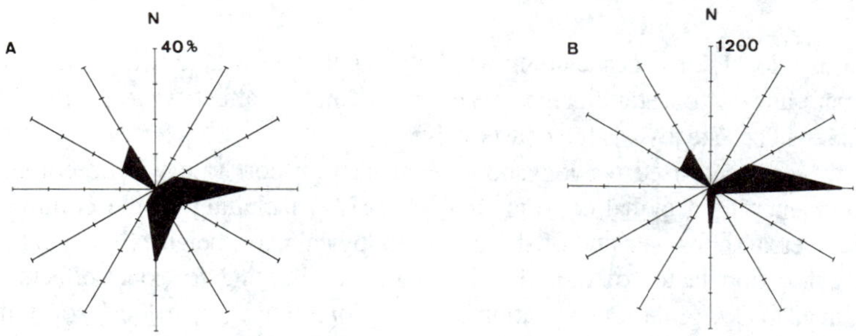

Figure 3. **(a)** Offshore directional wave distribution. **(b)** Offshore directional distribution of the yearly-integrated wave energy flux (calculated as $H_s^2 T_p f$) (Jiménez, 1996).

To select the *representative* long-term morphological wave angle the various wave energy fluxes associated with all effective wave directions (from N to S) were vectorially added, resulting in a "net" wave direction of 100°.

Once this *representative* wave angle was selected, the long-term transport coefficient, *Ksl*, was calculated using equation (4) fed by the estimated net longshore transport rates for the Ebro delta coast from measurements during the period 1988-1992 (Jiménez and Sánchez-Arcilla, 1993) and the coastal configuration for which these transport rates verify. The so obtained *Ksl* values along the Ebro delta coast will reflect all the terms affecting local net longshore sediment transport rates (such as the longshore variation in wave height and wave power along the coast, sediment grain size, etc). Fig. 4 shows the estimated long-term transport constants along the southern part of the Ebro delta coast and their variability as a function of the selected *representative* wave angle. As it can be seen, there are some stretches where the variability in *Ksl* values is high and, even, in some cases -such as the coastal stretch represented by profile 23- it presents a reversal in the transport direction for wave angles greater than the selected one (*i.e.* for waves arriving southwards of 100°).

This variability can also be reproduced by maintaining constant the *representative* wave angle but changing the local coastal orientation and, due to this, it can be used as a measurement of the sensitivity of the local longshore sediment transport to changes in effective wave angle (difference between *effective* wave angle and coastal orientation).

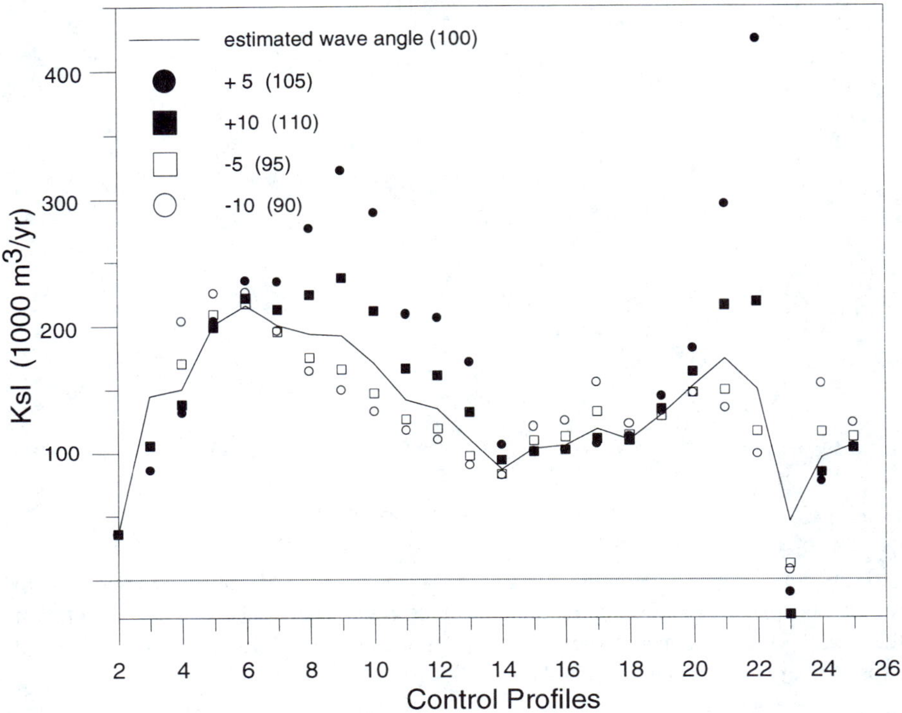

Figure 4. Long-term transport constant *Ksl* along the southern part of the Ebro delta coast as a function of the selected *representative* wave angle.

The coastline model

The effects induced by the net longshore sediment transport pattern along the Ebro delta coast have been modelled through a one-line coastal model.

This first sub-model of the generic conceptual one has been selected due to the good perfomance of this kind of models to reproduce such effects/changes (associated basically to a longshore sediment transport gradient). Moreover, this approach will permit to include other additional effects/processes to complete the ones already considered in the conceptual model by introducing them into the term representing local sediment sinks/sources and by expanding it to a kind of multi-line model.

Since the entire Ebro delta coastline can be considered as significantly curved (locally, with very strong curvatures as is the case of the apex of both spits), the use of an one-line model in Cartesian coordinates will not properly reproduce the coastline behaviour, specially for the most curved stretches. In order to solve this problem a curvilinear coordinate system (following to Le Blond (1972)) will be used. In this system the orientation of any point P of coordinates (x,y) can be written in terms of local coordinates (normal to and tangential to the shoreline), being θ the orientation between the local axes and the original ones (see Fig. 5).

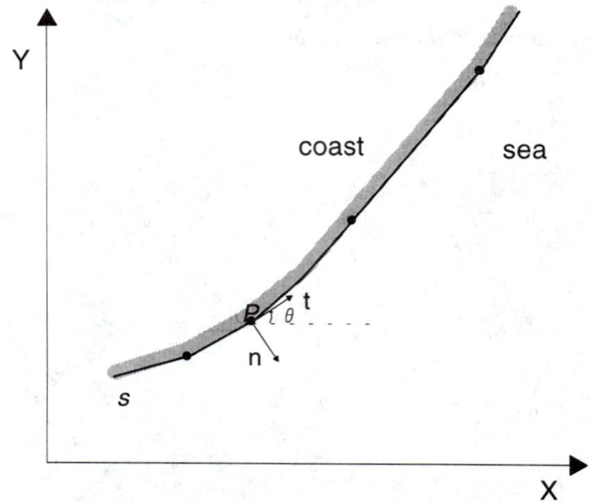

Figure 5. Curvilinear coordinate system.

Using the complex notation $P = x + iy$, in which $i = \sqrt{-1}$ and, assuming that the movement of any point $P(x,y)$ (induced by a longshore sediment transport gradient along the coast, s) will be perpendicular to the shoreline, the continuity equation can be expressed as (see *e.g.* Suh and Hardaway, 1994)

$$\frac{\partial P}{\partial t} = -\frac{1}{d_c}\left(\frac{\partial Sl}{\partial s} + q\right)\exp(i(\theta + \pi/2)) \qquad (5)$$

in which d_c is the closure depth and q a term to include local sediment sources and/or sinks.

Numerical model

The equation (5) is numerically solved by means of an explicit finite-differences scheme using a staggered grid in which shoreline positions are defined in the centre of cells and longshore transport rates are evaluated at cell edges (see *e.g.* Kraus, 1988).

In the selected approach (local coordinates) the cell length, s, is or can be variable along the coast and, to obtain the direction normal to a local point i, an average local orientation is defined considering the orientation upcoast and downcoast of the point of interest.

The finite-differences version of equation (5) for point i along the coast is thus

$$P_{i,j+1} = P_{i,j} - \frac{\Delta t}{d_c}\left(\frac{Sl_{i+1,j} - Sl_{i,j}}{0.5(s_{i+1,j} + s_{,ji})} + q_i\right)\exp\left[i\left(\frac{\theta_{i+1,j} + \theta_{i,j}}{2} + \frac{\pi}{2}\right)\right] \qquad (6)$$

where

$$s_i = \left[(x_{i+1} - x_i)^2 + (y_{i+1} - y_i)^2 \right]^{1/2}$$

The typical time step used in the simulations is of 1 year and the cell length is between 1 and 2 km. The selection of the time step will control the accuracy of the modelling, not only due to stability reasons, but also due to the fact that in any time step the transport rate is updated according to the final coastal orientation and it will be considered as steady during that time step. Thus, the longer the time step is, the smaller the capacity of transport updating will be.

The model requires boundary conditions at both limits of the domain, which in this specific case have been selected (based on the analysis of deltaic coastal behaviour) as *null* transport ($Sl_{net}=0$) in the southern boundary (apex of the spit) and *free beach* in the northern one.

Model verification

In order to verify the developed model, a hindcast of the Ebro delta coastal evolution has been performed. The selected time frame was from 1973 to 1989 in which detailed coastline configurations, obtained from aerial photographs, were available.

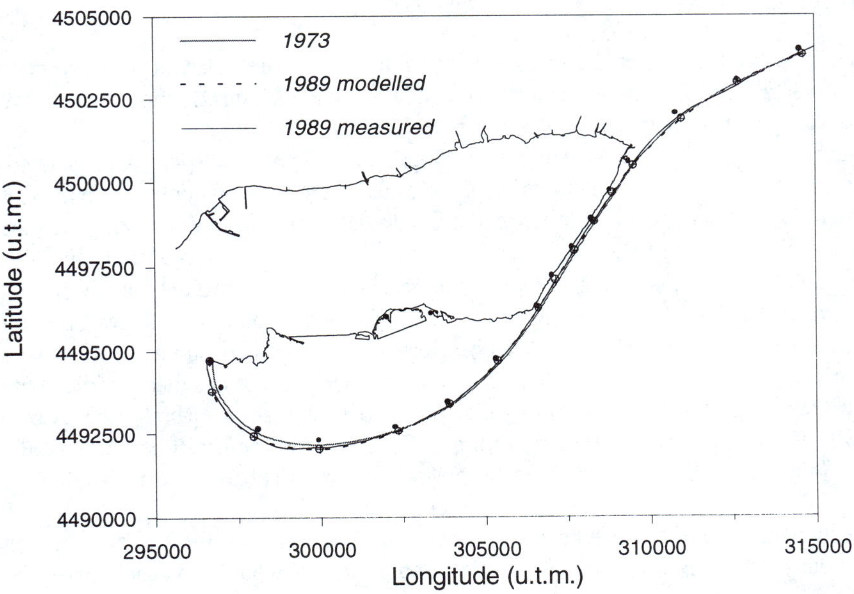

Figure 6. Comparison between modelled and measured coastline evolution between 1973 and 1989 (Jiménez, 1996).

The long-term coastal transport constant, *Ksl,* obtained from data measured during the period 1988-1992 were used together the Ebro delta coastal configuration corresponding to the year 1973. Fig. 6 shows the comparison between modelled and measured coastlines. As it can be seen, both coastlines are nearly coincident, with deviations measured in terms of the relative error ($|\Delta x_{mes} - \Delta x_{mod}| / |\Delta x_{mes}|$), giving an average relative error of about 15%. This magnitude can be considered as reasonable for the scale of modelled processes.

In any case, the comparison of measured and modelled deltaic coastlines cannot be considered as a model calibration in the sense that no parameter has been fitted. It should be stressed that, the values of the constant (*Ksl*) along the coast, have been obtained from the analysis of a period (1988-1992) not included in the verification interval (1973-1989).

This fact can be used as a clear indication that the net longshore transport pattern along the Ebro delta coast is the main driving term/process governing the outer coastal behaviour at this scale. The estimated net transport rates for a four year period, which according to the analysis of Jiménez and Sánchez-Arcilla (1993) are representative of the medium-term evolution (over a period of several years), will have a "residual" effect on the long-term/large scale coastal behaviour which is reflected in the observed coastal reshaping (although not contributing to the overall sediment budget). Thus, for the Ebro delta coast, if net longshore sediment transport rates at the medium-term scale are available, and the incident wave climate is steady, the large scale/long-term coastline behaviour can be reasonably predicted by considering the changes in coastal orientation.

Ebro delta coastline evolution under different scenarios

Fig. 7. shows the predicted coastline evolution for the southern part of the Ebro delta from 1989 to 2050 assuming that present conditions (river sediment supply and wave climate) are steady during all that period.

The predicted evolution, in which only longshore sediment transport is considered, shows a spit growth similar to the one observed under present conditions. Moreover, it also shows that the outer coast along the Trabucador Bar will shift landwards towards the present backbarrier coast. In this particular case, it must be taken into account that this implies that this barrier is "sandy rich", *i.e.* no breaching is considered and, in any case, it is assumed that behind the eroding barrier there is always enough sand available. Of course, this will only occurr if the effects of overwash processes which permit the barrier rollover are included. Although in this sub-model overwash is not included, existing data on the Trabucador Bar evolution during the last decades, have shown a similar behaviour. Fig. 8 shows thus the barrier evolution from 1957 to 1989 where it can be seen that the barrier outer coast in 1989 occupies the position of the backbarrier corresponding to 1957.

Additionally, the model results can also be used to characterise the barrier as a high vulnerability area in the sense that, if overwash is prevented, the barrier will be prone to be breached due to the existing positive gradient in the net longshore sediment transport pattern.

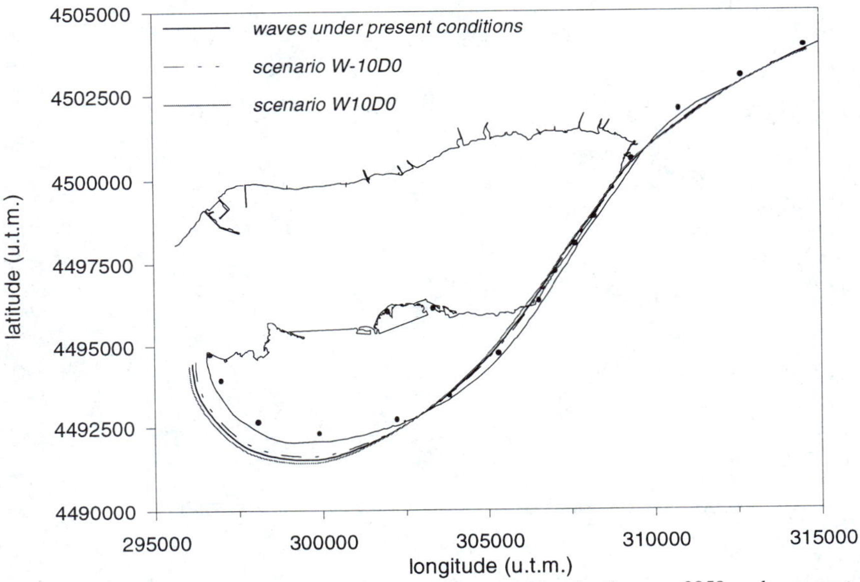

Figure 7. Prediction of the Ebro delta coastline evolution for the year 2050 under present conditions and assuming two scenarios of wave heigth change (*W10D0*: 10% increase in wave heigth, *W-10D0*: 10% decrease in wave heigth).

The potential of this kind of models to predict some climate effects on coastal evolution can be assessed from Fig. 7. Two climate scenarios, in terms of changes in wave height, have been defined: (i) scenario W10D0 where an increase of 10% in wave heigth is assumed and (ii) scenario W-10D0 with a decrease of 10% in wave heigth.

To account for these variations in wave heigth, the local transport constant, *Ksl,* has been modified for these two scenarios as follows:

$$Ksl_{\,(\mathrm{W10D0})} = Ksl\,(1.1)^{2.5}$$
$$Ksl_{\,(\mathrm{W\text{-}10D0})} = Ksl\,(0.9)^{2.5}$$

As it can be seen from Fig 7, the coastline evolution under both scenarios is similar same from a qualitative standpoint although differing in the magnitude of the predicted shoreline displacements according to the assumed variation in wave heigth (larger displacements for an increase in the wave heigth because it implies larger gradients in longshore sediment transport and viceversa).

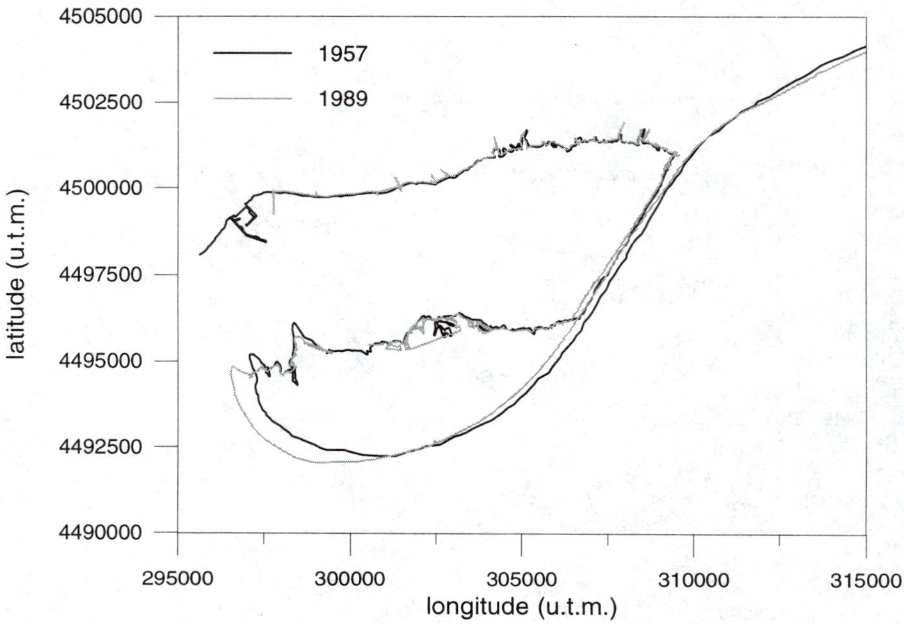

Figure 8. Evolution of the Trabucador Bar between 1957 and 1989 (modified from Sánchez-Arcilla and Jiménez, 1994b).

Summary and on-going works

A conceptual model to explain and simulate the long-term Ebro delta coast evolution has been presented as well as the approach to be followed to convert this conceptual model into a numerical one able to quantitatively predict the deltaic coast behaviour.

From all the considered modules, a first running sub-model which simulates most of the changes suffered by the outer coast of the Ebro delta, has been described in details. Although this sub-model only consideres coastal changes induced by the existence of longshore sediment transport gradients, it has been proven to reproduce most of the observed changes along the Ebro delta coast. This is due to the fact that longshore sediment transport is the main process governing the evolution of the Ebro delta coast at the long-term scale. However, other processes are also important, specially for determined coastal stretches in order to fully explain the coastal behaviour and not only the coastline evolution (*e.g.* overwash processes to explain barrier rollover).

In this way, other differents sub-models are being developed and included into a general model able to reproduce and simulate the large scale/long-term evolution of the Ebro delta coast. A first specific module to simulate the effects of overwash, which combined with the effects on the outer coast is able to reproduce different barrier behaviours is nearly operational. The inclusion of this process will permit to eliminate the assumption of a sandy-rich barrier (which in fact implies a quasi-infinity barrier width) as it has been used in the example shown here (Fig. 7). It will be then possible to simulate the evolution of both barrier coastlines (outer and inner) and this will permit identifying the barrier vulnerability conditions (leading to a barrier prone to breaching).

At present, all other processes considered in the conceptual model have been included in a parametric way (*i.e.* including them in the model as sediment sinks and sources), although a process-oriented approach is being followed to describe these terms in a dynamical manner.

Acknowledgements

This work was carried out as part of the MEDDELT and PACE projects. It was supported by the EU Environment and Marine Science and Technology (MAST-III) Programmes, under Contracts No. EV5V-CT94-0465 and No. MAS3-CT95-0002 respectively. This work was also co-sponsored by the Spanish Ministry of Education and Research through DGICYT (UE95-0033). JAJ was partly supported by a travel grant given by the DPTOP (Generalitat de Catalunya) through the ETSECCPB (UPC). We also thank to Marcel Stive and Michele Capobianco earlier fruitful discussions on some parts of this work.

References

De Vriend, H.J., Capobianco, M., Chesher, T., De Swart, H.E., Latteux, B. and Stive, M.J.F., 1993. Approaches to long-term modelling of coastal morphology: A review. *Coastal Engineering*, 21: 225-269.

Jiménez, J.A. 1996. *Evolución costera en el Delta del Ebro. Un proceso a diferentes escalas de tiempo y espacio.* Ph.D. Thesis, Catalonia Univ. of Technology, 274 pp.

Jiménez, J.A. and Sánchez-Arcilla, A., 1993. Medium-term coastal response at the Ebro delta, Spain. *Marine Geology*, 114: 105-118.

Jiménez, J.A., Sánchez-Arcilla, A., Valdemoro, H.I., Gracia, V. and Nieto, F. 1996. Forcing terms controlling reduction processes in the Ebro delta. *Marine Geology* (in review).

Kraus, N.C., 1988. Shoreline models. In: K. Horikawa (Ed.), *Nearshore dynamics and coastal processes.* Univ. of Tokyo Press, 321-366.

LeBlond, P.H., 1972. On the formation of spiral beaches. *Proc. 13th ICCE*, ASCE, pp. 1331-1345.

Sánchez-Arcilla, A. and Jiménez, J.A., 1994a. Ingeniería de playas (I): Conceptos de morfología costera. *Ingenieria del Agua*, 1(2): 97-114.

Sánchez-Arcilla, A. and Jiménez, J.A., 1994b. Breaching in a wave-dominated barrier spit: The Trabucador Bar (NE Spanish coast). *Earth Surface Processes and Landforms*, 19, 483-498.

Sánchez-Arcilla, A. and Jiménez, J.A., 1996. Physical impacts of climatic change on deltaic coastal systems (I): an approach. *Climatic Change*, (in press).

Suh, K.D. and Hardaway, C.S. 1994. Calculation of tombolo in shoreline numerical model. *Proc. 24th ICCE*, ASCE, 2653-2667.

CHAPTER 218

SEA DIKE EROSION AND COASTAL RETREAT AT NAM HA PROVINCE, VIETNAM

Ton That Vinh[1], G. Kant[2], Dr. Nguyen Ngoc Huan[3], Dr. Z. Pruszak[4]

ABSTRACT

Nam Ha Province is situated on the coast of the Red River Delta, immediately east of the mouth of the Red River. The coast of the delta is very dynamic in which large quantities of sediment are supplied by wet season run-off. However, in Nam Ha province the coast is retreating such that successive lines of dikes have been abandoned and a unique 'managed retreat' strategy is underway with double lines of dikes along large parts of the province's coast. This paper reviews and describes the possible causes of the erosion problem and highlights the unique way in which the local coastal engineers are adopting, out of necessity, the classical retreat strategy. Other impacts of the problem such as environmental and socio-economic impacts are also reviewed. Based on the qualitative first order analyses a strategy has been defined towards a sustainable development of this coastal area.

SITE DESCRIPTION

The Delta of the Red River is located in the northern part of Vietnam (Figure 1). The Thao River rises in the Dai Ly Lake in China. In Vietnam, near the town of

---

[1] Ministry of Agriculture and Rural Development, Department of Dike Management and Flood Control, 23 Hang Tri Street, Hanoi, Vietnam.

[2] Frederic R. Harris B.V., Badhuisweg 11, P.O. Box 87875, 2508 DG The Hague, The Netherlands.

[3] Marine Hydrometeorological Centre of HMS, Lang Trung, Dong Da, Hanoi, Vietnam.

[4] Polish Academy of Sciences, Institute of Hydroengineering, Koscierska 7, 80-953 Gdansk, Poland.

Viet Tri, the Thao River becomes Red River and receives flow from two other major contributories: the Da River and the Lo River. The Red River Delta has the classical triangular form downstream of Viet Tri. East of the Red River the Thai Binh River flows onto the Red River Delta. The Red River catchment area is about 155000 km^2 and reaches into China and Laos, resulting in an average annual discharge of $137 * 10^9$ m^3/year, and a total estimated sediment load of about $72 * 10^6$ t/year to the Gulf of Tongking (World Bank, 1994). The sediment is discharged through a large number of branches over a coastline with a length of some 165 km (as the crow flies). The enormous amount of sediment in general leads to accretion in the coastal areas for which the growth of the Red River Delta itself is the most outstanding proof and example. However, despite this fact there are local stretches of coast on which serious erosion problems are reported. Two provinces where such erosion problems exist are the provinces Thai Binh and Nam Ha. Surprisingly both provinces are directly located at Cua Balat, the mouth of the main branch of the Delta: the Red River itself (Figure 2).

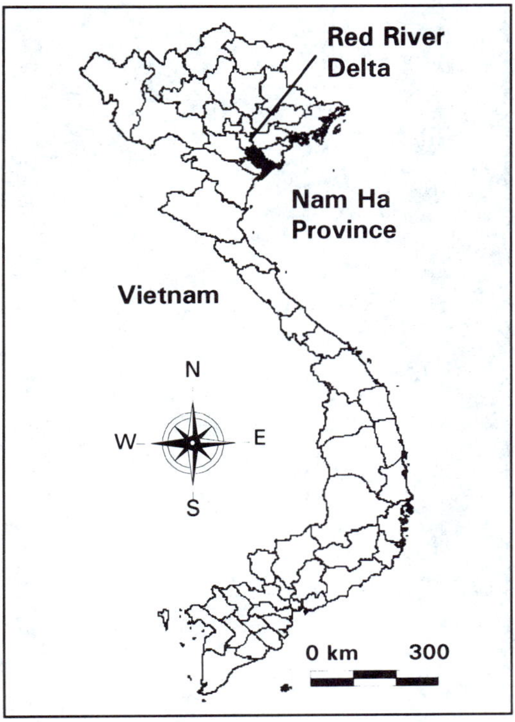

Figure 1 *Vietnam*

Nam Ha province is bound by the Red River in the north-east and by the Day River in the south west. The coast length is about 60 km and generally faces E to SE. The province has three coastal districts which are separated by branches of the Red River (Figures 2 and 3). One of the branches, Ngo Dong River was dammed in 1955. The middle district, Hai Hau district, experiences a serious consistent widespread erosion. About 75% of the total of 25 km coastline of Hai Hau district is retreating.

The coastline of Hai Hau district is being protected by dikes. In front of the dikes is a very gentle sloping beach. The slope of the beach varies along the coastline from 1:40 on eroding beaches to 1:200 in other places. The beaches consist of very fine sand with an average grainsize of about 80 µm. Measurements show an increase of the grainsize on the eroding sections of the coast.

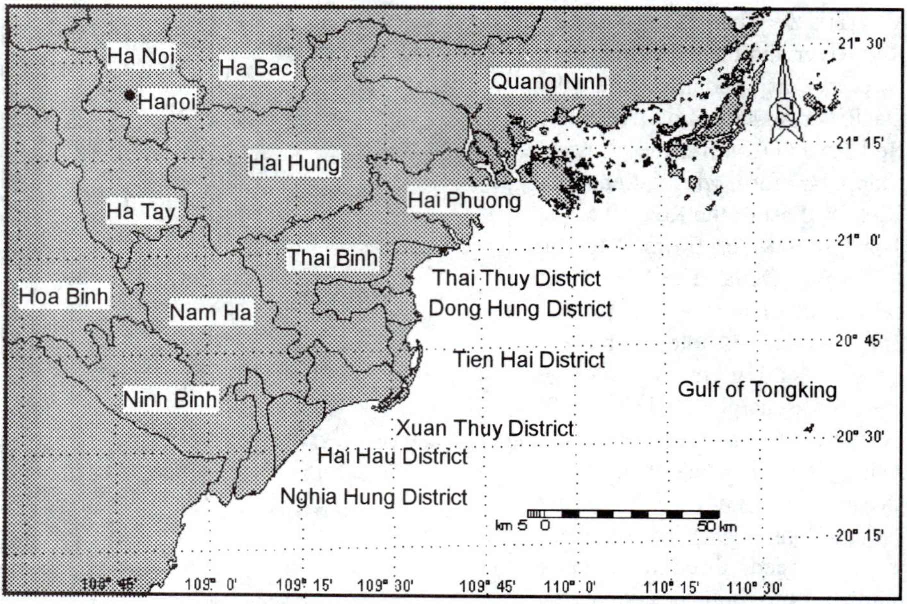

Figure 2 *Red River Delta Provinces and Coastal Districts Nam Ha Province*

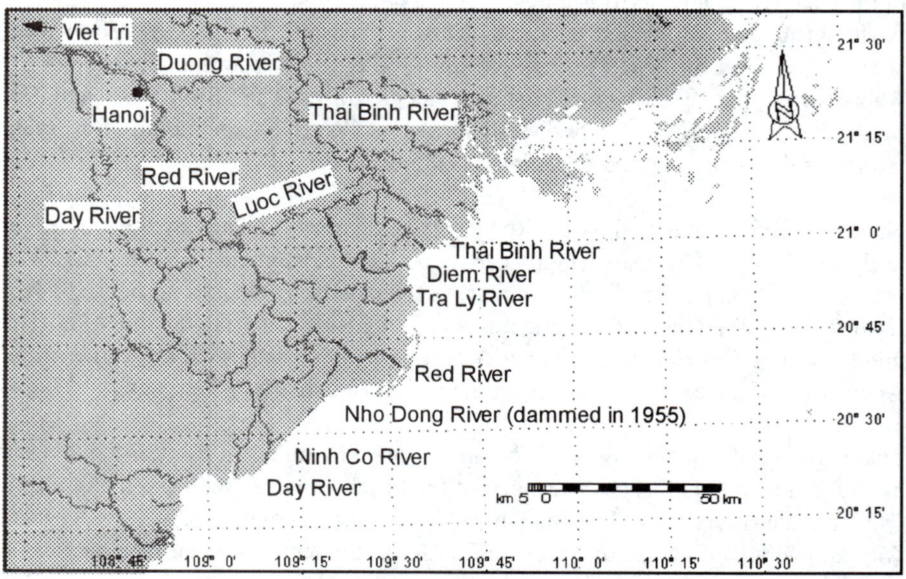

Figure 3 *Red River Delta River Branches*

HISTORIC SHORELINE DEVELOPMENT

The coastline development this century (from 1905 to 1992) in Nam Ha province is presented in Figure 4. It can be seen that the development of the coast is irregular. Near the mouths of the Red River (right hand side of the figure), the Ninh Co and the Day River (both on the left hand side of the figure) progressive sedimentation has led to accretion. This results in shoreline developments in these areas of 2.5 km to 6.5 km in seaward direction over the 87 year period. However, in Hai Hau district the development is just the opposite. Over the last 87 years a strip some 16 km long and about 2.5 km wide was lost to the sea. This is a time averaged shoreline retreat of about 29 m/year. It appears from this figure, which is based on maps, that the erosion at Hai Hau district started in the beginning of the century and seems to be slowing down after 1966. This indicates that the damming of the adjacent Ngo Dong River in 1955 is not the main reason of the consistent erosion problems.

LOCAL RESPONSE STRATEGY

The present coastal defence at Hai Hau district consists of a double dike system (Figure 5). These parallel dykes, with a separative distance of about 250 m, are constructed along 75 % of the Hai Hau coastline. The area between the dykes has been split into sections alongshore in the range of 500 to 3000 m long. The beach in front of the first dike reaches to about MSL. The slope of the dike is about 1:2 (v:h) and is protected by pitched limestone rock of about 30 cm diameter placed on a clay layer. The crest height is approximately 4 m above MSL. The dike body itself consists of clay and sand. On heavily attacked sections the toe of the dyke is reinforced with cement grouting.

Since the erosion is progressing continuously the first defence line is becoming weaker every year. During extreme events the first dike is therefore at some points not strong enough to withstand the attack. In

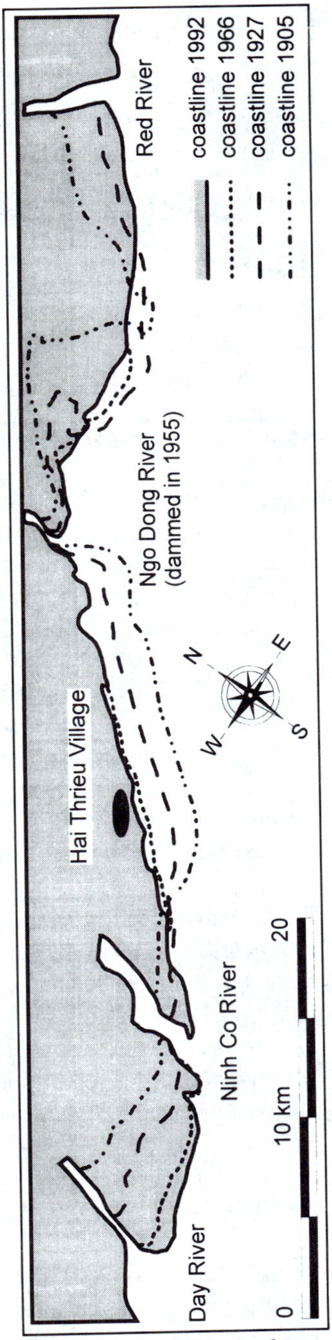

Figure 4 *Shoreline Development*

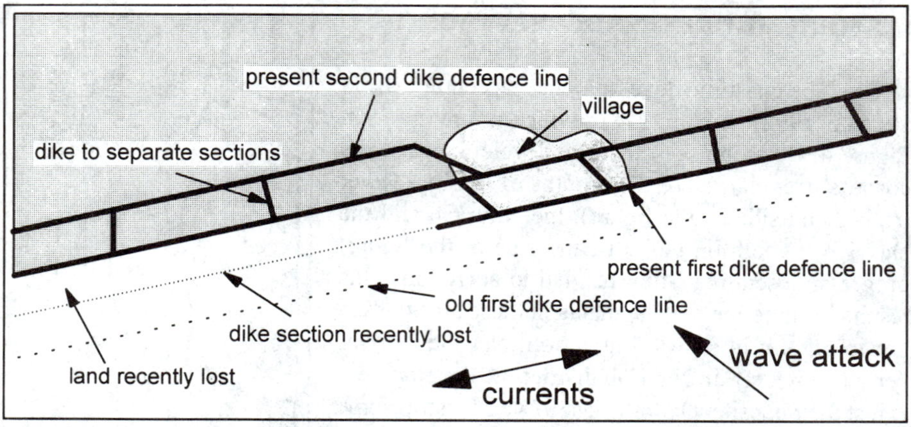

Figure 5 *Double Dike System*

order to prevent inundation the second defence line has been built. When the first dike fails the section behind the first dike is inundated and the second dike becomes the new first dike. Every time the first dike fails a section of land is lost to the sea since there are no resources to repair the old dike. In response to such a breakthrough a new second dike is built behind the new first defence line. On average every ten years a dike section is lost to the sea. This is a typical example of a managed retreat strategy although the Vietnamese authorities are forced by the circumstances. Beside the weakening of the dikes by the continuous erosion, the strength of the first defence line itself is in general insufficient due to problems such as funding and practical construction problems (such as accessibility of roads and availability of vehicles). This results in generally weak dykes, which are also too low, causing severe overtopping problems and revetment instabilities.

As the retreat of the coastline is known for many years evacuation plans are readily available. These plans are started on average every 2-3 years in case of high flooding risks. For maintenance and construction of the sea defences the authorities mobilise local inhabitants. As in the whole country the population has to invest a number of days per year in the community. The national average is about 10 workdays per year. In villages endangered by the erosion able bodied workers are required to work 40 days per year on dike repair, maintenance and construction.

ENVIRONMENTAL BOUNDARY CONDITIONS

The meteorological conditions of the area can clearly be divided into a summer (July and August) and winter (October to March) monsoon period. During the summer monsoon period moderate winds (average force: 3 Beaufort) from south western directions prevail. In winter the predominant wind direction is north east (average wind force about 4 Beaufort). The summer monsoon period is also known as the 'wet season'. During this season 80% of the yearly run-off is discharged through the Red River system. Peak discharges are to be expected towards the end of the wet season or at extreme storm or typhoon events. On average two typhoons per year hit the coastline of the northern provinces of Vietnam. The typhoon season starts in May and lasts till October with the highest activity in the months July and August.

Similarly, the wave climate is also strongly influenced by the monsoon periods which is typical for this region of the world. The average deep water wave height in the summer months is about 1.5 m (coming from SW directions) whereas the average deep water wave height in winter is about 2 m (NE directions). However, during storms and typhoons the deep water significant wave height can develop to 10 m.

The characteristic range of the diurnal tide at spring tide is about 2.4 m. Currents along the coast are mainly tidally driven with a significant (seasonal) influence of wind and waves. During the last 30 years water levels (astronomic tide plus wind set-up) up to 3.00 m above MSL have been reported in bays and inlets along the shallow coast.

QUALITATIVE MORPHOLOGICAL ANALYSIS

The morphological system of the Red River Delta is very dynamic. During the wet summer monsoon season 80% of the annual sediment load is discharged to the Gulf of Tongking resulting in huge deposits near the River mouths. As the prevailing wave and wind direction is south west most of the sediment will be deposited on the eastern side of the river mouths (Figure 6). Wave conditions in the summer season can

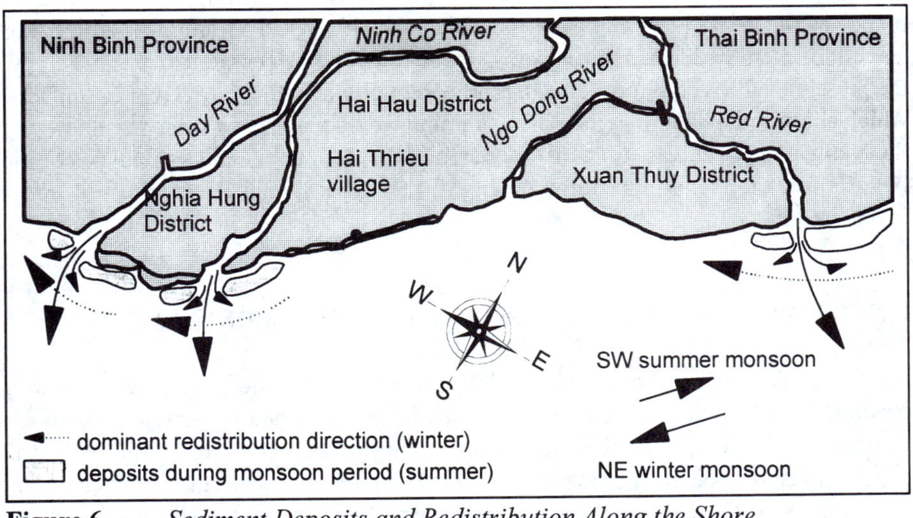

Figure 6 *Sediment Deposits and Redistribution Along the Shore*

be characterized as 'moderate' resulting in some redistribution of sediment in north-eastern direction. However, during the remainder of the year wind and wave conditions from north-eastern directions prevail. Compared to the summer monsoon the north-east monsoon is somewhat stronger and more persistent resulting in higher longshore sediment transports. During the winter season a substantial part of the fresh deposits will therefore be transported further offshore or be redistributed along the shore in south-western direction. The beach sections between the river mouths (beyond the direct influence of the river mouths) are therefore dependent on the redistribution mechanisms of deposited sediment from the nearest river branches.

Up to 1955 the Hai Hau district beaches were nourished by the Ngo Dong River in the north and the Ninh Co River in the south. The net sediment transport is directed to the south-west so the main source of sediment was the Ngo Dong River. With the damming of the Ngo Dong River the main sediment source was blocked resulting in erosion south of the River mouth. From then on the beaches of Hai Hau district had to be nourished from either the Red River or the Ninh Co River mouth (Figure 6). However, due to the shape of the Red River mouth hardly any nourishment took place from the north-east. Nourishment from the Ninh Co River could only develop during the summer months. From the coastline development graph (Figure 4) it can be seen that indeed some nourishment took place resulting in a stable or accreting development of the southern beaches of the Hai Hau district.

However, the erosion of the Hai Hau coastline did not start after the damming of the Ngo Dong River. The erosion started at the beginning of the century or even earlier. The reported erosion rates of 29 m/year over a length of 16 km indicate the Ngo Dong River must have been a major branch of the Red River. A possible explanation would be that at the end of the 19th century the Ngo Dong River lost its importance as a major branch of the Red River. This process, which can either be a natural development or a result of man made channelling works, redirected the discharge to, most probably, the present Red River mouth. The discharge through the Ngo Dong River diminished which allowed the damming in 1955. Damming would have never been possible when the Ngo Dong River was still a major branch of the Red River. This hypothesis of the historic development and importance of the Ngo Dong River is still under investigation.

QUALITATIVE PREDICTION OF THE SHORELINE DEVELOPMENT

Based on the above described hypothesis of the morphological system a prediction of the shoreline development can be made. When no action is taken the coastline will retreat at same pace for the coming years. On a longer timescale it is expected that the accreting coast of the Nghia Hung District will, due to the changing coastline orientation, develop an increasing nourishing effect on the Hai Hau coastline.

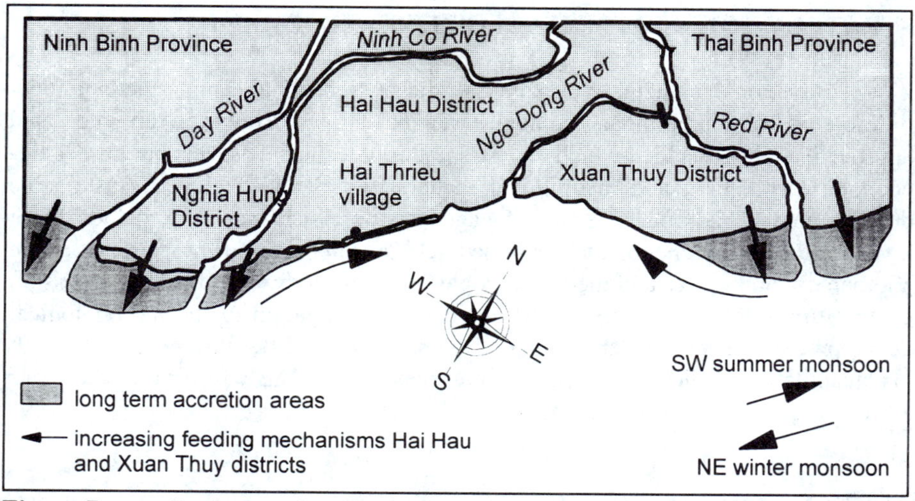

Figure 7 *Qualitative Prediction of Long Term Coastline Development*

The accretion near the mouths of the Ninh Co and Day River will slow down and the retreat at Hai Hau will decrease (Figure 7). This effect will develop along the coast starting in the south of the district.

SOCIO-ECONOMIC IMPACTS

Within Vietnam the Red River Delta (16,000 km^2) is the most densely populated region with 1,700 inhabitants per square kilometre. This is also among the highest population densities on arable land in the world (VVA, 1996). The erosion at Hai Hau district causes significant socio-economic impacts. The land in the coastal strip is completely in use for agriculture, aquaculture and salt production. Further, because of the progressing retreat, a number of villages are at risk being located directly behind the sea defence. Hai Thrieu for instance is a commune of which part of the 8000 inhabitants live between the first and second sea defence. At present the second dike is being reinforced as the authority reluctantly have to abandon the first dike in case of a breakthrough. Moving this part of the village requires resources and will include a social problem since practically all the land is presently in use. A more general problem of the coastal region is salt intrusion via overtopping of dikes. In this way storms and typhoons can impact a strip of 750 m wide along the coast with salt (eg. typhoon Wayne, 1986). It takes years after such an event before the original crop yield can be reproduced.

STRATEGY TOWARDS SUSTAINABLE DEVELOPMENT

Based on this qualitative analysis of the erosion at Hai Hau district a number of measures and preliminary strategies can be defined. However, an integrated approach is recommended which includes not only technical aspects but also socio-economics, planning and institutional matters in order to achieve a sustainable development. This Coastal Zone Management approach needs to include more investigations and studies on the mentioned fields in collaboration with the responsible Vietnamese authorities and the local people. The key study would be, in succession on this first qualitative analysis, the extensive investigation of the morphological development. A sound understanding of the processes and developments will result in a quantitative prediction of the shoreline development. This will form the basis for all further strategy decision making.

ACKNOWLEDGEMENTS

The authors gratefully acknowledge the support of DGIS (Netherlands Ministry of Foreign Affairs), sponsors for this work and the Vietnamese Government (Hydrometeorological Services, the local host for this project and the Ministry of Agriculture and Rural Development), for permission to publish these findings.

REFERENCES

VVA (1996); *Vietnam Coastal Zone Vulnerability Assessment, various reports*; Government of the Netherlands; Ministry of Foreign Affairs; Government of the Socialist Republic of Vietnam, Hydrometeorological Services.

World Bank (1994); *Red River Delta Master Plan*; Ministry of Science, Technology and Environment; UNDP; World Bank.

CHAPTER 219

BEACH PROFILE ANALYSIS AROUND INDIAN RIVER INLET, DELAWARE, U.S.A.

Kirk F. Bosma[1] and Robert A. Dalrymple[2]

ABSTRACT: The purpose of this study is to examine the recent shoreline history at Indian River Inlet, Delaware using beach profile data. Indian River Inlet was stabilized in the late 1930's with two parallel rubble mound jetties. The stabilization resulted in considerable modifications to the surrounding beach environment, leading to the construction of a sand bypassing system in 1990. The profiles adjacent to the inlet are first examined with the standard tools, such as shoreline change, volume change, and sand budget. Then, details are given on the use of Principal Component Analysis (PCA) in the coastal field, an explanation of its function, and the differences between the complex and non-complex versions. Results are presented for areas north and south of the inlet in both 2-mode and 3-mode versions of Complex Principal Component Analysis (CPCA). The results illustrate the ability of CPCA to detect moving features within the profile data, including its direction and speed, as bypassed sand is seen moving to the north.

INTRODUCTION

The Delaware Atlantic coastline, a sandy shore that spans approximately 24 miles, is an area of constant transformation. Significant littoral drift rates, man-made structures, remediation efforts, and occasional battering by large storms all effect the coastline. Thus, the beach profiles are 'dynamic' in character, changing continuously. These profile variations occur in both the subaqueous and subaerial elements of each profile, as wave energies constantly move sand on, off, or alongshore. In effect, changes in profiles can reveal a vast amount of information, both long and short term, about the coastline.

The focus of this study is the profiles adjoining Indian River Inlet (as shown in Figure 1), which is one of the most unique features along the Atlantic Coast of Delaware. After several failed attempts to keep the Indian River Inlet open by dredging alone, a 152 meter (500 foot) wide inlet was constructed in the late 1930's. The goal being to establish a stable passage way from the inner bays (Rehoboth and Indian River) to the Atlantic Ocean, increase bay salinity, reduce stagnation, and increase the tide range (Thompson and Dalrymple, 1976). The

[1]Graduate Student, Center for Applied Coastal Research, University of Delaware, Newark, DE 19716, USA.

[2]Professor and Director, Center for Applied Coastal Research, University of Delaware, Newark, DE 19716, USA.

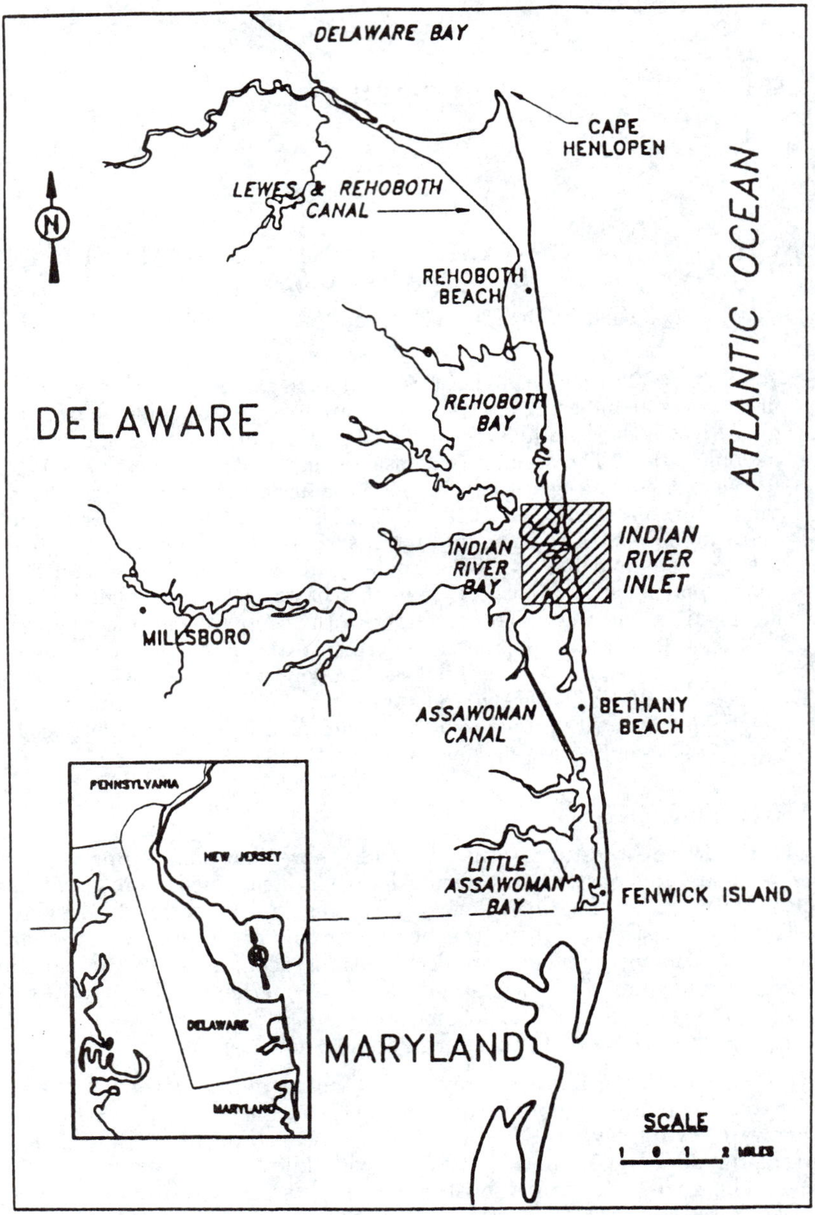

Figure 1: The Atlantic Coast of Delaware

inlet is stabilized by two parallel rubble mound jetties, originally extending a distance of 472 meters (1550 feet). 185 meters (600 feet) of this length extended seaward of the ocean shoreline at the time of construction. There have been significant problems with the engineered inlet over the years, including erosion of channel banks west of the jetties, accelerated scour along the jetties, and massive downdrift erosion, due to the predominant northward drift (Gebert *et al.*, 1992). The main concern related to the downdrift erosion is the threat it poses to the Route 1 highway traveling parallel to the shoreline. In fact, by 1954, a dune scarp had been created that was, in places, less than 60 meters (200 feet) from the roadway. From 1957 to 1990, mitigation of the beach erosion was accomplished by dredging of the flood tidal shoal and back barrier deposits. Approximately 380,000 m^3 (49,700 yd^3) of sand was placed on the north beach approximately every five years. Since February of 1990, however, a fixed sand bypassing system was constructed to pump sand from the southern shore and "slurry" it across the inlet to the northern shore. The system mines the south accretional fill by using an eduction unit deployed by a crane. Through May of 1995, approximately 350,000 m^3 (456,000 yd^3) have been pumped across the inlet at a cost of \$2.11 per m^3 (\$1.62 per yd^3). The system is performing well and is relatively inexpensive to run.

The objective of the present study is to investigate many aspects of the beach profile at Indian River Inlet through the use of the newest analytic methods and the most recent field data available. We want to determine what the profiles north and south of the inlet reveal about the region and to answer questions such as:

- What happens to the bypassed sand?
- Is enough sand being pumped?
- Can we identify moving forms or sandwaves?

To do this, we consider the early performance of the bypassing plant through shoreline changes, evaluation of littoral drift, and other standard analytical tools. We also explore the use of complex principal component analysis (CPCA), in both 2-mode and 3-mode versions, to evaluate propagating features that exist within a bathymetry.

FIELD DATA COLLECTION

Profile data sets are considered on both the north and south side of Indian River Inlet collected by the U.S. Army Corps of Engineers, Philadelphia District. The profiles span from 1984 to 1994, with an average of two surveys per year. 28 profile lines are taken in the range from 1524 meters (5000 ft) south of the inlet to 1524 meters (5000 ft) north of the inlet, as shown in Figure 2. 17 of the lines are located in the northern portion of the study area, while 11 are located in the south. As shown in Figure 2, some profile lines extend far offshore, while others only advance to the water line. The station numbers provided are the distance in hundreds of feet (30.5 m), from the respective jetty centerline. Survey points were taken randomly during each survey, thus requiring linear interpolation in both the alongshore and cross-shore directions for much of the analysis. Examples of the bathymetries north and south of the inlet are shown in Figure 3. Notice the scour hole that is evident at the tip of the northern jetty.

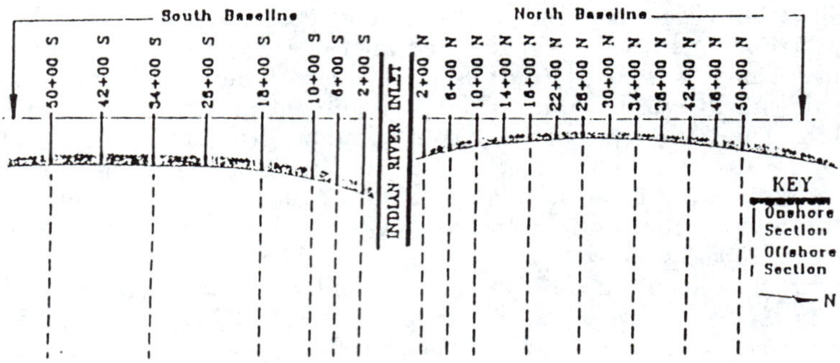

Figure 2: Profile lines north and south of Indian River Inlet

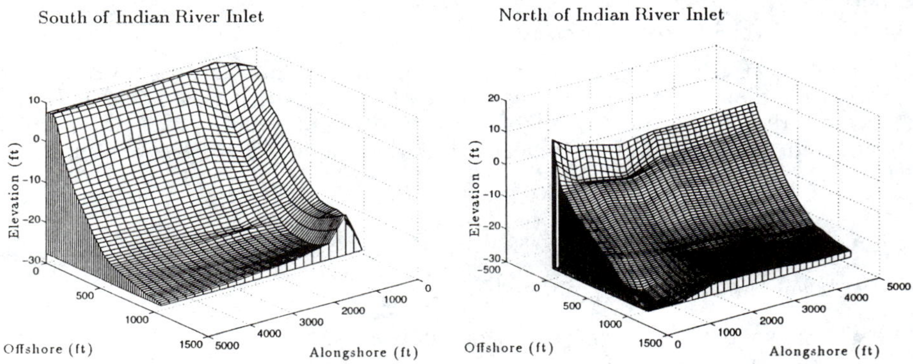

Figure 3: Sample bathymetries (October 1994) for north and south of Indian River Inlet. Offshore grid intervals are 152.4 meters (500 feet); alongshore, 304.8 meters (1000 feet); and elevation, 3.05 meters (10 feet).

SHORELINE CHANGE

The simplest way to examine what is occurring on a given region along the coast is to evaluate the change in shoreline position. This 1-line analysis quantifies beach behavior and allows for comparison of pre- and post-bypassing. Cumulative shoreline change plots for four profile stations in the range extending up to 305 meters (1000 feet) away from the inlet in both the north and south directions are shown in Figure 4. The figure illustrates the shoreline behavior at various stations through time. Watson *et al.* (1993) have computed similar results for a shorter interval of time. For this study, the pre-bypassing interval is from November of 1984 to October of 1989. The initial survey of November 1984 takes place after a large beach fill of 35,781 m^3 was placed between stations 0+00 to 30+00. Surveys measured after October of 1989 are considered post-bypassing. Since we only have a total of four years of data after bypassing start-up, only the short-term performance of the system can be determined.

North of the inlet the trend had been towards progressive retreat, as expected due to the influence of the inlet and the northward littoral drift. Only once, during the winter season from September 1987 to March 1988, did a substantial shoreline advance occur. This was most likely due to the net reversal of littoral drift in the winter months (response to "northeasters") and thus the trapping of some sediment in the shadow of the jetty. The general shoreline retreat was evident until approximately 1035 meters (3400 feet) from the jetty. At this distance, the effect of the inlet was lessened and larger seasonal variations tended to dominate. The bypassing operation was started with the aid of a 133,800 m^3 (175,000 yd^3) fill (evident in the March 1990 survey) obtained from the flood shoal . The initial increase in beach width was not retained due to the spreading of the beach fill, but the shoreline for the stations just north of the inlet seems to have stabilized since bypassing initiation.

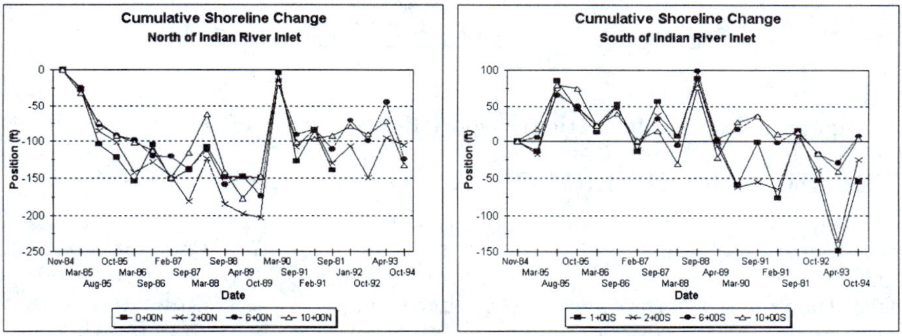

Figure 4: Cumulative shoreline plots showing relative distance from original shoreline position with time for the first four profile stations both north and south of Indian River Inlet.

South of the inlet, pre-bypassing, the trend had been overall stability with some slight accretion. The accretion, if any, was at a much lower rate than the erosion to the north. Again, at around 1035 meters (3400 feet), the effect of the inlet seemed to be minimized as the overall change in beach width was small. After bypassing start-up, Stations 1+00 and 2+00 exhibited immediate effects of the sand mining. Proceeding southward, the next two stations (6+00 and 10+00) show a slight lag in the response to the mining and a smaller shoreline retreat. Notice as well that all stations recover quickly from the effect of the mining. Sand bypassing influence is also typically seen to about 1035 meters (3400 feet) south of the inlet.

VOLUME CHANGE AND SAND BUDGET

Next, volume changes for the areas between profile stations were computed. From these volume changes, a standard sand budget analysis was calculated for the northern region by assuming that the only sediment entering the area was due to bypassing or beach fills. The results from this analysis, shown in Figure 5, yield a measure of the local transport rate. The littoral drift is found to be dominantly northward, as expected, at a rate of approximately 79,500 m^3 (104,000 yd^3) per year. This value was found to be consistent with values found in the past by other methods (Lanan and Dalrymple, 1977; U.S. Army Corps of Engineers, 1984).

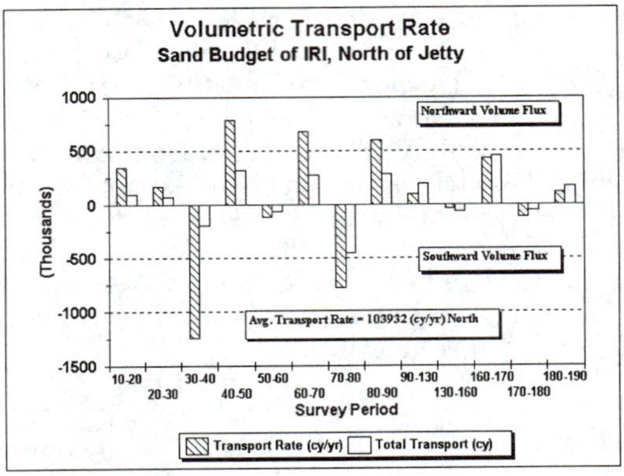

Figure 5: Local volumetric transport rate for north of Indian River Inlet.

THE HISTORY OF PRINCIPAL COMPONENT ANALYSIS

Many geophysical phenomena derive from interactions between traveling waves of different spatial scales and temporal frequencies. Principal component analysis (PCA) was developed to explore these spatial and temporal relations with the primary advantage of its ability to express the complicated variability of data into the fewest possible number of modes. Thus, applying this idea to beach profiles, beach changes can be described by linear combinations of space and time through the breakdown of data into spatial and temporal dependence (e.g. Winant *et al.*, 1975; Aubrey, 1979). Winant *et al.* (1975) found that most of the variation in a profile configuration is accounted for by the first three eigenfunctions, which corresponded to the "mean beach," "bar-berm," and "terrace" functions. However, PCA detects standing oscillations only, such as the standing phenomena of the shift from summer to winter profiles in seasonally sampled data (Winant *et al.*, 1975), not traveling waves. Therefore, PCA can not identify a coherent form moving through the data, such as a rapidly traveling bar as a sand wave alongshore.

Complex principal component analysis (CPCA) was developed for meteorological applications (e.g. Wallace and Dickinson, 1972; Barnett, 1983) and has been used to detect a fast moving sand bar by Liang and Seymour (1991). CPCA has also been shown to out-perform PCA by capturing more of the variance in fewer terms by Liang, White, and Seymour (1992). CPCA has considerable potential for being widely used to detect propagating features, yet its limitations as an analysis technique have not been well explored.

Both of the methods discussed above account for only one spatial direction when evaluating the temporal changes in the data set. This assumes the movement is directed in two independent directions and therefore, the analysis is limited to looking at only individual cross-shore or alongshore "lines." However, what if there is two dimensional movement of sand, as expected in response to a coastal structure or a beach nourishment? Then the 2-dimensional analysis may be rendered inadequate. So, in a further expansion of PCA, the analysis was carried into a third dimension allowing the break down of data into three separate components. This so-called 3-mode PCA was started for mathematical psychology applications, such as the evaluation of multiple personality patients

(Tucker, 1966; Kroonenberg and DeLeeuw, 1980). For the case of a coastal region, 3-mode PCA allows a bathymetric survey to be divided into two spatial directions (cross-shore and alongshore) and a temporal dependence. The technique was applied to a beach fill site in Spain by Medina *et al.* (1992).

In the next section, we apply the 2-mode CPCA tool to the beach profile data of Indian River Inlet and develop a 3-mode CPCA model to examine movement occurring in multiple directions within a bathymetry. The 2-mode CPCA is shown to not only detect the moving forms as well as standing forms, but also distinguish between them.

2-D COMPLEX PRINCIPAL COMPONENT ANALYSIS

To apply CPCA, the data field must first be augmented in a manner such that propagating features within it may be detected. This is done by deriving a complex data matrix, where the real part is simply the original data field and the imaginary part is the Hilbert transform, which represents a filtering operation upon the data in which the amplitude of each spectral component is unchanged, but each components phase is advanced by $\pi/2$. If g(t) is a real-valued function of time, we can define an analytic function

$$z(t) = g(t) + ih(t)$$

where h(t) is the Hilbert transform of g(t) given as:

$$h(t) = \mathcal{H}(g(t)) = \frac{1}{\pi}\int_{-\infty}^{\infty} \frac{g(t')}{t'-t} dt' \tag{1}$$

Using the complex data, we can compute complex eigenvectors (functional decompositions of the data) and eigenvalues (portions of the variation represented by each eigenvector). The goal is to expand the data, $z(x,t)$, into two dimensions (in this case offshore and time) as:

$$z(x,t) = \sum_{i=1}^{n} a_i \; g_i(t) \; e_i(x) \tag{2}$$

n=number of desired modes
a_i=normalizing factors (eigenvalues)
g_i=temporal eigenfunctions
e_i=spatial eigenfunctions

This is done through computation of the correlations between the spatial locations to develop a complex correlation matrix, as shown in Equation 3, where n_t is the total number of surveys and the large brackets denote a time average. The eigenvalues and eigenvectors are then determined from the complex correlation matrix.

$$A_{ij} = \langle \; z_j(t)^* \cdot z_i(t) \; \rangle_{\, n_t} \tag{3}$$

The 2-mode CPCA is used to look at an alongshore profile lines north of Indian River Inlet in the hope of identifying migrating features. Because of the random sampling of the data, many profile lines had to be discarded for various reasons (e.g. did not extend far enough offshore, a given survey was missing, etc.). The results shown below are for an alongshore profile line located approximately 76 meters (250 feet) from the baseline. The complex correlation between the complex time series at given alongshore grid points are shown in Figure 6a. Each complex correlation is plotted in vectorial format where the real portion (magnitude) is indicated in the vertical direction of the vector and the imaginary portion (phase) is indicated in the horizontal direction. A vector pointing upwards (downwards) indicates that the two time series are in-phase (out-of-phase). For example, a vector pointing to the right indicates a lag of 90 degrees. The complex correlation between the time series delineates the propagation of a moving "bump" through the domain. Figure 6b shows the resulting spatial eigenvectors for the alongshore line, plotted in vectorial format as in Figure 6a. The numbers correspond to the percent of variance retained by each eigenvector. The top panel is the first eigenvector which represents the mean alongshore profile and accounts for 98% of the variance. The eigenvector is almost entirely real valued, which signifies that no extensive movement is associated with it, and exhibits a depression in magnitude near the inlet entrance. The second eigenvector identifies a definite progressive feature, which represents 1% of the total variance. Imagine the movement as a spinning motion indicated by vectors "rotating" through space. The second eigenvector then represents a moving form that augments the mean alongshore profile. Similarly, the third eigenvector, which indicates no coherent movement, modifies the mean further.

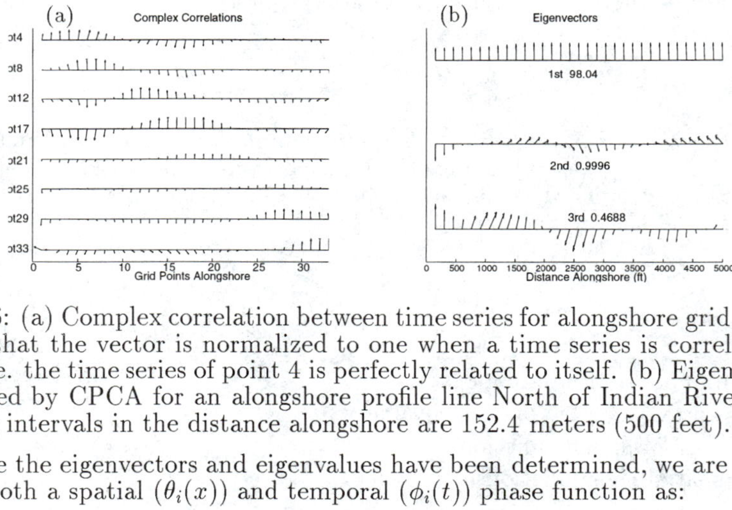

Figure 6: (a) Complex correlation between time series for alongshore grid points. Notice that the vector is normalized to one when a time series is correlated to itself; i.e. the time series of point 4 is perfectly related to itself. (b) Eigenvectors computed by CPCA for an alongshore profile line North of Indian River Inlet. Spacing intervals in the distance alongshore are 152.4 meters (500 feet).

Once the eigenvectors and eigenvalues have been determined, we are able to define both a spatial ($\theta_i(x)$) and temporal ($\phi_i(t)$) phase function as:

$$\theta_i(x) = arctan\left[\frac{\mathcal{I}m(e_i(x))}{\mathcal{R}e(e_i(x))}\right] \qquad \phi_i(t) = arctan\left[\frac{\mathcal{I}m(g_i(t))}{\mathcal{R}e(g_i(t))}\right] \qquad (4)$$

The spatial derivative of the spatial phase function then provides a measure of the "local" wavenumber. Similarly, the time derivative of the temporal phase function is directly proportional to the "instantaneous" frequency. Therefore, CPCA not only allows us to identify a moving form, but also determine the direction and the rate at which it is moving. The spatial and temporal phase functions for the same alongshore profile line, as presented in Figure 6, are shown in Figure 7. The numbers correspond to the approximate wavenumber

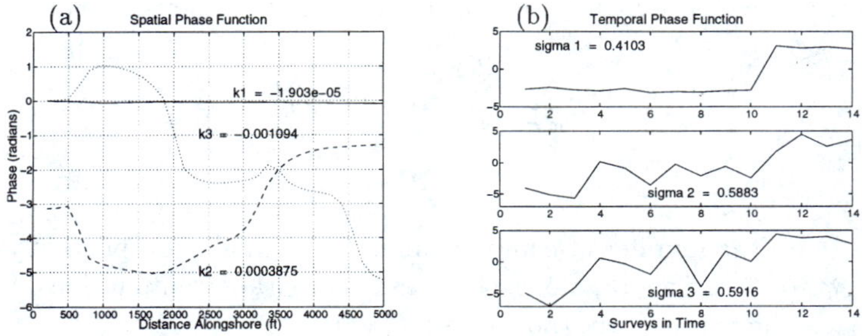

Figure 7: (a) Spatial phase function for alongshore location north of Indian River Inlet. Spacing intervals in alongshore distance are 152.4 meters (500 feet). (b) Temporal phase function for the same location.

and frequency of each eigenvector component. By using these values, the second eigenvector indicates a feature that is moving northward at the rate of around 1.68 meters (5.5 feet) per day. Therefore, over this time period, CPCA reveals that sand was relatively quickly moved out of the area to the north.

3-D COMPLEX PRINCIPAL COMPONENT ANALYSIS

The 2-mode CPCA appears to be very useful for many cases. However, what if 2-dimensional movement of sediment is expected, as mentioned earlier. The 3-mode CPCA is applied to the nearshore region, where significant movement is occurring, for both north and south sides of the inlet. In 3-mode CPCA, we begin with a set of data matrices or rather, a large 3 by 3 cube of data, which can also be thought of as a collection of 2-mode matrices (Figure 8).

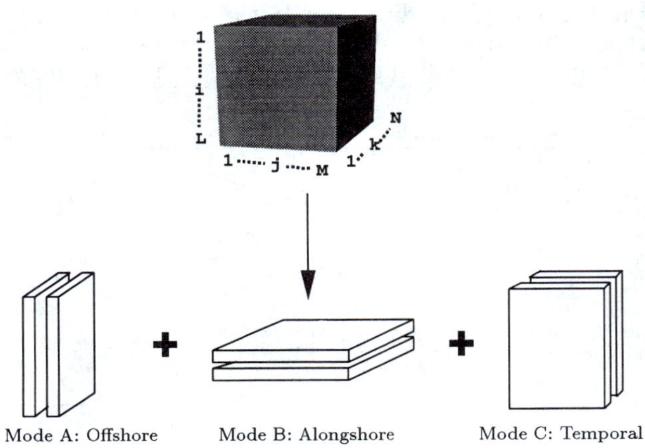

Figure 8: Schematic breakdown of a 3-mode data set into 2-mode submatrices.

Again, after Hilbert transforming the data in time to generate a complex data set, we seek to expand the data in two spatial dimensions and time by:

$$z(x,y,t) = \sum_{m=1}^{s} \sum_{p=1}^{u} \sum_{q=1}^{v} e_m(x)\, f_p(y)\, g_q(t)\, C_{mpq} \tag{5}$$

where

s, u, v are the number of components in the 3 modes, respectively
e, f, g are the offshore, alongshore, and temporal eigenfunctions, and
C_{mpq} is the core matrix (normalizing factors)

Rewriting Equation 5 in matrix form we arrive at

$$Z(x \times yt) = E(x \times s)C(s \times uv)[F(y \times u)' \otimes G(t \times v)'] \tag{6}$$

where $\otimes$ denotes the Kronecker product and the dimension of the matrices Z and C are augmented to be two dimensional arrays. The core matrix, C, is no longer a simple diagonal matrix of eigenvalues as in 2-mode analysis, but a complex combination of elements that describe the basic relations that exist between the various collections of variables as expressed through their components (Kroonenberg and DeLeeuw, 1980).

The cross-shore and alongshore eigenvectors for the nearshore region north of the inlet are presented in Figure 9. For simplicity, only the real part of each eigenvector is shown. To illustrate the relative importance between both the variables and components, the eigenvectors shown have been weighted. These eigenvectors can be thought of as the average form of all cross-shore or alongshore profile lines in the region. In the cross-shore direction, we see the mean is easily identifiable. The second component, typically referred to as the "bar-berm" function is significantly reduced in importance due to the larger fluctuations occurring in the alongshore direction. In the alongshore eigenvectors, the mean is characterized by a dramatic depression that occurs near the inlet. The second and third components highlight sizable changes in the alongshore direction.

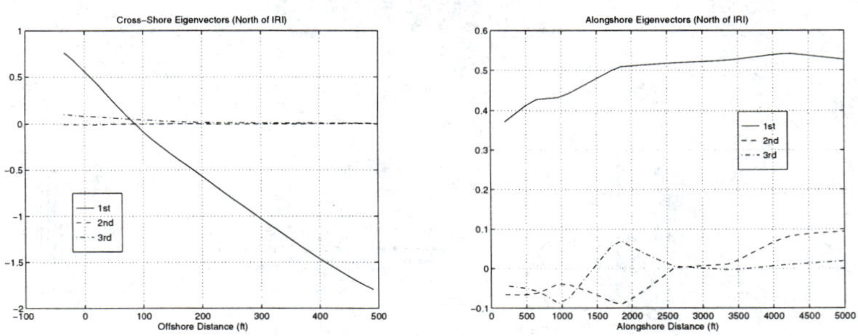

Figure 9: The real portion of cross-shore and alongshore eigenvectors computed from 3-mode CPCA. Offshore distance is spaced at 30.5 meters (100 feet), while the alongshore distance is spaced at 304.8 meters (1000 feet).

The eigenvector components can also be combined to represent various features of the bathymetry. These *eigenvector combinations* are products of the different alongshore, cross-shore, and temporal components. Similar to a large puzzle, if we have three components in each variable, then there are 27 combinations that make up a complete bathymetry. The three combinations that capture the highest amount of variance are shown in Figure 10. The real part of each combination is shown on the left and the imaginary part of the combination on the right. The top panel shows the combination A1O1, which is defined as the product of the first eigenvector in the alongshore direction (A) and the first eigenvector in the cross-shore direction (O). This combination represents the mean bathymetry and the real part is characterized by the erosion located at nearshore region of the inlet entrance. The middle panel shows the A1O3 combination, corresponding to the product of the first alongshore eigenvector and the third cross-shore eigenvector. This combination modifies the mean bathymetry most notably in the shoreline area where bypassing has a significant effect. The lower panel, which contains the A3O3 combination (same nomenclature as before), is perhaps the most intriguing. A large "bump" appears on the real portion of the bathymetry, while the imaginary portion depicts a wave-like phase rotation. This imaginary "hot spot" of movement indicates that the feature is in motion. Notice that the magnitude of the imaginary bathymetry decreases as we proceed offshore. Therefore, the alongshore movement is strongest at the nearshore area and becomes less severe further offshore.

The eigenvector combinations for the region south of Indian River Inlet are shown in Figure 11. The nomenclature remains the same as for the combinations north of the inlet. Again, the top panel represents the mean bathymetry and shows a build up of sediment in the area adjacent to the southern jetty. The A2O3 and A2O1 combinations illustrate changes that are occurring once again in the neighborhood of the inlet entrance. The difference between the two is that the imaginary portion of A203 identifies more of a movement in the alongshore direction, while the imaginary portion of A201 identifies more of a movement in the cross-shore direction. This may mean that, after bypassing occurs, the large eduction hole that remains recovers by receiving sediment from both the alongshore and offshore elements of the bathymetry.

CONCLUSIONS AND SUMMARY

An in-depth analysis of the profiles at Indian River Inlet, Delaware has been accomplished by using standard analysis tools and Complex Principal Component Analysis. Early returns of the sand bypassing system seem to be positive. The shoreline analysis reveals that the north shore has been stabilized by the bypassing and the south shore recovers quickly from the eduction process. The littoral drift indicates sand is moving to the north and that a larger amount of sand could be pumped each season. Generally, sand moves quickly to the north as evident from both the shoreline variation and CPCA analysis. As shown, CPCA has considerable potential, but needs further investigation on a more finely sampled data set.

ACKNOWLEDGMENTS

Special thanks to the State of Delaware Department of Natural Resources and Environmental Control, Beach Preservation Section for support and providing the profile data. The program TUCKALS3 for 3-mode real PCA was provided by R. Medina of Universidad de Cantabria.

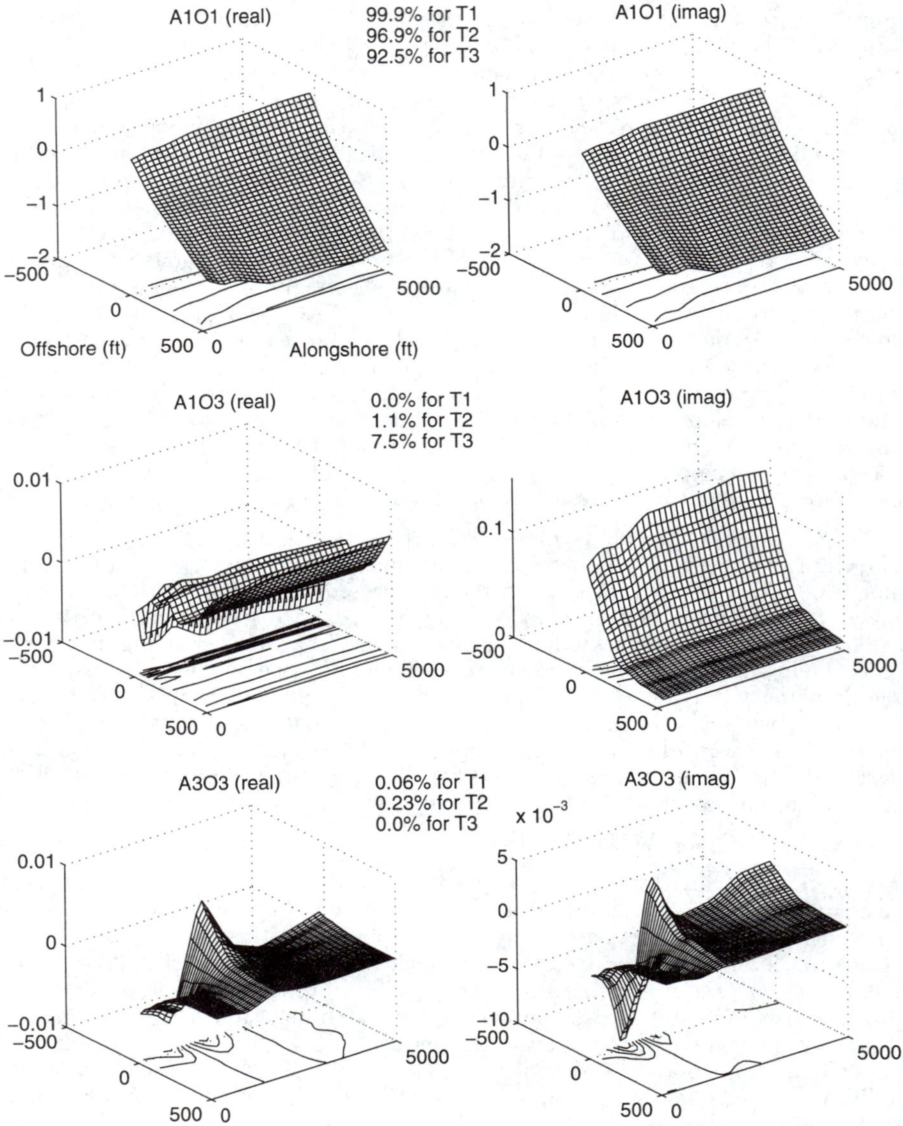

Figure 10: Eigenvector combinations north of Indian River Inlet. The numbers (or core matrix values) included represent the percent of variance captured by each eigenvector combination for a given temporal component. The offshore range is spaced at 152.4 meters (500 feet) and the alongshore range covers 1524 meters (5000 feet).

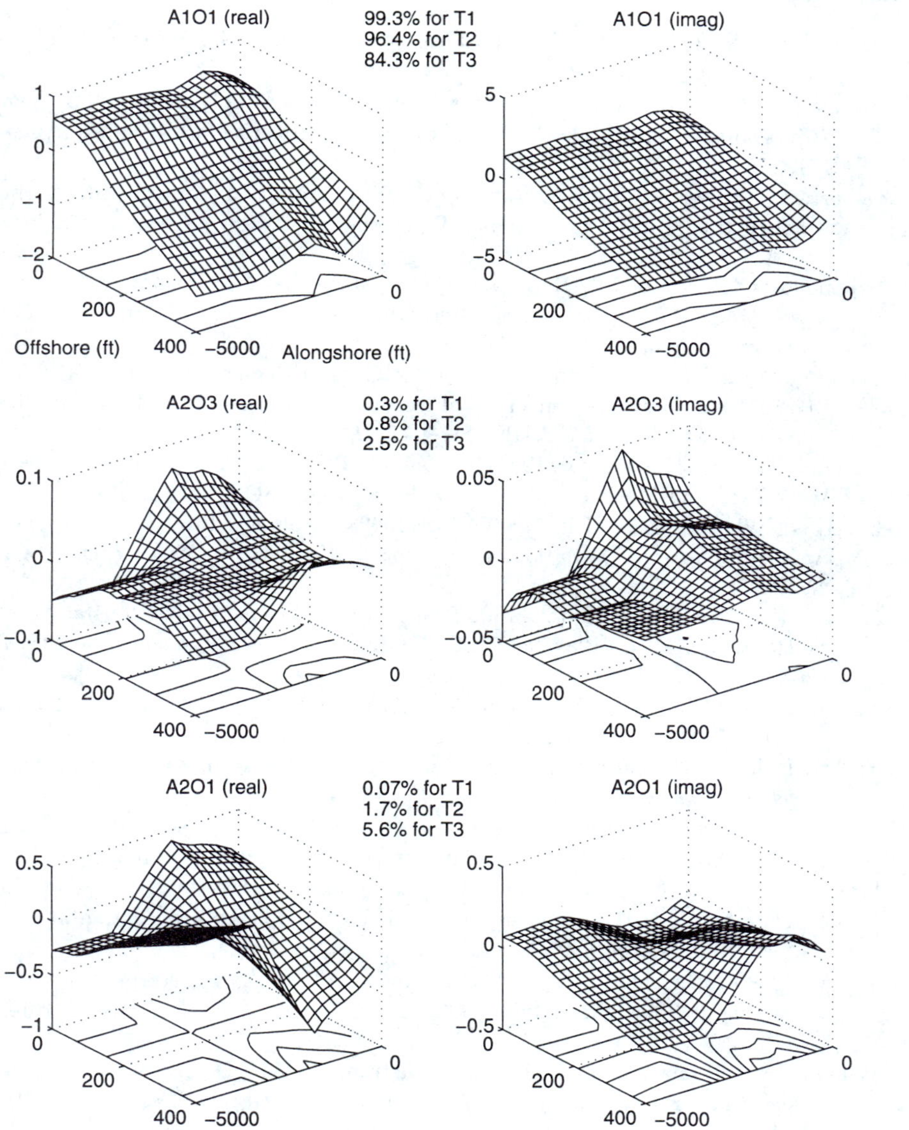

Figure 11: Eigenvector combinations south of Indian River Inlet. The numbers (or core matrix values) included represent the percent of variance captured by each eigenvector combination for a given temporal component. The offshore range is spaced at 61 meters (200 feet) and the alongshore range covers 1524 meters (5000 feet).

REFERENCES

Aubrey, D. G. (1979). "Seasonal Patterns of Onshore/Offshore Sediment Movement." *J. of Geophys. Res.*, 84, 10, 6347-6354.

Barnett, T. P. (1983). "Interaction of the Monsoon and Pacific Trade Wind Systems at Interannual Time Scale. Part I: The equatorial zone." *Mon. Weather Review*, 111, 756-773.

Gebert, J., Watson, K., and A. Rambo (1992). "57 Years of Coastal Engineering Practice at a Problem Inlet: Indian River Inlet, Delaware." *Coastal Eng. Practice*, ASCE, 1-17.

Kroonenberg, P. M., and J. DeLeeuw (1980). "Principal Components Analysis of Three-Mode Data by Means of Alternating Least Squares Algorithms." *Psychometrika*, 45, 69-97.

Lanan G. A. and R. A. Dalrymple (1977). "A Coastal Engineering Study of Indian River Inlet, Delaware." University of Delaware, Newark Col. of Marine Studies, MD Sea Grant, Report No. DEL-SG-5-77.

Liang, G. and R. J. Seymour (1991). "Complex Principal Component Analysis of Wave-like Sand Motions." *Coastal Sediments '91*, ASCE, 2175-2186.

Liang, G., White T. E., and R. J. Seymour (1992). "Complex Principal Component Analysis of Seasonal Variation in Nearshore Bathymetry." *Coastal Eng.*, ASCE, 2242-2250.

Medina, R. *et. al* (1992). "Three-Mode Principal Component Analysis of Bathymetric Data, Applied to 'Playa de Castilla' (Huelva, Spain)." *Coastal Eng.*, ASCE, 2265-2278.

Thompson, W. W. and R. A. Dalrymple (1976). "A History of Indian River Inlet, Delaware." *Shore and Beach*, Vol. 7, 24-31.

Tucker, L. R. (1966). "Some Mathematical Notes on Three-Mode Factor Analysis." *Psychometrika*, 31, 279-311.

U.S. Army Corps of Engineers, Philadelphia District (1984). "General Design Memorandum Sand Bypass Plant: Indian River Inlet, Delaware." Philadelphia, Pa.

Wallace, J. M. and R. E. Dickinson (1972). "Empirical Orthogonal Representation of Time Series in the Frequency Domain, Part I. Theoretical considerations." *J. Appl. Meteorology*, 11, 887-892.

Watson, K. D., Clausner, J. E., and R. D. Henry (1993). "Beach Response to Sand Bypassing at Indian River Inlet, Delaware." *Coastal Technology*, ASCE, Hilton Head S.C.

Winant, C. D., Inman, D. L., and C. E. Nordstrom (1975). "Description of Seasonal Beach Changes using Empirical Eigenfunctions." *J. Geophy. Res.*, 80, 15, 1979-1986.

CHAPTER 220

IMPROVED 3-D BEACH EVOLUTION MODEL COUPLED WITH THE SHORELINE MODEL (3D-SHORE)

Takuzo Shimizu [1], Takahiro Kumagai [1] and Akira Watanabe [2]

ABSTRACT

An improved 3-D beach evolution model coupled with the shoreline model, named "3D-SHORE", was newly developed to estimate both the spatial bottom topography change and the shoreline change. In calculation of the shoreline change, the total longshore sediment transport rate is estimated by integrating the local sediment transport rate in the direction parallel to the shoreline from the breaking point to the run-up point. The applicabilities of the model were verified through comparisons with both the results of the movable bed laboratory experiment and the actual beach evolution.

INTRODUCTION

In recent years, the 3-D beach evolution model which treats only the sediment transport due to nearshore currents have been applied to many practical problems in Japan. This model, based on the depth-averaged nearshore current model, is called "Medium-term 2DH Coastal Area Model" according to de Vriend et al.(1993). The authors have presented a few attempts to quantitatively verify its field applicabilities through comparisons with the actual medium-term topographical changes around a harbor entrance during 1 to 5 years(e.g. Shimizu et al., 1990, 1994). This conventional 3-D beach evolution model have reached the stage of being applicable for engineering use to estimating the volume of maintenance dredging around a harbor entrance and to investigating an effective countermeasure against harbor shoaling. It is, however, difficult to estimate the shoreline change, especially shoreline retreat, because the shoreline is used to be

[1]Penta-Ocean Construction Co., Ltd., 2-2-8 Koraku, Bunkyo-ku, Tokyo 112, Japan.
[2]Professor, Dept. of Civil Eng., Univ. of Tokyo, 7-3-1 Hongo, Bunkyo-ku, Tokyo 113, Japan.

treated as a fixed boundary in the model. Therefore, we cannot assess the impacts due to construction of a harbor on neighbouring beaches with good accuracy. In order to make the 3-D beach evolution model more practically useful, the model should be improved to treat the shoreline as a moving boundary and to properly estimate the shoreline change.

In this study, we developed an improved 3-D beach evolution model, named "3D-SHORE", which is coupled with the conventional shoreline model. The applicability of the model is verified through comparison with the result of movable bed laboratory experiment for the beach evolution around a detached breakwater. And we also tried to simulate the actual bottom topography changes around an offshore man-made island type fishing port during approximately a year.

IMPROVED BEACH EVOLUTION MODEL

Basic Model Concepts

Our interests are focussed on the impact assessment of neighbouring beaches due to construction of a maritime structure. It is demanded to evaluate long-term beach evolution during more than 10 years after the construction. For the beach evolution around a maritime structure, the sediment transport due to nearshore currents plays a predominant role and the contribution from cross-shore sediment transports due to waves and undertow is usually cancelled for a long period beyond approximately a year.

The shoreline model which describes only the longshore sediment transport is widely used in practise for such a case. We have recognized that the shoreline model is applicable to estimating beach evolution caused by the unbalance of sediment budget in the alongshore direction over long time scales and for broad spatial scales. However, in the vicinity of the harbor, where the nearshore circulations occur and the offshoreward sediment transport due to nearshore currents exist, the shoreline model approach has a limited applicability.

The 3-D beach evolution model which treats only the sediment transport due to depth-averaged nearshore currents, "Medium-term 2DH Coastal Area Model", is based on the same concept as that of the shoreline model. This medium-term predictive model can be, therefore, regarded as an improved version of the shoreline model, which has an advantage of estimating the spatial beach topography changes.

Another 3-D beach evolution model is a short-term predictive model which treat both longshore and cross-shore sediment transports. This model is based on the quasi-3D or fully 3D current model which can describe the vertical structure of nearshore current and undertow. In order to properly estimate the beach profile changes, the short interval iterations, probably every one or two hour repetitions are needed. The computations of the wave and current fields are, however, much time-consuming. At present, it is unrealistic to improve the short-term predictive model to properly estimate the shoreline change. We, therefore, tried to develop the 3D-SHORE for practical use by coupling the medium-term predictive model

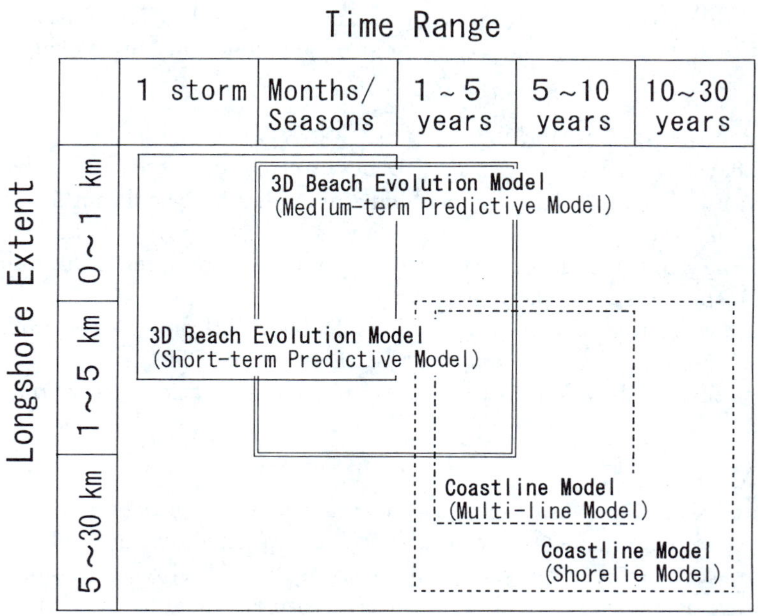

Fig. 1 Classification of numerical models of beach evolution.

Table 1 Comparisons of numerical models of beach evolution.

	Coastline Model	3D Beach Evolution Model (Coastal Area Model)	
	Shoreline Model/ Multi-line Model	Medium-term Predictive Model	Short-term Predictive Model
■Spatial Scale Temporal Scale	5~10 km 1~30 years	1~5 km 1~5 years	1~3 km 1 storm~1 year
■Current Calculation	×	2DH (Depth-averaged Nearshore Current)	· Quasi-3D : (Nearshore Current (2DH) +Undertow (1DVor2DV)) · Fully-3D
■Sediment Transport · Longshore Transport (due to nearshore currents) · On-offshore Transport (net transport due to waves + offshore transport due to undertow)	○ ×	○ ×	○ ○

with the shoreline model.

Fig.1 shows the classification of numerical models of beach evolution by spatial and temporal scales. The comparisons of the models are briefly summarized in Table 1.

Structure of 3D-SHORE

The flow chart of the improved 3-D beach evolution model (3D-SHORE) is shown in Fig.2. The model consists of four submodels for calculating 1)nearshore waves, 2)nearshore currents, 3)local sediment transport rate and spatial beach change, and 4)total longshore sediment transport rate by integrating local sediment transport rate and shoreline change.

The new bottom topography can be calculated by coupling the calculation results of shoreline change and spatial beach evolution. The new bottom topography is, then, fed back into the wave-current computations with appropriate intervals.

Wave Model

The waves in the swash zone play important roles on the shoreline change. In the nearshore wave model, the run-up height is estimated by the Hunt's formula, assuming the beach-face slope determined by the empirical relationship incorporating wave height, period and sediment grain size proposed by Sunamura(1984). The wave set-up and set-down are estimated by the approximate expressions for a plane beach according to Longuet-Higgins and Stewart(1962). The water depth in the swash zone is given virtually by decreasing linearly from the wave set-up height at the still water level to zero at the run-up point as shown in Fig.3. This virtual water depth is also used for calculation of the nearshore current field. Although this treatment of swash zone is not correct in a strict sense, it is a simple and effective method for engineering use in order to save computational time and take an important factor into consideration.

The parabolic-type equation model proposed by Isobe(1987) is employed in this study to properly estimate the wave field behind the offshore breakwater and man-made island and so on where combined diffraction and refraction occur. This basic equation is derived from the mild slope equation by using the wave ray-front coordinates. The energy dissipation term due to wave breaking is included. Random waves are described as a superposition of component regular waves with different frequencies and directions. The applicabilities of the model to the actual wave field were verified through comparisons with field measurement data (e.g. Shimizu et al.,1992).

Nearshore Current Model

In most of the previous computation of nearshore currents, the friction term is expressed as the general nonlinear form and a constant value has been used for the frictional coefficient C_f in the calculation domain. Its value, however, affects the magnitude of the nearshore current velocity and the resultant sediment transport rate. We, therefore, tried to directly estimate the local values of bottom frictional

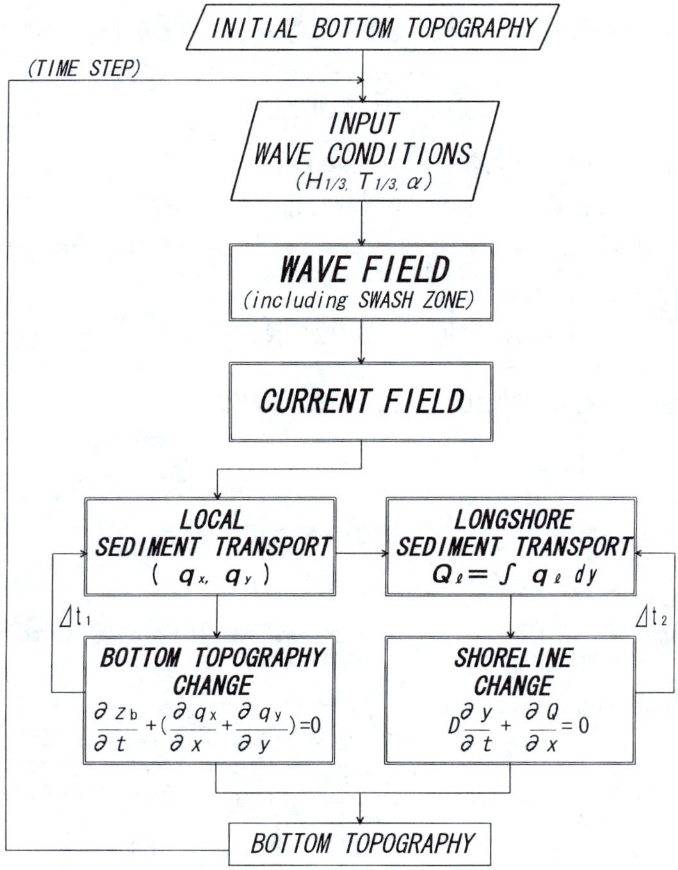

Fig. 2 Flow chart of the 3D-SHORE.

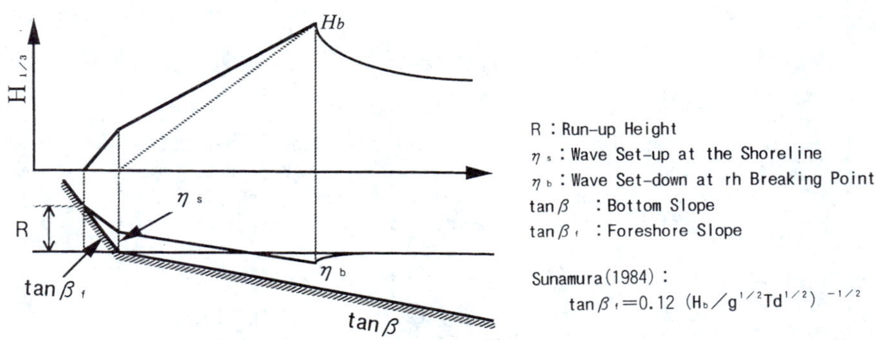

Fig. 3 Schematic illustration of the estimation method of the wave height and water level in the swash zone.

Table 2 Outline of the nearshore wave and current models.

【WAVE MODEL】

Parabolic Equation Model

using the curvillinear coordinates (Isobe, 1987)

- Combined Refraction, Diffraction, Shoaling and Wave Breaking
- Field verification : Shimizu et al. (1992, 1994)

【NEARSHORE CURRENT MODEL】

Depth-averaged Current Model

- Friction Term :

①Conventional Expression

$$\left.\begin{aligned}\tau_x &= \rho C_f (U+u_b)\sqrt{(U+u_b)^2+(V+v_b)^2}\\ \tau_y &= \rho C_f (V+v_b)\sqrt{(U+u_b)^2+(V+v_b)^2}\end{aligned}\right\} \quad (1)$$

where U, V: the mean currents, u_b, v_b: the orbital velocities, C_f : the friction coefficient.

②Improved Expression

directly estimate the local friction term in the mean current direction by using the friction law under combined wave and current action proposed by Tanaka and Thu (1994).

$$\tau_c = \rho u_{c*}^2 = \rho \kappa \alpha \sqrt{\frac{f_{cw}}{2}}\, U_w^2 \quad (2)$$

$$\alpha = \begin{cases} \dfrac{1}{\ln(z_h/z_0)-1}\,\dfrac{u_c}{U_w} & \text{: rough turbulent} \\[2ex] \dfrac{1}{\left[\ln\left(9.0\,\dfrac{R_c}{u_c/U_w}\sqrt{\dfrac{f_{cw}}{2}}\right)-1\right]}\,\dfrac{u_c}{U_w} & \text{: smooth turbulent} \\[2ex] \dfrac{1}{\kappa}\sqrt{\dfrac{f_{c(L)}}{2}} & \text{: laminar} \end{cases} \quad (3)$$

where u_c:the mean current, U_w:the amlitude of near-bottom orbital velocity, z_h:thewater depth, z_0:the roughness height, κ:the Karmanconstant, f_{cw}:the wave-current friction coefficient, $R_c = u_c z_h/\nu$, ν: the kinematic viscosity.

$$\tau_c = f_2[f_1 \tau_{c(L)} + (1-f_1)\tau_{c(S)}] + (1-f_2)\tau_{c(R)} \quad (4)$$

- f_1, f_2 : weight function [see Tanaka and Thu (1994)]
- Subscripts L, S, R describe Laminar, Smooth turbulent and Rough turbulent flows respectively.

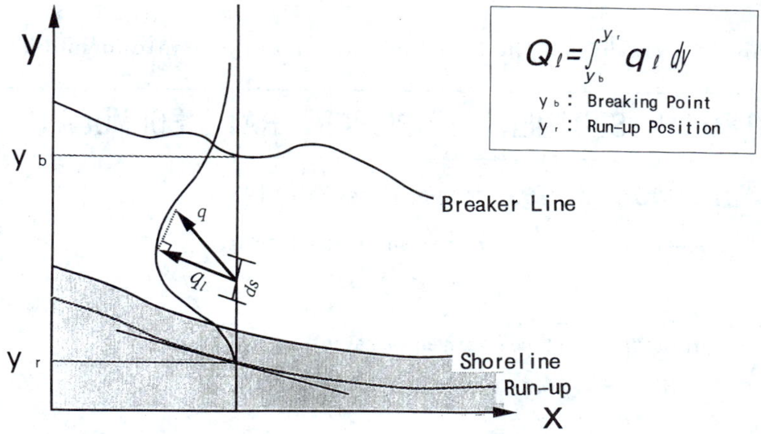

Fig. 4 Schematic illustration of the estimation method of the total longshore sediment transport rate by the cross-shore integration of the local sediment transport rate.

term by using the full-range explicit approximate expressions of frictional law for a wave-current coexistent system proposed by Tanaka and Thu(1994). The outline of the nearshore wave and current models is summarized in Table 2.

Sediment Transport and Beach Evolution Model

Fig.4 shows schematically the cross-shore integration of the local sediment transport rate in the direction parallel to the shoreline from the breaking point to the run-up point. The shoreline change is calculated using the mass conservation equation of the shoreline model based on the alongshore balance of the total longshore sediment transport rate.

In coupling the spatial bottom topography changes with the shoreline changes, both results are interpolated near the shoreline. In the area of shoreline retreat, the profile is determined by the smaller depth comparing between the profile extended offshoreward from the new shoreline position with the foreshore slope estimated by empirical formula by Sunamura(1984) and the profile extended onshoreward from the grid point adjacent to the old shoreline position with the local bottom slope. In the area of shoreline advance, in order to prevent the new shoreline calculated based on the mass conervation of total longshore sediment transport rate from advancing the new shoreline calculated by that of local sediment transport, the appropriate local sediment transport rate coefficient and duration of time-stepping are to be selected.

The local sediment transport rate is evaluated by the formula proposed by Watanabe et al.(1986). The outline of the model is shown in Table 3. The formula for local sediment transport rate under combined wave-current action was formulated so as to be consistent with previous studies on both longshore drift and cross-shore sediment transport. The total transport rate vector is divided

Table 3 Outline of the local sediment transport rate formula.

【LOCAL SEDIMENT TRANSPORT RATE FORMULA】

Watanabe Model (Watanabe et al., 1986)

- Seiment transport due to nearshore currentn q_c:

$$\vec{q}_c = A_c\,(\tau_m - \tau_c)\,\vec{u}_c / \rho g \tag{5}$$

- Seiment transport due to nearshore current q_w:

$$\vec{q}_w = A_w F_D\,(\tau_m - \tau_c)\,\vec{u}_b / \rho g \tag{6}$$

where τ_m : the maximum bottom shear stress under waves and currents,
τ_c : the critical shear stress,
u_c : the mean current, u_b : bottom orbital velocity
F_D : the direction function (+1 for onshore, −1 for offshore)

- **Relationships among the coefficients**
(Watanabe et al., 1991; Shimizu et al., 1994)

$$A_c = 10 A_w \tag{7}$$

$$A_w = w_0 \sqrt{0.5 f_w} \,/\, \{(1-\lambda)\, s' \sqrt{s' g D}\}\, B_w \tag{8}$$

where B_w : the nondimensional coefficient of the wave-induced sediment transport rate formula by Watanabe (1982),
f_w : the wave-current friction coefficient proposed by Tanaka and Shuto (1981),
w_0 : the fall velocity, D : teh grain diameter,
λ : the porosity, g : the gravity acceleration,
$s'\,(=\rho_s/\rho - 1)$ (ρ_s, ρ : the densities of sand and fluid).

- **Determination of local coefficients**

① Previous studies → B_w { = 3~5 for field; = 7 for laboratory experiment }

② H, T, θ, h, D, s' → f_w
→ A_w at each loacal point (eq. (8))

③ A_w → A_c (eq. (7))

into that due to mean currents including both nearshore current and undertow and that due to waves. This formula is based on the power model concept and assume that the sediments set in motion by the excess shear stress under combined wave-current action are transported with both mean currents and wave motion into the respective directions. In this study, only the sediment transport due to nearshore currents is taken into account. The value of the sediment transport rate coefficient is determined by using the empirical relations, according to the previous studies (Watanabe et al.,1991; Shimizu et al.,1994). The local transport rate coefficients depend on local wave conditions and properties of sea-bed material.

MODEL VERIFICATION

Outlines of Laboratory Experiment and Numerical Calculation

The applicability of the newly developed model is investigated on the basis of the movable bed laboratory experiment conducted by Mimura et al.(1983). The experiment was carried out using a wave basin 14m long and 7.5m wide. The sediment grain diameter was 0.2mm and the initial bottom slope was 1/20. After the initial beach was subjected to wave attack for approximately 12 hours, a 1.5m long detached breakwater was placed approximately at the breaker line. The incident wave height was 5.7cm and the period was 0.9s. The verification data used in this study are the beach topography change for about 6 hours after placement of the detached breakwater.

In the calculation of beach evolution, the new bottom topography is fed back into the hydrodynamic and sediment transport computations and the dynamic time-evolution of the seabed is calculated. The calculations of waves, currents and beach topography changes were repeated every 20 minutes and 18 time iterations in total were conducted. The change in the bottom shear stress caused by bottom elevation change is taken into account every 4 minutes in the calculation of beach evolution. The grid spacing is 15cm. The local sediment tarnsport rate coefficient B_w is 7 and the depth of closure D is set to 10cm.

Calculation of Nearshore Wave and Current Fields

Fig.5 shows the alongshore distribution of breaking wave height. The calculations show good agreement with the measurements.

Fig.6(a) and (b) show the examples of the calculated depth-averaged nearshore currents. Fig.6(a) is for initial bottom topography and Fig.6(b) is for that after 5 hours and 40 minutes. The shoreline advance behind the detached breakwater and the strong and sharp nearshore circulations are simulated.

Calculation of Beach Evolution

Fig.7 shows comparisons between the calculated and the measured positions of the shoreline and the depth contours of 2cm and 4cm. The shoreline position calculated by the shoreline model is also plotted. The shoreline change is reproduced better by the shoreline model than by the 3D-SHORE, because the

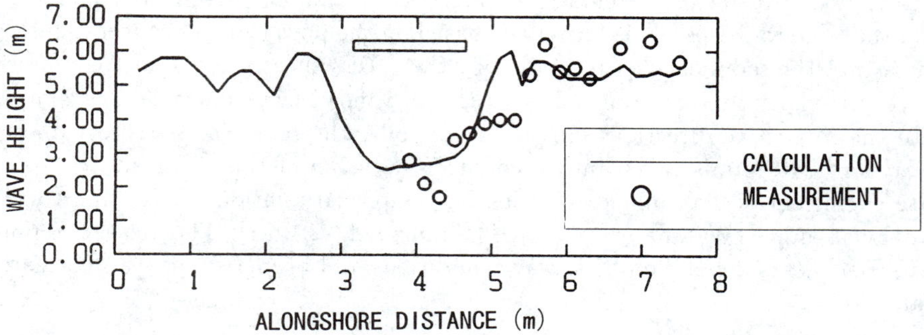

Fig. 5 Alongshore distribution of breaking wave height.

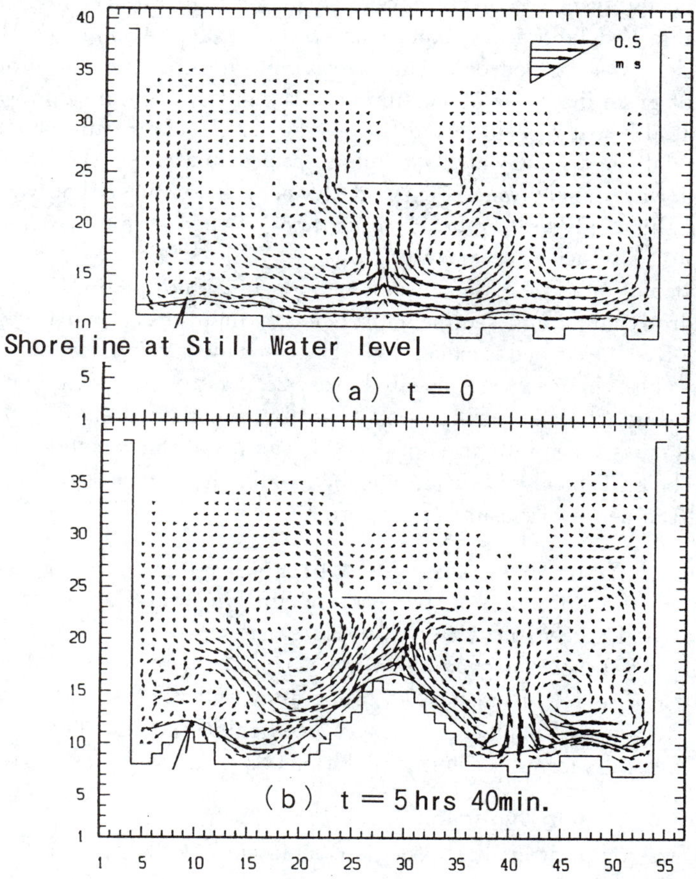

Fig. 6 Calculted nearshore current fields.

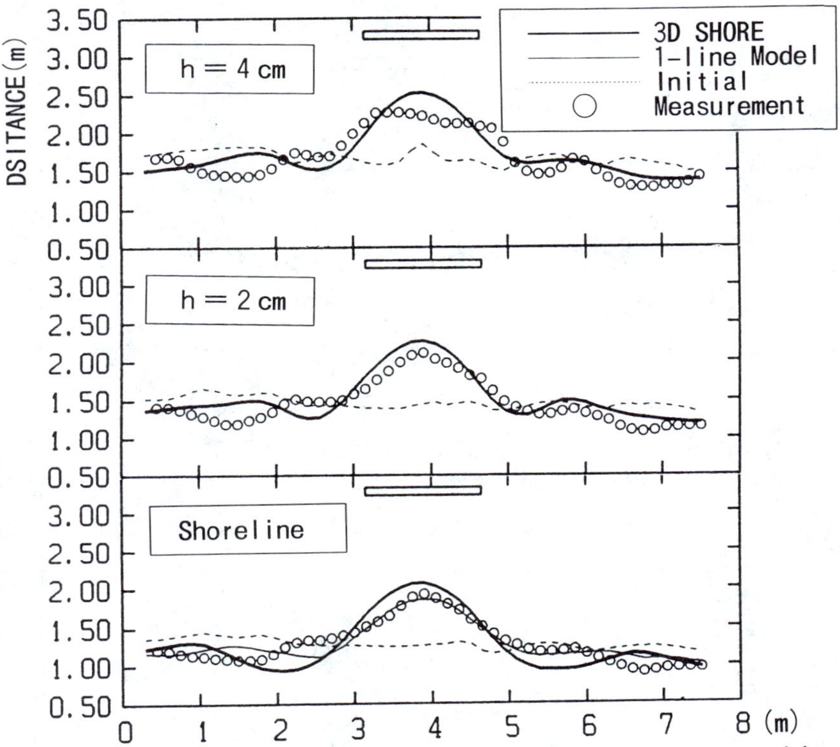

Fig. 7 Comparisons between the calculated and the measured positions of the shoreline and the depth contours of 2cm and 4cm.

computational time of the shoreline model is considerably shorter than that of the 3D-SHORE and it is, then, easy to adjust the parameters to agree with the measurement. The shoreline model, however, can calculate only the shoreline and cannot estimate the spatial bottom topography change. The present improved model, on the contrary, has good accuracy for estimating the changes in both the shoreline and the depth contours for approximately 6 hours.

Fig.8(a) and (b) show the time-evolutions of the shoreline and the 2cm depth contour calculated by the 3D-SHORE and (c) shows the time-evolution of the shoreline calculated by the shoreline model. The shoreline behind the detached breakwater advances gradually and reaches an equilibrium state after 5 hours according to the calculation result of the 3D-SHORE. The contour of 2cm advances faster than the shoreline.

According to the result calculated by the shoreline model, on the contrary, the shoreline advances rapidly and reaches an equilibrium state after only 3 hours. In the shoreline model, the longshore sediment transport rate becomes zero at the equilibrium state, because the breaking wave crest angle to the shoreline becomes zero. This shows that the shoreline model can describe only the static equilibrium. The improved 3-D beach evolution model, on the other hand, can describe

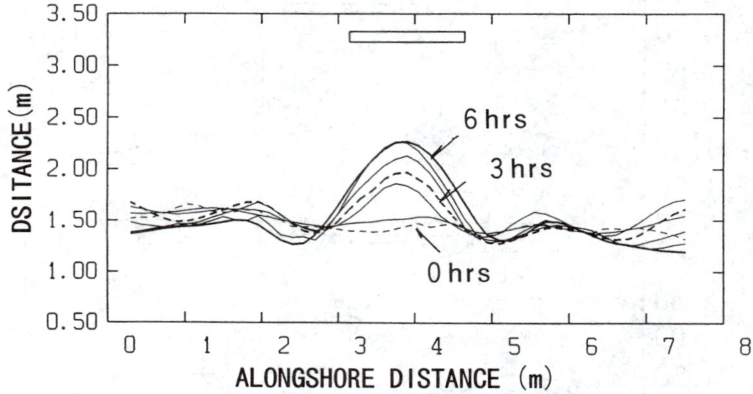

(a) h=2cm (3D-SHORE)

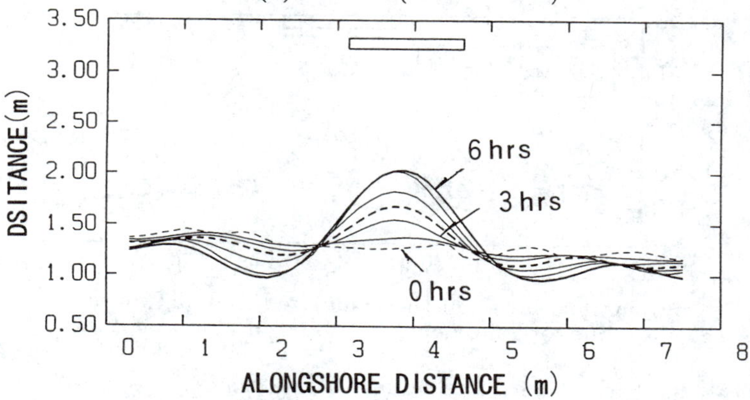

(b) Shoreline (3D-SHORE)

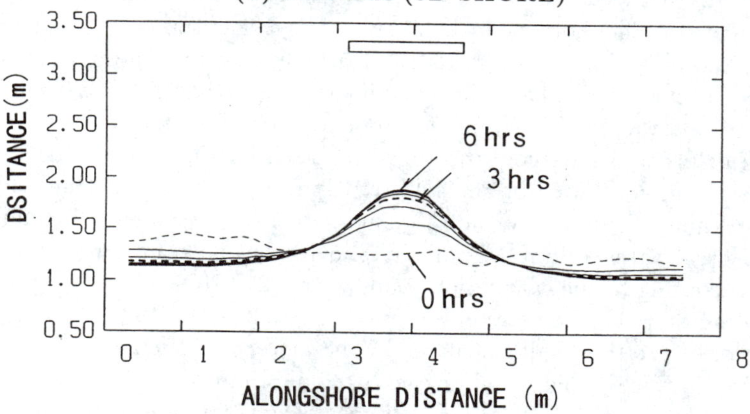

(c) Shoreline (1-LINE MODEL)

Fig. 8 Time-evolution of the calculated depth contours.

the dynamic equilibrium and is, therefore, more effective for properly predicting the time-evolution of bottom topography.

FIELD APPLICATION

In order to verify the field applicability, we try to simulate the beach evolution due to construction of the Kunnui Fishing Port, an offshore man-made island type fishing port, in Hokkaido, Japan (Kawaguchi et al., 1994). The man-made island is located about 200m offshore from the initial shoreline. The maximum width of the man-made island is about 180m, which approximately equals to the detached distance. Although the bottom contours are straight and parallel to the shoreline in 1985 before construction of the fishing port, they extend offshoreward like a tongue behind the fishing port owing to extreme accretion caused by nearshore circulations. The rapid and extreme beach evolution took place during only a year from 1989 to 1990 after the start of the construction of man-made island. At present, the tombolo is formed behind the fishing port.

The calculation was conducted with the area of 1.0km long in the alongshore direction and 0.8km long in the cross-shore direction. The grid spacing is 10m. The bottom slope is approximately 1/75. The grain diameter is about 0.2mm and the transport rate coefficient B_w is 4. The numerical simulation was performed under simply modelled two series of the storms by repeating the calculations of waves, mean currents and beach changes. The modelled series of waves have the same occurrence frequency in total as that of the observed wave climate data. The depth of closure D was expressed as a function of the incident wave conditions.

Fig.9 shows the comparisons between the calculated and the measured depth contours after a year from 1989 to 1990. The calculations show fairly good agreements with the measurements. The field applicability of the proposed model is also verified.

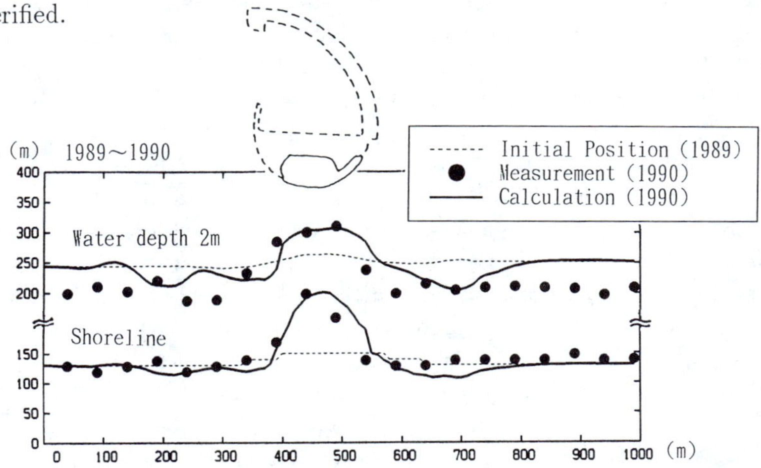

Fig. 9 Reproduction of the beach evolution around the Kunnui Fishing Port during a year.

CONCLUSIONS

In this study, an improved 3-D beach evolution model coupled with the shoreline model (3D-SHORE) is newly developed. And the applicabilities of the model were successfully verified through comparisons with both results of movable bed laboratory experiment and the actual beach evolution. The 3D-SHORE can describe the dynamic equilibrium of beach evolution and is, therefore, more effective for properly predicting the time-evolution of bottom topography than the conventional shoreline model which can only the static equilibrium. It is concluded that the 3D-SHORE has enough accuracy for practical use.

REFERENCES

De Vriend,H.J., J.Zyserman, J.Nicholson, J.A.Roelvink, P.Péchon and H.N.Southgate, 1993: Medium-term 2DH coastal area modelling, *Coastal Eng.*, Vol.21, pp.193-224.

Isobe, M., 1987: A parabolic equation model for transformation of irregular waves due to refraction, diffraction and breaking, *Coastal Eng. in Japan*, Vol. 30, No. 1, JSCE, pp.33-47.

Kawaguchi, T., O.Hashimoto, T.Mizumoto and A.Kamata, 1994: Construction of offshore fishing port for prevention of coastal erosion, *Proc. 24th Int. Conf. on Coastal Eng.*, ASCE, pp.1197-1211.

Longuet-Higgins, M.S. and R.W. Stewart, 1962: Radiation stress and mass transport in gravity waves, with application to 'surf beat', *J. Fluid Mech.*, Vol.13, pp.481-504.

Mimura, N., T. Shimizu and K. Horikawa, 1983: Laboratory study on the influence of detached breakwater on coastal change, *Proc. Coastal Structure '83*, ASCE, pp.740-752.

Shimizu, T., H. Nodani and K. Kondo, 1990: Practical application of the three-dimensional beach evolution model, *Proc. 22nd Int. Conf. on Coastal Eng.*, ASCE, pp.2481-2494.

Shimizu, T., A. Ukai and M. Isobe 1992: Field verification of numerical models for calculation of nearshore wave field, *Proc. 23rd Int. Conf. on Coastal Eng.*, ASCE, pp.590-603.

Shimizu, T., M. Tsuru and A. Watanabe, 1994: Field verification of a numerical model of beach topography change due to nearshore currents, undertow and waves, *Proc. 24th Int. Conf. on Coastal Eng.*, ASCE, pp.2610-2624.

Sunamura, 1984: Quantitative predictions of beach-face slopes, *Geol. Soc. Am. Bull.*, Vol.95, pp.242-245.

Tanaka, H. and N. Shuto, 1981: Friction coefficient for a wave-current coexistent system, *Coastal Eng. in Japan*, Vol.24, JSCE, pp.105-128.

Tanaka, H. and A. Thu, 1994: Full-range equation of friction coefficient and phase difference in a wave-current boundary layer, *Coastal Eng.*, Vol.22, pp.237-254.

Watanabe, A., 1982: Numerical models of nearshore currents and beach deformation, *Coastal Eng. in Japan*, Vol.25, JSCE, pp.147-161.

Watanabe, A., K. Maruyama, T. Shimizu and T. Sakakiyama, 1986: Numerical prediction model of three-dimensional beach deformation around a structure, *Coastal Eng. in Japan*, Vol.29, JSCE, pp.179-194.

Watanabe, A., T. Shimizu and K. Kondo, 1991: Field application of a numerical model of beach topography change, *Proc. of Coastal Sediments '91*, ASCE, pp.1814-1828.

CHAPTER 221

EVALUATION OF SHOREFACE NOURISHMENTS BY LINE MODELLING

Martin D. Groenewoud[1], Willem T. Bakker[2,3], Jan van de Graaff[1],
Ruud Spanhoff[3] and Christian Laustrup[4]

ABSTRACT

Shoreface nourishments are sometimes used as an alternative for ordinary beach nourishments. Many aspects of the behaviour of shoreface nourishments are still unknown. Shoreface nourishments affect the morphological behaviour of a coast. Line modelling might be used to study and to predict the behaviour of shoreface nourishments after execution. Although not every detail of the real behaviour can be dealt with, line modelling is still a rather simple but powerful tool. The recent application is discussed of the line modelling technique to study several shoreface nourishments carried out in the NOURTEC framework.

INTRODUCTION

Coastal zone managers may fight undesired structural erosion of coasts either by 'hard' or by 'soft' measures.
With 'hard' measures the basic idea is to interfere in the sediment transports involved in such a manner that the erosion in the stretch of coast under consideration is stopped, or at least reduced. With a well-tuned system of groynes or a number of shore-parallel detached breakwaters this aim can in principle be achieved. That often the erosion problem is shifted to the adjacent lee side beaches is a serious draw-back of these types of countermeasures.
With 'soft' methods (e.g. a beach nourishment [further: BN]) this adverse lee side effect is avoided. The basic idea of artificial nourishments is to accept the losses as observed (no attempts to interfere in the processes which cause the erosion) but replenish from time to time the apparent losses. Often life times in the range of 5 till 10 years are striven after. Although artificial BN's have to be repeated, it often turns out to be a very cost effective method in comparison with 'hard' alternatives.

Structural erosion of a stretch of coast means that the stretch loses sediments at a regular basis. Often a gradient in the longshore sediment transports is the main cause of the erosion problem. In a cross-shore profile the volume of sediments (m^3/m) in that profile in a predefined (fixed) area, diminishes as a function of time. All parts of such a profile (dunes; beach; shoreface) suffer eventually from this type of

---

[1] Delft University of Technology; Stevinweg 1; 2628 CN Delft; the Netherlands
[2] Netherlands Center for Coastal Research; Delft University of Technology; Stevinweg 1; 2628 CN Delft; the Netherlands
[3] Rijkswaterstaat, National Institute for Coastal and Marine Management; P.O. Box 20907; 2500 EX the Hague; the Netherlands
[4] Kystinspektoratet; Høbovej 1; 7620 Lemvig; Denmark

erosion. However, the upper part of the profile (beach and dunes) is often considered as the most important part of the entire profile; many important functions of a coast to mankind are concentrated in just this part. Because structural erosion manifests itself most clearly in the upper part of the profile, artificial nourishments are often applied just there. With a BN the 'health' of a beach is directly improved; one clearly can notice the improvements for the time being; e.g. the beach is widened which is favourable for recreational use.

Many methods exist to execute artificial BN's [see e.g. CUR (1987)]; most of them are, however, rather cumbersome in practice. The handling of the sediments from borrow area to fill site is often complicated and thus costly. Because of the recreational use of beaches often the summer period is excluded as execution time.

Instead of applying (cumbersome) artificial BN's, nourishing just the shoreface seems an (easy) alternative. Sediment handling is quite simple; the dredge sails from the borrow area and dumps the load at the shoreface and starts with a new cycle. Because of no (or less) hindrance to the recreational use of the beaches also the summer time can be used for execution. Large cost savings are expected to be achieved with shoreface nourishments [further: SN] in comparison with classical BN's.

In the scope of the NOURTEC project [Mulder *et al.* (1994)] at 3 sites (Terschelling, the Netherlands; Norderney, Germany; and Torsminde, Denmark; see Fig.1) sand has been supplied on the shoreface. The NOURTEC project implies evaluation and comparison of the coastal behaviour at those sites after the SN's.

Bakker *et al.* (1994) made a prediction of the expected coastal behaviour after the SN at the Dutch Wadden island of Terschelling; furthermore the coastal behaviour after the supply at the German Wadden island of Norderney was evaluated. They used the 3-line modelling technique. Later on the nourishments in Denmark and Terschelling have been evaluated with the same technique (Groenewoud 1996a and 1996b). The study has been finished with an overall evaluation of the 3-line modelling technique as a design and evaluation tool (Groenewoud 1996c).

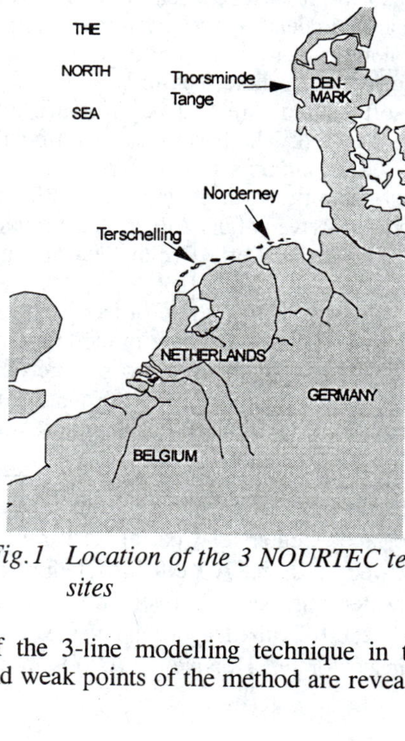

Fig.1 Location of the 3 NOURTEC test sites

In the present paper the application of the 3-line modelling technique in the 3 NOURTEC cases is discussed; strong and weak points of the method are revealed.

SHOREFACE NOURISHMENTS IN PRACTICE

Fig.2 shows in plan view a (shore parallel) SN along a sandy coast. Because a SN is meant to be a substitute of an ordinary BN it is expected that a part of the supplied volume is transported towards the coast. With the notion of the existence of equilibrium profiles one can indeed expect that because of cross-shore transports the extra volume of sand is eventually spread over the entire profile (in onshore as well in offshore direction). Consequently also a part of the supplied volume is thought to reach eventually the beach. For the judgment of the effectiveness of SN's in comparison to BN's it is of course important to know which part of the volume of SN's reaches the beaches, and which time scales are involved. With detailed cross-shore transport calculations these topics are to be evaluated.

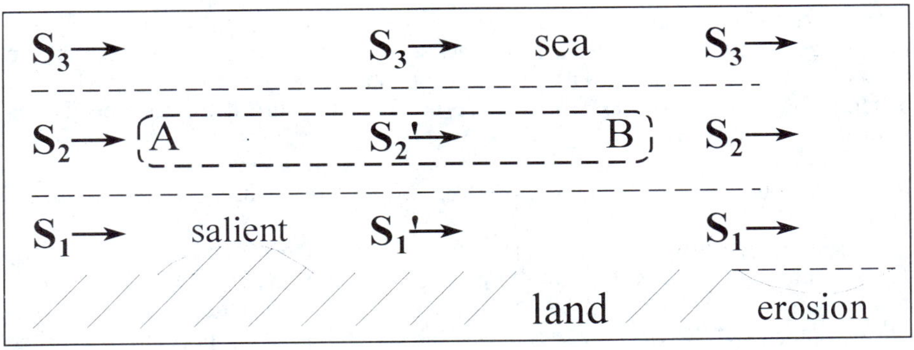

Fig.2 (Arbitrary) shoreface nourishment in plan view

In plan view (see Fig.2) a SN is limited; between points A and B the depths contours are just after the execution locally shifted in seaward direction. In the neighbourhood of the points A and B the orientation of the depth contours has changed. With a description of the (local) longshore transport depending on the orientation of the depth contours it is easily understood that near the points A and B a redistribution of the volume of sediments of the SN in longshore direction will take place. Also in the upper part of the profile, as soon as some sediment has reached that part of the profile by cross-shore transports, a similar reorientation of the local depth contours will occur with a consequent redistribution of sediments in longshore direction in the upper part of the profiles of the stretch of coast.

A SN acts, however, also as a (mobile) submerged breakwater. The presence of a SN will affect (reduce) the wave heights in the zone between the SN and the waterline. Consequently the longshore sediment transports in this zone are reduced compared to the original transports along the non-supplied coast at the left-hand side of A and at the right-hand side of B (see Fig.2). Assuming a net longshore transport in the zone behind the SN as indicated in Fig.2 from left to right, it is expected that behind the left side of the SN some accumulation of sediments will occur (salient formation). Right from B it is expected that (additional) erosion of the beaches will occur.

Especially the expected salient formation at the up-drift side behind the SN will confuse phenomenological studies based on bathymetric data. Is the observed accumulation of sediments in this area because of cross-shore sediment transport effects or because of longshore transport effects?

If a SN is applied in a zone with water depths where in the non-supplied situation still substantial yearly net longshore sediment transports do occur (either wave or tide driven or due to a combination of waves and tidal currents), the mere sudden uplift of the bottom because of the presence of the SN will cause increased sediment transports above the SN (see Fig.2). Induced gradients in the local longshore transport will cause erosion of the SN just right of point A and accumulation of sediments just right of point B. It seems that the entire SN is moving (within the schematization of Fig.2) from left to right along the coast.

In the brief discussion of the different morphological effects of a SN it was presumed that the borrow material of the SN is equal to the native material. In practical applications, however, this presumption might not be true; the borrow material may differ from the native material. It is for sure that consequently the sediment transports involved, will change.

In the previous part 4 morphological effects of a SN have been briefly discussed, viz.:

> i) (straightforward) cross-shore redistribution of the volume of a SN over the entire cross-section; ii) longshore redistribution of the volume of a SN due to reorientation of the depth contours in the zone of the SN itself (starts immediately after execution) and in the upper part of the profile (starts when after some time sediments have reached the upper part of the profile); iii) salient formation at the up drift side (and erosion at the down drift side) in the zone between SN and waterline because of reduction of the yearly net longshore sediment transport in this zone; iv) integral movement of the SN in longshore direction because of yearly net longshore sediment transports in the SN zone.

In the next section the 3-line modelling technique is discussed. It will turn out that the effects i) and ii) can in principle be properly modelled with the present technique; the effects iii) and iv) cannot be accounted for with the present model set up. In the present 3-line modelling approach it is also assumed that borrow and native material are the same. Additional effects due to differences in borrow and native material are not accounted for.

3-LINE MODELLING TECHNIQUE

General

In the present study 3 lines have been used to represent a cross-shore profile. More, or only 2 lines are in principle also possible. The 3-line modelling technique is a rather simple technique which can be used to study the behaviour of a morphological system with time. The approach relies on the assumption that the actual behaviour of a morphological system can be considered as a linear superposition of different sub-systems. In the present application for instance, the autonomous behaviour of the system is assumed to be the same before and after a SN application. Only the additional effect of the SN is modelled. So if the behaviour of the SN with time is modelled reliably, the eventual behaviour of the system with time is found by adding both sub-systems (autonomous behaviour + behaviour because of the SN).

Schematization of cross-shore profile and cross-shore transports

In a 3-line model the cross-shore profile is schematized by 3 zones with horizontal separation planes (see Fig.3). The volumes of sand in the zones (layers) are characterized by 3 lines which may, contrary to equilibrium profile approaches, develop 'freely' with time. It is assumed that in each zone the mutual distance between the upper and the lower limit remains the same and (thus) that the part of a profile coinciding with the zone moves only horizontally.

The characterizing lines each have a distance to an arbitrary vertical reference line which for L_1 can be defined as (see Fig.3):

$$L_1 = \frac{1}{h_1} \int_{-d_1}^{0} y(z)dz \qquad (1)$$

For lines L_2 and L_3 similar formulae can be derived.

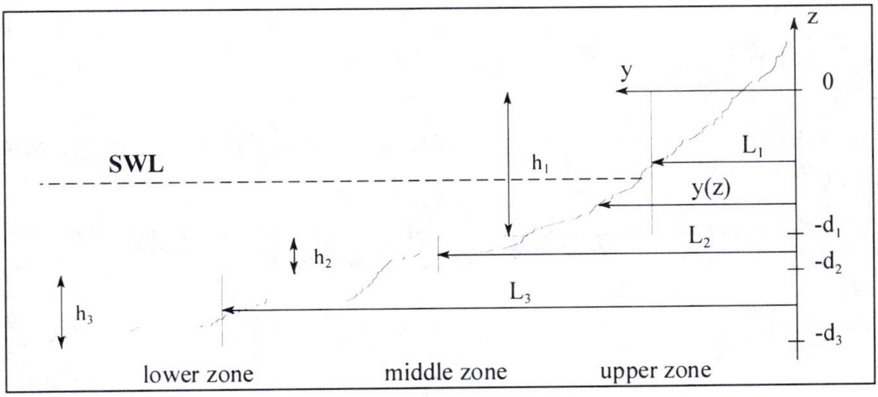

Fig.3 Schematization of cross-section

The parameters of Eq.1 and Fig.3 are:

d_1, d_2 = depth of separation plane between zone 1 and 2; zone 2 and 3;
d_3 = depth of lower limit of zone 3;
h_1, h_2, h_3 = height of zone 1; zone 2; zone 3;
L_1, L_2, L_3 = characterizing mean line of upper zone; middle zone; lower zone;
$y(z)$ = distance of point of profile at depth z to a reference line;
z = height above the y-axis.

The cross-shore sediment transport rate between the zones is assumed to be proportional to the difference between the actual distances and the equilibrium distance between the characterizing lines. This method of describing cross-shore sediment transports was originally proposed by Bakker (1968) and further developed by Swart (1974).

Since in the present application the profiles before the SN application are assumed to be equilibrium profiles, the use of L_1, L_2 and L_3 can be avoided; only perturbations from the initial equilibrium profile are considered. These perturbations are expressed as (see Fig.4) y_1, y_2 and y_3 respectively.

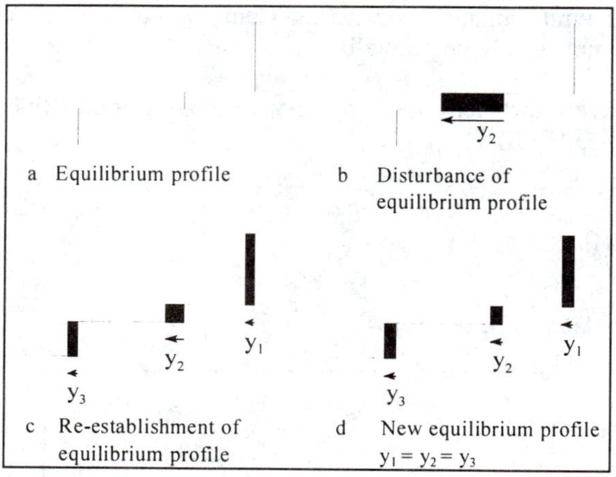

Fig.4 Schematized equilibrium profile (a); disturbance in middle layer (b) and behaviour (c and d)

If the transport rates S_{y1} [and S_{y2}] are assumed to be proportional to the differences (y_1-y_2) for S_{y1} [and (y_2-y_3) for S_{y2}], then the following equation for S_{y1} can be written:

$$S_{y_1} = s_{y_1}(y_1 - y_2) \tag{2}$$

(For S_{y2} similar formula.)

with: S_{y1} = cross-shore transport from the upper to the middle zone [m^3/m/year] (positive in seaward direction)
s_{y1} = cross-shore coastal constant [m/year]

The description of the cross-shore sediment transport relies on a proper estimate of the cross-shore coastal constants. It can be proved that these coastal constants can be rewritten as:

$$s_{y1} = \frac{1}{T_{01}} * \left(\frac{h_1 * h_2}{h_1 + h_2} \right) \tag{3}$$

with: T_{01} = time constant for diffusivity between upper and middle zone (year)

(For the time constant for diffusivity between the middle and lower zone T_{02} a similar formula can be derived.)

The T_0 period is the lapse of time in which a certain value of deviation from the equilibrium distance will decrease with a factor e. The problem of estimating proper values for the cross-shore constants is now shifted to the proper estimate of the T_0 periods.

Schematization of longshore transport

In each zone of the schematized cross-shore profile longshore sediment transports will occur. These longshore transports are indicated by S_1, S_2 and S_3 for the upper, middle and lower zone, respectively. In the following discussion S_x is generally used.

In the present NOURTEC cases the total longshore transport, and the distribution over the different zones, is calculated with the CERC formula; the possible contribution of tidal currents is consequently ignored. The method can, however, also be used if tidal currents are taken into account; the preparatory calculations are only somewhat more complicated in that case.

A gradient in the longshore transport results in the following equation of continuity:

$$\frac{\partial S_x}{\partial x} + h\frac{\partial y}{\partial t} = 0 \qquad (4)$$

with h being the thickness of the layer over which erosion or accumulation takes place. The gradient of the longshore sediment transport is due to changes in angle of wave approach.

The wave climate is assumed to be constant along the coast. With small changes of the angle, the longshore transport is assumed to depend linearly on the angle of wave approach.

$$\frac{\partial S_x}{\partial \varphi} = s_x \qquad (5)$$

with: s_x = longshore coastal constant

The chain rule gives:

$$\frac{\partial S_x}{\partial x} = \frac{\partial S_x}{\partial \varphi} * \frac{\partial \varphi}{\partial x} \qquad (6)$$

Substitution of Eqs.5 and 6 in Eq.4 and assuming small angles ultimately yields:

$$-s_x\frac{\partial^2 y}{\partial x^2} + h\frac{\partial y}{\partial t} = 0 \qquad (7)$$

The change of the position of the coastline ($\partial y/\partial t$) due to longshore transport is proportional to the curvature of the coastline. This also applies if the profile is divided into several zones. The equation for the upper zone then becomes:

$$\left[\frac{dy_1}{dt}\right]_{long} = \frac{s_1}{h_1} * \frac{\partial^2 y_1}{\partial x^2} \tag{8}$$

with: s_1 = longshore coastal constant

Combination of both cross-shore and longshore processes with linear addition gives:

$$\frac{dy_1}{dt} = \frac{s_1}{h_1} * \frac{\partial^2 y_1}{\partial x^2} - \frac{s_{y1}}{h_1}(y_1 - y_2) \tag{9}$$

$$\frac{dy_2}{dt} = \frac{s_2}{h_2} * \frac{\partial^2 y_2}{\partial x^2} + \frac{s_{y1}}{h_2}(y_1 - y_2) - \frac{s_{y2}}{h_2}(y_2 - y_3) \tag{10}$$

$$\frac{dy_3}{dt} = \frac{s_3}{h_3} * \frac{\partial^2 y_3}{\partial x^2} + \frac{s_{y2}}{h_3}(y_2 - y_3) \tag{11}$$

In these equations the longshore transport is determined by the constants s_1, s_2 and s_3 and the direction of the coast. The cross-shore transport is determined by the constants s_{y1} and s_{y2} and the deviation from the equilibrium position.
These three equations determine the development of y_1, y_2 and y_3 in time and position along the coast.

Numerical solution
The numerical method used to solve Eqs.9, 10 and 11 is the Euler Explicit Time Forward Central Space method. Time and space steps have to be carefully linked in order to fulfil the stability criterion.

Example
In Fig.5 the results of an arbitrary example are given. Along 2 km a shore-parallel SN is placed in the middle layer. The thicknesses of the layers are respectively: upper: 6 m; middle: 3 m and lower: 3 m. The cross-shore constants are s_{y1} = 1.33 m/year and s_{y2} = 0.3 m/year. The longshore coastal constants read: s_1 = $2.65*10^6$ m^3/year/rad; s_2 = $0.22*10^6$ m^3/year/rad and s_3 = 0 m^3/year/rad. The volume as nourished is 300 m^3/m; because of the nourishment just after the nourishment the y_2 values shifted with 100 m in seaward direction.

From Fig.5 it becomes clear that the SN diffuses in all directions. With the constants as used indeed a large part of the SN reaches the upper part of the profile.
The lower zone just seaward of the SN gets some sand directly from the middle zone because of cross-shore transports. In the adjacent lower zones also some sedimentation is noticed. Since the longshore coastal constant s_3 in the present

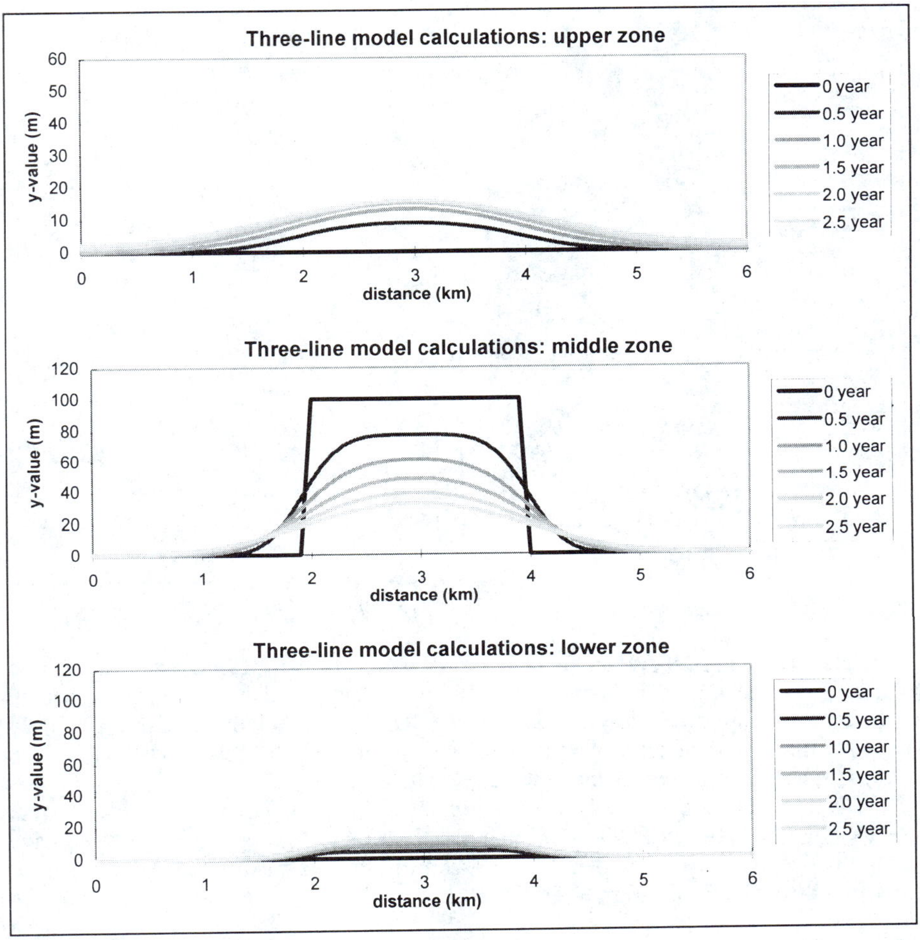

Fig.5 Development with time of a shoreface nourishment in middle layer

example is zero, this material has reached these zones first via the adjacent middle layers and next by cross-shore transport.

In the present example the autonomous behaviour of the coast is not taken into account; in real life problems the autonomous behaviour has to be added to the computation results. To determine the autonomous behaviour in real cases turns out to be a cumbersome task (see NOURTEC examples next section).

NOURTEC EXAMPLES

Torsminde

The Torsminde test side is located along the west coast of Denmark (see Fig.6). Fig.7 shows two typical cross-shore profiles; 1 to 3 offshore bars are present. The site is located in front of a sand dike. A typical value of D_{50} for the sand in the profile is 0.4 mm.

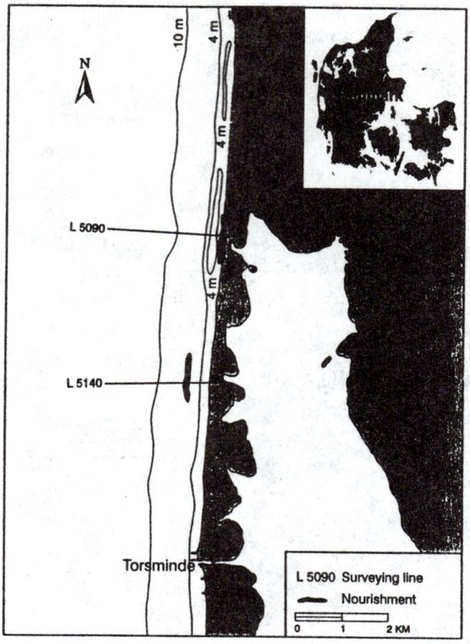

Fig.6 Location Torsminde test site

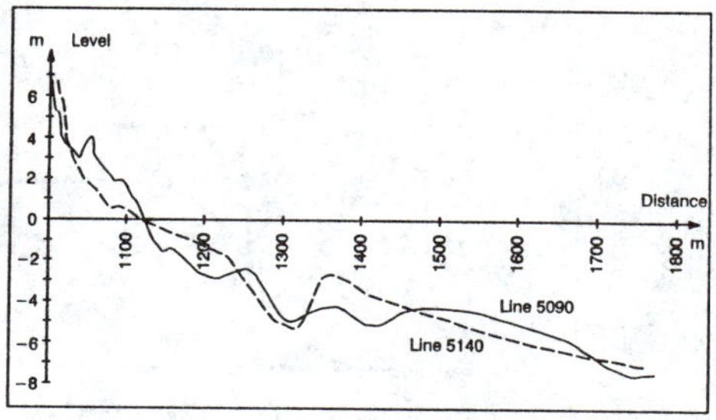

Fig.7 Typical cross-shore profiles

Separate beach and shoreface nourishments took place in 1993. Over longshore stretches of one km each, approximately 250,000 m^3 sand has been nourished in both cases; so approximately 250 m^3/m. The D_{50} of the borrow sand (D_{50} = 0.34 mm) for the BN was slightly less than the native sand; the borrow sand (D_{50} = 0.58 mm) for the SN was coarser than the native material.
The BN was placed between DNN +4 m and DNN -1 m (DNN = Danish Normal Zero); the SN between DNN -4 m and DNN -6 m. The cross-shore profiles have been divided in 3 layers: upper: DNN +4 m to DNN -4 m; middle: DNN -4 m to DNN -6 m; lower DNN -6 m to DNN -10 m.

Estimates of the longshore coastal constants have been calculated using the CERC formula; the distribution of the transport over the different zones have been found by using the Svašek and Bijker (1969) method.

In the modelling the determination of the cross-shore coastal constants is a vital item. In the Torsminde case different sets have been tested using the observed behaviour as criterion. A reasonable set was eventually found.

To judge the quality of the computation, a comparison with the observed behaviour was made. The autonomous behaviour was set on an average coastline retreat of 6.7 m/year. This value was based on measurements in the period 1978 - 1988.

In the example case of Fig.5 y-values are computed for each 100 m along the coast. In the Torsminde case also a 100 m spacing was used in the computations. In order to facilitate comparison of measurements with computations, averaged values for stretches of coast of 1 km length (so-called boxes) have been used.

The fluctuations of the measured y-values turned out to be relatively large compared to the calculated development of the y-values. It is not expected that these large fluctuations are mainly due to the two nourishments.

Overall the comparison between the calculated and the observed development showed in some cases similarities and in some other cases considerable differences. One of the shortcomings of the present model is that the model can not cope with the effects of the different grain sizes on the morphological development.

Norderney

As already discussed by Bakker *et al.* (1994), the Norderney case is in fact too complicated to be modelled properly with a 3-line modelling technique. The complicated bottom topography in front of the test site and the presence of a number of groynes made this site in fact unsuitable for a simple line-modelling technique.

Terschelling

In the Terschelling case (see Figs.8 and 9) the shoreface was nourished in 1993 with a volume of 2.1 million m^3 over a length of 4.6 km at a depth of Datum -7m to -5m. In fact the most seaward trough of the bar system was filled. The SN was located in the middle part of a 12 km long study area.

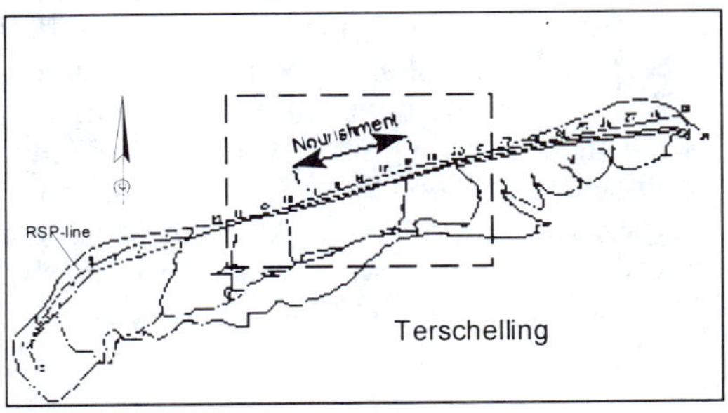

Fig.8 Location Terschelling test site

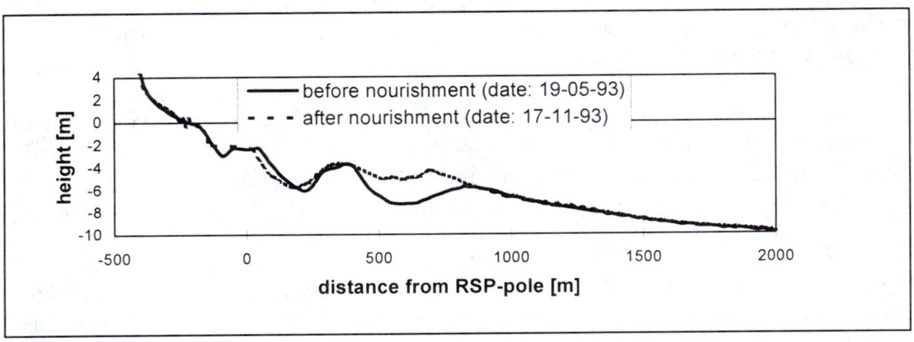

Fig.9 Typical cross-shore profiles

The native bed material shows a typical distribution over the cross-shore profile: D_{50}= 0.24 mm at the beach and gradually decreasing to D_{50}= 0.16 mm at the deeper shoreface. The D_{50} value of the borrow material was 0.20 mm.
Apart from the yearly profile measurements (since 1964; 200 m spacing), just before and after the execution of the SN many more profile measurements have been carried out. After the nourishment e.g. in a lapse of time of 2 years 11 series of measurements have been carried out.

The shape of the cross-shore profile was schematized in 3 layers. Normally (fixed) depth contours are chosen as layer limits. In the Terschelling case because of the bar system, the upper layer was defined from Datum +3 m to Datum -3 m (most landward contour); the lower layer was defined from Datum -6 m (most seaward contour) to Datum -9 m. All the sediment between the most landward Datum -3 m depth contour and the most seaward Datum -6 m depth contour was considered to belong to the middle layer. So even if a trough of the bar system is deeper than Datum -6 m or a crest is higher than Datum -3 m the sediments (or the lack of sediments) involved belong to the middle layer. With this procedure the entire (with time moving) bar system is confined in the middle layer.
The 3-line model is based on the hypothesis that an equilibrium profile exists in cross-shore direction. It might be questioned whether this concept holds as well in case of a bar system with a cyclic behaviour as for Terschelling.

The longshore coastal constants have been derived with the CERC formula. The cross-shore coastal constants have been derived by trial and error, taking the observed behaviour after the SN into account. Rather short T_0 (time constant for diffusivity) values had to be adopted (T_{01} = 1.5 years upper to middle zone; T_{02} = 5 years middle to lower zone). It is felt that the rather small value T_{01} of only 1.5 years is partly 'artificially'. As discussed in section 'Shoreface nourishments in practice' some salient formation landward of the SN is expected because of gradients in longshore transport. Although this phenomenon can not be properly modelled with the present technique, the effects are accounted for by the relatively small T_{01} value.

The study area was divided in several (8 in this case) boxes (1.2 km each).
Prior to the SN the autonomous behaviour of each box could be determined since many profile measurements are available. In the final judgment of the observed and calculated behaviour of the coast, the calculated autonomous behaviour of each box was taken into account.

The calculated development of the y-values of the upper zones was very similar to the measured development after the application of the SN. The large observed volume increase in the upper zone is, in the present simulation, almost entirely caused by cross-shore transports from the middle zones. It might be questioned whether this is physically correct.
The simulation results for the middle zones were less good. E.g. the observed general movement of the SN in eastward direction could not be properly modelled. The differences between the measured and modelled behaviour of the lower zone were in some cases large.

ESTIMATES OF COEFFICIENTS

In the preceding section the modelling results of the 3 NOURTEC cases have been briefly discussed. The Torsminde and Terschelling cases could be modelled more or less successfully. The Norderney case has been reported by Bakker *et al.* (1994); the complexity of that problem was demonstrated.

The quality of the modelling results depends to a large extent on a proper choice of the coastal constants. Reliable longshore constants could be determined with the CERC formula. The proper choice of the cross-shore constants turned out to be more difficult. Till now a sound theoretical basis is lacking for the determination of these constants. By trial and error (using the observed behaviour after execution of the SN's) useful estimates have been found. With these estimates at least some essential characteristics of the behaviour could be modelled. This procedure makes it difficult to use the model in an *a priori* predictive mode in other cases.
In Table 1 the constants as used in the Torsminde and Terschelling cases have been summarized.

	Torsminde	**Terschelling**
T_{01}	5.0 years (s_{y1} = 0.32 m/year)	1.5 years (s_{y1} = 1.33 m/year)
T_{02}	5.0 years (s_{y2} = 0.27 m/year)	5.0 years (s_{y2} = 0.30 m/year)
s_1	2.65 * 10^6 m^3/year/rad	3.63 * 10^6 m^3/year/rad
s_2	0.23 * 10^6 m^3/year/rad	0.96 * 10^6 m^3/year/rad
s_3	0.00 * 10^6 m^3/year/rad	0.00 * 10^6 m^3/year/rad

Table 1 Coastal constants Torsminde and Terschelling cases

The longshore constants for the upper zone for Torsminde and Terschelling are of the same order of magnitude; in the middle zone the constants for Torsminde are somewhat smaller than for Terschelling. This is partly caused by the difference of the height of the upper limit of the middle zone; DNN -4 m for Torsminde and Datum -3 m for Terschelling. Both in Torsminde and in Terschelling no longshore sediment transports are assumed to occur in the lower zone.

The time constants for diffusitivity in cross-shore direction are generally 5 years, except for the Terschelling case between upper and middle zone (T_{01} = 1.5 years). There is no reliable theory available yet to predict time constants. In the Terschelling case the observed cycle for the (offshore) movement of the bar system is 10 - 15 years. It is felt that the time constants should have the same order of magnitude; 5 years is in this respect not too strange. The rather small value of T_{01} = 1.5 years for Terschelling is probably because in this value also the (not modelled) effects of longshore sediment transport gradients are accounted for.

With the use of mathematical cross-shore transport (and morphological) models in principle estimates of cross-shore constants could be derived. Although this procedure has not yet followed in the present NOURTEC cases, promising results have already been achieved in other Dutch research cases. [See Steetzel (1996).]

CONCLUSIONS

The 3-line modelling technique has proved to be a powerful method to represent the most important characteristics of the behaviour of SN's after placement. Different design alternatives can be easily compared. It is certainly not a technique to answer all questions which might be raised related to the application of SN's.

A serious problem in analyzing the behaviour of SN's after placement is the distinction between the SN-induced behaviour and the autonomous behaviour. Much of the observed differences between model results (added to the estimated autonomous behaviour) and the measured behaviour, is caused by uncertainties about the estimated autonomous behaviour.

The present 3-line modelling technique is not able to cope with differences between borrow and native material.

Although the modelling results were in some respects not fully successful, the study has revealed that modelling attemps can serve as a very fruitful focus point for better understanding of the complex behaviour of shoreface nourishments.

ACKNOWLEDGEMENTS

The study was sponsored by partners in the NOURTEC project: Rijkswaterstaat, National Institute for Coastal and Marine Management/RIKZ; the Netherlands (Terschelling case), Coastal Research Station Norderney; Germany (Norderney case) and Danish Coastal Authority; Denmark (Torsminde case). The NOURTEC project was co-sponsored by the Commission of the European Union, Directorate General XII, Marine Science and Technology (MaST II) Programme under contract MAS2-CT93-0049.

REFERENCES

Bakker, W.T. (1968) The dynamics of a coast with a groyne system. Proc. 11th Int. Conf. on Coastal Eng., London, UK.

Bakker, W.T, N.F. Kersting and H.D. Niemeyer (1994) Line-modeling of Shoreface Nourishment. Proc. 24th Int. Conf. on Coastal Eng., Kobe, Japan.

CUR (Centre for Civil Engineering Research, codes and specifications) (1987) Manual on Artificial Beach Nourishment. Report 130; Rijkswaterstaat and DELFT HYDRAULICS; Gouda, the Netherlands.

Groenewoud, M.D. (1996a) Three-line modelling of the Torsminde supplies. Delft University of Technology, Fac. of Civil Eng., May 1996.

Groenewoud, M.D. (1996b) Three-line modelling of the Terschelling supply. Delft University of Technology, Fac. of Civil Eng., October 1996.

Groenewoud, M.D. (1996c) Summary of the three-line modelling of the NOURTEC supplies. Delft University of Technology, Fac. of Civil Eng., October 1996.

Mulder, J.P.M., J. van der Kreeke and P. van Vessem (1994) Experimental Shoreface Nourishment, Terschelling (NL). Proc. 24th Int. Conf. on Coastal Eng., Kobe, Japan.

Steetzel, H.J. (1996) Beach nourishment and rock structures; behaviour oriented modelling of the combination of beach nourishment and rigid shore parallel structures. DELFT HYDRAULICS, Delft (in preparation).

Svašek, J.N. and E.W. Bijker (1969) Two methods for determination of morphological changes induced by coastal structures. XXIInd Int. Nav. Congress Paris, section II, item 4.

Swart, D.H. (1974) Offshore sediment transport and equilibrium beach profiles. PhD thesis Delft University of Technology.

CHAPTER 222

BEACH MONITORING PROGRAM OF VALENCIA (SPAIN)

José Serra[1], and Josep R. Medina[1]

Abstract

The littoral processes and long-term shore protection plans are analysed for the coast of Valencia (Spain). The breakwaters built during this century by the Port Authorities have gratly affected the most Southern beaches by interrupting the natural longshore sand transport. The monitoring program of the beaches of Valencia (Spain) has been established for a precise estimation of the evolution in time of the beaches to the North and South by the Port of Valencia. Periodic topographic surveys provide the basic time-dependent beach description of the area. The two main objetives are the following: **a)** 3-D description of the beach surface, including estimation of reliability and measurement errors of bathymetrics, and **b)** stochastic description of wave climate.

High precision and cost-efficient beach surveying techniques have been developed during the monitoring program of the beaches of Valencia (Spain). Beach profiles and zero-shorelines have been obtained with systems and errors similar to the terestrial topography. The methods are simple enough to be applied by general land surveyors with a minimum training and may be extended to a variety of beaches for systematic and low-cost monitoring programs.

Introduction

Most beaches of the Gulf of Valencia in Spain suffer an erosion/acretion

---

1 **Department of Transportation, Universidad Politécnica de Valencia, Camino de Vera s/n, 46022 Valencia, SPAIN.**

processes induced by inappropiate constructions and unadequate coastal policies. The rivers and the natural littoral sand drift created more than a hundred miles continuos natural sand beach which has been altered in several points by breakwaters, jetties, comercial ports and marinas. According to Yepes (1995), economic, social and political presures and different kind of lobbies are interacting on the sand beaches, which are considered the most critical production factor of the touristic sector in Valencia (10% of the Spanish GNP).

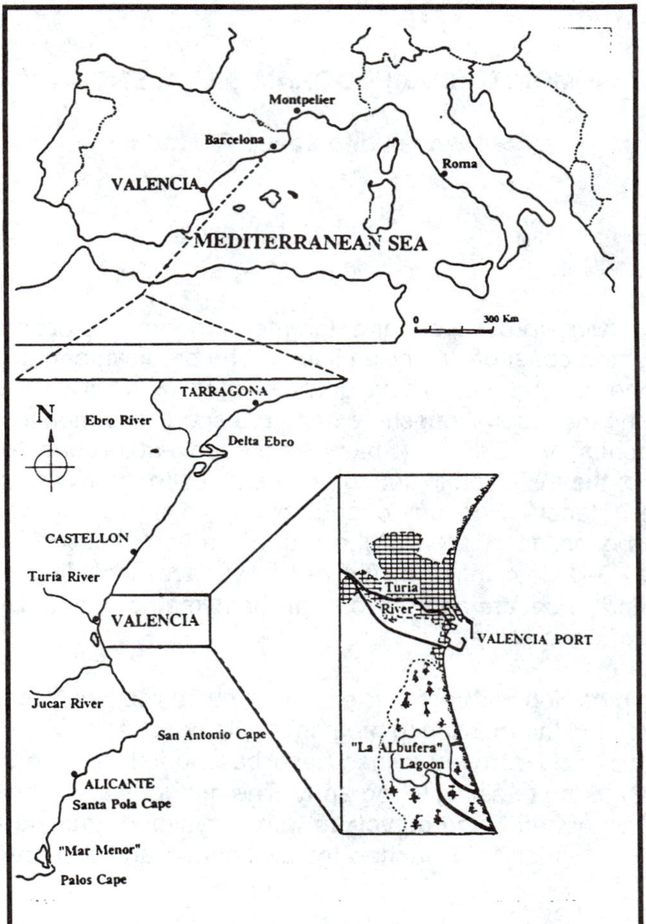

Figure 1. Gulf of Valencia and Area of Study

Cost-efficient beach monitoring programs appears as key elements in developing appropiate shore protection and beach nourishment plans. A high precision low-cost beach profile system developed during the monitoring program of Valencia (see Medina and Serra, 1993) is being applied for beach monitoring North and South of Valencia and may become a common low-cost system for autonomous and descentralized control of beaches along the Spanish Mediterranean coast

Environmental Characteristics

Figure 1 shows the location of the area of study in the Gulf of Valencia. Serra (1986) analysed the natural morphodynamic unit known as "Ovalo Valenciano" (Gulf of Valencia), located in the Westerm Mediterranean Sea, between the Delta Ebro (North), and the San Antonio Cape (South); it is almost a continuous sand beach of approximately 270 kilometers.

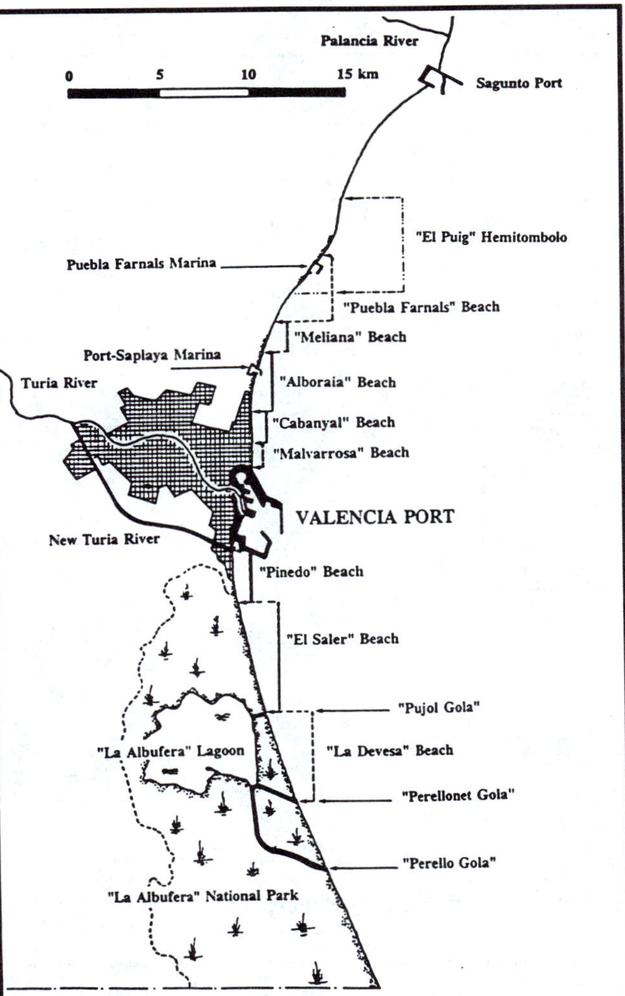

Figure 2. Area of Study

Figure 2 shows the monitorized area located in the "Ovalo Valenciano", between the "Puebla Farnals" marina on the Northern limit and de inlet of the "La Albufera" lagoon, known as "Gola del Perellonet" on the Southern limit, littoral stretch length approximately 30 kilometers. The longshore transport of the coast of Valencia is about 500.000 m^3/yr North to South and 200.000 m^3/yr South to North.

During centuries, there was not enough construction and dredging capacity to build up breakwaters and docks able to resist a sand current of half a million m^3/yr, and a variety of projects enlarging the ancient Port of Valencia failed because of the sand sedimentation processes. Nevertheless, modern construction techniques and the economic development of this century altered the traditional conditions.

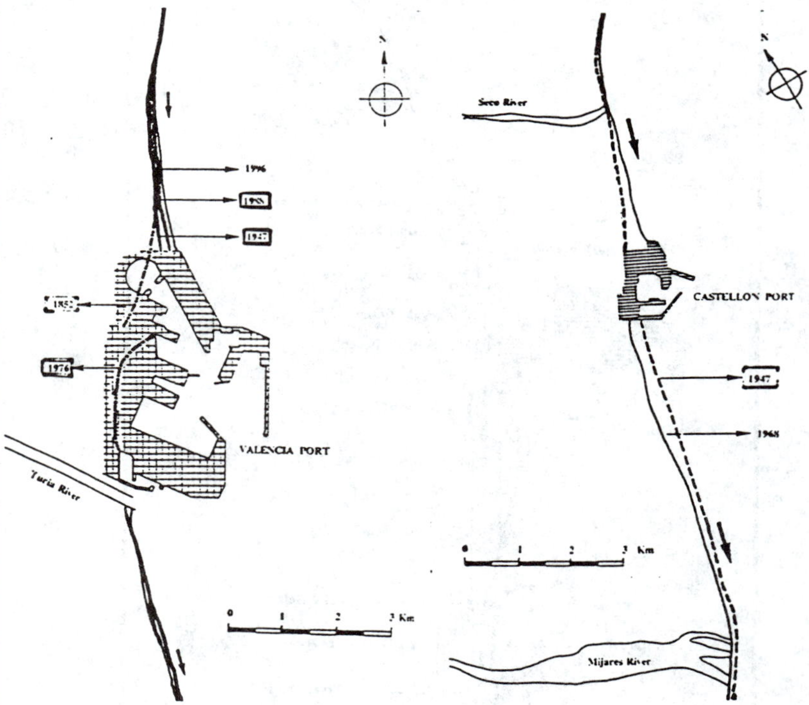

Figure 3. Modification of the Shoreline Around of the Ports of Valencia and Castellón

A barrier to the littoral drift was established on the sandy coast and an erosion process was generated since then on the beaches South of the harbor, were the shoreline is retreating about one meter per year. Not only the Port of Valencia, but also the dams built in the Turia river and marinas North of Valencia are contributing to the beach erosion process.

Unfortunately, neither the Port nor the dams are expected to be removed as barriers in the near future, because they are key elements to the economic development of the region. Additionally, the urban development of the Southern beaches (El Saler) has affected the natural mobility of dunes resulting in a significant retreat of the shoreline during the last three decades.

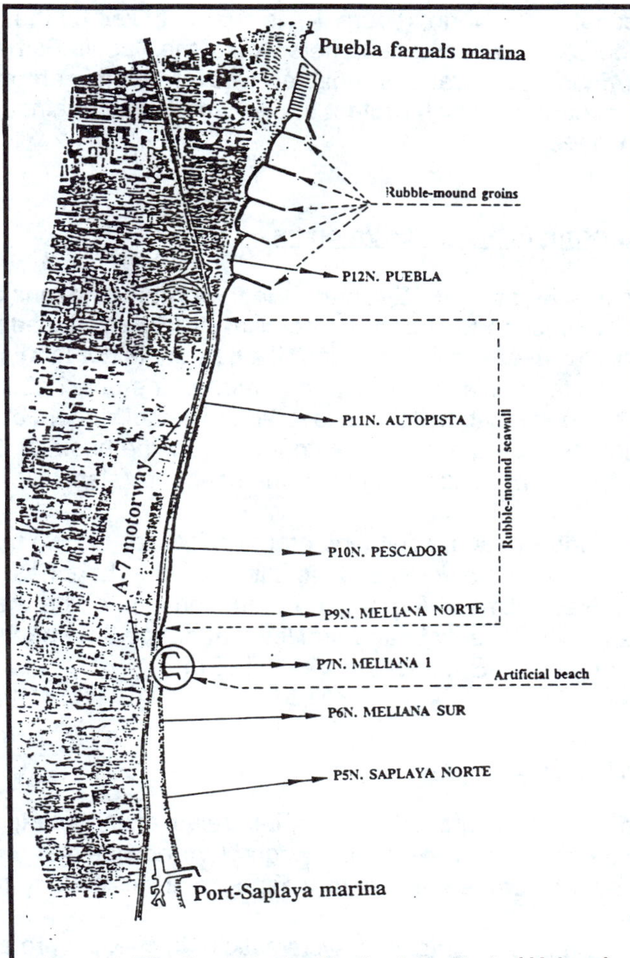

Figure 4. The Coast to the North of the Port of Valencia

The case of the Port of Valencia and the acretion on the Northern beaches and erosion of the Southern beaches is similar to other cases along theGulf of Valencia. The figure 3 shows the modification of the shoreline during this century around of the Ports of Valencia and Castellón.

The coast South of the Valencian Port has two different stretches. The one next to the Port is under erosion and the furthest from the Port is stable. On the other hand, the coast North of The Port of Valencia, presents streches in acretion and others in erosion. The erosion is located South of the barriers of the marinas; the areas in acretion are located North of the marinas, and on the beach North of the breakwater of the Port of Valencia.

The coast to the North, (figure 4), is characterized by the high number of constructions of coastal and nearshore works: the Puebla Farnals and Port-Saplaya marinas, and various coastal work fenders, emphasizing the longgshore revetments which protect the A-7 motorway against direct action from strong waves.

Beach Monitoring Program of Valencia

The sand deposits at "El Saler" is in the order of dozens of millions of cubic meters, and short term problems of vital structures caused by erosion is not expected. However, the construction of a new container terminal in the Port of Valencia and the social and economic value of the beaches in the area of Valencia (1.300.000 inhabitants) has favored the establishment of a systematic beach monitoring program. This monitoring will be the basis of the long-term planning and shore protection works on the coast of Valencia.

The Valencia beach monitoring program focuses on two main points: 1) the topographic and batymetric description of the beaches, and 2) the description of wave climate. Figure, shows the location of the wave recording equipment and in the twenty beach profiles that define the beach morphology to the North and to the South of the Valencia Port.

Bathymetric Survey

From the theorical point of view, it is relatively simple to describe the beach erosion process. The aerial topography and batymetry of the beach defines the surface of the sedimentary deposits or rocks along the coast.

The common equipment for systematic bathymetries are echosounders on boats with GPS which are affected by temperature, salinity, and calibration method. The bathymetries on boats must be related to a "zero level" which requires a high precision topographic network and a continuous measurement of the mean water level during the beach survey; it is a difficult to estimate the error level of the survey because the high precision measurement techniques, when available, are low-efficient.

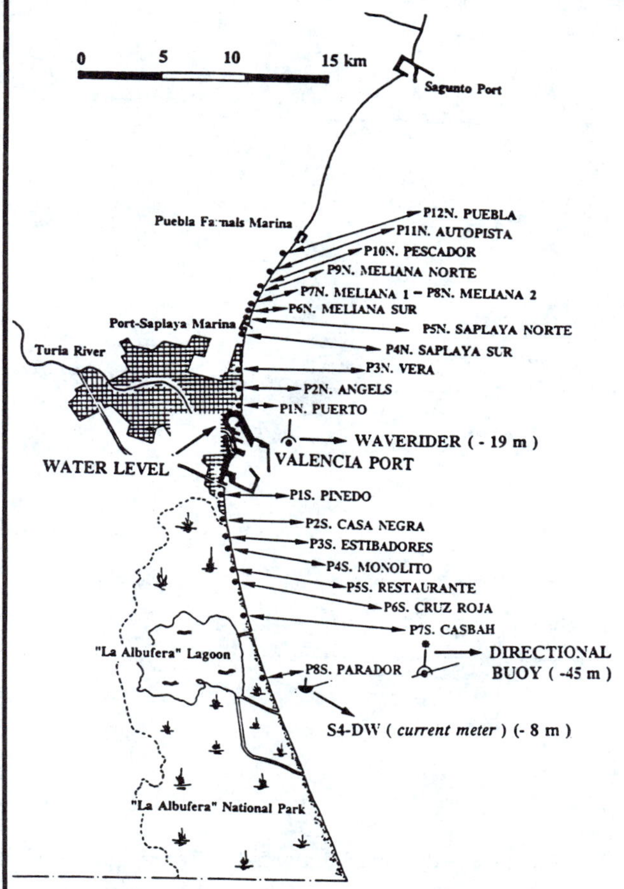

Figue 5. Beach Profiles and Wave Recording Equipment

Figure 6 shows a scheme of the three different surveying techniques commonly used in the beach monitoring programs of Valencia. The first technique is a systematic automated method based on an echosounder boat mounted with GPS. The second technique is a manual method based on an small dinghy and chain; and the third technique is a manual method based a land rod readapted for marine purposes, this technique is known as ***Beach Profiler (BP)***.

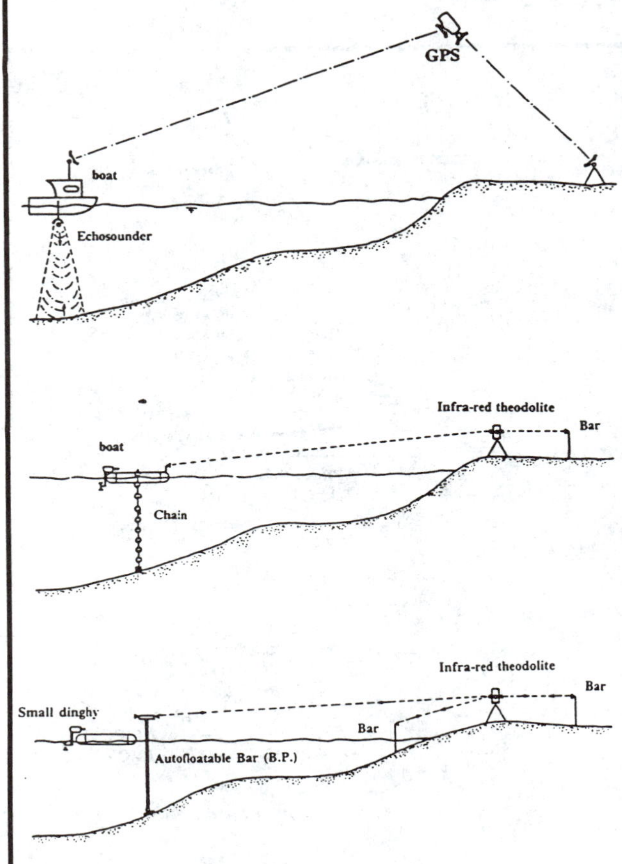

Figure 6. Scheme of Bathymetric Techniques

<u>**The Beach Profiler (BP)**</u> Within the last decade new measuring systems for bathymetric survey with a common criteria of moving land topography to marine topography have been considered by different research groups. Following this research line the Laboratory of Ports and Coast of Univerisidad Politécnica de Valencia has developed the denominated Beach Profiler (BP) represented in Figure 7. The unit has proved its effectivity as a precise unit for beach profile measuring in beach monitoring. A basic idea is that, in the long-term, cost-efficient beach monitoring techniques appear to be key elements in developing apropiate shore protection and beach nourishment plans.

BP is a high precision and low-cost beach profiler system developed during the frist monitoring program of "El Saler" which is being also applied for monitoring the beaches North of Valencia, and may become a standard low-cost system for regional and decentralized control of beaches along the Spanish Mediterranean coast.

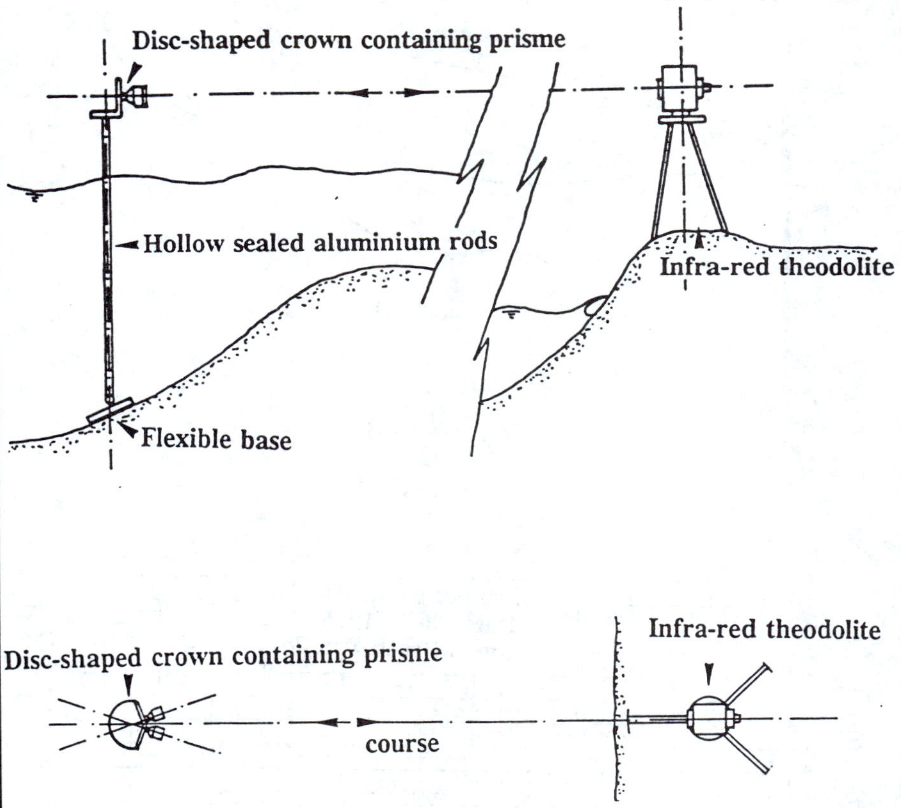

Figure 7. Beach Profiler (BP)

The BP system only requires a two man work team aboard to a small boat, and a land surveyor with a conventional infra-red equipment. The especial designed element to measure the level of the sea botton was a self-floatable aluminium bar, with a crow on the top on which the infra-red reflectors are fixed covering all directions and an articulated led plait in the botton. The BP shows the following characteristics: high precision (error < 2 cm), high efficiency (about 60 points/hour), low cost, and simplicity. Figure 8 shows a typical enveloped and standard deviation of the elevation of the measured profiles using BP.

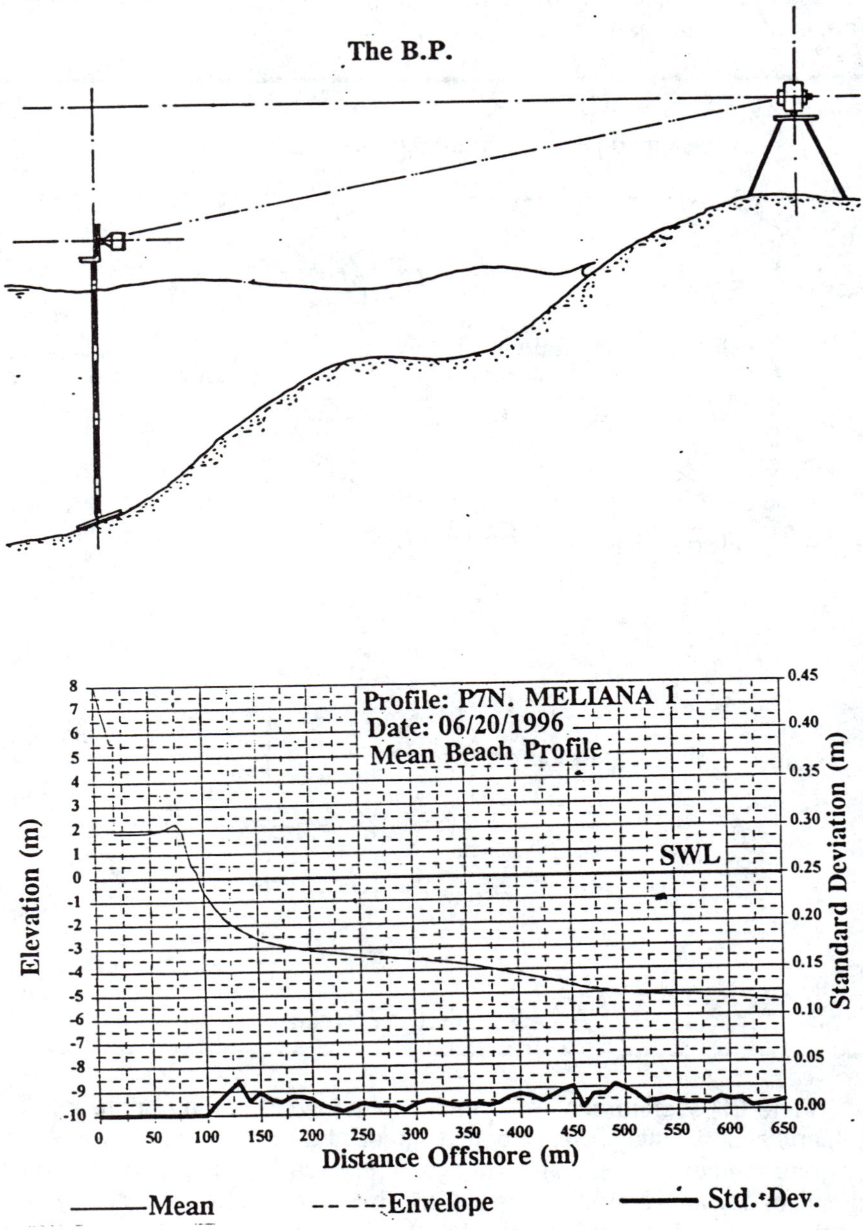

Figure 8. Profile Envelope and Standard Deviation in Elevation (P7N.MELIANA1) Using the BP

Beach Profiles: Errors

Esteban (1993) and Esteban et al. (1995) analysed the problems associated with the use of different bathymetric systems in beach monitoring. Table 1 shows a comparison of different bathymetric tecniques, incorporating the results obtained by the BP.

Bathymetric Tecnique	Sources of Error	Estimated Error	Cost
ECHOSOUNDER	⇨ Mean water level: *Tides* *Sstrong surges,...* ⇨ Calibration: *Temperature (° C)* *Salinity (g/l).* ⇨Boat movement: *Heave, Pitch, Roll.*	≈ 10 cm *(Mediterranean)*	High
CHAIN	⇨ Mean water level: *Tides* *Strong surges,...* ⇨ Manual operation	≈ 20 cm *(Mediterranean)*	Low
SLED *Stauble et al. (1993)*	⇨ Terrestrial topography ⇨ Positioning	σ ≈ 7 cm (Atlantic)	High
CRAB *Birkemeier et al. (1993)*	⇨ Terrestrial topography ⇨ Positioning	σ ≈ 1 cm (Atlantic)	Very High
BEACH PROFILER	⇨ Terrestrial topography ⇨ Positioning	σ ≈ 2 cm (Mediterranean)	Low

Table 1. Sources of Error Corresponding to Different Bathymetric Techniques

Conclusions

High precision and cost-efficient beach surveying techniques have been developed during the monitoring program of the beaches of Valencia (Spain).

The BP method is simple enough to be applied by a general land surveyor with a minimum training and may be extended to a variety of beaches for systematic and low-cost monitoring programs. The technical characteristics: are:

1.- *High precision (1 cm < error < 2 cm).*

2.- *High effiicency (about 60 points/hour).*

3.- *Maximun length profile: 1000 m.*

4.- *Maximun water depth: 10 m.*

5.- *It does not requires specially trained personal.*

6.- *The BP may be considered a simple terrestrial topography system adapted for monitoring beaches.*

Continuous mean water level measurements during surveys are unnecesary, neither is special equipment nor highly trained personell required.

Therefore, the BP is adequate for manual beach monitoring and the problems of the BP are similar to terrestrial topography.

References

Esteban, V (1993). Sistemas de Información y Medición Batimétrica. Proc. *II Jornadas Españolas de Ingeniería de Costas y Puertos*, Gijón (Spain), 10-11 May 1993, 163-174.

Esteban, V., Aguilar, J., Serra, J., and Medina, J.R. (1995). Levantamientos y Seguimientos Topo-Batimétricos en Ingeniería de Costas (in Spanish). *Ingeniería del Agua*, Vol. 2 (April 1995): 181-200.

Lee, G. And Birkemeier, W.A. (1993). Beach and Nearshore Survey Data: 1985-1991 CERC Field Research Facility. *Technical Report CERC-93-3*. U.S. Army Corps of Engineers.

Medina, J.R. , and Serra, J. (1993). Beach monitoring Program of El Saler (Spain). *Spain-U.S. Workshop on Natural Hazards*, Barcelona, 8-1 June, 1993, 279-291.

Stauble, D.K., Garcia, A.W., Kraus, N.C., Grosskopf, W.G. and G.P. Bass (1993). Beach nourishment Project Response and Design Evaluation: Ocean City, Maryland. *Technical Report CERC-93-13*. U.S. Army Corps of Engineers.

Serra, J. (1986). Procesos Litorales en las Costas de Castellón. (in Spanish) *Ph.D. Thesis* presented at the Universidad Politécnica de Valencia, Spain.

Yepes, V. (1995). Gestión Integral de las Playas como Factor Productivo de la Industría Turística. El Caso de la Comunidad Valenciana. (in Spanish). Proc. *III Jornadas Españolas de Ingeniería de Costas y Puertos*. Valencia, 3-4, May, 1995, 958-976.

CHAPTER 223

BAR MIGRATION AND DUNEFACE OSCILLATION ON DECADAL SCALES

Marcel J.F. Stive[1,2], Jorge Guillen[1,3] and Michele Capobianco[4]

Abstract

While structural erosion or accretion of a coastal stretch is of primary concern to coastal management, knowledge of coastline oscillations on larger time scales relative to the erosional or accretional trend would allow for a more efficient management practice. Our analysis of more than thirty years of observations of dune, beach and nearshore evolution reveals some of the typical oscillatory behaviour in time and space on a decadal scale. We have focused on the behaviour of the duneface and analyzed the possible relation between the duneface behaviour and that of the nearshore morphology. Amongst our results are the findings that the demeaned oscillatory duneface evolution on a decadal scale is not only correlated with the recurrence frequency of the migratory bar system, but also with a cumulative measure of episodic wave events.

Introduction

Moderate rates of structural erosion or accretion only become apparent on larger time and space scales, because of the dominance of shorter scale oscillations caused by natural processes. Most evident is that due to episodic wave events, which can cause strong erosion and associated steepening of the duneface, while the eroded sediment is deposited on the beach and nearshore profile, resulting in a more weakly sloping and wider beach. After these events a

---

[1] Netherlands Centre for Coastal Research, Delft University of Technology, PO Box 5048, 2600 GA Delft, The Netherlands
[2] Delft Hydraulics, PO Box 177, 2600 MH Delft, The Netherlands
[3] Instituto de Ciencias del Mar, Paseo Juan de Borbon, 08039 Barcelona, Spain
[4] Tecnomare SpA, 3584 San Marco, 30124 Venezia, Italy

restoration may occur due to aeolian and hydrodynamic processes. Our interest is typically into the scales intermediate to the short term, event related, behaviour and the long term, structural trend of erosion or accretion associated with gradients in cross-shore and longshore directions over a coastal stretch.

These intermediate scales concern the temporal behaviour on decadal scales, leading to duneface amplitude oscillations of magnitudes comparable to those of the structural trend. Our objective has been to develop an understanding of the mechanisms behind this behaviour, based on the "working-hypothesis" that a relation may be expected between the bar dynamics and the duneface behaviour. The JARKUS data-set, comprising more than thirty years of field observations of the duneface, beach and nearshore profile along the central Netherlands coast, has revealed typical and unexpected properties of subaqueous bar morphodynamics on a decadal scale. Amongst the most recent analyses is that of Wijnberg and Terwindt (1995) pointing towards the existence of four distinct morphodynamic regimes, each characterized by their own particular bar dynamics. These regimes occur in four coastal stretches, each constrained by coastal engineering structures. While the precise physical interpretation and explanation remains to be resolved, our objective (stimulated by the findings of Guillen and Palanques, 1993) has been to investigate and explain the possible relation between the above described bar behaviour and the duneface dynamics.

It is clear that the dune system and the bar system interact throught the dynamic processes of the coastal area; they are both subject to the same wave and tidal forcings and, at same time, they are part of an interconnected system. We recall the fact that not only tides have clearly defined astronomic cycles with a periodicity ranging from hours to decades, but also wave climate has at least a seasonal cycle and, on a long term basis, longer cycles linked to climate oscillations. These are still active subjects of investigation; think for instance to the so-called Bruckner cycles, basically consisting of the recurrence in northwester Europe of periods of cold and damp alternating with warm and dry years, the average interval between successive maxima being 34.8 years (as calculated by Bruckner in 1890), though individual cycles vary from 25 to 50 years. However there is an increasing evidence that, when looking at large scale and long term evolution of the coastal system, we need to highlight the long term character of the forcing factors.

Study area

Our study area is the Holland coast (Fig. 1), bounded by Rotterdam harbour and the port of Den Helder. It is characterized by an almost uninterrupted dune system, without barrier islands and tidal inlets. The main human interventions along this stretch of 120 km are the harbourmoles of Scheveningen and IJmuiden, the sea dike of Petten and the dike of Den Helder.

During the last three hundred years this coast is erosive in the north (approx. 1 m/yr), and the south (approx. 0.3 m/yr) and accretive in the centre (approx. 0.2 m/yr). Sediment grain sizes are 200 to 250 microns in the dunes and 250 to 300 microns on the beach. The tide ranges from 1.7 m near Scheveningen to 1.4 m near Den Helder. Mean wave heights and periods are 1.2 m and 5 s respectively, with approaches from south-west and north-northwest mainly.

Foredunes are regressive, stable or progressive over the last decades, with widths ranging from 500 to 2500 m. The subaerial beach averages 43 m in width, with a mean slope of 1:15 (Short, 1991). The bar-beach system commonly consists of a beach-bar, attached to the beach as a ridge and runnel system, an outer bar, highly rhythmic and rip-dominated and a longshore bar (Short, 1992). Their temporal and spatial behaviour has been analyzed by Wijnberg (1995). A common characteristic is that of a cyclic two or three bar system, initiated

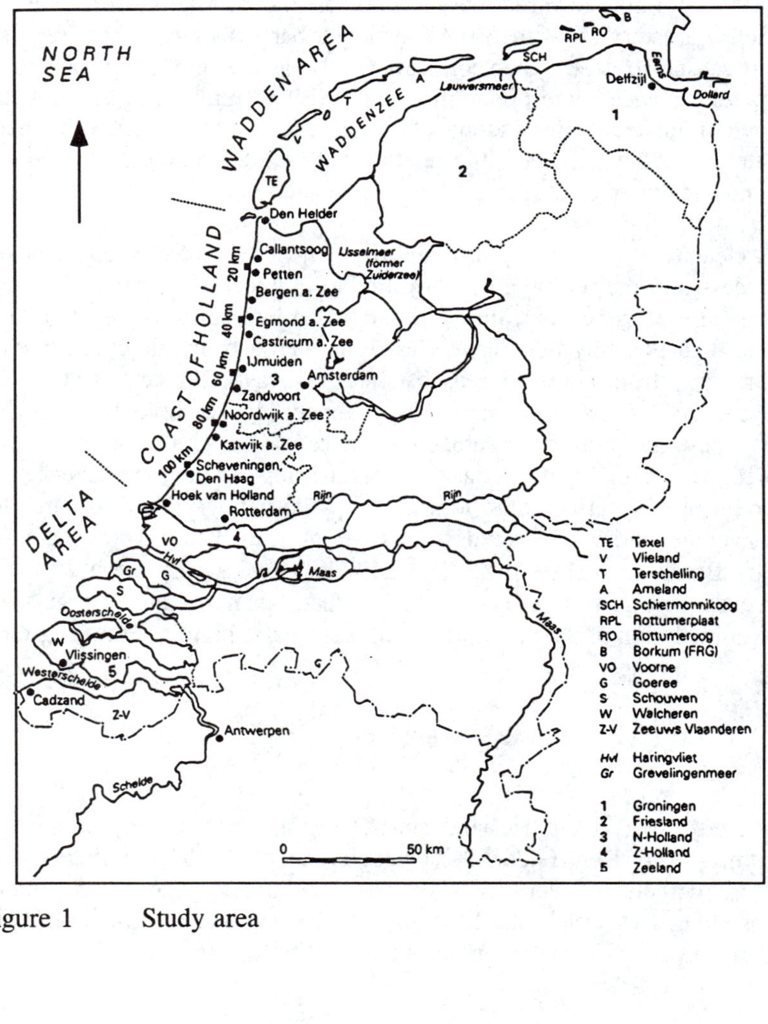

Figure 1 Study area

nearshore, migrating and growing through the surfzone and then decaying offshore. Apparent return periods range between 4 and 15 years.

Methodology

The medium-term dune, beach and nearshore profile evolutions were analyzed using the JARKUS data-set. It consists of (about 3,000) fixed measuring rays extending along the Dutch North Sea coast. The distance between the profiles is 200-250 m and yearly profiles approximately from the foredune to 1,000 m seaward are available since 1964. Profiles are usually surveyed between early April and late September. It is expected that the biased sampling does not affect a decadal analysis of the shapes of the profiles (Wijnberg and Terwindt, 1995).

Our morphometric analysis of the dune and beach system concerns first the determination of temporal and spatial oscillations on decadal scales in the dunefoot, and second the possible relation of this behaviour with the shoreface slope and the behaviour of bar and trough systems. Our parameters are the dunefoot position, the duneface slope and the beachface slope.

Qualitatively it is clear that the shoreface slope plays a fundamental role in the dynamics of the coastal system. Firstly, in the sense that in the steeper slope area, the bar system needs longer return period waves to be moved offshore. Second, in the steeper slope area, the dunefoot is subject to larger wave impact under storm conditions. In addition, residual tidal currents will determine a relatively larger speed of motion for the shallower part of the bar system with respect to the deeper part, which is a clear mechanism of introduction of a "structural modification" in the coastal system. If we consider two adjacent zones with different profile slope, this is clearly a mechanism for the introduction of a longshore mechanism of motion.

Following Wijnberg (1995) we distinguish four distinct spatial sectors (Fig. 1), viz. (I) km 8-20, (II) km 28-52, (III) km 63-95 and (IV) km 104-114. These sectors are bounded by structures, the Hondsbossche dike (km 23) and the harbours of IJmuiden (km 55) and Scheveningen (km 102), and have been found to exhibit a uniform bar behaviour within them. In our analysis we further concentrate on Zones (II) and (III), which contrary to (I) and (IV), are virtually without human interference.

Our initial effort has been to derive a representative duneface and dunefoot definition, such that it is not affected by local and instantaneous processes occurring immediately before the survey. A heuristic procedure, based on hydrodynamic and morphologic considerations, using two constant planar surfaces and a sediment volume balance was chosen (Fig. 2). By using this procedure the temporal changes in the calculated dunefoot position indicate the changes in the volume of sediment stored in the dune-beach system between +1 and +5 m NAP heights, allowing a comparison between all the profiles in a coherent way. The choice of the +5 m NAP height as the upper boundary is based on the storm surge design level height for dune erosion in Holland.

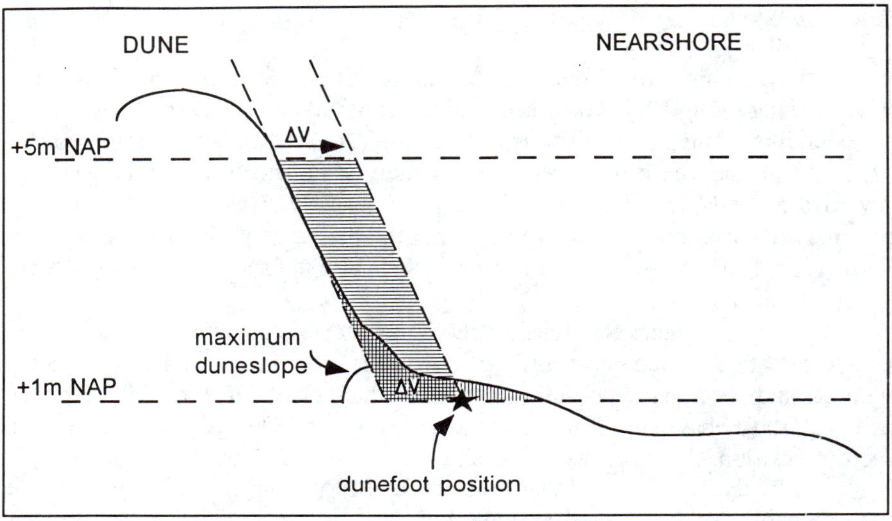

Figure 2 Dunefoot definition procedure

In the analysis the dunefoot and nearshore profile evolution has been based on space-time diagrams. Linear temporal trends were calculated in each profile and the "residuals" with respect to this trend were extracted. This approach has been complemented by application of Empirical Orthogonal Function (EOF) analysis techniques, the presentation of which has been omitted here because of space limitations. We have used the EOF's to derive periodicities more clearly.

In analyzing possible effects of wave forcing we have used a time history of 1,090 "storm events" during the period 1979-1991 (De Valk, 1994). They have been characterized by two parameters, viz.

$A = H_{m0}*SQRT(t)$, and

$B = SL*SQRT(t)$,

where H_{m0} is the average significant wave height during the storm event, SL the average surge level and t the duration of the storm event. Furtheron we have summed these parameters over the period in between soundings to account for the cumulative effect of storm events.

Results and analysis

Dunefoot data

Following our derivation of a morphologically equivalent and stable duneface profile, we find that the average maximum duneface slope is nearly constant along the different zones (tan b = 0.14 to 0.15). The average beach slope between the + 1 m NAP and the - 1 m NAP ranges from 0.025 to 0.030, implying a mean beachface width of 70 to 80 m.

The linear trend of the dunefoot position shows a high longshore variability (Fig. 3), but the average over several kilometres is less than 1 m/yr, while the accretional and erosional character is not very different from that over the last few hundred years. Standard deviations relative to the trend also show a high longshore variability (Fig. 4), while the average over several kilometres is between 5 and 10 m.

We now describe the behaviour of Zones II and III, showing a general erosive and accretive trend respectively, which are relatively untouched by human intervention.

The evolution of the residual dunefoot position in Zone II exhibits longshore and temporal oscillations (Fig. 5). Spatially, alternating accretional and erosional stretches of 2 to 3 km are observed alongshore. Temporally, this longshore rhythmicity may be viewed as a shoreline wave propagating towards the south with a propagation velocity of 150 to 200 m/yr. The amplitude of the oscillation is approximately 20 m and its periodicity 15 years.

In contrast the evolution in Zone III shows virtually no longshore oscillations (Fig. 6). Temporally, periods of relative accretion and erosion alternate simultaneously along the coast.

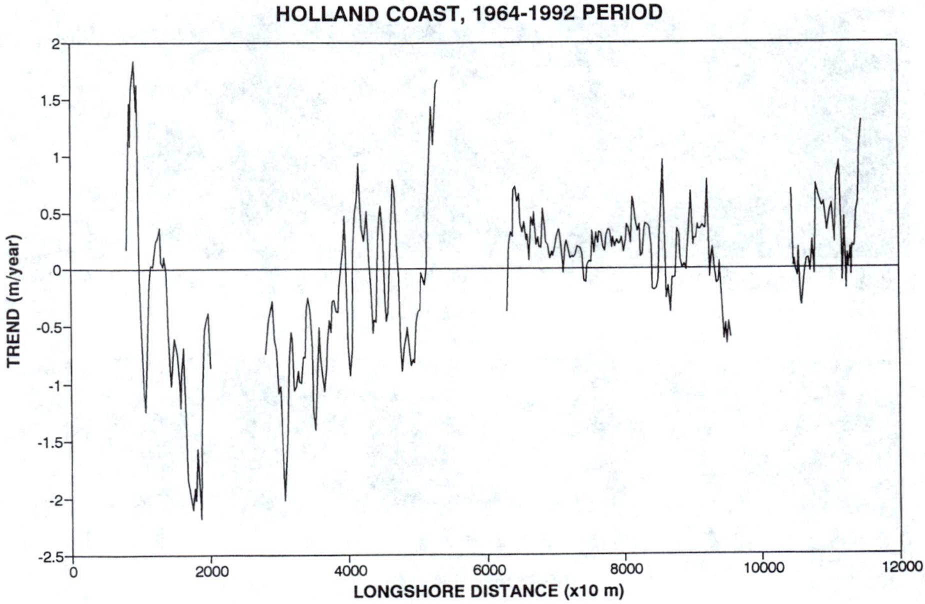

Figure 3 Trend in the dunefoot position (positive is seaward, negative is shoreward)

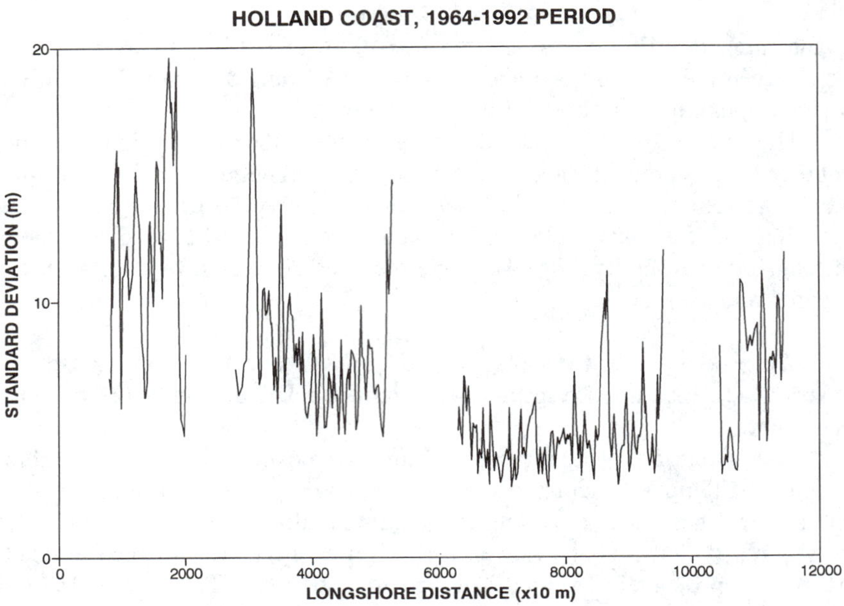

Figure 4 Standard deviation of the dunefoot position

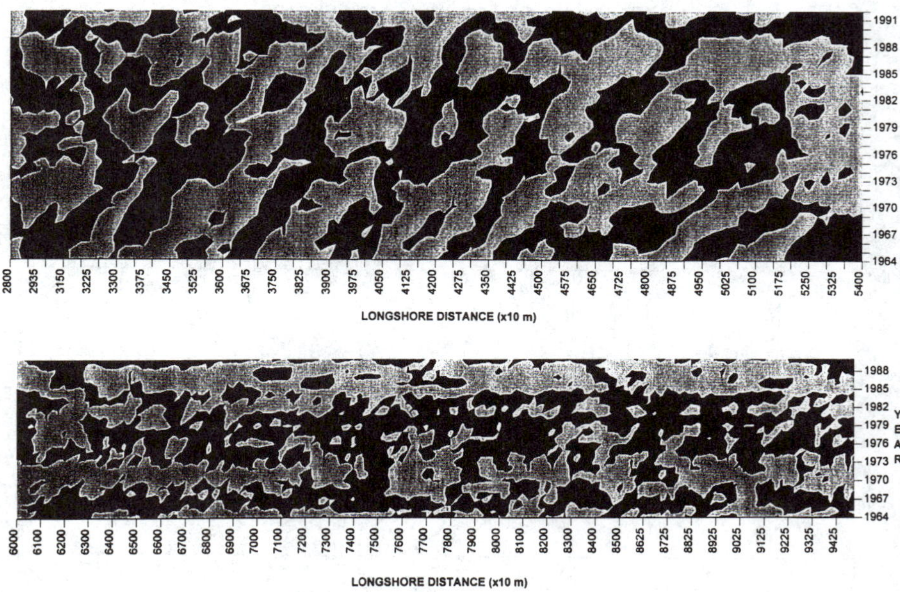

Figure 5 and 6 Residual dunefoot position relative to the trend for zones II and III respectively (dark shading is shoreward, grey shading is seaward)

Also, in spite of the above differences, inspection of Fig. 5 and 6 reveals some simultaneous occurrences of relative erosion and accretion. This may be more clear from Fig. 7, comparing the spatially averaged residual position over time in the two zones. Although our extreme event data only cover the period 1980 to 1991 (Fig 8), we note that we expect that these simultaneous occurrences are related to the decadal variation of the cumulative extreme events. For instance, the relative accretion between 1985 and 1988 appears to correspond to a relative minimum in the extreme events.

A preliminary analysis of the periodicity of the first empirical orthogonal function of the dunefoot position for the whole coast has also been performed. It consists of applying a simplified non-linear model as a reference concept. In practice the temporal evolution of the first and the second empirical orthogonal functions has been considered as determined by the sum of a linear (sinusoidal) term and two nonlinear terms related to the amplitude of the first one. Using empirical process-oriented considerations we assume that, under erosive conditions, the dunefoot is less mobile (composed by less mobile sediments), while under accretive conditions, the dunefoot is more mobile (composed by more mobile sediments). This way three fundamental periods have been identified: 51 years for the oscillation of the mean dunefoot position; 7.5 years for the oscillation of the dunefoot during erosive periods; 2 years for the oscillation of the dunefoot during the accretive periods. These results (see Figure 9) need to be further examined mainly because of the short duration of the dataset with respect to the longer period of oscillation. The results also need to be further examined with respect to the cycles in the forcing factors. We note that the dunefoot evolution is recovered by multiplying the "reconstructed" temporal orthogonal functions and the spatial temporal orthogonal functions. In case the results will be confirmed by further analyses, it is clear that, with a very simple modelling concept we are able to gain some predictive value.

Profile data

A main morphological feature of the profile along the Holland coast is the existence of migrating breaker bars. A comprehensive analysis of their behaviour was recently made by Wijnberg (1995), using EOF analysis. In each of the four coastal stretches the bar behaviour is found to be temporally and spatially different, but internally (i.e. within a particular stretch) coherent. The explanation of this is as yet undetermined.

Zone II and III both exhibit an offshore migratory behaviour of the bars. The average number of bars is constant in time, where the offshore decay of the outer bar is followed by the appearance of a new bar at the shoreline. The temporal behaviour is thus similar, only the recurrence frequency is different, viz. approximately 15 years in Zone II and 4 years in Zone III. Their spatial behaviour is significantly different, however. Whereas in Zone III the relative

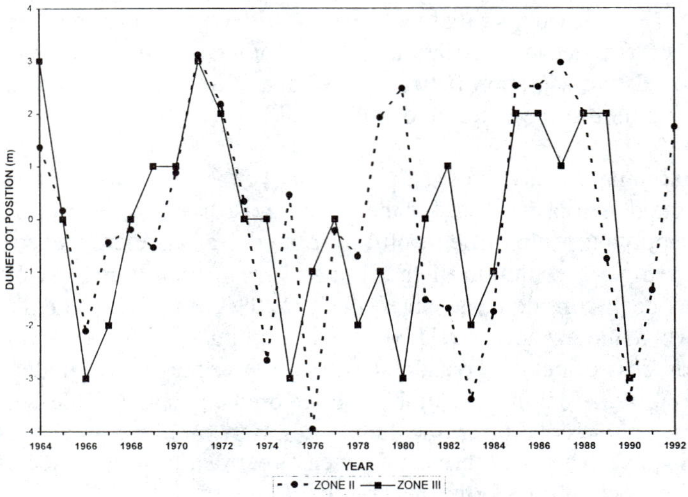

Figure 7 Longshore-averaged mean dunefoot position versus time

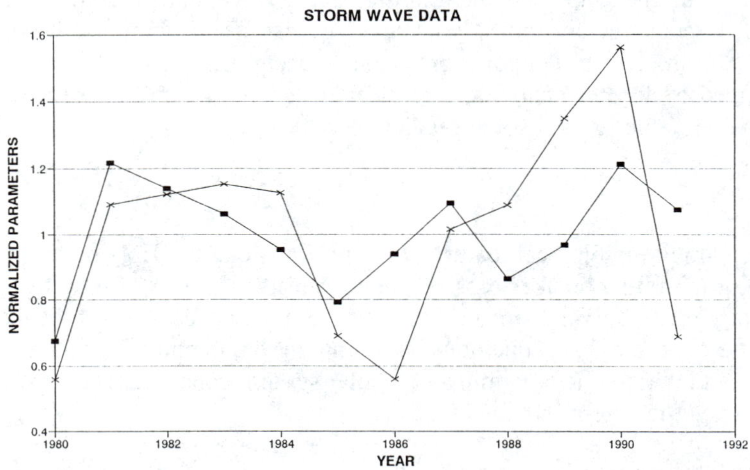

Figure 8 Normalized year-cumulative storm event parameters

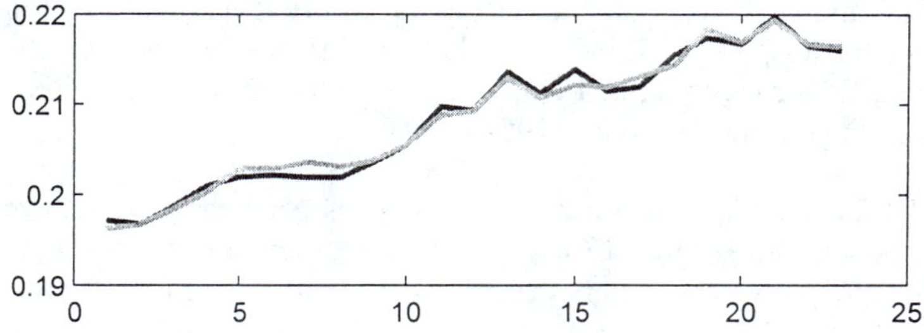

Figure 9 Normalized first temporal eigenfunction of the dunefoot displacement of all four zones (full line observed, grey line modelled)

position is homogenous alongshore, without any longshore rhythmicity, Zone II has two specific spatial features. First, at a large spatial scale, the bar migration is not in phase, exhibiting an obliquity relative to the shore. Apparently, the bar initiation first occurs in the south and this initiation propagates towards the north. The obliquity creates a discontinuity. Second, on a smaller spatial scale, longshore discontinuities (of a length of 2 to 3 km) in bar position exist, which are caused by the crescentic morphology of the bar systems here.

Discussion and conclusions

The standard deviation of the mean shoreline position has been termed beach mobility (Dolan et al., 1978). Short and Hesp (1982) found it to be a function of the morphodynamic beach state: dissipative, intermediate and reflective beaches correspond to low-moderate, moderate-high and low beach mobility respectively. On Australian shores beach mobility ranges between 5 and 14 m from data taken over 1 to 5 years. Shoreline mobility along the Holland coast, averaged over several km (as inferred from the + 1 m NAP depth contour) over the period 1964-1992 amounted to some 20 m, suggesting an increase of mobility over longer time scales. On the other hand, dunefoot mobility, such as defined here, is approximately half that of the shoreline. Apparently, the shoreline evolution exhibits more "noise" than that of the dunefoot. Further, it is noted that the dunefoot mobility on shorter time scales, e.g. due to extreme events, can be nearly an order-of-magnitude larger, i.e. 25 to 75 m of duneface retreat (Steetzel, 1993). While longer term erosive or accretive trends are relevant for long term nourishment policy, our findings of mobility are relevant to medium term (decadal) coastline management.

While the order-of-magnitude of dunefoot mobility has been found the same along the Holland coast, its spatial and temporal nature is quite different in the two, virtually unmanaged, Zones II and III. Our findings suggest that the dunefoot behaviour is strongly influenced by the subaqueous bar behaviour, confirming earlier suggestions by Bruun (1954).

In Zone III, both the dunefoot and the subaqueous bar behaviour behave spatially homogeneous alongshore. Erosional-accretive oscillations appear to correlate with the cumulative parameters describing the storm events.

The latter effect is also found in Zone II, but here most interestingly is the occurrence of the longshore rhythmicity and of the recurrence frequency. It exhibits spatially the same scale as that of the crescentic bar morphology (2 to 3 km) and temporally the same recurrence frequency as that of the migratory bar behaviour (15 yr). Clearly, there exists a strong signature of the bar dynamics in the shoreline and dunefoot behaviour.

A comparison of earlier observed temporal oscillations and -if present- their alongshore propagation speed is given in Table 1. We note the study of Verhagen (1989), who interpreted from long term shoreline observations the presence of 'sand waves' along the Holland coast, moving northwards.

Author	length (km)	migration rate (m/yr)	amplitude (m)	period (yr)
Bruun (1954)	0.5-3	0-1,000	60-80	-
Morton (1979)	5-7	-	-	-
	2.5-3	-	-	-
Dolan & Hayden (1981)	>1	-	-	-
Davidson-Arnott (1988)	0.5-2.5	150-300	50-90	10
Verhagen (1989)	5.5	65	40-60	75-100
Pelczar et al. (1990)	5-9	100-200	70-110	50-60
This study	2-3	150-200	20	15

Table 1. Observed shoreline oscillations

Our final conclusion is that our interpretation of shoreline mobility on decadal scales for the Holland coast indicates that the dunefoot, according to our definition, shows an oscillatory amplitude of some 20 m. It is quite moderate compared to other studies, and its behaviour is clearly related to the dynamics of bar behaviour on the one hand and to the cumulative effect of extreme wave events on the other hand.

Acknowledgements

Part of this work is based on work in the PACE- and SAFE-projects, in the framework of the EU-sponsored Marine Science and Technology Programme (MAST-III), under contract no. MAS3-CT95-0002 and MAS3-CT95-0004. The second author would like to acknowledge his grant from the Ministerio de Educacion y Sciencia of Spain, allowing for his post-doctoral stay at Delft University of Technology in 1995. We thank Mark Voorendt of Delft University of Technology for his technical support.

References

Bruckner E., Klimaschwankungen seit 1700 nebest Bemerkunge uber die Klimaschwankungen der Diluvialzeit, "Geographische Abbandlungen" B. IV. H. 2, Wien, 1890, pp. 153-184.

Bruun, P., 1954. Migrating sand waves or sand humps, with special reference to investigations carried out on the Danish North Sea Coast. Proc. 5th Int. Conf. Coastal Engineering, New York, ASCE: 269-295.

Davidson-Arnott, R.G.D., 1988. Temporal and spatial controls on beach/dune interaction, Long Point, Lake Erie. Journal of Coastal Research, Special Issue No 3: 131-136.

De Valk, C.F., 1994. Selection of storm events and estimation of exceedance frequencies of significant wave height for five North Sea locations. Delft Hydraulics, Report H 1931, 54 pp.

Dolan, R. and Hayden, B.P., 1981. Storms and shoreline configuration. Journal of Sedimentary Petrology, 51: 737-744.

Dolan, R.; Hayden. B.P. and Heywood, J., 1978. Analysis of coastal erosion and storm surge hazards. Coastal Engineering, 2: 41-53.

Guillen, J. and Palanques, A., 1993. Longshore bar and trough systems in a microtidal, storm-wave dominated coast: the Ebro Delta (NW Mediterranean). Marine Geology, 115: 239-252.

Morton, R.A., 1979. Temporal and spatial variations in shoreline changes and their implications, examples from the Texas Gulf Coast. Journal Sedimentary Petrology, 49: 1101-1111.

Pelczar, M.; Nejczew, P. and Mielczarski, A., 1990. Cartometric analysis of the shore-line changes on eastern part of the Polish Baltic coast in the last century (in Polish). Rozprawy Hydrotechniczne, 51.

Short, A.D., 1991. Beach morphodynamic systems of the central Netherlands coast, Den Helder to Hoek van Holland. Rapport GEOPRO 1991.01, University of Utrecht: 106 pp.

Short, A.D., 1992. Beach systems of the central Netherlands coast: Processes, morphology and structural impacts in a storm driven multi-bar system. Mar. Geol., 107: 103-137.

Short, A.D. and Hesp, P., 1982. Wave, beach and dune interaction in southeastern Australia. Marine Geology, 48: 259-284.

Steetzel, H.J., 1993. Cross-shore transport during storm surges. PhD. thesis, Delft University, The Netherlands: 242 pp.

Verhagen, H.J., 1989. Sand waves along the Dutch coast. Coastal Engineering, 13: 129-147.

Wijnberg, K.M., 1995. Morphologic behaviour of a barred coast over a period of decades. PhD. Thesis, University of Utrecht: 250 pp.

Wijnberg, K.M. and Terwindt, J.H.J., 1995. Quantification of decadal morphological behaviour of the central Dutch coast. Marine Geology, 126: 301-330.

CHAPTER 224

Morphological development of the Terschelling shoreface nourishment in response to hydrodynamic and sediment transport processes

P.Hoekstra[1], K.T. Houwman[1], A. Kroon[1], B.G. Ruessink[1], J.A. Roelvink[2] and R. Spanhoff[3]

Abstract

The Terschelling shoreface nourishment was studied over a period of about 2.5 years. The nourishment was originally designed to act as a feeder berm. Since its implementation, the nourishment has completely satisfied the design objectives. However, the nourishment only partly acts as a feeder berm and the actual success of the nourishment is also based on its breaker berm function. The creation of a wave shadow zone leads to an interception of part of the longshore drift and the development of a salient effect.

Introduction

The Dutch barrier island of Terschelling is part of the northern coastline of the Netherlands. The island is facing the North Sea and separates the North Sea from the backbarrier system of the Wadden Sea (Fig. 1). The North coast of the island consists of a sandy shoreface, flanked by sandy beaches and dunes. In the last decades the central part of this coastline has chronically suffered from erosion. Over an alongshore distance of 5 km, the average annual coastal retreat was about 2-3 m/yr. This corresponds with an estimated, average annual loss of sand of 110.000 m^3. In 1993 a shoreface nourishment was implemented along the Terschelling coast to stabilize the existing coastline. The main design objective of this nourishment was (and is) "to return the Transient Coastline to a position seaward of the Basal Coast Line (Dutch BKL concept; NOURTEC, 1995) and to ensure that this Transient Coast Line will not retreat landward of the Basal Coast Line during the next 8 years" (Biegel and Spanhoff, 1996). In terms of design dimensions in total 2.1 million m^3 of sediment was involved and this sediment was supplied to the nearshore zone, filling up the trough between the middle and outer breaker bar (Fig. 1).

----

[1]Institute for Marine and Atmospheric Research Utrecht, Utrecht University, P.O. Box 80115, 3508 TC Utrecht, The Netherlands.
tel: +31 302532753; fax: +31 302540604; e-mail: p.hoekstra@frw.ruu.nl
[2]Delft Hydraulics, P.O. Box 177, 2600 MH Delft, The Netherlands.
[3]National Institute for Coastal and Marine Management (RIKZ), P.O. Box 20907, 2500 EX 's-Gravenhage, The Netherlands.

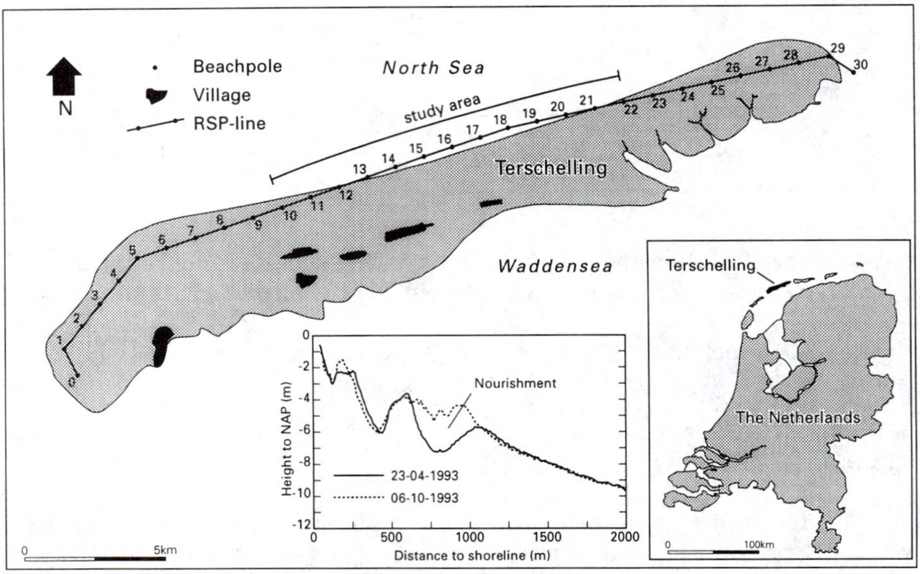

Figure 1. The North coast of the island of Terschelling with the study area. The beachpole (RSP)-line indicates the location of the cross-shore transects; the main measuring array is located in section 17

For the nourished zone, the amount of sediment supplied was equivalent to an average, vertical change in sea bed elevation of about 1 m and, in a longshore direction, the amount of nourished sediment per cross-section was in the order of 450 m^3/m. The nourishment, with a total length of about 4.4 km and extending from km section 13.7 to 18.1 (Fig. 1), is located in the depth interval between -5 and -7 m below NAP (Dutch Ordnance Datum; the zero - NAP level approximately equals mean sea level).
In 1993 the EC-MAST2 programme NOURTEC was launched to study and explain the behaviour of shoreface nourishments in different European coastal environments (Hoekstra et al., 1994, Knaack et al, 1997 and Laustrup et al, 1997). The final objective of the NOURTEC programme is to study, determine and explain the feasibility, effectiveness and optimum design characteristics of shoreface nourishment techniques for different environmental conditions.

The basic assumption underlying the design and implementation of the Terschelling shoreface nourishment is that eventually sand will be carried to the shore. The nourishment is expected to act as a feeder berm. In case of a feeder berm there has to be a significant net onshore directed sediment transport due to the presence of the nourishment.
Apart from being a potential feeder berm, a shoreface nourishment may also start acting as an offshore located breaker berm or as a combination of these two features. A shoreface nourishment acting as a breaker berm is able to reduce the erosional power of the waves by increasing the degree of wave energy dissipation, in particular due to wave breaking. The nourishment creates a "wave-shadow zone" landward from the

nourishment which will locally reduce the strength of the wave-driven longshore current and the littoral drift.

This paper will focus on the developments at the Terschelling site in the first 30 months after the implementation of the nourishment. The aim of the present study is to analyse and evaluate the potential role as both a feeder berm and submerged breakwater. First of all, the bathymetric data set is used to determine the main morphological evolution in the nearshore zone and to compute changes in sediment volumes for specific horizontal and vertical control sections. The second way of adressing the problem is by analyzing the available process measurements and model computations on (suspended) sediment transport in the nearshore zone. It comprises an analysis of the dominant role of wave- and flow-driven processes in generating cross-shore (and landward-directed) and/or longshore fluxes of sediment.

Methods of research

Bathymetry

Bathymetric surveys were frequently carried out with varying time intervals and the total number of surveys until January 1996 is 12. After the nourishment the time interval varied from 53 to 182 days between two consecutive soundings. The surveyed area North of Terschelling covers approximately 25 km^2 and measurements were performed along survey lines perpendicular to the coast. Inside the nourished area the longshore spacing of these survey lines varied from 25 to 100 m; outside the nourished area the longshore spacing was about 200 m. The supra- and intertidal part of the cross-shore survey line, starting at the dunefoot, was done by leveling. The subtidal section until a depth of at least 10-12 m was measured by a Rijkswaterstaat survey vessel using a digital acoustic depth sounder (ATLAS DESO, 210 kHz), in combination with an accurate positioning system (Syledis and dGPS).

Hydrodynamics

The hydrodynamic field measuring programme of Terschelling consisted of two main activities: a long-term, continuously recording monitoring network and a series of process-oriented measuring campaigns (4). The monitoring network has been operational for about 2.5 years and almost all measurements were more or less concentrated around cross-sectional transect 17 (Fig. 1), covering the central part of the nourishment.
Tidal water levels have been measured by a tide station, located in cross-section 14.60 and in a water depth of approximately 10 m. Every 10 minutes a mean value of the water level has been recorded. Offshore wave conditions were registered with a wave-directional buoy (WAVEC) positioned in the central measurement section in a water depth of about 15 m. The WAVEC measured continuously with a frequency of 1.28 Hz. Each block of 10 minutes was automatically processed using standard zero down-crossing and spectral methods. Information on the time-dependent variation in water levels (waves and tides) was also obtained from 2 measuring poles positioned in the surfzone (section 17.00; Fig.2). Both poles were equipped with a pressure transducer and a capacitance wire, measuring water levels for 40 minutes per hour with a frequency of 4 Hz. The fourth element of the monitoring network consisted of an instrumented tripod, equipped with 2 electro-magnetic flow meters at about 0.25 and 1.2 m. above the bed and with a pressure sensor. Occasionally, the tripod also carried two Optical Back Scatter (OBS) sensors to measure suspended sediment concentrations (0.15 and 0.25 m above the bed).

The instrumented tripod measured in a burst mode sampling scheme with a burst interval of one hour, a burst length of 2048 s. and a sampling frequency of 2 Hz. The tripod was located at the seaward margin of the nourishment, at a water depth of approximately 5.5 m.

During the process-oriented measuring campaigns the monitoring network was substantially expanded and consisted of the following configuration: 2 offshore located wave directional buoys, 6 instrumented tripods in the nearshore zone - including 3 tripods with OBS's - and 2 poles. Emphasis in the measuring programme was given to processes operating in a cross-shore direction and, consequently, a cross-shore array of instruments was installed in section 17 (Fig. 2). Four concentrated measuring campaigns were carried out in the period November 1993 until November 1995; each campaign lasted for approximately 5 to 6 weeks. Data calibration, validation and analysis is discussed in further detail by Houwman and Ruessink (1997) and Hoekstra et al (1996).

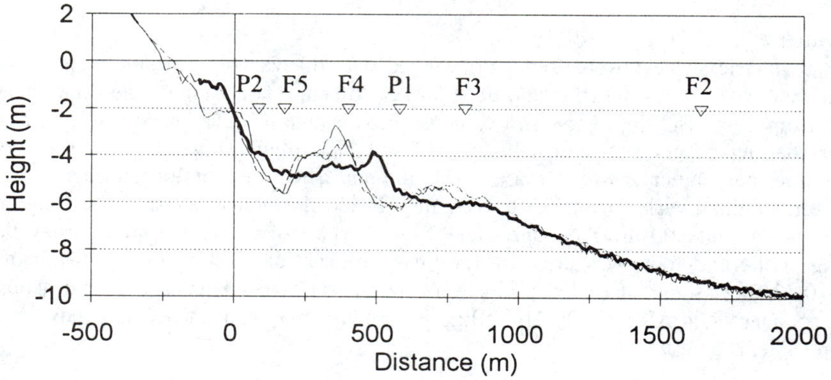

Figure 2. Measuring array in the cross-shore profile of section 17 (T2 campaign drawn line, T3 dashed line and T4 thick line); P = Poles and F = Frames or instrumented tripods

Hydrodynamic boundary conditions

Tidal conditions

Tides along the North coast of the island of Terschelling are semidiurnal and are characterized by a mixed micro- to mesotidal range. A maximum range of about 2.8 m is recorded at spring tide whereas the neap tide range is about 1.2 m. Associated tidal currents are basically flowing shore-parallel. The tidal flow during flood is in an ENE direction and the ebb tidal flow is directed towards the WSW. Especially at neap tide, the tidal ellipse is almost flat and rectilinear.

As expected, tidal flow is considerably modified by the presence of wind- and wave-driven currents (compare Whitford and Thornton, 1993). Earlier observations (Hoekstra et al., 1994) make clear that during moderate winds (< 8 Bf) wind-driven flow has the same order of magnitude as the tidal flow. During storm and heavy storm conditions, the wind- and wave-driven longshore flow, in combination with cross-shore (mean) flow patterns, will even fully dominate the nearshore flow regime.

Wind and waves

The wind climate of Terschelling shows a clear predominance of westerly winds with a prevailing wind force of 4 to 5 Bf. In the period 1984-1994 the predominant wind direction for each year is in the sector between 210 and 300 degrees. This is more or less parallel to the coastline in the nourished zone (Kruyt, 1995). More than 12% of the total number of observations is in the sector 210-240 degrees. In addition, wind forces between 3 and 6 Bf make up 79% of the total number of observations in the data set. The year 1994, just following the nourishment, appears to be a year with a relatively large number of storms with a considerable duration (Kruyt, 1995).
The wave climate of Terschelling has been analysed by Van Beek (1995). Nearly 65% of the total number of wave observations is related to obliquely incident waves from the West to North (270-360 degrees; Fig. 3). The most energetic wave fields are incident from the NW to NNW. This almost coincides with a shore-normal direction (Fig. 3). The average annual significant wave height ($H_{1/3}$) and significant wave period ($T_{1/3}$) for 1994 are 1.08 m (st.dev. $\pm$ 0.67 m) and 7.0 s (st.dev $\pm$ 1.4 s), respectively. For only 2% of the total distribution, wave heights are higher than 3 m (Van Beek, 1995). However, as already mentioned before, in the first three months following the nourishment a number of significant storm events with high seas and swell have been recorded in the study area: November 15, 1993 ($H_{1/3}$ = 5.50 m; data correspond to peak of the storm), December 10, 1993 ($H_{1/3}$ = 5.50 m), December 20, 1993 ($H_{1/3}$ = 5.00 m), January 24, 1994 ($H_{1/3}$ = 4.25 m) and January 28-29, 1994 ($H_{1/3}$ = 8 m; all data based on WAVEC observations offshore).

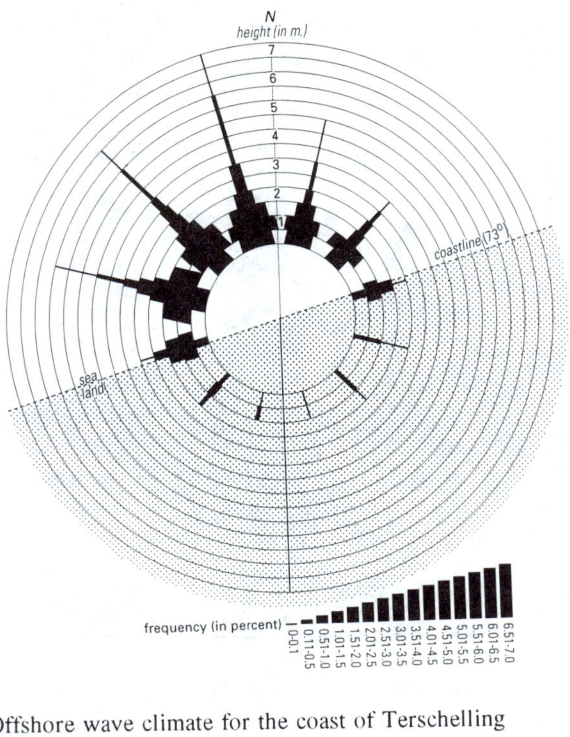

Figure 3. Offshore wave climate for the coast of Terschelling

Morphological evolution and sediment volumes

Coastal behaviour prior to the nourishment

The natural behaviour of the nearshore zone before the nourishment was studied in detail by Ruessink and Kroon (1994). Herein, both the alongshore and cross-shore tendencies of especially the long-term evolution of the nearshore breaker bars was discussed. In particular, the (geometrical) properties and dynamics of these bars, like their position and depth, were documented in time from 1965 to 1993. With the results of this study, the following conclusions were obtained:

- The cross-shore behaviour of bars is strongly influenced by an alongshore migration of bar attachment-points. Migration rates of these attachment-points are highly variable and can be in the order of 1200 m.year^{-1}.
- The cross-shore behaviour of a bar is schematized in *three phases*:
  1) generation close to the shore; 2) seaward migration from 300 to 1300 m offshore and 3) bar degeneration when the crest of the bar is at a depth of about -5.5 m NAP. There is a coupling mechanism involved: as soon as the outer bar disappears the inner is triggered, leaves phase 1 and enters phase 2.

This typical cyclic behaviour, also found along other parts of the Dutch coast (see for example Wijnberg, 1995), has a total period of about 12-15 years (Ruessink and Kroon, 1994).

Morphological response of the nourishment

The morphological behaviour of the nearshore zone in the post-nourishment conditions has been studied and reported by e.g. Kroon et al (1995) and Westlake (1995) and summarizing their results indicates the following trends (Hoekstra et al, 1996):

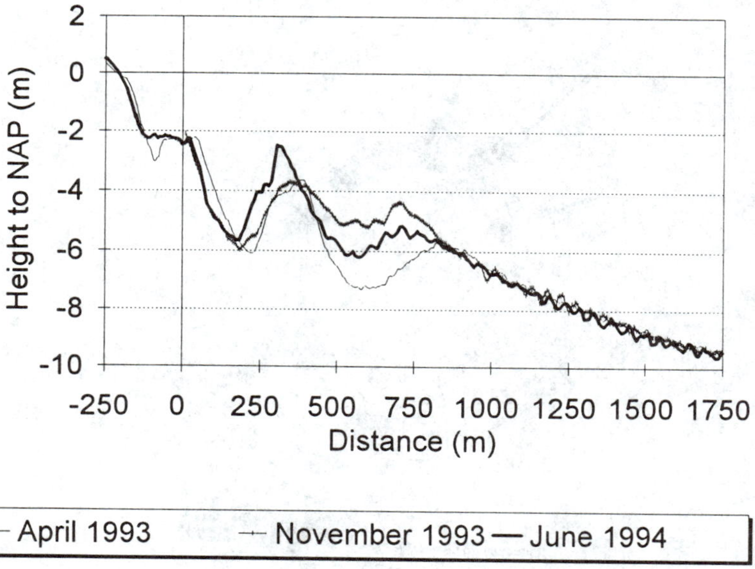

Figure 4. Morphological development of the cross-shore profile of section 17 after the nourishment

- the nearshore morphology inside and outside the nourished zone is becoming strongly 3-dimensional due to the presence of crescentic bars, the presence and migration of bar attachment points and drumstick-shaped nearshore breaker bars, separated by obliquely oriented rip channels. Such a pronounced development of a 3D morphological system has never been experienced in the past (Ruessink and Kroon, 1994);
- the disturbed bar-trough morphology caused by the nourishment is quickly adjusting itself and the former bar-trough pattern appears again (Fig. 4); the time of adjustment is rather short and the strongest response is visible in the first 150 days following the nourishment and the process slows down afterwards;
- the initial morphological response results in an extreme growth of the landward located middle bar and the adaptation of the profile is associated with a dominantly onshore movement of sediment (Fig.4);
- the western part of the nourishment is predominantly eroding whereas the eastern part shows accretion: the nourishment clearly migrates in an alongshore direction, towards the ENE (Fig. 1) at a variable rate of 280-320 m/yr (1994) or 400-420 m/yr (1995; Westlake, 1995).

As a matter of fact, (nourished) sediment is dominantly moving in an onshore and longshore (eastward) direction.

Sediment volumes

The consecutive bathymetric maps are now used in a more quantitative sense to identify areas of erosion and deposition and to compute the change in sediment volume in well-defined control sections. The computations have been reported previously by Westlake (1995). The selected control-sections are presented in Fig. 5 and are related to the nourished zone itself, areas landward and seaward of the nourishment and "reference" areas in both the East and West.

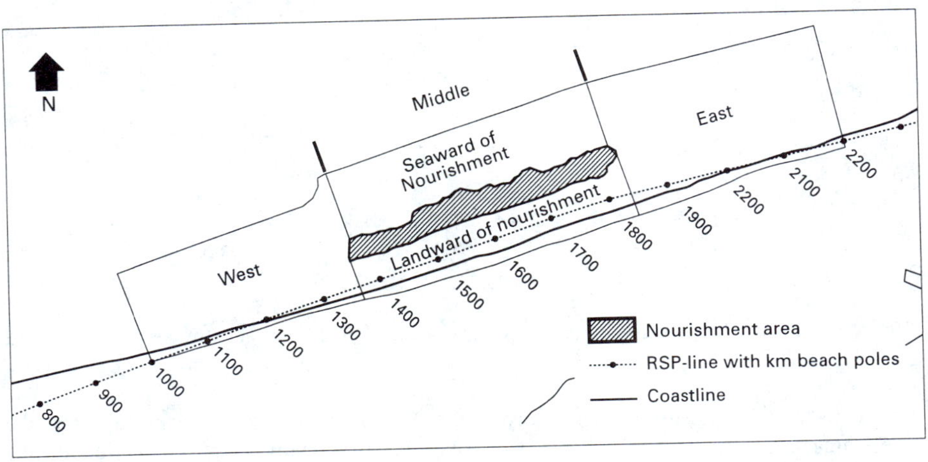

Figure 5. Control sections for volumetric computations of sediment budgets

The analysis of individual soundings occasionally shows the presence of systematic errors in the data sets and related bathymetric maps which may cause substantial inaccuracies in volume calculations. To reduce the effect of individual errors, results concerning the behaviour and effectiveness of the nourishment are based on the average linear trends in volumes, deduced from the total sequence of soundings, rather than the individual difference calculated for just two separate soundings (Westlake, 1995).

In total, the nourished volume of sediment is about 2.0 million m^3. In general it's rather difficult to assess whether there is an overall conservation of sediment volumes in the entire research area. For the most important sections though a number of trends are very consistent, as illustrated by Fig. 6. The nourishment area clearly shows erosion and the areas landward of the nourishment are definitely accreting. Remarkably though is the fact that the area landward of the nourishment is accreting at about twice the rate the nourishment is eroding. The nourished surface has lost about 560.000 m^3 of sediment. Part of this sediment - a conservative estimate is approximately 100.000 m^3 - is related to the alongshore migration of the nourishment. Meanwhile though, landward of the nourished zone a total gain of about 1.1 million m^3 of sediment is observed. And even if potential survey errors are taken into account, the order of magnitude of the difference suggests that the landward gain of sediment is not simply explained by a cross-shore redistribution of sediment. Significant longshore sediment transport gradients have to exist as well.

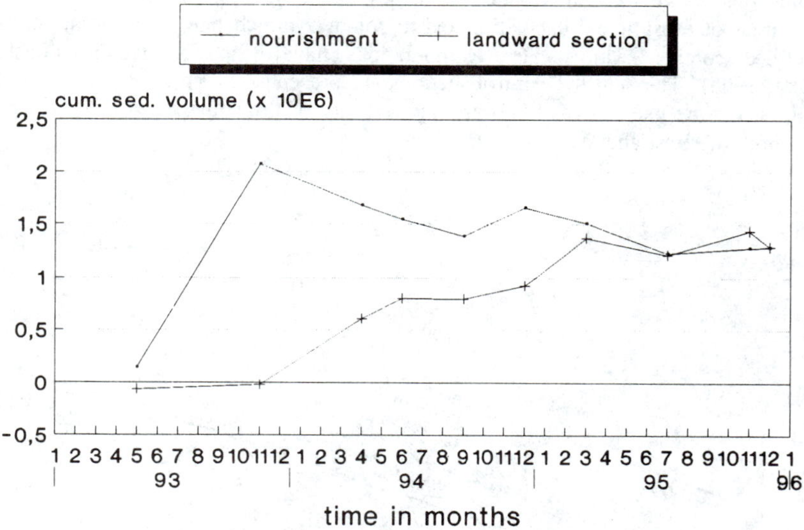

Figure 6. Cumulative sediment volumes for the nourished zone and the section landward of the nourishment

The frequent occurrence of storms and heavy storms in the period after the nourishment is predominantly held responsible for the relative rapid evolution of the nourished profiles and sediment volumes. This rapid response of the nearshore morphology is probably not fully representative for average conditions along the Terschelling coast.

The onshore directed redistribution of sediment certainly implies that the Terschelling shoreface nourishment has satisfied the design objective: the momentary or yearly computed coastline is estimated to have migrated seaward over a distance of about 16 m/yr (Biegel and Spanhoff, 1996). This totally reverses and exceeds the assumed retreat over the same period which should have been approximately 3 m/yr.
Based on the results of the present analysis, the Terschelling shoreface nourishment *only partly acts as a feeder berm*. Approximately 40% of the total gain of sediment in the inner nearshore can be explained by direct losses in the nourished zone. The success of the nourishment is not simply the result of the onshore movement of sediment from the nourishment. Accretionary processes in relation to longshore sediment transport gradients are probably equally important.

Cross-shore sediment transport processes

The partially, onshore-directed movement of sediment from the feeder berm has to be the result of cross-shore sediment transport processes. The transport study carried out in the framework of the Terschelling shoreface nourishment has focussed on both the dominant processes as well as dominant conditions that are mainly responsible for the onshore-directed fluxes of sediment. In this sediment transport analysis (Houwman and Ruessink, 1997) four main issues are adressed:

1) the contribution of oscillating and mean suspended transport to the net suspended transport;
2) the relative importance of high- and low-frequency suspended transport to the total oscillating suspended transport;
3) the ratio between bedload and suspended load;
4) the transport conditions under which most sediment is transported on the time scale of months to years.

Sediment transport measurements carried out in October and November 1995 (T4 campaign) at location F3 (Fig. 2) are selected to illustrate a number of features. Location F3 is one of the most interesting positions with respect to the hydrodynamic and sediment transport processes affecting the behaviour of the nourishment.

Some additional information is given here (Houwman and Ruessink, 1997):

- time series of u (cross-shore), v (longshore) and c (concentration) were divided in a mean and oscillating part; the net cross-shore suspended sediment transport, for example, is given by:

$$\langle u.c \rangle = \langle \bar{u}.\bar{c} \rangle + \langle u'.c' \rangle$$

  The net suspended sediment transport (left hand side) is based on the transport by mean currents (first term right hand side) and the oscillating transport (e.g the effect of wave asymmetry; 2nd term on the right). A positive cross-shore and longshore transport is onshore and eastward directed, respectively;
- the oscillating term is sub-divided into a high- and low-frequency part; the separation frequency between both was set to 0.04 Hz;
- cross-spectral analysis was applied to yield information about phase and coherence between u' and c' as a function of frequency.

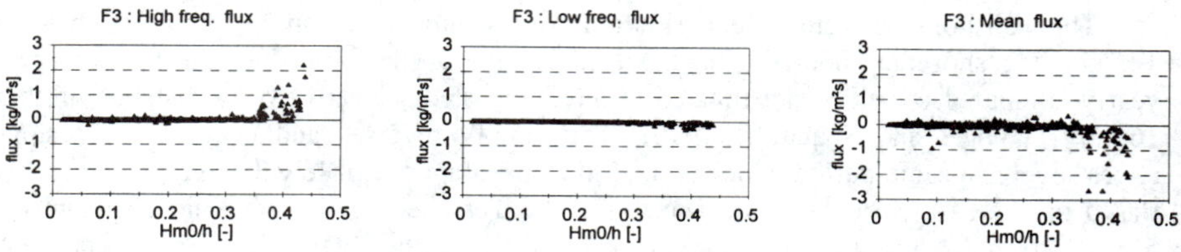

Figure 7. Cross-shore sediment fluxes measured at tripod F3 at 0.15 m above the bed (T4 campaign): high- and low-frequency (f < 0.04 Hz) fluxes and mean fluxes, respectively

Burst-averaged values (burst: measuring series of 34 minutes per hour with a sampling frequency of 2 Hz) of the high- and low-frequency oscillating fluxes in a cross-shore direction and at a height of 0.15 m above the bed are plotted in Fig. 7 as a function of local relative wave height (Hm0/h). The factor Hm0/h accounts for both variations in wave climate as well as (tide-induced) fluctuations in water depth. Non-zero fluxes only occured for Hm0/h larger than about 0.30-0.35, in other words, during situations that were just outside, or inside the surfzone. High-frequency fluxes were in an onshore direction and were associated with the horizontal asymmetry of the incident short waves, i.e. the difference in magnitude between onshore and offshore orbital velocity. Low-frequency fluxes were seaward directed and were caused by the dominance of bound long waves in the total long wave field (Ruessink, 1996). For this sensor height, the low-frequency sediment flux was only about 16% of the high-frequency one, averaged over the entire campaign.

Burst-averaged values of the mean flux for the T4 campaign are shown in Fig.7 as well, again as a function of relative wave height Hm0/h. The largest fluxes, which were in a seaward direction, occured for situations with the ratio Hm0/h above 0.3-0.35. This indicates that these negative fluxes were associated with wave-driven undertows. The mean flux was larger than the total oscillating flux (i.e. the sum of high- and low-frequency flux) averaged over the entire campaign. Consequently, the net flux was in an offshore direction. In addition, it is also clear that the (high-frequency) oscillating fluxes can not be neglected.

To estimate the long-term bedload and suspended load transport in the surfzone, the sediment transport is calculated with the Bailard model (Bailard, 1981) and the adapted Van Rijn/Ribberink model (Van Rijn, 1993). The suspended load component of the Van Rijn model does not include the oscillating suspended load transport. For that reason, Houwman and Ruessink (1997) incorporated an oscillating suspended load term in the model in a parameterized way.

Summarizing all measurements and modelling results (Houwman and Ruessink, 1997) it becomes obvious that the measured and computed cross-shore sediment transport fluxes (for Terscheling) are reasonably related to a dimensionless wave height (Hm0/h). This term can also be interpreted as a breaker coefficient and it turns out that sediment fluxes increase with Hm0/h. By making use of a probability density function for the term Hm0/h, it appears that the suspended sediment transport on a time scale of months to years is typically related to breaking wave conditions (Hm0/h > 0.33); in other words is limited to situations inside the surfzone. It also supports the view that the relative rapid evolution of the Terschelling shoreface nourishment is partly related to the frequent occurrence of storms in the period just following the nourishment.

Both measurements and modelling efforts also demonstrate that wave asymmetry is primarily responsible for the potential onshore movement of sediment; this sediment is mainly transported as suspended load by the oscillating flow related to seas and swell (high-frequency oscillating suspended load). Bedload, according to results obtained with the Van Rijn/Ribberink model, is of lesser importance.

However, the actual fate of the feeder berm is determined by a delicate balance of offshore directed mean fluxes and onshore directed oscillating fluxes. Paradoxically, in these conditions the contribution of some minor transport components, such as bedload and the effect of low-frequency waves, may eventually explain the success or failure of the nourishment. It's clear though that there is no significant and consistent net landward directed sediment transport in the nearshore zone. Therefore, it's not surprising that only part of the nourished sediment is moving in an onshore direction.

The observations make clear that it's not realistic to come up with a proper and reliable sediment balance for the area, based on sediment transport measurements and model computations.

Longshore sediment transport processes

Breaker berm and longshore sediment transport

The classic concept of a feeder berm is the idea that a cross-shore redistribution of sediment is mainly responsible for nourishing the upper part of the profile. Meanwhile, for the Terschelling coast, it's becoming quite evident that longshore processes are expected to be equally important. This is, for example, illustrated by the computations of sediment budgets inside and landward of the nourished zone, the longshore migration of the nourishment and the potential breaker berm function of the combination shoreface nourishment/nearshore breaker bar.

The breaker berm function is evaluated by using the UNIBEST-TC model. UNIBEST-TC is a morphodynamic coastal profile model (Roelvink et al, 1996). Application of the model required quite a lot of validation and calibration runs. During this process the model was further modified by using Terschelling data. Results of wave decay were obtained with the UNIBEST-TC model after considerable tuning of the model for wave heights using the breaker index (H/h) and the bottom friction factor f_w (Bakker, 1995 and Roelvink et al, 1996).

Model computations were carried out using measurements and conditions observed during the T2 campaign (May-June, 1994). The morphological boundary conditions are presented in Fig. 4. For all computations it is clear, that there commonly is a gradual reduction in wave heights across the seaward side of the profile and over the nourished zone. The wave dissipation is primarily due to bottom friction. A rapid decay in wave heights, though, is observed across the landward located middle bar (Fig. 1 and 4; Hoekstra et al, 1996). Here, wave breaking is the prime reason for a rapid decay in wave heights. In the landward located trough the wave field propagates without any major modifications. The model simulations definitely indicate that the design of the present shoreface nourishment not necessarily leads to an increase in wave energy dissipation: it is essentially the interaction of the shoreface nourishment with the nearshore breaker bars (Fig. 4) that is effective in reducing the landward propagating wave energy. Indirectly nourishing the middle bar is considered to be an effective method to construct or maintain a natural breaker berm.

The longshore transport module of the UNIBEST-TC model is further applied to study to what extent longshore sediment transport rates and gradients in these rates are initially affected
by the presence of the nourishment. The influence of the nourishment on longshore sediment transport patterns was determined by computing the cumulative longshore transports for a profile with (T1-condition) and without (T0-condition) a nourishment (Fig. 8). The measured wave climate for approximately the first 160 days since the implementation of the nourishment was used as input for the model, in combination with the recorded (tidal) water levels. The effect of tidal currents, however, was omitted. Every 3 hours a longshore transport computation was carried out; results are given in Fig. 8.

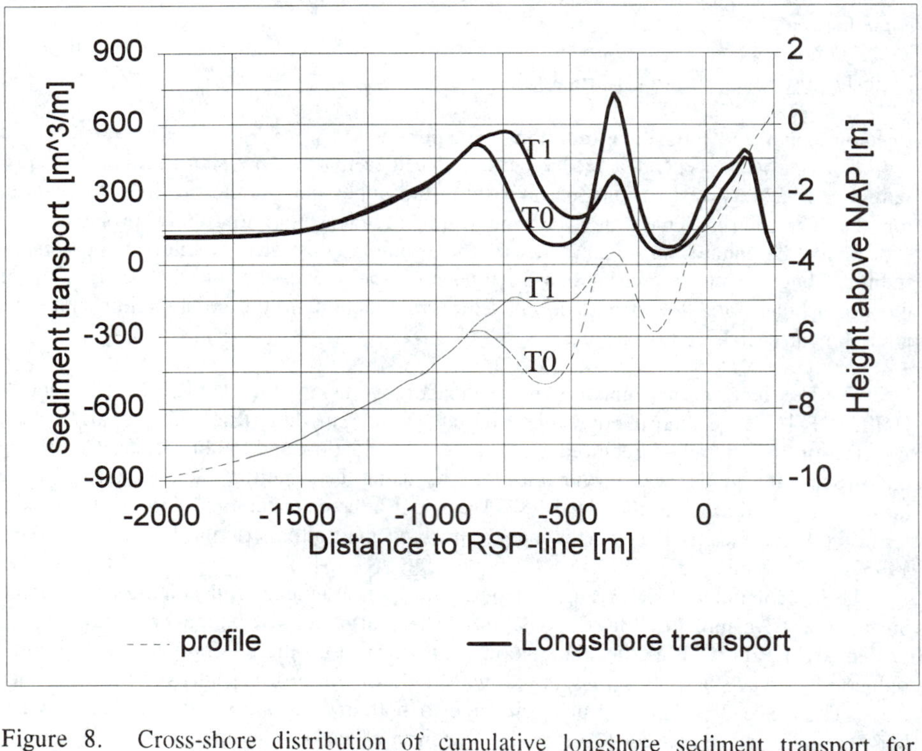

Figure 8. Cross-shore distribution of cumulative longshore sediment transport for conditions with and without a nourishment and for a period of about 160 days just following the nourishment

The net longshore sediment transport in the nearshore zone is predominantly towards the East (compare Fig. 3). Going from West to East - or from a non-nourished profile to a nourished one - longshore sediment transport in the trough is smaller than on the nourishment (Fig. 8). As a result, the West side of the nourishment will be eroding.

Sediment transport on the middle bar, by contrast, will be lower upon entering the nourished zone, which creates accretion on top of the bar (Fig. 8). On the nourished surface no transport gradients are expected to develop due to the longshore uniformity. On the downdrift side, however, longshore sediment transport patterns change again. The sediment transport on the nourished surface is larger than in the trough and the trough is gradually filled up with sediment. Simultaneously, the top of the middle bar shows erosion. The net effect is a gradual longshore migration of the nourishment towards the East, as observed for the Terschelling case.

The leeside deposition on the updrift side of the nourishment or breaker berm is commonly referred to as the salient effect. It probably partly explains the fact that the accretion in the cross-shore profiles landward of the nourished zone may exceed the losses from the nourishment. A salient is also frequently associated with erosion at a downdrift section of the coast. For the time being, no clear erosional patterns have developed yet.

In conclusion, longshore sediment transport gradients are important for understanding the behaviour of the nourishment. These gradients are responsible for the longshore migration of the nourishment and they can also partly explain the additional input of sediment in the inner nearshore zone.

Conclusions

The Terschelling shoreface nourishment has completely satisfied the design objectives in the first 2.5 years since the implementation. The nourishment not only compensates for the annual coastal retreat of about 3m/yr, but results in a net seaward migration of the momentary or yearly computed coastline of about 16 m/yr. The nourishment only partly acts as a feeder berm and the actual success of the nourishment is also based on the "interception" of longshore sediment transport due to the creation of a wave shadow zone and the salient effect. The combination of middle bar and shoreface nourishment effectively acts as a breaker berm.

Wave asymmetry is primarily responsible for the potential onshore movement of (nourished) sediment. This sediment is mainly transported in the form of a high-frequency oscillating suspended flux due to seas and swell. Bedload transport is of minor importance. In addition, the long-term suspended sediment transport from the nourished zone into an onshore direction is mainly related to breaking wave conditions ($Hm0/h > 0.33$). The net cross-shore suspended load flux, however, is determined by a very sensitive balance of two large components: a large offshore directed mean suspended flux due to currents is opposed by a large onshore directed oscillating flux due to waves. In these conditions the contributions of some minor transport components, such as bedload and the oscillating flux due to low-frequency waves, may eventually determine the behaviour of the nourishment.

Acknowledgements

This work was carried out as part of the project NOURTEC: Innovative Nourishment Techniques Evaluation. It was funded jointly by the Ministry of Transport, Public Works and Water
Management in the Netherlands and by the Commission of the European Communities, Directorate General for Science, Research and Development under the Marine Science and Technology programme contract no. MAS2-CT93-0049.

References

Bailard, J.A., 1981. An energetics total load sediment transport model for a plane sloping beach. Journal of Geophysical Research, 86 (C11), 10938-10954.

Bakker, R., 1995. Verifications of the UNIBEST-TC model. Masters thesis Delft University of Technology / National Institute for Coastal and Marine Management (RIKZ), OS-95.133x, Delft, 76 pp. with app.

Biegel, E.J. and Spanhoff, R., 1996. Effectivity of a shoreface nourishment, Terschelling, The Netherlands. National Institute for Coastal and Marine Management (RIKZ) and SEPRA bv.

Hoekstra, P., Houwman, K.T., Kroon, A., Van Vessem, P. and Ruessink, B.G., 1994. The NOURTEC experiment of Terschelling: process-oriented monitoring of a shoreface nourishment (1993-1996). Proc. Coastal Dynamics '94, ASCE New York, 402-416.

Hoekstra, P., Houwman, K.T., Kroon, A. and Ruessink, B.G., 1996. Morphodynamic Behaviour of the Terschelling Shoreface Nourishment. Institute for Marine and Atmospheric Research Utrecht (IMAU), Utrecht University, rep. R96.14, 64 pp. with appendix.

Houwman, K.T. and Ruessink, B.G., 1997. Cross-shore sediment transport mechanisms in the surfzone. Proc. 25th Int. Conf. Coast. Eng., ASCE New York (this volume).

Knaack, H., Kaiser, R. and Niemeyer, H.D., 1997. Effectiveness of a combined beach and shoreface nourishment. Proc. 25th Int. Conf. Coast. Eng., ASCE New York (this volume).

Kroon, A., Hoekstra, P., Houwman, K.T. and Ruessink, B.G., 1995. Morphological monitoring of a shoreface nourishment - NOURTEC experiment at Terschelling, the Netherlands. Proc. 24th Int. Conf. Coast. Eng., ASCE New York, 2222-2236.

Kruyt, N.M., 1995. Wind Climate of Terschelling. Institute for Marine and Atmospheric Research (IMAU), Utrecht University, rep. R95.19, 57 pp. with app.

Laustrup, C., Toxvig Madsen, H., Sørensen, P. and Brøker, I., 1997. Comparison of beach and shoreface nourishment Torsminde Tange, Denmark. Proc. 25th Int. Conf. Coast. Eng., ASCE New York (this volume).

NOURTEC, 1994. Design report of the three NOURTEC test sites. Coord. National Institute for Coastal and Marine Management (RIKZ), The Hague, The Netherlands.

Roelvink, J.A., Meijer, Th.J.G.P., Houwman, K.T., Bakker, R. and Spanhoff, R., 1996. Field validation of a coastal profile model. Proc. Coastal Dynamics '95, ASCE New York, 818-828.

Ruessink, B.G., 1996. On the origin of infragravity waves in the surf zone of a dissipative multiple bar system. Proc. Coastal Dynamics '95, ASCE New York, 93-104.

Ruessink, B.G. and Kroon, A., 1994. The behaviour of a multiple bar system in the nearshore zone of Terschelling, The Netherlands: 1965-1993. Marine Geology, 121, 187-197.

Van Beek, L.P.H., 1995. Wave climate of Terschelling. A statistical analysis for the NOURTEC project. Institute for Marine and Atmospheric Research (IMAU), Utrecht University, rep. R95.10, 48 pp. with app.

Van Rijn, L.C., 1993. Principles of sediment transport in rivers, estuaries and coastal seas. Aqua Publications, Amsterdam, The Netherlands.

Westlake, S.J., 1995. Behaviour of a Shoreface Nourishment, Terschelling, The Netherlands. Masters thesis International Institute for Infrastructural, Hydraulic and Environmental Engineering, Delft, the Netherlands, rep. HH249, 95 pp. with figs.

Whitford, D.J. and Thornton, E.B., 1993. Comparison of wind and wave forcing of longshore currents. Cont. Shelf Res., vol.13, 11, 1205-1218.

Wijnberg, K.M., 1995. Morphologic Behaviour of a barred Coast over a Period of Decades. PhD thesis, Utrecht University, 245 pp.

CHAPTER 225

JUSTIFICATION FOR BEACH NOURISHMENT

by John G. Housley, PE, M.ASCE [1]

ABSTRACT

Sand balance of the littoral system determines whether erosion or recession will occur. Adding sand to the system is the only engineered way to restore the balance and prevent erosion or recession. Justification of beach nourishment as a shore protection measure depends on the benefits criteria used in the analysis, and societal goals will decide the criteria.

INTRODUCTION

Mankind, in its love affair with the coast, has often encroached on sandy areas which are basically unstable - at least as far as lines on a map are concerned. Sandy coasts are particularly attractive as recreation sites, and in the last half century they have been the target of intensive development. In the United States, about 50 percent of the population lives within 30 miles of a coast. Erosion of those shores has become a national concern.

THE EROSION PROBLEM

Sand is in constant motion along the shores of this country due to waves, currents, and wind. As long as the sand leaving a stretch of beach is balanced by the same quantity of sand arriving on the beach, there is no erosion or recession. But if the balance is not maintained (either by the arriving sand being interrupted by a structure such as a jetty or by a lack of sand on the updrift beach), erosion/recession will result.

When the user of a beach wants to construct some fixed structure in the immediate vicinity, such as a road or a house, then, if the shoreline moves in the direction of the fixed structure, that is called a problem.

---

[1]Coastal Engineering Consultant, Michael Baker Corporation, Alexandria, VA 22304

There are three basic responses to an erosion threat: reduce the hazard, adjust human use, or accept the cost. Only the first one is amenable to an engineering approach; the others are either management options or the do- nothing alternative with its attendant increased costs. To reduce the hazard, the choices are to modify the physical process or to reduce the impact. From an engineering viewpoint, it is usual to pursue some combination of those hazard-reduction options.

A BEACH NOURISHMENT SOLUTION

For developed coasts, the serious problem of coastal erosion is caused by a deficit of sand in the littoral system. Adding sand to the system (beach nourishment) is the only engineering solution that remedies the basic problem. All other potential engineering solutions (groins, seawalls, breakwaters, etc.) may stop the erosion at a particular point, but the deficit of sand in the littoral system will cause erosion to occur at some other location. And the cost of mitigation of those adverse effects is one of the strong reasons for a beach nourishment approach.

When sand is first installed as part of a beach nourishment project, it is placed on the existing beach to raise it to a higher elevation and to widen the dry beach in the seaward direction; the seaward face is thus on a steep slope. The filling procedure is dictated by the least total cost of the operation. If the fill is made from the land side, say by truck haul, then the trucks dump near the waters edge and continually push that edge seaward. If the fill is by hydraulic dredge, then the pipes are positioned so that the effluent moves the edge of the fill seaward. If the fill is by offshore mounding, a split-hull hopper dredge or barge is maneuvered as close to shore as the draft of the vessel will allow, and the fill is placed in shallow water so that waves will move the sand toward the shore.

When first placed, the sand slope in the water is not at the equilibrium slope. As waves work on that slope, the sand tends toward the equilibrium position, that is, a slope which is stable under the wave regime present. Since the waves are not of constant height, period, or direction, each wave climate tries to put the beach slope in an equilibrium position for that instant. Seconds later, the regime has changed, thus the process of equilibration is continuous.

So as the sand pile is reshaped by waves, some of the dry beach sand is moved into the water to provide the material for a flatter equilibrium slope. And the perception that the beach was "lost" comes to those who only consider the subaerial beach. Without the subaqueous beach, surely the beach *is* lost.

BENEFITS

The recent study by a blue-ribbon committee of the Marine Board of the National Research Council found that, when used in appropriate situations, beach nourishment does indeed become a cost-effective measure to combat the effects of coastal erosion. The key is to determine what are the appropriate situations for the use of beach nourishment. First, a determination must be made of the benefits to be derived from such a project. Those benefits are not only in terms of dollars of increased revenues and decreased losses, but also social and environmental benefits. The costs associated with the project are not only the planning, design, construction, and maintenance costs, but also the social and environmental costs.

In an effort to stem what many consider unwise development on pristine shores in this country, the Coastal Barrier Resources Act precludes (with some exceptions) the federal government from expending funds on undeveloped coastal barriers. For developed shores, the rules for U.S. federal participation in shore protection projects are quite explicit about how and how much benefits can be claimed. The benefits in the category of storm damage reduction must be at least one half the costs of design and construction. After that benefit requirement is fulfilled, other benefits (such as recreation) can be added to bring the total benefits up to something greater than the total costs. However, these are not all the benefits that should be considered.

Justification usually has the connotation of financial break-even or better. Justification on the national, state, local, or personal levels uses different yardsticks and different elements of measure. Measures that would justify a project on the local level may not be counted on the national level, since benefit gains in one region may be offset by losses in another. And on the personal level, hard to quantify aesthetics may be the most important element for some people. And for environmentalists, the fill material as habitat for birds, turtles, and other creatures may be the important elements.

One of the major benefits not considered in beach nourishment cases involving federal participation are those benefits outside the project boundaries caused by sand being transported to property beyond those boundaries. In the recreation benefit category, the benefit to the non-federal sponsor is usually far greater than that allowable under U.S. federal rules. Those rules require the benefits to be national economic development (NED) benefits, but the local entities realize benefits in addition to NED benefits. An often overlooked cost is a navigation project that causes a sediment deficit in the littoral system by trapping sand in the channels and harbor areas, or in ebb-tide deltas that effectively divert sand away from a beach area.

CONSTRAINTS

The major problem with beach nourishment is not technical, but perception. The person who views a beach nourishment project has a certain expectation from that project. If it is a property owner in the vicinity, he looks for protection. If it is visitor, he looks for a recreation site. And if it is a taxpayer from a distant location, he looks for a return on his investment.

Beach nourishment mimics what nature does, so that sand is not held in one place but is allowed to move under the influence of the waves and current. Although beach nourishment is not appropriate for all locations, it is the preferred option in many problem areas. A beach nourishment project is sacrificial; that is, it is designed so that sand is removed from the beach rather than buildings being removed from the backshore. Sand replenishment restores the beach to the design dimensions upon which the benefits are derived, and if the project is well designed, constructed, and maintained, and if there is the political and financial will to continue the periodic renourishment program, the benefits will continue to accrue. While hard structures (groins, seawalls, etc.) require a large, up-front, construction-cost investment, beach nourishment involves a smaller, periodic expenditure for construction over the life of the project. In almost all cases, the benefits of beach nourishment outweigh those of hard structures.

It was once said that anyone could build a bridge, but it took an engineer to build one economically. The same can be said for beach fills.

Beach nourishment projects need to use state-of-the-art engineering principles, which includes a thorough understanding of shore processes. It is not enough that an adequate supply of sand is available to a project for it to be successful; there must also be an understanding of the goals of the project and the criteria for success. If the public does not understand how success is measured, and that beach nourishment is meant to be sacrificial, then when some of the subaerial beach is lost during a storm, the project may be labeled a failure, even when much of the sand is returned to the dry beach by natural processes in a few months, and the engineer has expected those results.

So beach nourishment (or renourishment) is the preferred solution to the sand deficit problem. To properly engineer the sand fill, the characteristics of the native sand and the proposed fill material must be known, as well as the forces acting on it, i.e., waves, current, wind, etc.

A recent (1995) legislative proposal by the Clinton administration would redefine the US Army Corps of Engineers' (the federal agency charged with shore protection) mission as "nationally significant missions." If enacted, the cost sharing for federally-funded water resources projects would change to 25/75 (federal/non-federal), and the benefit/cost ratio would be at least 2. Federal participation in storm damage reduction projects would be eliminated on the basis that these projects are local (not national or interstate), and should be paid for by non-federal dollars. As of the date of this paper, the proposal has been rejected by the Congress, but the administration is using the budget process to put shore protection in a low priority position.

CONCLUSION

The expense of a nourishment program (initial plus subsequent re-nourishments) is not inconsequential. However, the loss of property during storms and loss of revenue from tourism on a neglected beach can be much greater.

BIBLIOGRAPHY

Adams, J.W.R., "Florida's Beach Program at the Crossroads", *Shore and Beach*, v 49, nr 2, Apr 81, pp 10-14.

Adams, J.W.R., "The Politics of Beach Nourishment", *Shore and Beach*, v 50, nr 1, Jan 82, pp 2-5.

Basco, D.R., "Boundary Conditions and Long-Term Shoreline Change Rates for the Southern Virginia Ocean Coastline", *Shore and Beach*, v 59, nr 4, Oct 91, pp 8-13. Discussion by E.R. Thieler, R.S. Young, and O.H. Pilkey in *Shore and Beach*, v 60, nr 4, Oct 92, pp 29-30. Closure by D.R. Basco in *Shore and Beach*, v 60, nr 4, Oct 92, pp 31-34.

Bodge, K.R., "Damage Benefits and Cost Sharing for Shore Protection Projects", *Shore and Beach*, v 59, nr 2, Apr 91, pp 11-18.

Camfield, F.E., "Different Views of Beachfill Performance", *Shore and Beach*, v 61, nr 4, Oct 93, pp 4-8.

Campbell, T.J. and R.H. Spadoni, "Beach Restoration - An Effective Way to Combat Erosion on the Southeast Coast of Florida", *Shore and Beach*, v 50, nr 1, Jan 82, pp 11-12.

Davison, A.T., C.P. Ulrich, and R.J. Nicholls, "Accreditation of Beach Nourishment Projects: An Issues Discussion", *Shore and Beach*, v 61, nr 4, Oct 93, pp 9-15.

Denison, P.S., "The North Carolina Experience", *Shore and Beach*, v 49, nr 2, Apr 81, pp 15-16.

Domurat, G.W, "Beach Nourishment - A Working Solution", *Shore and Beach*, v 55, nr 3-4, Jul-Oct 87, pp 87-91.

Herron, W.J., "Sand Replenishment in Southern California", *Shore and Beach*, v 55, nr 3-4, Jul-Oct 87, pp 87-91.

Houston, J.R., "Beachfill Performance", *Shore and Beach*, v 59, nr 2, Jul 91, pp 15-24.

Houston, J.R., "Beach Nourishment", *Shore and Beach*, v 63, nr 1, Jan 95, pp 21-24.

Magoon, O.T., "A Letter to President Bill Clinton", *Shore and Beach*, v 62, nr 2, Apr 94, pp 2-3.

O'Brien, M.P., "Beach Stabilization by Sand Replenishment", *Shore and Beach*, v 52, nr 4, Oct 84, p 19.

Pilkey, O.H., "Another View of Beachfill Performance", *Shore and Beach*, v 60, nr 2, Apr 92, pp 20-25.

Seymour, R.J., "An Introduction to the Marine Board Study on Beach Nourishment and Protection", *Shore and Beach*, v 64, nr 1, Jan 96, p 3.

Seymour, J.R. et al., "Beach Nourishment and Protection: Executive Summary", *Shore and Beach*, v 64, nr 1, Jan 96, pp 5-10.

Stone, K.E. and B. Kaufman, "Sand Rights: A Legal System to Protect the 'Shores of the Sea'", *Shore and Beach*, v 56, nr 3, Jul 88, pp 7-14.

Stronge, W.B., "Beaches, Tourism and Economic Development", *Shore and Beach*, v 62, nr 2, Apr 94, pp 6-8.

Stronge, W.B., "The Economics of Government Funding for Beach Nourishment Projects: The Florida Case", *Shore and Beach*, v 63, nr 3, Jul 95, pp 4-6.

Smith, A.W.S., "Beaches and Tourism - An Example of the Results of a Dramatic Beach Erosion Episode: Gold Coast, Queensland, Australia", *Shore and Beach*, v 63, nr 3, Jul 95, pp 7-8.

Smith, A.W.S. and L.A. Jackson, "The Siting of Beach Nourishment Placements", *Shore and Beach*, v 58, nr 1, Jan 90, pp 17-24.

Sudar, R.A. et al., "Shore Protection Projects of the U.S. Army Corps of Engineers", *Shore and Beach*, v 63, nr 2, Apr 95, pp 3-16.

Wiegel, R.L., "Beaches - Tourism - Jobs", *Shore and Beach*, v 62, nr 2, Apr 94, pp 4-5.

Wiegel, R.L., "Beach Nourishment, Sand By-Passing, Artificial Beaches: Bibliography of Articles in the ASBPA Journal *Shore and Beach*", *Shore and Beach*, v 60, nr 3, Jul 92, pp 3-5.

CHAPTER 226

BEACH NOURISHEMENT VERSUS SHORE PROTECTION STRUCTURES

Alexander Khabidov[1], Yuri Bazhenov[2] , Marina Glodenis[2] , Nadya Ivanova[1], Liya Kaskevitch[1], Oleg Rybak[3] , Valery Savkin[2], Anatoly Trizno[2], Grigory Zelensky[2] , Veniamin Yedapin[2]

ABSTRACT

The problem of coastal erosion in Russia became especially urgent in 50th-60th during the extensive river damming, taking place mainly in the regions valuable for society. To prevent erosion, which spanned thousands of kilometres of new shoreline different shore protection techniques were used, including hard structures, shoreface and beach nourishment as well as hybrid methods. Most if not all of these techniques were previously applied on Novosibirsk Reservoir, one of the first Russian large man-made lakes. Long-term observations of structures state and recent studying of response of the coastal environment on the projects emplacing testify the greater efficiency of coastal stabilization by beach nourishment programs and hybrid structures.

INTRODUCTION

Novosibirsk Reservoir was created in Western Siberia in 1956 when the Ob River was dammed ca. 20 km upstream the city of Novosibirsk (latitude 55° N, longitude 83° E). It is a freezing reservoir which ice season lasts about 180-190 days from November till April or May. Like many man-made lakes seasonal changes in water level of Novosibirsk Reservoir occur because of the river flow fluctuations; they reach more than 5 m/yr in amplitude and include: (i) Water level rise in May-June: mean duration - ca. 50 at a rate of ca. 0.1 m per day, (ii) Water 00level stabilization: mean duration - ca. 120 days, and (iii) Water level subsidence in October-April: mean duration - ca. 195 days at a rate of ca. 0.02 m per day.

The fetch of Novosibirsk Reservoir is 220 km from northeast to southwest up to town of Kamen-on-Ob, its minimal, mean, and maximum width are 2, 10,

---

[1] Institute for Water and Environmental Problems, Siberian Branch of the Russian Academy of Sciences, 105, Papanintsev Street, 656099, Barnaul, Russial; e-mail: khabidov@iwep.altai.su
[2] Department of Natural Resourses of Russian Federation; 19, Geroev Truda St., 630055 Novosibirsk, Russia
[3] Scientific and Research Center of Marine Coasts; 1, Jan Fabritsius St., 354002 Sochy, Russia

and 22 km, respectively, water plane area is 1070 km^2, depth is up to 25 m, total storage is 8.8 km^3, effective storage is 4.4 km^3, and shoreline length is 550 km.

The principle basin morphometries of Novosibirsk Reservoir are: a shallow area in the south-western edge of the lake basin under fetch of about 60 km, an area of transitional depths, and a deep-water area with typical depths exceeding the average reservoir depth in the north-eastern part of the lake basin under fetch of 100-120 km. These areas correspond to main types of dynamical relief-forming and depositional sedimentary environments of the lake basin, namely: the deltaic/fluvial-dominated environment, the wave-dominated one with significant influence of the channel flows, and wave-dominated environment itself.

The wave heights are among the most significant features of dynamic relief-forming and depositional sedimentary environments of the reservoir. Within delta-like environment wave heights do not exceed 0.5 m even under severe storm. As for transitional zone the waves of up to 1.5-1.7 m height occur, while the waves of more than 3.5 m were observed in deep water area.

High rate of erosion of Novosibirsk reservoir coasts is primarily caused by wave. Two more factors contribute to erosion. First, the reservoir coasts are formed mainly of loose and soft grounds - sands, sandy loam, and loss-like loam. Second, like natural lakes and other man-made lakes, its beach berms are narrow or even absent at all. That is why the mean erosion rate reaches here 7-10 m/yr and more and the extention of eroding shores makes up 350 km.

Coastal erosion caused the most serious problems in wave-dominated area of Novosibirsk Reservoir. It is precisely here where the coast protection started before the reservoir filling. At present more than 40 km of its coasts have been protected already. As this takes place different shore protection techniques, including hard structures, shoreface and beach nourishment as well as hybrid methods were used.

Long-term observations of structures state showed considerably different reliability and effectiveness of hard structures, filled natural and artificial beaches as well as hybrid projects which combine a shoreface and/or beach nourishment with stabilization of sand by breakwaters, groins, artificial headlands, etc.

Recent studying of response of the coastal environment on the different projects emplacing allowed to reveal some causes of these phenomena.

METHODS

The work on the project began from the studying of data on long-term visual and tool supervision of seawalls, bulkheads, revetments, filled natural and artificial beaches as well as other coastal engineering sites of Novosibirsk Reservoir. This allowed to determine the key sites for special investigations. These investigations include repeated topographic surveys, continuous meteorological observations, synchronized with measurements of waves and studies of swash interactions by wire gauges, measurement of nearshore currrents by omni-directional and two-component impeller current meters, measurement of bedload and suspended sediment concentration by *in-citu* samplers, sand tracers movement observations and piezometric studies of beach groundwater interactions.

For comprehensive evaluation of the affect of protective measures on coastal environment, all observations are conducted not only on the coastal engineering sites, but on the upcoast and downcoast areas as well. Investigations on the key sites started a year ago. In this connection only the preliminary results of the observations are presented here.

BEACH MORPHOLOGY AND NEARSHORE PROCESSES NEAR A SEAWALL

As development increases along the shoreline and property is lost, engineers, developers as well as planners need quantitative descriptions and models of how coastal processes affect the shoreline (Dean, 1987) and how the projects, emplaced to protect the threatened structures, affect the surrounding coastal environment. In recent years, many of the investigations were devoted to the studying of the nearshore processes near a seawall and the resultant beach morphology under marine conditions (Kraus, 1988; Tait and Griggs, 1990; Plant and Griggs, 1992, and others). The definite severity of the environment occur at the man-made lakes, because of the long-term water level fluctuations from weeks to months.

Just as the seawalls, so bulkheads and revetments were built at the Novosibirsk Reservoir; as this took place, the first hard structure was emplaced before its initial filling. The concrete curved-face seawalls, concrete combination inclined and curved-face seawalls and rubble-mound seawalls as well as concrete and rubble-mound revetments had common occurance at the reservoir. Since 1956 more than 25 km of the seawalls were built here, but only about 12 km still persist to date.

For an understanding of the reasons of improper operation and destroying of the structures, beach morphology and nearshore processes were studied near seawalls of several types. The noted effects were both seasonal and long-term ones. According to observations, in the case typical for natural and man-made lakes, when a seawall is placed on the water edge or near it, and the adjacent beaches have a narrow berm or have not it at all, these effects are: (a) the bottom erosion rate increases after the seawall construction and on the fronting beach and its value is always higher than on the adjacent beaches without structures, (b) the bottom erosion in front of seawall before erosion on the protected beaches, (c) the coastal erosion increasing downcoast of seawalls. Near the seawall placed at the backshore, the erosion of beach berm in front of the structure had arised before equivalent erosion on the flanked beaches, because the erosion increses with the reflectivity of the seawall and the mobility of the fronting beach as the result of altering of the pattern of groundwater flux.

That is why, though under such sea-wall position the beach slows down the scour at the toe and ends of the structure, however as the result of the berm erosion the final response of the coastal environment will be equivalent to the first case.

In this manuscript the attention is focused on the effect of the nearshore currents on beach processes near a seawall that protrude into the surf zone. Dean (1976), Berkemeir (1980), and McDougal et al, (1987) noted, that structures of such type can effect longshore currents, trap sand upcurrent of wall and, perhaps, induce scour in front and downcoast of the seawall. However, the observations, conducted by Plant and Griggs (1992) in Monterey Bay, did not reveal this effect neatly.

The study site on Novosibirsk Reservoir consisted of a gently curving shorelines and a 350 m concrete combination inclined and curved-face seawall flanked by the continuous natural beaches at the upcoast and a 200 m beach at the downcoast. The flanked beaches, all backed by 3-5 m escarpment, had the berm the slope of which is 1:35, the width from 3-5 m (downcoast) up to 10 m (upcoast) and initial bottom slope was of order 1:25. The seawall protrude into the surf zone by 2:1 sloping revetment at its toe. The present investigation involved the pre-storm, storm, and post-storm beach surveys to capture morphological changes as well as measurement of the longshore and cross-shore currents at 10 cm level above the bottom in front of the seawall and in the flanked beaches. The storm surveys were carried out under the waves of Hsig 0.5 m, 1.2 m, and 2.3 m; the angle between wave crest and shoreline varied from 35o to 42o in all three events.

The beach surveys in the upcoast, seawall-backed, and downcoast sections of the site showed that the behavior of morphological changes within these sections differs greatly. Total losses of the bottom sediment per one metre of the shoreline made up 0.15, 2.75, and 4.25 m^3 within the upcoast section, 1.00, 5,14, and 8.00 m^3 within the seawall-backed section and 2.5, 4.25, and 6.25 m^3 for the 1st, 2nd, and 3rd storm events, respectively (Table 1).

Table 1.

Beach profile changes near the seawall and on the flanked beaches under different storm events

Beach profile area	Storm event	Net total volumetric changes in beach erosion and sediment accretion, m^3		
		Upcoast section	Seawall-backed section	Downcoast section
Beach Berm	1	+120	-	-150
Breacker Zone	1	-850	-1650	-1000
Offshore	1	+700	+1300	+1000
Beach Berm	2	-350	-	-450
Breacker Zone	2	-1500	-3900	-1900
Offshore	2	+1300	+2100	+1500
Beach Berm	3	-500	-	-650
Breacker Zone	3	-2450	-6700	-3400
Offshore	3	+2100	+3900	+2800
Note. Sign <+> corresponds to accretion, sign <-> corresponds to erosion				

Measurements of the nearshore currents during the storms in the front of seawall and on the flanked natural beaches correlate with beach surveys data, and to all appearances, explain the observed morphological changes. The absolute values of time-averaged longshore and cross-shore current velocities at the seawall-backed and downcoast sections of the study site were higher than the current velocities within the upcoast beach during the steady stage of all the storms (Table 2). Alongside, it is hard to state, that the observed rather higher rates of the nearshore currents at the downcoast beach are directly related to hydrodynamic processes in front of the seawall. It is stipulated by the fact that under small width of beach berm at the downcoast section, the backed cliff can induce nearshore processes, in principal similar to the processes caused by the seawall.

Table 2.

The range of the absolut values of time-averaged longshore (V) and cross-shore current (U) velosities near the seawall and on the flanked beaches during the steady stage of storms

Beach profile area	Storm event	Longshore and cross-shore current velosities, m/s		
		Upcoast section	Seawall-backed section	Downcoast section
Breacker Zone	1			
V		0.08-0.22	0.12-0.31	0.10-0.25
U		0.07-0.23	0.10-0.35	0.10-0.27
Offshore	1			
V		0.02-0.11	0.02-0.12	0.02-0.12
U		<0.03	<0.04	<0.04
Breacker Zone	2			
V		0.10-0.27	0.12-0.38	0.12-0.32
U		0.10-0.31	0.15-0.39	0.10-0.35
Offshore	2			
V		0.02-0.14	0.02-0.15	0.02-0.15
U		<0.05	<0.05	<0.05
Breacker Zone	3			
V		0.18-0.50	0.20-0.65	0.17-0.58
U		0.16-0.52	0.21-0.72	0.20-0.60
Offshore	3			
V		0.05-0.18	0.05-0.21	0.04-0.20
U		<0.10	<0.10	<0.10

Notes. 1)Time averaging interval by hardware is of 30 sec.
2)Time of continuous measurment is of 5 hours for all the storm events.

Visual observations of the seawall reliability started immediately after construction of Novosibirsk Reservoir. During the period of observation the most typical were the cases of seawalls destroy during several years or decades. At this time some events were noted, when the wall was completely destroyed under a prolonged severe storm with Hsig exceeding 3-3.5 m or under series of 2-3 such storms.

The survey of the walls destroyed showed that the main reasons of their destruction were not the construction failures but a scour at the toe and ends of the structures. No doubt, the scour was caused not only by the nearshore currents, but by wave reflection from a seawall, and the effect of groundwaters as well. However, in all cases of wall breakdown on the reservoir nearshore currents appeared to be of a prime consideration.

COASTAL ENVIRONMENT RESPONSE TO BEACH NOURISHMENT AND EMPLACING OF HYBRID PROJECTS

The first event of seawall damage on Novosibirsk Reservoir occured in 1959 near the Novosibirsk Academic Center. This wall was constructed in an effort to protect the shore site associated with federal highway and railway as well. Thus at this site renovation of shore defence, wherein the combined shoreface and beach nourishment were used for the first time in Russia, began.

At Novosibirsk Academic Center stretching 3 km alongshore, approximately 5,000,000 m^3 of sand have been nourished during three years. The constructed open artificial beach was located within the curved shoreline site where the longshore transport rate is not more than 20,000-25,000 m^3/yr. Owing to appropriate location the beach provided a powerful shore protection for 25 years without renourishment.

In spite of reasonable results of shore protection by beach nourishment, this technology became widely used on Novosibirsk Reservoir only at the end of 70th . At present there are more than 7 km of the open artificial beaches on the man-made lake.

The continuous monitoring of the beach morphology changes showed that under the environmental conditions of Novosibirsk Reservoir the equilibrium beach profile has shapes of several years. The process of profile formation usually begins from intensification of beach face erosion and is followed by cross-shore sediment transport. Superimposed terrace is formed in consequence of the sediment accretion at the outer border of the beach profile. As development progress of the profile the terrace edge is shifting to the seaward direction and the beach profile becomes level out. When two-three years elapse the terrace edge position has become fairly stable and the longshore sediment transport begins to dominate among the nearshore processes (Table 3). It seems likely that profile formation is completed.

The occurence of abnormal conditions with erosion of the accretive terrace at the outer border of the beach is mainly stipulated by severe storms effects during the lake's water level subsidence in autumn or water level rise in spring. It resulted in increaing of bottom slope (Table 3), that finally brought to intensification of berm beach erosion, cross-shore sediment transport and decreasing of longshore sediment rate.

Table 3.

Temporal morphological changes and prevailing trends in the nearshore processes on the nourished beach near the Novosibirsk Academic Center

Year	Total amount of the sediment losses			Total amount of the sediment accretion		Observed trends of the sediment transport
	Beach berm, m^3	Nearshore, m^3	Outer border of the beach, m^3	Near-shore, m^3	Outer border of the beach, m^3	
1	-82,000	-63,000	-	+21,000	+110,000	Cross-shore
2	-81,500	-52,000	-	+24,000	+97,000	Cross-shore
3	-78,000	-51,000	-	+54,000	+64,000	Cross-shore
4	-49,300	-51,000	- 5,000	+53,000	+35,000	Longshore
5	-46,500	-52,000	-35,000	+70,000	+15,000	Longshore
6	-54,500	-63,000	-	+38,000	+73,000	Cross-shore
7	-48,000	-54,000	-11,000	+49,000	+42,000	No clear trend
8	-46,700	-52,000	-27,000	+61,000	+21,000	Longshore
9	-39,200	-47,000	-	+34,000	+45,000	Cross-shore
10	-45,000	-54,000	- 4,000	+51,000	+34,000	Longshore
11	-40,500	-51,000	-31,000	+54,000	+19,000	Longshore
12	-46,100	-55,000	-23,000	+62,000	+20,000	Longshore
13	-59,200	-64,000	-16,000	+55,000	+49,000	No clear trend
14	-35,000	-40,000	-10,000	+54,000	+11,000	Longshore
15	-40,300	-53,000	-21,000	+59,000	+20,000	Longshore
16	-42,500	-55,000	-37,000	+53,000	+23,000	Longshore
17	-46,100	-57,000	-	+40,000	+47,000	Cross-shore
18	-45,400	-58,000	-9,000	+57,000	+25,000	Longshore
19	-45,500	-54,000	-10,000	+64,000	+18,000	Longshore
20	-53,000	-65,000	-23,000	+65,000	+30,000	Longshore
21	-62,500	-78,000	- 7,000	+68,000	+47,000	Longshore
22	-63,000	-83,000	-10,000	+64,000	+58,000	No clear trend
23	-70,300	-81,000	-14,000	+68,000	+62,000	No clear trend
24	-69,000	-86,000	-15,000	+71,000	+65,000	No clear trend

Within the wave-dominated region of Novosibirsk Reservoir the net longshore transport rate at the open shores varies from 25,000 m^3/ yr to 150,000-200,000 m^3/yr. In term of this, the priorities in policy of beach nourishment are established as following:

1. Initial beach nourishment and periodic replenishment after ten and more years existence of protective beach.
2. The accomplishment of hybrid projects, which combine beach nourish ment with the use of isolated groins, breakwaters or their systems.

Practically all attempts to prevent erosion along the Novosibirsk Reservoir shores by beach nourishment and hybrid structures were rather succesful and these projects offer numerous benefits. Firstly, unlike the seawall and similar hard structures the main part of nourishment-based projects are in good condition (Table 4). Secondly, groin systems and/or detached breakwaters retain the desired shape of beach berm. Thirdly, the longshore transport of borrow material from filled beaches allowed tangibly to mitigate the coastal erosion at downcoast areas and the costs on shore protection were reduced by virtue of this phenomenon.

Table 4.

General results of the Novosibirsk Reservoir shore protection

Shore Protection Type	Length km	Modern condition		
		Destroyed km	Indamage, km	In good condition, km
Seawalls and similar hard structures	11.88	5.51	4.50	1.87
Beach nourishement	7.50	–	0.50	7.00
Hybrid projects:				
protective beach + groins	9.00	–	1.50	7.50
protective beach + segmented breakwaters	5.20	–	1.20	4.00
protective beach + segmented breakwaters + groins	3.30	–	–	3.30
protective beach + artificial headlands	5.00	–	–	5.00

CONCLUSION

Preliminary data based on over one year of observations provide some information that can assist in future efforts to mitigate shoreline erosion. The construction of hard structures like seawalls under a man-made lake's environmental conditions marked by strong storms and long-term water level fluctuations had failed measure. This type of shore protection structure presumably will be admissible only in case of feeder beach creation on the upcoast area.

Beach nourishment and fulfilling of hybrid projects are considered to be the most efficient ones. There are numerous benefits providing these measures, among which are the significant mitigation of the shoreline erosion, fair reliability, and improvement of coastal environment on adjacent shore sites. It resulted in serious changes in shore protection policy in Russia, because Novosibirsk Reservoir is a study site for a full-scale testing of various ideas in applied coastal engineering.

ACKNOWLEDGEMENTS

The portion of this research was conducted under the support of Russian Foundation of Basic Research, the project # 96-05-64448.

REFERENCES

Birkemeier, W.A., 1980. The effect of structures and lake level on bluff and shore erosion in Berrigan County, Michigan, 1970-74. Coastal Eng. Research Center, Misc. Rept. No. 80-2, U.S. Army Corps of Engineers, 74p.

Dean R.G., 1976. Beach erosion: causes, processes, and remedial measures. CRS *Critical Review Environmental Control*, 6 (3), 259-296.

Dean R.G., 1987. Coastal sediment processes: Toward engineering solutions. *Coastal Sediments '87* (ASCE), pp. 1-24.

Kraus, N.C., 1988. The effects of seawalls on the beach: An extended literature review. Journal of Coastal Research , Special Issue No. 4, 1-28.

McDougal, W.G.; Sturtevant, M.A.,and Komar, P.D., 1987. Laborotory and field investigations of shoreline stabilization structures on adjacent propeties. In: Kraus, N.C. (ed.), *Proceedings of coastal sediments '87* (ASCE), pp. 961-973.

Plant, N.G., Griggs, G.B., 1992. Interaction between nearshore processes and beach morphology near seawall. *Journal of Coastal Research*, 8 (1), 183-200

Tait, J.F. and Griggs,G.B., 1990. Beach response to the presence of a seawall: A comparison of field observations. *Shore and Beach*, 58 (2),11-28.

CHAPTER 227

COMPARISON OF BEACH AND SHOREFACE NOURISHMENT
TORSMINDE TANGE, DENMARK

Christian Laustrup[1], Holger Toxvig Madsen[2], Per Sørensen[3], Ida Brøker[4]

Abstract

Within the framework of the EU–sponsored programme Marine Science and Technology (MAST), a full–scale test was implemented in 1993 in Denmark, Germany and the Netherlands with the aim of evaluating the use of shoreface nourishment.

In this paper the background, the set–up and the results of the Danish part of the NOURTEC project are presented.

1. Introduction

Nourishment has been used by the Danish Coastal Authority (DCA) on the Danish North Sea coast since 1974. The volume of sand has been gradually increased and has now reached a level of about 3 mill. m^3/year distributed along 150 km of the coast. Most of the nourishment has been beach nourishment but shoreface nourishment has also been used occasionally.

There has always been and there still is a considerable difference in the unit price of the two ways to protect the coast. The inevitable question is therefore which nourishment method has the best benefit/cost ratio.

Qualitatively, the following relative advantages are seen for the two nourishment methods:

Beach nourishment

– gives a better dune foot protection

---

1) Deputy Director, Danish Coastal Authority, Box 100, 7620 Lemvig, Denmark
2) Chief Engineer, Danish Coastal Authority, Box 100, 7620 Lemvig, Denmark
3) Project Engineer, Danish Coastal Authority, Box 100, 7620 Lemvig, Denmark
4) Chief Engineer, Danish Hydraulic Institute, Agern Allé 5, 2970 Hørsholm, Denmark

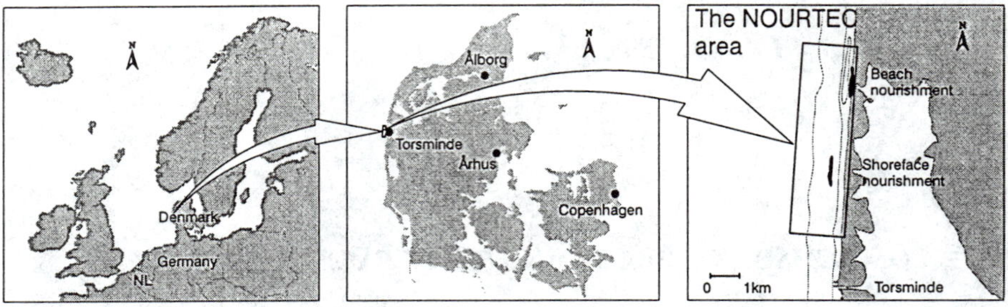

Fig. 1 Nourtec site, Torsminde Tange.

– gives a more well–defined local effect within the problem area

– is politically more attractive because a direct result of the nourishment can easily be seen

Shoreface nourishment

– is about 30% cheaper per m^3

– can be carried out without machinery on the beach

– is applicable to small nourishment projects because the mobilization costs are low

However, to our knowledge no quantitative methods exist which can be used to make a rational choice in the design situation. The design has been based on experience and on the use of rules of thumb.

The Danish Coastal Authority therefore decided to initiate a full–scale test on beach and shoreface nourishment, see Fig. 1. Together with similar projects in the Netherlands and Germany, the Danish project constitutes the NOURTEC project.

2. Research questions

In the planning and design of the test nourishment projects, the three NOURTEC partners agreed upon the following general design objectives [Niemeyer et al. 1995]:

– stabilization of the coastline

– coastal protection

– widening of the beach

Each of the design objectives is evaluated in terms of a corresponding design parameter, see section 4.2. The evaluation should end up giving the answers to these research questions:

– Does the nourishment fulfil the design objectives?

– What is the relative effect of the two nourishment projects? Here the benefit/cost ratio should be included.

– What is the importance of the grain size of the nourishment sand?

Besides, the project should give an improved understanding of the morphological changes induced by placing large volumes of sand in a coastal environment.

3. Experimental set–up

3.1 Test site description

Torsminde Tange is a 300 – 500 m wide barrier located north and south of Torsminde on the North Sea coast of Denmark, see Fig. 1. The barrier is a sand dune system on a clay layer. The top of the dunes is at +8 m DNN (Danish normal zero). The top of the clay layer is found at approximately –4 m DNN along the beach. The beach is 100 m wide and consists of sediment ranging from fine sand to pebbles. The beach slope is 1:20. A 2 – 3 bar system is found offshore and the sediment D_{50} range is from 0.3 to 0.8 mm. The closure depth is at –16 m DNN. The depth contours are straight and parallel to the coastline and the profile slope is 1:100. The average coastal retreat is 2.6 m/year.

3.2 Nourishment and survey programme

The shoreface nourishment is located on the outside of the outer bar in the southern sector and the beach nourishment is located from the dune foot to the shoreline in the northern sector. The nourishment volume in both nourishment projects was 250,000 m^3 of sand distributed with 250 m^3/m. Each of the stretches was 1 km long and 2 km apart from each other. The D_{50} of the beach nourishment sediment was 0.32 mm while the D_{50} of the shoreface nourishment was 0.57 mm. A red fluorescent tracer was mixed with the shoreface nourishment in a ratio of 1:100,000. Similarly, a blue tracer was mixed into the beach nourishment [Madsen et al. 1995].

The survey area is 7 km long and 1.5 km wide located symmetrically around the nourishment projects. The survey lines span from dune top to –12 m DNN with a mutual spacing of 100 m giving a total of 70 survey lines. A single beam echo sounder was used to survey the bathymetry and GPS was used to survey the topography and the zone from 0 to –2 m DNN. The survey accuracy was 0.04 m. A total of 17 survey campaigns were carried out during the 2–year NOURTEC period.

In every fifth of the survey lines 9 – 10 sediment samples were taken by hand on the beach and by a Van Veen grab offshore in predefined positions. All samples were sieved and analysed for content of tracer grains.

3.3 Boundary conditions

The water level is recorded at a groyne at Torsminde and at a groyne 5 km north of the survey area once every 15 min. The tidal range is 0.6 m and the 100 year water level is 3.40 m DNN. Wind speed and direction are recorded at Torsminde once every 15 min. Wind from the north west is dominant and the 100–year wind speed is 33 m/s. Wave height and direction are recorded by a waverider just north of the survey area at 18 m depth. Wave characteristics have also been recorded by a current meter, at first placed at 14 m depth just seaward of the NOURTEC area. Later it was placed at 4 m depth on the shoreface nourishment. Waves were recorded once every 3 hours. The dominant deep water waves come from NW and the 100–year significant wave height is 8.1 m. Current has been measured at the above described locations once every 15 min. At both locations the current meter was placed 1 m above the bottom. At 14 m depth the dominant current is the tidal current with an amplitude of 0.2 m/s. At 4 m depth the current is also significantly influenced by waves but still the tidal current is dominant.

4. Effectiveness analysis

4.1 Introduction

The surveys showed that the beach nourishment was eroded after about 3/4 year while the shoreface nourishment only lost about 40% of its volume in the whole monitoring period. The morphological description and interpretation of the development in the monitoring area is presented in section 6.

Here the results are presented of a strictly mechanistic method to analyse the survey data. This so–called effectiveness analysis gives objective results easy to compare with the results produced by the other two partners in the NOURTEC project.

The analysis is based on the use of a number of common design parameters. The parameters are calculated for each bathymetry and the development of the parameters is compared to the estimated development of the parameters in the absence of the nourishment projects. The difference between this so–called autonomous development and the actual development represents the net effect of the nourishment. Here it is assumed that the autonomous development and the nourishment effect can be superimposed.

4.2 Definition of the design parameters

In section 2, the general design objectives were presented. Each of these objectives is evaluated by means of one of the following three common design parameters:

- position of the coastline
- position of the upper part of the profile
- beach width

In Fig. 2 it is shown how the design parameters are calculated. Apart from the "beach width" the parameters are calculated on the basis of a volume divided by the height of the zone used in the volume calculation. The result is a horizontal distance to the reference line. The level DNN +4 m is considered to be representative of the beach level at the dune foot.

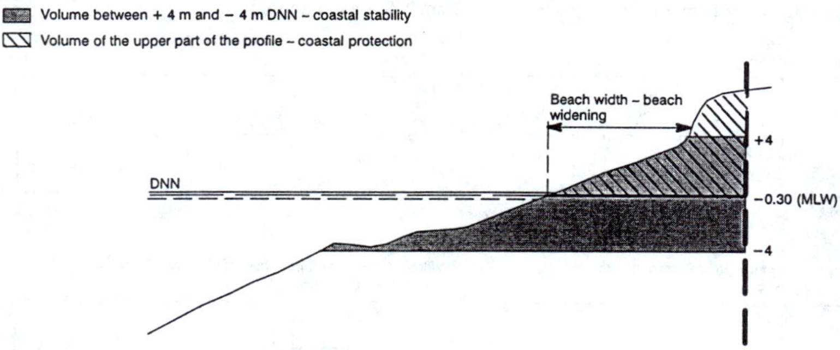

Fig. 2 Definition of the common design parameters.

4.3 The autonomous development

For practical reasons it was not possible to estimate the autonomous development in the monitoring area for the NOURTEC period 1993 – 95 from the development of a nearby reference stretch. So instead, historical data were used.

The wind rose for the monitoring period 1993 – 95 is nearly identical to the wind rose for the period 1985 – 92. This indicates that the meteorological boundary conditions are about the same for the two periods. On this basis it was decided to calculate the design parameters for the period 1985 – 92 and consider the slope of the corresponding trend lines as good estimates for the autonomous development in the monitoring period.

4.4 Results

The effect of the nourishment projects can be calculated for any section along the coast. Here it has been chosen to show the results for 3 km stretches located symmetrically around the two project areas. The reason for this choice is that the analyses have shown that the effect of the nourishment projects is not limited to the nourished 1 km stretches themselves.

The development of the three design parameters for the southern and the northern 3 km stretches is shown in Fig. 3. The autonomous development is shown as well.

From the figure it appears that in the 2–year monitoring period the beach nourishment stabilizes the coastline and improves the coastal protection level. On the other hand, the beach width has not been improved. For the shoreface nourishment there is a positive net effect on all three design parameters. Hence, the shoreface nourishment has fulfilled all the design objectives.

Quantitatively speaking the shoreface nourishment has an effect on the coastal stability of 10.4 m or 250,000 m^3 for the 3 km stretch while the similar effect of the beach nourishment is 2.8 m or 67,000 m^3. The effect on the coastal protection is 9.0 m or 189,000 m^3 for the 3 km stretch for the shoreface nourishment and 5.4 m or 113,000 m^3 for the beach nourishment.

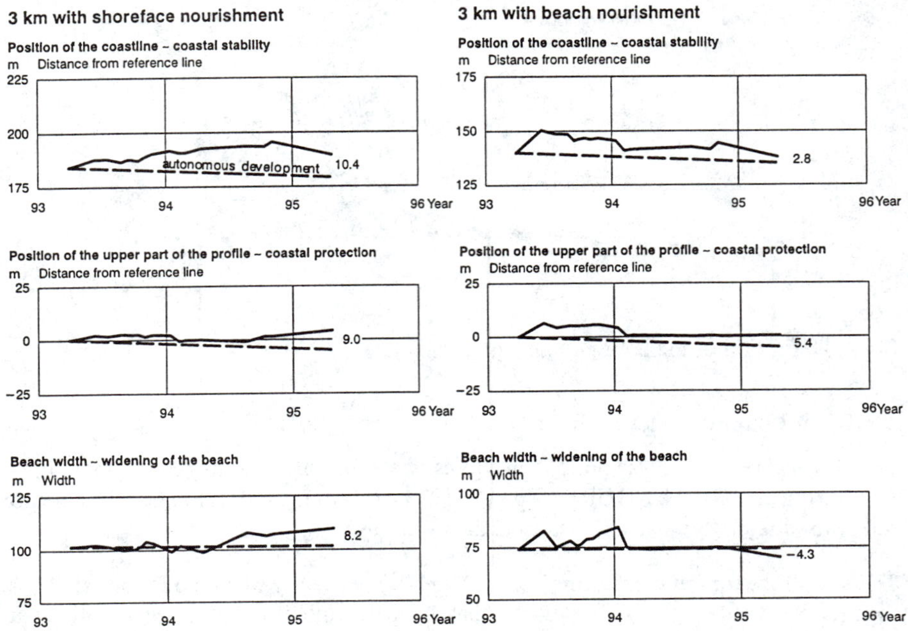

Fig. 3 Nourishment effect on the design parameters.

A direct comparison of the two nourishment projects is shown in Fig. 4. It appears that the shoreface nourishment gives the better stabilization of the coastline over the whole period. However, in the first 3/4 year the beach nourishment is the more effective. Concerning coastal protection the shoreface nourishment is the better option. Again the beach nourishment is the better option at the beginning of the period. For the beach width the result is similar. At the beginning, the beach nourishment is better but over the whole period the shoreface nourishment ends up being better.

Also in Fig. 4 the benefit/cost ratios are shown for the effect of the nourishment projects. Because it is about 30% cheaper to nourish on the shoreface than on the beach, the better relative effect of the shoreface nourishment mentioned before is increased when the costs are included in the evaluation.

5. Modelling

The 1–dimensional numerical model LITPACK and a 2–dimensional numerical model MIKE 21 have been used. Both models have been developed by the Danish Hydraulic Institute. The purpose of using the models is to analyse the effect of using coarse sand in the shoreface nourishment and to extrapolate the results to other boundary conditions.

5.1 1–dimensional modelling

LITPACK has been used to calculate the longshore sediment transport during the entire NOURTEC period and the results have been used in the calibration of MIKE 21. The

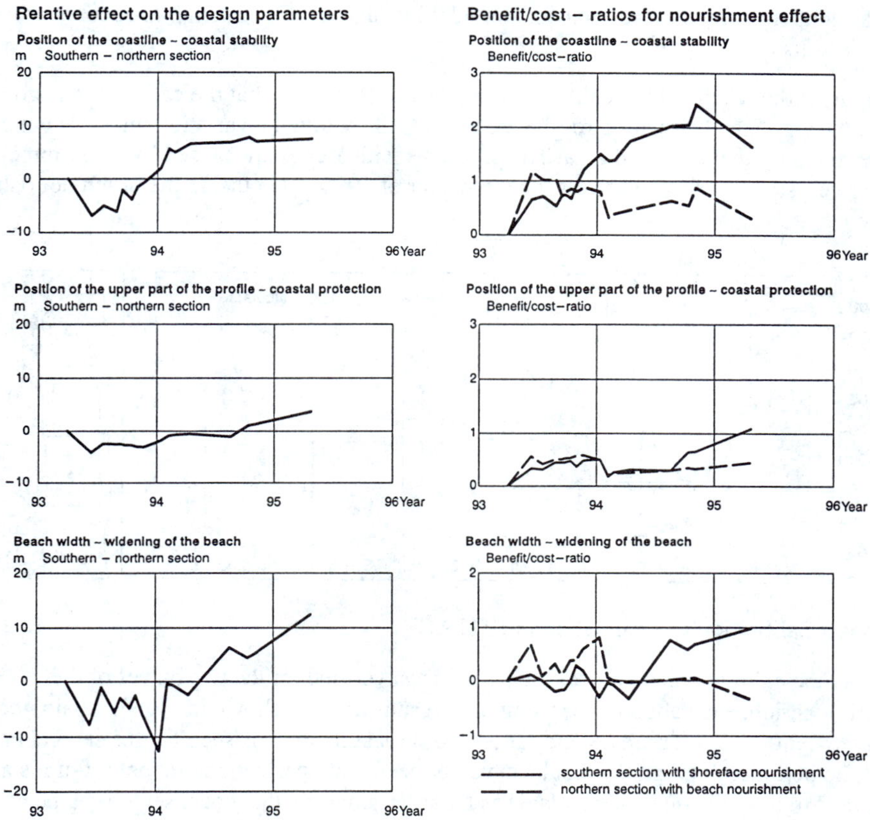

Fig. 4 Relative effect of the two nourishment methods and benefit/cost ratios for the nourishment effect.

nourishment projects are included in the two areas S(horeface nourishment) and B(each nourishment), see Fig. 5. For each area there is one survey line on both sides of the nourishment and 3 lines which include the nourishment.

First, LITPACK was calibrated to the annual littoral drift. The annual littoral drift has been calculated by accumulating the profile erosion from the nodal point for the littoral drift to the NOURTEC area by using three different approaches. Based on these calculations the south–going net littoral drift was estimated to be in the interval of 550,000 – 850,000 m^3/year. Secondly, LITPACK was calibrated to the measured volume changes in the areas S and B in the entire NOURTEC period, including the variation intervals due to survey inaccuracy.

In LITPACK, uniformity of the coastal profiles is a basic assumption. The applicability of LITPACK for quantification of erosion/accretion along a coastline depends entirely on the fulfillment of this assumption. This assumption was not completely met, which was confirmed by the 2–dimensional modelling, and must be taken into account when the results are analysed.

In Fig. 5 the results of the calibration are shown. It appears that the calibration is good for the area S and less good for the area B but still within the variation intervals which are given by the measured data. The problems with area B are caused by an extremely large erosion that took place during the winter of 1994 – 95 outside the beach nourishment.

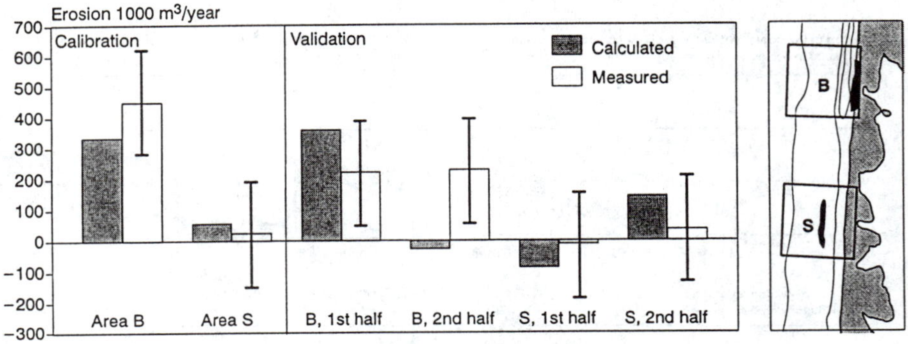

Fig. 5 Calibration and validation of LITPACK.

After the calibration, the model set–up has been verified for the two halves of the monitoring period, each about 1 year long. The results are also shown in Fig. 5. Again good agreement is seen for area S and for area B, less agreement is seen for the second half of the period. Although the results have not been quite perfect, the model set–up is assumed to be calibrated and validated as far as possible and the model set–up can be used in a descriptive and prognostic mode.

The cross–shore distribution of the longshore sediment transport in the NOURTEC period has been analysed and it is found to be very sensitive to the local shape of the shoreface nourishment on the outer bar. A decrease in the water depth above the nourishment causes extensive wave breaking and hence increased sediment transport on the bar and decreased transport near the coastline. The sediment transport distribution is not particularly affected by the the beach nourishment because only a small part of the nourishment is under influence of waves.

For both the shoreface and the beach nourishment area it is found that during the NOURTEC period waves with a significant wave height of 1 to 2 m cause most of the sediment transport and the net sediment transport direction is southward.

In order to quantify the effect of nourishment sediment characteristics on the sediment transport, the dependence of D_{50} is analysed. For the shoreface nourishment $D_{50} = 0.57$ mm and for the beach nourishment $D_{50} = 0.32$ mm. It is done by averaging the longshore sediment transport for the NOURTEC period in each of the areas S and B for each D_{50}. For the shoreface nourishment the ratio between $D_{50} = 0.32$ mm and $D_{50} = 0.57$ mm was found to be 2.3. Similarly for the beach nourishment, the ratio was found to be 1.2.

5.2 2–dimensional modelling

MIKE 21 is a 2–dimensional modelling complex. In the present case a parametrised wind wave model, a depth integrated hydrodynamic model and a deterministic, intra wave period sediment transport have been applied. The intention of using MIKE 21 is to calibrate the model so that it will reproduce the sediment transport in the NOURTEC period.

The simulated conditions had to be limited to 3 wave heights and corresponding water levels, 2 wave directions, 3 bathymetries and 4 different types of sediments due to computer time consumption and with that the costs. The wave heights are chosen from knowledge gained in the LITPACK study about which waves are significant for the sediment transport. Wave directions from south– and northwest were chosen as a minimum. Both are 40° off the coast normal. As a minimum, a bathymetry taken prior to the nourishing, one describing a "summer" bathymetry and one describing a "winter" bathymetry were chosen. Finally, the sediment range from natural sand to the coarse shoreface nourishment sand was described by the D_{50} parameter. The same sediment spreading was used to reduce the number of variables. These parameters were combined in 18 sets of data, each set representing a typical combination of boundary conditions.

The 2–dimensional modelling complex consists of a large wind wave model with a resolution of 20 by 100 m, which gives boundary conditions to the local model area. Here the resolution of the wave modelling is 4 by 20 m. The resolution of the hydrodynamic and sediment transport model is 20 by 20 m.

The results show that even small morphological changes are reflected in the sediment transport pattern so it was difficult to distinguish between the sediment transport caused by natural variations and the sediment transport caused by the nourishment volumes. An attempt has been made to eliminate all other changes but the ones caused by the nourishments by integrating the longshore transport in depth intervals of +4 m to –4 m ("coastline") and of –4 m to –8 m ("6 m depth"). It was found that the changes introduced by the nourishments are nearly insignificant compared to the natural variations both in space and time.

Fig. 6 shows examples of sediment transport fields calculated for 3 different bathymetries. Here the 2–dimensional effect of the nourishments are clearly seen. The longshore sediment transport components have been integrated within the intervals "coastline" and "6 m depth". The calculated integrated sediment transport rates are compared for a) the bathymetry just before nourishment and the "before nourishment bathymetry" supplemented with the "after nourishment bathymetry" within the area S and B, and b) for a "summer" bathymetry and a "winter" bathymetry.

Another result was that of migrating bed forms on deep water (15 – 20 m) influence the direction of the incoming waves and effect the longshore sediment transport. The effect is most pronounced in the most seaward part of the NOURTEC area.

The effect on the integrated longshore transport of using coarse sand for the shoreface nourishment was calculated. For the shoreface nourishment the ratio between the longshore sediment transport rates for D_{50}= 0.32 and 0.57 mm was found to be 1.17. Similarly for the beach nourishment, the ratio was found to be 1.11.

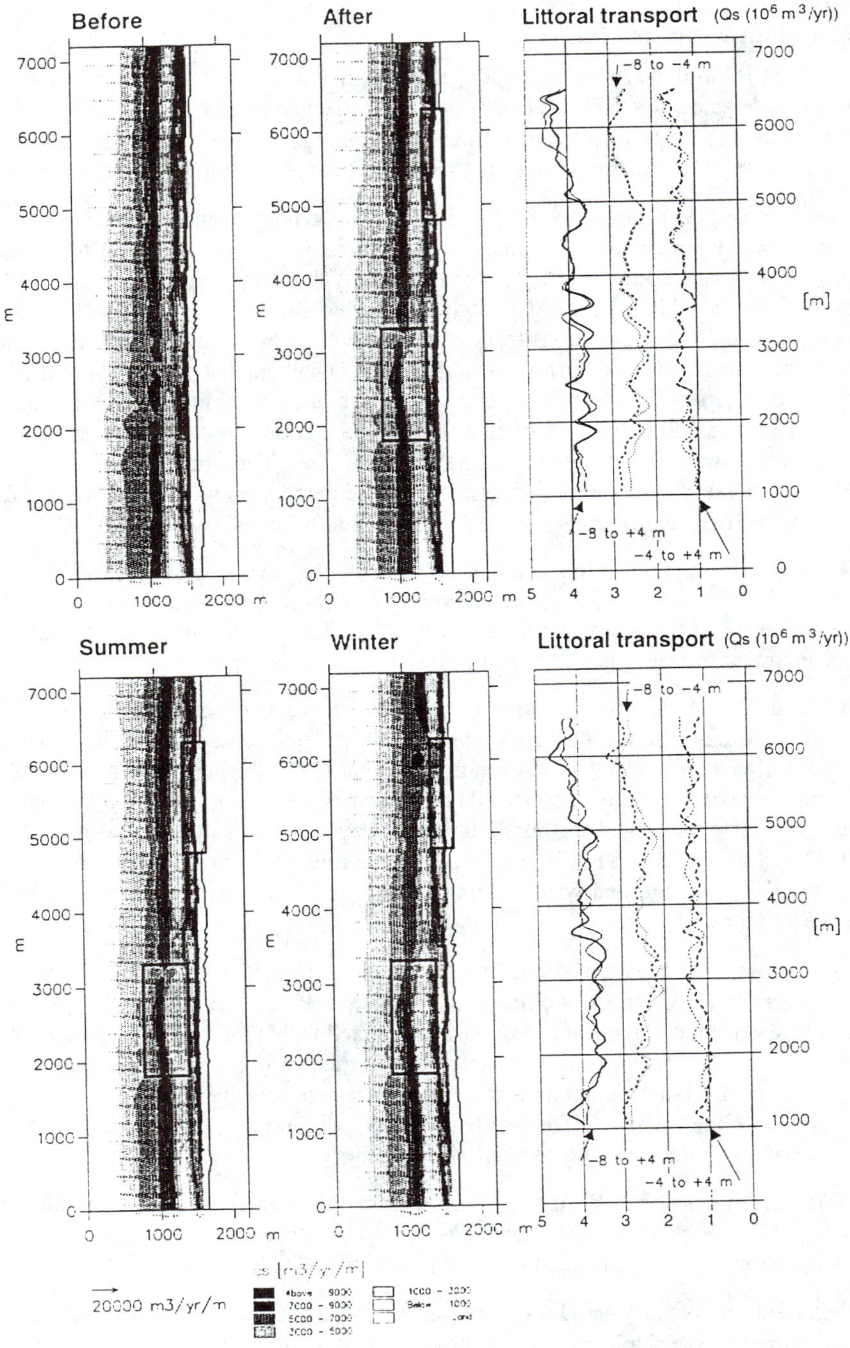

Fig. 6 Calculated, 2–dimensional sediment transport fields and integrated longshore sediment transport components for the intervals +4 m to –4 m, and –4 m to –8 m.

Finally, attempts were made to combine the MIKE 21 calculated volume changes in a system of 12 boxes by which the NOURTEC area is described to reproduce the development in the NOURTEC period. The longshore sediment transport variation with wave direction calculated by LITPACK was used to extrapolate the volume changes. The morphological changes were taken into account by using the "summer" or "winter" bathymetry. Fig. 7 shows the MIKE 21 weighted and measured accretion in 1 km long boxes. In 3 of the 6 boxes good agreement is seen but poor agreement is seen in the other boxes.

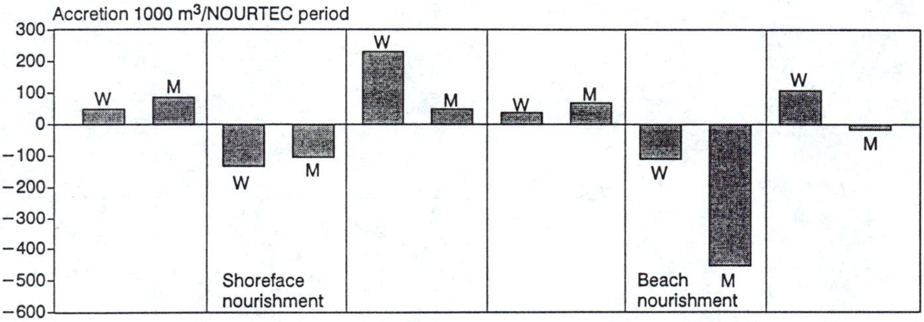

Fig. 7 Weighted Mike 21 and Measured volumes from south to north.

6. Morphological analysis

The volume changes in the monitoring area have been analysed by use of a system of cells with fixed horizontal boundaries. For each cell the volume development has been calculated.

In Fig. 8 the system of cells is shown together with the volume curves for selected cells and combinations of cells. The shoreface nourishment is located in cell H. For this cell there is a volume increase just after completion of the nourishment. From here the volume is stable with some fluctuations. In the cells A–B–C landward the shoreface nourishment there is a gradual volume increase over the whole monitoring period. For the beach nourishment in cell E a gradual volume decrease is seen over the monitoring period. After about 3/4 year the beach nourishment has been eroded. Tracer results indicate that part of this sand migrates towards south in the coastline zone.

Although the volume is constant in cell H, some morphological changes are seen. Over the monitoring period the nourishment migrates about 50 m onshore and the level of the top of the nourishment is reduced by about 0.5 m.

In Fig. 9 an overview is given of the horizontal development of the bar system. Just south of the shoreface nourishment the bar becomes continuous shortly after completion of the nourishment, probably because of the surplus of sand transported on the bar at the nourishment. It can also be seen that a weakening starts after a year at the northern end of the nourishment. Corings in the nourishment area show that about 60% of the nourishment sand is still in the initial position at the end of the monitoring period.

The 40% loss of the nourishment sand is only about 1/3 of the total accretion landward of the nourishment. It means that the accretion effect of the shoreface nourishment is

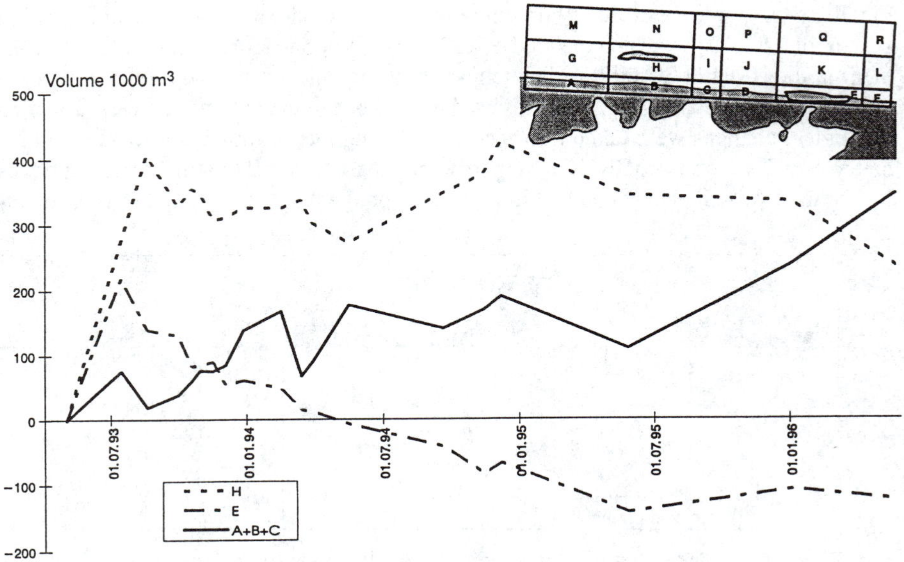

Fig. 8 Volume development in selected cells.

much larger than the nourishment volume itself. The reason is the breakwater effect of the nourishment. Like a breakwater, the nourishment reduces the longshore transport landward with an accretion as result.

An interesting result of this project is the better stability of the shoreface nourishment compared to the beach nourishment. Part of the explanation of this fact is the coarser sand in the shoreface nourishment. However, a reason which is even as important is the specific position of the shoreface nourishment. The position in the bar zone is a natural position for a perturbation so to say. Therefore the nourishment quickly becomes an integrated part of the bar system instead of being eroded away.

Outside the monitoring area migrating bed forms have been observed. However, it has not been possible to verify if the bed forms cause a longshore variation in the sand transport from the outer to the inner part of the profile.

7. Conclusions

The effects of a beach nourishment and a shoreface nourishment have been analysed and compared. Because the weather conditions in the monitoring period were about average, the results are considered to be valid for average weather conditions.

By use of a mechanistic analysis approach it has been determined if the nourishment projects fulfil three design objectives: stabilization of the coastline, coastal protection and widening of the beach. It was found that the beach nourishment stabilizes the coastline and improves the coastal protection but the beach width has not been improved. On the other hand, the shoreface nourishment fulfils all the design objectives.

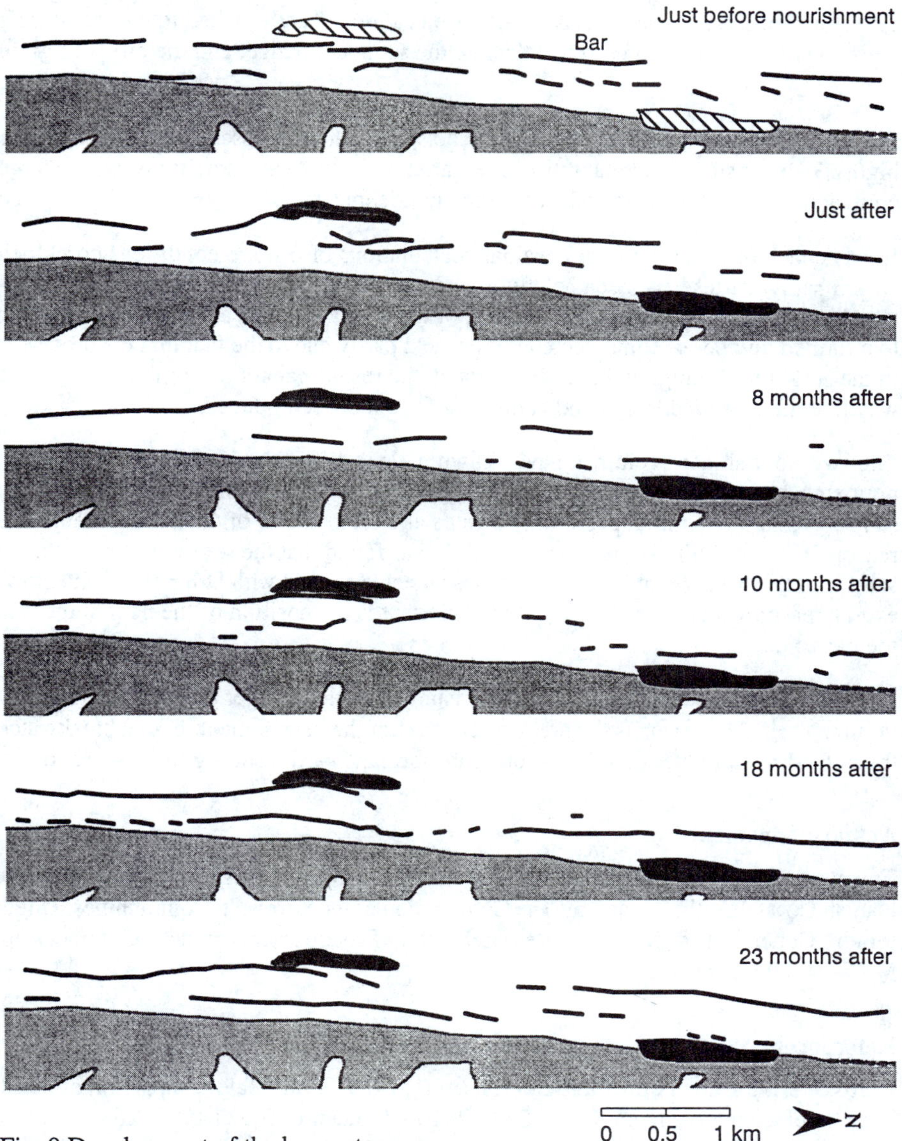

Fig. 9 Development of the bar system.

The relative effect of the two types of nourishment has also been analysed. For all three design objectives, shoreface nourishment is the better option at the end of the monitoring period but beach nourishment is the better option at the beginning. When the price of the nourishment sand is taken into account, the relative effect of the shoreface nourishment is improved because this nourishment type is cheaper.

The 1–dimensional model LITPACK has been calibrated and verified to the monitoring data. LITPACK was then used to calculate the sediment budget and the effect of grain size.

The 2–dimensional model MIKE 21 has shown that the processes in the surf zone are highly 3–dimensional and that the natural variations of the longshore transport are large compared to the changes introduced by the nourishment.

It was concluded that the 2–dimensional calculations of 6 wave conditions combined with 3 bathymetries have been helpful in understanding details of some of the processes which contribute to the natural re–shaping of the investigated area. However, partly due to a limited number of simulated conditions and partly due to the fact that the processes in the surf zone are highly three dimensional, the results cannot be combined in such a way that the observed losses and gains of sediment are reproduced.

The two nourishment volumes have shown a clear difference in stability. The beach nourishment was eroded after 3/4 year. For the shoreface nourishment about 60% of the sand was still in the initial position two years after completion of the nourishment. One reason for the stability is the coarser sand. It was found that the shoreface nourishment sand with D_{50} = 0.57 mm is about 2.3 times as stable as sand with D_{50} = 0.32 mm. However, a reason which is probably just as important is the position of the nourishment in the bar zone.

The accretion effect of the shoreface nourishment is much larger than the nourishment volume itself. The reason is the breakwater effect of the nourishment. Like a breakwater, the nourishment reduces the longshore transport landward resulting in an accretion.

Acknowledgement

This work was carried out as part of the NOURTEC project. It was funded jointly by The Danish Coastal Authority and by The Commission of the European Communities, Directorate General for Science, Research and Development under Contract no. MAS2–CT93–0049.

References

Madsen H.T., Laustrup C. and Soerensen P., 1995. A full scale comparison of beach and shoreface nourishment. COPEDEC IV Rio de Janeiro, Brazil 1995, 263–272.

Niemeyer H.D., Biegel E., Kaiser R., Knaack H., Laustrup C., Mulder J.P.M., Spanhoff R. and Toxvig H., 1995. General aims of the NOURTEC project– Effectiveness and execution of beach and shoreface nourishments, COPEDEC IV Rio de Janeiro, Brazil 1995, 311–325

CHAPTER 228

Mathematical and Physical Modeling of Beach Nourishment Projects

W. Erick Rogers[1] and Paul A. Work[2], Associate Member, ASCE

Abstract

This paper evaluates the utility of mathematical models for prediction of the planform movement of beach nourishment projects. The results of one-line models (analytical and numerical) are compared to the evolution of nine laboratory-scale model beachfills. Several methods of planform modeling are evaluated, varying in complexity. Comparison of planform modeling suggests that, in some circumstances, using a simple analytical treatment of refraction may be a feasible alternative to using more rigorous numerical wave modeling for shoreline modeling purposes. The relative ability with which different equations for net longshore sediment transport rate reproduce observed beachfill evolution is discussed. Information on the effect of wave climate and beachfill geometry on beachfill lifetime is extracted from the laboratory study. The observed effect of lower wave height, greater beachfill length, and the tapering of a beachfill is a greater beachfill lifetime, in agreement with conventional wisdom. Increasing beach slope and decreasing wave period are both observed to cause a slight decrease in beachfill lifetime.

Introduction

A beach nourishment project is an attempt to widen the dry portion of a beach and offer protection from storms to existing nearby structures by adding large quantities of sediment to some portion of the beach. They are sometimes appropriately referred to as "sacrificial beachfills." This emphasizes the fact that beach nourishment projects are rarely expected to stay in place for more than a decade without losing a large portion of the placed material. Any improvement in the ability to predict this redistribution of beachfill material, through the use of better modeling techniques, allows for more cost effective design and more efficient coastal zone management.

---

[1]U.S. Nav. Res. Lab. Contractor (Planning Systems Inc.); MSAAP Bldg.9121 Stennis Space Center, MS 39529 USA; e-mail:rogers@lincoln.nrlssc.navy.mil

[2]Asst. Prof., Dept of Civil Eng., 110 Lowry Hall, Clemson Univ., Clemson, SC 29634-0911 USA; e-mail: pwork@ces.clemson.edu

Laboratory Study

A laboratory study was conducted, with two primary purposes in mind: 1)to observe in detail longshore sediment transport patterns in the presence of beach nourishment projects, and 2) to provide prototype beachfills for validation of mathematical shoreline models. Similar studies have been conducted. Kamphuis et al. (1986), Kamphuis (1991), and Nielsen (1988) studied longshore sediment transport in a laboratory setting. Kamphuis and Meyer (1976) studied beachfills using physical models. Dean and Yoo (1994) conducted a laboratory study of beach nourishment projects in front of seawalls and compared the results to mathematical models.

Nine laboratory beachfills were monitored. A paddle-type wave-maker was used to generate nominally shore-normal, monochromatic waves. A set of parameters were compiled which were expected to have an effect on beachfill lifetime: project length, degree of beachfill tapering, breaking wave height, wave period, and beach slope. In a laboratory setting these parameters are easily controlled; each case was assigned a different set of values for these parameters. For each case, a beachfill "lifetime" was calculated based on the amount of beachfill remaining in place, by volume, after wave attack. Lifetime is defined as the time period required for a significant percentage of material to be eroded; loss of 5% of beachfill volume was used. Beachfill material which was dispersed in the longshore direction, outside the original location of the beachfill, was considered "lost" sediment volume. Based on these calculations, the effect of each of the important parameters on beachfill lifetime was inferred (see Table I). Comparison is made to mathematical models employing the "CERC equation" for longshore sediment transport rate (U.S. Army Corps of Engineers 1984).

Table I
Effect of Parameters on Beachfill Lifetime

Parameter	Mathematical Models (with CERC formula)	Physical Model Observations
H_b, wave height at breaking ⇑	Lifetime ⇓	Lifetime ⇓ (significant effect)
T, wave period ⇑	No effect	⇑ (slight)
ℓ project length ⇑	⇑	⇑ (moderate)
m, beach slope ⇑	No effect	⇓ (slight)
Tapering of fill	⇑	⇑ (moderate)

Figure 1 shows the location of two typical beach profiles from one of the laboratory cases. Figure 2 is a beach profile located at the "shoulder" of one of the beachfills; the profile experiences a net loss of sediment. Figure 3 shows a beach profile located in the adjacent, unnourished region; accretion occurs due to beachfill material deposited there via longshore transport.

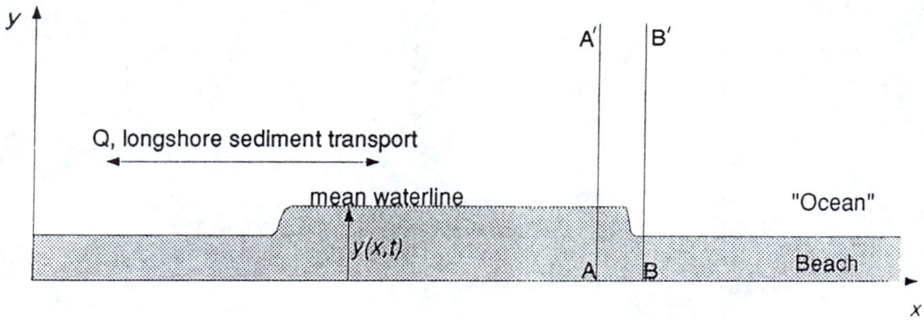

Figure 1. Locations of Laboratory Beach Profiles A and B

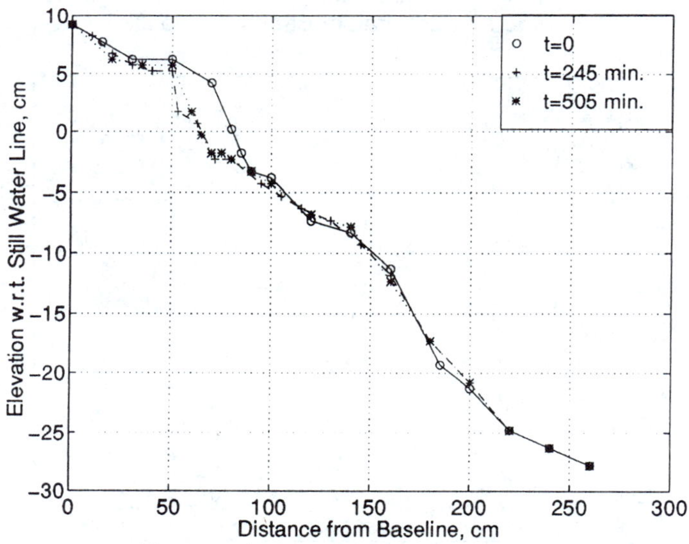

Figure 2. Laboratory Beach Profile A

With any physical model, scale effect is a major concern. Due to the relatively small scale of this laboratory study, it is unreasonable to expect accurate quantitative information. However, if the physical model behaves in a manner similar to full scale beachfills, useful qualitative information can be gained, like that shown in Table I. Figure 4 is a beach profile from the shoulder of a full-scale beach nourishment project (Perdido Key, FL). The similarity to Figure 2 is encouraging.

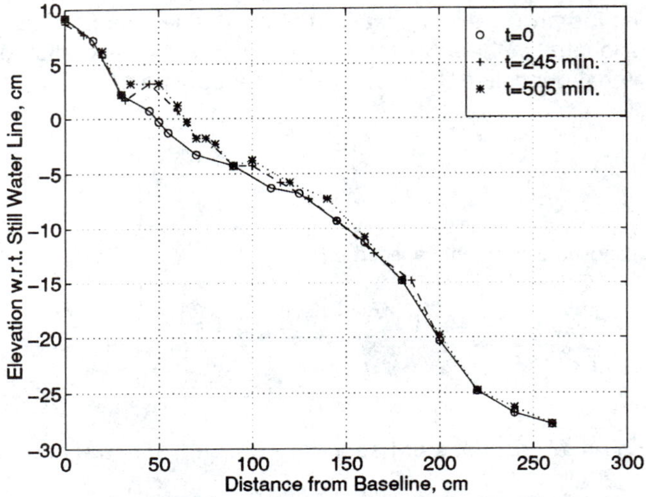

Figure 3. Laboratory Beach Profile B

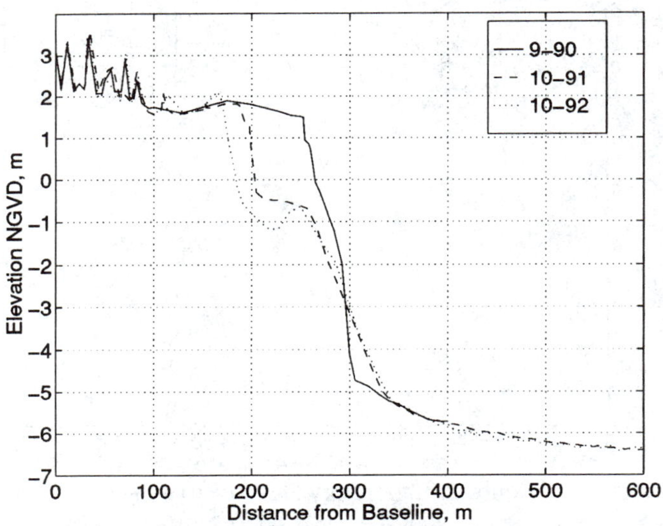

Figure 4. Beach Profile from Full-Scale Beachfill (Perdido, FL)

Shoreline Models

The shoreline model used in this study was a "one-line model", a tool which has been used extensively by researchers and engineers (e.g. Pelnard Considére 1956, Hansen and Kraus 1989). One-line models are used to predict the movement of a single elevation contour of a beach, typically the still water line. These models are governed by the sediment continuity equation,

$$\frac{\partial y}{\partial t}+\frac{1}{h_{*}+B}\frac{\partial Q}{\partial x}=0, \tag{1}$$

and an equation for longshore sediment transport. In equation (1), $h_{*}+B$ defines the vertical extent of the active beach profile.

If a one-line model is used to provide the shoreline location, $y(x,t)$, the assumption must be made that the shape of a beach profile does not change, but that the profile only shifts onshore or offshore as erosion or accretion occurs (implying zero net cross-shore sediment transport over the time period of interest). However, a recently constructed beachfill typically has an overly steep beach slope, resulting in significant offshore sediment transport. Shoreline models were used to calculate longshore distributions of longshore sediment transport gradients, $\partial Q/\partial x$. A longshore sediment transport gradient can be directly compared to the time averaged rate of change of beach profile area $\partial A/\partial t$, which can be inferred from laboratory and field beach profile measurements:

$$\frac{\overline{\partial Q}}{\partial x}=-\frac{\overline{\partial A}}{\partial t}=-\frac{\Delta A}{\Delta t}. \tag{2}$$

Here, the overbar denotes time averaging. This method of comparison removes the effect of cross-shore sediment transport from calculations, provided that the entire "active" portion of the beach profile is included in area calculations.

<u>Sediment Transport Equations used in Mathematical Shoreline Models</u>

Three basic equations for longshore sediment transport were tested with the shoreline change models of the laboratory beachfills: the "CERC equation," the "Kamphuis equation" (Kamphuis 1991), and the "GENESIS equation" (Hansen and Kraus 1989). The CERC equation states that longshore sediment transport is proportional to the longshore component of wave energy flux:

$$Q=\frac{KH_{sb}^{2}C_{gb}\sin2\theta_{b}}{16(SG-1)g(1-p)}. \tag{3}$$

Here K is an empirical coefficient.

The GENESIS equation is a slight modification of the CERC formula, with an added term to account for longshore sediment transport driven by longshore gradients in wave height caused by the presence of structures or (less significantly) irregular bathymetry:

$$Q=(H^2C_g)_b[a_1\sin(2\theta_b)-a_2\cos\theta_b\frac{\partial H}{\partial x}]_b\ , \qquad (4)$$

where

$$a_1=\frac{K_1}{16(SG-1)(1-p)(1.416)^{\frac{5}{2}}}\ , \qquad (5)$$

and

$$a_2=\frac{K_2}{8m(SG-1)(1-p)(1.416)^{\frac{7}{2}}}\ . \qquad (6)$$

The Kamphuis equation is an empirically-based equation, developed by a power fit analysis using data from lab and field studies:

$$Q=KH_{sb}^2T_p^{1.5}m_b^{0.75}D_{50}^{-0.25}\sin^{0.6}(2\theta_b). \qquad (7)$$

Inclusion of Refraction in the Numerical Shoreline Model

Longshore sediment transport rate is generally accepted to be dependent on the angle between the wave orthogonal and the local shore normal. This suggests that the inclusion of refraction is requisite for any numerical modeling of irregular shorelines, such as a coastline with a trapezoidal beachfill. In most beach nourishment scenarios, inclusion of refraction will lead to a lower predicted rate of evolution, as it tends to decrease wave obliqueness at breaking.

The numerical wave transformation model REFRACT (Dalrymple 1988) was used to calculate breaking wave conditions for use in the shoreline model. This wave model includes shoaling and refraction (but not diffraction).

As an alternate method for including refraction in a numerical one-line model, a modified form of the CERC equation for longshore sediment transport (Work and Dean 1995) was employed. This equation is based on a two-line, analytical approach to wave transformation, which is much simpler than using a numerical wave model to supply breaking wave input data. The equation is a generalized form of an equation proposed by Dean and Yoo (1992). It can be stated as:

$$Q=K\frac{H_R^{2.4}g^{0.4}\cos^{1.2}(\beta_R-\alpha_R)}{8(SG-1)(1-p)\kappa^{0.4}}\frac{C_{G_R}^{1.2}}{C_*}\left[\frac{C_*\sin(\beta_R-\alpha_R)}{C_R}-\beta_t\left(1-\frac{C_*}{C_t}\right)+\beta_s-\beta_R\frac{C_*}{C_t}\right]. \tag{8}$$

The notation is illustrated in Figure 5. Subscript t denotes the location of the "toe" of the beach nourishment.

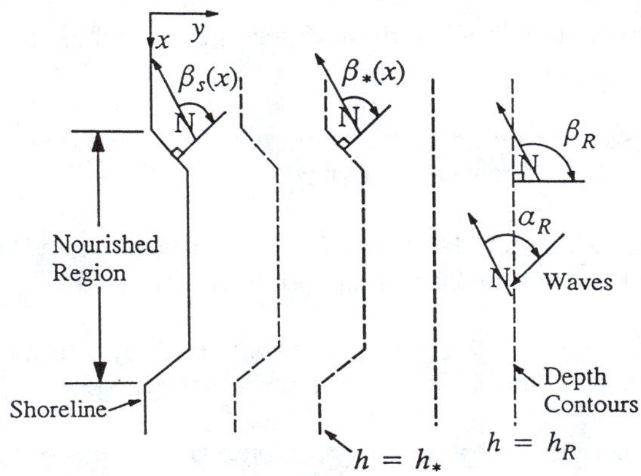

Figure 5. Notation used in Equation (8)

Inclusion of Diffraction in the Numerical Shoreline Model

Diffraction acts to reduce along-wave wave height gradients and thus tends to counter the effects of refraction. Without the inclusion of diffraction, a wave transformation model may tend to yield exaggerated focusing of wave energy near the "shoulders" of a beach nourishment project, which would compromise the predictive capability of the one-line model. The numerical wave model REF/DIF1 (Kirby and Dalrymple 1994) was used to generate breaking wave conditions (for use in the one-line model) including the effects of the shoaling, refraction, and diffraction of waves.

Comparison of One-Line Model Results to Laboratory Results

Laboratory data and numerical model results were compared based on longshore gradients of longshore sediment transport rate ($\partial Q/\partial x$), rather than shoreline evolution; thereby the effects of cross-shore sediment transport on shoreline movement were removed. Time-averaged longshore gradients of longshore sediment transport were inferred from beach profile measurements using equation (2).

Several shoreline modeling schemes were used, varying in complexity:

1. analytical calculation of shoaling (using conservation of wave energy flux), CERC equation for longshore sediment transport (equation (3));

2. analytical calculation of shoaling (using conservation of wave energy flux), Kamphuis equation for longshore sediment transport;

3. analytical treatment of the shoaling and refraction of waves in CERC equation (equation (8));

4. analytical treatment of shoaling and refraction in Kamphuis equation (analogous to equation (8));

5. shoaling and refraction determined by numerical wave model REFRACT, CERC equation for longshore sediment transport;

6. shoaling and refraction determined by numerical wave model REFRACT, GENESIS equation for longshore sediment transport; and

7. shoaling, refraction, and diffraction determined by numerical wave model REFDIF1, CERC equation for longshore sediment transport.

Figure 6 shows the longshore gradients of longshore sediment transport calculated by the one-line model for one of the laboratory cases. Note the similarity between the results using the numerical and analytical treatment of refraction.

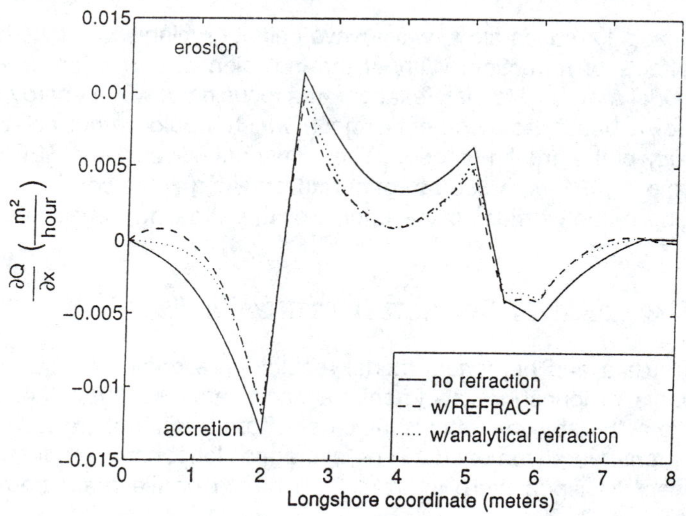

Figure 6. Comparison of Shoreline Modeling Schemes (1, 3, and 5)

Figures 7 compares longshore sediment transport gradients inferred from measurements for a laboratory beachfill to model output using option (1) above. This plot is representative of most comparisons of the one-line model results to lab results: though quantitative agreement is somewhat artificial due to calibration of the longshore sediment transport equation, qualitative agreement is good.

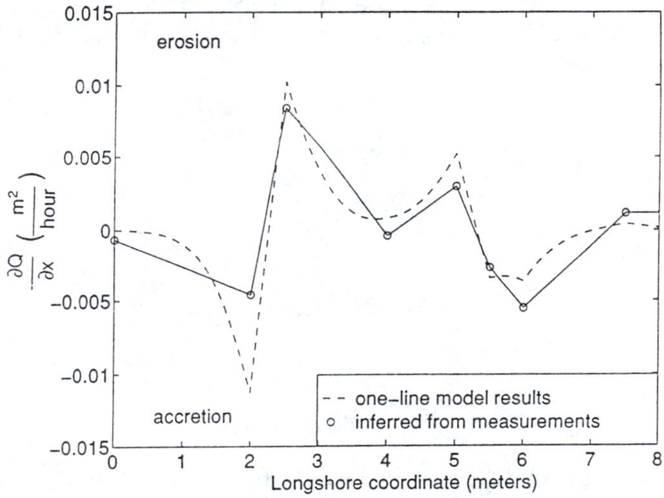

Figure 7. Comparison of Laboratory Results to One-Line Model Results

Using standard "field" values for coefficients of proportionality, none of the longshore sediment transport equations yielded good quantitative agreement. With standard coefficients, the equations typically greatly over-predicted longshore sediment transport. This is almost certainly due to scale effect, and has been observed by other researchers (e.g. Komar and Inman 1970).

The ability of the calibrated numerical one-line model to duplicate laboratory results was determined for each of the variations of the shoreline model. This comparison indicated that the use of the numerical wave models did not provide any added accuracy over method (1) above. On the other hand, use of the longshore sediment transport equation with an analytical treatment of refraction did yield a slight improvement in results.

In this comparison of methods, the Kamphuis equation yielded slightly better qualitative agreement than the other two basic equations for longshore sediment transport. The added term in the GENESIS equation did not result in any improvement over the CERC equation, probably because longshore gradients in wave height were not large in the laboratory.

Comparison to Field Data

Numerical one-line model results were compared to data from a well-monitored beach nourishment project at Folly Beach, South Carolina (Figure 8). The method for analytical treatment of refraction (method 3 above) was used with a standard calibration factor for the CERC equation (K=0.77 using H_{rms}). 2.1 million cubic meters of beachfill material were placed during the first four months of 1993 (Ebersole, et al., 1996). A 25-month period was investigated, from July, 1993 to August, 1995.

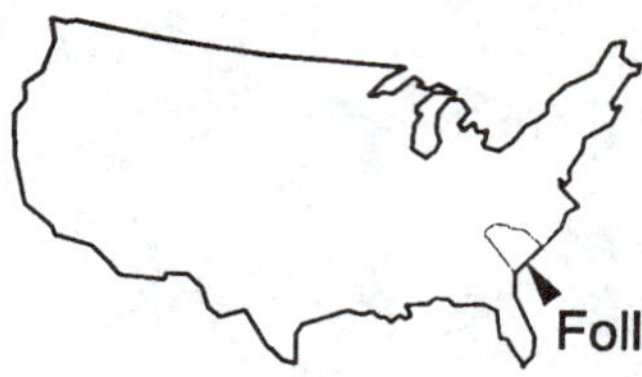

Figure 8. Location of Folly Beach

For the shoreline model initial condition, the "ad hoc transformation" method was used. This method is suggested by Dean and Yoo (1992) for situations "where substantial perturbations (human or natural) have placed the system out of balance." The initial planform used was simply the deviation from an assumed equilibrium planform caused by the placement of the beachfill.

Nine groins located northeast of the Holiday Inn (see Figure 9) were renovated at roughly the same time that the beachfill was placed. These groins were not included in the model results presented in this paper. For the time period modeled, the groins were thought to have a minor impact on shoreline movement; the groins are relatively short and were dry at low tide. Offshore wave data for 1993 and 1994 were obtained from a Wave Information Study (WIS) hindcast database (Brooks and Brandon 1995) for shoreline model input.

Comparison of the one-line model results to measured data is shown in Figure 10. The one-line model provides a reasonably good prediction of erosion and accretion, in both a quantitative and qualitative sense. Close inspection of Figure 10 suggests that the inclusion of the groins in the numerical model may have improved the model's accuracy--erosion was overpredicted by the model inside the groin field, and underpredicted downdrift (southwest) of the field.

Volume calculations indicate that 413,000 m^3, or 19% of the beachfill volume which existed in July 1993 was lost from the nourished region by August, 1995. Volume calculations include both the subaerial and submarine portions of Folly Beach; beachfill material deposited immediately offshore of the beach (above

-3m to -4m NGVD) was not considered a sediment loss, even though, by some measures, this cross-shore sediment transport would decrease the value of the beach. A majority of the 413,000 m^3 should therefore represent a loss of sediment due to longshore gradients in longshore sediment transport. However, other factors such as aeolian sediment transport, transport by tidal currents, and sand deposition well offshore of the beach could have contributed to the loss or gain of sediment. The shoreline model predicted a loss of 302,000 m^3 for the same time period. Had a larger empirical sediment transport coefficient been used, to account for the relatively fine sand at Folly Beach, this volumetric prediction would have been more accurate.

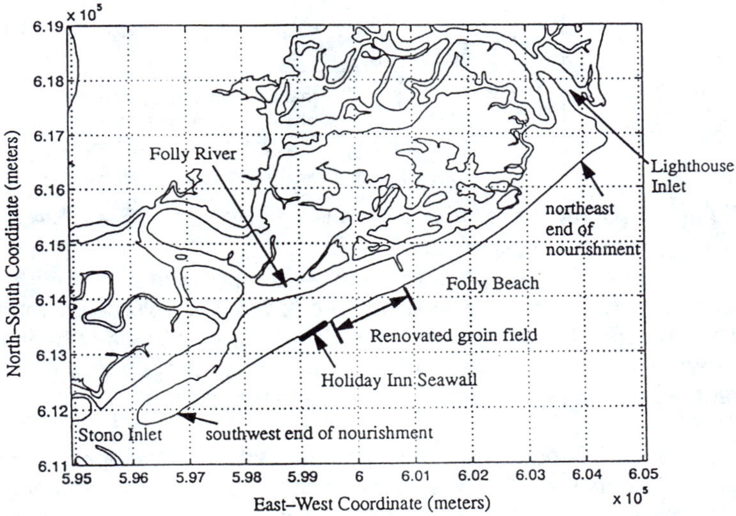

Figure 9. Map of Folly Island, S.C.

Conclusions

The laboratory observations regarding wave height and beachfill geometry shown in Table I agree with conventional wisdom on beach nourishment design. The observed weak dependence of sediment transport (and beachfill lifetime) on beach slope is not widely accepted, though it has been observed by other researchers, such as Kamphuis (1991). On the other hand, the observed positive correlation between wave period and beachfill lifetime is contrary to Kamphuis (1991). This discrepancy merits further study; though shorter waves impact on the shoreline with greater frequency, longer waves possess greater energy.

Results of the laboratory study suggest that relatively small-scale physical models of coastal sediment transport can yield useful qualitative information on shoreline and beach profile changes, though the magnitude of sediment transport rates may be greatly influenced by scale effects. The laboratory models cited in this study evolved similarly to prototype projects.

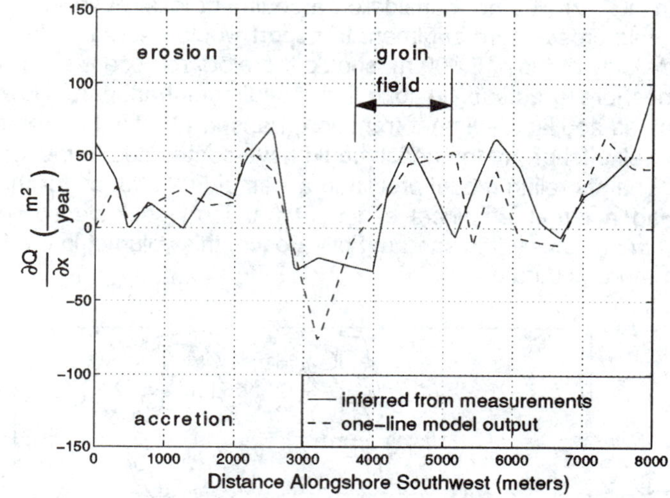

Figure 10. Comparison of One-Line Model Results to Field Measurements

The use of simple, one-line shoreline models in this study indicates that such models are adequate tools for predicting the volumetric redistribution of beachfill material. More rigorous, three-dimensional models of beachfill evolution are a worthy goal; however such models are equally, if not more, limited by present uncertainties in prediction of sediment transport.

The CERC equation for longshore sediment transport is adequate for predicting qualitatively the longshore trends of longshore sediment transport at both laboratory and field scale. The equation is a fair predictor of the magnitude of longshore sediment transport at field scale but requires calibration at small scales. Comparison to laboratory data suggests that the Kamphuis equation for longshore sediment transport is slightly more accurate than the CERC equation. Because of the possibilities of measurement error and scale effect with the laboratory study, this finding is inconclusive.

Based on limited laboratory and field-scale beach nourishment modeling, the analytical treatment of refraction seems to be an efficient method of including wave transformation in one-line models. This method may be considered by engineers for use in lieu of more rigorous, numerical approaches to wave transformation. However, in cases where a complex bathymetry exists (and accurate, high resolution survey data are available), a good numerical wave model is likely to give a better representation of the spatial variablilty of the wave climate. But again, the beachfill modeling as a whole is only as good as the "weakest link," so unless longshore sediment transport can be reliably predicted, extensive efforts at wave modeling may be unwarranted.

List of Symbols

The following symbols are used in this paper:

A=cross-sectional area of beach profile;
B=berm height;
C_g=wave group velocity;
C_t=wave celerity at toe of beachfill;
C_*=wave celerity at depth of closure;
D_{50}=representative grain size;
g=acceleration of gravity;
H_s=significant wave height;
h_*=depth to which longshore sediment transport affects beach profile;
K=empirical longshore sediment transport coefficient;
ℓ=length of beach nourishment;
m=beachface slope;
p=sediment porosity;
Q=volumetric longshore sediment transport rate;
SG=sediment specific gravity;
T=wave period;
T_p=wave period at peak of frequency spectrum
t=time;
x=longshore coordinate;
y=cross-shore location of shoreline;
β=local contour or shoreline orientation, relative to North;
κ=ratio of wave height to water depth at breaking;
θ=angle between wave orthogonal and local shore normal.

References

Brooks, R.M., and Brandon, W.A., 1995. *Hindcast Wave Information for the U.S. Atlantic Coast: Update 1976-1993 with Hurricanes.* U.S. Army Corps of Engineers. Waterways Experiment Station.

Dalrymple, R.A., 1988. "Model for refraction of water waves." *J. Waterway, Port, Coastal, and Ocean Engineering*, 114(4), ASCE, New York, N.Y., 423-435.

Dean, R.G., and Yoo, C., 1992. "Beach nourishment performance predictions." *J. Waterway, Port, Coastal, and Ocean Engineering* 118(6), ASCE, New York, N.Y.

Dean, R.G., and Yoo, C., 1994. "Beach nourishment in presence of seawall." *J. Waterway, Port, Coastal, and Ocean Engineering* 120(32), ASCE, New York, N.Y.

Ebersole, B.A., Neilans, P.J., Dowd, M.W. 1996. "Beach-fill performance at Folly

Beach, South Carolina (1 year after construction) and evaluation of design methods." *Shore and Beach,* Jan. 1996.

Hansen, H., and Kraus, N.C., 1989. *GENESIS: Generalized Model for Simulating Shoreline Change.* Coastal Engineering Research Center Technical Report CERC-89-19, U.S. Army corps of Engineers, Vicksburg, MS.

Kamphuis, J.W., 1991 "Alongshore sediment transport rate." *J. Waterway, Port, Coastal, and Ocean Engineering,* 117(6), American Society of Civil Engineers, New York, NY, 624-640.

Kamphuis, J.W., and Meyer, R.M., 1976. "Three dimensional tests on dynamic equilibrium and artificial nourishment." *Proc. 15th Coast. Engineering Conf.*, ASCE.

Kamphuis, J.W., Davies, M.H., Nairn, R.B., and Sayao, O.J. (1986). "Calculation of littoral sand transport rate." *Coastal Engineering,* 10, 1-21.

Kirby, J.T. and Dalrymple, R.A., 1994. *Combined Refraction/Diffraction Model: REF/DIF1.* Center for Applied Coastal Research. Department of Civil Engineering, University of Delaware, Newark, DE.

Komar, P.D. and Inman, D.L. 1970, "Longshore sand transport on beaches, *Journal of Geophysical Research,* v.75, No. 30, Oct. 20, 1970.

Nielsen, P., 1988. "Comparison of a numerical model for longshore sediment transport with measurements." *Computer Modeling in Ocean Engineering.* Schrefler & Zienkiewicz (eds.) Balkema, Rotterdam.

Pelnard-Considére, R., 1956. "Essai de théorie de l'evolution des formes de rivages en plage de sable et de galets." *4th Journées de la Hydraulique, les Energies de la Mer,* Question III, Rapport No. 1, pp.289-298 (in French).

U.S. Army Corps of Engineers, 1984. *Shore Prot. Manual,* Coastal Eng. Res. Center, Vicksburg, MS.

Work, P.A. and Dean, R.G. 1995, "Assessment and prediction of beach-nourishment evolution." *J. Waterway, Port, Coastal, and Ocean Engineering,* 121(3), ASCE, New York, N.Y., 182-189.

CHAPTER 229

COASTAL ZONE DYNAMICS DURING ARTIFICIAL NOURISHMENT

Elżbieta Zawadzka[1]

Abstract

The nearshore zone of the Hel Peninsula has a hydro-geomorphological system which, together with the influence of the port at Władysławowo, causes permanent erosion of the basal and central part of the peninsula's sea coast. After a period during which traditional methods of coastal defence (seawalls, groynes) were used, a massive artificial nourishment of the dune and beach was realised in the years 1989-1995. The borrow material used was of lagoon and marine origin. The progress of morpho- and lithodynamic processes was evaluated basing on quantitative analyses of changes in topography and granulometry. In the land part of the coastal zone, accretion of deposits resulted in rebuilding of the dunes and beaches. In the nearshore zone, accretion occured on the beach slope, and resulted in the development of shore terraces and breaker bars (especially the outer breaker bars). This was accompanied, by related with the accretion, processes of erosion in the form of U-shaped troughs. These processes proceeded at small changes in the erosion/accretion (e/a) system of the nearshore zone in the whole region. Smaller input of sandy material into the nearshore zone resulted in gradual return of the e/a system to initial state. If the safety of the peninsula is to be maintained, artificial nourishment must be executed every year. Frequency and volumes of renourishment will depend on the quality of used material. Therefore new sources of borrow material must be seeked.

Introduction

Artificial beach nourishment, considered as one of the most environment friendly methods of coastal defence, was used in Poland several times (Basiński, 1992; Basiński & Szmytkiewicz, 1991; Mierzyński, 1985). However, these were experimental pilot projects, realized in small temporal and spatial scale, using rather small amounts of material. In this paper are discussed results of investigations of changes in the coastal environment caused by a large-volume nourishment project implemented on the Hel Peninsula coast (Fig. 1). The decision to use artificial nourishment resulted from the cathastrophic state of the basal part of the peninsula, and from the imminent danger of flooding of the hinterland of the narrow and

[1]Dr., Department of Maritime Hydrotechnics, Maritime Institute, Abrahama 1, 80-307 Gdańsk, Poland.

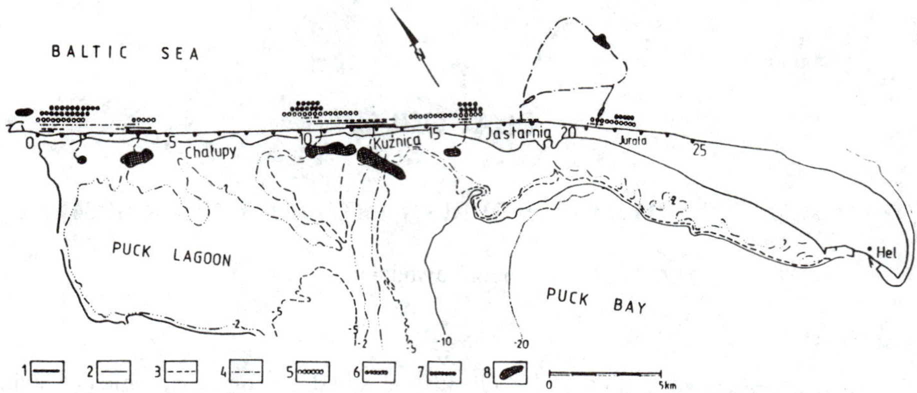

Fig. 1 Location of beach nourishment and borrow sites, Hel Peninsula 1989-1995. (1-7 - years of nourishment; 8 - borrow sites)

intensely eroded dune, at limited width of the spit. Strong storms in the autumn and winter of 1988/1989 resulted in a danger of breaking of the peninsula at 2-3 places.

The Hel Peninsula is an accumulative spit of 36 km length, which is an extension of the mainland to SE, and which shelters from the north the Gulf of Gdansk. In historical times it was broken several times. Analysis of maps from the period 1908-1978 (Zawadzka, 1994) has shown that the average rate of coastline retreat along the peninsula is -0.6 m/year (Fig. 2a). Processes of erosion in the period 1980-1987, directly preceding the nourishment project, are characterized by a similar spatial distribution as in the 70-year period 1908-1978 (Fig. 2b). The foredune became nearly completely eroded in the basal and central part of the peninsula along 17 km of coastline, resulting in a danger of breaking of the peninsula, especially at these stretches were the width of the spit did not exceed 100 m. In this situation, the decision was taken to use artificial nourishment as the most effective and least ingerentive method of coastal protection (Cieślak, 1994).

In the years 1989-1995 the largest project of artificial nourishment of a dune coast in Poland was realized. On the seaward coast of the Hel Peninsula, along a length of 23.4 km, 8 mln. m^3 of sand was supplied, gradually widening the range of sand suppletion works, which in total were executed along 14 km of the coastline. The material for nourishment was taken from the Puck Bay, from the sand trap at the port at Władysławowo, and from borrow sites at sea located to NE of the central part of the peninsula and to NE of Cape Rozewie (Zawadzka, 1993). In order to determine the effectiveness of the method, a comprehensive program of measurements and investigations was systematically realised. It included investigations of sediments and morphology of the seabed, beach and dune along the first 23 km of the peninsula's coastline. Each year 800-1000 sediment samples were analysed. Analyses of the variability of the dune, beach, beach slope, breaker bars, interbar troughs and of the deepwater slope to 10 m water depth were made. Comparison of maps of statistical grain size indicators (median diameter, sorting, skewness, steepness) on the background of proceeding changes in the shape of the nearshore bed and of its morphological forms, and taking into account the hydrodynamic conditions, has

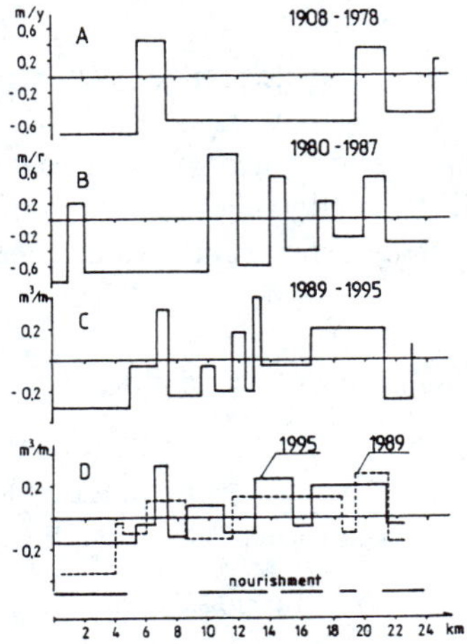

Fig. 2 Changes of coastline position, Hel Peninsula, 1908-1978, 1980-1987 and changes of the fill coefficient of the nearshore zone to -10 m MSL, 1989-1995. A-B - Zawadzka, 1990; C-D - Dubrawski, 1995.

shown that independtly of the local short term changes connected with artificial sand supply, in the coastal zone there is a hydro-geomorphological system, which forms relatively stable erosion/accretion trends in the coastal zone - also in conditions of artificial nourishment. Nourishment works carried out for 7 years have not resulted in a basic change of the general trend within the erosion/accretion system in the Hel peninsula region. Generally erosion has decreased, but its spatial distribution remained almost the same (Figs. 2c, 2d).

Characteristics of supplied material

For the nourishment material dredged from the Puck Bay and from sea borrow sites located at sea was used (Fig. 1). Deposits accumulated in the Puck Bay consist of fine and medium sand, well and medium sorted, and the median diameter (M_d µm) ranged between 139 and 237 µm. Marine material consists of medium sand of M_d 290 µm, medium sorted. Comaparative analysis of the "bay" borrow material and native material, using the R_j and R_A indicators (Manual, 1990), shows that these two sands do not fit satisfactorily. Highest stability of the supplied material may be attained when 2-3.5 larger quantities of material are nourished than the volumes resulting from calculations of deficiency of sand material in the basal part of the peninsula. In the central part, material borrowed from marine sources may be used in 1:1.5 (2) proportion. Grain size parameters of borrow material from the Puck Bay suggest that mass transport of that sand may proceed at speeds of 12-20 cm/s.

Nourishment started in 1989 at two sites: at Chałupy (km H3.5-4.5) and at Kuźnica (km H11.9-13.7). By 1993, about 5.85 mln. m^3 of sand were supplied, and the nourishment works were executed on 5 stretches of total length of 11.1 km. At the Chałupy nourishment field the resultant average volume of nourishment per metre of coastline was 510 m^3, at Kuźnica it was 730 m^3. On the other fields the average volume of nourishment per metre of coastline ranged between 100 and 330 m^3. In the next two years, until 1995, the length of directly nourished coastline grew to 14 km, and the total amount of sand supplied since 1989 increased to 8 mln. m^3. Most intensive nourishment was still realized at Chałupy and Kuźnica due to the highest defficit of sediments in the nearshore zone, and because of the limited stability of the nourished material.

Morphodynamics and lithodynamics of the region

The shore and the nearshore zone of the Hel Peninsula are characterized by a semistabile erosion/accretion system, with a strong predomination of erosional trend resulting from the hydrodynamic conditions in the region (Fig. 2).

The stable predisposition to erosion along given stretches of the coastline are caused by trough-like forms present in the sea bottom, which are obliquely oriented at an angle of 25-27^0 to the shoreline. In the western part of the peninsula they are of glacifluvial origin (Tomczak, 1994), in the eastern part they are remnants of the Yoldia Sea coastal systems (Musielak, 1989). Location of these forms influences the occurence of areas with permanent erosion of the sea bottom and shore. The stability of the erosion/accretion system is confirmed both by cartometric analysis of coastal change (Zawadzka 1990, 1994) and by means of modelling of hydrodynamic processes in the nearshore zone (Skaja & Szmytkiewicz, 1995). When wind blows from the west and north sectors, dangerous to the west and central part of the peninsula wave and current systems are generated. In extreme conditions current velocity may reach 2 m/s. Wind from the east results in higher energy supply to the east and central parts.

The variability of sediment transport characteristics in natural conditions, in the different zones of bottom and shore, influences the intensity with which the artificially nourished material is displaced in given meteorological conditions, resulting in different reactions of the erosion/accretion system of the coastal zone.

Transformation of morphological forms of the coastal zone

The caused by artificial nourishment increase of volume of the beaches resulted in adaptive transformation of the adjoining forms of the shore and nearshore zone.

Accretion of nourished material on the beach and dunes was accompanied by an appearance of erosion along stretches east (downstream) of the nourishment fields. Gradually these phenomena moved also into the nearshore zone.

After the first season of nourishment, along 14% of the observed coastline well developed shore terraces were present, and their average width was 75 m (Dubrawski, 1995). After 4 years terraces occured along 45%, and after 6 years - even along 76% of the coastline (Table 1).

The intensive development of inner and outer bars took place after four years of artificial nourishment, during which 5.8 mln. m^3 of sand were supplied. One of the most significant features of the transforming during the artificial nourishment seabed was the generation of rarely occuring in the natural bottom profile U-shaped troughs. Maximum development of the interbar U-troughs also was observed after 4 years of nourishment. The maximum number of shore U-troughs appeared after 6 years of nourishment, suggesting increased current velocities in the shallow part of the nearshore zone.

Table 1. Changes of parameters of morphological forms in the nearshore zone, 1989/1995 (Dubrawski, 1995)

Year	Shore terraces		Shore slope		shore U-trough		Inner bar lower class		interbar U-trough		Outer bar higher class	
	%	l [m]	l [m]	slope	%	l [m]	%	[m²]	%	l [m]	%	[m²]
1990	14	75	133	1:33	4	76	26	37	8	144	33	303
1993	45	76	141	1:37	11	80	12	43	21	108	52	332
1995	76	81	135	1:33	19	85	17	43	9	71	45	335

The decreased volumes of nourishment in the period 1994-1995 resulted in stronger erosion on the beach and in smaller differentiation of morphological forms in the nearshore zone and of grain size composition within these forms, which suggests a gradual return to the state before artificial nourishment (Dubrawski, 1995). Analysis of the average crosshore profile showed stabilisation of the volume of the inner bar after 4 years of nourishment, and a steady development of the outer bar where redeposition of fine sands occurs. The shore slope increased its volume by 4.8% of the volume of nourished material, attaining maximum after 4 years of nourishment. Similarly to shore terraces, in the next years it was subjected to gradual reduction.

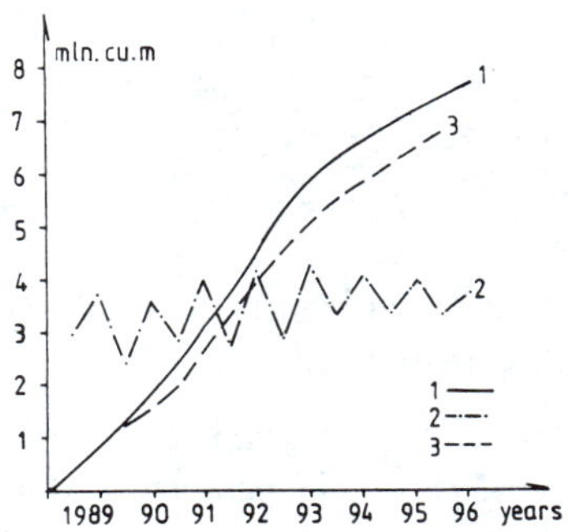

Fig. 3 Nourished stretches and net beach volume (1 - sum of annual suppletion; 2 - changes of dune and beach volume; 3 - sum of annual losses)

The balance of changes of the shore, evaluated basing on measurements carried out twice every year, shows that during the first 3.5 years, in spite of repeated nourishment, the volume of dunes and beaches remained at the level of 1989 (Fig. 3). When intense nourishment was carried out after strong storms of winter 1992/93, and after the relatively mild wind and wave conditions of the 1994/95 season with higher than average percentage of winds from the east, did result in stabilisation accretion on the beach and dune. After the annual amount of nourished material was reduced, the volume of material contained in the beach and dune decreased in spite of a consecutive mild winter season. In effect of 6.5 years of artificial nourishment, the volume of the shore (beach and dune) increased by about 1 mln. m^3.

Lithodynamics in conditions of artificial nourishment

Grain size analysis of bottom surface samples, taken each year from the seabed in the whole region before each annual nourishment works, show that the mean value of the median diameter has not changed significantly. Before the nourishment works began, the belt-like distribution of mean and fine sands was locally disturbed by zones of coarse sand and fine gravel, which indicated areas with higher wave and current activity. In these areas erosion of shore and bottom was observed (Fig. 4).

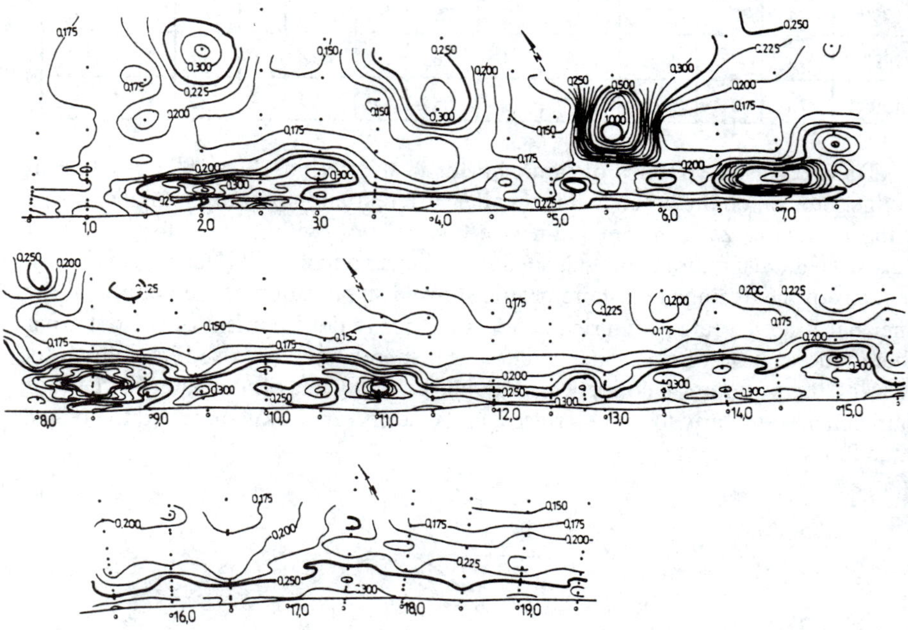

Fig. 4 Plan of median diameter (M_d mm) of bottom surface sediments in the coastal zone of Hel Peninsula, 1990

After 6 years of artificial nourishment, grain size composition shows changes of sediment parameters in the central part of the peninsula, which became supplemented with fine sands, at maintained erosional trend in the central part and at development of erosion along the basal part. The grain size composition of bottom surface sediments in the 10-12 m water depth zone was characterized, after 6 seasons of nourishment with sand from the Puck Lagoon (Puck Bay), by a slight decrease of the median diameter from 253 μm to 238 μm in 1993 and to 225 μm in 1995. The median diameter of beach deposits, which were successively supplemented with mean and fine sand, was after three years of artificial nourishment nearly the same as in prenourishment conditions.

The average grain size composition of deposits in the nearshore zone remained nearly unchanged through 3 years of nourishment. In the fourth year, due to erosion of nourishment fields during strong storms of the 1992/93 season, beach deposits became slightly coarser and the nearshore zone was supplemented with large amounts of fine sand, and the median diameter of bottom surface deposits in the nearshore zone decreased to 206 μm. It may be supposed that the stability of nearshore zone

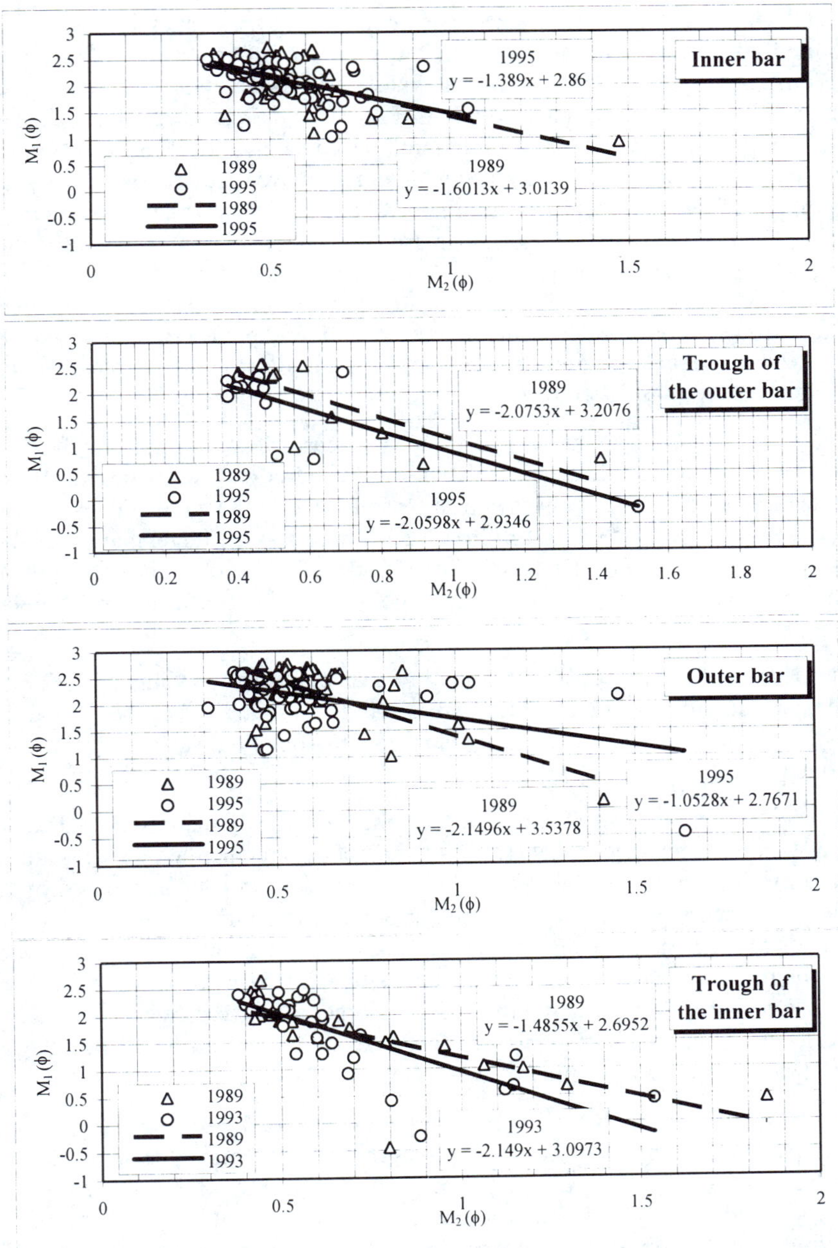

Fig. 5 Relationship between mean diameter (M_1) and standard deviation (M_2) of sediments in breaker bars and bar troughs, Hel Peninsula, 1989-1995

forms built of artificially supplied material, which was finer than the natural material, is smaller than in case forms built of material present before the nourishment project, and that thay will quicker undergo transformation.

Small changes in the grain size (evaluated using mean parameters) of the nearshore forms during the nourishment are proof of an inner stability of the hydrogeomorphological system and that the finest particles are displaced outside the active zone. However, at the same time within each of the morphological forms in the nearshore zone proceeded distinct granulometric changes, indicating a different distribution in the wave and current field (Basiński & Szmytkiewicz, 1991). Development of the shore terraces and the breaker bar zone was accompanied by deepening and widening of the troughs. Increased nearbottom current velocities in the troughs, which increase is largest in the trough between the inner and outer bar, suggest a transformation of the erosion/accretion system (Fig. 5). The development of interbar troughs is connected with the existence of gradient currents (Pruszak & Zeidler, 1995). Probably gradient currents increased in the nourished areas, and in effect U-shaped troughs developed.

The median diameter in the interbar troughs increased by 150 μm, from 280 μm in 1989 to 430 μm in 1995, in spite of relatively mild hydrodynamic conditions in the last winter season. After the strongest storm season 1992/93 the median diameter in the troughs was even up to 480 μm. In areas of most distinct development of U-shaped troughs, the median diameter exceeded 1000 μm, which could be related with current velocities of over 1 m/s (current velocities require confirmation by field measurements).

The decrease from 260 μm to 220 μm of median diameter of sediments building the breaker bars indicates that the accumulated in them material will be moved at lower initial current velocities. This fact may be the reason of quicker transformation of the outer breaker bars.

In conditions of a growing amount of supplied artificially material, of steadily growing length of nourished stretches, and of variable hydrodynamic conditions, the lithodynamic system of the Hel Peninsula nearshore zone was characterized by a variable spatial distribution of fields of erosion and accretion. Basing on analysis of the changes of median diameter of sediments and their mean values, calculated for each of the morphological zones (beach, shore slope, interbar troughs, breaker bar zone and deepwater slope), areas with finer or coarser sand were determined. This allowed to draw plans of of the surface erosion/accretion system of the nearshore zone, and to evaluate its dynamics.

Before artificial nourishment works started, there were 4 erosional areas along the basal part of the peninsula, covering all morphological zones: the shore slope, breaker bar zone and the deepwater slope. Along the central part of the peninsula, the shore slope and the breaker bar zone were eroded. Along the eastern part, alternatingly located zones of erosion and accretion occured, indicating routes of sediment transport to the deepwater slope and then out of the observed area.

In spring 1990, after the first artificial nourishment campaign, the initial system of erosion/accretion areas did not change along the basal and central part of the peninsula, though local accretion of the supplied material in the breaker bar zone and on the shore slope was observed. East of the fill field at Chałupy and Kuźnica (central part) the breaker bar zone was supplemented with fine sand material, at simultaneous erosion of the deepwater slope (Fig. 6). Influence of the fill material was observed in the interbar troughs and on the outer bar to 8 m water depth and up to 1.5 km from the fill field. On the other hand, segregation of the nourished material resulted in an increase of coarse sand and gravel on the beach and at the water line. No significant change of grain size was observed on the outer breaker bar.

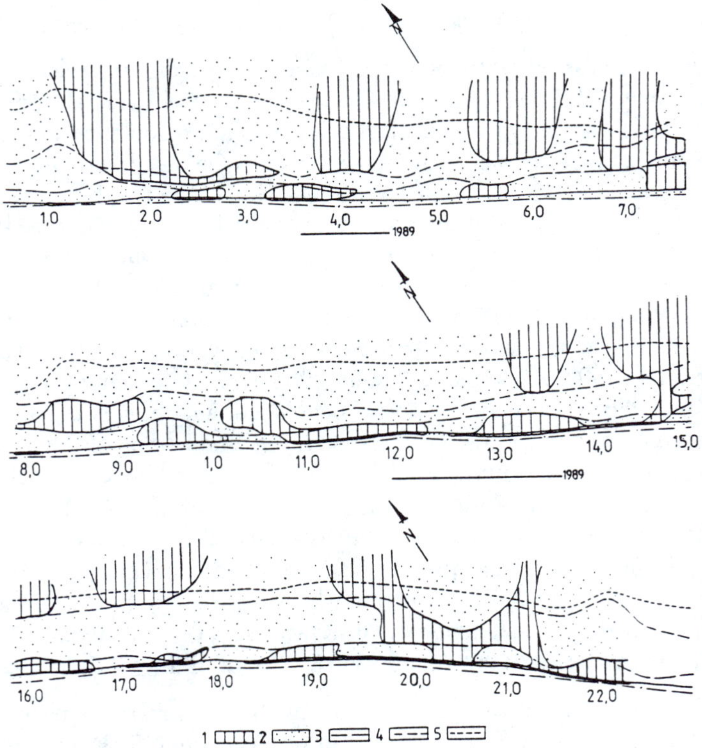

Fig. 6 Erosion/accretion system of surface sediments, Hel Peninsula, 1990 (1 - areas with erosive trend; 2 - areas with accretional trend; 3 - range of shore slope; 4 - range of breaker bar zone; 5 - 10m depth contour)

After two years of nourishment (in 1991), while distinct accretion occured in the shallow water and breaker bar zones, also development of areas of erosion of the breaker bars and deepwater slope was observed east of the fill field located in the basal part of the peninsula. In the central part, zonal accretion on the shore slope and on the breaker bars was observed, accompanied by erosion in the same zones east and west of the fill field. Rhythmically occuring erosion/accretion surfaces, encompassing the breaker bar zone occured in the eastern part, suggesting a development of circulation cells of 1 km order. In the Chałupy area, in effect of segregation of the material, the beach became saturated with mean and fine sand, while percentage of mean and coarse sand grew on the shore slope. The process of washing out of the shore slope proceeds at simultaneous enrichening of the breaker bar zone and of the deepwater slope with fine sediments. The influence of the fill material was observed in the breaker bar zone 2 to 3 km east of the fill field. Further migration of the fill material is limited by a convex bottom form at km H7.0-7.5, which is considered to be a natural boundary of a lithodynamic unit of the nearshore zone. Observations at Kuźnica showed that the fill material is transported over larger distances than in case of the Chałupy field. Within the shore slope and inner bar zones the material was transported 3 km, and in the outer bar zone - about 5 km to the east of the fill field. East of the Kuźnica fill field two independently occuring in 1990 areas of scour, one on the shore slope and the other on the deepwater slope, were transformed into an oblique to the coastline and cutting through the breaker bars zone of intense transport.

After the third year (in 1992) of artificial nourishment of the western stretch of coastline with material from the Puck Bay, supplemented with material from the sand trap at the port at Władysławowo, the inner bar and its inner trough became richer in fine sand. Deposition of sediments grew also locally in the outer bar zone. The large supply of material from the fill resulted in a shallowing of the interbar trough and in supplementing the balance of sediments up to about 1.5 km east of the fill field. Areas of bottom scour formed in 1991 became filled with sands of 225-250 μm diameter. Finer material (less than 150 μm) was deposited in the 8-10 m water depth zone about 3 km east of the fill field. Accretion in the area of transport of the fill material was always accompanied by erosion, which indicates a nonuniform character of the transport. In 1992, due to the continuing suppletion of the shore at Kuźnica, the system of shore terraces became distinctly developed, and the median diameter of sediments accumulated in them was 275-300 μm. The shallowing of the shore slope was accompanied by deepening of the trough before the inner bar and in its transformation from a V shape to a U shape. Suppletion of the inner bar occured at a distance of 1 km and of the outer bar - at a distance of 2.5 km east of the fill field. The change of zones of wave energy dissipation resulted in a local rebuilding of the double bar profile into a single bar profile, which was built of medium sands. The new system of eroded and accreting bottom surfaces suggests a distinct stabilisation of lithodynamic processes over a stretch of 5-6 km east of the fill field at Kuźnica.

Intensive development of large erosive forms, encompassing nearly all the morphological zones appeared after 4 years of nourishment (1993), pointing to the development of oblique to the coastline routes of sediment transport. The field at Chałupy still supplemented the coastal zone to a distance of 2.5-3.0 km east of the field, and to a water depth of 6 m. In the nearshore zone a strip of 350 m width seaward of the water line, and reaching to 4 m water depth, was supplemented at a length of about 3.5 km. Predominance of wind from the NW sector in autumn and winter of 1992/93 resulted in the generation of distinct "upstream" zones and of new routes of sediment transport out of the nearshore zone. The obliquely situated erosive areas were characterized by larger grain sizes than the average values determined for the whole observed area, and crossed all analysed nearshore zones. The larger rates of loss of fill material over the deepwater slope occur, among others, along the defined routes of intensive transport (Fig. 7).

In the last year of observations (1995), in the nearshore zone of the basal part of the Hel Peninsula accretion occured over large areas of the shallow water zone adjacent to the fill field. Locally in the central part erosive areas appeared, independently of the development of massive breaker bars, which are a transient zone

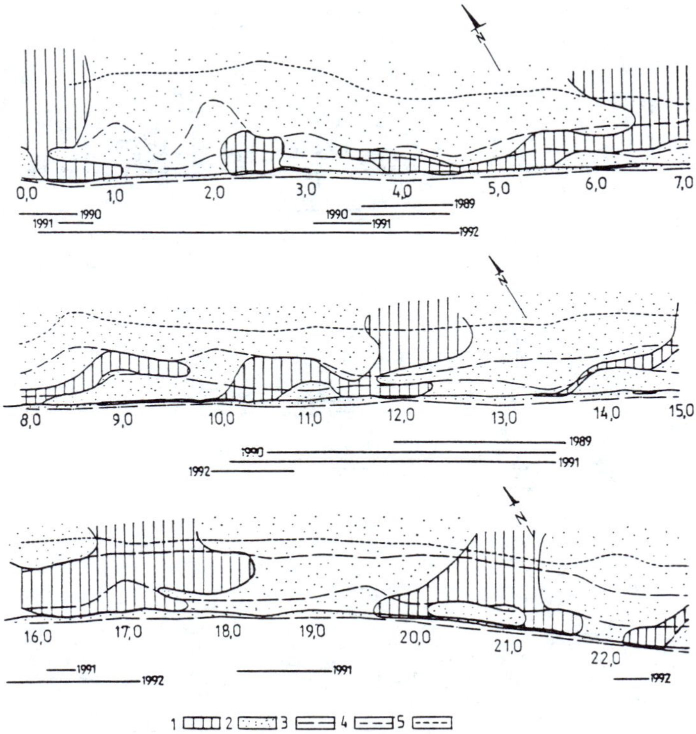

Fig. 7 Erosion/accretion system of surface sediments, Hel Peninsula, 1993 (explanations as in Fig. 6)

of deposition. Also distinct zones of higher energy appeared, which caused scouring of breaker bars east and west of the fill field at Kuźnica, formation of deep interbar troughs, and erosion on the deepwater slope. Hydrodynamic conditions at the break of 1994/95 resulted in the formation of two distinct circulation cells in the central and eastern region (Fig. 8). The direction of sediment transport out of the intensely artificially nourished region in the central part suggests an influence of a trough form situated obliquely to the coastline. A very similar arrangement of sediment transport routes (accretion in the central part of the system and erosion in the outer zones) was observed in the Jastarnia-Jurata region. The distribution of the eroded and accreting surfaces in the nearshore zone suggests that bottom forms farther offshore and deeper situated influence the nearshore zone, steering circulation at winds from the NW to NE sector.

The highest percentage of nearshore surface, subjected to processes of erosion in the breaker bar zone and on the shore slope, occured before artificial nourishment began (Table 2). The area of redeposition of the sediments was located on the deepwater slope below the breaker bar zone. Accumulation on the beach, shore slope and in the breaker bar zone caused increased erosion of the deepwater slope. After a period of strong storms at the break of 1992/93, processes of erosion increased again in the shallow part of the nearshore zone (shore terraces and breaker bars), but weakened below the breaker bar zone due redeposition of the sediments.

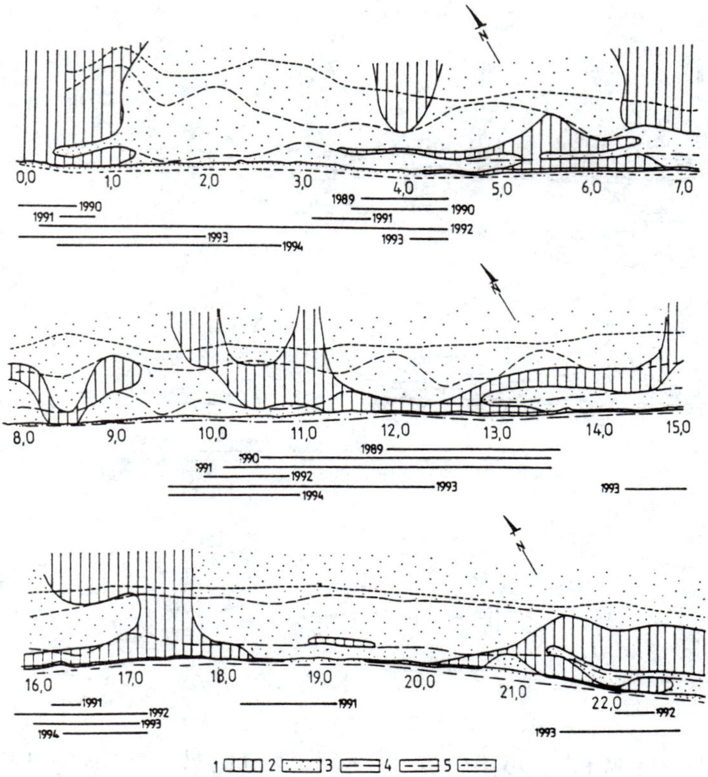

Fig. 8 Erosion/accretion system of surface sediments, Hel Peninsula, 1995 (explanations as in Fig. 6)

Table 2. Percentage of eroded nearshore zone surface, Hel Peninsula, period 1989-1995 (on the basis of granulometric parameters)

Year	Erosion (in %) of nearshore zone (to 10 m water depth)		
	Whole nearshore (0.0÷-10.0 m MSL)	Shore slope & breaker bar zone	Deepwater slope
1989	35.2	46.5	20.5
1990	21.0	16.9	33.7
1991	22.0	21.1	27.5
1993	29.0	36.4	19.4
1995	23.4	22.6	27.9

Obtained results show that in intensive hydrodynamic conditions the shallow part of the nearshore zone and the breaker bar zone was being destroyed over 36 to 46 % of its surface. Stabilisation on the shore slope and breaker bars is connected with development of processes of erosion on the deepwater slope, indicating that in the erosion/accretion system proceed compensatory processes.

Summary

The six years of artificial nourishment along the Hel Peninsula, due to the restoration of dunes and to the improvement of sediment balance in the coastal zone, resulted in higher safety of the coast and its hinterland. The total volumetric change in the strip from dunes to breaker bars inclusive was over the whole period 1.8 mln. m^3, i.e. about 28% of the total volume of the nourishment in the period 1989-1994 (Cieślak et al, 1995), in that about 12.5 % remained in the dunes and beach, and 15.5 % remained in the shore terraces and breaker bar zone. The rest was transported outside of the investigated area. Significant amounts of fill material were accumulated in the developing shore terraces, which at the end of the analysed period covered 3/4 of the length of the observed coastline. Terraces improve the stability of beaches and dunes because they dissipate wave energy.

Breaker bar height increased by 11.8%, and cross-section area by 22% in refrence to the 1989 state. Formerly present stretches with no breaker bars disappeared. A decisive change in the shape and volume of breaker bars took place. Participation of small breaker bar forms decreased. Massive breaker bar forms developed, especially after the fourth year of artificial nourishment. During the last two years of suppletion, when volumes of supplied fill material were reduced, the percentage of breaker bars of each of the volumetric classes began to stabilise at the levels of the period preceding the nourishment works, indicating a predominant influence of hydrodynamic processes. The grain size composition of fill material indicated that it may nourish especially zones with lower hydrodynamic activity. Gradual reduction of massive breaker bars caused by reducing the volume of annual artificial nourishment will result in the future in higher levels of danger to the coast. Development of shore terraces and breaker bars was connected with local changes in the erosion/accretion system, which appeared especially through the generation of U-shaped troughs and areas of scour downstream of the fill fields and terraces.

Observed morpho- and lithodynamic changes of the coastal zone during artificial nourishment are of adaptative character and proceed within an existing, well established hydro-geomorphological system of the Hel Peninsula. Gradual reduction of the volume of artificially supplied to the erosion/accretion system material will result in a slow return of the shore and seabed forms to the state present before artificial nourishment began. In the conditions of prevailing winds from the west sector and of generated by them waves and currents, and of the system of macroformations of the outer part of the coastal zone and of the breaker bar zone, there is a natural predisposition to transport a part of the sediments of the dynamic layer to east and north. This requires repeated artificial nourishment in the west and central part of the peninsula, in order to ensure relative stability of permanently eroded stretches, especially when too fine fill material is used. In periods when prevail winds from the east, morphodynamic processes in the erosion/accretion system become stabilised.

The basic recommendation, resulting from grain size and sediment transport analyses, is that borrow sites with grain sizes better fitted to open sea hydrodynamics should be seeked. Sand from the Puck Bay is very easily moved, and only for a short period supplement the areas with sediment defficit. Material with parameters ensuring 40-50% better stability in the whole strip from shore to outer bar than the presently used sand should be used. Maintaining the present state of the coast requires annual nourishment of the order of 0.6-1.0 mln. m^3.

The presented results show that artificial nourishment, as any other method, has a limited range of application, and - as any other technical ingerence into the coastal zone - causes changes in the wave and current system which result in a development of negative morpho- and lithodynamic effects.

References

Basiński, T. 1992. Experience from shore protection of the Hel Peninsula. *Proc. Int. Coastal Congress.* Kiel, p.639-647.

Basiński, T. & Szmytkiewicz M. 1991. Effectiveness of different artificial nearshore and beach nourishment technologies. *Proc. of the 3rd Int. Conf. on Coastal and Port Develop. Countries,* Mombasa.

Cieślak, A. 1994. Concept of Hel Peninsula coast protection. *Symp. on Changes of Coastal Zones, Polish Coast '94*, Gdynia p.57-59.

Cieślak, A., Semrau, I., Michowski, A. & Zawadzka, E 1989-1995 Stage. Complex protection of Hel Peninsula. (in Polish). Int. report IM WW4884, Gdańsk.

Dubrawski, R. 1995. Analysis of shore and offshore of Hel Peninsula during beach nourishment in 1989-1995. (in Polish). Int. Rep. IM WW 4864, Gdańsk.

Manual of artificial beach nourishment 1986. Center of Civ. Eng. Research. Codes and Specification, Report 130, Delft.

Mierzyński, S. 1985. Experimental artificial nearshore zone nourishment of the Hel Peninsula. (in Polish). *Materiały na sesję naukową*, IM, Gdańsk.

Musielak, S. 1989. Remarks on the Genesis of the Hel Peninsula in the Light of New Data. *Studia i Materiały Oceanologiczne 56*, p.311-322.

Pruszak, Z. & Zeidler, R. 1995. Sediment Transport in Various Time Scales. *ICCE '94, 24th Int. Conf. on Coastal Eng.* Kobe, p.570-571.

Skaja, M. & Szmytkiewicz, M. 1995. Calculations of waves and currents along the Hel Peninsula. (in Polish).Int Rep. of IBW-PAN, Gdańsk.

Tomczak, A. 1994. Hel Peninsula - relief, geology, evolution. *Symp. on Changes of Coastal Zones, Polish Coast '94*, Gdynia, p.45-49.

Zawadzka, E. 1990. Determination of coastline changes. (in Polish). Int. Rep. IM WW 4571, Gdańsk, pp.76.

Zawadzka, E. 1993. Sedimentological investigation of material from borrow sites in the Puck Bay. Int. Rep IM WW 4799, Gdańsk

Zawadzka, E. 1994. South Baltic coastal changes during the last 100 years. *ICCE'94 Int. Conf. on Coastal Eng.* Kobe. Book of Abstracts pp.568-569.

CHAPTER 230

Dune Damage Curves and Their Use to Estimate Dune Maintenance Costs

David R. Basco[1] and Cheol S. Shin[2]

Abstract

A beach profile numerical model is employed to calculate the change in dune cross-section for a wide range of increased, water level events. The volume loss in dune cross-section increases with relative increase in storm surge elevation above the design water level. These "damage curves" for dunes are analogous to rubble mound damage curves created by excessive wave energy above the design wave height.

The results are applied in the design of a protective, beach-dune structure (with buried seawall as a safety factor) at Dam Neck, Virginia for the US Navy. The damage curves are used to estimate annual dune maintenance costs and hence, total life-cycle costs for this "soft" alternative to shore protection. Interestingly, the dune-beach-buried seawall alternative costs less than a concrete structure using artifical units for armor.

Construction is complete and a three year monitoring project begun so that actual versus theoretical dune damage curves can be determined in the future.

1.0 Introduction

Dunes are "soft" coastal structures that quickly lose cross-sectional volume during elevated, storm surge events. Cross-shore sediment transport, beach profile models permit the development of dune "damage" curves showing the percent dune cross-sectional change, ΔV (damage) versus the relative storm surge level, S compared with the design storm surge, S_D. The dune damage curves can then be used in the classical, convolution integral method to compute annual dune maintenance costs. Life-cycle

---

[1] Professor of Civil & Environmental Engineering and Director, the Coastal Engineering Centre, Old Dominion University, Norfolk, Virginia 23529 USA

[2] Instructor, Department of Civil Engineering, Dongshin University, Daeho-Dong, Naju-City, Jeonranam-Do, 520-714 Korea

costs of the "soft" solution for coastal protection (dune-beach alternative) can then be compared with the "hard" alternatives (seawalls, dikes, etc.) on an equal, total annual cost per unit shoreline length basis.

The objective of this paper is to present the results of one analysis using the SBEACH model (Larsen and Kraus, 1989) for the design of a dune-beach restoration project to protect \$95 million of structures and property at the US Navy's, Fleet Combat Training Center (FCTC), Dam Neck, Virginia. Section 2 briefly reviews the available models and key independent variables. The dune damage curves are summarized in Section 3 including safety factors used in dune design. Maintenance costs are determined in Section 4 and a cost comparison with an armored seawall design presented in Section 5. Construction of the entire project has recently been completed and a three year, beach profile monitoring project begun so that the theoretical versus actual dune damage curves can be determined in the future.

2.0 Beach - Dune Erosion Models

Recently, ten cross-shore sediment transport models have been evaluated by Schoonees and Theron (1995). The model developed by Larsen and Kraus, 1989 called SBEACH was found to be "acceptable" on its theoretical basis and in the "best group" category based on the extent of verification with prototype data. It allows sand overwash during elevated water level events so that the total, cross-shore sectional volume is conserved.

The probabilistic design of dunes with examples from the Netherlands are presented by van de Graaff (1989) as illustrated in Figure 1. It was learned (Table 1) that storm surge elevation accounted for about 83 percent of the total variance associated with dune erosion. Wave height, particle size, initial profile shape, storm duration and other factors were far less important. The change in dune cross-sectional area (per unit width) was primarily related to elevated water levels during storm surge events.

Dune damage curves are thus analogous to rubble mound damage curves created by excessive *wave energy* above the design, wave height.

3.0 Dune Damage Curves for Dam Neck, Virginia

Figure 2 shows the location of the Navy's FCTC at Dam Neck, Virginia on the Atlantic Ocean below the City of Virginia Beach.

Using a 50 ft. (15.2m) crest width at elevation +22ft (6.7m) above the City of Virginia Beach datum and 2:1 side slopes produces a unit volume of 30cy per foot ($75m^3/m$) above the one percent chance, storm surge event (9.1 ft., 2.8m). This dune cross-sectional (Figure 3) is essentially that originally designed by Headland (1991) and contains fifty percent more sand than the minimum requirement of 540 ft^3/ft ($50m^3/m$) as established by the Federal Emergency Management Agency (FEMA) for breaching of natural dunes.

A nearshore survey taken in May, 1990 was considered representative for the existing bathymetric conditions in the model. The SBEACH model was run with eleven successively higher water level events to cover the return periods between 1

The objective of this paper is to present the results of one analysis using the SBEACH model (Larsen and Kraus, 1989) for the design of a dune-beach restoration project to protect $95 million of structures and property at the US Navy's, Fleet Combat Training Center (FCTC), Dam Neck, Virginia. Section 2 briefly reviews the available models and key independent variables. The dune damage curves are summarized in Section 3 including safety factors used in dune design. Maintenance costs are determined in Section 4 and a cost comparison with an armored seawall design presented in Section 5. Construction of the entire project has recently been completed and a three year, beach profile monitoring project begun so that the theoretical versus actual dune damage curves can be determined in the future.

2.0 Beach - Dune Erosion Models

Recently, ten cross-shore sediment transport models have been evaluated by Schoonees and Theron (1995). The model developed by Larsen and Kraus, 1989 called SBEACH was found to be "acceptable" on its theoretical basis and in the "best group" category based on the extent of verification with prototype data. It allows sand overwash during elevated water level events so that the total, cross-shore sectional volume is conserved.

The probabilistic design of dunes with examples from the Netherlands are presented by van de Graaff (1989) as illustrated in Figure 1. It was learned (Table 1) that storm surge elevation accounted for about 83 percent of the total variance associated with dune erosion. Wave height, particle size, initial profile shape, storm duration and other factors were far less important. The change in dune cross-sectional area (per unit width) was primarily related to elevated water levels during storm surge events.

Dune damage curves are thus analogous to rubble mound damage curves created by excessive *wave energy* above the design, wave height.

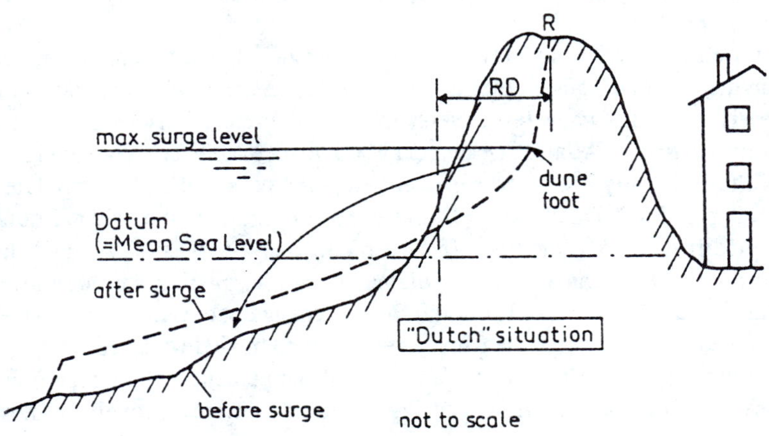

Figure 1 Schematic of Dune Erosion by Elevated Water Levels: Dutch Practice (from van de Graaff, 1986)

Table 1 Relative Importance of Independent Variables on Dune Erosion (from van de Graaff, 1986)

Contribution parameters to variance reliability function

Parameter	Contribution variance reliability function (%)
surge level	82.8
wave height	0.9
particle size	7.3
initial profile	1.3
surge duration	2.6
gust bump	0.0
accuracy computation	5.0

3.0 Dune Damage Curves for Dam Neck, Virginia

Figure 2 shows the location of the Navy's FCTC at Dam Neck, Virginia on the Atlantic Ocean below the City of Virginia Beach.

Using a 50 ft. (15.2m) crest width at elevation +22ft (6.7m) above the City of Virginia Beach datum and 2:1 side slopes produces a unit volume of 30cy per foot (75m^3/m) above the one percent chance, storm surge event (9.1 ft., 2.8m). This dune cross-sectional (Figure 3) is essentially that originally designed by Headland (1991) and contains fifty percent more sand than the minimum requirement of 540 ft^3/ft (50m^3/m) as established by the Federal Emergency Management Agency (FEMA) for breaching of natural dunes.

A nearshore survey taken in May, 1990 was considered representative for the existing bathymetric conditions in the model. The SBEACH model was run with eleven successively higher water level events to cover the return periods between 1 on Southeastern Coastline, Atlantic Ocean and 1000 years (Table 2). Wave heights associated with these water levels were established from measured and extrapolated values at the Corps' FRF, Duck, NC. As illustrated in Figure 3, the original dune section was modified by sand moving offshore $\Delta V3$, sand overwash $\Delta V1$, and the dune volume $\Delta V2$. Dune damage is the volume change (loss) $\Delta V2$ and here after simply ΔV as found in Table 2. The total volume change above mean low water (MLW) accounted for all but less than 1cy per foot which moved offshore.

Dune Volume loss in cy/ft (ΔV in Table 2) and the percent loss (damage) are plotted against the S/S_D ratio as shown in Figure 4a and 4b, respectively. These curves are conservative in that they do not include the renourished beach in front of the dune as discussed blow.

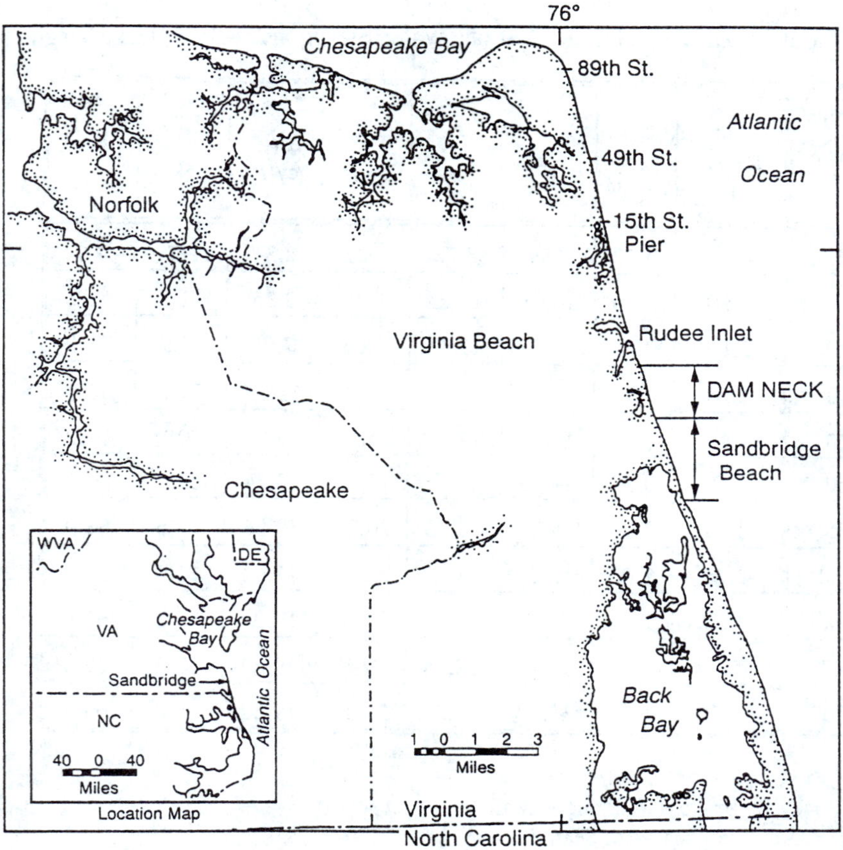

Figure 2 Location of US Navy, Fleet Combat Training Center, Dam Neck, Virginia on Southeastern Coastline, Atlantic Ocean

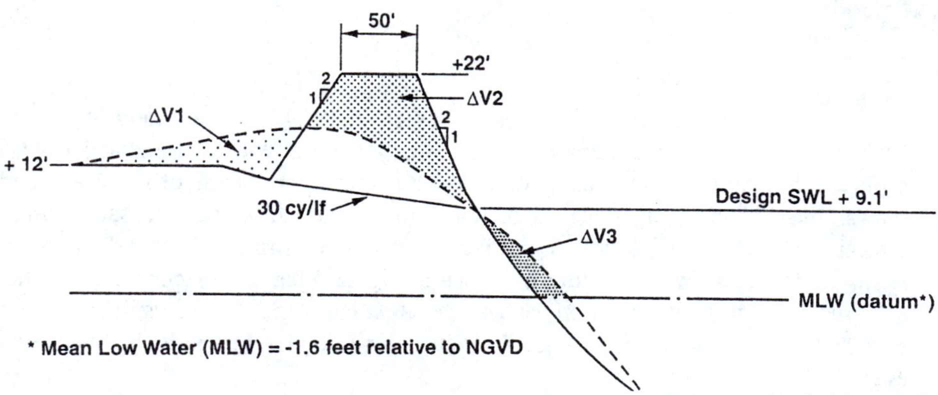

Figure 3 Dune Cross-Section Design for Dam Neck, Virginia and Definitions of Volume Change Regions, ΔV1, ΔV2 and ΔV3.

Table 2 Calculated ΔV Values for Various Water Elevations at Dam Neck, Virginia

Return Period in Year	Water Level, ft	S/S_D	H_{mo}, ft	T_p, sec	ΔV, cy/ft	ΔV_w, cy/ft	P, %	P_w, %
1	4.50	0.52	8.99	8.72	2.31	2.31	7.7	7.7
2	4.90	0.56	10.63	10.01	3.35	3.35	11.2	11.2
5	5.85	0.67	12.53	11.45	7.06	7.06	23.5	23.5
10	6.50	0.75	13.45	12.11	11.82	11.82	39.4	39.4
20	7.20	0.83	13.98	12.45	15.02	15.00	50.1	50.0
50	8.10	0.93	14.76	12.96	19.02	17.74	63.4	59.1
75	8.35	0.96	15.26	13.32	20.27	18.58	67.6	61.9
100	8.70	1.00	15.80	13.70	21.42	19.02	71.4	63.4
200	9.30	1.07	16.67	14.31	22.76	19.08	75.9	63.6
500	10.10	1.16	17.45	14.78	23.32	19.60	77.7	65.3
1000	10.70	1.23	18.18	15.25	23.40	19.68	78.0	65.6

ΔV = Volume Loss Without Seawall
ΔV_w = Volume Loss With Seawall
P = Percent Damage Without Seawall
P_w = Percent Damage With Seawall
$(P, P_w) = 100*(\Delta V, \Delta V_w)/V$, V=Total Volume of the Dune, 30 cy/ft

Dutch design practice for dunes (TAW, 1984) includes a safety factor or "remanent" dune volume remaining after a major storm event. Headland (1991) replaced this minimum remanent volume by an equivalent volume of rubble-mound seawall buried beneath the dune. Should a second major storm occur the same winter season, the buried seawall will be in place to protect the structures behind the dune. A similar buried seawall structure as shown in Figure 5 has been incorporated in the constructed project. Its effect on the damage curve is ΔV_w in Table 2 as illustrated in Figure 4b. The buried seawall acts to reduce dune volume loss at high water level events.

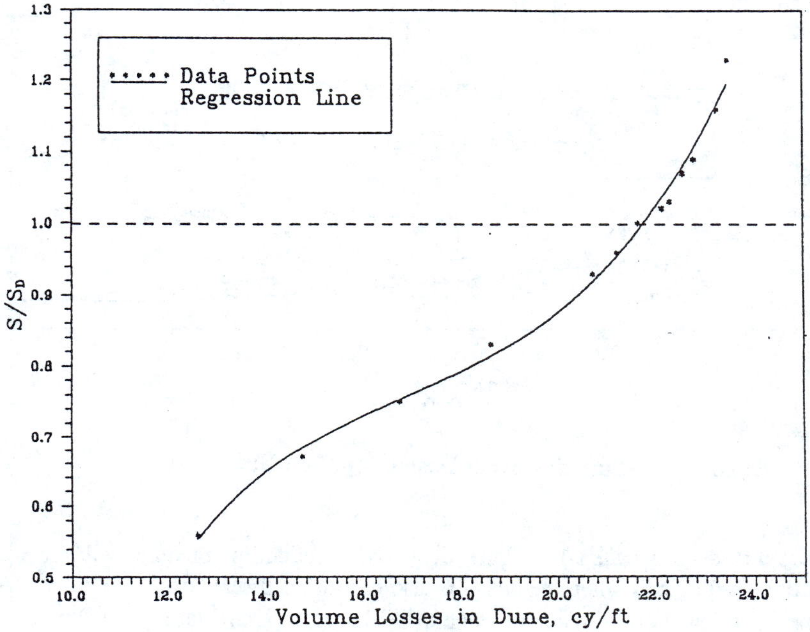

Figure 4a Dune Volume Loss (cy/ft) Versus S/S$_D$

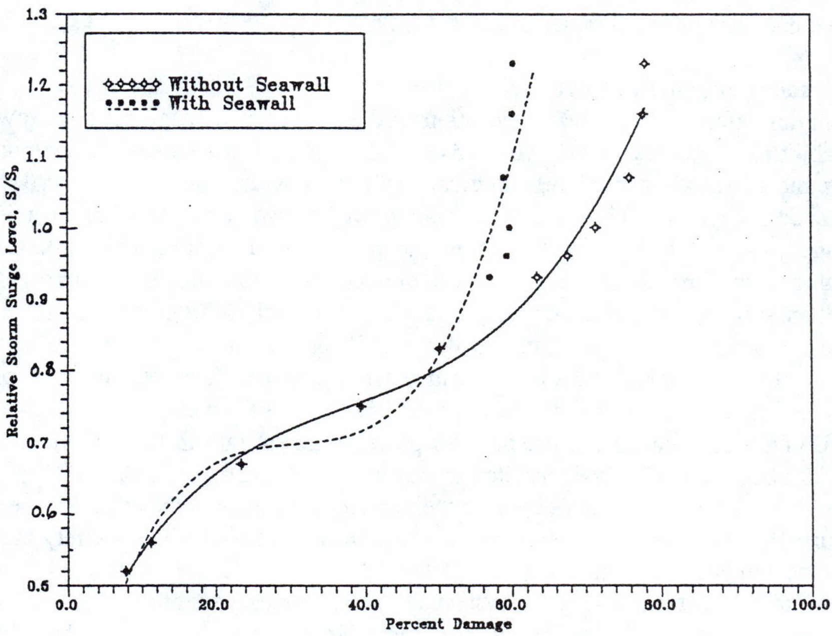

Figure 4b Dune Damage Curve: Percent Loss Versus S/S$_D$

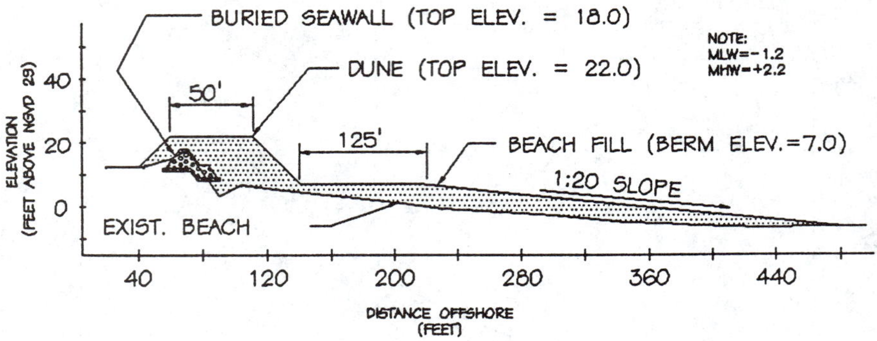

Figure 5 Dune-Beach-Buried Seawall Design for Dam Neck, VA

However, as illustrated in Figure 6, a major reduction in dune volume loss (percent damage) occurs when the beach is also renourished seaward of the dune. As an example, when the constructed beach width is 75ft (23m) damage curves for a design width after equilibration of 40 ft (12m) and with one-half life remaining of 20 ft (6m) are presented in Figure 6. Dune maintenance costs are obviously impacted by the volume of beach remaining over time to protect the dune system. The results of these computations are summarized in Table 3.

4.0 Dune Maintenance Costs

Simply stated, higher water elevations produce greater damage but have lower probabilities of occurrence each year. Annual dune maintenance costs are computed using the classical, convolution integral method of Kreeke and Paape, 1964 as illustrated in Figure 7. The dune damage curve was divided into eight subregions with damage greater than 70 percent requiring complete rebuilding of the dune structure. the storm surge probability curve was also divided into eight comparable subregions and then annual dune repair costs computed. As shown, subregion 2 for 10 - 20 percent damage from storms in the 0.2 - 0.35 probability of occurrence range accounted for the highest increment of maintenance expense. The total annual repair costs per unit foot of dune summed to about $74 per foot (Table 4).

Using a 25 year design life and 9.5 percent interest rate in the present worth method of economic analysis resulted in annual dune maintenance costs of $695 per foot. The value is conservative because the damage curve *without* the protective beach (Figure 4b) was used in the analysis. This can also be considered as a safety factor in the design and economic analysis.

Monitoring of the entire dune system and beach to closure depth will provide the requisite feedback information over a three year period to insure that dune maintenance is performed at appropriate intervals.

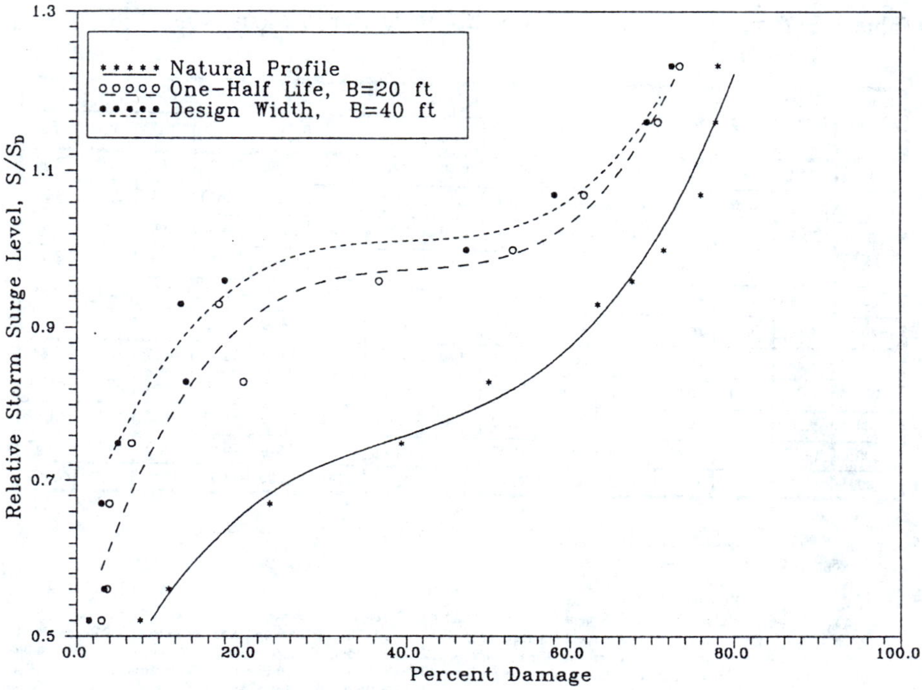

Figure 6 Effect of Beach Renourishment on Dune Damage Curves (Example for design beach width of 40 feet)

5.0 Cost Comparison With Armored Seawall

The dune damage curves used to compute dune repair costs permitted total, life-cycle cost estimates to be made for the dune-beach-buried seawall system (DBBS). Initial, unit costs for the dune of $15/cy or $450 per foot together with the maintenance costs of $695 per foot produced a total cost of $1145 per foot for the dune system. Using an offshore borrow area (Sandbridge shoal) for beach renourishment, one renourishment after 10 -12 years and realistic, unit cost estimates for sand and rubble materials (buried seawall), the "soft" alternative system totalled about $1820 per foot on a annual, life-cycle cost basis.

The original, engineering study (Cummings and Basco, 1995) also considered a conventional, large armor unit, revetment structure as the "hard" solution for storm damage mitigation.

Table 3 Summarized Computations of ΔV Values for Various Water Elevations at Dam Neck, Virginia

Return period in year	Water Level, ft	S/S_D	H_{mo}, ft	T_p, sec	Volume Loss, cy/ft					Percent Damage, %				
					ΔV	ΔV_{20}	ΔV_{30}	ΔV_{40}	ΔV_{65}	P	P_{20}	P_{30}	P_{40}	P_{65}
1	4.50	0.52	8.99	8.72	2.31	0.90	0.80	0.80	0.60	7.70	3.00	2.67	1.43	2.00
2	4.90	0.56	10.63	10.01	3.35	1.10	1.00	1.00	0.90	11.20	3.67	3.33	3.33	3.00
5	5.85	0.67	12.53	11.45	7.06	1.20	0.90	0.90	1.10	23.50	4.00	3.00	3.00	3.67
10	6.50	0.75	13.45	12:11	11.82	2.00	1.60	1.50	1.30	39.40	6.67	5.33	5.00	4.33
20	7.20	0.83	13.98	12.45	15.02	6.10	4.40	4.00	1.90	50.10	20.33	14.67	13.33	6.33
50	8.10	0.93	14.76	12.96	19.02	5.20	4.20	3.80	2.60	63.40	17.33	14.00	12.67	8.67
75	8.35	0.96	15.26	13.32	20.27	11.00	7.40	5.40	2.90	67.60	36.67	24.67	18.00	9.67
100	8.70	1.00	15.80	13.70	21.42	15.90	14.60	14.20	12.70	71.40	53.00	48.67	47.33	42.33
200	9.30	1.07	16.67	14.31	22.76	18.50	17.60	17.40	16.10	75.90	61.67	58.67	58.00	53.67
500	10.10	1.16	17.45	14.78	23.32	21.20	20.80	20.80	20.00	77.70	70.67	69.33	69.33	66.67
1,000	10.70	1.23	18.18	15.25	23.40	22.00	21.80	21.70	21.80	78.00	73.33	72.67	72.33	71.00

Notes :

- S_D = Design Storm Surge Level, ft
- S/S_D = Relative Storm Surge Level
- H_{mo} = Significant Wave Height, ft
- T_p = Spectral Peak Period, sec
- ΔV, P = Volume Loss and Percent Damage Without Any Beach Renourishment
- ΔV_{40}, P_{40} = Volume Loss and Percent Damage on Design Beach Width, B=40 ft under 75 ft Construction Beach Width
- ΔV_{20}, P_{20} = Volume Loss and Percent Damage on One-Half Life Width, B'=20 ft under 75 ft Construction Beach Width
- ΔV_{65}, P_{65} = Volume Loss and Percent Damage on Design Beach Width, B=65 ft under 125 ft Construction Beach Width
- ΔV_{30}, P_{30} = Volume Loss and Percent Damage on One-Half Life Width, B'=30 ft under 125 ft Construction Beach Width

$P=100*\Delta V/V$, V=Total Volume of the Dune, ≈ 30 cy/ft.

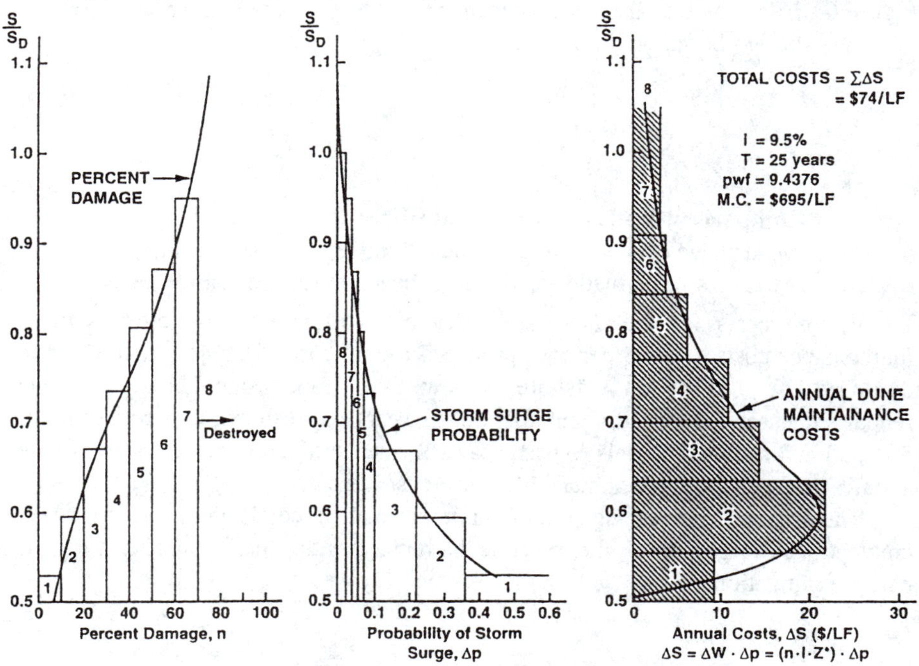

Figure 7 Probabilistic Design Method for Dune Maintenance Costs

Table 4 Dune Maintenance Costs

Region	Storm Surge Elev.*, S feet	Percent Damage (Average) n	Storm Probability Difference Δp	Dune Repair Costs $\Delta W = n.I.Z$ \$	Annual Dune Maint. Costs, \$ $\Delta S = \Delta p \cdot \Delta W$
1	4.57	5	0.42	22.50	9.45
2	5.18	15	0.32	67.50	21.60
3	5.83	25	0.125	112.50	14.06
4	6.39	35	0.068	157.50	10.71
5	7.00	45	0.030	202.50	6.08
6	7.57	55	0.015	247.50	3.71
7	8.27	65	0.012	292.00	3.51
8	> 8.7	> 100 destroyed	0.010	450.00 rebuild	4.50
				$\sum \Delta S$	73.62

* ZERO DATUM (NGVD)

Z = repair cost factor (say 1)

INITIAL COSTS, I.C.
IC = \$15/cy * 30 cy/LF = \$450/LF

MAINTENANCE COSTS, M.C.
i = 9.5%
T = 25 yrs
pwf = 9.4376
MC = \$73.62 * 9.4376 = \$695/LF

TOTAL COSTS, T.C.
TC = IC + MC
= \$450/LF + \$695/LF
TC = \$1145/LF

For a proper design study, the water depths at the toe of the stone revetment were estimated as those at the *end* of the 25 year design life. The historic, long term erosion rate, seasonal beach variations and toe scour during storms were all considered to determine the design water depth. In effect, at the end of 25 years, the beach would be gone in front of the seawall and the design water depth would be far different than today's conditions. The SBEACH model and others were used to calculate the surf zone wave conditions and design wave height for the armor layer. Artificial armor units (core-loc) were selected and resulted in total costs (minimal maintenance) of about \$2350 per foot.

Interestingly, the dune-beach-buried seawall alternative costs *less* than a concrete structure using artificial units for armor. This was primarily because the lowered beach elevations at the end of the design life permitted in large waves to directly attack the armor units with no beach remaining. Full details can be found in Cummings and Basco, 1997.

Costs are only one criteria in the decision matrix for choosing the "soft" or "hard" alternative for shore protection. Table 5 summarizes these decision criteria at Dam Neck. The only advantage of the armored revetment is lower annual maintenance costs. All others favored the dune-beach-buried seawall system. The primary benefits of this design are also that:

- a beach is present in the year 2020 at the end of the design life;
- the new sand spreads north and south to benefit all Dam Neck;
- the permitting agencies favor the "soft" alternative;
- the general public favors the "soft" alternative;
- the Navy's image will be enhanced by these efforts to protect/improve the environment.

6.0 Recommendations

Research is needed to develop generic damage curves and equations for use by coastal engineers for a wide range of practical applications. This dune design information could be developed through the systematic application of numerical models (e.g. SBEACH) plus confirmation of the results using large scale laboratory experiments and field data. The influence of the beach width should be quantified and design recommendations presented to aid the coastal engineer in dune design.

More case studies are needed comparing the life-cycle, costs of "soft" versus "hard" solutions for shore protection.

Table 5 "Soft" Versus "Hard" Alternatives for Storm Damage Mitigation

	ALTERNATIVE	
	"SOFT" Beach/Dune Buried Seawall	**"HARD"** Armored Revetment
• Shore Protection	✔	✔
• Economics - Initial Costs	✔	
Maintenance Costs		✔
Total Costs	✔	
• Environmental Consequences	✔	
• Recreation and Aesthetics	✔	
• Permit Application	✔	
• Public Perceptions	✔	
• Navy's Image	✔	
Combined Total	✔	

7.0 Acknowledgement

The authors wish to acknowledge the original contributions of John Headland (1991, 1992) for the buried seawall concept. Discussions with Professor van de Graaff, Delft Technical University are also appreciated. The project has been completed by Glenn and Sadler, consulting engineers, Norfolk, Virginia under the supervision of Mr. Robert E. Cummings, Jr.. The first author of this paper served as a special consultant on this project.

8.0 References

Cummings, R.B. and D.R. Basco (1997) "A Beach-Dune-Buried Seawall Shore Protection System for the Fleet Combat Training Center, Atlantic, Dam Neck, Virginia Beach, Virginia" (10th Annual, National Beach Preservation Tech. Conf., St. Petersburg, January)

Headland, J.R. (1991) "Erosion Study" Final Rept, NAVFAC, Alexandria, VA.

Headland, J.R. (1992) "Design of Protection Dunes at Dam Neck, Virginia," *Proceedings*, Coastal Engineering Practice '92, Specialty Conference, ASCE, Long Beach, CA.

Kreeke, J.v.d. and A. Paape (1964) "Europoort Breakwater Structures: A Cost Comparison," Ministry of Public Works (Rijkswaterstaat) and Delft Hydraulics, Final Rept W732M748.

Larsen, M. and N.C. Kraus (1989) "SBEACH: Numerical Model for Simulating Storm Induced Beach Change," Rept 1, Empirical Foundation and Model Development, Tech Rept, CERC-89-9, CERC, Vicksburg, MS.

Schoonees, J.S. and A.K. Theron (1995) Evaluation of Ten Cross-Shore Sediment Transport/Morphological Models," Coastal Engineering, Vol. 25, Nos. 1-2, pp.1-41.

TAW (1989) "Guide to the Assessment of the Safety of Dunes as a Sea Defense," Rept 140, CUR (Centre for Civil Engineering Research and Codes), Gouda, the Netherlands.

van de Graaff, J. (1986) "Probabilistic Design of Dunes: An Example from the Netherlands," Coastal Engineering, Vol. 9, pp.479-500.

CHAPTER 231

HURRICANE OPAL INDUCED CHANGES ON NATURAL AND NOURISHED BEACHES, WEST-CENTRAL FLORIDA

Richard A. Davis, Jr.[1] and Ping Wang[2]

ABSTRACT

Twenty-six beach profiles were surveyed immediately after the passage of storm conditions and were compared with pre-storm situations. They include 1) eleven locations spread throughout the entire 60 km reach of the Pinellas County coast including wide, narrow, natural and nourished sites, with and without seawalls; and 2) 15 locations confined to three adjacent nourishment projects along 14 km of Sand Key.

The overall behavior of the nourished and natural beaches along the 60 km reach of coast was similar, displaying a general trend of 1) shoreline erosion ranging from 2 to 10 m, 2) upward and landward migration of the nearshore bar, and 3) backbeach accumulation and increase in the berm height. Shoreline orientation and beach sand composition played no significant role in beach performance during the storm. The technique of dry beach replenishment using a dragline and conveyer belt may contribute to the more severe shoreline erosion at the Indian Shores nourishment project as compared to the traditional pumping technique used at Indian Rocks Beach and Redington Beach.

Temporary berm accumulation and shoreline accretion were recorded at two chronically eroding locations downdrift of structures. The shoreline accretion was caused by the landward sand transport induced by the storm waves. The storm accumulation was eroded by the normal-weather longshore sediment transport within three months after the storm.

The current version of SBEACH model (Larson and Kraus 1989) failed to reproduce the Opal-induced beach changes in the surf zone. The unsuccessful prediction was attributed to the uncertainties in offshore wave measurement, and morphological and computational complications caused by the exposure of hard bottom in the nearshore region.

1) Director and Distinguished Research Professor, Coastal Research Laboratory, Department of Geology, University of South Florida, Tampa, FL 33620, USA;
2) Research Associate, Coastal Research Laboratory, Department of Geology, University of South Florida, Tampa, FL 33620, USA.

INTRODUCTION

Hurricane Opal passed about 250 km to the west of the west-central Florida coast in its northerly path toward the Florida panhandle in early October, 1995. The speed of the hurricane center (Fig. 1) in the central Gulf of Mexico was relatively slow. The slow speed generated abnormally long-period, high waves in the Gulf. A storm surge of about 1 m was measured in the study area along the west-central Florida coast (Fig. 2). The energetic conditions and the storm surge lasted for approximately two days during spring tide conditions.

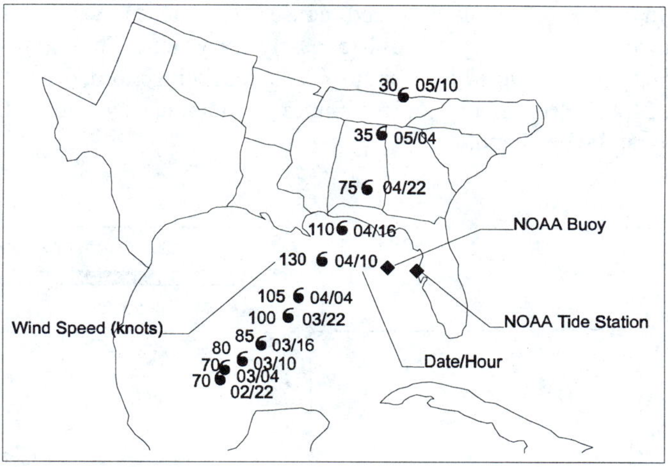

Figure 1. Storm track of Hurricane Opal and NOAA's wave buoy and tide gage.

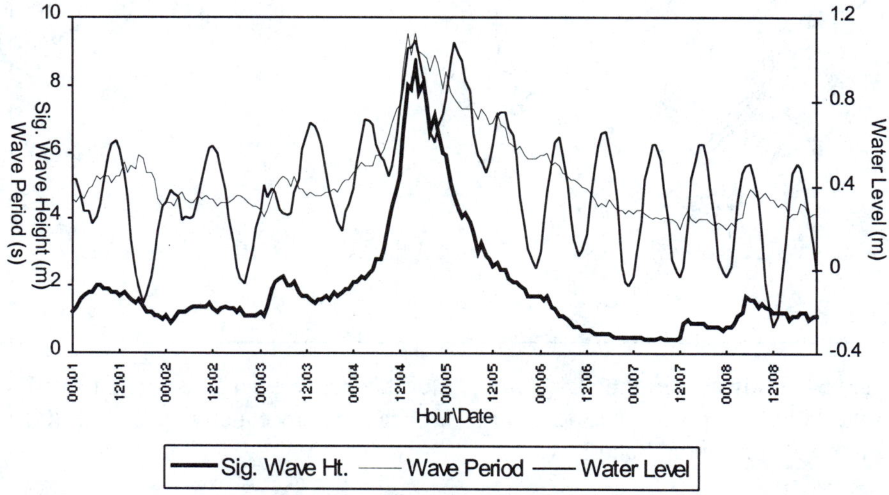

Figure 2. Measured significant wave height, average wave period, and storm surge by the NOAA buoy station 42036 and Clearwater tide station.

The significant wave height measured by the NOAA wave buoy (Fig. 1) reached a maximum over 9 m. Although the average wave period as shown in Figure 2 was less than 10 s, the dominant wave period reached 13 to 14 s during the peak of the storm. A storm surge of nearly 1 m was measured by the NOAA Clearwater tide station at the northern boundary of the study area (Fig. 2).

Twenty-six beach profiles were surveyed immediately after the passage of storm conditions and were compared with pre-storm profiles which were surveyed 1 to 2 months before the storm. Each profile was surveyed to a depth of 1.5 m below NGVD along a shore-normal transect. The 26 profiles are located along nearly 60 km of coast in Pinellas County, Florida and are part of an ongoing, long-term study of beach dynamics. Both natural and nourished beaches were included with construction ranging from 2 years old to nearly 10 years. The study area has a coastal orientation that ranges over about 40 degrees with a broad headland in the middle (Fig. 3). The shoreface gradient ranges from about 1:400 to 1:700 with the steepest being at the headland.

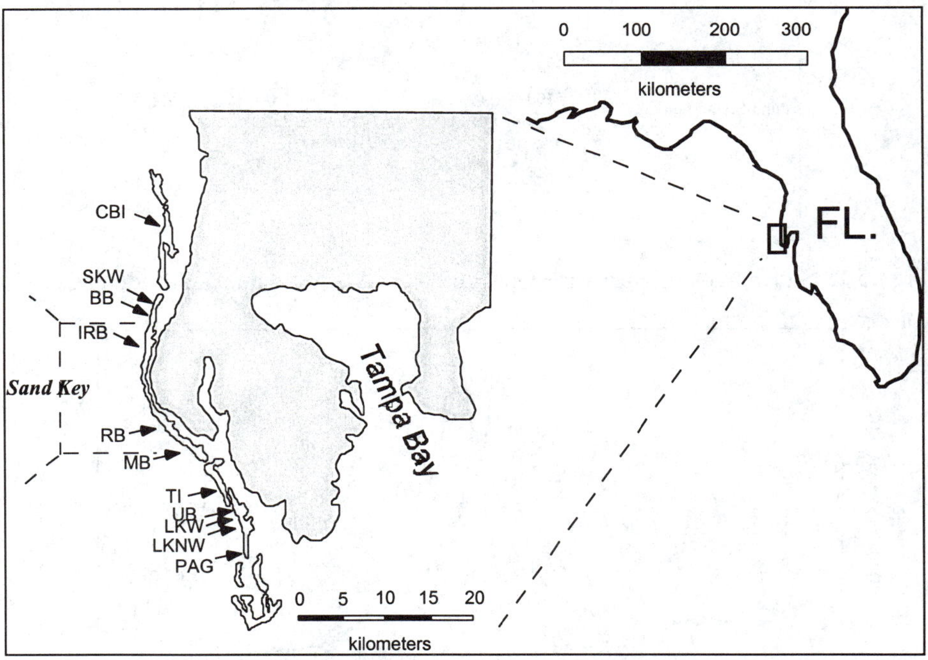

Figure 3. Study area in Pinellas County, Floirda. The eleven sites are indicated by the lettered arrows. The 15 nourished locations are equally spaced at 300 m intervals on central Sand Key.

The objective of this study was to examine the hurricane-induced beach morphology changes through the comparison of profiles surveyed before, immediately after, and 4 to 8 months after Hurricane Opal. Different types of beaches with different orientations and different degrees of human activities including natural and nourished, with and without seawalls, were examined and compared. Dominant direction of sediment transport during storm and normal-weather conditions is discussed. The application of the SBEACH model in predicting the storm induced beach changes along west-central Florida coast is also examined.

STUDY AREA

Two sets of profiles were surveyed: 1) eleven locations spread throughout the entire 60 km reach of coast including wide, narrow, natural and nourished beaches, with and without seawalls; and 2) 15 locations confined to three adjacent beach nourishment projects along 14 km of Sand Key (Fig. 3). The dominant longshore sediment transport along the entire coastal reach is toward the south, but there are local reversals. Two locations (SKW and UB), downdrift of structures, are experiencing severe beach erosion. SKW (Fig. 3) is currently protected by seawalls. The chronically eroding Upham Beach (UB in Fig. 3; Leonard et al. 1989, Dixon and Pilkey 1989) was protected by sand bags and renourished for the fifth time in the last 20 years in May, 1996.

Three adjacent beach nourishment projects were constructed on Sand Key. Five locations, R74, R75, R78, R80, and R81, were surveyed on Indian Rocks Beach which was nourished in 1990. The middle project at Indian Shores (R86, R87, R89, R91, and R92) which is located on the protruding headland was nourished in 1992 and the southern project at Redington Beach (R98, R99, R106, R107, and R108) was nourished in 1988. The nourishment at Indian Shores was constructed differently from the two adjacent projects. Instead of using the conventional pumping, the sand was replenished dry with a dragline and conveyer belt. The less expensive dry fill resulted in a looser packing than the wet pumping. The nearshore wave energy is usually higher at the Indian Shores headland due to the steeper shoreface gradient than at the adajcent Indian Rocks Beach and Redington Beach.

Sediment properties on natural beaches are different from those on the nourished beaches. Natural beaches or beaches that have not been nourished for the last decade or so are typically composed of well-sorted fine sand with less than 10% shell gravel. Nourished beaches, especially the two recently constructed at Indian Rocks Beach and Indian Shores, have significant amount of shell gravel, generally more than 20 %, inherited from the borrow material. Sediments in the swash zone have even higher shell-gravel concentration.

OPAL-INDUCED CHANGES ON NOURISHED BEACHES

All the nourished sites showed shoreline erosion of 2 to 10 m (Fig. 4A). The protruding 4-year old Indian Shores suffered the most shoreline loss, ranging from 8 m to over 10 m. The 8-year old Redington Beach lost the least shoreline, from 2 to 8 m. The 6-year old Indian Rocks Beach lost between 5 to nearly 10 m. The severe shoreline erosion at Indian Shores headland is believed to be caused by a combination of high wave energy and loose packing.

The sand-volume change above the -1.5 m NGVD datum was generally small, ranging from 15 m^3/m gain to 17 m^3/m loss (Fig. 4B). Although most of the locations lost 2 to 17 m^3/m sand, five of the 15 locations gained various amount of sand. The trend of volume change was not as apparent as the shoreline change. The reason for the less distinctive trend of volume change as compared to shoreline change was that the volume loss at the shoreline was compensated by the landward and upward migrations of the nearshore bar and the accumulation on the backbeach, forming a higher berm (Fig. 5).

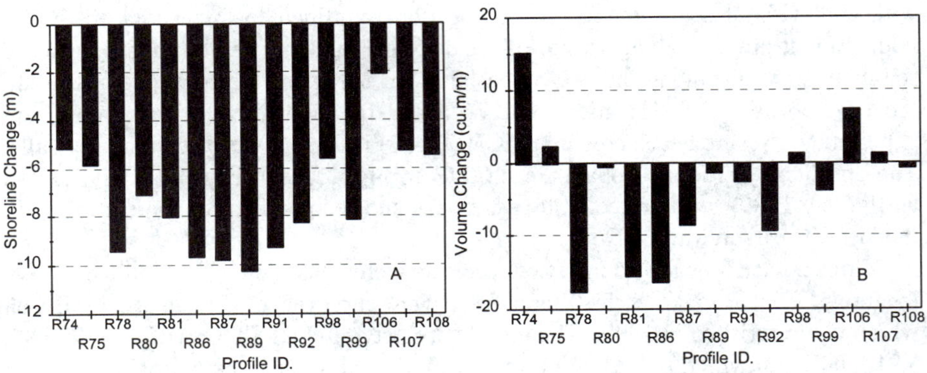

Figure 4. Shoreline (A) and volume (B) changes on the nourished beaches.

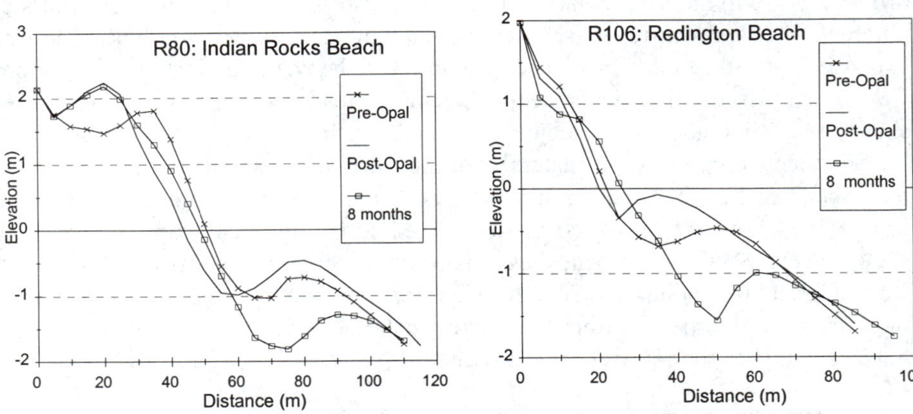

Figure 5. Beach profile changes: before, immediately after, and 8 months after.

It is generally assumed that large storm waves induce seaward migration of the breaker-point bar (e.g., Komar 1976, Dean 1995). This hypotheses which was developed from the original beach cycle study of Shepard (1950) has been used broadly in shoreface-development studies (e.g., Pilkey et al. 1991). Larson and Kraus (1994) documented landward sediment transport during 3 of the 4 examined storms in 1989 and 1991 at Duck, North Carolina. Lardward and upward migration of the nearshore bar induced by Hurricane Opal was measured on all the barred locations in the present study along the 60 km study area (Fig. 5). This landward migration of nearshore bar was also observed by Stone et al. (1996) on the Florida panhandle. Shoreline recovery and seaward migration of the nearshore bar was measured 4 and 8 months after Opal, indicating a seaward transport during the normal-weather conditions.

Seasonal beach cycles like those observed along the U.S. Pacific coast (e.g., Shepard 1950, Inman et al. 1993) are not observed along west-central Florida coasts. The landward migration of the nearshore bar observed in this study cannot be explained by the seaward shift of breaker point during high-energy storm wave conditions, which would result in seaward migration of the breaker-point bar. The unexpected upward and landward bar migration indicates that in addition to the wave steepness and the location of breaker point, the nearshore bar migration may be also controlled by other factors. Further study is needed to understand the mechanism of landward bar migration during storm conditions.

Another morphological change that was observed at all the profile locations except the "hot spot", R106 (Fig. 5B), was the accumulation on the backbeach. A large amount of sand was deposited landward of the previous berm crest. The berm crest was shifted landward and higher than before. The thickness of the wedge-shaped accumulation decreased landward and terminated, at most of the nourished locations, before the accumulation reached the seawalls. The backbeach accumulation was the thickest, up to 1 m thick at the storm-berm crest, on the high-energy Indian Shores headland, and the thinnest on the relatively low-energy Redington Beach.

Significant shoreline recovery and the seaward migration of the nearshore bar were observed 8 months after the storm (Fig. 5). The shoreline recovery resulted in more gentler beach than the storm beach, especially in the vicinity of the shoreline. The backbeach accumulation remained unchanged because the storm berm crest is beyond the reach of wave uprush under normal weather conditions.

Natural beaches or those that have not been nourished in the last decade or so (CBI, MB, TI, LKNW, and PAG in Fig. 3) showed a similar general trend, i.e., shoreline erosion, landward migration of nearshore bar, and backbeach accumulation, across the entire 60 km study area. The amount of shoreline retreat, ranging from 2 to 10 m, was similar to that measured on Indian Rocks Beach and Redington Beach and was less that that on Indian Shores (Fig. 6A) headland. Volume change above -1.5 m NGVD was generally small, mostly less than 15 m^3/m, and showed similar trend (Fig. 6B) as that observed on nourished beaches in Sand Key.

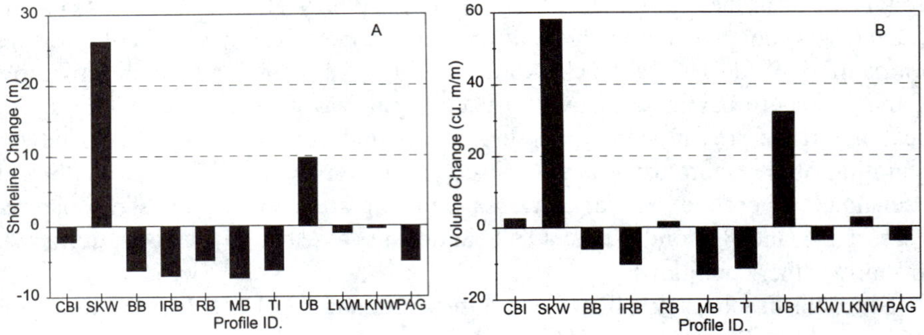

Figure 6. Shoreline (A) and volume (B) change of the 11 locations along Pinellas County, Florida.

OPAL INDUCED CHANGES ON THE CHRONICALLY ERODING STRUCTURED BEACHES

Two of the beaches that are historically erosional, SKW and UB (Fig. 3), are both downdrift of structures. Scour behind the seawall and damage to the adjacent residential buildings occurred due to storm wave over-topping at both locations. A significant amount of sand accumulation was measured in front of the seawall at SKW, and the sand bags at UB were buried by sand that was deposited on the beach (Figs. 6, 7). As much as 26 m of shoreline accretion with nearly 60 m^3/m volume gain was measured at SKW immediately after the storm. Ten meters of shoreline accretion and 35 m^3/m volume gain were measured at UB just south of the wave-dominated inlet, Blind Pass (Davis and Gibeaut 1990). Large carbonate rock fragments characteristic of the Tampa Limestone which is exposed offshore were found on the storm beach, indicating that the sediment was transported landward from offshore during the storm. It is believed that the seawalls and sand bags induced significant energy dissipation during the storm surge. The fast rate of energy dissipation resulted in backbeach accumulation above the normal wave uprush limit.

The temporary sediment accumulation induced by Opal was eroded completely within four months after the storm (Fig. 7). The longshore transport under normal weather and the lack of updrift sediment supply are believed to be responsible for the erosion. The southward transport is evident at Upham Beach (UB) from the most recent beach nourishment. A profile, LKW (Fig. 3), about 150 m south of the UB location, was surveyed as part of a long-term beach monitoring program. A large amount of sediment was replenished on the Upham Beach about 7 months after the storm. The LKW location was about 80 m south of the nourishment project. Significant shoreline erosion was measured at UB only 3 months after the nourishment (Fig. 8). Remarkable accumulation was measured at the LKW location during the same period. The shape and slope of the shoreface at Upham Beach remained fairly constant, suggesting that the shoreline retreat was not caused by the cross-shore profile adjustment. It is evident that the southward longshore transport is the cause of the long-term beach erosion at Upham Beach as well as the erosion of the storm accumulation from Opal.

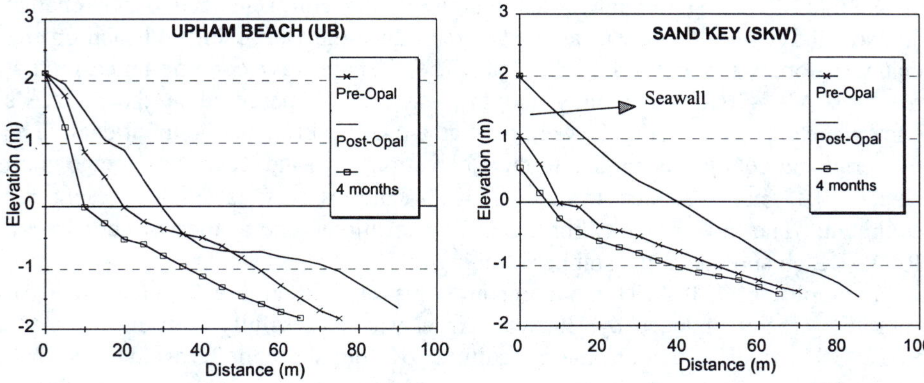

Figure 7. Temporary accumulation induced by Hurricane Opal on structured beaches.

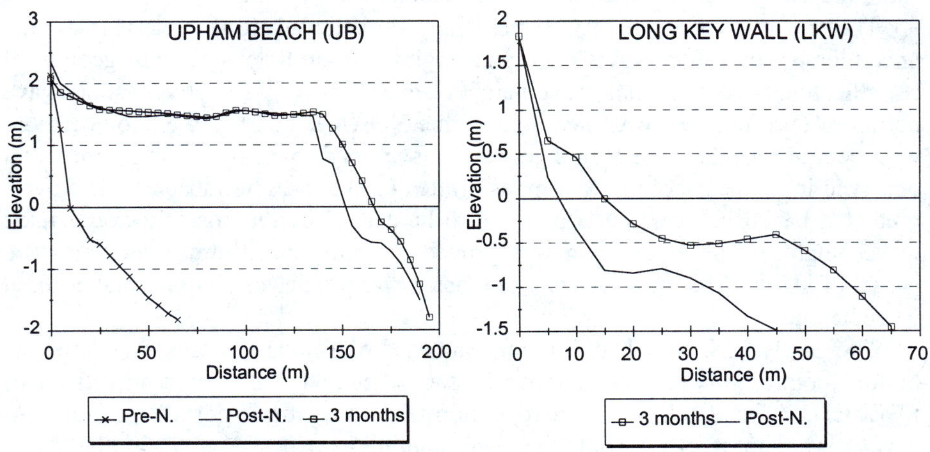

Figure 8. Shoreline erosion at the recently nourished Upham Beach and sand accumulation at the downdrift beach. Note the differences in scale.

The seawall and stablizing sand bags effectively protected the coast and dissipated a significant amount of wave energy during Hurricane Opal. The structures are not capable of controlling regional and long-term beach changes caused by longshore sediment transport. Beach nourishment not only provides a buffer for normal weather shoreline erosion caused mainly by longshore sediment transport but also protects the coasts against dramatic storm events.

SBEACH MODELLING OF THE BEACH PROFILE CHANGES

SBEACH, the numerical model for simulating storm-induced beach change (Larson and Kraus 1989), was applied to reproduce the Opal-induced beach profile changes along west-central Florida coast. The offshore wave condition measured at the NOAA's 42036 wave buoy and the water level measured at the NOAA's Clearwater tide station (Fig. 2) were used as the key hydrodynamic input data. The numerical modeling was applied to the 15 locations on Sand Key. An average grain size of 0.33 mm obtained from over 1100 samples was used to represent the sediment. The model was calibrated at two locations, one at the northern Indian Rocks Beach and one at the southern Redington Beach.

Examples of SBEACH modeling are illustrated in Figure 9. The trend of shoreline erosion induced by Hurricane Opal was successfully reproduced by the SBEACH model, although the magnitude of the shoreline erosion was over predicted. The backbeach accumulation and the unexpected upward and landward migration of the nearshore bar were not predicted. The bar/trough features were basically absent from the predicted profiles. The increased berm height and the wedge-shaped sediment accumulation on the backbeach were not predicted, on the contrary, significant berm erosion and a much gentler beach slope near the shoreline, as compared to the pre-storm situations, were predicted.

The unsatisfactory SBEACH modeling is believed to be caused by the uncertainties in input wave data and regional morphological and geological complications. A significant wave height of over 9 m with a relatively short wave period of less that 10 s was measured by the NOAA's 42036 wave buoy offshore west-central Florida. The 9 m significant wave height was much larger than that observed in the nearshore (3 to 4 m, as printed on the local newspaper). The over-predicted backbeach erosion is believed to be caused by the possibly exaggerated wave height. This assumption was proved by using an arbitrary smaller wave height of 3.5 m. Much less backbeach erosion was predicted with the smaller input wave height.

The nearly horizontal hard bottom composed of Tampa Limestone violated the bottom boundary condition of movable sand assumed in the current version of SBEACH. The depth of the hard bottom decreases from north to south. At northern Indian Rocks Beach, the hard bottom is exposed at 7 m, the depth decreases to about 3 m at the southern Redington Beach. Based on the wave breaking criterion used in the SBEACH model (Larson and Kraus 1989), the nearly horizontal hard bottom is within the breaker zone during the peak of the storm conditions with over 9-m waves. The current version of SBEACH is not capable of incorporating the non-erosible hard bottom, especially when the hard bottom is within the breaker zone. The above situation will be an ideal field test for the updated SBEACH model (N.C. Kraus, personal communication; available:

December 1996, Randy Wise, personal communication) which will incorporate the influences of non-erosible bottom.

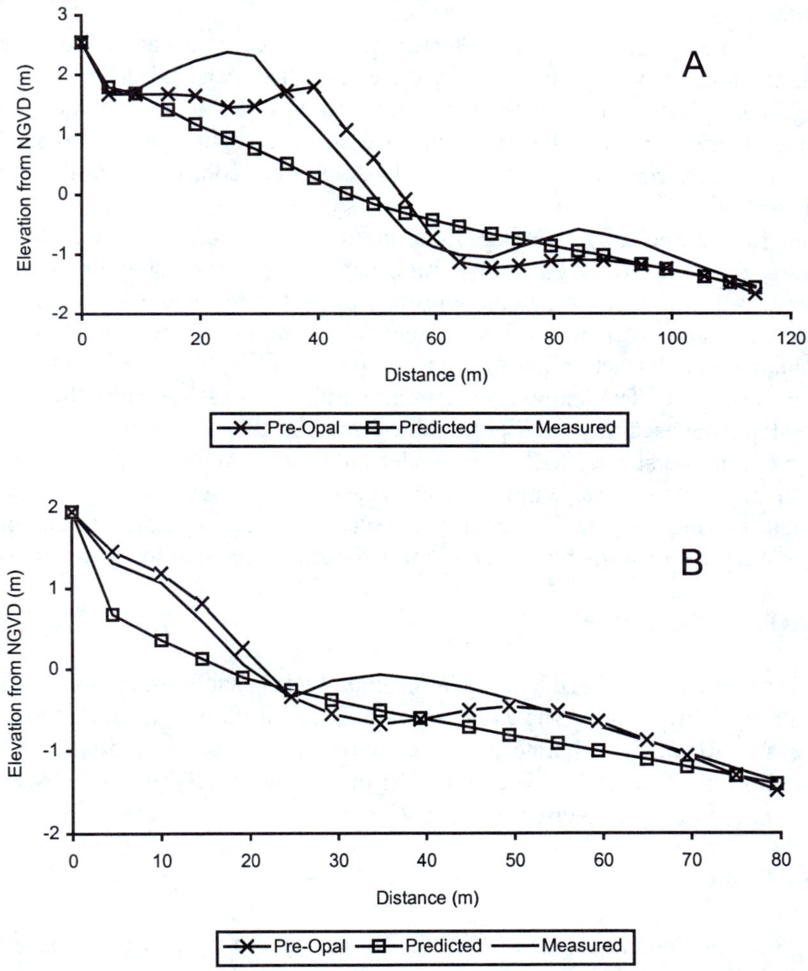

Figure 9. Measured and predicted beach profiles; A) R74, Indian Rocks Beach; and B) R106, Redingtion Beach.

SUMMARY AND CONCLUSIONS

The overall behavior of the nourished and natural beaches during the passage of Hurricane Opal is similar. All surveyed beaches along the 60 km reach of coast displayed a general trend of shoreline erosion, upward and landward migration of the nearshore bar, and accumulation on the backbeach resulting a higher berm. This trend was observed on beaches with different orientations and different sediment compositions, ranging from less than 10% shell gravel on natural beaches to greater than 30% shell gravel on nourished beaches. This indicates that shoreline orientation and sand composition played no significant role in beach performance under storm conditions.

It is apparent from the data collected on the three adjacent and differently constructed nourishment projects on Sand Key that they behaved differently. The oldest showed least change and the most recent showed the most. The most recent project was also constructed without the benefit of dredging and pumping, thus creating a loosely compacted beach. The loose packing contributed to the greater rate of erosion.

The two chronically eroding locations downdrift of structures behaved differently from the nourished and natural beaches. Significant shoreline accretion was measured at the two locations, with one protected by seawalls and the other protected by sand bags, immediately after the storm. A large amount of sand accumulated on the beach in front of the structure resulting from the storm-induced landward transport. The temporary storm accumulation was eroded by the normal-weather longshore sediment transport within 3 months after the storm.

The current version of SBEACH model failed to reproduce the Opal-induced beach changes along west-central Florida coast. The unsuccessful prediction is attributed to uncertainties in input wave data and the model's limitation in incorporating the non-erosible hard bottom in the study area into the computation.

ACKNOWLEDGEMENTS

This study was supported by Pinellas County, Florida and Florida Department of Environmental Protection. The authors would like to thank the graduate students in the Coastal Research Laboratory at the University of South Florida for their assistance in the field work. The SBEACH model was kindly provided by Randy Wise at the Coastal Engineering Research Center.

REFERENCES

Davis, R.A., Jr. and Gibeaut, J.C. (1990). "Historical morphodynamics of inlets in Florida." Florida Sea Grant, 81p.

Dean, R.D. (1995). "Cross-shore sediment transport processes." *Advances in coastal and ocean engineering*, P.L.-F., Liu. ed., World Scientific, New Jersey, 159-220.

Dixon, K., and Pilkey, O.H. (1989). "Beach nourishment along the U.S. coast of the Gulf of Mexico." *Proceedings of Coastal Zone'89*, ASCE, 2007-2020.

Inman, D.L., Elwany, M.H.S., and Jekins, S.A. (1993). "Shoreface and bar-berm proflies on ocean beaches." *Journal of Geophysical Research*, 98, 18,181-18,199.

Komar, P.D. (1976). "*Beach processes and sedimentation.*" Prentice-Hall, Englewood Cliffs, New Jersey, 429p.

Larson, M. and Kraus, N.C. (1989). "SBEACH: Numerical Model for Simulating Storm-Induced Beach Change." *Technical Report CERC-89-9*, U.S. Army Engineer Waterways Experiment Station, Vicksburg, MS.

Larson, M. and Kraus, N.C. (1994). "Temporal and spatial scales of beach profile change, Duck, North Carolina." *Marine Geology*, 117, 75-94.

Leonard, L., Clayton, T.D., Dixon, K., and Pilkey, O.H. (1989). "U.S. beach nourishment experience: a comparison of the Atlantic, Pacific, and Gulf coasts." *Proceedings of Coastal Zone'89*, ASCE, 1994-2006.

Pilkey O.H., Young, R.S., Riggs, S.R., Smith, A.W.S., Wu, H., and Pilkey, W.D. (1991). "The concept of shoreface profile of equilibrium: A critical review." *Journal of Coastal Research*, v. 9, p. 255-278.

Shepard, F.P. (1950). "*Beach cycles in southern California*." U.S. Army Corps of Engineers, Beach Erosion Board, Technical Memo., n. 15, 26p.

Stone, G.W., Grymes, J.M., Armbruster, C.K., Xu, J.P., and Oscar, K.H. (1996). "Researcher study impact of Hurricane Opal on Florida coast." EOS, May 7.

CHAPTER 232

Modelling Sand Transport and Profile Evolution on Macrotidal Beaches

Paul.R.Fisher[1] & Tim.J.O'Hare[2]

Abstract

An empirical model is presented which simulates the effect of tidal translation of cross-shore sediment flux patterns, in support of a hypothesis that wave height *and* tidal range are the key influences in the formation of characteristic macrotidal beach profiles. The energetics-based model is based upon field observations of cross shore currents and sediment fluxes from three high-energy macrotidal (tide range>4m) locations around the U.K, chosen as representations of dissipative, intermediate and reflective environments (after Wright and Short, 1984). The observed depth-dependant flux patterns are translated across linear beach profiles, with gradients modelled from sediment characteristics (Dean 1977, Kriebel et al 1991), the evolution of characteristic macrotidal beach profiles and the development of attendant features such as break-point bars, low tide terraces, and steepened foreshores are observed.

Introduction

The morphological significance of tidal translation of the swash, surf and shoaling wave zones across the beachface has received increasing attention over the last fifteen years (Wright et al (1982), Short (1991), Masselink (1993), Masselink and Short (1993)). It has been suggested that the continual variation in water depth by the tidal signal in a *macrotidal* environment is responsible for temporal variations in local dynamics. Furthermore, the intermittence of swash and surf processes in the high-tidal zone modifies the dynamics of that zone, whereas the mid and low-tidal zone dynamics are more similar to nearshore and offshore zones on micro-tidal beaches (Wright et al, 1982).

This contribution sets out to give some insight into the effects of asymmetric residence times of the water level on the beachface within macrotidal regimes.

[1]Research student, [2]Senior Lecturer, Institute of Marine Studies, University of Plymouth, Drake Circus, Plymouth, Devon, UK, PL4 8AA

According to the classification of Bird (1969) macrotidal regimes are those locations where the mean maximum tidal range during a month is greater than 4 metres, mesotidal regimes experience a mean maximum tidal range between 2 and 4 metres, and microtidal regimes are between 1 and 2 metres. While much work has been done relating the effects of a varying wave field on sediment transport processes in macrotidal regimes, (e.g. Wright et al (1982), Wright and Short (1984)), it is not clear how transport processes are modulated by a continual rise and fall of the water level by the tidal signal. This effect is pronounced within macrotidal environments where migration velocities and widths of swash, surf and wave shoaling zones are variant during the tidal cycle dependant on(a) the phase of

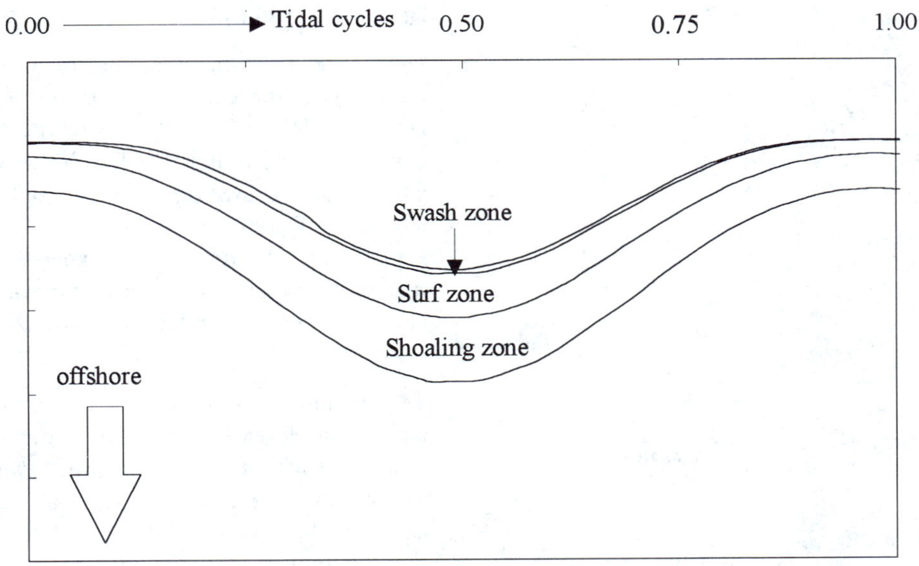

Figure 1. Tidal translation of swash, surf and shoaling zone processes across a beachface of exponential form. Because of the shoreward increasing beach gradient swash, surf and shoaling zones narrow significantly at high tide levels.

the tide and (b) the local gradient of the beach. The tidal variation in magnitude and position of these zones is illustrated in Figure 1 for the case of an exponential beachface. The approach taken in this study is to oscillate simple empirical models of cross-shore sediment transport across a beachface, allowing the profile to respond to the resulting patterns of erosion and deposition. The empirical model used is constructed from field observations of water and sediment dynamics at a variety of high energy macrotidal locations around the U.K during the British Beach and Nearshore Dynamics (B-BAND) experiment.

A useful overview of the B-BAND experiment is given by Davidson et al (1993). This three-year experiment investigated surf zone processes at three high energy macrotidal locations around the UK. The sites were chosen to represent dissipative, intermediate and reflective beaches following the morphodynamic classifications of Wright and Short (1984). The locations are illustrated in Figure 2 along with typical beach profiles and are described briefly in turn.

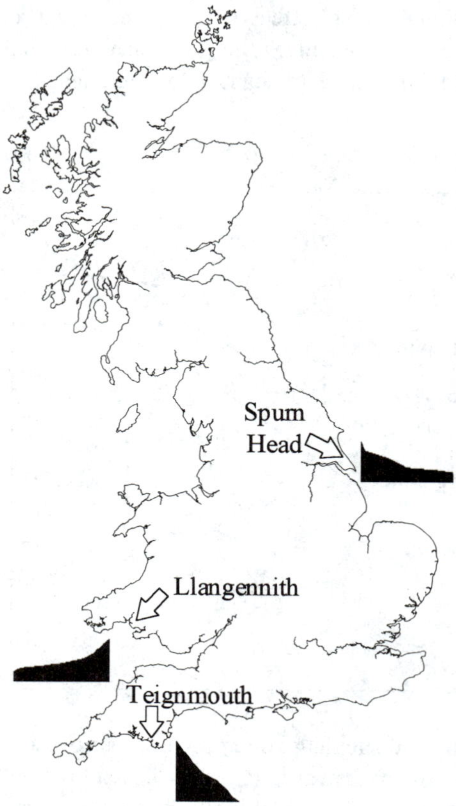

Llangennith: A dissipative high energy beach with a shallow beach gradient (0.014-0.020) with fine to medium quartz sands (D_{50}=0.21mm). This beach is west facing, has a tide range up to 9 metres, and broad surf (up to 350m) and intertidal (~ 500m) zones.

Spurn Head: An intermediate beach located near the end of a 5km long spit facing the North Sea. The beach profile comprises a steep high tide beach (tanβ ~ 0.0975) and a low tide terrace (tanβ ~ 0.023, and D_{50}=0.35mm). This beach has a tide range up to 6m and experiences strong longshore tidal currents which have a significant effect on surf zone dynamics.

Teignmouth: A reflective beach site facing south-east into the English Channel sheltered from Atlantic swell. The beach is backed by a sea wall, and has a gradient of 0.067-0.142. Cross - shore variation of sediment size at this location is significant, with a D_{50} value of 0.24mm. Tidal range is 6 metres.

Figure 2. Location of B-BAND field sites around the UK, showing typical cross-sectional profiles.

Shape Function Approach To Sediment Transport

Foote (1994), Foote et al (1994), and Russell and Huntley (1996) examined field measurements of cross-shore sediment transport from these three locations. They calculated velocity moment contributions to sand transport, by assuming the instantaneous velocity field to be composed of the following contributions:

$$u = \bar{u} + u_S + u_L \qquad (1)$$

where $\bar{u}$ is the mean flow, u_S is the incident wave component of velocity, and u_L is the long wave component. Using Bagnold's (1963, 1966) uni-directional stream flow total load sediment transport model, modified by Bailard (1981, 1987) for cross shore sediment transport under bi-directional flow, Guza and Thornton (1985) used field measurements to examine the relative importance of the velocity moment terms in Bailards model, and found the transport significant terms to be (in normalised form):

$$\overline{|u|^2 u} \Big/ \left(\overline{u^2}\right)^{3/2} \quad \text{for bedload transport, and} \quad (2)$$

$$\overline{|u|^3 u} \Big/ \left(\overline{u^2}\right)^{2} \quad \text{for suspended load transport} \quad (3)$$

Foote et al (1994) examined the cross-shore distribution of the significant moments for bedload and suspended load and found in both cases that the sum predicted a net onshore movement shoreward of the breakpoint and a net offshore movement seaward of the breakpoint. To clarify this pattern for each case they normalised the water depth by the depth of wave breaking. The position of wave breaking was determined from the position of maximum wave height from pressure transducer records and this was checked with visual observations in the field. Combining data from all three field sites they noted that not only was there a pattern of net onshore transport shoreward of wave breaking and a net offshore transport seaward of it, but that a curve could be reasonably fitted through the velocity moments calculated for the interim positions.

They subsequently proposed the existence of quasi-universal, cross-shore sediment transport spatial distribution curves for bed load and suspended load on high energy beaches. These sediment 'shape functions' are illustrated in Figure 3. Third order polynomials are fitted through each data set, with one root at x=0, y=0 (zero transport at the shoreline). O'Hare (1994) proposed that this 'shape function' approach be adopted to develop a tentative cross-shore sediment transport model, suggesting that the patterns observed in the B-BAND data are in fact required by the break-point bar hypothesis, and that characteristic macrotidal profiles may be constructed by the repeated tidal excursion of the shape function. This contribution presents the preliminary results from the development of such a model.

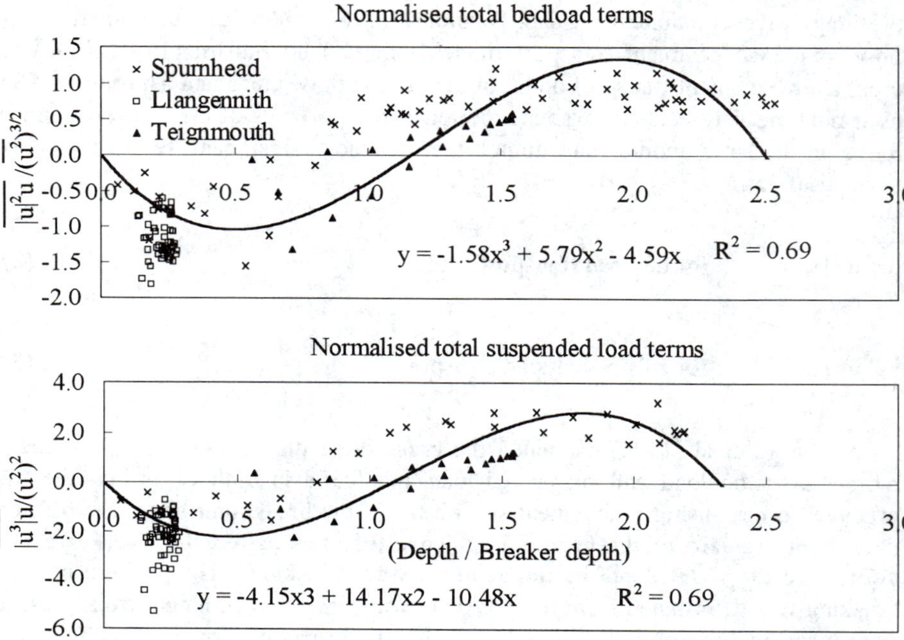

Figure 3. Sediment transport shape functions obtained from the B-BAND data.

Model Description

The simulation model migrates the proposed shape functions for bed and suspended load across a beachface. The input parameters for the model are breaker height H_b, grain diameter ϕ, tide range, wave period T, sediment and fluid densities ρ_s and ρ, gravity, kinematic fluid viscosity υ, bed drag coefficient C_d, and bed and suspended load efficiencies ε_b and ε_s. Related parameters such as sediment repose angle, $\tan\phi$, and sediment porosity, n, are evaluated using an empirical model (Allen, 1970).

The first stage for the simulation model is to evaluate the grain buoyancy and fall velocity of the beach sediment according to the CERC Shore Protection Manual procedures (US Army Corps of Coastal Engineers, 1984):

Grain buoyancy (A): $$\frac{(\rho_s - \rho)}{\rho}\frac{gD^3}{\upsilon^2} \quad (4)$$

Sediment fall velocity (w):

$$w = \frac{1}{18}\frac{(\rho_s - \rho)}{\rho}\frac{gD^2}{\upsilon} \qquad \frac{\left\{\frac{(\rho_s - \rho)g}{\rho}\right\}^{0.7} \cdot D^{1.1}}{6\upsilon^{0.4}} \qquad \left\{\frac{(\rho_s - \rho)gD}{0.91\rho}\right\}^{0.5} \tag{5}$$

(for A<39) (for 39<A<10⁴) (for A>10⁴)

Breaking depth is then calculated iteratively according to the model of Le Mehaute (1961) using an initial estimated value for breaking depth (taken as 5m):

$$d_b = \left[\frac{H_b\left(gd_bT^2\right)^{0.5} gT^2}{0.12\tanh 2\pi d_b}\right] \tag{6}$$

Next breaker coefficient (γ) and breaking wavelength (L_b) are evaluated from:

$$\gamma = \frac{H_b}{d_b} \qquad \text{(from Galvin, 1972)} \tag{7}$$

$$L_b = \sqrt{gd_bT^2} \qquad \text{(assuming shallow water)} \tag{8}$$

The preliminary (linear) gradient of the profile is evaluated at closure depth, d_c, (the outer root of the transport polynomial) from Dean (1977) and Kriebel et al (1991):

$$\tan\beta = \left(\frac{2.25}{d_c}\left(\frac{w^2}{g}\right)^{\frac{1}{3}}\right)^{\frac{3}{2}} x_c^{-1} \qquad \text{Where } x_c \text{ is the offshore distance to closure.} \tag{9}$$

The model starts at high tide with a vertical increment of 0.1 metres. Water depth is simply calculated by subtracting the profile level (initially linear) from the still water level. Next a simple model of shoaling wave height is used to determine velocity scales to apply to the polynomial coefficients. The model, again using shallow water approximations is linear from the shore to the breakpoint (depth = 0:gd), and offshore of the breakpoint, takes the form

$$H = \left(\frac{d_b}{d}\right)^{\frac{1}{4}} H_b \tag{10}$$

The shoaling wave heights are used to calculate cross-shore varying maximum orbital velocities from:

$$u_{max} = \frac{H}{2}\sqrt{\frac{g}{d}} \tag{11}$$

u_{max} is used to evaluate $\overline{u^2}$ by assuming a sinusoidal variation in u within the wave period. $\overline{u^2}$ is then used to scale the transport coefficients. Normalised depth, i.e. depth over breaking depth, is calculated cross-shore, so that bedload and suspended load velocity terms may be calculated from the B-BAND derived coefficients giving the $u^2|u|$ term for bedload transport and $u^3|u|$ for suspended load transport. The transport coefficients from the B-Band field data are then:

$$Ibed_{coeff} = \frac{\overline{|u^2|u}}{\left(\overline{u^2}\right)^{3/2}}\left\{-1.58\left(\frac{d}{d_c}\right)^3 + 5.79\left(\frac{d}{d_c}\right)^2 - 4.59\left(\frac{d}{d_c}\right)\right\} \quad \text{bed load} \tag{12}$$

$$Isus_{coeff} = \frac{|u^3|u}{\left(\overline{u^2}\right)^2}\left\{-4.15\left(\frac{d}{d_c}\right)^3 + 14.17\left(\frac{d}{d_c}\right)^2 - 10.48\left(\frac{d}{d_c}\right)\right\} \quad \text{suspended load} \tag{13}$$

The expression for bed and suspended load transport, i, from Bagnold (1966) and Bowen (1980) is:

$$i_{bed} = \frac{\varepsilon_b C_D \rho Ibed_{coeff}}{\tan\phi - \frac{u\beta}{|u|}} \tag{14}$$

$$i_{suspended} = \frac{\varepsilon_s C_D \rho Isus_{coeff}}{W - u\beta} \tag{15}$$

where ε_b is a bedload efficiency factor, C_d is the drag coefficient for the bed, ρ the density of fluid, tanϕ is the angle of repose of sediment and tanβ the beach slope. Similarly for suspended load transport ε_s is a suspended load efficiency factor, w is the sediment fall velocity and u is the instantaneous orbital velocity. Because our data provides a time averaged expression for $u^2|u|$ and $u^3|u|$ respectively, it is difficult to isolate the instantaneous orbital velocity to evaluate the slope terms in the denominator of the two expressions. In this contribution they are ignored (ie. uβ is set to zero in the denominators).

The total cross-shore immersed weight sediment transport rate is the sum of expressions (14) and (15), and the total volumetric transport rate, Q, is thus determined by:

$$Q = \frac{i_{bed} + i_{suspended}}{(\rho_s - \rho_w)g(1-n)} \tag{16}$$

where ρ_s and ρ_w are the sediment and fluid densities, g is gravity and (1-n) is the packing of settled grains (where n is the void ratio).

The sediment is then moved across the profile on the basis of the gradient of the volumetric transport rate, and the submerged part of the new profile is smoothed so that each point on the new profile is elevated or lowered to the mean of the height of its two neighbours. Having completed all these calculations for a particular tide level, the tide is moved at intervals of 1/100th of a tidal cycle through a simple sinusoid, and the process repeated.

Clearly the model works on several underlying assumptions, which may be summarised as follows:

1. The energetics model is valid.
2. Downslope and upslope transport are equal.
3. Airy wave theory is appropriate.
4. There is transmission of wave energy through bars.
5. Longshore transport is ignored.
6. Swash zone transport is ignored.
7. Sediment supply is unlimited.
8. No avalanching can occur.
9. There is a sinusoidal monochromatic tide signal.

Model Results and Discussion

Figure 4(a) shows the results of a model run with zero tidal range and a 2m breaking wave. Sediment is continually eroded at the shoreline and deposited offshore. The transport convergence point gradually moves offshore, the rate of movement being governed by the asymmetry of the transport profile in the surf zone, and the magnitude and asymmetry of the transport profile in the shoaling wave zone. As the profile evolves what started initially as an area of onshore erosion in the shoaling wave zone gets filled in by the offshore movement of sediment due to broadening of the surf zone. The plateau corresponds to the level of peak offshore transport, as modified by the wave energy profile. The onshore transport region quickly narrows into a spike as the beach gradient in the shoaling zone becomes steeper and steeper. Profile changes from cycle to cycle get smaller, suggesting that the model is approaching an equilibrium state.

Figure 4(b) shows what happens in the model when a tide range is introduced, so that the water surface is continually fluctuating and the transport zones migrate up and down the beachface. A pronounced high water bar forms, but

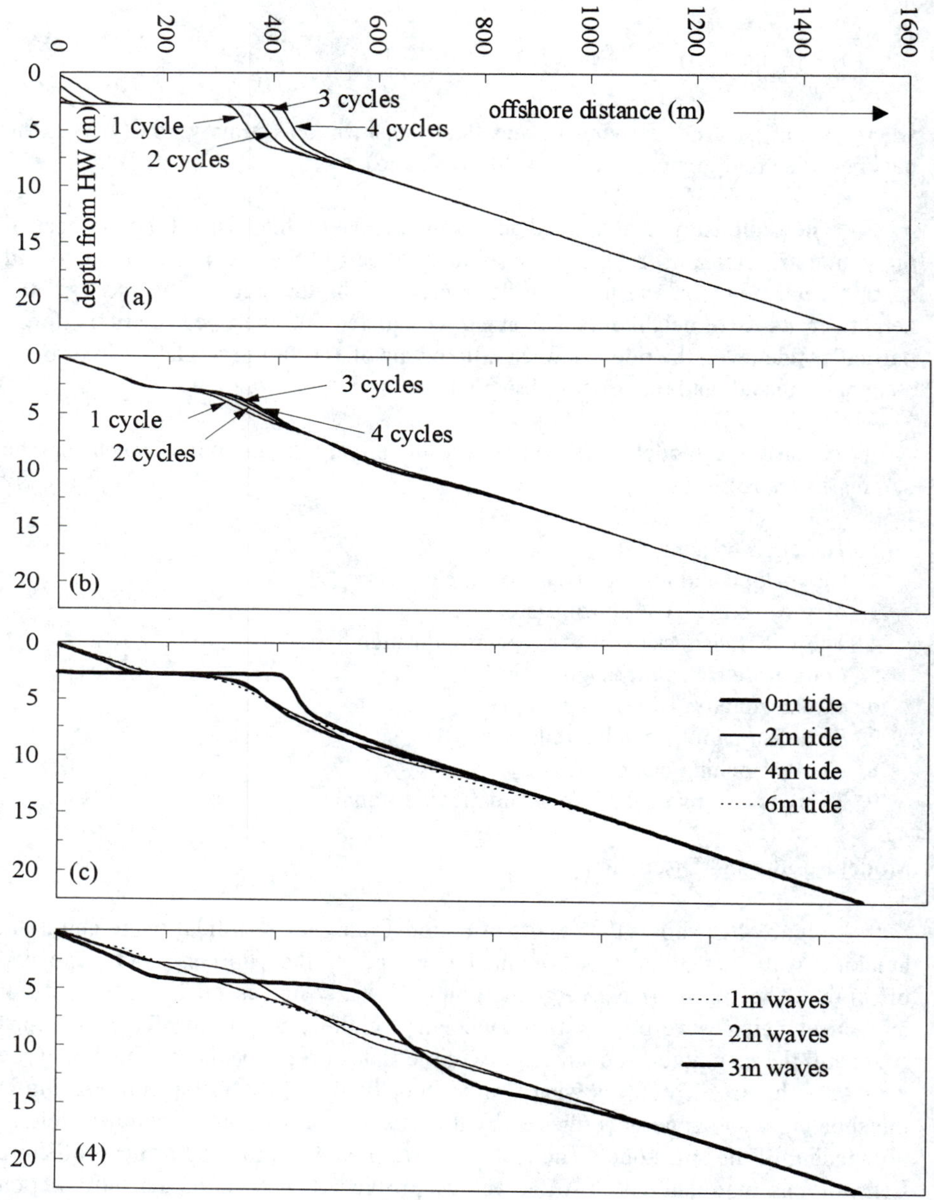

Figure 4. (a) Profile evolution through four tidal cycles with no tidal range and 2m wave height. (b) With 4m tidal range and 2m wave height. (c) Model sensitivity to tidal range 2, 4 and 6m tides (all with 2m waves). (d) Model sensitivity to wave height: 1, 2 and 3m breaking waves (all with 4m tide range).

the low water bar, corresponding to a depth from high water of approximately 6 metres, appears to be suppressed as a result of residence in both offshore and onshore transport zones. The area shoreward of the high water bar only ever resides in a zone of offshore erosion. Again the pattern of erosion and deposition is determined by the total effect of the transport profile, as modified by the shoaling wave energy profile. As the profile develops it appears to be approaching an equilibrium, and changes through time become progressively smaller. Because of the smoothing technique used, what is happening at the shoreline and in the offshore regions is that the elevation of cells outside of the transport zone is being modified by those inside, simulating a slumping effect. This model will only run for approximately 4 tidal cycles before it demands sediment from the very top, or very bottom of the profile, and fails a volume continuity test.

Figure 4(c) shows the evolution of the model profile under different tidal regimes after 4 tidal cycles, with 2 metre waves. The figure demonstrates the suppressing effect of larger tide ranges, with morphological change being the least for the 6m tidal range. In addition with the 6m range the high and low water bars become separated from each other. The position of the high water bar is similar for all cases except for the case of no tide range, where the amount of erosion has pushed it slightly further offshore.

Figure 4(d) shows the sensitivity of the model to wave height, it can be seen that the effect of increasing the wave height is similar to a reduction in tidal range - an increase in the magnitude of wave height leads to tendency for the profile to resemble the non-tidal case. 1m waves have a negligible effect on the profile even after 4 tidal cycles, while with 3m waves the erosion is so substantial, that, as with the zero tide case, the bar is moved offshore by the erosion, to the extent that it becomes indistinct from onshore accretion at low water.

Conclusion

At this stage in its development the model is not able to *quantitatively* simulate the cross-shore profile evolution on 2-dimensional beaches from a knowledge of the wave and tide regimes and sediment characteristics, although that is a possibility with further development. What has been shown is that, using quantitatively accurate flux profiles, the model produces realistic beach profiles within reasonable time-scales. Also, as has been observed by researchers examining field data from macrotidal beach sites (e.g. Wright et al (1982), Jago & Hardisty (1984), Short (1991), Masselink (1993), Masselink and Short (1993)), the model is able to confirm that wave height and tide range are the key influences on the form of the cross-shore profile.

The qualitative appearance of realistic features (offshore bar, intertidal terrace) indicates that the scale of macrotidal beach profile features can be predicted by the tidal translation of a transport shape function across the beachface. The model will now be used to explore more complex scenarios such as profile response to tide & wave field variations, and examine the effect of spring/neap variations within the tidal cycle.

Acknowledgements

This work was supported through funding by the Environmental and Physical Sciences Research Council, research grant reference 94305086.

References

Allen, J.R.L. (1970), *Physical Processes of Sedimentation*, George Allen and Unwin, London.

Bagnold, R.A. (1963), Mechanics of marine sedimentation, In: *The Sea, Ideas and Observations*, Wiley-Interscience Publishers, New York, **3**, 507-528.

Bagnold, R.A. (1966), An approach to the sediment transport problem from general physics, *U.S. Geol. Surv. Prof. Paper* 422-1.

Bailard, J.A. (1981), An energetics total load sediment transport model for a plane sloping beach, *Journal of Geophysical Research* **86**, 10938-10954.

Bailard, J.A. (1987), Surfzone wave velocity moments, *Proceedings of Coastal Hydrodynamics, ASCE*, 328-342.

Bird, A.C.F. (1969), *Coasts*, Australian National University Press, Canberra.

Bowen, A.J. (1980), Simple models of nearshore sedimentation; beach profiles and longshore bars, In: *The Coastline of Canada*, Ed. S.B. McCann, Geological Society of Canada, Paper 1-11.

Davidson, M.A., Russell, P.E., Huntley, D.A. and Hardisty, J. (1993), Tidal asymmetry in suspended sand transport on a macrotidal intermediate beach, *Marine Geology* **110**, 333-353.

Dean, R.G. (1977), Equilibrium beach profiles: U.S. Atlantic and Gulf coasts, *Ocean Engineering Technical Report #12*, 45pp.

Foote, Y.L.M. (1994), Waves, Currents and Sand Transport Predictors on a Macro-Tidal Beach, *Ph.D. Thesis*. Institute of Marine Studies, University of Plymouth.

Foote, Y.L.M., Huntley, D.A. and O'Hare, T.J. (1994), Sand transport on macro-tidal beaches, *Proceedings of the Euromech 310 colloquium*, Le Havre, 360-374.

Galvin, C.J. (1972), Waves breaking in shallow water, In: *Waves on Beaches*, Ed. R. Meyer, Academic Press, London, 413-455.

Guza, R.B. and Thornton, E.B. (1985), Velocity moments in the nearshore zone, *Journal of Waterways, Port, Coastal and Ocean Engineering*, **111**, 235-256.

Jago, C.F. and Hardisty, J. (1984), Sedimentology and morphodynamics of a macrotidal beach, Pendine Sands, SW Wales, *Marine Geology* **60**, 123-154.

Kriebel, D.L., Kraus, N.C. and Larson, M. (1991), Engineering methods for predicting beach profile response, *Proceedings of Coastal Sediments '91*, 557-571.

Le Mehaute, B. (1961), A theoretical study of waves breaking at an angle with a shoreline, *Journal of Geophysical Research* **66** no.2, 495-499.

Masselink, G. (1993), Simulating the effects of tides on beach morphodynamics, *Journal of Coastal Research* **15**, 180-197.

Masselink, G. and Short, A.D. (1993), The effect of tide range on beach morphodynamics and morphology: a conceptual model, *Journal of Coastal Research* **93**, 785-800.

O'Hare, T.J. (1994), What is the origin of the macro-tidal beach profile? *Unpublished Report,* Institute of Marine Studies, University of Plymouth.

Russell, P.R. and Huntley, D.A., (1996), The spatial distribution of cross-shore transport on high energy beaches, *(in prep.)* Institute of Marine Studies, University of Plymouth.

Short, A.D. (1991), Macro-meso tidal beach morphodynamics - an overview, *Journal of Coastal Research* **72**, 417-436.

U.S. Army Corps of Engineers, CERC (1984), *Shore Protection Manual*, 4th Ed, (2 vols), U.S. Printing Office, Washington DC.

Wright, L.D., Nielsen, P., Short, A.D. and Green, M.O. (1982), Morphodynamics of a macrotidal beach, *Marine Geology* **50**, 97-128.

Wright, L.D. and Short, A.D. (1984), Morphodynamic variability of surf zones and beaches: a synthesis, *Marine Geology* **56**, 93-11.

CHAPTER 233

BEACH EVOLUTION UNDER RANDOM WAVES.

Patrick Holmes[1], Thomas E. Baldock[2], Ray T. C. Chan[3] and M. Ahmad L. Neshaei[3]

ABSTRACT

This paper considers the evolution of steep mobile sediment beaches under random waves and results from a new experimental investigation are presented. Both hydrodynamic data obtained over fixed beds and the resulting profile evolution of fine, coarse and bimodal sediment beaches are discussed. Wave heights and undertow in the inner surf zone are found to be poorly predicted by commonly used numerical solutions. In addition, the undertow appears to be strongly influenced by wave grouping in the nearshore. The behaviour of the fine, coarse and bimodal sediment beaches are compared and contrasted. The fine sand beaches tend to erode in the inner surf and swash zones, with the sediment moving predominantly offshore to form a bar. In contrast, onshore sediment transport dominates over the coarse sand beaches, resulting in the formation of a berm above the initial still water level. The bimodal beaches show a similar evolution to the fine sand beaches. However, considerable sediment sorting occurs, with the swash zone largely denuded of fines and the coarser sediment deposited between the still water line and the bar. The data suggests that the stability of the coarse material is significantly reduced by the presence of fines, with little evidence of armouring effects under high incident energy conditions.

1) INTRODUCTION

The hydrodynamics within the nearshore region and the subsequent evolution of beaches under wave attack are important elements governing the stability of the coastal zone. However, numerical modelling of these processes has yet to result in consistent and realistic predictions of beach behaviour. The present

----

1) Professor.
2) Research Associate.
3) Ph. D. student.
Dept. of Civil Engineering, Imperial College, London, SW7 2BU, UK.

study addresses this issue and considers two key aspects of the problem; the undertow and the sediment particle size. The time mean flow, or undertow, is considered one of the dominant mechanisms in the erosion of beaches (Svendsen, 1984). However, particularly under random waves, significant difficulties remain in the modelling of the undertow. This may in part be due to the difficulties in successfully predicting the wave motion in the inner surf zone (Hamm et al., 1993). In the present study, a comprehensive series of measurements of both the wave heights and near bed horizontal velocity are compared to a deterministic numerical model of nearshore processes (Southgate and Nairn, 1993).

Sediment size has a significant effect on both the magnitude and direction of sediment transport under wave action but the mechanics of the process are still poorly understood. Recent work by Work and Dean (1991) considered the effects of varying grain size on equilibrium beach profiles and showed that steeper profiles were found on beaches with larger grain sizes. However, most beaches exhibit a range of grain sizes, with frequent cross-shore sorting of sediment sizes. Moutzouris (1991) found that the coarsest sediments on a beach were generally found just seaward of the shoreline, while the finest sediments tended to be deposited on offshore bars. Mechanisms for this process have been considered by a number of authors (see Horn, 1991), but numerical models have so far resulted in little success.

Bimodal sediment transport under unbroken waves over a flat bed was considered by Mansell (1992). This work indicated a transient process, whereby the winnowing of fine grains from the bed surface led to the formation of a coarse armour layer. Using model beaches, Quick and Dyksterhuis (1994) found some evidence of an armouring process at low energy conditions. In contrast, at higher energy conditions the coarser sediment had little effect on the profile evolution and the fines controlled the beach steepness. The beach permeability therefore appeared to have a controlling influence on the beach behaviour. However, previous experimental work has not considered the sediment sorting process on beaches and, in particular, the cross-shore distribution of sediment sizes. The present study considers this process and identifies key features that appear consistent with field observations.

2) HYDRODYNAMICS

The experiments were carried out in a large wave flume in the Civil Engineering Department at Imperial College. This flume is 60m long, 3m wide and was used with a working depth of 0.9m. Waves are generated by a hydraulically driven bottom hinged paddle with the facility to absorb wave components reflected from the far end of the flume. The end of the flume is subdivided into three sections to minimise cross-tank motions and consists of a composite beach, with an initial slope of approximately 1/20, rising to 1/10 at the shoreline. Jonswap spectra with three different significant wave heights and wave periods were used (table 1). Data was collected over 330s at 25Hz, giving 8192 points in each run. The water surface elevation was measured with standard surface piercing resistance type wave gauges and the wave heights calculated form the zeroth spectral moment ($Hrms = \sqrt{(8m_0}$). The horizontal velocity 6mm above the fixed bed was measured using LDA and an acoustic Doppler velocitymeter (ADV). The ADV was less sensitive than the LDA to noise from air bubbles generated by breaking waves, and allowed velocity data to be obtained in very shallow water.

Spectrum	*Hrms* (mm)	f_p (Hz)	T_z (s)
L	90	0.68	1.47
P	70	0.68	1.47
K	45	1.0	1.0

Table 1. Random wave spectral characteristics.

Wave heights

Figures 1a&b shows the *Hrms* wave height across the surf zone for cases P and K respectively. In the outer surf zone, the numerical solution (based on the Battjes and Janssen (1978) approach) gives good results. However, in the inner surf zone the variance and, consequently the wave height, does not decrease to zero at the shoreline. Indeed, the wave height only reduces to about half that of the initial offshore wave height. This is because the steep beach slope allows waves to travel close inshore before breaking and forming bores. The set-up just shoreward of the initial shoreline position is found to be small (of order 5mm) and this is therefore not the reason for the observed increase in wave height in the inner surf zone. It is interesting to note that calculations of the wave height based on the spectral approach and a zero crossing analysis give very similar results in the inner surf zone. This is despite the fact that the wave heights frequently exceed the water depth.

The measured spectra for case P are shown on figure 2, together with the smoothing interval and 95% confidence limits. Offshore of the breaker zone there is little energy in either the lower or higher harmonics, while just seaward of the initial shoreline the energy in the lower harmonics increases significantly. This is expected due to the generation of non-linearities by the shoaling process and the amplification of low frequency waves in shallow water. There is also a small frequency downshift of the spectral peak. However, the spectrum close to the shoreline is still dominated by energy close to the initial spectral peak frequency. The inner surf zone spectra are therefore significantly different from those observed on low slope beaches, where low frequency energy often dominates (e.g.

Holland et al., 1995). The present data are consistent with both the measured wave heights and the formation of bores with a frequency close to f_p.

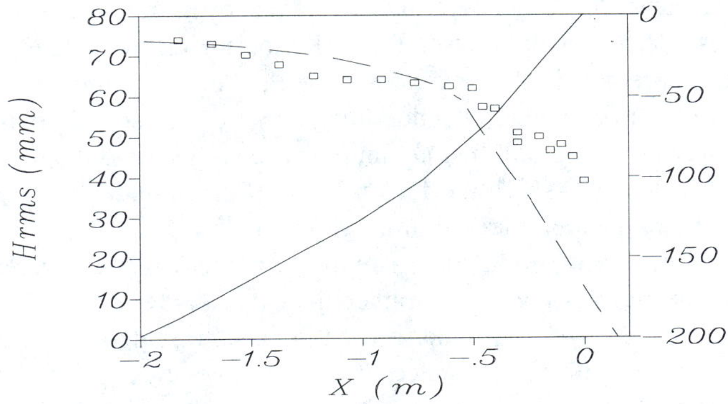

Figure 1a. RMS wave height across the surf zone, case P.
——— Depth , □ □ □ □ Data points, — — — Numerical solution.

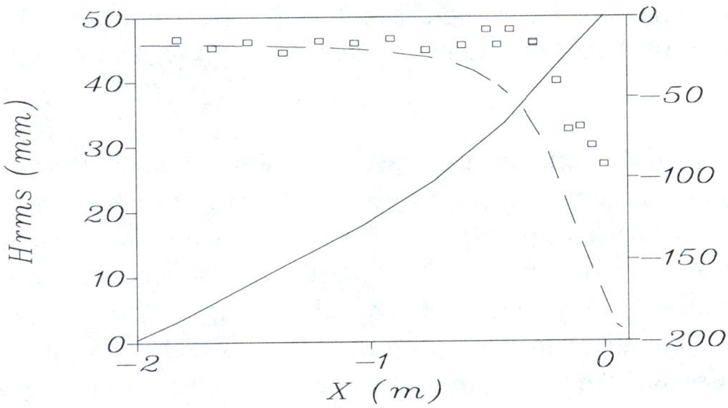

Figure 1b. RMS wave height across the surf zone, case K.
——— Depth , □ □ □ □ Data points, — — — Numerical solution.

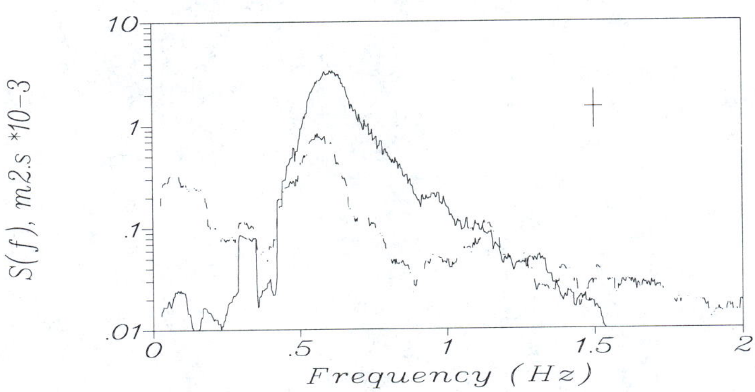

Figure 2. Wave spectra for case P outside the surf zone and close to the shoreline.
——— x=-2m (d=200mm), - - - - x=-0.05m (d=10mm).

Undertow

The time-mean flow, or undertow, averaged over the total run time is shown in figures 3a&b. For both cases P and K, the undertow in the outer surf zone is very well predicted. However, for case P, the maximum value of the undertow is over-estimated but the undertow close to the shoreline is under-predicted. The over-estimation of the maximum is probably due to the difficulty in determining the depth over which the forward mass flux should be averaged and the less well defined breakpoint (see figure 1a). In shallow water, and particularly with a significant degree of turbulence, it may be more realistic to average the mass flux over the mean water depth, rather than just below trough level as is usual (e.g. Svendsen, 1984). The under-estimation of the undertow in the inner surf zone is consistent with the presence of larger waves than predicted by the numerical solution. A similar effect may be observed for case K (figure 3b), where, in addition, the principal breakpoint is also closer to the shore than predicted (figure 1b). This is likely to increase both the maximum value of the undertow and the undertow close to the shoreline. It therefore appears that a more realistic model for the nearshore wave heights is required for beach slopes of this steepness.

Recent field measurements have suggested that suspended sand concentrations are strongly correlated with the occurrence of wave groups (e.g. Hay and Bowen, 1994). This prompted a closer examination of the velocity field beneath the random waves used in the present investigation. The measured Eulerian horizontal velocity, averaged over either one or three wave periods, shows significant variations in the "local mean" flow, which may be an order of magnitude greater than the long term mean. Figure 4 shows the velocity averaged over three wave periods, the wave height averaged over the same time period and the long term mean velocity.

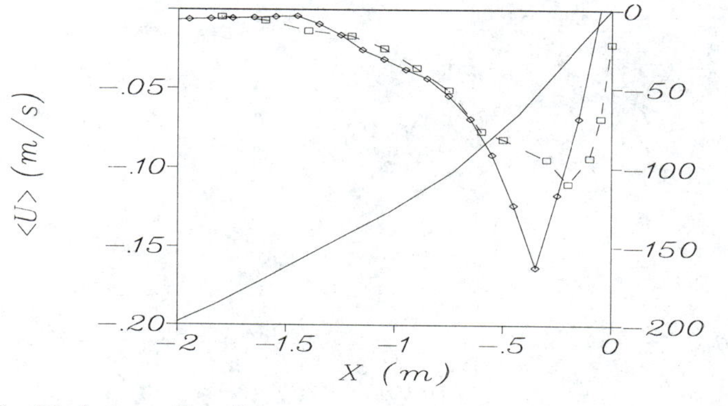

Figure 3a. Undertow. Case P.
——— Depth, – □ – data points, —◇— Numerical solution.

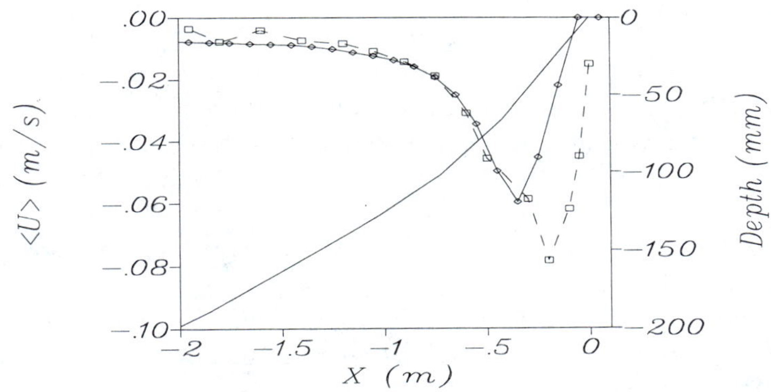

Figure 3b. Undertow. Case K.
——— Depth, – □ – data points, —◊— Numerical solution.

The variation in the "local mean" flow is strongly correlated with the occurrence of incident wave groups, which suggests that the groupiness of the wave field may be an important feature in suspended sediment transport. Note that the sign of the mean velocity has been inverted to show the correlation with wave height more clearly. The data show that large onshore mean velocities generally occur during the passage of smaller waves, while offshore mean velocities are found under the largest waves. If these variations in the mean flow are combined with coherent fluctuations in the concentration of suspended sediment, then a much large volume of sediment may be moved in suspension than a volume based on the product of the mean velocity and mean sediment concentration.

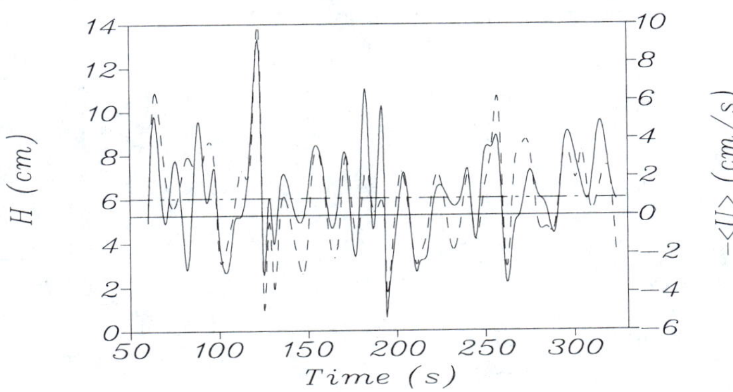

Figure 4. Temporal variations in the "local mean" flow. Case P.
——— Wave height, – – – – "local mean" flow (inverted),
— ·· — long term mean, <U> (inverted).

3) Beach evolution

Beach evolution experiments were carried out under random waves for a fine sand beach, a coarse sand beach and a bimodal sediment beach. In order to facilitate the experimental study, the sediment sizes needed to satisfy three criteria:

1. Both sediment sizes should be mobilised by the maximum velocity in the inner part of the surf zone.
2. The ratio of the two sizes should be large enough to result in significantly different profile evolution as well as allowing them to be readily separated from the bimodal mixture.
3. Each sediment should be well-sorted so as to produce the bimodal characteristics in the resulting grain size distribution of the mixture.

Using these criteria, the sediment size for the coarse and fine sand respectively were chosen to be D_c= 1.5 mm and D_f= 0.5 mm. The grading curves for the two sands and a 50:50 mix are shown in figure 5. Sediment beds with a thickness of 100 mm were laid over the existing solid beach and the water level in the flume raised by the same amount, resulting in the same initial beach profile as used while collecting the hydrodynamic data. The profile evolution experiments were conducted for a total duration of 240 minutes, using repeated 30 minute runs of the spectra shown in table 1. The flume was allowed to settle after each 30 minute run and there was no evidence of the build up of significant seiching due to wave reflections. The experiments were carried out on the coarse and fine grained beaches separately to serve as controls and to provide comparison with the evolution of the bimodal beach.

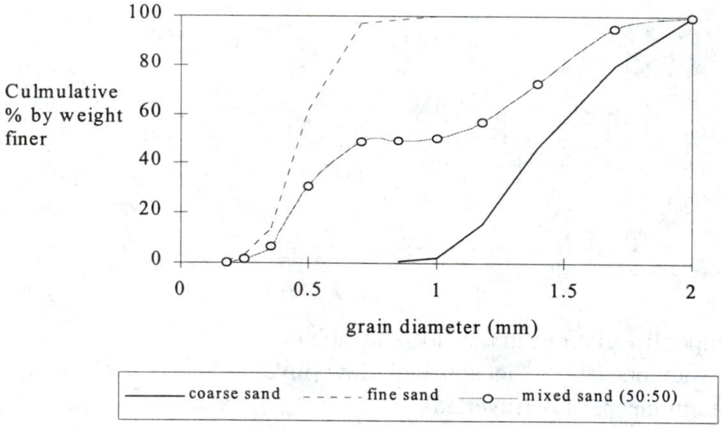

Figure 5. Grading curves.

Profile evolution with uniform sediment size

Several general observations apply to the beach profile evolution for each of the three spectra. Firstly, the coarse sand beaches tend to form a distinct berm shorewards of the initial SWL. The berm appears steepest for case P, which has the lowest deep water wave steepness. For the fine sand beaches, the formation a bar flattens the beach slope in the surf zone. Seaward migration of the bar was evident and the bar moves further offshore as the wave height increases. The SWL recedes shoreward in all cases, except for the coarse beach using case P. Examples of the cross-shore changes in bed level are shown on figures 6&7. A predominant onshore transport is observed in the case of coarse sand while the fine sand bed suffered much erosion around the shoreline, with sediment transported offshore to form a bar.

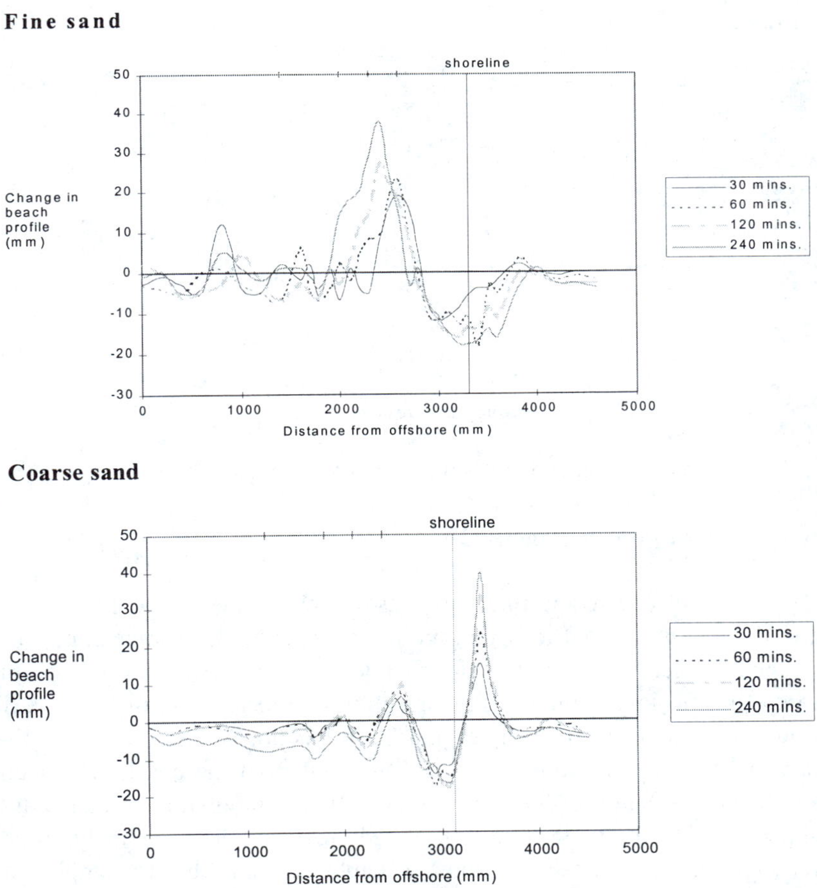

Figure 6. Change in bed elevation from original profile, case K.

Fine sand

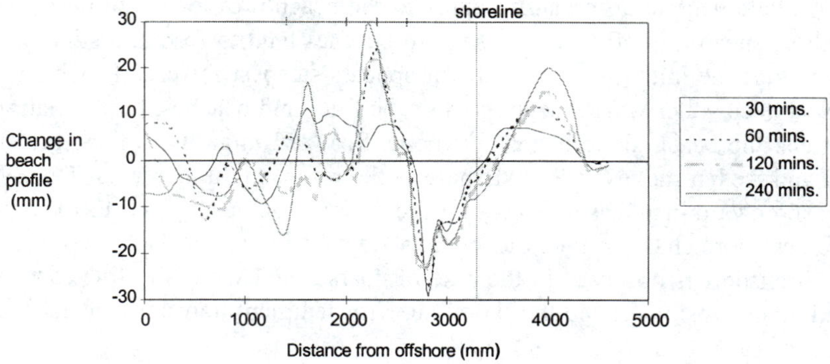

Coarse sand

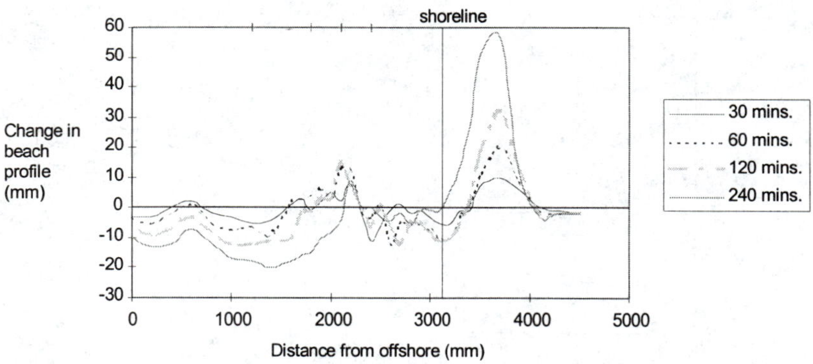

Figure 7. Change in bed elevation from original profile, case P.

Bimodal sediment sampling methods

A 50:50 mix of coarse and fine sand was chosen for the bimodal beach. Two sampling devices were used to determine the changes in the composition of the bimodal beach after wave action. The first was a 25mm square spot sampler, which was used to take sediment samples along the centre line of the bed. The second was an in-situ section sampler, placed in position before laying the beach and used to lift whole sections of the bed material from the beach. The section sampler allowed a section of the bed to removed with minimal disturbance of the sample and enabled the vertical section of sediment sample to be analysed layer by layer (with layers as small as 5 mm). A calibrated fall tube was employed to sort the sampled material, allowing the change with depth in the mix ratio of the beach material to be ascertained. Further details of the sampling method will be available shortly in the Ph.D. thesis of R. T. C. Chan. In order to establish the variability of initial bed composition a trial experiment was performed to sample the bimodal beach after it was laid. The sampling procedure included taking a section of the bed out (using the section sampler) at six different locations on the beach and 10 layers, each of 5 mm thick, were analysed at each location. The mean in-situ initial mix ratio was found to be 0.515 (coarse/total), with variations of less than 1% by volume between the different locations.

Profile evolution with bimodal sediment

Comparisons of the final beach profiles (after 240 minutes of wave action) for the fine, coarse and bimodal beaches are shown on figures 8a&9a for cases K and P respectively. The original profile in each case is also shown as a reference. The numbered circles at the top of each graph indicate the locations of the section samplers. Figures 8b&9b show the change in bed level from the original profile, where the positions of the section samplers are shown by the numbered triangles. The variation of the mix ratio (coarse/total) with depth at different sections of the beach is shown on figures 8c&9c. The experimental data show several features of particular interest.

In the inner surf zone and swash zone, the final profiles for the bimodal beach resemble the fine sand profiles much more closely than those for the coarse sand. This is the case for all three wave spectra and is confirmed by the bed level changes (figures 8b&9b). The bar positions on the bimodal beach are shoreward of those found in the fine sand beach. Recalling that a seaward migration of the bar was observed on the fine sand beach, it is apparent that there is a tendency for the coarse sand to curb the bar from migrating offshore. On the bimodal beach, the surface of the shoreward slope of the bar is dominated by coarse sediment (refer to section 4a on figures 8c and sections 4,5 and 6 on figure 9c). This is in marked contrast to the higher proportion of fines found on the seaward slope of the bar (section 4b on figure 8c, section 7 on figure 9c). Finally, the foreshore (just above the SWL in the lower swash zone) of the bimodal beach tends to be denuded of coarse sand (sections 2 and 3 on figures 8c&9c).

In summary, the results show that, although the bimodal sediment beach behaves in a similar fashion to the fine sand beach, in agreement with Quick and Dyksterhuis (1994), considerable sediment sorting has also been observed. The present data show that the region between the shoreline and the bar tends to be dominated by the coarse sediment. This appears consistent with the field data of Moutzouris (1990), which showed that the largest grain size often occurred just seaward of the shoreline. Within the swash zone, the fine sand within the mixture seems to destabilise the coarser sand. This facilitates the erosion of coarse sand on the foreshore, and is most conspicuous at the position of the steepest slope. A similar process is also evident on the seaward slope of the bar. This suggests that the mobility of the coarse sand in the mixture is substantially increased due to a greater exposure and a reduction in friction in the presence of fines. This appears consistent with the results of Miller & Byrne (1966), who found that there is a reduction of the angle of repose when large grains sit on a bed of smaller grains.

The difference in the evolution of the fine and coarse sand beds may be due to a combination of factors. The most dominant are probably a different threshold condition and a difference in the permeability of the beach material. A higher threshold condition for the coarse sand will tend to result in a greater net shoreward transport of sediment. The high permeability of the coarse sediment is also likely to lead to greater infiltration within the swash zone, reducing the backwash and leading to or increasing the asymmetry in the swash zone velocities. This hypothesis is consistent with the much larger berm formation observed on the coarse beaches. In contrast, the bimodal beach permeability will be largely controlled by the finest 10% of sediment within the mix (e.g. Hazen, 1892). Consequently, both the fine sand beaches and the bimodal beaches have a similar permeability. Since both beaches evolve similarly, the permeability appears to be the dominant controlling factor, as suggested by Quick and Dyksterhuis (1994). This has important implications for both numerical modelling of swash processes and beach recharge schemes. For example, if the fines control the beach permeability, and as a consequence the beach behaviour, the addition of coarser supposedly more stable sediment may have little effect on the overall stability of the beach. On the other hand, if the permeability is a controlling factor, beach drainage should be an effective method of beach stabilisation, although some questions remain as to its effectiveness.

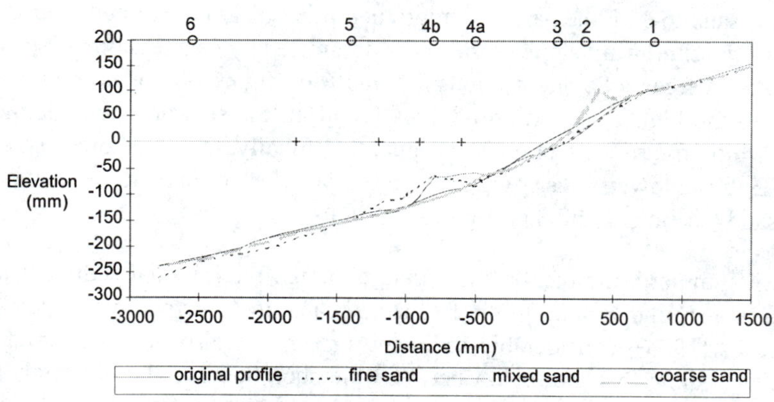

Figure 8a. Profile evolution after 240 minutes, case K.

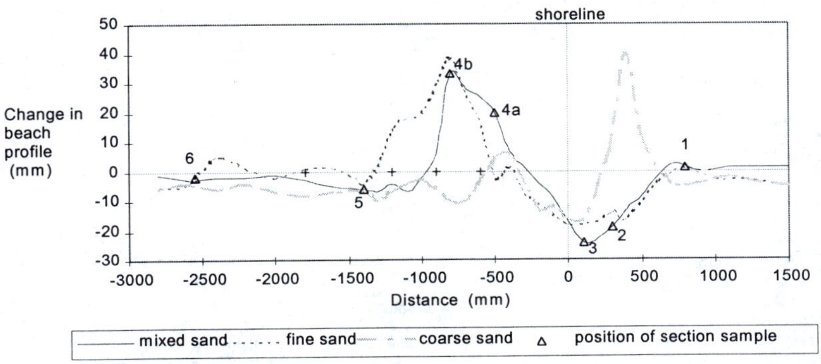

Figure 8b. Change in bed elevation from original profile, case K.

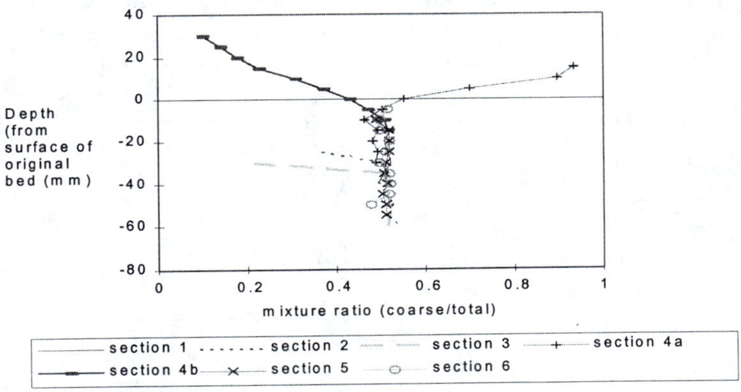

Figure 8c. Variation in mix ratio with depth at six cross-shore locations, case K.

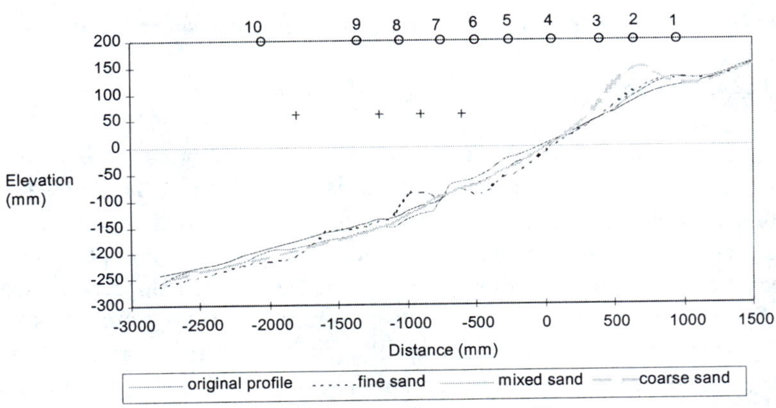

Figure 9a. Profile evolution after 240 minutes, case P

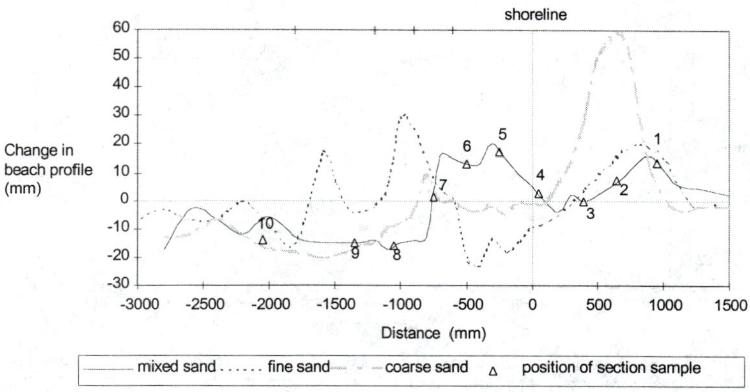

Figure 9b. Change in bed elevation from original profile, case P.

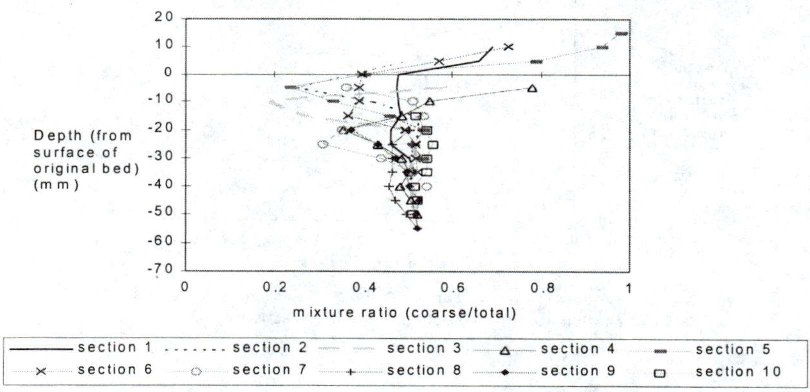

Figure 9c. Variation in mix ratio with depth at six cross-shore locations.

CONCLUSIONS

The results of a new experimental investigation have been presented. Hydrodynamic data obtained over a steep fixed beach show that the wave heights and undertow in the inner surf zone are under-estimated by the numerical solution. In particular, the wave heights in the inner surf zone are not depth limited to the expected degree. This is a consequence of the steep beach slope, with insufficient time for wave breaking fully to establish. Significant wave grouping is evident in the nearshore. This leads to large local fluctuations in the "mean flow", which are strongly correlated with the wave groups. The profile evolutions of fine and coarse sediment beaches were found to be very different, with the fine sand beaches tending to erode to form a bar, while the coarser beaches generally accreted, resulting in the formation of a berm. The profile evolution of bimodal beaches was found to be similar to that of fine sand beaches, although considerable sorting was observed. The data suggest that the permeability of beaches is of particular importance, in agreement with some previous studies. The presence of fines appeared to destabilise the coarser material within the bimodal beach, with little coarse material present on the bed surface in the swash zone and on the bar crest. In contrast, the coarse material appeared to have little effect on the movement of the fine sediment, with little evidence of armouring. Clearly there is a need further to investigate the influence of the fine/coarse ratio in bimodal sediments and to consider continuously-graded sediments.

REFERENCES

Battjes, J. A. and Janssen, J. P. F. M., 1978. Energy loss and set-up due to breaking of random waves. *Proc. 16th Int. Conf. Coastal Eng.*, ASCE, **1**, 569-587.

Hamm, L., Madsen, P. A. and Peregrine, D. H., 1993. Wave transformation in the nearshore: a review. *Coastal Engineering*, **21**, 5-39.

Hay, A. E. and Bowen, A. J., 1994. Coherence scales of wave induced suspended sand concentration fluctuations. *J. Geo. Phys. Res.*, **99**, C6, 12749-765.

Hazen, A., 1892. Discussion of : Dams in sand foundations, by A. C. Loenig. *Transactions ASCE*, 73:199.

Holland, K. T., Raubenheimer, B., Guza, R.T. and Holman, R. A., 1995. Run-up kinematics on a natural beach. *J. Geophys. Res.*, **100**, (C3), 4985-4993.

Horn, D. P, 1991. Computer simulation of shore-normal variations in sediment size. *Proc. Coastal Sediments' 91*. ASCE, 874-889.

Mansell, P., 1992. Transport of bimodal sediment, MSc. Dissertation, Department of Civil Engineering, Imperial College. London. pp 62.

Miller, R. L. and Byrne, R. L., 1966. The angle of repose for a single grain on a fixed rough bed. *Sedimentology*, **6**, 303-314.

Moutzouris, C. I., 1991. Beach profiles vs. cross-shore distributions of sediment grain sizes. *Proc. Coastal Sediments' 91,* ASCE, 860-874.

Quick, M. C. and Dyksterhuis, P., 1994. Cross-shore transport for beach of mixed sand and gravel. *Int. Symp.: Waves - physical and numerical modelling*, IAHR, Vancouver, 1443-1452.

Southgate, H. N. and Nairn, R. B., 1993. Deterministic profile modelling of nearshore processes. Part 1. *Coastal Engineering*, **19**, 27-56.

Svendsen, I. A., 1984. Mass flux and undertow in a surf zone. *Coastal Engineering*, **8**, 347-65.

Work, P. A. and Dean, R. G., 1991. Effect of varying sediment size on equilibrium beach profiles. *Proc. Coastal Sediments' 91*, ASCE, 890-904.

ACKNOWLEDGEMENTS

The authors gratefully acknowledge the support of the UK Engineering and Physical Sciences Research Council.

CHAPTER 234

BEACH PROFILE SURVEYS ALONG THE U.S. PACIFIC COAST 1945 – 1947

Nicholas C. Kraus[1], M. ASCE, Robert L. Wiegel[2], Hon. M. ASCE, Willard N. Bascom[3]

ABSTRACT: The Wave Project at the University of California at Berkeley was established in 1944 to develop a relationship between nearshore waves and the underlying topography for supporting amphibious military landings. The project was continued for scientific purposes until 1952 under the more general name of "Wave Observation and Beach Surveys," including the Amphibious Oceanography Project. These early field observations are a central part of the genesis of coastal engineering in the United States. Beach profiles were surveyed by stadia and by amphibious vehicle along the coasts of Washington, Oregon, and California, and the results were documented in University of California, Fluid Mechanics Laboratory, reports in the HE-116 series. The profile survey measurements, now 50 years old, are still the only such data available for many beaches along the U.S. Pacific Coast. The data set thus comprises a valuable baseline for documenting change in the coast, and it is a resource on the morphology and grain size for high-energy beaches that are extremely difficult to survey. The profile survey plots in the limited-circulation reports have been digitized and are available for general access in the present study. This paper reviews the early measurement program and the available data set.

INTRODUCTION

The Wave Project of the University of California at Berkeley, sponsored by the Navy Department, was established in 1944 to develop a relationship between nearshore waves and the underlying topography for aiding in WWII beach landings. The project was continued for scientific purposes until 1952 under the more general name of "Wave Observation and Beach Surveys (WOBS)," including the Amphibious Oceanography Project. These early field observations are the genesis of coastal engineering in the United States, and a typical objective was "... to relate wave characteristics to beach changes and to collect sufficient data on surf, beach, and meteorological conditions to make a comparison with the results of

1) Research Physical Scientist, U.S. Army Engineer Waterways Experiment Station, Coastal and Hydraulics Laboratory, 3909 Halls Ferry Road, Vicksburg, MS 39180.
2) Professor Emeritus, Department of Civil Engineering, University of California, Berkeley, 412 O'Brien Hall, Berkeley, CA 94720.
3) Scientist-Explorer (geophysics and underwater archeology), 5137 Vista Hermosa, Long Beach, CA 90815.

existing forecasting procedures and general wave theory" (Stump and Bascom 1947). Beach profiles were surveyed by stadia and by amphibious vehicle (Dukw – pronounced "duck") along the coasts of Washington, Oregon, and California, supplemented by measurement of numerous types of coastal processes. The Wave Project and WOBS are believed to have made the first systematic and comprehensive field measurements of the beach profile, sediment, waves, and water level. The results were documented in the HE-116 series of Laboratory Memorandums of the Fluid Mechanics Laboratory, Department of Civil Engineering, U. of California, Berkeley. In the early years, this pioneering program was planned, financed, and directed by Dean Morrough P. O'Brien (U. of California, Berkeley) and by John Isaacs (U. of California, San Diego). Isaacs (1947) made all the surveys in 1945, when he invented the system, and Willard Bascom made virtually all the rest, many at the risk of life. The profile survey measurements have been described authoritatively by Bascom (1964, 1980). The principals responsible for the work were M. P. O'Brien, J. D. Isaacs, W. N. Bascom, D. McAdam, D. Patrick, and R. L. Wiegel.

Surprisingly, the profile survey measurements, now 50 years old, are still the only data available for many beaches along North Pacific coast. In fact, Dean O'Brien had insisted that the northern beaches be surveyed in the winter when there was violent surf (see Bascom 1987 for further discussion), and it remains a challenge to repeat such measurements. The data set thus comprises a valuable baseline for documenting change in the beach that has occurred since that time, and it also serves as a rare resource on the morphology of high-energy beaches that are extremely difficult to survey. The data exist in limited circulation HE-116 reports as distance-elevation plots and have been little accessed, one exception being an article by Komar (1978), who discussed data from HE-116-229 (Isaacs 1947) and HE-116-247 (Bascom and McAdam 1947) for the Washington and Oregon coasts.

The objective of the present work was to capture and preserve the profile data taken on the U.S. West Coast over the period 1945-1952. However, only data for the period 1945-1947 could be recovered in the present effort. For our study, the available profile survey data were digitized manually from the original drawings and the information transferred to magnetic media. This paper presents an overview of the data set, including the institutions and personalities behind the WOBS projects. The paper was prepared as a contribution for the celebratory theme "Coastal Engineering Heritage" (Kraus 1996) of the Twenty-fifth International Coastal Engineering Conference. Further information about the history of coastal engineering in the United States can be found in Wiegel and Saville, Jr. (1996).

RESOURCES CONSULTED

For this study, the originals of ten HE-116 reports were borrowed from the Water Resources Center Library, Department of Civil Engineering, University of California, Berkeley. Of these, five reports contained beach profile survey data. Table A1 of the appendix lists by report the beaches, years, and ranges for which profile data are available from these reports. The data set developed in the present study contains 216 profile surveys from nine beaches in Washington, six in Oregon, and 16 in California.

Many of the reports contain black and white photographs, most taken from the ground, but with some from the air by military planes. HE-116-223 (Fluid Mechanics Laboratory 1946) includes pictures and captions with the location and date for "...various beaches which were not at the time subjects of intensive study." In order to preserve the photographic record of west-coast beaches in the late 1940s and early 1950s, for the present project more than 100 photographs compiled from the different reports were scanned to create digital files.

After making three sets of copies of the reports, including the oversize plots of the profile plots, the originals were returned to the Water Resources Center Library. A complete set was sent to the Coastal Engineering Archive at the University of Florida, Gainesville, Florida; the University Library at Oregon State University, Corvallis, Oregon; and the Technical Library at the U.S. Army Engineer Waterways Experiment Station, Vicksburg, Mississippi.

Many of the profile surveys extended from the upper beach to a depth of approximately 10 m, although surveys to wading depth were also made. Bascom (1980) gives a vivid first-hand account of the surveys made by amphibious vehicle (Dukw, Fig. 1). Horizontal position was measured by triangulation, and measurement points were marked through radio communication when a lead-line was heaved overboard. The sum of the water depth measured beneath the trough of a wave and one-third of the estimated wave height were reported by radio to a recorder on shore as an estimate of total water depth. The depths were converted to mean lower low water tidal datum by reference to either a tide gauge or predictions. Komar (1978) has summarized the measurement procedures and cites primary documents and reports not discussed here.

Other measurements or observations routinely made by the field parties included visual "surf observations" of the wave height and period, breaking wave type, and the nearshore current – both the longshore and rip currents. Measurements were sometimes made of the water table. Sediment samples were taken, but the authors were not able to locate the substantial information that should have been compiled. Other interested parties should continue in that search. An interesting aspect was documentation of the depth of tire impressions in the beach sediment made by the field-reconnaissance vehicles. Both military applications and relation of compaction to beach change were of interest, as was operation of the Dukws in future data collection.

In remainder of this paper, we focus on the beach profile survey data.

DIGITIZATION PROCEDURE

The original reports were unbound to mount the individual plots of the beach profiles on a digitization table. At first, it was planned to trace the profile lines by cross hair sighting on a high-resolution electronic digitization tablet. However, most of the pages containing the plots had become warped, making standard digitization mapping on to electronic axes inaccurate.

As an alternative to rubber sheeting through software, which would have introduced or distributed errors, distance-elevation pairs on the plotted profiles were located manually by placing a straight edge parallel to the grid lines on the plots and reading corresponding intersection values on the two axes. The locations of actual survey points on the plots were almost always unambiguously evident by discontinuities in the lines joining the points. Because of the presence of the grid lines on the plots, errors that might have been introduced by warping of the paper are believed to be minimal.

After manual digitization of the profile survey data, the distance-elevation pairs were entered into the computer and plotted to the same scale as the original drawings. Plots of the digitized data drawn on transparent paper were laid on top of the original drawings and visually compared for discrepancies, which were corrected. The resultant quality-controlled data set is believed to be an accurate representation of the original measurements. These data are available as ASCII files. The data are in the original American customary units (feet), but in this paper results are presented in metric units for the international audience.

12845
0+00 at Halfmoon Bay Municipal Beach
April 29, 1947 - 0930
Halfmoon Bay, California

12846
0+00 at Halfmoon Bay Municipal Beach
April 29, 1947 - 0835
Halfmoon Bay, California

Fig. 1. Survey preparations (from Wiegel 1947).

EXAMPLE RESULTS

In this section we give examples from Washington, Oregon, and California, from the data set compiled. Locations of many of the sites encompassed by the total data set are shown in Fig. 2. Appendix A gives the locations of all the beaches for which survey data were compiled.

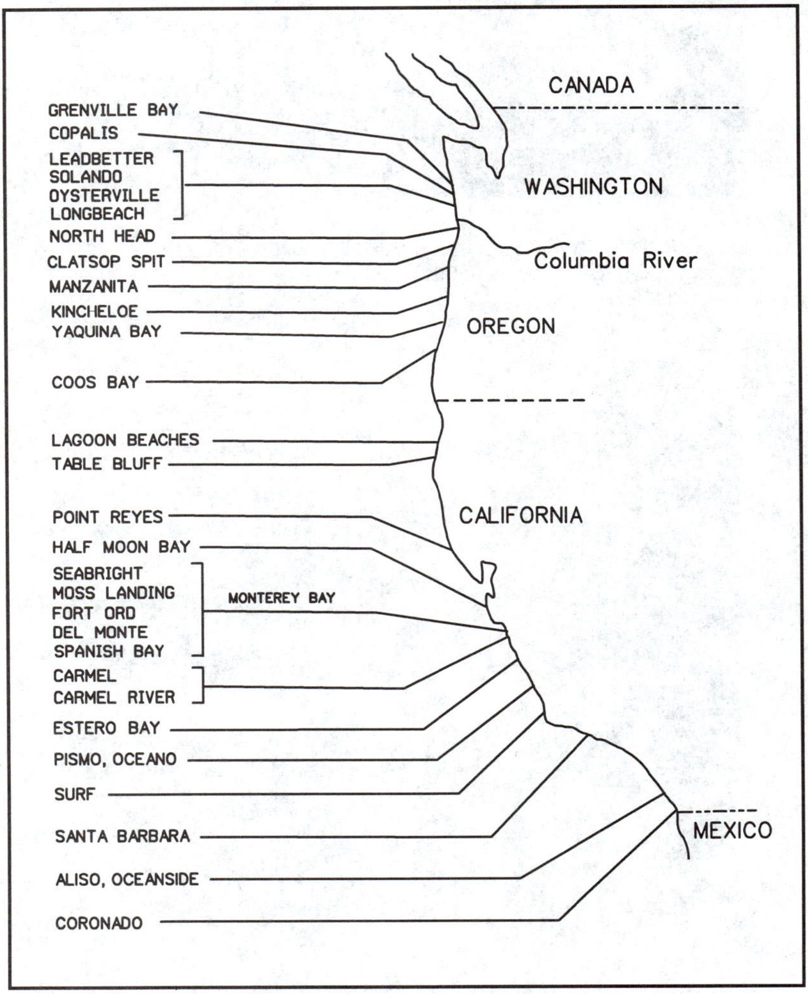

Fig. 2. Location of Pacific Coast beaches investigated (after Bascom 1951a).

Comparison to Echo Sounder

Depth soundings were typically made by lead line as described above, because echo sounders blanked out in the surf zone where air bubbles are present. Patrick and Bascom (1950) discuss technical details of the lead-line method and echo sounder measurement methods as applied in the Wave Project and subsequent studies. Many of the considerations are still valid today for making beach-profile surveys. As an example, we quote "After the ground control is established, the remaining field work is accomplished in two parts: water soundings during high tide and beach land profiles during low tide. The method gives *an overlap* (emphasis added here) of the data obtained and provides a check on the accuracy of the work." Lack of appreciation for the necessity of overlap of land and water surveys is found in some works today, which introduces ambiguity in collection of expensive data.

Patrick and Bascom (1950) compare the two methods "...made at Carmel (California, in 1947) and the results for one range line...," which is reproduced in Fig. 3 here. Although the lead-line method did not pick up detail of the bars on the profile, the comparison indicates remarkable skill in the lead-line survey that was consistent to almost 10-m depth, giving confidence in the reliability of the overall data set. The maximum difference in readings between the two methods is on the order of 1 m beyond the 6-m contour (MLLW).

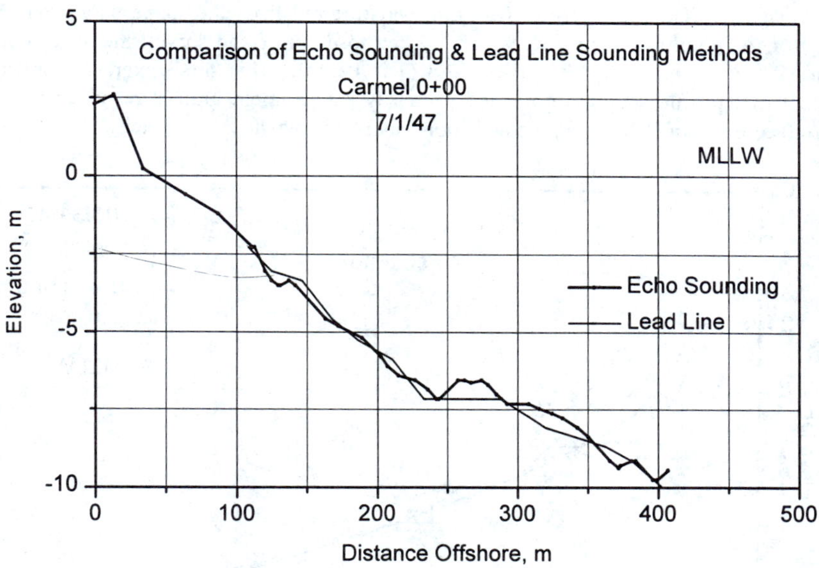

Fig. 3. Comparison of sounding methods (adapted from Patrick and Bascom 1950).

Point Grenville, Washington

This beach is located north of the mouth of the Columbia River and north of Grays Harbor, Washington. The two profiles plotted in Fig. 4 were surveyed approximately 1 year apart, with the 1945 survey made by a party led by Isaacs (1947) and the 1946 survey made by a party led by Bascom and McAdam (1947). Quoting from Isaacs: "(Between August 27 and September 27, 1945) …the field party of the University of California Wave Investigation Project made surveys of the beach and surf characteristics at a number of beaches of the Pacific Northwest. These beaches are noted for their gentle slopes and fine grained materials." Further in the report, it is stated for the Point Grenville, South Station: "About 2 miles (3.2 km) south of the Cape (Cape Grenville) a section of the beach existed where the beach face was comparatively steep. No berm existed and it was apparent that high water reached the low clay banks in this region. The profile there showed a slope of beach face of about 1:20. The breaker zone was quite irregular but the bottom offshore was smooth and of gentle slope." This description fits Fig. 4 well. Wave breaker heights tabulated by Bascom and McAdam (1947) at the approximate time of the 1946 survey were approximately 1.5, 2, and 3 m for the inner, middle, and outer bars, respectively.

Grain-size information is not given for Point Grenville in the two aforementioned reports. However, Bascom (1951a) indicates that a representative median grain size might be on the order of 0.17 mm. As an example of a possible use of the profile data, we fit an equilibrium profile following a "distance to the 2/3 power law" (Dean 1991) to the 1945 profile survey. The grain size of 0.14 mm was inferred from the shape parameter (Moore 1982) of the equilibrium profile, which is seen to describe the survey measurements well from MLLW shoreline to the survey limit (7-m depth). This small exercise confirms the concept that profile shape is governed primarily by sediment grain size on an open-coast beach (see Bascom (1951a) for a conclusions about a wave-sheltered beach).

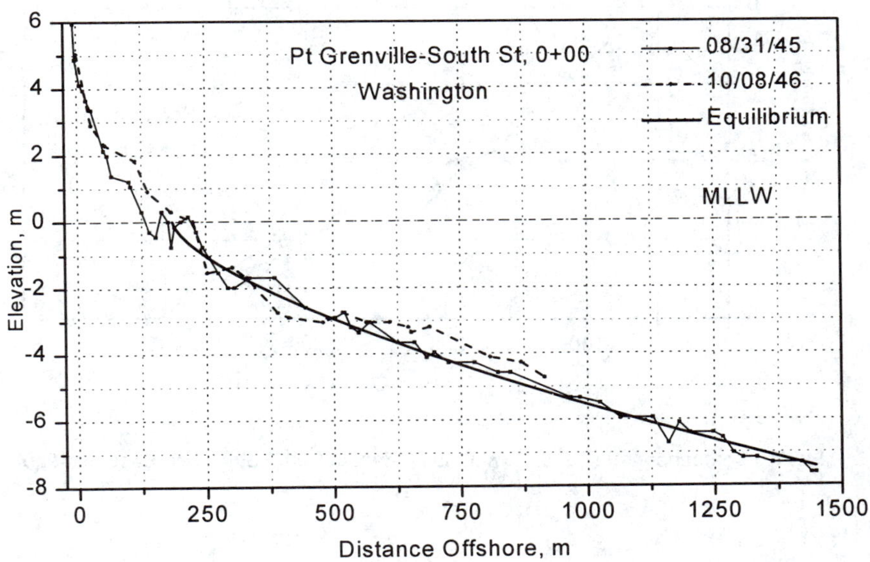

Fig. 4. Profile surveys approximately 1 year apart, Pt. Grenville, South Station, Washington.

Clatsop Spit Beach, Oregon

This beach is located south of the Columbia River, Oregon. The two profile surveys shown in Fig. 5 were taken only 3 days apart and within approximately 300 m of each other. A substantial bar-and-trough topography existed on the profiles. Komar (1978) speculated on mechanisms that might be responsible for the substantial trough. Clatsop Spit is presently experiencing moderate erosion, and data such as shown below will be useful for understanding the coastal processes acting at the site.

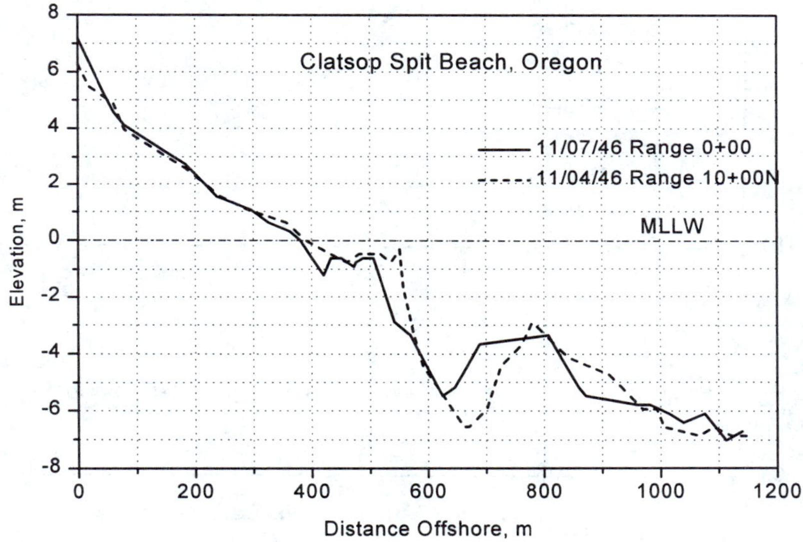

Fig. 5. Profile surveys on two range lines, Clatsop Spit Beach, Oregon.

Carmel Beach, California

The beach at Carmel, California, was studied intensively while the surveying parties "...would rest in the spring to recover from winter encounters with the northern surf" (Bascom 1980). Fig. 73 in Bascom (1980) illustrates seasonal shifts in the berm at Carmel, which were measured to be more than 60 m. Figs. 6 and 7, taken from data given by McAdam and Bascom (1947), reproduce some of the profiles in Bascom (1980) and display the remarkable seasonal changes at Carmel Beach. Bascom (1951a) indicates that a representative grain size for Carmel Beach is about 0.35 mm. This beach was one of those analyzed by Bascom (1951a) in a seminal paper on the relation between beach slope, grain size, and exposure to waves. In Fig. 6, it is interesting to note that little profile change occurs below approximately 4-m depth (MLLW) as compared to the elevation changes above that depth.

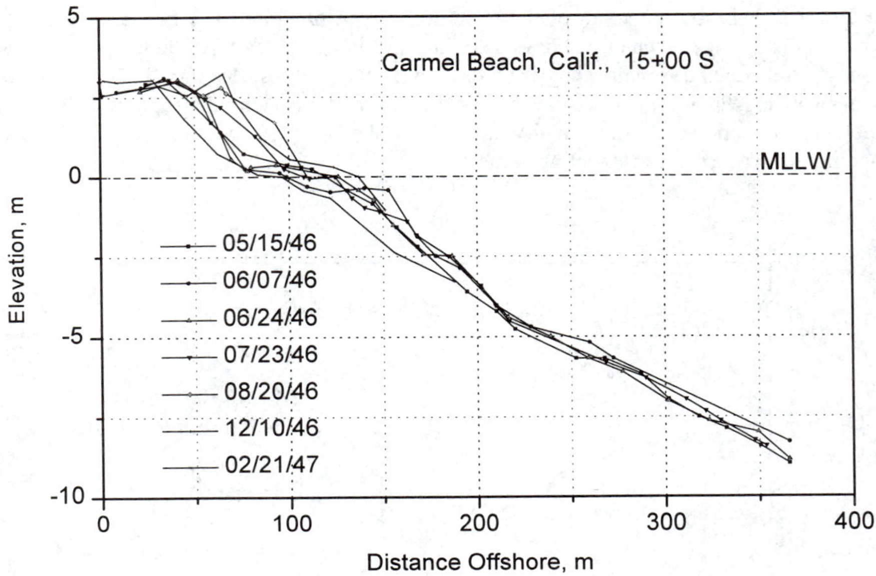

Fig. 6. Seasonal change in the berm and foreshore, Carmel Beach, California.

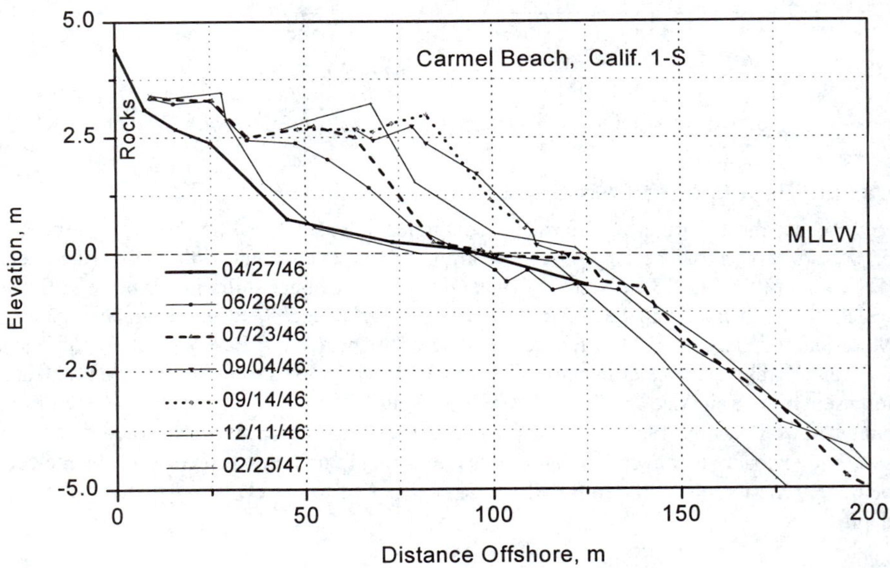

Fig. 7. Close-up view: seasonal change in the berm at Carmel Beach, California.

Miramar Beach, California

Discussion of the survey range lines along Halfmoon Bay, California, including Miramar Beach (Fig. 8) is given by Wiegel (1947). In the present study, it was originally intended to recover the locations of such range lines and reestablish survey monumentation. However, lack of funding prevented us from undertaking this task. Fig. 8 is provided to inform others that the HW-116 reports do contain information with which to reoccupy many of the survey ranges. Although recovery of the vertical datum may be impossible because of loss of benchmarks, location of many of the range lines along the beach will be straightforward because of the excellent photographic documentation.

CONCLUDING DISCUSSION

This study has attempted to capture and preserve, in digital form, beach profile survey data obtained on the Pacific Coast of the United States during the inception of the discipline of coastal engineering, the mid-1940s through mid-1950s. Ten original reports from that era were consulted, of which five contained plots of the survey data, which were digitized and quality checked. Approximately 100 photographs were also scanned to create digital copies. The basic product is a data base consisting of 216 profile surveys for 31 beaches in Washington, Oregon, and California, taken during the years 1945-1947.

The data set is expected to be valuable for coastal engineers, planners, and researchers because of its uniqueness – profile surveys extending to 10-m depth on high-energy beaches of various grain sizes. The data set allows study of fundamental aspects of beach morphology, including summer-winter profile change, time scales of profile change, bar size and movement, and beach slope versus grain size. Early analysis of the data (Bascom 1951a, 1951b) has yielded now-classic results on beach behavior. Many more insights and uses of the data can be expected.

The quest should not end for additional data from the WOBS project. Bascom (1951a) reports that approximately 500 beach profile surveys were made on some 40 beaches, so that the present study has located less than half of the available surveys. In addition, in this unfunded project we were not able to find much of the data for the approximately 600 sediment samples taken (see Bascom 1951a, 1951b for some data). Our goal of recovering all the beach profile survey ranges was not achieved, and we hope that this paper will serve as a stimulus for researchers to locate and preserve the remaining data, and for Federal, state, and local agencies in Washington, Oregon, and California, responsible for the coast to locate and reestablish the range lines.

ACKNOWLEDGEMENTS

This study would not have been possible without the cooperation of the Water Resources Center Library, Department of Civil Engineering, University of California, Berkeley, which permitted long-term loan of the HE-116 original reports. Digitization of the profile survey data was carried out at the Conrad Blucher Institute for Surveying and Science, Texas A&M University-Corpus Christi, while the first author served as Director of the Institute. The digitization and quality-control checks were performed by Ms. Karen Bridges and Ms. Margie Langely, both undergraduate students. The day-to-day work was supervised by Ms. Deidre Williams, graduate student in charge of the Blucher Institute Geographic Information Systems Laboratory. Permission was granted to N. C. Kraus by Headquarters, U.S. Army Corps of Engineers, to publish this information.

Fig. 8. Range lines at Miramar Beach, California (circa April 1947).

REFERENCES

Bascom, W. 1946. Report of Field Studies at Monterey, April to August 1946. Tech. Rep. HE-116-224, Dept. of Engrg., Univ. of Calif., Berkeley.

Bascom, W. N. 1951a. The Relationship Between Sand Size and Beach-Face Slope. *Trans. Am. Geophys. Union* 32(6): 866-874.

Bascom, W. N. 1951b. The Relationship Between Sand Size and Beach Face Slope. Series 14, Issue 2, Dept. of Engrg., Univ. of Calif., Berkeley, Beach Erosion Board Contract W49-055-eng-2, April, 1951.

Bascom, W. 1964. Revised and updated edition, 1980. *Waves and Beaches*. Anchor Books, Anchor Press/Doubleday, Garden City, New York, 367 pp.

Bascom, W. 1987. The Waves Project – An Illustrated Letter to Morrough P. O'Brien. *Shore & Beach* 55(3-4): 25-30.

Bascom, W. and McAdam, D. 1947. Beach and Surf Conditions on the Beaches of the Oregon and Washington Coasts Between October 9, 1946 and November 18, 1946. Laboratory Memorandum HE-116-247, Dept. of Engrg., Univ. of Calif., Berkeley.

Dean, R. G. 1991. Equilibrium Beach Profiles: Characteristics and Applications. *J. Coastal Res.*, 7(1): 53-84.

Fluid Mechanics Laboratory. 1946. Reconnaissance of Miscellaneous Pacific Beaches. Laboratory Memorandum HE-116-223, Dept. of Engrg., Univ. of Calif., Berkeley.

Isaacs, J. D. 1947. Beach and Surf Conditions on Beaches of the Oregon and Washington Coast Between August 27 and September 27, 1945. Laboratory Memorandum HE-116-229, Dept. of Engrg., Univ. of Calif., Berkeley.

Komar, P. D. 1978. Beach Profiles on the Oregon and Washington Coasts Obtained with an Amphibious Dukw. *Shore & Beach* 46(3): 27-33.

Kraus, N. C. (Editor). 1996. *History and Heritage of Coastal Engineering*. ASCE, N.Y., 603 pp.

McAdam, D. and Bascom, W. N. 1947. WOBS Field Party at Monterey, Supplement. Laboratory Memorandum HE-116-224 Supplement, Dept. of Engrg., Univ. of Calif., Berkeley.

Moore, B. 1982. Beach Profile Evolution Response to Changes in Water Level and Wave Height. M.S. Thesis, Dept. of Civil Engrg., Univ. of Delaware, Newark, Del.

Patrick, D. A. and Bascom, W. N. 1950. Methods for Surveying and Plotting Beach Profiles. Tech. Rep. 155-21, Dept. of Engrg., Univ. of Calif., Berkeley.

Stump, R. S. and Bascom, W. N. 1946 (revised 1947). Surf Observations and Photographic Data Obtained by Fieldd Party and Comparison to Hindcasts: Table Bluff Station, Dcember 1945 - January 1946. Laboratory Memorandum HE-116-205, Dept. of Engrg., Univ. of Calif., Berkeley.

Wiegel, R. L. 1947. Beach and Surf Conditions at Halfmoon Bay, April 28, 29, 1947. Tech. Rep. HE-116-256, Dept. of Engrg., Univ. of Calif., Berkeley.

Wiegel, R. L. and Saville, T., Jr. 1996. History of Coastal Engineering in the USA. In (N. C. Kraus, editor) *History and Heritage of Coastal Engineering*, ASCE, N.Y., 513-600.

APPENDIX A: DATA FOUND IN THIS STUDY

Table A1 lists profile data by source and state that was available for digitization in this study.

Table A1. Source and locations of available data

Report No.	Surveyed Beaches	Range	Date
		Washington	
HE-116-229	Grays Harbor to Columbia		
HE-116-229	Copalis to Cape Grenville		
HE-116-229	Longbeach to Leadbetter		
HE-116-229 HE-116-247	Point Grenville	0+00; 5+00S 10+00S	8/27/45, 10/8/46 8/28/45 8/31/45
HE-116-229 HE-116-247	Point Grenville, South St.	0+00	8/31/45, 10/8/46
HE-116-229 HE-116-247	Copalis Beach	0+00, 10+00N, 10+00S	8/28/45, 10/9/46
HE-116-229 HE-116-247	Ocean City	0+00	8/31/45, 10/11/46
HE-116-229 HE-116-247	Leadbetter Point	10+00S 20+00S 0+00 30+00S	8/31/45, 10/16/46 9/7/45, 10/16/46 9/8/45, 10/16/46 9/10/45, 10/16/46
HE-116-229 HE-116-247	Solando Wreck	0+00; 10+00N, 20+00N, 30+00N	10/16/46; 9/10/45, 10/16/46
HE-116-229 HE-116-247	Oysterville Beach	10+00S 0+00; 10+00N	9/14/45, 10/14/46, 10/17/46; 9/14/45, 10/17 - 11/18/46; 9/14/45, 10/14/46, 10/17/46
HE-116-229	Ocean Park	10+00N; 0+00; 10+00S	9/12/45; 9/14/45; 9/14/45
HE-116-229	Long Beach Portal	0+00	8/31/45
		Oregon	
HE-116-229 HE-116-247	Clatsop Spit Beach	0+00; 10+00N; 10+00S	9/24/45, 11/7/46; 9/24/45, 11/4/46; 9/24/45, 11/7/46
HE-116-247	Manzanita	0+00, 8+00N, 8+00S	10/18/46, 11/6/46
HE-116-229 HE-116-247	Cape Lookout Station (Lookout Beach)	0+00, 10+00S	9/27/45, 10/28/46
HE-116-229 HE-116-247	Cape Merriweather Station	0+00	9/27/45, 10/26/46, 11/12/46
HE-116-229 HE-116-247	Sand Lake Station	0+00; 10+00S; 10+00N	9/27/45, 10/26/46, 11/5/46, 11/12/46; 9/27/45, 11/12/46; 10/26/46
HE-116-247	Coos Bay (Empire Beach)	0+00, 10+00S&N	11/15/46

California			
HE-116-205	Table Bluff		10/27/45 - 2/10/ 46 3/24/46 - 3/30/46
HE-116-224	Seabright Beach	0+00; 4+47N; 1+50N	6/5/46, 7/30/46, 2/4/47; 6/12/46, 7/30/46, 2/4/47; 6/21/46
HE-116-224	Seacliff Beach	0+00; 5+00N; 2+85N	6/5/46, 7/30/46, 2/4/47; 6/13/46, 7/30/46, 2/4/47; 6/21/46
HE-116-224	Moss Landing Beach, North	0+00, 5+00N	6/6/46, 6/12/46, 8/14/46, 3/6/47
HE-116-224	Moss Landing Beach, South	0+00, 5+00N, 5+00S	8/14/46, 8/15/46, 3/6/47
HE-116-256	Halfmoon Bay Municipal Beach	0+00, 10+00N	4/29/47
HE-116-256	Coast Guard Beach	0+00	4/28/47& 4/29/47
HE-116-256	Princeton Beach	0+00, 10+00S	4/28/47& 4/29/47
HE-116-256	Miramar Beach	0+00, 6+00S, 6+00N; 0+00; 6+00S; 6+00N	4/28/47& 4/29/47; 7/17&18/45, 4/28 & 29/47; 7/17&18/45, 4/28 & 29/47; 7/17&18/45, 4/28 & 29/47
HE-116-224	Fort Ord (Soldier's Club)	0+00, 5+00S, 5+00N	6/13/46
HE-116-224	Fort Ord (Concrete Block)	0+00, 5+00S, 10+00S	6/13/46
HE-116-224	Fort Ord (Landing Barge)	0+00, 5+00S	6/13/46, 8/2/46, 2/24/47
HE-116-224	Spanish Bay	2+75N	5/24/46, 7/24/46
HE-116-224	Spanish Bay, South St.	2+75N	5/24/46, 7/24/46
HE-116-224	Carmel	1-S; 15+00S; 5+00S, 10+00S; 0+00 5+00N 10+00N 15+00N	4/27/46, 6/26/46, 7/23/46, 9/4/46, 9/14/46, 12/11/46, 2/25/47; 5/15/46, 6/7/46, 6/24/46, 7/23/46, 8/20/46, 12/10/46, 2/21/47; 5/15/46, 6/7/46, 6/24/46, 7/23/46, 8/20/46, 12/11/46, 2/21/47 4/24/46, 5/22/46, 6/24/46, 7/23/46, 8/21/46, 12/11/46, 2/21/47 5/15/46, 6/7/46, 6/24/46, 7/23/46, 8/21/46, 2/21/47 4/24/46, 5/15/46, 6/7/46, 7/23/46, 8/20/46, 2/21/47 5/15/46, 6/7/46, 6/26/46, 7/23/46, 8/20/46, 2/21/47
HE-116-224	Carmel River Bight	2+50N 0+00 2+50S 5+00S	6/11/46, 7/23/46, 2/25/47 6/11/46, 7/23/46, 2/25/47 6/11/46, 7/23/46, 2/25/47 6/11/46, 7/23/46, 2/25/47
HE-116-22	Point Sur	3+50S 0+00 5+00N	6/18/46, 7/31/46 6/18/46, 7/31/46 6/18/46, 7/31/46

CHAPTER 235

MECHANISM AND CALCULATION OF SAND DUNE EROSION BY STORMS

Ryuichiro Nishi[1] and Nicholas C. Kraus[2]

ABSTRACT: Dunes constitute a central element in shore-protection designs aimed at preventing inundation and erosion of the upland by storms. In the present study, the dune-erosion mechanism is investigated through field observations of the performance of sand dunes under storm action in Japan and in the United States. The observations are quantified by analysis of three test series carried out in the SUPERTANK project. The SUPERTANK data set includes tests on erosion of an uncompacted and a compacted near-vertical dune by random waves. The SBEACH numerical model of dune erosion and profile change is modified to erode dunes by the force of incident waves. This sediment can then supply the profile change model that demands offshore movement of sand in response to the occurrence of storm waves and elevated water level. The simulations show good agreement with the erosion measured at SUPERTANK for the uncompacted and the compacted dunes.

INTRODUCTION

Dune design is a central element in shore-protection projects aimed at preventing inundation and erosion of the upland by storms. In the United States and some other countries, the performance of protective dunes is often estimated with the Kriebel and Dean (1985) model or the Larson and Kraus (1989) (SBEACH) model of storm-induced beach erosion. These models operate under the assumption that erosion of the beach and dune complex is controlled by the demand for sand in the surf zone to satisfy establishment of an equilibrium profile under the impressed storm water level and waves. The demand, or, cross-shore sediment transport capacity, is estimated from the difference in wave energy dissipation between the existing profile and an assumed equilibrium profile shape. Conceptually, in such a demand-and-supply model (demand model), if a dune exists on an equilibrium beach profile for a given storm condition, little dune erosion is expected to occur. Clearly, however, a dune will erode if it is subjected to violent wave action almost independently of the equilibrium profile dynamics occurring offshore. Resolution of this problem is discussed here.

---

1. Assistant Professor, Department of Ocean Civil Engineering, Kagoshima University, 1-20-40 Korimoto, Kagoshima-shi 890, Japan.
2. Research Physical Scientist, US Army Engineer Waterways Experiment Station, Coastal and Hydraulics Laboratory, 3909 Halls Ferry Road, Vickburg, Mississippi 39180-6199, USA.

In the present study, the dune-erosion mechanism is investigated through field observations of the performance of sand dunes under storm action in Japan and the United States. These observations are quantified by analysis of test series from the SUPERTANK Data-Collection Project (Kraus et al. 1992, Kraus and Smith 1995, Smith and Kraus 1996). The data set includes two tests involving erosion by random waves of 0.8-m high, nearly vertical dunes composed of 0.23-mm median-diameter sand. One test involved an uncompacted dune (SUPERTANK Test ST_50) and the other a compacted dune (ST_60). Complete time series of the water-surface elevation are available from the offshore to the face of the dune, although only breaking waves were employed here to compute the impact parameter and related volume of dune erosion.

Dune erosion is calculated as a function of a wave-force parameter (Sunamura 1977, Fisher and Overton 1984). Material eroded from the dune is supplied to the nearshore where profile change is calculated through equilibrium-profile concepts. Therefore, we call the methodology a *supply-and-demand* dune erosion and profile change model. This paper describes different mechanisms of dune erosion and the new supply-and-demand model. The model is then tested to simulate dune erosion measured at SUPERTANK.

MECHANISM OF SAND DUNE EROSION BY STORMS

In this paper, we consider dune erosion produced by impact forces of waves incident nearly normal to the shore, a cross-shore transport process that is assumed to be two dimensional. This assumption is supported by the uniform dune recession commonly observed after large storms along kilometers of shore despite longshore variations in coastal structures and offshore bathymetry. However, we note the possibility of the action of shearing forces exerted on dunes by waves and associated currents passing tangentially to the shore. Such a situation occurs by waves generated by ships passing dunes in narrow channels. Shearing erosion is not considered further here.

Field Observations

Sunamura (1992) has described basal erosion and mass movement (failure or erosion) of cliffs on rocky coasts as four types: falls, topples, slides, and flows. In the present study, the authors have documented three types of erosion mechanisms of sand dunes by wave impacts during storms or strong wave action. The cross-shore dune-erosion mechanisms, schematized in Fig. 1, are classified as (a) layer separation, including layer separation and overturning, (b) notching and slumping, and (c) sliding and flowing.

Layer separation. Layer separation typically occurs if a near-vertical dune face is subjected to wave impact. Over the duration of a certain number of impacts, a vertical fault line (crack) develops, and this outer layer gradually separates (typically 30 to 50 cm thick) from the landward portion of the dune. As it separates, the outer layer detaches from the main body of the dune, becomes unstable, and either collapses suddenly (Fig. 1a) or tilts forward and overturns (Fig. 1a′).

Notching and slumping. Severe notching tends to occur if a dune slope is nearly vertical, permeated by roots, highly compacted, or composed of rocks such as a rocky cliff. Notching is limited to the elevation of wave attack and, after the notch is cut sufficiently deep into the base of the dune, the overlying sand column collapses. (Some notching may also occur during layer separation, but it is not the dominant factor in the collapse of the separated layer.) The material from the collapsed dune face is deposited in front of the new dune face.

The width of sand deposition at the foot of the dune face is less than that of layer separation, which involves the overturning and sliding of a layer.

Sliding and flowing. Sliding and flowing occur on uncompacted gently sloping dunes that have a face slope close to the angle of repose of the sediments forming them. In this situation, modest wave impact at the base of the dune or even pelting by rain or exposure to strong wind can cause a thin layer of sand to run down the slope. It is expected that this mode of dune erosion does not cause severe dune recession in a short period of time; however, this mechanism tends to steepen the dune face and a resultant steeper dune slope will probably trigger layer separation or notching and slumping under storm conditions.

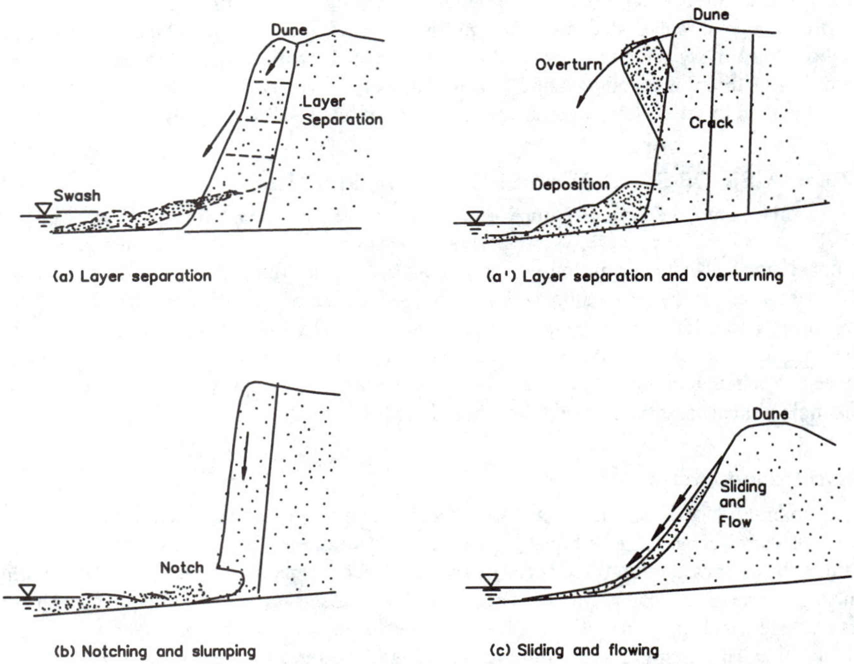

Fig. 1. Dune erosion mechanisms by cross-shore processes.

SUPERTANK Dune-Erosion Tests

The erosion mechanism was quantified by using profile response measurements made at the SUPERTANK project. The particular tests analyzed here concern profile steepening and dune erosion. These test series involved random wave incident to a near-equilibrium profile, and to uncompacted (loose sand) and compacted near-vertically faced dunes. Complete time series of the water surface elevation are available from the offshore to the dune face (Smith and Kraus 1995); however, in this study, as a preliminary and less calculationally intensive step, only wave-related quantities calculated at the breaker line were used.

The three SUPERTANK tests discussed in the present study were ST_10 (erosion of an equilibrium profile), ST_50 (uncompacted dune), and ST_60 (compacted dune). The waves were run in bursts of 10-40 min, and conditions were sometimes changed between bursts, for which representative waves are as follows:

Test Number	Significant Wave Height, m	Peak Spectral Period, sec
ST_10	0.8	3.0
ST_50	0.5 - 0.8	6.0 - 3.0
ST_60	0.5 - 0.7	6.0 - 3.0

Erosion above a near-equilibrium profile. For SUPERTANK Test ST_10, random waves and monochromatic waves were generated to act on an initial idealized sub-aqueous equilibrium beach profile. Evolution of the beach profile is shown in Fig. 2. Four series of slope steepening (scarping) events occurred in the test. The profile in the swash zone steepened and maintained a constant angle at the second through fourth wave-burst events, while the upper beach face receded. The slope of the upper beach face is $\tan\beta \approx 0.89$, slightly less than that of a one-to-one slope (45 deg) typically specified in the Kriebel and Dean (1985) model. The slope of the upper beach face appears to exert control on the speed and volume of upper beach erosion.

Fig. 3 shows the process of profile steepening and scarp generation at the first and second wave-burst events of the erosion processes shown in Fig. 2. The beach-face slope steepened, and the swash waves carried the sediment offshore while lowering the beach face. Once the slope of the upper beach face approached the angle of failure (avalanching), the upper portion of beach face collapsed and the sediment was deposited in front of the scarp. Thereafter, successive swash waves transported the sediment seaward that was supplied from the upper beach face. As the beach face was lowered by erosive swash waves, swash uprush intensely impacted the steep beach face, again causing avalanching. These erosion processes in the swash zone continued until the upper beach face was no longer vulnerable to swash waves. This test series demonstrates that the upper beach face behind a sub-aqueous near-equilibrium beach profile can be eroded by swash wave activity until it also achieves equilibrium with the water level and swash.

Dune erosion tests. Two dune-erosion tests were conducted at SUPERTANK, one for a dune formed of sand without compaction (ST_50) and the other for an artificially compacted dune (ST_60). The dune was compacted by applying a pavement vibrator for approximately 2 hr. Both dunes were subjected to short-period high waves. The water level was lower at the beginning of the tests and higher at the end of the tests (Kraus and Smith 1995, Smith and Kraus 1996). In the dune-erosion tests, the dune face tended to recede in parallel to itself, as shown in Fig. 4. Video records made during SUPERTANK indicate both the uncompacted and compacted dunes eroded primarily by layer separation. In all situations observed in the field and at SUPERTANK, dune faces tended to recede in parallel to themselves.

This coherent behavior of dune evolution illustrates the consistency and reproducibility of the dune-erosion mechanism. The sediment supply from the dune to the swash zone by either the layer-separation or notching-and-slumping erosion mechanisms was injected virtually instantaneously (order of one wave period) when a wave or backwash swept the eroded sediment toward the swash zone.

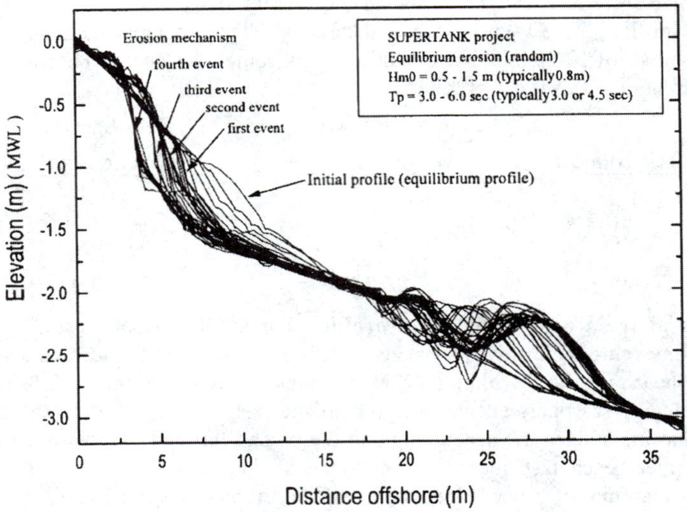

Fig. 2. Profile change on a near-equilibrium beach (ST_10).

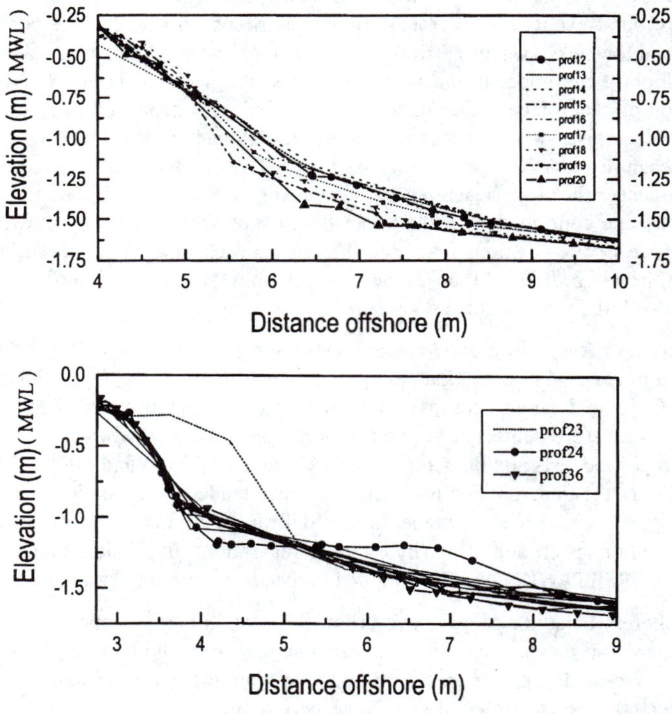

Fig. 3. Beach face steepening and scarp generation (ST_10).

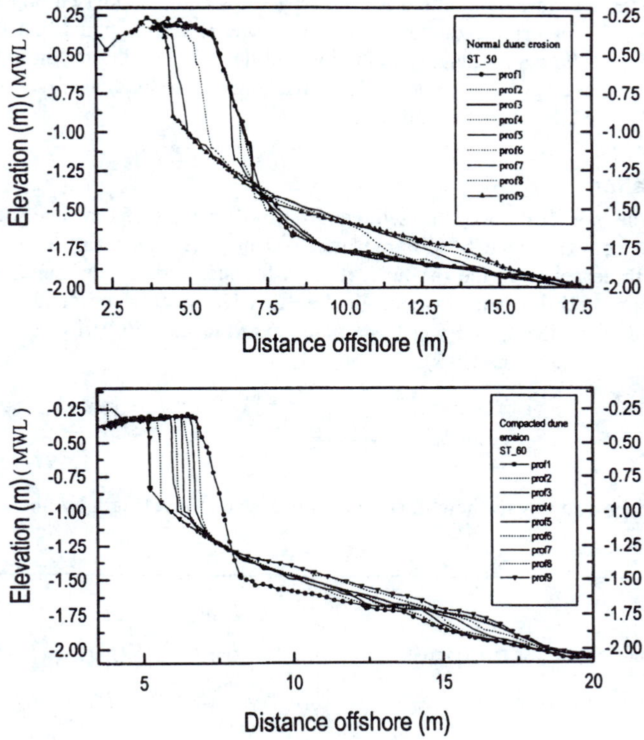

Fig. 4. Erosion of uncompacted (ST_50) and compacted (ST_60) dunes.

The volume of erosion of the compacted dune was less than that of the uncompacted dune, despite the slope of the compacted dune being steeper than that of the uncompacted dune during the erosion, as shown in Fig. 4. Compaction altered the strength of the dune and the volume of eroded material, thickness of layer separation, and angle of slope failure. It is clear that a compaction coefficient should be introduced in future studies of profile evolution.

DUNE EROSION MODEL

A numerical model of dune erosion was developed in the present study that simulates the impact force of individual waves on the dune face. Research conducted at Tuskuba University, Japan, for cliff erosion (see Sunamura 1977, 1992) and at North Carolina State University for sand dunes (see Fisher and Overton 1984, Fisher et al. 1986, and Overton et al. 1994) has advanced this approach. The impact-force and dune-erosion model was incorporated as a sub-model of SBEACH (Larson and Kraus 1989, Nishi et al. 1991) and employs the wave model of SBEACH supplemented by a swash model that includes bore velocity and height to estimate the impact parameter. The total model consists of a wave transformation model, cross-shore sediment transport model in the surf zone, and a sediment supply model from the dune.

In SBEACH, wave transformation in the surf zone is computed based on the Dally et al. (1985) model. For describing random wave incidence, it was assumed that the heights of individual waves follow a Rayleigh distribution and the wave period for an individual wave corresponds to an average wave period. The wave height of individual waves was computed by a Monte Carlo method (Larson and Kraus 1991).

Profile Zonation

Because the model computes the sediment transport in the surf zone and sediment supply and transport from the dune by different mechanisms, the beach and dune systems were divided into three zones as the (a) dune, (b) swash zone, and (c) surf zone (Fig. 5). This profile zonation modifies that of the original SBEACH model at the dune. The maximum runup or swash elevation Z_r is defined as a function of the surf-similarity parameter as given by Eq. (1) (Larson and Kraus 1989)

$$\frac{Z_r}{H_o} = 1.47\left(\frac{\tan\beta}{\sqrt{H_o / L_o}}\right)^{0.79} \tag{1}$$

where H_O = deep-water wave height, and L_O = deep-water wavelength by linear-wave theory.

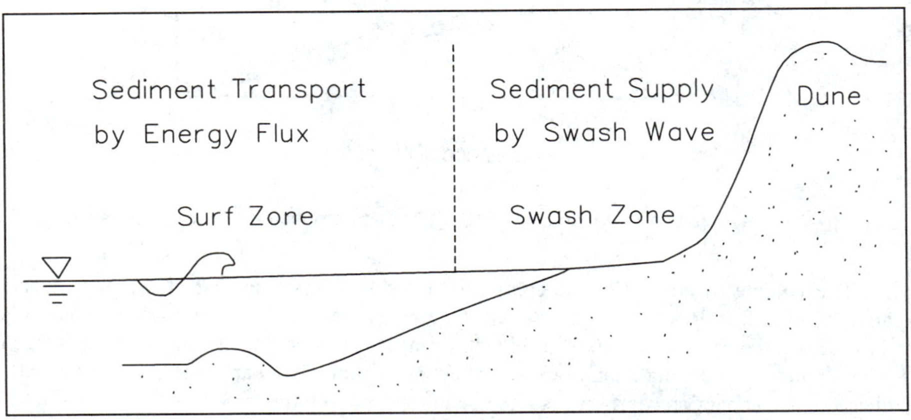

Fig. 5. Schematic diagram of dune and beach system.

Calculation of Sediment Supply from Dune

As storm waves approach the beach, bores impact the dune face and cause erosion. Thus, the volume of dune erosion during a storm or by waves that reach the dune face during times of elevated water level should be related to the force of wave impact. Sunamura (1977) parameterized this impact wave force by the incident wave height in study of rocky cliff erosion. A strong correlation between wave impact force and sand dune erosion was found by Fisher et al. (1986) for artificially constructed sand dunes in the field and by Overton et al. (1994), who obtained a correlation between the force of impacting waves and volume of dune erosion for the compacted dune at SUPERTANK.

The wave-force impact parameter is derived heuristically by considering the situation shown in Fig. 6 and the rate of change of wave momentum. The mass m of water per unit length of crest in a bore of height H and length L moving in shallow water of average depth h over the length of the wave is given by

$$m \approx \frac{1}{2}\rho_w HL = \frac{1}{2}\rho_w HT\sqrt{gh} \tag{2}$$

where ρ_w = density of water, T = wave period, g = gravitational acceleration, and $(gh)^{1/2}$ is the celerity of the wave. The wave impact force per unit length of dune is estimated by multiplying the mass m by the deceleration $(gh)^{1/2}/T$ resulting from the wave striking the dune and stopping in the time interval of the wave period T. This derivation suggests consideration of a cumulative wave-force impact parameter I defined by

$$I \equiv \rho_w gh H \frac{\Delta t}{T} \tag{3}$$

where Δt is the time interval or duration over which the waves impact, and the ratio $\Delta t/T$ is the number of waves. The ratio $\Delta t/T$ can be easily modified to describe random waves by summing the contributions of the individual waves of different height, period, and speed. The impact parameter has the dimensions of Newtons per unit wave crest or dune width.

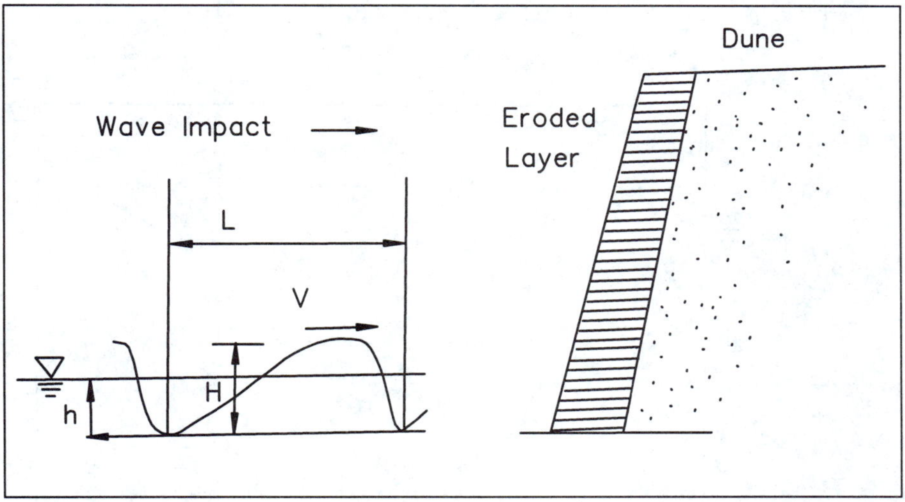

Fig. 6. Schematic diagram for the wave impact parameter.

For dimensional homogeneity it is convenient to work with the weight per unit width alongshore of the eroded material $W_E = \rho_S (1 - p) g V_E$, where ρ_S = density of the sand comprising the dune, p is the porosity of the sand (0.4 for uncompacted sand, 0.2 estimated for the compacted sand dune), and V_E is the volume of material eroded from the dune. The density and, therefore, the porosity, of the uncompacted and compacted dunes should be different.

Analysis of the data computed from the average of the impact forces indicated a linear relationship as shown in Fig. 7 and resulted in the empirical equation for the weight per unit width $(W_E)_U$

$$(W_E)_U = 0.81(I - I_{crit}) \tag{4}$$

for the uncompacted dune, and the weight $(W_E)_C$

$$(W_E)_C = 0.50(I - I_{crit}) \tag{5}$$

for the compacted dune. In the above, the critical wave impact parameter for inception of erosion I_{crit} is set to zero at the present time because of uncertainty in its value.

These equations and Fig. 7 show that waves with the same impact parameter erode a compacted dune less than a uncompacted or unconsolidated dune. An engineering lesson from this result indicates that greater erosion protection would be gained through construction of dunes by wetting the sand for consolidation and compacting them with vibrating compactors, rollers, and other heavy equipment. The above result was obtained from a limited number of tests available from the SUPERTANK project, and further verification needs to be done to refine the result.

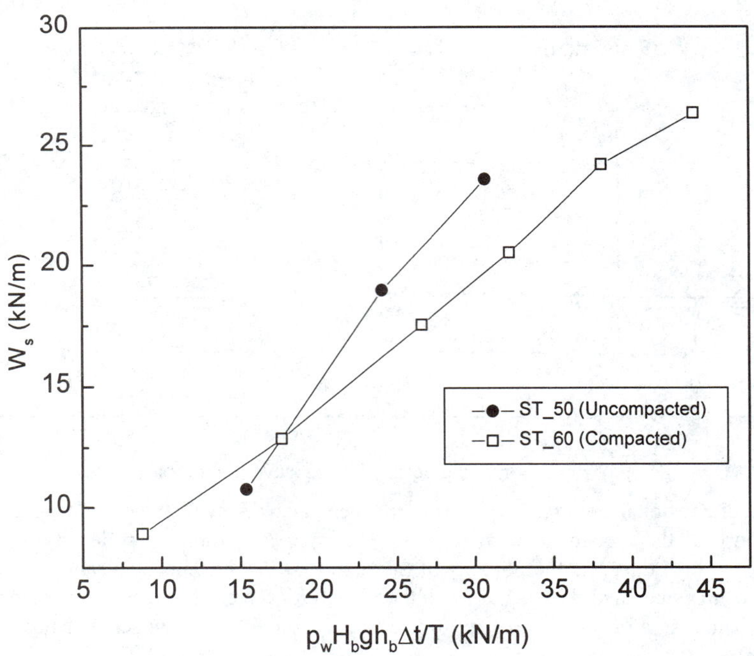

Fig. 7. Impact parameters for uncompacted and compacted dunes at SUPERTANK.

NUMERICAL MODEL OF DUNE EROSION

From the above discussion, the volume of sand eroded from a dune can be calculated as a function of time through knowledge of the incident waves and the wave-force impact parameter. Thus, the amount of sand supplied to the profile by the dune is known. As the next step, an assumption must be made as to how the sediment enters the swash zone or surf zone where other transporting mechanisms are operating. Here, it is assumed that sediment is introduced to the swash zone at a uniform rate within each time step of the model according to the supply available from the dune at that time step. A typical calculation time step in the model is 1 min, and the grid cell size on the foreshore is 0.1 m. Uniform distribution of sediment supply agrees with visual observations made at SUPERTANK and inferences from field observations, and it is also reasonable in a time-average sense. The structure of the numerical model does not preclude a more detailed description of the sediment-supply procedure to the swash zone and can be modified as understanding improves.

Dune Erosion Model (Supply)

We found above that a dune face recedes with a certain angle in accordance with the amount of compaction. Nearly vertical dune faces produced by erosion during storms are commonly observed in the field, in particular for well-established dunes presumably compacted by natural settling and wetting. In the SUPERTANK project, the angles of the dune face were approximately 68 deg for the uncompacted dune and 87 deg for the artificially compacted dune, as shown in Fig. 8. The irregularity in dune-face angle through time for the compacted dune may have been caused by uneven compaction close to the initial dune face. In the model simulations described below, it is assumed that the slope of the dune face is 68 or 87 deg depending on the amount of compaction.

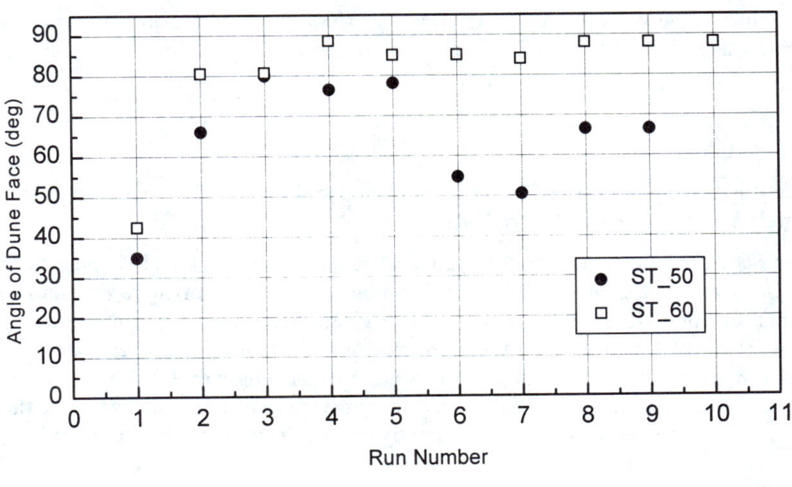

Fig. 8. Angle of dune face for uncompacted and compacted dunes at SUPERTANK.

As determined from the SUPERTANK data, the volume of dune erosion can be estimated by the duration and intensity of wave impact, which yields an eroded volume ΔV_E per unit longshore width of dune. A corresponding recession distance can be computed by assuming the dune face retreats a certain distance and at a certain angle to a baseline. The baseline is a vertical datum located at the toe of the dune. In the model, the volume of sand eroded from the dune is distributed uniformly as a thickness Δh_S over a distance X_S, which is taken to be an effective distance of the swash zone extending from the toe of the dune at the present time step to some depth defining the seaward limit of the swash zone, arbitrarily set to 0.3 m or a comparable value (Larson and Kraus 1989). Then we have

$$\Delta h_s = \frac{\Delta V_E}{X_S} \tag{6}$$

for the thickness of the sand layer.

Profile Change Model (Demand)

In the SBEACH model, sediment demand or the potential transport rate in the surf zone is calculated based on the dissipation of wave energy flux originally derived by Dean (1977) and implemented by Kriebel and Dean (1985) in dune-erosion modeling. In the surf zone, SBEACH computes the cross-shore sediment transport rate as

$$q = K\left(D - D_{eq} + \frac{\varepsilon}{K}\frac{\partial h}{\partial x}\right) \tag{7}$$

where K = empirical transport rate coefficient, ε = empirical coefficient controlling the strength of the slope-dependent transport rate term, and the energy dissipation per unit water volume D is

$$D = \frac{1}{h}\frac{\partial F}{\partial x} \tag{8}$$

in which F is the wave-energy flux. The dissipation for a profile in equilibrium with the existing waves and water level D_{eq} is given by (Dean 1977)

$$D_{eq} = \frac{5}{24}\rho_w g^{3/2}\left(\frac{H_b}{h_b}\right)^2 A^{3/2} \tag{9}$$

where A is an empirical "shape" parameter related to the form of the equilibrium profile and the grain size of the beach (Moore 1982).

In SBEACH, the transport rate in the surf zone as given by Eq. (7) is only calculated if $D > D_{eq} - \varepsilon/K\ \partial h/\partial x$, and the transport direction as onshore or offshore is determined by a separate function (Kraus et al. 1991). As originally developed (Larson and Kraus 1989), SBEACH calculates cross-shore transport rates in four zones, with a linear rate employed in the swash zone and a magnitude as determined by matching with Eq. (7) at the swash zone and surf zone interface. Material moved from the dune to the swash zone by the supply dune-erosion model is then moved offshore by the swash and surf zone transport.

Numerical Simulation of Dune Erosion at SUPERTANK

A numerical simulation was conducted to calculate dune erosion for Test ST_60 (compacted dune). The empirical dune-erosion predictor incorporated in the model was developed based on the SUPERTANK data and impact parameter, as described above. Therefore, the simulation is not a verification of the model; rather, it demonstrates the validity of the numerical scheme and behavior of the supply-and-demand procedure that connects the dune to the profile.

Fig. 9 shows the profile change simulated by applying the (standard) demand model, which can be compared to the profile change simulated by applying the supply-and-demand model, Fig. 10. The demand model underestimates the dune erosion for which the compacted dune is located on a beach profile in near equilibrium with the impressed waves.

We note in this paragraph independent contemporaneous work involving SUPERTANK Tests ST_10, ST_50, and ST_60. Wise et al. (1996) report comparisons of SUPERTANK measurements (and field measurements) and calculations performed with the most recent version of SBEACH operated by the US Army Engineer Waterways Experiment Station (WES), for which default calibration parameters (K and ε) were specified (the calibration parameters were not optimized for the individual tests). The most recent WES version of SBEACH incorporates sophisticated random-wave and cross-shore sediment transport models not applied in the version used in the present study. The WES version is still a demand model. For equilibrium Test ST_10, erosion of the foreshore was obtained by Wise et al. for random waves with the new WES version of SBEACH. For the dune erosion tests, the WES version well reproduced erosion of the uncompacted dune (ST_50) and overpredicted dune erosion for the compacted dune (ST_60). Wise et al. comment that "The difference in model predictions between the two dune cases might be expected due to greater erosion resistance associated with the compacted sediment which is not accounted for in SBEACH," This comment was verified in the present work.

CONCLUDING DISCUSSION

Three types of dune erosion mechanisms were identified through field observations as: (a) layer separation, (b) notching and slumping, and (c) sliding and flowing. The layer-separation mechanism was quantified by analysis of the SUPERTANK dune erosion tests by which the eroded volume could be related to the cumulative wave impact force. The degree of compaction was found to be a significant parameter that decreases the potential for dunes to erode. Therefore, an economic benefit might be gained by compacting artificially placed dunes to improve their performance as shore protection. The supply-and-demand model developed can simulate dune erosion and beach profile change based on representations of the hydrodynamics and sediment transport acting in each region and provides, in principle, a more accurate representation than existing demand models, especially in applications where the beach profile is approximately in equilibrium with the impressed storm waves and water level. The present study suggests ways through which geotechnical considerations might be incorporated in dune erosion modeling to both account for compaction and introduce supply and demand considerations.

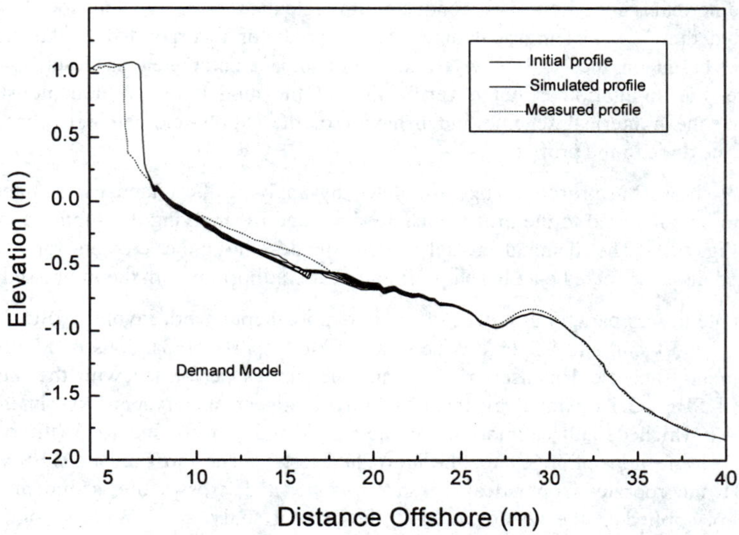

Fig. 9. Simulation for compacted dune (ST_60) by demand model.

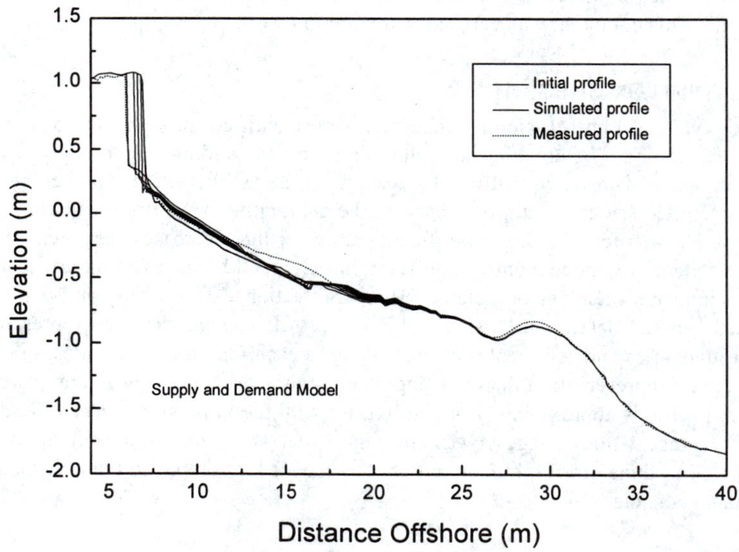

Fig. 10. Simulation for compacted dune (ST_60) by the supply-and-demand model.

ACKNOWLEDGEMENTS

Appreciation is expressed to all the individuals who participated in SUPERTANK and created the valuable data set. Bill McDougal, Oregon State University, compacted the dune for ST_60. Jane Smith, WES, Coastal and Hydraulics Laboratory (CHL), provided the SUPERTANK wave data used in this study. Randy Wise, CHL, reviewed an early draft of this paper. The second author received permission by Headquarters, U.S. Army Corps of Engineers, to publish this information.

REFERENCES

Dally, W. R., Dean, R. G., and Dalrymple, R. A. 1985. Wave height variation across beaches of arbitrary profile. *J. Geophys. Res.* 90(C6): 11917-11927.

Dean, R. G. 1977. Equilibrium beach profiles: U.S. Atlantic and Gulf Coasts. Dept. of Civil Engrg., Univ. of Delaware, Newark, Del.

Fisher, J. S. and Overton, M. F. 1984. Numerical model for dune erosion due to wave uprush. *Proc. 19th Coastal Engrg. Conf.*, ASCE, 1553-1558.

Fisher, J. S., Overton, M. F., and Chisholm, T. 1986. Field measurements of dune erosion. *Proc. 20th Coastal Engrg. Conf.*, ASCE, 1107-1115.

Kraus, N. C., Larson, M., and Kriebel, D. L. 1991. Evaluation of beach erosion and accretion predictors. *Proc. Coastal Sediments '91*, ASCE, 572-587.

Kraus, N. C. and Smith, J. M. 1994. SUPERTANK laboratory data collection project, Volume I: main text, Tech. Rep. CERC-94-3, US Army Engineer Waterways Experiment Station, Coastal Engrg. Research Center, Vicksburg, Miss.

Kraus, N. C., and Smith, J. M., and Sollitt, C. K. 1992. SUPERTANK laboratory data collection project. *Proc. 23rd Coastal Engrg. Conf*, ASCE, 2191-2204.

Kriebel D. L. and Dean, R. G. 1985. Numerical simulation of time dependent beach and dune erosion. *Coastal Engrg*, 9: 221-245.

Larson M. and Kraus N. C. 1989. SBEACH: Numerical model for simulating storm-induced beach change, Report 1, empirical foundation and model development. Technical Report CERC-89-9, US Army Engineering Waterways Experiment Station, Coastal Engrg. Research Center, Vicksburg, Miss.

Larson, M., and Kraus, N. C. 1991. Numerical model of longshore current over bar and trough beaches. *J. Waterway, Port, Coastal and Ocean Engrg.*, 117(4): 326-347.

Moore B. 1982. Beach profile evolution in response to changes in water level and wave height. Unpub. MS Thesis, Dept. of Civil Engrg., Univ. of Delaware, Newark, Del.

Nishi, R., Sato, M. and Wang, H. 1994. Field observation and numerical simulation of dune and beach scarping. *Proc. 24th Coastal Engrg. Conf.*, ASCE, 2434-2448.

Overton M. F., Fisher, J. S., and Hwang, K. 1994. Development of a dune erosion model using SUPERTANK data. *Proc.24th Coastal Engrg. Conf.*, ASCE, 2488-2502.

Smith, J. M. and Kraus, N. C. 1995. SUPERTANK laboratory data collection project, Volume II. Appendices A-I, Tech. Rep. CERC-94-3, US Army Engr. Waterways Experiment Station, Coastal Engrg. Research Center, Vicksburg, Miss.

Sunamura, T. 1977. A relationship between wave-induced cliff erosion and erosive force of waves. *J. Geol.*, 85: 613-618.

Sunamura, T. 1992. *Geomorphology of Rocky Coasts*. John Wiley & Sons, N.Y., 302 pp.

Wise, R. A., Smith, S. J., and Larson, M. 1996. SBEACH: Numerical model for simulating storm-induced beach change, Report 4: cross-shore transport under random waves and model validation with SUPERTANK and field data. Tech. Rep. CERC-89-9, US Army Engr. Waterways Experiment Station, Coastal Engrg. Research Center, Vicksburg, Miss.

CHAPTER 236

SIMULATION OF COASTAL PROFILE DEVELOPMENT USING A BOUSSINESQ WAVE MODEL

K.A. Rakha[1], R. Deigaard[2], P.A. Madsen[1], I. Brøker[3], and J.K. Rønberg[3]

ABSTRACT

A phase-resolving wave transformation module is combined with an intra-wave sediment transport module to calculate the on/offshore sediment transport rates. The wave module is based on the Boussinesq equations extended into the surf zone. The vertical variation of the instantaneous currents and concentrations are calculated. The net sediment transport rates are calculated, and the equation for conservation of sediment is solved to predict the beach profile evolution. The results of the present paper showed that the undertow contribution to the sediment transport rates dominated only at local areas even for eroding beaches, suggesting that other contributions should not be neglected.

1. INTRODUCTION

The study of beach profile evolution includes a large range of time and space scales. Process based morphology models (Roelvink and Brøker, 1993) usually include some averaging of the different space and time scales. This averaging is applied to either the hydrodynamic or the sediment transport calculations or both. The sediment transport calculations used in most of the present morphology models are based on phase-averaged calculations of the sediment concentrations or on the 'energetics approach'. Although models based on the energetics approach account for the intra-wave sediment transport rates, the intra-wave variation of the eddy viscosities and sediment concentrations are not calculated. Such models may be classified as semi-intra-wave sediment transport models. The model developed by Fredsøe et al. (1985) represents an example of a detailed intra-wave model for the sediment concentrations. Watanabe (1994) combined a wave model based on the Boussinesq equations with a semi-intra-wave sediment transport model. Such a model neglects a large part of the information provided by the wave module, and will require some approximations for irregular waves. The model described by Brøker et al. (1991) combines a detailed intra-wave sediment transport module with a phase-averaged wave module, where a wave theory is required to describe the intra-wave water motion. The extension of such a model to irregular waves will also require some approximations. In the present study a phase-resolving wave module is combined with a detailed intra-wave sediment transport module to study in more detail the process of sediment transport and morphological evolution.

---

[1]International Research Center for Computational Hydrodynamics (ICCH)
Danish Hydraulic Institute (DHI), Agern Allé 5, DK-2970 Horsholm, Denmark.
* Currently at Department of Irrigation and Hydraulics, Faculty of Eng., Cairo University, Egypt.
[2]Institute of Hydrodynamics and Hydraulic Eng. (ISVA).Technical University of Denmark, DK-2800 Lyngby, Denmark.
[3]Danish Hydraulic Institute (DHI), Agern Allé 5, DK-2970 Horsholm, Denmark.

2. MODEL DESCRIPTION

The morphological calculations are performed by updating the beach profile over the morphological time step Δt_{max}. For each morphological update the simulations are performed by four modules; a wave module, a hydrodynamic module, a sediment transport module and a bathymetry updating module.

2.1 Wave Module

The wave module simulates the wave conditions across the beach profile by a phase-resolving model based on the Boussinesq equations, with improved linear dispersion characteristics as explained in Madsen et al. (1991), and Madsen and Sørensen (1992). The effect of wave breaking is included by using the surface roller concept (Schäffer et al., 1993). Figure (1A) shows a sketch of the assumed velocity field under a wave with a surface roller. An extra term is included in the momentum equation to represent the momentum flux due to the rollers. This term extracts energy from the wave motion. Breaking is initiated when the local water surface slope exceeds the initial value of ϕ_b. The roller is defined as the water above the tangent slope tan(ϕ). Initially ϕ is equal to ϕ_b for each roller, which then decreases exponentially to ϕ_o, and breaking is assumed to cease when the maximum of the local slope becomes less than tan(ϕ). The resulting roller thickness δ is finally multiplied by the roller shape factor f_δ to compensate for the simple method of separating the roller from the rest of the flow.

The calculation proceeds into the swash zone using the slot-technique (Madsen et al. 1994), by extending the computational domain into an artificial permeable beach. Near the moving shoreline the water surface will intersect with the sea bed and continue into the porous beach. The instantaneous position of the shoreline is simply determined by this intersection.

2.2 Hydrodynamic Module

The hydrodynamic module consists of two parts, an oscillatory boundary layer model (Figure, 1B) and an undertow model (Figure, 1C). The boundary layer model calculates the vertical velocity distribution for the oscillatory wave motion u_o inside the boundary layer. The undertow model determines the vertical distribution of the mean undertow $\bar{U}_o$.

2.2.1 Boundary Layer Calculations

The oscillatory flow near the bottom is modelled by the momentum integral method developed by Fredsøe (1984). The shear stress is assumed to be zero at the top of the boundary layer, and the velocity distribution u inside the boundary layer is assumed to follow a logarithmic distribution. The boundary layer thickness is assumed to develop from zero at every zero-crossing of u_o outside the boundary layer. The values of the shear velocity U and the boundary layer thickness ζ are calculated for each time step. The streaming in the boundary layer and the wave asymmetry causes the time averaged bed shear stress to deviate from zero. To obtain this mean shear stress a constant drift velocity V_{st} is added to the near bed orbital velocity. The magnitude of V_{st} is found by iteration requiring that the time averaged bed shear stress in the boundary layer model is equal to the shear stress determined by the streaming.

2.2.2 Eddy Viscosity Calculations

The eddy viscosities are calculated by assuming that the total kinetic energy can be determined as the sum of three contributions (Brøker et al., 1991),

$$\epsilon = \sqrt{\epsilon_w^2+\epsilon_b^2+\epsilon_U^2} \quad \mathbf{(1)}$$

where, ϵ_w, ϵ_b, and ϵ_U are the eddy viscosities due to the bottom boundary layer, wave breaking, and the undertow, respectively. The eddy viscosity ϵ_w is calculated based on the variables calculated in the boundary layer, assuming a parabolic distribution inside the bottom boundary layer (Fredsøe et al., 1985). The eddy viscosity ϵ_U is calculated from the undertow velocity profile using a simple mixing length formulation. The eddy viscosity ϵ_b is calculated from the turbulent kinetic energy induced by wave breaking calculated from a one-equation turbulence model (Deigaard et al. 1991). The instantaneous production of turbulence due to wave breaking is calculated from the following equation proposed by Deigaard (1989),

$$P_r = \alpha_p Diss = \alpha_p \rho g c \delta \tan\phi_o \qquad (2)$$

where, Diss is the instantaneous energy dissipation due to wave breaking, and α_p is the fraction of the energy that is not dissipated immediately in the shear layer beneath the roller. α_p is assumed to be 0.33 as suggested by Deigaard et al. (1991). The production of turbulence is assumed to have a parabolic distribution over a distance of half the wave height (H/2) below the mean water depth as shown in Figure (2).

2.2.3 Undertow Calculation

The undertow is calculated by the following approximate equation,

$$\frac{\partial U_c}{\partial y} = \frac{1}{\rho \bar{\epsilon}} \bar{\tau} \qquad (3)$$

where an overbar denotes time averaging over a single wave. The time averaged shear stress distribution is assumed to vary linearly over the water depth as shown in Figure (2). The shear stress at the surface (mean water level, MWL) is calculated from the following formula,

$$\bar{\tau}_s = \frac{\overline{Diss}}{c} \qquad (4)$$

Eqn. (3) is solved with the no slip condition imposed at the bed, and the condition that the total flux compensates the wave drift (determined from the Boussinesq model) and the velocity V_{st} determined by the boundary layer model.

2.3 Sediment Transport Module

2.3.1 Bed Load and Sediment Concentrations

The instantaneous near bed concentration C_b is calculated from the formulation suggested by Zyserman and Fredsøe (1994). The value of C_b depends on the instantaneous value of the Shield's parameter. The instantaneous bed load q_b is calculated using the formulation by Engelund and Fredsøe (1976).

Neglecting the convective terms, the distribution of the sediment concentration C can be calculated from the diffusion equation,

$$\frac{\partial C}{\partial t} = w_f \frac{\partial C}{\partial y} + \frac{\partial}{\partial y}\left(\epsilon_s \frac{\partial C}{\partial y} \right) \qquad (5)$$

where, ϵ_s is the sediment diffusion coefficient assumed to be the same as the flow eddy viscosity ϵ, and w_f is the fall velocity. Eqn. (5) is solved with the boundary conditions $C = C_b$ at the bed ($y = 2D_{50}$), and zero sediment flux through the water surface.

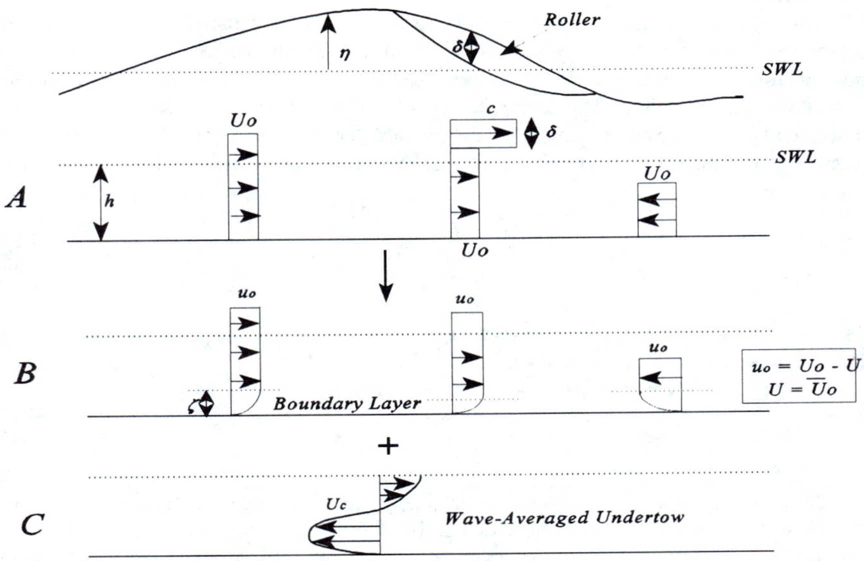

Figure (1): The elements of the hydrodynamic module; A- The wave description. B- The oscillatory current. C- The undertow.

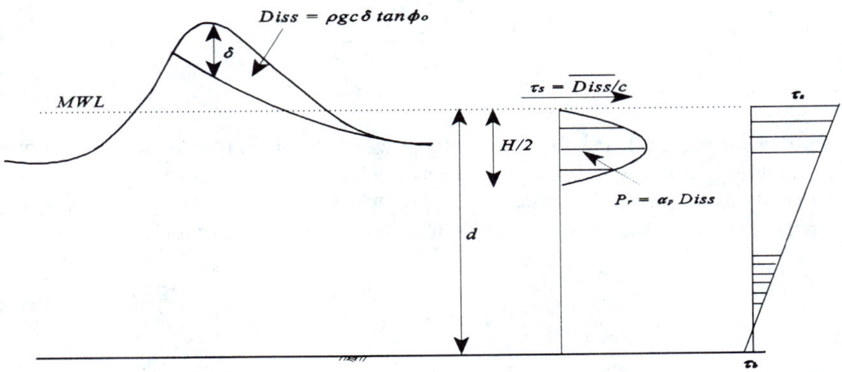

Figure (2): Vertical distribution of production of turbulence and shear stresses due to wave breaking.

2.3.2 Lagrangian Drift

For an Eulerian calculation a contribution due to the Lagrangian drift should be added as an approximation for the wave drift of suspended sediment (Brøker et al., 1991). This approximation is required because the convective terms were not included in the calculations for the sediment concentrations. The Lagrangian drift velocity U_l results from the fact that the water particles do not follow a closed path, but a net forward displacement exists. As an approximation it is assumed that the sediment on average follows the fluid motion. The drift current for each wave is calculated from,

$$U_l = \frac{1}{c}\left(\overline{u^2} - (\bar{u})^2\right) \tag{6}$$

where an overbar again denotes time averaging over a single wave.

2.3.3 Total Sediment Transport Rates

The total instantaneous sediment transport rates are obtained from,

$$q'_s = q_b + q_{sw} + q_{su} + q_{sl} \tag{7}$$

where the sediment transport rates due to the oscillatory motion q_{sw} are evaluated by integrating uC over the water depth. The contributions due to the undertow q_{su} and the Lagrangian drift q_l are evaluated by integrating U_cC and U_lC over the water depth respectively. The swash zone was included in an approximate manner by assuming that q'_s varies linearly from the last grid point to the location of the water line for each time step. Time averaging of q'_s over the time series provides the time averaged sediment transport rates q_s. Figure (3) shows the instantaneous sediment transport rates calculated for a regular and irregular wave of the same deep water rms wave height for Test 1c explained later. As shown in Figure (3) the time variation of the sediment transport rates under an irregular wave are quite different from a " representative " regular wave.

2.4 MORPHOLOGY MODULE

The bathymetry is updated by solving the conservation of sediment equation,

$$\frac{\partial Z_b}{\partial t} = -\frac{1}{(1-n)}\frac{\partial q_s}{\partial x} \tag{8}$$

where Z_b is the bed level, and n is the porosity of the sediment assumed to be 0.4. A Forward in Time Central in Space (FTCS) finite difference scheme was used to solve Eqn. (8), with an additional diffusive term for the numerical stability (Abbott and Basco, 1989). The resulting finite difference equation would correspond to the following differential equation,

$$\frac{\partial Z_b}{\partial t} + \frac{1}{(1-n)}\frac{\partial q_s}{\partial Z_b}\frac{\partial Z_b}{\partial x} = K\frac{\partial^2 Z_b}{\partial x^2} \tag{9}$$

where, K is a diffusion coefficient assumed to be proportional to q_s,

$$K = \epsilon_s |q_s| \tag{10}$$

which is similar to the additional gravitational term included by De Vriend et al. (1993) and

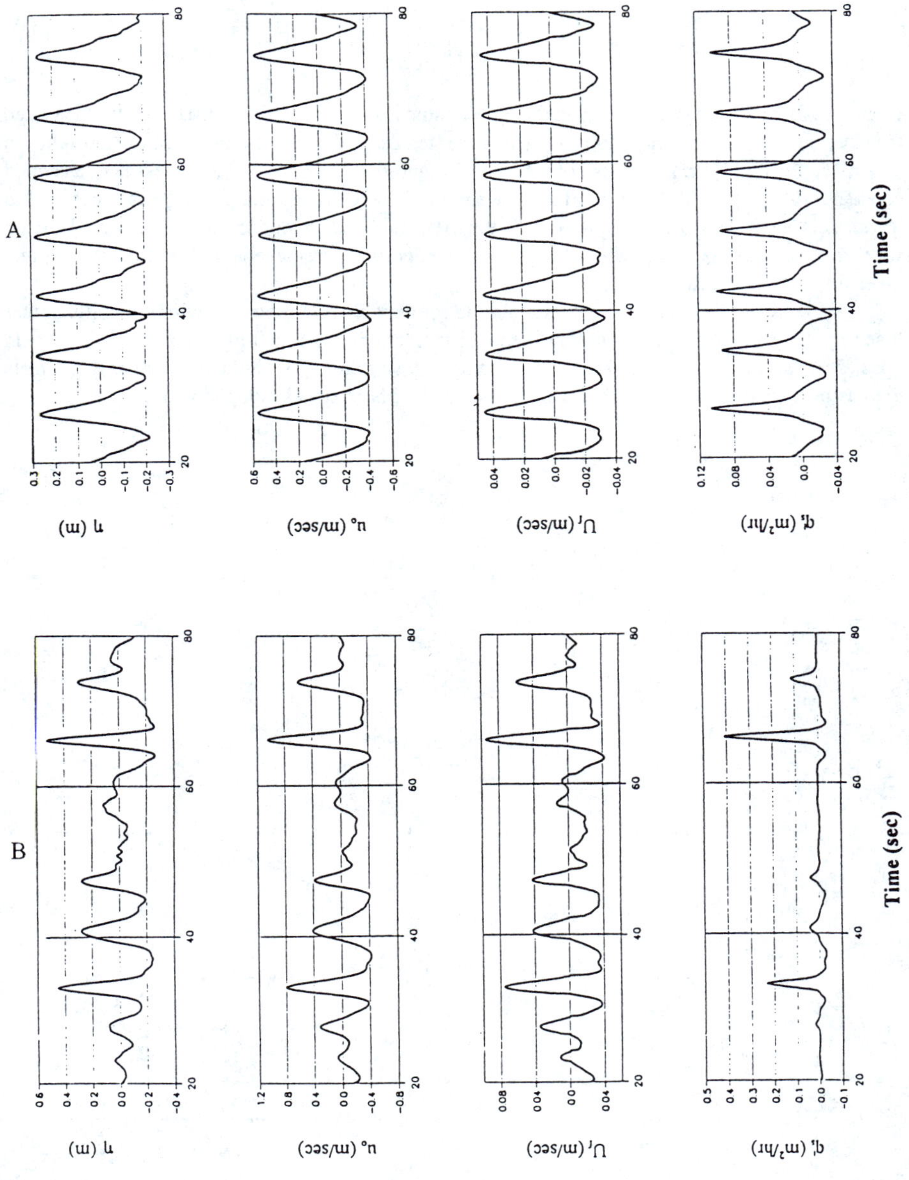

Figure (3): A- Time series of η, u_o, U_f, and q_s of Test 1c (regular wave).
B- Time series of η, u_o, U_f, and q_s of Test 1c (irregular wave).

Horikawa (1988). ϵ_s is an empirical coefficient. The following constraint on the morphological time step Δt must be satisfied (Abbott and Basco, 1989),

$$D_{cr} \leq \frac{1}{2} \ ; \ D_{cr} = \frac{K\Delta t}{\Delta x^2} \qquad \textbf{(11)}$$

where, Δx is the spatial grid spacing. In the present study two time steps are used; the first is called the inner time step Δt_i and the second the outer time step Δt_{max}. The maximum value of Δt_i is chosen to satisfy Eqn. (11). Eqn. (8) is solved over the duration Δt_{max} with the values of q assumed constant (Horikawa, 1988, and Rakha and Kamphuis, 1996). The value of Δt_{max} was specified to the model together with a criterion limiting the maximum change in bed level to 10% of the deep water average wave height (Rakha et al., 1996). A modified Lax-Scheme (Abbott, 1979) with no inner times steps was also tested.

Figure (4) shows a long term simulation for a highly erosive beach with a specified outer time step Δt_{max} = 0.2 hr. The actual time step was less than 0.1 hr resulting in nearly 80 wave field updates. As shown in Figure (4) the bars tended to move offshore with the depth over their crests increasing with time. Onshore of the previous bar, new bars also developed with time.

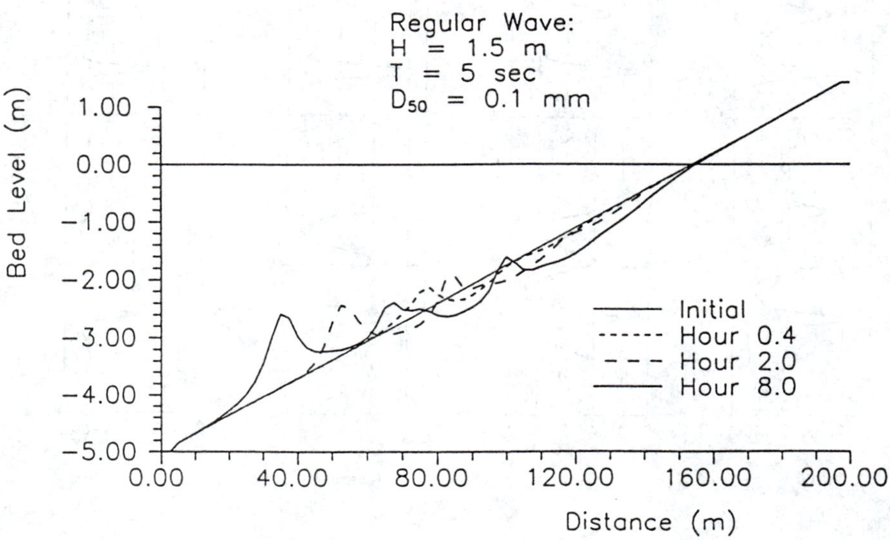

Figure (4): Long term morphology simulation for a regular wave.

3. MODEL VALIDATION

3.1 LIP 11D Delta Flume Tests

The Delta Flume '93 tests were performed at the Delft Hydraulics large-scale wave flume. These tests were supported by the "Large Installations Plan" (LIP) of the European Union. The main purpose of the tests was to provide high quality data to validate/calibrate numerical beach profile models. The details of the tests performed can be found in Arcilla et al. (1993) and Roelvink and Reniers (1995). Two tests (Tests 1b and 1c) were selected for the model verification. Test 1b represents a highly erosive wave condition, and Test 1c a strongly accretive wave condition. Table (1) provides a summary of the test conditions used in this paper.

Table (1): Summary of Test Conditions.

Test	H_{mo} (m)	T_p (sec)	D_{50} (mm)	Type
LIP 11D: Test 1c	0.60	8.0	0.2	Irregular
LIP 11D: Test 1b	1.40	5.0	0.2	Irregular
Shimizu et al. (1985): Test 3-2	1.05	6.0	0.27	Regular

The prediction of the characteristic wave height H_{mo} variation over the initial bathymetry for Tests 1c and 1b is shown in Figures (5) and (6) respectively. The Boussinesq model predicted the wave heights well for both Tests. It appears that the wave heights are slightly overpredicted inside the surf zone for Test 1c (Figure, 5). The results obtained using a phase-averaged wave model based on the Battjes and Janssen (1978) model are also shown in both figures, which shows that such a model provides a good estimate for H_{mo}.

Figures (7) and (8) show the time-averaged undertow for Tests 1c and 1b. At some locations measurements were performed twice at different hours of the test. Figures (7) and (8) show that the undertow is predicted well for most sections. The undertow however is overpredicted just before the bar (x = 138) for Test 1c and underpredicted in the trough of the bar (x = 145) for both Tests 1b and 1c.

The predicted on/offshore sediment transport rates for Test 1c are shown in Figure (5). Very good results are obtained outside the surf zone showing that the calculation of the sediment transport rates due to the oscillatory current is well predicted. Over the bar, the overestimation of the undertow (Figure, 8) results in the underestimation of the onshore sediment transport rates. Good predictions of the sediment transport rates for Test 1b are also shown in Figure (6), where the model follows the measured sediment transport rates well. The onshore sediment transport rates are overestimated inside the trough of the bar (x = 145) due to the underestimation of the undertow as explained earlier (Figure, 8). The use of the Boussinesq model provided much better results than the use of a phase-averaged wave model (Figures,5 & 6). Linear wave theory was used with the rms wave height assumed to be the representative wave for the phase-averaged wave model calculations.

Figure (9) shows a decomposition of the time averaged sediment transport rates into three of the four calculated components. The three contributions plotted are the Lagrangian drift, the integral of uC, and the contribution due to the undertow. The bed load was not included since it gives only a small contribution for these tests. From Figure (9) it can be seen that for Test 1c just before the bar where some of the waves start breaking, the undertow contribution increases rapidly. The measurements suggest that the drop in the sediment transport rates is delayed and occurs on the top of the bar. Figure (9) shows that for eroding beaches (Test 1b) the contributions due to the oscillatory current and the Lagrangian drift are important. The undertow contribution dominates only at the locations where a high percentage of waves are breaking.

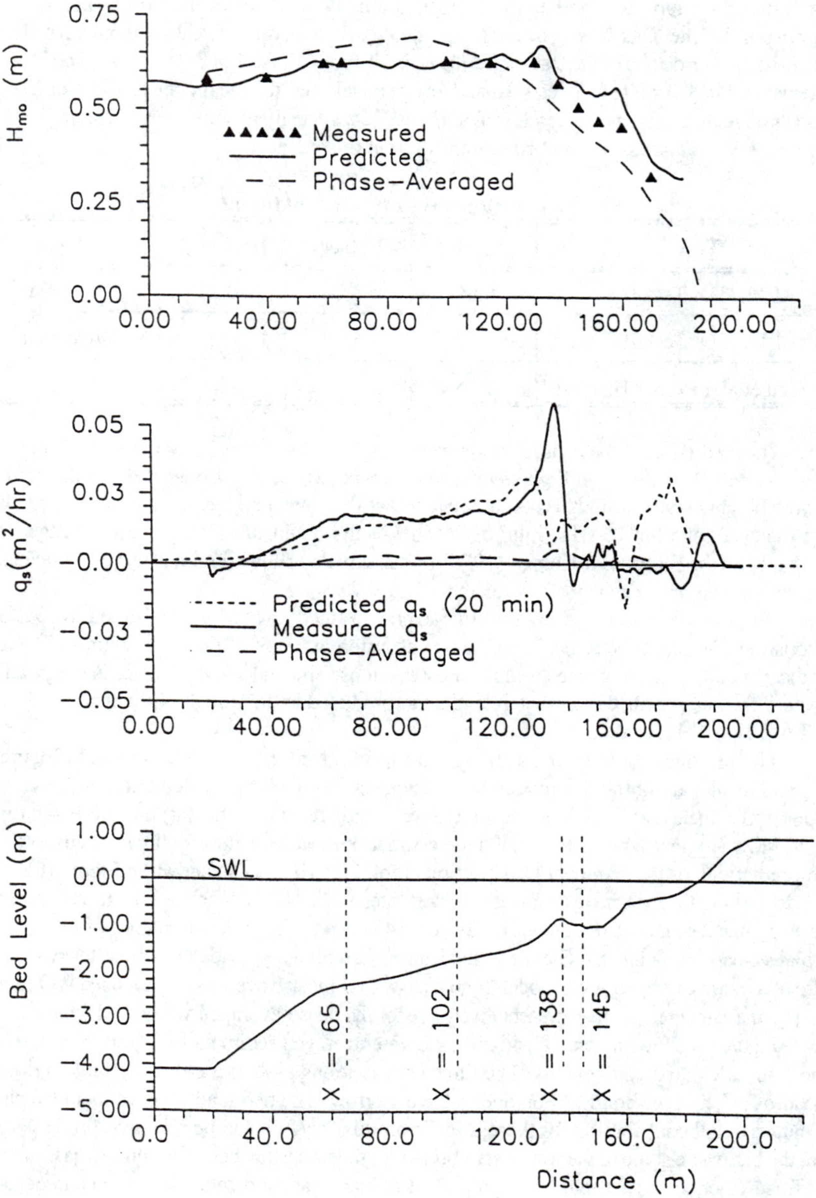

Figure (5): Model results for H_{mo} and q_s of Test 1c (LIP 11D).

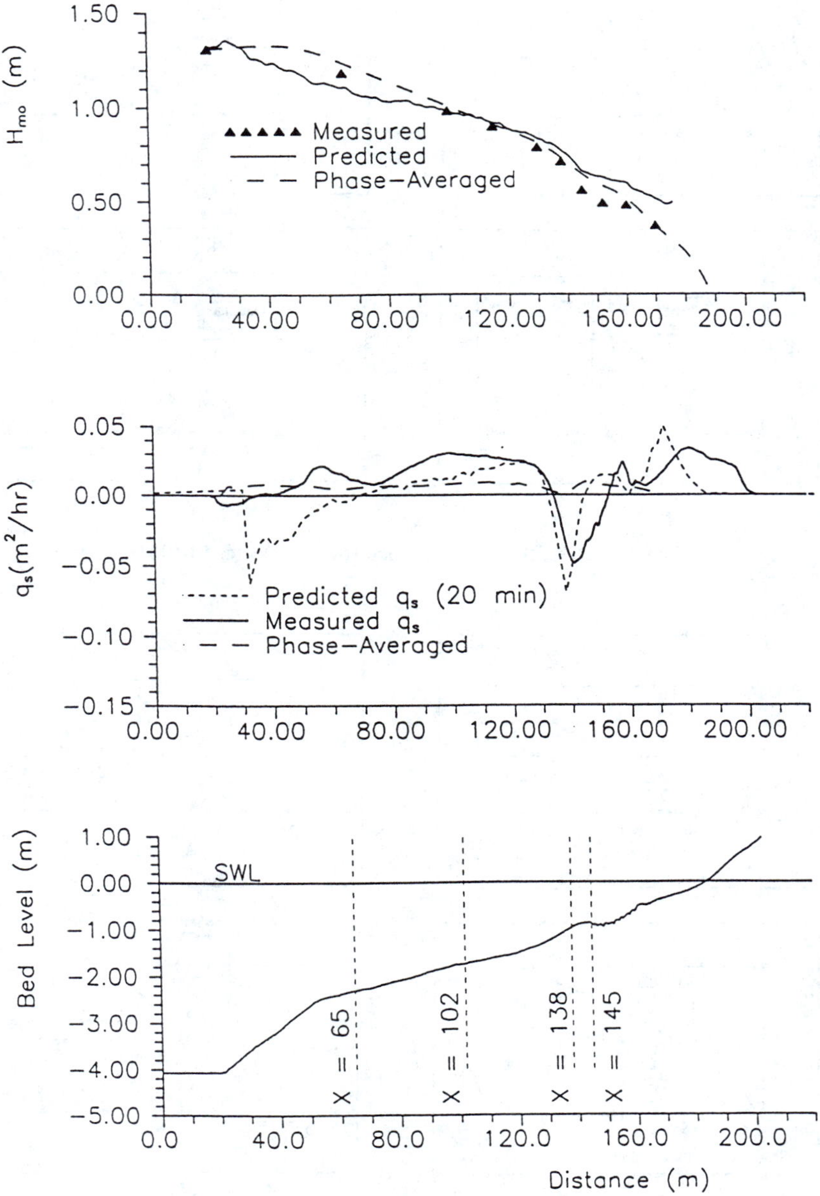

Figure (6): Model results for H_{mo} and q_s of Test 1b (LIP 11D).

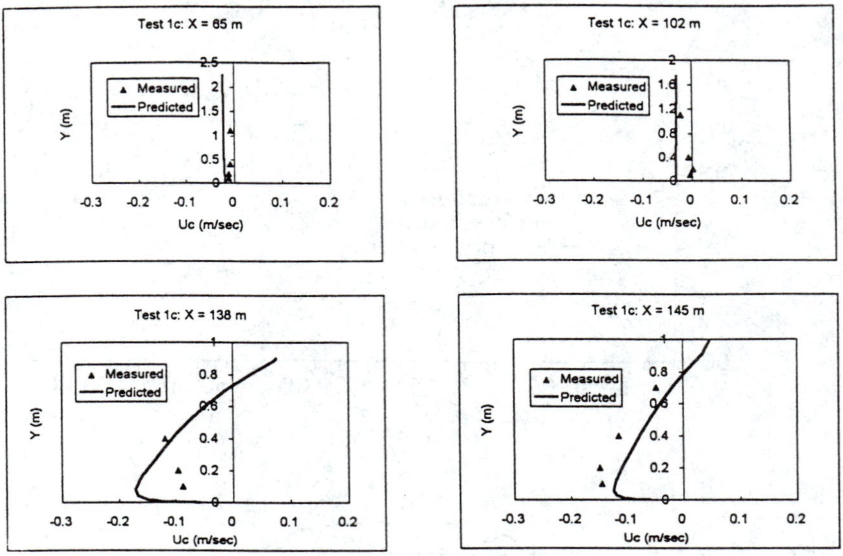

Figure (7): Model results for Undertow of Test 1c (LIP 11D).

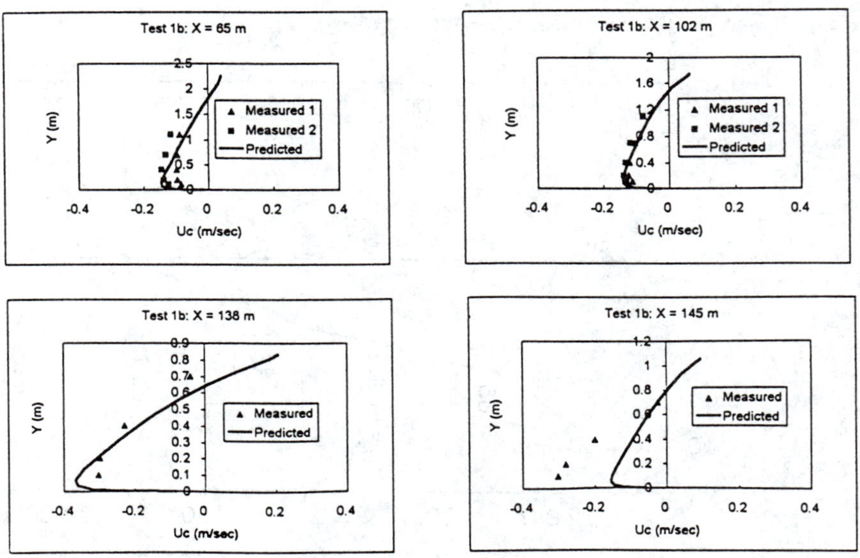

Figure (8): Model results for Undertow of Test 1b (LIP 11D).

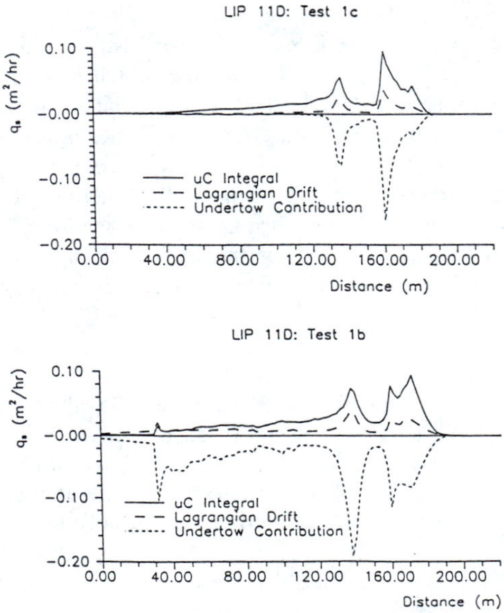

Figure (9): Sediment transport rate contributions for Tests 1c and 1b (LIP 11D).

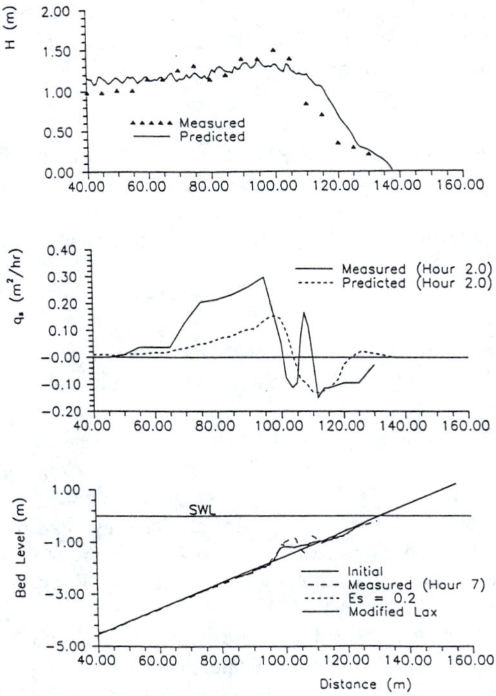

Figure (10): Model results for Test 3-2 (Shimizu et al., 1985).

3.2 Validation of Morphology Model

Figure (10) shows the morphology model predictions for Test 3-2 of Shimizu et al. (1985) with the test conditions provided in Table (1). Good results were obtained for the bar formation as shown in Figure (10) although the numerical model underpredicted the bar development due to the underprediction of the sediment transport rates outside the surf zone. Figure (10) shows that the Boussinesq model underpredicts the shoaling of the waves outside the surf zone. This underprediction of the wave shoaling could be responsible for the underprediction of the sediment transport rates. As shown in Figure (10) the modified Lax-scheme and the FTCS scheme with ϵ_s = 0.2 gave similar results. A sensitivity analysis on the effect of using a value of ϵ_s = 2.0 (as suggested by Watanabe, 1994) showed that the model was not sensitive to the value of ϵ_s for this case.

CONCLUSIONS

A phase-resolving wave transformation module was combined with an intra-wave sediment transport module to predict the on/offshore sediment transport rates and the resulting beach evolution. The roller geometry was used to determine the instantaneous production of turbulence and a mean shear stress at the water surface. The instantaneous sediment transport rates due to the oscillatory wave induced currents (bed load and suspended load), the undertow, and the Lagrangian drift were calculated.

The present model was verified against a few test conditions. The model predicted the undertow and the sediment transport rates well for both erosive and accretive conditions. The onshore sediment transport rates for accretive beaches were underpredicted before the bar due to the overprediction of the undertow, and the underprediction of wave shoaling.

The results of the present paper showed that even for eroding beaches the undertow contribution to the sediment transport rates dominated only at local areas, suggesting that other contributions should not be neglected for eroding beaches.

ACKNOWLEDGMENT

The Danish National Research Foundation financed the first and third authors and the Danish Technical Research Council under the program 'Marin Teknik' financed the fourth author, their support is greatly appreciated. The assistance of Ole Sørensen with the Boussinesq model is also appreciated. Dano Roelvink (Delft Hydraulics) helped in providing the data for the LIP experiments.

REFERENCES

Abbott, M.B. (1979). Computational Hydraulics. Pitman Books Ltd.

Abbott, M.B. and Basco, D.R. (1989). Computational Fluid Dynamics. John Wiley and Sons, Inc., New York. 425p.

Arcilla, A.S., Roelvink, J.A., O'Conner, B.A., Reniers, A., and Jimenez, J.A. (1994). The Delta Flume '93 experiment. Coastal Dynamics 1994, ASCE, Barcelona, pp. 488-502.

Battjes, J.A. and Janssen, J.P.F.M. (1978). Energy loss and set-up due to breaking of random waves. Proc. 16th Int. Conf. Coastal Eng., ASCE, Vol.1, pp. 569-589.

Brøker, H.I., Deigaard, R., and Fredsøe, J. (1991). On/offshore sediment transport and morphological modelling of coastal profiles. Proc. ASCE Specialty Conf. Coastal Sediments' 91, Seatlle, WA, pp. 643-657.

Deigaard, R., (1989). Mathematical modelling of waves in the surf zone. Progress Report 69, ISVA, Technical University, Lyngby, Denmark, pp. 47-59.

Deigaard, R., Justesen P., and Fredsøe, J. (1991). Modelling of undertow by a one-equation turbulence model. Coast. Eng., Vol.15, pp.431-458.

De Vriend, H.J., Zyserman, J., Nicholson, J., Roelvink, J.A., Péchon, P., and Southgate, H.N. (1993). Medium-term 2DH coastal area modelling. Coast. Eng., Vol.21, pp.193-224.

Engelund, F., and Fredsøe, J. (1976). A sediment transport model for straight alluvial channels. Nordic Hydrology, Vol. 7, pp. 293-306.

Fredsøe, J. (1984). The turbulent boundary layer in combined wave current motion. Journal of Hydraulic Eng., ASCE, Vol. 110, No.HY8, pp. 1103-1120.

Fredsøe, J., Andersen, O.H., and Silberg, S. (1985). Distribution of suspended sediment in large waves. Journal of Waterway, Port, Coastal and Ocean Eng., ASCE, Vol. 111, No. 6, pp. 1041-1059.

Fredsøe, J., and Deigaard, R. (1992). Mechanics of Coastal Sediment Transport. World Scientific. 369p.

Horikawa, K. (1988). Nearshore dynamics and coastal processes. University of Tokyo Press.

Madsen, P.A., Murray, R., and Sørensen (1991). A new form of the Boussinesq equations with improved linear dispersion characteristics. Coast. Eng., Vol.15, pp.371-389.

Madsen, P.A., Sørensen O.R. (1992). A new form of the Boussinesq equations with improved linear dispersion characteristics. Part 2. A slowly-varying bathymetry. Coast. Eng., Vol.18, pp.183-204.

Madsen, P.A., Sørensen O.R., Schäffer, H.A. (1994). Time domain modelling of wave breaking, runup, and surf beats. 24th Int. Conf. on coastal Eng., pp. 399-411.

Rakha, K.A., and Kamphuis, J.W. (1996). A morphology model for an eroding beach backed by a seawall. Submitted to Coastal Eng.

Rakha, K.A., Deigaard, R., Brøker, I. (1996). A phase-resolving cross shore sediment transport model for beach profile evolution. Submitted to Coastal Eng.

Roelvink, J.A. and Brøker, I. (1993). Cross-shore profile models. Coast. Eng., ASCE, Vol.21, pp.163-191.

Roelvink, J.A. and Reniers, A.J.H.M. (1995). LIP 11D delta flume experiments. Data Report, Delft Hydraulics, H 2130.

Schäffer, H.A., Madsen, P.A., and Deigaard, R. (1993). A Boussinesq model for waves breaking in shallow water. Coastal Eng., Vol. 20: 185-202.

Shimizu, T., Sato, S., Maruyama, K., Hasegawa, H., and Kajima, R. (1985). Modeling of cross-shore sediment transport rate distributions in a large wave flume, Report of Central Res. Inst. for Electric Power Industry, Rep. No. 384028, 60 p. (in Japanese).

Watanabe, A. (1994). A mathematical model of beach processes under sheet-flow condition using nonlinear wave theory. International Symposium Waves - Physical and Numerical Modelling, Vancouver, Canada, pp.1520-1529.

Zyserman, J.A. and Fredsøe, J. (1994). Data analysis of bed concentration of suspended sediment. Journal of Hydr. Engrg., ASCE, 120(9).

CHAPTER 237

Cross-shore Sediment Transport and Beach Deformation Model

Rattanapitikon Winyu [1]
Tomoya Shibayama [2]

Abstract

Based on a large amount of published laboratory results, reliable model is developed for computing beach profiles under regular wave actions. The sediment transport is separated into suspended load and bed load. The suspended load is computed as the product of the time-averaged suspended concentration and the time-averaged velocity. The bed load is developed following the similar step as Watanabe (1983) but the applied area is different. The wave model of Dally et al. (1985) is modified and used to compute wave height transformation. The beach profile change is computed from the conservation of sediment mass. The beach deformation model is verified with small scale and large scale experiments. Reasonably good agreement is obtained between measured and computed beach profiles.

I. Introduction

In the previous research works, most of the models were developed based on data with the limited experimental conditions. Therefore their validity is limited according to the range of experimental conditions which were employed in the calibration or examination. The evidence is that there are so many models exist. At this moment, the experimental results obtained by many researchers have been accumulated and a large number of experimental results have become available. It is a good time to develop a model based on the large amount and wide range of experimental results. The present model is developed based on 1138 data sets of 24 sources of published experimental results covering both small and large scale experiments, as shown in table 1.

----

[1] D. Eng., Dept. of Civil Engineering, Sripatum University, 61 Phahonyothin Rd., Jatujak, Bangkok, 10900, Thailand.

[2] D. Eng., Assoc. Prof., Dept. of Civil Engineering, Yokohama National University, Hodogaya-ku, 240, Yokahama, Japan.

Table 1. Sources and number of collected data for present sediment transport study.

Sources	$c(z)$	$u(z)$	H	$h(x,t)$	Others	Experiment
Bosman and Steetzel (1986)	3					SE
Deigaard et al. (1986)	6					SE
Dette and Uliczka (1986)	11					LE
Duncan (1981)					A,12	SE
Hansen and Svendsen (1984)		4	1			SE
Hayakawa et al. (1983)	4					SE
Horikawa and Kuo (1966)			213			SE
Horikawa et al. (1982)	7					SE
Irie et al. (1985)	27					SE
Kajima et al. (1983)	149	219	79	91		LE
Kraus and Larson (1988)				69		LE
Nadaoka et al. (1982)		11	2		h_{ot},2	SE
Nagayama (1983)			12			SE
Nakato et al. (1977)	3					SE
Nielsen (1979)	44					SE
Okayasu et al.(1988)		44	9		h_{ot},9	SE
Okayasu et al.(1989)					h_{ot},47	SE
Sato et al. (1988, 1989)			5			SE
Sato et al. (1990)	14					SE
Sawamoto et al. (1981)	4					SE
Shibayama and Horikawa (1985)			10	11		SE
Skafel and Krishnappan (1984)	8					SE
Sleath (1982)	4					SE
Vongvisessomjai (1986)	4					SE
Total	288	278	331	171	70	

where $c(z)$ is the time-averaged sediment concentration, $u(z)$ is the time-averaged fluid velocity, H is the wave height, $h(x,t)$ is the beach profile, A is the cross section area of surface roller, h_{ot} is the still water depth at transition point, SE is the small scale experiment, and LE is the large scale experiment.

1.1 Governing equation

The cross-shore change in local water depth, h, can be calculated by solving the conservation of sediment mass as the following

$$\frac{\partial h}{\partial t} = \frac{1}{(1-\Lambda)}\frac{\partial q_t}{\partial x} \qquad (1)$$

where t is the time, x is the horizontal coordinate in cross-shore direction, Λ is the porosity, and q_t is the total transport rate per unit width.

The sediment transport rate is usually expressed as

$$q_t = \int_{-\delta}^{h} c(z)u(z)\,dz \qquad (2)$$

where δ is the level above which there is no effective movement of sand particles, z is the vertical coordinate measured upward from the bed, $c(z)$ is the time-averaged sediment concentration, and $u(z)$ is the time-averaged fluid velocity.

In order to compute the transport rate, q_t, the sediment concentration and fluid velocity should be known first. The sediment concentration profile was described in Shibayama and Rattanapitikon (1993). The following sections will describe about velocity profile, sediment transport, wave model, and beach deformation model, respectively.

II. Time-Averaged Velocity Profiles

Following Okayasu et al. (1988, pp. 93-94), the vertical distribution of time-averaged velocity profile is calculated based on the assumption of eddy viscosity model. By considering time-averaged values, the eddy viscosity model can be expressed as

$$\tau = \rho \nu_t \frac{\partial u}{\partial z} \qquad \text{(3)}$$

where τ is the time averaged shear stress, ρ is the fluid density, ν_t is the time-averaged eddy viscosity coefficient, u is the time-averaged velocity, and z is the upward vertical coordinate from the bed.

To solve the eddy viscosity model, one boundary condition of velocity should be given and the expression of τ / ν_t should also be known. In this study, the vertical-averaged value of velocity, u_m, is used as the boundary condition for Eq. 3 and will be described in the following subsection.

2.1 Vertically averaged velocity

Using the concept of Svendsen (1984), the vertically averaged velocity, from bed to wave trough, u_m, consists of two components. One is caused by the wave motion, u_w, and the other is caused by the surface roller, u_r.

$$u_m = u_w + u_r \qquad \text{(4)}$$

Various formulas for computing u_w and u_r have been suggested by the previous researchers. From the previous studies, of Duncan (1981), Svendsen (1984), Stive and Wind (1986), Deigaard et al. (1991), and Fredsoe and Deigaard (1992), we can conclude that there are three formulas for computing u_w and three formulas for computing u_r (see in table 2).

Table 2. Formulas for computing u_m and u_r

	Formula 1	Formula 2	Formula 3
u_w	$\frac{B_o \sigma H^2 \coth(kh)}{h}$	$\frac{B_o c_w H^2}{h^2}$	$\frac{c_w H}{h}$
u_r	$\frac{H^3}{(0.7Th^2)}$	$\frac{0.9H^2}{(Th)}$	$\frac{0.1c_w H}{h}$

where $B_o = (1/T)\int_0^T (\xi/H)dt$, c_w is the phase velocity, H is the wave height, h is the mean water depth, T is the wave period, and ξ is the water surface elevation measured from mean water level.

The general form of u_m may be written as

$$u_m = au_w + bu_r \quad \text{(5)}$$

where a and b are the constants.

The proper combination of u_w and u_r and the constants a, b can be found from multi-regression analysis with the observed u_m.

After the regression analysis and include the effect of transition zone, the final equation for computing u_m can be written as

$$u_m = 0.77\frac{B_o \sigma H^2 \coth(kh)}{h} + b_1 0.1\frac{c_w H}{h} \quad \text{(6)}$$

where b_1 is the constant and expressed as

$$b_1 = \begin{cases} 0 & \textit{offshore zone} \\ \left(1/\sqrt{H} - 1/\sqrt{H_b}\right)/\left(1/\sqrt{H_t} - 1/\sqrt{H_b}\right) & \textit{transition zone} \\ 1 & \textit{inner zone} \end{cases}$$

where H_b and H_t is the wave height at breaking point and transition point.

Example of examination results of u_m along the cross-shore direction are shown in Fig. 1.

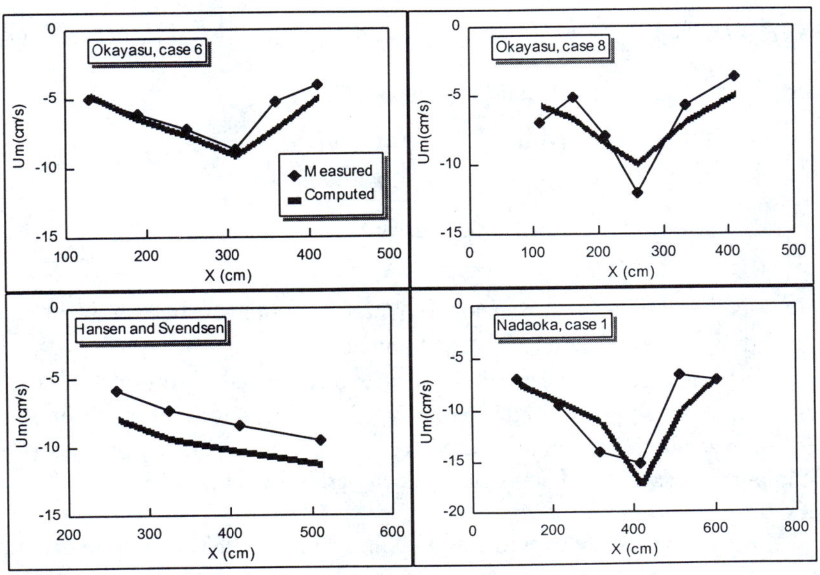

Figure 1. Example of cross-shore variations of measured and computed vertical averaged velocity, u_m.

2.2 Vertical distribution of shear stress and eddy viscosity coefficient

Based on dimensional analysis, Okayasu (1989, pp. 93-94) proposed a formulas for computing τ / ν_t as follows

$$\frac{\tau}{\nu_t} = \rho^{1/3} D_B{}^{1/3}\left[\frac{k_3}{d_t} + \frac{k_4}{z}\right] \quad \text{......} (7)$$

where k_3 and k_4 are the constants, d_t is the water depth at wave trough, and D_B is the energy dissipation. From the bore model (Thornton and Guza, 1983),

$$D_B = \frac{\rho g H^3}{4Th} \quad \text{......} (8)$$

By inserting Eq. 7 into Eq. 3 and then, after an integration of Eq. 3 and use u_m as the boundary condition, the analytical solution of u can be expressed as

$$u = \rho^{1/3} D_B{}^{1/3}\left[k_3\left(\frac{z}{d_t} - \frac{1}{2}\right) + k_4\left(\ln\frac{z}{d_t} - 1\right)\right] + u_m \quad \text{......} (9)$$

As mentioned in previous research works, the energy dissipation process at the breaking point is not in the same manner as in the inner zone (e.g., Okayasu, 1989). To incorporate this process, Eq. 9 may be written as follows

$$u = k_5 \rho^{1/3} D_B{}^{1/3}\left[k_3\left(\frac{z}{d_t} - \frac{1}{2}\right) + k_4\left(\ln\frac{z}{d_t} - 1\right)\right] + u_m \quad \text{......} (10)$$

where k_5 is the constant and equal to 1 at the inner zone.

From the multi-regression analysis and assume linear distribution of energy dissipation in the transition zone, Eq. 10 become

$$u = b_2 \rho^{1/3} D_B{}^{1/3}\left[b_3\left(\frac{z}{d_t} - \frac{1}{2}\right) - 0.22\left(\ln\frac{z}{d_t} - 1\right)\right] + u_m \quad \text{......} (11)$$

where b_2 and b_3 are the constants and expressed as,

$$b_2 = \begin{cases} 0.3 + 0.7\left(x_b - x\right)/\left(x_b - x_t\right) & \textit{transition zone} \\ 1.0 & \textit{inner zone} \end{cases}$$

$$b_3 = \begin{cases} \left(x_b - x\right)/\left(x_b - x_t\right) & \textit{transition zone} \\ 1.0 & \textit{inner zone} \end{cases}$$

where x is the position in cross-shore direction, x_b is the position at the breaking point, and x_t is the position of the transition point.

The comparison between measured u and computed u from Eq. 11, using measured u_m, are shown in Fig. 2. From Fig. 2 we can judge that the form of τ / ν_t in Eq. 7 are accurate enough to be used for computing the velocity profiles. Examples of measured and computed velocity are shown in Fig. 3.

2.3 Formula verification

The prototype scale experiment of Kajima et al. (1983) is used to examine the validity of present formula. By using the same formula as that used for small scale wave flume (Eq. 11), Fig. 4 shows examples of verification results in the term of averaged velocity of each section along the cross-shore direction.

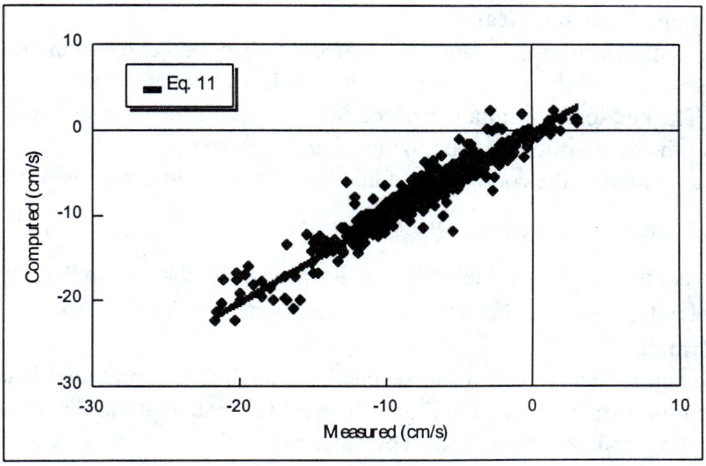

Figure 2. Comparison between measured and computed u (using measured u_m).

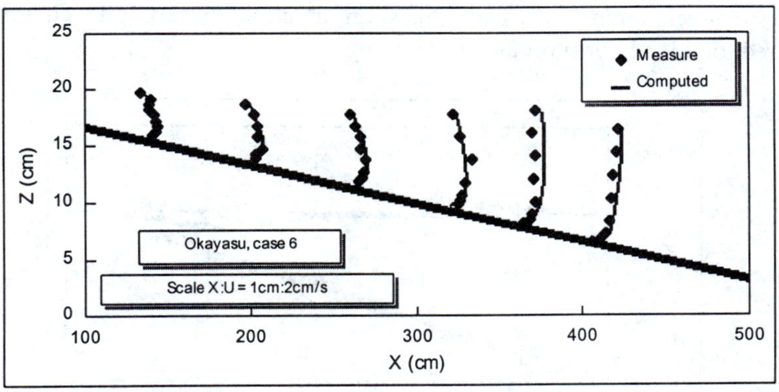

Figure 3. Comparison between measured and computed velocity profiles (laboratory data from Okayasu et al., 1988, case 6).

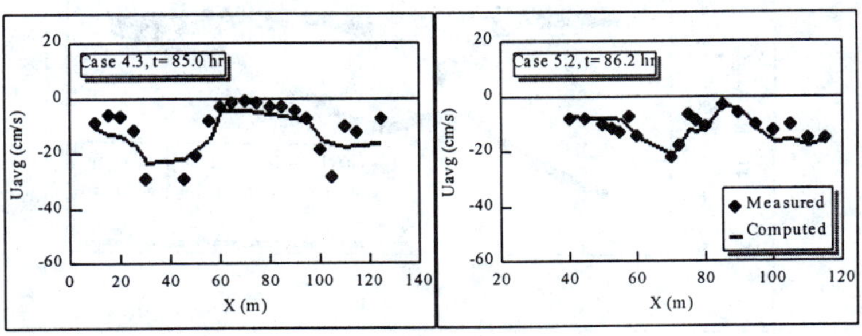

Figure 4. Example of verification results in term of averaged velocity along the cross-shore direction (laboratory data from Kajima et al., 1983).

III. Sediment Transport Rate

The sediment transport rate is separated into bed load and suspended load. The suspended load is expressed as the product of sediment concentration and fluid velocity. The bed load formula is derived in the similar procedure as Watanabe (1983) but the assumption on the time level and region of application are not the same. After calibrate the coefficient of bed load, the total load can be written as

$$q_t = \int_{\delta_s}^{d_t} c(z)u(z)\,dz + 2.0(\psi - \psi_c)\sqrt{\psi}\,w_s d \quad \text{.................................... (12)}$$

where ψ is the Shields parameter, ψ_c is the critical Shields parameter, w_s is the falling velocity, and δ_s is the bottom boundary layer which computed from Jonsson (1966) formula.

The comparison of computed suspended load, bed load and total load and the measured total load are shown in Fig. 5. It should be note that the separated equation for computing transport direction is not necessary in the present model. The transport direction is depended on the combination of bed load and suspended load. In the present model, bed load is the dominant transport for accretion beach (on-shore directed transport), and suspended load is the dominant transport for erosion beach (off-shore directed transport). These transport directions correspond well with the measured total load transport as shown in Fig. 5.

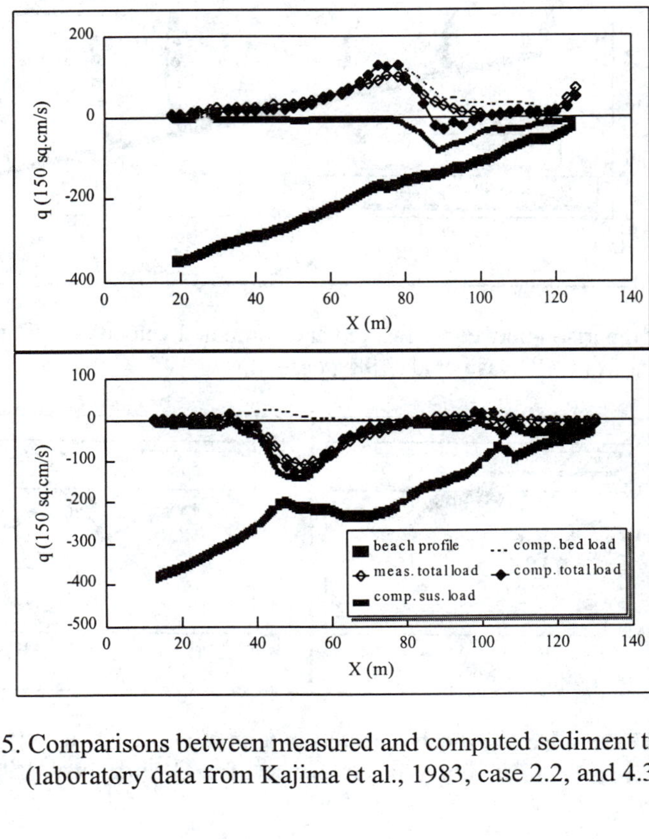

Figure 5. Comparisons between measured and computed sediment transport (laboratory data from Kajima et al., 1983, case 2.2, and 4.3).

IV. Wave Model

For computing beach transformation, the wave model should be kept as simple as possible because of the frequent updating of wave field for accounting the variability of mean water surface and the change of bottom profiles. In the present study, wave height transformation in cross-shore direction will be computed from the energy flux conservation.

$$\frac{\partial E c_g}{\partial x} = -D_B \quad \text{.......} \quad (13)$$

where E is the wave energy density, c_g is the group velocity, and D_B is the energy dissipation rate which is zero outside the surf zone.

4.1 Energy dissipation rate

Widely used formulas for computing energy dissipation rate are Bore model and Dally et al. (1985) model. The Bore model is shown in Eq. 8, and Dally model is

$$D_B = \frac{0.15 c_g \rho g}{8h}\left[H^2 - (\Gamma h)^2\right] \quad \text{.......} \quad (14)$$

The advantage of Dally model is that it is able to reproduce the pause (or stop breaking) in the wave breaking process at a finite wave height on a horizontal bed or in the recovery zone while the Bore model gives a continuous dissipation due to wave breaking. However, for the condition when waves continuously break both two models give nearly the same result of energy dissipation rates (for example see Fig. 6). Based on the above considerations, Dally model is selected for calculation in the present research.

The published experimental results of wave height transformation inside the surf zone have been collected to test the ability of Dally model as shown in the first column of table 3. Total 9 sources of published experimental results, covering 331 data sets, are used in this section. The experiments cover wide range of wave and bottom topography conditions, including both small scale and large scale wave flume experiments. Most of the experiments were performed under fixed bed conditions, except data of Kajima et al. (1983) and Shibayama and Horikawa (1985) which were performed under movable bed conditions.

The verification results are presented in term of error function, ER, as used by Dally et al. (1985), which is defined as

$$ER = 100\sqrt{\frac{\sum_{i=1}^{tn}\left(H_{ci} - H_{mi}\right)^2}{\sum_{i=1}^{tn} H_{mi}^{\ 2}}} \quad \text{.......} \quad (15)$$

where i is the wave height number, H_{ci} is the computed wave height of number i, H_{mi} is the measured wave height of number i, and tn is the total number of measured wave height. Small value of ER expresses a good prediction.

The verification results of Dally model are shown in the third column of table 3. From the third column of table 3, we can see that Dally model gives quite good estimation of wave height inside the surf zone.

Although, Dally model gives good estimation to the experimental results, it still has some error. We may be able to improve Dally model.

Considering Dally model, the measured Γ can be computed from the measured wave height and water depth by using the following formula (rewriting Eq. 14).

$$\Gamma = \frac{1}{h}\sqrt{H^2 - \frac{\partial\left(Ec_g\right)}{\partial x}\frac{8h}{0.15c_g\rho g}} \qquad \text{(16)}$$

The right hand side term of Eq. 16, RS, can be computed if we have the profile of wave height transformation. If we plot RS with any wave parameters, it will show a horizontal line (since Γ is constant).

Fig. 7 shows the relation between RS and the various dimensionless parameters, i.e., h/L_o, h/L, $h/\sqrt{LH}$. From Fig. 7 we can see that the parameter Γ is not a constant. Comparison among Fig. 7a-7c, the relation between Γ and $h/\sqrt{LH}$, in Fig. 7c, shows more consistent results than the others. A formula for parameter Γ, from Fig. 7c, can be expressed as

$$\Gamma = \exp\left[-0.36 - 1.25\frac{h}{\sqrt{LH}}\right] \qquad \text{(17)}$$

The next question is how much difference of computed wave height is resulted when we compare the results calculated by using constant Γ and calculated by using Γ from Eq. 17. Table 3 shows the error function, ER, when wave height is computed from Dally model by using either constant Γ or Γ from Eq. 17.

From table 3 we can see that the computed results show not much difference, but for most cases the results of computed wave height are improved. The selection of any type of model depends on each researcher. In the present study, the better estimation formula (Eq. 17) will be used.

Table 3. Verification results in term of error function parameter, ER.

Sources	No. of data sets	$\Gamma = 0.4$	Γ from Eq. 17
Hansen and Svendsen (1984)	1	16.12	6.95
Horikawa and Kuo (1966),slope=0	101	13.30	11.65
Horikawa and Kuo (1966),slope=1/80-1/20	112	20.58	17.67
Nagayama (1983)	12	9.19	8.62
Kajima et al. (1983)	79	18.36	16.37
Nadaoka et al. (1982)	2	11.97	10.81
Okayasu et al. (1988)	10	14.18	11.3
Sato et al. (1988)	3	11.36	7.74
Sato et al. (1989)	2	31.83	19.78
Shibayama and Horikawa (1986)	10	16.23	17.69

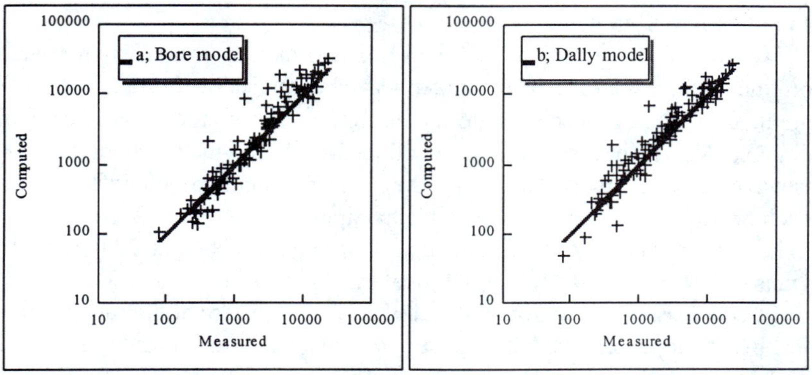

Figure 6. Comparison between measured and computed energy dissipation rate from a: Bore model, b: Dally model (laboratory data from Hansen and Svendsen, 1984, Okayasu et al., 1988, and Sato et al., 1988 and 1989).

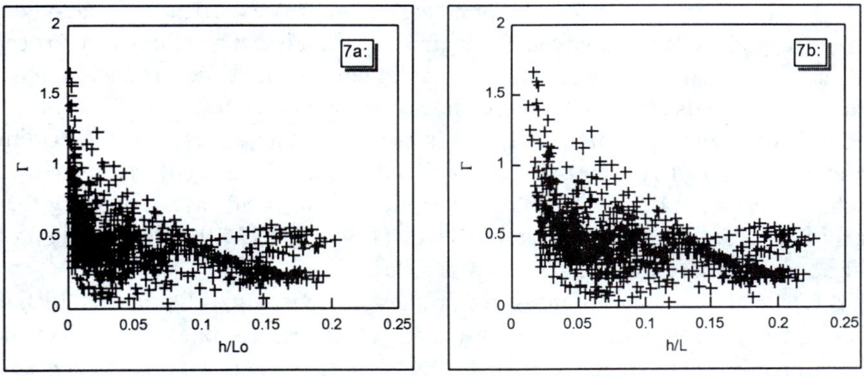

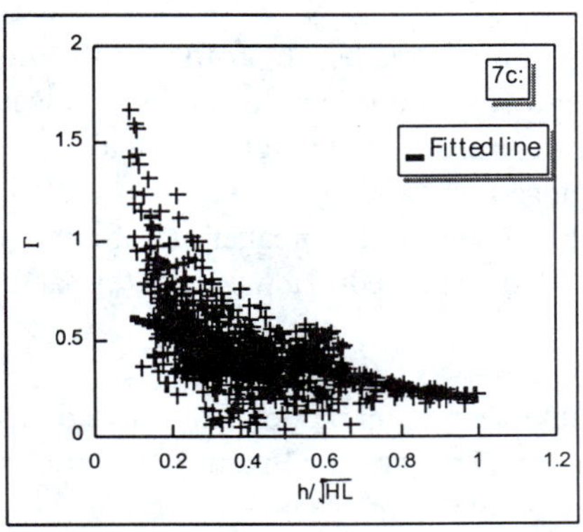

Figure 7. Relations between Γ and a: h/L_o, b: h/L, c: $h/\sqrt{LH}$ (laboratory data from Kajima et al., 1983).

V. Beach Deformation Model

Wave model is used to compute wave transformation from offshore boundary to shoreline boundary. From the computed wave height, sediment transport rate can be computed. Then new beach profile can be computed from the mass conservation equation (Eq. 1). The new beach profile will feed-back into the wave model and causes wave height changes. This yields the loop of dynamic beach deformation and the beach profile at any desired time can be computed.

The shoreward boundary is defined at the wave runup height. Following Larson and Kraus (1989, pp. 170-172), the sediment transport inside the swash zone is assumed to decrease linearly from the end of surf zone to zero at the runup limit. If the local beach slope exceeded the repose angle, avalanching concept will be used.

5.1 Model verification

A number of simulations are performed in order to examine the capability of the present model. All coefficients in the model are kept to be constant for all cases in the verification. Model results are compared with laboratory data of small scale experiment of Shibayama and Horikawa (1985) and large scale experiment of Kajima et al. (1983) and Kraus and Larson (1988).

The verification is performed for all cases, total 45 cases. The verification of these independent data sources and wide range of experiment conditions are expected to clearly demonstrate the accuracy, and stability of present model. The main input data of the model is the incident wave dimensions and initial beach profile. The model was run on workstations (HP 9000 series 700 computers); the total CPU time for simulation about 1000 hr (45 cases) is about 6 min. Examples of examination results are shown in Figs. 8-10. The examination results of all cases, including four profiles per case, are shown in Rattanapitikon (1995, pp. 173-196). Interesting points of the comparison results are described as follows

1. The global shapes of measured profiles are generally well predicted by the model. The predicted profiles are smoother than the measured ones. Small fluctuation of measured profiles can not be predicted by the present model (see Kajima, case 3.1 and Kraus, case 510 in Fig. 8).

2. The agreements of predicted beach profiles of the prototype scale are better than those of small scale wave flume. The fluctuation of measured beach profiles in small scale experiment are more than the prototype scale experiment and the present model can not predict those fluctuations.

3. The model is able to simulate a breaker bar of either the growth of breaker bar (see case 5.2 in Fig. 10) or the reduction of breaker bar (see case 1.2 in Fig. 9).

VI. Conclusions

Cross-shore beach deformation model is developed based on a large amount of published laboratory results. The model contains description of time-averaged concentration profile, velocity profile, sediment transport, wave height, and beach deformation. The validity of model is confirmed by small scale and large scale experiments. The main merit of this model is that it requires very short CPU time and the results are reasonably well.

Acknowledgment

The authors wish to express their gratitude to Prof. Yoshimi Goda of the Yokohama National University for his comments and fruitful discussions.

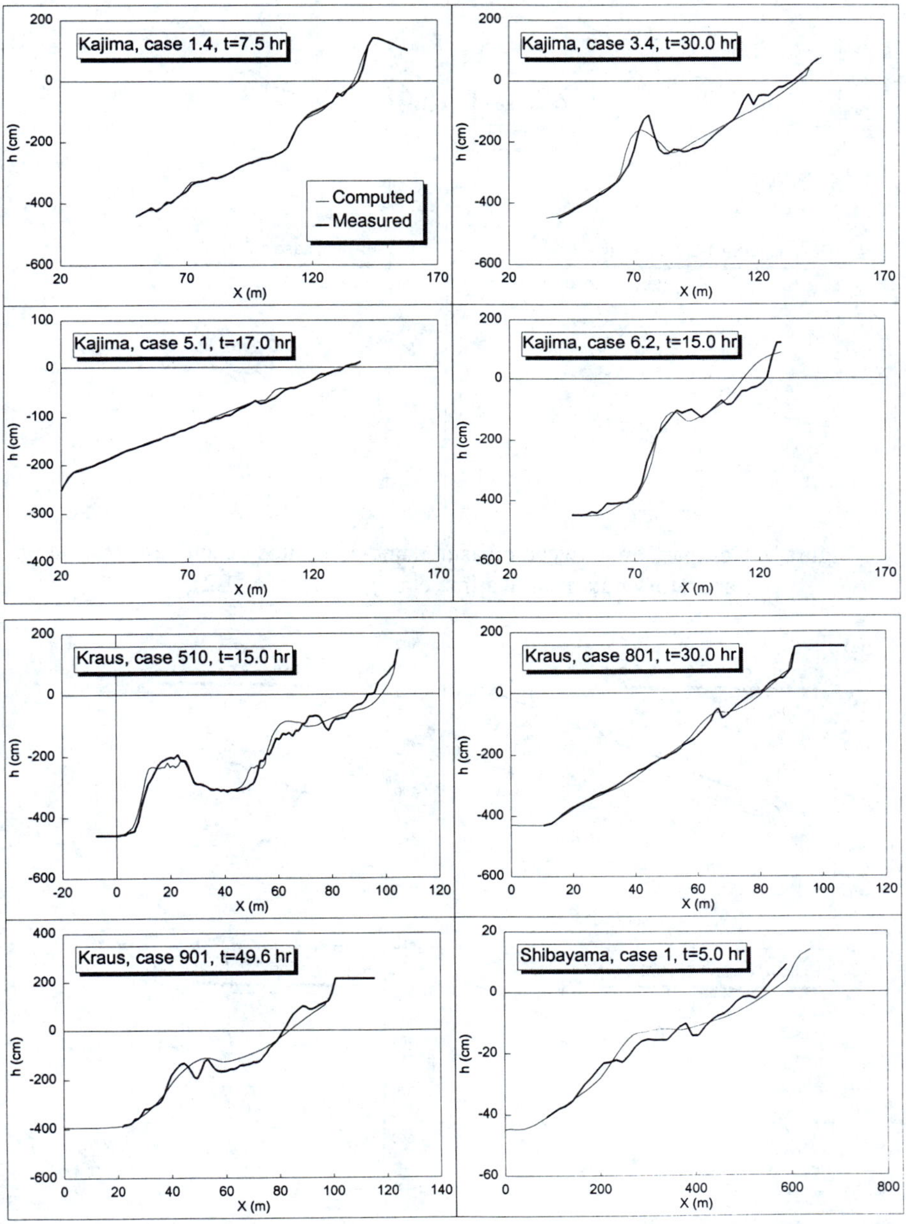

Figure 8. Examples of comparison between measured and computed beach deformation (laboratory data from Kajima et al., 1983, Kraus and Larson, 1988, and Shibayama and Horikawa, 1985).

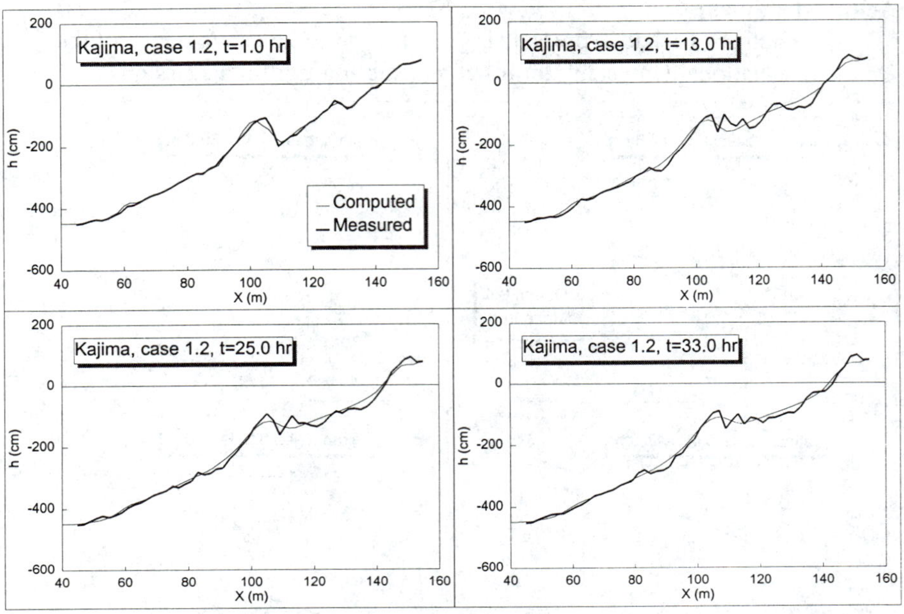

Figure 9. Comparison between measured and computed beach deformation (laboratory data from Kajima et al., 1983, case 1.2).

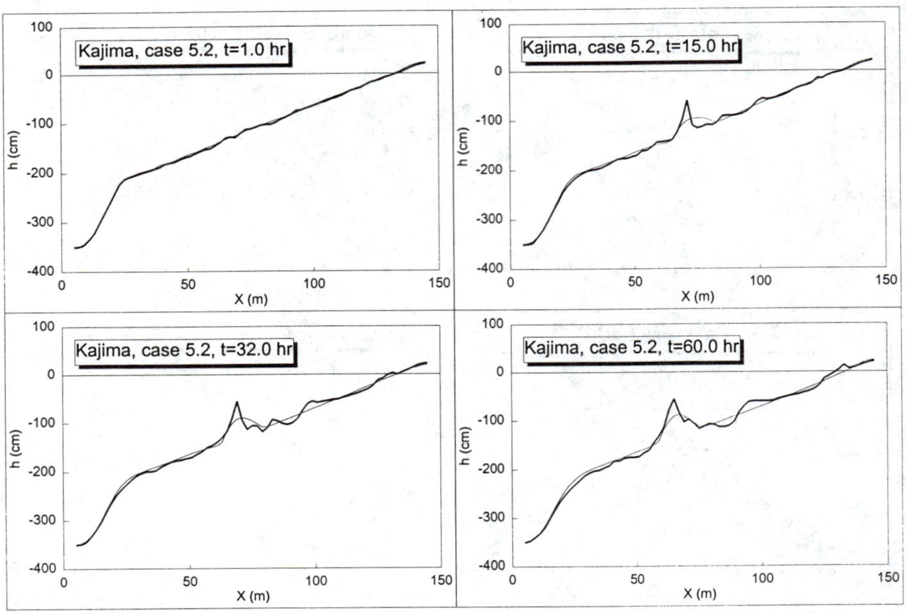

Figure 10. Comparison between measured and computed beach deformation (laboratory data from Kajima et al., 1983, case 5.2).

References

Dally, W.R., Dean, R.G., and Dalrymple, R.A. (1985): Wave height variation across beachs of arbitrary profile, J. of Geo. Res., Vol. 90, No. C6, pp. 11917-11927.

Deigaard, R., and Fredsoe, J. (1991): Modelling of undertow by a one-equation turbulence model, Coastal Engineering, No. 15, pp. 431-458.

Duncan J.H. (1981): An experimental investigation of breaking waves produced by a towed hydrofoil, Proc. R. Soc. London, Ser. A 377, pp. 331-348.

Horikawa, K., and Kuo, C.T. (1966): A study of wave transformation inside surf zone, Proc. 10th Coastal Engineering Conf., ASCE, pp. 217-233.

Kajima, R., Shimizu, T., Maruyama, K., and Saito, S. (1983): On-offshore sediment transport experiment by using large scale wave flume, Collected data No. 1-8, Central Research Institute of Electric Power Industry, Japan (in Japanese).

Kraus, N.C., and Larson, M (1988): Beach profile change measured in the tank of large waves, 1956-1957 and 1962., Technical Report CERC-88-6, Coastal engineering research center, US Army Eng. Waterways Exp. Station.

Larson, M. and Kraus, N.C. (1989): SBEACH: Numerical model for simulating storm-induced beach change, Report 1, Technical Report CERC-89-9, Coastal engineering research center, US Army Eng. Waterways Exp. Station.

Nadaoka, K., Kondoh, T., and Tanaka, N. (1982): The structure of velocity field within the surf zone revealed by means of laser-doppler anemometry, Report of The Port and Harbour Research Institute, Vol. 21, No. 2, pp. 50-102 (in Japanese).

Okayasu, A., Shibayama, T., and Horikawa, K. (1988): Vertical variation of undertow in the surf zone, Proc. 21st Coastal Eng. Conf., ASCE, pp. 478-491.

Okayasu, A. (1989): Characteristics of Turbulence Structure and Undertow in Surf Zone, D. Eng. Dissertation, Civil Eng., University of Tokyo, Japan.

Rattanapitikon, W. (1995): Cross-shore Sediment Transport and Beach Deformation Model, D. Eng. Dissertation, Civil Eng., Yokohama National University, Yokohama, Japan.

Sato, S., Isayama, T., and Shibayama, T. (1989): Long-wave component in near -bottom velocity under random waves on a gentle slope, Coastal Engineering in Japan, JSCE, Vol. 32, No. 2, pp. 149-159.

Shibayama, T., Horikawa, K. (1985): Numerical model for two-dimensional beach transformation, Proc. of JSCE, No. 357/II-3, pp. 167-176.

Shibayama, T., and Rattanapitikon, W. (1993): Vertical distribution of suspended sediment concentration in and outside surf zone, Coastal Engineering in Japan, JSCE, Vol. 36, No. 1, pp. 49-65.

Stive, M.J.F., and Wind, H.G. (1986): Cross-shore mean flow in the surf zone, Coastal Engineering, No. 10, pp. 325-340.

Svendsen, I.A. (1984): Mass flux and undertow in surf zone, Coastal Engineering, No. 8, pp. 347-365.

Thornton, E.B. and Guza, R.T. (1983): Transformation of wave height distribution, J. Geophys. Res., Vol. 88, pp. 5925-5983.

Watanabe, A. (1983): Lecture-1; Fundamental of littoral transport, Seminar on Sediment Problems in Beachs and Estuaries, Manila and Jakarta.

CHAPTER 238

Experimental Study on Sediment Transport in Surf and Swash Zones Using Large Wave Flume

Takao Shimizu[1], Masaaki Ikeno[2]

Abstract

In the present study, experiments were conducted using CRIEPI's Large Wave Flume (205m long, 3.4m wide and 6m deep) to investigate cross-shore hydrodynamics, sediment transport and beach change processes. On the experiments, suspended sediment transport in the surf and swash zones was directly measured under the condition of random waves. Moreover, the long wave effect to sediment transport was also investigated.

Introduction

Suspended sediment transport rate must be evaluated accurately to predict beach change, especially in a calm basin behind an artificial island, because suspended sediment in a surf zone is transported into the calm basin by circulating nearshore current. Recently, it was mentioned that low frequency component in a wave group influences remarkably suspended sediment transport(Sato et al.;1993). Sediment transport rate in a swash zone has to be estimated correctly to expect shoreline change, because erosion and accretion in a swash zone causes shoreline change directly. Ogawa et al.(1981) and Katori(1983) observed sediment transport in a swash zone by using sand traps.

In this study, large scale experiments on beach profile changes due to irregular waves were performed in the Large Wave Flume. Suspended sediment transport and sediment transport rate in a swash zone were measured directly. Effects of irregularity of waves on a sediment transport and influences of low frequency component in a wave group were discussed.

Experimental method

Uniform slopes of sand were piled up in the Large Wave Flume(Kajima et al; 1982) as shown in Fig.1. The bottom of offshore side 90m length of the flume is 1/15 slope. Horizontal coordinate X was defined as an seaward distance from the onshore side end of the flume. An origin of a vertical coordinate Z was defined on a still water level. Still water depth was 4m in cases of 1/10 and 1/20 slope of sand and 3.2m in a case of 1/50 slope of sand. A toe of the sand slope of 1/50 was cut

---

1) Research Fellow, Central Research Institute of Electric Power Industry, 1646 Abiko, Abiko-city, Chiba 270-11, JAPAN.

2) Senior Research Engineer, Central Research Institute of Electric Power Industry, 1646 Abiko, Abiko-city, Chiba 270-11, JAPAN.

off by a 1/10 slope, because the still water level can't be smaller than 3.2m for wave generating and the sand volume was limited.

A measuring vehicle that runs on the flume onshore-offshore was used for measurement in a surf zone. The measuring vehicle is equipped with a capacitance type wave gauge(W) and an installing arm going up and down just beside each other. Two sets of an electromagnetic current meter(C), an optical turbidity meter(T) and a pumping tube(P) for water sampling were assembled to the installing arm. Vertical distance between two sets is 40cm. Water sample including suspended sediment was sucked up by an engine pump, and a volume of sampled water was measured by a water counter. Between the pump and the counter, the suspended sediment was filtered out by a 97μ mesh plankton net and weighed. Median diameter of the sand was 1mm, and the grain size distributed from 0.2mm to 1.3mm as shown in Figure 2. Output of turbidity meter was calibrated individually by the time averaged sediment concentration based on the water sampling data. A wheel type sand surface profiler was loaded on a trolly dragged by the measuring vehicle. The profiler can trace the beach profile under the still water and in the air continuously, accurately and speedy.

In some cases, sediment transport due to uprush and sediment transport due to downrush of individual waves were measured respectively by sand traps(S) in a swash zone. The sand traps are original hand made with metal frame and 97μ mesh plankton net as shown in Photo 1. A mouth of the trap is 5cm wide and 20cm high. The length of the trap is 1m. Fifty traps were prepared and numbered. Two footings was built up in a swash zone, and sediment transport was trapped by three persons on the footings as shown in Photo 2. Right one of this side in the photo is trapping sediment transport due to downrush pressing the under edge of the mouth of the trap against the sand surface. Right one of back side is taking a trap away from the one of this side and going to stack it in a bucket. Left one is preparing the next trap. Trappings of sediment transport due to uprush were continued for fifty waves, and after the collection of trapped sand, sediment transport due to fifty downrushs were trapped. A trapping of sediment transport due to uprush was started when a front edge of an uprush reached the trapping point, and was finished when the water started to flow down. But if the front edge of the next uprush reached the trapping point before the water started to flow down, the trap was changed quickly to the next trap. A trapping of sediment transport due to downrush

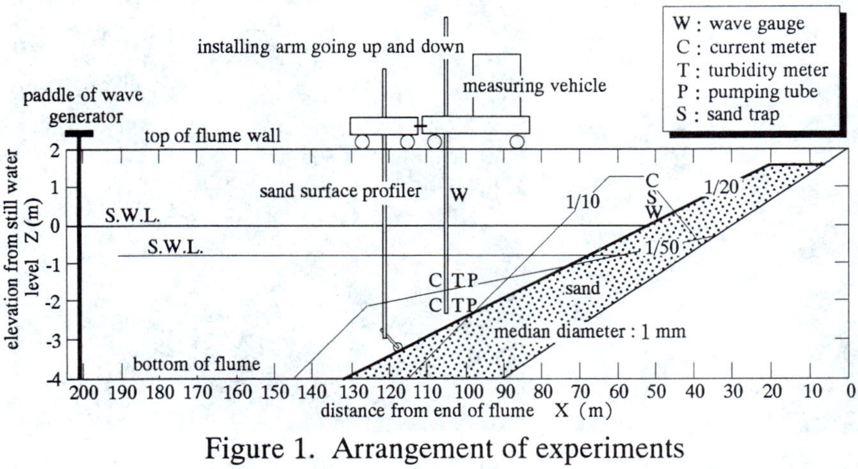

Figure 1. Arrangement of experiments

was started when the water started to flow down, and was finished when the front edge of the next uprush reached the trapping point or the down flow of the water sank into the sand. A wave gauge and a current meter was set up just beside the sand trapping. Bottom end of the wave gauge was stuck in a sand. Head of the current meter was set 1 or 2cm above the sand surface. The wave gauge on the measuring vehicle was set 8m offshore the trapping point. Outputs of wave gauges and a current meter were recorded in a memory of a personal computer and on a chart of linear-corder with trap numbers called by the trapper.

Irregular waves were generated by piston type wave generator. The wave train was repeated every hundred waves. Sampling of outputs of a wave gauge, current meters, turbidity meters and sampling water were continued for one hundred waves at every measuring points in a surf zone. Trappings of sediment transport in a swash zone was performed in the first half of the wave train and in the second half separately. In some cases, the free long wave generated at the paddle of the wave generator was canceled by the method after Ikeno et al.(1996).

Experimental cases and conditions

Table 1 shows experimental cases and conditions. In the case L1, sand slope "tan-β" was 1/20 and regular waves were generated. Offshore wave height "Ho" was 1m and wave period "To" was 5s. In the case L2, irregular waves were generated using JONSWAP spectrum. Significant wave height and significant wave period was same to the case L1. Sharpness parameter "γ" is 1 according to the wind wave condition. Conditions of the case L3 were same to the case L2 except for the sharpness parameter is 7 according to the swell condition. In the case L4, free long wave cancel method was adopted to the condition of the case L2, but the significant wave height had to be reduced to 0.6m because the stroke of the paddle of wave generator is limited up to 2.1m. Conditions of the case L5 were same to the case L3 except for the sand slope and the wave was steeper than the case L3. Condition of the case L6 was same to the case L5 except for the free long wave was canceled. Conditions of the case L7 was same to the case L3 except for the sand slope and the wave was more gentle than the case L3.

Photo 1. Sand trap

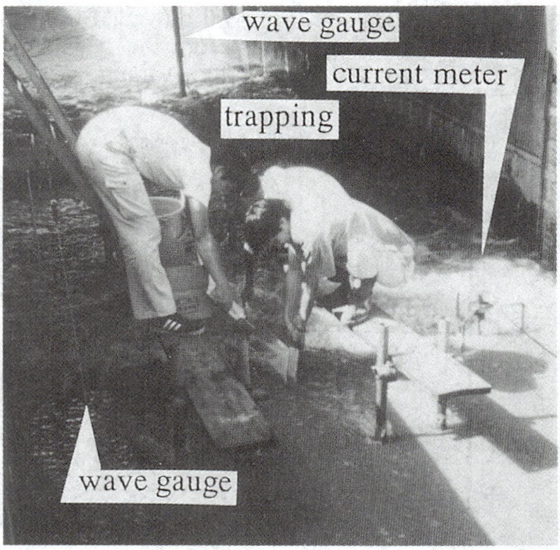

Photo 2. Observation in the swash zone

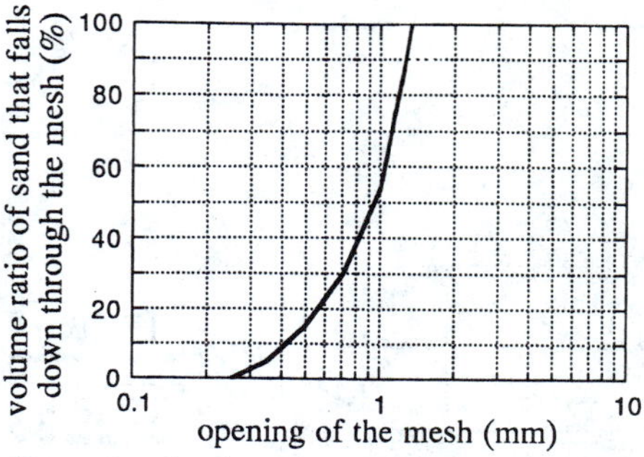

Figure 2. Sand particle diameter distribution

Table 1. Experimental cases and conditions

case	tan β	H_0 (m)	T_0 (s)	γ	wave generation
L1	1/20	1.0 m	5.0 s	r.w.	p.m.
L2	1/20	1.0 m	5.0 s	1	p.m.
L3	1/20	1.0 m	5.0 s	7	p.m.
L4	1/20	0.6 m	5.0 s	1	f.l.w.c.m.
L5	1/10	1.2 m	3.0 s	7	p.m.
L6	1/10	1.2 m	3.0 s	7	f.l.w.c.m.
L7	1/50	0.5 m	8.0 s	7	p.m.

γ : sharpness parameter of JONSWAP spectrum
r.w.: regular wave
p.m.: previous method
f.l.w.c.m.: free long wave cancel method

Experimental results

Figure 3 shows the initial beach profiles of each cases as dotted lines and final profiles as solid lines. Profile changes of the cases L1, L2 and L3 are bar and berm system. Profile change pattern is decided by significant wave characteristics, not by sharpness of spectrum or irregularity. Profile change due to swell is more rapid than due to wind waves of the same significant wave characteristics. Profile changes of the cases L5 and L6 are typical erosional pattern with remarkably developed bar. Profile change in the case L7 is accretional pattern with sediment transport from offshore zone to surf zone.

Figure 4 shows fluctuation of suspended sand concentration "c" due to wave motion near the breaking point in the cases L1 and L6 together with horizontal velocity "u" and water surface elevation "η". Suspension of sediment due to regular wave motion takes place at the phase when the horizontal velocity changes from onshore to offshore. Sediment is suspended due to irregular wave motion at the phase when horizontal velocity changes not only from onshore to offshore, but also from offshore to onshore.

Figure 5 to 8 show vertical distributions of suspended sediment transport flux in the cases. The graph at the top is cross-shore distribution of significant wave height and beach profiles. The graphs in the second row are vertical distributions of time averaged suspended sediment concentration. The graphs in the third row are vertical distributions of time averaged horizontal velocity. The graphs at the bottom is vertical distributions of time averaged product of suspended sediment concentration and horizontal velocity namely suspended sediment transport flux. In the cases L5 and L6, suspended sediment concentrations are higher than the other cases. In the almost cases, suspended sediment transport fluxes near the bottom are directed offshore, even in the accretional case L7.

Figure 9 shows low and high frequency components of suspended sediment transport flux in the cases L5 and L6. In the case L6 where the free long wave is canceled, long wave components of suspended sediment transport fluxes are directed offshore when the sort wave components are directed onshore. In the case L5, the above mentioned property is not clear.

Figure 10 shows synchronized data observed at the point A in a swash zone and at the point B in a surf zone of the case L5. The graph at the top of four is volume of sand par unit width trapped at the point A. Length of right side of each triangle is trapped volume. Left vertex points to the starting moment of trapping. Right

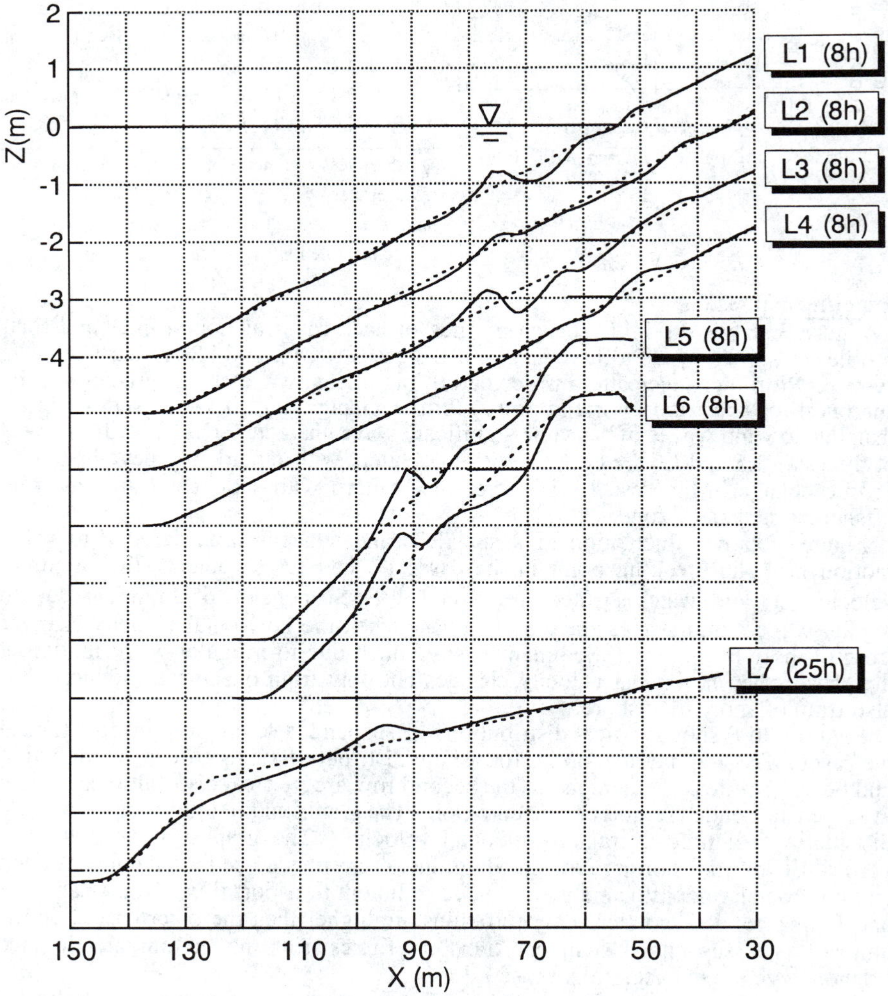

Figure 3. Beach profile change

vertices point to the finishing moment of trapping. For convenience' sake, sediment transport due to uprush and downrush are shown in one graph. The second graph is horizontal velocity 1 or 2cm above the sand surface at the point A. The curve is interrupted in the middle of downrush, because the head of current meter is exposed in the air. The third and the bottom graphs are water surface elevation at the points A and B. The waves corresponding to each other are connected by dotted lines. Number of waves reduces by half propagating from the point A to B. Waves in a swash zone are apt to be grouped and uprush velocity due to the first wave of the group is strong and uprush sediment transport is remarkable under the action of the first wave. Crest level of the last wave of the group in a swash zone is highest and downrush velocity due to the last wave is strong and the downrush sediment transport due to the last wave is striking and drags on.

Table 2 shows the uprush and downrush data of individual waves in a surf zone of the case L5. "H" is crest height measured from the trough just before the crest. "h" is water depth under the trough. "Uu" is horizontal velocity under the crest. Positive data is onshore velocity. Three data are negative, because the bottom velocity remains offshore when the front of the next uprush reached the trapping point. "Qu" is trapped volume of sediment transport due to uprush par unit width. "Ud" is trapped volume of sediment transport due to downrush par unit width. Total volume of Qu is 65.7 litter par meter and total volume of Qd is -84.4 litter par meter. Net volume of sediment transport during one hundred waves is -18.7 litter par meter. It is converted to the net sediment transport rate $-6.2\times10^{-5}m^3/m/s$.

Figure 11 shows sand transport rate counted back from profile change in the case L5. The net sediment transport rate at the trapping point $-6.3\times10^{-5}m^3/m/s$ is read

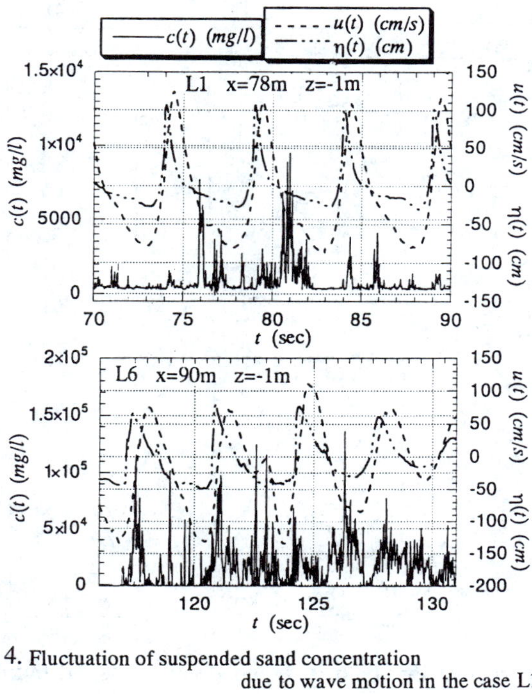

Figure 4. Fluctuation of suspended sand concentration due to wave motion in the case L1 and L6

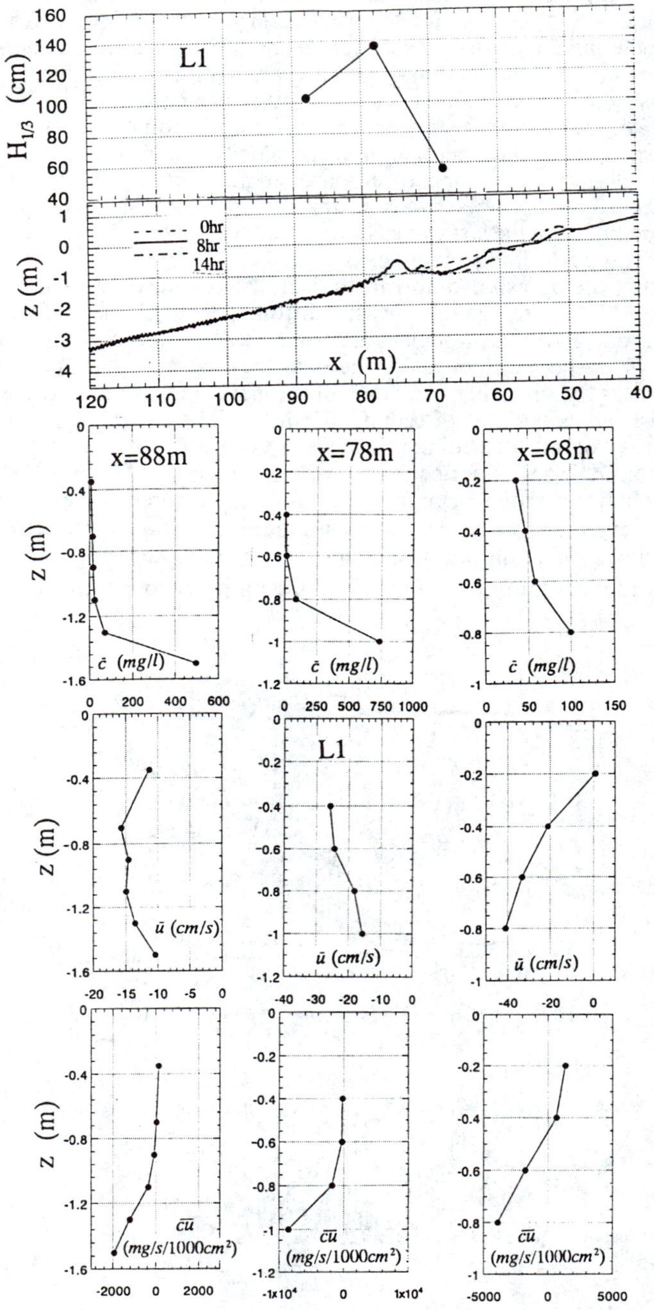

Figure 5. Virtical distribution of suspended sand transport flux in the case L1

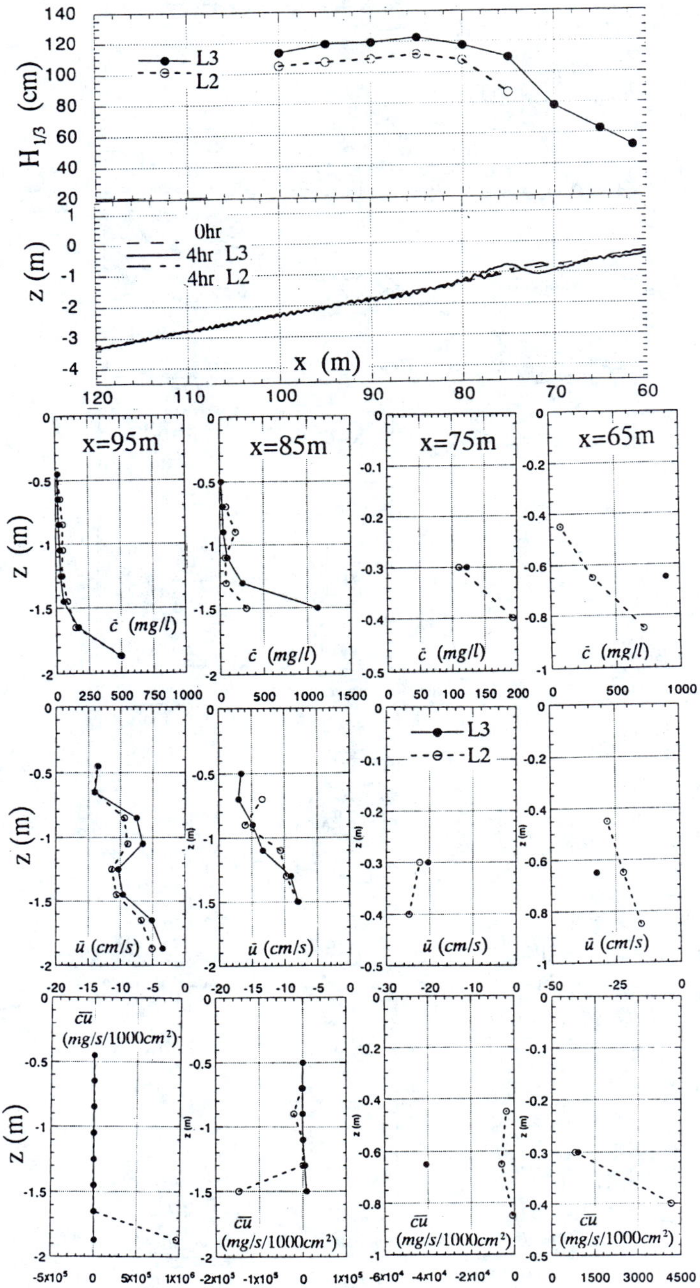

Figure 6. Virtical distribution of suspended sand transport flux in the cases L2 and L3

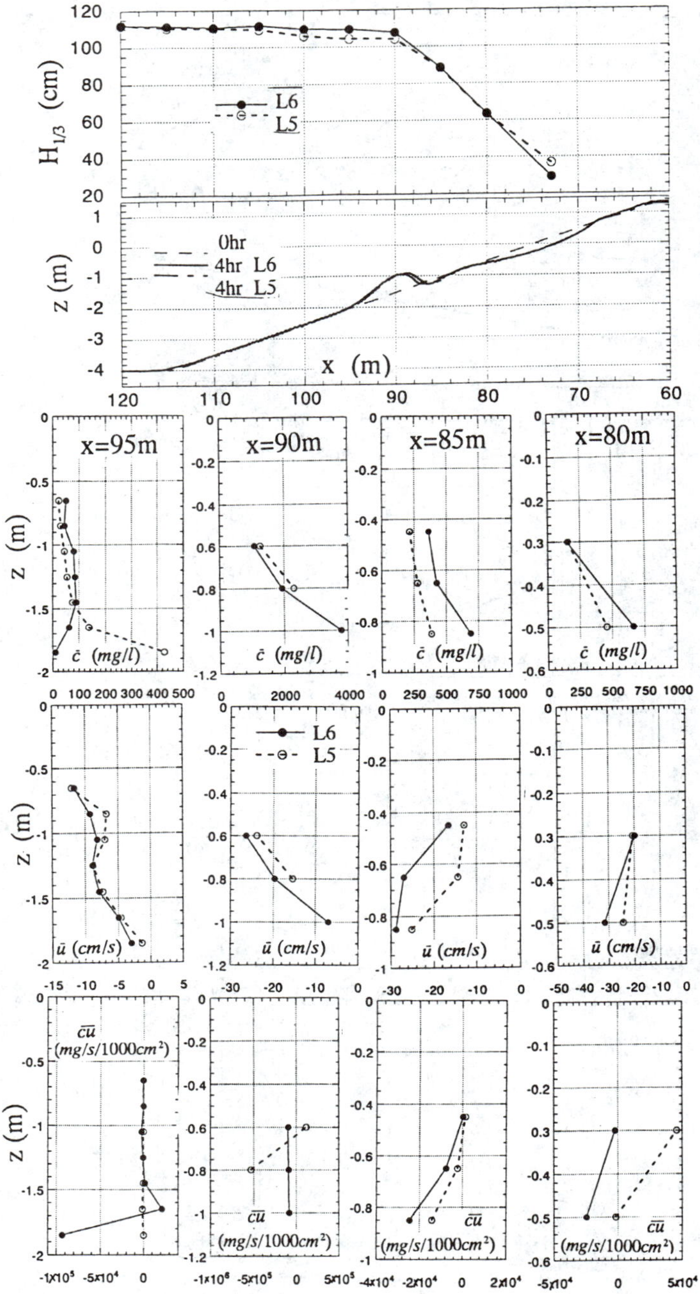

Figure 7. Virtical distribution of suspended sand transport flux in the cases L5 and L6

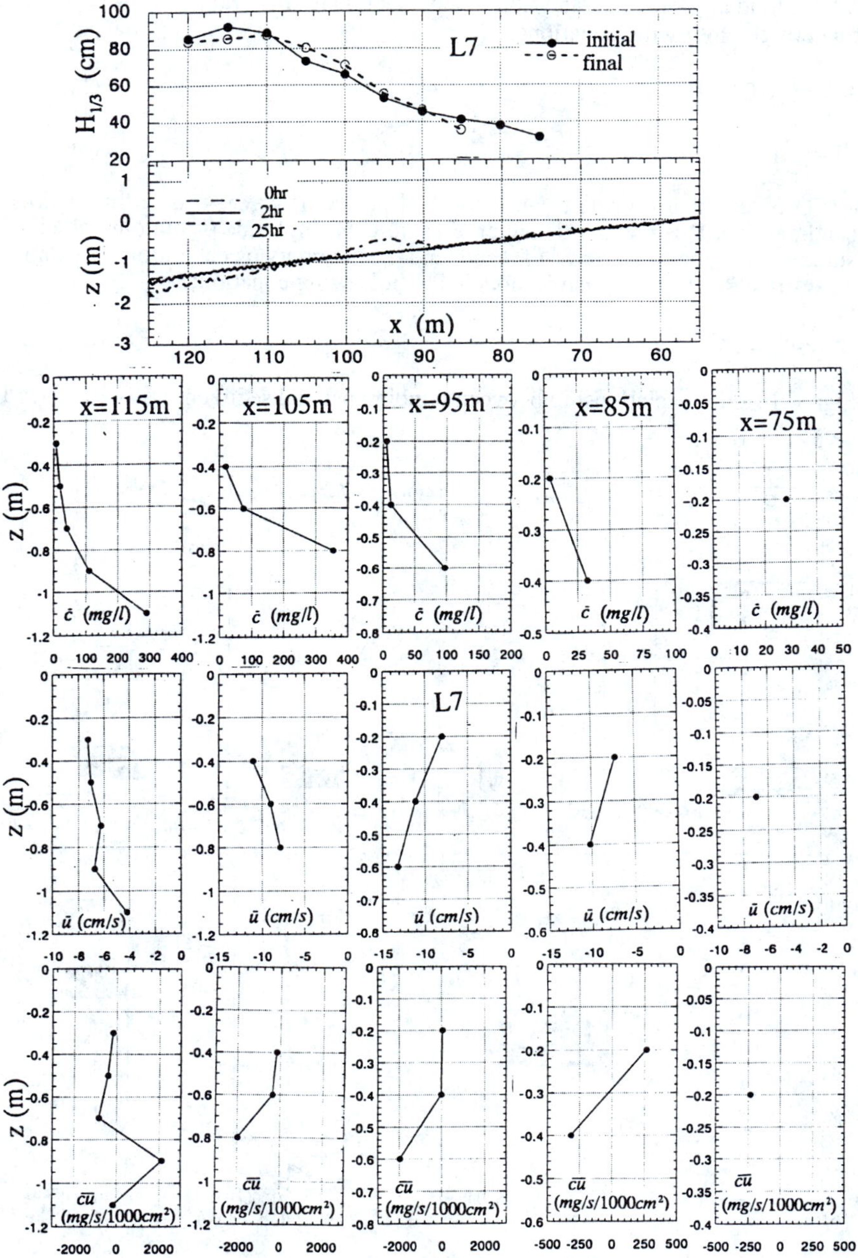

Figure 8. Virtical distribution of suspended sand transport flux in the case L7

from the figure. Good agreement of two independent values indicates high accuracy of the sand trapping method.

Figure 12 shows relationship between nondimensional sediment transport rate Φ' due to individual downrush and nondimensional bottom friction Ψ'. Φ' and Ψ' is defined by the following equations;

$$\Phi'=Q/(wD) \tag{1}$$

$$\Psi'=U^2/(s'gD) \tag{2}$$

where Q is sediment transport rate in the half period of wave, w is settling velocity of sand particle, D is median diameter of sand, U is amplitude of bottom velocity, s' is submerged specific gravity of sand and g is gravity acceleration. Sediment transport rate due to downrush is modeled by following equation;

$$\Phi'=-0.037\,\Psi'^{1.29} \tag{3}$$

Figure 13 shows relationship between nondimensional sediment transport rate Φ'

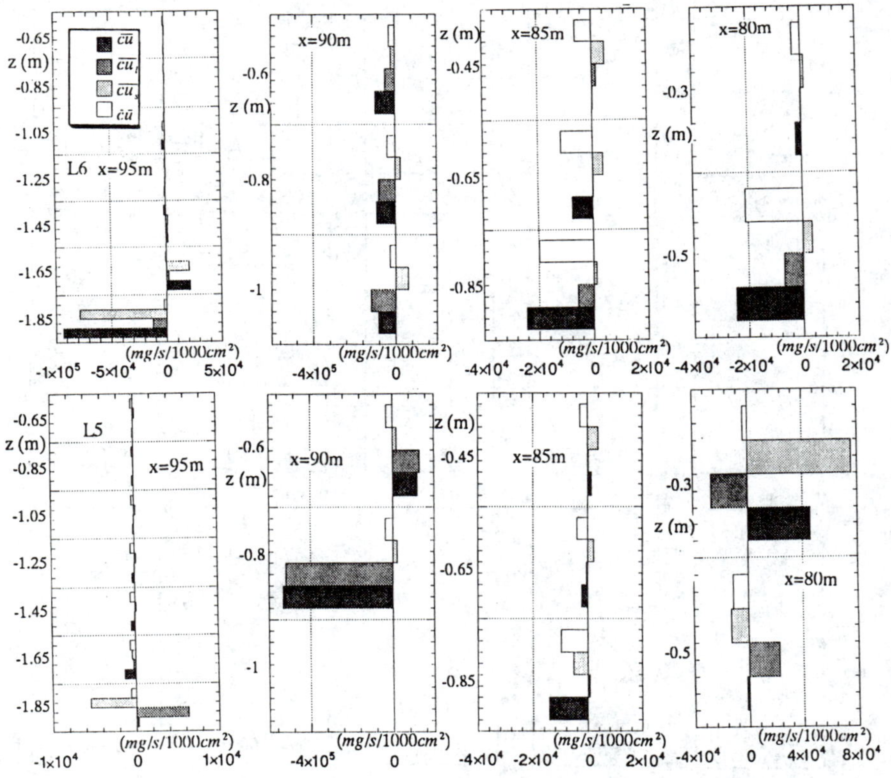

Figure 9. Low and high frequency components of suspended sand transport flux in the case L5 and L6

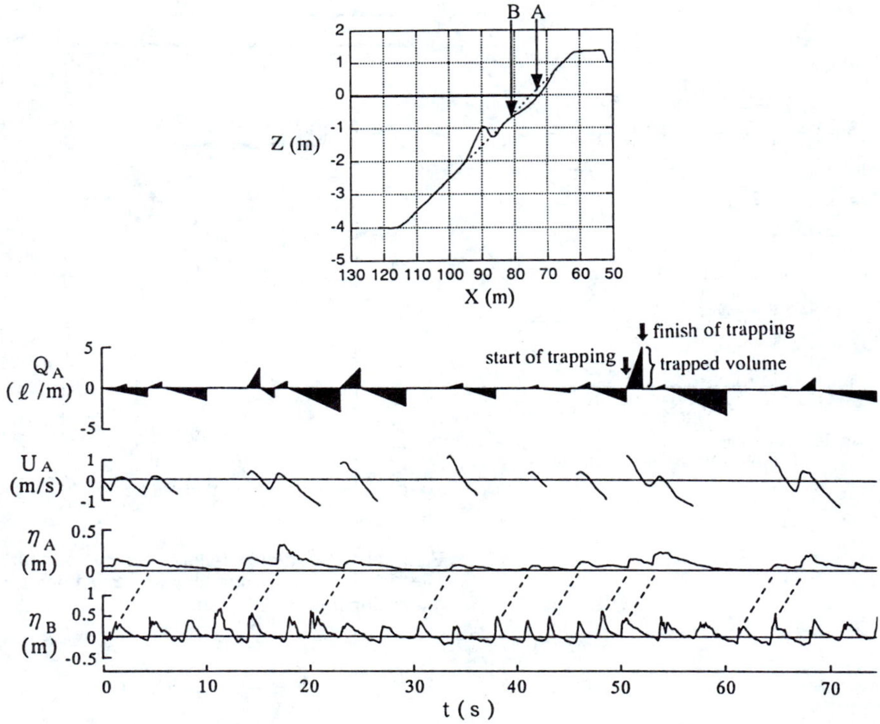

Figure 10. Synchronized data observed in the swash zone and the surf zone

Table 2. Uprush and downrush data of indivisual waves in the swash zone of case L5

H (m)	h (m)	Uu (m/s)	Qu (l/m)	Ud (m/s)	Qd (l/m)
0.225	0.000	0.90	3.15	-1.15	-2.74
0.150	0.000	0.90	1.43	-0.60	-0.23
0.100	0.025	0.40	0.29	-0.80	-0.70
0.175	0.013	0.65	1.69		
0.225	0.112	0.20	0.70	-1.30	-3.03
0.100	0.050	0.25	2.45	-1.00	-1.89
0.062	0.013	0.70	0.23	-0.60	-0.33
0.150	0.000	0.50	1.40	-1.00	-0.87
0.213	0.050	0.60	9.30		
0.200	0.125	0.35	0.23	-1.10	-3.21
0.138	0.000	0.60	1.05		
0.150	0.062	0.60	0.31	-1.00	-1.13
0.100	0.112	0.50	0.29	-0.60	-0.47
0.062	0.000	0.50	0.17	-0.30	-0.21
0.162	0.038	0.60	0.58		
0.213	0.087	0.00	0.64	-0.90	-1.54
0.150	0.000	0.60	2.45		
0.200	0.038	0.45	0.65		
0.125	0.138	-0.20	0.12	-0.80	-2.74
0.075	0.000	0.90	0.30		
0.138	0.038	-0.10	0.24	-0.50	-0.73
0.200	0.000	0.85	2.81	-1.10	-3.26
0.200	0.025	0.90	2.52	-1.10	-0.61
0.150	0.075	-0.20	0.45		
0.125	0.025	0.60	1.11	-0.90	-0.23
0.175	0.038	0.60	4.31	-1.20	-2.51
0.200	0.000	0.60	5.48		
0.138	0.100	0.00	0.12	-1.40	-1.07
0.062	0.000	0.80	0.27		
0.250	0.062	0.60	0.64	-1.30	-3.73
0.075	0.000	0.90	0.82	-1.20	-0.90
0.125	0.000	0.90	1.63	-1.20	-2.04
0.150	0.000	0.70	2.02	-1.00	-1.14
0.200	0.050	0.80	2.45		
0.350	0.125	0.90	3.21	-2.00	-15.51
0.087	0.000	0.85	0.97		
0.213	0.050	0.30	1.19	-1.30	-3.50
0.087	0.000	0.75	0.17		
0.125	0.000	1.25	0.58	-1.30	-3.11
0.200	0.025	0.90	0.12		
0.225	0.150	0.80	3.73	-1.80	-17.49
0.075	0.000	0.60	0.22		
0.275	0.050	0.70	1.14	-1.40	-4.90
0.125	0.000	1.00	1.35		
0.250	0.050	0.70	0.73	-1.35	-4.57
		Qu total	65.7	Qd total	-84.4

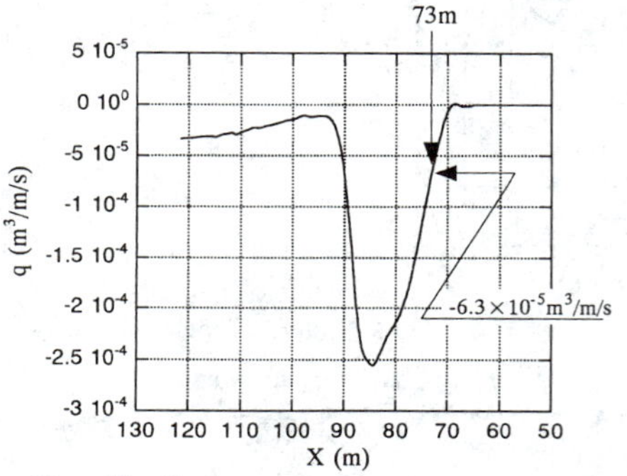

Figure 11. Sand transport rate counted back from profile change

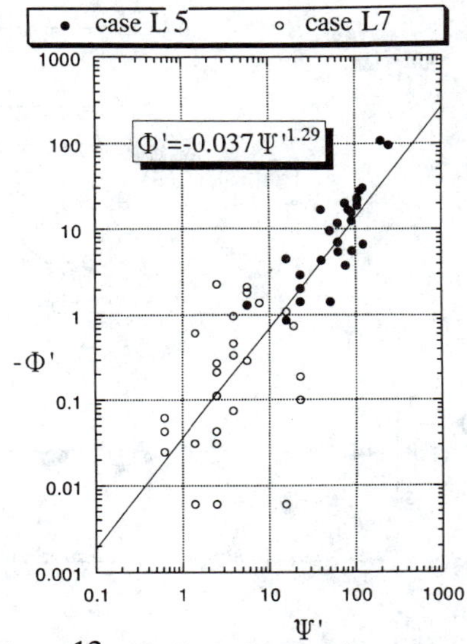

Figure 12. Nondimensional sediment transport rate Φ' due to individual downrush and non dimensional bottom friction Ψ'

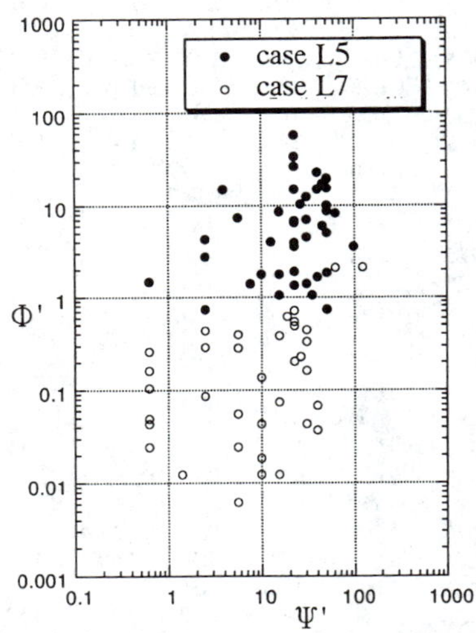

Figure 13. Φ' due to individual uprush and Ψ'

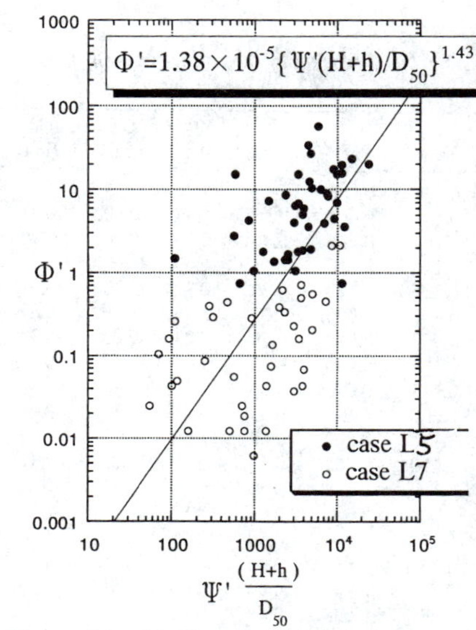

Figure 14. Φ' due to individual uprush and $\Psi'(H+h)/D_{50}$

due to individual uprush and nondimensional bottom friction Ψ'. Relationship between Φ' and Ψ' is bad, because uprushing bore involves suspended sediment according to not only bottom friction but also thickness of the bore.

Figure 14 shows relationship between nondimensional sediment transport rate Φ' due to individual uprush and

product of nondimensional bottom friction Ψ' and nondimensional thickness of the front bore of uprush. Sediment transport rate due to uprush is modeled by following equation;

$$\Phi'=1.38\times 10^{-5}\{\Psi'(H+h)/D\}^{1.43} \qquad (4)$$

Conclusions

❶Beach profile change due to swells is more rapidly than that due to wind waves.

❷Around the breaking point, there is the case that low frequency component of suspended sediment transport due to long waves is offshore, and high frequency one is onshore.

❸The number of waves decreases immediately in the swash zone, because a downrush prevents runup of the next wave in trough phases of long wave.

❹Waves in a swash zone are apt to be grouped and uprushing bottom flow due to the first wave of the group is strong. Hence, uprush sediment transport is remarkable under the action of the first wave.

❺Crest level of the last wave of the group in the swash zone is highest and downrush bottom flow due to the last wave is strong. So, downrush sediment transport due to the last wave is striking and drags on.

❻Sediment transport rate due to uprush and downrush in a swash zone was modeled relating to bottom friction and thickness of uprushing bore.

References

Ikeno, M. and Tanaka, H(1996): A new experimental system for reproduction of infragravity waves bounded by irregular wave groups and their wave deformation, Report of Central Research Institute of Electric Power Industry, No.U95042.(in Japanese)

Kajima,R., Shimizu, T., Maruyama, K. and Saito, S.(1982): Experiments on beach profile change with a large wave flume, Proc. 18th Coastal Eng. Conf., ASCE, pp.1385-1404.

Katori, S.(1983): Measurement of sediment transport by streamer sand trap, NERC, Rep. No. 17, TR-82-1, pp.110-117.(in Japanese)

Ogawa, Y. and Shuto, N.(1981): Field measurements of hydraulics and sediment movements in a swash zone, Proc. 28th Japanese Conf. on Coastal Eng., JSCE, pp.212-216.(in Japanese)

Sato, S., Shimatani, M. and Shimura, T.(1993): Analysis of suspended sediment transport rate in a surf zone of irregular waves, Proc. 40th Japanese Conf. on Coastal Eng., JSCE, pp.321-325.

CHAPTER 239

The Influence of Long Waves on Macrotidal Beach Morphology

David J. Simmonds[1], Tim J. O'Hare[2] & David A. Huntley[3]

Abstract

The aim of the following paper is to demonstrate that standing long waves can generate intertidal morphological features in a dissipative macrotidal setting. In this environment the tidal excursion can exceed the length scales associated with long waves. This extends the work of Simmonds et al (1995) in which it was shown that monochromatic long waves could generate sediment convergence patterns that bore a good similarity to observed topographic features, namely ridges and runnels from a field site on the Belgian coast. The simulation is modified to include a band of long waves and it is found that a good correspondence with real topography can still be achieved on a tidally averaged basis by allowing for a degree of uncertainty in the breaking criterion. It is also argued why the observed scales are more likely to correspond with the shorter end of the long wave spectrum. The simulation helps to shed light on the conditions in which such features are known to form.

Introduction

This paper examines the question "can a spectrum of long waves be capable of creating intertidal bedforms on a macrotidal beach?"

Despite the abundance of macrotidal coastlines Short (1991) has been led to remark that the "study of macro-tidal beaches has lagged considerably" in relation to the published material on micro– and meso–tidal coasts. Whilst numerical modellers are beginning to recognise the importance of long wave processes in the modeling of coastal morphology (Roelvink & Broker, 1993) very few of these models, even though applied to macrotidal coastal regions, appear to consider the further complication of a tidally modulated boundary. Although the concept of an equilibrium cross-shore profile is useful in a microtidal setting, the interpretation of equilibrium profile is questionable in the macrotidal setting when considered with

---

[1] Res. Fellow, School Civil Eng., Univ. Plymouth, Palace Court, Plymouth, UK PL1 2DE
[2] Senior Lecturer, Inst. Marine Studies, Univ. Plymouth, Drake Circus, Plymouth, UK PL4 8AA
[3] Reader, Inst. Marine Studies, Univ. Plymouth, Drake Circus, Plymouth, UK PL4 8AA

respect to timescales of the order of days, even when tidally averaged quantities are considered. The cross shore tidal excursion on a dissipative macrotidal beach is measured in hundreds of meters and is further modulated by the Spring–Neap cycle. The boundary is neither steady nor varying by the same amount from day to day. It seems more appropriate to speak of a dynamic equilibrium whereby the profile is constantly adjusting from minute to minute within limits determined by the tidal level and variations in the wave climate.

Standing long wave motion at the shoreline is often associated with the formation of submerged bars (Short, 1991, Sallenger & Holman, 1987) and support for this has been provided by field observations (e.g. Bauer & Greenwood, 1990). But what of the intertidal bedforms such as the bar-like features, referred to as ridges and runnels, that are evident on some macrotidal coastlines?

Wright et al (1982) suggested that macrotidal beach profiles are due to the combination of swash, surf and deeper water processes that operate over different regions of the beach face for different periods of time. It was also suggested that some wave-lain features may survive swash & surf zone excursion in low energy conditions — but that in high energy conditions any bedforms are erased by the more energetic processes.

These studies have been predominantly based on high energy exposed shorelines. In the European context, many dissipative macrotidal shorelines are to be found sheltering in fetch limited environments well away from the continental shelf breaks. In these situations, more subtle, lower energy mechanisms might be more evident.

It is tempting to suggest that standing long waves might be somehow responsible for the development of intertidal features such as ridges and runnels as the length scales of both are comparable. In addition it is widely recognised by field workers that long wave energy is present, to some degree, on all shorelines. In the simplest analysis, topographic features are believed to be generated under a stationary standing long wave. However on a macrotidal dissipative coast the reflection point of the long waves (whether considered at the shoreline or in the surfzone) can move cross–shore by several long wave wavelengths.

Simmonds et al (1995) showed how standing long waves might form bar-like features in the intertidal zone of a linear beach dominated by cross-shore long wave processes. They based their analysis upon a simple transport mechanism resulting from the correlation of the long waves with the incoming wave groups to demonstrate how intertidal bedform development might be initiated through a "tidal bar hypothesis". Using a monochromatic long wave envelope, a pattern for the *potential for erosion* across the beach face was shown to correspond well with observed scales of ridge and runnel morphology even around the mid–tide region, despite the inclusion of tidal modulation of the transport envelope across the beach face.

However, long waves observed in the field are generally broad-banded in nature (Huntley et al, 1993) and it might be predicted that the result of including of a realistic bandwidth of long waves would smear over the effect of the individual waves. This effect combined with tidal modulation of the reflection point would surely destroy any "memory" that the beach retains of the long wave structure – although Bowen and Huntley (1983) suggested that feedback, in the form of erosion of the bed and subsequent modification of long wave structure might result in a narrowing of the effective spectrum that shapes the beach through resonance.

A further criticism of the hypothesis is that given that there is a band of long waves why does there appear to be a stronger correlation at the shorter wavelength end of the measured band than at the longer scales?

We will investigate these criticisms by reference to an improved version of the original simulation outlined in Simmonds et al. (1995). It will be shown that under conditions recognised as precursory to ridge and runnel formation, rhythmic intertidal features are indeed predicted. We draw upon the geometry and field measurements of a particular example of ridge and runnel morphology. The aim is not to present a full sediment and hydrodynamic coupled model, but instead to show that the mechanism for forming intertidal bedforms involving standing long waves is feasible.

A Field Example of Rhythmic Intertidal Morphology

First of all it is useful to reintroduce the field example of ridge and runnel features (Simmonds et al, 1995). The macrotidal beach profile is shown in Figure 1 and is taken from a field site at Nieuwpoort on the Belgian coastline which was the

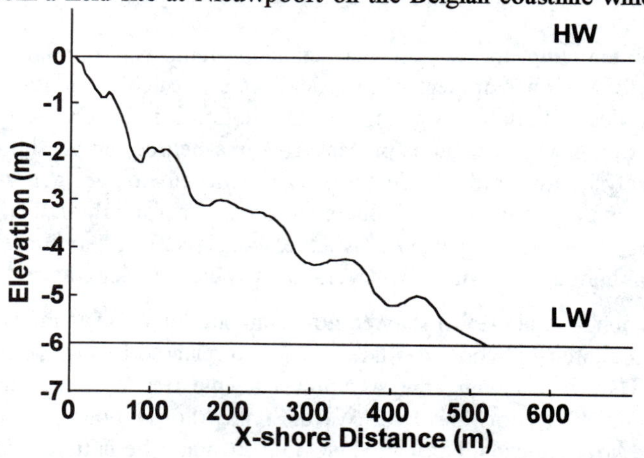

Figure 1. Ridge & Runnel profile at Nieuwpoort, Belgium

focus of the recent *MAST* project *Circulation and Sediment Transport Around Banks* (CSTAB —O'Connor et al, 1996). This illustrates the kind of intertidal features that are expected to be associated with long waves by virtue of their length scales. These ridges exist as low lying, shore–parallel bodies of sand with cross-shore scales of the order of 50*m* separated by runnels. Their longshore extent is interrupted by shore–normal channels spaced every 500*m* or so, down which water is drained from the runnels on the ebb of the tide. They are exclusive to dissipative macrotidal beaches of medium sand sized material and fetch limited wave climate.

It is hard to justify that these particular features are merely swash bars formed at low and high still stands at Spring and Neap tide levels (King & Williams, 1949). Five bars are clearly visible. Other documented cases of ridge and runnel beaches exhibit even greater numbers of ridges (Mulrennan et al, 1992). Indeed the number and spacing of these particular features on low energy beaches is reported to be strongly related to the beach slope and this is more reminiscent of long wave structure where the node–anti-node spacings are also dictated by the beach slope β. Therefore this swash bar hypothesis should be rejected.

The scales involved are certainly too large for comparison with gravity wavelengths. So the question "Could long waves be responsible?" arises. Simmonds et al (1995) previously reported the identification of standing long wave activity, even during low energy conditions at this particular beach in the frequency band of 0.01-0.04*Hz* (100*s* to 25*s* wave period). The scale of the features was shown to be consistent with shortest of these long wave scales (0.04*Hz*) measured at the beach which had a slope of around 1%. It therefore seems likely that long waves are indeed responsible.

Long Wave Transport Mechanisms

The mechanism by which a memory of the structure of a standing long wave can come to be imprinted upon a beach is now discussed. The cartoon (Figure 2) illustrates two mechanisms that can be considered.

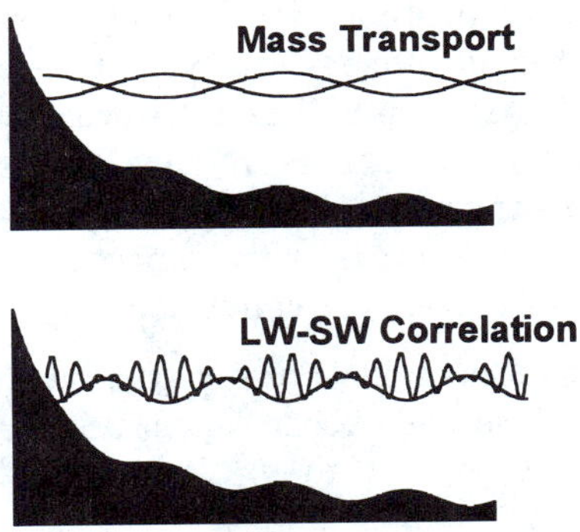

Figure 2 Transport mechanisms

The first involves sediment movement in response to the mass transport induced under a standing wave. This has been investigated in associated with sub–tidal bar formation (e.g. Carter et al. 1973). In this, the sediment is assumed to move in response to the drift velocities induced by the standing long wave structure. Sediment accumulates either at the nodes or anti-nodes according to different investigators though there is no general agreement as to which, although the answer is thought to depend upon the sediment grain size.

The second is a mechanism based upon the correlation of short-wave groups with a bound long wave. O'Hare & Huntley (1994) have discussed a model that furthers the concept of short waves acting as sediment stirrers and correlated long waves acting as transporters (pseudo-steady currents). In the case of a bound long wave the long wave is generated through the radiation stress variation under wave groups so that the largest short waves occur around the long wave troughs and the smaller at the long wave crests. Such a correlation creates a net offshore transport when integrated over a long wave cycle.

If a planar beach face is considered, in the absence of any other processes, this offshore sediment flux would be predominantly modulated by the simple depth dependence of the short wave orbital velocity. The inclusion of a phase–locked reflected long wave, however, introduces a rhythmic cross–shore spatial variability in the sediment flux. The function which is the gradient of this can be interpreted as regions of net erosion or accretion.

Both of these two mechanisms would produce an erosion pattern that would predict features of the same scale. However, the efficiencies of the two in transporting sediment will have different magnitudes. We argue here that the correlation mechanism is likely to be more important.

In the first case the transport mechanism is related through a power law to the "drift velocity". This second order parameter arises from the non-linearity in orbital motion under the standing waves. In the case of long waves this is likely to be very small in relation to the orbital velocity of the long waves.

In the second, the net transport is the result of averaging the correlation between the square of the heights of the (stirring) short wave and the velocity of the (transporting) long waves. Again, it is difficult to estimate the relative importance of this quantity since the resulting transport over a long wave cycle will depend upon, amongst other factors, the amplitude of the long wave orbital velocity and the groupiness of the short waves — that is, the net effect depends upon the difference between the onshore and offshore directed correlation's.

However, the magnitudes of the latter velocities are much greater than those of the drift velocities, and so the second mechanism has the potential to be significantly greater.

Bias to Shorter Scales

If it is accepted that there is a process for the imprinting of long wave scales on the beach face the next question to address is "why is a stronger correspondence with the shorter end of long wave spectrum observed in the field?" As we stated earlier, the observed correlation between the measured long wave spectrum and the observed morphological scales at Nieuwpoort indicated a good agreement but only at the shorter (0.04*Hz*) end of the long wave spectrum. This bias can be understood by considering the following arguments.

First, we should define what we understand by the description "broad banded" in relation to the long wave spectrum. If we take the naïve interpretation of "broad–bandedness" as implying equal wave heights at all frequencies in the spectrum, quite clearly the velocity (transporting) spectrum will be biased towards higher frequencies. This is because the velocity is obtained as the gradient of the time series. Put another way, a "white" amplitude spectrum corresponds to a "blue" velocity spectrum (predominance of higher frequencies).

We can also argue, on a geometric basis, that shorter long wave scales grow more rapidly. This is because the work done on the bed in changing from a flat to an undular bed is less if the undulations are of shorter wavelength (Figure 3). The distance that the centroid of sand has to move in order to create a depression and an adjacent hole is less for the shorter wavelength feature than the longer, although the

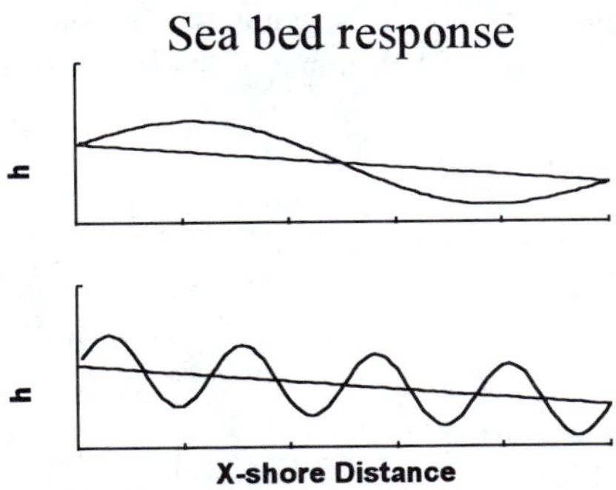

Figure 3 Geometric argument: mass centroid in (a) moves further than in (b)

amount of material that has moved is the same. It is therefore likely that the shorter scales are quicker to develop upon the beach and can be concluded that there is an inherent bias to short scales forming. This is especially important for a system in a state of dynamic equilibrium where features that require longer timescales in order to establish themselves will be discriminated against.

<u>Simulation</u>

Attention is now turned to the simulation reported in Simmonds et al. (1995). The simulation is used to predict the spatial cross–shore profile development. The basis for this is the assumption that the sediment transport is everywhere offshore directed due to the correlation mechanism outlined above. Obviously this is unrealistic— beaches do not simply disappear off–shore, (at least, in the short term), but this assumption is made so that the potential of this mechanism can be investigated in isolation. As such, other processes such as surf-zone and swash processes are not represented.

A bound long wave is assumed to interfere with a reflected free long wave to give a shallow water standing long wave. The sediment transport rate is assumed to be proportional to the product of the magnitude of the standing long wave orbital velocity and of the square of the short-wave envelope.

It is further assumed that the short wave envelope that stirs the sediment can be described by simple shoaling waves which break with a linear decrease in waveheight to the shoreline. The standing long wave envelope, calculated from the bessel function solutions of Lamb (op cite Kirby et al, 1981) defines the magnitude of the offshore flow. The product of these two represents an expression of the uncalibrated offshore sediment flux whose spatial representation is the sediment flux function. The gradient of this is the quantity that is of interest, and is referred to here as the *sediment flux convergence* (SFC). This is interpreted as an indication of zones of sediment accretion or erosion across the profile.

The bias to higher frequencies is incorporated in the model in terms of a "blue shift" of the velocity spectrum. This is generated by assuming uniform amplitude spectrum and hence a velocity spectrum biased linearly towards the high end of the frequency band.

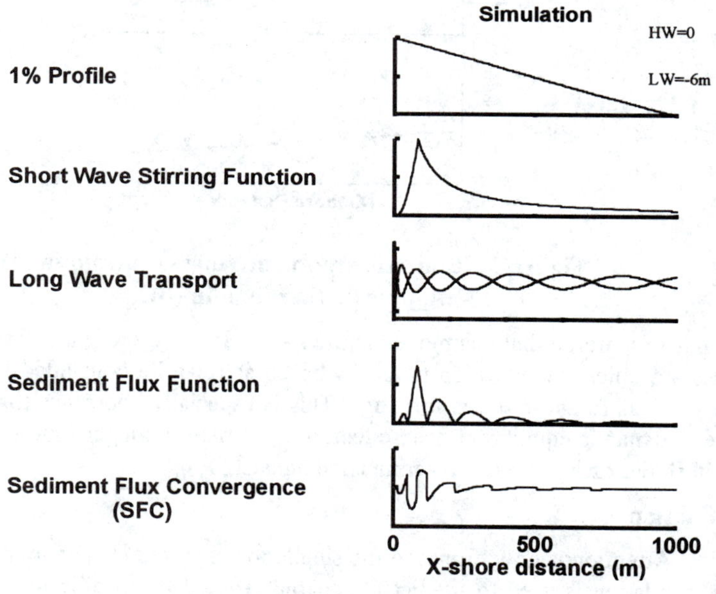

Figure 4. Components of the simulation

Taking parameters similar to those observed at the site at Nieuwpoort beach, a Spring tidal modulation of 6*m* height is introduced over a planar beach of 1% slope. Figure 4 shows how the simulation is constructed. This was implemented in the MATLAB language on a PC.

The simulation is also represented by the flow diagram (Figure 5). The model parameters specified include the short–wave height, the breaking depth, the long wave spectral amplitudes and the beach slope. From this information the standing long wave envelope (SLW env) is calculated and the short wave envelope (SW env). Multiplied together, these give the sediment flux function for a particular long wave frequency and reflection point on the profile.

The simulation has then been adapted to permit the summation of the sediment flux functions over a suitable spectrum of standing long waves at each position of the tide. The summation of the calculated sediment flux convergence functions is then performed to calculate the tidal average.

It is also possible to modify the short–wave envelope to allow for a degree of uncertainty in the short wave height and breaking criterion. This is desirable because, firstly, groupy waves do not all break at the same position, and, secondly, breaking is believed to be influenced by instabilities which determine that the nature of this process is chaotic (Longuet–Higgins, 1994).

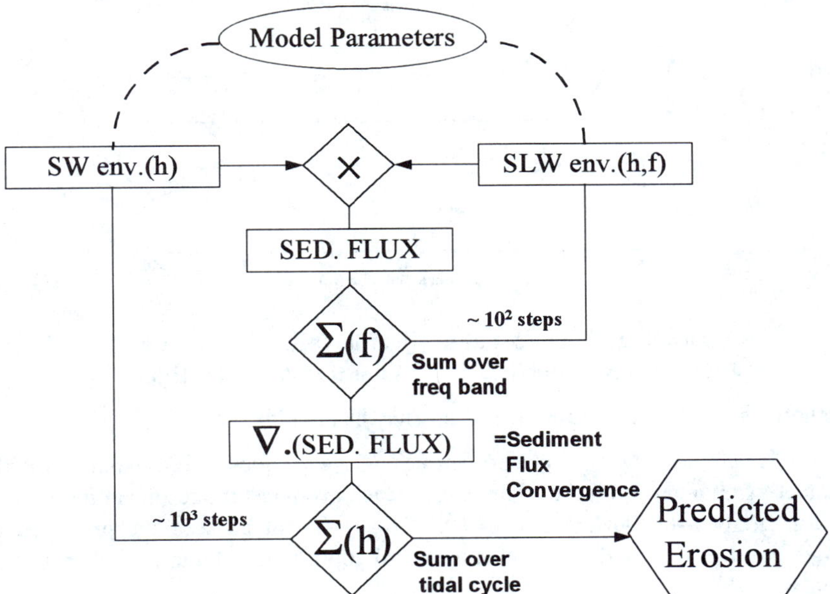

Figure 5 Flow–chart for improved simulation

As mentioned, the simulation involves two summations. Summation over frequency band and summation over all tidal amplitudes. An appropriate indication of the magnitude of the number of steps for these two summations is indicated in the figure. These have been selected to optimise processing time against numerical "noise".

It should also be pointed out that at each position of the tide the sum of the superimposed velocities is being calculated. It is therefore being assumed that all the different standing long wave modes occur simultaneously and, further, that their respective correlation's with the short waves also all occur together. Whether this is realistic is a matter for debate. Usually the statistical analysis of a wave record is interpreted to mean that the spectrum of frequencies is stationary. It may be that, in reality, there is a "wandering" of the peak band of frequencies from long wave to long wave.

The present simulation is not sensitive to the order in which the two summations are calculated. However to develop the model further, feedback, in the form of bed evolution, is to be introduced before summing over the next tidal

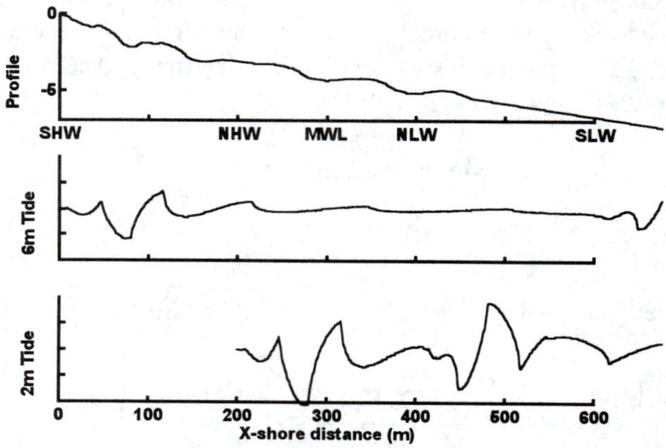

Figure 6(a) Ridge & runnel profile. Monochromatic calculations of the SFC for (b) 6m tide & (c) 2m tide.

amplitude after the summation over frequency bandwidth.

Tidally averaged runs for Spring & Neap tides using values for these parameters obtained from the Nieuwpoort beach produces the results in figure 6(b) & (c) for monochromatic long waves (0.04Hz). As can be seen the predicted SFC shows good correspondence with the actual topography, Figure 6(a), in terms of scale.

A Realistic Bandwidth of Long waves

The main query regarding the above simulation is: "what effect does a realistic, broad long wave bandwidth have on these predictions?" This question will now be addressed and the sensitivity of this simple model to other parameters will also be discussed.

At first sight, Figure 7 indicates that calculating the SFC for a band of long waves destroys the intertidal undulations, except in the case of a very narrow bandwidth indeed. However this can be shown to be misleading by studying the sediment flux functions. Figure 8(a) indicates that the flux function (the offshore transport) is dominated by the position of the breakpoint. This creates two relatively sharp peaks in the SFC near high and low water throwing the remaining details in to the shadows. The convergence here is emphasised by the fact that the breakpoint dwells at these two positions for the longest. As mentioned above, in reality the breaking of the short waves has a statistical variation caused by variations in the waveheight due to groupiness, return flows, interactions with preceding waves and an inherent chaotic instability in the breaking process.

To study the effect of this, an uncertainty in the breakpoint was modeled assuming a gaussian description of wave height and the breaking criterion. This was included in the description of the short wave envelope or "stirring function". The effect in the sediment flux function is evident in Figure 8(b). The breakpoint is rounded off and de-emphasised, allowing detail of the long wave structure to emerge. Figure 9(b) shows the effect of this in the predicted SFC which displays once again some undular detail across the profile, more obviously than in Figure 9(a). Again the scales correspond well with those observed in the field.

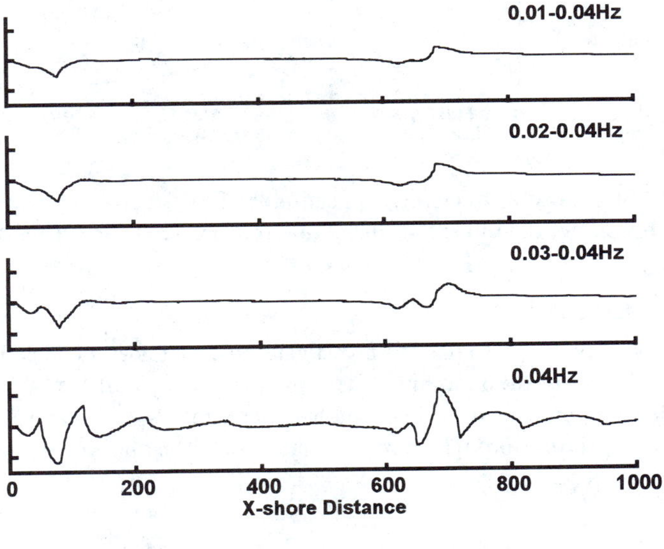

Figure 7 SFC calculated for different long wave bandwidths

The uncertainty in the breakpoint acts to widen the surf zone. Another way this can be achieved is to consider short waves with larger waveheight. As can be seen in Figures 8(c) and 9(c), the effect of this is to smooth off the breakpoint further and to bring out more intertidal profile detail. This remarkable result shows that despite summing over a band of long waves and averaging over a tide, the predicted SFC still shows details that correspond to the observed topography even in the mid tidal region. However, although larger wave heights are helpful in this model it is probably true, as Wright (1982) indicates, that in very high energy events other processes will dominate and the surf zone "bulldozer" will mask the effects of this subtle mechanism.

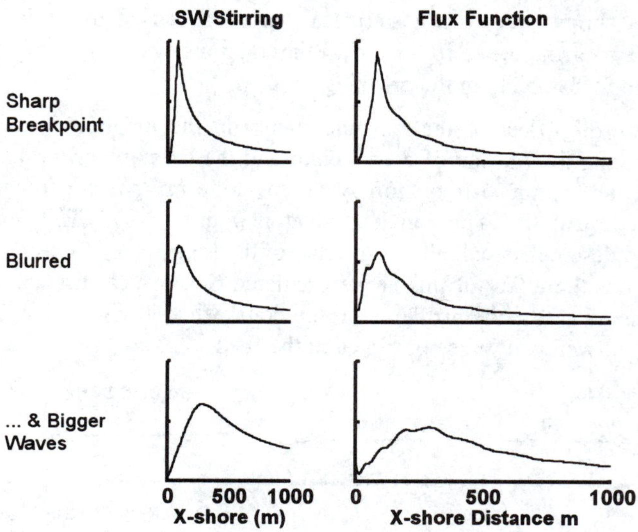

Figure 8 Short wave envelopes and sediment flux functions for (a) sudden breaking, (b) uncertain breaking, (c) uncertain breaking & larger wave height

Further Investigations

The simulation was also used to investigate other parameters, including shape of the tidal curve and of the spectrum. These predictably showed that a flatter tidal curve de-emphasised erosion at the high and low water marks, and that a "blue" long wave amplitude spectrum brought out even more detail in the undulations of the sediment flux convergence.

Time–scales

The results so far have all been calculated as averages over the tidal cycle. Figure 10 shows the sediment flux convergences calculated for averages over shorter fractions of a tide (1/6*ths*). As can be seen, the tidally averaged result shows no indication of some of the features that can develop during the middle of the tide. Two features at 200m and 500m are clearly visible in the mid tide SFC but these are cancelled by the SFCs approaching high and low water.

It is likely that the bed will respond over shorter time-scales than that of a tide. If the bed is modified, the long and short wave field are changed accordingly and the erosion patterns will be changed. It may be that the erosion pattern is capable of "tuning in" with the topography to favour growth of certain scales.

Southgate (1995) has remarked that since the interaction between hydrodynamics and the transport of sediment is non-linear, the chronology or sequencing of wave events is of direct influence. Further, in the macrotidal case, the erosion of the beach face will be dependent upon both the temporal variation of the long wave transport functions and the tidal variation.

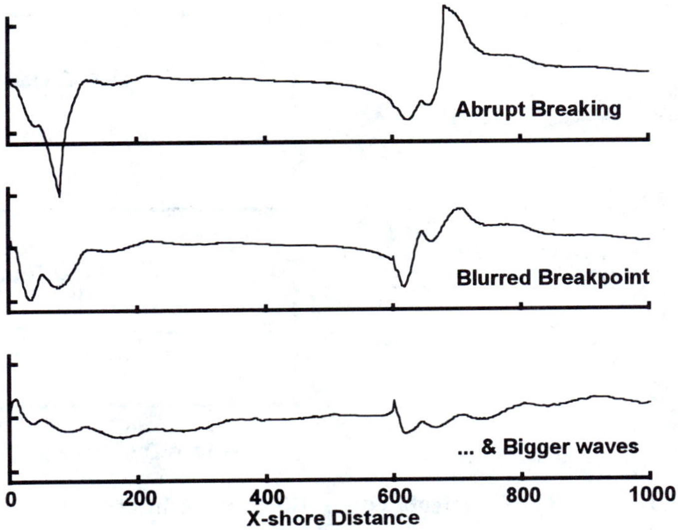

Figure 9(a) SFC for a band of long waves (0.01–0.04Hz), (b) inclusion of uncertain breaking & (c) bigger short waves.

Summary & Discussion

Simmonds et al (1995) demonstrated that a standing long wave mechanism such as the bound long wave–short-wave correlation is capable of generating intertidal bedforms at scales observed in nature. This idea has been extended in this paper to the consideration of a band of long waves. It has been argued that there exists an inherent bias towards the quicker development of shorter scales. It has been shown that a band of longwaves will still generate undular intertidal topography, even in the mid–tide region so long as a smoothing of the breakpoint and hence a widening of the surfzone is introduced. This was necessary to de-emphasize the erosion at the high and low water levels caused by the constancy of the breaking at these places. The smoothing is justified by considering the uncertainty in the breaking process. The resulting predicted regions of sediment convergence agree remarkably well with the observed topography.

It has also been suggested that the current approach to integrating results over tidal timescales needs re-examining since it is that the bed will respond over much shorter timesscales.

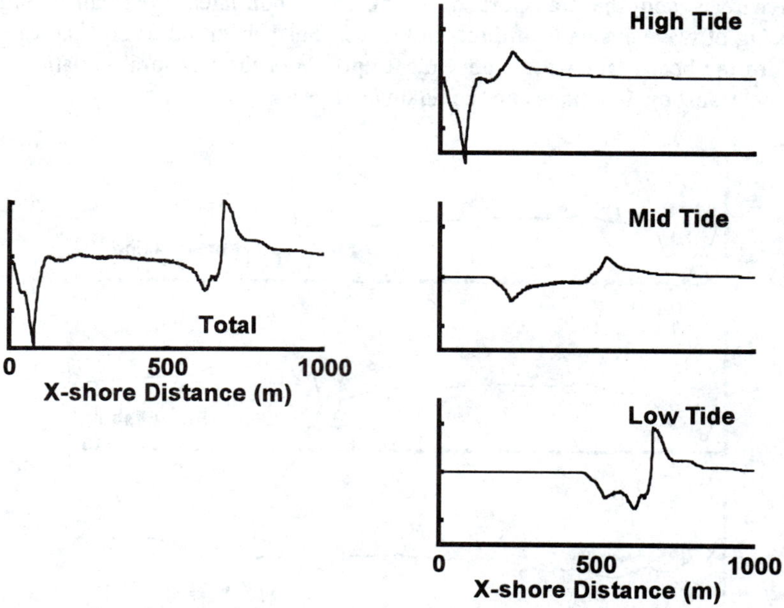

Figure 10 SFCs calculated for different sixths of the tide

The model helps us to explain the environment in which ridge and runnel features are found. For the mechanism to work, the cross shore tidal excursion must exceed the scale of the shorter long wave wavelengths otherwise no intertidal structure can be created. Broad, dissipative macrotidal beaches are ideal for this. The mechanism is, however relatively subtle, depending upon the long wave orbotal velocities to carry out the transporting of the sediment. and would probably be masked by other processes in a more energetic environment. This might explain why these features are only found on fetch limited macrotidal coastlines.

Future developments of the model are envisaged that allow erosion of the bed as a feedback mechanism which perturbs the structure of the sediment flux function. It is hoped that such a coupled sediment transport and hydrodynamic model can be used to study the parameters that control the behaviour of these intertidal features and the time-scales of their development.

Acknowledgments

This work was supported through funding by the European community under the Marine Science and Technology programme (EC contract MAS2-CT92-0024)

The authors would also like to acknowledge the contribution of those who assisted with collection of the original field data, especially Mike Wilkin, formerly of Southampton University, Dept. Oceanography.

References

Bauer, B.O. & Greenwood, B., (1990). Modification of a linear bar–trough system by a standing edge wave, *Marine Geology*, **92** :177–204.

Carter, T.G., Liu, P.L–F., Mei, C.C., (1973). Mass transport by waves and offshore sand bedforms, *J.Waterways Harbours and Coastal Eng.* **WW2** May 1973, :165–184.

Huntley, D.A., Davidson, M., Russell, P., Foote, Y., & Hardisty, J., (1993). Long waves and sediment movement on beaches: recent observations and implications for modeling, *J.Coastal.Res*, **SI 15**, :215–229

King C.A.M. and Williams W.W., (1949). The Formation and Movement of Sand Bars by Wave Action, *Geographical J.* **113**:70–85.

Kirby, J.T.,Jr, Dalrymple, R.A. and Liu, L.–F , (1981). Modification of edge waves by barred beach topography, *Coastal Engineering*, **5**: 35–49.

List, J.H., (1992). A model for the generation of two-dimensional surf beat, *J.Geophysical Research*, **97 C4**:5623–5635.

Longuet–Higgins, M.S., (1994). A fractal approach to breaking waves J.Phys.Oceanog, **24**: 1834–1838.

Mulrennan, M.E., (1992). Ridge and runnel beach morphodynamics: An example from the Central East Coast of Ireland, *J. Coastal Research* **8**(4): 906–918.

O'Connor *et al*, (1995). Circulation and sediment transport around banks —CSTAB final report. (held by MAST office, Brussels.)

O'Hare, T.J. and Huntley, D.A., (1994). Bar formation due to wave groups and associated long waves, *Marine Geology*, **116**: 313–325

Roelvink, J.A. & Broker, I, (1993). Cross–shore Profile Models. *Coastal Eng.*, **21**:163–191

Sallenger, A.H., Jr, and Holman, R.A., (1987). Infragravity waves over a natural barred profile. *J.Geophysical Res*., **92**, C9: 9531–9540.

Short, A.D., (1991). Macro–meso tidal beach morphodynamics– an overview. *J.Coastal.Res*. 7(2): 417–436

Simmonds, D.J., Voulgaris, G. & Huntley, D.A., (1995). Dynamic processes on a ridge and runnel beach. Paper presented at *Coastal Dynamics '95, Gdansk, ASCE*

Southgate, H.N., (1995). The effects of wave chronology on medium and long term coastal morphology *Coastal Engineering*, **26**: 251–270.

Wright, L.D., Nielsen, P., Short, A.D., Green, M.O. (1982). Morphodynamics of a macrotidal beach, *Marine Geology*, **50** :97–128

CHAPTER 240

MATHEMATICAL MODELS FOR WAVES AND BEACH PROFILES IN SURF AND SWASH ZONES

Akira Watanabe[1] and Mohammad Dibajnia[2]

Abstract

The paper presents mathematical models for wave deformation and for beach profile change in the surf and swash zones. Boussinesq-type equations including a breaker-induced energy dissipation term (Watanabe *et al.*, 1994) is extended to the swash zone by introducing a periodically moving shoreward boundary. Computation is made on the wave deformation on constant slope beaches, particularly on the runup and run-down heights as well as the wave heights at the still water shoreline, which are compared with existing formulas. Beach profile change is computed with the sediment transport rate formula proposed by Dibajnia and Watanabe (1992) after a further generalization, together with Lagrangian treatment for the sediment motion in the swash zone, using the near-bottom velocity obtained from the wave computation. The validity of the models is examined through comparisons with measurement data obtained in large wave flume experiments.

Introduction

The authors have presented a numerical model for profile change of sheet-flow dominated beaches based on Boussinesq-type wave equations and a sediment transport rate formula proposed by themselves (Watanabe *et al.*, 1994). It has been demonstrated that the model can predict well the cross-shore distributions of the wave height and the transport rate as well as the beach profile change in and around the surf zone. However, the model has a crucial limitation of the incapability of computing the wave behavior and the beach evolution in the surf zone, both of which are generally very important in practical applications and fundamental studies regarding

1 Professor, Department of Civil Engineering, University of Tokyo, Hongo-7, Bunkyo-ku, Tokyo, 113, Japan.

2 Associate Professor, ditto.

the coastal processes. A major effort in the present study is thus focused on the treatment of the swash zone dynamics both for waves and for sediment transport. Mathematical models are proposed for wave deformation and beach profile change and their validity is examined by comparing computations with measurements.

Mathematical Model for Wave Deformation

Watanabe *et al.* (1994) have presented a set of one-dimensional Boussinesq-type equations including a breaker-induced energy dissipation term as follows:

$$\frac{\partial \varsigma}{\partial t}+\frac{\partial Q}{\partial x}=0, \quad \varsigma=\varsigma(x,t), \quad Q(x,t)=\int_{-h}^{\zeta} u(x,z,t)\,dz \tag{1}$$

$$\frac{\partial Q}{\partial t}+\frac{\partial}{\partial x}\left(\frac{Q^2}{D}\right)+gD\frac{\partial \varsigma}{\partial x}-\frac{1}{3}\overline{D}^2\frac{\partial^3 Q}{\partial t \partial x^2}-M_D=0 \tag{2}$$

where

$$D(x,t)=h+\varsigma, \quad \overline{D}(x)=h+\overline{\zeta},$$

u is the horizontal velocity, and M_D corresponds to the momentum diffusion in the surf zone and has been expressed by Sato and Suzuki (1990) (See also Sato and Kabiling, 1994a, b) as

$$M_D=\frac{g\overline{D}}{\sigma^2}f_D\frac{\partial^2 Q}{\partial x^2} \tag{3}$$

in which σ is the angular frequency, and f_D is the following energy dissipation coefficient proposed by Watanabe and Dibajnia (1988).

$$f_D=\alpha_D\tan\beta\sqrt{\frac{g}{\overline{D}}}\cdot\sqrt{\frac{\hat{Q}-Q_r}{Q_s-Q_r}} \tag{4}$$

where α_D = 2.5, tan β is the bottom slope around the breaking point, $\hat{Q}$ is the amplitude of the flow rate Q, and Q_s and Q_r correspond to $\hat{Q}$ in the dissipation zone on a uniform slope and in the recovery zone of constant depth, respectively, defined in this study by

$$Q_s=0.4\left(0.57+5.3\tan\beta\right)C\overline{D}, \quad Q_r=\gamma_{\mathrm{B}}\cdot C\overline{D} \tag{5}$$

where C is the wave celerity, and γ_{B} is a coefficient proportional to the ratio of the wave amplitude to the total mean depth. In the present model, we don't use any breaking criterion that determines the location of the breaking point *a priori* as employed in most previous models. Instead, by setting an appropriate value of γ_{B} and by equating f_D to zero in regions where $\hat{Q}\le Q_r$, the model can automatically handle the problem of determining the breaking point and the dissipation zone as well as the recovery zone. This also makes it unnecessary to use the factitious increase of the value of α_D from 0 to 2.5 around the breaking point that is otherwise needed to avoid numerical wave reflection from there.

The Sommerfeld radiation condition is imposed on the offshore boundary of a computation domain in the same way as in the previous model (Watanabe *et al.*, 1994). On the other hand, the treatment of the shoreward boundary in the present model is completely different from the previous one that assumes the presence of a fictitious shoreward zone of constant depth and neglects the swash wave behavior. Here we adopt a periodically moving shoreward boundary (swash wave front) on which the condition $Q = 0$ is imposed all through the period of uprush and backwash.

Numerical computation is conducted by a finite difference method with a staggered grid scheme of the central difference except for the convection term that is treated with the upwind difference for the sake of stability.

Application of the Wave Model

Figure 1 shows one example of a time-series of wave profiles for one wave period in and near the swash zone computed by using the present model. The bottom consists of a uniform slope of 1/20 and a horizontal bed with a depth of 2 m. The time history of incident waves (H_I = 0.45 m, T = 8 s) has been calculated by the second-order cnoidal wave theory and given on the seaward boundary (x = 0 m). The swash wave behavior or runup/rundown process seems to be simulated reasonably.

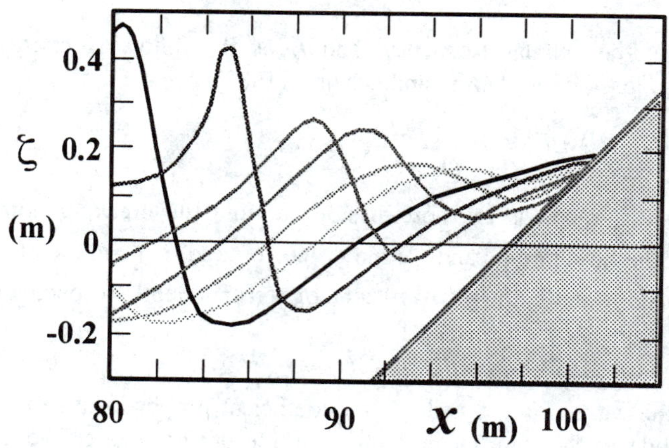

Fig. 1 Wave behavior near the shoreline; Runup and run-down.
(H_I = 0.45 m, T = 8 s, $\tan\beta$ = 1/20)

In order to examine the validity of the present wave model more quantitatively, swash-related parameters are computed and compared with existing formulas. Regarding the runup height R_u, Hunt (1959) proposed the following formula:

$$R_u / H_0 = \tan\beta / \sqrt{H_0 / L_0} \tag{6}$$

On the other hand, Ogawa and Shuto (1984) formulated the runup height Ru, on mild slope solid beds as the summation of the rundown height R_d and the swash zone height S_h as follows (See their paper for details):

$$R_u / H_0 = R_d / H_0 + S_h / H_0 \tag{7}$$

Comparisons of the runup height R_u are shown in **Fig. 2** between the computations by the present model and the above formulas. The agreement of the computations with Ogawa and Shuto's formula is remarkably good.

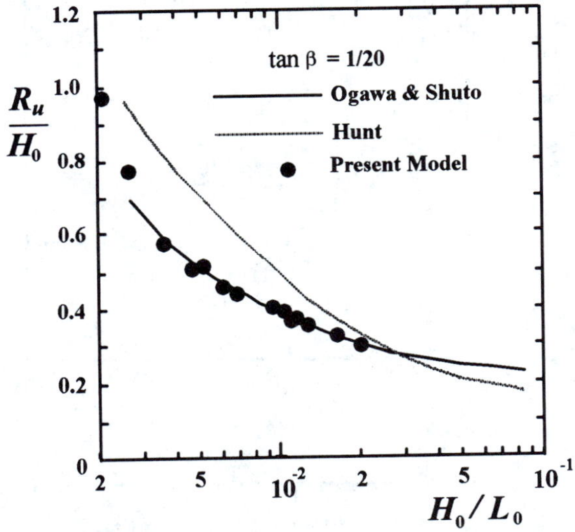

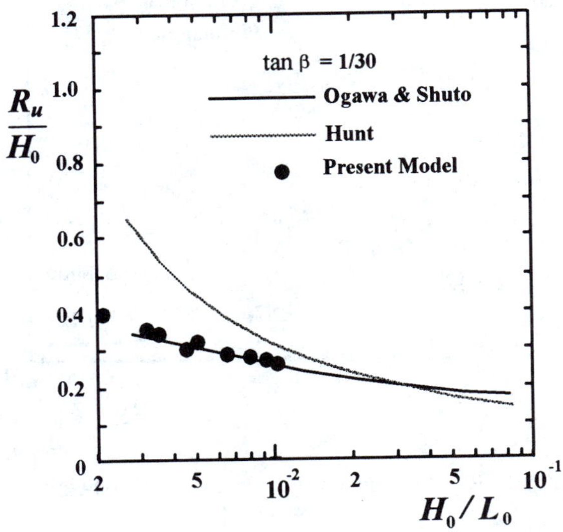

Fig. 2 Runup height R_u.

Similar comparisons are made in **Fig. 3** for the rundown height R_d and the vertical length of the swash zone S_h. The agreement with Ogawa and Shuto's formula is worse for R_d and thus for S_h than for R_u in particular for the small deepwater wave steepness, but still fairly good.

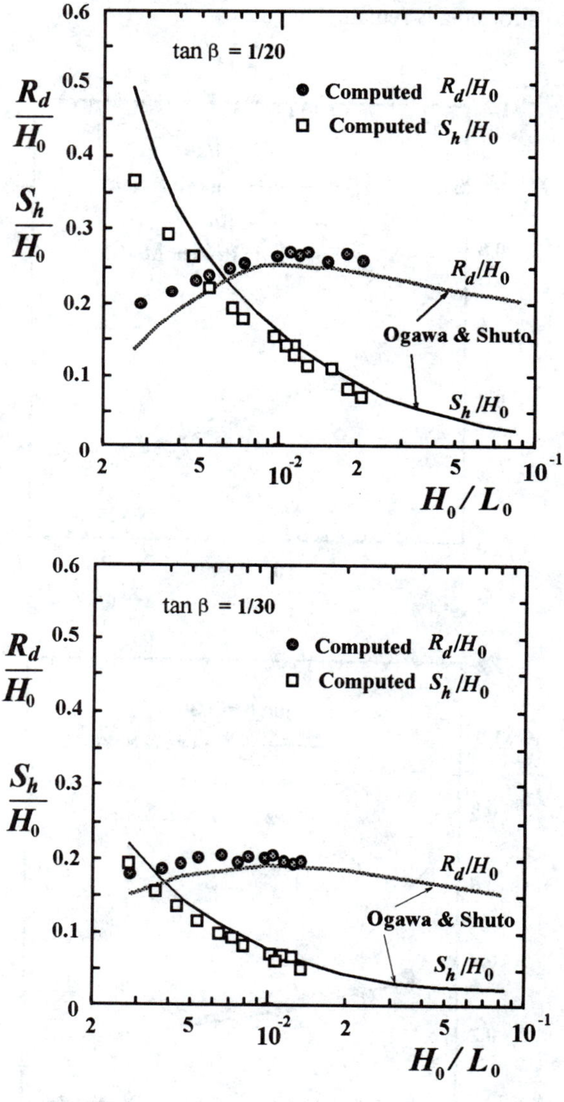

Fig. 3 Rundown height R_d and swash zone height S_h.

Now we will discuss the accuracy of the present model for evaluating the wave height H_S at the still water shoreline. Sunamura (1984) has reported the following empirical formula for H_S based on wide-range experimental data.

$$H_S / H_B = 2.5 \tan\beta \tag{8}$$

where H_B is the breaker height, which, according to Sunamura and Horikawa (1974), is related to the deepwater wave height H_0, wavelength L_0 and the bottom slope as

$$H_B / H_0 = (\tan\beta)^{1/5}(H_0 / L_0)^{-1/4} \tag{9}$$

Substituting Eq. (9) into Eq. (8), we obtain

$$H_S / H_0 = 2.5(\tan\beta)^{6/5}(H_0 / L_0)^{-1/4} \tag{10}$$

Figure 4 compares H_S between the computations and Eq. (10), also demonstrating the applicability of the present model to simulating the swash wave dynamics.

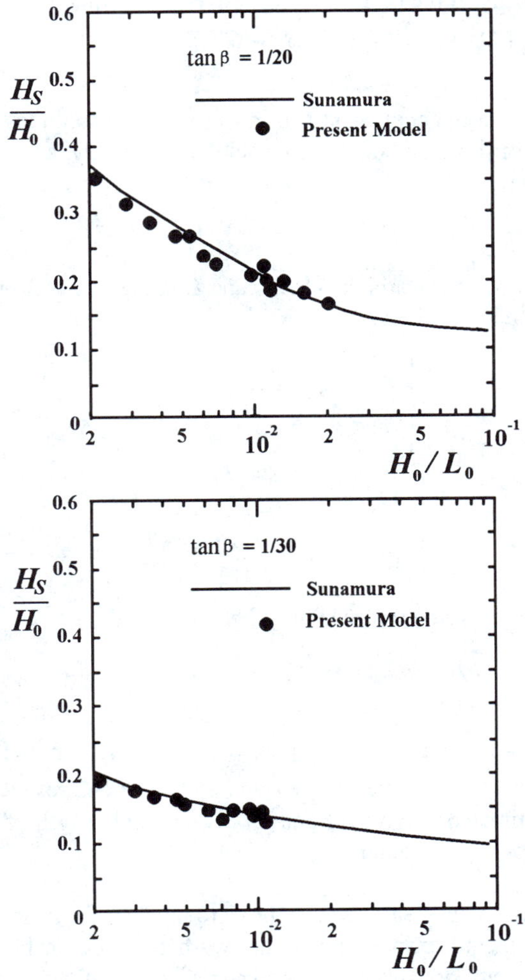

Fig. 4 Wave height H_S at the still water shoreline.

Mathematical Model for Beach Profile Change

The sediment transport rate formula proposed by Dibajnia and Watanabe (1992) is for the sheet flow in asymmetric oscillatory flow with superimposed steady flow, and has already been generalized by them for the bedload as well as for the suspended load over ripples (Dibajnia *et al.*, 1994). After a further slight modification, Dibajnia and Watanabe's sediment transport rate formula reads as follows:

$$\Phi \equiv \frac{q_{\text{net}}(1-\lambda_v)}{w_0 d} = 0.001 \cdot \text{sign}(\Gamma) \cdot |\Gamma|^{0.5} \quad (11)$$

where q_{net} is the net transport rate, and w_0, d, and λ_v are the settling velocity, grain diameter, and porosity of the sediment, respectively. The quantity Γ is defined by

$$\Gamma = \frac{u_c T_c(\Omega_c^3 + \Omega'^3_t) - u_t T_t(\Omega_t^3 + \Omega'^3_c)}{(u_c + u_t)T} \quad (12)$$

in which u_c and u_t are the equivalent root-mean-square amplitudes and T_c and T_t are the periods of the onshore and offshore velocity, respectively, and T (= $T_c + T_t$) is the wave period, namely,

$$u_c^2 = \frac{2}{T_c}\int_0^{T_c}(u_w + U)^2 dt, \quad u_t^2 = \frac{2}{T_t}\int_{T_c}^{T}(u_w + U)^2 dt \quad (13)$$

where u_w is the near-bottom orbital velocity and U is the steady flow velocity. Values of Ω_j are determined as follows:

$$\begin{cases} \text{if } \omega_j \le \omega_{\text{cr}} & \begin{cases} \Omega_j = \omega_j \cdot \dfrac{2w_0 T_j}{d} \\ \Omega'_j = 0 \end{cases} \\ \text{if } \omega_j > \omega_{\text{cr}} & \begin{cases} \Omega_j = \omega_{\text{cr}} \cdot \dfrac{2w_0 T_j}{d} \\ \Omega'_j = (\omega_j - \omega_{\text{cr}}) \cdot \dfrac{2w_0 T_j}{d} \end{cases} \end{cases} \quad (14)$$

where the subscript j is to be replaced by either c or t, and

$$\omega_j = \frac{1}{2} \cdot \frac{u_j^2}{s g w_0 T_j}, \quad s = \frac{\rho_s - \rho}{\rho} \quad (15)$$

$$\omega_{\text{cr}} = 1 - 0.97 \cdot \sqrt{\Lambda}, \quad \Lambda = \{1 - [(\Psi_{\text{rms}} - 0.2)/0.4]^2\} \cdot \min(1, 2\lambda / d_0) \quad (16)$$

in which ρ and ρ_s are the densities of the water and sediment, respectively, Ψ_{rms} is the Shields number estimated in terms u_c and d, λ is the pitch length of ripples if any, and d_0 is the near-bottom orbital diameter.

Now one very important issue is how to apply this generalized formula to computing the sediment transport rate in the swash zone reasonably, where the wave-induced orbital velocity drastically varies over a distance of the periodical movement of sediment mass. In the present model for beach profile change, we will adopt a Lagrangian method, namely, assuming that the velocity of every sediment mass in motion is given by the concurrent flow velocity, we trace the motion of a sediment mass over every wave period, memorize the velocity variation experienced by it, determine its mean position by taking the time-average, and evaluate its net transport rate at this position from the above formula with the velocity variation.

The undertow velocity, which is particularly large near the breaking point in the surf zone, is also incorporated in the computation of the transport rate (For details, readers are referred to Watanabe *et al.*, 1994) . Change in beach profiles is obtained from the following conservation equation of sediment mass including the effect of local bottom slope (Watanabe *et al.*, 1986) through alternate computation of the wave deformation, the net transport rate, and the beach transformation itself.

$$\frac{\partial z_b}{\partial t} = -\frac{\partial h}{\partial t} = -\frac{\partial}{\partial x}\left(q_{\text{net}} - \varepsilon_s \left|q_{\text{net}}\right| \frac{\partial z_b}{\partial x}\right) \tag{17}$$

where z_b is the bottom elevation, and a value of the coefficient ε_s is set equal to 2.0 in consideration of the repose angle of sand.

Application of Beach Profile Model

Now we will examine the validity of the model for the beach profile change. **Figures 5** and **6** show examples of comparisons of the wave height distribution and of the beach profile change between the computations and measurements. The measurement data are those obtained in large-wave-flume experiments by Shimizu *et al.* (1985). Figure 5 corresponds to an erosional condition, whereas Fig. 6 to a depositional case. For both the cases, the computed wave height distributions agree very well with the measurements except for the under-estimation near breaking points, which is attributable to the weak nonlinearity of Boussinesq-type equations. It is seen in Fig. 5 that the present model properly reproduces not only the formation of a bar but also the recession of the shoreline. Figure 6 also indicates high capability of the model in reproducing both the accretion near the shoreline and the berm formation.

Concluding Remarks

Mathematical models have been proposed in this paper both for the nearshore wave deformation and for the beach profile change in the surf and swash zones. The wave model is based on a Boussinesq-type equations including the breaker-induced energy dissipation and adopts a periodically moving shoreline boundary for the treatment of the swash zone dynamics. In the beach profile model, a Lagrangian method has been employed to evaluate the net rate of sediment transport in and near the swash zone. The validity and applicability of the proposed models have been verified through comparisons with the existing formulas and the experimental data both for the wave deformation and the beach profile change. Improvement of the computational efficiency and application to longer-term and H-2D conditions are left for future study.

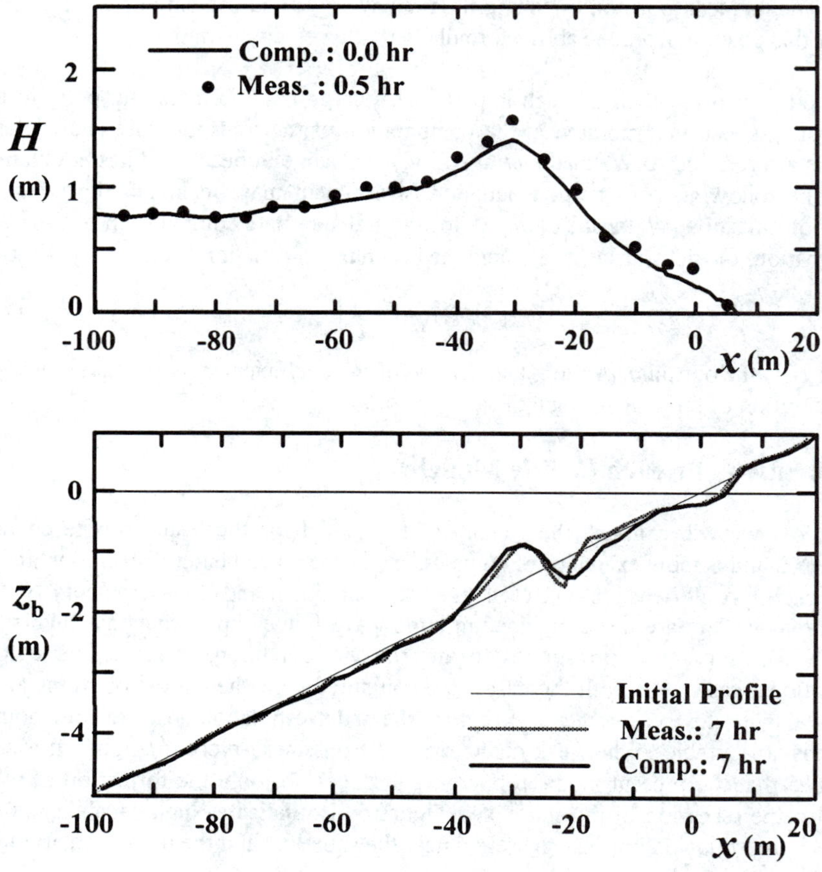

Fig. 5 Wave height distribution and beach profile change.
($H_0 = 0.81$ m, $T = 12$ s, $d = 0.27$ mm)

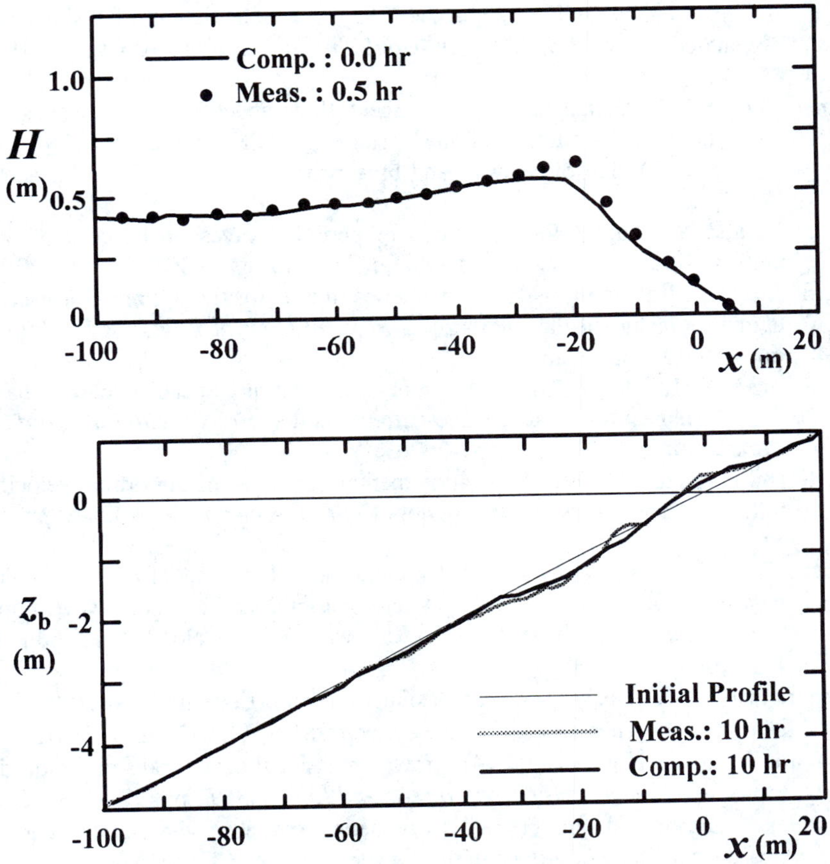

Fig. 6 Wave height distribution and beach profile change.
($H_0 = 0.46$ m, $T = 6$ s, $d = 0.27$ mm)

References

Dibajnia, M., T. Shimizu and A. Watanabe (1994): Profile change of a sheet flow dominated beach, *Proc. 24th Int. Conf. on Coastal Eng.*, ASCE, pp. 1946-1960.

Dibajnia, M. and A. Watanabe (1992): Sheet flow under nonlinear waves and currents, Proc. 23rd Int. Conf on Coastal Eng., ASCE, pp. 2015-2028.

Hunt, A.I. (1959): Design of seawalls and breakwaters, *Proc. ASCE*, Vol. 94, No. WW1, pp. 72-92.

Ogawa, Y. and N. Shuto (1984): Runup of periodic waves on beaches of non-uniform slope, *Proc. 19th Int. Conf. on Coastal Eng.*, ASCE, pp. 198-205.

Sato, S. and M.B. Kabiling (1994a): A numerical model for three-dimensional beach evolution including the swash zone, *Proc. of Coastal Eng.*, JSCE, Vol. 41, pp. 401-405. (in Japanese)

Sato, S. and M.B. Kabiling (1994b): A numerical simulation of beach evolution based on a nonlinear dispersive wave-current model, *Proc. 24th Int. Conf. on Coastal Eng.*, ASCE, pp. 2557-2570.

Sato, S. and H. Suzuki (1990): Estimation method for near-bottom orbital velocities in the surf zone, *Proc. 37th Japanese Conf. on Coastal Eng.*, JSCE, pp. 51-55.

Shimizu, T., S. Saito, K. Maruyama, H. Hasegawa and R. Kajima (1985): Modeling of cross-shore sediment transport rate distributions in a large wave flume, *Rep. of Central Res. Inst. for Electric Power Industry*, No. 384028, 60 p. (in Japanese)

Sunamura, T. (1984): Onshore-offshore sediment transport rate in the swash zone of laboratory beaches, *Coastal Eng. in Japan*, JSCE, Vol. 27, pp. 205-212.

Sunamura, T. and K. Horikawa (1974): Two-dimensional beach transformation due to waves, *Proc. 14th Int. Conf. on Coastal Eng.*, ASCE, pp. 920-938.

Watanabe, A. and M. Dibajnia (1988): A numerical model of wave deformation in surf zone, *Proc. 21st Int. Conf. on Coastal Eng.*, ASCE, pp. 578-587.

Watanabe, A., K. Maruyama, T. Shimizu and T. Sakakiyama (1986): Numerical prediction model of three-dimensional beach deformation around a structure, *Coastal Eng. in Japan*, JSCE, Vol. 29, pp. 179-194.

Watanabe, A., K. Shiba & M. Isobe (1994): A numerical model of beach change due to sheet-flow, *Proc. 24th Int. Conf. on Coastal Eng.*, ASCE, pp. 2785-2798.

CHAPTER 241

Comparisons of Erosion Models For Storms at Ocean City, MD

Jie Zheng[1] and Robert G. Dean[2]

Abstract

A new non-linear cross-shore sediment transport model called CROSS is developed based on equilibrium beach profile concepts and scaling relationships. CROSS and three other existing models, CCCL (Chiu and Dean 1984), EDUNE (Kriebel 1989) and two versions of SBEACH (version 2.0, Larson and Kraus 1989, and version 3.0, undocumented) are investigated by comparing erosion occurring from November 1991 to January 1992, at Ocean City, Maryland, the site of a major beach fill placed by the State of Maryland and Federal Government in 1988, 1990 and 1991. Among the four models, CCCL is the only one which overpredicts average dune erosion; the other three underpredict it. Overall CROSS and EDUNE yield better predictions than the other two models.

Introduction

Ocean City, Maryland, is located on Fenwick Island, a north-south oriented barrier island of the central Delaware-Maryland-Virginia coast. The Ocean City beach was nourished by the State of Maryland in 1988, and the Federal Government in 1990 and 1991 to protect the city against storm damage (Stauble et al. 1993). The project layout is shown in Figure 1. The entire project was finished in August 1991.

After the project, a series of storms occurred in late 1991 and early 1992. Among these 1991-1992 winter storms, the January 4, 1992 storm was very severe with a peak storm surge of 2 meters (Jensen and Garcia 1993). The initial pre-storm beach profiles were surveyed on November 2-4, 1991, and the post-storm profiles were surveyed on January 11, 1992. In this period, an additional storm occurred on November 11, 1991. For consistency with the measured pre-storm and post-storm profiles, both the November 1991 and January 1992 storms are included in the numerical simulations. A total of seven survey lines located from the southern (37th

---

[1] Coastal Engineer, Coastal Technology Corporation, 3625 20th St., Vero Beach, FL, 32960, USA.

[2] Professor, University of Florida, 336 Weil Hall, Gainesville, FL 32611, USA.

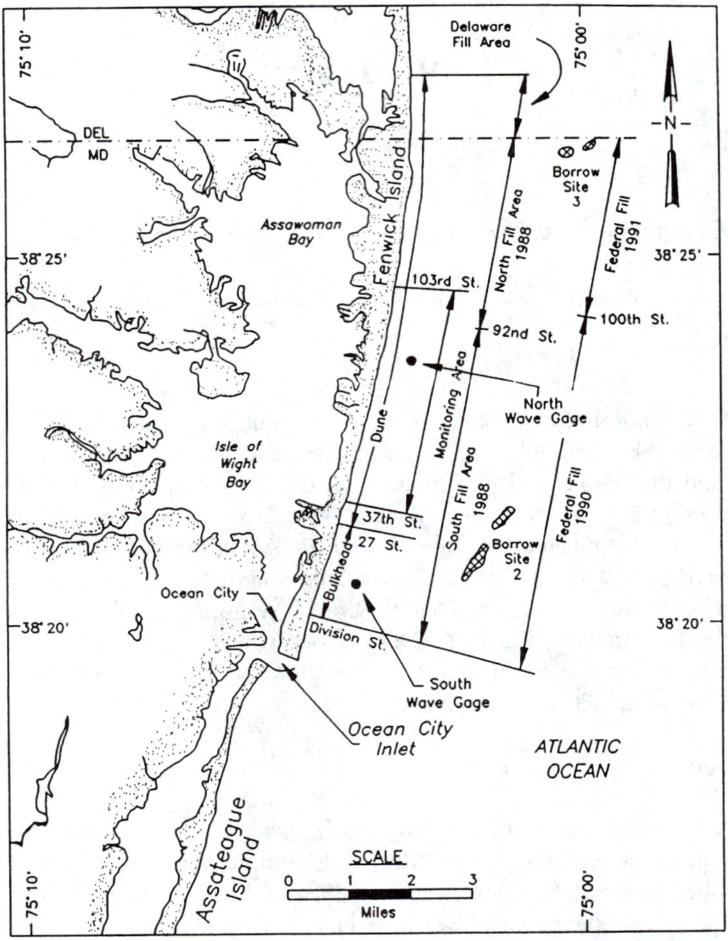

Figure 1. The Ocean City beach nourishment project layout.

Street) to northern (124th Street) portion of the project are available for both pre and post-storm surveys (Stauble et al. 1993, and Kraus and Wise 1993).

Four cross-shore sediment transport models are compared for the simulations of beach erosion at Ocean City during the November 1991 and January 1992 storms. These models are

- CCCL (Chiu and Dean 1984, 1986)
- EDUNE (Kriebel and Dean 1985, Kriebel 1986)
- SBEACH (Larson and Kraus 1989, Larson et al. 1989)
- CROSS (Zheng 1996, Zheng and Dean 1996)

Two versions (2.0 and 3.0) of SBEACH models will be included in the following comparisons. Version 3.0 was released in September 1994 by the U.S. Army Corps of Engineers after a study of storm erosion at Ocean City during the 1991-1992 winter storms. Three measures are presented for comparison of quantitative model performance.

Brief Description of Four Models

All four models discussed here are of the "closed loop" type and based on equilibrium beach profile concepts, which consider beach profile changes to be caused by deviations of a profile from its equilibrium. An equilibrium beach profile represents a dynamic balance of constructive and destructive forces acting on the beach. A change in the two competing types of forces will result in a disequilibrium. Considering wave energy dissipation per unit volume to represent the dominant destructive force, Dean (1977) proposed that a sediment of given size will be stable in the presence of a particular level of wave energy dissipation per unit volume, D_*, which is expressed as

$$D_* = \frac{1}{h}\frac{d(Ec_g)}{dy} \tag{1}$$

where h is the water depth, y is the shore-normal coordinate directed offshore, E is the wave energy density, and c_g is the wave group velocity. As a first approximation, D_* is assumed to dependent only on sediment size (Moore, 1982). With linear wave theory and shallow water assumptions, Eq (1) is integrated to

$$h = \left(\frac{24D_*}{5\rho g\sqrt{g}\,\kappa^2}\right)^{2/3} y^{2/3} = A\,y^{2/3} \tag{2}$$

where κ is the ratio of breaking wave height to water depth, and A is defined as a profile scale parameter.

CCCL Model

Under erosive water level and wave conditions, the time dependent beach recession, R(t) is given by

$$R(t) = R_\infty(1 - e^{K_R t}) \tag{3}$$

where R_∞ is equilibrium recession and K_R is a reciprocal time scale. At each time step, the equilibrium profile and recession are calculated by considering Eq. (3) and sand conservation. After the final computational time step, a factor of 2.5 is applied to the recession, in part to incorporate the alongshore variability of beach erosion.

EDUNE Model

The cross-shore sediment transport rate per unit beach width, Q, is considered as linearly proportional to the deviation of local wave energy dissipation per unit volume from the equilibrium,

$$Q = K_e(D - D_*) \tag{4}$$

where K_e is an empirical transport parameter and D is the local wave energy dissipation per unit volume. A beach steeper or milder than the equilibrium at a given depth will cause sediment transport offshore or onshore, respectively. Since the transport (Eq. (4)) has two variables, a continuity equation is used to close the system.

SBEACH Model

The cross-shore sediment transport rate is determined by

$$\begin{aligned} Q &= \pm K_s \left| D - D_* + \frac{\epsilon}{K_s}\frac{\partial h}{\partial x} \right| , \qquad && D > D_* - \frac{\epsilon}{K_s}\frac{\partial h}{\partial x} \\ &= 0 , && D \leq D_* - \frac{\epsilon}{K_s} \end{aligned} \tag{5}$$

where ϵ is an empirical constant and K_s is a transport coefficient. This relationship is similar to EDUNE except for the last term which incorporates the effect of local slope. The beach profile evolution is solved by combining the transport relationship with the continuity equation. The direction of transport (± in Eq. (5)) is determined according to

$$\begin{aligned} &\frac{H_0}{L_0} - 0.00070\left(\frac{H_0}{w_f T}\right)^3 < 0, \; + \\ &\frac{H_0}{L_0} - 0.00070\left(\frac{H_0}{w_f T}\right)^3 > 0, \; - \end{aligned} \tag{6}$$

CROSS Model

Based on scale analysis and the Froude relationship, the cross-shore sediment transport rate per unit beach width, Q, should scale as

$$Q_r = L_r^2 / T_r = L_r^{3/2} \tag{7}$$

where L_r and T_r are length and time scales, respectively. To satisfy this scale relationship (7) and ensure convergence to the equilibrium beach profile, the following sediment transport relationship is proposed

$$Q = K_c(D - D_*)^3 \tag{8}$$

where K_c is a dimensional constant. Time dependent profile response is determined by solving the transport equation (8) and sand conservation equation.

Storm and Beach Profile Characteristics

Water depth and profile elevations used here are referenced to NGVD (National Geodetic Vertical Datum), which lies 2 cm below mean water level for Ocean City. The wave height, wave period and storm surge time histories during the two storms were measured by two gages located directly offshore of Ocean City in a water depth of 10 meters. The measured significant wave height, peak spectral wave period and storm surge are shown in Figure 2 for the January 4, 1992 storm (Stauble et al. 1993).

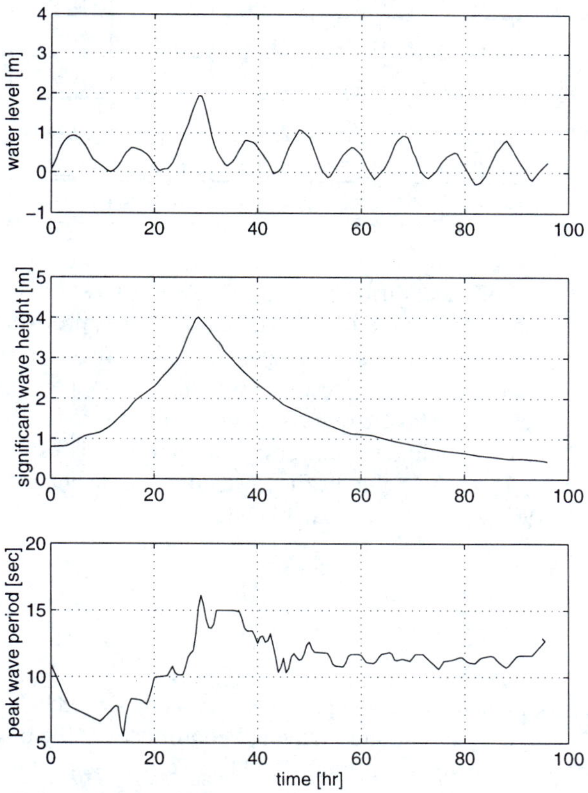

Figure 2. Water level, significant wave height and peak spectral wave period time history for the January 4, 1992 storm at Ocean City, MD (From Stauble et al. 1993).

Seven survey lines located at 37th, 45th, 56th, 74th, 103rd and 124th Streets (Stauble et al. 1993, Kraus and Wise, 1993) have both pre-storm and post-storm measured profiles available and are simulated. The mean grain sizes within each of 11 cross-shore morphologic zones are presented in Table 1. The seven measured pre- and post-storm profiles are shown in Figure 3 for the two surveys taken on November 2, 1991 and January 11, 1992, respectively. It appears that the storm-caused erosion is

Table 1 Average mean grain size in millimeters.

Sample Location	37th St.	45th St.	56th St.	63rd St.	74th St.	103rd St.	124th St.
Dune base	0.35	0.34	0.33	0.39	0.39	0.44	0.44
Berm crest	0.29	0.31	0.34	0.35	0.35	0.38	0.38
Mean-tide line	0.32	0.30	0.27	0.27	0.29	0.38	0.38
Swash zone	0.34	0.38	0.44	0.47	0.48	0.39	0.39
Nearshore trough	0.31	0.33	0.36	0.45	0.47	0.81	0.81
Nearshore bar	0.29	0.36	0.46	0.44	0.46	0.34	0.34
- 1.52 m contour	0.29	0.26	0.21	0.26	0.32	0.23	0.23
- 3.05 m contour	0.21	0.21	0.22	0.22	0.21	0.21	0.21
- 4.57 m contour	0.20	0.22	0.24	0.21	0.22	0.21	0.21
- 6.10 m contour	0.29	0.23	0.14	0.15	0.28	0.17	0.17
- 7.62 m contour	0.34	0.25	0.13	0.12	0.17	0.16	0.16

quite different for profiles at different locations. After two storms, the profiles at 37th and 56th Streets had almost no erosion, whereas the other profiles experienced quite severe losses in the dune area. Based on the measured profiles, the net volume changes are calculated for each profile and presented in Table 2. It appears that the net volume changes during the two storms are quite different from zero for most profiles due to the gradients in longshore sediment transport. To remove this effect, each post-storm profile is adjusted by shifting the profile horizontally a distance Δy to yield zero net volume change. The value of Δy is calculated by

$$\Delta y = \frac{1}{h_{total}} \int_{y_0}^{y_*} (h_{mb} - h_{ma})\, dy \tag{9}$$

where subscripts *mb* and *ma* denote profile elevation measured before and after storms, respectively, y_0 and y_* are offshore directed coordinates at the baseline and the limit of offshore profile change, respectively, and h_{total} represents the total elevation of the active storm profile. The sign of Δy is positive for a seaward translation. The profile retreat at the 3 meter contour and the eroded volume with and without the shifting adjustments are shown in Table 3. These parameters will be used as measures of the storm erosion and the performance of the numerical models.

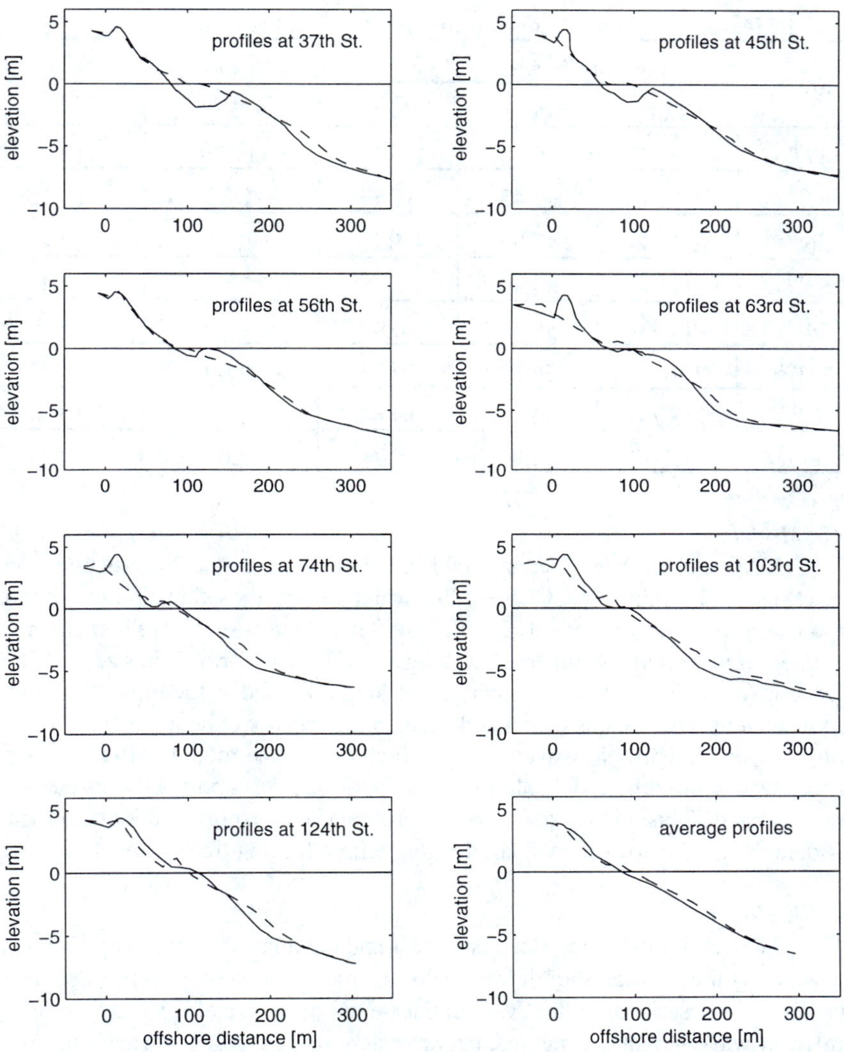

Figure 3 Measured pre-storm (solid line) and post-storm (dashed line) profiles.

Input Conditions for Each Model

The predictions of the four models are compared with the storm erosion measured at Ocean City, Maryland. The input parameters for each model in this study are selected to represent the conditions for which each model was calibrated. Among the four models, CROSS is the only one which can incorporate variable sediment size along a profile. The conditions run for each model are described briefly as follows.

Table 2 Measured volume change during the storm and adjustment Δy.

Street	Volume change (m^3 / m)			Adjustment Δy (m)	h_{total} (m)
	Gain	Loss	Net		
37th	158.77	20.78	137.99	-11.29	12.22
45th	81.01	49.42	31.59	-2.83	11.16
56th	27.71	47.96	-20.25	1.97	10.28
63rd	82.34	83.99	-1.65	0.16	10.31
74th	80.53	59.61	20.92	-2.19	9.55
103rd	136.06	56.33	79.73	-7.09	11.25
124th	75.74	56.10	19.64	-1.80	10.91
Average	55.71	24.97	30.74	-3.04	10.11

CCCL Model

This model has been used in establishment of the Coastal Construction Control Line (CCCL) in Florida. The CCCL is a line which depicts the landward limit of impact of a 100 year period storm event. The default dune slope is set to 1. Since this model cannot represent variable sediment size along a profile, a uniform grain size of 0.35 mm is applied. At each time, the significant wave height is used as the input wave height. The set-up at the shoreline is calculated based on linear wave theory as 0.23 times the incoming wave height. No wave run-up is included in the model. After running the storms, consistent with model calibration, a factor of 2.5 is applied to those contours which receded. This model does not incorporate a transport equation, but rather considers the profile to approach equilibrium with a folding time scale of 13 hours.

EDUNE Model

The default input dune slope is set to 1 and the input shoreline slope is taken as 0.05 which is the average shoreline slope of the measured pre-storm profiles. For the same reason as described for the CCCL model, a sediment size of 0.35 mm is used. The significant wave height is input at each time step. In EDUNE, the wave run-up is an input parameter held constant throughout the erosion simulation, and is used to control the location of the dune scarp above the peak water flood level. For the Ocean City storm erosion simulations, the run-up was fixed as 0.91 meters and 1.52 meters for the November 1991 and January 1992 storms, respectively, and was based on a combination of matching elevation of dune scarps and Hunt's equation (1958)

$$R = F_R H_b \frac{m}{\sqrt{H_b / L_0}} \tag{10}$$

where R is the run-up height measured vertically upward from the storm water level, F_R is a non-dimensional coefficient and is approximately 1, m is the average bed slope from the run-up limit to the breaking point, and L_0 is the deep water wave length. In the application here, it was found that the best agreement was provided by using the maximum significant wave height of each storm in Eq. (10). The transport coefficient, K_e, for this application is the program default value of 8.73×10^{-6} m^4/N.

SBEACH Model

Both versions (2.0 and 3.0) of SBEACH model are believed to incorporate wave run-up and set-up. The maximum slope that a predicted profile is allowed to achieve is required and is set to 17.5° as a default condition; this corresponds to a beach slope of 0.32. A constant sediment size of 0.35 mm is applied here, which was recommended by Kraus and Wise (1993) in their study. Both versions of the SBEACH model provide a choice of wave type (monochromatic or irregular). The option of irregular waves which requires an input of time history of significant wave height was chosen for this study. The default value for the transport coefficient, K_s, in both version is 1.50×10^{-6} m^4/N.

CROSS Model

The dune slope and the offshore slope are set equal to 1 and 0.5, respectively, as default conditions, and the shoreline slope is set to the average shoreline slope of the measured pre-storm profiles (0.05). Two different sediment size distributions are compared in this model. First, along each measured profile, the variable sand size listed in Table 1 is applied. Second, as a basis for comparison with the other three models, a uniform grain size of 0.35 mm is used. During each storm, a random wave series is generated according to the time history of the measured significant wave height and spectral peak wave period. The wave-by-wave set-up is calculated based on linear wave theory, and the wave run-up limit is established from Hunt's Equation (10). A transport coefficient value, K_c, of 7.14×10^{-10} m^8s^2/N^3 is used for this field application.

Numerical Results and Comparisons

The numerical results from the four models are quantified in terms of several parameters. A comparison of measured and predicted entire active profile changes is provided by the residual parameter, Res, defined in non-dimensional form as:

$$Res = \frac{\sum_{i=1}^{n} (h_{pi} - h_{mai})^2}{\sum_{i=1}^{n} (h_{mbi} - h_{mai})^2} \qquad (11)$$

where h is the profile elevation, the subscripts "p" and "m" denote predicted and measured, respectively, "b" and "a" indicate before and after storm conditions, respectively, "i" represents the ith location on the profile, and the sums extend across the entire active profile. The minimum possible value of Res is zero, which would correspond to a perfect simulation. If the numerical simulation predicts no changes, the value of Res is 1, which therefore should represent an upper limit of Res.

The dune erosion agreement between calculated and measured values is quantified by the eroded volume and the beach retreat at the 3 meter contour. To provided a measure of erosion and retreat, two different errors are presented: the mean square error, ERR_{ms}, and the algebraic average error, ERR_{ave}. These are expressed as

$$ERR_{ms} = \frac{\sum_{j=1}^{n} (S_{pj} - S_{mj})^2}{\sum_{j=1}^{n} S_{mj}^2}$$

$$ERR_{ave} = \frac{\sum_{j=1}^{n} S_{pj} - S_{mj}}{\sum_{j=1}^{n} S_{mj}} \qquad (12)$$

where S is an eroded volume or beach retreat, the subscripts "p" and "m" again represent the predicted and measured values, respectively, and "j" means the jth beach profile. ERR_{ms} represents a factor of simulation accuracy and ERR_{ave} provides a measure of over or under-prediction of erosion. Compared with the measure, Res, which is based on local differences across the entire active profile, ERR_{ms} and ERR_{ave} are based on average differences of eroded volume or beach retreat at a particular elevation.

Comparisons between predictions and measurements for the profiles at 45th Street are presented in Figure 4. The ridge features presented on the measured post-storm profiles suggest that there was some beach recovery even though the post-storm profiles were measured only a week after the second storm. The predictions of CROSS with both variable and uniform sand size are presented. It appears that the results predicted with variable sand size provide more reasonable simulations for dune erosion while the predictions with the uniform sand size fit the entire profile better. After applying the factor of 2.5 to those contours which receded, the CCCL model tends to overpredict erosion for most profiles. Since the application purpose of CCCL is to establish the Coastal Construction Control Line, the 2.5 factor is included to incorporate the variability of beach erosion at different locations. As mentioned above for EDUNE, the wave run-up is determined, in part, according to measured post-storm profiles used throughout the entire time of numerical simulation. Since beach erosion is quite sensitive to water level, the numerical results of EDUNE will depend greatly on the storm duration time and the input of wave run-up. Among all four models, SBEACH version 2.0 is the only model which represents the offshore bar feature. The newly modified version 3.0 of SBEACH presents substantially better agreement than version 2.0.

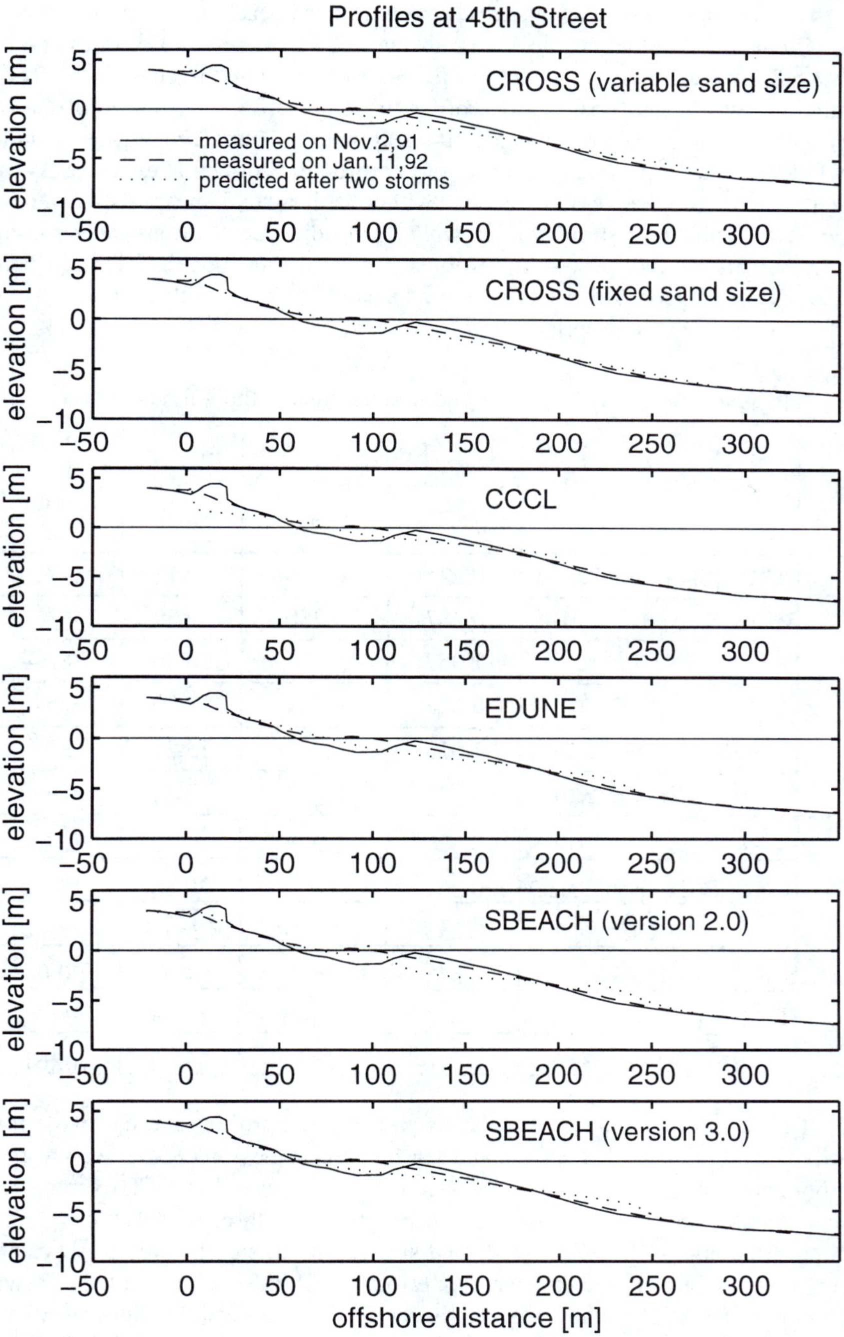

Figure 4. Comparisons of predicted and measured profiles at 45th Street.

The corresponding residuals, eroded volumes and beach retreat at the 3 meter contour for each individual profile were calculated. Two examples of profiles at 45th and 103rd Streets are shown in Table 4. Because the net volume changes between measured pre-storm and post-storm profiles are quite different from zero due to gradients of longshore sediment transport, in addition to the measured profiles, results are included based on shifting each post-storm profile the horizontal distance, Δy, to yield zero net volume change. In the table, "without adjustment" means data given by the original measured post-storm profiles, while "with adjustment" means data resulting from horizontally shifted post-storm profiles. Since all four models discussed here included only cross-shore sediment transport, the data presented "with adjustment" are considered to be more appropriate.

Table 4 The residuals, eroded volumes and beach retreat at the 3-meter contour.

Profile	Model	Residual		Eroded vol. (m^2)	Retreat at 3-m contour (m)
		w/o adjust.	With adjust.		
45th St.	CROSS(var. sand size)	0.525	0.496	20.13	8.71
	CROSS(fixed sand size)	0.287	0.280	19.39	8.36
	CCCL	1.067	0.851	60.56	18.34
	EDUNE	0.624	0.604	18.30	6.51
	SBEACH (ver. 2.0)	1.018	0.971	15.21	2.96
	SBEACH (ver. 3.0)	0.736	0.713	23.34	6.79
103rd St.	CROSS(var. sand size)	0.329	0.274	35.18	15.23
	CROSS(fixed sand size)	0.261	0.251	28.88	13.41
	CCCL	2.233	1.499	123.05	28.92
	EDUNE	0.548	0.413	54.66	20.04
	SBEACH (ver. 2.0)	0.596	0.590	19.54	7.03
	SBEACH (ver. 3.0)	0.510	0.433	35.20	10.61

The residuals averaged over the seven measured profiles and the two errors defined in Eq. (11) for eroded volume and beach retreat at the 3-m contour with respect to the horizontal shifted profiles are presented in Table 5. Overall, CCCL overpredicts the dune erosion during the two storms, whereas the other three models underpredict it. It appears that CROSS with fixed sand size results in significantly less average residual than the others. The dune erosion errors of CROSS with variable sand size, EDUNE and SBEACH version 3.0 are comparable. It is noticed that dune overwash occurred during the January 1992 storm. Among the four models, only EDUNE and SBEACH incorporate dune overwash processes. In the CROSS model, the profile shoreward of the dune crest is treated as a horizontal beach with the same elevation as the dune crest. Therefore, the numerical simulations of CROSS do not include the dune erosion part caused by overwash. It is expected that the underpredictions of CROSS could be improved by incorporating dune overwash processes in the model.

Table 5 The average residuals and errors of eroded volume and beach retreat at the 3-meter contour with respect to the horizontally shifted measured profiles.

Model	Averaged residual	Error of eroded volume		Error of retreat	
		ERR_{ms}	ERR_{ave}	ERR_{ms}	ERR_{ave}
CROSS (variable sand size)	0.725	0.227	-0.401	0.283	-0.359
CROSS (fixed sand size)	0.458	0.319	-0.511	0.327	-0.444
CCCL	1.331	0.707	0.847	0.480	0.561
EDUNE	0.792	0.182	-0.328	0.260	-0.324
SBEACH (ver. 2.0)	1.211	0.470	-0.663	0.623	-0.749
SBEACH (ver. 3.0)	0.752	0.183	-0.399	0.259	-0.479

Summary

This paper has presented comparisons of the predictions of the CROSS model and three other commonly used closed loop models (CCCL, EDUNE and SBEACH) for storm erosion at Ocean City, Maryland during the November 1991 and January 1992 storms. The "2.5" factor was applied in the CCCL model in this comparison. The wave run-up parameters used in the EDUNE model are determined according to the measured post-storm profiles and the maximum significant wave heights during two storms. Seven survey lines located from the southern (37th Street) to northern (124th Street) portions of the project are selected for evaluation of the four models. In most locations, the net volume changes in profiles are quite different from zero due to gradients in longshore transport. An adjustment is made by shifting the whole profile horizontally a distance Δy to yield a zero net volume change for each profile.

A non-dimensional mean square residual parameter is provided to evaluate the agreement between the entire measured and predicted profiles (including subaerial and subaqueous parts). Two kinds of error averaged with different methods (mean square and algebraic average) are provided to evaluate the prediction of dune erosion. It appears that the residuals are less affected by the shifting adjustment, while eroded volumes and beach retreat are affected more significantly by the shift. Among the four models, the CCCL model is the only one which overpredicts average dune erosion. CROSS yields the least average residual and presents the best prediction for an entire profile (including both subaerial and subaqueous parts). For errors of eroded volume and beach retreat, CROSS with variable sand size, EDUNE and SBEACH (version 3.0) provide reasonably comparable predictions. It is anticipated that by incorporating dune overwash, the under prediction of dune erosion by CROSS can be reduced.

References

Chiu, T.Y. and Dean, R. G. 1984. "Methodology on coastal construction control line establishment," *Tech. and Design Memorandum 84-6*, Beaches and Shores Resource Center, Florida State University, Tallahassee, FL.

Chiu, T.Y. and Dean, R. G. 1986. "Additional comparisons between computed and measured erosion by hurricanes," *Technical Report*, Beaches and Shores Resource Center, Florida State University, Tallahassee, FL.

Dean, R.G. 1977. "Equilibrium beach profiles: U.S. Atlantic and Gulf Coasts," *Ocean Engineering Report No. 12,* Department of Civil Engineering, University of Delaware.

Hunt, I.A. Jr. 1958. "Design of seawalls and breakwaters," *U.S. Lake Survey,* Corps of Engineers, U.S. Army, January 1958.

Kraus, N.C. and Wise R.A. 1993. "Simulation of January 4, 1992 storm erosion at Ocean City, Maryland," *Shore and Beach*, Vol. 61 No. 1, pp. 13-22.

Kriebel, D.L. 1989. "Users manual for dune erosion model EDUNE."

Kriebel, D.L. 1986. "Verification study of dune erosion model," *Shore and Beach*, Vol. 54, No. 3, pp13-21.

Kriebel, D.L. and Dean, R.G. 1985. "Numerical simulation of time-dependent beach and dune erosion," *Coastal Engineering,* Vol. 9, pp. 221-245.

Larson, M. and Kraus, N.C. 1989. "SBEACH: Numerical model for simulating storm-induced beach change, Report 1: Theory and model foundation," *Tech. Report CERC 89-9*, CERC, US Army WES, Vicksburg, MS.

Larson, M., Kraus, N.C. and Byrnes M.R. 1989. "SBEACH: Numerical model for simulating storm-induced beach change, Report 2: Numerical formulation and model tests," *Tech. Report CERC 89-9*, CERC, US Army WES, Vicksburg, MS.

Stauble, D.K., Garcia A.W. and Kraus N.C. 1993. "Beach nourishment project response and design evaluation: Ocean City, Maryland. Report 1, 1988– 1992," *Tech. Report CERC 93-13*, CERC, US Army WES, Vicksburg, MS.

Zheng, J. 1996. "Improved cross-shore sediment transport relationship and model," *Ph.D. Dissertation,* University of Florida, Gainesville, Florida, U.S.A.

Zheng, J. and Dean, R.G. "Numerical modeling and intercomparisons of beach profile evolution," to appear in *Coastal Engineering*.

CHAPTER 242

Measurements of Wave Generated Bedforms

Christopher D. Jetté[1] and Daniel M. Hanes[2]

Abstract

A 5 MHz multiple transducer array was used to measure wave generated bedforms near Duck, NC. The transducer array consists of 37 transducers with a center to center spacing of 1.2 centimeters. Seabed measurements were made with approximately 1 millimeter vertical and 2 centimeter horizontal resolution. Measured ripple dimensions are compared to three popular ripple prediction models. It is shown that errors of over 100 percent were found between measured and predicted ripple heights and lengths for all ripple models compared. Errors in predicting ripple steepness ranged from 37 to 55 percent.

Description Of Experiment

Field measurements of bedforms were made during an experiment at the Army Corps of Engineers Field Research Facility in Duck, NC during August 23-25, 1995. The instruments were deployed using a Sensor Insertion System (SIS) on the research pier. The SIS was positioned at multiple locations along the cross-shore profile during the experiment.

Bedform measurements were made using a multiple transducer array (MTA) developed at the University of Florida. The MTA consists of 37 ultra-sonic transducers operating at a frequency of 5 MHz. The center-to-center spacing of the transducers is 12 mm. Such configuration allows for O(1 mm) vertical resolution and O(2 cm) horizontal resolution. The MTA can scan a profile in approximately 4 seconds. The distance to the seabed is calculated from the elapsed time between the pulsing of the transducer and the time at which the return exceeds a software selectable threshold.

---

1) Research Assistant, Department of Coastal and Oceanographic Engineering, University of Florida, PO Box 116590, Gainesville, FL 32611, USA.
2) Associate Professor, Department of Coastal and Oceanographic Engineering, University of Florida, PO Box 116590, Gainesville, FL 32611, USA.

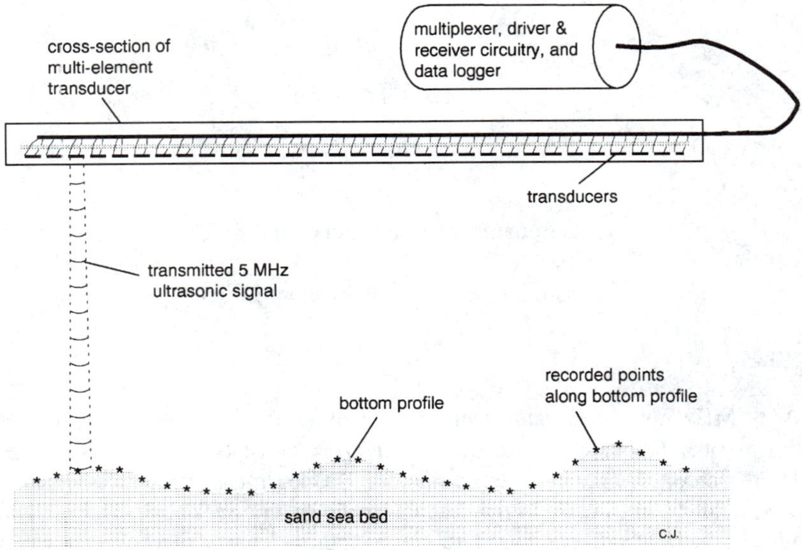

Figure 1. Schematic of the MTA

In addition to the MTA, an under-water video camera was deployed in order to document bedform orientation. Electro-magnetic current meters and pressure transducers were deployed to measure the wave and current conditions. Measurements of the vertical distribution of suspended sediment by acoustical backscatter methods were also made.

The data was collected in 13 minute bursts. Each burst contained 131 profiles taken with the MTA. Current, pressure, and concentration profile data were collected at 4 Hz.

<u>Conditions During Experiment</u>

Exploiting the mobility of the SIS, measurements were made under a variety of wave and sediment conditions. During the experiment, an off-shore bar was present at approximately 210 meters from the high water line. Measurements were made at 8 separate cross-shore locations in the regions offshore of the bar, on the bar, in the trough, and on the nearshore beach-face. Depths where measurements were made ranged from 1.6 to 6.8 meters. The vertical lines in figure 2 indicate cross-shore locations where measurements were obtained.

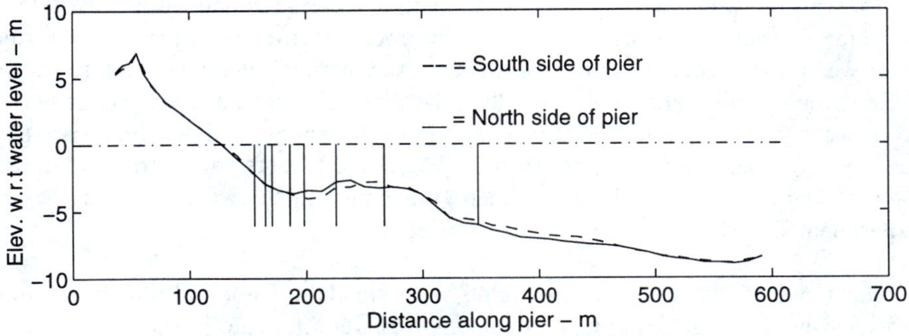

Figure 2. Cross-shore profiles from North and South sides of pier during experiment.

The sediment characteristics varied greatly over the cross-shore profile. D_{50} values of the mostly quartz sediment ranged from 0.19 to 1.67 mm. The range of D_{50} values versus run number are plotted in Figure (3) along with H_{mo} (a), peak wave period (b), and water depth (c).

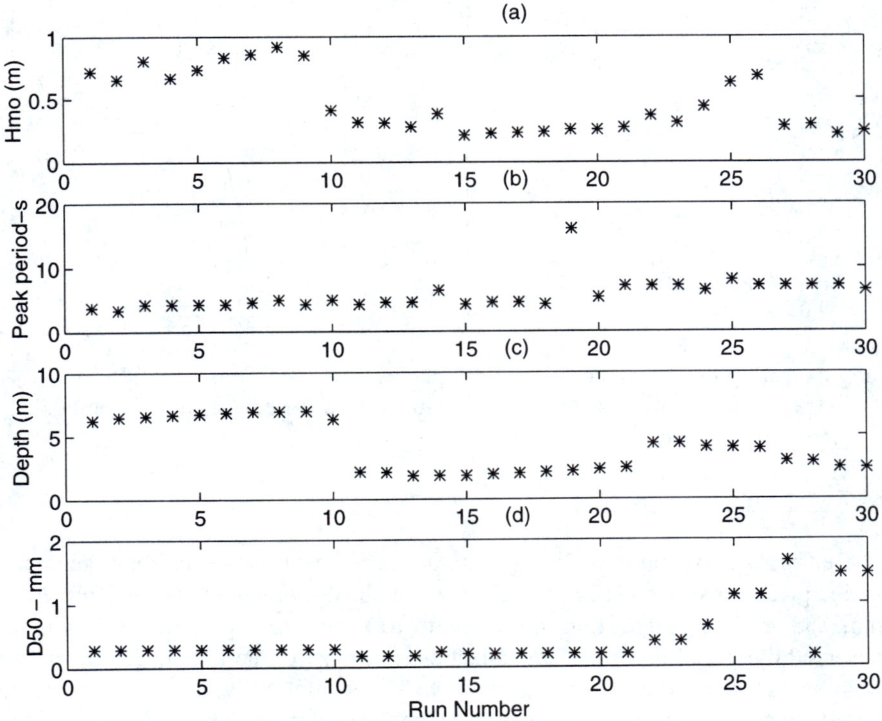

Figure 3. H_{mo} (a), peak wave period (b), water depth (c),and D_{50} (d) vs. Run number for SIS experiment

Most of the wave data taken over the three day period contained at least two, if not three frequency peaks in the surface elevation spectrum. For many of the runs there was a low frequency component (12-15 sec. period) generated from hurricane Felix, which was several hundred miles off-shore, as well as a moderate frequency component (6-8 sec. period), and generally in the afternoon, a higher frequency (4-5 sec. period) locally generated component. H_{mo} wave heights varied from 0.2 to 0.9 meters. During most of the experiment the predominate wave direction at the experiment site was directly on-shore.

For most of the runs measurements were made offshore of the breakpoint, however during runs 11 through 21 some waves were noted to break in the vicinity of the instruments. It is also noted that, after transforming the pressure spectrums into surface elevation spectrums, in the majority of cases the higher frequency component became the peak frequency. Consequently, the wave period used in model prediction was the lowest period component in the wave profile.

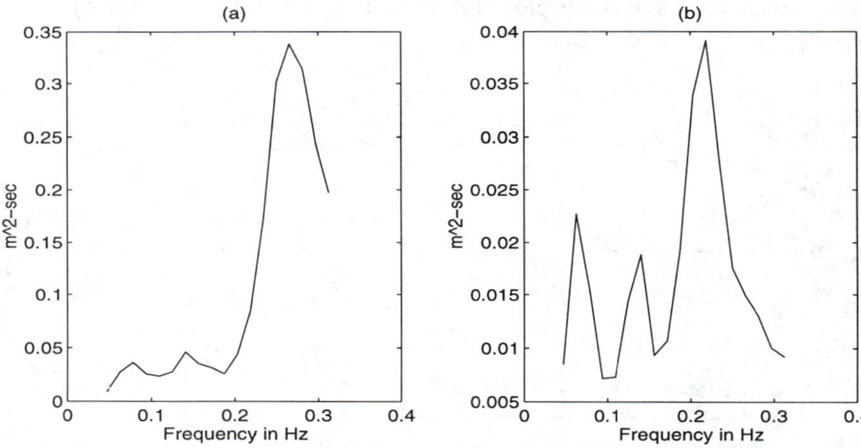

Figure 4. Representative corrected surface elevation spectrums from SIS experiment for runs 1 (a) and 13 (b) with Hmo wave heights of 0.72 and 0.28 meters respectively

Measurements

Bedforms were present throughout the experiment, however their geometries varied greatly over the beach profile. Measured bedform heights ranged from 7 to 50 mm, and bedform lengths ranged from 80 to 1000 mm. Active ripple conditions were observed during all times of data collection. For most of the files under-water video was available to document near-bed sediment motion and ripple orientation. Significant ripple migration was not observed during any of the 13 minute runs.

Bedform profiles were collected at a rate of ten profiles per minute. For the measurements presented herein, a single representative profile was found for each minute of data (10 scans). An example of the thirteen representative profiles for an entire run is shown in figure 5 (a-c). In these figures, each profile has been offset by -3 mm from the previous profile for comparative display. The top profile represents the first minute of data and contains circles at each of the 37 measured data points.

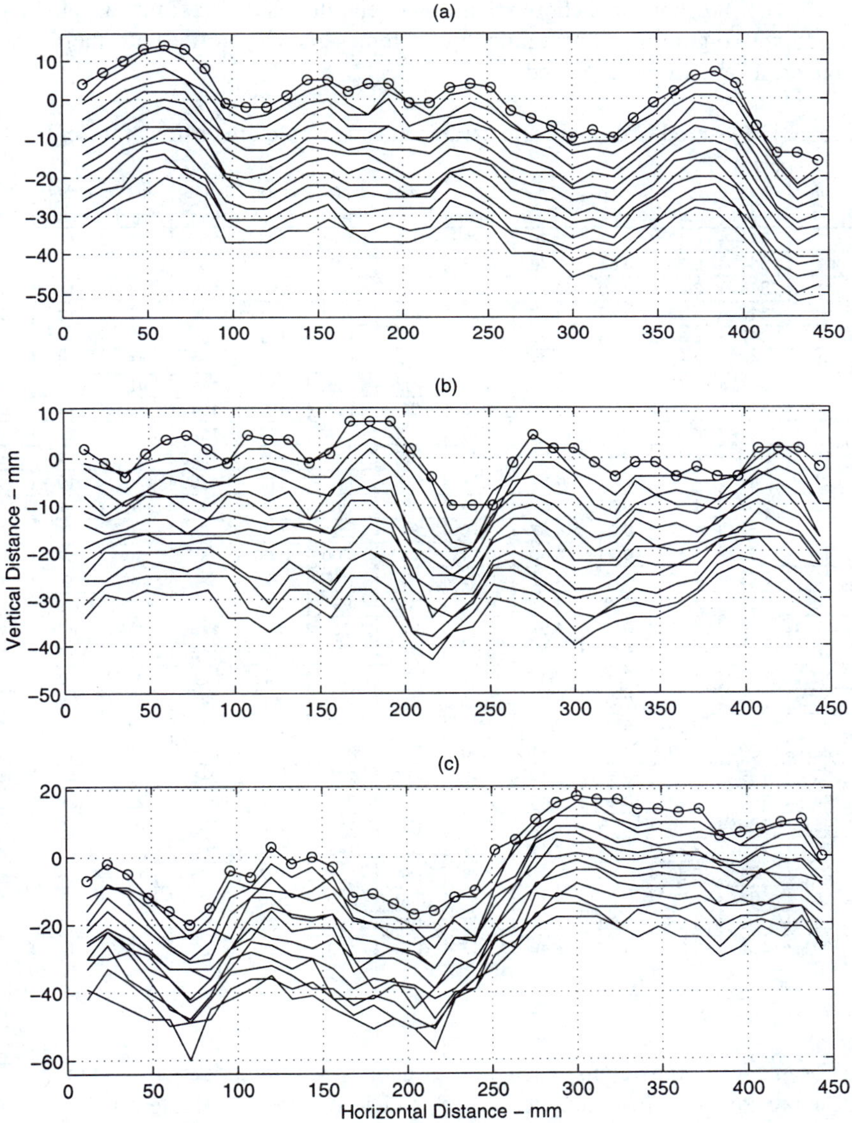

Figure 5. Representative profiles for each minute of a 13 minute run from runs 1(a), 13(b), and 26(c). Data from first minute of run is plotted with circles.

In order to compare these data to models, the bedforms need to be characterized in terms of a wavelength and height. In previous work this has typically been accomplished by visual inspection. We therefore chose a threshold method to determine ripple height and length dimensions for each representative ripple profile. This method first found the peaks and troughs along a profile. Then, if the distance between adjacent peaks and troughs met a certain threshold criterion, they were recorded. From these peaks and troughs average ripple heights and lengths were found. The threshold was chosen conservatively, that is, to maximize ripple heights and lengths. Figure (6) is an example of a representative profile with the threshold ripple crests and troughs marked.

Table 1. Measured hydrodynamic parameters and ripple dimensions for field experiment.

run	loc. (m)	depth (m)	Hmo (m)	Tp (s)	D50 (mm)	ltm(1) (mm)	ltm(4) (mm)	ltm(8) (mm)	ltm(13) (mm)	htm(1) (mm)	htm(4) (mm)	htm(8) (mm)	htm(13) (mm)
1	347	6.10	0.71	3.76	0.28	103	81	103	107	13	11	13	14
2	347	6.33	0.65	3.76	0.28	103	103	82	105	13	13	12	13
3	347	6.42	0.80	4.27	0.28	100	106	101	106	14	14	12	16
4	347	6.53	0.66	4.27	0.28	108	107	101	154	13	13	14	19
5	347	6.58	0.73	4.27	0.28	107	103	101	105	14	14	15	14
6	347	6.69	0.83	4.27	0.28	152	150	147	149	18	19	16	17
7	347	6.78	0.85	4.57	0.28	103	146	144	105	12	16	16	12
8	347	6.79	0.92	4.92	0.28	98	136	95	104	12	14	10	12
9	347	6.82	0.85	4.27	0.28	158	164	154	152	18	17	17	17
10	347	6.16	0.41	4.92	0.28	102	102	101	102	14	14	15	14
11	267	1.94	0.31	4.27	0.18	112	112	113	115	10	10	9	11
12	267	1.90	0.31	4.57	0.18	117	86	89	113	10	11	10	12
13	267	1.62	0.27	4.57	0.18	84	110	110	105	10	11	14	13
14	267	1.65	0.38	6.40	0.24	1000	1000	1000	1000	50	50	50	50
15	226	1.63	0.21	4.27	0.21	77	101	97	94	8	9	11	10
16	226	1.76	0.22	4.57	0.21	102	99	106	124	13	13	14	16
17	226	1.82	0.23	4.57	0.21	94	90	88	113	11	12	11	11
18	226	1.91	0.24	4.27	0.21	107	139	97	113	8	11	10	12
19	226	1.99	0.26	16.00	0.21	70	86	84	82	9	10	8	10
20	226	2.14	0.25	5.33	0.21	83	83	83	83	7	7	7	7
21	226	2.25	0.27	7.11	0.21	101	101	101	101	11	11	11	11
22	198	4.19	0.37	7.11	0.40	275	275	275	275	40	40	40	40
23	198	4.21	0.31	7.11	0.40	275	275	275	275	40	40	40	40
24	186	3.90	0.44	6.40	0.64	240	232	224	240	26	28	28	26
25	186	3.86	0.63	8.00	1.12	136	134	138	131	23	20	22	20
26	186	3.80	0.69	7.11	1.12	146	147	154	154	23	25	23	23
27	171	2.80	0.28	7.11	1.66	144	136	127	139	15	15	16	15
28	165	2.72	0.29	7.11	0.18	500	500	500	500	45	45	45	45
29	155	2.25	0.22	7.11	1.47	197	108	103	120	27	17	17	17
30	155	2.20	0.24	6.40	1.47	117	128	106	125	18	17	16	17

For model comparisons, each run of data was summarized into four profiles. These profiles were from minutes 1, 4, 8, and 13 of each 13 minute run. The reasons for this were to minimize the amount of clutter in model comparisons, while at the same time retain the range of ripple geometries present during each run. Measured ripple length (ltm) and height (htm) along with summaries of hydrodynamic and sediment conditions are presented in table 1 for each run.

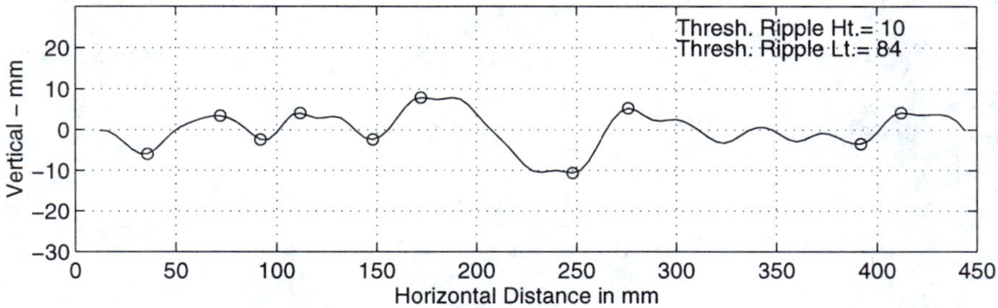

Figure 6. Representative ripple profile with threshold method results for first minute of data from run number 13.

Model Comparisons

Measured bedform dimensions were compared to three predictive models; Nielsen (1981), Wiberg and Harris (1994), and Grant and Madsen (1981). For each run measured values of water depth, Hmo wave height, peak period, and sediment properties were entered into the predictive models. Measured bedform dimensions were plotted with the model curves, if possible, in order to better illustrate the relationships between measured values and predicted values.

Nielsen (1981) derived different predictive models for regular (laboratory) and irregular (field) conditions. Comparisons included herein refer to the irregular or field ripple models. The Nielsen (1981) bedform model characterizes flow conditions using sediment mobility number (Ψ), grain roughness Shield's parameter ($\theta_{2.5}$), and the wave friction factor "$f_{2.5}$" given by Swart (1974). These parameters are defined as follows

$$\theta_{2.5} = 0.5 f_{2.5} \psi \qquad \text{equation 1}$$

$$f_{2.5} = \exp\left[5.213\left(\frac{2.5 D_{50}}{A}\right)^{0.194} - 5.977\right] \qquad \text{equation 2}$$

$$\psi = \frac{(a\omega)^2}{(s-1) g D} \qquad \text{equation 3}$$

where "D" is the mean grain diameter, "a" is the near-bed wave semi-excursion, "s" is the sediment specific gravity, "g" is the acceleration of gravity, and ω is the angular frequency of the waves. The Nielsen (1981) irregular wave model for non-dimensional ripple steepness, height, and length are respectively

$$\eta/\lambda=0.342-0.34(\theta_{2.5})^{0.25} \qquad \text{equation 4}$$

$$\eta/a=21\Psi^{-1.85} \qquad \text{for } \Psi>10 \qquad \text{equation 5}$$

$$\eta/a=0.275 - 0.022\Psi^{0.5} \qquad \text{for } \Psi<10 \qquad \text{equation 6}$$

$$\lambda/a = \exp\left[\frac{693-0.37\ell n^{8}\psi}{1000+0.75\ell n^{7}\psi}\right] \qquad \text{equation 7}$$

where η and λ represent ripple height and length respectively.

Relationships between measured ripple dimensions and the Nielsen (1981) model curves are illustrated in figure (8). As can be seen in figure 8(a), for all but 5 out of the 30 runs the Nielsen (1981) irregular wave ripple model over-predicted ripple height. The largest amount of over-prediction was found to occur at low values of sediment mobility number (Ψ<20). The Nielsen (1981) ripple length model agreed best with measured dimensions for higher values of mobility number (Ψ>12), however for lower values of mobility number (Ψ<12), the model over-predicted ripple length for all runs. Measurements of ripple steepness had better agreement with the Nielsen (1981) model curves than did ripple height or ripple length. Best agreement was found at higher values of the grain roughness Shield's parameter ($\theta_{2.5}$>0.08). For lower values of the Shield's parameter the Nielsen (1981) model over-predicted ripple steepness.

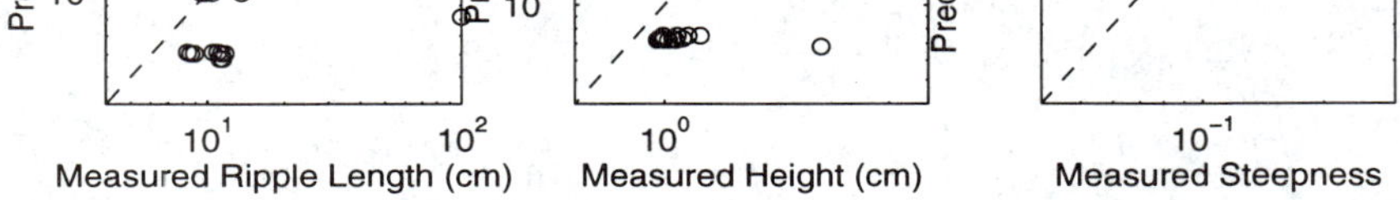

Figure 7. Nielsen (1981) irregular wave ripple comparison with measurements

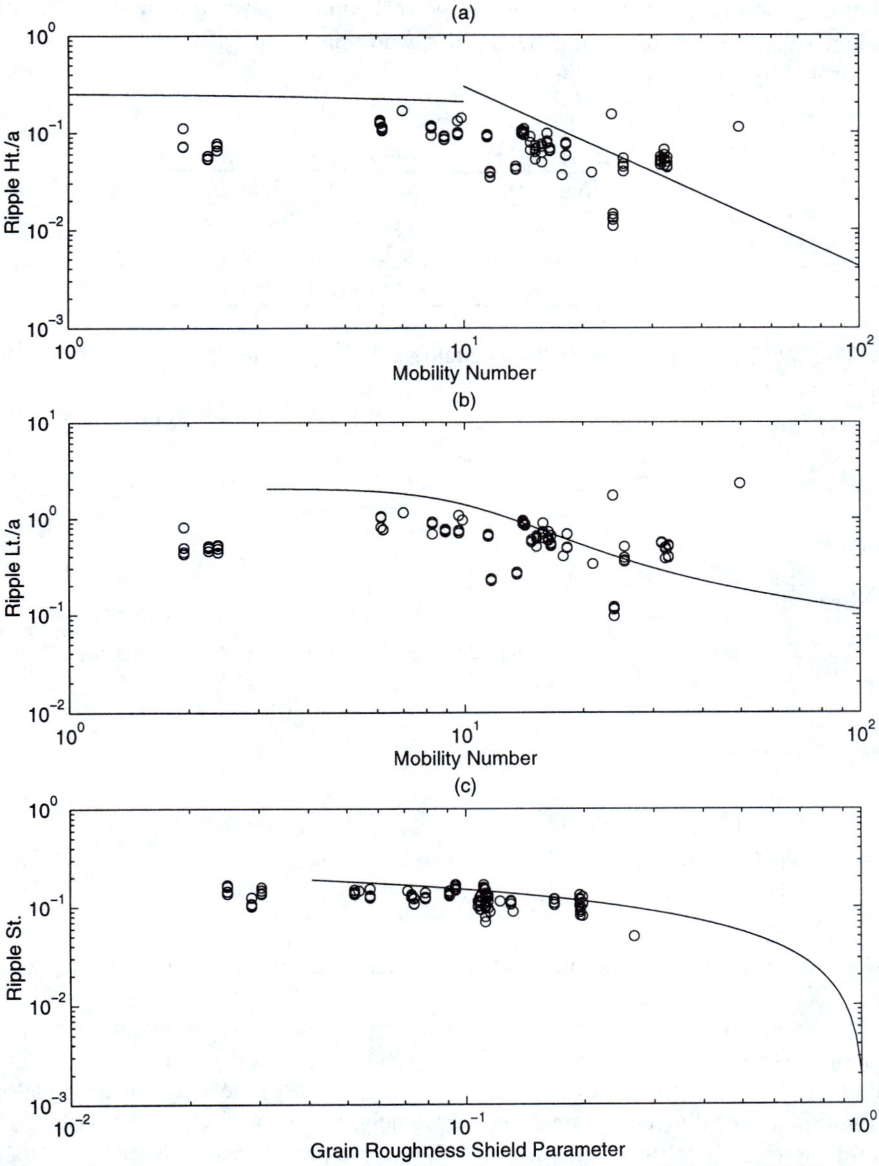

Figure 8. Comparison of measured field ripple dimensions and Nielsen (1981) model curves for height (a), length (b), and steepness (c).

Wiberg and Harris (1994) reexamined existing ripple data from oscillatory flows in both flume and field studies to construct a model to predict ripple geometry for all types of oscillatory flow environments. Wiberg and Harris (1994) classified bedforms according to the ratio of the wave boundary layer thickness and ripple height (δ_ω/η). From this ratio, the ripples were classified as orbital, anorbital, or suborbital by the following criteria

Table 2. Wiberg and Harris (1994) ripple classification

flow conditions	ripple classification
$d_o/\eta_{ano} < 20$	orbital ripples
$20 < d_o/\eta_{ano} < 100$	suborbital ripples
$d_o/\eta_{ano} > 100$	anorbital ripples

The Wiberg and Harris (1994) equations for ripple length and steepness for orbital ripples are:

$$\lambda_{orb} = 0.62 d_o \qquad \text{equation 8}$$

$$(\eta/\lambda)_{orb} = 0.17 \qquad \text{equation 9}$$

for anorbital ripples

$$\lambda_{ano} = 535\ D \qquad \text{equation 10}$$

$$\frac{\eta}{\lambda} = \exp\left[-0.095\left(\ell n\frac{d_o}{\eta}\right)^2 + 0.442\ln\frac{d_o}{\eta} - 2.28\right] \quad \text{for } d_o/\eta>10 \qquad \text{equation 11}$$

and for suborbital ripples

$$\lambda_{sub} = \exp\left[\left[\frac{\ln\left(d_o/\eta_{ano}\right) - \ln 100}{\ln 20 - \ln 100}\right]\left(\ln\lambda_{orb} - \ln\lambda_{ano}\right) + \ln\lambda_{ano}\right] \qquad \text{equation 12}$$

where "d_o" is the near bottom orbital diameter, and "D" is the mean sediment size.

In figure (9), measured non-dimensional ripple dimensions are plotted versus the parameters used in the Wiberg and Harris (1994) ripple model. Where possible, the model curves were included. It is noted that in figure (9d) the x-axis is near bottom orbital diameter divided by measured ripple height. For these plots, the dashed and solid curves refer to the orbital and anorbital ripple models respectively. The data in figure (9) is also plotted according to the Wiberg and Harris (1994) ripple classification scheme. Each data point is plotted as an "o", "+",or "*" indicating ripple types of orbital, suborbital, or anorbital ripples, respectively. The Wiberg and Harris (1994) model also over-predicted ripple length and height for most of the runs. It is also interesting to note that only 2 of the 30 runs were classified as anorbital ripples.

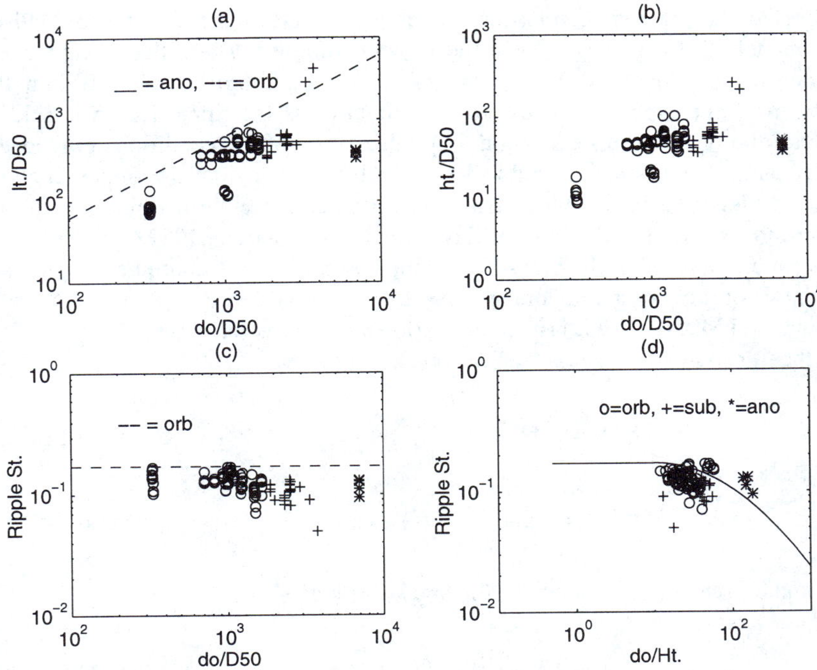

Figure 9. Wiberg and Harris (1994) ripple models with measured values.

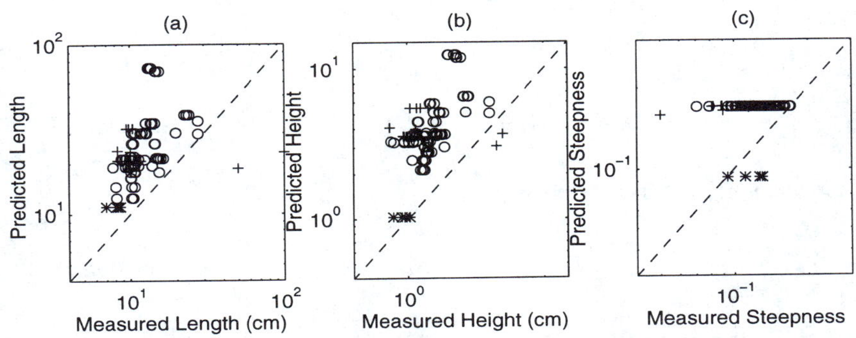

Figure 10. Wiberg and Harris (1994) model ripple comparison with measurements

The Grant and Madsen (1982) ripple model uses the ratio of the maximum value of the skin friction (shear stress) under the wave (τ_{*sf}) to the critical value of shear stress for the initiation of sediment motion ($(\tau_*)_{cr}$. Also a dimensionless sediment parameter (S_*) is used. This ripple model was based primarily on the Carstens et al. (1969) laboratory oscillatory ripple study.

Two separate regions of transport stages were used in the Grant and Madsen (1982) model. The first stage, when the flow conditions are less than the breakoff point defined by equation (18), is known as the equilibrium stage. Within the equilibrium range, ripple steepness is a maximum and the ripple length is said to scale with the near-bottom excursion amplitude. When flow conditions exceed the breakoff point, conditions are said to be in the breakoff region; the ripple length is said to no longer be in equilibrium and a decorrelation between ripple length and near-bottom excursion amplitude occurs (Grant and Madsen (1982)). As the flow intensity increases within the breakoff region, ripple steepness continues to decrease to a point where ripples are no longer present.

The Grant and Madsen (1982) ripple prediction model is as follows:
when the transport stage is less than the breakoff point

$$\eta/\lambda = 0.16\left(\tau_{*sf} / (\tau_*)_{cr}\right)^{-0.04} \qquad \text{equation 13}$$

$$\eta / a_o = 0.22\left(\tau_{*sf} / (\tau_*)_{cr}\right)^{-0.16} \qquad \text{equation 14}$$

and at higher transport stages, above the breakoff range

$$\eta/\lambda = 0.28 S_*^{0.6}\left(\tau_{*sf} / (\tau_*)_{cr}\right)^{-1.0} \qquad \text{equation 15}$$

$$\eta / a_o = 0.48 S_*^{0.8}\left(\tau_{*sf} / (\tau_*)_{cr}\right)^{-1.5} \qquad \text{equation 16}$$

where

$$S_* = (d / 4v)\left[(S-1)gD\right]^{0.5} \qquad \text{equation 17}$$

and the breakoff point is defined as

$$\left[\tau_{*sf} / (\tau_*)_{cr}\right]_B = 1.8 S_*^{0.6} \qquad \text{equation 18.}$$

where ‘v’ is the kinematic viscosity of water, ‘D’ is the grain diameter, ’S’ is the specific gravity of the sediment, and ‘g’ is the acceleration of gravity.

A comparison of Grant and Madsen (1982) predicted ripple heights and lengths versus measured values is shown in figure 11. It can be seen that the Grant and Madsen (1982) model overp-redicted ripple length and height for almost every run.

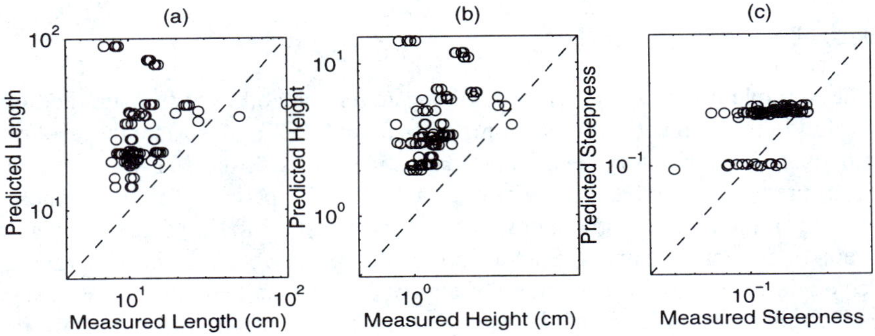

Figure 11. Grant and Madsen (1982) model ripple comparison with measurements

In order to evaluate the performance of the predictive models, the relative error was computed between measured and predicted values. The relative error is defined as:

$$\Delta = \exp\left\{\left[\frac{1}{n}\sum_{1}^{n}\left(\ln(y)-\ln(\hat{y})\right)^{2}\right]^{1/2}\right\} \qquad \text{equation 19}$$

where y is the measured value and $\hat{y}$ is the predicted value. This quantity is a multiplicative factor that indicates the possible variation about the predicted value (Wikramanayake, 1993). For example if Δ equals 1.34, the average error is equal to 34 percent.

Table 3 contains values of the relative error for each of the different predictive models. It can be seen that the predictive models performed much better at predicting ripple steepness than ripple height or length independently. The Nielsen (1981) and Grant and Madsen (1982) ripple steepness models performed the best at predicting ripple steepness with an average error of 37 percent, the Wiberg and Harris (1994) model had and average error of 55 percent. The Nielsen (1981) model predicted ripple height the best, however the average error was found to be 153 percent. The Wiberg and Harris (1994) and Grant and Madsen (1982) models had even greater error. Ripple length comparisons showed the Wiberg and Harris (1994) model having the least error with an average error of 125 percent. The Nielsen (1981) and Grant and Madsen (1982) ripple length models had slightly higher relative errors with 150 and 154 percent respectively.

TABLE 3. Relative Error for model predictions and measured values.

	Ripple Length	Ripple Height	Ripple Steepness
Nielsen (1981) Field	2.50	2.53	1.37
Wiberg & Harris (1994)	2.25	2.84	1.55
Grant and Madsen (1982)	2.54	3.03	1.37

Conclusions

The multiple transducer array (MTA) has proven the ability to measure bedforms in the field environment with O(1 mm) vertical and O(2 cm) horizontal resolution. Comparisons of oscillatory field ripple data collected with the MTA and the most popular predictive models show errors of over 100 percent for all models in predicting ripple heights and lengths. Average errors of 37 to 55 percent were found in predicting ripple steepness. Such errors could be due to errors in the previously collected data sets used to construct these models, or to the current understanding of the mechanics of ripple formation and geometric equilibrium with the flow field. More investigation is necessary in this field in order to improve bedform prediction models.

Acknowledgments

The authors wish to ackowledge the financial support of the U.S. Office of Naval Research, Coastal Sciences Program, the field assistance provided by the Field Research Facility, U.S. Army Corps of Engineers, and the help of Chuck Broward, Eric Thosteson, Mike Krecic, and Carl Miller.

References

Bagnold, R. A., Motion of waves in shallow water: Interaction between waves and sand bottoms, Proc. R. Soc. London Ser. A, 187, 1-15, 1946.

Carstens, M. R., R. M. Neilson, and H. D. Altinbilek, Bed forms generated in the laboratory under oscillatory flow: Analytical and experimental study, Tech. Memo. 28, U.S. Army Coarps of Eng., Coastal Eng. Res. Center, June 1969.

Dingler, J. R., and Inman D. L., Wave-formed ripples in near-shore sands, Proc. Fifteenth Conf. Coastal Engng., Amer. Soc. Civil Eng., 1976.

Grant, W. D., and O. S. Madsen, Movable bed roughness in unsteady oscillatory flow, J. Geophys. Res., 87, 469-481, 1982.

Nielsen, P., Dynamics and geometry of wave-generated ripples, J. Geophys. Res., 86, 6467-6472,1981.

Nielsen, P., Field measurements of the time-averaged suspended sediment concentration under waves. Coastal Engn., 8,51-72, 1984.

Sleath, J. F. A., Sea bed mechanics, Wiley Interscience, New York, 1984.

Swart, D. H., Offshore sediment transport and equilibrium beach profiles. Delft. Hydr. Lab. Publ. No. 131, 1974.

Vincent, C. E., and P. D. Osbourne, Bedform dimensions and migration rates under shoaling and breaking waves, Unpublished manuscript (1992).

Wiberg, P.L., and C. K. Harris, Ripple geometry in wave dominated environments, J. Geophys. Res., 99, 775-789, 1994.

Wikramanayake, P. N., Velocity profiles and suspended sediment transport in wave-current flows. Ph.D. Thesis. Dept. of Civil Eng., MIT, Cambridge, Mass., 1993.

CHAPTER 243

Field Measurement of Bed Roughness for Waves on an Off-Shore Reef

Raymond C. Nelson[1]

Abstract

This paper takes advantage of fortuitous, near ideal conditions which existed for a subset of field data, to obtain the hydraulic roughness of the surface of a coral reef platform. This was accomplished by observing the transformation of natural, unbroken, oscillatory wave trains, propagating over the reef top in water of constant depth, where wave bed friction damping was the only active wave transformation process. The paper also demonstrates that while coral reefs may be perceived as being much rougher than sandy beds, they may in fact be no rougher in hydraulic terms, than many sandy beds experiencing the same wave and water depth conditions.

Introduction

The bed friction factor (f) associated with wave motion, is a function of a Reynolds number and a relative roughness.

$$f = \phi_1\left(R_e, \frac{r}{a}\right) \qquad (1)$$

R_e is the Reynolds number associated with the oscillatory water movements at the bed as estimated from linear wave theory.

$$R_e = \frac{aU}{v} \qquad (2)$$

a is the amplitude of the oscillatory bed movements, U is the amplitude of the periodic bed velocity, and v is the kinematic viscosity of the fluid. r/a is the relative roughness where r is the hydraulic roughness. For all turbulent flow conditions, the friction factor can be expressed in the form of Eq. 1. However,

[1] Senior Lecturer, School of Civil Engineering, University of New South Wales, University College, Australian Defence Force Academy, Canberra, ACT, 2600, Australia

for all prototype wave conditions of engineering significance, the oscillatory flow at the bed is always fully developed rough turbulent flow (Jonsson, 1963 and 1966; Swart, 1974). In this regime the friction factor is independent of Reynolds number, reducing Eq. 1 to Eq. 3 for field conditions.

$$f = \phi_2\left(\frac{r}{a}\right) \tag{3}$$

On a movable bed, both r and a are variables, depending on the wave and water depth conditions, the changes in r being caused by the variability of bed form (Nelson, 1989 and 1995).

On fixed beds, such as coral reef platforms, a still varies with wave and water depth conditions, but r remains constant. A measured value of this fixed roughness is far more useful than a measured value of friction factor. The latter is only applicable to wave and water depth conditions that produce the same value of a existing when the measurement was made, while the hydraulic roughness value can be used to obtain wave friction factors for all wave and water depth conditions.

This paper presents values of reef top hydraulic roughness derived from the observed transformation of natural random waves, propagating in water of constant depth, over the fixed bed of a platform coral reef. The resulting values are to some extent site specific. However, from an engineering viewpoint, the use of a measured roughness that can be varied by visual comparisons and reasoning, is far more desirable than the use of a *guesstimate.*

Experiment Description

The experimental data were collected from John Brewer Reef, located inside the line of the Barrier Reef approximately 70 km north east of Townsville, Australia. The reef is elliptical in shape (6 km by 3 km - see Fig. 1) with a major axis approximately normal to the prevailing south easterly winds. The windward reef platform is a continuous reef flat 200m to 300m wide and uniform in elevation at a level approximating Lowest Astronomical Tide. High tide water depths over the reef platform seldom exceed 3m. Seaward of the windward edge, the reef drops rapidly to a depth of about 50m below the reef platform, while the lagoon enclosed by the whole of John Brewer Reef averages about 10m in depth.

Wave and current measuring instruments were deployed on the windward side of the reef (Fig. 1). WR1 was a Waverider buoy located about 500m seaward of the reef front and transmitted incident wave data to a receiver located in the floating hotel. ZP1, ZP2, ZP3 and ZP4 were 6m high, surface piercing wave transducers known as Zwarts Poles (Zwarts, 1974). These were aligned normal to the reef front and parallel to the prevailing south east winds. ZP1 and ZP4 are of special interest to this study. ZP1 was 27m from the reef front while ZP4 was located at a further interval of 141m. Most of the time the line of Zwarts Poles was located along the line of the reef top wave orthogonal but there were some exceptions at high tide during incident wave events with very low energy.

Each wave sample collected from WR1 contained 2048 readings at intervals of 0.391 seconds (sample length about 14 minutes). Each wave sample from each Zwarts pole contained 4800 readings at an interval of 0.25 seconds (sample length about 20 minutes). Data sets consisted of wave samples collected concurrently from all 5 wave instruments with the recording commencement times for each instrument being identical. The interval at which data sets were recorded was sometimes 1 hour and at other times 2 hours.

CM1 was an S4 electromagnetic XY current meter that measured reef top currents. Every 10 minutes it took 120 readings at 0.5 second intervals of both the easterly and northerly velocity components and logged the average.

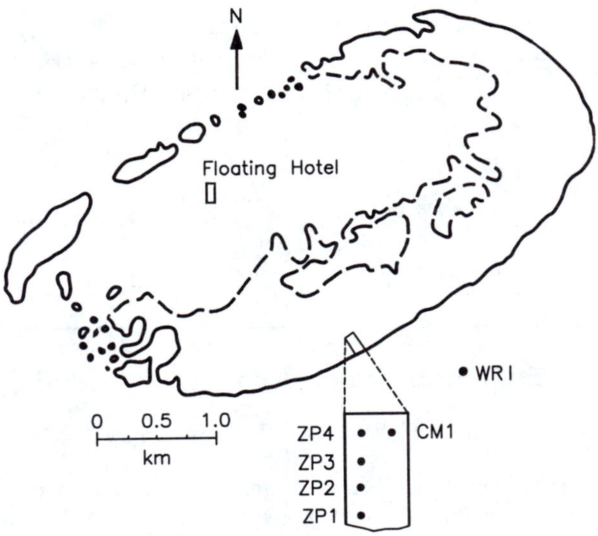

Fig. 1 John Brewer Reef and Instrument Locations

Data Selection

The extraction of bed friction effects from other interacting sub-processes, increases the uncertainties incorporated in the results, the preferred option being to have a prototype location and wave motion where bed friction damping is the only active sub-process. Some of the John Brewer Reef Zwarts Pole data conformed with this requirement. The water depths were, for all practical purposes, uniform eliminating the effects of shoaling and refraction. There were no barriers around which diffraction could occur, and careful selection of data sets ensured that the effects of wave breaking were all but eliminated.

The data base used in this study contained 72 consecutive data sets, some times recorded at one hourly intervals, and at other times, two hourly intervals. They spanned a sustained period of high incident wave energy and some periods of low incident wave energy, both coupled with large tide ranges. This provided a wide range of reef top wave and water depth combinations. These data are presented as a time series in Fig. 2.

Incident high wave energy events were dominated by sea states with peak energy periods of between 4 and 6 seconds with direction determined by the dominant south east wind. During low energy events, when the sea state was minimal, longer peak energy periods were observed when small amounts of wave energy from distant sources became significant relative to the local sea state. The direction of this longer period wave energy was unknown and it was necessary to eliminate it from the data records. All characteristic spectral wave parameters have been restricted to the wind wave component by only considering wave frequencies greater than or equal to 0.16 Hz.

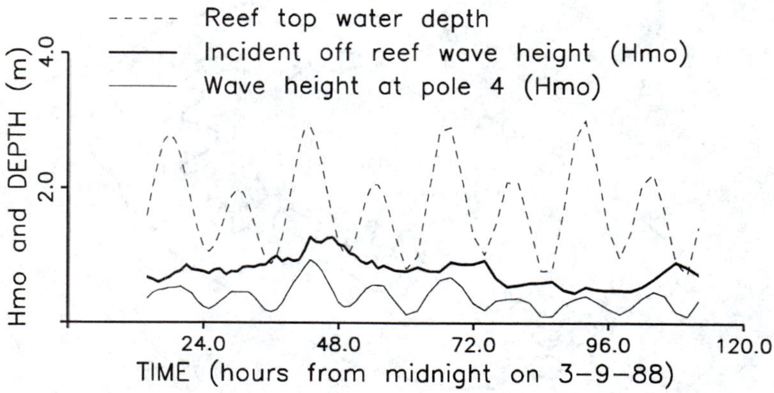

Fig. 2 Wave and Tide Summary

The selected subset of wave records were to be free of wave breaking influences, with the wave energy dissipation over the flat bed between ZP1 and ZP4 (Fig. 1) due only to bed friction damping. However, for most data sets at John Brewer Reef, waves broke at the reef edge due to depth limiting effects and dissipated energy, primarily through the turbulence associated with the breaking process, until stable wave heights compatible with reef top water depths were attained. The waves then reformed into stable oscillatory waves. It was established that even the largest waves in any of the recorded incident wave trains, reformed into oscillatory waves before reaching ZP4 (Nelson, 1994). However, this was not the case at ZP1 making it necessary to delineate which data sets were largely free of breaking waves at this station. These will obviously be those biassed towards lower incident wave energy and larger reef top water depths.

This screening was made using the criterion of Nelson (1994) for the maximum stable oscillatory wave heights sustainable over horizontal beds, namely Eq. 4, and the fact that this criterion was shown to be applicable to the individual waves of a naturally occurring random wave train.

$$\frac{H}{h} = \frac{F_c}{22 + 1.82F_c} \tag{4}$$

F_c is a non-linearity parameter after Swart and Loubser (1979), as defined in Eq. 5. F_c values less than 10 indicate deep water waves, those between 10 and 500 indicate transitional water depth waves, and values greater than 500, shallow water waves when H/h approaches a value of 0.55.

$$F_c = \frac{g^{1.25} H^{0.50} T^{2.50}}{h^{1.75}} \tag{5}$$

Waves with larger H/h ratios can exist on horizontal beds but these would be turbulent breaking waves losing height rapidly by the turbulent dissipation of energy.

A wave by wave analysis of all 72 wave records at ZP1 determined the H/h population for each record and the value of F_c associated with each H/h value. This took into account the effect of tide and infragravity wave activity in varying the effective water depth on which the wind waves were superimposed, since the maximum wave height that could be sustained varied with time because local water depth was time dependent. The F_c value assigned was based on the wave height and water depth associated with each wave and the wave period equal to the lapsed time between the crest of this wave and the crest of the previous wave (see Nelson, 1994 for greater detail).

The results for all 72 wave records were compared with the criterion of Eq. 4. Wave records with little or no wave breaking were those for which none (or at least very few) of the observed values of H/h exceeded the value given by Eq. 4 when using the observed value of F_c. Based on this criterion 20 wave records were available for further consideration.

A further 6 of these 20 records were rejected because the reef top wave orthogonal was not sufficiently aligned to the line of wave poles. There remained 14 acceptable data sets.

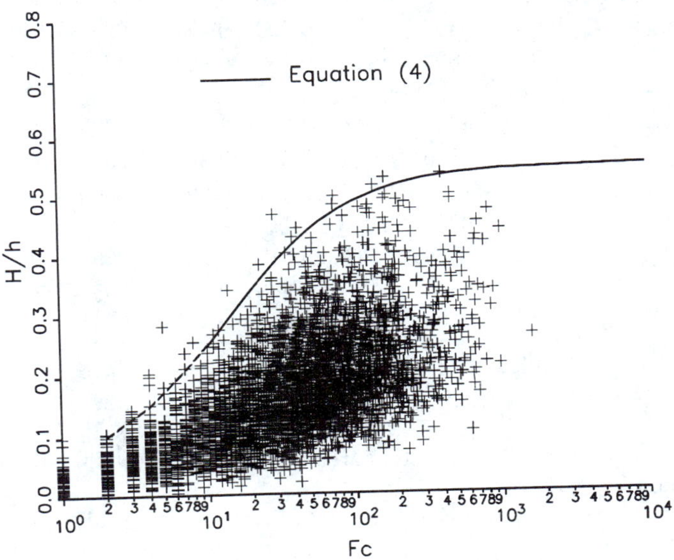

Fig. 3 ZP1: H/h versus F_c for all Waves in all 14 Data Sets

Fig. 3 shows the results for all waves in all 14 data sets plotted on the H/h versus F_c plain. The solid line is Eq. 4 and the dotted line is the deep water wave steepness limit at F_c values less than 10. The average number of waves per wave record is 357, and on average, only 1 or 2 waves per record significantly exceeded the envelope curve. While it is possible some waves did break at H/h values less than the envelope curve (Nelson 1987), it is reasonable to assume that this impact was slight, and that for the 14 selected data sets, all wave energy dissipation between ZP1 and ZP4 was due to bed friction damping only.

Table 1 summarises the characteristic parameters associated with the adopted 14 data sets. The data set number is the location of the data set within the consecutive order of the original 72 data sets. The day and time indicate the common instrument commencement time for the recording of each data set. H_1 and H_4 are respectively the spectral estimates of significant wave height (H_{mo}) of the wind wave component at ZP1 and ZP4 and H is the mean of these two values. T_p is the mean of the four peak energy periods at ZP1, ZP2, ZP3, and ZP4 for the wind wave component. h is the mean, reef top water depth during the data set.

Table 1
Observed Reef Top Parameters - September 1988

Data set	Day	Time (hrs)	$^*H_1$ (m)	$^*H_4$ (m)	H (m)	$^*T_p$ (s)	h (m)
02	4	1500	0.544	0.418	0.481	4.280	1.865
03	4	1600	0.536	0.446	0.491	3.443	2.273
04	4	1700	0.521	0.461	0.491	4.455	2.555
07	4	2000	0.570	0.496	0.533	4.265	2.343
17	5	0600	0.559	0.429	0.494	4.025	1.843
49	6	1800	0.625	0.582	0.604	3.975	2.693
50	6	2000	0.641	0.599	0.620	4.265	2.758
55	7	0600	0.379	0.319	0.349	4.932	1.948
56	7	0800	0.361	0.315	0.338	5.200	1.935
60	7	1600	0.337	0.257	0.297	3.880	1.505
65	8	0200	0.253	0.125	0.189	4.020	0.828
66	8	0400	0.323	0.210	0.266	4.265	1.208
67	8	0600	0.407	0.349	0.378	3.595	1.905
68	8	0800	0.539	0.456	0.498	3.710	2.078

* Based on all frequencies $\geq$ 0.16 Hz

Reef Top Wave Energy Dissipation Factors

Two wave related friction factors are in common use. The wave friction factor (f_w) is defined by Eq. 6.

$$\tau_{max} = 0.5 \rho f_w U^2 \tag{6}$$

τ_{max} is the maximum bed shear stress due the horizontal oscillatory water particle velocities at the bed and ρ is the water density. The determination of f_w requires the direct or indirect measurement of bed shear. The wave energy dissipation factor (f_e) is defined in terms of the time averaged rate of wave energy dissipation due to bed friction as given in Eq. 7.

$$\frac{dE_f}{dx} = (\tau u)_{mean} = \frac{2}{3\pi}\rho f_e U^3 \qquad (7)$$

E_f is the wave energy flux, τ is the instantaneous bed shear stress, u is the instantaneous bed velocity, and U is the amplitude of the of the bed velocity variations. The determination of f_e requires the measurement of wave energy loss over a known distance.

The second of the above procedures was adopted at John Brewer Reef, but in so doing it is important to note the following. The two wave related friction factors are different according to how they are defined and measured. However, it has been shown that for fully developed rough turbulent oscillatory flows, $f_w \approx f_e$ (Jonsson, 1963; Swart, 1974; Nielsen, 1992; Raudkivi, 1988). The oscillatory flow at the bed is fully developed rough turbulent flow for all prototype wave conditions of engineering significance. Therefore, for all practical purposes, the field estimates of f_e can be interchanged with f_w. This will be important when estimates of the reef top hydraulic roughness are made.

f_e can be evaluated two ways. The first considers the attenuation of the total energy using only characteristic spectral parameters. The second considers the attenuation of individual frequency components in the spectrum, based on the assumption that these components travel independently of all other spectral components and that there are no non-linear interactions. Experimental limitations dictated the use of the first method. Despite the difficult and hostile environment, the field measurements of small total energy deficits between poles 1 and 4 displayed good consistency and trends. To deal with still smaller energy deficits within individual frequency components introduces anomalies due to the lesser confidence associated with these components relative to that of the total wave energy measurement.

In water of constant depth, the loss rate of wave energy per unit surface area (left hand side of Eq. 7) can be expressed as,

$$\frac{dE_f}{dx} = \frac{\rho g C_g}{8}\frac{dH^2}{dx} \qquad (8)$$

where C_g is the group velocity and g is the acceleration due to gravity. The right hand sides of Eqs. 7 and 8 can be equated, integrated, and re-arranged to yield the solution for f_e shown in Eq. 9, and can be applied directly to the John Brewer Reef data shown in Table 1. Δx is the distance between ZP1 and ZP4 namely 141m.

$$f_e = \frac{3g}{8\pi^2}\frac{(H_1 - H_4)}{H_1 H_4}\frac{C_g}{\Delta x}(T_p sinh\,(kh))^3 \qquad (9)$$

The resulting values of f_e are shown in Table 2. Also shown are the associated values of a, U, and R_e obtained using the values of H, h, and T_p given in Table 1.

A plot of f_e versus R_e is shown in Fig. 4. At first glance the scatter appears considerable with f_e values of between 0.1 and 0.2 over a Reynolds number range

Table 2
Derived Reef Top Parameters - September 1988

Data set	Day	Time (hrs)	a (m)	U (m/s)	R_e (-)	f_e (-)
02	4	1500	0.323	0.474	153000	0.164
03	4	1600	0.206	0.376	77000	0.218
04	4	1700	0.281	0.396	111000	0.150
07	4	2000	0.304	0.449	137000	0.132
17	5	0600	0.308	0.480	148000	0.162
49	6	1800	0.279	0.441	123000	0.090
50	6	2000	0.314	0.463	145000	0.082
55	7	0600	0.273	0.348	95000	0.152
56	7	0800	0.283	0.342	97000	0.120
60	7	1600	0.202	0.327	66000	0.177
65	8	0200	0.194	0.303	59000	0.207
66	8	0400	0.234	0.344	80000	0.190
67	8	0600	0.196	0.342	67000	0.144
68	8	0800	0.253	0.428	108000	0.143

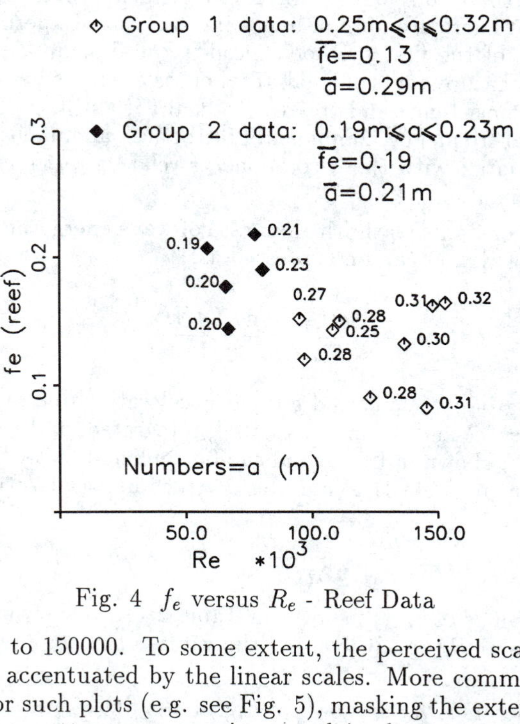

Fig. 4 f_e versus R_e - Reef Data

of between 50000 to 150000. To some extent, the perceived scatter is more apparent than real, accentuated by the linear scales. More common practice is to use log log axes for such plots (e.g. see Fig. 5), masking the extent of the scatter. While much of the scatter must remain unexplained, some can be attributed to the different characteristic values of relative roughness associated with each data set, and the dependence of f_e on that value (see Eq. 3). The values shown beside each point are the a values given in Table 2. These alone are an indicator of relative roughness ($= r/a$) because the reef top roughness remains unchanged for all data sets. Based on this indirect measure of relative roughness the results can be divided into two groups. Group 1 data has a values between 0.25m and 0.32m (mean=0.29m) and a mean f_e value of 0.13. Group 2 data has a values between 0.19m and 0.23m (mean=0.21m) and a mean f_e value of 0.19.

Analyses demonstrated that the small prevailing pole line current that existed on the reef (average value of 0.06 m/s for the 14 data sets) would have had little influence on the computed values of f_e shown in Table 1.

Reef Top Hydraulic Roughness

Kamphuis (1975) used an oscillating water tunnel to measure the maximum shear stresses on a smooth bed and 5 different artificially roughened beds. The shear stress values, measured using a shear plate incorporated in the bed, enabled the computation of wave friction factor, f_w (Eq. 6). Kamphuis summarised the results in a friction factor diagram reproduced here in Fig. 5. The results are consistent with the friction factor being independent of Reynolds number for fully rough turbulent flow. In this region, f_w and f_e (the wave energy dissipation factor) are for all practical purposes, interchangeable as indicated on the vertical axis of Fig. 5. The John Brewer Reef data in Fig. 4 fall well within the fully rough turbulent region of Fig. 5 ($0.1 < f_e < 0.2$ and $50000 < R_e < 150000$) and can be used to estimate values of reef top hydraulic roughness in conjunction with Fig. 5.

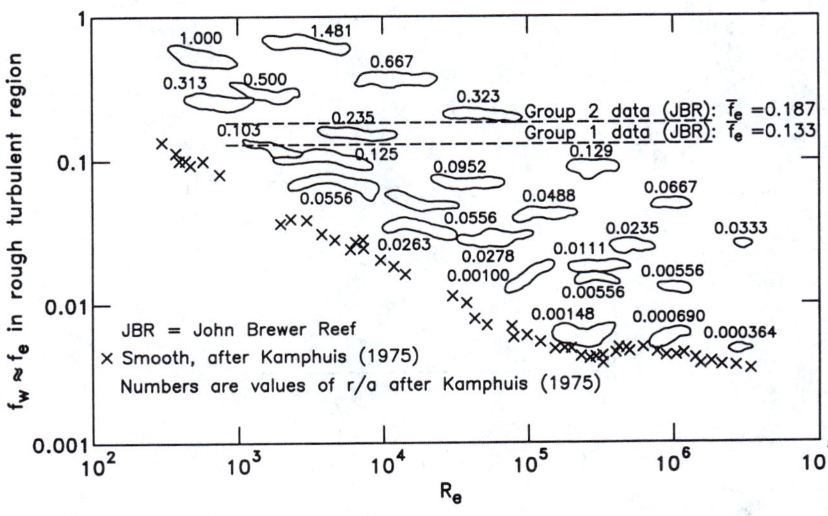

Fig. 5 f_e versus R_e (after Kamphuis, 1975)

The group 1 data has a mean f_e value of 0.130, corresponding to a relative roughness (r/a) value in Fig. 5 of about 0.2. Based on the mean a value of 0.29m for the group 1 data, the hydraulic roughness (r) of the reef top is estimated as $0.2 \times 0.29 = 0.06$m. Similarly for group 2 data, the mean values of $f_e = 0.183$ and $a = 0.21$m lead to values of $r/a \approx 0.3$ and $r = 0.3 \times 0.21 = 0.06$m. Hence, one estimate of the reef top roughness is 0.06m.

A second independent estimate of hydraulic roughness can be made using the work of Jonsson (1963) who developed the following implicit equation for fully rough turbulent oscillatory flow over a fixed immobile bed.

$$\frac{1}{4f_w} + log\left(\frac{1}{4\sqrt{f_w}}\right) = -0.08 + log\left(\frac{a}{r}\right) \tag{10}$$

Swart (1974) presented Eq. 11 as an explicit alternative to Eq. 10. The author has replaced f_w in Swart's equation with f_e as has been previously justified in this paper for fully rough turbulent flow.

$$f_e = exp\left(5.213\left(\frac{r}{a}\right)^{0.194} - 5.977\right) \tag{11}$$

Fig. 6 shows the John Brewer Reef data over-plotted with curves corresponding to fixed values of r computed from Eq. 11. These show that the best estimate of reef top hydraulic roughness is 0.07 m, and this closely approximates the previous estimate of 0.06m.

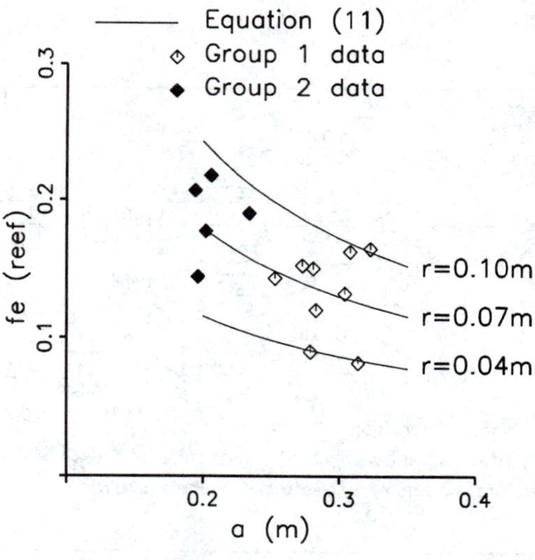

Fig. 6 f_e versus a - Reef Data

Movable Sand Bed Equivalents

An interesting comparison has been made between the influence of this fixed reef surface on wave damping and that of a movable bed of sand using the dimensionless function of Nelson (1995). This expresses the wave energy dissipation factor (f_e) and the physical state of a movable bed as a function of two dimensionless parameters namely $T_c = T\sqrt{g/D}$ and $D_c = D/a$. A range of bed particle sizes were used in the function for each reef top wave and water depth condition listed in Table 1. It was found that movable bed f_e values, comparable with the observed fixed bed reef top values, occurred most consistently when the bed particle size was equal to 0.6 mm. The comparative results are shown in Fig. 7.

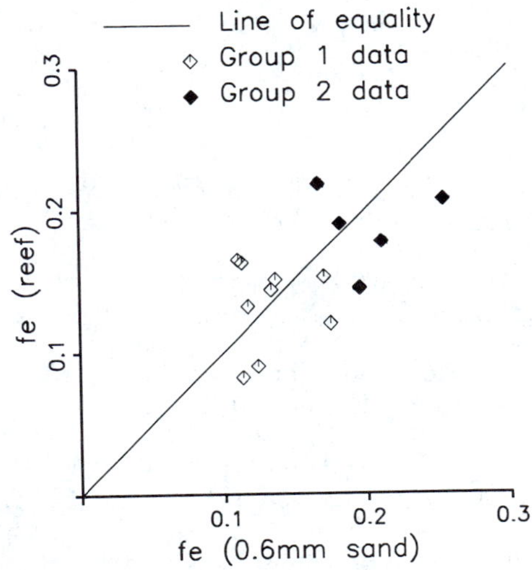

Fig. 7 Movable Bed versus Reef Top Friction Factors

The scatter in the plot results from the fact that a movable bed of given particle size no longer has a constant roughness because the bed form varies with the prevailing wave and water depth conditions. Nevertheless, the comparison demonstrates that while coral reefs may be perceived as being much *rougher* than sandy beds, they may in fact be no rougher, in hydraulic terms, than many sandy beds.

Comments and Conclusions

Field observations have enabled the hydraulic roughness for the fixed bed of a coral reef platform to be determined. In so doing, advantage has been taken of fortuitous and near ideal conditions, where natural trains of random, stable and unbroken oscillatory waves were propagating in water of constant depth, such that wave bed friction damping was the only active wave transformation process present. The roughness estimate of 60mm to 70mm was based on small total energy deficits measured over a known length of wave orthogonal.

The measured fixed bed roughness value is of more use to wave transformation modelers than any of the measured values of wave friction factor. The latter change with wave and water depth conditions while the former remains unchanged, so that the applicable relative roughness and wave friction factor can be determined no matter what wave and water depth conditions prevail.

The measured hydraulic roughness value is transferable to other similar reefs. If differences between reefs are perceived, the measured roughness is a datum value that can be varied based on visual comparisons and reasoning.

The paper has demonstrated that, while coral reefs may be perceived as being much rougher than sandy beds, they may in fact be no rougher in hydraulic terms than many sandy beds experiencing the same wave and water depth conditions.

Acknowledgements

The author wishes to acknowledge the efforts of Dr T. Hardy (James Cook University, Queensland) during the field data collection program.

Appendix 1. References

Jonsson, I.G., 1963. Measurements in the turbulent wave boundary layer. In: Proc. 10th IAHR Congress, London, Vol. 1, pp. 85-92.

Jonsson, I.G., 1966. Wave boundary layers and friction factors. In: Proc. 10th Coastal Eng. Conf., Tokyo. ASCE, New York, Vol. 1, pp. 127-148.

Kamphuis, J.W., 1975. Friction factors under oscillatory waves. J. Waterways Harbors and Coastal Eng., ASCE, Vol. 101, pp. 135-144.

Nelson, R.C., 1987. Design wave heights on very mild slopes. Civil Eng. Trans., Inst. Eng. Australia, Vol. CE29, pp. 157-161.

Nelson, R.C., 1989. A simplified function for wave energy dissipation factor over mobile beds. Coastal Engineering, Vol. 13, pp. 149-159.

Nelson, R.C., 1994. Depth limited design wave heights in very flat regions. Coastal Engineering, Vol. 23, pp. 43-59.

Nelson, R.C., 1995. Wave bed friction damping over shoaling movable beds. Coastal Engineering, Vol. 25, pp. 65-80.

Nielsen, P., 1992. Coastal bottom boundary layers and sediment transport. World Scientific, Singapore.

Raudkivi, A.J., 1988. The roughness height under waves. J. Hydraulic Research, IAHR, Vol. 26, pp. 569-584.

Swart, D.H., 1974. Offshore sediment transport and equilibrium beach profiles. Delft Hydr. Lab., Publication No. 131.

Swart, D.H. and Loubser, C.C., 1979. Vocoidal wave theory. Vol. 2: verification. CSIR (South Africa), NRIO, Research Report No. 360.

Zwarts, C.M.G., 1974. Transmission line wave height transducer. In: Proc. Symp. on Ocean Wave Measurement and Analysis, New Orleans. ASCE, New York, Vol. 1, pp. 605-620.

CHAPTER 244

Effect of Wave-Induced-Pressure on Seabed Configulation

Tetsuo Sakai[1], M. ASCE and Hitoshi Gotoh[2]

Abstract

In almost all of the previous studies on the seabed configuration, the seabed configuration is classified based on the bottom shear stress, or Shields number. While, in the wave field, the wave-induced-water-pressure change can have a secondary effect on the sediment transport. The laboratory experiment on the seabed configuration under the coexistence of the oscillatory flow and the water-pressure change is conducted in this study, to investigate the effect of water-pressure change on the seabed regime classification.

The existing region of various bed configurations, such as no-motion, bed-lad, suspension over ripple and sheet flow, changes with the change of the amplitude of water-pressure change. The ripple geometry, or flatness of ripple, are also affected by the water-pressure change. The mechanism of theses changes are discussed related to the lift force due to the porewater-pressure distribution in sand layer.

Introduction

Seabed configuration, which is formed as the result of sediment transport, is quite different under the various conditions of wave and current. On the other hand, the mode of sediment transport depends on the seabed configuration.

Although there is an interactive structure among the wave-current field, the sediment transport and the seabed configuration, the time scale of the development of seabed configuration is sufficiently longer than that of others. Therefore, to describe

---

1 Professor, Department of Civil Engineering., Kyoto University
Yoshida Honmachi, Sakyo-ku, Kyoto, 606, Japan
2 Lecturer, Department of Civil Engineering., Kyoto University

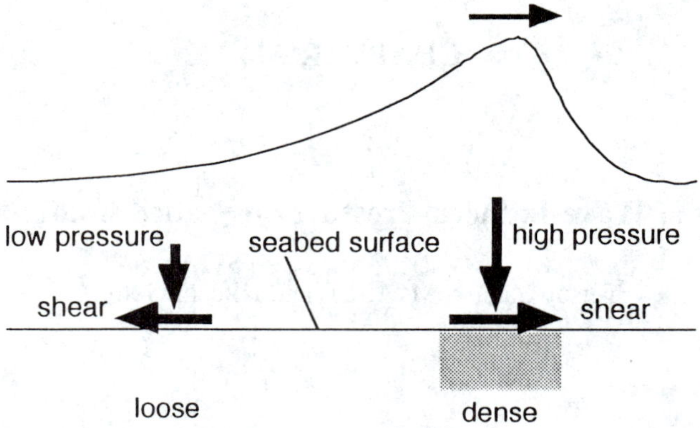

Figure 1. Forces acting on seabed under the wave propagation

the sediment transport process based on the mechanics of the motion of sediment particles, it is essential to know the relation between the seabed configuration and the bottom shear stress, or Shields number, due to wave and current.

These relation has been investigated through some experimental studies; while the effect of wave-induced-water-pressure change on the seabed configuration was not examined in detail. Fig. 1 shows the schematic expression of external forces acting on seabed due to the wave propagation. High water pressure acts on the seabed under the crest of a wave, then the particle density of the bed becomes to be dense. On the other hand, low water pressure acts on the seabed under the trough of a wave, then the particle density of the bed becomes to be loose. In both situations, the bottom shear stress is the dominant driving force of sediment motion. To discuss the additional effect due to the water-pressure change by wave propagation in detail, porewater pressure distribution should be investigated.

In this study, the experiment on the seabed configuration under the coexistence of the oscillatory flow and the water-pressure change is performed by using an oscillating water tunnel with cylinder system for the control of the water pressure.

Laboratory experiment

experimental apparatus

Figure 2 shows the oscillating water tunnel with the water-pressure-control system. Both sides of the rectangular water tank are connected to the water circulating pipe, at the top of which a water-circulating-propeller system, or an oscillatory-flow generator,

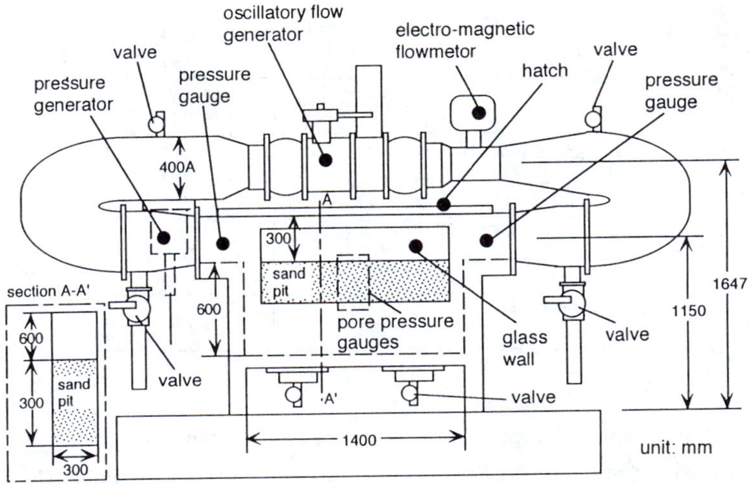

Figure 2. Oscillating water tunnel with water-pressure-changing system

is installed. Lower part of the water tank is a sand pit, the depth of which is 600 mm. The front and back sides of the water tank is made by glass wall for the observation of the motion of sediment particles. In the back side wall, porewater-pressure sensors are flash mounted. In order to generate the water-pressure change, a oil-hydraulic-pressure-driven piston system is connected to the water-circulating pipe. The electromagnetic current meter and the water-pressure sensors are equipped to monitor the discharge of current and the changing water pressure, which are used as the feedback signal of the personal-computer based current-water-pressure control system. Although various kinds of the time series of current-velocity and water-pressure can be generated by using this apparatus, the in-phase sinusoidal time series of current-velocity and water-pressure, which correspond to the condition of a progressive small amplitude wave, are generated.

procedure of experiment and experimental conditions

The transition of bed configuration occurs with the increase of the bottom shear stress in the following order: (i) no motion; [critical state for sediment movement]; (ii) bed-load motion; (iii) suspended load over ripples; and sediment transport in sheetflow regime. In the experiment, the flow-velocity amplitude is gradually increased with keeping the amplitude of water-pressure change, to observe the transition of bottom

Table 1. Experimental conditions

	oscillatory flow	oscillatory flow and pressure
velocity amplitude u_b (cm/s)	3.0 - 78.0	3.0 - 78.0
total pressure amplitude p_0 (m)	0.0	0.5, 1.0, 1.5
period T (s)	6.0	6.0

configuration mentioned above. The series of the experiment were conducted with changing the magnitude of the amplitude of the water-pressure change, to investigate the effect of the water-pressure change on the transition of the bed configuration.

The ripple geometry affected by the amplitude of water-pressure change is examined based on the detailed observation of sediment motion over ripples, and the mechanism to determine the ripple geometry is considered.

Table 1 shows the experimental conditions. The test particle are the two kinds of the uniform sand, the diameter of which are d=0.025 cm and d=0.035 cm. The specific density of the sand is 2.65.

Classification of the seabed configuration

Figure 3 shows the experimental results of the bed configuration under the action of oscillatory flow for the cases of d=0.025 cm and d=0.035 cm. The lines in this figure shows the division of the bed configuration proposed by Shibayama and Horikawa (1982). In the figure, u_b=amplitude of the flow velocity in the neibourhood of the bottom; w=terminal fall velocity of sediment particle; and Ψ=Shields number defined as $\psi = f_w u_b^2 / 2(\sigma/\rho - 1)gd$ (f_w=friction coefficient proposed by Jonsson (1966); σ=density of sediment; ρ= density of water; and g=graviational acceleration). The flow velocity in the neighborhood of the bottom is estimated by extrapolating the measured velocity near-bottom-wall region by supposing the logarithmic velocity profile. The friction coefficient was calculated by using the explicit expression proposed by Tanaka (1990). The thresholds of four modes, such as no motion, bed-load motion, suspension over ripples and sheetfolw, estimated from the present experiment show fairly good agreement with the empirical relation proposed by Shibayama and Horikawa, which was determined based on the existing experimental results.

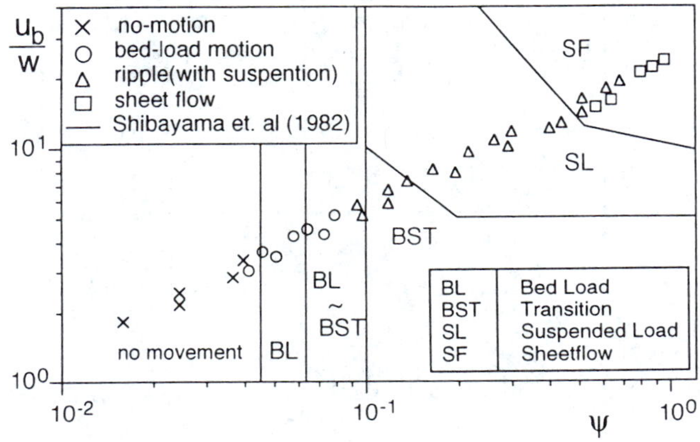

Figure 3. Classification of seabed

Figure 4 shows the classification of the bed configuration, namely, no motion, bed-load, suspension over ripple and sheetflow, under the various combination of flow-velocity amplitude, u_b, and total water-pressure-change amplitude, p. In this figure, the experimental results of the cases of the particle diameter d=0.025 cm are shown. The critical shear stress of the sediment movement decreses with the increase of the amplitude of water-pressure change. In other words, the no-motion region decrease with the increase of the amplitude of water-pressure change. The supression of the ripple generation increases with the increase of the amplitude of water-pressure change, consequently, the be-load region increase with the increase of the amplitude of water-pressure change. In contrast with the ripple genaration, the generation of the sheetflow is promoted by the increase of the amplitude of water-pressure change.

The changing porewater pressure is induced in the sandy bed constituted by fine sediment under the action of water-pressure change on the bed surface. Because of the phase lag between the porewater pressure and the water-pressure acting on the sand surface, an additional lift force acts on the sediment layer near the sand surface. This additional lift force depends on the porewater-pressure profile, and the positive lift force acts on the sediment layer when the changing water-pressure is smaller than the average, namely in wave-trough phase. The decrease of the critical bottom shear stress of sediment movement is caused by the action of this additional lift force.

Figure 5 shows the classification of the bed configuration under the various combination of flow-velocity amplitude, u_b, and water-pressure-change amplitude, p,

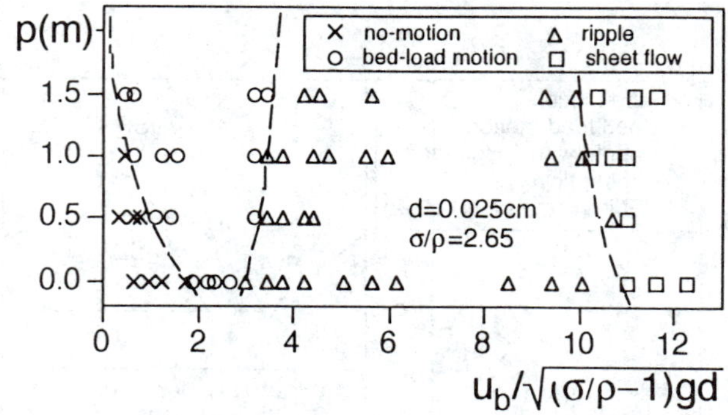

Figure 4. Classification of the bed configuration (d=0.025 cm)

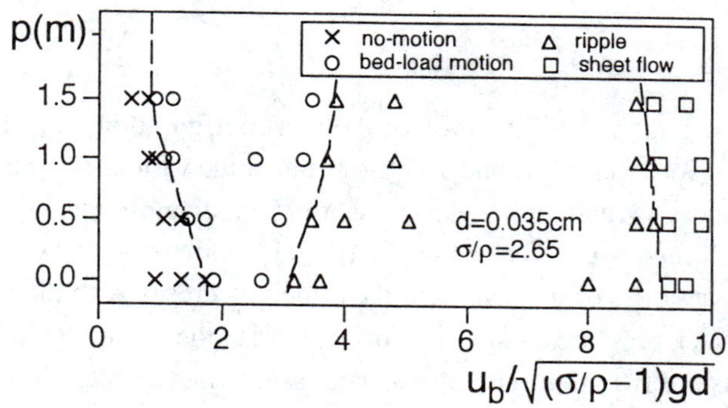

Figure 5. Classification of the bed configuration (d=0.035 cm)

for the particle diameter d=0.035 cm. Although the tendency of the transition of bed configuration shown in Fig. 5 is the same as in Fig. 4, the effect of the amplitude of the water-pressure change to the transition of the bed configuration is stronger in Fig. 4 than that in Fig. 5. To clarify this point, the bottom neighboring flow velocities, u_{cp}, at three threshold such as the transition point from no motion to bed load, from bed-load to suspension and from suspension to sheetflow for the change of the amplitude of water-pressure change are shown in Fig. 6. The bottom neighboring flow velocities, u_{cp}, are normalized by that of oscillatory flow condition, u_c. Fig. 6 shows that the transition points of the bed configuration clearly depend on the magnitude of the

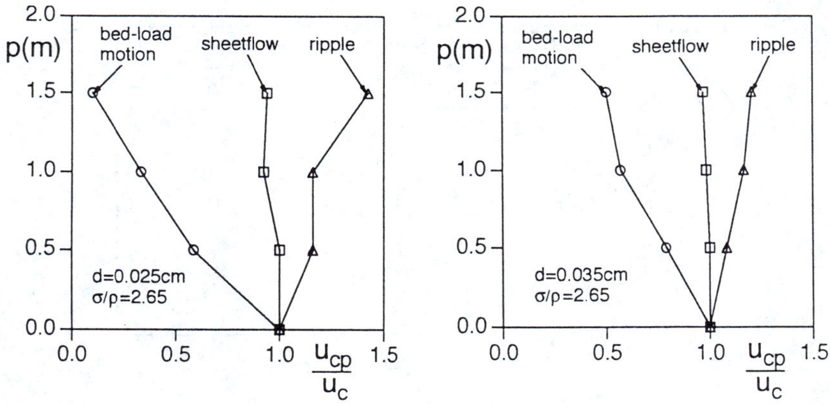

Figure 6. Transition points of the bed configuration

amplitude of water-pressure change. This tendency is most clearly detected in the transition point from no motion to bed load. By comparing the cases of d=0.025 cm and the cases of d=0.035 cm, the change of the threshold of bed configuration against the change of the amplitude of water-pressure change in small size sediment is more significant than that in large size sediment.

Under the action of critical bottom shear stress of sediment movement, hydrodynaminc force, or driving force, acting on the sediment particle balances with the bottom frictional force, or resistant force. Therefore the sediment motion is keenly affected by the additional lift force due to the changing porewater pressure in sand layer. While, in the cases of the transition from suspension to sheetflow, the amplitude of the bottom shear stress are sufficiently larger than that of no-motion-bed-load transition. Hence the bottom shear stress plays a significant role in the transition from suspension to sheetflow, and the change of the threshold from suspension to sheetflow against the change of the water-pressure amplitude is smaller than that from no motion to bed load.

Geometry of ripples under coexistence of shear and water-pressure changes

The geometry of ripples in suspension mode is also affected by the change of the amplitude of water-pressure change. Maeno et al. (1989) conducted the experiment on the ripple geometry under the action of porewater pressure in sandy bed. They mainly discussed the energy dissipation due to the ripples based on the measurement of the porewater pressure change. While in this study, the role of the porewater pressure

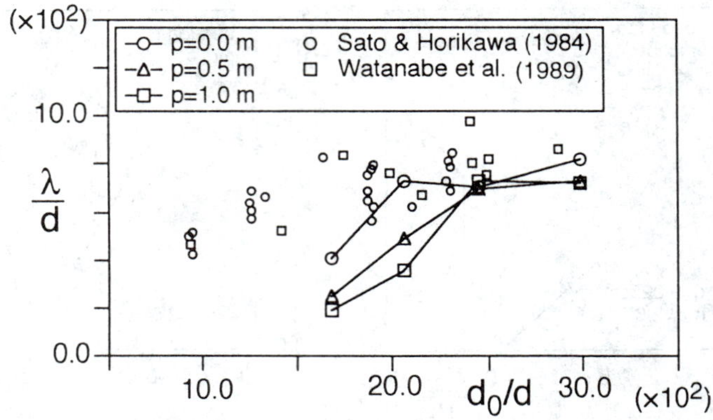

Figure 7. Relation between ripple wave length and excursion length of water particle

change in sandy bed is investigated as a part of the mechanism to determine the ripple geometry.

Figure 7 shows the relation between ripple wave length, λ, and excursion length of water particle, d_0. In this figure, the cases of three kinds of total water-pressure amplitude, p=0.0, 0.5 and 1.0 m are shown with the experimental result by Sato et al. (1984) and Watanabe et al. (1989). In the region of $15.0<d_0/d<25.0$, the ripple wave length decreases with the increase of amplitude of water-pressure change; while, in the region of $25.0<d_0/d$, the dependence of ripple wave length on the amplitude of water-pressure change becomes less significant, and the ripple wave length is regarded to be constant.

Figure 8 shows the relation between ripple wave steepness, η/λ, and Shields number, Ψ (η=ripple wave height) . In this figure, the cases of three kinds of total water-pressure amplitude, p=0.0, 0.5 and 1.0 m are shown with the experimental result by Sato et al. (1984) and Watanabe et al. (1989). In the region of small Shields number, or in the region of ripple generation, the ripple wave length decreases with the increase of amplitude of water-pressure change. In other words, ripples become flatter with the increase of amplitude of water-pressure change. This tendency becomes less significant with the increase of Shields number. In the neighborhood of Ψ=0.3, the dependence of the ripple wave length on the amplitude of water-pressure change is not clear.

Figure 9 shows the change of the transition from two-dimensional ripples to three-dimensional ripples with the change of the amplitude of water-pressure change. The dashed lines in this figure shows the threshold of two- and three-dimensional

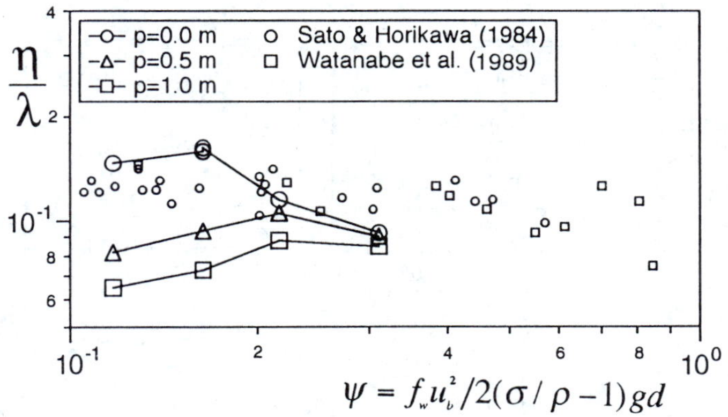

Figure 8. Relation between ripple wave steepness and Shields number

ripples proposed by Watanabe et al. (1989). In the pure oscillatory cases, p=0.0 m, the experimental results agree fairly well with the proposal by Watanabe et al. The transition point from two-dimensional ripple to three-dimensional one gradually decreases with the increase of d_0/d. In the coexistence cases of oscillatory flow and water-pressure change, the transition point from two-dimensional ripple to three-dimensional one shifts upward, and the dependence of the transition point to the excursion length of water particle becomes smaller.

Two-dimensional ripples are dominantly observed when the ripple wave length and the ripple wave steepness is affected by water-pressure change. The motion of sediment in the neighborhood of the crest of ripple was recorded by a CCD video camera, to investigate the flatting process of ripple due to the water-pressure change in detail. Table 2 shows the experimental condition. Figure 10 shows the behavior of sediment at every $3/\pi$ intervals on the symmetric ripple, which is generated in the pure oscillatory flow condition, p=0.0 m. On the other hand, Fig. 11 shows the behavior of sediment on the asymmetric ripple, which is generated in the oscillatory flow and water-pressure change coexisting condition (p=1.0 m).

For the simplicity, direction of the current form left to right is taken positive, and the opposit one is taken negative. In this study, the oscillating flow is in phase with the water-pressure change, hence the pressure differnce from the average is positive when the current direction is positive; while the pressure difference from the average is negative when the current direction is negative.

In the pure oscillatory flow case, the time series of the current velocity follows the sinusoidal curve, then the flow velocity is symmetric between the positive-current phase and negative-current phase. Two ripple geometry with the phase lag π are

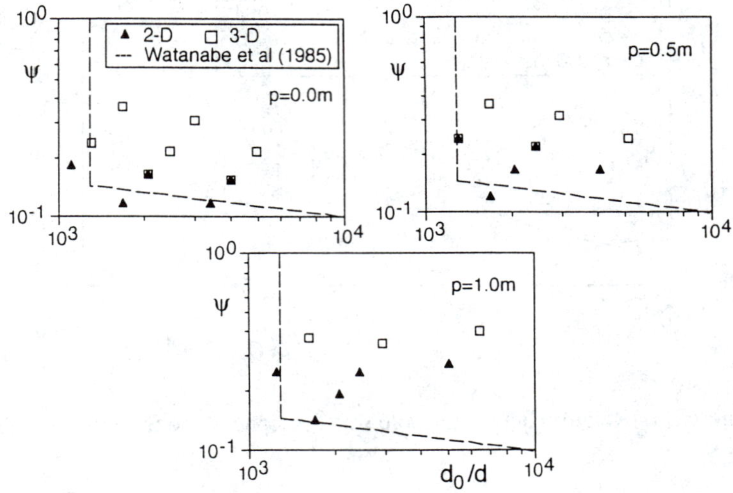

Figure 9. Transition from two-dimensional ripples to three-dimensional ripples

Table 2. Experimental conditions of the visualization

	oscillatory flow	oscillatory flow and pressure
velocity amplitude u_b (cm/s)	29.0	29.0
total pressure amplitude p_0 (m)	0.0	1.0
period T (s)	6.0	6.0
diameter of sand d(cm)	0.025	
specific density of sand σ/ρ	2.65	
ripple wave length λ(cm)	20.0	23.0
ripple wave height η(cm)	2.5	2.0

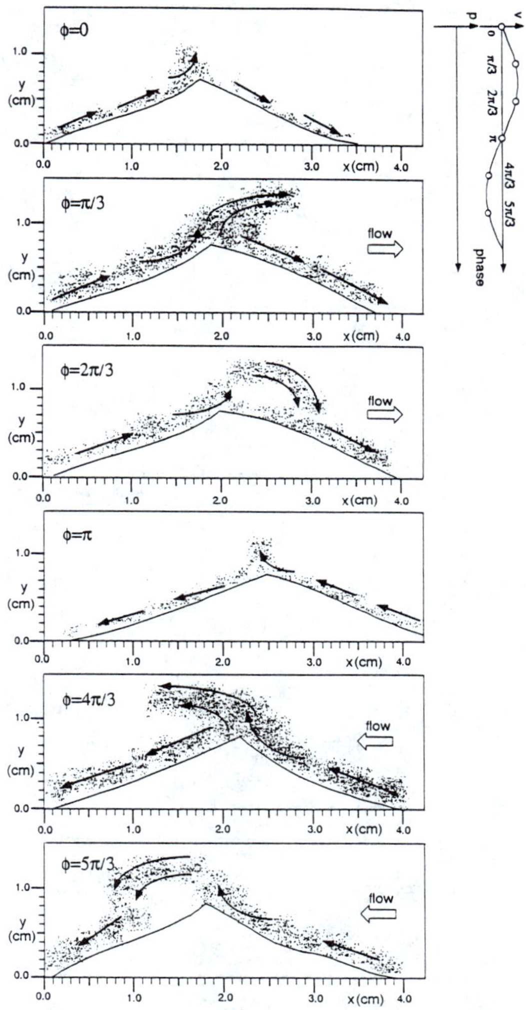

Figure 10. Behavior of sediment at every 3/π intervals on the symmetric ripple

symmetric. On the contrary, under the coexistence of oscillatory flow and water-pressure change, a significant asymmetric ripple geometry is detected between the positive-current phase and the negative-current phase. Suspended sediment transported along a backward slope of the ripple in the negative-current direction at the phase ϕ=4π/3 is significantly larger than that in the positive-current direction at the phase

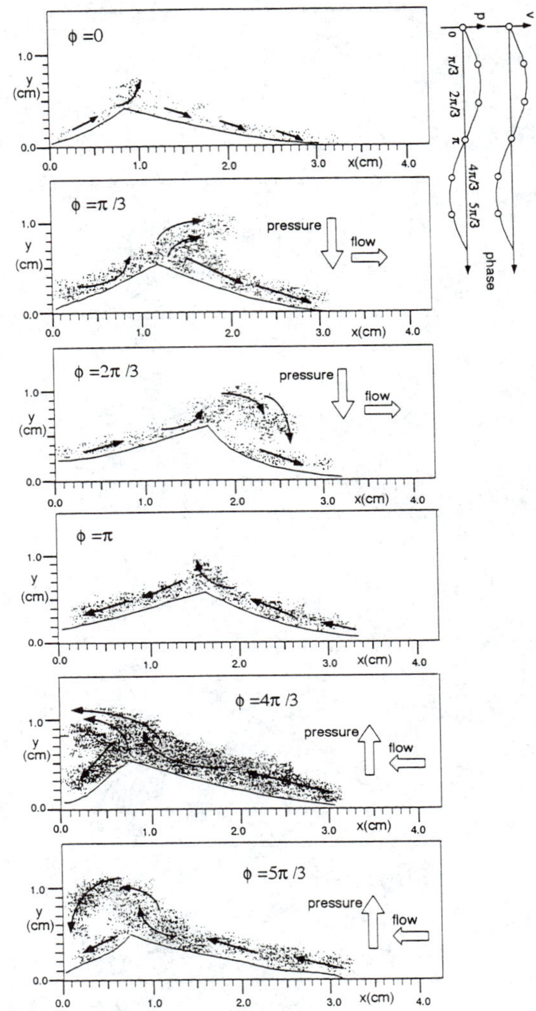

Figure 11. Behavior of sediment at every 3/π intervals on the asymmetric ripple

$\phi=\pi/3$. When the pressure difference from the average is negative, or the current direction is negative, the additional lift force due to the porewater pressure in sand layer acts on the sediment particles near the surface of sand layer. Because of this additional lift force, sediment particles are promoted to be dislodged, consequently the sediment transport rate increases. On the other hand, when the pressure difference

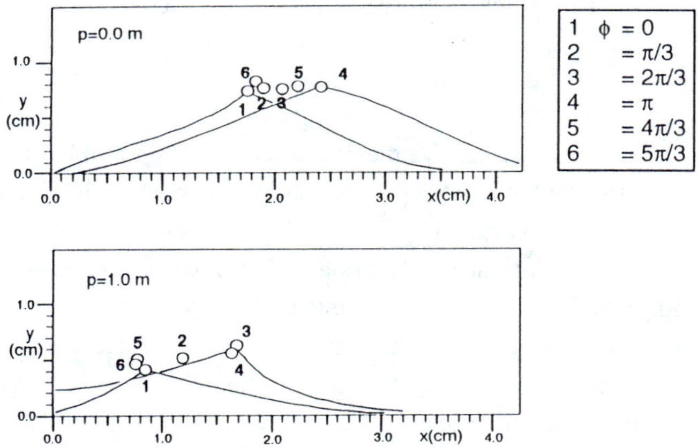

Figure 12. Movement of the crest of ripple

from the average is positive, or the current direction is positive, the negative lift force, or the additional force acting in downward direction, due to the porewater pressure in sand layer acts on the sediment particles near the surface of sand layer. This force increases the resistance of the sediment particle against the dislodgment and, consequently the sediment transport rate decreases.

Figure 12 shows the movement of the crest of ripple during one oscillating cycle. In the symmetric ripple, which is generated under the pure oscillating flow (p=0.0 m), the moving path of the crest toward right and that toward left overlap with each other. In other word, the vertical displacement of the crest of ripple is sufficiently small compared with the horizontal motion of the crest of ripple. On the contrary, in the case of the coexistence of the oscillatory flow and water-pressure change, the moving path of the crest during the positive-current-direction phase and that during the negative-current direction phase is significantly different. The moving velocity of the crest is slower during the positive-current-direction phase (1-3 in Fig. 12) than during the negative-current-direction phase (4-6 in Fig. 12). The crest of ripple moves upward during the positive-current-direction phase and moves downward during the negative-current-direction phase. Because the ripple wave length is approximately constant in all phases, this characteristics of the ripple wave height, or the movement of the crest in vertical direction, are interpreted as the change of the ripple wave steepness. The ripple wave steepness is larger during the positive-current-direction phase than during the negative-current-direction phase. The additional force due to porewater pressure induced by water-pressure change acts in upward direction during the negative-current-direction phase. This additional force brings the decrease in the apparent angle of repose of sand, and the resultant flatting of ripple geometry.

Conclusion

In this study, the experiment on the seabed configuration under the coexistence of the oscillatory flow and the water-pressure change is performed by the oscillating water tunnel with cylinder system for the control of the water pressure.

The existing region of various bed configurations, such as no-motion flat bed mode, bed-load mode, suspension over ripple bed and sheet flow, changes with the change of the amplitude of water-pressure change. Furthermore, the ripple geometry, such as ripple wave length and wave steepness, or flatness of ripple, are also affected by the amplitude of water-pressure change.

Lift force due to the porewater-pressure distribution in sand layer acts in both upward and downward direction. The upward lift force promotes a sediment motion; while the downward force, or negative lift force, suppress a sediment motion. These difference of the sediment motion is the reason of the change of the existing region of the dominant bed configuration and the asymmetric characteristics of ripple geometry.

References

Jonsson, I. G.(1966): Wave boundary layer and friction factors, Proc. of 10th Conf. on Coastal Eng., pp.127-148.

Maeno, Y, Matsuoka, Y, Hayashida, H. and Mase, H.(1989): Effects of ripple geometry on wave-induced pore pressure, Proc. of Coastal Eng., JSCE, Vol. 36, pp. 789-793(in Japanese).

Sato, S. and Horikawa, K.(1984): Laboratory study on sand transport over ripples due to asymmetrical oscillatory flows, Proc. of Coastal Eng., JSCE, Vol. 31, pp. 286-290(in Japanese).

Shibayama, T. and Horikawa, K.(1982): Sediment transport and beach transformation, Proc. 18th ICCE, pp.1439-1458.

Tanaka, H.(1990): An explicit expression of a friction coefficient for a wave-current coexisting motion, Proc. of JSCE, No. 417/II-13, 285-288 (in Japanese).

Watanabe, A., Sakinada, M. and Isobe, M.(1989): Ripple formation and sand transport rate in a wave-current coexistent system, Proc. of Coastal Eng., JSCE, Vol. 36, pp. 299-303(in Japanese).

CHAPTER 245

The Measurement of Bed Form Shapes in Hydraulic Models

Ian E Shepherd[1]

Abstract

Mobile bed hydraulic models have been used in the laboratory for a very long time. Attempts to predict the consequences of river engineering works, dams and barrages, flood protection schemes and dredging methods have led to wide ranging scientific research in material scaling. The design of breakwaters, beach protection systems and sea walls have long been studied in physical models. All of these require a method for measuring the effects of water movements and the resultant erosion, deposition and scouring of materials. This paper reviews some methods that have been used on laboratory models, and attempts to suggest ways forward in the future.

Introduction

Laboratory models of mobile bed properties require measurements of the topography resulting from water moving over the surfaces. This ranges from sedimentation effects in rivers to wave action on beaches. Measurements can be on the macro scale (ie, methods for producing 3D contours over the complete surface) or local 2D tracks of surface elevation (eg, scour around pile structures). It is a difficult measurement, as a wide range of materials and circumstances is encountered. Many experimental laboratories have developed specific techniques to suit particular projects, so it has always been difficult to design equipment and methods that are easily transportable between investigations.

---

[1]HR Wallingford Ltd, Howbery Park, Wallingford, Oxon, OX10 8BA, UK

The durability of this style of investigation is confirmed by the fact that they still occur, and the long history (Hydraulic Models, 1942). In this reference, a description is given of a mechanical system for lowering a small probe onto the surface of the bed from a bridge over the model, and a scale to indicate its elevation. It is a source of surprise and regret that little real progress has been made in the last 50 years. This is partly due to the reluctance of civil engineering laboratories in all areas (government, private and university) to adequately invest in instrumentation development. Recent work by Jetté (1996) has shown what can be achieved by individual innovation, in this case to solve a specific field measurement problem.

This paper reviews some options, but concludes that a simple, adaptable, universal method has still only been possible using old technology, with the advantage of PC control. There must be better ways but time and resources have not permitted their development, underlining the comments above about lack of investment.

In recent years there has been a continuing interest at HR Wallingford in the effects of flowing water and waves over mobile beds. The range of work has included:

(a) Under-mining of sea walls
(b) Loss of sand from amenity beaches
(c) Prevention of siltation in navigation channels
(d) Research on the combined effects of waves and currents on bed load transport
(e) Embankment erosion and protection
(f) Scour around bed mounted structures

Most of this work requires the measurement of bed form disturbance to assess the effects of the existing regime and the consequences of engineering changes to the environment. Some researchers have used remote optical techniques (for example Oesch et.al., 1985) but most of these are time consuming to set-up, require sophisticated and expensive equipment, and cannot record what is happening underwater.

To overcome these objections a number of different mobile probe measurement methods have been developed by several laboratories.

Sensor Method	Use in Air	Use Under Water	Typical min. depth mm	Use in Air-Water Transition	Typical horizontal resolution at the bed mm	Typical vertical resolution mm
Conductivity	X	√	20	X	5	0.2
Optical retro-reflective	√	√	10	X	10	0.2
Ultrasonic	√	√	100	X	20	1
Laser distance sensor	√	√	60	X	2	<0.1
Touch Sensitive	√	√	Zero	√	2	0.1
Remote photographic	√	X	(Surface reflection)	X	?	?
Image analysis	√	?	?	?	?	?

TABLE 1 - COMPARISON OF METHODS

Table 1 shows some sensing methods that have been used and typical performance comparisons. There is only one universally applicable method, that of the touch sensitive, incremental technique. HR Wallingford has used optical servo probes for many years (HRS 1968), and the Delft conductivity probe has been popular (Delft 1985, and Villanueva, 1989). They are all subject to the limitations in the table. Models in flumes, for example, can be measured either by draining the water out or by flooding; it is always inconvenient, and a method which will traverse the (shallow) air-water boundary is better, even if the measurement process itself is slower. Early designs of touch probe were used by SOGREAH (Limnidyn, 1985) for a mobile bed river model, and by KENEK in Japan. The design requirements for an instrument include:

(a) Must operate in air, underwater, and through the air-water boundary in both directions.

(b) Must be suitable for sand, mud, coal dust, rocks, plastics, saw dust and in fresh or saline water.

(c) Vertical resolution to be better than 1mm, horizontal resolution to be better than 5mm in a working range of 10m. Horizonal range ideally unlimited.

(d) Options for use in flowing water to be available.

(e) Should be easy to move from one location to another, and be cheap to build.

(f) Expandable from 2D to 3D operation.
(g) Must be able to operate over steep rock faces.
(h) Automatic pc control and data logging.
(i) Good vertical range.
(j) Remote control options

There is an obvious choice of sensor operation, either

(a) Remote sensing probe with a large gap between the probe tip and the bed, or
(b) Servo operated probe following the bed surface with a small gap.

Option (a) is ideal for flowing water (to prevent scour around the probe tip) but has a poorly defined horizontal "footprint".

Option (b) provides good horizontal and vertical resolution but is unsuitable for use in flowing water or under waves.

The well-equipped experimental hydraulic laboratory should have the capability to use any of the techniques described above, so that the varied demands of different experiments can be accommodated. Each method is described in more detail below.

Methods Available

Most of the techniques depend on a probe that is servo-driven so that its tip is near to the measured surface. The error between a reference value and the value of the measured variable is used to maintain a 'constant' following distance of the probe above the surface. Exceptions to this are the photographic, laser distance and image analysis methods.

Conductivity

This option has been available for a long time (Wilkie, 1954), and had been used even earlier using the water surface as a reference (Anon 1952). Essentially it consists of a probe that can measure the bulk conductivity of the water surrounding a probe tip, and variations that depend on the proximity of non-conducting surfaces. Figure 1 (a) shows the principle. Since the water properties affect the measured values, some method must be used to eliminate the problems caused by water conductivity variations.

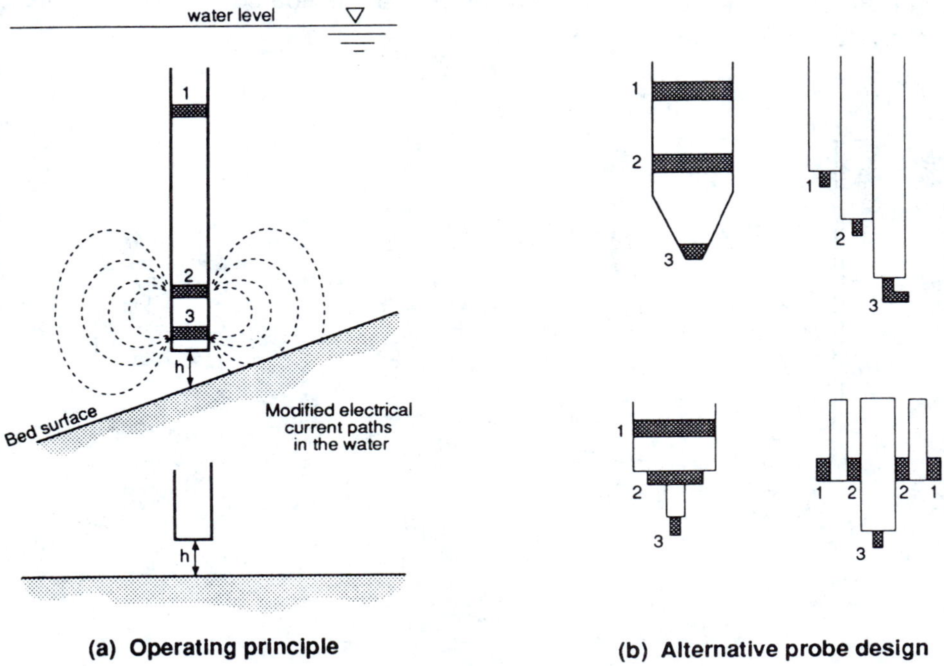

Figure 1 Conductivity Probes

The usual approach is to use an additional electrode (1) which is not affected by the surface proximity, and to use the bulk conductance measured between (1) and (2) as a form of compensation in a 'bridge'-type of arrangement.

At HR Wallingford we have experimented with a range of different probe tip designs, some of which are shown in figure 1(b). We have found that the following height H in 1(a) varies depending on the steepness of the surface. This would, of course, be expected, as the modification to the conductive path depends on the shape of the surface as well as the spacing. This is a very subjective problem, and care must be taken in the interpretation of results. The retro-reflective optical probe also has the same problem.

Optical retro-reflection

Again, this method has been used for a very long time (HRS 1968). It can be used in air (whereas the conductivity method cannot) but must be carefully optimised for the type of material over which it is working. It is also independent of water conductivity variations. The basic principle is to shine pulsed (infra-red) light on to the surface and detect the back scattered light. A convenient way to make a small diameter probe without active electronics in the water is to use fibre optic guides so that the source and detector are above water level, as in figure 2 (a).

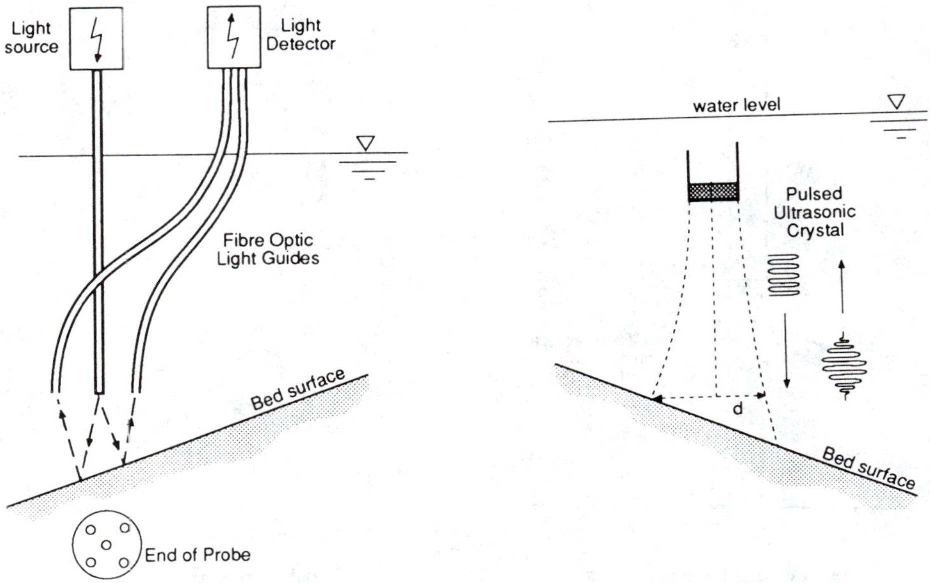

(a) Optical Retro-Reflective **(b) Ultrasonic**

Figure 2 Optical and ultrasonic methods

The use of pulsed infra-red and synchronous detection helps to minimise the effects of ambient lighting. For effective operation, the surface must be a good homogeneous reflector such as sand or light plastics. It will not work satisfactorily over coal or gravel of mixed size and colour. Although it will operate in air or in water, the following height is different due to changes of refractive index. It can be adjusted to operate through the air-water boundary (with a step change of level at the boundary), but not the other way round. We have found that the probe collects water droplets at its tip as it emerges from the water, and this causes internal light reflections.

During some experiments, where the water has been drained away from a rippled sand bed, problems again occur even though the probe is always in air. This is because small pools of water are trapped between the ripples and cause surface reflection. The optical probe method has been used successfully over an eroded mud bed in a greenhouse to measure the effects of irrigation sprays.

Both the conductivity probe and the optical probe can also be adjusted to detect the water surface location, which is useful for datum location. The HR optical probe has been mounted on a reversible carriage and used to traverse a rippled sand bed by ourselves for many projects, and also by others (Richards, 1986), using computer based logging.

Ultrasonic

The use of ultrasonics for hydrographic survey work in the field has been developed to an extremely sophisticated level, and impressive imaging is achieved with side scan sonar. Sensors have also been used for sand ripple measurements (Jetté, 1996). Hydraulic modelling is not in the same price/performance environment, so the same intense development has not occurred.

Some laboratories have used ultrasonic probes, but there is a minimum water depth for satisfactory operation. This depends on the sensor frequency (hence the transmitted wavelength) and the accuracy of the time resolution of the returned pulse within the electronics. Figure 2 (b) shows the usual principle. It is clear that there must be enough time for the transmitted pulse to be detected back at the source before the next pulse is transmitted. Theoretically, bed slope could be measured by driving the probe at a constant horizontal speed, and then detecting the Doppler shift of the backscattered acoustic signal.

A potentially more important limitation is the transmitted beam width. The beam pattern of an acoustic source is complex, and the reader is referred to Stansfield for information. Simply, if the beam angle is 10°, then at a distance of 160mm above the bed, the horizontal projection 'd' will be about 28mm. A narrowed beam angle will improve this, but a larger spacing will degrade it. This sort of horizontal resolution is often inadequate for ripple measurements in models. A typical result is shown in figure 3. Steep ripples and edges of small stones are not measured accurately. This performance has been achieved with a 2.5 MHz probe mounted about 160mm above the bed. ideally, a vertically driven servo probe (as with the HR systems) would be used, but has not yet been tried. In addition, the vertical resolution is limited by the wave length (0.6mm at 2.5 MHz).

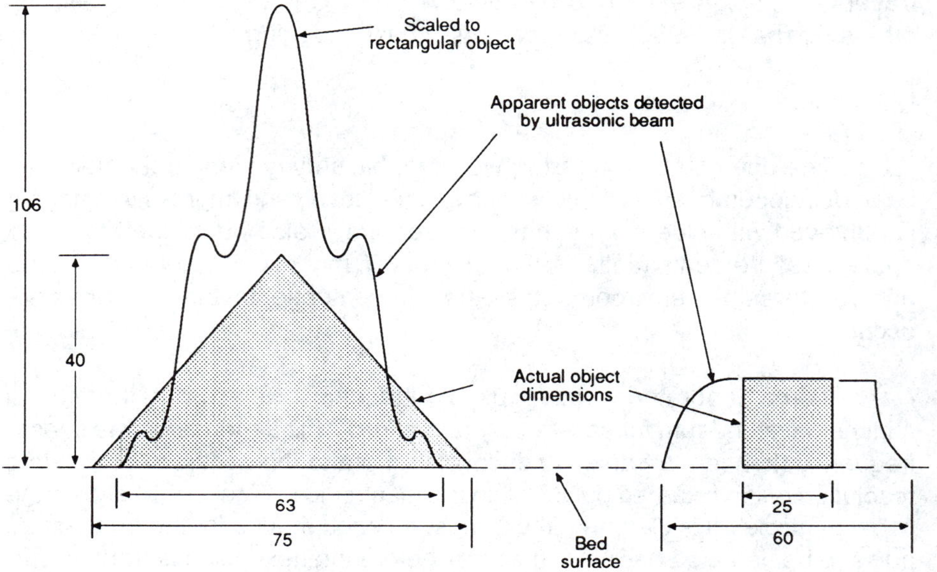

(a) Result from 'shiny' surfaces (mm)

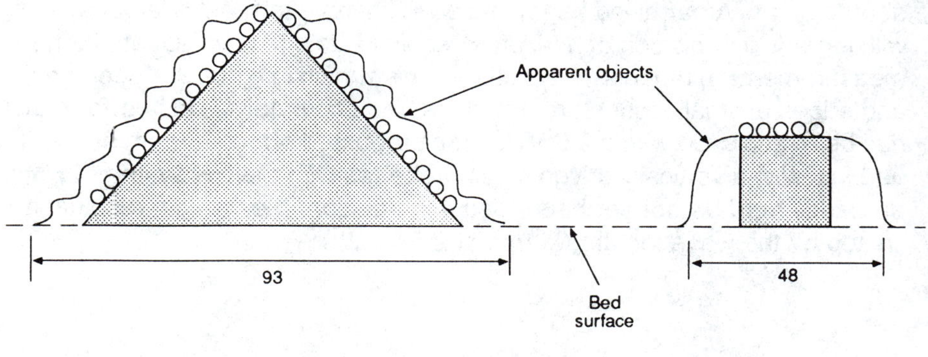

(b) Objects as (a) covered with fine coal dust (mm)

Figure 3

Objects detected by a 10° ultrasonic beam with a source 160mm above the bed f = 2.5 MHz

Laser distance sensors

A wide range of solid state laser displacement sensors have become available in recent years. Some of these have been described by Shepherd (1994). At HR we have used the 'Keyence' type of sensor for bed measurements. The probe was packaged in a waterproof perspex housing and used on a fixed support underwater. A great advantage of this type of sensor is that it can be used with a good "stand off" (distance between the probe and the surface) so that flowing water does not cause scour of the bed. Used in this way, it is not necessary to operate the probe on a vertical servo drive unless the total range required is beyond the range of the sensor.

The method again relies on scattered reflected infra-red light, and will work well on sand but generally requires a good homogeneous light scattering surface. It is not satisfactory on shiny reflecting surfaces. The typical optical path is shown in figure 4, and the calibration is different in water than in air. The horizontal resolution is very good as the projected light spot has a diameter of less than 1mm. Calibration in air was linear but underwater it was non-linear.

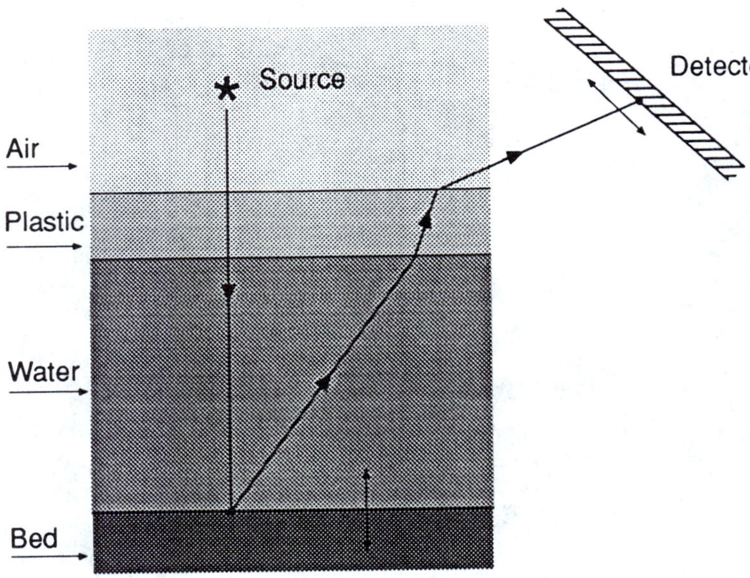

Figure 4 Laser Position Detector

There are other longer range optical 'time-of-flight' laser distance detectors and scanners that are potentially useful, but we have not yet investigated these. They are generally much more expensive than the short range distance detectors.

Touch sensitive detector

This is an extension of the retro-reflective probe described above, and as can be seen from Table 1, the only universal technique. This relies on the use of a 'diving bell' type of attachment as shown in figure 5, attached to the bottom of the optical probe. The very light weight (< 5g) piston has a good reflecting surface on its top, so that the intensity of the collected light is independent of the type of bed material.

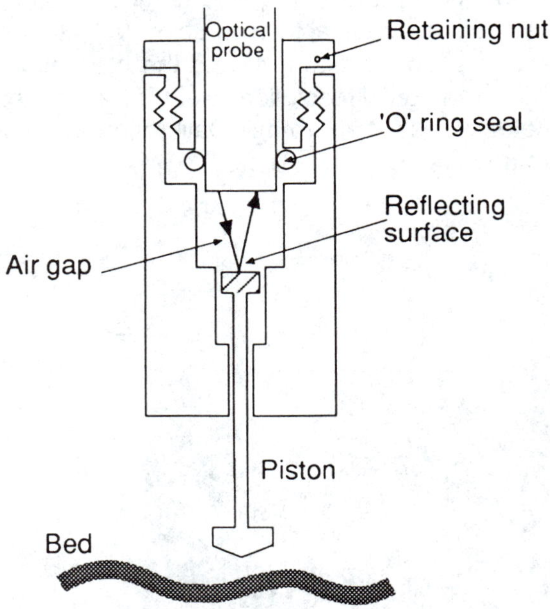

Figure 5 Universal Touch Probe

In operation, the probe is driven down to the bed, the piston tip contacts the bed and deflects by an amount which produces a repeatable level of light reflection at its top. The light level is compared with a fixed threshold which generates the error signal for the probe drive. The air gap is retained when the probe is underwater, so that the optical detection path is unchanged. Use through the air-water boundary has no effect on operation, and any type of bed material can be detected. The weight of the piston does not produce significant bed deformation.

Remote photography

An early example of remote measurements used Moiré fringe techniques (Oesch et al 1985), and a method used for height detection has potential for bed profiling (Nagle et al ca 1988). Early attempts to use digital image analysis on a rock beach at HR was not very successful due to the variations of surface reflection from wet and dry rock faces.

More recently Lowe and Steele at HR have been developing a method suitable for wide area beach contour measurement.

In principle, it consists of arranging suitable lighting to illuminate the edge of the water at the beach so that it produces adequate contrast to provide a good photographic image from an overhead camera. The water level is changed in known increments, and successive photographic images of the water edge produced, which can then be digitised as two coordinates. The values are then subject to numerical calibration based on the location of the data point and the measured projections of vertical calibration rods located at known positions on the model. Future plans include automatic image analysis of the photographic or video based images. Image processing has already been used for surface float tracking in hydraulic models, and it is probable that these techniques will be adopted for beach movement measurement.

Typical Equipment and Results

Application examples were presented at the Conference, together with results. Figure 6 shows a typical result of a a changing profile resulting from wave action. We have used profile measuring equipment for beach erosion studies, sea wall armour research, sediment transport, sand ripple propagation and mud erosion measurements. It is difficult to design universally applicable equipment, as detailed requirements vary with the type of experiment. However, it is possible to make available a range of methods that can be adopted for differing needs.

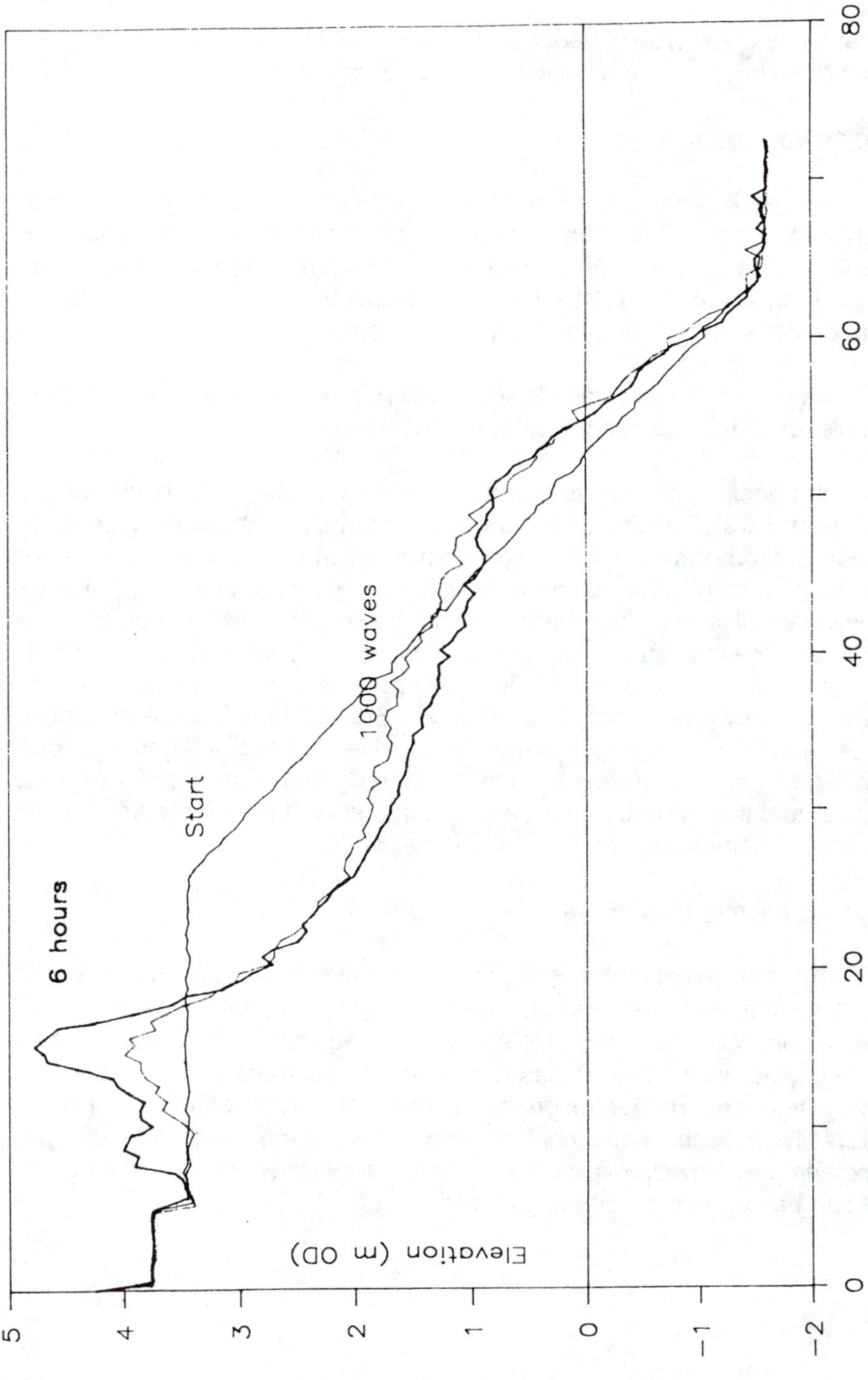

Figure 6 Profile Measurements on a Model Beach

Typical problems include the sag of instrument beam supports, either due to the weight of a moving carriage, or due to the weight of the beam itself. This can be remedied in several ways, for example:

(a) Use of a water surface datum location
(b) Software correction for measured values which are dependent on carriage position
(c) Overhead fixed datum reference compensation.

In most of our applications adequate support for traversing beams can be provided so that vertical errors due to sag effects are less than 1mm.

Conclusions

Measurements of bed profiles in hydraulic modelling have been very popular in recent years. Project specific studies and basic research both demand this type of measurement. The advances in numerical techniques also require experimental data to help prove computational models. Various methods have been used, but we have found that the touch sensitive technique is the only method completely independent of bed material, slope steepness and air-water boundaries. Photographic and image analysis methods are still in development, and solid state lasers will be important in the future.

Acknowledgements

I am greatly indebted to all my colleagues at HR Wallingford, many of whom have contributed to the successful instrument developments. In particular I want to thank Les Smith and Ian Payne.

APPENDIX

References

Anon. Houille Blanche, 7, p 289, 1952.

Delft Hydraulics. PV-07 Profile Indicator, 1985.

Hydraulic Models. ASCE Manual of Engineering Practice No 25, p53, 1942.

Hydraulics Research Station. The Bed Level Plotter. 1968.

Jetté, C.D., Hanes, D. M. Measurements and Model Comparisons of Wave Generated Bedforms. Paper No 196, 25th Int. Conference on Costal Eng. ASCE, 1996.

Limnidyn. SOGREAH, 1985.

Nagle, T. J., Lindenmuth, W. T., Baldwin, B. M., Snyder, G. I. Measurement of the wave pattern generated by a ship model in a towing basin using close range photogrammetry. David Taylor Naval Ship Research and Development Center Bethesda, Maryland, USA.

Oesch, S., Burry, J-M., Schneiter, L. Moiré Topography of Hydraulic Models. IAHR 21st Congress Vol 5, pp 41-46. Melbourne, August 1985.

Richards, K. S., Robert, A. Laboratory Experiments with the HR Multipurpose Profile Follower on a Rippled Sand bed. Dept. of Geography, University of Cambridge, 1986.

Shepherd, I. E. Optical methods for movement measurements (To be published).

Stansfield, D. Underwater Electroacoustics Transducers. Pub. Bath University Press and the Institute of Acoustics. 1991.

Villanueva, L. F. A Novel Technique of Automatic Beach Profile Measurements. IAHR Workshop on Instrumentation for Hydraulics Laboratories. pp 267-275. Canada, August 1989.

Wilkie, M. J., King, R. F. J. A bed level indicator for detecting the boundary of a body of water. Jnl.Sci.Inst. Vol 31, pp 8-9, January 1954.

CHAPTER 246

Asymmetric boundary layer flow above sand ripples under progressive waves

Takao Toue[1], Kazuo Nadaoka[2] and Hidehiro Katsui[3]

Abstract

Boundary layer flow above sand ripples under both oscillatory flow and progressive waves is examined by two dimensional numerical simulation. The simulation shows the difference of the flow between under an oscillatory flow and progressive waves. The boundary layer flow under oscillatory flow is symmetric, on the other hand, the flow under progressive waves is asymmetric.The mechanism of the asymmetric flow is explained by the convection effects of progressive waves.

Introduction

To understand the boundary layer flow above sand ripples is important for the mass transport rate and the sediment transport.The boundary layer flow above sand ripples is characterized by vortices which are generated at ripple crests.The generation, development, movement and dissipation of the vortices should be carefully examined for the understanding of the boundary layer flow.

Many experimental and numerical studies for the boundary layer flow above ripples have been carried out. There are several types of numerical simulation such as the vortex method, the turbulent simulation by k-ε and the direct numerical simulation (DNS) . The discrete vortex method is performed by Longuet-Higgins (1981) , Smith and Stansby (1985) and Ismail and Syuto (1985) . Sato, Uehara and Watanabe (1985) simulated the boundary layer flow using the k-ε model, and Penasantana, Watanabe and Isobe (1990) examined the accuracy of the three types of k-ε turbulent model. Both Blondeaux and Vittori (1990) and Hamanaka and Sato (1992) simulate the boundary layer flow by DNS using the spectral method or the pseudo spectral method, and examine the characteristics of the vorticity.

In most of the numerical studies, the main flow,i.e,the flow above boundary layer, is an oscillatory flow. Nadaoka and Yagi (1988) report the difference of the

---

1 M.S, Senior Research Engineer, Dep. of Coastal and Hydraulics Eng.,Tech. Res. Center,Taisei Corp.
2 Dr. of Eng. ,Professor, Department of Mechanical and Environmental Informatics,Graduate School of Information Science and Engineering, Tokyo Institute of Technology
3 Dr. ,Manager Dep. of Coastal and Hydraulics Eng.,Tech. Res. Center,Taisei Corp.

boundary layer flow between under an oscillatory flow and progressive waves over a horizontal bed. Takigawa, Kim and Gonotani (1989) , and Toue , Nadaoka and Katsui (1993) also show the differences above ripple beds. According to Nadaoka et al (1988) ,the difference are caused by the vertical convective effects of progressive waves. Takigawa et al (1990) showed the differences of the vortex formation between at left side of a ripple and at right under progressive waves, but the shape of the ripples is asymmetric .

The vertical convective effects also exist above symmetric ripples, and the boundary layer flow can be asymmetric under progressive waves.In this report , the boundary layer flow under progressive wave above symetric ripples is numerically simulated , and then the difference of the flow under an oscillatory flow and progressive waves are shown , and finally the mechanism of the asymmetry is also stated.

Methodology of Numerical Simulation

The method of the numerical simulation used is based on the method of Blondeaux et al (1990) or Hamanaka et al (1992) . The basic equations are followings.

$$\omega_t = \{-\Psi_\eta \omega_\xi + \Psi_\xi \omega_\eta + \nu(\omega_{\xi\xi} + \omega_{\eta\eta})\} / J \tag{1}$$

$$\Psi_{\xi\xi} + \Psi_{\eta\eta} = -J\omega \tag{2}$$

where ω is the vorticity, Ψ is the stream function, t is the time, ξ and η are the boundary fitted coordinate system and J is the Jacobian. Sub-scripts denote the differentiation. ξ, η and J are expressed as below.

$$\begin{aligned} x &= \xi + ae^{-\eta}\cos\xi \\ y &= \eta - ae^{-\eta}\sin\xi \end{aligned} \tag{3}$$

$$J = 1 - 2ae^{-\eta}\cos\xi + a^2 e^{-2\eta} \tag{4}$$

where x and y are the horizontal and vertical coordinate in Cartesian coordinate system and a is the ripple height. All valuables are non-dimensionalized by the ripple wave number, $k^* = 2\pi / l^*$ and σ^* is the angular frequency, where l^* is the ripple length.Furthermore, to generate finer meshes near the bottom, the new valuables, ς, is introduced as below.

$$\eta = \eta_T \frac{e^{b\zeta} - 1}{e^b - 1} \tag{5}$$

where η_T is the non-dimensional height of the calculation region, and b is the coefficient to adjust the mesh size.

The lateral boundary condition is the periodic condition, and at the top of the boundary,

$$\omega = 0 \tag{6}$$

$$\Psi = u_\infty n_r \sinh\frac{\eta_T}{n_r}\sin(\frac{\xi}{n_r} - t) \qquad \text{for progressive waves} \qquad (7)$$

$$\frac{d\Psi}{d\varsigma} = u_\infty \sin(t) \qquad \text{for an oscillatory flow} \qquad (8)$$

where u_∞ is the horizontal velocity amplitude at the top of the boundary layer and n_r is the number of ripples in one wave length.At the bottom, the following boundary conditions are used ;

$$\Psi = 0 \qquad (9)$$

$$\omega = -J^{-1}\left(\frac{\partial\varsigma}{\partial\eta}\right)^2 \frac{2\Psi(\Delta\varsigma)}{\Delta\varsigma^2} \qquad (10)$$

Toue et al (1992) analyze the boundary layer flow by this method. They apply the method to high-Reynolds number flows (Re=3400) .The calculation can not reach the steady condition, and the time-averaged flow can not be symmetric under an oscillatory flow above a symmetric ripple.Exact two dimensional simulation can express the two dimensional high Reynolds-number flow, but not the three dimensional flow that we usually observe.The method, however, can simulate the boundary layer flow exactly as long as the Reynolds number is small, around 600,and is quite useful to understand the difference of the boundary layer.

In this study the calculation condition is limited to the low Reynolds number, but the flow has vortex sheddings. According to Hamanaka et al (1992) , the phenomena are controlled by the non-dimensional ripple height, a, Reynolds number, Re, and Strohal's number, S_t, being expressed as below.

$$\text{Re} = 2\pi\frac{u_\infty}{\nu} \qquad (11)$$

$$S_t = \frac{2\pi}{u_\infty} \qquad (12)$$

where ν is the non-dimensional viscous coefficient.

Nadaoka et al (1988) showed the ratio of the representative horizontal velocity, u, to wave celerity , C, u/C , is another control parameter for progressive waves. If u/C is large, the vertical convective effects is large. u/C is also expressed as

$$u/C = u_\infty / n_r \qquad (13)$$

in our formulation.The calculation condition is listed in Table-1.

Table-1 Condition of Caculation(p-1 and p-2 are for verification)
O.F=Oscillatory Flow,P.W=Progressive Waves

	a	u_∞	b	ν	η_T	Re	St	Ψ_o	nr	u/C	
case p-1	0.00	5.0	3.0	0.05	5.0	628					O. F
case p-2	0.00	0.08	3.0	0.0023	5.0	221		0.37	10	0.01	P.W
case 1−0	0.50	5.0	3.0	0.05	5.0	628	1.26				O. F
case 1−1	0.50	5.0	3.0	0.05	5.0	628	1.26	23.11	10	0.50	P.W
case 1−2	0.50	5.0	3.0	0.05	5.0	628	1.26	19.04	5	1.00	P.W
case 1−3	0.50	5.0	3.0	0.05	5.0	628	1.26	24.11	15	0.33	P.W
case 2	0.80	8.0	3.0	0.05	5.0	1005	0.79				O. F
case 3	0.70	7.0	3.0	0.05	5.0	880	0.90				O. F
case 4	0.60	5.0	3.0	0.05	5.0	628	1.26				O. F
case 5	0.75	3.0	3.0	0.03	5.0	628	2.09				O. F

$$\Psi = \Psi_o \sin\left(\frac{\xi}{n_r} - t\right)$$

Verification of Method

To verify the the numerical method, the boundary layer flow above a flat bed are calculated. Fig.1 is the velocity distribution under oscillatory flow. The theoretical value is the 1st order solution. The agreement is quite well. Fig.2 shows the distribution of the vorticity under progressive waves. Fig.2 (a) is the analytical value

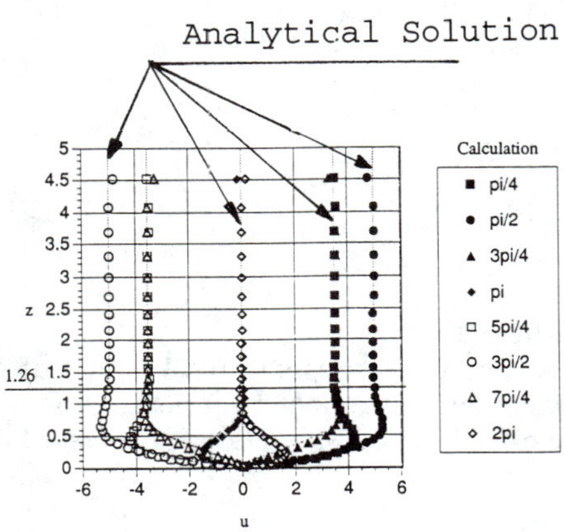

Fig.1 Comparison of Vertical Velocity Profile in Laminar Oscillatory Boundary Layer over Flat Bed between Numerical Simulation and Analytical solution.

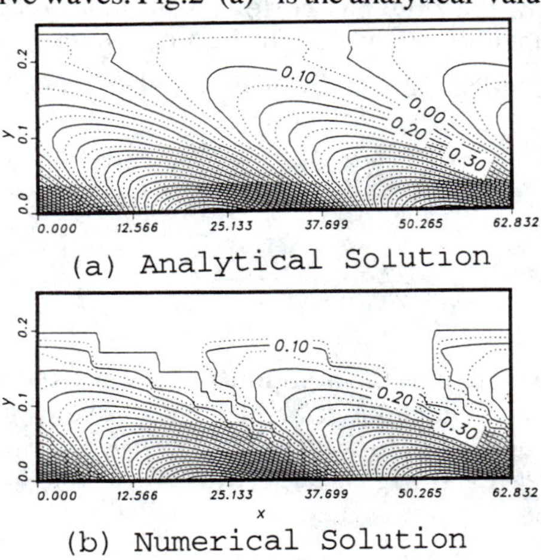

Fig.2 Comparison of Vorticity Profile in Laminar Wave Boundary Layer over Flat Bed between Numerical Simulation and Analytical solution.

, and Fig.2 (b) is the numerical one . Both have quite good agreements.

Time Averaged Vorticity and Stream Function

The time averaged vorticity is shown in Fig.3. A pair of the symmetric vorticity cell can be seen in the case of oscillatory flow, i.e. case1-0, case4 and case5. On the other hand, the profile of the vorticity is asymmetric in case1-1, case1-2 and case1-3 which are the cases of progressive waves.When the flow become high Reynolds number, the calculation could not obtain steady state. Those cases are case2 and case3, and they are not analyzed further.

Asymmetry of the time averaged vorticity means the existence of the residual components of the circulation in a ripple, and the residual components cause steady currents. Fig.4 is the time averaged stream function in case 1-1.The profile of the stream function above all ripples should have been the same for each ripple, but since the calculation has not reached the complete steady condition, there are small differences among the stream functions.In spite of the minor differences, unidirectional and uniform steady currents can be seen near the top boundary.

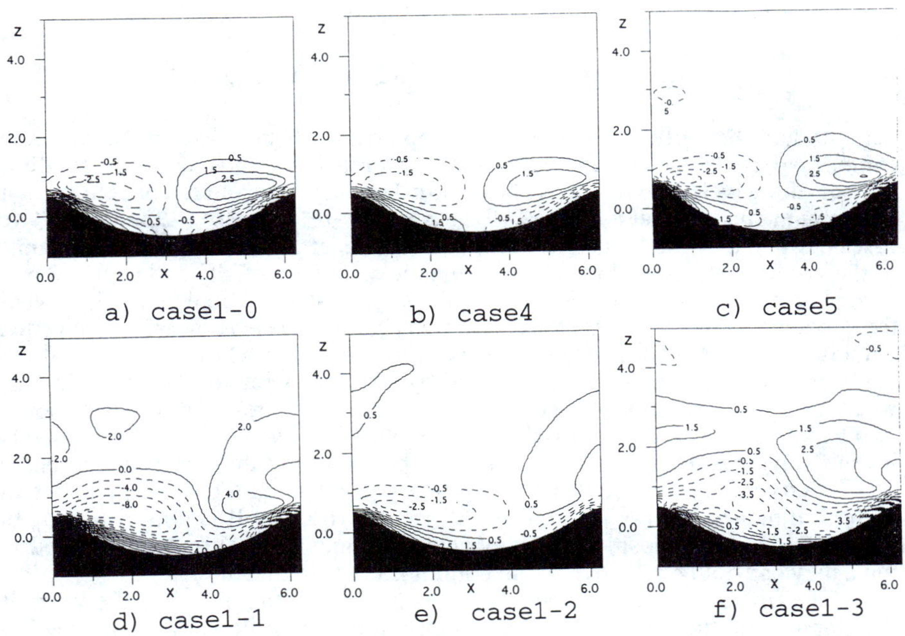

Fig.3 Time Averaged Vorticity(a) case1-0 b) case4 c) case5 d) case1-1 e)case1-2 f) case 1-3)

General Feature of Vorticity Development under Progressive Waves

Fig.5, Fig.6 and Fig.7 are the instant vorticity and the stream function for case1-1,case1-2 and case1-3 respectively in one wave length for $t = \pi/4$. In the figures, waves propagates from the left to the right, and when t=0, the horizontal

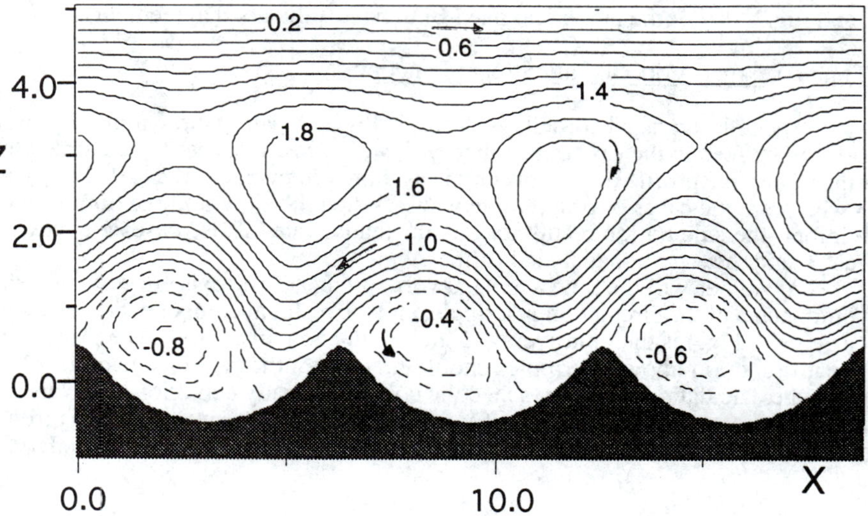

Fig.4 Time Averaged Stream Function for case1-1

velocity and the vertical velocity on the top boundary are 0 and maximum at the center respectively. Since the calculation has reached almost a steady condition after the third cycle , the figure can be regarded as the time evolution of both vorticity and stream function at every $\pi/5$ for case1-1, $2\pi/5$ for case1-2 and $2\pi/15$ for case1-3 when they are looked from the right to the left. It is clear that the dynamics of the vorticity are quite different between the positive vorticity and negative one.

The positive vorticity, denoted by A, is generated at Ripple 9 in Fig.5 where the horizontal velocity is negative.The vorticity A is convected upward by the vertical velocity of waves, and finally disperse during a half period of the waves. On the contrary, the negative vorticity, denoted by B, is generated at Ripple 5 in Fig.5 where the horizontal velocity is positive.The vorticity B is pressed down to the bottom also by the vertical velocity of waves. The another positive vorticity can be found at Ripple 3 on the bottom. That is generated by the velocity of the vortex which consists of vorticity B.The vorticity B and the positive vorticity on the bottom interact and then diminished each other. In other word, the vorticity B is diminished by the bottom friction. The positive vorticity convected upward is at far from the bottom, thus, they can not be affected by the bottom friction so much.

The differences mentioned above is related to the phase of the vorticity developing and the direction of the velocity of the main flow. When the positive vorticity is developing, the horizontal velocity is negative and the vertical velocity being directed upward. On other hand, the negative vorticity is developing, the horizontal velocity is positive and the vertical velocity being directed downward.

The phase when the vorticity stop developing is not clear, but it would be the phase during the velocity decreasing in its amplitude.These processes seem to occur at the phase between ripple 2 and 3 for the positive velocity, and 7 to a8 for the negative velocity.

The similar phenomena can be seen in case 1-2 and 1-3 in Fig.6 and 7. In case 1-2, the convection is stronger than case 1-1 and case 1-3. Thus, the positive vorticity A disperse and the negative vorticity B dissipate more quickly. In case 1-3, the

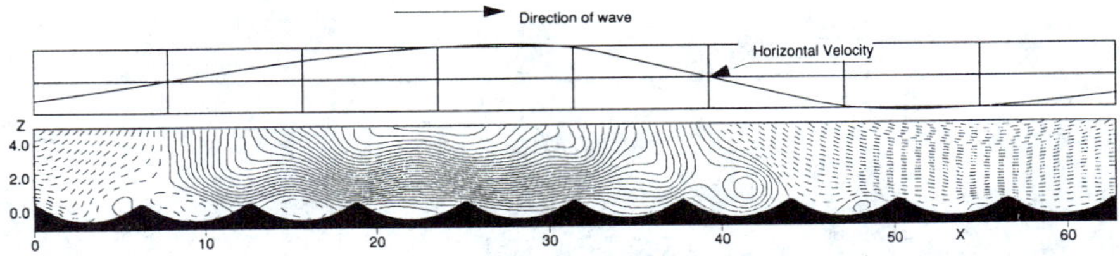

a) Stream Function

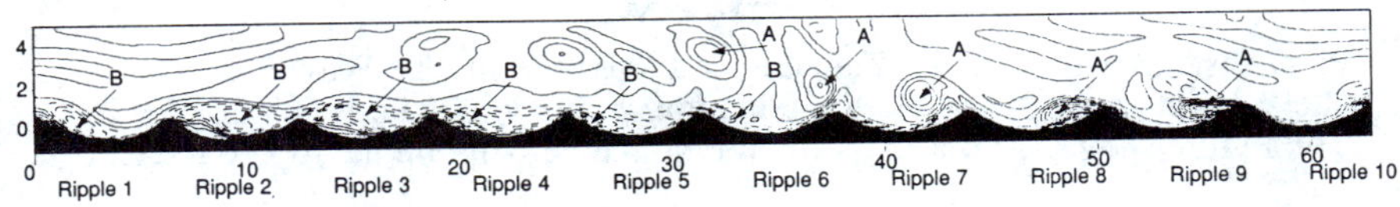

b) Vorticity

Fig.5 Instantaneous Stream Function and Vorticity in One Wave Length at t= π /4 for case1-1 (a) Stream Function $\Delta\ \psi$ =1.0 b) Vorticity $\Delta\ \omega$ =2.0 , Top Figure shows the phase of the horizontal velocity on the top boundary)

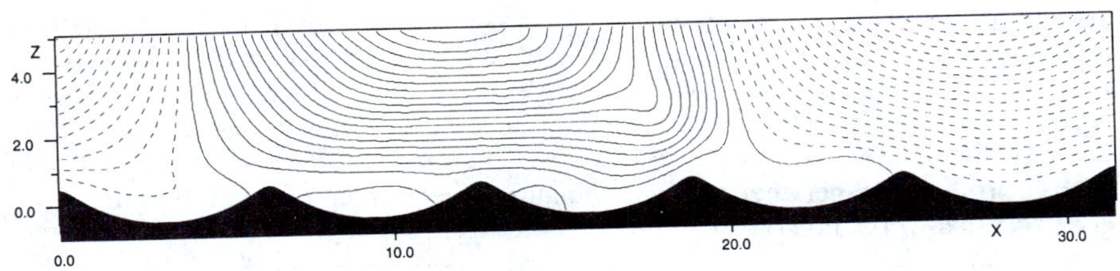

a) Stream Function

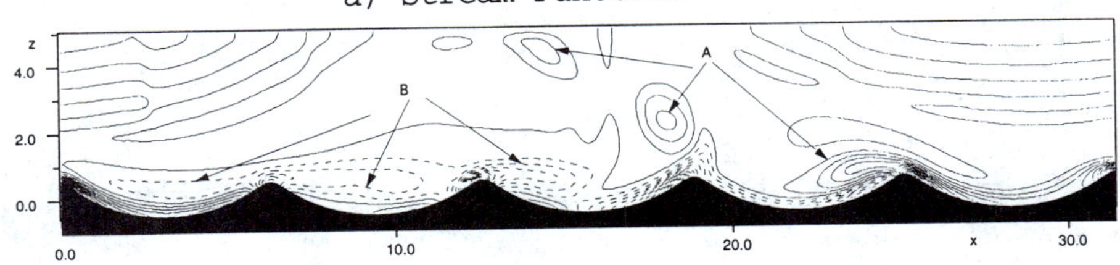

b) Vorticity

Fig.6 Instantaneous Stream Function and Vorticity in One Wave Length at t= π /4 for case1-2 (a) Stream Function $\Delta\ \psi$ =1.0 b) Vorticity $\Delta\ \omega$ =2.0 , Top Figure shows the phase of the horizontal velocity on the top boundary)

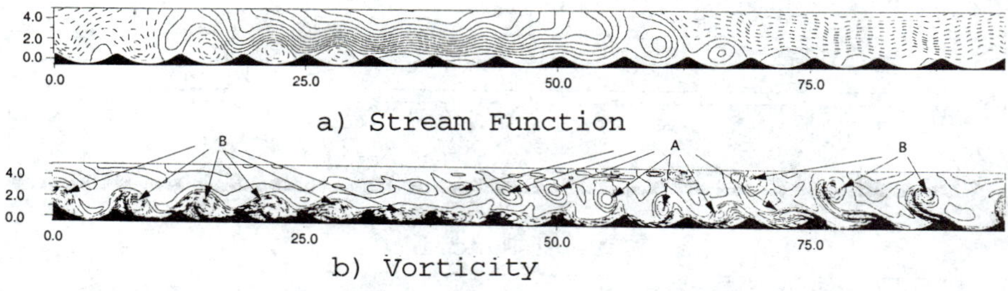

Fig.7 Instantaneous Stream Function and Vorticity in One Wave Length at t= π /4 for case1-3 (a) Stream Function $\Delta\ \psi$ =1.0 b) Vorticity $\Delta\ \omega$ =2.0 , Top Figure shows the phase of the horizontal velocity on the top boundary)

convection effects is week, thus, the negative vorticity B still remain when the positive velocity change the direction. The survived negative vorticity B and the part of the generating positive vorticity interact, and then they goes upward by their own self-propelling velocity.

Vorticity Flux over Ripple Crest

To examine this more quantitatively, the vorticity flux, $u\omega$, in case1-1 that goes across a ripple crest is examined next.In Fig.8, the vertical profiles of $u\omega$ at the left crests of ripples are shown. Fig.8 (a) is the vorticity flux when the velocity is positive, and (b) is for negative velocity. From Fig.8 (a) , when the horizontal velocity is positive, the negative vorticity enters near the bottom and the positive vorticity enters near the top boundary. On the contrary, in the case of the negative velocity, the positive vorticity goes into near the bottom. By comparing the strength of the vorticity flux between the positive and the negative, it is clear that $u\omega$ which goes across a ripple crest is also quite asymetric.

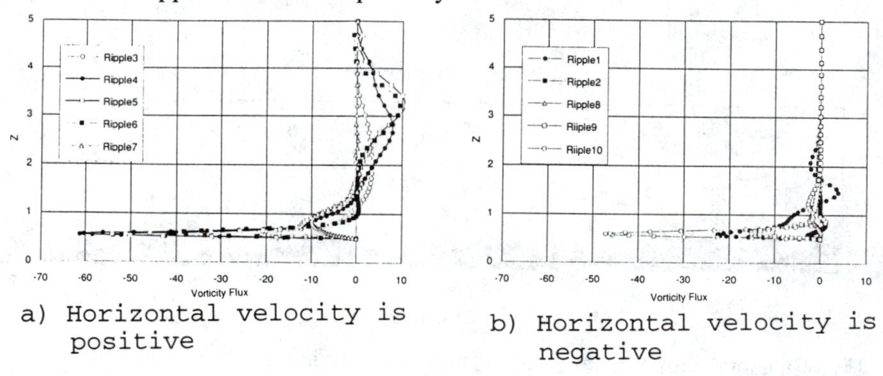

Fig.8 Vertical Distribution of Vorticity Flux at Ripple Crests (a) The horizontal velocity is positive, b) The horizontal velocity is negative.)

Circulation in A Ripple

The time evolution of the circulation that a ripple obtain for case1-1 is shown in Fig.9 from t=0 to t=π of the first wave cycle. The time evolution of the circulation for case 1-0 is also shown in Fig.9. The reason why the results of the first cycle is examined is to eliminate the effect of the vortex that is formed at a half cycle before and the effects of the steady currents. The circulation are divided into four layers which are shown in Fig.10 to understand how the vorticity which is generated at the bottom transfer to upword. The circulation of Fig.9 (b) is the value for Ripple 6 in Fig.5, and that of Fig.9 (c) is the value for Ripple 1. Fig.9 (a) is the circulation under an oscillatory flow, (b) and (c) are the circulation under progressing waves. In Fig.9 (a) and (b), the horizontal velocity is positive, and the horizontal velocity is negative in Fig.9 (c).

In Fig.9 (b) and (c), the circulation in the lowest layer (Layer 1) has some value from the beginning, because even though the main flow is exactly zero at t=0 at the crest, the velocity is not zero above the other part of the ripple.The circulation in Layer 1 develop first since the source of the vorticity exists on the bottom. The value of circulation monotonously decreases in (b) and that monotonously increase in (c).The value has a peak before $\pi/2$, since the opposite sign of vorticity is generated on the bottom beneath a vortex. This opposite sign of vorticity is caused by the velocity of the vortex that has oppsite sign to the main flow. Before the peak of the circulation in Layer 1, the circulation in Layer 2 and Layer 3 start to develop, because the vorticity near the bottom disperses and the vortex at a ripple begins to form. The characteristics of the circulation mentioned above are common in Fig.9 (a) to (c).

The circulations of (b) and (c) in Layer 1 are the same up to around t = 0.5, though the sign is different. After this stage, the differences between (b) and (c) can be seen. The circulation of (c) develops more than that of (b), because the negative vorticity remains near the bottom, thus, the positive vorticity beneath the negative vorticity develops. In Fig.9 (c), the positive vorticity is convected upward

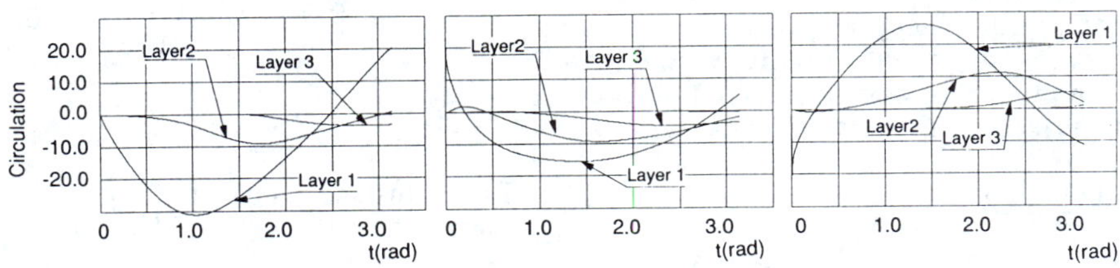

a) Under Oscillatory Flow(case1-0)
b) Under Progressive Waves; Horizontal Velocity is positive(case 1-1)
c) Under Progressive Waves; Horizontal Velocity is negative(case 1-1)

Fig.9 Evolution of Circulation in a Ripple (a) under Oscillatory Flow b)under Progressive Waves and Positive Horizontal Velocity c) under Progressive Waves and Negative Horizontal Velocity)

, thus, the negative vorticity on the bottom beneath the vortex is small. Also comparing the circulation in Layer 2 and 3 between (b) and (c), the circulation in (b) has a peak earlier than (c) has a peak, and the peak vaule of (b) is smaller than that of (c). This means that the positive vorticity goes upward more easily than the negative vorticity as mentined before.

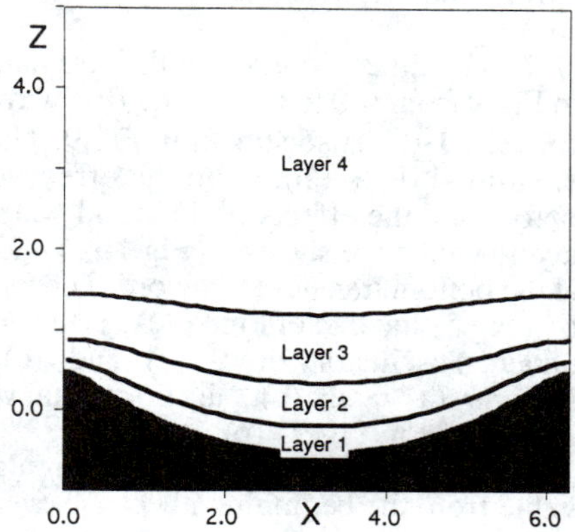

Fig.10 Explanation of Layers

Difference of Phase in Separation

From the difference of the time development of the circulation among Fig.9 (a), (b) and (c), there might be a phase difference in vorticity separation. Fig.11 is the vorticity distribution at $t = \pi/4$ of Ripple 5 and 1 in case1-1 in the first cycle. When the velocity is negative, the vorticity forms along the bottom topography, but the vorticity starts to separate when the velocity is positive. The separation time (phase) of the vorticity is earliest when the velocity is positive under progressive waves (see the right side of the bottom figure of Fig.11). The separation time under oscillatory flow is next and that is latest when the velocity is negative under progressive waves. This is caused by the difference of the attack angle of the velocity against the ripple crest. In the vortex evolution, i.e., t=0 or $\pi/2$, the positive velocity of the progressive waves attack with a steep angle, and the angle becomes milder in oscillatory flow, and much milder for negative velocity. This is also effects of the vertical velocity.

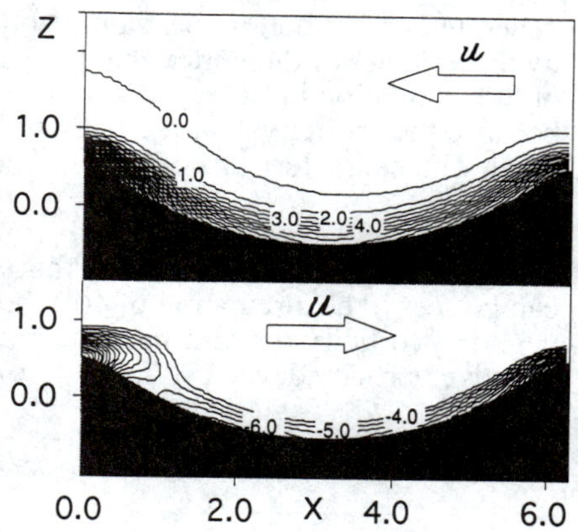

Fig.11 Instantaneous Vorticity Distribution at t= π /4 of The First Wave Cycle in case1-1(Top figure is for Ripple 5 where the horizontal velocity is negative and the bottom is for Ripple 1 where the horizontal velocity is positive.

Concluding Remarks

Boundary layer flow above ripples under both an oscillatory flow and progressive waves is examined by two dimensional numerical simulation. The simulation shows the differences between the flow under an oscillatory flow and progressive waves. The boundary layer flow under an oscillatory flow is symmetric, on the other hand, the flow under progressive waves is asymmetric. The mechanism of the asymmetric flow is explained by the vertical convection of waves. Furthermore, there are the differences of the phase of the vortex separation.

The Reynolds number of the flow is small and the relation of main flow and the dimension of ripples are neglected in this simulation because this calculation is to examine the mechanism of the asymmetric flow. It is necessary to calculate the high Reynolds number flow by including the appropriate turbulent model for further quantitative discussion.

References

Blondeaux,P. and G.Vittori (1991) : Vorticity dynamics in an oscillatory flow over a rippled bed,J.Fluid Mech.,Vol.226,pp.257-289

Ismail Aydin and Nobuo Shuto (1988) : Flow field over sand ripples due to combined wave-current motion, Proceeding of Coastal Eng. in Japan ,Vol.32,pp.278-282 (in Japanese)

Nadaoka ,K., and H.Yagi (1988) : Numerical Simulation of Wave Boundary Layer by Vortex Method, Proceedings of Coastal Eng. in Japan ,Vol.35,pp.16-20 (in Japanese) .

P.G.Pena-Santana,A.Watanabe and M.Isobe (1990) : Numerical Simulation of Oscillatory Turbulent Boundary Layer Flow and Suspended Sediment Movement,Proceedings of Coastal Eng. in Japan ,Vol.37,pp.264-268 (in Japanese)

Sato, S., H. Uehara and A. Watanabe (1986) : Numerical simulation of the oscillatory boundary layer flow over sand ripples by a k-e turbulence model, Coastal Eng. in Japan, Vol. 29, pp.65-78

Smith,P.A. and Stansby,P.K. (1985) : Wave-induced bed flows by a Lagrangian vortex scheme. J.Comput. Phys.60, 489-516

Takigawa,K.,N.H.Kim and S.Gonotani (1989) : Numerical Analysis of Turbulent Boundary Layer Flow over Arbitrary bottom Topography under Wave Motion,Proceedings of Coastal Eng. in Japan ,Vol.36,pp.1-5 (in Japanese) .

Toue,T., K.Nadaoka and H.Katsui and K.Ito (1993) : Two dimensional boundary layer flow under progressive waves above sand ripple,Proceedings of Coastal Eng. in Japan ,Vol.40,pp.1-5 (in Japanese) .

CHAPTER 247

Undertow Profiles in the Bottom Boundary Layer under Breaking Waves

Daniel T. Cox[1] and Nobuhisa Kobayashi[2]

ABSTRACT: The vertical distribution of the mean shear stress inside the surf zone is compared to the terms in the time-averaged horizontal momentum equation using one set of laboratory measurements of the free surface elevations and fluid velocities u and w induced by regular waves spilling on a plane slope. The vertical distribution of the eddy viscosity is estimated directly from the measured mean shear stress and velocity. The shear stress distribution in the surf zone is shown to vary linearly with depth until the bottom boundary layer where it reached a nearly constant, negative value. The shear stress variation in the transition region differs distinctly from the inner surf zone. The vertical variation of $\overline{uw}$ is shown to be small outside the surf zone except near the bottom. Inside the surf zone, it is shown that the $\overline{uw}$ term of the horizontal momentum equation is likely to be important in the transition region and that its importance diminishes in the inner surf zone. The vertical distribution of the eddy viscosity has a form which is small near trough level, increases to a maximum value about one-third of the depth below trough level, and then decreases toward the bottom. The eddy viscosity in the middle of the bottom boundary layer is two orders of magnitude less than the eddy viscosity in the interior.

INTRODUCTION

Detailed cross-shore sediment transport models require accurate prediction of near-shore currents, particularly an accurate description of the flow in the bottom boundary layer under breaking waves. The horizontal component of the mean cross-shore flow, or undertow, is driven by the vertical imbalance of the depth-varying momentum flux and the depth-uniform pressure gradient due to the setup (Dyhr-Nielsen and Sørensen, 1970). In the last decade, several undertow models have been developed based on this concept and vary in the degree of empiricism, choice of boundary conditions, specification of eddy viscosity, and treatment of the bottom boundary layer (e.g., Dally and Dean, 1984; Svendsen, 1984; Stive and Wind, 1986; Svendsen *et al.*, 1987; Deigaard *et al.*, 1991; Stive and de Vriend, 1994). Cox *et al.* (1994) presented for the first time detailed laboratory measurements of the instantaneous velocities and shear stresses in the bottom boundary layer of about 1 cm thickness under breaking waves. In the present paper, these measurements are used to compare the vertical distribution of the mean shear stress to the terms of the time-averaged horizontal momentum equation and to estimate the vertical distribution of the eddy viscosity used to express the mean shear stress in terms of the mean horizontal velocity.

The time-averaged horizontal momentum equation may be written

$$\frac{\partial}{\partial z}\left(\frac{\overline{\tau}}{\rho}\right) = g\frac{\partial \overline{\eta}}{\partial x} + \frac{\partial}{\partial x}\left(\overline{u^2} - \overline{w^2}\right) + \frac{\partial \overline{uw}}{\partial z} = \alpha(x, z) \qquad (1)$$

[1]Asst. Prof., Ocean Engrg. Prog., Dept. of Civil Engrg., Texas A&M Univ., College Station, TX 77843-3136, USA; Email: dtc@arlo.tamu.edu; Tel: +1 409 862 3627; Fax: +1 409 862 1542.

[2]Prof. and Assoc. Dir., Center for Applied Coastal Res., Univ. of Delaware, Newark, DE 19716, USA; Email: nk@coastal.udel.edu; Tel: +1 302 831 8044; Fax: +1 302 831 1228.

where x is the horizontal coordinate, positive onshore; z is the vertical coordinate, positive upward with $z = 0$ at the still water level (SWL); τ is the shear stress; ρ is the fluid density; η is the free surface elevation; u and w are the horizontal and vertical components of the fluid velocity; g is the gravitational acceleration; α is the vertical gradient of $\overline{\tau}/\rho$; and the overbar denotes time-averaging over the wave period. The subscripts x and z are used later to denote differentiation with respect to the horizontal and vertical coordinates. The mean shear stress $\overline{\tau}$ is generally related to the vertical gradient of the mean horizontal velocity $\overline{u}$ through an eddy viscosity ν_t in the following form:

$$\frac{\overline{\tau}}{\rho} = \nu_t \frac{\partial \overline{u}}{\partial z} \tag{2}$$

In this paper, each of the terms in (1) are analyzed using the laboratory measurements of Cox *et al.* (1994). The eddy viscosity is estimated using (2) with the measured values of $\overline{\tau}/\rho$ and $\overline{u}$.

The data of Cox *et al.* (1994) consists of free surface and velocity measurements for the case of regular waves spilling on a rough, impermeable 1:35 slope. The roughness consisted of a single layer of sand grains glued on the plane slope. The median diameter of the sand grains is $d_{50} = 0.10$ cm. The velocity profiles were measured at six vertical lines in the cross-shore direction to include the shoaling region seaward of breaking (denoted L1), the break point (L2), the transition region (L3) and the inner surf zone (L4, L5, and L6). Each of the measuring lines include measuring points at a fraction of the grain height above the rough, fixed bottom. Table 1 lists basic statistics of the data set where x is the cross-shore coordinate with $x = 0$ at L1 and $x = 980$ cm at the still water shoreline, H is the wave height, $\overline{h}$ is the local water depth including the setup, k is the wavenumber, η_{min} is the trough level, and $\overline{\eta}$ is the setup. The wave period is $T = 2.2$ s, and the free surface statistics are based on the phase-average of 50 waves.

Table 1: Measuring line locations and free surface statistics for L1 to L6.

Line. No.	x (cm)	H (cm)	$\overline{h}$ (cm)	$k\overline{h}$	η_{min} (cm)	$\overline{\eta}$ (cm)	$[\overline{\eta}_x]_m$ $\times 10^3$	$[\overline{\eta}_x]_B$ $\times 10^3$	Q_s (cm^2/s)
L1	0	13.22	27.60	0.4982	-3.88	-0.30	-0.6	-0.5	-88
L2	240	17.10	20.64	0.4265	-3.60	-0.44	0.0	—	-99
L3	360	12.71	17.56	0.3917	-2.82	-0.05	3.3	4.6	-148
L4	480	8.24	14.38	0.3529	-2.33	0.20	3.3	3.1	-114
L5	600	7.08	11.51	0.3144	-1.60	0.75	3.3	3.3	-70
L6	720	5.05	8.50	0.2690	-0.82	1.13	3.3	3.5	-45

ANALYSIS OF HORIZONTAL MOMENTUM EQUATION

The shear stress appearing in (1) is estimated from the measurements assuming $\overline{\tau}/\rho = -\overline{\sigma_{uw}}$ where $\overline{\sigma_{uw}}$ is the time-average of the phase-averaged covariance of the measured horizontal and vertical velocity components (Cox *et al.*, 1995). Fig. 1 shows the vertical variation of $\overline{\tau}/\rho$ below trough level at the four measuring lines inside the surf zone. Outside the surf zone, $\overline{\tau}/\rho \simeq 0$ even in the bottom boundary layer (Cox *et al.*, 1995).

The mean water level and trough level corresponding to $\overline{\eta}$ and η_{min} of Table 1 are indicated in the figure with the long and short horizontal lines, respectively. The vertical variation of $\overline{\tau}/\rho$ below trough level increases from the transition region (L3) to the inner surf zone (L4). For the three measuring lines in the inner surf zone, the vertical variation of $\overline{\tau}/\rho$ is fairly linear and decreases with depth, as shown in Okayasu *et al.* (1988); but $\overline{\tau}/\rho$ is negative and almost constant in the bottom boundary layer. In the transition region, $\overline{\tau}/\rho$ also decreases linearly with depth, but the thickness for which $\overline{\tau}/\rho$ is negative is much larger.

The variations of $\overline{\tau}/\rho$ in the bottom boundary layer are shown in detail in the bottom panel of Fig. 1. A second vertical coordinate z_b is introduced and is related to the vertical coordinate z by $z = (z_b - d)$ where d is the depth below the SWL. This figure clearly shows that $\overline{\tau}/\rho$ is non-zero and negative near the bottom in the inner surf zone and could be modeled as depth-invariant near the bottom. The solid circle at $z_b = 0$ indicates $\overline{\tau}_b/\rho$ computed by a linear regression analysis assuming a logarithmic profile of the mean velocity in the bottom boundary layer (Cox and Kobayashi, 1996). The thick line indicates the 95% confidence interval from the regression analysis. These points are added as a check of the $\overline{\tau}/\rho$ estimates near the bottom. The temporal variations of the bottom shear stress τ_b were discussed in Cox *et al.* (1996).

The solid line in the upper and lower panels is a piecewise continuous line approximated by a constant shear stress in the bottom boundary layer and a linear best-fit line through the interior points. The constant shear stress was computed using a number of measured points, j, above the bottom as listed in Table 2. The number j was determined at each measuring line by choosing all of measured values for which $\overline{\tau}/\rho < 0$ near the bottom. The best-fit line through the interior points was estimated using the interior points from the jth point to the trough level. Because of the large scatter in $\overline{\tau}/\rho$ in the interior and because of a regression analysis minimizing the squared error, it was necessary to use a weighting function which emphasized the jth point by counting it ten times. The jth point is indicated in the figure with an open circle.

Table 2 lists the average shear stress in the bottom boundary layer, denoted $\overline{\tau}_b$. These values are reasonably close to the values obtained by Cox and Kobayashi (1996) as indicated in Fig. 1. Table 2 lists the vertical gradient of $\overline{\tau}/\rho$, denoted α, for the interior points using the weighted best-fit line. The uncertainty using a 95% confidence interval is indicated in parenthesis. The value of α without the weighting, denoted α_*, is also indicated in Table 2 with the 95% confidence interval. The values of α and α_* are similar, and the effect of the weighting is to change the intercept of the best-fit line rather than its slope. The confidence intervals, however, were affected since the weighting function effectively doubled the number of points in the regression analysis; hence, the uncertainties based on the unweighted regression analysis are adopted hereafter as realistic estimates. For completeness, Table 2 lists an estimate of the boundary layer thickness δ defined here as the location where the shear stress changes sign from negative to positive. The boundary layer thickness is approximately $\delta \simeq 0.9$ cm in the inner surf zone, similar to the values obtained by Cox *et al.* (1996). At L3, the shear stress changes sign at a much higher elevation, $\delta \simeq 5$ cm, which may not give a good indication of the actual boundary layer thickness in the transition region.

The setup gradient $\overline{\eta}_x$ is the dominant term on the right-hand-side of (1) in the surf

Table 2: Shear stress estimates and gradients for L3 to L6.

Line. No.	j	$\overline{\tau}_b/\rho$ (cm^2/s^2)	α (cm/s^2)	α_* (cm/s^2)	δ (cm)
L3	14	-1.1	1.1 (0.2)	1.1 (0.6)	4.5
L4	10	-0.8	2.6 (0.5)	2.3 (1.0)	1.0
L5	9	-0.5	1.9 (0.3)	1.7 (0.5)	0.7
L6	8	-0.6	2.8 (0.3)	2.6 (0.5)	0.8

zone. The estimates of $\overline{\eta}_x$ are listed in Table 1, where the m subscript is used to denote the gradient estimated from the measured $\overline{\eta}$. For L1, $\overline{\eta}_x$ is estimated by a forward difference of the measured values at L1 and L2. For L2, no estimate is given, although $\overline{\eta}_x \simeq 0$ at the break point (e.g., Bowen *et al.*, 1968). For L3 to L6, $\overline{\eta}_x$ is estimated using a linear regression which includes an additional measuring value of $\overline{\eta} = 1.45$ cm at $x = 840$ cm (Cox *et al.*, 1995). The regression analysis indicated

$$\overline{\eta}_x = (3.3 \pm 0.6) \times 10^{-3}, \quad \gamma_{xy}^2 = 0.988$$

where the uncertainty is estimated by a 95% confidence interval and γ_{xy}^2 is the square of the correlation coefficient. To check the accuracy of $\overline{\eta}_x$ estimated from the measurements, comparisons were made with $\overline{\eta}_x$ estimated using the analytical expression in Bowen *et al.* (1968). These analytical values are listed in Table 1 and are indicated with the B subscript. The agreement is good except at L3 where the analytical $\overline{\eta}_x$ may be overpredicted.

The second term on the right-hand-side of (1) involves the horizontal and vertical velocities, u and w, which can be expressed as the sum of the orbital velocity, denoted with a tilde, and the mean velocity, denoted with an overbar,

$$u = \tilde{u} + \overline{u} \quad \text{and} \quad w = \tilde{w} + \overline{w} \tag{3}$$

The time-average of the squared velocities may be written

$$\overline{u^2} = \overline{\tilde{u}^2} + \overline{u}^2 \quad \text{and} \quad \overline{w^2} = \overline{\tilde{w}^2} + \overline{w}^2 \tag{4}$$

Undertow models based on (1) generally assume that $u \simeq \tilde{u}$ and $w \simeq \tilde{w}$. As a further simplification, it is often assumed that $\overline{u^2}$ is depth-invariant and that $\overline{w^2}/\overline{u^2} << 1$. Fig. 2 shows the vertical variation of $\overline{u^2}$, $\overline{u}^2$, and $\overline{w^2}$ for L1 to L6. This figure indicates that $\overline{u^2}$ is nearly constant over depth, except very near the bottom and that $\overline{w^2}/\overline{u^2} << 1$ is a reasonable assumption, except near the trough level. Outside the surf zone, the approximation of $u \simeq \tilde{u}$ is also quite good since $\overline{u}^2 \simeq 0$ everywhere below trough level. Inside the surf zone, however, the magnitude of $\overline{u}^2$ is sufficiently large compared to $\overline{u^2}$. Table 3 lists the depth-averaged values of $\overline{u^2}$, $\overline{\tilde{u}^2}$, and $\overline{u}^2$ where the a subscript denotes averaging from the trough level to the bottom. Outside the surf zone, $[\overline{u}^2]_a/[\overline{u^2}]_a \simeq 0.05$, and inside the surf zone, $[\overline{u}^2]_a/[\overline{u^2}]_a \simeq 0.25$. Table 3 also indicates that the vertical velocity component is negligible both outside and inside the surf zone, $[\overline{w^2}]_a/[\overline{u^2}]_a \simeq 0.05$, except near trough level where $[\overline{w^2}]_{tr}/[\overline{u^2}]_{tr} \simeq 0.15$.

Table 3: Depth-averaged values and horizontal gradients of $\overline{u^2}$, $\overline{\tilde{u}^2}$, and $\overline{u}^2$ for L1 to L6.

Line	$\left[\overline{u^2}\right]_a$	$\left[\overline{\tilde{u}^2}\right]_a$	$\left[\overline{u}^2\right]_a$	$\frac{\partial}{\partial x}\left[\overline{u^2}\right]_a$	$\frac{\partial}{\partial x}\left[\overline{\tilde{u}^2}\right]_a$	$\frac{\partial}{\partial x}\left[\overline{u}^2\right]_a$	$\frac{\left[\overline{w^2}\right]_a}{\left[\overline{u^2}\right]_a}$	$\frac{\left[\overline{w^2}\right]_{tr}}{\left[\overline{u^2}\right]_{tr}}$
No.	$(cm/s)^2$	$(cm/s)^2$	$(cm/s)^2$	(cm/s^2)	(cm/s^2)	(cm/s^2)		
L1	473	460	13	0.09	0.01	0.07	0.080	0.20
L2	494	464	30	-0.4	-0.9	0.4	0.071	0.19
L3	410	307	103	-0.8	-1.0	0.2	0.054	0.16
L4	309	220	89	-0.5	-0.4	-0.2	0.036	0.11
L5	260	202	58	-0.5	-0.4	-0.2	0.031	0.09
L6	183	135	48	-0.5	-0.4	-0.2	0.022	0.07

Fig. 3 shows the cross-shore variation of the depth-averaged values of $\overline{u^2}$, $\overline{\tilde{u}^2}$, and $\overline{u}^2$ with the horizontal gradients. The estimates of the horizontal gradients are listed in Table 3. At L1, L2 and L3, the gradients were computed using finite differences as follows: L1, forward difference with $(L2 - L1)/(\Delta x)$ with $\Delta x = 240$ cm from Table 1; L2, central difference with $(-L1-3L2+4L3)/(6\Delta x)$ with $\Delta x = 120$ cm; and L3, central difference with $(L4 - L2)/(2\Delta x)$ with $\Delta x = 120$ cm. The horizontal step size is large given that the vertical resolution of the measuring points is only a few centimeters in the interior and a few millimeters near the bottom. The gradients for L4, L5 and L6 were computed using linear regression which gave

$$
\begin{aligned}
(\overline{u^2})_x &= -0.5 \pm 0.3 \text{ cm/s}^2, \quad \gamma_{xy}^2 = 0.98 \\
(\overline{\tilde{u}^2})_x &= -0.4 \pm 0.5 \text{ cm/s}^2, \quad \gamma_{xy}^2 = 0.90 \\
(\overline{u}^2)_x &= -0.2 \pm 0.2 \text{ cm/s}^2, \quad \gamma_{xy}^2 = 0.93
\end{aligned}
$$

Although there is large uncertainty in the horizontal gradient estimates, the analysis based on the depth-averaged values of $\overline{u^2}$, $\overline{\tilde{u}^2}$, and $\overline{u}^2$ indicates that the approximation of $u \simeq \tilde{u}$ is reasonable for estimating the horizontal gradient of the $\overline{u^2}$ term in (1) in the surf zone.

Deigaard and Fredsøe (1989) indicated the importance of $\overline{uw}$ appearing in the third term on the right-hand-side of (1). Fig. 4 shows the temporal variation of uw with u and w at five elevations $z = -7.9$, -15.9, -23.9, -26.9, and -27.8 cm, indicated (a) to (e), respectively, for L1. In this figure, uw is reduced by a factor of 10 to facilitate plotting, and the phases have been adjusted such that the zero upcrossing of the free surface elevation is at $t = T/4 = 0.55$ s. The figure shows that the variation in uw is large, especially in the upper portion of the water column $z = -7.9$ cm (a) where uw reaches a maximum value of $uw \simeq 500$ cm^2/s^2. Near the bottom at $z = -27.8$ cm (e), uw is small since the vertical velocity component is approximately zero.

Fig. 5 shows the vertical variation of the time-averaged value of uw below trough level for L1 to L6. The time-averaged values for the five elevations in Fig. 4 are indicated by a solid circle for L1. The magnitude of $\overline{uw}$ is 10 cm^2/s^2 which is an order of magnitude less than the fluctuations of uw shown in Fig. 4. Outside the surf zone, $\overline{uw}$ is nearly depth-invariant except near the bottom. Inside the surf zone, the variation of $\overline{uw}$ is noisy. Nevertheless, crude estimates of the vertical gradients of the $\overline{uw}$ at L3 and L4 may be made. At L3, $(\overline{uw})_z \sim (-2.7)$ cm/s^2 in the lower portion of the water column, and $(\overline{uw})_z \sim 2.3$ cm/s^2 in the upper portion. These estimates were made "by eye" using the curves of Fig. 5 re-plotted on a separate sheet of graph paper. The uncertainties are roughly ±50%. For L4, $(\overline{uw})_z \sim (-2.5)$ cm/s^2 in the lower portion of the water column. In the upper portion, $(\overline{uw})_z$ is approximately zero. For L5 and L6, $(\overline{uw})_z$ is small. Again, these are very crude estimates based on time-averaged quantities that are an order of magnitude less than the fluctuating quantities.

Considering (3), the $\overline{uw}$ term can be expressed as

$$\overline{uw} = \overline{\tilde{u}\tilde{w}} + (\overline{u}\,\overline{w}) \tag{5}$$

and the assumption of a weak current gives $\overline{uw} \simeq \overline{\tilde{u}\tilde{w}}$. Fig. 5 indicates that the $(\overline{u}\,\overline{w})$ term is indeed small below trough level and in the bottom boundary layer outside the surf zone. Inside the surf zone, the $(\overline{u}\,\overline{w})$ term appears to give a substantial contribution to the $\overline{uw}$ term.

Table 4 summarizes the quantification of the terms in the horizontal momentum equation (1) based on the measurements of Cox *et al.* (1994) inside the surf zone. The uncertainties are indicated in parenthesis. The table indicates that for L4 and L6 in the inner surf zone, the shear stress gradient is balanced by the sum of the first two terms on the right-hand-side of (1). The relative error is less than 5% for both L4 and L6 which is well within the uncertainty of the estimates. For L5, the measured shear stress gradient is smaller than for L4 and L6, but the gradient is still balanced by the first two terms within the uncertainties of the measurements. At L3 however, the measured shear stress gradient is smaller than the sum of the first two term by a factor of two. Therefore, it is likely that the $\overline{uw}$ term plays an important role in the transition region and that its importance decreases in the inner surf zone. Outside the surf zone, the gradient of $\overline{uw}$ is small, except in the bottom boundary layer.

Table 4: Summary of terms in horizontal momentum equation for L3 to L6.

Line. No.	$(\overline{\tau}/\rho)_z$ (cm/s^2)	$g\,\overline{\eta}_x$ (cm/s^2)	$\left(\left[\overline{u^2}\right]_a\right)_x$ (cm/s^2)	$(\overline{uw})_z$ (cm/s^2)
L3	1.1 (0.6)	3.2 (0.6)	-0.8 (0.5)	-2.7 to 2.3
L4	2.6 (1.0)	3.2 (0.6)	-0.5 (0.3)	-2.5 to 0
L5	1.9 (0.5)	3.2 (0.6)	-0.5 (0.3)	small
L6	2.8 (0.5)	3.2 (0.6)	-0.5 (0.3)	small

ESTIMATE OF EDDY VISCOSITY

The use of (2) in undertow models requires the specification of an eddy viscosity, ν_t. Several forms have been proposed including depth-invariant, two-layer, parabolic, and piecewise continuous. The eddy viscosity is specified empirically, with a velocity scale related to the wave celerity or to the local turbulent kinetic energy and a length scale related to the local water depth.

In this paper, ν_t is estimated from (2) using the measured values of $\overline{\tau}/\rho$ and $\overline{u}$. The left panel of Fig. 6 shows the vertical variation of the measured $\overline{u}$ with a best-fit cubic spline at L4. The values of the seaward volume flux, Q_s, per unit width listed in Table 1 were calculated by integrating this curve from the bottom where $\overline{u} = 0$ to just above trough level where again $\overline{u} = 0$. The middle panel of Fig. 6 shows the measured values of $\overline{\tau}/\rho$ below trough level and is the same as that plotted at L4 in Fig. 1. The right panel shows the eddy viscosity at L4 estimated from

$$\nu_t = \frac{\overline{\tau}/\rho}{(\overline{u})_z} \tag{6}$$

where $\overline{\tau}/\rho$ is given by the piecewise continuous best fit curve and the vertical gradient of the horizontal velocity is estimated from the best-fit cubic spline using a finite difference approximation with a vertical resolution of $\Delta z = 0.01$ cm. The figure shows that from trough level, the eddy viscosity increases downward to a maximum value about one-third of the depth below trough level and then decreases. The variation in the eddy viscosity is unrealisticly large as the velocity gradient approaches zero at $z \simeq (-12)$ cm. This is the major limitation of the eddy viscosity approach. Near the bottom, the eddy viscosity is small as the velocity gradient becomes large.

Fig. 7 shows the vertical variation of the measured eddy viscosity at the four measuring lines inside the surf zone similar to the right panel of Fig 6. Only those portions for which (6) gave reasonable estimates are shown. The vertical distribution of the eddy viscosity is qualitatively similar for L4, L5, and L6 in the inner surf zone. For L3, reasonable estimates could not be obtained over most of the depth. The magnitude of ν_t in the bottom boundary layer for the four measuring lines is

$$\nu_t = 0.10 \pm 0.05 \text{ cm}^2/\text{s} \quad \text{at} \quad z_b = 0.5 \text{ cm}$$

and is about two order of magnitude smaller than ν_t estimated in the interior. This large decrease is consistent with the assumption made by Svendsen *et al.* (1987).

CONCLUSIONS

The following conclusions are supported in this paper based on the analysis of one set of laboratory measurements of regular wave spilling on a rough, plane slope.

- The mean shear stress distribution in the inner surf zone varies linearly with depth until the bottom boundary layer where it reaches a nearly constant, negative value. The bottom shear stress is also negative. The shear stress variation in the transition region differs distinctly from the inner surf zone in that the negative shear stress extends much higher in the water column.

- The $\overline{w^2}$ term is negligible compared to $\overline{u^2}$ except near the trough level.
- Outside the surf zone, $\bar{u}^2/\overline{u^2} << 1$. Inside the surf zone, $\bar{u}^2$ contributes significantly to $\overline{u^2}$. The approximation of $u \simeq \tilde{u}$, however, is reasonable for estimating the horizontal gradient of $\overline{u^2}$ inside the surf zone.
- The vertical variation of $\overline{uw}$ is small outside the surf zone except near the bottom.
- The shear stress gradient can be balanced by the first two terms of the horizontal momentum equation (1) in the inner surf zone.
- The $\overline{uw}$ term is likely to be important in the transition region and its importance diminishes in the inner surf zone.
- The vertical distribution of the eddy viscosity in the surf zone estimated from the measured mean shear stress and horizontal velocity has a form which is small near trough level, increases to a maximum value about one-third of the depth below trough level, and then decreases toward the bottom.
- The eddy viscosity in the middle of the bottom boundary layer is two orders of magnitude less that the eddy viscosity in the interior.

Dynamic undertow models based on (1) and (2) are difficult to apply with confidence because of the many uncertainties involved in estimating the terms in (1) and the eddy viscosity in (2). An alternative kinematic model is proposed by Cox and Kobayashi (1996). This model relates the horizontal velocity, bottom shear stress, and boundary layer thickness in a simple but general manner.

ACKNOWLEDGMENTS

This work was sponsored by the U.S. Army Research Office, University Research Initiative under contract No. DAAL03-92-G-0116 and by the National Science Foundation under grant No. CTS-9407827.

REFERENCES

Bowen, A., Inman, D., and Simmons, V. (1968) "Wave 'set-down' and setup." *J. Geophys. Res.*, 73, 2569–2577.

Cox, D.T. and Kobayashi, N. (1996) "A kinematic undertow model with a logarithmic boundary layer." *J. Waterway, Port, Coastal, and Ocean Engrg.*, (submitted).

Cox, D. T., Kobayashi, N., and Okayasu, A. (1994) "Vertical variations of fluid velocities and shear stress in surf zones." *Proc. 24th Coast. Engrg. Conf.*, ASCE, 1, 98–112.

Cox, D.T., Kobayashi, N., and Okayasu, A. (1995) "Experimental and numerical modeling of surf zone hydrodynamics." *Rpt. No. CACR-95-07*, Center for Applied Coastal Research, University of Delaware, Newark, Delaware, 293 p.

Cox, D.T., Kobayashi, N., and Okayasu, A. (1996). "Bottom shear stress in the surf zone." *J. Geophys. Res.*, 111 (C6), 14337–14348.

Dally, W.B. and Dean, R.G. (1984). "Suspended sediment transport and beach evolution." *J. Waterway, Port, Coastal, and Ocean Engrg.*, 110 (1), 15–33.

Deigaard R. and Fredsøe, J. (1989). "Shear stress distribution in dissipative water waves." *Coast. Engrg.*, 13, 357-378.

Deigaard, R., Justesen, P., and Fredsøe, J. (1991). "Modeling of undertow by a one-equation turbulence model." *Coast. Engrg.*, 15, 431-458.

Dyhr-Nielsen, M. and Sørensen, T. (1970). "Sand transport phenomena on coasts with bars." *Proc. 12th Coast. Engrg. Conf.*, ASCE, 182–192.

Okayasu, A., Shibayama, T., and Horikawa, K. (1988) "Vertical variation of undertow in the surf zone." *Proc. 21rd Coast. Engrg. Conf.*, ASCE, 478–491.

Stive, M. J. F. and de Vriend, H. J. (1994) "Shear stresses and mean flow in shoaling and breaking waves." *Proc. 24th Coast. Engrg. Conf.*, ASCE, 1, 594–608.

Stive, M.J.F. and Wind, H.G. (1986) "Cross-shore mean flow in the surf zone." *Coast. Engrg.*, 10, 325–340.

Svendsen, I. A. (1984) "Mass flux and undertow in a surf zone." *Coast. Engrg.*, 8, 247-365.

Svendsen, I. A., Schäffer, H. A., and Hansen J. B. (1987) "The interaction between the undertow and the boundary layer flow on a beach." *J. Geophys. Res.*, 92(C11), 11845–11856.

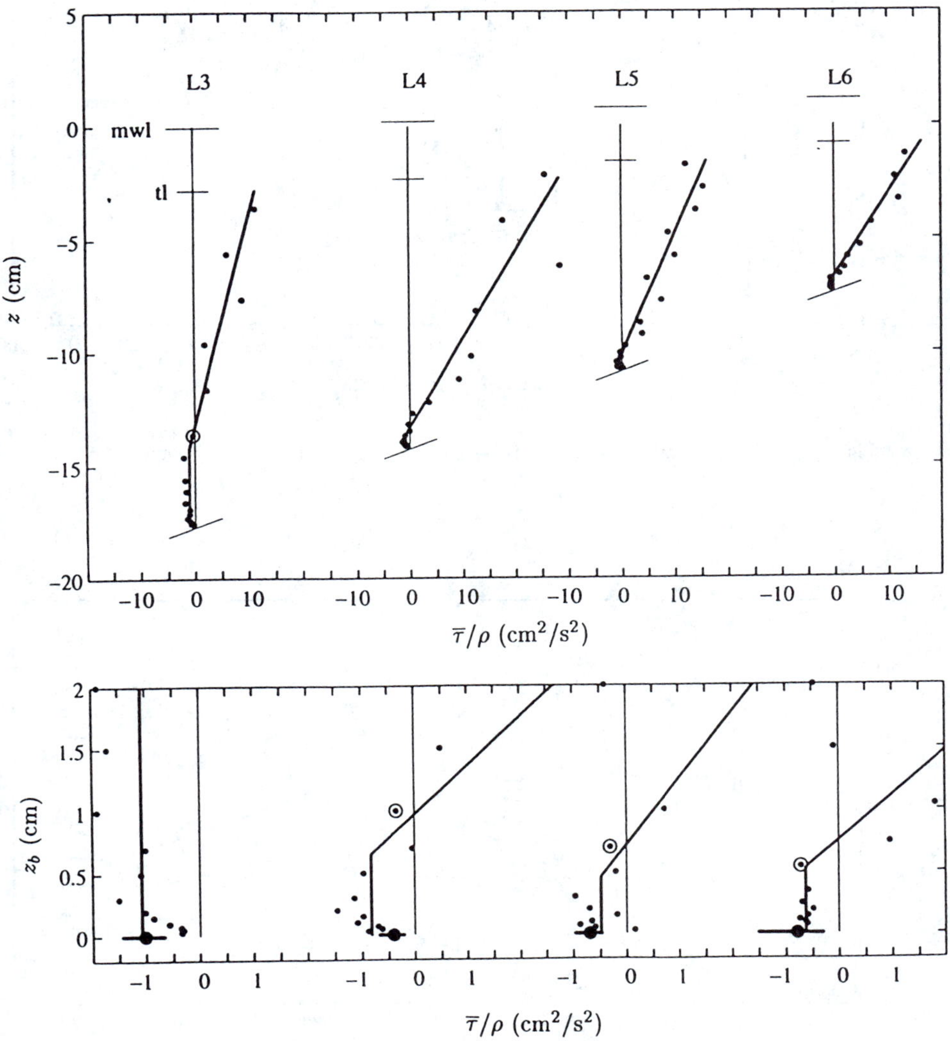

Figure 1: Vertical variation of measured shear stress $\overline{\tau}/\rho$ (•) with piecewise continuous best-fit line (——) (top) and detail in bottom boundary layer with $\overline{\tau}_b$ from linear regression (●) and 95% confidence interval (——) for L3 to L6. Circle indicates jth elevation listed in Table 2.

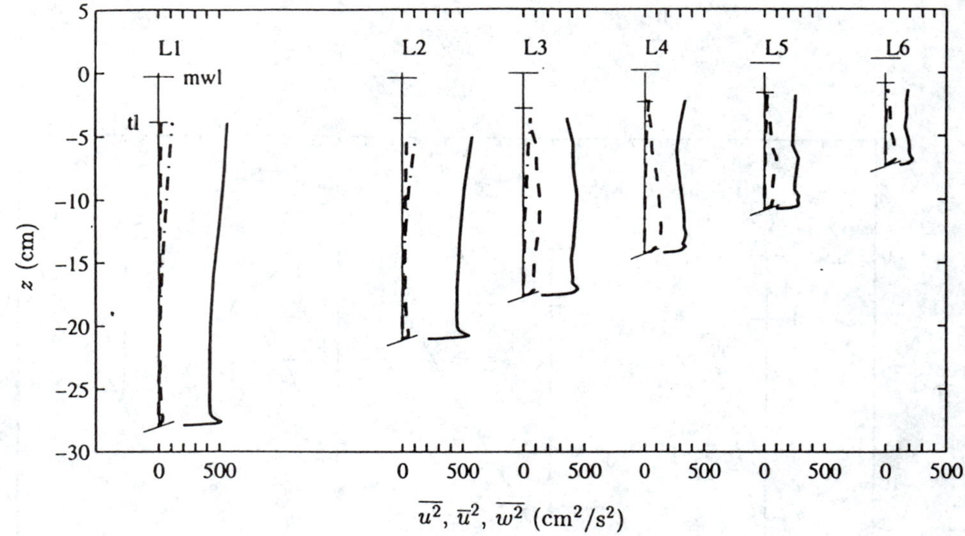

Figure 2: Vertical variation of $\overline{u^2}$ (——), $\overline{u}^2$ (– –), and $\overline{w^2}$ (–·–) for L1–L6.

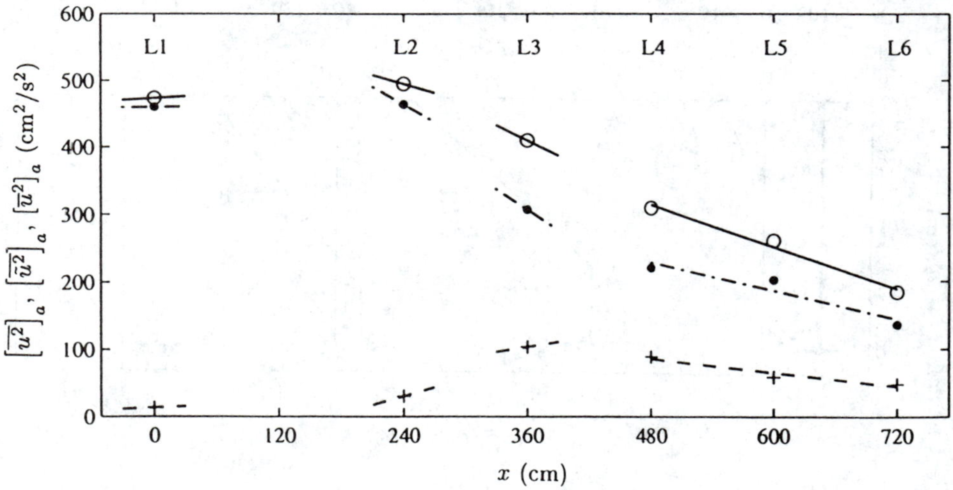

Figure 3: Cross-shore variation of $\left[\overline{u^2}\right]_a$ (○ ——), $\left[\overline{\tilde{u}^2}\right]_a$ (• –·–), and $\left[\overline{u}^2\right]_a$ (+ – –) with gradients for L1–L6.

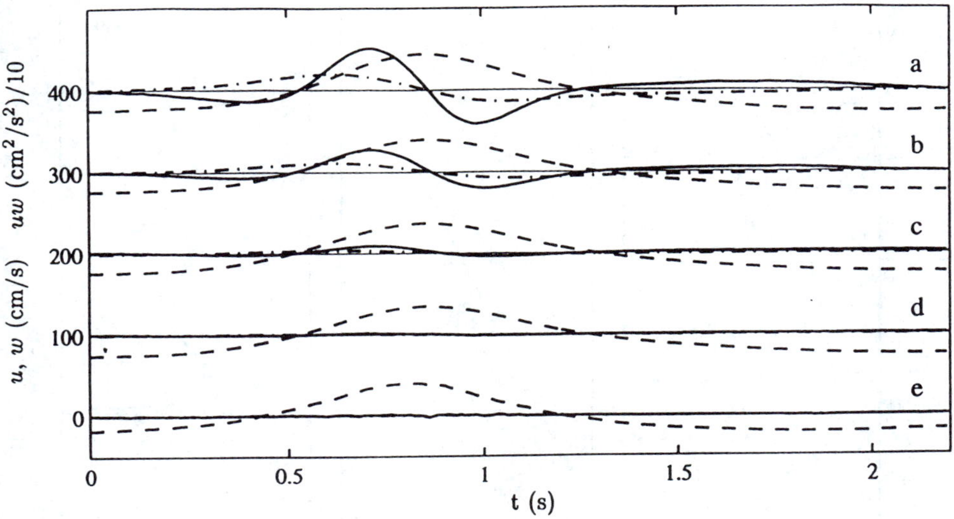

Figure 4: Temporal variation of u (– –), w (–·–), and uw (——) at five elevations for L1. To facilitate plotting, uw is reduced by a factor of 10, and there is an offset of a factor of 100 in the ordinate.

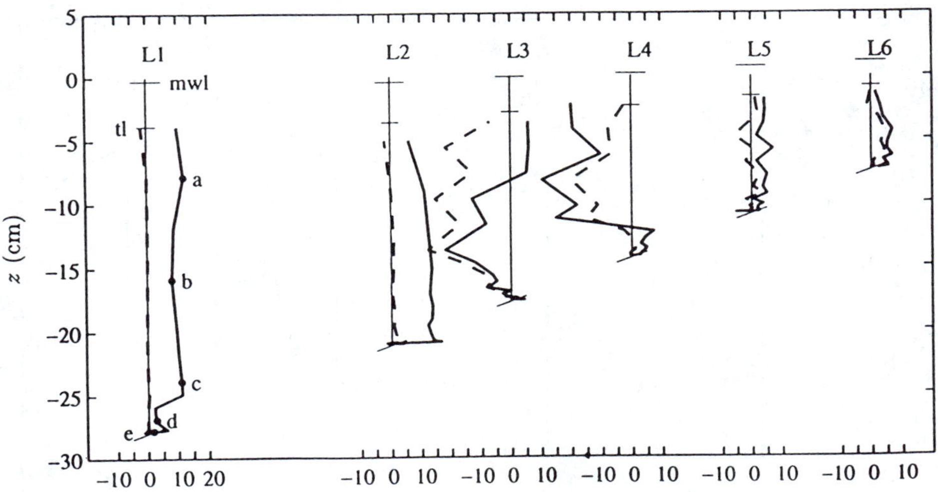

Figure 5: Vertical variation of $\overline{uw}$ (——) and $(\overline{u}\,\overline{w})$ (– –), for L1–L6. Five elevations of Fig. 4 at L1 indicated by solid circle (•).

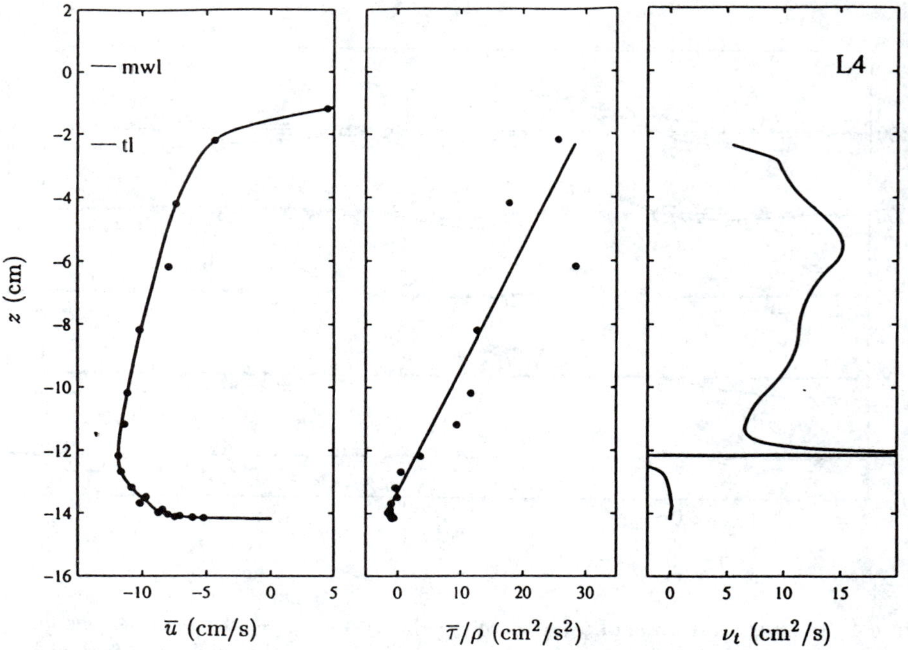

Figure 6: Vertical variation of measured $\overline{u}$ (•) with cubic spline (——) (left); measured shear stress $\overline{\tau}/\rho$ (•) with best-fit curve (——) (middle); kinematic viscosity ν_t (——) (right) for L4.

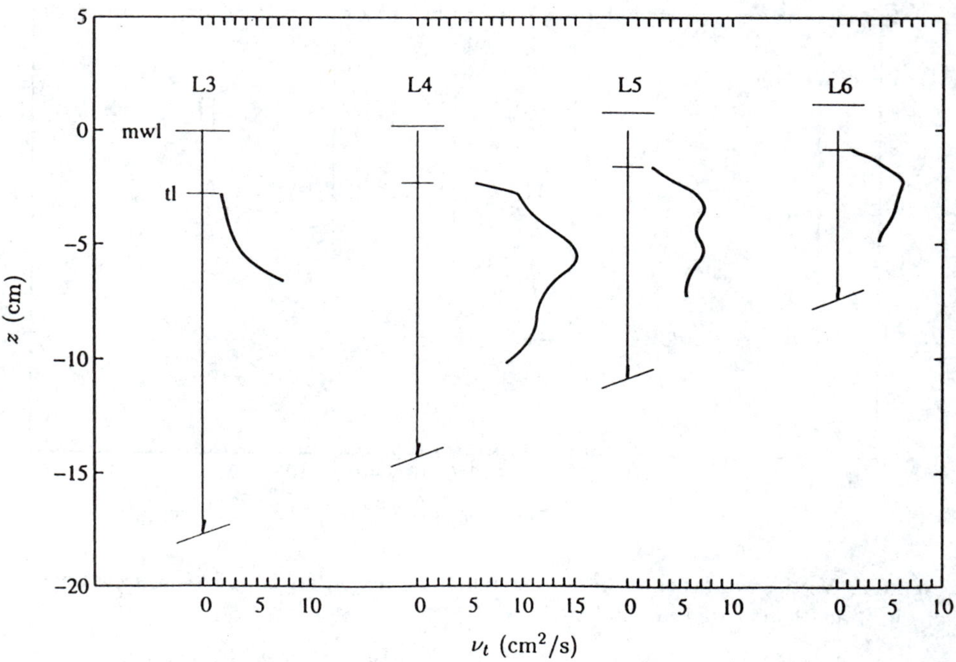

Figure 7: Vertical variation of measured eddy viscosity ν_t (——) for L3 to L6.

CHAPTER 248

A Comparison of Field Observations and Quasi-Steady Linear Shear Instabilities of the Wave Bottom Boundary Layer

D. L. Foster[1], A. J. Bowen[1], R. A. Beach[2] and R.A. Holman[2]

Abstract:

Field observations of near-bed high frequency velocity fluctuations and suspended sediment are compared with predictions of momentary flow stability by a quasi-steady linear instability model. Field observations were made as part of the Duck94 cooperative field experiment and consisted of a vertical array of four hot-film anemometers and 19 fiber-optic backscatter sensors located within the wave bottom boundary layer. Predicted instabilities for the quasi-steady model occur over length scales ranging from 2 cm to 1 m and therefore may be assumed to be a plausible mechanism for the generation of turbulence and suspension of sediment. However with the quasi-steady model, the instabilities are at the same phase as the sediment suspension events, but are occuring too late in the wave phase for them to be generating the observed turbulence.

[1] Department of Oceanography, Dalhousie University, Halifax, Nova Scotia, Canada, B3H 4J1.
internet: diane.foster@dal.ca

[2] Department of Oceanography, Oregon State University, Ocean Admin Bldg 104, Corvallis, Oregon, USA 97331.

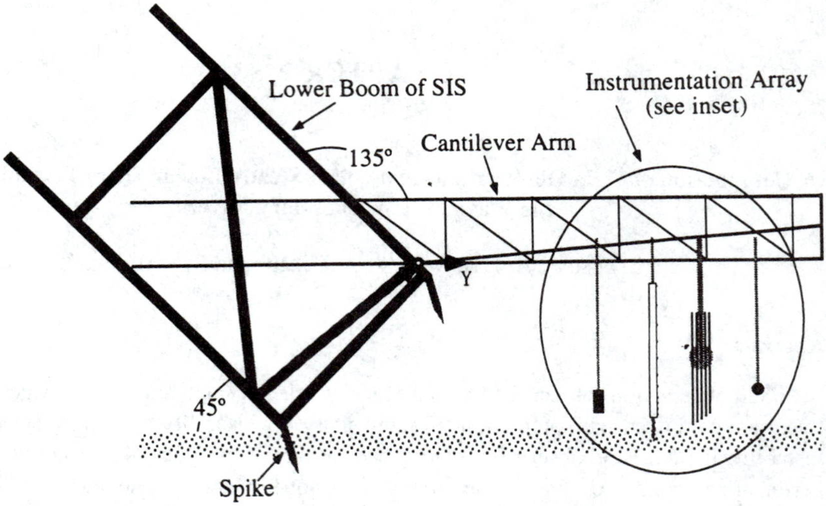

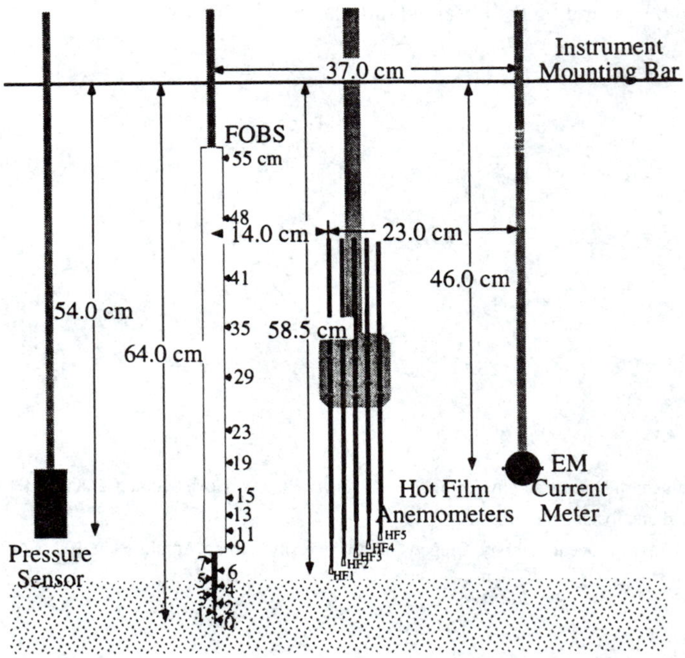

Figure 1: A schematic showing the instrument cantilever as deployed from the Sensor Insertion Sytem on the FRF pier(top panel) and positions of the pressure sensor, FOBS probe, hot film anemometer array, and electromagnitic current meter (bottom panel) (Foster, et al., 1996).

1 Introduction

The seemingly random occurrence of sediment suspension events in the surf zone has long been the topic of research for coastal oceanographers and engineers. These short-lived events are the first step in morphologic evolution and consequently hold significance in our understanding of coastal dynamics. Intuition suggests that suspension events are a result of rapid increases of turbulence levels in the near-bed region. The wave bottom boundary layer is the region of fluid bounding the seabed which responds to the oscillatory surface waves. Within this relatively small region ($\delta \sim$ O(5 cm)), the velocity transitions from the free stream value (u $\sim$ O(100 cm/s)) to zero. Thus far, most of the previous research on the introduction of turbulence to the wave bottom boundary layer has focussed on bottom shear generation. An additional and equally plausible mechanism is the generation of turbulence from within the wave bottom boundary layer through a shear instability. Using a piecewise continuous linear instability model, Foster, et al. (1994) showed that at particular wave phases, the high shears in this region may lead to shear instabilities. With laboratory observations of oscillatory boundary layer flow, Hino et al. (1983) concluded that increased levels of turbulence during the decelerating phase of the flow were triggered by a shear instability.

The objective of this paper is to compare Duck94 field observations of near-bed high frequency velocity fluctuations and suspended sediment with predictions of momentary flow stability by a quasi-steady linear instability model. This paper is organized in the following manner. An overview of the field experiment and data collection technique is given in section 2. The theoretical formulation and solution method of the shear instability problem is presented in section 3. Comparisons between the theory and observations are made in section 4. Finally, the conclusions are given in section 5.

2 Observations

The experiment was conducted at the Army Corps of Engineers, Field Research Facility (FRF) in Duck, NC (Figure 1) on August 17, 1994, as part of the Duck94 field experiment. The significant offshore wave height, angle, and period measured in 8 m water depth were 0.83 m, 50 from the southeast, and 4.54 s, respectively. The observations were made in 2 m water depth on the crest of the bar over flat bed conditions. The instruments were deployed from a cantilever arm attached to the lower boom of the sensor insertion system on the FRF pier, see Figure 1. Near bed velocity observations were made with an array of 5 hot-film anemometers, sampled at 256 Hz, separated with a 1 cm vertical spacing, Figure 1. The velocity outside the wave bottom boundary layer was measured with an electromagnetic current meter. The bed elevation and suspended sediment were measured with a vertical stack of 19 fiber-optic backscatter sensor probe (FOBS), sampled at 16 Hz. The bed elevation is determined by examining each independent sensor for burial. A thorough presentation of field techniques and instrument calibration may be found in Foster et al. (1996).

3 Theoretical Formulation

As formulated in Foster, et al. (1994), let the total velocity be partitioned into perturbation and wave components and given by

$$\begin{aligned} u^T(x,z,t) &= u(x,z,t) + U(z,t) \\ w^T(x,z,t) &= w(x,z,t), \end{aligned} \tag{1}$$

where x and z are the horizontal coordinates (x being positive offshore from the shoreline and z being positive up from the seabed), t is time, u and w are the perturbation velocities, and U is the known horizontal oscillatory background velocity. In a linear instability analysis, it is assumed that perturbation velocities are significantly smaller than the background velocity, simply $u, w << U$. Substituting (1) into the continuity equation, subtracting the linearized wave bottom boundary layer equation and neglecting terms of $O(u^2)$, we obtain the governing equation:

$$\begin{aligned} u_t + Uu_x + wU_z &= -\frac{1}{\rho}P_x + \nu(u_{xx} + u_{zz}) \\ w_t + Uw_x &= -\frac{1}{\rho}P_z + \nu(w_{xx} + w_{zz}), \end{aligned} \tag{2}$$

where P is the pressure and ν is the kinematic viscosity. By assuming conservation of mass and two-dimensional flow, it is possible to represent u and w in terms of the stream function, Ψ, such that

$$\begin{aligned} u &= \Psi_z \\ w &= -\Psi_x. \end{aligned} \tag{3}$$

Substitute the stream function definitions into (2), cross differentiate, and subtract to eliminate the pressure reducing to one equation

$$\left(\frac{\partial}{\partial t} + U\frac{\partial}{\partial x} + \nu\nabla^2\right)\left(\nabla^2\Psi\right) - U_{zz}\Psi = 0. \tag{4}$$

For this investigation, we neglect the effect of viscosity.

The quasi-steady model assumes that the time-varying background profile may be represented with a series of time-independent profiles, such that $U(z,t) \equiv U_n(z)$. At each instant in time, t_n, we assume a solution of the form

$$\Psi_n(x,z,t_n) = \phi_n(z)e^{i(k_n x - \sigma_n t_n)} \tag{5}$$

where k is the real wavenumber and $\sigma = \sigma_r + \imath\sigma_i$ is the complex frequency. The substitution of (5) into (4) results in the well known Rayleigh equation

$$(U_n - c_n)\left(\phi_{n_{zz}} - {k_n}^2\phi_n\right) = U_{n_{zz}}\phi_n. \tag{6}$$

The boundary conditions are specified by assuming that there is no mass flow through the bottom and that the perturbation must decay as $z \Rightarrow H$, where H is the water depth, and are given by

$$\begin{aligned} \phi(z = z_o) &= 0 \\ \phi_z(z = \delta) &= -k\phi(z = \delta). \end{aligned} \quad (7)$$

(6) may be solved with a finite centered difference approximation scheme. For this model, we assume that the background flow may be approximated with a time- and depth-dependent eddy viscosity model (Foster, 1996).The wave bottom boundary layer background flow model was forced with a 256 second record of velocity measured at 13 cm above the bed. The system is discretized into 100 nodal points and solved numerically as an eigenvalue problem (Dodd, et al., 1992), where the wave celerity, $c = \sigma/k$, is the eigenvalue and the stream function, Ψ, is the eigenfunction.. The momentary flow stability is examined by searching for the fastest growing mode at each instant in time over a given wavenumber range of interest.

The concept of momentary stability (Shen, 1961) implies that when $\sigma_\imath > 0$ the disturbances are growing relative to the background flow and the flow may be considered to be "momentarily unstable". Conversely, when $\sigma_\imath < 0$ the disturbances are decaying relative to the background flow and the flow may be considered to be "momentarily stable".

4 Results

Flow stability was determined at 1/8 second intervals using the eddy viscosity model-generated background velocity profile, $U(z,t)$. An example of this can be seen in Figure 2. Notice that at times a and d the flow is stable and that at times b and c, when the free stream flow is decelerating and there exists an infection point in the profile, the flow is unstable. The peak growth rate occurs prior to flow reversal when the near-bed internal shear is largest. The oscillatory perturbation frequency, σ_r, is nearly zero in the two unstable cases, b and c, indicating that there will exist a fixed spatial pattern which grows exponentially in time. This may be a possible explanation for ephemeral ripples which form over a wave phase following a suspension event. An example of the eigenfunction amplitude and phase structures for 3 wavenumbers is given in Figure 3. For cases b and c, the peak phase shift is as large as $\pi/2$.

A 60 second time series of near bed cross shore velocity, turbulent variance, concentration, and predicted growth rates is given in Figure 4. The perturbation growth rates, concentration, and turbulent variance all have 'event-like' structures. Although instabilities are predicted during each 1/2 wave and more often than concentration or turbulent variance events, the larger perturbation growth rates are correlated to the highly correlated concentration and turbulent variance events. The concentration (increasing with increasing proximity to the bed) and turbulent variance (decreasing with increasing proximity to the bed) magnitudes vary over the vertical (Foster, 1996), however the temporal event-like structure is uniform throughout this near-bed region. For further comparisons, we use one FOBS sensor located at 3 cm above the bed and the vertically averaged turbulent variance.

To examine the evolution of the growth rates, concentration, and turbulent variance

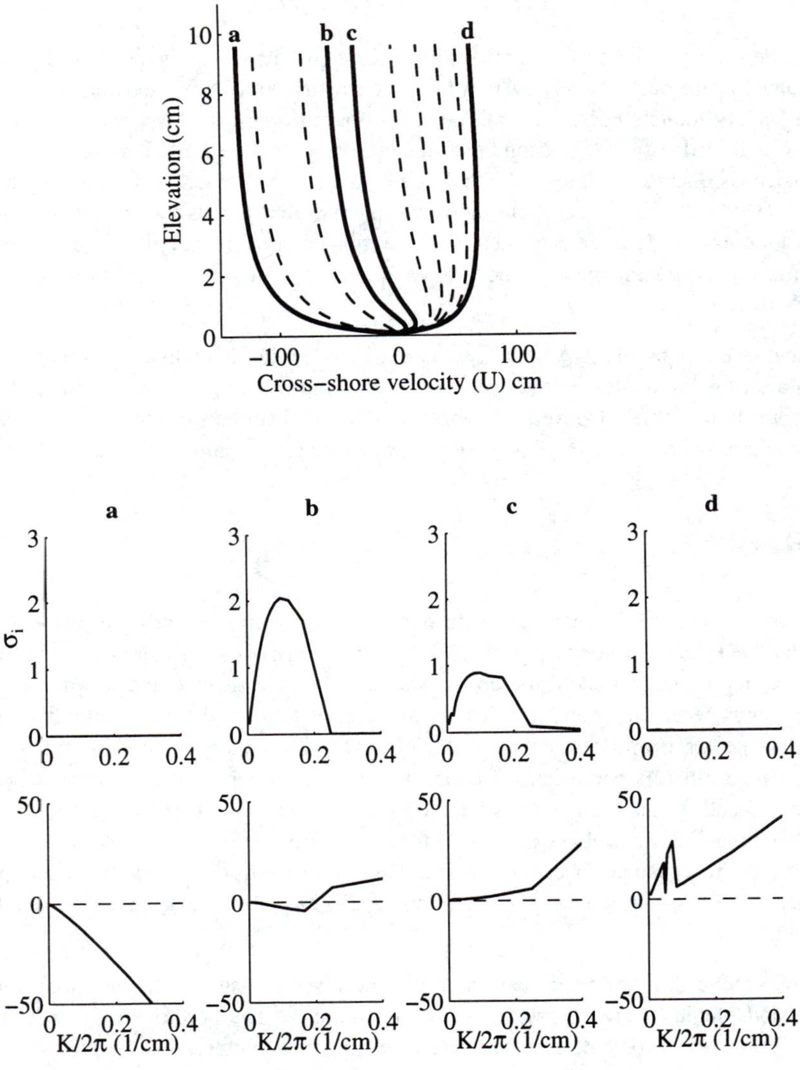

Figure 2: The top panels shows the vertical structure of $U(z,t)$ at 1/4 sec intervals over a 2.5 second record. At 4 of these times, a-d, we have shown the predicted perturbation growth rates, $\sigma_i(k)$ (Hz) (middle panels), and the real frequencies, $\sigma_r(k)$ (Hz) (bottom panels).

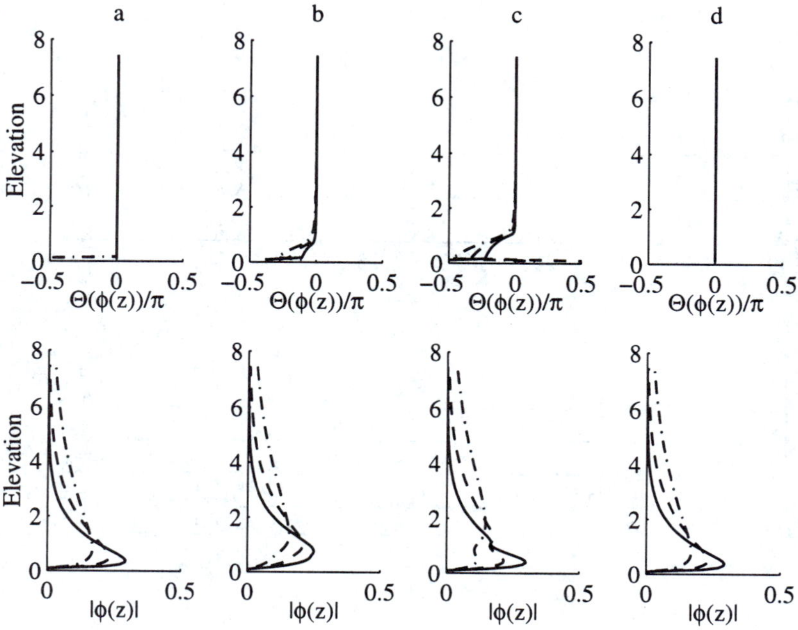

Figure 3: The top panels show the eigenfunction phase, $\Theta\phi(z)$, and the bottom panels show the amplitude, $|\phi(z)|$, for each of the 4 cases (a-d) for 3 selected wavelengths (straight line, 4 cm; dashed line, 10 cm; and dashed-dotted line, 20 cm).

over the course of the wave, we use a phase space averaging (PSA) technique. The PSA technique averages quantities of similar velocity and accelerations bins over a 256 second record. The acceleration is determined in the frequency domain with a tapered 1 Hz low pass cutoff frequency. This technique allows us to evaluate the phase averaged temporal event structure as it corresponds to the magnitude and phase of the background wave. The PSA for the growth rates, concentration, and turbulent variance are given in Figures 5, 6, and 7, respectively. The growth rate PSA shows two peaks, one prior to each of the flow reversal on the larger waves (the outer part of the wave ellipse), with the largest peak occuring after the wave crest when the internal shears are largest. The concentration PSA shows one peak during the flow reversal period following the wave crest prior to flow reversal at relatively the same phase as the growth rate. This peak also occurs at the outer part of the wave ellipse, indicating, not surprisingly, that concentration events are closely associated with the larger waves. The second peak in the concentration PSA occurs following the flow reversal and may be a result of the sediment plumes being advected back through the instrument array. The turbulent variance PSA shows a peak intensity at and following the large wave crest, leading to the concentration and predicted growth rates peaks. This indicates that the turbulence is occuring too early in the wave phase for it to a result from the instabilities predicted here.

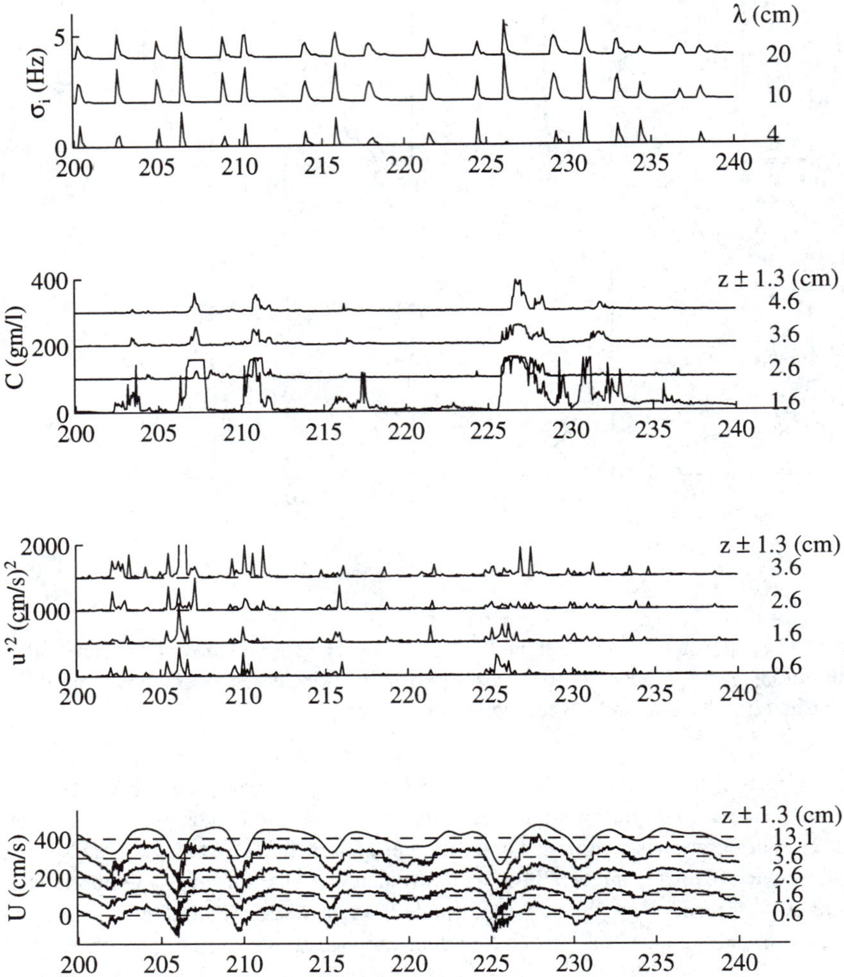

Figure 4: This example 60 sec time series shows near bed velocity as measured by four hot film anemometers and the free stream velocity as measured by an EMCM (bottom panel); high frequency VITA velocity variance at four elevations above the bed (lower middle panel); and concentration at 4 elevations above the bed (upper middle panel). The average distance of each sensor as measured by the FOBS is listed in the right column. The predicted growth rates at three wavelengths (4, 10, 20 cm) listed in the right column are given in the upper panel. Each velocity , variance, concentration, and predicted growth rate time series is offset by 100 cm/s, 500 $(cm/s)^2$, 100 g/l, and 2 Hz, respectively. Offshore directed flow is defined with positive velocities.

5 Conclusions

Predicted instabilities for this quasi steady model occur over the appropriate length scales for them to be a plausible mechanism for the generation of turbulence and suspension of sediment. With this quasi-steady model, the instabilities are at the same phase as the sediment suspension events, but are occuring too late in the wave phase for them to be generating the observed turbulence. Work in progress indicates that considering the time-dependent nature of the background flow may yield growth rates which are larger and earlier in the wave phase.

Acknowledgments

This work was graciously funded by the Office of Naval Research, Coastal Sciences Program, and the Andrew Mellon Foundation. The first author wishes to thank Keith Thompson for his insight into the numerical computations world.

References

Dodd, N., J. Oltman-Shay, and E.B. Thornton (1992), *Shear Instabilties in the longshore current: a comparison of observation and theory*, Journal of Physical Oceanography, 22.

Foster, D.L., R.A. Beach, and R.A. Holman (1994), *Correlation sedminet suspension events and shear instabilities in the bottom boundary layer of the surf zone*, Coastal Dynamics '94, ASCE, Barcelona, Spain.

Foster, D.L., R.A. Beach, and R.A. Holman (1996), *Field Observations of the Wave Bottom Boundary Layer*, submitted to Journal of Geophysical Research.

Foster, D.L. (1996), *Dynamics of the nearshore wave bottom boundary layer*, Ph.D. thesis, Oregon State University.

Hino,M., M. Kashiwayanagi, A. Nakayama, and T. Hara (1983), *Experiments on the turbulent statistics and the structure of a reciprocating oscillatory flow*, Journal of Fluid Mechanics, 131, 363-400, .

Rayleigh, L., and J.W. Strutt (1880), *On the stability, or instability of certain motions*, Proceedings of the London Mathematical Society, 11, 57-70.

Shen, S.F.(1961), *Some Considerations on the laminar stability of incompressible time-dependent basic flows*, Journal of Aerospace Science, 28, 397-404 and 417.

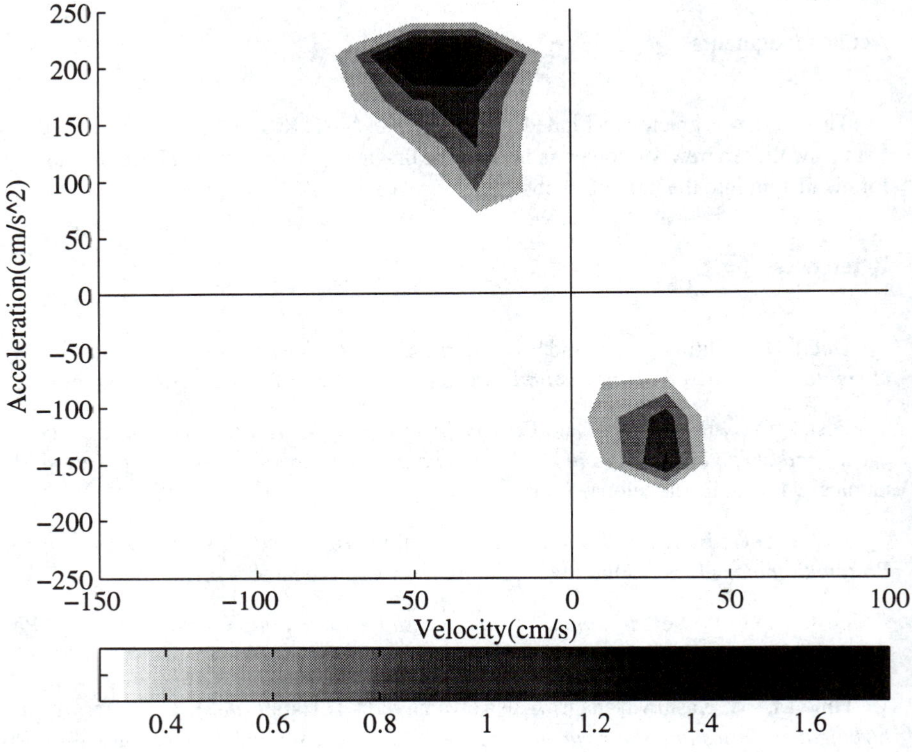

Figure 5: The phase space average of perturbation growth rate shows the background flow phases where the flow is momentarily unstable. The velocity and acceleration are binned at 20 cm/sec and 40 cm/sec ^{2} intervals, respectively. Note that the flow moves clockwise around the figure, and that zero acceleration and maximum negative velocity corresponds to the wave crest. The units of the grayscale colorbar are in Hz.

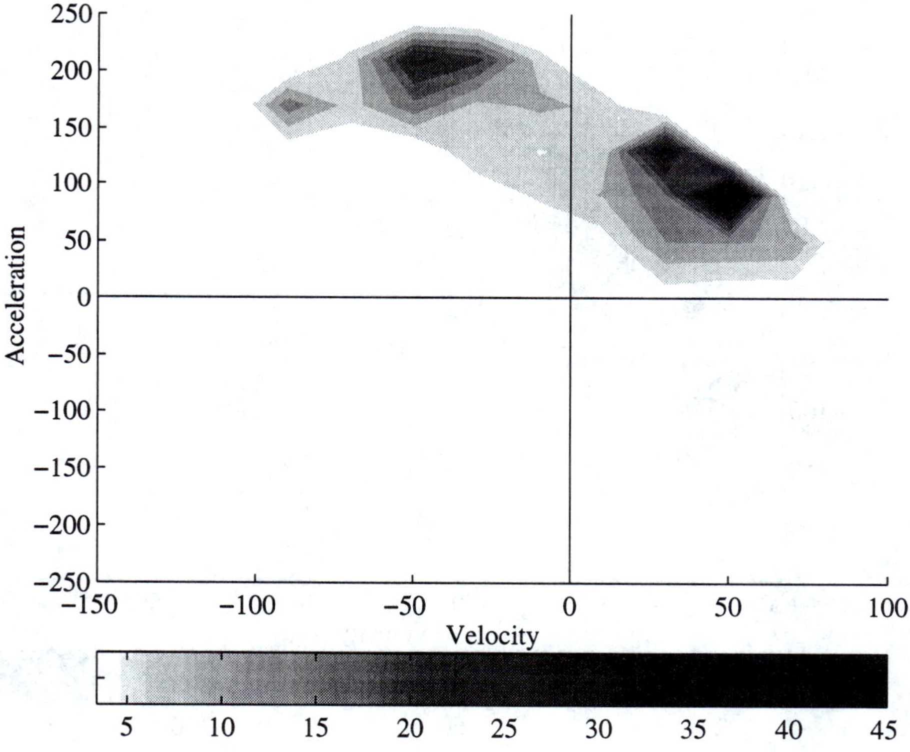

Figure 6: The phase space average of concentration at 3 cm above the bed shows the background flow phases where the suspension events occur. The velocity and acceleration are binned at 20 cm/sec and 40 cm/sec ^{2} intervals, respectively. Note that the flow moves clockwise around the figure, and that zero acceleration and maximum negative velocity corresponds to the wave crest. The units of the grayscale colorbar are in Hz.

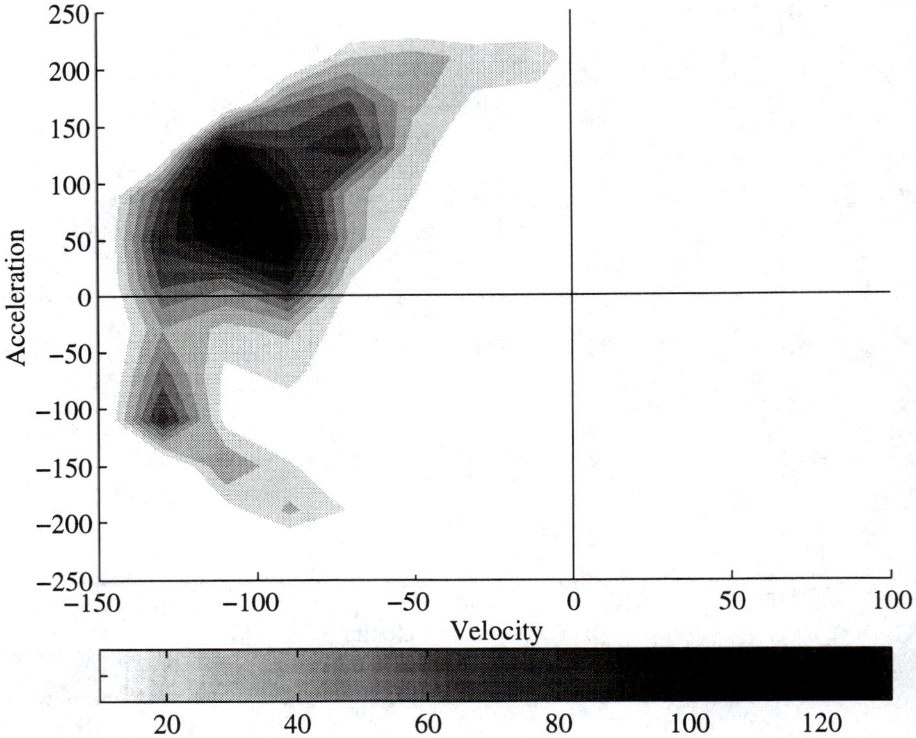

Figure 7: The phase space average of average turbulent variance over the 4 hot-film sensors shows background flow phases where the turbulent variance events occur. The velocity and acceleration are binned at 20 cm/sec and 40 cm/sec ^{2} intervals, respectively. Note that the flow moves clockwise around the figure, and that zero acceleration and maximum negative velocity corresponds to the wave crest. The units of the grayscale colorbar are in Hz.

CHAPTER 249

Wave Boundary Layer Flows and Pore Pressures in Permeable Beds

H. H. Hwung[1] K. S. Hwang[2] B. H. Lee[3]

Abstract

The boundary layer flows and wave-induced pore pressures inside porous beds are both investigated in wave tank. Two kinds of porous beds are composed of two different grades of quartz sands which the porosities are 0.437 and 0.472 respectively. According to the experimental results, it is found that the larger pore pressure in permeable bed is induced by the longer wave period, and the pore pressure has an exponential attenuation with the wave steepness increasing.

From the elaborate measurements in the boundary layer, the overshooting induced by wave motion is effected by bed porosity and it becomes smooth. In the experimental cases, the maximum horizontal overshooting occurs at $\tilde{Z}/\delta \approx 1.2$,which is lower than the impermeable case of $\tilde{Z}/\delta \approx 2.7$ obtained by Hwung and Lin (1989). Furthermore, the vertical overshooting has also been found in our experiments.

Introduction

The natural sea beds are usually permeable, however, most investigations about the bottom wave boundary layer flow were conducted on rigid impermeable beds. Longuet-Higgins (1953, 1958), Collins (1963), Jonsson (1966), Kajiura (1968) and so on, analyzed the characteristics of boundary layer flows with linear or nonlinear wave theory in uniform water depth. Furthermore, Du Toit and Sleath (1981), Sleath (1982, 1984, 1987), Hwung and Lin (1989) and other investigators detected the velocity profile of boundary layer on rigid bottom.

---

[1] Professor of Hydraulic and Ocean Engineering Department, Director of Tainan Hydraulic Laboratory, National Cheng Kung University, Tainan, Taiwan
[2] Associate Researcher of Tainan Hydraulic Laboratory, National Cheng Kung University, Taiwan
[3] Graduate Student of Hydraulic and Ocean Engineering Department, National Cheng Kung University, Taiwan

Concerning the pore pressure induced by wave motion in permeable bed, many researchers have paid more attention on the theoretical studies, such as Putnam (1949), Sleath (1970), Liu (1973) and so on. However, only a few of experimental studies has been done in the past years. Therefore, the elaborate experiments including the pore pressure and boundary layer flow in permeable beds have been carried out in this paper.

Experimental set-up

Two different grade of quartz sand of which the porosities are 0.437 and 0.472 respectively, are used in the experiments. The wave flume and related facilities are schematically illustrated in Figure 1. Within the wave flume, a section which 120 cm x 30 cm x 26 cm is filled with quartz sands as the testing section. Four pore pressure transducers are buried inside the porous bed to measure the pore pressure, while a Laser-Doppler velocimetry is used to detect the velocities above the porous bed. The enlargement testing section with the experimental installation are shown in Figure 2.

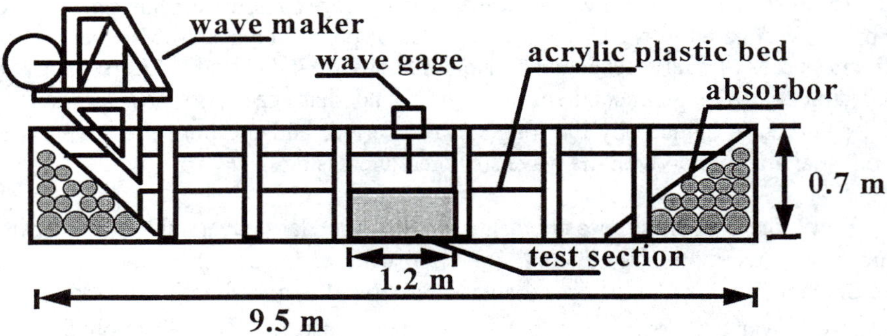

Fig. 1 The layout of wave flume

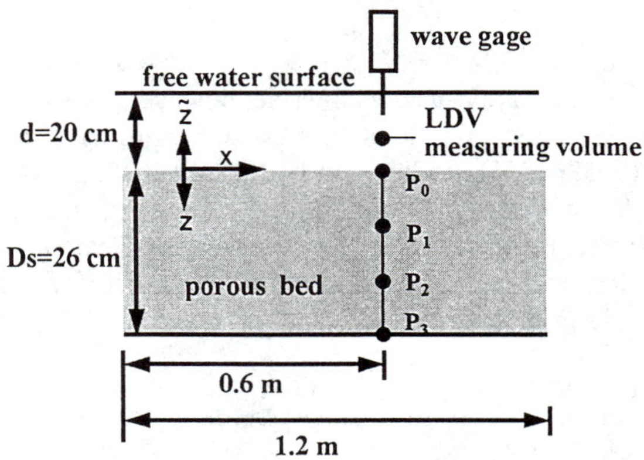

Fig. 2 The installation of test section

Herein, there are three kinds of experimental wave conditions listed in Table-1.

Table-1 The experimental wave conditions

Test Cases	T (sec)	d (cm)	H (cm)	H/L
Case 1	1.00	20	5.0	0.041
Case 2	0.95	20	7.0	0.062
Case 3	1.20	20	5.0	0.033

Results and Discussions

According to the experimental results, Figure 3 shows the original measurements of water surface elevation and horizontal, vertical velocities. Due to the wave reflection and boundary turbulent effect, we can see the fluctuations are existed in the velocity measurements. And Figure 4 is the original measurements of water surface elevation and the corresponding dynamic pore pressure. It is obviously to see that the dynamic pore pressure is decreasing from surface to bottom.

From the above original measurements, it shows that the length of each wave cycle in wave flume is not exactly the same due to wave reflection. This phenomena will lead to phase time shifting and create a false image of wave component by the phase average method. Therefore, according to the modified phase average method proposed by Hwung *et al.* (1988), the quantities of phase average can be express as

$$\langle Q \rangle = \frac{1}{N}\sum_{j=1}^{N} \tilde{Q}\left(\frac{T_j}{T} \cdot t_i\right) + \overline{Q} \qquad i = 1 \sim M$$

where $\langle Q \rangle$: the quantities of phase average,
$\tilde{Q}$: the quantities of wave component,
$\overline{Q}$: the quantities of time average,
N : numbers of testing wave cycle,
M : numbers of phase time within one wave cycle.

After using the modified phase average method, the water surface elevation, velocities and dynamic pore pressure measurements are shown in Figure 5 and Figure 6 respectively, and the quantities of velocities and dynamic pore pressure at any phase time can be calculated from the above figures.

From the above measurements and analysis, we obtained the relationship between dimensionless pore-pressure attenuation and wave period for the two different quartz sands in Figure 7. It shows that the pore pressure increases linearly as

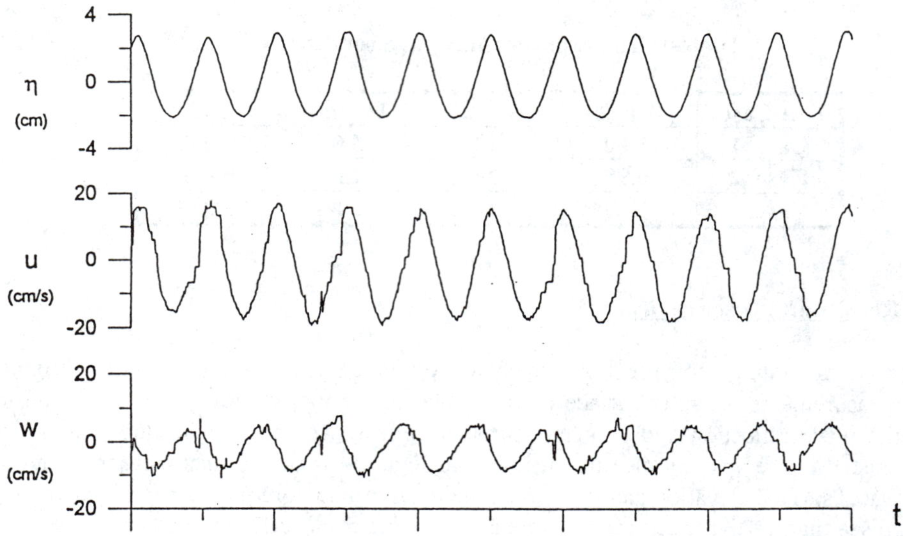

Fig. 3 The surface elevation and velocity measurements

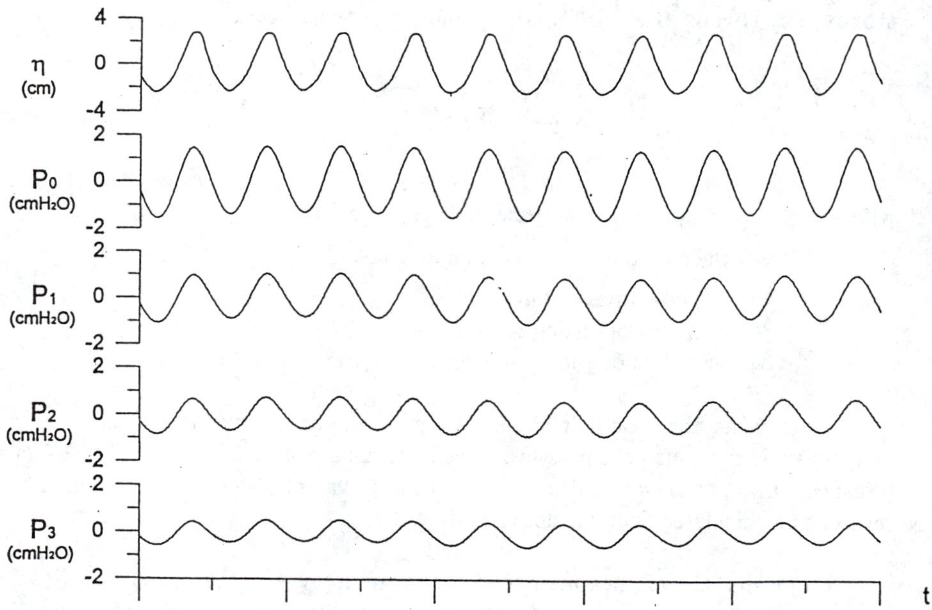

Fig. 4 The surface elevation and pore pressure measurements

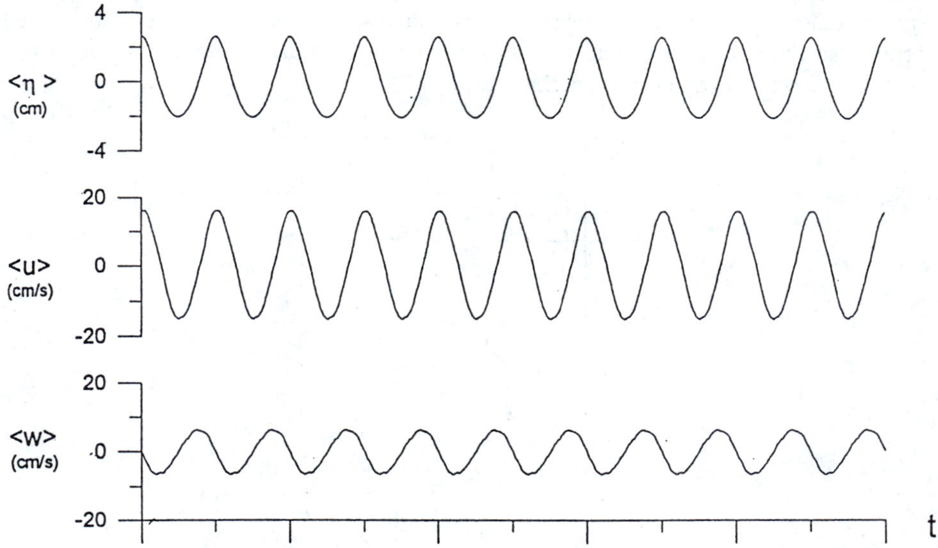

Fig. 5 The surface elevation and velocity measurements of the modified phase average

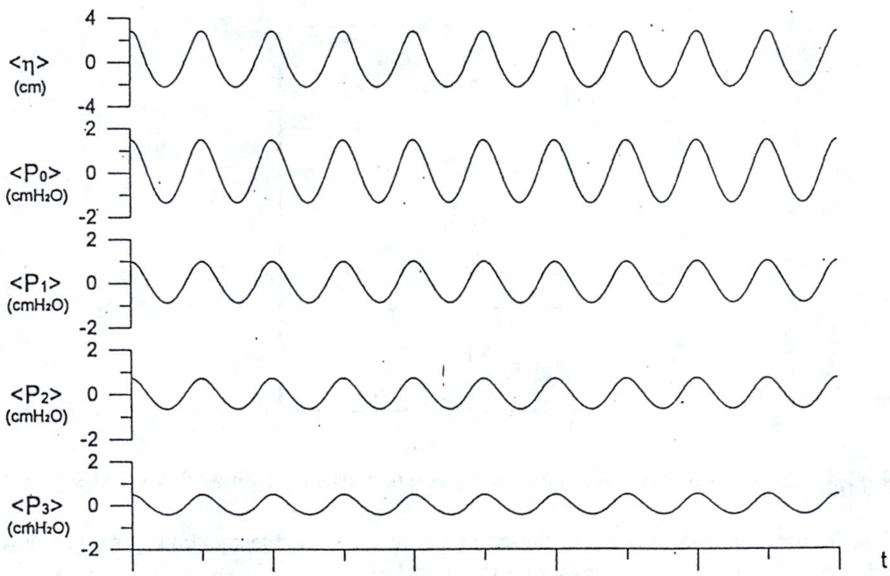

Fig. 6 The surface elevation and pore pressure measurements of the modified phase average

wave periods increasing, and the larger porosity bed has larger pore pressure at the same permeable depth. Besides, Figure 8 shows the relationship between dimensionless pore pressure attenuation and wave steepness. From the results, we found that the pore pressure decreases as wave steepness increasing, and it seems to have an exponential attenuation in this figure.

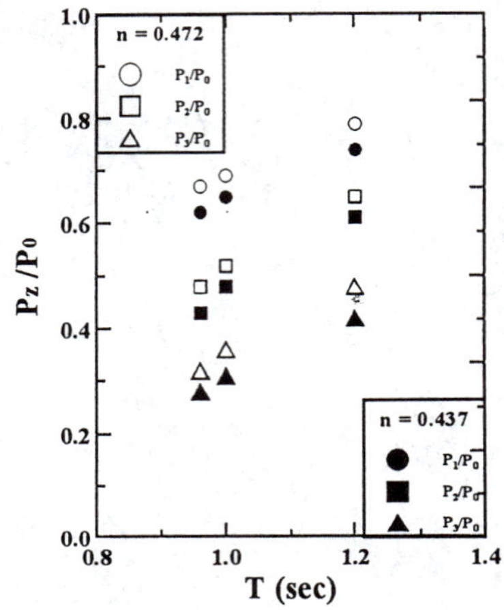

Fig. 7 The relationship between pore pressure attenuation and wave period

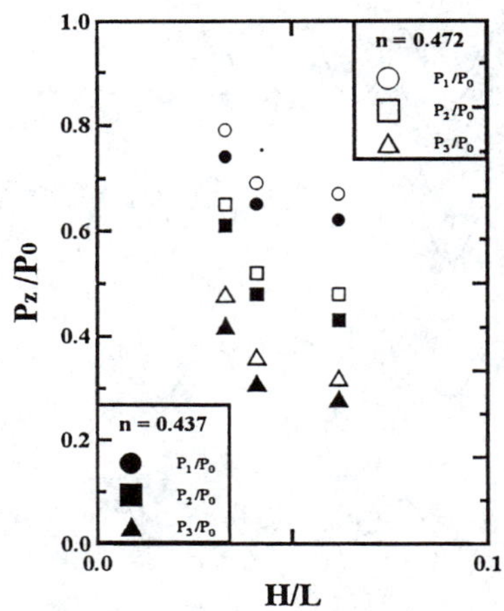

Fig. 8 The relationship between pore pressure attenuation and wave steepness

In addition, we also made the comparison between the dynamic pore pressure and previous theoretical studies as shown in Figure 9. We see that it has similar variation for the measurements and theoretical values, however, there is a little difference at the bottom due to the reflection of rigid bed.

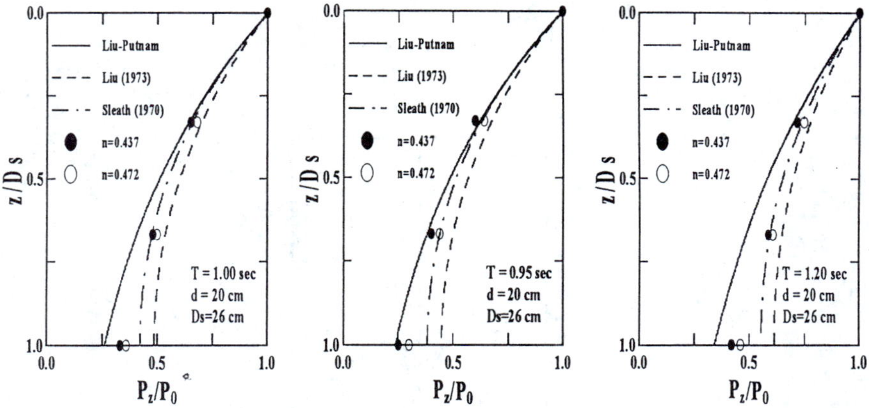

Fig. 9 The comparison between pore pressure measurements and theoretical values

In order to understand the boundary flows on permeable beds, a two-dimensional LDV system was employed to detect the horizontal and vertical velocities simultaneously. Figure 10 shows the horizontal velocity distributions on the permeable bed of porosity is 0.437 at seven different time phases. We can see that it also has overshooting in the boundary layer at $\widetilde{Z}/\delta \approx 1.2$, and it is less than on rigid smooth bottom which the overshooting is at $\widetilde{Z}/\delta \approx 2.7$ obtained by Hwung and Lin (1989). Figure 11 shows the horizontal velocity distributions on the permeable bed of porosity is 0.472 at different time phases. And the overshooting phenomena disappears in this larger porosity case.

Further, the extreme horizontal velocity distributions of all experiments are plotted in Figure 12, we can see that the overshsooting phenomena in the boundary layer is clearer in shorter wave period and the overshsooting disappears in longer wave period. The another cases of the extreme horizontal velocity distributions on the permeable bed of porosity is 0.472 as shown in Figure 13. From the comparison with Figure 12, we see that the boundary layer flows are more uniform in larger porosity bed.

Finally, the extreme vertical velocity distributions in bottom boundary layer on the permeable beds are shown in Figure 14 and Figure 15 respectively. It is interesting to see that the velocity in lower layer are larger than in upper layer within the boundary layer, and the velocities are more scattering due to the effect of vertical low passing through the porous material. From the vertical velocity distribution, we found the overshooting phenomena also occurs in porous bottom boundary layer.

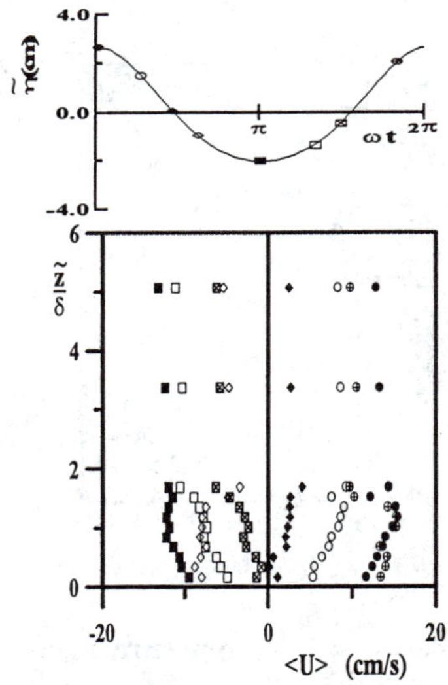

Fig. 10 Case 1, The horizontal velocity distributions on permeable bed of n=0.437

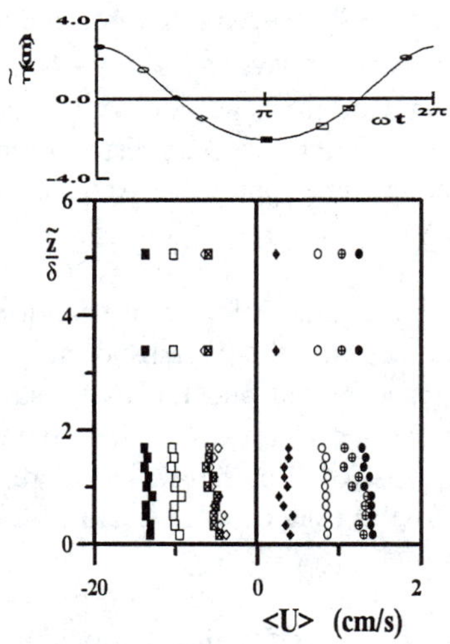

Fig. 11 Case 1, The horizontal velocity distributions on permeable bed of n=0.472

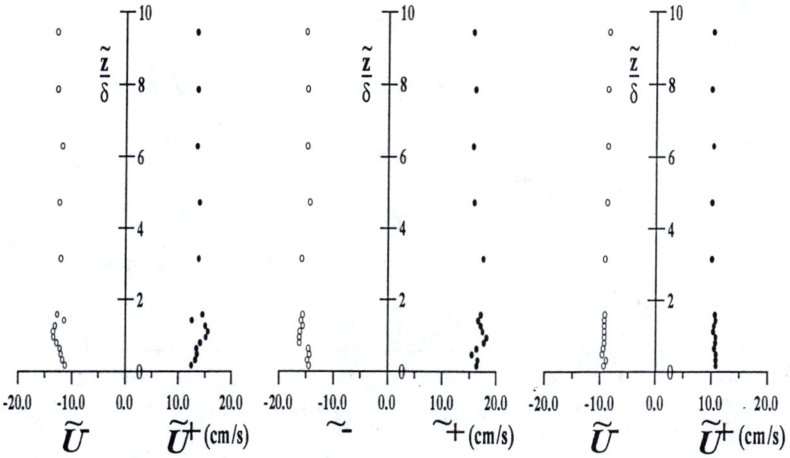

Fig. 12 The extreme horizontal velocity distributions on permeable bed of n=0.437

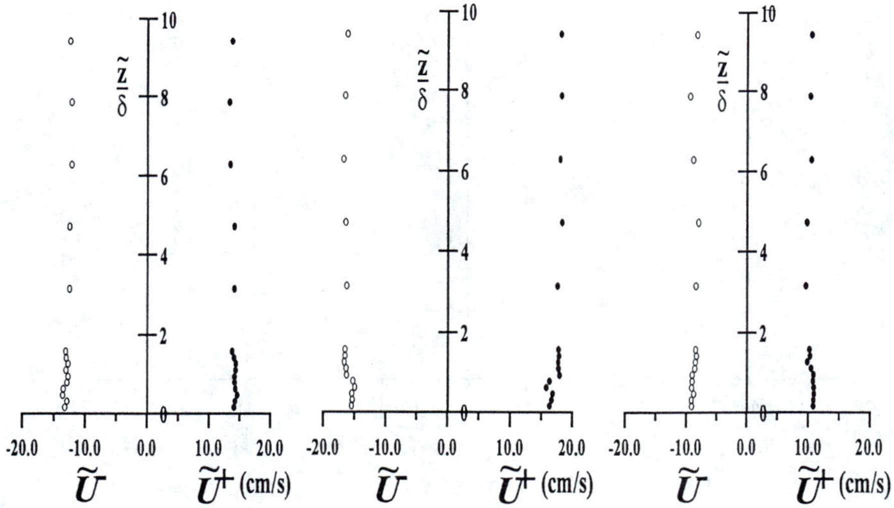

Fig. 13 The extreme horizontal velocity distributions on permeable bed of n=0.472

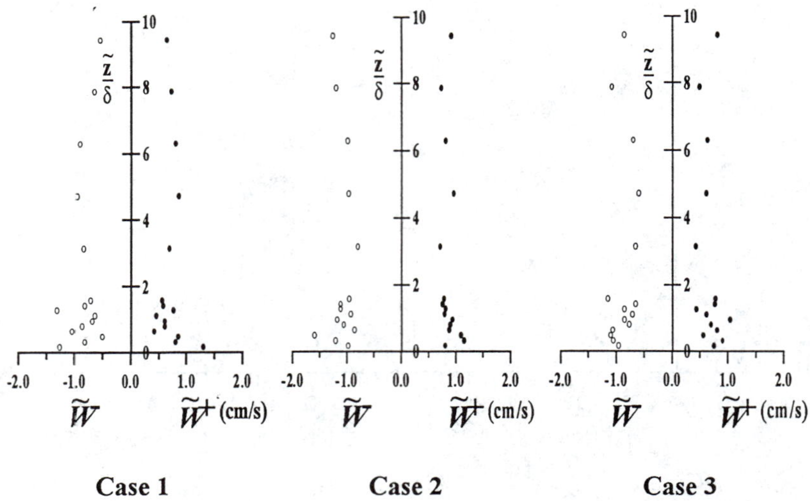

Fig. 14 The extreme vertical velocity distributions on permeable bed of n=0.437

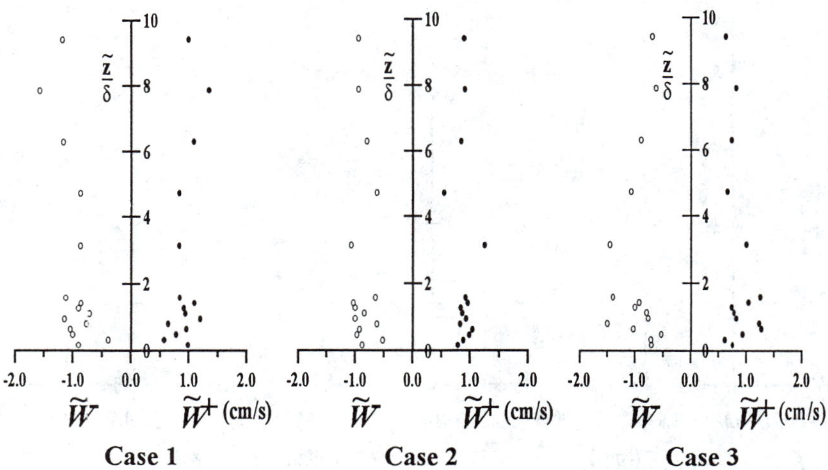

Fig. 15 The extreme vertical velocity distributions on permeable bed of n=0.472

Conclusions

The results summarized above indicate that the permeability of the bottom would result in interesting phenomena in wave boundary flows and dynamic pore pressure. Some remarkable conclusions would be described as follows:

1. In permeable beds, the larger pore pressure is induced by the larger wave period, and the dynamic pore pressure decreases exponential with the wave steepness increasing.

2. On the same wave condition, the larger porosity bed has larger dynamic pore pressure.

3. In the boundary layer, the horizontal overshooting induced by wave motion is influenced by bed porosity and becomes smoother. In our experimental cases, the maximum horizontal overshooting occurs at $\tilde{Z}/\delta \approx 1.2$, which is lower than the impermeable case of obtained by Hwung and Lin (1989).

4. From the vertical velocities measurements, we found that the vertical overshooting phenomena also occurs within bottom boundary layer.

References

1. Longuet-Higgins, M.S. (1953): Mass transport in water waves, Phil. Trans. Royal Society of London, Series A, No. 903 Vol. 245, pp. 535-581.

2. Longuet-Higgins, M.S. (1958): The mechanics of the boundary-layer near the bottom in a progressive waves, Proceedings of the 6th International Conference of Coastal Engineering, pp. 184-193.

3. Collins, J.I. (1963): Inception of turbulence at the bed under periodic gravity waves, J. Geophysical Research, Vol. 68, pp. 6007-6014.

4. Jonsson, I.G. (1966): Wave boundary layers and friction factors, Proceedings of the 10th International Conference of Coastal Engineering, pp. 127-148.

5. Kajiura, K. (1968): A model of the bottom boundary layer in water waves, Bull. Earthquake Research Inst., Vol. 46, pp. 75-123.

6. Du Toit and J.F.A. Sleath (1981): Velocity measurements close to rippled beds in oscillatory flow. Journal of Fluid Mechanics, Vol. 112, pp.77-96.

7. Putnam, J.A. (1949): Loss of wave energy due to percolation in a permeable seabed, Trans. Am. Geo. Un., 30, pp. 349-356.

8. Sleath, J.F.A. (1970): Wave-induced pressure in beds of sand. J. Hyd. Dvi., ASCE, 96, HY2. pp.367-378.

9. Liu, P.L.F. (1973): Damping of water waves over porous beds. J. Hyd., Div., ASCE, 99, HY12, pp.2263-2271.

10. Sleath, J.F.A. (1984): ***Sea Bed Mechanics.*** John Wiley & Sons, Inc. Press.

11. Hwung H.H., S.C. Wang and C. Lin (1988): The investigation of turbulent characteristics in the surf zone by LDV, Proceedings of the 10th Conference on Ocean Engineering in Republic of China, pp. 17-34.

12. Hwung, H.H. and C. Lin (1989): The experimental study on the boundary layer flow of waves propagating on a sloping bottom, The Chinese Journal of Mechanics, Vol.5, No.1, pp.67-76. (in Chinese)

CHAPTER 250

Backfilling Of Trenches Exposed to Waves

Jacob Hjelmager Jensen[1] and Jørgen Fredsøe[2]

Abstract

This paper treats the numerical prediction of initial and long-term morphology of small pipeline trenches. For this purpose a refined flow and sediment transport description is applied such that the entire mathematical problem is formulated and solved on a curvilinear grid using a $k-\epsilon$ turbulence-closure.

The backfilling process of trenches exposed to either waves or a steady current is of importance in relation to the implementation of pipelines in the marine environment. With respect to the sedimentation of trenches, the non-dimensional Trench-Keulegan-Carpenter number, $KC = a/L$, where a is the excursion length of a particle in waves and L the trench length, is investigated in detail, and an optimal KC-number for the backfilling rate is found. Coherent structures in the non-uniform unsteady trench-flow are shown to dominate the deposition rate when present.

The importance of a detailed description of the flow is further justified by comparing predictions of a very simple flow-model neglecting convective terms with the sophisticated flow-model.

Introduction

The morphodynamics of trenches or other man-made excavations in the marine environment involves the process of natural backfilling. In some cases backfilling is intended, as is the case with pipeline trenches, yet, in other cases, such as navigation channels, self-maintainance is striven for. For scientific as well as for economic reasons it is desirable to fully understand and control the mechanisms responsible for this specific morphological process.

A number of investigations of flow and sediment transport processes in a trench or a navigation channel have been conducted during the years. Bijker (1980), van Rijn (1986) and the analytical work of Fredsøe (1979) focus on the sediment transport processes. Other papers such as Alfrink's (1983) and Basara's (1995) focus on the sophisticated modelling of the hydrodynamics. These papers are all confined to the steady current case, and treat waves, if modelled, as a stirring mechanism for sediment only.

[1]Ph.D.-student, ISVA, Tech. Univ. of Denmark, DK-2800 Lyngby, Denmark.
[2]Professor, ISVA, Tech. Univ. of Denmark, DK-2800 Lyngby, Denmark.

Deposition of sediment takes place whenever a suspended particle advected by the flow starts adapting to an environment with less capacity. This might happen when the velocity or the turbulent kinetic energy level of the fluid decreases. A change in fluid capacity is exactly what a particle travelling with the flow across a trench will experience. A net deposition of sediment within the boundaries of the trench may therefore take place. On top of that the bottom concentration of the suspended sediment decreases, which means that less sediment will erode from the bed. Another important contribution to the backfilling process is the action of gravity on bed-load particles moving on a sloping bed that makes particles go downhill more easily than uphill.

For particles carried by a steady stream past a trench the maximum change in capacity (depth-averaged velocity for instance) is obviously some function of the maximum expansion in depth. However, in the case of an oscillatory flow crossing the trench a suspended particle might not experience the maximum expansion in depth. For a given wave situation the expansion felt by a particle advected with the flow generally becomes a function of both the Trench-Keulegan-Carpenter-number and the maximum expansion in depth.

Problem

The trench is initially sinusoidal and characterized by a length, L, and a depth, h, see figure (1). The water depth is called D.

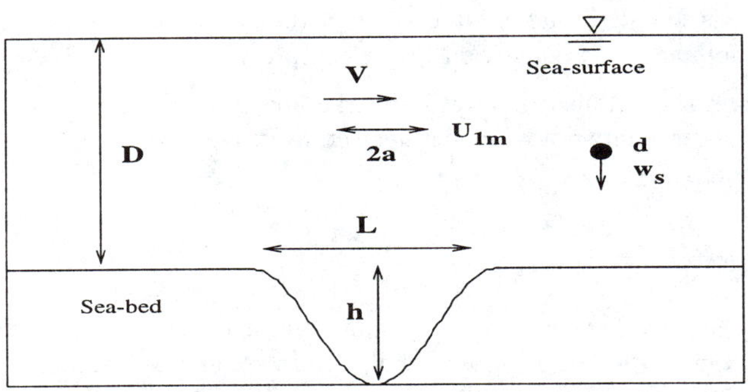

Figure 1: Sketch defining some of the key parameters used in the text.

The bed sediment is characterized by a fall velocity, w_s, and a diameter, d. The flow is generally combined by waves and currents (co-directional) and these are separately characterized by the fluxes, V and U_{1m}. The excursion length for a particle in a wave motion is symbolized by a. The wave is simulated by an oscillating pressure gradient neglecting any deformations of the water surface.

The initial backfilling rate is generally a function, Π, of all above-mentioned quantities, i.e.:

$$\Delta q = \Pi(L, h, D, d, w_s, a, V, U_{1m}, g)$$

where Δq is the amount of trapped sediment. A natural selection of non-dimensional parameters from these quantities become:

$$\Delta\phi = \Psi(\frac{h}{L}, \frac{h}{D}, \theta, Z, KC, \frac{V}{V+U_{1m}}, \frac{k_N}{D})$$

The parameters are in order of appearance referred to as: steepness, expansion, Shields, ($\theta = U_f^2/((s-1)gd)$) and Rouse parameter ($Z = U_f/(\kappa w_s)$), Trench-Keulegan-Carpenter number (a/L) and the wave/current parameter, where U_f is the friction velocity at a reference point far from the trench. The spatial coordinates, x (horizontal) and z (vertical), are non-dimensionalized by D and the time, t, by $(U_{1m}+V)/D$. For the present case the wave/current parameter is either 0 or 1, that is for waves and currents, respectively. All intermediate points define a combined wave/current flow. Furthermore, we will confine ourselves to an analysis based on the expansion parameter and the KC-number.

Solution of flow and sediment-transport equations

The finite volume concept (see Patankar (1980)) formulated in curvilinear orthogonal multi-block coordinates is applied, solving the transformed continuity equation and the fully 2D transformed Reynolds-averaged Navier-Stokes equations employing the standard $k-\epsilon$ closure of turbulence. A thorough description of the present model is reported by Tjerry (1997).

The partial differential equations of the turbulence model in Cartesian form, i.e. turbulent kinetic energy, k, and dissipation of turbulent kinetic energy, ϵ, are given by:

$$\frac{\partial k}{\partial t} + \frac{\partial u_j k}{\partial x_j} - \frac{\partial}{\partial x_j}(\frac{\nu_T}{\sigma_k}\frac{\partial k}{\partial x_j}) = \nu_T \frac{\partial u_i}{\partial x_j}(\frac{\partial u_j}{\partial x_j} + \frac{\partial u_j}{\partial x_i}) - \epsilon \qquad (1)$$

$$\frac{\partial \epsilon}{\partial t} + \frac{\partial u_j \epsilon}{\partial x_j} - \frac{\partial}{\partial x_j}(\frac{\nu_T}{\sigma_\epsilon}\frac{\partial \epsilon}{\partial x_j}) = C_{\epsilon 1}\frac{\epsilon}{k}\nu_T \frac{\partial u_i}{\partial x_j}(\frac{\partial u_j}{\partial x_j} + \frac{\partial u_j}{\partial x_i}) - C_{\epsilon 2}\frac{\epsilon^2}{k} \qquad (2)$$

Here, t is time, x_i the spatial coordinates, u_i the velocities and ν_T is the eddy-viscosity, calculated from:

$$\nu_T = C_\mu \frac{k^2}{\epsilon} \qquad (3)$$

The set of constants that appears in the $k-\epsilon$ model is standard choice and originally proposed by Launder (1974):

C_μ	σ_k	σ_ϵ	$C_{\epsilon 1}$	$C_{\epsilon 2}$
0.09	1.0	1.3	1.44	1.92

The transport of sediment, q, is divided into a bed and a suspended load contribution and corrected with the porosity, n, implicitly. The non-dimensional bed load, ϕ_B, is calculated using the Meyer-Peter formula modifying tractive stresses with slope effects as suggested by Fredsøe (1978):

$$\phi_B = \frac{q_B/(1-n)}{\sqrt{(s-1)gd^3}} = \frac{8}{(1-n)}(\theta - \theta_c - 0.1\frac{\partial h}{\partial x})^{3/2} \qquad (4)$$

where q_B is the bed load, $\partial h/\partial x$ the slope of the bed, g the acceleration of gravity, s the relative density and θ_c the critical Shields parameter (~ 0.05).

The concentration of suspended sediment is derived by solving the transport equation:

$$\frac{\partial c}{\partial t} + u\frac{\partial c}{\partial x} + v\frac{\partial c}{\partial z} = w_s\frac{\partial c}{\partial z} + \frac{\partial}{\partial x}(\nu_T\frac{\partial c}{\partial x}) + \frac{\partial}{\partial z}(\nu_T\frac{\partial c}{\partial z}) \tag{5}$$

The non-dimensional suspended transport, ϕ_S, is found as:

$$\phi_S = \frac{q_S/(1-n)}{\sqrt{(s-1)gd^3}} = \frac{\int_{-h}^{1} cudz/(1-n)}{\sqrt{(s-1)gd^3}} \tag{6}$$

where q_S is the suspended transport, and u and v are the longitudinal and vertical velocities, respectively.

Intra-wave calculation is applied for the sediment-transport model in order to take into account non-linear and phase-lag effects.

In the present model the water surface is taken to be plane, and the waves are introduced as an oscillatory motion caused by an oscillatory pressure gradient. At the water surface the flow equations apply symmetry conditions while the c-equation applies a no-flux condition. At the bed boundary a no-slip condition is used for the momentum equations, and a Dirichlet condition is used for the c-equation as the Engelund-Fredsøe formula (1976) is applied, i.e. $c_b = c_b(\theta)$, where c_b is the concentration at bed level.

Furthermore the k-equation uses a no-flux condition at the bed, and the bed condition for the ϵ-equation reads: $\epsilon = (30C_\mu^{0.75})/(\kappa k_N/D)k^{3/2}$, where k_N/D is the relative roughness of the sea-bed and κ the von Karman mixing length constant. On the in- and outlets Dirichlet and Neumann are applied, respectively, if not, periodic conditions are used.

The total flux of sediment is used in a continuity equation, which governs the morphodynamics. This reads:

$$\frac{\partial h}{\partial t} + \frac{\partial q}{\partial x} = 0 \tag{7}$$

stating that the divergence of depth-integrated sediment flux, $q = q_B + q_S$, is equal to the rate of change in bed elevation, h. This equation is discretized with a second order accurate scheme minimizing numerical diffusion. This is crucial when predicting diffusive processes such as backfilling of trenches.

If the entire flow, sediment, and morphological modules are run together, the dynamics of a mobile bed can be described and the backfilling tracked.

Backfilling in currents

In addition to the $k-\epsilon$ model an alternative flow description of the depth-averaged type is employed for the steady state case to emphasize the effect of including a more refined flow description. The depth-averaged approach uses a prescribed logarithmic velocity profile matched by the flux of water and a parabolic eddy-viscosity distribution scaled with the friction velocity. This is identical to a flow solution where convective terms have been ignored, i.e. local equilibrium.

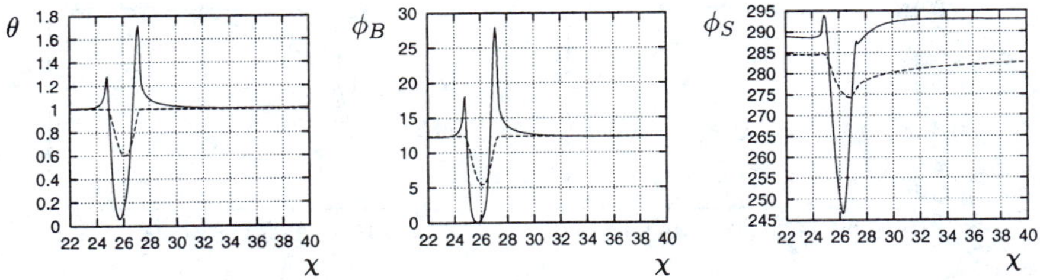

Figure 2: The Shields parameter along with the non-dimensional bed and suspended load distributions across a trench with the deepest section of the trench at $x = 26$ predicted by a $k - \epsilon$ model and a depth-averaged model. Reference flow: $\longrightarrow$. Parameters are: L/h=10, θ=1, $h/D = 0.25$ and Z=0.6. $k - \epsilon$ model: ——, Depth-averaged model: – – –.

The depth-averaged approach has earlier been applied in the field of sediment transport for analytical purposes, and it is therefore of interest to compare the performance of these flow models with respect to morphology (cf. e.g. Fredsøe (1979)).

The shear stress along with the non-dimensional bed and suspended load predicted by the two flow models typically vary as illustrated in figure (2). It is obvious from the shear-stress variation (i.e. $\theta(x)$) that the depth-averaged approach cannot in detail capture the complex behaviour of flow over a negative bump, which involves separation (or retardation), a phase shift between shear stress and bed form, speed-up and the relaxation of speed-up.

It is seen from figure (2) that both models predict erosion and deposition on the down- and upstream trench side, respectively. The additional shear stress features uncovered by the $k - \epsilon$ model relate in a morphological sense to a faster migration, a slower backfill (partly due to the above mentioned phase shift) and a more violent downstream deformation. The progressing asymmetrical shape of the bed, $h(x)$, is captured with both models as seen in figure (3), where bed forms at $T = tV/D$=3000, 9000, 18000 and 28000 are shown.

It is surprising how well the depth-averaged flow model performs in general and with respect to backfilling in particular. One reason for this is that the sediment transport mechanics such as the response length of suspended sediment is the governing parameter for the backfilling, and that very detailed flow mechanics to some extent is unnecessary. However, for $h/D \rightarrow 0$ (deep-water case) the depth-averaged model must fail since the gradient in capacity as well as in bottom concentration approaches zero in the limit. In this case the depth-averaged mechanism is not capable of explaining the deposition process, and a detailed description of the flow past a trench, including separation or even flow retardation, becomes necessary.

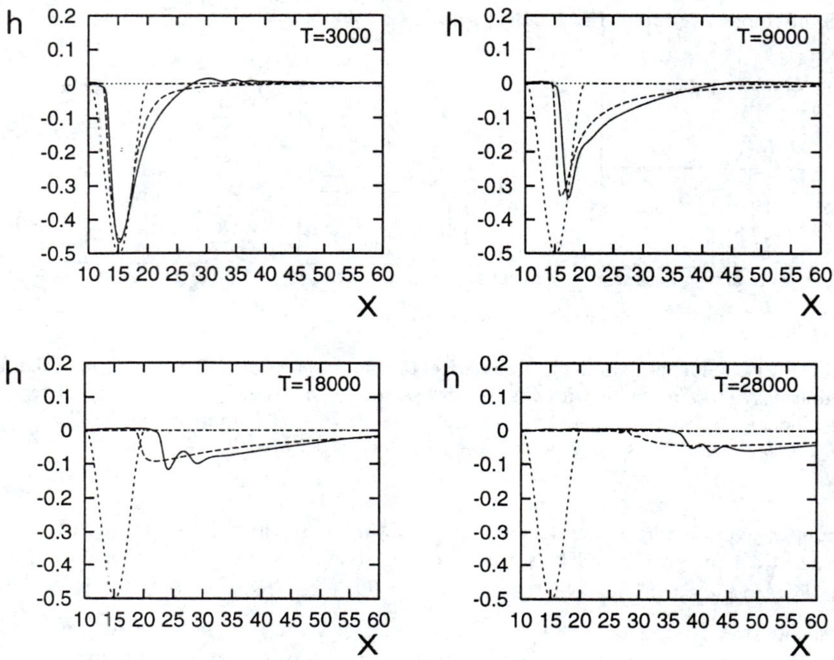

Figure 3: Shape of trench at four stages of its morphological evolution, for $k_N/D = 0.001$, $w_s/V = 0.02$ and $\theta = 0.25$. Reference flow (current alone): $\longrightarrow$.
$k-\epsilon$ model: ——, Depth-averaged model: – – –, Initial trench shape: - - - -.

Backfilling in waves

If wave effects are included matters become more complicated. The waves introduce an additional parameter that in non-dimensional form is described by a KC-number. This is the Keulegan-Carpenter number well-known from the offshore hydrodynamics where it is defined as:

$$KC = \frac{2\pi a}{D_c}$$

in which D_c = the diameter of the cylinder. For small KC-numbers (< 1-2) flow separation will not occur, while at larger KC-numbers (> 10-20), the flow is fully separated. In trenches, the flow may also separate if the slope is steep enough, and the equivalent KC-number is sufficiently large. In the following we define this number by:

$$KC = a/L$$

With respect to the sedimentation of trenches the KC-number involves at least one important aspect other than flow-separation. For $KC \sim 1$ it is seen that a sand particle picked up on the plane bed may reach the deepest part of the trench where it may settle due to a reduced capacity. The capacity is reduced on the upstream slope as the depth increases (non-uniformity) and as the flow decelerates (unsteadiness), which gives rise to smaller flow velocities as well as a decrease in intensity of turbulent kinetic energy.

So this is what happens: As suspended particles adapt to the local flow capacity by settling, they might get trapped in the separation bubble growing simultaneously on the upstream slope. If the vortex (former separation bubble) maintains its rotation and is able to keep the particles 'alive' when ejected at flow reversal, particles will be convected out of the trench and into the outer flow with the vortex.

The situation described is captured in figure (4) where time series of contours of concentration and turbulent kinetic energy are shown side by side for a short span of time around flow reversal. The phase in the wave period is symbolized as a dot on the sine-curve (positive values indicate left- to right-going flow).

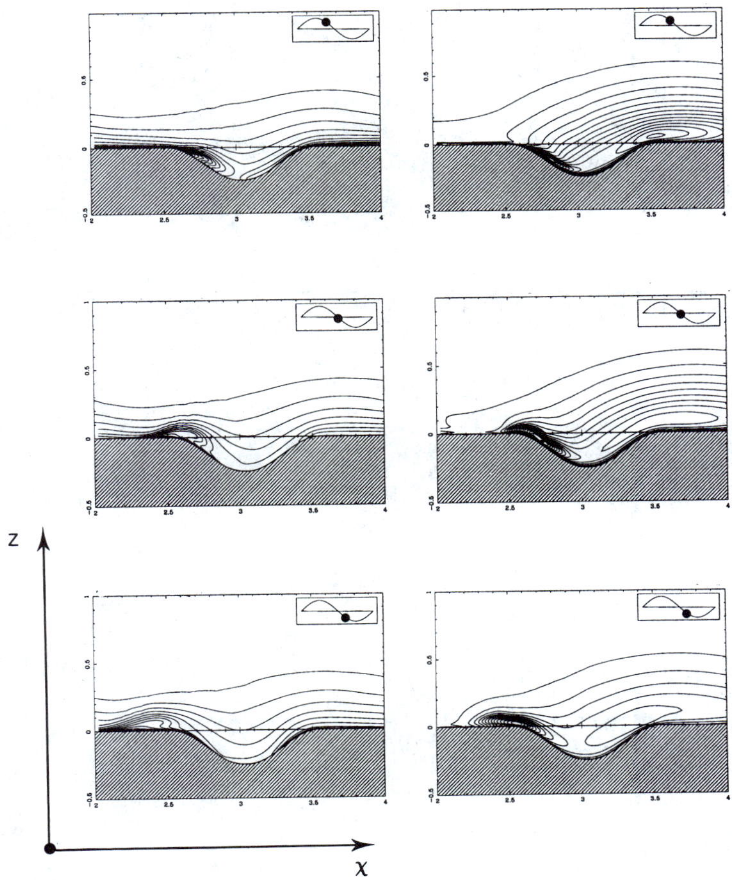

Figure 4: Time series of contours of concentration (left hand side of figure) and turbulent kinetic energy (right hand side of figure) around flow reversal. For L/h=5, θ=1, h/D=0.25, w_s/U_{1m}=0.02 and KC=1.

While the separation develops on the upstream slope, the speed-up of fluid on the downstream trench-shoulder creates high shear-stress levels, which causes a large amount of sediment to suspend. This is seen in figure (5) where time series of profiles for horizontal velocities and contours of concentration are placed side by side at a phase close to flow reversal. The speed-up process in the converging section is visualised by sediment being bursted off the sloping trench bed. This results in an overloaded cloud of sediment that returns with the reversing flow to deeper parts of the trench where it most likely will be deposited. This process, captured in figure (5), may intensify the backfilling.

The time series of concentration contours shown in figure (5) are created by shutting off the bed erosion shortly before the reversal of the flow occurs.

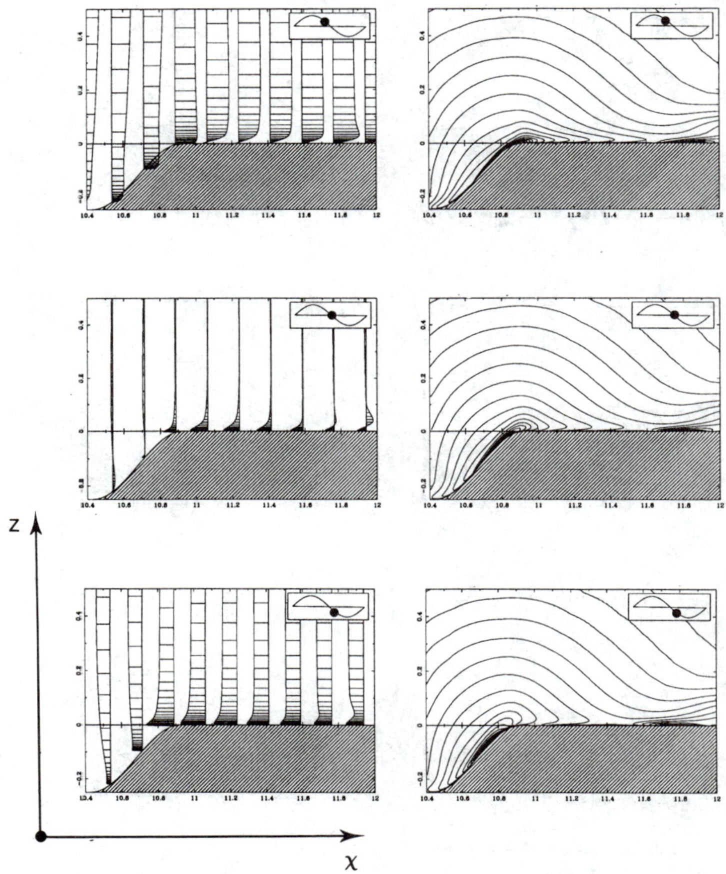

Figure 5: Time series of profiles of horizontal velocities (left hand side of figure) and contours of concentration (right hand side of figure). For L/h=5, θ=1, h/D=0.25, w_s/U_{1m}=0.02 and KC=1.

The non-uniform and unsteady flow separates and will speed-up succeedingly, which in the period-averaged sense will create a near bed flow directed towards the banks. A return flow, required for reasons of continuity, is therefore directed towards the centre of the trench so that the time-averaged streamline pattern, depicted in figure (6), shows two cells rotating clock- and anti-clockwise on the left and right hand side, respectively. It is also worth noticing that two additional rotating cells will develop on the shoulders of the trench.

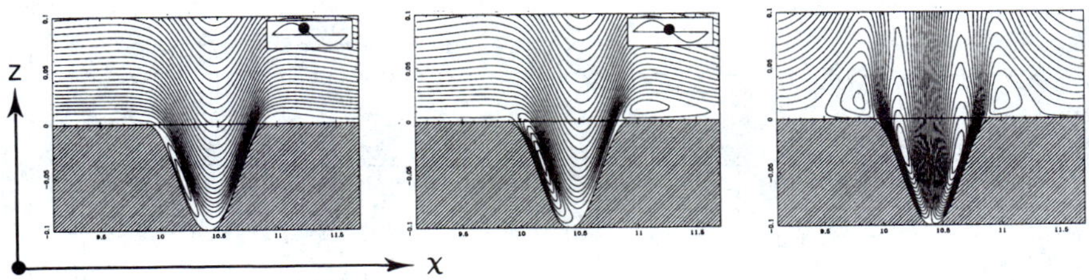

Figure 6: Two patterns of instant streamlines just before flow reversal occurs (in both cases the direction of main flow: $\longrightarrow$) along with the corresponding period-averaged streamline pattern. KC=0.5, h/D=0.1, L/h=10, k_N/D=0.001.

The result seen on the right hand side of figure (6) is a time-integrated effect from using a detailed flow-description.

Distributions of the time-averaged bed and suspended loads are shown in figure (7). With the streaming pattern in mind it is interesting to study the period-averaged bed load distribution in the neighbourhood of the trench. The resulting transport rate is an outcome of a battle between the action of gravity and the time-averaged shear stress. It is seen that the gradient, $\partial q_B/\partial x$, which determines whether the sinusoidal trench is being eroded (positive) or sediment is being deposited (negative) in some cases may shift sign at the centre of the trench, such that the deepest regions of the trench is eroded. The cases where the deepest regions are eroded are, for this particular span of parameters, found (see left hand side of figure (7)) as the KC-number is increased and the steepness parameter decreased, thus the time-averaged hydrodynamical forces are strengthened and the gravitational forces weakened.

The period-averaged rotating cells which appear on top of the shoulders are reflected in the bed transport, $\bar{\phi}_B$, at that particular location. This is shown on the left hand side of figure (7). The bed transport here are, however, damped by the vortices (increases in strength with the KC-number) which pass through the shoulder region shortly after flow reversal.

The influence of the vortex on $\bar{\phi}_B$ and $\bar{\phi}_S$ at a given location on the flat bed is recognized as a departure from $\bar{\phi}_B = 0$ and $\bar{\phi}_S = 0$. The vortex disturbs the flat bed region at a distance which is of the same order as the excursion length and is for obvious reasons amplified as h/L is increased. On the flat bed the sediment transport is, in the time-averaged sense, directed away from the trench by the strength of the advected vortex.

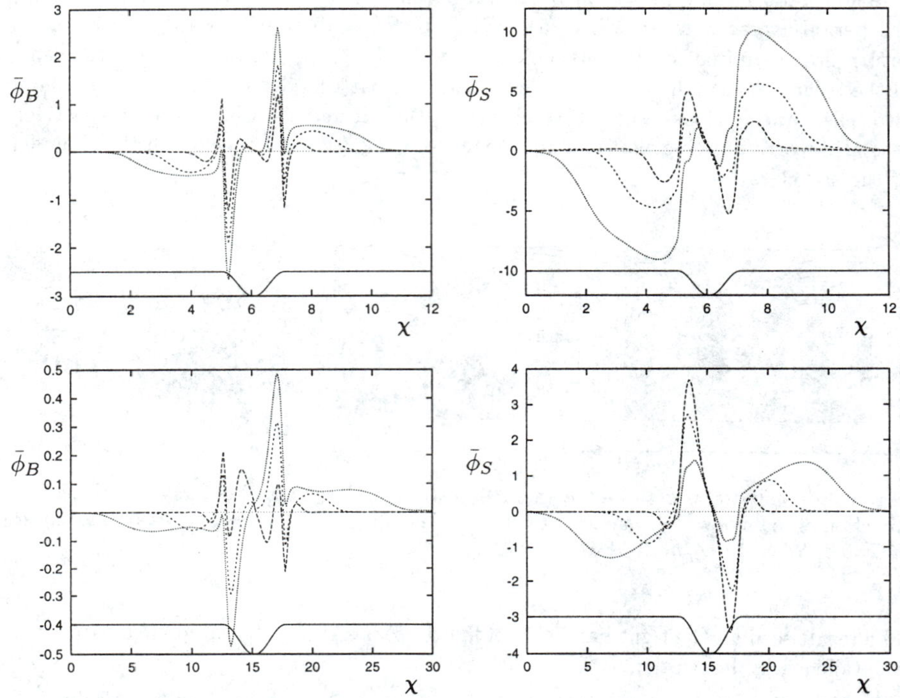

Figure 7: Distribution of period-averaged bed (left-hand side) and suspended sediment transport (right hand side) for a steep ($L/h = 4$ for the top of figure) and a moderate trench slope ($L/h = 10$ for the bottom of figure). $KC = 1.0$: $\cdots\cdots$, $KC = 0.5$: - - - -, $KC = 0.25$: — — —, with $h/D = 0.5$, $\theta = 0.65$ and $w_s/U_{1m} = 0.02$ The trench height depicted is highly distorted.

At the deeper regions of the trench the suspended sediment transport, $\bar{\phi}_S$, is directed towards the centre (see right hand side of figure (7)), implying that suspended material will be deposited here as opposed to erosion of the shoulder region.
More generally the backfilling process can be analysed by looking at period-averaged trapped sediment within control boxes (CB). Figure (8) shows the amount of trapped sediment as a function of the KC-number for three expansion parameters in two control boxes; an inner and an outer. The outer control box surrounds the entire trench from shoulder to shoulder, and the inner control box is bounded by the two steepest sections of the trench.

The trapped sediment is furthermore divided into contributions from bed and suspended load and where deposition is assigned with positive values. The Shields parameter in these calculations are: $\theta = 1$, and the steepness: $h/L = 0.1$.

An interesting feature of figure (8) is that an optimal backfilling rate takes place at $KC \sim 0.2$ for the suspended sediment. This is before the vortex mechanism becomes dominant. At this point the backfilling rate drops as the separation bubble grows in size and strength. However, beyond a certain KC-number the strength of the separation bubble will not increase any more, which is evident in figure (8) where a small increase in deposition of suspended sediment is seen for large KC-numbers.

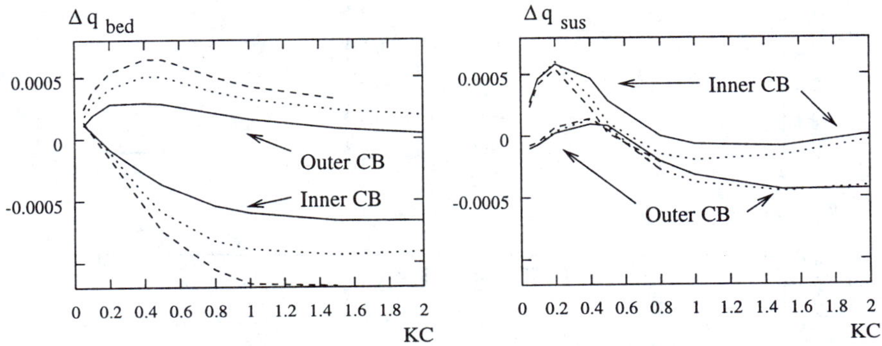

Figure 8: Trapped bed, (left hand side) and suspended load (right hand side) within control boxes (CB) for $h/D = 0.05$: - - - -, $h/D = 0.10$: · · · · ·· and h/D=0.25: ——-, and $L/h = 10$, $\theta = 1$, $Z = 0.6$.

The trapped sediment originating from bed load is of the same order of magnitude but negative as the trapped suspended load (at least when $\theta = 1$ and $L/h = 10$) even though $q_{S,max} \gg q_{B,max}$. It is seen from the left hand side of figure (8) (inner CB) that the action of gravity for this particular steepness parameter is merely capable of reducing the forces from the time-averaged shear-stress.

The time-averaged rotating cells on the shoulders are also seen to affect the in-fill of bed load through the outer box. This in-fill rate decreases as the KC-number increases because the separation bubble is washed over the shoulder thus counteracting this time-averaged motion.

It is clear from figure (8) that the typical morphological evolution is a net erosion of the trench shoulders together with a deposition in the centre of the trench. This is, however, exactly the way a diffusion process works.

The morphological development of a trench exposed to waves has been calculated using both flow models discussed previously. Four selected results are shown in figure (9), which clearly demonstrates the diffusion-like process.
As opposed to the steady current case, the depth-averaged model performs poorly in the pure wave case. This is not surprising since a detailed flow description is required to resolve the significant effect from the separation bubble dynamics on the backfilling process in waves. This causes the depth-averaged model to predict a faster backfill.

One interesting, but not surprising, consequence of the change in bed shape was the slow vanishing of the separation bubble as the trench slope becomes gentler. This involves a similar decrease in the vortex mechanism. One would immediately guess that this would lead to an increase in the backfilling rate. However, the trench widens and with this a decrease in the effective KC-number was found. After $T = tU_{1m}/D$=1000 the change in bed form was extremely weak.

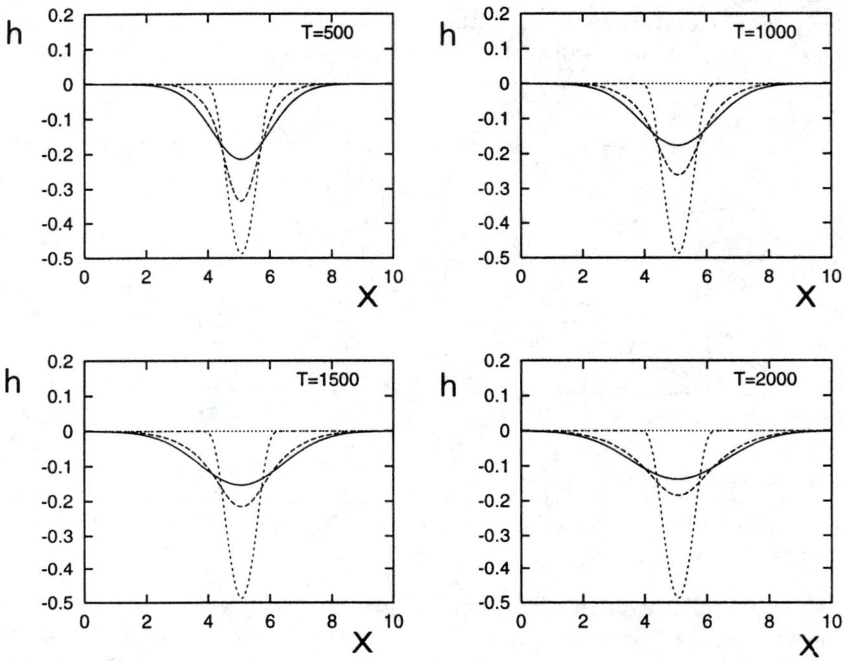

Figure 9: Shape of trench at four stages of its morphological evolution, for $k_N/D = 0.001$, $w_s/U_{1m} = 0.02$, $\theta = 0.25$ and $KC = 0.2$. $k-\epsilon$ model: – – –, Depth-averaged model: ——, Initial trench shape: - - - -.

Conclusion

The dynamics of vortices originally born as separation bubbles seem to dominate the backfilling process in waves whenever they are present. This is because the separation bubble traps incoming sediment particles on the upstream slope and keeps some of these 'alive' by the massive level of turbulent kinetic energy found in separating regions. When it detaches from the surface at flow reversal, a cloud of sediment will be advected out of the trench with the vortex.

It has been shown that the vortex mechanism is important and that models which neglect this process may over-predict the backfilling rate.

The optimal backfilling rate in waves was found to be $KC \sim 0.2$.

Furthermore, depth-averaged steady state flow models give good results when applied in morphological calculations particularly in shallow waters which include navigation channels etc.. However, these simple flow models fail to describe more complicated KC-effects found in waves as the effect of separation and vortex shedding is not taken into account.

Acknowledgments.

The computer-code applied in the present investigation is originally developed at ISVA by Ph.D. Søren Tjerry.

This work is jointly supported by "Numerical and Experimental Fluid Mechanics" and "Marine Technique" both funded by the Danish Technical Research Council (STVF).

References

Alfrink, B.J., and van Rijn, L.C. (1983). Two-equation turbulence model for flow over trenches. Journal of Hydraulic Engineering (ASCE), Vol. 109, No.7, pp. 941-958.

Basara, B. and Younis B.A. (1995). Prediction of turbulent flows in dredged trenches. Journal of Hydraulic Research, Vol. 33, No. 6.

Bijker, E.W. (1980). Sedimentation in channels and trenches. Proc., 17'th ICCE'80, Vol 2. pp. 1708-1718, Sydney, Australia.

Engelund, F. and Fredsøe J. (1976). A sediment transport model for straight alluvial channels. Nordic Hydrology 7:293-306.

Fredsøe, J. (1978). Sedimentation of river navigation channels. Journal of the Hydraulics Division (ASCE), Vol. 104, No. HY2, pp.223-236.

Fredsøe, J. (1979). Natural backfilling of pipeline trenches. Journal of Petroleum Techn. Vol. 31, pp. 1223-1230.

Launder, B.E. and Spalding, D.B. (1974). The numerical computation of turbulent flows. Comput. Meths. Appl. Mech. Eng., 3:269-289.

Patankar, S.V. (1980). Numerical Heat Transfer and Fluid Flows. Hemisphere, New York.

Tjerry, S. (1997). Morphological Calculations of Dunes in Alluvial Rivers. Ph.D.-thesis ISVA, Technical University of Denmark. (In print).

van Rijn, L.C. (1986). Sedimentation of Dredged Channels by Currents and Waves. Delft Hydraulics Laboratory. Delft Hydraulics Communication No. 369.

CHAPTER 251

A CLOSED-FORM SOLUTION FOR TURBULENT WAVE BOUNDARY LAYERS

Magnus Larson[1]

ABSTRACT: A general, closed-form solution is presented to the linearized equation describing the velocity in a turbulent boundary layer. The solution is valid for any type of time-varying free-stream velocity, although the focus of the paper is on oscillatory flows generated by surface waves propagating over the sea bottom. A primary objective is to employ the solution for waves with an asymmetric velocity field, where nonlinear effects are small enough to be neglected in the linearized turbulent boundary layer equation. The general solution is developed to more convenient forms for a few special cases, including a sinusoidally varying free-stream velocity and a velocity given by stream function theory. The solution is tested against two different data sets from oscillatory tunnels; one set involves a sinusoidal and the other set a cnoidal free-stream velocity.

INTRODUCTION

The oscillatory boundary layer that develops when surface waves propagate over the sea bottom has many important implications for flow-dependent phenomena in the coastal zone. Examples of such phenomena are wave energy dissipation due to bottom friction and the initiation and transport of sediment (Grant and Madsen 1986). The velocity changes with elevation at a high rate in the wave boundary layer, which typically causes large shear stresses, high dissipation rates, and strong turbulence intensities. An oscillatory free-stream velocity generates a boundary layer with a characteristic thickness that is much smaller than the corresponding boundary layer under a uni-directional current. Thus, the near-bottom velocity gradients and the associated shear stresses are much larger under oscillatory waves than under a uni-directional current, which suggests that the waves dominate the flow close to the

[1] Associate Professor, Department of Water Resources Engineering, University of Lund, Box 118, S-221 00 Lund, SWEDEN. (Email: magnus.larson@tvrl.lth.se)

bottom in many cases. However, because the oscillatory wave motion often does not give rise to any net transport of for example sediment, but primarily has a stirring effect, a superimposed current may have a marked influence on the net transport rate, although the current itself may be to weak to mobilize and transport the sediment on its own.

In nature the boundary layer under waves will almost always be turbulent (Nielsen 1992). Most simple approaches to calculate the velocity in the wave boundary layer employ the eddy viscosity concept to model the turbulence, where the viscosity is taken to be a function of the elevation above the bottom (Kajiura 1968, Grant and Madsen 1979, Brevik 1981, Myrhaug 1982). These models mainly differ in the formulation of how the eddy viscosity ν varies with the elevation z; the most simple models are based on a linear variation in ν with z (Grant and Madsen 1979), whereas more complex models assume several different layers, each having a separate equation to relate ν and z (Kajiura 1968). In reality, ν should also depend on time and Trowbridge and Madsen (1984a) developed a model where a time-varying eddy viscosity was employed. However, in most models the prediction of the velocity in the wave boundary layer is not overly sensitive to the formulation of ν, and a simple model such as the one suggested by Grant and Madsen (1979) often yields satisfactory results. This is especially true if the velocity calculations are performed with the aim of computing sediment transport rates; existing equations for calculating the sediment transport rate include large uncertainties that do not warrant excessively detailed flow computations (Madsen and Wikramanayake 1991).

The main objective of the present study is to develop a simple, analytical model of the flow in an oscillatory boundary layer under rough turbulent conditions that may be employed for situations where the free-stream velocity is not purely sinusoidal. Such velocity fields are generated by nonlinear shoaling waves, which could produce a net transport of sediment due to the asymmetry in the wave velocity. In deriving the analytical model, it will be assumed that the effects of the nonlinear terms in the momentum equations are negligible, implying that the linearized boundary layer equation may be used (Nielsen 1992). A simple eddy viscosity formulation in accordance with Grant and Madsen (1979) is employed to model the turbulent stresses. The proposed model is tested with data from Jonsson (1980) for a sinusoidal free-stream velocity and with data from Nadaoka et al. (1994) for an asymmetrical free-stream velocity of cnoidal type.

The present study focusses on boundary layer development under oscillatory waves, but the general solution presented for the velocity profile in the boundary layer is valid for any time-varying free-stream velocity. A first-order approach is taken (Grant and Madsen 1979) where the governing equation is linearized in order to obtain a closed-form solution. Thus, for cases where the free-stream velocity is generated by nonlinear waves second-order effects, such as the mass transport in the boundary layer (Trowbridge and Madsen 1984b), are not included in the solution.

Even though the mass transport may be small it could still be of significance for calculating the sediment transport.

THEORETICAL CONSIDERATIONS

General Solution

Employing the simple eddy viscosity model by Grant and Madsen (1979), the linearized turbulent boundary layer (TBL) equation may be written (Nielsen 1992),

$$\frac{\partial}{\partial t}(u_w - u_b) = \frac{\partial}{\partial z}\left\{\kappa u_{*m} z \frac{\partial}{\partial z}(u_w - u_b)\right\} \tag{1}$$

where $u_w(z,t)$ is the velocity in the TBL, $u_b(t)$ the free-stream (wave) velocity, t time, z a vertical coordinate, κ von Karman's constant (=0.40), and u_{*m} a constant, representative bottom shear velocity. With the boundary conditions $u_w=0$ for $z=z_o$, where z_o is the characteristic height of the bottom roughness, and $u_w=u_b$ for $z \rightarrow \infty$, Equation 1 has the following general solution,

$$u_w = \int_o^t \frac{\partial}{\partial t}(u_b(t-\xi))\, I_u(\xi,z)\, d\xi \; + \; u_{bo} I_u(t,z) \tag{2}$$

where u_{bo} denotes u_b at $t=0$ and,

$$I_u(t,z) = \frac{2}{\pi}\int_o^\infty e^{-\frac{1}{4}y^2 \kappa u_{*m} t} \frac{J_o(y\sqrt{z_o})Y_o(y\sqrt{z}) - J_o(y\sqrt{z})Y_o(y\sqrt{z_o})}{J_o^2(y\sqrt{z_o}) + Y_o^2(y\sqrt{z_o})} \frac{dy}{y} \tag{3}$$

in which J_o and Y_o are zero-order Bessel functions of the first and second kind, respectively, and y is a dummy integration variable. Equation 3 corresponds to the solution for a time-independent free-stream velocity employed at $t=0$ (that is, $u_b(t)=u_{bo}$=constant); thus, the solution for any $u_b(t)$ is obtained through the superposition of the response from an infinite number of temporal changes in u_b, as expressed by the convolution integral in Equation 2. Figure 1 illustrates I_u in non-dimensional form, and the curves may be interpreted as velocity profiles at different times indicating the TBL growth through the water column.

The shear stress at any elevation z may be derived from,

$$\frac{\tau}{\rho} = \kappa u_{*m} z \frac{\partial u_w}{\partial z} = \kappa u_{*m} z \left\{\int_o^t \frac{\partial}{\partial t}(u_b(t-\xi)) I_s(\xi,z)\, d\xi \; + \; u_{bo} I_s(t,z)\right\} \tag{4}$$

where,

$$I_s(t,z) = \frac{1}{\pi\sqrt{z}} \int_o^\infty e^{-\frac{1}{4}y^2\kappa u_{*m}t} \frac{-J_o(y\sqrt{z_o})Y_1(y\sqrt{z})+J_1(y\sqrt{z})Y_o(y\sqrt{z_o})}{J_o^2(y\sqrt{z_o})+Y_o^2(y\sqrt{z_o})} dy \tag{5}$$

in which J_1 and Y_1 are first-order Bessel functions of the first and second kind, respectively. The integral $I_s \rightarrow \infty$ as $t \rightarrow 0$, but the approach towards infinity is slow and the singularity is easy to handle within the convolution integral in Equation 4. However, if $u_{bo} \neq 0$ an infinite shear stress will be obtained at $t=0$, because $u_w=0$ at $z=z_o$ simultaneously as $u_w=u_{bo}$ an infinitesimal distance above. The integrals in Equations 3 and 5 are time-consuming to evaluate, so in order to speed up the calculation of u_w and τ in Equations 2 and 4, respectively, it is convenient to derive look-up tables for I_u and I_s.

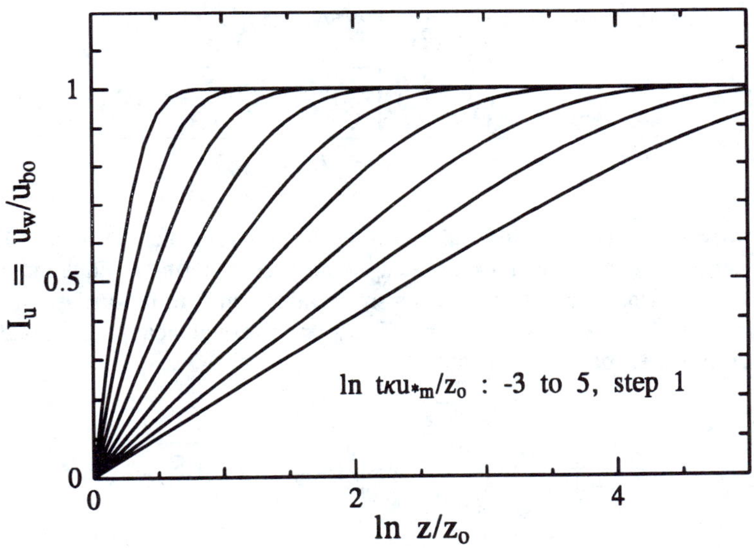

Figure 1. The integral I_u as a function of non-dimensional distance and time.

Equation 2 is a general solution of Equation 1 for any type of free-stream velocity, although from a physical point of view the solution only makes sense for a wave boundary layer where the assumptions behind Equation 1 are applicable. There is only one free parameter in the solution, namely the roughness length scale z_o, which in the case of rough turbulent flow over a flat bed is typically set to $k_n/30$, where k_n is the equivalent Nikuradse sand grain roughness (Grant and Madsen 1979). After it has been defined, the representative shear velocity u_{*m} may be obtained implicitly from the solution. Grant and Madsen (1979) studied the TBL under a sinusoidal free-stream velocity and used the maximum bottom shear stress τ_{bmax} during a wave period to define $u_{*m}=(\tau_{bmax}/\rho)^{1/2}$, where ρ is the water density. For more complex variations in the free-stream velocity other choices to define u_{*m} may be more appropriate, such as the mean absolute value of the bottom shear stress during a cycle (τ_{bav}).

Sinusoidal Free-Stream Velocity

Equation 2 was derived using Laplace transform technique and for elementary $u_b(t)$ more convenient forms than Equation 2 may be obtained. If the free-stream velocity is sinusoidal and described by $u_b=u_o\cos\omega t$, where u_o is the velocity amplitude at the bottom and ω is the angular frequency, the following solution satisfies Equation 1 and the boundary conditions,

$$u_w = u_o\cos\omega t - u_o\frac{N_o(2\sqrt{f\frac{z}{z_o}})}{N_o(2\sqrt{f})}\cos\left(\omega t+\Phi_o(2\sqrt{f\frac{z}{z_o}})-\Phi_o(2\sqrt{f})\right)$$
$$-\frac{2u_o}{\pi}\int_0^\infty e^{-\frac{1}{4}\frac{\omega t}{f}y^2}\frac{y^3}{y^4+(4f)^2}\frac{J_o(y\sqrt{\frac{z}{z_o}})Y_o(y)-Y_o(y\sqrt{\frac{z}{z_o}})J_o(y)}{J_o^2(y)+Y_o^2(y)}dy \qquad (6)$$

where N_o and Φ_o is the modulus and phase, respectively, of the zero-order Kelvin function $ker_o x + i\ kei_o x$, and $f=\omega z_o/\kappa u_{*m}$. The second term in Equation 6 is a transient that is dampened out quickly for small values on f; in most cases this term is negligible already after a wave period. The shear stress at steady-state conditions is given by (see first part of Equation 4),

$$\frac{\tau}{\rho} = -\kappa u_o u_{*m}\sqrt{f\frac{z}{z_o}}\frac{N_1(2\sqrt{f\frac{z}{z_o}})}{N_o(2\sqrt{f})}\cos\left(\omega t-\frac{\pi}{4}+\Phi_1(2\sqrt{f\frac{z}{z_o}})-\Phi_o(2\sqrt{f})\right) \qquad (7)$$

where N_1 and Φ_1 is the modulus and phase, respectively, of the first-order Kelvin function. Grant and Madsen (1979) previously derived the steady-state part of the solution for a sinusoidal free-stream velocity (see also Kajiura 1968).

To permit rapid evaluation of the maximum bottom shear stress it is convenient to define a wave friction factor f_w from $\tau_{bmax}=0.5\rho f_w u_o^2$ (Jonsson 1966, 1980). In accordance with Grant and Madsen (1979), who defined u_{*m} based on τ_{bmax}, f_w may be obtained through the expression for τ_{bmax} given by Equation 7. An alternative approach would be to use τ_{bav} in u_{*m} instead. This would yield a smaller τ_{bmax}, because the turbulent mixing is less if τ_{bav} is used to define u_{*m}, also implying that f_w is smaller than if the conventional definition of u_{*m} is employed. Figure 2 displays f_w as a function of the bottom excursion amplitude A_b $(=u_o/\omega)$ over the grain size k_n for a u_{*m} defined based on τ_{bmax} and τ_{bav}.

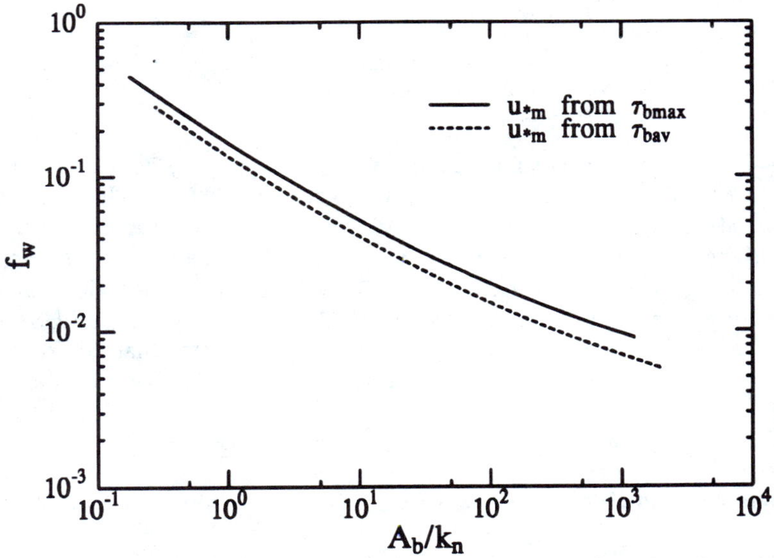

Figure 2. Friction factors for a sinusoidal wave velocity, where the friction velocity is defined based on τ_{bmax} respective τ_{bav}.

Free-Stream Velocity from Stream Function Theory

Stream function theory (Dean 1965) is convenient for describing nonlinear wave properties, because the theory is valid from deep water up to wave breaking. The bottom orbital velocity at a point under a wave described by stream function theory may be expressed as,

$$u_b = -\frac{2\pi}{L}\sum_{n=1}^{N} nX(n)\cos(n\omega t) \tag{8}$$

where L is the wavelength, $X(n)$ stream function coefficients, and N the order of the theory employed. The velocity given by Equation 8 is a sum of sinusoidal components and the steady-state solution to Equation 1 with this u_b is,

$$u_w = \sum_{n=1}^{N} u_n \left\{ \frac{N_o\left(2\sqrt{nf\frac{z}{z_o}}\right)}{N_o(2\sqrt{nf})} \cos\left(n\omega t + \Phi_o\left(2\sqrt{nf\frac{z}{z_o}}\right) - \Phi_o(2\sqrt{nf})\right) - \cos(n\omega t) \right\} \tag{9}$$

where $u_n = 2\pi nX(n)/L$. The corresponding shear stress is given by,

$$\frac{\tau}{\rho} = \kappa u_{*m}\sqrt{f\frac{z}{z_o}}\sum_{n=1}^{N} u_n\sqrt{n}\,\frac{N_1(2\sqrt{nf\frac{z}{z_o}})}{N_o(2\sqrt{nf})}\cos\left(n\omega t-\frac{\pi}{4}+\Phi_1(2\sqrt{nf\frac{z}{z_o}})-\Phi_o(2\sqrt{nf})\right) \tag{10}$$

A wave described by stream function theory is uniquely defined by the two ratios h/L_o and H/L_o (Dean 1990), where h is water depth, H wave height, and the subscript o denotes deepwater conditions. Waves with identical values on h/L_o and H/L_o yield the same dimensionless velocity $u_b/(H/T)$; thus, the quantity H/T appears as a normalizing "velocity". A friction factor derived for a stream function wave will depend not only on the normalized roughness k_n/H, but also on h/L_o and H/L_o. The friction velocity may be computed by using τ_{bav}, which is obtained from time integration of the absolute shear stress over a wave period T.

RESULTS

Data From Jonsson (1980)

Measurements by Jonsson (1980) of u_w in a water tunnel for a sinusoidally varying u_b with the amplitude u_o was employed as a first step to validate Equation 1 for describing the velocity in the TBL. It was verified that Equation 2 produced identical results to Equation 6, and the steady-state portion of the solution was used for the comparison with the data. Jonsson (1980) presented data for two cases: 1) $u_o=2.11$ *m/s*, $T=8.39$ *s*, $k_n=2.3$ *cm*, and 2) $u_o=1.53$ *m/s*, $T=7.20$ *s*, $k_n=6.3$ *cm*. Comparison between the analytical solution and the measurements was performed for the phases t/T = *1/2, 5/8, 3/4, 7/8,* and *1*. The Reynolds number *Re* for Cases 1 and 2 were 6.0 10^6 and 2.7 10^6, respectively, based on A_b and u_o. The roughness values given by Jonsson were employed and there were no free calibration parameters. Two different definitions of u_{*m} were used in the comparison, namely $u_{*m}=(\tau_{bmax}/\rho)^{1/2}$ and $u_{*m}=(\tau_{bav}/\rho)^{1/2}$.

Figures 3 and 4 display the comparison between the analytical solution and the measurements for Cases 1 and 2, respectively. In general, the difference between the two formulations for u_{*m} is small, although using τ_{bav} seems to consistently produce somewhat better agreement with the data. Some of the overshoot effect in the data is not entirely captured by the analytical solution, especially for Case 2. The computed wave friction factors f_w for Cases 1 and 2 were 0.019 and 0.033, respectively, when τ_{bmax} was employed, and 0.014 and 0.025 when τ_{bav} was used. Using different formulations for u_{*m} will not change u_w as much as it will affect the calculation of the shear stress.

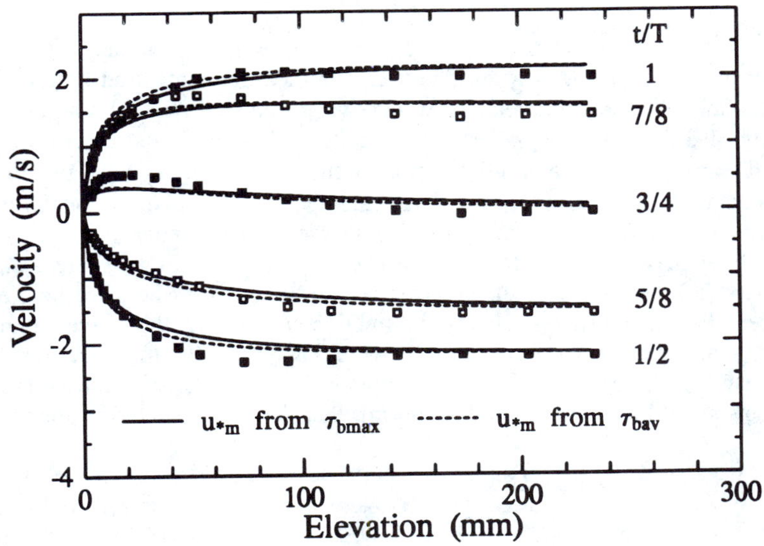

Figure 3. Calculated and measured velocity in the turbulent boundary layer for Case 1 from the data by Jonsson (1980).

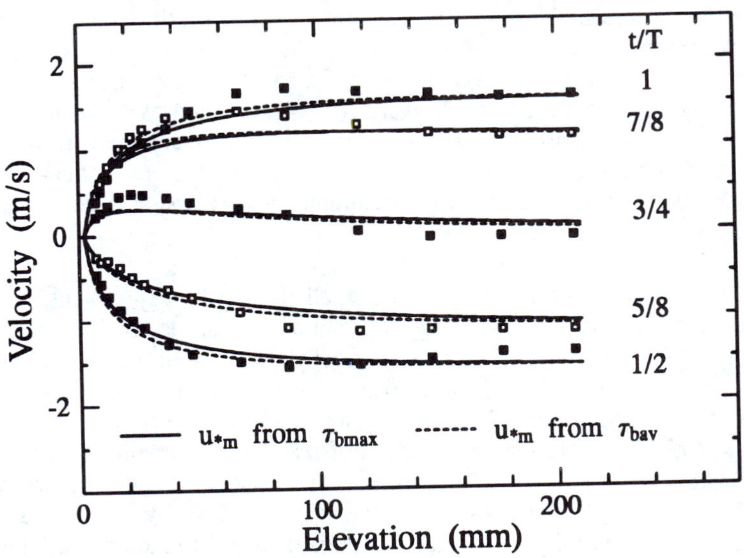

Figure 4. Calculated and measured velocity in the turbulent boundary layer for Case 2 from the data by Jonsson (1980).

Data From Nadaoka et al. (1994)

Nadaoka et al. (1994) measured u_w in an oscillatory tunnel using air for a free-stream velocity that was asymmetric. The measurements used here to evaluate the TBL model involved a velocity u_b that was of cnoidal type with a positive peak velocity of 2.50 m/s, a negative peak velocity of 1.05 m/s, and a period of 5 s. A cnoidal wave producing a non-dimensional time variation in u_b corresponding to the experimental conditions implies an Ursell number of $U_r=57.8$, although during the experiment u_b was generated to agree with the velocity induced by a hyperbolic wave. In general, such a strongly nonlinear wave provides a severe test for the linearized TBL equation; neglecting the nonlinear terms in the governing equation assumes that the particle velocity is small compared to the wave phase speed (Madsen and Wikramanayake 1991), which may not be the case for strongly nonlinear waves. However, for data obtained in oscillatory tunnels the spatial gradients should be small enough to permit that the nonlinear terms are neglected.

Instead of using a cnoidal or hyperbolic wave to describe u_b in the solution given by Equation 2, u_b was approximated using a wave described by 20-order stream function theory ($H=5.26$ m, $T=10.3$ sec, and $h=8.22$ m; the resulting velocity as a function of non-dimensional time is shown in Figure 5 together with the generated hyperbolic velocity variation during the experiment). Stream function, cnoidal, and hyperbolic theory produced essentially identical variation in u_b with time, but the former theory allows direct calculation of u_w from Equation 9 for steady-state conditions without having to compute for the transient phase, which is necessary if the general solution in Equation 2 is employed. The bed consisted of spray-painted aluminum and was judged to be hydraulically smooth during the experiments. Thus, the length scale z_o is independent of the boundary roughness and may be calculated from $z_o=(3.3\nu_a/u_{*m})/30$ (Madsen and Wikramanayake 1991), where ν_a is the kinematic viscosity for air. The air temperature was about 10 deg during the experiment and the corresponding Reynolds number was $Re=2.8\ 10^5$. The origin of the vertical axis in the measurements was assumed to approximately coincide with z_o.

Since smooth turbulent flow prevailed during the experiment, z_o could be obtained from ν and u_{*m} and no calibration was needed to estimate the bed roughness. The representative shear velocity was based on τ_{bav}, which was determined through time integration over T. A value of $u_{*m}=0.065$ m/s was thus calculated implying $z_o=0.024$ mm. Figure 6 displays measured and calculated velocity profiles for selected phase values of t/T. The maximum positive peak in u_b occurred at about $0.18t/T$, the maximum negative peak at $0.68t/T$, and zero velocity at $0.36t/T$. The model captures the overall features of the velocity variation in the boundary layer, but the overshoot effect is not well predicted by the model, especially during the phase of flow reversal in the boundary layer in connection with large gradients in the wave velocity. The simple eddy viscosity model employed in the linearized TBL equation is most likely the reason for the disagreement between the model and the measurements (Sleath 1987), although lack of detailed information on z_o and the use of stream function theory to describe u_b may also contribute to the discrepancy.

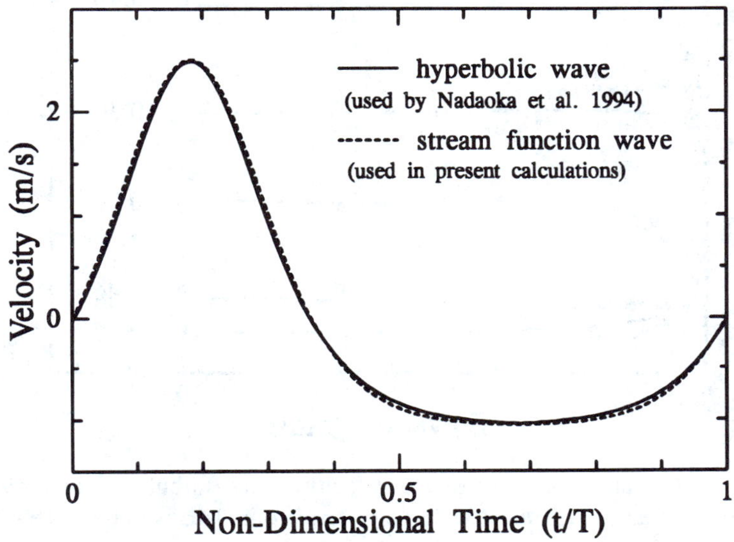

Figure 5. Hyperbolic wave velocity generated by Nadaoka et al. (1994) in their experiment on turbulent boundary layers and the velocity from a stream function wave that approximates the hyperbolic wave.

Figure 7 shows the computed normalized shear stress as a function of time at selected elevations based on the experimental conditions from Nadaoka et al. (1994). The time variation in the shear stress differs significantly from u_b and τ is completely asymmetric. This feature of the shear stress under nonlinear waves is important to include in for example detailed sediment transport calculations that employ the instantaneous shear stress to compute the transport rate.

CONCLUDING REMARKS

The analytical solution presented in Equation 2 is valid for any time-varying free-stream velocity, although from a physical point of view the solution is only interesting when the conditions underlying the linearized TBL equation are fulfilled. For example, the solution will describe the temporal growth of a TBL under a uni-directional current (compare Figure 1); however, the boundary layer will grow infinitely large and at some time the assumption that the layer thickness is small compared to the water depth will be violated and the solution lacks physical meaning. Also, for strongly nonlinear waves where the convective terms in the momentum equations are significant, the linearized TBL equation will yield results that are not

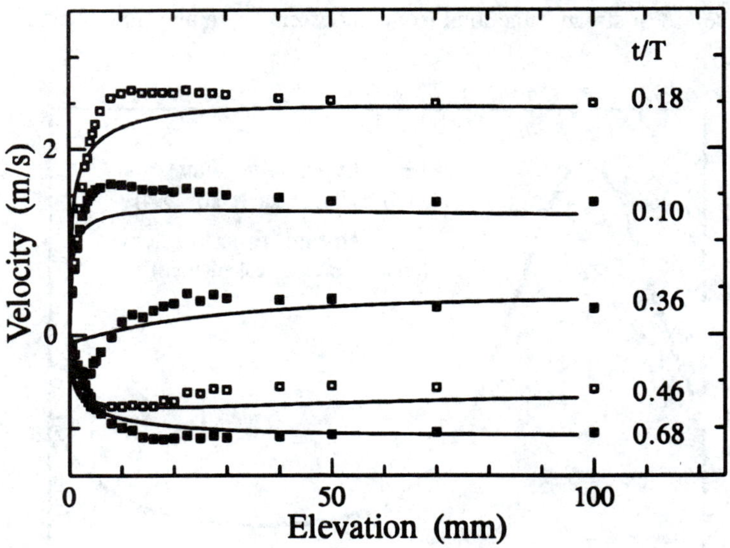

Figure 6. Calculated and measured velocity in the turbulent boundary layer using the asymmetric velocity case from Nadaoka et al. (1994).

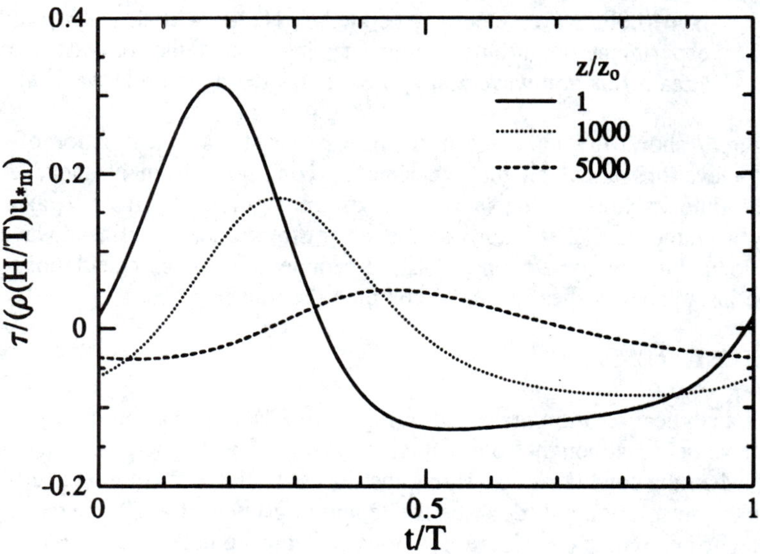

Figure 7. Calculated time variation in the shear stress at selected elevations for the experimental conditions given by Nadaoka et al. (1994).

reliable. For the case tested from the data by Nadaoka et al. (1994) the linearized equation failed to produce accurate predictions, especially during the phase of flow reversal in the boundary layer in connection with large gradients in the wave velocity. This is mainly attributed to the simple eddy viscosity model approach employed which fails to realistically capture the variation in the turbulence during a wave cycle (Sleath 1987). A time-varying eddy viscosity would most likely improve the agreement with the data, but such a formulation would not permit an analytical solution of the governing equations.

However, the simple eddy viscosity model of Grant and Madsen (1979) seemed to produce acceptable results for a sinusoidal wave velocity, even if ν increases without limit with distance from the boundary. This observation pertains mainly to the velocity profile, whereas the shear stress is more sensitive to the eddy viscosity formulation because of the dependence on the velocity gradient. Similarly, the velocity profile is not overly sensitive to the choice of representative friction velocity, in contrast to τ which is more affected by the u_{*m} used. It was found in the present study that a u_{*m} based on the mean absolute shear stress during a wave cycle produced somewhat better agreement with data for a sinusoidal wave than using the maximum absolute shear stress.

ACKNOWLEDGEMENTS

The author is grateful to Professor Akira Watanabe, University of Tokyo, for valuable comments on the paper. Professor Masahiko Isobe, University of Tokyo, developed the program for calculating waves by stream function theory. Professor Kazuo Nadaoka and his coworkers at Tokyo Institute of Technology kindly provided the data from their experiments. The support from the Japan Society for the Promotion of Science for the research visit of the author to the University of Tokyo is gratefully acknowledged, as well as the assistance from all members of the Coastal Engineering Laboratory during his stay.

The research presented in this paper was partly conducted under the SAFE Project of the Marine Science and Technology Program (Contract No. MAS3-CT95-0004) funded by the Commission of the European Communities, Directorate for Science, Research, and Development.

REFERENCES

Brevik, I. 1981. "Oscillatory Rough Turbulent Boundary Layers," Journal of Waterways, Port, Coastal and Ocean Engineering, Vol 103, pp 175-188.

Dean, R.G. 1965. "Stream Function Representation of Nonlinear Ocean Waves," Journal of Geophysical Research, Vol 70, No. 18, pp 4561-4572.

Dean, R.G. 1990. "Stream Function Wave Theory and Applications," In: Handbook of Coastal and Ocean Engineering, Herbich, J.B (ed), Gulf Publishing Company, Vol 1, pp 63-94.

Grant, W.D. and Madsen, O.S. 1979. "Combined Wave and Current Interaction with a Rough Bottom," Journal of Geophysical Research, Vol 84, No. C4, pp 1797-1808.

Grant, W.D. and Madsen, O.S. 1986. "The Continental-Shelf Bottom Boundary Layer," Annual Review of Fluid Mechanics, Vol 18, pp 265-305.

Jonsson, I. 1966. "Wave Boundary Layers and Friction Factors," Proceedings of the 10th Coastal Engineering Conference, American Society of Civil Engineers, pp 127-148.

Jonsson, I. 1980. "A New Approach to Oscillatory Rough Turbulent Boundary Layers," Ocean Engineering, Vol 7, pp 109-152.

Kajiura, K. 1968. "A Model of the Bottom Boundary Layer in Water Waves," Bulletin of the Earthquake Research Institute, Vol 46, pp 75-123.

Madsen, O.S. and Wikramanayake, P.N. 1991. "Simple Models for Turbulent Wave-Current Bottom Boundary Layer Flow," DRP-91-1, Coastal Engineering Research Center, US Army Waterways Experiment Station, Vicksburg, MS.

Myrhaug, D. 1982. "On a Theoretical Model of Rough Turbulent Wave Boundary Layers," Ocean Engineering, Vol 9, No. 6, pp 547-565.

Nadaoka, K., Yagi, H., Nihei, Y., and Nomoto, K. 1994. "Characteristics of the Structure of Turbulence Under Asymmetrical Oscillatory Flow," Proceedings of the Japanese Coastal Engineering Conference, Japan Society of Civil Engineers, pp 141-145. (in Japanese)

Nielsen, P. 1992. "Coastal Bottom Boundary Layers and Sediment Transport," World Scientific, Singapore.

Sleath, J.F.A. 1987. "Turbulent Oscillatory Flow Over Rough Beds," Journal of Fluid Mechanics, Vol 182, pp 369-409.

Trowbridge, J. and Madsen, O.S. 1984a. "Turbulent Wave Boundary Layers. 1. Model Formulation and First-Order Solution," Journal of Geophysical Research, Vol 89, No. C5, pp 7989-7997.

Trowbridge, J. and Madsen, O.S. 1984b. "Turbulent Wave Boundary Layers. 2. Second-Order Theory and Mass Transport," Journal of Geophysical Research, Vol 89, No. C5, pp 7999-8007.

CHAPTER 252

Observed Suspended Sediments in Storm Conditions

J.J. Williams[1], C.P. Rose[2], P.D. Thorne[1], L.E. Coates[2], J. R. West[2], P.J. Harcastle[1], J.D. Humphery[1], S.P. Moores[1] & D. J. Wilson[1]

Abstract

High resolution synchronous measurements of turbulence, waves and vertical suspended sediment (S_{sed}) concentration profiles have been obtained during a storm above a rippled bed comprising coarse/medium sand located close to the end of a large offshore sand bank in a water depth of approximately 20m. These data are utilised to study the temporal characteristics of sediments in suspension. Sediment resuspension 'events', exhibiting vertical coherence up to approximately 30cm above the sea bed, are observed to span a range of temporal scales encompassing half wave, wave and wave group periods. The vertical concentration of S_{sed} is shown to be enhanced significantly by wave groups. Average S_{sed} profiles are described using a simple Rouse-type and an new 'convective' model.

Introduction

At the most fundamental level, a fluid motion directed away from the bed with a velocity exceeding the *in situ* grain settling velocity is the basic requirement for the suspension of sediments. In marine situations, such fluid motion can arise in three ways (*Pykhov et al., 1995*): (a) turbulent fluid '*bursting*' (*Kline et al., 1967; Hino et al., 1983; Sleath, 1987*); (b) vortices formed by interactions between waves, currents and the bed (*Mogridge & Kamphuis, 1972; Kaneko, 1981*); and (c) vortices formed under breaking waves which penetrate to the bed (*Nadaoka & Kondoh, 1989*). Spatial and temporal scales determined by the hydrodynamics and the physical properties of the sediment makes simple descriptions of the instantaneous physical characteristics of suspended sediment difficult and most approaches consider only time-average conditions. Further, in field situations, waves at angles to tidal currents frequently lead to the generation of complex and interacting bedform assemblages (*Sleath, 1984*) adding further complexity to the modelling of sediment dynamics.

[1] Proudman Oceanographic Laboratory, Bidston Observatory, Birkenhead, Merseyside, L43 7RA, UK, (*jjw@pol.ac.uk*).

[2] School of Civil Engineering, The University of Birmingham, Edgbaston, Birmingham, B15 2TT, UK.

Recently, the use of acoustic instruments has enabled acquisition of high temporal resolution data pertaining to the vertical distribution of suspended sediment in marine conditions (e.g. *Thorne et al., 1993*; *Lee & Hanes, 1996*) and progress has been made in understanding and modelling of sediment resuspension and suspension dynamics. However, the majority of experiments examining sediment processes in marine situations have been conducted just outside the breaker zone in moderate conditions (e.g. *Osborne et al., 1994*) and few field data relate to the mobilisation of sediments during storm conditions at offshore locations (e.g. *Green et al., 1995*). As a consequence, data necessary for validation of numerical models of sediment transport in extreme conditions remains scarce. The present paper presents field observations of suspended sediments and hydrodynamic conditions in the first metre of the water column above a rippled coarse/medium sandy bed in a water depth of 20m during a moderate storm.

Field site and measurements

The study site was located close to the northern end of Middelkerke Bank in the Flemish sand banks, southern North Sea, Europe (*Figure 1a*). Middelkerke Bank is approximately 10km long, 1-2km wide and is located approximately 10km offshore. Bottom sediments consisted of medium/coarse sand (D_{50} = 450μm). Approximately 15% of the sediment by weight was composed of platy angular shell fragments (D_{50} = 650μm). Data relating to hydrodynamic conditions (turbulence and waves) and vertical suspended sediment concentration (S_{sed}) profiles have been obtained using the autonomous benthic lander *STABLE* (*Figure 1b*, *Humphery & Moores, 1994*).

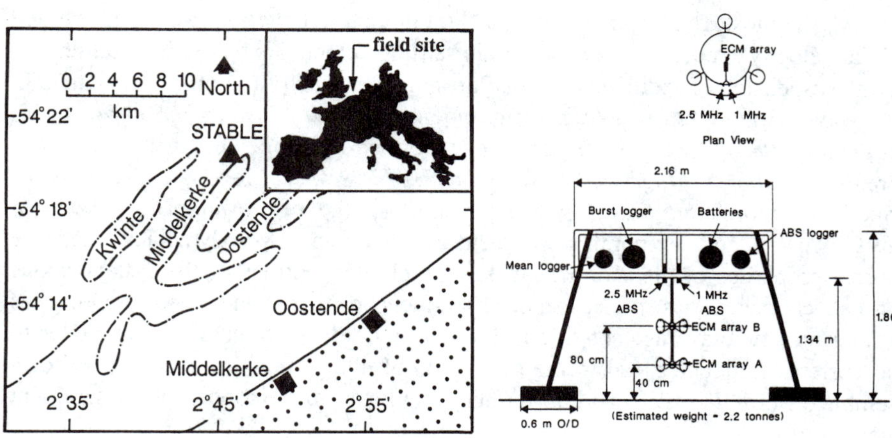

(*a*) (*b*)

Figure 1. (a) Location of the field site at Middelkerke Bank, Belgium; (b) *STABLE* (Sediment Transport And Boundary Layer Equipment).

Instantaneous near-bed flow in orthogonal vertical planes (x-z, y-z) was measured using pairs of electromagnetic current meters (*ECM*, diameter 10cm) at heights (z) of approximately 36cm and 76cm above the sea bed (*Figure 1b*). A pressure sensor at z = 1.63m was used to measure waves. *ECM* and wave data were sampled at 8Hz for approximately 19 minutes every hour in '*burst*' mode. The analysis methods used to obtain hydrodynamic parameters referred to below are described by *Hannay et al., (1994).* A vertical profile of S_{sed} perpendicular to the *ECM*'s over the range 1cm < z < 100cm with a vertical resolution of 1cm was derived from two acoustic backscatter sensors (*ABS*) operating at 1.0MHz (*ABS1*) and 2.5MHz (*ABS2*), (*Thorne et al., 1993*). These sensors were separated horizontally by a distance of 10cm along a line normal to the *ECM* axis (*Figure 1b*) and were synchronised precisely with *ECM* and pressure sensor data. *ABS* data were recorded independently in *burst* mode at 4Hz for approximately 9 minutes every hour. *ABS* data were calibrated in the laboratory using samples of the bed sediments from the *STABLE* deployment site (*Wylie et al., 1994*). For the range of instantaneous and average S_{sed} values observed in the present study ($\approx 10^{-4}$ - 10^{1}g/l), close agreement was found between *ABS1* and *ABS2*. Estimates of ripple wavelength ($\lambda \approx$ 50cm) were obtained in the field using the *ABS* instruments and ripple height (a) was estimated using an empirical relationship between ripple steepness and grain Shields number (*Nielsen, 1981*) and approximated to 7cm. The physical roughness (k_s) of the ripples was calculated using $k_s = 8a^2/\lambda$, (*Nielsen, 1992*).

Experimental conditions

In total 50 *burst* records encompassing calm and storm conditions were obtained. A detailed description of the hydrodynamic conditions occurring during the periods preceding and following the storm is given by *Williams et al., (1996).* In the present paper attention is focused upon a period of 5 hours spanning the peak of storm activity (*bursts* 34-39) when the average S_{sed} value reached a maximum. *Burst* average hydrodynamic conditions and S_{sed} values observed during this period are summarised in *Figure 2*.

During *bursts* 34-39, wave direction ($\psi_w \approx 330^o$) and current direction ($\psi_c \approx 200^o$) remained approximately constant and *burst* average S_{sed} values increased in response to increasing tidal current speed and wave orbital speed (*Figure 2*). Following *burst* 38, S_{sed} values decreased as the storm subsided. During the storm, waves were observed to be long-crested and groupy in nature and the vertical position of the bed was observed to change relative to *ABS* sensors. Since the motion sensors on *STABLE* indicated that the rig remained static during the experiment it is considered likely that the observed changes in bed elevation are attributable to the migration of bedforms beneath *STABLE*.

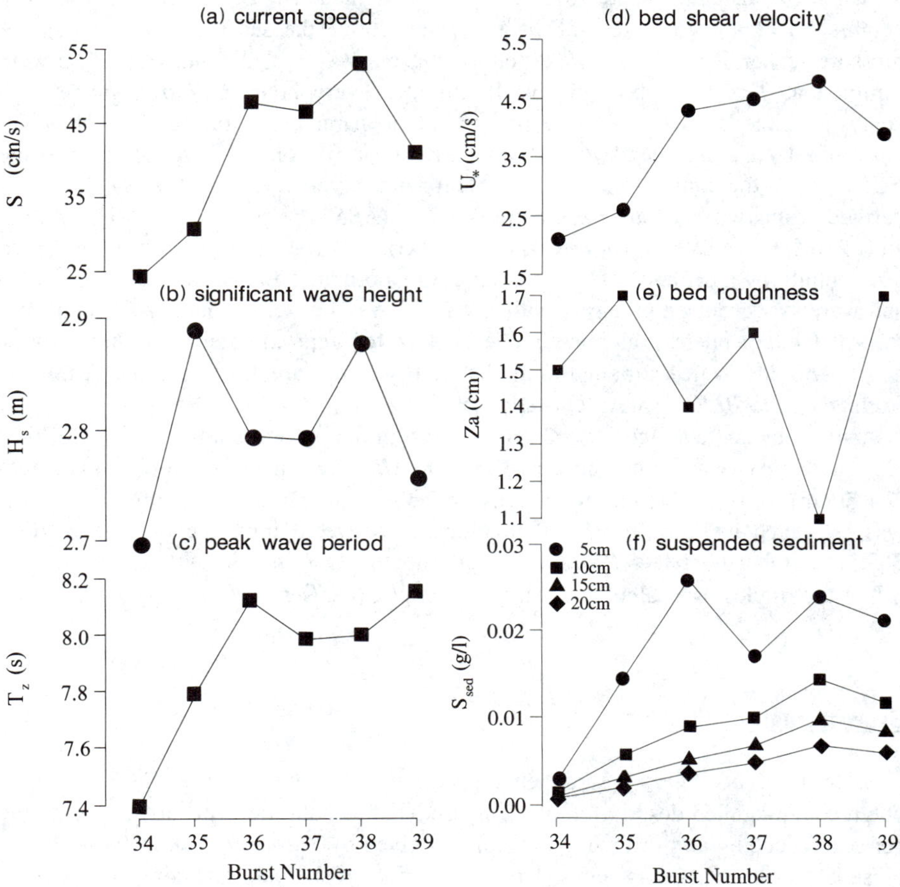

Figure 2. (a) *burst* average current speed (S) at z = 76cm ; (b) significant wave height (H_s); (c) peak wave period (T_p); (d) *burst* average bed shear velocity (currents) obtained using the turbulent kinetic energy (t.k.e.) method (*cf. Hannay et al., 1994*); (e) apparent bed roughness (z_a); and (f) S_{sed} at z = 5-20 cm.

Results and discussion

Observed temporal and spatial structure of sediment resuspension events

Typical time series measurements of S_{sed} at z = 4cm, 6cm, 10cm and 40cm and surface wave amplitude are shown in *Figure 3*. S_{sed} time series in this *figure* comprise a 'background' population of suspended material ($\approx$0.02g/l at z = 4cm) and discrete resuspension *events* in which instantaneous S_{sed} peaks frequently exceed 1.0g/l. Whilst *Figure 3* shows strong visual correlation between waves and S_{sed} *events* and marked vertical coherence in S_{sed}, coherent vertical fluid motion necessary to suspend sediment was not detected by the *ECM* sensors. This is attributed to the dissipation of coherent vertical fluid motion by $z = 40$ cm and to the physical size of the *ECM* measuring volume.

When examining *Figure 3* it is important to remember that during the storm, long-crested waves were approximately normal to the tidal currents and that the location of the *ABS* sensors changed in relation to ripple crests through time as ripples migrated. However, if it is assumed that resuspension *events* occurred along a significant length of the rippled bed parallel to the wave crests, (i.e. vortices formed by wave action are also long-crested), then material in suspension will be advected past the *ABS* sensors by the combined action of waves and the tidal current as coherent clouds approximately in phase with the waves.

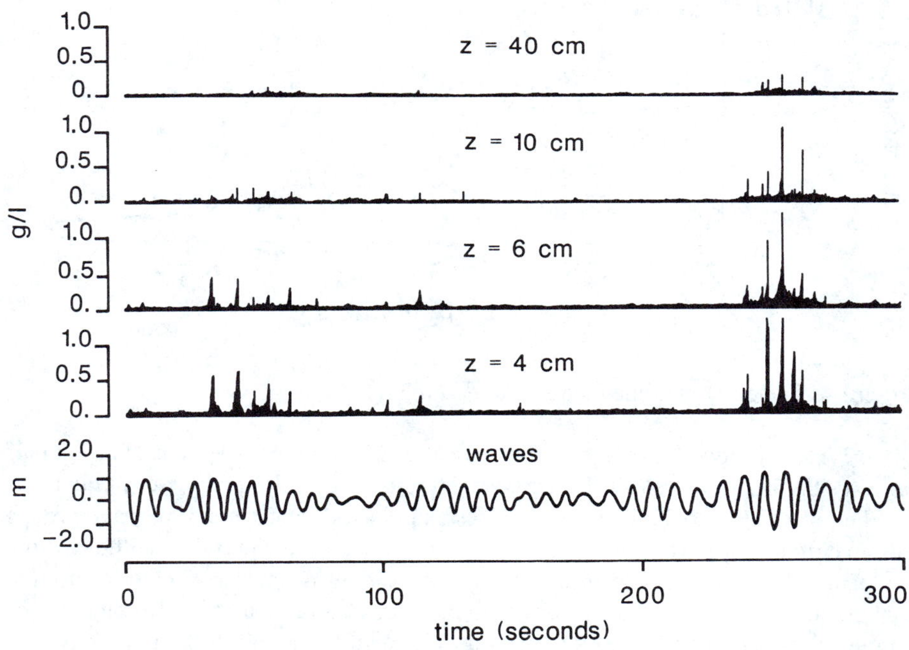

Figure 3. Time series of observed suspended sediments and waves.

Surprisingly, given the strong visual correlation between waves and resuspension *events* illustrated in *Figure 3*, periodic components at wave frequencies were not identified in power spectra of S_{sed} time series irrespective of the observation height above the sea bed. This is attributed to two problems: (i) the S_{sed} time series are 'spiky' in nature; and (ii) the temporal changes in spatial location between *ABS* sensors and resuspension *event* sources (i.e. ripple crestlines) and spatial variation in horizontal S_{sed} values combine with advected populations of suspended material to results in complex phase relationships between observed surface waves and S_{sed} time series. However, velocity^2-sediment cospectra (*Figure 4*) show statistically significant correlation's between the horizontal flow components squared (u^2 and v^2) and S_{sed}.

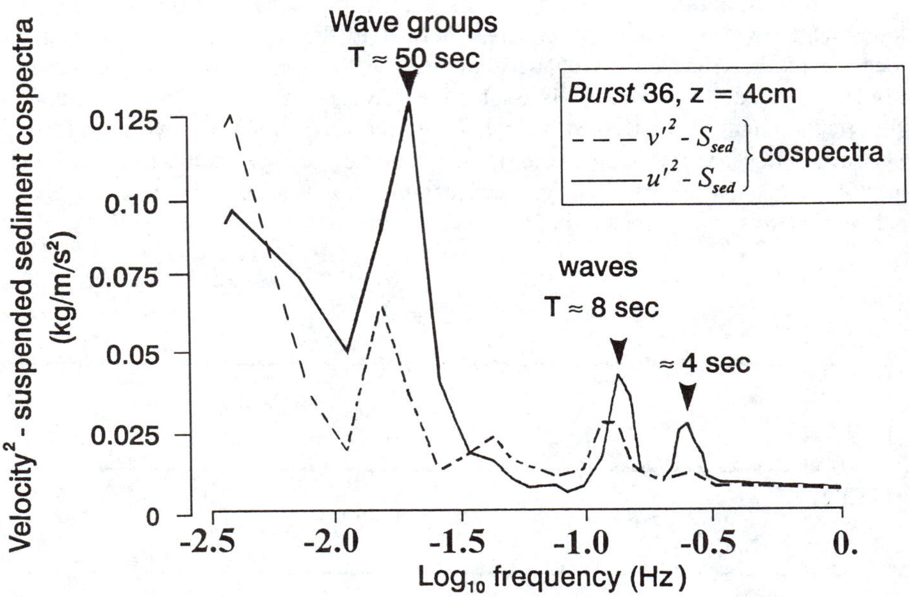

Figure 4. Velocity^2-sediment cospectra, *burst* 38.

Figure 4 shows correlation's between u^2 and S_{sed} and between v^2 and S_{sed} at half wave (≈0.25Hz), wave (≈0.125Hz), wave group (≈0.01Hz) and lower frequencies (in the range 0.01 - 0.003Hz). In all cases examined during the storm, wave groups dominate the cospectra. This is also illustrated visually in *Figure 3*. Although low frequency fluid motion is recorded by *ECM* sensors, it is unclear whether or not the low frequency peak in the velocity^2-sediment cospectra is a real phenomenon or simply an artefact of the analysis method. If shown to be real, further work is then required to quantify the role of this low frequency component in suspended sediment transport.

Waves recorded during *bursts* 34-39 exhibited well developed groupy structure (*Figure 3*). Initial visual inspection of correlation's between the wave groups and modulation in suspended sediment concentration indicated strongly that successive waves in a group resulted in a progressive increase in average S_{sed} values up to $z \approx$ 30 cm consistent with the idea of *wave pumping*. It was also determined that average

S_{sed} values associated with the last wave in a given group of more than 6 waves were a factor of approximately 2 times larger than values associated with the first wave in the same group. Following passage of a group, average S_{sed} values were observed quickly to decline to their former value.

In order to investigate further the wave-by-wave 'pumping' effect, the Goda run-length was utilised (*Goda, 1985*). Here, the start of a wave group was defined by the passage of a wave of height (h_j) that was observed to mobilise bottom sediments. For the group to continue, successive waves must be equal to or exceed h_j. In all cases an observed linear increase in average S_{sed} values with j demonstrated the enhancement of the sediment resuspension process by the passage of a successive series of larger than average waves (*Figure 5*). At present suspended sediment transport formulae do not include explicitly this important effect and thus are likely to underestimate the sediment transport rate in combined wave-current situations.

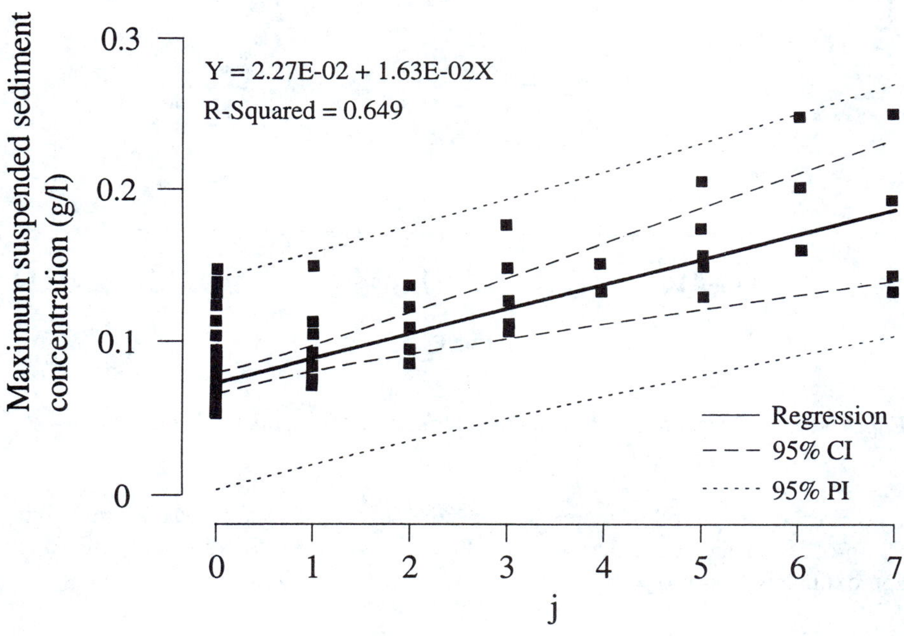

Figure 5. Goda wave group run length (*j*) versus observed suspended sediments.

<u>*Burst* average suspended sediment concentration profiles</u>

Burst average suspended sediment concentration profiles observed over the range 1cm < z < 50cm are shown in *Figure 6*. Since ripples are observed to migrate beneath the *ABS* sensors during a *burst* measurement period, these profiles also represent horizontally spatially averaged S_{sed} values in the direction of ripple migration. The observed profiles show a progressive increase in the *burst* average suspended sediment concentration over the range 1cm < z < 50cm from a minimum during *burst* 34 to a maximum during *burst* 38. At z > 50cm, S_{sed} values are small (0.001g/l to 0.005g/l) and slowly decay with height above the bed. In all cases the profiles are smooth, continuous and are Rouse-like above z ≈ 20cm. In the bottom 10cm, S_{sed} profiles for *bursts* 37-39 exhibits a significant concave curvature which develops through the storm.

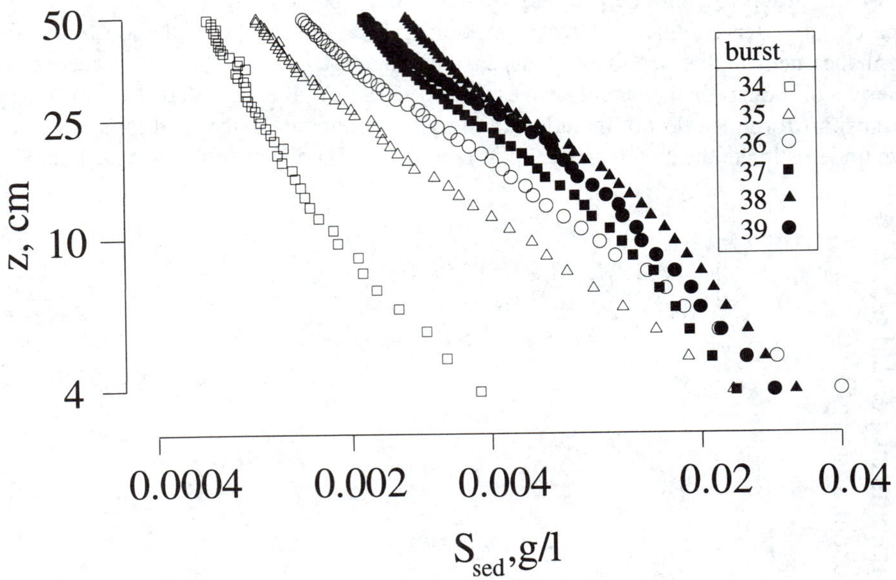

Figure 6. Suspended sediment concentration profiles, *bursts* 34-39.

During the period of increasing wave height preceding the peak storm activity, S_{sed} profiles in the range 3cm < z < 40cm for *bursts* 34-36 were found to follow approximately Rouse-type profiles defined by *Smith (1977)* as

$$C(z) = C(a)(z/a)^{-\alpha} \quad (1)$$

where α is the Rouse parameter (i.e. $\alpha = w_s / \beta\kappa U_{*wc-max}$), *C(z)* and *C(a)* are S_{sed} values at heights z and a, respectively, above the sea bed, w_s is the settling velocity (estimated to be 3.5 cm/s, *van Rijn, 1989*), β is the ratio of the eddy diffusion coefficients for the fluid and the sediment (assumed to be 1.0), κ is the von Kármán coefficient (0.4) and $U_{*wc-max}$ is the maximum wave-current bed shear velocity.

Using a least squares fit to the straight line portion of the observed S_{sed} profiles, values for α were obtained from the gradient of the straight line fit, (*Table 1*).

Burst	S (cm/s)	σ_w (cm/s)	α	$U_{*wc-max}$ *(cm/s)*
34	24.2	15.5	1.27	-
35	30.9	20.7	1.32	6.6
36	47.8	22.7	1.33	6.6

Table 1. *S*, RMS wave orbital speed σ_w, α and $U_{*wc-max}$, values derived from observed S_{sed} profiles (*Eq. 1*), *bursts* 34-36.

Utilisation of the observed α values in *Eq. 1* gave predicted S_{sed} values that matched well observed S_{sed} values for *bursts* 34-36 ($R^2 > 0.95$). $U_{*wc-max}$ values for *bursts* 35 and 36 agree broadly with values predicted by a number of wave-current models (e.g. *Soulsby, 1994*). Although S and σ_w values measured during *burst* 34 are significantly lower than those pertaining during *bursts* 35 and 36, a larger than expected value for $U_{*wc-max}$ was obtained from the *burst* 34 S_{sed} profile data (6.8cm/s). This may imply that the assumed settling velocity of 3.5 cm/s is incorrect for this *burst* and for this reason the value for $U_{*wc-max}$ for *burst* 34 is not included in *Table 1*.

In the case of *bursts* 37, 38 and 39 a new expression is required to simulate the concave curvature in the bottom 10cm of the observed profiles (*Figure 6*). In these cases it was found that the whole S_{sed} profile for 1 cm < z < 40 cm could be matched closely to the "convective" empirical profile proposed by *Nielsen (1992)* for natural sand ripples composed of coarse grains in oscillatory flows. Thus

$$C(z) = C(0)\left(\frac{z}{L} + 1\right)^{-\alpha'} \qquad (2)$$

where L is a characteristic length scale. *Nielsen (1992)* suggests a value of 2 for α'. In contrast to *Eq. 1*, *Eq. 2* tends towards a Rouse-type profile when $z/L >> 1$ and also tends toward $C(0)$ at the bed. The gradients of S_{sed} profiles depart from the form of *Eq. 2* at z > 30-40cm where finer material in suspension might be anticipated. Present data indicates that the vertical distribution of S_{sed} in the range 35cm < z < 100cm can be modelled well using a second Rouse-type profile with a smaller α value. However, for z > 50cm, S_{sed} values are observed to be 3-4 orders of magnitude less than those observed at z < 10cm and can therefore be neglected for practical purposes when estimating rates of sediment transport.

In the present study, *Eq. 2* was obtained using a semi-analytical approach by assuming that the total wave-current sediment diffusivity is the sum of the Rouse-type diffusivity defined in *Smith (1977)* and a constant diffusivity (ε_c) found in oscillatory only flow over ripples (*Nielsen, 1992*). This gives

$$-w_s C = \left(\kappa U_{*wc-max} z + \varepsilon_c\right)\frac{\partial C}{\partial z} \tag{3}$$

and hence

$$C(z) = C(a)\left(\frac{z+L}{a+L}\right)^{-\alpha'} \tag{4}$$

where α' is a Rouse-type parameter and $L = \varepsilon_c / \kappa U_{*wc-max}$. When $a = 0$, this profile is identical to *Eq. 2*.

The values of the parameters used in *Eq. 4* to obtain the best fit between observed and predicted S_{sed} profiles for *bursts* 37-39 are shown in *Table 2*. These were obtained by firstly estimating the gradient of the straight line portion of a given S_{sed} profile (α') on log-log axes (typically this fell within the range 15cm < z < 25cm). Using the suspended sediment concentration values at z = 4cm as a 'reference' concentration, values of L were obtained using $C(z)$ values up to z = 25cm. Finally, taking the average value of L (*Table 2*), $C(0)$ was calculated using observed concentration values at z = 4cm and $U_{*wc-max}$ values were derived from α' assuming w_s = 3.5cm/s (*van Rijn, 1989*). A comparison between computed and observed S_{sed} profiles is illustrated in *Figure 7*. In all cases the fit between observed and predicted S_{sed} profiles is statistically significant (R^2 >0.97).

Burst	S (cm/s)	σ_w (cm/s)	$C(0)$ (g/l)	α' –	L (cm)	k_s (cm)	$U_{*wc-max}$ (cm/s)
37	46.6	21.8	0.058	1.27	4.2	7.3	6.9
38	53.1	22.1	0.090	1.48	5.6	10.1	5.9
39	41.0	23.5	0.079	1.29	3.7	13.3	6.8

Table 2. S, σ_w and *Eq. 4* parameter values derived from observed hydrodynamics and S_{sed} profile data, *bursts* 37-39.

Table 2 shows that values for α' are lower than the value suggested *by Nielsen (1992)* and are similar to the α values shown in *Table 1* suggesting α' may be the same parameter. Values of the parameter L in the range 3.7-5.6 are only a factor of two to four times lower than k_s values. A relationship between these two parameters is therefore suggested. Work to improve the fit between observed and predicted S_{sed} profiles for these and other field observations of suspended sediment profiles in marine situations is currently being undertaken by the authors.

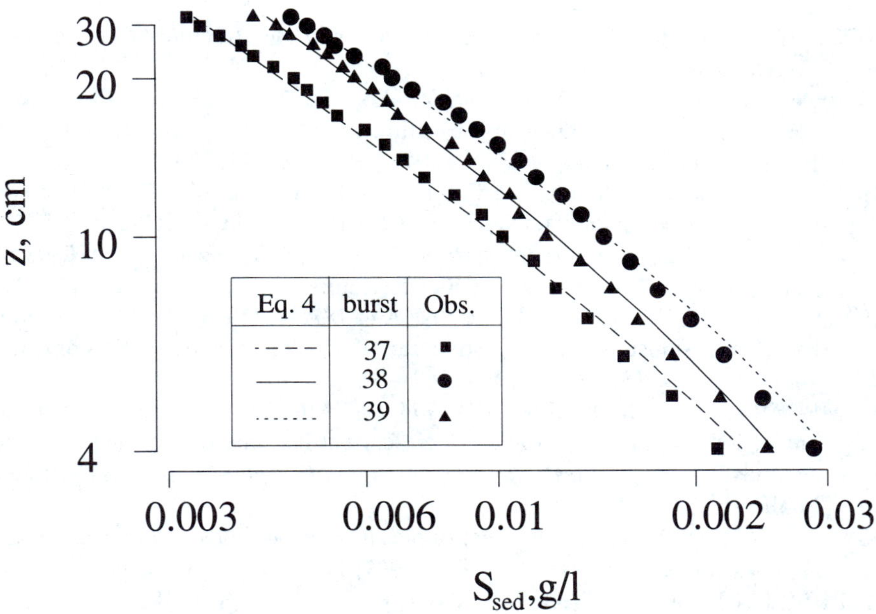

Figure 7. Comparison between and computed S_{sed} profiles, *bursts* 37-39.

Conclusions

(i) Field data have been obtained in the bottom 1m of the water column during a storm.

(ii) Present analysis has examined the temporal response of medium/coarse sandy sediments to waves and tidal currents.

(iii) Sediment resuspension has been shown to be enhanced significantly by the passage of groups of waves.

(iv) Average S_{sed} profiles observed during storm conditions have been represented using a 'simple' Rouse-type profile (*Smith, 1977*) and a new wave-current profile based upon the "convective" empirical profile proposed by *Nielsen (1992)*. The new profile allows accurate determination of *C(0)*.

Acknowledgements

This work was funded jointly by NERC and by The Commission of the European Communities Directorate General for Science and Education, Research and Development under contract number MAS2-CT-0024 (CSTAB).

References

GODA Y (1985) Random sea and design of maritime structures. University of Tokyo Press.

GREEN M.O., VINCENT C.E., McCAVE I.N., DICKSON R.R., REES J.M. & PEARSON N.D. (1995) Storm sediment transport: observations from the British North Sea shelf. *Continental Shelf Research*, **15**, 889-912.

HANNAY A., WILLIAMS J.J., WEST J.R. & COATES L.E. (1994). A field study of wave:current interactions over a rippled sandy bed. *EUROMECH 310: Sediment Transport Mechanisms in Coastal Environments and Rivers*, M. Belorgey, R.D. Rajaona & J.A.F. Sleath (editors), World Scientific, 345-359.

HINO M., KASHIWAYANAGI M. & NAKAYAMA A. (1983) Experiment on the turbulence statistics and the structure of a reciprocating oscillatory flow. *Journal of Fluid Mechanics*, **131**, 363-400.

HUMPHERY J.D. & MOORES S.P. (1994) STABLE II - An improved benthic lander for the study of turbulent wave-current-bed interactions and associated sediment transport. *Electronic Engineering in Oceanography*, IEE Conference Publication No. **394**, 170-174.

KANEKO A. (1981) Oscillation sand ripples in viscous fluids. *Proceedings of the Japanese Society of Civil Engineering*, **307**, 113-124.

KLINE S.J., REYNOLDS W.C., SCHRAUB F.A. & RUNDSTADLER P.W. (1967) The structure of turbulent boundary layers. *Journal of Fluid Mechanics*, **30**, 741-773.

LEE T.H. & HANES D.M. (1996) Comparison of field observations of the vertical distribution of suspended sand and its prediction by models. *Journal of Geophysical Research*, **101**, 3561-3572.

MOGRIDGE G.R. & KAMPHUIS J.W. (1972) Experiments on bedform generation by wave action. *Proceedings 13th. Conference on Coastal Engineering*, Vancouver, Canada, 1123-1142.

NADAOKA K. & KONDOH T. (1989) Turbulent flow field structure of breaking waves in the surf zone. *Journal of Fluid Mechanics*, **204**, 359-387.

NIELSEN P. (1981) Dynamics and geometry of wave-generated ripples. *Journal of Geophysical Research*, **86**, 6467-6472.

NIELSEN P. (1992) *Coastal bottom boundary layers and sediment transport*, World Scientific Publishing Co., River Edge, N.J, USA.

OSBORNE P.D., VINCENT C.E. & GREENWOOD B. (1994) Measurement of suspended sand concentrations in the nearshore: field intercomparisons of optical and acoustic backscatter sensors. *Continental Shelf Research*, **14**, 159-174.

PYKHOV N.V., KOS'YAN R.D. & KUZNETSOV S. YU. (1995) Time scales of sand suspending by irregular waves. *Proceedings of the Second International Conference on the Mediterranean Coastal Environment*, Tarragona, Spain, Vol. **2**, 1073-1092.

SLEATH J.P.A. (1984) *Sea Bed Mechanics.* John Wiley & Sons, 335 pp.

SLEATH J.P.A. (1987) Turbulent oscillatory flow over rough beds. *Journal of Fluid Mechanics*, **182**, 369-400.

SMITH J.D. (1977) Modelling of sediment transport on continental shelves. In: *The Sea*, 6, E.D Goldberg, I.N. McCave, J.J. O'Brien & J.H. Steele (Eds.), Wiley, New York, pp. 539-578.

SOULSBY R.L. (1994) *Manual of Marine Sands.* Report **SR 351**, HR Wallingford, UK, 70pp.

THORNE P.D., HARDCASTLE P.J & SOULSBY R.L. (1993) Analysis of acoustic measurements of suspended sediments. *Journal of Geophysical Research*, **98**, 899-910.

VAN RIJN L.C. (1989) *Handbook Sediment Transport by Currents and Waves.* Delft Hydraulics, Report **H 461** (2nd. edition).

WILLIAMS J.J., THORNE P.D., O'CONNOR B.A., HUMPHERY J.D., HARDCASTLE P.J., MOORES S.P. & COOKE J.A. (1996) Interactions between currents, waves and sediments in calm and storm conditions. *Continental Shelf Research*, (submitted).

WYLIE T., TAYLOR K. & BORN A. J. (1994) Design and calibration of the sediment tower. *POL Internal Document* Number **60**, 15pp, (*unpublished manuscript*).

CHAPTER 253

EBB TIDAL DELTA EVOLUTION OF COASTAL INLETS

Michael R. Dombrowski[1], Associate Member, ASCE
Ashish J. Mehta, Ph.D[2], Member, ASCE

ABSTRACT

Previous investigations have established the dependence of ebb delta volumes on wave energy and tidal energy at sandy inlets. In this study, this dependence was examined with respect to the rate of delta growth and the final equilibrium delta volume starting with the opening of a new inlet when no delta is present. A diagnostic model was developed for examining the influence of the ratio of wave energy to tidal energy on delta growth. Model sensitivity tests showed that increasing the suspended sediment concentration in the littoral zone caused the delta to approach equilibrium faster, but did not affect the equilibrium volume. Increasing the wave height increased the time of approach to equilibrium but decreased the volume. Finally, increasing the sand size increased the growth rate as well as the equilibrium volume. The model was applied to five Florida inlets. It was shown that the delta may never attain a true equilibrium volume, and the actual volume may fluctuate about a "quasi"-equilibrium volume consistent a wave energy to tidal energy ratio representative of the long-term wave and tidal conditions at the entrance.

INTRODUCTION

At the seafloor in the immediate vicinity of a coastal inlet the interrupted littoral sediment tends to accumulate and raise the floor, leading to the formation of an ebb delta. The ebb tidal delta grows due to the supply of littoral sediment and ultimately reaches an equilibrium volume when the condition of no net deposition is attained. At new inlets, or ones which have been closed for a period of time, the rate at which the seafloor is modified by deltaic formation depends on the prevailing physical

---

1) Coastal Engineer, Coastal Tech, 1234 Airport Road, Suite 104, Destin, Florida 32541, USA.

2) Professor, Coastal & Oceanographic Engineering, 336 Weil Hall, University of Florida, Gainesville, Florida 32611, USA.

conditions, availability of littoral sediment, and geologic setting. Previous investigations have established the dependence of the delta volume on wave energy and tidal energy at sandy inlets. In this study, this dependence was examined with respect to the rate of delta growth and final equilibrium delta volume starting with the opening of a new inlet when no delta is present. The aim of this study was therefore to examine the inter-dependence between significant physical parameters governing sediment transport and the rate of delta formation at coastal inlets.

PROCEDURE

To examine the influence of the effects of current and waves on the growth rate of ebb deltas, a diagnostic approach is developed. The growth process of the delta will have an initial condition of a new inlet with no delta present (Figure I.A). The opening of an inlet the ebb delta volume increases as the inlet tidal current deposits material derived from the littoral system and ultimately reaches an equilibrium volume when the condition of no net deposition is attained. Delta accumulation height will be simulated by modeling tidal currents and superimposed waves to determine the combined shear stress, τ_b (Pa). The seafloor will continue to rise on the condition that the combined shear stress is smaller than the critical shear stress, τ_{cr} (Pa), for deposition (Figure I.B). The model must then determine the delta volume when the seafloor reaches an equilibrium elevation (Figure I.C) due to a balance of shear stresses (ie. $\tau_b = \tau_{cr}$), and the estimate the time for the equilibrium to occur.

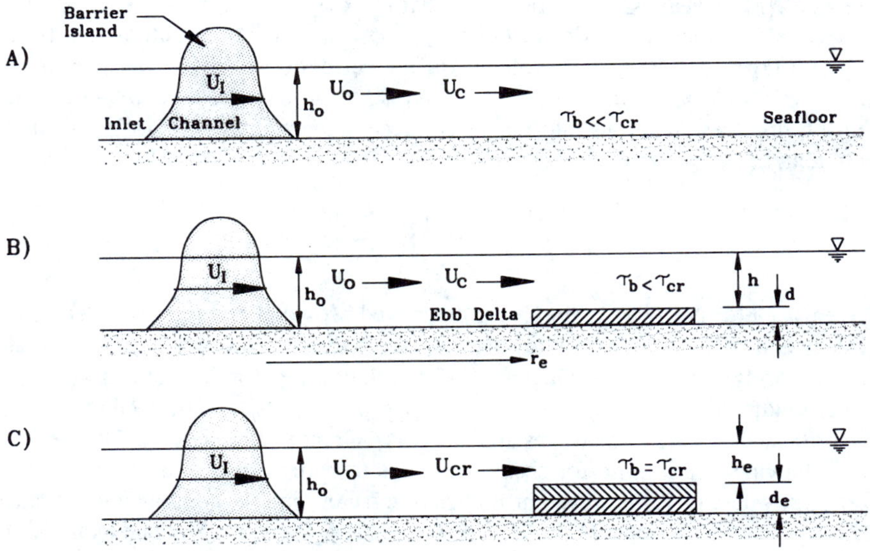

Figure I. Idealized ebb delta growth: A) initial condition; B) growth period; and, C) equilibrium condition

DIAGNOSTIC EXAMINATION OF SEAFLOOR EVOLUTION

The net decrease of suspended sediment mass per unit delta bed area, m, with respect to time, t (s), is related to sediment settling flux by

$$\frac{dm}{dt} = F_s \tag{1}$$

Substituting $F_s = -pW_sC_s$, where W_s = particle settling velocity (m/s), C_s = depth-averaged suspended sediment concentration (kg/m^3), and p = probability of deposition which is given as $p = (1 - \tau_b/\tau_{cr})$ (Krone, 1962) which can range between 1 (total deposition) and 0 (no deposition). The settling velocity can be expressed as $W_s = [(4gd_{50}/3C_D)(\rho_s\text{-}\rho_w/\rho_w)]^{1/2}$, where ρ_s = particle density (kg/m^3), ρ_w = seawater density (kg/m^3), d_{50} = median particle size (mm), g = acceleration due to gravity (m/s^2), and C_D = drag coefficient (Schiller and Naumann, 1933). The value of C_D outside the Stokes range (Reynolds number < 1) decreases rapidly then levels off and becomes nearly constant (e.g., 0.43 for spheres) in the fully turbulent flow regime considered.

Equation 1 can thus be expressed as

$$\frac{dm}{dt} = -H_f\left(1 - \frac{\tau_b}{\tau_{cr}}\right)W_sC_s \tag{2}$$

where H_f [x] = heavyside function such that H_f [x>0]=x, and H_f [x≤0]=0. Next, $\rho_d = A_Dm/V$, where ρ_d = dry bed density, A_D = ebb delta deposition area (m^2), m = mass (kg), and V = delta volume (m^3). Furthermore, $dV = dhA_D = d(d)A_D$, where dh = change in water depth and $d(d)$ = change in ebb delta height. Substituting these relations into Equation 1 results in an expression for the change of ebb delta height over time:

$$\frac{d(d)}{dt} = -H_f\left(1 - \frac{\tau_b}{\tau_{cr}}\right)\frac{W_sC_s}{\rho_d} \tag{3}$$

Given W_s, C_s, and ρ_d, Equation 3 can be solved provided τ_b and τ_{cr} are determined. Komar and Miller (1974) found that data for sediment threshold under oscillatory flows closely agreed with Shields' (1936) relationship for incipient grain motion under unidirectional flows. Thus, $\tau_{cr} = 0.058(\rho_s - \rho_w)gd_{50}$, can be used to determine the critical shear stress for waves and currents. Grant and Madsen (1978) prescribed the following relationship for shear stress due to both current and waves, $\tau_b = 0.5\,\rho_w f_{cw} U_t^2$, where f_{cw} is the wave-current friction factor, and the combined wave-current velocity near the bottom, $U_t = (U_{wb}^{\ 2} + U_{cb}^{\ 2} + 2U_{cb}\,U_{wb}\cos\phi)^{1/2}$. The quantity, U_{cb}, is the near-bed current velocity over the bottom, U_{wb} = near-bed orbital velocity due to waves (m/s), and ϕ = angle between the current and wave direction. During flood flow, when $\phi = 0$, the waves are able to penetrate over the shoal and into the inlet channel thereby causing more bottom scour at the delta during flood flow than ebb. Therefore, $\phi = 0$ will be assumed for this study.

For calculating the bottom stress, τ_b, the friction factor due to the combined current and waves is given by $f_{cw} = (|U_{cb}|\, f_c + |U_{wb}|\, f_w)/(|U_{cb}| + |U_{wb}|)$, where f_c = friction factor due to current, and f_w = friction factor due to waves. The near-bed orbital velocity due to waves can be obtained from linear wave theory, $U_{wb} = [H\pi \cosh(kh)/T \sinh(kh)]$ (Dean and Dalrymple, 1984), where k = wave number equal to $2\pi/L$ (1/m), L = wave length (m), T = wave period (s), and h = water depth at the delta (m). As a wave train propagates from offshore into shallower water, the wave height changes as the depth changes. According to linear wave theory, the shoaled wave height (m), $H = [H_o(C_o/2C)^{1/2}(b_o/b_s)^{1/2}]$, where b_o = distance between two adjacent deep water wave rays (m), b_s = distance between two adjacent nearshore wave rays (m), C_o = deep water wave celerity (m/s) equal to $gT/2\pi$, C = shallow water wave velocity (m/s) equal to $(gh)^{1/2}$, and H_o = deep water wave height (m). The model assumes the contours are to remain straight and parallel. Thus, the refraction coefficient, $(b_o/b_s)^{1/2} = 1$.

The initial step to determine the near-bed current velocity over the delta, U_{cb}, is to obtain the maximum velocity (m/s) through the inlet for a spring tide is given by $U_{max} = (0.86\pi P/\,T_t\,A_c\,)$, where P = spring tidal prism (m^3), T_t = tidal period (s), and A_c = throat cross-sectional area of the inlet (m^2) (O'Brien, 1969). The average inlet velocity at the mouth of the inlet channel over one-half tidal cycle (m/s), $U_I = (2U_{max}/\pi)$. As the flow exits the inlet channel it is considered to spread out from the inlet mouth. To obtain a characteristic velocity (m/s), U_o, at the shoreward end of the deposition area, this velocity is assumed to occur along an arc, one-half the distance (m), r_e (Figure I.B), from the entrance mouth to the outer edge of the tidal prism based ebb delta area (m^2), A_P, where $r_e = (2A_P/\pi)^{1/2}$ is obtained from continuity. Thus, $U_o = (2U_I w/\pi r_e)$, where w = width of the entrance (m).

As the seafloor rises, the water depth decreases with respect to the initial water depth (m), h_o, whereby to maintain the continuity of flow, the current velocity over the delta (m/s), U_c, must increase. The current velocity, U_c, decreases with distance from the entrance as the flow spreads out over the delta from its inner to outer limit. For this study, U_c will be defined as its value at the inner limit of the delta. It should also be noted that the velocity profile of U_c is vertically uniform, it is therefore necessary to apply a correction factor to obtain the near-bed velocity (m/s), U_{cb}.

From the logarithmic velocity profile (Mehta, 1978), the ratio of the near-bed velocity to the depth averaged current velocity, $U_{cb}/U_c = \ln(z_b/z_o)/\ln(h/z_o)\text{-}1$, in which z_o = theoretical origin of the logarithmic profile (m), and z_b = distance above profile origin (m) and is set here equal to 0.05 m. The virtual origin of the profile is obtained from the Manning-Strickler formula, $z_o = 10^7 n^6$.

Mehta and Özsoy (1978) noted that a representative Manning's n value of 0.028 can be used for sandy inlets with a typical initial water depth of 4.0 m. Thus, the current velocity obtained by continuity is multiplied by a correction factor of 0.40, $U_{cb} = 0.40U_o(h_o/h)$. Note that when the equilibrium delta volume is attained, $U_c = U_{cr}$, hence $U_{cb} = U_{cr} = 0.40U_o(h_o/h_e)$, where h_e = equilibrium water depth (m).

Inserting equations above into Equation 3 results in the governing equation for ebb delta height variation with time, and is expressed as

$$\frac{d(d)}{dt} = -\frac{W_s C_s}{\rho_d}\left(1 - \frac{\rho_w f_{cw}}{2\tau_{cr}}\left[\frac{(H\sigma)^2\cosh^2 kh}{4\sinh^2 kh} + 0.16\left(\frac{U_o h_o}{h}\right)^2 + \frac{H\sigma\cosh kh}{2.5\sinh kh}\frac{U_o h_o}{h}\right]\right) \tag{4}$$

In the finite difference form, the left hand side of Equation 4 becomes $\Delta d/\Delta t$, and was solved using the fourth order Runge-Kutta iteration method for the incremental change in delta accumulation, Δd, for $\Delta t = T_t$ (tidal period). The incremental change in delta accumulation, Δd, can then be multiplied by the depositional area, A_D, to obtain the incremental ebb delta volume, ΔV. The cumulative volume change is then plotted to illustrate the effects of waves and currents on ebb delta growth rate and estimate the duration to achieve an equilibrium volume.

MODEL PARAMETERS

Ebb delta area, A_D. It is necessary to identify the ebb delta depositional area, A_D, over which deposition occurs. This was achieved by empirically correlating the tidal prism based ebb delta area, A_P, with A_D using measurements of 21 ebb delta areas of Florida's lower Gulf Coast inlets (Davis and Gibeaut, 1990). The equation of the regression line relating the tidal prism based delta surface area and the delta depositional area, $A_D = 2.34A_P^{0.81}$, corresponds to the coefficient of regression of $r^2 = 0.65$ which shows an acceptable relationship. The surface area, $A_P = P/2a_{os}$, is characterized by spring tidal prism, P and spring sea tidal amplitude (m), a_{os}.

Suspended sediment concentration, C_s. Downing (1984) presented a time-series of sediment concentrations at three locations across the surf zone at Twin Harbor Beach, Washington. The investigator found two distinct types of vertical concentration profiles. The first occurred between resuspension events ranged from 0.0002 to 0.0004 kg/m^3, when the sediment concentration had vertical uniformity. While during resuspension events a concentration gradient, 0.0015 to 0.0100 kg/m^3, occurred within 0.10 m above the bed in a total column water depth of 0.25 m. These concentration ranges will be assumed to apply for this study.

Sediment grain size diameter, d_{50}. Mehta and Özsoy (1978) noted that for sandy inlets the median grain size at most inlets range between 0.2 and 0.4 mm. This range will be considered in the present study.

Deep water wave height and period, H_o. The deep water wave height has a significant effect on the growth rate of the ebb delta and its equilibrium volume. By adjusting the wave height, the model generated delta volume-time curve can be made to pass through the appropriate smallest and largest measured delta volumes at a given inlet. A characteristic wave period of 8 seconds will be used for all model runs.

Friction factors, f_w, f_c. The friction factor due to current, $f_c = 8gn^2/h^{1/3}$, where Manning's n and h = water depth (Mehta, 1978). Mehta and Özsoy (1978) noted that a typical mean Manning's n value of 0.028 can be used for sandy inlets. The initial water depth used to model the evolution of the ebb deltas averaged 4 m (Dombrowski, 1994), resulted in a characteristic friction factor due to current of 0.039. It should be noted that Mehta (1978) determined friction factors for three inlets on the Gulf Coast of Florida ranging between 0.021 to 0.050.

The friction factor due to waves, f_w, was obtained from the wave friction factor diagram developed by Jonsson (1965) which plots the friction factor against the wave Reynolds number. Given, h = 4 m, H_o = 0.4 m, and wave period of 8 seconds, $R_e = 1.7 \times 10^4$, corresponds to the fully turbulent flow range (Figure 6 in Jonsson, 1965). Given the typical variation of R_e in the present study, a representative value of $f_w = 0.005$ in the fully turbulent flow range was chosen.

Tidal inlet characteristics. The tidal inlet characteristics used in the analysis are derived from the database found in Dombrowski (1994). The characteristics include: inlet throat width, throat depth, tidal prism, and spring tidal range.

EFFECTS OF IMPORTANT PARAMETERS ON DELTA GROWTH

The effects of important parameters on the rate of delta formation at coastal inlets is examined. The three selected parameters are 1) suspended sediment concentration, C_s; 2) median sediment grain size, d_{50}; and 3) deep water wave height, H_o. The influence of varying these parameters on the volume growth curves are shown in plots of ebb delta volume versus time, beginning with a new inlet with no delta. The range of values of these three parameters are found in Dombrowski (1994).

Suspended sediment concentration, C_s. In the Equation 4 for the change rate of ebb delta height, $d(d)/dt$, is proportional to the suspended sediment concentration, C_s. Figure II plots the ebb delta volume, $V = A_D(h_o-h)$, versus time (years) for three suspended sediment concentrations. A characteristic that is evident from the growth curves is that the equilibrium ebb delta volumes are the equal (1.4×10^6 m^3) for the three concentrations. However, it is evident that as C_s increases the rate of deposition becomes more rapid.

Sediment grain size diameter, d_{50}. Two physical parameters are dependent on the median grain size diameter, d_{50}, the settling velocity, W_s, and the critical shear stress

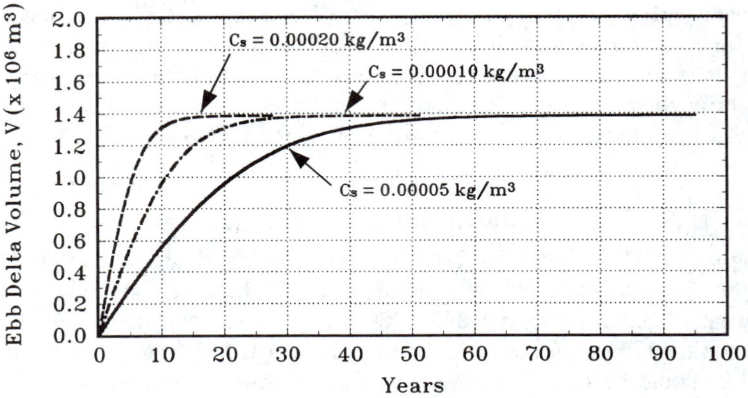

Figure II. An illustration of the influence of suspended sediment concentrations on calculated delta growth rate.

for sediment transport, τ_{cr}. The ebb delta volume versus time plot for varying sediment diameters (Figure III) is characterized by three different growth rates and equilibrium volumes. The increase in the sediment diameter increases the rate of deposition, due to the dependence of particle fall velocity on sediment size. An increase in the sediment size also increases the critical shear stress, allowing the sediment bed to remain more stable as compared to a bed composed of smaller grain size under the same flow conditions. This effect results in an increase in the equilibrium volume for increasing grain diameters.

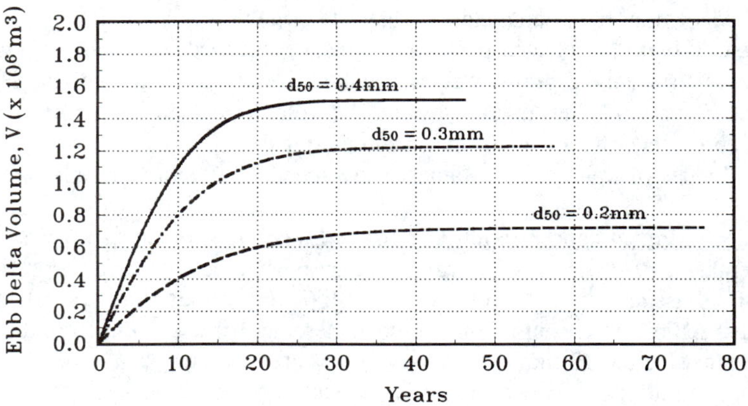

Figure III. An illustration of the influence of sediment grain size diameters on calculated delta growth rate.

Deep water wave height, H_o. As the waves approach the shoreline, its height increases as the water depth decreases. This increase in wave height in turn increases the near-bed orbital velocity, U_{wb}, hence reduces the rate of deposition. Figure IV plots the ebb delta volume versus time illustrating delta growth due to current alone, 0.0 m wave height, and two additional deep water waves heights of 0.4 and 0.8 m. During sea conditions when the deep water wave height is equal to 0.0 m, the rate of deposition is observed to be relatively rapid compared to the other two wave conditions. Note the drastic decrease in the equilibrium volume with increasing H_o.

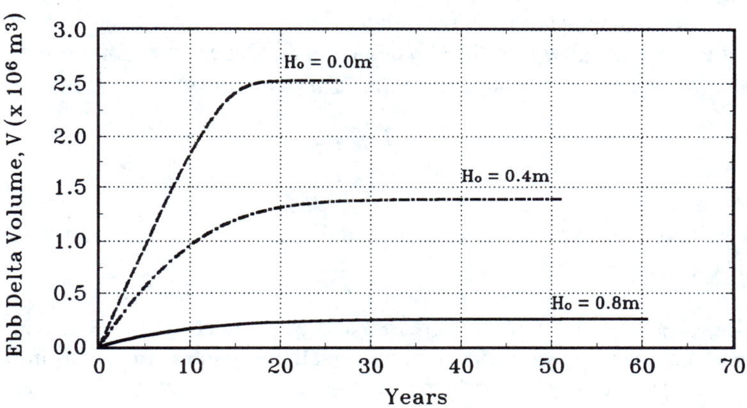

Figure IV. An illustration of the influence of deep water waves on calculated delta growth rate.

Model sensitivity tests showed that increasing the suspended sediment concentration in the littoral zone caused the delta to approach equilibrium at a greater rate, but did not affect the equilibrium volume. Increasing the wave height increased the time the delta approached equilibrium but decreased the equilibrium volume. Finally, increasing the sand size increased the growth rate as well as the equilibrium volume.

TIME-EVOLUTION OF DELTA VOLUMES

The time-evolution of sand volumes of five selected deltas along the east coast of Florida including those at 1) Jupiter Inlet; 2) South Lake Worth Inlet; 3) Boca Raton Inlet; 4) Bakers Haulover Inlet; and 5) Sebastian Inlet were analyzed. These inlets were chosen because 1) the date when each inlet was opened was available, and 2) four or more data points were available per inlet to represent the time-variation of ebb delta volumes. A plot of the measured delta volumes versus date of survey with the corresponding volume ranges for Jupiter Inlet obtained from the model is presented. The theoretical volume curves were derived from the model using the specific characteristics of the respective inlet. These data including 1) spring tidal prism, P; 2) inlet throat width, w; 3) inlet depth, h_o; and 4) spring tidal range, $2a_{os}$

are summarized in Dombrowski (1994). By adjusting the deep water wave height, H_o, and the suspended sediment concentration, C_s, the delta volume-time curves were made to pass through the appropriate smallest and largest measured delta volumes.

Wave energy to tidal energy ratio, α. The inlet stability parameter, α, was introduced by O'Brien (1971) and was later expanded by Mehta and Hou (1974) to provide an indicator of the stability of inlets. An inlet in equilibrium is due to a balance between the wave energy which tends to close an inlet and the tidal energy which maintains the opening. For the present study, the (non-dimensional) stability coefficient, α, defined as the ratio of longshore wave energy to the tidal energy is used to provide an indication of the relative effect of waves and tidal current in governing the rate of growth of the ebb delta. In a reduced form for a representative deep water wave height H_o, the relationship can be expressed as:

$$\alpha = \frac{H_o^2 T w T_t}{64 \pi a_{os} P} \qquad \textbf{(5)}$$

where T = wave period (s), T_t = tidal period, a_{os} = spring tidal range, and P = spring tidal prism).

As the deep water wave height is increased at a given inlet, α increases and reflects a tendency to drive material toward the inlet and the nearshore area, thus limiting the delta volume. Conversely, if the deep water wave was set to zero, the corresponding α would equal zero indicating a current-determined delta. This condition results in a larger ebb delta volume as compared to a higher α-value for the same inlet when waves are present.

Jupiter Inlet. Nine delta volumes were available for Jupiter Inlet since this entrance was re-opened for navigation in 1947 (Figure V). The near linear delta growth rate from 1947 to 1967 is characteristic of the high initial growth of the delta following the opening of the inlet. This high growth rate is consistent with the occurrence of a deep sea floor at the time of the entrance opening when the incipient influence of wave action is low. The non-zero delta volumes range between 0.23 x10^6 m^3 and 0.77 x10^6 m^3, for the years 1981 and 1993, respectively. This variability may, in part, be due to the method used in estimating the volumes which could be on the order of 15% for Jupiter Inlet. We infer that this variability is primarily influenced by wave action and its seasonal as well as year-to-year variation.

The model was used to simulate the growth curves matching the volume range. The α-value of 0.17 resulting from a H_o = 0.54 m yielded a volume of 0.77 x10^6 m^3 in 1993. The higher α-value of 0.27 was calculated for a H_o = 0.68 m to modify the growth curve to achieve a volume of 0.23 x10^6 m^3 in 1981. It is therefore surmised that the relative wide range in the delta volumes between the two curves is the result of waves relative to current. Larger delta volumes correspond to lower values of α and vice versa. The maximum range of α being 0.17 to 0.27 for this inlet.

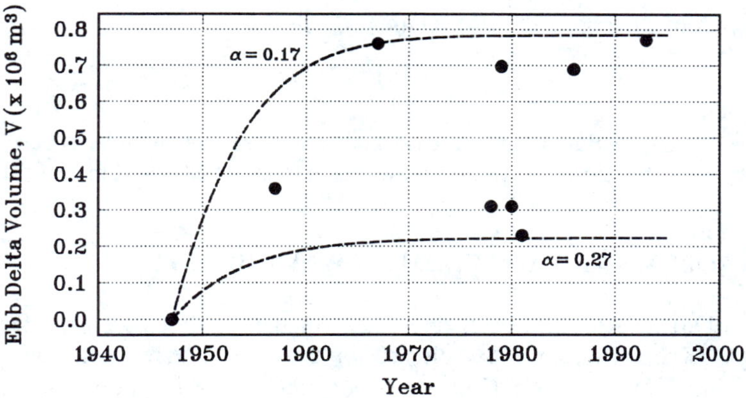

Figure V. Ebb delta volume versus year with model-calculated volume ranges for Jupiter Inlet.

INFLUENCE OF α ON DELTA GROWTH

A comparison of the ebb delta volume ranges and the corresponding wave to tidal energy ratio, α, for each of the five inlets is illustrated in Figure VI.

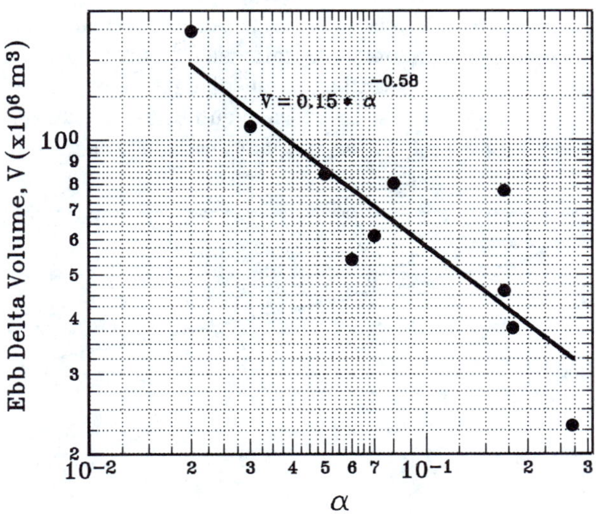

Figure VI. Ebb delta volume against wave to tidal energy ratio, α

As α increases, the ebb delta volume has a tendency to decrease, and vise versa. Although there is data scatter and a r^2-value of 0.58, which is low, the regression line does show that there in an inverse relationship between the delta volume to α.

The equation of the regression line relating α and V is

$$V = 0.15\alpha^{-0.58} \tag{6}$$

Note that V (delta volume) in this case may not represent the actual equilibrium value, but may be close to it, given the manner in which the curve fitting was conducted.

DELTA VOLUME VERSUS MAXIMUM WAVE HEIGHT

Devine (1996) related historical ebb delta volumes with episodic extreme sea conditions. The calculated delta volumes for three inlets were plotted against the corresponding maximum WIS (Hubertz *et al.*, 1993) wave height of the preceding year which showed an inverse relationship between the volumes and wave heights. For one inlet, there was no district relationship presumably because the ebb delta has not reached a equilibrium condition.

For this study, 13 surveys performed between 1974 and 1995 of Sebastian Inlet ebb delta were estimated and plotted versus the maximum WIS (Hubertz *et al.*, 1993 and Brooks, unpublished) wave height of the preceding year (Figure VII). A "best-fit" line through the data shows an inverse relationship of wave conditions on the ebb delta volumes. The average ebb delta volume for this time period, based on the available data, is 1.48 x10^6 m^3. The inlet has not undergone any major modification since before 1974, therefore the only one parameter controlling the delta volume is the changing sea conditions. The fluctuation of data about the average may be indicative of the ebb delta being in a "quasi"-equilibrium condition due to: 1) the variation of sea conditions; and, 2) the sink not available to accumulate more sand.

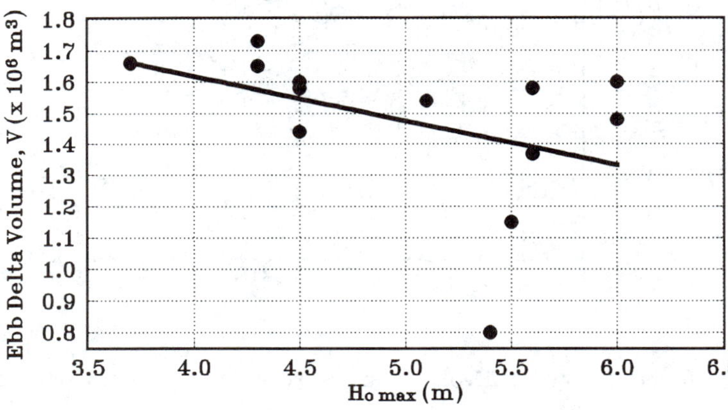

Figure VII. Ebb delta volume versus maximum WIS wave height for Sebastian Inlet.

CONCLUSIONS

Effects of Significant Physical Parameters on Delta Growth. Three parameters, namely the suspended sediment concentration, sediment grain size, and the deep water wave height, were varied in the diagnostic model developed for delta growth to determined their effects on the rate of delta formation at coastal inlets. It was found that an increase in the suspended sediment concentration increases the rate of approach to equilibrium, but does not result in a change in the equilibrium volume. On the other hand, a change in the sediment grain size and the deep water wave height effect both the rate of growth and the equilibrium volume. Thus, an increase in the sediment diameter increases the rate of growth due to the dependence of the particle fall velocity on sediment size, and increases the critical shear stress resulting in an increase in the equilibrium volume. An increase in the deep water wave height increases the near-bed orbital velocity at the site of the delta, hence decreases the rate of growth. The equilibrium delta volume likewise decreases.

Effects of α on Equilibrium Delta Volumes. It was shown through the application of the model to five Florida inlets that there is a dependence between the ebb delta volume and the wave to tidal energy ratio, α. The growth of the delta is determined by the rate at which the sand, supplied by the littoral system, is deposited by the ebb tidal flow. As wave action increases, thus increasing α-value, the delta growth rate decreases as wave and current induced bottom shear stresses scours sand deposited on the delta.

The dependence of delta volume on α partly explains the observed fluctuations in the delta volume at many inlets, since α tends to vary seasonally as well as annually. This relationship was further illustrated by plotting the ebb delta volume of Sebastian Inlet versus the maximum wave height of the preceding year. Another cause of variation of the delta volume at a given inlet is that even under constant sea conditions, the equilibrium volume often occurs after several decades following the opening of an inlet. Thus the delta volumes measured during the early years of evolution will be lower than the equilibrium volume. It was shown that the delta may never attain a true equilibrium volume, and the actual volume may fluctuate about a "quasi"-equilibrium condition. This would be consistent with a value of the wave energy to tidal energy ratio representative of the long-term wave and tidal conditions at the entrance.

ACKNOWLEDGMENTS

The authors would like to extend their appreciation to Rebecca Brooks of CERC for processing the additional wave information and to Ray LaRoux and Kathy FitzPatrick of the Sebastian Inlet Taxing District for providing the historic survey data of the inlet.

REFERENCES

Brooks, R.M., unpublished. Maximum WIS wave heights 1994 to 1995.

Davis, R.A. and Gibeaut, J.C., 1990. Historical morphodynamics of inlets in Florida: models for coastal zone planning, *Florida Sea Grant College Technical Paper 55*. University of South Florida, Department of Geology, Tampa, Florida, 81p.

Dean, R.G. and Dalrymple, R.A., 1984. *Water Wave Mechanics for Engineers and Scientists*, P.L. Liu (ed.), World Scientific Publishing Company, Singapore.

Devine, P.T., 1996. Modulation of inlet tidal deltas by severe sea. *M.S. Thesis, UFL/COEL-96/010*, University of Florida, Coastal and Oceanographic Engineering Department, Gainesville, Florida, 113p.

Dombrowski, M.R., 1994. Ebb tidal delta evolution and navigability in the vicinity of coastal inlets. *M.S. Thesis*, *UFL/COEL-94/010*, University of Florida, Coastal and Oceanographic Engineering Department, Gainesville, Florida, 95p.

Downing, J.P., 1984. Suspended sand transport on a dissipative beach. *Proceedings of the 19th Coastal Engineering Conference*, Houston, Texas, pp. 1765-1781.

Grant W.D. and Madsen, O.S., 1978. Bottom friction under waves in the presence of a weak current. *NOAA Technical Report ERL-MESA*, National Oceanographic and Atmospheric Administration, Washington D.C., 150p.

Hubertz, J.M., Brooks, R.M., Brandan, W.A. and Tracy, B.A., 1993. Hindcast wave information for the U.S. Atlantic Coast. *WIS Report No. 30*, Coastal Engineering Research Center, U.S. Army Engineer Waterways Experiment Station, Vicksburg, Mississippi, 20p., plus appendices.

Jonsson, I.G., 1965. Friction factor diagrams for oscillatory boundary layers. *Basic Research Report No. 10*, Coastal Engineering Laboratory, Technical University of Denmark, Copenhagen, pp. 10-21.

Komar, P.D. and Miller, M.C., 1974. Sediment threshold under oscillatory waves. *Proceedings of the 14th Coastal Engineering Conference*, Copenhagen, Denmark, pp.756-775.

Krone, R.B., 1962. Flume studies of the transport of sediment in estuarial shoaling processes. *Final Report*, Submitted to the U.S. Army Corps of Engineers, San Francisco District. Hydraulic Engineering Laboratory, University of California, Berkeley, California, 110p.

Mehta, A.J., 1978. Bed friction characteristics of three tidal inlets, *In: Coastal Engineering*, Elsevier Scientific Publishing Company, Amsterdam, The Netherlands, pp. 69-83.

Mehta, A.J. and Hou, H.S., 1974. Hydraulic constants of tidal entrances II: stability of Long Island inlets. *Technical Report No. 23*, University of Florida, Coastal and Oceanographic Engineering Department, Gainesville, Florida, 88p.

Mehta, A.J. and Özsoy E., 1978. Inlet hydraulics. *In: Stability of Tidal Inlets*, P. Bruun (ed.), Elsener Scientific Publishing Company, Amsterdam, The Netherlands, pp. 83-161.

Mehta, A.J., Dombrowski, M.R., and Devine, P.T., 1996. Role of waves in inlet ebb delta growth and some research needs related to site selection for delta mining. *Journal of Coastal Research*, Special Issue 23, Winter 1996, Fort Lauderdale, Florida, pp. 121-136.

O'Brien, M.P., 1969. Equilibrium flow areas of inlets on sandy coasts. *Journal of Waterways and Harbors Division*, American Society of Civil Engineers, 95(1), pp. 43-52.

Schiller, L. and Naumann, A., 1933. Über die grundlegenden Berechnungen bei der Schwerkraftaufbereitung, *Z. VDI*, Volume 77.

Shields, A., 1936. Anwendung der Ähnlichkeitsmechanik und der Turbulenzforschung auf die Geschiebebewegung, *Versuchsanstalf für Wasserbau und Schiffbau*.

CHAPTER 254

Adjustments toward Equilibrium of a Large Flood-Tidal Delta after a Major Dredging Program, Tauranga Harbour, New Zealand

Terry Healy, Joseph Mathew, Willem de Lange, and Kerry Black*

Abstract

The Port of Tauranga is located within a tidal inlet estuarine lagoon system which has been dredged to improve navigation for shipping since 1968. In 1992 a major dredging program deepened the entrance navigation channel through the ebb-tidal delta from 10 to 14 m, and the inner harbor channels to ~13 m. As a condition of the consent to dredge a detailed monitoring program was required which included annual hydrographic surveys over the flood and ebb-tidal delta as well as recording current meter records from morphodynamically sensitive locations to compare with records taken before the dredging. Results of the hydrographic surveys showed significant morphodynamic change of the flood-tidal delta had occurred essentially by the time the 6 month dredging program was completed. Comparison of the S4 current meter recordings showed current changes consistent with expectations of the EIA. Changes on the ebb-tidal delta were not expected but have occurred although not as rapidly, and seem to be ongoing.

Introduction: The Pre-dredging Assessment of Impacts

The Port of Tauranga was established last century within a meso-tidal estuarine lagoon enclosed by a Holocene barrier island and tombolo system (Fig.1). The inlet exhibits many morphological features typical of a meso-tidal inlet as presented by Boothroyd (1985), including a flood and ebb-tidal delta system and a narrow inlet gorge. Prior to dredging, channel depths reached 7-8 m through the delta systems with the narrow inlet gorge attaining 30 m.

---

* Coastal Marine Group, Department of Earth Sciences, University of Waikato, Hamilton, New Zealand. Email: trh@waikato.ac.nz

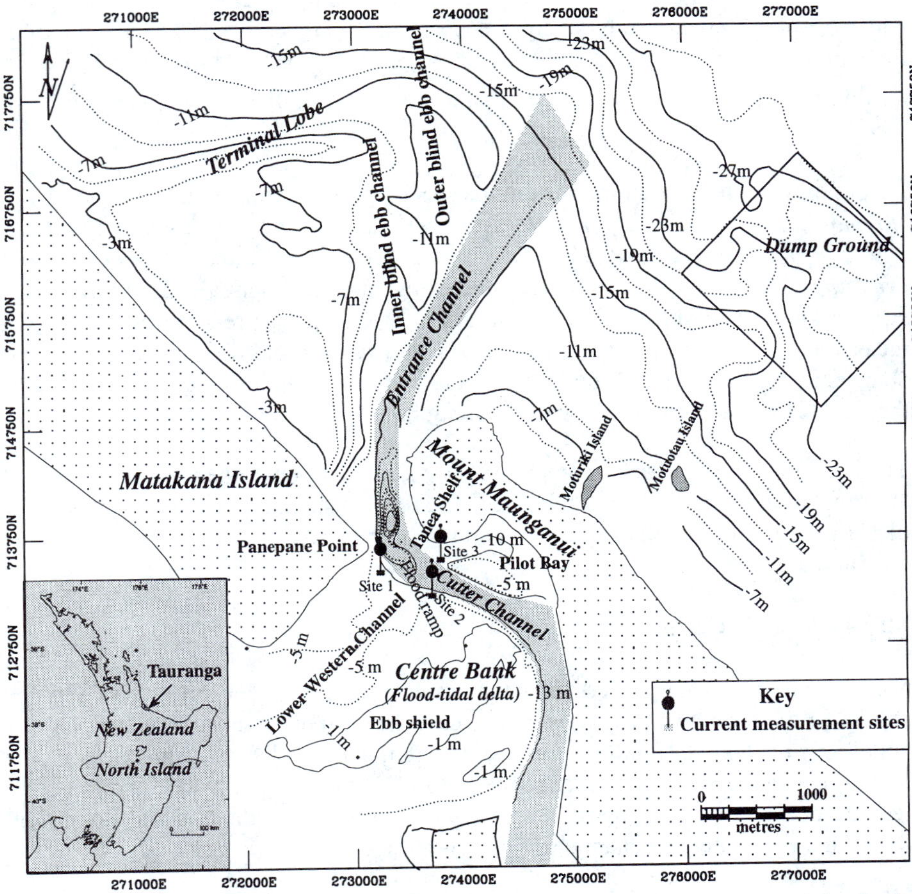

Fig. 1. Location map of Tauranga Harbour flood and ebb-tidal deltas and dredged shipping channels.

The port was expanded in the 1950s, and in the 1960s Wallingford Hydraulics Laboratory of the UK was commissioned to undertake a study investigating improvements for navigation approaches to the port. To improve port operations for ocean-going vessels, dredging of a navigation channel to depths of 10 m through both the flood and ebb-tidal deltas was carried out in the late 1960s. As was to be expected in a high current flow sandy tidal inlet located on a littoral drift shoreline, the navigation channels required significant maintenance dredging. Since the initial navigation channel construction through the tidal delta system, the average maintenance dredging has been ~70,000 m^3 per year, which was taken primarily from specific zones of channel deposition located either side of the inlet gorge, i.e., in the Entrance Channel through the ebb-tidal delta, and in the Cutter Channel which transects the flood-tidal delta.

In 1992, as part of further port expansion, a major shipping channel deepening and widening program was undertaken to deepen the channels from 10 to 13 m within the harbor, and to 14 m outside of the entrance. This involved dredging some 5.5 million m^3 of predominantly sandy material with some shell gravel (Healy et al., 1991).

A condition of the consent permitting the development, required the port company "*to monitor all aspects of the dredging programme and its impact upon the environment, and take such action as necessary to mitigate adverse effects that the dredging shall have.*" Accordingly, monitoring included annual full hydrographic surveys over the flood-tidal delta and adjacent channels, pre- and post-dredging measurement of current flows using RCMs, and surveys of barrier island shoreline change, in conjunction with biological monitoring. The monitoring continued until the end of 1995, some 4 years after the dredging.

The aims of this paper are to report on the morphodynamic changes that occurred after the dredging in this large tidal inlet and delta system, and make an assessment of the time it took for the inlet-tidal delta system to readjust to changed hydraulic conditions.

Morphodynamic Change

Morphodynamic change refers to change in bathymetry or landform created by the actions of the formative processes, in this case the currents and waves. Changing the conditions of the tidal inlet dynamic equilibrium, such as by inducing current flow hydraulic change will likely change the sediment transport pathways, and therefore lead to morphological change. The most sensitive areas are the ebb and flood-tidal deltas and the associated channels.

The projected impact on morpho- and hydrodynamics was studied at the EIA stage using 2-dimensional hydrodynamic modelling (Bell, 1991). The modelling indicated a substantial alteration to the current flow regime over the flood-tidal delta and shipping channel (the Cutter Channel), but no detectable impact on tidal currents over the ebb-tidal delta. Within the harbor the major impact predicted was that the shipping channel of ~13 m depth (c.f. original natural pre-dredged depths of 5-6 m) would act as a tidal "sink" on the ebb flow which would substantially alter the tidal cycle residual flow vectors over the flood-tidal delta (Fig.2). Accordingly, Healy et al. (1991; 1993) predicted as a result of the dredging that parts of the flood-tidal delta sediment transport pathways would be subject to change which would likely induce some morphodynamic change. What was not known was how long it would take for the flood-tidal delta to come to equilibrium after the dredging.

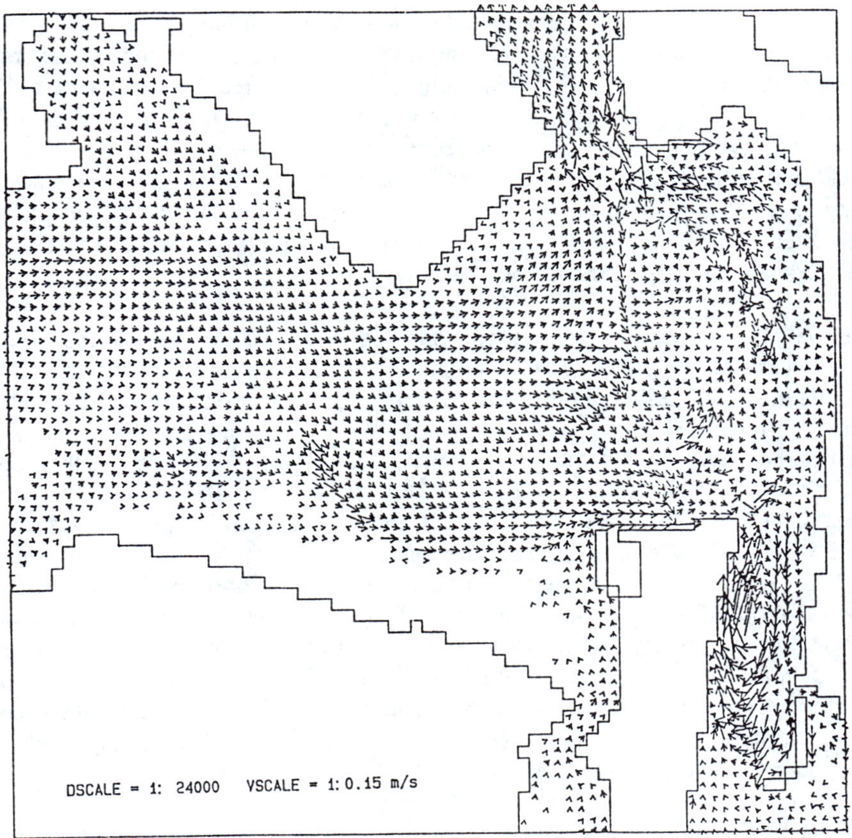

Fig. 2. Residual velocity vector differences showing the changes due solely to the deepened shipping channel in a spring tide (after Bell, 1991).

Sedimentation Within The Dredged Channels

Deepening an entrance channel to an inlet creates a greater hindrance to littoral drift and a more effective sediment trap. The long term effectiveness of the sediment trap is measured by the maintenance dredging requirement. For this case, since the channel deepening of 1992, the maintenance dredging has increased from a long term average of ~70,000 m^3 per year to ~110,000 m^3 per year for each of 1993, 1994, and 1995. An increase of this magnitude was expected, and to date has been remarkably uniform in amount per year. The sediment has accumulated on both the seaward and the harbor side of the inlet gorge.

The Flood-Tidal Delta (Centre Bank)

The flood-tidal delta is perhaps the most sensitive to morphodynamic change because it is formed largely as an interplay between the tidal current hydraulics and sediment supply, whereas the ebb-tidal delta may also be influenced by wave action. Centre Bank shows some features typical of the Boothroyd (1985) "horseshoe crab" morphological model of flood-tidal delta, but is distorted by alignment of the harbour in relation to the inlet entrance, and the consequent predominant ebb-tidal discharge flowing laterally across the feature rather than being diverted by the ebb shield around the delta. For Centre Bank the EIA showed that deepening of the shipping channels within the harbor would be expected to induce a much greater west-to-east component of flow across Centre Bank, and this would likely have an impact on the sediment transport pathways.

The availability of earlier hydrographic surveys over Centre Bank, and the regular soundings since 1989, allow a close monitoring of the changes. The changes in tidal channel and bank outline bathymetry for the 1 m and 5 m depths relative to chart datum (extreme low water spring tide level) are presented in Fig.3. The plots show that between 1982 and 1989 the Lower Western Channel where it crossed the flood-tidal delta, markedly reduced in depth with a greater proportion of the tidal flow being diverted around and to the south of Centre Bank, at which time the disparate, shallower small southern shoals were scoured away. In July 1992 the major dredging program was just being completed, and as a result of the greater west-to-east tidal flows across Centre Bank, the Lower Western Channel was already opening up again. By September 1993, the Lower Western Channel had re-established itself, coincident with shallow shoals also redeveloping on the southeastern margins, and increased area of shoaling on the eastern margin adjacent to the shipping channel. Since 1993 the flood-tidal delta morphology has not changed in broad outline (Mathew et al., 1995).

The net bathymetric differences exceeding 0.5 m between April 1990 and April 1994 are presented in Fig. 4. A broad zone of scour is evident along the Lower Western Channel, resultant upon the deepening of the channel. A zone of minor scour occurred along the outer ebb shield, with a parallel band of minor accretion, indicating a relocation of the distal ebb shield. Minor scour also occurred along the southeastern flank of Matakana barrier island. Accretion was marked on the flood ramp due to channel realignment, as well as an additional capping on the highest part of the inner ebb shield, and along the northeastern rim of the Centre Bank adjacent to the shipping channel.

J. Mathew (in prep.) has undertaken detailed analysis of the sounding data including detailed cross sections. The profiles showing the most change are presented in Fig 5. The sounding sections show post-dredging infilling and scour of parts of the Lower Western Channel, with only minor change on the other parts of the Centre Bank. In particular the shallowing of the ebb shield, interpreted as a strong possibility in the EIA (Healy et al., 1991), has partly occurred, but only up to 0.5 m vertical accretion over an area of ~ 600 x 300 m^2 , and this change likewise occurred rapidly after the 1992 dredging.

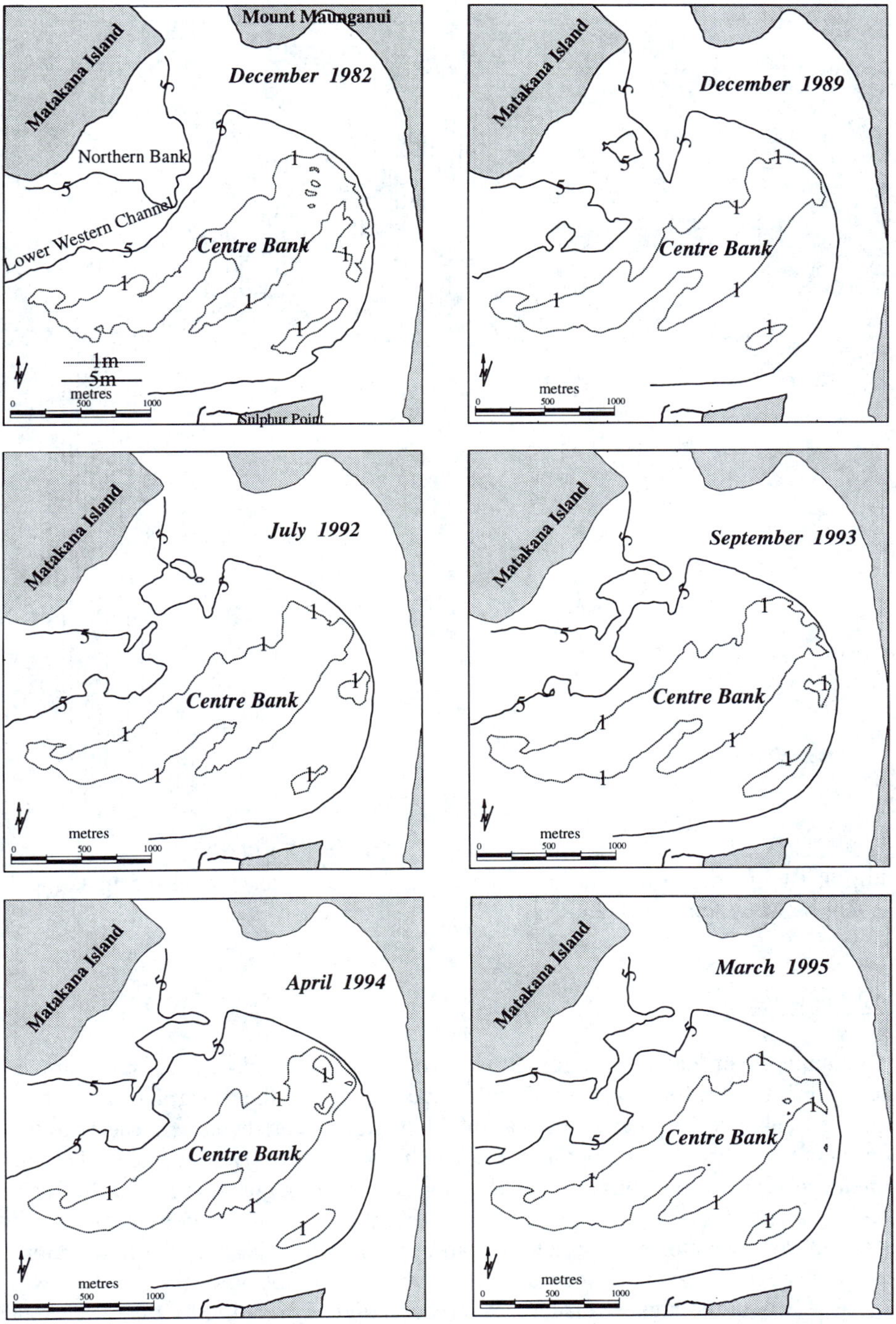

Fig. 3. Bathymetric changes in the Lower Western Channel between December 1982 and March 1995.

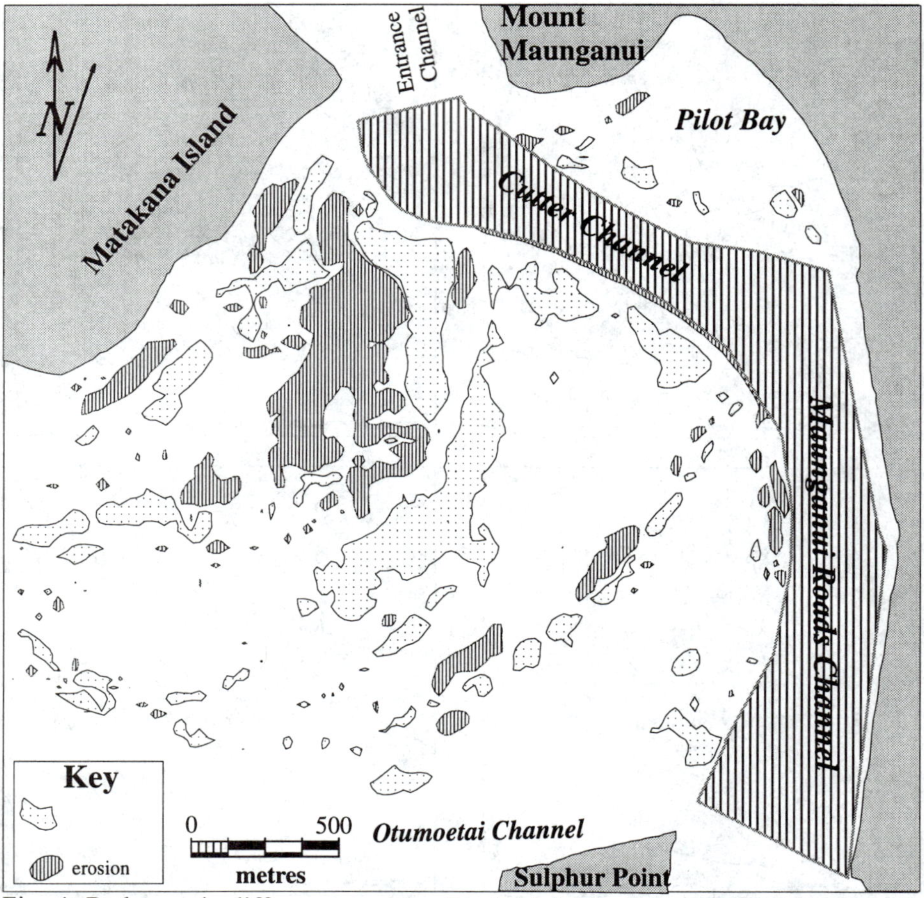

Fig. 4. Bathymetric differences on the flood tidal delta exceeding 0.5 m vertical extent between April 1990 and April 1994.

Air Photo Record

The annual vertical air photos flown since the major dredging have shown realignment of the direction of the sand waves which cap the topographical high of Centre Bank, geomorphically the ebb shield. Noticeable on the eastern sector of the ebb shield the sand wave bedforms have shown an easterly movement. The air photos also show a response of the flood ramp to the dredging and removal of part of a rocky shelf which had been protruding into the inlet gorge and was considered a hazard to navigation. This resulted in an expanded flood jet onto the flood ramp, and created a new area of scoured shell-gravel lag. Expansion of the flood jet was consistent with the higher current flows recorded after dredging by the S4 current meters, discussed below.

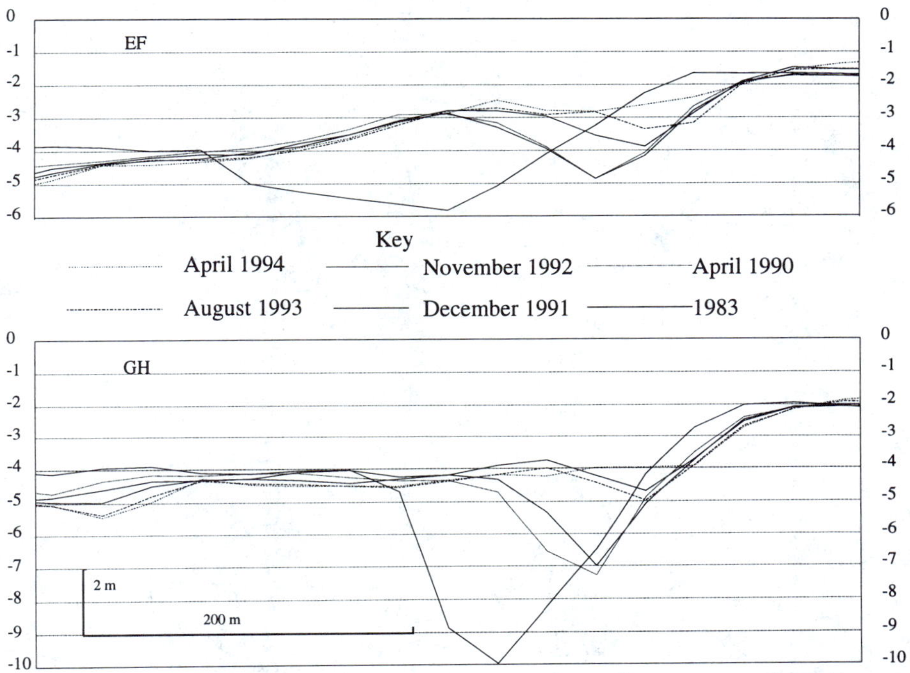

Fig. 5. Cross-sections across Western Channel showing deposition of up to 3 m in Lower Western Channel (EF) and up to 6 m in the blind flood channel (GH) between 1983 and 1994.

Changes in Current Flows Near Panepane Point

Current flows were monitored at 3 locations adjacent to the barrier island side of the inlet using S4 electromagnetic current meters. These were located in the Lower Western Channel, Pilot Bay Channel and on the edge of the Cutter Channel. The resulting data (Figs 6a-b) show some changes in velocities for the Cutter Channel site with greater flood tide dominance rising from 0.85 m/s to 1.20 m/s after the 1992 major dredging. In the Lower Western Channel near Panepane Point there was an increase in current magnitude from 1.00 m/s peak flood flow and 0.75 m/s peak ebb flow before the dredging to 1.25 m/s peak flood flow and 1.10 m/s peak ebb coincident with a directional change after the dredging. However a later deployment at the same site in 1994 showed that the current direction had resumed its pre - 1992 dredging alignment.

The observed increase in current speed at points measured is consistent with a re-aligned and laterally extended tidal jet as a result of removal of the Tanea Shelf from the inlet gorge. That, as expected, has induced some change in the affected sea floor, in particular the Lower Western Channel and flood ramp area.

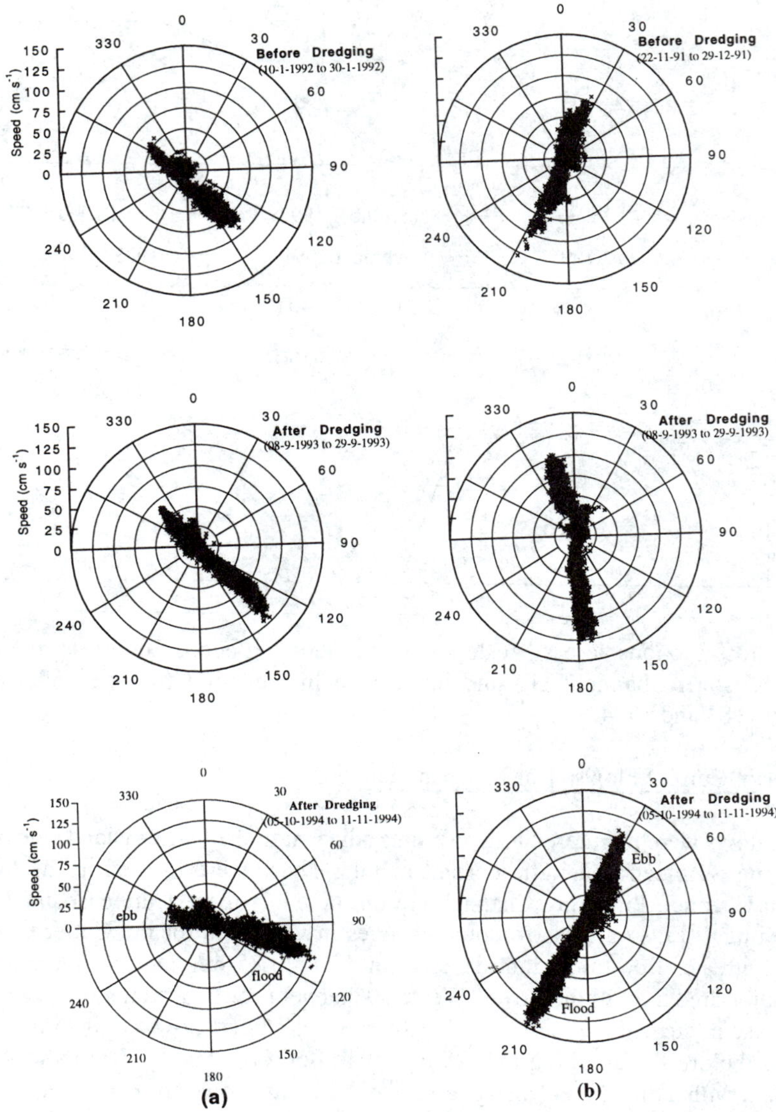

Fig. 6. Current velocities at 1 m above the bed before and after dredging at: (a) the edge of Cutter Channel and (b) at Lower Western Channel near Panepane Point.

The Ebb-Tidal Delta

Annual bathymetric surveys were also carried out over the ebb-tidal delta since 1989. Comparison of the surveys indicates that the ebb delta has been largely stable in terms of its gross morphology, for example the terminal lobe has not changed in location between 1989 and 1995 (Fig.7). The major identifiable change has been that a proximal (inner) blind ebb channel northwest of Mount Maunganui has tended to infill from about 13 m depth in 1989 to 10 m depth in 1995. A distal (outer) blind ebb channel has increased in size.

Overall there has been no sudden or substantial change to the ebb delta morphology which can obviously be linked to the entrance shipping channel deepening in 1992. This is consistent with the numerical model residual vector patterns carried out in the EIA which showed that there was unlikely to be identifiable difference from deepening the Entrance navigation channel from 11 m to 14 m.

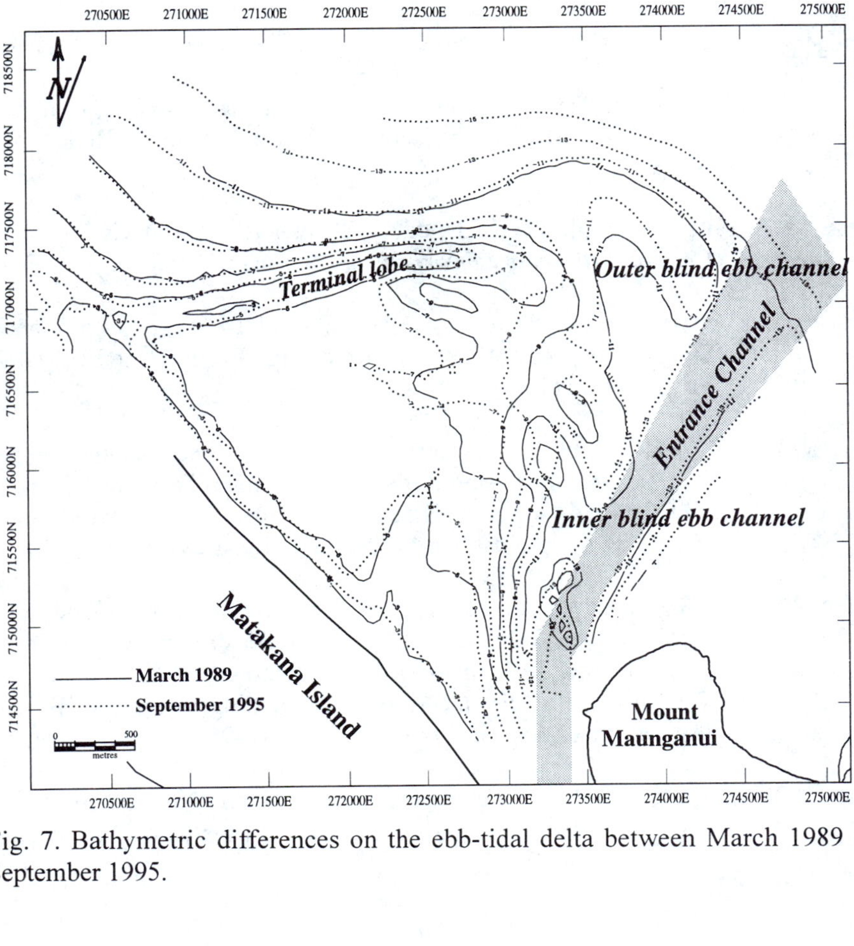

Fig. 7. Bathymetric differences on the ebb-tidal delta between March 1989 and September 1995.

Conclusions

Since the 1992 major channel dredging, the hydrographic surveys have shown that some areas of the flood ramp have accreted by up to 3 m in vertical extent while up to 2.5 m of scour associated with increased currents have occurred in the non-dredged semi-major Western Channel. Alignment of large 'dune' bedforms on the topographic high of the flood delta, morphologically the ebb shield, has changed significantly, but the depths have changed only by about 0.5 m, and overall the broadscale morphology of the flood-tidal delta has retained its coherency.

In terms of how rapidly the tidal inlet and delta system has adjusted to new conditions after the substantial hydraulic change to the tidal delta system consequent upon the 1992 dredging, it is clear that major readjustment for the flood-tidal delta was very rapid, and essentially occurred by the time the dredging was completed - over a 6 month period. On the other hand, the ebb-tidal delta adjustment has been much slower and seems to be ongoing.

References

Bell, R.G., (1991). Port of Tauranga Model Study (Deepened Shipping Channel Proposal). DSIR Water Quality Centre Consultancy Report No 6127/1. 22p.

Boothroyd, J. (1985). Tidal inlets and tidal deltas, chapter 7 in R.A.Davis Jnr.(ed.): *Coastal Sedimentary Environments*, Springer-Verlag, pp 445-532.

Healy, T., McCabe, B., and Thompson, G., (1991). Port of Tauranga Ltd. Channel Deepening and Widening dredging Programme 1991-92 Environmental Impact Assessment. 122p.

Healy, T., Bell, R., and de Lange, W., (1993). Predicting morphodynamic change from tidal residual vectors at a large tidal inlet, Tauranga Harbour, New Zealand. In List, G.H., ed., Large Scale Coastal Behaviour - 1993, U.S. Geological Survey Open File Report 93 - 381,pp64-67.

Mathew, J., Healy, T.R. and W.P. de Lange, (1995). Changes to a large flood-tidal delta system following major channel dredging, Tauranga Harbour, New Zealand. Proc. Fourth International Conference on Coastal and Port Engineering in Developing Countries, Rio de Janeiro, Brazil, September 1995. pp 1464-1478.

Mathew, J. (in prep.) Morphodynamic change in response to dredging and dumping, Tauranga, New Zealand. D.Phil. Thesis, University of Waikato.

CHAPTER 255

LABORATORY MOBILE BED MODEL STUDIES ON EBB TIDAL SHOAL EVOLUTION

Xu Wang[1], Lihwa Lin[2] and Hsiang Wang[3]

Abstract

Laboratory mobile bed experiments were conducted to study the ebb tidal shoal evolution process under storm wave conditions. An idealized inlet configuration was chosen for the experiment representing a typical median-sized inlet on the east coast of Florida, USA. Six different cases were tested including a natural inlet and a jettied inlet with different jetty length and jetty type. Formation of ebb tidal shoal was observed in all cases; the rate of growth and location of ebb tidal shoal were different for the cases. Inlet channel shoaling and beach erosion next to the inlet are far more severe in the case of natural inlet than that in the case of jettied inlet. In general, the established tidal shoal tends to grow during the ebb cycles and deteriorate during flood cycles. And partial removal of ebb tidal shoal has shown to increase downdrift beach erosion and reduce the rate of ebb shoal growth, though the rate of change of the mined case rapidly approached that of the case without mining.

1. Introduction

Ebb tidal shoal is a common feature associated with tidal inlets in coastal area. It is created by the combined deposition of littoral material diverted from adjacent beaches together with the alluvial material carried out from inlet by the tidal current. When inlets are stabilized with training structures, ebb tidal shoal can become more prominent as littoral material is diverted further offshore into deeper water. As a consequence, the ebb shoal volume also increases. This causes additional disruption of the normal longshore sediment transport and often results in severe downdrift shoreline recession. In Florida, over 85% of the shoreline erosion is considered to be related to inlets, particularly to those with training structures.

---

[1]Engineer I, Division of Natural Resources Management of Lee County, FL 33902, [2]Research Scientist, [3]Professor, Coastal and Oceanographic Engineering Department, Univeristy of Florida, Gainesville, FL 32611,USA.

Since ebb tidal shoal is formed mainly by material diverted from the updrift beach it is a tempting source, and reasonably so, to tap for downdrift beach nourishment. However, such practice is not common because the formation of ebb tidal shoal is part of the natural process and disturbing an ebb tidal shoal environment so close to shoreline without knowing clearly its effects is unsettling.

This paper is aimed at investigating the evolution process of ebb tidal shoal on one hand and finding the possibility of utilizing the sand from ebb tidal shoal on beach nourishment on the other by means of movable-bed physical modeling in the laboratory. To achieve this goal, an inlet model testing is designed and carried out to investigate ebb tidal shoal evolution process and corresponding shoreline responses for a natural and unimproved inlet, an inlet with jetty structures and with the ebb tidal shoal partial removal.

2. Design Of Inlet Model Experiments

The inlet-beach physical model was designed with considerations on experimental constraints and modeling laws. The model was tested under simulated storm wave conditions to insure turbulent flow and suspended sediment transport mode. The modeling law adopted in the present study is shown in Table 1, which is for an inlet-beach system based on the analyses of the experimental results of a series of 2-D and 3-D laboratory model study (Wang, *et al.*, 1994: Wang, *et al.*, 1995).

Table 1: Modeling Law

Geometric Distortion*	Wave Height Distortion	Hydrodynamic Time Scale	Morphological Time Scale
$N_\delta = N_w^{0.4} N_\lambda^{0.8}$	$N_H = \sqrt{N_\lambda}$	$N_T = \sqrt{N_\lambda}$	$N_t = \sqrt{N_\lambda}$

* N_w, N_δ, N_λ are fall velocity, vertical and horizontal length ratios, respectively.

The model experiments were carried out in a wave basin located in Coastal and Oceanographic Engineering Laboratory at the University of Florida. The wave basin has a physical size of 25m wide, 30m long, and 1m deep as shown on Figure 1. The inlet-beach model and a wave maker were located at two long ends of the basin. An ideal inlet, of straight, rectangular channel, with uniform width and depth of 1.75m and 0.2m, respectively, was constructed cutting through a plane beach made of a natural quartz sand (D_{50}=0.19mm). The overall length of the beach from updrift to downdrift ends is 19m. The model is laterally bounded on two wave guides formed by concrete blocks. The wave guides are perforated to allow flows in and out of the test section. The downdrift wave guide is open at the beach end to allow downdrift littoral transport to leave the test section and to be collected in the catch channel. The plane beach consists of a flat back shore segment, a steep-slope foreshore segment, and a mild-slope offshore profile which extends seaward to about 7m form the shoreline beach face before merging with the concrete floor. The beach profile approximates an equilibrium shape of $h = Ax^{0.8}$, which h is water depth, x is seaward distance from shoreline, A is a scale factor.

A schematic of experiment setup

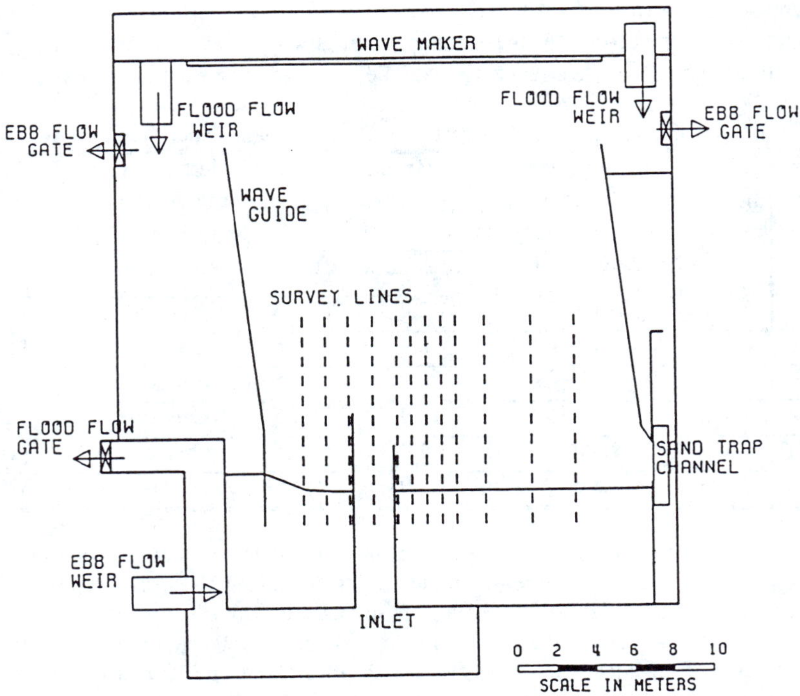

Figure 1: The Schematic Setup for Movable-Bed Inlet Physical Model.

The inlet is offset from the center towards the updrift with the updrift beach length of 5m and downdrift beach length of 12m. The wave generator is located about 27m form the shoreline based on an average water depth about 0.35m. Tidal currents are generated by recirculating water through the circulation channels connected with the wave basin. The flow discharge is controlled by the flood and ebb flow weir boxes. Water is supplied form the upper basin weir boxes (flood flow weirs) for flood current and from the lower basin weir box (ebb flow weir) for ebb current. A curved feeder beach section at the updrift end allows for continuous and more uniform sediment supply to the downdrift which is purely due to wave-induced transport.

3. Experimental Conditions

Six different experimental cases and associated conditions are summarized in Table 2. Tidal currents are simulated in the experiment by alternating ebb and flood cycles at equal interval of 40 minutes, which is equivalent to a semi-diurnal tidal period at 1:80 geometrical scale ratio according to Froud number criterion. A constant discharge of 0.04 m^3/sec is used for both ebb and flood cycles. The cross-sectional averaged currents in the inlet are maintained to be 0.12 m/s and 0.14 m/s for the flood and ebb cycles, respectively. The tidal range is 3 cm in the experiment.

Table 2: Test Conditions of Inlet-Beach System Experiment.

Case	Wave height	Incident wave condition (Wave period: 1sec) Wave angle	Beach slope		Jetties (type) e: even length u:uneven length	Model time (min)
			Foreshore	Offshore		
C1	8 cm	15 deg	1:2.4	1:14.5	none	480
C2	8 cm	15 deg	1:2.9	1:14.5	Riprap(u)	1600
C3	8 cm	7.5 deg	1:2.9	1:14.5	Caisson (u)	3200
C4	7 cm	7.5 deg	1:2.9	1:14.5	Caisson (e)	3200
EC1	7 cm	7.5 deg	1.2.9	1:14.5	Caisson (e)	3200
EC2	7 cm	7.5 deg	1:2.9	1:14.5	ebb shoal mining	3200

Experiment C1 is to simulate a natural inlet; C2 is to simulate a jettied inlet with riprap type jetties; C3 and C4 are to simulate a jettied inlet with caisson type jetties. In Case C2 and C3, the inlet consists of an updrift jetty of 1.5m and a downdrift jetty of 0.7m, both straight and perpendicular to initial shoreline. The uneven updrift and downdrift jetty geometry of an inlet is common in Florida. In C4, the updrift and downdrift jetty jetties have the same length of 1m. Experiment EC1 and EC2 were conducted to investigate the effects of ebb tidal shoal removal. Experiment EC1 is the case without removal of ebb tidal shoal and EC2 is the case with removal of ebb tidal shoal. Both experiments have the same test conditions as that of C4, except C4 has a slight larger width of the inlet. The jetty elevation is about 5cm above the flood tide water surface and jetty width is 20cm. Figure 2 shows the initial bathymetry for natural inlet case. Also, the major difference between riprap and caisson type jetties is that the riprap is porous and not sand tight whereas the caisson is impervious.

4. Experimental Procedures

The model experiments is conducted according to the following procedures: (1) Prepare initial inlet model bathymetry, (2) Conduct initial profile survey at selected cross-sections as shown in Figure 1, (3) Adjust water level and discharge to the specified ebb conditions, (4) Start wave maker to generate storm waves, (5) Run the ebb cycle experiment for 40 minutes, (6) Readjust water level and tidal flow for the flood condition, (7) Start wave generator and run the flood cycle for 40 minutes, (8) Repeat steps from (3) to (7) until an prominent ebb tidal shoal is established, and (9) Reshape the model to its initial bathymetry for the next experiment. Eleven bottom profile surveys were conducted at irregular time intervals, shorter in the early stage and longer later in the experiment. Sediment accumulated inside the inlet and outside of the downdrift boundary was collected at the same time when bottom surveys were conducted. Both dye and sand tracer studies were conducted from time to time. The dye study was for current pattern observation and was documented by video recordings. Sand tracers were used for visual examination on sediment transport pattern. No quantitative measure of sand tracer movement was attempted.

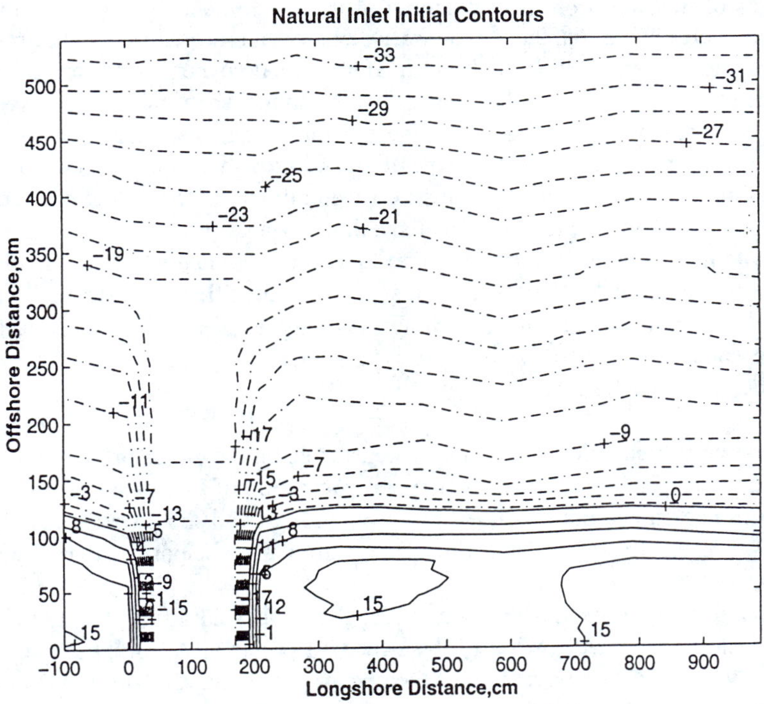

Figure 2: Initial Topographic Contours for Natural Inlet Experiment

The initial topography of Case EC2 was prepared by modifying the final topography (3200 min) of EC1, which included mining of ebb tidal shoal and downdrift beach nourishment. Sand was removed from the ebb shoal and inlet channel areas which is indicated by the dashed rectanger in Figure ?. This sand was used in all for the downdrift beach nourishment in the preparation of initial topography of EC2.

5. Experimental Results

Formation of ebb tidal shoal was observed in all cases; the location of ebb tidal shoal and the rate of growth were different. In the laboratory, + 2cm above the initial bottom profile was chosen as the reference plane to present the results of ebb tidal shoal growth.

In the natural inlet experiment, a small shoal formed immediately near the entrance in the first ebb and flood cycles as the beach material was rapidly carried seaward to form nearshore breaking bars. Inside the breaking line, sediment from the updrift beach was seen to move towards the inlet by strong wave action. Outside the breaking line, sediment was carried across the channel and deposited at downdrfit side of the channel outside the surf zone. Beach erosion was severe at both sides of the inlet from the beginning of the experiment. In the following ebb and flood cycles, the initial shoal built near the inlet entrance continued to grow and expand to form channel shoal and ebb tidal shoal. The experiment was stopped at 480 minutes or six complete tidal cycles as both channel shoaling and beach erosion became excessively severe. The shoreline erosion pattern was nearly symmetrical with respect to the inlet center. The generation and growth of the ebb tidal shoal and also the shoreline patterns in Experiment C1 using the net +2cm as the base contour are exhibited in Figure 3. It is seen that shoaling began at the channel entrance and grew in both directions towards offshore and into the channel. At 120 minutes, channel shoal and ebb tidal shoal can be separately identified. The ebb shoal began to shift towards downdrift after 120 minutes. At the end of 480 minutes, a drastic ebb tidal shoal was establish while the channel shoaling was seen to extend and reconnect with the ebb shoal.

In the jettied inlet experiments, general sediment transport patterns were similar in the beginning ebb and flood cycles. Accretion of sediment occurred at the tips of both updrift and downdrift jetties. Generation of ebb shoal was not evident in this early stage. In subsequent time, the sediment transport patterns became different,

which then influenced the development of the ebb tidal shoal. In C2, the updrift jetty tended to attract sediment owing to the structural porosity. Accordingly, sediment was heavily deposited on both sides of the updrift jetty around its tip. In C3 and C4, on the other hand, more updrift sediment was seen to bypass the jetty. In these jettied inlet experiments, beach erosion was significant only at the downdrift side, particularly in C2 owing to the larger incident wave angle. In C4, the ebb tidal shoal was generated more closer to the inlet than in C2 and C3 due to the small incident wave height. Compared with C4, EC1 had more centered ebb tidal shoal and much less channel shoaling because of the smaller width of the inlet which tended to transport more sediment bypassing the jetties.

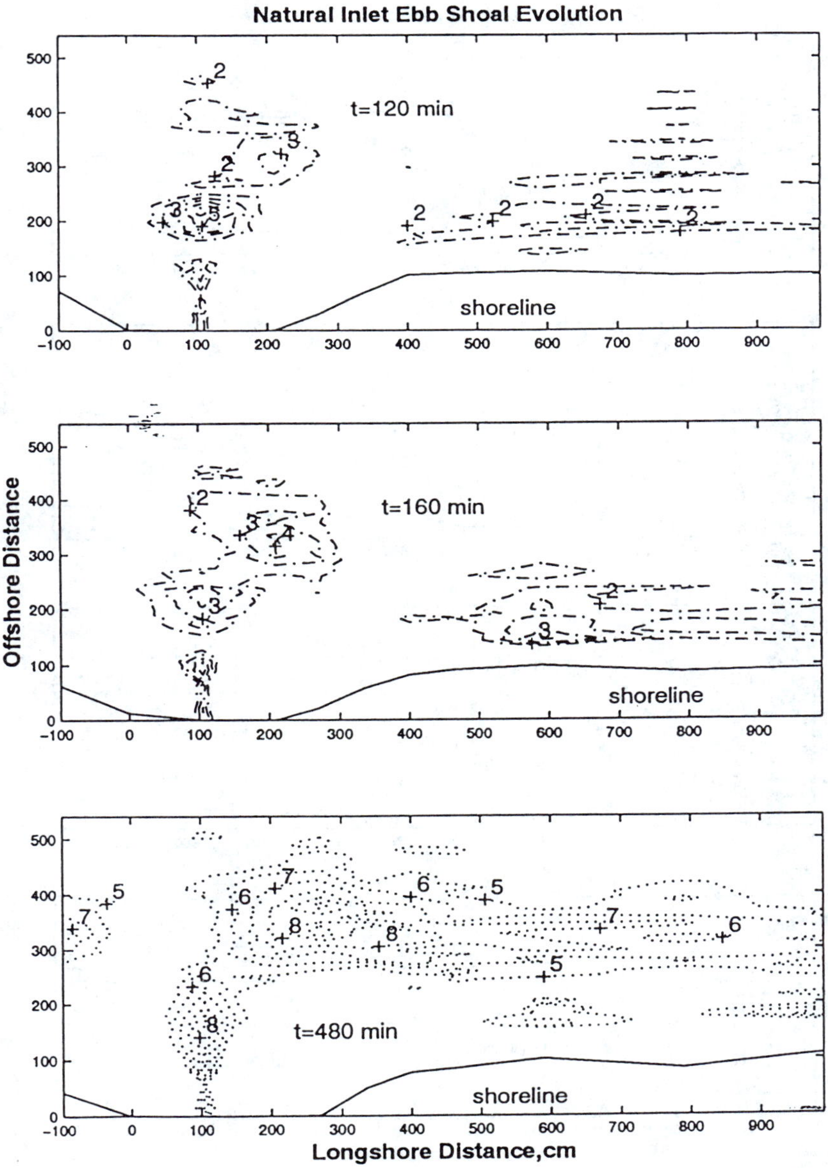

Figure 3: Ebb Tidal Shoal Evolution in Experiment C1.

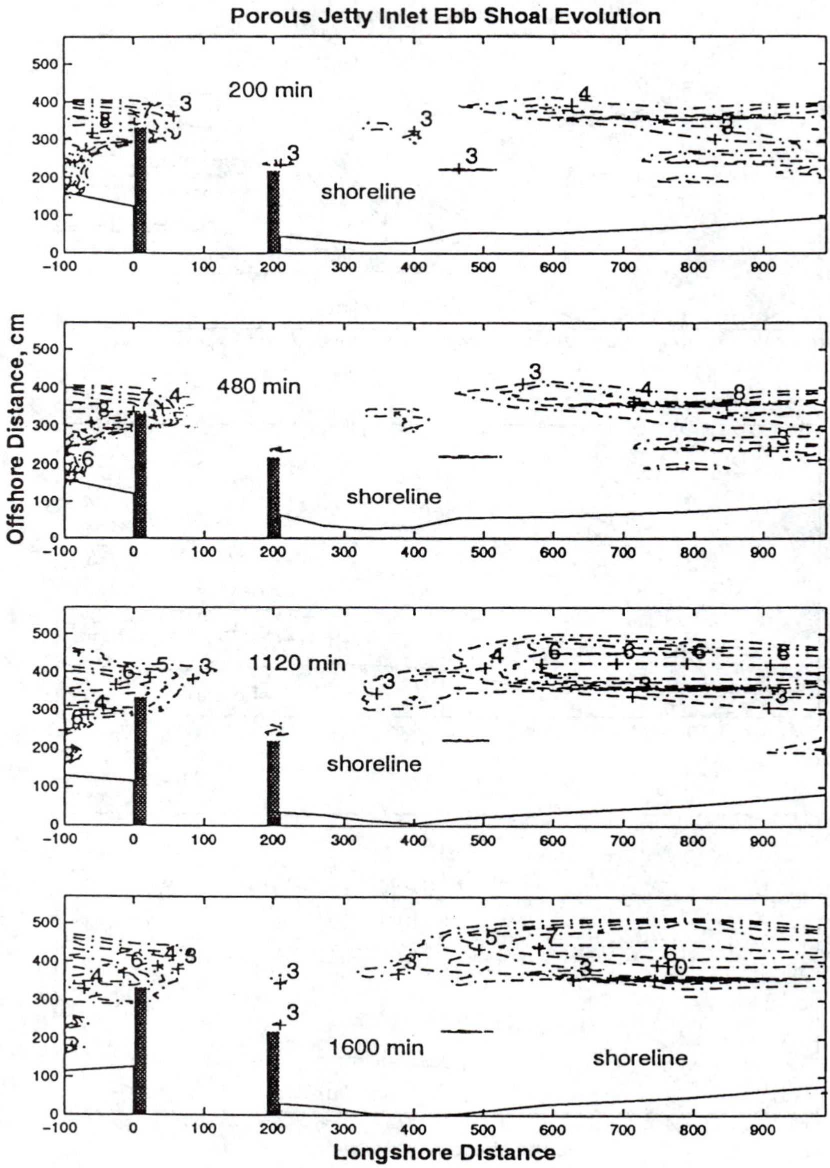

Figure 4: Ebb Tidal Shoal Evolution in Experiment C2.

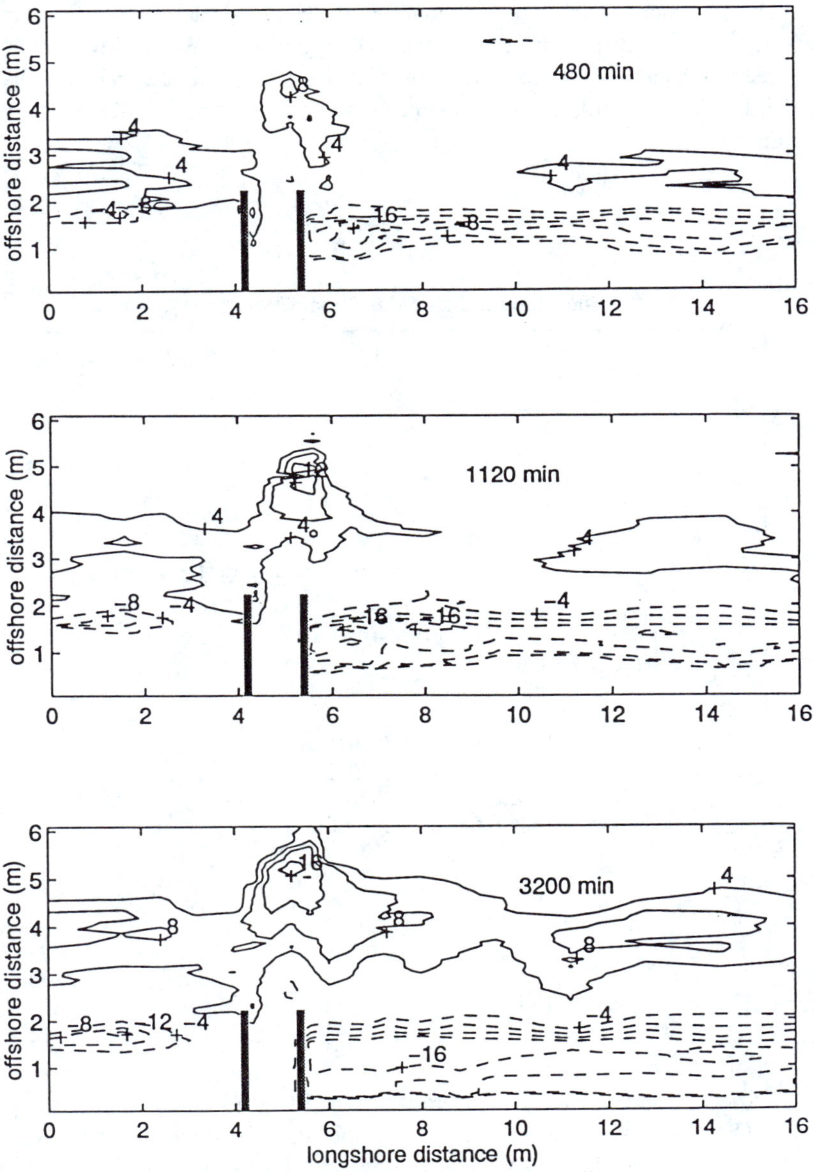

Figure 5: Ebb Tidal Shoal Evolution in Experiment EC1.

Figure 4 to 5 present the evolution pattern of ebb tidal shoal in C2 and C3 , respectively. Figure 6 compares the ebb tidal shoal volume for C1,C2, C3 and EC1. It is seen that the growth of ebb tidal shoal was unsteady in the early stage in these cases. The ebb shoal simply grew during ebb cycles but shrunk during flood cycles. After the first few cycles, the ebb tidal shoal began to grow steadily, almost in a linear fashion. The process was much rapid in the case of the natural inlet than the jettied inlet. The rate of growth apparently slowed down at a later stage of the experiment.

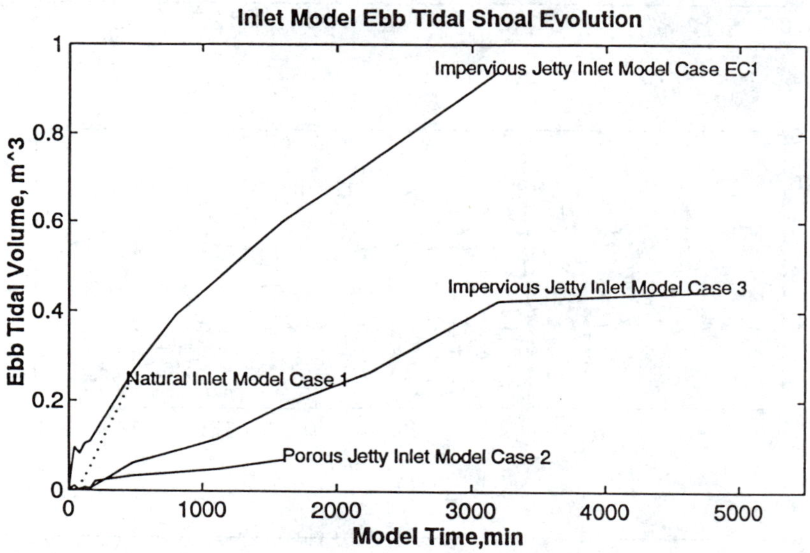

Figure 6: Comparison of Ebb Tidal Shoal Volume Changes for C1, C2, C3 and EC1

Experiments EC1 and EC2 were conducted to study the effect of the partial ebb tidal shoal removal. Figure 7 shows the ebb tidal shoal evolution process after part of the ebb tidal shoal sand was removed and used as the downdrift beach nourishment in EC2.

The effects of ebb tidal shoal removal on downdrift beach were evaluated by comparing the total volume of sand eroded from the beach between the downdrift jetty and downdrift boundary in experiments EC1 and EC2. This comparison of downdrift beach erosion and erosion rate versus the elapsed time is shown in Figure 8. It is seen that the downdrift beach erosion is overall significant in EC1 and EC2, with a greater erosive rate in EC2 than in EC1, though at the later stage, this difference became much smaller.

The degree of restoration of the ebb tidal shoal was evaluated by comparing the volumes of the ebb shoal in EC1 and EC2 as shown in Figure 9. At 800 min, the ebb tidal shoal in EC2 has less volume than the shoal in EC1 implying a slower rate of growth in EC2. However, from 800 min to 1600 min, the ebb shoal growth approaches a steady rate in both experiments. The rate of growth fluctuates greatly in the first 160 min reflecting the effects of the short time intervals corresponding to individual ebb and flood tides.

6. Conclusion

Based on the experimental results, inlet channel shoaling and beach erosion in the cases of jettied inlet experiment were not as severe as compared to the case of the natural inlet. Apparently, the presence of jetties slows down the inlet shoaling and beach erosion near the inlet. It is evident that jetty structures are necessary for tidal inlets under strong wave environment.

Formation and growth of ebb tidal shoal were observed in all the inlet experiments. The location and rate of growth of ebb shoal were different, though.The ebb tidal shoal were created and expanded during ebb cycles but deteriorated and diffused during flood cycles. The porous jetty tended to attract sediment deposition near the inlet entrance whereas impervious jetty caused more sediment bypassing the inlet.

Mining an ebb tidal shoal has shown to increase downdrift beach erosion and reduce the rate of ebb shoal growth at certain degree. However, the rate of change of the mined case rapidly approached that of the case without mining, which implies the feasibility of using sediment from ebb tidal shoal in downdrift beach nourishment. The experiment was successful in reproducing ebb tidal shoals observed in nature.
However, more research work is needed to gain the insight of dynamics of the system.

Acknowledgement

The work is partially supported by the Coastal Engineering Research Center, US Army Waterway Experimental Station under Contract: DACW39-95-K-0084. Mr. Jack Davis of CERC is the project monitor. Both their financial support and personnel assistance are sincerely acknowledged here.

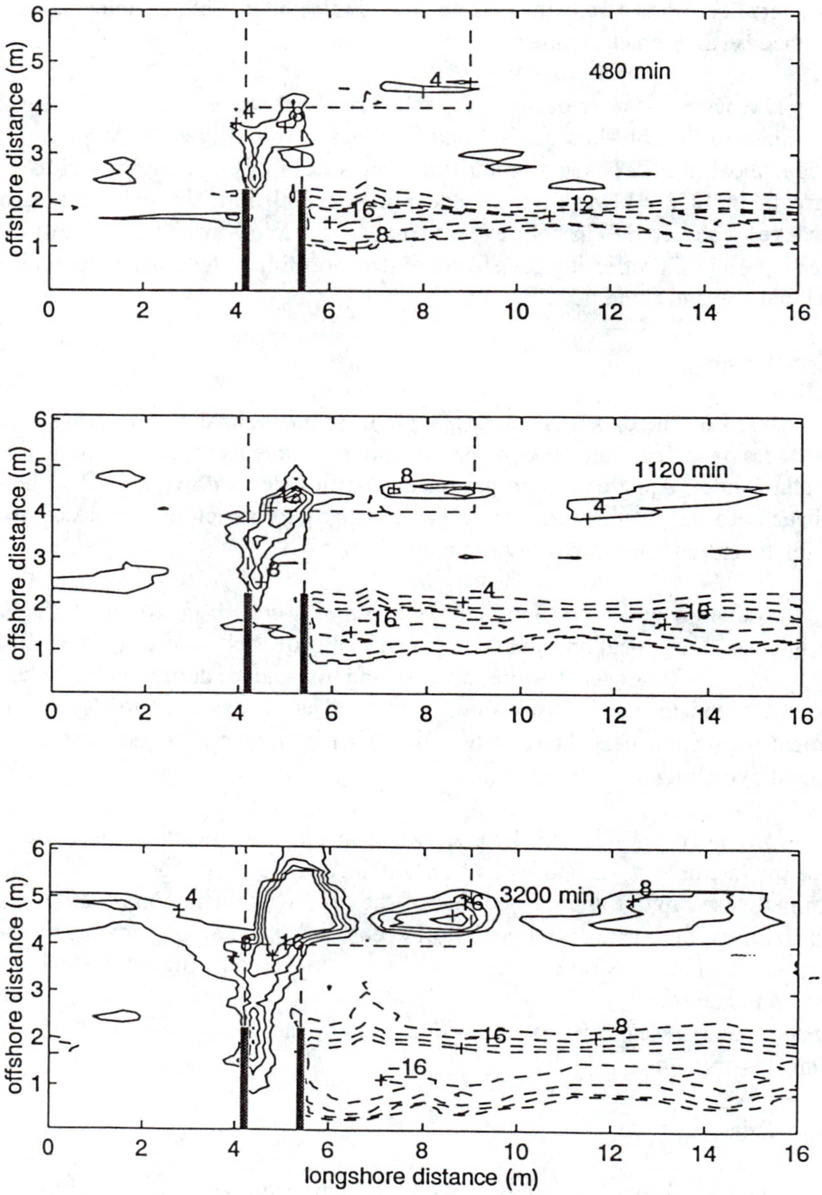

Figure 7: Ebb Tidal Shoal Evolution in Experiment EC2.

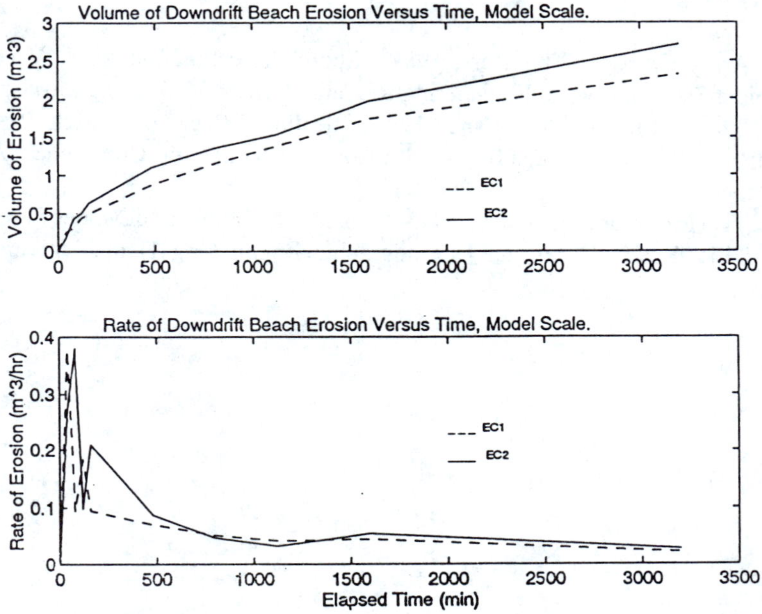

Figure 8: Comparison of Downdrift Beach Erosion in Experiments EC1 and EC2.

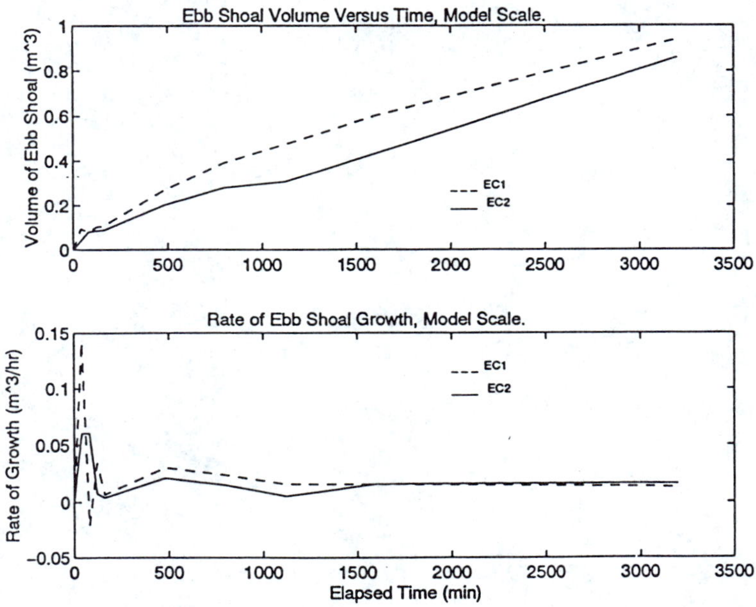

Figure 9: Comparison of Ebb Tidal Shoal Evolution in Experiments EC1 and EC2.

REFERENCE

1. U.S. Army Corps of Engineers, 1984. "Shore Protection Manual," Waterways Experiment Station, Coastal Engineering Research Center. Vicksburg, Mississippi.

2. X. Wang, L.Lin, and H. Wang, 1994. "Scaling Effects on Beach Response Physical Model," Prodeedings of the 24th Coastal Engineering Conference, ASCE, Kobe, Japan.

2. Wang, H, L.Lin, and X. Wang, 1995. "Laboratory Mobil Bed Model Studies on Inlet;," UFL/95-021, Coast and Oceanographic Engineering Dept., University of Florida.

CHAPTER 256

INTERACTION OF THE COLORADO RIVER PROJECT, TEXAS, WITH LONGSHORE SEDIMENT TRANSPORT

Daniel J. Heilman[1], M. ASCE and Billy L. Edge[2], M. ASCE

ABSTRACT

In 1985, a weir jetty system was constructed at the mouth of the Colorado River, Texas, and river discharge was diverted from this mouth in 1992. An evaluation of project impacts on longshore sediment transport was performed to determine the effectiveness of the project at preserving an open, navigable channel while preventing excessive erosion along adjacent beaches. This evaluation included both physical and numerical analysis of pre- and post-project conditions at the Colorado River mouth. Analysis of site data revealed that under the dredging and mechanical bypassing schedule followed during the first 8 years after jetty construction, the project resulted in significant trapping and sorting of sediment transported alongshore. Numerical modeling of the design maintenance dredging plan confirmed the plan to be sufficient for achieving project objectives. However, the entrance channel has continued to shoal considerably more rapidly than was anticipated during the project design.

INTRODUCTION

General Setting

The Colorado River discharges into the Gulf of Mexico through Matagorda Peninsula on the Texas Gulf Coast about midway between Galveston and Corpus Christi (Fig. 1). With a generally linear WSW-ENE orientation, the Matagorda Peninsula outer shoreline is dominated primarily by Gulf waves from the northeast (Fig. 2), and it is subjected to a resulting longshore current that transports sediment in a southwestward direction (McGowen and Brewton 1975). The peninsula ranges in width from about 1.2 to 1.6 km with an average elevation of approximately 2 m and faces the Gulf with a moderately wide sandy beach. The tide is diurnal with a range on the order of 0.6 m (U.S. Army Corps of Engineers (USACE) 1977). As reported by Morton *et al.* (1976), the beach varies in composition and texture from fine sand to shell and rock-fragment gravel.

---

1) Research Scientist, Conrad Blucher Instit. for Surveying and Science, Texas A&M Univ.-Corpus Christi, 6300 Ocean Drive, Corpus Christi, TX 78412. Formerly M.S. student, Ocean Eng. Program, Dept. of Civil Eng., Texas A&M Univ., College Station, TX 77843.

2) Head, Ocean Eng. Program, Dept. of Civil Eng., Texas A&M Univ., College Station, TX 77843.

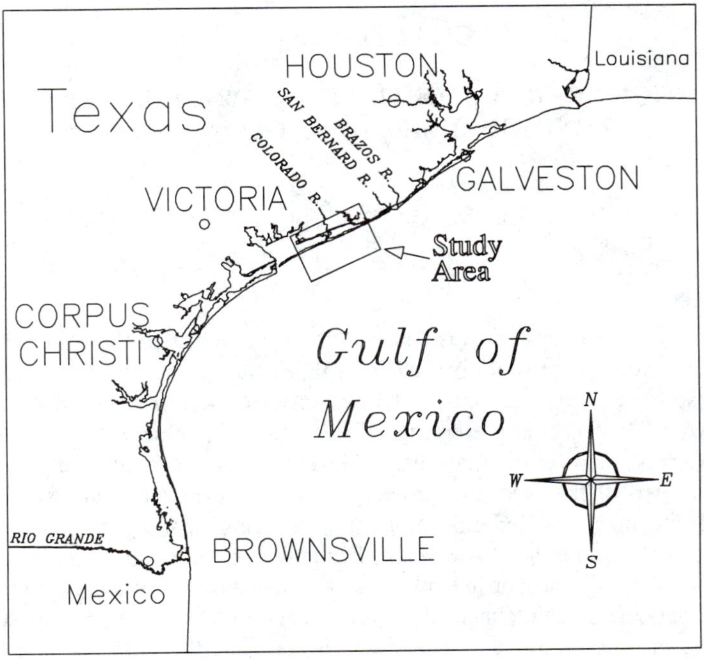

Figure 1. Site location map.

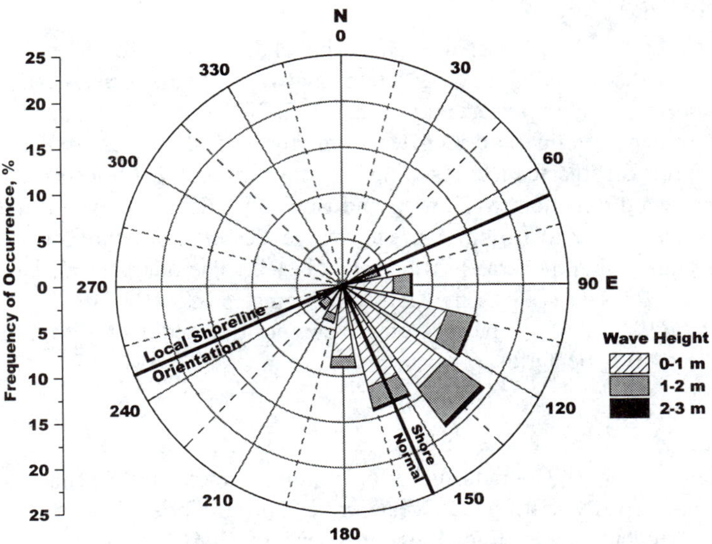

Figure 2. Wave rose for study site (data collected by USACE, CERC, May, 1990 - August, 1992).

The Colorado River Mouth

Originally discharging into a large estuary shared with the nearby Brazos River, the Colorado River began discharging into Matagorda Bay near the mainland shore at Matagorda, Texas, around 1,000 years ago (McGowen and Brewton 1975). Sediment transported by the river to the bay was restricted in the mid and late 1800s by a jam consisting of tangled masses of logs and brush embedded in silt, which was characteristic of the Lower Colorado River (Morton *et al.* 1976). By 1926, the log mass backed up in the river 74 km above Matagorda and created a lake, further restricting river sediment discharge by impounding course sediment. As shown in Fig. 3, which is a digitized rendering of a survey map prepared in January, 1839, the lake and log mass effectively prevented the river from creating a delta in the bay. In the late 1920s, a channel was dredged through the log mass, resulting in the release of trapped sediment and the rapid growth of the Colorado River delta into Matagorda Bay (Fig. 4) (Bouma and Bryant 1969). By 1935, the delta extended across Matagorda Bay to Matagorda Peninsula. In search of a direct path connecting the river to the Gulf, the Matagorda County and Reclamation District dredged a straight channel from one of the river outlets in the delta, across the bay, and through Matagorda Peninsula to the Gulf of Mexico (Morton *et al.* 1976, Ralston 1987).

The Colorado River Project

The Gulf entrance of the Colorado River has historically been subject to heavy shoaling, primarily due to longshore transport, and was often restricted enough to prevent the passage of small boats or flood flows. The major portion of the bed load material transported by the river was deposited above the GIWW in a silting basin (USACE 1977). The inlet was occasionally re-opened by high river discharges and by periodic maintenance dredging (Ralston 1987).

The shoaling which occurred at the Gulf entrance of the river continued to interrupt the capability of the river to serve as a dependable small-craft channel. In 1968, federal authorization was given for the USACE Mouth of the Colorado River Project in response to a need for a dependable, navigable channel connecting the Gulf Intracoastal Waterway (GIWW) to the Gulf of Mexico near the town of Matagorda, Texas. The project included a jettied entrance at the Gulf of Mexico, a navigation channel along the existing Colorado River through the delta, a harbor and turning basin, recreational facilities, and full, uncontrolled diversion of the Colorado River into Matagorda Bay (Fig. 5) (USACE 1977). The primary benefit from the project was dependable navigation between the Gulf and the GIWW.

The jetty system for the project was completed in 1985 and included a weir jetty on the northeast side of the inlet and an impoundment basin adjacent to the weir to accumulate southwesterly moving longshore sediment. The river diversion was completed in 1992, and construction of a 4.6-m deep and 61-m wide entrance channel and a 3.7-m deep and 30.5-m wide navigation channel extending to the GIWW

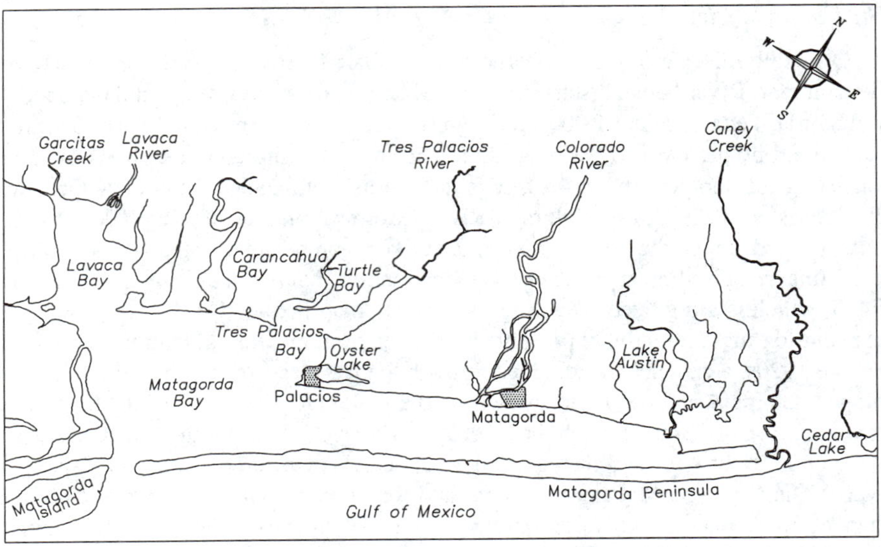

Figure 3. Survey map of Matagorda Bay, dated January 16, 1839 (original map archived by the Texas General Land Office).

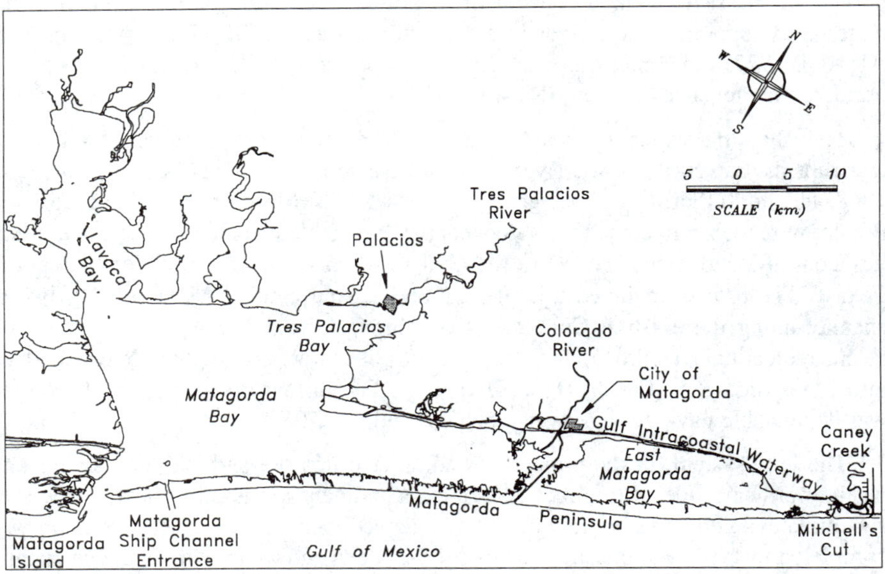

Figure 4. Map showing Colorado River delta which extended across Matagorda Bay to Matagorda Peninsula by 1935.

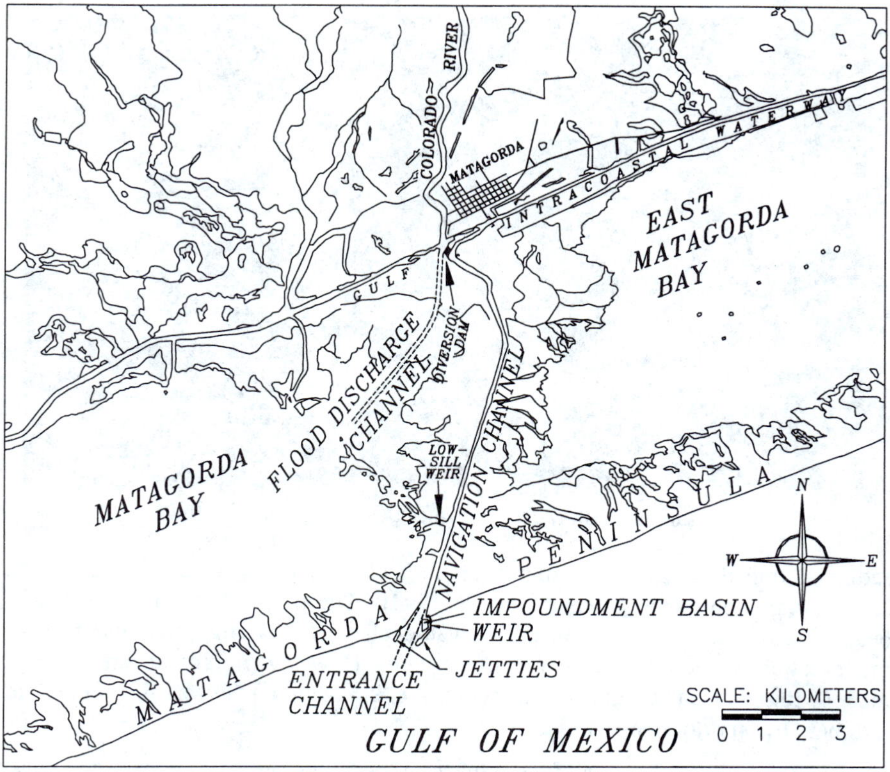

Figure 5. Features of the USACE Colorado River Project.

near Matagorda was completed by 1990 (USACE 1977). The jetties and entrance channel were aligned along the pre-existing flood discharge channel into the Gulf, with the seaward ends of the jetties spaced about 400-m apart. The weir section in the northeast jetty extended 305-m seaward from the shoreline at a crown elevation of 0 mean low tide (MLT). Past the weir section the jetty extended to the 3.7-m depth contour and was aligned to serve as a breakwater for the impoundment basin and entrance channel. The southwest jetty extended seaward to the 1.5-m depth contour.

The impoundment basin was initially designed to hold approximately 460,000 m^3 of littoral sediment (USACE 1977). Based on estimated rates of longshore sediment transport, it was assumed the basin would require dredging on average once every two years. Material dredged from the basin was to be mechanically bypassed by pipeline and discharged in the surf zone approximately 610 m downdrift of the southwest jetty (Fig. 6).

STUDY OBJECTIVES

Because the entrance to the navigation channel and the impoundment basin have accumulated sediment more rapidly than was anticipated, assumptions made about the

Figure 6. Colorado River mouth, December 12, 1992.

sediment transport rate at the mouth of the Colorado River have been in question since completion of the jetties (Rozsypal 1994). The frequent maintenance that has been required to retain an open, navigable channel has suggested the need for modification of the jetties or bypassing system. However, improvements cannot be made without accurate understanding of the interaction between the project and sediment transport processes.

To evaluate the influence of the project on longshore sediment transport, a coastal processes analysis was conducted using historical and recent data, and numerical analysis was conducted using the GENESIS shoreline change model (Hanson and Kraus 1989). The updrift and downdrift limits of inlet impact on shoreline migration were estimated, a summary of predicted longshore sediment transport rates was developed, and pre- and post-project beach profiles were used along with dredging records to quantify sediment trapping. Present and future project impacts on shoreline stability and longshore sediment transport were estimated through the use of GENESIS. Future project effectiveness was tested based on the anticipated pre-project maintenance dredging plan.

PROCEDURE

Historic Inlet Impact on Adjacent Shorelines

An evaluation of historical volumetric shoreline changes along Matagorda Peninsula was conducted using shoreline position data obtained from the University of Texas, Bureau of Economic Geology (Morton *et al.* 1976, Paine and Morton 1989). The data were compiled from topographic charts and aerial photographs and cover the period from 1855 to 1982. Based on these data, Morton *et al.* (1976) and Paine and Morton (1989) concluded that the Matagorda Peninsula Gulf shoreline is historically erosional.

Beach volume change trends were estimated by application of the equilibrium beach profile concept as presented by Moore (1982) and Dean (1991). The beach volume change trends were evaluated to a distance of 33.6 km downdrift and 11.5 km updrift. Using methods developed by Bodge (1995), whereby shoreline change data representing pre- and post-introduction of a potential sediment sink are compared, the differences between the average of the pre- and post-inlet beach volume change rates were considered to identify the net effects of the inlet (which was cut in 1935) on the adjacent beaches. In the method, all beach volume change updrift and downdrift of the inlet is attributed to the interruption of the longshore sediment transport by the inlet (i.e., the effects of hurricanes, cross-shore transport, wind-blown sand transport, etc., are neglected). Based on this assumption, the maxima in a curve plotting the difference between the averages of the pre- and post inlet volume change rates, calculated cumulatively as a function of distance from the inlet, represents the alongshore terminus of the beach segment within which the inlet influences beach volume change. As plotted in Figs. 7 and 8, the un-jettied Colorado River mouth (as existed from 1935 to 1982) may have impacted shoreline stability as far downdrift as 18.4 km and as far updrift as 5.4 km along the Gulf-facing shoreline of Matagorda Peninsula.

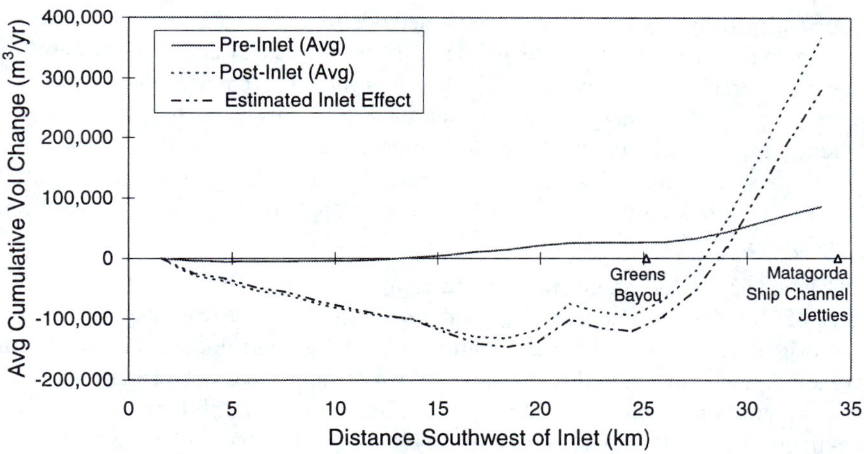

Figure 7. Cumulative annualized volume changes downdrift of the un-jettied Colorado River mouth. The pre-inlet average is based on 1855 to 1937 shoreline position data, and the post-inlet average is based on 1937 to 1982 shoreline position data.

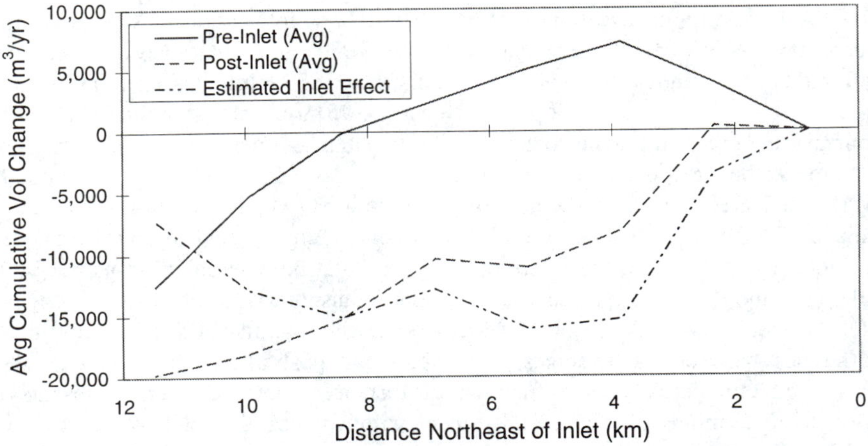

Figure 8. Cumulative annualized volume changes updrift of the un-jettied Colorado River mouth. The pre-inlet average is based on 1855 to 1937 shoreline position data, and the post-inlet average is based on 1937 to 1982 shoreline position data.

Sediment Transport

As outlined by USACE (1984), four methods of estimating longshore sediment transport exist, including (in order of decreasing preference) calculations based on: (1) A known rate at a nearby site; (2) Measured historical changes in bathymetry or topography; (3) Wave height and direction data; (4) Only wave height data. These prediction methods can be combined to compare estimated rates or to supplement a lack of data that precludes the complete application of any one method. The traditional methods outlined by the USACE were applied to the study area, and the results are summarized in Table 1.

Calculated net longshore transport rates for the Texas Gulf Coast which are in excess of 200,000 m^3/yr appear to the authors to represent over-estimates, particularly if comparison is made to other calculations of net transport documented for harsher wave climates. For example, Bodge (1995) determined the net average longshore transport rate to be about 184,000 m^3/yr for a beach on the north Florida coast facing the Atlantic Ocean; the transport rate along Texas beaches should be less because the average wave climate is less severe. Heilman and Kraus (1995) calculated the net average transport rate along South Padre Island, Texas, to be 115,000 m^3/yr based on a sediment budget, and Kraus *et al.* (1996) calculated the net rate along Mustang and north Padre Islands to be 34,000 m^3/yr and the gross rate to be 150,000 m^3/yr based on 8 years of wave hindcast data. Also, as noted by Mason (1971), Mason and Sorensen (1971), and Kraus and Militello (1996), small cuts have historically existed near the eastern end of East Matagorda Bay. These cuts, which are not subjected to strong tidal flows, would be expected to rapidly close under a strong longshore transport rate.

Table 1. Summary of longshore transport rate estimations for Matagorda Peninsula. Net rates are to the southwest.	
Method	**Rate (m^3/yr)**
1. Estimate based on hindcast wave data, spit growth, and beach erosion at Brown Cedar Cut, Matagorda Peninsula (Mason and Sorensen 1971).	38,000 *net* 460,000 *gross*
2. Estimation derived from sediment budget developed for nearby Sargent Beach (Seelig and Sorensen 1973).	31,000 *net*
3. Average rate based on 47 years of historical shoreline position data (Heilman 1995).	152,000 *net*
4. Estimation derived from evaluation of sediment impoundment at nearby Matagorda Ship Channel jetties (USACE 1977).	150,000 *net*
5. Estimation based on inspection of historic data at the project site, including survey data, aerial photography, and dredging records (USACE 1977).	170,000 *net*
6. Design rate used for the project based on combination of methods 2 and 3 above (USACE 1977).	230,000 *net*
7. Post-project rate based on inspection of beach profile survey data and dredging records (revised after Heilman 1995) (data obtained from USACE).	250,000 *net* 420,000 *gross*
8. Potential rate calculated from a 22-month time series of directional wave data using $K = 0.77$ (Heilman 1995) (data obtained from USACE, CERC).	310,000 *net* 580,000 *gross*

Over-prediction of sediment transport rates based on bathymetric changes at the study site may be due to use of the assumption that sediments are predominantly introduced to the system from alongshore. A significant volume of sand may be transported onshore as the historic Colorado River ebb delta (if it exists) is eroded by waves (Kraus 1995). The cause of the suspected over-prediction of the longshore transport rate based on collected directional wave data is unknown, but may have occurred if the sampling period (May, 1990 to August, 1992) was not representative of the typical, long-term wave climate. A comparison between wave and wind data for the study site indicated that the measured wave directions were accurate, with the predominant direction being from the southeast and sediment transport reversals occurring only 20% of the year. The average mean wave height at a depth of 5.2 m was 0.77 m, the average wave period was 5.3 sec, and the average incident wave angle was 84 deg measured clockwise from the local shoreline orientation.

Inspection of aerial photographs and observations made at the site indicate that wind-blown sand also plays a significant role in channel infilling, as sand along the beach is transported across and around the jetties to constrict the channel opening.

This process has been noted to play an important role at other jettied Texas inlets (Kieslich 1977, Kraus and Heilman 1996).

Sediment Sorting and Trapping

Sediment grain-size data collected in August, 1994, and October, 1995, by the USACE, Galveston District, were examined to determine that finer sand (0.02 to 0.04 mm) collects in the impoundment basin than in the portion of the entrance channel adjacent to the basin (0.14 to 0.18 mm). The existence of finer sand in the impoundment basin is probably a result of: 1) the sorting of suspended-load sediments from bed-load sediments by the weir; 2) the lack of sediment sorting by currents in the deep (low-energy) basin; and 3) the transport of fluvial silts and clays to the inlet from the navigation channel during ebb flows (these fine sediments may be trapped in the basin). Analysis of survey data collected by the USACE revealed that the channel and impoundment basin fill from the weir side in a southwestward direction and that the heaviest shoaling occurs close to the shoreline.

As shown in Table 2, a comparison was made between pre- and post-project beach volume change rates near the jetties. The change from pre-project beach erosion to post-project beach accretion was considered to represent a reduction in the net longshore transport of sand past the inlet. The comparison indicated an average increase in the beach volume change rate of about 53,700 m^3/yr (the difference between -9,700 m^3/yr and +44,000 m^3/yr) between 1984 and 1992 along 3.9 km of shoreline updrift and downdrift of the channel. Based on a longshore transport rate of approximately 230,000 m^3/yr, this impoundment of sand at the inlet is equivalent to a 23% reduction in the transport of sand to beaches downdrift of the inlet.

Table 2. Comparison of pre- and post-project beach volume change rates.

	Rate of Beach Volume Change (m^3/yr)		
Period (Years)	**Updrift Beach (from 0.8 to 3.6 km from inlet)**	**Downdrift Beach (from 1.6 to 2.7 km from inlet)**	**Total (along 3.9 km of shoreline)**
1857-1935 (Pre-Inlet)	6,700	-2,800	3,900
1935-1956	13,800	22,000	35,800
1956-1965	-58,000	60,400	2,400
1965-1974	-15,400	-109,000	-124,400
1974-1982	8,100	-13,000	-4,900
1935-1982 (Average)	-7,100	-2,600	**-9,700**
1984-1992	31,000	13,000	**44,000**

Numerical Simulation of Shoreline Change

As a means of supplementing knowledge gained through evaluation of field data collected at the project site, numerical modeling was conducted to simulate shoreline change based on theory and approximation. The GENESIS model, developed by the USACE, Coastal Engineering Research Center (CERC) (Hanson and Kraus 1989) was applied to the study site to test the effectiveness of the design maintenance dredging schedule at preventing channel shoaling. Shoreline reaches updrift and downdrift of the jettied inlet were modeled separately. Details of the model setup, input, and calibration are presented in Heilman (1995).

Model results indicated that the design maintenance dredging volume of 460,000 m^3 consistently bypassed every other year is sufficient for preserving the transport of sand past the inlet and for minimizing channel shoaling. However, the model results also suggested that accelerated shoreline erosion will occur updrift of the weir due to the unidirectional transport of sediment over the weir. Based on inspection of aerial photography and visits to the site, this shoreline recession has not distinctly occurred updrift of the weir, although a series of beach profile surveys conducted by the USACE, CERC indicate beach erosion occurred within about 500 m updrift of the weir between 1984 and 1992 (Liang 1995). Model results also revealed a tendency for sediments to be temporarily stockpiled in the dredged material discharge zone (approximately 600 m downdrift) due to wave sheltering by the jetties. The sediment stockpiling process may increase rates of channel shoaling as the effective length of the downdrift jetty is reduced and northeastward-flowing sand enters the channel.

CONCLUSIONS

The Colorado River Project was a response to the need for a dependable, navigable channel connecting the GIWW to the Gulf of Mexico. Jetties were constructed in 1985 including provisions for the mechanical bypassing of littoral sediments past the entrance from an impoundment basin filled by functioning of an innovative weir. Through utilization of the bypassing system, whereby sediments are regularly dredged from the impoundment basin and discharged into the surf along the downdrift beach, the potential for accelerated beach erosion has been reduced. However, the efficiency of the system has been decreased by the unexpectedly-high infilling of the impoundment basin and resulting excessive channel shoaling, and by the sorting of littoral sediments.

Possible causes for the rapid shoaling which occurs at the Colorado River mouth include:

1. A higher than anticipated net longshore transport rate.
2. The breakup of an assumed historic Colorado River shoal.
3. The reduction in effective length of the downdrift jetty (which originally extended only to the 1.5-m depth contour) by the growth of a shoreline fillet; this process results in a nearly unimpeded transport of sediment around the jetty during longshore transport reversals.

4. The diversion of river discharge to Matagorda Bay and associated reduction in flushing of channel sediments from the channel entrance.
5. The growth of shoals across the channel adjacent to open, unvegetated areas of the beach by wind-blown sand.

Main results of this study are:

1. Historical records indicate that the un-jettied Colorado River inlet influenced migration of the Matagorda Peninsula Gulf shoreline as far updrift as 5.4 km and as far downdrift as 18.4 km. The Colorado River jetties and channel are expected to exert control on long-term shoreline migration at least as far as these extents.
2. The Colorado River jetties trap sediment despite mechanical bypassing. Some sorting occurs as the weir separates suspended and bed-load sediments.
3. A varying degree of sediment is lost from the mechanical bypassing process when more sediment reaches the impoundment basin than can be stored. Overflowing sediment may be either lost to shoals (causing navigation hazards) or transported farther downshore (starving the local beach). Numerical simulation indicates that this sediment overflow can be reduced with a consistent bypassing schedule.
4. Numerical simulation predicts that a unidirectional transport of sediment occurs across the weir, resulting in a sediment deficit directly northeast of the weir during periods when waves arrive from the southwest. The effects of this process have not been observed at the study site.

ACKNOWLEDGEMENTS

Special thanks is given to Dr. Nicholas Kraus of the Coastal Engineering Research Center (CERC) for providing technical review of this report. The authors also give thanks to Mr. T. Neil McLellan at the U.S. Army Corps of Engineers, Galveston District and to Dr. David King and Dr. Tom White, both of CERC, for providing technical assistance and data. Recognition is given to Dr. Robert Morton at the University of Texas, Bureau of Economic Geology, and Dr. Richard Seymour at Texas A&M, College Station, for their patience and cooperation in providing other needed information.

REFERENCES

Bodge, K.R. 1995. The Extent of Inlet Impacts Upon Adjacent Shorelines. *Proc. 24th International Conf. Coastal Engin.*, ASCE, 2943-2957.

Bouma, A. H., and Bryant, W. R. 1969. Rapid Delta Growth in Matagorda Bay, Texas, Lagunas Costeros Un Sinposio, Nov., 1967, 171-189.

Dean, R. G. 1991. Equilibrium Beach Profiles: Characteristics and Applications. *J. Coastal Res.*, 7(1): 53-84.

Hanson, H. and Kraus, N. C. 1989. GENESIS: Generalized Model for Simulating Shoreline Change, Report 1: Tech. Ref. Technical Report CERC-89-19. U.S. Army Coastal Eng. Res. Center, Vicksburg, Mississippi.

Heilman, D. J. 1995. The Effects of the Colorado River Project on Longshore Sediment Transport at Matagorda Peninsula, Texas. Unpublished M.S. thesis, Texas A&M Univ., Dept of Civil Eng., College Station, Texas, 105 p.

Heilman, D. J., and Kraus, N. C. 1995. Beach Fill Functional Design, Town of South Padre Island, Texas. Tech. Rep. TAMU-CC-CBI-95-14, Conrad Blucher Instit. for Surveying and Science, Texas A&M Univ.-Corpus Christi, Corpus Christi, Texas, 35 p.

Kieslich, J. M. 1977. A Case History of Port Mansfield Entrance Channel, Texas. GITI Report 12. U.S. Army Coastal Eng. Res. Center, Vicksburg, Mississippi, 66 p.

Kraus, N. C. 1995. *Personal Communication.* Director, Conrad Blucher Instit. for Surveying and Science, Texas A&M Univ.-Corpus Christi, Corpus Christi, Texas.

Kraus, N. C., Heilman. 1996. Packery Channel Feasibility Study: Inlet Functional Design and Sand Management, Report 1 of a Two-Part Series. Tech. Rep. TAMU-CC-CBI-96-06, Conrad Blucher Instit. for Surveying and Science, Texas A&M Univ.-Corpus Christi, Corpus Christi, Texas, 112 p.

Kraus, N. C., and Militello, A. 1996. Hydraulic Feasibility of Proposed Southwest Cut, East Matagorda Bay, Texas. Tech. Rep. TAMU-CC-CBI-96-03, Conrad Blucher Instit. for Surveying and Science, Texas A&M Univ.-Corpus Christi, Corpus Christi, Texas, 80 p.

Liang, G. 1995. Shoreline and Sediment Volume Changes at the Colorado River Mouth, Texas. Contract Report CERC-95-2, U.S. Army Coastal Eng. Res. Center, Vicksburg, Mississippi, 31 p.

Mason, C. 1971. Properties and Stability of a Texas Barrier Beach Inlet. M.S. thesis, Texas A&M Univ., Dept. of Oceanography, College Station, Texas, 105 p.

Mason, C. and Sorensen, R. M. 1971. Properties and Stability of a Texas Barrier Beach Inlet. TAMU-SG-71-217, COE No. 146, Depts. of Oceanography and Civil Eng., Texas A&M Univ., College Station, Texas.

McGowen, J.H. and Brewton J.L. 1975. *Historical Changes and Related Coastal Processes, Gulf and Mainland Shorelines, Matagorda Bay Area, Texas.* The Univ. of Texas at Austin: Bureau of Economic Geology, in Cooperation with The General Land Office of Texas, 72 p.

Moore, B. D. 1982. Beach Profile Evolution in Response to Changes in Water Level and Wave Height. Unpub. M.S. thesis, Univ. of Del., Newark, Delaware.

Morton, R. A., Peiper, M. J., and McGowen, J. H. 1976. Shoreline Changes on Matagorda Peninsula (Brown Cedar Cut to Pass Cavallo): An Analysis of Historical Changes of the Texas Gulf Shoreline. *Geological Circular 76-6*. The Univ. of Texas at Austin: Bureau of Economic Geology, 37 p.

Paine, J.G. and Morton, R.A. 1989. Shoreline and Vegetation-Line Movement, Texas Gulf Coast, (1974-1982). *Geological Circular 89-1*. The Univ. of Texas at Austin: Bureau of Economic Geology, 50 p.

Ralston, D. O. 1987. *Summary for Mouth of Colorado River.* 47th Meeting of the Coastal Eng. Res. Board, Corpus Christi, Texas, 19-20 May, 1987, U.S. Army Eng. District, Galveston, Texas.

Rozsypal, J. 1994. *Personal Communication.* Area Engineer, Fort Point Area Office, U.S. Army Eng. District, Galveston, Texas.

Seelig, W. N., and Sorensen, R. M. 1973. Investigation of Shoreline Changes at Sargent Beach, Texas. TAMU-SG-73-212, COE No. 169, Department of Civil Engineering, Texas A&M Univ., College Station, Texas.

U.S. Army Corps of Engineers. 1977. *General Design Memorandum (Navigation Features), Mouth of Colorado River, Texas, Phase I.* U.S Army Eng. District, Galveston, Texas, 122 p.

_________. 1984. *Shore Protection Manual.* (4th ed), 2 vols, Coastal Eng. Res. Center, U.S.A.E. Waterways Expt. Sta., U.S. Govt. Printing Office, Washington, D.C.

CHAPTER 257

A Comprehensive Field Study of Tidal Inlet Processes at Ponce de Leon Inlet, Florida

Gary L. Howell[1]

1 Introduction

Traditional coastal engineering concerns at tidal inlets include navigation and channel maintenance. Today coastal engineers are faced with new concerns such as adjacent shoreline impacts, water quality, and wetlands preservation. The development of new engineering tools to address these problems depends on improved understanding of tidal inlet processes. A necessary component of this research is comprehensive field measurement of the complex hydrodynamic and sedimentation processes at tidal inlets (Mehta, 1995).

The field measurement work unit of the Coastal Inlets Research Program (CIRP) of the US Army Corps of Engineers is a coordinated effort to acquire a comprehensive data set of physical inlet processes. The measurements will be used to evaluate and refine existing process models and motivate new predictive methods. A wide range of approaches must be served by common measurements. These include physical, numerical, analytical, empirical, semi-empirical, and hybrid models. Field data are required for development, calibration, and validation.

This paper describes the planning, site selection, and design of the study. Preliminary results from the first year of data collection are presented. It is intended as a reference for investigators using the data. Problems as well as successes are described so that future studies will hopefully improve on the efforts described here.

2 Planning

2.1 Measurements

The Ponce study is the first major long term field study of a tidal inlet since the General Investigations of Tidal Inlets program (G.I.T.I.) of the 1970s. It is also the first to benefit from the major improvements in instrumentation and computer technology since that time. However a tidal inlet remains a difficult environment for field studies. The same conditions that cause hazardous navigation impact the installation and survivability of instrumentation. Rapid changes in bedforms

[1]Research Engineer, US Army Engineer Waterways Experiment Station, Coastal and Hydraulics Laboratory, 3909 Halls Ferry Road, Vicksburg, Mississippi 39180-6199 USA.

make siting of in-situ gages difficult and mandate frequent surveys. The variety of measurements required to describe all of the interacting processes still exceeds todays measurement capabilities.

Careful planning of field experiments is required to match limited resources with conflicting needs. Approaches to field experiments are affected by the technical discipline and research interests of various investigators. For example, there is not a consensus on whether it's best to study a single inlet intensively or study several inlets with limited measurements.The planning of the Ponce experiment reflects the evolution of research from traditional tidal hydraulics problems to problems of decadal or longer time scales such as impacts on adjacent shorelines from dredging and other engineering activities. Today there is more emphasis on these longterm geomorphologic responses and the sediment transport processes that drive them. Measurement technology has improved in the areas of hydrodynamics but unfortunately, progress in sediment transport measurements has been much slower.

The approach which led to the Ponce study was based on applying an engineering design method to the field studies. Study proposals and measurement requirements documents were developed by each investigator. The requirements focussed on the end use of the data, rather than specific gages or surveys. Based on requirements, experiment plans were developed which attempted to meet as many requirements as possible. Gages and surveys were prioritized by the number and technical value of requirements they satisfied. The decision to initially concentrate on comprehensive measurements at a single inlet was based on cost/benefit analysis. There remains a need to examine variability between different types of inlets, but this will be left to future studies. The planning resulted in the following types of measurements:

- Water levels at the open coast, outer throat, inner throat, and back bay open boundaries.
- Directional wave measurements in the offshore, ebb shoal, and throat areas.
- Current measurements in the throat and on the ebb shoal.
- Wind velocity and barometric pressure.
- Bathymetry
- Beach profiles on the ocean side
- Sediment samples
- Long term monitoring of shoreline, breaker lines, and the shoals.
- Regular and post-storm aerial photography.

2.2 Site selection

Since the CIRP is a national program, there were many possible field study sites. The selection was based on technical merit by using a formal analysis that matched technical and operational requirements with the characteristics of candidate inlet field sites. Each investigator developed requirements for hydrodynamic and sedimentary characteristics of the study inlet. Operational and cost factors were also considered for each inlet.

The set of all U.S. inlets was reduced to 12 candidate inlets by establishing minimum requirements. From the research viewpoint, availability of historical data was a major factor. Relevance as demonstrated by real world problems representative of those faced by the Corps was important. Finally the inlet should have a manageable physical size for field operations.

A weighted requirements matrix analysis was employed to compute separate metrics for technical and operational characteristics of the candidate inlets. In the case of the technical requirements, the field study requirements were used to compute the weights. For operational and cost factors, weights were assigned based on a consensus of senior field technicians and project engineers.

The candidate inlets were ranked by operational score, technical score, and a combined score which was normalized to provide equal weight to the operational and technical factors. Separately each candidate inlet was evaluated to determine it's suitability for physical modeling. The major factor in this evaluation was the scale factor necessary to represent the inlet in available facilities. For some inlets, factors such as tide range and wave generation requirements were also important.

The results of the formal analysis ranked Ponce de Leon Inlet, Fl. highest. The recommendation was reviewed and accepted by all investigators, program managers, and sponsors. Experience during the first year of the field study has supported the selection. Ponce is a tidal inlet typical of the east coast of the U.S. It has a jettied entrance with a visible north jetty and a south jetty covered with sand. The design channel is $5m$ deep and $70m$ wide. Historically, the inlet has undergone significant evolution due to engineering activity and natural response. A substantial set of historic data is available (USAE Jacksonville District, 1993). Present engineering problems include dangerous navigation, potential breach locations, severe scour along the inside of the north jetty, inner bank erosion, and jetty sand bypassing problems. Many unique operational advantages at Ponce have been important in achieving a successful monitoring effort. These include undeveloped land south of the inlet, the Coast Guard Station close by, and a single tall condominum building near the inlet which was ideally sited for video monitoring. Ponce is a dynamic inlet which can be expected to undergo measureable changes during the lifetime of the study.

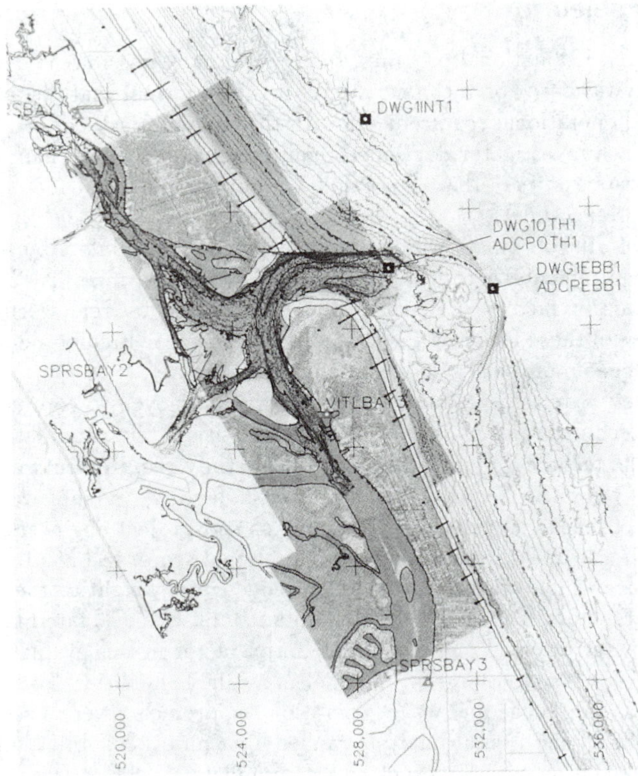

Figure 1: Ponce Inlet long term field measurements.

3 Instrumentation

3.1 Hydronamic gages

Figure 1 shows the layout of the long term instrumentation at Ponce. There are directional wave gages at three sites on the ocean side of the inlet. The site **DWG1INT1** is intended as a reference for incident waves. The gage is located in $14m$ of water north of the inlet in an area where the bottom contours are relatively parallel. The site **DWG1EBB1** is located on the face of the ebb shoal in approximately $7m$ of water. The site in the outer throat of the inlet is **DWG1OTH1**. Each of these sites has a DWG-1 (Howell, 1992) directional wave gage. The DWG-1 is a short-baseline pressure gage array. The pressure gages are Paros Digiquartz which provide accurate water level information. *PUV* type directional wave gages using electro-magnetic current meters were not used because of the high cost and difficulty of the regular cleaning. The Ponce area is regularly visited by shrimp trawlers, making the trawler resistant design of the

DWG1EBB1	Ebb Shoal	29deg 04.565min N, 80deg 53.902min W
DWG1INT1	North of Inlet	29deg 05.462min N, 80deg 54.620min W
DWG1OTH1	Outer Inlet Throat	29deg 04.605min N, 80deg 54.489min W
SPRSBAY1	Deepwater Marina	29deg 05.453min N, 80deg 56.432min W
SPRSBAY2	Utility Company (W)	29deg 03.944min N, 80deg 56.332min W
SPRSBAY3	Riverview Restaurant (S)	29deg 02.266min N, 80deg 54.299min W
VITLBAY3	Coast Guard Station	29deg 03.776min N, 80deg 54.909min W

Table 1: Map coordinates of the principle gage sites.

Figure 2: DWG-1 directional wave gage modified with a colocated ADCP (top right). Continuous digital time series data from both instruments are transmitted on a single cable.

Gage type	Sample period	Resolution
Directional Wave Gages	0.2-sec	2-mm
Current Profiler	1-sec	$1 - cm/sec$
Bay Water Levels	360-sec	1-cm

Table 2: Sampling periods for the hydrodynamic measurements. Record lengths are approximately continuous.

DWG-1 advantageous.

The ebb shoal site and the inlet throat site also have RDI [2] broadband 1200kHz Acoustic Doppler Current Profilers (ADCP) located on the same mounting pod as the DWG-1. Figure 2 shows the right angle upward looking style ADCP used. The ADCP is configured for continuous real time output at a $1Hz$ sample rate and $1m$ vertical bins. The combined DWG-1 and ADCP share the same cable for both power and data output. All data recording is continuous, real-time via the cable. The internal recording modes of the instruments are not used.

Water level data in the inner bay is provided by Paros Digiquartz pressure sensors. The sensors are mounted in a steel protective sleeve sleeves are attached to a fixed length pipe which is used to mount the sensors to a pile. The top of the pipe has a reference mark which can be surveyed to establish the datum for the water levels. Both the offshore water levels (from the DWG-1) and the inner bay water levels use absolute pressure sensors which require compensation for barometric pressure. Another Paros sensor is used as the principle barometric pressure measurement and is recorded simultaneously with the water level sensors. A backup barometric measurement is available from the meteorological station installed on the north beach[3]The meteorological station is provided by the Battelle institute. A standard stilling well tide gage is located at the Coast Guard Station (**VITLBAY3** in Figure 1). This gage uses an air acoustic sensor and provides a check on the air pressure corrections and datums for the pressure water level sensors.

Datums for the offshore water levels obtained from the DWG-1 wave gages were not surveyed like the inner bay gages. Instead data analysis of six months of data was used to estimate the datums. The technique is based on an iterative minimization of the residuals between the Coast Guard Station tide gage (**VITLBAY3** and the offshore water levels. The technique selects data windows at slack water when the inner and outer water levels approach equality. The method produces accuracies comparable to that obtained with conventional surveys. Details of the method will be provided in a future publication.

Table 1 lists the locations of the principle measurement sites.

---

[2]RD Instruments, see `http://www.adcp.com/` for additional information.

[3].

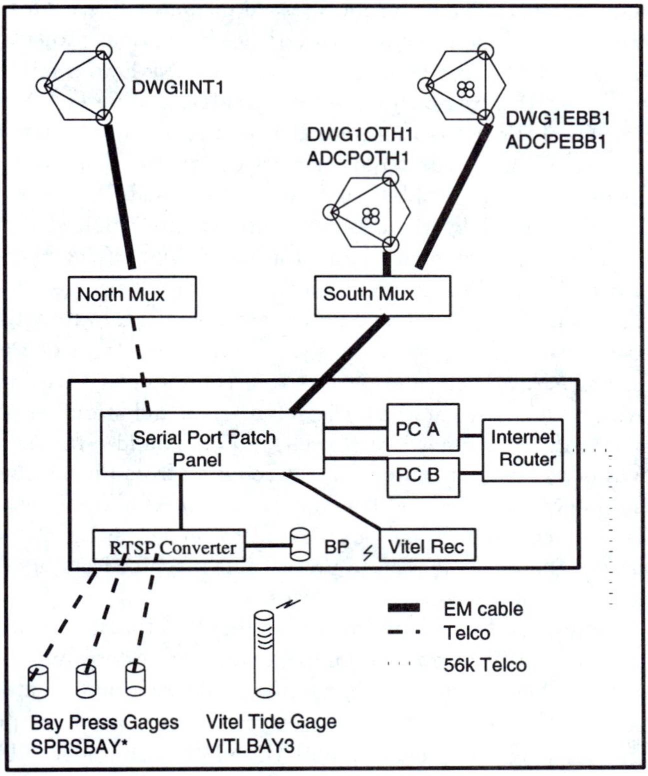

Figure 3: Diagram of the Ponce hydrodynamic gage data acquisition system.

3.2 Time series data acquistion

Table 2 gives the sample rates and resolution of the measurements. All data are acquired continuously. However some of the routine analysis use shorter record lengths. For example directional wave spectra are computed using 34 minute records windowed from the available 59 minute records.[4] All data analyses are performed at least hourly. Data from all of the time series gages are logged by a single computer located in a leased trailer at the Coast Guard Station. The trailer is connected to the Internet via a 56k Baud leased line to the Jacksonville COE District office in Jacksonville. Figure 3 shows the layout of the connections from the gages to the central logging computer. Most of the connections are through

[4]A variety of analysis products are produced from the Ponce data. Users should check the analyses parameters used for the particular product rather than relying on the general description here.

cable installed for the experiment or local leased telephone lines. This design was used instead of radio telemetry because experience in previous projects has shown that wireline installations provide higher data return at a lower maintenance cost than telemetry. Another reason for use of a central computer facility was the desire to preposition the infrastructure that would be required for a larger short term experiment. Without this requirement a design based on several outdoor Remote Transmission Units (RTUs) could have been used.

All of the data logged by the central computer are acquired through serial ports. The computer is a standard industrial rackmount PC computer. A multiport serial card is installed to provide extra ports. There are three identical computers installed. Two of the computers are operational all of the time and have the serial data arriving continuously. Logged data from one of the computers is transferred over the Internet link to a computer in Vicksburg which is configured with a standard data acquisition, analysis and storage system known internally as PMAS (McKinney and Howell, 1996). Should a failure of the principal computer occur, data are automatically acquired from the backup computer. The third computer is installed and tested, but disconnected from power, network, and serial lines. It can be manually configured to replace either of the two operating computers. This mode is used to protect this machine from lightning surges which overpower the surge protection devices.

The software for the PC compatible computers logs the data from each serial port to files. The Posix compliant Linux[5] operating system provided most of the building blocks for the system. Customized code was implemented primarily with shell scripts. Procedures too complex for shell scripts were implemented as Ada programs using the freely available GNAT Ada compiler[6]. The software is configured to operate completely unattended. Status information is available at any time on-site or remotely using the *finger* command. Automated email messages are sent to advise of any unusual situation or error. A status monitor implemented as a shell script runs continuously on the computer in Vicksburg. It uses a combination of the *finger* and *traceroute* commands to detect problems at the site and network outages. Automated email messages are also sent by the Vicksburg monitor computer to warn of problems.

3.3 Problems and performance

The general performance of the hydrodynamic monitoring system has been very good. The data acquisition system has achieved over 99% reliability. The ebb shoal directional wave gage and ADCP experienced a cable failure during the March 1996 storm. The failure was caused by incomplete disassembly of the instrument by unauthorized divers. The instrument located in the throat of the

[5]A freely available Unix like operating system which runs on PCs. See `http://www.li.org/` for additional information.

[6]GNAT is a the Free Software Foundation Ada 95 compiler. See `http://www.gnat.com/` for additional information.

Figure 4: The SHOALS helicopter Lidar bathymetric system. The Pod on the bottom of the helicopter contains the laser, scanner, and optics. The computers and operator's consoles are located inside the helicopter.

inlet has experienced periodic burial by sand. The directional wave gage continued to yield useful data. However the sand cover which ranged from 1-2m rendered the ADCP data unuseable. Relocation of these instruments is under consideration. Some problems have been encountered with the Vitel tide gage in setting its range to accomodate the extreme highs and lows of the water levels. These problems have been easily corrected but compromise the peak tide data until correction. After an initial period of working with the telephone company to debug the leased telephone lines, the bay pressure tide gages have been very reliable.

Data analysis of the wave and water level gages has been generally routine. However the Ponce system is our first experience with continuous time series data from multiple ADCPs. Automated analysis and quality control software was not available at the beginning of the study. Development of this software is underway, but will require extensive effort before its performance is comparable with existing wave and tide measurement software. ADCPs have the potential to provide details of both mean and orbital velocities, as well as suspended sediment concentrations. As better analysis tools become available these measurements should become feasible.

4 Spatial monitoring and surveys

4.1 Bathymetry

The Ponce study is fortunate to have available a powerful new tool for coastal engineering work, the helicopter Lidar bathymetric system. The system developed by the Corps of Engineers is known as SHOALS. Irish *et al.*(1996) in this Proceedings provides a detailed description of the SHOALS system and its use for monitoring volume changes at inlets. Figure 1 shows bathymetry obtained from the SHOALS system. The detailed and rapid coverage provided by the laser system allow the details of the complex bottom topography of the inlet system.

A Lidar system operates by measuring the time difference between the reflection of a laser light pulse from the water surface and the bottom. Scanning the laser allows coverage of a $100m$ swath as the helicopter moves across the survey area. Figure 4 shows a photograph of the SHOALS system mounted on the survey helicopter. The accuracy of SHOALS is continually being evaluated and improved. Experience at Ponce shows that accuracies are already comparable with that of traditional fathometer surveys. Improved GPS technology now allows SHOALS to survey the above water portions of the beach profile. The accuracies of this capability are currently under evaluation.

The main limitation of Lidar technology is dependence on water clarity. At Ponce there have been periods where surveys could not be performed because the water was too turbid. Turbidity problems were most severe near the throat and in the back bay. Turbidity can be caused by a variety of weather conditions. These include storms which suspend large quantities of sediment. The rapid survey capability of SHOALS makes it ideal as a post-storm survey tool, however turbidity will prevent its use during the storm and the recovery period immediately afterwards.

4.2 Shoreline response

Monitoring of the long term shoreline response is performed using previously developed techniques. Data are acquired from

- Color digital aerial photography
- Wading beach profiles
- Continuous video monitoring

The aerial photography and beach profiles are combined with the SHOALS bathymetry in a Geographic Information System (GIS) database. Once in the GIS, standard tools such as differencing and volume change mapping are applied. As longer records become available, the GIS will be used as a tool to help manage the spatial data and facilitate correlations with the hydrodynamic forcing.

Figure 5: Video photo looking out from the top of the condominium toward the inlet channel and jetties. The photo is a result of a timed exposure which causes the breaker lines to appear as white streaks.

Potential development of inlet response models require long term data of shoreline position that is sampled more frequently than annual or semiannual beach surveys. One approach that has been applied to beaches on the open coast is video monitoring. Konicki and Holman (1996) describe the technique in a paper in this Proceedings. For the Ponce experiment, five video cameras were mounted on top of a condominium on the south side of the inlet. The cameras provide views of the up and down coast beach, the throat, ebb shoal, and flood shoal. Figure 5 shows an unrectified view from one camera of the channel and jetties. The photo is the result of a six minute time averaged exposure. Averaging highlights wave breaking zones as white bands. By processing many such photos, shoreline, bar, and shoal positions may be tracked.

The usefulness of the video data has not yet been demonstrated. Long term records and substantial efforts in data analysis are required. If successful, results from this monitoring will appear in future publications.

5 Sample storm data

Figure 6 shows waves from both the offshore gage and the wave gage in the throat of the inlet. This storm occurred in March 1996 and was unusual in that the wind

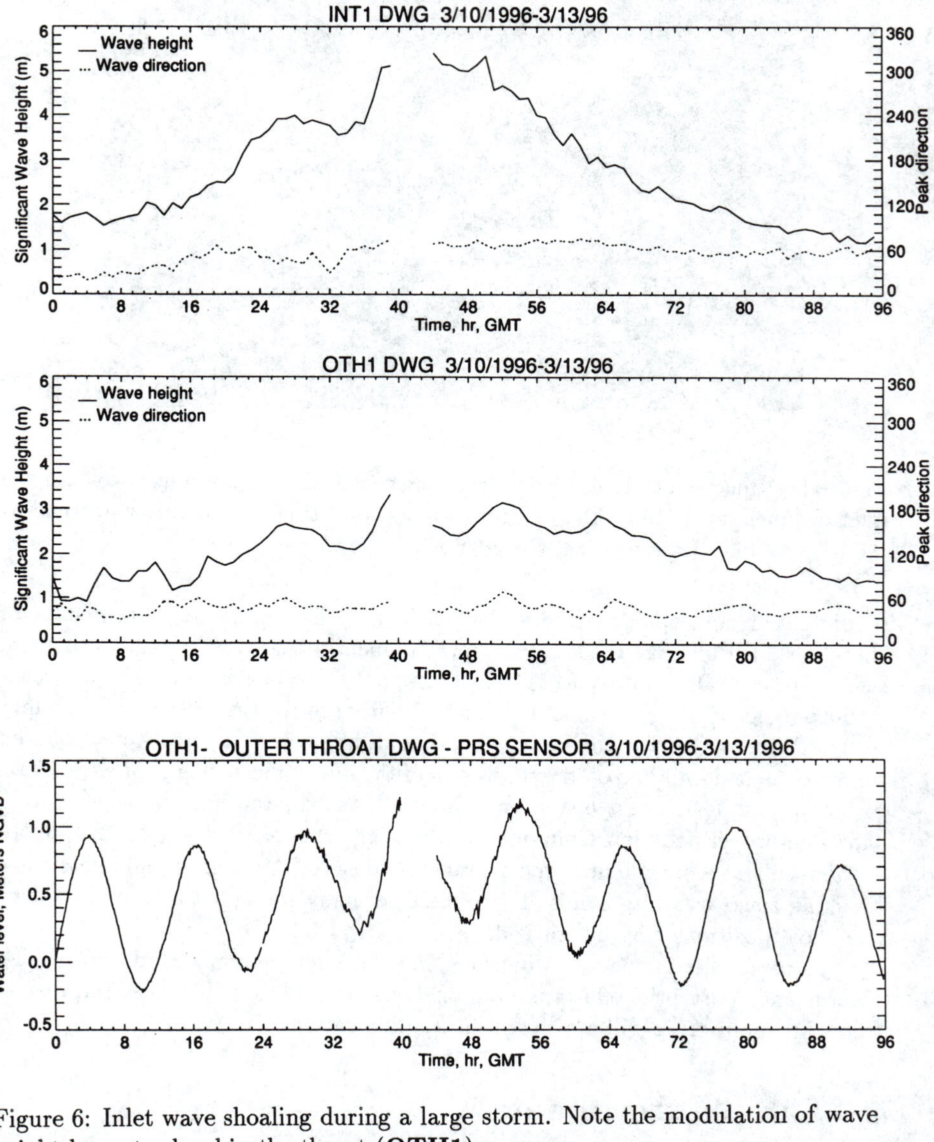

Figure 6: Inlet wave shoaling during a large storm. Note the modulation of wave height by water level in the throat (**OTH1**).

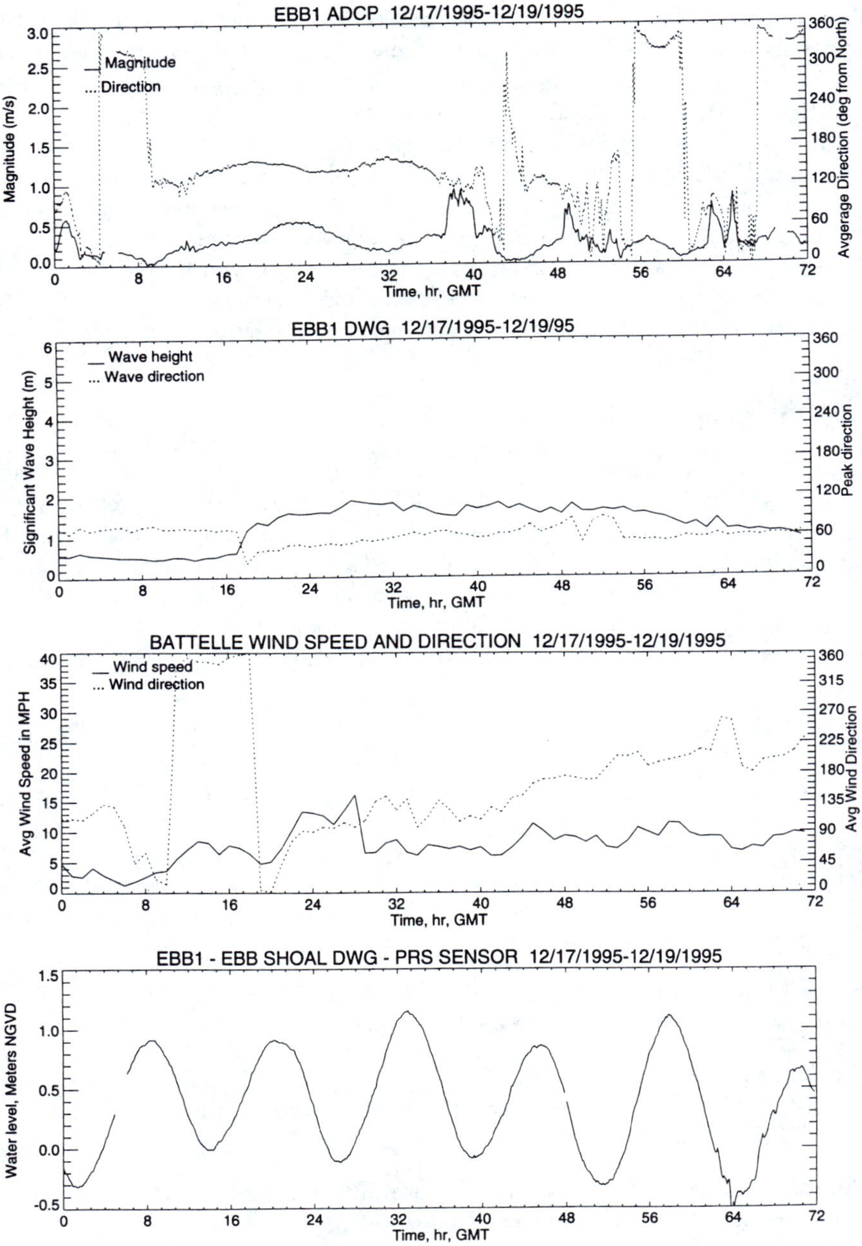

Figure 7: Ebb shoal waves and currents. Note the wind and waves driven currents as well as tidal currents.

direction stayed fairly constant as can be seen by the wave direction. The throat waves show both the effects of shoaling and the sheltering of the north jetty. The modulation of the wave height by the tide confirms the importance of water levels and tidal currents in predicting waves in inlets.

Figure 7 shows vertically averaged mean currents on the ebb shoal during a moderate Nor'easter. The beginning and end of the record show tidal currents dominant as is the ambient condition. As the wind and wave directions change, the tidal component of the mean current is strongly influenced.

Additional data analysis is currently underway and will be reported by the program investigators. The data already acquired provide confirmation of the complexity of the hydrodynamic and sedimentary processes at Ponce inlet.

6 Acknowlegements

Many people made important contributions to this study. Bill Grogg, Trap Puckette, Sam Corson, Pat McKinney, Rhonda Lofton, Kent Hathaway, Jeff Lillycrop, and Tom Martin all played key roles. The U.S. Coast Guard Station, the Inlet Condominium, the Battelle Paint testing facility, and many other public and private land owners made their property and facilities available. This work was supported by the Coastal Inlets Research Program of the US Army Corps of Engineers. Permission for publication was granted by the Chief of Engineers.

References

Howell, G.L. "A New Nearshore Directional Wave Gage," Proc. 23rd ICCE, Venice, Italy, ASCE, October, 1992.

Irish, J.L., Lillycrop, W.J., and Parson, L.E. "Accuracy of Sand Volumes as a Function of Survey Density," Proc. 25th ICCE, Orlando, Fl., ASCE, September, 1996.

Konicki, K.M. and Holman, R.A. "Transverse Bars in Duck, North Carolina," Proc. 25th ICCE, Orlando, Fl., ASCE, September, 1996.

McKinney, J.P. and Howell, G.L. "A Data Acquisition and Analysis System for Coastal Ocean Measurements," Proc. Oceans 96, Ft. Lauderdale, Fl., IEEE, September, 1996.

Mehta, A.J. ed. "Understanding Physical Processes at Tidal Inlets," SI 23 *Journal of Coastal Research,* Winter 1996.

USAE Jacksonville District "Navigation Study for Ponce de Leon Inlet, Reconnaisance Report," U.S. Army Corps of Engineers, Jacksonville District, Jacksonville, Fl., January 1993.

CHAPTER 258

Transport of Fluid Mud generated by Waves on Inclined Beds

Thijs van Kessel*, C. Kranenburg† and J.A. Battjes‡

Abstract

Liquefaction of a freshly deposited mud bed may occur if shear stresses in the bed, generated by wave-induced pressure gradients, locally exceed the yield strength. In this way fluid mud is generated, which has a concentration close to the original bed concentration (generally a few hundreds kg m^{-3}). If fluid mud is displaced, very high transport rates are to be expected. This may be an explanation for rapid mud accumulations often observed in navigation channels after storm periods. In this paper wave-induced liquefaction of a sloping mud bed is investigated with laboratory experiments. A yield strength profile of the bed is obtained with a small-scale sounding test. Rheological properties of the mud after liquefaction are determined with independent rheological experiments. It is shown that calculated shear stresses in the bed at the onset of liquefaction just exceed the measured yield strength. The combination of waves and a slope turns out to be very effective for the transport of fluid mud.

1 Introduction

Harbours and coastal areas are affected by the accumulation of mud in several respects. Resulting bathymetric changes often hamper navigation and make expensive dredgeing operations unavoidable in order to maintain navigable depth. Moreover, mud is often heavily polluted as it easily adsorbs many pollutants such as heavy metals and pesticides because of its high specific area and surface properties. This negatively affects water and sea bed quality and makes disposal of dredged material expensive. In order to prevent the undesired accumulation of mud, the relevant transport mechanisms have to be identified, thus providing a tool for coastal management.

*PhD Student, Hydromechanics Group, Dept. of Civ. Engrg., Delft Univ. of Technol., P.O. Box 5048, 2600 GA Delft, The Netherlands. E-mail: T.vanKessel@ct.tudelft.nl

†Assoc. Prof., Hydromechanics Group, Dept. of Civ. Engrg., Delft Univ. of Technol., P.O. Box 5048, 2600 GA Delft, The Netherlands.

‡Prof., Hydromechanics Group, Dept. of Civ. Engrg., Delft Univ. of Technol., P.O. Box 5048, 2600 GA Delft, The Netherlands.

Mud may accumulate gradually by sedimentation from the water column during slack tide. This is a slow process, as the sediment concentration in the water column generally does not exceed 1 g l^{-1}, and the settling velocity of the mud particles is low. The freshly deposited sediments may also be resuspended during ebb and flood tides, which reduces the net accumulation rate during a tidal cycle.

Another transport mechanism is the transport of concentrated, near-bottom mud layers—often referred to as 'fluid mud'—under the influence of waves and gravity. This may result in very high transport rates during a short period of time, as sediment concentrations in fluid mud may be up to a few hundreds kg m^{-3}. This transport mechanism may be an explanation for the rapid, event-like mud accumulations often observed after storm periods.

In this paper the latter mechanism is studied by means of laboratory experiments. In §2 the theory underlying the generation and transport of concentrated mud layers is discussed in greater detail; in §3 the experimental results are presented and discussed. Conclusions are drawn in §4.

2 Theory

2.1 Generation of fluid mud

Several erosion mechanisms of mud are possible. Surface erosion will occur if the bed shear stress exerted by the orbital motion of water under waves or by current, exceeds the critical shear strength for erosion (Mehta et al. 1989). Bulk erosion will occur if the shear stress inside the bed generated by wave-induced pressure gradients on the bed surface, exceeds the yield strength of the bed. For the experiments reported in this paper the latter mechanism prevailed, as the frictional bed shear stress remained small because of the absence of current.

Because of the very low permeability of a mud bed, water flow inside the bed caused by wave-induced pressure gradients is negligible (undrained conditions). Erosion caused by fluidisation or swell will not occur in this case. The behaviour during and after undrained failure of a mud layer will depend on its state of consolidation. Freshly deposited and consolidated beds tend to decrease in volume when subjected to shear, as they are loosely packed. A (positive) excess pore water pressure is generated, resulting in decreases in effective stress and strength. As the bed liquifies without the flow of water, sediment particles become mainly supported by the pore water instead of the grain skeleton. A liquified bed is easily transported if net forces are acting on it (De Wit 1995).

However, if the bed is closely packed because it has been exposed to high effective stresses in the past—by burial and subsequent exposure, for example—it tends to expand under shear, resulting in the generation of negative pore pressures. No liquefaction occurs in this case; on the contrary, the bed 'breaks'. Shear stresses needed for this type of failure to occur are generally much higher than those for the liquefaction of loosely packed mud beds.

The sediment beds used in the experiments described in the present paper were freshly consolidated and loosely packed. Therefore, a positive excess pore water pressure generated upon failure, and liquefaction of the bed are to be expected.

The yield stress of the sediment bed is the key parameter for possible bed failure for the present experiments. It is essential that yield strength is measured *in situ*, as remoulded yield strength tends to be much lower than *in situ* strength. Measurements with a rheometer using a standard geometry in which a bed sample is injected gives erroneous results. A better option would be to use a vane geometry that inserted into the undisturbed bed, as the shear surface during insertion is different from the shear surface when measuring yield strength.

Another technique, which has been applied to the beds used in this study, is a sounding test, during which the force exerted on a small geometry slowly penetrating into the sediment bed is measured with a sensitive balance. With bearing capacity theory (Terzaghi 1943) the yield strength profile can be calculated from the force as a function of vertical position (Van Kessel 1996b).

As was remarked before, undrained failure occurs if the wave-induced shear stresses inside the mud bed exceed the local yield strength. Shear stresses in the bed generated by the pressure gradients caused by waves can be calculated from prescribed boundary conditions assuming a certain constitutive behaviour of the bed. For stresses well below the yield stress the bed will behave as an elastic material, whereas for stresses close to the yield stress the bed will behave as an visco-elastic material. Stresses beyond the yield stress result in liquefaction and concurrently viscous behaviour. If the constitutive behaviour of the bed is purely elastic, then the applied shear stress and the resulting strain are in phase. If the behaviour is purely viscous, then the shear stress and shear rate are in phase.

The wave-induced shear stresses in the bed during the present experiments were calculated assuming an elastic constitutive behaviour, which is in accordance with the concept of a yield strength as long as it is not exceeded. Horizontal and vertical displacements at the interface between sediment bed and fixed bottom were assumed to be zero (no-slip condition), whereas at the interface between sediment bed and overlying water the shear stress was assumed to be zero. The latter boundary condition can be justified if the stresses inside the bed caused by the pressure gradients are much larger than the frictional shear stress on the bed surface caused by the orbital motion of water under waves, which is shown to be true in §3.2. The final boundary condition needed to obtain a closed set of equations is the wave-induced pressure gradient exerted on the bed surface, which can be calculated from linear surface wave theory, for example. Details of the calculations are not presented herein for reasons of brevity. An analytical solution can be obtained (Van Kessel et al. 1997), which is a special case of the Yamamoto et al. (1978) model.

2.2 Transport of fluid mud

After the yield strength has been exceeded, the liquified bed starts to flow if subjected to a net force, which is the gravity force in the present study. As failure of the bed is undrained, the sediment concentration in the liquified layer equals the original bed concentration, which generally is a few hundreds kg m^{-3}. Because of the high viscosity, the flow of these layers tends to be laminar. Mixing with overlying water hardly occurs, because of the large density gradient stabilizing the interface.

In order to predict the flow of fluid mud layers, it is essential to know their rheological behaviour. The rheology of concentrated suspensions of cohesive sediment particles is quite complex; for a detailed discussion the reader is referred to Van Kessel (1996a), Toorman (1993), James et al. (1987) and Williams (1984). Fluid mud subjected to simple shear flow can be rheologically characterized by a (residual) yield strength and a shear rate dependent viscosity (non-Newtonian behaviour). A complication is that the viscosity is a function of the shear rate history rather than the actual shear rate, because structural changes that take place within fluid mud if the shear rate is changed, need time to reach an equilibrium situation. The constitutive model used to describe the fluid mud flow in the present experiments includes thixotropic behaviour (Toorman 1997):

$$\tau_{xz} = \lambda\tau_y + \left(\eta_\infty + \frac{c_1}{1+\beta|\dot{\gamma}|}\right)\dot{\gamma} \quad (\tau_{xz} > \lambda\tau_y) \tag{1}$$

where τ_{xz} is the shear stress parallel to the bed, $\dot{\gamma}$ is the shear rate, τ_y is the yield stress, η_∞ is the viscosity as infinite shear rate and c_1 and β are coefficients incorporating shear-thinning behaviour. The structure parameter λ decreases with time from 1 to 0 as a result of structural break-up of the sediment caused by wave action (Moore 1959):

$$\frac{d\lambda}{dt} = -c_2\lambda|\dot{\gamma}| \tag{2}$$

Structural recovery is not taken into account for the present experiments, because the time scale of recovery is much larger than that of the experiments. In §3.3 the calibration of the model coefficients τ_y, η_∞, c_1, β and c_2 for the sediment type and concentration used is discussed.

The equation of motion for the fluid mud layer can be solved together with the rheological model and boundary conditions. The equation of motion can be simplified into:

$$\rho\frac{\partial u}{\partial t} = -\frac{\partial p}{\partial x} + \frac{\partial \tau_{xz}}{\partial z} + (\rho - \rho_w)g\sin\theta \tag{3}$$

where ρ is the density of fluid mud, ρ_w is the density of water, u the local velocity, t is time, $\partial p/\partial x)$ is the the wave-induced pressure gradient, z the coordinate normal to the bed, g is the acceleration of gravity and θ is the angle of the slope to the horizontal. The convective terms that should appear in (3) have been neglected, as for the present experiments they are very small compared to the other terms. Additionally, it has been assumed that the motion of the overlying water layer is not influenced by the mud layer, and that the slope is small. The pressure gradient at the bed caused by wave action is expressed by:

$$-\frac{\partial p}{\partial x} = -k\frac{\rho_w g a}{\cosh kh}\cos(\omega t - kx) \tag{4}$$

where k is the wave number, a is the wave amplitude, h is the water depth, ω is the circular wave frequency and x the coordinate parallel to the water surface.

Boundary conditions are:

$$\left.\begin{array}{lcl} z = D_1 & : & u = 0 \\ z = 0 & : & \dot{\gamma} = 0 \end{array}\right\} \tag{5}$$

Here $z = D_1$ is the elevation of the interface between non-liquified and liquified mud determined from observations and $z = 0$ is the interface between liquified mud and water. Initial conditions are:

$$t = 0 \ : \ u = 0 \text{ for all } z \tag{6}$$

These equations, which describe the motion of the mud layer after liquefaction, were solved numerically.

3 Experimental results and discussion

3.1 Experimental methods

An experimental setup was built to study wave-induced liquefaction and transport on a slope experimentally (Figure 1). Sediment beds were prepared by sedimentation and consolidation of a suspension of China clay in tap water in which 0.5% NaCl was dissolved. The initial sediment concentration was 275 kg m^{-3}. The suspension was mixed for 2 weeks in order to reach physicochemical equilibrium. China clay—mainly consisting of kaolinite—was used as an artificial mud because of its reproducible properties and easy handling. After 1 week of consolidation the bed height remained constant and the bed was titled to its desired angle (0.05 rad). The bed height after consolidation was 0.12 m, its width 0.65 and its length 4.67 m.

At $z = 0$, 0.02, 0.06 and 0.10 m from the bed surface pore pressure transducers were mounted to observe the liquefaction behaviour. These transducers were located at measuring station 3. The sediment concentration profile in the bed was measured with a conductivity probe. Velocities both in the bed (after liquefaction) and above the bed were measured with two electromagnetic flow meters (EMF) mounted on traversing units to obtain vertical profiles. These traversing units were located at measuring stations 1, 2 and 3. Sediment concentrations in

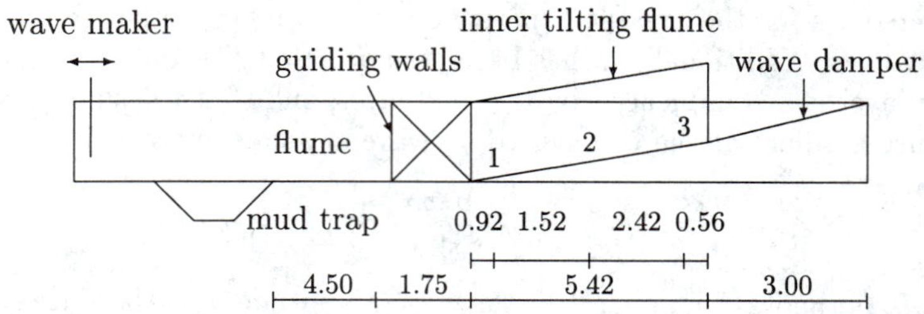

Figure 1: Experimental set-up (not to scale); measuring stations 1, 2 and 3 are indicated; lengths in m

the water column were measured with turbidity meters in order to estimate the importance of interfacial mixing. Also three wave height meters were installed. The signals of these instruments were logged onto a PC.

3.2 Wave-induced liquefaction

The experiments were started by generating sinusoidal waves with a period of 1.65 s and an initial wave amplitude of 0.005 m. During the experiments the wave amplitude was increased in steps to 0.008, 0.010, 0.013, 0.016, 0.021, 0.027, 0.036 and finally 0.042 m and the pore pressure response was measured (Figure 2). Initially no changes in the wave-averaged pore pressure were observed. However, after an increase in wave-height from $a = 0.005$ to $a = 0.008$ m, suddenly excess pore pressures were generated and the bed started to liquefy. Bed flow does not yet occur at this wave-height, as is illustrated in Figure 3, where the velocity histories at several levels inside and above the bed are shown.

The bed did flow if the wave amplitude was increased to 0.010 m. This can be derived from Figure 2, where a sudden drop in pore pressure can be attributed to the removal of the liquified mud layer. The thickness of this layer can be calculated from the observed pore pressure drop (approximately 150 Pa at $z = 0.06$ m), as the initial sediment concentration is known from the conductivity probe measurements ($\rho = 1340$ kg m^{-3}). The thickness thus estimated ia 0.045 m. This result is consistent with Figure 4, where the velocity histories shown indicate that the bed is eroded up to the level $z = 0.065$ m.

A comparison of the wave-induced stresses in the bed—calculated from the elastic model—with the yield strength profile obtained with the sounding test (§2) is shown in Figure 5. At $a = 0.005$ m, when no liquefaction is observed, wave-induced shear stresses just remain below the yield strength. At $a = 0.008$ m, however, onset of liquefaction was observed, which is consistent with Figure 5, where, for this wave amplitude, the shear stress exceeds the yield strength in the

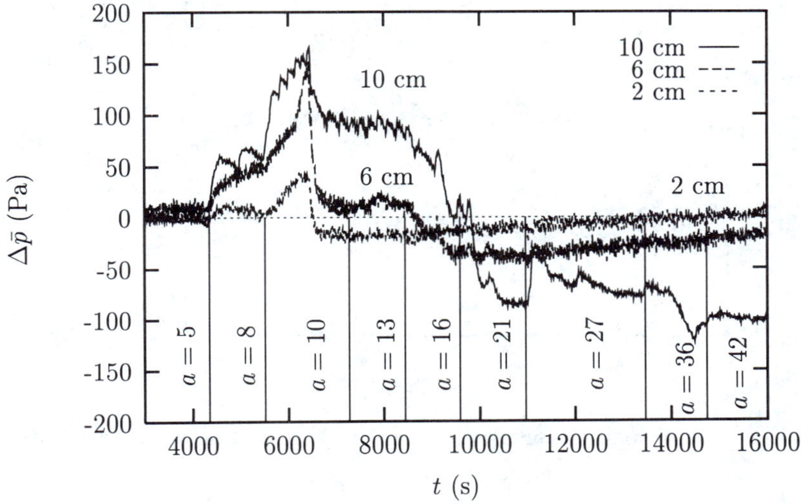

Figure 2: Changes in wave-averaged pore pressure $\Delta\bar{p}$; wave amplitudes in mm

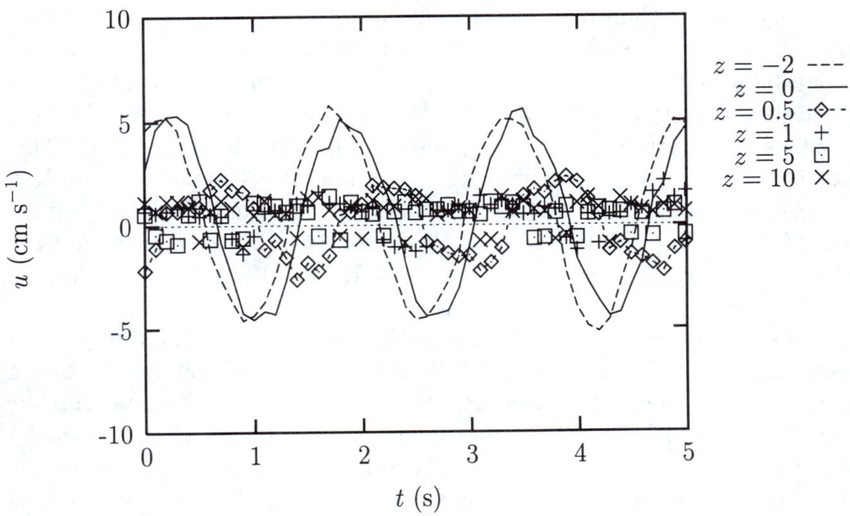

Figure 3: Velocities versus time for $a = 5$ mm; z in mm; station 3

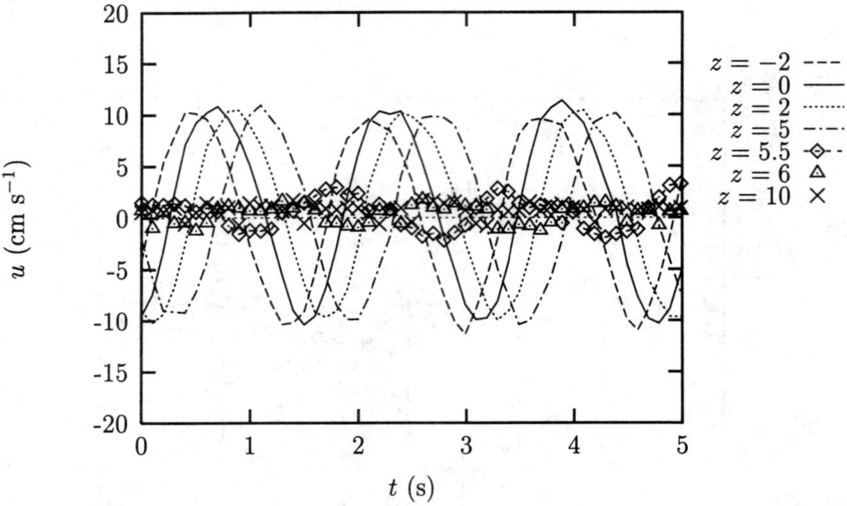

Figure 4: Velocities versus time for $a = 26$ mm; z in mm; station 3

upper 8 cm of the bed. The onset of liquefaction can therefore be well predicted if the yield strength profile of the bed is known. The yield strength of the bed increased with depth in a similar fashion as the effective stress, wich was also observed by Bowden (1988).

3.3 Fluid mud flow

After liquefaction the fluid mud starts to flow. Rheological properties of China clay have been investigated independently to calibrate the rheological model applied (§2). In Figure 6 the equilibrium flow curve of China clay with $C = 467$ kg m^{-3} suspended in tap water with 0.5% NaCl is shown. It was measured with a Carrimed controlled stress rheometer equipped with a cone-plate geometry. This flow curve has been used to calibrate the Toorman (1997) model. In addition to measuring the flow curve, also thixotropic experiments have to be performed to measure the time-scales of structural break-up. These measurements are not discussed herein, the reader is referred to Van Kessel (1996a). Values for the model parameters used are listed in Table 1.

With the rheological model and the equation of motion presented in §2 fluid mud flow can be calculated numerically. Results of these calculations are shown in Figure 7. Peak velocities are approximately 0.02 m s^{-1} and increase in time because of structural break-up of fluid mud. Plug flow can be explained by the residual yield strength of the bed; negative velocities do not occur for the conditions during the experiments because of the finite yield stress. The model shows that the combination of wave-loading and the shear-thinning behaviour of mud results in a decrease in its effective viscosity, which markedly enhances

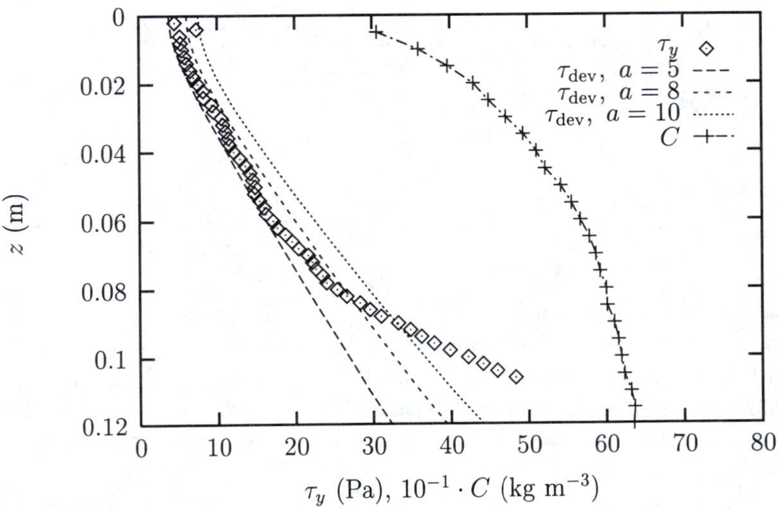

Figure 5: Yield strength, deviator stress and concentration profiles of a bed of China clay, obtained from sedimentation of a suspension of 275 kg m^{-3}

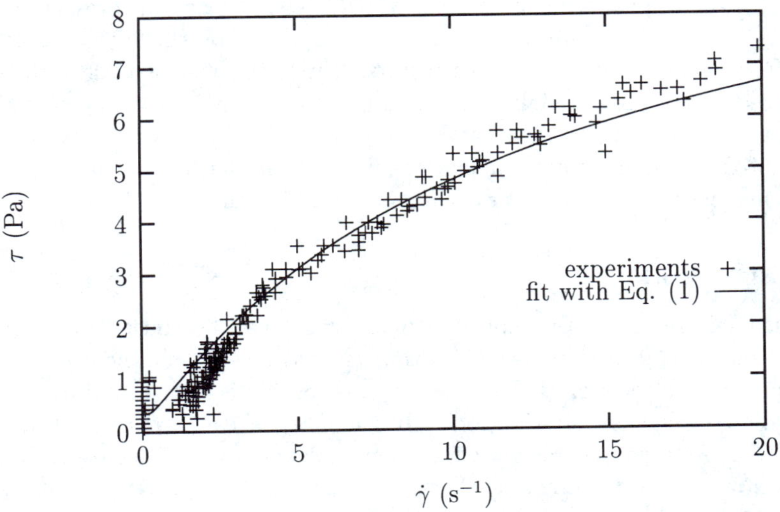

Figure 6: Equilibrium flow curve for China clay; $C = 467$ kg m^{-3}; with Carrimed controlled stress rheometer; cone-plate configuration used

Table 1: Calibrated values of the parameters in (1) based on Figure 6 and τ_y shown in Figure 5

parameter	value	dimension
τ_y	$3.4 + 235z$	Pa (z in m)
η_∞	0.0066	Pa s
c_1	0.86	Pa s
β	0.081	s
c_2	10^{-2}	-

transport under the influence of net forces, such as gravity.

A displacement of 5 m (approximately the length of the test section) is reached within a few minutes, therefore fluid mud flow can be observed only for a short period of time, which is confirmed by visual observations. However, quantitative comparison between the flow model and measured velocities proved difficult. The accuracy of the flow meters in the velocity range of 0.01 m s^{-1} is low, too.

After removal of the fluid mud layer, a new non-liquified layer became exposed, which would only liquefy if the wave amplitude was increased. At $a = 0.027$ m, when the bed is clearly completely eroded at $z = 0.06$ m, a pore pressure increase at $z = 0.10$ m was observed. Erosion of this layer occurred at $a = 0.027$ and 0.036 m.

Even at the largest wave amplitude, $a = 0.042$ m, not all sediment was liquified. After the experiments a layer of 0.02–0.03 m was still present in the upper part of the tilting flume, which is not influenced by the sill at the lower end. This can be explained by two factors. First, the yield strength close to to the fixed bottom is the highest as the mud has been most compacted by the weight of the overlying sediment, and secondly, the (high) pressure gradients at the end of the experiments are much less effective, as the bed becomes thinner and thinner, which leads to lower shear stresses inside the bed.

4 Conclusions

Undrained failure of freshly deposited mud layers may be caused by pressure gradients on the bed surface, if the resulting shear stresses inside the layer locally exceed the yield strength. If the yield strength profile is known, failure can be well predicted using an elastic model to calculated shear stresses in the mud layer. For loosely packed layers, which have not yet been subjected to higher effective stresses than the actual effective stress, positive excess pore pressure are generated upon failure and liquefaction occurs. The strength of the layer is then much reduced, and a transition from predominantly elastic to predominantly viscous behaviour takes place.

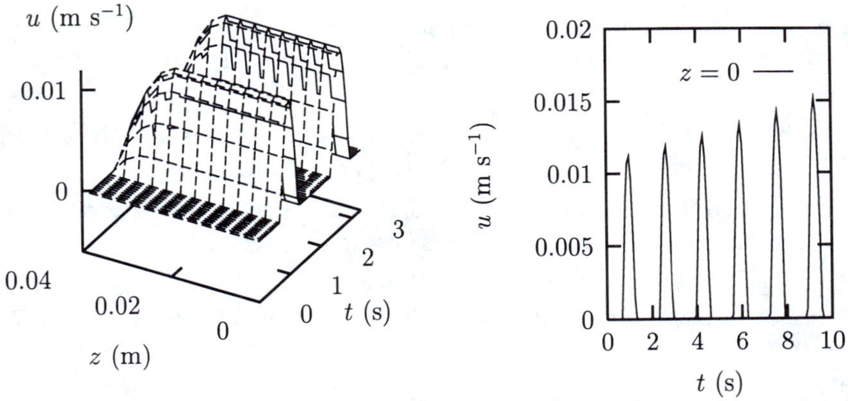

Figure 7: Velocity profile $u(z,t)$ resulting from wave-induced liquefaction on a slope. $D_1 = 0.04$ m, $\rho = 1346$ kg m^{-3}, $\theta = 0.05$ rad, $h = 0.30$ m, $a = 10$ mm, $k = 2.49$ m^{-1}, $\omega = 3.93$ s^{-1}; rheological coefficients used are listed in Table 1

If the fluid mud layer generated by liquefaction starts to flow under the influence of a net force, e.g. gravity, very high transport rates in a short period of time are possible, as the concentration of the liquified mud equals the original bed concentration (generally a few hundreds kg m^{-3}). This may partly explain the large mud accumulations often observed in the field after a storm period. In order to model fluid mud flow, its rheological properties have to be studied first, taking into account shear rate dependent and thixotropic behaviour.

The transport mechanism described in this paper may be important for the redistribution of cohesive sediments in depositional areas, where layers of freshly deposited, still consolidating mud are present. In erosional areas, however, the surface layer has generally been buried in the past and is therefore more compacted. Also ageing effects will contribute to the strength which tends to be much higher than for surface layers in depositional areas. Liquefaction of these old layers is therefore unlikely.

Acknowledgments

This work was carried out as part of the G8 Coastal Morphodynamics program. It was funded jointly by the Delft University of Technology and the Commission of the European Communities, Directorate General for Science, Research and Development under contract no. MAS2-CT-92-0027. It was carried out in the framework of the Dutch Center for Coastal Research. Special thanks go to

M. Reijnders, who assisted in the measurements reported.

References

Bowden, R. (1988). *Compression behaviour and shear strength characteristics of a natural silty clay sedimented in the laboratory.* Ph. D. thesis, Oxford University.

De Wit, P. (1995). *Liquefaction of Cohesive Sediments caused by Waves.* Ph. D. thesis, Delft University of Technology, the Netherlands.

James, A., D. Williams, and P. Williams (1987). Direct measurement of static yield properties of cohesive suspensions. *Rheol. Acta 26*, 437–446.

Mehta, A., E. Hayter, W. Parker, R. Krone, and A. Teeter (1989). Cohesive sediment transport. I: Process description. *J. Hydraul. Eng. 115*(8), 1076–1093.

Moore, F. (1959). The rheology of ceramic slips and bodies. *Trans. Brit. Ceram. Soc. 58*, 470–494.

Terzaghi, K. (1943). *Theoretical Soil Mechanics.* John Wiley and Sons, New York.

Toorman, E. (1993). *Rheological properties.* In: On the methodology and accuracy of measuring physico-chemical properties to characterize Cohesive Sediments. MAST-I Report. pp. 79–123.

Toorman, E. (1997). Modelling the thixotropic behaviour of dense cohesive sediment suspensions. *Rheologica Acta*, (to be publ.).

Van Kessel, T. (1996a). Rheological properties of cohesive sediment suspensions. Report 1–96, Hydromechanics Group, Delft University of Technology, the Netherlands.

Van Kessel, T. (1996b). Small-scale sounding tests on soft saturated cohesive soils. Report 7–96, Hydromechanics Group, Delft University of Technology, the Netherlands.

Van Kessel, T., C. Kranenburg, and J. Battjes (1997). Wave-induced liquefaction and transport of mud on inclined beds. (to be publ.).

Williams, D. (1984). *Rheology of cohesive suspensions*, Volume 14 of *Lecture notes on coastal and estuarine studies.*

Yamamoto, T., H. Koning, H. Sellmeijer, and E. Van Hijum (1978). On the reponse of a poro-elastic bed to water waves. *Journal of Fluid Mechanics 87*(1), 193–206.

CHAPTER 259

Wave on pebble beach and deformation of pebble beach

Ichiro Deguchi[1)], Masanobu ONO[1)] and Toru SAWARAGI[2)]

Abstract

Hydraulic function of pebble are investigated by carrying out 2-D experiments in laboratory. Numerical model for predicting wave height and wave run-up are developed by applying a Boussinesq type equation and a non-linear unsteady Darcy's law for the fluid motion on the pebble beach and in the pebble layer. The applicability of the model are examined using experimental results. Deformation of pebble beach are also measured in the field and laboratory to investigate the applicability of Dean's profile and to examine the shoreline change by a single-line theory.

Introduction

Urban coastal region in Japan have been developed in various ways for various purposes. Especially, after the World War II, through the post-war industrial reconstruction, Japan experienced a rapid and high economic growth. During that period, a large part of natural shoreline around big cities in Japan disappeared by the reclamation for heavy industries. As a result, public access to the shoreline decreased and natural coastal environment was lost. Recently, there are strong demands for restraining lost natural coastal environment and creating new pro-water front structures to increase public access to the coast and coastal amenity.

Target structure of this study is the pebble beach constructed as a permeable gentle slope seawall to increase public access and improve coastal view. It is usually permitted for such kind of pebble beaches to deform their profiles until they lose their originally expected function. The aims of this study are to examine hydraulic function of pebble beach and to establish numerical model for predicting wave transformation and run-up on pebble beach through carrying out two dimensional experiments. Characteristics of deformation of pebble beach is also investigated by using experimental results and field data.

2-D experiments on the hydraulic function of pebble beach

Experimental set-up and conditions:

Hydraulic function of pebble beach was examined by carrying out two-dimensional experiments in a laboratory. Figure 1 shows a rough sketch of the experimental set-up.

1) Department of Civil Engineering, Osaka University,, Suita-city, Osaka 565, Japan
2) Department of Civil Engineering, Osaka Sangyo University, Daito-city Osaka, Japan

We examined two slopes 1/10 and 1/5 constructed by pebbles of different sizes in a wave tank of 27m long, 1,9m high and 1,5m wide. To investigate the effect of permeability of the slope on incident wave, wave run-up, reflection, we also conducted the same measurements on the impermeable slope. The mean diameter D of the larger one was about 4.1cm and that of smaller one was 1.8cm. Table 1 shows the permeability k_p, turbulent drag coefficient C_f, added mass coefficient C_m and void ratio λ of these two pebbles obtained from the unsteady permeability tests. Depth at the horizontal bottom was varied from 25cm to 55cm and incident wave height was in the region of 6cm to 14cm with the period of 1.2s to 2.0s.

Surface displacement was measured between 0.5m landward from the shoreline and horizontal bottom. When deformation of pebble beach took place, we recorded the profile.

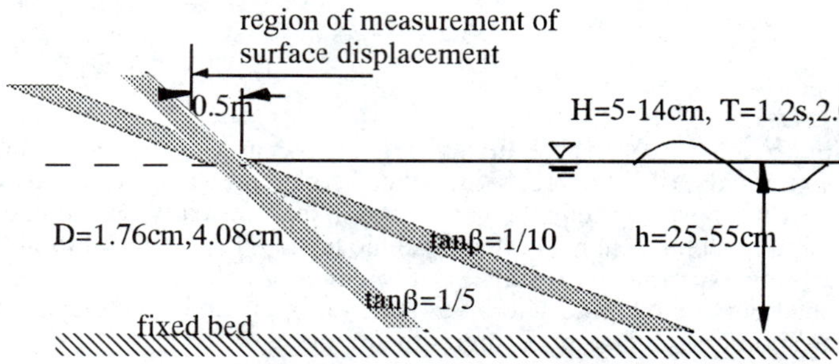

Figure 1 Experimental set-up

Table 1 Characteristics of pebbles

D (cm)	k_p (cm^2)	C_f	C_m	λ
4.08	0.0038	0.151	1.0	0.47
1.76	0.0015	0.255	1.0	0.46

Wave height distribution on pebble beach:

Figures 2 and 3 show measured cross-shore distribution of wave height on pebble beach and on impermeable slope of the slope 1/5 and 1/10. Incident wave period and height on horizontal bottom in both figure are 1.6s and 14cm and water depth at horizontal bottom was 50cm. Closed and open circles are the wave height and set-up measured on the impermeable slope and another symbols are the results obtained on the pebble beaches.

Wave height on pebble beach decreases significantly when compared with that on the impermeable slope. Especially the decay of wave height on the slope of 1/10 is large because incident waves have to travel for long distance on the permeable layer when compared with the case of the slope 1/5. On the other hand, there is little difference between the wave height measured on the pebble beaches of different sizes.

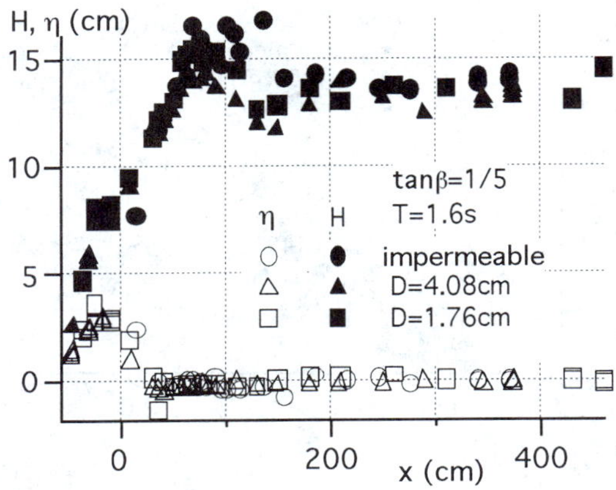

Figure 2 Wave height distribution on 1/5 slope

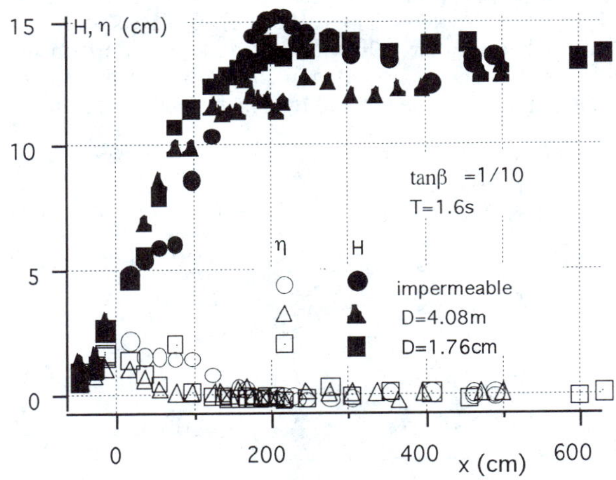

Figure 3 Wave height distribution of 1/10 slope

Wave run-up and reflection coefficient:

Figure 4 shows the non-dimensional run-up height normalized by the incident wave height R/H. The horizontal axis is the surf similarity parameter. Measured run-up height on pebble beaches shown by closed triangle and rectangle are 40 to 50% smaller than that measured on impermeable slope shown by the closed circles. Solid line in the figure show the relation $R/H=\xi$ that is usually applied to the run-up height on impermeable slope. Broken line in the figure is the empirical result for the non-dimensional run-up height on rubble mound breakwater obtained by Losada and Kurto(1981).

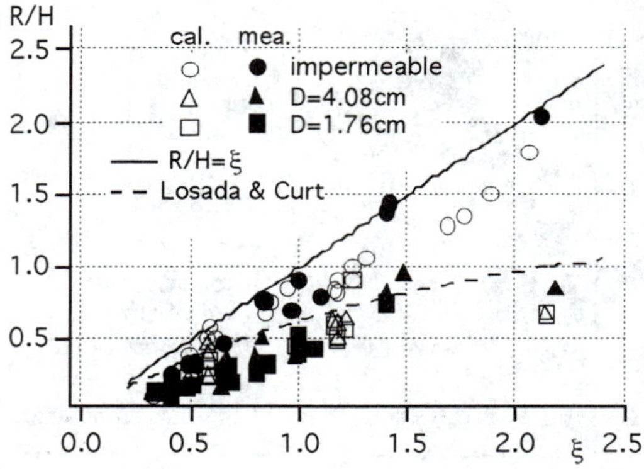

Figure 4 Wave run-up height

Figure 5 illustrates measured reflection coefficient K_r by Goda's method using two time series of measured surface displacements on the horizontal bottom. Horizontal axis is the incident wave steepness. The value of K_r in all cases are less than 30% and significant difference between the value of K_r measured on impermeable slope and pebble beaches can not be seen. K_r plotted in the region of wave steepness smaller than 0.005 is the result of long wave generated by a bichromatic waves.

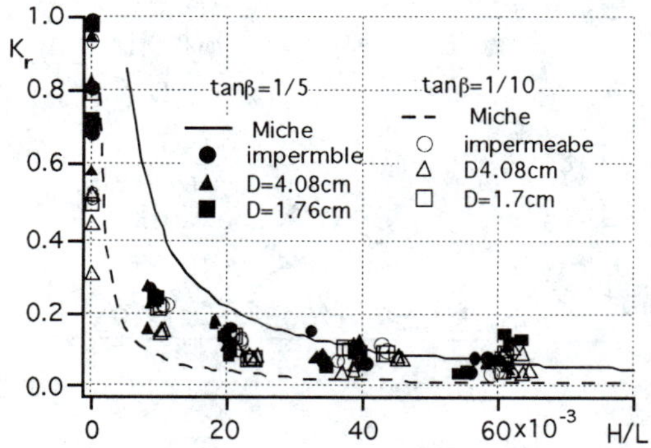

Figure 5 Reflection coefficient

Solid and broken line in the figure are the reflection coefficient on the impermeable slope of 1/5 and 1/10 evaluated by the Mich's formulae.

Numerical model for predicting wave height distribution on pebble beach

We have already proposed a simple model for estimating wave height on the permeable slope (Deguchi et al., 1995). Definition of variables and coordinate system used in the model is illustrated in Fig.6. h is the depth on pebble beach and d is the thickness of the pebble layer. For evaluating wave height on the sloping beach, we apply nonlinear shoaling model proposed by Shuto(1974) with the energy dissipation after wave breaking E_b, energy loss in the permeable layer E_p and surface drag on the pebble beach E_{fb}. These expression are shown below:

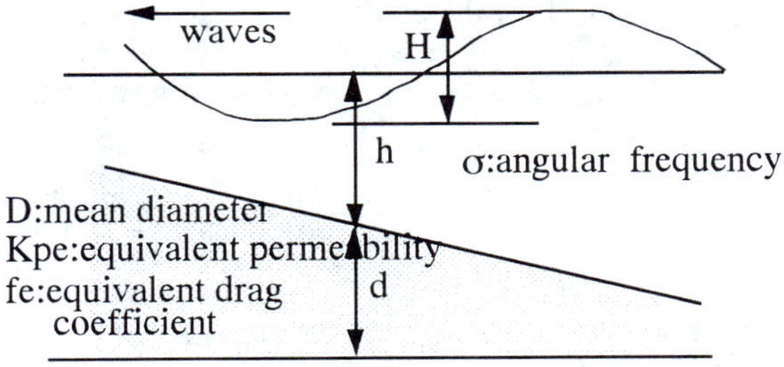

Figure 6 Definition of variables

In the calculation, wave height on the sloping beach was firstly calculated from Eq.(1) by giving incident wave height at offshore. When the wave height became grater than wave breaking height, wave height was reduced according to the energy loss by wave breaking that is given by Eq.(2) (Sawaragi, et al., 1984).

$$gHT^2/h^2 \leq 30: \textit{small amplitide wave theory}$$
$$30 \leq gHT^2/h^2 \leq 50: Hh^{2/7} = const. \quad (1)$$
$$50 \leq gHT^2/h^2: Hh^{5/2}\left(\sqrt{gHT^2/h^2} - 2\sqrt{3}\right) = const.$$

$$E_b = \alpha_1\left(5.3 - 3.3 I_r - \frac{0.07}{\tan\beta}\right)\left(\frac{\left(\rho g H^2/8\right)^3}{\rho h^3}\right)^{1/2} \quad (2)$$

where ρ is the density of water, g is the gravitational acceleration, β is the slope, ξ is the surf similarity parameter and α_1 is the empirical constant of the order of one.

When waves propagate on pebble beach, wave height is also reduced by energy dissipation on the permeability and the surface drag that are expressed by Eqs.(3) and (4) (Sawaragi et al., 1992).

$$E_p = \frac{\rho g H^2}{4} C_g k_i \quad (3)$$

$$E_{fb} = \frac{2}{3}\pi^2 \frac{\rho f_w H^3}{T^3 \sinh^3 kh}(1 - \lambda) \quad (4)$$

where C_g is the group velocity, f is Jonsson's friction factor, and k_i is the imaginary wave number determined by the following dispersion relation on the permeable layer:

$$\sigma^2 = g\mathbf{k}\frac{(S/f_e + i)\sinh \mathbf{k}h \cosh \mathbf{k}d + (1/f_e)\cosh \mathbf{k}h \sinh \mathbf{k}d}{(S/f_e + i)\cosh \mathbf{k}h \cosh \mathbf{k}d + (1/f_e)\sinh \mathbf{k}h \sinh \mathbf{k}d} \quad (5)$$

in which **k** is the complex wavenumber and f_e is the equivalent drag coefficient defined by using equivalent permeability k_{pe} as follow:

$$1/f_e = K_{pe}\sigma/\nu \text{ , } f_e = 0.1 + 1.8(u/\sigma D_m) \quad (6)$$

Figure 7 is the comparisons of measured and calculated wave height and set-up on permeable and impermeable slope of 1/10. The mean diameter of the pebble on the permeable slope was D=1.76cm. Shift of the location of breaking point and decrease in wave height are reproduced well by the model.

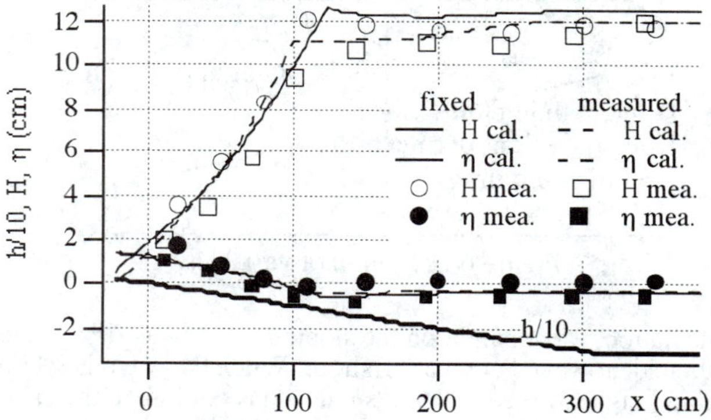

Figure 7 Comparison of measured and calculated wave height

Numerical model for predicting wave run-up on pebble beach

Some numerical model have already developed to predict wave run-up on the permeable slope (fro example, Kobayashi and Wurjanto, 1990). However, it is not sufficient to explain various non-linear phenomena including wave breaking, wave propagation into the permeable layer , wave run-up on the slope and so on. Here, we developed a numerical procedure for predicting surface displacement on the slope in a coordinate system shown in Fig.8. To construct numerical model, we neglected a vertical water particle velocity and applied Boussinesq type equation and a non-linear unsteady Darcy's law for the fluid motion on and in the pebble beach. Driving force of the fluid motion on and in the pebble beach is the pressure gradient. Although pressure on the pebble beach calculated from the Boussinesq equation is not hydrostatic, driving force on fluid motion in the pebble layer is assumed to be determined by the gradient of surface displacement. Definition of variables are shown in Fig.8.

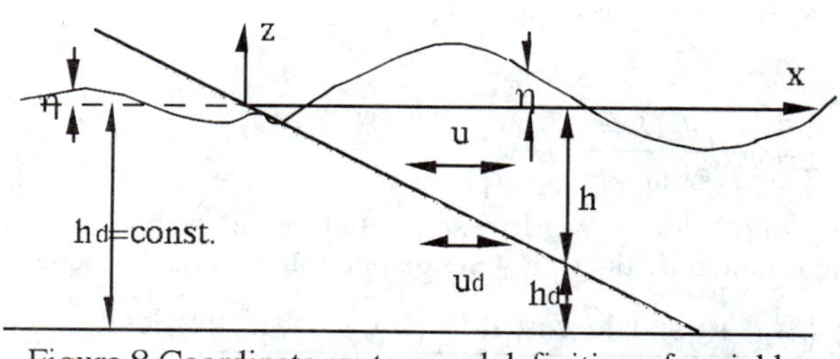

Figure 8 Coordinate system and definition of variables

Equation of motion of water particles on and in the pebble beach are expressed as follows:

$$\frac{\partial u}{\partial t}+u\frac{\partial u}{\partial x}+g\frac{\partial \eta}{\partial x}=\frac{h}{2}\frac{\partial}{\partial x}\left[\frac{\partial}{\partial x}\left(h\frac{\partial u}{\partial t}\right)\right]-\frac{h^2}{6}\frac{\partial}{\partial x}\left[\frac{\partial}{\partial x}\frac{\partial u}{\partial t}\right]-D_l \tag{7}$$

$$D_l=\alpha_2\frac{\partial}{\partial x}\left[\kappa g(h+\eta)\left(\frac{u}{h}\right)^2\right]+\frac{\alpha_3}{h+\eta}fu|u| \tag{8}$$

$$S\frac{\partial u_d}{\partial t}+u_d\frac{\partial u_d}{\partial x}+g\frac{\partial \eta}{\partial x}=-\frac{\nu}{K_p}u_d-\frac{C_f}{\sqrt{K_p}}u_d|u_d| \tag{9}$$

$$S=\{1+(1-\lambda)C_m\}/\lambda$$

where u and u_d are the water particle velocity on the pebble beach and in the pebble layer, ρD_l is the momentum dissipation, κ is Karman's constant, α_2 and α_3 are the empirical constants.

Equation of continuity of the fluid motion on the horizontal bottom and on the pebble beach are expressed by the following two equations.

$$\frac{\partial \eta}{\partial t}+\frac{\partial}{\partial x}[(\eta+h)u]=0 \tag{10}$$

$$\frac{\partial \eta}{\partial t}+\frac{\partial}{\partial x}[(\eta+h)u+h_d u_d]=0 \tag{11}$$

In the region shoreward of the wave front on pebble beach, h in Eq(11) becomes zero.

These equations are transformed into finite difference equations and are solved by giving time variations of surface displacement and water particle velocity at the offshore boundary. Continuity condition of surface displacement inside and on the pebble beach is imposed. A so-called moving boundary condition is used to determine the location of wave front on the pebble beach.

Figure 9 illustrates a comparison of calculated and measured surface displacements in the breaker zone on the pebble beach of the slope 1/5 of large grain size.

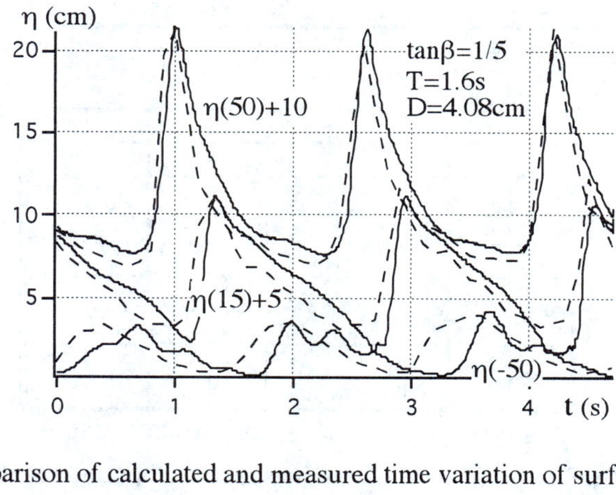

Figure 9 Comparison of calculated and measured time variation of surface displacement

In the figure, η(-50), η(15) and η(50) are the surface displacements at -50cm, 15cm and 50cm from the initial shoreline, solid and broken lines are the calculated and measured surface displacement. Calculated and measured surface displacements in front of the pebble beach, η(15) and η(50), coincide well with each other. Calculated amplitude of the surface displacement in pebble beach, η(-50), also reproduces the measured one. However, the time variation of measured surface displacement is smooth when compared with the calculated result and there is a small phase lag between them. Until now, we can not explain the reason of these discrepancy.

Figure 10 is an example of calculated surface profiles drawn at time interval of 0.2s on impermeable slope of 1/5. Incident wave period is 2.0s.

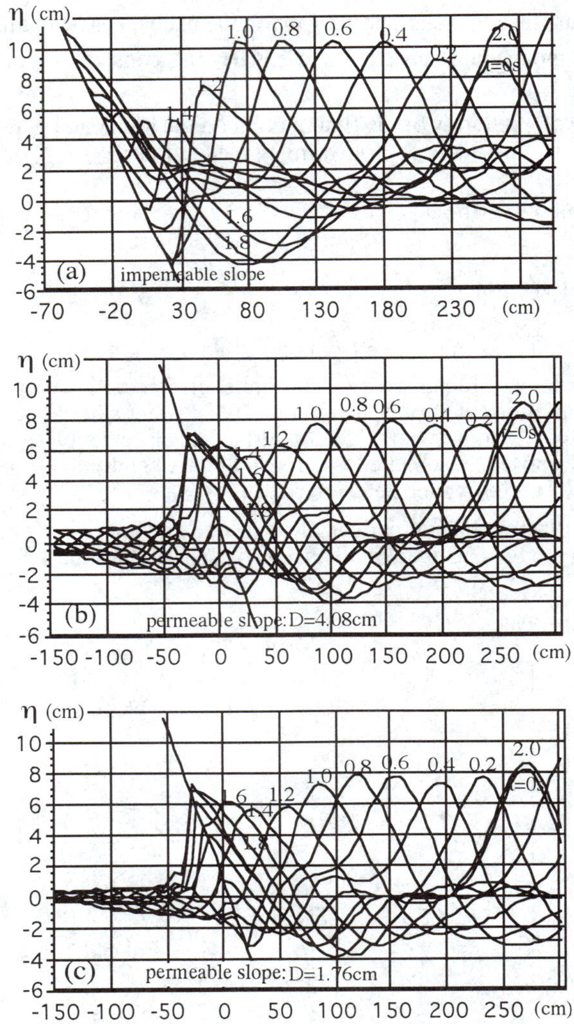

Figure 10 Surface profile in front of impermeable slope and pebble beach

Figure(a) is the result on the impermeable slope and Figures(b) and (c) are the results on the pebble beach of different materials. The maximum run-up height on the impermeable slope is almost the same as the incident wave height. Run-up height on the pebble beach of large materials under the same wave conditions shown in Fig,(b) is smaller than the incident wave height and even in the permeable layer, surface displacement of the amplitude of more than 1cm exists. Run-up height on the pebble beach of small materials shown in Fig.(c) is almost the same as that in the former case but the amplitude of the surface displacement in the permeable layer is smaller then the former case.

Open symbols in Fig.4 are the run-up height determined from the calculated surface profiles examples of which are shown in Fig.10. Although the predicted run-up height on both pebble beach and impermeable slope are a little bit smaller than those of the measured results, the decrease in the run-up height on a pebble beach is expressed by the numerical model.

Characteristics of topography change of pebble beach

Field and laboratory experiments:

We examined the characteristics of the deformation of the pebble beach based on the bottom topographies measured in two dimensional experiments and field measurements. In the experiments, significant deformation took place only in the case of pebble beach of small materials of the slope of 1/5. Field measurement were carried out on the artificial pebble beach constructed as a gentle slope seawall of the reclamation just landward of the Kansai International Airport in Osaka Bay as shown in Fig.11. The total length of the beach is about 3km and we measured bottom topography in one section of the beach surrounded by two groins. The length of the section is 300m. Construction of the beach finished in March 1992. In this paper, characteristics of the deformation are discussed based on the measured results on October 1995 about 3.5years after the construction of the beach.

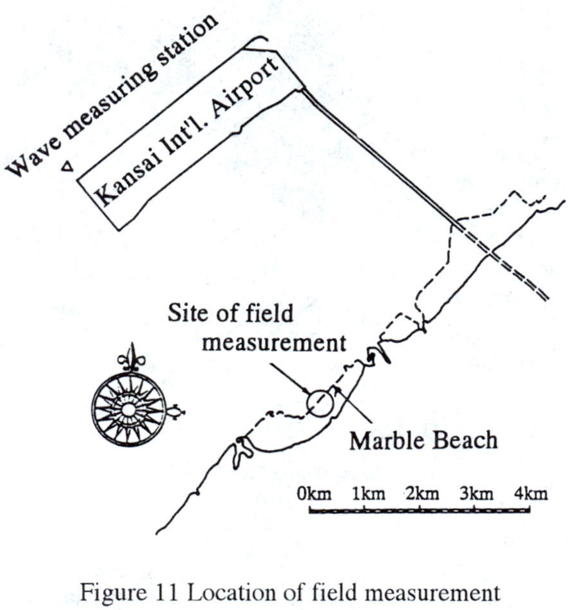

Figure 11 Location of field measurement

The representative planned cross-section of the beach is shown in Fig.12. Average depth at the toe is about 5m and the slope of the beach is 1/15. The surface of the beach is covered by pebbles of marble of the diameter 4-10cm. The thickness of the cover layer is 1m. We call this beach as Marble Beach.

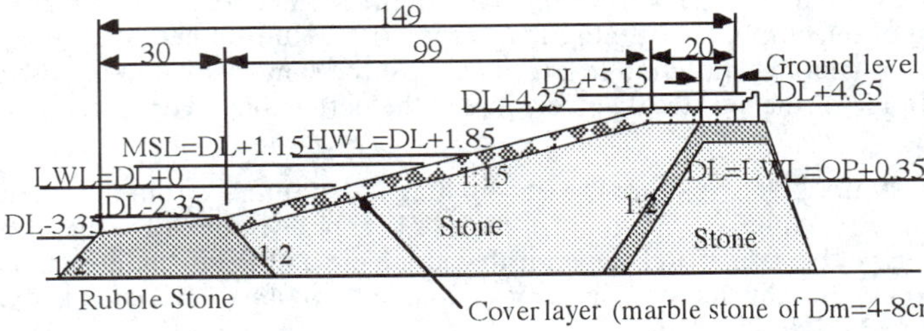

Figure 12 Representative cross-section of the Marble beach

Bottom topography was measured along 15 measuring lines set at an interval of 20m. Figure 13 illustrates the bottom topography when we see the beach from the land. There are groins at both side of this section.

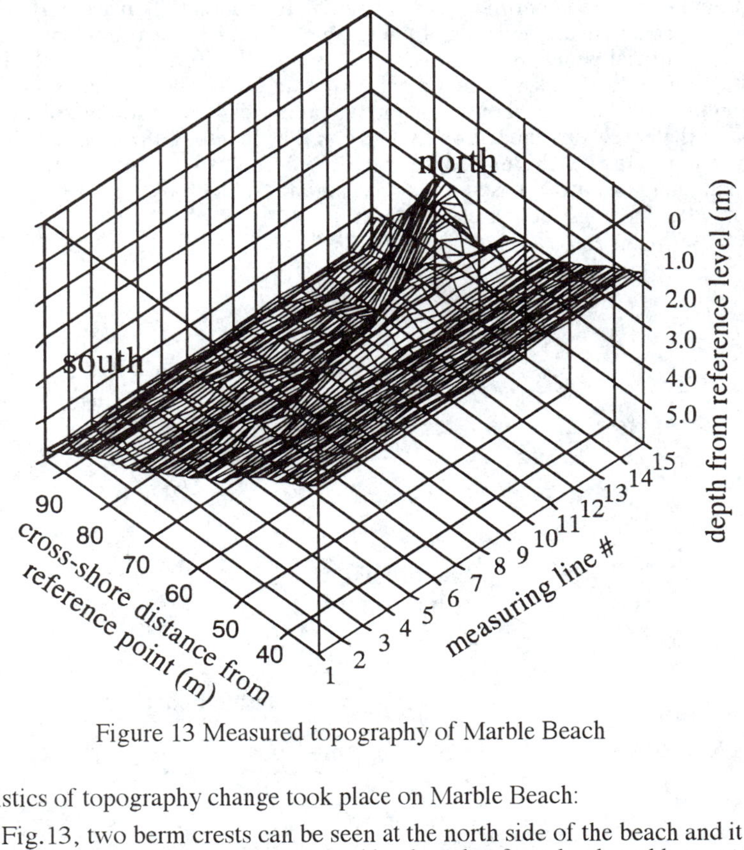

Figure 13 Measured topography of Marble Beach

Characteristics of topography change took place on Marble Beach:

In Fig.13, two berm crests can be seen at the north side of the beach and it is easy to imagine that pebbles were transported landward to form landward berm at first stage of the deformation and then they were transported from south to north to form the second berm. According to the empirical eigenfunction analysis, it is found that about 60% of the total deformation was caused by the longshore pebble transport and about 30% was caused by the net onshore pebble transport. Measuring line #15 is the north end of the beach and #8 is almost the neutral section for the topography change due to longshore transport.

Applicability of 2/3-power law:

Figures 14 and 15 are the representative cross-sections measured on the Marble Beach and in the laboratory.

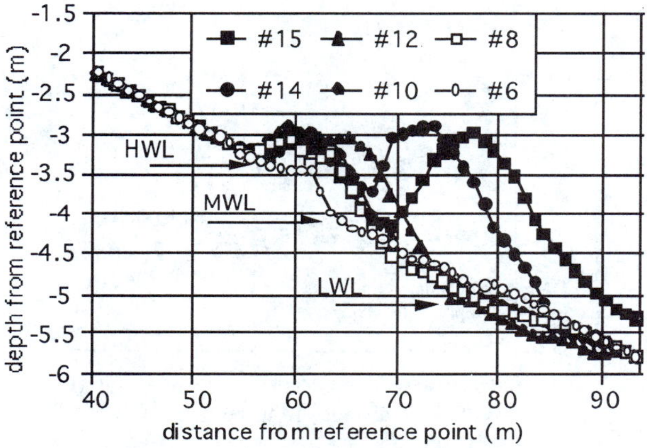

Figure 14 Representative measured cross-sections of Marble Beach

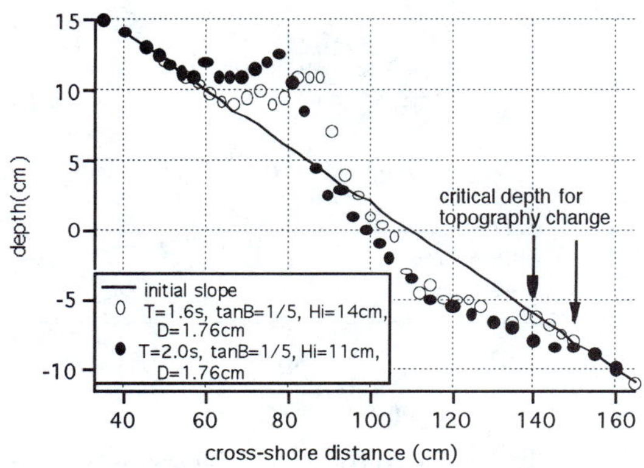

Figure 15 Measured cross-sections in experiments

Both sections shown in the figure, as well as those shown in Fig.14, are the typical accretion-type profile. We examined the applicability of the 2/3-power-law to these sections. Bruun(1962) and Dean(1991) proposed the equilibrium beach profile that is expressed by Eq.(12).

$$h = Ax^{2/3} \qquad (12)$$

where h is the depth of the equilibrium beach profile, x is the cross-shore distance and A is the empirical constant.

Dean (1991) gave physical meaning to Eq.(12) from the view point of energy dissipation in the surf zone and proposed the following empirical expression for the value of A as a function of settling velocity of the bed material W_f.

$$A = 0.067 W_f^{0.44} \tag{13}$$

We examined the applicability of the expression of Eq.(12) to the measured profile of pebble beach in the laboratory and in the Marble Beach. Figures 16 and 17 are the results. Profiles of both laboratory and field roughly coincide with the predicted profiles with the value of A 0.34 and 0.4, respectively.

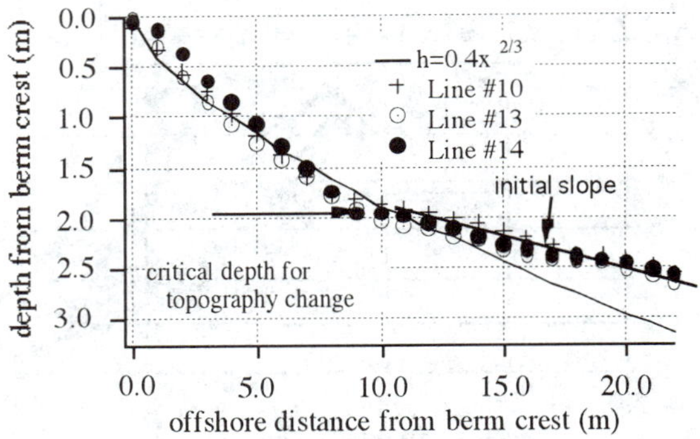

Figure 16 Comparison of Dean's profile and that measured on Marble Beach

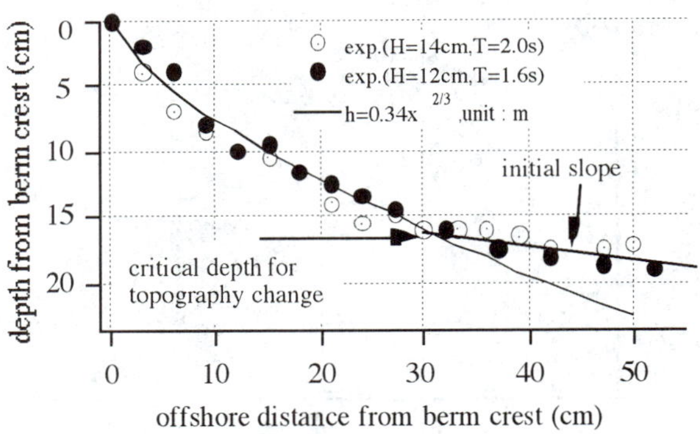

Figure 17 Comparison of Dean's profile and that measured in experiments

Shoreline change:

To evaluate total energy flux to the beach, we used wave records measured at observation station near the Kansai International Airport where various quantities concerning with the sea state are measured at one hour interval.

The total incident wave energy flux are evaluated from south to north and from north to south separately. The result is shown in Table 2. Energy flux from south to north is larger than that from north to south. Using this result together with the calculated volume of pebbles that were transported from south to north, we determined the longshore pebble transport rate and one representative wave to calculate location of contour line by a so-called one line theory.

Table 2 Incident wave energy flux in longshore direction (ton)

	1992	1993	1994	1995	total
from south to north	107280	178740	136656	68076	490752
from north to south	45720	61128	56988	32580	196416

It is found that the total longshore pebble transport rate Q_y is related to the longshore energy flux of incident waves in deep water by the following relation:

$$Q_y\left(m^3 / day\right) = 0.036\left(EC_g\right)_o \sin\theta\cos\theta\left(t/m/day\right) \qquad (14)$$

where $(EC_g)_o$ is the incident wave energy flux in deep water and θ is the incident wave direction.

Figure 18 is the comparison of the calculated shift of the -4m contour line and measured location of the same contour line. As can be seen from the figure, -4m contour line almost becomes equilibrium 4 years after the construction and the measured location coincides calculated location fairly well. The measured contour line locates a little shoreward than the location of the calculated contour line. This is because the advancement of the contour line caused by the net onshore pebble transport is not taken into account in the calculated location of the -4m contour line.

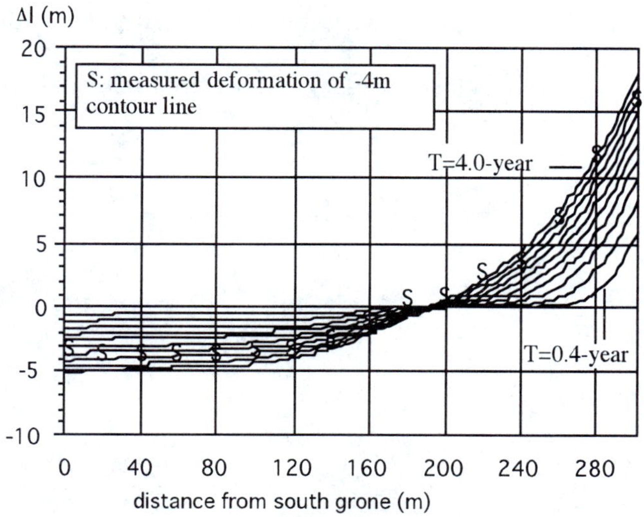

Figure 18 Comparison of measured and calculated deformation of -4m contour line

Conclusions

In this paper, numerical models for predicting wave height and run-up on pebble beach are proposed and the applicability of the model are examined through the laboratory experiments. Although the proposed model is too simple to reproduce the phenomena perfectly, we can predict rough figure of the wave tun -up on pebble beach.

Characteristics of the deformation of the pebble beach are also investigated using measured bottom topography in the field and laboratory. It is found that we can apply a so-called 2/3-power law and a single line theory for the cross-sectional profile and shoreline change of pebble beach.

Acknowledgment

This study were carried out by as a part of the research sponsored by the Grant-in-Aid for Scientific Research of the Ministry of Science, Culture and Education to DI (0605065)

Reference

Bruun, P., 1962, Sea level rise as a cause of shore erosion, J. Waterways and Harbor Div., ASCE, 88(WW1), pp.117-130

Dean, R. G.,1991, Equilibrium beach profiles, Characteristics and applications, J. Coastal Res., 7(1), pp.53-84.

Deguchi, I. and K. Moriwaki, 1995, Wave transformation on pebble beach, Proc. Coastal Engineering, JSCE, Vol.41, pp.41-45.

Kobayashi, N. and A. Wurjanto, 1990, Numerical model for waves on rough permeable slopes, J. Coastal Res., No.7, pp.149-166.

Losada, M. and L. Gimenez-Kurto, 1981, Flow characteristics on rough, permeable slopes under wave action, Coastal Engineering, 4, pp.187-206

Sawaragi, T., I. Deguchi and K. Kim, 1984, Energy loss and wave set-up due to breaking waves, Tech. Rept., Osaka University, Vol.34, No.,1779, pp.329-338.

Sawaragi, T. and I. Deguchi, 1992, Waves on permeable layer, Proc. 23rd ICCE, pp.1531-1544.

Shuto, N., 1974, Nonlinear long waves in a channel of variable section, Coastal Eng. in Japan, Vol.17, pp.1-12.

CHAPTER 260

Wind-Induced Waves and Currents in a Nearshore Zone

Nobuhiro Matsunaga[1], Misao Hashida[2] and Hiroshi Kawakami[3]

Abstract

Characteristics of waves and currents induced when a strong wind blows shoreward in a nearshore zone have been investigated experimentally. The drag coefficient of wavy surface has been related to the ratio u_{*a}/c_p, where u_{*a} is the air friction velocity on the water surface and c_p the phase velocity of the predominant wind waves. Though the relation between the frequencies of the predominant waves and fetch is very similar to that for deep water, the fetch-relation of the wave energy is a little complicated because of the wave shoaling and the wave breaking. The dependence of the energy spectra on the frequency f changes from f^{-5} to f^{-3} in the high frequency region with increase of the wind velocity. A strong onshore drift current forms along a thin layer near the water surface and the compensating offshore current is induced under this layer. As the wind velocity increases, the offshore current velocity increases and becomes much larger than the wave-induced mass transport velocity which is calculated from Longuet-Higgins' theoretical solution.

1. Introduction

When a nearshore zone is under swell weather conditions, the

1 Associate Professor, Department of Earth System Science and Technology, Kyushu University, Kasuga 816, Japan.
2 Professor, Department of Civil Engineering, Nippon Bunri University, Oita 870-03, Japan.
3 Graduate student, Department of Earth System Science and Technology, Kyushu University, Kasuga 816, Japan.

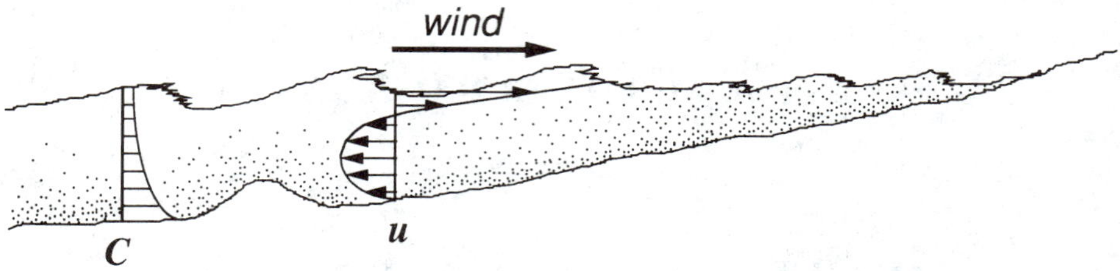

Fig.1 Sketch of sediment transport process in a nearshore zone under a storm.

sediment transport process depends strongly on the wave-induced mass transport velocity. The process is relatively slow. On the other hand, a large amount of sediment is suspended and transported under a storm. The transport process under storm weather conditions is very different from that under swell weather conditions. Shepard (1950) observed the change of beach profiles along Scripps Pier, La Jalla, California. He revealed that a beach profile with longshore bars forms under storm weather conditions and a profile with pronounced berms develops under swell weather conditions. The former has been referred to as the winter profile, and the latter as the summer profile. Komer (1976) claims the use of terms 'storm profile' and 'swell profile' to be preferable.

Many researchers (e.g., Johnson (1949), Rector (1954), Iwagaki and Noda (1963)) investigated seasonal variations of beach profiles and obtained a critical wave steepness at which they change from the storm profile to the swell one. In their studies, the steepness of storm waves was regarded as the most important factor to determine the beach profile. However, it seems to be difficult to explain the sediment transport process under storm weather conditions without considering the wind effect. In the case when a strong wind blows shoreward, a strong onshore wind-driven current forms along a thin layer near the water surface, and the compensating offshore current along the bed (see figure 1). The offshore current may transport a large amount of sediment seaward because the concentration of suspended sediment increases to the seabed. After a storm, in fact, we can often see a beach being eroded remarkably and floating matters such as seaweeds and pieces of wood being cast ashore.

In this study, waves and currents formed in a nearshore zone under storm weather conditions have been investigated experimentally in order to understand the wind effect on the onshore-offshore sediment transport.

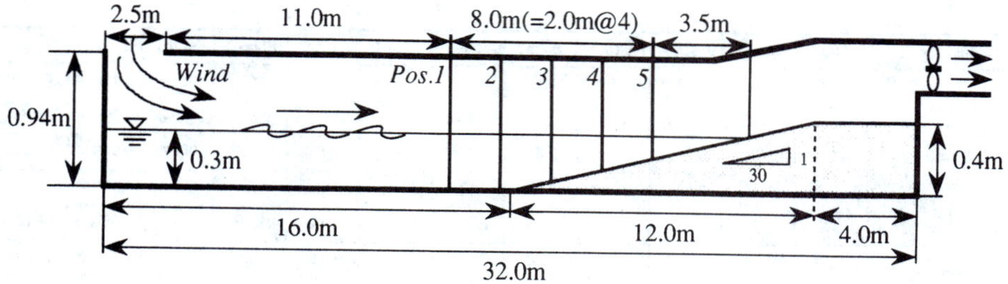

Fig.2 Experimental apparatus.

2. Experimental set-up

Experiments were carried out by using a water tank equipped with an inhalation-type wind tunnel. Figure 2 shows a schematic diagram of the experimental apparatus. The tank was 32 m long, 0.6 m wide and 0.94 m high. A sloping bed was attached to the end of the tank as a beach model. Its gradient was fixed at 1/30. The mean water depth was 0.3 m at the horizontal bed section. Wind waves were generated by the shoreward wind. Measurements of the wind velocity, wave height and wind-induced current velocity were made at positions 1 to 5. The distance from the intake of the wind to Position 1 was 11 m. The intervals between the adjacent measuring positions were 2.0 m. Positions 1 and 2 were in the horizontal bed section and positions 3 to 5 on the sloping bed. The wind velocity was measured by using a propeller-type current meter. In the wave height measurements, two capacitance-type wave gauges were used in order to obtain the phase velocity. They were set 28 cm away. The wave signals were digitized at the intervals of 1/50 s and 16,384 data were sampled. Horizontal and vertical components of the wind-induced currents were obtained by using an electromagnetic current meter. The sampling rate of the velocity signals was 1/20 s and the number of sampled data was 2,048.

Table 1 shows the wind parameters and the wave ones. Five tests in all were carried out by varing the wind velocity. The cross-sectionally averaged wind velocity U_m was varied from 7.60 m/s to 21.8 m/s. F is the fetch and h the mean water depth. The air friction velocity on the wavy surface is denoted by u_{*a} and the mean wind velocity at a 10 m height by U_{10}. H is the mean wave height. The periods, lengths and phase velocities of predominant waves are

Table 1 Experimental parameters.

U_m(m/s)	Pos.	F(m)	h(cm)	u_{*a}(m/s)	U_{10}(m/s)	H(cm)	T_p(s)	L_p(m)	c_p(m/s)	E(cm^2)	u_{*a}/c_p	C_D	U_r
	1	11.0	30.0	0.335	10.9	2.20	0.467	0.400	0.856	0.688	0.391	9.45E-4	0.130
	2	13.0	30.0	0.259	10.4	2.43	0.467	0.417	0.893	0.836	0.290	6.20E-4	0.157
7.60	3	15.0	25.0	0.274	10.4	2.63	0.474	0.433	0.914	0.977	0.300	6.94E-4	0.316
	4	17.0	18.4	0.238	10.7	2.83	0.515	0.493	0.956	1.17	0.249	4.95E-4	1.10
	5	19.0	11.9	0.294	11.1	2.99	0.535	0.483	0.903	1.26	0.326	7.02E-4	4.14
	1	11.0	30.4	0.682	18.1	4.68	0.585	0.637	1.09	2.88	0.626	1.42E-3	0.676
	2	13.0	30.2	0.568	17.3	5.06	0.595	0.685	1.15	3.31	0.494	1.08E-3	0.862
11.6	3	15.0	25.2	0.723	19.1	5.30	0.658	0.743	1.13	3.53	0.640	1.44E-3	1.83
	4	17.0	18.7	0.823	20.6	5.33	0.662	0.834	1.26	3.59	0.653	1.60E-3	5.67
	5	19.0	12.4	0.658	19.1	4.46	0.699	0.741	1.06	2.57	0.621	1.19E-3	12.8
	1	11.0	29.9	1.18	26.4	5.58	0.662	0.768	1.16	4.01	1.02	2.00E-3	1.23
	2	13.0	30.1	1.29	28.0	6.41	0.714	0.971	1.36	5.07	0.949	2.10E-3	2.22
15.5	3	15.0	25.5	1.19	27.3	6.55	0.709	0.851	1.20	5.26	0.992	1.91E-3	2.86
	4	17.0	19.3	1.32	29.1	6.11	0.746	0.896	1.20	4.86	1.10	2.06E-3	6.82
	5	19.0	13.4	1.49	31.1	4.55	0.775	0.791	1.02	3.01	1.46	2.30E-3	11.8
	1	11.0	28.5	1.10	29.0	6.05	0.699	0.881	1.26	4.71	0.873	1.45E-3	2.03
	2	13.0	30.0	1.63	34.8	6.39	0.719	0.971	1.35	5.45	1.21	2.20E-3	2.23
18.9	3	15.0	25.6	1.82	36.9	6.69	0.746	0.955	1.28	6.06	1.42	2.44E-3	3.64
	4	17.0	19.6	2.10	40.3	5.87	0.787	1.00	1.27	4.95	1.65	2.72E-3	7.80
	5	19.0	14.0	2.57	45.6	4.81	0.840	0.830	0.988	3.43	2.60	3.19E-3	12.1
	1	11.0	27.8	1.99	39.4	6.30	0.719	0.971	1.35	5.38	1.47	2.54E-3	2.76
	2	13.0	28.1	2.04	43.3	6.49	0.741	1.02	1.38	5.92	1.48	2.21E-3	3.04
21.8	3	15.0	24.1	2.02	42.9	6.91	0.794	1.04	1.31	6.41	1.54	2.21E-3	5.34
	4	17.0	18.2	2.19	45.1	5.76	0.794	1.02	1.28	4.64	1.71	2.36E-3	9.94
	5	19.0	13.5	2.40	45.9	4.19	0.885	1.04	1.18	3.09	2.03	2.72E-3	18.4

represented by T_p, L_p and c_p, respectively. E denotes the total wave energy. The drag coefficient C_D is defined by $(u_{*a}/U_{10})^2$, and U_r is an Ursell parameter defined by HL_p^2/h^3.

3. Experimental results and discussion

3.1 Drag coefficient of wavy surface

Vertical profiles of the mean wind velocity U in the case of U_m = 15.5 m/s are shown in figure 3, where z is the vertical coordinate taken upward from the mean water level. Though the wind velocity near the ceiling of the wind tunnel decreases due to the boundary layer, a logarithmic profile is formed near the water surface. The wind set-up increases the wind velocity and the velocity gradient near the water surface in the leeward direction. The values of u_{*a} were calculated by

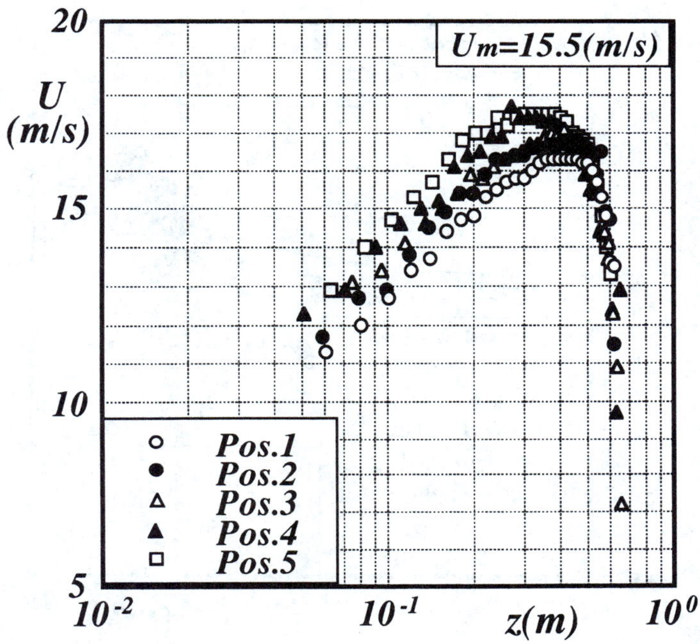

Fig.3 Vertical profiles of mean wind velocity.

fitting the logarithmic law

$$U = \frac{u_{*a}}{\kappa} \ln \frac{z}{z_0} \tag{1}$$

to the wind velocity profiles, where κ (= 0.4) is von Karman's constant. It is read from table 1 that the values of u_{*a} increase with increase of U_m and with increase of F. The drag coefficient C_D is defined by

$$C_D = \left(\frac{u_{*a}}{U_{10}}\right)^2 \tag{2}$$

The relation between C_D and U_{10} has been investigated until now by many researchers. Some of the empirical expressions and our experimental data are shown in figure 4. The data include ones obtained through other experiments in which the wind blew on swells made by a wavemaker. Though the U_{10}-dependence for the wind waves is different from that for the swell and wind waves, the data approach gradually to Kondo's empirical curve when $U_{10} \geq 25$ m/s. Some of our data are under the values to which the empirical curves

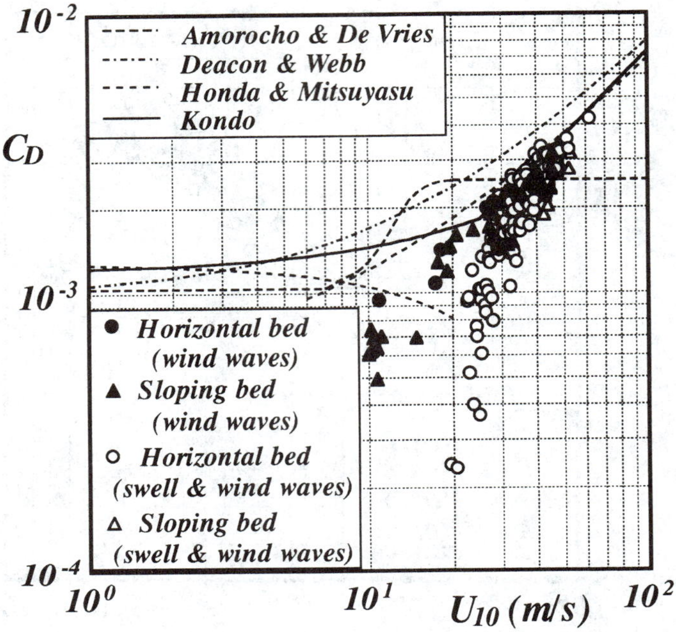

Fig.4 U_{10}-dependence of C_D.

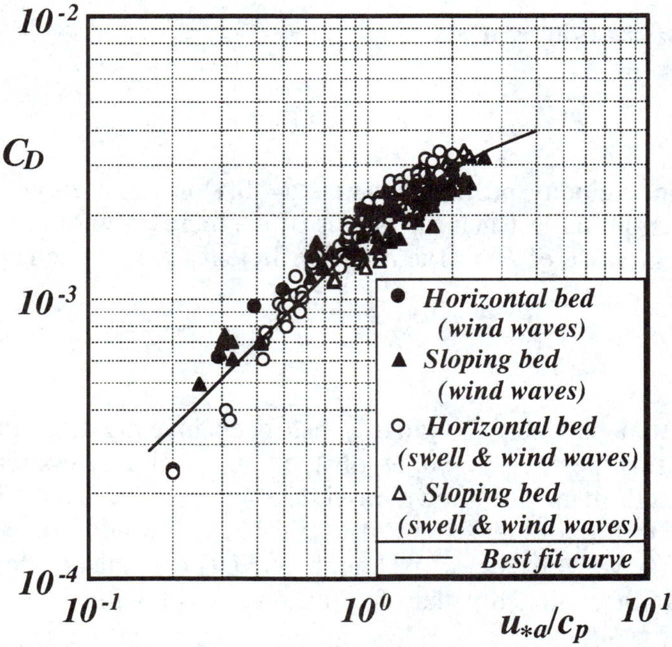

Fig.5 Relation between C_D and u_{*a}/c_p.

approach with increase of U_{10}. The validity of these data remains to be proved.

The relation of C_D and U_{10} has been investigated from the standpoint of practicality. However, if we try to obtain a universal form for C_D, relations between C_D and dimensionless parameters should be examined. Figure 5 shows a relation between C_D and u_{*a}/c_p. The solid line is drawn by the least-square fit method. The discrepancy between the data for the wind waves and ones for the swell and wind waves becomes much smaller than that shown in figure 4. Increasing linearly with u_{*a}/c_p when $u_{*a}/c_p \leq 1$, C_D becomes constant for a large value of u_{*a}/c_p.

3.2 Wind waves in shallow water

It is well-known that the energy of wind waves in deep water and the periods of the predominant wind waves increase with increase of u_{*a} and F. The empirical fetch-relations proposed by Mitsuyasu (1968) are

$$\frac{gE^{1/2}}{u_{*a}^2} = 1.31\times10^{-2}\left(\frac{gF}{u_{*a}^2}\right)^{0.504} \tag{3}$$

and

$$\frac{u_{*a}\,f_p}{g} = 1.00\left(\frac{gF}{u_{*a}^2}\right)^{-0.330} \tag{4}$$

where $f_p = 2\pi/T_p$. As read from table 1, the values of T_p in shallow water increase with increase of F but the values of E do not increase monotonically with F because of the wave breaking. We can also read that the increase of F corresponds to that of U_r. It means that the wind waves progress into a shallow region with increase of F. If equations (3) and (4) are rewritten by using the wave energy E_0 and the frequency f_{p0} at a standard point,

$$\frac{E}{E_0} = \left(\frac{F}{F_0}\right)^{1.008} \tag{5}$$

and

$$\frac{f_p}{f_{p0}} = \left(\frac{F}{F_0}\right)^{-0.330} \tag{6}$$

are obtained, where F_0 is fetch at the standard point. Equations (5) and (6) give the increasing rate of the wave energy to E_0 and the decreasing rate of the predominant wave frequency to f_{p0}, respectively. The

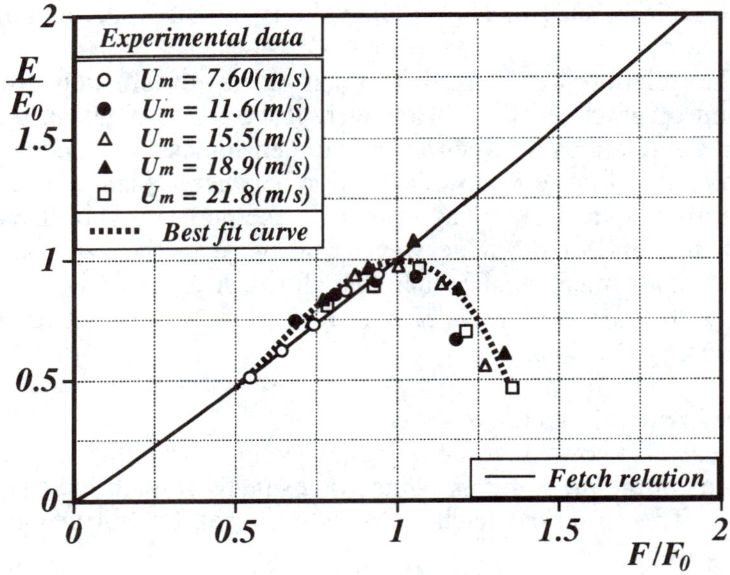

Fig.6 Comparison of fetch relation of E in shallow water with that in deep water.

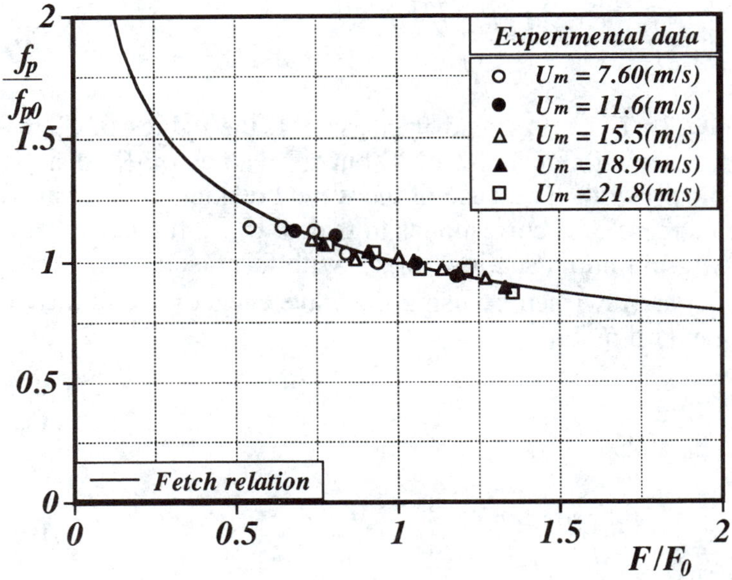

Fig.7 Comparison of fetch relation of f_p in shallow water with that in deep water.

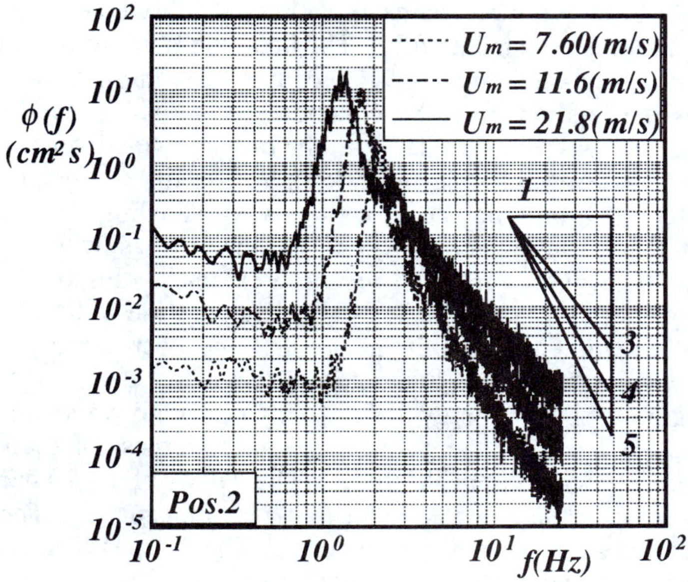

Fig.8 Variation of energy spectra of wind waves with increase of U_m.

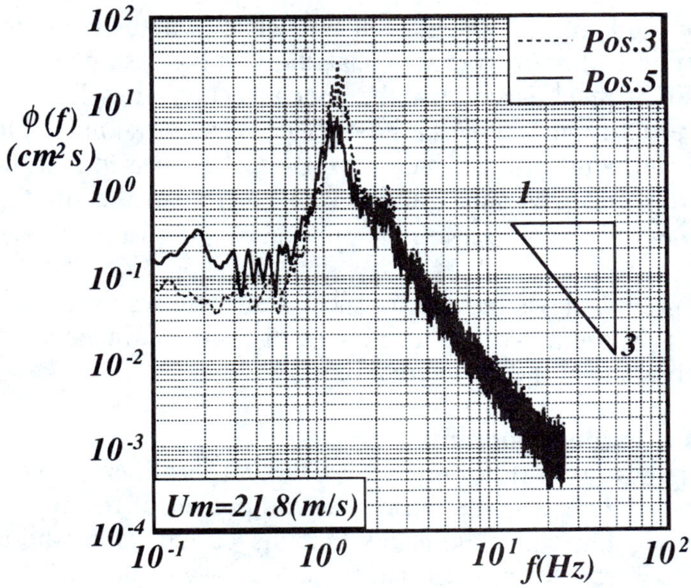

Fig.9 Comparison between energy spectra before and behind breaking point.

values of E/E_0 are plotted against F/F_0 in figure 6, where the fetch to the breaking point is selected as F_0. The breaking point of the wind waves is defined as a position at which E becomes maximum. The solid line expresses the relation given by equation (5). The dashed line is the best fit curve based on the data. In the offshore side from the breaking point ($F/F_0 < 1$), the increasing rate of E is a little larger than that in deep water. It may be caused by the wave shoaling. On the other hand, in the onshore side ($F/F_0 > 1$), the increasing rate decreases rapidly with fetch because of the wave breaking. Figure 7 shows the relation between f_p/f_{p0} and F/F_0. The data collapse well onto the curve given by equation (6). It means that the decreasing rate of f_p in shallow water agrees well with that in deep water.

Figure 8 shows energy spectra $\phi(f)$ of wind waves measured at Position 2. The total energy increases with increase of U_m because no wave breaking occurs at Position 2. The values of $\phi(f_p)$ become large with the U_m-increase and the values of f_p become small. These are the same features as in deep water waves. The f-dependence of $\phi(f)$ in the high frequency region changes from f^{-5} to f^{-3} as the wind velocity increases. The spectral form in an equilibrium region is given by

$$\phi(f) \propto g^{\alpha} u_{*a}^{2-\alpha} f^{-3-\alpha} \tag{7}$$

with the aid of a dimensional analysis, where g is the gravity acceleration and α an arbitrary constant. In the case when the effect of the gravity is much larger than the wind effect, α takes 2. At that time, $\phi(f)$ is proportional to f^{-5} in the high frequency region. On the other hand, as the wind velocity increases, it can be guessed that the effect of the gravity becomes small and $\phi(f) \propto f^{-3}$ in the limit. These dimensional considerations are supported by the results shown in figure 8.

Figure 9 shows the energy spectra of wind waves at positions 1 to 5. It is seen that the energy of the predominant waves decays remarkably due to the wave breaking.

3.3 Wind-induced currents

Figures 10 (a) to (e) show vertical profiles of wind-induced currents. Here, u is the horizontal component of the current velocity and the negative value indicates that the current is offshore. The vertical axis z is normalized using the local water depth h. Offshore wind-induced currents are formed in the range of $-1.0 \le z/h \le -0.1$. This suggests that an onshore strong current is generated in a thin layer near the water surface. As a natural result, the offshore currents

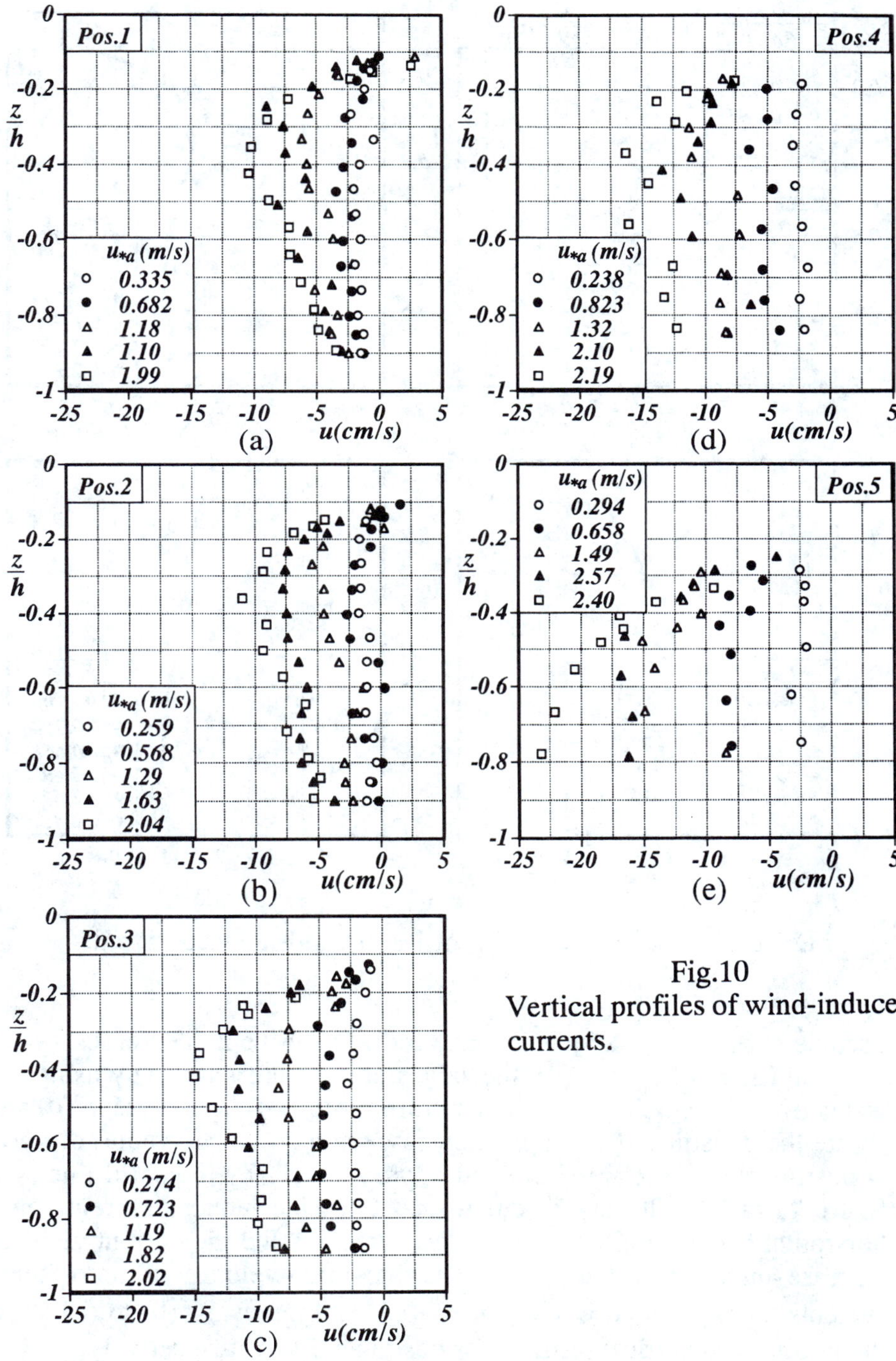

Fig.10
Vertical profiles of wind-induced currents.

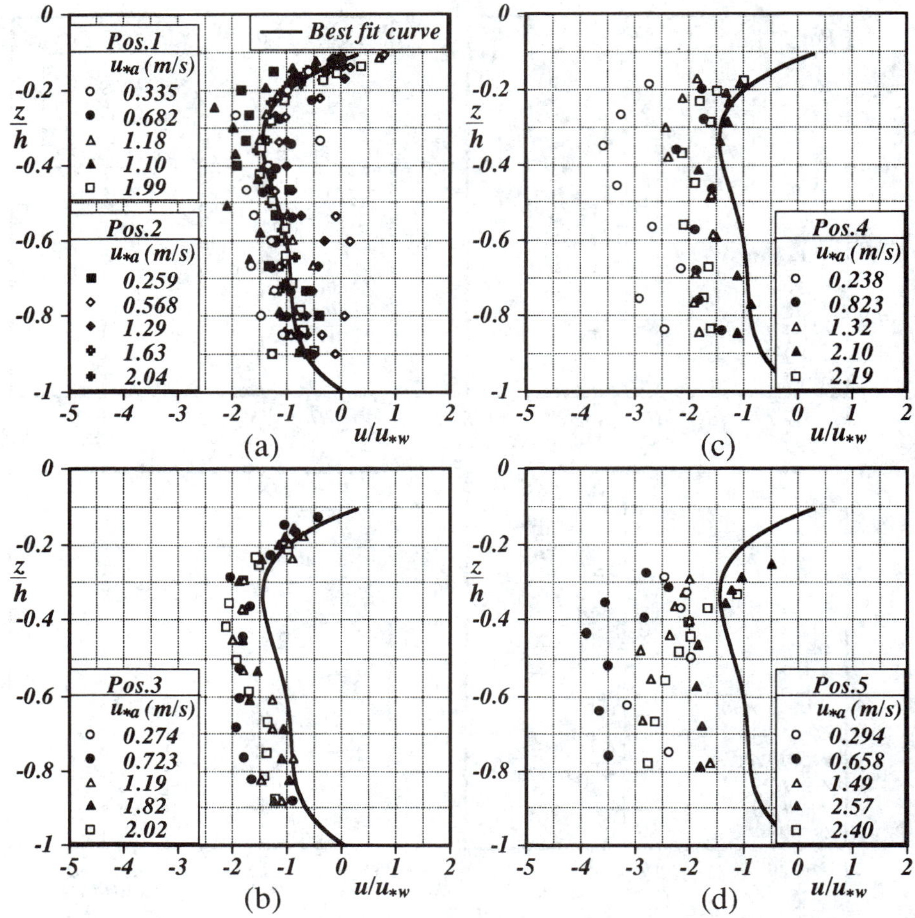

Fig.11 Normalized vertical profiles of wind-induced currents.

become stronger as u_{*a} increases and as the water depth decreases.

In figures 11 (a) to (d), the values of u are normalized by using the water friction velocity u_{*w} calculated from $(\rho_a/\rho_w)^{1/2}u_{*a}$. Here ρ_a and ρ_w are the densities of air and water, respectively. The values of u/u_{*w} at positions 1 and 2 are expressed approximately by the solid line (see figure 11 (a)). It may be due to that the wind-induced currents on a horizontal bed are uniform in the flow direction and the current velocity increases in proportion to u_{*a}. The maximum velocity of the offshore currents takes about $1.5u_{*w}$ at z/h = -0.3. Figures 11 (b) to (d) show the normalized vertical profiles for positions 3 to 5, respectively. The solid lines in these figures are the one drawn in figure 11 (a). It is

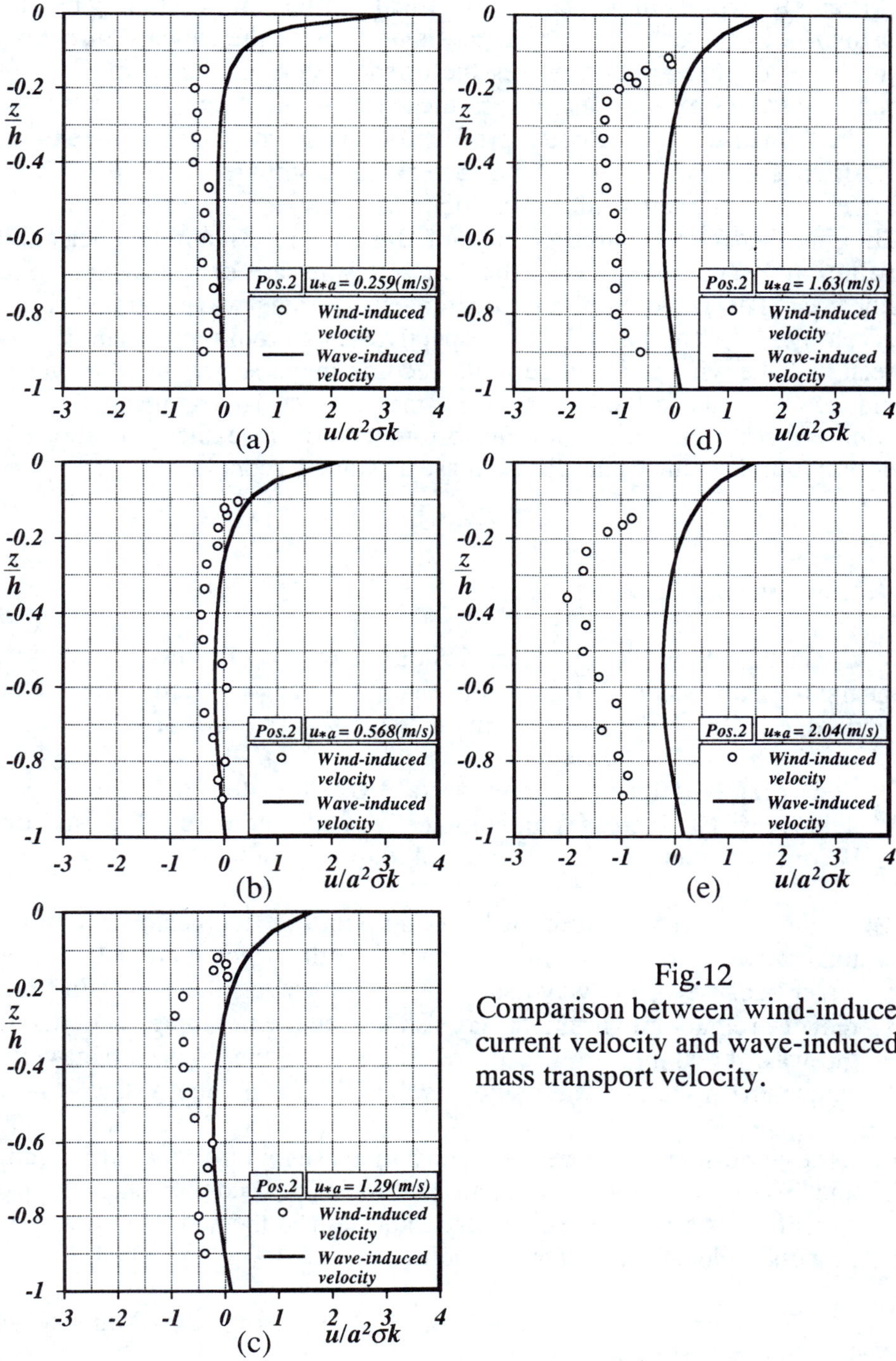

Fig.12
Comparison between wind-induced current velocity and wave-induced mass transport velocity.

difficult to express universally the vertical profiles on the sloping bed by using u_{*w} and h, because even if u_{*w} is uniform in the leeward direction, the water depth variation makes the offshore currents accelerate. In fact, the values of u/u_{*w} increase in the leeward direction.

Dimensionless velocity profiles of the wind-induced currents at Position 2 and the wave-induced currents are compared in figures 12 (a) to (e). The wave-induced velocity is estimated by using Longuet-Higgins' theoretical solution (Longuet-Higgins (1953)) and the measured values at Position 2. The theoretical results are drawn by the solid lines. The wave amplitude a, frequency σ and wave number k are given by $H/2$, $2\pi/T_p$ and $2\pi/L_p$, respectively. From these figures, it is seen that the wind-induced currents become much larger than the wave-induced currents as the values of u_{*w} increase. This suggests that the wind effect on the sediment transport under storm weather conditions is very important rather than the increase of wave steepness.

4. Conclusions

In this study, the caracteristics of waves and currents formed in a nearshore zone under storm weather conditions have been investigated experimentally. The obtained main results are as follows.

1) The drag coefficient of wavy surface C_D is related to u_{*a}/c_p. The values of C_D increase monotonically with increase of u_{*a}/c_p but become constant for a large value of u_{*a}/c_p.

2) In the offshore side from the breaking point, the increasing rate of the total wave energy in shallow water is a little larger than that in deep water because of the wave shoaling. However, the increasing rate reduces remarkably in the onshore side from the breaking point due to the wave breaking. The decreasing rate of the predominant wave frequency in shallow water agree well with that in deep water.

3) The wind-induced current velocity increases with the wind velocity and becomes much larger than the wave-induced current velocity. Therefore, the wind effect is very important in the sediment transport process under storm weather conditions.

The authors are grateful for a grant from the Ministry of Education, Science and Culture.

References

Iwagaki, Y. and Noda, H. (1963) Laboratory study of scale effects in two-dimensional beach processes, Proc. 8th Conf. on Coastal Eng., ASCE, 194 - 210.

Johnson, J.W. (1949) Scale effects in hydraulic models involving wave motion, Trans. Am. Geophys. Union **30**, 517 - 525.

Komar, P.D. (1976) Beach processes and sedimentation, Prentice-Hall, Englewood Cliffs, N.J., 288 - 300.

Longuet-Higgins, M.S. (1953) Mass transport in water waves, Phil. Trans. Roy. Soc. (London), series A, **245**, 535 - 581.

Rector, R.L. (1954) Laboratory study of the equilibrium profiles of beaches, U.S. Army Corps of Engrs., Beach Erosion Board Tech. Memo., **41**.

Shepard, F.P. (1950) Beach cycles in Southern California, U.S. Army Corps of Engrs., Beach Erosion Board Tech. Memo., **20**.

CHAPTER 261

PREDICTING LARGE-SCALE, CROSS-SHORE SEDIMENT MOVEMENT FROM ORBITAL SPEEDS

Edward B. Hands[1], John P. Ahrens[2], and Donald T. Resio[3]

ABSTRACT: The ratio U of near-bed peak orbital speed to grain threshold speed should express the competence of coastal waves to agitate loose seafloor sediments. Using Stream Function Wave Theory, U can be evaluated separately under the wave crest U_C and trough U_T to produce a pair of parameters whose relative magnitudes indicate the direction of the displacing force. The authors tested this simple parametrization in two distinctly different, large-scale coastal transport situations. One situation involved relative stabilities and displacements of submerged dredged-material mounds outside the normal surf zone. These mounds contained 10s to 100s of thousands of cubic meters of sandy material. Predicted wave responses match well with measurements made over months and years at 11 such mounds widely scattered around the United States. The second situation involved predicting whether beaches accrete or erode during single storms. Comparison between U-based predictions and 99 beach responses, compiled from the published literature, provided good confirmation in the second situation.

Critical values of U are surprisingly skillful in predicting both types of cross-shore movement. Where extreme U_C's exceeded U_T's by more than about 5, mounds migrated shoreward; where waves were more linear, mounds remained stationary. Beaches eroded significantly where $U_T < -2$; and accreted otherwise regardless of the degree of wave linearity.

Research Phy Scientist [1] and Senior Scientist [3] USAE Waterways Experiment Station, Coastal and Hydraulics Laboratory, 3909 Halls Ferry Road, Vicksburg, MS 39180

Specialist [2], Coastal Processes, NOAA, Sea Grant, 1335 East-West Highway, Silver Spring, MD 20910

INTRODUCTION

Purpose

Two very simple critical conditions "explain" different types of coastal sediment dispersion. Either the tested large-scale laboratory and field studies fail to capture some significant class of conditions or the relative speed of bottom wave oscillations and sediment thresholds is the dominant factor controlling coastal profile response from the shoreline to well offshore. Broad-scale predictors, even if not precise, are useful to engineers who must often work with limited input data. Accordingly, the profile adjustment predictors developed here depend on the height and period of the wave, size and density of the sediment, and depth and density of the water.

Background

Artificial beach nourishment is a widely popular form of storm damage reduction. Good uses for dredged sands that improve cost-to-benefit ratios of inlet channel maintenance are of keen interest to coastal managers. Much of the sand dredged continually to maintain navigation can be used to reduce coastal storm damages. Several new uses involve designing submerged mounds to either shelter adjacent shorelines from erosive waves or cost-effectively augment the natural sediment supply to the coast (Bodge 1994 a 1994b; Foster, Healy, and de Lange 1996; Hands and Resio 1994; Landin, Davis, and Hands 1995; Stive et al. 1992). Motivated by the need for an easy method to determine which conditions move sand onshore and offshore, we found two velocity parameters to be effective predictors in widely disparate transport situations.

DEFINITION OF PARAMETERS

Both parameters are ratios of near-bed oscillatory peak speeds (NOPS) to the sediment threshold speed, i.e., $U \equiv u_{\text{-d max}} \div u_{\text{crit}}$, where $u_{\text{-d max}}$ is the NOPS and u_{crit} is the threshold speed required to initiate motion of selected grain sizes. As waves approach shore, orbital speeds increase under narrowing crests while decreasing under widening troughs. A pair of parameters results if a nonlinear theory is used to evaluate U separately under the wave crest (U_C) and trough (U_T). Differences (U_C-U_T) may be crucial especially in contrasting nearshore transport effects of steep storm waves versus gentle swell.

METHODS OF APPLICATION

NOPS were determined from Dean's (1974) Stream Function Wave Theory (SFWT). SFWT contains the crucial nonlinearities and is easy to apply because extremely accurate regression equations were developed in this study for a suitably wide range of conditions ($0.002 \leq d/L_o \leq 0.20$ and $H_b/4 \leq H \leq H_b$, where d is water depth, L_o is the deepwater wave length, H is local wave height, and H_b is the breaking wave height).

Under the wave crest,

$$u_{-d\ max\ c} = \left(\frac{H}{T}\right)\left(\frac{d}{L}\right)^{-0.579} e^{0.289-0.491(H/d)-2.97(d/L_o)} \tag{1}$$

where T is the wave period and L is the local wave length. Under the trough,

$$u_{-d\ max\ t} = -\left(\frac{H}{T}\right) e^{1.996-1.73(H/d)-6.70(d/Lo)+5.58(H/Lo)} \tag{2}$$

More information on fitting near-bed speeds to Stream Function Wave Theory will be provided in Ahrens and Hands 1997.

If the representative grain size $d_{50} \leq 2\ mm$, threshold speeds come from Hallermeier (1980)

$$u_{crit} = \sqrt{8\gamma g d_{50}} \tag{3}$$

where γg is the grain to fluid ratio of unit submerged weights. If $d_{50} > 2\ mm$, threshold speeds come from Komar and Miller (1974)

$$u_{crit} = \left[0.47\gamma g T^{1/4}\ (\pi d_{50})^{3/4}\right]^{4/7} \tag{4}$$

For application to long-term fates of submerged mounds, where wave conditions fluctuated over a wide range, NOPS were calculated for an arbitrary, but common representation for wave extremes: the 12-hr/year exceedance value. Threshold speeds were calculated based on median grain sizes as sampled soon after mound placements.

For discrimination between beach erosion and accretion, the threshold speeds were calculated for the reported typical beach grain sizes. The breaker depth was selected as a reasonable standard location at which threshold ratios (U_C and U_T) are compared. This reference depth was obtained by fitting a breaker depth index to the SFWT data to obtain

$$d_b = 0.68 \left(\frac{L_o}{2\pi}\right) \ln\left(\frac{1+s}{1-s}\right) \quad (5)$$

where $s = H_b / 0.171\ L_o$ and H_b comes from Kaminski and Kraus's (1993) expression for the breaker height index

$$\frac{H_b}{H_o} = 0.46 \left(\frac{H_o}{L_o}\right)^{-0.26} \quad (6)$$

RESULTS

The skill of these new U parameters is determined by applying them to a number of published laboratory and field studies. The single parameter U_T shows considerable and unsuspected skill in predicting erosion or accretion of beaches. Surprisingly, the seaward-directed component better discriminates between eroding and accreting beaches. Combining both parameters, U_C and U_T, explains observed movement of the 11 test mounds.

Response of Shore Profiles from Large Wave Tank Tests

Larson and Kraus (1989) published results from two sets of large-scale wave tank tests. One set was run with monochromatic waves at the Coastal Engineering Research Center (CERC) (Saville 1957). The other set was run at the Central Research Institute of Electric Power Industry in Japan (CRIEPI) (Kajima et al. 1982). Deep-water wave heights were in the range of $0.30 < H_o < 1.78$ m, wave periods were in the range of $3.0 < T < 16.0$ sec., and sediment sizes were in the range of $0.22 < d_{50} < 0.47$. These laboratory conditions thus cover a wide range of prototype conditions.

Erosional profiles had no berm above uprush and at least one pronounced bar offshore; *accretional* profiles had a prominent berm and no bar formations. Kraus et al. (1991) has shown that H_o/L_o and $H_o/w_f T$, where w_f is the sediment fall velocity, correctly categorize these two types of storm profile changes. Dalrymple (1992) combined Kraus's two variables into a single profile change predictor.

Figure 1 shows *accretional* (A) and *erosional* (E) profile responses as functions of U_C and U_T. Profile transitions during the storms were *erosional* if U_T was less than a critical value near -2. In other words, if magnitude of NOPS under the trough was greater than twice the grain threshold speed, the storms ended with *erosional* profiles. Otherwise profiles became *accretional*. Even events with high U_C values remained *erosional* so long as the critical value of U_T remained less than -2.

Skill is a simple statistical measure for quantifying the performance of a categorical predictor on a given set of data (Seymour and Castel 1989). Skill equals the ratio of correct predictions to total observations. Using U_T= 1.8 as a threshold level, there is one miscategorized erosion and one miscategorized accretion in the 32 tank results, for a predictive skill of 0.94.

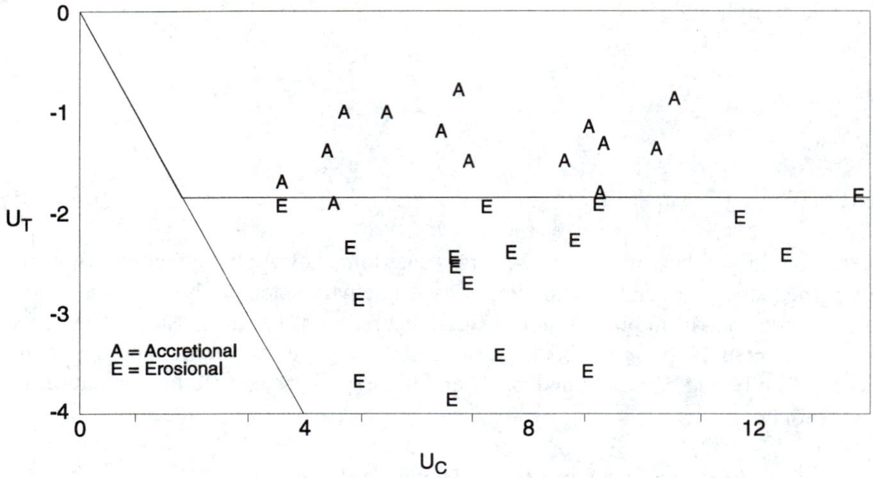

Figure 1 Large wave tank data

Storm Response of Shore Profiles from Field Measurements

Kraus and Mason (1991) compiled and standardized 99 cases of storm profile change from field studies published by many researchers worldwide. Seventy-two cases were *erosional*. Twenty-seven were *accretional*. To qualify as *accretional* the storm had to have resulted in a notable seaward advance of the shoreline, a buildup of the subaerial berm, or a landward movement of the longshore bar. Kraus et al. (1991) used this data set, along with the previously presented results from large wave tank tests, to develop discriminators between *erosional* and *accretional* storms. In the field data set, deepwater significant wave heights ranged from 0.08 to 7.90 m, wave periods from 2.0 to 15.3 sec, and sediment sizes from 0.17 to 3.5 mm. Wave periods were associated with either deepwater significant wave height or the spectral peak.

Accretional and *erosional* type profiles are denoted in Figure 2 as functions of U_C and U_T. U_T discriminates well between *erosional* and *accretional* profiles at a value around -2, just as for the laboratory data. And if $U_T < -2$, *erosional* profiles occur even if U_C is quite large.

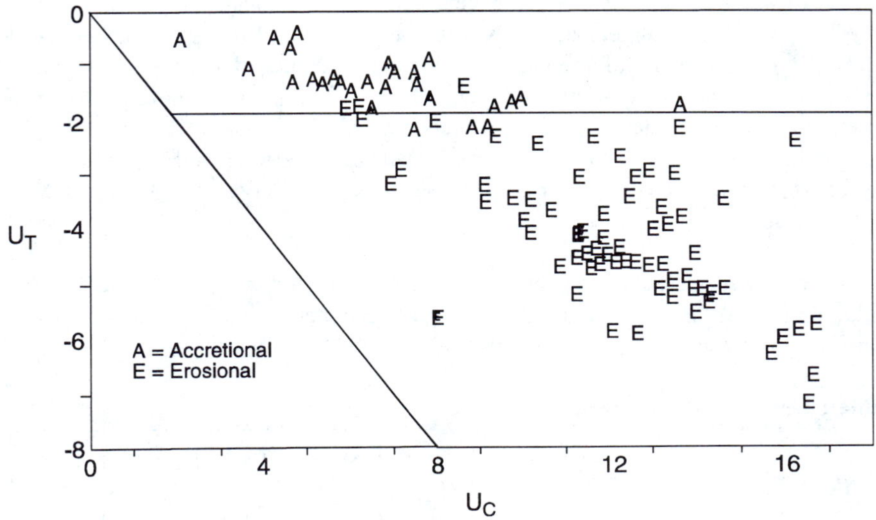

Figure 2 Field beach profile data

Using U_T = -1.90 as the discriminator, three cases are miscategorized as *erosional* and three are miscategorized as *accretional* in this field data set of 99, for a predictive skill (0.94) similar to that of previously proposed criteria. The skill of Dalrymple's (1992) criterion on this field data is 0.94 and the favored pair one of eight criteria examined by Kraus et al. (1991) had a skill of 0.91. Largely by coincidence, the skill of U_T on field and laboratory data are identical. More importantly, the critical U_T threshold levels are essentially equal for both field and large wave tank results.

Long-Term Response of Submerged Mounds

Nearshore placement of feeder deposits is an increasingly attractive form of erosion protection offering a variety of environmental, social, and economic benefits (Bodge 1994a and b; Bruun 1988; de Lange and Healy 1994; Foster, Healy, and de Lange 1994; Foster, Healy, 1996; Mulder, van de Kreeke, and van Vessem 1995; Roelvink and Stive, 1988; Russell, Robinson, and Soward 1994; Uda, Naito, and Kanda 1991; Zenkovich and Schwartz 1987). Hands and Allison (1991) compiled results from feeder mound tests and developed criteria to distinguish between stable nearshore deposits and others that moved promptly shoreward. Those mound criteria were used to identify conditions where dredged material mounded outside the surf zone acts not only as a temporary wave dissipator, but also gets pushed shoreward to nourish the surf zone.

Eleven reference mounds monitored in the United States have been categorized as *active* or *stable* depending on whether repeated surveys indicated significant loss of material from the placement area. All but one of the *active* cases showed clear shoreward displacement of mound centroids. Evidence of change was always identifiable within months. S*table* mounds remained stationary without evidence of dispersion or displacement for years (Hands 1991). Wave forces appear to be the dominant factor in moving mounds landward (Hands and Resio 1994, Douglass, Resio, and Hands, 1995). There has never been strong evidence of any seaward movement at the test sites.

Table 1 lists the locations of all 11 field test mounds, premound water depths, median grain sizes of placed material (d_{50}), wave parameters, and indicators of which mounds were *active* (A) and which *stable* (S).

Table 1 Dredged Mound Data

Site Location	Mound Depth (m)	Smaller of $H_{99.863}$ or H_b (m)	$\bar{T}_A$ (sec)	Grain Size d_{50} (mm)	Stable or Active	U_C	U_T
Long Island, NY	4.6	2.4	8.0	0.40	A	7.20	-3.27
Long Branch, NJ	11.6	2.9	8.0	0.23	S	6.25	-5.38
Atlantic City, NJ	5.8	2.5	8.0	0.35	S	7.16	-3.98
Dam Neck, VA	10.4	3.4	10.0	0.08	S	15.39	-10.06
Dam Neck, VA (crest)	7.6	3.1	10.0	0.08	A	16.97	-9.36
New River, NC	2.1	1.6*	7.0	0.50	A	6.62	-1.63
Sand Island, AL (berm)	5.8	2.2	9.1	0.20	A	8.96	-4.63
Sand Island, AL (mound)	5.6	2.2	9.1	0.22	A	8.63	-4.34
Brazos, TX	8.1	2.8	10.0	0.13	A	11.88	-6.79
Silver Strand, CA	5.8	2.3	16.7	0.22	A	10.70	-2.82
Santa Barbara, CA	6.7	1.4	15.0	0.20	S	6.66	-2.65
Humboldt, CA	15.8	7.6	14.3	0.23	A	16.29	-8.01

* = only case with $H_b > H_{.99863}$

To test the velocity ratios, the full time series of waves were transformed to each mound from the nearest offshore Wave Information Study hindcast site. Velocity ratios were evaluated using transformed spectral peak heights and average associated peak periods, T_A, i.e., the average period of all waves having a height within 0.1 m of the 99.863 percentile nonexceedance wave height ($H_{99.863}$). This wave height was chosen to match Hallermeier's (1980) selection of the 12-hr/ year exceedance wave height as the determinate for beach profile zonation. Use of $H_{99.863}$ to calculate the velocity ratio is also consistent with its use in the Empirical BERM model (Hands and Resio 1994). In only one case would this extreme wave have broken before passing over the mound. It seems reasonable in such a case to use the H_b estimated from SFWT. Both approaches gave identical results, however, to within the two significant figures used here.

Mound responses are denoted in Figure 3 as functions of U_C and U_T. In the mound situation, U_T alone is a poor discriminator. U_C and U_T are both needed. *Stable* mounds have smaller values of U_C than active mounds for approximately equivalent values of U_T. The $U_C = U_T$ line indicates the limiting condition for NOPS which is reached only by linear waves. Above this line a curve follows the trend between *active* and *stable* mounds.

99

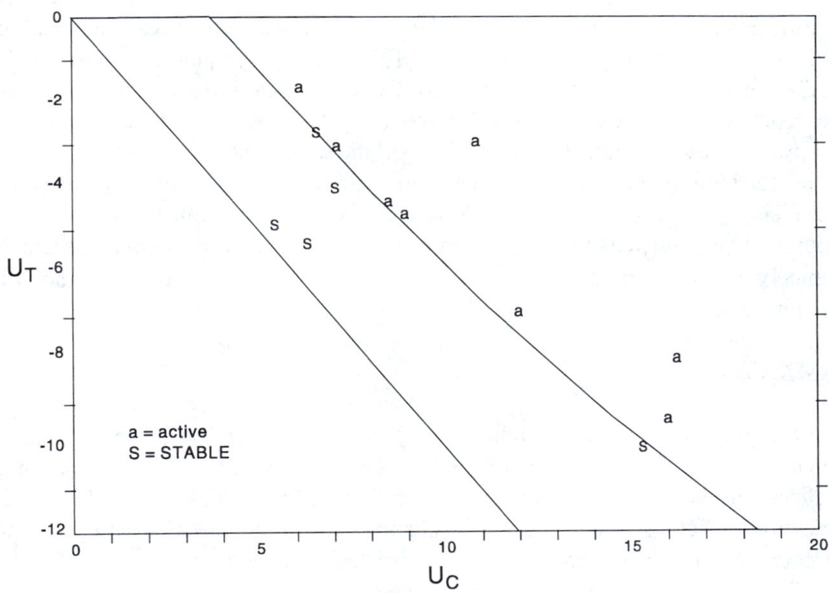

Figure 3 Field mound response data

Bivariate classification of the reference mounds does not imply one should expect unambiguous behavior at future mounds. An ill-defined zone of uncertainty separates the two classes. Occurrence of a severe storm or extended periods of unusual calm will affect a mound's response and the accuracy of any climatological-based prediction. And the role of unusual storms should be most critical for the mounds in the transition zone separating expected *stable* and *active* regions.

Earlier Approach for Mound Predictions

Hallermeier (1977) proposed a sediment entrainment parameter to characterize fluid motion at the onset of intense bed agitation. This parameter had the form of a Froude number which Hallermeier (1980) simplified using linear wave theory to obtain his two profile zonation limits. Hands (1991) used dimensionless ratios of these profile limits to mound depths as feeder-berm citing criteria. Linear theory was also used to show that the distributions of predicted near-bed oscillatory speeds from hindcasted waves could be used to distinguish between *active* and *stable* mound sites (ibid.). The new approach, presented here, is the first attempt we know to improve mound predictions by advancing beyond linear wave theory.

Revised Approach for Evaluation of Nonlinear Oscillatory Speeds

Threshold ratios presented here are very much works in progress. The form of the threshold ratios, the theory and procedure for evaluating oscillatory speeds, and alternative methods for summarizing distributions have not been thoroughly explored. We briefly examined the impact of basing the criteria on deepwater, local, and breaking wave heights; fitting $u_{-d\,max}$ instead of U to SFWT, and optimizing fit in terms of U_T instead of to both U_T and U_C because only the single criterion seems necessary for erosion prediction. These variations led to considerable differences in the spread of U_T and U_C values, yet, each of these versions support the same conclusions except for small adjustments as to the best critical values. With different versions, the critical U_T for beach erosion field data ranged from -1.8 to -2.3. Present uncertainties about sediment transport and the limited available prototype data do not support refinement of details. Fortunately threshold ratios seem to be robust with respect to tested methods of evaluation.

SUMMARY

Simple threshold ratios indicate tendencies for waves to drive cross-shore sediment fluxes. These ratios combine oscillatory peak speeds with initiation of movement criteria. The given equations are applicable over a wide range of indicated wave conditions and sediment sizes. Results from testing these ratios against known laboratory and field data produce trends consistent with present understandings of onshore and offshore sediment movement under waves.

The single parameter U_T shows considerable and unexpected skill (0.94) in distinguishing between beach erosion and accretion. Surprisingly, the speed under the trough is more diagnostic than under the crest. In fact U_C seems to be unimportant for predicting the shore profile change. Both parameters, U_T and U_C , are necessary, however, to predict responses of nearshore mounds built of sand- to silt-sized dredged material. While experiences with mounds is limited and many cases cluster close to the discrimination boundary, the present approach has an advantage over previous methods given by Hands (1991), because U captures the inherent nonlinearity of waves nearshore and provides a logical explanation for the noted preferential shoreward movement of *active* mounds.

CONCLUSIONS

Admittedly, this presentation does not investigate the mechanics of sediment motion and thus does not offer a general solution to cross-shore coastal transport. Nevertheless, we hope our results focus more attention on what certainly seems to be the dominant effect of nonlinear wave motion in some important coastal situations. The success of U criteria over the demonstrated range of wave conditions, water depths, and grain sizes seems to justify their adoption as simple decision criteria needed now to help manage coastal sediment resources. For long-term mound movement, the option of using a more encompassing parameter to represent ranges of near-bed fluid motion seems promising as an easy improvement that should characterize the net effect of the range of waves impacting mounds over their months and years of migration. On a spatial scale, some broader measure of surf conditions may offer improvements over the simple default to breaker conditions tested first.

Obviously, however, wave effects are more complex than can be represented by peak bottom speed and many other factors affect sediment motion (e.g., bottom slope, bed forms, cohesive particle forces, and other currents). Net effects of wave groupiness, undertow, grain inertia, and intra-wave phase lags between stress and sediment concentration and a time-varying eddy viscosity are simply lumped in a single critical ratio. The fact that such a simple characterization as U, even with reduction of NOPS distributions to a default percentile, produced a more than acceptable correlation with laboratory and field data, will hopefully spark theoretically based improvements.

ACKNOWLEDGMENTS

Support from the NOAA Sea Grant Office and the U.S. Army Engineer Waterways Experiment Station is acknowledged, especially the recently concluded Dredging Research Program for collecting mound response data and the new Dredging Operations and Environmental Research Program for follow-on analyses. Comments and suggestions from Mr. Thomas W. Richardson and Drs. Nicholas C. Kraus and Andrew Morang were very helpful and are greatly appreciated. The first author received permission from Headquarters, U.S. Army Corps of Engineers, to publish this information.

REFERENCES

Ahrens, J. P., and Hands, E. B. 1997. "Velocity Parameters, for Predicting Cross-Shore Sediment Movement," in preparation for *Journal of Waterway, Port, Coastal and Ocean Engineering*, ASCE, NY.

Bodge, K. R. 1994a. "Performance of Nearshore Berm Disposal at Port Canaveral, Florida," *Proceedings 2nd International Conference Dredging and Dredged Material Placement, Dredging '94*, ASCE, NY, 1182-1191.

Bodge, K. R. 1994b. "Performance Of 1992/93 Nearshore Berm Disposal At Port Canaveral, Florida," *Proceedings Beach Preservation Technology 1994*, L.S. Tait, ed., Florida Shore & Beach Preservation Association, Tallahassee, FL, 285-301.

Bruun, P. 1988. "Profile Nourishment: Its Background And Economic Advantages," *Journal of Coastal Research*, 4 (2) 219-288.

Dalrymple, R. A. 1992. "Prediction of Storm/Normal Beach Profiles," *Journal of Waterway, Port, Coastal, and Ocean Engineering*, 118 (2) 193-199.

Dean, R. G. 1974. "Evaluation and Development of Water Wave Theories for Engineering Application," Volumes I and II, Coastal Engineering Research Center Special Report No.1, Fort Belvoir, VA.

de Lange, W., and Healy, T. 1994. "Assessing The Stability Of Inner Shelf Dredge Spoil Mounds Using Spreadsheet Applications On Personal Computers," *Journal of Coastal Research* 10 (4) 946-958.

Douglass, S. L., Resio, D. T., Hands, E. B. 1995. "Impact of Near-Bottom Currents on Dredged Material Mounds Near Mobile Bay," Dredging Research Program Technical Report DRP-95-6, U.S. Army Engineer Waterways Experiment Station, Vicksburg, MS.

Foster, G. A., Healy, T. R., and de Lange, W. P. 1996. "Presaging Beach Renourishment from a Nearshore Dredge Dump Mound, Mt. Maunganui Beach, New Zealand," *Journal of Coastal Research* 12 (2) 395-504.

Foster, G. A., Healy, T. R., and de Lange, W. P. 1994. "Sediment Budget and Equilibrium Beach Profiles Applied To Renourishment of an Ebb Tidal Delta Adjacent Beach, Mt-Maunganui, New Zealand," *Journal of Coastal Research* 10 (3) 564-575.

Hallermeier, R. J. 1977. "Calculation of a Yearly Limit Depth to the Active Beach Profile," Technical Paper 77-9, U.S. Army Engineer Coastal Engineering Research Center, Ft. Belvoir, VA.

Hallermeier, R. J. 1980. "Sand Motion Initiation by Water Waves: Two Asymptotes," *Journal of Waterway, Port, Coastal, and Ocean Engineering* 106 WW3 299-318.

Hands, E. B., and Resio, D. T. 1994. "Empirical Guidance for Siting Berms to Promote Stability or Nourishment Benefits," *Proceedings 2nd International Conference.,*

Dredging and Dredged Material Placement, Dredging '94, Buena Vista, FL, ASCE, NY, 220-228.

Hands, E. B. 1991. "Unprecedented Migration Of a Submerged Mound Off the Alabama Coast," *Proceedings 12th Conference Western Dredging Association and 24th Annual Texas A&M Dredging Seminar*., ed., Center for Dredging Studies, College Station, TX, 245-269.

Hands, E. B., and Allison, M. C. 1991. "Mound Migration in Deeper Water and Methods of Categorizing Active and Stable Depths," *Proceedings Specialty Conference on Quantitative Approaches to Coastal Sediment Processes, SEDIMENTS '91*, Seattle, WA, ASCE, NY, 2, 1985-1999.

Kajima, R., Shimizu, T., Maruyama, K., and Saito, S. 1982. "Experiments of Beach Profile Change with a Large Wave Flume," *Proceedings 18th International Coastal Engineering Conference*, ASCE, NY, 1385-1404.

Kaminski, G. M., and Kraus, N. C. 1993. "Evaluation of Depth-limited Wave Breaking Criteria," *Proceedings Ocean Wave Measurement and Analysis Conference*, New Orleans, LA, ASCE, NY, 180-193.

Komar, P. D., and Miller, M. C. 1974. "Sediment Threshold under Oscillatory Waves," *Proceedings 14th International Coastal Engineering Conference*, ASCE, NY, 756-775.

Kraus, N C., Larson, M., and Kriebel, D. L. 1991. "Evaluation of Beach Erosion and Accretion Predictors," *Proceedings Coastal Sediments '91*. N. C. Kraus, K. J. Gingerich, and D. L. Kriebel, eds., ASCE, NY, 572-587.

Kraus, N. C., and Mason, J. G. 1991. "Field Data Set for Testing Beach Erosion and Accretion predictive Criteria," Memorandum for Record, Coastal and Hydraulics Laboratory, U.S. Army Engineer Waterways Experiment Station, Vicksburg, MS.

Larson, M., and Kraus, N. C. 1989. "SBEACH: Numerical Model for Simulating Storm-Induced Beach Change," Report 1: Empirical Foundation and Model Development, TR CERC 89-9.U.S. Army Engineer Waterways Experiment Station, Vicksburg, MS.

Landin, M. C., Davis, J. E., and Hands, E. B. 1995. "Wetland Restoration, Underwater Features, and Other Beneficial Uses, A Phase I Report on the Delaware Bay Deepening Project," U.S. Army Engineer District, Philadelphia, PA.

Mulder, J. P. M., van de Kreeke, J., and van Vessem, P. 1995. "Experimental Shoreface Nourishment, Terschelling (NL)," *Proceedings 24th International Coastal Engineering Conference*, B.L. Edge, ed., ASCE, NY, 2886-2899.

Roelvink, J. A., and Stive, M. J. F. 1988. "Large Scale Tests of Cross-Shore Sediment Transport on the Upper Shoreface," *Proceedings of the Symposium on Mathematical Modelling of Sediment Transport in the Coastal Zone*, Copenhagen, Denmark, IAHS, Washington, DC, 137-147.

Russell, R. J., Robinson, D. A., and Soward, C. L. 1994. "Southern Gold Coast Beach Nourishment Project: Implementation, Results, Effectiveness," *Terra et Aqua* 66, 12-23.

Saville, T. 1957. "Scale Effects in Two Dimensional Beach Studies," *Transactions 7th General Meeting International Association of Hydraulic Researchers*, 1 A3.1-A3.10, IAHS, Washington, DC.

Stive, M. J., de Vriend, H. J., Nicholls, R. J., and Capobianco, M. 1992. "Shore Nourishment and the Active Zone: a Time Scale Dependent View," *Proceedings 23rd International Conference Coastal Engineering*, B. L. Edge, ed., ASCE, NY, 2464-2473.

Uda, T., Naito, K., and Kanda, Y. 1991. "Field Experiment On Sand Bypass Off the Iioka Coast," *Coastal Engineering in Japan* 34 (2) 205-221.

Zenkovich, V. P., and Schwartz, M. L. 1987. "Protecting the Black Sea-Georgian SSR Gravel Coast," *Journal of Coastal Research* 3 (2) 201-209.

CHAPTER 262

Small-scale Morphology Related to Wave and Current Parameters Over a Barred Beach

A.F. Garcez Faria[1], E.B. Thornton[1], T.P. Stanton[1]

ABSTRACT: Small-scale morphology (< 5 m horizontal wavelength) was measured by combining CRAB surveys with bed elevations acquired from a 1 MHZ sonic altimeter mounted on the CRAB during the DUCK94 experiment in October 1994. Bedform plan views were recorded simultaneously using a 500 KHz side-scan sonar mounted on the CRAB. Waves and currents were measured at the same time. Three cases are examined in detail: mild waves and weak longshore currents resulting in wave ripples throughout the study area; narrow-band, normally incident waves with a strong rip current resulting in a planar bed except in the throat of the rip where mega-ripples were measured; and storm waves with strong longshore currents resulting in lunate and straight-crested mega-ripples in the trough of the barred beach. During these strong current days, bed shear stresses calculated from logarithmic velocity profiles are equated to a quadratic bottom shear stress formulation. The associated bed shear stress coefficients vary by more than an order of magnitude across the surf zone (0.0006-0.012). Bottom roughness was obtained by calculating the wavenumber spectra of the bed. The bed shear stress coefficients are positively correlated with bottom roughness (linear correlation coefficient, 0.68). A higher linear correlation coefficient (0.80) is obtained by subtracting skin friction from the total bed shear stress.

INTRODUCTION

The nearshore bed on a barred beach is composed of complex bedforms caused by varying wave and current regimes. With the exception of wave ripples in deeper water, few quantitative observations of bedforms are available, especially in the nearshore, and simultaneous quantitative wave and current measurements are almost non-existent. The lack of data is due to difficulties in making measurements in the nearshore. Observations historically have been made by divers who are limited by visibility and the harsh nearshore environment, which can become unfavorable for observations even in the mildest of storms.

Earlier field measurements were reported by Dingler and Inman (1976), which profiled wave ripples using a moveable acoustic altimeter placed on a track on the bottom outside the surf zone, but were limited to a single track of approximately 2 m in length. Other methods have included using waxed combs pushed into the bed by divers to record ripple profiles. Miller and Komar (1980) found ripple wavelength to be related to the orbital diameter of the waves.

The measurements described in this paper were made in shallow water (<8m) where shoaling and breaking waves are typically asymmetric, resulting in asymmetric bedforms. A conceptual view of wave-induced small-scale morphology on a barred beach is summarized by Clifton (1976) based on diver observations, which describes the asymmetric transition of bedforms with increasing velocity from long-crested

----

[1]Oceanography Department, Naval Postgraduate School, Monterey, CA, 93943-5000, USA; E-mail: faria@oc.nps.navy.mil; Tel: +1 408 656 2379; Fax: +1 408 656 2712.

ripples, to irregular ripples, to cross-ripples, to lunate mega-ripples on the seaward slope of the bar, to a planar bar crest where the waves break, and to wave ripples again in the trough where wave intensity decreases after waves cross the bar. These are the least observed nearshore bedforms owing to difficulties in obtaining quantitative measurements.

The objective of this paper is to correlate small-scale morphology measured over a barred beach with waves and currents. In the following sections, a description of the DUCK94 experiment, the methodology of the data analysis, the description of the small-scale morphology observed on Oct. 8, 12, and 20, and the relationship between measured bottom roughness and bed shear stress are presented.

DUCK94 EXPERIMENT

The measurements described here are part of the comprehensive nearshore DUCK94 experiment conducted during October 1994 at the U.S. Army Corps of Engineers Field Research Facility (FRF), Duck, North Carolina. The weather during October was climatologically characterized by three distinct phases: weak currents and winds from the north (4-9 Oct.), relatively strong currents from the north (0.6-1.0 m/s) caused by a storm with predominant winds and waves from the north (10-12 Oct.), and variable currents and winds from the north/south (13-21 Oct.).

The morphology of the bottom (bathymetry) was measured at various scales using the Coastal Research Amphibious Buggy (CRAB). The surveys were performed daily during the experiment by driving the CRAB in a series of cross-shore survey lines with alongshore separation of 25 to 30m. Large-scale variations of bathymetry were obtained by recording the CRAB position using a laser ranging/auto-tracking system approximately every meter (order 3 cm rms vertical and horizontal accuracy). Three-dimensional, large-scale morphology mapped by the CRAB during the three periods of interest (8, 12, and 20 Oct.) are shown in Fig. 1.

Small-scale vertical bottom variations relative to the CRAB, including ripples and megaripples, were measured with a 1 MHZ sonic altimeter mounted on the CRAB 70 cm from the bed. The altimeter has a nominal sampling rate of 25 Hz, which resulted in a sample spacing of 2-4 cm (dependent on CRAB speed) with mm vertical resolution and accuracy less than 2 cm (Gallagher *et al*., 1995). The decrease in accuracy relative to resolution is due to the changing reflective surface owing to the bed dialating or sediment transported along the bed as waves pass overhead. The CRAB survey and altimeter measurements were combined to obtain a high resolution description of the bottom (Thornton *et al.*, 1996).

A 500 KHz side-scan sonar suspended in the center of the CRAB approximately 1.25 m from the bed generates high-resolution plan views of the morphology. The nominal horizontal range of the sonar was approximately 10 to 15 m to each side of the sonar, which allowed overlap of sonar data from adjacent survey lines, giving a relatively complete view of the geometry and spatial representation of the ripple features.

Corroborating wave and current data were acquired using an instrumented sled. A vertical stack of eight Marsh-McBirney two-component electromagnetic current meters (em's, hereafter) with 2.5 cm diameter spherical probes were mounted on a 2.5 m mast to measure vertical profiles of longshore and cross-shore currents. The sled was oriented with the vertical stack of current meters placed on the up-current side to prevent the sled structure from contaminating flow measurements. Waves and mean water level were measured using an array of five pressure sensors configured in a 3 m square with sensors at the corners and one at the center. For data collection, each morning the sled was towed by the CRAB to its furthest position offshore, dependent upon wave conditions, for the first data run. A forklift on the beach pulled the sled shoreward approximately 10 to 30 m for each subsequent run. Data were acquired at each location for approximately one hour at four to eight locations. Additionally, directional wave spectra were acquired using a linear array of 10 pressure sensors in 8 m depth offshore of the survey area.

METHODOLOGY

The bottom roughness is examined by calculating wavenumber spectra of the bed. To calculate wavenumber spectra, the unevenly spaced data from the combined CRAB surveys and altimeter measurements are linearly interpolated to evenly spaced 2 cm increments of the cross-shore distance. The small-scale morphology in general shows large cross-shore variation; as a consequence, the condition of spatial homogeneity (stationarity) required for calculating averaged spectra is not met. Therefore, continuous wavenumber spectra are calculated for 20 m cross-shore segments at increments of 1 m across the surf zone. Lowest wave numbers are filtered by subtracting a third-order polynomial best-fit curve from each 20 m section. A 10 percent cosine-taper data window is applied. The spectra are summed over three wavelength bands (0.4 - 0.83 m, 0.83 - 1.67 m, and 1.67 - 5 m) plus the total band (0.4 - 5 m), resulting in 52, 24, 16, and 92 degrees of freedom for each band respectively. The wavelength bands chosen are based on examination of individual spectra that were generally broad, indicating that several ripple wavelengths coexisted as a result of newly formed ripples combining with residual ripples from the past to form a complex series of ripple patterns. The *rms* height of each band is calculated as the square root of the variance within each band.The general trend is that the bottom was smoothest offshore and over the bar where wave ripples were planed-off due to higher near-bottom velocities, with increased roughness within the trough associated with mega-ripples (see Thornton, *et al.*, 1996, for details).

Linear shear stress gradients occur in flows driven by constant hydrostatic pressure gradients, such as in steady open channel flows, are well described by a logarithmic velocity profile. Therefore it is hypothesized that a steady, uniform, turbulent boundary layer flow over a rough surface in the alongshore direction can be described by a logarithmic profile:

$$V(z) = \frac{v_*}{\kappa} \ln \left(\frac{z + h}{z_o}\right) \quad (1)$$

where z is positive upwards from the surface, h is the mean water depth, κ is the Von Karman constant (0.4), v_* is the alongshore shear stress velocity and z_o is the physical roughness height, determined by bottom topography and sediment grain size. When waves are present, nonlinear interactions between waves and currents within the bottom boundary layer increase the bottom shear stress. Following Grant and Madsen (1979), this additional stress can be modeled by an apparent roughness height z_a, that is analogous to, but larger than z_o. Logarithmic profiles are fit to the data based on a linear-regression least-squares method. The value of z_a is calculated from the z intercept of the linear-regression on a semi-log plot of z versus V(z), and the shear stress velocity v_* is calculated from the slope.

Mean along shore bottom shear stress ($\overline{\tau_y(-h)}$) is related to the alongshore shear stress velocity (v_*) through (overbar indicates time averaging)

$$\overline{\tau_y(-h)} = \rho v_*^2 \tag{2}$$

where ρ is water density. In addition, the bed shear stress coefficient, C_f, can be calculated assuming a quadratic bed shear stress relationship

$$\overline{\tau_y(-h)} = \rho C_f \overline{(u^2 + v^2)^{1/2} v} \tag{3}$$

and combining with (2), gives

$$C_f = \frac{v_*^2}{\overline{(u^2 + v^2)^{1/2} v}} \tag{4}$$

C_f values are calculated using measured velocities (u,v) at the elevation of 1 m above the bed, with v_* determined by least-squares log fit (Garcez Faria *et al.*, 1996).

DISCUSSION

October 8 Morphology

The large-scale morphology on Oct. 8 at first glance appears to be relatively homogeneous. The offshore bar is nearly parallel to the shore and is located approximately 230-250 m offshore (Fig. 1). Therefore, it would be expected that the small-scale morphology would also be homogenous alongshore throughout the survey area given the same wave and current conditions. However, close examination shows that the small-scale morphology is not homogeneous and that ripple patterns vary both in the cross-shore and alongshore directions.

Two representative lines (710 and 940 m along shore coordinate) of cross-shore combined CRAB and altimeter profiles are selected to depict the differences present in the small-scale morphology for the survey area (Fig. 2). The primary differences in the ripple dimensions between the two representative lines are observed in the trough and the seaward slope of the bar. The small-scale morphology in the trough of line 940 is predominantly long wavelength mega-ripples (1.67-5 m) with shorter wavelength wave ripples superimposed on them. The trough region of line 710 has a more homogeneous ripple pattern in which no wavelength band predominates. The seaward slope of the bar on line 940 shows the ripple wavelength bands of 0.83-1.67 m and 1.67-5 m are dominant and the rms height is on the order of 3 cm. The seaward slope of the bar on line 710 shows relatively homogeneous ripple wavelengths with rms heights of approximately 2 cm.

October 20 Morphology

The most striking observation on the 20th is the hole in the bar with the bar bowed seaward due to the rip current between 800 and 1100 m in alongshore coordinates (Fig. 1). The bar is parallel to the beach and is located approximately 210-225 m offshore north of the rip area, shifting to greater than 300 m offshore in the rip area. The narrow trough, north of the rip area, acts as a rip-feeder channel and has the same small-scale morphological features as the rip area. The dominant small-scale morphology observed in the rip area are seaward-facing lunate and straight-crested mega-ripples. The mega-ripples were generally asymmetric with the steep slope facing seaward indicating the ripples were formed primarily as a result of the rip current. However, some ripples were symmetric while others were asymmetric facing toward the beach, which may infer a more complex formation due to a combination current and wave interactions. The area immediately north and south of the rip current is void of any discernable small-scale morphology with the exception of the rip-feeder channel. Typical rms roughness values in the rip area range from 5 to 15 cm, which are an order of magnitude larger than rms roughness values away from the rip. A contour plot of rms roughness values for the survey area is shown in Fig. 3, which displays the extreme variability of roughness in the rip current area compared with that of the area away from the rip current. The rip current, as qualitatively observed with dye on the 20th, is superimposed on the accompanying contour plot of the macro-scale bathymetry. The bed is essentially smooth away from the rip current with rms roughness values being less than 2 cm, which appears to be the result of planing action of the long-period swell.

October 12 Morphology

The large-scale morphology of Oct. 12 is similar to that of Oct. 8. The offshore bar is nearly parallel to the beach, but migrated 30 m offshore to approximately 260-280 m offshore (Fig. 1). Although the large-scale morphology is similar for the two days, the small-scale morphology is dramatically different owing to differences in the wave and current conditions associated with the arrival of a storm on Oct. 10. Large amplitude mega-ripples are now present in the trough and no regular bedform patterns are observed seaward of the bar for the entire survey area. The strong longshore current that appeared on the 10th is believed to be the dominant process that formed the small-scale morphology observed in the trough on the 12th.

The small-scale morphology across the surf zone is relatively homogeneous alongshore, and therefore, the survey area is represented by the rms roughness from the combined CRAB and altimeter profile for line 940 (Fig. 4, lower panel), which is close to the position occupied by the sled. The small-scale morphology in the trough compared with the region seaward of the bar crest are distinctly different. Seaward of the bar the bed is essentially plane, due to intense breaking associated with large waves. In the trough, the rms roughness is primarily associated with the larger ripple wavelengths of 1.67-5 m. Side scan plots showed the mega-ripples in the outer trough to be lunate, facing in the direction of the longshore current; while straight crested in the inner trough.

Bottom Roughness and Bed Shear Stress

The more energetic period (Oct 10-12) is used to relate small-scale morphology to wave and current forcing. The data are qualitatively sorted by location into the two regions of over the bar and in the trough. This sorting allows a better identification of the possible correlations among variables, as wave breaking which is a major controlling factor within the surf zone is significantly different for these regions.

All 22 vertical profiles of longshore currents obtained during these three days are analyzed. The profiles are based on the measurements by three to seven em's over the vertical. The em closest to the sea bed was not used because of malfunction. The observed and logarithmic predicted velocity profiles at successive offshore positions (runs) that the sled occupied during 12 Oct. is shown in the upper panel of Fig 4. The data agree well with the model.

A high correlation coefficient for the linear regression, is commonly accepted as an indicator of the validity of the logarithmic approach (Gross *et al.*, 1994; Li, 1994). The linear correlation coefficients for all profiles ranged from 0.95 to 0.99, with an average value of 0.98, and the largest deviations occurring over the bar, where wave breaking was strong.This can be related to the increase of turbulent mixing due to wave breaking producing a more uniform vertical profile of the current for a given shear stress, compared with profiles in the absence of breaking. Therefore, larger discrepancies between observations and logarithmic profile predictions (lower correlation coefficients) would be expected for increased turbulent mixing caused by wave breaking, as the logarithmic profile presumes no surface turbulence source.

The bottom shear stress coefficient (C_f) varied by an order of magnitude across the surf zone, with the values offshore and over the bar in the order of 10^{-3}, while the values in the trough were in the order of 10^{-2}. An attempt was made to find empirical relationships between C_f with measured physical parameters commonly used throughout the literature such as $|\vec{u}_b| / V$ (ratio of near-bottom wave velocity magnitude to mean current speed), and the *rms* bottom roughness (*r*). Surprisingly, no statistically significant correlation was found between C_f and $|\vec{u}_b| / V$. The strongest correlation was found between C_f and the bottom roughness, with C_f increasing with bottom roughness (Fig. 5, upper panel), with a linear correlation coefficient of 0.68, which is statistically significant at the 99.75 percent confidence level. Theoretically this is expected as larger roughness implies larger bottom stress due to form drag, and consequently larger C_f values.

The scatter of data observed in Fig. 5 (upper panel) may be due to correlating the non-synoptic velocity measurements with bottom roughness measured once in the morning, prior to the positioning of the sled for the first station. Wave forcing changed during the period of observation due to both wind and tidal variations. Changes in wave forcing have a direct effect in the measured velocities and an indirect effect in bottom roughness due to modification of wave ripples associated with variations in wave height.

Another reason for scatter of data arises from C_f values being calculated from the total bottom shear stress, which has stress contributions from skin friction related to sediment grains, waves-current nonlinear interactions within the bottom boundary layer as well as form drag related to bed forms. Smith and McLean (1977) linearly partitioned the total bed shear stress into skin friction and form drag and found good agreement with the data from the Columbia River. Extending this concept to the surf zone environment requires adding a component due to non-linear interactions between waves and currents within the bottom boundary layer to the total bed shear stress. Assuming that the linear stress partition is valid within the surf zone, the skin friction contribution can be removed from the total bottom stress and a new bed shear stress coefficient $C_f^{/}$ is defined

$$C_f^{/} = \frac{v_{*_d}^{2} + v_{*_{wc}}^{2}}{(u^2 + v^2)^{1/2} v} \quad (5)$$

where v_{*_d} and $v_{*_{wc}}$ are the form drag and waves-current interactions shear stress velocities. The relationship between $C_f^{/}$ and the total bed shear stress coefficient (C_f) can be determined from (4) and (5)

$$C_f^{/} = [1 - (\frac{v_{*_s}}{v_*})^2]\, C_f \quad (6)$$

where v_{*_s} is the skin friction related shear stress velocity. As skin friction was not measured during the DUCK94 experiment, an attempt is made to isolate its contribution from the total bottom shear stress by applying a stress partitioning model. The probabilistic approach used to quantify bottom roughness does not allow the adjustment of the two empirical coefficients C_D and a_1 necessary to apply the Smith and McLean (1977) model. Therefore, the empirical relationships obtained by Li (1994) are used to estimate skin friction shear stress velocity from the total shear stress velocity obtained from the logarithmic profile

$$\frac{v_{*_s}}{v_*} = 0.125(\frac{v_*}{R}) + 0.373 \qquad for \quad \frac{v_*}{R} < 2.3\ s^{-1} \quad (7)$$

$$\frac{v_{*_s}}{v_*} = 0.107\left(\frac{v_*}{R}\right) + 0.266 \qquad for \quad \frac{v_*}{R} \geq 2.3\ s^{-1} \tag{8}$$

where R is ripple height, assumed here to be equal to the measured *rms* bottom roughness ($R = r$) . The recalculated bed shear stress coefficients $C_f^{/}$ (Equation 6) are also positively correlated with bottom roughness (Fig. 5, lower panel), but show a higher linear correlation coefficient (0.80). Theoretically this is expected as the removal of the skin friction component from the total stress should enhance the form drag contribution, and consequently increase the correlation between bed shear stress and bottom roughness.

CONCLUSIONS

Quantitative measurements of small-scale morphology on a barred beach were acquired during the DUCK94 experiment. Three separate sequences are examined in which distinctly different ripple dimensions and bedform types were observed owing to varying wave and current regimes.

The measurements of small-scale morphology on Oct. 8, when mild waves and weak longshore currents prevailed, shows surprising differences within the survey area. The absence of significant variation associated with wave and current forcing suggests that the mega-ripples in the trough of line 940 may be residual from some earlier time of formation, such as on Oct. 3 when a strong longshore current was present.

Oct. 20 is characterized by a rip current with two distinct areas of small-scale morphology. In the throat of the rip current, straight-crested and lunate mega-ripples were observed. The area away from the rip current was void of any ripple patterns as a result of the planing action of the long-period swell.

Oct. 12 is characterized by the occurrence of a strong longshore current. The small-scale morphology of the inner trough region is characterized by relatively straight-crested mega-ripples (1.67-5 m) generated by the waves, superimposed by smaller wavelength ripples O(1 m). In the outer trough, lunate mega-ripples in the direction of the longshore current predominated. Seaward of the bar, the bed was essentially void of ripples.

The wavenumber spectral results infer a spatial and temporal variability of ripple dimensions in both the cross-shore and alongshore directions. The wavenumber spectra were generally broad, indicating that several ripple wavelengths coexisted as a result of newly formed ripples combined possibly with residual ripples from the past to form a complex series of ripple patterns. This broadening is also due to non-cross-shore (cross-track) orientation of the morphological features.

The vertical structure of mean longshore currents on a barred beach is well described by a logarithmic profile for the three strong longshore current days examined. This hypothesis works better in the trough where turbulent bottom boundary layer processes are more dominant than over the bar, where breaking-wave induced turbulence generated at the surface modifies the profile.

A relationship between bed shear stress coefficients (C_f) and bottom roughness was found. While the relatively small amount of data used (three days) are not enough to establish a definitive empirical relation from the available data, it can be concluded that C_f is directly proportional to bottom roughness, and therefore, the latter is an important parameter to characterize the bottom boundary layer. Surprisingly, only poor correlations of C_f and z_a were found with wave parameters.

The empirical relationships obtained by Li (1994) were used to remove the skin friction contribution from the total bottom shear stress. The improved correlation between bottom shear stress and bottom roughness obtained, although not conclusive to validate these simple expressions, indicates that the linear stress partition concept introduced by Smith and McLean (1977) can be extended to the surf zone environment.

ACKNOWLEDGMENTS

This research was funded by the Office of Naval Research, Coastal Sciences Program, under contract N00114-95-AF-002. The authors wish to express their appreciation to all those who participated in DUCK94 experiment, particularly the staff of the U.S. Army Field Research Facility. In addition, special appreciation is expressed to R. Wyland, Jim Stockel, and Mary Bristow, Naval Postgraduate School, for their role in the acquisition and processing of wave and current data.

REFERENCES

Clifton, H.E. (1976) "Wave-formed sedimentary structures- A conceptual model." *Beach and Nearshore Sedim., SEPM, Special Publication*, 24, 126-148.

Dingler, J.R., and Inman, D.L. (1976) "Wave formed ripples in the seashore sands." *Proceedings 15th Coastal Engineering Conference, ASCE*, pp. 2109-2126.

Gallagher, E.L., Boyd, W., Elgar, S., Guza, R.T., and Woodward, B. (1996). "A performance of a sonar altimeter in the nearshore." *Marine Geology*, 133, 241-248.

Garcez Faria, A.F., Thornton, E.B., Stanton, T. P., and Soares, C.M.C.V. (1996) "Vertical profiles of longshore currents and related bed shear stress and bottom roughness." *J. of Geophysical Research*, (submitted).

Grant, W.D., and Madsen, O.S. (1979) "Combined wave and current interaction with a rough bottom." *J. of Geophysical Research*, 84 (C4), 1797-1808.

Gross, T.F., Williams III, A.J., and Terray, E.A. (1994) "Bottom boundary layer spectral estimates in the presence of wave motions." *Continental Shelf Research*, 14, 1239-1256.

Li, M.Z. (1994) "Direct skin friction measurements and stress partitioning over movable sand ripples." *Journal of Geophysical Research*, 99(C1), 791-799.

Miller, M.C., and Komar, P.D. (1980) "A field investigation between oscillation ripple spacing and the near-bottom water orbital motions." *J. Sedimentary Petrology*, 50, 180-191.

Smith, J.D., and McLean, S.R. (1977) "Spatially averaged flow over a wavy surface." *Journal of Geophysical Research*, 82(C12), 1735-1746.

Thornton, E.B., Swayne, J.L., and Dingler, J.R. (1996) "Small-scale morphology related to waves and currents across the surf zone." *Mar. Geology* (accepted).

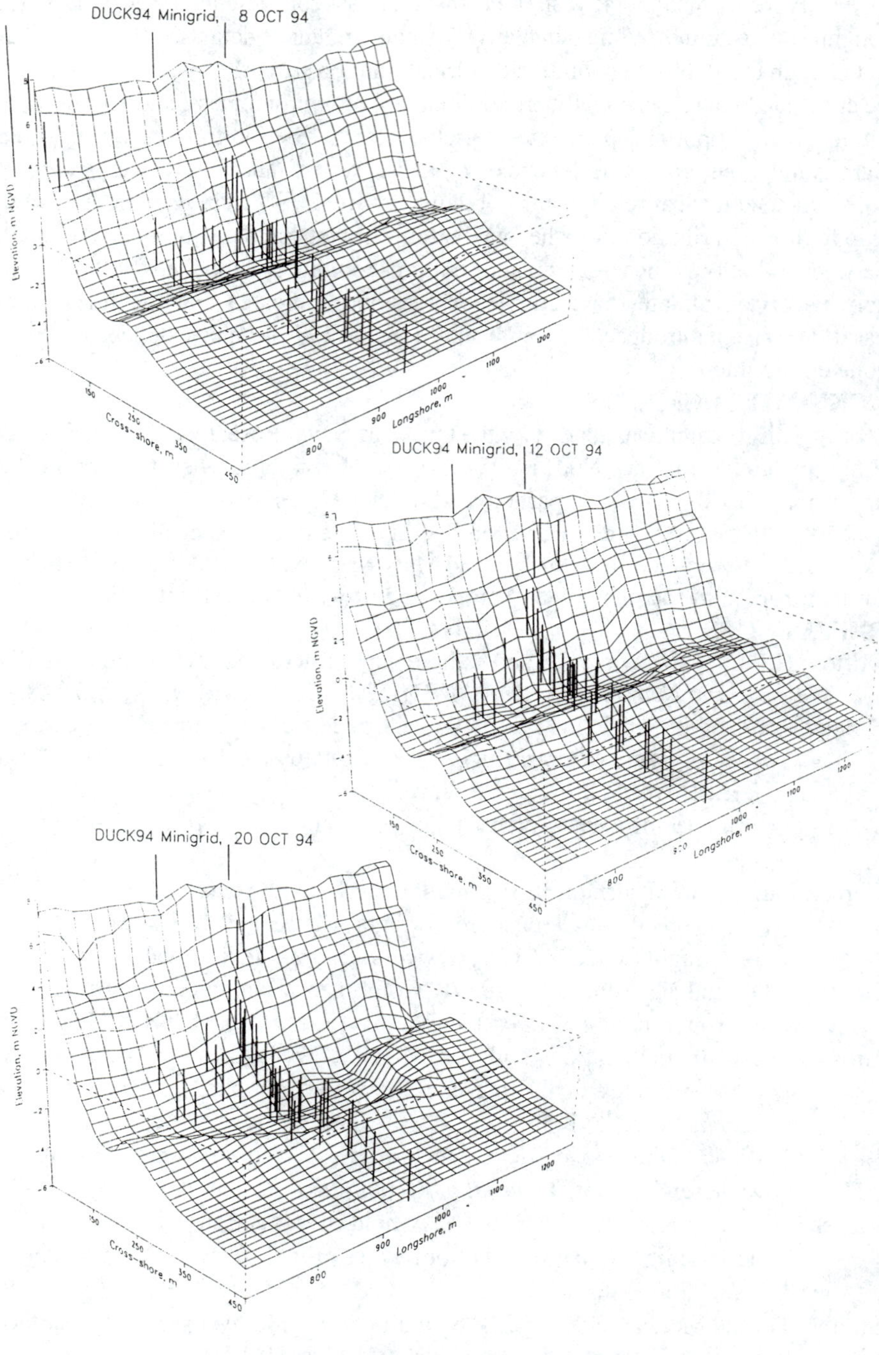

Figure 1. Three-dimensional bathymetry profiles of the survey area as mapped by the CRAB for the three days considered (Oct. 8, 12, and 20).

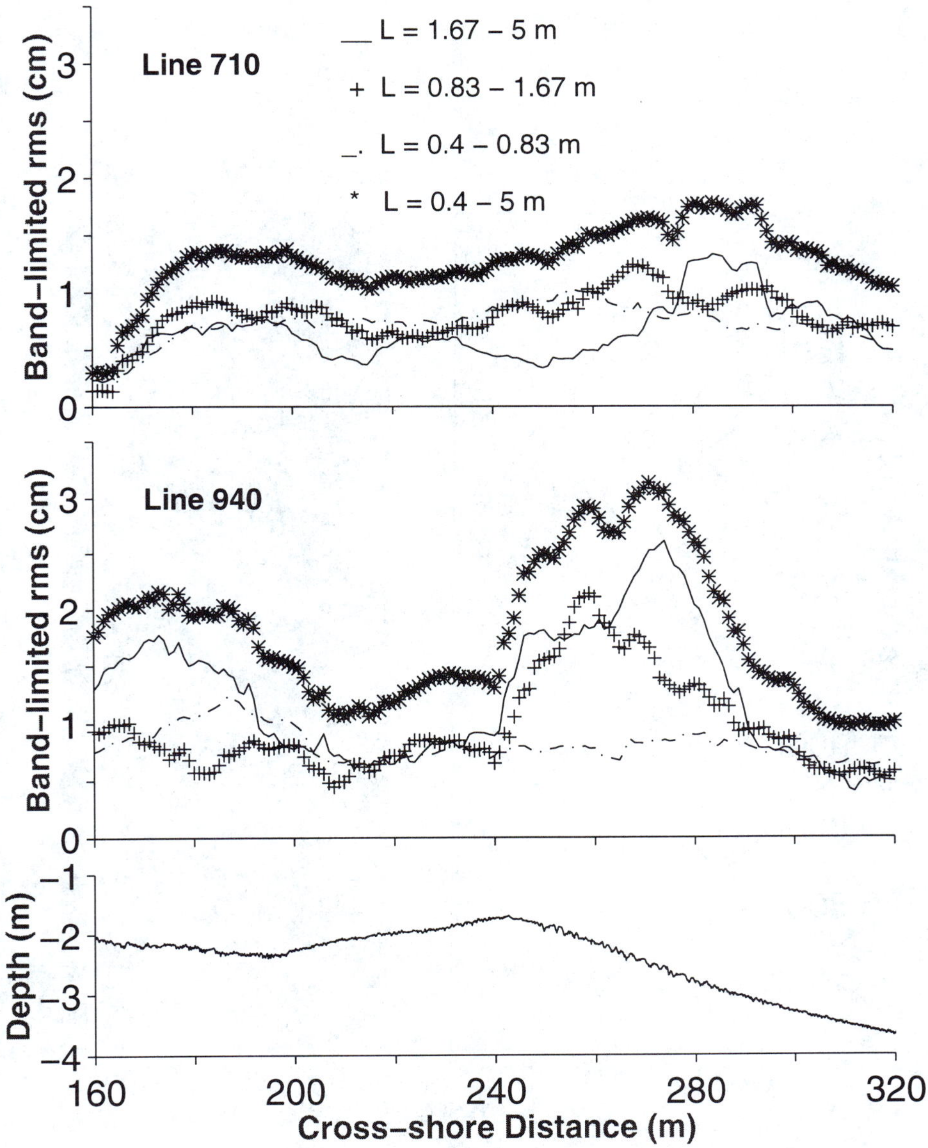

Figure 2. Band limited rms heights calculated across lines 710 (upper panel) and 940 (middle panel) for Oct. 8. Combined CRAB and altimeter bathymetry profile of line 940 for Oct. 8 (lower panel).

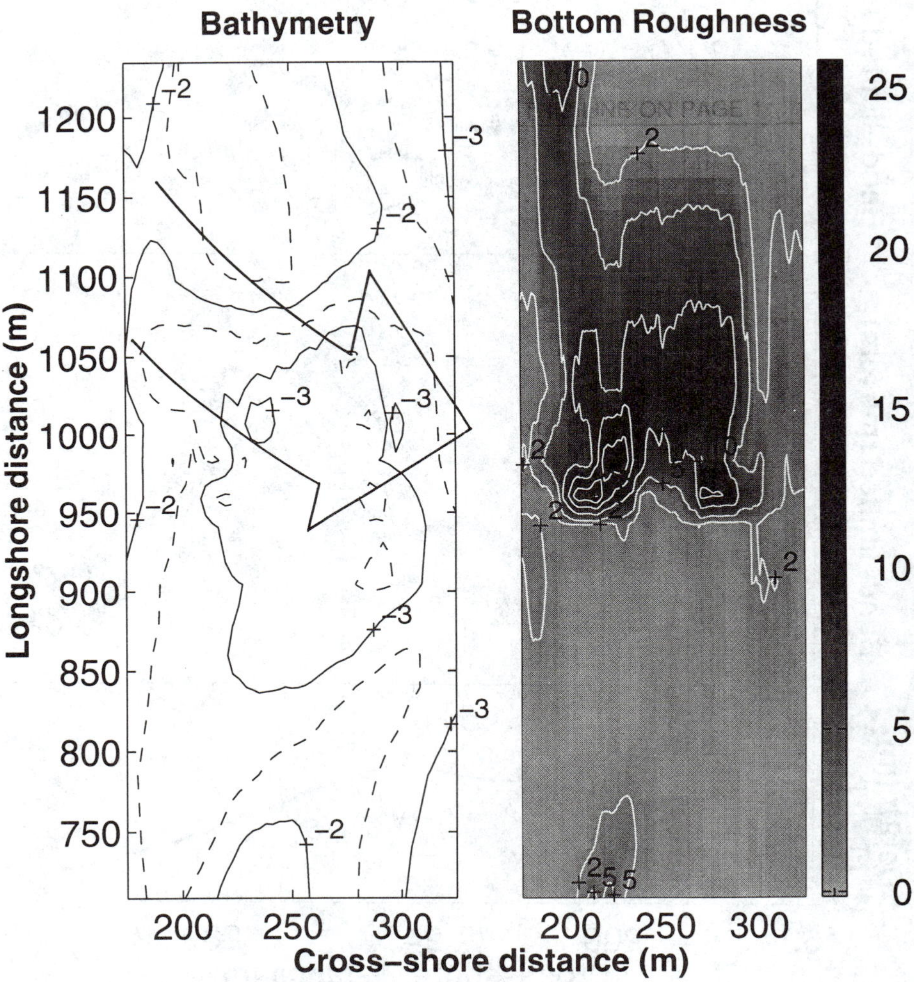

Figure 3. Bathymetry contour plot of DUCK94 survey area for Oct. 20 (left plot). Contour levels are in meters. Rip current is qualitatively indicated based on dye observations. Bottom rms roughness contour plot of survey area for Oct. 20 (right plot). Contour levels are in centimeters.

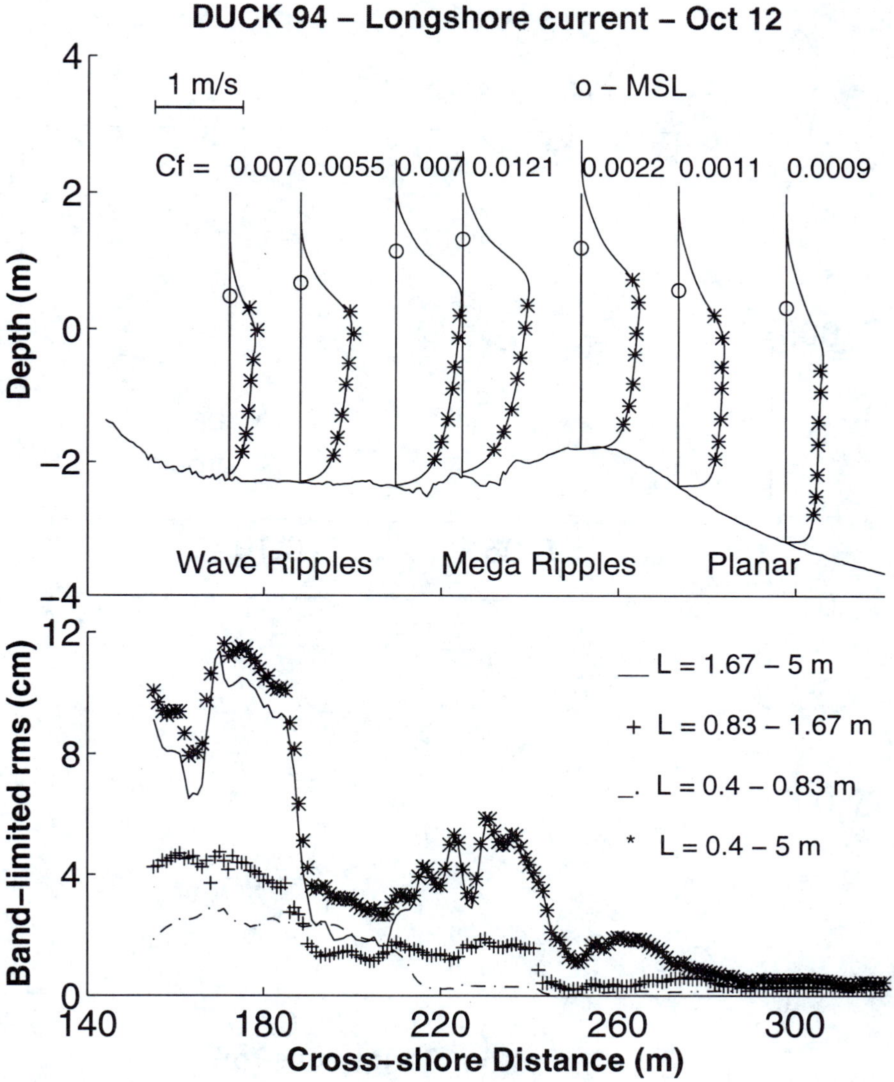

Figure 4. Vertical profiles of mean longshore currents superposed on bottom profile with tide elevation indicated by (o) and measured C_f values (upper panel). Variation of band-limited rms bottom roughness with the cross-shore distance for Oct. 12 (lower panel).

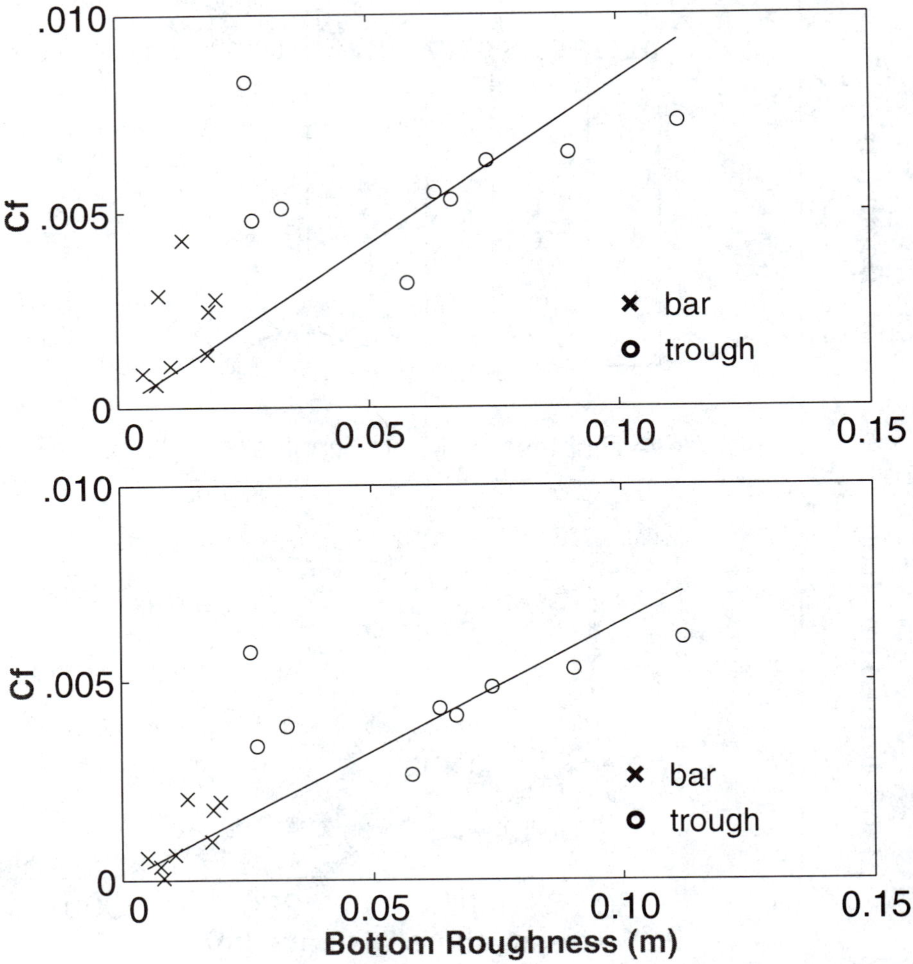

Figure 5. Total bed shear stress (C_f) versus bottom roughness (upper panel). The line represents a linear regression with a regression coefficient of 0.68. Bed shear stress coefficient without skin friction contribution (C_f') versus bottom roughness (lower panel). The line represents a linear regression with a regression coefficient of 0.80.

CHAPTER 263

PREDICTION OF SHORELINE CHANGE CONSIDERING CROSS-SHORE SEDIMENT TRANSPORT

Yoshimichi Yamamoto[1], Kiyoshi Horikawa[2], Katsutoshi Tanimoto[3]

ABSTRACT

Relations of cross-shore sediment transport rate with the grain size of sediment, the sea bottom slope in a surf zone and others were investigated by using data of field observation and large scale model experiments. The results are as follows :

(1) the coefficient of a cross-shore sediment transport rate varies inversely as the 1.31th power of the grain size. Then, the steeper a initial bottom slope is, the faster a beach profile reaches a state of equilibrium.

(2) The amount of a shoreline change is roughly proportional to the square root of the cross-shore sediment transport rate.

(3) The stabilized bottom slope in the surf zone increases with the grain size and the wave period, and it decreases as the breaking wave height increases.

Then, new equations to predict a beach profile change induced by cross-shore sediment transport were introduced from this investigation. Moreover, the adequate applicability of these equations to actual coasts was confirmed.

1. INTRODUCTION

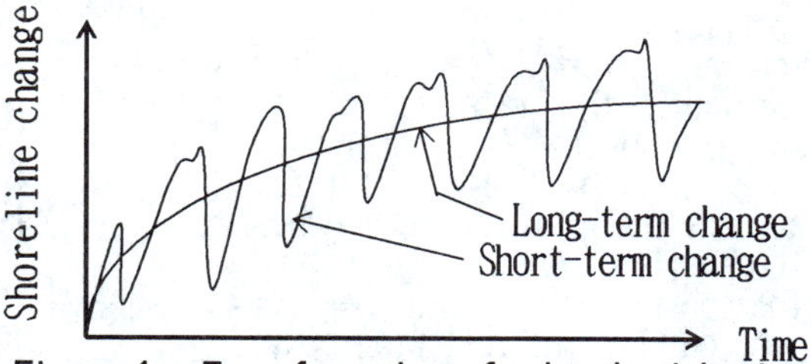

Figure 1 Transformation of a beach with time.

As practical models for predicting long-term transformation of long beaches, a shoreline change model and Uda et al.'s contour change model (1991, 1996) were proposed. However, these numerical models do not take cross-shore sediment transport rate into consideration. In designing measures to control coastal erosion and wave overtopping, it is necessary to take account of short-term beach transformation under stormy weather condition in addition to the long-term transformation as shown in Figure 1. The short-term transformation, which cannot be

---

[1] Dr.-Eng., Coastal Eng. Dept., INA Corporation, 1-44-10 Sekiguchi Bunkyo-ku, Tokyo, 112, Japan.
[2] Dr.-Eng., President, Saitama University, 255 Shimo-okubo, Urawa, 338, Japan.
[3] Dr.-Eng., Professor, Ditto.

determined without considering the cross-shore sediment transport, can be predicted by means of 2 or 3 - dimensional beach transformation models combined with the computation of waves and currents (e.g., Horikawa, 1988; Sato, 1994; Shibayama et al., 1994). However, it is difficult to apply these models to long-term prediction of long beaches, because long computational time is required and numerous coefficients introduced in sand transport rate formulas have not yet been generalized.

In this study, an attempt was made to generalize the coefficient of cross-shore sediment transport rate formulas. Moreover, a convenient beach evolution model using this result was proposed and applied to actual coasts.

2. CROSS-SHORE SEDIMENT TRANSPORT NEAR SHORELINE

Sunamura (1984) proposed the following formula for calculating cross-shore sediment transport rate, Q near the shoreline per unit time and unit beach width :

$$Q = K U_r^{0.2} \phi(\phi - 0.13 U_r) w d \qquad (1)$$

where K : a coefficient of sediment transport rate, U_r : Ursell parameter [$= gHT^2/h^2$], ϕ: Hallermeier parameter [$= H^2/shd$], w : the settling velocity of sediment, d : the median grain size of sediment, g : the acceleration of gravity, h : the wave setup height at shoreline against the still water level [= (1.63 tanα + 0.048) H_b, Sasaki and Saeki (1974)], H : the wave height at shoreline against the still water level [= 2.4(tan α)$^{0.3}$ h Yamamoto(1988)], T : the wave period, s: the specific gravity of sediment in water, tanα : the initial bottom slope in the surf zone, H_b : the breaker height.

Now, let us generalize the coefficient K of Eq. (1). As long as external forces remain constant, the rate of cross-shore sediment transport decreases with the lapse of time, and the beach profile approaches the equilibrium state. Therefore the coefficient K can be expressed by the following equation with the elapsed time t:

$$K = A \cdot e^{-Bt/T} \qquad (2)$$

where A and B are coefficients.

Then, we assume that the coefficients A and B are dominated by tanα and d/H_0 (H_0 is the wave height in deep water), and investigate relations of these coefficients with tanα

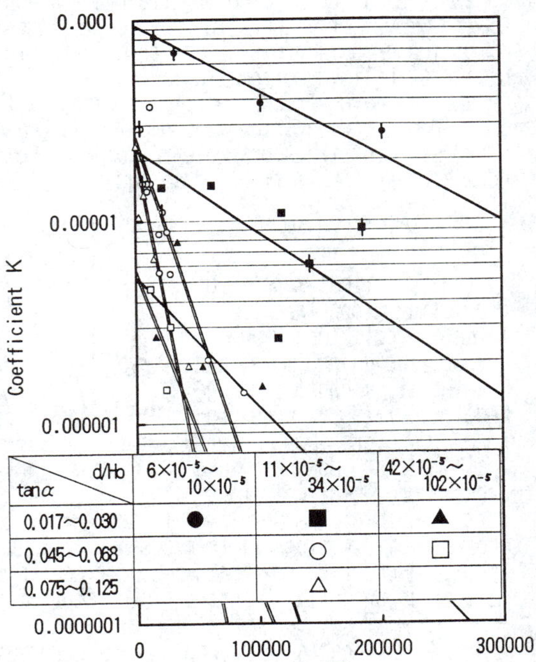

Figure 2 Relationship of the coefficient K with t/T.

and d/H_0 by using data of field observation and large scale model experiments given in Tables 1 ~ 4 in the appendix.

Figure 2 is a semilogarithmic graph of the relationship between K and t/T based on these data. Symbols with a vertical segment mean field data and other symbols mean experimental data. Then, average values of significant waves are used for the wave height and the wave period of irregular waves, because data of significant waves are used in many countries. However, experimental data of regular waves are used in order to supplement lack of data. Each straight line in this figure shows a tendency of data of each group. When t/T is 0, K (= A) varies greatly depending on the value of d/H_0, but changes very little depending on the value of $\tan\alpha$. Thus, the coefficient A is dominated strongly by d/H_0. Moreover, the slope of the straight lines in this figure varies widely depending on the value of $\tan\alpha$, while it is little affected by the value of d/H_0. This means that the coefficient B is strongly dominated by $\tan\alpha$.

$\tan\alpha$ \ d/Ho	6×10^{-5}~10×10^{-5}	11×10^{-5}~34×10^{-5}	42×10^{-5}~102×10^{-5}
0.017~0.030	●	■	▲
0.045~0.068		○	□
0.075~0.125		△	

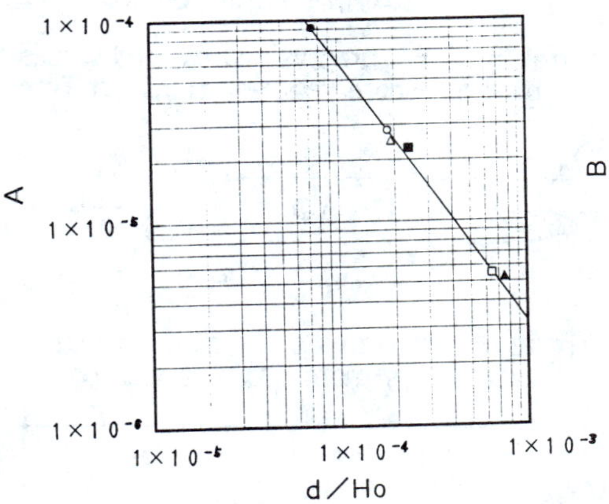

Figure 3 Relationship of A with d/H_0.

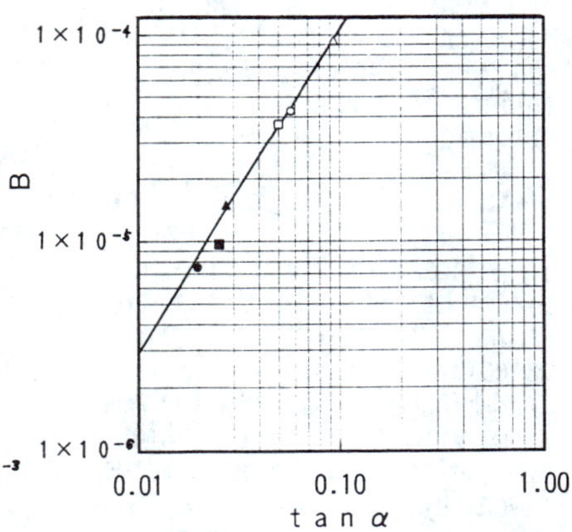

Figure 4 Relationship of B with $\tan\alpha$.

Figure 3 illustrates the relationship of the coefficient A with d/H_0 and Figure 4 shows the relationship of the coefficient B with $\tan\alpha$. As these figures indicate clearly that the coefficient A increases as d/H_0 falls and the coefficient B increases with $\tan\alpha$. Namely, the smaller the grain size of sediment is the larger the coefficient A is and the steeper the sea bottom slope is the faster the beach profile reaches the equilibrium state. These relations can be expressed by the following equations :

$$\left.\begin{array}{c} A = 3.61\times10^{-10}(d/H_0)^{-1.31}, \quad B = 4.20\times10^{-3}(\tan\alpha)^{1.57} \\ [\tan\alpha = 0.017\sim0.125, \quad d/H_0 = 0.00006\sim0.00102] \end{array}\right\} \quad (3)$$

The rate of cross-shore sediment transport near the shoreline can be obtained from Eqs. (1), (2), and (3). Figure 5 shows the comparison between measured values and calculated values obtained from data shown in Tables 2 and 4. This figure shows that the calculated values agree fairly well with the measured values.

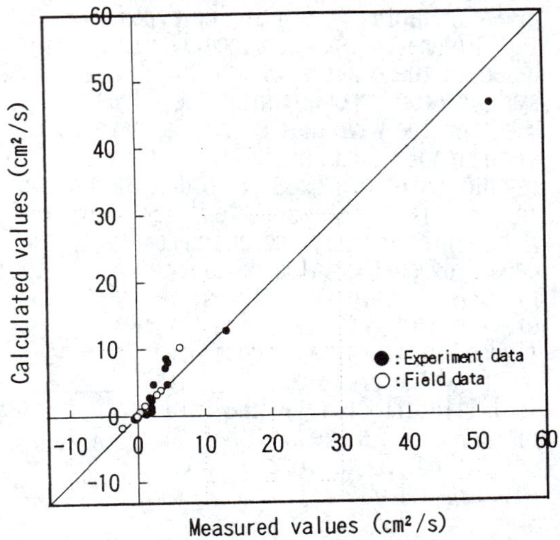

Figure 5 Measured vs. calculated values of cross-shore sediment transport rate.

3. SHORELINE CHANGE DUE TO CROSS-SHORE SEDIMENT TRANSPORT

Let us consider a simplified pattern of beach profile change, as shown in Figure 6, due to cross-shore sediment transport. Transforming slightly the continuity equation of cross-shore sediment transport, we can obtain the shoreline displacement as $\Delta y \propto (\int_0^t Q\,dt)^{0.5}$. Moreover, by using the data shown in Tables 2 and 4, the following equation can be obtained:

$$\Delta y = 2.7\left(\int_0^t Q\,dt\right)^{0.5} \quad (4)$$

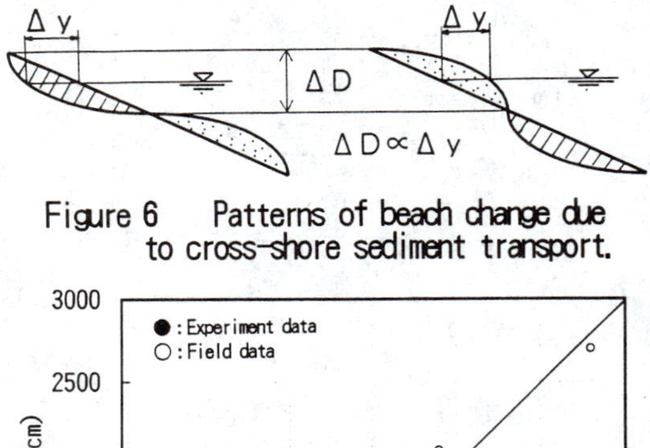

Figure 6 Patterns of beach change due to cross-shore sediment transport.

Therefore, the shoreline displacement due to cross-shore sediment transport can be calculated by using Eq. (4).

Figure 7 compares the measured values shown in Tables 2 and 4 with the calculated values given by Eq. (4). The data marked with + mean cases that initial bottom slopes above the still water level are steep by cliffs or steps. This figure shows that the calculated values agree well with the measured values. However, since Eq. (4) is intended for the simple beach profile change due to cross-shore sediment transport, Application of Eq. (4) to shores undergone complex beach changes should be preceded by careful study.

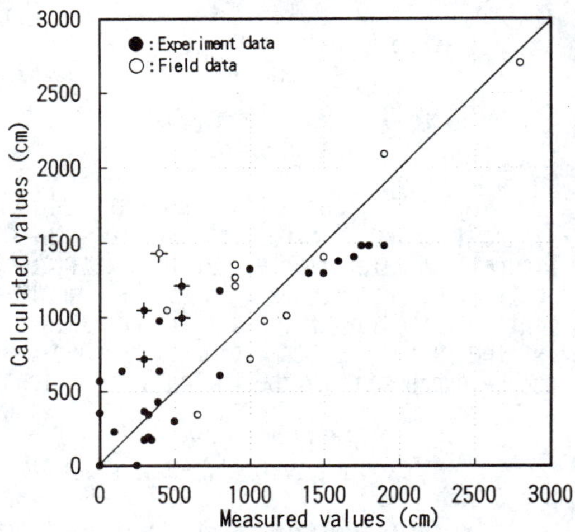

Figure 7 Measured vs. calculated values of Δy.

By combining the above equations with a formula for calculating the stable slopes of sea bottoms, the beach profiles after transformation can be determined.

From Eq. (1), when the beach profile reaches the equilibrium state, the following relation can be obtained :

$$\phi = 0.13\,U_r \qquad (5)$$

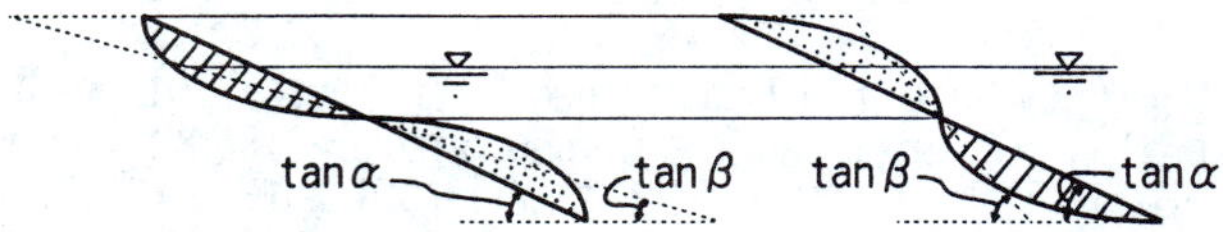

Figure 8 Image of tan β (sea bottom slope under equilibrium state in the surf zone).

Substituting Hallermeier parameter, Ursell parameter and Eq. (6) (Yamamoto, 1988) to Eq. (5), and transforming slightly, we can obtain Eq. (7).

$$H = 2.4(\tan\beta)^{0.3}\,h = 1.9(\tan\beta)^{0.9}H_b \qquad (6)$$

$$\tan\beta = \left(\frac{0.0864\,s g d T^2}{H_b^{\,2}}\right)^{2/3} \qquad (7)$$

where $\tan\beta$ is the sea bottom slope under the equilibrium state in the surf zone, and H_b is the breaking wave height.

Then, assuming that the rate of time change of the sea bottom slope equals $e^{-Bt/T}$, the sea bottom slope of the arbitrary elapsed time t in the surf zone, $\tan\theta$, can be expressed by the following equation :

$$\tan\theta = \tan\beta + \frac{\tan\alpha - \tan\beta}{e^{-Bt/T}} \qquad (8)$$

Moreover, substituting the data shown in Tables 5 and 6 in the appendix, we can obtain Figure 9. The figure indicates that the calculated values agree well with the measured values.

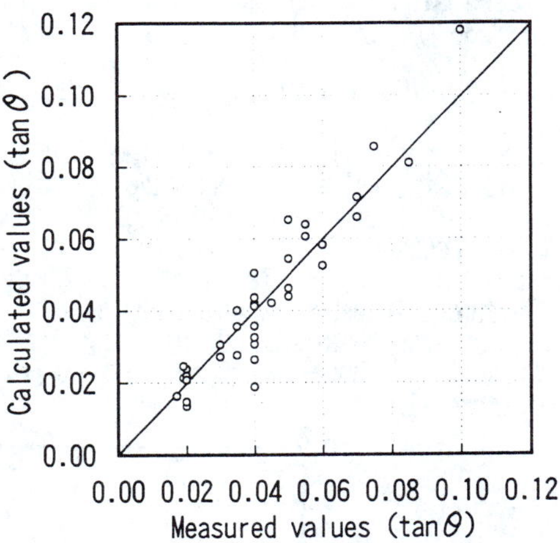

Figure 9 Measured vs. calculated values of tan θ.

4. APPLICATION OF THE PROPOSED EQUATIONS TO ACTUAL COASTS

We performed the following two beach change simulations.

(1) SHORELINE CHANGE AT HAZAKI COAST

The first one is the time series hindcast for 18 days at Hazaki coast in Ibaragi Prefecture, Japan, based on the data given in Katoh and Yanagishima's paper (1988). In their paper, time series data of the daily mean wave energy flux, and limited data of the maximum significant wave height $[(H_{1/3})_{max}]$ and period on stormy days were given. Therefore, time series data of the significant wave height $(H_{1/3})$ from the data of the square root of the daily mean wave energy flux $(E^{1/2})$ were calculated by using the following empirical relations :

$$H_{1/3} = (H_{1/3})_{max}/1.5 \quad (9), \qquad E^{1/2} = (H_{1/3})_{max} \quad (10)$$

Eqeation (10) can be obtained from Figure 10 drawn by limited data under stormy weather. However, as the offshore bars exist and the mean water depth at the bar crown is about 2.9 m, the significant wave height of waves acting on the shoreline is less than approximately 2.2 m due to wave breaking. Therefore, the breaking wave height heigher than 2.2 m is reduced to 2.1 m. Then, the time series data of the wave period are calculated by using Eq. (11) obtained empirically in their paper.

$$H_{1/3}/L^{1/3} = 0.25 \times E^{0.37} \quad (11)$$

where L is the wavelength.

However, as larger waves are diminished in this case, the breaking wave height is cut down, the wave period in this case should be shortened by using the following equation based on Bretschneider's formula [$T = 3.86\,(H_{1/3})^{0.5}$]:

$$T_a/T_b = C\,(2.1/H_{1/3b})^{0.5} \quad (12)$$

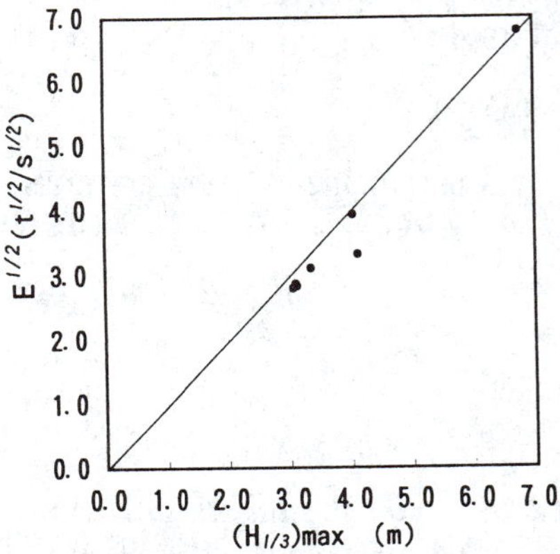

Figure 10 Relation between maximum significant wave height and square root of daily mean wave energy flux.

where the suffix a means the values after wave breaking, while the suffix b means the values before wave breaking, and C is a proportional coefficient (= 1.1 from a few field observation data).

Because the shoreline of Hazaki coast is straight and no coastal structure like a groin exists along this coast, the shoreline change due to longshore sediment transport can be neglected. Therefore, the shoreline change on this coast can be simulated by using Eqs. (1) ~ (4), (7), and (8). The calculated result of shoreline change agrees well with the measured result as shown in Figure 11.

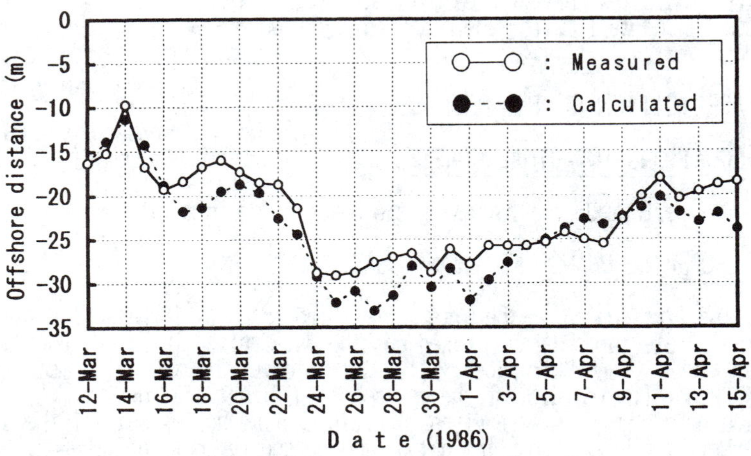

Figure 11 On-offshore changes of shoreline position (D.L.+ 1.4 m).

(2) SHORELINE CHANGE AT MISAWA PORT COAST

Another example is the simulation for one year at Misawa Port coast in Aomori Prefecture, Japan, based on data of Hashimoto and Uda's paper (1979).

A remarkable shoreline change occured during the period of one year from 1976 to 1977, when the offshore breakwater at Misawa Port became long enough to bring about remarkable diffraction effect (refer to Figure 12). It is likely that the diffraction effect of the breakwater made the wave height small in the water area sheltered by the breakwater, thereby the rate of onshore sediment transport increased in this area as shown in Figure 13. Hashimoto and Uda applied an empirical eigenfunction expansion method to predict shore transformation at and around Misawa Port and pointed out the existence of cross-shore sediment transport. Thus, the probable mechanism of this shoreline change was that alongshore transport sediment entered the port area due to influence of the diffraction effect of the breakwater, then the waves in the port transported the sediment onshore, therefore the shoreline advancement occured.

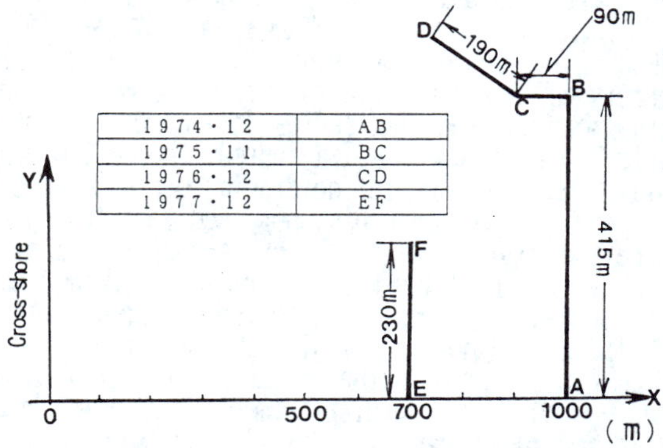

Figure 12 Configuration of breakwaters at Misawa port.

The hindcast of the shoreline change in this area was performed by combining Eqs. (1) ~ (4) with the shoreline change model.

First, the shoreline change model was applied under the following conditions on the basis of the Hashimoto and Uda's paper :

(a) The height of the longshore sediment transport zone was 11 m .

(b) Ozasa and Brampton's formula (1979) was used to calculate the longshore sediment transport rate, and the figure 0.2 was selected as a coefficient in

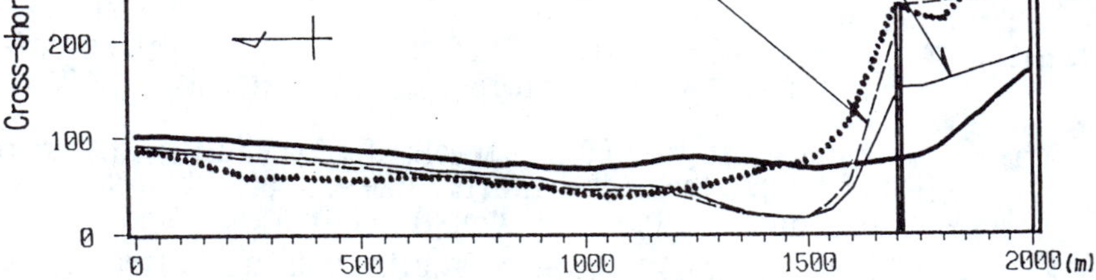

Figure 13 Result of shoreline change simulation.

the formula.

(c) The wave height, period and direction are shown in Table 7 in the appendix.

(d) The sea bottom slope was 1/50.

The result of this simulation was plotted in Figure 13 by a solid line. This result, which was the same as that calculated by Hashimoto and Uda, did not agree with the measured result shown as a dotted line.

Then, the shoreline change due to the cross-shore sediment transport was taken into account by using Eqs. (1) ～ (4). The influence of the wave hysteresis on the shore transformation seems to be great. However, the available data were not the time series data of waves but the statistical data shown in Table 7. Therefore, we adjusted the median grain size of sediment so as to obtain reasonable shoreline displacement. Because the natural beach change after one year induced by cross-shore sediment transport is regarded as small, the shoreline change due to the cross-shore sediment transport near the 0 m point, which is far from the breakwater, can be deemed small. Namely, the median grain size of sediment must be selected so as to obtain a small shoreline change due to the cross-shore sediment transport near the 0 m point. When the figure 0.43 mm was selected as the median grain size from usual grain size at Misawa Port coast, the shoreline displacement near the 0 m point became small as shown in Table 7. Therefore, the calculation by Eqs. (1) ～ (4) was performed under the condition of the median grain size 0.43 mm. By combining this result with the result obtained by the shoreline change model, the significantly improved shoreline displacement was obtained as shown in Figure 13 by a broken line.

5. CONCLUSIONS

Main conclusions are as follows :

(1) The relations among $\tan\alpha$, d/H_0 and the coefficient K of cross-shore sediment transport rate (in Sunamura's formula) can be expressed by Eqs. (2), (3) based on the analyses of data obtained by field observation and large scale experiments.

(2) The formulas [Eqs. (4), (7) and (8)] were proposed for calculating the shoreline displacement and the bottom slope change in the surf zone due to the cross-shore sediment transport. Then the effectiveness of these formulas against actual coasts was demonstrated by the simulation results of shoreline change on two actual coasts.

REFERENCES

Hashimoto, H. and Uda, T. (1979). "The application of empirical shoreline change model to Ogawarako coast", Proc. 26th Japanese Conf. on Coastal Eng., JSCE, pp.215-219 (in Japanese).

Horikawa, K. (1988). "Nearshore Dynamics and Coastal Processes" Univ. of Tokyo Press, 522pp.

Katoh, K. and Yanagishima, S. (1988). "Predictive model for daily changes of shoreline", Proc. 21st International Conf. on Coastal Eng., ASCE, pp.1253-1264.

Ozasa, H. and Brampton, A. H. (1979). "Models for predicting the shoreline evolution of beaches backed by seawalls", Report of The Port and Harbour Research Institute, Vol.18 - no.4, Ministry of Transport, pp.77-104.

Sasaki, M. and Saeki, K. (1974). "Study on wave transformation after wave breaking (2)", Proc. 21st Japanese Conf. on Coastal Eng., JSCE, pp.39-44 (in Japanese).

Sato, S. and Kabiling, M. B. (1994). "Numerical model of 3-dimensional shore transformation including swash zone", Proc. 41st Japanese Conf. on Coastal Eng., JSCE, pp.401-405 (in Japanese).
Shibayama, T., Yamada, M., and Kobayashi, A. (1994). "Shore transformation model around mouth of a river and its verification", Proc. 41st Japanese Conf. on Coastal Eng., JSCE, pp.466-470 (in Japanese).
Sunamura, T. (1984). "Study on cross-shore sediment transport in surf zone including swash zone", Proc. 31st Japanese Conf. on Coastal Eng., JSCE, pp. 316-320 (in Japanese).
Uda, T., Yamamoto, K., and Kouno, S. (1991). "Prediction method of 3-dimensional shore transformation due to longshore sediment transport", Proc. 38th Japanese Conf. on Coastal Eng., JSCE, pp.386-390 (in Japanese).
Uda, T., Yamamoto, Y., Itabashi, N., and Yamaji, K. (1996). "Field observation of movement of sand body due to waves and verification of its mechanism by numerical model", Proc. 25th International Conf. on Coastal Eng., ASCE, pp. 774-775.
Yamamoto, Y. (1988). "On the wave run-up height after wave breaking on a complicated nearshore profile", Proc. of Civil Engineering in the Ocean, Vol.4, JSCE, pp.295-300 (in Japanese).

APPENDIX

Ho : significant wave height in deep water.
T : wave period.
Lo : wavelenght in deep water.
Ho/Lo: wave steepness in deep water.
$\tan\alpha$: initial mean bottom slope in the surf zone.
Hb : breaking wave height.
h : wave setup on shoreline against the still water level.
H : wave height on shoreline against the still water level.
d : median grain size of sediment.
ϕ : Hallermeier parameter.
Ur : Ursell parameter.
Ar : cross sectional area of erosion part near shoreline.
t : observation time.
Q : rate of cross-shore sediment transport.
w : settling velocity of sediment.
K : coefficient in the cross-shore sediment transport rate formula.
Δy : shoreline displacement due to cross-shore sediment transport.
$\tan\beta$: mean bottom slope under the equilibrium state in the surf zone.
$\tan\theta$: mean bottom slope of the arbitrary elapsed time in the surf zone.
α : wave direction.
Tp : frequency of incoming waves.

Suffix c means the calculated value.

Table 1 Field observation data (No.1).

Ho(cm)	T(s)	Lo(cm)	Ho/Lo	tan α	Hb(cm)	h(cm)	H(cm)	d(cm)	Φ	Ur	Ar(cm²)	t(s)	Researcher
100	13.5	28431	0.004	0.040	216	24.4	22.3	0.030	440	6684	-200000	88620	Sonu, 1960
170	6.0	5616	0.030	0.045	219	26.6	25.2	0.030	515	1256	150000	133260	Sonu, 1960
120	8.5	11271	0.011	0.025	179	15.9	12.6	0.030	216	3544	-270000	1209600	Uda, 1982
250	9.0	12636	0.020	0.020	305	24.6	18.2	0.020	439	2398		172800	Sunamura et al., 1983
250	9.0	12636	0.020	0.020	305	24.6	18.2	0.027	325	2398		172800	Sunamura et al., 1983
300	10.0	15600	0.019					0.076			280000	1036800	Takeda et al., 1985
280	9.5	14079	0.020					0.026			130000	259200	Takeda et al., 1985
290	10.0	15600	0.019	0.017	346	26.0	18.3	0.018	463	2646	380000	1036800	Kuriyama, 1991
180	8.5	11271	0.016	0.017	223	16.8	11.8	0.018	298	2965	-220000	1728000	Kuriyama, 1991
130	7.0	7644	0.017	0.017	159	11.9	8.4	0.018	212	2828	-70000	1296000	Kuriyama, 1991
230	9.0	12636	0.018	0.045	337	40.9	38.7	0.026	913	1839	275000	43200	Nairn, 1991
300	10.5	17199	0.017	0.023	388	33.2	25.7	0.020	644	2520	600000	172800	Katoh et al., 1992
300	10.5	17199	0.017	0.023	388	33.2	25.7	0.020	644	2520	1000000	345600	Katoh et al., 1992

Table 2 Field observation data (No.2).

Q(cm²/s)	W(cm/s)	K	d/Ho	t/T	Kc	Qc(cm²/s)	Δyc(cm)	Δy(cm)	Researcher
2.257	4.490	0.0000153	0.00030	6564	0.0000125	-1.846	1207	900	Sonu, 1960
1.126	4.490	0.0000111	0.00018	22210	0.0000146	1.479	1046	450	Sonu, 1960
0.223	4.490	0.0000061	0.00025	142306	0.0000030	-0.111	1403	1500	Uda, 1982
	2.178		0.00008	19200	0.0000707	0.812	1011	1250	Sunamura et al., 1983
	3.558		0.00011	19200	0.0000477	0.093	342	650	Sunamura et al., 1983
0.270	14.793		0.00025	103680			1429	400	Takeda et al., 1985
0.502	3.087		0.00009	27284			973	1100	Takeda et al., 1985
0.367	2.008	0.0000382	0.00006	103680	0.0000580	0.556	1664		Kuriyama, 1991
-0.127	2.008	0.0000274	0.00010	203294	0.0000158	-0.074	1266	900	Kuriyama, 1991
-0.054	2.008	0.0000092	0.00014	185143	0.0000117	-0.068	714	1000	Kuriyama, 1991
6.366	3.087	0.0000287	0.00011	4800	0.0000458	10.158	1416		Nairn, 1991
3.472	2.178	0.0000816	0.00007	16457	0.0000887	3.773	2091	1900	Katoh et al., 1992
2.894	2.178	0.0000680	0.00007	32914	0.0000737	3.136	2700	2800	Katoh et al., 1992

Sunamura et al.'s Δy is the calculated value by using Qc.
Nairn's data is based on information by Kriebel and Dean's paper.

Table 3 Experimental data in large wave tanks (No.1).

Ho(cm)	T(s)	Lo(cm)	Ho/Lo	tan α	Hb(cm)	h(cm)	H(cm)	d(cm)	Φ	Ur	Ar(cm²)	t(s)	Researcher
141	11.3	19920	0.007	0.066	200	31.1	33.0	0.022	1034	4270	300000	144000	Saville, 1957
108	11.3	19920	0.005	0.066	170	26.4	28.1	0.022	879	5024	300000	144000	Saville, 1957
48	11.3	19920	0.002	0.070	110	17.8	19.3	0.022	614	7585	0	144000	Saville, 1957
171	5.6	4892	0.035	0.066	220	34.2	36.3	0.022	1137	953	300000	144000	Saville, 1957
46	6.0	5616	0.008	0.050	100	13.0	12.7	0.047	170	2662	-18000	72000	Simizu et al., 1984
95	9.0	12636	0.008	0.050	128	16.6	16.2	0.047	218	4679	-130000	250200	Simizu et al., 1984
85	3.0	1404	0.061	0.050	88	11.4	11.1	0.047	150	756	0	75600	Simizu et al., 1984
176	6.0	5616	0.031	0.030	199	19.3	16.2	0.047	187	1534	-7000	126000	Simizu et al., 1984
73	9.0	12636	0.006	0.030	147	14.2	11.9	0.047	138	4671	-50000	140400	Simizu et al., 1984
71	3.1	1499	0.047	0.030	82	7.9	6.7	0.047	77	993	-5000	105840	Simizu et al., 1984
96	9.1	12918	0.007	0.050	190	24.6	24.0	0.027	564	3222	55000	255600	Simizu et al., 1984
110	6.0	5616	0.020	0.050	180	23.3	22.8	0.027	534	1479	55000	353160	Simizu et al., 1984
65	12.0	22464	0.003	0.050	160	20.7	20.2	0.027	475	6654	-44000	288000	Simizu et al., 1984
162	3.1	1499	0.108	0.050	160	20.7	20.2	0.027	475	444	25000	273960	Simizu et al., 1984
34	3.5	1911	0.018	0.030	47	4.6	3.8	0.027	77	2209	-4000	360000	Simizu et al., 1984
106	4.5	3159	0.034	0.030	132	12.8	10.7	0.027	216	1300	17000	280800	Simizu et al., 1984
161	3.1	1499	0.107	0.030	159	15.4	12.9	0.027	260	512	16000	360000	Simizu et al., 1984
30	5.8	5248	0.006	0.020	62	5.0	3.7	0.027	66	4896	-12000	313920	Simizu et al., 1984
80	3.1	1499	0.053	0.020	91	7.3	5.4	0.027	97	953	-4000	370800	Simizu et al., 1984
178	5.0	3900	0.046	0.100	191	40.3	48.5	0.027	1400	731	270000	212400	Simizu et al., 1984
110	7.5	8775	0.013	0.100	137	28.9	34.8	0.027	1004	2294	230000	109800	Simizu et al., 1984
150	6.0	5616	0.027	0.075	221	37.6	41.5	0.033	900	1035	70000	15480	Dette et al., 1986
151	5.4	4549	0.033	0.075	211	35.9	39.6	0.023	1258	879	150000	36000	Vellinga, 1986
200	5.6	4892	0.041	0.068	230	36.5	39.1	0.022	1235	901	230000	54000	Kraus et al., 1988
200	5.6	4892	0.041	0.068	230	36.5	39.1	0.022	1235	901	260000	108000	Kraus et al., 1988
200	5.6	4892	0.041	0.064	230	35.0	36.9	0.040	628	923	240000	54000	Kraus et al., 1988
200	5.6	4892	0.041	0.064	230	35.0	36.9	0.040	628	923	190000	108000	Kraus et al., 1988
140	6.0	5616	0.025	0.085	215	40.1	46.0	0.022	1552	1007	200000	15120	Nairn, 1991
150	6.0	5616	0.027	0.125	245	61.6	79.3	0.033	2002	736	135000	2590	Southgate, 1991

Wave heights of Saville's, Simizu et al's, Dette et al's, Kraus et al's, and Southgate's data are not significant values but mean values, because of experiments carried out by using regular waves.

Table 4 Experimental data in large wave tanks (No2).

Q(cm²/s)	W(cm/s)	K	d/Ho	t/T	Kc	Qc(cm²/s)	Δyc(cm)	Δy(cm)	Researcher
2.08	2.354	0.0000153	0.00016	12743	0.0000165	2.25	1479	1750	Saville, 1957
2.08	2.354	0.0000369	0.00020	12743	0.0000117	0.66	1479	1900	Saville, 1957
0.00	2.354	0.0000000	0.00046	12743	0.0000037	-0.26	0	250	Saville, 1957
2.08	2.354	0.0000089	0.00013	25714	0.0000099	2.33	1479	1800	Saville, 1957
-0.25	7.947	0.0000046	0.00102	12000	0.0000019	-0.10	362	300	Simizu et al., 1984
-0.52	7.947	0.0000030	0.00049	27800	0.0000027	-0.46	973	400	Simizu et al., 1984
0.00	7.947	0.0000000	0.00055	25200	0.0000026	0.03	0	0	Simizu et al., 1984
-0.06	7.947	0.0000146	0.00027	21000	0.0000121	-0.05	226	100	Simizu et al., 1984
-0.36	7.947	0.0000027	0.00064	15600	0.0000042	-0.55	604	800	Simizu et al., 1984
-0.05	7.947	0.0000079	0.00066	34142	0.0000029	-0.02	191	330	Simizu et al., 1984
0.22	3.558	0.0000055	0.00028	28088	0.0000056	0.22	633	400	Simizu et al., 1984
0.16	3.558	0.0000021	0.00025	58860	0.0000021	0.16	633	150	Simizu et al., 1984
-0.15	3.558	0.0000015	0.00042	24000	0.0000039	-0.40	566	0	Simizu et al., 1984
0.09	3.558	0.0000014	0.00017	88374	0.0000011	0.07	427	390	Simizu et al., 1984
-0.01	3.558	0.0000015	0.00079	102857	0.0000007	-0.01	171	300	Simizu et al., 1984
0.06	3.558	0.0000149	0.00025	62400	0.0000064	0.03	352	0	Simizu et al., 1984
0.04	3.558	0.0000026	0.00017	116129	0.0000044	0.07	342	330	Simizu et al., 1984
-0.04	3.558	0.0000019	0.00090	54124	0.0000022	-0.04	296	500	Simizu et al., 1984
-0.01	3.558	0.0000109	0.00034	119613	0.0000043	0.00	171	350	Simizu et al., 1984
1.27	3.558	0.0000019	0.00015	42480	0.0000003	0.20	1403	1700	Simizu et al., 1984
2.09	3.558	0.0000065	0.00025	14640	0.0000037	1.18	1295	1500	Simizu et al., 1984
4.52	4.709	0.0000105	0.00022	2580	0.0000185	7.95	714	300	Dctte ct al., 1986
4.17	2.459	0.0000135	0.00015	6667	0.0000230	7.11	1046	300	Vellinga, 1986
4.26	2.354	0.0000153	0.00011	9643	0.0000305	8.52	1295	1400	Kraus et al., 1988
2.41	2.354	0.0000086	0.00011	19286	0.0000168	4.70	1377	1600	Kraus et al., 1988
4.44	6.349	0.0000140	0.00020	9643	0.0000147	4.68	1323	1000	Kraus et al., 1988
1.76	6.349	0.0000055	0.00020	19286	0.0000086	2.73	1177	800	Kraus et al., 1988
13.23	2.354	0.0000290	0.00016	2520	0.0000278	12.68	1207	550	Nairn, 1991
52.12	4.709	0.0000235	0.00022	432	0.0000208	46.27	992	550	Southgate, 1991

Table 5 Field observation data (No3).

tan β (-)	tan θ c(-)	tan θ (-)	Researcher
0.062	0.044	0.040	Sonu
0.021	0.032	0.040	Sonu
0.043	0.040	0.035	Uda
0.017	0.020		Sunamura et al.
0.021	0.020		Sunamura et al.
			Takeda et al.
			Takeda et al.
0.016	0.016	0.017	Kuriyama
0.023	0.021	0.019	Kuriyama
0.028	0.024	0.019	Kuriyama
0.018	0.041	0.040	Nairn
0.015	0.022	0.020	Katoh et al.
0.015	0.021	0.020	Katoh et al.

Table 6 Experimental data in large wave tanks (No3).

tan β (-)	tan θ c(-)	tan θ (-)	Researcher
0.044	0.054	0.050	Saville
0.054	0.060		Saville
0.097	0.085	0.075	Saville
0.015	0.026	0.040	Saville
0.079	0.061	0.055	Simizu et al.
0.097	0.081	0.085	Simizu et al.
0.037	0.042	0.045	Simizu et al.
0.031	0.030	0.030	Simizu et al.
0.081	0.042	0.040	Simizu et al.
0.043	0.036	0.035	Simizu et al.
0.040	0.044	0.040	Simizu et al.
0.025	0.028	0.035	Simizu et al.
0.073	0.064	0.055	Simizu et al.
0.012	0.013	0.020	Simizu et al.
0.073	0.065	0.050	Simizu et al.
0.026	0.027	0.030	Simizu et al.
0.012	0.015	0.020	Simizu et al.
0.098	0.050	0.040	Simizu et al.
0.026	0.024	0.020	Simizu et al.
0.018	0.019	0.040	Simizu et al.
0.048	0.058	0.060	Simizu et al.
0.022	0.066	0.070	Dette et al.
0.016	0.052	0.060	Vellinga
0.014	0.044	0.050	Kraus et al.
0.014	0.031	0.040	Kraus et al.
0.021	0.046	0.050	Kraus et al.
0.021	0.036	0.040	Kraus et al.
0.017	0.072	0.070	Nairn
0.019	0.118	0.100	Southgate

Table 7 Calculated shoreline position at Misawa Port coast.

α	Ho(cm)	T(s)	Lo(cm)	Ho/Lo	Tp(%)	t(s)	t/T	$\tan\alpha$	Hb(cm)	h(cm)	H(cm)	d(cm)	Φ	Ur	d/Ho	Kc	W(cm/s)	Qc(cm /s)	Δyc(cm)
ENE	150	9.0	12636	0.012	25.0	7884000	876000	0.020	208	16.7	12.4	0.043	139	3517	0.00029	0.000000006	7.689	-0.00043	158
ENE	250	9.0	12636	0.020	7.0	2207520	245280	0.020	305	24.6	18.2	0.043	204	2398	0.00017	0.000003362	7.689	-0.11586	1365
ENE	350	9.0	12636	0.028	2.8	883008	98112	0.020	392	31.6	23.5	0.043	263	1863	0.00012	0.000019746	7.689	0.15718	-1006
ENE	450	9.0	12636	0.036	0.2	63072	7008	0.020	474	38.2	28.3	0.043	317	1543	0.00010	0.000062500	7.689	3.31069	-1234
E	150	9.0	12636	0.012	28.0	8830080	981120	0.020	208	16.7	12.4	0.043	139	3517	0.00029	0.000000002	7.689	-0.00017	104
E	250	9.0	12636	0.020	6.0	1892160	210240	0.020	305	24.6	18.2	0.043	204	2398	0.00017	0.000004615	7.689	-0.15901	1481
E	350	9.0	12636	0.028	1.9	599184	66576	0.020	392	31.6	23.5	0.043	263	1863	0.00012	0.000026254	7.689	0.20899	-955
E	450	9.0	12636	0.036	0.1	31536	3504	0.020	474	38.2	28.3	0.043	317	1543	0.00010	0.000064510	7.689	3.41716	-886
ESE	150	7.5	8775	0.017	20.0	6307200	840960	0.020	190	15.3	11.3	0.043	127	2676	0.00029	0.000000008	7.689	-0.00036	128
ESE	250	7.5	8775	0.028	3.0	946080	126144	0.020	278	22.4	16.6	0.043	186	1824	0.00017	0.000009864	7.689	-0.13885	979
ESE	350	7.5	8775	0.040	0.1	31536	4205	0.020	358	28.9	21.4	0.043	240	1417	0.00012	0.000046121	7.689	0.86454	-446
																		Σ	-313

CHAPTER 264

THE INFLUENCE OF ROLLERS ON LONGSHORE CURRENTS

Daniel A. Osiecki[1] and William R. Dally[2]

Abstract

A computational model is developed for depth-averaged cross-shore and longshore currents which includes the effects of the surface roller generated by wave breaking. The creation and evolution of the roller itself is modeled explicitly (Dally and Brown, 1995), and convective acceleration terms are included in both the cross-shore and longshore momentum equations. Lateral mixing is parameterized in terms of the local cross-shore current and a turbulent eddy viscosity, as proposed by Svendsen and Putrevu (1994); however, a new model for eddy viscosity is proposed which contains contributions from both the roller-induced and bed-induced turbulence. The laboratory measurements of quasi-uniform longshore currents reported by Visser (1991) are used to calibrate and verify the model. For driving the model, it is shown that using stream function wave theory produces significantly better results than linear wave theory. Also, comparisons of longshore current distributions with and without the roller terms included show that the roller plays an essential role in faithfully modeling the longshore current. The calibrated model also produces accurate results for the set-up/set-down using stream function theory, for the limited data available from Visser (1991).

Introduction

Recent investigations of the roller have demonstrated its importance in mean cross-shore currents (undertow); e.g., see Svendsen (1984b), Okayasu et al. (1988), Deigaard et al. (1991), and Dally and Brown (1995). We now focus attention on

---

[1] Research Engineer and [2]Associate Professor, Ocean Engineering Program, Florida Institute of Technology, 150 W. University Blvd., Melbourne FL 32901, USA.

the potentially important influence of the aerated roller on longshore currents, both in driving longshore flows as well as in cross-shore mixing.

For shore normal waves, Dally and Brown (1995) showed that the roller momentum flux (R_{xx}) was comparable to the Radiation Stress (S_{xx}) associated with the organized wave motion, and that their gradients tend to balance in the outer surf zone. This balance is responsible for the landward shift in the initiation of setup, as well as the landward shift in the location of the peak of the undertow, that have been observed in laboratory data (Bowen, et al., 1968; Nadaoka and Kondoh, 1982). In the present study it will be shown that with obliquely incident waves, the gradients of the cross-product momentum terms (R_{xy} and S_{xy}) also tend to balance in the outer surf zone. This results in a landward shift in the peak of the longshore current, which is also evident in laboratory data (Visser, 1991).

Although lateral mixing of the longshore current has been an important modeling issue since the 1970's, it was not until recently that a physically realistic mechanism for observed cross-shore mixing, i.e. residual convective acceleration induced by the vertical structure of the cross-shore and longshore currents, was proposed by Svendsen and Putrevu (1994). In the present study the influence of the roller, which enhances vertical mixing and consequently *reduces* lateral mixing, will also be modeled and explored.

General Governing Equations for Longshore and Cross-shore Currents

Assuming steady currents, longshore uniformity, and no flow through the shoreline, a simple equation for the mean, depth-integrated, period-averaged, cross-shore current below the mean water level (U) is developed by integrating the conservation of mass equation. This relation is given by

$$U = \frac{-(Q_{wx} + Q_{rx})}{(h + \overline{\eta})} \tag{1}$$

The Stokes Drift term (Q_{wx}) is provided by the wave theory selected to drive the model, the roller volume flux (Q_{rx}) is provided by an energy-based roller model, and the still water depth (h) is known. The elevation of the mean water level ($\overline{\eta}$) is as yet unknown.

Again, assuming steady state conditions, applying longshore uniformity, and neglecting wind stress, the period-averaged, depth-averaged momentum equations for the cross-shore and longshore directions are

$$-\frac{\partial}{\partial x}\left[U^2(h + \overline{\eta})\right] + \frac{1}{\rho}\left(\frac{\partial}{\partial x}S_{xx} + \frac{\partial}{\partial x}R_{xx}\right) + g(h + \overline{\eta})\frac{\partial \overline{\eta}}{\partial x} = \frac{1}{\rho}\left(\overline{\tau_{zx}\big|_{-h}}\right) \tag{2}$$

and

$$\frac{\partial}{\partial x}\left(D_c h \frac{\partial V}{\partial x}\right) - \frac{1}{\rho}\left(\frac{\partial}{\partial x} S_{xy} + \frac{\partial}{\partial x} R_{xy}\right) = \frac{1}{\rho}\left(\overline{\tau_{zy}}\Big|_{-h}\right) \tag{3}$$

where S is the Radiation Stress, R is the roller momentum flux, D_c characterizes a horizontal mixing mechanism, τ is the time-averaged bed stress, and V is the depth-averaged longshore current. The first term in Eq. 3 represents a lateral mixing mechanism due to the vertical structure of the cross-shore and longshore currents (Svendsen and Putrevu, 1994) with D_c approximated by,

$$D_c = \frac{0.5\, h^2 U^2}{\nu_t} \tag{4}$$

where ν_t is a turbulent eddy viscosity. The eddy viscosity model developed herein contains vertical mixing introduced by both the aerated surface roller and the bottom, as described below.

A quadratic bed stress model is adopted:

$$\overline{\tau_{zx}}\Big|_{-h} = \rho \frac{c_f}{8} \overline{u_b \left(u_b^2 + v_b^2\right)^{\frac{1}{2}}} \tag{5}$$

$$\overline{\tau_{zy}}\Big|_{-h} = \rho \frac{c_f}{8} \overline{v_b \left(u_b^2 + v_b^2\right)^{\frac{1}{2}}} \tag{6}$$

where u_b and v_b are the total instantaneous velocities at the bed. Following Smith et al. (1993) c_f is the Darcy-Weisbach friction factor, related to Manning's n_m and the total water depth by

$$c_f = \frac{8 g\, n_m^{\ 2}}{\left(h + \overline{\eta}\right)^{\frac{1}{3}}} \tag{7}$$

Manning's resistance coefficient has units of ($s{*}m^{-1/3}$) and was determined experimentally by Chow (1959) for a wide variety of channels.

To solve the system of equations, the initial conditions of wave height, wave angle, and bottom topography must be specified. The selected wave theory is then used to determine the mass and momentum fluxes while shoaling and refracting the wave to the next grid. The cross-shore mass and momentum equations are next solved for the undertow and the wave induced set-up/set-down. By iterating between the mass equation, the cross-shore momentum equation, and the roller model, the undertow and the set-up/set-down are determined across the entire transect. The present model employs the wave height decay model of Dally, Dean, and Dalrymple (1985), hereafter referred to as the D^3 model.

Once the cross-shore hydrodynamics are computed, the longshore momentum equation is solved with a forward difference scheme. The longshore momentum equation is solved from onshore to offshore because the boundary condition of V=0 must be applied at the shoreline. Equation (3) is solved across the entire transect using an implicit tridiagonal solution method. Because there is a quadratic dependence on V in the longshore mean bed stress, the longshore momentum equation is iterated until V converges to a selected tolerance for all points across the transect. The solution method outlined in this section is the same regardless of the wave theory used to drive the model.

The Roller Model

The roller model recently developed by Dally and Brown (1995) is used to predict the development and decay of the aerated surface roller in the surf zone, and is based on a depth-integrated and period-averaged energy balance equation. For obliquely incident waves, the roller model, under the condition of longshore uniformity, is expressed by

$$\frac{\partial \overline{F_w} \cos\theta}{\partial x} + \frac{\partial}{\partial x}\left(\tfrac{1}{2} M_R c^2 \cos^2\theta\right) = -M_R g \beta_d \cos\theta \tag{8}$$

where $\overline{F_w}$ is the depth-integrated, time-averaged, wave-induced energy flux, x is the cross-shore coordinate positive in the shoreward direction, M_R is the roller mass flux, θ is the local wave angle (computed from Snell's law), g is gravity, and β_d is the roller dissipation coefficient. β_d was calibrated by Dally and Brown (1995) using a number of laboratory data sets, and a value of 0.1 yields good agreement when stream function theory is used to compute the driving terms. The volume flux of the roller, used in the conservation of mass equation (Eq. 1), is calculated from

$$Q_{rx} = \frac{M_R}{\rho}\cos\theta \tag{9a}$$

$$Q_{ry} = \frac{M_R}{\rho}\sin\theta \tag{9b}$$

The momentum flux terms for the roller in used in Eqs. 2 and 3 are calculated from

$$R_{xx} = \rho Q_{rx} c \cos\theta \tag{10a}$$

$$R_{xy} = \rho Q_{rx} c \sin\theta = \rho Q_{ry} c \cos\theta \tag{10b}$$

The Eddy Viscosity Model

It is suggested that two sources of turbulence, i.e. roller-induced turbulence and bottom-induced turbulence, should be accounted for in the eddy viscosity model. The measurements of Battjes and Sakai (1981) and Nadaoka and Kondoh (1982) suggest that as a wave dissipates across the surf zone, the turbulence it creates and leaves behind is similar to a turbulent wake. Consequently, an analogy is made that the turbulent wake produced by the roller of a breaking wave is similar to the turbulent wake produced by a cylinder submerged in a moving fluid. From the theory of turbulent wakes, the eddy viscosity in a turbulent wake (ν_T) can be described by (Tennekes and Lumley, 1977)

$$\nu_T = \frac{\kappa}{R_T} U_o \Theta \tag{11}$$

where κ is Von Karman's constant (equal to 0.4), R_T is the turbulent Reynolds number (equal to 12.5 based on experimental observations), U_o is the free stream velocity, and Θ is the momentum thickness of the wake. For a coordinate system moving with the breaking wave, it would seem that the wave celerity (c) is analogous to U_o. Tennekes and Lumley (1977) show that the momentum thickness for a circular cylinder is approximately half the frontal height of the cylinder, for Reynolds numbers between 10^3 and 3×10^5. Thus, if the roller is equated to a cylinder, a logical estimate of the momentum thickness, or '*radius*', of the roller (Θ_R) is given by

$$\Theta_R = \frac{1}{\sqrt{\pi}} \sqrt{M_R \frac{T}{\rho}} \tag{12}$$

Completing the cylinder analogy, the roller contribution to the eddy viscosity is given by

$$\nu_R = \frac{0.4}{12.5} c \Theta_R \tag{13}$$

where ν_R is uniform over depth, which also follows from turbulent wake theory.

With the upper layer turbulence represented, the bottom-induced turbulent eddy viscosity is now developed. Nielsen (1985) suggests that the length scale of the near-bed eddies should be that of the oscillatory water particle excursion, whereas the velocity scale should be that of a friction velocity. We also suggest that the friction velocity is well-represented by the maximum water particle velocity multiplied by a friction factor. Consequently, it appears that the turbulent eddy viscosity contribution from the bed might be modeled by

$$\nu_B = \sqrt{\frac{c_f}{2}} \left| u_{max} \right| \xi_{max} \tag{14}$$

$$u_{max} = \frac{H\sigma}{2 \sinh kh} \tag{15}$$

$$\xi_{max} = \frac{H}{2} \frac{1}{\sinh kh} \tag{16}$$

where, c_f is the friction factor discussed in the previous section, u_{max} is the maximum orbital velocity given by linear wave theory in equation (15), ξ_{max} is the maximum water particle displacement given by linear wave theory in equation (16), H is wave height, σ is angular frequency, and k is wave number. With the bottom-induced eddy viscosity component modeled, the combination of mixing induced by the roller of the breaking wave and mixing induced by bottom turbulence is given by

$$\nu_T = \left(\nu_R^2 + \nu_B^2 \right)^{\frac{1}{2}} \tag{17}$$

A comparison of the new eddy viscosity model with the eddy viscosity model proposed by Svendsen and Putrevu (1994) (Figure 1) shows an order of magnitude agreement. The new eddy viscosity model does provide a better behaved transition across the breakpoint, with the point of maximum turbulence located inside the surf zone, whereas the Svendsen and Putrevu (1994) eddy viscosity model has a maximum at the breakpoint.

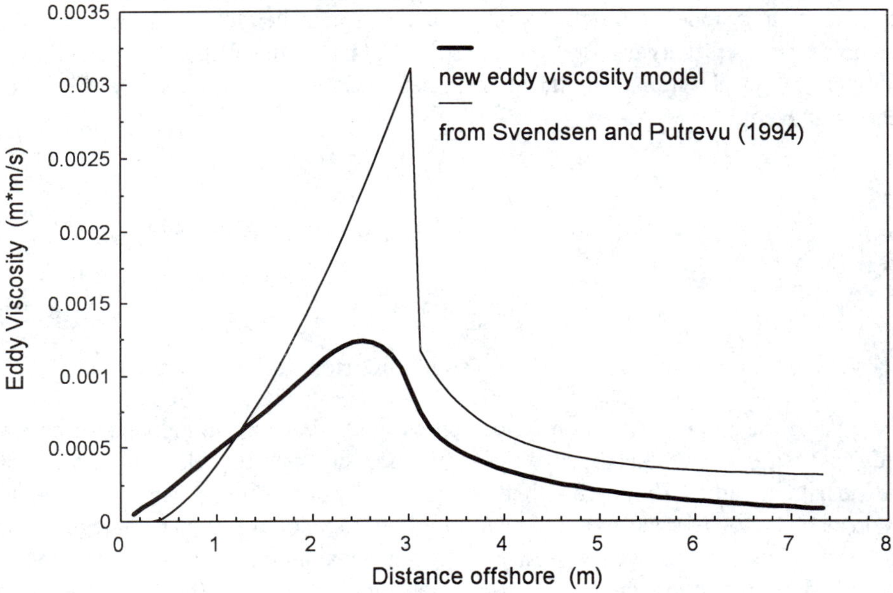

Figure 1 Comparison of the new eddy viscosity model to Svendsen and Putrevu's (1994) eddy viscosity model for Visser's (1991) Case 4 conditions.

Calibration

The numerical model presented here is compared to the laboratory data set of Visser (1991) for calibration and verification. This data set is unique because it is the only experiment to minimize the recirculation present in enclosed basins by utilizing a longshore pumping method, to develop a uniform longshore current. Calibration of the longshore current model entailed the selection of a Manning's friction factor from a range of values presented by Chow (1959) for open channel flow conditions. The empirical decay coefficient (κ) and the stable wave height coefficient (Γ), in the D^3 wave decay model are also calibrated to minimize the error in wave height prediction. Manning's friction factor and the D^3 coefficients are selected based on their relative least squares error to the measured data defined by equation (18) where B represents the variable in question:

$$E_v = \left(\frac{\sum \left(B_{meas} - B_{calc} \right)^2}{\sum \left(B_{meas} \right)^2} \right)^{\frac{1}{2}} \tag{18}$$

Calibration results for the longshore current model are presented for two wave theories in tables 1 and 2. The numerical model results with stream function theory are presented in contrast to the model results for linear wave theory.

	D^3 Coefficients		Least squares error in wave height	
Case #	κ	Γ	*Stream Function Theory*	*Linear wave theory*
4	0.15	0.2	7.89 %	5.80 %
5	0.15	0.4	12.45 %	16.70 %
6	0.2	0.3	4.09 %	4.51 %
7	0.2	0.2	9.06 %	7.03 %

Table 1 Selected D^3 wave decay model coefficients and relative least squares error from Visser's (1991) measured data.

Least squares error in longshore current velocity			
Case #	$\mathbf{n_m = 0.01}$	$\mathbf{n_m = 0.011}$	$\mathbf{n_m = 0.013}$
Stream Function Theory			
4	9.42 %	6.85 %	17.13 %
5	19.10 %	13.05 %	16.92 %
6	25.35 %	26.65 %	31.92 %
Linear Theory			
4	19.02 %	20.16 %	25.29 %
5	10.85 %	8.84 %	16.71 %
6	33.13 %	34.97 %	39.59 %

Least squares error in longshore current velocity			
Case #	$\mathbf{n_m = 0.015}$	$\mathbf{n_m = 0.017}$	$\mathbf{n_m = 0.02}$
Stream Function Theory			
7	59.32 %	43.89 %	25.75 %
Linear Theory			
7	43.27 %	31.35 %	17.01 %

Table 2 Least squares percent error for the range of Manning's friction coefficients (n_m) from Chow (1959) for Visser's (1991) cases 4 through 7.

Because Visser did not measure the cross-shore current distribution, it is impossible to asses the validity of the cross-shore velocity predictions. However, based on the work of Dally and Brown (1995), the cross-shore current is expected to be reasonably represented. Sensitivity testing of the longshore current model shows that the magnitude of the longshore current is not overly dependent on Manning's friction factor. For the smooth bottom, cases 4-6, the normal friction factor value provides accurate results; however, for the gravel bottom in case 7 a value larger than Chow's (1959) suggested maximum provides better results.

Visser's experiments 1-3 were performed on a beach with a slope of 1/10. Because stream function theory is limited by the breaker height to water depth ratio of 0.78, cases 1-3 were not investigated because the breaking waves in these experiments exceeded this ratio. Because case 8 was reported without wave height measurements, this case was also discarded.

Results and Discussion

The most obvious way to asses the effect of the roller on the longshore current is to contrast the results to a model without the roller term. Figure 2 shows the calibrated longshore current model results for Case 4, with and without the roller term in the cross-shore and longshore momentum equations. It is noted that the roller contribution to turbulence is still included. The model that neglects the roller term in the momentum equations shows a dramatic decrease in the longshore current velocity inside the surf zone and the location of the maximum current is shifted offshore. A similar effect occurs in the model predictions for all the cases studied.

To asses the importance of the roller contribution to turbulence in the surf zone, results are presented that neglect the roller contribution to the eddy viscosity model. It is noted that the roller contribution to the longshore momentum equation is included in this comparison so that the effect of the roller turbulence on lateral mixing may be examined exclusively. Figure 3 clearly shows that the roller-induced turbulence has a significant effect on the overall longshore current distribution. The absence of the roller turbulence in the eddy viscosity model decreases the maximum velocity in the longshore current and shifts this point landward. Also, outside the breakpoint, a slight increase in the longshore current is produced without the roller-induced turbulence included.

(a)

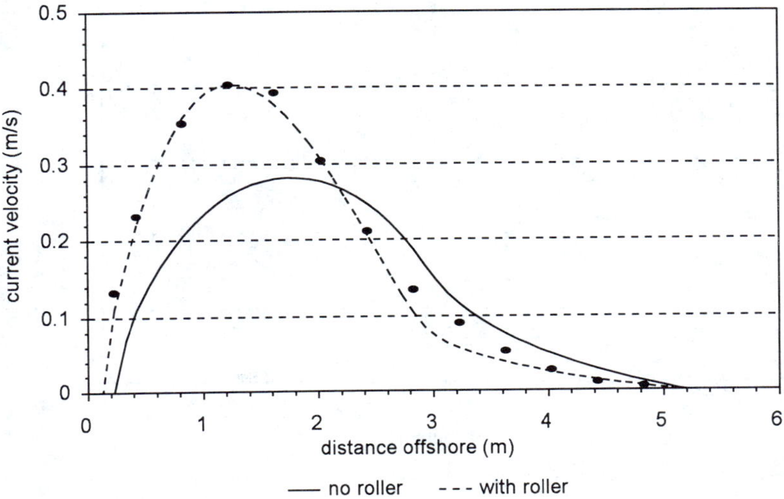

(b)

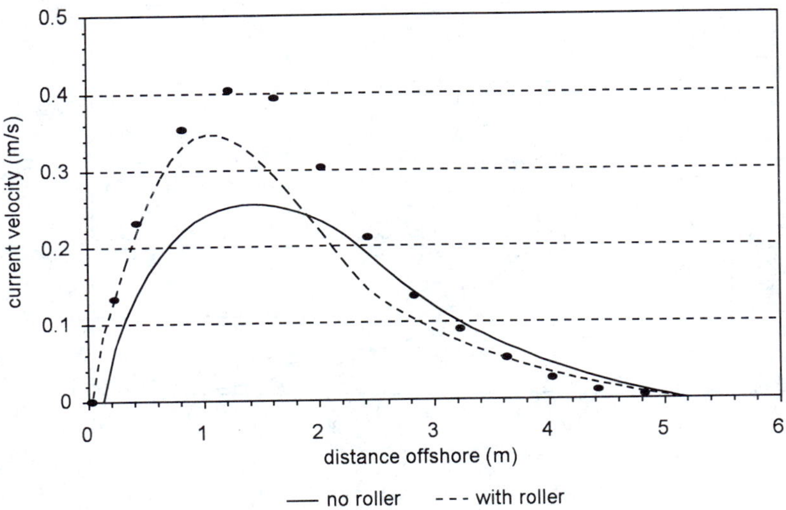

Figure 2 Visser's (1991) Case 4 model comparisons with and without the roller contribution to the longshore momentum equation for (a) Stream Function theory and (b) Linear Wave theory.

(a)

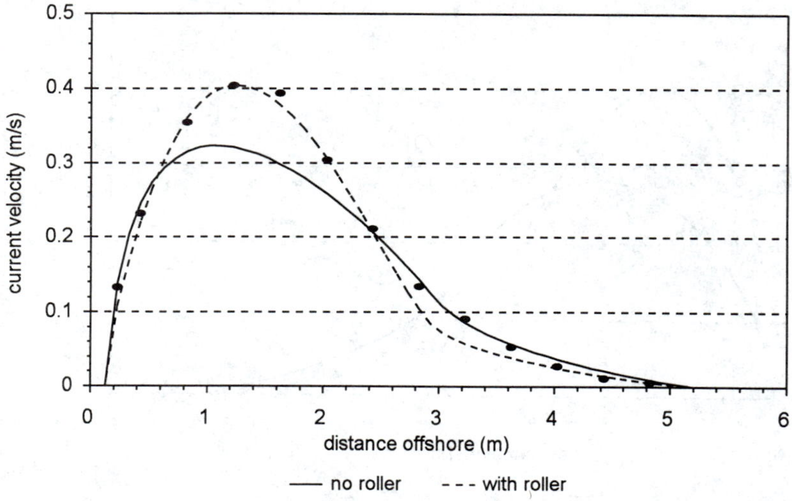

(b)

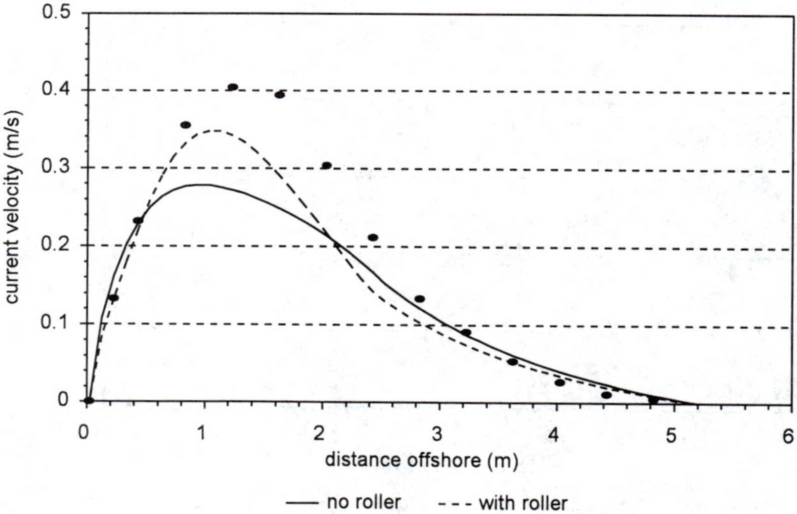

Figure 3 Visser's (1991) Case 4 model comparisons with and without the roller contribution to the eddy viscosity model for (a) Stream Function theory and (b) Linear Wave theory.

It is important to note that the predicted cross-shore current driven by linear wave theory is much greater than the stream function theory cross-shore current. According to Dally and Brown (1995) linear theory in general overpredicts the Stokes drift, whereas stream function theory provides a better prediction of this wave-induced mass transport. Although Visser (1991) did not measure the cross-shore velocity component, it is reasonable to conclude that the hydrodynamic model that uses stream function theory is the better model for calculating the cross-shore and longshore currents.

Conclusions

The major findings of this investigation are as follows: (1) The nonlinear mixing term described by Svendsen and Putrevu (1994) provides an adequate amount of lateral mixing, with realistic eddy viscosity values, in comparison to the laboratory data for longshore currents collected by Visser (1991). (2) The addition of roller momentum to the longshore equation of motion is necessary for accurate longshore velocity predictions. (3) The contribution of turbulence from the roller is an important aspect of realistically modeling the turbulent eddy viscosity. These findings affirm the premise that the roller plays a significant role in driving the longshore current and in shifting the maximum longshore velocity inside the breakpoint.

Acknowledgments

Funding for this research was provided in part by the Sebastian Inlet Tax District Commission.

References

Battjes, J.A., and T. Sakai, Velocity field in a steady breaker, *J. Fluid Mech.*, 111, 421-437, 1981.

Bowen, A.J., D.L. Inman, and V.P. Simmons, Wave 'set-down' and set-up, *J. Geophys. Res.*, 73(8), 2569-2577, 1968.

Chow, V.T., *Open channel hydraulics*, McGraw-Hill, New York, 1959.

Dally, W.R. and C.A. Brown, A modelling investigation of the breaking wave roller with application to cross-shore currents, subm. to *J. Geophys. Res.*, 100(C12), 24,873-24,883, 1995.

Dally, W.R., R.G. Dean, and R.A. Dalrymple, Wave height variation across beaches of arbitrary profile, *J. Geophys. Res.*, 90(C6), 11917-11927, 1985.

Dean, R.G., Evaluation and development of water wave theories for engineering applications, Special Report No. 1, U.S. Army Waterw. Exp. Stn., Vicksburg, Miss., 1974.

Deigaard, R., P. Justesen, and J. Fredsøe, Modeling of undertow by a one-equation turbulence model, *Coastal Eng.*, 15, 431-458, 1991.

Nadaoka, K. and T. Kondoh, Laboratory measurements of velocity field structure in the surf zone by LDV, *Coastal Eng. Jpn.*, 25, 125-145, 1982.

Nielsen, P., A short manual of coastal bottom boundary layers and sediment transport, Public Works Dept. of New South Wales, TM 85/1, 1985.

Okayasu, A., T. Shibayama, and K. Horikawa, Vertical variation of undertow in the surf zone, *Proc. 21st Conf. Coast. Engr.*, 478-491, 1988.

Putrevu, U., and I.A. Svendsen, A mixing mechanism in the nearshore region, *Proc. 23rd Conf. Coast. Engr.*, 2758-2771, 1992.

Smith, J.M., M. Larson, and N. Kraus, Longshore current on a barred beach: field measurements and calculation, *J. Geophys. Res.*, 98(C12), 22,717-22,731, 1993.

Svendsen, I.A., Mass flux and undertow in a surf zone, *Coastal Eng.*, 8: 347-365, 1984b.

Svendsen, I.A. and U. Putrevu, Nearshore mixing and dispersion, *Proc. R. Soc. Lon.*, 445, 561-576, 1994.

Tennekes, H., and J.L. Lumley, *A first course in turbulence*, MIT Press, 1977.

Visser, P.J., Laboratory measurements of uniform longshore currents, *Coastal Eng.*, 15, 563-593, 1991.

Visser, P.J., Uniform longshore current measurements and calculations, *Proc. 19th Conf Coast. Eng.*, 2192-2207, 1984.

CHAPTER 265

Velocity Field Measurements in a "Coastal Buffer Zone"

Hiroshi Yagi[1], Hirofumi Hinata[1] and Kazuo Nadaoka[2]

Abstract

To investigate the characteristics of flow field in the coastal buffer zone (C.B.Z.), velocity measurements in a field were performed both in non-stratified and stratified conditions. The results of the data analysis have shown that in the non-stratified condition wind action, together with tidal motion, may induce unexpectedly large velocity even near the bottom with the magnitude beyond the threshold of the sediment movement, suggesting that the appreciable sediment transport may be caused by these wind-driven currents in C.B.Z.. In the stratified condition, it is revealed that internal tides with baroclinic velocity fluctuation may dominate the cross-shore exchanging motion of water mass.

INTRODUCTION

In coastal current system, the surf zone and the far-offshore region may be regarded as two extremes in a sense that the former is dominated exclusively by nearshore currents, while the latter is governed mostly by the ocean currents (Fig.1). The intermediate-depth zone, on the other hand, may be affected by both the factors as well as tidal currents, wind-driven currents, density currents, river flow, and hence may be called "coastal buffer zone (C.B.Z.)". Therefore for developing a reliable tool to describe the flow field in this zone, more integrated knowledge on these factors and, if any, the interactions among them are necessary. However, still limited parts

---

[1] Department of Civil Engineering, Tokyo Institute of Technology, 2-12-1, O-okayama, Meguro-ku, Tokyo 152, Japan.

[2] Department of Mechanical and Environmental Informatics, Graduate School of Information Science and Engineering, Tokyo Institute of Technology, 2-12-1, O-okayama, Meguro-ku, Tokyo 152, Japan.

of the actual physical process of these factors are known for this purpose (Sato,1995; Yasuda et al.,1995). Hence, in the present study, field measurements in a C.B.Z. have been performed to clarify the fundamental characteristics of its velocity field.

OUTLINE OF FIELD OBSERVATION

The observation was curried out in the Kashima coast, as a typical open coast in Japan, which has a straight sandy beach facing the Pacific Ocean (Fig.2(a)). To clarify the characteristics of the current system in the C.B.Z. both in the non-stratified and stratified conditions, field observations were performed in the periods from Nov. 6 to 28 in 1994 and from July 19 to Aug. 20 in 1995.

In the field measurements, three observation stations (St.1-3) were set to be aligned in the cross-shore direction as shown in Fig.2(b). At these stations electromagnetic current meters, pressure-type wave gauges and thermometer were installed. The location of observation stations and the arrangement of measuring instruments were slightly changed between the observations in the non-stratified and stratified conditions as shown in Fig.2(b).

RESULTS AND DISCUSSTION

Non-stratified condition

Figure 3 shows the time histories, at the three stations, of the mean water surface elevation and the shore-parallel component of the mean velocity close to the bottom (0.7-1.7m above the bottom). These mean values were obtained by a moving time averaging over one hour. The regular diurnal tidal fluctuations appear in the mean surface elevation. On the other hand, the shore-parallel bottom velocity shows appreciable irregularity and its maximum magnitude attains almost 40cm/s at St.1. This is an unexpectedly large value, and such a large near-bottom velocity is observed also by Sato (1995) at the Ishikawa coast facing the Sea of Japan. The velocity of 40cm/s is beyond the threshold for the sediment movement, which is estimated as about 25cm/s by the Shields experimental curve for the sand with the mean diameter of 0.2mm. This fact suggests that in C.B.Z. appreciable sediment transport may be caused by these currents, although conventionally significant sediment transport has been believed to occur mostly in the surf zone and its vicinities.

To find the causes of the large velocities near the bottom, the non-tidal irregular components were extracted from the near bottom velocity data by a moving time average over 24 hours. The results are shown in Fig.4, in which the long-period fluctuations with the period of 6 or 7 days appear at each station. Although the velocity amplitude at St.1 in the shallower region is larger than that at St.3 in the deeper region, their fluctuation patterns are very similar each other. This fact suggests that there exists a common mechanism to cause these long-period velocity fluctuations in the C.B.Z.. To find the mechanism, the wind speed, which is indicated with a thick line in Fig.4 (wind-speed data are partly lacked because of trouble in the measurement), is compared with these long-period velocities.

Obviously there is a close correlation among them. Hence, it is suggested that the wind may induce the long-period fluctuating currents with an appreciable magnitude even near the bottom.

To obtain more quantitative understanding of the wind-driven currents in the C.B.Z., its numerical simulation has been done. The governing equations of the computations are 2-DV momentum equations with the coriolis force and the continuity equation. The k-ε model described below has been employed to evaluate the vertical eddy viscosity ν_t.

$$\frac{\partial k}{\partial t} + u\frac{\partial k}{\partial x} + w\frac{\partial k}{\partial z} = \frac{\partial}{\partial z}\left(\frac{\nu_t}{\sigma_k}\frac{\partial k}{\partial z}\right) + P_k - \varepsilon\ , \qquad (1)$$

$$\frac{\partial \varepsilon}{\partial t} + u\frac{\partial \varepsilon}{\partial x} + w\frac{\partial \varepsilon}{\partial z} = \frac{\partial}{\partial z}\left(\frac{\nu_t}{\sigma_\varepsilon}\frac{\partial k}{\partial z}\right) + C_{\varepsilon 1}\frac{\varepsilon}{k}P_k - C_{\varepsilon 2}\frac{\varepsilon^2}{k}, \qquad (2)$$

$$\nu_t = C_\mu \frac{k^2}{\varepsilon}, \qquad (3)$$

where x and z represent the cross-shore and vertical coordinate, u and w the velocity components in the x- and z-direction, k the kinetic energy of turbulence, ε the dissipation rate of the turbulence kinetic energy, and P_k the production rate of k. C_μ, $C_{\varepsilon 1}$, $C_{\varepsilon 2}$, σ_k and σ_ε are the turbulence model parameters, for which the standard values (C_μ=0.09, $C_{\varepsilon 1}$=1.44, $C_{\varepsilon 2}$=1.92 and $\sigma_t = \sigma_\varepsilon$=1.0) are adopted here. In the present computation, the shore-parallel uniformity is assumed and the topography of the computational area is simplified as shown in Fig.5. The spatially uniform wind shear stress at the water surface was given, being based on the wind data obtained at the nearshore observation station (as shown in Fig.4) and the friction factor at the water surface (C_d=0.002).

Figure 6 shows the computed vertical profiles of the shore-parallel velocity for the three cases of the water depth; i.e. h=20, 60, 100m. The indicated results are those after 9.5 days from the beginning of the computation. In the case of h=100m, the vertical gradient is relatively large in the surface layer, while it becomes small beneath it. This fact suggests that the effect of wind shear stress is limited to the surface boundary layer, and the geostrophic current under the balance between the coliolis force and the cross-shore gradient of the water surface elevation is formed beneath the surface layer. On the other hand, in the case of h=20m, the influence of wind shear stress reaches the bottom so that the large vertical gradient of velocity is induced through the entire depth. This means that in the present observation area, because of its shallow water depth less than 20m, the large bottom velocity may be induced by the wind shear stress. These facts indicate that the ratio between the surface Ekman boundary layer width δ_E and the local water depth h, δ_E/h, is one of the principal factors for the overall understanding of the velocity field of a C.B.Z..

Figure 7 represents the computed time histories of shore-parallel bottom velocities at St.1 to 3, which show the long-period fluctuation similar to those found in Fig.4. However, the magnitude of the computed velocity, except for St.3, is smaller than that of the measured, and the velocity differences among the three stations are not well reproduced. One of the possible causes of these discrepancies is the difference between actual and simulated turbulence structure of the wind-driven currents. Besides the offshore wave breaking due to wind may act to augment the bottom velocity (Yasuda et al., 1995).

Stratified condition

Figure 8 (a) and (b) show respectively the shore-parallel and cross-shore components of the bottom velocity fluctuation in the case of the stratified condition at St.2, where the water depth is 13m and the distance from the shoreline is 1.5km. The long-period fluctuation as found in the non-stratified condition dose not appear in this case. However, more conspicuous feature in the time variation characteristics can be found; namely, as compared with those in the non-stratified condition, these velocity data show high frequency fluctuation with appreciable irregularity, especially in the cross-shore component. To examine the frequency characteristics of these velocity fluctuations, the frequency spectra have been computed as shown in Fig.9, which indicate that in the shore-parallel component the diurnal tidal fluctuation has dominant contribution to the spectrum both in the surface and bottom velocities. In the cross-shore velocity component, on the other hand, the diurnal tidal fluctuation prevails in the surface velocity, while in the bottom velocity the semi-diurnal fluctuation arises with appreciable magnitude.

These diurnal and semi-diurnal velocity fluctuations were further examined with the water surface tidal fluctuation by extracting them from the raw data by applying a band-pass filtering. The results are shown in Fig.10 and indicate that the diurnal components of the shore-parallel velocity in the surface and bottom layer are nearly in phase. This means that the shore-parallel velocity is governed mainly by the *barotropic* tidal motion. On the other hand, the semi-diurnal components of the cross-shore velocity in the surface and bottom layer are mutually 180 degrees out of phase. These facts suggest that the *baroclinic* tidal motion is dominant in the cross-shore direction.

To examine the cross-shore barotropic motion in more detail, the water temperature variations both in time and space has been investigated. The field data for this purpose was obtained with the vertical arrays of several thermometers at the three stations. Figure 11 represents a typical example of the time sequence of the spatial distribution of the water temperature. It can been seen that the bottom water body with relatively low temperature runs up and down on the sloping sea bottom with the excursion length of about 1 km for the front movement. Corresponding to this bottom water motion, the upper water body with higher temperature shows cross-shore oscillating motion in the opposite phase from the bottom water motion.

From these results, the velocity field of the C.B.Z. in the stratified condition may be summarized as follows; i.e., in the shore-parallel direction, the barotropic tidal motion is dominant, while in the cross-shore direction the baroclinic tidal fluctuation is significant. This difference may be explained by the role of the shore boundary; i.e. it prevents the barotoropic tidal motion exclusively in the cross-shore direction. However, the baroclinic cross-shore tidal motion appears as running up and down motion near the shore with large excursion length. Since this baroclinic tidal motion results in appreciable cross-shore exchange of water mass, it may have a dominant role in the cross-shore mass transfer in C.B.Z..

CONCLUSIONS

The major conclusions obtained in the present study for C.B.Z. are summarized as follows:

1. In the non-stratified condition, wind action may induce unexpectedly large velocity even near the bottom with the magnitude beyond the threshold of the sediment movement. This fact suggests that the wind-driven current may cause appreciable sediment transport in the C.B.Z..
2. The ratio between the surface Ekman boundary layer width δ_E and the local water depth h is one of the principal factors for overall understanding of the velocity field in C.B.Z..
3. In the stratified condition, internal tide with baroclinic velocity fluctuation may dominate the cross-shore exchanging motion of water mass.

References

Sato, S. (1995): Characteristics of wave and current induced by low pressures in Japan Sea, Journal of Hydraulic, Coastal and Environmental Engineering, JSCE, No.521, 113-122. (*in Japanese*)

Yasuda, T., Mori, N., Kato, S. and Sato,S. (1995): Currents off the Ishikawa coast and their breaker effects, Proc. of Coastal Engineering, JSCE, Vol.42, 431-435. (*in Japanese*)

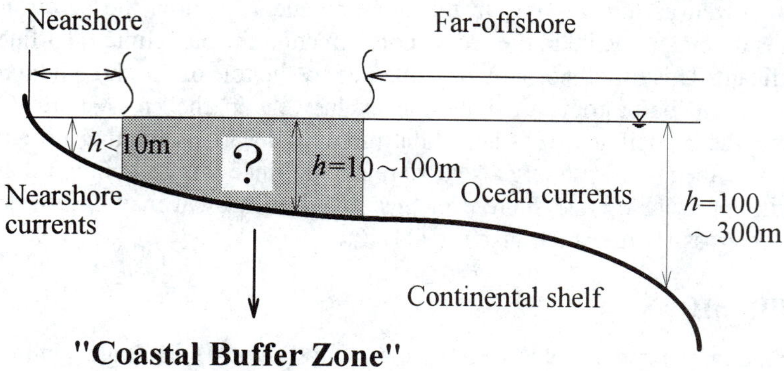

Fig.1 Conceptual illustration of coastal buffer zone in an open coast.

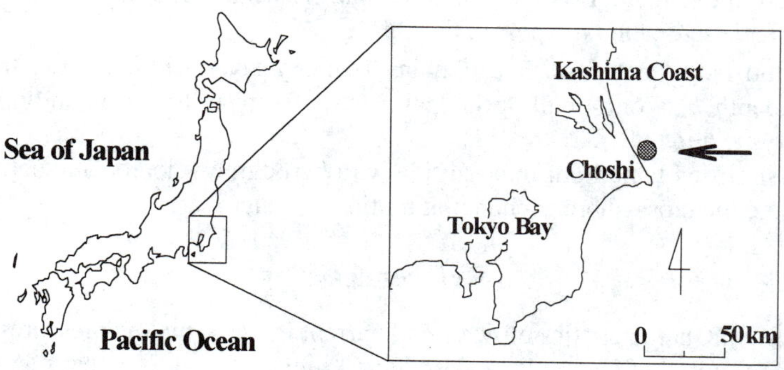

Fig.2(a) Location of study.

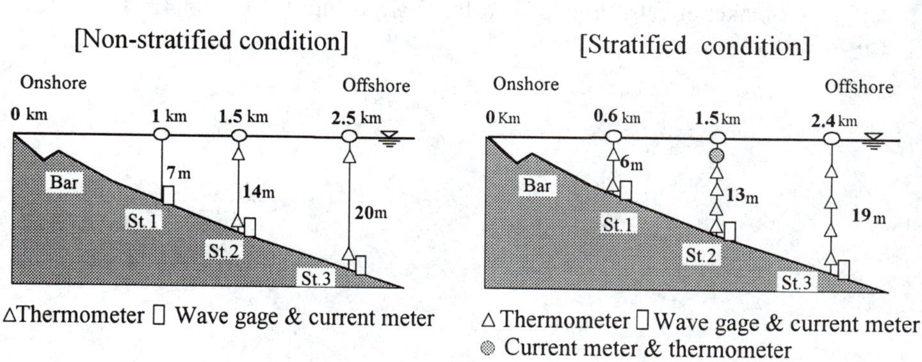

Fig.2(b) Location of observation stations and arrangement of measuring instruments.

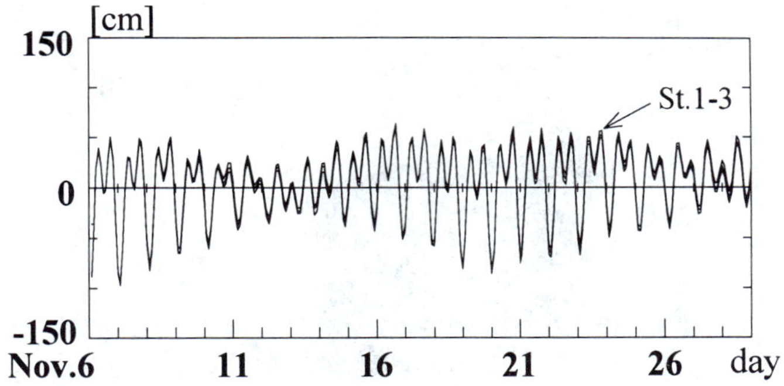

Fig.3 (a) Mean water surface elevations.

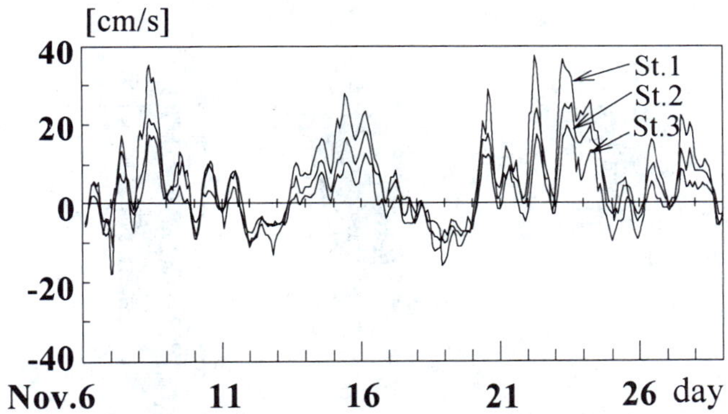

Fig.3 (b) Shore-parallel component of bottom velocities.

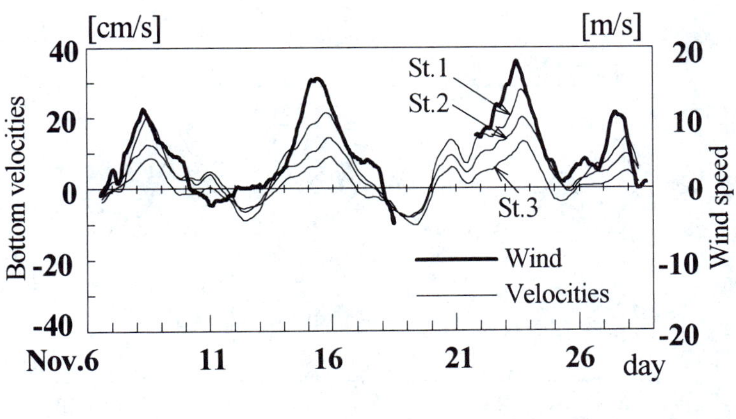

Fig. 4 Long-period fluctuating components of shore-parallel bottom velocities and wind speed.

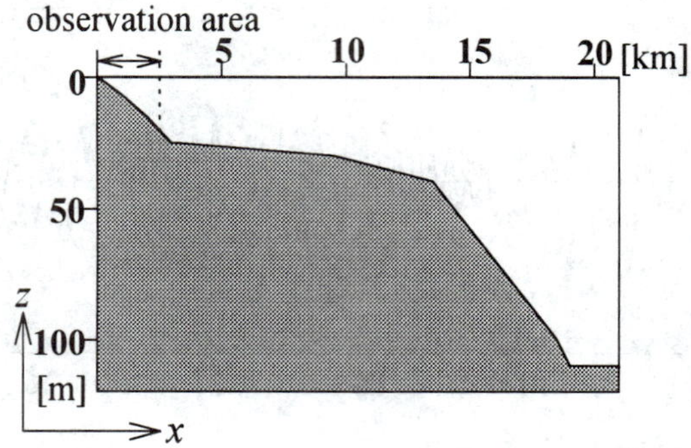

Fig.5 Computational area.

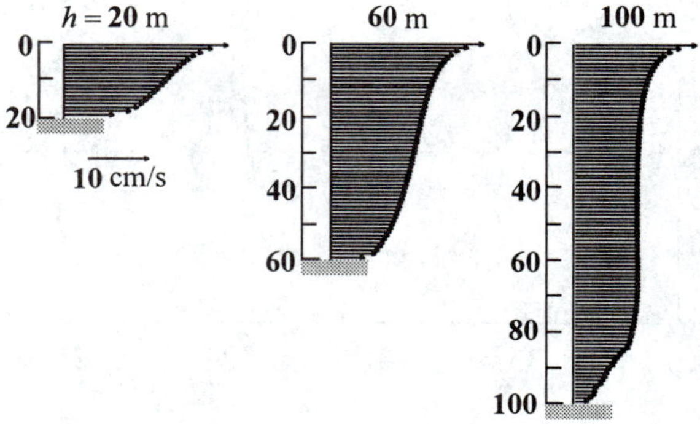

Fig.6 Computed vertical profiles of shore parallel velocity.

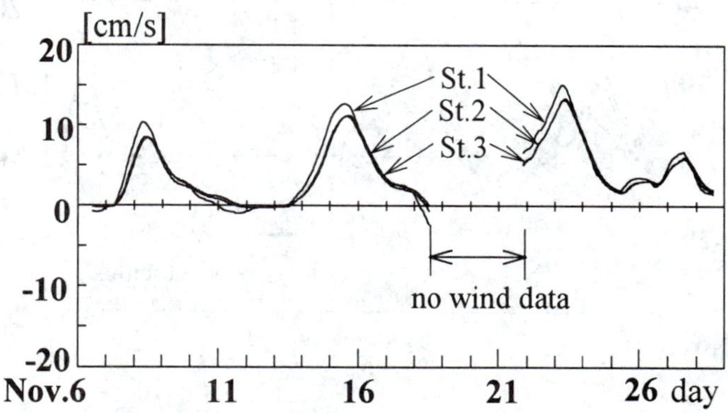

Fig.7 Computed shore-parallel velocities at St.1to 3.

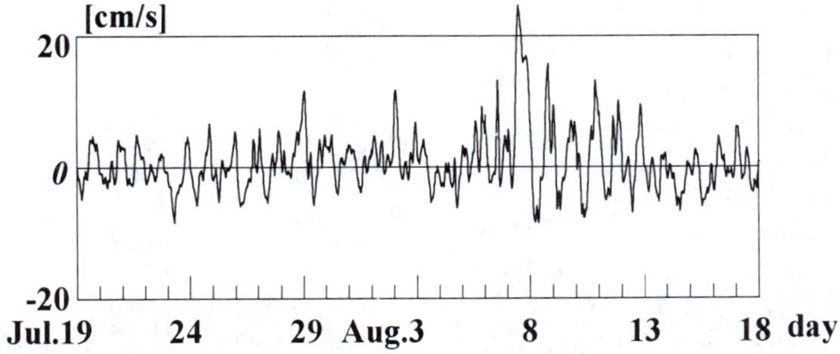

Fig.8(a) Shore-parallel component of bottom velocity at St.2.

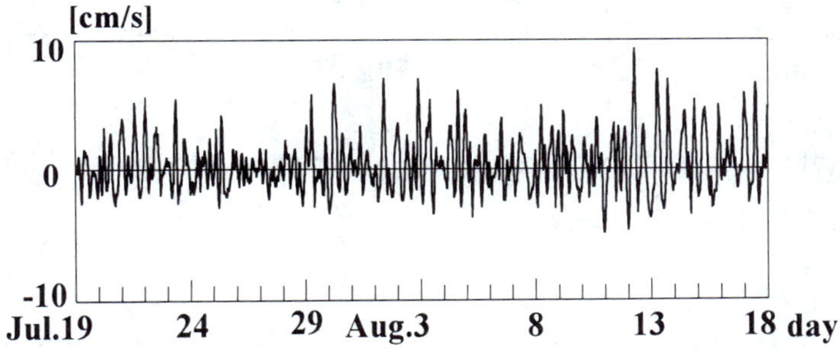

Fig.8(b) Cross-shore component of bottom velocity at St.2.

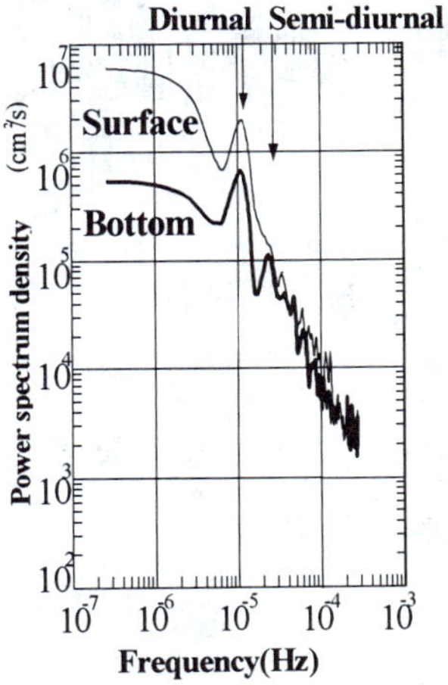

(a) Shore-parallel component.

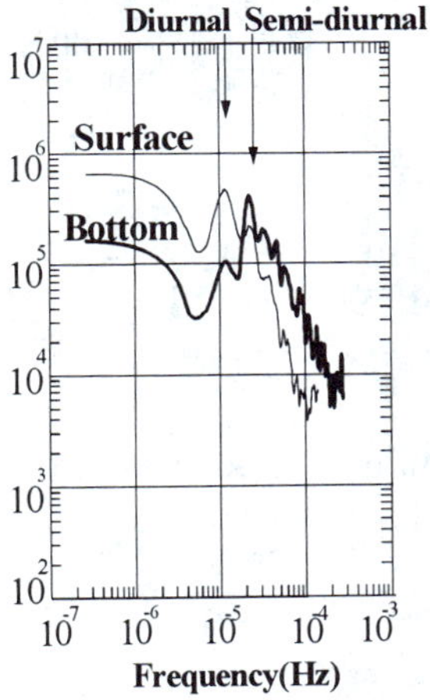

(b) Cross-shore component.

Fig.9 Power spectra of velocity fluctuations at St.2.

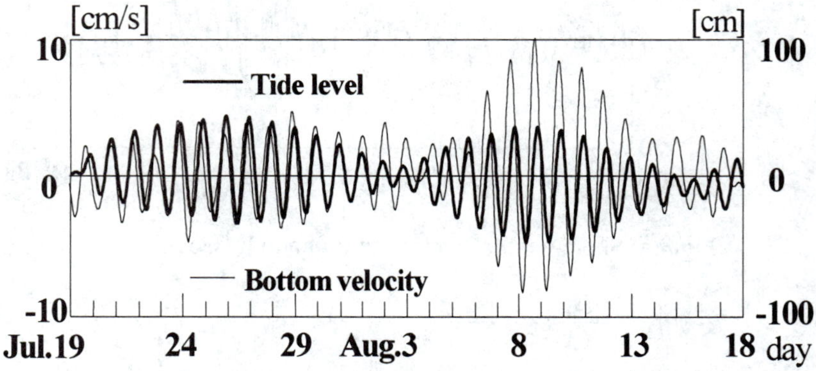

(a) Diurnal component of shore-parallel bottom velocity and tide level at St.2.

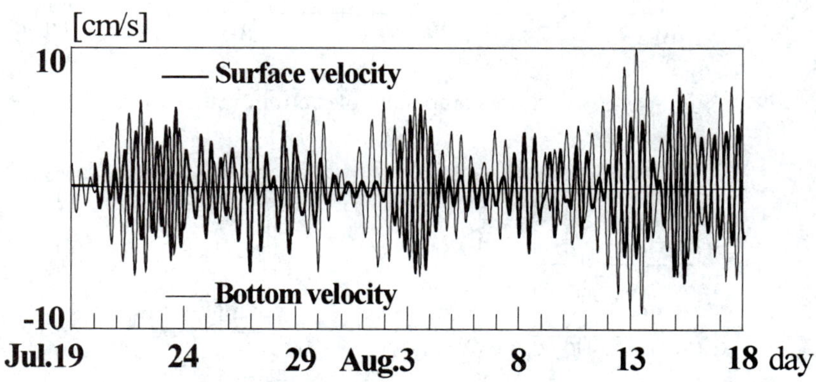

(b) Semi-diurnal component of cross-shore bottom velocity and tide level at St.2.

Fig.10 Band-pass filtered velocities.

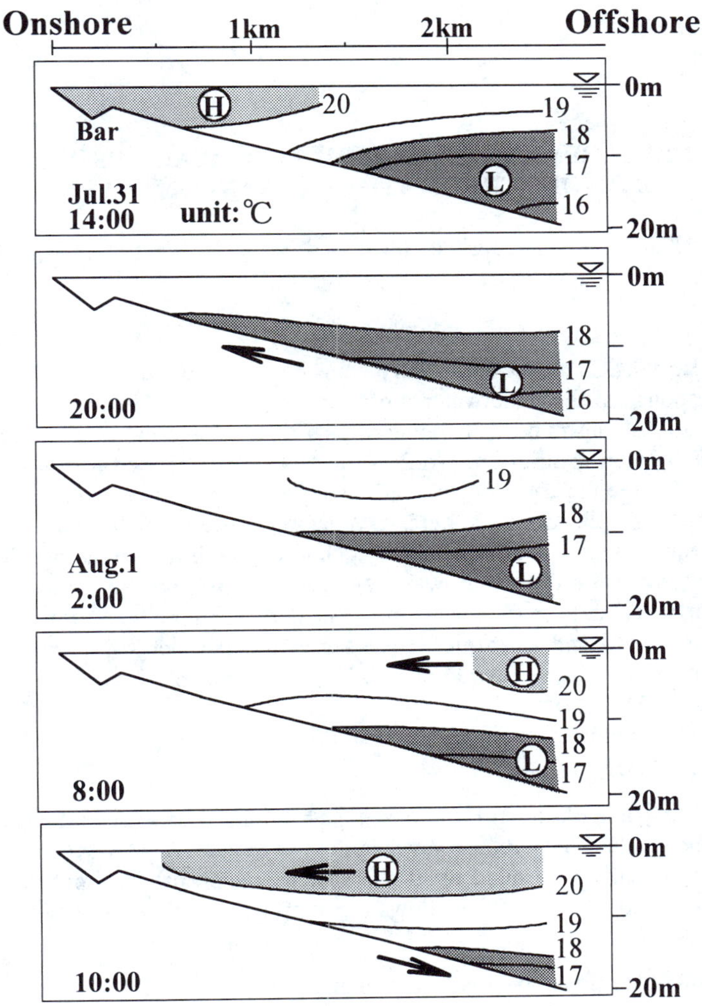

Fig. 11 Time sequence of spatial distribution of water temperature in diurnal time scale.

CHAPTER 266

DEVELOPMENT OF UNDERWATER BEACH PROFILE BY MONOCHROMATIC AND RANDOM WAVES

Karsten Peters[1], Jürgen Newe[1] and Hans-H. Dette[2], M.ASCE

ABSTRACT: Central to all problems involving alluvial coasts is the response of the underwater profile to given wave conditions. It has been demonstrated in laboratory tests that a constant wave pattern leads to an equilibrium profile. BRUUN (1954) defined this profile by a parabolic equation and various refinements have been proposed since. Applications to field data show, however, that the fit is frequently less than satisfactory. Noteworthy is also that profiles generated in the large wave flumes tend to have significantly narrower surf zones than similar sand beaches in the nature. To answer this and other questions a series of experiments were carried out in the Large Wave Flume (LWF) in Hannover.

INTRODUCTION

The Large Wave Flume in Hannover is 324 m long, 5 m wide and has a depth of 7 m and is capable of generating waves up to 2 m height. These waves are generated by an electronically controlled hydraulic system, which drives a piston-type wave board. A wave absorption device allows the minimization of wave reflections during the tests.

The main objective in the 1993 experiments was the investigation of beach profile response under erosional and accretional wave conditions. The experiments included tests with monochromatic waves and spectral waves at constant water level and spectral waves with superimposed tidal simulation with 1m tidal range.

---

[1] Research Assistant [2] Academic Director, Department of Hydrodynamics and Coastal Engineering, Leichtweiss-Institute for Hydraulics, Technical University Braunschweig, Beethovenstr. 51a, 38106 Braunschweig, Germany.

EXPERIMENTAL SET-UP

The initial profile geometry consisted of a dune, a 1:30 beach slope and a 70 m long toe berm (Fig. 1). A sand volume of 1.700 m^3 was necessary to build up the initial profile. The used sand had a grain diameter of D_{50} = 330 µm with a fairly narrow grading. This is representatively for the beaches along the west coast of the high energy coast of the Island of Sylt/North Sea. Data were collected on bed profile development, water levels and waves, wave energy dissipation, velocity distributions and suspended sediment distributions, only aspects relating to profile development will be discussed here.

The experiments in the Large Wave Flume can be devided into three major parts:

- In part I monochromatic waves were used to generate an equilibrium profile under erosional conditions (Test Series 0 and 2) and accretional conditions (Test Series 1), while the water level was kept constant.
- In part II (Test series 3 and 4) random waves (TMA-Spectra) were used to investigate the profile response under erosional and accretional test conditions with a constant water level.
- In part III (Test Series 5 and 6) random waves (TMA-Spectra) were run together with water level variations in order to simulate a profile development under tidal conditions. The tide had a range of 1 m with the previous constant water depth of 4.50 m representing the high water level. The 12 hours 'sinusoidal' tide is simulated as step function in 25 cm steps of 60 minutes, except high and low water phases with 90 minutes duration.

Table 1 summarizes information on wave parameters, water level, test duration and profile response for the Test Series' 0 to 6. The final profile of each single Test Series was used as initial profile for the following series.

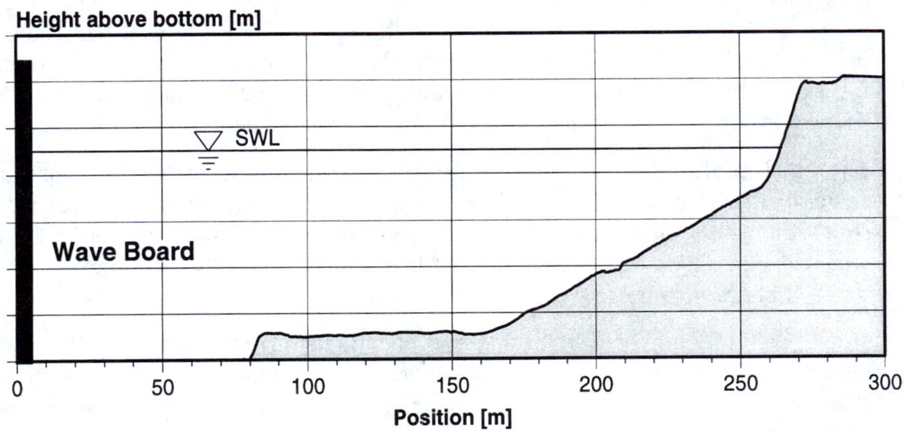

Figure 1. Initial profile built up in the Large Wave Flume in Hannover.

Table 1: Test conditions for the flume experiments 1993.

Test Series	Initial Geometry	Wave Parameters			Water Level [m]	Duration [hrs]	Profile Response
		Type	H, H_{m0} [m]	T, T_p [sec]			
0	1:30 slope	M	1.20	5	4.50	17:00	Erosion
1	result of 0	M	1.20	10	4.50	15:45	Accretion
2	result of 1	M	1.20	5	4.50	07:20	Erosion
3	result of 2	S	1.20	5	4.50	15:00	Erosion
4	result of 3	S	1.20	10	4.50	10:30	Accretion
5	result of 4	S+T	1.10	5	3.5 - 4.5	45:00	Erosion
6	result of 5	S+T	1.05	10	3.5 - 4.5	39:00	Accretion

with M - monochromatic waves with H and T
S - TMA-spectrum with H_{m0} and T_p
S+T - TMA-spectrum with water level variations (tidal simulation)

RESULTS AND DISCUSSION

Profile Shape

- In part I monochromatic waves in Test Series 0 with a height of 1.2 m and a period of 5 sec. were used to generate an equilibrium profile under erosional conditions. The initial 1:30 beach slope and the dune (Fig. 2a) were transformed during 17 hours of erosion in Test Series 1 (Fig. 2b) into a typical bar-trough-profile, which can be characterized by the relatively steep and narrow bar and the following trough. The slope of the seaward bar face had a magnitude of 1:5. Arrows in Fig. 2 indicate the direction of the bar movement.

  The final profile form in Test Series 0 was stabilized under 16 hours of accretional waves with a period of 10 sec. in the following Test Series 1 (Fig. 2c), where ripple structures were formed seaward and landward of the widened bar. The surf zone has been filled with sand so that the profile level inside the surf zone raised.

  Additional 7 hours of erosional wave conditions in Test Series 2 (Fig. 2d) widened the surf zone and formed a steeper bar. The inner surf zone was raised again and rebuild like a foreshore terrace, while the ripple structures at the sea floor disappeared.

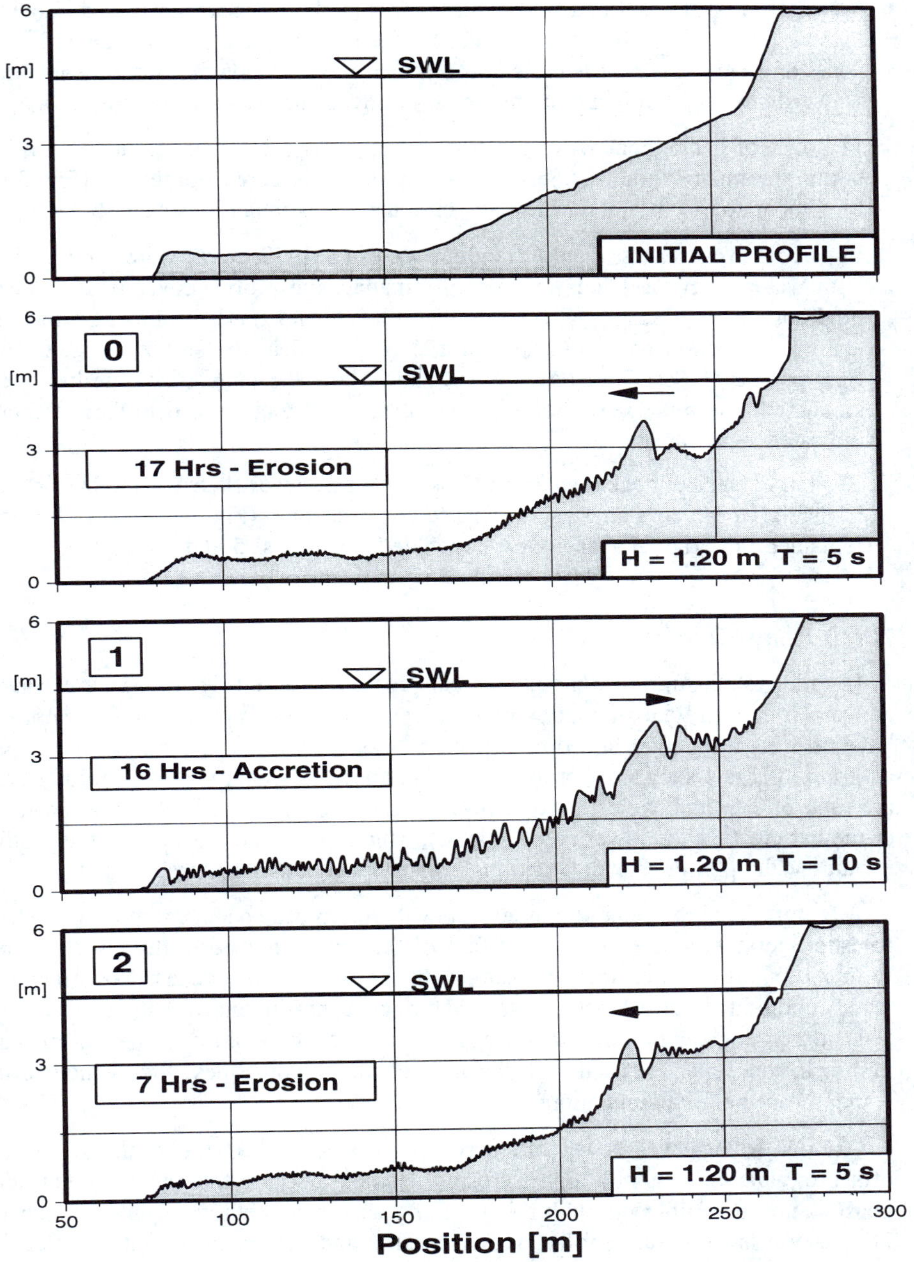

Figure 2. Initial and final beach profiles under monochromatic waves and constant water level (Test Series 0 - 2).

- In part II (Test series 3 and 4) irregular waves (TMA-Spectra) were used in order to investigate the profile response under erosional and accretional waves. In 15 hours of erosion the profile has smoothed inside and outside the surf zone and the seaward slope of the foreshore terrace has moved further offshore (Fig. 3a).

  11 hours of accretional wave conditions formed ripple structures at the profile bottom (compare with Test Series 1) and rounded the foreshore terrace (Fig. 3b). In comparison to the monochromatic case the bar is much lower and broader.

- In part III (Test Series 5 and 6) random waves (TMA-Spectra) were run together with water level variations in order to simulate the profile development under tidal conditions. Caused by different water levels and spatially moving zones of energy dissipation an equilibrium profile developed in the surf zone area. The final profile of Test Series 5 (Fig. 3c) shows how the inner surf zone has been smoothed, while the seaward front face of the surf zone moved further offshore indicated by an arrow.

  39 hours of accretional wave conditions in Test Series 6 flattened the bar profile in the surf zone area and a short terrace could develop (Fig. 3d). The wet beach remained the original shape over the whole Test Series' 5 and 6.

2D/3D Profile Development

In order to illustrate the morphological processes caused by the spectral waves and superimposed tidal simulation more in detail the single profiles of Test Series 5 have been joint together into time-dependent profile evolution diagrams. The upper diagram in Fig. 4 shows the profile evolution under erosional wave conditions over the time of 3½ tidal cycles. Two arrows indicate the starting and the final position of the bar crest. In the lower graph (Fig. 4), which shows the top view of the profile evolution, the phases after low and high water level conditions are marked.

It is significant to see in both graphs how the bar profile changed into a smoothed foreshore terrace, while the seaward face of the terrace moved in rhythmic manner further offshore. The offshore movement of the bar, which is caused by erosional wave conditions, is overlapped by the rhythmic bar movement over the tidal cycles. This morphological rhythm, which can be observed between position 200 m and 220 m in the upper diagram with the help of the contour lines, shows the profile response on the tidal simulation.

At low water level zones of energy dissipation are spatially placed further offshore, which results in a higher erosion potential because the profile in that tidal phase is not in equilibrium with the incoming waves. At high water level the zones of energy dissipation are shifted further onshore and the waves do not break at the front face of the forshore terrace as at low water levels. The front face of the foreshore terrace moves onshore, by which means the crest height of the bar is increased.

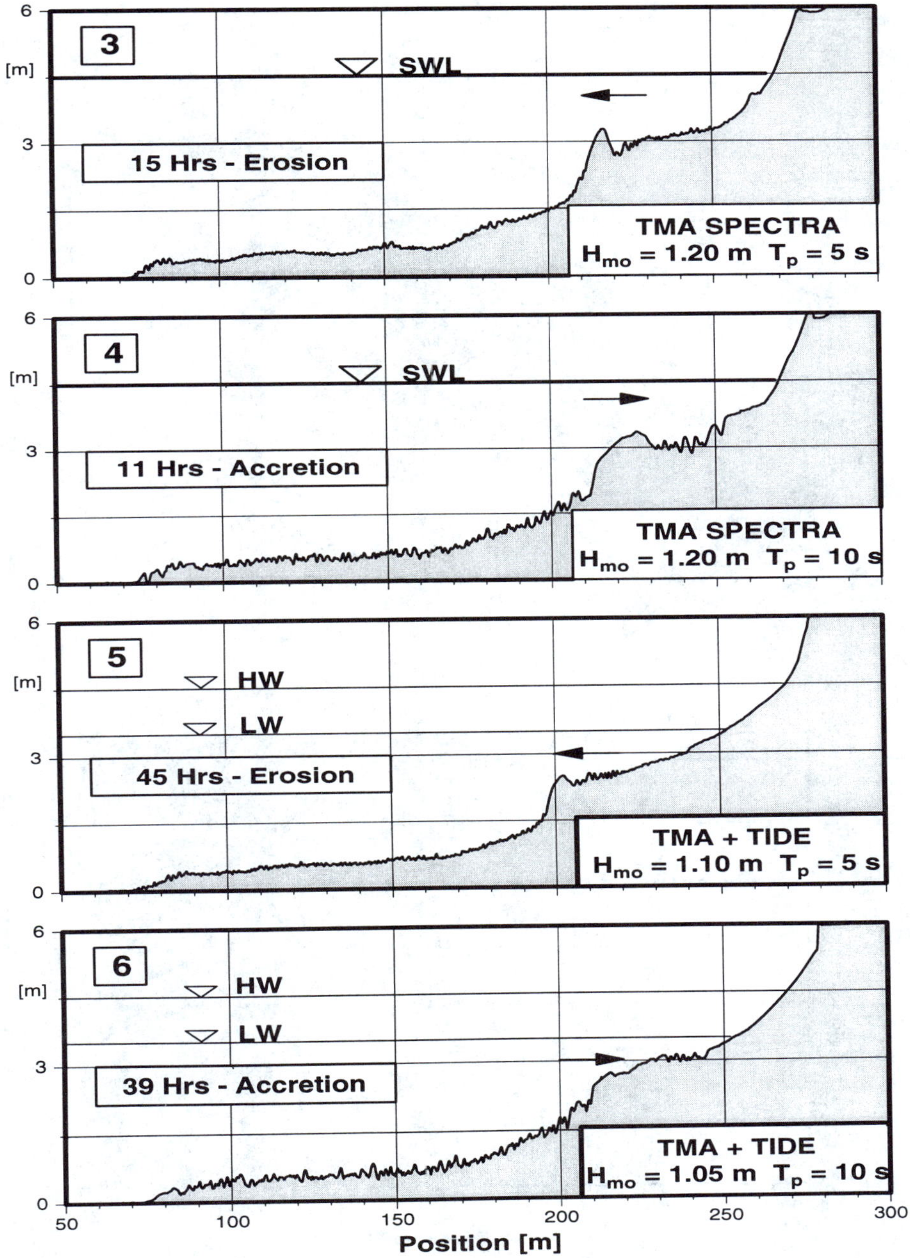

Figure 3. Initial and final beach profiles under spectral waves with constant water level (Test Series 3 and 4) and spectral waves with superimposed tidal simulation (Test Series 5 and 6)

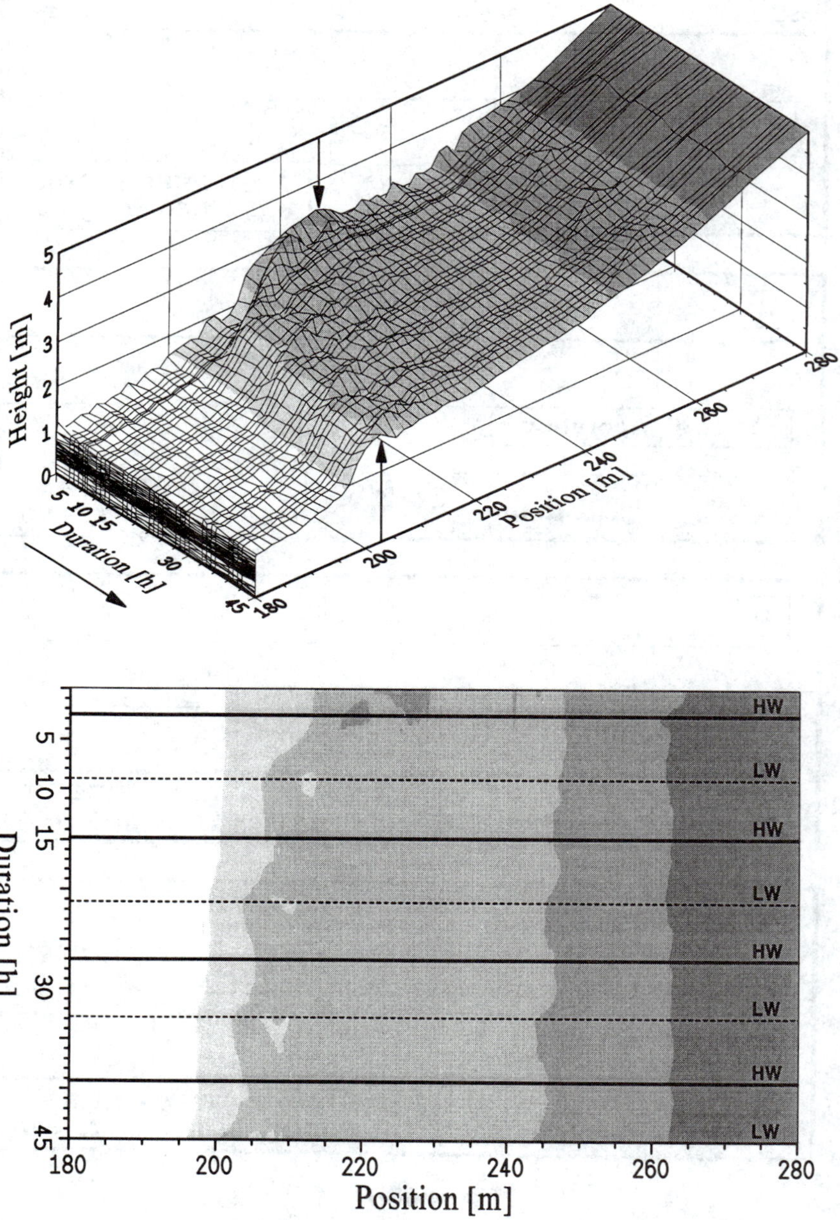

Figure 4. Profile evolution under erosional wave conditions (TMA spectra) with superimposed tide as a 3D-diagram (above) and as a top view (2D diagram below).

By tidal simulation the bar movement has been intensified. This was also observed for Test Series 6 (not illustrated here) under accretional wave conditions, where the rhythmic bar movement caused by the tide overlapped the onshore movement of the bar system.

Bar Position and Bar Height

All profile data for each wave condition can be analysed as a time history over the whole test duration. In the upper graph of Fig. 5 the definitions of the bar height, the bar position and the surf zone width are shown. The bar crest in this case is defined as crest height of the seaward shore face.

In the lower graph of Fig. 5 the surf zone and the enclosed beach with the adjacent shoreline is shown as a top view and the bar height with the corressponding water level as a side view. It can be seen how the bar position at the seaward end of the surf zone developed, and how the crest height of the bar corresponds to this. Additionally the water level is indicated. For Test Series 5 and 6 the manner of tidal simulation is illustrated.

For the Test Series' 1 to 6 it can be summarized that erosional wave conditions (marked with 'E' in the graph) widen the surf zone and lower the crest of the bar, while accretional wave conditions (marked with 'A' in the graph) cause a reduction of the surf zone width and connected with an increase of the bar crest level. The superimposed tide and the tidal range intensify the bar movement, compared to the Test Series with constant water level. These two components, the bar movement at a constant water level and the tidal cycles itself, lead to a rhythmic bar movement. After 2-3 tides the morphological changes during each tide became stable, which means that the morphological changes on the profile over one tidal cycle reached already a minimum.

Ripple Structures

Accretional monochromatic waves (see Test Series 1) have caused large bed forms at the profile bottom. This is demonstrated in Fig. 3b. As a time history the plotted profiles in Fig. 6 show a considerable ripple movement seaward of the bar. Starting with the initial profile at the bottom the intermediate profiles are plotted above in temporal order. It can be seen how the bed forms grow and develop at the seafloor and how they propagate in landward direction until they reach the toe of the bar. The bed forms were not able to propagate into the surf zone, so that they stopped at the toe of the bar and increased the width of the bar. Because the bar crest remained nearly at its position the seaward slope of the bar became flatter.

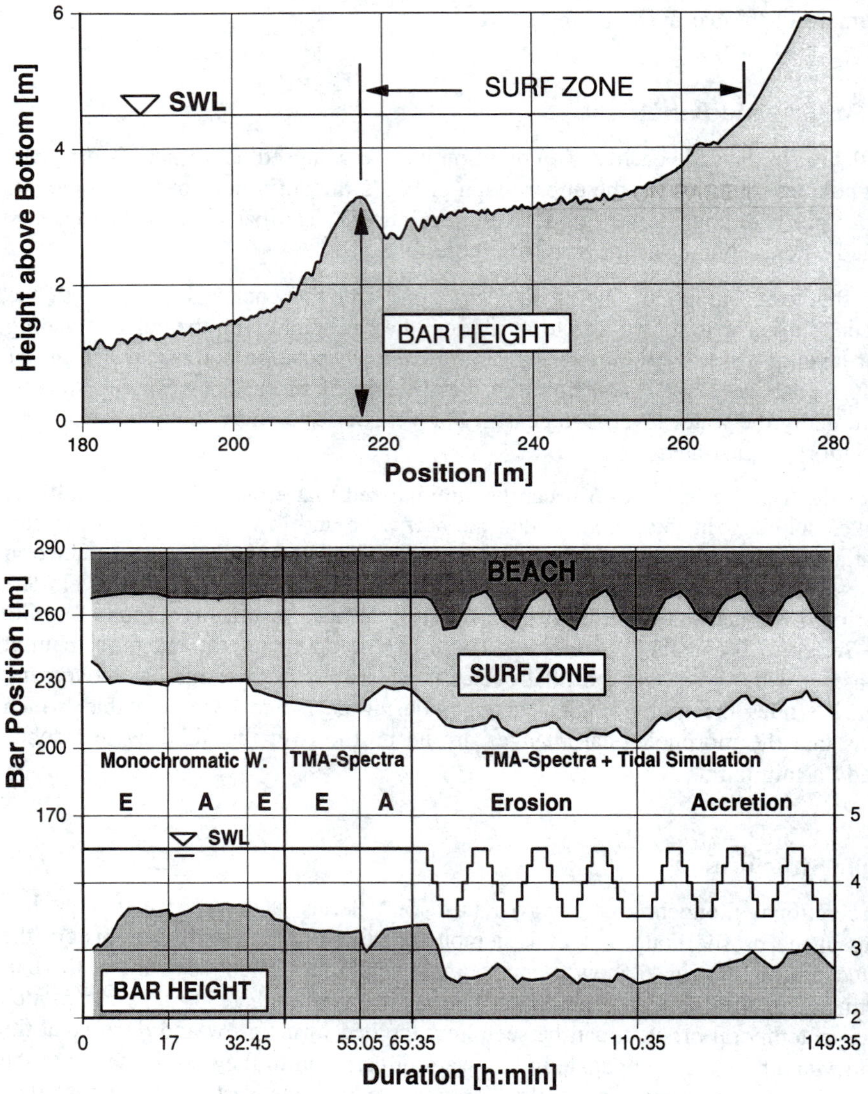

Figure 5. Definition of the surf zone (above) and position and height of bar over all Test Series' (below)

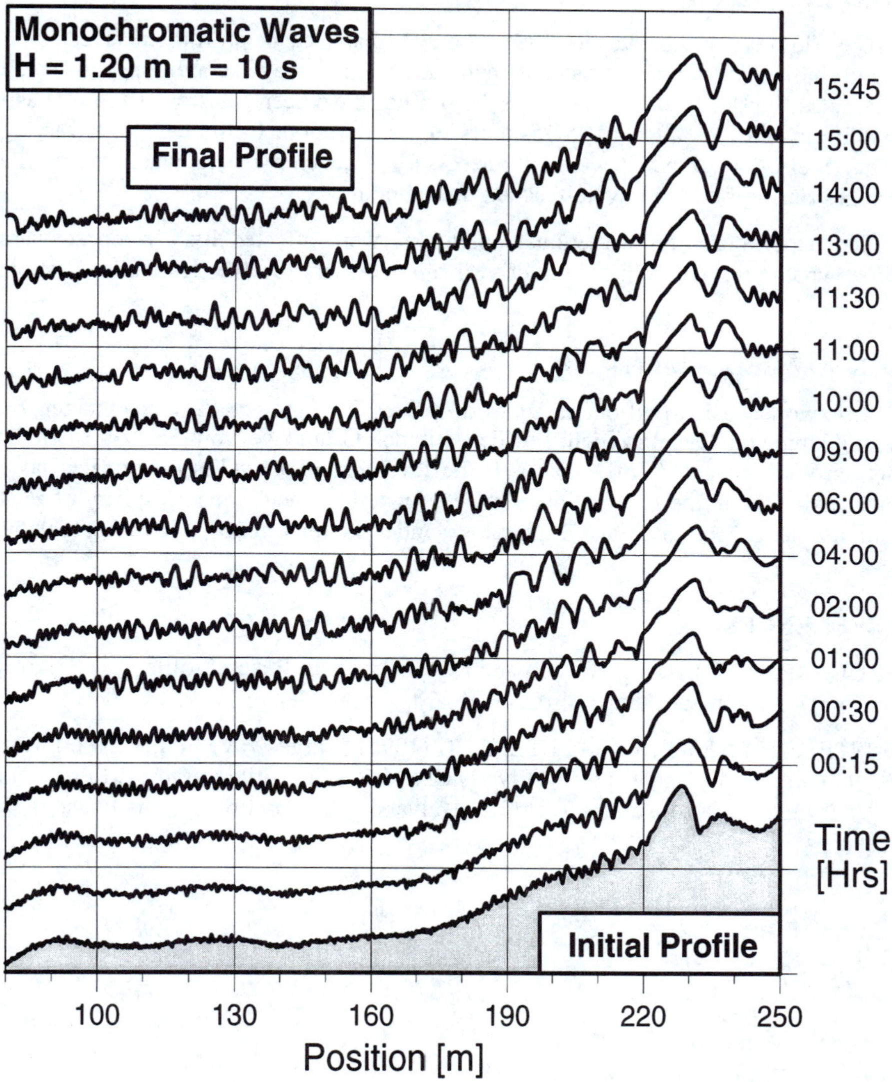

Figure 6. Development of bed forms during accretional wave conditions (Test Series 1).

CONCLUSIONS

The flume experiments in 1996 showed that beach profiles reached their equilibrium state under erosional and accretional wave conditions in small timescales. For a constant water level the profiles have been transformed into quasi-equilibrium profiles after 10 to 15 hours of test duration. In the case of a varying water level 2-3 tidal cycles were only necessary to develop a state, where the profile changes during one tidal cycle reached a minimum.

The experiments for accretional conditions were affected by a steep seaward slope, what is not naturelike. Further tests are necessary to investigate this question.

ACKNOWLEDGEMENT

This work was carried out as a part of the research project "Untersuchungen zur Ausbildung von Gleichgewichtsprofilen auf der Grundlage von Experimenten im Großen Wellenkanal" ("Investigation in the development of equlibrium profiles based on experiments in the Large Wave Flume") funded by the German Ministry of Education, Science, Research and Technology under contract no. 515-3892-MTK 0559.

REFERENCES

BRUUN, P. (1954) "Coast Erosion and Development of Beach Profiles", U.S. Army Corps of Engineers, Beach Erosion Board., Tech. Memo. No. 44.

DETTE, H.H.; NEWE, J. and PETERS, K. (1995) "Large Wave Flume Experiments '93. Volume I: Data Report - Wave Data and Beach Profile Surveys", Report No. 787, Leichtweiss-Institute, Technical University Braunschweig (unpublished).

CHAPTER 267

TURBULENT STRESSES IN THE SURF–ZONE: WHICH WAY IS UP?

John W. Haines[1] and Guy Gelfenbaum[1]

Abstract

Velocity observations from a vertical stack of three-component Acoustic Doppler Velocimeters (ADVs) within the energetic surf-zone are presented. Rapid temporal sampling and small sampling volume provide observations suitable for investigation of the role of turbulent fluctuations in surf–zone dynamics. While sensor performance was good, failure to recover reliable measures of tilt from the vertical compromise the data value.

We will present some cursory observations supporting the ADV performance, and examine the sensitivity of stress estimates to uncertainty in the sensor orientation. It is well known that turbulent stress estimates are highly sensitive to orientation relative to vertical when wave motions are dominant. Analyses presented examine the potential to use observed flow-field characteristics to constrain sensor orientation. Results show that such an approach may provide a consistent orientation to a fraction of a degree, but the inherent sensitivity of stress estimates requires a still more restrictive constraint. Regardless, the observations indicate the degree to which stress estimates are dependent on orientation, and provide some indication of the temporal variability in time-averaged stress estimates.

Introduction

Few observations of turbulent stresses in energetic surf–zone conditions exist. This observational void is in marked contrast to modeling efforts describing surf–zone circulation. Published models include details of the stress distribution which have yet to be constrained by field observations (*Svendsen et al.*, 1987, *Deigaard et al.*, 1991). Moreover, investigations of nearshore circulation have suggested that simpler (constant eddy viscosity) models of the stress distribution may be inconsistent with field observations of the mean flow field (*Haines and Sallenger,* 1994).

---

[1]U.S. Geological Survey, Center for Coastal Geology, 600 4th St. South, St. Petersburg, FL, 33701

In an attempt to better constrain modeling efforts, a field program was designed to collect data on the vertical and temporal distribution of turbulent stresses within the surf-zone. As part of the Duck94 field program a vertical array of SonTek Acoustic Doppler Velocimeters (ADVs) was deployed and 24 days of nearly continuous data collected (Figure 1). The principle aim of the deployment was to return observations of the high-frequency intermittent phenomena associated with bottom and surface boundar–layer processes. The objective was to accurately describe the turbulent statistics of the flow-field adjacent to both boundaries under breaking and non-breaking wave conditions.

A critical requirement for the calculation of stress estimates is the precise definition of an appropriate coordinate system. Because of the relatively large horizontal flows associated with shallow–water incident waves, stress estimates are inherently sensitive to the choice of a vertical coordinate. Previous work has suggested that, under representative incident wave conditions, a precision of 0.03 degrees is required in definition of the tilt angle to assure stress estimates are correct within $1cm^2s^{-2}$ (*Lohrmann et al.*, 1995). The appropriate coordinate system may depend on the modeling approach taken. For example, a gravitational coordinate system may produce results significantly different than a coordinate system aligned with the local bed slope. In the absence of field observations of the magnitude and variability of the stresses, the precision required for array design is speculative. Observed variability also provides an indication of the sensitivity of results to the choice of coordinate frame.

The ADV array deployed was instrumented with a continuously recording tilt-meter. Subsequent analyses suggest that the tilt–meter lacked the required accuracy for definition of a reliable coordinate system. Given the shortcomings of the tilt measurements, and the sensitivity of stress estimates to tilt, we are investigating the potential for using the flow data itself to determine an appropriate coordinate system. Regardless of the success of this effort, the data also provides a means for investigating the sensitivity of field observations to deviations from the vertical. The observations will also provide some indication of the magnitude of the temporal and spatial variability of surf-zone stresses.

Data

As part of the Duck94 experimental program, a vertical stack of Sontek ADVs was deployed at the US Army Corps of Engineers (ACOE) Field Research Facility (FRF) in Duck, NC. The deployment consisted of 7 ADV sensors in a vertical array, with supplementary observations of water level, optical backscatterance, array tilt and rotation, and bed elevation from a sonic altimeter (Figure 1). All sensor outputs were digitized *in situ*, synchronized via GPS clock, and written to shore-based storage via a fiber optic cable. The ADVs were sampled at 25 Hz. All other sensors were sampled continuously at $64Hz$. The ADV distribution was intended to span the bulk of the water column and return observations of near-bed and near-surface turbulent fluctuations.

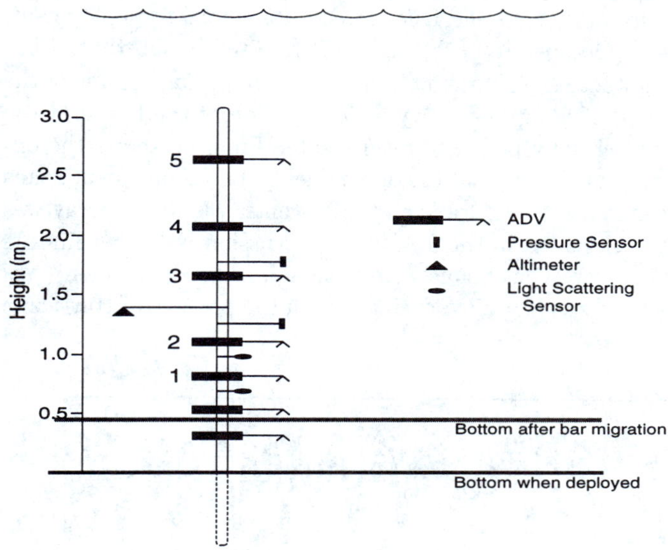

Figure 1: Array configuration. Bed height determined from altimeter is approximately $1m$ from ADV position. ADV sensing volumes are $15cm$ below the sensor heads shown.

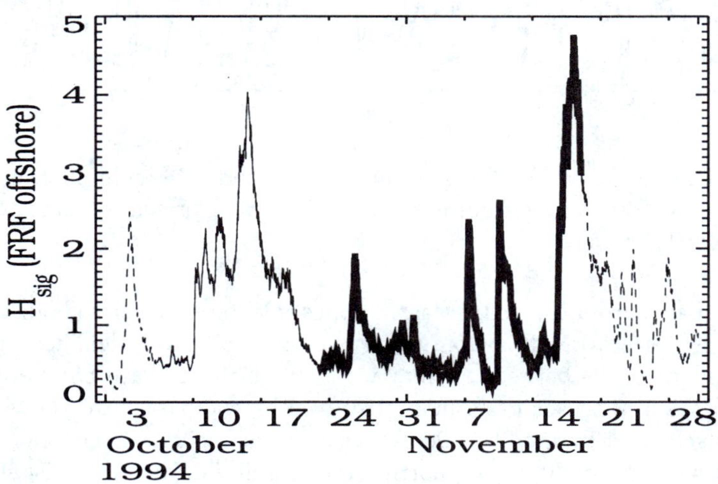

Figure 2: Offshore significant wave heights from the Field Research Facility pressure array in $8m$ water depth. Solid lines indicate period of sensor deployment, thick solid line shows period of data recovery.

The array was initially deployed in approximately 4.5m water depth, outside the primary bar normally present at the Duck site. Subsequent to deployment, an energetic storm event (October 16-18th, Figure 2) resulted in substantial bar migration, burying the lowermost ADV and obscuring the sampling volume of the next highest sensor. The following weeks saw a slow erosion of the bar, and the upper impacted sensor eventually began to return data. Figure 2 shows the offshore wave conditions present throughout the deployment. The deployment spans 3 substantial storm events, and only data from these events, where the array was within the surf zone, are discussed in the following. Discussion is further limited to those 5 sensors which were continuously operational. The deployment was violently terminated by surf-zone debris associated with the passage of Hurricane Gordon.

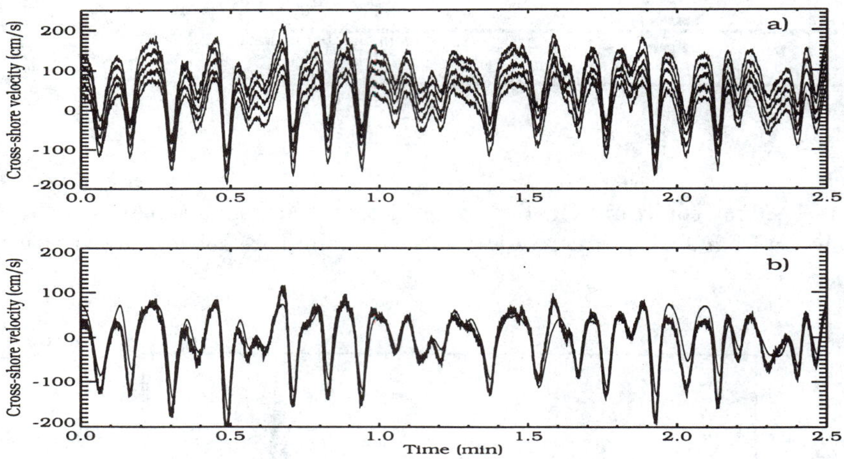

Figure 3: a) Time series from ADV sensors, offset to indicate increasing height above the bed. b) Time series from uppermost ADV and pressure sensor (normalized by velocity variance).

The data quality from the ADV sensors appears to be uniformly good. Figure 3a shows representative data from the 5 operational ADV's, demonstrating extremely high coherence between sensors across the incident wave band. Independent verification of sensor performance is given by comparison to the water level data (as shown in Figure 3b.) Depth decay of the incident-band velocities are consistent with linear theory, supporting the overall data quality. Visual inspection of the data suggests that high-frequency variance "events" occur with greater frequency high in the water column. This may be suggestive of breaking–produced turbulence, and is an area for further investigation.

The data are further described in Figure 4a, showing cross-shore, u, and vertical, w, velocity spectra (Figure 4a). Above incident–band frequencies the

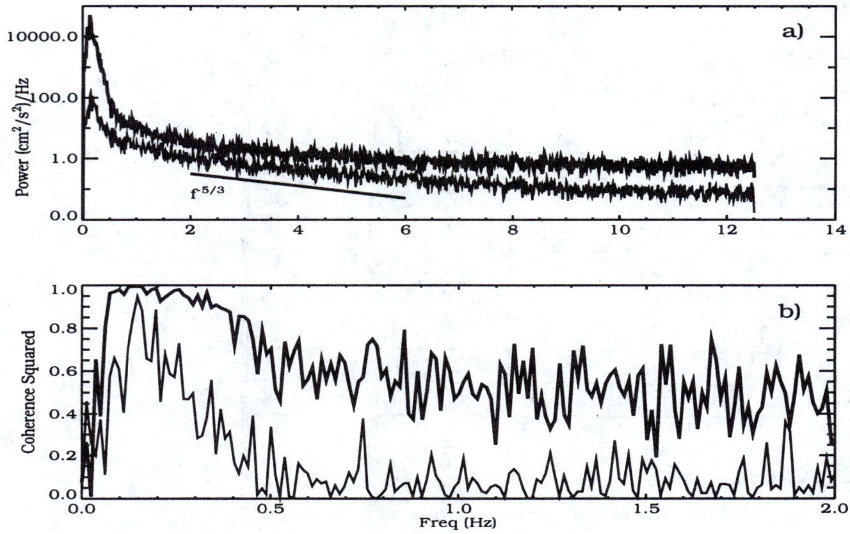

Figure 4: a) u (higher energy) and w spectra. b) pu (thick line) and pw squared coherences. Frequency scales are not equivalent for the two panels. Data shown are post–rotation as determined by the analysis described herein.

spectra show an $f^{-5/3}$ decay. The horizontal velocities generally show enhanced energy levels relative to the vertical velocities at all frequencies. The leveling of spectral energy at high frequencies for the horizontal velocity is due to the noise-floor of the sensor, which is substantially reduced for the vertical component. Coherences between pressure observations, p, and horizontal velocities are high across the incident band and uniformly higher than pw coherences.

Analysis

While the data appear reliable, calculation of the turbulent stresses requires accurate constraint of the coordinate frame. In the following we will attempt to determine the sensor orientation from the observed flow characteristics. In order to maximize the signal-to-noise ratio of the data, further analysis is restricted to high-energy conditions. Selected one–hour data segments, described in Table 1, were analyzed.

The underlying assumption for the following analysis is that the flow field, as described by the observations, contains information about the sensor orientation. A plausible first assumption might be that, following linear wave theory over a horizontal bed, within the incident-wave band, vertical and horizontal velocities are in quadrature. It follows that the time-averaged stress, $\langle uw \rangle$, should be zero. We might then choose to rotate the observations into a coordinate frame where this criteria is met. For a sloping bed, the result is rotation into a bed-normal

Run #	Date	Time UTC	Depth m	H_{rms} m	T_p s	bed ht. m
0	10/26	13:00	4.49	.81	4.9	.25
1	10/26	15:00	4.78	.92	5.3	.26
2	10/26	19:00	4.54	1.21	6.1	.27
3	10/26	21:00	4.20	1.21	6.4	.26
4	11/7	08:00	3.50	.55	12.8	.35
5	11/7	09:00	3.65	.68	11.6	.36
6	11/7	10:00	3.77	.79	11.6	.37
7	11/7	11:00	4.18	1.14	11.6	.38
8	11/7	12:00	4.52	1.29	11.6	.39
9	11/7	13:00	4.80	1.38	6.4	.41
10	11/7	16:00	4.90	1.60	7.5	.42
11	11/7	17:00	4.69	1.35	7.5	.37
12	11/7	22:00	3.80	.99	11.6	.38
13	11/8	14:00	4.73	.62	11.6	.48

Table 1: Summary statistics for analyzed runs. All quantities are from array observations of surface water level and bed elevation. Bed height is distance beneath lowermost operational sensor. Sensor height for remaining sensors are, relative to lowermost sensor, 20.8 *cm*, 86.4 *cm*, 130.1 *cm*, 189.4 *cm*.

coordinate system, or more correctly, rotation into a wave-defined coordinate system. Alternative criteria for determination of the coordinate system might include rotation to zero mean vertical velocity, $\langle w \rangle = 0$, or minimization of the vertical variance, $\langle ww \rangle$. The $\langle w \rangle$ criteria would follow on assuming that mean flows were constrained to be bed–parallel. The validity of this assumption may vary with distance from the bed. In all cases the resulting coordinate frame may be expected to reflect the influence of a sloping bed. It is of interest to examine whether variations on the order of the bed slope significantly change the estimated stress quantities.

The rather simplified view outlined supports a variety of approaches for determining a best rotation. The assumption that relevant information is contained in the wave–driven flow requires application of any methodology across some subset of the entire frequency range sampled. Any of the above criteria might be applied, singularly or in a weighted combination. The assumption of uw–quadrature further supports minimization of a number of co–spectral quantities across the frequency band of interest. In fact, any of the variance properties ($\langle uw \rangle$, $\langle ww \rangle$) may be minimized as band-averaged quantities, or minimized frequency band–by–band in a least–squares sense across the entire frequency range of interest.

A number of candidate criteria were tested. The "best" criteria were determined by the consistency of the resulting coordinate system. This is based on the necessary assumption that the array was stationary throughout the data examined. Here the deposition of approximately $50cm$ of sand associated with the bar migration may be viewed as a benefit. This deposition, we presume, further stabilized the array, reinforcing the stability properties provided by a deep central jetted pipe and three anchored guy wires. A further assumption is that the coordinate frame, as defined by the flow, is stationary. This assumption is invalidated by any significant change in the underlying bed topography.

Among all the criteria tested the most consistent results were achieved by constraining the mean vertical velocity to be zero ($\langle w \rangle = 0$) and minimizing the vertical variance subject to this constraint. The minimization requires selection of a frequency band of interest, and it is the band–averaged variance which is ultimately minimized. Variations arising from this selection will be discussed in the following.

The constraint of $\langle w \rangle = 0$ defines a functional relationship between the two tilt angles given by $\gamma = f(\beta)$, where γ is the tilt angle in the alongshore/vertical (yz) plane, and β is the tilt angle in the cross-shore/vertical (xz) plane. The search for $\langle ww \rangle = \min$ may then be restricted to the line in $\gamma\beta$ space defined by the $\langle w \rangle$ constraint.

The shore–normal horizontal coordinate frame chosen approaches, for the nearly shore–normal wave conditions examined here, a frame oriented with the principle component of the wave velocities. As a result, as will be shown, the tilt angle, β, in the shore–normal direction is strongly constrained by the energetic wave motions, while the γ solution, subject to less energetic wave motions, is less well defined. Conversely, small variations in β have a large (relative to γ) impact on the estimated stresses. The alongshore component of the system is dominated by the strong alongshore currents present during most of the runs analyzed. The mean currents influence the rotation through the $\langle w \rangle = 0$ constraint.

This approach to determining a coordinate frame implies a further, less obvious constraint. The minimization procedure ($\langle w \rangle = 0, \langle uw \rangle = \min$) is equivalent to constraining the stress terms, $\langle uw \rangle$, to be zero. We do not expect this constraint to be strictly valid, due to system noise and the influence of "non–wave" motions. Nonetheless, the results provide the greatest consistency in determination of sensor tilt. While it might be hoped that averaging the results over several data segments might reduce the effects of such contamination, it is equally (or more) likely that this procedure has a systematic bias in the tilts determined.

The minimization procedure was applied to the data subsets described in Table 1. Stable estimates resulted only when conditions were energetic, suggesting the wave signal must exceed some "noise" level in the data (where noise may include a variety of non–wave motions as well as system or random noise). Further examination indicated that the stability of the estimates was enhanced when the incident wave field was narrow–banded in direction. The stability of the tilt estimates was determined from the variance of the estimates from segmented runs. While variability in the results increased with decreasing wave height,

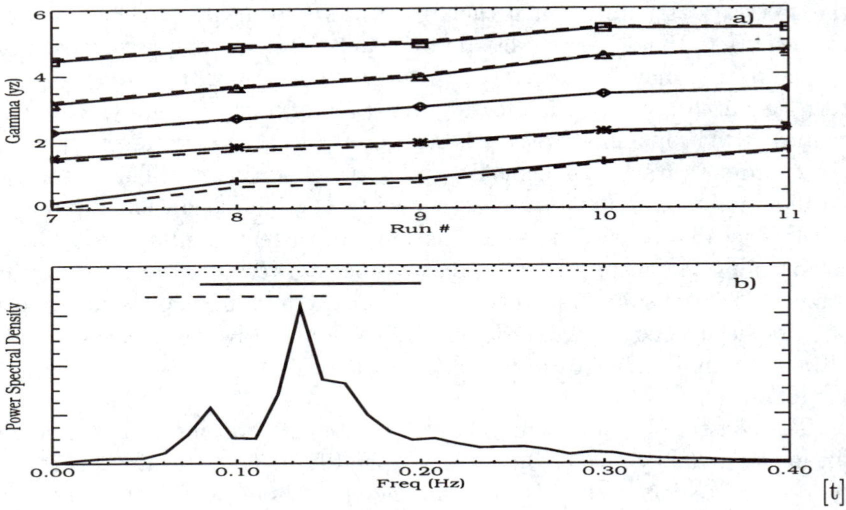

Figure 5: a) Resulting best γ (tilt in yz plane) values for 1 hour data runs. Values offset to indicate sensor location. Line types indicate the frequency range of the solution. b) Representative cross-shore velocity spectrum and frequency range of solutions (shown by horizontal lines).

there was no significant change in the mean values determined. The final values determined are consistent with the data from the tilt sensor, though this provides constraint within only a few degrees.

Determination of the best rotation in the following is restricted to data Runs $7-11$ (Table 1) which are relatively continuous in time and subject to energetic incident wave conditions. In fact, all runs listed are well within the surf–zone. Only beneath breaking waves were conditions energetic enough to result in stable estimates. The impact of breaking wave conditions on the approach can not be assessed as non–breaking conditions with large waves were not observed.

Results

Results for the γ (yz plane) tilt determination are shown in Figure 5a. The minimization was applied over three separate frequency bands as shown in Figure 5b. For determination of γ the results are insensitive to the frequency band chosen. The best γ values resulting show significant variance and a marked trend with time. While the variance in γ is large, the relatively modest amounts of incident wave energy in the along-shore direction results in little effect on calculated stress quantities.

Determination of β, in the plane of maximum wave orbital motions, has a far greater impact on stress calculations. Figure 6 shows that while the overall variance in the estimated orientation is reduced relative to γ, the sensitivity to

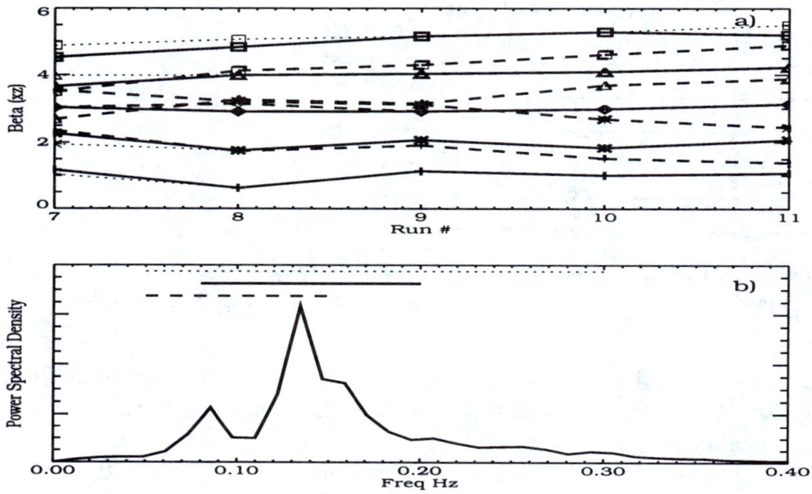

Figure 6: a) Resulting best β (tilt in xz plane) values for 1 hour data runs. Values offset to indicate sensor location. Line types indicate frequency range of solution. b) Representative cross-shore velocity spectrum and frequency range of solutions (shown by horizontal line).

frequency band selection is increased. Stable estimates require a choice which encompasses the entire incident band, and stability is further enhanced by inclusion of the higher harmonics. This is a somewhat problematic finding. Ideally we would like to determine orientation using an approach independent of information in the higher turbulent frequencies.

The standard deviations of the β values found for the 5 runs are shown in Table 2. The variance reduction with increasing frequency range is clear. There is also an indication that the upper and lower sensors are less well constrained. This is consistent with enhanced turbulence generation at the boundaries (reducing the wave signal relative to other motions) or may suggest some deviation from linear-wave behavior near the free-surface and bed. Overall, consistent tilt estimates are found to within half a degree (2 standard deviations) for all sensors.

Discussion

The mean tilt values from the 5 runs described were applied to correct all the data. The resulting vertical velocity spectra and uw coherences are shown in Figure 4. Vertical velocity variances are substantially reduced relative to the unrotated data, as are uw coherences; suggesting a removal of wave fluctuations from the vertical component. The data still show marked spectral and coherence peaks in the incident wave band. The method applied minimizes the w–variance across the entire band. Peak removal is somewhat enhanced if the variance mini-

Std. deviation β (degrees)			
Sensor #	Low Freq.	Incident	Broad band
5	.53	.32	.23
4	.47	.29	.10
3	.10	.09	.09
2	.44	.20	.14
1	.38	.22	.20

Table 2: Standard deviation of β values determined over 5 runs analyzed. Values are shown (in degrees) for frequency ranges indicated in Figure 5b.

mization is applied in a least–squares sense, band–by–band, across the frequency range selected. Results from this approach are comparable, though somewhat more variable.

The approach followed here is based on some general assumptions about the frequency contribution to the stress. Figure 7 describes this contribution, and the effect of rotation. The figure shows the cumulative stress as higher frequencies are included in the stress calculation. Shown are the results for the best orientation determined previously (averaged over 5 runs), and rotations representive of ± 0.2 and ± 0.4 degrees in γ and β. The solid horizontal line shows the frequency range for which the tilt angles are determined. The best rotation for this run (as opposed to the best average rotation) would result in the intersection of the cumulative stress curve and the high–frequency end of the frequency range. The case shown is a worst case result, serving to illustrate the sensitivity of the stress to relatively small changes in tilt.

The total stress is given by the high–frequency end of the cumulative stress curve (flat beyond the limits shown). It is clear that the resulting estimates are highly sensitive to the tilt. Furthermore, the stress is completely specified to a high degree by frequencies lower than $0.5Hz$. This result is independent of uncertainties in the orientation. Also of note, the resultant stresses change in a near–linear fashion with orientation, with the rate of change fairly constant for all sensors. This implies that, though absolute values of the stress may be highly sensitive to orientation, temporal and spatial trends (between-sensor variability) may be more amenable to observation.

For the sensor and data run shown the "best" average rotation results in a flattening of the cumulative stress curve across the incident band. In contrast, the best solution specific to this run and sensor (approximated by the lowermost cumulative stress curve) shows stress contributions which cancel within the incident band. This cancellation of regions (in frequency) of opposing stress is common to all the runs examined. A more rigorous approach will require a more complete consideration of this structure and the indicated physics of incident–band motions.

Overall the message from Figure 7 is rather disheartening. The suggestion

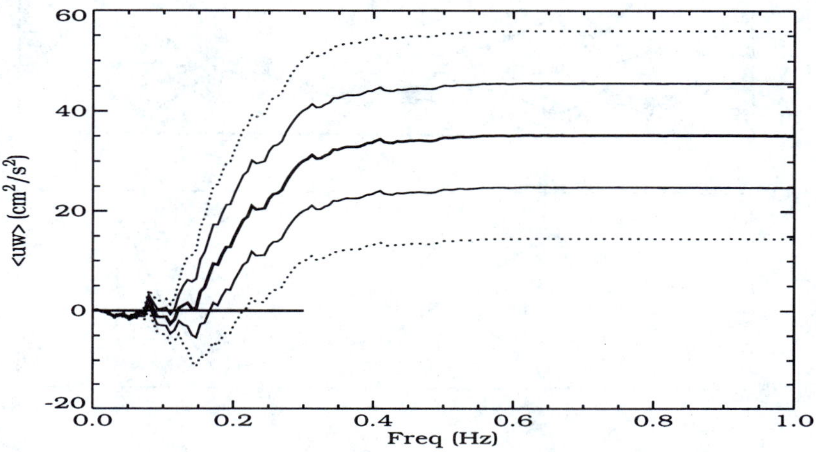

Figure 7: Cumulative stress plots for representative time series. Results from best rotation (thick line), plus (minus) 0.2 degrees in γ and β are given by upper (lower) solid lines, plus (minus) 0.4 degrees in γ and β are given by upper (lower) dotted lines.

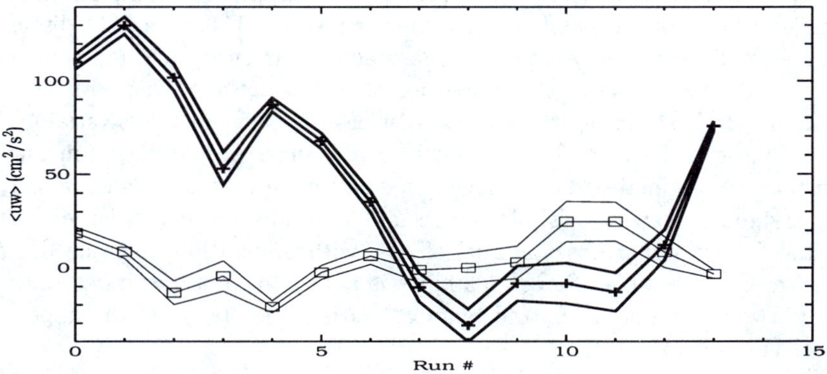

Figure 8: Stress estimates for all runs for uppermost (thin) and lowermost (thick) sensors. Error bounds indicated represent $\pm$ 2 standard deviations in γ and β.

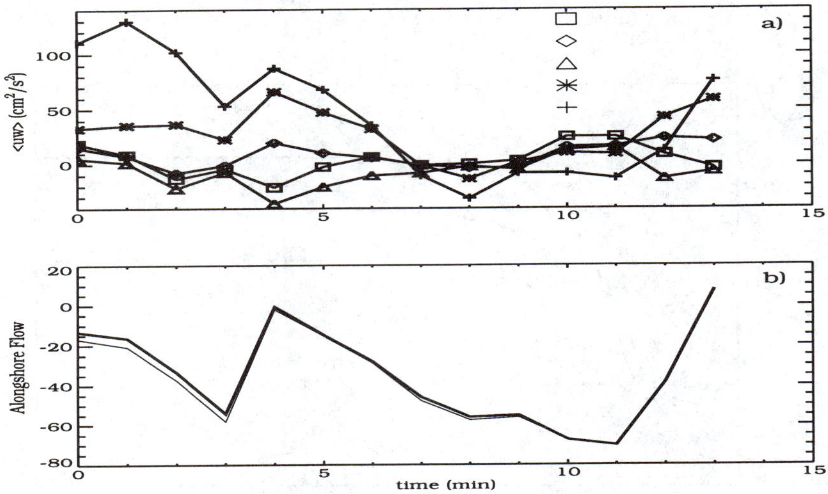

Figure 9: a) Estimated stresses for all sensors. Sensor location indicated by symbols. b) Variation in alongshore mean current for uppermost (thin line) and lowermost (thick line) sensors.

is that the most consistent results require utilization of flow information across the entire frequency band responsible for the observed stresses. There is no clear indication that the wave band is, in fact, independent of some turbulence band which contributes significantly to the stress.

The stresses resulting from the top and bottom sensors are shown in Figure 8. Calculations have been extended to incorporate all the data from Table 1, applying the best rotation as found for the 5 runs (7 – 11) previously discussed. While we believe the array was stationary across the entire period, we can not demonstrate this to the level required and the data must be viewed with some skepticism. Indications are that the near surface stress is much less variable than that near the bed. The effects of significant changes in bed elevation may be contributory. Estimates show changes in sign for both sensors, and changes in relative sign between sensors. The degree of temporal variability is large near the bed and suggests that observational efforts, with orientation precision of order 0.2 degrees are feasible. Some credibility is added to these results as both the bottom two sensors show marked temporal variability relative to the uppermost sensors (Figure 9a).

The temporal variability noted may be related to both changes in bed elevation and changes in the fluid forcing. The mean alongshore flow was highly variable across the runs examined. Figure 11b shows this variability, which is qualitatively similar to that observed for the stress estimates in the lower water column. The temporal trend in γ previously shown (Figure 5) is strongly correlated with the

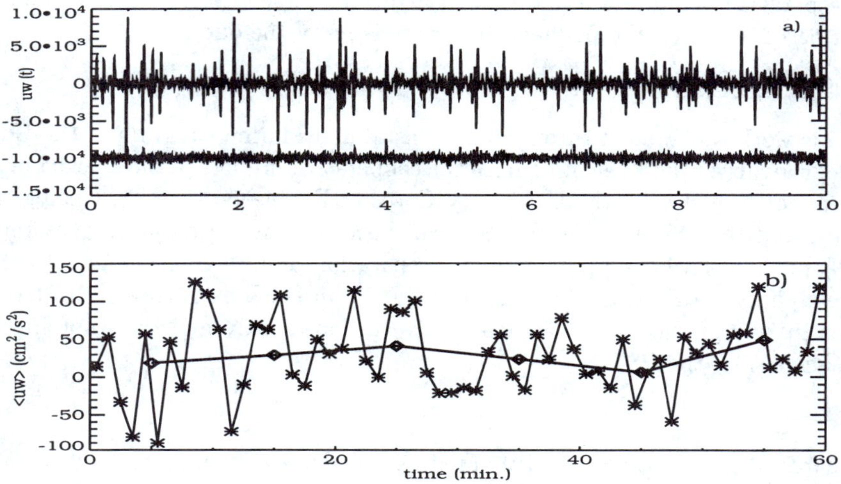

Figure 10: a) Instantaneous stress (uw) for uppermost and lowermost (offset by -10^{-4}. b) One and ten minute averages of stress from uppermost sensor.

variability in the alongshore current.

As a final note we leave the rotation problem behind to examine the variability of the stress estimates on short time scales. Figure 10a shows time series of stress from the uppermost and lowermost (offset by -10^{-4}) sensors. The degree of variability in the instantaneous stresses is elevated near the free surface, in constrast to the behaviour previously shown for time–averaged values. Averaged values (1 and 10 minute averages, Figure 10c) from the uppermost sensor clearly show that, over 10 minute periods, the stress estimates display substantial temporal variation.

Conclusions

First and foremost, precise sensor orientation is critical to estimating the absolute magnitude of turbulent stresses. Observed sensitivity to variations in tilt suggests that a precision of order 0.1 degrees in the vertical is required. Flow derived orientations are consistent to roughly 0.4 degrees. This level of precision appears to be sufficient for investigations of the temporal and spatial variability in stress estimates in at least some cases. The lack of a clear physical underpinning for the resulting coordinate frame limits further application of the results. The sensitivity of the results to orientation further suggests that consideration of bed slope effects may be required.

The observations demonstrate that stress estimates are highly variable across a variety of temporal scales. The bulk of the stress is generated by motions at frequencies approaching the incident band fundamental and harmonic frequencies.

The separation of motions into turbulent and wave–driven components is dubious when based solely on the frequency characteristics of the data.

Acknowledgements

This work was funded through the Coastal and Marine Program of the U.S. Geological Survey, and benefited from the cooperative support of the Duck94 field experiment provided by the U.S. Army Corps of Engineers, the U.S. Geological Survey, and the Office of Naval Research. Special thanks are due to the many USGS personnel who supported us in preparation and deployment and to the entire staff of the Field Research Facility (FRF) in Duck, NC. Any use of trade names in this publication is for descriptive purposes only and does not imply endorsement by the USGS.

References

Deigaard, R., P. Justesen, and J. Fredsoe, Modelling of undertow by a one-equation turbulence model, *Coastal Eng.*, 11, 565–602, 1987.

Haines, J. and A. H. Sallenger, Vertical structure of mean cross-shore currents across a barred surf zone, *J. Geophys. Res.*, 99, 14,223–14,342, 1994.

Lohrman, A., R. Cabrera, G. Gelfenbaum, and J. Haines, Direct measurements of Reynolds stress with an acoustic doppler velocimeter, *Proceedings IEEE Fifth Workshop on Current Measurements*, 205–210, 1995.

Svendsen, I., A. Hemming, and J. Buhr Hansen, The interaction between the undertow and the boundary layer flow on a beach, *J. Geophys. Res.*, 93, 11,845–11,856, 1987.

CHAPTER 268

Asymmetric and Irregular Wave Effects on Bedload: Theory versus Laboratory and Field Experiments

Leszek M. Kaczmarek & Rafał Ostrowski[1]

1. INTRODUCTION

To parameterise the sediment transport in terms of the wave spectrum and a few other parameters, it is necessary to start from the surface elevation spectrum which can be transformed into orbital velocities at the bed and then to bed shear stress. The bed shear stress so derived is used in a new sediment transport formula put forth by the Authors. The problem is highly non-linear and at a few points in the process it is necessary to make non-linear transformations. This route: surface elevation - sediment transport and the importance of the non-linearity are central to this study.

The theoretical model is based on the concept proposed by Kaczmarek & O'Connor (1993) who used the procedure for matching the solutions of equation of motion in the turbulent flow above the theoretical bed level and in the collision-dominated granular-fluid region. This concept, first used for regular linear waves, has recently been developed for random waves by Kaczmarek et al. (1994) and for non-linear waves by Kaczmarek (1995).

Then, on the above basis, the first attempt was made by Kaczmarek et al. (1995) to formulate bedload theory and verify it using available laboratory data and IBW PAN radio-tracer field results.

This paper presents the results of experiment carried out at IBW PAN laboratory. The verification of the theory for regular and irregular waves using own experimental data is the major goal of this study. There are few experimental data sets for ripple regime, especially in the range $\theta_{2.5}$=0.1-0.4. Therefore the laboratory survey covered this range, particularly taking account of the identification of non-linear effects. This range of small $\theta_{2.5}$ is extremely important as one can expect the equivalence of bedload and total sediment transport in this regime (small suspended load rate). Thus,

[1] both Polish Academy of Sciences' Institute of Hydro-Engineering, *IBW PAN*,
7 Kościerska, 80-953 Gdańsk, Poland

only in this regime bedload theory can be precisely verified while in more severe hydrodynamic conditions bedload is a minor part of total sediment transport.

The various aspects of the non-linearity in respect to moveable bed roughness including the asymmetry of waves (described by 2nd and 3rd Stokes theory) and the wave-current interactions as well as wave irregularity are also discussed. A methodology for predicting the bed shear stress time series for irregular waves is applied to a large number of data sets of bottom velocity, some of which are simulated from field free surface elevation spectrum and some taken from IBW PAN laboratory data. The resulting bed shear stress is used in the bedload transport formula to predict sediment transport rate series for each data set. Hence, some questions concerning the equivalent regular waves generating the same shear stress root-mean-square values and the same bedload root-mean-square values as the irregular waves are answered.

2. MOVEABLE BED BOUNDARY LAYER THEORY

2.1. FORMULATION OF THE PROBLEM

The nearbed dynamics is examined and modelled for the flow regions above and beneath the original static bed line, see Figure 1. The collision-dominated granular-fluid region I stretches below the nominal static bed while the wall-bounded turbulent fluid region II extends above it. Since both water and sand grains are assumed to move in both regions, there must be a certain transition zone between I and II, in which the velocity profiles of I and II would merge and preserve continuity of shape.

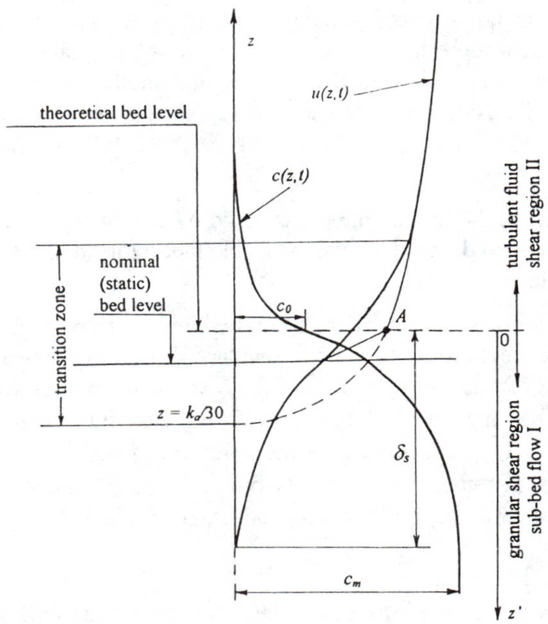

Figure 1. Definition sketch

The velocity distribution about a porous rough bed is controlled by various features of roughness and bed permeability. At first it is assumed that the velocity is determined by roughness geometry and outer flow parameters, such as the free-stream wave velocity.

The iteration procedure permits matching of velocities in the regions I and II. The velocity profile in the upper turbulent layer, which is linked with identification of roughness k_a, is determined first and then passed to the lower collision-dominated layer. The intersection of the two velocity profiles is marked as point A in Figure 1.

2.2. BASIC EQUATIONS

The flow in the turbulent upper region for regular waves is described by the integral momentum model based on the solution proposed by Fredsøe (1984):

$$\frac{dz_1}{d(\omega t)} = \frac{30\kappa^2 U}{k_a \omega e^{z_1}(z_1 - 1) + 1} - \frac{z_1(e^{z_1} - z_1 - 1)}{e^{z_1}(z_1 - 1) + 1}\frac{1}{U}\frac{dU}{d(\omega t)} \tag{1}$$

$$\frac{dz_1}{d(\omega t)} = \frac{30 z_1^2\left[\left|\frac{\kappa U}{z_1} \mp u_{f0}\right|\left(\frac{\kappa U}{z_1} \pm u_{f0}\right) \pm u_{f0}^2\right]}{k_a \omega U\left[e^{z_1}(z_1 - 1) + 1\right]} - \frac{z_1(e^{z_1} - z_1 - 1)}{e^{z_1}(z_1 - 1) + 1}\frac{1}{U}\frac{dU}{d(\omega t)} \tag{2}$$

for pure wave motion and wave with/against current, respectively.

The solutions of Equations (1) and (2) are achieved by the Runge-Kutta second-order method. As a result for a particular case, the function $z_1(t)$ is obtained and the time distributions of the friction velocity $u_f(t)$ and boundary layer thickness $\delta(t)$ are calculated thereafter using the following equations:

$$z_1 = \frac{U\kappa}{u_f \mp u_{f0}} \tag{3}$$

$$\delta = \frac{k_a}{30}\left(e^{z_1} - 1\right) \tag{4}$$

It should be emphasised that the free stream velocity $U(t)$ can be described as linear or nonlinear, thus Fredsøe's model can be adapted to nonlinear (asymmetric) wave motion, cf. Kaczmarek & Ostrowski (1992).

The solutions of Equations (1) and (2) enable the values of $u_{f\max}$ to be determined, if the quantities k_a are specified. To evaluate the roughness parameter k_a an iterative procedure is proposed for finding the matching point A.

The approach for irregular waves incorporates a time-invariant, two-layer eddy viscosity and bottom boundary layer thickness. Bed roughness is calculated using the iteration procedure taking account of the reduction of this parameter due to irregularity of wave motion. More detailed discussion on modelling of bed roughness is given in section 3.1.

In the sub-bed flow region, the sediment concentration is high and chaotic collisions of grains are the predominant mechanism. Particle interactions are assumed to produce two distinct types of behaviour. The Coulomb friction between particles give rise to rate-independent stresses (of the plastic type) and the particle collisions bring about stresses that are rate-dependent (of the viscous type). The use of the mathematical description by Sayed & Savage (1983) for determination of the stress tensor was made and the balance of linear momentum according to Kaczmarek & O'Connor (1993) leads to the following equations:

$$\alpha^0\left(\frac{c-c_0}{c_m-c}\right)\sin\varphi\sin 2\psi+\mu_1\left(\frac{\partial u}{\partial z'}\right)^2=\rho u_f^2 \quad (5)$$

$$\alpha^0\left(\frac{c-c_0}{c_m-c}\right)(1-\sin\varphi\sin 2\psi)+(\mu_0+\mu_2)\left(\frac{\partial u}{\partial z'}\right)^2= \left.\left(\frac{\mu_0+\mu_2}{\mu_1}\right)\right|_{c=c_0}\rho u_f^2+(\rho_s-\rho)g\int_0^{z'}c\,dz' \quad (6)$$

in which:

ρ_s and ρ are the densities of the solid and fluid, respectively;

α^0 is a constant;

c_0 and c_m are the solid concentrations corresponding to fluidity and the closest packing, respectively;

μ_0, μ_1 and μ_2 are functions of the solid concentration c:

$$\frac{\mu_1}{\rho_s d^2}=\frac{0.03}{(c_m-c)^{1.5}} \quad (7)$$

$$\frac{\mu_0+\mu_2}{\rho_s d^2}=\frac{0.02}{(c_m-c)^{1.75}} \quad (8)$$

The value φ in Equations (5) and (6) is the quasi-static angle of internal friction, while the quantity ψ is equal to:

$$\psi=\frac{\pi}{4}-\frac{\varphi}{2} \quad (9)$$

For the calculations the following numerical values are recommended:

$$\frac{\alpha^0}{\rho_s g d}=1 \qquad c_0=0.32 \qquad c_0=0.32 \qquad \varphi=24.4° \quad (10)$$

where d denotes the diameter of grains.

3. MODELLING OF BEDLOAD UNDER IRREGULAR WAVES

3.1. BED ROUGHNESS

The apparent bed roughness parameter k_a is a central quantity in the theoretical model. Both the turbulent and sub-bed velocity profiles depend on k_a, which is not known a priori. Therefore an iteration procedure is proposed for finding the matching point A between these profiles. The matching is assumed to take place at the phase of maximum shear stress, although at any other phase of oscillatory motion there must be some transition between the two profiles. Under monochromatic waves the maximum shear stress is the maximum value of shear stress during a wave period, while for spectral waves it becomes the maximum value of the random shear stress time series. To calculate this value it is proposed to use, the following simple relation, Kaczmarek et al. (1994):

$$\tau_{max} = \frac{3\tau_{rms}}{\sqrt{2}} = 3\delta_{\tau} \tag{11}$$

The choice is fairly arbitrary, however, as it was shown by Kaczmarek et al. (1994), it yields the best agreement of the calculations of wave friction using Madsen et al.'s (1990) data.

For engineering purposes it is useful to approximate the theoretical results by a curve expressed in terms of skin friction. To determine the skin friction one can follow Nielsen's (1992) description:

$$\theta_{2.5} = \frac{1}{2} f_{2.5}\,\psi_1 = \frac{1}{2} f_{2.5} \frac{(a_{1m}\omega)^2}{(s-1)gd} \tag{12}$$

$$f_{2.5} = \exp\left[5.213\left(\frac{2.5d_{50}}{a_{1m}}\right)^{0.194} - 5.977\right] \tag{13}$$

in which $s=\rho_s/\rho$.

The iteration procedure proposed for finding the matching point A between the velocity profiles (Fig. 1) was run for a wide range of small and large scale conditions and for a few diameters of sandy bed. Then the results were approximated yielding the following formulae for regular and irregular waves, respectively:

$$\frac{k_a}{d} = 47.03\theta_{2.5}^{-0.66} \tag{14}$$

$$\frac{k_a}{d} = 26.64\theta_{2.5}^{-0.71} \tag{15}$$

For irregular waves k_a is calculated using Eq. (15) taking root-mean-square wave height (or free-stream velocity) and peak period for determination of $\theta_{2.5}$.

The moveable bed roughness under sinusoidal and asymmetric waves with/versus a steady current has also been considered. The numerical solution of Eq. (2) is used with the free stream velocity $U(\omega\ t)$ expressed using linear or nonlinear approximation. The considerations on nonlinearity have been limited to the 2nd Stokes wave propagating without current and with/against weak and strong currents. For wave-current motion the ratio U_{mean}/U_{1m} was kept constant and amounted to 0.2 and 0.5 for weak and strong current, respectively. The quantity U_{mean} is the steady current, known as an input, averaged over water depth in the outer region (outside the bed boundary layer). The mean slip velocity (at the top of the bed boundary layer) is calculated from U_{mean} with the assumption of the logarithmic velocity profile in the outer region, using the model of Kaczmarek & Ostrowski (1992). This model also provides the value of shear stress u_{f0} resulting from the presence of steady current.

Figure 2 depicts the relative differences in roughness computed for regular waves (linear and nonlinear) and currents, as well as irregular wave motion, with respect to pure sinusoidal cases. The results imply that the greatest reduction of the roughness parameter k_a (about 40%) is observed when the waves are irregular and in a certain range of Shields parameter - when the nonlinear waves are accompanied by a steady current. Smaller reductions of k_a occur for sinusoidal waves superimposed on currents. It can be seen that for nonlinear cases the reduction of k_a is more distinct for waves propagating with the current than for waves versus the current.

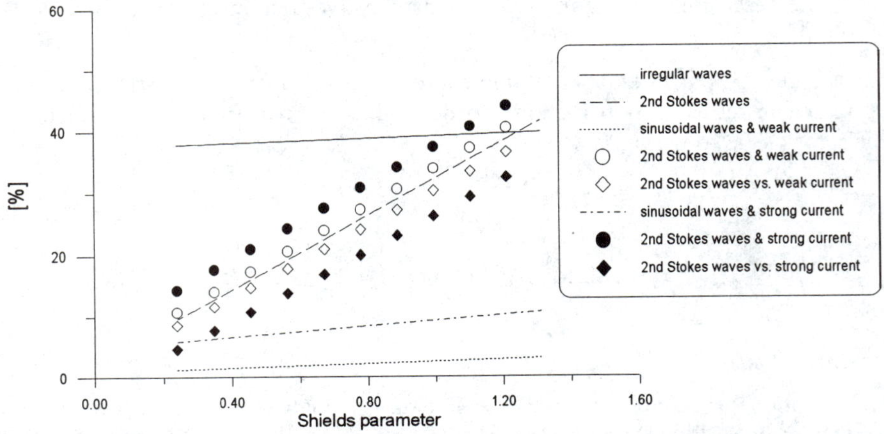

Figure 2. Relative differences (smoothed values) in k_a with respect to sinusoidal case

Hence, the irregularity of waves is the most important factor causing the reduction of moveable bed roughness.

3.2. SHEAR STRESS AND BEDLOAD

It is convenient to introduce complex notation which determines the dependence of velocity on time and will ensure the analytical solution of the equation of motion. The free stream velocity will be written as $U(t)=U\exp(i\omega\, t)$, in which U = real velocity amplitude. Similarly, one may write:

$$u(z,t) = u(z)e^{i\omega t} \qquad \text{and} \qquad u_d(z,t) = u_d(z)e^{i\omega t} \tag{16}$$

in which $u(z)$ and $u_d(z)$ are in general complex due to the phase shift relative to the free stream velocity. The equation of motion may now be written as:

$$\frac{d}{dz}\left[\nu_t(z)\frac{du_d(z)}{dz}\right] - i\omega\, u_d(z) = 0 \tag{17}$$

It is fortunate that this equation is explicitly solvable both when $\nu_t(z)$ varies linearly with z and when it is a constant. Thus, following Brevik (1981), the two-layer eddy viscosity model is proposed:

$$\nu_t(z) = \kappa u_{fr} z \qquad \text{for} \qquad \frac{k_a}{30} < z \le \frac{\delta_r}{4} + \frac{k_a}{30} \tag{18}$$

$$\nu_t(z) = \kappa u_{fr}\left(\frac{\delta_r}{4} + \frac{k_a}{30}\right) \qquad \text{for} \qquad z > \frac{\delta_r}{4} + \frac{k_a}{30} \tag{19}$$

where $u_{fr} = \sqrt{\tau_r / \rho}$ and δ_r are the representative friction velocity and representative bottom boundary layer thickness, respectively.

To solve the Eq. (17) the free stream velocity is specified as that associated with a wave spectrum, i.e.:

$$U(t) = \sum_n U_n e^{i\omega_n t} \tag{20}$$

in which the index n denotes summation over frequencies. With such a representation of $U(t)$ the velocity amplitudes U_n are related to the near-bottom orbital velocity spectrum and to the surface amplitude spectrum through:

$$U_n = \sqrt{2S_U(\omega_n)d\omega} = \frac{\omega_n}{\sinh(k_n h)}\sqrt{2S_\eta(\omega_n)d\omega} \tag{21}$$

in which ω_n and k_n are related to each other by linear dispersion relationship.

The linearity of Eq. (17) combined with the assumed time-invariant eddy viscosity concept (Equations (18) and (19)) suggests a solution in the form of:

$$u(z,t) = \sum_n u_n(z)e^{i\omega t} \tag{22}$$

in which $u_n(z)$ represents the complex velocity component amplitudes and only the real part of Eq. (22) constitutes the solution sought.

Introducing Equations (18) and (19) into Equation (17), one can obtain the equation for each velocity component n. Introducing the dimensionless variable:

$$\xi_n = \left(4\omega_n \frac{z}{\kappa u_{fr}}\right)^{\frac{1}{2}} \tag{23}$$

Eq. (17) reduces to the standard differential equation for the Kelvin functions of zeroth order, with ξ_n as the independent variable.

On the basis of the velocity solution one can obtain the shear stress at $z=z_0$ (complex value) which can be expressed as:

$$\tau(t) = \tau_n \exp[i(\omega_n t + \varphi_{\tau n})] \tag{24}$$

where τ_n and $\varphi_{\tau n}$ are the bed shear stress amplitude and phase, respectively, corresponding to nth harmonic component U_n of the input free stream velocity random series $U(t)$.

The solution obtained for the turbulent flow in the wave boundary layer involves the representative friction velocity u_{fr} and the representative thickness of the boundary layer δ_r, which are yet to be specified. Although each harmonic component of wave motion is described by the same equation, i.e. Equation (17), there is a coupling between the components incorporated in the eddy viscosity. This appears in the modelling of the representative values u_{fr} and δ_r by iterative procedure which is schematically shown in Figure 3. The quantities u_{fr} and δ_r are determined by using Eq. (1) and specified as:

$$u_{fr} = u_{f\max} \tag{25}$$

$$\delta_r = \delta_m = \max(\delta_1, \delta_2) \tag{26}$$

where $u_{f\max}$ is the maximum value of bed shear velocity during the wave period, that is $\max[u_f(\omega\ t)]$ and δ_1 and δ_2 the boundary layer thicknesses at the moments corresponding to maximum and minimum velocity at the top of the turbulent boundary layer. For a given by free stream irregular series the iterative procedure of Fig. 3 is used to determine the representative period T_r, friction velocity u_{fr} and the boundary layer thickness δ_r. The exemplary shear stress results vs. Gdańsk lab data, collected by Ostrowski (1993), are shown in Figure 4.

< 1 > Fourier decomposition of the free stream velocity input $U(t)$

$$U(t) = \sum_n U_n \sin(n\omega t + \varphi_n) + \tfrac{1}{2}U_0$$

< 1a > *Alternatively* Fourier decomposition of the water surface elevation input $\eta(t)$

$$U(t) = \sum_n \eta_n \frac{n\omega}{\sinh(k_n h)} \sin(n\omega t + \varphi_n) + \tfrac{1}{2}U_0$$

< 2 > Calculation of the input root mean square value:

$$U_{rms} = \sqrt{\sum_n U_n^2}$$

< 3 > Computation of bed roughness for irregular waves

< 4 > Assumption of representative period T_r

< 5 > Determination of parameters of representative eddy viscosity distribution: u_{fr} & δ_r (running Fredsœ's (1981) model with U_{rms} & T_r as an input)

< 6 > Computation of representative shear stress amplitude ρu_{fB}^2 (using Brevik's (1981) approach with U_{rms}, T_r & eddy viscosity distribution from step < 5 > as an input)

< 7 > Computation of bed shear stress components τ_n & $\varphi_{\tau n}$ using Brevik's approach with U_n, $n\omega$ (from step < 1 > or < 1a >) and representative eddy viscosity (determined in step < 5 >) as an input

< 8 > Calculation of bed shear stress root mean square value:

$$\tau_{rms} = \sqrt{\sum_n \tau_n^2}$$

< 9 > Checking whether ρu_{fB}^2 (step < 6 >) = τ_{rms} (step < 7 >)

if NO ⟶ correction of T_r and going to step < 5 >

if YES ⟶ going to step < 10 >

< 10 > Calculation of output time series (bed shear stress):

$$\rho u_f^2(t) = \tau(t) = \sum_n \tau_n \sin(n\omega t + \varphi_n + \varphi_{\tau n})$$

< 11 > Calculation of bedload time series using present model with the boundary conditions $u_f|_{z'=0} = u_f(t)$

Figure 3. Computation of bed shear stress and bedload under irregular waves

Once the bed shear stress is determined the velocity distributions can be computed for the regions I and II (Fig. 1), as well as the concentration of grains in bedload layer. Basing on Bagnold's (1956) definition, according to which the bedload is a part of sediment transport subject to inter-granular forces, the bedload layer can be represented as the region II. Knowing the instantaneous shear stress one can calculate the instantaneous bedload rate from velocity and concentration profiles in this layer:

$$Q_B = \int_0^{\delta_s} u(z',t) \cdot c(z',t)\, \mathrm{dz'} \tag{27}$$

and the dimensionless bedload rate defined as:

$$\phi_B = \frac{Q_B}{d\sqrt{(s-1)gd}} \tag{28}$$

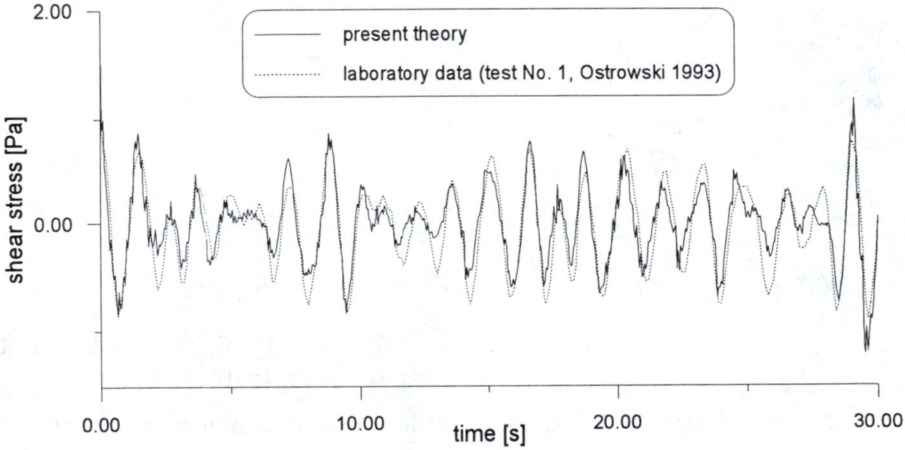

Figure 4. Theoretical and experimental irregular shear stress

The methodology for predicting the bedl oad time series for irregular waves was applied to a number of laboratory data sets. These results are discussed in the next section. The presented concept was also used for modelling of the bedload series for the field free surface elevation series taken at IBW PAN Coastal Research Facility in Lubiatowo. The exemplary results of the bedload simulation are given in Figures 5 and 6.

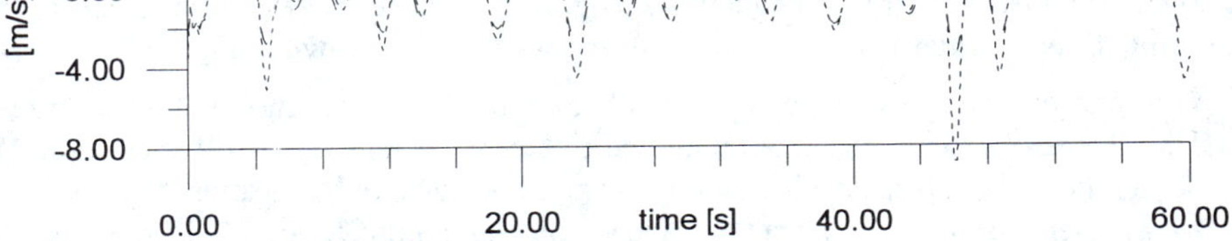

Figure 5. Time series of shear stress and bedload simulated for field wave input

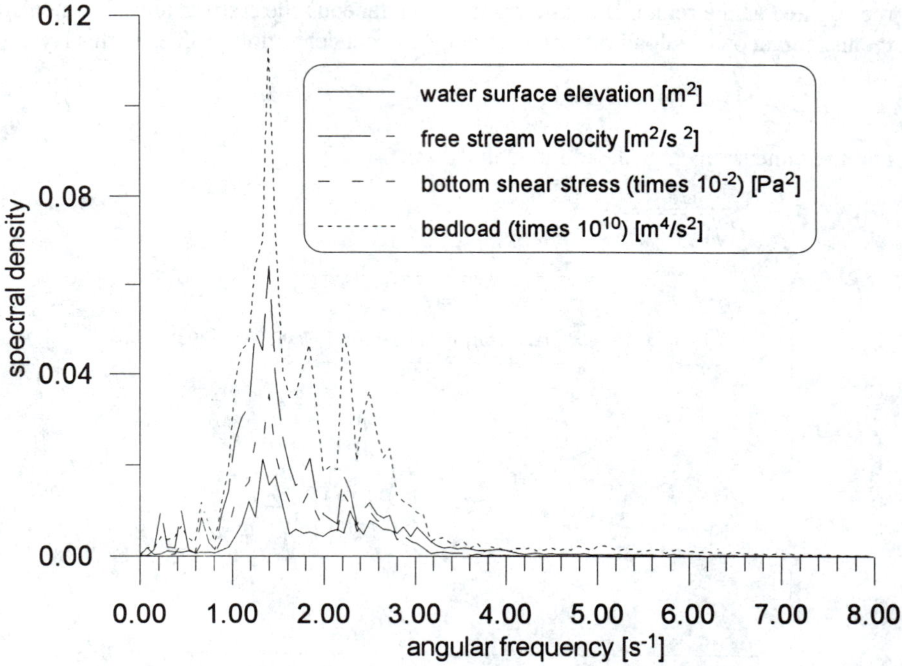

Figure 6. Spectra of shear stress and bedload simulated for field wave input

4. LABORATORY BEDLOAD SURVEY

4.1. EXPERIMENTAL SETUP

The measurements were carried out in the IBW PAN wave flume. The wave flume, 0.5 m wide and 22.5 m long, is equipped with a programmable wave maker and can be filled with water up to 0.7 m. Reinforced concrete slabs 8 cm thick were placed on the bottom with a sandy measuring section (also 8 cm thick) 7 m long situated about the middle of the flume. Natural sand was used in the experiments with a grain diameter d_{50}=0.22 mm.

For each test, free surface elevation was registered at three points along the flume. The horizontal component of free stream velocity was measured at one point in the measuring section, using a micro-propeller, together with one of the wave gauges which was located above the sand trap. The other two wave gauges were spaced 1/4*L from each other (L being a wave length) to estimate the reflection effects in the flume. Experimental setup, together with the sand trap, is shown in Figure 7.

The sand trap was covered by a lid and buried in the sandy section before each test. Then the waves were generated until bed ripples were fully formed, which took 25-60 minutes. The lid, suspended on four strings, was removed thereafter, together with a thin layer of sand on it. Then the wave action was continued for 1.2-15 minutes and sediment grains were being accumulated in the sand trap. Finally, the grains were siphoned from the trap and weighed to determine the sedimentation quantity.

The sand trap had two cells to ensure the determination of onshore and offshore bedload components.

Together with the sedimentation, bed forms geometry was measured and their shape was assessed after each test.

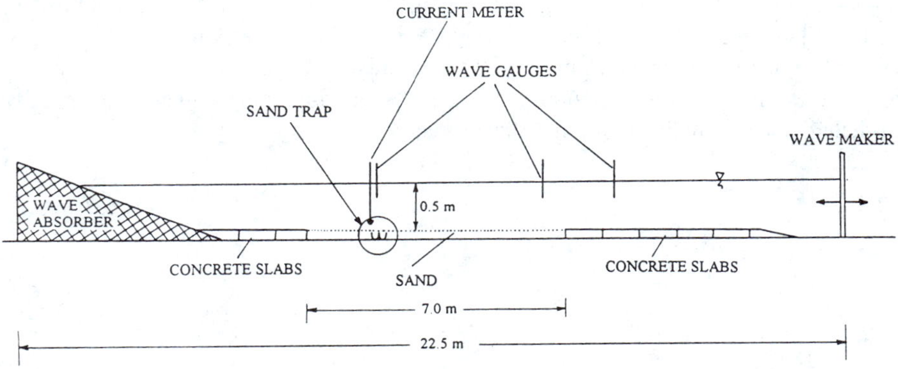

Figure 7. Experimental setup

A constant water depth of h = 0.5 m above the measuring section was maintained during each test. The experiments were carried out for six sets of parameters for regular waves (Tests 1, 2, 3, 4, 11 and 12) and for six irregular wave series (Tests 5, 7, 8 for JONSWAP and Tests 6, 9, 10 for Pierson-Moskowitz spectrum generated by the wave maker). As a total, 141 series were run, among which 103 ones were regular.

The conditions expressed in terms of $\theta_{2.5}$ varied roughly from 0.1 in Test 3 to 0.4 in Test 12. This range of rippled bed (cf. Fig. 4b) is extremely important as one can expect the equivalence of bedload and total sediment transport in this regime (small suspended load rate). Thus, only in this regime bedload theory can be precisely verified while in more severe hydrodynamic conditions bedload is a minor part of total sediment transport.

4.2. RESULTS OF MEASUREMENTS VS. PRESENT THEORY

The experimental bedload data in comparison with theoretical results obtained by using present theoretical linear approach are shown in Figure 8.

The waves registered in the flume had a slightly asymmetric shape. Therefore, for better representation of actual laboratory conditions, the theoretical results have been produced using present approach adapted for non-linear waves. The free stream velocity in Equation (1) was described using 2nd order Stokes theory, as indicated by Kaczmarek & Ostrowski (1992). Then, for known distributions of u_f in time, the instantaneous bedload transport rate has been found from Equations (5) and (6) for the entire wave period, integrated over time and averaged over wave period. These results are also shown in Figure 8.

The conformity of theoretical evaluations and experimental data can be seen in Figure 8, especially while using non-linear approach (except for Test 2). It should be pointed out that for long and highly asymmetric waves, represented by Test 11 (Ursell parameter of 40), the experimental data fit the present non-linear theory while they differ significantly from the linear approach. The most severe shear stress conditions generated in the flume without wave breaking had a $\theta_{2.5}$=0.4 and were achieved in Test 12. However, the Ursell number (equalled to 31) was less than in Test 11. In Test 12, a very distinct concentration of suspended sediment was observed which could result in increased accumulation in the sand trap, bigger than theoretically modelled, using both linear and non-linear theory.

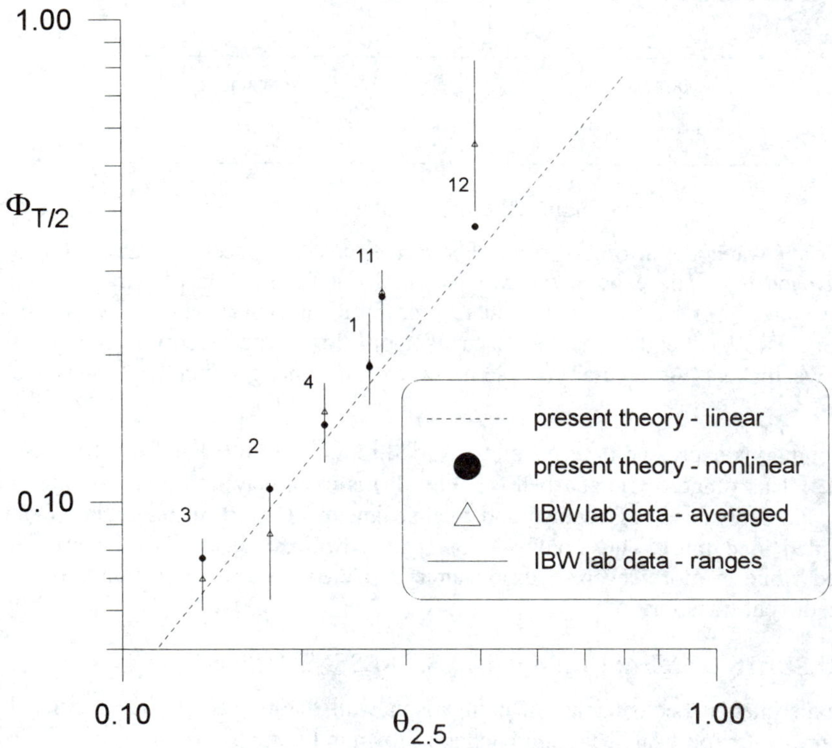

Figure 8. Regular bedload IBW PAN lab data vs. present linear and non-linear theory

The laboratory bedload data for irregular waves in comparison with theoretical results are depicted in Figure 9. Computed values have been modelled using the present computational procedure (Fig. 3) for the full-time (15 minutes) water surface elevation series registered at the measuring section. The agreement between theoretical and experimental results appears to be very good.

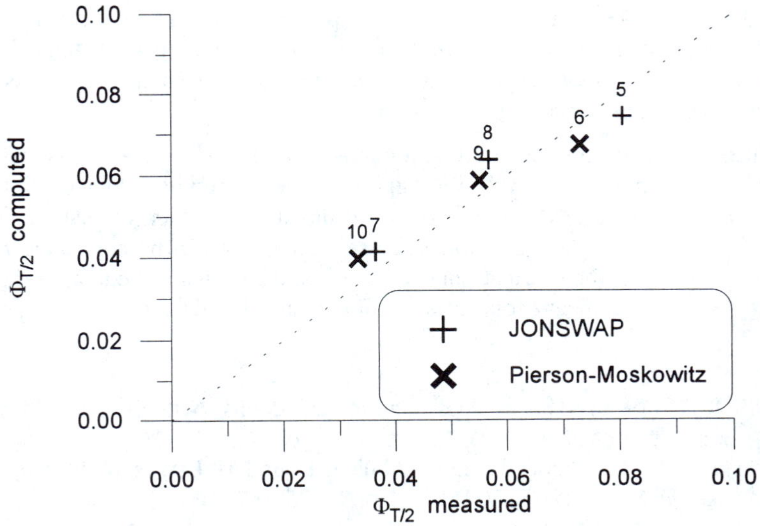

Figure 9. Bedload IBW PAN lab data vs. present theory for irregular waves

For the sake of full comparison the laboratory bedload data for regular and irregular waves are plotted together in Figure 10, against linear theory. Additionally, the bedload results simulated for field-recorded waves are included in Fig. 10. All results for irregular waves are presented as a function of dimensionless stress $\theta_{2.5}$, calculated using root-mean-square wave height and representative wave period T_r.

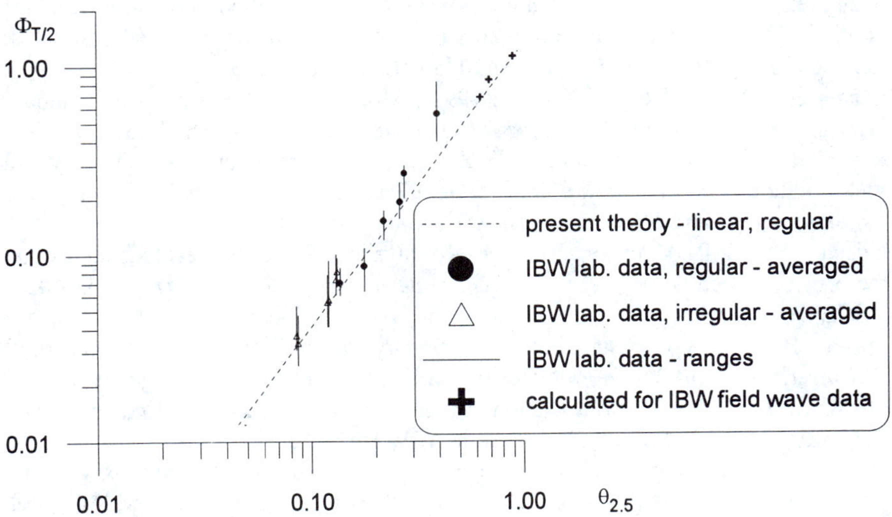

Figure 10. Bedload under regular and irregular waves vs. present linear theory

Figure 10 implies that the bedload rate for irregular cases can be successfully modelled with the use of present linear theory taking root-mean-square wave height

H_{rms} and representative wave period T_r as an input. It should be noted that in analysed cases the computed values of T_r are close to those of peak period, thus the peak period (T_p) can be used as representative. However, further studies are necessary to find out whether this conclusion is valid for all types of spectra.

It can also be seen from Figure 10 that for weak and moderate shear stress conditions the regular laboratory bedload data lie slightly above the results obtained using linear theory, thus for precise determination of bedload rate under regular-asymmetric waves the non-linear approach should be used. Finally, it can be concluded that for high shear stresses the present linear theory underestimates bedload rate, most probably - due to significant concentration of suspended sediment.

REFERENCES

Bagnold, R.A. (1956). The flow of cohesionless grains in fluids, *Phil. Trans.*, R. Soc., London, Ser. A, 249 (964), 235-297

Brevik, I. (1981). Oscillatory Rough Turbulent Boundary Layers, *J. Wtrway., Port, Coast. and Oc. Eng.*, ASCE, Vol. 107, No. 3, 175-187

Fredsøe, J. (1984). Turbulent boundary layer in combined wave-current motion, *J. Hydraulic Eng.*, ASCE, Vol. 110, No. HY8, 1103-1120

Kaczmarek, L.M. (1995). Nonlinear Effects of Waves and Currents on Moveable Bed Roughness and Friction, *Archives of Hydro-Engineering and Environmental Mechanics*, Vol. 42, No. 1-2, pp. 3-27

Kaczmarek, L.M., J.M. Harris & B.A. O'Connor (1994). Modelling moveable bed roughness and friction for spectral waves, *Proc. 24th ICCE*, 300-314, ASCE, New York

Kaczmarek, L.M. and B.A. O'Connor (1993a, b). A new theoretical approach for predictive evaluation of wavy roughness on a moveable flat/rippled bed, Parts I/II, *Reps. CE*/14-15/93, Dept. Civ. Engng., Univ. Liverpool

Kaczmarek, L.M. and Ostrowski, R. (1992). Modelling of wave-current boundary layer in the coastal zone, *Proc. 23rd Intern. Conf. Coast. Eng.*, ASCE, 350-363

Kaczmarek, L.M., R. Ostrowski & R.B. Zeidler (1995). Boundary Layer Theory and Field Bedload, *Proc. International Conf. Coast. Res. in Terms of Large Scale Experiments (Coast. Dynamics '95)*, ASCE, 664-675

Madsen, O.S., P.P. Mathison & M.M. Rosengaus (1990). Moveable bed friction factors for spectral waves, *Proc. 22nd Intern. Conf. Coast. Engng.*, 420-429, ASCE, New York

Nielsen, P. (1992). Coastal bottom boundary layers and sediment transport. *Advanced Series on Ocean Engineering*, Vol. 4, World Scientific, Singapore

Ostrowski, R. (1993). Bottom friction induced by surface waves and currents with nonlinear effects (in Polish), *Ph. D. thesis*, IBW PAN, Gdańsk

Sayed, M. and Savage, S.B. (1983). Rapid gravity flow of cohesionless granular materials down inclined chutes, *J. Applied Mathematics and Physics* (ZAMP), Vol. 34, 84-100

CHAPTER 269

KINEMATICS AND SHEAR STRESSES FROM COMBINED WAVES AND LONGSHORE CURRENTS IN THE UK COASTAL RESEARCH FACILITY

Richard R.Simons[1], Ruairi D.MacIver[2], Wameidh M.Saleh[2]

Abstract

The paper describes a series of experiments involving regular and random waves propagating over an orthogonal longshore current in the UK Coastal Research Facility at Wallingford. Mean and wave-induced velocities have been measured above a horizontal fixed rough bed using three acoustic velocimeters, and simultaneous measurements of mean and oscillatory bottom shear stress have been made with a UCL shear cell. Wave-induced shear stresses have been expressed as friction factors and compared with widely used empirical formulae. Other results have been compared with predictions for shear stress and apparent bed roughness from eight wave-current theories. The predicted enhancement of these parameters is judged against the relative strengths of the waves and currents tested, and against the practical range of conditions achievable in such laboratory basins.

Introduction

In trying to understand the effects of waves and currents on the coastal environment it is important to be able to predict the forces exerted on the seabed by the fluid motion and, conversely, what effect the seabed has on that motion. Such boundary layer processes are included in the wave-current interaction elements of the new generation of coastal numerical models, but there remains insufficient reliable data against which predictions from these models can be validated. Researchers at UCL are involved in a number of projects aimed at filling this gap.

In an earlier series of tests performed by the UCL group [Simons *et al.* (1992), (1994)], bottom shear stresses were measured directly using a novel shear cell device under conditions including regular and random waves propagating across an orthogonal current. This work suggested that waves have a significant effect on the current-induced mean shear stresses but that an additional current makes little difference to the wave-induced stresses. However, the relatively small basin used for those tests meant that the flows were not all fully rough turbulent. To produce

---

1) Senior Lecturer, Civil & Environmental Engineering Department, UCL
2) Research Assistant, Civil & Environmental Engineering Department, UCL
University College London, Gower Street, London WC1E 6BT, UK

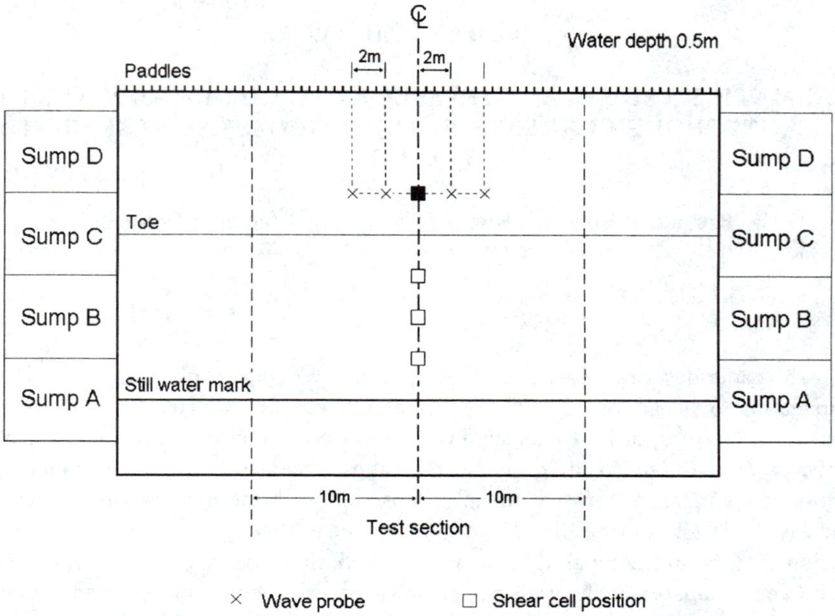

Figure 1 Schematic layout of UK Coastal Research Facility

results which can be applied to field-scale conditions and provide a valid test for the models, the experiments reported here have been carried out in the UK Coastal Research Facility (UKCRF) at Reynolds numbers well beyond transition and over a large fixed bed roughness.

The number of models in the literature which offer possible solutions to the problem of wave-current interaction in a turbulent coastal flow field is too extensive for detailed consideration here. All such models require empirical assumptions about the physical structure of the combined flow, and many also lead to difficult calculations in achieving a solution. In their review of the subject, Soulsby *et al.* (1993) put forward a simplified method for applying these theories using a set of standard formulae to relate the significant parameters. Each model is characterised by a unique set of 26 coefficients - found by fitting curves through model solutions for a wide range of input parameters - and these are used in the standard formulae to determine mean and maximum shear stresses. These "parameterised" versions of certain models are compared with the present experimental data later in this paper, both in terms of the enhanced bottom shear stress and also the apparent increase in bed roughness when waves are superimposed.

<u>UK Coastal Research Facility</u>

The UKCRF has been designed to provide a controlled environment in which various coastal processes can be simulated at relatively large scale. It measures 54m by 27m overall, with a central test region 20m by 15m **(fig.1)**, and is designed for water depths between 0.3m and 0.8m - although the present tests have been performed with 0.5m depth. The wave generation system consists of 72 individually controlled wave boards (each 0.5m wide) mounted along one wall of the basin. This can produce orthogonal or oblique-incidence regular or random waves up to 30^0 from shore-normal, with periods between 0.8s and 3.5s and heights up to 0.25 metres.

Currents are circulated by 4 independent reversible flow pumps and are introduced into the basin through 40 inlet flumes, each controlled by its own undershoot weir, with a matching set of flumes at the outlet end. One of the novelties of the UKCRF is that this system allows wave-driven currents to be circulated and at the same time allows for simulated wind- or tidally-driven longshore currents to be superimposed. Another is that the pumps act under programmable control, thus allowing time-varying (tidal) longshore currents with a user-defined period to be superimposed onto a controlled sequence of waves.

The facility is equipped with a wide range of instrumentation. For the present tests, a UCL shear cell was used to make direct measurements of the wave- and current-induced bottom shear stresses on the horizontal region of the basin. Briefly, the shear cell consists of a thin 250mm diameter plate supported level with the bed on four thin needles and deflecting laterally under the action of any horizontally applied shear stress at the bed. The movement (less than 0.5 mm) is recorded by two orthogonal eddy-current sensors and converted into an analogue voltage - as described in Grass *et al.* (1995).

A 3-d fibre-optic laser Doppler velocimeter (restricted here to 2-d operation only) provided detailed information within the bottom boundary layer; three Sontek 3-d acoustic Doppler velocimeters (ADV) were mounted, one above the other **(fig.2)**, on the z-axis of the instrument deployment bridge, thereby speeding up the process of measuring full vertical profiles of mean and wave-induced velocities; and 4 wave probes were deployed close to the shear cell to provide measurements of wave height and period.

To control the ADV's when they are operating near the centre of the basin, it is necessary to deploy a dedicated computer on the instrument carriage. The system is then operated from the control room via long cables linking the keyboard, screen and data logger to the remote processor.

For the present set of tests, the bed was roughened with nominal 10mm diameter granite chippings stuck to the concrete base of the basin and also to the surface of the UCL shear cell **(fig.3)**. The bottom roughness was stuck as a single layer, with a thin coating of adhesive painted onto the bed, the chippings rolled into it and left to set, and the surplus swept off some time later. The observed Nikuradse roughness was 18.7 mm.

Long-crested waves were generated in the offshore region and propagated across

Figure 2 **Photograph of the three Sontek acoustic velocimeter probes**

Figure 3 Granite roughness on the surface of the UCL shear cell

the horizontal bed before moving on to a 1-in-20 sloping plane beach. The general design and capabilities of the basin are described by Simons *et al.* (1995).

Experiments

The experiments reported here were performed with waves propagating orthogonally across a turbulent current above a fixed bed roughness. They involved:

a) 2 wave sequences generated from Jonswap spectra with peak frequencies of 0.4 Hz and significant wave heights of 0.15m and 0.18m; and

b) 5 regular wave conditions, with periods in the range 1.7s to 3s and heights between 0.18m and 0.26m.

Each test condition was repeated with the velocity measuring system positioned at different heights above the bed to allow vertical velocity profiles to be determined. For the tests on regular waves, between 150 and 200 wave cycles were recorded at each position - from which ensemble averages were processed; for the random waves, the sequences lasted 20 minutes.

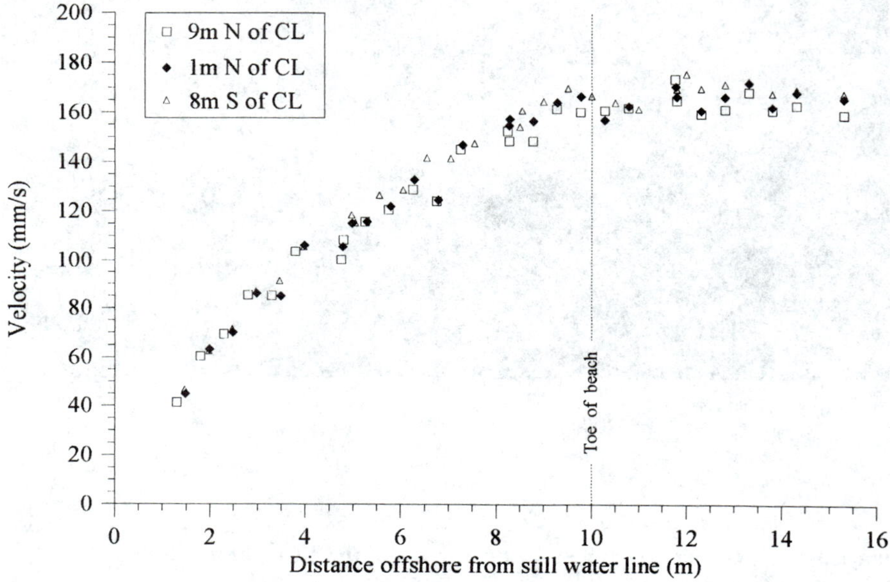

Figure 4 **Graph showing the distribution of longshore current across the beach in the central test region of the basin.**

The longshore current superimposed on the waves had a maximum velocity in the deep water region of approximately 0.14 m/s, reducing parabolically up the plane beach - see **fig.4**. At each position on the basin centreline, the vertical profile was logarithmic and demonstrated that the boundary layer was fully developed. This particular current setting is one of the standard reproducible conditions for the UKCRF established during initial evaluation work on the Facility. Test conditions and results are listed in **Table 1**.

MEAN CURRENT PARAMETERS

Current Alone

TEST	T (s)[a]	H (m)[b]	U_{bar}[c] (m/s)	Z_0[d] (m)	τ_c (N/m^2) Log layer	τ_c (N/m^2) Shear cell
0410953	--	--	0.119	6.24 E-4	0.086	0.079

Regular Waves & Current

TEST	T (s)[a]	H (m)[b]	U_{bar}[c] (m/s)	Z_a[d] (m)	τ_m (N/m^2) Log layer	τ_m (N/m^2) Shear cell	WCI[k] parameter $y = \tau_m/(\tau_c+\tau_w)$
0310951	1.70	0.213	0.122	3.33 E-3	0.254	--	0.047
0610953	2.10	0.219	0.125	4.52 E-3	0.364	0.264	0.057
0210951	2.45	0.256	0.119	6.19 E-3	0.395	0.365	0.064
0610951	2.70	0.262	0.121	3.48 E-3	0.258	0.263	0.058
0410951	3.00	0.260	0.116	4.73 E-3	0.284	--	0.064

WAVE PARAMETERS

Regular Waves Alone

			Crest				Trough			
TEST	T (s)[a]	H (m)[b]	U_{bed}[e] (m/s)	a/Z_0[f]	τ_w[g] (N/m^2)	WCI[k] parameter $x = \tau_c/(\tau_c+\tau_w)$	U_{bed}[e] (m/s)	a/Z_0[f]	τ_w[g] (N/m^2)	WCI[k] parameter $x = \tau_c/(\tau_c+\tau_w)$
0310952	1.70	0.199	0.331	131	6.41	0.013	0.253	100	5.31	0.016
0610954	2.10	0.177	0.409	204	7.34	0.012	0.326	162	6.30	0.013
0210952	2.45	0.275	0.406	194	7.52	0.011	0.233	111	6.09	0.014
0610952	2.70	0.254	0.435	203	9.45	0.009	0.228	107	4.36	0.019
0310953	2.99	0.243	0.451	220	10.01	0.008	0.269	131	4.38	0.019

Irregular Waves Alone

TEST	T (s)[a]	H (m)[b]	U_{bed}[h] (m/s)	a/Z_0[i]	τ_w[h] (N/m^2)
2909952	2.24	0.152	0.130	74	1.98
2809951	2.19	0.187	0.149	83	2.32

Regular Waves & Current

			Crest				Trough			
TEST	T (s)[a]	H (m)[b]	U_{bed}[e] (m/s)	a/Z_0[f]	τ_{max}[g] (N/m^2)	WCI[k] parameter $Y = \tau_{max}/(\tau_c+\tau_w)$	U_{bed}[e] (m/s)	a/Z_0[f]	τ_{max}[g] (N/m^2)	WCI[k] parameter $Y = \tau_{max}/(\tau_c+\tau_w)$
0310951	1.70	0.213	0.310	124	6.08	0.937	0.237	94	4.97	0.921
0610953	2.10	0.219	0.358	176	5.70	0.768	0.290	142	5.86	0.919
0210951	2.45	0.256	0.416	187	7.97	1.048	0.187	84	4.78	0.774
0610951	2.70	0.262	0.439	197	8.76	0.919	0.210	94	4.14	0.930
0410951	3.00	0.260	0.464	237	9.69	0.960	0.266	136	4.09	0.916

Irregular Waves & Current

TEST	T (s)[a]	H (m)[b]	U_{bed}[h] (m/s)	a/Z_0[i]	τ_w[h] (N/m^2)
2909953	2.23	0.153	0.128	73	1.97
2909951	2.20	0.186	0.156	88	2.37

[a] T_{13}, established from phase locking wave probe 3. (Data file B)
[b] H_{13}, established from wave probes 2 & 3. (Data file B)
[c] Depth averaged velocity.
[d] Established from log layer.
[e] At z'=35mm. (Ensemble average data file C)
[f] *a* calculated as $U_{bed}/(\pi/\text{half period})$.
[g] Established from shear cell.
[h] RMS of time series. (Data file A)
[i] *a* calculated as $U_{bed}/(2\pi/\text{wave period})$.
[k] WCI parameters from Soulsby *et al*. (1993)

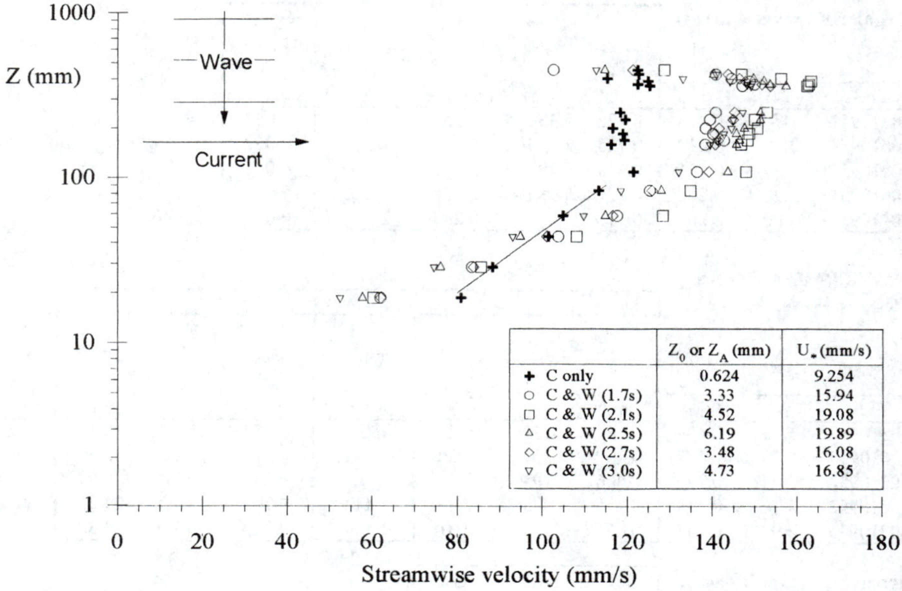

Figure 5 **Longshore current mean velocity profiles: u v $\log_e z$**
Current alone and with waves superimposed.

<u>Results</u>

Analysis of the logarithmic mean velocity profiles for the current alone suggested a bed roughness, ks, of 18.7mm. When orthogonal (or oblique) waves were superimposed, the logarithmic profiles **(fig.5)** showed the expected increases in mean shear stress and apparent bed roughness from their values for current alone. It was also noted that the average-over-depth flow rate was slightly higher at the offshore measuring position when the waves were present. This was attributed to the non-linear wave-current enhancement of mean shear stress being greater in shallow water than in the deep water region, thereby producing a greater resistance to the longshore current inshore on the 1-in-20 beach and thus redirecting the flow out towards the deeper water - where the present measurements were made.

Table 2 shows a comparison between the observed mean longshore shear

Test Data			Model predictions for τ_{wc} Nm^{-2}							
T s	τ_c Nm^{-2}	τ_{wc} Nm^{-2}	F 84	MS 90	HTT 91	DSK 88	OY 88	CN 86	S 91	N 92
1.7	0.09	0.25	0.16	0.14	0.22	0.16	0.15	0.20	0.20	0.20
2.1	0.09	0.36	0.19	0.17	0.26	0.18	0.15	0.22	0.25	0.23
2.5	0.09	0.39	0.17	0.16	0.23	0.17	0.14	0.19	0.22	0.20
2.7	0.09	0.26	0.18	0.17	0.24	0.17	0.13	0.19	0.23	0.21
3.0	0.09	0.28	0.19	0.18	0.25	0.19	0.13	0.20	0.24	0.18

Table 2: Mean bed shear stress τ_m: comparison of measurements from regular wave tests with predictions from 8 theories.

stresses τ_{wc} deduced from the logarithmic velocity profiles and predictions from a number of wave-current models. τ_c was measured for the current on its own. The values from Davies *et al.* (1988), Fredsoe (1984), Huynh-Thanh & Temperville (1991), Myrhaug & Slaattelid (1990), and O'Connor & Yoo (1988) were all deduced from the parameterised versions of those models presented by Soulsby *et al.* (1993) and discussed above; this was particularly helpful for the fully numerical models which would otherwise have been inaccessible. The values from Coffey & Nielsen (1986), Sleath (1991), and Nielsen (1992) were calculated (more-or-less) directly.

Test data			Model predictions for za							
T s	z0 mm	za mm	F 84	MS 90	HTT 91	DSK 88	OY 88	CN 86	S 91	N 92
1.7	.63	4.73	3.82	3.09	6.92	3.65	3.45	6.04	9.12	5.7
2.1	.63	3.48	4.63	3.81	8.29	4.43	3.04	6.47	12.3	6.68
2.5	.63	6.19	4.96	4.35	8.13	4.75	3.13	5.55	11.7	6.24
2.7	.63	4.52	4.92	4.35	8.00	4.71	2.81	5.40	12.4	6.34
3.0	.63	3.33	6.31	5.67	9.89	6.06	3.20	5.93	14.5	6.97

Table 3: Apparent bed roughness k_a: comparison of measurements from regular wave tests with predictions from 8 theories.

However, these three models are intended primarily to predict the enhanced bed roughness assuming an appropriate value for wave-current shear stress. To deduce the predicted shear stress in these cases, it has been assumed that the mean-over-depth velocity is as measured for the wave-current tests, and also that the velocity profile remains logarithmic across the full flow depth. Then, starting with an initial guess for shear stress, it is possible to calculate a first estimate for apparent bed roughness from an integrated form of the logarithmic boundary layer equation. An improved estimate for shear stress can then be found, and repeated iteration used to produce solutions both for apparent roughness and for mean shear stress.

Table 3 gives a similar comparison to that described above, but now looking at the apparent bed roughness. The predictions have been calculated, again assuming fully logarithmic velocity profiles and also that the mean-over-depth velocity is as measured in the combined wave-current flow. This was done to allow flows to be compared "like-for-like" and to overcome the problem of the local flow rate having been altered by the redirection of the longshore current (as discussed above). In order to obtain these data for each case, it was necessary to integrate the velocity profiles manually - taking into account the additional Eulerian mean flow taking place above Still Water Level through the wave crest. A typical example, showing a curve-fit through the scattered ADV data from which the mean-over-depth velocity was calculated, is shown in **fig.6**.

The first thing to note from these two tables is that the apparent bed roughness and mean shear stress have, as the models all predict, both increased when the waves are superimposed. That there is not a steady increase with relative wave strength can be attributed to the observation that the longer period waves have become non-linear with significant secondary crests. However, even when looking at the 1.7s and 2.1s period tests, it can be seen that there is a wide range of predictions and that no single model stands out as ideal.

Turning to the wave-induced stresses, friction factors for the regular wave tests calculated from ensemble averaged wave-induced velocities and shear stresses showed no visible change when the current was superimposed, so confirming UCL's earlier results.

Continuous time-series of velocities and shear stresses from the random wave tests were analysed half a wave cycle at a time to produce some hundreds of independent values of friction factor through each sequence of irregular waves **(fig.7)**. Values were calculated for a specific half-wave period, determining the amplitude between consecutive trough and crest - of wave-induced shear stress from the shear cell, and of velocity from ADV measurements outside the wave boundary layer.

Again, these showed that for the range of test conditions possible in the UKCRF

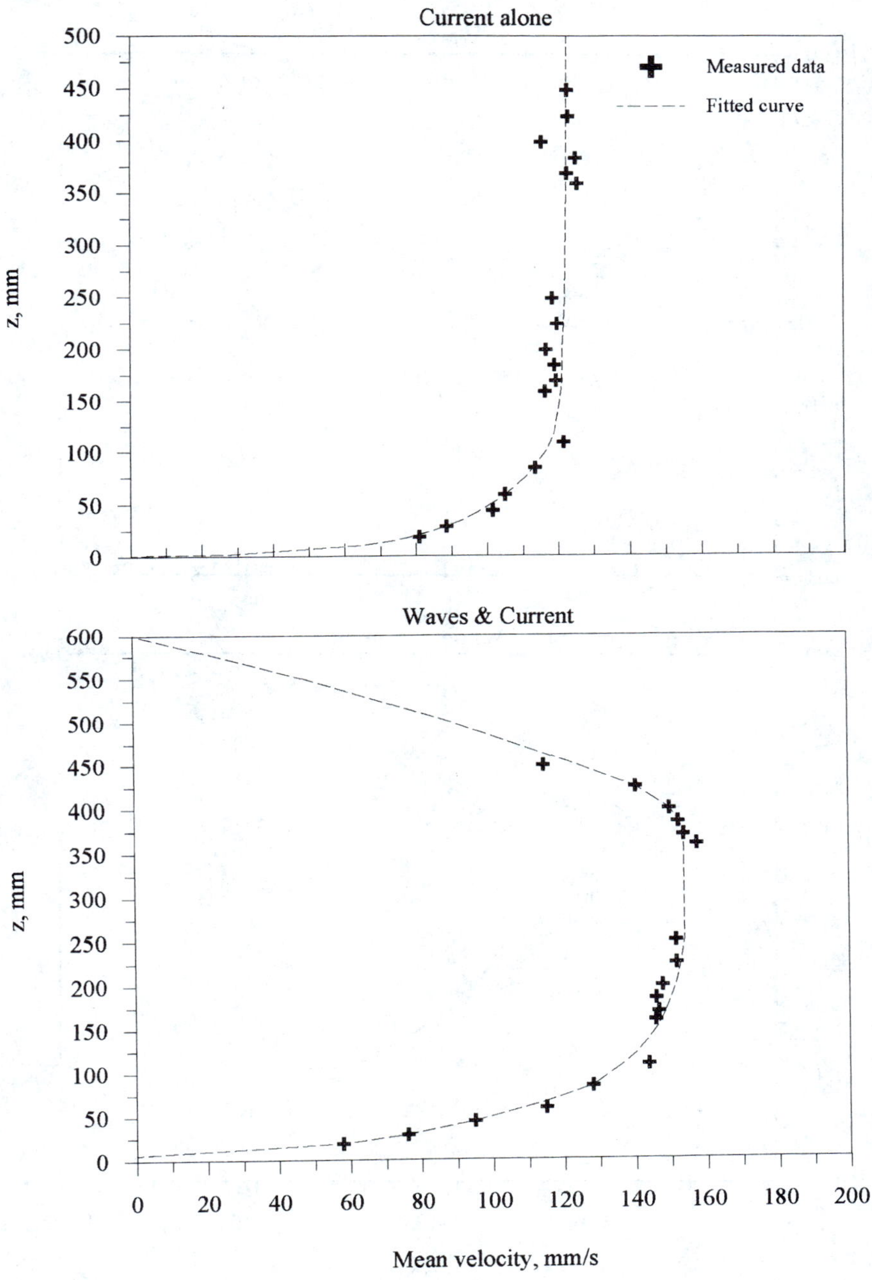

Figure 6 **Mean velocity profiles for:**
a) current alone, b) current + wave (T=2.5s, H=0.2m)

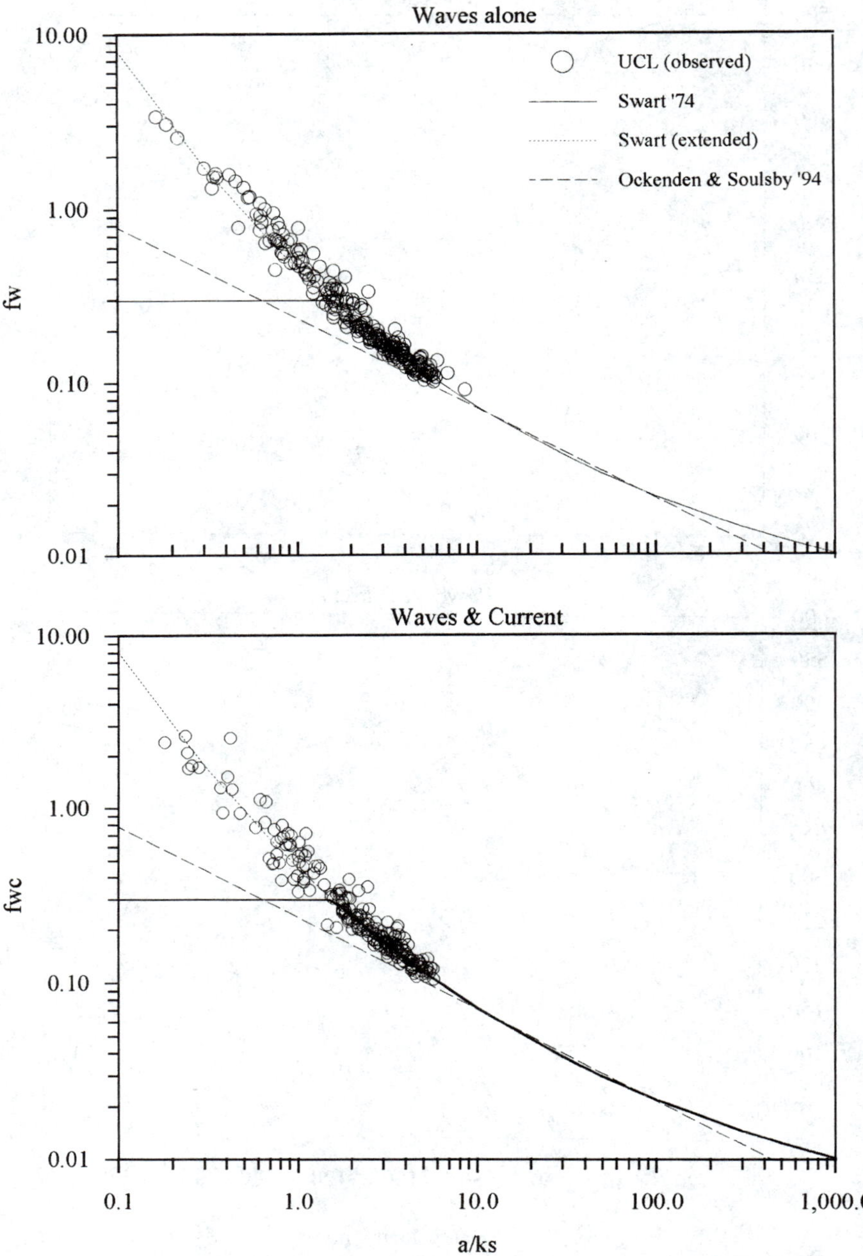

Figure 7 **Friction factor calculated from each half wave cycle: Random wave sequence: Tp=2.5s, Hs=0.15m**

there is no discernable increase in friction factor (and hence no effect on the wave-induced bottom shear stress) when a current is superimposed. However, they did confirm that Swart's (1974) formula for friction factor gives a good estimate, for waves alone or with an orthogonal current added, even at very low a/k where a constant value has sometimes been proposed. Comparison with the power law approximation proposed in Ockenden and Soulsby (1994) is also encouraging for the practical range 5<a/k<100 although this appears less accurate at low values of a/k and cannot be correct at high a/k where the friction factor should tend to a constant value for a quasi-steady current.

If we now consider what enhancement of wave-induced shear stress (or friction factor) is to be expected when a current is superimposed, it seems that none of the models actually predicts any significant enhancement under the conditions being investigated. This is because the wave boundary layer only starts to be modified when the current-induced mean shear stress is of a similar magnitude to the wave-induced stress, and this can only be achieved in the laboratory if mean flow velocities are impractically high or the waves are so small as to be dominated by viscosity and surface tension. So the consistent behaviour of the waves whether with and without the longshore current is indicating that this aspect of wave-current theory is valid.

Conclusions

For a longshore current, mean shear stress and apparent bed roughness are both increased significantly by the addition of relatively strong orthogonal waves. Predictions from the wave-current models vary significantly, although the Davies *et al.* (1988) and Fredsoe (1984) theories appear most consistent with the present data.

Swart's (1974) formula for wave friction factor gives excellent agreement with the values measured by the shear cell for both regular and irregular waves.

The addition of a relatively weak longshore current has no effect on the bottom shear stress generated by waves on a rough bed.

The UCL shear cell is capable of measuring mean and oscillatory bottom shear stresses in a large-scale three-dimensional flow field.

Acknowledgements

The authors would like to express their gratitude to the UK Engineering and Physical Sciences Research Council for their financial support of this project, and also to HR Wallingford Ltd for their co-sponsorship of the UKCRF.

References:

Coffey, F.C. and Nielsen, P. 1986 The influence of waves on current profiles. Proc. 20th Int.Conf. on Coastal Engng., Taipei, pp.82-96

Davies, A.G., Soulsby. R.L. and King, H.L. 1988 A numerical model of the combined wave and current bottom boundary layer. J.Geophysical Res., 93(C1), pp.491-508

Fredsoe, J. 1984 Turbulent boundary layer in wave-current motion. J.Hydraulic Engng., ASCE 110: pp.1103-1120

Grass, A.J., Simons, R.R., MacIver, R.D., Kalopedis, A. 1995 Shear cell Proc.IARH Congress, London, 11-15 Sept. 1995, Vol.1, 1C8, pp.415-420

Huynh-Thanh, S. and Temperville, A. 1991 A numerical model of the rough turbulent boundary layer in combined wave and current interaction. In: R.L.Soulsby and R.Bettess (Editors), Sand Transport in Rivers, Estuaries and the Sea. Balkema, Rotterdam, pp.93-100

Myrhaug, D. and Slaattelid, O.H. 1990 A rational approach to wave-current friction coefficients for rough, smooth and transitional turbulent flow. Coastal Engineering, 14, pp.265-293

Nielsen, P. 1992 Coastal bottom boundary layers and sediment transport. Advanced series on Coastal Engineering, Vol.4. World Scientific Publishing Co., Singapore.

O'Connor, B.A. and Yoo, D. (1988) Mean bed friction calculation of combined wave-current flow. Coastal Engineering, 12, pp. 1-21.

Ockenden, M.C. and Soulsby, R.L. 1994 Sediment transport by currents plus irregular waves. HR Wallingford Report SR 376, February 1994.

Simons R.R., Whitehouse R.J.S., MacIver R.D., Pearson J., Sayers P., Zhao Y. and Channell A. 1995 Evaluation of the UK Coastal Research Facility. Coastal Dynamics '95. Proc. Int.Conf. - Large Scale Experiments, Gdansk, Poland, 4-8 September 1995, pp.161-172

Simons, R.R., Grass, A.J., Saleh, W.M. and Mansour-Tehrani, M. (1994) Bottom shear stresses under random waves with a currents superimposed. Proc. 24th Int. Conf. on Coastal Engng, Kobe, Japan, Ch.42, pp.565-578

Simons, R.R., Grass, A.J. and Mansour-Tehrani, M. (1992) Bottom shear stresses in the boundary layer under waves and currents at right angles. Proc. 23nd Int. Conf. on Coastal Engng, Venice, Italy, Ch.45, pp.604-617

Sleath, J.F.A. 1991 Velocities and shear-stresses in wave-current flows. J.Geophysical Res., Vol.96 (C8), pp.15237-15244

Soulsby, R.L., Hamm, L., Klopman, G., Myrhaug, D., Simons, R.R., and Thomas, G.P. (1993) Wave-Current Interaction within and outside the bottom boundary layer. Coastal Engineering, 21, pp.41-69.

CHAPTER 270

Movable Bed Roughness in the Flow of Irregular Waves and Currents over Movable Beds

Zai-Jin You[1] Peter Nielsen[2]

Abstract

The bed roughness k_s in the flow of irregular waves and currents over a movable bed is studied on the basis of measured current profiles and the model of You (1994). It is found that the movable bed roughness k_s is affected by both waves and currents and that the existing formulae derived in purely oscillatory flow generally fail to predict k_s in the flow of waves and currents. A new formula is proposed to calculate k_s in the flow of irregular waves and currents over a movable bed. The present bed roughness formula together with You's (1994) model gives better prediction of current shear velocity $\overline{u}_{cw}^*$ and especially apparent roughness z_1 than the existing formulae derived in purely oscillatory flow.

Introduction

The bed roughness k_s is an important input parameter in the modelling of coastal processes, but usually unknown in the coastal zone where the seabed is often movable owing to irregular waves and currents. The movable bed roughness k_s in purely oscillatory flow has been studied by many investigators (Van Rijin, 1982, Grant and Madsen, 1982; Nielsen, 1983; Raudkivi, 1988), but little investigated in the flow of irregular waves and currents. Consequently, the formulae derived in purely oscillatory flow have been often applied to estimate the movable bed roughness k_s in

----

[1]Victorian Institute of Marine Sciences, Locked Bag 7, Collins Street East Post Office, Melbourne 8003, AUSTRALIA.

[2]Department of Civil Engineering, University of Queensland, Brisbane 4072, AUSTRALIA.

in the flow of irregular waves and currents. For example, Coffey (1987) used the formulae of Nielsen (1983), Van Rijin (1982), and Grant and Madsen (1982) to calculate k_s at $\theta = 0^o$, 90° and 180°, but did not explain why the different formulae were applied for the different angles, where θ is the angle between the wave propagation and a current.

Van Kampen and Nap (1988) found that the movable bed roughness k_s estimated at $\theta = 0^o$ or 180° from their experimental data was very large and could not be predicted by the existing formulae. In the sequential experiment of Havinga (1992), however, k_s estimated at $\theta = 60^o$ or 120° was found to be very small and could not be predicted by the existing formulae either.

Drake *et al.* (1992) used the model of Grant and Madsen (1979) to study the movable bed roughness k_s in the flow of irregular waves and currents on the basis of measured current profiles. A simple empirical formula was suggested to be $k_s = 28\eta^2 / \lambda - 0.14\theta$, where $0 \le \theta \le 90^o$. Unfortunately, the model of Grant and Madsen (1979) together with the empirical formula still failed to predict the apparent roughness z_1 as shown in Figure 4 of Drake *et al.* (1992).

More recently, Mathisen and Madsen (1996) studied the fixed bed roughness in steady flow, oscillatory flow and combined wave-current flow, respectively. It was found that k_s experienced by a pure current was equal to that by waves alone or by waves and currents. However, this is not valid for a movable bed simply because the movable bed form, which is suggested to a main contributor to k_s, is found to be quite different in steady flow compared to that in a combined wave-current flow (Arnot and Southard, 1990; Havinga, 1992).

In the present study, the movable bed roughness k_s in the flow of irregular waves and currents over a movable bed is studied on the basis of measured current profiles and the calibrated model of You (1994). A practical application is also given to calculate k_s in the coastal zone.

Model of Wave-Current Flows over Fixed Beds

In steady flow over a fully rough fixed bed, the current profile is logarithmic

$$\overline{u} = \frac{\overline{u}_c^*}{\kappa} \ln \frac{z}{z_o} \tag{1}$$

in which $\overline{u}_c^*$ is the current shear velocity, κ is the von Karman constant, $k_s = 30 z_o$ and k_s is the bed roughness. When waves are superimposed on a current, a logarithmic current profile near the bed can still be found and similarly expressed as

$$\overline{u} = \frac{\overline{u}^*_{cw}}{\kappa} \ln \frac{z}{z_1} \tag{2}$$

but with $\overline{u}^*_{cw} > \overline{u}^*_c$ and $z_1 > z_o$ as first studied experimentally by Bijker (1967) and recently by Havinga (1992) and Klopman (1994).

Many models of combined wave-current flows over fixed beds have been developed to calculate the current shear velocity $\overline{u}^*_{cw}$ and the apparent roughness z_1 in Eq.(2). A review of existing models of combined wave-current flows refers to You (1992). You (1994, 95a) compared the four analytical models of Christoffersen and Jonsson (1985), Coffey and Nielsen (1986), Sleath (1991) and You (1994) with the laboratory measurements of van Doorn (1991, 82) and Kemp and Simons (1982). It was concluded that the model of You (1994) was simpler and gave better agreement with the experimental data than the others. In the present study, the model of You (1994) is chosen to study the movable bed roughness k_s in the flow of irregular waves and currents.

In the model of You (1994), the input parameters were chosen to be the wave parameters (A, ω), a reference current velocity $\overline{u}_r$ at an arbitrary level, and the bed roughness k_s. For irregular waves, significant wave height H_s and period T_s were suggested to calculate A and ω using linear wave theory. The current shear velocity $\overline{u}^*_{cw}$ in Eq.(2) was explicitly calculated by

$$\overline{u}^*_{cw} = 0.5 u^*_w \left\{ \log^b_a + \left[\left(\log^b_a \right)^2 + 1.6 \frac{\overline{u}_r}{u^*_w} \log^e_a \right]^{0.5} \right\} \tag{3}$$

in which the wave shear velocity u^*_w was evaluated as

$$u^*_w = \sqrt{0.5 f_w}\, A\omega \quad \text{and} \quad f_w = 0.108 \left(\frac{k_s}{A} \right)^{0.343} \tag{4}$$

and the parameters a and b were defined by

$$a = \frac{30 e \delta_1}{k_s}, \quad b = \frac{2\delta_1}{z_r} \quad \text{and} \quad \delta_1 = \frac{0.5 \kappa u^*_w}{\omega} + 0.2 z_o \, . \tag{5}$$

The apparent roughness z_1 in Eq.(2) was also calculated explicitly by

$$z_1 = 2\delta_1 \left(\frac{30 e \delta_1}{k_s} \right)^{-\frac{\overline{u}^*_{cw}}{u^*_w}} . \tag{6}$$

The derivations of Eqs.(3)-(6) refer to You (1994). Alternatively, when the current shear velocity $\overline{u}_{cw}^{*}$ and the apparent roughness z_1 are measured, the movable bed roughness k_s in the flow of waves and currents can be determined by the model.

Movable Bed Roughness

The formulae derived in purely oscillatory flow have been often used to calculate the movable bed roughness k_s in the flow of waves and currents (Grant Williams, 1984; Coffey, 1987; Drake et al, 1992)

$$k_s = C\eta^2/\lambda + f(\Theta) \tag{7}$$

in which η and λ are the ripple height and length, $f(\Theta)$ is an additional bed roughness owing to the near-bed sediment transport, and C is constant and was chosen to be 8, 16 and 28 by Nielsen (1983), Raudkivi (1988) and Grant and Madsen (1982), respectively. Since no direct measurements of k_s are available in the flow of waves and currents over movable beds, a direct comparison of Eq.(7) with experimental data on k_s becomes impossible. When the current shear velocity $\overline{u}_{cw}^{*}$ and apparent roughness z_1 are measured, however, Eq.(7) may be verified indirectly by using a calibrated model of combined wave-current flows. That is, if the calibrated model together with Eq.(7) can agree well with the measured $\overline{u}_{cw}^{*}$ and z_1, it may be concluded that Eq.(7) gives good estimation of k_s.

In the present study, the laboratory experimental data from Kaaij and Nieuwjaar (1987), Van Kampen and Nap (1988) and Havinga (1992) are used to study the movable bed roughness k_s in the flow of irregular waves and currents. The irregular waves in the three sequential experiments were generated by a directional wave generator. The wave spectrum was JONSWAP. The significant wave heights H_s ranged from 7.5 to 18cm, the peak wave period T_p was about 2.5s, and the depth-averaged current velocity $<\overline{u}>$ varied from 10 to 40cm/s. The current profiles and bed forms in the absence and presence of irregular waves were measured. The movable roughness k_s calculated from Eq.(7) was found to be generally dominated by the first term in Eq.(7).

In Figure 1, the models of Sleath (1991) and You (1994) together with Grant and Madsen's (1982) formula of $k_s = 28\eta^2/\lambda$ are used to calculate the apparent roughness z_1 and compared with the experimental data. It can be seen that the models generally underestimate z_1 at $\theta = 0^o$ and 180°, but overpredict z_1 at $\theta = 60^o$ and 120°. This indicates that the bed roughness k_s used in the models has been underestimated at $\theta = 0^o$ and 180°, but overcalculated at $\theta = 60^o$ and 120°. The models would not give better prediction of z_1 by adjusting the constant C or the second term in Eq.(7).

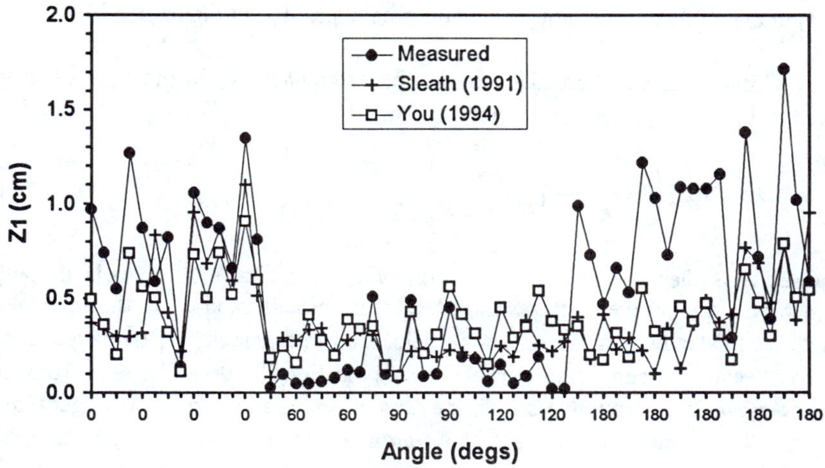

Figure 1. Comparison of the apparent roughness z_1 measured by Kaaij and Nieuwjaar (1987), Van Kampen and Nap (1988) and Havinga (1992) with those calculated by the models of Sleath (1991) and You (1994) with $k_s = 28\eta^2/\lambda$. The ripple height η and length λ were directly measured in the experiments.

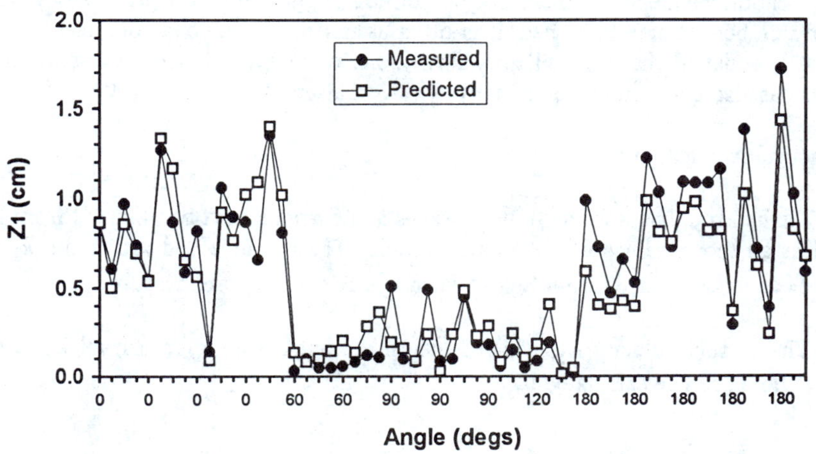

Figure 2. Comparison of the apparent roughness z_1 measured by Kaaij and Nieuwjaar (1987), Van Kampen and Nap (1988) and Havinga (1992) with those calculated by You's (1994) model with the input of k_s estimated from Eq.(8). The bed roughness z_o in steady flow alone, the depth-averaged current velocity $<\overline{u}>$ and the nearbed wave velocity amplitude $A\omega$ in Eq.(8) were measured in the experiments.

In other words, the use of the other former formulae (eg Nielsen, 1983; Raudkivi, 1988) in the models would not give better prediction of z_1 in Figure 1.

As a first approximation, the movable bed roughness k_s in the flow of irregular waves and current flow may be simply structured as

$$k_s = 30 z_o \left(1 + \frac{A\omega}{<\bar{u}>}\right) \tag{8}$$

in which A is the nearbed semi-excursion of wave orbital motion, ω is the angular frequency, $<\bar{u}>$ is the depth-averaged current velocity, and z_o is the movable bed roughness in steady flow alone. In the field, a current velocity $\bar{u}_A$ averaged from the bed to a level h_A is suggested to replace $<\bar{u}>$ in Eq.(8). It can be seen from Eq.(8) that the movable bed roughness k_s in the flow of waves and currents is larger than that in steady flow alone. The increase of k_s in the presence of waves may be understood to result from the appearance of wave-generated sand ripples and the increase of the nearbed sediment transport. It is shown in Figure 2 that the model of You (1994) together with Eq.(8) gives satisfactory agreement with the measured apparent roughness z_1. The three variables z_o, $<\bar{u}>$ and $A\omega$ in Eq.(8) were directly measured in the experiments. It may be concluded here that the new formula expressed by Eq.(8) gives good estimation of k_s in the flow of waves and currents.

It should be mentioned here that the formula suggested by Eq.(8) is only valid for a movable bed. For a fixed bed, it is often assumed that the bed roughness in steady flow is equal to that in oscillatory flow alone or in a combined wave-current flow. This has also been studied quantitatively by Mathisen and Madsen (1996).

Practical Applications

The bed roughness in the flow of waves and currents can't be obtained from Eq.(8) unless all the variables in Eq.(8) are known. The movable bed roughness k_s in the field can be determined from Eq.(8) when $\bar{u}^*_{cw}$, z_1 and $A\omega$ are measured.

The current velocity $\bar{u}_A$, which is depth averaged from z_o to a level h_A in steady flow alone, can be calculated by

$$\bar{u}_A = \frac{\bar{u}^*_c}{\kappa} \ln \frac{h_A}{e z_o}. \tag{9}$$

In the presence of waves, $\bar{u}_A$ can also be approximately estimated from

$$\bar{u}_A = \frac{\bar{u}^*_{cw}}{\kappa} \ln \frac{h_A}{e z_1} \tag{10}$$

in which h_A is an arbitrary level and is chosen to be about 1.0m above the seabed in the present study. On the other hand, the relationship between $\bar{u}^*_c$ and $\bar{u}^*_{cw}$ was derived by You (1995b) as

$$\bar{u}^*_{cw} = \bar{u}^*_c + K \times A\omega \tag{11}$$

in which K is constant and was determined to be 0.026 and 0.020 for the laboratory and field data, respectively. For irregular waves, significant wave height H_s and period T_s were suggested to calculate $A\omega$ in Eq.(11) using linear wave theory. The comparison of Eq.(11) with available experimental data is shown in Figure 3. Therefore, the movable bed roughness k_s in the coastal zone can be explicitly calculated from Eqs.(8)-(11). The depth-averaged current velocity $<\bar{u}>$ in Eq.(8) should be replaced by $\bar{u}_A$ in Eq.(10).

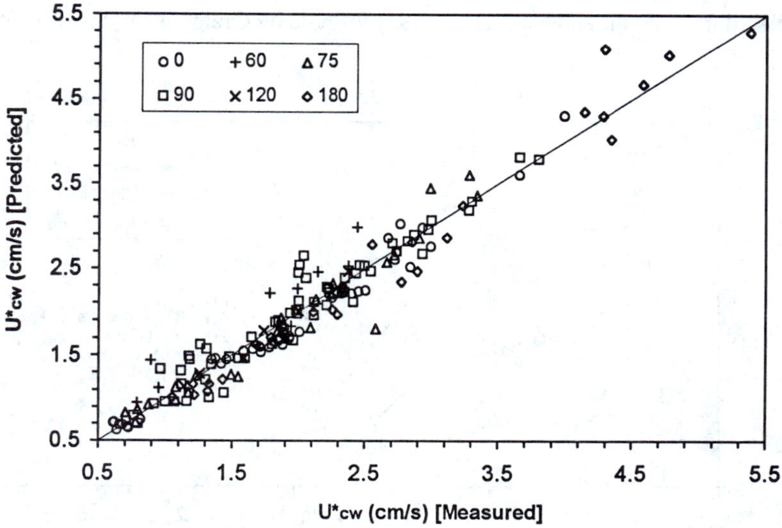

Figure 3. Comparison of Eq.(11) with the available experimental data from Bijker (1967) [θ = 75° and 90°, regular waves], Kemp and Simons (1982, 1983) [θ = 0° and 180°, regular waves], Visser (1986) [θ = 90°, regular waves], Kaaij and Nieuwjaar (1987) [θ = 0° and 180°, irregular waves], Van Kampen and Nap (1988) [θ = 0° and 180°, irregular waves], Simons *et al.* (1988) [θ = 0°, regular waves], and Havinga (1992) [θ = 60°, 90° and 120°, irregular waves]. After You (1995b).

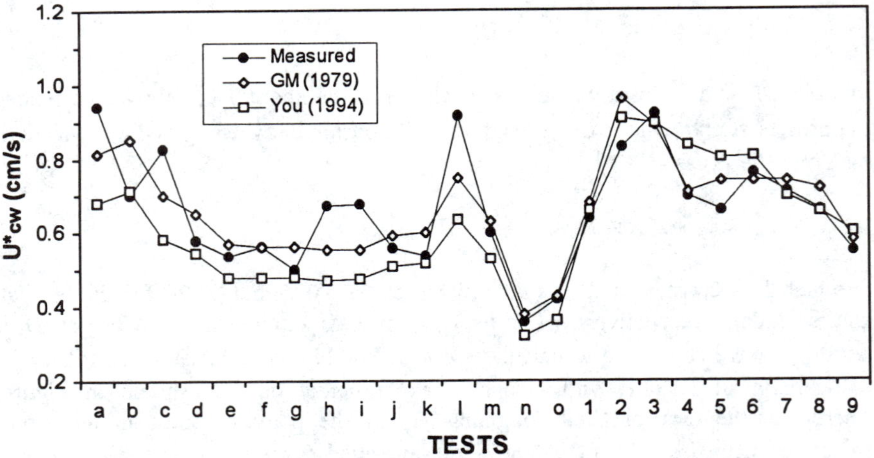

Figure 4. Comparison of the current shear velocities $\overline{u}^*_{cw}$ calculated by the models of Grant and Madsen (1979) and You (1994) with the field measurements of Grant and Williams (1984, 85). The movable bed roughness of k_s=6.0cm used in the models was calculated from Grant and Madsen's (1982) formula by Grant and Williams (1984, 85).

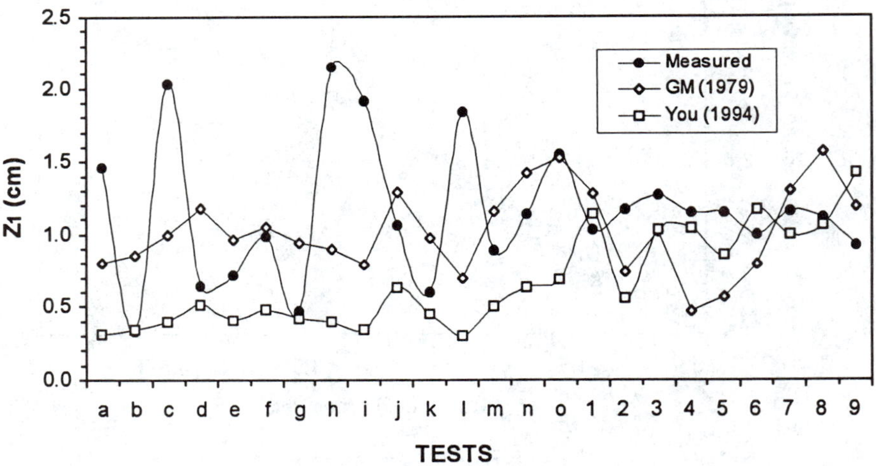

Figure 5. Comparison of the apparent roughness z_1 calculated by the models of Grant and Madsen (1979) and You (1994) with the field measurements of Grant and Williams (1984, 85). The bed roughness of k_s=6.0cm used in the modes was estimated from Grant and Madsen's (1982) formula by Grant and Williams (1984, 85).

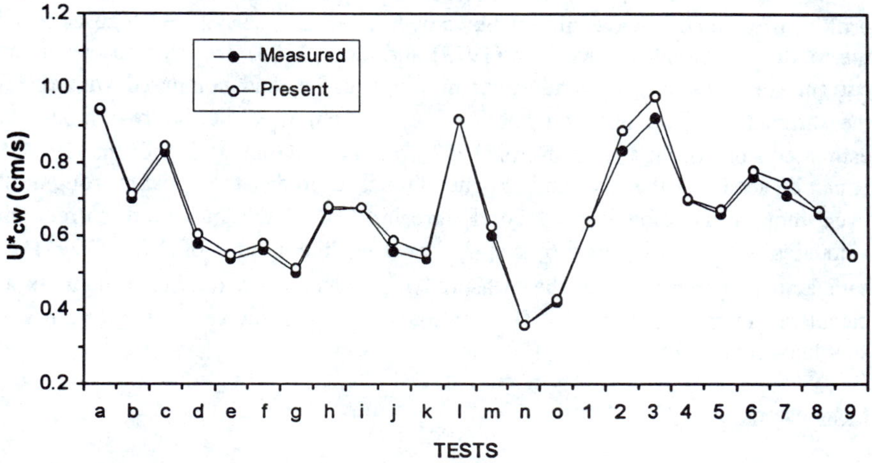

Figure 6. Comparison of the current shear velocities calculated by the model of You (1994) with the field measurements of Grant and Williams (1984, 85). The bed roughness k_s used in the model is calculated from Eqs.(8)-(11).

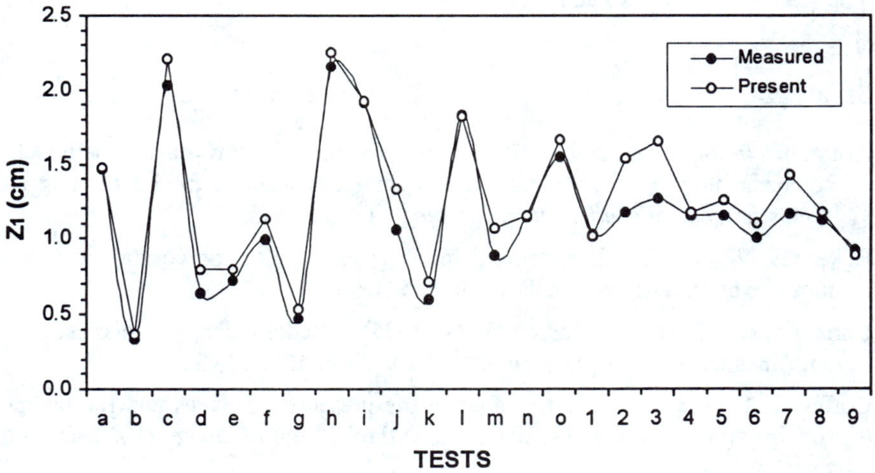

Figure 7. Comparison of the apparent roughness z_1 calculated by the model of You (1994) with the field measurements of Grant and Williams (1984, 85). The bed roughness k_s used in the mode is calculated from Eqs.(8)-(11).

Grant and Williams (1984, 85) provided high-quality field measurements of current profiles of combined wave-current flows over a movable seabed. In Figures 5 and 6, the models of Grant and Madsen (1979) and You (1994) are used to calculate the current shear velocity $\overline{u}^*_{cw}$ and apparent roughness z_1 and compared with the field measurements. The bed roughness of $k_s = 6.0\text{cm}$ used in the two models was estimated from Grant and Madsen's (1982) formula by Grant and Williams (1984, 85). It can be seen that the two models generally fail to predict the apparent roughness z_1 even though the models give good agreement with the measured current shear velocities $\overline{u}^*_{cw}$. In Figures 6 and 7, however, the model of You (1994) gives satisfactory agreement with the measured $\overline{u}^*_{cw}$ and z_1 when the bed roughness k_s is calculated from Eqs.(8)-(11). The depth-averaged current velocity $<\overline{u}>$ in Eq.(8) should be replaced by $\overline{u}_A$ in Eq.(10), and $h_A = 1.0\text{m}$.

Conclusions

The movable bed roughness in the flow of irregular waves and currents are studied based on the simple model of You (1994) and measured current profiles. A new formula expressed by Eq.(8) is proposed to calculate k_s in the flow of irregular waves and current. The present bed roughness formula together with the model of You (1994) gives better agreement with the measured apparent roughness z_1 than the former formulae derived in purely oscillatory flow. The field measurements of current profiles in the presence of irregular waves can be also predicted well with the model of You (1994) and the new bed roughness formula.

References

Arnot, R. W and Southard, J. B (1990). Exporatory flow duct experiments on combined flow bed configurations and some implications for interpreting storm even stratification. *Sedimentary Petrology*, 60: 211-219.

Bijker, E. W (1967). Some considerations about scales for coastal models with movable beds. Delft Hydraulics Lab, Pub No. 50.

Christoffersen, J. B and Jonsson, I. G (1985). Bed friction and dissipation in a combined current and wave motion. *Ocean Eng.*, 12: 387-423.

Coffey, F. C (1987). Current profiles in the presence of waves and the hydraulics roughness of natural sand beaches. Ph.D thesis, Dept of Geography, University of Sydney, Australia.

Coffey, F. C and Nielsen, P (1986). The influence of waves on current profiles, *Proc 20th Int Conf Coastal Eng.*, Taipei, p. 82-96.

Drake, D. E, Cacchinone, D. A and Grant, W. D (1992). Shear stress and bed roughness estimates for combined wave and current flows over a rippled bed. *J Geophys Res.*, 97: 2319-2326.

Grant, W. D and Madsen, O. S (1979). Combined wave current interaction with a rough bottom. *J Geophys Res.*, 84: 1797-1808.

Grant, W.D and Madsen, O.S (1982). Movable bed roughness in unsteady oscillatory flow. *J Geophys Res.*, 87: 469-481.

Grant, W. D and Williams, A. J (1984). Bottom stress estimates and their prediction on the Northern California Continental Shelf during CODE-1: the importance of wave-current interaction. *J Phys Oceanogr.*, 14: 506-727.

Grant, W. D and Williams, A. J (1985). Reply to Huntley's Comments. *J Phys Oceanogr.*, 15: 1219-1228.

Havinga, F. J (1992). Sediment concentrations and sediment transport in case of irregular non-breaking waves with a current. Draft Thesis, Faculty of Civil Engineering, Delft University of Technology.

Kemp, P. H and Simons, R. R (1982). The interaction between waves and a current: waves propagating with the current. *J Fluid Mech.*, 116: 227-250.

Kemp, P. H and Simons, R. R (1983). The interaction between waves and a current: waves propagating against the current. *J Fluid Mech.*, 130: 73-89.

Klopman, G (1994). Vertical structure of the flow due to waves and currents, Delft Hydraulics Laboratory, Report H840, Part II.

Mathisen, P. P and Madsen, O. S (1996). Waves and currents over a fixed rippled bed: 1. Bottom roughness experienced by waves in the presence and absence of currents. *J Geophys Res.*, 101: 16,533-16,542.

Mathisen, P. P and Madsen, O. S (1996). Waves and currents over a fixed rippled bed: 2. Bottom and apparent roughness experienced by currents in the presence of waves. *J Geophys Res.*, 101: 16,543-16,550.

Raudkivi, A, J (1989). The roughness height under waves. *J. Hydraulic Res*, 26: 569-584.

Simons, R. R, Kyriacou, A, Soulsby, R. L, Davis, A. G (1988). Predicting the nearbed turbulent flow in waves and currents. *IAHR Symp Mah Modelling of Sediment Transport in the Coastal Zone*, Copenhagen, pp.33-47.

Nielsen, P (1983). Analytical determination of nearshore wave height variation due to refraction, shoaling and friction. *Coastal Eng.*, 7: 233-251.

Sleath, J. D (1991). Velocities and shear stress in wave-current flows. *J Geophys Res.*, 96: 15237-15244.

van Doorn, Th (1981). Experimental investigation of near-bottom velocities in water waves without and with a current. Delft Hydraulics Laboratory, Report No M143.

van Doorn, Th (1982). Experimental investigation of the velocity field in the turbulent bottom boundary layer in an oscillatory water tunnel. Delft Hydraulics Laboratory, Report No M1562-b (in Dutch).

van Kampen, H. F. A and Nap, E. N (1988). Sediment concentrations and sediment transport in case of irregular non-breaking waves with a current. Faculty of Civil Engineering, Delft University of Technology.

van Rijin, L.C (1982). The prediction of bedforms and alluvial roughness. *In Mechanics of Sediments Transport*, Proc Euromech 156, pp. 133-135.

You, Z. J (1992). Oscillatory boundary layers with and without currents. Ph.D thesis. School of Civil Engineering, University of New South Wales.

You, Z. J (1994). A simple model for current velocity profiles in combined wave-current flows. *Coastal Eng.*, 23: 289-304.

You, Z. J (1995a). A simple model for current velocity profiles in combined wave-current flows: reply to the comments by P. Nielsen. *Coastal Eng.*, 26: 101-104.

You, Z. J (1995b). Increase of current bottom shear stress due to waves. *Coastal Eng.*, 26: 291-295.

CHAPTER 271

Seabed and Foundation Response to Wave Loading

Michael H. Davies[1]

Abstract

The interactions between wave loading and soil response in both open waters and around the foundations of coastal structures are observed experimentally. Experimental results for wave action over sand beds are compared to analytical predictions based on both Sleath's porous bed model and on Biot's consolidation theory. Experiments on wave action in the vicinity of a rigid caisson, in conjunction with Mei's boundary layer theory, allow for the development of guidelines for interpreting the relative importance of drainage on the effective stress response of foundations to wave action.

Introduction

The geotechnical response of seabeds and foundations to wave loading can be a critical factor in evaluating the stability of proposed coastal and offshore structures. This paper reviews some of the techniques available for analysis of seabed response to wave action and presents a new heuristic approach for evaluating the relative importance of soil drainage in the response of coastal foundations.

Bea and Aurora (1981) and Wright and Dunham (1972) proposed total stress analysis for seabeds under wave loading. These techniques view the seabed loading as a harmonic tractive stress on the upper boundary, total stress analysis is used to calculate seabed response. Such analysis can use elastic, elasto-plastic or visco-elastic constitutive models and is commonly undertaken within a finite element framework. These approaches are particularly suitable to conditions where the soil is relatively soft and impermeable. These techniques have been successfully applied in analyzing the behaviour of soft soils such as Mississippi Delta muds in the Gulf of Mexico.

The pore pressure response within a seabed was originally explored by Putnam (1949) and later expanded by Sleath (1970). This analysis assumes that pore fluid flows are independent of soil stresses. This assumption is generally valid for stiff, permeable beds such as coarse sands and gravels. Yamamoto (1978) and Madsen (1978) independently proposed the use of 'poro-elastic' analysis for seabed response to wave action using Biot's (1941) linearized equations of consolidation. This allows the treatment of the coupled response to wave-seabed interactions, that is the porous

---

[1] Canadian Hydraulics Centre, Bldg.M-32, National Research Council, Ottawa, Canada, K1A 0R6 (e-mail: Michael.Davies@NRC.CA)

media response coupled with the elastic soil response. This provides a technique suitable for a wide range of soil and wave conditions ranging from soft, impermeable silts and clays through to stiff sands. Since the initial work by Madsen and Yamamoto, a wide range of papers has been written proposing expansions and adaptations of the work, e.g. Finn et al (1983) and Okusa (1985). The development of full finite element solutions of time-dependent effective stress problems has provided another analysis tool. Models such as those of Shen et al (1986) provide a finite element framework within which the coupled effective stress response of soils can be investigated.

In experimental studies of soil-wave-structure interaction, measurements of soil stress or strain are difficult. Pore pressure response is, however, readily measured and is thereby often used as the main indicator of soil behaviour. Other techniques tend to be highly intrusive and fail to provide a time history of soil response (e.g. cone penetration testing). One promising technique for measurement of soil response is measurement of the acoustic emissions from the soil mass (AE). This technique was successfully employed in this research program and was described in Davies, et al (1990). The present paper focuses on the use of pore pressure response as a measure of soil behaviour.

Pore pressure in seabeds

In a porous seabed under harmonic wave loading (wave frequency, ω) the pore pressure, U of a fluid with unit weight, γ, has three components:

$$U = \gamma z + \gamma P(\cos \omega t) + u_r \quad (1)$$

Where γz is the hydrostatic component of the pore pressure at depth z below the free surface, P is the wave-induced component of the pore pressure fluctuation (expressed as a piezometric head), and u_r is the residual component of the pore pressure (due to consolidation and or shearing action). It is important to draw the distinction between the wave-induced fluctuation in the pore pressure, P and u_r, the residual pore pressure in excess of hydrostatic. The wave-induced pore pressure fluctuation is an indicator of the steady-state harmonic fluctuations in the effective stress state of the soil caused by wave loading. The residual pore pressure, u_r is a more gradual change in the pore pressure (non-harmonic) possibly caused by shearing action of the soil which results in soil volume change and consequent changes in the residual pore pressure.

Experimental Set-up—Flat Bed Testing

A series of flume tests was conducted on a sand bed consisting of 15 m^3 of a fine Ottawa sand (D_{50}=0.07mm). The sand was hydraulically placed using a positive displacement slurry pump. The resulting berm had a uniform crest height of 0.91m over a 10 m width. This uniform section was flanked by 1:10 side slopes. A vertical array of pore pressure transducers had been installed in the flume prior to construction of the berm. This consisted of 14 Druck PDCR81 pore pressure transducers (7 kPa capacity) at 10 cm vertical spacing. A capacitance-type wave gauge was located directly above the pressure transducers.

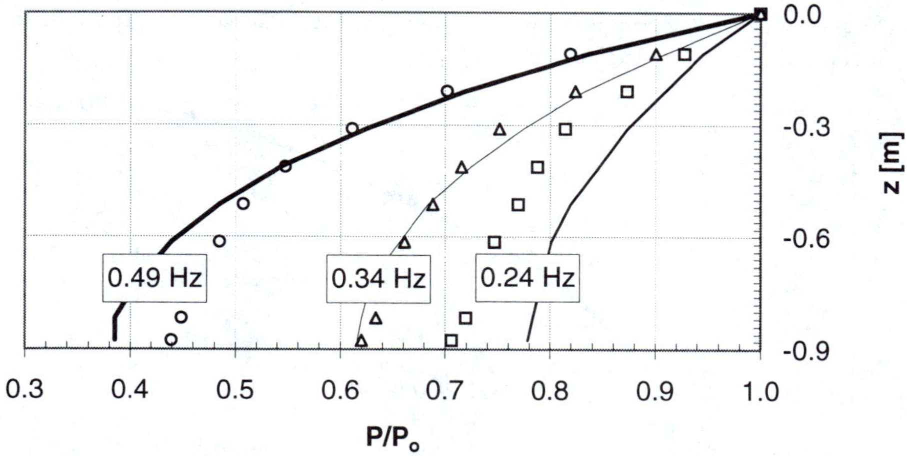

Figure 1 Vertical distribution of pore pressure as per Sleath (1970) for k_x=1.8k_z

Sand densities were determined using a 12 cm diameter brass sampling tube. Soil relative density was determined to be 47% (soil unit weight was 15.5 kN/m^3). Constant head permeability tests yielded a coefficient of permeability of k=5.0 x 10^{-5} m/s. The experimental setup is described in detail in Davies (1992).

Steady-state harmonic response

The pore pressure response measured in these experiments was compared to Sleath's rigid, porous bed model and to that predicted by poro-elastic theory as per Yamamoto, and Finn.

In general, the Sleath formulation was seen to provide a good estimate of the vertical distribution of the magnitude of the pore pressure fluctuation if the effects of anisotropic permeability are considered (see Figure 1). Here the amplitude of the harmonic pore pressure, P is normalized by P_o, the pressure at the top of the seabed. Since Sleath's model assumes a porous media flow within a rigid bed, this solution provides no information about the stress response of the soil. Poro-elastic analysis is required to examine the coupled response of the soil and the pore fluid.

In comparing the pore pressure response to that predicted by poro-elastic theory, initial comparisons were made to the model of Yamamoto, 1978. This formulation does not consider the effect of anisotropic permeability and it was not possible to obtain good agreement between the model and theory. To match the observed vertical attenuation of the pressure magnitude, the predicted phase lags were far too large – on the order of 30°. Measured phase lags in the model were not seen to exceed 15°. Consequently, Finn's Stabmax routine was employed since it allows for anisotropic permeability. Using Stabmax, good agreement could be obtained for both the amplitude and phase of the pore pressure response. This agreement is shown in Figures 2 and 3. It should be noted here that the poro-elastic models were seen to be extremely sensitive to the volumetric degree of soil saturation, S_r. Sensitivity tests showed that varying the soil saturation ratio from 98% to 99.8%

resulted in predicted phase lags at the bottom of the bed varying from 28° down to 3°, the trend being that phase lag reduces with reducing air content. These findings are described in more detail in Davies (1992).

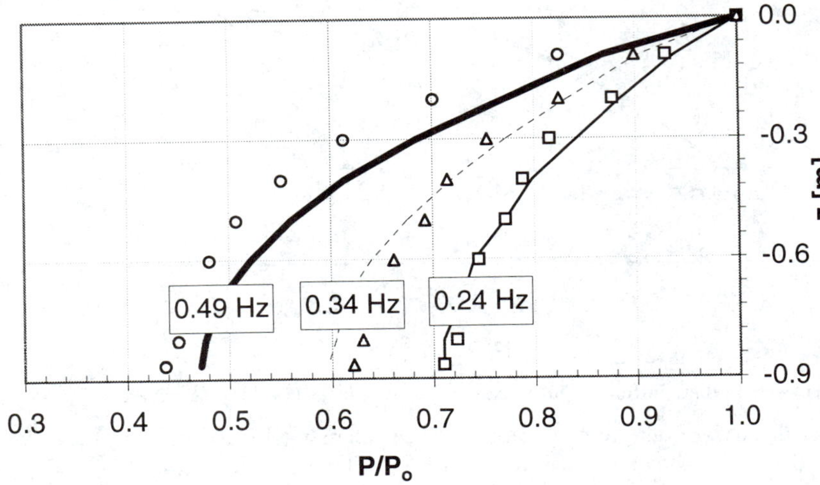

Figure 2 Measured pore pressure response vs poro-elastic theory (Stabmax, Finn, 1982)

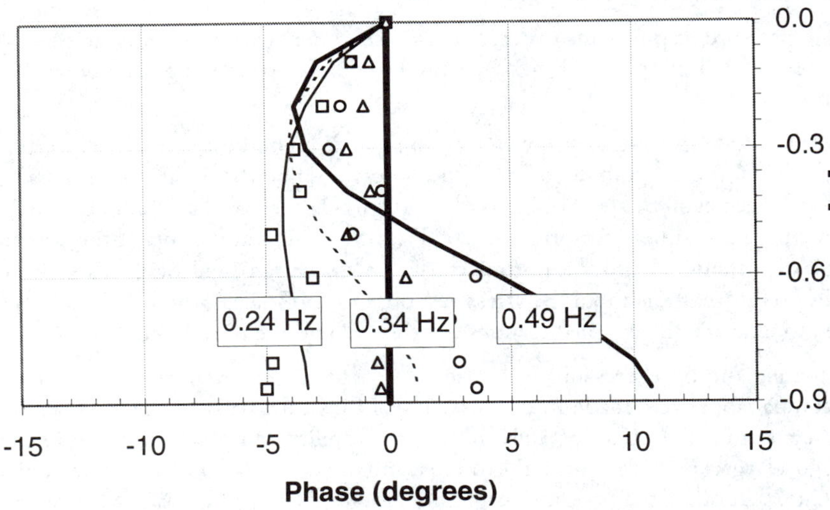

Figure 3 Phase lag of pore pressure response vs poro-elastic theory (Stabmax, Finn, 1982).

Experimental Set-up—Caisson Testing

The second experimental test program in this study was designed to examine the interactions between seabeds and rigid coastal structures under the influence of wave action. Large-scale flume tests were undertaken to study the effective stress response of seabeds beneath caissons, and to examine the combined effective stress, pore pressure and scour response of the toes of caisson-type coastal structures. This experimental study was also used to explore the validity of the Tsai et al (1990) analytical poro-elastic approach for the prediction of foundation response.

The Coastal Wave Basin of the National Research Council in Ottawa was used for the caisson testing. This basin is 85m long and 14m wide. The maximum water depth available for testing is 1.2m (allowing 0.3m of freeboard for waves). Two parallel 2.4m wide channels were built within the basin. This layout enabled the simultaneous testing of two different caisson foundations. Berms were built in each of these channels to serve as foundations for two identical steel caissons (see Figure 4). The berms were composed of two different sands; a coarse sand with D_{50}=0.38mm, and a fine sand, with D_{50}=0.10mm. Due to the large amount of fill required to build the foundations for the caisson (roughly 25 tonnes of sand in each flume) it was not possible to rebuild the berm between tests. The test sequence employed gradually exposed the caissons to irregular waves of increasing amplitude and period in a manner simulating the building of a storm. For all tests, the hydrodynamics acting on the structure and the response of both the soil and the caisson were measured in detail.

One of the results of the test program was a series of measurements of the pore pressure response within the foundations beneath the leading edge of the caissons. The following section gives a review of some of these pore pressure measurements.

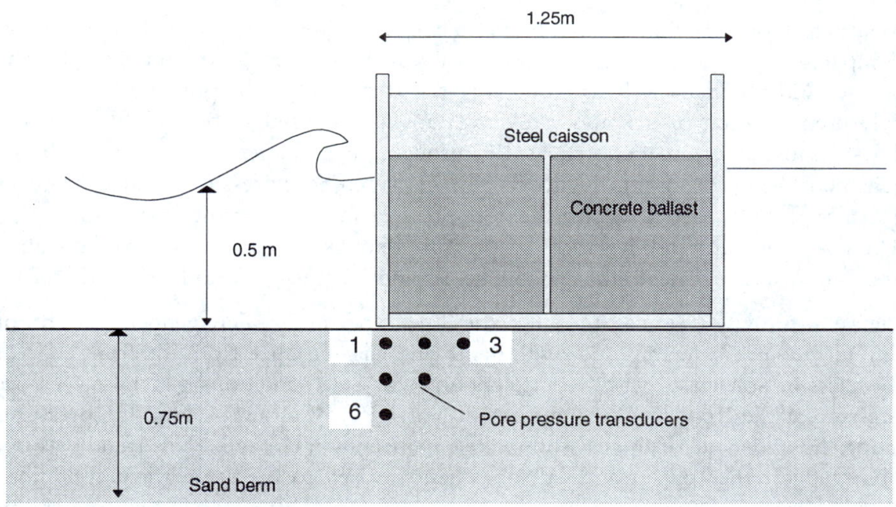

Figure 4 Schematic cross-section of caisson tests.

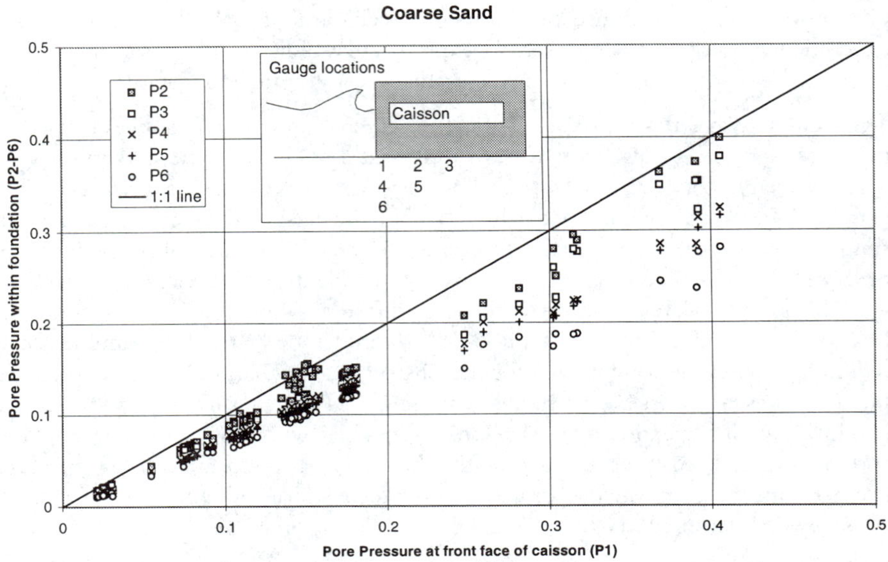

Figure 5 Pore pressure response - coarse sand.

Pore pressure response within coastal foundations

Figure 5 shows how the pore pressure response measured beneath the caisson compared to that at the outside front corner of the caisson (at the mudline) for the coarse sand foundation. This plot shows the amplitude of the pressure fluctuations measured by each of six pore pressure transducers installed beneath the caisson. These transducers are denoted as P1 through P6 and their locations are shown schematically in the inset in Figure 5. The measured pore pressure at each of gauges P2 through P6 is plotted against the pore pressure measured at the front corner (P1). There is a general trend that the pore pressure diminishes with depth into the seabed and with distance beneath the caisson, i.e. P3 and P6 are significantly lower than P1.

Figure 6 shows a similar plot for the fine sand test. The general pattern of diminishing pressure amplitude with depth and with distance from the front edge of the caisson still exists with one exception: for test conditions generating a pore pressure at the front face of the caisson (gauge P1) of around 0.1 m, the pressures along the underside of the caisson were significantly increased. This trend was not observed for the coarse sand caisson when exposed to the same wave conditions with the same caisson geometry and caisson ballast. This suggests that, under this combination of wave conditions, the finer sand caisson was experiencing complex wave-soil-structure interactions which were not observed with the coarser sand. The following sections describe a boundary layer approach to provide some insight to these interactions.

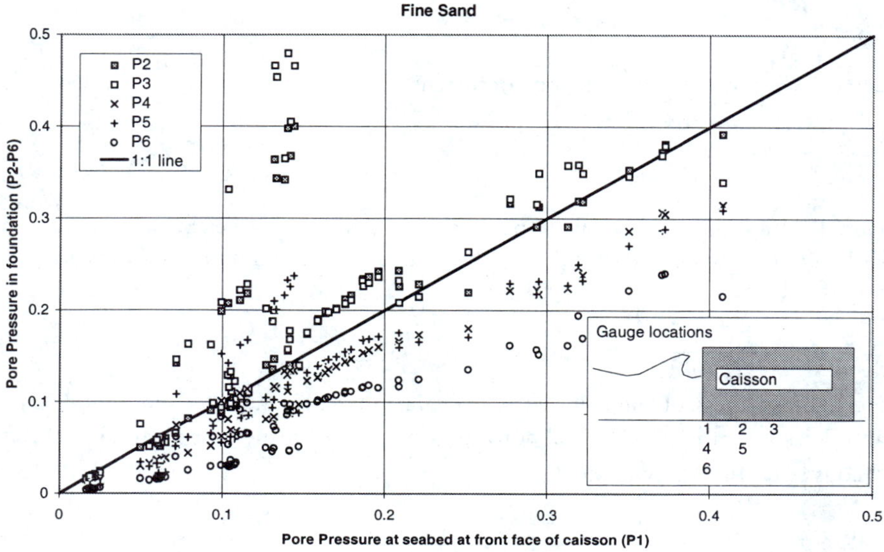

Figure 6 Pore pressure response - fine sand

The Boundary Layer Approach

The boundary layer approach of Mei and Foda (1981) shows that, in the treatment of a poro-elastic solid using Biot's linearized theory of consolidation, for the frequencies of typical ocean waves, a Stokes' type of boundary layer exists near the top of the bed. The soil and fluid move together as a single phase in the region below this boundary layer. Within the boundary layer, the effects of the free draining upper boundary are evident. This allows a greatly simplified analysis where the outer region (far below the mudline) can be treated as a single phase elastic solid for which analytical solutions to the soil's behaviour are readily found. Mei and Foda show that in the boundary layer, a one-dimensional analysis using Terzaghi's theory of consolidation can be applied (provided that the boundary layer thickness is small relative to wavelength). Solutions to a variety of poro-elastic problems using this boundary layer technique have been presented in the literature (see Mei and Foda, 1981 and Mei, 1982)

Intuitively, one would expect that for a soil of permeability, k, which is small relative to the frequency of loading, $\omega=2\pi f$, there is little fluid motion relative to the soil skeleton. For such a case the soil and fluid move together and the medium can be treated as a single phase as described by the equations of elastic dynamics. Near the seabed there will be a free-draining region extending to some depth, δ. For $\delta/L << 1$, where L is the wavelength, then the upper boundary region becomes essentially one-dimensional. Mei and Foda show that the first order approximation of the boundary layer is the Terzaghi equation for one-dimensional consolidation.

In the boundary layer $\partial/\partial z >> \partial/\partial x$. The thickness of the boundary layer, δ is given by

$$\delta = \sqrt{\frac{c'_v}{\omega}} \tag{2}$$

where c'_v is the soil consolidation coefficient

$$c'_v = \frac{k\,\gamma}{m'_v} \tag{3}$$

Here k is the soil's coefficient of permeability, γ is the unit weight of water, and m_v' is the effective modulus of volume compressibility for soil with a partially saturated pore fluid.

$$m'_v = n\beta' + \frac{1-2\nu}{2(1-\nu)G} \tag{4}$$

Here the stiffness of the soil matrix is characterized by the bulk shear modulus, G, and Poisson's ratio, ν. The soil porosity is n and the stiffness of a partially saturated fluid is β' (as per Verruijt, 1969),

$$\beta' = \beta + \frac{1-S_r}{P_{tot}}. \tag{5}$$

Here β is the compressibility of pure water (β=4.3 x 10^{-10} Pa^{-1}), and S_r is the volumetric degree of saturation of the soil. P_{tot} is the absolute pressure at the point of interest.

The exact total solution for dynamic stresses and pore pressures is determined through the solution of the dynamic elastic equations governing the outer region and subsequent application of a boundary layer correction:

$$(\ )_{exact} = (\ )^{o} + (\ )^{b} \tag{6}$$

where the superscripts o and b denote outer region and boundary layer region terms, respectively.

The boundary layer formulation provides an accessible and heuristically appealing treatment of the soil-wave interaction problem. The boundary layer thickness, δ quantifies the relative depth of influence of wave action in a seabed.

Foundations - Drained vs Undrained

The problem of wave action in the vicinity of coastal structures can be considered in terms of the permeability and stiffness of the soil mass relative to the size of the structure. Wave loading on a rigid coastal structure such as a caisson causes wave stresses to be transferred from the caisson to the seabed. At the same time wave-induced pressure fluctuations at the seabed cause uplift pressures along the underside of the structure. The complex interactions between a rigid structure and the seabed under wave loading have been examined in the literature by Lee and Focht (1985), Lindenberg et al, (1982) and Tsai et al (1990). Analytic solutions of the wave-soil-structure interaction problem have been formulated by Mei (1982) as well as by Tsai et al (1990). These analytic solutions use elastic solutions for a rigid block resting on a flexible base to solve the total stress state in the foundation. A boundary layer correction is then applied to take account of the presence of the free-draining upper boundary. These solutions rely on an assumption that the boundary layer thickness, δ is small relative to the wavelength, L and furthermore that the boundary layer thickness is small relative to the caisson size, C (for some solutions the half-width of the caisson is used as the representative caisson dimension).

These calculations are quite complex but yield some interesting insight into the behaviour of wave-soil-structure interactions. For practical problems, a finite element effective stress model is more adaptable (e.g. Shen, 1990).

One of the insights provided by boundary layer theory is the concept of relative boundary layer thickness. In assessing the appropriate analysis to be undertaken for a specific structure, simple calculations of boundary layer thickness relative to wavelength and relative to caisson geometry can allow the classification of the problem in terms of the influence of pore fluid flow on soil response.

There are two bounding conditions often considered in analysing foundation response. These are **undrained** and **drained** behaviour:

1) For **undrained** analysis, it is assumed that the rate of application of loading is rapid relative to the soil's permeability. Consequently the pore fluid and soil matrix move together. At a lower bound, for example, the case of a structure resting on clay, the role of pore pressures in generating uplift forces is negligible. For undrained analysis the total stress state is used, and the concept of effective stresses is not applicable.

2) For **drained** analysis, it is assumed that the rate of application of loading is slow relative to the soil's permeability. Consequently, the pore fluid is free to move relative to the soil matrix. At an upper bound, the Shore Protection Manual (1984) and Goda's models of wave-induced uplift pressures are reasonable, (a triangular pressure distribution acting along the underside of the structure). For drained analysis, an effective stress approach is employed.

In reality, these limiting cases rarely exist. The true soil response lies somewhere between fully undrained and fully drained. What is needed is a set of practical guidelines to evaluate the relative importance of drainage to the structure.

Foundation response guidelines

When the boundary layer thickness, δ is small, free-drainage occurs only close to the mudline. When δ is large the free-drainage zone extends further into the seabed.

The pore pressure response within the seabed in the vicinity of a caisson will be related to two dimensionless parameters:

1. f/f_n - describes the frequency response of the system and how close the loading frequency, f is to the natural frequency of the caisson, f_n. For a single degree of freedom oscillator (such as the caisson response in pure pitch motion), the natural frequency of the caisson-soil system is related

to the foundation stiffness and the polar mass moment of inertia of the caisson as follows:

$$f_n = \frac{1}{2\pi}\sqrt{\frac{E'B^3}{12\,I_p}} \qquad (7)$$

Here, E′ is the equivalent foundation stiffness, and B is the caisson width.

2. δ/C - the thickness of the boundary layer, δ relative to the caisson size gives a measure of how large the free-draining boundary is relative to the size of the caisson.

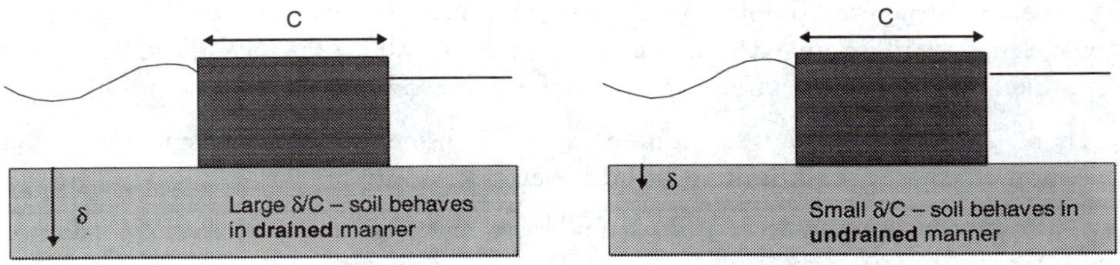

Figure 7 Illustration of boundary layer thickness concept for caissons

For a given caisson geometry (and hence given f/f_n), the pore pressure response could be expressed as a function of the ratio of boundary layer thickness to caisson width alone:

$$P/P_o = \phi\left(\delta/C\right) \qquad (8)$$

Three conditions can then exist:

1) If $\delta \ll C$ then the boundary layer effect will be small, uplift pressures will only exist very close to the outer edge of the caisson and undrained analysis will be sufficient.

2) If $\delta \gg C$ then the entire caisson will lie within the boundary layer, uplift pressures will be large and drained analysis will be appropriate. Here solution of the pore pressure response (and uplift) can be 'de-coupled' from the soil stress response.

3) Where δ is of the order of C, the response will be partially drained – the interactions between the pore fluid and soil matrix must be considered. Figure 7 provides a schematic of these concepts.

Consider the behaviour of the two caisson test series shown in Figure 5 and Figure 6. The major differences between these two caissons were the soil's coefficients of consolidation. This is reflected in different boundary layer thicknesses for the two caisson datasets. Figure 8 shows a plot of the relative amplitude of the pore pressure response beneath the caisson (normalized by P1) against the ratio of boundary layer thickness to caisson size, δ/C. This figure shows that for test conditions corresponding to δ/C around 0.1, amplification of the pore pressure response beneath the caisson is observed (e.g. P3 in caisson 1). This trend is not observed for sensors mounted deep in the bed at the front face of the caisson (vis P6). For test conditions of δ/C much greater than 0.5, the amplitude of the pore pressure response starts to approach the linear distribution suggested by Goda and others for caissons resting on very coarse rubble beds.

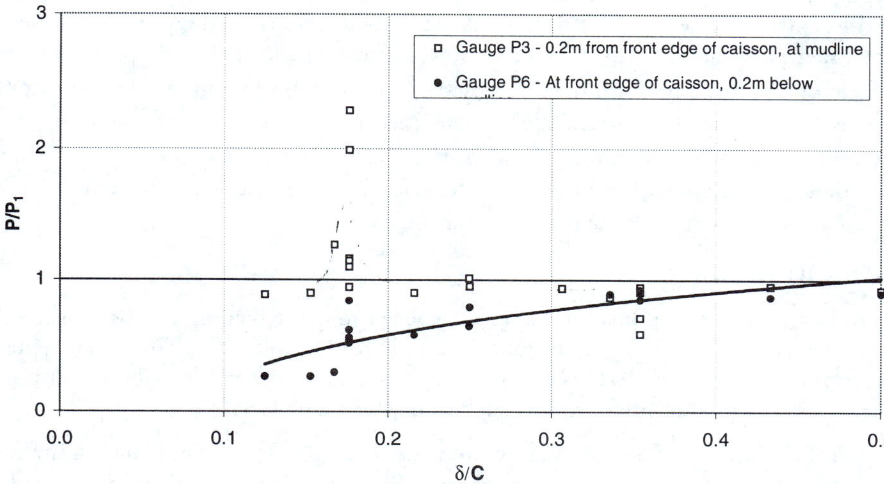

Figure 8 Pore pressure response vs drainage ratio, δ/C (combined results for both coarse and fine sands).

Table 1 shows typical ranges of foundation conditions and wave conditions for some typical coastal structure geometries

Wave period, T= 15 s			ω=2π/T= 0.42	
Influence of foundation type	c_v '(m^2/s)	δ [m]	C [m]	δ/C
Rubble fill	1000	48.86	15	3.26
Hard sand	100	15.45	15	1.03
Soft sand	10	4.89	15	0.33
Stiff clay	0.1	0.49	15	0.03
Soft clay	0.001	0.05	15	0.00
Influence of caisson size	c_v '(m^2/s)	δ [m]	C [m]	δ/C
Rubble fill	1000	48.86	*3*	16.287
Rubble fill	1000	48.86	*10*	4.886
Rubble fill	1000	48.86	*100*	0.489
Soft clay	0.001	0.05	*3*	0.016
Soft clay	0.001	0.05	*10*	0.005
Soft clay	0.001	0.05	*100*	0.000

Table 1 Relative boundary layer thickness for typical field conditions

For cases where the foundation material is stiff and permeable (such as rubble fill), Table 1 shows that, except in the case of exceptionally wide structures, the ratio δ/C is large and the free-draining condition can be expected to be observed along the entire underside of the structure. This is in accordance with the design techniques presented by Goda and the Shore Protection Manual where, for caissons resting on rubble foundations, a triangular uplift pressure distribution is assumed along the underside. For cases where either the foundation material is soft and impermeable (soft clay) or where the structure is very large (large C), the ratio δ/C can be seen to be quite small. In these cases, the uplift pressures due to wave action will be negligible and the undrained analysis is most appropriate (e.g. as described by Lee and Focht). In cases such as this, skirts are sometimes installed around the base of the structure to act as cut-off walls to further reduce the likelihood of uplift pressures existing. The structures are then designed for a no-uplift condition. The ratio δ/C serves to delineate how foundations in intermediate conditions may behave. For example if a structure rests on soft sand or stiff clay, the ratio δ/C becomes useful in combining the effects of wave period, consolidation coefficient and structure size to give a sense of the relative size of the drainage path, δ.

Conclusions

The pore pressure response of a seabed exposed to wave loading can be reasonably well described by the rigid, porous bed solution of Sleath (1970). To obtain information about the effective stress state in the soil, however, it is better to use a poro-elastic seabed response model such as that of Finn (1982).

Often the response of a coastal structure is neither fully drained nor fully undrained but somewhere in between. For engineering analysis, practical guidelines are needed to delineate the extent of influence of the free-draining boundary. Through boundary layer theory, it is possible to interpret this influence through δ, the boundary layer thickness. Preliminary analysis of test results indicates that the ratio, δ/C might be useful here. For small values of δ/C (less than 0.05, say), the soil can be treated as fully undrained and a total stress analysis can be used. As δ/C becomes large (greater than 0.3, say), soil behaviour starts to become fully drained and analysis techniques such as those proposed by Goda and the Shore Protection Manual become more appropriate. Soil conditions in the intermediate range ($0.05 < \delta/C < .3$) require particularly close attention since the response of the soil and pore fluid is closely coupled.

The idea of using the relative boundary layer thickness, δ/C as a predictor of seabed response is hypothetical at this stage. Further verification and extension of this idea requires examination of a wider range of experimental conditions. Variation of the ratio of f/f_n (the frequency of wave loading relative to the natural frequency of the system) should also be further explored.

Acknowledgements

This work was made possible by funding from Program on Energy Research and Development of Natural Resources Canada and the National Research Council of Canada. The continuing support of my colleagues at CHC greatly appreciated, in particular the technical assistance of Dave Watson.

References

Bea, R.G. and Aurora, R.P. (1981): *A simplified evaluation of seafloor stability.* Proc. Offshore Technology Conference, paper no. 3975, pp. 223-240.

Been, K., and Jefferies, M.G. (1985): *A state parameter for sands.* Geotechnique, Vol. 35, No. 2, pp. 99-112.

Biot, M.A. (1941): *General theory of three-dimensional consolidation.* Journal of Applied Physics, Vol. 12, pp. 155-164.

Davies, M.H. (1986): *Wave-induced liquefaction of sands.* Proc. of the Ocean Structure Dynamics Symposium, Oregon State University, Corvallis, Ore.

Davies, M.H., Chichibu, A., and Garga, V.K. (1990): *Study of AE response of coastal structures under storm wave loading.* Proc. of the 10th International Acoustic Emission Symposium, Sendai, Japan, Publ. by the Japanese Society for Non-Destructive Testing, pp. 153-159.

Davies, M.H. (1992): *Soil-wave-structure interaction in coastal engineering*, Ph.D. thesis, Queen's University at Kingston, January, 1992, 301 pp.

Finn, W.D.L., Siddhartan, R., and Martin, G.R., (1983): *Response of seafloor to ocean waves.* J. of Geotechnical Engg., ASCE, Vol.109, No.4, April 1983, pp.556-573.

Goda, Y. (1985): *Random seas and design of maritime structures.* University of Tokyo Press, Japan 323 pp.

Lee, K.L., and Focht, J.A.. (1985): *Liquefaction potential at Ekofisk tank in North Sea.* Journal of Geotechnical Engg., ASCE, Vol. 101, No. GT1, pp. 1-18.

Lindenberg, J., Swart, D.H., Kenter, L., and den Boer, W. (1982): *Wave-induced pressures beneath caisson: Comparison between theory and large scale tests.* BOSS '82, Cambridge, Mass., pp. 337-357.

Madsen, O.S. (1978): *Wave-induced pore pressures and effective stresses in a porous bed.* Geotechnique, Vol. 28, No. 4, pp. 377-393.

Mei, C.C. (1982) *Analytical theories for the interaction of offshore structures with a poro-elastic seabed.* Proc. of BOSS '82, Cambridge, Mass., August 1982, Published by Hemisphere Publishing, Washington, D.C., USA pp. 358-370.

Mei, C.C. and Foda, M.A. (1981): *Wave-induced responses in a fluid-filled poro-elastic solid with a free surface – A boundary layer theory.* Geophysical Journal of the Royal Astronomical Society, Vol. 66, pp. 597-631.

Okusa, S. (1985): *Wave-induced stresses in unsaturated submarine sediments.* Geotechnique, Vol. 35, No. 4, pp. 517-532.

Putnam, J.A. (1949): *Loss of wave energy due to percolation in a permeable sea bottom.* Transactions of the American Geophysical Union, Vol. 30, No. 3.

Seed, B.H., and Rahman, M.S. (1978): *Wave-induced pore pressures in relation to ocean floor stability in cohesionless soils.* Marine Geotechnology, Vol. 3, pp.123-150.

Seed, B.H., Idriss, I.M., and Arango, I. (1983): *Evaluation of liquefaction potential using field performance data.* Proc. J. of Geotechnical Engg., ASCE, Vol. 109, No. 3, March 1983, pp. 458-482.

Shen, C.K., Wang, Z.L., and Herrmann, L.R. (1986) *Final report on ocean wave - induced effective stresses in an elasto-plastic seafloor.* Calif Sea Grant College Program Proj. No. R/OT-13, Dept. of Civil Engg., University of California, Davis, November 1986, 52 pp.

Sleath, J.F.A. (1970): *Wave-induced pressures in beds of sand.* Journal of Hydraulics Division, ASCE, Vol. 96, No. HY2, pp. 367-378.

SPM (1984) *Shore Protection Manual,* Coastal Engineering Research Centre, U.S. Army Corps of Engineers, 3 volumes.

Tsai, Y.T., McDougal, W.G., and Sollitt, C. (1990) *Response of finite depth seabed to wave and caisson motion.* J. of Waterway, Port, Coastal and Ocean Engg., ASCE, Vol. 116, No. 1, January 1990, pp. 1-20.

Wright, S.G., and Dunham, R.S. (1972): *Bottom stability under wave-induced loading.* Proc. Offshore Technology Conference, Paper no. 1603, pp. 853-862.

Yamamoto, T., Koning, H.L., Sellmeijer, J., and van Hijum, E. (1978): *On the response of a poro-elastic bed to water waves.* J. Fluid Mechanics, Vol. 87, Part 1, pp. 193-206.

CHAPTER 272

Interannual Shoreline Variations at Duck, NC, USA

Nathaniel Plant and Rob Holman

ABSTRACT

We have analyzed 3 years of daily observations of the shoreline position along a 2 km segment of the coast at Duck, NC. Shoreline variations were found to be dominated by length scales of order 1 kilometer, with an amplitude that was modulated by a seasonal cycle, and a spatial structure that progressed alongshore at an average rate of 1 m/day. Surprisingly, the alongshore-averaged shoreline position did not display an annual cycle. Instead, the most pronounced alongshore-averaged shoreline changes occurred over several years. The dominance of large spatial and temporal scales suggests that short scale studies at Duck (typical duration < 1 month and lengths < 500 m) seriously under-resolve an important component of shoreline variability.

INTRODUCTION

To describe general beach behavior, we have often referred to a "summer-winter" conceptual model (e.g. Komar, 1974), in which beaches erode and sand bars form or move offshore during large storms (winter), followed by beach accretion and onshore bar movement during calm periods (summer). Recently, however, observations of nearshore bathymetry have revealed interannual and decadal morphologic behavior, which are not described by a seasonal model. Examples include offshore progressive sand bars at Duck, NC, USA (Birkemeier, 1985; Lippmann and Holman, 1990; Kraus and Larson, 1992) and along the Holland Coast (Ruessink and Kroon, 1994; Wijnberg and Terwindt, 1995) over periods of several years to over a decade. In addition, there is evidence that beach response over these long time periods is not alongshore-uniform. Wijnberg and Terwindt (1995) showed examples of alongshore progressive variations at the shoreline, with an alongshore wavelength of about 2 km and periodicity matching that of the bars. It is appears likely that, with increasing time scales, beach variability may be dominated by large length scales. Quantifying the contribution of various alongshore length scales to beach variability, and determining characteristic time scales associated with each length scale are clearly important to understanding (or simply describing) the long term behavior of beaches.

In this paper, we focus on the behavior of a particular beach elevation contour near the shoreline at Duck, NC. One reason to focus on the shoreline is that it

---

Ocean Admin. Bldg 104, Oregon State University, Corvallis OR 97331 delineates the width of the subaerial beach, which is of practical significance to those who live on or visit the beach. Also, the shoreline is a morphologically active (and therefore interesting) region of the beach, which responds to variations in wave conditions, tides, and (perhaps) interannual forcing. We seek to determine the relative contribution of different length scales to shoreline variability, and to describe the relationships between length and time scales. First, we will describe observations of the cross-shore position of a particular shoreline contour, obtained from analysis of video images. Then, we will present an analysis of the spatial and temporal variability of this data set. Since the dominant variability is associated with large length scales, we quantify this variability by focusing on two scales: (1) the alongshore-averaged shoreline position and (2) variability with a wavelength of about 1 km. The observed variability is partitioned into amplitude time series for both scales, as well as a time series of the spatial phase shift of the 1 km scale, which describes alongshore propagation. Finally, we describe the correlation between the shoreline variability and the wave field.

DATA COLLECTION

Video images of the surf zone at Duck, NC have been collected since 1986. Originally, the images were obtained once per day; however, they have been obtained every hour since 1993. The type of images that we will use in this study are time exposures (Lippmann and Holman, 1989, 1990). A time exposure (Figure 1) represents the average image brightness over, in this case, a 10 minute period. In relatively shallow regions where waves tend to break (e.g. over sand bars and at the shoreline), the average brightness is relatively high (white) due to the persistent production of foam, hence, morphologic features are often easily identified. At Duck, time exposures have been used successfully to locate the position of sand bars as well as the shoreline (Lippmann and Holman, 1989, 1990).

Plant and Holman (in press) described the methods that we used to locate and map the shoreline position. To summarize: we assumed that the shore-parallel band of high image intensity at the shoreline, also called the shoreline intensity maximum (SLIM), visually marked the still water level shoreline. The SLIM was identified in each image and its cross-shore position at a series of alongshore sample locations was determined from simple photogrammetric relationships (Lippmann and Holman, 1989). Since the cross-shore position of the SLIM varied with tidal elevation, we selected images corresponding to tidal elevations within 0.25 m of the 0.50 m level. We attempted to locate the shoreline once per day, within a 2 km long region, sampling every 2 m. The shoreline position is overlain on each image in Figure 1, and is plotted in map coordinates on the adjacent panels.

This shoreline estimation technique ought to be most accurate when applied to simple intertidal bathymetry (i.e. slowly varying in both alongshore and cross-shore directions) and when the wave height is low, so that the SLIM is a narrow band located near to the still water shoreline. The beach slope and wave height at the shoreline scale the errors in the estimates of the cross-shore position of the shoreline. For a barred beach, the error is approximately $\frac{1}{2}\frac{H}{\beta}$, where H is the (potentially depth-limited) rms wave height shoreward of the sand bar and β is the beach slope at the shoreline. The intertidal beach slope at Duck is typically about 0.1 and the shoreline wave height is limited by breaking over the bar to 1.0 m, resulting in a maximum error of about 5 m. Finite resolution of the digital images also contributed to measurement errors that ranged from 0.50 m nearest to the camera to 2.0 m farthest from the camera. The alongshore resolution of the shoreline position estimates varied from 1 m to 50 m .

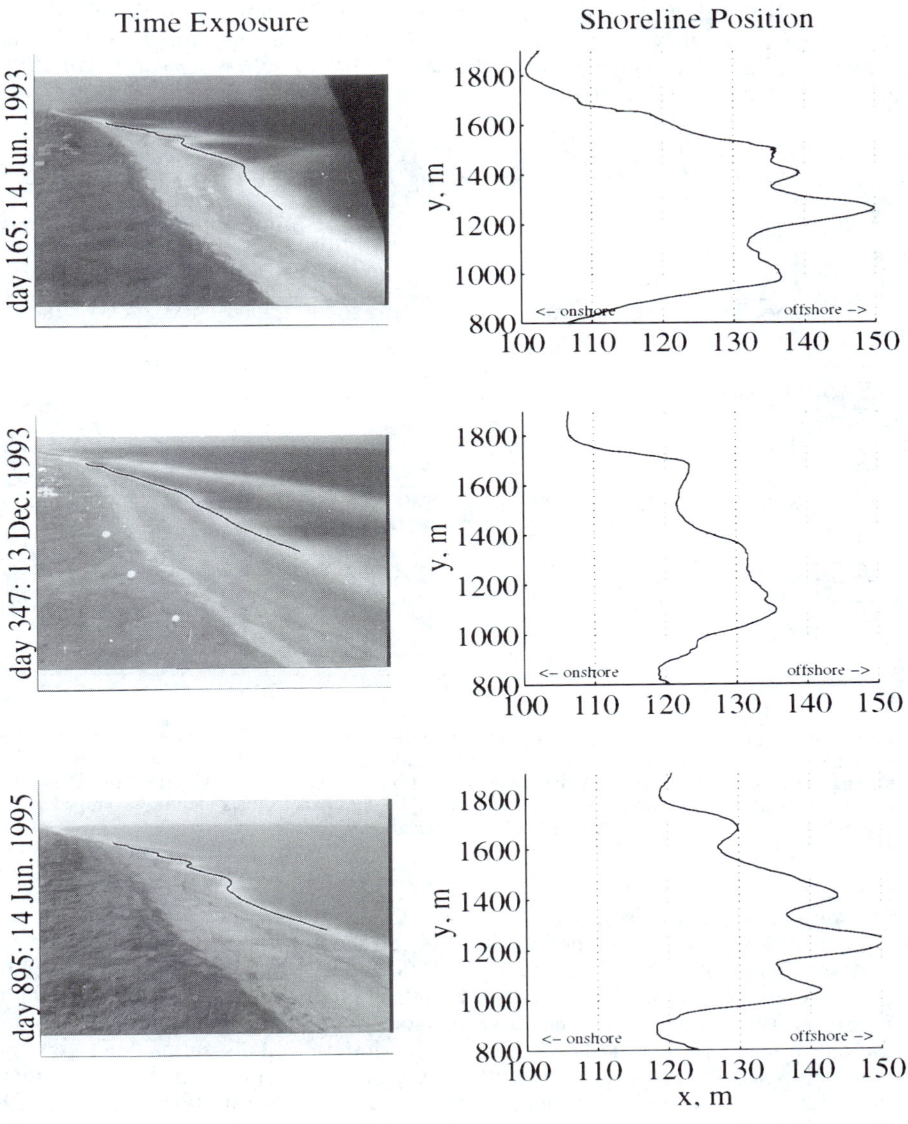

Figure 1. Examples of time exposure images, from Duck, NC. The location of the shoreline intensity maximum (SLIM) line is identified on each image and the map projection of that line is shown adjacent to each image. On the map projection, the ordinate is the alongshore range; positive is to the north (away from the camera). The cross-shore scale is exaggerated 20x.

RESULTS

Of the 980 days in the study period, 748 images fell within the desired tidal level range (0.25 ≤ tide ≤ 0.75). Due to problems with image quality (rain on the lens, fog, equipment failure, and very small wave heights that did not produce a SLIM), only 578 images yielded shoreline positions. Of these, some shoreline position estimates did not cover the entire 2 km sample region. Only 336 images yielded data fully spanning a 1 km subset of the sample region (900 m ≤ y ≤ 1900 m). This subregion was sampled at an average rate of about 1 sample/3 days. We will focus much of the analysis on the 1 km sub region.

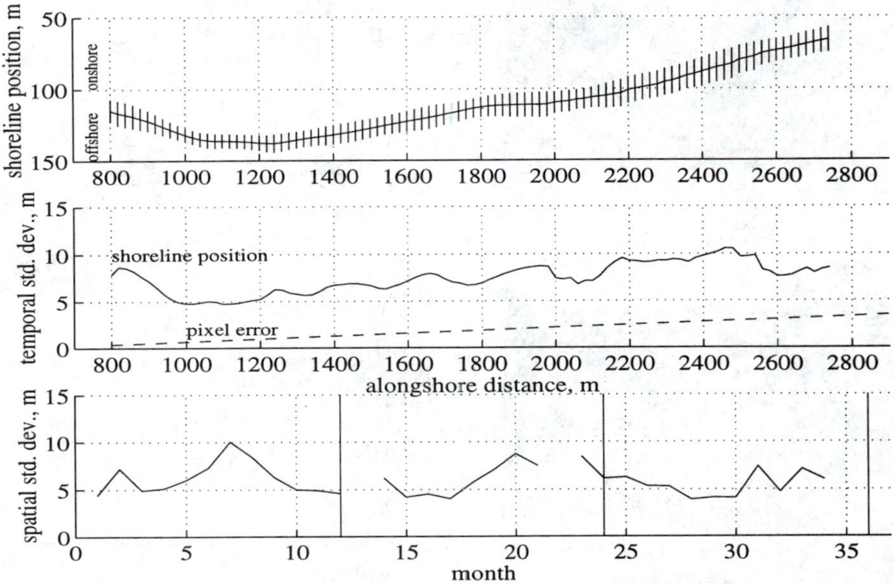

Figure 2. Mean and standard deviation of the shoreline position. The time-averaged shoreline position is shown in the upper panel, where error bars are the standard deviation, plotted at 20 m intervals. In the second panel, the standard deviation is plotted again, along with the cross-shore error due to the pixel footprint dimensions (dashed line). The third panel shows monthly-averaged spatial variability. January of each year is marked by a vertical line.

Shoreline Mean and Variance

Over the 3 year study period, the observed shoreline position varied by as much as 45 m. The spatial structures of the shoreline and its variability are summarized by plots of the mean and standard deviation as a function of alongshore location, shown in Figure 2. The mean shoreline position was not straight, but curved smoothly through the entire study area. The bend in the mean shoreline likely stems from geologic constraints. The temporal variability of the shoreline, indicated by its standard deviation, was not uniform either. The minimum variance occurred near the most seaward protrusion of the mean shoreline (y = 1100). The observed shoreline variance increased with distance from the camera, which may be due, in part, to the increasing size of the pixel footprint. The expected cross-shore pixel error is plotted for reference. Our choice of coordinate system was not shore-normal everywhere, which also contributed to increased apparent variability.

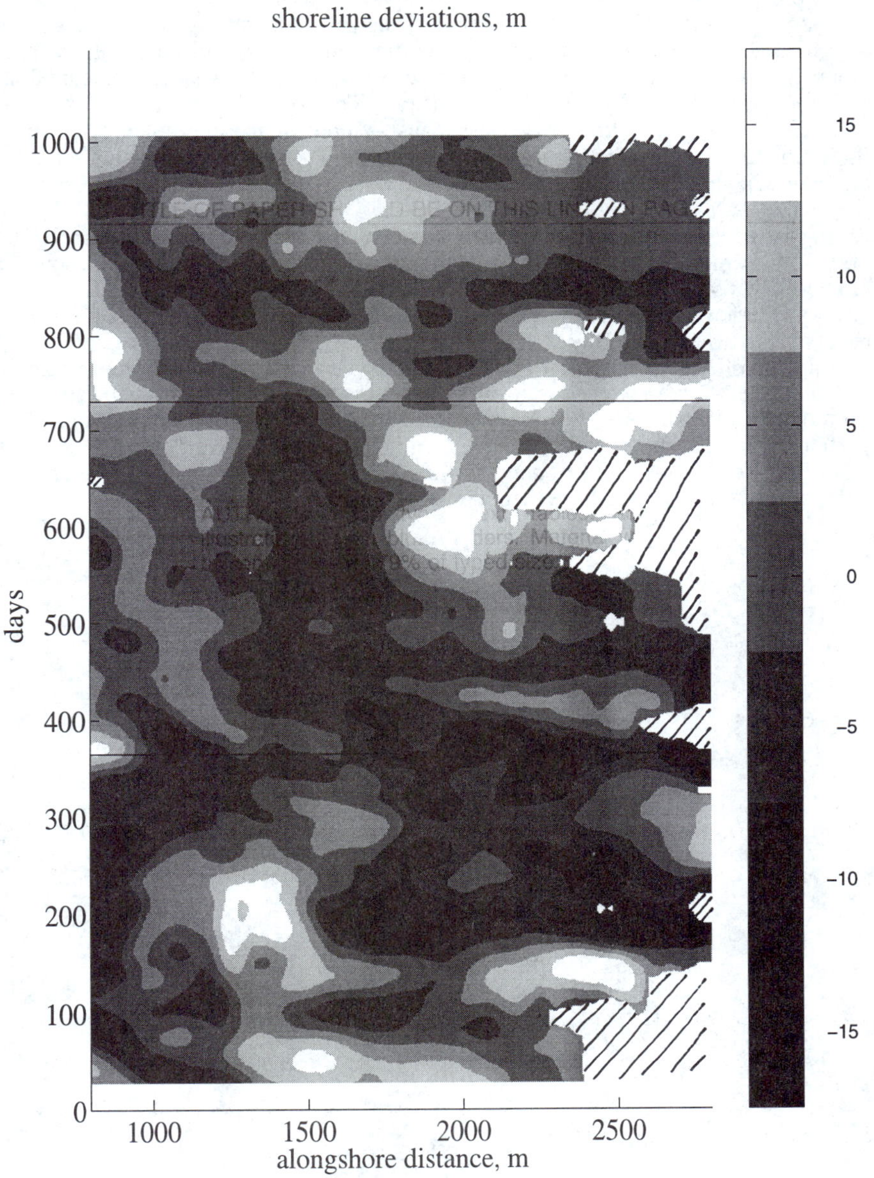

Figure 3. Gray scale plot of shoreline deviations. The scale bar is in meters. Positive deviations (whiter) correspond to displacement of the shoreline seaward of it mean position. Missing data are filled by hatching, and January 01 of each year is marked by a horizontal line.

Time series of shoreline deviations were obtained by subtracting the mean shoreline position from each observation, at each alongshore location. The final panel in Figure 2 shows the monthly-averaged, spatial variation from the mean shoreline position, as a function of time. In order to visualize the variation of the Duck shoreline, Figure 3 displays low-pass-filtered deviations of the shoreline position at each alongshore location and each time. Positive deviations (light shades) indicate that the shoreline was seaward of its mean position. The low pass filter (Ooyama, 1989) damped variations with time scales less than 1-2 months, and interpolated missing data if the time gaps were shorter than the filter cutoff. The remaining missing data values are filled with hatching.

In Figure 3, the most striking features are lineations that generally trend obliquely from the lower right towards the upper left (traveling toward the south) and have length scales of about 1 km. The behavior of these large scale features was not always characterized by simple southward propagation. For example, their direction reversed at times (e.g. day 200 to day 300). Also, shorter scales contributed to prominent coherent behavior. For example, between days 850 to 900, megacusps with approximately 200 m wavelengths were well developed. The megacusps appeared to propagate at the same speed as the large scale features. An annual cycle is not particularly apparent in Figure 3. We turn now to quantitative descriptions of these shoreline deviations, partitioned into an alongshore-averaged component and a component with an alongshore length scale of 1 km.

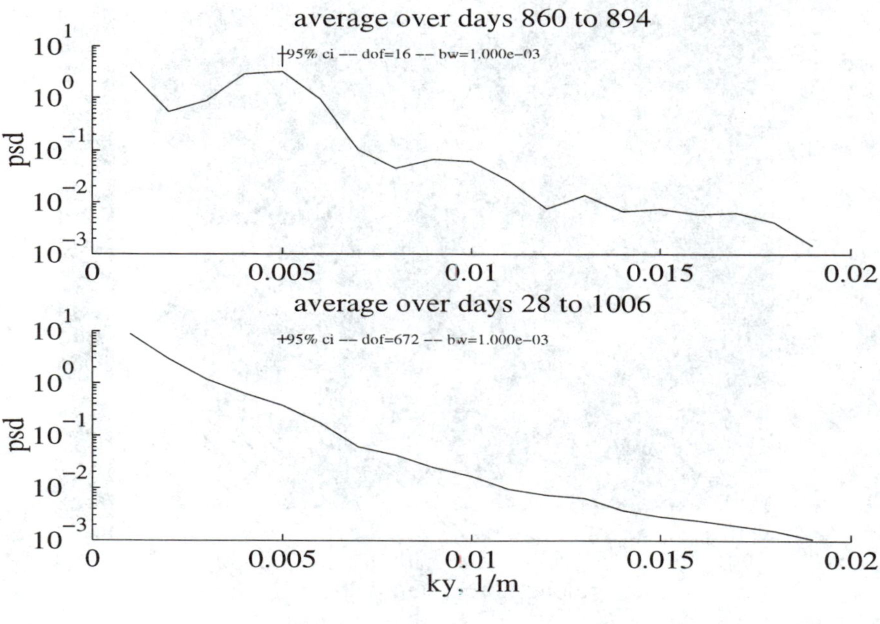

Figure 4. Shoreline variance spectra as a function of alongshore wavenumber. The upper panel shows sample spectral estimates averaged over a 34 day period (containing only 8 images) when megacusps were present with 200 m wavelengths. The lower panel is the spectrum averaged over the entire 3 year study period.

Length Scales

The contribution of different length scales to the observed shoreline variability was quantified using wavenumber spectra of the shoreline deviations. A discrete spectrum was estimated on each day, using data within the 1 km subregion, $900 \leq y \leq 1900$ m. Spectral estimates corresponded to the Fourier wavenumbers, $k_y = 0, \pm\frac{1}{1000}, \pm\frac{2}{1000}, \ldots,$ and $\frac{250}{1000}$ m^{-1}. The alongshore-averaged shoreline position was removed at each time before estimating the spectra, and was the estimate of the first component, $k_y = 0$. The variance spectrum, $S(k_y)$, on any particular day describes the contribution of each length scale to the spatial variability on that day. As an example, we show several spectra, averaged over time to improve the reliability of the spectral estimates (Figure 4). For instance, megacusps, which were obvious in the space-time map of deviations (days 850 to 900), also appeared as a well defined and significant spectral peak. However, the dominance of large length scales (low wavenumbers) was the most typical characteristic of these spectra. The spectrum averaged over the entire 3 year study period (Figure 4) emphasizes this point. In fact, the first two Fourier components, which are the alongshore-averaged shoreline position ($k_y = 0$) and the 1 km wavelength ($k_y = \frac{1}{1000}$ m^{-1}), explain nearly 80% of the total shoreline variance. We will focus the remaining analysis on these two components.

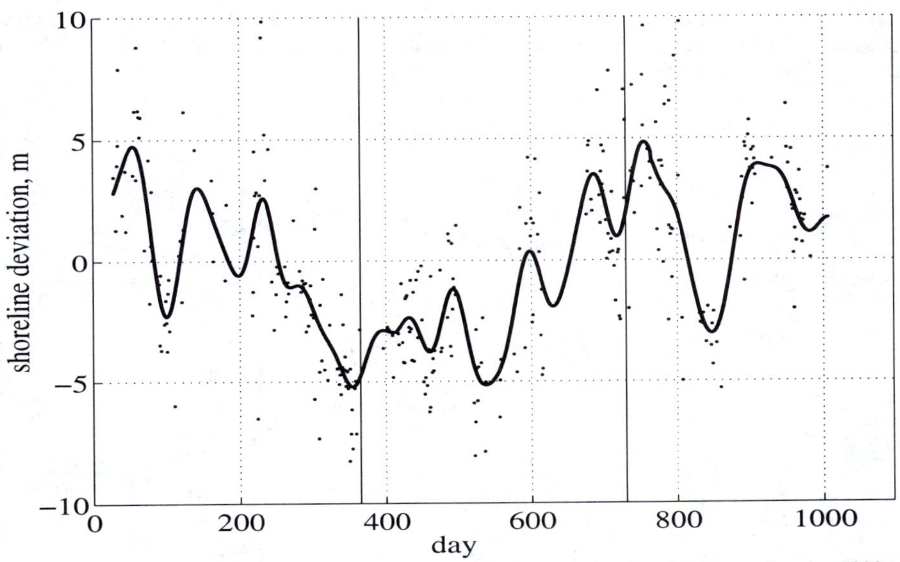

Figure 5. Time series of the alongshore-averaged shoreline position over the 1 km subregion (900 m ≤ y ≤ 1900 m). The dots correspond to the actual samples, while the solid line is the filtered time series. January 01 is of each year is marked by a vertical line.

Alongshore-averaged shoreline variability explained 36% of the total, observed shoreline variance. This component represented net onshore or offshore change of the shoreline (Figure 5). Apparent, high frequency fluctuations due to measurement errors associated with rapid variation of the wave height were removed by filtering using a 30-60 day cutoff. The filtered time series of the alongshore-averaged shoreline deviations shows that dominant variability occurred over time periods greater than one year (interannual behavior). More rapid variations likely corresponded to storm/calm cycles. The absence of an annual cycle, as might be predicted with the summer/winter model, is conspicuous.

The second component that we will consider corresponds to the 1 km wave length ($k_y = \frac{1}{1000}$ m^{-1}), and accounted for 40% of the total shoreline variance. This component may be represented by time series of both amplitude, A, and phase, ϕ.

$$A(k_y,t) = \left| a(k_y,t) + i\, b(k_y,t) \right| \quad (1)$$

and

$$\phi(k_y,t) = \mathrm{Tan}^{-1}[-b(k_y,t)/a(k_y,t)], \quad (2)$$

where $a(k_y,t)$ and $b(k_y,t)$ are the real and imaginary parts of the Fourier amplitudes. The amplitude time series had a strong annual cycle (Figure 6). This is the quantitative representation of typical observations that the shoreline is relatively straight during the winter (largest waves) and develops increased alongshore variation during calmer periods, also evident in Figure 2. The phase time series describes alongshore shifts of a coherent morphological structure. A simple phase ramp (constant slope) is indicative of a progressive wave form, while phase jumps of 180° suggest standing motions. The net phase shift over the study period corresponded to an average southward propagation of about 1 m/day. The estimated celerities are shown in the lower panel of Figure 6. We expect the amplitude and celerity for this component to be correlated to wave height and direction.

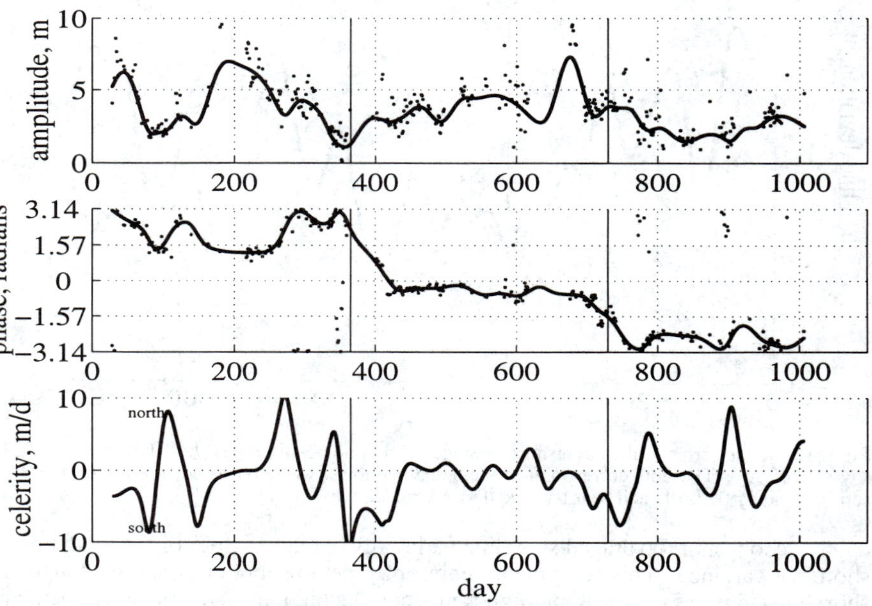

Figure 6. Time series of the amplitude (upper panel), phase (middle panel), and celerity (lower panel) of the 1 km wavelength component. The trend in the phase (ramp) corresponds to an average alongshore celerity of -0.95 m/day.

Correlation to Waves

We move now to determining how the two components of variability (ky = 0, $\frac{1}{1000}$ m^{-1}) were related to variations in the wave field. In particular, we expect that the amplitudes of the two components should be correlated to the wave height, while we expect that the alongshore propagation of morphologic features should be related to the product of the wave height and angle of incidence. The alongshore component of wave energy flux ($Pl \sim H_s^2 \, Sin[-2\,\alpha]$) will be used to test this hypothesis. Figure 7 shows a time series of the wave height, wave angle, and wave energy flux over the study period. The wave directions are relative to shore normal and positive wave angles correspond to waves approaching from the north, consequently driving sediment transport to the south. High frequency fluctuations dominate the wave time series, so time series were low-pass filtered with a 30-60 day filter cutoff. The filtered time series contain seasonal cycles in which wave heights tended to be high in the winter months and low in the summer months. Waves tended to approach from the north during winter and south during summer. Comparing the filtered wave time series to the filtered morphologic time series allows us to test for correlation between low-pass variations.

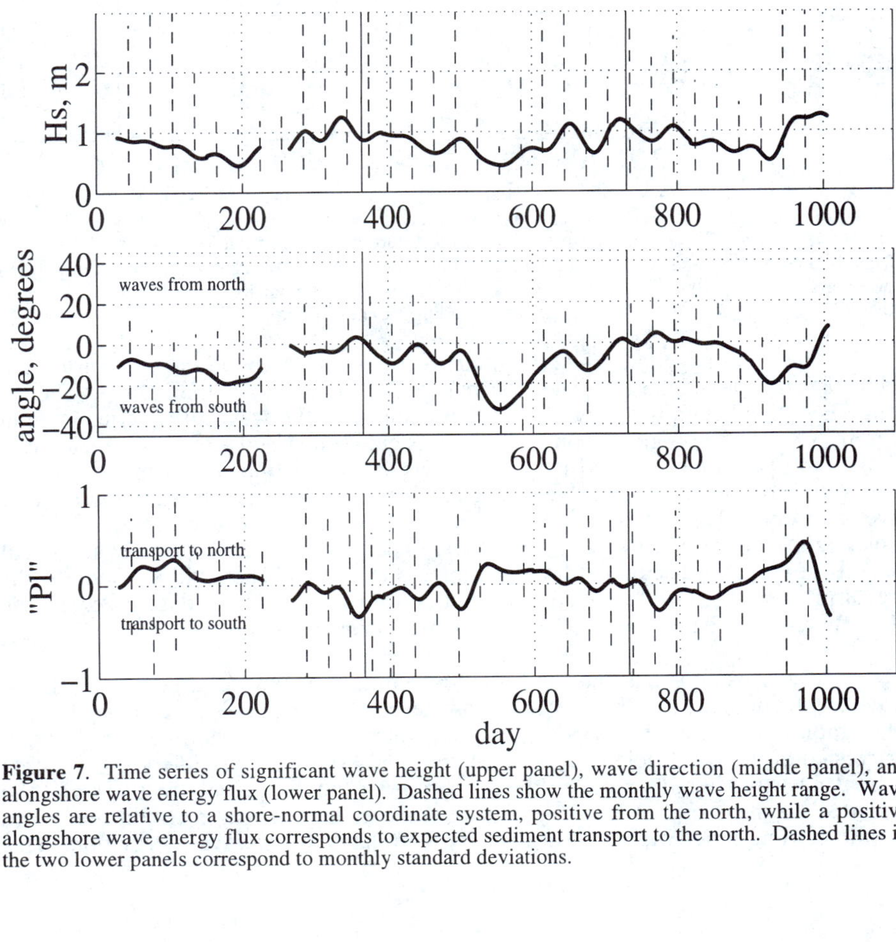

Figure 7. Time series of significant wave height (upper panel), wave direction (middle panel), and alongshore wave energy flux (lower panel). Dashed lines show the monthly wave height range. Wave angles are relative to a shore-normal coordinate system, positive from the north, while a positive alongshore wave energy flux corresponds to expected sediment transport to the north. Dashed lines in the two lower panels correspond to monthly standard deviations.

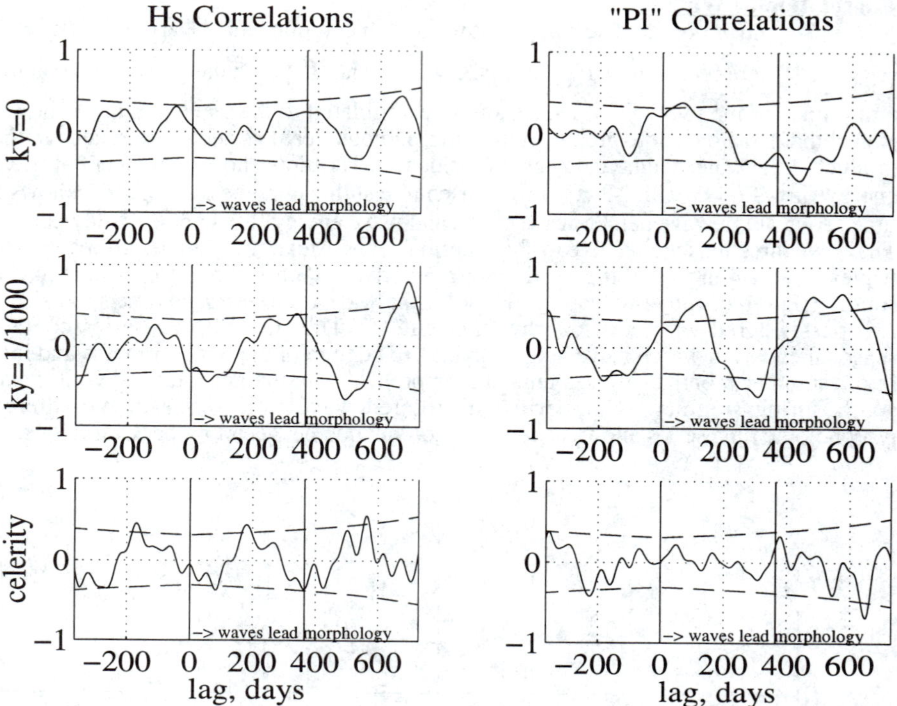

Figure 8. Cross-correlations between wave parameters and morphologic parameters. The wave parameters (H_s and Pl) are labeled at the top of the figure and the morphologic parameters (k_y=0, k_y=1/1000 amplitude, and celerity) are labeled to the left of each pair of figures. In each plot, dashed lines mark the 95% significance level of the correlation and vertical lines mark lags of ±1 year. Positive lags correspond to the wave time series leading the morphology.

The time-lagged, cross-correlations between the morphologic time series and both wave height and the alongshore wave energy flux are shown in Figure 8. The lag range is defined such that positive lags correspond to the wave time series leading (forcing) the morphologic response. Time lags between -1 and +2 years are shown in all the correlation plots. The first panel shows the correlations between alongshore-averaged shoreline position and wave height, which were less than the 95% significance level at nearly all lags. In the adjacent panel, the correlation to the wave energy flux was most significant at a lag of about 60 days. The positive correlation suggests that periods of northward sediment transport lead to increased beach width. Both results are surprising, since we expected a correlation to wave height but did not expect the alongshore-averaged response to be a function of wave direction.

The cross-correlations of both the amplitude and celerity of the 1 km wavelength component with the wave parameters contained strong seasonal signatures. The amplitude of the 1 km wavelength component was significantly and negatively correlated to the wave height, at a lag of about 60-90 days. We infer from this that increasing wave heights reduced the amplitude of the 1 km scale features and the response of the morphology lagged behind the change in wave conditions. The amplitude time series was also, surprisingly, correlated to the wave energy flux. This may stem from the strong correlation between wave height and direction, rather than a direct forcing-response relationship of wave direction on these amplitudes.

The celerity and wave height were correlated significantly at a lag of about 90 days. The negative correlation indicates that when the wave height increased, the celerity increased towards the south. Surprisingly, the correlation between the celerity and wave energy flux was relatively poor, which was not expected. Interpretation of cause and effect relationships between the wave field and morphology is potentially impeded by the possibility that the celerity of the 1 km features was partly a function of their amplitude, which was modulated by the wave height.

DISCUSSION

Source of alongshore propagating waves

Sand waves with various amplitudes and celerities have been observed in several nearshore environments (e.g. Sonu, 1969). Clearly, large scale, progressive, morphologic features exist at Duck. We can suggest a few explanations for the existence of these features. We can also reject some possible explanations, based on characteristic time scales. For example, oblique sand ridges are large scale features, recently discussed by Trowbridge (1995). These features extend to deep water (~10 m), and have characteristic, long-term, alongshore celerities of $0(10^{-2}$ to $10^{-1})$ m/day. The forcing mechanism for them is probably storm driven currents, which, as Trowbridge showed, may cause an initially alongshore uniform coastline and bathymetry to become unstable. Hulscher (1996) presents theoretical descriptions of another type of sand wave, which is an instability driven by tidal currents. However, both oblique sand ridges and tidal sand waves, which are O(0.1 to 1) km in length, have much slower celerities than those observed at Duck. Since the length scale of these deeper water features matches the scales we have observed, it is possible that there is some interaction between the two.

Another sand wave formation mechanism, treated recently by Thevenot and Kraus (1995), is ejection of sand plugs from tidal inlets. These plugs may weld to the shore and perturb the longshore current field. Then, they are expected to propagate alongshore in the direction of the longshore current. Thevenot and Kraus observed sand waves with wavelengths of about 1 km and celerities of order 1 m/day. The alongshore-progressive behavior was modeled by Thevenot and Kraus. However, the modeled sand waves tended to diffuse more rapidly than those observed in the field. The addition of wave refraction in the model reduced the tendency of the modeled waves to dissipate, and appears to be an important sand wave maintenance mechanism. Although there are no nearby inlets at the Duck field site to allow formation of sand waves as ejected sand plugs, the kinematics of the features we have observed seem consistent with the Thevenot-Kraus model.

Potential Pier Effects

A possibly important physical characteristic of our study site is the presence of a pier, which extends 500 m offshore (6 m depth). Since we observed the shoreline behavior only north of the pier, we have no information about what happens near to and south of the pier. If sand waves propagate past the pier, how are they affected by its presence? Does the pier generate sand waves? We hope to extend analysis to observations from both sides of the pier, as there is an extensive collection of surveyed profiles at Duck, spanning over a decade.

CONCLUSIONS

Observations of the shoreline position at Duck, NC indicated that alongshore progressive features with a length scale of order 1 km were responsible for a large portion (40%) of shoreline variability at this site. The amplitude of these features (~10 m) was modulated by the annual variations in wave height. The amplitude was reduced during periods of increased wave height (winter) and the amplitude increases during calm conditions (summer). Although these features propagated to the south, variations in celerities were not well correlated to the direction of expected wave-driven sediment transport, and were more correlated to changes in the wave height. Surprisingly, the alongshore-averaged shoreline behavior was not well correlated to wave height. Although the wave field varied on an annual cycle, the alongshore-averaged shoreline position did not, in contrast to expectations from a simple conceptual model of summer-winter behavior. Instead, this component varied most dominantly over an interannual period, in which one cycle was completed in about two years.

The observations and analysis presented here describe the contribution of various alongshore length scales to beach variability at Duck, NC. The results of this analysis indicated the importance of large length and time scales to observed nearshore variability. Extending analysis to observations that cover larger lengths and longer time periods will improve our resolution of the phenomena presented here.

ACKNOWLEDGMENTS

The data used here comes from a large image database collected as part of an ever expanding research program called ARGUS. If John Stanley had not organized all the data so well, the pilot study for this project would not have been completed in the two days available to do so and the first author's scientific career would have been ruined. The research staff at the Army corps of Engineers Field Research Facility are similarly appreciated for their maintenance of the wave data. Thanks, also, to those people who discussed this work with us at the ICCE '96 meeting. This work was funded by the Office of Naval Research, Coastal Dynamics Program (grant # N00014-960237).

REFERENCES

Birkemeier, W.A., Time scales of nearshore profile change, in 19th International Conference on Coastal Engineering, pp. 1507-1521, ASCE, New York, 1985.

Hulscher, S.J.M.H., Tidal-induced large-scale regular bed form patterns in a three-dimensional shallow water model, Journal of Geophysical Research, 101 (C9), 20,727-20,744, 1996.

Komar, P.D., Beach processes and sedimentation, 429 pp., Prentice-Hall, Englewood Cliffs, N.J., 1974.

Larson, M. and N.C. Kraus, Dynamics of longshore bars, in 23rd International Conference on Coastal Engineering, pp. 2219-2232, ASCE, New York, 1992

Lippmann, T.C., and R.A. Holman, Quantification of sand bar morphology: A video technique based on wave dissipation, Journal of Geophysical Research, 94 (C1), 995-1011, 1989.

Lippmann, T.C., and R.A. Holman, The spatial and temporal variability of sand bar morphology, Journal of Geophysical Research, 95 (C7), 11,575-11,590, 1990.

Ooyama, K.V, Scale-controlled objective analysis. Monthly Weather Review 115(10): 2479-2506, 1987.

Plant N.G. and R.A. Holman, Intertidal beach profile estimation using video images. Marine Geology, in press.

Ruessink, B.G., and A. Kroon, The behavior of a multiple bar system in the nearshore zone of Terschelling, the Netherlands, 1965-1993, Marine Geology, 121, 187-197, 1994.

Sonu, C.J., Collective movement of sediment in the littoral environment, in Proc. of the 11th Conference of Coastal Engineering, pp. 373-400, ASCE, 1969.

Thevanot M.M. and N.C. Kraus, Longshore sand waves at Southampton Beach, New York; observations and numerical simulation of their movement. Marine Geology, 126(1-4), 249-269, 1995.

Trowbridge, J.H., A mechanism for the formation and maintenance of shore-oblique sand ridges on storm-dominated shelves, Journal of Geophysical Research, 100 (C8), 16,071-16,086, 1995.

Wijnberg, K.M., and J.H.J. Terwindt, Extracting decadal morphological behavior from high-resolution, long-term bathymetric surveys along the Holland coast using eigenfunction analysis, Marine Geology, 126, 1995.

CHAPTER 273

Long Term Behaviour of the Sediment Volume inside a Tidal Basin after Poldering

Ernst Schroeder[1], Roland Goldenbogen[2], Hans Kunz[3]

Abstract

The development of morphological reactions over time as a result of polderings has been investigated for the Dithmarscher Bucht (Bay of Dithmarsch), a tidal basin which is located in the North Sea/Germany. Two methods of parametrization have been used: the first method is based on volume-differences and the second one refers to the vertical differences of the topography. The results have been used for an estimation of the stability of the tidal basin with respect to an accelerated rise of the relative sea level.

Introduction

Tidal basins are transition areas between the mainland and the sea. They exist under special hydrological and sedimentological boundary conditions. On account of these special conditions, tidal basins have been formed along the Southern North Sea coast during the last 10,000 years (Bantelmann 1966 and Streif 1990). In most of the areas, the actual situation has been substantially influenced by human interferences, especially by land reclamation.

Besides coastal engineering aspects the Wadden Sea areas are an unique ecological system, which is essential as breeding and feeding ground for migratory birds on their East Atlantic Flyway (Lozán et al. 1994).

---

[1] Dipl.-Ing., Coastal Research Station (CRS) of the Lower Saxonian Central State Board for Ecology, An der Mühle 5, D-26548 Norderney/Germany

[2] Dipl.-Ing., Coastal Research Station

[3] Director Dr., Coastal Research Station

An accelerated rise of the relative sea level is a realistic prospective; therefore the question about its effects on the long term behaviour of the Wadden Sea-tidal basins along the North Sea coast is important. Only if the morphology of the tidal basin is able to adapt to the changing hydrological conditions, is it ensured, that the whole system keeps its stability.

A tidal basin is a feed-back system which tends towards an equilibrium stage, determined by the hydrological and morphological boundary conditions. Changes in the relation between these boundary conditions lead to reactions of the morphology. The morphology compensates the non-equilibrium stage by more or less intensive accumulation or erosion processes. These adjustments proceed with a phase lack, depending on the quantity of the disturbance in relation to the system itself.

Comparable processes occur after cutting off (poldering) parts of a tidal basin. The daming reduces the catchment area of the tidal basin and alters the relationships between the hydrological and the morphological boundary conditions. Investigations of the morphological development after cutting off parts from a tidal basin led to calculations over time of the adjustment-ability.

The morphological reaction of a tidal basin, forced into a non-equilibrium stage by cutting off parts, can be compared with the effects of the changing hydrological boundary conditions as a result of sea level rise. Thus, the adaptability can be estimated qualitively for different values of sea level rise.

PARAMETRIZATION

The description of the morphological development has been based on two different methods for the parametrization. The parameters for the first method has been determined by calculating the difference between volumes. The second method is based on the averaged heights difference between bathymetric surveys of the morphology for successive stages. The introduced parameters are named "transpose-rates", for the first method as h_I and for the second one as h_A or h_E, with the dimension m/a. The transpose-rates which are presented and discussed in the paper represent mean yearly values.

Method N°. 1

The sediment volume V_I has been calculated above a fixed datum as horizontal reference level (Fig. 1). The reference level has been set below the deepest point of the topography for which morphological changes are recorded. For the total tidal basin

Dithmarscher Bucht, the reference level has been fixed at -22.0 m NN[4] and for the defined subareas at -12.0 m NN. The upper limit is 2.0 m NN[5]. The volume difference ΔV_I (Fig. 2) of two successive surveys are divided by the time period between the surveys Δt and by the area of the cross section A_I (Schroeder et al.1995b). This parameter is named "transpose rate h_I" (m/a):

$$h_I = \Delta V_I / (\Delta t_I \cdot A_{I,i+1}) = (V_{I,i+1} - V_{I,i}) / ((t_{i+1} - t_i) \cdot A_{I,i+1}). \qquad (1)$$

The division by the cross section A_I provides a standardization of the parameter, which allows the comparison of catchment areas A_b with different sizes.

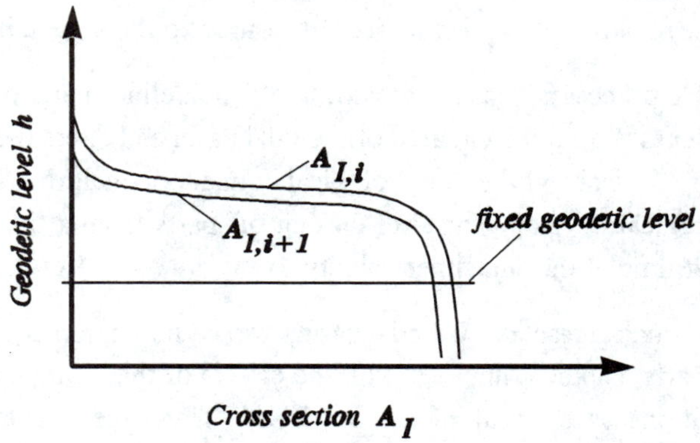

Figure 1. Idealized graphs of the geodetic level h over the cross section A_I

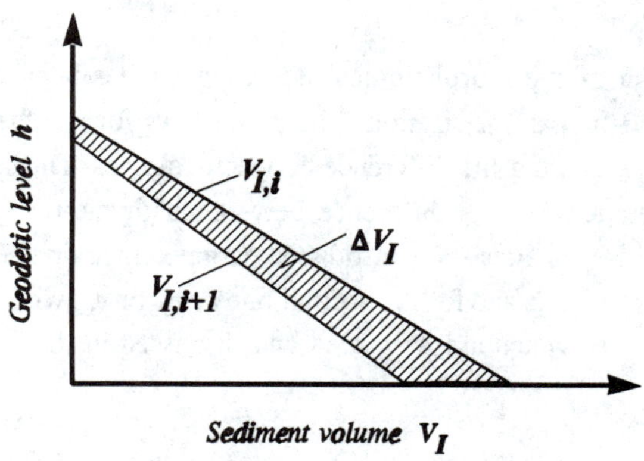

Figure 2. Idealized graphs of the geodetic level h over the sediment volume V_I

[4] NN = Normal Null = German ordnance-datum (app. MSL)
[5] MHW app. 1.55 m NN (min. HW/max. HW = 1.46/1.63 m NN; 1942 - 1990) at the tide-gauge Büsum (Fig. 6)

Method N°. 2

The second method leads to "transpose-rates" which have been determined separately for accumulation as "h_A" and for erosion "h_E". The calculation of these parameters is based on the heights of a standardized area-element (vertical difference). Therfore, the topography has been transformed into square grids (Fig. 3). The resolution of the grid has been fixed as 25 m in x- and y-direction, hence, one grid-element covers an area of 625 m².

The horizontal reference level of one grid yields as mean value of the "average node" H_i (mNN) of the heights of each grid node (see A, B, C and D in Fig. 4):

$$H_i = (A+B+C+D)/4. \quad (2)$$

The difference of the parameter H of two successive surveys (H_i, H_{i+1}) is the accumulation or erosion-rate ΔH_i in m for each grid element:

$$\Delta H_i = H_{i+1} - H_i . \quad (3)$$

This rate, divided by the time interval between the successive surveys, leads to the yearly transpose-rate of each grid element, either as accumulation ($\Delta H_{A,i}$) or as erosion ($\Delta H_{E,i}$). The sum of all elements with accumulation or erosion inside the defined investigation areas (total basin or subarea) yields to the "transpose-rate", which is indicated as h_A (accumulation) and h_E (erosion) with the dimension m/a:

$$\Sigma \Delta H_{A,i} / \Delta t = h_A \quad (4)$$

$$\Sigma \Delta H_{E,i} / \Delta t = h_E . \quad (5)$$

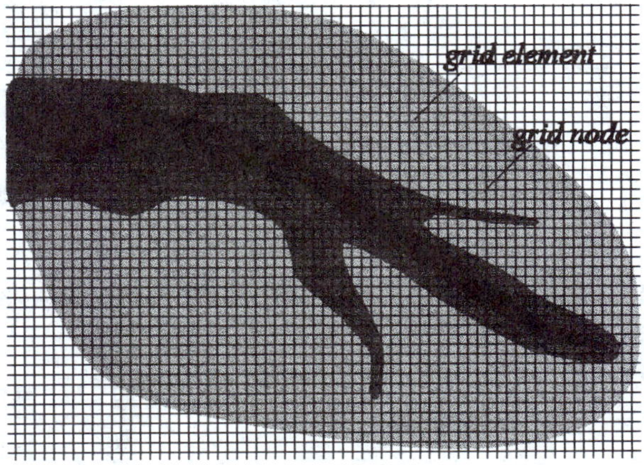

Figure 3. Idealized investigation area with grid

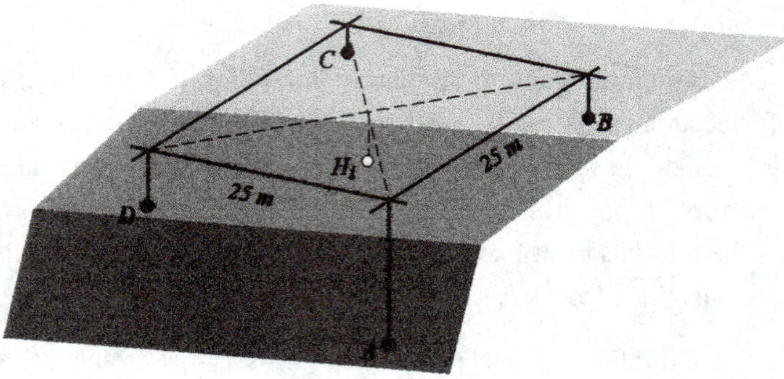

Figure 4. Idealized grid element with the grid nodes A, B, C and D and the average node H_i

INVESTIGATION AREA

The investigation area is called the *Dithmarscher Bucht* (Bay of Dithmarsch). It is located on the German North Sea coast on the West side of the German federal state of *Schleswig-Holstein* (Fig. 5).

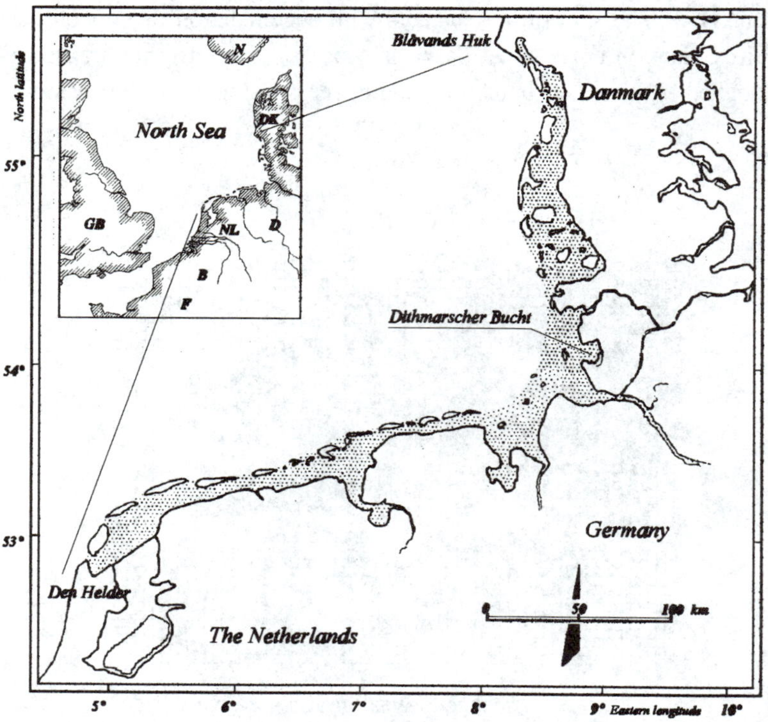

Figure 5. Location of the investigation area

The shape represents a single bay which opens towards the main wind- and wave direction. Inside the bay, the mean tidal range (semidiurnal tides) is 3.25 m (1942-1990); the dominate wind direction scatters around West. The mean offshore waveheight is about 1.00 m. The geological structure of the bay, developed during the Holozän-time-period, consists mainly of marine sediments (Fischer 1955).

The tidal basin had been reclaimed two times by poldering. The bay covered an area of approximately 157 km² before the polderings had been carried out. The first poldering (1972) reduced the bay-area (A_b) by about 16 km² and the second one (1978) by 26 km² (see dotted lines in Fig. 6).

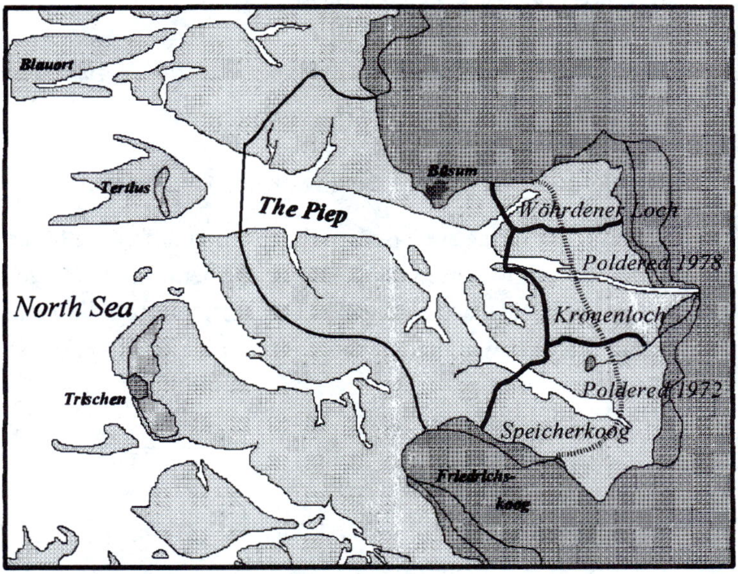

Figure 6. Investigation area *Dithmarscher Bucht* with poldered areas and the investigated subareas

With regard to the reaction of the bay as result of the polderings, the tidal basin has been split up into subareas according to the application of an applied empirical model (Goldenbogen 1994, Niemeyer et al. 1995 and Schroeder et al. 1995a). The subareas which include the poldered parts are shown on fig. 6: *Speicherkoog*, *Kronenloch* and *Wöhrdener Loch*. The total bay-area is represented by *The Piep*. During the investigation period from 1942 up to 1990 the *Dithmarscher Bucht* had been surveyed 9 times (1942, 1956, 1969, 1973, 1976, 1979, 1982, 1985 and 1990). The bathymetric surveys of 1942, 1956 and 1969 decribe an undisturbed topography. The surveys after the year 1973 are influenced by the poldering-means.

The morphological data for the *Dithmarscher Bucht* and parametrizations for "equilibrium stages" have been described by Schroeder 1994.

RESULTS

The following diagrams show the calculated transpose-rates h_I. The data (ordinate) has been plotted over the half period of an investigation intervall Δt_i (abscissa, Fig. 7).

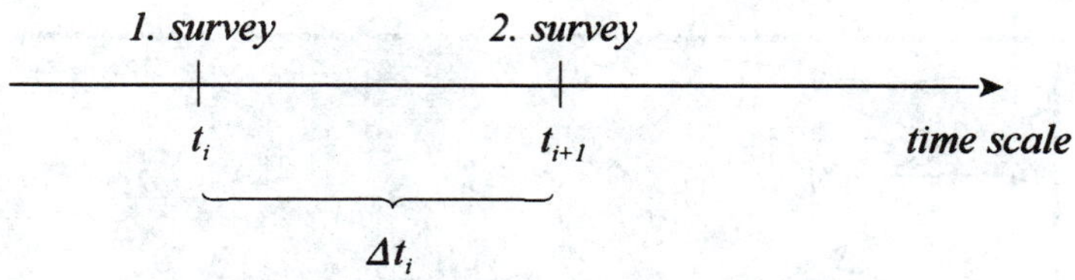

Figure 7. Definition of the investigation intervall Δt_i

Figure 8 shows data which are based on the the total difference of the sediment volume in accordance with the first method. During the period from 1942 up to 1969 the anthropogenic influences are negligible. The scattering of the results can be explained by natural effects and, additionally, in the case of the year 1956, by the quality of the bathymatric survey.

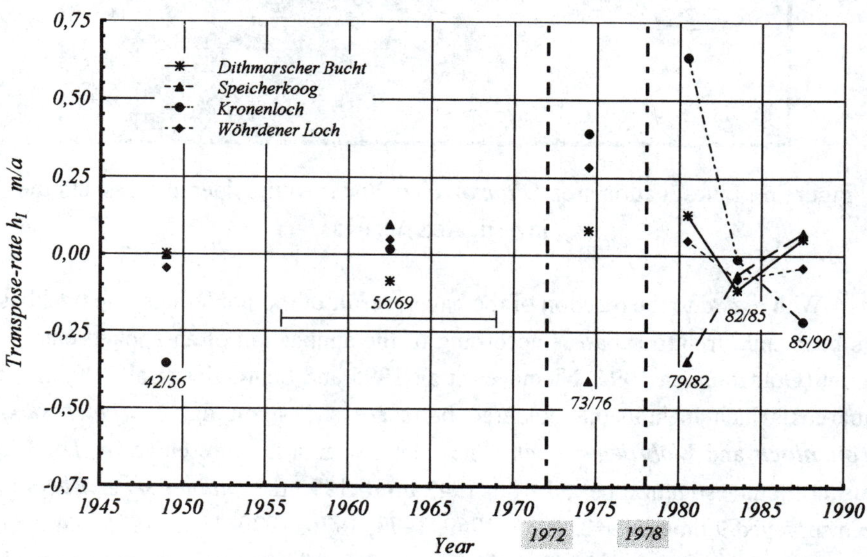

Figure 8. Transpose-rate h_I versus investigation intervall Δt_i

The first poldering was carried out inside the subarea *Speicherkoog* in 1972 (Fig. 6). The value of the subarea *Speicherkoog* shows a distinct tendency of erosion as a result of the first poldering, whereas the other subareas, as well as the total tidal basin, are characterized by accumulation.

The second poldering was carried out in 1978. This poldering extends over the subareas *Kronenloch* and *Wöhrdener Loch* (Fig. 6). After the second poldering, the tendency of erosion continued inside the subarea *Speicherkoog*. The transpose-rate inside the subarea *Wöhrdener Loch* decreased in the range of the average values, as monitored during the undisturbed period before 1969. Only the transpose-rates of the subarea *Kronenloch* increased significantly. However, already during the second investigation intervall from 1982 to 1985, the values for this subarea also decreased. All values indicate a low erosion-tendency. The erosion effects continue for the subarea *Kronenloch* during the last interval (1985 to 1990). The results for the last interval scatter within a range which is comparable to the values during the undisturbed investigation period before 1969.

A similar development display the results for the accumulation rate h_E gained by the second method (Fig. 9).

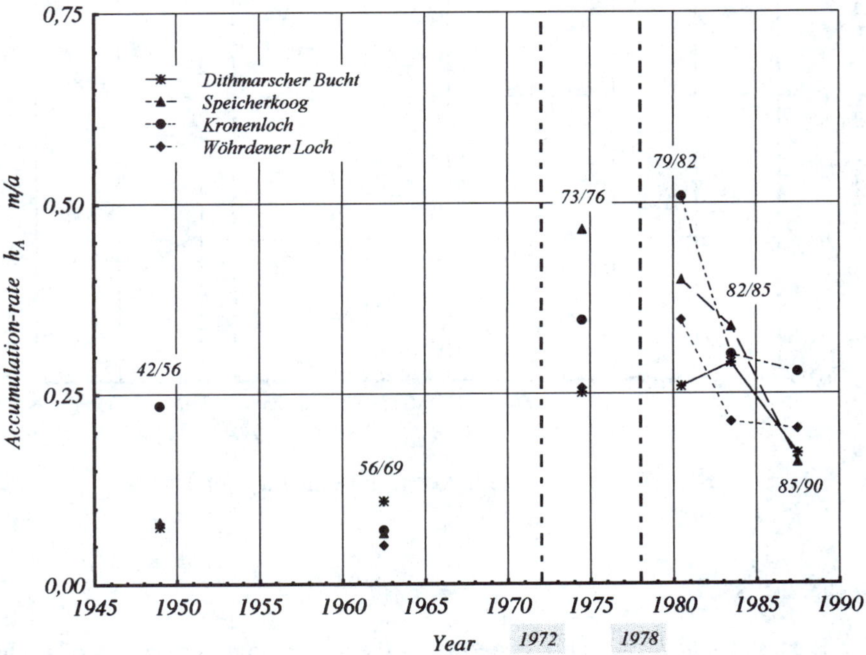

Figure 9. Transpose-rate h_A versus investigation interval Δt_i

The value increased after the first poldering. The subarea *Speicherkoog*, where the poldering was carried out, shows the highest value. This value decreased after the second poldering, whereas the transpose-rates of the subarea *Kronenloch* and *Wöhrdener Loch* increase. The next investigation-interval indicates a reduction of the values for all subareas. During the following interval, this tendency continued for the subarea *Speicherkoog* and also for the whole tidal basin, whereas the subareas *Kronenloch* and *Wöhrdener Loch* show more or less constant accumulation-rates.

The data of the erosion rate also indicate clear trends (Fig. 10). The undisturbed period is followed by an increase of the values except for *Kronenloch.* Over the last two investigation intervals the values decrease, including those for the *Kronenloch.* During the last interval (1985-1990), the value scatter already in the range of the undisturbed period before 1969.

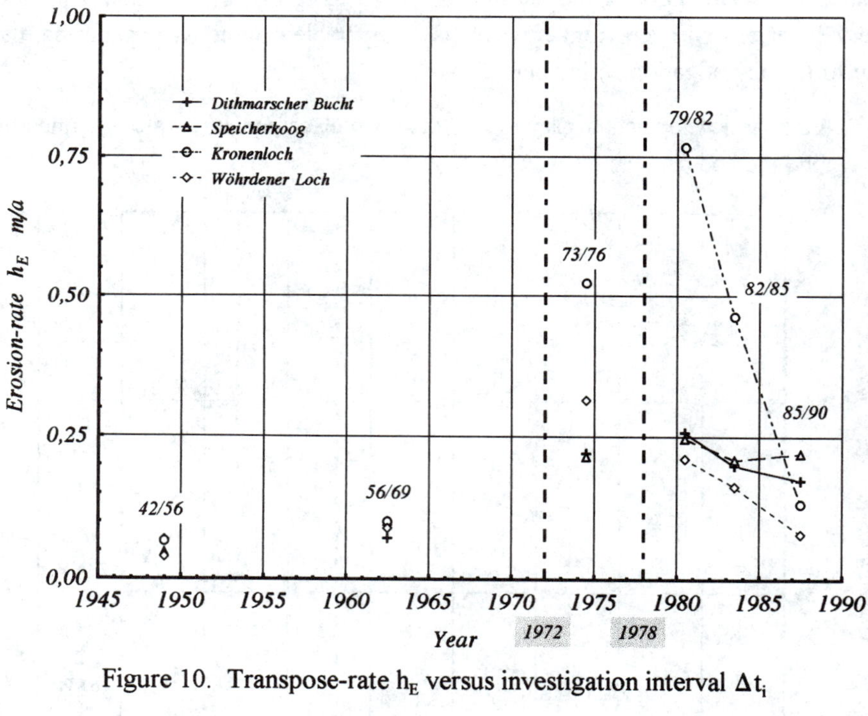

Figure 10. Transpose-rate h_E versus investigation interval Δt_i

CONCLUSIONS

The results allow estimations concerning the morphological reactions of a tidal basin on the changes of geographic boundaries by poldering (reduction of the bay-area leading to a disturbance of the equilibrium condition with respect to the tidal volume as hydrographical boundary condition and the cross-section/area-size as morphological boundary condition). The discussed results show comparable reactions of the described parameters over the whole investigation-period from 1942 to 1990.

The results concerning the development of the *Dithmarscher Bucht* over the last three investigation intervals from 1979 up to 1990 (time period after the polderings) showed, that after approximatly 5 years, the parameter-values had decreased, in part, more than 50% of the maximum values after the polderings. The assumption can be drawn that after 5 more years there will be no significant distinction between the anthropogenic and the natural influences of the transpose-rates. With respect to the strongly decreasing tendency of the investigated transpose-rates, the prediction seems to be reliable, that after one to two decades, the influences of the anthropogenic impacts (disturbance of the "equilibrium stage" by poldering) will fade away and the adaptation to the anthropogenically-created new boundary conditions will reach the new equilibrium.

The results indicate the *Dithmarscher Bucht* as an area with strong accumulation. Comparable investigations, decribed in literature, came up with similar results (Wieland 1984, Siefert 1987, Niedoroda et al. 1994), and Unsöld 1974 with respect to the regeneration after dredging.

With regard to a tranfer of the described results to other tidal basin-areas, the available sediment capacity of the area is one of the main restricting factors that has to be taken into account.

The investigation have shown that, for the tidal basin *Dithmarscher Bucht*, there will be no danger concerning the stability of the morphology with respect to changes of the forcasted or feared hydrological boundary conditions, even if the acceleration of the relative mean sea level rise should reach the IPCC-best-estimate-values (IPCC 1990) of about 60 cm in hundred years.

ACKNOWLEDGEMENT

The findings result from the German-Dutch research project *Wadden Sea Morphological Development - WADE* (Niemeyer et al. 1995). The German part has been funded by the Federal Ministry for Science, Education, Research and Technology (BMBF), Bonn under contract MTK 0508.

REFERENCES

Bantelmann, A. (1966): Die Landschaftsentwicklung an der schleswig - holsteinischen Westküste, eine Funktionschronik durch fünf Jahrtausende. Die Küste, Heft 2, KFKI, Boyens & Co., Heide i. Holst., , (in German).

Fischer, O. (1955): Landgewinnung und Landerhaltung in Schleswig-Holstein. Teil III: Das Festland. Bel. 5: Dithmarschen. Verlag Dietrich Reimer, Berlin, (in German).

Goldenbogen, R. (1994): Erste Ergebnisse empirischer Modellierung der Morphodynamik eines Watteinzugsgebiet am Beispiel der Dithmarscher Bucht. Die Küste, Heft 56, KFKI, Boyens & Co., Heide i. Holst., (in German).

IPCC (1990): Strategies for adaption to sea level rise. Intergovernmental Panel on Climate Change (IPCC), Response Strategies Working Group. Edit: Dutch Ministry of Transport and Public Works, Rijkswaterstaat, Tidal Water Division, Den Haag.

Lozán, J.L.; Rachor, E.; Reise, K.; Westernhagen, H. v. & Lenz, W. (Ed.) (1994): Warnsignale aus dem Wattenmeer. Blackwell Wissenschaftsverlag, Berlin, (in German).

Niedoroda, A.W.; Reed, Chr.W.; Swift, D.J.P.; Arato, H. & Hoyanagi, K. (1994): Modeling shore-normal large-scale coastal evolution. In: Mulder, J.P.M.; Vriend, H.J. de & Heeming, A.W. (Ed.): Validation on large scale morphological models. Nederlands Centrum voor Kustonderzoek, Workshop, Noordwijk.

Niemeyer, H.D.; Goldenbogen, R.; Schroeder, E. & Kunz, H. (1995): Untersuchungen zur Morphodynamik des Wattenmeeres im Forschungsvorhaben WADE. Die Küste, Heft 57, KFKI, Boyens & Co., Heide i. Holst., (in German).

Schroeder, E. (1994): Parametrisierungen morphodynamischer Strukturen von Watteinzugsgebieten für empirisch-konzeptionelle Modellierungen. Die Küste, Heft 56, KFKI, Boyens & Co., Heide i. Holst., (in German).

Schroeder, E.; Goldenbogen, R. & Kunz, H. (1995a): Parametrization for conceptual morphodynamic models of the Wadden Sea areas. Proc. 24th Int. Conf. Coast. Eng. (1994), Kobe/Japan, ASCE, New York.

Schroeder, E.; Goldenbogen, R. & Kunz, H. (1995b): Parametrization of intertidal areas and volumes. Proc. 4th Int. Conf. on Coast. and Port Eng. Develop. Countr. - COPEDEC IV, Rio de Janeiro/Brazil, Brazilian Water Resources Association (ABRH), Rio de Janeiro.

Siefert, W. (1987): Umsatz- und Bilanz-Analysen für das Küstenvorfeld der Deuschen Bucht.Grundlagen und erste Auswertungen (Teil 1). Die Küste, Heft 45, KFKI, Boyens & Co., Heide i. Holst., (in German).

Streif, H. (1990): Das ostfriesische Küstengebiet - Nordsee, Inseln, Watten und Marschen. Sammlung Geologischer Führer 57, Gebr. Borntraeger, Berlin/Stuttgart, (in German).

Unsöld, G. (1974): Jahreslagen und Aufwachsraten in Schlicksedimenten eines künstlichen, gezeitenoffenen Sedimentationsbeckens (Wattgebiet südlich Nordstrand/Nordfriesland). Meyniana, Bd. 26, (in German).

Wieland, P. (1984): Untersuchungen über geomorphologische Veränderungen in der Dithmarscher Bucht. Die Küste, Heft 40, KFKI, Boyens & Co., Heide i. Holst., (in German).

CHAPTER 274

WAVE GROUPS IN A BARRED NEARSHORE

B. Boczar-Karakiewicz[1], W. Romańczyk[1], J.L. Bona[2] and B. Greenwood[3]

ABSTRACT

Observations of waves and sediment movement in coastal zones have shown evidence of groupiness in the wave field associated with episodic events of high sediment concentration above the sandy bed. These observations raised the question about the role of wave groups on large-scale coastal sand bars. The present study addresses this question. Analysis is based on data from a lacustrian experimental study site located in Nottawasaga Bay, Lake Huron, Ontario, Canada. The results of field measurements provided initial and boundary conditions for numerical simulations of wave-seabed interactions under waves and wave-groups. Model predictions are compared with observations. The conclusions suggest that nonlinearly modulated wave trains may potentially form and modify sand bars in coastal zones due to large-scale spatial variation in the nonlinear structure of these waves. However, strong irregularities in observed wave groups and related episodic sediment concentration events suggest that these groups do not represent an effective factor in the 'slow' process of sand bar formation.

1. INTRODUCTION

In deep ocean, with limited breaking, wave groups can be described by superposition of linear wave components or by applying nonlinear modulation instabilities. Medina and Hutspeth (1990) have shown that linear algorithms describe observed wave groups in a water depth greater than 10 m with satisfactory agreement. Battjes and Vledder (1984) supported this conclusion for the surf zone.

[1]INRS-Océanologie (Université du Québec), Rimouski, Québec, Canada
[2]University of Texas, Applied Math, Austin, Texas, USA
[3]Scarborough College Coastal Research Group, University of Toronto, Toronto, Ontario, Canada

However, in the linear approach, waves are assumed to be a Gaussian process, with independent components of the wave field. In earlier studies, it also has been assumed that upon initial breaking, wave groups vanish, except for a small zone surrounding the mean breakpoint.

These assumptions were shown not to be valid in shallow coastal zones. List's quantitative study (1991, 1992), using the concept of wave group envelope, has confirmed that groupiness decreases rapidly through the surf zone, and is sensitive to initial wave breaking. However, he also proved that the degree of groupiness does not become negligible, even in a saturated surf zone. His analysis revealed that groupiness of unbroken waves remains constant over a wide range of incident wave conditions from fairweather swells to storm seas. That observation contradicts the intuitive concept of groupiness resulting from the superposition of sinusoidal waves of similar frequency. According to the linear concept, an increase of spectral bandwidth should result in decreased groupiness.

Laboratory tests and numerical simulations carried out with wave shoaling over a beach of constant slope allowed experimentations with nonlinear wave groups in a simpler environment than a natural coast. A set of such experiments has been conducted in the wave basin of the Hydraulics Laboratory of the National Research Council of Canada (Nwogu 1993). Bichromatic linear waves were generated at the deep-water end of a 1:25 sloping concrete beach. The time series of water surface elevation were measured at two locations: at the toe of the slope and at the initial breaking location. A numerical model based on Boussinesq equations simulated hydrodynamics in the experiment. Comparisons of model results and observations showed good agreement and explained the nature of nonlinear transformations in the initial modulated train composed of sinusoidal waves. When these waves approach shallow water, their shape changes into a complex nonlinear form. The nonlinearity in an individual wave profile increases with decreasing water depth, due to a transfer of energy from the two fundamental frequencies of the incident biharmonic wave train into respective higher harmonic modes. A small amount of energy is also transferred to a low-frequency group-bound wave. That energy increases with decreasing water depth.

The results of a similar numerical study based on a KdV equation with variable coefficients describe the nonlinear evolution of a single group of modulated sinusoidal waves propagating over a gently-sloping beach (Talipova *et al* 1995). Nonlinear and dispersive effects in this group lead to recurrent energy transfer between the fundamental and the higher harmonic components. The initial wave group evolves with increasing distance into a strongly nonlinear wave packet. At a distance of several tens of wavelengths from the deep-water end, the initial modulation changes shape. Higher waves are shifted to the front of the packet due to frequency dispersion. Nonlinearity causes the appearance of secondary crests in the individual waves of the packet. Therefore, it is not

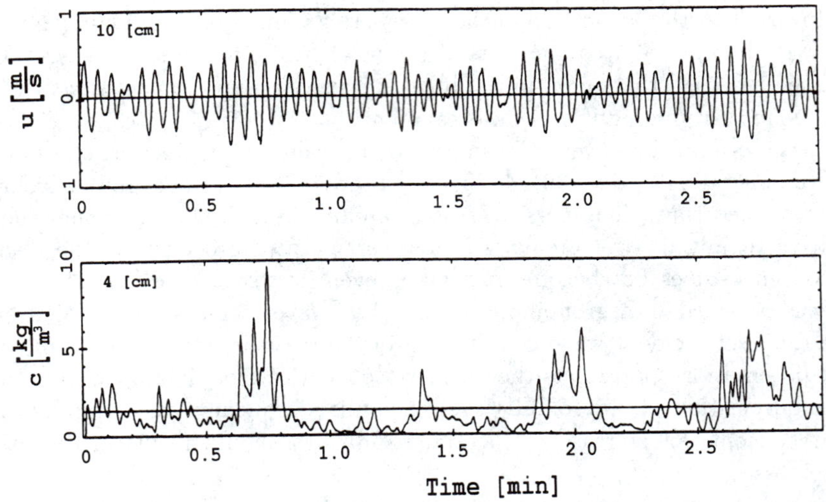

Figure 1: (a)Cross-shore velocity (at 0.10 m above the bed), (b) suspended sediment concentration (at 0.04 m above the bed).

clear how to relate the number and position of waves in the packet with the number of sinusoidal waves in the initial group.

Examples of numerical simulations and laboratory experiments with idealized wave groups show how their properties result from nonlinearity and dispersion of the shallow-water environment. On a natural coast, wave group behaviour is even more complex because of the presence of several flow components.

Observations of waves and sediment movement in coastal sites (Greenwood *et al.* 1992) show evidence of groupiness in the wave field associated with episodic events of high sediment concentration above the sandy bed. A typical record of nearbed velocities and suspended sediment concentration is shown in Figure 1. These observations raised the question about the role of wave groups on large-scale coastal sand bars.

The present study addresses this question and concentrates on the role of wave groups in sand bar formation and dynamics. Our analysis is based on observations from an experimental study site presented in Section 2. Numerical simulations of wave-seabed interactions under waves and wave-groups are presented in Section 3. Model calculations are carried out with initial data from the field experiment. Predictions are compared with observations from the experimental site. The conclusions are formulated in Section. 4.

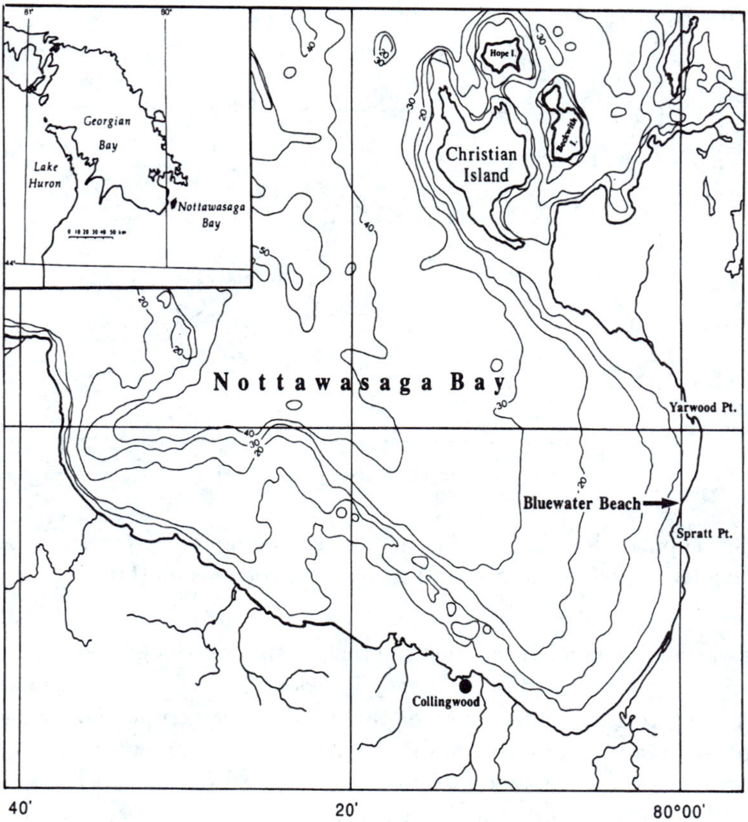

Figure 2: Field site, Bluewater Beach, Nottawasaga Bay, Ontario, Canada.

2. THE EXPERIMENTAL SITE AND FIELD OBSERVATIONS

The study site (Bluewater Beach, Nottawasaga Bay, Lake Huron, Ontario, Canada) is located in a narrow bay, as shown in Figure 2. Incident wind-generated storm waves arrive to the nearshore zone from a narrow window and impinge on the beach from the shore-normal direction. Under such conditions, we may apply our two-dimensional morphological model in calculations presented in Section 3. In the chosen site, surface hydrodynamics are simpler than on ocean coasts as they are not affected by tidal flows.

The study site has a barred shoreface (see Fig. 3 for the location of sensors used in measurements). The typical multibar shoreface from Figure 3 is not static. In 1987, a beach survey revealed a one-bar profile (Fig. 4a), which transformed into a multi-bar profile in the following year (Fig. 4b).

Field measurements of waves and sediment concentrations at Bluewater Beach analyzed by Osborne and Greenwood (1992) have shown that the lo-

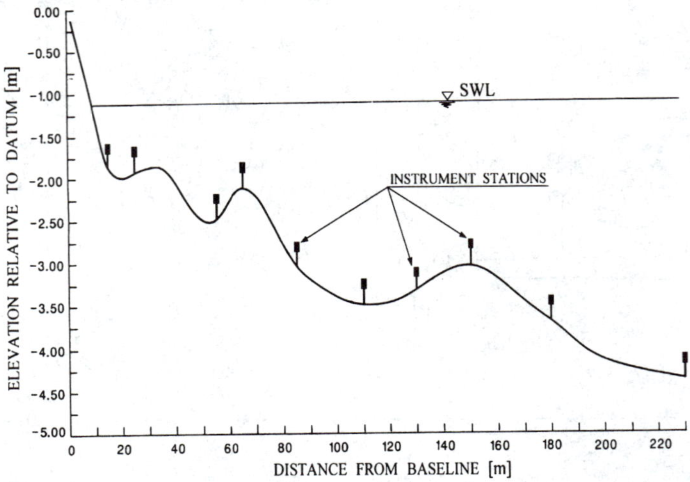

Figure 3: Typical shoreface profile and sensor deployment (pressure sensors, electromagnetic current meters, optical backscatter sensors (OBS)).

cal time-varying suspended sediment flux is a response to high-frequency wind-wave oscillatory currents, low-frequency waves and quasi-steady currents (undertow). The relative importance of these transport components varies spatially and temporally in association with variability in incident wave energy. In contrast to a non barred shoreface, where variations of transport components are strongly controlled by monotonically-decreasing local water depth, across a barred shoreface sediment fluxes are constrained by position with respect to the bars.

Co-spectral analyses of cross-shore velocities and sediment concentrations reveal that grouped, nonlinear shoaling waves induce large onshore sediment transport rates at wind-wave frequencies. However, simultaneously, net offshore-directed transport at low frequencies was also observed. This offshore sediment flux compensates onshore currents induced by grouped waves. During extreme storms, the group-bound long wave could potentially dominate the net oscillatory component of suspended sediment transport and, consequently, displace sediment in the offshore direction.

Additional complexity at a barred foreshore results from waves breaking over a bar crest and waves reforming in the bar trough. Observations show that, typically, outside the surf zone, the re-suspension of sediment under groups of large waves is coherent with the offshore phase of low-frequency modulation of the velocity field due to the presence of wave groups and the group-bound, forced long wave.

However, landward of the zone of wave breaking, on the bar crest and on the upper lakeward slope, the wave-group structure was observed to be diminished.

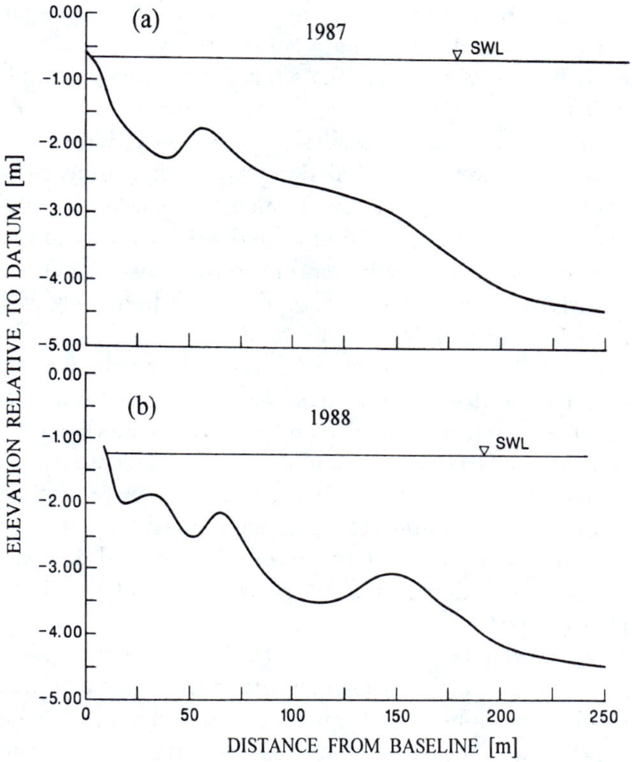

Figure 4: Shoreface profiles at Bluewater Beach in 1987 and 1988.

Here, low-frequency waves were still evident, which suggests the presence of edge waves or reflected long waves. These waves may contribute to observed high sediment concentrations and to the observed morphodynamic response under highly energetic conditions.

Observations have also shown that bar topography exerts an important constraint on both the magnitude and direction of transport by oscillatory and mean currents. The bar crests experienced the largest oscillatory currents and greatest levels of turbulence induced by wave breaking. Surprisingly, however, the net suspended sediment transport rates were minimal, near balance, between the mean transport rate directed offshore by undertow and the net onshore oscillatory transport rate controlled by asymmetric waves at or close to breaking. The oscillatory transport rates at wind-wave frequencies decreased with increasing water depth both landward and lakeward of the bar crest.

According to the interpretation of reported spatial changes in suspended transport rates, it has been concluded that spatial variation in sediment transport does concur with bar morphodynamics that resulted from the monitored storm event. After the storm, the inner bar migrated landwards. However, the lower lakeward slope did not show any erosion despite large and persistent sediment flux gradients and net offshore transport.

The reported results of observations analyzed by Osborne and Greenwood (1992) were concluded by a statement that the net oscillatory sediment transport rate induced by low-frequency waves increases both landward and lakeward of the breaker zone on the bar crest, as does the mean sediment transport rate. Additionally, it was shown that the low-frequency group-bound wave transports sediment offshore. However, it could only be inferred that the secondary low-frequency wave may transport sediment onshore because observations did not provide quantitative data on the structure of the low-frequency wave and about its role in the maintenance of the bar form.

The results of field data analysis presented above show that surface hydrodynamics and related sediment fluxes are extremely complex in a barred foreshore. With available technology and analytical methods applied in the study of Osborne and Greenwood (1992), it is, at present, impossible to discriminate between the contribution of wave groups to sediment transport from the contribution of the remaining transport components active in the nearshore. In their conclusion, Osborne and Greenwood also pointed out that the spatial variation of suspended sediment transport rates correlated only partially with observed changes in bar morphology.

The numerical modeling presented in Section 3 allowed us to investigate the influence of wave groupiness on bar dynamics in idealized, simple conditions. Wave and topography data from field measurements at Bluewater Beach were used in initial and boundary conditions required in the model, and in the experimental verifications of model predictions.

3. METHODOLOGY OF NUMERICAL SIMULATIONS

We present three sets of numerical experiments investigating wave-bed interactions in a coastal environment. In these experiments, surface hydrodynamics are induced by wave trains at wind-wave frequencies characterizing the wave climate at Bluewater Beach. In the first set of calculations, model predictions were carried out with incident nonlinear waves without any group-like modulation. In the remaining calculations (set 2 and set 3), wave-bed interactions were induced by modulated wave trains with some properties of groupiness observed in the coastal wave field.

In the analysis presented, quasi-steady flows (undertow) and low-frequency group-bound waves are not considered. The first simplification is sustained by observations from Bluewater Beach (Osborne and Greenwood 1992). The results of these observations show that despite evidence of steady flow contribution to sediment transport, the changes seen in bar morphology correlated only partially with measured transport rates. The second assumption is based on conclusions from the DUCK'85 field experiment. In this experiment, low-frequency group-bound waves and other secondary, low-frequency flows were

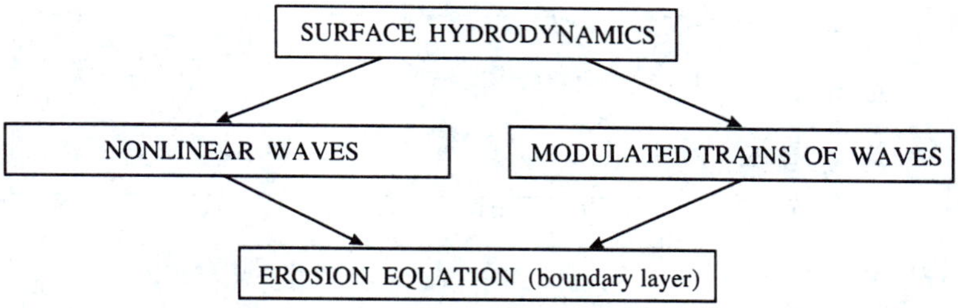

Figure 5: Schematical presentation of modeling non-modulated and modulated wave trains.

shown to contain lower energy levels than oscillations at wind-wave-induced frequencies (List 1991).

To predict the spatial and temporal evolution of the bed morphology under waves, we applied our two-dimensional morphological model (see, e. g., Boczar-Karakiewicz *et al.* 1995). In this modular model, the surface wave induces a nearbed flux of sediment, resulting in a deformation of the underlying seabed. For nonlinear incident, non grouped waves, surface hydrodynamics are described by a simplified Bussinesq equation (Boczar-Karakiewicz *et al.* 1987). To simulate incident trains of wave groups we have modified the original description of the surface wave. In this simulation, group-like modulations in the incident wave train result from superposition of two waves of similar frequency. The original and modified approaches are shown schematically in Figure 5. In all calculations, incident wave conditions are given at the deep-water end of a two-dimensional coastal zone. The initial topography, a uniform featureless slope, corresponds to mean slopes observed at Bluewater Beach (broken line in Fig. 6c).

In the first set of calculations, we analyzed the morphological response to incident nonlinear waves without any group-like modulation (Fig. 6). The spatial evolution of the surface wave (Fig. 6a and 6b) and the temporal change of the underlying bed from the initial featureless slope (broken line in Fig. 6c) to a final multibar profile (heavy line in Fig. 6c) are predicted for an extreme storm event. This event is simulated in the model by a train of regular waves with period T ($T = T_p = 6.9$ s), where T_p denotes the peak period of extreme storms at Bluewater Beach. (In the results presented, the amplitudes of harmonic wave components are non-dimensionalized by the amplitude of the incident wave.)

The temporal evolution of waves and morphology is shown in more detail in Figure 7. The morphological time scale τ (Fig. 7a) characterizing the formation of a multibar morphology from an initially featureless slope is of the order of hundreds of thousands of wave periods. The comparisons presented in Figure 7b show satisfactory agreement between the predicted and observed number of bars and bar-crest locations.

For a sequence of incident waves simulating a typical season (extreme winter

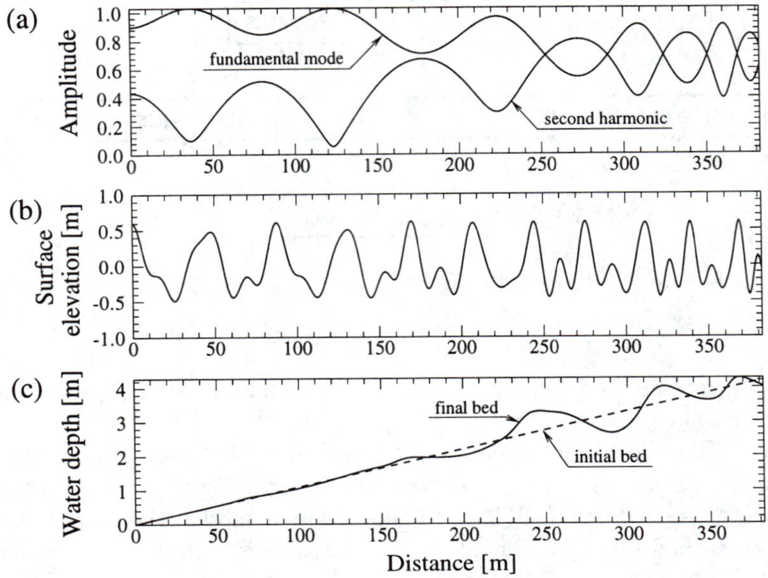

Figure 6: Prediction of wave and morphology evolution under nonlinear waves without group-like modulation (simulation of an extreme storm event with T_p = 6.9 s): (a) amplitude of the two first harmonic modes, (b) free-surface elevation, (c) initial and final bed.

storms followed by moderate summer wave conditions, with T_p = 6.9 s and T_p = 3.5 s, respectively), the model correctly reproduces the observed seasonal variability of the shoreface morphology that transforms from a multiple bar profile into a single bar, as observed at Bluewater Beach (Fig. 3; see also Boczar-Karakiewicz and Jackson 1991, Boczar-Karakiewicz *et al.* 1995).

In the second set of calculations, we simulated the linear effects of groupiness on coastal morphology. In this calculation presented in Figure 8, the group-like modulated wave train (Fig. 8a) propagates over the featureless mean slope of Bluewater Beach (Fig. 8b) without interacting with the underlying bed. A similar result is obtained for an initial multibar bed profile which remains unchanged under a train of linear wave groups (see again Fig. 8b). The amplitude modulation in the train of linear wave groups at x = 100 m is shown in Figure 9. Modulation results from a superposition of two waves of similar frequency, derived from the peak period, T_p = 6.9 s, of extreme storm events observed at Bluewater Beach.

In the third set of calculations shown in Figure 10, surface hydrodynamics are again induced by a group-like modulation (in the frequency band of extreme storms at Bluewater Beach). In contrast to previous experiments with linear wave groups (set 2), we have imposed spatial, nonlinear energy flow between

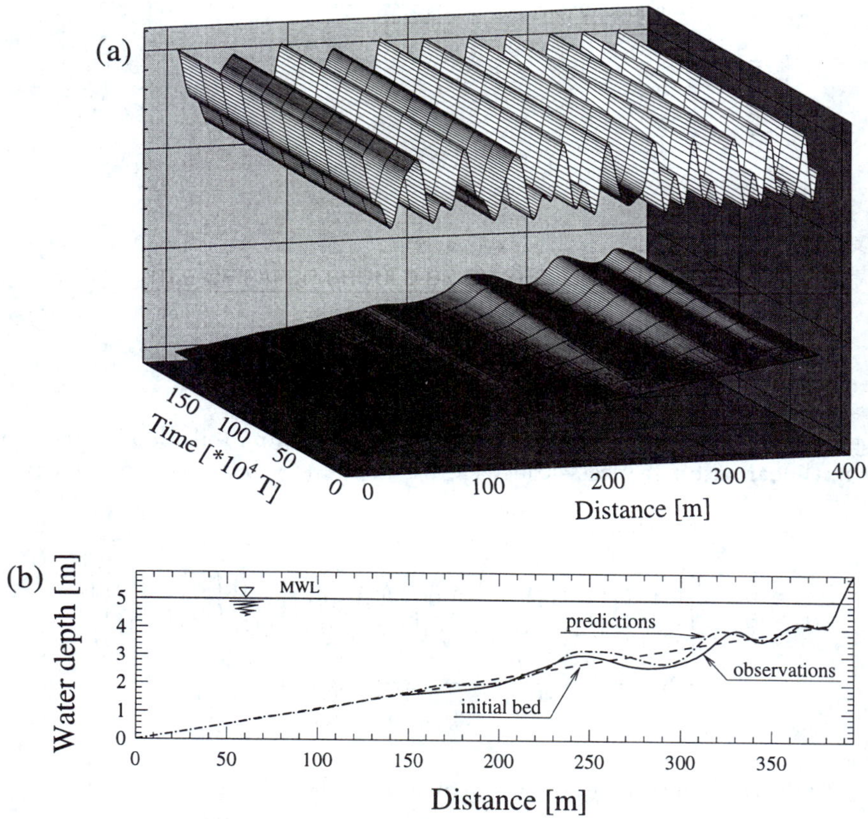

Figure 7: Prediction of morphology evolution under nonlinear waves without group-like modulation (simulation of an extreme storm event with $T_p = 6.9$s): (a) temporal evolution of waves and morphology, (b) predicted and observed multibar morphology.

fundamental modes of the incident wave train in the present simulation (see Fig. 10a). This pattern results from the solution of approximated Boussinesq equations (Fig. 5; Boczar-Karakiewicz *et al.* 1987). The spatial view of the resulting surface elevation is shown in Figure 10b. The velocity field of simulated wave groups generates nearbed sediment fluxes that gradually deform the underlying bed. Eventually, at time τ ($\tau = 10^4 T_p$), the initially-featureless, uniform slope (broken line in Fig. 10c), evolves into an equilibrium configuration (solid line in Fig. 10c). In the deeper part of the shoreface (at 0-260 m, Fig. 10c), the initial featureless slope remains nearly intact. In the shallower part (at 260-380 m in Fig. 10c), the initial bed deforms into multibar morphology. The spatial distribution of bar crests correlates with the nonlinear pattern of the imposed energy exchange among fundamental modes of the wave group

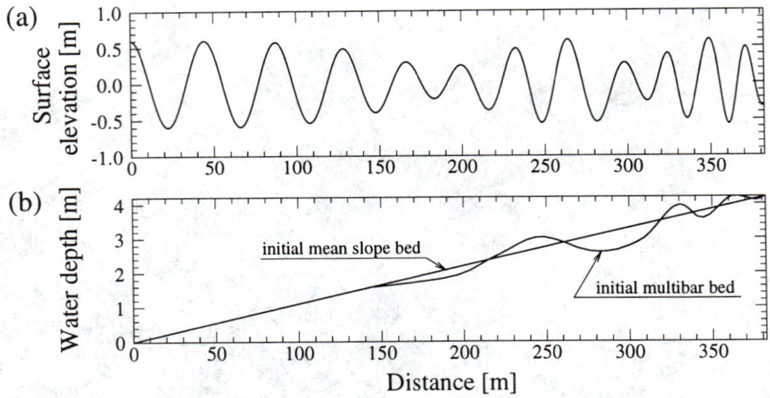

Figure 8: Evolution of waves and morphology under group-like modulated wave train (linear case): (a) free-surface elevation, and (b) shoreface morphology.

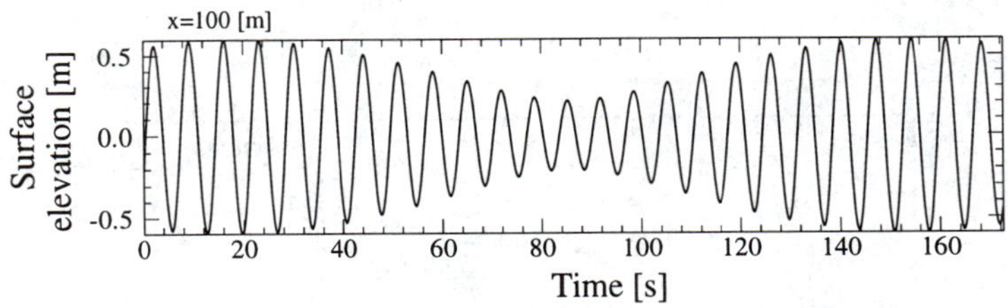

Figure 9: Time history of the incident bimodal wave train (linear case)

(Fig. 10a).

Figure 11 shows, at three chosen locations indicated in Figure 10c, the time history of surface elevation in nonlinear simulation. These records indicate strict periodicity in the time domaine, T. In contrast, velocity records from field measurements show evidence of pronounced temporal irregularity in observed wave groups (see again Fig. 1a). The reported results of model calculations suggest that large-scale morphology in coastal zones responds to the passage of nonlinear waves. In both nonlinear cases considered (set 1 and set 3), a rhythmical bed configuration is generated by waves characterized by large-scale nonlinear modulation in their amplitudes. The process of bar formation (Fig. 6 and Fig. 10) and bar transformation (Fig. 4) lasts over time denoted by τ in Figure 6. This morphological time scale τ is several orders of magnitude lower than the hydrodynamical scale T, characterized by the wave period of the incident wind wave.

In contrast, the large-scale coastal morphology (featureless uniform slope and a multibar profile) does not respond to the passage of trains of linear wave groups. This conclusion explains observations reported from the study

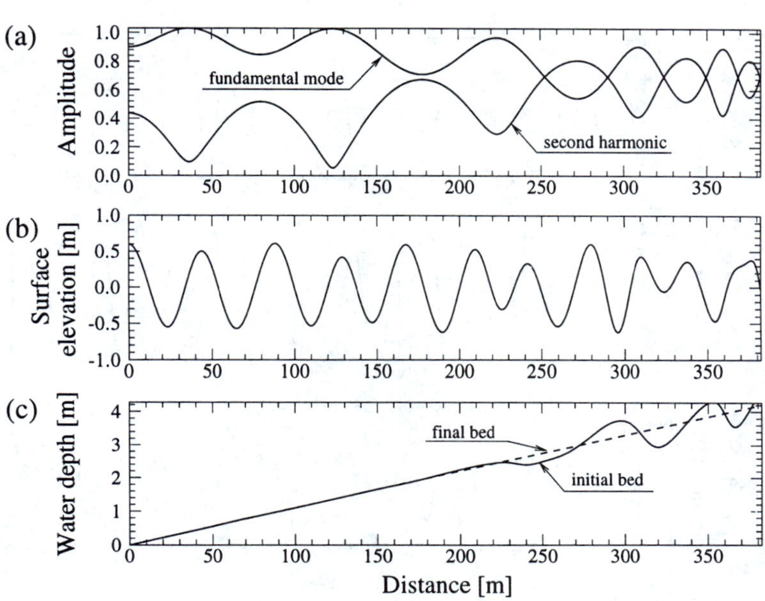

Figure 10: Evolution of waves and morphology under a group-like modulated wave train (nonlinear case): (a) initial spatial modulation of wave amplitudes, (b) free-surface elevation, and (c) shoreface morphology.

site according to which the deeper lying outer bar and the lakeward slope of the inner bar were unaltered during the monitored moderate summer storm (with peak period waves $T_p = 3.5$ s). During this storm, part of the unaltered underwater beach was located in a deep-water region for both the incident peak period wave and for observed wave groups related to summer wave conditions (see, e.g., Fig. 8).

4. CONCLUSION

1. Model predictions show that nonlinearly-modulated wave trains may potentially form and modify sand bars in coastal zones. These morphological changes are correlated with large-scale spatial variation in the nonlinear structure of these waves.

2. The predictions presented with strictly periodic nonlinearly-modulated waves suggest that sand bars are formed over time scales in the order of hundreds of thousands of wave periods. Velocity records from Bluewater Beach reveal strong irregularities in observed wave groups and related episodic sediment concentration events. Therefore, on natural coasts, these irregular wave groups do not represent an effective factor in sand bar generation.

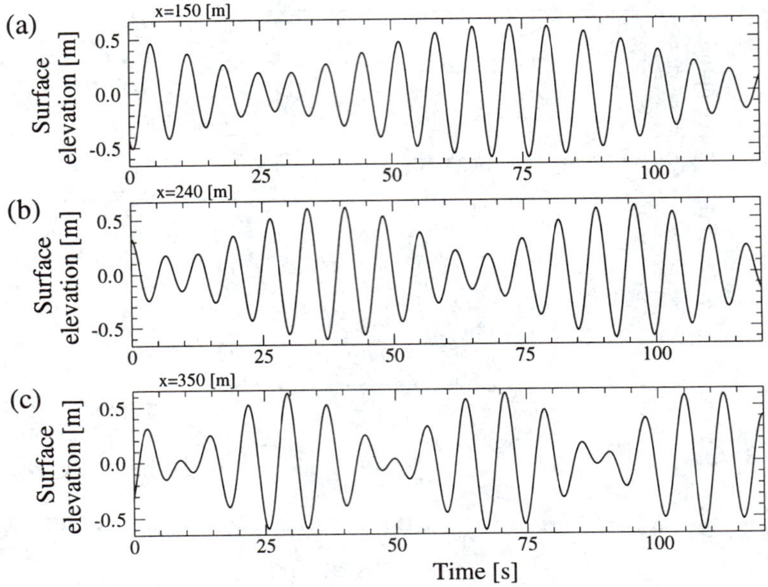

Figure 11: Time history of the shoaling group-like modulated wave train (in three chosen cross sections of the shoreface indicated in Fig. 10).

3. Predictions and observations show that trains of linear wave groups do not interact with the large-scale nearshore morphology.

5. REFERENCES

Battjes, J.A. and V. Vledder. 1984. Verification of Kimura's theory for wave group statistics. *Proceedings of the 19th International Conference of Coastal Engineering, American Society of Civil Engineers, Cape Town* : 642-648.

Boczar-Karakiewicz, B., D.L. Cohen and J.L. Bona. 1987. Interaction of shallow water waves and bottom topography. *Dynamical problems in continuum physics. Mathematics and its applications.* J.L.Bona, ed., Springer Verlag, New York, NY, 4: 131-176.

Boczar-Karakiewicz, B., D.L. Forbes and G. Drapeau. 1995. Formation and stability of coastal bars in the Gulf of St. Lawrence. *J. Waterways, Port, Coastal and Ocean Eng.* American Society of Civil Engineers, 129 : 49-60.

Boczar-Karakiewicz, B. and L.A. Jackson. 1991. On the protection of a beach system, Gold Coast, Queensland, Australia. *Proceedings 22nd International Conference of Coastal Engineering,* American Society of Civil Engineers, Delft: 2265-2278.

List, J.H. 1991. Wave groupiness variations in the nearshore. *Coastal Engineering*, 15: 475-496.

List, J.H. 1992. A model for the generation of two-dimensional surf beat. *J. Geophys. Res.*, 97, (C4): 5623-5635.

Medina, J.R. and R.T. Hudspeth. 1990. A review of the analyses of ocean wave groups. *Coastal Engineering*, 14: 515-542.

Nwogu, O. 1993. Alternative form of Boussinesq equations for nearshore wave propagation. *J. Waterways, Port, Coastal and Ocean Eng.* American Society of Civil Engineers, 129 : 618-638.

Osborne, P.D. and B. Greenwood. 1992. Frequency dependent cross-shore suspended sediment transport. 2. A barred shoreface. *Marine Geol.*, 106: 45-51.

Talipova, T., E. Pelinovsky and E. Kit. 1995. Wind wave simulation in coastal zone. *Proceedings of the International Conference Coastal Dynamics'95*, American Society of Civil Engineers, Gdańsk, Poland: 105-127.

CHAPTER 275

COASTAL MORPHODYNAMIC INSTABILITIES

Albert Falqués, Amadeu Montoto and Vicente Iranzo [1]

Abstract

An initially plane beach is considered with waves incoming obliquely. For this basic steady state the Longuet-Higgins solution is used to give set-up and longshore current. The beach is assumed to be erodible. Then, a stability analysis is performed considering infinitesimal disturbances in the current, the free surface and the sea bottom. The basic undisturbed state is found to be unstable giving rise to topographic features similar to oblique or transverse bars and to a meandering in the longshore current. Some comparison with the erlier work by Christensen et al., 1994, is made.

1. Introduction

Beach topography often show tridimensional patterns with some kind of recurrence or rhythmicity in the alongshore direction. Typical examples are oblique/transverse bar systems, crescentic longshore bars, ridge and runnel systems, etc. The explanation for rhythmic topography has followed to main theoretical approaches: i) the infragravity wave influence on sediment transport and distribution (see for instance, Holman and Bowen, 1982) and ii) the morphodynamic instability of the surf-zone under the action of the incoming waves and the longshore current (see Hino, 1974 and Christensen et al., 1994). The basic idea in this latter approach is as follows. Given an initial beach topography, the incoming waves produce some currents. This currents carry sediments and this sediment transport can produce changes in the topography which in turn affects the incoming wave field and the current. In this way there is a feedback and if some perturbation within this loop produces a positive feedback the perturbation will start to grow. Then an instability arises leading to topographic features. In earlier works (Falqués et al., 1996a, Falqués et al., 1996b) the instability of the current-sea bed system

---

[1] Departament de Física Aplicada, Universitat Politècnica de Catalunya, 08034 Barcelona, Spain

keeping fixed the incoming wave field was analyzed (bed-flow instability). The aim of the present contribution is to include the effect of the perturbations in the incoming wave field. In this way we reproduce and extend the results of Hino, 1974, and Christensen et al., 1994, with a different numerical model. Special attention is paid to the bar shape and orientation, the systematic trends with respect to bottom friction and other parameters and the influence of lateral momentum diffusion.

2. Model equations

Since our aim is to look at large 2D horizontal patterns in the near shore, a 2D shallow water model with time-average over incoming wave period is considered. As governing equations we take the momentum equations:

$$\frac{\partial \vec{v}}{\partial t} + \vec{v} \cdot \nabla \vec{v} + g \nabla z_s - \frac{\vec{\tau}_b}{\rho \zeta} - \vec{\mathcal{V}} = \frac{1}{\rho \zeta} \nabla \cdot S \quad (1)$$

the mass conservation equation

$$\frac{\partial \zeta}{\partial t} + \nabla \cdot (\zeta \vec{v}) = 0 \quad (2)$$

and the sediment conservation equation

$$\frac{\partial z_b}{\partial t} + \nabla \cdot \vec{q} = 0 \quad (3)$$

A cartesian coordinate system is assumed with x cross-shore, y alongshore and z vertical upwards (see Fig.1). $\zeta = z_s - z_b$ is the total depth, $\vec{\tau}_b$ the bottom friction and S is the radiation stress tensor. The lateral momentum diffusion terms read:

$$\mathcal{V}_i = \frac{1}{\zeta} \sum_{j=1}^{2} \frac{\partial}{\partial x_j} [\nu_t \zeta (\frac{\partial v_i}{\partial x_j} + \frac{\partial v_j}{\partial x_i})] \qquad i = 1, 2 \quad (4)$$

Here, $x_1 = x$ and $x_2 = y$. The sand transport is parametrized by the formula:

$$\vec{q} = \nu |\vec{v}|^m (\frac{\vec{v}}{|\vec{v}|} - \gamma \nabla h) \quad (5)$$

where h is any bottom perturbation with respect to equilibrium and where the term $\gamma \nabla h$ takes account of the tendency of the sand to move downslope. The exponent m has been set to 3 which is suitable for bed-load transport and $\gamma \sim 1$. Now we are going to perform a linear stability analysis for the system of equations 1, 2, 3. First, let us consider a basic steady equilibrium. In this state we assume waves incoming obliquely on the beach (see Fig. 2). The equilibrium consists of a set-up/down in the mean water level and a longshore current. For simplicity so that to be able to use analytical expressions, we will use the Longuet-Higgins solution (Horikawa, 1988). Thus, we assume a saturated surf zone, $H = \gamma_b \zeta$, $x \leq X_b$. The bottom friction (weak current and small angle) is given by

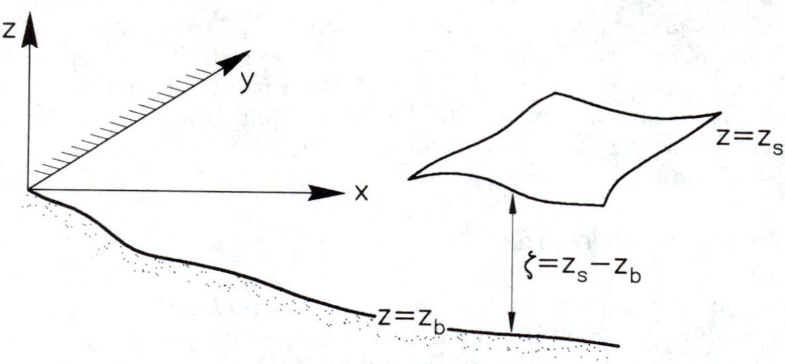

Figure 1: coordinate system.

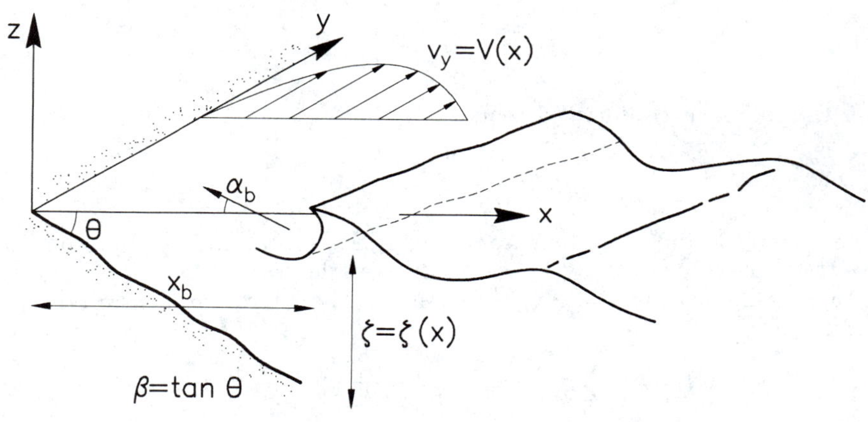

Figure 2: basic steady equilibrium.

$$\tau_{bx} = -\frac{4}{\pi}\rho c_f u_0 v_x \qquad , \qquad \tau_{by} = -\frac{2}{\pi}\rho c_f u_0 v_y \tag{6}$$

and the eddy viscosity reads $\nu_t = Nx\sqrt{g\zeta}$.

Consider now a small perturbation on the basic equilibrium of the form:

$$\vec{v} = (0, V(x)) + (\hat{u}(x), \hat{v}(x))e^{\sigma t + iky}$$

$$z_s = z_s^0(x) + \hat{\eta}(x)e^{\sigma t + iky}$$

$$z_b = z_b^0(x) + \hat{h}(x)e^{\sigma t + iky}$$

where $\lambda = 2\pi/k$ is the wavelength or the alongshore spacing of the growing bedforms. Then, by linearizing equations 1, 2, 3 we obtain:

$$\frac{\partial u}{\partial t} + V\frac{\partial u}{\partial y} + g\frac{\partial \eta}{\partial x} = [\frac{\tau_{bx}}{\rho\zeta}]^{\mathcal{L}} + [\mathcal{V}_x]^{\mathcal{L}} + [\frac{1}{\rho\zeta}(\frac{\partial S_{xx}}{\partial x} + \frac{\partial S_{xy}}{\partial y}]^{\mathcal{L}} \quad (7)$$

$$\frac{\partial v}{\partial t} + V\frac{\partial v}{\partial y} + \frac{dV}{dx}u + g\frac{\partial eta}{\partial y} = [\frac{\tau_{by}}{\rho\zeta}]^{\mathcal{L}} + [\mathcal{V}_y]^{\mathcal{L}} + [\frac{1}{\rho\zeta}(\frac{\partial S_{yx}}{\partial x} + \frac{\partial S_{yy}}{\partial y}]^{\mathcal{L}} \quad (8)$$

$$\frac{\partial \eta}{\partial t} - \frac{\partial h}{\partial t} + [\nabla \cdot (\zeta\vec{v})]^{\mathcal{L}} = 0 \quad (9)$$

$$\frac{\partial h}{\partial t} + [\nabla \cdot \vec{q}]^{\mathcal{L}} = 0 \quad (10)$$

These equations are too long to be written here in detail in such a way that we abbreviate 'linear part of' by

$$[.]^{\mathcal{L}}$$

The perturbation in the radiation stress terms will be called bed-surf terms since they describe the perturbation in the incoming wave field due to growing bedforms. The perturbation in the wave refraction should be also included in the bed-surf terms but this has not been done in the present work. The remaining terms in the linearized equations if bed-surf effect is neglected will be called bed-flow terms since they describe the perturbation in the current due to the growing bedforms if the forcing by the waves is kept fixed.

Before solving the linear problem, a scaling and some non-dimensional parameters will be introduced. As horizontal lengthscale, the width of the surf zone, $L_H = X_b$, is choosen and $L_V = \beta X_b$ will be the vertical scale where β is the beach slope (see Fig.3). The velocity scale is the scale for the Longuet-Higgins model,

$$U = \frac{5\pi}{16}\frac{\gamma_b}{c_f}\sqrt{g\zeta_b}\beta' \sin\alpha_b$$

where β' is the effective beach slope including set-up and α_b is the wave angle with rescect to the cross-shore at breaking. Two time scales appear , $T_h = L_H/U$, $T_m = L_H L_V/Q$, where $Q = \nu U^m$ is the scale for sediment transport. The second one arises in a natural way from the sediment conservation equation (10) and will be called morphological time scale. It is the scale at which bedforms are expected to grow. The first one will be called hydrodynamical time scale and it is much shorter than the other one.

The variables are made nondimensional by means of:

$$(x, y) = L_H(x', y') \quad , \quad z_b = L_V z_b' \quad , \quad z_s = \frac{U^2}{g}z_s'$$

$$(u, v) = U(u', v') \quad , \quad t = T_m t'$$

Then, we deal with non-dimensional linear equations with the following dimensionless parameters. First a nondimensional wavenumber and growth rate, $k' = kL_H, \sigma' = \sigma T_m$. Also, the breaking index, γ_b, the wave angle at breaking, α_b and the viscosity parameter, N, appear. The parameter related to the tendency of the sedimend to go downslope now become $\gamma' = \gamma\beta$. An important parameter is the characteristic Froude number of the longshore current, $F = U/\sqrt{gL_V}$ and the frictional parameter, $r = c_f/\beta$. Finally, the ratio between both time scales, $\epsilon = T_h/T_m$, is very small but not set to 0 (which is usually called quasi-steady hypothesis. See 'response time concept' in Christensen et al, 1994). In this way, the model allows for the computation of purely hydrodynamic instabilities like shear waves or even for the possible interaction between both kind of instabilities. However, we have focussed only in morphological instabilities for the present contribution. Hereinafter, nondimensional quantities will be handled dropping accents for simplicity.

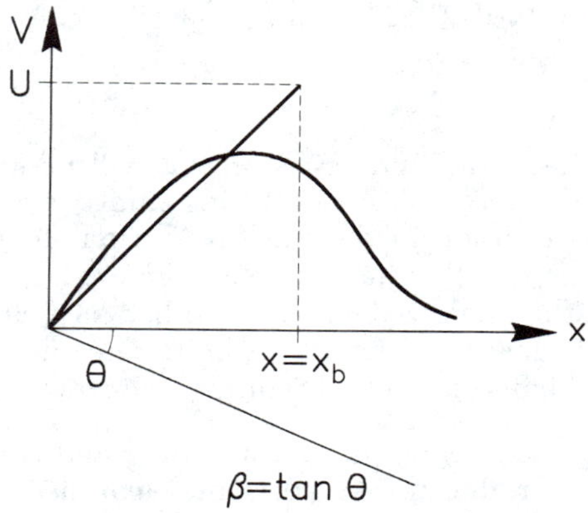

Figure 3: scaling.

Finally, equations (7),(8),(9) and (10) result in an eigenproblem where the eigenvalue is the growthrate, σ, and the eigenfunctions the perturbations $(\hat{u}(x), \hat{v}(x), \hat{\eta}(x), \hat{h}(x))$. This problem is discretized by a numerical spectral method. In the next section, an indication is given of the solution procedure.

3. Numerical method

For simplicity, we will ilustrate the numerical method by means of a simpler eigenproblem than (7)-(10). Consider the differential equation

$$p(x)u''(x) + q(x)u'(x) + r(x)u(x) = \sigma u(x) \qquad 0 < x < \infty \tag{11}$$

with boundary conditions $u(0) = u(\infty) = 0$. Here, σ is the eigenvalue and $u(x)$ the eigenfunction.

The method uses an expansion in Chebyshev polynomials and a

transformation from $[0, \infty)$ to $[-1, 1)$. First, this map is given by

$$x = \phi(z) = l\frac{1-z}{1+z} \tag{12}$$

(see Falqués and Iranzo, 1994).

Second, an expansion in Chebyshecv polynomials is assumed:

$$u(x) = \sum_{n=0}^{N} \hat{u}_n T_n(z) = \sum_{n=0}^{N} \hat{u}_n T_n(\phi^{-1}(x)) \tag{13}$$

The following step is to find a combination of Chebyshev polynomials which verify the boundary conditions. The N-1 functions which satisfy this requirement are:

$$g_n(x) = T_n(\phi^{-1}(x)) - \frac{1}{2}(1+(-1)^n)T_0(\phi^{-1}(x)) - \frac{1}{2}(1+(-1)^{n+1})T_1(\phi^{-1}(x)) \tag{14}$$

where $n = 2, 3, ..., N$. Then, expansion (13) is substituted by

$$u(x) = \sum_{n=2}^{N} \hat{u}_n g_n(x) \tag{14}$$

where the unknowns (eigenvector) are the $N-1$ coefficients $\hat{u}_2, ...\hat{u}_N$.

Now, to discretize equation (11) we will apply collocation at the Gauss-Lobatto nodes $z_i = cos(\pi i/N)$ transformed by the map (12) (Falqués and Iranzo, 1994). For this purpose, the first step is to know the values of g_n at the nodes. This is given by the martix $G_{ij} = g_j(x_i)$, so that

$$u(x_i) = \sum_{n=2}^{N} G_{in}\hat{u}_n$$

In a similar way, the first derivative of u will be given by a matrix G' :

$$u'(x_i) = \sum_{n=2}^{N} G'_{in}\hat{u}_n$$

where $G'_{in} = g'_n(x_i)$ can be computed by using the derivatives of the Chebyshev polynomials and the map (12). Similarly, the second derivative will be computed by

$$u''(x_i) = \sum_{n=2}^{N} G''_{in}\hat{u}_n$$

where $G''_{in} = g''_n(x_i)$ can also be computed by means of the derivatives of the Chebyshev polynomials and the map (12). Explicit expressions of the derivatives of Chebyshev polynomials combined with (12) can be found in Falqués and Iranzo, 1994.

Finally, after performing collocation at x_i, $i = 1...N-1$ the discretized eigenproblem reads:

$$p(x_i)\sum_{k=2}^{N} G''_{ik}\hat{u}_k + q(x_i)\sum_{k=2}^{N} G'_{ik}\hat{u}_k + r(x_i)\sum_{k=2}^{N} G_{ik}\hat{u}_k = \sigma\sum_{k=2}^{N} G_{ik}\hat{u}_k$$

where the eigenvector is $(\hat{u}_2...\hat{u}_N)$.

4. Results

One of the drawbacks of Hino model, 1974, without the quasisteady hypothesis, i.e., hydrodynamic instabilities out of control, has been solved. All hydrodynamic instabilities could be identified as shear waves with a clear maximum in the $\sigma - k$ curve. Shear waves have growthrates an order ϵ^{-1} larger than bed wave growthrates. By comparison with other models (Falqués and Iranzo, 1994) they could be recognized. Usually, however, when a realistic friction and viscosity parameters are choosen, no hydrodynamic instabilities appear. Figure 4 shows the nondimensional growth rate $Re(\sigma)$ as a function of the wavenumber k, for three different cases. In one set of curves only bed-flow terms were kept in the equations while in the other set the full equations (bed-flow + bed-surf) were considered. It can be seen that bed-surf, that is, the effect of the wave field perturbation enhances significantly the instability. This was also found by Christensen at al., 1994. Also, bed-surf gives shorter spacing between the bedforms. The imaginary part of σ is not shown here. It is negative and of order one. This means that the topographic waves migrate downcurrent with a speed of the order L_H/T_m.

Regarding the shape of the topographic features, two kind of bedforms were found. First what we call current dominated bedforms. In this case, the shape of the bars looks very similar when only bed-flow terms were taken or when bed-flow and bed-surf were considered. The bars are upcurrent rotated, very oblique (a small angle with the shoreline of the order of 10°). These topographic waves come out in the case of a relatively high characteristic Froude number, say $F > 0.6$ (note that this is not the maximum local Froude number which can be smaller). According to

$$F = \frac{U}{\sqrt{g\zeta_b}} = \frac{5\pi}{16}\frac{\gamma_b}{\sqrt{1+0.375\gamma_b^2}}\frac{sin\,\alpha_b}{r} \tag{15}$$

this corresponds to small frictional parameter, r, and large angle, α_b. Figure 5 shows how the bottom perturbations with only bed-flow and with bed-flow and bed-surf look quite similar. Figure 7 shows the full equation bedform when the basic slope is added to the perturbation.

Second, we found what we call wave-dominated bedforms. In this case the shape of the bars with only bed-flow or with bed-flow and bed-surf looks quite different. With only bed-flow they are similar to alternate bars in a river and in the other case they are similar to transverse bars slighly upcurrent rotated (see Fig.6). Figure 8 shows, in this case, the total topography with bed-flow and bed-surf. Wave-dominated bedforms appear for small Froude number, say $F < 0.6$, that is, for large frictional parameter and small angle.

An important issue of morphological models for rhythmic features is the alongshore spacing or wavelength. This is found by looking at the fastest growing wavenumber, i.e., the maxima in the σ_r, k curves for different values of the parameters. From experimental data (usually from the spacing between rip currents assumed to be related to rhythmic topography) the wavelength to surf zone width ratio is in the order $\lambda/X_b \sim 1.5 - 8$. The mean value is between 3 and 4 (Sasaki and Horikawa, 1975). Christensen et al., 1994 found a ratio about 6. In our modelling this ratio ranges between 1 and 7, depending on the frictional parameter, r, and on the wave angle, α_b. It decreases with increasing r and with decreasing α_b (see Fig. 9).

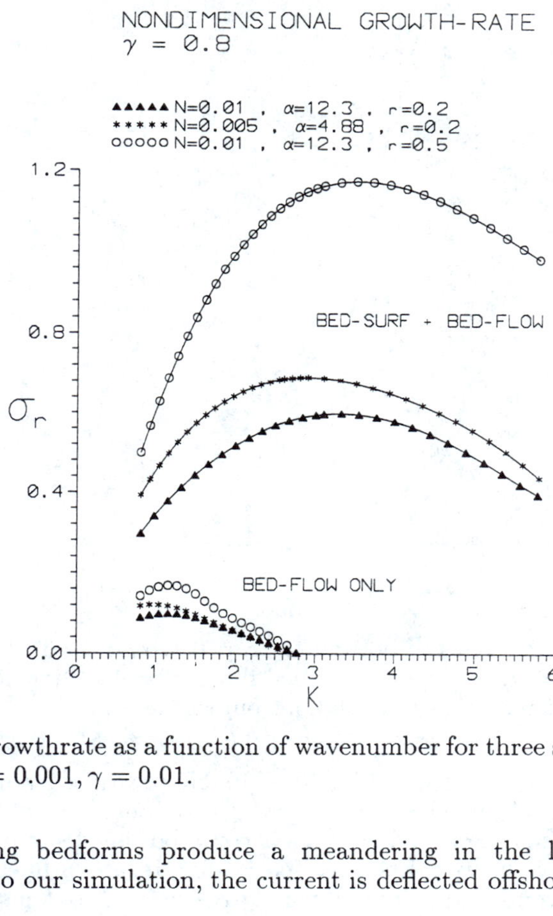

Figure 4: growthrate as a function of wavenumber for three sets of parameters. $F = 0.3, \epsilon = 0.001, \gamma = 0.01$.

The growing bedforms produce a meandering in the longshore current. According to our simulation, the current is deflected offshore over the shoals

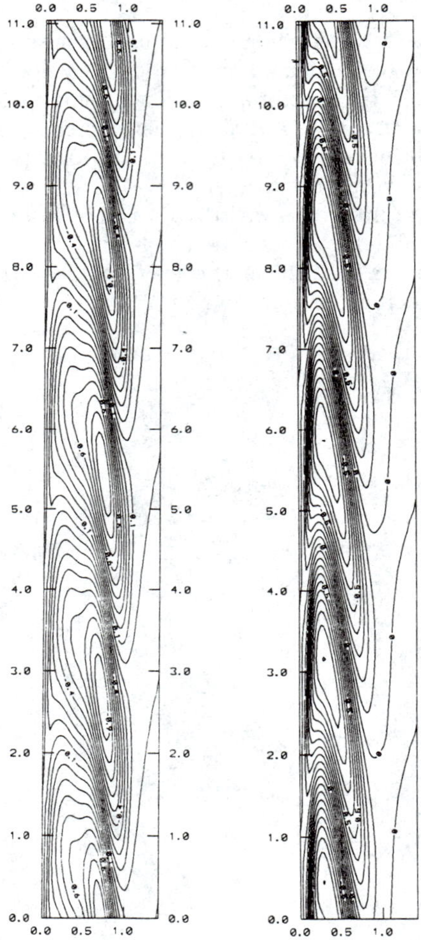

Figure 5: perturbation in the bottom for $r = 0.1, \alpha_b = 12.3^o, F = 0.8, N = 0.005, \gamma_b = 0.4, k = 1.14$. Right: only bed-flow, left= bed-flow and bed-surf. In this plot the current runs from the bottom to the top.

and inshore over the pools (see Fig.10). Thinking of rip currents, this is in contrast with many observed rip currents. But two things have to be taken into account. First this is the initial growth. It could be that for finite amplitude features this behaviour would be reversed leading to a non-linear saturation of the growth. Second, these are not exactly rip currents but just the

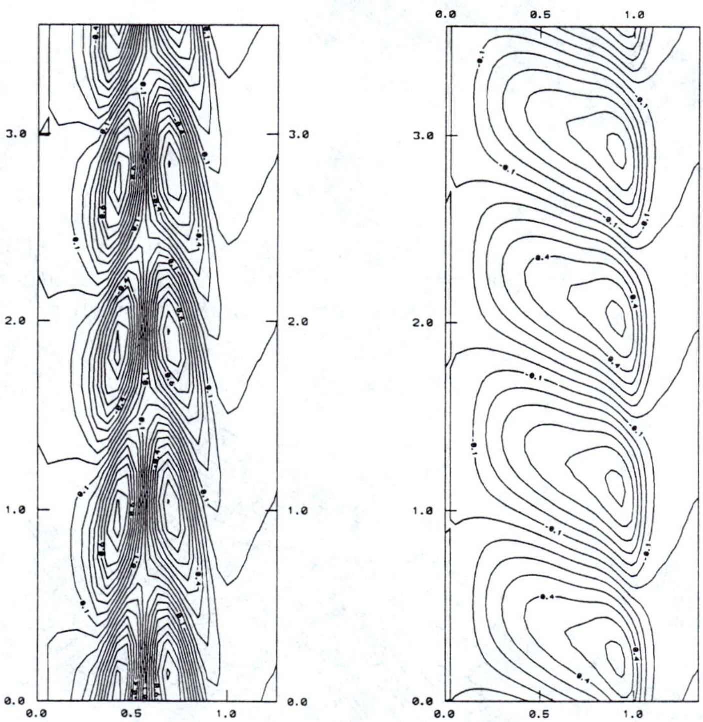

Figure 6: perturbation in the bottom for $r = 0.5, \alpha_b = 12.3^o, F = 0.16, N = 0.005, \gamma_b = 0.4, k = 3.5$. Right: only bed-flow, left= bed-flow and bed-surf. In this plot the current runs from the bottom to the top.

deflection of the longshore current. Also Christensen et al., 1994, found the same behaviour.

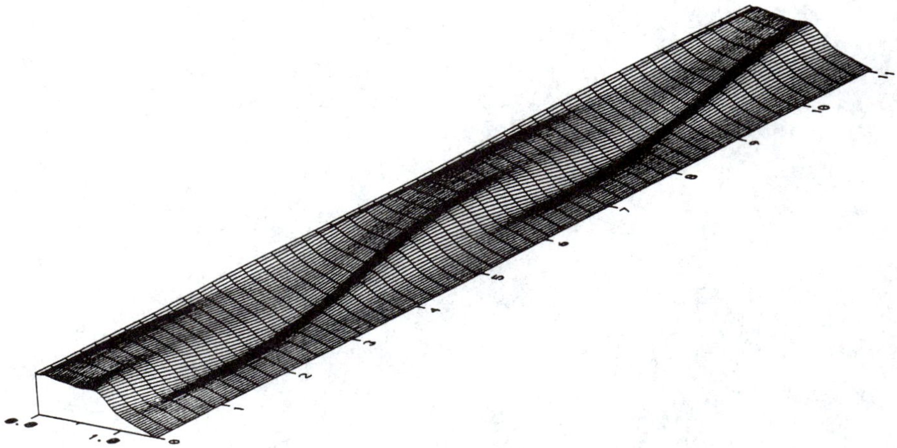

Figure 7: bottom topography for $r = 0.1, \alpha_b = 12.3^o, F = 0.8, N = 0.005, \gamma_b = 0.4, k = 1.14$. Basic slope plus perturbation (with arbitrary amplitude since we are dealing with a linear problem). Current running from left to right.

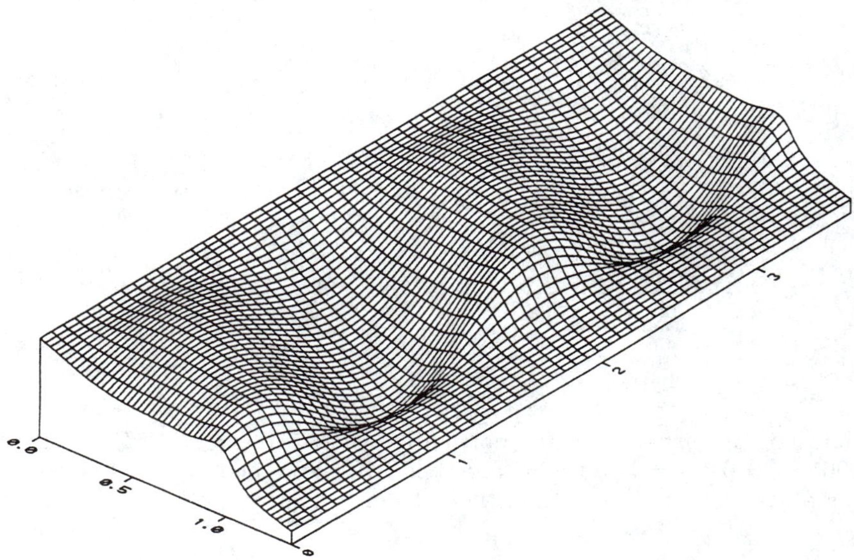

Figure 8: bottom topography for $r = 0.5, \alpha_b = 12.3^o, F = 0.16, N = 0.005, \gamma_b = 0.4, k = 3.5$. Basic slope plus perturbation. Current running from left to right.

5. Conclusions

It has been shown that the topography of a plane beach can be unstable due to the interaction with the incoming waves and the longshore current. The effect of the perturbation on the waves (bed-surf) turns out to be the most de-stabilizing. This is in line with Christensen et al., 1994. The instability produces the growth of upcurrent rotated bars which is also in agreement with the earlier work of Christensen et al. However, as these authors pointed out, this is sometimes in agreement with the observed bars in the field and sometimes is in contrast. We found two kind of topographic features: i) 'current dominated' (large angle, small friction) and ii) 'wave dominated' (small angle, large friction). Instead of fixed values of the alongshore spacing

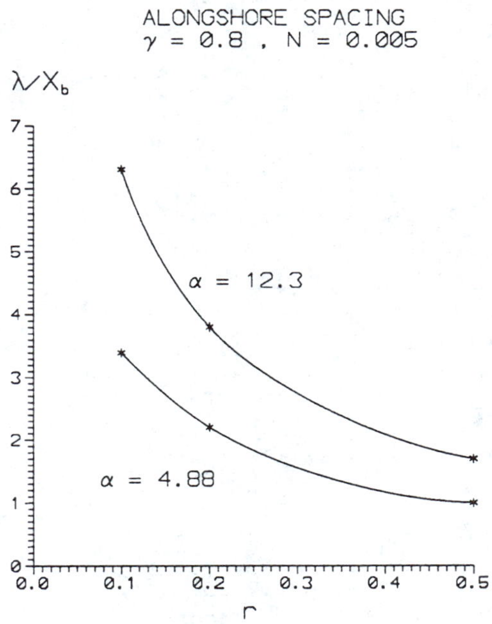

Figure 9: Alongshore spacing of the topographic waves.

λ/X_b (4 in the case of Hino and 6 in the case of Christensen et al.,) our alongshore spacing ratio ranges from 1 to 7, decreasing for increasing r, and depending on α_b and N. Finally, the growing bedforms produce a small meandering of the longshore current with an offshore deflection over the shoals. This can be a little surprising but it is also in line with earlier results of Christensen et al. Clearly, more research is needed to clarify the mechanism which produces this perturbed flow in combination with the perturbation in the set-up.

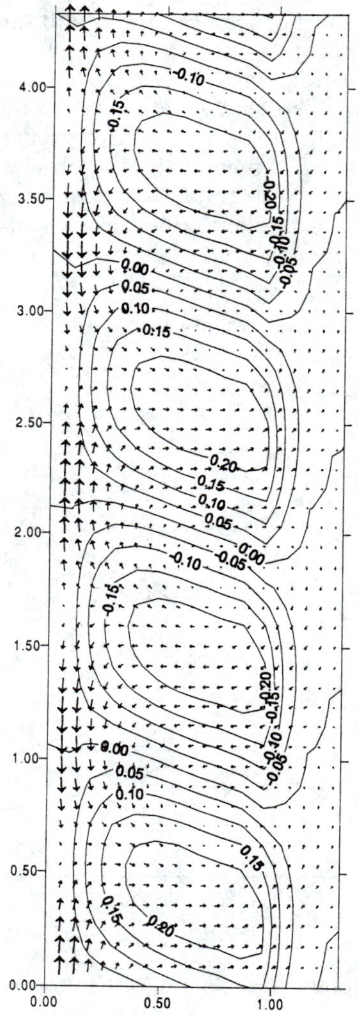

Figure 10: wave-dominated bedform and the perturbation produced on the current. Note the offshore deflection over the shoals. $r = 0.2, \alpha_b = 4.88^o, F = 0.3, N = 0.005, \gamma_b = 0.8$. Basic current running from bottom to top.

Acknowlodgements

This work has been funded by the Spanish government (DGICYT) under contract PB93-0948. Part of the work of A.Falqués has been done in Utrecht University (IMAU) funded by RIKZ (Netherlands Institute of Coastal Management).

References

Christensen,E.,Deigaard,R.and Fredsoe,J.,1994. Sea bed stability on a long straight coast 24th Int. Conf. Coastal Engineering, 1865-1879.

Falqués,A. and Iranzo,V.,1994. Numerical simulation of vorticity waves in the nearshore. J.Geophys. Res., 99, 825-841.

A.Falqués, A.Montoto and V.Iranzo, 1996a. Bed-Flow instability of the longshore current. Cont.Shelf Res. (in press).

A.Falqués, A.Montoto, V.Iranzo, 1996b. Bed-flow instability of longshore currents. XXI European Geophysical Society General Assembly, The Hague, 6-10 May 1996.

Hino,M.,1974. Theory on formation of rip-current and cuspidal coast 14th Int. Conf. Coastal Engineering, 901-919.

Holman,R.and Bowen,A.,1982. Bars, bumps, and holes: models for the generation of complex beach topography J.Geophys. Res.,87,457-468.

K.Horikawa(editor), 1988. Nearshore Dynamics and Coastal Processes. University of Tokio Press.

Sasaki,T. and Horikawa,K., 1975. Nearshore current system on a gently sloping bottom. Coastal Eng. in Japan, 18, 123-143.

CHAPTER 276

On validation of a sand waves and sand banks model

S. J. M. H. Hulscher *

Abstract

A morphological model is described for the interaction between tidal motion and an erodible bed. This model is able to distinguish between a flat bed, and the generation of tidal sand banks and/or sand waves, based on physical parameters. The physical meaning of the turbulence parameters A_v and S, as used in this model is subject of discussion in the present paper, therefore a simpler system is investigated. A simple flow is described by a logarithmic profile model (a generally accepted turbulence model), as well as by a partial slip model, using A_v and S. It is shown that the two models can be calibrated such that they produce the same bed shear stress, depth-averaged velocity and depth-averaged eddy viscosity. This leads to expressions for A_v and S as function of the roughness height and a shape parameter in the logarithmic profile model. Application of this information to two locations shows that the bed form prediction model gives encouraging results.

1 Introduction

The offshore seabed of shallow seas is covered with rhythmic patterns on a large scale, see e.g. *Off* [1963]. Tidal sand banks have wavelengths of about five kilometers and reach heights up to 40 meters. Sand waves, see figure 1 are smaller, wavelengths of 500 meters, heights up to 10 meters. Both patterns are shown in a schematic way in figure 2. When these two patterns overlap, the crests are oriented differently (angle between 60°- 90°), which suggests that these two patterns

*Civil Engineering & Management, University of Twente, P.O. box 217, 7500 AE, Enschede, The Netherlands.

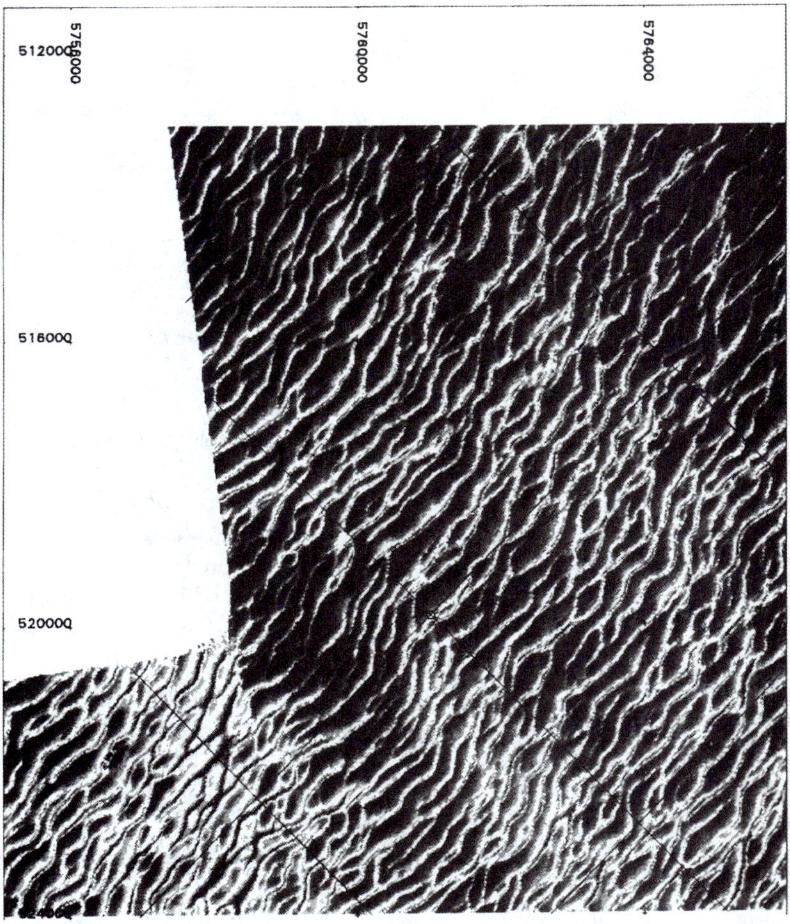

Figure 1: *Visualization of the North Sea bed near the Eurogeul, from Van Goor & Andorka Gal, 1996.*

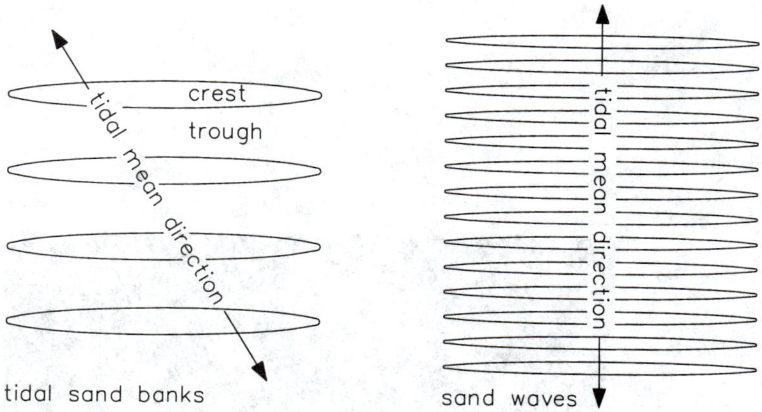

Figure 2: *Sketch of tidal sand banks and sand waves, the relation to the main tidal current axis is indicated.*

are different. Therefore different mechanisms are responsible for their formation and behavior.

A model to describe formation of these large-scale sea-bed patterns due to tide-topography interactions is given in *Hulscher* [1996]. In this model the tide is described by a three-dimensional shallow water model. This model uses a very simple turbulence closure scheme, see also *Engelund* [1970]: constant eddy viscosity A_v in combination with a partial slip parameter at the bottom. The latter introduces a so-called resistance parameter S.

Application of this turbulence model often raises questions how these results compare with more realistic turbulence-closure schemes. Unfortunately it is not completely known how turbulence should be modeled, therefore the answer to this question can never be definite. A further, even more important question is how these, rather abstract, parameters can be determined from observations. Illustrative issue is whether observations are to be collected either on the top and in the trough of the bank or at the adjacent sides.

A start to answer the first question is made studying a simplified system: a two-dimensional horizontal model without Coriolis force and neglecting inertia effects. This system is analyzed in two ways: using a realistic turbulence closure scheme and the simple scheme based on eddy viscosity and the resistance parameter. By comparing the results of these two methods one finds explicit relations between at one hand more accepted model parameters an at the other hand, the eddy viscosity A_v and the resistance parameter S. Furthermore this enables to investigate how the latter two variables are connected, and if they represent two degrees of freedom.

These results can be used to make estimates for the three dimensional model in which Coriolis effects and inertia effects are included. These lead to choices for the parameters which are following Hulscher [1996] crucial to determine the bed structure: a flat bed or sand waves or tidal sand banks or a combination of the latter two patterns. This procedure enables the estimate of these parameters without analyzing an network of current meter recordings. For two cases explicit estimates are given: an average North Sea location and the Middelkerke bank.

The outline of this note is as follows. A short description of the three-dimensional bed form model is given in section 2. Next, the simplified system and subsequent modeling is given in section 3. The analyses in particular for the logarithmic profile and the $A_v - \hat{S}$ model are given in sections 3.3 and 3.4, respectively. Matching of these two models is discussed in section 3.5. These results are transferred to the three dimensional model, which is shown and discussed in section 4. In this section two physical locations are treated as examples. Lastly, conclusions are presented in section 5.

2 Model for sand waves and tidal sand banks

The generation of large-scale bed form patterns is studied based on the idea that these structures might be free instabilities of the coupled morphological system: sea water and sea bed. Therefore a suitable model of this system is formulated and use for analysis.

The morphological shallow water is based on tidal flow described by three-dimensional shallow water equations, bed load transport and conservation of sediment. Tidal averaging and application of a linear stability analysis lead to prediction of the dominant bottom mode starting from a flat bottom. Translating the measures of these fastest growing sinusoidal sea bed waves into physical quantities shows that the pattern is similar to either similar to sand waves or to tidal sand banks; patterns which are significantly different, see figure 2. This leads to the qualitative result in figure 3; the derivation of this figure is described in detail in *Hulscher* [1996].

Figure 3 shows that the bed structure prediction depends strongly on the Stokes number E_v of the tidal flow and the bed resistance parameter $\hat{S}$. These are defined as

$$E_v = \frac{2A_v}{H^2\sigma} \qquad \text{and} \qquad \hat{S} = \frac{2S}{H\sigma}, \tag{1}$$

in which H is the local mean depth and σ the frequency of the tidal motion. The constant turbulent eddy-viscosity, A_v (m^2s^{-1}), quantifies the way in which horizontal momentum is transferred in vertical direction. The quantity S (ms^{-1}) is used to model the partial slip near the bed boundary. This overcomes the problem that the constant eddy viscosity is near the bottom too simple to model both velocity and shear stress in a realistic way. After *Engelund* [1970] here A_v and S are considered as two constants which describe together the vertical profile of the vertical flow profile.

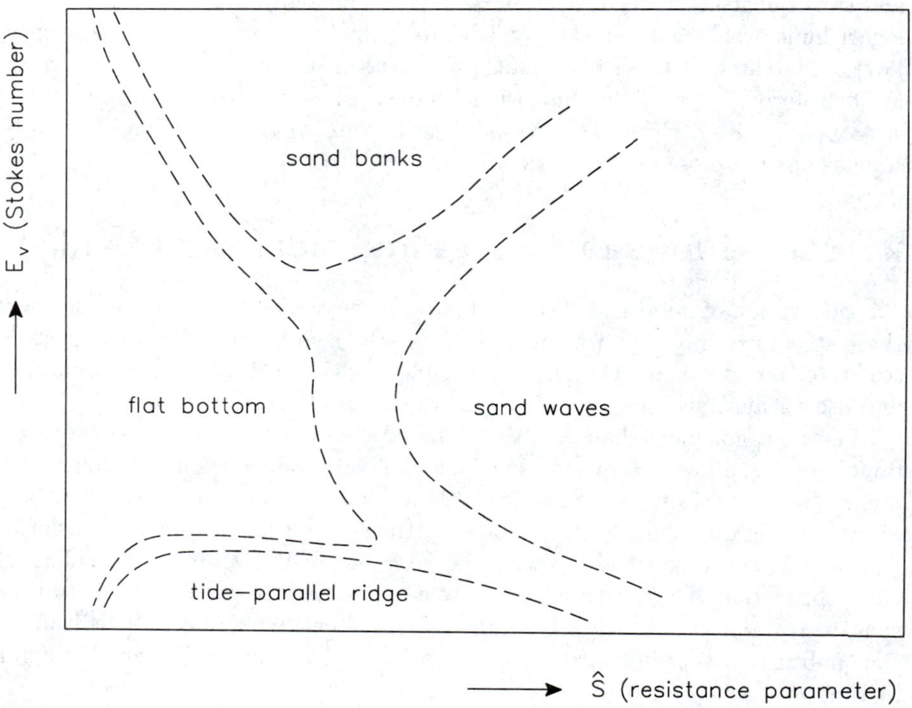

Figure 3: *Characteristic bed forms predicted by the three-dimensional shallow water model as a function of the bottom slip parameter* $\hat{S}$ *and the Stokes number* E_v. *Quantification depends on values of the Coriolis parameter, local depth, bed-slope effects on sediment transport and the nonlinearity in the bedload transport formula.*

3 Modelparameter derivation

3.1 Simplified situation model

To illustrate the procedure a very simplyfied model and model-situation is sufficient. Starting from the three-dimensional shallow water model as in *Hulscher* [1996] the simplifications are as follows:

- Coriolis effects are neglected
- tidal flow, $\tilde{u}(z,t)$, is horizontally uniform
- friction dominates inertia effects: the eddy turn-over scale is much smaller than the scale on which the flow varies, such that eddies can reach a statistical equilibrium

Here the tidal flow is driven by pressure gradient, oscillating at the tidal frequency σ and having amplitude P_x. The flow is opposite the pressure gradient; the equation of motion becomes

$$0 = -\frac{P_x}{\rho}\sin\sigma t + \frac{\partial}{\partial z}\left[\nu_t(z,t)\frac{\partial\tilde{u}}{\partial z}\right], \tag{2}$$

in which ν_t is the turbulent eddy viscosity coefficient. The boundary conditions near the sea bed and surface are as follows

$$\tilde{u} = 0 \qquad \text{at} \quad z = z_0\,, \tag{3}$$

$$\nu_t(z,t)\frac{\partial\tilde{u}}{\partial z} = 0 \qquad \text{at} \quad z = H\,, \tag{4}$$

in which z_0 is the roughness height. The z-axis is here directed upwards, from the sea bottom $z = 0$ till the sea surface $z = H$. Here the surface stress component is chosen zero. The bottom shear stress is denoted by $\tilde{\tau}_b(t)$ and is time-dependend:

$$\tilde{\tau}_b = \rho\nu_t(z)\frac{\partial\tilde{u}}{\partial z} \equiv \rho|\tilde{u}_*|\tilde{u}_* \qquad \text{at} \quad z = 0\,, \tag{5}$$

where $\tilde{u}_*$ is the time-dependent friction velocity.

3.1.1 Linearization of the shear stresses

Definition (5) shows that, in general, the friction velocity $\tilde{u}_*$ depends on time, so that the momentum equation is difficult to solve. To deal with this one usually replaces one friction velocity factor $\tilde{u}_*$ by a representative constant $\bar{u}_*$. Now the bed shear stress condition becomes

$$\tilde{\tau}_b = \rho\nu_t(z)\frac{\partial\tilde{u}}{\partial z} \equiv \rho\bar{u}_*\tilde{u}_* \qquad \text{at} \quad z = 0\,, \tag{6}$$

A possible way to derive such a representative value for $\bar{u}_*$ is requiring that the tidal-averaged shear stresses of both formulations are equal:

$$\rho \left\langle |\tilde{u}_*|\tilde{u}_*^2 \right\rangle = \rho \left\langle \bar{u}_* \tilde{u}_*^2 \right\rangle , \tag{7}$$

in which <> denotes the tidal averaging. Further assuming that the friction velocity oscillates with the tidal frequency σ (like the forcing pressure gradient) and that it has an amplitude $\hat{u}_*$, the searched constant becomes [*Zimmerman*, 1981, and *Mei*, 1989]

$$\bar{u}_* = \frac{8}{3\pi} \hat{u}_* . \tag{8}$$

3.2 Model analysis: general part

In this model the tidal flow is forced by the pressure gradient and as the latter oscillates at frequency σ, so in absence of inertia effects the tidal flow can simply be decomposed as follows

$$\tilde{u}(z,t) = u(z) \sin(\sigma t) . \tag{9}$$

Substitution of (9) into the equation (2) leads to the following equation of motion for $u(z)$:

$$0 = -\frac{P_x}{\rho} + \frac{\partial}{\partial z} \left[\nu_t(z) \frac{\partial u}{\partial z} \right] . \tag{10}$$

By integration of (10) using the upper boundary condition (4) and the lower as given in (6), the following equality is obtained

$$\frac{P_x}{\rho} H = \bar{u}_* \hat{u}_* = \frac{8}{3\pi} \hat{u}_*^2 . \tag{11}$$

Up till this point no simplifications are made regarding the choice of a specific turbulence closure scheme. In the next two sections equation (10) will be solved using two different turbulence closure schemes.

3.3 Logarithmic profile model and analysis

Viscosity parametrization From turbulence modelling is known that the turbulent eddy viscosity increases from the boundaries in which the distance to the boundary is a measure for the length scale of the turbulent eddies. So at the fixed sea bottom the eddy viscosity equals zero and it increases with the distance from the bottom z:

$$\nu_t(z) \sim z . \tag{12}$$

The sea surface can act as a less stringent boundary. Many physical processes lead to a more effective mixing in the upper part of the water column, e.g. action of wind waves, swell, stratification. Here this fact is parametrized by a parameter ϵ, which value is here between the limits 1/2 (little influence of surface on turbulence) and 1 (rigid surface). This is chosen such that the viscosity at the sea surface is:

$$\nu_t(H) \sim H\,(1-\epsilon)\;. \tag{13}$$

And near the surface $H-\zeta$ (in which ζ positive) the length scale of the turbulent eddies increases following

$$\nu_t(H-\zeta) \sim \nu_t(H) + \zeta\,(2\epsilon - 1)\;. \tag{14}$$

The previous considerations motivate a parabolic function as parametrization for the dimensional turbulent eddy viscosity ν_t

$$\nu_t(z) = k\hat{u}_* z \left(1 - \epsilon \frac{z}{H}\right), \tag{15}$$

in which κ is the Von Karman constant $\kappa \simeq 0.41$.

Velocity profile Solving equation (10), using boundary conditions (3), (4) and expressing it in terms of $\bar{u}_*$ using (6) yields

$$u(z) = \frac{\bar{u}_*}{\kappa}\left[\ln\left(\frac{z}{z_0}\right) + \frac{(1-\epsilon)}{\epsilon}\ln\left(\frac{1-\epsilon\frac{z}{H}}{1-\epsilon\frac{z_0}{H}}\right)\right]. \tag{16}$$

Depth-averaged quantities The depth-average value of the turbulent eddy viscosity can easily be computed from equation (15) and yields

$$\bar{\nu}_t = k\hat{u}_* H \frac{(3-2\epsilon)}{6}\,. \tag{17}$$

In figure 4 the normalized viscosity profile is shown. The depth-averaged velocity follows from equation (16 and is given by

$$\bar{u} = \frac{\bar{u}_*}{\kappa}\left[\ln\left(\frac{H}{z_0}\right) - \frac{1}{\epsilon} + \frac{1-\epsilon}{\epsilon}\ln\left(\frac{1-\epsilon}{1-\epsilon\frac{z_0}{H}}\right) + \frac{\epsilon-1}{\epsilon^2}\ln\left(1-\epsilon\right)\right]. \tag{18}$$

3.4 $A_v - S$ model

In this section the highly simplified turbulence model is chosen, here called S and A_v model, which is applied in *Hulscher* [1996]). The model has a vertically constant eddy viscosity, here denoted by A_v. This constant eddy viscosity model lacks variations which lead to describe the correct velocity and shear stress both in the interior as well as at the bottom. However, the aim here is to use this model for studying sediment transport, therefore it has to produce the correct bed shear stress, rather than the exact horizontal velocity near the sea bed. This can be achieved by choosing a partial slip condition, instead of condition (3), at the sea bed. In this context the partial slip condition is formulated as follows

$$\bar{u}_*\hat{u}_* = Su \qquad \text{at} \qquad z = 0\,, \tag{19}$$

in which S is the so-called resistance parameter.

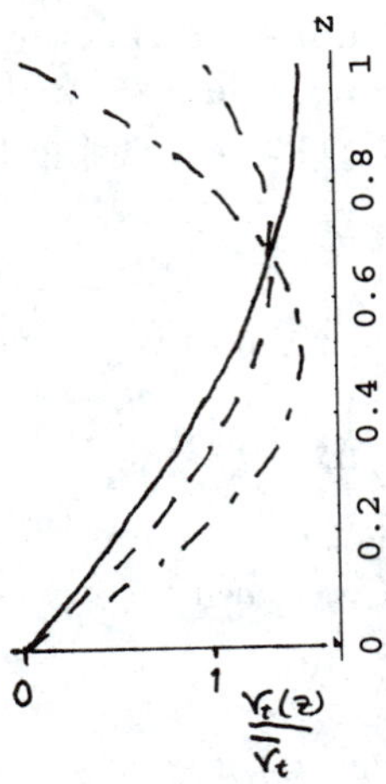

Figure 4: *The dimensionless viscosity profile $\nu_t(z)/\bar{\nu}_t$ for three values of ϵ (straight $\epsilon = 0.5$, dashed/dotted $\epsilon = 0.75$, dashed $\epsilon = 1.0$).*

Velocity profile The solution of equation (2), using (4) and (19) becomes

$$u(z) = \bar{u}_* \hat{u}_* \left[\frac{z}{A_v} \left(1 - \frac{z}{2H} \right) + \frac{1}{S} \right] . \tag{20}$$

Depth-averaged quantities The depth-averaged viscosity in this model is simply the constant

$$\bar{\nu}_t = A_v . \tag{21}$$

The depth-averaged velocity $\bar{u}$ in this model yields

$$\bar{u} = \bar{u}_* \hat{u}_* \left(\frac{H}{3A_v} + \frac{1}{S} \right) . \tag{22}$$

3.5 Matching

Now the models, discussed in the previous two sections, section 3.3 and section 3.4, will be matched by appropriate requirements. First remark is that sediment transport is a function of the bed shear stress. This motivates the first requirement: the $A_v - S$ model has to produce the same bottom shear stress as in the logarithmic model. Based on this, the bottom boundary condition in the $A_v - S$ model has already been adapted such that expression (6) is valid in both models.

The second condition is that the water discharge is equal in both models, This condition is here transferred into the requirement of equal depth-averaged velocities resulting from both turbulence formulations. Using equations (18) and (22) this means

$$\ln\left(\frac{H}{z_0}\right) - \frac{1}{\epsilon} + \frac{1-\epsilon}{\epsilon} \ln\left(\frac{1-\epsilon}{1-\epsilon\frac{z_0}{H}}\right) + \frac{\epsilon - 1}{\epsilon^2} \ln(1-\epsilon) = \kappa \hat{u}_* \left(\frac{H}{3A_v} + \frac{1}{S}\right) . \tag{23}$$

The third condition is that the depth-averaged value of the eddy viscosity $\bar{\nu}_t$ equals the constant eddy viscosity A_v. Using equation (17) this leads to

$$A_v = k\hat{u}_* H \frac{(3-2\epsilon)}{6}. \tag{24}$$

4 Discussion

Before the discussion here starts, it is worthwhile to remark that all values for turbulent eddy viscosity coefficients and bottom shear stress are results of the application of a turbulence closure scheme. The logarithmic model, used as a reference here, is a often applied and produces satisfactory results. Therefore this model is generally accepted.

In the present model this leads to the following Stokes number E_v by using equations (11) and (24)

$$E_v = \frac{\kappa\sqrt{\frac{3\pi}{8}\frac{P_x}{\rho}H}}{H\sigma}\left(1-\frac{2}{3}\epsilon\right). \tag{25}$$

So if P_x, H, σ are already determined, the value of E_v is still a function of ϵ.

Using equations (11), (24) and the matching condition (23) the following relation for $\hat{S}$ and E_v is obtained

$$\frac{E_v}{\hat{S}} = \frac{3-2\epsilon}{6}\left[\ln\left(\frac{H}{z_0}\right) - \frac{1}{\epsilon} + \frac{1-\epsilon}{\epsilon}\ln\left(\frac{1-\epsilon}{1-\epsilon\frac{z_0}{H}}\right) + \frac{\epsilon-1}{\epsilon^2}\ln(1-\epsilon)\right] - \frac{1}{3} \tag{26}$$

This relation shows that $\hat{S}$ is a function of the roughness height z_0, if the other parameters are specified. As z_0 usually is between $5*10^{-5}$ m and $2*10^{-1}$ m and $\frac{1}{2} \leq \epsilon \leq 1$, this will restrict the values of $E_v, \hat{S}$ which are physically relevant.

4.1 Physical discussion

In general, the pressure gradient P_x can be determined from observations. Here the pressure gradient P_x is roughly estimated from the M_2 tidal (spring) range Δ, following

$$P_x = \rho * g * \frac{2\Delta}{L}, \tag{27}$$

in which L is the tidal wavelength and g the gravity acceleration. Assuming that Δ is 2 meters and the tidal wave follows from shallow water theory (so L= 770 kilometers on a depth of 30 meters) one derives $P_x = \rho * 5.1\,10^{-5} ms^{-2}$; the tidal frequency is $\sigma = 1.4\,10^{-4} s^{-1}$.

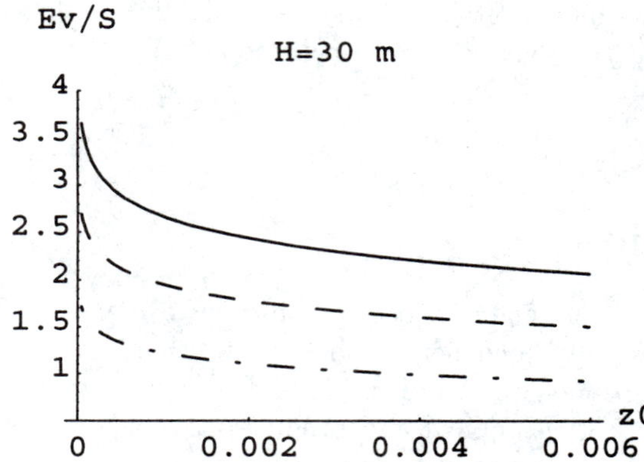

Figure 5: *(a) The ratio $E_v/\hat{S}$ and (b) $\hat{S}$ for three values of ϵ (straight $\epsilon = 0.5$, dashed/dotted $\epsilon = 0.75$, dashed $\epsilon = 1.0$). Here avarage North Sea conditions are chosen, the depth $H = 30$ meters.*

Offshore North Sea A suitable North Sea average depth is 30 m. Using equation (25) gives that $1.4 \leq E_v \leq 2.8$ due to the range of $0.5 \leq \epsilon \leq 1$. The correct value of ϵ has to be estimated based on current profiles. Using many field measurements *Sousby* [1990] has found that the roughness height z_0 in the North sea usually is between $5\,10^{-5}$m and $6\,10^{-3}$ m, in which the larger values are found for rippled sand. Using (26) the ratio $E_v/\hat{S}$ and subsequently the resistance parameter itself can be evaluated. Figure 5(a) shows the ratio $E_v/\hat{S}$ and in figure 5(b) the resistance parameter $\hat{S}$ is shown.

Middelkerke Bank The Middelkerke Bank is part of the Flemish Banks system; these large tidal sand banks are partly covered with sand waves. At the Middelkerke Bank the mean depth is significantly smaller, being about 15 meters. So here the expected range for the Stokes number becomes $2.0 \leq E_v \leq 3.9$. *Vincent & Stolk* [1993], p 217, reported that the roughness height z_0 at two stations around the Middelkerke Bank is between $9\,10^{-4}$m and $3.3\,10^{-2}$ m. For these values the ratio $E_v/\hat{S}$ and the resistance parameter $\hat{S}$ are shown in figure 6.

General The areas in parameter space of both discussed locations are shown in figure 7. In this figure also the expected bed structure based on the bed form prediction model *Hulscher* [1996] is also indicated. For the offshore North Sea bed a part of the possible parameter combinations predict a flat sea bed, also a part of these combinations predicts tidal sand banks. Around the Middelkerke bank a part of the parameter combinations predicts tidal sand banks, for a different part the slowly growing tidal sand banks are dominated by sand waves, which grow faster. Comparison between prediction and the actual bed structure show that these results are not unrealistic. So the ranges in the parameters E_v and $\hat{S}$

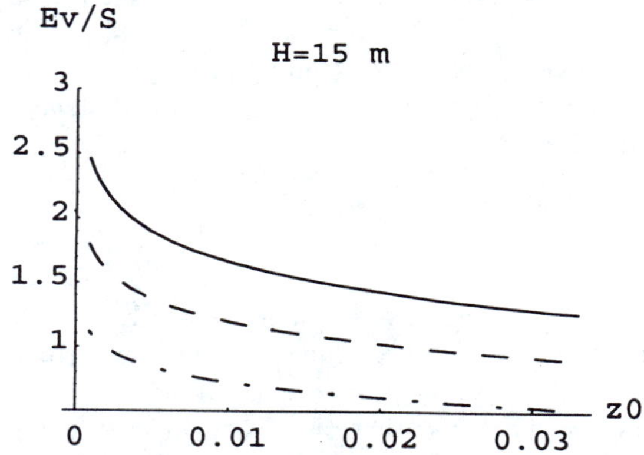

Figure 6: *(a) The ratio $E_v/\hat{S}$ and (b) $\hat{S}$ for three values of ϵ (straight $\epsilon = 0.5$, dashed/dotted $\epsilon = 0.75$, dashed $\epsilon = 1.0$.). Here Middelkerke Bank conditions are chosen, the depth $H = 15$ meters.*

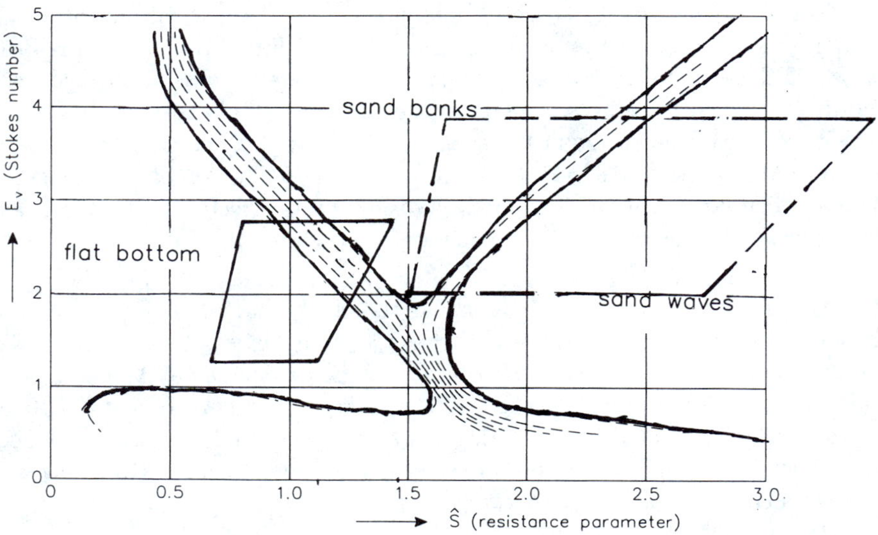

Figure 7: *Parts of the parameter space E_v and $\hat{S}$ possible in the offshore North Sea bed and around the Middelkerke bank. Also the expected bed behavior as found by Hulscher [1996] is indicated.*

based on natural variations in the quantities roughness height z_0 and turbulence characteristics ϵ cover different bed structures. These results show that it will be worthwile to determine the value of z_0 and ϵ more precise and subsequently compare the model prediction with the observed local bed structure, for several locations in the North Sea.

5 Conclusions

A way to investigate the implication of a potenetial sea-bed prediction model is presented. To quantify the parameters which describe the turbulence processes, first a simpler system, based on steady flow, is analyzed.

In the simpler system is shown that the S, A_v-turbulence model is able to produce the same bed shear stress as the accepted logarithmic model. The comparison between these models gives expressions for S and A_v in terms of the model parameters in the logarithmic model: ϵ and z_0. So based on this comparison between the logarithmic model and the S, A_v model can furthermore be concluded that there are two basic degrees of freedom to choose S and A_v left, here expressed as ϵ and z_0. One might argue that these two degrees of freedom are restricted, however these restrictions are the result of a calibration of the S, A_v-model with a logarithmic model. Boundaries limiting the physical realistic values of $\hat{S}, E_v$ will certainly be different if the S, A_v-model is calibrated using another turbulence closure scheme than the one used in this note.

The results of the steady flow case are used to estimate the model parameters in the tidal morphological model. For two cases the prediction of the model and the local bed structure are compared and show good agreement. This indicates that it is worthwhile to continue this line of validation.

Acknowledgements The author thanks H.J. De Vriend and H.E. De Swart for the ideas and discussions which have led to this paper. This note is based on work both in the PACE-project, in the framework of the EU-sponsored Marine Science and Technology Programma (MAST-III), under contract no. MAS3-CT95-0002, as well as the project of The UK Ministry of Agriculture, Fisheries and Food under contract no. CSA3051-Offshore sand banks: Basic processes and effects on long-term coastal morphodynamics.

References

Hulscher, S. J. M. H., Tidal-induced large-scale regular bed form patterns in a three-dimensional shallow water model, *Journ. of Geoph. Res* Vol 101 No C9, 20,727-20,744, 1996.

Mei, C.C., *The applied dynamics of ocean surface waves.* World Scientific, 740pp, 1989.

Off, T., Rhythmic linear sand bodies caused by tidal currents, *Bull. of the Am. Ass. of Petroleum Geologists* 47, 324-341, 1963.

Soulsby, R. L., Tidal-current boundary layers, in *The Sea, Vol. 9. Part A*, edited by le Mehaute, B. and Hanes, D. M., pp 523- 566, Ocean Engineering Science, John Wiley and Sons, New York, 1990.

Van Goor, S. and Andorka Gal, J.H., Bodem (in Dutch), document RIKZ/OS-96.109X, 28 pp, 1996.

Vincent, C. and Stolk, A., Hydrodynamics and suspended load: measurements by total sediment-load sampler and acoustic backscatter, in *Sediment Mobility and Morphodynamics of the Middelkerke Bank*, edited by De Moor, G., and Lanckneus, J., pp 211-227, Final Report MAST Project, Gent, 1993.

Zimmerman, J.T.F., On the Lorentz linearization of a quadratically damped forced oscillator, *Phys. Lett.* 89A, 123-124, 1982.

CHAPTER 277

Transverse Bars in Duck, North Carolina

Kathryn M. Konicki[1] and Rob A. Holman[1]

Abstract

In 1985, unusual streaks were first seen in time exposure images of nearshore wave breaking. These features were persistent from image to image and ran perpendicular or oblique to the shoreline. They have been observed both between the shoreline and inner bar and offshore of the inner bar, and have a typical longshore wavelength of ~40-130 m, substantially shorter than normal sand bar length scales. Comparison with bathymetric data collected by the CRAB showed the bands to be bathymetric ridges with a relief of 0.3-0.5 m.

Both intensity transects as well as geometrically rectified images were used to determine the length scales of the bars. Longshore length scales of offshore bars (mean spacing of 131 m) were typically three times the spacing of trough bars (mean spacing of 41 m). The offshore limit of the visible features in the trough varied from as short as 5 m to a maximum of 35 m.

Frequency of occurrence statistics for the bars were determined by viewing Argus data from 1987-1995. Trough bars appeared in the images an average of 62 days/year while offshore bars appeared 41 days/year although there was substantial interannual variability. Offshore bar occurrence is more frequent in the winter, and trough bar occurrence is more frequent in the summer.

Introduction

The classic model of a sandbar is a large-scale feature aligned parallel to the shoreline. Sandbars that are oriented slightly oblique to the shoreline have also been observed and are called welded bars. In 1952, Shephard described a feature oriented perpendicular or at a high angle to the shoreline with a relatively short length scale that he called a transverse bar. Only a few studies have been done on these transverse bars, including Barcilon and Lau's (1973) and Niedoroda and Tanner's (1970) work on transverse bars off the coast of Florida.

In 1985, similar features were first observed in time exposure images off the coast of Duck, North Carolina. Figure 1a shows an oblique snapshot from January 10, 1994. The vegetated dunes, dry beach, and shoreline are evident, as is the pattern of breaking waves in the nearshore. By contrast, Figure 1b shows an oblique ten-minute time exposure. The time-averaged breaking waves now show up as smooth bands, clearly delineating the shoreline (from the shorebreak) and the

[1] College of Oceanic and Atmospheric Sciences, Oregon State University, Oregon State University, Corvallis, OR 97331

Figure 1. Images from Duck, NC. a). Snapshot from January 10, 1994. b). 10-minute time exposure from January 10, 1994. Transverse features are evident in the trough as well as offshore of the shore-parallel bar. Box shows area rectified in Figure 2.

offshore bar (Lippmann and Holman 1989). Intermediate-scale white bands, oblique to the shoreline, are also obvious in the trough, as well as offshore of the shore-parallel bar. Research has been carried out on these features to first determine whether or not the oblique bands in the video images are real bathymetric features or simply an artifact of the imaging technique. Statistics of the features, such as their frequency of occurrence and longshore and cross-shore lengths were then calculated.

Methodology

The data were taken at the Field Research Facility (FRF) at Duck. Duck is a dune-backed beach located on one of the barrier islands of the Outer Banks. It is a low-energy, intermediate to reflective beach that is barred. An inner bar is located at ~125 m offshore and an outer bar is also present at times. The spring tide range is 1.6 meters and the mean water level is 0.35 m above NGVD. The average annual breaker height is 88 cm (Birkemeier1981). The study area for our research was a 1700 meter stretch along the beach.

The data were collected automatically as part of the Argus program. An Argus station is a video camera operated by a personal computer. Every hour, both a snapshot and a ten-minute time expose are taken of the nearshore region. These images are sent back to the Coastal Imaging Lab at Oregon State University every night. A data set of images sampled with Argus is available for Duck spanning 10 years (1986-1996).

A great deal of information can be extracted from time exposures such as the one shown in Figure 1b. Since the geometries of the video camera are known, the exact location on the ground for each pixel in the image is also known. The coordinate system that is used for the images is the FRF coordinate system. The video camera is located at x=32.54, y=585.78, with x and y increasing positively offshore and away from the camera, respectively.

One method of determining length scales and positions of the features is to use geometrically rectified images (plan views) of the study area. Knowledge of the camera geometries allows a rectification of a region on an oblique image to be made. (Figure 2). Measurements such as longshore length spacing, cross-shore position and cross-shore length can be manually made from such rectifications. Transects can also be taken along the rectification, or along the oblique images, to measure intensity. Intensity data were compared with bathymetric data that were collected using the FRF's Coastal Research Amphibious Buggy (CRAB) to determine if the features have a bathymetric nature.

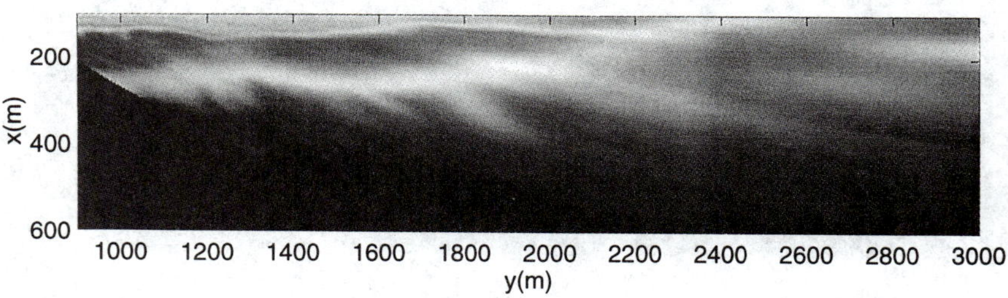

Figure 2. Rectification of boxed area in Figure 1b. Measurements can be made manually from the rectification.

In order to determine the frequency of occurrence of the bars, 8 years of Argus images (1987-1994) were viewed. One image from each day was studied for a total of 2600 images There were some days when no data was available due to equipment failure, bad weather, or other reasons that an image could not be obtained. In order to quantify the presence/absence of transverse bars, each image was given a rating on a scale of 0-5, with 0 being no evidence of transverse bars and 5 being a clear indication of multiple transverse bars. A rating was given both for the *trough bars* (transverse bars found in the trough) as well as *offshore bars* (transverse bars found offshore of the shore-parallel bar). Figure 3 shows examples of images that were given ratings of 1, 3, and 5.

All the images were separated into two categories. An image that was given a rating of 3, 4, or 5 was considered a day in which bars were present. Images that received a rating of 0, 1, or 2 were considered as days in which bars were not present.

Results

The first objective of this study was to determine the persistence of the intensity streak features to insure that they are not random processing features. Longshore intensity transects were taken along a line at x=145 m, from y=790 to 1060 m, for a series of 4 hours (5 images) for January 10, 1994 (Figure 4a). From Figure 4b, it is evident that there was a peak in intensity where there was a white band in the time exposure. This figure also shows that intensity peaks were persistent over substantial periods, including tidal variations, suggesting that these features are not a random artifact of the image processing. Figure 5 shows intensity transects that were taken across a set of offshore transverse bands for 3 consecutive days in November of 1993. Transects were taken at x=300 m, from y=1200 m to 1700 m. The intensity maxima and corresponding white bands in the video images persisted for a period of days, supporting the fact that these features are real, not artifacts.

On July 18, 1995, a detailed CRAB survey was carried out to determine if the white bands in the images are real bathymetric features. Figure 6a shows a transect along which intensity data were taken and also shows the transect along which bathymetric data were collected. Intensity and depth along the transects in Figure 6a are shown in Figure 6b. It is evident that intensity maxima correspond to bathymetric highs and intensity minima correspond to bathymetric lows, proving that the oblique bands in the video images are ridges. The bathymetric data showed the relief of these bars to range from 0.3-0.5 m.

Length scales and positions for both the trough and offshore bars were calculated from intensity plots and rectified images. The mean longshore length spacing (distance between two consecutive bars), distance offshore (offshore extent of bars as measured from the shoreline), and cross-shore length of the bars, were determined. Longshore length spacing was determined using 18 images for trough transverse bars and 33 images for offshore transverse bars. Five images (up to 3 bars per image) were used to determine the distance offshore and cross-shore length for trough bars, and 7 images (up to 3 bars per image) were used to calculate those statistics for the offshore bars. These data are summarized in Tables 1 and 2 for trough and offshore bars, respectively.

Figure 3. Examples of ratings on a scale of 0-5 that were given to images. a). Time exposure from November 13, 1993. Trough bars were given a rating of 1. b). Time exposure from November 11, 1994. Trough bars were given a rating of 3. c). Time exposure from February 9, 1993. Both trough and offshore bars were given a rating of 5.

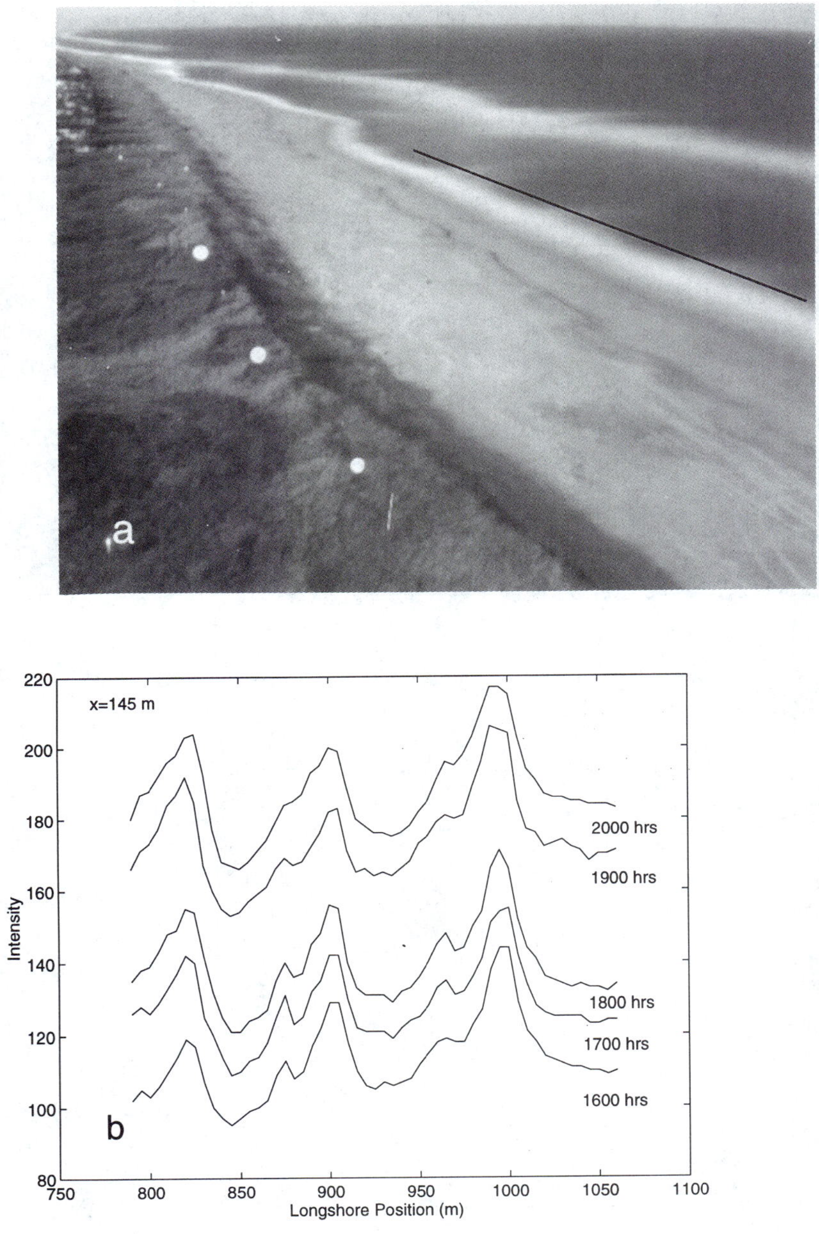

Figure 4. a). Time exposure from January 10, 1994 showing transverse bars in the trough. Transects along which intensity data were collected for 5 consecutive images are also shown. b). Intensity vs. longshore length from 1600-2000 hours along transect shown in a). This graph shows that the bars remain fixed over a period of 4 hours.

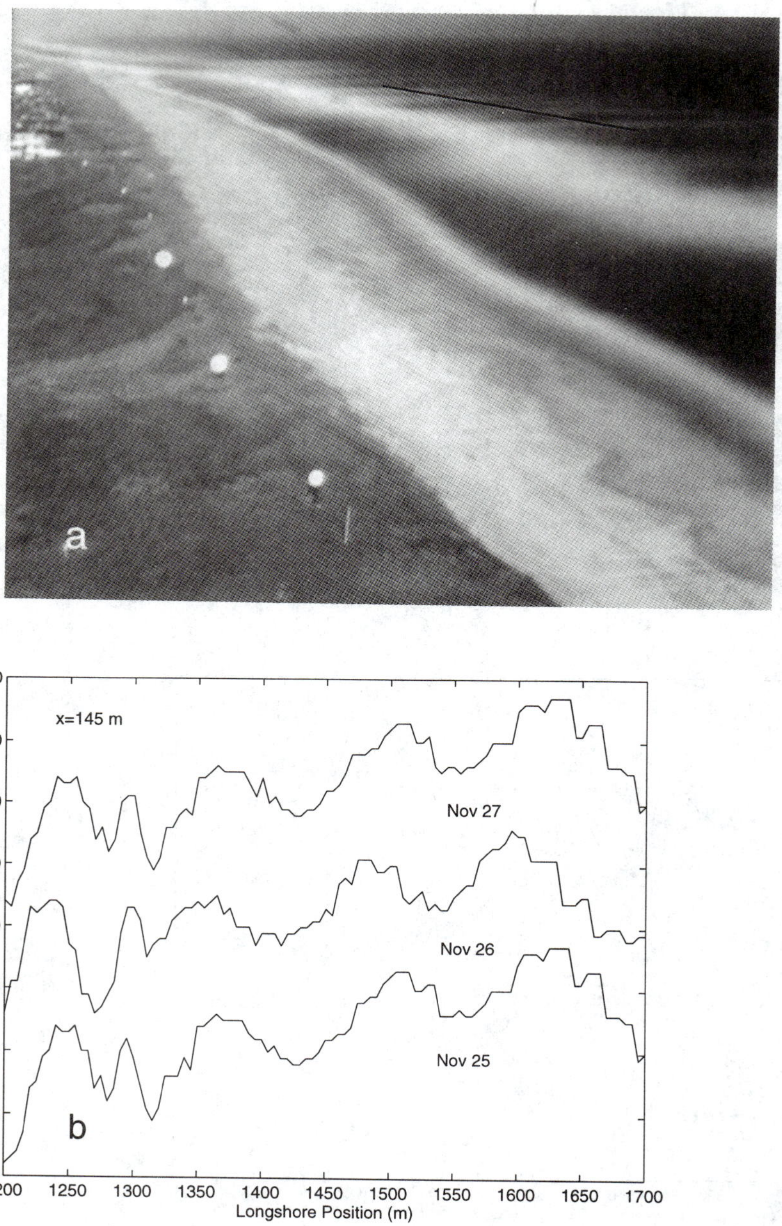

Figure 5. a). Time exposure from November 27, 1993 showing transverse bars offshore of the inner bar. Transects along which intensity data were collected are also shown. b). Intensity vs. longshore position for November 25, 26, and 27, 1993 along transects shown in a). This graph shows that most of the bars remain fixed over a period of days.

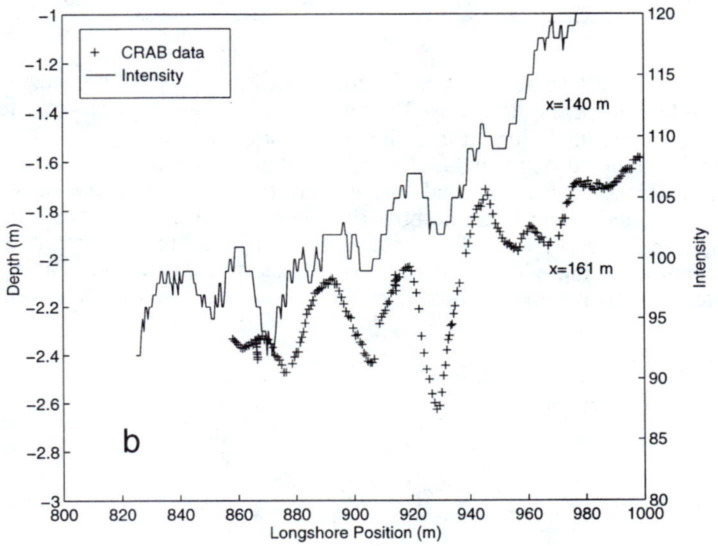

Figure 6. a). Time exposure from July 18, 1995 showing transverse bars in the trough. Transects along which intensity and bathymetric data were collected are shown. b). Comparison between intensity vs. longshore position and depth vs. longshore position for transects in a). This figure clearly shows that there is a correlation between the high pixel intensity in the video images and topographic highs.

Table 1. Length scales and locations for ***trough bars.***

	longshore length spacing (m)	*distance offshore (m)*	*cross-shore length (m)*
mean	45.0	20.0	23.8
std dev.	21.4	9.2	15.5
range	12-87	5-35	5-50

Table 1. Length scales and locations for ***offshore bars.***

	longshore length spacing (m)	*distance offshore (m)*	*cross-shore length (m)*
mean	130.7	249.1	180.0
std dev.	62.9	37.4	117.9
range	45-325	175-300	75-400

Offshore transverse bars are larger scale features than trough transverse bars. Their longshore length spacing is almost 3 times as large and their cross-shore length is up to an order of magnitude longer than the trough bars.

From the extensive Argus database, presence/absence data for the transverse bars were determined. We calculated the number of days transverse bars were observed annually and the average number of days transverse bars were observed per month. One problem that was encountered when calculating the statistics was the fact that there were many days for which no image was available. Taking this into consideration, the frequency of occurrence of transverse bars was determined as follows:

For annual statistics: $$N_{annual} = \frac{N_{obs}(yr)}{N_{images}(yr)} * 365$$

For monthly statistics: $$\overline{N}_{month} = \frac{N_{obs}(month)}{N_{images}(month)} * N_{days}(month)$$

where N_{obs} is the number of images in a year or month in which transverse bars were present, N_{images} is the number of images available for that month or year, N_{annual} is the number of days transverse bars were observed annually, and $\overline{N}_{month}$

is the average number of days transverse bars were observed per month. If less than 20 images were available for a given month, that month was excluded from the monthly statistics because a representative average could not be determined.

The average number of days trough and offshore transverse bars appeared in the video images each month for the 8 years varies greatly (Figure 7). Trough bars were observed from 3 to 7 days/month and offshore bars were observed less frequently (~1 to 5 days/month). Trough bars were most frequently observed in the summer while offshore bars seem to follow a seasonal cycle, appearing in the video images more frequently in the winter than in the summer.

Each year, trough bars were observed more frequently than offshore bars (an average of 62 and 41 days/year, respectively), and there was an interannual variability in the statistics as well (Figure 8). There is a positive correlation between trough and offshore bars for most years and there seems to be an increasing trend for both sets of bars.

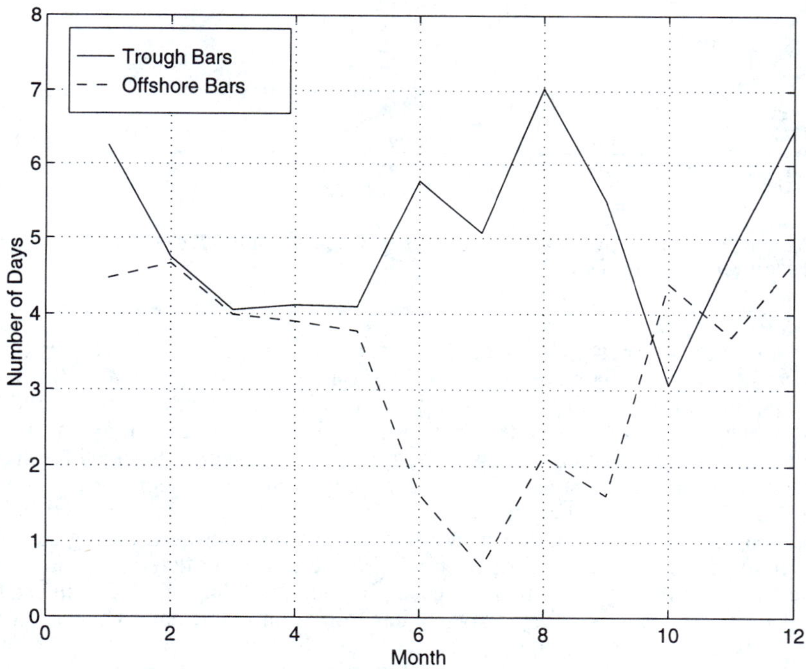

Figure 7. Average number of days transverse bars are observed per month.

Discussion

Transverse bars seem to be a common feature in the nearshore at Duck. They persist in the video images for periods up to days. There were, however, many instances when they were seen for consecutive days, "disappeared" from the images for a few days and then returned to what seems to be the exact location in which they

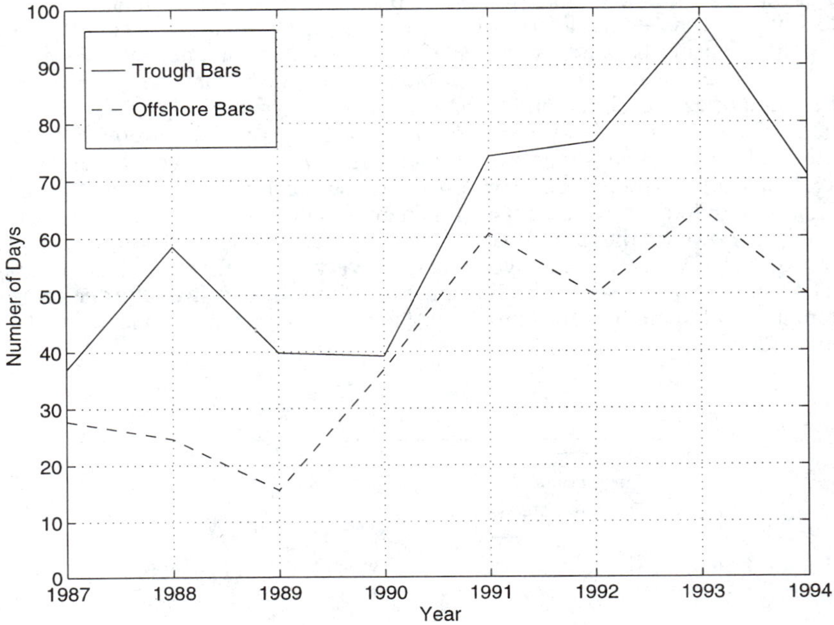

Figure 8. Number of days transverse bars are observed annually. Trough bars are observed more frequently than offshore bars.

were previously seen. This appearance/disappearance may be due to fluctuations in wave height. The bars are only seen in the images on days in which the waves are high enough to be breaking. Although they may "disappear" from the images when the wave height is lower and no breaking is taking place, they may still be present, but not observable. Wave height has not yet been taken into consideration in the calculation of bar statistics. Images in which no bars were seen were given a rating of 0 even when wave height was low. Wave height data from 1987-1995 will be studied to determine whether or not bars were always present during conditions of high wave height. This information will be used to better determine the statistics of the bars, as only days in which waves are breaking will be included in the statistics.

The mechanism for transverse bar formation is unclear. Their presence may be attributed to strong longshore currents and they may be formed in the same way that bedforms in a river are created. While longshore currents could perhaps be the mechanism for trough bar formation, it is not clear that they will be strong enough to locally form offshore bars.

Research will continue on the transverse bars at Duck. More complete statistics will be calculated for bar lengths and positions, and images from 1995 and 1996 will be viewed to augment the existing database. The movement, if any, of the bars will be determined, and an attempt will be made to understand the environmental conditions that are conducive to transverse bar formation. Finally, the Argus database for other beaches will be searched to determine if this process is unique to Duck.

Conclusions

Time exposures from Duck frequently show white bands perpendicular or oblique to the shoreline. These bands, which have been observed both in the trough and offshore of the shore parallel bar, persist for periods of hours up to days. Ground truthing has shown that they are transverse ridges with a relief of 0.3-0.5 m. Trough transverse bars (ridges in the trough) are a smaller scale feature than offshore transverse bars (ridges offshore of the shore-parallel bar). Their mean longshore length spacing is 3 times smaller than that of the offshore bars (44 m and 131 m, respectively). Trough bars obliquely extend up to 35 m from the shoreline while their actual cross-shore lengths may be up to 50 m. Offshore bars obliquely extend up to 300 meters from the shoreline with actual lengths up to 400 m. It appears that the cross-shore length of trough bars is constrained by the shore-parallel bar.

Trough transverse bars appear in the video images an average of 62 days/year and offshore transverse bars appear in the images an average of 41 days/year. There is an interannual variability in the statistics and a seasonal cycle. Offshore bars appear more frequently in the winter and trough bars appear most often from June-August and in December and January.

ACKNOWLEDGMENTS

We would like to thank Mike Leffler and Bill Birkemeir of the FRF for carrying out the CRAB survey. Thanks also to John Stanley for help with the wave data, and Eric Gilbert, Hilary Stockdon, Nathaniel Plant, and Peter Ruggiero for their continual help. Funding for this research is made available from The Office of Naval Research, Coastal Dynamics Program (grant number N00014-95-10753).

REFERENCES

Barcilon, A.I. and J.P. Lau (1973). A model for formation of transverse bars: *Journal of Geophysical Research*, 78 (15): 2656-2664.

Birkemeier, W.A. *et al*. (1981). A User's Guide to CERC's Field Research Facility US Army Corps of Engineers, CERC, Miscellaneous Report No. 81-7, p. 36.

Lippmann, T.C. and R.A. Holman (1989). Quantification of sand bar morphology: A video technique based on wave dissipation: *Journal of Geophysical Research*, 94: 995-1011

Niedoroda, A.W. and W.F. Tanner (1970). Preliminary study of transverse bars: *Marine Geology*, 9:41-62.

Shephard, F.P. (1952). Revised nomenclature for depositional coastal features: *Bull. Am. Assos. Petrol. Geologists*, 36 (10): 1902-1912.

CHAPTER 278

On the systematic offshore decay of breaker bars

Kathelijne M. Wijnberg [1,2]

ABSTRACT

Multiple breaker bar systems along the Holland coast exhibit long term cyclic behaviour. Over periods of many years, all bars migrate in a net seaward direction; the outer bar decays offshore, and a new bar is generated near the shoreline. This behaviour is essentially a cross-shore redistribution of sediment; the sediment of the offshore decaying outer bar is transported back onshore.

This paper presents a tentative evaluation of a part of the hypothesis formulated by Wijnberg (1995) to explain the cyclic bar system behaviour. The systematic offshore decay of the outer bar is considered to be the key to explaining the cyclicity in this behaviour. The offshore decay is assumed to occur due to a change in the long term balance of bar maintaining conditions (breaking waves) and bar diminishing conditions (very asymmetric waves). This balance will change as a result of offshore migration of the outer bar to deeper water during single storm events.

The calculations presented in this paper support the hypothesis, but field observations of the hydrodynamic conditions during the proces of bar decay are required for a more rigorous verification.

1. INTRODUCTION

Along the Dutch coast, multiple bar systems of nearshore breaker bars exist that exhibit systematic, cyclic behaviour on a time scale of many years (Edelman, 1974; Bakker and De Vroeg, 1988; Ruessink and Kroon, 1994; Wijnberg and Terwindt, 1995). This behaviour consists of a net seaward migration of all bars, with the outer bar decaying offshore and a new bar being generated near the shoreline (Fig. 1). This

---

[1] Dept. of Physical Geography, Institute for Marine and Atmospheric research Utrecht, Utrecht University, P.O. Box 80115, 3508 TC Utrecht, The Netherlands.

[2] College of Oceanic and Atmospheric Sciences, Oregon State University, Ocean Admin Bldg # 104, Corvallis, OR 97331-5503, USA.

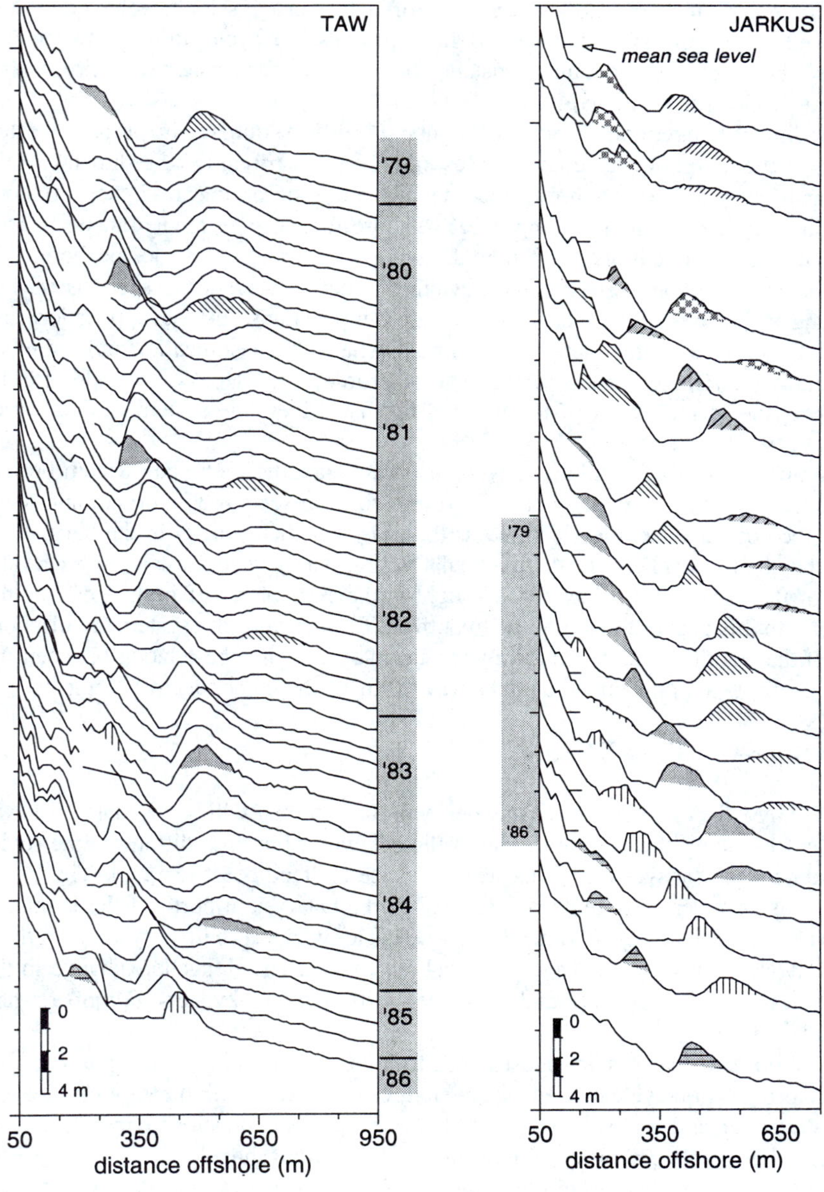

Fig. 1: Illustration of the offshore progression cycle for one profile line (line 8425). Left-hand figure based on TAW data base (several surveys per year, 1979-1986); right-hand figure based on JARKUS data base (one survey per year, 1965-1989). Bars are shaded to emphasise the offshore progression cycle; each bar has its own shading pattern through time.

behaviour implies that every so many years the same general cross-shore bar system configuration re-appears (i.e. when one offshore progression cycle has been completed). The time scale for the offshore progression cycle varies between bar systems, but seems to be quite constant for a given bar system (Wijnberg and Terwindt, 1995; Ruessink and Kroon, 1994).

Sediment budget considerations (Wijnberg, 1995; Wijnberg, in prep.) revealed that this cyclic morphologic behaviour is essentially a redistribution of sediment within the nearshore zone. So, although bar shapes are progressing in a net offshore direction, they are not representing a net offshore directed sediment transport. This implies that the sediment of the offshore decaying bar is transported back onshore.

The observed long term cyclic behaviour of multiple breaker bar systems is very intruiging in its own right, but may also have implications for the application and design of shoreface nourishments. For example, what is the best location and timing of a shoreface nourishment on a barred coastal profile? (eg. Hoekstra et al., 1994). Therefore, the identification of the mechanism that drives this systematic behaviour deserves further attention.

Wijnberg (1995) formulated a hypothesis to explain the cyclic bar system behaviour. In this hypothesis, which is briefly summarized in section 3, morphologic feedback in the coastal system is identified as the cause for the cyclicity in the bar behaviour rather than a cyclicity in the external forcing. The key to explaining the cyclic nature of the bar system behaviour is sought in the systematic offshore decay of the outer bar. In this paper, the part of the hypothesis dealing with the systematic offshore decay of the outer bar will be tentatively evaluated by studying the relation between the development stage of the outer bar and a wave climate integrated 'decay parameter'.

2. THE STUDY AREA

The hypothesis on the systematic decay of the outer bar will be evaluated for two multiple bar systems that exhibit similar cyclic behaviour but on a different time scale. The two multiple bar systems are located along the Holland coast (Fig. 2). These two bar systems have very different cycle times, viz. the bar system north of the IJmuiden breakwaters exhibits an approximately 15 year cycle, whereas the bar system south of the breakwaters has a typical cycle time of about 4 years. The alongshore change in the cycle times is rather abrupt, and coincides with the breakwaters (Wijnberg and Terwindt, 1995).

The northern bar system consist of 2 to 3 bars, depending on the phase in the offshore progression cycle (i.e. either 2 well developed bars or, in the case of 3 bars, a gentle outer bar and a small new inner bar, with a well developed middle bar in between). The whole bar system extents about 30 km alongshore.

In the southern bar system the number of bars changes alongshore: in the very south the number varies between 1 and 2, going north this changes into 2 to 3 bars, then 3 to 4 bars, and finally in the very north back to 2 to 3 bars. This bar system has a total length of about 40 km. The alongshore change in the number of bars has no obvious effect on the time span of the offshore progression cycle (Wijnberg and Terwindt, 1995).

Along the Holland coast, the mean wave height (H_{m0}) is about 1.2 m and the mean wave period (T_{m01}) is about 5 s; the mean tidal range is about 1.6 m. The grain size of the sediment in the breaker bar zone is generally in the fine sand range. There seems to be no significant difference in grain size between the two bar systems (Wijnberg, 1995)

Except for the large breakwaters that separate the two studied bar systems, the considered coastal stretch is almost free from engineering structures. In the northern bar system groins are present along about 5 km of beach at the northern end of the region (so still 25 km of beach has no groins). In the southern bar system a small discharging sluice is present (Fig. 2) which discharges some fresh water into the North Sea.

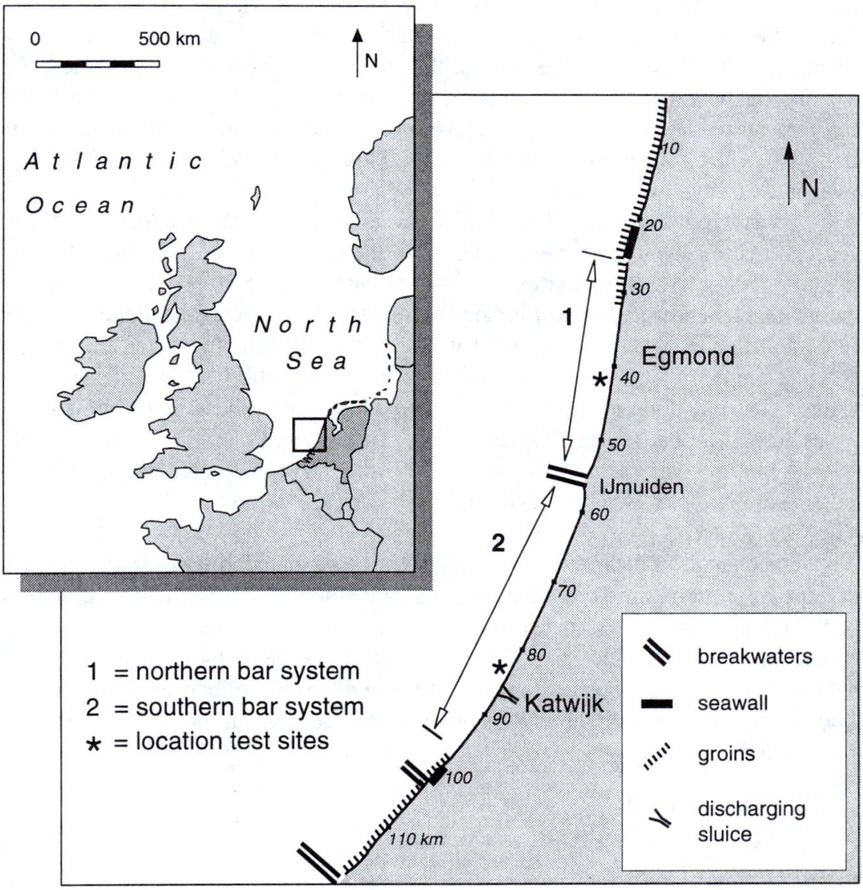

Fig. 2: Location of the study area

3. A CONCEPTUAL MORPHODYNAMIC MODEL

The multi-year cyclic behaviour of breaker bar systems can not readily be explained by existing theories on bar behaviour. These theories focus on the formation and maintenance of bar topography and pay little attention to the degeneration of bars. Our lack of understanding of the decay of bar features is illustrated by one of the findings of Roelvink et al. (1995) who mimicked long term cyclic bar behaviour with a cross-shore profile model. Roelvink et al. had to prescribe a reduced angle of repose for the sediment in the outer nearshore zone to force the outer bar to decay offshore.

According to Wijnberg (1995; in prep.), the cyclicity in the long term bar behaviour is most likely explained by morphologic feedback in the breaker bar system rather than by cyclicity in the external forcing. To explain the observed cyclic bar system behaviour, a central role was attributed to the behaviour of the outer bar. As long as the outer bar remains well developed at a certain position, the inner bar(s) do not move net offshore and will only move to and fro within a limited cross-shore range. Only as the outer bar disappears can the inner bar migrate net offshore (cf. Lippmann et al., 1993). This key role of the outer bar is plausible because the outer bar will obviously affect the hydrodynamic conditions in a down-wave direction, and therefore systematically affects the wave and current climate in the inner nearshore zone. Therefore, attention is further focused on mechanisms that may explain the systematic decay of the outer bar.

It was hypothesized by Wijnberg (1995; in prep.) that the systematic decay of the outer bar occurs due to a changing balance in the occurrence of conditions that favour decay of the outer bar topography (%DCAY) and the occurrence of conditions that favour maintenance of the bar topography (%MAIN). The bar-maintaining conditions were identified as those with breaking waves across the bar, favouring the occurrence of break-point or long-wave mechanisms for the formation and maintenance of bar features. The bar-diminishing conditions were identified as those with very asymmetric waves across the bar (so the highest waves in the wave field will just start to break), allowing for an onshore directed sediment transport. A more complete justification for the choice of these two process 'regimes' can be found in Wijnberg (1995) or Wijnberg (in prep.)

The change in balance between the two types of conditions is induced by the offshore migration of the outer bar on a seaward sloping bed. The latter happens when the bar is migrating into deeper water during individual storm events, i.e. when waves are breaking on the outer bar. The deeper the bar gets, the less likely that bar-maintaining conditions occur. Consequently, the relative dominance of bar-diminishing conditions will increase. The rate of increase will depend on the characteristics of the wave climate.

4. METHODS

To evaluate the proposed mechanism for the systematic offshore decay of the nearshore bars, the balance of %DCAY and %MAIN will be calculated for various development stages of the outer bar (Fig. 3). The balance is expressed in terms of the decay parameter I_{decay}:

$$I_{decay} = \frac{\%DCAY}{\%DCAY + \%MAIN} = \frac{\%DCAY}{\%ACT} \quad (1)$$

%DCAY = mean annual percentage of occurrence of bar-diminishing conditions on the outer bar

%MAIN = mean annual percentage of occurrence of bar-maintaining conditions on the outer bar.

%ACT = mean annual percentage of occurrence of morphologically active conditions on the outer bar.

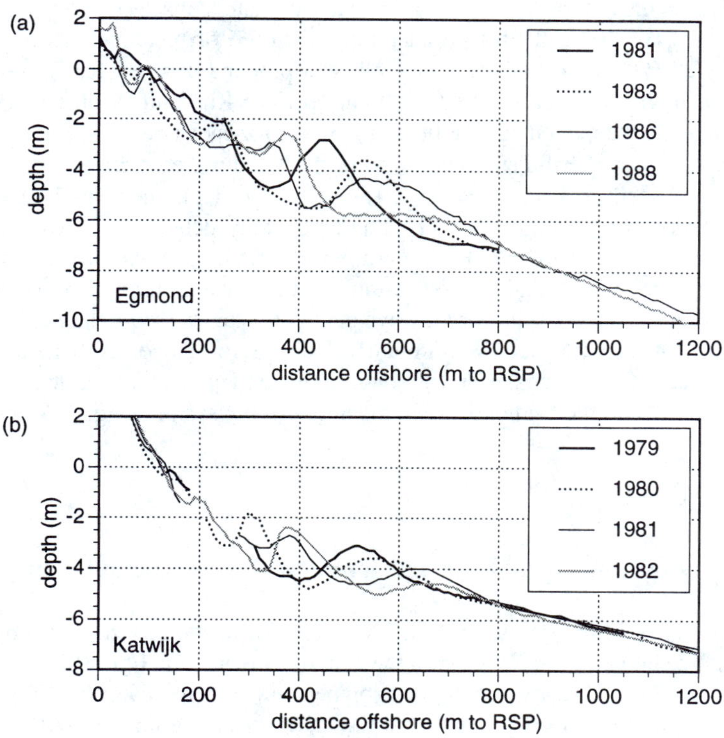

Fig. 3: Nearshore profiles used in the model computations at (a) the Egmond site, and (b) the Katwijk site. (All profiles extend down to about 20 m water depth.)

'Morphologically active conditions' (%ACT) are those conditions during which the outer bar is changing its morphology. This does not necessarily imply that during morphologically *in*active conditions the sediment is completely immobile; it is just so that any sediment transport under those conditions does not result in noticeable morphological change. The outer bar has been observed to be morphologically inactive during the calm summer months (Wijnberg, 1995).

Bar-maintaining conditions are defined (Wijnberg, 1995; Wijnberg, in prep.a) as those conditions that produce more than 5% of breaking waves on the bar. Bar-diminishing conditions are defined as those which produce only 0.1% to 5% of breaking waves on the bar. %DCAY and %MAIN are estimated from the mean annual offshore wave climate.

The data on the offshore hydrodynamic climate along the Holland coast are of the following type: a mean annual percentage of occurrence, $P(H_{m0},\alpha)$, and a mean water level elevation, $h(H_{m0},\alpha)$, are available, with wave height bands of 0.5 m and wave direction bands of 15 degrees. The mean water level represents the average set-up or set-down that occurs during given wave conditions. H_{m0} and α are the central values of the wave height class respectively wave directional class that they represent.

To estimate %DCAY and %MAIN one needs to know the percentage of breaking waves occurring on the outer bar for each of the wave climate 'cells' (H_{m0}, α). Adding all $P(H_{m0},\alpha)$ for all (H_{m0}, α) cells that produce more than 5% of breaking waves gives %MAIN; adding all $P(H_{m0},\alpha)$ for all (H_{m0}, α) cells that produce between 0.1% and 5% of breaking waves gives %DCAY. The WAVIS model (Van Rijn and Wijnberg, 1996) was used to calculate the percentages of breaking waves for each (H_{m0}, α) cell.

Since WAVIS is a probabilistic type of wave model, a representative wave field (H_i,T_i, p_i, α_i) was generated for each (H_{m0},α)-cell, where: H_i is the central value of the ith (0.25 m wide) wave height class; T_i is an empirically determined wave period that goes on average with H_i ($T_i = 6*H_i^{0.333}$ (Wijnberg, 1995)); p_i is the probablility of occurrence of the ith wave height class assuming the wave heights are Rayleigh distributed; α_i was chosen constant for all wave height classes, viz. equalling α.

The values of %DCAY and %MAIN are calculated for the outer bars at two locations, viz. near Katwijk and near Egmond. The Katwijk site is located in the southern bar system, and the Egmond site in the northern bar system (Fig. 2).

5. RESULTS

5.1 The decay parameter and the development stage of the outer bar

The values of %ACT, %DCAY, %MAIN, and I_{decay} in relation to the depth of the outer bar crest are shown in Fig. 4, for both the Katwijk site and the Egmond site. The depth of the outer bar crest represents the development stage of the outer bar (Fig. 3). The depth of the bar crest is given relative to the Dutch ordnance datum NAP which approximates mean sea level.

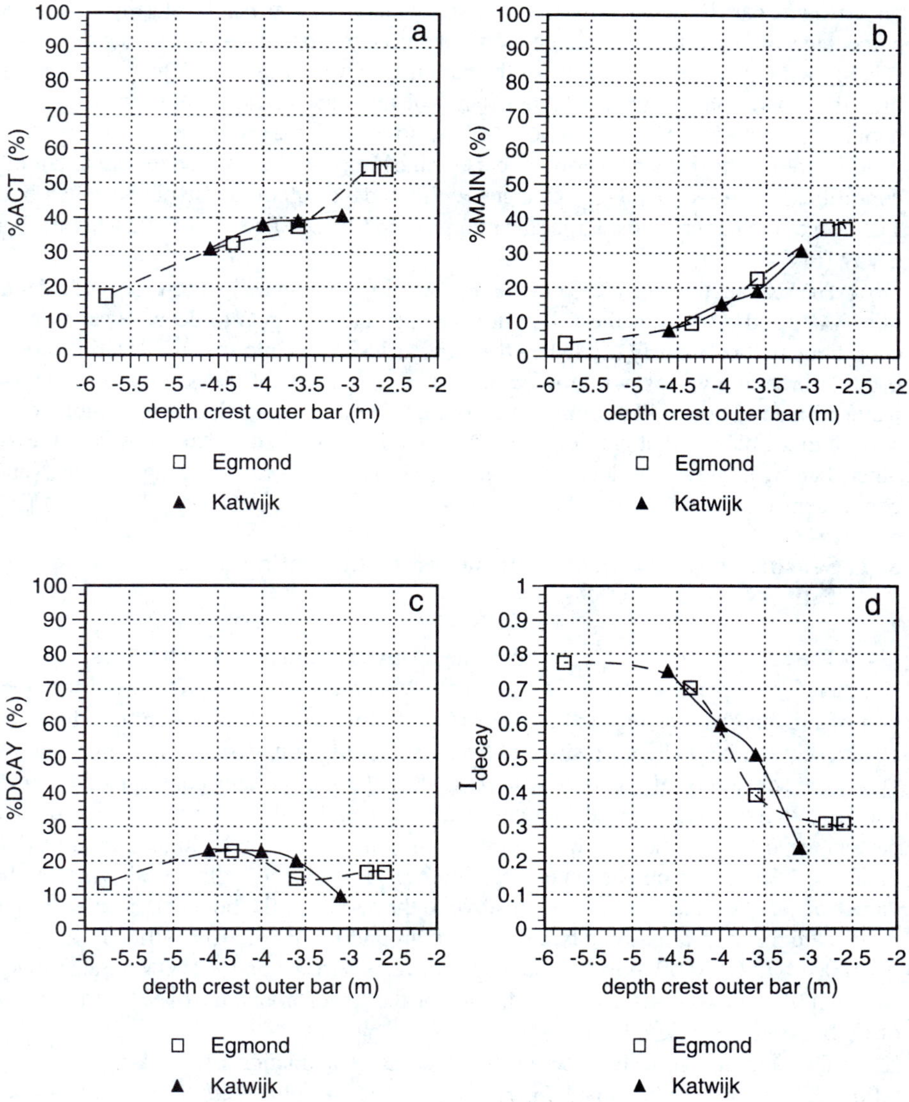

Fig. 4: Nearshore wave climate parameters as a function of the development stage of the outer bar. (a) %ACT, (b) %MAIN, (c) %DCAY, and (d) I_{decay}.

As expected, the percentage of time that morphologically active conditions occur on the outer bar decreases with the increase of depth of the bar crest (Fig. 4a). The percentage of time that bar-maintaining conditions occur on the outer bar decreases steadily with the increase of depth of the bar crest as well (Fig. 4b). The percentage of time that conditions occur that favour decay of the outer bar appears to vary between about 15% and 25% (Fig. 4c). %DCAY seems to reach a maximum around 4 to 4.5 m water depth. The balance between bar-diminishing conditions and bar-maintaining conditions, expressed by I_{decay}, changes with the depth of the crest of the outer bar. The deeper the bar crest, the larger the relative importance of conditions that favour bar decay (Fig. 4d).

The explanation for the fact that %DCAY does not steadily decrease with depth like %ACT and %MAIN, is the following. As the outer bar gets to deeper water, wave conditions that previously (i.e. when the bar was located shallower) produced just over 0.1% of breaking waves, will now produce less than 0.1% of breaking waves. Those wave climate classes will no longer add to the value of %DCAY. However, some of the wave climate classes that previously produced more than 5% of breaking waves on the outer bar will now produce between 0.1% and 5% of breaking waves, and consequently will add to the value of %DCAY.

5.2 Sensitivity of the decay parameter to the definition of %DCAY and %MAIN

The decay parameter I_{decay} reflects the relative importance of %DCAY vs. %MAIN and therefore depends on the definition of the two process regimes in terms of the fraction of breaking waves on the outer bar (Q_b). In this section, the extent to which the results shown in Fig. 4 are sensitive to the choice of the boundaries will be evaluated.

The sensitivity of the values of %DCAY and %MAIN, and consequently %ACT and I_{decay}, to the definition of the process regimes is analysed by plotting the cumulative percentage of occurrence of wave conditions with increasingly higher fractions of breaking (Q_b) waves on the outer bar (Fig. 5). These plots will reveal whether the values of %DCAY and %MAIN are sensitive to the choice of the boundaries between the process regimes in a reasonably close range around the previously defined values of Q_b=0.001 and Q_b=0.05. For the boundary between %DCAY and %MAIN a value of Q_b between 0.04 and 0.06 is considered, and for the lower boundary of %DCAY a value of Q_b between 0.001 and 0.01

The cumulation in the plots in Fig. 5 starts at a fraction of breaking waves of 0.001. Therefore, the cumulative percentage in Fig. 4 only refers to wave conditions that produce at least 0.1% of breaking waves on the outer bar. For example, on the bar crest at 3.1 m -NAP (Fig. 5a) about 22.5% of the mean annual wave conditions will produce between 0.1% and 8.3% of breaking waves on the outer bar (near Katwijk).

It appears that for the bar crest at 3.1 m -NAP, the variables %DCAY and %MAIN will be quite sensitive to the location of the boundary at Q_b=0.05. For the bar crest at 3.1 m -NAP about 11 % of the wave climate is enclosed in the discretely defined wave climate cells that produce a fraction of breaking waves of 0.055 on the outer bar (Fig. 5a). The value of %ACT will be hardly affected at this position of the bar crest.

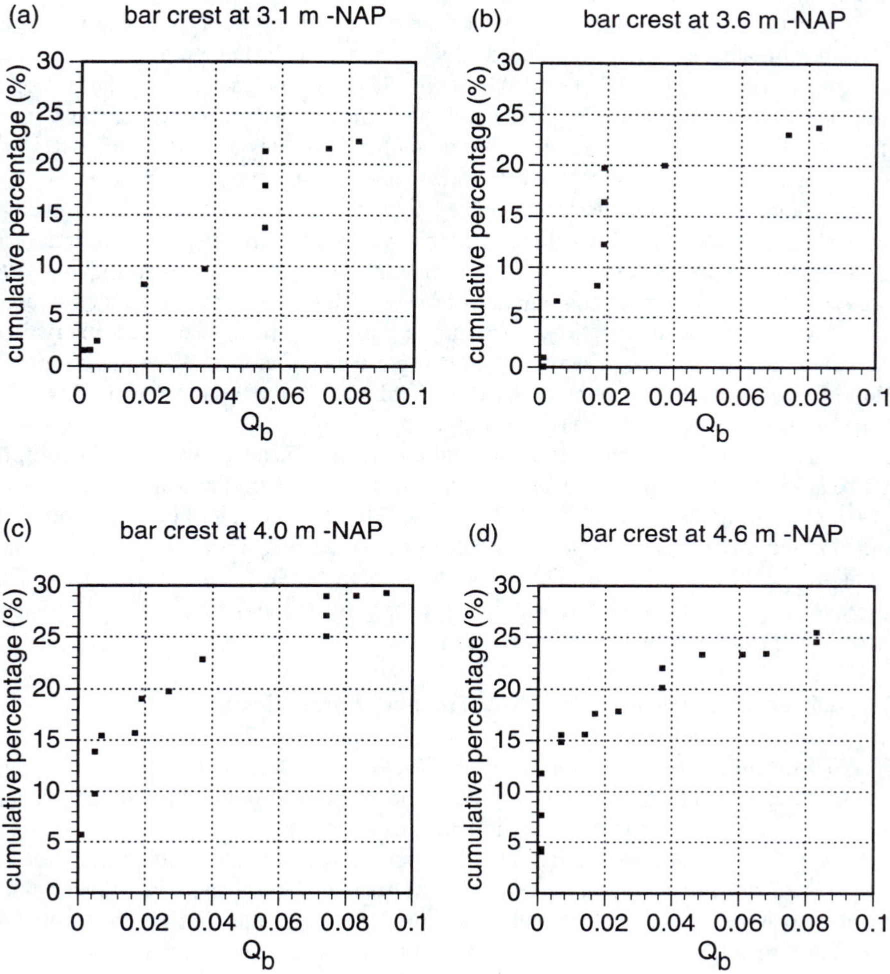

Fig. 5: Cumulative percentage of occurrence of wave conditions with increasingly higher fraction of breaking waves on the outer bar, for Qb≥0.001 (Katwijk site). (a) Bar crest at 3.1 m -NAP, (b) bar crest at 3.6 m -NAP, (c) bar crest at 4.0 m -NAP, (d) bar crest at 4.6 m -NAP

A comparable sensitivity to the Q_b=0.05 boundary exists at the Egmond site for the bar crests at 2.6 m depth and 2.8 m depth (not shown).

The values of %DCAY and %MAIN on the bar crest located at 3.6 m -NAP will be insensitive to a shift between Q_b=0.04 and Q_b=0.06 (Fig. 5b). However, %DCAY will be somewhat affected by a raise of the lower boundary to Q_b=0.01.

For the positions of the bar crest at 4 m -NAP and 4.6 m -NAP, about 15% of the wave climate appears to be enclosed in wave climate cells that produce a fraction of breaking waves between 0.001 and 0.01 (Fig. 5c,d). Consequently, for these deeper bars the choice of the boundary between morphologically inactive conditions and morphologically active conditions is quite important for the values of %ACT and %DCAY.

The sensitivity of %DCAY and %MAIN - and consequently of %ACT and I_{decay} - to the choice of the boundaries is related to the stepwise increase of cumulative "curves" shown in Fig. 4. This shape is probably related to the discrete description of the wave climate and the the discrete description of the input wave fields in the WAVIS model. Only a limited set of combinations of wave heights and wave directions is used for the wave climate description. Each single wave height represents an interval of wave heights and each single wave direction represents a sector of wave directions. In addition, the discrete description of the wave field input in the WAVIS model allows the value of Qb to change in discrete steps only.

The effect of the discrete description of the wave climate and wave field on Fig.4 can be smoothed by averaging over a set of calculations where the lower boundary of %DCAY was varied between Q_b=0.01 and Q_b=0.001, and the boundary between %DCAY and %MAIN was varied between Q_b=0.04 and Q_b=0.06. The result of the averaging is shown in Fig. 6. This figure appears to be very similar to Fig. 4. The smoothed figures, however, reveal more clearly the simmilarities and differences between the Egmond site and the Katwijk site.

5.3 Differences between the Katwijk and Egmond site

To discuss differences between the Katwijk and Egmond site the smoothed results are analysed (Fig. 6), because these are less affected by the discrete representation of the wave information (see section 5.2).

A difference between the Egmond site and the Katwijk site in the occurrence of morphologically active conditions (Fig. 6a) only occurs for the shallower positions of the outer bar (bar crests shallower than 3.5 m -NAP). At both sites, those shallower outer bars are still in the non-decaying stage of their development. In that non-decaying stage, the percentage of time that morphologically active conditions occur on the outer bar seems to be slightly larger at the Egmond site than at the Katwijk site.

The bar-maintaining conditions occur about equally often at both sites for the deeper positons of the outer bar (Fig. 6b). However, for bar crest positions shallower than 4 m -NAP the bar-maintaining conditions occur somewhat more often near Egmond than near Katwijk.

Regarding the bar-diminishing conditions (Fig. 6c), an obvious difference only occurs for the bar crest position at 3.6 m -NAP, where the bar-diminishing conditions occur more frequently near Katwijk than near Egmond.

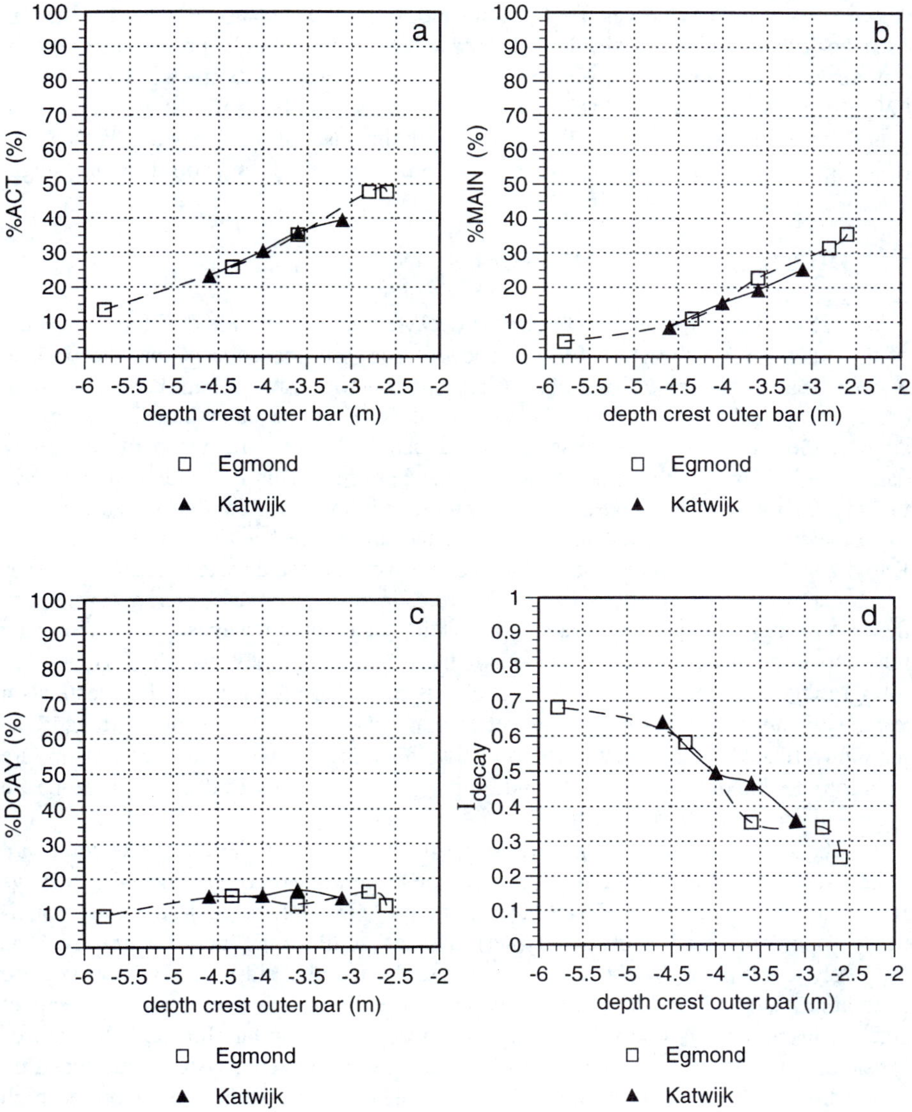

Fig. 6: Smoothed values of the nearshore wave climate parameters as a function of the development stage of the outer bar.
(a) %ACT, (b) %MAIN, (c) %DCAY, and (d) I_{decay}.

The slight differences between the %MAIN and %DCAY 'curves' for Egmond and Katwijk result in I_{decay} 'curves' for Egmond and Katwijk that clearly deviate for the bar crest position at 3.6 m -NAP. This discrepancy was consistently present irrespective ofthe exact definitions of the boundaries between the process regimes in terms of Q_b (Wijnberg, 1995). The relatively frequent occurrence of bar-diminishing conditions near Katwijk around 3.6 m -NAP coincides with the onset of bar decay at this location. Near Egmond, an outer bar with its crest at 3.6m -NAP is still in a non-decaying stage.

6. DISCUSSION AND CONCLUSIONS

The relationship found between the decay parameter and the development stage of the outer bar, as shown in Fig. 4d and 6d, supports the concept that the balance between bar-diminishing conditions (very asymmetric waves) and bar-maintaining conditions (breaking waves) is important for the fate of the outer bar. These results support the bar-decay hypothesis for two reasons. Firstly, the value of the decay parameter increases with the depth of the outer bar crest. This indicates an increase in the relative importance of the bar-diminishing conditions with depth, as was expected from the formulated hypothesis. Secondly, the difference between the Egmond and Katwijk site in the depth of onset of bar decay is well indicated by the decay parameter.

The results presented in Section 5 further indicate that the chosen definitions of the two process regimes in terms of the fraction of breaking waves (Q_b) enclose the relevant processes for respectively maintaining bar topography and diminishing bar topography. It is emphasized that no choice is made here for a particular mechanism that maintains the bar topography other than that this mechanism is directly or indirectly related to the occurrence of wave breaking. Long-wave mechanisms are therefore not excluded here (eg. bound long waves can be released in the surfzone by the breaking of the short waves).

The suggested mechanism for the systematic decay of the outer bar could not directly be verified with field observations. The currently existing bathymetric surveys in the studied area are to widely spaced in time to pinpoint the hydrodynamic conditions responsible for the observed (net) changes in the outer bar topography. The relatively wide spacing of the surveys is due to the fact that ships can only survey the nearshore bathymetry during relatively calm conditions, and consequently cannot survey during the conditions that are hypothesized to be important for the behaviour of the outer bar. Remote-sensing systems like the video-monitoring system introduced by Lippmann and Holman (1989) might be a valuable tool for obtaining the required high resolution morphologic information during the more severe wave conditions.

To conclude, the presented conceptual model for the systematic offshore decay of breaker bars seems viable. However, the results presented in this paper are considered to be supporting evidence for the hypothesis rather than a real proof of it. For a more rigourous verification of the conceptual model field observations are needed of the process of bar decay itself.

ACKNOWLEDGEMENTS

This paper is based on work in the PACE-project, in the framework of the EU-sponsored Marine Science and Technology Programme (MAST-III), under contract number MAS3-CT95-0002. The work was co-sponsored by the Andrew Mellon Foundation.

The *Rijkswaterstaat* (Dutch Ministry of Transport, Public Works and Water Management) is thanked for providing the wave climate data and the bathymetry data.

REFERENCES

Bakker, W.T., and H.J. De Vroeg, 1988. Is de kust veilig? Analyse van het gedrag van de Hollandse kust in de laatste 20 jaar. Nota GWAO 88.017, Rijkswaterstaat, Den Haag, The Netherlands, 42 p.

Edelman, T., 1974, Bijdrage tot de historische geografie van de Nederalndse kuststrook. Rijkswaterstaat / Directie Waterhuishouding en Waterbeweging, Den Haag, The Netherlands, 84 p.

Hoekstra, P., K.T. Houwman, A.Kroon, P. van Vessem, and B.G. Ruessink, 1994. The Nourtec experiment of Terschelling: process-oriented monitoring of a shoreface nourishment (1993-1996). Proceedings Coastal Dynamics '94, Barcelona, ASCE: 402-416.

Lippmann, T.C., and R.A. Holman, 1989. Quantification of sand bar morphology: a video technique based on wave dissipation. Journal of Geophysical Research 94 (C1): 995-1011.

Lippmann, T.C., R.A. Holman, and K.K. Hathaway, 1993. Episodic, nonstationary behaviour of a double bar system at Duck, North Carolina, U.S.A., 1986-1991. Journal of Coastal Research 15 Special Issue: 49-75.

Roelvink, J.A., Th.J.G.P. Meijer, K. Houwman, R. Bakker, and R.Spanhoff, 1995. Field validation and application of a coastal profile model. Proceedings Coastal Dynamics '95, Gdansk, ASCE: 818-828.

Ruessink, B.G. and A. Kroon, 1994. The behaviour of a multiple bar system in the nearshore zone of Terschelling, the Netherlands: 1965-1993. Marine Geology 121: 187-197.

Van Rijn, L.C. and K.M. Wijnberg, 1996. One-dimensional modelling of individual waves and wave-induced longshore currents in the surf zone. Coastal Engineering 28: 121-146.

Wijnberg, K.M., 1995. Morphologic behaviour of a barred caost over a period of decades. PhD thesis Utrecht University, The Netherlands. Netherlands Geographical Studies 195, KNAG, 245 p.

Wijnberg, K.M., in prep. Long term behaviour of multiple breaker bar systems along the Holland coast.

Wijnberg, K.M., and J.H.J. Terwindt, 1995. Extracting decadal morphological behaviour from high-resolution, long-term bathymetric surveys along the Holland coast using eigenfunction analysis. Marine Geology 126: 301-330.

CHAPTER 279

LONGSHORE BED-LOAD TRANSPORT

Jesper S. Damgaard [1] and Richard L. Soulsby [1]

Abstract.

An analytical formula for longshore bed-load sediment transport is derived. The derivation is based on sediment physics and the formula is therefore less dependent on calibration. The formula is compared to the CERC formula, a numerical model and field data.

1 Introduction.

Around the UK coastline shingle and gravel beaches are common and considerable resources are spent on preserving and/or predicting the behavior of these beaches. Therefore, it is important for coastal engineers to have an easily applicable tool with which to predict the littoral drift of sediments consisting of larger grain sizes.

The most widely used longshore sediment transport formula is probably the CERC -formula (Coastal Engineering Research Center, 1977) which is based on the concept that the longshore sediment transport rate is dependent on the longshore component of wave energy flux $E_b(C_g)_b \cos(\alpha_b) \sin(\alpha_b)$, where subscript $'b'$ denotes values at the breaking point, E is the wave energy, C_g is the group velocity and α is the angle of wave incidence. There exist numerous extensions and variations of the CERC formula. Despite its popularity there are three major shortcomings of this type of formulae: i) it relies heavily on calibration, ii) it does not contain a threshold term, iii) it contains no dependence on grain or beach properties. The necessity of a calibration limits the predictive quality of the formula. Furthermore the calibrations of CERC type formulae are often performed using field data sets with different levels of variance within those data sets and with considerable systematic errors (Greer and Madsen, 1978). The inclusion of a threshold term is especially important for the determination of transport rates for large grain size sediments, as will be shown in the present paper.

[1]Marine Sediments Group, HR Wallingford Ltd.,Wallingford, Oxon OX10 8BA, UK.

2 The present study

It is the purpose of the present study to derive an analytical cross-shore integrated bed-load transport formula. The derivation is based on a bed-load transport formula for combined waves and currents developed by Soulsby (1996), which relates the wave-averaged value $\langle \ldots \rangle$ of the non-dimensional bed-load transport rate vector $\vec{\Phi}$

$$\vec{\Phi} = \frac{\vec{q_b}}{\sqrt{(s-1)gd^3}} \tag{1}$$

to the non-dimensional bottom shear-stress vector due to skin friction induced by waves plus currents (Shields parameter) $\vec{\theta}$

$$\vec{\theta} = \frac{\vec{\tau_b'}}{\rho g(s-1)d} \tag{2}$$

where q_b is the volumetric bed-load transport rate per unit width and time, s is the relative density of the sediment, g is the acceleration of gravity, d is the grain size and τ_b' is the skin friction at the bed. The bed-load transport formula is based on a separation of the shear stress vector into a mean and an oscillatory part:

$$\vec{\theta}(\omega t) = \vec{\theta}_m + \vec{\theta}_w \cos(\omega t) \tag{3}$$

where ω is the frequency of the incoming waves. The coordinate system (x, y) is chosen such that $\vec{\theta}_m$ is parallel to the x-axis (see figure 1).

The formula has been extended to cover asymmetric waves (Damgaard *et al*, 1996) but since the asymmetry will have no significant impact on long-shore sediment transport calculations all higher order harmonics are set to zero for the present application. The bed-load transport formula of Soulsby (1996) reads:

$$\langle \vec{\Phi} \rangle = \begin{pmatrix} \Phi_x \\ \Phi_y \end{pmatrix} \tag{4}$$

with the components given as

$$\Phi_x = \text{sign}\{\Phi_{x1}, \Phi_{x2}\} \max\{\Phi_{x1}, |\Phi_{x2}, |\} \tag{5}$$

$$\Phi_{x1} = \begin{cases} 12\sqrt{\theta_m}(\theta_m - \theta_{cr}) & \text{for } \theta_m > \theta_{cr} \\ 0 & \text{for } \theta_m \leq \theta_{cr} \end{cases} \tag{6}$$

$$\Phi_{x2} = 12(0.95 + 0.19\cos 2\varphi)\theta_m\sqrt{\theta_w} \tag{7}$$

$$\Phi_y = \frac{12(0.19\theta_m\theta_w^2 \sin 2\varphi)}{\theta_w^{3/2} + 1.5\theta_m^{3/2}} \tag{8}$$

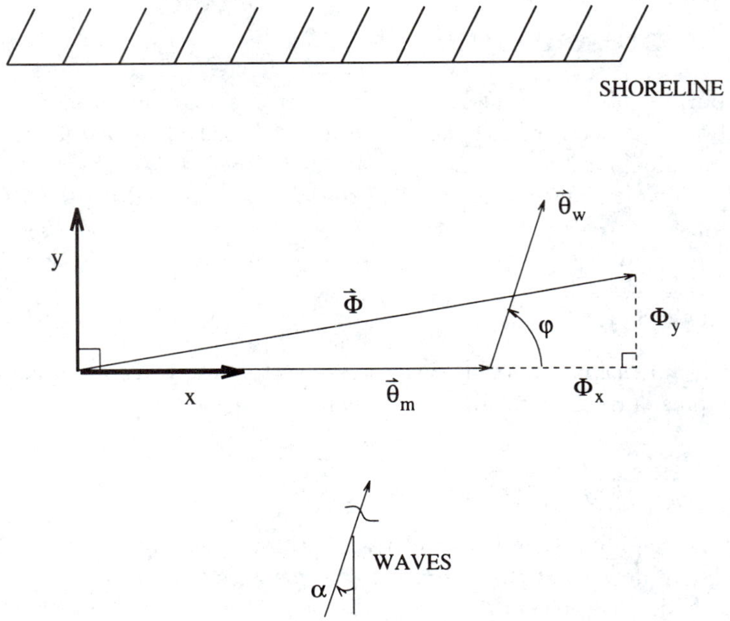

Figure 1: Definitions.

subject to the threshold condition

$$\Phi_x = \Phi_y = 0 \text{ for } \theta_{max} \leq \theta_{cr} \tag{9}$$

where φ is the angle between $\vec{\theta_m}$ and $\vec{\theta_w}$ and θ_{max} is the maximum Shields parameter for waves + currents within a wave cycle. Current dominated conditions correspond to Φ_{x1} and wave-dominated conditions to Φ_{x2}. The threshold condition is necessary as no threshold term appears explicitly in the formula for Φ_{x2}. It is important to note the fact that the formula is based on dimensionless parameters which makes it applicable to model as well as prototype scale.

The procedure of derivation for the analytical formula is as follows: Approximate expressions for the mean bottom shear-stresses, required as input to the bed-load transport formula are derived directly from the depth-integrated momentum equation. This avoids the need to calculate the speed of the longshore current. In order to solve the momentum equation analytically a number of assumptions have been made.

(i) The ratio of wave-height, H_b, and water depth, h_b, at the breaker point is constant, $H_b/h_b = \gamma_b = 0.8$.
(ii) After breaking the 1 decays linearly, so that $H = \gamma_b h$.
(iii) No further refraction takes place beyond the breaking point, i.e. the angle of wave incidence, α, is constant in the surf zone, $\alpha(y) = \alpha_b$.
(iv) All waves break at the same location.

(v) Shallow-water linear wave theory is used in the surf zone.
(vi) The effect of a lateral bed slope on the sediment transport is neglected.
(vii) The effect of cross-shore currents (e.g. undertow) on the longshore transport is neglected.

Furthermore the variation in mean water level (set-down and set-up) is ignored and the beach is assumed to be plane with a constant slope, $tan\beta$. Ignoring the lateral transfer of turbulent momentum the depth-integrated momentum equation for a longshore uniform beach reduces to a balance between the cross-shore gradient of the shear component of the radiation stress and the bottom friction:

$$\frac{\partial S_{xy}}{\partial y} = -\tau_{b,x} \tag{10}$$

where S_{xy} is the shear component of the radiation stress and $\tau_{b,x}$ is the x-component of the bed shear stress. The (x, y)-coordinate system used here is chosen such that the x-axis is parallel to the coast, and y is directed on-shore with origin shoreward of the breaker line. The relation between φ and α_b is $\varphi = \frac{\pi}{2} - \alpha_b$. The radiation stress can be expressed in terms of the momentum flux, F_m, in the direction of wave propagation:

$$S_{xy} = F_m \cos(\alpha)\sin(\alpha) \tag{11}$$

in which α is the angle of wave incidence and F_m is determined according to linear wave theory as:

$$F_m = \frac{1}{8}\rho g H^2 \tag{12}$$

Combining equations (10), (11) and (12) yields an expression for τ_b

$$\tau_b = -\frac{1}{16}\rho g \frac{\partial}{\partial y}(\sin(2\alpha)H^2) \tag{13}$$

According to the assumptions employed regarding refraction and wave height decay, (13) reduces to:

$$\begin{aligned}\tau_b &= -\frac{1}{16}\rho g \sin(2\alpha_b) 2H \frac{\partial H}{\partial y} \\ &= \frac{\gamma_b}{8}\rho g H \sin(2\alpha_b) tan\beta\end{aligned} \tag{14}$$

in which h is the depth and γ_b is the constant ratio between wave height and depth in the surf zone. Thus the shore parallel component of the mean Shields parameter can be expressed as:

$$\theta_{m,x} = \frac{\gamma_b H \sin(2\alpha_b) tan\beta}{8(s-1)d} \tag{15}$$

The amplitude of the oscillatory component of the bottom friction is calculated using a quadratic friction law

$$\tau_w = \frac{1}{2}\rho f_w U_w^2 \tag{16}$$

where f_w is the wave friction coefficient. The wave orbital velocity is estimated using shallow water linear wave theory i.e. $kh \ll 1$, $c = \sqrt{gh}$:

$$\begin{aligned} U_w &= \frac{\pi H}{T} \frac{1}{\sinh(kh)} \\ &\simeq \frac{1}{2}\sqrt{\frac{g}{h}} H = \frac{1}{2}\sqrt{g\gamma_b H} \end{aligned} \tag{17}$$

where T is wave period and k is the wavenumber. Two different friction factors were used. For rough turbulent flows Soulsby (1994) obtained an expression for the wave friction factor $f_{w,r}$ based on the analysis of a large data set:

$$\begin{aligned} f_{w,r} &= 1.39 \left(\frac{A}{z_0}\right)^{-0.52} \\ &\simeq \sqrt{\frac{d}{U_w T}} \\ &\simeq (g\gamma_b H)^{-1/4} \sqrt{\frac{2d}{T}} \end{aligned} \tag{18}$$

in which A is the amplitude of the near bed orbital motion, and the roughness length $z_0 = d/12$. The above equation is valid for weak sediment motion or immobile beds. For sheet flow conditions ($\theta >$ say 1) Wilson (1989) derived a friction coefficient, $f_{w,sf}$:

$$\begin{aligned} f_{w,sf} &= 0.114 \left(\frac{A}{g(s-1)T^2}\right)^{2/5} \\ &\simeq 0.0655 \left(\frac{\gamma_b H}{g}\right)^{1/5} (\pi(s-1)T)^{-2/5} \end{aligned} \tag{19}$$

The actual friction coefficient for a particular wave condition was taken as the maximum of $f_{w,r}$ or $f_{w,sf}$. In accord with experimental observations (Simons *et al.*, 1994), the friction factor for the waves is not enhanced by the presence of a near-normal incidence current.

The oscillatory component of the Shields parameter can be expressed using equations (2), (16), (17) and (18) and (19) respectively

$$\theta_{w,r} = \frac{\sqrt{2}}{8} g^{-1/4} \frac{(\gamma_b H)^{3/4}}{(s-1)\sqrt{dT}} \tag{20}$$

and

$$\theta_{w,sf} = 8.187 \cdot 10^{-3} g^{-1/5} (\pi(s-1)T)^{-2/5} \frac{(\gamma_b H)^{6/5}}{(s-1)d} \tag{21}$$

Now the current-dominated case of the dimensionless bed-load transport rate is calculated by inserting equation (15) into equation (6)

$$\Phi_{x1} = 0.530 \left(\frac{\gamma_b H \tan\beta}{(s-1)d}\right)^{3/2} (\sin(2\alpha_b) - \theta_{cr}^*) \sqrt{|\sin(2\alpha_b)|} \tag{22}$$

where

$$\theta^*_{cr} = \theta_{cr}\frac{8(s-1)d}{\gamma_b H \tan\beta} \tag{23}$$

and the wave-dominated case (subject to flow regime) is determined by inserting equation (15) and (20) or (21) into equation (7)

$$\Phi_{x2,r} = (0.602 + 0.120\cos(2\varphi))\frac{(\gamma_b H)^{11/8}\tan\beta\sin(2\alpha_b)}{g^{1/8}d^{5/4}T^{1/4}(s-1)^{3/2}} \tag{24}$$

or

$$\Phi_{x2,sf} = (0.129 + 0.0259\cos(2\varphi))\frac{(\gamma_b H)^{8/5}\tan\beta\sin(2\alpha_b)}{g^{1/10}d^{3/2}(\pi T)^{1/5}(s-1)^{17/10}} \tag{25}$$

In order to transform to a bulk transport formula, equations (22), (24) and (25) are integrated across the surf zone, using assumptions (i) and (ii) above:

$$\int_{\text{surf zone}} \Phi_x\, dy = \int_{H_b}^{0} \Phi_x \frac{1}{dH/dy}\, dH = \frac{-1}{\gamma_b \tan\beta}\int_{H_b}^{0}\Phi_x\, dH \tag{26}$$

The cross-shore integrated volumetric sediment transport rate $Q'_{b,l}$ $[m^3/s]$ is given as

$$Q'_{b,l} = \sqrt{g(s-1)d^3}\int_{\text{surf zone}} \Phi\, dy \tag{27}$$

and the resulting analytical formula for $Q_{b,l}$ is expressed as a combination of a current dominated transport (Q_{x1}) and a wave dominated transport (Q_{x2}):

$$Q'_{b,l} = \text{sign}\{\alpha_b\}\max\{|Q_{x1}|, |Q_{x2},|\} \tag{28}$$

$$Q_{x1} = \begin{cases} 0.21\dfrac{\sqrt{g\gamma_b\tan\beta}H_b^{5/2}}{(s-1)}\left(\sin(2\alpha_b) - \frac{5}{3}\theta^*_{cr}\right)\sqrt{|\sin(2\alpha_b)|} & \text{for } \sin(2\alpha_b) > \frac{5}{3}\theta^*_{cr} \\ 0 & \text{for } \sin(2\alpha_b) \le \frac{5}{3}\theta^*_{cr} \end{cases} \tag{29}$$

$$Q_{x2} = \begin{cases} (0.25 + 0.051\cos(2\varphi))\dfrac{g^{3/8}d^{1/4}\gamma_b^{3/8}H_b^{19/8}}{T^{1/4}(s-1)}\sin(2\alpha_b) & \text{for } f_{w,r}/f_{w,sf} > 1 \\ (0.050 + 0.010\cos(2\varphi))\dfrac{g^{2/5}\gamma_b^{3/5}H_b^{13/5}}{(\pi T)^{1/5}(s-1)^{6/5}}\sin(2\alpha_b) & \text{for } f_{w,r}/f_{w,sf} \le 1 \end{cases} \tag{30}$$

subject to the threshold condition

$$Q'_{b,l} = 0 \text{ for } \theta_{max} \leq \theta_{cr} \tag{31}$$

where

$$\theta_{max} = \sqrt{(\theta_m + \theta_w \cos\varphi)^2 + (\theta_w \sin\varphi)^2} \tag{32}$$

In general the current dominated term, Q_{x1}, is used for large wave heights, small grain sizes and large beach slopes. In most cases $f_{w,r} > f_{w,sf}$. This is especially true for larger grain sizes which are rarely transported as sheet flow. A study of a field data set with approximately 25000 wave conditions sampled throughout a 4 year period (see section 4) showed that Q_{x1} was applied for approximately 60 % of the wave conditions, Q_{x2} with $f_{w,r}$ for the remaining 40 % of the conditions and Q_{x2} with $f_{w,sf}$ was not applied at all for that particular data set.

3 Comparison with the CERC formula.

The CERC-formula is probably the most widely used longshore sediment transport formula. In the course of the time since the CERC-formula was derived, several 'mutations' have emerged; only some of them will be mentioned here. Van Hijum and Pilarczyk (1982) analysed laboratory experiments on shingle transport and, using dimensional analysis, came up with an expression for littoral shingle drift. Their expression did contain a threshold term. Bailard (1987) derived an energetics based sediment transport formula which had the ratio of amplitude of orbital velocity to fall velocity of the sediment as an additional parameter. Kamphuis (1991) derived a longshore sediment transport formula based on dimensional analysis. The formula was calibrated using laboratory results obtained by Kamphuis and co-workers at Queens University. The drawback of these formulations is that they are not founded on sediment transport physics but on a relation between overall dissipation of wave energy and transport of sediments in a heuristic way.

Since the present formula (equations (28) to (32)) claims to be an improvement compared to the CERC- formula it is relevant to compare them. Using linear shallow-water wave theory and a constant breaking criterion the CERC-formula can be written as

$$Q_{CERC} = \frac{K_{CERC}}{16}\sqrt{\frac{g}{\gamma_b}}\frac{H_b^{\frac{5}{2}}\sin(2\alpha_b)}{(s-1)} \tag{33}$$

in which Q_{CERC} is the longshore sediment transport rate in solid volume per second. The constant K_{CERC} is equal to 0.77 for regular waves and for irregular waves with $H = H_{rms}$. Figure 2 shows a comparison between the analytical model (equations (28) to (32)) and the CERC formula (equation (33)). Figures 2(a) to

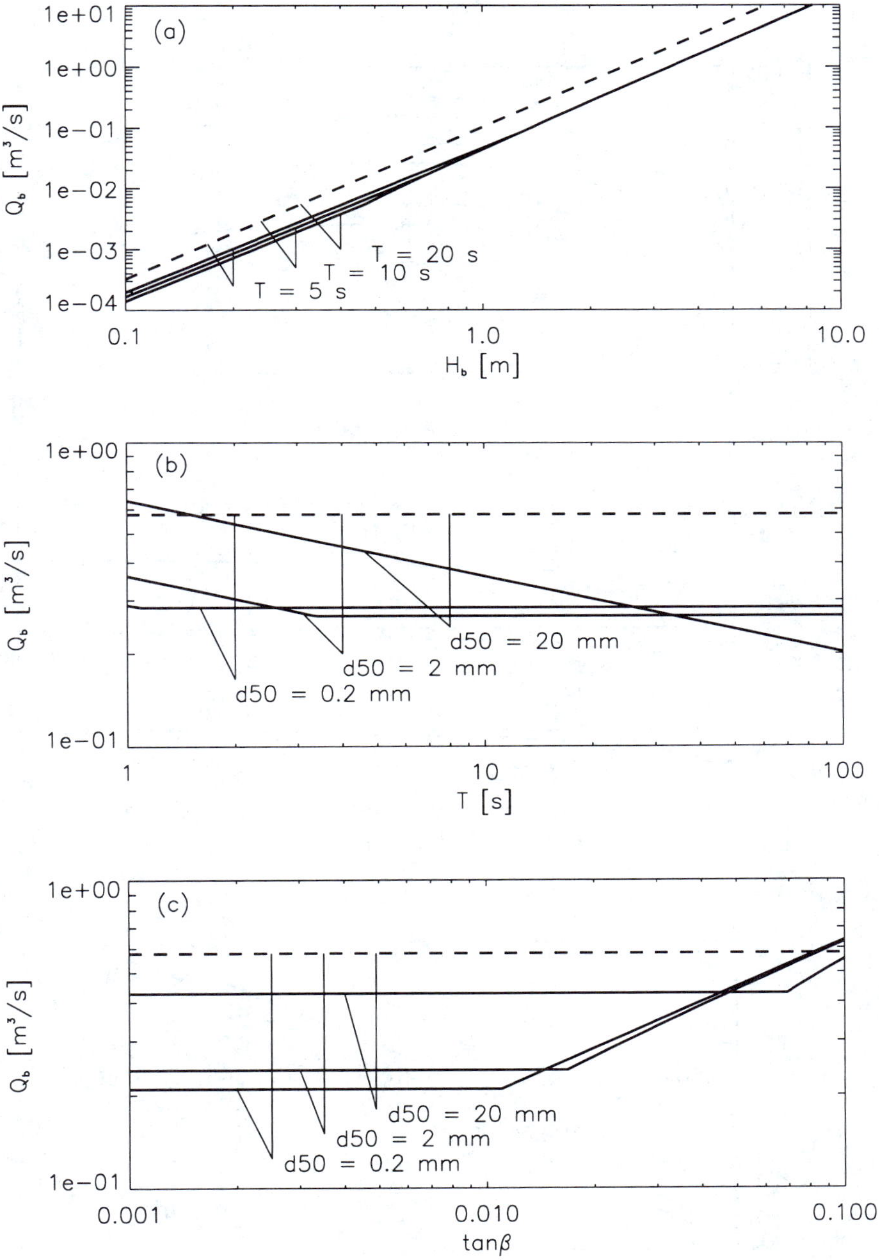

Figure 2: Comparison between the CERC formula (dashed line) and the analytical model (solid lines).

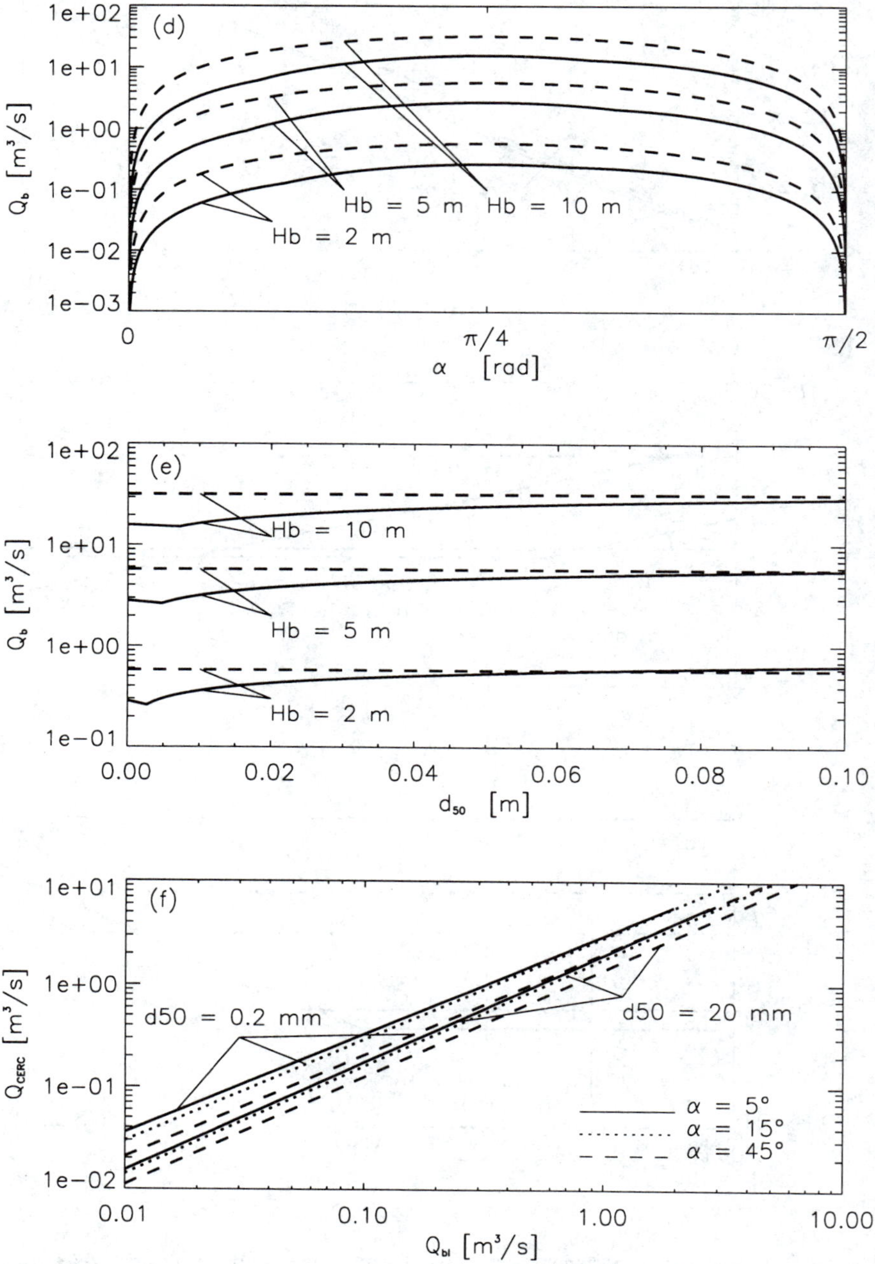
(d)
1e+02
1e+01
1e+00
1e−01
1e−02
1e−03
Q_b [m³/s]
Hb = 5 m
Hb = 10 m
Hb = 2 m
0
π/4
π/2
α [rad]
(e)
1e+02
1e+01
1e+00
1e−01
Q_b [m³/s]
Hb = 10 m
Hb = 5 m
Hb = 2 m
0.00
0.02
0.04
0.06
0.08
0.10
d50 [m]
(f)
1e+01
1e+00
1e−01
1e−02
Q_CERC [m³/s]
d50 = 0.2 mm
d50 = 20 mm
α = 5°
α = 15°
α = 45°
0.01
0.10
1.00
10.00
Q_bl [m³/s]

2(e) depicts the respective values of the littoral drift for the parameter range of the governing variables. The basic values of the parameters are: $H_b = 2.0m$, $T = 5s$, $d_{50} = 2.0mm$, $tan\beta = 1/50$, $\alpha_b = 45$ and $\theta_{cr} = 0.05$. Each plot uses mainly the basic values, and varies the values of two parameters over a range. The parameter ranges are supposed to represent realistic values plus a bit more at each end of the range. Figure 2(f) shows a direct comparison between the analytical model and the CERC formula. The new method follows a broadly similar pattern to the CERC formula, but differs by up to a factor of 3 (generally smaller), depending on the value of T, d_{50}, $\tan\beta$, and α.

One of the shortcomings of the CERC formula is its lack of a grain size dependency. The sediment transport rate might actually not be to sensitive to modest changes in grain size but since the grain size on natural beaches can easily vary by two orders of magnitude (from fine sand $\simeq 100\mu m$ to gravel $\simeq 10mm$) the grain size dependency can be important. Thus, when it is taken into account that the CERC formula was calibrated using data from sandy beaches with values of d_{50} less than $500\mu m$, it is not surprising that it does not perform well for large grain sizes. For use on shingle or gravel beaches it is customary to multiply the CERC formula with a factor $\simeq \frac{1}{18}$ (see Brampton and Motyka, 1984) however, it is not clear what the procedure is for grain sizes in the intermediate range and this highlights one of the shortcomings of the CERC formula *viz.* the CERC-constant is not constant throughout the range of naturally existent grain sizes.

4 Comparison with data.

In order to quantify the deviation caused by the hydrodynamic assumptions employed, the results of the analytical formula were compared to the results of a numerical profile model using a state-of-the-art hydrodynamic module to propagate the waves inshore and calculate the bed shear-stress distribution (Southgate and Nairn, 1993). Bed-load transport was predicted by the formula of Soulsby (1996) given by equations (4) to (9) at a number of grid points across the profile and then cross-shore integrated numerically. The field data were provided by an extensive field observation programme of shingle beach re-nourishment at Seaford Beach on the south coast of England including more than three years of wave and sediment data. The wave data were obtained by a 0 located offshore at 10 m depth. Data were sampled every third hour and from the spectra the significant wave height, $H_s = \sqrt{2}H_{rms}$, the peak period, T_p, and the direction, α, were derived. For the analytical model and the CERC formula, the waves were propagated to the breaking point using Snel's law. Shingle transport volumes were determined through analysis of the beach profile surveys taken every 2-3 months during the period in question, at transects positioned along the beach at approximately $100m$ intervals. The results of the comparison are shown in figure 3, for the analytical model (equations (28) to (32), the numerical model and the CERC formula. The full numerical model in conjunction with the bed-load trans-

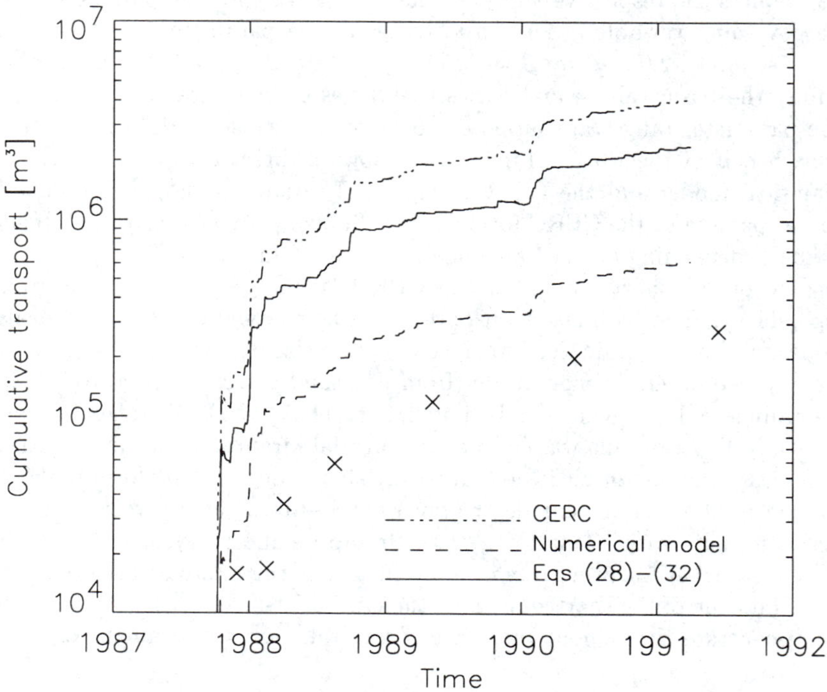

Figure 3: Comparison between the CERC formula, the numerical model, and the analytical model (equations (28) to (32)) for the Seaford data.

port formula gave relatively good results: it predicted the littoral drift within a factor 3, whereas the analytical model consistently over predicted by a factor 12, and the CERC formula over predicted by a factor 20. The fact that the hydrodynamically more advanced numerical model performed better was taken as an expression of the relevance of the sediment transport formulation and the factor 12 was interpreted as a quantitative representation of the errors made in applying the before mentioned hydrodynamic assumptions to the analytical model. The formula has therefore been corrected through dividing by this constant factor. Thus the corrected analytical formula gives the longshore transport rate, $Q_{b,l}$, as $Q_{b,l} = \frac{1}{12}Q'_{b,l}$ with $Q'_{b,l}$ given by equations (28) to (32).

The existence of a threshold term is very important particularly for larger grain sizes. The study of the Seaford data showed that a decrease in θ_{cr} from 0.05 to zero, increased the shingle transport by a factor 3.

The corrected analytical formula has been applied to a number of shingle field and laboratory data sets: Van Hijum and Pilarczyk (1982) did laboratory experiments at Delft Hydraulics (DH) with shingle transport under the action

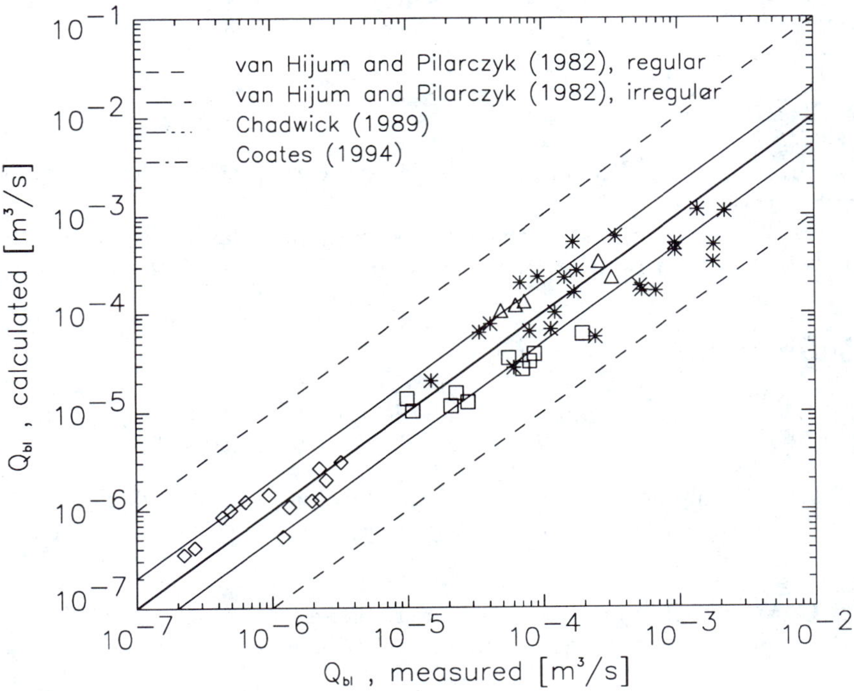

Figure 4: Comparison of the corrected analytical formula with existing field and laboratory data.

of regular and irregular waves, Chadwick (1989) collected field data of longshore shingle transport near Brighton, UK, and Coates (1994) did physical model studies of shingle transport at HR Wallingford. The results are shown in figure 4: Most transport rates are predicted within a factor 2 and all data sets are predicted within a factor 10.

5 Conclusions.

An analytical longshore bed-load transport formula has been developed from a bed-load transport formula developed at HR. Thus, the analytical formula is based on sediment transport physics and has been verified against a state-of-the-art numerical model and field data. This indicated the need to introduce a correction factor arising from grossly simplified assumptions about the hydrodynamics in the analytical model.

A major improvement on the CERC formula is dependencies on the grain size, beach slope and wave period, and the inclusion of a threshold term, which has proven to be very important for the quantitative prediction of littoral shingle drift.

Acknowledgements.

The UK Ministry of Agriculture, Fisheries and Food provided the financial support for this work.

The authors would like to thank Stuart Stripling, Marine Sediments Group, HR Wallingford for running the numerical model.

References

[1] J.A. Bailard. A simplified model for longshore sediment transport. In *Proc. 19th Int. Conf. Coastal Eng.*, pages 1454–1470, Houston, 1984.

[2] A.H. Brampton and J.M. Motyka. Modelling the plan shape of shingle beaches. In *POLYMODEL7 Conference*, pages 219–233, Sunderland Polytechnic, 1984.

[3] Coastal Engineering Research Center. Shore protection manual. Technical report, U.S.Army Corps of Engineers, Washington, 1977.

[4] A.J. Chadwick. Measurements and numerical model verification of coastal shingle transport. In Palmer M.H., editor, *Advances in Water Modelling and Measurement.* British Hydromechanics Research Association, Cranfield, 1989.

[5] T.T. Coates. Effectiveness of control structures on shingle beaches. Report SR 387, HR Wallingford, Wallingford, 1994.

[6] J.S. Damgaard, S. Stripling, and R.L. Soulsby. Numerical modelling of coastal shingle transport. Report TR 4, HR Wallingford, Wallingford, 1996.

[7] M.N. Greer and O.S. Madsen. Longshore sediment transport data: A review. In *Proc. 16th Int. Conf. Coastal Eng.*, pages 1563–1576, Hamburg, 1978.

[8] J.W. Kamphuis. Alongshore sediment transport rate. *J. Waterway, Port, Coastal and Ocean Eng.*, 117(6):624–640, 1991.

[9] R.R. Simons, A.J. Grass, W.M. Saleh, and M.M. Tehrani. Bottom shear stress under random waves with a current superimposed. In *Proc. 24th Int. Conf. Coastal Eng.*, pages 565–578, Kobe, 1994.

[10] R.L. Soulsby. Manual of marine sands. Report SR 466, HR Wallingford, Wallingford, 1996.

[11] H.N. Southgate and R.B. Nairn. Deterministic profile modeling of nearshore processes. *Coastal Eng.*, 19:27–56, 1993.

[12] E. van Hijum and K.W.Pilarczyk. Equilibrium profile and longshore transport of coarse material under regular and irregular wave attack. Publication no. 274, Delft Hydraulics, Delft, 1982.

[13] K.C. Wilson. Friction of wave induced sheet flow. *Coastal Eng.*, 12:371–379, 1989.

CHAPTER 280

Design Capacity of a Longshore Current Recirculation System for a Longshore Sediment Transport Laboratory Facility

David G. Hamilton[1], Julie D. Rosati[2], Jimmy E. Fowler[2], and Jane M. Smith[2]

ABSTRACT

A new longshore sediment transport laboratory facility for conducting three-dimensional moveable-bed experiments is being developed by the U.S. Army Engineer Waterways Experiment Station, Coastal and Hydraulics Laboratory. This paper focuses on the problem of designing the pumping capacity for an external longshore current recirculation system, designed to minimize adverse laboratory effects created by the updrift and downdrift lateral boundaries of the new facility. A review of longshore current recirculation systems used in other laboratory facilities is presented. The numerical model NMLONG is used to predict the magnitude and cross-shore distribution of the wave-driven longshore current that will be generated during experiments in the new facility. This paper concludes with a performance curve that defines the required pumping capacity of each of the individual pumps in the longshore current recirculation system.

1. INTRODUCTION

In performing its mission to maintain navigable waterways along U.S. coasts, the U.S. Army Corps of Engineers (USACE) regularly applies analytical and numerical models to estimate the total longshore sediment transport (LST) rate. Accurate prediction of LST rate is essential when predicting beach response in the vicinity of coastal structures, designing artificial beach nourishment projects, and approximating sedimentation rates in navigation channels. For design applications with adequate field measurements, the commonly used CERC formula (Shore Protection Manual, 1984) can be calibrated and applied to estimate total LST rates with reasonable confidence. However, for design applications without calibration data, the CERC formula provides only order-of-magnitude accuracy.

The present work is part of a research program intended to improve the USACE's capabilities to predict local and total LST rates and to evaluate errors associated with these predictions. The first goal of this research is to develop a world-class Longshore Sediment Transport Facility (LSTF) for conducting three-dimensional moveable-bed experiments. The LSTF will simulate nearshore hydrodynamic and sediment transport processes at a relatively large geometric scale. Further information on the objectives of this research program and general planning and design considerations for this new laboratory facility can be found in Rosati et al. (1995).

---

1 Coastal Engineer, C5-310 Cain Ridge Rd., Vicksburg, Mississippi, USA, 39180,
Tel: 601 634-3029, Fax: 601 634-4314, Email: D.Hamilton@CERC.WES.ARMY.MIL

2 Research Hydraulic Engineer, USAE Waterways Experiment Station, Coastal and Hydraulic Laboratory, 3909 Halls Ferry Road, Vicksburg, Mississippi, USA, 39180-6199

This paper focuses on the problem of calculating the required pumping capacity of a longshore current (LSC) recirculation system being designed to minimize the adverse laboratory effects created by the updrift and downdrift lateral boundaries of the LSTF.

The following sections review other LSC recirculation systems, provide a brief description of the LSTF, and discuss results obtained using the numerical model NMLONG to estimate the magnitude and cross-shore distribution of LSC that will be generated in the LSTF. Subsequent sections discuss limitations of the numerical analysis and other design parameters required to determine the design capacity of the LSC recirculation system. The final section provides a performance curve that defines the required pumping capacity for each of the individual pumps in the LSC recirculation system.

2. REVIEW OF OTHER LSC RECIRCULATION SYSTEMS

Early longshore current laboratory experiments were conducted without a LSC recirculation system that would have minimized the adverse laboratory effects created by the updrift and downdrift boundaries of each facility. Putman, Munk and Taylor (1949) used the sidewalls of the tank as wave guides to train the waves onto the beach. The LSC generated on the beach became part of the internal circulation pattern in the wave basin, which is considerably different then the situation on a long straight coast. Brebner and Kamphuis (1963) terminated the wave guides near the break point allowing the LSC to passively enter and exit the beach. Galvin and Eagleson (1965) and Mizuguchi and Horikawa (1978) terminated the downdrift wave guide at the break point, and completely closed the updrift wave guide. Apparently the longshore flux, in both of these experiments, was entering the test area underneath the wave generator, which would cause a non-uniform LSC distribution along the beach. Galvin and Eagleson (1965) were the first to measure the cross-shore distribution of LSC at several transects along the beach. Their data revealed the adverse effects created by the updrift and downdrift boundaries of the facility, in that the LSC was not able to reach its equilibrium magnitude and cross-shore distribution.

Visser (1982) conducted what appears to be the most detailed LSC experiments conducted to date. The wave basin was relatively large with a beach length between the wave guides of approximately 20 m. Visser's experiments were the first to use a LSC recirculation system to recirculate water from the downdrift to the updrift end of the beach. For most of Visser's experiments, the cross-shore distribution at the updrift boundary was controlled using 12 flow channels and control gates. Each flow channel was 0.2 and 0.4 m wide for tests with beach slopes of 1:10 and 1:20, respectively. Visser found it impossible to optimize the recirculation procedure from measurements of the mean water level in the longshore direction. He developed an alternative method where the optimum rate of external recirculation was determined by minimizing the internal circulation between the wave guides.

Simons et al. (1995) described some of the evaluation tests conducted in the recently developed Coastal Research Facility at HR Wallingford, in the United Kingdom. This facility was constructed in a large wave basin (27 x 54 m) and has a 36-m long directional wave generator. The most impressive component of the facility is the LSC recirculation system that has 4 independent reversible axial flow pumps with a total pumping capacity of

1.2 m³/s. Currents are introduced into the basin at the updrift boundary through 40 flow channels, each controlled by its own undershoot weir, with a matching set of flow channels at the downdrift end of the facility. All of the flow channels are 0.5 m in width and the fixed-bed beach has a slope of 1:20. This system allows the wave-driven LSC to be externally recirculated to establish longshore uniformity of waves and wave-driven currents. Some technical specifications on the facility were found in HR Wallingford (1994).

3. PHYSICAL DESCRIPTION OF THE NEW LABORATORY FACILITY

The LSTF occupies a 30-m cross-shore by 50-m longshore by 1.4 m deep wave basin (Figure 1). Monochromatic and random waves can be generated with four digitally controlled, piston-type, servo-electric wave generators. The wave generators are synchronized to create unidirectional long-crested waves up to 0.5 m in height for wave periods up to 3.0 s, and can be oriented at various angles ranging from 0 to 20 deg with respect to shore normal. For oblique angles of wave attack, the wave generators are phase-shifted to maximize the cross-shore dimension of the testing area. End-baffles, located between two adjacent wave boards, are used to guide the waves and prevent the formation of spurious waves caused from wave diffraction from the ends of each wave board and from wave energy reflected from the wave absorber located behind the wave generators. Wave guides that follow the refracted wave ray will be designed to facilitate current and sediment input and output at the lateral boundaries of the beach. Wave, current, and bathymetric data will be collected using a suite of sensors located on a custom-designed instrumentation bridge, as described by Rosati et al. (1995).

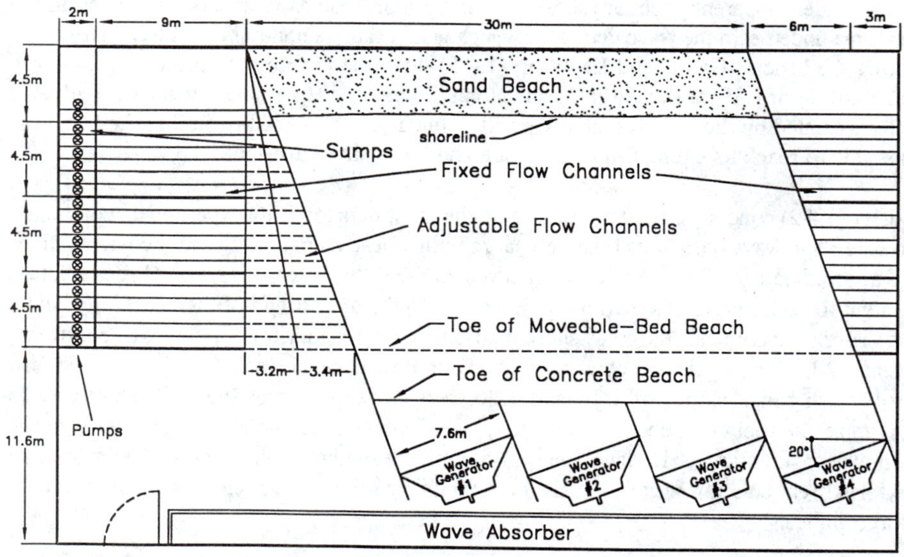

Figure 1: Plan View of the Longshore Sediment Transport Facility

The fixed-bed (concrete) beach has a longshore dimension of 30 m and a cross-shore dimension of 21 m, and was accurately constructed (tolerance of ± 2.0 mm) with parallel contours. The main section of the concrete beach has a constant slope of 1:30, and the toe of the beach slopes down to the basin floor at a slope of 1:18. During future LST experiments, a 0.3 m thick moveable-bed of quartz sand will be placed on top of the fixed-bed beach and will extend 18.0 m offshore (see toe of moveable-bed beach in Figure 1). The toe of the beach will taper down to the basin floor at a slope of 1:6.5.

A LSC recirculation system is being designed to minimize the adverse laboratory effects resulting from the updrift and downdrift boundaries, and to maximize the length of beach over which longshore uniformity of waves and wave-driven currents exist in the facility. A LSC recirculation system creates a closed-loop system that continuously recirculates LSC from the downdrift to the updrift lateral boundary of the beach, while waves are being generated. As shown in Figure 1, the LSC recirculation system will extend 18.0 m offshore corresponding to the offshore limit of the moveable-bed beach which will be constructed for future LST experiments. A series of pumps will be located at the downdrift end of the facility and will recirculate the wave-driven LSC back to the updrift end of the facility through a series of pipes.

4. NUMERICAL SIMULATIONS OF LSC

The first part of this section provides a description of the numerical model used to estimate the magnitude and cross-shore distribution of the longshore current that will be generated in the LSTF. Subsequent sections describe the setup of the numerical model, validate the results obtained from the numerical model with dye measurements in the LSTF, and show the influence of wave height, period and direction on the estimated magnitude and cross-shore distribution of the longshore current.

4.1 Description of the Numerical Model

Kraus and Larson's (1991) numerical model NMLONG (Numerical Model of the LONGshore current) was used to estimate the wave-driven longshore current in the LSTF. NMLONG is a PC-based model that calculates wave transformation, mean water surface elevation, and longshore current for a 1-D cross-section of beach. The major assumptions in NMLONG are longshore homogeneity and linear wave theory.

Wave transformation in NMLONG includes shoaling, refraction, breaking with energy dissipation, and wave reformation. Random wave transformation is simulated by assuming a Rayleigh distribution of wave heights offshore (waves are assumed to be narrow banded in frequency and direction, so randomness enters only in variability in wave height). NMLONG randomly selects wave heights from the Rayleigh distribution and transforms them as individual waves. The individual wave heights are used to calculate root-mean-square wave height at each point across the profile. This approach neglects wave-wave interactions. The mean water surface elevation (setup and setdown) produced by waves is calculated from the cross-shore momentum balance. The wave forcing (radiation stress) is calculated from linear wave theory.

Wave-driven longshore current is calculated from the longshore momentum balance. The model includes wave forcing (gradients in radiation stress), lateral mixing, and bottom friction. Lateral mixing is modeled with an eddy viscosity approach. Bottom friction is nonlinear (in the unknown current velocity) and requires time-averaging of nonlinear terms over the wave period. The offshore boundary condition for the current is zero wave-driven velocity. For random wave simulation, longshore currents from individual waves are averaged at each profile point.

Using the data from Visser (1982), Kraus and Larson (1991) illustrated that NMLONG can be calibrated to reproduce the LSC measured in the laboratory with reasonable success. In particular, the magnitude and cross-shore location of the peak of the LSC distribution was reproduced with reasonable accuracy. This suggests that NMLONG can be used to provide a reasonable estimate of the magnitude and cross-shore distribution of LSC that can be generated in the LSTF. However, in the Visser cases, NMLONG predicted that the magnitude of the offshore tail of the LSC distribution was higher than measured by Visser, even after the numerical model had been calibrated.

4.2 Numerical Model Setup

NMLONG requires the following input parameters: offshore wave height, period, and direction, specification of regular or random waves, offshore water depth, and beach profile elevation relative to mean water level. Random waves are characterized in NMLONG using the root-mean-square wave height, H_{rms}. Values of H_{rms} were converted to significant wave height, H_s assuming $H_s = 1.414 \times H_{rms}$. Nonlinear bottom friction with a friction coefficient equal to 0.01 was used in the LSTF simulations. For one wave condition, the sensitivity of results to this value was evaluated by reducing and increasing the coefficient to 0.005 and 0.02, respectively. All other empirical parameters were set to the default values: incipient breaking-wave-height-to-water-depth-ratio equal to 0.8, stable wave-height-to-water-depth-ratio equal to 0.4, energy flux dissipation rate equal to 0.15, and lateral mixing coefficient equal to 0.3.

NMLONG was set up to calculate the depth-averaged LSC velocities at each cell spaced 1 m across the beach profile. These depth-averaged velocities were multiplied by the corresponding mean water depth and integrated across the entire profile to estimate the total longshore volume flux for each wave condition.

Two series of numerical simulations were conducted to represent two different experimental configurations planned for the LSTF. The first series was conducted with an offshore water depth equal to 0.6 m and represents the configuration that will be used in the LSTF during the hydrodynamic testing phase. The second series was conducted with a water depth equal to 0.9 m and represents the configuration that will be used for the moveable-bed LST experiments. In this configuration a 0.3 m thick uniform layer of sand will be placed on top of the existing concrete beach. Increasing the water level from 0.6 to 0.9 m translates the shoreline directly upwards (i.e., there is no cross-shore translation of the nearshore zone). However, the resulting magnitude and cross-shore distribution of the LSC is slightly different for the two different configurations, due to the slight difference in wave transformation caused by the steeper toe of the beach in the case with the moveable-bed.

Numerical simulations in both test series covered the following range of hydrodynamic conditions: significant wave height, H_s = 0.1 to 0.4 m, peak wave period, T_p = 1.0 to 2.5 s, and angle of wave incidence, θ = 5 to 20 deg relative to shore normal. For these test conditions total longshore volume flux ranged from 0.04 m^3/s for H_s = 0.1 m, T_p = 2.5 s, and θ = 20 deg to 1.3 m^3/s for H_s = 0.4 m, T_p = 2.5 s, and θ = 20 deg.

The sensitivity of the bottom friction coefficient was evaluated for a relatively energetic wave condition; namely H_s = 0.3 m, T_p = 2.5 s and θ = 20 deg. The bottom friction coefficient was varied over a range of two times the default value of 0.01 (i.e., from 0.005 to 0.02). Results obtained using the lower friction coefficient (0.005) resulted in a total longshore volume flux of 1.1 and 1.2 m^3/s for the 0.6 and 0.9 m water depth cases, respectively. The higher friction coefficient (0.02) reduced the total longshore volume flux to 0.4 m^3/s for both water depth cases. Therefore, the results obtained using NMLONG to predict the total longshore volume flux in the LSTF are sensitive to the value of the bottom friction coefficient.

4.3 Validation with Dye Measurements in the LSTF

Preliminary estimates of the magnitude of the LSC were obtained in the LSTF by injecting dye into the surf zone while generating monochromatic waves. These measurements were then used to assess the validity of the numerical simulations. However, the following facility components had not been installed in the LSTF at the time these measurements were required: (1) lateral wave guides were not available to train the incident waves from the wave generators to the surf zone, (2) none of the LSC recirculation system components were available to minimize the laboratory effects caused by the lateral boundaries of the facility (i.e., pumps and flow channels), and (3) only monochromatic waves could be generated with the wave makers, As a result, strong adverse reflection patterns (from the vertical walls at the two ends of the basin) and circulation cells developed throughout the facility as the first 5-10 waves were generated. In addition, no flow velocity measurement sensors were available to accurately measure the wave driven LSC in the surf zone.

Three conclusions were made based on these limited dye measurements. First, NMLONG produced reasonable estimates of the peak LSC magnitude and cross-shore location, using the default bottom friction coefficient of 0.01. However, insufficient data were collected to calibrate this coefficient. Secondly, we were unable to verify the cross-shore distribution of the LSC predicted using NMLONG due to the physical limitations and resulting adverse laboratory effects discussed above. Therefore, the default lateral mixing coefficient (0.30) was used. Thirdly, observations made while these dye measurements were being carried out strongly reinforced the expectation that a properly designed external LSC recirculation system would be required to maintain longshore uniformity of waves and wave-driven currents in the facility, especially for energetic wave conditions.

4.4 Influence of Hydrodynamic Variables

The influence of wave height, period and direction on the magnitude and cross-shore distribution of the LSC was investigated. All numerical simulations were conducted using random waves. Figure 2 shows the LSC distribution for H_s = 0.2, 0.3, and 0.4 m, with T_p = 2.5 s and θ = 20 deg at the wave generator. Three general trends can be seen in this figure. First, as H_s increases, the magnitude of the LSC at the peak of the distribution significantly increases. The magnitude of the peak LSC equals 0.21, 0.28 and 0.36 m/s for H_s = 0.2, 0.3 and 0.4 m, respectively. This is a relative increase of approximately 30% for each 0.1 m increase in wave height. Secondly, the peak of the LSC distribution moves offshore as H_s increases, because the incident waves begin to break further offshore. Thirdly, the width of the LSC distribution increases as the H_s increases, since the width of the surf zone increases.

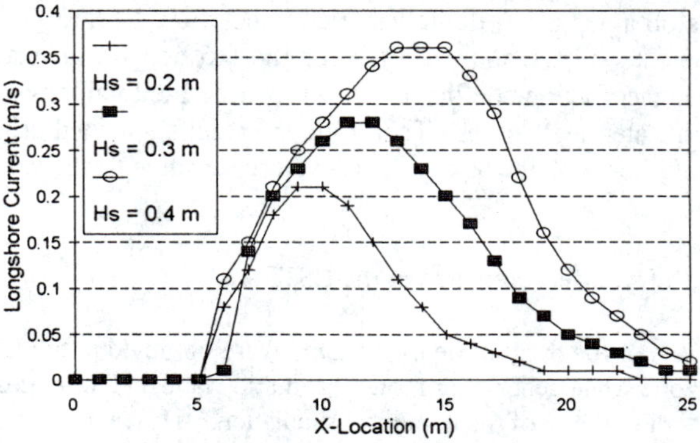

Figure 2: Influence of Significant Wave Height

Figure 3 shows the LSC distribution for T_p = 1.0, 1.5, 2.0 and 2.5 s with H_s = 0.3 m and θ = 20 deg at the wave generator. Three general trends can be seen in this figure. First, as T_p increases, the magnitude of the LSC at the peak of the distribution increases slightly. The magnitude of the peak LSC equals 0.25, 0.26, 0.27 and 0.28 m/s for T_p = 1.0, 1.5, 2.0 and 2.5 s, respectively. This is a relative increase of only 4% for each 0.5 s increase in T_p. Secondly, the peak of the LSC distribution moves slightly further offshore as T_p increases, since the incident waves begin to break slightly further offshore. Thirdly, the width of the LSC distribution increases slightly as T_p increases, since the width of the surf-zone increases slightly.

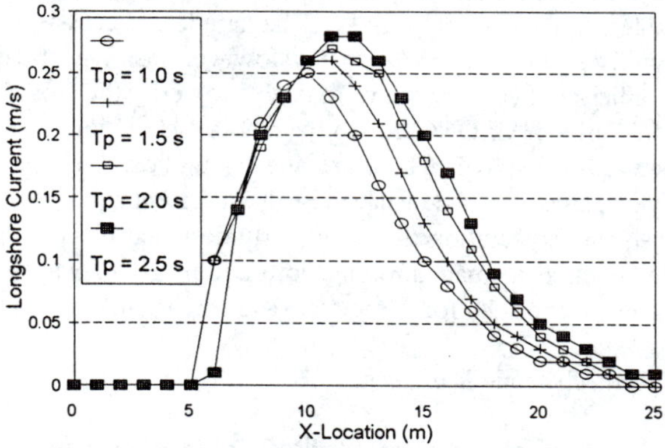

Figure 3: Influence of Peak Wave Period

Figure 4 shows the LSC distribution for θ = 5, 10, 15 and 20 deg with H_s = 0.3 m and T_p = 2.5 s. Three general trends can be seen in this figure. First, as θ increases, the magnitude of the LSC at the peak of the distribution significantly increases. The magnitude of the peak LSC equals 0.10, 0.18, 0.23 and 0.28 m/s for θ = 5, 10, 15 and 20 deg, respectively. This is a relative increase of about 80%, 28%, and 22%, for each consecutive increase in θ. Secondly, the cross-shore location of the peak of the LSC distribution remains constant as θ increases. Thirdly, the width of the LSC distribution remains reasonably constant as θ increases.

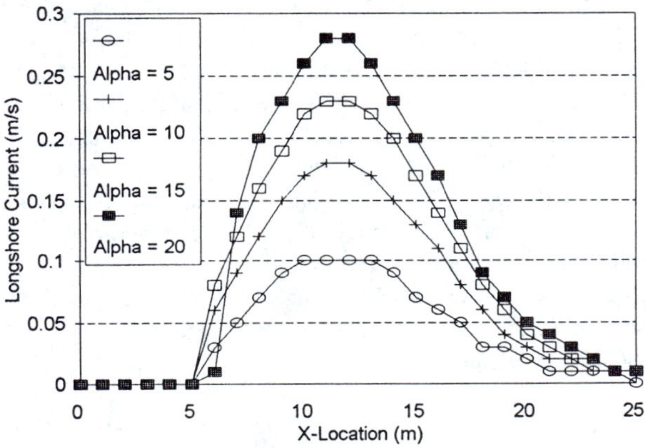

Figure 4: Influence of Incident Wave Direction

In summary, increasing H_s has a strong influence on increasing the magnitude and cross-shore location of the peak of the LSC distribution, whereas, increasing T_p has a much milder effect. Increasing θ has a very strong influence on increasing the peak magnitude, but no influence on the cross-shore location of the peak of the LSC distribution.

5. DESIGN PARAMETERS FOR THE LSC RECIRCULATION SYSTEM

The first part of this section describes the design wave condition used to determine the required pumping capacity of the LSC recirculation system. Subsequent sections discuss a factor of safety applied to the numerical modeling results, and incorporate the requirements of a moveable-bed beach and a variable operating water level into the design.

5.1 Design Wave Condition

To determine the design capacity of a LSC recirculation for the LSTF, a maximum design wave condition was selected. After observing the location of the break point for a number of different wave height and period combinations in the LSTF (and comparing these results with the numerical simulations), it was decided that H_s = 0.3 m and T_p = 2.5 s is probably the most energetic wave condition that will be generated during the moveable-bed LST experiments. As mentioned previously, the maximum angle of wave incidence (measured at the wave generators) to be tested in the LSTF is 20 deg. Therefore, the "design wave condition" used to determine the maximum required pumping capacity of the LSC recirculation system was characterized by $H_s = 0.30$ m, $T_p = 2.5$ s and $\theta = 20$ deg with an offshore water depth of 0.9 m. This is the LSC distribution shown previously in Figure 2 (middle curve) and Figure 3 and 4 (upper curve). Results obtained using NMLONG to predict the magnitude and cross-shore distribution of LSC for this "design wave condition" indicated a total longshore flux of 0.76 m^3/s within the surf zone.

5.2 Determining a Factor of Safety

A factor of safety was incorporated into the design to compensate for inaccuracies in calculating the magnitude and cross-shore distribution of LSC that will be generated in the LSTF. Our concerns included: (1) having inadequate data to calibrate the bottom friction coefficient and the lateral mixing coefficient in the numerical model, and (2) changes in the bottom roughness between the initial hydrodynamic tests on a fixed-bed concrete beach and multiple grain sizes used for the moveable-bed beach for the LST experiments. The surface of the concrete beach was broom-finished to simulate the roughness of a course grained sand. However, because bed-forms will develop in the LST experiments it is likely that the moveable-bed beach will have a higher friction coefficient for a given wave condition. Due to these uncertainties, the magnitude of the LSC at each cross-shore location was increased by 10% (i.e., a Factor of Safety = 1.1) for the purpose of determining the required pumping capacity of the LSC recirculation system.

5.3 Allowance for Maximum Depth of Erosion of the Moveable-Bed Beach

At this point in the design process, predictions of the LSC magnitude and cross-shore distribution to be generated in the LSTF are based on a 1:30 plane sloping fixed-bed beach having parallel contours. However, for future moveable-bed experiments in the LSTF, the beach profile will respond to incident wave conditions, creating a shore-parallel bar and trough feature near the location of initial wave breaking. Assuming longshore uniformity in the beach bathymetry, the capacity of the LSC recirculation system must be increased wherever the beach profile erodes, due to the increased longshore flux at the cross-shore location. For example, in the trough located shoreward of the offshore bar.

To investigate this problem, 2-D flume tests were conducted to estimate the maximum depth of erosion during future moveable-bed LST experiments in the LSTF. Each test started with a 1:30 plane sloping sand beach, with $D_{50} = 0.11$ mm. This is the same initial beach slope and sediment size to be used for the fine-grained sediment test series in the LSTF. In each experiment, the maximum depth of erosion was located just offshore of the still-water shoreline. For the design wave condition the maximum depth of erosion did not exceed 0.1 m, however, this conclusion is based on 2-D flume tests, and not 3-D tests in the LSTF.

For design purposes, it was assumed that the entire 18-m width of the moveable-bed beach will erode by 0.1 m. This is a fairly crude assumption, although it is conservative in that it allows for the maximum scour depth to occur at any location across the beach profile. In particular, this is a very conservative assumption near the offshore end of the beach where accretion should occur (as opposed to erosion) as the offshore bar develops. However, future plans for the LSTF include a test with a terminal groyne located at the downdrift end of the moveable-bed beach. If this physical model behaves as expected, a scour hole will develop at the head of the groyne and the longshore flux (shoreward of the offshore end of the groyne) will be deflected offshore and bypass the end of the groyne. Therefore, over designing the LSC recirculation system near the offshore region of the beach will give the system more flexibility in the future when conducting experiments with coastal structures.

Note that this analysis assumes that the depth-averaged LSC at any cross-shore location does not change substantially (from the case with a plane sloping fixed-bed beach) as the offshore bar and trough feature develop in the moveable-bed experiments. Although this may not be the case, we feel that changes to the depth-averaged LSC (as the beach profile develops) will be relatively small since the initial beach slope is relatively gentle (1:30) and the total profile adjustment should be relatively small. Therefore, this assumption should be adequate for design purposes.

5.4 Allowance for Variable Operating Water Levels in the Facility

The design of the LSC recirculation system also accounted for the requirement to conduct experiments with variable water levels in the facility. As mentioned previously, we plan to conduct the majority of the moveable-bed LST experiments with an offshore water depth of 0.9 m. However, if we decide to increase the water level in the future, the LSC distribution will be translated shoreward on the beach.

A number of physical constraints in the vertical dimension of the facility dictate that the maximum operating water level in the facility will not exceed 1.0 m (at least for energetic wave conditions). Lower water levels can be used, however, the wave generation capability of the wave makers will decrease with decreasing water level. Therefore, the critical design parameter is the maximum operating water level in the facility. The first curve in Figure 5 shows the longshore flux distribution for the design wave condition, with an offshore water depth of 0.9 m. This curve includes the increased capacity required by using a factor of safety of 1.1 and assuming the moveable-bed erodes 0.1 m across the entire width of the beach. The abrupt change in cross-shore gradient at X = 18 and 21 m is caused by the sudden change in beach slope at these locations. The second curve was obtained by increasing the offshore water depth to 1.0 m, for the same design wave condition. Essentially, this 0.1 m increase in operating water level translates the LSC distribution 3.0 m shoreward, because the slope of the concrete beach is 1:30. The third curve in the figure encompasses both the first and second curve, and represents the required pumping capacity of the LSC recirculation system for water levels ranging from 0.9 to 1.0 m. As shown in the figure, incorporating a variable water level into the design essentially broadens the width of the LSC distribution that needs to be recirculated within the facility.

6. OTHER DESIGN CONSIDERATIONS

The following sections describe uncertainties related to how far the LSC distribution extends offshore, the magnitude of the longshore flux of water flowing in the swash zone, and the issue of what percentage of the wave-driven LSC needs to be recirculated.

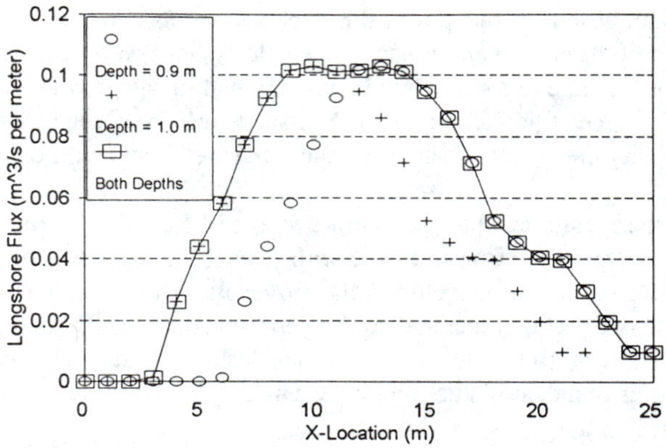

Figure 5: Influence of Offshore Water Depth

6.1 Offshore Tail of the LSC Distribution

Another design consideration is the possibility that a small longshore flux may exist offshore of the toe of the moveable-bed beach (X > 18.0 m), for very energetic wave conditions. Approximately 17% of the area under the curve in Figure 5 is offshore of X = 18 m, which is the location of the offshore end of the recirculation system, as shown in Figure 1. In Section 4.1, it was discussed that NMLONG may over predict how far the LSC distribution extends offshore. Unfortunately, this was impossible to verify due to problems encountered making dye measurements in the partially constructed LSTF, as discussed in Section 4.3. If the numerical simulations are correct, this offshore tail of the LSC distribution may cause adverse internal circulation patterns within the facility.

If this problem exists in the future, there are a number of possible solutions. First, we could decrease the magnitude of the maximum wave condition that is generated in the facility. Second, we could increase the offshore water depth to 1.0 m for very energetic wave conditions, which would shift the LSC distribution 3.0 m shoreward. The area under the curve (offshore of X = 18 m) associated with an offshore water depth of 1.0 m, is only 9% of the total area. Third, it may be possible to oversize the furthest offshore pump, and use the additional capacity to recirculate some of the LSC which exists offshore of X = 18 m. Based on these considerations, it was decided that the LSC recirculation system would initially be constructed to extend offshore to X = 18.0 m. If the numerical predictions turn out to be correct, one or more of the options discussed above will be employed.

6.2 Longshore Flux of Water in the Swash Zone

Another design feature to be considered is the longshore flux of water flowing downdrift within the swash zone. The swash zone was defined as the area between the still-water shoreline and the maximum point of wave uprush, including any setup in the mean water level. Unfortunately, it is impossible to accurately calculate the longshore flux in the swash zone due to our limited understanding of swash zone processes. One of the interesting results obtained from the 2-D moveable-bed tests, discussed previously in Section 5.3, is that the swash zone was approximately 4 m in width for the design wave condition. Therefore, it is believed that it will be necessary to recirculate the longshore flux in the swash zone, due to the physical size of the swash zone that develops at this relatively large geometric scale.

Figure 5 indicates that at X = 5.0 m, the longshore flux of water will be approximately 0.045 m^3/s per meter width of beach. In addition, we know that the longshore flux of water at the landward boundary of the swash zone will be zero. Therefore, we can make a somewhat crude assumption and draw a straight line between these two points to represent the time-averaged longshore flux of water flowing downdrift in the swash zone. Although we expect the longshore flux of water to have a nonlinear distribution across the swash zone, this assumption provides a first-order estimate for design purposes.

6.3 Percentage of Longshore Flux to be Externally Recirculated

Intuitively one would expect that 100% of the longshore flux within the surf zone should be externally recirculated back to the updrift end of the facility with a LSC recirculation system. However, laboratory facilities are not ideal systems and in many cases an internal circulation pattern may exist, in which case it may not be necessary (or possible) to externally recirculate 100% of the wave-driven longshore flux without adding too much momentum into the system. Visser (1982) developed a method to obtain a uniform LSC distribution along the finite length of a straight beach in a 3-D wave basin. Visser determined the "proper" rate of external recirculation by adjusting the pumping rate and inflow distribution until the internal circulation between the wave guides was minimized. However, after reviewing Visser's work, Svendsen (1991) concluded that in some of Visser's experiments the internal circulation was still disturbingly large (as much as 40-50% of the wave-driven longshore flux within the surf zone), even though Visser's data indicate that he was able to establish a LSC distribution that was relatively uniform alongshore.

Based on Svendsen's conclusion and the principle of continuity, Visser's external recirculation flow rate must have been substantially less than the total longshore flux in the surf zone. This indicates that it may not be necessary to recirculate 100% of the longshore flux in the LSTF, to obtain longshore uniformity. This problem will be investigated in the LSTF during the initial hydrodynamic experiments on a fixed-bed beach. However, for the purpose of estimating the required pumping capacity of the LSC recirculation system, it was prudent to assume that 100% of the longshore flux will need to be externally recirculated.

7. DESIGN CAPACITY FOR EACH INDIVIDUAL PUMP

The last step in this phase of the design process was to determine the design capacity of each of the individual pumps that make up the LSC recirculation system. Required pump capacity is a function of the width of the flow channels at the downdrift end of the facility, which is directly related to the cross-shore resolution of the LSC recirculation system. Intuitively, one would expect that the higher the cross-shore resolution, the longer the "working beach region" where longshore uniformity of waves and wave-driven LSC's exist. After reviewing the design of LSC recirculation systems used in other laboratory facilities (see Section 2) it was decided that the flow channels at the downdrift and updrift end of the LSTF should be 0.75 m in width. This was a compromise between cross-shore resolution and the estimated cost of construction of the entire LSC recirculation system.

Knowing the required pumping capacity of the LSC recirculation system, shown in Figure 5 (third curve), and given a flow channel width of 0.75 m, the capacity of each individual pump could be readily calculated. Figure 6 shows a bar graph expressed in m^3/s for every 0.75 m of beach width, and the magnitude of each bar represents the required pumping capacity of each individual pump. This figure also includes the required pumping capacity to recirculate the longshore flux in the swash zone (see Section 6.2). Figure 6 shows 23 pump capacities. However, to reduce the number of pumps near the shoreline, where the longshore flux is relatively small, one large pump will be used as opposed to four smaller pumps, as illustrate in Figure 1. Therefore, there will be a total of 20 pumps in the system.

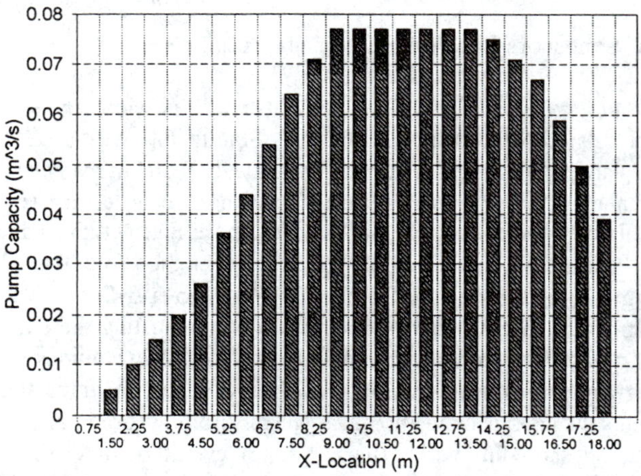

Figure 6: Design Capacity of the LSC Recirculation System

The total design capacity of the LSC recirculation system was calculated to be approximately 1.25 m^3/s. To put this pumping capacity into perspective, HR Wallingford (1994) states that the LSC recirculation system in the Coastal Research Facility has a design capacity of 1.2 m^3/s (i.e., four axial flow pumps each with a capacity of 0.3 m^3/s). The LSC recirculation system in the Coastal Research Facility is used to generate not only wave-driven LSC within the surf zone, but also tidal currents offshore of the toe of the concrete beach. Therefore, although the purpose of the two systems in not exactly the same, the total design pumping capacities are very similar.

8. SUMMARY

This paper documents the design process used to determine the required pumping capacity of an external longshore current recirculation system for the development of a longshore sediment transport laboratory facility. A design wave condition of $H_s = 0.3$ m, $T_p = 2.5$ s, and $\theta = 20$ deg, with an offshore water depth of 0.9 m was used to estimate the magnitude and cross-shore distribution of the longshore current that will be generated in the new facility. By including a factor of safety and considering a number of functional design requirements, the total required design pumping capacity was determined to be 1.25 m^3/s. It was estimated that twenty independent pumps and flow channels would be required to provide adequate cross-shore resolution for the system and to provide the required capability to recirculate a broad range of LSC distributions for a wide range of incident wave conditions, water levels and moveable-bed beach profiles.

9. ACKNOWLEDGEMENTS

The authors would like to thank the staff at HR Wallingford, in the United Kingdom for generously providing technical information about the Coastal Research Facility. This paper benefited from the review comments of Bruce A. Ebersole. This research is funded by the "Large-Scale Laboratory Investigation of Longshore Sediment Transport" work unit of the Coastal Sediment and Dredging Program, under the Coastal Navigation and Flood Damage Reduction Research Area. Permission to publish this paper was granted by the Chief of Engineers, U.S. Army Corps of Engineers.

10. REFERENCES

Brebner, A., and Kamphuis, J.W. 1963. Model Tests on the Relationship Between Deep-Water Wave Characteristics and Longshore Currents. Report No. 31, Department of Civil Engineering, Queen's University at Kingston, Canada.

Galvin, C.J., and Eagleson, P.S. 1965. Experimental Study of Longshore Currents on a Plane Beach. Technical Memorandum No. 10, US Army Waterways Experiment Station, Coastal Engineering Research Center, Vicksburg, MS, USA.

HR Wallingford (1994). Understanding the Nearshore Environment. A publication describing the Coastal Research Facility at HR Wallingford, 4 p.

Kraus, N.C., and Larson, M. 1991. NMLONG: Numerical Model for Simulating the Longshore Current, Report 1: Model Development and Tests. Technical Report DRP-91-1, US Army Engineer Waterways Experiment Stations, Vicksburg, MS.

Mizuguchi, M. And Horikawa, K. 1978. Experimental Study on Longshore Current Velocity Distribution, Bulletin 21, pp 123-150, Facul. Sci. & Eng., Chuo University.

Putnam, J.A., Munk, W.H., and Taylor, M.A., 1949. The Prediction of Longshore Currents. Transactions of the American Geophysical Union, Vol. 30 (3), pp 337- 345.

Rosati, J.D., Hamilton, D.G., Fowler, J.E. and Smith, J.M. 1995. Design of a Laboratory Facility for Longshore Sediment Transport Research. Proc. of Coastal Dynamics '95, pp 771-782, Gdansk, Poland.

Shore Protection Manual 1984. USACE Waterways Experiment Station, Coastal Engineering Research Center, Vicksburg, MS, 2 Vols.

Simons, R.R., Whitehouse, R. JS., MacIver, R.D., Pearson, J., Sayers, P.B., Zhao, Y., and Channell, A.R. 1995. Evaluation of the UK Coastal Research Facility. Proc. of Coastal Dynamics '95, pp 161-172, Gdansk, Poland.

Svendsen, I.A. 1991. Development of a Comprehensive Plan for Modeling Longshore Current Generation in the Laboratory. Unpublished Contract Report for the Waterways Experiment Station, Coastal Engineering Research Center, Vicksburg, MS.

Visser, P.J. 1982. The Proper Longshore Current in a Wave Basin. Report No. 82-1, Department of Civil Engineering, Delft University of Technology, The Netherlands.

CHAPTER 281

INFLUENCE OF NEARSHORE HARDBOTTOM ON REGIONAL SEDIMENT TRANSPORT

Paul C.-P. Lin, Ph.D., P.E.[1] and R. Harvey Sasso, P.E.[2]

ABSTRACT

The influence of nearshore hardbottom on longshore and cross-shore sand movement along a 21.2 km long southeast Florida coastline is evaluated. This study correlates a variation of longshore sediment transport rate within the region with the extent of nearshore hardbottom. The nearshore sand movement within this region was determined based on analysis of the hydrographic survey data over a 15-year period. The study area includes three major beach renourishment sites, which were renourished a combined total of seven (7) times with 5.5 million cubic meters of sand. The objectives of this study are to obtain a greater understanding of the causes of severe beach erosion and a more accurate estimate of nearshore sand movement in the region, thus providing a better design tool for future beach nourishment projects and inlet sand bypassing practices.

INTRODUCTION

The area of study extends from Port Everglades Inlet in Broward County to Bakers Haulover Inlet in Dade County. This region of coastline encompasses six (6) municipalities including, from north to south, John U. Lloyd, Dania, Hollywood/Hallandale, Golden Beach, Sunny Isles, and Bakers Haulover Park, with their major beach nourishment projects illustrated in Figure 1.

This study region is bounded by Port Everglades Inlet to the north and Bakers Haulover Inlet to the south. Port Everglades Inlet was opened in 1926 and stabilized by north and south jetties in 1931. Its 190-meter wide and 15 m deep inlet channel acted as a complete sand barrier and allowed no sediment bypassing

---

[1]Senior Engineer, Coastal Systems International, Inc., 464 South Dixie Highway, Coral Gables, Florida

[2]President, Coastal Systems International, Inc.

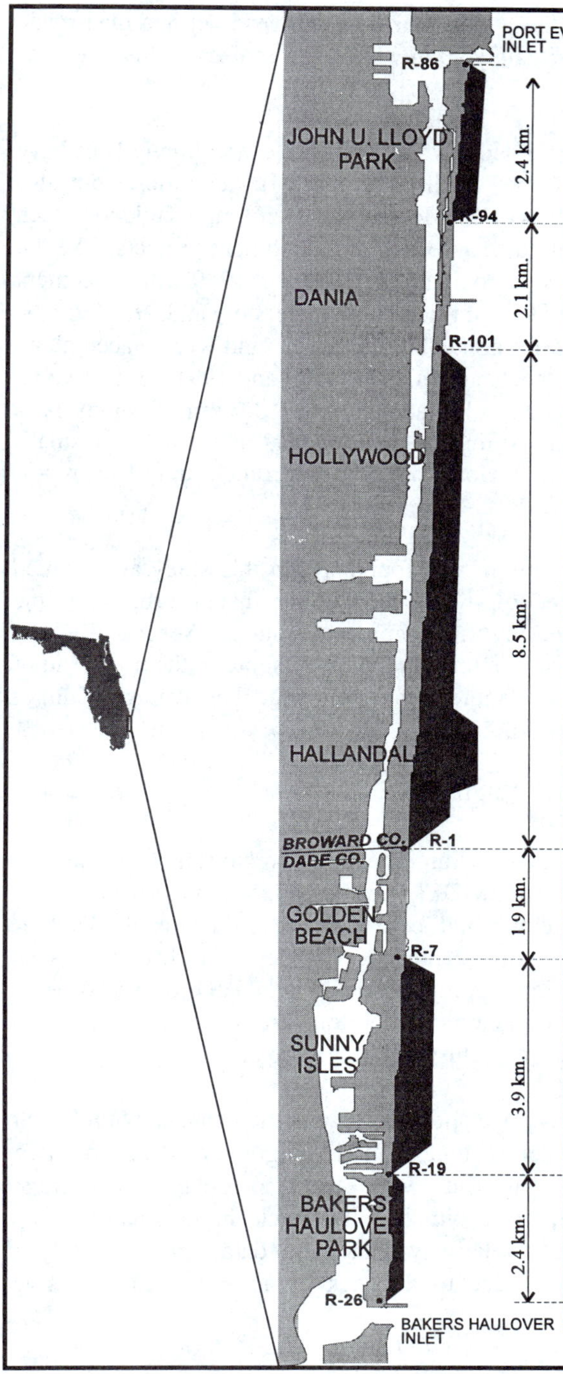

PROJECT	YEAR	VOLUME (Cubic Meters)
JOHN U. LLOYD PARK NOURISHMENT	1977 1989	833,000 461,000
HOLLYWOOD/ HALLANDALE BEACH NOURISHMENT	1971 1979 1991	268,000 1,514,000 847,000
SUNNY ISLES BEACH NOURISHMENT	1988 1990	1,009,000 23,000
HAULOVER BEACH NOURISHMENT	1960 1978 1987	138,000 230,000 180,000

Figure 1. Study Area and Historic Beach Nourishment Projects

across the inlet (Coastal Tech, 1994). The Bakers Haulover Inlet is a man-made improved inlet measuring approximately 90 m wide and 5 m deep. This Inlet was opened in 1925 (CP&E, 1992a).

In the region, John U. Lloyd, Hollywood/Hallandale, and Sunny Isles have been periodically renourished. Some small-scale beach fill activities were also conducted in the Bakers Haulover Park. Meanwhile, Dania and Golden Beach have been accreting as a result of adjacent beach renourishment projects. The 2.4 km beach of John U. Lloyd Park was nourished in 1977 and 1989 with placement of approximately 0.83 million and 0.46 million cubic meters of sand, respectively. An estimated 0.27, 1.5, and 0.85 million cubic meters of sand were placed along the 8.5 km of Hollywood/Hallandale beach in 1971, 1979 and 1991, respectively. A major beach nourishment project took place along the 3.9 km of Sunny Isles beach in 1988 with placement of 1.0 million cubic meters of sand. Other small-scale beach fill projects within this region were detailed in the Sunny Isles report (Coastal Tech, 1993).

Extensive nearshore hardbottom was found within this study area, which aligned in a shore-parallel direction with an offshore distance between the hardbottom and shoreline increasing from north to south. Specifically, the nearshore hardbottom is located approximately 200 m offshore at the north end of the region and increases to 900 m offshore at the south end. The average width of the nearshore hardbottom area is 2,400 m.

REGIONAL SEDIMENT TRANSPORT

The overall sand movement within the study region is evaluated by examining volumetric changes at each DEP (formerly DNR) monument. The volumetric changes occurring above and below the -2 m National Geodetic Vertical Datum (NGVD) elevation are calculated. The depth of closure for sand movement in this region varies from approximately -4 m NGVD to -6 m NGVD, depending on the location of nearshore hardbottom. A detailed variation of volumetric changes along the region is illustrated in Figure 2.

Table 1 summarizes the average and total volumetric changes within each coastal municipality. In general, long-term hydrographic data for two consecutive beach nourishment projects were used to obtain an average volumetric change rate after a beach renourishment project. In the area of Sunny Isles and Bakers Haulover Park, relatively short-term data (approximately a three-year time period) was used due to the lack of accurate and extensive hydrographic survey data.

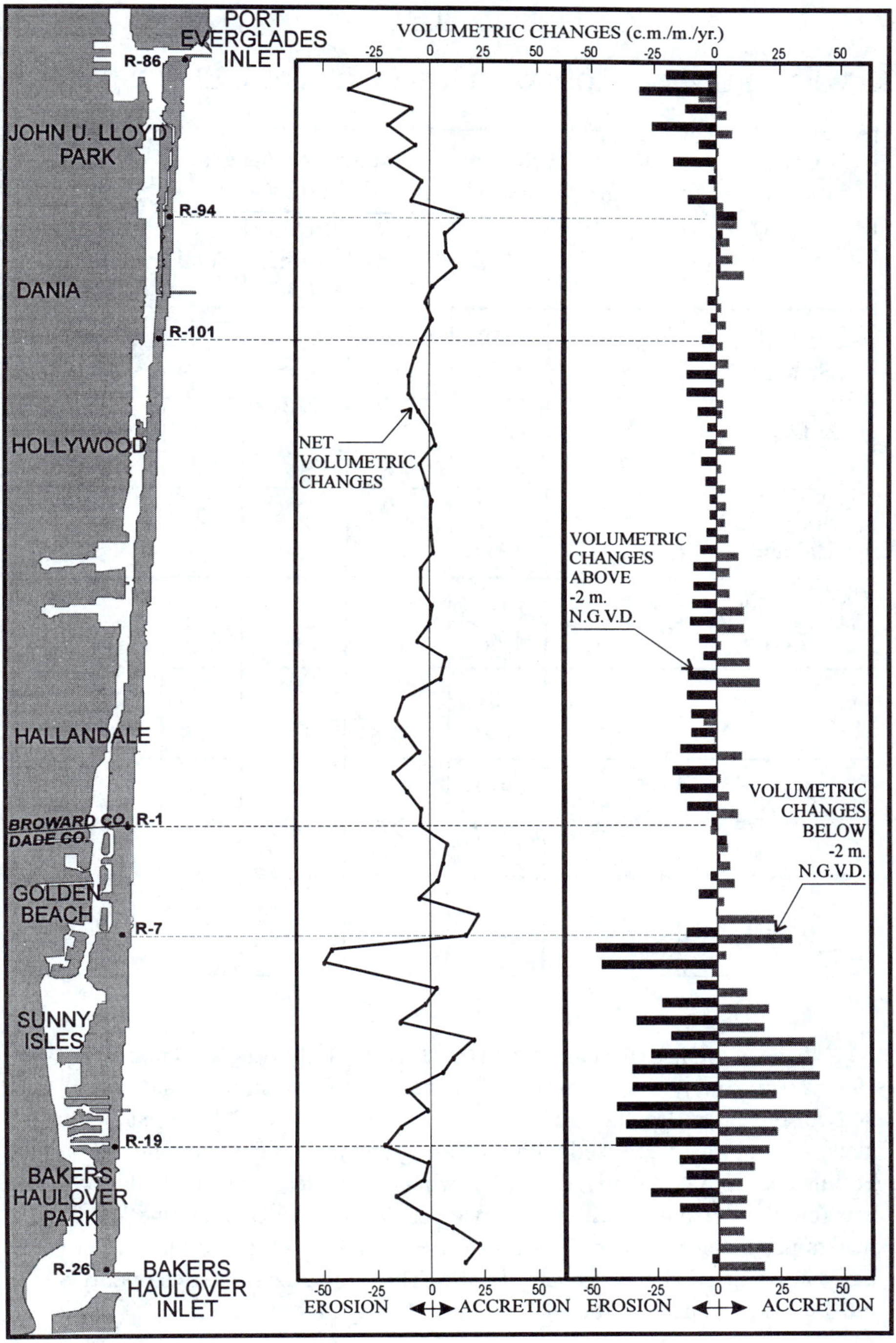

Figure 2. Net Volumetric Changes and Cross Shore Sand Movement

TABLE 1

VOLUMETRIC CHANGES WITHIN EACH COASTAL MUNICIPALITY

Coastal Municipality	Distance (km)	Survey Period	Volumetric Changes (c.m./m/yr)			Total Volumetric Changes (c.m./yr)
			Above -2 m NGVD	Below -2 m NGVD	Total	
John U. Lloyd State Park	2.4	1978-1989	-14.5	1.0	-13.5	-33,000
Dania	2.1	1978-1993	0.8	4.3	5.1	11,000
Hollywood/ Hallandale	8.5	1979-1991	-8.0	4.0	-4.0	-34,000
Golden Beach	1.9	1980-1991	-2.3	9.5	7.2	14,000
Sunny Isles	3.9	1989-1991	-30.6	22.1	-8.5	-34,000
Bakers Haulover Park	2.4	1989-1991	-8.8	14.0	5.2	13,000
Total	21.2	—	—	—	—	-63,000

The nourished beach at John U. Lloyd, Hollywood/Hallandale (CP&E, 1992b) and Sunny Isles was found to be eroding at similar rates of 33,000, 34,000 and 34,000 cubic meters per year, respectively. The erosion rate per meter of shoreline decreased with increasing length of the renourishment project. For instance, Hollywood/Hallandale, having the greatest project length of 8.5 km, was found to have the least erosion rate per linear meter of shoreline at 4.0 cubic meters per meter per year. The greatest erosion of 13.5 cubic meters per meter per year occurred at John U. Lloyd Park. This high erosion rate is mainly due to the downdrift impacts associated with Port Everglades Inlet.

Non-nourished areas adjacent to the renourished beaches were accreting, as illustrated in Table 1. Dania, Golden Beach and Bakers Haulover Park were accreting at a rate of 11,000, 14,000 and 13,000 cubic meters per year, respectively. The accretion rates in these areas were relatively uniform, ranging from 5.1 to 7.2 cubic meters per meter per year.

A sediment budget, as illustrated in Figure 3, is developed for the region based on a known boundary condition and the results of the above volumetric changes. This sediment budget encompasses a region that extends beyond the area that is conventionally studied under inlet management plans and beach renourishment projects. The well-defined northern boundary condition of Port Everglades Inlet has improved the accuracy in the derivation of this regional sediment budget. A number of studies have reported that Port Everglades Inlet has interrupted 100 percent of longshore sediment transport and resulted in no sediment being naturally bypassed from north of the inlet to John U. Lloyd Park (COE, 1990; Coastal Tech, 1988 and 1994).

The regional sediment budget indicates that the longshore sediment transport rate at the north end of the region is less than the rate at the south end. This is contradictory to the trend experienced along the east coast of Florida, where longshore sediment transport rates usually decrease toward the south. The wave-induced longshore transport rate is found to be 33,000 cubic meters per year at John U. Lloyd/Dania, as compared to 76,000 cubic meters per year at Sunny Isles/Bakers Haulover Park. A net southern longshore transport rate of 77,000 cubic meters per year at Bakers Haulover Park was predicted by COE (1995) using the GENESIS numerical model.

The results of volumetric changes show that beach erosion generally occurred above the -2 m NGVD elevation and accretion occurred below the -2 m NGVD. Among all the beach renourishment projects, the amount of erosion that occurred above -2 m NGVD elevation was greater than the accretion in the area below -2 m NGVD.

Substantial beach erosion occurred above -2 m NGVD elevation in Sunny Isles, where a great amount of sand that was placed on the beach moved offshore. More substantial cross-shore sand movement occurred in Sunny Isles than that in other areas, as illustrated in Figure 2. This is due to the nearshore hardbottom in these other areas that have restricted sand movement in a cross-shore direction.

IMPACTS OF HARDBOTTOM ON SAND MOVEMENT

The relationship between the net longshore sediment transport rate and offshore location of hardbottom is presented in Figure 4. In general, the longshore sediment transport rate increases from north to south, as does the offshore

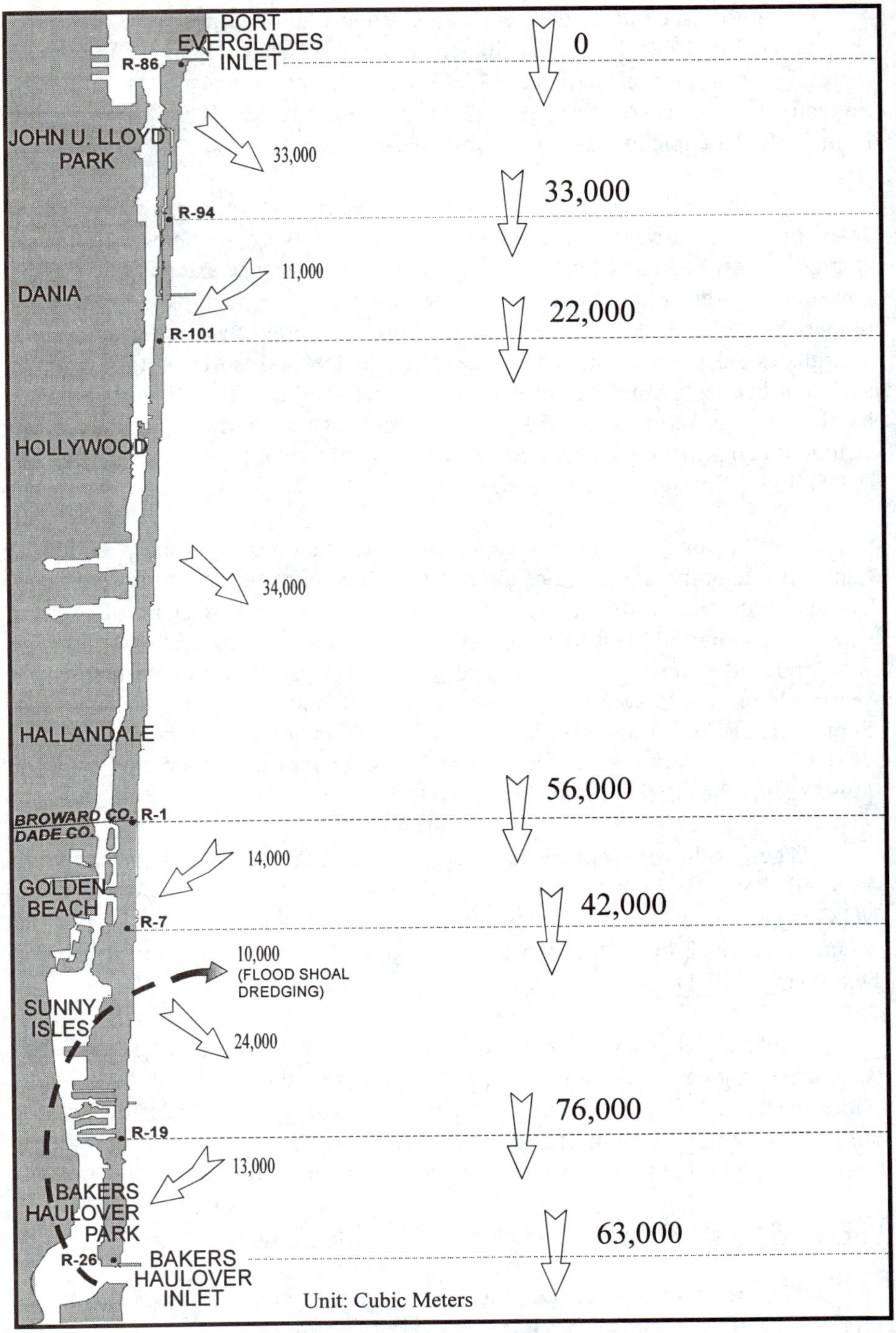

Figure 3. Regional Sediment Budget

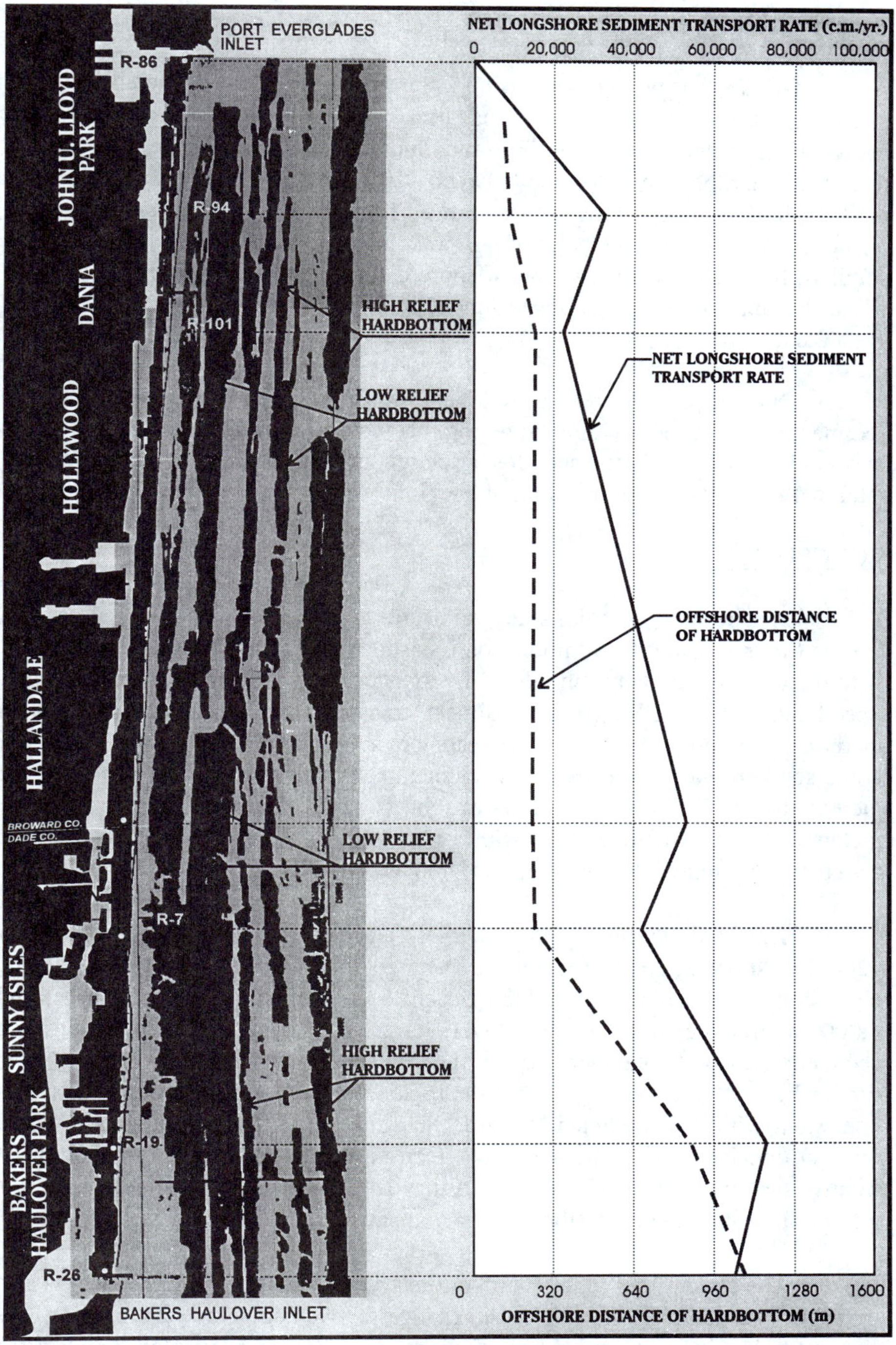

Figure 4. Longshore Sediment Transport Rate vs. Offshore distance of Hardbottom

distance of hardbottom.

Offshore wave climates along this study area are generally similar (Stauble, 1993). Along the north end, nearshore hardbottom has reduced wave energy and resulted in less longshore sand movement. In the northern half of the region, the hardbottom is located between 120 m and 300 m offshore, while the net longshore sediment transport rate ranged from 22,000 to 33,000 cubic meters per year. However, the hardbottom was located further offshore along the southern half of the area with the offshore distance between 300 m to 1,000 m, while the annual net longshore sediment transport rate ranged from 42,000 to 76,000 cubic meters.

Due to the differential sediment transport rate, there is an inherent sediment supply deficit within the region that would result in beach erosion. This may explain historically the need for beach renourishment projects for a combined total of 14.8 km of shoreline within the region.

CONCLUSIONS

The results of this study indicate that the existence of nearshore hardbottom has a significant impact on longshore and cross-shore sand movement within the region between Port Everglades and Bakers Haulover Inlet. Specifically, the net longshore sediment transport rate within the study area increased from north to south as the nearshore hardbottom aligned further offshore at the southern part of the area. The annual net longshore sediment transport rate increased from 33,000 cubic meters at John U. Lloyd State Park to 63,000 cubic meters at Bakers Haulover Inlet with a maximum of 76,000 cubic meters at the area between Sunny Isles and Bakers Haulover Park.

The hardbottom remains relatively close to the shore at approximately 120 m offshore at John U. Lloyd, then increases to 300 m offshore at Dania, Hollywood, Hallandale and Golden Beach, 600 m offshore at Sunny Isles, and 1,000 m offshore at Bakers Haulover Park. It is evident that the longshore sediment transport rate increases as the protection of nearshore hardbottom is diminished. At the southern end of the region, the net longshore sediment rate is consistent with that predicted by the COE using the GENESIS numerical model without consideration of hardbottom effect. Hence, the offshore hardbottom at Sunny Isles and Bakers Haulover Park has little or no effect on reducing wave energy and subsequently on the sand movement.

The effect of nearshore hardbottom on cross-shore sand movement is also evident. By examining the volumetric changes above and below -2.0 m NGVD as illustrated in Figure 2, less active offshore sand movement was observed in the area from John U. Lloyd to Hallandale. It suggests that the nearshore hardbottom may act as a perch structure preventing sand from moving offshore.

The use of the GENESIS numerical model to predict shoreline changes and sediment transport, may not provide reasonable results in regions with extensive nearshore hardbottom, such as occurs in southeast Florida. The effects of these nearshore hardbottom areas on longshore and cross-shore sand movement should be considered in developing a sediment budget for inlet or beach management plans.

ACKNOWLEDGEMENT

Assistance from the Department of Natural Resources Protection (DNRP) in Broward County, Florida, in providing valuable data is greatly appreciated.

REFERENCES

Coastal Planning and Engineering, Inc. (CP&E) , "Bakers Haulover Inlet Management Plan," Interim Report No.1, Dade County, Florida, 1992a.

Coastal Planning and Engineering (CP&E), Inc., "Hollywood/Hallandale First Periodic Beach Renourishment Project," One-year Beach Monitoring Study and Post-Storm (Hurricane Andrew) Analysis, 1992b.

Coastal Technology Corporation, "Port Everglades Inlet Management Plan," Department of Natural Resources Protection, Broward County, Florida, 1994.

Coastal Technology Corporation, "Sunny Isles Beach Restoration Project, 18-month Monitoring Study," Department of Environmental Resources Management, Dade County, Florida, November, 1993.

Coastal Technology Corporation, "Port Everglades Sand Bypassing Study," Department of Natural Resources Protection, Broward County, Florida, 1988.

Stauble, D.K., "An Overview of Southeast Florida Inlet Morphodynamics," Beach/Inlet Processes and Management: A Florida Prospective, Journal of Coastal Research, Special Issue No. 18, Fall, 1993.

U.S. Army Corps of Engineers (COE), "Broward County, Florida, Port Everglades to the South County Line (Segment III), Shore Protection Project Reevaluation Report, Section 934 study with environmental assessment," Jacksonville District South Atlantic Division, 1990.

U.S. Army Corps of Engineers (COE), "Coast of Florida Erosion and Storm Effects Study, Region III" May, 1995.

CHAPTER 282

IMPROVEMENT OF THE MOST ACCURATE LONGSHORE TRANSPORT FORMULA

by

J S Schoonees[1] and A K Theron[1]

ABSTRACT

The ability to predict the longshore sediment transport rate accurately is essential for many coastal engineering applications. Because of the existance of a large number of existing longshore transport formulae, it is important to know which formula to use/apply. Thus, the most universally applicable formula was identified and tested against a comprehensive data set. This formula (Kamphuis formula) was also re-calibrated and guidance is given regarding its use.

INTRODUCTION

The ability to predict the time-averaged longshore sediment transport rate accurately is essential for the design of breakwaters at harbour entrances, navigation channels and their dredging requirements, beach improvement schemes incorporating groynes, detached breakwaters and beach fill as well as for the determination of the stability of inlets and estuary mouths.

Because of the large number of existing longshore transport formula it is important to know which formula to apply in practice. The aim of this paper is therefore to identify the most universally applicable formula and to test this formula against a comprehensive field data database. Finally, this formula is re-calibrated and guidance is given regarding its use.

The data considered in this paper are only for particulate (non-cohesive) sediment (including sand, gravel and shingle) being transported alongshore from the swash zone across the surf zone to deep water. Bulk (total rate calculated perpendicular to the shoreline) as well as local (at a specific point)

---

[1]Research Engineers, CSIR, P O Box 320, Stellenbosch, 7599, South Africa

point) transport rates are considered. These bulk rates include both the bedload and the suspended load. Only field data are used because laboratory investigations often contain possible scale effects and/or use regular waves. Furthermore, the ultimate aim is to be able to predict longshore transport accurately in the *field* (Komar, 1988).

It is assumed that if a longshore transport formula is capable of predicting transport rates accurately for the wide ranging data sets described herein, it can be used with reasonable confidence at similar sites to determine the long-term longshore sediment transport budget if representative wave and other input parameter data are available. It would of course be preferable to have site-specific calibration data before calculating average long-term transport rates at a specific site.

Previous studies where longshore transport formulae have been tested against data include Swart (1976), Fleming *et al.* (1986) and Kamphuis *et al.* (1986). These studies entailed a relatively small number of formulae and limited data. Schoonees (1996) evaluated 52 formulae against an extensive data base. This paper reports some of the findings of the above-mentioned study, with specific regard to the Kamphuis formula (Kamphuis, 1991).

FIELD DATA ANALYSIS

Schoonees and Theron (1993) compiled and reviewed almost all the available field data on longshore transport (as recorded up to 1993).

The data were collected at a wide variety of sites around the world, yielding a large number of data-points, of which 273 points give bulk transport rates. (This is considerably more than the 41 data points used in the Shore Protection Manual by US Army, Corps of Engineers, 1984). Included in the database are also 184 points which give local transport rates.

A point rating system was devised whereby the quality of the data could be assessed. The recording method and the accuracy thereof as well as the representativeness of the data were taken into account. It was found that this evaluation was done reasonably objectively and consistently (Schoonees and Theron, 1993).

According to this evaluation, the data sets were assigned to three categories, namely, the lower, middle and higher quality categories. Most of the data sets fell in the middle category which exhibited a very gradual increase in the overall accuracy of the data within this category. Distinguishing between short- and long-term bulk transport data yielded similar trends in the accuracy of the data. The highest score achieved in the data evaluation was only 71%, thus reflecting the difficulty of measuring longshore transport accurately.

EVALUATION OF LONGSHORE TRANSPORT FORMULAE

The method used in this study to evaluate the longshore transport formulae, was to compare the predicted longshore transport rates to the measured rates and to calculate the relative standard error of estimate (σ). (For a definition of σ, see Schoonees and Theron (1994)). The lower σ is, the better the predictions by the particular formula. In addition the residuals (e_i = measured transport rate - predicted rate) and the distribution of the discrepancy ratio (r_d = predicted/measured rate) were also determined. (The residuals were plotted against the predicted rates to check whether there are systematic trends in the residuals - these are not shown here (Schoonees, 1996)). The longshore transport formulae were also tested under as many different conditions and at as many sites as possible.

From the above-mentioned field data database, Data Set 1 containing 123 points was extracted (see Schoonees and Theron, 1994 for a full description of Data Set 1). In Data Set 1 all the parameters required for testing the transport formulae are available. This same data set was used to evaluate existing longshore transport formulae as well as a newly derived formula (Schoonees, 1996) based on the applied wave power concept.

It is important to note that the data ranges of Data Set 1 are:

0,058	<	H_{bs} (significant breaker height, m)	<	3,400
2,32	<	T_p (peak wave period, s)	<	16,60
0,30	<	θ_b (breaking wave angle, °)	<	35,00
0,007 (=1/143)	<	beach slope	<	0,138 (=1/7,2)
0,154	<	D_{50} (median grain size, mm)	<	15,000
600	<	S (longshore transport rate, m^3/year)	<	14 793 000

From the above values it is clear that the data ranges of this data set are quite wide. Most conditions encountered on natural beaches are covered and the data were collected on beaches from a variety of sites from around the world. These factors give credibility to the conclusions drawn in this comparison of predicted versus measured transport rates.

In total, 52 different longshore transport formulae were evaluated (Schoonees, 1996). These formulae were classified into different categories with regard to the theories on which they are based. The following three formulae were found to be the most accurate as tested against Data Set 1:

Order of accuracy	Name of the formula	Relative standard error of estimate (σ)	Category
1	Kamphuis (1991)	0,393	Dimensional analysis
2	Van Hijum, Pilarczyk and Chadwick (1989)	0,417	Energetics (energy flux)
3	Van der Meer (1990)	0,447	Empirical

These formulae are all bulk (total rate) as opposed to detailed predictors. Figures 1, 2 and 3 illustrate the fit of the predicted transport rates against the measured rates (log-log scales).

The formulae were also ranked according to the highest percentage of discrepancy ratios (r_d) between 0,5 and 2 (i.e. under or over prediction by a factor of 2; $r_d = 1$ indicates perfect agreement). A similar ranking of the "best" five formulae was found. However, it was found that σ provides a better way to judge the accuracy of a formula than using the percentage of r_d between 0,5 and 2. This is because, when applying a transport formula to determine a longshore transport budget at a site, a single badly predicted transport rate can distort the calculated budget greatly. At the same time, however, the above-mentioned percentage of r_d can still be very high compared to σ which would be affected greatly by a single badly predicted rate. Therefore σ is a better yardstick.

Dimensional analysis incorporating all relevant variables ensured that the Kamphuis formula contains the most important parameters. The three top formulae (Kamphuis; Van Hijum, Pilarczyk and Chadwick and Van der Meer) are relatively simple. These "simpler" methods performed well probably because a lower degree of inaccuracy is or can be introduced by having fewer (but all the most important) parameters. It is very difficult to acquire accurate input data; and the more parameters incorporated in a formula, the more input data is required (thereby potentially increasing the noise).

It is common practice to compare the predictions from different longshore transport formulae when computing the annual longshore transport regime at a site. Swart and Fleming (1980) advocated the use of a so-called package deal approach. In this approach, the highest and lowest transport rates predicted by six formulae were ignored and the median of the remaining values was determined. The question then remains whether better results can be achieved by means of this or a related method. Three approaches were tried

(Schoonees, 1996). Firstly, by considering the median of the predictions by the five best formulae; secondly, by determining the mean of the three middle values after discarding the highest and lowest predictions; and thirdly, by computing a weighted mean transport of the five predictions. The variation in the transport rates predicted by the five best formulae was investigated. It was found (Schoonees, 1996) that these predictions are reasonably consistent; that is, the individual formulae do not yield excessive outliers. It can therefore be concluded that none of the package deal approaches yield better answers than the best formula (the Kamphuis method) and as such , are not worth pursuing if the above-mentioned five best formulae are used. The reason for this probably lies in the consistency (reliability) of the five best formulae.

RECALIBRATION OF THE KAMPHUIS FORMULA

The Kamphuis formula can be written as follows:

$$
\begin{aligned}
S &= (31\ 557\ 600.1.3.10^{-3})\ x_{Kamphuis} \\
&= 41\ 024{,}88\ x_{Kamphuis}\ (m^3/yr)
\end{aligned}
\qquad \textbf{(1)}
$$

$$
\begin{aligned}
x_{Kamphuis} = \frac{1.}{(1-p)\ \rho_s} &\cdot (\rho / T_p)\ L_o^{1{,}25}\ H_{bs}^2\ (\tan\alpha_K)^{0{,}75} \\
&\cdot (1/D_{50})^{0{,}25} (sin\ 2\theta_b)^{0{,}6}
\end{aligned}
\qquad \textbf{(2)}
$$

where p = porosity
ρ_s = density of the sediment grains
ρ = density of sea water
L_o = deep-water wavelength
$\tan\alpha_K$ = beach slope to the breaker line

See Kamphuis (1991) for a more comprehensive definition of all the parameters.

Equation (1), the original Kamphuis formula, is plotted *on linear scales* in Figure 4a and b. Note that Figure 4b shows the detail of Figure 4a for $x_{Kamphuis}$ values up to 80 (instead of 200). The 80%, 90% and 95% confidence intervals for the predicted responses of the original Kamphuis formula are also shown in Figures 4a and b. Despite the fact that the Kamphuis formula fares the best of the 52 formulae tested, it is immediately apparent that the confidence intervals are very wide. For example, at the 80% confidence level, the predicted transport rate for $x_{Kamphuis}$ = 8,7 varies between -1 290 000 m^3/year and + 2 004 000 m^3/year (predicted rate = +357 000 m^3/year) - Figure 4b. The coefficient of determination (R^2) is 0,284.

Illustrated in Figure 5 is the best-fit straight line through all the data (called $S_{Kamphuis}$ recalibrated, 1):

$$S = 88\ 248 + 61\ 892\ x_{Kamphuis}\ (m^3/year) \tag{3}$$

If Equation (3) is used, it is evident that the lowest transport rate that can be predicted, is 88 248 m^3/year, which is when $x_{Kamphuis} = 0$. This is clearly unacceptable, because the transport rate must be zero if $x_{Kamphuis} = 0$. Furthermore, it can be seen from Figure 4a that there are three main outliers (which fall beyond the 95% confidence limit) and that Equation (3) fits the higher transport rates better than the original Kamphuis formula. Therefore, disregarding the three main outliers and fitting the line through the origin, the following equation is found:

$$S = 75\ 549\ x_{Kamphuis}\ (m^3/year) \tag{4}$$

This relationship, $S_{Kamphuis}$ recalibrated 2, is shown in Figure 5a and b (the latter Figure 5b again presents the detail of Figure 5a). Although this formula fits the high transport rates well, it over predicts significantly for $x_{Kamphuis}$ values below 10 (Figure 5b). This is caused by two influential points where transport rates higher than 3 x $10^6 m^3$/year were measured (Figures 5a and b). (Remember that the three main outliers, although shown, have not been used in this regression).

To eliminate this problem all the data points (123) were again considered to yield the third regression line, the $S_{Kamphuis}$ recalibrated, 3:

$$S = 63\ 433\ x_{Kamphuis}\ (m^3/year) \tag{5}$$

This formula fits the data reasonably well over the whole range (Figures 5a and b). It gives virtually the same answers at high $x_{Kamphuis}$ values than the first recalibrated formula Equation (3). It also fits the data at lower transport rates quite well. R^2 is 0,620 and thus Equation (5) explains 62% of the variance in the data (which is a 118% improvement compared with Equation (1)). However, the standard error of estimate (σ) for this formulation is 0,405, which is slightly poorer than the 0,393 of the original Kamphuis equation. The reasons for this apparent contradiction are:

- The least squares approach (Equations (3), (4) and (5)) minimizes $\sum (S_{m,i} - S_{p,i})^2$ while the standard error of estimate (σ) uses $(\log S_{p,i} - \log S_{m,i})^2$. (Subscripts m and p denote "measured" and "predicted" respectively while i is the number of the data point).

- The effect of a few data points with high transport rates act as influential points in the least squares approach. On the other hand, the number of data points at low transport rates play an important role in the value of the standard error of estimate (σ).

To investigate the effect of such low rates on the standard error of estimate, certain data points below a cut-off transport rate were temporarily disregarded and σ re-calculated. The result was the following:

Cut-off measured transport rate (m^3/year)	Number of data points	σ for $S = 63433\ x_{Kamphuis}$ (Equation 5)	σ for the original Kamphuis formula (Equation 1)
0	123	0,405	0,393
5 000	115	0,392	0,393
10 000	106	0,368	0,380
25 000	103	0,365	0,383
50 000	87	0,324	0,377
100 000	68	0,299	0,374

These values have been plotted in Figure 6. It is clear from the above table that except when very low transport rates of less than 5 000 m^3/year are included, Equation (5) is superior to the original Kamphuis formulation. Figure 6 shows that the original Kamphuis formula is relatively insensitive to the cut-off transport rate. On the other hand, the standard error of estimate decreases significantly (to only 0,299 compared to 0,374 of the original formula, which is a 20% improvement) for Equation (5), if the cut-off transport rate increases. If a cut-off rate of 50 000 m^3/year is applied, σ reduces from 0,377 to 0,324, a 14% improvement. This finding is important because relatively few storm conditions at any site usually contribute the major part of the longshore sediment transport budget. It is therefore important that the higher transport rates are predicted accurately.

In order to obtain an indication of which wave conditions would cause such cut-off transport rates, the following typical values were chosen:
T_p = 10s, θ_b = 2°, D_{50} = 0,3 mm and tanα = 1/25 (= 0,04). It was also assumed that these wave conditions will occur throughout the year.

Using the *original* Kamphuis formula, the longshore transport was then computed for a range of wave heights:

H_{bs} (m)	Longshore transport rate ($S_{Kamphuis}$) (m^3/year)
0,1	2 007
0,3	18 067
0,5	50 187
0,7	98 366

For this particular (typical) case, it is clear that relatively low wave heights of about 0,5 m and 0,7 m will already cause transport rates of about 50 000 m^3/year and 100 000 m^3/year respectively. For these cut-off rates, the standard errors of estimate are 0,299 and 0,324 respectively (Figure 6). Equation (5) is therefore judged to be "good".

Instead of using the least squares approach, the question may also be asked: What value of K will cause the minimum standard error of estimate (σ) when using *all* the data points? That is, K in:

$$S = Kx_{Kamphuis} \quad (m^3/year) \tag{6}$$

Computations with different K values yielded the following:

K	σ
30 000	0,438
41 025 (Equation (1))	0,393
45 000	0,388
48 000	0,387
49 000	0,387
50 000	0,387
55 000	0,391
64 433 (Equation (5))	0,405
70 000	0,420

These data have been plotted in Figure 7. From this figure it is evident that the minimum standard error of estimate(s) is 0,387. It can also be seen that σ is not very sensitive with regard to the value of K near the turning point: σ = 0,387 even if K varies from 48 000 to 50 000. Taking the accuracy of predicted transport rates during storms into account, the preferred (rounded off) equation is:

$$S = 50\,000\ x_{Kamphuis}\ (m^3/year) \qquad \textbf{(7)}$$

$$\text{with } R^2 = 0{,}397$$

FINAL RECOMMENDATIONS

The least squares as well as the minimum standard error of estimate approaches were used to recalibrate the Kamphuis formula, yielding Equations (5) and (7) respectively. Finally the recommended procedure for calculating longshore sediment transport is as follows:

For obtaining bulk longshore transport rates, it is recommended that **Equation (5)** be applied at sites where the significant wave heights normally exceed say, 0,3 m and where the sediment grain size is usually less than 1 mm; that is, at partially protected and exposed sites (i.e. relatively high transport rates). Only at sites where very calm conditions prevail and/or where the sediment is coarse, is **Equation (7)** expected to yield better answers. The considerably higher R^2 for Equation (5) compared with the corresponding value for Equation (7) supports the preference for Equation (5). A significant improvement of 118% according to R^2 and up to 20% in σ in the predicted transport rates was obtained.

REFERENCES

Chadwick, A J (1989). Field measurements and numerical model verification of coastal shingle transport. *Advances in Water Modelling and Measurement*, Palmer, M H (ed), BHRA, Cranfield, Bedford.

Fleming, C A, Pinchin, B M and Nairn R B (1986). Evaluation of models of nearshore processes. *20 Intern. Conf. on Coastal Eng.*, ASCE, Taipei, Taiwan, Vol 2: 1116-1131.

Kamphuis, J W (1991). Alongshore sediment transport rate. J. Waterways, *Port, Coastal and Ocean Eng.*, ASCE, Vol. 117(6): 624-640.

Kamphuis, J W, Davies, M H, Nairn, R B and Sayao, O J (1986). Calculation of littoral sand transport rate. *Coastal Engineering*, Vol 10: 1-21.

Komar, P D (1988). Environmental controls on littoral sand transport. *21st Intern. Conf. on Coastal Eng.*, ASCE, Malaga. Vol 2: 1238-1252.

Schoonees, J S (1996). Longshore sediment transport in terms of the applied wave power concept. *PhD thesis*, University of Stellenbosch, Stellenbosch (in preparation).

Schoonees, J S and Theron, A K (1993). Review of the field data base for longshore sediment transport. *Coastal Engineering* Vol. 19: 1-25.

Schoonees J S and Theron, A K (1994). Accuracy and applicability of the SPM longshore transport formula. *24 Intern. Conf. on Coastal - Eng.* ASCE, Kobe, Japan. Vol 3: 2595-2609.

Swart, D H (1976). Predictive equations regarding coastal transports. 15th *Intern. Conf on Coastal Eng.* ASCE, Honolulu, Hawaii. Vol 2: 1113-1132.

Swart, D H and Fleming, C A (1980). Longshore water and sediment movement. *17th Intern. Conf. on Coastal Eng.*, ASCE, Sydney. Vol 2: 1275-1294.

U S Army, Corps of Engineers (1984). *Shore Protection Manual*, Volumes I and II, Coastal Engineering Research Center, Vicksburg.

Van der Meer, J A W (1990). Static and dynamic stability of loose materials. In: *Coastal protection*, Pilarczyk K W (ed), Balkema, Rotterdam.

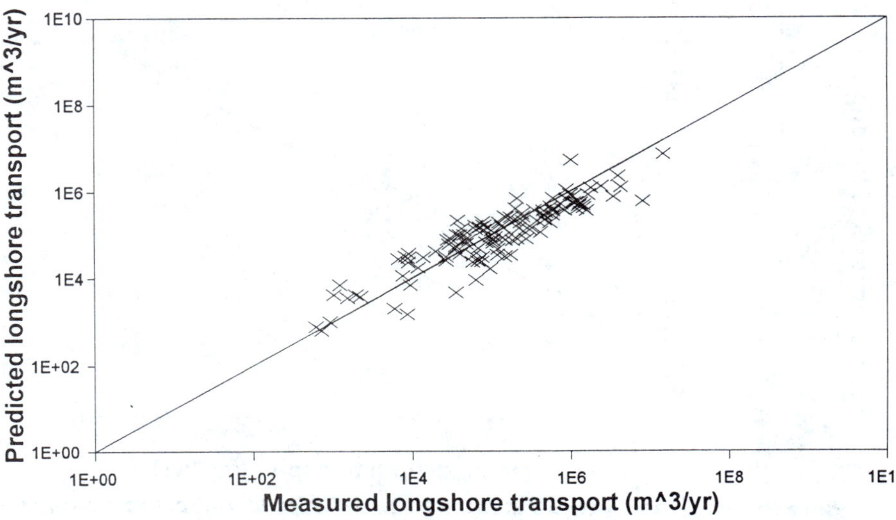

Figure 1: Predicted versus measured longshore transport rates for the Kamphuis formula

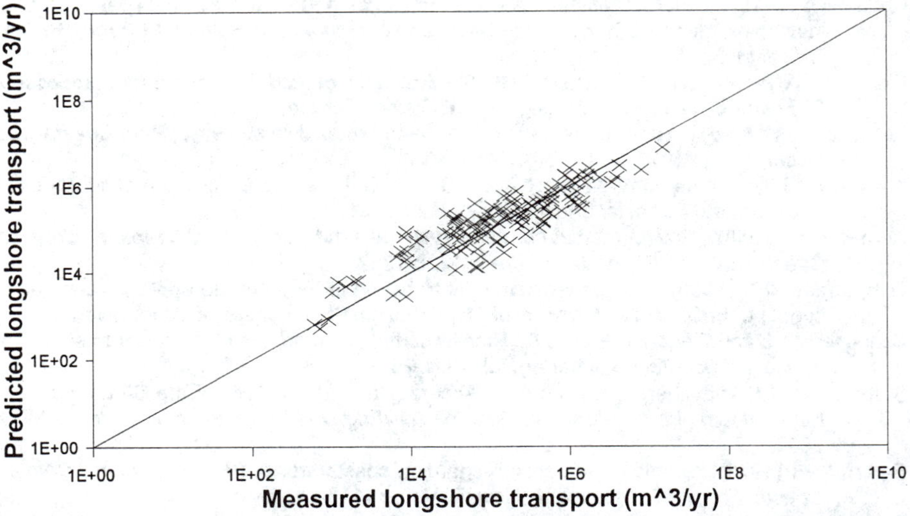

Figure 2: Predicted versus measured longshore transport rates for the Van Hijum, Pilarczyk and Chadwick formula

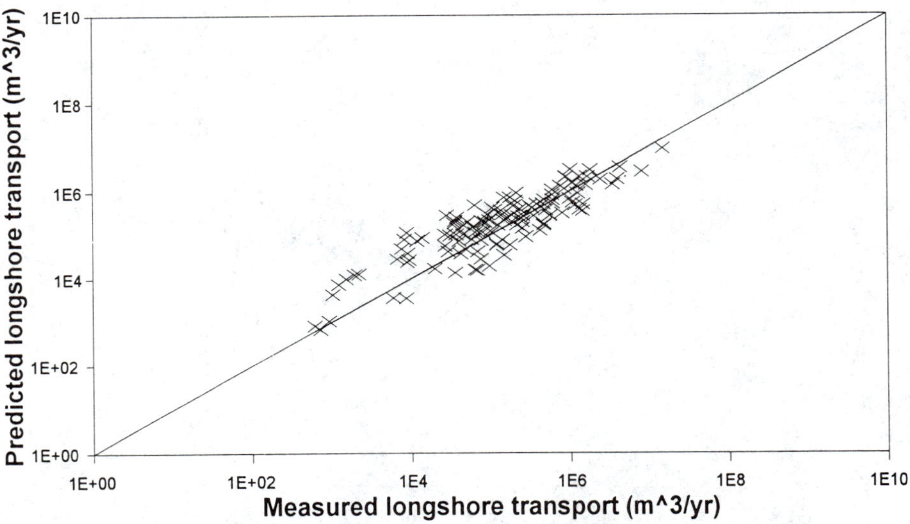

Figure 3: Predicted versus measured longshore transport rates for the Van der Meer formula

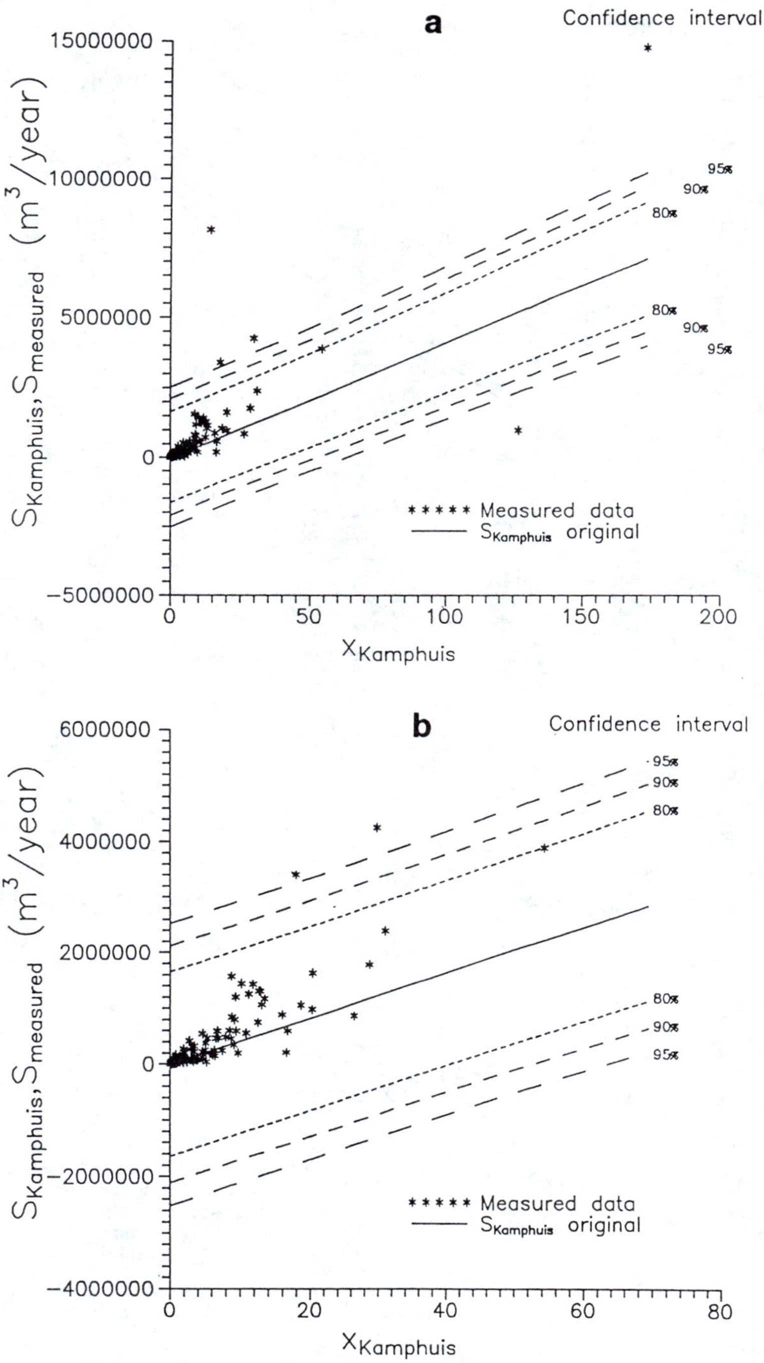

Figure 4: Confidence intervals for the original Kamphuis Formula

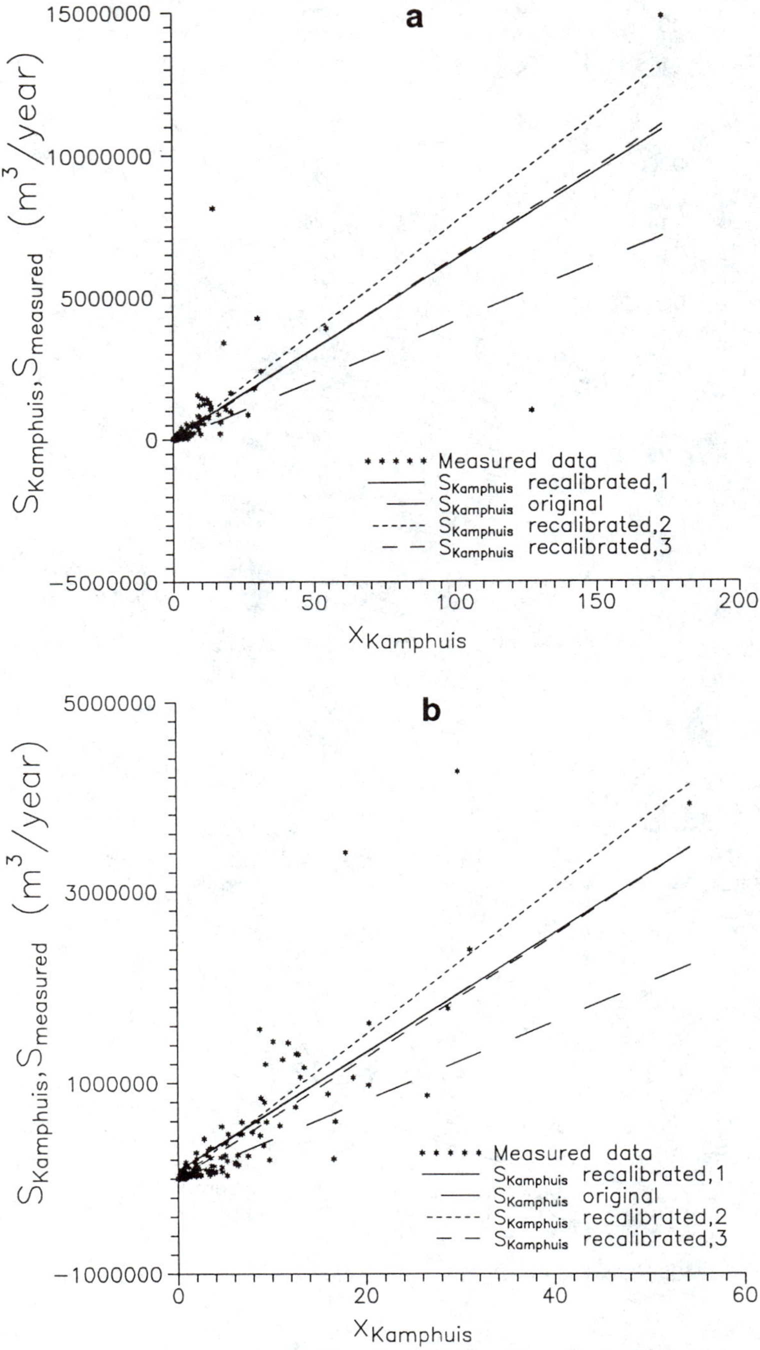

Figure 5: Recalibration of the kamphuis formula

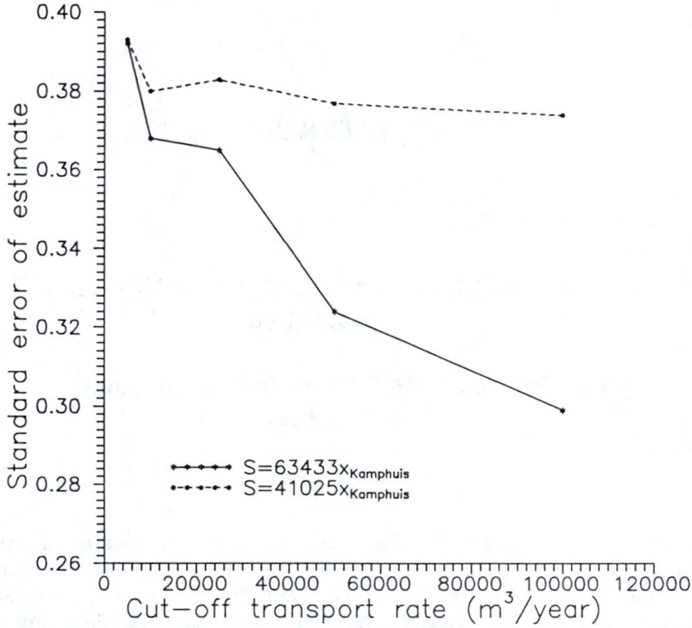

Figure 6: Standard error of the estimate for different cut-off rates

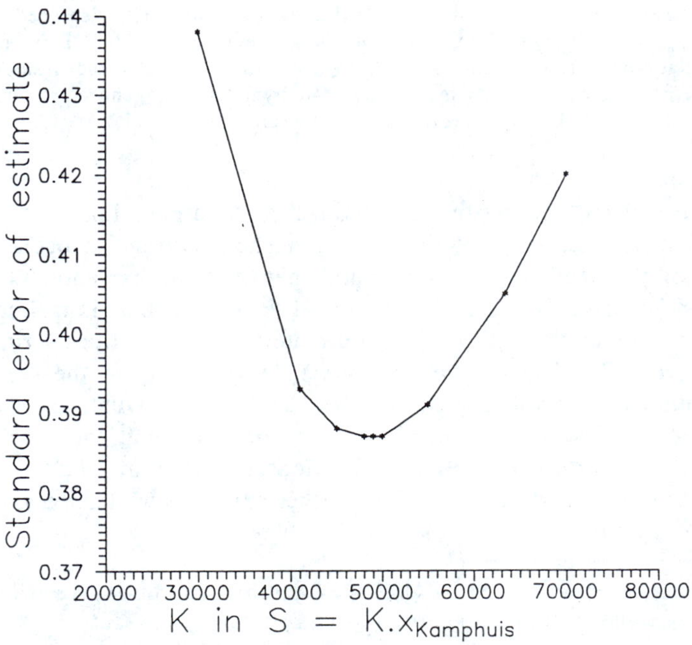

Figure 7: Minimizing the standard error of estimate for the Kamphuis formula

CHAPTER 283

Cross-shore Structure of Longshore Currents during Duck94

Falk Feddersen[1], R.T. Guza[2], Steve Elgar[3], and T.H.C Herbers[4]

Abstract

The cross-shore structure of the mean longshore current on a barred beach is investigated with observations from the Duck94 field experiment. Maxima of the hourly-averaged longshore current are most frequently located either slightly inshore of the bar crest or near the shoreline. At low tide the longshore current maxima are located closer to the bar crest, and the current is stronger and narrower than at high tide. The tidal cross-shore displacement of the longshore current maximum is qualitatively consistent with the observed radiation stress, S_{xy} , although the maximum currents are displaced shoreward of the maximum S_{xy} gradients at both high and low tide. This spatial lag suggests that some mechanism (such as wave rollers) delays the transfer of momentum from waves to the mean low. Bathymetric longshore inhomogeneities may also affect the cross-shore structure of the longshore current.

Introduction

Selected observations from the DELILAH field experiment at Duck N.C. suggest that the maximum of the mean longshore current, $\bar{v}$, occurs between the crest of the sand bar and the trough between the bar and the shoreline (Figures 8-11 of *Church and Thornton*, (1993)), and that longshore current variability is coherent with the approximately 1 m semidiurnal tidal fluctuations in water level (*Thornton and Kim*, 1993). However, the generality of these results over the wide range of wave (*Long*, 1996), wind, and bathymetric (e.g. pronounced sandbars and alongshore inhomogeneities (*Lippmann and Holman*, 1990; *Gallagher*, 1996)) conditions observed at Duck is unknown. Here the cross-shore structure of $\bar{v}$ is further explored with observations from the Duck94 field experiment.

---

[1] Graduate Student, Scripps Institution of Oceanography, 0209, La Jolla Ca,92037-0209, falk@coast.ucsd.edu

[2] Professor, Scripps Institution of Oceanography, 0209, La Jolla Ca, 92037 -0209, rguza@ucsd.edu

[3] Professor, Washington State University, Pullman Wa, 99164-2752, elgar@eecs.wsu.edu

[4] Assistant Professor, Naval Postgraduate School, Monterey Ca, 93943, herbers@zee.oc.nps.navy.mil

Duck94 Observations

The experiment site is located on a long straight barrier island exposed to the Atlantic Ocean. Directional properties of sea and swell (including the wave radiation stresses, S_{xy}) were estimated with data from a 2-dimensional array of 15 bottom-pressure sensors located in 8 m water depth operated by the Field Research Facility (FRF) of the U.S. Army Corps of Engineers (*Long*, 1996). Longshore currents, wave-induced bottom pressures, and the location of the sea floor were observed with colocated bidirectional electromagnetic current meters, pressure sensors, and sonar altimeters (*Gallagher et al.*, 1996) deployed on a cross-shore transect extending 750 m from the shoreline to 8 m water depth (Figure 1). The sensors were sampled at 2 Hz for approximately two months. At each pressure sensor-current meter pair, hourly values of S_{xy} were crudely estimated using linear theory.

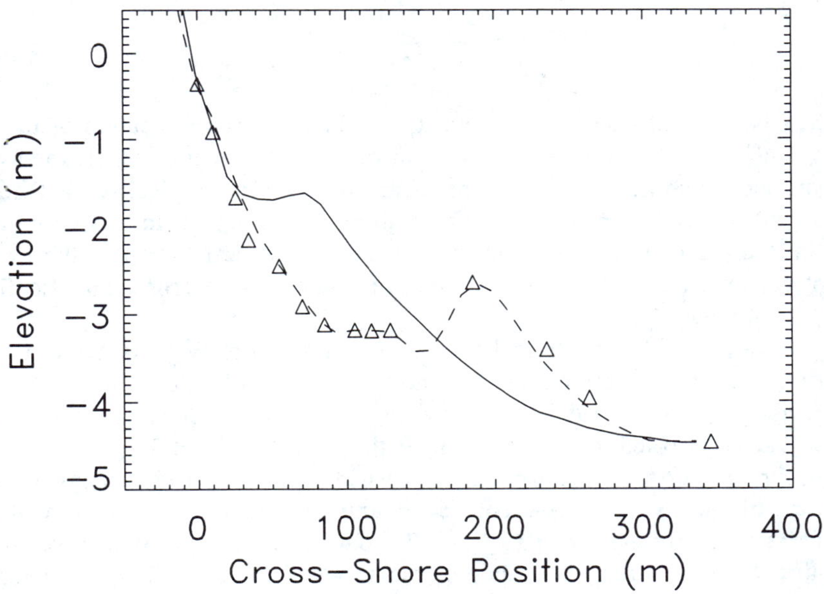

Figure 1. The cross-shore location of current meters (Δ), and measured bathymetry relative to sea level on August 25 (solid line) and October 26 (dashed line). An additional current meter at cross-shore location 749 m in 8 m water depth is not shown.

Conditions during Duck94 are shown in Figure 2. In 8 m water depth, the significant wave height (H_{sig}) ranged between 0.2 m to 4 m and the mean wave angle between $\pm 50^{o}$, so that the total (e.g. frequency-integrated) incident wave radiation stress S_{xy}/ρ (where ρ is the water density) in 8 m depth ranged from -0.7 to 0.5 m^3/s^2 (figures 2a-c). The mean (e.g. centroidal) wave frequency ranged from about 0.08 to 0.2 Hz (not shown). Maximum mean longshore currents (in each hour-long record, $\bar{v}_{max}$) ranged from 0.1 to 1.4 m/s (Figure 2d). The bar crest, originally located 80 m from the mean shoreline, gradually migrated 120 m offshore (Figure 2e, after *Gallagher*, 1996). Fluctuations in mean water level were about 1 m at spring tide. The slope of the beach foreshore is about 1/10 (Figure 1), so tidal fuctuations in the location of the mean shoreline are about 10 m. Tidal currents in depths less than 8 m are less than 0.03 m/s (S. Lentz, personal communication 1996).

Results

The sand bar is expected to strongly effect the longshore current, so a normalized cross-shore coordinate is defined as,

$$x' = \frac{x - x_{02}}{x_{bar} - x_{02}}$$

where x is the cross-shore coordinate, x_{bar} is the cross-shore location of the bar crest, and x_{02} is the cross-shore location of the most nearshore sensor uv02. The distance from sensor uv02 to the mean shoreline was typically less than 10 m. Depths at uv02 ($x' = 0$) ranged from about 0.2 to 1.2 m. In this coordinate system, the bar crest is always located at $x' = 1$, but the location of the trough is not constant because of variability in the shape of the cross-shore seafloor profile (e.g. Figure 1).

The location of the longshore current maximum, $\bar{v}_{max}$, is broadly distributed and roughly bimodal in the normalized coordinate system (1) (Figure 3, after *Gallagher*, 1996). Maxima most often occur either slightly inshore of the bar crest ($0.65 \leq x' \leq 1.2$) or near the shoreline ($x' < 0.3$). Rarely does $\bar{v}_{max}$ occur seaward of the bar crest, even when large waves were breaking well seaward of the bar ($x' > 2$). The few maxima located well seaward of the bar crest typically are weak ($\bar{v}_{max} \sim 0.3$ - 0.4 m/s) and approximately correspond to times of strong buoyancy driven flows (Rennie and Largier, personal communication 1996). The stronger longshore currents ($\bar{v}_{max} > 0.8$ m/s) are associated with large offshore S_{xy} (Figures 2c and 2d) and occur near the bar crest ($0.65 < x' < 1.2$). Weaker maxima (0.25 - 0.7 m/s) occur typically near the shoreline ($0 < x' < 0.3$) or near the bar crest, with few maxima in the region in between. Many of the larger $\bar{v}_{max}$ (0.4 - 0.6 m/s) in the region $0 < x' < 0.3$ occur after the sandbar migrated far offshore in mid-October (Figure 2e).

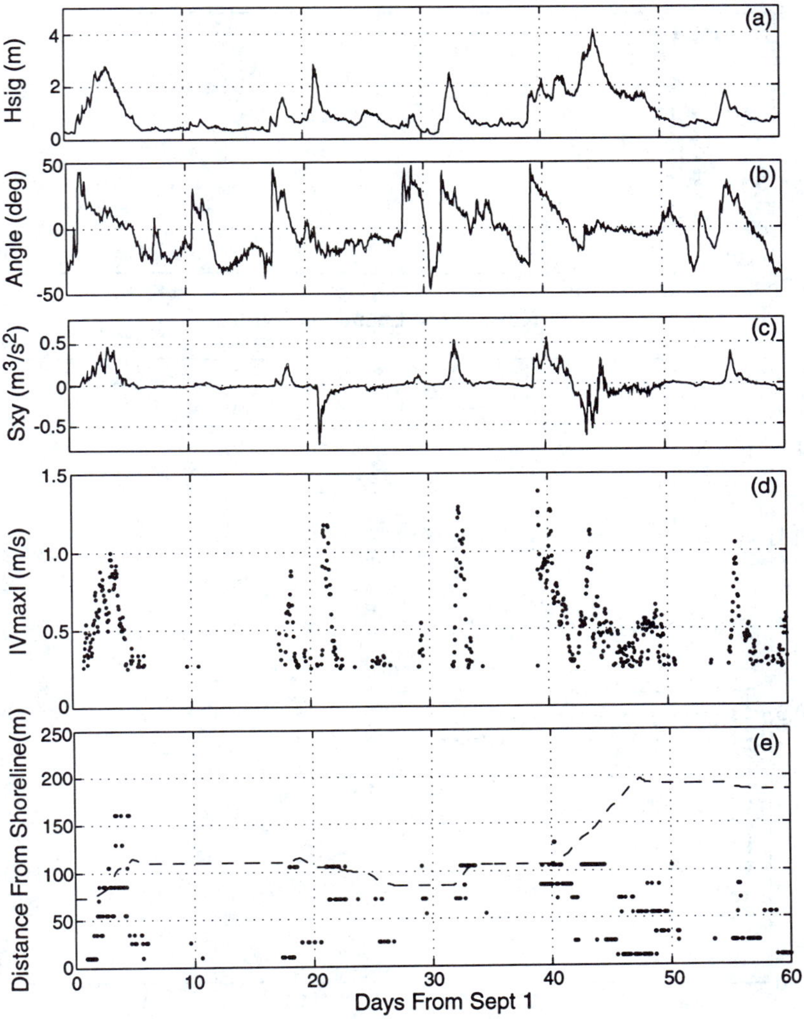

Figure 2. Hourly values of (a) significant wave height H_{sig} in 8 m depth (b) deviation of the mean incident wave angle from shore normal (positive angles correspond to waves from the northern quadrant) (c) S_{xy}/ρ in 8 m depth (d) maximum hour-averaged longshore current $\overline{V}_{max}$ (e) bar crest location (dashed) and the cross-shore location of $\overline{V}_{max}$. $\overline{V}_{max}$ is plotted only if there were at least five active current meters and $\overline{V}_{max} \geq 0.25$ m/s. Out of 1464 possible values, 572 hourly maxima pass these criteria. The few maxima occurring more than 250 m from shore are not shown in (e).

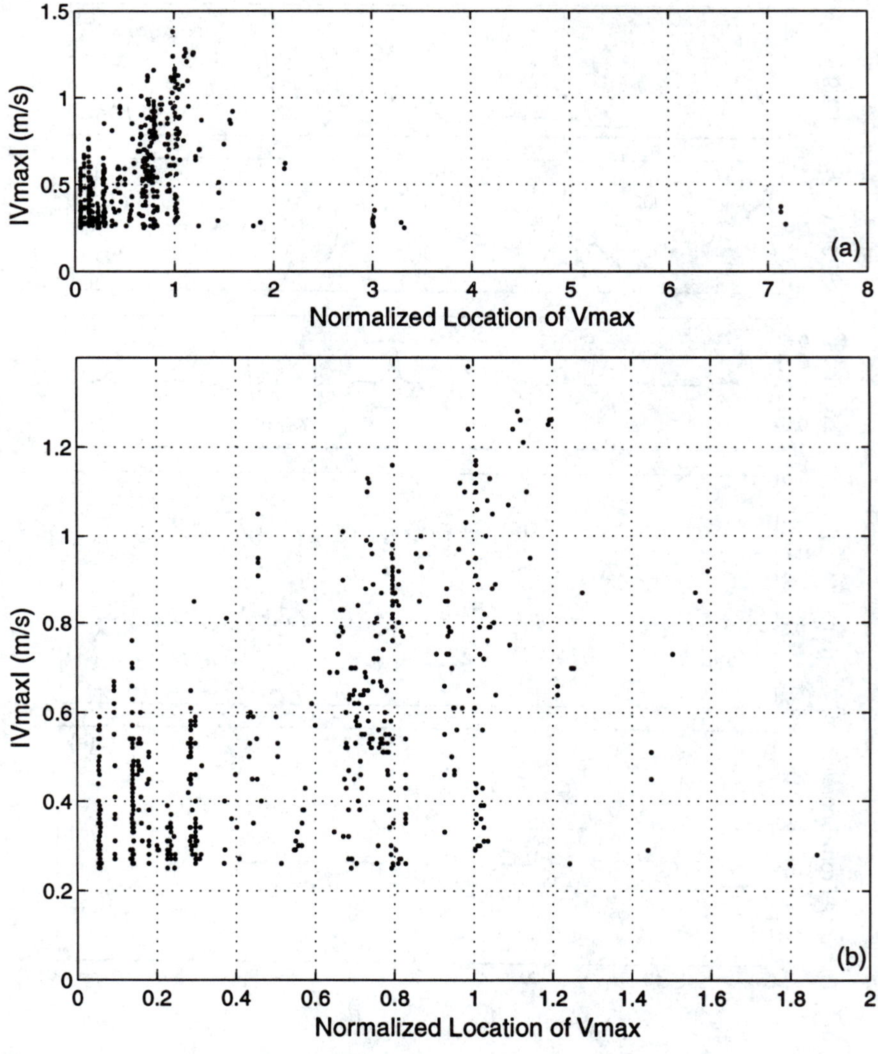

Figure 3. Magnitude of the maximum of the hourly averaged longshore current in normalized cross-shore coordinate, x' (1). the upper panel shows the entire cross-shore region. The region $0 < x' < 2$ is enlarged in the lower panel.

Tides dominate the variability of the local water depth, and thus affect wave shoaling, breaking, and the longshore current. When $\bar{v}_{max}$ is near the bar crest ($0.65 \leq x' \leq 1.2$), the location of $\bar{v}_{max}$ is tidally modulated (Figure 4). At high tide, $\bar{v}_{max}$ is rarely located seaward of $x' = 0.8$, whereas at low tide $\bar{v}_{max}$ is near $x' = 1$. Tidal effects on the location of $\bar{v}_{max}$ are smaller when $\bar{v}_{max}$ is near the shoreline ($x' < 0.3$). Tidal differences in the cross-shore structure of $\bar{v}$ are further illustrated in Figure 5. The current profiles at low tide exhibit a number of similar features; $\bar{v}_{max}$ is located close to $x' = 1$, and the current falls off rapidly shoreward of the maximum. Close to the beach, $\bar{v}$ is approximately 1/4 of $\bar{v}_{max}$ In contrast, at high tide $\bar{v}_{max}$ is farther shoreward (around $x' = 0.7$, consistent with Figure 4) and weaker. The current profile is broader, and near the shoreline $\bar{v}$ is about 1/2-2/3 of $\bar{v}_{max}$. The observed $\bar{v}$ tidal variation is consistent with the phase relationships between $\bar{v}$ and sea-level at tidal frequencies found by *Thornton and Kim* (1993). Wave-breaking induced gradients in significant wave heights are greater during low tide than high tide ($0.8 < x' < 1.5$ in Figure 6), resulting in smaller waves shoreward of the sandbar at low tide ($H_{sig} \sim 0.7$ m) relative to high tide ($H_{sig} \sim 1.0$ m).

The differences within the low (Figure 5a) and high (Figure 5b) tide cross-shore structure of $\bar{v}$ are primarily owing to differences in the conditions over the two days spanned by the observations. The bar was relatively stationary moving 7 m, H_{sig} in 8 m depth ranged from 1 to 2 m (Figure 6), the mean incident wave angle from 15 to 45 degrees, and S_{xy}/ρ from 0.2 to 0.5 m^3/s^2. Similar qualitative features in the tidal variation of $\bar{v}$ were observed at other times when the longshore current was strong for several tidal cycles (e.g. Sept 2-5 and Oct 10-16).

The modeled tidal variation of H_{sig} and $\bar{v}$, for waves and bathymetry representative of Figures 5 and 6, are shown in Figure 7. The wave heights are modeled using *Church and Thornton* (1993), with free model parameters selected to best fit the observed wave heights. The qualitative features of the observed H_{sig} distributions (Figure 6) are reproduced by the model (Figure 7a), except close to the shoreline. The longshore current predictions are made using the modeled H_{sig} variation and observed (in 8 m depth) directional wave properties in the *Thornton and Guza* (1986) longshore current model. A drag coefficient of 0.015 results in similar magnitudes for the modeled (Figure 7b) and observed (Figure 5) currents. Similar to previous results (e.g. *Church and Thornton*, 1993 and others), the modeled longshore currents have the familiar problem of predicting a flow with two maxima, one seaward of the bar crest and one near the shoreline, with no flow in the bar trough. The modeled low and high tide maxima occur at $x'=1.25$ and $x'=1.1$ respectively, farther offshore than observed (Figure 5). In physical units, the displacement of the maximum is about 30 m.

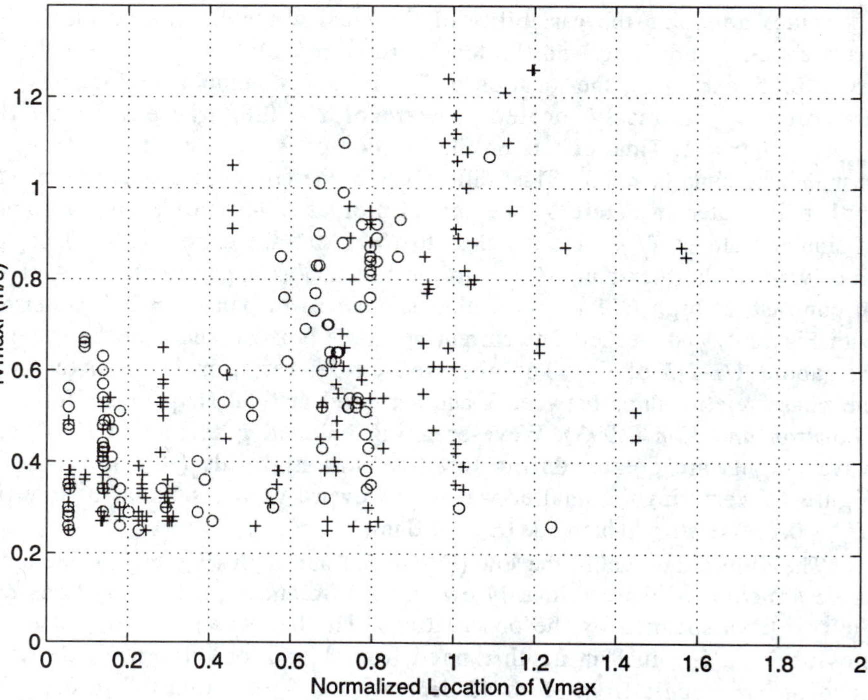

Figure 4. Location and magnitude of $\overline{V}_{max}$, observed within about 1.5 hours of tidal extrema in normalized cross-shore coordinate. There are 157 low tide (+) and 108 high tide (o) values.

Two commonly given reasons (among many) for the model failure are: (1) There may be a spatial lag in the transfer of momentum from the waves to the mean flow, possibly associated with wave rollers (*Svendsen*, 1984; *Dally and Brown*, 1995; and many others). (2) Longshore inhomogeneities in the bathymetry and wave field may result in nonlinear terms or longshore pressure gradient terms in the longshore momentum equation (e.g. *Putrevu et al.*, 1995). Some of the observations are consistent with the lag hypothesis, and there are other examples where alongshore inhomogeneities are likely dominant.

The cross-shore variation of S_{xy} for the successive low and high tide cases (Figure 5 and 6) is shown in Figure 8. At low tide, strong S_{xy} gradients are observed seaward of the bar ($1 \leq x' \leq 1.5$, Figure 8a), whereas shoreward of the bar crest, S_{xy} is relatively constant. The shoreward displacement of the observed longshore current maxima (Figure 5a) relative to strong S_{xy} gradients (Figure 8a) is consistent with a spatial lag in the transfer of momentum to the mean longshore current. At high tide, the region of strong S_{xy} gradients is slightly seaward of the bar crest ($0.9 \leq x' \leq 1.25$, Figure 8b), and seaward of the location of $\overline{v}_{max}$ ($x' \sim 0.7$) again indicating a spatial lag in the transfer of momentum to the mean flow. The tidal differences in the cross-shore structure of $\overline{v}$ (i.e. the the broading of $\overline{v}$ at high tide, Figure 5) may be related to tidal variations in the lagging mechanism.

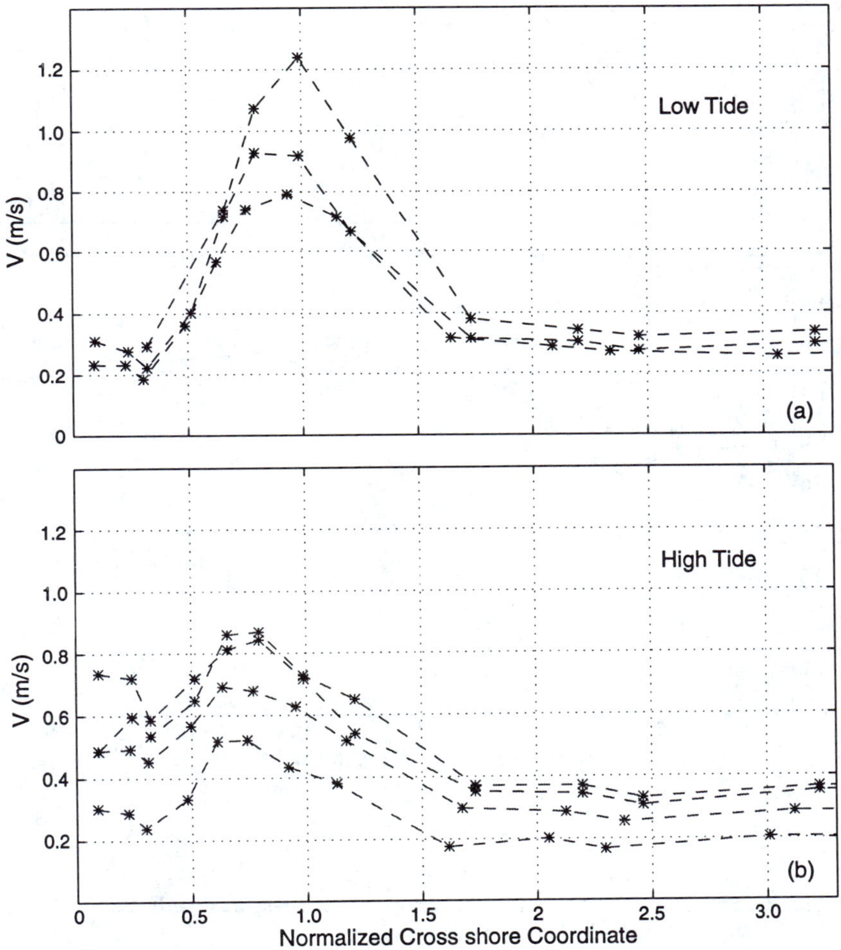

Figure 5. Mean longshore currents in the normalized cross-shore coordinate x' (1) during October 10 and 11. (a) 3 low tides (b) 4 high tides.

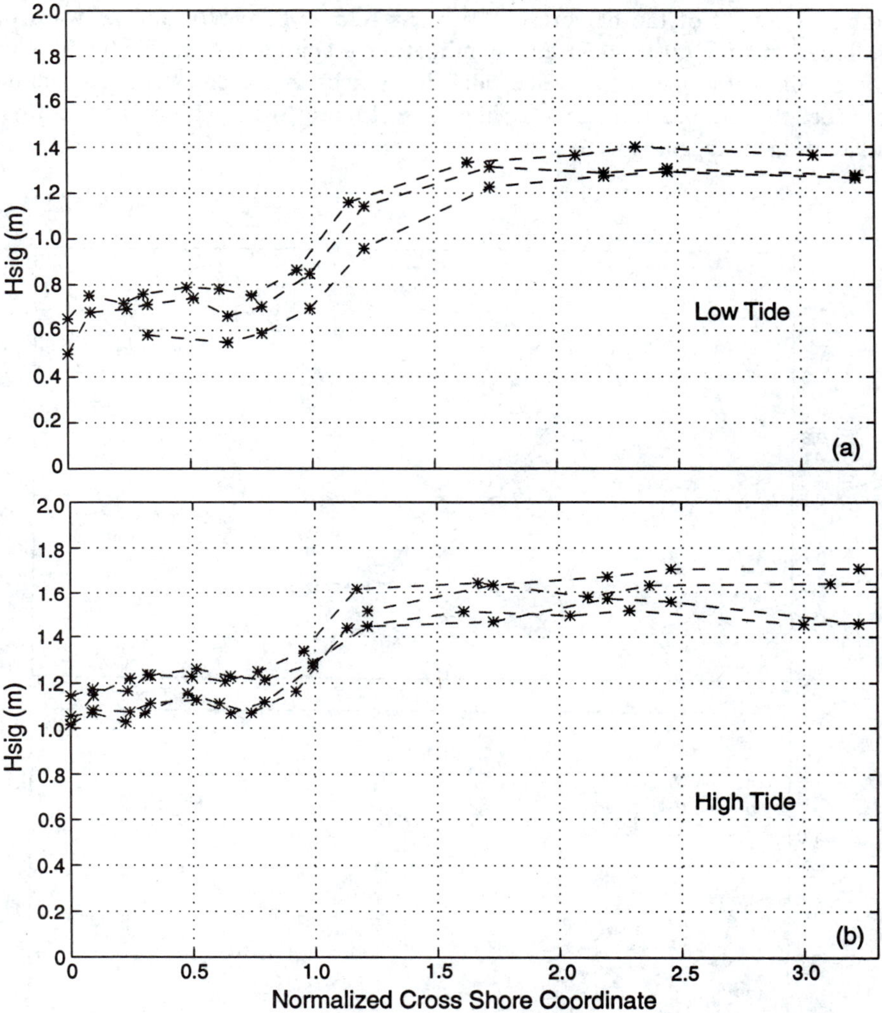

Figure 6. Significant wave height in the normalized cross-shore coordinate x' (1) during October 10 and 11. (a) 3 low tides (b) 4 high tides.

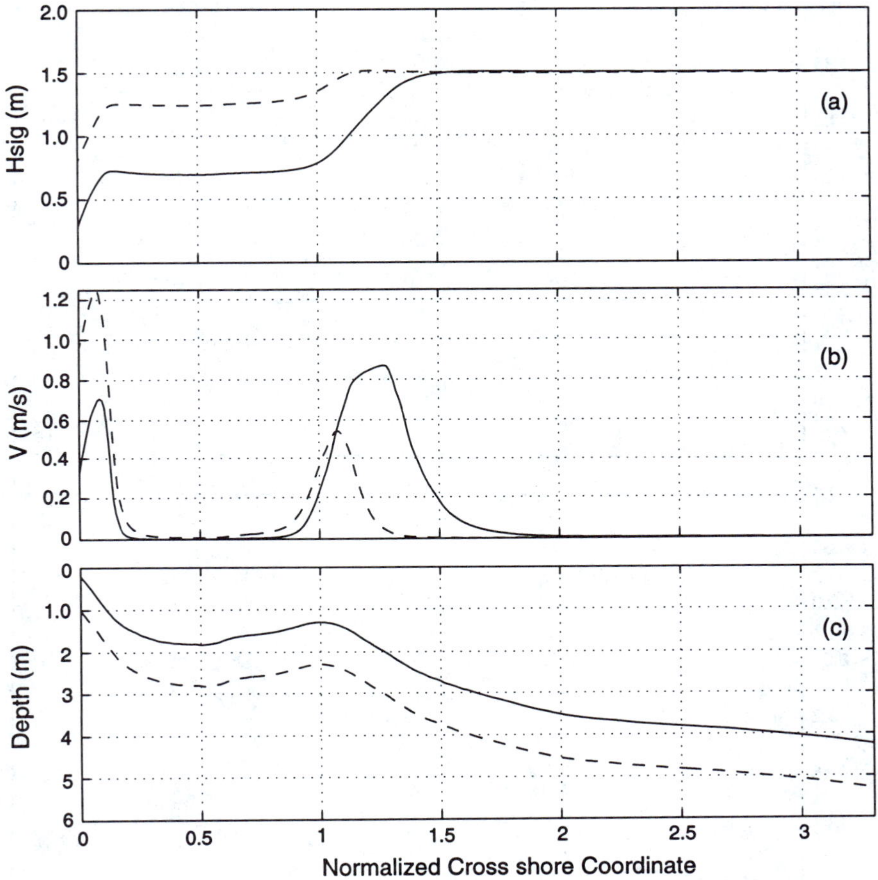

Figure 7. Modeled (a) significant wave heights (b) mean longshore currents (c) model bathymetry in normalized cross-shore coordinate (1). Solid curves represent low tide and dashed curves are high tide. The incident wave parameters are representative of conditions in Figure 5 and 6: H_{sig} = 1.5 m and θ = 25°.

An example of longshore currents likely affected by longshore bathymetric inhomogeneities is shown in Figure 9. Although the waves were energetic (H_{sig} = 3 m in 8 m depth), they were nearly normally incident (the mean incident wave angle in 8 m depth was 2°) and thus S_{xy} in 8 m depth was small (about 0.1 of S_{xy} with less energetic but more obliquely incident waves in Figure 8). The observations suggest that S_{xy} and S_{xy} gradients are small everywhere (Figure 9b), and that wave breaking began (x >200 m, Figure 9a) far

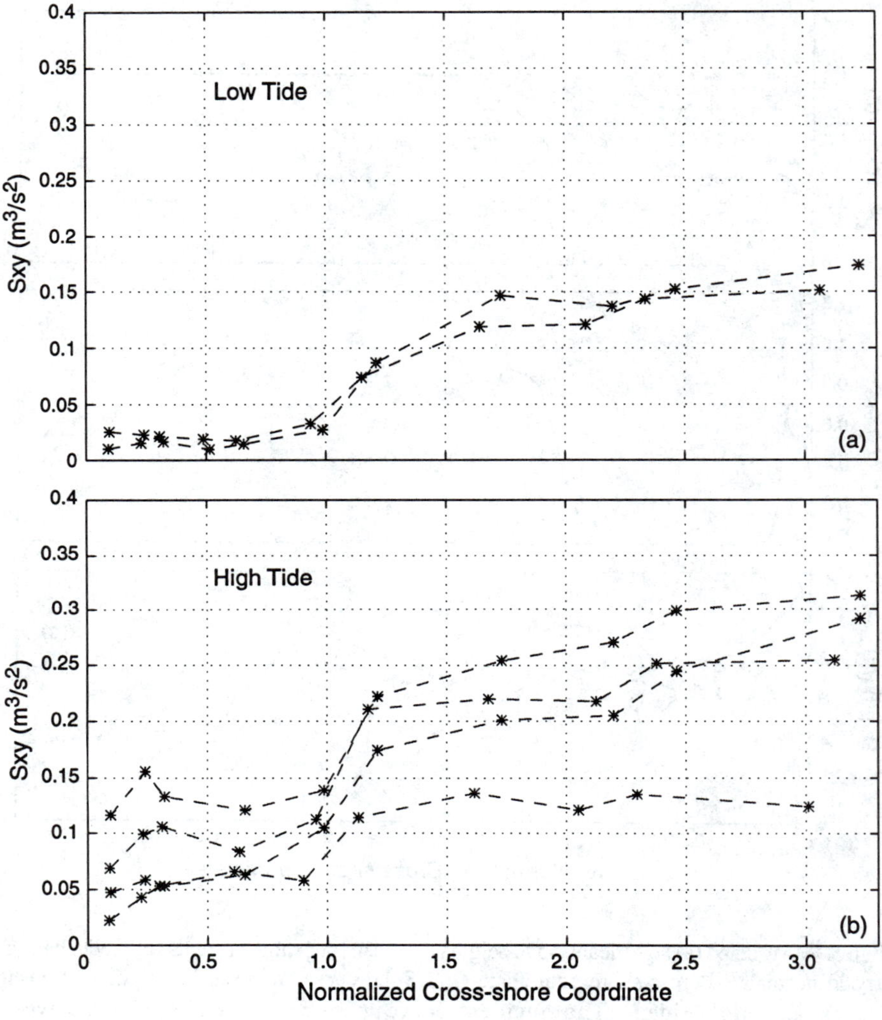

Figure 8. S_{xy}/ρ for the cases (October 10 and 11) shown in Figures 5 and 6 in the normalized cross-shore coordinate x' (1). (a) low tide (b) high tide. The accuracy of S_{xy} estimates is limited because of uncertainties in current meter response and orientation, and because errors in linear theory may be significant in the surfzone. Only two low tide profiles are shown because of unavailable data.

offshore of the location of $\overline{v}_{max}$ ($x \approx 25$ m, Figure 9c). The observed longshore current jet (Figure 9c) near the shoreline is inconsistent with model predictions (based on *Thornton and Guza*, 1986) of negligibly small currents (not shown). Time elapsed video images (R.A. Holman, personal communication, 1996) suggest that the bathymetry was longshore inhomogeneous. Three dimensional bathymetric surveys were not available because of the stormy conditions, however, the first post-storm survey (October 20) did reveal strong longshore inhomogeneity. *Sancho et al.* (1996) demonstrated that this inhomogeneous bathymetry can cause pressure gradients and nonlinear terms to become important in the longshore momentum balance.

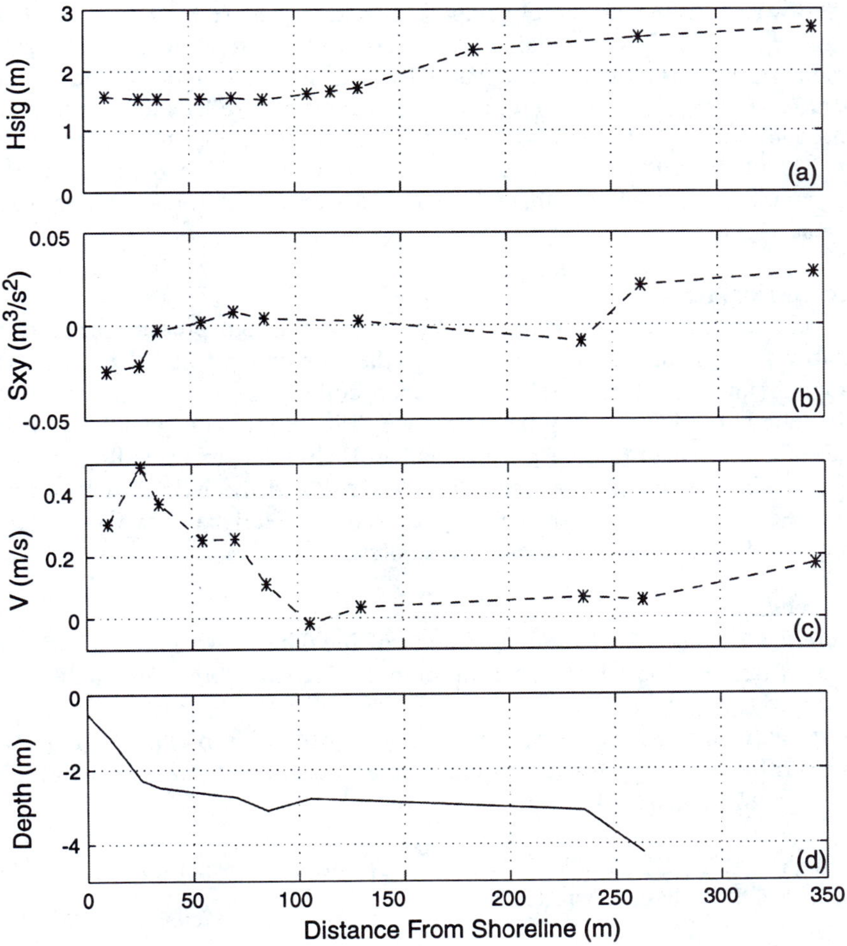

Figure 9. Observed cross-shore variation of (a) significant wave height (b) S_{xy}/ρ (c) longshore current $\overline{v}$ (d) bathymetry on 0500 October 10.

Conclusions

Observations spanning a wide range of bathymetric and wave conditions show that the probability distribution of cross-shore location of longshore current maxima is bimodal. Maxima typically occur near the crest of the sand bar or near the shoreline (Figure 3). At both high and low tide, the observed currents (Figure 5) are qualitatively consistent with the observed S_{xy} (Figure 8), although with a spatial lag, perhaps associated with wave rollers. The 1 m tidal fluctuations in sea level affect the cross-shore structure of the longshore current (Figure 4). At low tide, the current jet is stronger, narrower, and located closer to the bar crest than at high tide (Figure 5). The reason for the tidal changes in the longshore current structure is unknown, however tidal changes in the wave forcing are surely important. Clearly more quantitative work is necessary. The observed variation in the cross-shore structure of the longshore current across a tidal cycle is a robust feature (see also Figure 12 of *Thornton and Kim*, (1993)), and one step in the verification of a roller model would be to reproduce qualitatively the observed tidal variation of the longshore current. Finally, although the magnitude of the effects of longshore bathymetric inhomogeneities is not known, there are cases where they dominate the longshore current (Figure 9).

Acknowledgments

The array of current meters, sonar altimeters, and pressure sensors was deployed and maintained by staff from the Center for Coastal Studies. Britt Raubenheimer and Edith Gallagher helped collect the data. Excellent logistical support was provided by the FRF. We thank Edith Gallagher for providing sonar-altimeter based depth profiles and Rob Holman for providing the video images. This research was supported by ONR and the NSF CoOP program. Falk Feddersen received trainee support from the National Sea Grant College Program, NOAA, U.S. Department of Commerce.

References

Church, J.C. and E.B. Thornton, Effects of breaking wave induced turbulence within a longshore current model, *Coastal Eng.*, 20, 1-28, 1993.

Dally, W.R. and C.A. Brown, A modeling investigation of the breaking wave roller with application to cross-shore currents, *J. Geophys. Res.*, 100, 24,873-24,883, 1995.

Gallagher, E.L., Observation of sea oor evolution on a natural barred beach. SIO PhD thesis, 62pp. 1996.

Gallagher, E.L., W. Boyd, S. Elgar, R.T. Guza, B. Woodward, Performance of a sonic altimeter in the nearshore, *Marine Geology*, 133, 3, 241-248, 1996.

Lippmann, T.C., and R.A. Holman, The spatial and temporal variability of sand bar morphology, *J. Geophys. Res.*, 95, 11,575-11,590, 1990.

Long, C.E., Index and bulk parameters for frequency-direction spectra measured at CERC Field Research Facility, June 1994 to August 1995, *Miscellaneous Paper CERC-96-6*, U.S. Army Engineer Waterways Experiment Station, Vicksburg, MS, 1996.

Putrevu, U., J. Oltman-Shay, and I.A. Svendsen, Effect of alongshore nonuniformities on longshore current predictions, *J. Geophys. Res.*, 100, 16,119-16,130, 1995.

Sancho, F.E., I.A. Svendsen, A.R. Van Dongeren, and U. Putrevu, Longshore nonuniformities of nearshore currents, *Coastal Dynamics 1995*, 425-436, 1996.

Svendsen, I.A., Wave heights and set-up in a surf zone, *Coastal Eng.*, 8, 303-329, 1984.

Thornton, E.B. and R.T. Guza, Surf zone longshore currents and random waves: field data and models, *J. Phys. Oceanogr.*, 16, 1165-1178, 1986.

Thornton, E.B., and C.S. Kim, Longshore current and wave height modulation at tidal frequency inside the surf zone, *J. Geophys. Res.*, 98, 16,509-16,519, 1993.

CHAPTER 284

THREE-DIMENSIONAL HYDRODYNAMICS ON A BARRED BEACH

Tae-Myoung Oh[1] and Robert G. Dean[2]

ABSTRACT

Results are described from fixed bed wave tank tests to examine the generation and existence of three-dimensional flows on a barred beach. All tests were carried out with the use of two-dimensional and three-dimensional bar, which represented periodic forms parallel to the coast superimposed on a linear sloping beach. For these tests, wave heights and wave direction fields within and adjacent to the surf zone were documented to calculate the wave-induced horizontal torque based on mean vorticity and momentum equations. The circulation pattern clearly demonstrated that rip currents were stable on a three-dimensionally barred beach. It was found that the individual component including the depth gradients in the longshore direction, i.e., the effects of the bar morphology, tended to stabilize the bar-induced cellular circulation on the three-dimensional beach morphology although increased momentum fluxes of waves in the area with opposing currents induced by the bar tended to exert a counter force against them. It appeared that the mass transport over the more elevated portions of the bar tended to return to offshore through the deeper portions of the bar, where the depth gradients in the longshore direction existed. The cause is believed to be due to the greater hydraulic efficiency of flows through deeper portions of the profile.

[1]Research Director, Woobo Space Co., Ltd., Seongkok Bldg. 7FL., #823-22 Yeoksam-Dong, Kangnam-Ku, Seoul, Korea, Zip-Code 135-080

[2]Professor, Coastal and Oceanographic Engineering Department, University of Florida, Gainesville, Fl 32611

INTRODUCTION

There are several proposed mechanisms to explain the presence of three- dimensional forms that are fairly regular and generally associated with rip currents, which range from the relatively simple to the complex, including: (1) Instability between the hydrodynamics and sedimentary systems (Hino, 1974; Deigaard, 1990), (2) Interaction beween waves and currents (Dalrymple and Losano, 1978), (3) Interaction between synchronous edge waves and incident waves (Bowen, 1969; Bowen and Inman, 1969), (4) Intersecting wave trains (Dalrymple, 1975), and (5) Structural controls by natural and/or constructed features (Gourlay, 1976; Wind and Vreugdenhil, 1986). Some of those mechanisms (Hino, 1974; Dalrymple and Losano, 1978) contended that rip currents in nature could be explained by the reinforcement of three-dimensional flows over two-dimensional morphology. However, previous laboratory investigations (Dean and Oh, 1994) suggested that rip currents were neither self-reinforcing nor stable on two-dimensional topography due to the stabilizing wave-induced torque associated with the increased momentum fluxes in the vicinity of seaward mass return flow; hence, substantial rip currents could not occur without the presence of a rip channel.

This paper comprises the results of a series of fixed bed wave tank experiments conducted on a barred beach to examine the effects of three-dimensional bar morphology on the tendency for celluar circulation inside the surf zone, thus discussing the questions including: (1) Rip currents of significant strength can form on a barred beach, and (2) Whether the interaction between rips and incident waves causes increased or decreased wave heights at the location of the rip current with the presence of the rip channel.

Before presenting the experiments, it is helpful to discuss theoretical background governing three-dimensional circulation inside the surf zone. Primary emphasis is focused on the driving forces (torques) of the three-dimensional flows rather than the details of the hydrodynamics.

DRIVING FORCES OF THE THREE-DIMENSIONAL FLOWS

Consider a Cartesian coordinate system, where the y-axis is located at the still water line and the x-axis is directed seaward. Assuming horizontal flows, wave-induced circulation inside the surf zone is usually formulated based on depth-integrated, time-averaged continuity and momentum equations (Phillips, 1977). After neglecting the lateral shear stress coupling,

Continuity Equation

$$\frac{\partial \overline{\eta}}{\partial t} + \frac{\partial}{\partial x}[U(h+\overline{\eta})] + \frac{\partial}{\partial y}[V(h+\overline{\eta})] = 0 \quad (1)$$

Momentum Equations

$$\frac{\partial U}{\partial t} + U\frac{\partial U}{\partial x} + V\frac{\partial U}{\partial y} = -g\frac{\partial \overline{\eta}}{\partial x} + D_x + B_x \tag{2}$$

$$\frac{\partial V}{\partial t} + U\frac{\partial V}{\partial x} + V\frac{\partial V}{\partial y} = -g\frac{\partial \overline{\eta}}{\partial y} + D_y + B_y \tag{3}$$

where, t is the time, (U, V) the mean horizontal velocity components in the x- and y-direction respectively, $\overline{\eta}$ the mean water level measured from the still water level, h the still water depth, g the gravitational acceleration, and (B_x, B_y) are the shear stress components. The driving force components (D_x, D_y) are expressed in terms of the gradients of the radiation stresses.

$$D_x = -\frac{1}{\rho(h+\overline{\eta})}\left(\frac{\partial S_{xx}}{\partial x} + \frac{\partial S_{xy}}{\partial y}\right) \tag{4}$$

$$D_y = -\frac{1}{\rho(h+\overline{\eta})}\left(\frac{\partial S_{xy}}{\partial x} + \frac{\partial S_{yy}}{\partial y}\right) \tag{5}$$

in which,

$$S_{xx} = E[n(\cos^2\theta + 1) - \tfrac{1}{2}] \tag{6}$$

$$S_{xy} = E\, n \sin\theta \cos\theta \tag{7}$$

$$S_{yy} = E[n(\sin^2\theta + 1) - \tfrac{1}{2}] \tag{8}$$

where, E is the total wave energy per unit surface area and n is the ratio of the group velocity C_g to the wave celerity C.

Mean Vorticity Equations

Mean vorticity equation is obtained by cross-differentiating Equations (2) and (3) such that the set-up terms are eliminated.

$$\frac{\partial \overline{\omega}}{\partial t} + \frac{\partial}{\partial x}(U\overline{\omega}) + \frac{\partial}{\partial y}(V\overline{\omega}) = \frac{\partial D_y}{\partial x} - \frac{\partial D_x}{\partial y} + \frac{\partial B_y}{\partial x} - \frac{\partial B_x}{\partial y} \tag{9}$$

where, the mean vorticity, $\overline{\omega}$, is defined as

$$\overline{\omega} \equiv \frac{\partial V}{\partial x} - \frac{\partial U}{\partial y} \tag{10}$$

Sometimes, this time-averaged and depth-integrated vorticity equation is helpful since the contribution of set-up term to wave-induced current system is considered only indirectly through the total mean water depth $(h + \overline{\eta})$, thus making it possible to neglect $\overline{\eta}$ by assuming that $\overline{\eta}$ is small relative to the still water depth.

Vorticity Driving Torque based on Mean Vorticity Equation

The local vorticity driving torque, T_V, can be obtained directly from the mean vorticity equation presented in Equation (9):

$$T_V = \frac{\partial D_y}{\partial x} - \frac{\partial D_x}{\partial y} \tag{11}$$

in which D_x and D_y are given in Equations (4) and (5). Since D_x and D_y include the contributions by S_{xx}, S_{yy} and S_{xy}, the local torque T_V can also be expressed as the sum of three contributions: (1) $(T_V)_1$ by S_{xx}, (2) $(T_V)_2$ by S_{yy}, and (3) $(T_V)_3$ by S_{xy}.

$$T_V = (T_V)_1 + (T_V)_2 + (T_V)_3 \tag{12}$$

where,

$$(T_V)_1 = -\frac{1}{\rho(h+\overline{\eta})^2}\frac{\partial(h+\overline{\eta})}{\partial y}\frac{\partial S_{xx}}{\partial x} + \frac{1}{\rho(h+\overline{\eta})}\frac{\partial^2 S_{xx}}{\partial x \partial y} \tag{13}$$

$$(T_V)_2 = \frac{1}{\rho(h+\overline{\eta})^2}\frac{\partial(h+\overline{\eta})}{\partial x}\frac{\partial S_{yy}}{\partial y} - \frac{1}{\rho(h+\overline{\eta})}\frac{\partial^2 S_{yy}}{\partial x \partial y} \tag{14}$$

$$(T_V)_3 = \frac{1}{\rho(h+\overline{\eta})^2}\left[\frac{\partial(h+\overline{\eta})}{\partial x}\frac{\partial S_{xy}}{\partial x} - \frac{\partial(h+\overline{\eta})}{\partial y}\frac{\partial S_{xy}}{\partial y}\right] - \frac{1}{\rho(h+\overline{\eta})}\left[\frac{\partial^2 S_{xy}}{\partial x^2} - \frac{\partial^2 S_{xy}}{\partial y^2}\right] \tag{15}$$

Assuming that waves encounter the coast with nearly normal incidence (i.e., $\theta \cong \pi$), and using the shallow water asymptotes ($n = 1$), then the radiation stress terms can be written as

$$\begin{aligned} S_{xx} &= \tfrac{3}{16}\rho g H^2 \\ S_{xy} &= \tfrac{1}{8}\rho g H^2(\theta - \pi) \\ S_{yy} &= \tfrac{1}{16}\rho g H^2 \end{aligned} \tag{16}$$

Assuming further that $\overline{\eta}$ is small relative to still water depth, i.e., $(h+\overline{\eta}) \rightarrow h$ (this assumption will not be valid near the still water shoreline where $h = 0$), then vorticity driving torque components $(T_V)_1$, $(T_V)_2$ and $(T_V)_3$ can be written as follows:

$$(T_V)_1 = \frac{g}{16h}[-\frac{3}{h}\frac{\partial h}{\partial y}\frac{\partial H^2}{\partial x} + 3\frac{\partial^2 H^2}{\partial x \partial y}] \tag{17}$$

$$(T_V)_2 = \frac{g}{16h}[\frac{1}{h}\frac{\partial h}{\partial x}\frac{\partial H^2}{\partial y} - \frac{\partial^2 H^2}{\partial x \partial y}] \tag{18}$$

$$(T_V)_3 = \frac{g(\theta - \pi)}{8h}[\frac{1}{h}\frac{\partial h}{\partial x}\frac{\partial H^2}{\partial x} - (\frac{\partial^2 H^2}{\partial x^2} - \frac{\partial^2 H^2}{\partial y^2})] + \tag{19}$$

$$\frac{g}{8h}[\frac{1}{h}\frac{\partial h}{\partial x}H^2\frac{\partial \theta}{\partial x} - H^2(\frac{\partial^2 \theta}{\partial x^2} - \frac{\partial^2 \theta}{\partial y^2})] - \frac{g}{4h}[\frac{\partial \theta}{\partial x}\frac{\partial H^2}{\partial x} - \frac{\partial \theta}{\partial y}\frac{\partial H^2}{\partial y}]$$

Hence, if the distributions of wave height and local wave directions are known, the local vorticity driving torque induced on the surf zone can be calculated. It should be noted here that the torque T_V has dimensions of $[1/sec^2]$, which is not actual dimensions of torque, [force × length].

Vorticity Driving Torque based on Momentum Equations

Another expression for the vorticity driving torque, T_M, can be obtained by integrating local torque about a reference point over the entire surf zone.

$$T_M = \int\int \rho(h + \overline{\eta})[(x - x_o)D_y - (y - y_o)D_x]\, dxdy \tag{20}$$

where, (x_o, y_o) represent the coordinate of the reference point about which the total driving torque is to be calculated. Here, T_M has dimensions of torque, [force × length].

Using the radiation stress expressions given in Equation (16) with the assumption that $\overline{\eta}$ is small, T_M also can be written as the sum of three contributions: (1) $(T_M)_1$ by S_{xx}, (2) $(T_M)_2$ by S_{yy}, and (3) $(T_M)_3$ by S_{xy}, as follows:

$$T_M = (T_M)_1 + (T_M)_2 + (T_M)_3 \tag{21}$$

where,

$$(T_M)_1 = \int\int (y - y_o)\frac{\partial S_{xx}}{\partial x}dxdy = \frac{3\rho g}{16}\int\int (y - y_o)\frac{\partial H^2}{\partial x}dxdy \tag{22}$$

$$(T_M)_2 = -\int\int (x - x_o)\frac{\partial S_{yy}}{\partial y}dxdy = -\frac{\rho g}{16}\int\int (x - x_o)\frac{\partial H^2}{\partial y}dxdy \tag{23}$$

$$(T_M)_3 = \int\int [-(x - x_o)\frac{\partial S_{xy}}{\partial y} + (y - y_o)\frac{\partial S_{xy}}{\partial x}]dxdy \tag{24}$$

$$= \frac{\rho g}{8}\int\int [(y - y_o)\frac{\partial\{H^2(\theta - \pi)\}}{\partial y} - (x - x_o)\frac{\partial\{H^2(\theta - \pi)\}}{\partial x}]dxdy$$

Hence, T_M also can be calculated if the distributions of the radiation stresses, i.e., the distributions of wave heights and wave incidence angle, are known.

EXPERIMENTS

All experiments were conducted in a wave tank of the Coastal and Oceanographic Engineering Laboratory of the University of Florida. This tank is approximately 20 *m* long, 1.5 *m* high and 0.6 *m* wide, and is equipped with a piston type wavemaker and has glass panels on both sides of the tank. The water depth was 0.40 *m* in the horizontal section of the tank. Two types of regular waves of 3 *cm* and 9 *cm* heights were used while the wave period was fixed at 1.43 *sec* during the experiments. The bathymetry was configured to represent a bar morphology by superimposing two-dimensional and three-dimensional wood bars parallel to the coast on a uniform slope of 1:20. The two-dimensional bar extended to the whole width of the tank, while the three-dimensional bar extended from one wall to two-thirds of the tank width. These bars were installed approximately at the surf line, which depends on wave height condition. All four tests (two waves and two bars) were carried out to document the wave heights and wave direction distribution within and adjacent to the surf zone. The details are described in Oh (1994).

Wave hights and direction fields were measured to document and interpret the interaction of the waves and bar-induced currents. Figures 1 and 2 present the isolines of wave heights in the presence of the three-dimensional bar for the small waves and for the large waves, respectively. For the small waves, as shown in Figure 1, wave crest lines were almost uniform in the longshore direction even in the presence of the three-dimensional bar, which implied that the return flow due to the three-dimensional bar was weak to modify the wave fields. It is seen from Figure 2 that waves were much higher in the vicinity of the seaward return flow than those near the bar area. Much higher waves appeared to be caused by increased return current induced by the three-dimensional bar. However, it should be noted that the breakerline was not significantly influenced by the return flow. [It was observed by Dean and Oh (1994) that the breakerline was significantly displaced seaward near the return current during the circulation tests with the jet present on a prismatic beach.] Uniform breakerline with the three-dimensional bar morphology appeared to be due to both relatively weak return current induced by the three-dimensional bar and deeper water depth at the return current area than the area over the bar, and it was agreed with the field observations by Sonu (1972).

RESULTS AND DISCUSSIONS

Vorticity Driving Torque based on Mean Vorticity Equation

Considering that waves encounters the coast with normal incidence ($\theta = \pi$), Equations (17), (18), and (19) can be expressed as the combinations of different individual components depending on the wave height gradients and water depth near the bar area, as follows:

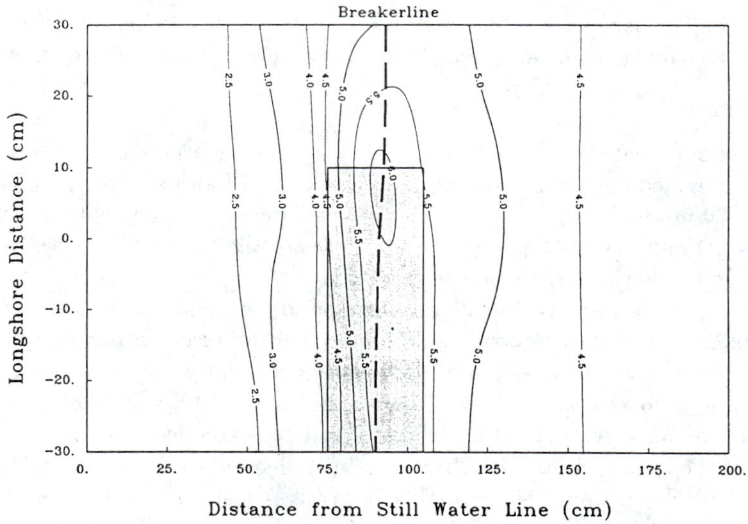

Figure 1: Wave Height Distributions in the Presence of Three-Dimensional Bar when $H_o = 3cm$. Wave heights are given in cm. A dark area represents the area covered by the three-dimensional bar. Note that wave fields are not greatly influenced by the return flow induced due to the presence of the three-dimensional bar, resulting in more or less uniform breakerline.

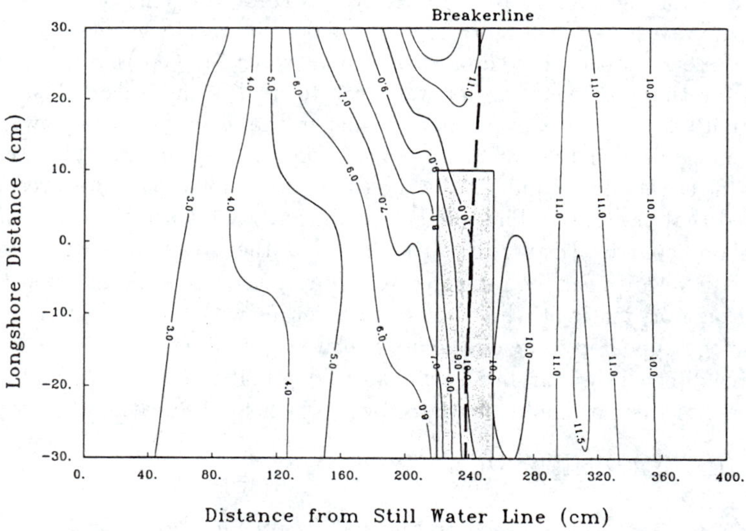

Figure 2: Wave Height Distributions in the Presence of Three-Dimensional Bar when $H_o = 9cm$. Wave heights are given in cm. A dark area represents the area covered by the three-dimensional bar. Note that wave fields are substantially influenced by the return flow induced due to the presence of the three-dimensional bar. Note also almost uniform breakerline due to deeper water depth at the area of return flow than the area over the bar.

$$\begin{aligned} (T_V)_1 &= 3\ (II - VI) \\ (T_V)_2 &= I\ -\ II \\ (T_V)_3 &= 0 \end{aligned} \tag{25}$$

where,

$$I = \frac{g}{16h^2}\frac{\partial h}{\partial x}\frac{\partial H^2}{\partial y} \tag{26}$$

$$II = \frac{g}{16h}\frac{\partial^2 H^2}{\partial x \partial y} \tag{27}$$

$$VI = \frac{g}{16h^2}\frac{\partial h}{\partial y}\frac{\partial H^2}{\partial x} \tag{28}$$

The marginal distributions of individual components are presented in Figure 3 for the large waves. When the three-dimensional bar is installed, the vorticity driving torque is mainly contributed by Component VI, which includes the gradients of water depth in the y-direction, $\frac{\partial h}{\partial y}$, and is confined in the vicinity of the three-dimensional bar area. With the two-dimensional bar present, in which case $\frac{\partial h}{\partial y}$ is zero, the magnitude of the Components I and II are negligible. It is noted for the large wave tests in the presence of the three-dimensional bar that Component I increases up to a comparable magnitude to Component VI, as shown in Figure 3(b). This effect might be due to the increase in the gradients of wave heights in the y-direction. Due to space limitations, the results for the small wave condition will not be presented here; however, it was found that the results were consistent with those for the large waves.

The marginal distributions of the total torque are shown in Figure 4 for the large wave condition in the presence of the three-dimensional bar. Positive circulations occur within the surf zone and in the area seaward of the bar region but the magnitudes of these circulations are smaller than and overshadowed by the negative circulation near the bar area. Hence, it would be necessary for future experiments to simulate the more realistic and smooth bar morphology which could avoid these exaggerated effects of the depth gradients in the y-direction. However, these results clearly demonstrate that rip currents are stable on a three-dimensional barred beach although increased momentum fluxes of waves in the area with opposing currents induced by the bar tend to work against rip currents. It appeared that the water mass transported over the bar tended to return to offshore through deeper area in the bar morphology.

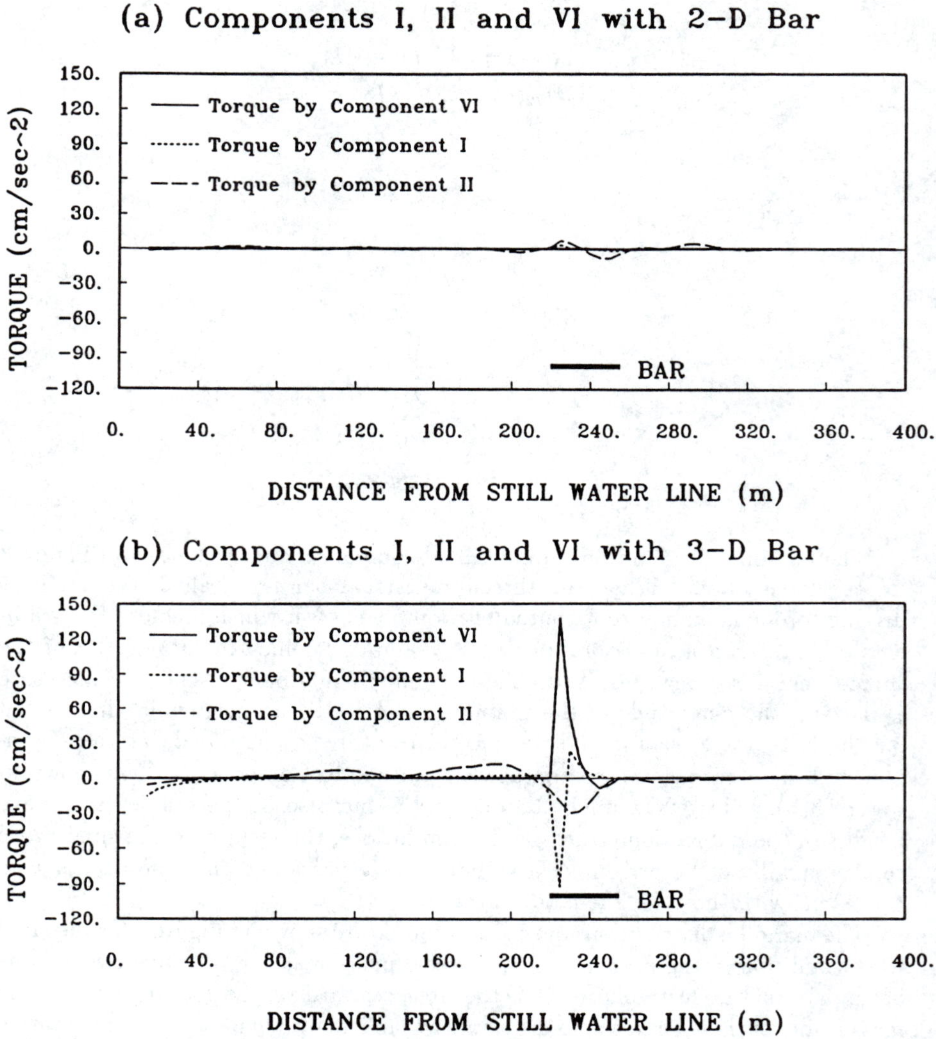

Figure 3: Marginal Distributions of Individual Components along x-axis in cm/sec^2 when $H_o = 9cm$: (a) Components I, II and VI in the presence of the two-dimensional (2-D) bar and (b) Components I, II and VI in the presence of the three-dimensional (3-D) bar

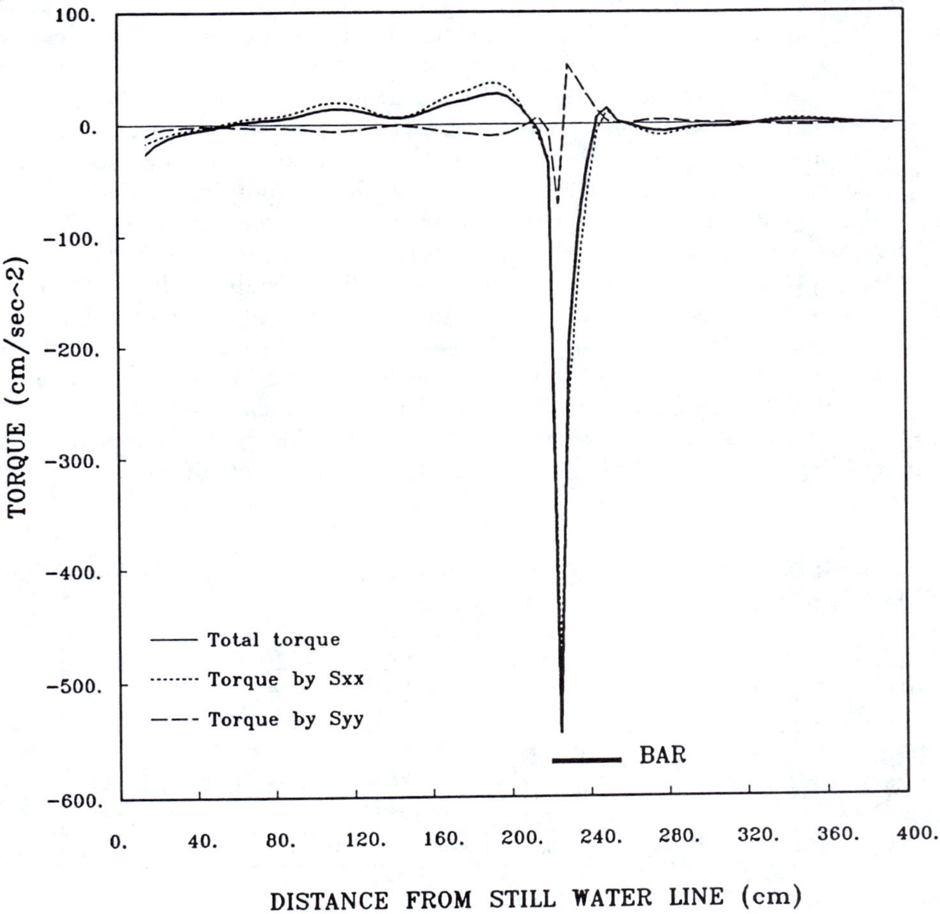

Figure 4: Marginal Distributions of Total Torque and the Contributions of Radiation Stresses Components along x-axis in cm/sec^2 when $H_o = 9cm$ in the Presence of the Three-Dimensional Bar. Note that torque by S_{xx} is dominant over the torque contributed by S_{yy}. Note also the positive values within the surf zone and in the area seaward of the bar. Hence, positive circulation occurs within the surf zone, which tends to suppress the bar-induced circulation.

Vorticity Driving Torque based on Momentum Equation

Marginal distributions of the total torque and the contributions by S_{xx} and S_{yy} were calculated based on the momentum equations, as presented in Equations (21) to (24), and the results are presented in Figure 5 for the large waves ($H_o = 9cm$) in the presence of the three-dimensional bar. For these calculations, the reference point was selected as the centerline point at the same x-location as the location of the bar crest. The contribution by S_{yy} dominates over the contribution by S_{xx} within the whole area of interest except the area over the bar, where the contribution by S_{xx} is dominant since $(x - x_o)$ is small in this area. The total torque changes from negative values near the shoreline to positive values within the surf zone and again to negative values over the bar, and demonstrates oscillating patterns beyond the breakerline. This figure clearly exhibits the opposing effects of waves to the bar-induced circulation, thus tending to stabilize the surf zone.

Table 1 summarizes the total torque induced on the area of interest from the shoreline to offshore. The positive values of the total torque T_M demonstrate that the wave-induced torque is in opposition to the three-dimensional bar-induced circulation (flowing from the bar area, i.e. shoals, to the base of return current, i.e. the embayments). The opposing wave effects appeared to be due to the increased momentum fluxes associated with higher waves in the vicinity of the return flow.

Table 1: Summary of Total Vorticity Driving Torque based on the Momentum Equation.

Item	Unit	$H_o = 3cm$		$H_o = 9cm$	
		2-D Bar	3-D Bar	2-D Bar	3-D Bar
$\sum\sum D_x$	N	-0.98	-1.29	-8.08	-8.40
$\sum\sum D_y$	N	0.049	0.011	0.23	-4.04
$(T_M)_1$ by S_{xx}	$N-m$	-0.002	-0.005	-0.001	0.035
$(T_M)_2$ by S_{yy}	$N-m$	0.015	0.036	0.376	1.103
Total torque, T_M	$N-m$	0.013	0.031	0.375	1.138

SUMMARY AND CONCLUSIONS

A series of fixed bed tests was carried out to investigate the generation and existence of three-dimensional flows on a barred beach. The bar morphology represented periodic forms parallel to the coast superimposed on a linear sloping beach. For these tests, wave heights and wave direction fields within and adjacent to the surf zone were documented to calculate the wave-induced horizontal torque.

It was found for the large wave condition ($H_o = 9cm$) that three-dimensional circulations occurred due to the effect of the three-dimensional bar morphology and flowed from the bar area (shoals) to the return channel (embayments), thus

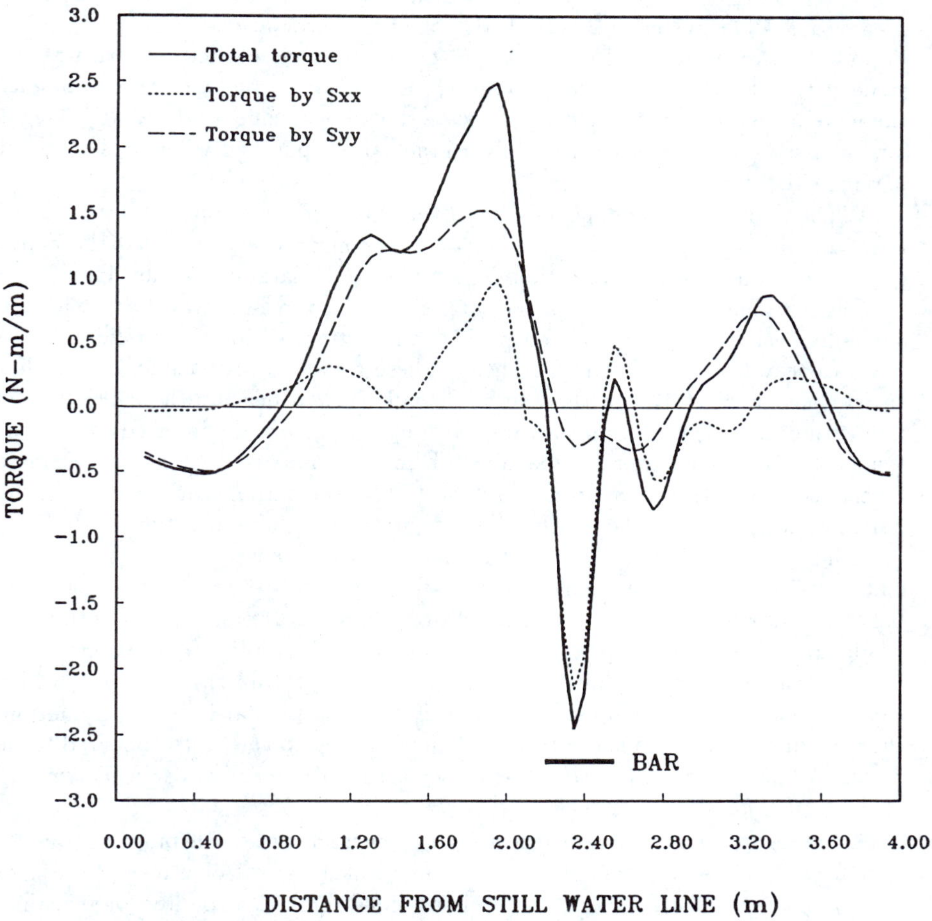

Figure 5: Marginal Distributions of Total Torque and the Contributions of Radiation Stresses Components based on Momentum Equations along x-axis in $N-m/m$ when $H_o = 9cm$ with the Three-Dimensional Bar. Note that the contribution by S_{yy} is dominant term over the whole area of interest except over the bar area, where the contribution by S_{xx} is dominant. Positive circulation occurs within the surf zone, resulting in opposing effects of waves on the three-dimensional bar-induced circulation.

modifying the wave fields such that waves were higher in the vicinity of the return current. However, the breakerline was not significantly influenced by the return flow since the return current induced by the three-dimensional bar was weak and the water depth in the vicinity of the return current was deeper than that in the area over the bar, which was agreed with the field observations.

When waves were small ($H_o = 3cm$), the resulting circulation pattern was the same as that for the large waves; however, the circulation was too weak to modify the wave fields. These results could indicate that rip currents need the wave height larger than a lower limit to form even with the bar morphology, which was observed in the field.

When individual components of the vorticity driving torque was calculated based on the mean vorticity equation, the component which included the depth gradients in the longshore direction demonstrated the largest contribution.

The marginal distributions of the total torque for the large wave tests with the three-dimensional bar present show positive circulations within the surf zone and in the area seaward of the bar region; hence, these results clearly demonstrated that rip currents were stable on three-dimensionally barred beach although increased momentum fluxes of waves in the area with opposing currents induced by the bar tended to exert a counter force against them. It appeared that the water mass transported over the bar tended to return to offshore through deeper area in the bar morphology, where the depth gradients in the longshore direction existed.

When the vorticity driving torque was calculated based on the momentum equations for the large wave condition tests in the presence of the three-dimensional bar, the contribution by S_{yy} dominated over the contribution by S_{xx} within the whole area of interest except the area over the bar. The total torque changed from negative values near the shoreline to positive values within the surf zone, which clearly demonstrated the opposing effects of waves to the bar-induced circulation, thus tending to stabilize the surf zone. The opposing wave effects appeared to be due to the increased momentum fluxes associated with higher waves in the vicinity of the return flow.

In aggregate, these studies suggest that rip currents are stable on three- dimensionally barred beach although increased momentum fluxes of waves in the area with opposing currents induced by the bar tended to exert a counter force against them. It appeared that the water mass transported over the bar tended to return to sea through deeper area in the bar morphology. This cause is believed to be due to the greater hydraulic efficiency of flows through deeper portions of the profile, which leads to a hydraulic/sedimentary instability which causes development of a vestigial channel.

REFERENCES

Bowen, A.J. (1969), "Rip Currents. Part 1: Theoretical Investigation", J. of Geophysical Research, Vol. 74, No. 23, pp. 5467-5478.

Bowen, A.J. and D.L. Inman (1969), "Rip Currents. Part 2: Laboratory and Field Observations", J. of Geophysical Research, Vol. 74, No. 23, pp. 5479-5490.

Dalrymple, R.A. (1975), "A Mechanism for Rip Current Generation on an Open Coast", J. of Geophysical Research, Vol. 80, No. 24, pp. 3485-3487.

Dalrymple, R.A. and C.J. Lozano (1978), "Wave-Current Interaction Models for Rip Currents", J. of Geophysical Research, Vol. 83, No. C12, Paper No. 8C0556, pp. 6063-6071.

Dean, R.G. and T.M. Oh (1994), "Three-Dimensional Morphology in a Narrow Wave Tank: Measurement and Theory", Proc. 24th Int. Conf. on Coastal Eng., Kobe, Japan, ASCE

Deigaard, R. (1990), "The Formation of Rip Channels on a Barred Coast", Progress Report No. 72, Institute of Hydrodynamics and Hydraulic Eng., ISVA, Tech. Univ. Denmark, pp. 65-74.

Gourlay, M. R. (1976), "Non-Uniform Alongshore Currents", Proc. 15th Int. Conf. on Coastal Eng., Honolulu, Hawaii, ASCE, pp. 701-720.

Hino, M. (1974), "Theory on Formation of Rip-Current and Cuspidal Coast", Proc. 14th Int. Conf. on Coastal Eng., Copenhagen, Denmark, ASCE, pp. 901-919.

Oh, T.M. (1994), "Three-Dimensional Hydrodynamics and Morphology Associated with Rip Currents", Ph.D. Dissertation, Dept. of Coastal and Oceanographic Eng., Univ. of Florida, Gainesville, Florida, 210 pp.

Phillips, O.M. (1977), "The Dynamics of the Upper Ocean", Cambridge Univ. Press, New York, 363 pp.

Sonu, C.J. (1972), "Field Observation of Nearshore Circulation and Meandering Currents", J. of Geophysical Research, Vol. 77, No. 18, pp. 3232-3247.

Wind, H.G. and C.B. Vreugdenhil (1986), "Rip-Current Generation near Structures", J. of Fluid Mech., Vol. 171, pp. 459-476.

CHAPTER 285

Rip Current Generation on a Plane Beach

D.A. Suriamihardja[1] and Yoshito Tsuchiya[2], Member, ASCE

ABSTRACT: In order to establish a wave-current interaction model for rip currents, a set of first-order linearized governing equations describing the 2-D nearshore circulation is derived from the equations of nearshore currents by the perturbation method. Wave refraction induced by the generated currents is fully considered. The solutions to the field equations of rip currents both in the surf zone and shoaling zone are obtained, and expressed in the form of Gaussian hypergeometric functions and modified Bessel functions, respectively. The matching condition of the solutions at the breaking point determines the spacing of the rip currents. Comparison of the computed rip current spacing, circulation pattern, and rip current discharge, with laboratory and field data shows a satisfactory agreement especially in the so-called instability region of the surf similarity parameter.

INTRODUCTION

The change in momentum flux of incoming waves can be described by the offshore and longshore gradients of the radiation stresses which act as the driving forces in the nearshore circulation system. Therefore to formulate the driving forces in the momentum equations for the generation of nearshore currents is of importance (Dingemans, Radder and De Vriend, 1987). Two main causes of the driving forces are the so-called wave-current interaction, and structural interaction (Dalrymple and Lozano, 1978).

The first investigation of nearshore currents as 2-D horizontal circulation generated by the interaction between the incoming waves and the resulting rip currents was by Le Blond and Tang (1974), whose theory applies to a plane beach and normally incident waves. Similar to their work, Iwata (1976) developed a theory which assumed the nonuniformity of bottom friction between surf and shoaling zones. However, these two theories still failed in obtaining the nondimensional alongshore spacing of rip currents as an eigenvalue of the governing equations. Iwata further attempted to find

---

[1]Lecturer, Department of Physics, Faculty of Mathematics and Natural Sciences, Hasanuddin University, Ujung Pandang 90245, Indonesia.

[2]Professor Emeritus, Kyoto University, and Professor, Meijo University, Kani 509-02, Japan.

currents as an eigenvalue of the governing equations. Iwata further attempted to find an asymptotic solution for both the surf and shoaling zones. Finally, he obtained the real eigenvalue from the characteristic equation which is derived by matching conditions at the breaking point, and found that its value is a function of a parameter determined by the bottom friction coefficient, the surf zone width and the breaker height.

Mizuguchi (1976) also attempted to obtain the eigenvalues from the characteristic equation without using any approximation such as asymptotic solution as in Iwata's work, by considering both the uniformity and nonuniformity of the bottom friction. However, no eigenvalue was obtained. He concluded that this failure occurred due to the exclusion of lateral mixing in the governing equation, and because the contribution of the bottom friction was not sufficient to represent the dissipative effect in the nearshore current. Consequently, he reformulated the bottom friction term to be a function of the distance from the shoreline, similar to the lateral mixing. Dalrymple and Lozano (1978) argued that no reason exists to justify this formulation, so that the real eigenvalue obtained as a function of a parameter proportional to the so-called surf zone similarity parameter and inversely proposional to the bottom friction was invalid.

Dalrymple and Lozano (1978) presented two wave-current interaction models. One was similar to the theory of Le Blond and Tang (1974), in which changes in local wave length due to currents were considered. However, their assumption of an extremely small refraction angle, which implied that no longshore variation in wave orthogonals was allowed, resulted in no rip-current formation. In the second model, the wave-current interaction effect was considered through wave refraction due to current, and the formation of longshore periodic nearshore circulation cells was calculated numerically. The obtained eigenvalue was a function of a parameter expressed by the ratio of beach slope to bottom-friction coefficient. The relationship between the eigenvalue and this parameter showed that the rip-current spacing increases as the parameter increases and vice-versa. Comparison with the rip current spacing of Balsillie (1975) obtained by field measurement showed a good agreement in the region of small values of the parameter, which implied that the theoretical eigenvalue of Dalrymple and Lozano was partly suitable for the prediction of rip-current spacings generated by the incidence of infragravity waves, as it was stated by Balsillie that almost all of the data are catagorized into this type of waves. However, when compared with the field measurements collected by Sasaki (1977), which were made in wider regions such as the regions of instability and edge waves, the theoretical spacing showed a lower value than that of field measurements.

The present study investigates the steady-state nearshore circulation on a plane beach based on the interaction between normally-incident waves and the resulting rip currents. A mathematical formulation of the governing field equations for rip currents on a plane beach is made by means of the wave-current interaction model, including the formulation of the driving forces. The governing equations of 2-D nearshore currents on the plane beach are first established by employing the conservation laws for mass, momentum and wave action. The so-called mild slope equation (MSE), which is able to calculate wave transformation due to interaction with the nearshore currents, such as wave refraction, is applied to formulate the driving forces in the momentum equations. Nevertheless, to simplify the analysis without sacrificing the generality of the theory, the lateral mixing terms are omitted. Using a perturbation method for a small parameter of beach slope, the field equations of rip currents are formulated. At the first order of approximation, the field equations of rip currents are obtained in both the surf and shoaling zones, where the wave refraction due to currents is fully considered. With some additional assumptions, solutions to the field equations are obtained both in the surf and shoaling zones in terms of the Gaussian hypergeometric function and the modified Bessel function, respectively. The matching condition for these solutions at the breaking point makes it possible to determine the eigenvalues of the characteristic equation of the derived field equations.

The theoretical results of the rip current characteristics such as the rip current spacing and flow patterns, are compared with both the previous theoretical results, laboratory and field data. The comparison of the theoretical rip current spacings and those of field data showed a satisfactory agreement. Current patterns in a nearshore circulation cell are also calculated numerically through the determination of integration constants in the stream function based on energy budgeting at the breaking point, which resulted in a reasonable value of rip discharge, both at laboratory and field scales.

THE BASIC EQUATIONS OF NEARSHORE CURRENTS

The MSE derived by Kirby (1984) which includes an additional wave energy dissipation term, $i\omega_r W\Phi$, is employed to formulate the driving forces. The terms in the MSE of Kirby are as follows: W is the ratio of the wave energy dissipation rate D to the total wave energy E, C is the wave celerity, C_g is the group velocity, where both values are assumed to be nearly equal due to the shallow water approximation, and D/Dt is the Lagrangian derivative. By introducing velocity potential of linear wave theory, $\Phi = \phi \exp(-i\omega t)$, and $\phi = (ga/i\omega_r)\exp(iS)$, where S is the phase function, into the MSE of Kirby in the steady state condition, the MSE is written as:

$$\left\{\frac{\partial}{\partial x_j}\left(C C_g \frac{\partial}{\partial x_j}\right) + 2i\omega U_j \frac{\partial}{\partial x_j}\right\}\phi = -\left\{i\left(\omega\frac{\partial U_j}{\partial x_j} + \omega_r W\right) + \left(\omega^2 + k^2 C C_g - \omega_r^2\right)\right\}\phi \qquad (1)$$

Following lengthy algebraic procedures using the expressions for the radiation stresses by Dingemans, Radder and De Vriend (1987), and for the depth-integrated mass transport by Crapper (1984) (see Tsuchiya and Suriamihardja, 1989), the driving force F_i can be obtained as:

$$F_i = \frac{\rho\omega_r}{2g}\frac{\partial S}{\partial x_i}W\phi\phi^* - \frac{\rho}{4g}d\frac{\partial}{\partial x_i}\left(\phi\phi^*\frac{\omega_r^2}{d}\right) + \frac{\rho\omega_r}{2g}\phi\phi^*\frac{\partial S}{\partial x_j}\left(\frac{\partial U_i}{\partial x_j} - \frac{\partial U_j}{\partial x_i}\right) - \frac{\rho}{4g}C C_g\frac{\partial}{\partial x_i}\left(\frac{\phi\phi^*}{A}\frac{\partial^2 A}{\partial x_j^2}\right) \qquad (2)$$

where ϕ^* is the complex conjugate of ϕ. The first term in the right side of (2) is the rotational term, and is the contribution from the dissipation of wave energy. The second is the irrotational term, and the third term describes the interaction between wave-induced mass transport and currents. The fourth term represents the additional effect resulting from the diffraction of waves. Dingemans, Radder and De Vriend (1987) demonstrated that only these rotational terms are able to generate non-zero depth-averaged current velocities. For simplicity, the horizontal components of the nearshore current are assumed to be independent of depth. Therefore the conservation laws to be employed in this study are presented in depth-integrated form (For example, Phillips, 1966; Dolata and Rosenthal, 1984; and Crapper, 1984). Neglecting the wave diffraction effect and assuming the shallow water condition, the steady state mass and momentum equations can finally be reduced to

$$\frac{\partial}{\partial x_j}\left(\rho U_j d\right)=0 \tag{3}$$

$$U_j\frac{\partial U_i}{\partial x_j}+R_i=-g\delta_{ij}\frac{\partial \bar{\eta}}{\partial x_j}+\frac{D}{\rho\omega_r d}\frac{\partial S}{\partial x_i} -\frac{\partial}{\partial x_i}\left(\frac{E}{2\rho d}\right)+\frac{E}{\rho\omega_r d}\frac{\partial S}{\partial x_i}\left(\frac{\partial U_j}{\partial x_i}-\frac{\partial U_i}{\partial x_j}\right) \tag{4}$$

where $$R_i=\frac{1}{\rho d}\left(\frac{\partial}{\partial x_j}\left(\rho d\tau_{ij}\right)+f_{ij}U_j\right) \qquad E=\frac{1}{2}\rho g\omega_r^2\phi\phi^*$$

in which U_i are the vertically-averaged horizontal current velocity components, τ_{ij} are the lateral mixing terms, η is the mean water level, ρ is the density of water, d is the water depth, g is the acceleration of gravity, t is time, x_j are the horizontal coordinates x and y for $j=1$ and 2, and f_{ij} are the bottom friction coefficients following Iwata's (1976) formulation in the case of normal incidence and are written as

$$\left.\begin{aligned}
\begin{pmatrix} f_{xx} & 0 \\ 0 & f_{yy}\end{pmatrix}\begin{pmatrix} U \\ V\end{pmatrix}&=\frac{2}{\pi}K_*\sqrt{gd}\begin{pmatrix} 2 & 0 \\ 0 & 1\end{pmatrix}\begin{pmatrix} U \\ V\end{pmatrix}, \quad x\le x_B \\
\begin{pmatrix} f_{xx} & 0 \\ 0 & f_{yy}\end{pmatrix}\begin{pmatrix} U \\ V\end{pmatrix}&=\frac{2}{\pi}K_*\sqrt{gd_B}\begin{pmatrix} 2 & 0 \\ 0 & 1\end{pmatrix}\begin{pmatrix} U \\ V\end{pmatrix}, \quad x\ge x_B
\end{aligned}\right\} \tag{5}$$

where $K_* = 1.41\,(\gamma / k.\, k_e\,)^{-2/3}$ in which k_e is the bottom roughness, k is the wave number and γ is the ratio of wave amplitude to local water depth, and suffix B stands for the position of wave breaking and indices x and y represent the seaward x and alongshore y directions. The conservation of wave action, which can be derived directly from (1), and wave number conservation which is equivalent to the irrotationality condition for wave number, are written as:

$$\frac{\partial}{\partial x_j}\left\{\frac{E}{\omega_r}\left(U_j+C_j\right)\right\}+\frac{D}{\omega_r}=0; \qquad \frac{\partial k_j}{\partial x_i}=\frac{\partial k_i}{\partial x_j} \tag{6}$$

THE FIELD EQUATION OF RIP CURRENT

Perturbation Scheme

Before ordering the equations using the perturbation method, it is convenient to nondimensionalize them. The water depth d_B at the breaking point is selected as the representative length to facilitate the nondimensionalization. The process is defined as:

$$\left.\begin{aligned}&(x, y, \bar{\eta}, d) = d_B(x^*, y^*, \bar{\eta}^*, d^*) \qquad (U, V, C) = \sqrt{g d_B}(U^*, V^*, C^*)\\&(\omega, \omega_r) = \sqrt{g/d_B}(\omega^*, \omega_r^*) \qquad (k, k_x, k_y) = d_B^{-1}(k^*, k_x^*, k_y^*)\end{aligned}\right\} \tag{7}$$

where an asterisk represents the dimensionless quantities.

. The beach slope s is selected as a parameter of perturbation in ordering the equations, and the series expansion, in which asterisks have been dropped for convenience, are given as:

$$d = s(d_0 + s\zeta_1 + s^2\zeta_2 + \ldots) \quad ; \quad a = s(a_0 + sa_1 + s^2 a_2 + \ldots) \tag{8a}$$

$$U = s^{\frac{1}{2}}(0 + sU_1 + s^2 U_2 + ..) \quad ; \quad V = s^{\frac{1}{2}}(0 + sV_1 + s^2 V_2 + ..) \tag{8b}$$

$$\mathbf{k} = (k_{0x} + sk_{1x} + s^2 k_{2x} + ..)\,\mathbf{i} + (0 + sk_{1y} + s^2 k_{2y} + ..)\,\mathbf{j} \tag{8c}$$

$$\omega_r = s^{\frac{1}{2}}\{k_{0x}C_0 + s(k_{1x}C_0 + k_{0x}C_1) + ..\}; \; S = (S_0 + sS_1 + s^2 S_2 + ..) \tag{8d}$$

$$C = s^{\frac{1}{2}}(d_0 + s\zeta_1 + s^2\zeta_2 + \ldots)^{\frac{1}{2}}\sqrt{(\tanh kd_B)/(kd_B)} \tag{8e}$$

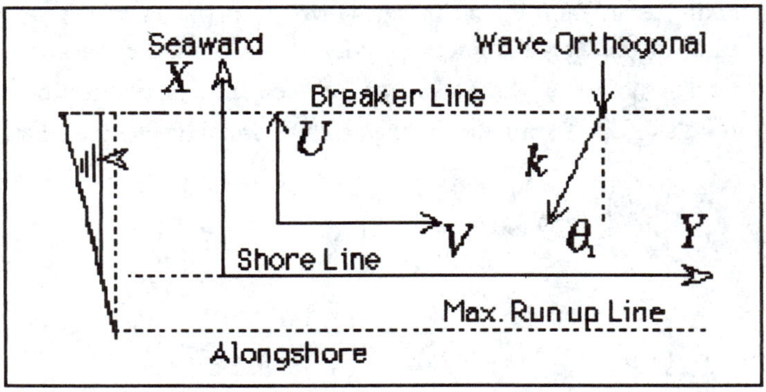

Figure-1 The coordinate system and geometry for nearshore current vectors U, V, and wave number k.

The Conservation of Wave Number

The normally-incident waves refract due to their interaction with rip currents. This may mean that the wave number direction intersects the beach obliquely. Therefore, based on Figure 1 the wave number can be given as:

$$\mathbf{k} = -(|\mathbf{k}|\cos\theta_1)\;\mathbf{i} - (|\mathbf{k}|\sin\theta_1)\;\mathbf{j} \tag{9}$$

where $|\mathbf{k}|$ is the magnitude of vector $\mathbf{k}$. Equation (9) can be expanded with respect to the perturbation parameter s, and as was mentioned above that the group velocity is to be nearly equal to the wave celerity in the surf zone, it was assumed to be in the same direction as the wave number vector. Consequently, it can also be expanded by the perturbation parameter. The conservation equation for the wave number is finally formulated for each order of the perturbation parameter (see Tsuchiya and Suriamihardja, 1989) as:

$$\left.\begin{aligned} O(s^0) &\equiv \frac{\partial k_{0x}}{\partial \tilde{y}} = 0 \\ O(s^1) &\equiv \frac{\partial k_{1x}}{\partial \tilde{y}} = \frac{\partial k_{1y}}{\partial \tilde{x}} \Rightarrow \frac{\partial}{\partial \tilde{y}}(C_1 - U_1) = -\frac{\partial}{\partial \tilde{x}}(V_1) + \frac{1}{\tilde{x}} V_1 \end{aligned}\right\} \quad (10)$$

where $\tilde{x} = m(x + x_s) = d_0;$ $\tilde{y} = my;$ $m = (1 + \mathrm{d}\zeta_0 / \mathrm{d}\tilde{x})$

m is the slope of mean water level, and x_s is the maximum run-up position.

Zeroth-order equations and their solutions

The leading terms of the governing equations produced by the perturbation expansion are given for the surf zone and shoaling zone, respectively. In the surf zone, the zeroth order equations of momentum and wave action conservation are given respectively as:

$$d_0\left(\frac{\mathrm{d}\zeta_0}{\mathrm{d}x}\right) = \frac{D_0}{\omega_{0r}}\frac{\mathrm{d}S_o}{\mathrm{d}x} - d_0\frac{\mathrm{d}}{\mathrm{d}x}\left(\frac{E_0}{2d_0}\right); \qquad -\frac{\mathrm{d}}{\mathrm{d}x}(E_0 C_0) + D_0 = 0 \quad (11)$$

In order to solve the above equations, the wave amplitude a_0 in the surf zone is assumed to be proportional to the local depth to leading order: $a_0 = \gamma\, d_0$ where γ is an empirical constant. This formulation is employed in surf zone models. The wave energy dissipation term D is hypothesized as:

$$D = (5\gamma^2/4)\left\{md_0^{\frac{3}{2}} + s\left(3md_0^{\frac{1}{2}}\zeta_1/2 + d_0^{\frac{3}{2}}(\partial\zeta_1/\partial x) + \ldots\ldots\right)\right\} \quad (12)$$

where D_0 in (11) is equal to $(5\,\gamma^2 m/4)\, d_0^{\,3/2}$

The wave amplitude in the shoaling zone is obtained from the leading order of the equation of wave action conservation as $a_0 = B_0^{\,1/2} d_0^{-1/4}$, where B_0 is an integration constant. The wave set-up is obtained from (11) (see Tsuchiya and Suriamihardja, 1989) as:

$$\zeta_0(x) = -\left\{\left(\tfrac{3}{2}\gamma^2\right)x + x_s\right\}m \quad (13)$$

where $m = (1 + \frac{3}{2}\gamma^2)^{-1}$; $x_s = \gamma^2\left\{\frac{3}{2} - \frac{1}{4}\left(1 + \frac{3}{2}\gamma^2\right)/\left(1 - \frac{3}{2}\gamma_B^2\right)\right\} x_B$, and $\gamma_B = B_0\, x_B^{-5/4}$

and x_B is the distance from the shoreline to the breaking point.

The first-order equations of rip currents in the surf zone

The mass conservation equation is given from the first equation of (3), considering (8) for its expansion and using variables defined from mass conservation, of which the stream function can be defined as follows:

$$\frac{\partial}{\partial \tilde{x}}(\tilde{x} U_1) + \frac{\partial}{\partial \tilde{y}}(\tilde{x} V_1) = 0 \Rightarrow (\tilde{x} U_1) = \frac{\partial}{\partial \tilde{y}}(\Psi) \text{ and } (x V_1) = -\frac{\partial}{\partial \tilde{x}}(\Psi) \quad (14)$$

The approximation may be made to the wave dispersion relation, to evaluate C_1. It is necessary, of course, to obtain a simple relation between C_1 and ζ_1 in order to reduce higher-order differentation terms. The simple relation can be derived from the established set of four equations. In order to approximate the relation along with the four equations, however, the x-direction of the first order of momentum equation (4) can be used by considering the most effective terms, thus:

$$C_1 = q\tilde{x}^{-\frac{1}{2}}\zeta_1; \qquad q = \frac{1}{2}\sqrt{(\tanh kd_B)/(kd_B)} \quad (15)$$

if the other non-leading terms are neglected, q may equal to 1/2 which is equivalent to that of Darlymple and Lozano's approximation. However, based on this approximation, the value of q might be somewhat less than 1/2.

When the horizontal mixing terms remain in (4), the field equation of rip currents may become a fourth-order partial differential equation. Without loss of generality in the formation of rip currents, as previously treated in theoretical approaches to the formation of rip currents, the horizontal mixing terms are neglected, but the bottom friction terms are included. By use of the four equations using (15) to replace C_1 the first-order equation of momentum can finally be obtained (see Tsuchiya and Suriamihardja, 1989 for the detailed reduction) as:

$$\begin{pmatrix} M^{\kappa}_{xx} & 0 \\ 0 & M^{\kappa}_{yy} \end{pmatrix}\Psi = \begin{pmatrix} \beta & 0 \\ 0 & \alpha \end{pmatrix}\frac{\partial^2}{\partial \tilde{x}\partial \tilde{y}}\left(\tilde{x}^{\frac{3}{2}}\zeta_1\right) - \frac{5\gamma^2}{4}\begin{pmatrix} (1+2q) & 0 \\ 0 & q\kappa \end{pmatrix}\frac{\partial}{\partial \tilde{y}}\left(\tilde{x}^{\frac{1}{2}}\zeta_1\right) \quad (16)$$

where

$$\begin{cases} M^{\kappa}_{xx} = \left\{\frac{1}{4}\gamma^2\frac{\partial}{\partial \tilde{x}}\left(\tilde{x}\frac{\partial^2}{\partial \tilde{y}^2}\right) - \frac{15}{8}\gamma^2(1-\kappa)\tilde{x}^{-1}\frac{\partial}{\partial \tilde{x}} - \frac{7}{8}\gamma^2\left(1+\frac{20\kappa}{7}\right)\frac{\partial^2}{\partial \tilde{y}^2}\right\} \\ M^{\kappa}_{yy} = \left\{\frac{1}{4}\gamma^2\frac{\partial}{\partial \tilde{x}}\left(\tilde{x}\frac{\partial^2}{\partial \tilde{y}^2}\right) - \frac{5}{2}\gamma^2(1-\kappa)\tilde{x}^{-1}\frac{\partial}{\partial \tilde{x}} + \frac{3}{8}\gamma^2\left(1-\frac{10\kappa}{3}\right)\frac{\partial^2}{\partial \tilde{y}^2}\right\} \end{cases}$$

and $\alpha = \left\{1+\left(\frac{3}{8}+\frac{1}{4}q\right)\gamma^2\right\}$; $\beta = \left\{1+\left(\frac{13}{8}+\frac{1}{4}q\right)\gamma^2\right\}$; $\kappa = \frac{4K}{5\gamma^2 ms}$

By eliminating ζ_1 from (16), the first-order field equation of rip currents can finally be obtained as:

$$\left(\hat{P}\frac{\partial^2}{\partial \tilde{x}^2} - \hat{Q}\frac{\partial^2}{\partial \tilde{x}^2}\frac{\partial^2}{\partial \tilde{y}^2} - \hat{R}\frac{1}{\tilde{x}}\frac{\partial}{\partial \tilde{x}} - \hat{S}\tilde{x}\frac{\partial}{\partial \tilde{x}}\frac{\partial^2}{\partial \tilde{y}^2} - \hat{T}\frac{\partial^2}{\partial \tilde{y}^2}\right)\Psi = 0 \quad (17)$$

where
$$\hat{P}=\left\{1+2q\left(1-\frac{\beta\kappa}{2\alpha}\right)\right\}\left\{q(1-\kappa)+\left(q-\frac{1}{2}\right)\right\}+\frac{3q}{2}\left(\frac{4\beta}{3\alpha}-1\right)(1-\kappa)$$
$$\hat{Q}=\frac{q}{5}\left(\frac{\beta}{\alpha}-1\right)$$
$$\hat{R}=\left\{1+2q\left(1-\frac{\beta\kappa}{2\alpha}\right)\right\}\left\{2\left(q-\frac{1}{2}\right)+\frac{3q}{2}(1-\kappa)\right\}+\frac{9q}{4}\left(\frac{4\beta}{3\alpha}-1\right)(1-\kappa)$$
$$\hat{S}=\frac{3q}{5}\left(\frac{2}{3}+\frac{\beta}{\alpha}\right)+2\kappa q\left(1-\frac{\beta}{2\alpha}\right)$$
$$\hat{T}=\left\{1+2q\left(1-\frac{\beta\kappa}{2\alpha}\right)\right\}\left\{q(1-2\kappa)-\frac{1}{2}\right\}+\frac{q}{4}\left(1+\frac{\beta}{\alpha}\right)+q\kappa\left(1-\frac{\beta}{2\alpha}\right)$$

The first-order equation of rip currents in the shoaling zone

In the shoaling zone the governing equations differ from the surf zone equations principally in the wave energy dissipation term and in the formulation of the bottom-friction stresses. In this zone, the driving forces are presented in irrotational form, which may be incapable of generating currents of the first order. The remaining terms in the cross-differentiated equations originate from the friction terms. Consequently, the momentum equations in the shoaling zone play an important role in the decay of currents produced in the surf zone. As was previously indicated in (14), a stream function in the shoaling zone can be equally defined. The momentum equations in the x and y directions and cross-differentiated momentum equations are given respectively by

$$\begin{pmatrix}\frac{2\kappa}{s^{3/2}\tilde{x}} & 0\\ 0 & \frac{2\kappa}{s^{3/2}\tilde{x}}\end{pmatrix}\begin{pmatrix}U_1\\ V_1\end{pmatrix}=-\left(1-\frac{3\gamma_B^2}{8}\right)\begin{pmatrix}\frac{\partial}{\partial\tilde{x}}\\ \frac{\partial}{\partial\tilde{y}}\end{pmatrix}\left(\zeta_1-\frac{B_0}{4}\tilde{x}^{-\frac{5}{2}}\zeta_1+\frac{B_0^{\frac{1}{2}}}{2}\tilde{x}^{-\frac{5}{4}}a_1\right) \tag{18}$$

$$\left(2\frac{\partial^2}{\partial\tilde{y}^2}+\frac{\partial^2}{\partial\tilde{x}^2}-\frac{2}{\tilde{x}}\frac{\partial}{\partial\tilde{x}}\right)\Psi=0 \tag{19}$$

THE SOLUTIONS OF THE FIELD EQUATIONS OF RIP CURRENTS

The solution in the surf zone

The solutions to (17) for the surf zone and (19) for the shoaling zone can be obtained by means of the method of separation of variables. The stream function Ψ to be solved for can be expressed by $\Psi(\tilde{x},\tilde{y})=\Xi(\tilde{x})Y(\tilde{y})$. Substituting this expression into (17), the equation can be transformed into two equations as

$$\left.\begin{aligned}&\left[\left\{1+\frac{\hat{Q}}{\hat{P}}(\lambda\tilde{x})^2\right\}\frac{d^2}{d\tilde{x}^2}+\left\{\frac{\hat{S}}{\hat{P}}(\lambda\tilde{x})^2-\frac{\hat{R}}{\hat{P}}\right\}\frac{1}{\tilde{x}}\frac{d}{d\tilde{x}}-\frac{\hat{T}}{\hat{P}}\lambda^2\right]\Xi(\tilde{x})=0\\&\left(\frac{d^2}{d\tilde{y}^2}+\lambda^2\right)Y(\tilde{y})=0\end{aligned}\right\} \tag{20}$$

where λ is the separation constant which is the eigenvalue corresponding to the number of rip currents. Introducing the new variable

$$\tilde{\xi} = \left\{1 + \hat{Q}(\lambda\tilde{x})^2 / \hat{P}\right\}^{-1} \tag{21}$$

into the first equation of (20) yields

$$\left[\tilde{\xi}^2(1-\tilde{\xi})\frac{d^2}{d\tilde{\xi}^2} - \left\{(\hat{a}+\hat{b}-1)\tilde{\xi} + (2-\hat{c})\tilde{\xi}^2\right\}\frac{d}{d\tilde{\xi}} + \hat{a}\hat{b}\right]\Xi(\tilde{\xi}) = 0 \tag{22}$$

where $(\hat{a},\hat{b}) = \left\{(\hat{S}/\hat{Q}-1) \pm \sqrt{(\hat{S}/\hat{Q}-1)^2 + 4\hat{T}/\hat{Q}}\right\}/4\,;\ \hat{c} = (\hat{S}/\hat{Q} + \hat{R}/\hat{P})$

The general solution of this equation can be expressed as a Gaussian hypergeometric function in the form

$$\Xi_{surf} = A_1\tilde{\xi}^{\hat{a}}F(\hat{a},1+\hat{a}-\hat{c};1+\hat{a}-\hat{b};\tilde{\xi}) + A_2\tilde{\xi}^{\hat{b}}F(\hat{b},1+\hat{b}-\hat{c};1+\hat{b}-\hat{a};\tilde{\xi}) \tag{23}$$

where A_1 and A_2 are the integration constants to be determined, and the suffix "surf" represents thesurfzone.To solve for regular rip-current spacing as an eigenvalue problem of the equation, a maximum point for (23) must exist in the surf zone. The solutions should have a finite value and a maximum point within $0<\tilde{\xi}<1$. This behaviour can be examined through the following conditions:

$$\Xi(\tilde{\xi})\Big|_{\tilde{\xi}=0} = 0 \quad ; \quad \frac{\partial}{\partial\tilde{\xi}}\Xi(\tilde{\xi})\Big|_{\tilde{\xi}_B \le (\tilde{\xi}=\tilde{\xi}_m) \le 1} = 0 \quad ; \quad \frac{\partial^2}{\partial\tilde{\xi}^2}\Xi(\tilde{\xi})\Big|_{\tilde{\xi}_B \le (\tilde{\xi}=\tilde{\xi}_m) \le 1} < 0 \tag{24}$$

where B indicates the breaking point. It is also confirmed by numerical calculation, that for A_2=0, the conditions will be fulfilled.

The solution in the shoaling zone

The solution in the shoaling zone can be obtained by solving (19) using the method of separation of variables. Introducing the new variables

$$\varphi(\tilde{x}) = \tilde{x}^{-\frac{3}{2}}\,\Xi(\tilde{x}) \quad ; \quad \tilde{x} = \lambda\tilde{x}\sqrt{2} \tag{25}$$

Equation (19) is reduced to

$$\left\{\frac{\partial^2}{\partial\tilde{x}^2} + \frac{1}{\tilde{x}}\frac{\partial}{\partial\tilde{x}} - \left(1 + \frac{9}{4\tilde{x}^2}\right)\right\}\varphi(\tilde{x}) = 0 \quad ; \quad \left(\frac{\partial^2}{\partial\tilde{y}^2} + \lambda^2\right)Y(\tilde{y}) = 0 \tag{26}$$

The solution to (26) can be expressed using modified Bessel functions of the first and second kinds. By use of (25) the solution can finally be written as

$$\Xi_{shoal}(\tilde{x}) = B_1\tilde{x}^{\frac{3}{2}}I(\lambda\tilde{x}\sqrt{2}) + B_2\tilde{x}^{\frac{3}{2}}K(\lambda\tilde{x}\sqrt{2}) \tag{27}$$

where B_1 and B_2 are integration constants to be determined, and the suffix "shoal" represents the shoaling zone. To fulfill the condition that $\Xi_{Shoal} \to 0$ far offshore, therefore, B_1 should be zero.

The boundary conditions and characteristic equation

The eigenvalues can be determined by the matching condition at the breaking point, that is:

$$\frac{d}{d\tilde{x}}\Xi_{surf}(\tilde{x})\Bigg|_{\tilde{x}=\tilde{x}_B} = \frac{d}{d\tilde{x}}\Xi_{shoal}(\tilde{x})\Bigg|_{\tilde{x}=\tilde{x}_B} \quad ; \quad \Xi_{surf}(\tilde{x})\Big|_{\tilde{x}=\tilde{x}_B} = \Xi_{shoal}(\tilde{x})\Big|_{\tilde{x}=\tilde{x}_B} \tag{28}$$

where B indicates the breaking point.

The real and positive eigenvalues λ which satisfy the matching conditions (28) can be determined if the stream functions in both the surf and shoaling zones are decreasing monotonically in the offshore direction, the characteristic equation is given by

$$\frac{\hat{F}\left(-\hat{b},\hat{c}-\hat{b};1+\hat{a}-\hat{b};\xi_B\right)}{\hat{F}\left(1-\hat{b},\hat{c}-\hat{b};1+\hat{a}-\hat{b};\xi_B\right)} = \left(\frac{\hat{P}}{\hat{a}\hat{Q}}\right)\left\{\frac{(1-\xi_B)/\xi_B}{1+\sqrt{2\hat{P}(1-\tilde{\xi}_B)/\hat{Q}\tilde{\xi}_B}}\right\} \tag{29}$$

By numerical solution of this characteristic equation, the eigenvalues λ can be determined. The eigenvalues λ directly correspond to rip spacing along the shore line.

Determination of the integration constants in the stream functions

The integration constant of the stream function A_{surf} in the surf zone can be determined from the steady-state wave power conservation in the shoaling and surf zones. The wave energy dissipation rate immediately after breaking is different from that in the inner region. This is revealed by the different wave-elevation decay rates. Therefore, the ratio of wave amplitude to local water depth should be taken into consideration in wave energy budgeting. This difference in the rate of wave energy dissipation is responsible for the structure of the nearshore current. Consequently, the wave power contribution to the generation of the nearshore current initiates at the breaking point. The change in wave power occurring in the region from just before to just after the breaking point can be expressed by:

$$\left(\frac{1}{2\pi}\right)^{\frac{1}{4}}\left(\frac{H'_\infty}{d_B}\right)\left(\frac{L'_\infty}{d_B}\right)^{\frac{1}{4}}\left(\frac{d_B}{d_{0B}}\right)^{\frac{5}{4}} = s^{\frac{5}{4}}\left[H^{*2}_{0B}C^*_{0B} + s\left\{2H^*_{0B}H^*_{1B}C^*_{0B} + H^{*2}_{0B}\left(C^*_{1B}-U^*_{1B}\right)\right\}+\ldots\right]^{\frac{1}{2}} \tag{30}$$

Kimura, Goto, and Seyama's (1988) experimental work suggests the condition $\gamma_B > \gamma$. Based on this fact, the first-order approximation of (30) is reduced to:

$$\frac{1}{s}\left(\frac{\gamma_B}{\gamma}-1\right)\tilde{x}_B \cong \left\{\frac{a_{1B}}{\gamma}+\frac{1}{2}\left(C_{1B}-U_{1B}\right)\tilde{x}_B\right\} \Rightarrow \gamma_B = \frac{1}{2}\left(\frac{1}{2\pi}\right)^{\frac{1}{4}}\left(\frac{H'_\infty}{d_B}\right)\left(\frac{L'_\infty}{d_B}\right)^{\frac{1}{4}} \tag{31}$$

Using (31), the integration constant A_{surf} can immediately be evaluated, and the stream function Ψ can finally be expressed as

$$\left.\begin{aligned}\Psi(\xi,\tilde{y}) &= A_{surf}\,\xi^{\hat{a}}\,\hat{F}\left(\hat{a},1+\hat{a}-\hat{c};1+\hat{a}-\hat{b};\xi\right)\cos\left(\lambda\tilde{y}\right)\\ \Psi(\tilde{x},\tilde{y}) &= \mathrm{Re}\left\{(-1)^{-\hat{a}}\frac{\Gamma(\hat{c})\Gamma(\hat{b}-\hat{a})}{\Gamma(\hat{c}-\hat{a})\Gamma(\hat{b})}\right\}\Psi(\xi,\tilde{y})\end{aligned}\right\} \tag{32}$$

The depth-integrated rip current velocity at first-order is now ready to be expressed numerically as:

$$Q(\tilde{x},\tilde{y}) = \mathbf{i}\ dU(\tilde{x},\tilde{y}) = \mathbf{i}\ (s\tilde{x})s^{\frac{1}{2}}\left\{0 + sU_1(\tilde{x},\tilde{y}) + s^2U_2(\tilde{x},\tilde{y}) + \ldots\right\} \tag{33}$$

Theoretical Results And Comparison With Field Data

A theory of the steady nearshore horizontal circulation cells induced by normally-incident waves on a plane beach has been proposed in this paper. The theoretical results predict two main characteristics of rip currents, i.e. the alongshore spacing and the depth-integrated velocity distribution in the seaward direction. The obtained rip current spacing is compared with the theoretical curve of Dalrymple and Lozano (1978) and the field data of Sasaki (1977) and other investigators where the surf similarity of the waves are categorized into instability region. To compare the theoretical results with the field data of Sasaki (1977) and other investigators, their values of the parameter q at the breaking point should be evaluated using their wave characteristics data, and their values of κ should be evaluated using bottom roughness k_e, bottom slope s, and wave length data at the breaking point, then the results are plotted over the theoretical curves. In the evaluation of κ, Kajiura's expression for bottom friction coefficient was used, as Dalrymple and Lozano used and assumed the bottom roughness as $k_e = 0.4$ mm.

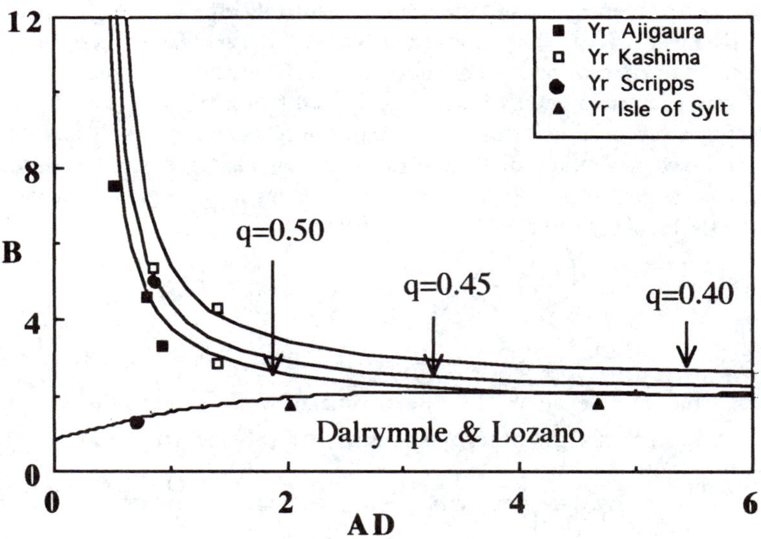

Figure 2. Theoretical curves of dimensionless rip spacing (Y_r / x_B) in terms of Dalrymple number A_D for q-values of 0.40, 0.45, and 0.50 including Dalrymple & Lozano's curve (1978) and rearranged field data of rip currents generated by waves having surf similarity parameter of the instability region.

In comparing the results of this study with those of Dalrymple and Lozano, the parameter κ, which has been previously defined, is transformed into Dalrymple and Lozano's number $A_D = 0.4/\kappa$. Figure 2 shows the curve of dimensionless rip spacing as a function of Dalrymple parameter A_D. The present curves (q=0.5, 0.4, and 0.3) rapidly increase as A_D becomes less than 1, while Dalrymple's curve decreases as A_D goes to 0. The present theoretical results and those of Dalrymple and Lozano differ particularly at small values of A_D. In the course of deriving the characteristics equation in the present eigenvalue problem, the dimensionless rip spacing depends on the values of κ and q. It was defined that q corresponds to wave characteristics and beach slope s, and κ depends on the bottom roughness and wave length at the breaking point. Therefore, we conclude that the dimensionless rip spacing is determined by incoming wave characteristics, i.e.wave height H and wave length L, and morphological characteristics, i.e. beach slope s and bottom surface roughness k_e.

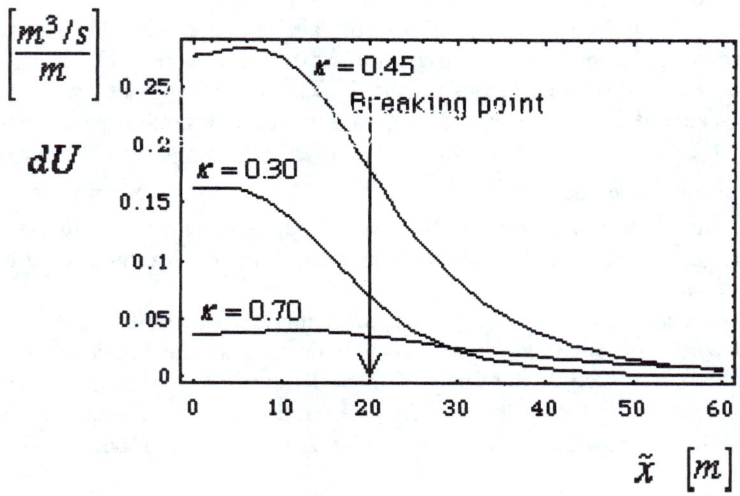

Figure 3. Seaward distribution of rip discharge, where beach slope s = 0.05 and dimensionless rip spacings of 3.12 (κ = 0.30), 4.21 (κ = 0.45), and 7.70 (κ = 0.70) are used.

Figures 3 illustrate the distribution of depth-integrated rip current velocity or seaward rip discharge as a function of seaward distance. The illustrations have dimensionless rip spacings of 7.70, 4.21, and 3.12, a beach slope of 0.05, and q of 0.50.The breaking point is located at 20 meters from run-up line. By using γ_B of 0.50, the numerical value of A_{surf} gives values of rip discharges with reasonable values at both laboratory and field scales.

CONCLUSIONS

In this study a new mathematical model for steady-state horizontal nearshore circulation has been developed, in which the wave-current interaction is taken into consideration. The basic equations consist of the depth-integrated equations for conservation of mass, momentum, wave action, and wave number. The wave-current interaction includes not only wave refraction due to currents but also the work done by radiation stresses against the mean current.

The basic equations are decomposed into the leading order and the instability order using the beach slope as a pertubation parameter. The leading order gave the solution for wave set-up in the surf zone and wave set-down in the shoaling zone. The instability order gave linearized field equations of nearshore currents. It is found that the nearshore circulations were generated by the rotational driving forces in the momentum equation. The solutions were characterized by boundary conditions at a shore line, a breaking line, and far offshore. In the surf zone, the solution was represented by the Gaussian hypergeometric function. In the shoaling zone, the solution was represented by the modified Bessel function. The eigenvalues which correspond to the dimensionless rip spacing are obtained from the characteristics equation extracted from the matching condition at the breaking point.

The present theoretical results for the dimensionless rip spacing were consistent with the scatter of the field data of Sasaki in terms of surf similarity parameter of the region of instsbility. The curves suddenly increase in the range $A_D < 1.0$, and tend to be a constant value for larger values of A_D. The field data on dimensionless rip spacings caused by infragravity waves are difficult to predict by this theory. Outside of this range, the theory gives reasonable predictions for rip spacing caused by wave-current interactions.

Rip discharge is illustrated with exact numerical values, using an integration constant which is obtained from energy budgeting at the breaking point. This budgeting was applied at the suggestion of Kimura et.al, that the wave energy dissipation rate immediately after breaking is larger than the dissipation rate in the inner region. The resulting value is within the range of data at both laboratory and field scales.

Acknowledgments

This work was completed while the first author was a fellow at the Coastal Disaster Section, Disaster Prevention Research Institute, Kyoto University, Japan. Most of this work was completed while the first author was a postgraduate student under the supervision of the second author at the same university. The authors' appreciations are expressed to Professor T. Takayama, Head of the Section, for his support in executing this work, Associate Professor T. Yamashita for his crucial advice and arragement to make this work was possible, Dr. Gary Watson, Visiting Research Fellow, for his kindness in critical correction of the draft, and Dr. H.Yoshioka for his cooperation during the execution this work at the Institute.

References

Balsillie, J.H.(1975)." Surf observation and longshore current prediction", Technical Memorandum no.58, CERC, 39.

Bowen, A.J.(1969)."Rip currents,I, Theoretical investigation", Jour. Geophys. Res.,74, 5479-5490.

Bowen, A.J. and Inman, D.L.(1969)." Rip currents, II, Laboratory and field observation", Jour. Geophys. Res., 74, 5479-5490.

Crapper, G.D.(1984)." Introduction to water waves", John Wiley & Sons, New York, 93-106.

Dalrymple, R.A., and Lozano, C.J. (1978)." Wave-current interaction model for rip currents", Jour. Geophys. Res., 83, C12, 6063-6071.

Dingemans, M.W., Radder, A.C. and De Vriend, H.J. (1987)." Computation of the driving forces of wave induced currents", Coastal Engineering,11, 539-563.

Dolata, J.B., and Rosenthal,W. (1984)." Wave set-up wave-induced currents in coastal zones", Jour. Geophys. Res., 82, C2, 1973-1982.

Horikawa, K., and Mizuguchi,M. (1975)." Experiment on nearshore currents on a plane beach", Proc. 22nd Japanese conf. on Coast. Eng., 141-153.(in Japanese).

Iwata, N. (1976)." Rip Current Spacing", Jour. of the Ocean. Soc. Japan, 32, 1-10.

Kimura, A., Goto, K., and Seyama, A. (1988)." On the wave height variation of irregular waves in shallow water", Proc. 35th Japanese conf. on Coast. Eng., 168-172. (in Japanese).

Kirby,J.T. (1984)."A note on linear surface wave-current interaction over slowly varying topography", Jour. Geophys. Res., 89, C1, pp. 745-747.

LeBlond,P.H., and Tang,C.L.(1974)."On energy coupling between wave and currents", Jour. Geophys. Res., 79, 6, 811-816.

Noda, E.K.(1974)." Wave-induced Nearshore Circulation", Jour. Geophys.Res.,79, 27, 4097-4106.

Phillips, O.M. (1977)." The dynamics of the upper ocean", second edition, Cambridge University Press, 23-81.

Sasaki, T. (1977)." Field Investigation of nearshore currents on gently sloping bottom", PhD Disertation, NERC, Tokyo University, Japan.

Sonu, C.J. (1972)." Field observation of nearshore circulation and meandering currens", Jour. Geophys. Res., 77, 18, 3232-3247.

Tsuchiya, Y., Kawata, K., Shibano, T., Suriamihardja, D.A., and Shishikura, T.(1986). "Generation of rip currents on a plane beach with reflective boundaries", Proc.33rd Japanese conf. on Coast. Eng., 36-40. (in Japanese).

Tsuchiya, Y., and Suriamihardja, D.A. (1989)." Nearshore circulation: Theory of rip current generation on a plane beach", Ann. DPRI, Kyoto Univ., 32, B-2. (in Japanese).

CHAPTER 286

NEARSHORE PLACEMENT OF SAND

By Scott L. Douglass[1], Member ASCE

ABSTRACT:
A methodology that evaluates the influence of depth on the rate of onshore migration of sand placed in the nearshore is presented. The results appear appropriate for the Atlantic and Gulf coasts of the United States. Migration rate is found to be extremely dependent, to the 4th or 5th power, on the water depth. This implies that doubling the rate of migration requires placement in only about 15% less depth. It is argued that accounting for the net shoreward bottom stress due to the velocity asymmetry under finite-amplitude waves is the appropriate way to model the fate of nearshore placed sands.

BACKGROUND

Nearshore placement of sand is becoming a more popular option in two related types of coastal engineering projects; beach nourishment and inlet dredging. "Nearshore" is defined rather loosely here as beyond the day-to-day surf zone but within the depths that are disturbed by storm waves. Nearshore placement is an alternative to direct placement on the beach that has been called "profile nourishment" or "shoreface nourishment" since the placement is farther out on the beach profile than the dry beach.

1

Associate Professor, Civil Engineering Department, University of South Alabama, Mobile, Alabama, USA (334) 460-6174

Placing the sand in the nearshore instead of directly on the beach can reduce costs and improve the overall cost-benefit ratio of a beach nourishment project (Laustrup, et al. 1996). There may also be a perception that construction-related aesthetic and environmental impacts to the beaches are improved with nearshore placement instead of direct beachfill because the sand moves ashore in the form of migrating sand bars instead of being placed there by construction equipment.

Nearshore placement of sand is also an option in navigation dredging projects for similar reasons. First, it may be less expensive than other dredged material disposal options such as hauling to offshore disposal. Second, there may be a recognition that inlet sand transfer is required for the long-term maintenance of the downdrift beaches. The goal of nearshore placement of sand in this case would be to keep the sand resource in the littoral system for the least cost.

The primary question is, "what is the fate of sands placed in the nearshore?" More specific questions are:

1. does the sand move?
2. If so, which way does it move?
3. How fast does it move?
4. What is the influence of depth on this fate?

Depth is one of the primary parameters that the design engineer can specify.

Several tools to answer these questions have been developed in the United States as part of the Corps of Engineers Dredging Research Program in the last decade. Hands and Allison (1991) and Hands (1991) developed an empirically based answer to the first question concerning movement. The stability of bathymetric features (mounds or bars) constructed with dredged sands at eleven locations discussed in the American literature is summarized. A method based on combining extreme wave statistics and Hallermeier's estimates of the limits of sand movement successfully discriminates between mounds that moved and mounds that did not move.

Larson and Kraus (1992) suggest using empirical estimates of surf zone sand bar migration and beachface accretion/recession for estimating the direction of movement of sand placed in the nearshore. A design option is to place the sand in a long, linear, shore-parallel bathymetric feature that is similar to natural sand bars. They validate their method by comparison to the onshore migration of a constructed bar feature at Silver Strand, California (Andrassy 1991).

Scheffner (1991, 1996) suggests combining a hydrodynamic model with a sediment transport model for evaluating the fate of sand placed in the nearshore. The influence of waves on the bottom friction factor is considered, but mean currents provide the transport mechanism for the sand movement. The migration of a mound south of Dauphin Island, Alabama (Hands and Allison 1991) is used for validation.

The answer to the primary question concerning the fate of sands placed in the nearshore is tied to a broader, underlying question that has been the focus of much research in the nearshore oceanography community for the past decade: "What moves sand in these depths beyond the day-to-day surf zone but within the influence of storm waves?" Considering the rich variations in sediment transport processes such as bed and suspended load, ripple and dune migration, sheetflow, and bed-ventilation; combined with the varying time-scales of the hydrodynamic forcing conditions and our limited understanding of the near-bottom boundary layer of the fluid column under real waves, the answer is only marginally understood at this time. Solving applied problems requires the use of that part of the available science that is considered to be the most important of the geophysical processes for the problem.

One concept for modeling surf zone sand bars is a balance between the cross-shore movement of sand due to the velocity asymmetry of waves and the undertow (Stive 1986, Roelvink and Stive 1989). Simplistically, the waves drive the sand bar landward and the undertow drives it seaward. Considering that the location of constructed mounds is offshore of the location of the typical sand bar, the undertow should be reduced and the wave asymmetry mechanism should dominate. This is the fundamental concept used in this paper.

Douglass (1995) briefly proposed a model for estimating migration of sand mounds constructed in the nearshore. The model assumes that landward migration is due primarily to the net landward transport by the velocity asymmetry of finite-amplitude waves. Conservation of sand considerations in the cross-shore direction lead to the classical convection-diffusion model equation. The "convection" and "diffusion" coefficients are functions of the wave conditions and depth and are based on the wave and transport theory, not traditional concepts of convection and diffusion. The model produced results which were in reasonable agreement with measured migration directions and rates at both Silver Strand and Dauphin Island. The implication of the diffusion is that mounds should spread out while moving landward. This type of behavior has been noticed to different degrees in many of the monitoring reports on mounds (Hands and Allison 1991; Andrassy 1991; Bodge 1994; Foster, et al. 1996; Mesa 1996) and is shown schematically in Figure 1.

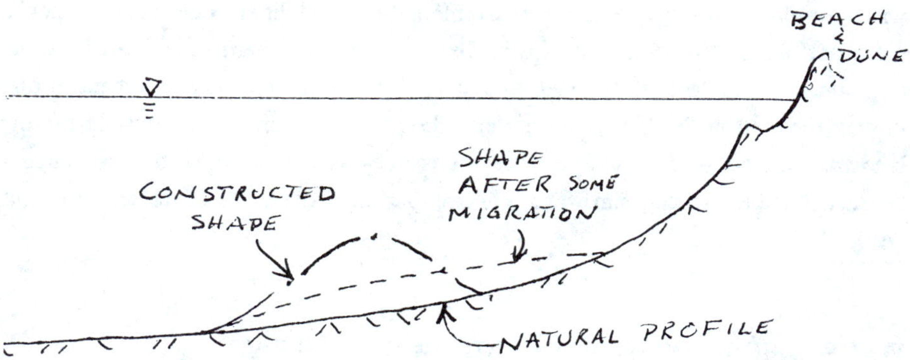

Figure 1. "Typical" migration of sand placed in the nearshore.

This present paper briefly discusses the results and implications of a study of the near-bottom current field at a nearshore sand placement site. The Douglass (1995) model is further developed. Modifications include the use of a calibration coefficient and a simple adjustment for wave breaking when using WIS wave climatology data as input. Results for a few locations on the Atlantic and Gulf coasts of the United States are presented. The implications of the dependence of migration on depth of placement are discussed. Some of the practical limitations of the model as well as the theoretical problems are briefly discussed. The implications concerning the important geophysics dominating the problem are discussed. The conclusion is that some form of modeling based on the influence of wave asymmetry is probably appropriate concerning the fate of sand placed in the nearshore.

DAUPHIN ISLAND, ALABAMA NEAR-BOTTOM CURRENT STUDY

The fate of sand mounds constructed in about 6 m depths south of Dauphin Island, Alabama in 1987 was monitored extensively by the US Army Corps of Engineers. A number of reports by Mr. Edward Hands and other co-authors documented the movement of these nearshore mounds (e.g. Hands and Bradley 1989, Hands and Allison 1991). Repetitive bathymetric surveys showed that the mounds moved landward.

Instantaneous, near-bottom currents were measured, along with waves, at the site. These near-bottom currents were examined in detail by Douglass, et al. (1995).

A number of possible mechanisms for mound migration; including mean currents, large wave correlation with mean currents, and mound-induced differences in current fields; were not able to satisfactorily explain the observed shoreward direction of mound migration. Douglass, et al. (1995) concluded that some unknown mechanism that moves the bottom back towards its equilibrium shape may be responsible for the landward migration of mounds and that the velocity asymmetry of storm waves might be part of such a mechanism. These finding played a role in the development of this work.

WAVE-BASED MODEL OF MOUND MIGRATION RATE

Assuming that wave-induced near bottom velocity asymmetries are the dominant mechanism for the net transport of sand in the depths of interest, net sand transport can be expressed as (Douglass 1995),

$$Q=\frac{9\pi^2}{64}\frac{\rho g C_f \epsilon_b}{(\rho_s-\rho)\,a'\tan^2\phi}(\tan\phi-\tan\beta)\frac{H^4 T}{L^3}\operatorname{sech}^2\left(\frac{2\pi h}{L}\right)\operatorname{csch}^4\left(\frac{2\pi h}{L}\right)$$

where Q=net volumetric transport rate, ρ = density of water, ρ_s=sand density, C_f=a friction coefficient, ϵ_b = Bagnold's efficiency of transport coefficient, ϕ = angle of internal friction of the sediment, β = angle of slope of the bottom, a'=ratio of total volume to volume of solids, H=wave height, T=wave period, L= wave length, and h=water depth. This is an expression for cross-shore bedload transport rate using Stokes wave theory in Bailard and Inman's (1981) form of Bagnold's (1963) bedload transport equation assuming all motion is in the cross-shore direction.

It can be assumed that transport due to irregular waves with a spectral significant wave height, H_{mo} , with a period corresponding to the peak of the energy density spectrum, T_p, will be proportional to transport due to the corresponding monochromatic conditions. Also, given that several of the empirical coefficients, in particular ϵ_b , have typical values based on limited tests, it seems logical to modify this transport equation with an empirical coefficient that can be used to fit the results to observed data specific to the migration problem. Therefore another empirical coefficient, A, has been added to the above transport equation modified to use the spectral estimates of wave climate:

$$Q=A\frac{9\pi^2}{64}\frac{\rho g C_f \epsilon_b}{(\rho_s-\rho)a'\tan^2\phi}(\tan\phi-\tan\beta)\frac{H_{mo}^4 T_p}{L_p^3}sech^2\left(\frac{2\pi h}{L}\right)csch^4\left(\frac{2\pi h}{L}\right)$$

where L_p =wave length corresponding to T_p computed with linear wave theory. The A will be used to fit the results to observed migration rate data below.

Considering conservation of sand in the cross-shore direction,

$$\frac{\partial h}{\partial t}-\frac{\partial Q}{\partial x}=0$$

where t= time and x = the cross-shore direction.

Substitution of the above transport equation into the conservation equation results in a form of the convection-diffusion equation (Douglass 1995):

$$\frac{\partial h}{\partial t}+C\frac{\partial h}{\partial x}-D\frac{\partial^2 h}{\partial x^2}=0$$

where

$$C=A\frac{9\pi^3}{16}\frac{\rho g C_f \epsilon_b}{(\rho_s-\rho)a'\tan\phi}\frac{H_{mo}^4 T_p}{L_p^4}\left\{2csch^5\left(\frac{2\pi h}{L}\right)sech\left(\frac{2\pi h}{L}\right)\right.$$

$$\left.+csch^3\left(\frac{2\pi h}{L}\right)sech^3\left(\frac{2\pi h}{L}\right)\right\}$$

$$D=A\frac{9\pi^2}{64}\frac{\rho g C_f \epsilon_b}{(\rho_s-\rho)a'\tan^2\phi}\frac{H_{mo}^4 T_p}{L_p^3}sech^2\left(\frac{2\pi h}{L}\right)csch^4\left(\frac{2\pi h}{L}\right)$$

C has the role of the "convection" coefficient and D has the role of the "diffusion" coefficient. The expressions for C and D originated from the consideration of conservation of sand with the wave theory and transport model and not on physical concepts of convection or diffusion. The equation implies that the form of a submerged mound will "convect" or move at speed C while it diffuses. Similar behavior of constructed bottom mounds has been observed (Hands 1991, Bodge 1994, Foster et al. 1996, Mesa 1996).

The equations outlined above can be used as a planning and design tool when combined with an estimate of wave climate at a specific site. The equations predict the behavior of a mound under a given, specified sea state. To apply it to a non-constant sea state, either the time history of sea must be estimated or a statistically valid estimate of the overall wave climate must be evaluated in terms of an “expected value of migration.” Given a tabular estimate of the onshore portion of the wave climate, the expected value of mound movement in any given depth, E[C(h)], can be estimated with C, the "convection" coefficient as

$$E[C(h)] = \sum p(H,T)\, C(H,T,h)$$

where the summation is across all (H,T) bins, p(H,T)= probability or percentage of time that the wave height and period is of that magnitude.

APPLICATION

The migration rate equation above can be roughly calibrated by matching the predictions to the observed migration of constructed mounds. The “expected value” of the migration rate can be estimated with an estimate of the “expected” wave climate such as the US Army Corps of Engineers Wave Information Study (WIS) wave hindcast data for the Atlantic and Gulf coasts of the United States (Hubertz, et al. 1993). Using WIS data provides a consistent estimate of wave climatology when comparing different locations.

WIS data is given at specific depths. Since this methodology evaluates migration rate at varying depths, some form of transformation of wave climatology is needed. For the purposes of this study, a very simple, depth-induced, wave height reduction was used. It was assumed that the maximum spectral significant wave height in any depth of water was half of the water depth, i.e. $(H_{mo})_{max} = 0.5$ h. Wave heights were reduced to that level to evaluate migration for the shallower depths. A more sophisticated wave transformation scheme could be used.

The first comparison is for the mounds constructed in around 6 m depth off the Alabama coast in 1987-88. Hands and Allison (1991) showed survey data that indicated the mounds moved roughly 30 m/yr initially. The methodology outlined above with the WIS climatology for that location matches this expected value when the calibration coefficient is A=0.5.

The second comparison is for a disposal mound created in about 6 m depth off Cocoa Beach in the Cape Canaveral area in 1992. Bodge (1994) found that mound moved roughly 30 m to 45 m landward in the first six months. Most of this movement occurred during and immediately after construction. By the time of the second post-construction survey at six months, the peak of the constructed mound had been flattened to the extent that there was not a clear crest, i.e. the bottom elevation profile had no significant decrease in depth in the offshore direction. Apparently the top of the mound moved about 60 m/yr to 100 m/yr during the first 6 months and then slowed down as the mound or bar-like feature was lost. The methodology presented here with the WIS data for Cape Canaveral, matches this expected migration rate when the calibration coefficient is A=0.2 to 0.33.

Based on these two comparisons, A (using WIS data) can be set to roughly A=0.3 with a reasonable expectation that it may vary from $0.1 < A < 1$. The use of observed migration rates for comparison to expected migration rates is not optimum since the observations are based on much less time than the overall WIS estimate of wave climate, 20 years. The waves would be expected to be larger or smaller than typical for any 6-month to one-year time spans because of the natural variability in wave climate. The observed migration rates were due to the actual waves for that time period and not the long-term wave climate.

Any calibration of this methodology is probably dependent on the wave climatology estimate used. For example, wave climatology based on long-term, in-situ wave gages will probably have different calibration coefficients.

Figure 2 shows the results of this methodology at three locations with very different wave climates. The calibration coefficient was set to A=0.3 for Figure 2. Alabama is on the Gulf of Mexico and has a relatively mild wave climate. Cape Canaveral on the Atlantic coast of Florida has a moderate wave climate. Cape Hatteras has one of the more severe wave climates on the Atlantic coast of the United States. Figure 2 shows that the model estimates migration rates that vary significantly, up to an order of magnitude, for these three locations. For a given depth, migration rate estimates

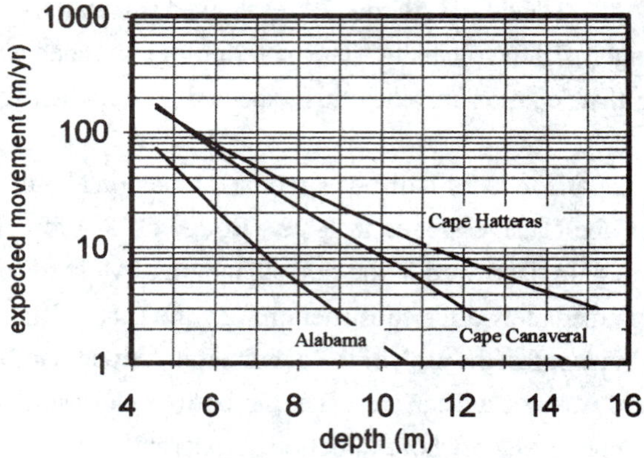

Figure 2. Expected migration rate at three U.S. locations

are higher for the Atlantic sites than for Alabama. Such variation would be expected for any process based on wave climate. In particular, wave lengths are typically significantly longer for the Atlantic Ocean than the Gulf of Mexico. For deeper depths, Cape Hatteras migration rates are higher than Cape Canaveral. The estimates for the two Atlantic sites become close to each other in shallower depths. This closeness is due to the depth-induced, wave height reduction employed to transform the data into shallower water as discussed above. The infrequent, largest waves at Hatteras are more influential in the deeper depths but are less so after breaking as the depths decrease.

Interestingly, Figure 2 shows a similar general shape and dependence with depth at the three locations. This similar shape apparently is a function of the model equations. The same general shape has been found using wave climatology based on wave gages for the Santa Anna River, California nearshore placement of sand (Mesa, personal communication, 1996). This dependence with depth has practical value for planning and design. The results shown on figure 2 can be fit to a simple power relation:

$$E[C] = ah^{-b}$$

where E[C] = expected value of migration rate (m/yr), a=coefficient, b=exponent, and h=depth (m). For most of Figure 2 and all of the deeper depths at each location, the best fit for the exponent is b=5. These are the depths where the wave height decay due to depth ($(H_{mo})_{max}$ = 0.5 h) was small. This was all of the depths shown for the Alabama site, most of the depths for Cape Canaveral, and the deeper depths for Cape Hatteras. In the portion of Figure 2 where the depth reduced the largest waves in the climatology, the depth dependence exponent is b=4. Thus, this methodology implies that migration rate is dependent on depth to the 4th or 5th power.

DISCUSSION

Implications of influence of depth on migration

The depth dependence of migration rate outlined above provides some useful concepts for planning and design of nearshore placement. The above equation implies that migration is extremely sensitive to depth at any specific site. Doubling the depth of placement will decrease the rate of migration by a factor of 16 to 32. Considering it another way, to double the migration rate, it is necessary to place the sand in only 13-16% shallower depth. This guidance should be of value in evaluating the cost vs. depth-of-placement issues typical of nearshore placement of sand.

The extreme sensitivity to depth implied by this methodology has not been verified by observations. Verification would require placement at different depths at the same general location and time with adequate monitoring.

Theoretical limitations

There a number of theoretical limitations to the methodology presented here. First, many of the physical mechanisms that influence sediment transport in these depths including the effects of suspended sediment load, wave groups, ripple and dune migration, sheetflow, bed-ventilation, wave grouping, undertow, wind, bottom streaming and other bottom boundary layer complexities have been ignored. Second, the one physical mechanism modeled, the influence of wave asymmetry on net bedload sand transport could be modeled in a more sophisticated manner. A primary limitation is the use of Stokes 2nd order theory - particularly in shallow water. It was used because it is the lowest order theory with asymmetry. Better representations of the actual velocities should be used. Thus, this methodology can be considered as a rough first attempt at applying such a model to the problem at hand. Its abilities, in spite of this serious limitation, are encouraging in that they indicate the proper physics may be included.

Practical limitations

One serious practical limitation of this methodology is the use of a single depth for modeling migration rate. Nearshore placement in a mound-like feature implies that different parts of the feature are in different depths. More importantly, the equations given above are extremely sensitive to depth. This presents a practical problem in specifying the single, appropriate depth for the use of this model. For this work, that single depth was chosen as the pre-existing depth at the middle of the feature. Conceptually, the top, or shallower portion, of the feature should move fastest and be flattened out by landward movement of sand into the deeper depth shoreward. There is evidence at many observed mounds that this has occurred (Bodge 1994, Mesa 1996, Hands and Bradley 1990, Foster et al. 1996).

Appropriate physics for modeling migration

The implication of the observations and the results presented here is that a net sand transport equation that is dependent on depth could be used with the conservation of sand equation to model the fate of nearshore disposed sands. More specifically, the use of a model that accounts for the influence of velocity asymmetry on sand transport is recommended for modeling the fate of nearshore placed material. Such models have been proposed by Stive (1986) and several of the investigators in the European NOURTEC project (Houwman and Ruessink 1996, Roelvink, et al. 1996, Hoekstra, et al. 1996). The results presented in this paper and the detailed simulation of the behavior of the Silver Strand, California mound shown in Douglass (1995) imply that such modeling is an appropriate approach. The use of some form of Bailard's model (Bailard and Inman 1981) is common to all of these approaches. This present work does not endorse the use of Bailard's model specifically. Other approaches that explicitly account for the effect of net onshore bottom stress due to wave asymmetry should also be considered.

Other existing American methods

The other existing American methods for estimating the fate of nearshore placed sand are not explicitly based on this physical mechanism (wave asymmetry) although it may be implicitly included through the empiricism. The Hands and Allison (1991) method for movement or non-movement is essentially empirical and does not attempt to explain migration direction. The Hands and Allison method is based on extreme wave heights and a sand movement initiation criterion, and thus is really is not incompatible with the wave asymmetry mechanism proposed here. The Hands and Allison method may be successful because it adequately captures the wave asymmetry mechanism through the empiricism.

The Larson and Kraus (1994) method is purely empirical so the same argument could be made. However, that method is based on empirical expressions for surf zone sand bars and beaches. It does not seem likely that the dominant physical mechanisms in the inner surf zone occur in the same relative proportions in the offshore depths for a number of obvious reasons. Any skill shown by methods based on inner surf zone empiricism may be due to fortune as much as physics.

The physical mechanism modeled by Scheffner (1991, 1996), the sediment response to the mean currents with transport rate modified by the presence of waves, seems relatively unimportant to the migration of nearshore placed sand. The evidence from the Mobile, Alabama site demonstrates the fundamental problem with a model based on mean currents. Douglass, et al. (1995) found that the measured mean near-bottom currents in the area were shore-parallel during the time that the mounds migrated shoreward. Shore-parallel mean currents are consistent with expectations for tidal currents near the coast, i.e. along the bathymetric contours. The cross-shore component of the mean near-bottom current was offshore during the time the mounds migrated landward! This offshore dominance was more pronounced during times of large waves, presumably due to downwelling due to onshore wind stress during storms. It appears that Scheffner (1991, 1996) was only able to verify his model at the Mobile, Alabama site by the explicit specification of shoreward mean currents. This was apparently based on a numerical, tidal simulation of a selected seven month period. Given that the actual measured, mean near-bottom currents at the site were in the other direction, Scheffner's method seems incorrect.

Diffusion concept

The diffusion coefficient, D, presented above has also been investigated. It is notable that the diffusion coefficient does not have any free parameters. It has the same coefficient, A, that is in the C equation. No results are presented here for two reasons. One, the methodology estimated more diffusion was was observed at Cape Canaveral. Other mounds; such as Mobile, Alabama and Perdido Pass, Florida; have shown less diffusion than the one at Cape Canaveral. The primary reason for not focussing on the diffusion portion of the model is conceptual. The "diffusion" term results directly from the way that bottom slope, $\tan \beta$, is included in the Bagnold/Inman and Bailard formulation for quasi-instantaneous transport. It accounts for influence of gravitational effects to pull sand down-slope. None of the observed mounds have shown a tendency for sand to move offshore down the slope, only onshore down the slope.

CONCLUSIONS

A design and planning tool for the nearshore placement of sand that estimates the expected onshore migration rate is presented. The results appear appropriate for the Atlantic and Gulf coasts of the United States. For any specific wave climate, the results indicate an appropriate way to evaluate the influence of depth on migration rate. Specifically, doubling the depth of placement will decrease the rate of migration by a factor of 16 to 32. Doubling the rate of migration requires placement in 13-16% shallower depths.

The physical basis for the model is the velocity asymmetry of large waves. This is consistent with observations reported in the literature and the findings of a field study off the Alabama coast that eliminated several other potential physical mechanisms for controlling mound migration. This model is limited by several theoretical and practical shortcomings. It abilities, however, indicate that a wave-based model that accounts for the net onshore transport or shear stress under finite-amplitude wave is the appropriate way to model the fate of nearshore placed sands.

REFERENCES

Andrassy, C. J. (1991). "Monitoring of a nearshore disposal mound at Silver Strand State Park." *Proc., Coast. Sediments 91 Conf.*, ASCE, New York, N.Y., Vol 2, 1970-1984.

Bagnold, R. A. (1963) "Mechanics of marine sedimentation," in *The Sea*, vol. 3, Interscience, New York.

Bailard, J.A. and Inman, D. L. 1981 "An Energistics Bedload Model for a Plane Sloping Beach: Local Transport" *J. Geophysical Res.*, 86(C3):2035-2043.

Bodge, K. R. (1994) "Performance of 1992 nearshore berm disposal at Port Canaveral, Florida." Olsen Associates, Inc. report to the Canaveral Port Authority, Jacksonville, FL, 55 p.

Douglass, S. L. (1995) "Estimating Landward Migration of Nearshore, Constructed, Sand Mounds," *J. Wtrwy, Port, Coast. and Oc. Engrg.*, ASCE, 121:5:247-250.

Douglass, S. L., Resio, D. T., and Hands, E. B. (1995) "Impact of near-bottom currents on dredged material mounds near Mobile Bay," *Tech. Rep. DRP-95-6*, US Army Engrs. Wtwys. Exper. Sta., Vicksburg, MS.

Foster, G. A., Healy, T. R., and DeLange, W. P. (1996) "Presaging beach renourishment from a nearshore dredge dump mound, Mt. Maunganui Beach, New Zealand." *J. Coastal Res.*, 12:2:395-405.

Hands, E. B. (1991) "Unprecedented migration of a submerged mound off the Alabama coast." Proc., 12th Annu. Conf. Western Dredging Assoc. And 24th Annu. Texas A&M Dredging Seminar, Oc.Engr. Program, Texas A&M Univ., College Station, Tx.

Hands, E. B. and Allison, M. C. (1991) "Mound migration in deeper water and methods of categorizing active and stable berms," *Proc. Coast. Sediments 91*, ASCE, 2:1985-1999.

Hands, E. B. and Bradley, K. P. (1990) "Results of monitoring the disposal berm at Sand Island, Alabama; report 1: construction and first year's response." *Tech. Rep. DRP-90-2*, US Army Engrs. Wtwys. Exper. Sta., Vicksburg, MS.

Hoekstra, P., Houwman, K. T., Kroon, A., Ruessink, B. G., Roelvink, J. A., and Spanhoff, R. (1996) "Morphological Development of the Tershelling shoreface nourishment in response to hydrodynamic and sediment transport processes," *Pre-Conf. Paper Summaries*, 25th Inter. Coastal Engrg. Conf., Orlando, Fl., Amer. Soc. of Civil Engrs., pp. 548-549.

Hubertz, J. M., Brooks, R. M., Brandon, W. A., and Tracy, B. A. (1993) "Hindcast Wave information for the US Atlantic coast," *WIS Report 30*, US Army Corps of Engrs. Wtwys. Exper. Sta., Vicksburg, MS, 502 p.

Larson, M. and Kraus, N. C. (1992) "Analysis of cross-shore movement of natural longshore bars and material placed to create longshore bars," *Tech. Rep. DRP-92-5*, US Army Engrs. Wtwys. Exper. Sta., Vicksburg, MS, 115 p.

Laustrup, C., Madsen, H. T., Sorensen, P., and Broker, I. (1996) "Comparison of Beach and Shoreface Nourishment Torsminde Tange, Denmark" *Pre-Conf. Paper Summaries*, 25th Inter. Coastal Engrg. Conf., Orlando, Fl., Amer. Soc. of Civil Engrs., pp. 550-551.

Mesa, C. 1996. Paper presented at 25th Inter. Coastal Engineering Conf., Orlando, Fl.

Roelvink, J. A., and Stive, M. J. F. (1989) "Bar generating cross-shore flow mechanisms on a beach," *J. Geophysical Res.*, 94:C4:4785-4800.

Roelvink, J. A., Meijer, Th., Spanhoff, R., and Hoekstra, P. (1996) "Process-based modelling of the behavior of barred coastal profiles" *Pre-Conf. Paper Summaries*, 25th Inter. Coastal Engrg. Conf., Orlando, Fl., Amer. Soc. of Civil Engrs., pp. 546-547.

Scheffner, N. W. (1991) "A systematic analysis of disposal site stability." *Proc., Coast. Sediments 91 Conf.*, ASCE, New York, N.Y., Vol. 2, 2012-2025.

Scheffner, N. W. (1996) "Systematic analysis of long-term fate of disposed dredged material." *J. Wtrwy, Port, Coast. and Oc. Engrg.*, ASCE, 122:3:127-133.

Stive, M. J. F. (1986) "A model for cross-shore sediment transport." *Proc., 20th Inter. Coastal Engrg. Conf.*, ASCE, New York, N. Y., pp. 1550-1564.

CHAPTER 287

Influence of Nearshore Berm on Beach Nourishment

Paul A. Work[1], A.M. ASCE, and Emre N. Otay[2]

Abstract

The influence of a large (3.0×10^6 m^3) nearshore berm on a large (4.1×10^6 m^3) beach nourishment project at Perdido Key, FL, is addressed via a monochromatic, numerical wave transformation model. Two years of post-placement survey data indicate that the nearshore berm, placed at a depth of 6 m with a relief of 1-1/2 m, did not migrate during this time period. Wave model results suggest that the berm influences the beach nourishment project via refraction, but breaking and diffraction effects are not significant. The berm causes wave transformation and leads to zones of wave energy focusing (and corresponding de-focusing) that affect longshore sediment transport rates in the lee of the berm. For a typical wave condition, predicted high-energy regions correlate well with observed erosional areas. The findings indicate the relative importance of project-induced wave transformation that should be considered during the design process for nearshore berm projects.

Project Background

The beach nourishment project at Perdido Key, FL, adjacent to Pensacola Pass, was initiated in late 1989. Pensacola Pass serves as an access channel to a major naval port; the beach nourishment material was generated by dredging of the entrance channel to 19 m. The region is characterized by diurnal tides with a very small range (mean tide range = 34 cm; spring range = 58 cm) and a relatively mild wave climate. The mean value of the significant wave height measured over 4 years

---

[1] Asst. Prof., Dept. of Civil Eng., 110 Lowry Hall, Clemson University, Clemson, SC 29634-0911. (864) 656-3000. pwork@eng.clemson.edu.
[2] Asst. Prof., Civil Eng., Boğaziçi Univ., Bebek 80815 Istanbul, Turkey. otay@boun.edu.tr.

as a part of this study near the 6 m contour is 0.6 m. The mean value of the wave period at which the peak of the energy spectrum is found is 6 sec for the same time interval.

The beach nourishment aspect of the project involved widening the dry beach by 140 m over a length of 7 km and required a volume of 4.1 million m^3. This was completed in September, 1990. Construction of an offshore berm then commenced. The offshore berm was completed in 1991. The placed berm was 4 km (longshore) by 300-800 m (cross-shore) and included 3.0 million m^3 of sand. The bathymetry in the region where the berm was placed was quite flat (slope of 1:1000). Water depth averaged 6 m, and the berm relief is nominally 1.5 m, leaving 4.5 m of water above the berm. Figure 1 indicates the site location, and Figure 2 shows cross-sections through the berm at different times.

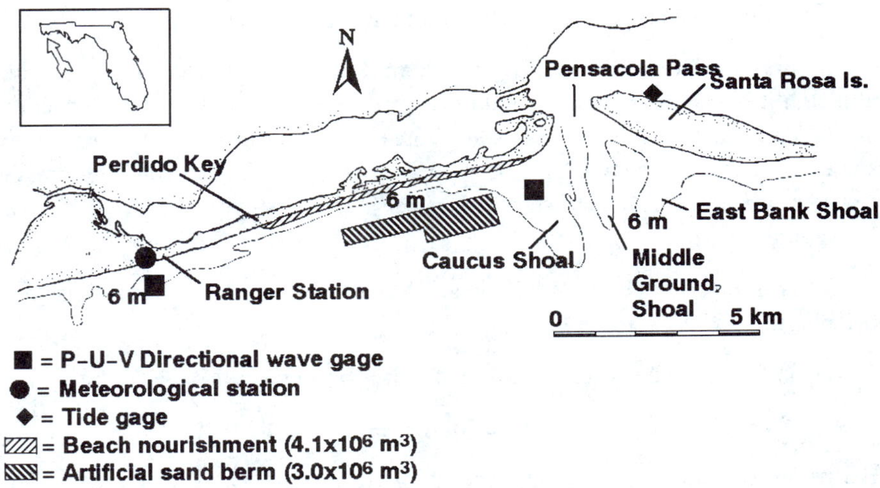

Figure 1. Site location and monitoring plan.

A monitoring program from 1989-1994 yielded a data set that is useful for assessment of hydrodynamics and sediment transport at Perdido Key, including annual bathymetric survey data and long-term wave and current data. Details of the monitoring program may be found in Work (1992), Otay (1994), and Work and Dean (1995).

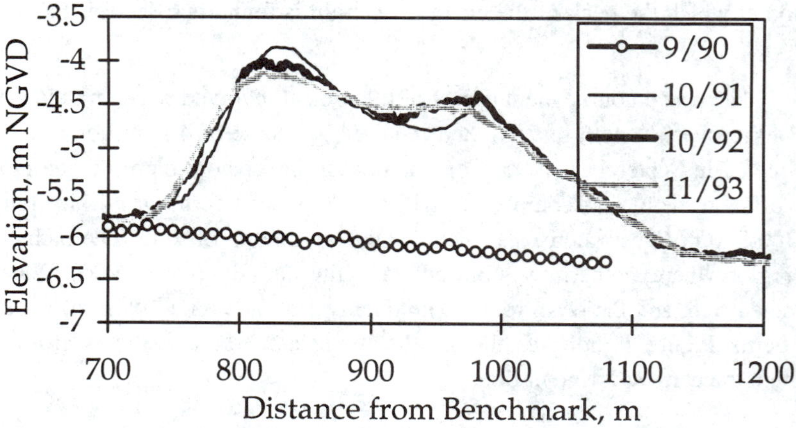

Figure 2. Cross-section through berm near midpoint.

Survey data indicate that the berm has not migrated since placement, although it has been smoothed slightly (Figure 2). But due to the 25% reduction in water depth that it introduces over a large area, it is expected that the berm will exert some influence on the waves that pass over it, and therefore modify nearshore sediment transport. The purpose of this paper is to investigate this effect and assess its significance.

A nearshore berm can provide at least two benefits, depending on its behavior:

1) Nourish the beach as sediment moves onshore.

2) Shelter the beach in the lee of the berm.

Berms are sometimes favored over beach nourishment because of the reduction in cost due to the relative ease of placement. But correct assessment of the benefits requires accurate prediction of both any migrational tendency and any effect on adjoining areas.

<u>Bathymetry</u>

A numerical grid of depths at the site was created for wave modeling. The grid represents a combination of measurements taken as a part of the monitoring study and digitized nautical charts. Depths at the offshore boundary are not uniform and the grid does not extend to deep water, posing some problems for wave modeling that are addressed below.

A second grid was created by artificially stripping out the berm from the first grid. A detail of the two grid files is shown in Figure 3. Some "scalloping" of the nearshore contours is evident in the no-berm plot. This is due to limitations of the contouring routine when there are "stripes" of data (in this case due to surveying beach profiles at 300 or 600 m intervals along the beach), a common problem. This will lead to some overestimation of the effects of wave transformation.

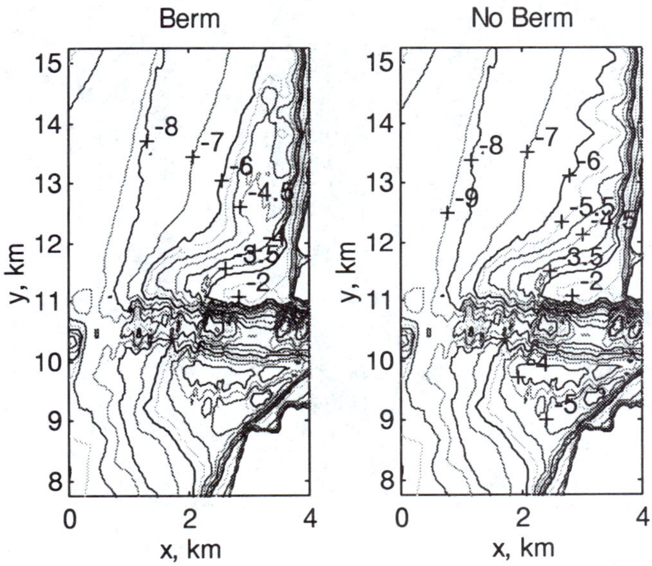

Figure 3. Bathymetry with and without offshore berm. Contour interval 1 m, with the addition of 3.5, 4.5, and 5.5 m depth contours. x-axis points onshore, y-axis in longshore direction. Pensacola Pass at right near y = 10 km.

Hydrodynamic Modeling

Measurements of wave energy spectra and mean currents are available at two locations at the site (Figure 1). A finite element hydrodynamic model (RMA-2, Norton, King, and Orlob, 1973, Thomas and McAnally, 1990) was employed to predict tidal currents. The friction factor in the model was adjusted until modeled currents agreed with published values for predicted tidal currents in the inlet throat (U.S. Dept. of Commerce, 1994). Results were then compared to measurements at the two wave gage locations for maximum ebb and maximum flood conditions. The model typically underpredicted currents slightly, not surprising due to the omission of wind- and wave-driven currents in the model. Figure 4 illustrates the results for maximum ebb tide.

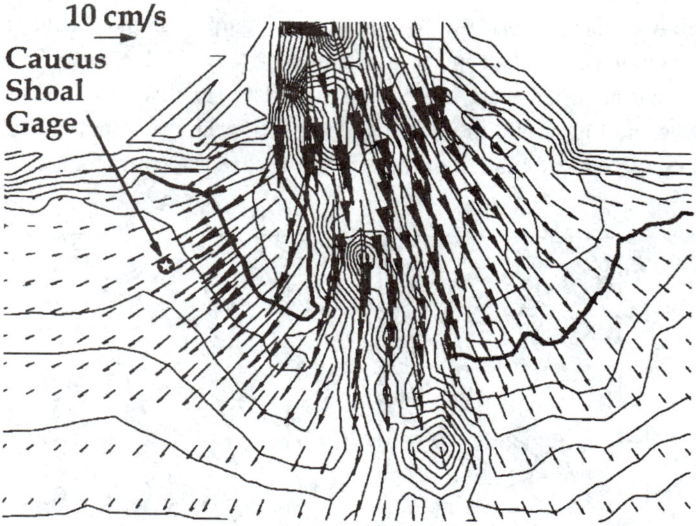

Figure 4. Predicted tidal currents (from RMA-2 model) for maximum ebb tide through Pensacola Pass. Note that flow is concentrated between shoals to either side of navigation channel.

Wave Modeling

A monochromatic, finite difference wave transformation model (REF/DIF 1, Kirby and Dalrymple, 1994) was used to investigate wave transformation at the site. The monochromatic model was selected for two reasons: 1) ease of use, compared to a spectral model, and 2) to assess the viability of a monochromatic model for such a situation. Ease and speed of use become major factors when long-term simulations (years) will be performed.

Since neither bathymetry grid extended to deep water for most of the wave conditions to be modeled, an iterative procedure was developed to estimate wave conditions at the offshore boundary, given wave conditions at the western wave gage, where the bathymetric contours are relatively straight and parallel (Work and Kaihatu, in press). This allows one to drive the model with data from one nearshore wave gage (a common situation) and predict wave conditions anywhere else in the domain.

A large number of model runs have been performed with the real (berm) bathymetry at the site, using data from the western wave gage to drive the model. The monochromatic model, on average, yields a reasonable prediction of wave height at the other gage, although there is a large amount of scatter (the average value of the ratio of model to measured wave heights is 0.99; standard deviation of this ratio is 0.34). The model is not a good predictor of wave direction at the other gage, in part because of the fact that the contours run nearly perpendicular to the nominal shoreline orientation at the eastern wave gage, which sits near a shoal. Figure 5 shows results for one simulation near this shoal.

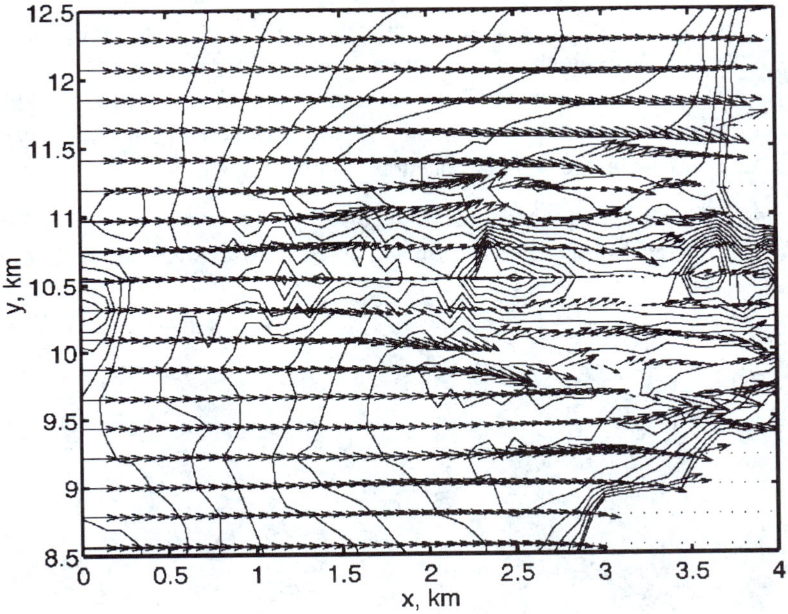

Figure 5. Wave vectors near Caucus Shoal, immediately west of Pensacola Pass. Conditions at western wave gage: H_{mo} = 0.6 m, T_p = 6 sec, shore-normal incidence.

Wave model results indicate that tidal currents at the site do not exert any significant influence on wave conditions outside of the inlet channel. Thus bathymetric gradients are the primary cause of wave transformation. This simplifies modeling efforts substantially.

Hypothetical cases were simulated to isolate the influence of wave period, tide stage, and incident wave height and direction, on wave conditions in the lee of the nearshore berm. Figure 6 compares the wave heights for the two bathymetric grids for the "typical" case (T_p = 6 sec, shore-normal waves with height H_{mo} = 0.6 m at western wave gage). Figure 7 shows the difference between the two results, normalized by the incident wave height. Some findings:

- Tide stage has minimal influence on the wave height differences (berm vs. no berm). This is largely due to the very small tide range at the site.

- Dependence on wave period is as expected: as period increases, refraction becomes more pronounced, and differences between the two results increase.
- Dependence on wave height is weak until the incident wave height is increased to the point where waves begin to break on the nearshore berm (H_{mo} > 3.5 m). The largest significant wave height measured during the 4-year wave monitoring program was 2.9 m, in August, 1992, while Hurricane Andrew was in the Gulf of Mexico. This event also yielded the largest measured wave period (T_p = 13 seconds). In summary, wave breaking on the berm has occurred rarely, if at all.
- Dependence on incident wave direction is relatively strong. Greatest differences between the berm and no-berm cases were observed for nearly shore-normal waves.

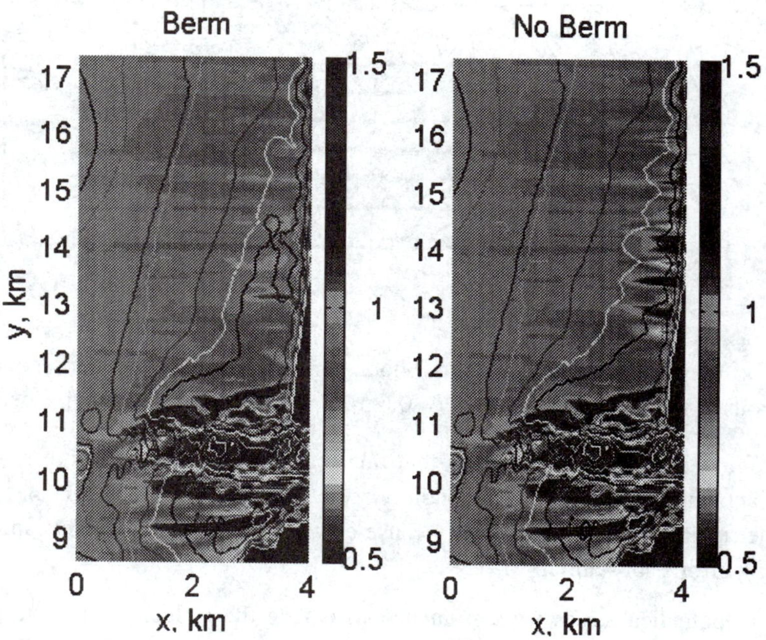

Figure 6. Wave model results with typical wave conditions (H_{mo} = 0.6 m, T_p = 6 sec, shore-normal incidence, at western wave gage), with and without berm. Contours indicate depth at 1 m intervals. Shade indicates wave height relative to wave height at western wave gage.

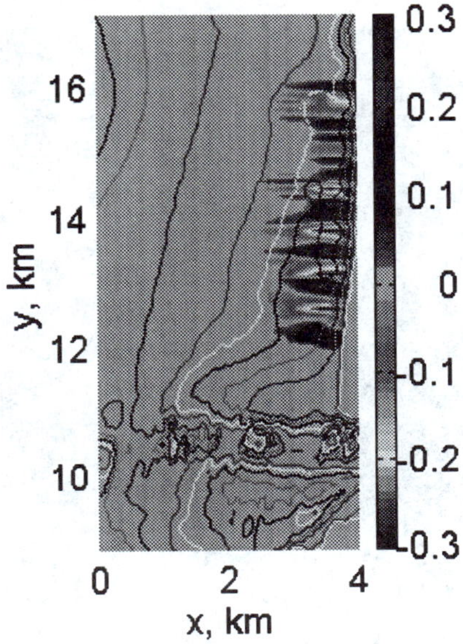

Figure 7. Difference between no-berm and berm wave model results shown in Figure 6. Normalized by no-berm results. Shade indicates percent difference in wave height. Positive values indicate percent increase in wave height in absence of berm, negative indicate percent reduction.

The maximum difference between the berm and no berm wave height fields, for the region behind the berm, is 30%. Because there is typically a minimal difference in energy dissipation between the berm and no berm cases, the total energy reaching the shoreline is the same whether or not the berm is present. The berm acts to redirect energy and leads to zones of focusing and de-focusing. If depths over the berm were less, breaking would occur on the berm and the shoreline behind the berm would be more sheltered. Figure 8 compares the wave height fields for large vs. small incident wave heights, both over the berm bathymetry, to illustrate this effect.

Limited tests were also performed with a spectral version of the same wave model (REF/DIF S, Ozkan and Kirby, 1993). This model essentially consists of superimposing many results from the monochromatic model for different frequencies. Results with the spectral model were similar to the monochromatic model results, but more muted. The presence of additional frequency components leads to less pronounced focusing of wave energy. One example is provided in Figure 9.

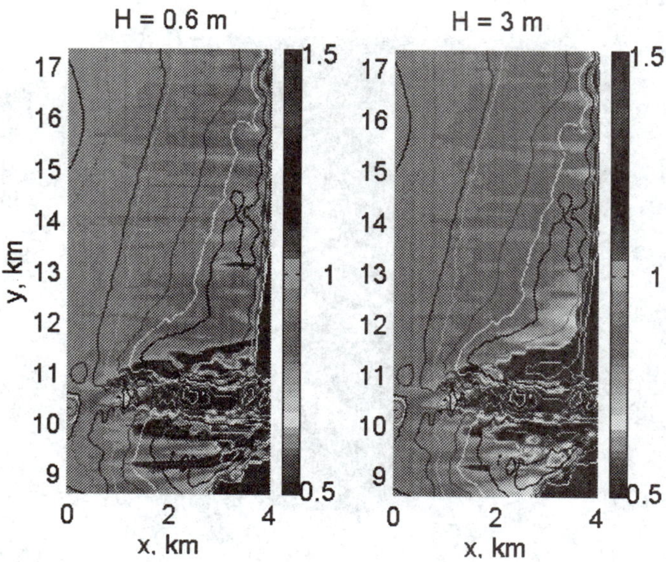

Figure 8. Small wave height (H_{mo} = 0.6 m, T_p = 6 sec, shore-normal at western wave gage) vs. large (H_{mo} = 3.0 m) wave height results over berm bathymetry. Shade indicates wave amplification. Contours indicate depth at 1 m intervals.

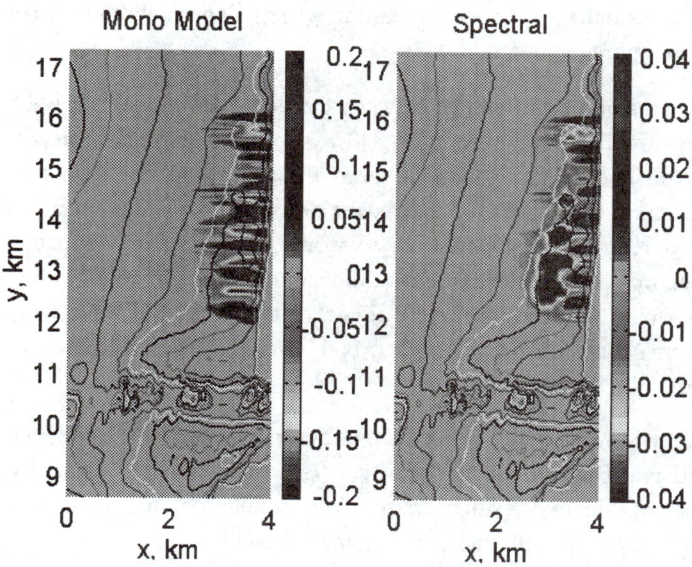

Figure 9. Monochromatic model (REF/DIF 1) vs. spectral model (REF/DIF S) results for conditions corresponding to Figure 6. Shade indicates percent difference in wave height. Positive values indicate percent increase in wave height in absence of berm, negative indicate percent reduction.

Shoreline Change

Redistribution or reduction of wave energy reaching the shoreline will affect sediment transport. Figure 9 shows the measured changes in shoreline planform, relative to the pre-nourishment condition. This is a classic signature for a beach nourishment project: the waterline moves back, rapidly at first, due to cross-shore sediment transport, since the as-built beach profile slope is much steeper than natural slopes. The "shoulders" (ends) of the project erode more rapidly due to the high rate of sediment diffusion in the longshore direction induced by strong planform gradients.

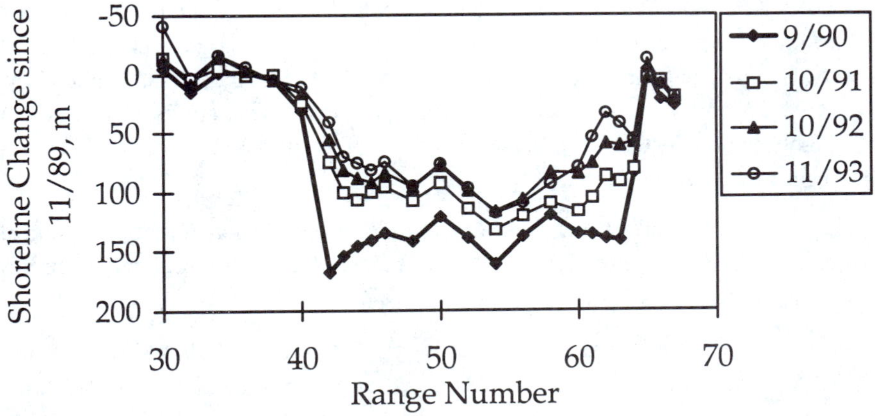

Figure 10. Measured change in shoreline position, relative to pre-nourishment condition. Nominal distance between survey ranges is 300 m, Pensacola Pass at right limit of figure.

Another means of investigating the evolution of a nourishment project is to calculate the volumetric changes per unit length of beach, i.e. the change in the cross-sectional area of the beach profile. Doing so allows one to integrate out the effects of cross-shore sediment transport: all observed changes are due to longshore redistribution of sediment arising from longshore gradients of longshore sediment transport. This requires assumption of a depth beyond which there is no cross-shore sediment transport. For the Perdido Key beach nourishment project, it is quite clear that there is negligible sediment transport beyond the 5.5 m contour (Work and Dean 1995).

Figure 11 shows the computed longshore gradient of longshore sediment transport, which is equal in magnitude (but opposite in sign) to the change in cross-sectional area of the beach profile per unit time. Positive values indicate erosion and negative indicate accretion. The question is then whether these zones of erosion and accretion can be predicted by the numerical wave model. A definitive answer to this question would require that the wave model be run for every measured wave event to calculate a four-year time series of breaking wave height and direction. These time series could then be used to drive a shoreline change model.

Unfortunately it is not possible to model wave transformation for every measured wave event. There are many measured wave directions which are highly oblique (> 45 degrees) to the shoreline. The chosen wave model (and any other similar model which employs a parabolic approximation to the elliptic governing equation) has a limitation on the incident wave angle. If the incident wave angle exceeds a certain value (the value of which depends on details of the model formulation), the model will yield erroneous results or no results at all. In addition, the method developed to estimate wave conditions at the offshore boundary of the grid, given wave conditions at one nearshore wave gage, similarly will not work for highly oblique waves. In practice, it was possible to model only about one-half of the measured wave conditions.

A "typical" wave condition was therefore chosen for comparison of wave height fields and observed shoreline changes. Noteworthy regions are indicated by the arrows on Figure 11. Several features stand out:

- There is consistent focusing predicted (both with and without the berm, for nearly all wave conditions) at the eastern limit of the nourishment project. This is indicated by the arrow at the extreme right of Figure 11. A strong erosional trend is also observed there. This is erosion of the end of the beach nourishment project, enhanced by the focusing of wave energy.

- Placement of the berm reduced, for the "typical" wave condition, the wave heights near the midpoint of the beach nourishment project, where a switch from erosional to stable behavior is observed (middle arrow in Figure 11). A similar switch occurred near the western end of the nourishment project, where placement of the berm reduced wave heights, and a switch from a stable shoreline to accretion occurred (left arrow).

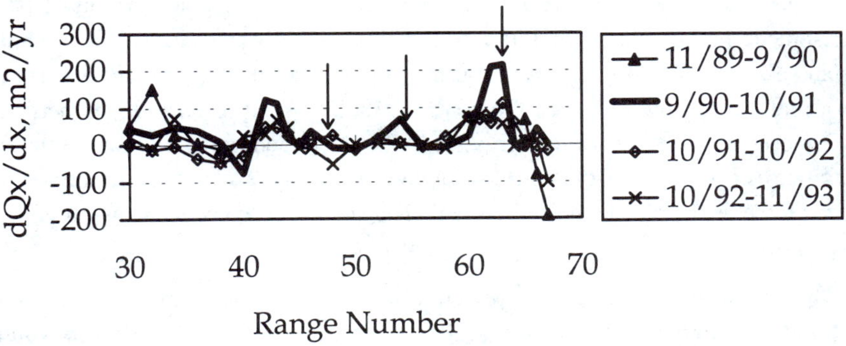

Figure 11. Longshore gradient of longshore sediment transport, calculated from surveyed beach profiles. Equivalent in magnitude to change in cross-sectional area of beach profile. Nominal spacing between ranges is 300 m, Pensacola Pass to right.

Project Performance

Volume of sand remaining within the nourished region was computed as a measure of project performance. Three years after placement of the beach nourishment material, 84% of the initial volume remains (Figure 12). If the entire monitored area is considered, the value is 82%. Some of the nourishment material has been deposited into the inlet, and some has moved westward due to the predominant wave conditions.

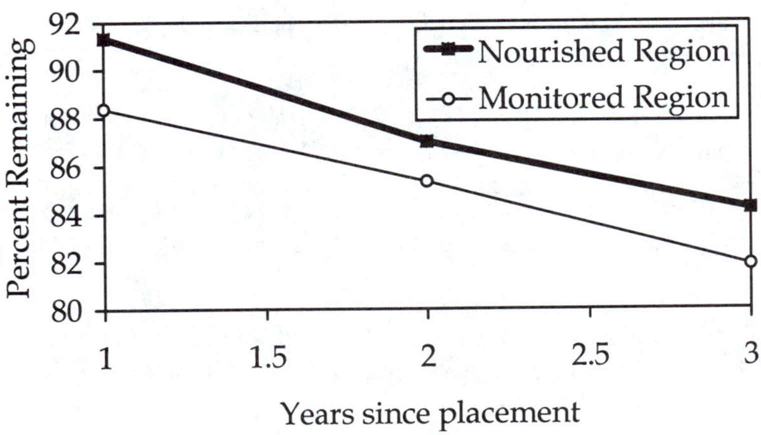

Figure 12. Volume changes for project. Nourished region is a subset of the monitored region.

Conclusions

Nearshore berms can and have been placed with the goal of gradual beach nourishment through onshore migration. Existing predictive capability with regard to rate and even direction of migration (onshore or offshore) is still limited. A berm can also exert a sheltering effect on the beach in its lee. The presence of the berm can either merely redistribute wave energy, or, if it induces additional dissipation through breaking over the crest of the berm, reduce wave energy reaching the shoreline. The effect of the berm on the shoreline behind it should be considered as part of the design process.

Data for two years after placement of a berm at Perdido Key, FL, indicate the berm did not migrate during this time. It has been slightly smoothed but its volume unchanged. Numerical wave model results indicate that the berm does exert some influence on the breaking wave climate by redistributing energy alongshore. This occurs even in the absence of waves breaking on the berm. Changes in nearshore wave heights which occur as a result of the berm are reflected in the measured shoreline signature, for a typical incident wave.

It was not possible to model all wave events with the chosen numerical wave transformation model. Waves which deviate significantly from shore-normal occur frequently at the site but could not be modeled. Improvements in wave modeling techniques will eventually remove this constraint.

References

Kirby, J.T., and Dalrymple, R.A., 1994. REF/DIF 1: Combined refraction/diffraction model. Documentation and user's manual. CACR Report No. 94-22, Dept. of Civil Eng., Univ. of Delaware, Newark, DE.

Norton, W.R., King, I.P., and Orlob, G.T., 1973. A finite element model for Lower Granite Reservoir. Water Resources Engineers, Inc., Walnut Creek, CA.

Otay, E.N., 1994. Long-term evolution of nearshore disposal berms. Ph.D. Dissertation, Coastal and Oceanographic Eng. Dept., Univ. of Florida, Gainesville, FL.

Ozkan, H.T., and Kirby, J.T. 1993. Evolution of breaking directional spectral waves in the nearshore zone. Proc. Ocean Wave Measurement and Analysis, New Orleans, LA, ASCE, 849-863.

Thomas, W.A., and McAnally, W.H., Jr., 1990. Users manual for the generalized computer program system: Open-channel flow and sedimentation, TABS-2. U.S. Army Engineer Waterways Expt. Station, Vicksburg, MS.

U.S. Department of Commerce, 1994. *Tidal Current Tables, 1995, Atlantic Coast of North America*. Natl. Oceanic and Atmos. Admin., Natl. Ocean Serv., Silver Spring, MD.

Work, P.A., and Dean, R.G., 1995. Assessment and prediction of beach-nourishment evolution. J. Wtrwy, Port, Coast., and Ocean Eng. 121(3), ASCE, 182-189.

Work, P.A., and Kaihatu, J.M., in press. Wave transformation at Pensacola Pass, FL. Accepted for J. Wtrwy, Port, Coast., and Ocean Eng., ASCE.

CHAPTER 288

ACCURACY OF SAND VOLUMES AS A FUNCTION OF SURVEY DENSITY

Jennifer L. Irish[1], W. Jeff Lillycrop[2], and Larry E. Parson[3]

ABSTRACT

Hydrographic surveys are the primary tool for calculating beach nourishment project volumes. With costs for beach-quality sand as high as $30 US per cubic meter, miscalculating project design volumes can result in significant cost differentials. Because conventional bathymetry and topography of beach projects are collected along shore-normal profile lines spaced anywhere from 30 m to 300 m, calculation of project volumes relies heavily on the assumption that there is little along-shore variability from one profile to another. In most cases, however, the beach and nearshore are highly irregular and this assumption is violated. With the development of high-resolution bathymeters, such as the SHOALS airborne lidar system, it is now feasible to collect accurate, high-density beach surveys. These types of data sets create a highly accurate, quantitative measurement of beach and nearshore conditions. This paper describes the SHOALS system and lidar technology and presents a comparison of volumes calculated using high-density lidar data and conventional nearshore profile surveys. Volumes are calculated to compare differences for beaches on the Atlantic Ocean, the Gulf of Mexico, and the Great Lakes.

1.0 INTRODUCTION

Cost for a cubic meter of sand placed on a beach ranges from $5 US to $30 US, depending on many geographic and engineering factors. Beach nourishment projects range in size from several thousands of cubic meters, such as the 1993 project at St. Joseph, Michigan of 39,000 m^3, to millions of cubic meters, such as the project at Miami Beach, Florida of 12 million m^3. Underestimated project design volumes can result in cost overruns or a reduced amount of sand being placed on the beach, while overestimated project design volumes can result in excessive budgeting and planning. Hydrographic surveys are the primary tool for calculating project

---

1) Hydraulic Engineer, 2) Research Physical Scientist, 3) Research Hydraulic Engineer, US Army Engineer Waterways Experiment Station, Coastal and Hydraulics Laboratory, 3909 Halls Ferry Road, Vicksburg, MS 39180-6199, USA

volumes. Conventional survey techniques use shore-normal wading depth surveys matched with offshore acoustic surveys, spaced at intervals along the beach ranging from 30 m to 300 m.

Calculation of project volumes relies on the assumption that there is little topographic or bathymetric variability from one profile line to the next, or that if there is variability, it averages out over the project limits. However, beach and nearshore topography are highly three-dimensional as a result of sub-aerial sand dunes, nearshore bars, hard-bottom outcrops, seawalls, and groins. At profile spacings typically on the order of 300 m, these assumptions are often violated. With the development of lidar (LIght Detection And Ranging) hydrographic systems, such as the US Army Corps of Engineer's (USACE) Scanning Hydrographic Operational Airborne Lidar Survey (SHOALS) system, it is now economically feasible to collect accurate, high-resolution beach and nearshore surveys.

The following discusses the SHOALS system and its technology and discusses the benefits of high-resolution bathymetry and topography. Four beach projects surveyed with SHOALS are presented and analyzed to determine sand volume computation sensitivity to survey density.

2.0 SHOALS SYSTEM

In 1994, the USACE introduced a new airborne hydrographic survey system capable of remotely collecting high-resolution, accurate bathymetry. The SHOALS system is an airborne-lidar system operating from a Bell 212 helicopter (Figure 2.1). The system uses state-of-the-art lidar technology to measure water depths (Guenther, 1996). A laser-transmitter/receiver is housed inside a pod mounted underneath the aircraft. As the laser pulses at 200 Hz, it is scanned in an arc producing a swath width equal to approximately one-half the aircraft altitude. This yields a uniform sounding spacing, nominally 4-m by 4-m. The laser pulse travels from the airborne platform to the water surface where part of the energy reflects back to the receiver (Figure 2.2). The remaining energy penetrates the water column and reflects off the sea bottom. The time differential between these two returns indicates the water depth (Lillycrop et al., 1996).

Figure 2.1 The SHOALS system

Table 2.1 gives SHOALS current performance specifications. SHOALS has a vertical accuracy of ±15 cm and a horizontal accuracy of ±3 m and meets the USACE Class I, dredge payment, performance specifications (Riley, 1995). Operating at a nominal speed and altitude of 30 m/s and 200 m, SHOALS surveys 8 km^2 per hour. In optically clear water, SHOALS records to depths of 40 m. Since early 1995, SHOALS also records accurate topographic

Table 2.1 SHOALS performance

maximum depth	40 m
minimum depth	< 1 m
vertical accuracy	±15 cm
horizontal accuracy	±3 m
sounding density	4 m
operating speed	30 m/s
swath width	110 m

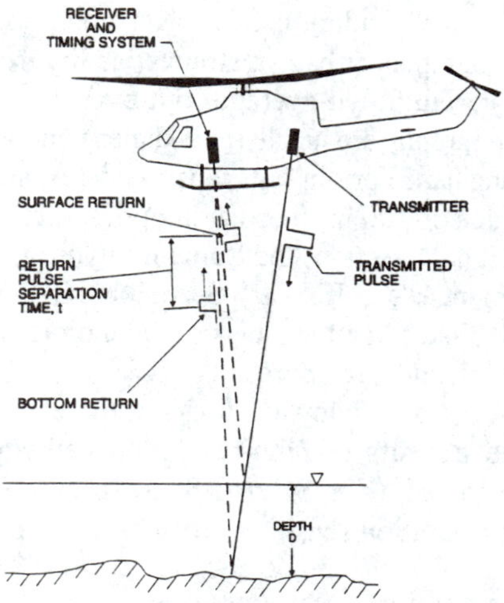

Figure 2.2 SHOALS operating principle

elevations of adjacent beaches allowing full mapping of both the beach and nearshore. To date, SHOALS surveyed over 80 USACE projects totaling 2,000 km^2.

3.0 SITE DESCRIPTIONS

Because SHOALS collects very dense bathymetry, typically 120,000 soundings per square kilometer, it is an ideal tool for monitoring the beach and nearshore in dynamic, irregular areas. For a particular stretch of beach, SHOALS data provides detailed bathymetry and adjacent beach topography allowing accurate identification of high-erosion areas and complex nearshore features. The four data sets presented herein include Longboat Key on the Gulf of Mexico, Island Beach State Park on the Atlantic, and St. Joseph and Presque Isle on the Great Lakes. Each of these sites is characterized by irregular cross-shore and/or along-shore variation in topography.

3.1 Longboat Key

Longboat Key, Florida is on the east shore of the Gulf of Mexico and is situated between Longboat Pass, to the north, and New Pass, to the south. Over 2 million cubic meters of beach-quality sand were placed on the southern-most 8.5 km of the key in 1993 to protect the shoreline from further erosion. SHOALS surveyed the area 5 times since March 1994 to monitor the nourishment project (Irish and Truitt, 1995). The high-resolution SHOALS bathymetry collected in November 1995 reveal a complex sand-bar system in the nearshore (Figure 3.1). The dual-bar system merges together and separates as it parallels the shoreline. At the southern end of the key, the seaward bar diverges into the ebb shoal of New Pass. Conventional profile data are regularly collected along profiles spaced 300 m apart and outline the dual-bar system; however,

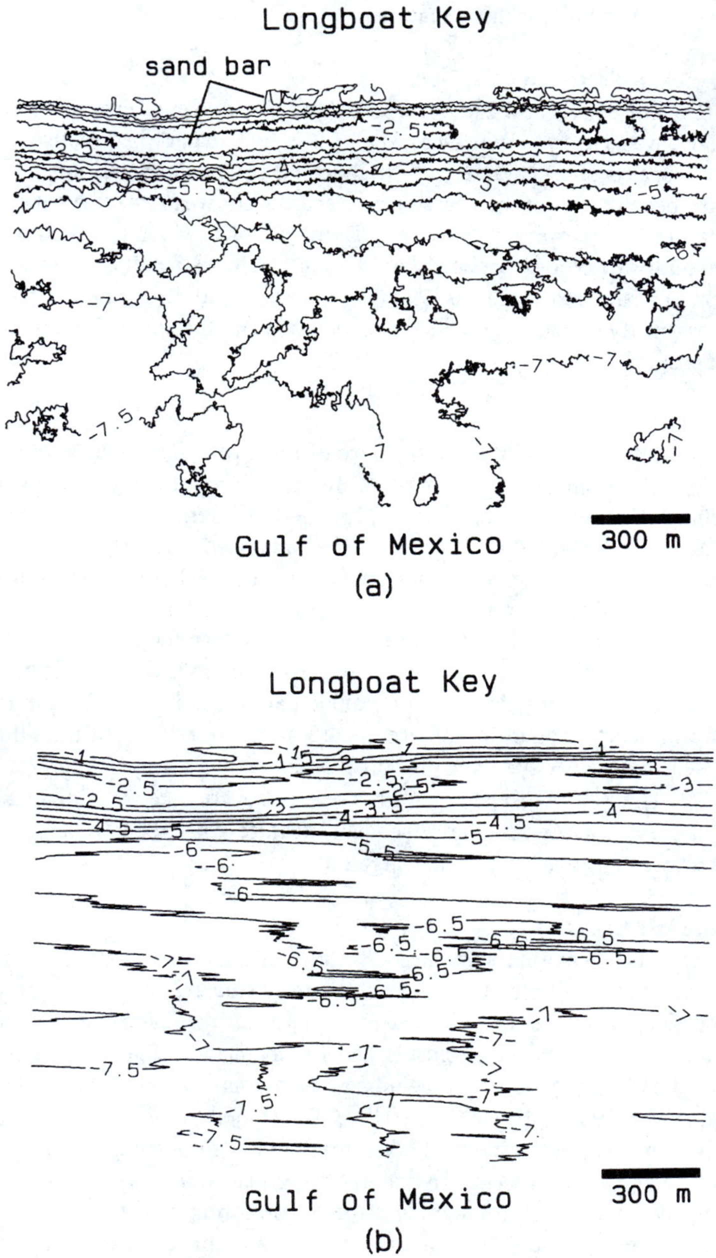

Figure 3.1 Longboat Key, Florida, November 1995: a) representative section of SHOALS bathymetry, b) simulated profile bathymetry at 300-m spacing. All depths are in meters referenced to NGVD.

these data sets do not reflect local complexities in the nearshore. The southern-most 3.4 km were analyzed in this investigation.

3.2 Island Beach State Park

A 2.5-km stretch of Island Beach State Park, New Jersey, just north of Barnegat inlet was surveyed with SHOALS in 1994. The park is directly exposed to the Atlantic Ocean and is characterized by shore-perpendicular sand ridges stretching from the dry beach through the nearshore. Because the area is a state park, no man-made alterations are permitted. However, Barnegat inlet, which is jettied on both its north and south sides, does impact Island Beach. SHOALS surveyed the southern-most 2.5 km of Island Beach in June 1994 (Figure 3.2). The survey details the beach's three-dimensionality quantifying the sand formations and the shoaled areas formed by inlet processes.

3.3 St. Joseph

St. Joseph, on the southeastern shore of Lake Michigan, was authorized as a Federal beach nourishment project in 1976. In 1903, two jetties were constructed to stabilize the St. Joseph River entrance. These jetties interrupt the natural southerly longshore transport of 84,000 m^3 per year, and as a consequence, the downdrift beach experienced erosion and the lake bed suffered downcutting (Parson and Smith, 1995). Since 1976, the USACE annually places dredged material from the maintenance dredging of St. Joseph Harbor south of the entrance to form a feeder beach to replenish 6 km of shoreline. Additionally, coarser material from an upland source is periodically placed. Typically, bathymetric and topographic data are collected along survey lines spaced 152 m apart through the fill area and 800 m apart south of the fill area. In August 1995, SHOALS surveyed the project collecting nearly 400,000 soundings (Figure 3.3). The SHOALS data quantify areas of severe lake-bed downcutting and identified a previously undiscovered headland feature with a 2-m relief. The entire 6-km project was analyzed in this investigation.

3.4 Presque Isle

Presque Isle Peninsula is on the south shore of Lake Erie at Erie, Pennsylvania. The peninsula historically tends to migrate easterly causing erosion of the lake-side beach. Occasionally the area breaches causing dangerous navigation conditions in Erie Harbor, situated between the peninsula and the mainland. The USACE has taken several measures to prevent erosion including beach nourishment and the construction of groins (Grace, 1989). In 1992, the USACE installed 55 breakwaters offshore of Presque Isle. Each breakwater is 47.5-m long and is separated by a 106.7-m gap. Additionally, 285,000 m^3 of beach fill material were placed initially and renourishment occurs annually (Mohr, 1994). Ongoing project monitoring of the performance of the breakwater system includes bathymetry and topography collected annually along profiles spaced 300 m to 600 m apart. In August 1995, SHOALS surveyed this breakwater system detailing the salient and tombolo formations shoreward of the structures and the fairly uniform bottom seaward of the structures (Figure 3.4). The entire 9.9-km long project was used in this investigation.

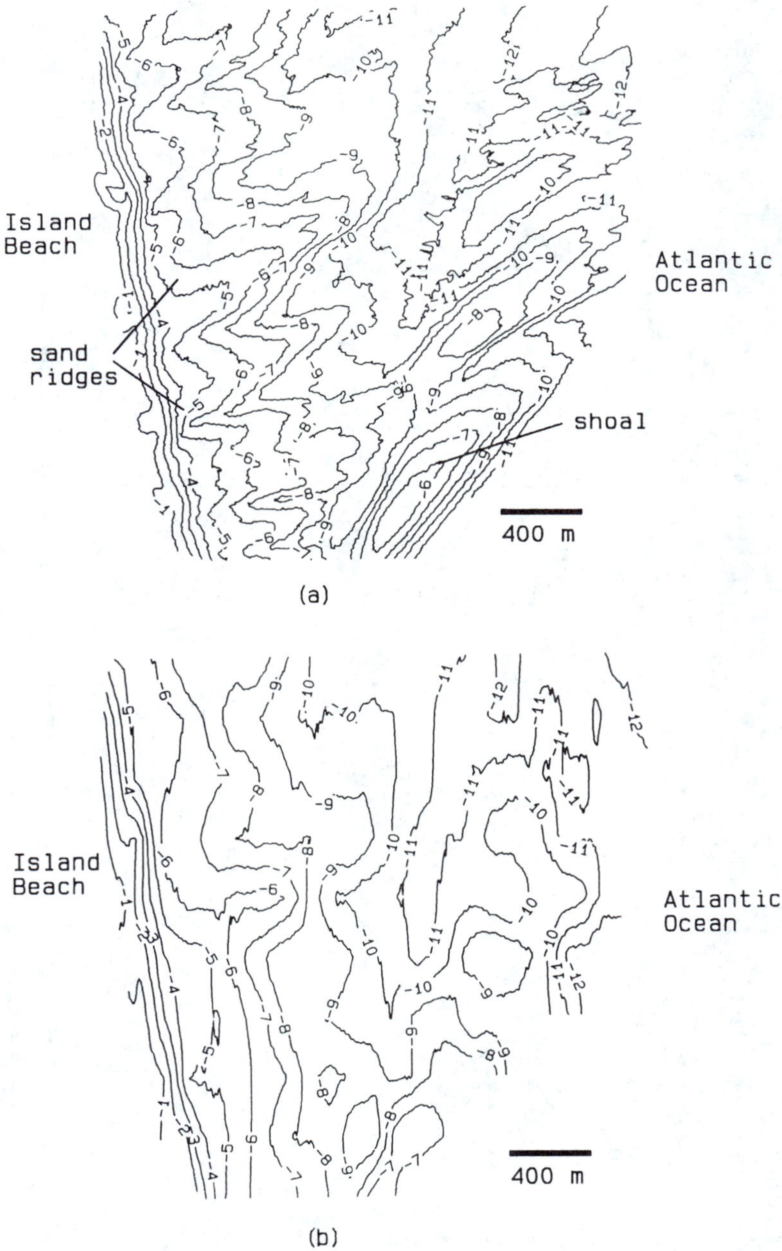

Figure 3.2 Island Beach State Park, New Jersey, June 1994: a) SHOALS bathymetry, b) simulated profile bathymetry at 300-m spacing. All depths are in meters referenced to NGVD.

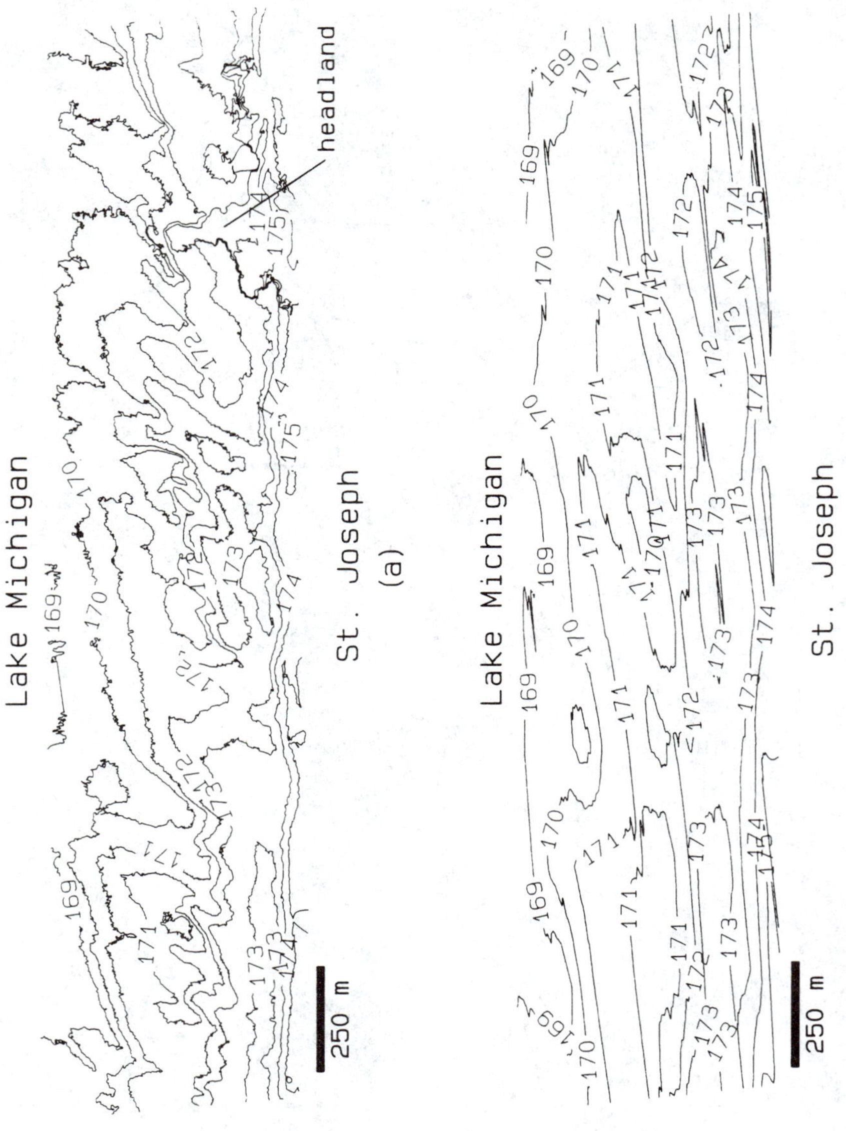

Figure 3.3 St. Joseph, Michigan, August, 1995: a) representative section of SHOALS bathymetry, b) simulated profile bathymetry at 300-m spacing. All depths are in meters referenced to IGLD.

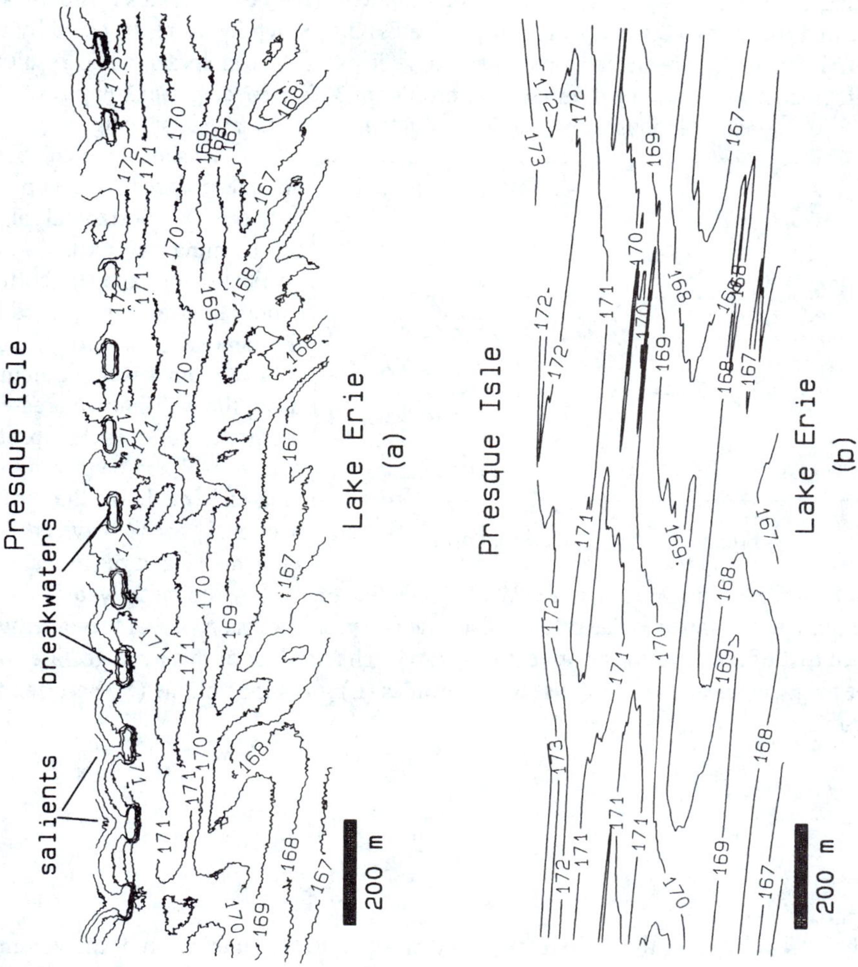

Figure 3.4 Presque Isle, Pennsylvania, August, 1995: a) representative section of SHOALS bathymetry, b) simulated profile bathymetry at 300-m spacing. All depths are in meters referenced to IGLD.

4.0 COMPARISON PROCEDURE

To evaluate the importance of data density on beach nourishment volume calculations, profile data along spacings varying from 5 m to 300 m were simulated from the SHOALS bathymetry using a commercially available CAD and engineering package, TERRAMODEL™. First, profile alignments spaced every 5 m were created along the project length. A digital terrain model (DTM) of the SHOALS data was created to represent it mathematically. The SHOALS depths, as represented by the DTM, were then projected onto the profile lines at 4-m intervals. The resulting bathymetric contours for the simulated profiles at 300-m spacing are in Figures 3.1b, 3.2b, 3.3b, and 3.4b illustrating the loss of detail with such a wide spacing.

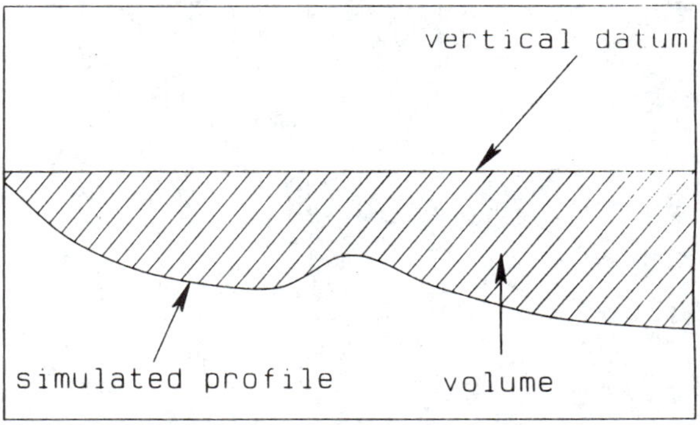

Figure 4.1 Cross-sectional area

To simplify calculations, volumes were computed between the horizontal plane representing a mean water vertical datum and the bottom topography as represented by the simulated profiles (Figure 4.1). Volumes were computed using the well-known area-end method for each profile spacing. When employing the area-end method, the cross-sectional area between the bathymetry and mean water at each profile location was first calculated. (Only the area below the mean water vertical datum was computed.) The calculated area at one profile location (A_1) is then averaged with that of the next consecutive profile (A_2). The product of this averaged area and the length between the two consecutive profiles (L) gives the volume (V) between the profiles:

$$V = \frac{A_1 + A_2}{2} \cdot L$$

The total volume for the entire project is then equal to the summation of the volumes. The computed results are presented in Table 4.1 as volume difference between the highest resolution set, 5-m spacing, and the stated spacing, in cubic meters per meter length of beach. Positive differences indicate that the stated spacing resulted in a volume larger than the 5-m spacing volume while negative differences indicate a smaller volume.

5.0 DISCUSSION

The results, in general, show that as the spacing between survey profiles increases, so too does the error in computed volumes. Figure 5.1 gives a plot of

Table 4.1 Computed volume differences

Profile spacing	Volume difference (m³/m)			
	Longboat Key	Island Beach	St. Joseph	Presque Isle
5	0.0	0.0	0.0	0.0
10	-0.4	0.1	0.0	1.2
25	0.0	-0.1	-0.4	1.9
30	-1.8	0.1	0.2	0.8
50	0.2	-0.3	-1.7	3.3
60	-3.1	0.2	2.3	0.9
100	-1.6	-1.9	-3.3	6.1
150	-0.7	-0.3	-6.5	-10.4
200	2.2	5.3	-18.5	0.8
250	-6.6	-15.9	8.4	4.2
300	12.3	-19.8	-5.3	-9.8

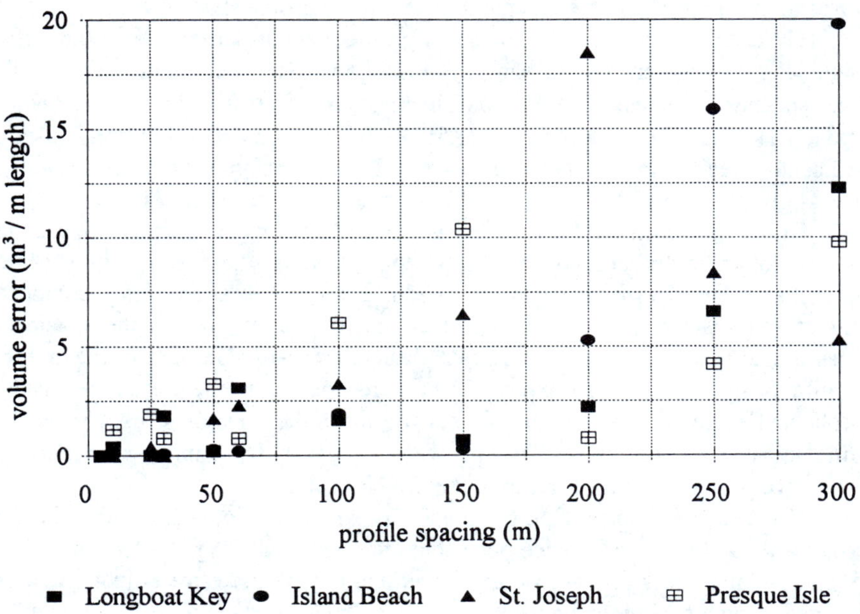

Figure 5.1 Measured volume error

absolute deviation from the 5-m volume (volume error) versus profile spacing for all four sites. In general, the volume error becomes more random and larger in magnitude as profile spacing increases. This is consistent with the findings of Saville and Caldwell (1952). In their investigation, the use of average profiles at spacings varying from 120 m to 2,800 m to represent lengths of a fairly uniform beach were analyzed for accuracy in evaluating engineering volumes. The spacing error, defined as the accuracy measurement of a particular profile in representing a section of beach, was evaluated by comparing the selected profile with the average profile for that section. The spacing error was then translated into a volume error, defined as the total volume difference over the project length. Their conclusions state that the volume error increases nearly linearly as profile spacing increases. Saville and Caldwell's results do not indicate a randomness in volume error; however, the results from this investigation do. This is most probably attributed to the irregularities in the bathymetry evaluated herein.

Of the four sites, the bathymetry at Longboat Key shows the least along-shore variation. This is reflected in the results where the volume error at Longboat Key is within 5 m^3/m for profile spacings as great as 200 m. However the area is still highly three-dimensional, and the volume error when using 200-m, 250-m, and 300-m spacings continually increases and is as much as 12.3 m^3/m. The calculations for Island Beach indicated similar findings. Here, the volume error is small for profile spacings less than 200 m, and as the spacing increases beyond 200 m, the measured volume error increases. In contrast to Longboat Key, the bathymetry at Island Beach is highly variable along-shore with shore-normal sand ridges occurring every 400 m to 500 m. This investigation indicates that bathymetric variations associated with these large features do not impact volume computations until the profile spacing exceeds 200 m.

The analysis at St. Joseph shows that profile spacings larger than 100 m result in significant volume error. Differing from Longboat Key and Island Beach, as the profile spacing is increased at St. Joseph, the volume error does not continually increase. The measured volume error at St. Joseph when a 250-m spacing is used is significantly lower than that measured when a 200-m spacing is used. However, all spacings greater than 100 m result in volume errors larger than 5 m^3/m.

Of the four projects, Presque Isle is most affected by survey spacing changes between 5 m and 50 m, and volume computations deviate significantly with spacings greater than 50 m. Of the spacings evaluated at Presque Isle, the volumes computed using 150-m spacing yielded the largest error, 10.4 m^3/m. When the spacing is increased to 200 m, the volume error dramatically decreases. Because of the uniform repetition of salient formations every 155 m corresponding to the breakwater field, it is probable this decrease in volume error is a result of these features. By chance, the profile locations at the 200-m spacing were such that the volume error between consecutive profiles averaged out over the project length.

The economic impact of profile spacing is evident in Figure 5.2 where the cost error, or absolute cost difference per cubic meter per unit length of beach, for the Presque Isle project is displayed. The lower and upper cost error boundaries were calculated using a cost per cubic meter of beach-quality sand equal to \$5 US and \$30 US, respectively. Even at profile spacings as small as 50-m, the cost error is as great as \$100 US/m. As the profile spacing increase to 100 m or greater, the cost error

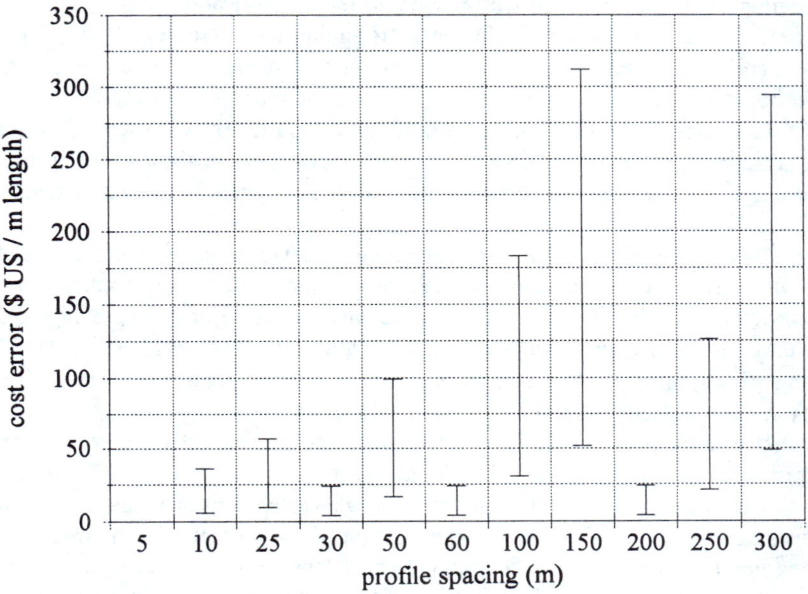

Figure 5.2 Cost error at Presque Isle, Pennsylvania

becomes as great as $325 US/m, translating to a total project cost difference well over $3 million US. With renourishment at $22 US to $26 US per cubic meter occurring annually at Presque Isle, higher density surveys are certainly warranted. Typical monitoring at profile spacings of 300 m or greater may result in cost differentials around $2.5 million US over the project length.

Similar cost impacts may be observed at the other three projects. At Longboat Key, a 300-m profile spacing results in cost differentials between $ 200,000 US and $1.2 million US over the 8.5-km project length. At Island Beach and St. Joseph, profile spacings greater than 150 m may result in cost errors in excess of $75 US/m. At all four projects, the economic benefits of higher density bathymetry are obvious.

Hydrographic surveys serve as the base for engineering planning and design of beach nourishment projects. Usually, a fixed budget for a particular project is developed from volume calculations between the design profile and the collected bathymetry. If design volumes were computed using sparse data, a fixed budget may result in too little, or too much, sand placement at the site. Ultimately, project performance is affected: an under-designed project will not adequately protect against further shoreline erosion.

6.0 CONCLUSIONS

This investigation clearly indicates the economic and planning benefits of high-resolution bathymetry for beach nourishment along irregular beaches. Low density data may result in gross miscalculation of fill volumes ultimately impacting project performance and financial management. The conclusions of this analysis are based on engineering volumes computed with respect to the below-water portion of the profile and only reflect underwater bathymetric irregularities. However, initial design, maintenance, and monitoring of beach nourishment projects also encompasses the above-water portion of the profile. There are commonly occurring irregularities on the above-water portion as well including sand dunes and features associated with man-made structures. Based on the findings herein, inclusion of the above-water portion of the profile should result in even greater inaccuracies in engineering volumes when wide profile spacings are used.

Data density for a particular hydrographic survey mission should be governed by the degree of along-shore variability and the intended engineering application of the data set rather than the capability of the survey instrument. For example, high-density data may not be required when monitoring long-term regional coastal evolution; however, high-density data are required when preparing plans and specifications for a beach nourishment project in an area with irregular bathymetry. High-density bathymetry are also valuable when evaluating the performance of erosion-control structures such as groins or detached breakwaters.

The findings herein prompted several new investigations including the expansion of this study to include the above-water portion of the project. Furthermore, studies are ongoing to optimize profile locations to best represent a project while minimizing data collection requirements. Finally, similar investigations are evaluating the impacts of data density on other engineering computations, such as dredging volumes.

With complete coverage, accurate bathymetry now available at costs nearly two-thirds that of conventional methods, engineers can more accurately and cost-effectively plan, design and monitor beach nourishment projects. Furthermore, future research will provide new guidance for determining necessary survey densities for accurate engineering computations.

ACKNOWLEDGMENTS

The authors would like to thank Ms. Jennifer McClung and Mr. Wesley McCleese for their help in the analysis of these data. Results and information, unless otherwise noted, are based on work funded by Headquarters, US Army Corps of Engineers. The use of trade names does not constitute an endorsement in these products by the US government. Permission to publish this paper was granted by the Chief of Engineers.

REFERENCES

Grace, P.J., 1989. "Investigation of Breakwater Stability at Presque Isle Pennsylvania Erie, Pennsylvania", TR-CERC-89-3. US Army Corps of Engineers, Vicksburg, MS.

Guenther, G.C., Thomas, W.L., and LaRoque, P.E., 1996. "Design Considerations for Achieving High Accuracy with the SHOALS Bathymetric Lidar System," *SPIE Laser Remote Sensing of Natural Waters: From Theory to Practice*. Feigels, V.I. and Kopilevich, Ed. SPIE, Bellingham, WA, Vol. 2964, pp. 54-71.

Irish, J.L. and Truitt, C.L., 1995. "Beach Fill Response at Longboat Key, Florida," *Proc. National Conference on Beach Preservation Technology*. Florida Shore and Beach Preservation Association, Tallahassee, FL.

Lillycrop, W.J., Parson, L.E., and Irish, J.L., 1996. "Development and Operation of the SHOALS Airborne Lidar Hydrographic Survey System," *SPIE Laser Remote Sensing of Natural Waters: From Theory to Practice*. Feigels, V.I. and Kopilevich, Ed. SPIE, Bellingham, WA, Vol. 2964, pp. 26-37.

Mohr, M.C., 1994. "Presque Isle Shoreline Erosion Control Project," *Shore and Beach*. American Shore and Beach Preservation Association, Berkeley, CA, Vol. 62, pp.23-28.

Parson, L.E., and Smith, J.B., 1995. *Assessment of Native Beach Characteristics for St. Joseph, Michigan -Southeastern Lake Michigan*. US Army Corps of Engineers, Washington, DC.

Riley, J.L., 1995. "Evaluating SHOALS Bathymetry Using NOAA Hydrographic Survey Data," *Proc. 24th Joint Meeting of UINR Sea Bottom Survey Panel*. Tokyo, Japan.

Saville, T., and Caldwell, J.M., 1952. "Accuracy of Hydrographic Surveying In and Near the Surf Zone," *Proc. 1952 Coastal Engineering Conference*. American Society of Civil Engineers, New York, NY.

CHAPTER 289

Shoreline Analysis Using Digital Photogrammetry

Margery F. Overton[1]
John S. Fisher[2]

Abstract

This paper discusses the application of digital photogrammetry to the analysis of shoreline change with examples from Holden Beach, NC. Beach cross sections are generated from a three dimensional stereo image. These profiles are then used to determine shoreline position and erosion rates. The paper compares these rates of shoreline change with those determined from the more traditional methodology based on the two dimensional interpretation of the wet-dry line. The paper also illustrates the improved accuracy obtained by using survey grade GPS as opposed to USGS topographic maps to establish ground control.

Introduction

Coastal engineers are always seeking better tools to analyze shoreline changes. These changes include the result of long-term erosion or accretion and short-term impacts of severe storms. The two most frequently used methods to record shoreline change are directly from field surveys and remotely from aerial photographs. Both methods have advantages and disadvantages.

---

[1] Associate Professor, NCSU-Kenan Natural Hazards Mapping Program, Department of Civil Engineering, NCSU, Raleigh, NC, USA, 27695.

[2] Professor, NCSU-Kenan Natural Hazards Mapping Program

Field surveys provide detailed information regarding the topography and bathymetry of the beach and shoreline. These surveys are relatively expensive. In addition, they can not be used to determine long-term change if no historical surveys have been conducted.

Alternatively, the interpretation of shoreline change from aerial photographs permits one to perform a historical analysis when such photographs are available. Unlike field surveys, traditional methods for analyzing shoreline change from aerial photographs do not generally furnish the three dimensional data provided by a field survey. Three dimensional data can be obtained from aerial photographs, however, if one utilizes photogrammetric techniques including the analysis of pairs of overlapping photographs.

Until recently, few coastal engineering projects have included the application of three dimensional photogrammetric techniques to analyze shoreline change, simply because traditional photogrammetry relies upon complex analytical stereoplotters. These devices do not lend themselves to routine use outside of the highly specialized photogrammetry laboratory.

The difficulties of using photogrammetry for shoreline change analysis have been significantly reduced by the introduction of the new computer based digital techniques. Digital photogrammetry replaces the analytical stereoplotter with a three dimensional image that has been generated by a computer coupled with a high resolution scanner. A pair of photographs is scanned, the two images merged and viewed on the terminal as a single three dimensional image with the use of special glasses worn by the operator. In addition to reducing the difficulty in generating a three dimensional image, the technique supplies an image in digital format. This greatly expands the opportunities for data manipulation and analysis.

This paper describes our experience with the application of digital photogrammetry to the analysis of shoreline change along the coast of North Carolina. While we are still learning, it is already clear that this technology has greatly expanded our abilities to make accurate measurements of erosion rates and the impacts of severe storms on the beach.

Project Study Area

The Holden Beach project described in the paper is located in Brunswick County, on the southernmost portion of the North Carolina coast, Figure 1. The study area is a typical barrier island separated from the mainland by a narrow estuary (the Intracoastal Waterway). Most of the beaches in Brunswick County are developed with low density single family houses as is typical for the mid-Atlantic coast. Long-term shore erosion is on the order of 2 to 4 ft/yr. The occasional severe storm, including both hurricanes and northeasters, coupled with the effects of long-term erosion has led to the loss of oceanfront houses and a general need to be able to make reasonable estimates for the future changes in shoreline position.

The analysis of the rates of shoreline change at Holden Beach is part of a larger study of the area being undertaken for the North Carolina Department of Emergency Management through a grant from the Federal Emergency Management Agency. The larger study is intended to develop new techniques for natural hazard identification through the use of digital photogrammetry.

Project Description

The analysis presented in this paper focuses on the measurement of the change in position of the shoreline. In this context the shoreline is defined as the interface between the wet and dry beach as seen from aerial photographs. This interface has been used by many investigators (e.g., Dolan, et al. 1978) as the shoreline, and is often referred to as the high water line in the literature. Strictly speaking it is not the high water line, but rather a time average of this line. This line is not fixed in space; it moves up and down the beach face depending upon the tide, beach slope, and to some extent the wave climate (Fisher and Overton, 1994).

Our laboratory utilizes both hardware and software developed by Intergraph Corporation. The photogrammetric grade scanner with a maximum scanning capability of 3386 dpi is a key feature of the system. It provides the digital resolution needed to achieve the accuracy desired in the data analysis. Figure 2 illustrates a typical digital photogrammetric project work flow. Two or more photographs are scanned (at a typical scanning resolution of 1693 dpi) and used to build a stereo image.

Standard photogrammetric procedures including interior and relative orientation correct for camera distortion and tilt.

One of the most important steps in this work flow is to establish an accurate set of ground control points. These points determine the absolute position of objects seen on the photographs. In order to maximize the ultimate accuracy of the project we have elected to use survey grade GPS equipment for this ground control.

Ideally the ground control should be established at the time of the aerial photographs. However, in many cases this is not possible, and we frequently have to establish ground control after the fact. For example, control points can be established from features that are visible in the photography and which can be surveyed accurately. Using photography from the last five years, this procedure has allowed us to establish control points with approximately 3 ft accuracy.

Once the control has been established, the computer can then generate a three dimension image of the beach. As noted above, the operator must be wearing a special set of glasses to view this image, much like one would do in a theater when viewing a 3-D movie.

The operator, using the stereo image, can next proceed to create a digital terrain model, or DTM. This model is developed by the operator "placing" the computer cursor on ground at multiple locations. This process requires a certain amount of training and experience, and therefore the quality of the DTM depends upon the skill of the photo analyst. Once the DTM has been created, a number of different digital products can be produced, depending upon the specific application and interest. Examples include orthophotos, triangulated irregular networks (TINs), grids, and draped images.

Shoreline Change Measurements

Figure 3 is a digital orthophoto for a portion of the study area for June 1992. A digital orthophoto is a computer generated 2-dimensional image which has been corrected for distortion. Accurate measurements can be made directly from this orthophoto. Because of the digital format it is possible to zoom in on a portion of the photo as shown by the inset in this figure. The shoreline (wet-dry line) has been identified by the operator as indicated by the line drawn on the beach. As noted above, this shoreline is dependent upon the waves and tide at the site at the instant in time when the photograph was taken.

When analyzing shoreline change it is necessary to look at two dates. In addition to the 1992 date shown in Figure 3, we chose a set of photographs taken in 1955 as the second date. Figure 4 shows a pair of orthophotos at the same site for 1992 and 1955. The development that has taken place between these dates is clearly evident. In addition, the 1955 orthophoto was taken less than 12 months after Hurricane Hazel struck this coastline. Hazel (prior to Hurricane Fran in 1996) was the most severe storm to impact this shoreline in memory. The extremely wide overwash penetration shown on this photograph is due to that single storm. It is important to note that the absence on the 1955 photograph of most of the man-made features shown on the 1992 photograph makes it difficult to establish reasonable control for the earlier date.

Once we have identified the wet/dry line for the two different dates it is a relatively easy task to determine shoreline change. We use a technique developed by Dr. Robert Dolan (Dolan, et al. 1978). This technique uses an imaginary, shore parallel off-shore baseline as a reference line. The distance from this baseline to the shoreline is measured along an orthogonal drawn between the shoreline and the baseline. In our work we set these orthogonals at 150 ft intervals. Once the distance between the baseline and the shoreline is known for the two dates in question, it is a simple matter to compute the erosion/accretion rate.

Prior to our use of the digital orthophotos we used the more widely practiced techniques described by Dolan and others which depend upon less accurate corrections for the aerial photographs. In this prior method, ground control points for the photographs were identified using USGS topographic maps. Figure 5 illustrates the difference in erosion rates we found between these two methods at the study site. The difference is about 1 ft/yr. At this site, where the average erosion rate is about 3 ft/yr, this is a significant difference. Depending upon the site, this difference can be important when one attempts to use these data for shoreline management and construction permit programs.

The method to determine shoreline change described above is an essentially 2-dimensional technique. Since the digital image is stored in the computer as an 3-dimensional object, it is interesting to consider how one could determine shoreline change from this latter data. In this case we work from a computer generated TIN. The TIN is a three dimensional wire diagram of the stereo image, as shown in Figure 6. The triangles connect points of known elevation.

From the TIN the software can generate cross-sections normal to the shoreline at any location. Figure 7 illustrates two such profiles, for 1955 and 1992. As an alternative to measuring the change in position in the wet/dry line, one can now measure the change in position of a specific elevation contour. For comparison, the wet/dry lines are also shown on this figure. Of course, the use of beach profiles to determine rates of shoreline change is not new. This is the normal procedure when working from field survey data. However, the ability to generate these profiles from digital historical aerial photographs is a relatively new procedure, and one worthy of future development.

Figure 8 illustrates the difference in erosion rates determined by these two methods. For this site the change in position of the wet/dry line between 1955 and 1992 yields a higher erosion rate than the method which is based upon the change in position for the 3 ft contour for these same dates. Note that the general trend in the data is consistent between the two methods. All things being equal, we feel that the erosion rate based upon the contour is preferable because it reduces the influence of the waves and tides on the data.

We are continuing to explore the pros and cons of the two methods to measure shoreline change. A rigorous test with field survey data is needed prior to making any conclusions regarding the preference of one over the other.

Draped Images

The discussion above has focused on the use of digital photogrammetry for measuring shoreline change. Another important use for this technology is the ability to generate realistic looking computer images of the beach. One way to generate such an image is to "drape" the orthophoto over the TIN, as shown in Figure 9. This image (shown here with a 5x vertical distortion) enables us to visualize the topography and see clearly the dunes, roads, houses, and vegetation. Such pictures are valuable both to the analyst as well as the general public. In the latter case these images can be used to help explain a problem, proposed solution, or probable impact. Again, since this is a digital image, it is possible to easily change the scale, angle of viewing, and distortion as needed. We are currently working with these draped images to prepare a report which identifies areas most vulnerable to damage from severe storms.

Conclusions

While the application of digital photogrammetry to coastal engineering is new, it is already clear that this technology will provide important improvements to the field. In particular, the ability to generate orthophotos from historical aerial photographs will allow us to make more accurate determinations of rates of shoreline change. In addition, the computer generated images will create new tools for both shoreline managers as well as the general public to understand the nature of shoreline change.

Acknowledgments

The authors are greatly for the support this research has received from the Kenan Institute for Science and Technology, the NC Division of Emergency Management, the NC Division of Coastal Management, the NC Department of Transportation, the Federal Emergency Management Agency, and the Intergraph Corporation. In addition, we greatly appreciate the help from our research support staff, including Cheryl Hapke, Roger Grenier, Rob Hardy, Brian Joyner, and Beth Judge. Dr. Robert Dolan at the University of Virginia has been a valuable colleague in our work on this and related projects.

References

Anders, F. J. and Byrnes, M. R., 1991. Accuracy of shoreline change rates as determined from maps and aerial photographs. Shore and Beach, 59 (1), 17-26.

Dolan, R., Hayden, B. and Heywood, J., 1978. A new photogrammetric method for determining shoreline erosion. Coastal Engineering, 2:21-39.

Fisher, J. S. and Overton, M. F., 1994. Interpretation of shoreline position from aerial photographs. Coastal Engineering 1994: Proceedings of the twenty-fourth international conference. Vol. 2, 1998-2003.

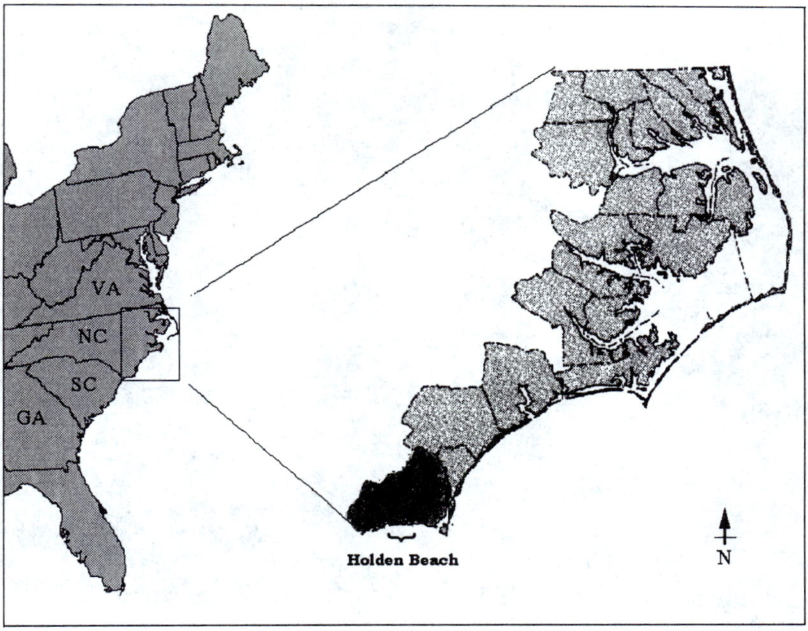

Figure 1. Project Study Area

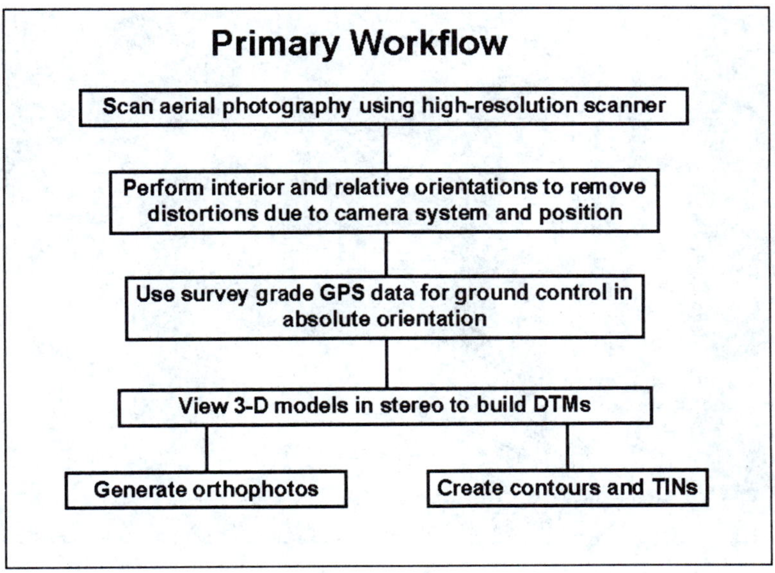

Figure 2. Project Workflow

Figure 3. Holden Beach Orthophoto

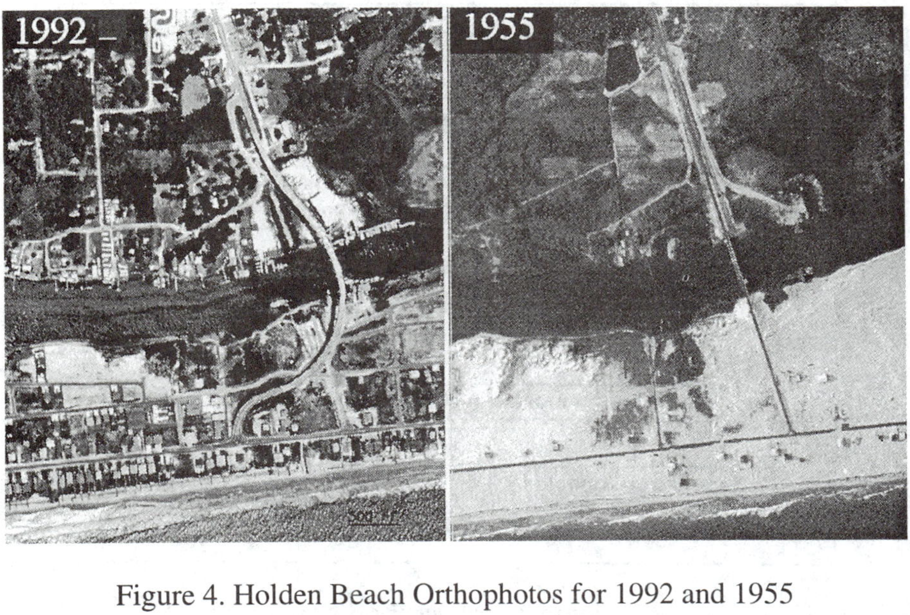

Figure 4. Holden Beach Orthophotos for 1992 and 1955

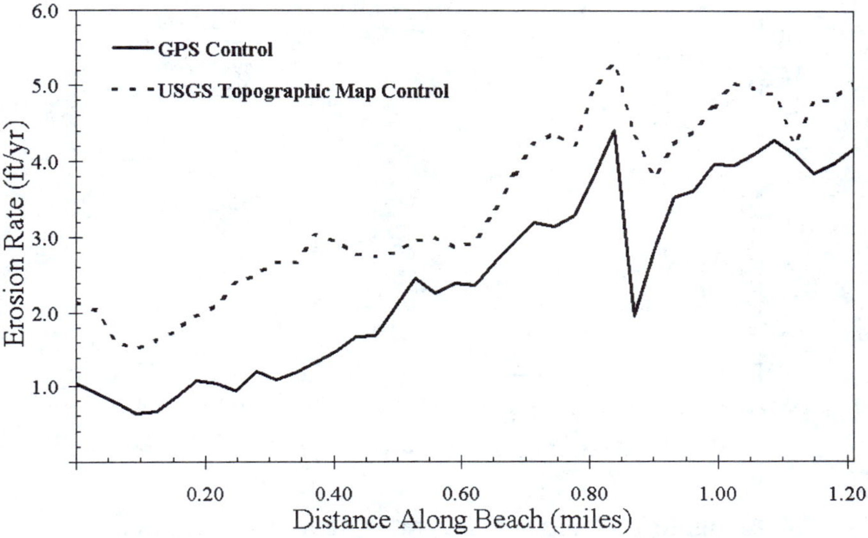

Figure 5. Comparison of Erosion Rates: Mapping Error

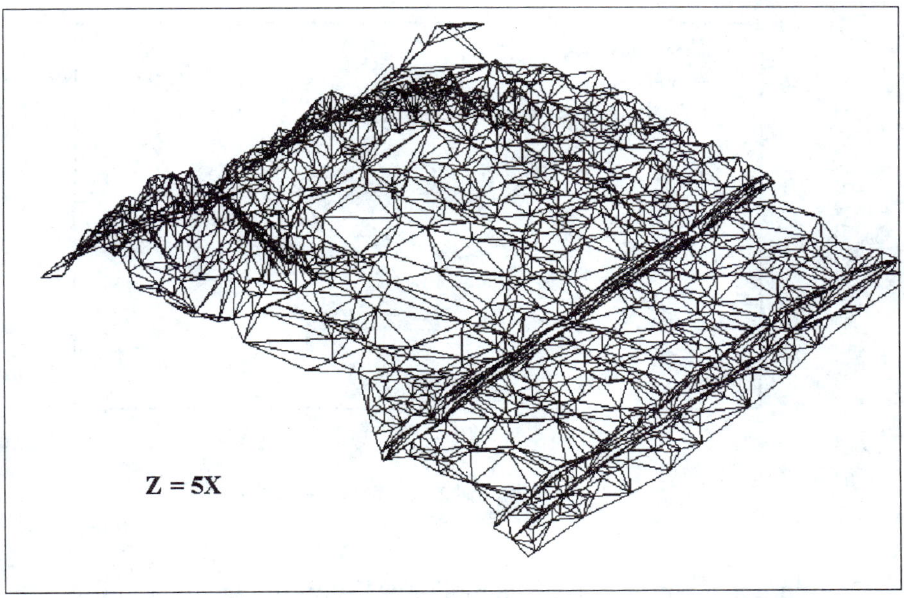

Figure 6. 1955 TIN for Study Site

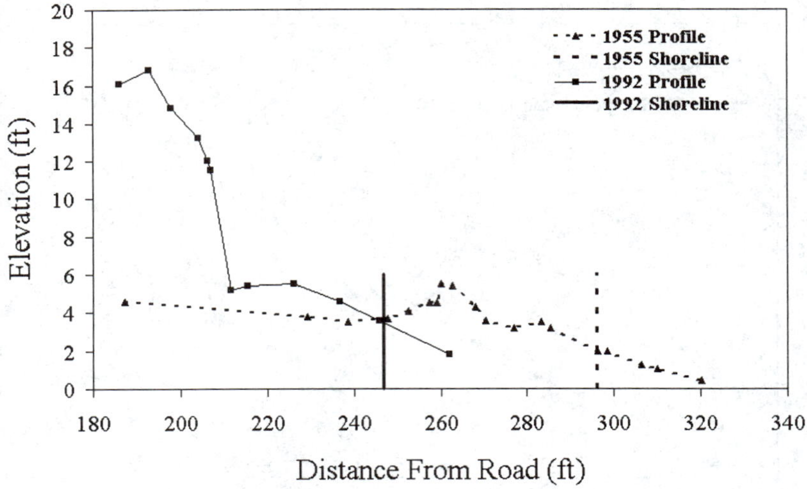

Figure 7. Computer Generated Shoreline Profiles

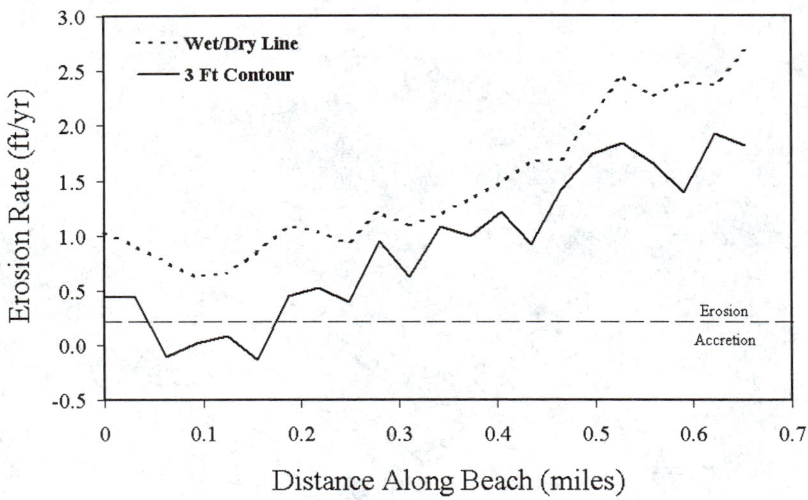

Figure 8. Comparison of Methods for Determining Erosion Rate

Figure 9. 1955 Orthophoto Draped Over Computer Generated Grid

CHAPTER 290

Application of Satellite Images to the Detection of Coastal Topography

Ji Wu[1] and Jea Tzyy Juang[2]

Abstract

Taking advantage of the high resolution visible images from the remotely sensed data of SPOT, the phenomenon of waves in the coastal area of Taichung was analyzed and its changing characteristics were studied. Changes of wave direction are affected by coastal topography; therefore, the changes of water depth can be detected by waves characters. In this study, the P band image data of SPOT of Taichung coast taken on 8th December 1993 was used. After dividing the area from deep to shallow into several blocks, each with 128x128 pixels, the wave images were transformed into two dimensional wave number spectra by ways of Fourier Transform. After that, the wave direction and the wave length were calculated. Results show that the shallower the water depth, the shorter the wave length; and the wave rays tend to approach shore normally. The phenomenon coincides with the theorem of wave dispersion. At last, based on the Snell's Law and the theory of shoaling, the average depth of each block was calculated. The results are encouraging.

Introduction

Remote sensing techniques have been developing rapidly since 1970. Since then, the LANDSAT, SEASAT and the SPOT satellites, have been launched and operated. The earth resources satellite receiving station in Taiwan was also installed in 1993.

---

[1] Assistant Researcher, Institute of Harbour & Marine Technology No.83, Lin-Hai Rd., Wuchi, Taichung District, Taiwan, R.O.C.

[2] Deputy Director, Institute of Harbour & Marine Technology No.83, Lin-Hai Rd., Wuchi, Taichung District, Taiwan, R.O.C.

In recent years, many researchers are using this new techniques to solve problems in the field of Oceanography, J. Populus (1991) used SPOT data to do the wave analysis. P.N. Bierwrith (1993) tried to detect depths on shallow water by studying the radiance of sea bottom reflectance. SPOT satellite is a sun-synchronous salelite with an average orbit height of 832 km. It passes Taiwan at 10:45 A.M. and reappears at the same track every 26 days. Since SPOT has the ability of inclined scan, we are able to get information of the same area every 3~5 days. The XS band image has a spatial resolution of 20 meters, and P band image has a spatial resolution of 10 meters. This study try to detect water depth at intermediate depth zone by using the characteristics of wave propagation.

Background Theory

The Dispersion equation states that

$$C^2 = \frac{g}{k} \tanh kh \tag{1}$$

where
C : wave celerity
σ : the angular frequency
k : wave number
g : gravitational acceleration
h : depth

During the propagation of waves from deep sea to shallow water zone, wave height, wavelength and wave direction undergo changes due to changes of water depth. In region with intermediate water depth, we have

$$L = L_0 \tanh kh \quad ; \quad C = C_0 \tanh kh \tag{2}$$

where L_0 , C_0 = wavelength & wave celerity in deep water

The irrotationality condition of wave number k yields

$$\frac{\partial(k \sin\theta)}{\partial x} - \frac{\partial(k \cos\theta)}{\partial y} = 0 \tag{3}$$

where θ is the angle between wave ray and the normal of coastline.

For a coast with straight and parallel contours, the changes of the wave direction obey the Snell's law :

$$\frac{\sin\theta}{c} = \frac{\sin\theta_0}{c_0} \tag{4}$$

By using all the above equations and the data from satellite image, the variations of the water depthes at different locations can be estimated.

Techniques Used

Although the wave patterns are visible roughly in the satellite image, its real direction and wavelength can not be measured by eyes. Therefore, certain techniques of wave-number spectrum analysis using the two-dimensional Fourier transform are required. From the satellite image, we obtained the wave-number spectrum $S(k) = S(k_x , k_y)$. If $I(x , y)$ stands for the spatial intensity of the image, then the wave-number spectrum can be obtained by using the two-dimensional Fourier transform as

$$S(k_x , k_y) = \iint I(x , y)\, e^{-jk_x x - jk_y y}\, dxdy \tag{5}$$

When the spectrum peak was calculated, the wavelength and wave direction in each subregion can be determined as follows ：

$$\frac{2\pi}{L} = k = \sqrt{k_x^{\ 2} + k_y^{\ 2}} \quad ; \quad \theta = \tan^{-1}(k_x / k_y) \tag{6}$$

where $k_x , k_y =$ the wave number components of the peak position on the spectrum diagram; $\theta =$ the angle between the wave direction and the Y-axis.

The above method is suitable for cases where wave phenomenon is simple and significant. If there is noise on the image scene, for example, a boundary of different water colors exists. We'll get a confused spectrum. In such cases we have to use special techniques, such as top-hat transform, to reduce the influence of noise. Fig.1 & Fig.2 show the comparison of the original image scene and the image after using Top-hat transformation, a detail flow chart of processing is shown in Fig.3.

Fig. 1 Original image scene of the sea

Fig. 2 Image scence after using Top-hat transformation

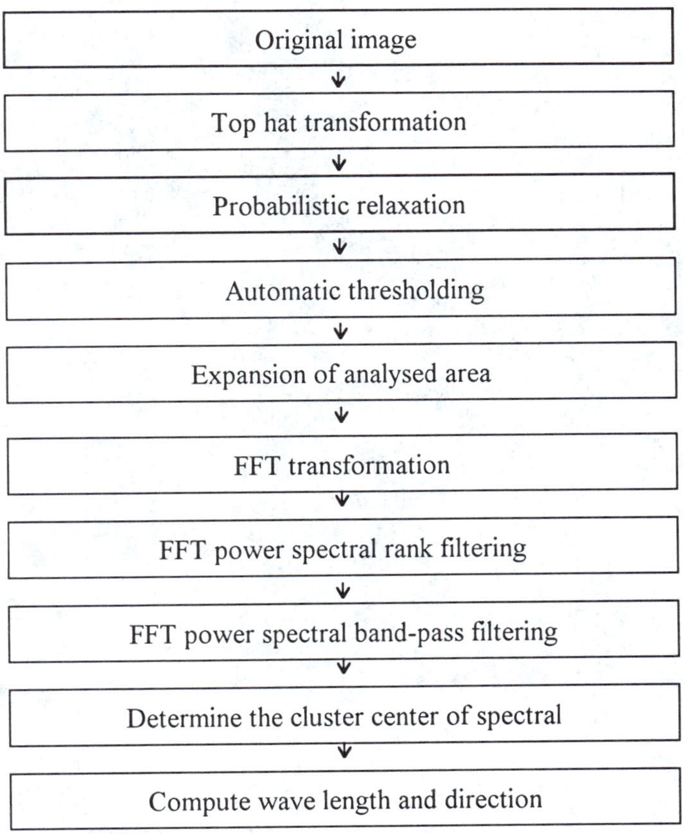

Fig. 3 Detail flow chart of processing

Fig. 4 Satellite images off Taichung Coast

Results

In this study, the Pan Band image data off Taichung Harbour on 8th December, 1993 were used (Fig.4), it covers an area of 18km x 12.5km. Taichung Harbour was builded on 1970's, locate at the central western coast of Taiwan. From September to March, the winter monsoon season, the strong wind always blows from north. Therefore, the waves travel from the upper to the lower in Fig.4. The upper left corner is 9 kms away from coast and the depth is about 50 meters. The lower left corner is the outlet of Da-Du River and is quite shallow. Therefore we can observe the phonomenon of waves shoaling. After selecting and dividing some scope in the left part area from deep to shallow water and labelling them as A, B, ... S, with each block having 128x128 pixels, the wave image is then transformed into the wave-number spectrum.

Two example diagrams of the wave-number spectrum at block E (in deep water) and S (in shallow water) are shown in Fig.5 and Fig.6 respectively. In these figures, the longer the distance of the bright spot to the original point, the larger the wave number, which means shorter the wavelength.

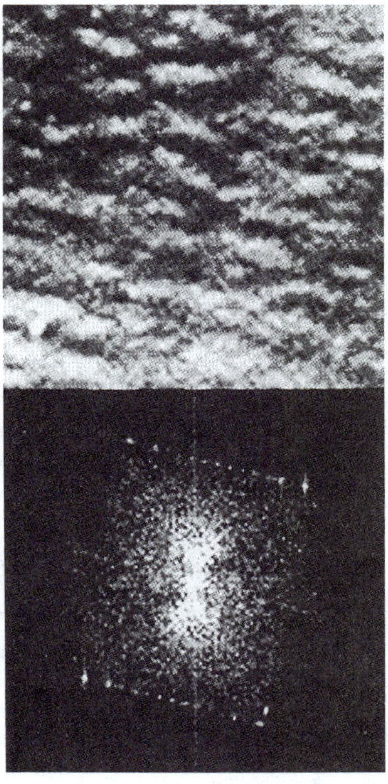

Fig. 5 Wave image & wave number spectrum at block E

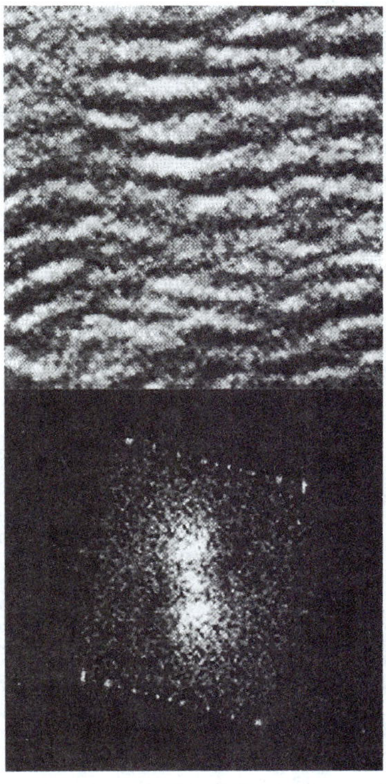

Fig. 6 Wave image & wave number spectrum at block S

From Table 1, we can see that the wavelength is decreasing and wave direction changes from 15° at block E to 350° at block S also. Fig.7 shows the variation of wavelength with wave direction from block E to block S.

Table 1 Results of water depth computation

Block	Wavelength	Wave direction 0° for N: clockwise	c/c_0	k	dp(m)	dr(m)
C	107	16	0.99	0.059	44	---
E	101	15	0.98	0.062	41	43
G	101	14	0.97	0.062	40	46
I	92	6	0.94	0.068	26	30
K	88	3	0.92	0.071	22	20
M	83	0	0.90	0.076	19	19
O	81	356	0.86	0.078	17	17
Q	75	353	0.83	0.083	14	15
S	68	350	0.80	0.092	12	10

Finally, the Snell's law of wave refraction and the small amplitude wave formula were used to compute the ratio of wave celerity from deep to shallow water and to predict the water depth at each block individually. Comparison of the predicted water depth (dp) and the measure's (dr) was also shown in Table 1 and Fig.8. The results are encourging.

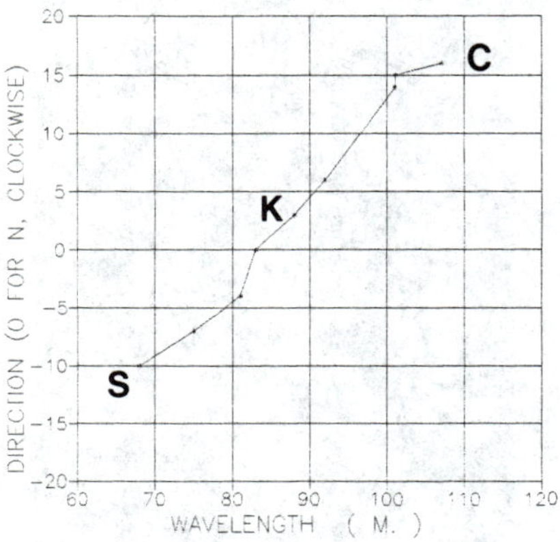

Fig. 7 Variation of wavelength with wave direction

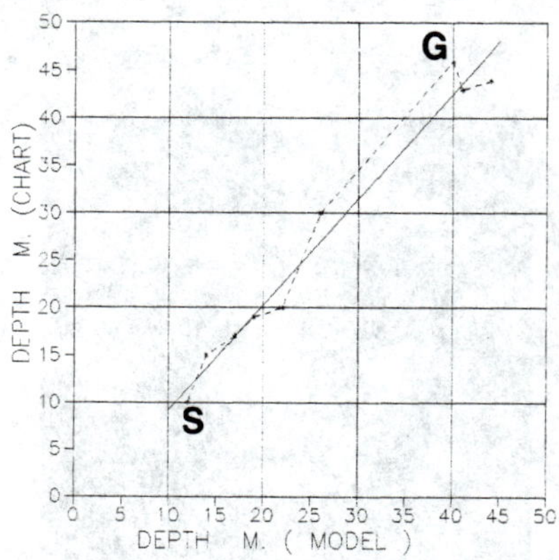

Fig. 8 Comparison of the predicted water depth and the measure's

Conclusions and Discussions

1. From the satellite image of the sea, we can get direct information of wave propagating from deep to shallow water, such as refraction, diffraction and reflection. It has the advantages of wide spatial information and lower coast, when compared to traditional field survey.

2. By combining Snell's law, dispersion relationship and spectral results, the bottom topography are easily derived. The results are encouraging.

3. The method used here takes the assumption of straight and paralled offshore contours; therefore the range of water depths can be estimated is within intermediate depth, i.e., from $L/2$ to $L/20$, where L is the wave length.

4. Block size influences the analyses. Larger block size provides better spectral resolution while causing more uncertainty due to shoaling effect. On the other hand, small block size has the problem of insufficient details. The present study uses a 128x128 pixel block size.

5. Effects of water color on the image may yield confusing in the detection of spectra. The top-hat transformation is used to reduce the influence of this factor.

6. SPOT Pan band images are used in the present spectral analyses. In comparison with most of the earlier works where SAR images were used, these images generally provides higher space resolution and lesser non-linear effects.

References

1. Bierwirth P. N., Lee T. J., and Burne R. V. "Shallow Sea-Floor Reflectance and Water Depth Derived by Unvixing Multispectral Imagery", Photogrammetric Engineering & Remote Sensing, Vol.59, No.3, pp.331-338, 1993.

2. J. Populus, C. Aristaghes, L. Jonsson, J. M. Augustin, and E. Pouliquen (1991) "The Use of SPOT Data for Wave Analysis", Remote Sens. Environ. 36:55-65.

3. J. Serra, "Image Analysts and Mathematical Morphology", London, Academic, 1982.

CHAPTER 291

SEDIMENT TRANSPORT IN SWASH ZONE UNDER OBLIQUELY INCIDENT WAVES

Toshiyuki Asano[1]

ABSTRACT

Time dependent calculations on two-dimensional sediment transport are carried out. The swash wave front is treated as a moving boundary which can reproduce non-vanishing sediment transport rate even for landward region of the still water shoreline. Meanwhile, experimental investigations on this subject are conducted by florescent sand tracer method. By comparing the present and existing experimental results with the numerical ones, the sediment transport mechanism in the swash zone is discussed.

1. INTRODUCTION

Recently, swash zone has received much attention because the sediment process in this zone provides an important boundary condition for the entire beach evolution. And also, recent studies have reported that the alongshore sand transport takes bimodal distribution with maxima at the swash zone as well as near the breaking point(White-Inman (1986), Bodge-Dean (1987)). Under obliquely incident waves, sediment near shoreline moves in a zig-zag way which results in the inherent alongshore transport in the swash zone. Intensive turbulence generated in uprush and backwash waves causes a large volume of sediment in suspension. However, these mechanisms have not been sufficiently understood.

The author proposed a horizontally two-dimensional shallow water wave model for predicting the free surface elevation and fluid velocities in the swash and surf zone under obliquely incident waves (Asano, 1994). The

---

[1] Associate Professor, Dept. of Ocean Civil Engrg., Kagoshima Univ., 1-21-40, Korimoto, Kagoshima, 890, JAPAN

sediment movement in swash zone should be described in a time dependent way because the bed is immersed and dried alternatively corresponding to run-up and run-down motion. The objective of this work is to improve the understanding of sediment transport in the swash zone by numerical analysis as well as experimental investigations focusing on the alongshore transport rate.

2. NUMERICAL ANALYSIS

(a) Two dimensional velocity fields in swash zone

The author has presented the numerical model for the two-dimensional velocity field in a swash zone (Asano, 1994). The outline of the calculations is as follows:

Two dimensional coordinate system is considered where x- and y- axis is chosen to be in the normal and the parallel direction to the shoreline, respectively. The z-axis is taken positive upward with z=0 at the still water shoreline. The beach slope S is herein restricted to be uniform and its contour is straight and parallel to the shoreline. The incident monochromatic waves with straight parallel crests are assumed to arrive at the seaward boundary with an angle θ_B.

The basic equations are two-dimensional shallow water equations with a moving boundary at the front of wave runup. Ryrie's(1983) method is applied in order to obtain simplified equations. That is, the wave crest is straight parallel and bottom topography does not vary in y-direction, the observed wave motion moving along the alongshore direction at the speed C'/sinθ becomes independent in the y-direction. By introducing pseudo-time $\hat{t} = t - (\sin\theta_B / C_B)y$ to unify two independent variable t and y, and assuming the incident wave angle θ_B is small enough to be used as a small parameter, the basic equations are decoupled into independent equations both for longshore motion and cross shore motion.

(b) Particle trajectory in swash zone

Instantaneous behavior of a single spherical particle placed on a uniform slope is simulated under two-dimensional swash motion calculated in the above mentioned model. The momentum equation for the particle is given by

$$\begin{aligned}(\rho_s/\rho + C_A)A_3 d^3 \dot{\boldsymbol{u}}_s &= \frac{1}{2} C_D A_2 d^2 |\boldsymbol{u} - \boldsymbol{u}_s|(\boldsymbol{u} - \boldsymbol{u}_s) \\ &+ (1 + C_A)A_3 d^3 \dot{\boldsymbol{u}} - (\rho_s/\rho - 1)A_3 g d^3 \sin S \\ &- (\rho_s/\rho - 1)A_3 g d^3 \tan\phi \cos S \cdots\cdots\cdots\cdots (1)\end{aligned}$$

in which, g: gravitational acceleration, ρ and ρ_s: specific gravity of water and sediment, d: sediment particle diameter, A_2, A_3: 2- and 3- dimensional shape factor of sediment particle, ϕ : internal friction angel, C_A, C_D: added mass and drag coefficient, u, u_s : fluid and sediment particle velocity vector, respectively.

Since the fluid velocity u has been obtained at every calculation grid, interpolation is needed to calculate u at the point sediment particle locates. The instantaneous position of the particle can be calculated by integrating sediment particle velocity u_s with a time increment of 1/1000 of the wave period.

Fig. 1 shows one example of the calculated trajectories of a single particle placed on a relatively steep slope S=0.4 to emphasize the gravitational effect. Here, x- and y- axis is chosen to be in the normal and parallel direction to the shoreline, respectively. The results reproduce the zig-zag transport under which a particle is conveyed onshore with a certain angle to the shoreline, then transported offshore with greater angle due to the gravity.

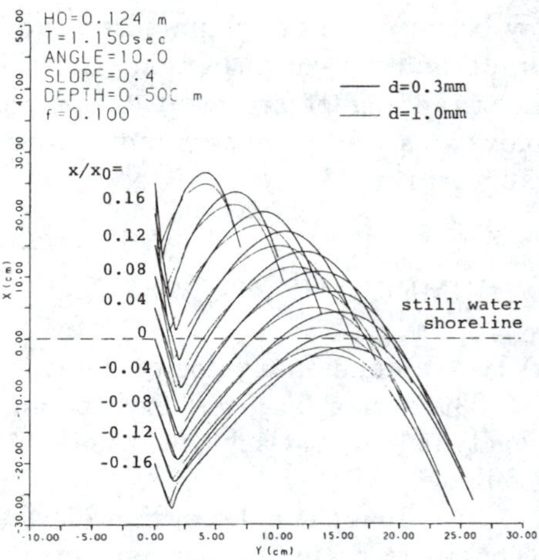

Fig. 1

Trajectories of a sediment particle in a swash zone

(c) Sediment Transport Rate

Instantaneous sediment transport rate was calculated based on Kobayashi's(1982) bed load formula, which accounts for the effect of a slope explicitly. In the model, the composing forces acting on a spherical particle are the submerged weight, frictional force, fluid drag and lift forces. An inertia force acting on the particle was neglected in the model for simplicity.

To illustrate the external force of a sediment particle, the temporal variations of on-offshore velocity at the still water shoreline x=0 are shown in Fig. 2. The parameter of Fig.2 (a) and (b) is the wave period T and the slope steepness S, respectively. The results show that each velocity fluctuation in swash zone rapidly increases at a passage of bore-like crest, and has long lasting backwash motion. Even though the incident wave height is same in the figures, the breaking wave height H_b will change by the parameter T and S, and that leads to different set-up height at the still water shoreline. The magnitude of the backwash flow is found to be large because the precedent onshore mass flux should be returned through a thin backwash flow. These figures show greater integrated areas in the offshore direction than those in the onshore direction.

The variations of sediment transport rate reflect the velocity variations shown above. Fig. 3(a),(b) illustrates the instantaneous on-offshore transport rates q_x/wd (w: sediment fall velocity) at the still water shoreline x=0. At the limit of horizontal bed, Kobayashi's formula yields that the sediment transport rate is proportional to the Shields number $\Psi(t)$ to the power of 1.5 (3 rd power of the instantaneous fluid velocity). As a result, the curves of sediment transport rate in Fig. 3 show steeper peaks than those in Fig. 2. Fig. 3(b) shows the results with a slope steepness S as a parameter. The steeper the bottom slope, the sharper the onshore transport fluctuations and the greater and the longer the offshore transport fluctuations.

Fig. 4 depicts the instantaneous alongshore transport rate q_y. The alongshore transport is generated mainly during wave up-rush under mild slope condition. Meanwhile, under steep slope condition, a large amount of alongshore transport is observed during down-rush.

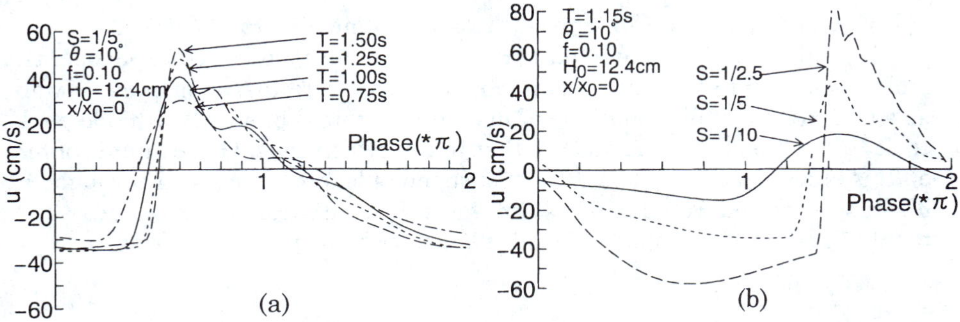

Fig. 2 Temporal variations of on-offshore velocity at x/x0 =0

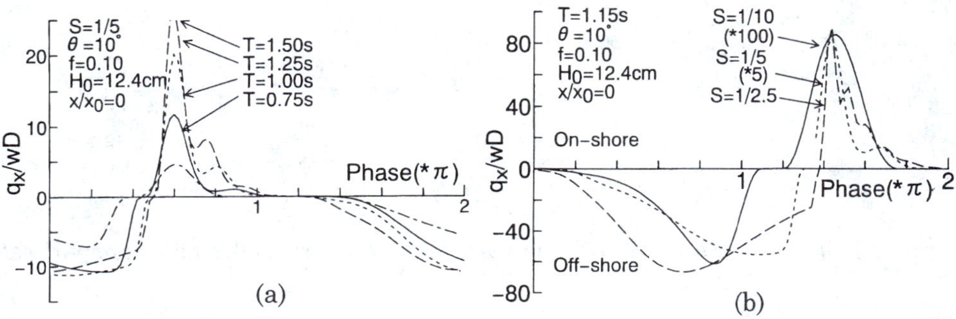

Fig. 3 Temporal variations of on-offshore sediment transport rate at x/x0 =0

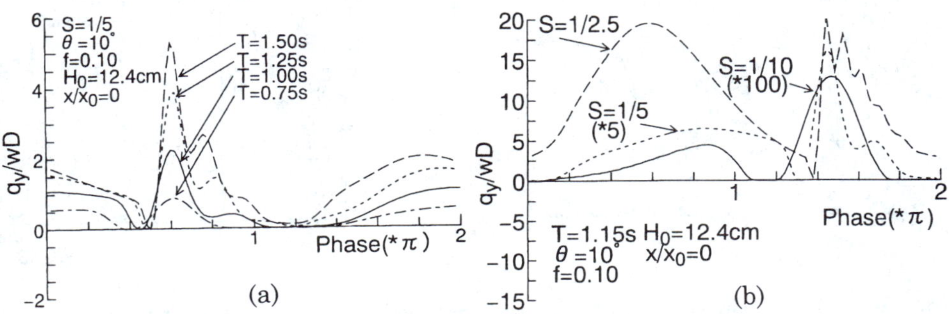

Fig. 4 Temporal variations of alongshore sediment transport rate at x/x0 =0

Fig. 5 and Fig. 6 depicts spatial variations of time averaged on-offshore and alongshore sediment transport rates Qx, Qy, respectively. The positive direction of the abscissa is taken in the onshore direction from S.W.S.L.. It is recognized that substantial values both for Qx and Qy are observed even in the onshore region of the shoreline. The time dependent calculation treating the swash wave front as a moving boundary is able to reproduce non-vanishing transport rates both Qx and Qy in the region $x/x_0>0$. These figures show that Qx are offshore direction except for the mildest case S=1/20, and Qy decreases monotonously toward the onshore direction, hence, no bi-modal peak around the shoreline is obtained.

(d) Comparison with Conventional Experimental Data

Kamphuis(1990) conducted three-dimensional mobile bed experiments, and investigated alongshore sediment transport rate in swash zone as well as surf zone. Most distributions of his data exhibited bimodal with one peak close to the breaking zone and another in the swash zone. One example of his data is shown in Fig. 7. The ratio of suspended load and bed load throughout his tests showed roughly constant for the entire region, even near the breaking point where suspend load is expected to be predominant.

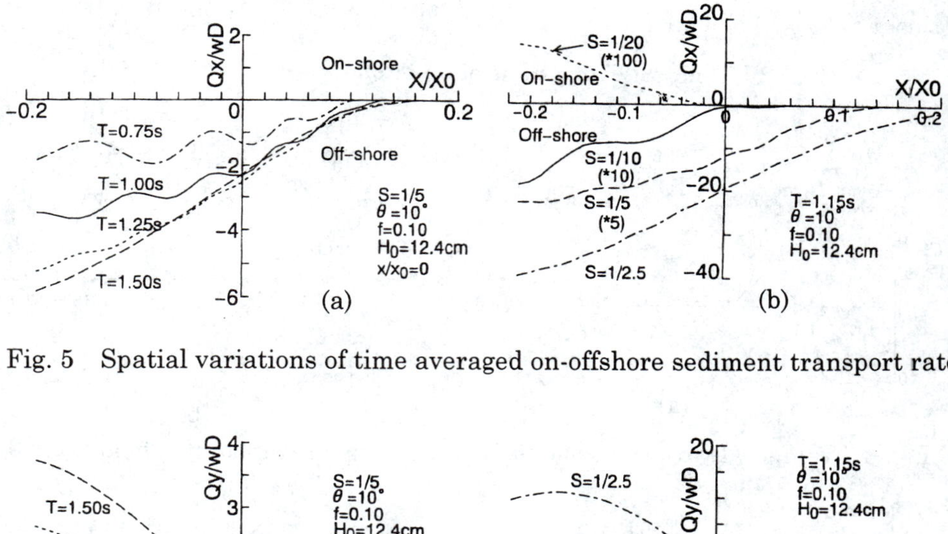

Fig. 5 Spatial variations of time averaged on-offshore sediment transport rate

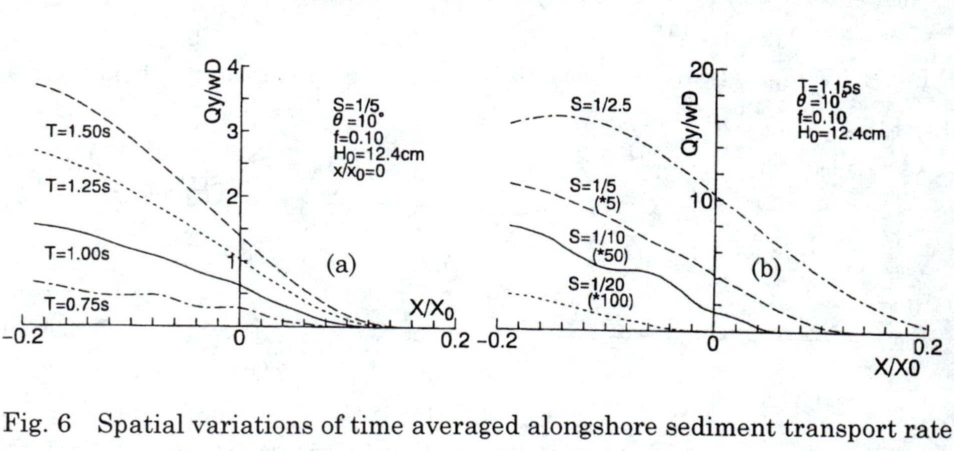

Fig. 6 Spatial variations of time averaged alongshore sediment transport rate

Comparisons with the present calculation under the same input condition of Kamphuis's experiment have revealed that his results were 2 order greater than the calculated results. Fig. 7 shows one of the results where the transport rate Qy at still water shore line x=0 for the bed load was around 17kg/m/hr in its immersed weight. Converting into a non-dimensional transport rate Qy/wd, this amount corresponds to 1.06, while the calculated rate is 0.029.

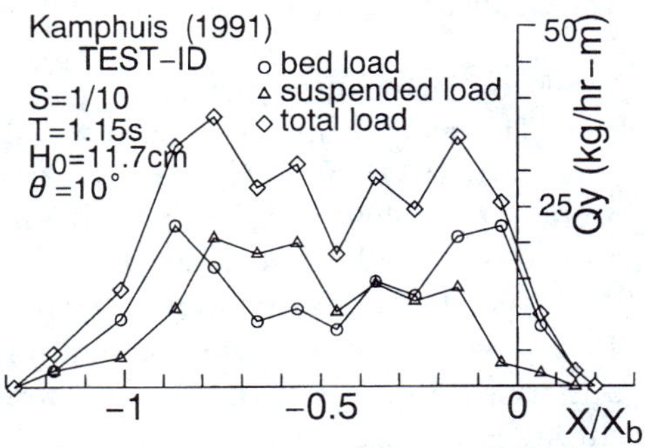

Fig. 7 Distributions of alongshore sediment transport rate (Kamphuis, 1991)

Sunamura(1984) measured on-offshore sediment transport rate focusing on the swash zone in a two-dimensional wave tank. Based on the results, he proposed a transport rate formula for the swash zone. Comparisons between the calculated on-offshore sediment transport rates Qx at x=0 with those of Sunamura's empirical formula are shown in Fig. 8, in which the parameter H_0, T, d is the incident wave height, wave period and sediment diameter, respectively. Qualitative agreements are fairly good in all the figures, but some discrepancies are found. In Fig. 8(a), Sunamura's formula predicts positive Qx values for small wave height, which implies onshore sediment transport corresponding accretional beach formation. Meanwhile the calculated Qx also decreases to zero with decreasing of wave tractive force but the direction is still offshoreward.

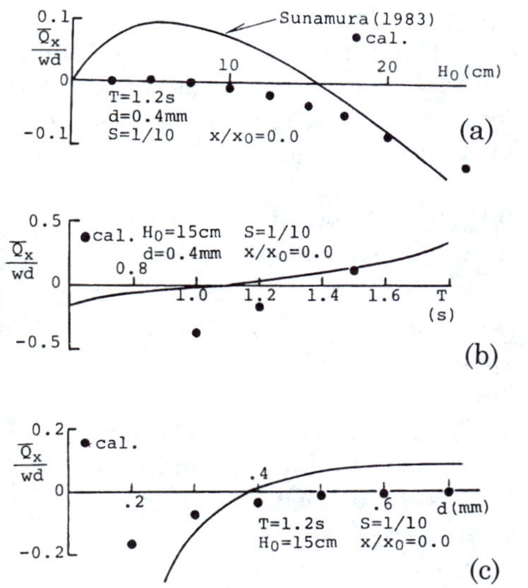

Fig. 8 Comparison of calculated on-offshore sediment transport rate with Sunamura's(1984) formula

The present calculations deal only with bed load transport under non-breaking waves. Large amount of sediment will be suspended by the vortex when the backwash collides with the incoming bore. The discrepancy from the empirical formula suggests the significant contribution of the generated vortex.

3. EXPERIMENTAL STUDY

(a) Experimental Apparatus

The wave tank used here was 26.7m long, 13m wide and 1.2m deep wave basin in Dept. of Ocean Civil Engrg., Kagoshima Univ.(Fig. 8). In this section, the x-axis is taken positive in the offshore direction with the origin x=0 at the still water shoreline. The y-axis is taken in the alongshore direction. A uniform concrete slope of gradient S=1/7.5 was set up at an angle θ=15° with a wave generator equipped at the other end of the basin. The uniform water depth in the offshore region was kept at constant 79cm throughout the experiments.

Ten capacitance type wave gauges were used, one for the offshore waves, and nine were set in an array on the slope from 210cm offshore to 40cm onshore of the still water shoreline(S.W.S.L.), which covered the surf and swash zone. The wave gauges were inserted into slit holes, which were ditched in order to maintain small water depth even when the bed was dried up at the downrush phase.

A 10mm thick sand layer extending 1.0m width by 3.2m length was laid on the mortar bed. The sand used here was clastic sediment particle originated from igneous rock, of which diameter d was 0.9mm and specific gravity s was 2.96.

(b) Experimental Procedures

Measurements using the florescent sand tracer were conducted at 16 points with 15cm intervals in the range between x=150cm offshore and x=-75cm onshore of S.W.S.L. First, natural undyed sand was laid and leveled its surface accurately. Then, using a core sampler of 36mm inner-diameter and a thin spatula, a cylindrical sand mass was removed. Instead of this, a cylindrical plug of florescent colored sand was buried. After one minute wave action on the sand bed, core samples were collected at 16 points each for two concentric circles of 20cm and 40cm radius.

The samples were evaporated in a drying oven, then, weighed on a precision balance. Meanwhile, the weight

Table-1 Test conditions

	H_i (cm)	T (s)	x_b (cm)	W_u (cm)	ξ
CASE-5	6.4	2.5	-70	85	1.65
CASE-6	7.9	2.0	-60	70	1.19
CASE-7	3.3	3.0	-70	100	2.75
CASE-8	7.5	1.5	-65	65	0.91

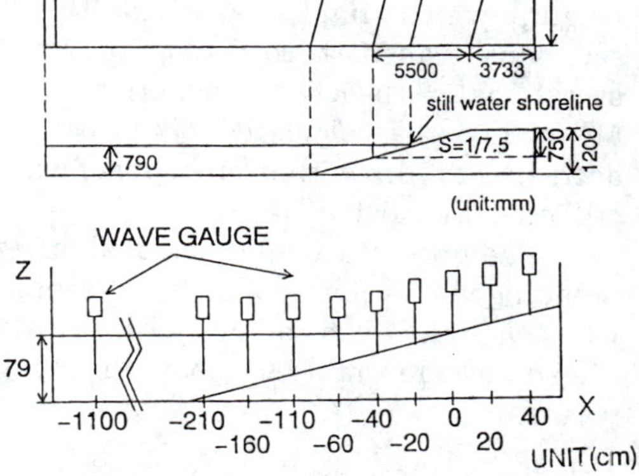

Fig. 8 Wave tank

of dyed sand was estimated by the number which was determined by visual counting under ultraviolet illumination. For this conversion, a calibration curve between the number of particles and its weight was determined beforehand. The final results were arranged in volumetric concentration; the occupation ratio of dyed sand against the whole sand.

Assuming advection is predominant, the sediment transport velocity was estimated in the following way. If a particle initially located at the center moves to the position (r, θ) after Δt time, the traveling velocity for on-offshore and alongshore direction u_s, v_s will be given by $r\cos\theta/\Delta t$, $r\sin\theta/\Delta t$, respectively. The representative sediment transport velocity was estimated by spatial averaging as follows:

$$u_s = \frac{\iint c(r,\theta)\frac{r\cos\theta}{\Delta t}dS}{\iint c(r,\theta)dS} = \frac{\int_0^\infty\int_0^{2\pi} c(r,\theta)\frac{r\cos\theta}{\Delta t}rd\theta dr}{\int_0^\infty\int_0^{2\pi} c(r,\theta)rd\theta dr}$$

$$v_s = \frac{\iint c(r,\theta)\frac{r\sin\theta}{\Delta t}dS}{\iint c(r,\theta)dS} = \frac{\int_0^\infty\int_0^{2\pi} c(r,\theta)\frac{r\sin\theta}{\Delta t}rd\theta dr}{\int_0^\infty\int_0^{2\pi} c(r,\theta)rd\theta dr} \tag{2}$$

However, the number of the measuring points was restricted as 16 for each concentric circle, 32 in total, thus the estimation for u_s, v_s was actually carried out by the following equations.

$$u_s = \frac{\sum_{i=1}^{16} c_i\frac{r_1^2\cos\theta}{\Delta t} + \sum_{j=1}^{16} c_j\frac{r_2^2\cos\theta}{\Delta t}}{\left(\sum_{i=1}^{16} c_i\right)r_1 + \left(\sum_{j=1}^{16} c_j\right)r_2}$$

$$v_s = \frac{\sum_{i=1}^{16} c_i\frac{r_1^2\sin\theta}{\Delta t} + \sum_{j=1}^{16} c_j\frac{r_2^2\sin\theta}{\Delta t}}{\left(\sum_{i=1}^{16} c_i\right)r_1 + \left(\sum_{j=1}^{16} c_j\right)r_2} \tag{3}$$

The test conditions are summarized in Table-1, in which Hi is the incident wave height in uniform depth region, x_b is the location of breaking wave, and W_u is the maximum run-up distance on the slope obtained by eye-measurements.

(c) Wave characteristics on a slope

Fig.9 shows the water surface fluctuations at each measuring point from offshore to onshore. From the fluctuation records, the following parameters were defined to describe the sediment transport; wave height, skewness (asymmetry in respect of horizontal level), asymmetry (asymmetry in respect of center line).

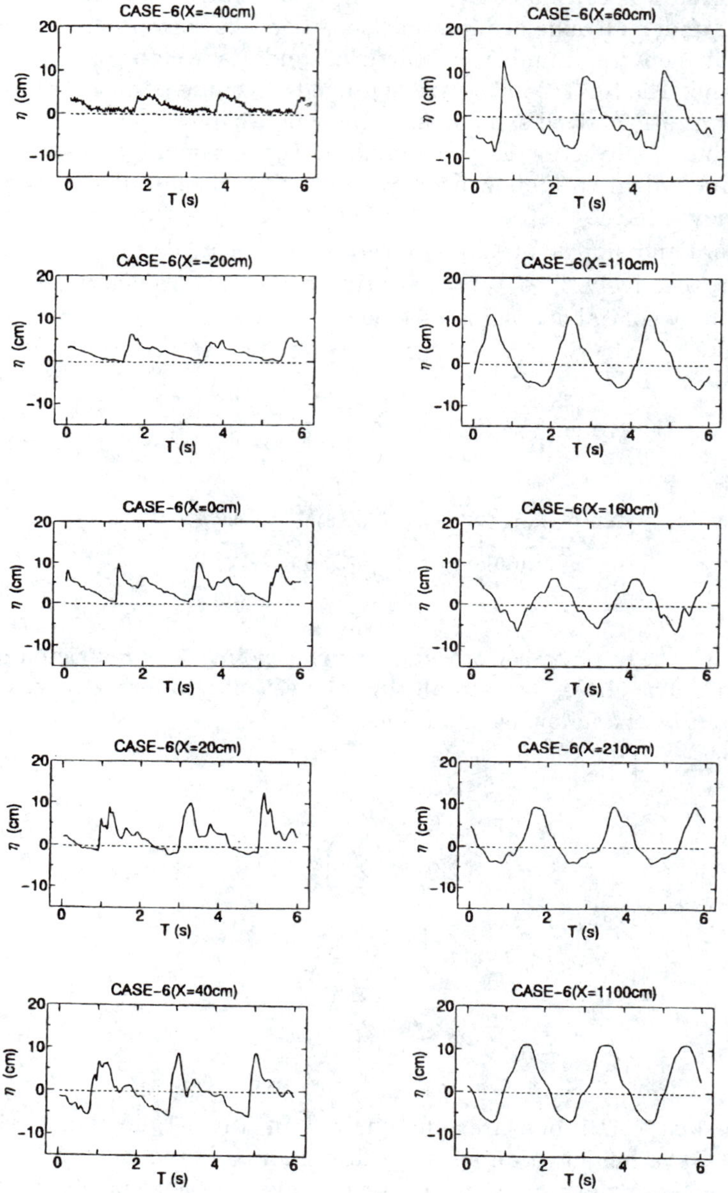

Fig. 9 Measured water surface fluctuations

The definitions of skewness and asymmetry are given by

$$skewness = \frac{1}{\eta_{rms}^3}\overline{(\eta - \overline{\eta})^3}$$

$$asymmetry = (t_{peak} - t_0)/T$$

in which, η : water surface fluctuation, η_{rms}: root mean square of η 、t_0: time at the lowest water surface fluctuation, t_{peak} : time at the succeeding fluctuation peak.

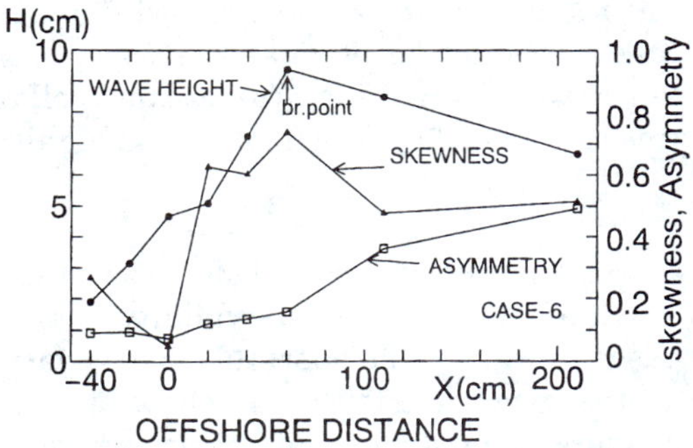

Fig. 10 Spatial variations of wave profile properties

Fig. 10 illustrates that *skewness* becomes small in the swash zone where the water surface fluctuation shows bore-like variation. Meanwhile, *asymmetry* monotonously decreases toward onshore. This sudden increase of wave profile will cause significant onshore-ward sediment transport in the swash region.

(d)Transport velocity determined by tracers

Fig. 11 shows an example of directional distribution of florescent sand concentration. Onshore transport predominance in the offshore region of the breaking point(in this case;x=70cm), alongshore transport in the surf region, and onshore transport around still water shoreline (x=0) are observed. Large scattering was found in the surf zone, where not only advection but diffusion contributes the tracer transport.

The sediment transport velocities estimated from Eq. (3) are plotted in a vector form(Fig. 12). In the figure, the solid vector indicates the data where more than 10 florescent sands were sampled, the dotted vector points out the data where 6 ～ 10 florescent sands were found (less reliable). Following characteristics are obtained:

In the offshore region of the breaking point(B.P.) sediment is transported onshore because the velocity fluctuations are skewed in the onshore direction.

In the region between the breaking point and still water shoreline, alongshore transport is predominant. This is due to generated longshore current in this region. For on-offshore transport, the onward sediment transport due to the skewness of velocity fluctuation may be almost canceled by offshore transport due to an undertow generated near the bottom.

In the swash zone, the transport direction is offshore (except for CASE-7). Since water mass running up on a slope will be returned in a thin layer flow, the velocity of the return flow becomes large enough to transport the sediment offshore in averaging over a period.

The result of CASE-7 is slightly different from others especially in the swash zone , it could be because the wave period was the longest among all the cases, so the breaking type became collapsing (surf similarity parameter in this case ξ =2.75) which breaks suddenly close to the shoreline.

(e) Comparison with the calculated transport velocity

In order to check the validity of the experimental results, calculations based on Kobayashi's formula are performed under the same input condition as the experiments. The results are illustrated in Fig. 13. All of the on-offshore components of transport velocity vector direct offshore. This directional property agrees with the experiment in the swash zone, but disagree in the surf zone and offshore zone.

The comparison also shows the discrepancy in the magnitude of transport velocity. For the experiment, most of the results ranges 0.3 ~ 0.4cm/s, and maximum is 0.66cm/s, which is so small compared to the calculation. In terms of a ratio to the generated longshore current velocity, the data were not so small, but agreed with existing field data. Nadaoka et al.(1981) reported that the sand transport velocity was less than 1.0cm/s, which corresponded to 1 ~ 2 % of the observed longshore velocity.

Several reasons can be considered to explain the difference. One would be due to the large grain size. In this experiment, relatively coarse sand d=0.9mm was used. This might be beyond the applicability of Kobayashi formula which calibrated with existing data using medium and fine sand. The run-up waves propagate on a slope with generating much turbulence at the front. In the numerical model, the velocity field is treated as uniform and no-turbulence, the predictablity of sediment transport in the swash zone is severely restricted.

4. CONCLUSIONS

(1) Analysis on single particle trajectory reproduces the zig-zag sediment movement, which is inherent transport mechanism in the swash zone.

(2) The time dependent analysis treating the swash wave front as a moving boundary enable the calculation to reproduce the swash zone sediment transport. Non-vanishing transport rates both for on-offshore and alongshore directions have been obtained even for the onshore region of the still water shoreline.

(3) From the experiments using florescent sand tracers, the sediment transport in the swash zone is comparable in the magnitude with those in the surf zone, and to be directed generally offshore. The offshore transport can be explained due to the strong backwash flow prevailing over the onshore flow generated at a passage of bore-like front.

(4) The present model postulates uniform velocity distribution and no turbulence. Quantitatively, the model could not reproduce the sediment transport rate in the existing and present experiments. For improving model to describe sediment transport in a swash zone, vortex and turbulence roles for sediment stirring up mechanism should be included.

REFERENCES

Asano, T. H. Suetomi and J. Hoshikura(1994): Measurements of current velocity generated in swash zone by obliquely incident waves, Coastal Engrg., in Japan, Vol. 37, No. 2, pp.125-136.

Asano, T. (1994): Swash motion due to obliquely incident waves, Proc. of 24 th I.C.C.E., ASCE, Vol.1, pp.27-41.

Bodge, K. R. and R. G. Dean(1987): Short-term impoundment of longshore transport, Proc. of Coastal Sediment '87, pp.468-483.

Kamphuis, J.W. (1991): Alongshore sediment transport rate distribution, Proc. of Coastal Sediment'91, pp.170-183.

Kobayashi, N. (1982): Sediment transport on a gentle slope due to waves, J. Waterways, Port, Coastal and Ocean Div., Vol. 108, No. WW3, pp.254-271.

Ryrie, S. C. (1983): Longshore motion generated on beaches by obliquely incident bores, J. Fluid Mech., Vol. 129, pp.193-212.

Sunamura, T. (1984): Onshore-offshore sediment transport rate in the swash zone of a steeply inclined laboratory beach, Ann. Rep., Inst. Geosci., Univ. Tsukuba, No.10, p.69-72.

White, T.E. and D. L. Inman(1989): Measuring alongshore transport with tracers, Nearshore Sediment Transport, Edited by R. J. Seymore, Plenum Press, pp.287-312.

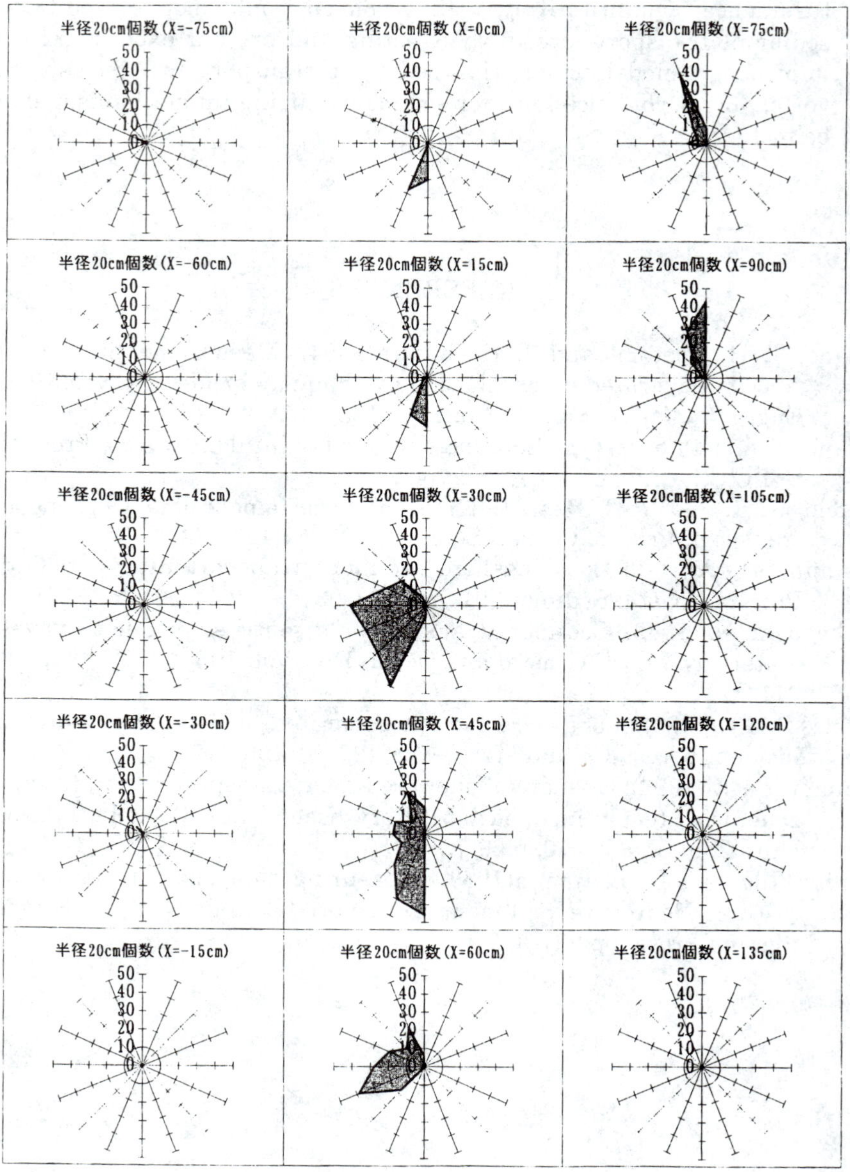

Fig. 11 Directional distributions of florescent sand tracers

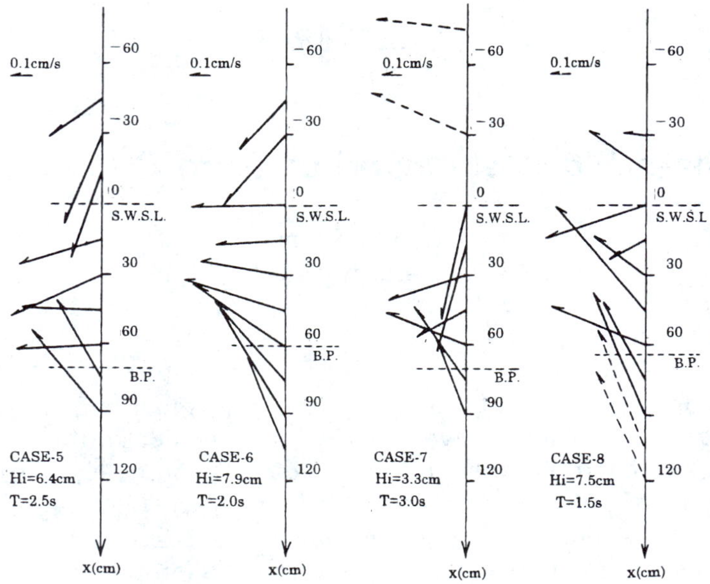

Fig. 12 Measured sediment transport velocity vectors

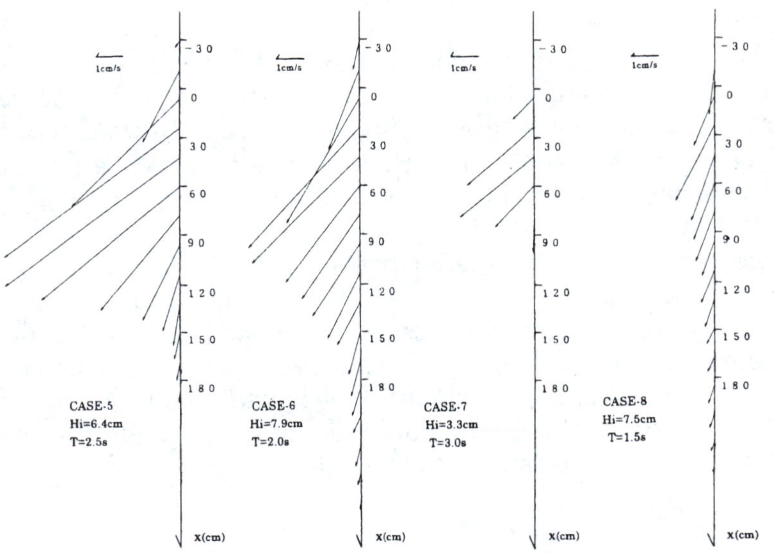

Fig. 13 Calculated sediment transport velocity vectors

CHAPTER 292

Sheet Flow Modelled as Pure Convection

Stephen Clark[1]
Peter Nielsen[2]

Abstract

A considerable amount of experimental data regarding the transport of sand under violent wave conditions (specifically the sheet flow regime) has been gathered in recent years. Historically, modelling of the concentration distributions of the suspended sediments immediately above the sheet flow layer has been undertaken using the diffusion equation. A physically more appropriate model is the combined convection diffusion approach. However, analysis of the experimental results to date indicate that a pure convection which is far simpler than either of the other two is adequate.

Introduction

The transport of sand under violent wave conditions with Shield's parameter values greater than one is often described by the term sheet flow because no bedforms survive through the wave cycles and a layer (sheet) of sand, many grain diameters thick, can be seen to move back and forth with the wave motion.

A considerable amount of experimental details of the process have been gathered in recent years by Ribberink and co-workers (1992). However, modelling of the process is still inadequate. This paper concentrates on the "upper" or suspended part of the sediment distribution. Traditionally, concentration distributions have been modelled using forms of the diffusion equation. As will be presented below, analysis of experimental results indicates that a combined

[1] Senior Engineer, Lawson & Treloar, PO Box 115, Ashgrove, Queensland, Australia 4060.
[2] Reader, Dept of Civil Engineering, University of Queensland, Australia, 4072.

convection diffusion model or in some cases a pure convection model appears to be adequate.

Experimental Data

A series of experiments have been conducted in the Large Oscillating Water Tunnel (LOWT) of Delft Hydraulics which have focussed on both time averaged and time dependant near bed and within bed sediment concentrations and sediment transport rates.

Flow conditions investigated have ranged from purely sinusoidal to regular, asymmetric waves over both rippled and plane beds. Sediments investigated have been predominantly sands with a D_{50} = 0.2 mm. An additional set of experiments have been conducted using a finer sand D_{50} = 0.12 mm.

Figure 1 provides an example of the results obtained through this procedure (Ribberink et al, 1992).

Figure 1: Experiment C8 Results

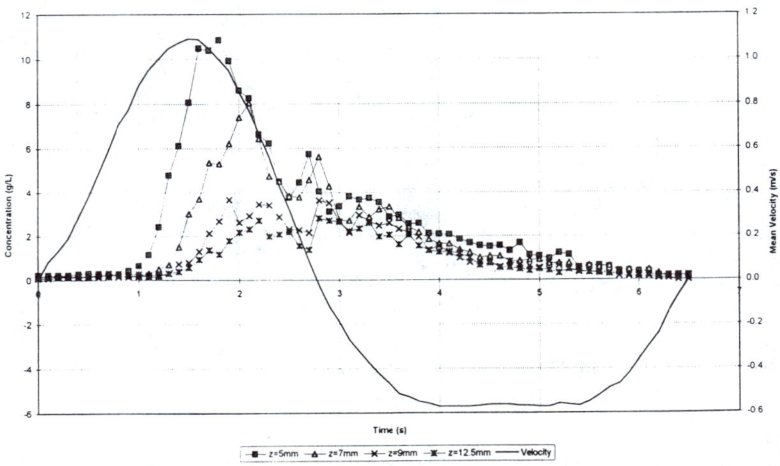

Suspended Sediment Distribution Models

Natural suspended sediment processes are a combination of both convective and diffusive processes. The process by which sediment (heavier than water) is driven upward has traditionally been described in terms of pure gradient diffusion. This process is due to driven by vertical mixing and consists of a large number of small, random vertical steps. Alternatively, the scale over which this mixing takes place is much smaller than the scale of the concentration profile.

Analysis of recent experimental data indicates that the pure gradient diffusion model is inadequate and that in many cases, a larger scale or convective process is reflected by many suspensions as well. These results reflect a process whereby the mixing scale is of the same order as the concentration profile.

Considering a horizontally uniform sediment concentration field $c=c(z,t)$ and a correspondingly uniform sediment velocity field $u_s=u_s(z,t)$, the vertical sediment flux $q=cu_s$ is considered to consist of three components. These components are a downward component $-w_0c$ due to gravitational settling and an upward flux which can be of convective (subscript C) or diffusive (subscript D) nature or a combination of both. This process is illustrated in Figure 2 (Nielsen, 1992).

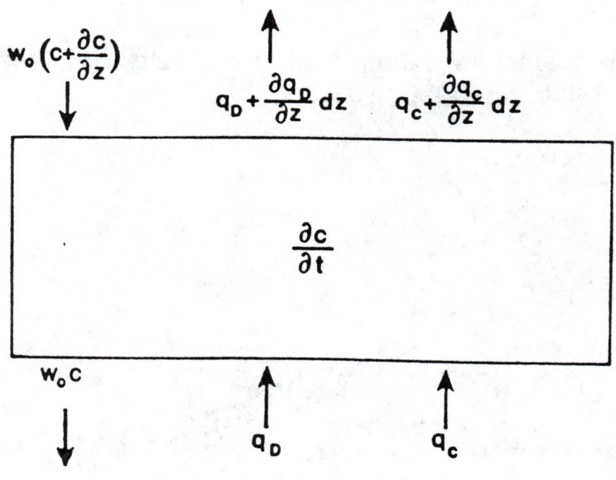

The total vertical flux is thus written as:

$$q_z=-w_0c+q_D+q_C$$

enabling the conservation equation to be written as

$$\frac{\partial c}{\partial t}=w_0\frac{\partial c}{\partial z}-\frac{\partial q_D}{\partial z}-\frac{\partial q_C}{\partial z}$$

The forms of convective and diffusive functions considered for this paper are as below:

Convective Flux, $q_c(z,t) = p(t - \frac{z}{w_c})F(z)$

Diffusive Flux, $q_D(z,t) = -\varepsilon_s \frac{\partial c}{\partial z}$

The pickup function $p(t)$ is a non-negative function describing the instantaneous pick up rate at the bed and the dimensionless convective distribution function $F(z)$ determines the fraction of the entrained sand which travels (convectively) beyond the level z above the bed.

Analytical Solution - Pure Convection

The vertical convective sediment flux, q_c is generally written in the following form

$$q_c(z,t) = p(t - \frac{z}{w_c})F(z)$$

where w_c is the average vertical velocity with which the sand is convected upwards.

Assuming that the distribution function $F(z)$ is an exponential of the form

$$F(z) = e^{-z/L}$$

the general form of the time dependant sediment concentration equation may be expressed in terms of periodic functions as:

$$c_n(z,t) = C_n f(z)\, e^{in\omega t}$$

becoming

$$c_n(z,t) = C_n e^{-\beta_n z/L}\, e^{in\omega t}$$

where

$$C_n = \frac{P_n}{w_0} \frac{1 + in\omega L / w_c}{1 + in\omega L / w_c + in\omega L / w_0}$$

and

$$\beta_n = 1 + \frac{in\omega L}{w_c}$$

If the real portion of c_n is considered, have

$$Re\{c_n(z,t)\} = Re\{C_n\}\, e^{-z/L}$$

Analytical Solution - Pure Gradient Diffusion

As with the pure convection description, with the general form of the time dependant sediment concentration equation as

$$c_n(z,t) = C_n f(z) e^{in\omega t}$$

for the pure gradient diffusion solution, this becomes

$$c_n(z,t) = C_n e^{-\alpha_n z w_0/\varepsilon_s} e^{in\omega t}$$

where

$$C_n = \frac{P_n}{w_0 \alpha_n}$$

and

$$\alpha_n = \frac{1}{2} + \sqrt{\frac{1}{4} + \frac{in\omega\, \varepsilon_s}{w_0^2}}$$

Note that for the pure gradient diffusion solution, the length scale, L of the convective solution is replaced by the sediment diffusivity over the still water settling velocity ε_s / w_0.

There are two important differences between the analytical solutions for pure gradient diffusion and pure convection. Firstly, all the harmonic concentration components decay as $e^{-z/L}$ for the convective solution and secondly, the phase lag relative to the pick up function grows at the same rate for all frequencies. This indicates that in a process dominated by convection, defined concentration peaks will travel upwards through the water column at a constant rate. However, for a process dominated by diffusion, the phase lags will grow with elevation above the bed, leading to "blurring" of successive peaks.

Data Analysis

The experimental results previously illustrated may be expressed in terms of simple periodic functions via a Fourier Analysis. This has been carried out on the available data to the tenth harmonic. It has been found to date that six harmonics provide an appropriate level of accuracy.

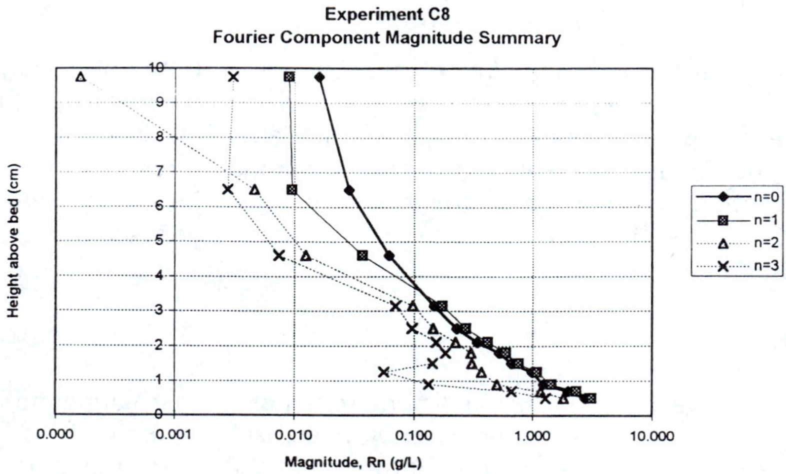

Figure 3 provides an example of the relative magnitudes of the primary harmonics obtained for the data set presented in Figure 1. From these results, parameters representing the mixing length (pure convection solution) and diffusivity (pure gradient diffusion solution) can be derived.

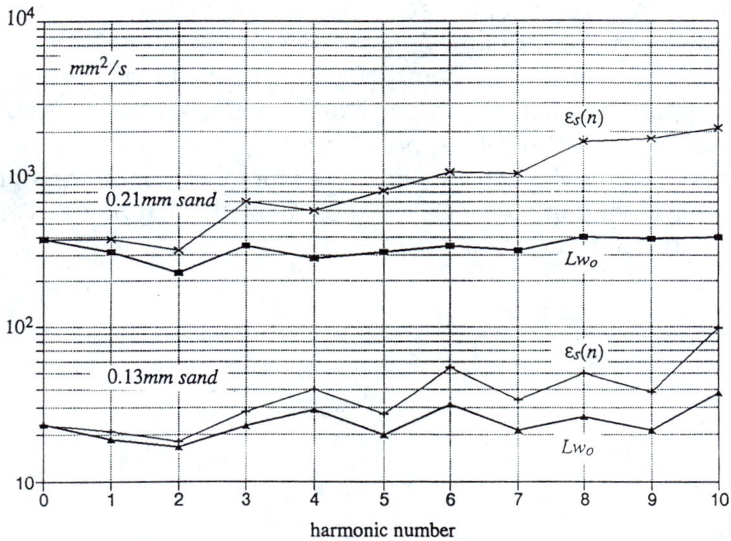

Figure 4 presents the results of this process. Note that the length scale (corresponding to pure convection) is approximately constant across all harmonics. However, the diffusivities show an increasing "drift" as the harmonic number increases. As could be expected, this is more pronounced for the sand of larger grain size.

Conclusions

The primary conclusion which may be drawn from this analysis is that the experimental results analysed are more appropriately represented by a purely convective model. This is convenient as computationally, the solution of the equations corresponding to pure convection is much more straightforward than for the second order partial DE for the pure diffusion (and combined convection diffusion) equation.

References

1. AA Al-Salem (1993), "Sediment transport in oscillating boundary layers under sheet flow conditions", Delft Hydraulics.
2. Katepodi, I et al (1994) "Intra-wave sediment transport in an oscillatory flow superimposed on a mean current", Delft Hydraulics Technical Report no. 1684 Part III
3. Nielsen, P (1992), "Coastal Bottom Boundary Layers and Sediment Transport", Advanced Series on Ocean Engineering, World Scientific, Singapore.
4. Nielsen, P (1991), "Combined Convection and Diffusion: A new framework for suspended sediment modelling", Proc Coastal Sediments '91", Seattle, pp 418-431.
5. Ribberink, JS & A Al-Salem (1992), "Time dependent sediment transport phenomena in oscillatory boundary layer flow" Delft Hydraulics Technical Report no 840.20 Part VI.
6. Ribberink, JS & Z Shen (1993), "Sediment transport of fine sand under asymmetric oscillatory flow", Delft Hydraulics Technical Report no. 840.20 Part VII.

CHAPTER 293

A TRANSPORT RATE FORMULA FOR MIXED-SIZE SANDS

Mohammad Dibajnia [1] and Akira Watanabe [2]

Abstract

Experiments were conducted on the transport of mixed grain size sands due to nonlinear waves, over both rippled and flat beds. The sand mixture was composed of a fine sand with median diameter of 0.2 mm and a coarse sand with median diameter of 0.87 mm. Conditions for the initiation of sheet flow were investigated and transport rates were measured. It was found that the transport rate of fine sand is significantly reduced by the existence of the coarse sand, whereas the coarse sand behaved almost as if there were no other sand present. A transport rate formula for mixed sands was presented and verified by using the measured data.

1. Introduction

Sandy beaches are the focus of many attentions because of their role in preserving the coastal environment, their ability in reducing disasters, and their attractiveness for recreational activities. They are, however, subjected to erosion problems due to the action of natural forces or artificial impacts. As a countermeasure, beach nourishment is one of the most acceptable methods for protecting sandy beaches. The grain size of the sand used for filling the beach is, however, decided by the available sources of material and in many cases does not coincide with the grain size of the native sand. Also, using a sand coarser than the native sand is in general more desirable because it can better stabilize the beach.

[1]Associate Professor, Dept. of Civil Eng., Univ. of Tokyo, Bunkyo-ku, Tokyo 113, Japan.
[2]Professor, ditto.

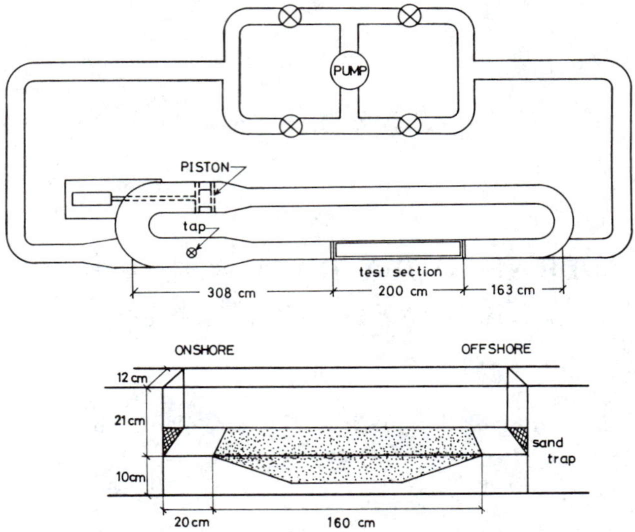

Figure 1: **Oscillatory/steady flow water tunnel and its test section.**

Needless to mention that even the grain size of the native sand itself may be nonuniform. Therefore, estimation of the transport rates of mixed-size sands is of crucial importance in predicting the behavior of both natural and artificial beaches.

In this study experiments were conducted on the transport of mixed grain size sands due to nonlinear waves over rippled and flat beds. The transport rate formula of Dibajnia and Watanabe (1992), which has originally been derived for 0.2 mm sand, was first generalized to cover the transport of granular sediments with other grain sizes and densities. A method was then presented to apply this formula to estimating the transport rate of mixed size sands. Conditions for the initiation of sheet flow were also investigated.

2. Experiments

The present experiments were carried out in a loop-shape oscillatory flow water tunnel at the University of Tokyo. The tunnel and its test section are shown in Fig. 1. Two kinds of sand, fine and coarse, were mixed together. The fine sand had a median grain size of 0.2 mm and the fall velocity of 2.3 cm/s. For the coarse sand these values were 0.87 mm and 8.7 cm/s, respectively. The percentage of the weight of each sand in the total weight of a mixed sample was defined as the percentage of that sand, P_i. Three types of mixed sands with percentages of fine sand, $P_{0.2}$, equal to 25%, 50%, and 75% were prepared. The oscillatory velocity profiles generated in the tunnel were calculated by the first

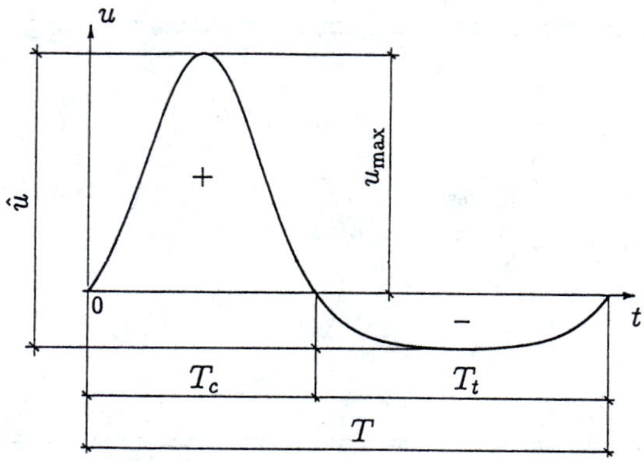

Figure 2: **Typical velocity profile in the experiments.**

Table 1: Experimental conditions for transport rate measurements.

T (s)	3	3	3	3	3	3	3	3	3	5	5	5
d_0 (cm)	23	28	35	38	43	48	50	55	71	62	78	98
$P_{0.2} = 100\%$				$\odot$	$\odot$						$\odot$	
$P_{0.2} = 75\%$	$\odot$	$\odot$	$\odot$	$\odot$	$\odot$		$\odot$	$\odot$	$\odot$			
$P_{0.2} = 50\%$	$\odot$	$\odot$	$\odot$	$\odot$	$\odot$	$\odot$	$\odot$	$\odot$		$\odot$	$\odot$	$\odot$
$P_{0.2} = 25\%$	$\odot$	$\odot$	$\odot$		$\odot$		$\odot$		$\odot$			
$P_{0.2} = 0\%$				$\odot$		$\odot$						

$\odot$: Points where transport rate is measured.

order Cnoidal wave theory. Two wave periods, $T = 3$ and 5 s and one nonlinearity index $u_{\max}/\hat{u} = 0.7$ (see Fig. 2) were selected. By changing the velocity amplitude, transport of sands over rippled beds as well as under sheet flow conditions were observed and net transport rates were measured. The experimental conditions for which transport rates were measured are summarized in Table 1. In this table, d_0 and T are the excursion length of water particle and the priod of oscillation, respectively. Conditions for initiation of sheet flow were also investigated. The present experimental data have originally been reported in Suzuki *et. al* (1994) and Inui *et. al* (1995).

3. Initiation of Sheet Flow

For a uniform sand, the transport mode usually changes from the bedload transport over an initial flat bed to the suspended load transport over the ripples and, as the velocity amplitude continue to increase, to the sheet flow transport over the reflattened bed. Several experiments were performed to observe the

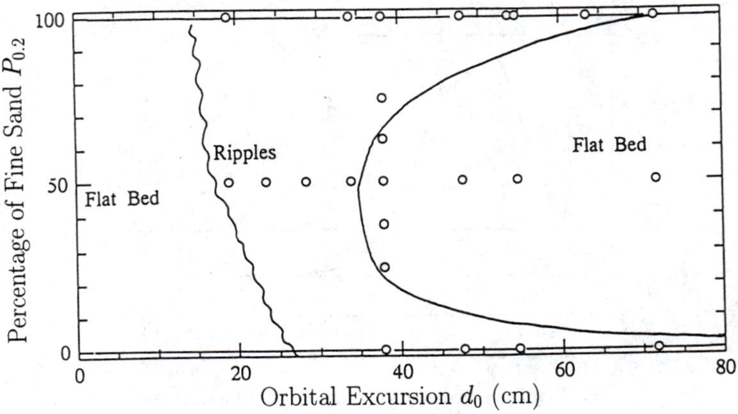

Figure 3: **Occurrence of different transport modes.**

behavior of mixed sand beds. The procedure was to start each experiment from a flat bed and see whether ripples appear or not. The results for $T = 3$ s were summarized in Fig. 3. It is interesting to see that for mixed sands, sheet flow occurs under lower velocities than those required for uniform sands. However, this was the case when the bed has initially been flat. Additional few experiments were performed with an initially rippled bed. As shown in Fig. 4, it was found that a higher velocity than that in the case of uniform sand is required to wash out the ripples from the mixed sand bed. It was thus concluded that the mixture of fine and coarse sand provides a firm structure which is not deformed as easily as uniform sands are.

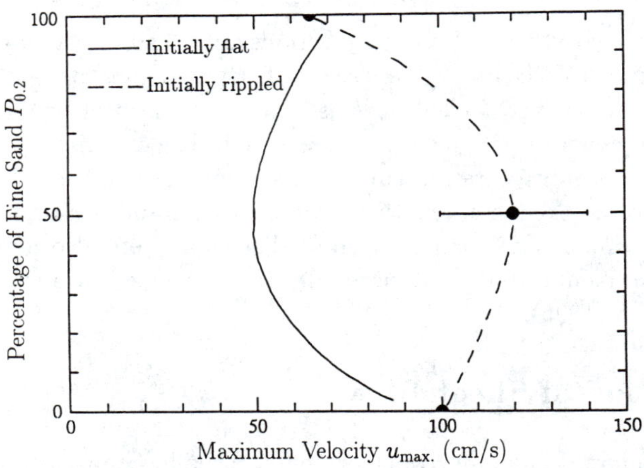

Figure 4: **Dependence of the initiation of sheet flow on the initial bed form.**

4. Generalized Transport Rate Formula

A transport rate formula involving unsteady aspects of the sand transport phenomenon has been presented by Dibajnia and Watanabe (1992) for estimating the sheet flow sand transport rates. The formula has later been extended by them (Dibajnia *et. al*, 1994) to cover suspended load over ripples as well as bed load transports. Their formula, however, is based on the experimental data obtained by using a sand with median grain size of 0.2 mm only, and needs further verification for sands of other sizes before it is applied to the transport of mixed sands.

In their experiments on sheet flow, Dibajnia and Watanabe (1992) noticed that once there was a positive velocity (Fig. 2) large enough to raise up sand particles to such a level that they could not reach the bottom before the negative velocity occured, then these particles tended to be carried to the negative direction. Dibajnia and Watanabe (1992), therefore, defined a dimensionless parameter, $\varGamma$, as

$$\varGamma = \frac{u_c T_c(\varOmega_c^3 + \varOmega_t'^3) - u_t T_t(\varOmega_t^3 + \varOmega_c'^3)}{(u_c + u_t)T} \tag{1}$$

and assumed the net transport rate to be a function of $\varGamma$. For the dimensionless net transport rate, $\varPhi$, Dibajnia (1995) proposed

$$\varPhi = \frac{q_{\rm net}(1-\epsilon)}{Wd} = 0.0023 \cdot {\rm sign}(\varGamma) \cdot |\varGamma|^{0.5} \tag{2}$$

In the above relations $q_{\rm net}$ is the net volumetric transport rate, ϵ the sediment porosity, d the sediment grain size, W the sediment fall velocity, and u_c and u_t are representative velocity amplitudes for the positive and negative portions of the velocity profile, respectively, as

$$u_c^2 = \frac{2}{T_c}\int_0^{T_c} u^2 dt, \quad u_t^2 = \frac{2}{T_t}\int_{T_c}^{T} u^2 dt \tag{3}$$

In Eq. (1) $\varOmega_c'$ represents the amount of suspended sand remaining from the positive half cycle, to be carried by the negative velocity. Similarly $\varOmega_t'$ stands for the amount of sand still in suspension after the negative half cycle, which will be transported to the positive direction. On the other hand, $\varOmega_c$ represents that amount of sand which is entrained and carried only by the positive velocity; and $\varOmega_t$ indicates the amount of sand entrained, transported, and settled during the negative half cycle. Values of $\varOmega$ were estimated in terms of another parameter, ω, equal to the ratio of the time required for a sand particle, suspended during a half cycle, to reach the bottom to the time duration of that half cycle. The reader is referred to the original paper for more details.

Now consider the simple case of sheet flow sand transport under half of a sinusoidal cycle. In this condition, Eq. (1) reduces to

Table 2: Transport rate data of Sawamoto and Yamashita (1986)

sediment type	specific gravity	d (cm)	W (cm/s)	T (s)	$u_{\max}$ (cm/s)	Q (cm^2/s)
sand A	1.65	0.02	2.5	3.8	125.3	2.013
sand A	1.65	0.02	2.5	3.8	114.6	1.575
sand A	1.65	0.02	2.5	3.8	101.5	1.180
sand A	1.65	0.02	2.5	3.8	88.7	0.876
sand A	1.65	0.02	2.5	3.8	74.4	0.642
sand B	1.65	0.07	10.5	3.8	125.3	1.308
sand B	1.65	0.07	10.5	3.8	114.6	1.063
sand B	1.65	0.07	10.5	3.8	101.5	0.641
sand B	1.65	0.07	10.5	3.8	88.7	0.500
sand B	1.65	0.07	10.5	3.8	74.4	0.353
sand C	1.65	0.16	22.5	3.8	125.3	1.240
sand C	1.65	0.16	22.5	3.8	114.6	1.001
sand C	1.65	0.16	22.5	3.8	101.5	0.587
sand C	1.65	0.16	22.5	3.8	88.7	0.357
sand C	1.65	0.16	22.5	3.8	74.4	0.271
coal dust	0.58	0.15	7.6	3.8	125.3	4.750
coal dust	0.58	0.15	7.6	3.8	114.6	3.215
coal dust	0.58	0.15	7.6	3.8	101.5	2.405
coal dust	0.58	0.15	7.6	3.8	88.7	1.835
coal dust	0.58	0.15	7.6	3.8	74.4	1.080
coal dust	0.58	0.15	7.6	3.8	56.2	0.441
coal dust	0.58	0.15	7.6	3.8	44.3	0.158

$$\Gamma = \Omega^3 \tag{4}$$

which means that a method to evaluate Ω can best be verified by using averaged transport rates measured under half cycle of sinusoidal oscillations. Such data have been reported by Sawamoto and Yamashita (1986) for sheet flow sand transport, and are shown here in Table 2. The data are for quartz sands of three different sizes (0.2, 0.7, and 1.8 mm) and for coal dust with density of 1.58. According to Dibajnia and Watanabe (1992), Ω may be evaluated as

$$\begin{aligned}\Omega_j &= \omega_j \cdot \frac{WT_j}{d} = \frac{1}{2}\frac{u_j^2}{sgWT_j} \cdot \frac{WT_j}{d} \\ &= \frac{1}{2}\frac{u_j^2}{sgd} = \Theta \quad : \quad \text{Mobility number}\end{aligned} \tag{5}$$

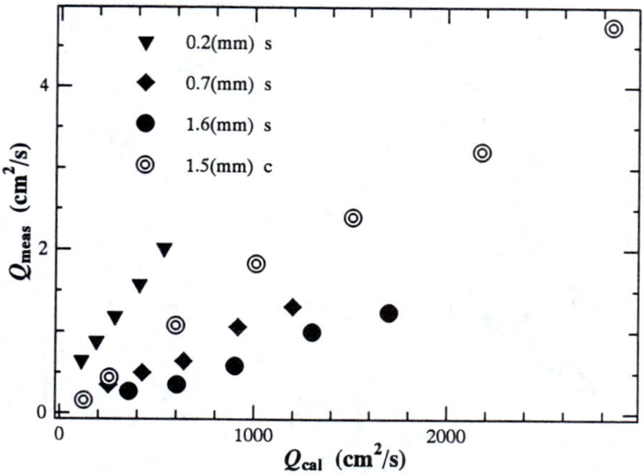

Figure 5: **Comparison of calculated half cycle averaged transport rates with measurements (previous method).**

where $s = (\rho_s - \rho)/\rho$ in which ρ and ρ_s are the densities of water and sediment, respectively, g is the acceleration of gravity, and the subscript j should be replaced by either c or t. Figure 5 shows the comparison between measured and calculated transport rates when Eq. (5) was applied. In this figure $Q = q(1-\epsilon)$ where q is the half cycle averaged transport rate. The proportionality coefficient in Eq. (2) is not considered. It is seen that each sediment type defines a different line of agreement. This indicates that the proportionality coefficient should depend on the grain size and density of sediment whenever the transport rate is assumed to be a function of either Mobility or Shields numbers.

Sawamoto and Yamashita (1986) found that the half cycle averaged sheet flow transport rate is proportional to the third power of the ratio of the friction velocity to the sediment fall velocity. Considering this, a new way to evaluate Ω is as follows.

$$\begin{aligned}\Omega_j &= \omega_j \cdot T_j\sqrt{\frac{sg}{d}} = \frac{1}{2}\frac{u_j^2}{sgWT_j}\cdot T_j\sqrt{\frac{sg}{d}} \\ &= \frac{1}{2}\frac{u_j^2}{W\sqrt{sgd}} \propto \left(\frac{u_j}{W}\right)^2\end{aligned} \tag{6}$$

Figure 6 shows the results after applying Eq. (6). All the data follow the same line of agreement, indicating that the new method for calculating Ω is valid for granular sediments with different grain sizes as well as different densities. The proportionality constant in Fig. 6 is about 0.0035, and is valid for half cycle averaged transport rates. For the net transport rate, original data of Dibajnia and Watanabe (1992) were used ($d = 0.2$ mm) to obtain the value of the constant.

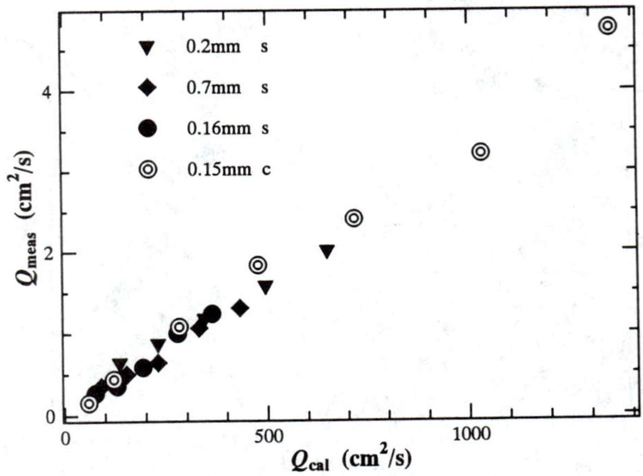

Figure 6: **Comparison of calculated half cycle averaged transport rates with measurements (new method).**

The generalized transport formula was thus obtained (Takasawa *et. al*, 1996) as

$$\Phi = \frac{q_{\text{net}}(1-\epsilon)}{Wd} = 0.0015 \cdot \text{sign}(\Gamma) \cdot |\Gamma|^{0.5} \tag{7}$$

$$\begin{cases} \text{if} \quad \omega_j \leq \omega_{\text{cr}} & \begin{cases} \Omega_j & = & \omega_j \cdot T_j\sqrt{\frac{sg}{d}} \\ \Omega'_j & = & 0 \end{cases} \\ \text{if} \quad \omega_j > \omega_{\text{cr}} & \begin{cases} \Omega_j & = & \omega_{\text{cr}} \cdot T_j\sqrt{\frac{sg}{d}} \\ \Omega'_j & = & (\omega_j - \omega_{\text{cr}}) \cdot T_j\sqrt{\frac{sg}{d}} \end{cases} \end{cases} \tag{8}$$

$$\omega_j = \frac{1}{2}\frac{u_j^2}{sgWT_j} \tag{9}$$

where Γ should be obtained as before from Eq. 1. In order to verify the above formulation, net sand transport rate data for grain sizes other than 0.2 mm are required. The only such a data set available to the authors was that reported by Watanabe and Isobe (1990), giving the net sand transport rate over ripples under nonlinear waves for sands of 0.2 and 0.87 mm median diameter. Value of ω_{cr} in Eq. (8) is equal to 1 for sheet flow. In presence of ripples, ω_{cr} is equal to 0.03 for 0.2 mm sand and equal to 0.05 for 0.87 mm sand. Threshold velocity should also be considered. Comparison between measured and calculated nondimensional net transport rates for the experiments of Watanabe and Isobe (1990) is shown in Fig. 7. In this figure, the transport rates of the 0.87 mm sand are multiplied by a factor of ten to be more visible. Satisfactory agreement is observed.

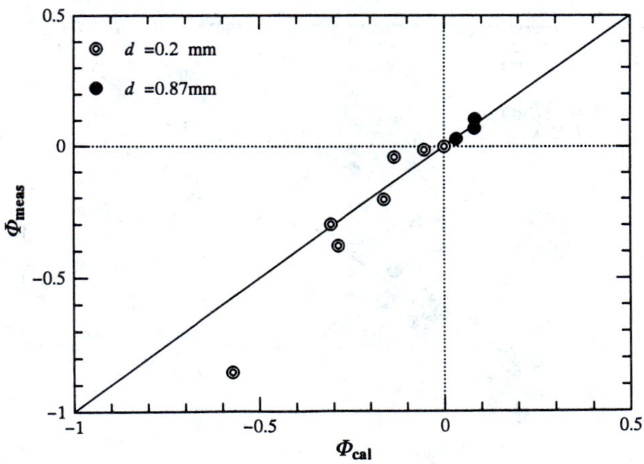

Figure 7: **Comparison of calculated net transport rates over ripples with measurements.**

5. Transport Rate of Mixed Sands

Let us first assume that the transport of each of the sands in a mixture, is not affected by the existence of the other sand. Then the transport rate of each sand could be obtained by using the following formulation.

$$\Phi_i = \frac{M_i}{(\rho_s - \rho)W_i d_i} = 0.0015\ \mathrm{sign}(\Gamma) \cdot |\Gamma|^{0.5} \cdot P_i \tag{10}$$

$$\begin{cases} \text{if} \quad \omega_j \le \omega_{cr} & \begin{cases} \Omega_j = \omega_j \cdot T_j \sqrt{\dfrac{sg}{d_i}} \\ \Omega'_j = 0 \end{cases} \\ \text{if} \quad \omega_j > \omega_{cr} & \begin{cases} \Omega_j = \omega_{cr} \cdot T_j \sqrt{\dfrac{sg}{d_i}} \\ \Omega'_j = (\omega_j - \omega_{cr}) \cdot T_j \sqrt{\dfrac{sg}{d_i}} \end{cases} \end{cases} \tag{11}$$

$$\omega_j = \frac{1}{2} \frac{u_j^2}{sgW_iT_j} \tag{12}$$

In the above relations M is the immersed weight of sand corresponding to Φ, subscript i is to address each grain size, and it is assumed that sands have the same density ρ_s. The above formulation was applied to the present experimental data. The results were compared with the measurements as shown in Fig. 8. The dash line in this figure is to separate sheet flow data from those obtained over ripples, and the solid line is the line of perfect agreement. The coarse sand data are again multiplied by a factor of ten for clarity. It is noticed that the fine sand

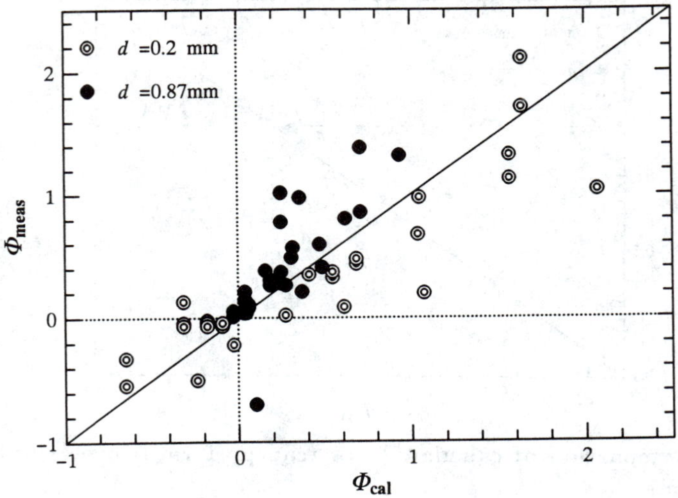

Figure 8: **Calculated net transport rates of mixed sand (Eq. 10) compared with measurements.**

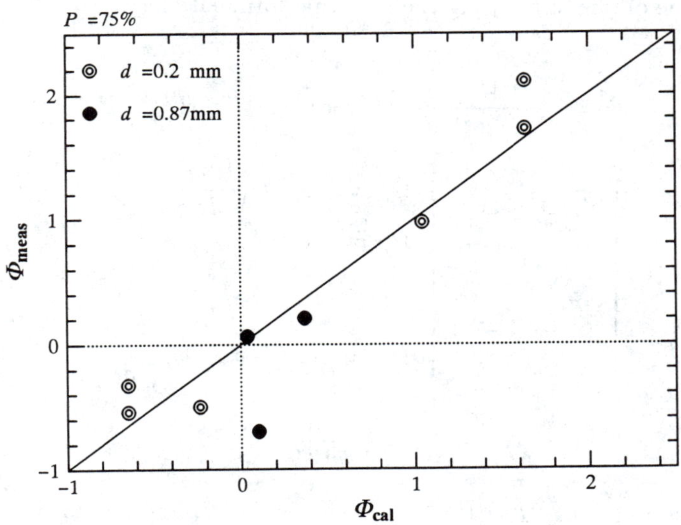

Figure 9: **Calculated net transport rates of mixed sand (Eq. 10) compared with measurements for** $P = 0.75\%$.

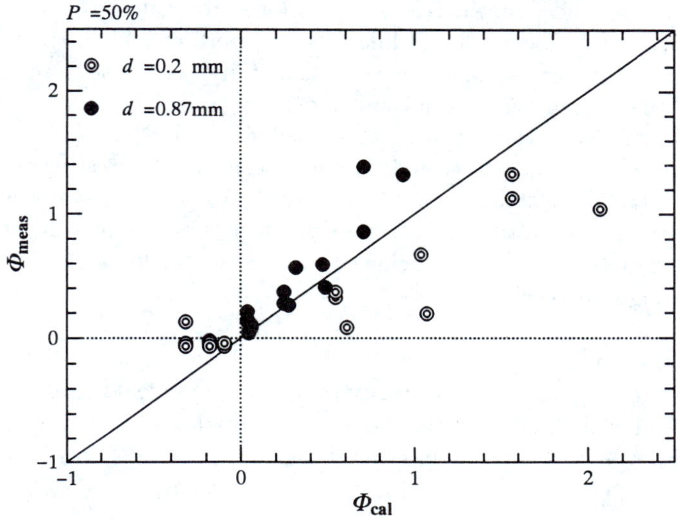

Figure 10: **Calculated net transport rates of mixed sand (Eq. 10) compared with measurements for** $P = 0.50\%$.

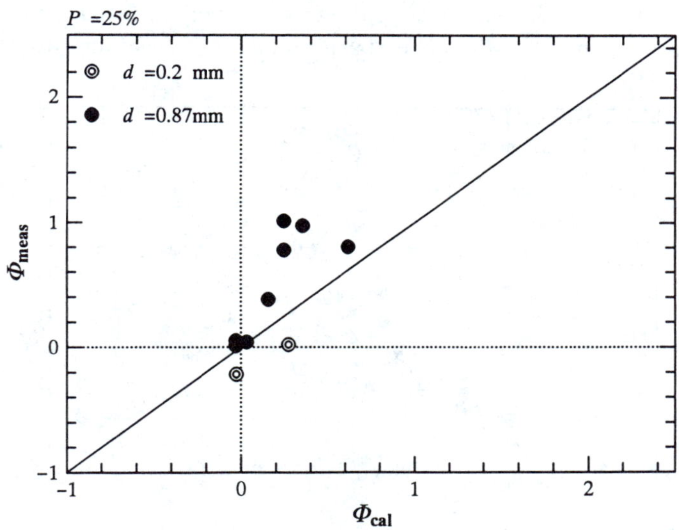

Figure 11: **Calculated net transport rates of mixed sand (Eq. 10) compared with measurements for** $P = 0.25\%$.

data lay below the solid line, whereas the data for coarse sand are located above this line. This simply means that while the transport rate of fine sand has in fact been reduced, the coarse sand has kept its activity. The data were separated according to the percentage of sand and plotted again in Figs. 9, 10, and 11. For $P = 75\%$, *i.e.* when each sand has occupied the major part of the mixture, the data more or less follow the solid line of agreement. Interesting results are observed for $P = 25\%$, where measured transport rates of coarse sand show much larger values than those estimated. Despite the fact that coarse sand amounted only to 25% of the mixture, it has behaved as if it were alone; a clear indication of the armouring phenomenon.

The reduction in the transport rate of fine sand was explained by assuming that a percentage of the total flow energy equal to the percentage of fine sand is consumed to carry the fine sand. In the present formulation, ω is the parameter involving the flow kinetic energy. In general, Eqs. (10) and (12) may be rewritten as

$$\Phi_i = \frac{M_i}{(\rho_s - \rho)W_i d_i} = 0.0015\ \mathrm{sign}(\Gamma) \cdot |\Gamma|^{0.5} \tag{13}$$

$$\omega_j = \frac{1}{2} \frac{u_j^2}{sgW_i T_j} \cdot P_i^{\frac{1}{\alpha_i}} \tag{14}$$

where α is expected to depend on the ratio of the mean diameter of each sand to that of the other sand. For the present sand mixture it was found that $\alpha_{0.2} = 1$

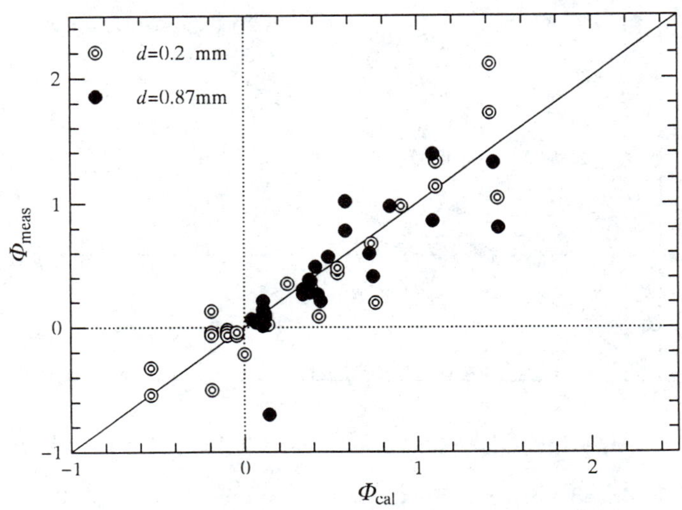

Figure 12: **Comparison of net transport rates of mixed sand, calculated by the presented method, with measurements.**

and $\alpha_{0.87} = 4$ are appropriate. Equations (1), (3), (11), (13), and (14) were thus applied to estimating the net transport rates of the present experiments. Calculated transport rates reasonably agreed with the measurements as shown in Fig. 12.

6. Concluding Remarks

The transport rate formula of Dibajnia and Watanabe (1992) was generalized to cover the transport of uniform granular sediments of any size and density.

Experiments were performed on transport of mixed-size sands. It was shown that fine and coarse sands when mixed together may make a firm structure and thus the bed may not be deformed as easily as in the case of uniform sands. Measurements of transport rate showed that armouring of fine sand by the coarse sand causes a siginificant reduction in the transport rate of the fine sand. But the transport rates of the coarse sand itself were almost unaffected by the presence of the fine sand. A method to apply the above generalized transport formula to estimating the transport rate of mixed sands was presented. Application of this method to beach profile simulation will be presented in a separate paper.

In the present experiments, the oscillatory velocity amplitude had to be limited because of the limited length of the test section (maximum velocity was about 95 cm/s) and higher velocities could not be tried. Therefore, further experiments under sheet flow conditions but with higher velocities are required to confirm whether the armouring function of the coarse sand will persist or not.

References

Dibajnia, M. (1995): Sheet flow transport formula extended and applied to horizontal plane problems, Coastal Eng. in Japan, Vol. 38, No. 2, pp. 179-194.

Dibajnia, M. and A. Watanabe (1992): Sheet flow under nonlinear waves and currents, Proc. 23rd Int. Conf. on Coastal Eng., ASCE, pp. 2015-2028.

Dibajnia, M., T. Shimizu and A. Watanabe (1994): Profile change of a sheet flow dominated beach, Proc. 24th Int. Conf. on Coastal Eng., ASCE, pp. 1946-1960.

Inui, T., M. Dibajnia, M. Isobe and A. Watanabe (1995): A transport rate formula for mixed-size sands and its application, Proc. 42nd Japanese Annual Conf. on Coastal Eng., JSCE, to be published (in Japanese).

Sawamoto, M. and T. Yamashita (1986): Sediment transport rate due to wave action, J. Hydrosc. Hydraul. Eng., Vol. 4, No. 1, pp. 1-15.

Suzuki, K., A. Watanabe, M. Isobe and M. Dibajnia (1994): Experimental study on transport of sediment with mixed-grain size due to oscillatory flow, Proc. 41st Japanese Annual Conf. on Coastal Eng., JSCE, pp. 356-360 (in Japanese).

Takasawa, S., M. Dibajnia, and A. Watanabe (1996): A generalized sheet flow transport formula, Proc. 51st Annual Civil Eng. Conf., JSCE, Vol. B2, pp. 124-125 (in Japanese).

Watanabe, A. and M. Isobe (1990): Sand transport rate under wave-current action, Proc. 22nd Int. Conf. on Coastal Eng., ASCE, pp. 2495-2507.

SUBJECT INDEX

Page number refers to the first page of paper

AUTHOR INDEX

Page number refers to the first page of paper

Volume 4

Coastal

e

ASCE American Society of Civil Engineers
345 East 47th Street
New York, New York 10017-2398

Abstract:

This proceedings contains over 370 papers presented at the 25th International Conference on Coastal Engineering which was held in Orlando, Florida, 2-6 September, 1996. The book is divided into six parts: 1) Characteristics of coastal waves and currents; 2) long waves and storm surges; 3) coastal structures; 4) coastal processes and sediment transport; 5) coastal, estuarine and environmental problems; and 6) case studies. The individual papers include such topics as the effects of wind, waves, storms and currents as well as the study of sedimentation, erosion and beach nourishment. Special emphasis is given to case studies of completed engineering projects. With the inclusion of both theoretical and practical information, these papers provide the civil engineer and professionals in related fields with a broad range of information on coastal engineering.

Library of Congress Cataloging-in-Publication Data

Coastal engineering 1996 : proceedings of the twenty-fifth international conference, September 2-6, 1996, The Peabody Hotel, Orlando, Florida / edited by Billy L. Edge.

p. cm.

"Held under the auspices of the Coastal Engineering Research Council of the American Society of Civil Engineers."

ISBN 0-7844-0242-6

1. Coastal engineering--Congresses. 2. Ocean waves--Congresses. I. Edge, Billy L. II. Coastal Engineering Research Council (U.S.) III. International Conference on Coastal Engineering (25th : 1996 : Orlando, Fla.)

TC203.5.C6184 1997 97-10664

627'.58--dc21 CIP

Library of Congress Catalog Card No: 97-10664
ISBN 0-7844-0242-6
Manufactured in the United States of America.

Cover: One year after nourishment at Palm Beach, Florida. Photo courtesy of Applied Technology & Management, Inc. and Aerial Photography, Inc.

CONTENTS

Part I Characteristics of Coastal Waves and Currents

Part II Long Period Waves, Storm Surges and Wave Groups

Part III Coastal Structures

Part IV Coastal Processes and Sediment Transport

Part V Coastal, Estuarine and Environmental Problems

Part VI Case Studies

CHAPTER 294

MODELLING OF 3D SEDIMENT TRANSPORT IN THE SURF ZONE

by

Berry Elfrink [1], Ida Brøker[1], Rolf Deigaard [2], Erik Asp Hansen[1] and Peter Justesen[1]

ABSTRACT

The three-dimensional sediment transport in the surf zone has been investigated using two different approaches for modelling of the flow pattern. The first approach is based on the integrated momentum concept for the turbulent wave-current boundary layer, see Fredsøe (1984), the second approach is based on the k-model, as described in Deigaard et al. (1991). The k-model allows for a more consistent description of the time and space varying eddy viscosity than the integrated momentum approach, but demands considerably more computation effort. The driving forces are calculated according to the formulations of Deigaard et al. (1991) and Deigaard (1993). The model based on the integrated momentum equation is able to reproduce the details of the flow satisfactorily. The presence of a longshore current increases the turbulence near the bed. This results in a decreased offshore directed flow velocity near the bed and an increase in the sediment concentration. Comparisons with field measurements show good agreement.

INTRODUCTION

When waves break turbulent kinetic energy is produced at the water surface. This energy is partly dissipated in the surface roller and partly transported downwards into the water column. The radiation stress gradients, associated with the wave breaking, are balanced by the pressure gradient originating from the wave setup and the mean bed shear stress.

In case of uniform conditions along the shore, the net discharge in the mean cross shore flow must balance the fluid mass transported in the surface roller and the discharge associated with the asymmetric wave motion and streaming in the bottom boundary layer. The zero mass flux in the cross shore direction is achieved by a slope of the mean water level. The vertical variation of the cross shore flow velocity is therefore charaterized by offshore directed velocities near the bed (the undertow) and onshore velocities further away from the bed.

---

[1] Danish Hydraulic Institute, Agern Allé 5, DK-2970 Hørsholm, Denmark

[2] Institute of Hydrodynamics and Hydraulic Engineering, Technical University of Denmark, 2800 Lyngby, Denmark

In the longshore direction, the forces associated with the wave breaking can only be balanced by a mean bed shear stress which drives the wave induced longshore current.

The interaction of the turbulence originating from wave breaking with the turbulence generated in the bottom boundary layer is of great importance in the description of the sediment transport in the surf zone, see Justesen et al. (1986). Due to the higher level of turbulence more sediment can be kept into suspension under breaking waves than under non-breaking waves.

In the present work, the flow conditions along the shore are assumed to be uniform, see figure 1. The waves approach the coast under an angle α and are assumed to be long crested and monochromatic.

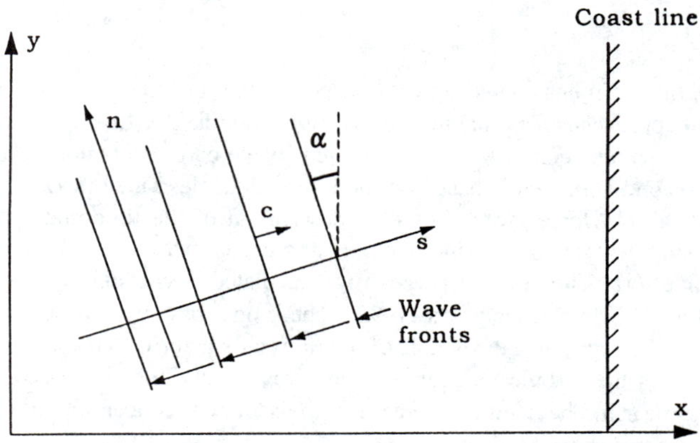

Figure 1 : plan view

VERTICAL VARIATION OF THE DRIVING FORCE

Deigaard and Fredsøe (1989) determined the time mean shear stress in normally incident breaking waves. It was found that the vertical transfer of horizontal momentum gives a significant contribution to the momentum balance. Deigaard (1993) derived expressions for the time mean shear stress in case of oblique waves by considering the momentum balance for the surface rollers. The surface shear stress is given as :

$$\frac{\overline{\tau_s}}{\rho} = \left(-\frac{g}{8c}\frac{d\,H^2 c}{ds} - \frac{d}{ds}\frac{Ac}{T} \right) \begin{pmatrix} \cos\alpha \\ \sin\alpha \end{pmatrix} \tag{1}$$

where H=wave height, s=direction of wave propagation, A=surface roller cross section, c=wave celerity, α=wave angle, g=gravitational accelleration and T=wave period.

The cross shore and longshore components of the time mean bed shear stress are determined by the gradients in the radiation stress and the mean water surface slope, the wave set-up, in the cross shore direction, Justesen et al. (1994) :

$$\frac{\bar{\tau}_{b,x}}{\rho} = -\frac{g}{8}\frac{dH^2}{dx}\cos^2\alpha - \frac{g}{16}\frac{dH^2}{dx} - \frac{d}{dx}\left(\frac{Ac}{T}\right)\cos^2\alpha - gDs_s \qquad (2)$$

$$\frac{\bar{\tau}_{b,y}}{\rho} = -\frac{g}{8}\frac{dH^2}{dx}\cos\alpha\,\sin\alpha - \frac{d}{dx}\left(\frac{Ac}{T}\right)\cos\alpha\,\sin\alpha \qquad (3)$$

Here, s_s is the slope of the water surface, and x and y are the cross shore and longshore coordinate, respectively. The shear stress varies linearly across the water column. The vertical distribution of the time mean shear stess is illustrated in figure 2.

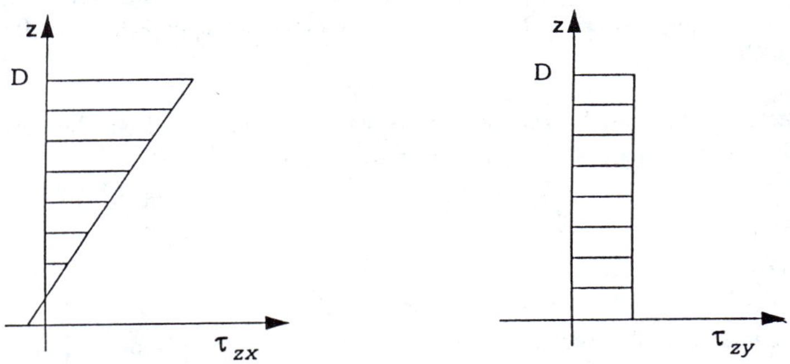

Figure 2 : Illustration of the time mean shear stress in the cross shore and longshore direction.

In both models, the wave orbital motion was calculated from linear wave theory, but any potential wave theory could be used here. If a non-linear wave description is used, the mean bed shear stress associated with the wave motion is non-zero. In the model based on the integrated momentum equation, the wave motion outside the boundary layer is described by potential theory. In order to fulfil the requirements of potential theory, the mean bed shear stress due to the wave motion must be zero. Brøker-Hedegaard et al. (1991) showed that a perfect force balance can be obtained by adding a small constant velocity to the orbital motion. This additional velocity is found iteratively by the requirement of zero mean bed shear stress, see Elfrink et al. (1993). In the K-model, this velocity is found automatically.

In case of progressive waves, a small net shear stress is generated due to the non-uniformity of the wave boundary layer. Brøker-Hedegaard (1985) showed that this streaming induced shear stress at the bed can be expressed as :

$$\tau_{str.} = \frac{\rho}{c} \overline{U_o U_f |U_f|} \tag{4}$$

where U_o= wave orbital velocity and U_f= instantaneous friction velocity.

THE INTEGRATED MOMENTUM APPROACH

In this model the instantaneous velocity profiles and eddy viscosity profiles inside the boundary layer are assumed to be given by a logarithmic and a parabolic distribution, respectively. The interaction of the undertow, the longshore current and the waves is taken into account by superposing the wave by a steady current and solving the turbulent boundary layer for the combined wave/current motion. The total mean eddy viscosity has contributions from the wave breaking, from the bottom layer and from the mean current. The eddy viscosity due to the mean current outside the boundary layer is calculated by applying the mixing length concept. The equations are solved by the assumption that the net flux perpendicular to the shoreline is zero. It is noted that the integrated momentum approach is by far the most computationally efficient.

The time-averaged flow velocity distribution is calculated from the vertical distribution of the driving forces (eq. 1 - 4) and the mean eddy viscosity :

$$\frac{d\vec{U}}{dz} = \frac{\vec{\tau}}{\rho \nu_t} \tag{5}$$

The three contributions to the eddy viscosity are calculated independently from each other. The contribution from the wave boundary layer is calculated from the boundary layer model of Fredsøe (1984). The contribution from the mean flow is modelled by applying a mixing length concept, see Justesen et al. (1994) and the eddy viscosity due to wave breaking is calculated from the transport equation for turbulent kinetic energy, see Deigaard et al. (1986).

THE K-EQUATION MODEL

The k-model approach in the present work is an extension of the model developed by Deigaard et al. (1991) for the case of normally incident waves.

The driving force in the unsteady motion is composed of a steady part as given in eqs. 1 - 4 and an oscillatory part which is given by the periodically varying horizontal pressure gradient associated with the wave motion. The flow equations, neglecting convective terms, read :

$$\frac{\partial u}{\partial t} = -\frac{1}{\rho}\frac{\partial p}{\partial x} + \frac{1}{\rho}\left[\frac{\partial \tau_{zx}}{\partial z} - \frac{\partial \bar{\tau}_{zx}}{\partial z}\right] \tag{6}$$

$$\frac{\partial v}{\partial t} = -\frac{1}{\rho}\frac{\partial p}{\partial y} + \frac{1}{\rho}\left[\frac{\partial \tau_{zy}}{\partial z} - \frac{\partial \bar{\tau}_{zy}}{\partial z}\right] \tag{7}$$

where u=cross shore flow velocity, v=longshore flow velocity, u=cross shore coordinate, y=longshore coordinate, z=vertical coordinate, p=pressure gradient associated with the wave motion, ρ=density of water, τ_{zx} and τ_{zy} = cross shore and longshore components of the shear stress respectively, t = time.

$$\tau_{zx} = \rho \nu_t \frac{\partial u}{\partial z} \quad ; \quad \frac{1}{\rho}\frac{\partial p}{\partial x} = -\frac{\partial U_o}{\partial t} \tag{8}$$

$$\tau_{zy} = \rho \nu_t \frac{\partial v}{\partial z} \quad ; \quad \frac{1}{\rho}\frac{\partial p}{\partial y} = -\frac{\partial V_o}{\partial t} \tag{9}$$

The eddy viscosity, ν_t, is calculated from the diffusion equation for turbulent kinetic energy, see Justesen et al. (1994) :

$$\nu_t = l\sqrt{k} \quad ; \quad \frac{\partial k}{\partial t} = -\frac{\partial}{\partial z}\left(\frac{\nu_t}{\sigma_n}\frac{\partial k}{\partial z}\right) + \frac{1}{\rho} PROD - c_l \frac{k^{\frac{2}{3}}}{l} \tag{10}$$

PROD = production term, which consists of contributions from wave breaking , as expressed through the gradient in wave energy flux, E_f, and from the shear stress in the model.

$$PROD = \tau_{zx}\frac{\partial u}{\partial z} + \tau_{zy}\frac{\partial v}{\partial z} - \frac{\partial E_f}{\partial x} \tag{11}$$

SEDIMENT TRANSPORT

The sediment transport is calculated according to Fredsøe et al. (1985) and Deigaard et al. (1986). Here the bed load transport model of Engelund and Fredsøe (1976) is used, where the bedload transport is calculated from the instantaneous bed shear stress. The vertical variation of the suspended sediment concentration is calculated from the vertical diffusion equation for suspended sediment.

The omission of the convective terms in the diffusion equation is alleviated by adding the lagrangian flow velocity times the mean concentration to the time averaged product of instantaneous velocity and concentrations.

COMPARISON OF THE 2 MODELS

Figs 3 and 4 show the mean cross shore and longshore flow velocities for the 2 models for varying wave angles. The wave height in this example was taken as 0.6 m and the period as 6 s. The water depth was 1 m and the bed roughness 0.5 mm. The energy dissipation, the surface roller volume and the cross shore gradient of the surface roller were calculated by using a wave height gradient of -0.015 and a bed slope of 0.033. The sediment grain size was taken as 0.2 mm.

It can be seen that the agreement between the two models is good. In the lower part of the water column, the gradients in the velocity profiles are somewhat steeper for the k-model than for the integrated momentum approach, which indicates slightly higher values of the eddy viscosity for the latter.

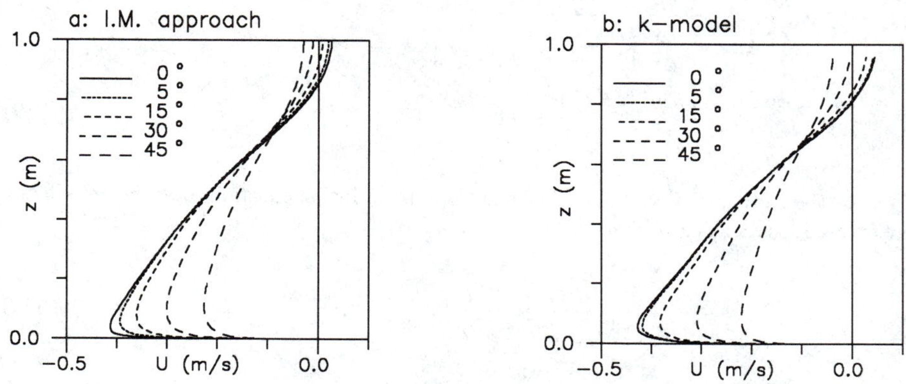

Figure 3 : Time averaged cross shore flow velocities for the two models. a = integrated momentum approach , b = k-model.

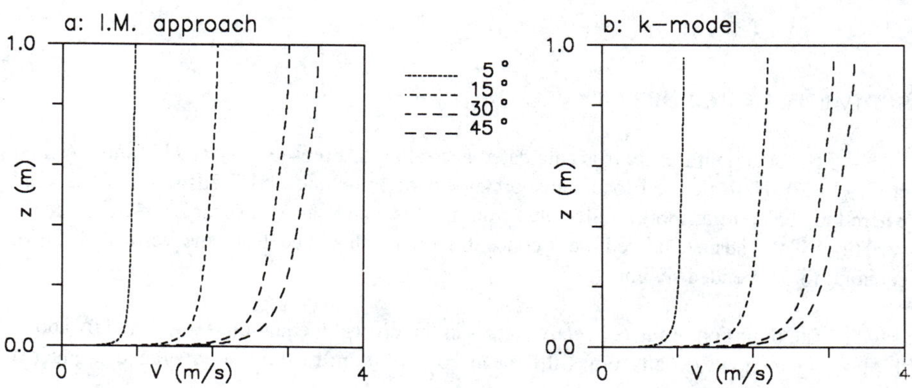

Figure 4 : Time averaged longshore flow velocities for the two models. a = integrated momentum approach, b = k-model.

The time variation of the cross shore and longshore components of the friction veloctiy are shown in figs. 5 and 6. For low angles of wave incidence, the cross shore friction velocities are slightly lower under the wave crest for the integrated momentum approach than for the k-model, but generally the agreement is good.

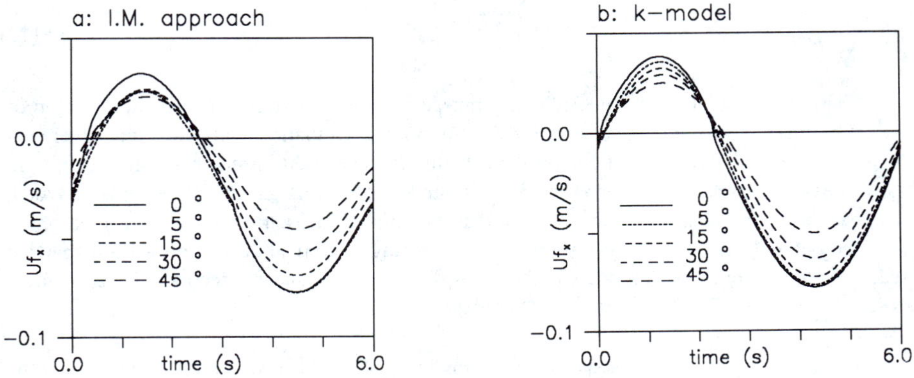

Figure 5 : Time variation of the cross shore component of the shear velocity. a = integrated momentum approach, b = k-model.

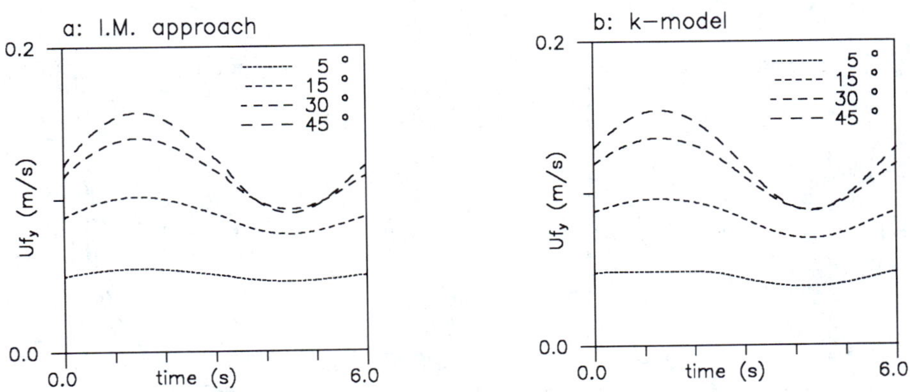

Figure 6 : Time variation of the longshore component of the shear velocity. a = Integrated momentum approach, b = k-model.

In this example, the cross momentum exchange of horizontal momentum, associated with the gradients in the longshore current velocity across the shore has been neglected. This causes unrealistically high values of the calculated longshore current velocites and friction velocities. Usually, the mean bed shear stress is calculated from a surf zone model which includes this momentum exchange. However, this has no importance in the context of intercomparing the two different models.

In both models, the suspended sediment transport is calculated as the product of the instantaneous flow velocities and the instantaneous sediment concentration :

$$q_s = \frac{1}{T}\int_0^T\int_0^D (uc)\, dzdt \tag{12}$$

In the calculation of the sediment transport in the surf zone, the vertical variation of the suspended sediment concentration is important. Deviations in the time averaged suspended sediment concentrations may occur due the phase difference between the flow in the wave boundary layer and the outer flow. In the k-model, this phase difference is included, whereas the boundary layer in the integrated momentum approach is assumed to be in phase with the outer flow. Figs. 7 and 8 show the time averaged vertical variation of the suspended sediment fluxes for the two models in the cross shore and longshore direction. The agreement is seen to be good for all examined wave angles.

A number of comparisons was performed were the hydrodynamic parameters (e.g the wave height, period, angle of incidence and energy dissipation) were varied systematically. Also different grain sizes were analyzed. The integrated momentum approach tends to give slightly higher transport rates for high levels of turbulence, associated with wave breaking. This is due to the different techniques in determining the turbulence originating from the boundary layer, the mean flow and wave breaking. Generally the agreement is good.

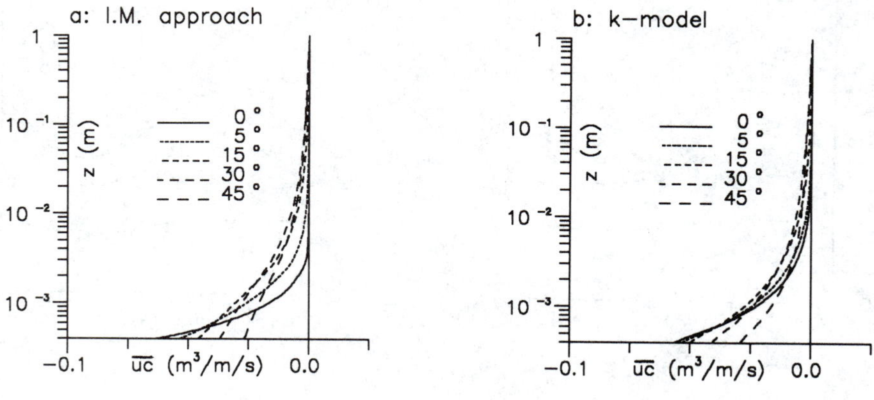

Figure 7 : Vertical variation of the time averaged cross shore sediment flux. a = Integrated momentum approach, b = k-model.

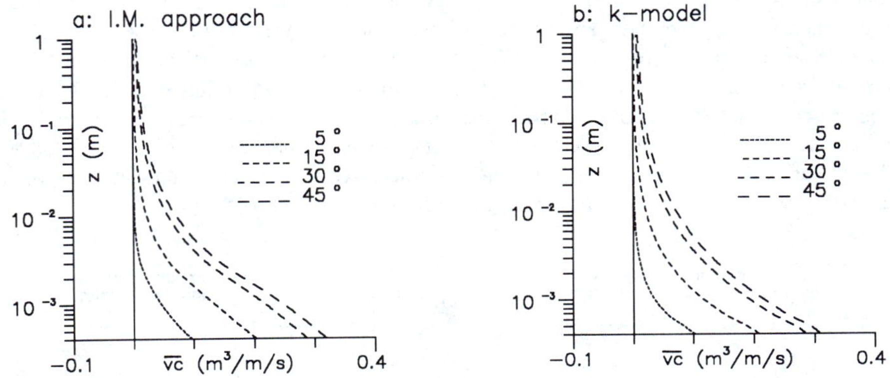

Figure 8 : Vertical variation of the time averaged longshore sediment flux.
a = Integrated momentum approach, b = k-model.

It can be concluded that the integrated momentum approach gives results very similar to the more complete k-model in determining the mean flow, the time variation of the bed shear stess and the sediment transport.

THE INTERACTION OF UNDERTOW AND LONGSHORE CURRENT

The presence of a longshore current will affect the undertow and the resulting cross shore sediment transport. The interaction of the longshore current with the undertow is important in the determination of the cross shore profile evolution. The longshore current causes higher levels of turbulence near the sea bed, this affects the vertical shape of the cross shore velocity as the vertical velocity gradients decrease due to the higher eddy viscosity. This results in a reduction of the near bed cross shore flow velocity. At the same time, the higher levels of turbulence allow higher concentrations of suspended sediment.

A test was performed with the model based on the integrated momentum equation for a situation with and without the presence of a longshore current of 0.5 m/s. In the present example, a wave height of 0.7 m and a period of 5 s was used. The water depth is 1.5 m and the energy dissipation corresponds to the energy loss in a hydraulic jump. Figure 9a shows the simulated time averaged cross shore velocity profile with and without a longshore current. In both cases, the mean flow under the wave crest balances the mass transport in the surface roller and the wave drift. It can be seen that the near bed offshore directed velocities are lower in case of a longshore current. Consequently, the velocities in the vicinity of the water surface are slightly higher.

The bed concentrations for both cases are shown in figure 9b. The main difference is that the bed concentration does not vanish at the flow reversal (t/T ≈ 0.05, and t/T ≈ 0.45) in case of a longshore current. The bed concentrations are lower under the wave crest (t/T < 0.5) than under the trough (t/T >0.5) because the wave orbital motion and the undertow counteract each other, whereas they both are directed offshore under the wave trough. The maximal bed concentration is slightly higher in case of a longshore current. The vertical variations of the time-mean sediment concentration for the two examples are shown in figure 9c. The increased eddy viscosity and the higher mean bed concentration result in a higher mean concentration in case of a longshore current. The combined effect of lower offshore velocities near the bed and the higher sediment concentrations result in an increased cross shore sediment transport in case of a longshore current in this example as shown in figure 9d.

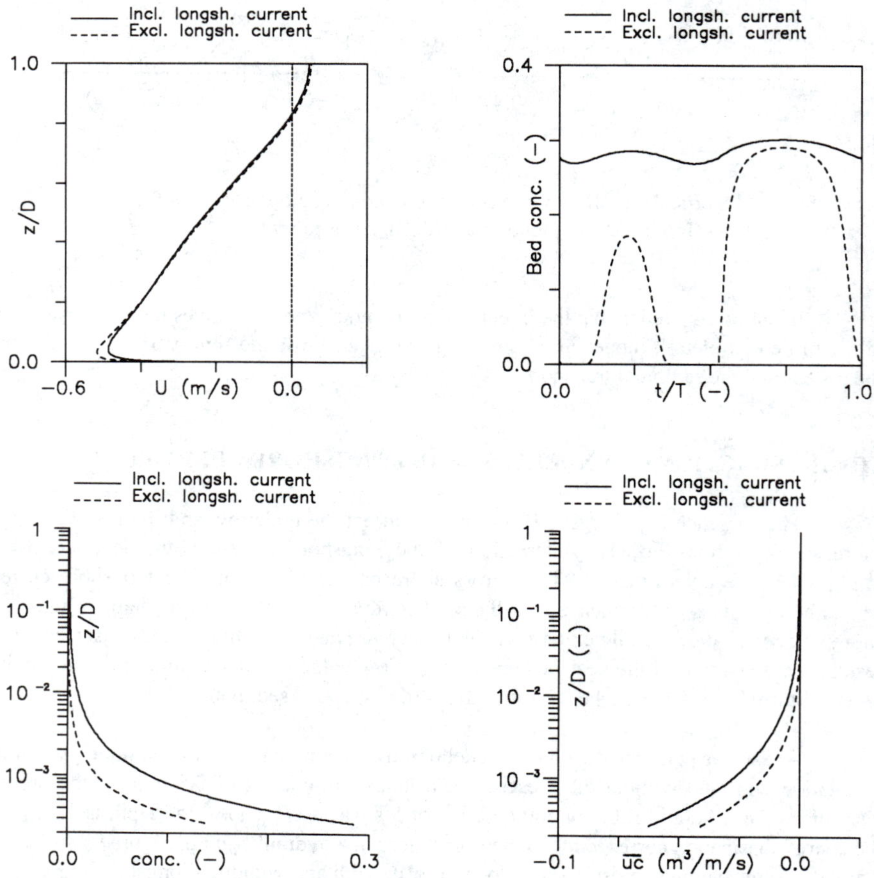

Figure 9 : Interaction of the undertow with a longshore current. a: Mean cross shore flow, b: time variation of the bed concentration, c: time-mean sediment concentration d: time-mean sediment flux.

COMPARISON WITH FIELD MEASUREMENTS

A comparison was made with measured flow velocity profiles at the Ebro delta, Spain, which were reported in Rodriguez et al. (1994). The comparison was made for three levels of calibration. First, the mean cross shore flow velocities were calculated without any calibration. In these simulations, a fully developed roller, where the associated energy dissipation corresponds to the energy loss in a stationary hydraulic jump was assumed.

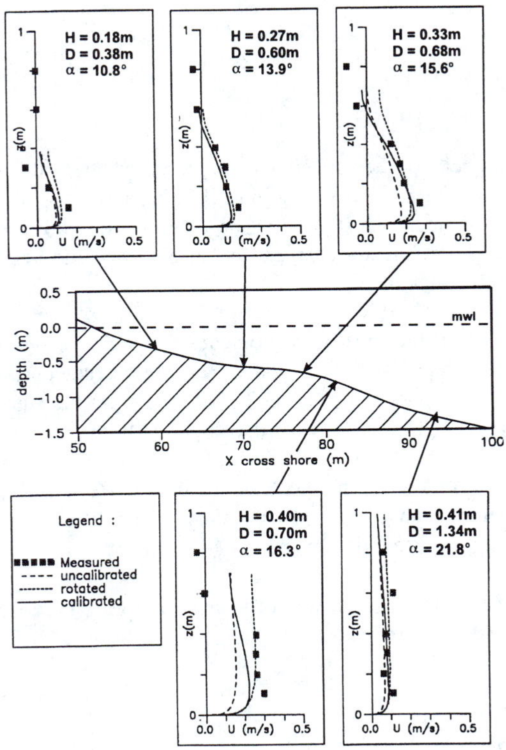

Figure 10 : Comparison of the present model with field measurements from Rodriguez et al. (1994). Three levels of calibration were applied : 1 - uncalibrated, 2 - rotated, 3 - adjusted roller volume.

secondly, the uncalibrated flow velocity profiles were rotated such that the calculated depth averaged velocity matched the measured values. This adjustment accounts for possible 2D effects, e.g. the coast is not perfectly uniform in the longshore direction. Thirdly, a calibration was performed by determining the volume of the surface rollers from the measured mean cross shore flow velocity and the estimated wave drift, which was calculated from sinusoidal wave theory. In all cases the waves were assumed to be unidirectional and regular. The wave orbital motion was calculated by using 1st order Stokes theory. The results for the mean cross shore flow are shown in figure 10. Generally the agreement is satisfactorily. The uncalibrated model underestimates the measured velocities near the bed slightly. The rotation of the coordinate system improves the agreement to a limited degree. The agreement is worst for the sample point located at a cross shore coordinate of approximately 80 m. For the present test, this corresponds to the outer surf zone. The calibrated velocity profiles match the velocity gradients better than the uncalibrated model.

CONCLUSIONS

A numerical model was developed for the flow and the sediment transport under oblique incident waves in the surf zone. The model is based on the integrated momentum equation for the wave boundary layer of Fredsøe (1984). The eddy viscosity consists of contributions from the wave boundary layer, the mean flow and wave breaking. These contributions are calculated separately.

The model based on the integrated momentum equation gives very similar results compared to a more complete turbulence model. For engineering purposes, the integrated momentum approach is recommended because it is able to reproduce the details of the flow and the sediment transport and is very efficient in computing time.

In case of a longshore current the level of turbulence near the bed is increased compared to a situation without current. This results in lower near bed velocities of the undertow and an increased concentration of suspended sediment.

Comparison of the simulated mean cross shore and longshore flow velocities show good agreement with field measurements.

ACKNOWLEDGEMENT

This work was supported by the Danish Technical Research Council (STVF) under the program “Marin Teknik”.

The authors would like to express their gratitude to Andres Rodriquez for supplying the data of the field measurements at the Ebro delta.

REFERENCES

Brøker Hedegaard, I. (1985). Wave generated Ripples and resulting Sediment Transport in Waves. Series paper no 36. Institute of Hydrodynamics and Hydraulic Engineering (ISVA). Technical University of Denmark.

Brøker Hedegaard, I. , R. Deigaard, and J. Fredsøe (1991). Onshore/Offshore sediment transport and morphological modelling of coastal profiles. Proc. Coastal Sediments '91, ASCE, Seattle, pp. 643-657

Deigaard, R. , J. Fredsøe and I. Brøker Hedegaard (1986). Suspended sediment in the surf zone. J. Waterway, Port, Coastal and Ocean Engineering, ASCE, vol. 112 (1), pp 115-128

Deigaard, R and J. Fredsøe (1989). Shear stress distribution in dissipative water waves. Coastal Engineering, ASCE, vol. 13, pp 357-378

Deigaard, R. (1993). A note on the 3-dimensional shear stress distribution in a surf zone. Coastal Engineering, vol. 20, pp 157-171.

Deigaard, R., P. Justesen, J. Fredsøe (1991). Modelling of the undertow by a one equation turbulence model. Coastal Engineering, vol. 15, pp 431-458.

Elfrink, B., I.Brøker Hedegaard, R. Deigaard and J. Fredsøe. (1993). Net sediment transport in non-breaking waves. Progress report no 74. Institute of Hydrodynamics and Hydraulic Engineering (ISVA). Technical University of Denmark. pp 1-10.

Engelund, F and J. Fredsøe (1976). A sediment transport model for straight alluvial channels. Nordic Hydrology, vol 7, pp 293 - 306

Fredsøe, J. (1984). Turbulent boundary layers in wave-current motion. Journal of Hydraulic Engineering, ASCE, vol. 110 (HY8), pp 1103-1120.

Fredsøe, J., O.H. Andersen and S. Silberg. (1985) . Distribution of suspended sediment in large waves. J. of Waterways, Port, Coastal and Ocean Engineering, ASCE, vol. 111 (6), pp 1041 - 1059.

Justesen, P., J. Fredsøe and R. Deigaard. (1986). The bottleneck problem for turbulence in relation to suspended sediment in the surf zone. Proc. 20th Coastal Engineering Conference, Taipei, Vol 2, pp 1225 - 1239.

Justesen, P., E.A. Hansen, I. Brøker and R. Deigaard. (1994). Longshore and cross shore velocity profiles in spilling breakers with an oblique angle of incidence. Progress report no 75. Institute of Hydrodynamics and Hydraulic Engineering (ISVA). Technical University of Denmark. pp 41-54.

Rodriguez A., A. Sánchez-Arcilla, F.R. Collado, V. Gracia, M.G. Coussirat, J. Prieto. Waves and currents at the Ebro delta surf zone: Measurements and modelling. Proc. 24th Int. Conf. on Coastal Eng. 1994. pp. 268-269

CHAPTER 295

A Numerical Model of Sheet Flow Sediment Transport

Sadakazu Katori[1], Masaru Mizuguchi[2] and Akira Watanabe[3]

Abstract

Governing equations are derived for a numerical model to simulate the sediment motion under sheet flow condition. They consist of equations for the conservation of the horizontal momentum, the turbulence energy and the sediment mass. Values of coefficients in the governing equations are determined using the experimental results of Horikawa *et al.* (1982). The mean sediment transport rates during half a period of sinusoidally oscillatory flow are calculated and the computed values agree well with the experimental results of Sawamoto and Yamashita (1986). Computations are also made for the net transport rate in unsinusoidally oscillatory flow with and without superposed steady flow, showing a good agreement with the measurements of Dibajnia and Watanabe (1992). Finally the thickness of the sheet flow moving layers is computed and related to grain-flow parameters.

Introduction

The first author conducted a detailed experiment on the sheet flow sediment transport in an oscillatory flow tank with sinusoidal velocity variations, measured the velocity as well as concentration of the sediment, and evaluated the mean transport rate during half a period of sinusoidal flow (Horikawa *et al.*, 1982). After this experiment that remarkably advanced our understanding of the sheet flow transport, many studies have been performed until now. According to recent studies on the sediment transport, it has been widely recognized that the sheet flow predominates in the surf zone not only during storms but also even under moderate waves in the field (e.g., Watanabe *et al.*, 1991, Dibajnia *et al.*, 1994).

[1]Director, Coastal Engineering Consultants Co., Ltd., Tamura Bldg., 5-23-13, Hongo, Bunkyo-ku, Tokyo, 113 Japan.

[2]Professor, Department of Civil Engineering, Chuo University.

[3]Professor, Department of Civil Engineering, University of Tokyo.

Hence not a few numerical models for the sheet flow sediment movement have been presented so far, but the validty and applicability of most of them are not sufficient enough.

This study aims to develop governing equations and a numerical model that can reproduce time-space changes in the velocity and concentration of the sediment as well as the net transport rate under the sheet flow condition.

Governing Equations Derived for A Numerical Model

Equations (1) to (3) are newly proposed to describe the sheet flow sediment motion.

For the conservation of the horizontal momentum:

$$\partial u/\partial t = (\rho_w/\rho)\left[\partial U/\partial t + \partial\left(K_u \partial u/\partial z\right)/\partial z\right] \tag{1}$$

$$K_u = \begin{cases} \nu + \kappa_{ul} d k^{1/2} + \kappa_{uc}\nu F(c) & (-\infty < z < z_5) \\ \nu + \left[\kappa_{ul} d + \kappa_{uh}(z - z_5)\right] k^{1/2} + \kappa_{uc}\nu F(c) & (z_5 < z < \infty) \end{cases}$$

$$F(c) = \left[1 + 0.183c/(1 - 0.878c)\right]^2 \text{ - - - - - - - - - - - - - (Eilers, 1941)}$$

For the conservation of the turbulence energy:

$$\partial k/\partial t = K_{kp}\left(\partial u/\partial z\right)^2 + \partial\left(K_{kd}\partial k/\partial z\right)/\partial z - \varepsilon \tag{2}$$

$$K_{kp} = \begin{cases} 0 & (-\infty < z < z_5) \\ \left[\nu + \kappa_{kp}(z - z_5) k^{1/2}\right](c_5 - c) & (z_5 < z < \infty) \end{cases}$$

$$K_{kd} = \begin{cases} 0 & (-\infty < z < z_{10}) \\ (\nu + \kappa_{kdl} d k^{1/2})(1 - c) & (z_{10} < z < z_5) \\ \left\{\nu + \left[\kappa_{kdl} d + \kappa_{kdh}(z - z_5)\right] k^{1/2}\right\}(1 - c) & (z_5 < z < \infty) \end{cases}$$

$$\varepsilon = \kappa_{k\varepsilon} g F(c) k^{1/2}$$

For the conservation of the sediment mass:

$$\partial c/\partial t = \partial\left(K_c \partial c/\partial z\right)/\partial z + \partial(wc)/\partial z \tag{3}$$

$$K_c = \begin{cases} \nu + \kappa_{cl} d k^{1/2}|_{z=z_5} & (-\infty < z < z_5) \\ \nu + \left[\kappa_{cl} d + \kappa_{ch}(z - z_5)\right] k^{1/2} & (z_5 < z < \infty) \end{cases}$$

$$w = w_0(1 - c)$$

Here U is the main flow velocity, u the velocity of the fluid-sediment mixture, k the turbulence energy, c the volumetric sediment concentration$(0 \le c \le 1.0)$, t the time, z the vertical distance measured upward from an initial bed surface, ρ_w the fluid density, ρ the density of the fluid-sediment mixture, ν the fluid kinematic viscosity, d the sediment grain size, w_0 the free fall velocity of sediment particles, z_5 and z_{10} are z where $c = c_5 = 0.5$ and $c = 0.99$, respectively, and g is the acceleration of gravity. All the coefficients κ with subscripts are constants.

Equation (1) is a turbulence model for the conservation of the horizontal momentum in the bottom boundary layer. Here it is assumed that the fluid and sediment move together with the same horizontal velocity u. The density of the fluid-sediment mixture $\rho(z,t)$ is defined as

$$\rho = (1-\lambda)c\rho_s + [1-(1-\lambda)c]\rho_w \tag{4}$$

where λ is the porosity of the fluid-sediment mixture at c=1.0, ρ_s the density of the sediment particles. A main point of Eq. (1) is how to express the eddy viscosity K_u. According to Kolmogorov (1941) and Prandtl (1945), K_u is determined as $K_u \sim \ell k^{1/2}$, where ℓ is the length scale of turbulence and k is the tubulence energy. As observed in one of the experimental results of Horikawa *et al.* (1982) in Fig. 1, the vertical distribution of concentration has an inflection point around the vertical distance z_5 where concentration c is c_5. This implies that the length scale in the inner layer is very different from that in the outer layer. Then we assume that the length scale in the inner layer is governed by the grain size d, because the concentration in this layer is very high. On the other hand, the length scale in the outer layer will be set as $z - z_5$, because more free turbulence field is formed in this layer. The grain size $d, z - z_5$, and the turbulence energy k are thus included in the expression of the eddy viscosity K_u. In addition, we employ an equation proposed by Eilers (1941) to describe the chaotic collision of grains.

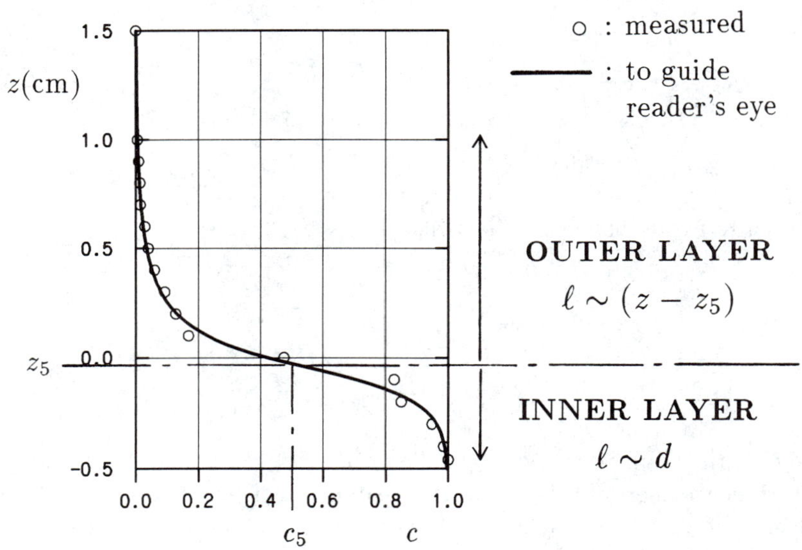

Fig. 1 Distribution of sediment concentration.
(d=0.2mm, T=3.64s, $\hat{U}$=127cm/s, $\theta = 90°$,i.e., $U=\hat{U}$)

Equation (2) shows that time change in the turbulence energy k is determined by three terms; the production, diffusion and dissipation. Regarding coefficients K_{kp} and K_{kd} in Eq. (2), similar formation is made in terms of $\nu, z - z_5, k$ and d for the same reasons mentioned above. A main difference is that there are layers where no production or diffusion of the tuburence energy takes place. This assumption is acceptable, because the concentration in the vicinity of the bed surface is very high and it is nearly impossible for the fluid-sediment mixture with high concentration to form the strong turbulence field. Another point is that the effect of concentration is simply expressed by $(c_5 - c)$ or $(1 - c)$ and that the dissipation rate ε is described in a very simple way. On these assumptions, Eq. (2) becomes a one-equation model of the turbulence energy which is much simpler to solve than two-equation models such as k-ε models.

Equation (3) for the concervation of sediment mass takes similar formulation except that the diffusion coefficient K_c for the inner layer is determined by the turbulence energy k at $z = z_5$, because sediment particles are expected to diffuse even in the inner layer where the magnitude of turbulence energy is very small. In nother words, the reduction of concentration in the inner layer is assumed to take place at the phase of acceleration of the main flow through loosening of the grain arrangement due to shear stresses around $z = z_5$. The reduction rate of fall velocity w with the increase of sediment concentration can be approximately expressed as $(1 - c)$ according to Katori and Homma (1984).

Boundary conditions are

$$\left.\begin{matrix} u = 0 \\ k = 0 \\ c = 1 \end{matrix}\right\} \quad \text{at} \quad z < z_{10} \qquad \left.\begin{matrix} u = U \\ k = 0 \\ c = 0 \end{matrix}\right\} \quad \text{at} \quad z = z_\infty \tag{5}$$

Here z_{10} is the vertical distance where $c = 0.99$, and z_∞ is that of the outer edge of the bottom boundary layer. It has been reported that the thickness of the bottom boundary layer becomes much larger when fluid contains granular sediment particles. For this reason, z_∞ should be set as large as possible in computation. According to trial and error, z_∞ = 5cm is large enough for grain diameter smaller than several milimeters. In addition it has been found that the concentration around $z = z_{10}$ sometimes exceeds 1.0 at the phase of sediment falling and in the worst cases the computaion fails by oscillation or divergence of solutions. We employ the following procedure to avoid such computaional failure. At each time step of calculation of the sediment concentration by Eq. (3), we detect the top vertical distance z_B where the concentration becomes or exceeds 1.0 by searching downward from z_∞, set the concentration below z_B as 1.0, and make the following adjustment for the upper region so as to maintain the conservation of total sediment mass.

$$\begin{aligned} c &\Leftarrow c\ [(V_o - z_B)/(V_i - z_B)] \qquad \text{at } z > z_B \\ V_i &= \textstyle\int_{z-\infty}^{z\infty} c\ dz \end{aligned} \tag{6}$$

where V_o is the volume of sediment per unit horizontal area at the begining of calculation, and $z_{-\infty}$ is the vertical distance of the lower boundary of a calculation domain.

Determination of Coefficients κ

In the governing eqations, there are several coefficients κ with subscripts to be determined. These coefficients are determined using the experimental data presented by Horikawa *et al.* (1982). Figure 2 shows another example of their measurements. The experimental conditions are: d=0.2mm, ρ_s=2.66, T=3.64s and $\hat{U}$=127cm/s. In Fig. 2 the phase θ of the main flow is defined as $\theta = 0°$ for $U = 0$ and $\theta = 90°$ for U=$\hat{U}$. Open circles in Fig. 2 are the measurement data of the horizontal velocity u and concentration c. As for the experimental data of the turbulence energy k, Yamashita *et al.* (1985) measured the velocity by

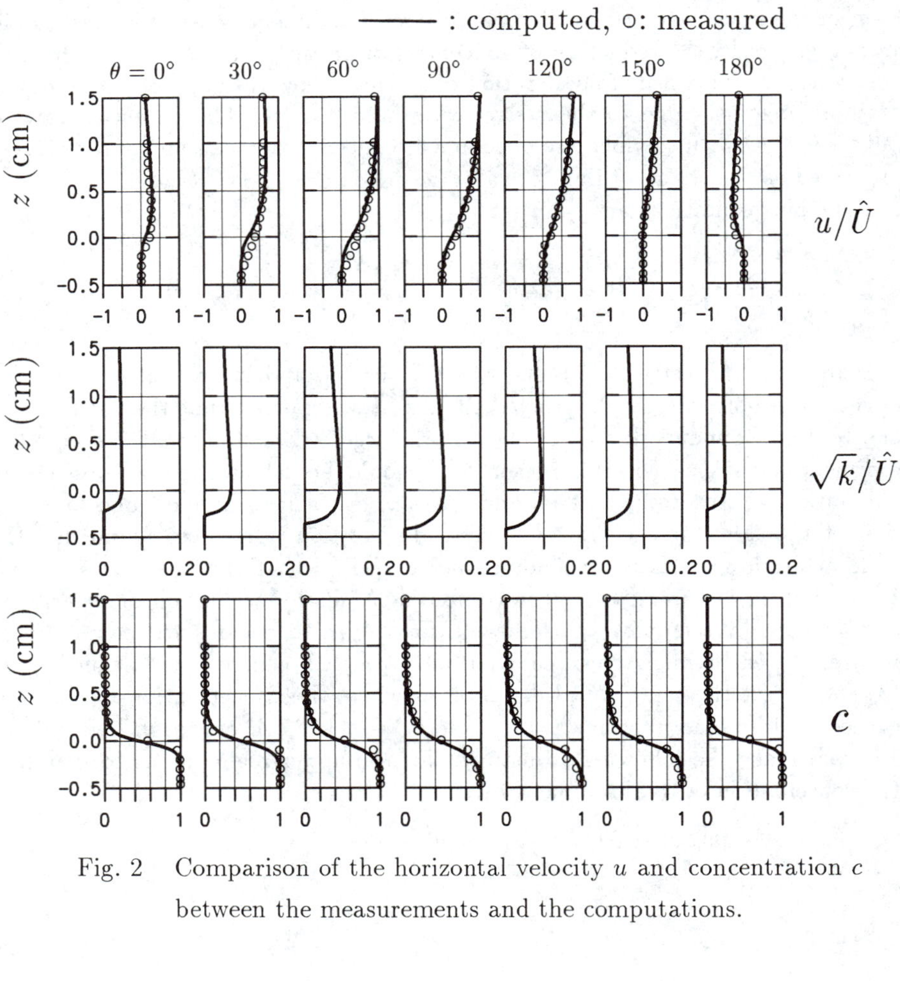

Fig. 2 Comparison of the horizontal velocity u and concentration c between the measurements and the computations.

using an optic fiber velocity meter and reported that the maximum magnitude of square root of the turbulence energy is about 1/10 of the main flow velocity. The computations shown by solid lines in Fig. 2 show a remarkably good agreement with the measurments when we set values of the coefficients κ as follows:

$$\begin{aligned} &\kappa_{ul} = 1.8, \quad \kappa_{uh} = 0.045, \quad \kappa_{uc} = 10.0 \\ &\kappa_{kp} = 0.984, \quad \kappa_{kdl} = 45.0, \quad \kappa_{kdh} = 18.0, \quad \kappa_{k\varepsilon} = 0.006 \\ &\kappa_{cl} = 0.833, \quad \kappa_{ch} = 0.09 \end{aligned} \tag{7}$$

It should be emphasized that these constants have given a good agreement between the computations and the measurements also for all the other cases of Horikawa *et al.* (1982), and that they will be commonly used in the following calculations.

In the numerical computation the initial conditions are u=0, k=0, and

$$\begin{aligned} &c=0 \qquad z > 0 \\ &c=1 \qquad z \leq 0 \end{aligned} \tag{8}$$

The time interval Δt and the vertical grid interval Δz are set as follows:

$$\begin{aligned} &\Delta t = T/2400 \\ &\Delta z = \min(\sqrt{2q\Delta t}, d) \end{aligned} \tag{9}$$

where q is the mean sediment transport rate during half a period estimated by the following formulas presented by Sawamoto and Yamashita (1986).

$$\Phi \equiv (1-\lambda)q/w_0 d = 2.2\ (u_*/w_0)^3 \tag{10}$$

in which Φ is the nondimensional mean transport rate, u_* the friction velocity, and w_0 the free fall velocity. To estimate the friction velocity u_*, we employ the following wave friction factor f_w proposed by Jonsson (1966).

$$\begin{aligned} \frac{1}{4\sqrt{f_w}} + \log\frac{1}{4\sqrt{f_w}} &= -0.08 + \log(a_m/\xi) \qquad (a_m/\xi > 1.57) \\ f_w &= 0.3 \qquad (a_m/\xi < 1.57) \end{aligned} \tag{11}$$

where a_m is the amplitude of main flow orbital movement, and ξ is the bottom roughness, which is set as ξ=d in this study.

Sediment Transport Rate

Computations of the sediment transport rate are conducted to verify the validity and applicability of the present model. Sawamoto and Yamashita (1986) have carried out a series of experiments in a U-shape tube with a rectangular cross section. In the tube, quasi-sinusoidal flow is generated by releasing the initial inbalance of water heads between the tube ends. The period of the oscillatory flow is fixed to be 3.8s owing to the natural frequency of the fluid motion in the tube. The range of the amplitude of main flow velocity is 44.3-125.3 cm/s. They have used three kinds of quartz sands and a coal powder as the bed materials to measure their transport rates. Imitation pearl pellets have also been used to observe the sediment motion through a side glass wall. Figure 3 shows the comparison between $\Phi_{\rm cal}$ and $\Phi_{\rm meas}$, where $\Phi_{\rm cal}$ is the calculated nondimensional mean transport rate during half a period and $\Phi_{\rm meas}$ is that of the measurement data by Sawamoto and Yamashita (1986), Horikawa *et al.* (1982) and Abou-Seida (1965). Although a little scatter is observed especially for the case of Abou-Seida, the overall agreement between $\Phi_{\rm cal}$ and $\Phi_{\rm meas}$ is fairly good.

In addition Sawamoto and Yamashita (1986) proposed a mean transport rate formula during half a period of sinusoidal flow as already shown by Eq. (10). Figure 4 shows that the computed values of Φ shown by symbols agree well with thier formula except for the range of small u_*/w_0, where the sediment movement approaches to its threshold of sediment motion. Here the computation conditions are almost the same with those of Sawamoto and Yamashita, while the range of main flow velocity amplitude $\hat{U}$ has been extented to 0.2-4.0m/s.

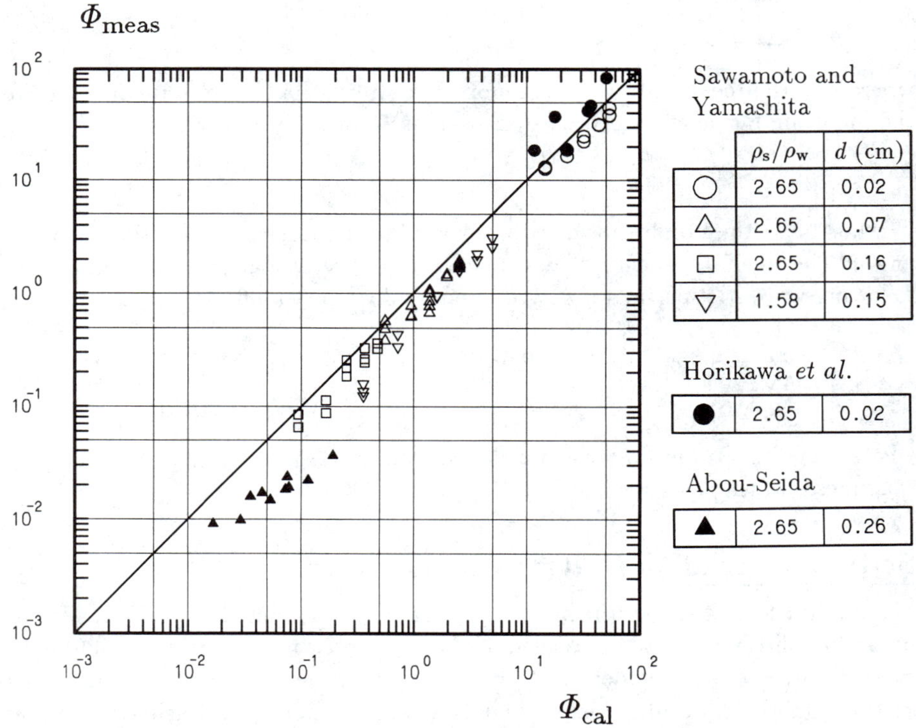

Fig. 3 Comparison between $\Phi_{\rm cal}$ and $\Phi_{\rm meas}$ in sinusoidally oscillatory flow.

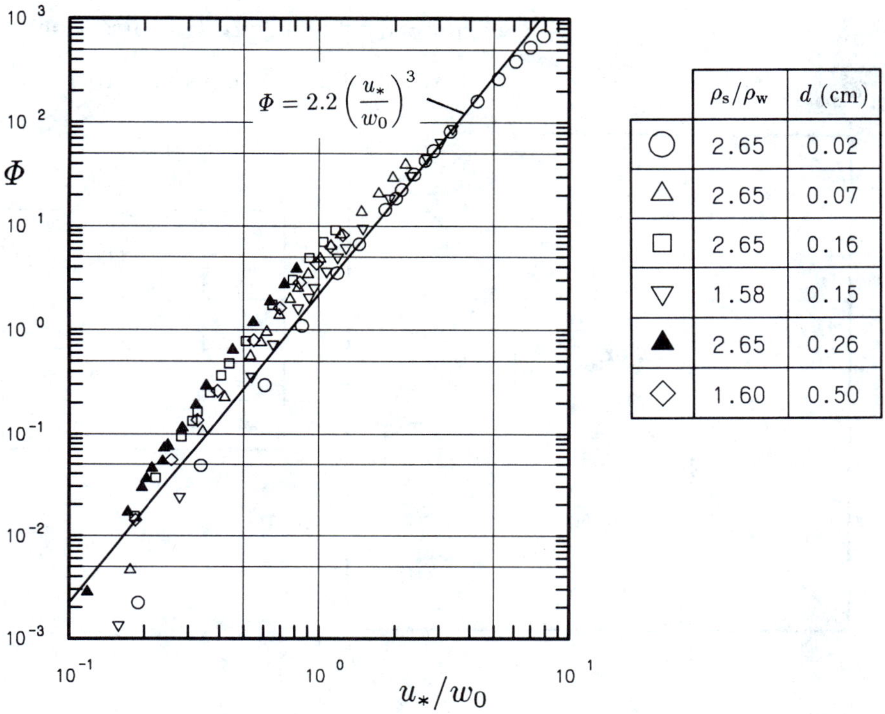

Fig. 4 Relation between Φ_{cal} and u_*/w_0 in sinusoidally oscillatory flow.

Dibajnia and Watanabe (1992) have measured the net rate of sheet flow sediment transport in unsinusoidally oscillatory flow with and without superposed steady flow. The experiments have been carried out in a loop-shape oscillatory/steady flow water tunnel. The flow is driven by a piston to be oscillation with arbitrary temporal variations and by a pump to superpose a steady current on the oscillation. The bet material is the sand with a grain diameter of 0.2mm. They have selected five oscillation periods ranging from 1s to 4s and four conditions of the asymmetry index $U_{max}/\hat{U}$=0.5 to 0.8, where U_{max} is the maximum value of the oscillatory flow velocity U and $\hat{U}$ is the total amplitude of U. In addition the steady flow with four velocities ($\bar{V} \simeq$ -20, -10, 10, 20cm/s) have been superposed on each of the oscillations. In Fig. 5 their data are compared with computations, which shows a good agreement for both the positive and the negative transport rate. Here the results are plotted for the two oscillation periods of 3s and 4s, because the other experimental results for T= 1.0s, 1.5s and 2.0s show large scatter. The reason for such large scatter is to that Dibajnia and Watanabe have not measured the flow velocities directly in coxistent oscillatory

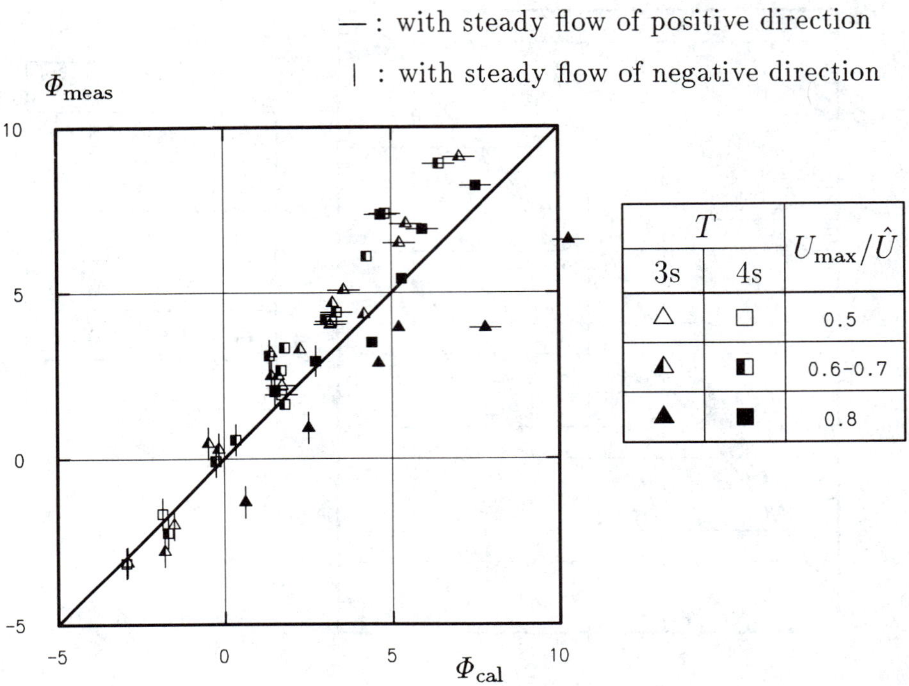

Fig. 5 Comparison between Φ_{cal} and Φ_{meas} in unsinusoidally oscillatory flow with and without superposed steady flow.

and steady flow. The expected flow conditions may not have been obtained when the period of oscillatory flow is short and the interaction between oscillatory flow and steady flow becomes significant in their water tunnel.

Thickness of Moving Layers

The thickness of moving layers is another major concern about the sheet flow sediment transport. In this study, the moving layer thickness δ_B is defined as the maximum change in the elevation of the bed surface, where the sediment particles never move during the whole period of the oscillatory flow. The quantity δ_B is not only important to be used as a boundary condition for analytical treatment of the sheet flow transport but also useful for rough evaluation of the transport rate. It is possible to make such evaluation by multiplying δ_B by the mean concentration and the mean near-bed velocity. Although experimental studies on δ_B have been performed (e.g., Li *et al.*, 1993, Yamashita *et al.*, 1993), most of them are observations through side glass walls where the movement of sediment particles is greatly affected by the side boundary layer. For this reason, we only show computations without comparison with the measurements. Figure 6 shows the computed results, indicating a high correlation with grain-flow parameters. The solid line fitted for the computations reads

$$\delta_B/\sqrt{d\nu/\hat{U}} = 230\ \Psi^{0.891} \tag{12}$$

where d is the grain size, ν the fluid kinematic viscosity, $\hat{U}$ the amplitude of main flow velocity, $\Psi(=u_*^2/sdg)$ is the Shield number, and $s = (\rho_s/\rho_w - 1)$ is the immersed specific gravity of sediment.

Conclusions

This paper has presented a numerical model for the sheet flow sediment transport. The model consists of governig eqations for the conservation of horizontal momentum, the turbulence energy and the sediment mass. Comparisons have been made between the computations and laboratory measurements for vertical distributions of the horizontal velocity and sediment concentration as well as the

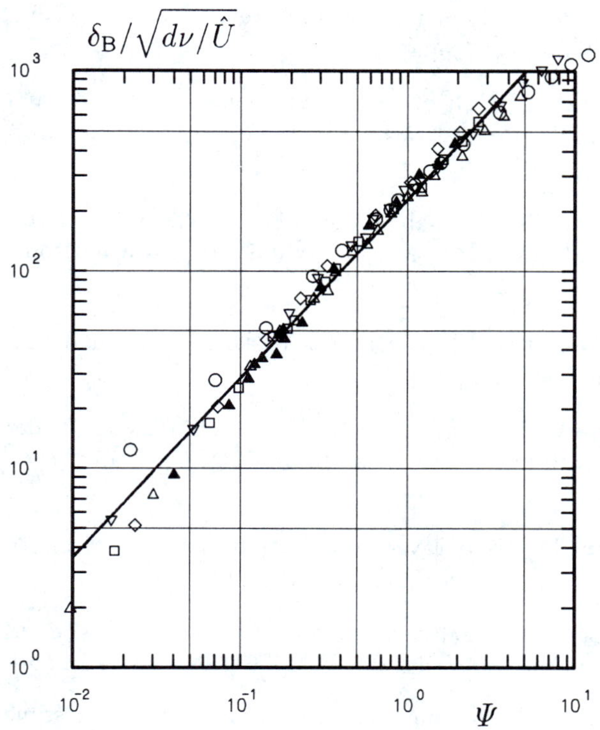

	ρ_s/ρ_w	d (cm)
○	2.65	0.02
△	2.65	0.07
□	2.65	0.16
▽	1.58	0.15
▲	2.65	0.26
◇	1.60	0.50

Fig. 6 Relation between $\delta_B/\sqrt{d\nu/\hat{U}}$ and Ψ in sinusoidally oscillatory flow.

sediment transport rate under a variety of main flow conditions, showing a remarkably good agreement for any of them. It should be emphasized that every coefficient in the governing equations has been set to be a constant respectively common for all the computations. These results indicate the validity and applicability of the present model for estimating the net transport rate of the sheet flow as well as time-space changes in its velocity and concentration over a very wide range of conditions.

Acknowledgement

The authors would like to express their appreciation to Assoc. Profs. M. Dibajnia and Y. Kawahara of University of Tokyo, Japan for their helpful discussion and to Assoc. Prof. T. Yamashita of Hokkaido University, Japan for his greatful kindness to provide us with his experimental data.

References

Abou-Seida, M. M. (1965): Bed load function due to wave action, Univ. Calif., Hydraulic Eng. Lab. Rep., HEL-2-11, 78pp.

Dibajnia, M. and A. Watanabe (1992): Sheet flow under nonlinear waves and currents, Proc. 23rd Int. Conf. on Coastal Eng., ASCE, Venice, Italy, pp. 2015-2028.

Dibajnia, M., T. Shimizu and A. Watanabe (1994): Profile change of a sheet flow dominated beach, Proc. 24th Int. Conf. on Coastal Eng., ASCE, Kobe, Japan, pp. 1946-1960.

Eilers, H. (1941): Die Viskositat von Emulsionenhochviskoser Stoffe als Funktion der Konsentration, Koloid. Z.: 97, pp. 313-321.

Horikawa, K., A. Watanabe and S. Katori (1982): Sediment transport under sheet flow condotions, Proc. 18th Int. Conf. on Coastal Eng., ASCE, Cape Town, South Africa, pp. 1335-1352.

Jonsson, I.G. (1966): Wave boundary layers and friction factors, Proc. 10th Int. Conf. on Coastal Eng., ASCE, Tokyo, Japan, pp. 127-148.

Katori, S. and M. Homma (1984): Fall velocity of sediment in high concentrations, Proc. 28th Japanese Conf. on Hydr., JSCE, pp. 641-646 (in Japanese).

Kolmogorov, A. N. (1941): The local structure of turbulence in incompressible viscous fluid for very large Reynolds numbers, C.R. Acad. Sci., U.R.S.S., 30, pp. 301-305.

Li, L., M. Sawamoto and T. Sato (1993):Experimental study on sediment transport in sheet flow regime, Proc. 40th Japanese Conf. on Coastal Eng., JSCE, pp. 341-345 (in Japanese).

Prandtl, L. (1945): Über ein neues formelsystem der ausgebildeten turbulenz, Nachr. Akad. Wisis, Göttingen, pp. 6-19.

Sawamoto, M. and T. Yamashita (1986): Sediment transport rate due to wave action, Jour. Hydroscience and Hydr. Eng., Vol. 4, No. 1, April, pp. 1-15.

Watanabe, A., T. Shimizu and K. Kondo (1991): Field application of a numerical model of beach topography change, Proc. Coastal Sediment '91, pp. 1814-1828.

Yamashita, T., M. Sawamoto, H. Takeda and G. Yokomari (1985): A study on oscillatory flow and sediment transport at sheet-flow condition, Proc. 32nd Japanese Conf. on Coastal Eng., JSCE, pp. 297-301 (in Japanese).

Yamashita, T., T. Someya and K. Nozaki (1993): The effect of unsteadiness of oscillatory flow and specific gravity of sediment on the mechanism of sheet flow motion, Proc. 40th Japanese Conf. on Coastal Eng., JSCE, pp. 331-335 (in Japanese).

CHAPTER 296

Numerical Modeling of Nearshore Morphological Changes under a Current-Wave Field

Taerim Kim[1] and Hsiang Wang[2]

Abstract

This paper presents a time-dependent 3-D nearshore morphological response numerical model which includes the slope effect, undertow current, wave-borne transport, and transition zone effect. These effects are very important but often not estimated in other sediment transport models. By including the cushioning effect in the water column, which prevents the advection of turbulence generated by wave breaking from the surface to the bottom, this model is shown to be able to approach an equilibrium state of sediment transport. The computed cross-shore and longshore sediment transport rates are calibrated based on the large wave tank data, 3-D basin experiments, and available sediment transport formulas. The predicted changes of bottom topography near structures and inlet in physical model are compared for model verification.

Introduction

Prediction of beach response to the engineering activities is important for assessing the impact of coastal structures and to improve their design. It is also useful for evaluating remedial and mitigation measures. Numerical modeling is clearly an attractive alternative for this purpose and is becoming increasingly viable with the advancement of computational facilities and improved understanding on wave mechanics and sediment transport processes.

There have been two types of approach to predict the 3- dimensional morphological changes in the nearshore zone. The so called 3DBEACH (3-dimensional decoupled model of beach change) by Larson et al. (1989) utilizes the profile change model, SBEACH, for calculating storm-induced beach erosion and recovery and the shoreline change model, GENESIS, for calculating long-term change in shoreline position. Since the profile model is based on equilibrium profile, the combined model also maintains

1 Senior Research Engineer, Ocean Engineering Division, KORDI, Ansan, P.O. Box 29, SEOUL 425-600, KOREA

2 Professor, Department of Coastal and Oceanographic Engineering, University of Florida, Gainesville, FL 32611, USA

the ability to approach an equilibrium bottom configuration under steady wave condition. This is difficult to attain in the models of other type. However, since the submodels are all wave driven, combined current wave effect can not be directly addressed. The other approach computes nearshore topographic change based on local sediment flux balance. This approach usually links a hydrodynamic submodel with a sediment transport submodel. Models of this kind with varying degrees of sophistication have been developed. van Rijn et al. (1989) estimated the sediment transport rate by multiplying the wave-averaged mean vertical sediment concentration by the wave-averaged local horizontal velocity. In their model, the wave-borne transport mechanism is not directly addressed. A model developed by Ohnaka and Watanabe (1990), on the other hand, computes the flow field with considerations of current and wave interaction based on coupled mild-slope wave equation and depth-averaged circulation equations. It calculates the rate of sediment transport as the summation of two energetic mechanisms, one due to the mean current and the other due to waves (Watanabe et al., 1986). Recently, the nearshore circulation model was improved by a 3-D approach, which employed a combined depth-integrated current model and a vertical profile model(De Vriend and Stive, 1987). This improvement when incorporated into the sediment transport model enables one to more realistically represent the 3-D nature both inside and outside the surfzone. Briand and Kamphius (1993) multiplied the time averaged quasi-3D velocities to a time averaged exponential sediment concentration profile to achieve a 3-D sediment flux. This technique of combining a quasi-3D velocity profile with a vertical distribution of suspended sediment concentrations is a promising step to lead to full 3-D models in the future.

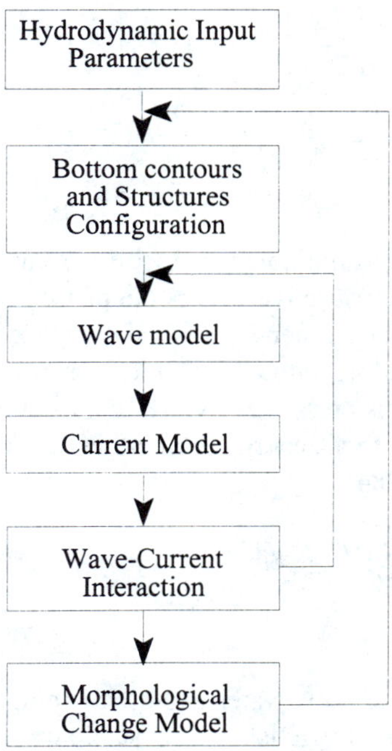

Figure 1 Structure of Nearshore Morphodynamic Model.

The model developed in this study consists of three submodels for calculation of waves, nearhsore currents, and beach changes. At the first step, the initial beach topography and the geometry of the structures for the study area are given as input data. Next, the wave model determines the spatial distributions of radiation stresses and near-bottom orbital velocities for a given incident wave condition. Then, the circulation model computes the mean water surface level and the depth-averaged mean currents using depth-averaged momentum and continuity equations. Inside the surf zone, mean undertow current, transition zone length, and cushioning effect are calculated based on the wave-current model results. Spatial distribution of sediment transport fluxes are computed in the domain of interest. Finally, bottom topography changes are computed based on sediment mass conservation. The change in bottom topography will modify the flow field. Therefore the hydrodynamic model needs to be updated from time to time. Figure 1 shows the computational flow chart of the model.

Wave and Circulation Models

The flow field in the nearshore is computed by depth integrated 2-D hydrodynamic model developed by Winer(1988). The hydrodynamic model consists of wave and circulation models which are fully coupled through interaction terms. First, the wave model determines the spatial distributions of radiation stresses and near-bottom orbital velocities based on the following parabolic wave equation,

$$(C_g\cos\theta + U)A_x + \frac{\sigma}{2}\left(\frac{C_g\cos\theta + U}{\sigma}\right)_x A + VA_y + \frac{\sigma}{2}\left(\frac{V}{\sigma}\right)_y A$$
$$-\frac{i}{2}kC_g(1-\cos^2\theta)A - \frac{i}{2}\left[CC_g\left(\frac{A}{\sigma}\right)_y\right]_y + \frac{W}{2}A = 0 \qquad (1)$$

where, U and V are the depth averaged horizontal velocities in the x (cross-shore) direction and y (longshore) direction, respectively, C is the phase velocity, C_g is the group velocity, σ is the intrinsic angular frequency, A is the complex amplitude , θ is the angle of the wave propagation relative to the x axis, the subscripts x and y denote derivatives in the x and y directions, respectively, W is the energy dissipation coefficient. The coefficient W is related to the energy dissipation due to wave breaking following the work of Dally et al., (1984).

$$D_E = \frac{K}{D}[(E-E_s)C_g] \qquad (2)$$

where, D_E is the energy dissipation rate, D is water depth, K is empirical parameter (K=0.17) , E is the local wave energy density, E_s is the local stable wave energy density that the breaker is striving to attain.

The circulation model computes the mean water surface level and the depth averaged mean currents using depth averaged momentum and continuity equations with the radiation stresses imported from the wave model as the driving force.

The governing equations are given by (Ebersole and Dalrymple, 1979)

$$\frac{\partial U}{\partial t}+U\frac{\partial U}{\partial x}+V\frac{\partial U}{\partial y}+g\frac{\partial \bar{\eta}}{\partial x}+\frac{1}{\rho D}\tau_{bx}-\frac{1}{\rho D}\tau_{sx}$$

$$+\frac{1}{\rho D}\left(\frac{\partial S_{xx}}{\partial x}+V\frac{\partial S_{xy}}{\partial y}\right)+\frac{1}{\rho}\frac{\partial \tau_l}{\partial y} = 0$$

$$\frac{\partial V}{\partial t}+U\frac{\partial V}{\partial x}+V\frac{\partial V}{\partial y}+g\frac{\partial \bar{\eta}}{\partial y}+\frac{1}{\rho D}\tau_{by}-\frac{1}{\rho D}\tau_{sy} \tag{3}$$

$$+\frac{1}{\rho D}\left(\frac{\partial S_{xy}}{\partial x}+V\frac{\partial S_{yy}}{\partial y}\right)+\frac{1}{\rho}\frac{\partial \tau_l}{\partial x} = 0$$

and the continuity equation

$$\frac{\partial \bar{\eta}}{\partial t}+\frac{\partial}{\partial x}(UD)+\frac{\partial}{\partial y}(VD) = 0 \tag{4}$$

where, t is the time; x and y are the Cartesian coordinates in a horizontal plane; U and V are the corresponding velocity components of the nearshore current; $D = h + \bar{\eta}$; h is the still water depth; $\bar{\eta}$ is the elevation of the mean water level due to wave set up/set down; τ_l is the lateral shear stress due to turbulent mixing; τ_{bx} and τ_{by} are the bottom shear stresses; τ_{sx} and τ_{sy} are the surface shear stresses; and S_{xx}, S_{xy}, and S_{yy} are the radiation stress components which arise from the excess momentum flux due to waves. These equations are obtained by integrating the local x and y momentum equations and the continuity equation over the depth of the water column and then time-averaging the results. The governing equations in the circulation model are solved by a matrix analysis using the alternating direction implicit (ADI) scheme (Winer, 1988). In order to treat the wave-current interaction, waves and currents are calculated alternatively.

Inside the surf zone, the 2-D model is inadequate as the model will yield null current in the cross-shore direction, consequently, zero net cross-shore transport. In reality, the current inside the surfzone is highly three dimensional. Field and laboratory observations of surf zone flow show the existence of current that is directed offshore on the bottom, balanced with the onshore flow of water carried by the breaking waves. This offshore-directed steady current near the bed, commonly referred to as undertow, is known to be the most important mechanism causing profile erosion and bar. The depth-integrated discharge of x component by undertow current, Q_u is expressed as

$$Q_u = \frac{\rho g H^2 k_x}{8\sigma} \tag{5}$$

The mean undertow velocity is simply estimated as $\bar{u} = Q_u/(\eta_c + h)$, which is directed offshore.

Sediment Transport Model

The computed current-wave field is used to drive the sediment model which, in turn, computes the morphological changes by utilizing the equation of conservation of sediment material. In the present study, the sediment transport formula contains two parts, bed load and suspended load. The bed load transport is based on an energetic approach driven by mean current and bottom wave orbital velocity. Owing to the asymmetric wave bottom orbital velocity in a wave cycle, this bed load transport by wave orbital velocity has a net onshore component. The suspended load transport which dominates in surfzone is built upon an undertow current. Here, the suspended sediment concentration is related to breaking wave energy dissipation and the transport velocity is the mean undertow current. This component is always directed offshore. The total transport Q is the sum of bed and suspended loads as followings

$$
\begin{aligned}
Q &= q_b + q_s \\
q_b &= A_{bc}(\tau_m - \tau_{cr})U_c/\rho g + A_{bw}(\tau_m - \tau_{cr})U_w/\rho g \\
q_s &= A_s \tau_{turb} Q_u
\end{aligned}
\qquad (6)
$$

, where ρ is the density of water, g is gravity coefficient, U_c is the integrated depth mean wave induced-current, U_w is the maximum orbital velocity at the bottom, Q_u is the discharge by the undertow, τ_m is the maximum bottom shear stress generated by wave and current, τ_{turb} is the turbulent shear stress generated by waves and mean current, τ_{cr} is the critical shear stress under waves and mean current, A_{bc}, A_{bw}, and A_s are calibrated coefficients.

Sediment transport is influenced by the bottom slope as downslope reduces resistance and upslope increases resistance. The down-slope gravitational transport is the most important mechanism to keep the bed from growing indefinitely and , to enable the coastal profile to reach a dynamic equilibrium state. A basic formulation to express the slope gravitational transport component proposed by Horikawa (1988) assumes the following functional form,

$$
\begin{aligned}
q_x &= q_x^1 + \varepsilon_x |q_x^1| \frac{\partial h}{\partial x} \\
q_y &= q_y^1 + \varepsilon_y |q_y^1| \frac{\partial h}{\partial y}
\end{aligned}
\qquad (7)
$$

and in which q_x and q_y are the transport components in x and y directions, respectively, q_x^1 and q_y^1 are their equivalents for horizontal bed, h is the water depth and ε are empirical coefficients

The zone between the incipient breaking and the plunging point is defined as transition zone. The transition zone effect on the wave setup, undertow current, and sediment transport is a subject of increased research over recent years. It showed that the inclusion of the transition zone in profile modeling has an important effect on the predicted profiles, particularly for the bar features. Several approaches have been proposed to evaluate the distance and influence of the transition zone based on wave tank experiments(Galvin (1969)). The present model uses the following formula analyzed from the experimental data(O'Shea et al. (1991)).

$$l_t = 0.556 \tan\beta \; L_b \xi^{-1.465} \qquad \textbf{(8)}$$

in which, $tan\beta$ and L_b are respectively the bed slope and wave length at breaking, and ξ is the surfzone parameter.

Cushioning Effect

Most of the laboratory results as well as field evidence seems to support the concept of beach equilibrium. The profile equilibrium can be reached either when the tangential thrust by fluid is not strong enough to overcome the sand resistance or when the local landward sediment transport is balanced by the local seaward sediment transport. In the latter case, although there could be active sediment movement, the profile remains stable. Including this feature into the model is a difficult subject. One popular approach is to predesignate an equilibrium profile shape. The sediment transport formulas should then be consistent to lead the profile evolution to this equilibrium profile. Kriebel's (1985) storm profile model was of this kind .This type of models can be classified as closed loop model. On the other hand, most sediment transport models do not have a targeted equilibrium profile therefore often fail to reach an equilibrium state. This type of models is known as open loop model. For the open loop model to reach equilibrium, mechanisms must be devised to locally balance the on/offshore transport. The present model is an open loop model. In the formulas presented earlier, inside the surf zone the offshore transport is much too strong to be balanced by the onshore transport component. The reason is that in the formulation, once wave breaks wave breaking-induced turbulence immediately produces a strong bottom stress and the resulting offshore transport is much larger than the onshore transport computed by the bed load transport equation. Clearly, the offshore transport equation needs to be modified. Dally and Dean (1984) pointed out that the expression developed for the sediment concentration profile apparently lacks the ``cushioning" effect which the water column provides in reducing the amount of sediment entrainment as breaking wave form and trough evolve. This cushioning effect is especially important in the stabilization of the bar and trough formation. Almost no work has been done for this cushioning effect. The present model attempted to incorporate this cushioning effect into the model based on the ratio of the depth and wave height at the plunging point as follows,

$$C_{us} = \tanh(7\frac{H}{D})^{700} \qquad \textbf{(9)}$$

where, C_{us} is the cushioning effect, H is the wave height, D is the water depth. Figure 2(a) plots this cushioning coefficient and its effect on sediment transport. As can be seen, when relative water depth increases, the cushioning equation limits the turbulence penetration to the bottom. When wave height is larger than water depth, there is no cushioning effect. However, as the water depth at the trough zone becomes larger, cushioning effect becomes stronger. Finally when water depth reaches 2.5 times the wave height breaking induced turbulence will not reach the bottom. This equation is, of course, purely empirical at this stage with no supporting data. Figure 2(b) shows the effect on sediment transport rate. With the inclusion of this effect, the open loop model presented in this study could reach profile equilibrium without a predetermined profile shape.

Calibration of Sediment Transport Rate

The nearshore sediment transport characteristics under storm wave conditions as predicted by the model are illustrated here. In order to validate the 3-D morphological response model, cross-shore transport rate and longshore transport rate are calibrated separately by using available experiment data or empirical formulas.

Cross-Shore Transport Rate

For the calibration of cross-shore transport rate, two sets of 2-D tank experiment data were used. These included the case of a sand beach backed by a sloping dike tested in the German Big Wave Flume (GWK) and case CE 400 from Saville's large wave tank tests (CE). Comparisons were made between computed values and experimental results both in profile changes and transport rates.

The GWK experiment(Dette and Uliczka, 1986) used sand with a median diameter of 0.33 *mm* and the test profile was subjected to regular wave attack (H=1.5 *m*, T=6 *sec*). Figure 3 shows the comparison between the predicted profiles and the experimental results at run times of 62, 111, and 273 minutes, respectively. The computed wave height distribution across shore

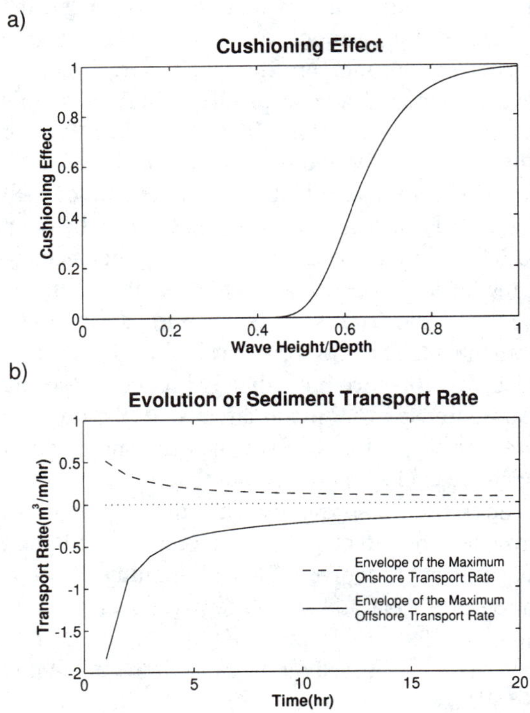

Figure 2 a)Distribution of Cushioning Effect. b)Evolution of Sediment Transport Rate.

is also shown. The waves cut back the foreshore to produce a vertical scarp and a bar formed shoreward of the breaking point which grew and moved offshore with continued wave action. The numerical model satisfactorily reproduced the observed foreshore erosion and main breakpoint bar development. Simulated shoreline retreat and bar growth were initially rapid and gradually slowed as the bar moved offshore to reach a location close to that of the observed bar at the end of the run (20 hr). However, the bar trough is less well reproduced. All smaller features inshore of the main breakpoint bar were not reproduced in the simulations. Figure 4 shows the comparison of the cross-shore transport rate between the profile changes in Figure 3. Initially, a high peak appears near the shoreline as beach material slumps down slope into the foreshore zone in this experiment. These material was transported offshore to form the bar. The seaward changes of the peak transport rate explain the offshore movement of the bar. With the exception at the initial stage, the predicted transport rate distribution is in good agreement with the laboratory data. At the initial stage the experimental values were larger than predicted. There are many factors that could contribute to the difference such as slumping effect mentioned earlier. It was also found later in small scale experiments conducted at UF that the profile erosion particularly in the dune region was much more severe in the initial stage if the beach is dry and loosely compacted as opposed to wet and well compacted. The best overall fit of both profile

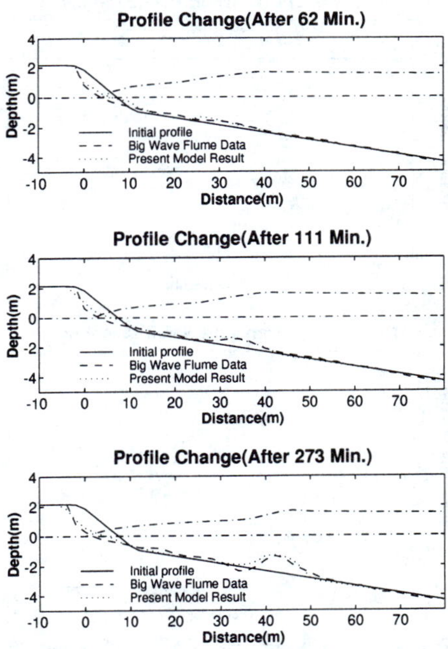

Figure 3 Comparison between Computed and Measured Beach Profiles for t= 62, 111, 273 minutes. Data from Dette and Uliczka (1986).

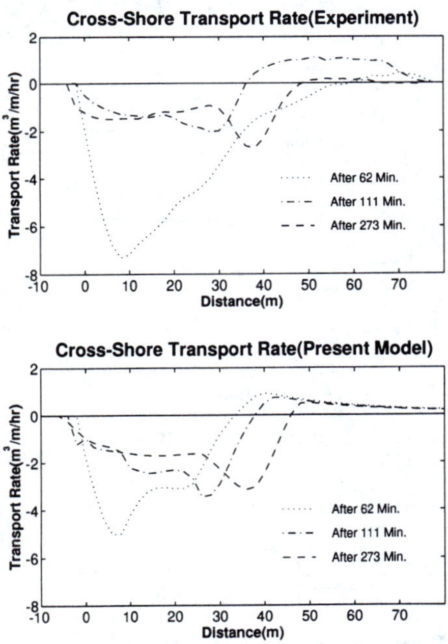

Figure 4 Comparison of Cross-Shore Transport Rates between Experiment and Present Model for t=62, 111, 273 minutes. Data from Dette and Uliczka(1986)

and transport rate was obtained by using A_{bw}

in a range from 0.05 to 0.07 and A_s from 1.0 x 10^{-5} (m^2/N) to 1.5 x 10^{-5} (m^2/N). With these determined A_{bw} and A_s values, the model was used to simulate the CE case CE400. Figure 5 shows the comparison of measured and calculated profiles. The test conditions were: initial uniform slope = 1/15; grain size = 0.22 *mm*; wave height and period of 1.62 *m* and 5.6 *sec*. The numerical and test results are shown for simulation times of 1, 3, and 5 hours. In this case, the bar development was also well predicted except at the initial stage. However, the profile change near the shoreline shows very different results between the experiment and the model. In experiment, there was strong erosion at the foreshore region above the water level, but this erosion was limited near the shoreline in the numerical model. The spatial distribution of the transport rate from the CE tests were very different from that of the GWK tests. Here in the CE case, peak transport initially occurred near the breaking point and did not show any onshore transport outside the breaking point (The numerical model results did show a very small onshore transport component). Also, in the CE experiments, the transport in swash zone apparently played an important role to cause shore face erosion. The numerical model does not have an appropriate swash zone transport mechanism. Therefore, it was unable to reproduce the large erosion near the shoreline in the experiment.

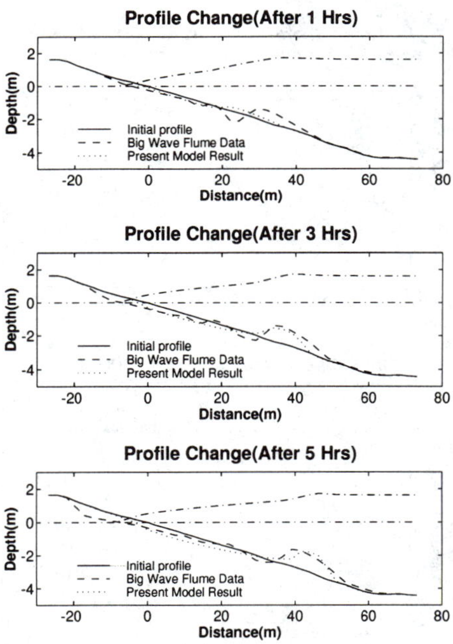

Figure 5 Comparison between Calculated and Measured Beach Profiles for t= 1, 3, 5 hours. Data from Saville(1957).

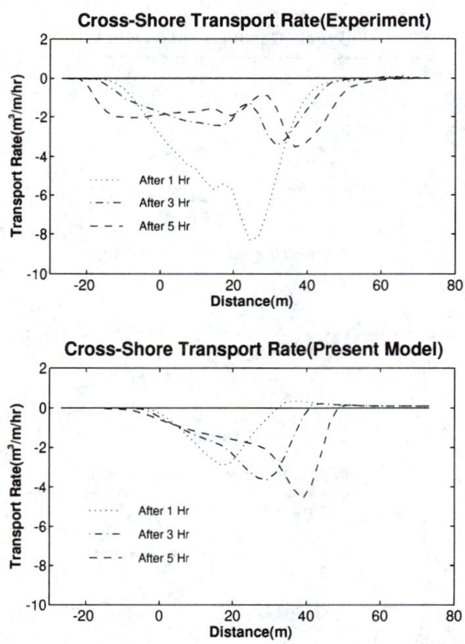

Figure 6 Comparison of Cross-Shore Transport Rates between Experiment and Present Model for t=1,3,5 hours. Data from Saville(1957)

Longshore Transport Rate

Longshore sediment transport plays a very important role particularly in long term beach evolution. The longterm evolution of many coastlines is the result of slight gradients in the longshore transport rate. Available data suitable for calibration and verification of longshore transport rate are scarce and mostly limited to 2-D cases. The 2-D data will not yield information on A_{bc}. In this section, the characteristics of longshore transport rate are investigated by comparing with CERC formula and the coefficient A_{bc} was calibrated based on a 3-D basin experiment. The CERC's littoral drift formula (Shore Protection Manual, 1984) has the following form as,

$$Q = \frac{K(EC_g \cos\theta \sin\theta)_b}{\rho g(s-1)(1-p)} \tag{10}$$

where Q is the volumetric longshore transport rate, E is wave energy, C_g is group velocity, s is specific gravity, p is porosity and K is an empirical transport coefficient. The formula is based on energetic concept with empirically determined coefficient based on field evidence.

In the present model, the sediment transport formulas as presented are of very different nature. However, for lack of laboratory and/or field data for verification, CERC's formula is used here as a bench mark to test the model. A case of 2-D beach that has an initial profile of a uniform slope (1/20) was used here to compare the characteristics of the present model and CERC's formula. Figure 7(a) shows the longshore transport distribution across the surf zone as calculated by the model at different wave angles under the same wave height of 1.5 m. The transport patterns are all bell-shaped with a uni-modal peak inside the breaking point. This shape is consistent with observations on beaches of more or less uniform slopes excluding the swash zone (dual peaks were common if swash zone is included). It is also shown that the longshore transport rate increases drastically as

a) b)

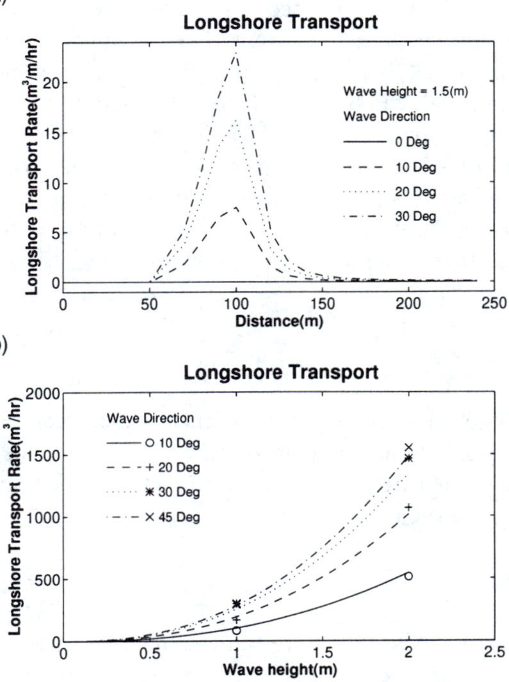

Figure 7 a)Distributions of Longshore Transport Rate in Different Wave Directions
b)Comparison of Longshore Transport Rates between CERC Formula and Present Model in Different Wave Directions.

wave angle increases. The total transport rate is obtained by integrating the volume under the curves. These total transport rates are compared with the CERC's formula in Figure 7(b) which plots the transport rate *vs.* wave height for different wave angles. The comparisons are good for wave up to 20°. For larger wave angles, the model yields values slightly higher than the CERC's formula. The coefficient A_{bc} in the transport equation given by Equation (6) is equal to unity and 0.77 was used for the K value of CERC formula in this comparisons. As discussed earlier, most experiments and field measurements were conducted for cross-shore transport, and there is very little information on longshore transport rate. Therefore, a plain beach movable bed physical model was constructed in a wave basin to investigate the sediment transport patterns , specially longshore transport rate. The experiments were performed in the 16x23 *m* 3-D wave basin of the Coastal Engineering Laboratory of the Department of Coastal and Oceanographic Engineering at University of Florida. The designed initial beach profile consists of a flat backshore, a steep-sloped foreshore, and a mild-sloped offshore, and has simple straight shoreline and parallel offshore contours.
After 80 minutes run time, a total of 0.28 m^3 sand was collected at the sediment trap channel located in the downdrift boundary. This longshore transported material was used to calibrate the A_{bc} value in the numerical model and to check the K value in CERC's formula. By applying wave height, and angle used in the experiment to the CERC's formula, it was found that K=0.23 from the results. This value of the proportionality coefficient obtained in this study is smaller than 0.77 in Komar and Inman's formula and is rather close to the value in Sato and Tanaka (1966)'s formula. It is noted here that many one line models utilizing CERC's formula the K coefficient was adjusted downward to the range of 0.05 - 0.4 based on model calibrations with physical experiment data. Here, the numerical longshore sediment transport is calibrated with measured total longshore transport quantitatively. The best fitting value was found to be A_{bc} = 0.1. This value is also smaller than the previous value which gave good agreement with CERC's formula.

Performance Test

The model performance is demonstrated here with two types of man-made structures including shore perpendicular structures(groins) and offshore breakwaters. The purpose is not for model calibration or verification as there is no available data in hand. Rather, model utilizations are illustrated with realistic coastal structures.

Groins

Groins are shore perpendicular structures which are built to intercept littoral transport from updrift. They are used to trap sand locally or as end structures to stabilize sand placed in conjunction with beach nourishment projects. For trapping sand, a series of groins is often used. The presence of groins is generally known to cause accretion on the updrift end and erosion on the downdrift end. The ability to be able to predict morphological changes associated with groin structure is undoubtedly useful in coastal engineering. Perlin and Dean (1985) developed an N-line model using a simple wave refraction and diffraction scheme and longshore sediment transport equation to modify the topography. Kraus et al.(1994) investigated parameters governing beach response to groins and incorporated them into the so-called GENESIS model that can be used to predict shoreline changes in the presence of groins. In the present model, the presence of a groin is reflected in the boundary conditions. The boundary conditions for the groin in the sediment transport model are given as,

$$q_y(I,\ J_{groin}+1) = 0, \quad \text{if} \quad q_y(I,\ J_{groin}+1) > 0$$
$$q_y(I,\ J_{groin}-1) = 0, \quad \text{if} \quad q_y(I,\ J_{groin}-1) < 0 \qquad \textbf{(11)}$$

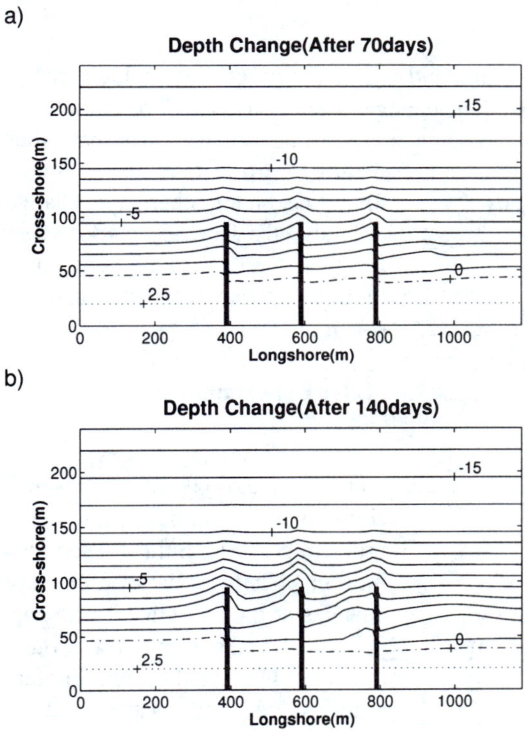

Figure 8 Depth Changes after a)70 days, and b)140 days on the Plain Beach with Three Groins for 0.5 m, 8 sec, and 10 degree Incident Waves.

where q_y is the transport rate in y direction and J_{groin} is the Jth grid column which has the groin.

Figure 8 shows the morphological changes for the case of a three-groin after 70 days and 140 days, respectively, when wave of 0.5 m, 6 sec approaches the beach at 15 degree angle. It is shown that shoreline and bottom contours advances at the updrift of first groin and retreats at the down drift of the last groin. In each compartment, erosion occurs at the updrift end and accretion takes place at the downdrift end. At present, there is a lack of reliable data especially on the morphological effects of groins to compare with numerical results.

Breakwater

An offshore breakwater is generally a shore parallel structure designed to protect the beach behind the structure against severe erosion. Behind the breakwater wave height is reduced and circulation cells are generated which draw sediment into the sheltered area. A salient feature soon appears which grows into a tombolo and sometimes becomes attached to the breakwater. Many numerical models, some based on rather artificial mechanisms are able to produce the described topographic changes. Hanson and Kraus (1990) employed a numerical model (GENESIS) to investigate the various depositional types in the lee of a single detached breakwater. Suh and Hardway (1994) developed a one-line numerical model for predicting shoreline change in the vicinity of multiple breakwaters and compared with the field data. The boundary condition for the breakwater in the present sediment transport model is as follows,

$$q_x(I_{break}+1,\ J) = 0, \quad \text{if} \quad q_x(I_{break}+1,\ J) > 0$$
$$q_x(I_{break}-1,\ J) = 0, \quad \text{if} \quad q_x(I_{break}-1,\ J) < 0 \qquad \textbf{(12)}$$

where q_x is the transport rate in x direction and I_{break} is the Ith grid row which has the breakwater. Figure 9 shows the topographic changes after 2 days and 7 days under the normal incident wave condition with 1 m height and 8 sec period. It is clear that salient feature grows rather rapidly. Shoreline, on the other hand, changes only slightly. The absence of swash transport mechanism might be one of the reasons that shoreline change is so slight.

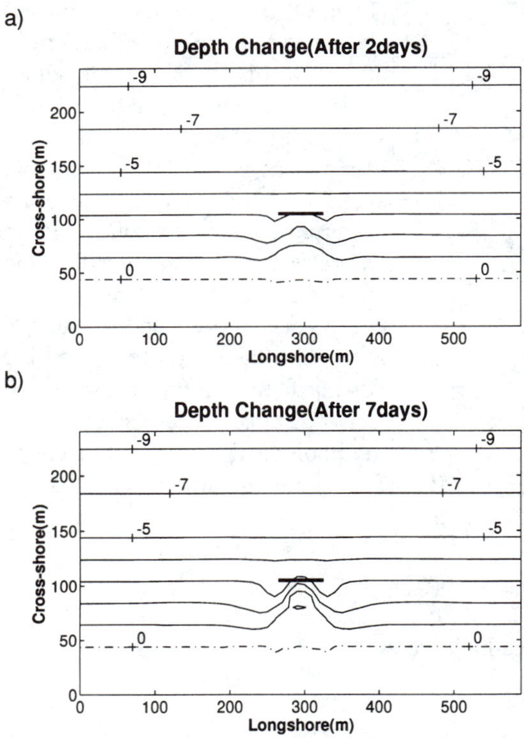

Figure 9 Depth Changes after a)2 days, and b)7 days on the Plain Beach with Breakwater for 1 m, 6 sec, and Normal Incident Waves.

Inlet Experiment

The sedimentary processes in the vicinity of a tidal inlet present a complex dynamic interaction problem between fluid and sediment motion. There exist water level changes at the shoreline by tide and periodical tidal currents in and out of the inlet. This tidal current interacts with wave and wave induced longshore current. Under the combined effects of waves, longshore current, tidal current and coastal structures, the behavior of the sediment movement and morphological evolution in the inlet region is a very complicated process and the current knowledge is extremely limited. Laboratory modeling using movable bed-material is still a viable tool despite its limitations such as expensive cost, time consuming, and scale effects (Wang et al., 1992). Numerical model on inlet morphology is at its infancy. An attempt is made here to apply the present model to an inlet beach system. On a separate study(Wang et al., 1995), an inlet-beach physical model was constructed in the wave basin facility in the Coastal and Oceanographic Engineering Laboratory at the University of Florida. The inlet beach model consists of an idealized inlet on a plain beach with profile identical to that in the plain beach model. The inlet has two parallel jetties extending offshore with the updrift jetty twice as long as the downdrift jetty. These jetties are made of concrete blocks and impervious to sediment transport. The experiments were carried out under the condition of 7.5 degree oblique waves. The test conditions are as follows:

Wave condition: 8 cm wave height, 1 sec wave period, 7.5° wave direction
Current condition: 0.14 m/sec ebb current, 0.1 m/sec flood current
Tidal range(between flood and ebb): 3 cm

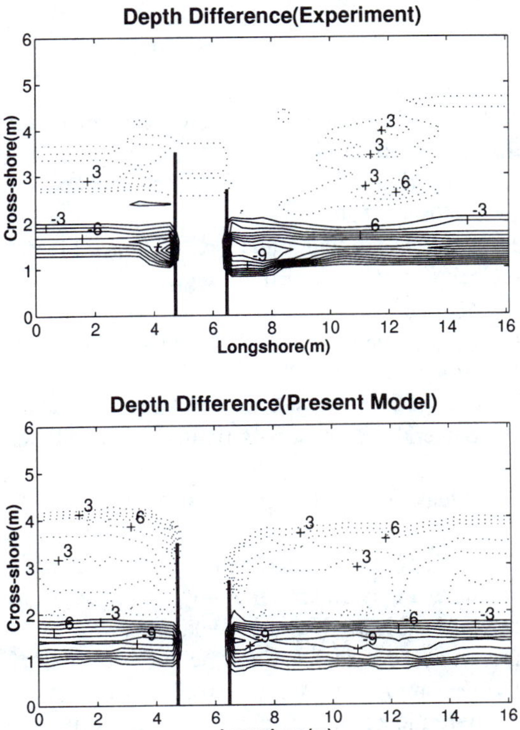

Figure 10 Comparison of Bathymetric Changes after 160 minutes between Experiment and Numerical Model for 8 cm, 1 sec, and Oblique Incident Waves.

The flood and ebb tidal conditions were simulated alternatively in the model at every 40 minute intervals by holding the high and low water levels, respectively, and reversing tidal currents in the inlet. Figure 10 shows the changes of the topographies by plotting the difference between the initial contours and the contours after 160 minutes. The numerical model appears to be able to reproduce reasonably well some of the general features found in the physical model, as the locations of erosion and accretion, the position and size of breakpoint bars and the shoreline offset. Detailed topographic comparisons are still difficult and may not be too meaningful owing partly the three dimensional effects produced in the physical model.

Results

1) The prediction of cross-shore transport from model was compared to the evolution of beach profiles and sediment transport rate distributions in the 2-D German's Big Wave Flume tests and CERC's large tank tests. The model is capable of predicting the growth and movement of main breakpoint bars and beach processes with reasonable reliability.
2) The comparison of longshore transport rates between present model and CERC formula shows close agreement for different wave angles when the coefficient A_{bc} =1 is adopted in the model as compatible with K=0.77 in CERC's formula. A different calibration using 3-D basin experiments yields A_{bc} =0.1 and K=0.23.
3) An application of the model to different coastal structures including groins and breakwaters shows the model is capable of producing the general scouring, accretion and erosion features found in nature. The model was tested with inlet beach movable bed physical model experiment. The comparison of depth changes shows reasonable agreement between numerical model and physical experiment.

References

Briand, M.-H.G. and Kamphuis, J.W., 1993a. ``Waves and currents on natural beaches: a quasi 3-D numerical model", Coastal Eng., 20:101-134.
Briand, M.-H.G. and Kamphuis, J.W., 1993b. ``Sediment transport in the surf zone: a quasi 3-D numerical model", Coastal Eng., 20:135-156.
Dally, W. R. and Dean, R. G., 1984. ``Suspended sediment transport and beach profile evolution", Journal of Waterway, Port, Coastal and Ocean Engineering, Vol.109, No. 4. pp.401-405.
Dally, W. R., Dean, R. G. and Dalrymple, R. A., 1984. ``A model for breaker decay on beaches", Proc. 19th Coastal Eng. Conf., Houston, pp.82-98.
Dette, H.H. and Uliczka, K., 1986.``Seegangserzeugte Wechselwirkung Zwischen Vorland und Vorstrand und Kustenschutzbauwerk", Technisher Bericht SFB 205/TPA6, University Hannover.
De Vriend, H.J. and Stive, M.J.F., 1987. ``Quasi-3D modeling of nearshore currents", Coastal Eng., 11;pp.565-601.
Ebersole, B.A. and Dalrymple, R.A., 1979. ``A numerical model for nearshore circulation including convective accelerations and lateral mixing", Ocean Engineering Report No. 21, Dept. of Civil Eng., Univ. of Delaware, Newark, Delaware.
Glavin, C.J., 1969. ``Breaker travel and choice of design wave height", Journal of Waterway, Port, Coastal and Ocean Engineering, ASCE, Vol.95, No. 2. pp.175-200.
Hanson, H. and Kraus, N.C. 1990. ``Shoreline response to a single transmissive detached breakwater", Proc. 22th, ICCE, ASCE pp. 2034-2046.
Horikawa, K., 1988., ``Nearshore dynamics and coastal processes ", University of Tokyo Press, pp. 522.
Kraus, N.C., Hanson, H., and Blomgren, S.H. 1994., ``Modern functional design of

groin systems", Proc. 24th, ICCE, ASCE, pp.1327-1342.
Kriebel, D. L., 1985. ``Numerical simulation of time-dependent beach and dune erosion",Coastal Eng., Vol 9, pp. 221-245.
Larson, M., Kraus, N. C. and Byrnes, M. R., 1989. ``SBEACH: Numerical Model for Simulating Storm-Induced Beach Change", Technical Report CERC-89-9, Coastal Engineering Research Center, US Army Corps of Engineers.
Ohnaka, S. and Watanabe, A., 1990. ``Modeling of Wave-Current Interaction and Beach Change", Proc. 22nd Coastal Eng. Conf., Delft, The Netherlands, pp. 2443-2456.
O'Shea, K.F., Nicholson, J. and O'Connor, B.A., 1991., ``Analysis of existing laboratory data to find expression for the transition zone length", Report MCE/1, Dept. Civil Eng., University of Liverpool, UK.
Perlin, M., and Dean, R.G., 1985. ``3-D model of bathymetric response to structures," Journal of Waterway, port, coastal and ocean engineering. Vol.111, No.2, pp.153-170.
Sato, S. and Tanaka, N., 1966. ``Field investigation on sand drift at Port Kashima facing the Pacific Ocean", Proc. 10th Coastal Eng. Conf., pp. 595-614.
Saville, T., 1957. ``Scale Relationship for Equilibrium Profiles", Int. Assoc. of Hydraulic Research, 1957
Shore Protection Manual, 1984(4th edition). Coastal Engineering Research Center, U.S. Army Corps of Engineers, Vicksburg.
Suh, K.D. and Hardaway C.S. 1994., ``Calculation of Tombolo in Shoreline Numerical Model", Proc. 24th, ICCE, ASCE, pp. 2653-2667.
van Rijn, L.C. and Meijer, K., 1989. ``Three dimensional mathematical modeling of suspended sediment transport in currents and waves", IAHR Symp., Copenhagen, Denmark.
Watanabe, A., Maruyama, K., Shimizu, T., Sakakiyama, T., 1986. ``Numerical prediction model of three dimensional beach deformation around a structure", Coastal Eng. in Japan. 29, pp.179-194.
Wang, H., Lin, L., Zhong, H., and Miao, G., 1992, ``Sebastian Inlet Physical Model Studies, Part II --- Movable Bed Model", Coastal and Oceanographic Engineering Department, University of Florida. UFL/COEL--91/014.
Wang, H., Lin, L., Zhong, H., and Wang Xu, 1995, ``Laboratory Mobile Bed Model Studies on Inlet , Part II---Ebb Tidal Shoal Evolution Process", Coastal and Oceanographic Engineering Department, University of Florida. UFL/COEL--95/021.
Winer, H.S., 1988. ``Numerical Modeling of Wave-Induced Currents Using a Parabolic Wave Equation", Ph.D. Dissertation, Coastal and Oceanographic Engineering Dept. Univ. of Florida.

CHAPTER 297

An Analysis of Particle Saltation Dynamics

Michael R. Krecic[1] and Daniel M. Hanes[2]

Abstract

A two-dimensional particle saltation model for unidirectional flow is applied to simulate the motion of single particles. The equations of motion include added mass, gravity, drag, shear lift, Basset history, and Magnus or spin lift forces. A sensitivity analysis is performed on the forces, initial lift-off speeds and angles, and for different size particles. The Magnus lift force is found to have a significant effect on a particle's trajectory for coarse sand sized and larger particles. The shear lift and Basset history forces cause particles to saltate farther. Most of the forces vary with particle size. The model predictions compare favorably to observations if appropriate initial conditions are assumed.

Introduction

The term saltation was first used by Gilbert (1914) and comes from the Latin word "saltare" meaning to leap or dance. Saltation is analogous to a ballistic trajectory in the sense that trajectories are smooth and not strongly influenced by turbulent fluctuations. However, hydrodynamic forces such as lift and drag significantly influence the particles' trajectory (Nino, Garcia, and Ayala, 1992).

Fernandez Luque and van Beek (1976) performed experiments in a flume with different bed slopes. They were able to measure the mean critical bed shear stress for the initiation of motion, rate of bedload transport, average particle velocity, and the average length of individual saltations. This was accomplished through the use of high speed photography. Two different sediment types, gravel and magnetite, were studied. When they compared their model results, they concluded that a lift force was needed to explain the observed saltation characteristics.

---

[1] Research Assistant, Coastal and Oceanographic Engineering Department, University of Florida, Gainesville, FL 32611

[2] Associate Professor, Coastal and Oceanographic Engineering Department, University of Florida, Gainesville, FL 32611

There is some debate as to what angles the saltating particles leave the bed. Owen (1964) examined particle saltation in air. Owen suggested that those particles that leave the bed nearly vertically with a certain initial speed will saltate to a higher elevation than those with other initial angles. White and Schulz (1977), in contrast, observed that particles in air eject at angles ranging from 30 to 70 degrees.

Other researchers such as Murphy and Hooshiari (1982), van Rijn (1984), Wiberg and Smith (1985), Nino and Garcia (1992), and Lee and Hsu (1994) have tried to derive a set of equations to describe the motion of a particle in saltation from bed ejection to bed impact based on the fluid forces. This paper focuses on developing a complete equation of motion from a Lagrangian perspective and provides a detailed analysis of the model parameters.

Particle Saltation Model

This model simulates the trajectory of a single saltating sphere in a steady state, unidirectional flow. An equation of motion is developed and evaluated using previous experimental observations. Models based on only drag and gravity proved insufficient, so other forces have been included, as will be described below. Some of these other forces are found to be significant while others can be ignored in certain situations. What follows is a brief description of the relevant forces.

The effect of gravity is usually written as a submerged weight,

$$\mathbf{F}_G = (\rho_s - \rho) g V \tag{1}$$

where V is the volume of the particle. Obviously, this force increases with increasing particle size because of its dependence on the volume of a particle.

The added mass force arises from the relative accelerations of the particle and the fluid. A submerged body induces accelerations on a fluid if the body is moving with an acceleration relative to the surrounding fluid. The particle can be thought of as having an 'added mass' of fluid attached to its own mass when it accelerates relative to the surrounding fluid (Patel, 1989). The added mass force is given by Auton et al. (1988) as

$$\mathbf{F}_A = -\rho C_M V \left(\frac{d\mathbf{u}}{dt} - \frac{Du_f}{Dt} \right) \tag{2}$$

where **u** is the particle velocity, u_f is the fluid velocity, and C_M is the added mass coefficient. The added mass coefficient is defined as a ratio of the additional mass of fluid that is accelerated with the particle to the mass of the displaced fluid by the particle. For a sphere, C_M equals 0.5.

A drag force is a net force in the direction of the fluid relative to the body due to pressure and viscous forces on the body. The drag force on a particle may be written as

$$\mathbf{F}_D = C_D A \frac{1}{2} \rho \mathbf{V}_r^{\,2} \tag{3}$$

where C_D is the coefficient of drag, A is the cross-sectional area of the particle normal to the force, and $\mathbf{V}_r$ is the relative velocity (Fredsoe and Deigaard, 1992). The drag coefficient is a strong function of Reynolds number and shape; thus, it is generally not constant. There have been many empirical formulas for C_D such as those found by Graf (1984) and Morsi and Alexander (1972) which provide C_D as a function of particle Reynolds number. The following formula is used for this analysis:

$$C_D = \frac{24}{\mathrm{Re}} + \frac{7.3}{1+\sqrt{\mathrm{Re}}} + 0.25 \tag{4}$$

where Re is the Reynolds number, $\mathrm{Re} = \dfrac{\mathbf{V}_r d}{\nu}$

In shear flow the particle develops a pressure gradient across it which results in a lift force, which is commonly called the shear lift force. This phenomenon can be attributed to the Bernoulli effect where the lift force acts in the direction of the velocity gradient. The shear lift force may be written as

$$F_L(shear) = C_L A \frac{1}{2} \rho \left(u_{\Delta Top}^2 - u_{\Delta Bot}^2\right) \tag{5}$$

where C_L is the lift coefficient and A is the cross-sectional area of the particle normal to the force. The $u_{\Delta Top}$ and $u_{\Delta Bot}$ are the relative velocities evaluated at the top and bottom of the particle, respectively (Wiberg and Smith, 1985).

The lift coefficient has been related to the drag coefficient by Chepil (1958). He conducted experiments involving evenly-spaced hemispheres and allowed wind to flow over them. He then proceeded to calculate drag, lift, and the ratio of lift to drag for different wind speeds. It was determined that the ratio was approximately equal to 0.85; therefore we assume here that C_L equals 0.85 times C_D.

A. B. Basset (1888) first acknowledged that a particle's history had a role in the present particle path; hence, the Basset history force bears his name. Mei (1995) described the force as

> ... derived from the diffusion of vorticity generated at the surface of the particle at a rate proportional to the particle's relative acceleration. Since the diffusion rate is finite, this means that the force is dependent on the history of the particle motion.

Consider a sphere in a steady fluid and then instantaneously increasing the flow to some higher value. It takes some time for the boundary layer on the sphere to adjust to the new flow intensity. This time is accounted for through the Basset history force. The Basset force is defined as

$$\mathbf{F}_B = 6\rho\sqrt{\pi\nu}\left(\frac{d}{2}\right)^2 \int_0^{t_s} \frac{\frac{d\mathbf{u}}{dt} - \frac{d\mathbf{u}_f}{dt}}{\sqrt{t_s - \tau}} d\tau \tag{6}$$

where $\mathbf{u}_f$ is the fluid velocity, ν is the kinematic viscosity, t_s is time, and τ is a dummy variable (Mei, 1994). This force has both a vertical and horizontal component. For steady, shear flow, the fluid acceleration is zero, so the term can be simplified to be a function of the particle acceleration.

In addition to the shear lift force, the Magnus lift force also results from the velocity gradient across the particle. This force, named for Heinrich Magnus who first discovered the phenomenon in 1853, is a pressure force due to the circulation around a spinning sphere (Murphy and Hooshiari, 1982). As a result of viscous effects, angular momentum is supplied to the particle. This force accounts for the different types of pitches in baseball such as the curveball and the slider (Munson, Young, and Okiishi, 1990). The shear flow induces a rotation which causes a particle to saltate higher and further than without the inclusion of this term. If the velocity gradient is positive, the particle will rotate clockwise (cw); hence, an upward lift. If the velocity gradient is negative, the particle will rotate counterclockwise (ccw); hence, a downward lift. For a moving sphere in a shear flow, the force is expressed as

$$\mathbf{F}_L(Magnus) = \frac{\pi}{8} d^3 \rho V_r \left(\Omega - \frac{1}{2}\frac{d\mathbf{u}_f}{dz} \right) \tag{7}$$

with $\mathbf{\Omega}$ as the angular velocity of the particle with units in rad/s (White and Schulz, 1977). This force was developed from the work of Rubinow and Keller (1961). Rubinow and Keller (1961) looked at rotating a moving sphere in a still viscous fluid with low Reynolds numbers only. Based on their analysis, the Magnus force was independent of viscosity. From this they derived the form of the Magnus force and the moment acting on the sphere. The force has been described previously. The moment has the form

$$Moment = I\frac{d\Omega}{dt} = -\pi\mu d^3 \left(\Omega - \frac{1}{2}\frac{d\mathbf{u}_f}{dz} \right) \tag{8}$$

where I is the particle's moment of inertia. This moment equation is solved simultaneously with the equations of motion to constantly adjust the particle rotation to that induced by the fluid. Note that the dynamic viscosity appears in (8). It acts to dampen the effects of the initial particle rotation to that of the fluid. It must be said that the Magnus force and moment equations were derived for small Reynolds numbers. We consider the Magnus force in this shear flow analysis for the purpose of determining whether the effect improves the agreement between experiments and theory.

Applicability of Forces For Different Reynolds Number Ranges

The drag, added mass, shear lift, Magnus lift, and Basset history forces have been formulated as previously published in the literature. The drag force, as defined here, is valid for the entire range of Reynolds numbers encountered. The form of the shear lift force was originally verified for turbulent flow. Also, theoretically a solid sphere in an inviscid fluid has an added mass coefficient value of 0.5; so, the added mass force defined herein is applicable to large Reynolds number flows, too. From Rubinow and Keller (1961), the Magnus force was derived for low Reynolds number flows only. The Basset history force is not well known for low Reynolds number flow (Nino and Garcia, 1992). For analysis purposes, the forces are extended to the entire range of Reynolds numbers. This follows the previous research procedures as done by Wiberg and Smith (1985) and Nino and Garcia (1992) which have yielded satisfactory results.

Equation of Motion

The equation of motion for a saltating particle may be divided into longitudinal and vertical components. The forces acting on a saltating particle are described by the following equations for a steady, horizontal flow.

$$\rho_s V \frac{du}{dt} = -C_D A \frac{1}{2} \rho \sqrt{(u-u_f)^2 + w^2}\,(u-u_f) - \rho V C_M \frac{du}{dt} - \frac{6\pi\mu\left(\frac{d}{2}\right)^2}{\sqrt{\pi \upsilon}} \int_0^{T_s} \frac{\frac{du}{dt}}{\sqrt{T_s - \tau}} d\tau$$
$$+ \frac{\pi}{8} d^3 \rho \sqrt{(u-u_f)^2 + w^2} \left(\Omega - \frac{1}{2}\frac{du_f}{dz} \right) \frac{w}{\sqrt{(u-u_f)^2 + w^2}} + \rho(s-1) g V \sin\beta \tag{9}$$

and

$$\rho_s V \frac{dw}{dt} = -C_D A \frac{1}{2} \rho \sqrt{(u-u_f)^2 + w^2}\,(w) - \rho V C_M \frac{dw}{dt} - \frac{6\pi\mu\left(\frac{d}{2}\right)^2}{\sqrt{\pi \upsilon}} \int_0^{T_s} \frac{\frac{dw}{dt}}{\sqrt{T_s - \tau}} d\tau$$
$$- \frac{\pi}{8} d^3 \rho \sqrt{(u-u_f)^2 + w^2} \left(\Omega - \frac{1}{2}\frac{du_f}{dz} \right) \frac{u-u_f}{\sqrt{(u-u_f)^2 + w^2}} \tag{10}$$
$$+ \frac{\rho}{2} A C_L \left(u_{\Delta Top}^2 - u_{\Delta Bot}^2 \right) - \rho(s-1) g V \cos\beta$$

where

$$u_{\Delta Top}^2 = \left(u - u_{fTop}\right)^2 + w^2 \tag{11}$$

and

$$u_{\Delta Bot}^2 = \left(u - u_{fBot}\right)^2 + w^2 \tag{12}$$

The variables, u_{fTop} and u_{fBot}, are the fluid velocities evaluated at the top and bottom of the particle, respectively. The quantities, u and w, are the particle's translational speed parallel and normal to the bed, respectively. All these assume no w-component of fluid velocity. Figure 1 provides a sketch of the particle forces. β is the bed slope as defined in Figure 2.

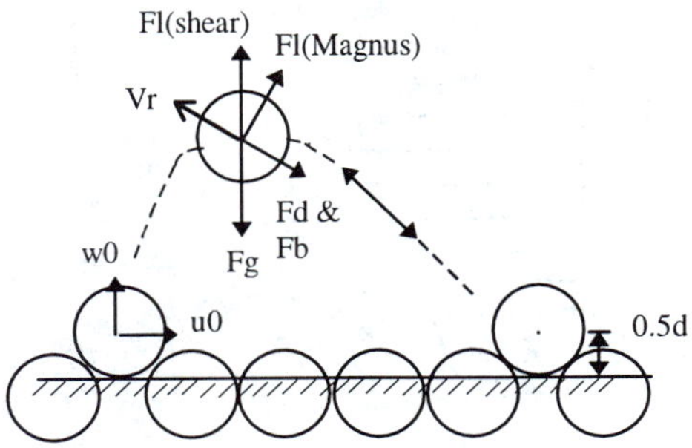

Figure 1 Force definition sketch of particle saltation

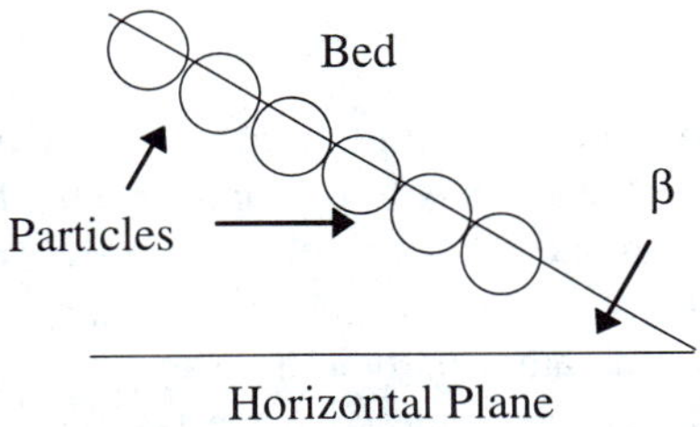

Figure 2 Definition of bed slope

The fluid velocity profile used in the model is the "law of the wall" profile,

$$\frac{\bar{u}}{u_*} = \frac{1}{\kappa} \ln\left(\frac{29.7z}{k_s} + 1\right) \tag{13}$$

where $\bar{u}$ is the mean fluid velocity, u_* is the fluid friction velocity, κ is Karman's constant, z is height above the "theoretical bed", and k_s is the bed roughness. The "theoretical bed" is located at z = 0. Note that the fluid velocity has only a horizontal, or bed-parallel, component.

Boundary Conditions

The "theoretical bed" is assumed to be 0.2 times the particle diameter below the top of the particles as is shown in Figure 3. The initial position of the saltating particle is at one-half a particle diameter above the "theoretical bed". In order to solve the equations of motion given previously, the particle's initial vertical and horizontal velocity components are needed. The values of these components come from the equations of White and Schulz (1977). They found that the speeds varied from u_* to $2u_*$. In addition, they found that lift-off angles varied from 30 to 70 degrees. When the Magnus force is considered, an initial particle angular velocity is needed. This value is adjusted to yield a best match to a known trajectory.

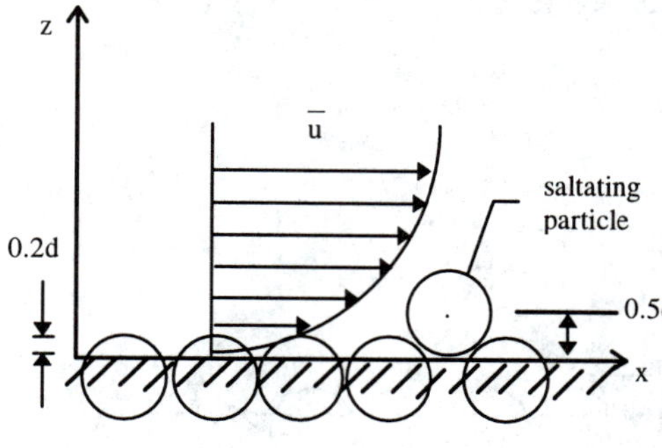

Figure 3 Saltating particle initial position

Method of Solution

The equations of motion along with the moment equation are defined as first order ordinary differential equations. A fourth order Runge Kutta approach is used to yield a vertical and horizontal particle velocity and an angular velocity. The vertical and horizontal velocity components are numerically integrated with a simple Simpson's Rule approach to obtain a particle trajectory. The model includes some adjustable parameters. The bed roughness can be taken to be any multiple of the particle diameter. The initial velocities, angular velocity and the lift-off angle also may be adjusted.

Sensitivity Analysis

This section looks at the relative effect force combinations and initial conditions have in determining the trajectory of a saltating grain. The parameters that are held constant unless otherwise specified through the force sensitivity test are grain diameter, specific gravity, initial particle velocity, initial angle, and bed roughness. They are 0.18 cm, 2.65, $2u_*$, 45 degrees, and 2d, respectively. The coefficient of lift was taken to be 85% of the coefficient of drag. These are the values that were either used or observed by Fernandez Luque and van Beek (1976). In addition, it needs to be stated that the bed slope was taken to be zero in this case.

The most basic form of the model contains the drag, added mass, and gravity forces. The shear lift force developed by Wiberg and Smith (1985) and the history force from Basset (1888) are successively added to the model. Figure 4 shows the results. For this case, $u_* = 4$ cm/s with $u_0 = w_0 = 2u_*$. The shear lift force increases the length of the trajectory by about one grain diameter while the Basset history force appears to have a lesser effect on the trajectory. The results with the shear lift force included are consistent with our intuition. Only a small velocity gradient develops across the grain because of its size. As a result, the shear lift has a relatively small effect. The height is increased by less than one-tenth of a grain diameter while the length increased by approximately one-half grain diameter when the Basset term is included. The increase in length is due to the fact that the grain travels higher into the fluid column and thus attains a greater velocity.

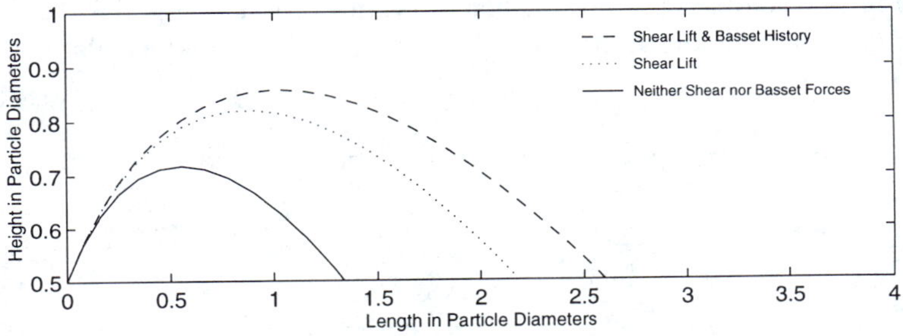

Figure 4 Effect of shear lift and Basset history forces on small particle

We then ran the model with a larger particle size. A diameter of 3.1 cm was chosen, roughly in the gravel regime. The other parameters maintained their same values with the exception of $u_* = 22.84$ cm/s and $\beta = 0.07$ to match the experimental conditions of Nino et al. (1992). Figure 5 shows that the shear lift force had more of an effect than it did with the smaller grain. This result is expected because the velocity gradient is large on a larger particle. The Basset force, however, did not posses the same significance as it did in the other case. This force had little effect on the particle trajectory. The Basset history force had a greater effect on the smaller particle than on the larger particle. This same result was obtained by Nino and Garcia (1992).

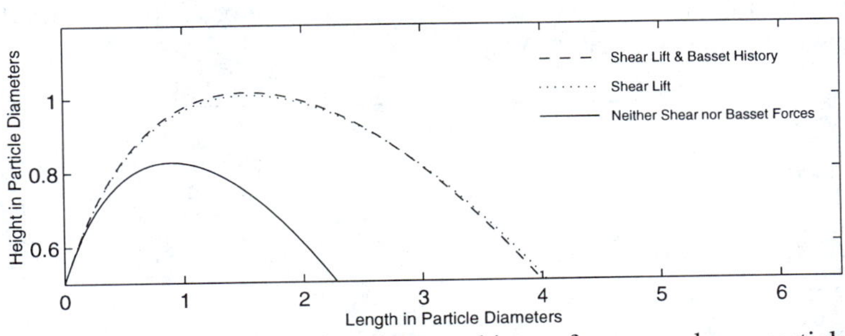

Figure 5 Effect of shear lift and Basset history force on a large particle

The next force to consider adding to the formulation is the Magnus lift force. To review, this force results from grain rotation which is caused in two ways. First, a grain may be transferred an angular velocity from the shear flow. Since the fluid velocity is greater at the top of a grain than at the bottom of a grain, a net torque may be exerted upon the grain. Second, a grain on the bed may obtain an angular velocity from a collision with another grain. The magnitude of the rotation is dependent on the speed and the placement of the blow that the grain striking the bed delivers. Figure 6 displays the trajectories that result by varying the initial angular velocity. The diameter of the particle is 0.18 cm with k_s=2d, C_L=0.85C_D, u_*=4 cm/s, u_0=2u_*, and w_0=2u_*. It is obvious and expected that the trajectories should both increase in length and height with increasing initial angular velocity. The effect of the Magnus force on a particle's path is quite significant. This is a somewhat discouraging result because little is known about the angular velocities of saltating grains. A common practice is to match a known trajectory by "tweaking" the initial angular velocity of the particle.

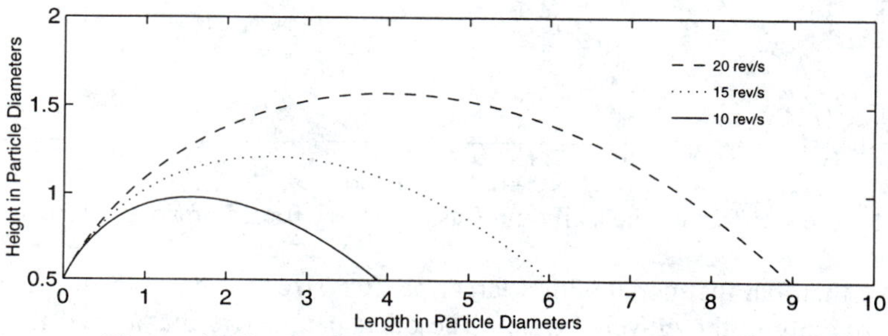

Figure 6 Effect of Magnus force on particle trajectories

The initial angle and velocity of the grain is the next area of interest. Again the grain size used is 0.18 cm. Figure 7 displays the results of varying the take-off angles and maintaining a constant initial velocity. The initial angles are 30, 45, and 60 degrees. The initial longitudinal velocity and initial vertical velocity are both 2u_*. Figure 8 shows what happens if the velocities are varied and the angles are held constant at 45 degrees. As the lift-off angles increased, the grains saltated farther. Also, the higher the initial velocity, the longer the trajectory. These results are obviously from the fact that the grain attains a higher velocity from the fluid with a larger angle and higher initial velocity.

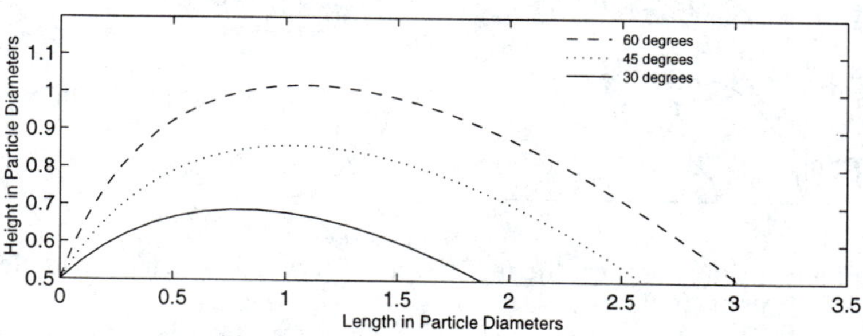

Figure 7 Trajectory sensitivity to particle take-off angle

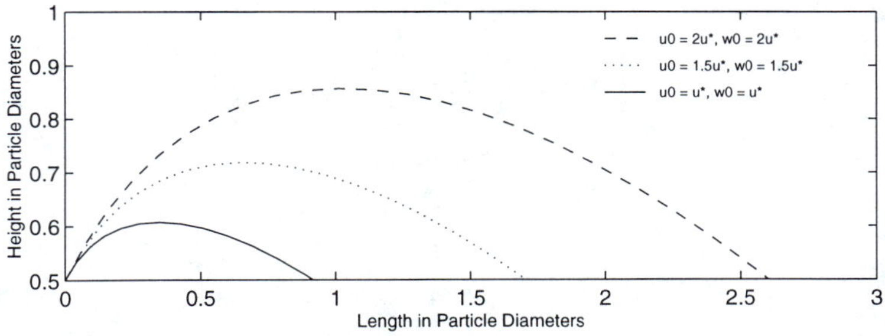

Figure 8 Trajectory sensitivity to particle take-off speed

To summarize, the forces acting on the particle vary significantly with grain size. The shear lift force and the Basset history force vary the greatest of the forces examined. The Magnus effect increases saltation length and height greatly as initial angular particle velocity increases. Finally, the larger the initial velocity and angle, the greater the saltation length and height.

Comparison of Model to Data

The Magnus Effect is neglected for the first comparison with a given data set to see if the shear lift force is sufficient. The trajectory data used for the model comparison is provided by Fernandez Luque and van Beek (1976). The grain diameter was 0.18 cm with a friction velocity of 4 cm/s. The initial conditions for the model were u_0 = 8 cm/s, w_0 = 8 cm/s, $\beta = 0$, $C_L = 0.85C_D$, and $k_s = 2d$. The trajectory is shown as the solid line in Figure 9. It is easily seen that the shear lift force alone does not adequately describe the motion of the particle as observed by Fernandez Luque and van Beek (1976).

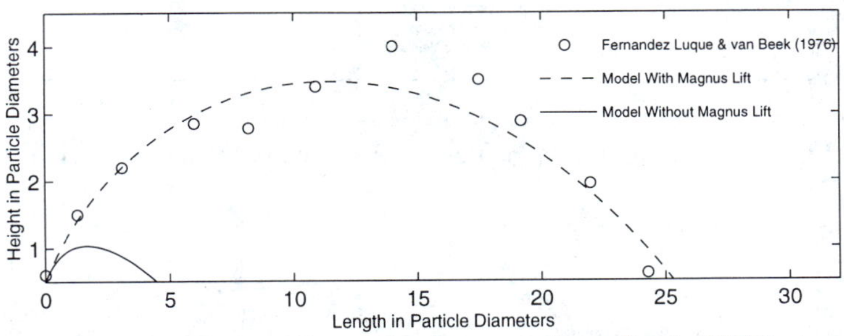

Figure 9 Model comparison to Fernandez Luque and van Beek (1976) observations

Figures 10 through 14 show some of the relevant velocities and forces as predicted by the model. The fluid velocity is always in the rough, turbulent range as shown by the Reynolds number. The drag coefficient was between 0.60 and 0.85. Those quantities are graphically shown in Figure 10 and Figure 11. Figure 12 shows how the particle, fluid, and relative velocities varied over the trajectory length of the saltation. The plot reveals that the particle velocity is continuously increasing

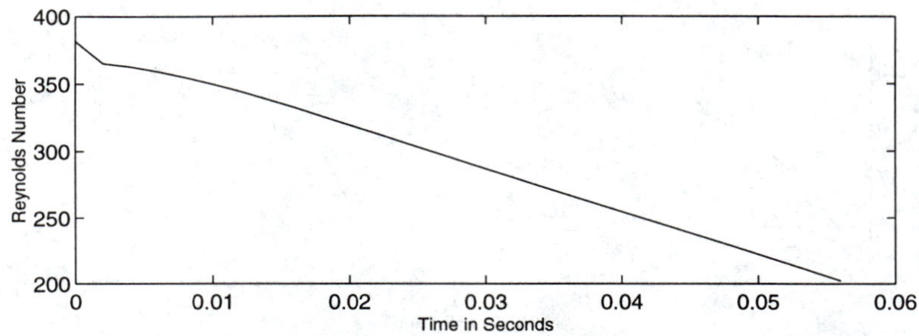

Figure 10 Reynolds number simulated saltation trajectory

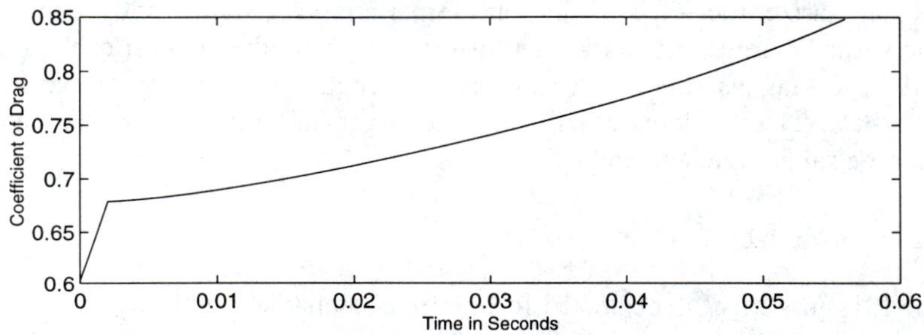

Figure 11 Drag coefficient for simulated saltation trajectory

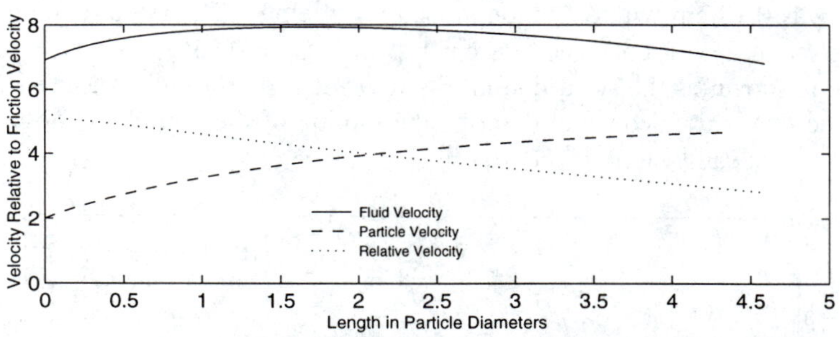

Figure 12 Strength of velocities of simulated particle trajectory

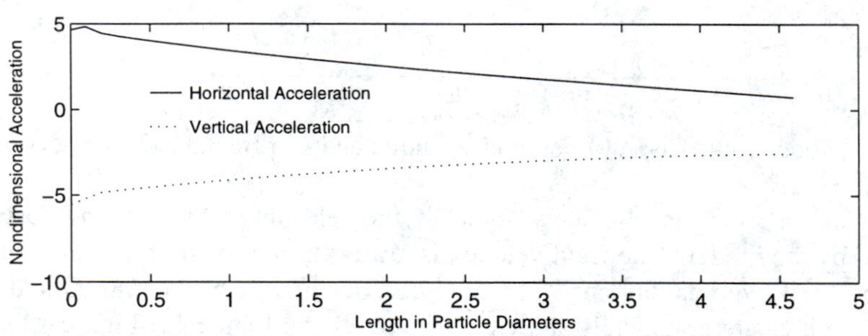

Figure 13 Strength of particle accelerations for a simluated particle trajectory

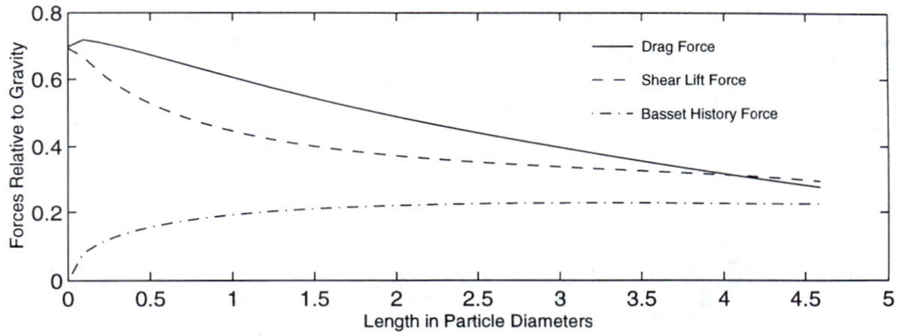

Figure 14 Strength of relevant forces for a simulated particle trajectory

toward that of the fluid velocity. So as a result, the relative velocity is decreasing. This means that the drag and lift forces are both decreasing over the grain trajectory. This is shown in Figure 14. The drag force is represented by the solid line and the shear lift force is represented with a dashed line. The Basset history force is smaller than both the lift force and the drag force for much of the saltation also in Figure 14. These forces in this figure are given relative to gravity. Note that the Basset force continually increases as the particle saltated as a result of the integral nature of the force.

The horizontal acceleration of the particle decreases as the particle velocity approaches the fluid velocity as shown in Figure 13. The vertical acceleration becomes less negative as the particle saltates. The accelerations are non-dimensionalized by multiplying the acceleration by T_s/u_*. The particle decelerates significantly in the rising part of the trajectory and continues to do so until the particle approaches the bed. It accelerates again near the bed because of the aforementioned lift force. This shear lift force is significant when compared to the drag force. It is evident from Figure 9 that the lift force is not providing enough upward thrust for the particle to saltate as observed by Fernandez Luque and van Beek (1976).

The Magnus force was added to the equation of motion to improve the agreement with observations. In addition, the moment equation was added and solved simultaneously with the equation of motion. The initial conditions were maintained with one exception. An initial angular velocity, Ω_0, of 30 rev/s was added. The model comparison with the data range is displayed in Figure 9. The model now predicts the grain trajectory fairly well and certainly much better than previously. The Magnus force significantly increases the overall lift effect on the particle especially in the rising part of the trajectory according to Figure 15. For this particular case, the Magnus force increased the saltation height by 2.3 grain diameters and the length by 21 grain diameters. Wiberg and Smith (1985) reached a similar conclusion that the Magnus force should be included.

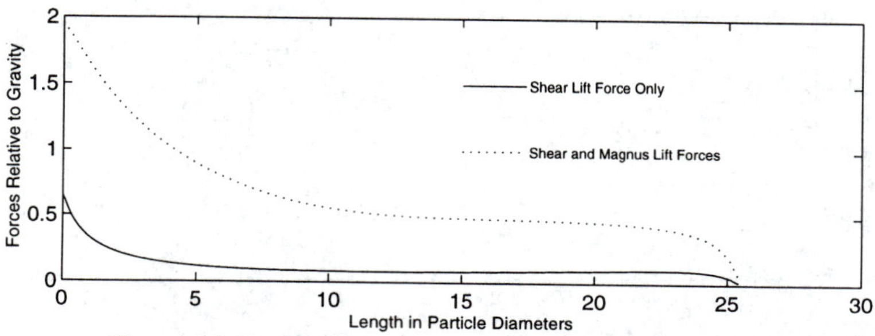

Figure 15 Comparison of shear lift and Magnus forces

Conclusion

The saltation model considered particles from incipient motion through a trajectory and returning to the bed. The saltation model included gravity, added mass, drag, shear lift, Magnus lift, and Basset history forces. The shear lift force is the same as defined by Wiberg and Smith (1985). When the Magnus lift force is applied, the moment equation from White and Schulz (1977) is used to continually update the particle's rotation.

It was found that a particle will saltate farther with increasing take-off speed and angle. The shear lift and Basset history forces vary significantly with particle size. The shear lift force has a greater effect on large particles and Basset history has a greater effect on small particles. The Magnus lift force affects trajectories significantly when included in the model formulation.

The model is able to match observed trajectories from Fernandez Luque and van Beek (1976). It is necessary to include the Magnus lift force to best match their observations. The shear lift force may need further examination. It appears to underpredict the velocity gradient that is expected at the end of a particle's trajectory near bed. These results suggest experiments be conducted to measure the rotation of saltating grains.

Acknowledgments

This work was supported by the U.S. Office of Naval Research, Coastal Sciences Program. Discussions with Renwei Mei are appreciated.

References

Auton, T.R., Hunt, Jc.R., and Prud'Homme, M., The force exerted on a body in inviscid unsteady nonuniform rotational flow, J.Fluid Mech., 197, 241-257, 1988.

Abbott, J. E., and Francis, J. R. D., Saltation and suspension trajectories of solid grains in a water stream, Phil. Trans. R. Soc. London, Ser. A, 284, 225-254, 1977.

Bagnold, R. A., The nature of saltation and of 'bed-load' transport in water, Proc. R. Soc. London, Ser A, 332, 473-504, 1973.

Basset, A. B., A Treatise on Hydrodynamics Volume Two, Dover Publications, New York, 1888.

Chepil, W. S., The use of evenly spaced hemispheres to evaluate aerodynamic forces on a soil surface, Trans. AGU, 39 (3), 397-404, 1958.

Einstein, H. A., The bed-load function for sediment transportation in open channel flow, U.S. Dept. Agric., Tech. Bull. no. 1026, 1950.

Fernandez Luque, R., and van Beek, R., Erosion and transport of bed-load sediment, J. Hydraulic Res., 14 (2), 127-144, 1976.

Gilbert, G. K., Transportation of debris by running water, U. S. Geol. Survey Prof. Paper, 86, 1914.

Graf, W.H., Hydraulics of Sediment Transport, Water Resources Publications, Littleton, Colorado, 1984.

Krecic, M. R., Evaluation and development of particle saltation Models, thesis presented to the University of Florida, Gainesville, Fl, in 1995, in partial fulfillment of the requirements for the degree of Master of Science.

Lee, H. Y., and Hsu, I. S., Investigation of saltating particle motions, J. Hydr. Engrg., 120 (7), 831-845, 1994.

Mei, R., Flow due to an oscillating sphere and an expression for unsteady drag on the sphere at finite Reynolds number, J. Fluid Mech., 270, 133-174, 1994.

Morsi, S. A., and Alexander, A. J., An investigation of particle trajectories in two-phase flow systems, J. Fluid Mech., 1972 (part 2), 193-208, 1972.

Murphy, P. J., and Hooshiari, H., Saltation in water dynamics, J. Hydr. Div., ASCE, 108 (HY11), 1251-1267, 1982.

Nino, Y., and Garcia, M., Gravel saltation 2. modeling, Water Resour. Res., 30 (6), 1915-1924, 1994.

Nino, Y., M. Garcia, and L. Ayala, Gravel Saltation 1. Experiments, Water Resour. Res., 30 (6), 1907-1914, 1994.

Owen, P. R., Saltation of uniform grains in air, J.Fl. Mech., 20(Pt 2), 225-242, 1964.

Patel, M. H., Dynamics of Offshore Structures, Butterworth & Co. Ltd, NY, 1989.

Rubinow, S. I., and Keller, J. B., The transverse force on a spinning sphere moving in a viscous fluid, J. Fluid Mech., 447-459, 1961.

Van Rijn, L. C., Sediment transport, Part I: Bed load transport, J. Hydr. Engrg., 110 (10), 1431-1456, 1984.

White, B. R., and Schulz, J. C., Magnus effect in saltation, J. Fluid Mech., 81 (part 3), 497-512, 1977.

Wiberg, P. L., and Smith, A theoretical model for saltating grains in water, J. Geophys. Res., 90, 7341-7354, 1985.

CHAPTER 298

Sediment Movement and Stress Condition in Sea Bed

Yoshihiko Maeno[1], Motoyasu Ishikawa[2], Kozo Bando[3], Yoshinobu Akiyama[4] and Kouichi Yabe[5]

ABSTRACT

The stress condition in the sea bed such as effective stress and hydrodynamic condition were measured in the surf zone to understand the detailed mechanism of local and temporal sediment movement. Analysis of the measured data suggests that the large effective stress drop and strong flow in the onshore direction when the wave crest passes are the possible cause for the sand movement in the surf zone. It is also found that the sea bottom elevation change can be evaluated from the effective stress in the sand by a simple equation. The effective stress sensor used in the field measurement was proved to be useful in estimating local and temporal sediment transport.

INTRODUCTION

Sediment transport is usually studied in terms of wave condition and characteristics of bed material. However, in order to understand the detailed mechanism of local and temporal sediment movement, it is important to associate the stress condition in sea bed with wave condition and then to relate the stress to sediment transport. For example, Nago(1982) pointed the relation between the pore pressure and fluidization of the bed.

---

[1]Professor, Dept. of Civil Engineering, Nihon University, 7-24-1 Narashinodai Funabashi, Chiba 274, Japan.
[2]Research Associate, Dept. of Civil Engineering, Nihon University.
[3]Chief, Offshore and Hydraulic Engineering Sec., Kajima Technical Research Institute, 19-1 Tobitakyu 2-Chome Chofu, Tokyo 182, Japan.
[4]Senior Research Engineer, Kajima Technical Research Institute.
[5]Manager of Patents, Engineering Dept., Kyowa Electronic Instruments Co., Ltd., 3-5-1, Chofugaoka Chofu, Tokyo 182, Japan.

Maeno and Hasegawa(1985) examined the effects of wave-induced pore pressure in sand bed on liquefaction. Maeno and Tokutomi(1989b) indicated the strong correlation between the liquefaction of sand bed and the mechanisms of sediment transport. Maeno (1992) introduced the experimental results on wave-induced liquefaction and sediment transport, and indicated the needs of the further verification by field observation.

Maeno *et al.*(1993) measured the effective stress, pore pressure and wave conditions outside the surf zone, and concluded that liquefaction occurs by the delayed response of the pore pressure to the surface elevation change. In this research, the similar approach was taken, and the bottom elevation change was tried to relate to the effective stress, pore pressure, wave height and current velocity, which were measured in the surf zone.

This research also attempts to examine the performance of a newly developed effective stress sensor (Maeno *et al.*,1992), which may give the local and temporal information that are hard to obtain by the conventional methods; sand trap, surveying, etc.

FIELD MEASUREMENT

Measurement Site

Measurement was made in a beach at Hazaki-cho, Ibaragi Prefecture, Japan (see Figure 1). This beach is a part of 80 km straight sandy coast facing the Pacific Ocean. There is a 427 m long observation pier on the beach owned by Port and Harbour Research Institute, Ministry of Transport, to study sediment transport. This study made use of the pier, from which various data were taken. Two measurement locations were chosen in the surf zone along the pier: one is 230 m offshore from the shoreline and the other is 260 m offshore. Figure 2 illustrates the measurement locations and the bottom profiles under the pier during the measurement period. The water depth at those locations fluctuated in the range of 1 m, but both the locations remained on a large bar.

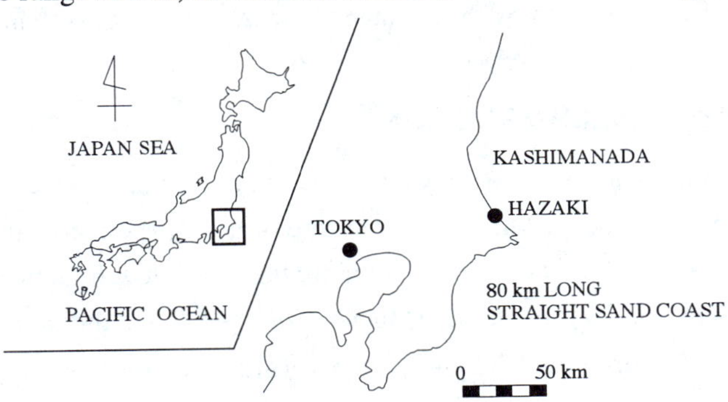

FIG. 1. MEASUREMENT SITE

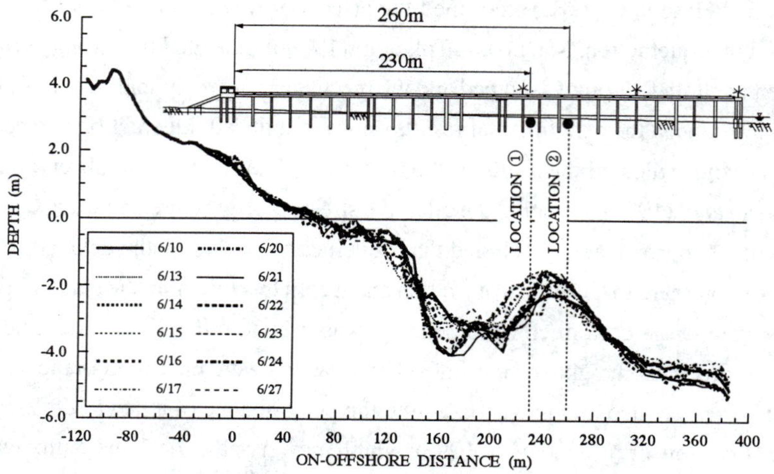

FIG. 2. MEASUREMENTS OF SEA BOTTOM PROFILE

Instruments

The instruments deployed in the measurement locations are listed in Table 1. To install these instruments, a frame structure was put in sea bed at each location as shown in Figure 3. It held an effective stress sensor and a dynamic pore pressure gauge in the sea bed, a pressure gauge on the sea bed, a current meter, and a wave gauge. The effective stress sensor and dynamic pore pressure gauge were the differential type made of the flexible filter (Maeno and Tokutomi,1989a). Additional data such as sea surface elevation by acoustic wave gauge, wind data, water depth and ground water level were supplied by Port and Harbour Research Institute, who takes these data at the pier on a regular basis. The measured data were recorded at 10 Hz from June 12th to 26th, 1994 for the location 1 and from June 12th to 15th for the location 2. Data acquisition was terminated by the damage to the measurement system due to the high seas.

MECHANISM OF LIQUEFACTION

From a soil mechanical point of view, the sand can be mobilized when the stress acting on it exceeds its capacity. The sand in the zero or negative effective stress condition, which may be called liquefaction, is certainly the case. How liquefaction takes place outside the surf zone was explained (Maeno *et al.*, 1993), and this study focuses on the condition in the surf zone. The mechanism of liquefaction outside the surf zone is reviewed first, and the mechanism in the surf zone found by analyzing the data measured in the present study is followed.

TABLE 1. SUBJECTS FOR MEASUREMENT

MEASURING SUBJECTS	MEASURING EQUIPMENTS	REMARKS
DYNAMIC PORE WATER PRESSURE	DYNAMIC PORE WATER PRESSURE TRANSDUCER	1.0m BELOW SEABED SURFACE
EFFECTIVE EARTH PRESSURE	EFFECTIVE EARTH PRESSURE TRANSDUCER	1.0m BELOW SEABED SURFACE
WATER PRESSURE	WATER PRESSURE TRANSDUCER	1.0 m ABOVE SEABED SURFACE
WAVE PROFILE	ULTRASONIC-TYPE WAVE GAUGE	AIR BEAM TYPE
	CAPACITANCE-TYPE WAVE GAUGE	
VELOCITY	ELECTROMAGNETIC VELOCIMETER	1.0 m ABOVE SEABED SURFACE
WIND DIRECTION AND WIND VELOCITY	PROPELLER-TYPE WIND VANE AND ANEMOMETER	385m OFF SHORELINE
SEA BOTTOM PROFILE	SOUNDING	DAILY MEASURING

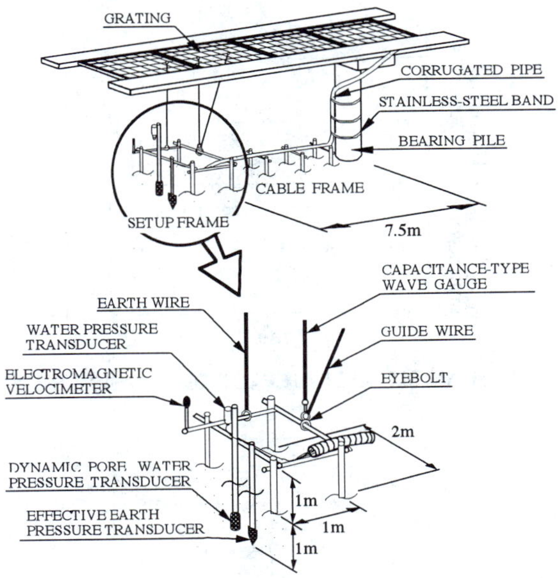

FIG. 3. INSTALLATION OF INSTRUMENTS

Outside Surf Zone

Generally, the dynamic pore pressure and effective stress in the sea bed follow the sea surface elevation change. However, it was found from close observation of these data taken outside the surf zone (Maeno *et al.*, 1993) that the dynamic pore pressure is slightly lagged behind the change of the sea surface elevation. This yields excessive positive effective stress at crest and negative value after the crest passes to compensate for the insufficient pore pressure response. The negative part of the effective stress gives rise to the liquefaction of the sand, which can be swept away by the currents near the bed. This process is illustrated in Figure 4. Figure 5 shows the time histories of the stresses measured at a depth of 6 m (Maeno *et al.*, 1993). Sharp peaks are observed in the effective stress record due to the late response of the pore pressure to the surface elevation. The magnitude of the negative peak indicates the thickness of the sand layer that was liquidized, and the difference in the stress level before and after the wave is equal to the thickness of the sand layer that was washed away. In this example, the sand in the top 47 cm thick layer was liquidized and the sand layer in the thickness of about 17 cm was carried away.

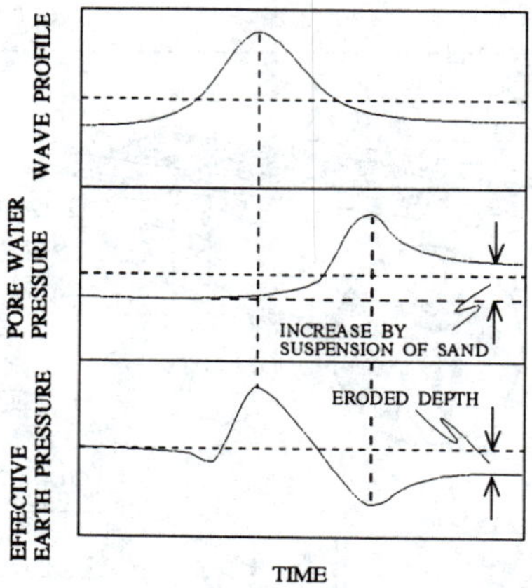

FIG. 4. MECHANISM OF LIQUEFACTION OUTSIDE SURF ZONE

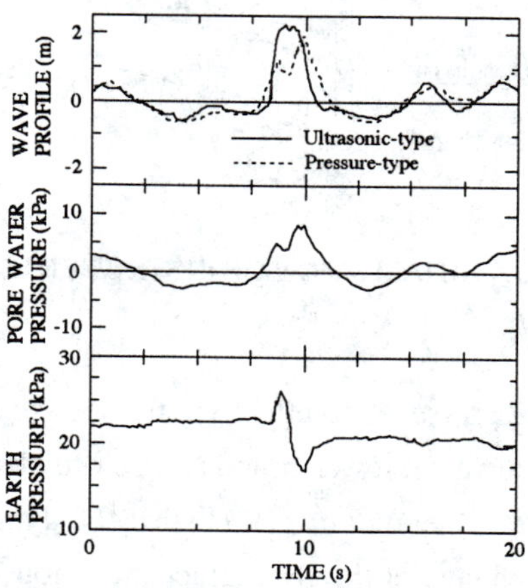

FIG. 5. TIME HISTORIES FOR STEEP AND LARGE WAVE HEIGHT

Inside Surf Zone

In this study the similar kinds of data were collected in the surf zone. Figure 6 shows the responses of those stresses collected at location 2, where waves were steep enough to break. In this case there is little time lag in the response of the pore pressure, which is different from the stress response outside the surf zone. The wave action in the surf zone is severer and the top layer of the sand is more likely to be loose so that the pressure exerted on the sea bottom is transmitted instantaneously to inside.

The response of the effective stress is characterized with the sharp and large drops when the wave crests pass. These drops become larger as the wave height increases. The magnitude of this drop is much smaller than that of the pore pressure, and the negative effective stress may be due to the overshooting response of the pore pressure.

The current data at the sea bottom in Figure 6 show the strong flow in the onshore direction when the crests pass and the effective stress is very low. It is surmised that the loose sand in the top layer where the effective stress is nearly zero may be carried

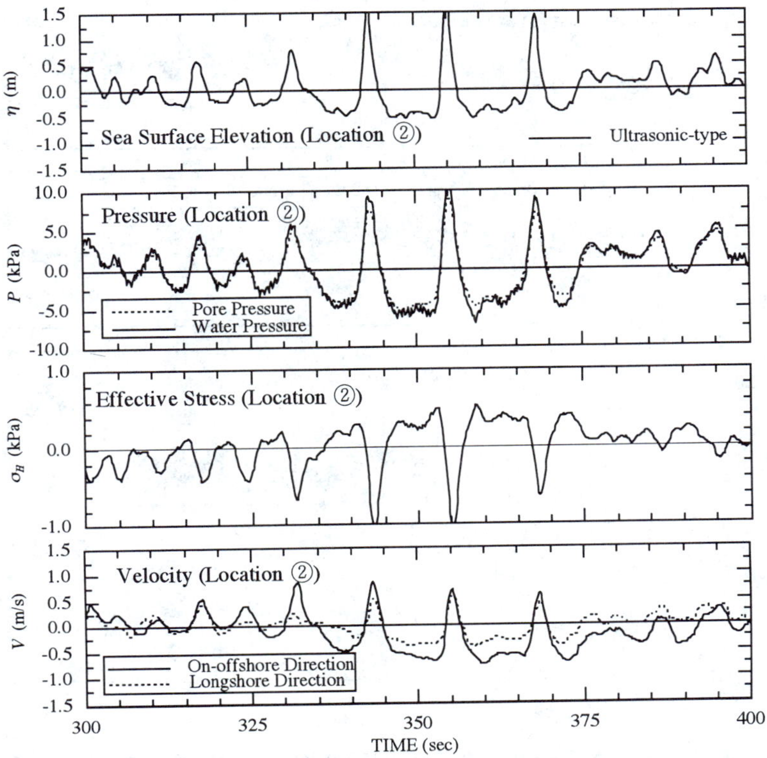

FIG. 6. TIME HISTORY DATA AT LOCATION 2 (6/15/1994 01:17:16)

away by the strong onshore flow. Compared with the condition outside the surf zone, the larger negative effective stress and stronger current due to severer wave action are expected in the surf zone. The condition in the surf zone that large negative effective stress and strong current occur at crest is one of the mechanisms for the sediment movement in the surf zone

ANALYSIS OF FIELD DATA

Wave Condition

The significant wave height and period, and wind information are computed based on 50 minutes-long data to represent the condition at that hour, and are shown in Figure 7. There were two moderate storms with approximately 2 m significant wave height coming in the measuring period: on 15th to 16th and on 19th to 20th. The swells were dominant at the first event, since wind was weak and the wave period was relatively long. On the other hand, the second event was cause by the strong easterly winds.

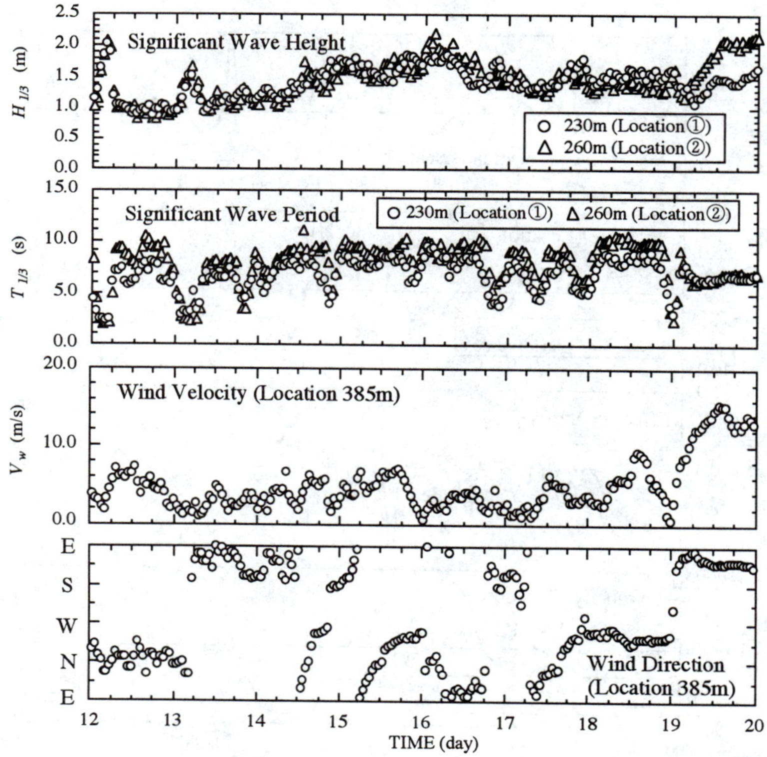

FIG. 7. SIGNIFICANT WAVE HEIGHT AND PERIOD, AND WIND INFORMATION (50 MINUTES-LONG AVERAGED DATA)

Response of Stress in Sand to Wave Action

To compare the stress condition in the sand with the wave condition, the stress data were also processed in the same way as the waves were. Figure 8 illustrates the time histories of the significant pressure, stresses and velocity data. The pressure data can be linearly converted to the sea surface elevation. It is seen that the amplitude of the effective stress change in one wave period becomes larger with the significant wave height. It implies that the effective stress is directly controlled by the wave action.

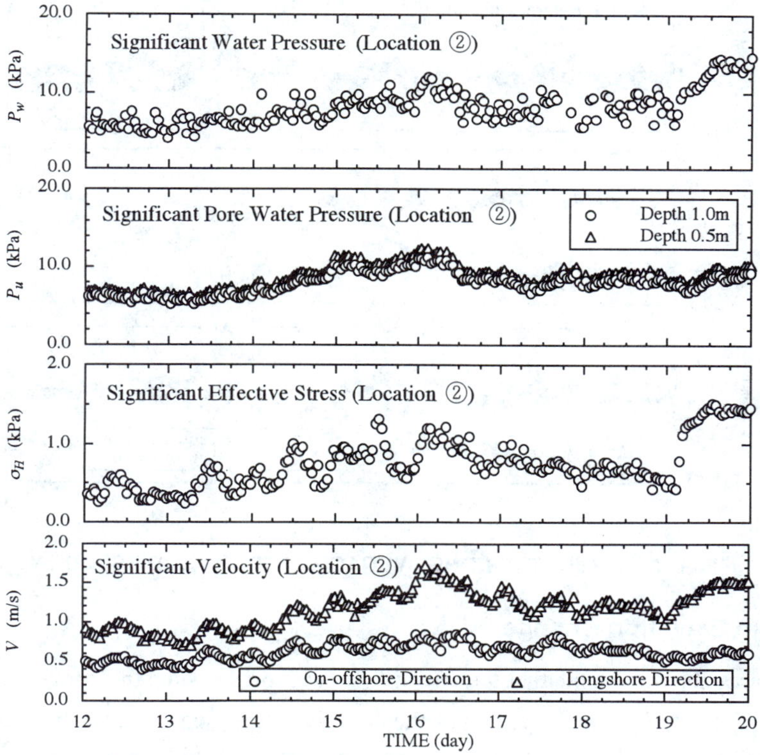

FIG. 8. SIGNIFICANT STRESS AND VELOCITY (50 MINUTES-LONG AVERAGED DATA)

The averaged data are shown in Figure 9. There is a sharp depression everyday in the effective stress data, and it becomes larger with time until 17th. The pore pressure exhibits the similar response but in the positive direction. Then adding them together makes the variation at the time become negligible. It is also seen that the drop in the averaged effective stress occurs at the lowest elevation of the sea surface in each day. Clear cause for these events are not understood, but the non-linearity associated with shallower water depth may be attributed. Correlation with the wave height is not clear either, since high waves came from 15th evening to 16th morning.

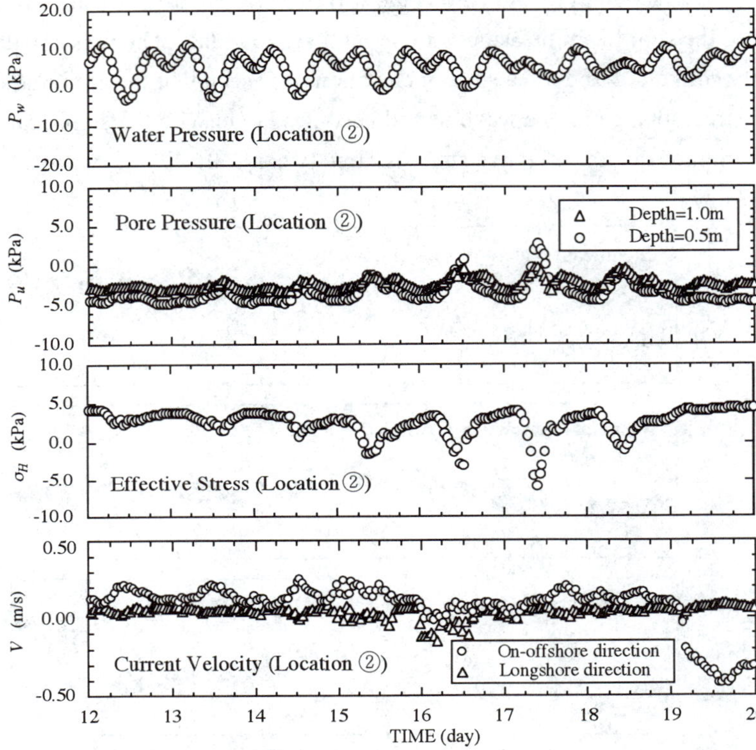

FIG. 9. 50 MINUTES-LONG AVERAGED STRESS AND VELOCITY

Bottom Elevation Change

It is reasonable to assume that the effective stress represents the thickness of the sand above. Then, the vertical location of the effective stress sensor in relation to the sea bottom can be estimated by the following equation (1), and consequently the sea bottom elevation, Δh is evaluated.

$$\Delta h = \frac{\sigma_H}{K_0\{(1-n)(\rho_s - \rho_w)\,g\}} \tag{1}$$

where K_0, n, ρ_s, ρ_w and g are coefficient of earth pressure at rest, porosity, density of soil particle, density of sea water and gravity acceleration. The averaged effective stress, σ_H was used in the equation. Note that since there are large drops in the averaged effective stress series as mentioned above, those data was unconsidered. If these data were converted to the thickness based on equation (1), it turned out to be about 3 m depth change, which is unrealistic. The effective stress change is assumed to be solely by the change of sand layer thickness in equation (1), and those large stress drops are more likely to be caused by other physical factors.

The estimated sea bottom elevation is shown in Figure 10 along with the measured elevation. The estimate appears to oscillate about the measurement. But, considering that the measurement was made only once a day and that the elevation possibly changes within a day, the estimated changes are compared well with the measured data at the corresponding time. This trend is observed both at location 1 and at location 2. Agreement between the measured elevation and the estimate by the effective stress implies that the effective stress is directly related to the sand movement and the sea bed elevation change can be estimated from the information on the effective stress. From this point of view the effective stress sensor employed in this study is found to be one of the most efficient instrument to provide the information on the sediment movement. It also suggests that the bottom elevation may fluctuate in a day and may have to be measured more than once in such a day to understand the actual movement.

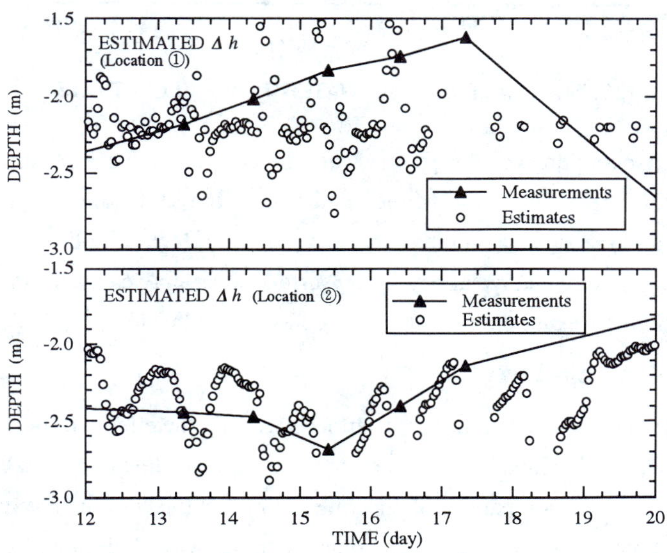

FIG. 10. ELEVATION OF BOTTOM SURFACE

Effect of Ground Water

There are some studies indicating connection between ground water and sediment transport (*e.g.*, Grant, 1948). Since the ground water level near the shoreline was continuously measured by Port and Harbour Research Institute, its effect was attempted to examine. The sea surface level and ground water table changes are compared in Figure 11. There is a long-period variation in the time history of the ground water level, which is similar to the wave height record in the same period; *i.e.*, the ground water level rose as the wave became higher. There is also oscillations with short periods, responding to the tides. Thus, it is clear that the ground water near the shore is influenced by the sea conditions.

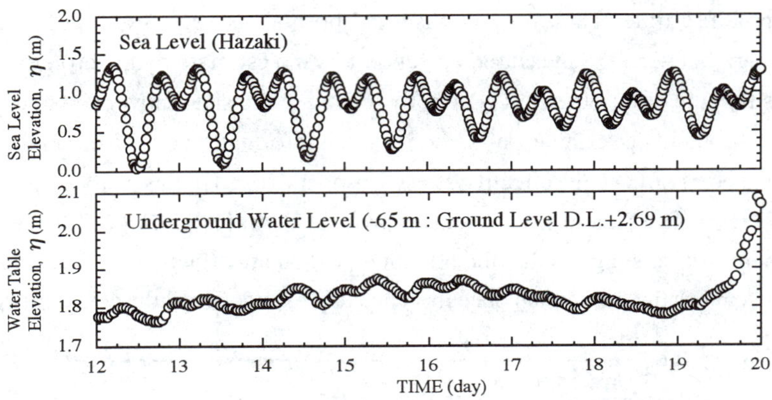

FIG. 11. SEA LEVEL AND WATER TABLE ELEVATIONS

Further, this rise and drop of the ground water level due to the sea condition change may affect the stress condition in the sand bed in the surf zone. Especially, the liquefaction and the sand movement on a bar may be affected by the ground water, because rapid pressure change at bottom by surface wave may cause the discontinuity between that pressure and the pressure defined by ground water.

Effects of Non-Linearity

To examine the non-linear effects of the surface waves on the effective stress, skewness and kurtosis were calculated from the measured sea surface elevation data. When those values in Figure 12 are compared with the time histories of the wave height and effective stress, the skewness becomes slightly bigger with the wave height and the kurtosis remains constant. From these observations clear relation between the non-linearity of the surface waves and the effective stress is not recognized.

Effects of Long-Period Component of Waves

From the spectrum computed from the sea surface data, the long period waves are estimated and shown in Figure 13. In this study, only the components with the period longer than 30 s are picked. The long-period wave had larger height on 15 th and 16th, the duration of which coincides with that of the higher significant wave. The trend of long-period wave height is also found in the time history of the ground water level. It implies that the ground water near the coast can be affected by the longer waves.

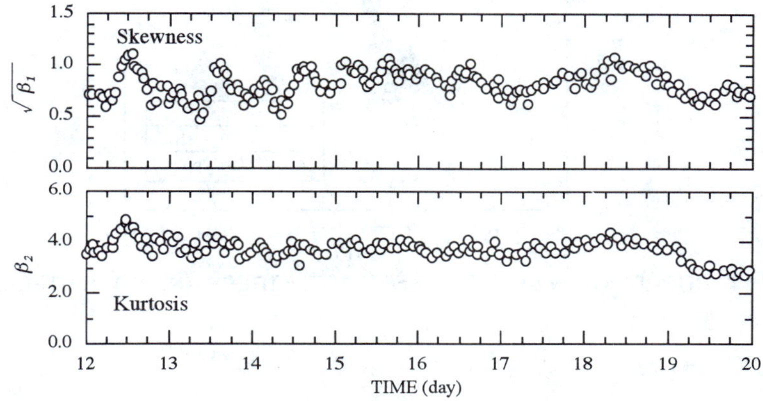

FIG. 12. EFFECTS OF NON-LINEARITY

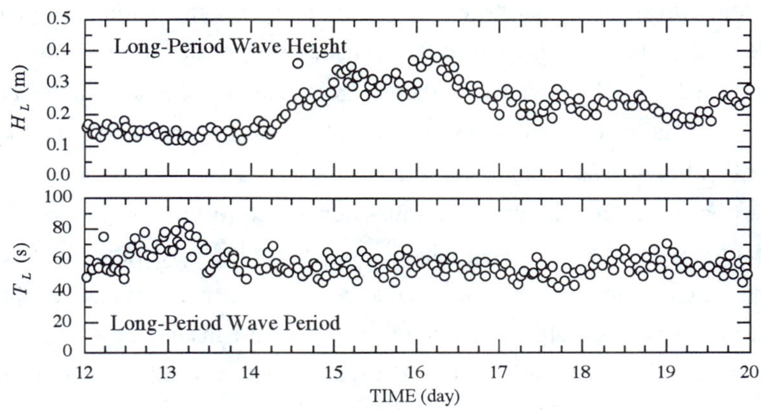

FIG. 13. EFFECTS OF LONG-PERIOD WAVES

Relation between Stresses and Wave Steepness

Using the significant wave height and period, and the corresponding effective stress change, their relation is examined in Figure 14, which exhibits the strong linear relation between them.

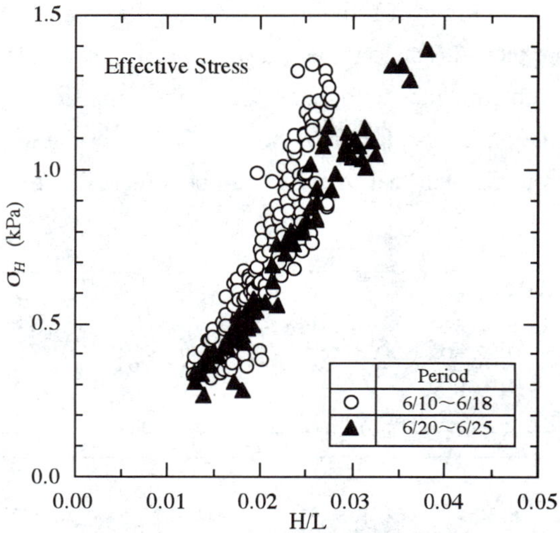

FIG. 14. CORRELATION BETWEEN EFFECTIVE STRESS AND WAVE STEEPNESS

CONCLUSIONS

The effective stress and pore pressure in the surf zone were directly measured to study the relation with the stress, and the wave condition and other factors, and to evaluate the bottom elevation change from the effective stress data. Though the data were collected only 10 days because of difficult installation of the instruments in the surf zone and the damage by high waves, there were two moderate storms in the measuring period. From the careful observation and analysis of the measured data during this period the followings are found.

(1) The pore pressure and effective stress appears to respond to the water surface elevation without any time lag. The measured area is always attacked by breaking waves and the sand there is anticipated to be loose. Then, the pressure on the bottom transmits almost simultaneously to the pore pressure inside the soil.

(2) The mechanism of liquefaction in the surf zone seems to be different from that outside the surf zone. The large effective stress drop and strong flow in the onshore direction when the wave crest passes are the possible cause for the sand movement in the surf zone.

(3) The water depth change estimated from the effective stress records agrees well with the measured depth change. This leads to the following three conclusions.

(a)The effective stress in sea bed is closely related to liquidization of sand and then to sand movement.

(b)The sea bottom elevation change can be evaluated from the effective stress in the sand and equation (1).

(c)The effective stress sensor was proved to be useful in estimating local and temporal sediment transport.

(4) Both the pore pressure and effective stress show larger variation when the short waves with large wave steepness pass.

(5) The currents going offshore were observed when the sea level was low. The effective stress varied corresponding to these currents.

(6) The averaged effective stress over 50 minutes drops sharply at low tide. The shallower water depth may be a possible cause, and the exact reason for this is still being studied.

(7) The ground water level near the coast is likely to affected by the long waves and tidal changes.

ACKNOWLEDGMENTS

The writers gratefully acknowledge the Port and Harbour Research Institute, Ministry of Transport, Japan, for permission to use the observation pier of Hazaki Oceanographical Research Facility.

REFERENCES

Grant, W. S. (1948)."Influence of the Water Table on Beach Aggradation and Degradation." *J. Marine Res.*, 7(3), 655-660.

Maeno, Y. and Hasegawa, T. (1985)."Evaluation of Wave-Induced Pore Pressure in Sand Layer by Wave Steepness." *Coastal Engineering in Japan*, JSCE, Vol.28, 31-44.

Maeno, Y. and Tokutomi, K. (1989a)."Characteristics of Flexible Filter for Pore Pressure Transducer." *IAHR Workshop on Instrumentation for Hydraulics Laboratories*, Ontario, 105-119.

Maeno, Y. and Tokutomi, K. (1989b)."Measurements of Wave-Induced Pore Pressure by Flexible Filter." *IAHR Workshop on Instrumentation for Hydraulics Laboratories*, Ontario, 121-134+1.

Maeno, Y. (1992)."Mechanism of Wave-Induced Liquefaction and Densification in Seabed." (Discussion)*Soils and Foundations*, Japanese Society of Soil Mechanics and Foundation Engineering, Vol.32, No.3, 177-181.

Maeno, Y., Karauchi, H., Yabe, K., and Koketsu, K. (1992)."Development of Transducers for Measuring Effective Stresses and Pore Water Pressure in Seabed." *Proc. of Techno Ocean '92* , Vol.1, 79-85 (in Japanese).

Maeno, Y., Karauchi, H., Yabe, K., and Suzuki, Y. (1993). "Field Measurements of Wave-Induced Liquefaction and Suspension of Sediments." *Proc. of Coastal Engineering* , JSCE, Vol.40, 576-580 (in Japanese).

Nago, H. (1982)." Liquefaction of Highly Saturated Sand Layer under Oscillating Water Pressure.", *Proc. of the 26th Conference on Hydraulics*, JSCE, 589-594 (in Japanese).

CHAPTER 299

APPLICATION OF THE DEPTH OF CLOSURE CONCEPT

Robert J. Nicholls, ASCE Associate Member[1],
William A. Birkemeier, ASCE Member[2], and Robert J. Hallermeier[3]

ABSTRACT

Using data from Duck NC (a wave-dominated, microtidal, sandy oceanic beach), depth of closure is critically evaluated. A meaningful closure is observed for most erosional events, and annual to 30 month time intervals, supporting the application of this concept within coastal engineering. However, the magnitude of depth of closure is sensitive to the definition and analysis approach utilized and estimates of closure need to be explicitly linked to this information.

The limit depth d_ℓ (Hallermeier, 1981) is found to define a conservative bound to the observations of closure during erosional events and in those annual cases where we have data. This confirms the ability to compute a meaningful limit depth simply using extreme wave conditions. At longer time scales there is evidence of a decoupling of the relationship between d_ℓ and observed depth of closure, the observations increase less rapidly than the predictions of d_ℓ. Understanding how closure evolves from individual erosional events to annual and longer time intervals improves the interpretation of sparse surveys and can assist engineering judgement when applying closure predictions.

INTRODUCTION

Depth of closure (DoC) is a widely-used concept within coastal engineering which describes the seaward limit of appreciable depth change (Hallermeier, 1978; 1981; Nicholls *et al.*, in review). It is based on the observation that repetitive beach-nearshore profiles show a decline in vertical variability with increasing depth. Empirically, a closure depth is observed in most high-quality profile data, corresponding to a pinch-out depth

---

1. School of Geography and Environmental Management, Middlesex University, London EN3 4SF, U.K.
2. US Army Engineer Waterways Experiment Station, Field Research Facility, 1261 Duck Road, Kitty Hawk, NC 27949.
3. Dewberry & Davis, 8401 Arlington Boulevard, Fairfax, VA 22031.

below which depth changes become small (Figure 1). DoC is often used to infer a seaward limit to significant cross-shore sediment transport, leading to applications such as: (1) estimating coastal sediment budgets (*e.g.*, Hands, 1983); (2) numerical modeling of coastal change (*e.g.*, Kraus and Harikai, 1983) and (3) beach nourishment design (*e.g.*, Stive *et al.*, 1991).

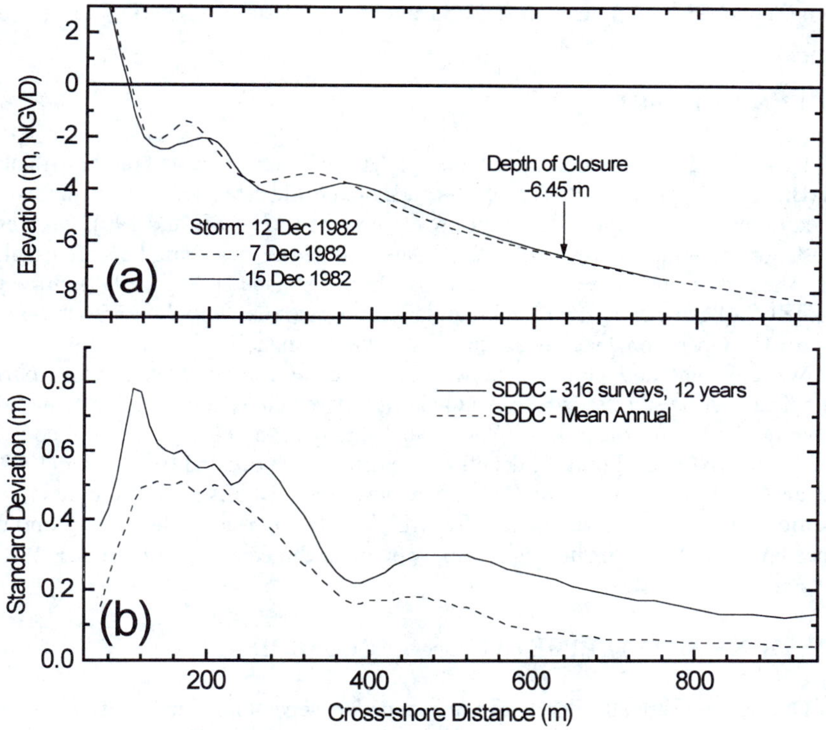

Figure 1. Example of (a) depth of closure produced by single storm, and (b) standard deviation of depth change (SDDC) over a typical year and 12 years. Variability decreases offshore, but increases with time scale. (Profile line 188, Duck, NC).

At sites with repetitive, beach-nearshore profile data, empirical estimates of DoC may be made directly. However, such sites are unusual and DoC is normally estimated using a range of possible indicators (USACE, 1984). The only analytical method to estimate DoC has been proposed by Hallermeier (1978; 1981). Annual DoC is the seaward limit of the littoral zone (depth — d_ℓ) which can be estimated using extreme wave conditions. In a generalized time-dependent form:

$$d_{\ell,t} = 2.28H_{e,t} - 68.5(H_{e,t}^2/gT_{e,t}^2) \qquad (1)$$

where $d_{\ell,t}$ is the predicted DoC over t years, referenced to Mean Low Water; $H_{e,t}$ is the non-breaking significant wave height that is exceeded 12 hours per t years, $(100/730t)$%

of the time); $T_{e,t}$ is the associated wave period; and g is the acceleration due to gravity.

The detailed characteristics of DoC, including the validity of Equation 1, have been evaluated (Nicholls *et al.*, in review) using 12 years of profile data collected at the Field Research Facility (FRF) of the U.S. Army Engineer Waterways Experiment Station's, Coastal and Hydraulic Laboratory (formerly the Coastal Engineering Research Center), located in Duck, NC, U.S.A. This paper extends previous analyses, develops additional conclusions, and examines implications for coastal engineering design and practice.

THE DUCK DATA SET

Duck, NC is a wave-dominated, microtidal sandy beach located on the Atlantic Ocean (Birkemeier *et al.*, 1985). The beach-nearshore profile data used in this study were collected to about 8-m depth along four profile lines (58, 62, 188, and 190). On these lines, offshore contours are relatively straight and one or two nearshore bars are usually present. These bars tend to be three-dimensional (Lippmann and Holman, 1990). Surveys are typically collected every two weeks and after most storms, providing pre- and post-storm profiles. Operational survey accuracy is <3 cm (Lee and Birkemeier, 1993).

Wave height and period data were collected every six hours, with hourly measurements during storms. Tidal elevation is also measured on site, with the mean tide range being about 1 m. Mean Low Water (MLW) is 0.42 m below National Geodetic Vertical Datum (NGVD). For more details on the data set see Lee and Birkemeier (1993), Larson and Kraus (1994), Lee *et al.* (1995; in review) and Nicholls *et al.* (in review). All observations of DoC are given relative to MLW as this appears to be an appropriate reference level for DoC estimates based simply on wave dimensions (Hallermeier, 1981; Nicholls *et al.*, in review).

DEFINITION AND INTERPRETATION OF CLOSURE

DoC is a fundamental morphodynamic boundary separating a landward active zone from a seaward less active zone *over the period defined by the profile observations used to define closure*. It is not an absolute cross-shore boundary and some depth change, and hence some cross-shore transport is expected to occur at DoC and further seaward (*cf.* Hallermeier, 1981). The position of DoC depends on several factors, including definition. Time scale is also significant as profile activity increases with time scale and a fixed closure criterion will typically move offshore as time scale increases (*e.g.*, Figure 1(b)). (Note that data accuracy often influences our ability to resolve DoC, but at Duck this is a minor issue). The processes which are observed to control DoC at Duck are the typical annual to decadal storm-accretion processes of the beach-nearshore zone (Lee *et al.*, in review), but at longer time scales, shoreface processes will probably control closure (*cf.* Stive and DeVriend, 1995). Therefore, while the concept of DoC is relatively simple, stating its magnitude at any site as x meters is meaningless without qualification concerning the definition utilized and the pertinent time scale. In some cases, closure may not be susceptible to a practical empirical definition, as observed by Inman *et al.* (1993) in samples of surveys at Oceanside, CA which include major storms in 1982/83.

DoC can be defined for (1) events, such as storms, (2) time-interval (or endpoint) change, which ignore intermediate changes (or in practical terms is equivalent to analysis of a typical sporadic survey program), or (3) time-integrated (or cumulative) change. While events can be related to specific energetic wave conditions and processes (*e.g.*, erosion), time-interval and time-integrated changes are controlled by the balance of erosional processes (rapid offshore transport during high energy wave events) and accretional processes (slow, but continuous onshore transport between erosional events) (Lee *et al.*, in review).

To illustrate and examine the variability of DoC due to different definitions at a site with frequent surveys, annual DoC values from profile lines 62 and 188 at Duck are presented in Table 1. Annual time-interval DoC using three fixed depth change criteria: (6 cm, 10 cm and 15 cm) are considered; In this analysis, profiles are compared proceeding seaward to determine where the profile change consistently declines below the depth change criterion, hence defining closure. Annual time-integrated DoC is measured using the standard deviation of depth change (SDDC) which avoids bias from outliers —

Table 1. Annual closure (m, MLW) derived from three fixed change criteria and from an SDDC analysis, including the annual elevation range shown at closure. Each annual period is July to July (fixed criterion) and July to end June (SDDC). "*" indicates non-closing case.

Start Year	Profile Line 62					Profile Line 188				
	Fixed Depth -Change Criteria			SDDC		Fixed Depth -Change Criteria			SDDC	
	6 - cm	10- cm	15- cm	Depth (m)	Range (m)	6- cm	10- cm	15- cm	Depth (m)	Range (m)
1981	7.2	6.5	4.4	5.4	0.16	7.9	7.6	7.4	5.8	0.16
1982	8.0	6.1	5.9	6.7	0.12	6.4	6.2	6.0	6.6	0.19
1983	6.4	6.2	6.0	6.2	0.14	7.7	5.9	5.7	6.0	0.13
1984	6.3	5.0	3.8	6.6	0.12	7.3	6.6	6.2	5.1	0.21
1985	8.0	7.5	6.8	5.1	0.20	*	8.3	6.6	4.8	0.20
1986	*	*	8.0	6.9	0.23	*	*	7.7	8.3	0.18
1987	6.3	6.1	5.9	6.1	0.13	6.0	5.8	5.6	6.0	0.12
1988	*	*	*	*	*	*	*	*	7.7	0.43
1989	*	*	*	7.3	0.13	*	*	5.1	7.2	0.20
1990	*	*	5.2	5.7	0.16	*	*	*	6.8	0.24
1991	*	8.1	5.3	5.9	0.22	7.9	7.4	4.8	5.6	0.22
1992	6.9	6.3	6.0	6.8	0.16	5.2	5.0	4.1	6.0	0.14

the DoC then corresponds to the start of a non-zero tail (Kraus and Harikai, 1983). Note that depth change at DoC is a variable defined by the standard deviation envelope and hence there is more potential for an irregular criterion using this approach. Annual SDDC-defined closures from Profile line 62 are illustrated in Figure 2.

The first observation concerning Table 1 is that annual DoC is not always defined within the surveys (≤8-m depth): the 6-cm fixed criterion defines 58% of closures, while SDDC defines 96% of closures. The largest annual DoC is usually given by the 6-cm change criterion: the 10-cm and 15-cm changes are on average about 90% and 80% of the

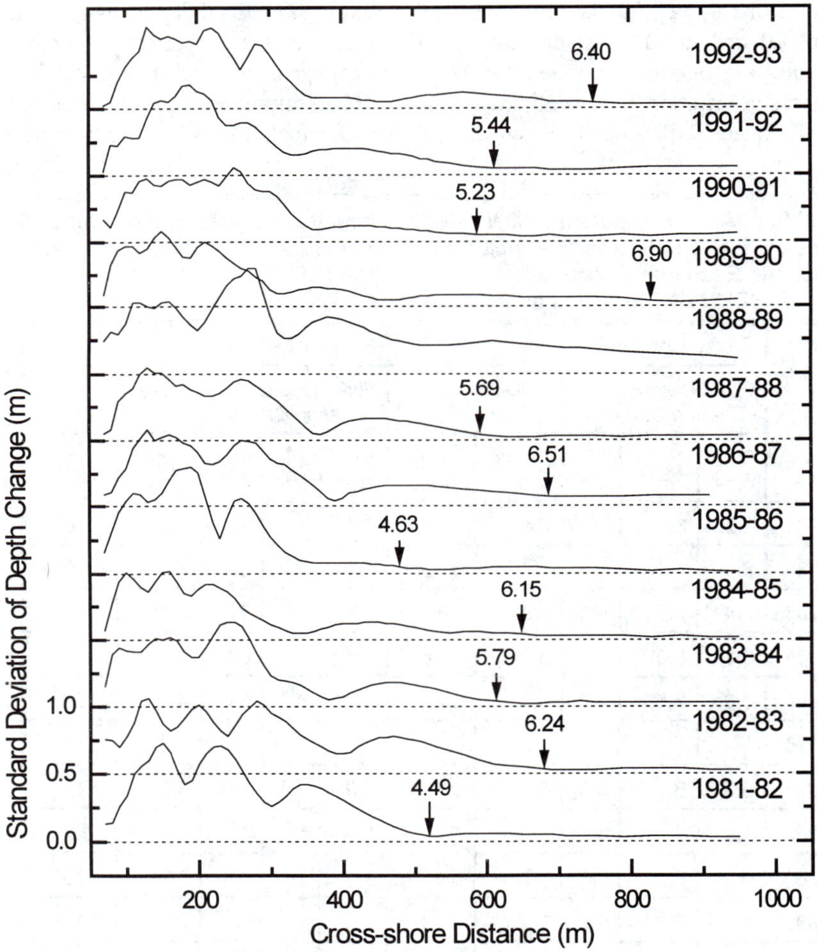

Figure 2. Standard deviation of depth change envelopes for 12 annual periods from July 1981 until end June 1993 (Profile line 62). Arrows mark SDDC-derived closure and are labeled with depths in m below MLW.

6-cm changes, respectively. Clearly, DoC defined with any constant depth change criterion is just one of a family of possible DoCs, depending on the depth change criterion selected. The smaller the depth change criterion, the greater the proportion of the active profile which will be defined.

The SDDC-defined boundary defines most annual DoCs, including periods of significant profile accretion when all the fixed change criteria fail. However, these cases are associated with a larger non-zero tail, indicating larger uniform changes to the seaward limit of the profile envelope than in typical years. On average, the SDDC DoC lies between the 10-cm and 15-cm time-interval DoC, but this relationship is not apparent for the non-closing cases using a fixed change criterion. Therefore, it is concluded that while the SDDC approach is a reasonable empirical method to define DoC, at Duck it shows bias towards smaller annual closure when there are uniform profile changes at the seaward ends of profiles. A fixed, standard deviation criterion (*e.g.,* 10 cm) would provide results more consistent with the fixed change criterion.

It is also noted that these annual SDDC analyses differ from Larson and Kraus (1994). Using all the Duck data from 1981 to 1991, they discriminated a boundary at about 3.6 m below MLW (400 m offshore). However, comparing Figures 1 (b) and 2 shows that while the minimum at 400 m is an intriguing aspect of the survey data, it is part of a larger declining trend to greater depths. Therefore, this position should not be considered the start of a non-zero tail in the sense of the results in Table 1.

Clearly, different definitions for DoC yield different depth estimates, so any stated empirical result needs to make the definitions explicit. Closure at Duck due to erosional events and over annual and longer time scales is now examined in more detail.

DEPTH OF CLOSURE DURING EROSIONAL EVENTS

Predictions Using Hallermeier (1981)

DoC during erosional events, defined by consistent offshore bar migration, was measured for 68 events with high waves (>2 m height) (Nicholls *et al.*, in review) and is henceforth referred to as erosional event-dependent (E-D) DoC. Following the rationale behind Equation 1, which defines a depth where most waves will not have experienced shallow water breaking, analysis of DoC during erosion at Duck shows that a 6-cm change criterion is the best indicator of closure. It is also most consistent with the documented survey accuracy. A 6-cm change criterion is used in all subsequent analysis

Equation 1 is found to define a conservative bound to the observations (Figure 3), agreeing with Hallermeier's (1981) original recommendations concerning input parameters. Below this limit, the observations show considerable scatter which is partly explained by the pre-event bar configuration — the largest DoC appear to be associated with the most dissipative profile morphology, namely, single outer bars (bar crest about 300 m offshore). Three erosional events produced observed DoCs which were near the survey limit (8-m depth), while three erosional events did not close within the measured profile. These results do not contradict Equation 1 as d_ℓ>8 m for those cases which close beyond the survey range.

Using an empirical best-fit approach to the data in Figure 3, a typical erosional event realizes 69% of the change predicted by Equation 1 (>95% confidence) (Figure 3):

$$d_E = 0.69\ d_\ell \qquad \textbf{(2)}$$

where d_E = observed E-D DoC. However, Equation 2 is affected by the fact of missing observations and its validity for other sites is uncertain.

Probability Analysis

The 68 available observations of erosional E-D DoC approximately conform to a lognormal probability distribution, *i.e.* plot near a straight line in Figure 4. The observations have been ranked in ascending order from i = 1 to 68, then assigned evenly-spaced probabilities of (i-0.5)/N, where N = (68+3); this procedure allows for the three presumably higher missing observations noted in the previous section. A lognormal probability distribution arises "when many random quantities cooperate multiplicatively so that the effect of a random change is in every case proportional to the previous value of the quantity" (Sachs, 1984, p. 107). For these nearshore DoC, the controlling variables include wave height and period and pre-event profile geometry (Nicholls *et al.*, in review).

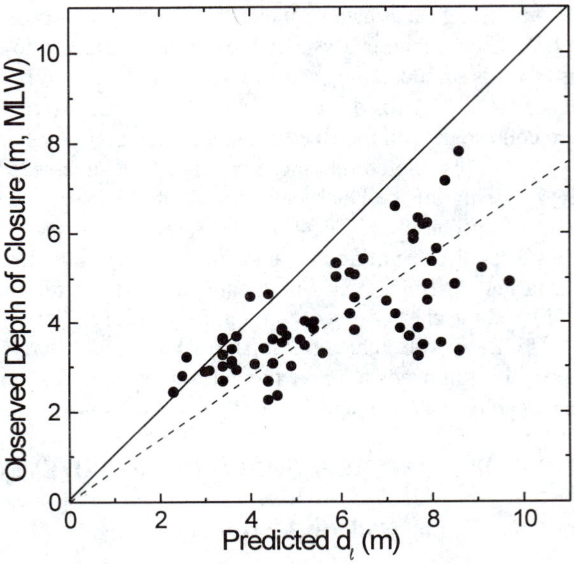

Figure 3. Observed versus predicted depth of closure (averaged alongshore) during 68 storms The best-fit line (Equation 2) is dashed.

Such well-behaved observations suggest a coherent sample of erosional events at Duck, so that extrapolation to rarer events than those yet recorded at the site maybe meaningful. One approach to extrapolation is to use annual maxima of the erosional E-D DoC, giving observations a direct basis in annual probabilities. Figure 4 also displays the nine observed annual maxima (three missing) in the same basic format, with these values showing fair conformance to a shifted lognormal probability distribution. The median annual extreme corresponds to 50% annual probability, or a recurrence interval of 2 years. Note that a lognormal distribution yields lower extremes than a exponential distribution, which would typically describe wave heights (USACE, 1984).

Empirical evidence in Figure 4 consistently supports the order-of-magnitude

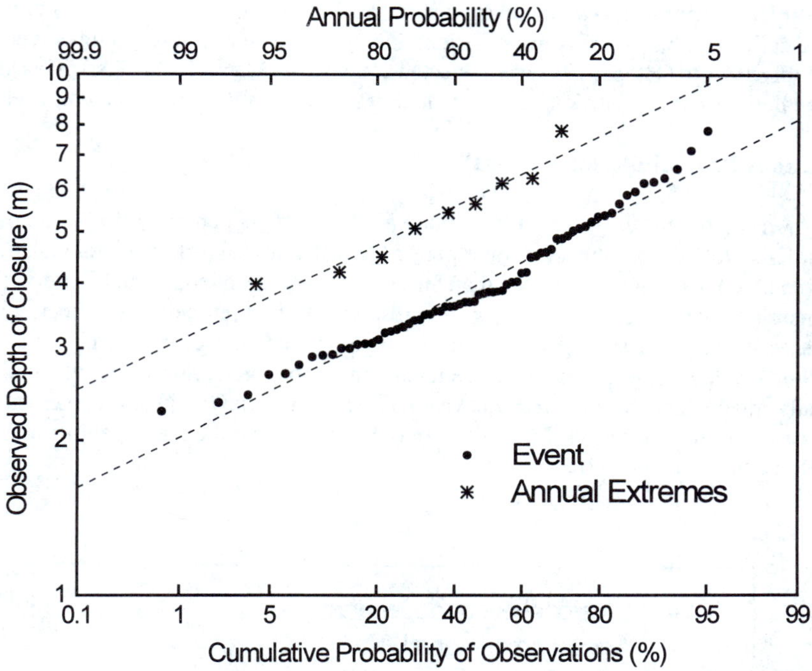

Figure 4. Lognormal distributions of all observed erosional event-dependent cases, comprising all observations (Event), and the nine observed annual extremes

estimates for extreme DoC in Table 2. These projections suggest that during more extreme erosional events, advances of DoC into water depths beyond 8- to 9-m depth below MLW are rather rare. This behavior is broadly consistent with (1) the predictions of Equation 1 which suggest that over 12.5 years, DoC during erosional events only exceeded 9-m on four occasions (or 6% of cases), and (2) conclusions about observed DoC in a probabilistic context, derived from an independent analysis of all the Duck survey data from 1981 to 1991 (Larson and Kraus, 1994).

Table 2. Approximate recurrence interval for the annual maximum erosional event-dependent closures based on a probability analysis (see text for more details).

Closure Depth (m below MLW)	7	8	9	10	11
Approximate Recurrence Interval (years)	2.5	5	12	30	70
Annual Probability (%)	40	20	8	3	1.4

DEPTH OF CLOSURE OVER ANNUAL AND LONGER TIME INTERVALS

This section examines the magnitude of DoC over a range of time intervals (1, 1.5, 2, 2.5, 4 and 8 years), including the applicability of Equation 1. DoC was evaluated using a time-interval approach (*i.e.*, comparing surveys separated by a fixed time interval, as defined earlier) with a 6-cm change criterion. Surveys three months apart in time over the period July 1981 to October 1993 were selected as the basic data set for analysis. This gave 46 time periods for annual closures, 44 annual time periods for 18-month closure, etc.

Predictions Using Hallermeier (1981)

An important result is that time-interval (T-I) DoC is generally deeper than the largest E-D DoC during the same time period. For the annual case, the residuals (d_ℓ - observed DoC) are about 1 m smaller than the erosional events, as shown in Figure 5. This demonstrates that T-I DoC is an integrated response to both erosional and accretional processes, rather than simply indicating the biggest erosional event during a period. Equation 1 is known to fail under accretional conditions, producing estimates that are generally smaller than the observed DoC (Nicholls *et al.*, in review). Therefore, if erosional processes dominate a period, Equation 1 may have more predictive capability than if accretion dominates the period.

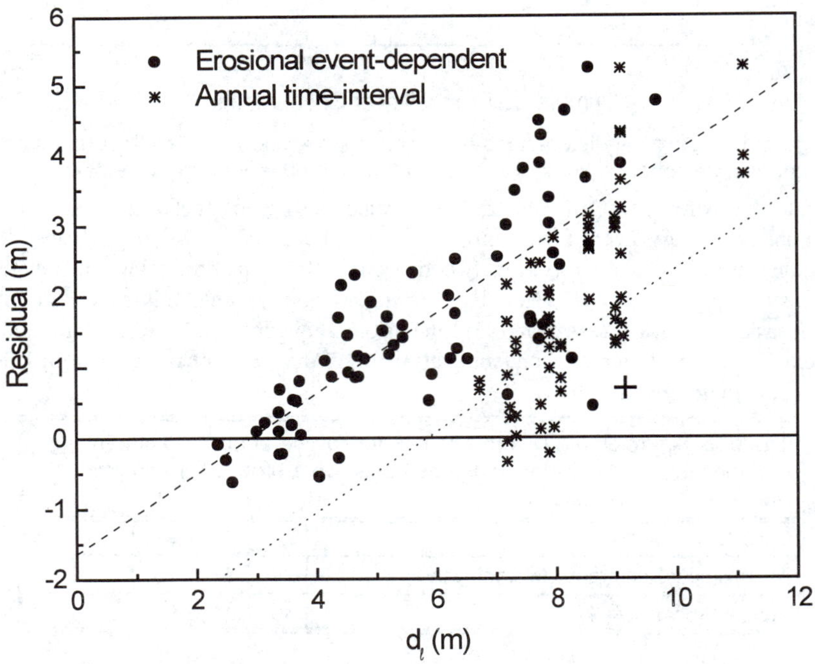

Figure 5. Residuals (d_ℓ - observed DoC) versus d_ℓ for erosional event-dependent closures and annual time-interval closure using a 6-cm change criterion. The dashed line is the best-fit to the erosional data, while the dotted line is the same line offset to pass through the approximate median of the annual time-interval data.

As the time interval between surveys increases, fewer cases close within the surveyed profile range at Duck (≤ 8-m depth below MLW) — 65% of annual interval cases, 63% of 18-month interval cases, 61% of 2-year interval cases, 51% of 30-month interval cases, 44% of 4-year interval cases and only 3% of 8-year interval cases close. The 8-year intervals always include significant profile movement and hence 6-cm closure is lost, except in one case. In all the non-closing cases, net profile change (>6 cm) is occurring seaward of 8-m depth.

Equation 1 is found to provide a useful conservative bound to those annual DoC which are defined within the survey range, see Figure 6. Using an empirical best-fit approach to the data, over a typical year, 76% of the change predicted by Equation 1 occurs (>95% confidence):

$$d_{T,1} = 0.76\ d_{\ell,1} \qquad (3)$$

where $d_{T,1}$ = observed annual T-I DoC. Note that Equation 3 has a numerical coefficient 10% larger than Equation 2.

Most of the annual cases which do not close are consistent with Equation 1 as $d_{\ell,t}$>8 m. However, there are five cases (8%) which are predicted to close but did not within the survey extent. This characteristic is also observed in the 18 month, 2 year and 30 month interval data, comprising 9%, 8% and 17% of cases, respectively. All the time periods where this behavior is observed were characterized by slow near-continuous onshore feed of sand from the upper shoreface (>5-m depth) (*cf.* Lee *et al.*, in review). Under these conditions, Equation 1 might be expected to be inapplicable, as already discussed. The large number of missing observations and the scatter in results makes objective assessment of the predictive value of Equation 1 more difficult to assess than for the erosional events.

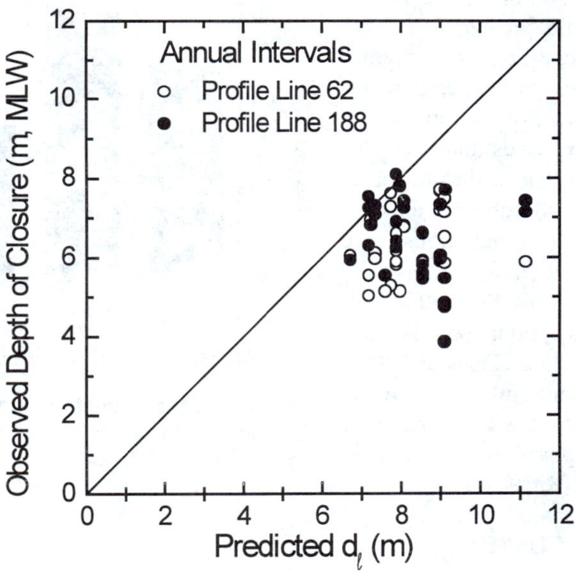

Figure 6. Observed versus predicted annual depth of closure for individual profiles. The dashed line is the best-fit (Equation 3). Of the 32 cases which are not observed to close, $d_{\ell,1}$> 8 m for 27 cases.

Sensitivity Analysis

The data were examined considering median values to circumvent the missing values, and annual up to 30 month time intervals. (Longer intervals have too few observations to define medians). The cases with d_ℓ of 11.1 m due to the most energetic wave event of record occurring on 31 October 1991, the "Halloween storm," are discarded as outliers not likely to be observed. The remaining data is divided as near equally as possible into a low and high bin, while retaining the capacity to define observational medians: *i.e.*, a meaningful central tendency for measurements. The median for each bin was then determined, treating all missing values as higher than available observations. The sensitivity of observations to predictions is simply the ratio of the difference in observation medians between the two bins, to the difference between the prediction medians. The sensitivity results are presented in Figure 7. Annual observations appear more strongly related to predictions than the erosional event observations. Above the annual time scale, there is a significant decline in sensitivity to negligible levels at 24- and 30-month time intervals. This indicates that at time intervals longer than one year, the observed closures are increasing much less rapidly than Equation 1 would suggest — *i.e.* the observations and predictions are decoupling. (Of course, Equation 1 still provides estimates of closure that tend to be larger than the observations, so that its predictions have value as a conservative bound.) From this evidence, Equation 1 is interpreted at Duck as being meaningful up to annual time scales, as originally proposed by Hallermeier (1978; 1981).

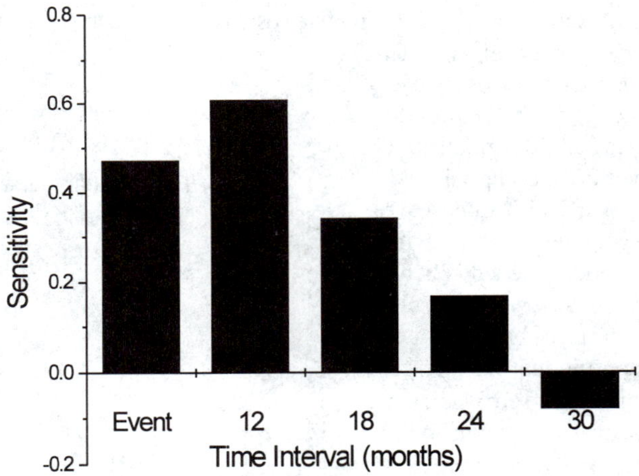

Figure 7. Sensitivity of observation to $d_{\ell,t}$ as a function of time interval: from individual events to a 30-month interval.

DISCUSSION

The time-interval behavior of closures in this extensive database appears to validate a common empirical definition of a useful DoC: such analyses usually group all available surveys of a profile, regardless of the time interval between data (*e.g.*, Inman *et al.*, 1993). The results documented here indicate that a closure estimate from such analysis is meaningful for a site, provided that the time interval of available data includes at least one seasonal profile translation. However, Figure 2 shows that profile variation in any one year may be a poor description of the statistical properties of a profile and more than one annual cycle of profile data is preferred. Further, a practical DoC may not be evident if the time span included exceptionally energetic wave sequences yielding a permanent profile translation (Inman *et al.*, 1993).

Some previous analyses may appear to be a little misleading about the DoC implications of the Duck profile database. Our analyses focusing on isolated energetic episodes have revealed many cases of resolvable and volumetrically significant changes at water depths well beyond the 4-m or 6-m deduced as limits by Larson and Kraus (1994). In addition, our analysis, focusing on intervals of one year clearly established that, within survey limits, an annual DoC is the prevalent situation at Duck, despite an absence of definite DoC over some longer time spans (Inman *et al.*, 1993). For each type of observation, the seaward limits of profile change are related to extreme wave conditions by Equation 1. Previously repeated conclusions about DoC at Duck are empirically supported by particular analyses addressing certain longer time intervals, but the present findings reflect more detailed and intensive examination of the same database. Additions to and analysis of that database continue at present, and a more coherent picture is expected to emerge progressively.

As Equation 1 provides a robust conservative bound for significant cross-shore sediment transport at Duck for individual erosional events up to the annual time scale, simply knowing the extreme wave conditions will allow estimates of closure to be made at similar microtidal, wave-dominated sites. Interestingly, both Equation 3 and the sensitivity analysis indicate a stronger relationship between annual DoC and Equation 1 than erosional DoC and Equation 1. Given that Equation 1 is an event-based formulation (*i.e.*, based on the biggest event in a period), this is a surprising result. Above an annual time scale, the available evidence suggests that DoC increases less rapidly than Equation 1 would predict. This is an important result which disagrees with earlier work (*e.g.*, Stive *et al.*, 1992) and has important implications for application. While we are unsure about applying Equation 1 to a *t*-year time interval, we are more confident about applying Equation 1 as a conservative bound with known wave conditions during an *t*-year erosional event. An estimate of how far sand might be carried offshore during extreme events can assist in analysis of a range of problems. In intervening periods between erosional events when Equation 1 may not be applicable, sand is generally being slowly transported onshore at open-coast sites such as Duck (Lee *et al.*, in review) with beneficial effects for the beach-nearshore sediment budget. In more sheltered sites with more limited swell, this may not be the case.

A key question is the specification of the wave information to use in Equation 1. The predictions presented here are based on 12-hour exceeded wave height and period measurements at a waverider buoy, 6-km offshore at a depth of about 18-m. Therefore, the measurements are shallow water, but well seaward of the breaker zone. In many cases such data are unavailable, although other wave measurements or hindcast wave information may be available. Improved guidance on optimum wave inputs to Equation 1 need to be developed.

Another problem is the variation in $d_{\ell,1}$ at any site due to variations in annual extreme waves. Twelve years of wave data at Duck show that $d_{\ell,1}$ can vary by more than 2 meters depending on the annual period selected: the 25th percentile is 7.7 m, the median is 8.7 m and the 75th percentile is 9.2 m below MLW. Given the apparent tendency for overprediction of closure depth, most attention might be focused on the two lower values. Taking beach nourishment as an example, selecting a smaller DoC will result in a lower initial project cost, but it may be prudent to plan for more frequent renourishment. This type of analysis will help to identify the trade-offs when selecting from the range of legitimate DoC estimates.

Despite the apparent decoupling of Equation 1 and observations with increasing time scale, it is apparent that application of DoC requires a more explicit consideration of time scale. At Duck, longer time scales are associated with greater profile variability at depth (*cf.* Figure 1(b)), implying larger DoC for a constant depth change criterion. The surveys define DoC for 95% of erosion episodes and for 65% of annual intervals, but over the 12-year window they have failed to define a seaward limit of the profile envelope. An important lesson is that when designing long-term, high accuracy, beach monitoring programs to define the profile envelope, the distribution of $d_{\ell,1}$ helps to define a realistic *minimum* target depth for routine surveys. At Duck, the surveys considered in this paper typically only reach the 25th percentile of $d_{\ell,1}$, with the maximum depths attained being less than the median $d_{\ell,1}$. With our present knowledge of Duck, occasional surveys to the 75 percentile of $d_{\ell,1}$ would appear a reasonable target depth for Duck and similar sites.

CONCLUSIONS

A meaningful depth of closure is observed at Duck under erosional events and a range of time intervals from 1 year to 30 months. The depth limit $d_{\ell,t}$ is found to be robust and useful for erosional events and annual time intervals. Therefore, with limited environmental information (the event or annual 12-hour exceeded wave height and period), a reliable closure estimate apparently can be made. However, Equation 1 is a conservative bound to the observed closures and results should be applied with this in mind. At longer time scales, the observations and Equation 1 decouple as observed closure appears to increase more slowly than the predictions of $d_{\ell,t}$ would suggest.

Knowing the nature of Equation 1 and how closure evolves with increasing time scale, coastal engineers can make better judgements when interpreting sparse survey data. They can also optimize application of a closure value by considering the trade-offs in selecting from a range of legitimate estimates for a site. Finally, the Duck data indicates our limited understanding of cross-shore sand exchange between the beach and the upper shoreface. To capture more of the change for extreme cases, routine measurements are required to greater depths. Collection and investigation of the Duck data set continues.

ACKNOWLEDGMENTS

This study is a collaboration between the PACE-project in the framework of the EU-sponsored Marine Science and Technology (MAST-III) under contract no. MAS3-CT95-002 and the Geologic Analysis of Shelf/Beach Sediment Interchange Work Unit under the Coastal Research and Development Program of the US Army Corps of Engineers. Permission was granted by the Chief of Engineers to publish this information.

REFERENCES

Birkemeier, W.A., Miller, H.C., Wilhelm, S.D., DeWall, A. E. and Gorbics, C. S., 1985. A user's guide to the coastal engineering research center's (CERC's) Field Research Facility. *Instruction Report CERC-85-1*, U.S. Army Corps of Engineers, Waterways Experiment Station, Vicksburg, MI, 121 pp.

Hallermeier, R. J., 1978. Uses for a calculated limit depth to beach erosion. *Proceedings 16th Coastal Engineering Conference,* ASCE, 1493-1512

Hallermeier, R. J., 1981. A profile zonation for seasonal sand beaches from wave climate, *Coastal Engineering*, 4, 253-277.

Hands, E. B., 1983. The Great Lakes as a test model for profile responses to sea level changes. In: Komar, P.D.(ed.), *CRC Handbook of coastal processes and erosion*, CRC Press, Boca Raton, pp.167-189

Inman, D.L., Hany, M., Elwany, S. and Jenkins, S.A., 1993. Shorerise and bar-berm profiles on ocean beaches. *Journal of Geophysical Research*, 98 (C10), 18181-18199.

Kraus, N.C. and Harikai, S., 1983. Numerical model of the shoreline change at Oarai beach. *Coastal Engineering*, 7, 1-28

Larson, M. and Kraus, N.C., 1994. Temporal and spatial scales of beach profile change, Duck, North Carolina, *Marine Geology*, 117, 75-94.

Lee, G. and Birkemeier, W.A., 1993. Beach and nearshore survey data: 1985-1991, CERC Field Research Facility. *Technical Report-93-3,* US Army Corps of Engineers, Waterways Experiment Station, Vicksburg, MS., 26 pp.

Lee, G., Nicholls, R. J., Birkemeier, W.A. and Leatherman, S.P., 1995. A conceptual fairweather-storm model of beach-nearshore profile evolution at Duck, North Carolina, U.S.A. *Journal of Coastal Research*, 11, 1157-1166.

Lee, G., Nicholls, R. J. and Birkemeier, W.A., in review. Storm-driven variability of the beach-nearshore profile at Duck, North Carolina., 1981-1991. *Marine Geology*.

Lippmann, T. C. and Holman, R.A., 1990. The spatial and temporal variability of sand bar morphology. Journal of Geophysical Research, 95(C7), 11575-11590.

Nicholls, R. J., Birkemeier, W. A. and Lee, G. In review. Evaluation of depth of closure using data from Duck, NC, USA. *Marine Geology*.

Sachs, L. 1984. *Applied Statistics*, 2nd Edition, Springer-Verlag, New York.

Stive, M. J. F., Nicholls, R. J. and DeVriend, H. J. 1991. Sea-Level Rise and Shore Nourishment: A Discussion. *Coastal Engineering*, 16, 147-163.

Stive, M. J. F., DeVriend, H. J., Nicholls, R. J. and Capobianco, M., 1992. Shore nourishment and the active zone; a time scale dependent view. *Proceedings 23rd Coastal Engineering Conference*, ASCE, New York, pp. 2464-2473.

Stive, M. J. F and DeVriend, H. J. 1995. Modelling shoreface profile evolution. *Marine Geology*, 126,235-248.

USACE, 1984. *Shore Protection Manual*, 4th Ed. U.S. Army Corps of Engineers, Waterways Experiment Station, Vicksburg, MI (2 vol.).

CHAPTER 300

Numerical modeling of sediment transport for various mode

Masanobu ONO[1)], Ichiro DEGUCHI[2)] and Toru SAWARAGI[3)]

Abstract

There are various mode of sediment transport on sandy beach. They are bed load, suspended load and sheet flow. Until now, various kinds of numerical model for analyzing these sediment transport have been developed based on the assumption of each mode of sediment transport. Consequently, it is difficult to apply them to different mode of sedimentation.

In this study, we propose a relatively simple numerical model for simulating sediment motion of various mode based on the semi-multi-phase flow model. The applicability of the model to sheet flow transport and suspended load transport is examined by using existing experimental data.

The results show that for the suspended sediment, it is necessary to take into the effect of the vortex to lift up sediment to high position. To construct numerical model for simulating sediment transport phenomena of the various mode , we have to introduce generation , development and disappearance process of bed ripple in this model.

Introduction

Mode of sediment transport continually changes from bed load, suspended load to sheet flow according to the increase in magnitude of agitation force. Until now, various kinds of numerical model for analyzing these sediment transport have been developed. An advection-diffusion equation has been widely used to analyze suspended sediment concentration. Suspended load is expressed as the product of concentration and migration speed of suspended sediment that is assumed to be equivalent to water particle velocity when the concentration is not so high. Concentration of bed load and sheet flow has also been analyzed by solving the advection-diffusion equation and the migration speed is estimated by solving a momentum equation of multi-phase flow. When we analyze sediment motion by these model, we have to know the mode of sediment transport before we determine what kind of model we should use.

[1)] Research Assoc., and [2)]Assoc. Prof., Dept. of Civil Engineering, Osaka University, Yamada-oka, Suita, Osaka, 565, Japan

[3)]Prof., Dept. of Civil Engineering, Osaka sangyo University, Japan

In this study, we propose a relatively simple numerical model for simulating sediment motion of various mode based on the semi-multi-phase flow model. The applicability of the model to sheet flow transport and suspended load transport is examined by using existing experimental data. The effect of vortex shedding from the rippled bed on the Schmidt number is also investigated numerically.

Numerical simulation method for analyzing sediment transport

(1)Basic equations

Numerical model consists of a horizontal momentum equation of one-phase flow and equations of mass conservation of two-phase flow. There are generally six equations, that is, conservation of mass and conservation of horizontal and vertical momentum flux for sediment phase and fluid phase.

The continuity equations for sediment and water phase are shown by Eq.(1) and Eq.(2).

$$\frac{\partial \rho_S c}{\partial t} + \frac{\partial \rho_S c u_S}{\partial x} + \frac{\partial \rho_S c w_S}{\partial z} = 0 \tag{1}$$

$$\frac{\partial \rho_f (1-c)}{\partial t} + \frac{\partial \rho_f (1-c) u_f}{\partial x} + \frac{\partial \rho_f (1-c) w_f}{\partial z} = 0 \tag{2}$$

where c is volumetric concentration of sediment, ρ_S and ρ_f are the density of sediment and water, us and uf are horizontal velocity of sand and water particle, ws and wf are vertical velocity of sand and water particle.

The momentum equations for sediment and water phase are shown by Eqs.(3),(4) and Eqs.(5),(6).

$$\frac{\partial \rho_S c u_S}{\partial t} + \frac{\partial \rho_S c u_S^2}{\partial x} + \frac{\partial \rho_S c u_S W_S}{\partial z} = -c\frac{\partial p}{\partial x} + \frac{\partial r_{xx}}{\partial x} + \frac{\partial r_{zx}}{\partial z} + f_x \tag{3}$$

$$\frac{\partial \rho_S c W_S}{\partial t} + \frac{\partial \rho_S c u_S W_S}{\partial x} + \frac{\partial \rho_S c W_S^2}{\partial z} = -c\frac{\partial p}{\partial z} + \frac{\partial r_{xz}}{\partial x} + \frac{\partial r_{zz}}{\partial z} - \rho_S c g + f_z \tag{4}$$

$$\frac{\partial \rho_f (1-c) u_f}{\partial t} + \frac{\partial \rho_f (1-c) u_f^2}{\partial x} + \frac{\partial \rho_f (1-c) u_f w_f}{\partial z} = -(1-c)\frac{\partial p}{\partial x} - f_x \tag{5}$$

$$\frac{\partial \rho_f (1-c) w_f}{\partial t} + \frac{\partial \rho_f (1-c) u_f w_f}{\partial x} + \frac{\partial \rho_f (1-c) w_f^2}{\partial z} = -(1-c)\frac{\partial p}{\partial z} - \rho(1-c)g - f_z \tag{6}$$

where p is pressure, fi(i=x,z) is the i-th component of interaction force per unit volume between sediment and water, γ_{ij} is inter granular stress tensor.

There are six unknowns, p, uf, wf, us, ws, and c. However, it is very difficult to get stable numerical results of these equations due to very strong non-linear interaction of each equations, Therefore in our model we use the horizontal momentum equations and mass conservation equations and apply empirical relation for determining vertical velocity of sediment phase.

(2)Numerical simulation method

To construct numerical model we made some assumptions and simplification. One is that the horizontal velocity of moving sand is the same as the horizontal fluid velocity. This implies that we apply single layer model for the sediment larded water and Eqs.(3) and (5) become Eq.(7) as shown below:

$$\frac{\partial}{\partial t}\left\{\rho u-\left(\rho_s-\rho_f\right)k_x\frac{\partial c}{\partial x}\right\}+\frac{\partial}{\partial x}\left\{\rho u^2-2u\left(\rho_s-\rho_f\right)k_x\frac{\partial c}{\partial x}-\left(\mu_e+\rho\Omega_x\right)\frac{\partial u}{\partial x}\right\}$$
$$+\frac{\partial}{\partial z}\left[\left\{(1-c)\rho w_f+c\rho_s w_s\right\}u-\left(\mu_e+\rho\Omega_z\right)\frac{\partial u}{\partial z}\right.$$
$$\left.-u\left(\rho_s-\rho_f\right)k_z\frac{\partial c}{\partial z}-\left(\rho_s w_s-\rho_f w_f\right)k_x\frac{\partial c}{\partial x}\right]=-\frac{\partial p}{\partial x} \tag{7}$$

where u ($=u_s=u_f$) is the horizontal velocity of sand and fluid particle and μ_e is apparent viscosity of sediment laden water (Savage-McKeown ,1983).

Sediment concentration C is calculated from the continuity equation for the sediment phase Eq. (1) by transforming it into an advection-diffusion equation with diffusion coefficient k_i as follows:

$$\frac{\partial c}{\partial t}+\frac{\partial cu}{\partial x}+\frac{\partial cw_s}{\partial z}=\frac{\partial}{\partial x}\left(k_x\frac{\partial c}{\partial x}\right)+\frac{\partial}{\partial z}\left(k_z\frac{\partial c}{\partial z}\right) \tag{8}$$

Eq.(9) is derived from Eq.(2) by performing the same transform.

$$-\frac{\partial c}{\partial t}+\frac{\partial(1-c)u}{\partial x}+\frac{\partial(1-c)w_f}{\partial z}=-\frac{\partial}{\partial x}\left(k_x\frac{\partial c}{\partial x}\right)-\frac{\partial}{\partial z}\left(k_z\frac{\partial c}{\partial z}\right) \tag{9}$$

The diffusion coefficient is assumed to be proportional to the kinematic eddy viscosity Ω_i by using Schmidt number Sm as follow.

$$k_x=\Omega_x/Sm \qquad k_z=\Omega_z/Sm \tag{10}$$

$$\mu_e=1.2\lambda^2\mu_0 \qquad \lambda=1/\left\{\left(c_{max}/c\right)^{1/3}-1\right\} \tag{11}$$

where μ_0 is water viscosity, λ is the Linear concentration, c_{max}(=0.65) is the maximum possible static concentration of uniform spheres. We used the expression for kinematic eddy viscosity proposed by Nadaoka et al.(1990).

$$\Omega_x=l_{(z)}^2\left|\frac{\partial u}{\partial x}\right| \qquad \Omega_z=l_{(z)}^2\left|\frac{\partial u}{\partial z}\right| \qquad l_{(z)}=\kappa\int_{-\infty}^{z}\left(c_{max}-c\right)/c_{max}dz \tag{12}$$

where l(z) is mixing length and κ is Karman constant.

To evaluate the vertical velocity of sediment phase, an empirical expression to relate settling velocity of sand particle and concentration is used.

$$w_s=-w_{s0}\left(1-c/c_{max}\right)^{2.3} \tag{13}$$

where w_{s0} is the settling velocity of sand particle in clear water.

Vertical velocity of water particle is calculated by the continuity equation for two phase flow. Eq.(14) is derived from Eqs.(8) and (9).

$$\frac{\partial u}{\partial x}+\frac{\partial}{\partial z}\left\{cw_s+(1-c)w_f\right\}=0 \tag{14}$$

Equations (7),(8),(13) and (14) were simultaneously solved by transforming them into finite difference equations. In this model, horizontal phenomena is assumed to be homogeneous in the direction of wave propagation.

(3)Initial and boundary conditions

Z axis is taken upward from the initial sand surface as shown in Fig.1. The up-ward boundary conditions are given by Eqs.(15) and (16) at the boundary layer whose thickness is given by Eq.(17).

$$cw_s + k_z \frac{\partial c}{\partial z} = 0 \qquad (15) \qquad\qquad u = U_0 \sin(\sigma t) \qquad (16)$$

$$\max(D_z) = 4\kappa \max(u^*)/\sigma \qquad u^* = l_{(z)} \frac{\partial u}{\partial z} \qquad (17)$$

The pressure gradient in Eq.(7) is given by the pressure gradient outside the boundary layer using the expression of Eq.(16).

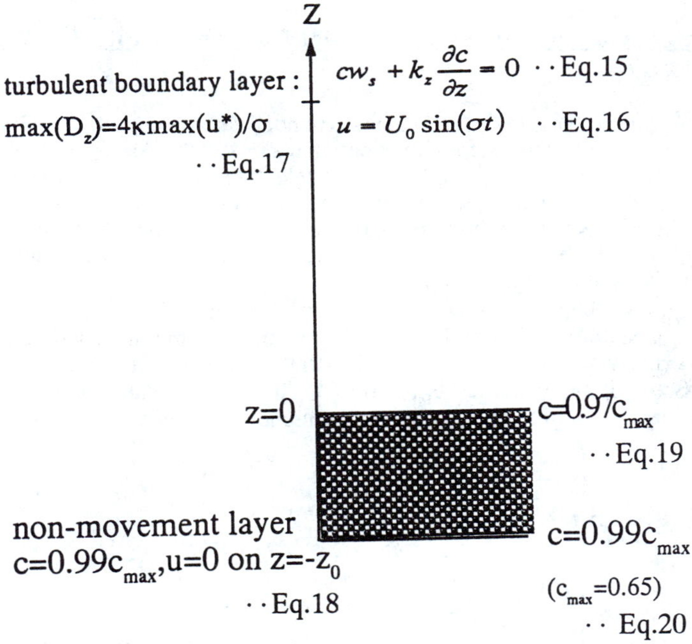

Fig. 1 Coordinate system

Two boundary conditions at the bottom are applied. One is the velocity of sediment laden water is 0 and the other is sediment concentration is $0.99c_{max}$. These are given at assumed initial thickness of sand layer $z=-z_0$. To prevent dispersion of vertical gradient of apparent viscosity near sand surface, we distribute sediment concentration in sand layer from $0.97c_{max}$ to $0.99c_{max}$ between $z=-z_0$ and $z=0$ linearly. Calculations are started from the still water and there is no suspended sediment concentration in the region of z is greater than 0.

The thickness of moving sand layer is determined by using Bagnold's condition as shown by Eq.(21).

$$\mu_e \frac{\partial u}{\partial z} \geq \int_z^{+\infty} (\rho_s - \rho_f) gc \tan\phi dz \qquad (21)$$

The friction angle ϕ in Eq. (21) also depends on sediment concentration. When the linear concentration is smaller than 1, there is no resisting force against shear. We assume that the value usually used for ϕ is applicable when the linear concentration is larger than 14. It is said that fluid act as a Newtonian fluid when the linear concentration is smaller than 14. We change the value of ϕ continuously in this region as shown by Eq. (23).

$$\mu_e \frac{\partial u}{\partial z} \geq \int_z^{+\infty} (\rho_s - \rho_f) g c \tan(\alpha\phi) dz \qquad (22)$$

$$\alpha = \lambda / \lambda_{14} \qquad \lambda \leq 14 (c \leq 0.53)$$

$$\alpha = 1 \qquad 14 \leq \lambda (0.53 \leq c) \qquad (23)$$

Applicability of the model under sheet flow and suspended sediment conditions

We carried out a series of calculation under various conditions. Here, the calculated results are compared with experimental results under sheet flow condition reported by Horikawa et al.(1982) and experimental results under pure suspension condition given by Nakato et. al.(1977) to examine the applicability of the numerical method. There was not any bottom configuration in the former experiments and there were ripples on the bottom in the latter experiments. Both of the experiments were carried out in oscillatory flow tunnels.

Figures 2 and 3 show the comparisons of calculated and measured horizontal velocity, concentration profile. Fig. 4 is the thickness of moving sand layer under the same experiment (Horikawa et. al.(1982)). In these calculation we assumed that the value of Schmidt number was 10. Measured velocity , concentration and thickness of moving layer are roughly reproduced by the numerical simulation.

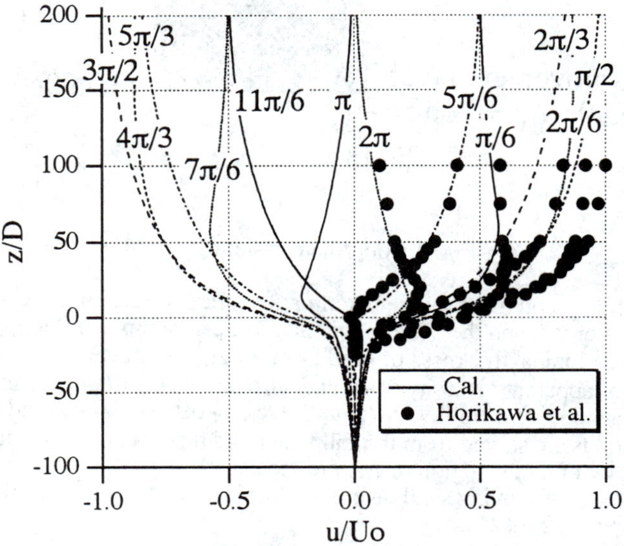

Fig. 2 Comparison between Horikawa's experimental results and calculated results under sheet flow condition (Horizontal velocity profile, Schmidt number :Sm=10)

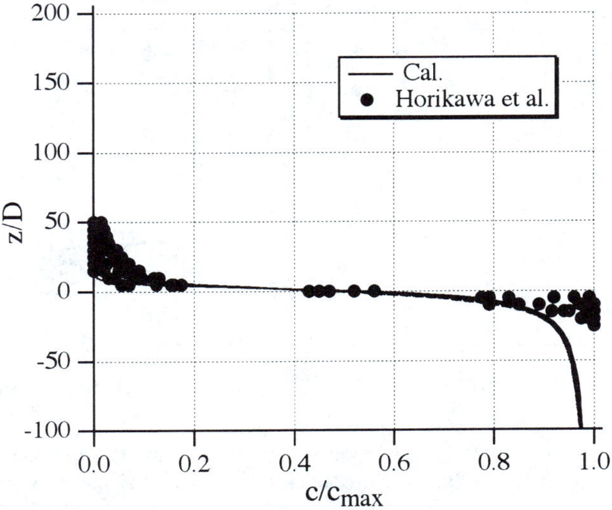

Fig. 3 Comparison between Horikawa's experimental results and calculated results under sheet flow condition (Sediment concentration profile , Sm=10)

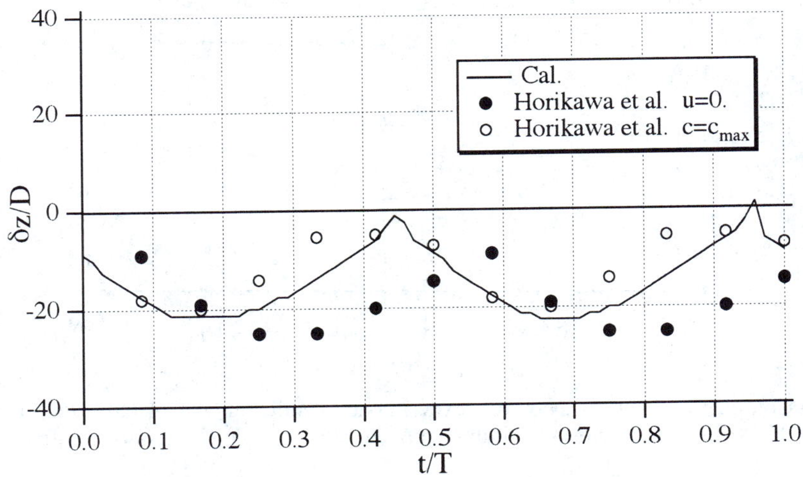

Fig. 4 Comparison between Horikawa's experimental results and calculated results under sheet flow condition (Thickness of moving layer , Sm=10)

We calculate the vertical distribution of purely suspended sediment concentration given by Nakato and others by using the same Schmidt number, that is Sm=10. In Figs.5 and 6, some examples of calculated results (shown by solid line) are compared with the measured results. As can be seen from this figure, we could not reproduce the measured distribution by the calculation.

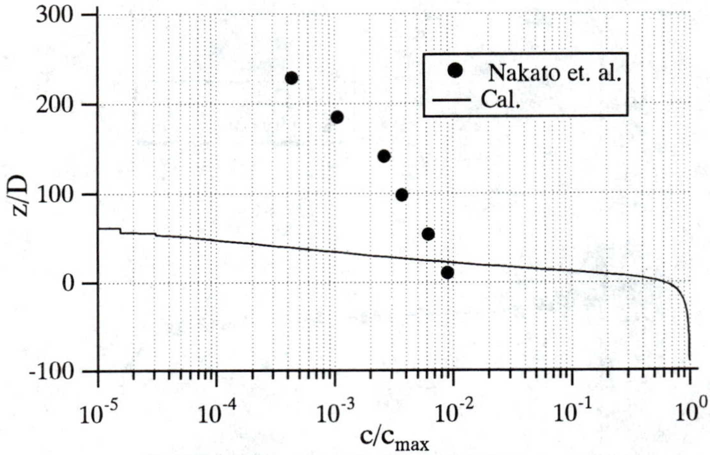

Fig. 5 Comparison between Nakato's experimental results and calculated results under suspended sediment condition (sediment concentration profile ,Case1, Sm=10)

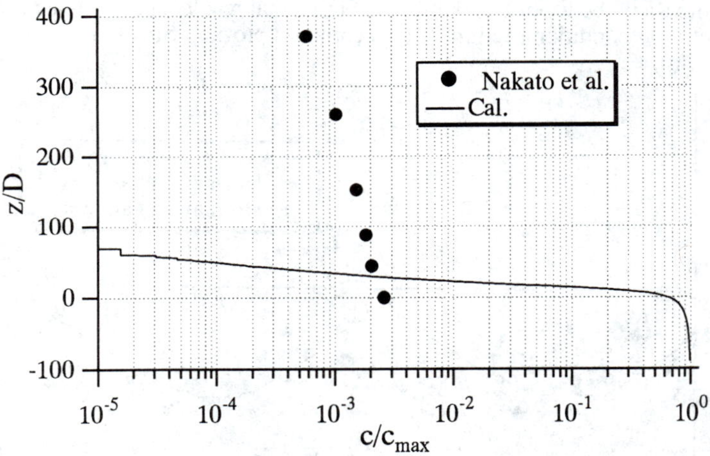

Fig. 6 Comparison between Nakato's experimental results and calculated results under suspended sediment condition (sediment concentration profile ,Case2, Sm=10)

The reason for this discrepancy was simply because the kimematic eddy viscosity did not diffuse from the horizontal bottom and diffusion coefficient outside the boundary layer become almost 0. To improve our numerical model, we changed the value of Schmidt number from bed load layer to suspended load layer. In the bed load layer, the same value of Schmidt number as the sheet flow is used and for the suspended load layer, the value is reduced to be 0.5 to increase diffusion coefficient. We defined the bed load layer where the concentration is lower than the reference concentration given by Sawaragi et al.(1985).

The calculated results are shown by broken lines in Figs. 7 and 8. The degree of agreement between measured and calculated concentration is improved. This implies that the flow field, especially structure of turbulence in the sheet flow region and suspended load layer is perfectly different from each other and systematic vortices generated by the ripples play an important role in the purely suspended load layer.

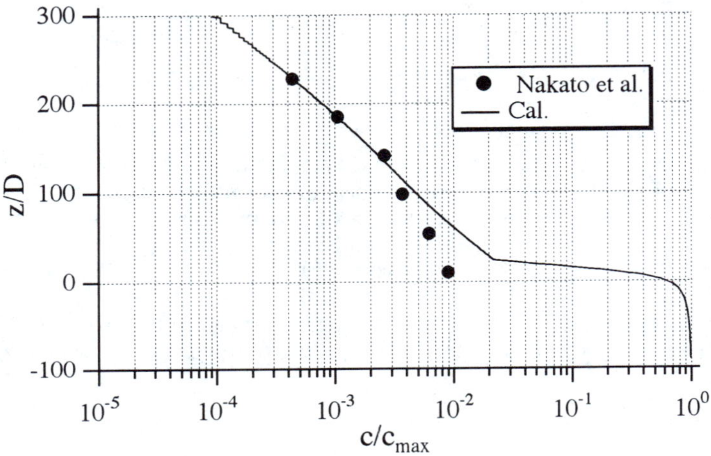

Fig. 7 Comparison between Nakato's experimental results and calculated results under suspended sediment condition (Case1, Suspended load layer :Sm=0.5, Bed load layer :Sm=10)

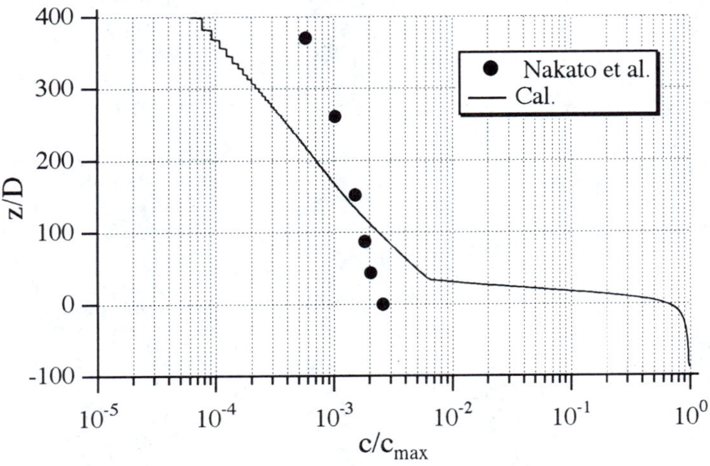

Fig. 8 Comparison between Nakato's experimental results and calculated results under suspended sediment condition (Case2, Suspended load layer :Sm=0.5, Bed load layer :Sm=10)

Diffusion coefficient of suspended sediment on the ripple bed

To investigate flow field and the effect of the systematic generation of vortex on rippled bottom, we calculated vortex flow field on ripple bed by modified vortex filament method originally proposed by Longuet-Higgins(1981). We used circular vortex instead of vortex filament to avoid velocity from going infinity. Eqs.(24) and (25) are the complex velocity potential of circular vortex used in the analysis.

$$|W - W_n| > Gr \Rightarrow \frac{\Gamma_n}{2\pi i}\ln(W - W_n) \tag{24}$$

$$|W - W_n| < Gr \Rightarrow \frac{\Gamma_n}{2\pi i}\frac{1}{Gr}(W - W_n) \tag{25}$$

where W is the arbitrary point in the interior of the unit circle on the complex number plane, W_n is the position of circular vortex, Γ_n is the circulation of circular vortex, Gr is the radius of circular vortex that will be determined from the comparison of measured and calculated turbulence intensity of horizontal velocity on the ripples. When the distance between W and Wn is longer than Gr, the complex velocity potential becomes the same equation of vortex filament. But when Wn approaches to W ,the complex velocity potential converges to 0.

Decay of circulation of separated vortex is reproduced by decreasing circulation of each circular vortex by assuming the following time variation with the damping coefficient α_G :

$$\Gamma_n = \Gamma_{0n}\exp(-\alpha_G t_{0n}/T) \tag{26}$$

where Γ_{on} is the initial circulation of n-th circular vortex, t_{0n} is the elapsed time since n-th circular vortex has been released. The damping coefficient will also be determined from the measured time variation of vortex circulation.

Solid lines in Fig.9 show the calculated time variation of circulation under various values of damping coefficient. In the figure, measured time variation of circulation of separated vortex (Horikawa et al.,1992) is also shown. Fig. 11 shows the comparison of measured turbulent intensity of horizontal velocity on the ripples (Sawamoto et al.,1981) and calculated result by giving various values of radius for the circular vortex. The vertical axis z is the distance upward from the ripple crest.

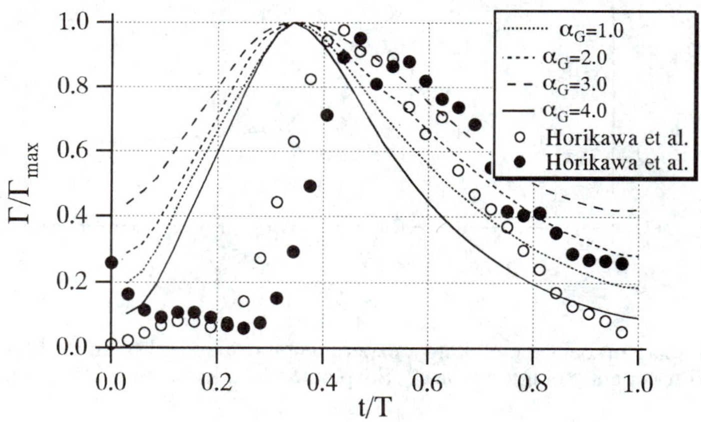

Fig. 9 Time variation of circulation of separated vortex

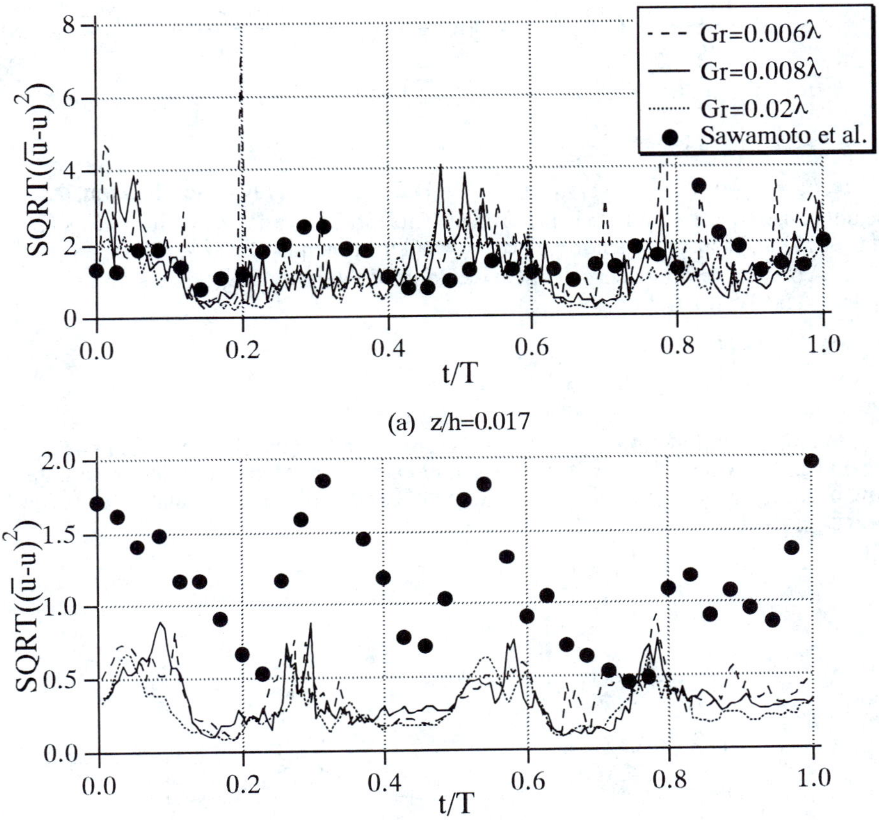

(a) z/h=0.017

(b) z/h=0.050

Fig. 10 Vertical distribution of turbulent intensity of horizontal velocity

By comparing calculated and measured result of time decay of circulation and turbulent intensity, circular vortex radius is determined to be 0.8% of ripple length and vortex circulation damping coefficient to be 4.

Calculation of diffusion coefficient of suspended sediment

The motions of suspended sediment are calculated in the flow field obtained on the ripples by an above-mentioned procedure. The governing equations of sediment motion are horizontal and vertical equations of motion as shown in non-dimensional forms by Eqs.(27) and (28) and are solved by using the forth-order Runge-Kutta method.

$$\frac{\partial u_s^*}{\partial t^*} = \frac{3}{4}\frac{1}{\rho_s^* + C_M} C_D K_C \left|u_f^* - u_s^*\right|\left(u_f^* - u_s^*\right) + \frac{\partial u_f^*}{\partial t^*} \tag{27}$$

$$\frac{\partial w_s^*}{\partial t^*} = \frac{3}{4}\frac{1}{\rho_s^* + C_M} C_D K_C \left| w_f^* - w_s^* \right| \left(w_f^* - w_s^* \right) + \frac{\partial w_f^*}{\partial t^*} - \frac{\rho_s^* - 1}{\rho_s^* + C_M} g^* \tag{28}$$

$$t^* = t\sigma, \left(u_s^*, w_s^*\right) = \left(u_s, w_s\right)/w_{s0}, \left(u_f^*, w_f^*\right) = \left(u_f, w_f\right)/w_{s0},$$
$$\rho_s^* = \rho_s/\rho_f, g^* = g/\left(w_{s0}\sigma\right) \tag{29}$$

where $R_e = w_{s0}D/\nu$, $K_C = w_{s0}/D\sigma$, $C_M = 1/2$, g is the gravity acceleration, σ is the angular frequency and C_D is the drag coefficient. The variables with ' in Eqs.(27) and (28) are dimensionless normalized by relation shown by Eq.(29).

To evaluate drag coefficient C_D we adopted the following expression of Molerus and Werther(1968):

$$C_D = \frac{24}{R_e}\left(\frac{1}{\left| w_f^* - w_s^* \right|} + \frac{0.152}{\sqrt{\left| w_f^* - w_s^* \right|}} + 0.0151 R_e \right) \tag{31}$$

Initial positions of sand particles are distributed on the ripples as shown Fig.11. and they are released four cycles after calculation of the flow field where the flow becomes steady and stable. The motion of sand particles is calculated for seven cycles.

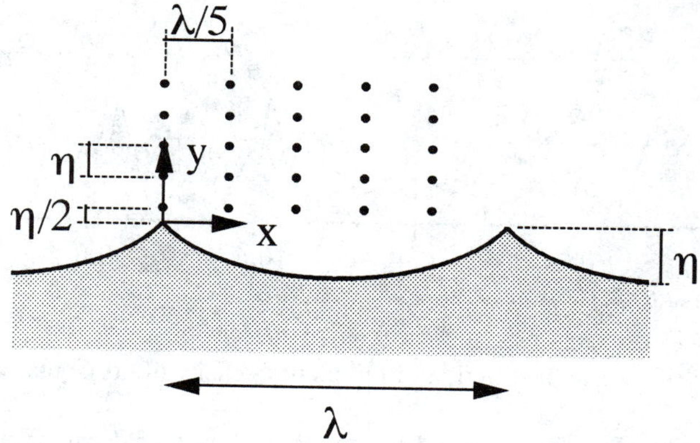

Fig. 11 Coordinate system and initial arrangement of sand particles

Diffusion coefficient of suspended sediment is evaluated by the calculated displacement of sand particles by Eq.(32).

$$k_z(t) = \frac{1}{2}\frac{d\left\{\sum_{i=1}^{25} l_i(t)^2/25\right\}}{dt} \tag{32}$$

where l(i) is relative position vector of sand particles evaluated by Eq.(33).

$$l_i(t) = s_i(t) - S(t) \tag{33}$$

where $s_i(t)$ is the position of i-th sand particle, S(t) is the average position of all sand particles.

Fig.12 shows the comparison of the calculated and measured diffusion coefficient of suspended sediment. Calculated results show good agreement with the measured diffusion coefficient. This means that systematic vortex shedding from the ripple bed plays very important role in sediment suspension. It is not necessary for the sediment movement under the sheet flow condition that takes place on a flat bed to exist such kind of systematic vortex. For the suspended sediment, it is necessary to take into the effect of the vortex to lift up sediment to high position.

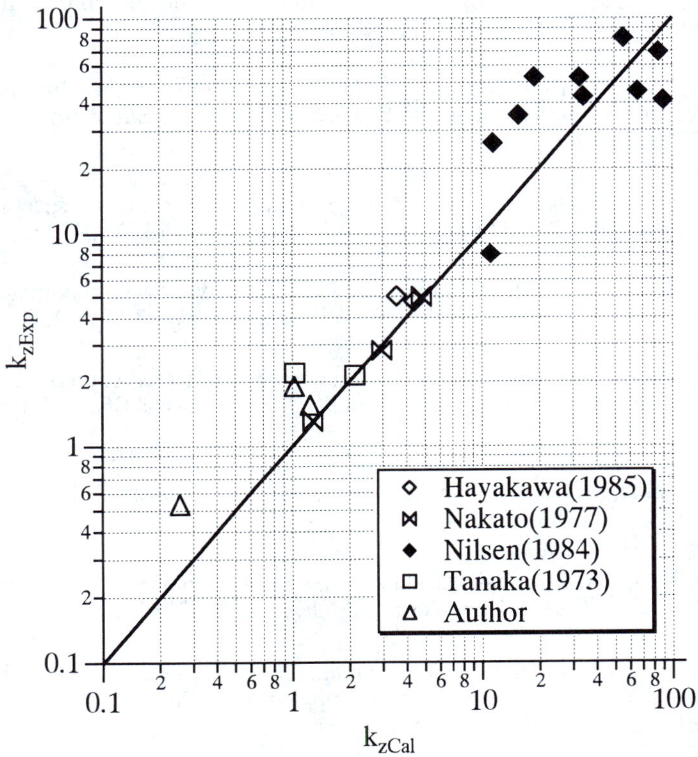

Fig. 12 Comparison of measured and calculated diffusion coefficient

Conclusions

We propose a relatively simple numerical modeling of sediment transport for various mode based on the semi-multi- phase flow model and examine the applicability of the model by using existing experimental results of sheet flow and suspended sediment. Although calculated results roughly reproduce the experimental results under sheet flow condition, we could not predict vertical distribution of the concentration of the suspended sediment.

To investigate this reason, we examine the suspended sand particle motion in vortex flow on the ripples. The diffusion coefficient that is evaluated by analyzing sand particle motions on ripples is coincide with the experimental results. This result implies that systematic vortices generated by the ripples lift up suspended sediment to high position. Therefore to construct numerical model for various mode of sediment transport, we have to introduce generation, development and disappearance process of bed ripple in the model.

Reference

Savage S. B. and S. McKeown : Shear stresses developed during rapid shear of concentrated suspensions of large spherical particles between concentric cylinders. J. Fluid Mech. ,Vol. 127, pp.453-472,1983.

Nadaoka, K. and H. Yagi: Single-phase fluid modeling of sheet flow toward the development of "Numerical mobile bed", Proc. 22nd Coastal Eng. Conf., ASCE, pp.2346-2359, 1990.

Horikawa, K., A. Watanabe and S. Katori: Sediment transport under sheet flow condition, Proc. 18th Coastal Eng. Conf., ASCE, pp.1335-1352, 1982.

Nakato T., F. A. Locher, J. R. Glover and J. F. Kennedy : Wave entrainment of sediment from rippled beds, A.S.C.E., WW1, Vol. 103, pp.83-99, 1977.

Sawaragi T., J. S. Lee, and I. Deguchi: A new model for a prediction of beach deformation around a rivermouth, Proc. Int'l Sympo. on Ocean Space Utilization '85, pp.229-236, 1985.

Longuet Higgins, M. S.: Oscillating flow over steep sand ripples, J. Fluid Mech., Vol. 107, pp.1-35, 1981.

Horikawa, K. and S. Mizutani: Oscillatory flow behavior in the vicinity of ripple models, Proc. 23rd Coastal Eng. Conf., ASCE, pp.2122-2135, 1992.

Sawamoto M., T. Yamashita, and T. Kitamura: Distributions of turbulence intensity and suspended sediment concentration over rippled beds, Proc. 29th Japanese Conf. Coastal Eng., pp.232-236, 1981. (in Japanese)

Molerus, O. and J. Werther: Berechnung der Sinkbewegung kugeliger teilchen in einem vertikal pulsierenden stromungsfeld, Chemie-Ingenieur-Technik, Vol. 40, pp.522-524, 1968.

Hayakawa, N., G. Tsujimoto and H. Hashimoto, Velocity distribution and suspended sediment concentration over large scale ripple, Coastal Eng. in Japan, Vol. 26, pp.91-100,1983.

Nielsen, P.: Field measurements of time averaged suspended sediment concentration under waves, Coastal Eng., Vol. 8, No. 1, pp.51-72, 1984.

Tanaka, N., H. Ozawa and A. Ogasawara, Experiments on sand movement by waves and currents, Rept. Port and Harbour Res. Inst., Vol. 12, No.4, pp.2-22,1973,(in Japanese).

CHAPTER 301

Nearbed sediment concentration from tracer studies

Zbigniew PRUSZAK[1], Ryszard WIERZCHNICKI[2], Andrzej OWCZARCZYK[2] & Ryszard B. ZEIDLER[1]

ABSTRACT

This paper reports on field investigations, in which the nearbed layer was one of major foci. The study was carried out at the IBW PAN Coastal Research Facility Lubiatowo. The radioactive sand tracer used in the study consisted of iridium glass beads with medium grain diameter D_{50} from 0.2 to 0.25 mm and having physical properties close to their counterparts of natural sand on the Baltic coast of Lubiatowo. Some 150 core samples, each 30 cm long, were taken from the areas where the tracer was set in motion, i.e. bar crests, troughs between bars, and other regions subject to the action of waves and currents. Data analysis shows that sediment movement is characterized by high intermittence reflected in a stratified structure of bedlayer, whereby each sublayer is governed by different forcing factors. For identical external conditions the vertical tracer profiles are different at bar crests and troughs, showing clear diversification of sediment mechanics across the surf zone. Fairly thin sediment sheets, a few millimeters thick, travel at bar crests and troughs, due to high sand speed in one-directional (longshore) movement. The aggregated vertical profile of tracer concentration stems from the superposition of a few quasi-parabolas (close to the injection point) or thin uniform laminas (futher away from the tracer source).

1. INTRODUCTION

The surficial layer of seabed is affected by continuous dynamic motion of sea water due to waves and currents. As a result, sediment moves in an unstable and variable moveable layer. Adequate description of the active layer of sediment transport in the sea bed becomes now an urgent but yet controversial problem of coastal dynamics. The thickness of that layer has been assessed in a number of studies to range from a few grain diameters, see Kraus (1984) and Galvin (1987) to several grain diametes (Ingle 1966 and Horikawa et al. 1982) or as much as tens of centimetres (Drapeau et al. 1990). These estimates originate from analyses and investigations having different

---

[1] Polish Academy of Sciences' Institute of Hydro-Engineering, IBW PAN, Kościerska 7, 80-953 Gdańsk, Poland

[2] Institute of Nuclear Chemistry, 16 Dorodna, 03-195 Warsaw, Poland

degree of accuracy, scales and site conditions. Among the conditions bringing about those different estimates of bed layer thickness one can enumerate:

a) ambiguity of the distinction between bedload and suspended load;

b) difficulties in determination of real sand grain speed profiles within the entire active layer of sea bed due to variable waves and currents, especially under random field conditions;

c) problematic delineation of the vertical extension of the movement of single sand grains in sea bed during storms, and controversial singling out individual storm events from aggregated bed thickness due to storm spectrum;

d) diversified conditions of investigations, analysis and their time and space scales (laboratory versus prototype etc.).

Because of the sophisticated patterns of the transmission of random energy of waves and currents to sea bed and its constituent grains, the problem is complex and highly dependent on scales of the phenomenon. Undistorted relationships between the forces controlling sediment transport stemming from the prototype seem to provide the most reliable and valuable information on the processes investigated here.

This study has aimed at analysing and estimating the bedload thickness in the sea, and its dynamic variability due to random waves and currents.

2. FIELD STUDIES

Our field studies were carried out by the use of radioactive tracers (Ir-192) in the form of iridium glass having geometric and physical properties identical with those of natural sand in the area of investigations (Pruszak & Zeidler 1994).

Some 150 core samples taken from various locations of the cross-shore profile (Fig. 1) were taken all over the area of characteristic coastal features such as sand bars at crests and troughs, under various circumstances created by random waves and currents.

Waves and two components of nearshore currents were measured continuously at two stations D1 and D3 on the cross-shore transect (Fig. 1) every three hours. In addition, wind and sea level oscillations were also recorded every three hours. The results described here stem basically from two measuring series of 23th September to 4th November 93 and 11th May to 30th June 94. Graphical distributions of some hydrodynamic parameters representative of the hydrological background of those situations are shown as examples in Figure 2. The times of core sampling over the area of tracer movement are also indicated on the drawing. Each of the core samples was about 30 cm thick and was subsequently subject to analysis in 2-cm layers with a sensitive apparatus for measurement of low radioactivity.

A special casting box 0.5×0.5×0.25 m filled up with the radioactive tracer was deployed at the point P1 (Fig. 1). Core samples were taken from that box at intervals of a few hours, in the course of wave and current action. This section of the study was intended as monitoring of the dynamic variability of sea bed in its linkage to changes in the thickness of the active sand layer as a function of time and external factors.

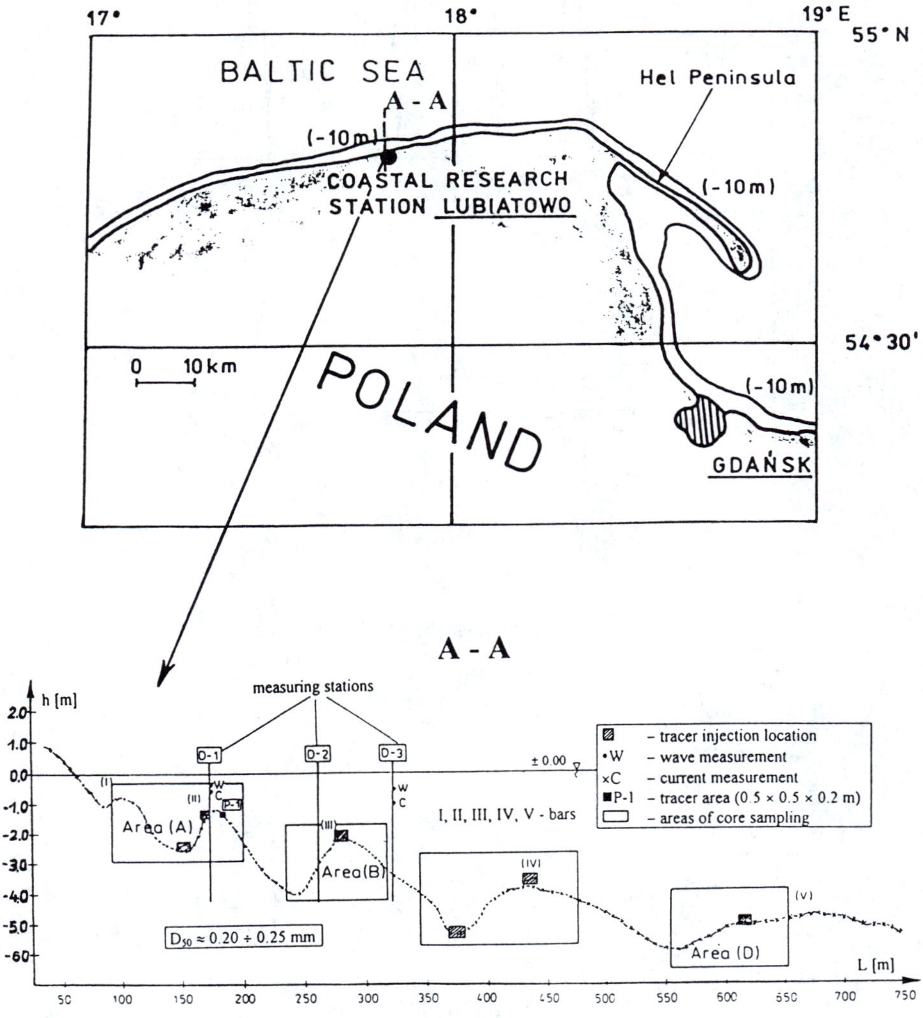

Figure 1. Study area

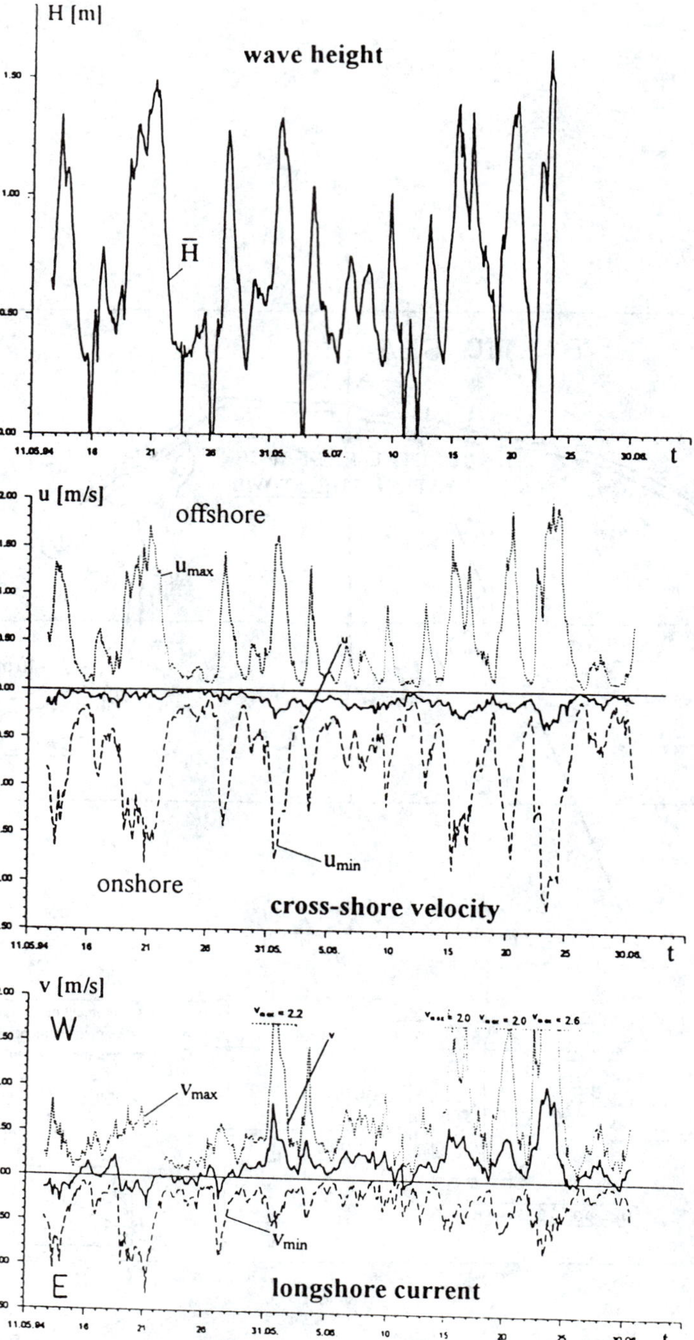

Figure 2. Distribution of wave height and currents at point D-3 during the second series of measurements (1994)

3. ANALYSIS

3.1 TIME AND SPACE CHARACTERISTICS OF THE BEDLOAD LAYER

Geometric characteristics of the variable active layer (primarily its thickness), in the time and space domain, have been analysed by reference to the core samples taken at different locations of the tracer plume. One of the situations tested in area B is illustrated in Figure 3, while the vertical distribution of the tracer concentration, providing insight into the thickness of the active layer with increasing distance from the point of injection, is depicted in Figure 4.

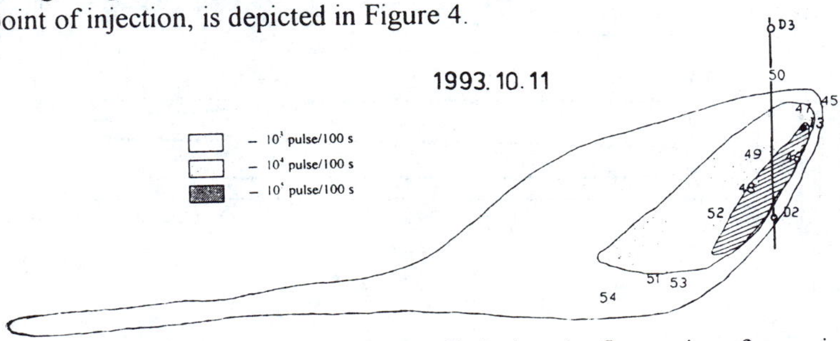

Figure 3. Location of tracer plume in area B during the first series of experiments

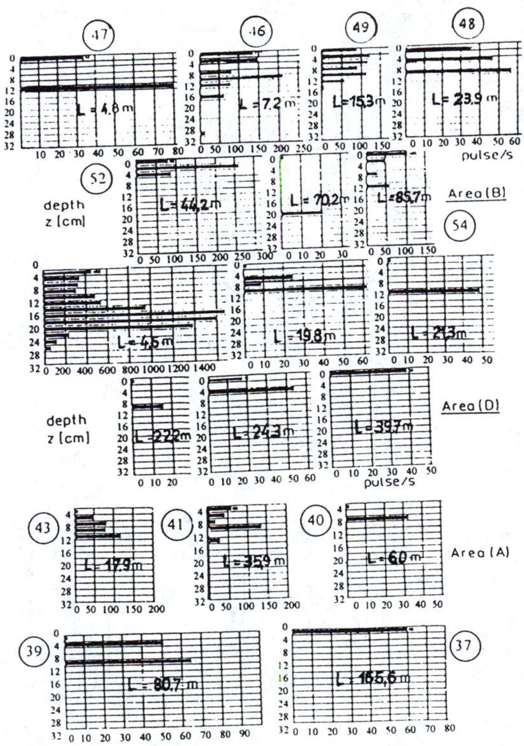

Figure 4. Vertical tracer concentration profiles at various location as a function of spacing L from injection point

Some other vertical profiles, illustrating the wide range of diversity at other locations of the cross-shore transect, are shown in Figure 5. All these results, on the background of the controlling factors have been employed in the determination of characteristic descriptors of the bedload layer.

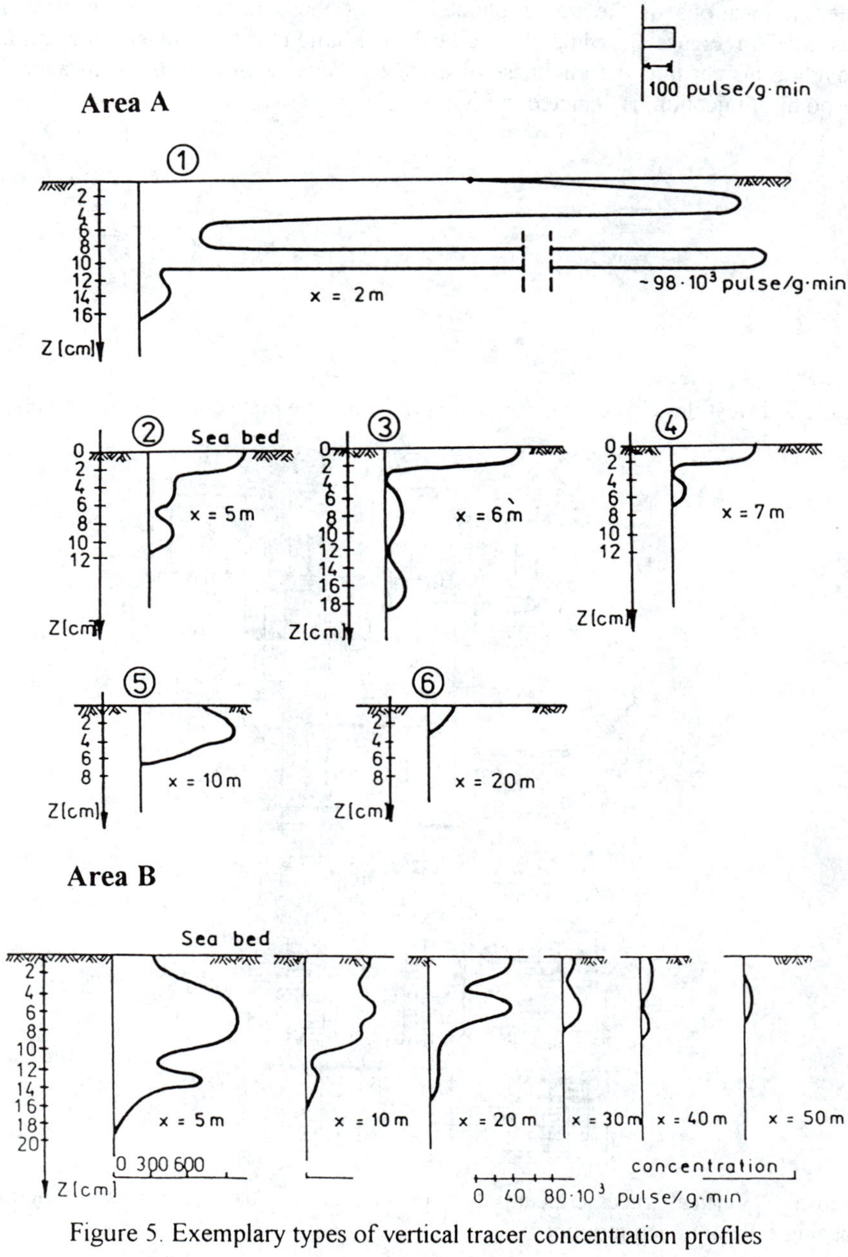

Figure 5. Exemplary types of vertical tracer concentration profiles

As a result of our analysis of the core samples taken in both measuring series it has turned out that there is no prevailing type of vertical concentration within the bedload layer. Among a few possible general types of distribution one can suggest a quasi-parabolic profile resulting from the superposition of two or even three „moving" parabolas and one or a few thin surface layers on top of those parabolas. The schematic representation of that profile is illustrated in Figure 6. The number of components, thus complexity of the profile, depends on the intensity of changes in wave and current climates at a given location. Since every sublayer of the core sample represents a certain instantaneous random hydrodynamic situation cut out of the complex series of events during the measurements, the storm history is obviously reflected in the configuration of the tracer within the core sample. The higher the diversity of the storm history the more complex the core sample.

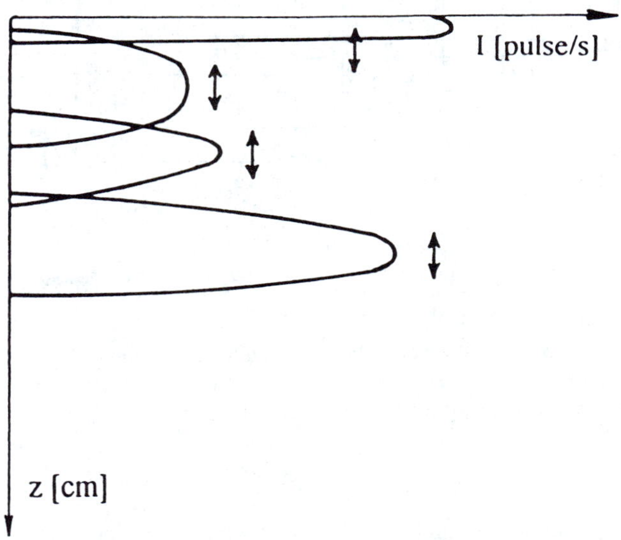

Figure 6. Schematic decomposition of the vertical profile of tracer concentration

Clear-cut variability of the vertical tracer profiles in time and space (distance from the injection point) is visible. In the near field of the injection point, the vertical distribution of the tracer is close to the parabolic type, sometimes with an additional thin surficial sublayer. As one moves away from the injection point, the thickness of the tracer layer decreases and the higher activity „moves" from deeper depth to the sea bed surface.

The shape of the vertical bedload profile depends very much on the shape of sea bed itself and its macroforms such as sand bars. Under the same external conditions, at both crests and troughs the tracer travels in a thin surface layer (Fig. 7). The picture is different at bar slopes and in areas without large bed forms; the vertical profiles of tracer concentrations at those locations are less regular, more variable and close to the types depicted in Figures 4, 5 and 6.

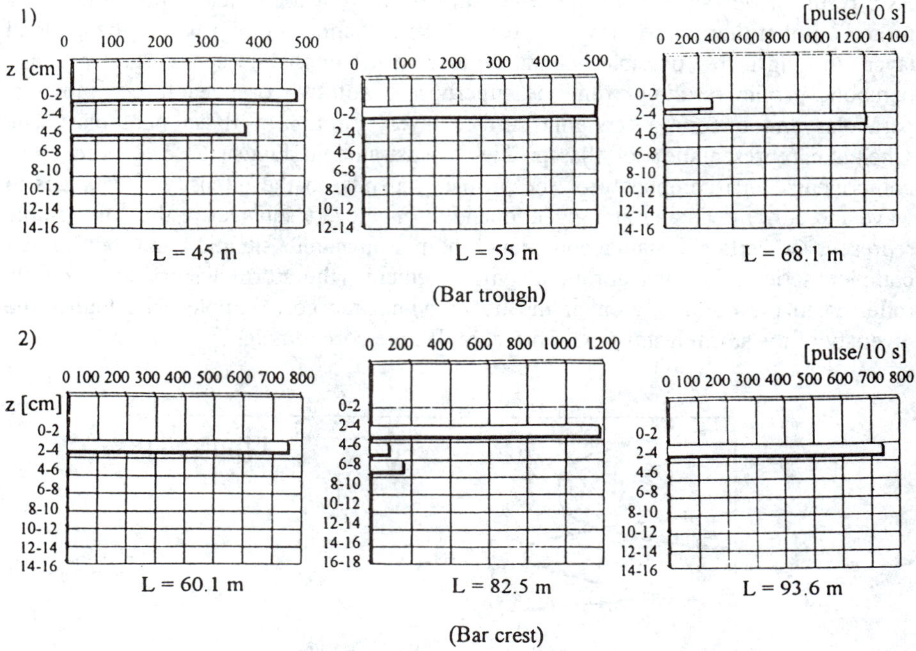

Figure 7. Vertical profile of tracer concentration at bar crests and bar troughs

The core samples tested display a strong nonlinear variability of the bedload thickness as a function of waves and currents. The thickness of each sublayer in the vertical profile, created by various sequences of waves and currents, depends on the intensity and duration of wave motion. Very roughly, it can be described by the general function:

$$a = F(H,t) \approx f(H) \cdot g(b \cdot t/T_{st}) \qquad (1)$$

in which T_{st} = time after which stabilisation of the controlling factors takes place and some equilibrium is attained for a given wave situation. This time is estimated as 10 to 100 wave periods T.

The function $f(H)$ stands for the strength of forcing, thus the intensity of sediment transport, while the function $g(b \cdot t/T_{st})$ represents the effect of the time necessary for generation of the bedload layer. After a sufficiently long time $t = T_{st}$ one should get $g(b \cdot t/T_{st}) = 1$ and accordingly $a \approx f(H)$, for a given combination of waves and currents. The vertical profiles of tracer concentration shown in Figures 4 and 5 (or schematically in Figure 6) are thus integrals of the quantity a due to individual segments of the wave and current history.

The thickness a can be estimated from the sediment transport rate $q = (\rho_s - \rho) \cdot v_s \cdot a$ and independent assessments of the average grain speed v_s, both resulting from our analysis of the tracer movement and core samples. By such procedure one can claim that the thickness *a* is in the range of 2-4 centimetres during moderate storms.

Theoretical investigations of the bedload layer (Kaczmarek et al. 1995) suggest that the bedload layer thickness *a* should be smaller. A numerical simulation belonging to those studies is depicted in Figure 8 for the grain diameter D_{50} and wave period T typical of the area of investigation; the thickness *a* should obviously be a function of wave height and the depth of water.

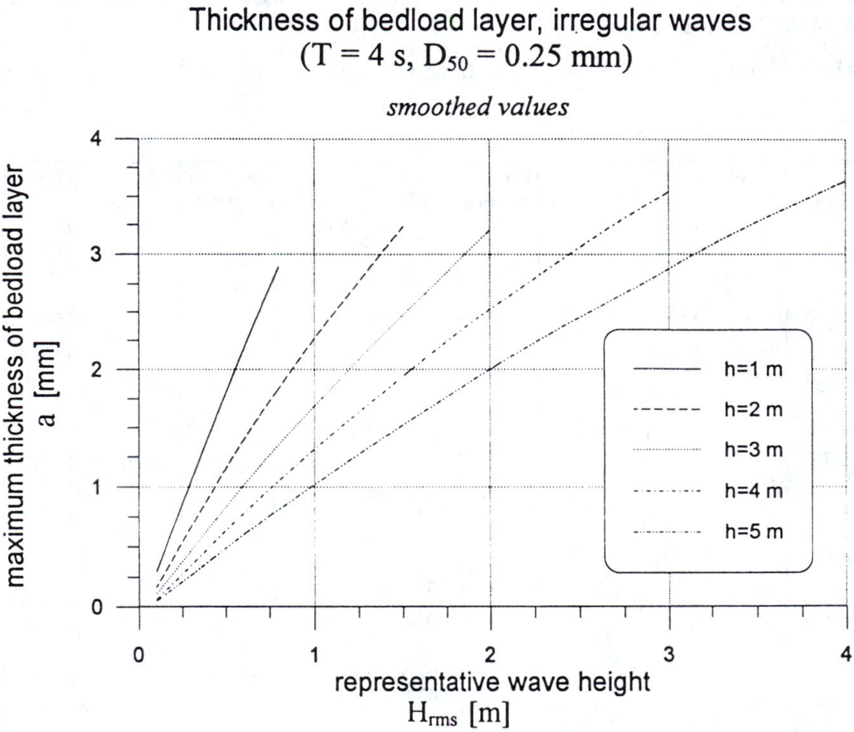

Figure 8. Theoretical bedload layer thickness for wave period T=4 s and variable wave height (H_{rms}) and depth of water h (Kaczmarek et al. 1995)

Upon comparison of the simulated quantities and their prototype counterparts it can be inferred that the bedload layer is about a few millimetres thick, thus by one order of magnitude smaller theoretically than under prototype conditions. This disparity can partly be associated with the laboratory origin of the theoretical outcome. More important, the definitions of the bedload layer are different — Kaczmarek et al. (1995) adhere to the classical few grain diameters above and beneath bedline, while in this study the movable bed layer encompasses sheet flow as well.

The function $f(H)$ can be approximated by $f(H) \approx 0.027H_{br}$, as postulated by Kraus et al. (1982) and Pruszak & Zeidler (1994). On the other hand, the function $g(b \cdot t/T_{st})$ can be estimated once T_{st} is given more accurately, which is however beyond the scope of this study.

3.2 LOCAL VARIABILITY OF ACTIVE SAND LAYER

In order to shed light on local variability of the active sand layer as a function of time and variable forcing by waves and currents, the second series of measurements included a special experiment under prototype conditions. The test box denoted by P1 was situated on a depth h=1.2 m, on the seaward slope of the sand bar, some 6 m from the bar crest (Fig. 1). The experiment was initiated at 12:00 on 26th May 1994 at calm weather. The tracer concentration on a depth about 25 cm below original sea bed was constant c=1500 pulse/s (Fig. 9). Moderate waves were noted between 12:00 and 14:00 (Table 1).

Table 1. Parameters of water and sediment movement at box P1 (depth = 1.2 m, Fig.1)

Time (t)	Wind		Wave (m)		Current (m/s) Average/ Max		Sea level change (cm)	Depth of mixing (m)	Rate of depth changes	Remarks
	φ	v (m/s)	H	H_{max}	Long shore	Cross-shore				
26.05.94										
13^{00}	272°	7.5	0.12	0.44	0.2/0.6	0.01/0.4	0.0	0.04		
12^{00}-14^{00}									few D/ 2 h	One-point breaking
16^{00}	260°	2.2	0.11	0.45	0.2/0.6	0.01/0.4	-2.5	0.04		
14^{00}-18^{00}									-1cm/4 h	
19^{00}	274°	6.8	0.16	0.48	0.2/0.6	0.01/0.4	+5.0			
22^{00}	328°	2.4	0.2	0.54	0.2/0.5	0.01/0.4	0.0			
27.05.94										
1^{00}	19°	4.9	0.25	0.96	0.3/1.1	0.01/0.8	+1.0			Multiple-
4^{00}	38°	7.5	0.35	1.0	0.4/1.2	0.02/0.9	+6.0	a few	mean value =	point wave
7^{00}	33°	6.5	0.40	1.17	0.4/1.4	0.02/1.1	+20	cm	10cm/10h	breaking
10^{00}	21°	3.0	0.35	1.0	0.4/1.4	0.02/1.0	+20			

Waves broke in a close vicinity of shoreline, far away from the area of investigations. Although quite small, the waves clearly affected the upper sand layer, reaching some 4 cm below original sea bed line. For similar waves and currents between 14:00 and 18:00 the activity of sand movement in the bedload layer was similar, but its intensity was growing, with the maximum at the surface and linear decrease to the depth h_d= 4 cm below sea bed line. The latter value can be regarded as the depth of wave effect on sea bed grains; thus it somehow defines the thickness of the active sea bed layer at given external forcing. Substantially thinner bedload thickness, about a few grain diameters, is typical of that period for the grains belonging to the sea bed and being in motion at the same time. As a result of such wave and current forcing the bed was eroded by 1 - 2 centimetres over four hours, (Fig. 9). By the time of the next measurement at 9 o'clock on the next day, waves and currents grew considerably (Tab. 1). Multiple wave breaking was observed at the time of the next measurements. About the area of the experiment, where one of the wave breaking lines was noted, the breaker height was about $H_{br}=1.2$ m and the maximum current velocity exceeded 1 m/s. Tracer concentration in the vertical profile decreased substantially and the concentration profile became irregular, close to the parabola shifted by a certain value below sea bed line. The maximum concentration of a tracer some 6-10 cm below sea bed line was much smaller than its initial counterpart (about five times). From thorough observations by scuba divers it can be reported that ripples were not present and strong sediment flux, somehow resembling a moveable sheet a few centimetres thick occurred. It should be stressed that the thickness of sediment movement layer estimated by scuba divers was about 2-4 cm (depending on the accuracy of estimation), but not a few millimetres. Further erosion by 9-10 cm was noted from 18:00 on 26th May to 9:00 on 27th May; the most intensive changes due to growing waves and currents took place most probably about midnight (Tab. 1).

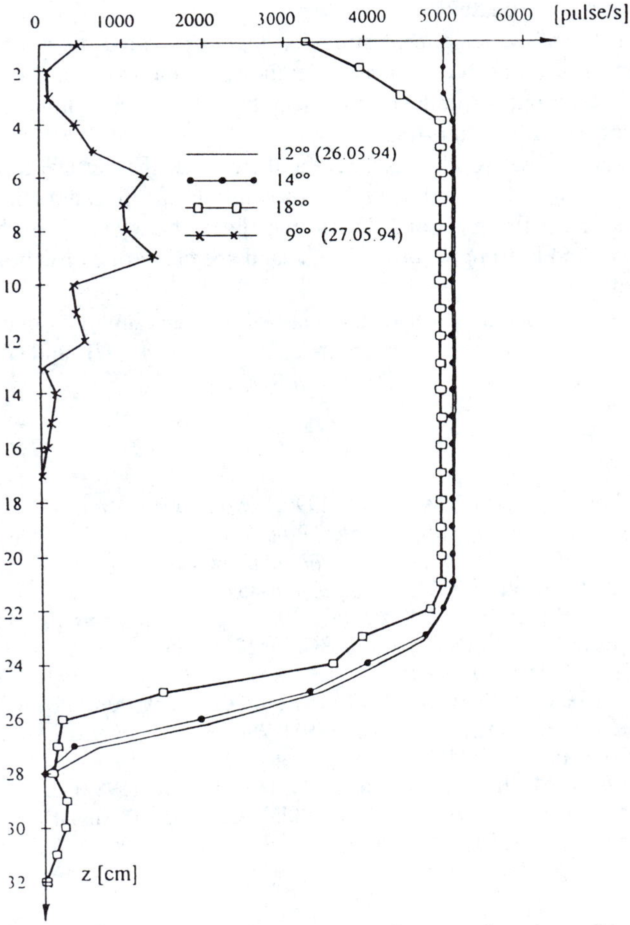

Figure 9. Variability of vertical concentration profile as a function of hydrodynamic parameters

Subsequent measurements were impossible because of the growing storm. After several hours of that medium storm it was realised that the test box was washed away and no tracer was found in its area.

4. CLOSING REMARKS

From our analysis of the tracer distribution and its coupling with waves and currents one can draw some conclusions concerning the bedload layer. The following points should be made:

- sediment movement under dynamic surf zone conditions is characterized by high intermittence reflected in a stratified structure of bedlayer, whereby each sublayer is governed by different forcing factors. The entire vertical profile of tracer concentration represents then a sum of various sublayers, which are correlated with subsequent wave climate changes,
- vertical profile of tracer concentration displays a pronounced variability in time and space, with predominance of quasi - parabolic profile close to the tracer injection point and a thin quasi - uniform one away from the tracer source; being often the superposition of a few thin `foils',
- for identical external conditions (wave and currents) the vertical tracer profiles are different at bar crests and troughs, showing clear diversification of sediment mechanics across the surf zone. Fairly thin sheets, below 2 cm in thickness, travel at bar crests and in troughs, due to high sand speeds in one directional (longshore) movement,
- the thickness of each sublayer of the nearbed load is a highly nonlinear function of wave and current velocity, depending on intensity (H) and duration (t) of subsequent wave action, $a = f(H,t) \approx f(H) \cdot \log\,(b \cdot t/T_{st})$.

5. REFERENCES

Drapeau G., Long B., Kamphuis J., W. (1990), *Evolution of radioactive sand tracers to measure longshore sediment transport rates*, Proc. 22nd ICCE, Delft, ASCE, p.2710-2723

Galvin C. (1987), *Vertical profile of littoral sand tracers from a distribution of waiting times*, Proc. Coastal Sediment's 87, ASCE, p. 436-451

Horikawa K. A., Watanabe A., Katori S. (1982), *Sediment transport under sheet flow condition*, Proc. 18th ICCE, Cape Town, ASCE, p.1335-1352

Ingle J., C. (1966), *The movement of beach sand: An analysis using fluorescent grains*, Developments in Sedimentology, No. 5, Elsevier Pub. Company, Amsterdam

Kaczmarek L., Ostrowski R., Zeidler R. (1995), *Boundary layer theory and field bedload*, Proc. Coastal Dynamics'95, Gdańsk, ASCE, p. 664-675

Kraus N., C., Isobe M., Igarashi H., Sasaki T., Horikawa K. (1982), *Field experiments on longshore sand transport in surf zone*, Proc. 18th ICCE, Cape Town, ASCE, p.969-988

Pruszak Z., Zeidler R. (1994), *Sediment transport in various time scales*, Proc. 24th ICCE, Kobe, ASCE, p.2513-2526

CHAPTER 302

A SLOPING DUCT FOR THE STUDY OF SEDIMENT TRANSPORT

Jesper S. Damgaard [1] , Richard J.S. Whitehouse [1] and Richard L. Soulsby [1]

Abstract.

The details of a large sloping duct at HR Wallingford Ltd. are presented. The duct is capable of sloping to the angle of ±33° which exceeds the angle of repose of normal quartz sand. The duct can be tilted laterally as well as longitudinally so that; 1) the effect of combined slopes can be investigated; 2) the aspect ratio can be inverted. At present the duct operates with a steady flow but it can accommodate a wave piston.

1 Introduction.

An important aspect in modelling coastal morphology is the ability to describe the effects of sloping surfaces on sediment transport rates, for example due to longshore bars, dune and ripple formations, dredged channels and trenches and beaches. The research carried out to date is limited, and to remedy this HR Wallingford Ltd has constructed a unique facility to permit the measurement of sediment transport at steep slopes in a controlled environment.

2 Description.

The sloping sediment duct has been purpose-designed for conducting tests on sediment transport on horizontal beds, and shallow and steep slopes up to the angle of repose for the sediment. It has a maximum longitudinal tilt of ±33° and can be tilted, or turned, laterally up to this angle for sediment transport studies where a transverse slope is required. In figure 1 the duct is shown at

[1]Marine Sediments Group, HR Wallingford Ltd.,Wallingford, Oxon OX10 8BA, UK.

Figure 1: Sloping duct at maximum downslope setting.

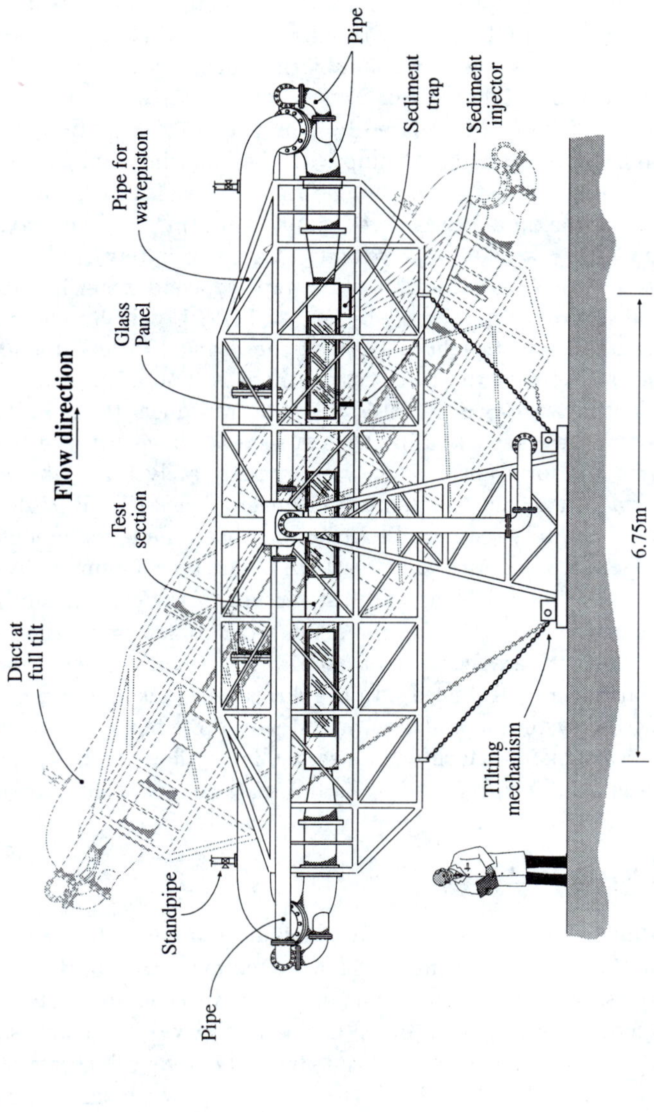

Figure 2: Drawing of sloping duct (not to scale).

full downslope (in the direction of the flow) tilt. The duct consists of a support cage, a test section and a central support on which the support cage pivots. See figure 2.

The dimensions of the test section are 6.75 m long by 0.6 m wide by 0.25 m high with three sections fabricated with glass viewing windows. The ends of the working section have rotating joints, and by tilting the duct laterally through 90° the aspect ratio can be either 1:2.4 or 2.4:1. The downstream end of the duct has a sediment trap which can be emptied by removing a cover plate.

The pump is a P76 centrifugal pump with the following characteristics: maximum head = $13m$ @ zero flow, design rating = $9m$ @ $0.17m^3/s$ and head corresponding to the maximum discharge =$5.3m$ @ $0.21m^3/s$. The maximum discharge corresponds to a mean flow velocity of approximately $1m/s$ in the duct. The system is a closed conduit with a stand-pipe to avoid water hammer effects.

The flow discharge is measured by an ALTOFLUX K380 electromagnetic flowmeter situated in a straight portion of the $8inch$ ($0.20m$) diameter feeder pipework. In each end of the test section there is a small pressure outlet. A Sandhurst Scientific $20mbar$ differential pressure transducer (GA 64/20E) is used to determine the total head loss in the test section. The distance between the outlets is $5.57m$ and for a mean velocity of approximately $1m/s$, the head loss is of the order $1cm$. An absolute pressure transducer (Druck PDCR 930) is installed in the middle of the test section. Provision has been made for mounting a two component Laser Doppler Anemometer (LDA). The laser mounting system rests on rails and can be moved along the test section. An electronically indicating point gauge is installed on the laser mount in order to determine the vertical position of the LDA with an accuracy of $\pm 0.01mm$. Seeding material for boosting the LDA signal can be introduced into the duct while it is operating.

At present the system operates with a steady flow, but it is designed to accommodate a wave piston which will allow oscillatory flows to be generated with periods and amplitudes corresponding to full-scale near-bottom wave motions.

3 Hydrodynamics.

The hydrodynamic properties of the duct have been investigated as a part of the ongoing research. Of particular interest for the sediment transport research is the wall shear stress. For all the experiments that have been conducted so far the floor of the duct has been covered by PVC plates onto which a quartz sand with a median grain diameter of $208\mu m$ has been glued. The sidewall correction method of Vanoni and Brooks (1957) has been used to partition the total pipe friction into wall and bed shear stress. The shear stress has been estimated through head loss measurements obtained by the differential pressure transducer and through LDA measurements in the constant stress layer. It turned out that the variance on the shear stress estimate obtained via head loss measurements was much larger than that of the estimates derived from LDA measurements. Therefore the results

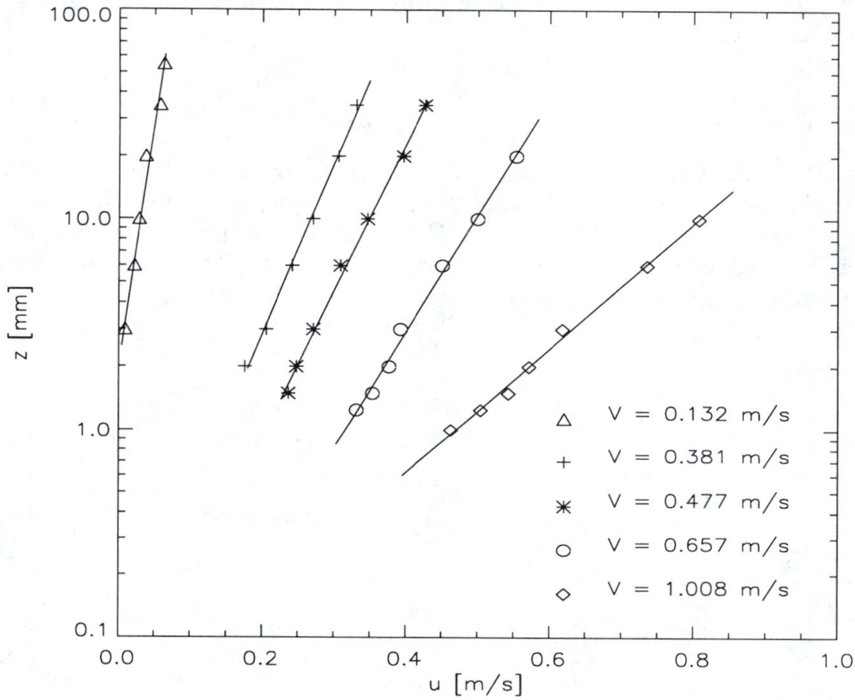

Figure 3: Vertical profiles of mean horizontal velocity measured with the LDA

obtained from the LDA measurements were used to establish the particular pipe friction relation for the duct. The procedure is as follows. First a vertical velocity profile is measured for a range of different discharges. The measuring points in the vertical profile should lie above the viscous sublayer but beneath the core region. See Schlichting (1960) for details. For each discharge the bed friction velocity, u_{*b}, is determined assuming a logarithmic velocity profile

$$\frac{u}{u_{*b}} = \frac{1}{\kappa} \ln\left(\frac{z}{z_0}\right) \qquad (1)$$

where u = mean horizontal velocity at distance z above the bed, κ = Von Karman's constant and z_0 = the bed roughness length. A number of velocity profiles are shown in figure 3 for different values of V, mean flow speed, $V = Q/A$, where Q = discharge and A = cross-sectional area.

The bed friction values obtained by equation 1 can be used to estimate the overall pipe friction by applying the sidewall correction method of Vanoni and Brooks (1957). These results have been compared with the Colebrook and White formula (Colebrook, 1939) and a modified version of the C&W formula where the theory of Christoffersen and Jonsson (1985) was used to account for the effect of uniform grain roughness and where a different value of the constant was used (1.94 rather than 1.74):

$$\frac{1}{\sqrt{\lambda}} = 1.94 - 2\log\left(\frac{k_s}{R}(1 - e^{-\mathbf{Re_g}/27}) + \frac{18.7}{\mathbf{Re}\sqrt{\lambda}}\right) \tag{2}$$

where λ is the resistance coefficient, $\lambda = 8(u_*/V)^2$, $\mathbf{Re_g} = u_* k_s/\nu$ is the grain Reynolds number, k_s = the Nikuradse equivalent sand roughness, equal to the median grain diameter of bed material, $\mathbf{Re} = 4RV/\nu$ is the pipe Reynolds number and R = hydraulic radius = A/p, p = wetted perimeter. The results are shown in figure 4 using axes based upon Colebrook's original paper.

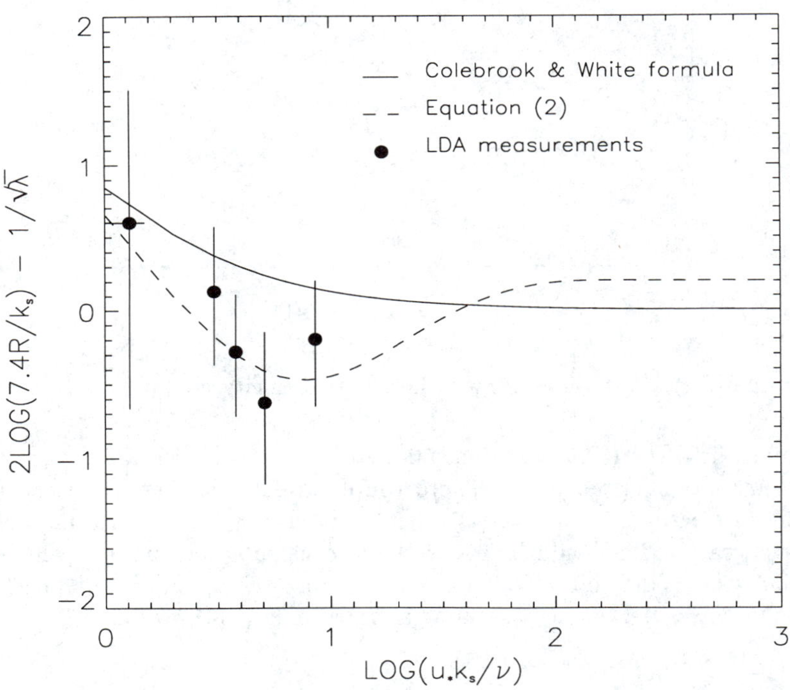

Figure 4: Friction diagram for the sloping duct with a rough floor.

4 Current research.

A study of bed-load transport (erosion rate) on sloping beds in steady flow has been conducted as a part of the MAST G8M Coastal Morphodynamics Project. It is for that purpose that the rough bottom mentioned previously was installed.

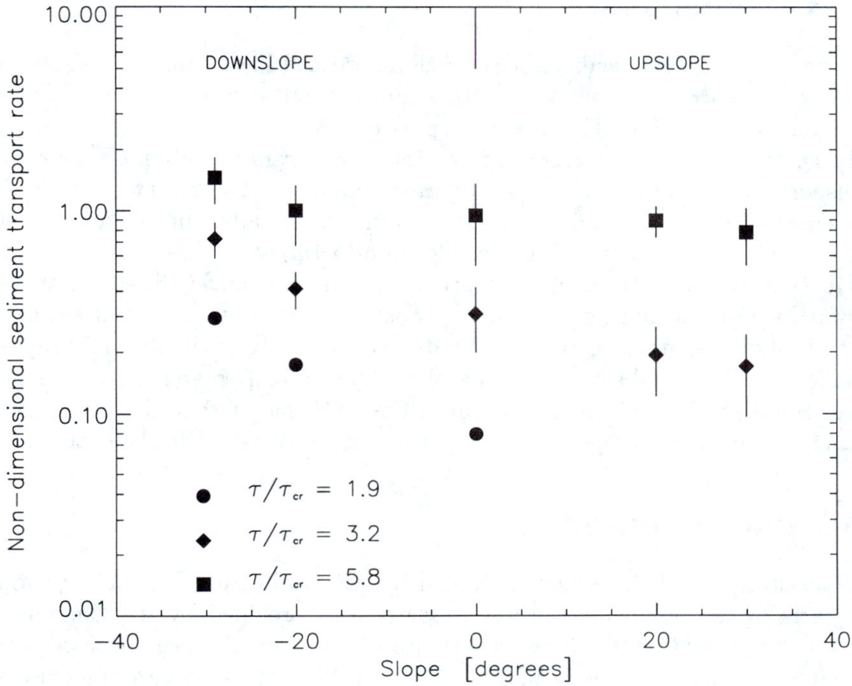

Figure 5: Experiments on slope effect on bed-load sediment transport. The values of τ_{cr} are for a horizontal bed.

An electronically driven sediment injector was installed downstream from the centre of the duct with an inner diameter of 20 mm. This device was used to inject sand into the flow to determine the equilibrium pick -up rate for sediment by the flow. The speed of the piston was adjusted until the rate of delivery of sediment matched the rate of removal by the flow. A porthole in the roof of the duct above the injector could be opened to charge the injector with sand and to make observations. At the downstream end of the test section the sediment trap captured the sediment injected into the flow. Figure 5 contains some results obtained in the duct for a range of bedslopes and shear stresses. The bed shear stress, τ, is given as $\tau = \rho u_{*b}^2$, and τ_{cr} is the critical shear stress for initiation of sediment motion on a horizontal bed. The results of this research are discussed further in Damgaard et al. (1996).

5 Conclusions.

The new sloping duct facility at HR Wallingford Ltd has proven to be suitable for large-scale experimental tests with sediments and can also be used for hydrodynamical research. The unique sloping capability of the facility allows the influence of gravity due to a sloping bed (of an arbitrary angle) on the sediment transport to be investigated for all types of sediment. The recent research has concentrated on a fine sand but the duct would be equally suitable for working with gravels as the flow speed achievable is up to $1m/s$.

It is expected that the duct will continue to be used for steady flow tests and, in the future with the addition of a wave piston, for wave or wave-current sediment or hydrodynamic investigations. The duct is currently available to European researchers through the EC Training and Mobility of Researchers - Access to Large-Scale Facilities - programme and it is anticipated that the facility will become available for wider usage by UK and other researchers in the near future.

Acknowledgements.

The construction of the duct was funded by HR Wallingford Ltd. The funding for parts of the sediment transport research was provided by the Commision of the European Communities Directorate General for Science, Research and Development as part of the MAST2 G8M Coastal Morphodynamics collaborative research programme, under Contract Number MAS2-CT92-0027, and for other parts by the U.K. Ministry of Agriculture, Fisheries and Food (Flood and Coastal Defence Division).

References

[1] J.B. Christoffersen and I.G. Jonsson. Bed friction and dissipation in a combined current and wave motion. *Coastal Eng.*, 12(5):387–423, 1985.

[2] C.F. Colebrook. Turbulent flow in pipes, with particular reference to the transition region between smooth and rough pipe laws. *Proc. Inst. Civ. Eng.*, 11:133–156, 1939.

[3] J.S. Damgaard, R.J.S. Whitehouse, and R.L. Soulsby. Bed-load sediment transport on steep slopes. *accepted for J. Hydr. Eng.*, 1996.

[4] H. Schlichting. *Boundary layer theory.* McGraw-Hill, New York, Fourth edition, 1960.

[5] V.A. Vanoni and N.H. Brooks. Laboratory studies of the roughness and suspended load of alluvial streams. Technical Report Report No. E68, California Inst. Tech., Pasadena, California, 1957.

CHAPTER 303

SEDIMENT DYNAMICS AND PROFILE INTERACTIONS: DUCK94

Donald K. Stauble[1] and Mary A. Cialone[2]

ABSTRACT

Beach profiles and sediment samples were collected on a daily basis along three cross-shore transect lines during the DUCK94 nearshore experiment lasting for 18 days in October 1994. Conditions ranged from near calm during the first week of the experiment to full storm conditions during the second and third weeks, with a two day initiation of beach recovery at the end of the experiment. The profiles responded with similar elevation change, with little morphologic variability during the calm period. During the storm, the bar migrated seaward 70 to 100 m, but the foreshore exhibited little change. The bar began to migrate shoreward at initiation of recovery. Sediment grain-size distributions vary in the cross-shore direction, with medium size grains on the upper foreshore, coarse gravel deposits on the lower foreshore and progressively finer sands in the offshore direction. After the storm, the foreshore and bar/trough samples were coarser with little change in the nearshore sediment distributions.

INTRODUCTION

To quantify sediment distribution response to forcing functions on the foreshore, bar/trough and nearshore along a three-dimensional beach area, a sediment sampling and analysis experiment was conducted as part of the DUCK94 nearshore field experiment (see Birkemeier et al., 1997 for overview of DUCK94 experiment). This experiment was designed to examine the three-dimensional natural sediment distribution and its relationship to profile change at the U.S. Army Engineer, Waterways Experiment Station, Field Research Facility (FRF) at Duck, N.C. (Figure 1). Sediment distribution changes were measured over the short-term (18-day period) during October, 1994. To increase our understanding of sediment distribution, the experiment focused on our knowledge of the short-term 3-D sediment variations of the entire active profile to document relationships between beach morphology and sediment dynamics from the high water line, seaward to closure depth.

Sediment grain-size distributions for different beach environments change as the beach erodes and accretes in response to changes in wave and tidal forcing. A previous long-term study at the FRF along one profile line characterized a cross-shore variability pattern in

---

[1] Res. Phy. Sci. [2] Res. Hyd. Eng., U.S. Army Engineer Waterways Experiment Station, Coastal and Hydraulics Laboratory, 3909 Halls Ferry Road, Vicksburg, Mississippi 39180

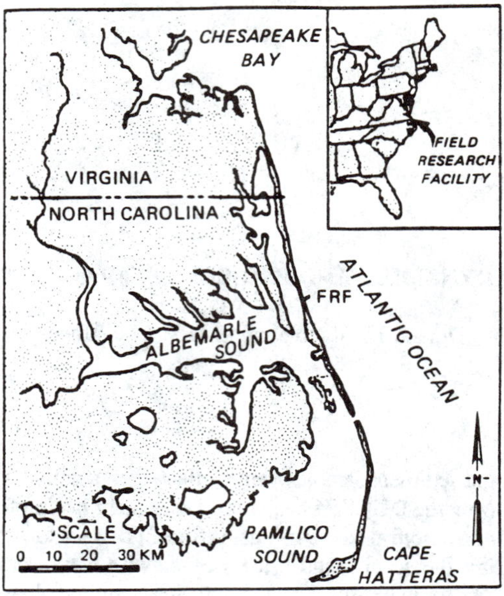

Figure 1. Location map.

grain-size distributions (Stauble, 1992). During the SUPERDUCK (1986) experiment, a short-term storm erosion and recovery study showed a high degree of three-dimensional grain-size variability but was limited to the foreshore and hinted at nearshore control of the foreshore sedimentation (Stauble et al., 1993). A lack of knowledge exists in relating three-dimensional sediment movement on the entire active beach profile during both fair weather and storm periods to the processes that cause the movement. These processes include a) the swash processes on the foreshore, b) wave breaking, longshore currents, and/or cell circulation (rip currents) in the bar and trough area, and c) wave, tidal and wind driven circulation on the nearshore slope out to closure depth and beyond. The variation in grain sizes in each of these environments is indicative of the different active processes.

EXPERIMENT DESIGN

The experiment design included collection of profile data and sediment samples along three lines, approximately 100 m apart, extending from the dune base to the 6-m depth contour (Figure 2). Surface sediment samples were collected at the dune base, mid-berm, berm crest (area around high water), mid-tide, swash (area around low water), trough, bar crest, and the 3-m, 4-m, 5-m, and 6-m depth contours. Shallow surface sediment samples were collected daily with a hand scoop from the foreshore to wading depth and with a grab sampler on subaqueous portion of each profile using the Coastal Research Amphibious Buggy (CRAB), during the beach profile survey. Samples were collected around the time of low tide on the foreshore. The sampling schedule included daily profile and sediment collection of the foreshore and alternate days for the nearshore area sediment collection for a duration of 18 days in October, 1994. The data sampling time period covered physical conditions ranging from near calm to storm conditions.

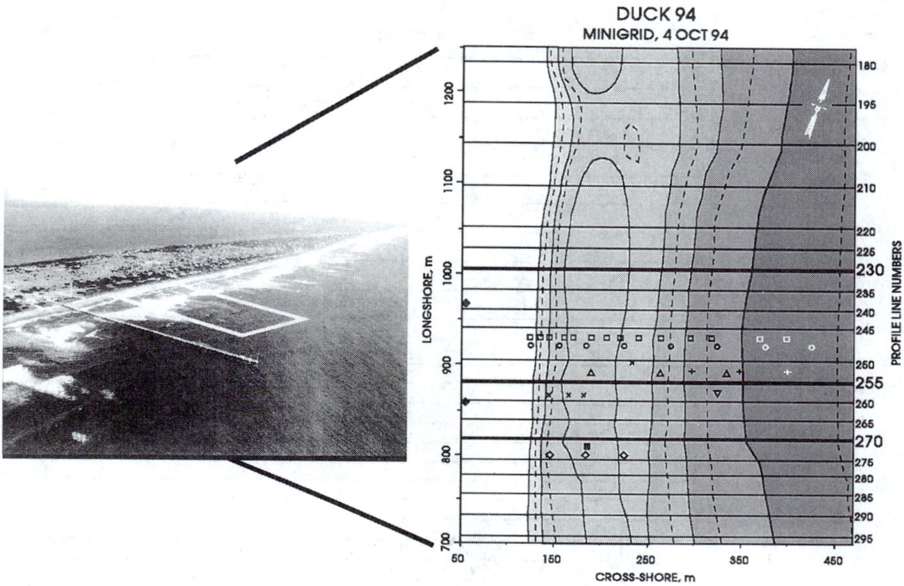

Figure 2. Location of profile transects samples (Lines 230, 255, and 270).

DATA ANALYSIS

Physical data: During the experiment several arrays of wave gauges and current meters were deployed within the study area. An offshore wave gauge array was located in 8 m of water, just offshore of the data collection area. Figure 3 shows the mean wave height (H_{mo}) and peak period (T_p) recorded from 2 to 22 October 1994. Profile and sediment data used in this paper were collected from 4 to 21 October. The experiment began just after a period of high wave activity, as the waves diminished to an average H_{mo} of around 0.5 m and a T_p between 3 and 7 sec. This relatively calm period lasted until 9 October. Currents recorded in the trough area landward of the bar between transects 245 and 250 (Figure 2), indicated that the longshore velocity was near zero and the cross-shore velocity ranged from 0 to 0.2 m/s in the offshore direction during this time period (data courtesy of S. Elgar). At the onset of the storm (10-13 October), H_{mo} rapidly increased to around 2 m and T_p increased from a low of 3 sec to around 7 sec. The longshore currents reached their maximum velocity (around 1.3 m/s) to the south, with a steadily increasing offshore component. The wave gauge recorded its maximum H_{mo} of around 4 m on 15 October as T_p increased to around 11 sec. The longshore currents reversed during this time reaching a peak velocity (1 m/s) to the north as the storm progressed up the coast and wave approach angles switched from a northeasterly direction to a more easterly to southeasterly direction. The offshore velocity component in the trough continued to increase to a maximum of around 0.64 m/s on 19 October. The wave height decreased to around 1.5 to 1.0 m as the experiment ended, but the wave period remained around 14 sec.

Profile Data: Beach profile data were collected from a shore-parallel baseline landward of the dune out to a depth of 6 m, some 700 m seaward of the baseline. All profiles are referenced to the 1929 National Geodetic Vertical Datum (NGVD). Profile data were

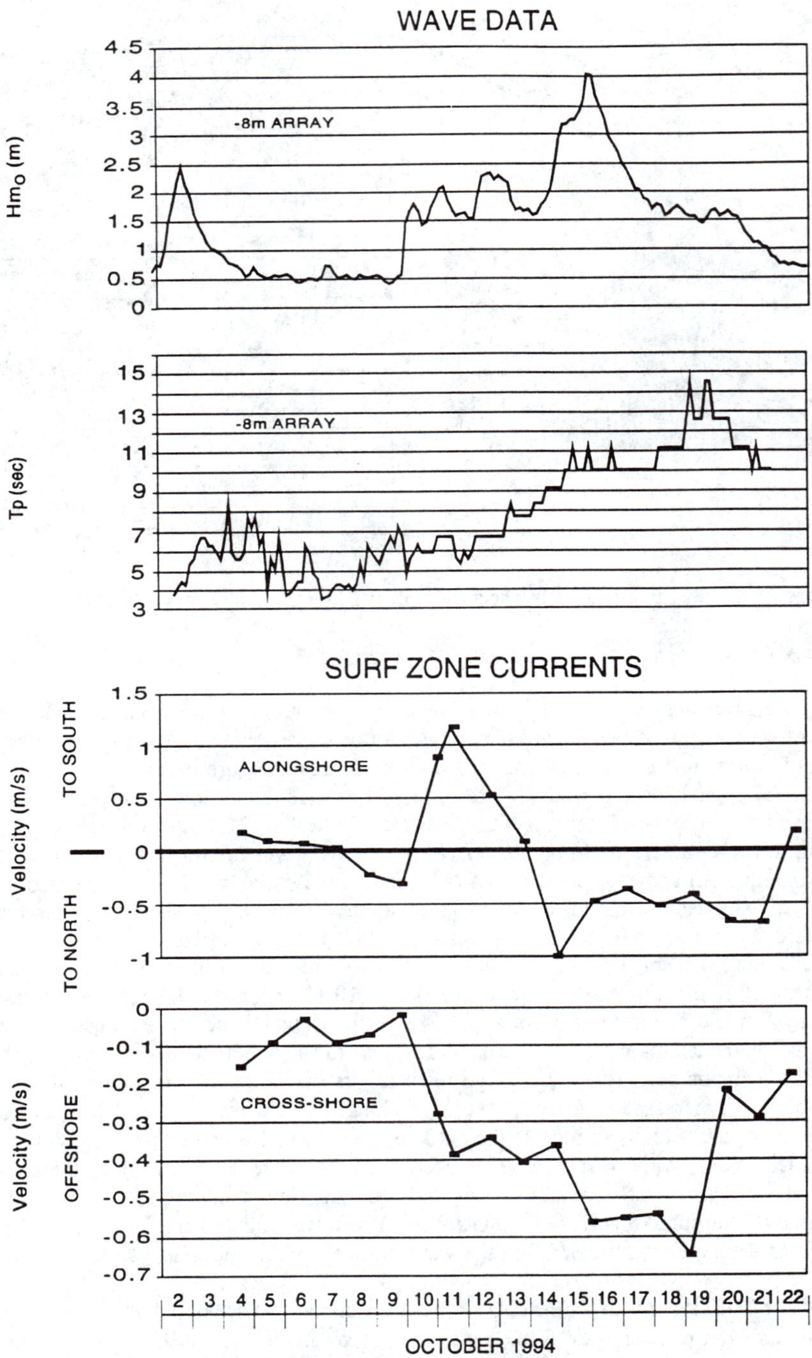

Figure 3. Plot of nearshore wave parameters and surf zone currents in trough.

analyzed and plotted using the Interactive Survey Reduction Program (ISRP) as described by Birkemeier (1984). Plots of the pre-storm beach profile (5 October) at all three transects showed that the bar crest was about 250 m from the baseline (Figure 4). With the increase in wave height beginning on 10 October, the bar migrated seaward. By 15 October the bar had moved to around 280 m seaward of the baseline. Just one day later, at the height of storm wave activity, the bar had moved to between 310 and 350 m offshore. With steadily decreasing wave heights and long period swell (as the storm moved out of the area), the bar migrated back onshore at profile lines 255 and 270. Profile line 230 was the location of a rip current and the bar remained in its seaward position through 21 October The foreshore did not change significantly in elevation over the study. During the low wave period at the beginning of the experiment the foreshore was planar, but beach cusps developed on 20 and 21 October as the waves subsided.

Sediment Data: A total of 256 sediment samples were collected and sieved using a sonic sifter at quarter-phi (¼ ϕ) intervals, ranging from -3 ϕ (8.0 mm) to 4.25 ϕ (0.53 mm), and weight percentages for each interval were computed. Statistical data were calculated using the method of moments (described in Friedman and Sanders, 1978). High variability was found in the cross-shore grain-size distributions, with the foreshore exhibiting the highest variability. This area between the berm crest and low tide swash contained a bi-modal gravel component along with sand size fractions. A localized source of coarse relict sediment has been identified in the area of the FRF by previous investigators (Calliari, 1994). The highest variability in profile elevation and sediment distribution occurred in the lower foreshore and trough/bar area, however the gravel component was restricted to the beach foreshore. The grain-size distribution was much more well-sorted and more uniform in mean size in the nearshore, both temporally and spatially. Figure 5a shows an example of the cross-shore sediment distribution on profile line 270 collected just at the beginning of the storm, representative of the sediment distributions deposited during the period of low wave activity. In contrast, at the end of the high wave period on 20 October, the bar had migrated offshore and the trough had widened. The grain-size distributions (Figure 5b) show a coarser and more poorly-sorted sample from the foreshore to the 3-m depth. Little change in nearshore sand distributions were found between low and storm wave conditions, while an increase in coarse material was measured on the foreshore. The finer-sized sediments were removed from the foreshore, trough and bar crest area, leaving behind a coarse lag deposit.

ANALYSIS AND DISCUSSION

Alongshore and cross-shore sediment distribution variability was related to beach morphology changes as a function of wave and current fluctuation. During the October 1994 experiment, the alongshore beach shape was uniform for the low wave period (first six days) and a "linear" beach was the prevalent form. The cross-shore profile elevation change and sediment grain-size distribution exhibited more variability, with a zone of medium size sands on the upper foreshore and coarse material in the lower foreshore. The size distribution became progressively finer in the offshore direction, through the nearshore trough, over a single bar feature, and the sloping nearshore region.

On 10 October, an extratropical storm system developed over the northern Gulf of Mexico and migrated northeastward into the Atlantic Ocean off of Cape Hatteras on 15 October. During this "northeaster", longshore currents increased to the south and offshore

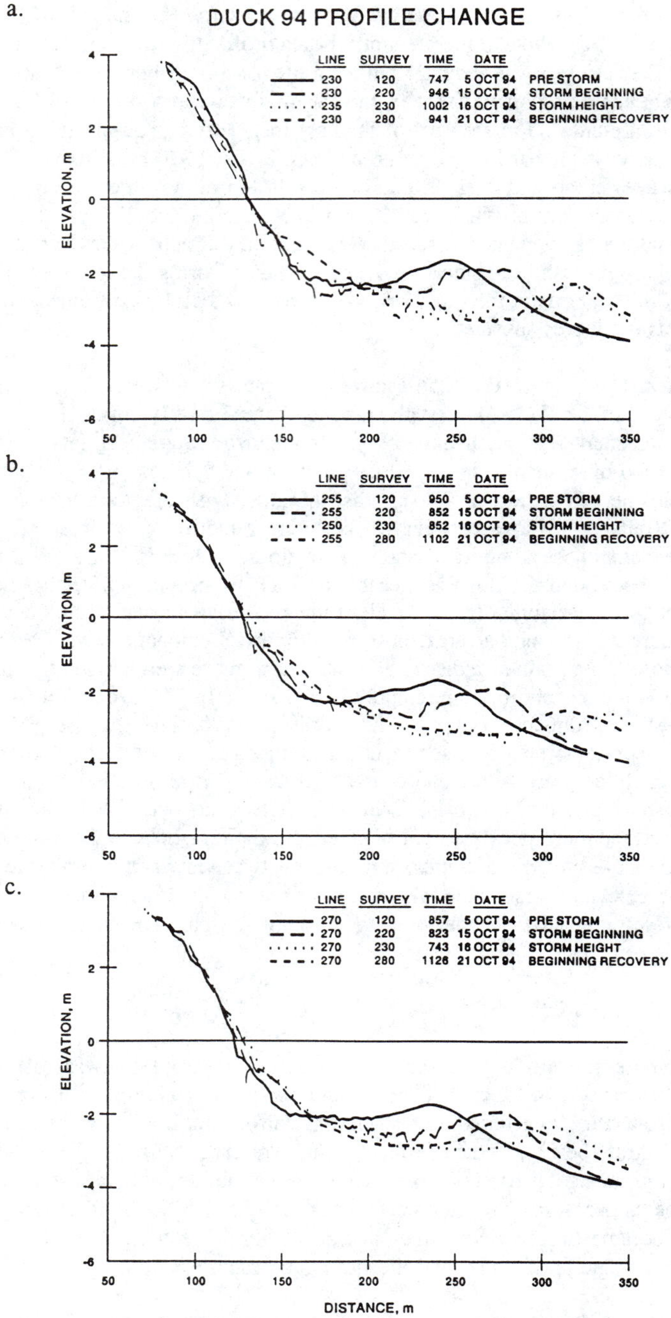

Figure 4. Plot of selected profiles showing bar migration due to storm. a) line 230, b) line 255, and c) line 270

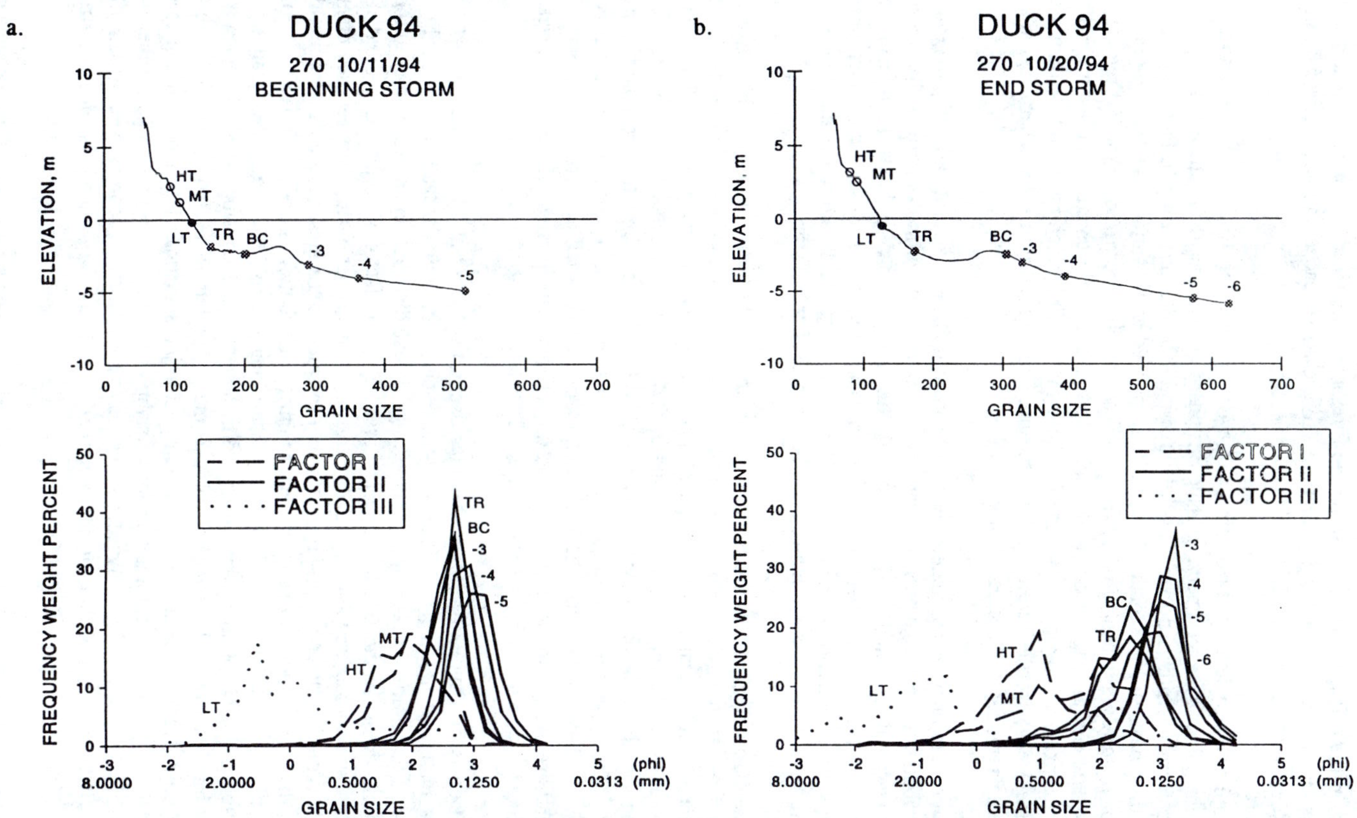

Figure 5. Sediment sample locations and grain-size distributions on line 270 a) pre-storm and b) post-storm
HT = High tide, MT= Mid tide, LT = low tide, TR= trough, BC = bar crest, -3 = 3 meter depth sample, etc...

flow increased in the trough area. During this phase of the experiment, the bar feature migrated in the offshore direction, while the foreshore maintained a uniform planar shape along the study area. Wave heights continued to grow and reached a maximum of 4 m at the 8-m depth wave gauge on 15 October.

High pressure developed over the eastern United States on 16 October and continued through the end of the experiment. During this phase of the experiment, the longshore drift switched from a southward to northward direction. In the surf zone, offshore currents reached their peak on 19 October. The profile responded with continued seaward migration of the bar feature and widening of the trough. Wave heights decreased on the last two days of the experiment, as long period swell conditions prevailed. Beach cusps formed and a rip current was present at the northern end of the study area. The beach began to take on more three-dimensional profile features with the bar migrating onshore in the southern region as offshore currents rapidly decreased in velocity.

In order to improve our understanding of both spatial and temporal variability in grain-size distribution during high energy events and compare them with calm periods, foreshore samples were collected and analyzed along all three lines, on a daily basis throughout the experiment. Nearshore sediment data collection was planned on a two-day cycle, but large waves precluded collection in the nearshore during the height of the storm event. However, nearshore samples were taken immediately before and after the storm.

Standard statistical techniques were used to analyze the sediment distribution of each beach and nearshore environment. Mean and standard deviation values were calculated for each sample. In general the coarser samples tended to have poorer sorting. The widest range in mean and sorting values was found in the low tide sediments, where the samples contained coarse shell and gravel components, as well as medium to fine quartz sand material. The upper foreshore (high and mid tide) samples had a coarse to medium-sized sand that was better sorted than the low tide (less of a coarse shell and gravel component). The nearshore (trough to the 6-m depth) had a relatively narrow range of mean grain sizes in the fine sand range, with little shell and no gravel size components.

Analysis of a suite of sediment samples using just the mean and standard deviation values is somewhat limiting. The use of Q-mode factor analysis (Klovan, 1966) provides a method to determine the relationship between grain-size distribution and variability in the 3-D sedimentation of the beach and nearshore. Q-mode factor analysis, as applied to sediment investigation, involves the determination of interrelationship between sediment samples. With this method, a group of sediment samples can be arranged into a meaningful order so that the relationship between each sediment distribution is deduced (Davis 1973). One of the main advantages of Q-mode factor analysis is that the entire grain-size distribution is considered in the analysis, yielding a detailed relationship especially when ¼ ϕ sample intervals are used. Using an analytical method to determine statistical relationships is more objective because it does not require arbitrary statistical descriptors or a-priori knowledge of the environment and location of samples (Klovan, 1966). A large number of samples can be objectively analyzed without having to manually compare each pair of curves. This reduces the "human interpretation" in relating large numbers of grain size distributions.

Q-mode factor analysis relies on how similarity between samples is defined (Reyment et al., 1993). In this application, the technique of Imbrie and Purdy (1962) for defining similarity was used. They defined an index of proportionality, or cosine theta function, to determine the degree of similarity in weight percent in each size class between each pair of samples. The cosine theta matrix shows all the information on the relationship between the sample vectors, but it is difficult to interpret (Klovan, 1966). Factor analysis provides a means of analyzing the cosine theta matrix to determine the minimum number of mutually orthogonal "factor axes" needed to account for most of the information in the cosine theta matrix. The first axis accounts for the majority of the information in the cosine theta matrix, the second axis accounts for most of the remaining information in the matrix, and so forth. Thus, the problem is to determine the number of eigenvalues needed to account for most (95-99%) of the information in the cosine theta matrix. Eigenvalues and eigenvectors were determined for the sediment sample data set. Factor loadings, a measure of each sample's weighting or correlation to each factor, can then be determined from the eigenvectors and eigenvalues. This provides the coordinates of the samples in a space of reduced dimensionality (Syvitski, 1991). Lastly, a varimax method of factor rotation is used to rotate the factor axes to maximize the variance of the factor loadings on each factor axes, while retaining their orthogonality.

The sample's weight percent in each phi-class interval is used to represent the sample distribution and can be used to "place" the sample relative to other samples in the Q-mode analysis. For example, if a sample is sieved into twenty-seven phi intervals, the sample can be defined as a vector in 27-dimensional space whose position is uniquely determined by the amount of sediment in each of the 27 classes (Klovan, 1966). With this technique, similarities and differences between samples can be determined and comparisons from day to day can also be made. Factor analysis relates the sediment distribution curves of similar shape, and dominant grain-size distribution peaks. Three factors accounted for 88.7% of the variance in the sample distributions. Factor I accounted for 57.8% of the variance and represented the medium sands between 1 and 2 ϕ (0.5 and 0.25 mm). Factor II accounted for 18.3% of the variance and represented fine sands with a peak frequency of occurrence between 2 and 3 ϕ (0.25 and 0.125 mm), and Factor III accounted for 12.6% of the variance and represented coarser sands, with a peak between -1 and 0 ϕ (2.0 and 1.0 mm). Figure 6 is a triangular diagram, which illustrates the distribution of the sediment samples within the three factors from the Q-mode analysis. Samples that are at the corners of the triangle represent the "end-members" of each factor group and depict a particular sediment distribution (Klovan 1966). The other samples within the triangle can be considered as a mixture of these three sediment distributions. The sediments from the high and mid tide area were strong in Factor I and embody medium grain-size distributions. The samples collected from the nearshore (trough to -6 m) covered a range from strong Factor II through near Factor I and comprise fine to medium grain-size distributions. The low tide samples were strongest in Factor II and include predominantly coarse sands.

To better understand the temporal changes over the duration of the 18 day experiment, factor analysis was run on each individual group (i.e. all high tide or all 3-m depth samples) within the cross-shore distribution of sediment data set to investigate the differences in grain-size distributions at each profile position over time. The factors will change depending on which sediment distributions are more dominant within each data set. Each cross-shore sediment data set was compared to differentiate the change over the low to high wave activity period. Little change was found in the nearshore sediment groups, with each data set grouping around Factor I (representing the fines) over the study period. A somewhat

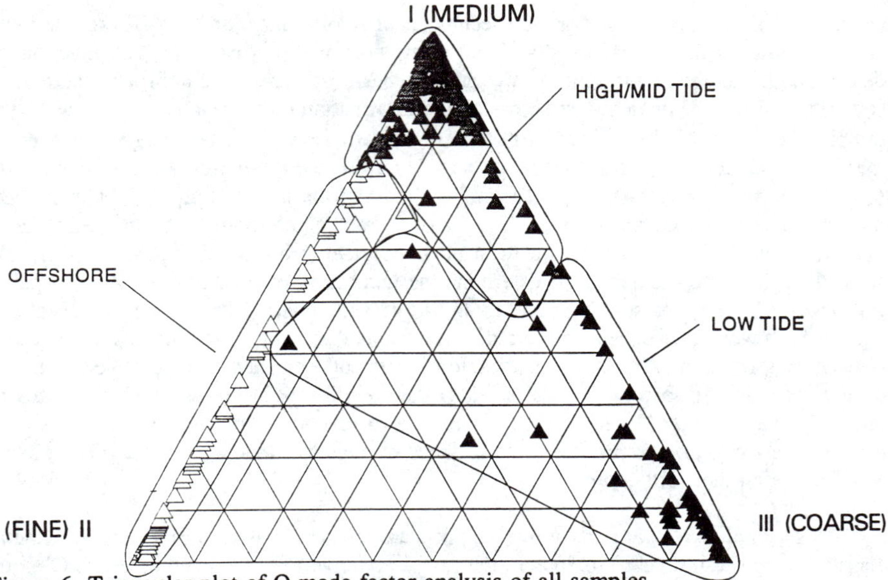

Figure 6. Triangular plot of Q-mode factor analysis of all samples.

more complex grouping was found for the individual foreshore samples over the 18 day study, with the high and mid tide samples grouping more commonly between Factor I (finer) and Factor III (medium), and the low tide samples around Factor II (coarser).

To remove some of the high spatial variability in the foreshore samples, a mathematical composite was made of the high, mid and low tide samples to allow a more clear picture of the temporal change due to storm activity. The 54 composite samples mathematically derived from each of the three profiles per day were analyzed and the grain size variability of the composites was described by three factors representing 94.2% of the variance. Figure 7a shows the triangular plot of the sample distributions within the three factors and representative frequency plots of "end member" samples. Factor I accounted for 79.2% of the variance and represented the finer distributions with peaks around 2 ϕ (0.25 mm). Factor II accounted for 12% of the variance and represented coarser grain-size distributions with peaks around -1 ϕ (2.0 mm). There was no strong Factor III end member, which accounted for only 3.1% of the variance. Samples that plotted in this area of the diagram had peaks in the medium size range between 1 and 2 ϕ (0.5 and 0.25 mm).

To give physical significance of the results of this analysis requires inferring what the end member samples indicate and how the samples group together around these end members. Foreshore composite samples collected from the period of low wave activity (4-10 October) all have a strong grouping toward Factor I. This grain-size distribution is represented by the composite of 9 October on profile line 230 (Figure 7b). All three profile lines exhibited similar composite grain-size distributions. During the period of higher waves (11-19 October) the composites plotted in two groups, one between Factor I and Factor III representing a shift to more medium grain sizes and the second strongly associated with Factor II, the coarse grains. A bi-modal sample from profile line 230 collected on 19 October is representative of this group. Except for the line 230 composite of 20 October

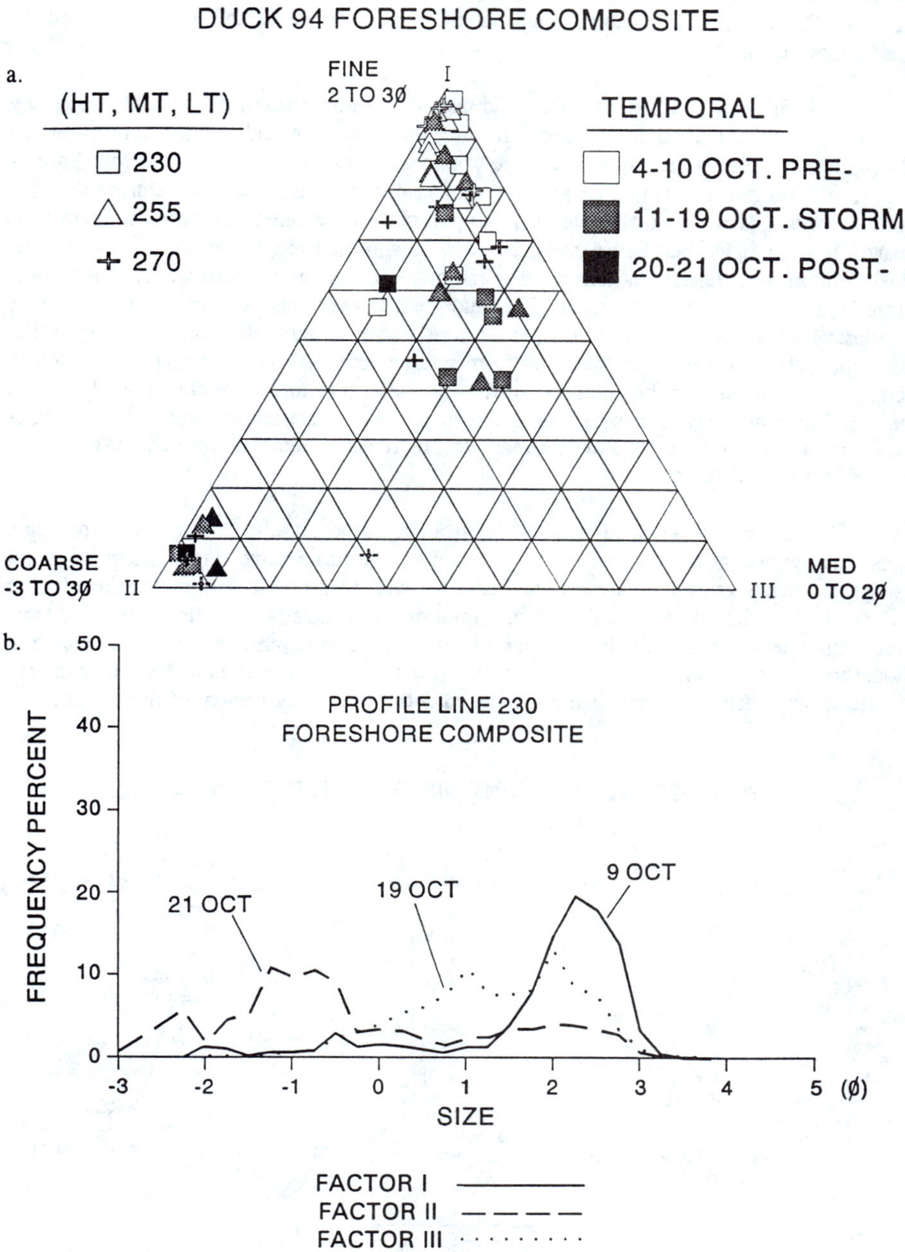

Figure 7. a) Foreshore composite factor analysis and b) examples of grain-size distributions representing Factor I, II and III groups.

all post-high wave samples (20-21 October) plotted strongly as coarse Factor II distributions. The 21 October line 230 sample distribution is an example of this coarse distribution resulting after the storm.

A picture emerges that foreshore and nearshore sediment dynamics are controlled by the wave and current input. During the time of low wave and current activity at the beginning of the experiment, little change was measured on the profile of the foreshore or bar/trough positions and little change in sediment distributions occurred. During the time of high wave and current activity at the latter part of the experiment, the foreshore remained basically unchanged, but the bar migrated seaward and the trough expanded in width. The sediment on the foreshore became coarser and more poorly sorted, particularly in the lower foreshore. Samples from the trough, bar and the 3-m depth also became coarser. Figure 8 illustrates the general trend of the pre- to post-storm sediment distribution change using the mid tide and 3-m depth samples from profile line 270 as an example. A coarse component is present in the grain-size distribution after the storm, possibly a lag deposit of underlying coarse layers exposed as the surficial finer material was removed. Further seaward at the 4-m, 5-m and 6-m depth samples, there was a minimal change in distributions as a result of the storm.

Correlation of sediment data in the trough, bar crest, 3-m, 4-m, and 5-m depth positions, with near real-time physical data such as significant wave height and mean cross-shore and mean longshore currents were possible with an array of sensors positioned 45 m to the north of the center line of sediment sampling. This array of 9 stations extended from the trough seaward over the bar to approximately the 5-m depth contour. The change in longshore current from flow to the south to flow to the north maintained the trough area while strong offshore currents were associated with seaward movement of the bar and the

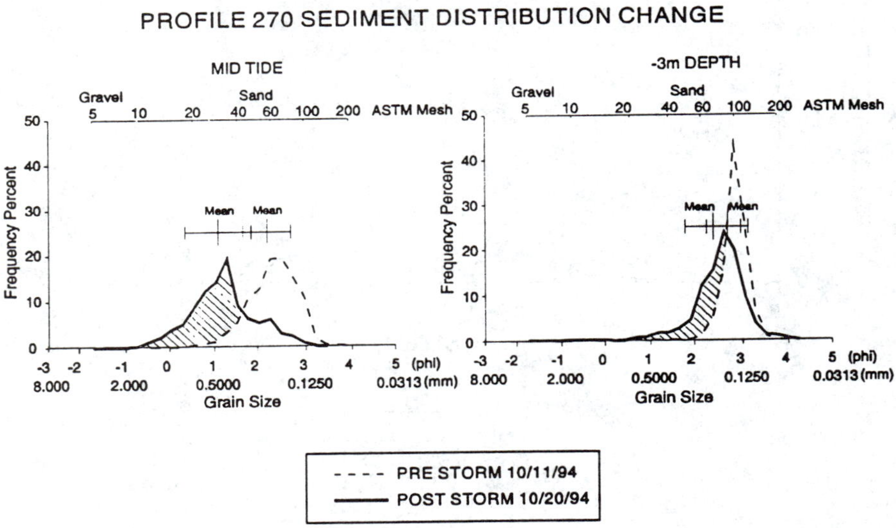

Figure 8. Post-storm coarsening of sediment at mid tide and -3 m depth at profile line 270.

development of a rip current. On the last two days of the experiment the strong currents dissipated and the bar moved landward. The 3-D component of beach morphology was observed with the formation of beach cusps and the rip current at the northern end of the study area. Sediments became much coarser on the foreshore, and a coarser component was present even in the trough, bar and at the 3-m depth position.

CONCLUSION

The dynamics and evolution of sedimentation patterns on beaches and their link to hydrodynamics are poorly understood and present problems in effectively managing erosion control and storm damage reduction. Present research, using sediment statistical data analysis, is proving beneficial to characterize beach sediment distributions and their spatial and temporal deposition patterns. This study afforded a unique opportunity to couple the beach profile evolution, sediment deposition patterns and their resulting grain size distributions with the physical processes active at that coast.

This research examines the interaction of sediment along a three-dimensional active profile during both erosional and accretional events. Previous sediment studies at the FRF were a long term (17.8 month) study limited to one profile (corresponding to Line 230 in this study) (Stauble, 1992) and a 3-D study of a small 50 m wide area of the foreshore during SANDYDUCK (Stauble et. al, 1993). The DUCK 94 profile response covering one storm event was typical of long-term profile response where the most active part of the profile (bar/trough area) alternately moved seaward during storms and landward during fair weather conditions. From the 3-D perspective, the erosion during high wave and strong offshore surf zone currents moves the bar seaward uniformly with a linear foreshore planform. The recovery phase was more three-dimensional. The bar remaining in a fixed seaward position in the presence of a rip current (north end of the study at Line 230), while the bar migrated landward as the wave and surf zone current energy decreased outside of the rip area (southern area of the study at Lines 255 and 270). The foreshore was also very three-dimensional with the formation of beach cusps during this recovery period.

A better understanding of the dynamic processes of sediment deposition and interaction with profile elevation change on a natural beach was documented. The zonation of sediment characteristics over the entire active beach profile provides a picture of cross-shore grain-size data variability, with highest variability on the foreshore. The low tide samples were the most coarse and poorly sorted, with decreasing grain size and better sorting in the offshore direction. Q-mode factor analysis indicated that the foreshore was finer during the calm period at the beginning of the experiment and became coarser during the storm and recovery period. A coarse lag component contributed to the coarsening of the foreshore. The present study provided more detail to similar findings of Stauble (1992) and Stauble et al. (1993). The nearshore became slightly coarser over the bar and out to the 4-m depth. The seaward most sample distributions (5-m, 6-m depth) remained unchanged throughout the study.

Future research needs should focus on the zonation of coarse material and the interaction of the vertical distribution of layered beds on the foreshore/trough area. Conservation of grain-sizes within the three-dimensional beach (cross-shore and alongshore) is still not well understood. A better understanding of sediment processes can provide input into beach profile numerical models. These studies will ultimately help to understand the fate of beach fill material placed anywhere on the beach profile.

ACKNOWLEDGEMENTS

This research was funded through the Coastal Research Program, Geological Analysis of Shelf/Beach Sediment Interchange work unit. The authors would like to thank J.B. Smith, Andrew Morang and the many volunteers who assisted in the field data collection effort. William Birkemeier provided helpful discussion and review, and along with Michael Leffler and the staff of the FRF provided many hours of assistance in collecting and compiling the profile and sediment data. Dr Steven Elgar provided hydrodynamic data in the trough area, Dr Charles Long provided the 8-m array wave data, Kent Hathaway provided video of the beach and Dr Rob Holman provided detailed topographic maps of the foreshore. Sediment analysis was performed by Robert Jones, Chris Murhead and Amy Hampton. Sediment analysis, data reduction and graphics was performed by Johnnie Davenport, with graphics assistance by Robert Chain. Permission to publish was granted by the office Chief of Engineers

REFERENCES

Birkemeier, W.A., 1984. A User's Guide to ISRP: The Interactive Survey Reduction Program. Instructional Report CERC-84-1, U.S. Army Engineer Waterways Experiment Station, Vicksburg, MS., 114 p.

Birkemeier, W.A., Long, C.E., and Hathaway, K.K., 1997. DELILAH, DUCK94, SandyDuck: Three Nearshore Field Experiments. *Proceedings 25th International Conference on Coastal Engineering*, Orlando, FL, ASCE, New York, N.Y., (this issue).

Calliari, L.J., 1994. Cross-shore and Longshore Sediment Size Distribution on Southern Currituck Spit, North Carolina: Implications for Beach Differentiation. *Journal of Coastal Research*, Vol. 10 (2), pp. 360-373.

Davis, J.C., 1973. Statistics and Data Analysis in Geology. John Wiley and Sons, New York, N.Y., 550 p.

Friedman, G.M. and Sanders, J.E., 1978. Principals of Sedimentology. John Wiley and Sons, New York, N.Y., 792 p.

Imbrie, J., and Purdy, E.G., 1962. Classification of Modern Bahamian Carbonate Sediments. *In*, Ham, W.E. (ed.), Classification of Carbonate Rocks, a Symposium, American Association of Petroleum Geologists Memoir 1, pp. 253-272.

Klovan, J.E., 1966. The Use of Factor Analysis in Determining Depositional Environments from Grain-Size Distributions. *Journal of Sedimentary Petrology*, Vol. 36 (1), pp. 115-125.

Reyment, R.A., Joreskog, K.G., and Marcus, L.F., 1993. Applied Factor Analysis in the Natural Sciences, 2nd Ed. Cambridge University Press, Cambridge, England, 371 p.

Stauble, D.K., 1992. Long Term Profile and Sediment Morphodynamics: Field Research Facility Case History. Technical Report CERC-92-7, U.S. Army Engineer Waterways Experiment Station, Vicksburg, MS., 252 p.

Stauble, D.K., Holem, G.W., Byrnes, M.R., Anders, F.J., and Meisburger, E., 1993. SUPERDUCK Beach Sediment Sample Experiment: Beach Profile Change and Foreshore Sediment Dynamics. Technical Report CERC-93-4, U.S. Army Engineer Waterways Experiment Station, Vicksburg, MS., 41 p.

Syvitski, J.P.M., 1991. Factor Analysis of Size Frequency Distributions: Significance of Factor Solutions Based on Simulation Experiments. *In*, Syvitski, J.P.M. (ed.), Principals, Methods, and Application of Particle Size Analysis, Cambridge University Press, Cambridge, England, pp. 249-263.

CHAPTER 304

A STUDY ON FLOW STRUCTURE AND SUSPENDED SEDIMENT CONCENTARTION OVER SEAWEED BED

Gozo Tsujimoto

ABSTRACT

Numerical calculation of flow structure and the suspended sediment concentration over seaweed bed under surface wave is conducted using the k- ε turbulence model. Velocity distribution and suspended sediment concentration over seaweed bed, which is made of a plastic sheet, are measured by use of the electromagnetic current meter and a turbidity meter, respectively.

The characteristic phenomenon such as the vortex formation over seaweed bed is simulated and the calculated vertical profile of suspended sediment concentration is similar to that of experiment. The value of drag force coefficient of seaweed is a few times larger than that of rigid seaweed.

I. INTRODUCTION

Seaweed bed in the sea is an important ground for fishes and shellfishes as nursery ground. So method for artificial seaweed bed creation and technique for selecting appropriate site have been studying. But there are many problems to be resolved. One of them is to understand the mechanism of sediment transport induced by the wave motion and the seaweed motion. The seaweed acts against the water body in motion as the drag force, and the flow mechanism is so complicated by that.

Also a study on wave attenuation and control of sediment transport by using the flexible artificial seaweed a like plastic sheet have been investigated recently. This paper presents an experimental approach and a numerical approach on a study of the flow mechanism and the concentration profile of suspended sediment over seaweed beds.

Dr. Eng., Associate Professor, Dept. of Civil Eng., Kobe City College of Tech., 8-3, Gakuen-Higashimachi, Nishiku, Kobe, 651-21, Japan

II. EXPERIMENTAL METHOD

The experiment was performed in a water tank which was 18m long, 0.6m wide and 1.0m high. Since the seaweed motion in the field is very complicated, an artificial flexible roughness element like a plastic sheet is used as the seaweed in this experiment(from now on, it is called seaweed). The plastic sheet is cut at space of 1mm to 10cm in the length, being placed at the bed. Number of seaweed and its settling space were made to change and eight test runs were performed as summarized in Table 1.

For each of the eight test run, capacitance wave gages were used for the measurement of the free surface elevations, the electromagnetic current meter for that of velocity and the turbidity meter for that of suspended sediment concentration. Sand with diameter of 0.014mm was used as bed materials. The bed material was placed in a uniform thickness of 8cm at the tank bottom, about 2.5m long and 0.4m wide. Also the natural frequency of used seaweed is measured by using the oscillating plate, and its value is between 4.5 and 5.5Hz. The output voltages of their measured values were transmitted to personal computer. In table 1, "u" stands for horizontal velocity, "w" for vertical velocity, "c" for concentration, "H" for wave height and "S" for settling space of seaweed. In the follow figures, the settling position of seaweed is marked as "↑" and the wave phase is set to be zero when the flow direction over the seaweed on the offshore side is changed from offshore to onshore.

Table-1 Experimental Conditions

Run	*H* (cm)	*Depth* (cm)	*Period* (sec)	*S* (cm)	*Number Seaweed*	*Item*
1	4.1	30	1.22		1	u, w
2	3.6	21	1.22		1	u,w,c
3	7.3	21	1.22		1	c
4	4.8	21	1.22	10	2	u,w
5	1.1	21	1.72	10	2	u,w,c
6	3.9	21	1.22	10	3	u,w
7	4.6	21	1.22	20	2	u,w
8	4.3	21	1.22	20	3	u,w

III. EXPERIMENTAL RESULTS

3.1 Velocity vector

Figure 1(a) and 1(b) shows the measured velocity over the seaweed without its motion at the wave phase of t/T=0.25 for run No.2 and over the seaweed with its motion at that of t/T=0.5. A blank space in fig.1(b) shows no data because of the limitation in measurement system. It is thought that formation of vortex over the vertical plate(like fig.1(a)) under the wave motion is seen at the wave phase of a wave crest. Meanwhile the vortex in fig.1(b) with the seaweed motion is formed over seaweed beds at the wave phase when the flow direction change from onshore to offshore. There is a phase lag of $\pi/2$ on the formation of vortex between both fig.1(a) and (b) and the location of vortex formation in fig1(b) is lower than that of fig.1(a), and after the vortex in fig.1(b) is expected, following the motion of the free surface.

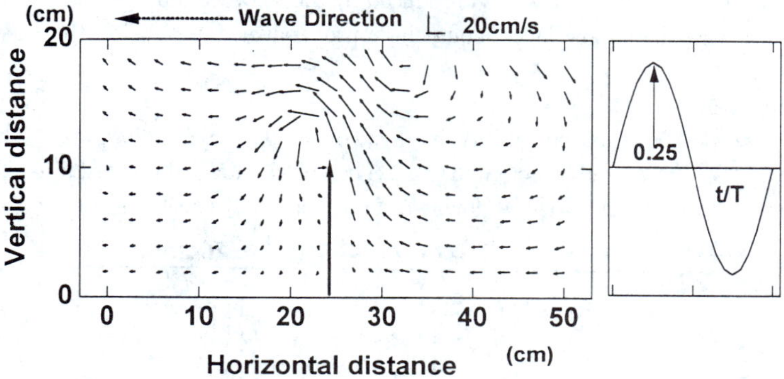

Fig.1(a)Velocity profile over seaweed without motion for run 2

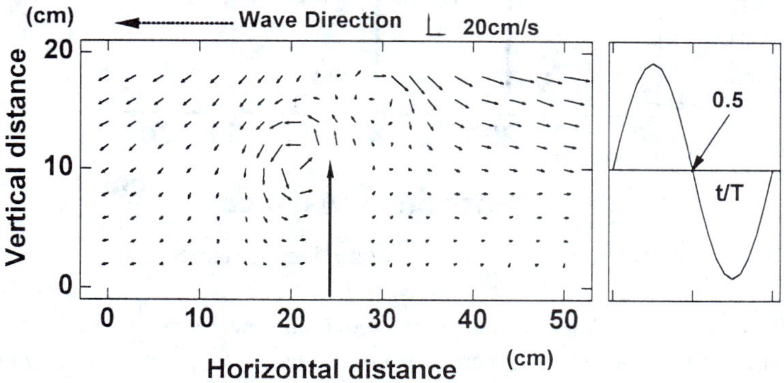

Fig.1(b) Velocity profile over seaweed with motion for run 2

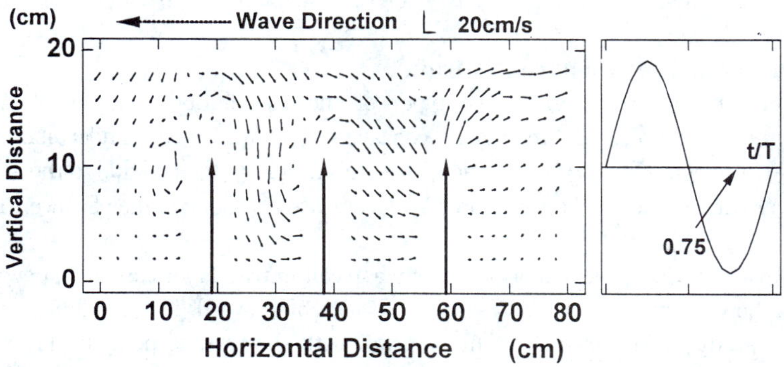

Fig2.Velocity profile over three seaweed with motion for run 8

Figure 2 show the velocity profile for run No.8. The vortex formation is not seen clearly near the top of seaweed. The counter clockwise vortex, however, is formed over the seaweed bed in the onshore side when the flow direction changes from onshore to offshore and its scale is smaller than that of fig.1(b). Also the other clockwise vortex around there is seen. It is thought that the flow filed produced by the seaweed motion before the half period forms a wavy boundary layer along the top of seaweed, producing this clockwise vortex.

3.2 Steady velocity

Figure.3 shows the measured velocity averaged over a wave period for No.8. The flow profile near the top of seaweed is oscillated wavy. The velocity toward upward direction just over seaweed and the velocity toward downward direction between the seaweed are seen,

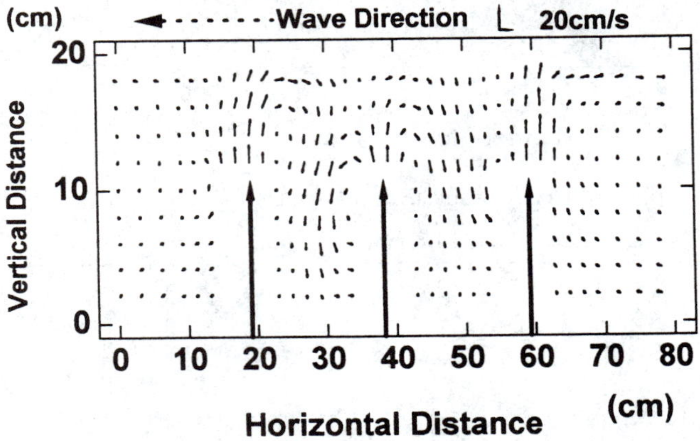

Fig.3 Steady velocity over seaweed for run 8

respectively, so that these flows form a set of circulation flow, having a large scale which compares with the scale of depth. It, therefore, has a effect on the suspension mechanism of sand particles close to the bed among the seaweed. These phenomena are seen in the case of run No.7,too. No vortex, however, is formed for run 4 and 5 in which the distance between both seaweed is half of run 7 or for run No.1,2 and 5 in which the number of seaweed is only one.

3.3 Suspended sediment concentration

It is known that there are three types of the distribution patterns of sediment concentration in the vertical direction, i.e. concave, convex and linear. Under non-breaking wave field, a region of high concentration is produced close to the bed due to the separated vortex over sand ripples and away from the bed surface the mean value of concentration is decreasing.

Figure 4 shows the vertical distribution of mean concentration for run 3 and in which the numbers show the distance from the settling position of seaweed. The distribution patterns of present study, as shown in fig.4, are concave with the turning point in the middle of depth, away from there all profiles reach a constant value. Those are not seen under non-breaking wave condition. Because no turbulent kinematics energy is supplied from the upper layer unlike the breaking wave and the flow filed by wind shear stress(e.g., Nielsen 1985; Tsujimoto et al.1995). It is thought that the turbulent kinematic energy generated by separated vortex near the top of seaweed can picks up the sand particles from the bed. Especially the separated vortex on the onshore side is conspicuous, having a large scale in space, so that it exerts the major effects on the suspension mechanism of sand particles. Also sand ripples have been formed at the bottom through eight runs.

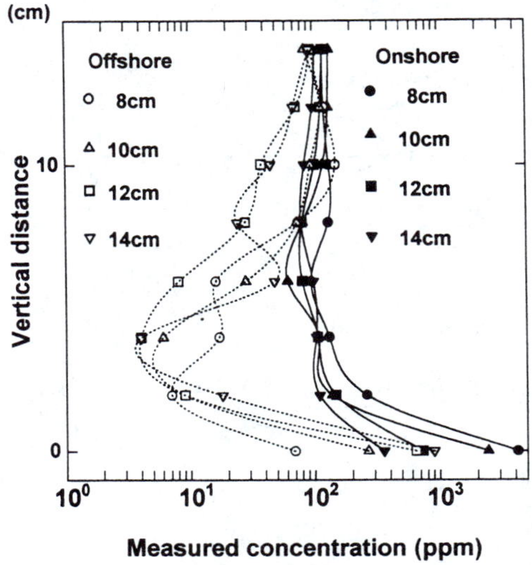

Fig.4 Measured mean concentration of suspended sediment

IV. NUMERICAL APPROACH

4.1 Previous study

Although there are many studies on the flow field with the substance in motion like seaweed, few have considered the mechanism of turbulent flow and sediment transport under wave motion. Asano et al.(1988) investigated the wave damping by using the artificial vegetation. Tsujimoto(1992) studied the mechanism of turbulent flow over an artificial seaweed field using the k- ε turbulent model. The agreement between the calculation and experiment, however, is not satisfactory.

These approaches are characterized by considering the effects of seaweed in the momentum equation as drag force. To calculate the turbulent flow filed over the seaweed bed, the turbulent productions and its dissipation by the seaweed motion should be modeled. There, however, is less knowledge on the production mechanism of turbulence over the seaweed bed under the wave motion and it is not easy for accurate determination on drag coefficient when drag force is adapted in the momentum equation.

4.2 Basic equation

The basic equations describing fluid motion consist of the continuity equation, momentum equations, diffusion equation of suspended sediment and conservation for

turbulent kinematic energy k and its dissipation rate ε. These equations are the same equations as author's used(1991) on calculation of suspended sediment over sand rippled and the effects of seaweed motion includes directly in the calculated grid without estimating that as drag force. These equations are written as follows:
For the continuity

$$\frac{\partial u}{\partial x}+\frac{\partial w}{\partial z}=0 \tag{1}$$

For the momentum equation in x-direction

$$\frac{\partial u}{\partial t}+\frac{\partial u^2}{\partial x}+\frac{\partial uw}{\partial z}=-\frac{1}{\rho}\frac{\partial p}{\partial x}+\nu\left(\frac{\partial^2 u}{\partial x^2}+\frac{\partial^2 u}{\partial z^2}\right)+\frac{\partial}{\partial x}\left(2\nu_t\frac{\partial u}{\partial x}\right)+\frac{\partial}{\partial z}\left(\nu_t\left(\frac{\partial u}{\partial z}+\frac{\partial w}{\partial x}\right)\right) \tag{2}$$

For the momentum equation in z-direction

$$\frac{\partial w}{\partial t}+\frac{\partial uw}{\partial x}+\frac{\partial w^2}{\partial z}=-\frac{1}{\rho}\frac{\partial p}{\partial z}+\nu\left(\frac{\partial^2 w}{\partial x^2}+\frac{\partial^2 w}{\partial z^2}\right)+\frac{\partial}{\partial z}\left(2\nu_t\frac{\partial w}{\partial z}\right)+\frac{\partial}{\partial x}\left(\nu_t\left(\frac{\partial u}{\partial z}+\frac{\partial w}{\partial x}\right)\right) \tag{3}$$

For the diffusion equation of the suspended sediment

$$\frac{\partial c}{\partial t}+\frac{\partial uc}{\partial x}+\frac{\partial wc}{\partial z}=-\frac{\partial}{\partial x}\left(\left(\nu+\frac{\nu_t}{\sigma_t}\right)\frac{\partial c}{\partial x}\right)+\frac{\partial}{\partial z}\left(\left(\nu+\frac{\nu_t}{\sigma_t}\right)\frac{\partial c}{\partial z}\right)+\frac{\partial w_o c}{\partial z} \tag{4}$$

For k-equation

$$\frac{\partial k}{\partial t}+\frac{\partial uk}{\partial x}+\frac{\partial wk}{\partial z}=-\frac{\partial}{\partial x}\left\{\left(\nu+\frac{\nu_t}{\sigma_k}\right)\frac{\partial k}{\partial x}\right\}+\frac{\partial}{\partial z}\left\{\left(\nu+\frac{\nu_t}{\sigma_\varepsilon}\right)\frac{\partial k}{\partial z}\right\}+P-\varepsilon \tag{5}$$

For ε-equation

$$\frac{\partial \varepsilon}{\partial t}+\frac{\partial u\varepsilon}{\partial x}+\frac{\partial w\varepsilon}{\partial z}=-\frac{\partial}{\partial x}\left\{\left(\nu+\frac{\nu_t}{\sigma_\varepsilon}\right)\frac{\partial \varepsilon}{\partial x}\right\}+\frac{\partial}{\partial z}\left\{\left(\nu+\frac{\nu_t}{\sigma_\varepsilon}\right)\frac{\partial \varepsilon}{\partial z}\right\}+C_{1\varepsilon}\frac{\varepsilon}{k}-C_{2\varepsilon}\frac{\varepsilon^2}{k} \tag{6}$$

$$P=2\nu_t\left\{(\partial u/\partial x)^2+(\partial w/\partial z)^2\right\}+\nu_t(\partial u/\partial z+\partial w/\partial x)^2 \tag{7}$$

$$\nu_t=c_\mu\frac{k^2}{\varepsilon} \tag{8}$$

where x and z are the horizontal axis and the vertical axis, respectively, u and w are velocity in x and z direction, p is the pressure, ρ is the fluid density, ν and ν_t the molecular and eddy viscosities, c is the suspended sediment concentration, w_o is the fall velocity of the sediment particle and the constants in these equations are set at the values proposed by Rodi(1980) as given as follows:

$$C_\mu = 0.09, C_{1\varepsilon} = 1.44, C_{2\varepsilon} = 1.92, C_{3\varepsilon} = 1.00, \sigma_k = 1.00, \sigma_\varepsilon = 1.3, \sigma_t = 1.00 \qquad (9)$$

The natural frequency of seaweed is different from that of the experimental wave, so that the oscillation period of seaweed was set to be the same as wave's. The seaweed motion should be coupled from the wave motion in a strict sense, but the experimental value of velocity of seaweed motion are used herein.

There is no relative velocity in the normal direction to the surface of seaweed, and it is given by the following equation;

$$Vn = V'n \qquad (10)$$

where **n** is a unit vector in the normal direction to the seaweed surface, V is the velocity of water particle and V' is the velocity of seaweed motion. The values of V' are given by the displacement of seaweed per a second in the horizontal direction. The displacement of seaweed in the vertical direction is assumed so small that the vertical velocity can be neglected. The boundary conditions at seaweed surface on the turbulent kinematic energy k and its dissipation rate ε are the same as that at the bottom and the concentration is set to be zero there. Other boundary conditions are the same as author's(1991).

4..3 Method of calculation

The conditions of calculation are equivalent to that of the experiment as given in Table 1. Forty elements in the vertical direction are used and the number of element in the horizontal direction is changed from forty to one hundred twenty depended on number of seaweed and a fine mesh scheme is employed near the seaweed. The calculation was conducted to investigate the effect to the flow structure when the phase lag between the wave motion and the seaweed motion and the height of seaweed were changed.

V. CALCULATED RESULTS

5.1 Steadv flow

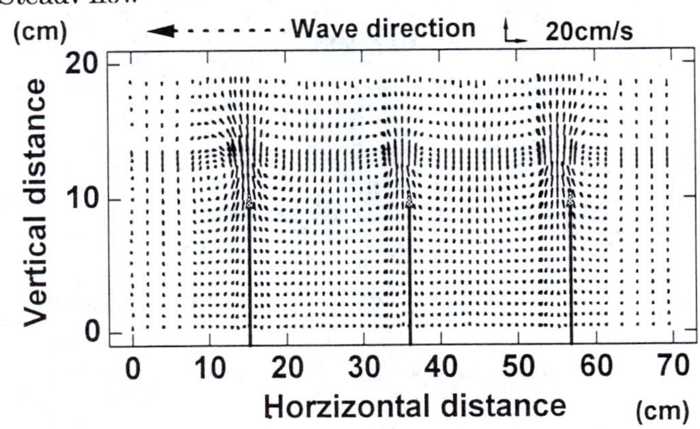

Fig.5 Calculated steady velocity

Figure 5 shows a steady velocity for run 8. The wavy flow near the top of seaweed and the vertical circulation flow between the seaweed are seen as shown in the experimental results(e.g. figure 3). These values are about 20cm/sec and approximately the same value as the experimental one. The calculated values of vertical downward direction, however, are smaller than the experimental values.

5.2 Velocity profile and suspended sediment concentration

Figure 6(a)and(b) show the calculated results on velocity and concentration for run 6. The separated vortex is not seen at the wave phase when a wave crest passes over the seaweed and the wavy flow is only formed near the top of seaweed. So the value of suspended sediment concentration is still low. The suspended sediment cloud is transported from the offshore side to the top of seaweed, but it has little sand particle. A small separated vortex is formed near the top of seaweed for fig.6 (b), but the suspended sediment between the seaweed could not be transported to the upper layer.

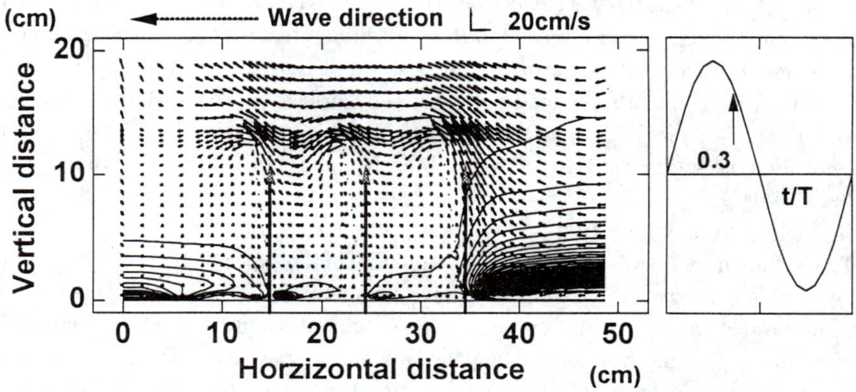

Fig.6(a) Calculated velocity and suspended sediment concentration for run 6

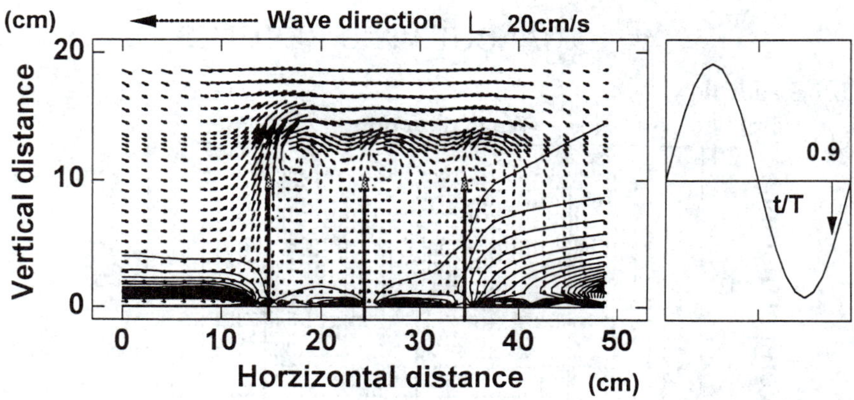

Fig.6(b) Calculated velocity and suspended sediment concentration for run 6

Figure 6(c) shows the calculated results at the same wave phase as that of figure 6 (b) for run 8. The settling space of seaweed in this figure is two times longer than that of figure 6(b). Since the separated vortex grows easily in size, much suspended sediment are transported to upward from the bed. The results of this figure gives an evidence that the settling space of seaweed is very important for suspended sediment transport.

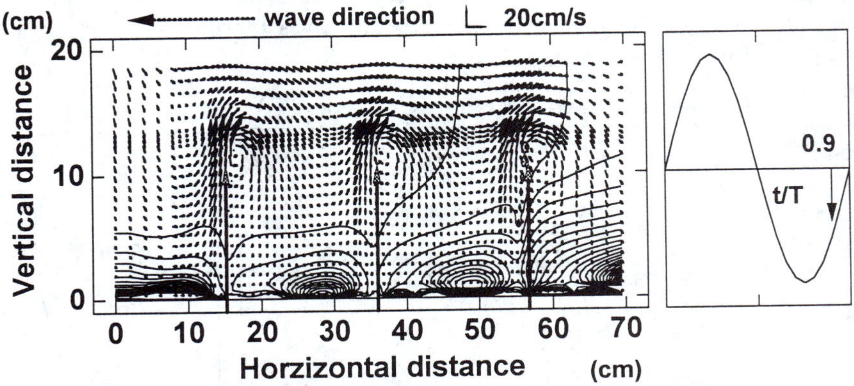

Fig.6(c) Calculated velocity and suspended sediment concentration for run 8

5.3 Suspended sediment concentration

Figure 7 shows the calculated vertical distribution of suspended sediment concentration averaged over a wave period, in which the number shows the distance from the position where the seaweed is located. The value of suspended sediment concentration close to the bed is lower than that of experiment, as the effects of sand ripples are not considered in the present calculation. The calculated results around the top of seaweed are close to the experimental values.

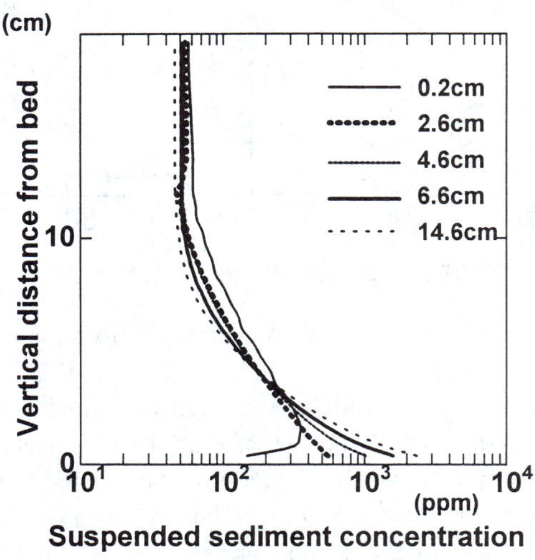

Fig.7 Calculated mean suspended sediment concentration for run 2

The suspended sediment concentration decreases lineally with increasing distance from the bed, whereas away from there it attains a constant value and the distribution pattern is concave. Also a certain level where the slope of the vertical distribution of the sediment concentration changes is approximately equal to the height of seaweed.

5.4 Vorticity

The experimental distribution of vorticity is shown in figure 8 for run 2. The vortex with negative sign at the preceding half period and the vorticity with positive sign at the next half period are seen near the top of seaweed.

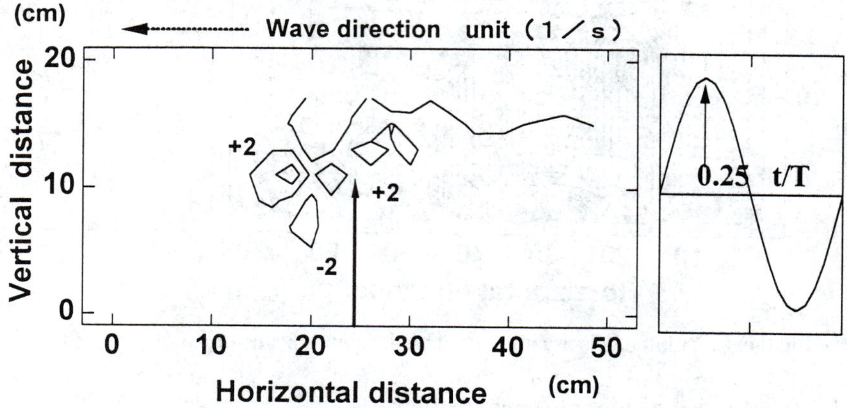

Fig.8 Contour lines of measured vorticity for run 2

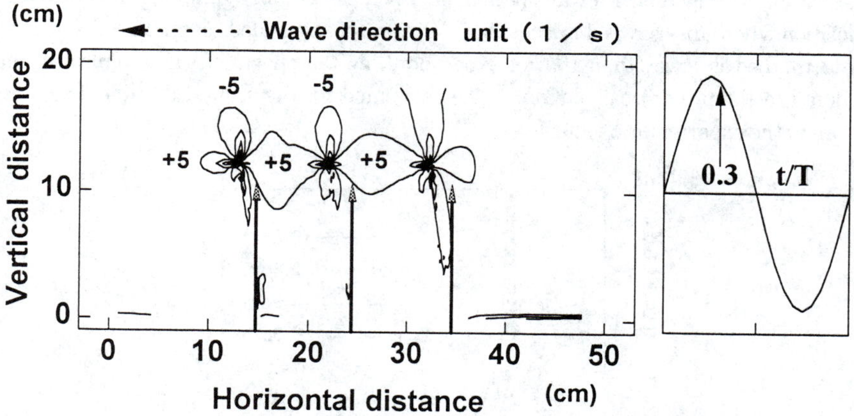

Fig.9 Contour lines of calculated vorticity for run 6

It is thought that the vorticity with positive sign is dominant at this phase; but the seaweed is oscillated at the same frequency of that of water particles toward the same direction, and the vorticity with positive sign are produced late. The similar results are obtained in the present calculation(no figure). The calculated results are shown in figure 9 for run 6. Each vorticity are combined among the seaweed, and they does not grow in size.

5.5 Coefficient of drag force

In the present study the effect of seaweed to the flow mechanism is considered by given the velocity of seaweed motion to calculated mesh. The seaweed field should be considered as the drag force averaged in space in the field, and the estimation on the coefficient of drag force are very important.

Hydrodynamic force F(t) at the each wave phase is calculated by integrated the calculated hydraulic pressure along the seaweed surface, being separated into the drag force and the inertia force as shown eq.(11) .

$$F(t) = \frac{1}{2} C_D \rho A u(t)|u(t)| + C_M \rho V \frac{\partial u(t)}{\partial t} \tag{11}$$

in where u(t) is the horizontal velocity just over seaweed close to the free surface, C_D is the coefficient of drag force, C_M is the coefficient of inertia force, A and V are the reference area and volume, respectively. As shown by Ishida(1988), the values of C_D and C_M are assumed to be constant between the adjacent phase and eq.(11) is solved by using the simultaneous equations.

Figure 10 shows the variation of the horizontal velocity and the coefficient of drag force over both the seaweed and the fixed seaweed for run 2. The coefficient of drag force reaches the biggest value at the wave phase when the flow direction changes from onshore to offshore and constant value at other wave phase. Also its value over the seaweed is bigger than that over the fixed seaweed. The drag coefficient C_D is changeable at the each phase, varying from the used value of horizontal velocity. The drag coefficient C_D is generally estimated by the number of the Keulegan-Carpenter number at the wave phase when the wave crest or the wave trough passed over the reference point, when the drag force become maximum.

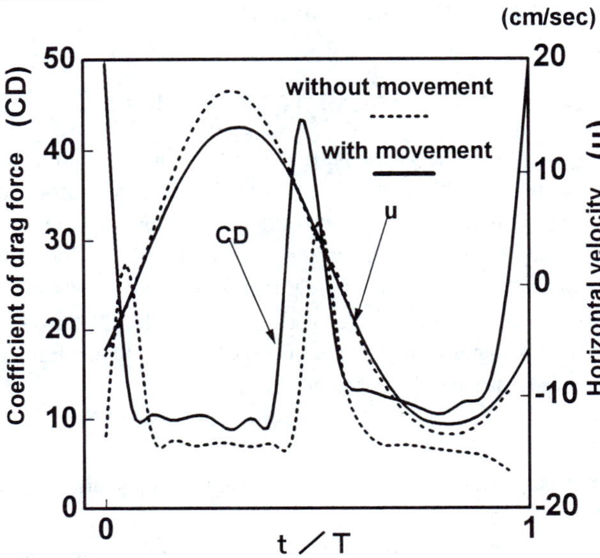

Fig.10 Time history of drag force and horizontal velocity

It is shown through the experiment that the wave phase lag between the seaweed motion and the wave motion is $\pi/2$. The wave phase lag is varied from the wave condition and kinds of seaweed. Figure 11 shows the relationship between the drag coefficient C_D and K.C. number when the wave phase lag between the seaweed motion and the wave motion is changed from 0 to $3\pi/2$. With increasing the value of KC number, the value of KC number attains to about 5. That is similar to Ishida's results, who measured the wave force acting on a pile structure.

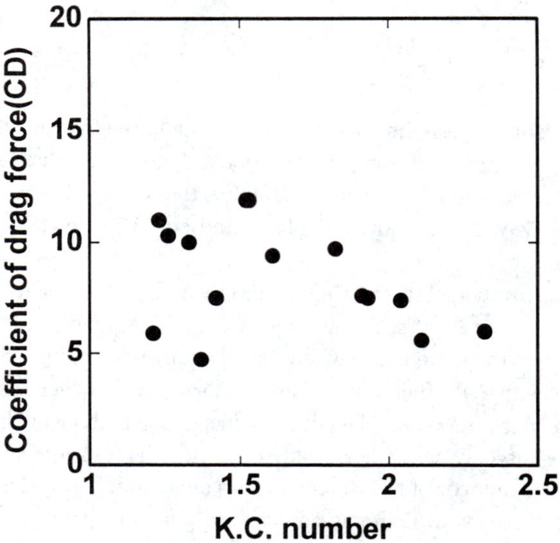

Fig. 11 Drag coefficient and KC number

VI. CONCLUSIONS

The seaweed bed is made of a plastic plate which is oscillated symmetrically with the wave motion and the flow mechanism and suspended sediment concentration over the seaweed bed were investigated by an experimental approach and a numerical simulation. The wavy flow near the top of seaweed and the vertical circulation flow which compared with the scale of water depth are produced by the seaweed motion, respectively. As the separated vortex is formed at the middle of depth, the value of suspended sediment concentration reaches a constant one or increase by the turbulence kinematics energy. That can not be seen under the non-breaking wave.

The turbulent calculation considered the existence of seaweed in the calculated mesh showed as follows:

1) The separated vortex over the seaweed is formed later than that over the fixed seaweed.
2) The value of suspended sediment concentration attains a constant value away from the bed.
3) The settling space between the seaweed and the wave phase lag between wave

motion and seaweed motion have a major effect on the flow mechanism.

4) The value of drag coefficient C_D of seaweed estimated as the drag force is several times lager than that of a plate(2.01).

ACKNOWLEDGEMENT

This experiment work was conducted at Kobe City College of Technology. The author wish to express thanks to graduate students; Y.Nagao, H.Arima, Y.Ito and Y.Masuda. A part of this study is supported by the Grant in Aid for Science Research Ministry of Education.

Reference

Asano T., S. Tsustui, and T. Sakai(1988):Wave damping characteristics due to seaweed,Proc.35th Coastal Eng. Conference in Japan,pp.138-142(in Japanese).

Ishida, H. and Tamura K.(1988):Coefficient of drag force of a pile structure under separated flow ,Proceeding. of JSCE,No.393,II-9,pp.120-130(in Japanese).

Nilesen P.(1984):Field measurement of time-averaged suspended sediment concentration under waves, Coastal Engineering,Vol.8,pp.51-72 .

Tsujimoto G., N. Hayakawa, M. Ichiyama, Y. Fukushima and Y. Nakamura(1991) :A study on suspended sediment concentration and sediment transport mechanism over rippled sand bed using a turbulence model, Coastal Engineering in Japan,Vol.34,No.2,pp. 177-189.

Tsujimoto G.(1992):Flow and suspended sediment concentration over seaweed field, Proceeding Coastal Eng. of JSCE,Vol.39,pp.276-280(in Japanese).

Tsujimoto G. and H.G.Stefan(1994):Simulation of sediment resuspension transport by wind driven current in a shallow lake, Research Reports of Kobe City College of Technology No.32,pp.65-72.

CHAPTER 305

An Energetics Approach for Suspended Sand Transport on Macrotidal Ridge and Runnel Beaches

George Voulgaris[1], Travis Mason[2] and Michael B. Collins[2]

Abstract

High frequency observations of hydrodynamics and suspended sediment concentration were acquired from the intertidal zone of a macrotidal ridge and runnel beach. The hydrodynamic data from the electromagnetic current meters were used to calculate sediment transport rates according to Bailard's (1981) energetics model. Additionally, suspended sediment fluxes were calculated for the period of data collection. The contribution of long waves, short waves and mean currents to suspended sediment transport was examined. Application of the energetics model suggested that suspended sediment is the dominant mode of transport. Further, it showed that the waves define net sediment transport rate and direction. However, the measurements revealed that the transport due to mean currents was dominant. The discrepancy between the model and measurements was attributed to vortex ripples which were developed in the intertidal zone. Bailard's model failed to predict the oscillatory mode of transport but it was found to predict accurately the mean mode of suspended sediment transport assuming an efficiency factor value of 0.021 is used in the calculations.

Introduction

Sand resuspension under the action of waves and currents is the primary process controlling seabed erodibility. Successful parameterization of this mechanism is the key element for the development of accurate and predictive morphological models. Presently, Bailard's (1981) sediment transport formula is considered one of the best theoretical models for time-dependent, cross-shore sediment transport (Schoonees and Theron, 1995) and is widely used for the prediction of changes in the coastal zone (*e.g.* Scott *et al.*, 1991). The model is based on the importance of various moments of the fluid velocity in determining transport rates. Previous studies examined the various velocity moments

---

[1]Woods Hole Oceanographic Institution, Woods Hole, MA 02543, U.S.A.

[2]Department of Oceanography, University of Southampton, Southampton Oceanography Centre, European Way, Southampton, SO14 3ZH, U.K.

under either wave (*i.e.,*Guza and Thornton, 1985; Russell *et al.,* 1995) or steady current dominating conditions (*i.e.*, Thornton *et al.*, 1996). Ridge and runnel beaches, however, are found throughout areas with moderate wave conditions (short-fetched seas) and high tidal ranges (King and Williams, 1949; Orford and Wright, 1978). The combination of high tidal currents and moderate wave activity invalidates the assumptions of the previous investigators regarding the dominant forces (*i.e.* waves or steady currents).

In this contribution, high frequency measurements of instantaneous currents and suspended sediment concentration were obtained on a macrotidal ridge and runnel beach. The data collected are used to: (i) examine determine the relative contributions of long waves, short waves and mean currents to suspended sediment transport; (ii) evaluate the energetics sediment transport model (Bailard, 1981) and velocity moments for macrotidal conditions; and (iii) calibrate the energetics model for suspended sand transport.

Energetics Model

Bagnold's (1966) energetics-based total load sediment transport model for streams has been used as a basis for the development of a model of time-varying transport over a plane sloping bed. The idealised equation for the total (i.e bedload and suspended load, $<I_t>$) time-averaged immersed weight sediment transport is given by Bailard (1981):

$$<\vec{I}_t> = \rho\, C_f \frac{\epsilon_b}{\tan\phi}\left[<|\vec{U}_t|^2\vec{U}_t> + \frac{\tan\beta}{\tan\phi}<|\vec{U}_t|^3>\vec{i}\right] + \rho\, C_f \frac{\epsilon_s}{W}\left[<|\vec{U}_t|^3\vec{U}_t> + \frac{\epsilon_s}{W}\tan\beta<|\vec{U}_t|^5>\vec{i}\right] \quad (1)$$

where W is the sediment fall velocity; ϵ_b and ϵ_s are the bed load and suspended load efficiency factors, respectively; β is the local bed slope; ϕ is the angle of internal friction of the sediment; i is the unit vector in the down-slope, offshore (-x) direction; and U_t the total instantaneous velocity.

The above general model has been analysed into the cross-shore ($<I_x>$) and longshore ($<I_y>$) components for application with random waves (Guza and Thornton, 1985):

$$<I_x>=\rho C_f U_m^3\{\frac{\epsilon_B}{\tan\phi}\cdot[\psi_1\cos\alpha_1+\delta_u^3+\delta_u(\frac{1}{2}+\cos^2\alpha_2+\delta_u^2)+\delta_u\sin\alpha_3\cos\alpha_3 + \frac{\tan\beta}{\tan\phi}u3^*] + \frac{U_m}{W}\epsilon_s[\psi_2\cos\alpha_5+\delta_u\cdot u3^*] + \frac{U_m^2}{W^2}\epsilon_s^2\tan\beta\cdot u5^* \quad (2)$$

$$<I_y>=\rho\cdot C_f\cdot U_m^3\cdot\{\frac{\epsilon_B}{\tan\phi}\cdot[\psi_1\sin\alpha_1+\delta_v^3+\delta_v(\frac{1}{2}+\sin^2\alpha_2+\delta_u^2) + \delta_u\sin\alpha_3\cos\alpha_3] + \frac{U_m}{W}\epsilon_s[\psi_2\sin\alpha_5+\delta_v\cdot u3^*] \quad (3)$$

where U_m is the oscillatory velocity (rms) amplitude. δ_u and δ_v are the steady current strength for the cross-shore and longshore directions defined as $\bar{u}/U_m$ and $\bar{v}/U_m$, respectively. The most important normalised velocity moments are:

$$\psi_1 = \frac{<|U_t|^2(u_{osc},v_{osc})>}{U_m^3} \tag{4}$$

$$\psi_2 = \frac{<|\vec{U}_t|^3(u_{osc},v_{osc})>}{U_m^4} \tag{5}$$

where the symbols <> denote a time-averaged quantity. The velocity components are decomposed into an oscillatory (u_{osc} , v_{osc}) and a mean ($\bar{u}$, $\bar{v}$) part, so that:

$$|\vec{U}_t| = \sqrt{u_{osc}^2+v_{osc}^2+\bar{u}^2+\bar{v}^2+2(u_{osc}\bar{u}+v_{osc}\bar{v})} \tag{6}$$

The definitions of the remaining parameters can be found in Guza and Thornton (1985, see page 251, eqns 35 and 36). Briefly, αn (n=1 to 5) are the angles associated with the directional characteristics of the various flow moments, while $u3^*$ and $u5^*$ are the odd moments of the total instantaneous flow, integrated over the whole wave period. Application of the above model requires information on the moments of the instantaneous velocity field.

In order to obtain a quantitative prediction of immersed weight sediment transport rate, values for the drag coefficient (C_f) and the efficiency factors are required. Bowen (1980) proposed a value of 0.005 for the drag coefficient. Here, the drag coefficient is determined by the flow conditions as function of the ratio z_0/h (Soulsby *et al.*, 1993). The efficiency factors used in this model are known to depend on factors such as flow conditions (laminar or turbulent, steady or unsteady, etc), and the grain size characteristics of the sediments. A list of various suggested values is presented in Table 1.

Author	Bedload	Suspended load
Bagnold (1966), steady flow	0.13	0.010
Bailard (1981,1982) longshore	0.21	0.025
Bailard (1981,1982), cross-shore	0.10	0.020
Bowen (1981), cross-shore	0.15 tanϕ	0.010

Table 1. Efficiency parameter values proposed by various investigators.

Guza and Thornton (1985) applied field data to Bailard's cross-shore sediment transport model to determine the dominant terms in the bedload and suspended load modes. They concluded that suspended load dominates sediment transport processes and

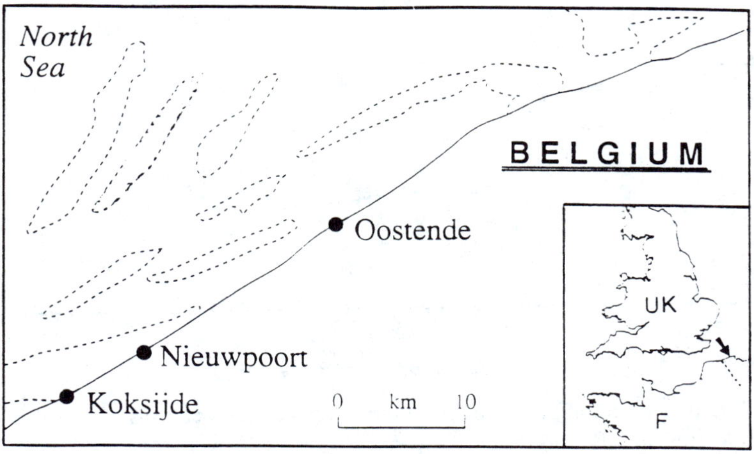

Figure 1. Location of experimental site.

that the normalised velocity moments ψ_2 and $\delta_u u3^*$ are the most important terms, representing the contribution of the oscillatory and mean flow to suspended sediment transport, respectively.

Bowen (1980), Roelvink and Stive (1989), and Russell et al (1996) have extended Guza and Thornton's (1985) work by examining the relative importance of the long, short period oscillations and mean flow to cross-shore suspended sediment transport. Their analysis was based on the assumption that the longshore currents are much smaller than the orbital velocities ($\bar{v} << U_m$). On the other hand, Thornton et al. (1996) presented an expansion of equation (1) which is based on the assumption that mean currents dominate the flow.

Macro-tidal ridge and runnel beaches are found though, throughout areas with moderate wave conditions (short fetched seas); this, combined with the high tidal currents invalidates the assumptions used by the above investigators. In the present analysis the velocity terms of eqns (2) and (3) are examined, assuming that both waves and currents are of equal importance. In addition, almost all of the above studies concentrated on the cross-shore direction only. Due to the importance of the longshore tidally-induced currents found on ridge and runnel beaches it is considered appropriate that both horizontal components (cross-shore and longshore) should be included in the velocity moments analysis.

Experimental Site - Data Collection

The field measurements were obtained at Nieuwpoort-aan-Zee, situated some 13km southwest of Oostende (Fig. 1), from 20th February to 4th March 1994 (Julian days 51-63). The experimental site was outside the influence of any man-made obstructions, such as groynes which are typical elsewhere. The sediment is well sorted, fine/medium sand (mean grain size 0.183mm); at the time of the observations the average beach slope ($\tan\beta$) was 0.012 with an intertidal zone of between 250 and 500m wide. The

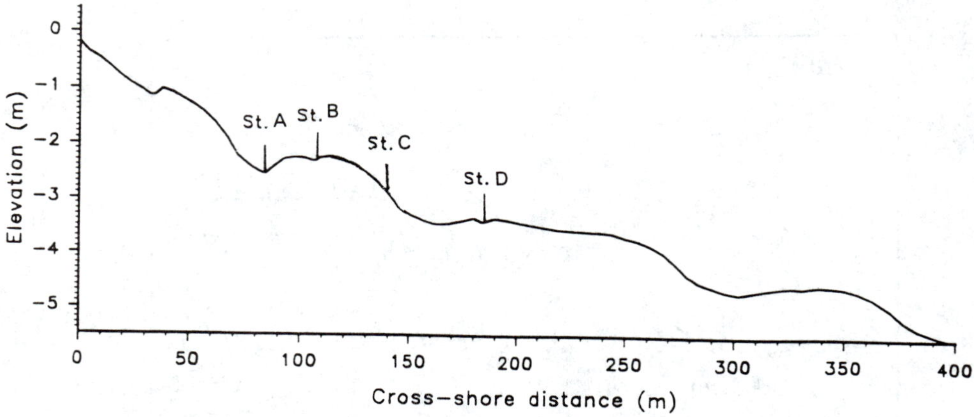

Figure 2. Cross-sectional profile of the intertidal zone and locations of instrument stations (A, B, C and D).

experimental site exhibited typical ridge and runnel beach characteristics (Fig. 2) with up to five well defined ridges exposed at low tide.

Tides in this area are semi-diurnal with a large neap to spring variation of 2.7 to 6.5m during the deployment period. Offshore currents are characterised by elongate and asymmetrical tidal current ellipses, with a northeasterly directed residual flood displacement controlling the sub-tidal sediment transport pathways (De Moor, 1991).

The wave climate of the area is typical of short-fetched seas with wave periods 3 to 6s. However, typical wave periods recorded during the deployment period were between 4 and 12 seconds with offshore wave heights of the order of 0.5 to 1m. Dominant offshore wave direction, extracted from wave hindcasting modelling (U.K. Meteorological Office), is primarily from the west-southwest (23% of the time) and secondarily from the north-northeast (10% of the time).

Four instrumented stations were deployed in the intertidal zone (Fig. 2). The instruments deployed consisted of electromagnetic current meters (Stations A, B, C and D), pressure transducers (Stations B, C and D) and optical backscatter sensors (OBS) for measuring instantaneous suspended sediment concentration (SSC, Stations C and D). The OBS data obtained from station C were of too poor quality to be used for analysis. Therefore, the results presented here are based on the data collected at Station D which was located in the upper mid-tidal zone and on the top of a ridge with a relatively flat crest (Figure 2).

Four 2-axis electromagnetic current meters, 3 OBS sensors and a pressure transducer were deployed at station D. All instruments were cantilevered from a stainless steel pipe, set vertically into the beach sand during low water. Three of the electromagnetic current meters were arranged to measure instantaneous horizontal currents at three levels above the sea bed (21.5, 48 and 108.5cm, respectively). The fourth sensor was orientated to measure vertical currents at 70cm above the sea bed. The OBS sensors were installed at elevations 9, 19 and 34.5cm above the sea bed, respectively. Data were collected when the instruments were fully submerged and for consecutive time intervals of approximately 17min, spanning flood, high water and ebb tide. The sampling frequency within each burst was 8 Hz.

Sediment from the deployment area was used for the calibration of the OBS sensors. A temporally variable background level of "turbidity" was observed during the data collection period. This was due to fine material in suspension, referred t hereafter as "washload", coming from Nieuwpoort Harbour. To maintain uniformity in data analysis, the offset due to this washload was determined as the 10th percentile of the sample values as suggested by Aagaard and Greenwood (1995). This was the magnitude of concentration which corresponded approximately to a significant break in the cumulative frequency of signal values and best represents the separation between the background signal and sediment resuspension events. As Aagaard and Greenwood (*op. cit.*) noted such an approach may introduce some error in computing absolute magnitudes of sediment transport rates. Laboratory experiments showed that this error is less than 10%.

Hydrodynamic and suspended sediment concentration measurements were obtained during 3 tidal cycles on 3rd and 4th March, 1994 (Julian days 62 and 63, respectively). In particular, 6 data sets (runs) were collected during the flood of the morning flood tide of Julian day 62; 9 runs during the evening of Julian day 62; and finally 12 runs were collected during the morning of Julian day 63.

Time-series of cross-shore and longshore velocity (u,v) and sediment concentration (c) were plotted to check data quality. Statistical descriptions of velocity and sediment concentration were computed (i.e mean and standard deviations). Wave characteristics were computed using the moments of the spectra of the pressure transducer signal after correcting for depth attenuation (Voulgaris, 1992). Significant orbital velocities were estimated from the standard deviation of each time-series ($U_m=(2(u_\sigma^2+v_\sigma^2))^{1/2}$).

The total fluxes have been analysed into mean and oscillatory modes with the latter being decomposed further into fluxes due to incident short (gravity, subscript $_S$) waves and long period (infragravity, subscript $_L$) flow.

$$\overline{u\cdot c} = \bar{u}\cdot\bar{c} + \overline{u_{osc}\cdot c_{osc}} = \bar{u}\cdot\bar{c} + \overline{u_S\cdot c_S}+\overline{u_L\cdot c_L} \tag{7}$$

The mean component was calculated (by averaging the entire 17 minute record), and then subtracted from the record, to obtain the oscillatory component. The infragravity (long) component of the signal (u_L, v_L and c_L) was calculated by applying a low-pass filter (cut-off frequency 0.06Hz) to the oscillatory component. The infragravity component was subtracted from the total oscillatory component, thus giving the gravity contribution.

Local fluxes were calculated for each data run according to eqn. (7), for both the cross-shore and longshore directions and for the three elevations above the sea bed where SSC measurements were obtained. Measurements from the current meter installed at 21.5cm above the sea bed were used for the local flux calculations. Oscillatory currents were assumed to be constant through the water column whilst a logarithmic distribution was assumed for estimating the fluxes due to mean currents.

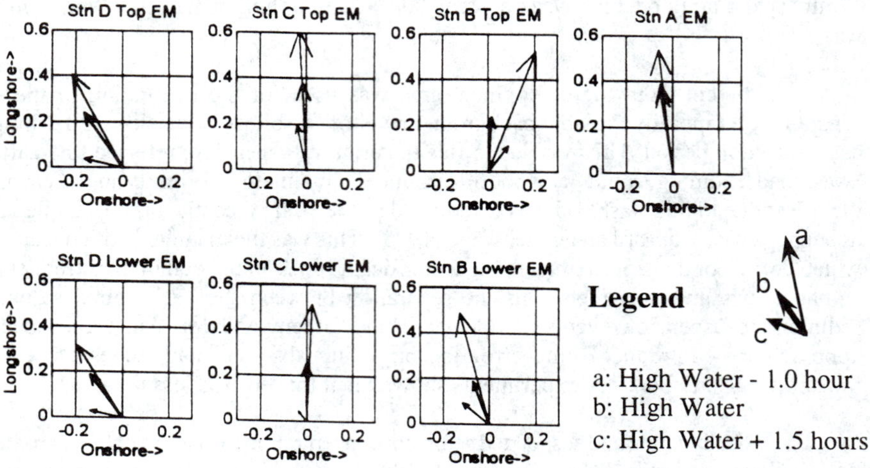

Figure 3. Vector diagrams of mean flow in the intertidal zone (for locations see Fig. 2) at 1 m (top row) and 0.2m (bottom row) above the sea bed (units in m/s).

Results

Overview of Hydrodynamics

In this section an overview of the hydrodynamic conditions in the intertidal zone is presented. This is based on observations from the whole of the experimental period. Typical mean flow patterns in the intertidal zone were found to be tidally-induced. Figure 3 shows the magnitude and direction of the mean current, for a typical tide, before (-1 hour) during and after (+1.5 hours) high water. It is evident that the dominant flow direction is consistently alongshore in the positive (northeast) direction. Maximum longshore flow occurs during the flood stage of the tide and the magnitude of the longshore current reduces as the tide progresses. Preliminary analysis of the hydrodynamic data collected indicates the this behaviour is due to bed friction which causes the longshore currents to lead the tidal elevation inshore. The mean cross-shore flow is generally offshore for all stations. An exception to this is Station B where the currents at 1 m above the bed are generally directed onshore. This station is located at the top of a ridge (see Fig. 2) where the waves exhibit their maximum asymmetry, thus this onshore flow is attributed to onshore mass transport which is balanced by an offshore flow near the seabed.

The wave climate in the intertidal zone exhibits a dependence on tidal stage (ebb, flood, see Figure 4). Local wave height is larger during the flood than during the ebb stage of the tide. This is more enhanced at the offshore stations C and D). This wave height dependence on tidal stage is the result of the interaction of the waves with the tidally-induced longshore currents. During flood, the currents are strongest and are directed toward the northeast. This results in the currents opposing the surface waves which, during the time of observation, were propagating at an angle towards the southwest. This

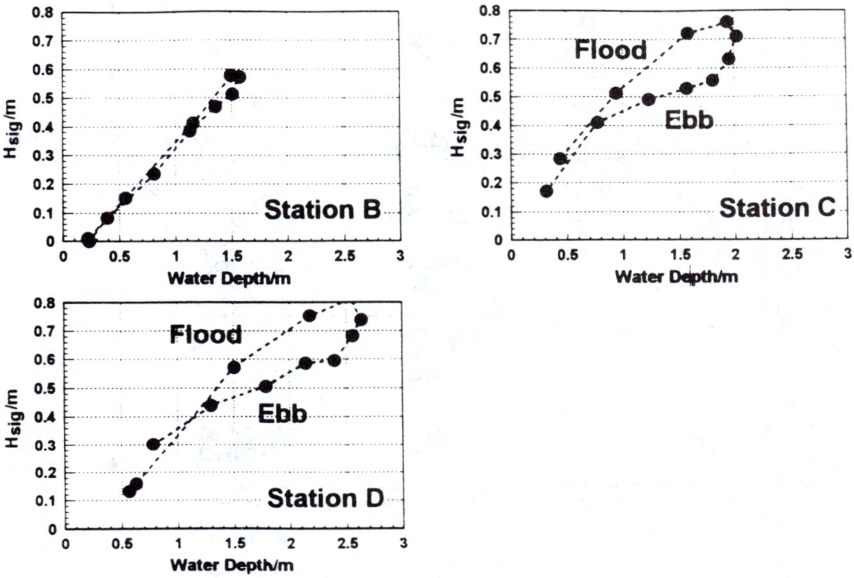

Figure 4. Tidal modulation of the local wave height (H_{sig}) for stations B, C and D (for location see Fig. 2).

opposition of waves and currents resulted in an increased wave steepness. During the ebb stage of the tide, the current strength is significantly reduced (see Figure 3), so that the interaction is much lesser, leading into less steep waves.

Velocity Moments

The relevant flow moments to sediment transport are examined in order to identify the most important terms for macrotidal beaches. Following Guza and Thornton's (1985) approach, the field flow data collected at station D from the lowest current meter are used to estimate the terms of eqns. (2) and (3).

The values of the individual terms such as ψ_1, ψ_2, $u3^*$, $u5^*$, δ_u, δ_v and the angles α_1,α_2, α_3, and α_5 have been evaluated for each run during the data collection period and the results are shown in Figure 5. The nonlinear scalar moments ψ_1 and ψ_2 are linked in the transport eqns (2) and (3) to the angles α_1 and α_5. The moments ψ_1 and ψ_2, (Fig. 5a) are associated with the wave asymmetry and this is why their variation through the tidal cycles corresponds to the variation in the wave orbital velocity field, as discussed in the previous section (see Fig. 4). However, in the case of a dominant cross-shore oscillatory component, the angles α_1 and α_5 in the case of a dominant cross-shore oscillatory components are expected to by around zero or 180degs (Guza and Thornton, 1985). In the present data set though α_1 varies between -10 and 10 degs with an average value of 5 degs (Figs 5d and e). The angle α_5 is around zero for the first and the last tidal cycles but achieves values of up to 20 degs during the second tidal cycle implying that a significant part of the oscillation occurs in the alongshore direction as well. However these angles are linked to the velocity

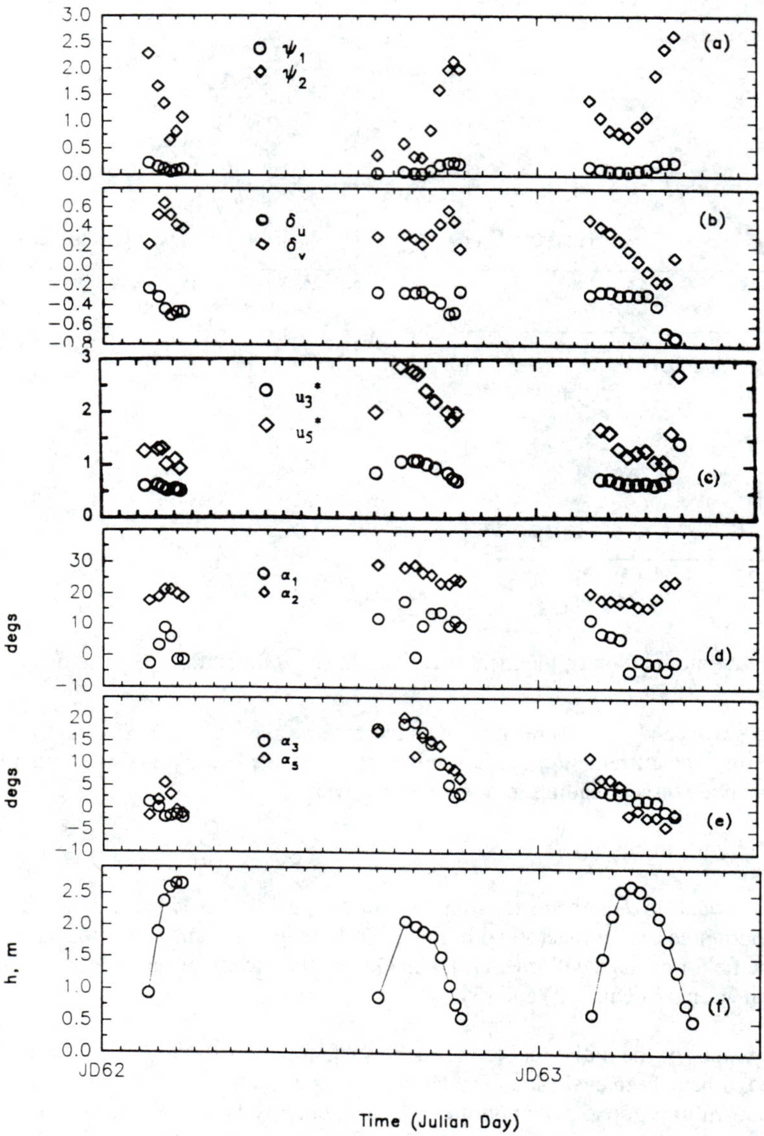

Figure 5. Velocity moment terms used in Bailatd's (1981) energetics sediment transport model (eqs (2) and (3)) as calculated from the measured current velocities from Station D.

moments through their cosine which for the extreme values of 10 and 20degs for the α_1 and α_5 are 0.98 and 0.94 respectively; this implies that omission of these terms will involve an error of 2 and 6%, respectively in the sediment transport rate attributed to these particular terms.

The relative current strength terms δ_u and δ_v vary in accordance with the mean currents observed at the measuring station. It is characteristic that both terms are very important for ridge and runnel beaches reaching absolute values of up to 0.8 which is significantly higher than the average value of 0.01 observed by Guza and Thornton (1985).

The terms $u3^*$ and $u5^*$ express the contributions of the total currents (mean and oscillatory flow) averaged over the whole wave period. In Guza and Thornton's (1985) studies, these were dominated by contributions from the oscillatory component of the cross-shore currents only. In the present study, both oscillatory and mean currents appear to define these moments. It is characteristic that $u3^*$ is almost constant throughout the data collection period, with an average value of 2.5 (standard. deviation 0.34). This value is larger than the theoretical one, derived either assuming a sinusoidal wave ($u3^*$=0.42) or a Gaussian distribution of the wave field ($u3^*$=0.56) with the latter being in good agreement with the field data collected presented by Guza and Thornton (1985). The increased value in the present data set is due to the strong mean flows present which produce an asymmetry of instantaneous flow vector during a wave cycle. The moment $u5^*$ exhibits significant variations throughout the three tidal cycles.

The relative magnitudes of the bedload and suspended load sediment transport and of each individual term of equations (2) and (3) has been examined, assuming that ϵ_s=0.025 and ϵ_b=0.20. Each sediment transport term in eqns (2) and (3) was calculated and then multiplied by the appropriate factor ($\epsilon_b/\tan\phi$, $U_m\epsilon_s/W$). The contribution of each term to the total sediment transport was expressed as a percentage of the total (bedload and suspended load) gross sediment transport rate, with the sign indicating direction of transport. In order to identify any differences in processes under shoaling and breaking waves the analysis was undertaken separately for broken (H_{rms}/h > 0.20) and shoaling waves(H_{rms}/h < 0.20).

Component	Mode	Term	Contribution
Cross-shore	Suspended	$\psi_2\cos\alpha_5$	+55%
		$\delta_u u3^*$	-37%
	Bedload	$\delta_u(0.5+\cos^2\alpha_2)$	+15%
		$\psi_1\cos\alpha_1$	+ 4%
		Others	< 4%
Longshore	Suspended	$\delta_v u3^*$	+70%
		$\psi_2\sin\alpha_5\cos\alpha_5$	+17%
	Bedload	$\delta_v(0.5+\sin^2\alpha_2)$	+16%
		δ_v^3	+ 7%
		Others	< 4

Table 2. Largest terms in energetics sediment transport eqs (2).and (3).

The results suggested that the most important mode of sediment transport is suspended load both in the longshore and cross-shore; this applies to both shoaling and breaking wave conditions. The most important terms within each mode of transport and for each direction, together with the percentage contribution to the total gross sediment transport rate are shown in Table 2. From these results, it can be seen that suspended load accounts for more than the 80% of the gross sediment transport rate. Thus it can be argued that equations (2) and (3) can be simplified with significant predictive accuracy to:

$$\langle I_x \rangle_s = \frac{\rho \cdot C_f \cdot U_m^4 \cdot \epsilon_s}{W} \cdot (\psi_2 \cos\alpha_5 + \delta_u \cdot u3^*) \tag{8}$$

and

$$\langle I_y \rangle_s = \frac{\rho \cdot C_f \cdot U_m^4 \cdot \epsilon_s}{W} \cdot (\delta_v \cdot u3^*) \tag{9}$$

for the cross-shore and longshore directions, respectively.

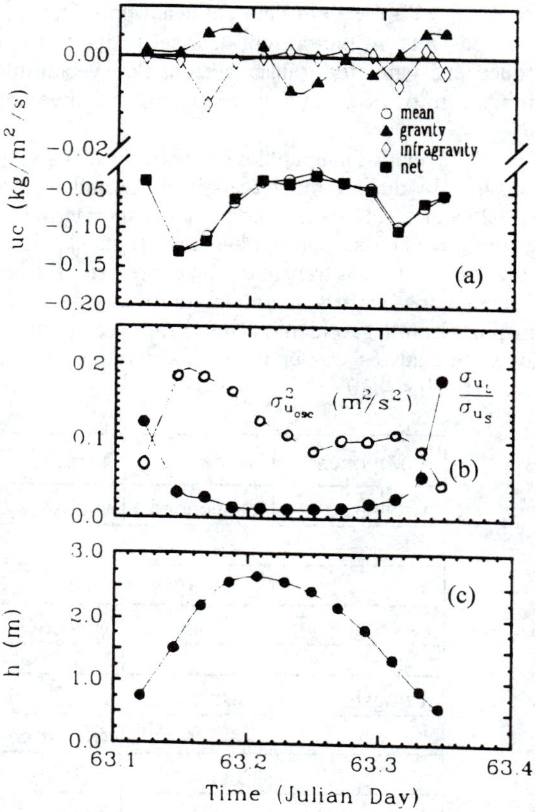

Figure 6. (a) Mean, gravity, infragravity and net local cross-shore suspended sediment flux for the last tidal cycle (Julian day 63) measured at 9 cm above the sea bed. (b) Total variance of oscillatory currents (○) and ratio of infragravity over gravity kinetic energy (•). (c) Local mean water depth (h).

<u>Suspended Sediment Fluxes</u>

Local suspended sediment fluxes from station D and for the lower elevation (9 cm) of SSC measurement are shown in Fig 6 for the cross-shore direction. Mean cross-shore, suspended sediment fluxes dominate; they are directed consistently offshore with maxima occurring at the beginning and the end of the tidal cycle, in shallow water depths where the local resuspension processes are more significant. The absolute values of the mean fluxes are an order of magnitude greater than the oscillatory ones due to gravity and/or infragravity motions. Gravity fluxes are inconsistent; they are directed offshore during the first data cycle, whilst during the second tidal cycle they are contributing to an onshore flux. Finally, during the last tidal cycle they contribute to an onshore

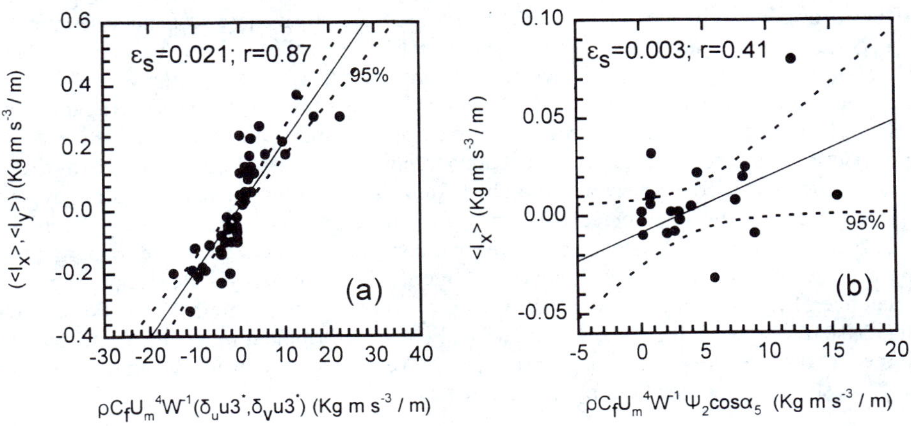

Figure 7. Comparison of measured, depth-integrated immersed weight suspendended sediment transport rats with the terms of eqs (2) and (3). (a) Mean mode of suspended sediment transport (cross-shore, $<I_x>$ and longshore $<I_y>$ data superimposed). (b) Oscillatory mode of transport (cross-shore only).

sediment transport during the flood; during high water the flux is directed offshore and turns again to onshore at the last stage of the ebb tide. The contribution of the infragravity motions is comparable to that of the gravity waves, with the main characteristic being that when the infragravity flux is of any significance it is always directed offshore, in other cases its direction is variable.

Longshore local fluxes (not shown here) are dominated also by the mean fluxes. Both mean and net fluxes are northeasterly (*i.e.* towards the Netherlands). Gravity fluxes although an order of magnitude smaller than the mean, contribute to a northeasterly sediment transport in the direction of wave approach. However, infragravity motions, which result in a southwesterly transport during the first tide, exhibit a variability in direction during the second tidal cycle whilst during the last tidal cycle contribute to a northeasterly transport during the flood and a southwesterly flux during the ebb.

Modelling Suspended Sediment Transport

Depth-integrated values of immersed weight suspended sediment transport rates were obtained by box integrating the measured local suspended sediment fluxes (Voulgaris and Collins, 1996). The lower and upper limits for the integration were the sea bed and 0.5m above the sea bed, respectively. The general patterns observed in the local fluxes analysis were present in the depth-averaged sediment transport rates.

The terms describing sediment transport due to mean currents for both the cross-shore (2nd term in eq. (8)) and longshore directions (eq. (9)) have been combined and plotted against the predicted mean depth-integrated flux in Figure 6a. A very good agreement was found with a correlation coefficient of 0.87. The efficiency factor (ϵ_s) was determined by regression analysis and was found to be 0.021. The oscillatory component (which is dominated by the gravity mode) of the measured cross-shore suspended sediment transport is plotted against the 2nd term of eq (8) in Figure 7b. There, it can be seen that the correlation is poor; the model always predicts onshore transport due to the waves while the measurements revealed periods of offshore sediment transport. This discrepancy is believed to be due to the presence of vortex ripples which can reverse the sediment transport direction near the bed (Vincent and Green, 1990). Visual observations, during low water, identified ripples with height 0.7cm and wavelength 8.5cm. These were asymmetrical in shape, directed offshore, with curved crests characteristic of ripples occurring under the co-existence of waves and currents. Application of the Wiberg and Harris (1994) model for the period when the intertidal zone was inundated suggested the existence of ripples with height of the order 3 to 5 cm and wavelength 10 to 20cm. The inconsistency between the modelled and visually observed ripple dimensions is attributed to the fact that the latter observations were undertaken during low water when the ripples were washed and flattened out by the swash zone.

Conclusions

In the mid-tidal zone of a macrotidal ridge and runnel beach, under mild conditions, both steady (tidally-induced) and oscillatory currents are equally important for the resuspension of sediment. Their importance is expressed not only in terms of their absolute magnitude but also in terms of wave-current interaction.

Suspended sediment transport is dominated by the mean currents

Bailards (1981) energetics model appears to predict suspended sediment transport due to mean currents quite accurately. However, it fails to predict suspended sediment transport due to wave-induced orbital velocities; this is probably due to the presence of vortex ripples.

The efficiency factor for suspended sediment transport and for the mean mode of transport was found to be 0.021.

Acknowledgments

The authors would like to acknowledge M.P. Wilkin and D.Simmonds for their invaluable assistance. The study was funded by the Commission of the European Community (DG XII) under Contract No. MAS2-CT92-0024-C (C-STAB). This contribution was prepared while G.Voulgaris was at Woods Hole Oceanographic Institution with a Post-Doctoral Award funded by the Mellon Foundation.

References

Aagaard, T. and Greenwood, B.M., 1995. Longshore and Cross-shore Suspended Sediment Transport at Far Infragravity Frequencies in a Barred Environment. Continental Shelf Research, 15(10): 1235-1250.

Bagnold, R.A., 1966. An Approach to the Sediment Transport Problem from General Physics. U.S. Geol. Sur. Prof. Pap. 422-I.

Bailard, A., 1982. Modelling On-Offshore Sediment Transport in the Surfzone. Proc. ASCE 18th Conf. Coastal Eng.: 1419-1438.

Bailard, J.A., 1981. An Energetics Total Load Sediment Transport Model for a Plane Sloping Beach. Journal of Geophysical Research, 86:10,938-10,954.

Bowen, A.J., 1980. Simple Models of Nearshore Sedimentation: Beach Profiles and Alongshore Bars. In: The Coastline of Canada. McCann, S.B. (Editor). Geological Society of Canada, Paper 80-10, Ottawa, pp. 1-11.

De Moor, G., 1991. The Beach Nourishment of Bredene-De Haan and Its Impact on the Beach Morphology and the Coastal Evolution of the Belgian Coast East of Oostende. Proceedings IGU Symposium "Coastal Protection". Nantes, October, 1991.

Guza, R.T. and Thornton E.B., 1985. Velocity Moments in the Nearshore. J. Waterway, Port, Coastal & Ocean Eng., 111(2): 235-256.

King, C.A.M. and Williams, W.E., 1949. The Formation and Movement of Sand Bars by Wave Action. Geographical Journal, 113: 70-85.

Orford, J.D. and Wright, P., 1978. What's In a Name? - Descriptive or Genetic Implications of "Ridge and Runnel" Topography. Marine Geology, 28: M1-M8.

Roelvink, J.A. and Stive, M.J.F., 1989. Bar-Generating Cross-Shore Flow Mechanisms on a Beach. Journal of Geophysical Research, 94(C4): 4785-4800.

Russell,P, Foote, Y. and Huntley, D.A., 1995. An Energetics Approach to Sediment Transport on Macrotidal Beaches. In: Coastal Dynamics '95, ASCE:

Schoonees, J.S. and Theron, A.K., 1995. Evaluation of 10 Cross-Shore Sediment Transport/Morphological Models. Coastal Engineering, 25: 1-41.

Soulsby, R.L., Hamm, L., Lopman, G., Myrhaug, D., Simons, R.R. and Thomas, G.P., 1993. Wave-Current Interaction Within and Outside the Bottom Boundary Layer. Coastal Engineering, 21: 41-69.

Thornton, E.B., Humiston, R.T. and Birkemeier, W., 1996. Bar/Trough Generation on a Natural Beach. Journal of Geophysical Research. 101(C5): 12097-12110.

Vincent, C.E. and Green, M.O., 1990. Field Measurements of the Suspended Sand Concentration Profiles, and of the Resuspension Coefficient γ over Rippled Bed. Journal Geophysical Research, 95: 15591-15601.

Voulgaris, G., 1992. Suspended Sediment Transport in the Nearshore Zone. Department of Oceanography, University of Southampton. Unpublished Ph.D. Thesis, 425pp.

Voulgaris, G. and Collins, M.B, 1996. Suspended Sediment Fluxes: *In:* B.A. O'Connor (Editor). Project MAS2-CT92-0024, Circulation and Sediment Transport Around Banks, CSTAB. Vol 2. Report No CE/05/96. Department of Civil Engineering, University of Liverpool, pp.495-545.

Voulgaris, G. and Simmonds, D., 1996. Incident Waves. *In*: B.A. O'Connor (Editor) Project MAS2-CT92-0024. Circulation and Sediment Transport Around Banks, CSTAB. Vol. 2. Report No CE/05/96. Department of Civil Engineering, University of Liverpool, pp. 451-460..

Wiberg, P.L. and Harris, C.K., 1994. Ripple Geometry in Wave-Dominated Environments. Journal of Geophysical Research, 99(C1): 775-789.

CHAPTER 306

Field Tests of Suspended-Load Transport Theories Used in Numerical Models

Thomas E. White [1]

Abstract

Surfzone optical-backscatter measurements were conducted at a long, straight sandy beach near Colorado River, Texas, in order to test velocity-based theories for predicting suspended-load sediment transport. Statistical methods were applied that provided estimates of theory performances, independent of any calibration or tuning of the model coefficients. The models of Ackers and White, Bowen, and Roelvink and Stive showed trends opposite that of the data (increasing transport estimates as measured transport decreased). The Bailard model performed very well in the cross-shore, explaining both the trends in transport and most of the variance, as determined by correlation coefficients. Bailard's model did not explain most of the variance in the longshore data, but comparison of the model's performance with other datasets show good correlation in the longshore. When the original coefficients in the unidirectional-flow-based Ackers and White model are used, very low transport estimates are produced. This suggests that this empirical model that was calibrated for unidirectional flows should not be used in coastal areas without considerable recalibration of the coefficients using high-quality datasets.

Introduction

A series of tests of sediment-transport theories has been performed with different data sets for different transport types:

- Bedload theories were tested with small-scale tracer experiments (White, 1987).

[1]Coastal Engineering Research Center, USAE Waterways Experiment Station, Attn: CEWES-CD-P, 3909 Halls Ferry Road, Vicksburg, MS 39180-6199, USA

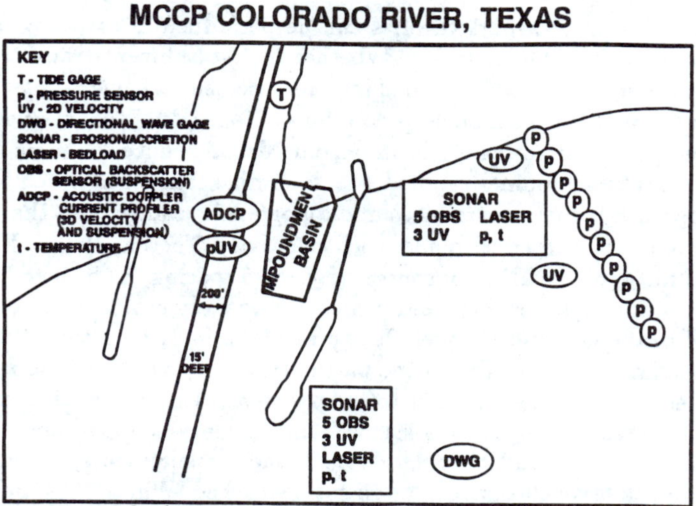

Figure 1: Areas of deployment of each gage type. Gage platforms are not to scale, but for visualization are drawn very large relative to the jetties. The lasers were replaced by micro-OBS gages for bedload.

- The Shore Protection Manual (1984) equation for total longshore transport was tested with large-scale tracer and suspension-trap experiments (White and Inman, 1989).
- Both approximate and exact versions of the SPM longshore-transport formula were tested against each other with surfzone optical-backscatter experiments (White, 1994).
- This paper tests several suspended-load transport equations with optical-backscatter experiments.

Comprehensive arrays of 43 sensors were deployed to measure waves, currents, tides, and sediment transport at a long straight beach with gently-sloping plane parallel contours near a river outlet, Colorado River, Texas (Figure 1).

This project of monitoring the behavior of the jetty system had several objectives. One was to identify and quantify the sediment budgets for the different parts of the system, in order to evaluate the performance of the sediment-management system of jetties, weir, and dredging (King and Prickett, 1997). Another was to evaluate longshore-transport formulas in the Shore Protection Manual (1984) by comparing the measured longshore sediment transport with the transport estimated from the offshore wave data and the SPM formulas (White, 1994). The topic of this paper is testing suspended-load transport theories at a point, in order to find the best ones to use in numerical models. This paper will describe only the results of this final objective.

Tests of sediment-transport theories can be performed in two very different means on two vastly different scales. Whether to test sediment-transport equations used in numerical models by making measurements at one point ("micro-scale") or by comparing the models' predictions of topographic change with measured topography ("macro-scale") is an ongoing debate. Since the main use of a transport theory is in a numerical model to determine changes in beach topography, one approach is to compare measured topographic change with the changes predicted from each numerical model (Schoonees and Theron, 1995). However, this method incorporates **all** error sources (e.g., grid spacing, coefficients' calibration, "skill" of the modeler, etc.) and cannot **separately** evaluate the transport theory used in the numerical model. It is possible in individual tests that these other factors have more effect than the particular theory that is in the model.

When detailed tests are made of total-load (macro) transport theories, such as the Shore Protection Manual's (1984) well-known longshore-transport formula, the error bars are huge and often larger than the measurements, so that high-confidence objective conclusions can often not be reached (White, 1994; Schoonees and Theron, 1994).

A third problem encountered in testing models with the "macro-scale" method is that individual vertification tests invariably test only one model against one set of data. Schoonees and Theron (1995) point out: "Without direct comparative prototype tests the final conclusion as to which are the better models in practice cannot be given."

In this paper, such a direct comparative prototype test is outlined. This approach is to test each theory against transport measured at a single horizontal location. The theory used in the numerical model, not the model itself, is tested.

Experimental Methods

Several state-of-the-art instruments were deployed: an offshore trawler-resistant Directional Wave Gage (DWG), a puv gage of colocated pressure and current sensors, Optical Backscatter Sensors for suspended-sediment concentrations, electromagnetic current meters for velocities, a new cable with internally imbedded pressure sensors for surfzone waves, and Acoustic Doppler Current Profilers for inlet currents and sediment flux.

The surfzone transport was measured by cross-shore integration of the point measurements from three synoptic multi-sensor platform arrays of Optical Backscatter Sensors (OBSs) (Downing, Sternberg, and Lister, 1981) and Electromagnetic Current Meters (ECMs), sonars, and pressure sensors. Figure 1 shows a plan view of the site with various platforms of gages drawn in the general areas where they were deployed. The sediment concentration was measured with OBS2 optical backscatter sensors, which were each calibrated in a flow tank over a wide range of transport values. The sand from the site was sampled prior to each experiment and generally had a median size of 130 microns, with a size distribution that was always single-peaked. The sand from the site was used in the laboratory to calibrate the sensors over a wide range of transport values (from 1 to 300 grams per liter), both prior to and after each of the three experiments.

The currents were measured with standard 1-inch diameter Marsh-McBirney electromagnetic probes, calibrated just prior to use. Pressure sensors were used to determine the range of vertical integration for each platform, and also times when uppermost sensors were out of the water and thus not to be used.

The sediment transport is the direct integration of the product of these two types of calibrated-sensor outputs, as measured in the surfzone:

$$\vec{\imath} = (\rho_s - \rho) g N_o \int_0^{\eta} \vec{u}(x,z) c(x,z) dz \qquad (1)$$

where $\vec{\imath}$ is the sediment transport at one horizontal location, ρ_s is the sediment density, ρ the fluid density, g the acceleration of gravity, N_o the solids density (one minus porosity), $\vec{u}$ the fluid velocity, c the sediment concentration, and the integral is taken over the vertical surf depth.

The above integration is first accomplished in the vertical at each platform location by vertically integrating the values of concentration c, longshore velocity v, and vertical range z. See White (1994) for more details. It is important to note that the vertical integration is performed *at each time step of 0.2 seconds*, and is thus unaffected by phenomena with longer time periods, such as bursts of sediment out of the boundary layer. Each sensor measures at one instant, and all the instantaneous products of c, v, and z are vertically summed before moving to the next 0.2s time step.

Theoretical Methods

Theories for computing suspended-sediment transport were obtained via an extensive literature search of several hundred modeling documents (Drake and White, 1995). Five were found to be unique formulas of point transport. Other theories are either variations on these five types, or are not testable with "point" transport measurements. Such other models include (1) diffusion-based theories such as Smith (1977), Grant and Madsen (1979), and Glenn and Grant (1987); (2) the two-layer model of Dally and Dean (1984), and (3) the eddy-viscosity model of Deigaard et al (1986).

The following variables are found in common to many suspended-load theories:
$\vec{\imath}$ is the local immersed-weight sediment transport,
ρ is fluid density, ρ_s is sediment density,
c_f is a friction coefficient,
e_s is a suspended-load efficiency, e_b is a bedload efficiency,
W is sediment fall speed,
$\vec{u}$ is the fluid velocity,
ϕ is the sand's angle of internal friction (related to grain size),

h is depth, H is wave height, T is wave period,
g is the acceleration of gravity,
β is the beach slope angle,
$\hat{i}$ is the unit vector in the offshore direction,
x is the cross-shore spatial coordinate, and y is the longshore coordinate.

The value always used for the suspension efficiency was the 0.025 best-fit value found from several experiments, as published by Bailard (1981). A consensus in the literature was reached for a friction coefficient c_f value of 0.007, since to one significant figure, this was the value determined by both Thornton (1970) and Bailard's (1981) experiments.

Three of the models we tested are based on Bagnold's (1966) derivation for both bedload and suspended load in streams. Bagnold's basic concept was that the transport uses a fixed fraction of the available stream power. The derivation of Bagnold's transport relation, along with all of the assumptions may be found in Drake and White (1995).

(1) Bowen's (1980) suspended-load model expands Bagnold's basic stream-transport relation using two velocities, an oscillatory and a steady component. Assuming the oscillatory velocity dominates, the formula is expanded in a series and the terms retained yield:

$$\left\langle \vec{i} \right\rangle = \rho c_f \frac{e_s}{W} \left[\left\langle |\vec{u}|^3 \, \vec{u} \right\rangle + \frac{\beta}{W} \left\langle |\vec{u}|^5 \right\rangle \, \hat{i} \right] \tag{2}$$

(2) Bailard (1981) worked on this problem at the same time as Bowen and independently derived a similar suspended load model. Again expanding in a series and retaining terms yields:

$$\left\langle \vec{i} \right\rangle = \rho c_f \frac{e_s}{W} \left[\left\langle |\vec{u}|^3 \, \vec{u} \right\rangle + \frac{e_s \tan \beta}{W} \left\langle |\vec{u}|^5 \right\rangle \, \hat{i} \right] \tag{3}$$

As opposed to Bowen, the reason for the efficiency e_s appearing again in the second term is based on interpretations of Bagnold's claims regarding whether power is contributed through an efficiency factor. For most applications, there is no practical import as to whether β or $\tan \beta$ is used, since the slope is usually small.

(3) Roelvink and Stive's (1989) transport model contains transport from two different mechanisms. The stream-power based transport follows the work of Bagnold, Bowen, and Bailard. But they also include transport caused by turbulence from breaking waves:

$$\left\langle \vec{i} \right\rangle = \rho c_f \frac{e_s}{W} \left[\left\langle |\vec{u}|^3 \, \vec{u} \right\rangle + K_1 \hat{i} \right] \tag{4}$$

where $K_1 = \beta_d \{ k_t [\exp(h/H_{rms}) - 1]^{-1} \}^{3/2}$

(4) Larson and Kraus' (1989) cross-shore transport model extends the energy-dissipation model of Kriebel and Dean (1985) to predict total cross-shore transport. They add an empirical criterion that determines the **sign** of the transport as either bar growth (offshore transport) or berm growth (onshore transport). Various datasets were used to determine an empirical dividing line between these cases. The resulting model, which forms the basis for CERC's SBEACH cross-shore numerical model is:

$$\langle i \rangle = sgn\left\{ M\left[\frac{\langle H_0 \rangle}{WT}\right]^3 - \frac{2\pi \langle H_0 \rangle}{gT^2}\right\}(\rho_s - \rho)gK\left[E - E_{eq} + \frac{\varepsilon_\beta}{K}\frac{dh}{dx}\right] \quad (5)$$

when the rate of energy dissipation is greater than that for an equilibrium beach: $E > E_{eq} - \frac{\varepsilon_\beta}{K}\frac{dh}{dx}$, and zero otherwise.

H_0 is the deep wave height, T the period, W the grain fall velocity, and dh/dx is the beach slope. M, K, and ε_β are empirical constants. The quantity in the first set of brackets is used solely to determine the **direction** of transport, with M the empirical constant dividing their cases of onshore and offshore transport.

The energy dissipation is follows the work of Dean. The dissipation in general and the value for the equilibrium beach are:

$$E = \frac{1}{h}\frac{d}{dx}\left[\frac{1}{8}\rho g H_s^2\sqrt{gh}\right]; \; E_{eq} = \frac{5}{24}\rho g^{3/2}\gamma_b^2 A^{3/2}$$

where γ_b is the ratio between wave height and depth, and A is the profile shape parameter.

Note that the Larson and Kraus model requires substantially different measured quantities than the other models. Not only are wave heights and periods used, but their variation across the surf zone must be known.

(5) Ackers and White's (1973) river-transport model was not originally developed for or intended for use in an oscillatory flow. However, it has been adapted for use by the Coastal Engineering Research Center for use in modelling the fate of offshore dredge dumps. The fate of these dumps is thought to initially be governed primarily by currents rather than waves. Hence this model was used, and is now being applied to other situations where there is a strong unidirectional (current) component, such as at inlets. Since this model is seeing so many uses in the coastal region, it was included in this testing program.

The unidirectional transport in the original paper is modelled as:

$$\langle i \rangle = 10^{-6}\frac{\rho_s - \rho}{\rho_s}C_g\left(\frac{F_*}{A_g} - 1\right)^m\left(\frac{\rho_s D\bar{u}}{\rho h u_*}\right)(\rho \bar{u} h) \quad (6)$$

Some of the final parameters, such as depth h and density ρ cancel. What remains are sediment parameters (density ρ_s and size D) and fluid parameters (ρ, a velocity in the direction of transport $\bar{u}$, and a friction velocity $u_* = \sqrt{c_f}u_T$). The velocity u_T is the **total** velocity, and the velocity $\bar{u}$ is the velocity in the direction of transport. So in the cross-shore case, the appropriate instantaneous velocities are the one that appears twice in the numerator, $\bar{u} = u$ (u being the instantaneous cross-shore velocity) and $u_* = \sqrt{c_f}(u^2 + v^2)^{0.5}$ in the denominator.

The remaining parameters are four 'numbers' used in the model. C_g is a function of grain size determined by fitting to Ackers and White's original data set:

$$\log C_g = 2.86 \log D_* - (\log D_*)^2 - 3.53; \; D_* = D\left[\frac{g(\rho_s - \rho)}{\rho \nu^2}\right]^{1/3}$$

where D_* is a dimensionless grain diameter. The exponent m is another empirically fit number $m = 9.66D_*^{-1} + 1.34$, as is $A_g = 0.23D_*^{-1/2} + 0.14$. For fine to medium sediments (suspension) the mobility number F_* is the square root of Shields' number:

$$F_* = \sqrt{\theta} = \frac{\sqrt{\rho} u_*}{[gD(\rho_s - \rho)]^{1/2}}; \; G_* = C_g\left[\frac{F_*}{A_g} - 1\right]^m$$

G_* is the so-called transport parameter into which are lumped all the empirical numbers.

Statistical methods

Methods were developed to objectively judge theory performance (White, 1987). The degree of success of the developed theories can be judged by computing correlations between predicted and measured variables by objective statistical techniques. A simple linear regression is performed between measured transport (y) and transport predicted from the models (x):

$$y = mx + b; \; r = m\frac{\sigma_x}{\sigma_y} \tag{7}$$

where r is the correlation coefficient and σ is the standard deviation using N weighting. Negative values represent anti-correlation. The r^2 values are the fraction of the total variance in the data explained by the method.

For a selected level of confidence, say 90%, confidence limits can be computed. Since the correlation r is not a Gaussian function, it must first be translated into a variable that is Gaussian in order to compute confidence limits. The upper confidence limit is: $w^+ = w + Z/\sqrt{N(N-3)}$ where Z comes from the cumulative normal (Gaussian) distribution tables. The lower limit is computed likewise, except there is a + instead of − before the fraction. The quantity w is from a theorem that determined w is a Gaussian function of the correlation r: $w = \tanh^{-1} r$. The confidence limits allow us to state how confident we are that one theory performs better than another theory. Once translated back into r-space via the inverse of the above equation, we obtain a **range** of correlations for each theory. If the ranges in correlations r for one theory's performance do not overlap the range computed for another theory's performance, then we may conclude at the specified level of confidence that the theory with the higher r values performs better than the one with lower values. The reason for this exercise is to show that random variation in small amounts of experimental data have not, by chance, provided an incorrect conclusion. The percent confidence specifies the probability that such incorrect conclusions have not occurred.

The statistical methods separate the issues of calibration and testing. Correlation coefficients determine the degree of success of each theory **without calibrating** to the data; confidence intervals determine which theories are better than which other theories; regression slopes determine appropriate coefficients for each of the theories, independently of the quality or goodness-of-fit issue.

Results

Experiments were performed over a two-week period in January 1992 over greatly varying wave conditions. Each experiment consisted of measurements taken each fifth of a second. The vertically integrated transports were then averaged over the recording time for the experiment, typically one to three hours. This dataset consisted of 81 such cross-shore transport and 36 longshore transport experiments.

Unfortunately, we encountered practical problems when using two of the five theories. The **Larson and Kraus** theory requires somewhat different data than the others: considerable information about wave height and period, and also cross-shore variation in these quantities. The original project had the purpose of measuring transport rates and of testing velocity-based theories. Thus such detailed wave information was not computed in the original dataset. Future work is planned to compute these quantities in order to test this theory.

The **Ackers and White** formula was used but produced extremely small values of transport, by several orders of magnitude. Numerous independent checks of the testing program failed to reveal any programming errors. Another possibile explanation is that we are incorrectly surmising how to compute the different velocities required by this unidirectional-flow theory. Of course, it is also quite possible that the results are correct and suggest that such a unidirectional-flow theory was not intended for and should not be used in oscillatory flow. Since we encountered such low values in these cross-shore computations, we did not continue testing this theory with longshore data. Thus their theory does not appear in Table 3.

Correlations between theories and measurements were computed over each time period and are listed in Tables 1 and 2 for cross-shore transport and Table 3 for longshore transport. Confidence limits were also computed in order to determine how confident one can be of conclusions that one theory performs better than another.

Modifications to the theoretical transport equations may be extracted from Tables 1, 2, and 3. The slope m could be used to modify the theory's empirical coefficient or the friction coefficient c_f, *provided* a y-intercept were also added to the equation. So one might compute i from the original formula, and then $i_{new} = mi_{old} + b$.

Table 1: Correlation of Theoretical and Measured CROSS-SHORE Transports for the OUTER Surf Zone

Method being correlated with measured transport:	Measured Transport	Bowen (1980)	Bailard (1981)	Roelvink and Stive (1989)	Ackers and White (1973)
	Outer surf, top water column (number of experiments, N = 15)				
μ Mean, dynes/(cm-s)	-3833.	6178.4	-291.9	2062.5	.00003096
σ Standard Deviation, dynes/(cm-s)	8388.	16373.6	776.2	3395.1	.00008103
m, Regression Slope		-.4612	10.2	-2.2	-93015000.
b, Intercept, dynes/(cm-s)		-983.7	-854.1	778.9	-953.7
r, Correlation Coefficient		-0.900	0.945	-0.905	-0.899
r^2		0.810	0.893	0.819	0.808
Lower 90% Confidence Limit for r		-0.921	0.930	-0.925	-0.920
Upper 90% Confidence Limit for r		-0.874	0.957	-0.880	-0.873
	Outer surf, mid water column (number of experiments, N = 15)				
μ Mean, dynes/(cm-s)	-3833.	181.5	-83.5	303.9	.000000982
σ Standard Deviation, dynes/(cm-s)	8388.	221.1	162.7	301.2	.000001599
m, Regression Slope		-19.9	50.7	-12.7	4192780000.
b, Intercept, dynes/(cm-s)		-219.3	397.8	13.8	1008.9
r, Correlation Coefficient		-0.525	0.983	-0.455	-0.799
r^2		0.276	0.966	0.207	0.638
Lower 90% Confidence Limit for r		-0.608	0.978	-0.547	-0.352
Upper 90% Confidence Limit for r		-0.430	0.987	-0.352	-0.750
	Outer surf, bottom water column (number of experiments, N = 15)				
μ Mean, dynes/(cm-s)	-3833.	40.4	-21.8	151.0	.000000219
σ Standard Deviation, dynes/(cm-s)	8388.	55.2	40.5	157.9	.000000289
m, Regression Slope		-129.7	203.4	-45.3	27350800000.
b, Intercept, dynes/(cm-s)		1411.8	595.5	3005.1	1168.6
r, Correlation Coefficient		-0.854	0.982	-0.852	-0.942
r^2		0.729	0.964	0.726	0.887
Lower 90% Confidence Limit for r		-0.884	0.977	-0.882	-0.954
Upper 90% Confidence Limit for r		-0.817	0.986	-0.814	-0.926

The computed statistics are reported with far more significant digits than the method justifies, in order to retain digits until the rounding of the final result.

Table 2: Correlation of Theoretical and Measured CROSS-SHORE Transports for the MIDDLE and INNER Surf Zone

Method being correlated with measured transport:	Measured Transport	Bowen (1980)	Bailard (1981)	Roelvink and Stive (1989)	Ackers and White (1973)
	Mid surf, top water column (number of experiments, N = 8)				
μ Mean, dynes/(cm-s)	-24000.	3229.4	-2456.2	1173.7	
σ Standard Deviation, dynes/(cm-s)	12177.	4172.0	3252.4	1303.7	
m, Regression Slope		-1.674	2.038	0.525	
b, Intercept, dynes/(cm-s)		-18595.2	-18994.6	-24616.5	
r, Correlation Coefficient		-0.573	0.544	0.056	
r^2		0.328	0.296	0.003	
Lower 90% Confidence Limit for r		-0.723	0.335	-0.202	
Upper 90% Confidence Limit for r		-0.372	0.702	0.307	
	Mid surf, bottom water column (number of experiments, N = 8)				
μ Mean, dynes/(cm-s)	-24000.	3220.2	-2092.6	1498.7	
σ Standard Deviation, dynes/(cm-s)	12177.	3521.5	2986.0	1072.3	
m, Regression Slope		-2.208	2.096	0.932	
b, Intercept, dynes/(cm-s)		-16890.8	-19614.5	-25397.4	
r, Correlation Coefficient		-0.638	0.514	0.284	
r^2		0.407	0.264	0.081	
Lower 90% Confidence Limit for r		-0.768	0.298	0.680	
Upper 90% Confidence Limit for r		-0.457	0.680	0.503	
	Inner surf, mid water column (number of experiments, N = 20)				
μ Mean, dynes/(cm-s)	1275.	1024.4	44.9	1188.4	
σ Standard Deviation, dynes/(cm-s)	8076.	1212.8	532.1	1235.5	
m, Regression Slope		2.775	13.800	3.575	
b, Intercept, dynes/(cm-s)		-1567.7	655.9	-2973.4	
r, Correlation Coefficient		0.417	0.909	0.547	
r^2		0.174	0.826	0.299	
Lower 90% Confidence Limit for r		0.341	0.892	0.482	
Upper 90% Confidence Limit for r		0.488	0.923	0.606	

The computed statistics are reported with far more significant digits than the method justifies, in order to retain digits until the rounding of the final result.

Table 3: Correlation of Theoretical and Measured LONG-SHORE Transports

Method being correlated with measured transport:	Measured Transport	Bowen and Bailard (same model in the longshore)
	Mid surf, top water column (number of experiments, N = 8)	
μ Mean, dynes/(cm-s)	-48350.	-3810.0
σ Standard Deviation, dynes/(cm-s)	18563.3	3714.2
m, Regression Slope		2.205
b, Intercept, dynes/(cm-s)		-39950.3
r, Correlation Coefficient		0.441
r^2		0.194
Lower 90% Confidence Limit for r		0.209
Upper 90% Confidence Limit for r		0.626
	Mid surf, bottom water column (number of experiments, N = 8)	
μ Mean, dynes/(cm-s)	-48350.	-3796.4
σ Standard Deviation, dynes/(cm-s)	18563.	3370.2
m, Regression Slope		2.549
b, Intercept, dynes/(cm-s)		-38674.7
r, Correlation Coefficient		0.463
r^2		0.214
Lower 90% Confidence Limit for r		0.235
Upper 90% Confidence Limit for r		0.642
	Inner surf, mid water column (number of experiments, N = 20)	
μ Mean, dynes/(cm-s)	11,505.	736.9
σ Standard Deviation, dynes/(cm-s)	10,792.	761.7
m, Regression Slope		9.417
b, Intercept, dynes/(cm-s)		4564.8
r, Correlation Coefficient		0.665
r^2		0.442
Lower 90% Confidence Limit for r		0.613
Upper 90% Confidence Limit for r		0.712

The computed statistics are reported with far more significant digits than the method justifies, in order to retain digits until the rounding of the final result.

Table 4: Correlations between Model Predictions and Measured Transport

	Bowen (1980)	Bailard (1981)	Roelvink & Stive (1989)	Ackers & White (1973)
January 1992: 81 Experiments for Cross-Shore Transport				
r for outer surf, upper water column	-0.900	0.945	-0.905	-0.899
r for outer surf, mid water column	-0.525	0.983	-0.455	-0.799
r for outer surf, bottom water column	-0.854	0.982	-0.852	-0.942
r for mid surf, upper water column	-0.573	0.544	0.056	
r for mid surf, lower water column	-0.638	0.514	0.284	
r for inner surf	0.417	0.909	0.547	
January 1992: 36 Experiments for Long-Shore Transport				
r for mid surf, upper water column	0.441	0.441		
r for mid surf, lower water column	0.463	0.463		
r for inner surf	0.665	0.665		

All the correlations between models and data are summarized in Table 4. Note that many correlations are negative. This means that the **trend** in the theoretical predictions are opposite the trend in the data. (As measured transport increases, predicted transport decreases.) It does **not** mean that the direction of the predicted transport is opposite that of measured transport.

Conclusions

Conclusions about model performances can be made from the correlations in Table 4. For the 6 groups of 81 surfzone experiments of 1 to 3 hours length each:

1. The Bailard model performs very well in the cross-shore, predicting most of the variation in cross-shore transport.

   Mean r = 0.81 and mean $r^2 = 0.70$ (Explains **70% of the variance**)

2. The Bowen model and the Roelvink & Stive model do not predict most of the variation in cross-shore transport, and frequently predict the wrong trend (e.g., predicted transport increases as measured transport decreases). The only significant difference between the Bowen and Bailard models is a appearance of the suspended-load efficiency coefficient in the second term, caused by different interpretations of Bagnold's original work on this point by Bowen and Bailard. The results of this study are strong evidence that Bailard's interpretation is correct.

   Bowen: mean r = - 0.52 and mean $r^2 = 0.45$ (**45% of the variance**)

Roelvink & Stive: mean r = - 0.22 and r^2 = 0.36 (**36% of the variance**)

3. The Ackers and White model performs strangely. Using the coefficients provided in their original publication and making our stated assumptions about how to compute the velocities, results in extremely low values of transport. Results show strong **anti-correlation** with the data. The most likely explanation of the anti-correlation is some local circulation pattern at the outer surf platform, since some of the other models also exhibit this anti-correlation result at that site.

   Mean r = 0.88 and mean r^2 = 0.77 (Explains **77% of the variance**)

4. The Bowen model and the Bailard model (identical in the longshore) model do not predict most of the variation in longshore transport.

   Mean r = 0.52 and mean r^2 = 0.28 (Explains **28% of the variance**)

   Knowledge of the performance of this model in the longshore is by no means conclusive. Thornton reports in meetings for the Delilah/Duck94 surfzone experiments, that performance levels are similar to what we found here. But we have also tested this model in the longshore against numerous other Shore Protection Manual-type formulas and also other velocity-based formulas in another study using Nearshore Sediment Transport Study sand-tracer data at Santa Barbara, California (White and Grandon, draft). In that study this model performed excellently, with correlations around 0.9.

In another study, the main factor not included in these equations that was found to improve *local* estimates of transport, was inclusion of a sediment threshold-of-motion criterion (White, 1987; White, 1989). In comparing predictions of different local **bedload** transport theories with transport measured by sand tracer, it was found that agreement on direction of transport improved from 70% to 100% of the experiments, once a threshold criterion was added to the theory. We have included threshold subroutines in our testing program and plan to apply such criteria to these suspension models and report the results in a followup paper.

Acknowledgments

The considerable efforts in obtaining the large field data set at Colorado River were performed by the capable staff of the Prototype Measurement and Analysis Branch of CERC, including Mike Carpenter, Bill Grogg, Doug Lee, Rhonda Lofton, Bill Kucharski, Chuck Mayers, Debbie Shafer, and Kerry Taylor. The data-collection systems were designed and built by Ralph Ankeny, Bill Grogg, Gary Howell, and J. Rosati. The analysis of the surfzone time-series data and computation of measured transport were completed by Joon Rhee. Conversion of the author's testing programs to use with OBS data was performed by Rhonda Lofton. Without her considerable programming and data-organization efforts, this paper would not have been possible.

The work presented herein was performed at the Waterways Experiment Station of the US Army Corps of Engineers. Permission was granted by the Chief of Engineers to publish the results.

References

Ackers, P. and W.R. White, 1973, "Sediment transport: new approach and analysis," *Proc. Hydraulics Div.*, ASCE, v 99, n HY11, p 2041-2060.

Bailard, J.A., 1981, "An energetics total load sediment transport model for a plane sloping beach," *J. Geophys. Res.*, v 86, n C11, p 10938-10954.

Bowen, A.J., 1980, "Simple models of nearshore sedimentation: beach profiles and longshore bars," *The Coastline of Canada*, Geological Survey of Canada Paper 80-10, ed., S.B. McCann, p 1-11.

Dally, W.R., and R.G. Dean, 1984, "Suspended sediment transport and beach profile evolution," *J. Waterway, Port, Coastal, and Ocean Engr.*, v 110, n 1, pp. 15-33.

Deigaard, R., J. Fredsoe, and I.B. Hedegaard, 1986, "Suspended sediment in the surf zone," *J. Waterway, Port, Coastal, and Ocean Engr.*, v 112, n 1, pp. 115-128.

Downing, J.P., R.W. Sternberg, and C.R.B. Lister, 1981, "New instrumentation for the investigation of sediment suspension processes in the shallow marine environment," *Marine Geology*, v 42, p 19-34.

Drake, T.G. and T.E. White, 1995, "Compilation of suspended-load point-transport theories," USAE Waterways Experiment Station, Technical Report CERC-95-18, 40pp.

Glenn, S.M. and W.D. Grant, 1987, "A suspended sediment stratification correction for combined wave and current flows," *J. Geophys Res.*, v 92, n C8, pp. 8244-8264.

Grant, W.D. and Madsen, O.S., 1979, "Combined wave and current interaction with a rough bottom," *J. Geophys. Res.*, v 84, pp. 1979-1808.

King, D.B., Jr., and T.L. Prickett, 1997, "Mouth of the Colorado River, Texas, monitoring program," USAE Waterways Experiment Station Technical Report.

Kriebel, D.L., and R.G. Dean, 1985, "Numerical simulation of time-dependent beach and dune erosion," *Coastal Engr.*, v 9, n 3, pp. 221-245.

Larson, M. and N.C. Kraus, 1989, "SBEACH: numerical model for simulating storm-induced beach change," Technical Report CERC-89-9, USAE Waterways Experiment Station, Coastal Engineering Research Center.

Roelvink, J.A. and M.J.F. Stive, 1989, "Bar-generating cross-shore flow mechanisms on a beach," *J. Geophys. Res.*, v 94, n C4, p 4785-4800.

Schoonees, J.S. and A.K. Theron, 1994, "Accuracy and applicability of the SPM longshore transport formula," *Proc. 24th ICCE, Kobe*, ASCE, p 2595-2609.

Schoonees, J.S. and A.K. Theron, 1995, "Evaluation of 10 cross-shore sediment transport-morphological models," *Coastal Engr.*, v 25, p 1-41.

Shore Protection Manual, 1984, Vol. I, U.S. Army Corps of Engineers' Coastal Engineering Research Center, U.S. Gov't Printing Office.

Smith, J.D., 1977, "Modeling of sediment transport on continental shelves," in *The Sea*, E.D. Goldberg (Editor), John Wiley, New York, v. 6, pp. 539-577.

Thornton, E.B., 1970, "Variation of longshore current across the surf zone," *Proc. 12th ICCE*, ASCE, p. 291-308.

White, T.E., 1987, "Nearshore sand transport," *University of California*, Ph.D. dissertation, 210 pp.

White, T.E., 1989, "Using sediment-threshold theories in waves and currents," *Proc., Sediment Transport Modeling*, Intl. Symposium, HY Div., ASCE, p 248-253.

White, T.E., 1994, "Field tests of radiation-stress estimators of longshore sediment transport," *Proc. 24th ICCE, Kobe*, ASCE, p 2799-2812.

White, T.E. and J. Grandon, draft, "Longshore sediment transport: a comparison of energetics-based total-load models," Technical Report, USAE Waterways Experiment Station, Coastal Engineering Research Center.

White, T.E. and D.L. Inman, 1989, "Measuring longshore transport with tracers," Chapter 13 of *Nearshore Sediment Transport*, R.J. Seymour, editor, Plenum Pub., New York, p 287-312.

CHAPTER 307

Shore Protection Studies for Ras-Elbar, Egypt

John B. Herbich[1], Fellow ASCE, Abd-Elfattah Elfiky[2], Abd-Elmohsen Elmongy[3]
Ali Elbahrawy[4], Gamal Elsaeed[5], Member ASCE

Introduction and Background of the Coastal Changes Along the Northern Coast of the Nile Delta

The northern coastline of the Egyptian Nile Delta extends some 280 km from a point 30 km west of the City of Alexandria to 30 km east of the City of Portsaid. The coastline is considered to be in a state of continuous change under the action of sea waves and currents.

Over the years the River Nile and its two main branches, Damietta and Rosetta, supplied large volumes of sediment from the mouth of these branches to the sea. This large volume of sediment exceeded the loss of sediment caused by wave and current action; thereby providing natural beach protection and excess sediment to the Nile Delta.

Construction of the Aswan Nile Dam significantly reduced the volume of sediment deposited along the Nile Delta shoreline causing severe erosion. The erosion started after construction of the first Aswan dam, the development of other dams, and the increasing diversion of the river water for irrigation purposes. After closing the Aswan High Dam in 1966, erosion along the Delta coastline accelerated considerably, resulting in loss of several beaches and blockage of estuaries and navigation channels because of accreting sediment and flooding of coastal villages. The erosion had a severe effect on the nearby agricultural areas because of salt water intrusion affecting these areas.

---

[1]Professor Emeritus, Ocean Engineering Program, Department of Civil Engineering, Texas A&M University, College Station, TX 77843, USA.
[2]Professor, Department of Civil Engineering, Ain Shams University, Cairo, Egypt.
[3]Professor, Department of Civil Engineering, Ain Shams University, Cairo, Egypt.
[4]Assistant Professor, Department of Civil Engineering, Ain Shams University, Cairo, Egypt.
[5]Researcher, Department of Civil Engineering, Ain Shams University, Cairo, Egypt.

A successful shore protection management plan must be developed for the next twenty years.

Available Data and Description of the Ras-Elbar Area

Ras-Elbar is located at the Damietta branch mouth of the Nile. This popular resort beach has declined severely over the past twenty years due to extensive erosion. Many hectares of beachfront property have been lost since 1965 and erosion is continuing.

The Ras-Elbar area is only five kilometers away from the new Port at Damietta. The development of the port facilities, related industries and new town development will have a significant economic impact upon the coastal erosion at the Ras-Elbar area, and will affect tourism and industrial development in the area.

The Damietta branch of the River Nile was developed during the tenth century. The Nile sediments, directed toward the west and east sides of this branch, developed this area by wave action. The old maps show that the Ras-Elbar-Damietta promontory advanced about three kilometers in the period between 1800 and 1900. At present, the coast is retreating due to sediment supply cut off since 1966 and the permanent closures of the Faraskour Dam some 20 km south of the mouth. Since 1966 erosion has been quite active in this area. The average retreat of the Ras-Elbar Peninsula had been about 30 m/yr before the construction of the western jetty in 1941. This jetty has stabilized the northern part of the Ras-Elbar shoreline, although the rest of the coast continued to erode with a diminishing rate for a distance of 3 km west of the jetty.

In 1970 a series of three short groins were constructed to protect the eastern part of the resort. In 1982 two Dolosse embankments between groins 1 and 2, and 2 and 3 were constructed. The Ras-Elbar nearshore region showed continuous deepening of the water depth and steepening of its underwater shore slope.

Mathematical Studies and Procedure

Computer hardware. The computer used in this study was an IBM 486 DX2 with a 66 MHz clock speed, a 16 MB RAM memory, a 240 MB hard drive memory and a math co-processor. A Hewlett Packard Laser Jet IV P printer and 1 7475A plotter were used. A Numonics 1.0 x 1.20 m digitizer was used to digitize the Ras-Elbar area drawings. The above equipment was proven to be reliable within the range of this study. The average CPU time was 2 hours.

Phases of this study. This study was conducted for the purpose of establishing and examining protection plans for the Ras-Elbar area coastline and to determine their long-term impact on the shoreline and the volume of sediment transported.

The study was conducted in four main phases. The first phase established the behavior of the initial shoreline subject to the design wave and served as a base reference.

The second phase studied the effect of a system of detached breakwaters on the behavior of the shoreline under the same wave climate conditions. The system proposed by Delft Hydraulics in 1987 was examined for comparative purposes.

The third phase studied the effect of a system of groins on the behavior of the shoreline having the same wave climate conditions.

Finally, the fourth phase evaluated the effect of a combined system of detached breakwaters and groins on the behavior of the shoreline under the same wave climate conditions. All computer runs were conducted for three wave directions.

<u>Simulation Procedure and System Setup</u>. The procedure was as follows:

1. The shoreline data were digitized using the Autocad release 11.0 with a digitized accuracy of 4.16 m. The digitized data were analyzed and extracted using the Quick surf-Cad based program, and the shoreline digitized data were prepared as DXF file. Ras-Elbar shoreline was digitized using Drawing No. 2 prepared by Tetra Tech Inc.
2. The wave data were prepared using the Coastal Research Institute (CRI) and the Suez Canal Authority (SCA) data recorded over various periods of time.
3. A computational program (GENESIS) (Gravens and Kraus 1992) was used to conduct the required simulation. This program was developed by Hanson and Kraus (1982).

Two major tests were conducted:

1. Conducting sensitivity and error analysis runs:
   The changes in the output resulting from the intentional change in the input were examined to determine the limitations of different variables. To produce quality results, a knowledge of the sources of errors is important.
2. Conducting sensitivity analysis runs:
   Compromise between the efficiency and accuracy of the numerical code while maintaining its stability. To achieve the stability analysis test, several pre-runs were conducted with various conditions and various shore-protection structure alternatives. As a result of these

runs, two parameters were fixed to maintain the stability (without loosing either the efficiency or the accuracy of the numerical scheme). The time step was chosen to be 6 hours; 18,992 time steps were calculated. The number of grid cells within the simulation area reach was 60 with a grid spacing of 50 meters.

The start and setup files were selected for each run of the proposed shore plan. If the proposed plan was not successful due to the formation of tombolos, diffracting structures interaction or ineffective structure position, the file had to be redesigned until a successful run was reached. A simulation time of thirteen years was selected for each run to establish a case of relatively-stable state, undergoing minor or moderate changes. The simulation starting date was November 30, 1986, and the ending date was November 30, 1999. The results were calculated and plotted once a year. Figure 1 shows a general layout of the study area.

The following numerical model arrangement was used for all cases:

a. the total length of simulated coast was 3 km. This length was divided into sixty equal cells, each 50 m in width.
b. simulation starting date, November 30, 1986 and ending November 30, 1999,
c. time step used was 6 hours with a total of 18,992 time steps per run,
d. shoreline position was calculated and presented once a year,
e. longshore sediment transport coefficients K_1 and K_2 selected were 0.5 and 0.25 respectively (these values are typically $0.5\ K_1<K_2<1.5\ K_1$ for sandy beaches from experiments and experience),
f. the depth of the offshore wave input was 10.0 m,
g. the number of incoming calculated wave cells was four,
h. only one wave direction was assumed,
i. the average effective sand diameter was 0.25 mm,
j. the average height of land above mean water level was 1.5 m, and the limiting depth of profile movement seaward was -6.0 m,
k. average beach slope was one on twenty,
l. it was assumed that the breakwaters and groins were impermeable structures, and
m. it was assumed that there was no wave reflection from the structures.

Analysis

The series of runs conducted were very illustrative in showing the long-term effect of different structures on the shoreline position. The results of the simulation performed can be divided into four different groups:

1. no protection measures,

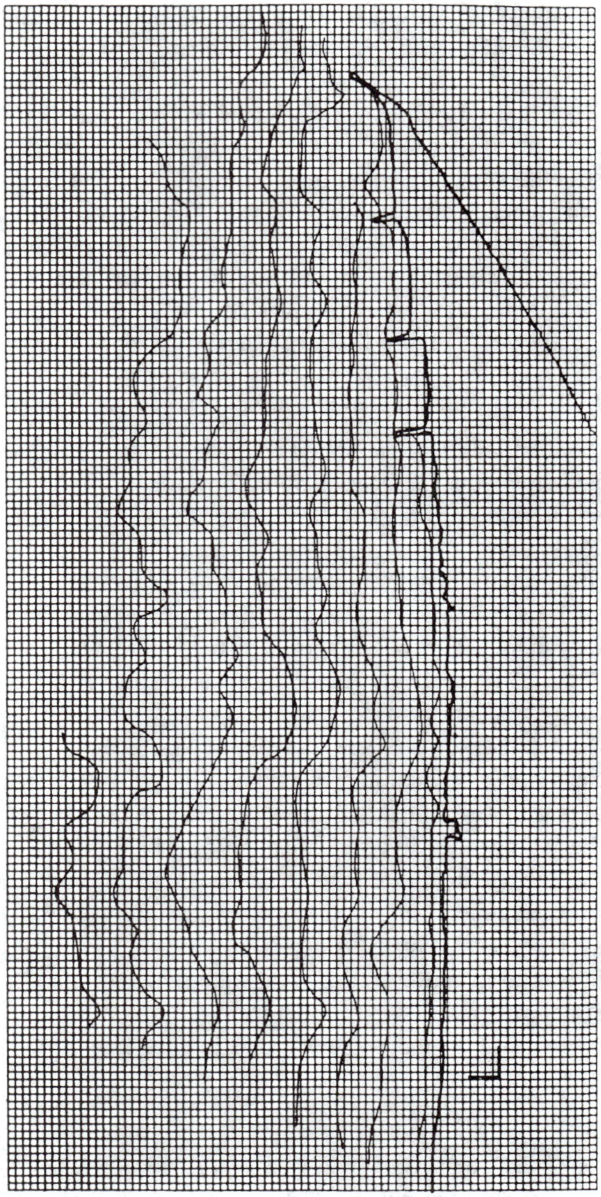

Figure 1. Ras-Elbar Study Area

2. effect of a system of detached breakwaters on the shoreline including the systems proposed by Tetra Tech, Inc. and Delft Hydraulics,
3. effect of a system of groins on the shoreline, and
4. effect of a combined system of detached breakwaters and groins on the shoreline.

For all groups, the final shoreline positions were plotted for each year until reaching a semi-equilibrium state. Relations representing the shoreline change and the average net sediment transport rate for all runs were also provided.

The breaking wave heights and angles were computed for each cell point on the grid. A figure representing the net volumetric change of sediment was prepared to indicate whether a region was gaining or losing sediment.

The different shoreline positions, longshore transport rates, and the shoreline changes from measured values were prepared for each run to illustrate the effect of different parameters and protection plans on the shoreline.

Conclusions

The results of this study were found to be reliable and applicable to a wide range of coastal engineering problems. For the Ras-Elbar area the following conclusions can be made:

1. Large fluctuations in littoral drift were noted.
2. The advantage of being able to vary wave conditions and structural protection plans was clearly apparent throughout this study.
3. A system of offshore detached breakwaters has proven to be the most efficient and reliable plan as a long-term shore protection for long reaches of the coastline.
4. Shore protection planning for a whole region was proven to be useful in providing a continuous interaction plan for the whole coastline.
5. This study has shown that increasing the number of breakwaters (from 4 to 7) for the whole region provided better results than employing a two- or four-breakwater system (Figure 2).
6. The ratio of G = 1.5 B proved to provide a good estimate of the length and the gap of a system of detached breakwaters (where G = distance between two breakwaters, B = length of the breakwater). Four locations of detached breakwaters were studied (165, 294, 445 and 545 m) from the shoreline. At a distance of 294 m from shore, the breakwaters provided minimum shoreline changes. For every case there is an optimum design for the detached breakwater system that will provide minimal or moderate shoreline changes, or cause the formation of tombolos.

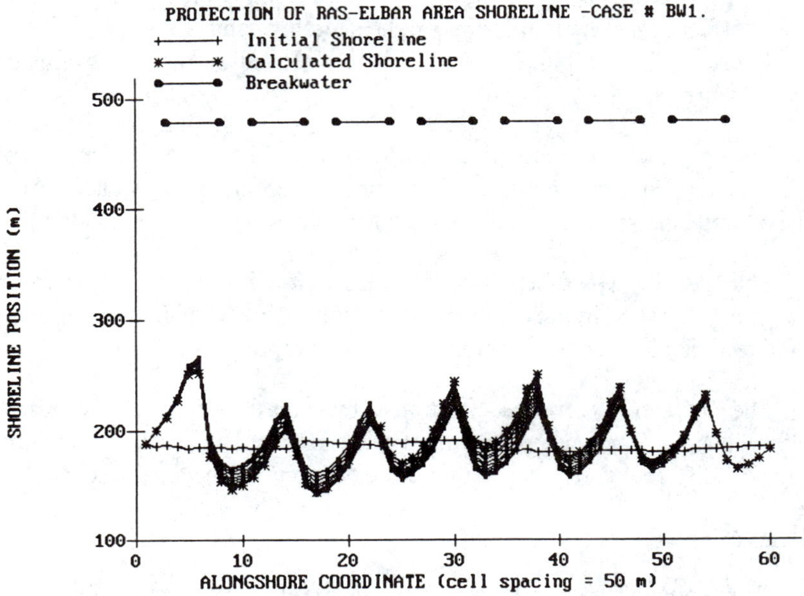

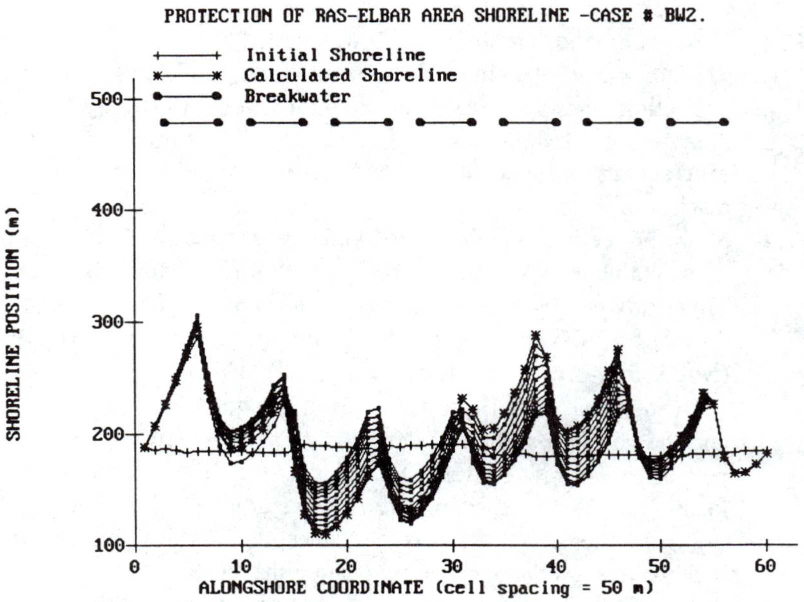

Figure 2. Protection of Ras-Elbar Area Shoreline - Case Numbers BW1 and BW2

7. A system of inclined and staggered breakwaters behaved in a similar manner to the parallel and continuous systems. The higher cost of constructing staggered or inclined offshore structures would make such a system uneconomical.
8. A combined system of offshore detached breakwaters and non-diffracting groins had a fairly similar result to that of using groins only. However, the higher cost involved in construction of two protection systems make it economically unfavorable. Using diffracting groins (which is more realistic) created more problems due to an overlap of the diffracted wave patterns from the groins and the breakwaters.
9. It is concluded that a complete system of detached breakwaters permits the continuation of shore sediment transport and will not affect the coast downstream of the system after equilibrium is reached. At the same time, this system breaks the energy of the incoming waves causing minimum shoreline change in the shadow zones. Sand nourishment should be considered in conjunction with a system of detached breakwaters to accelerate reaching equilibrium and assuring supply of sediment downstream of the system. Moreover the detached breakwater system tends to change the shoreline in a time-dependent manner. A fairly long time will take place to affect and cause changes in the shoreline.

Recommendations

Recommendations for future work are summarized as follows:

1. A uniform numerical grid of 50 m spacing was used, a finer grid spacing less than 50 m should be considered for refinement of the numerical plan.
2. A one-dimensional program employing constant grid spacing and time-step representations was used in this study and proved to be a flexible, reliable and economical method for shoreline change simulation. Higher order two- and three-dimensional schemes with variable grid spacing and time steps should be considered employing a mainframe computer. It should be noted in representing the physics of the flow and the sediment transport process that a compromise between the overall accuracy and the computation costs must be made.
3. As the shore protection structures employed in this study were assumed to be impermeable, consideration of the use of different coefficients of transmissions and permeability are strongly recommended.
4. Use different wave periods and wave heights to study their effect on the shoreline (only one wave height and one wave period were used in this study).

Appendix - References

Arafa, F. (1981) "Coastal Protection Structures Effect on Northern Coast of Egypt," Report No. 1, Cairo University, MIT.

Boer, S., De-Vriend, H. and Wind, H. (1984) "System of Mathematical Models for the Simulation of Morphological Processes in the Coastal Area," *Proceedings, Coastal Engineering Conference*, ASCE, pp. 1437-1453.

CERC (1990) "Shoreline Change and Storm Induced Beach Erosion Modeling," Report No. CERC-90-2 WES, U.S. Army Corps of Engineers, Vicksburg, MS.

CERC (1984) ***Shore Protection Manual***, Volumes I and II, WES, U.S. Army Corps of Engineers, Vicksburg, MS.

Delft Hydraulics (1987) "Ras-Elbar Shore Protection," Report No. 11500.

Elsaeed, Gamal H. (1996) "Study of Protection of the Northern Coast of Nile Delta Using Mathematical Models," Ph.D. Dissertation, Faculty of Engineering, Ain Shams University, Cairo.

Fanos, A., Khafagy, A. and Sharaf-Eldin, S. (1989) "Coastal Changes Along the Egyptian Mediterranean Coast," International Seminar, Egypt.

Fletcher, C. (1991) ***Computational Techniques for Fluid Dynamics***, Volumes 1 and 2, Springer-Verlag.

Frihy, O., Fanos, A., Khafagy,A. and Komar, P. (1991) "Patterns of Nearshore Sediment Transport Along the Nile Delta Egypt," *International Journal of Coastal Engineering*, Vol. 15, The Netherlands, pp. 409-429.

Gravens, M. and Kraus, N. (1992) "GENESIS Workbook," Report No. 2, WES U.S. Army Corps of Engineers, Vicksburg, MS.

Hanson, H. and Kraus, N. (1982) "GENESIS," Technical Reference Report No. 1, WES, U.S. Army Corps of Engineers, Vicksburg, MS.

Herbich, J.B. (Ed.) (1991-92) ***Handbook of Coastal and Ocean Engineering***, Volumes 1, 2 and 3, Gulf Publishing Company, Houston, TX.

Inman, D. and Jenkins, S. (1984) "The Nile Littoral Cell and Man's Impact on the Coastal Zone of the Southeastern Mediterranean," *Proceedings, Coastal Engineering Conference*, ASCE, pp. 1282-1290.

Inman, D., Golik, A. and Carmel, Z. (1984) "Transport of Nile Sand Along Southeastern Mediterranean Coast," *Proceedings, Coastal Engineering Conference*, ASCE, pp. 124-173.

Kadib, A., Shak, A., Mazen, A. and Nader, A. (1986) "Shore Protection Plan for the Nile Delta Coastline," *Proceedings Coastal Engineering Conference*, ASCE, pp. 2530-2544.

Kraus, N. (1989) "Beach Change Modeling and Coastal Planning Process," *Proceedings, Coastal Zone 1989 Conference*, ASCE, pp. 553-567.

Kraus, N. and Harikai, S. (1983) "Numerical Model of the Shoreline Change of Oarai Beach," *International Journal of Coastal Engineering*, Vol. 7, The Netherlands.

Perlin, M. and Dean, R. (1983) "A Numerical Model to Simulate Sediment Transport in the Vicinity of Coastal Structures," WES Report No. 83-10, U.S. Army Corps of Engineers, Vicksburg, MS.

Tetra Tech, Inc. (1984-85) "Shore Protection Masterplan for the Nile Delta Coast," Reports 1 and 2.

United Nations Development Program (UNDP) (1973) "Coastal Erosion Studies," Technical Report, Project No. Egy/70/581.

CHAPTER 308

Stabilizing Beaches Downcoast of Harbor Extensions

John R C Hsu[1] and Richard Silvester, M. ASCE[2]

Abstract

In geomorphological terms, the sandy shoreline of a bay downcoast of a harbor may be stable or in static equilibrium, or could be in dynamic condition, if sediment is still being supplied from upcoast or from downcoast to form a salient predicted by a static bay shape equation. Should commercial expansion demand a larger port, the general solution is to run a breakwater from the headland or existing structure. This has the potential to create a new static equilibrium beach, often with accretion in the lee which is at the expense of beach erosion downcoast. It is strongly recommended that geomorphic approach be incorporated to stabilize downcoast beaches early in the planning stage of a harbor, or as remedial measures. By creating bay beaches in static equilibrium, the potential beach erosion downcoast of a harbor will be kept to a minimum or may be prevented completely.

Introduction

As Inman (1974) has recalled, man has been interfering with river and coastal environments dated back in 1500 BC when the Minoans constructed wharves in Crete and by 480 BC as the Phoenicians built harbors along the coasts of Lebanon and Israel. Fleming (1992) has also discussed the modern history of harbor developments for fishing ports and major maritime trade. Oliveira et al (1982) has reported disappearances of towns in Portugal, as beaches eroded towards their stable shape.

Despite our current understanding of coastal processes there is a missing link as instanced by the many stories of beach erosion still reported in the technical literatures (for example, Ozaki 1964, Dunham 1965, Inman and Frautschy 1965, Herron and Harris 1966, Jordaan 1970, Sato and Irie 1970, Moni 1972, Lepetit 1976, Tanaka and Sato 1976, Nir 1982, Oliveira et al 1982, Komar 1983, Saxena 1983, Kraus et al 1984, Uda et al 1986, Gonzalez et al 1988, Kuo 1988, Moutzouris 1990).

[1] Senior Lecturer, Department of Environmental Engineering, University of Western Australia, Nedlands, WA 6907, Australia E-mail: hsu@cwr.uwa.edu.au

[2] Honorary Research Fellow, Department of Civil Engineering, University of Western Australia, Nedlands, WA 6907, Australia

The present status of man's abilities in coastal engineering and shore protection has been well summed up by Inman (1974): "In view of man's present extensive intervention in the coastal zone, mostly based on 'brute force' technology, a careful study of the ancients' ability to work with Nature provides valuable insight for today's problems."

In spite of our great knowledge of wave kinematics and sediment movements in marine situations, engineers have not appreciated the macroscopic view of the coast as imposed by Nature. She has transported millions of tonnes of material to sea and back again in a matter of days and also moved it alongshore. More observations are necessary to ascertain the long-term stability of waterlines produced by persistent swell and the aberrations that occur during storm sequences. It is the former which is the concern of this paper.

Many researchers have recognized the effect of harbor construction on sediment transport downcoast, but without adequate quantitative prediction. Herron and Harris (1966) believed that "harbor works are the principal offenders" that have interrupted the balance of natural littoral drift existing over hundreds or thousands of years. Even so, the question of how to predict a stable shoreline in the lee of such structures remains unanswered. It is hoped this situation is overcome with the current submission and other works by Hsu et al (1993) and Silvester and Hsu (1993). In terms of the beach erosion downcoast of harbor extensions, these authors believe it is not so much a need to look back on our achievements but a need to observe natural processes in geomorphological terms and apply them.

Beaches Downcoast of Harbors

In geomorphological terms the stability of a sandy shoreline can assume a bay shape that is *stable* or in *static equilibrium* if the littoral drift is negligible or non-existent, or it can be in *dynamic equilibrium* if drift is still being supplied with sand from upcoast or from within the bay. Whilst this supply is constant the bayed beach can remain in place for decades, but should it decrease the shoreline will recede towards the static equilibrium shape, the final limit to which it will go.

If on a straight sandy length of coast a breakwater or other structure is run out to sea, the immediate result will be interception of littoral drift and the formation of a bay downcoast of the impediment. This will become further indented until sediment bypasses the new structure and dynamic equilibrium is restored. This sudden loss of land instigates the insertion of protective measures such as seawalls or groins, that are usually ineffective since man is fighting with Nature. The walls reflect waves obliquely and so generate short-crested waves (Silvester 1972; Hsu 1990) which expedite the longshore transport. On the other hand, straight groins accelerate drift by forming rip currents during storms that carry vast volumes of material offshore, there to be transmitted downcoast by subsequent swell.

Similar to the eroding beaches on straight coasts, discontinuity or interruption of sediment supply from upcoast has been considered as the main cause of beach erosion downcoast of harbors (Herron and Harris 1966, Komar 1983, Uda et al 1986). This may be attributed to the construction and extension of long breakwaters and dredging of deep access channels.

An additional factor is the potential of salient/bay formation in the lee of the breakwaters, caused by an improper alignment of a new breakwater and the positioning of its tip. This implies that a large volume of sand is needed, often from the beach downcoast, to build up a salient in the form of a bay. In addition to the evolution of shoreline, it also causes change to the bottom topography in the vicinity of the structures (Sato and Irie 1970). The slopes of beach faces downcoast were found to become steeper with armouring units subsided in front of seawalls. (Uda et al 1986; Hsu et al 1994)

Universal Bay Shape Equation

Crenulate-shaped bays are ubiquitous, occurring as they do on oceanic margins, coasts of enclosed seas, lakes, and even river shorelines. They are indicative of Nature's way of balancing the wave energy with the sediment load to be carried. Bays have thus been maintained in the same position for thousands of years inspite of the fierce storms that hit them infrequently. But for the rocky headlands spaced along the coast, vast indentations would have been formed with the downcoast extremity almost normal to the orthogonals of the most persistent swell (Silvester 1976).

Geographers and geologists have published on bay shapes (Halligans 1906) with the first recognition of this as a stable feature by Jennings (1955), unfortunately without full knowledge of the wave action involved. Davies (1958) observed the importance of wave refraction but not diffraction. Langford-Smith and Thom (1969) noted the zeta-shaped beaches of the New South Wales coasts of Australia, without scientific analysis of their profiles. Although the crenulate shaped bays have received the attention of many scientists (Bird 1984, Carter 1988, Davis 1985, Phillips 1985, Shepard 1973, Zenkovich 1967), their stability has only been examined by engineers (Silvester 1960; Le Blond 1979; Silvester et al 1980). This geomorphic feature has existed for some thousands of years (at least since the last still stand of sealevel), but have become more indented during the past century due to the dearth of sediment for many reasons (Silvester and Hsu 1993). In this relatively short period in geologic time rivers have been harnessed, which results in less supply of sediment to the coast for spreading by wave action. This has resulted in bays approaching their static equilibrium condition, causing greatest erosion at the most indented position of the bay with practically no change at either end.

Yasso (1965) suggested the logarithmic spiral for predicting the shape of the crenulate-shaped bays, which Silvester (1974) adopted for many years. But this was found to be inaccurate for the downcoast region where the beach straightens (Hsu and Evans 1989). A polynomial form resulted from data of a multitude of bays examined as follows:

$$R/R_o = C_0 + C_1\,(\beta/\theta) + C_2\,(\beta/\theta)^2 \tag{1}$$

with the variables R, R_o, β and θ as given in the definition sketch of Figure 1. The values of constants C_0, C_1 and C_2 are available in Hsu and Evans (1989) and Silvester and Hsu (1993), in tabular and graphical form. It is noted that the wave obliquity (β) is measured between the control line (R_o) and the downcoast beach tangent. For a stable or near stable beach this is normal to the orthogonal of the persistent swell which sculptures the shape. The straight crests of these incoming waves at the point of diffraction, at the other end of the control line, are also angled β to it, from which arcs (R) are drawn at angle θ to it. Thus, by knowing R_o and β the value of R is determined for a specific θ, so that the complete periphery of the static equilibrium bay can be drawn. It has been shown that the logarithmic spiral fits only the curved portion of the actual bay, whereas the parabolic formula fits the complete periphery exceedingly well (Hsu and Evans 1989).

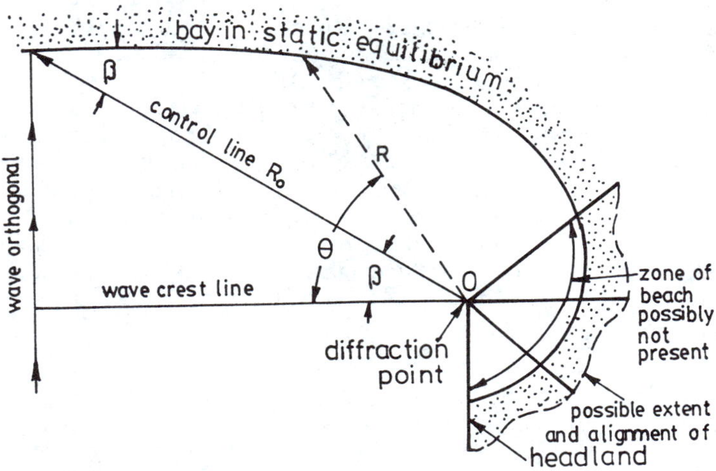

Figure 1. Definition sketch of a new empirical equation for static bay shape.

Besides the shaping of bays in the lee of headlands, equation (1) can be applied to salients behind offshore breakwaters (Hsu and Silvester 1990), as seen in Figure 2. These accretions are not triangular but assume the parabolic curve either side of the apex, as dictated by the diffracting and refracting waves and the associated longshore current circulation. This added land is at the expense of beaches beyond the extremities of the breakwater, if beach nourishment is not carried out.

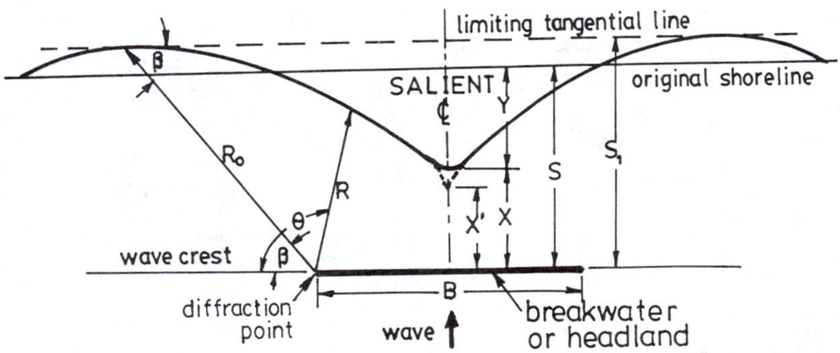

Figure 2. Definition sketch of a salient formed in lee of an offshore breakwater, with normal wave approach, based on the static bay shape equation (1).

Influence of Upcoast Headland Conditions

A natural headland is depicted in Figure 3 a & b as *A* and the downcoast limit as *B* in which control line AB ($= R_{01}$) angled to the downcoast tangent by wave obliquity β_1. When breakwater *AC* is built the new diffraction point *C* gives a new control line *CB* and new angle β_2, which results in a changed static equilibrium or stable shape. This natural reshaping will take place inspite of any modest attempts with groins or seawalls to prevent its fruition.

Looking first at Figure 3a, the existing bay may be in dynamic equilibrium, with littoral drift still passing through, as seen by the full line in the figure. The dotted curve denotes the stable shape for control line *AB*, to which the beach would erode if all sand supply around point *A* should cease. Now, when the breakwater *AC* is inserted its length and orientation could be so chosen that diffraction at point *C* and control line *CB*, with the static bay for *CB* being the same as the dynamic for *AB*. This can be achieved close enough for erosion at all points around the periphery to be negligible. If, however, sand were finally to bypass point *C* the resulting beach may become dynamic once more with a beachline seawards of this stable shape. But it should be designed for the static condition in case this additional drift ceases in the future.

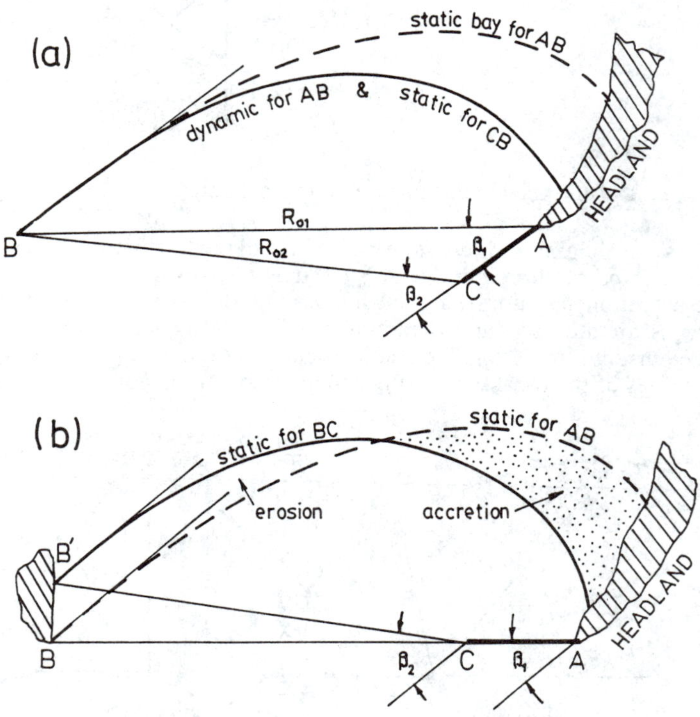

Figure 3. Effect of positioning a breakwater tip (a) in a dynamic bay, and (b) in a static bay, for harbor planning and beach erosion control.

Now turning to Figure 3b, the dotted curve represents the static equilibrium shape for control line *AB* which is the existing beach. When breakwater *AC* is added the new control line *CB* creates a new static beach which involves massive accretion in the lee of headland *A*. The material for this has to be supplied from the region of *B* because the beach was originally stable. This causes a new downcoast tangent to operate that determines a new β_2 for equation (1). The situation would detract from the calm water area serving as an anchorage but could be attractive for tourism purposes.

It must be remembered that the stable waterline being predicted applies to swell conditions or calm weather. When storm waves arrive, the beach berms will be removed offshore to form a bar, which take some days to be returned. In the case of static equilibrium the same waterline will result, but all infrastructure should be kept landward of the receded storm limit.

Case Studies of Stabilization

Conventional defense solutions, such as seawalls, groins and offshore breakwaters, have been used to stabilize eroding beaches downcoast of harbors without much success. Sand bypassing and nourishment schemes have also been tried, with some degree of success in the short term. However, the geomorphological approach, involving the construction of stable bay beaches, will be attempted in this paper. This may be regarded as a permanent solution.

Like the medical profession, which prefers to make sick people better by the administration of drugs or surgery, coastal engineers have concentrated on rectifying problems rather than preventing them. Some of these are due to their own mismanagement. Many medicos are now working holistically in looking at the whole person (body, mind and emotions) to find causes rather than remedies as such. In approaching a coastal situation from a geomorphological viewpoint the final trends can be examined which point to a permanent rather than a temporary solution.

As will be seen in many of the cases to be presented, the macroscopic view of the whole coastline gives a better picture of a problem, than looking at a limited section of it. Erosion can be predicted long before it is likely to take place and hence means can be devised to prevent it. In some of the Japanese harbors the breakwater extensions are mammoth, with lengths in terms of kilometers. The incentive may be large current expenditure to generate a larger budget next year. These structures certainly produce many subsidiary problems such as siltation and erosion which demand more money.

1. Durban harbor, South Africa

The erosion of beaches north of Durban harbor on the Indian Ocean of South Africa has been presented by Jordaan (1970). This involved siltation at the port entrance, which is continually dredged and dumped out to sea or along the eroding beaches. The suggested solutions entailed Y or L shaped groins plus two offshore sand breakwaters, each 300 m in length and spaced apart the same amount, and 600 m from the beach. Whether any of these have been implemented does not affect the this exercise to apply the principle of static bay shape.

Using the tip of the outer breakwater as the diffraction point (*A*) the resulting stable bay *A* is shown in Figure 4. This indicates substantial erosion even if minor attempts are made to prevent it. The need is for a new diffraction point to the north in order that a beachline seaward of the present can be established. Such a point (*B*) can be provided by an island breakwater *BD* from which bay *B* is formed. The space between structure *BD* and the existing inner breakwater will permit some wave action to round-off this waterline to the existing (old) breakwater as shown. The beach between it and the main port structure will be reoriented as indicated by the dotted line.

The accretion anticipated for bay beach *B* could be supplied from dredging of the entrance channel that appears to be continuous. Otherwise the siltation would come from beaches to the north. It is expected that the groins will be removed and their stone etc used in the new breakwater *BD*. The offshore sand mounds (if built) could also be removed and perhaps utilized as fill for the beach *B*.

To help overcome the necessity for dredging at the port entrance the outer breakwater could be extended as shown (*AC*) which would reflect waves obliquely. The short-crested system so established would transmit material across the slightly diverted navigation channel and create a shoal near the tip *B*. It would finally finds its way to the beaches. This new structure would prevent swell entering the gap between breakwater *B* and the existing structure so that beach *B* would extend to the tip of the old breakwater.

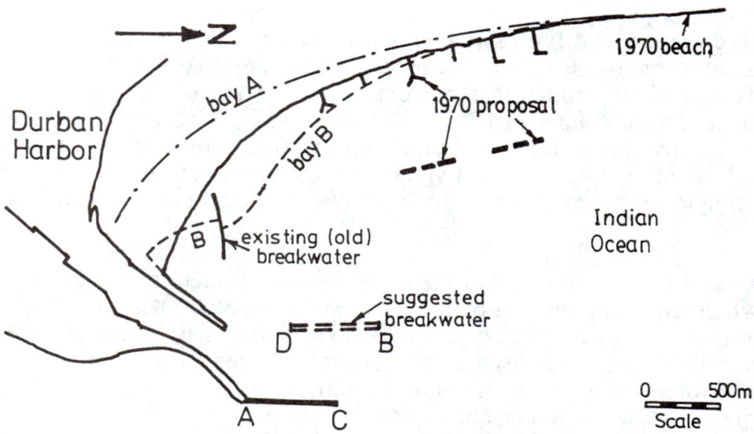

Figure 4. Durban harbor, South Africa

2. Esperance port, Western Australia

Esperance Bay is part of a continuous chain of embayments facing the Southern Ocean in the south of Western Australia. The curved shoreline, in geomorphological terms, was in dynamic equilibrium, implying its stability was maintained by the uninterrupted supply of littoral drift. However, the construction of a breakwater and dredging of an entrance channel for the Port of Esperance between 1962-65, with breakwater extension in 1973-75, has cut off the littoral drift. This has forced the shoreline to recede back to a potential static equilibrium, which is destined by Nature. This limit is landwards of the existing shoreline (see Figure 5). A total of 573,000 m^3 of dredged material was placed onto the beach.

The engineering solutions taken to combat with the eroding beach at Esperance has been a low-cost rock groin field with beach renourishment, which has been an ongoing task ever since 1969. Between 1977 and 1995, sand renourishment at an average rate of 23,100 m^3/year has been placed on sections of beaches suffering the worst erosion. Two headlands in the shape of a walking sticks with fronting beaches are proposed to achieve the overall stability with one-off beach nourishment to form a bay, rather than a field of short groins or offshore breakwaters with continual renourishment.

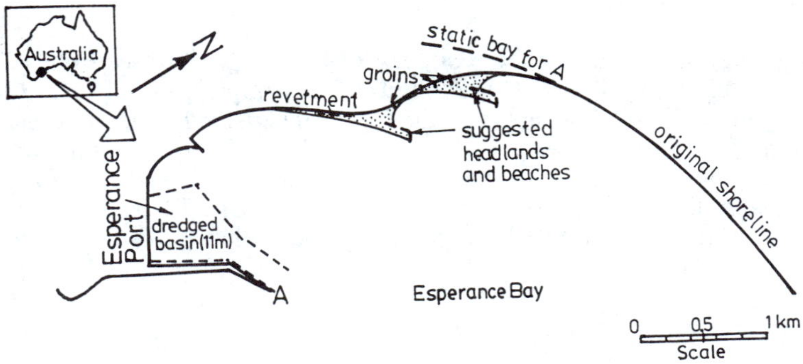

Figure 5. Esperance Bay, Western Australia

3. Hua-Lien harbor, Taiwan

Hua-Lien harbor is located on the Pacific Ocean coast of Taiwan at latitude 24°N. Two breakwaters (*AB* and *CD*) were constructed in 1987, as noted in Figure 6 (Hsu et al 1994). Prior to this extension, erosion of the coast south of the harbor entrance had been experienced, resulting in lengthy seawalls protected by concrete armour units which have steepened the offshore area. The stable beach for breakwater *CD* is shown dotted (bay for *D* in the figure) which is landward of the majority of the protected area, except at the northern tip where accretion would inhibit the discharge of the stream in the lee of breakwater *AB*.

One solution to these two problems is to extend the breakwater from *CD* to *DE*. The new diffraction point at *E* will produce the waterline shown as salient for *DE*, which is seaward of the current shoreline protected by the seawall. The remainder of the beaches will be stabilized, so not requiring other protection such as groins or seawalls. An offshore breakwater *FG* may be constructed which will intercept material in the lee of the main breakwater *DE* to form the secondary curve in the form of a salient to ensure that the stream outlet will not silt up.

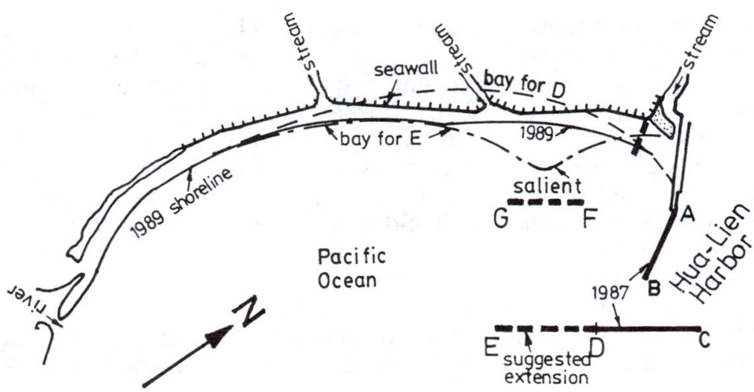

Figure 6. Hua-Lien harbor, Taiwan

4. Ohtsu harbor, Japan

The fishing harbor at Ohtsu on the east coast of Honshu, Japan (see Figure 7 for location), as seen in Figure 8, had a small breakwater complex initially in 1977. It was added to point *A* in 1984, as noted in the figure. The island structure created a new diffraction point *A* that caused accretion west of the groin at Edokomi river and between that and the Satone river outlet. The static equilibrium bay shape for point *A* is seen to match this erosion some distance from the port. To overcome this a revetment was built which suffered undermining and collapsed in places during 1983.

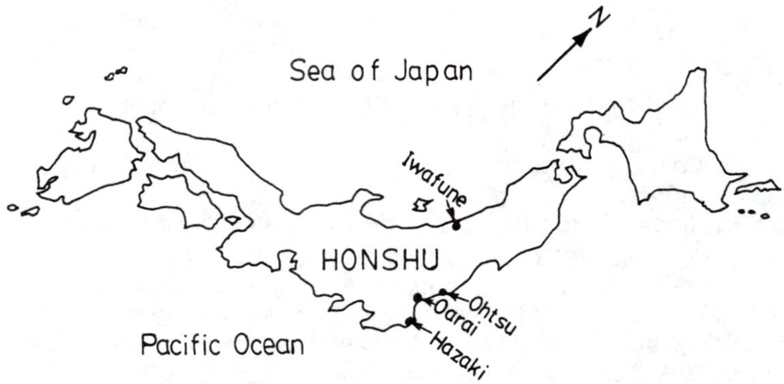

Figure 7. Map of Japan, showing locations of harbors discussed.

Offshore, in the region of the island breakwater, accretion was measured between 1977 and 1984. This was probably due to the oblique reflection of waves from this 600 m long structure that transmitted scoured material to this area. Uda et al (1986) believed these measurements were erroneous soundings. They were aware that the island breakwater had caused the beach accretion but stated: "the current induced the sand movement toward the lee of the breakwater". These authors believe it is the wave direction of approach due to diffraction that generates the new waterline orientation.

Two alternatives are shown in Figure 8 to rectify this problem, namely, either to extend from *A* to *B* or to provide a separate structure *CD*. The new diffraction points *B* and *D* result in stable beaches as shown, which indicate substantial accretion adjacent to the Edokomi river groin but less erosion further west. The additional material required for beach stability could be provided from dredging the outlets of these two rivers, to give the stable beach shown between the two groins. To deal with the slight erosion to the west an offshore breakwater could be constructed from the revetment that should be removed.

5. Oarai harbor, Japan

Oarai harbor on the Honshu coast of Japan (see Figure 7) has had many researchers analyze it (Sato and Irie 1970, Mizumura 1982, Kraus et al 1984 and Uda et al 1986). It was a small port comprising two breakwaters at 45° to the coast which was silted up quickly between 1911 and 1916. By 1976 a new breakwater was built to point *A*, in Figure 9, for which the predictable bay shape (not shown) matched the existing beach at the time (Hsu et al 1993; Silvester and Hsu 1993). In

1981 it was extended to point *B*, and to point *D* by 1985, of which the stable bays for *B* and *D* are shown. This involves massive accretion, which had partially been achieved by 1988. The sediment needed was at the expense of erosion downcoast.

In order to achieve a stable shape more in keeping with the 1988 shoreline, or even the current profile, two extensions to *E* or *F* are suggested, together with their resulting beachlines. It is seen that *F* matches the 1988 shoreline at the groin-type structure, even though it is landward of it towards the south. It is envisaged that the two small groins would be removed, which would permit the sand to find its way back to renourish beaches to the south that have been eroded. The extension of *AB* to *F* infers that breakwater *CD* would be removed as it is no longer needed for wave attenuation.

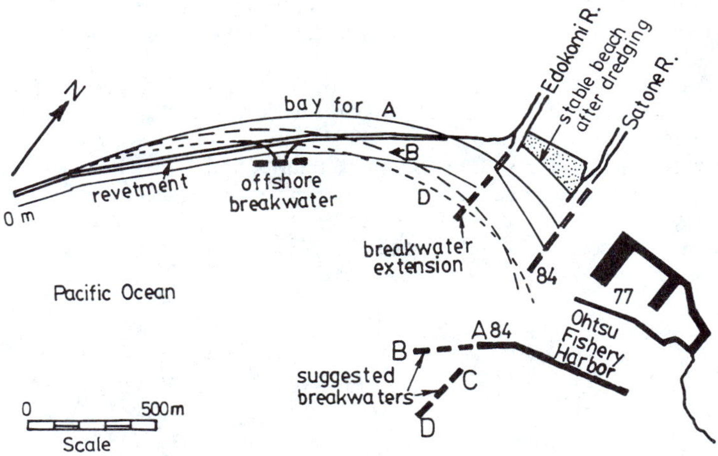

Figure 8. Ohtsu harbor, Japan

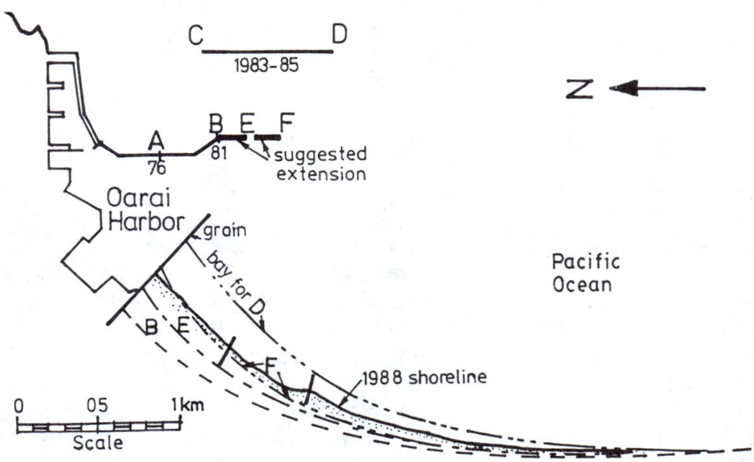

Figure 9. Oarai harbor, Japan

6. Hazaki harbor, Japan

Hazaki fishing harbor is located at the outlet of the Tone river in Honshu, Japan (see Figure 7 for location). As seen in Figure 10, a breakwater was commenced in 1974 and extended in 1976 with another started nearer the mouth. Additions were made to both in 1978 and to the SE one in 1981. By this time erosion had commenced some 600 m from the NE breakwater with accretion adjacent to it. In 1983 a shore parallel section 350 m in length was inserted to calm waters in the harbor, which became silted very quickly. Uda et al (1986) believed this was due to sediment discharge from the Tone river but the actual mechanism was not discussed. These authors believe it was due to waves being reflected obliquely from this shore parallel arm. These transported sand along it, depositing it at the entrance, which subsequently was moved into the harbor. The leg added in 1984 (see Figure 10) did little to prevent this as the mass-transport current of the short-crested wave system would have still formed a shoal at the entrance.

An extension of this leg to point *A* in the figure would achieve two important objectives. Firstly, continued oblique reflection will transmit sediment from the Tone river in a NW direction to form a shoal beyond the scoured area in proximity to its tip where an access channel can be maintained. This shoal will continually be fed to the downcoast beach. Secondly, the new diffraction point *A* will create a stable bay shape, as indicated in the figure, which will require no further structures to keep it in place. The accretion already there will thus remain for all time.

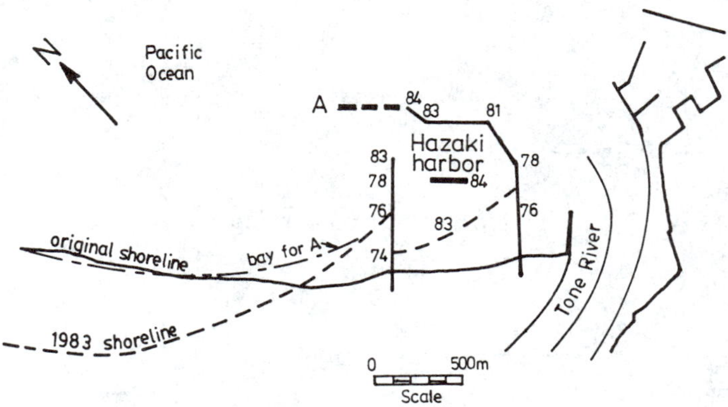

Figure 10. Hazaki harbor, Japan

7. Iwafune harbor, Japan

The shoreline evolution at Iwafune harbor (see Figure 7 for location) has been discussed by Hsu et al (1993) and Silvester and Hsu (1993). As seen in Figure 11 a large triangular land mass had been accreted up to 1988, from having attained to a stable bay shape for *A* in 1965. The static bay for diffraction point *A* matched extremely well the waterline at that time. The extensions of the main breakwater from *A* to *B* and then to *C* (by means of a detached structure) relocated the diffraction points for the incoming swell waves. The resulting bay shapes are shown which predict much more siltation if sand is available from the southern beaches. To overcome the present erosion many groins and an offshore breakwater system have been installed with little effect. The reason for this is that littoral drift still occurs offshore, just after each storm, as the storm bar is being moved back to the beach.

To prevent the transposition of further material from southern beaches it is recommended that two offshore breakwaters be installed (see Figure 11), with concomitant groin removal. The outer structure would have stable bay *C* pass through its southern tip, with the tombolo curved bay stretching to the tip of the inner breakwater. its curved tombolo would be tangential to the 1988 or present shoreline. The fill required in the lee of and south of each new breakwater could be transported by truck or pipeline from the large triangular landmass, in order to reduce siltation at the harbor mouth.

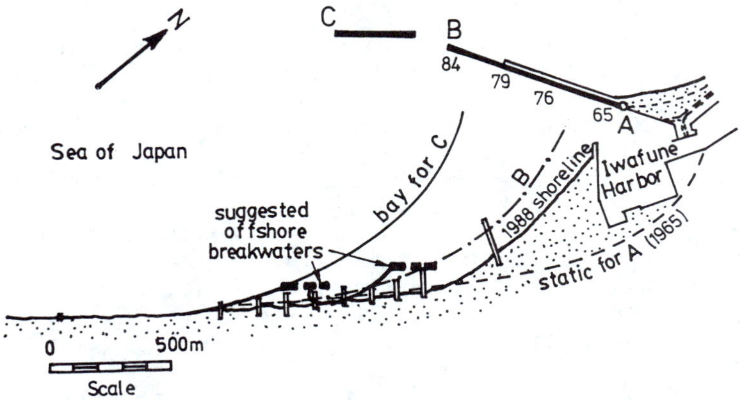

Figure 11. Iwafune harbor, Japan.

Conclusion

1. Bay beaches form downcoast of headlands or man-make structures that can be in dynamic equilibrium (continued littoral drift) or static equilibrium (zero littoral drift).
2. Should the diffraction point for persistent swell waves be altered, the bay as in (1) above will change, which may involve siltation in the lee of the structure and/or erosion downcoast.
3. A bay initially in dynamic equilibrium can be stabilized by a judicious selection of breakwater location, which then does not require any other defense measures.
4. Many major ports have suffered siltation as noted in (2) above after extension of breakwaters, together with erosion of downcoast beaches.
5. There is a dire need for coastal engineers to emulate Nature, who has maintained crenulate-shaped bays for aeons, by taking a geomorphic approach to their problems, both at the planning stage and in remedial measures later.

References

Bird, E.C.F. (1984). *Coasts: An Introduction to Coastal Geomorphology.* 3rd e d., Aust. National Univ. Press, Canberra.

Carter, R.W.G. (1988). *Coastal Environments: An Introduction to the Physical, Ecological and Cultural Systems of Coastlines*. Academic Press, London.

Davies, J.L. (1958). Wave refraction and the evolution of shoreline curves. *Geogr. Stud.,* 5: 1-14.

Davis, R.A. (ed.) (1985). *Coastal Sedimentary Environments.* 2nd rev. expanded ed., Springer-Verlag, New York.

Dunham, J.W. (1965). Use of long groins as artificial headlands. *Coastal Eng. Santa Barbara Specialty Conf., ASCE*, 755-762.

Fleming, C.A. (1992). The development of coastal engineering. In M.G. Barrett (ed.), *Coastal Zone Planning and Management*, Thomas Telford, London, 5-20.

Gonzalez, J.J.D., Paredes, V.A.E. and Herrando, J.A. (1988). Ria De Foz: Solution for a harbour and coastal problem. *Proc. 21st Int. Conf. Coastal Eng., ASCE,* V. 3, 2896-2907.

Halligan, G.H. (1906). Sand movement on the New South Wales coast. *Proc. Limn. Soc. N.S.W.*, 31: 619-640.

Herron, W.J. and Harris, R.L. (1966). Littoral bypassing and beach restoration in the vicinity of Port Hueneme, California. *Proc. 10th Int. Conf. Coastal Eng., ASCE*, V.1, 651-675.

Ho, S.K. (1971). Crenulate shaped bays. *Master Eng. thesis* No. 346, Asian Inst. Tech., Bangkok.

Hsu, J.R.C. (1990). Short-Crested Waves. Chapter 3 In *Handbook of Coastal and Ocean Engineering*, (ed. J.B. Herbich), Gulf Publishing Co., Texas, 1: 95-174.

Hsu, J.R.C. and Evans, C. (1989). Parabolic bay shapes and applications. *Proc. Instn. Civil Engrs.*, London, Part 2, 87: 557-570.

Hsu, J.R.C. and Silvester, R. (1990). Accretion behind single offshore breakwater. *J. Waterway, Port, Coastal & Ocean Eng.,* ASCE, 116 (3): 362-380.

Hsu, J.R.C., Silvester, R. and Xia, Y.M. (1989). Generalities on static equilibrium bays. *Coastal Eng.*, 12: 353-369.

Hsu, J.R.C., Uda, T. and Silvester, R. (1993). Beaches downcoast of harbours in bays. *Coastal Engineering*, 19: 163-181.

Hsu, T.W., Ou, S.H. and Chang, H.K. (1994). Countermeasures against beach erosion in Hua-Lien coast. *Proc. 16th Ocean Eng. Conf.,* Taiwan, ROC, 186-206. .

Inman, D.L. (1974). Ancient and modern harbors: a repeating phylogeny. *Proc. 14th Int. Conf. Coastal Eng., ASCE,* V. 3, 2049-2967.

Inman, D.L. and Frautschy, J.D. (1965). Littoral processes and the development of shorelines. *Coastal Eng. Santa Barbara Specialty Conf., ASCE,* 511-536.

Irie, I., Sakai, M., Nagyu, T. and Katoh, K. (1990). Development of shore protection engineering in Japan. *Proc. 27th Intel. Nav. Congr.*, PIANC, Paper S.II-2-1, 29-45.

Jennings, J.N. (1955). The influence of wave action on coastal outline in plan. *Aust. Geogr.*, 6: 36-44.

Jordaan, J.M. Jr. (1970). Study of Durban Harbor silting and beach erosion. *Proc. 12th Int. Conf. Coastal Eng., ASCE,* V. 2, 1097-1116.

Kraus, N.C., Hanson, H. and Harikai, S. (1984). Shoreline change at Oarai beach: past, present and future. *Proc. 19th Int. Conf. Coastal Eng., ASCE*, V. 2, 2107-2124.

Kuo, S.D. (1988). Shore protection of Kaohsiung harbour outer breakwater. *Proc. 21st Int. Conf. Coastal Eng., ASCE*, V. 3, 2882-2895.

Komar, P.D. (1983). Coastal erosion in response to the construction of jetties and breakwaters. In: P.D. Komar (Ed.), *Handbook of Coastal Processes and Erosion*. CRC Press Inc, Boca Raton, Florida, 191-204.

Langford-Smith, T. and Thom, B.G. (1969). New South Wales coastal geomorphology. *J. Geol. Soc., Aust.*, 16: 672-580.

Le Blond, P.H. (1979). An explanation of the logarithmic spiral plan shape of headland bay beaches. *J. Sed. Petrology*, 49: 1093-1000.
Lepetit, J.P. (1976). Impact of yachting marinas on beaches. *Proc. 15th Int. Conf. Coastal Eng., ASCE*, V. 2, 1844-1867.
Mizumura, K. (1980). Littoral drift of sand near Port of Oarai. *Proc. 17th Int. Conf. Coastal Eng., ASCE*, V. 3, 2159-2173.
Moni, N.S. (1972). Systematic study of coastal erosion and defence works in the southwest coasts of India. *Proc. 13th Int. Conf. Coastal Eng., ASCE*, V. 2, 1427-1450.
Moutzouris, C.I. (1990). Effect of harbour works on the geomorphology of three Greek coasts. *Proc. 27th Inter. Nav. Congr.*, PIANC, 17-21.
Nir, Y. (1982). Offshore artificial structures ad their influence on the Israel and Sinai Mediterranean beaches. *Proc. 18th Int. Conf. Coastal Eng., ASCE*, V. 3, 1837-1856.
Oliveira, I.B.M., Valle, A.J.S.F. and Miranda, F.C.C. (1982). Littoral problems in the Portuguese west coast. *Proc. 18th Int. Conf. Coastal Eng., ASCE*, V. 3, 1950-1969.
Ozaki, A. (1964). On the effect of an offshore breakwater on the maintenance of a harbour constructed on a sandy beach. *Proc. 9th Int. Conf. Coastal Eng., ASCE*, 323-345.
Phillips, D.A. (1988). Headland-bay beaches revisted: an example from Sandy Hook, New Jersey. *Marine Geol.*, 65: 21-31.
Sato, S. and Irie, I. (1970). Variation of topography of sea-bed caused by the construction of breakwaters. *Proc. 12th Int. Conf. Coastal Eng., ASCE*, V. 2, 1301-1319.
Saxena, P.C. (1983). Role of coastal morphology in the design of layouts of major harbours along the eastern coast of India. *Proc. 8th Inter. Harbour Congr.*, Antwerp, Belgium.
Shepard, F.P. (1973). *Submarine Geology*. 3rd Ed., Harper & Row, New York.
Silvester, R. (1960). Stabilization of sedimentary coastlines. Paper 4749, *Nature*, 188: 467-479.
Silvester, R. (1972). Wave reflection at seawalls and breakwaters. *Proc. Instn. Civil Engrs.*, 51:123-131.
Silvester, R. (1974). *Coastal Engineering, II*. Elsevier Science, Amsterdam.
Silvester, R. (1976). Headland defense of coasts. *Proc. 15th Int. Conf. Coastal Eng., ASCE*, V. 2, 1394-1406.
Silvester, R., Tsuchiya, Y. and Shibano, Y. (1980). Zeta bays, pocket beaches and headland control. *Proc. 17th Inter. Conf. Coastal Eng., ASCE*, V. 2: 1306-1319.
Silvester, R. and Hsu, J.R.C. (1993). *Coastal Stabilization: Innovative Concepts.* Prentice Hall, Englewood Cliffs, New Jersey, 578 pages.
Tanaka, N. and Sato, S. (1976). Topographical change resulting from construction of a harbor on a sandy beach: Kashima Port. *Proc. 15th Int. Conf. Coastal Eng., ASCE*, V. 2, 1824-1843.
Uda, T., Sumiya, M. and Kabayashi, Y. (1986). Analysis of beach erosion around large-scale coastal structures. *Proc. 20th Int. Conf. Coastal Eng, ASCE*, V. 3, 2329-2343.
Vichetpan, N. (1969). Equilibrium shapes of coastline in plan. *Master Eng. thesis* No. 280, Asian Inst. Tech., Bangkok.
Yasso, W.E. (1965). Plan geometry of headland bay beaches. *J. Geol.*, 63: 702-714.
Zenkovich, V.P. (1967). *Processes of Coastal Development*. Ed. J.A. Steers, trans. D.G. Fry, Oliver & Boyd, Edinburgh.

CHAPTER 309

OBSERVATION OF NEARSHORE CURRENTS AND BEACH CHANGES AROUND HEADLANDS BUILT ON THE KASHIMANADA COAST, JAPAN.

Saito,K.*, T.Uda**, K.Yokota*, S.Ohara*, Y.Kawanakajima* and K.Uchida***

ABSTRACT

Nearshore currents and beach changes around headlands built on the Kashimanada coast facing the Pacific Ocean were investigated through the field observation. In order to confirm the effectiveness of headland system to stabilize sandy beach, nearshore current field and shoreline change around headlands were measured as well as the comparison of measured value with the results of numerical simulation of nearshore currents. Comparison of development of a rip current around a completed, anchor-type headland and a groin-type headland under construction was also made and superiority of anchor-type headland to suppress the development of a rip current in the vicinity of the structure was clarified.

I. INTRODUCTION

In recent years, a method to stabilize sandy beach by using headlands has been applying on several coasts in Japan in place of the detached breakwaters and the effectiveness has been confirmed. One of the examples is given on the Kashimanada coast as shown in Photo 1, where headland system was introduced to stabilize coastline(Uda et al., 1988). Now, construction of 40 headlands is underway with 1km intervals. The headland system to stabilize sandy shoreline was originally proposed by Silvester(1976). In his study headland system in which a headland of asymmetric shape is built on the sandy beach, was considered under the obliquely incident wave conditions. On many coasts, however, wave direction varies from season to season. In this condition, symmetric type of headland is more effective to stabilize the sandy beach. On the Kashimanada coast, therefore, headlands of a symmetric shape were built.

In the planning of headland system, due considerations must be paid to the shape and its effect to surrounding nearshore current field. For example, under

---

* River Dep., Ibaraki Pref. Government, Sannomaru 1-5-38, Mito, Ibaraki 310, Japan(FAX 0292-28-0904).

** Dr. Eng., Res. Coordinator for Hydraulic Structures, Public Works Research Inst., Ministry of Construction, Tsukuba, Ibaraki 305, Japan(FAX 0298-64-1168).

***INA Consultant Co., Ltd., Sekiguchi 1-44-10, Bunkyo-ku, Tokyo 112, Japan (FAX 03-3268-8256).

the obliquely incident wave conditions, a rip current will be developed along the groin-type headland, whereas in case of the T-shape headland, development of a rip current is suppressed because of the sheltering effect of head-part of the headland. These differences deeply relate to the stabilization of sandy beaches between headlands. However, poor experience of constructing headland system on the real coasts prevents these understandings.

Photo 1 Aerial view of headlands on the Kashimanada coast (Dec.1993).

This study aims at the investigation of nearshore currents, beach changes around these headlands, and the confirmation of their effectiveness. To this purpose, the comprehensive field observations and the numerical simulation of nearshore currents were conducted. The study is composed of three contents as shown in Fig.1. First, the results of the field observation on nearshore currents around headlands are described. Field observations were carried out at three times from 1991 to 1993, in which aerial pictures were taken at a constant time interval from a helicopter to investigate diffusion of dyed water. In 1991, pre-observation was done in the vicinity of headlands, aiming at studying the difference between a groin-type and an anchor-type headland. In 1992, observations were done not only near the headland but also in the extensive area between two completed headlands in order to investigate nearshore currents and rip currents. In these observations, however, photographs were taken from a relatively low altitude. Therefore, in 1993, in order to simultaneously investigate current field in extensive area, high altitude photographs were taken from 1,000 m, as well as the bottom sounding.

In addition to these observations, beach stabilizing effect was investigated around a group of headland on the basis of several photographs and soundings, which have been carried out recurrently. Moreover, numerical simulation of nearshore currents was carried out and results of numerical simulation were compared with the observed nearshore current pattern. Current pattern was well agreed with the observed one.

Field Observation of Nearshore Currents

① **Dec.17, 1991** : Pre-observation in the vicinity of headlands
② **Dec.15, 1992** : Individual observation between two completed headlands
③ **Dec. 7 , 1993** : Simultaneous observation from high altitude

Beach Stabilizing Effects By Headlands

① Effectiveness of shoreline stabilization by a group of headland
② Cyclic shoreline changes around headland
③ Accretion in profile changes around headland
④ Vertical distribution of depth changes based on bottom soundings

Numerical Simulation of Nearshore Currents

① Calculation of nearshore currents under the condition observed in 1993
② Trial by both measured and smoothed topographies
③ Usage of irregular wave model
④ Calibrating wave direction by comparing with observed one

Fig.1 Outline of this study.

II. FIELD OBSERVATION OF NEARSHORE CURRENTS

2.1 GENERAL

Figure 2 shows the location of study site. The Kashimanada coast is located in the southern part of Ibaraki Prefecture facing the Pacific Ocean, where construction of 40 headlands with 1 km intervals is now underway. Construction works were begun in 1984, and by 1995, six sets of headlands were completed. Figure 3 shows the plane shape of the headland. Headland has an anchor-shape, of which offshore length is about 150 m and the head-part length is about 100 m. The head-part has a semi-circular shape and its offshore slope is as mild as 1/3 to reduce wave impact against the structure. The offshore surface of the headland is covered by armor units of 4 tons and the main part is built by natural stones of 1 ton. Field observations were carried out mainly between No.10 and No.11 headlands on the Ono-kashima coast located in the middle of the Kashimanada coast. The Ono-kashima coast is a sandy beach next to Kashima Port and beach slope is about 1/80, and the coastline length is about 12.5 km.

2.2 NEARSHORE CURRENT PATTERN AROUND THE COMPLETED HEADLAND

Photo 2 shows the nearshore current pattern around the completed headland measured on Dec. 17th, 1991, ① : right after the injection of dyestuffs to the foot part of a completed anchor-type headland No.10, ② : 6 minutes after the injection, and ③ : oblique aerial photo after ten minutes. In case of a anchor-type headland, it is realized from the comparison of successive aerial photographs that a nearshore circulation cell was formed on the south side of headland because of the wave dissipating effects of head-part. This is considered to be effective to contain the nearshore currents between headlands.

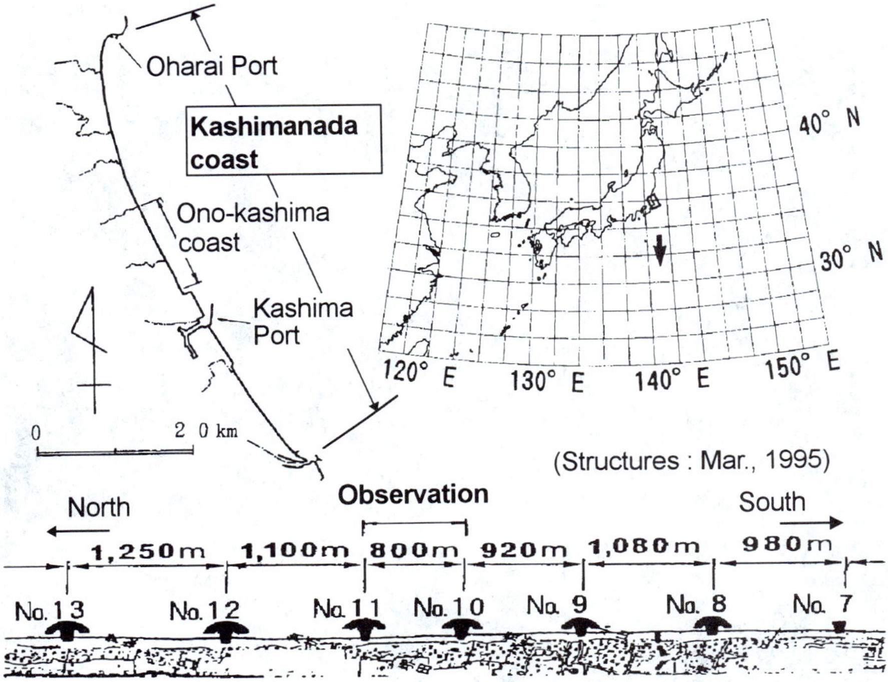

Fig.2 Location of study site.

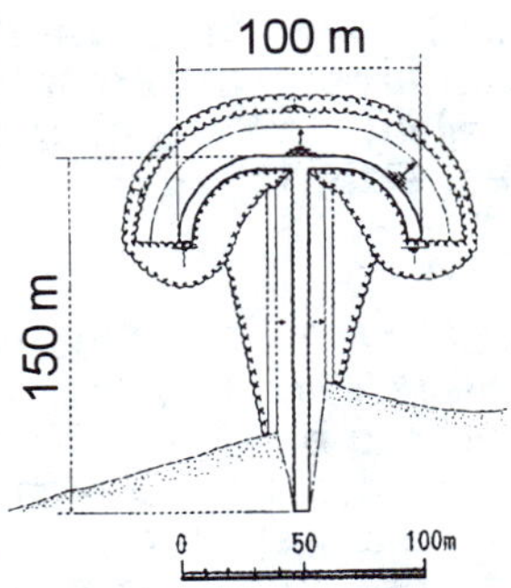

Fig.3 Plane shape of anchor-type headland.

2.3 NEARSHORE CURRENT PATTERN AROUND THE GROIN-TYPE HEADLAND

Photo 3 shows the rip currents developed along the groin-type headland (No.9) under construction observed on Dec. 17th, 1991. In this case, dyed water injected to the foot of the groin-type headland was carried fast along the groin by the rip current. Then, dyed water spread out offshoreward and was diffused, implying that sand will be carried away to downcoast to accelerate the beach erosion on upcoast of the headland.

2.4 RESULT OF NEARSHORE CURRENT OBSERVATIONS IN 1992

Figure 4 shows the nearshore current field observed on Dec. 15th in 1992 in the extensive region covering between No.10 and No.11 headlands . The figure shows changes in dyestuff clouds, their moving velocity and wave direction, which were measured from aerial photographs. Diffusion of injected dyed water was traced by the aerial photographing from the helicopter at 15 seconds interval, and the diffusion pattern of dyed water is traced at 1 minute interval.

① RIGHT AFTER THE INJECTION OF DYESTUFFS TO THE FOOT PART.

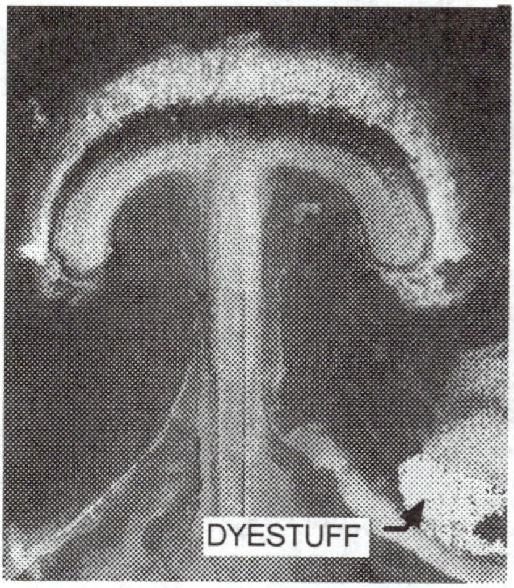

② 6 MINUTES AFTER THE INJECTION.

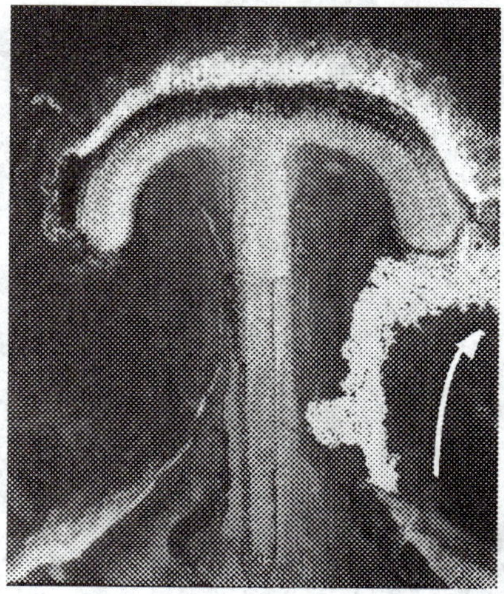

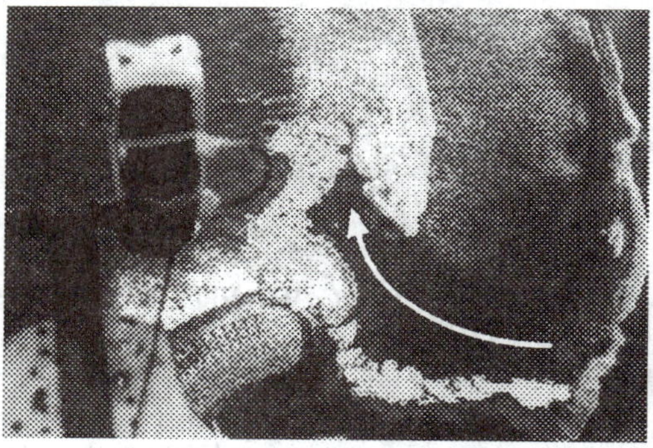

③ OBLIQUE AERIAL PHOTO AFTER 10 MINUTES.

Photo 2 Advection of dyed water around the completed headland of anchor-type (Dec. 17th, 1991).

During the observation, significant wave height was 1.3 m, and the period was 9.1 s due to the wave record measured off Kashima Port by an ultrasonic wave gauge. And also incident wave angle was read to be about 15 degree counterclockwise to the normal to the shoreline from the aerial photograph.

Because of the oblique wave direction, southward longshore currents are predominant and the circulation cell is formed on the north side of No.10 headland. On the other hand, on the north side of No.11 headland, a rip current occurs, which passes over head-part and flows out in the offshore direction. Furthermore, a rip current is occurred at a central part of the beach between No.10 and 11. The position of the rip current is located approximately 500m south of No.11 headland and is a little south from the center of headlands.

When mass flux by waves is considered, it is considered that the flux is mainly heading shoreward from between No.11 headland and the position of the rip current. And it returns to offshore by this strong rip current at central part of the beach. On the contrary, the longshore distance between the rip current in the central part and No.10 headland is short, so that the shoreward mass flux in this area is considered to be small and therefore the strength of the rip current just north of No.10 headland is weak.

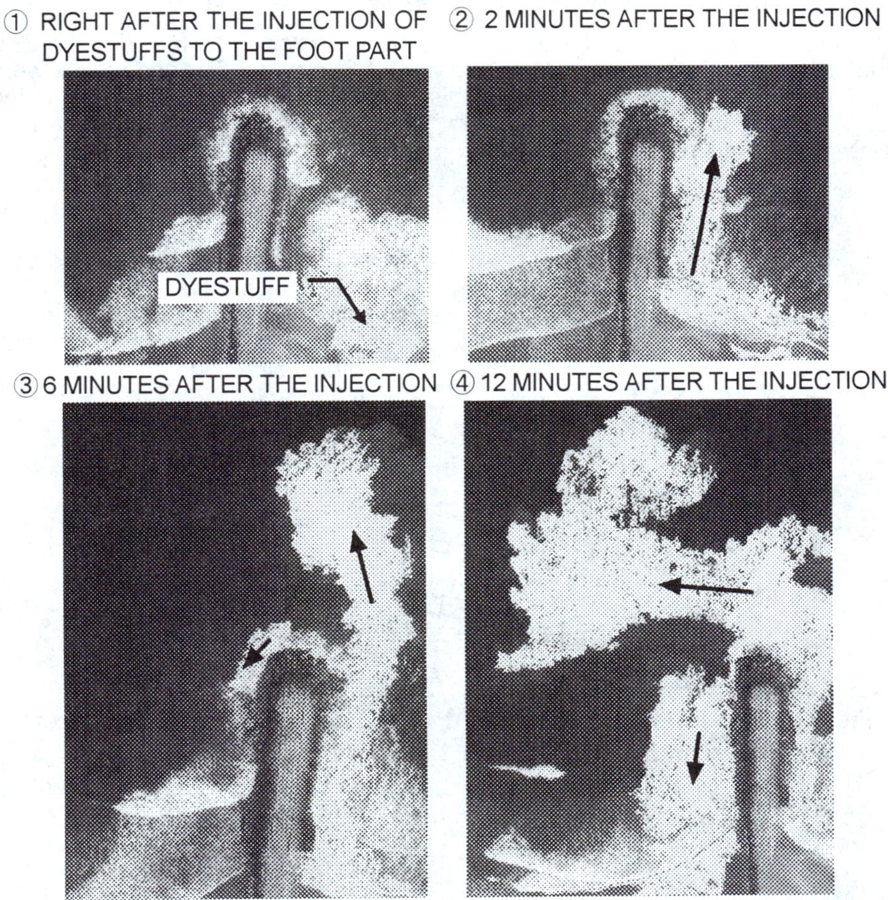

Photo 3 Advection of dyed water around the groin-type headland (Dec. 17th, 1991).

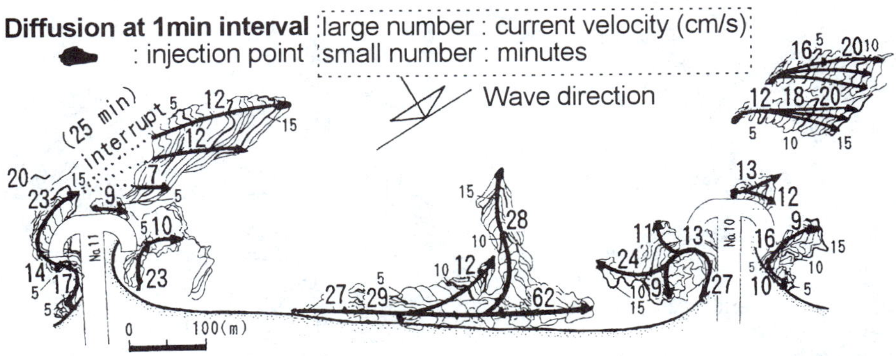

Fig.4 Diffusion of injected dyed water between No.10 and No.11 headlands (Dec. 15th, 1992).

2.5 RESULT OF THE HIGH ALTITUDE PHOTOGRAPHING IN 1993

Figures 5 and 6 show the diffusion process measured on Dec. 7th in 1993. The significant wave height and period were 1.6 m and 7.6 s, and waves were incident from clockwise direction in this observation period as shown in the figure. The condition of this incident wave angle contrasts to the wave characteristics in 1992.

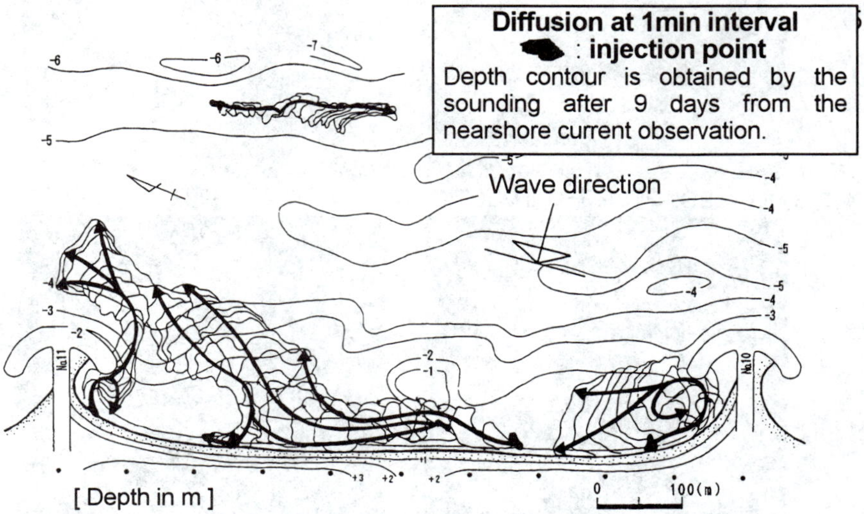

Fig.5 Diffusion of injected dyed water between No.10 and No.11 headlands(Dec. 7th, 1993).

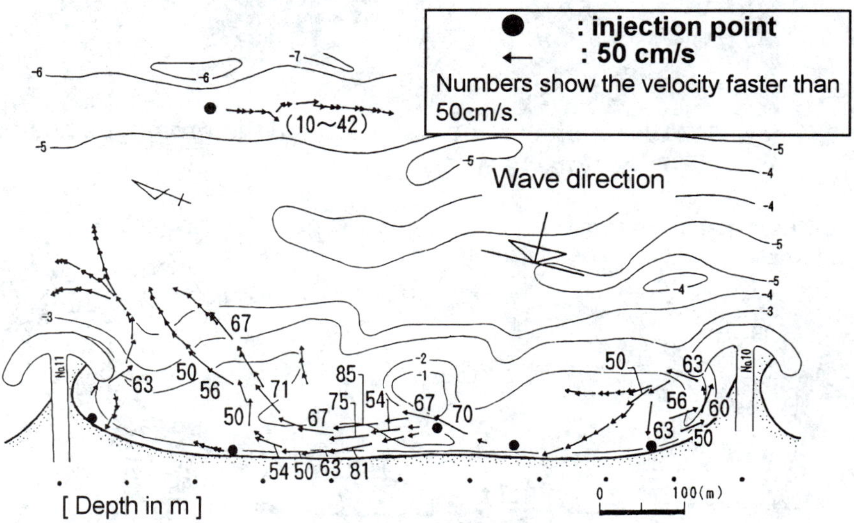

Fig.6 Velocity field of nearshore current determined from advection rate of dyed water front (Dec. 7th, 1993).

As a whole, northward longshore currents were induced, and counterclockwise circulation cell was observed in the vicinity of the north side of No.10. Furthermore, northward longshore currents developed next to this circulation cell. This longshore current is reduced to a rip current at a location of 300m distant from No.11 headland. On the south side of No.11 located on the downcoast, an another clockwise circulation cell was formed in the lee side of the headland. A part of this current reduces to the rip current and it superimposes the main rip current induced in the central part of the beach. In addition, about 400m off the shoreline, weak southward longshore currents were measured.

These results mean the formation of a large-scale clockwise circulation cell between headlands. Development of this circulation cell is considered to be effective to confine sediment between headlands, and to stabilize the sandy beach.

III. TOPOGRAPHIC CHANGES

3.1 SHORELINE CHANGE

Sea bottom surveys and aerial photographing for monitoring shoreline changes have been recurrently carried out on this coast. Figures 7 and 8 show the shoreline changes around headlands determined from the aerial photographs. As shown in Fig.7, the shoreline retreats in the southern part between headlands and a stepwise shoreline was formed due to the southerly wave incidence, which is the predominant wave direction on this coast. Near the headlands as shown in Fig.8, wide foreshore (tombolo) was formed behind the headland. All these show the effectiveness of headland system to contain sand between headlands, and to form dynamically stable beach between headlands. In this headland system, the interval between them is about 1 km, which is wide enough to keep the coastal scenic beauty on the natural sandy beach. Detached breakwaters are effective against beach erosion. But they have killed scenic beauty of the beach by the dense installation. Usually in Japan, the length is 100m and the opening width is 50m. In the headland system, however, the number of the headlands per unit distance is much less than that of detached breakwaters, thus this system is much more economical.

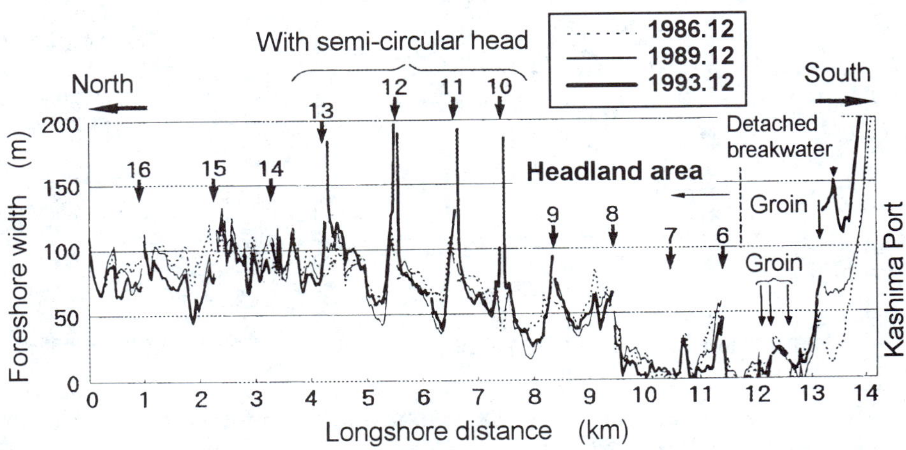

Fig.7 Changes in foreshore width between 1986 and 1993.

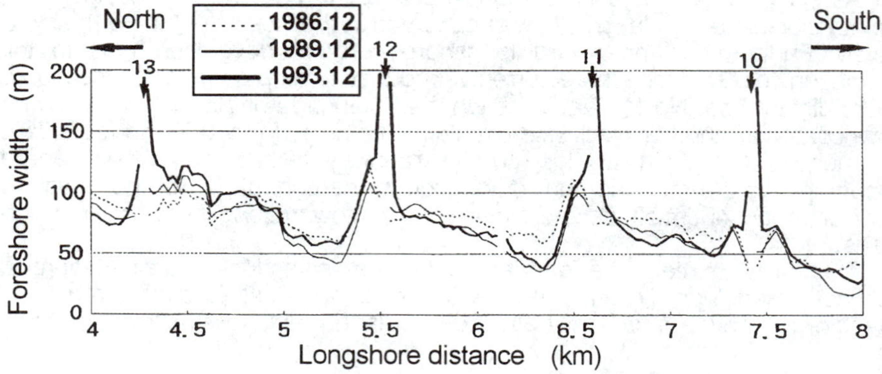

Fig.8 Shoreline changes around completed headlands between 1986 and 1993.

Figure 9 shows the temporal shoreline changes from 1986 to 1993 in the vicinity of No.11 headland. Surveys have been carried out in the same month of December in each year. In this figure, completion time of groin-part and head-part of the headlands are also shown. As shown in this figure, shorelines on the south side (S1~S3) of the headland No.11 change with the same phase, whereas on the north side (N2~N3) of the headland, the opposite mode of shoreline changes occur except the change at N1 located very close to headland and at the concave location. After the construction of the head-part in 1989, shoreline changes turn to the accretion tendency. It is understood that cyclic shoreline change occurs because of the seasonal change in wave direction in the zone next to the structure obstructing longshore sand movement. In short, the north side of headland is eroded and accretion occurs on the south side by southerly waves in summer, and then in winter, opposite phenomenon appears. Furthermore, it is realized that the accretion effect of headland overlaps on the above seasonal changes.

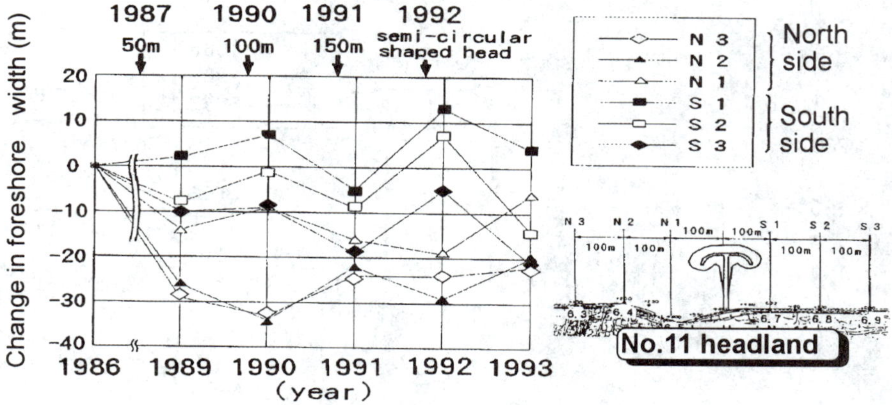

Fig.9 Temporal shoreline changes around headland No.11 between 1986 and 1993.

3.2 PROFILE CHANGES

Regarding the accretive effect of headland, it is also investigated by the bottom soundings along the survey lines arranged with 25m interval. Figure 10 shows the typical cross-shore profile changes around No.10 headland between July, 1987 and August, 1993. On the line of CN100, development of bar-trough topography was seen before the construction, but since 1992 when the head-part had been completed, bar-trough development becomes weak. On CN50 approaching 50m to headland further, because of the wave sheltering effect of the headland, accretion is seen at nearby the shoreline. Then, on CN0 nearest to the headland, a large amount of sand is accumulated to form a tombolo. The thickness of accreted sand at a distance of 100m offshore reaches to approximately 3m. On the other hand, a local scouring is not seen off the head, and a gentle slope of approximately 1/100 is formed. Thus, it is confirmed that accretive effect of headland is remarkable around No.10. because of the wave sheltering effect by its head.

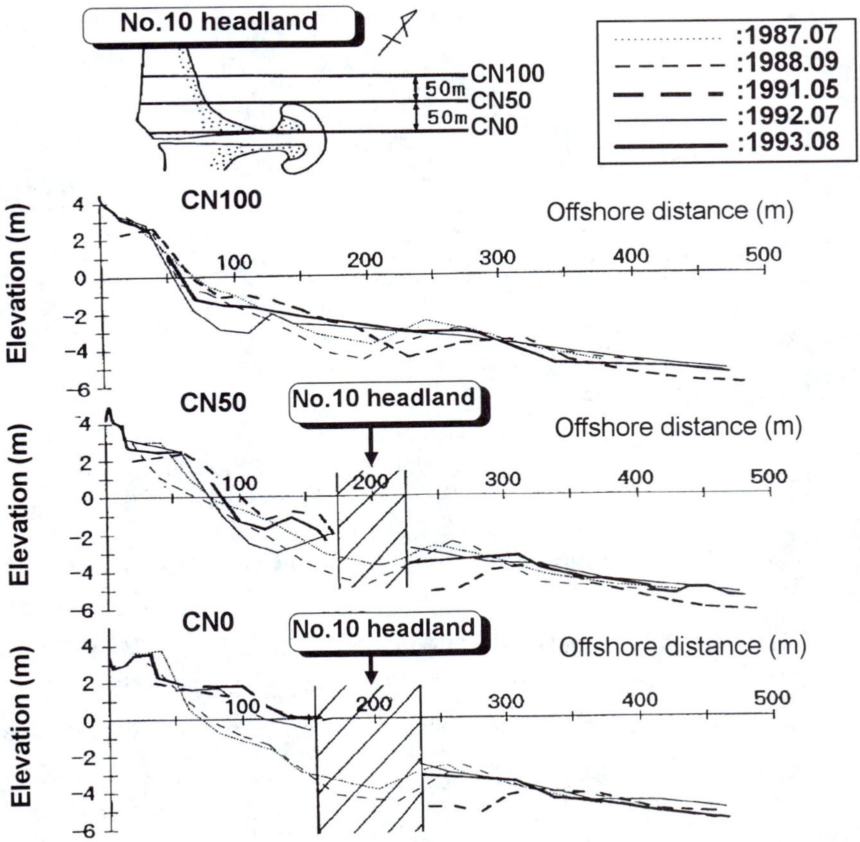

Fig.10 Typical profile changes in a period between July, 1987 and August, 1993.

3.3 VERTICAL DISTRIBUTION OF DEPTH CHANGES

On the basis of the sounding data, a vertical distribution of the standard deviation of the water depth change was examined. Fig. 11 shows the result. The vertical axis indicates the mean depth in meter. They were examined between No.10 and 11 headlands from 1987 to 1993.The significant topographic changes is occurred between -6m to +3m with two peaks at the shoreline and -3m. The peak of -3m is considered to be caused by the changes of bar-trough topography. This distribution shows a range of the remarkable topographic change. And the tip depth of the headland is approximately 3m depth. Therefore, the headland is located at a place which almost contains the range of the remarkable topographic change. This fact means the headlands on this coast is located at a offshore limit for demonstrating its effect.

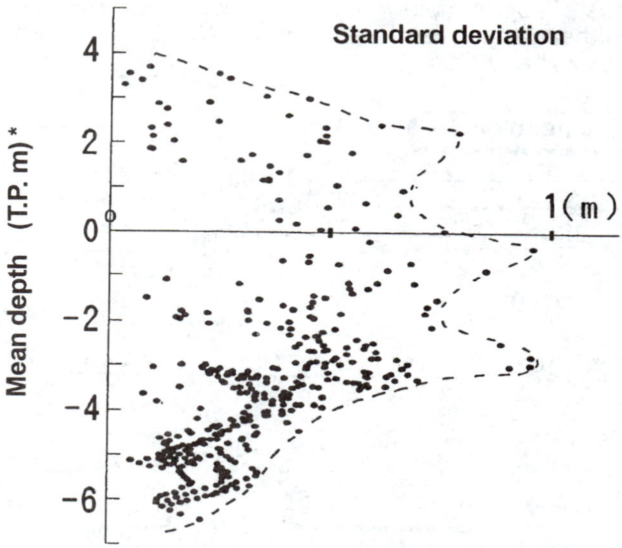

* : T.P. is a datum level frequently used in Japan and equal to annual mean sea level.

Fig.11 Vertical distribution of depth change between No.10 and No.11.

IV. NUMERICAL SIMULATION OF NEARSHORE CURRENTS

4.1 NUMERICAL MODEL AND COMPUTATIONAL CONDITION

Numerical simulation for the nearshore currents was carried out to compare field data observed in 1993. Table 1 shows the computational condition. In the nearshore current simulation, wave field was first calculated by the irregular wave model Then, nearshore current field was calculated on the basis of the above wave field and with given some coefficients. On the other hand, regarding the sea bottom topography, two cases of the calculations as described in Table 1 and shown in Figs.12 (Chart 1) and 13 (Chart 2) were carried out. Chart 1 is the original topography data sounded on Dec. 16th, 1993, and Chart 2 is the smoothed one. The reason using also Chart 2 is why the bottom topography would be changed by the storm continued during 9 days since the above-mentioned nearshore current observation.

Table 1 Computational conditions.

Term	Condition
Computation area	Longshore 2,000 m × Offshore 1,000 m
Mesh size	$\Delta x = \Delta y = 10$ m
Initial topography	Chart 1 : Original bathymetric chart (Dec. 1993) Chart 2 : Smoothed bathymetric chart
Structures	Present (headland No.10 and No.11)
Sea wave conditions	Height and Period : decided from observation at Kashima Port on Dec. 7th, 1993. Direction : Trial in the range of observed direction
Tidal level	T.P.+0.806m
Computation time	1 day
Irregular wave computation	Dividing number of frequency : 5 Dividing number of direction :11 Directional concentration coefficient : $S_{max} = 25$
Nearshore current computation	Friction factor : $C_f = 0.01$, Lateral eddy viscosity : 25m^2/s Time step : $\Delta t = 0.5$ s, Boundary condition : Open

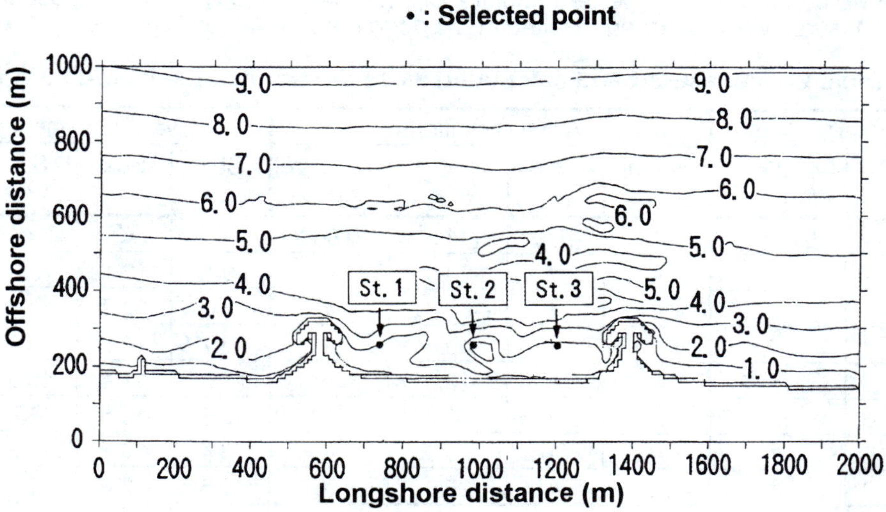

Fig.12 Original bathymetric chart (Chart 1) and selected point for verification of wave direction.

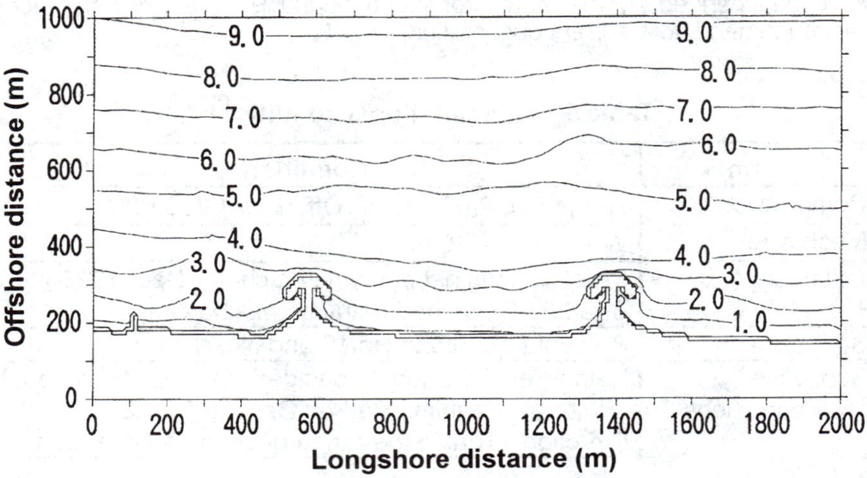

Fig.13 Smoothed contour chart (Chart 2).

4.2 VERIFICATION OF WAVE DIRECTION

First, the wave direction was verified. Table 2 shows the trial cases and results. Besides, as a representative point for verification, St.2 in Fig.12 was selected because it is located in the center, and therefore the influence due to the structures is small enough. As shown in Table 2, in total 8 cases of trials were conducted, and the best fit of the offshore wave direction to reproduce the measured wave direction resulted in -25 degree as case 7 or 8.

Table 2 Measured and calculated wave directions at selected points.

No.	Topography	Wave conditions				Wave directions		
		Height (m)	Period (s)	Direction	Tide level (T.P. m)	St.1	**St.2**	St3
1	Chart 1	1.57	7.6	-14.5°	0.806	-1.4°	**2.0°**	-7.4°
2	//	//	//	-20.0	//	-5.1	**-2.1**	-9.5
3	//	//	//	-35.0	//	-11.6	**-8.2**	-14.3
4	//	//	//	-40.0	//	-13.7	**-9.5**	-16.2
5	Chart 2	//	//	-20.0	//	-11.6	**-11.4**	-10.5
6	//	//	//	-30.0	//	-15.8	**-16.6**	-15.4
7	//	//	//	-25.0	//	-13.4	**-13.6**	-12.8
8	//	2.39	//	-25.0	//	-12.6	**-12.8**	-11.7
Mean angle measured						-13.7	**-13.3**	-5.1

4.3 RESULTS OF COMPUTATION

Figure 14 compares calculated current pattern with the observed one. Nearshore circulation cell on the downcoast of No.10, a rip current at the head of No.11, and a rip current at a location of longshore distance 900 m agreed considerably well. On the other hand, in case of using Chart1, a very strong onshore current occurred and discrepancy was large on the shoal located in the middle of the sandy beach.

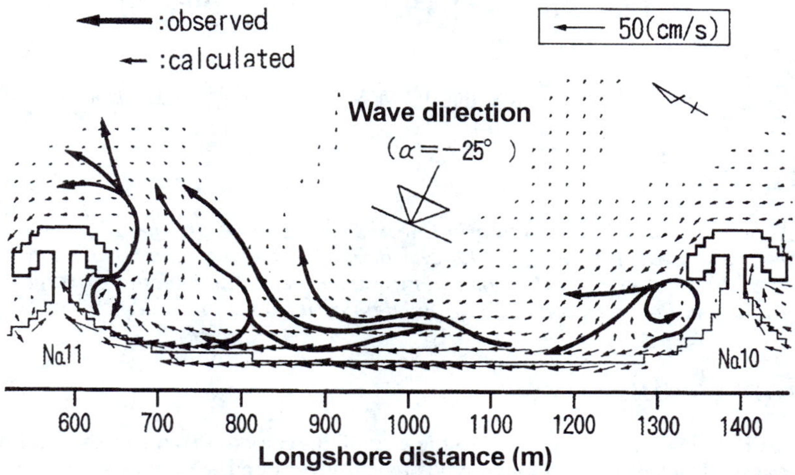

Fig.14 Comparison of observed and calculated nearshore current field.

V. CONCLUSIONS

Main conclusions in this study are summarized as follows :

① Nearshore current field with a longshore stretch of about 800m was measured. Development of nearshore circulation cell in the vicinity of headlands was confirmed. And, a rip current induced from the shoreward mass flux were also confirmed.

② Due to the shoreline changes around headlands observed by aerial photographs, the shoreline was stabilized to form a step-wise configuration responding to the oblique wave incidence. Furthermore, it was found that seasonal shoreline variation was superimposed to this shoreline to form dynamically stable shoreline.

③ Numerical simulation of wave field and nearshore currents were carried out. Results of numerical simulation agreed well with the observed one to show the effectiveness of the wave and current model.

④ This study summarizes field experiences regarding headland system in Japan. And this information is thought to be important for further application on the other coast in place of the detached breakwater.

REFERENCES

Silvester, R. (1976) : Headland defense of coasts, Proc. 15th Coastal Eng. Conf., ASCE, pp.1394-1406.

Uda,T., M.Sumiya and H.Sakuramoto (1988) : Stabilization of coast by construction of headlands on the Kashimanada coast, Japan, Proc. 21st ICCE, pp.2791-2805.

CHAPTER 310

NOURISHED BEACH CONTROL BETWEEN BALIS AND ARENYS HARBOURS (SPAIN)

F. Javier Escartín [1], Alfonso Vidaor[2], A.M. Castañeda[3] and Nuria Lupón [4]

ABSTRACT

A description of the monitoring works carried out in the Maresme beaches between Arenys de Mar and Balis Harbour after the completion of the regeneration works consisiting on the demolition of the existing groins and the execution of an artificial nourishment is presented.

INTRODUCTION

Maresme beaches are located close to the city of Barcelona (40 km in the northwest) and because of this they are frequently visited all year round.

The zone was suffering an important regression that couldn´t be restrained by any attempts of stabilization. So in February of 1992, a regeneration project of this coastal zone was carried out as a part of the works that the Spanish Public Works Ministery has been doing on the Catalan Coast.

The nourishment works began in August 1993 and the first phase of the regeneration was finished in October of the same year, in March of 1994 the second phase ended.

Because of the high investment dedicated to this work, an accurate monitoring of the beaches was planned in order to obtain a better understanding of the beach behaviour with respect to the littoral processes and thus to implement appropiate coastal management strategies.

---

(1) Senior Engineer; Europrincipia, sl., Centre Tecnològic Europroject, Parc Tecnològic del Vallès, 08290 Cerdanyola, Barcelona, SPAIN; Tel: 34-3-5820187; Fax: 34-3-5824394; e-mail: fjeg@europroject.es

(2) Director; Europrincipia, sl., Centre Tecnològic Europroject, Parc Tecnològic del Vallès, 08290 Cerdanyola, Barcelona, SPAIN; Tel: 34-3-5820187; Fax: 34-3-5824394; e-mail: alfonso@europroject.es

(3) Projects and Works Service Chief; Catalunya Coasts Demarcation, General Directorate of Coasts, Environtment Ministery, Ronda Litoral Mar 36, 08005 Barcelona, SPAIN.

(4) Senior Engineer; Europrincipia, sl., Centre Tecnològic Europroject, Parc Tecnològic del Vallès, 08290 Cerdanyola, Barcelona, SPAIN; Tel: 34-3-5820187; Fax: 34-3-5824394; e-mail: nuria @europroject.es

The monitoring has been divided in two different parts:

- A field systematic monitoring of the beaches
- A shoreline evolution analysis using a numerical model of simulation

Following the predicted evolution, the possible actions will be analized in order to guarantee the adequate maintenance of the desired stability conditions and quality.

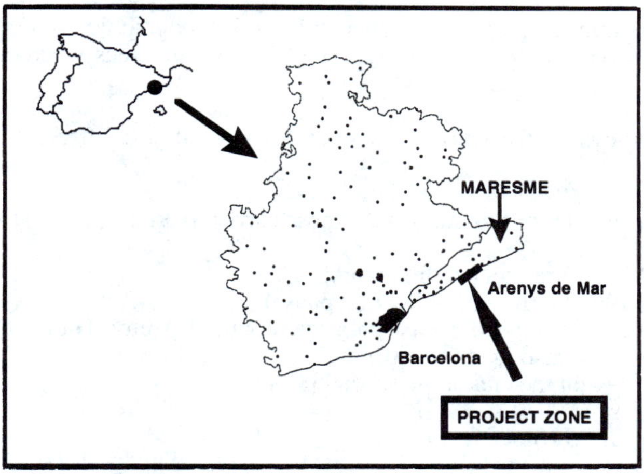

Figure 1: Project location

HISTORY

Maresme littoral is a fisiographic unit which is spread out between Tordera River mouth and Barcelona Harbour.

From the littoral dynamics point of view the working scheme of the stretch, is simple: This area has a NE-SW orientation with a net littoral transport from east to west, and net rates around 40.000 and 100.000 m^3/year.

After Arenys Harbour was built in the late forties, littoral dynamics of the beaches in the southwest of the Harbour were strongly modified suffering a regression that could not be restrained by any of the attempts of stabilization carried out. Mainly seawalls and transversal groins.

In the closest zone to Arenys Harbour, a seawall about 2 km long was built to protect the railway line, and far away to the south, for 2.5 km, a set of eleven transversal groins were constructed.

The nourishment project was designed with the double objective of:

- obtaining a great stable beach area
- establishing an important nourishment source for the Maresme beaches in the southwest of Arenys Harbour as far as Barcelona.

The total length of the regeneration was 4.5 km, with a beach width of 70 to 110 m and an elevation of 2.5 m. The mean slope was 1/10.

The total volume of sand was 3.8 million cubic meter. Ten of the eleven transversal groins were demolished.

MONITORING WORKS

Monitoring program consists in bathymetric and topographic works which are performed every six months with the purpose of identifying areas where accretion and erosion occur.

Bathymetric surveys of the whole area were carried out just after the works were completed.

The field data available for this study were those carried out in:

July 1993 (before regeneration works)
October 1993 (at the end of the first phase)
November 1993 (specific survey for the mouth of Arenys Harbour)
January 1994 (also in the mouth)
March 1994 (at the end of the second phase)

February 1995 (first survey)
July 1995 (second survey)

With these data and employing a digital terrain model, comparative shoreline and isopaches planes, and comparative diagrams of control profiles are obtained.

Some of the results of these works are shown in figure 2 where shorelines corresponding to every bathimetric survey are drawn.

Since the nourishment ends until the february of 1995's monitoring survey, shoreline stepped back around 30 meters along all the study zone except the both ends, westward Balis Harbour and eastward Arenys Harbour. It has been a sediment accumulation which means a shoreline advance, in some points a very significant advance (70 meters close to Arenys Harbour counterdike).

It has also been a shoreline advance in the central part of the stretch, downstream of the Hotel Colon groin.

From February to July 1995, shoreline position has hardly modified what indicates a relative tendency to stabilization. A light tilting takes place at westward beach of Arenys Harbour with an important accretion in the counterdike due to wave diffraction effect.

The control profile analysis also indicates (fig. 3) the local advance and receed of the shoreline. Different studied profiles show how the shoreline has receed 30 meter in profile 82 or has advanced in the other points. It can be seen in any profile that July 95's shoreline tilt around February 95 position with 3 meters variation.

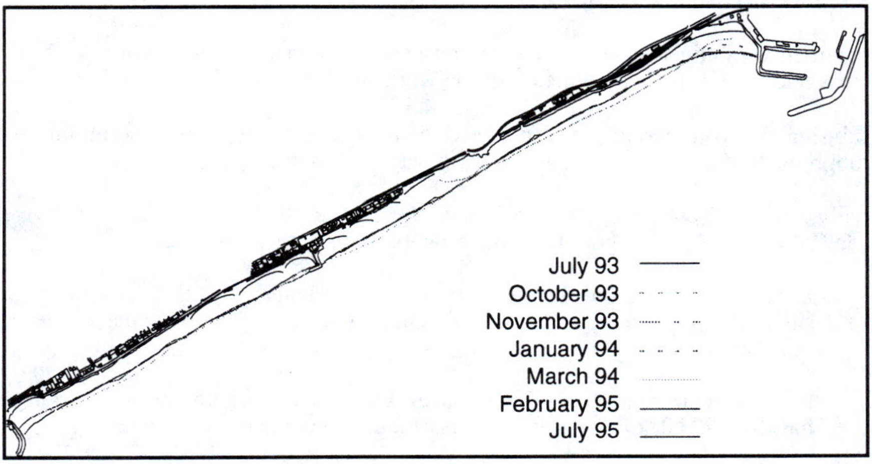

Figure 2: Shoreline evolution

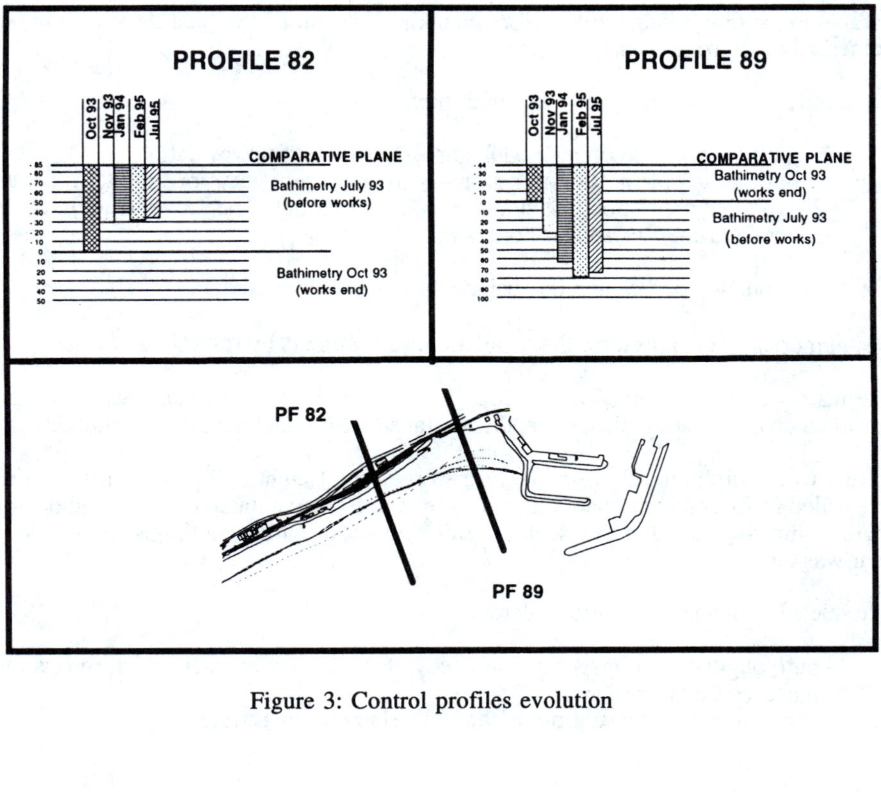

Figure 3: Control profiles evolution

NET SEDIMENT TRANSPORT

A sedimentary budget was done with the purpose of estimating sand volume mobilized by wave action since the completion of the works until July 1995.

Analysing the volumes in the considered cells, the sand quantity lost in this area through the SW extreme during this period was 46,923 m^3.

Finally, from these volumes, estimate longshore transport calculations were made.(Figure 4). At this figure two zones can be clearly distinguished:

- In the first zone, from Balis Harbour, and for a length of 3,000 meters, the net sand transport has the NE-SW direction with a maximum value of 89,307 m^3/year.

- In the second zone, from this point to Arenys Harbour, the net sand transport has the SW-NE direction with a maximum value of 81,190 m^3/year.

In the first zone a relative minimum exists as a consequence of the sand accumulation due to the Hotel Colon groin that acts as a partial barrier.

SHORELINE EVOLUTION NUMERICAL MODELIZATION

Parallel to the monitoring work performed taking account of the field data, a shoreline numerical modelization was executed.

This study was structured in two different parts:

- calibration of numerical model employing the first survey data
  In order to obtain good transport coefficients necessary for long term coastal evolution prediction, a calibration was carried out. Then a verification of the obtained coefficiens was done.

- simulation of the shoreline behaviour in a 10 year term

The numerical model used for the simulation was GENESIS by CERC.

The main variables involved in one-line shoreline evolution model are wave climate, sediment characteristics, closure depth, coastal structures and boundary conditions.

In this work, different sets of wave data were used. Height and period of wave data proceeded from Tordera escalar buoy, and directions were obtained from visual data of routing ships registered in this sector. With both set of data a continous register for a year was built up.

The selected boundary conditions were:

- null longitudinal transport in the Arenys Harbour breakwater which means that it acts as a total barrier
- permeable transversal groin in the Balis Harbour breakwater

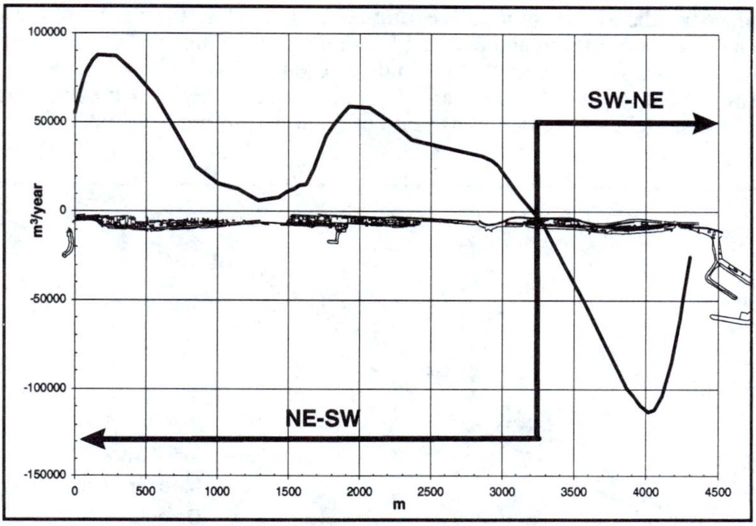

Figure 4: Measured net transport

Some simulations for the period end of the works - February of 1995 period were made, by varying transport parameters k1 and k2, in order to obtain a proper adjustement between the numerical values of the longitudinal transport and shoreline evolution in comparison with the measured values. Values found to agree more accurately were k1=0,8 and k2=0,5.

The net longshore tranport distribution along shoreline is shown in figure 5, which can be compared with that obtained from measured values: Both distributions present similar global behaviour with identical peaks of transport.

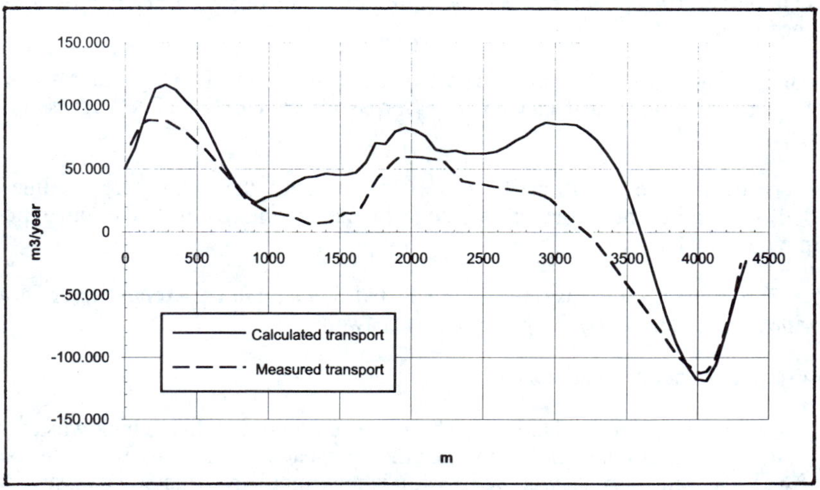

Figure 5: Net transport comparison

The obtained prediction of shoreline evolution is shown in figure 6. A reasonable agreement between measured and predicted shoreline positions can be observed. Both lines present identical accumulation and erosion areas. There is a sediment accumulation close to Balis Harbour and to Arenys Harbour, and sediment erosion in the rest of the stretch, specially downstream of Arenys Harbour.

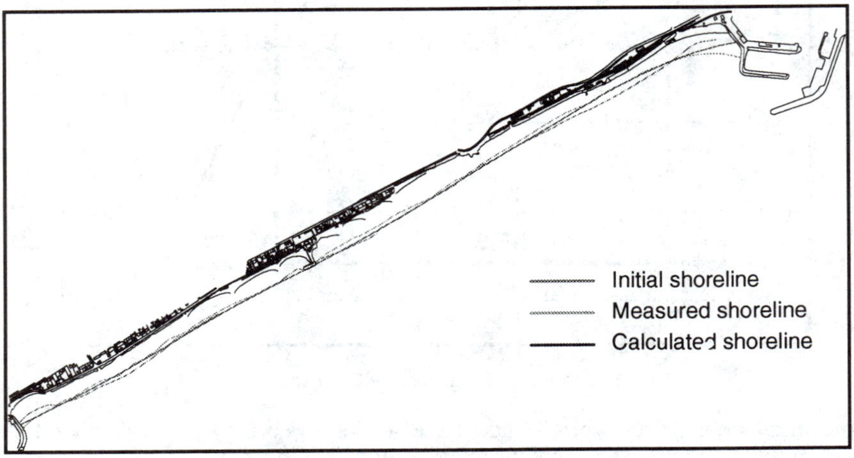

Figure 6: Shoreline prediction

10 YEAR SHORELINE MODELIZATION

Afterwards, with parameters obtained previously, a long term simulation was made in order to find out the beach evolutive trend for the next ten years. In this case a mean climate obtained from visual data is used. Shoreline from July of 1995 is employed as initial position.

Boundary conditions are the same. The existence of a seawall which represents the edge of urbanized area, that can not be overpassed by the shoreline regression, is considered as well.

Figure 7 shows the result. It can be observed that since the 7th year, shoreline has receeded, in some points, reaching the edge of the urbanized area. The other points remain rather stable.

Points where regression is maximum are situated dowstream of Arenys River Mouth. This zone has been traditionally protected by seawalls.

The most important observations are:

- Westward of Arenys Harbour, shoreline tends to stability but downstream a strong erosion takes place, so after the 6th year less than 10 m of the beach width remains. Points were regression is maximum are situated downstream of Arenys stream outlet. This zone has been traditionally protected by seawalls.

- Hotel Colon Groin doesn´t represent, a total barrier to longshore transport and its presence has no influence in the beach behaviour.

- But in the long term, erosive trend of the coast situated downstream of the Arenys stream outlet could move downstream, causing the beach to lie in the Hotel Colon groin.

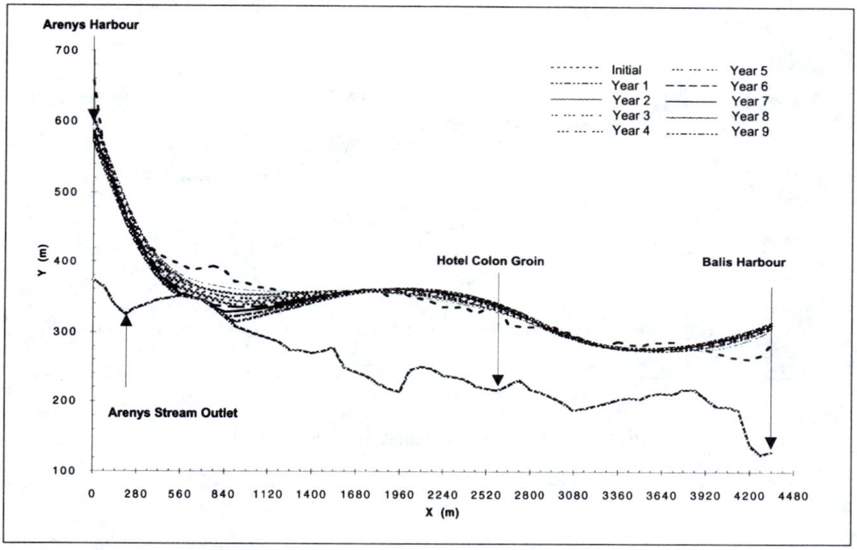

Figure 7: 10 year shoreline evolution

ACTIONS

To restrain the erosive trend of the stretch, two possible actions have been suggested and studied:

- A periodical beach fill, upstream of the stretch (westward of Arenys Harbour)
- Construction of a detached breakwater

Figure 8 shows the first action modelization. As it can be seen in the diagram, an anual beach fill of 40,000 m^3 of sand, after the third year, produces stability of the entire beach.

Secondly, with a detached breakwater a salient is generated (figure 9), what means an important increment in the beach surface in this area.

Since there is a possibility of an erosion later on in this point, beach fills are also studied (figure 10). Subsequently the beach would be totally stabilized.

Both solutions satisfy the stabilization objective of the regenerated beach.

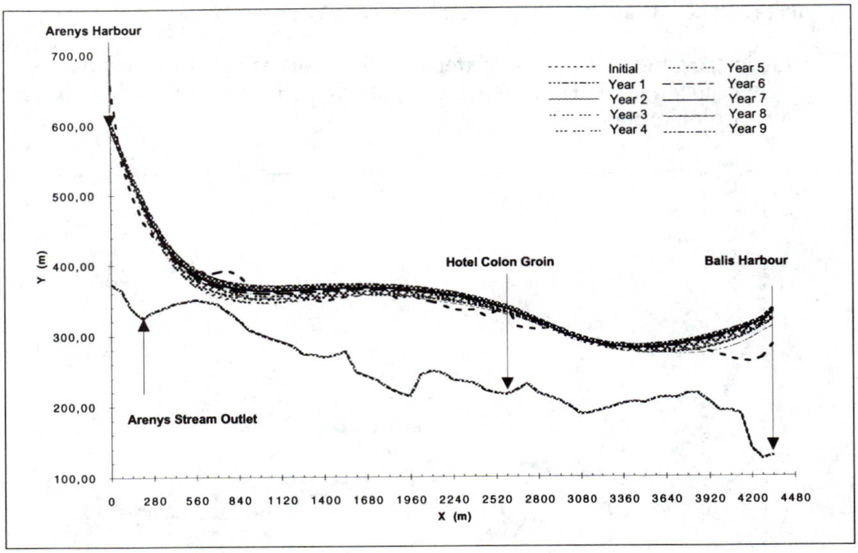

Figure 8: Periodical beach fill modelization

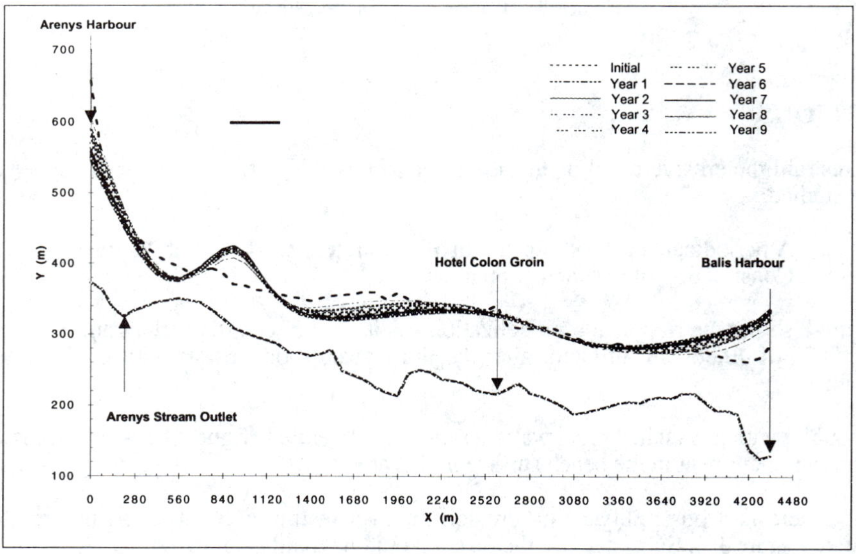

Figure 9: Detached breakwater modelization

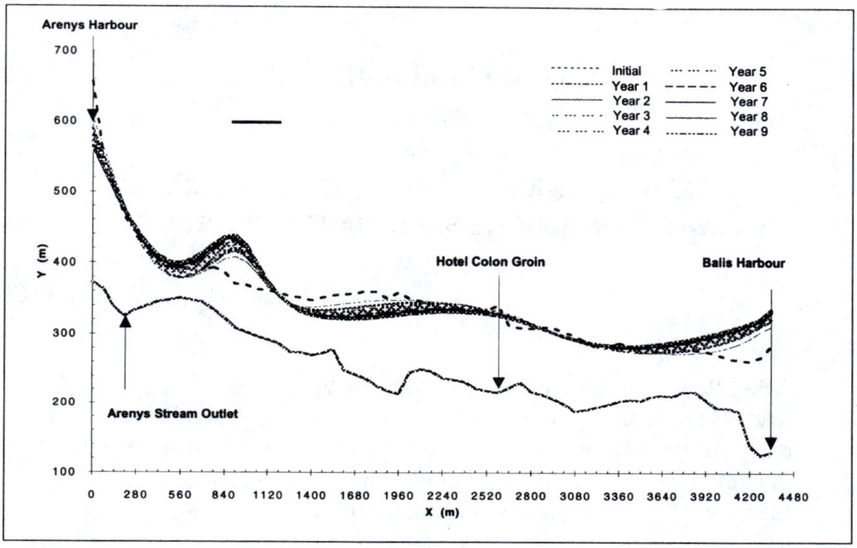

Figure 10: Detached breakwater with periodical beach fill modelization

CONCLUSIONS

- To guarantee extensive coastal zone regeneration, it is very important to carry out monitoring of the shoreline evolution.
- The monitoring and control processes should be completed with simulation models in order to:
  - improve the understanding of littoral processes
  - predict the evolutive trends
  - the necessary actions, their evolution and cuantificatioon could therefore be analyzed
- Monitoring continuity will allow the feed back of the calibration of the model and a better adjustment of both, the regenerated beach evolution simulation and the effect of the suggested actions.

CHAPTER 311

NUMERICAL SIMULATION OF SHORELINE CHANGE WITH LONGSHORE SAND WAVES AT GROINS

Hans Hanson[1], M. ASCE, Michelle M. Thevenot[2], Nicholas C. Kraus[3], M. ASCE

ABSTRACT: Longshore sand waves (LSWs) are macro-morphologic features that maintain form while migrating along the shore with speeds on the order of kilometers per year. LSWs can dominate shoreline evolution by causing both apparent long-term erosion and accretion seemingly unrelated to the calculated or estimated net and gross longshore sand transport rates. This paper explores three possible mechanisms, wave asymmetry, form advection, and surf-zone contraction, hypothesized to maintain and translate LSWs. The mechanisms are implemented within the framework of a shoreline change numerical model. Simulations implementing the LSW evolution mechanisms are tested with observations made of LSWs at Southampton, Long Island, New York. Consideration is also given to movement of LSWs at and through groins. It is concluded that fundamental questions remain on processes governing the behavior of LSWs.

INTRODUCTION

Shoreline change can be calculated by one-line numerical simulation models for a wide range of coastal structures, beach fills, waves, and boundary conditions. Such models are based on the continuity equation and a transport rate formula for the particulate movement of sand, such as the CERC formula (SPM 1984). It is well known that the one-line model, with a particulate transport rate formula dependent upon the incident wave angle, reduces to the diffusion equation. The result is that perturbations in

---

1) Associate Professor, Department of Water Resources Engineering, University of Lund, Box 118, S-221 00 Lund, SWEDEN. (Email: Hans.Hanson@tvrl.lth.se)

2) Deceased. Formerly, Research Hydraulic Engineer, US Army Engineer Waterways Experiment Station, Coastal and Hydraulics Laboratory, 3909 Halls Ferry Road, Vicksburg, MS 39180-6199, USA.

3) Research Physical Scientist, US Army Engineer Waterways Experiment Station, Coastal and Hydraulics Laboratory, 3909 Halls Ferry Road, Vicksburg, MS 39180-6199, USA.

shoreline position will tend to be smoothed, unless controlled or sustained by a boundary condition or another constraint. Particulate transport rate formulas pertain to micro-scale or meso-scale motion (minutes to hours or days) and are stepped through time, typically at 3- to 6-hr intervals, for calculation cell widths typically on the order of 50 to 500 m.

Engineers are becoming aware of morphologic features in the nearshore having much longer time and space scales that may impact project performance. Such features maintain their identities for months to years and move while preserving form. One such collective movement (Sonu 1968) of consequence is that of longshore sand waves (LSWs) (see Thevenot and Kraus (1995) for a literature review of LSWs), large wave-like features that migrate alongshore with a characteristic speed of kilometers per year. Verhagen (1989) examined a 100-year record of LSWs present along 20 km of Dutch coast and concluded that periodic accretion observed in a groin field coincided with the passage of LSWs and not to trapping of littoral (particulate) drift by the groins. LSWs have been associated with intermittency in sand supply, such as the discharge of river sediments, sediments discharged from inlets, artificial injection of a large quantity of sand, and welding of shoals on to the shore.

Recently, Thevenot and Kraus (1995) presented a one-line model that includes both particulate transport and representation of LSWs as a collective motion, and they successfully simulated LSW migration observed at Southampton Beach, Long Island, New York. The purpose of the present paper is to investigate possible mechanisms that maintain and translate LSWs and to extend the one-line model to examine LSW movement in groin fields. The model is tested with data from Southampton Beach.

THEORETICAL CONSIDERATIONS

Here, we discuss three mechanisms we hypothesize might act singly or jointly in the nearshore to preserve LSWs and translate them alongshore. First, basic concepts underlying longshore sand transport and shoreline change calculation are introduced.

General Relations

Shoreline change models have become a standard technique for calculating the long-term evolution of sandy beaches under impressed waves, boundary conditions, and coastal engineering activities (Hanson and Kraus 1989). In these models, the longshore sand transport rate is usually represented by an expression of the form

$$Q = Q_o \sin 2\theta_b = Q_o \sin\left[2\left(\theta_o - \arctan\left(\frac{\partial y}{\partial x}\right)\right)\right] \tag{1}$$

where Q_o = amplitude of longshore sand transport rate, θ_b = angle of breaking wave crests relative to the shoreline, θ_o = angle of breaking wave crests relative to an axis set parallel to the trend of the shoreline, y = shoreline position, x = distance alongshore, and $\partial y/\partial x$ = local shoreline orientation. For beaches with gentle offshore slopes, it can be assumed that the breaking wave angle, θ_b, relative to the shoreline and the shoreline orientation, $\partial y/\partial x$, are small. For such a situation, if the amplitude of the longshore sand transport rate and the incident breaking wave angle are constant (independent of x and time t), the equation for the change of shoreline position y reduces to the one-dimensional (1D) diffusion equation (Larson *et al.* (1997)

$$\frac{\partial y}{\partial t} = \epsilon \frac{\partial^2 y}{\partial x^2} \tag{2}$$

in which

$$\epsilon = \frac{2Q_o}{D} \tag{3}$$

where D = sum of depth of profile closure and elevation of the berm. Eq. (2) describes diffusion or spreading of perturbations that might be located along the shoreline, the diffusion acting to obliterate the distinct and persistent shoreline sand forms which are the subject of this study.

In this study, three possible mechanisms are postulated and investigated, by means of a one-line model, which may be responsible for the preservation of the form and translation of the center of mass of LSWs. These mechanisms are (1) wave asymmetry, (2) form advection, and (3) surf-zone contraction.

Wave Asymmetry Mechanism

There are a wide range of expressions for calculating the amplitude of longshore sand transport rate Q_o. For example, the SPM (1984) gives the relation

$$Q_o = \frac{\rho g}{16} H_b^2 \, C_{g_b} \frac{K}{(\rho_s - \rho)\lambda} \tag{4}$$

where H_b = significant breaking wave height, C_{gb} = wave group velocity at breaking, ρ (ρ_s)= density of water (sediment), g = acceleration due to gravity, K = non-dimensional empirical coefficient (approx. 0.5 - 0.8), and λ = porosity of sand (approx. 0.4). Eqs. (1) and (4) show that breaking wave height and direction are the dominant parameters determining the magnitude and direction of the longshore sand transport rate.

Thevenot and Kraus (1995) postulated that a LSW would refract waves toward it, similar to wave convergence at a headland, and this wave asymmetry would tend to "pack" the sand in place on the LSW. Their calculations with a simple 1D wave model showed this phenomenon to be potentially valid. To further evaluate the possible influence of wave asymmetry on LSW migration, breaking wave input for the shoreline change model was calculated with a full 2-dimensional (2D) wave transformation model (Larson and Hanson 1996) that was executed every time step.

The mild-slope equation (Berkhoff 1972) is often employed to describe the transformation of linear water waves. A simplified form of this equation was used in the present study, where diffraction was neglected, but energy dissipation due to depth-limited breaking was included. The dissipation was incorporated as a sink term in the equation for conservation of wave energy flux, and the magnitude of the dissipation was estimated according to the procedure of Dally et al. (1985). The mean water elevation was computed from the cross-shore momentum equation. The wave calculations involved in these simulations were, therefore, much more rigorous that what is normally done for shoreline-change model simulations. A practical limitation in accuracy is the finite grid size, as the breaking wave height and angle are calculated at discrete points across-shore.

Advective Form Mechanism

As shown by Inman (1987) and by Larson and Kraus (1991), the longshore migration of LSWs may be incorporated into Eq. (3) by including a morphologic form-advective term *V(∂y/∂x)* to yield the advection-diffusion equation for a conservative substance

$$\frac{\partial y}{\partial t} + V\frac{\partial y}{\partial x} = \epsilon\frac{\partial^2 y}{\partial x^2} \tag{5}$$

where V = the migration speed of the LSW. Thevenot and Kraus (1995) related V to a longshore water discharge parameter R defined as (Kraus & Dean 1987)

$$R = \frac{1}{2}d_b y_b v_{ls} \tag{6}$$

in which d_b = depth at wave breaking, y_b = the distance from the shoreline to the break point, and v_{ls} = mean velocity of the longshore current. The discharge parameter was considered as an appropriate means to express the form speed because R is a macro-scale quantity, as is V.

The longshore current velocity may, in turn, be calculated using an accepted empirical relation (Komar & Inman 1970)

$$v_{ls} = \frac{1.35}{2}\gamma\sqrt{gd_b}\sin(2\theta_b) \tag{7}$$

in which γ = ratio of wave height to water depth at breaking. Based on these parameters, the volume rate of transport Q_{LSW} may be calculated as (Thevenot and Kraus 1995)

$$(Q_{LSW})_{calc} = \alpha(R - R_{crit}) \tag{8}$$

where α = empirical proportionality coefficient, and R_{crit} = threshold value of R. Both of these values must be determined from field measurements or from inferences made through modeling of shoreline change. On the other hand, based on geometric properties, the volume rate of transport may be estimated as

$$(Q_{LSW})_{est} = \eta D V \tag{9}$$

where η = amplitude of the LSW measured from the ambient shoreline. Elimination of Q_{LSW} between Eqs. (8) and (9) gives an expression for the migration speed of the LSW to be substituted in Eq. (5) as:

$$V = \frac{\alpha (R - R_{crit})}{\eta D} \tag{10}$$

Contracted Surf Zone Mechanism

This analysis is based on the assumption that the presence of a LSW does not alter the alongshore location of depth contours beyond the depth of closure. Under this assumption, because the LSW protrudes seaward from the ambient shoreline, the slope of the profile on the LSW must be steeper than that of the adjacent beach.

The beach profile is assumed to follow a Bruun - Dean equilibrium ($y^{2/3}$) shape. Along the ambient shoreline (unaffected by the LSW), the horizontal distance from the shoreline to D_c is the active width of the surf zone y_c. The equilibrium shape relation for y_c may be expressed as

$$y_c = (D_c/A)^{3/2} \tag{11}$$

where A = empirical scale parameter. Along sections of shore occupied by a LSW, the active width of the surf zone will decrease. If an equilibrium profile is assumed to exist along these sections as well, the steeper profile will be characterized by a modified scale parameter A^* given by

$$A^* = D_c/(y_c^*)^{2/3} \tag{12}$$

where y_c^* is the modified active width, $y_c - \eta$ (Fig. 1). Similarly, the surf-zone width, defined as the distance y_b from the shoreline to the breaker line at depth D_b, may be written as

$$y_b = (D_b/A)^{3/2} \tag{13}$$

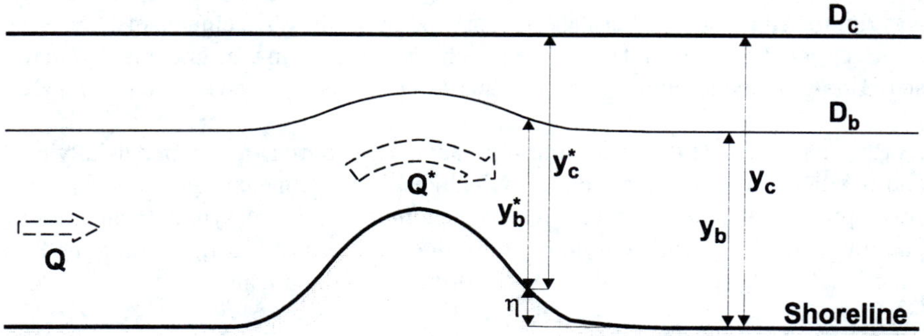

Fig 1. Definition sketch for the contracted surf zone approach.

along ambient beach sections, whereas along the beach at the LSW, the corresponding relation reads:

$$y_b^* = (D_b / A^*)^{3/2} \tag{14}$$

By expressing the longshore sand transport rate Q as an Inman-Bagnold-type relation and applying the continuity equation for the longshore transport of water yields

$$Q \sim v_{ls}\, H_b^2, \quad Q^* \sim v_{ls}^*\, H_b^{*\,2}$$
$$v_{ls}\, Area_b = v_{ls}^*\, Area_b^* \tag{15}$$

where $Area_b$ = cross-sectional water area between the shoreline and breaker depth, and the superscript * denotes values in the region of the LSW. Based on the unmodified longshore sand transport rate distant from the LSW, the corresponding rate along the LSW is given by

$$\frac{Q^*}{Q} = \frac{A}{A^*}\left(\frac{y_c}{y_c^*}\right)^{5/3} \tag{16}$$

CALCULATION RESULTS

In the following, predictions of the three approaches for maintaining and translating LSW are evaluated through numerical simulation.

Wave Asymmetry Calculation Approach

Starting with a representative single LSW with a length of 5.5 km and an amplitude of 34.2 m (Fig. 2), a standard shoreline change model (particulate transport rate only) was run for 2,000 hr with constant offshore (20-m depth) wave climate of $H = 1.0$ m, wave period $T = 6$ sec, and $\theta = 10$ deg. Other parameters in the simulation were time step $\Delta t = 4$ hr, cell spacing $\Delta x = 50$ m, and $K = 0.7$. As has been shown previously

(Hanson and Larson 1987), wave transformation modules run on assumed locally plane and parallel contours that are typically included in shoreline change models produce wave properties alongshore that results in a high degree of symmetrical transport patterns around a protruding symmetric feature such as a LSW. As a result, the LSW experiences little advection, although the diffusion is significant (Fig 2).

In the 2D wave modeling simulations, the number and size of alongshore calculation cells were the same as in the 1D simulation. The calculation area was divided into 50 sections, each of 20-m spacing across shore. Time step and duration of the simulation were the same as in the 1D case. As seen in Fig. 2, the LSW evolution produced by the 2D wave modeling displays a much greater degree of asymmetry, but there was no significant advective LSW motion, although diffusion was significant.

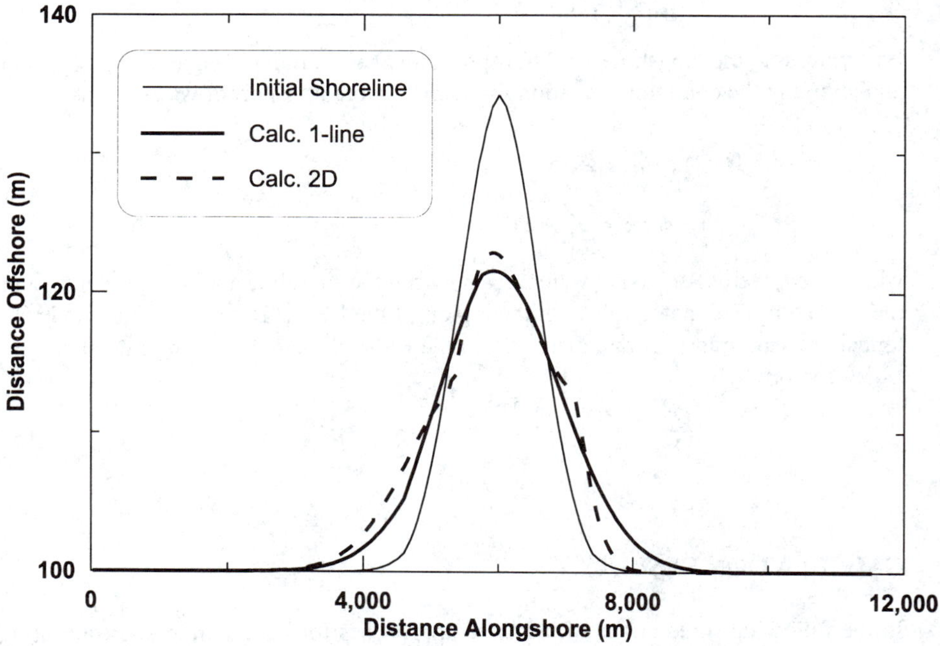

Fig 2. Shoreline change calculated using 1D and 2D wave transformation modeling.

Based on this and other calculations, it is concluded that a 2D wave calculation scheme as run in typical engineering studies will not create enough wave- and associated sand-transport asymmetry around a LSW to cause the LSW to migrate.

Advective Form Calculation Approach

As seen from Eq. (10), the advective speed of any section of a LSW is inversely proportional to its displacement from the baseline. Thus, the center of the initially symmetric LSW in Fig. 3 moves faster than the flanks, resulting in a flatter up-drift plan shape of the LSW and a steeper down-drift flank, in agreement with field observations. From a modeling point of view, the advective speed concept offers significant flexibility. The migration speed is controlled by α and R_{crit} in Eq. (10), whereas the diffusion is controlled by K in Eq. (4).

The initial LSW in Fig. 3 is the same as in the previous figure. In this case, the wave conditions were $H = 1.0$ m, $T = 8$ sec, and $\theta = 20$ deg, and the simulation duration 720 days. The LSW migrates with an average speed of 3 km/yr with a decrease in amplitude of about 50%. Variation of input wave characteristics in several similar simulations (not shown) indicated that the LSW would move in either direction, always migrating in the direction of net sand transport, in agreement with field observations. At the same time,

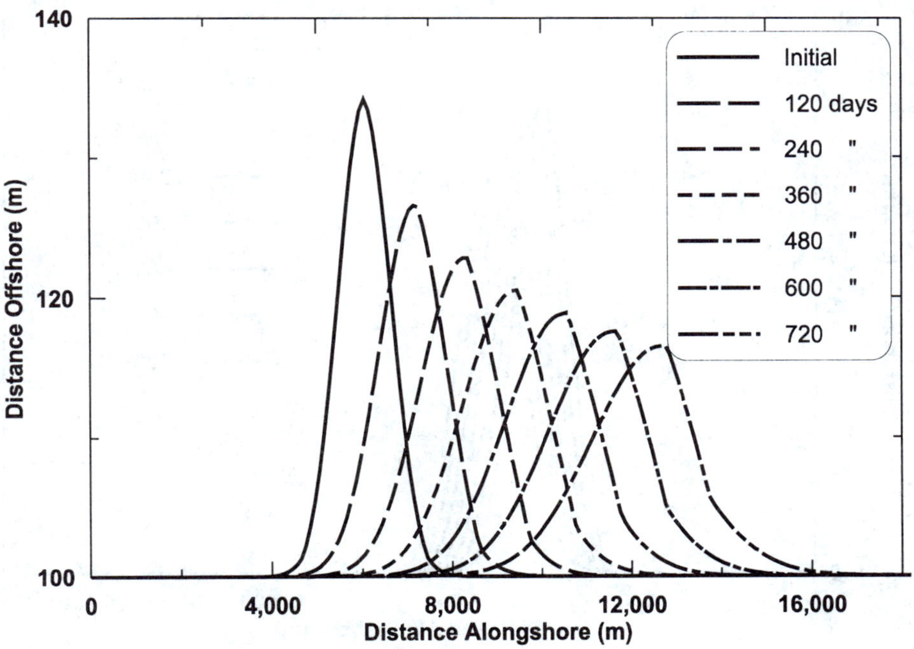

Fig 3. Calculated LSW migration using the advective form approach.

there was always an increase in length of the calculated LSW with time, while its amplitude decreased due to diffusion. Available field observations seem to indicate that LSWs have a more stable amplitude and do not increase in length.

Contracted Surf Zone Calculation Approach

A simulation was performed with the contracted surf zone approach implemented and with identical conditions as in the previous cases. As seen from Fig. 4, the calculated LSW does indeed move downdrift, but with considerably less speed and diffusion than in the previous case. During the simulated 2 years, the LSW moves with an average speed of 0.6 km/yr while maintaining 82% of its amplitude. At this stage, it is not possible to determine which of the two very different results is correct, because both are reasonable.

Distinguishing Advection and Diffusion

One problem both approaches have in common is, however, that any kind of perturbation in the shoreline will move downdrift, which is not reasonable. For example, the local erosion and accretion associated with the presence of a groin are expected to remain in the vicinity of the groin and not to move alongshore as a coherent form.

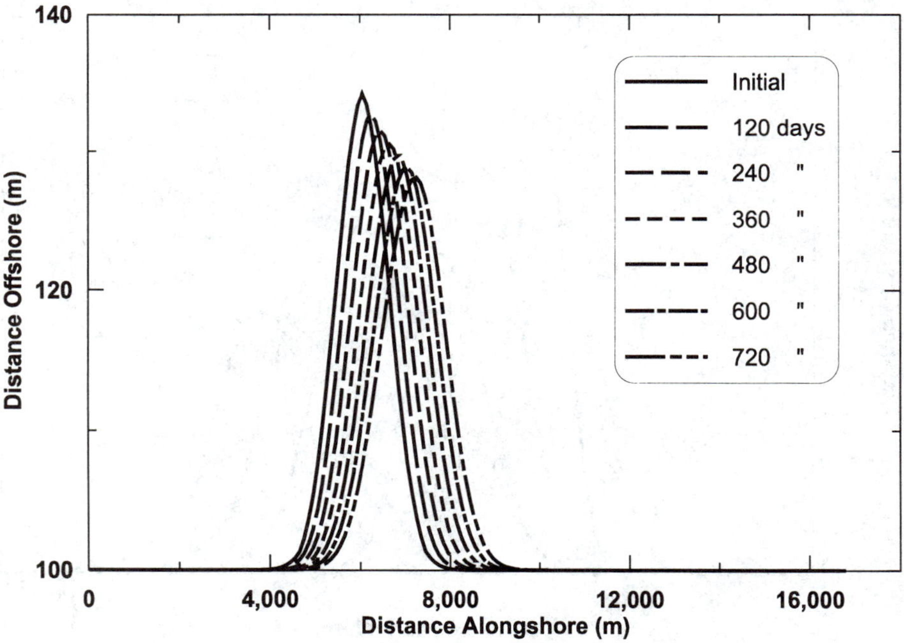

Fig 4. Calculated LSW migration using the contracted surf zone approach.

Fig. 5 shows the situation of an initial LSW identical to that in Fig. 3, calculated with the advective-form approach. The simulation was driven with a Corps of Engineers hindcast wave time series for Southampton Beach, Long Island, New York, that covers the period 1956-75. The simulation started on Jan. 1, 1958, and continued for 1,600 days, chosen to represent a period of unusually strong longshore transport wave conditions and a persistent sediment transport to the west (right). With a groin blocking the sediment transport, impoundment or accretion are expected on its updrift side and corresponding erosion on the opposite side of the structure. However, as these features start to develop, they, too, begin to migrate, resulting in the wavy shoreline shape downdrift of the groin. This behavior is unrealistic.

One *ad-hoc* means of eliminating advection of all perturbations is to *a-priori* identify and specify which features are to migrate. These features will subsequently evolve by advection as well as diffusion. Other features that appear spontaneously in response to constraints or placement of beach fills, such as shoreline displacements near groins, will not be subject to the advective mechanisms and not migrate alongshore. Such an *ad-hoc* specification might be implemented in an engineering application if the results are monitored with caution, but it is not satisfactory because it does not address why some

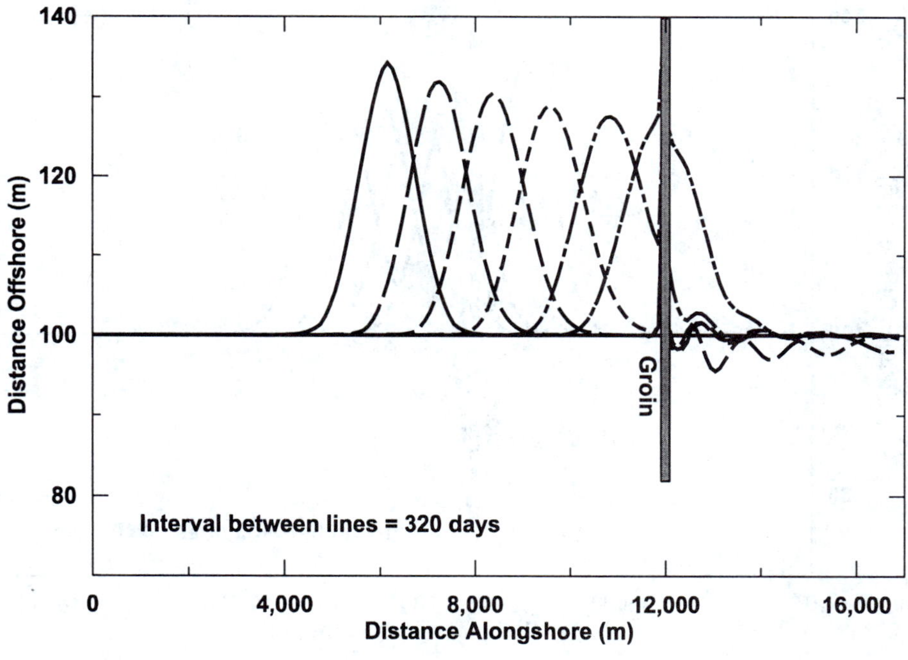

Fig 5. Calculated LSW interaction with a groin using the advective form approach.

shoreline features move and others do not, that is, the physical processes associated with cause and effect of the LSW migration are omitted.

A result of a simulation with such a distinction between advection and diffusion in the advective-form approach is displayed in Fig. 6, representing the same wave conditions as in the previous case, but now only run for 1,200 days. With the LSW distinguishing capability active, accretion as well as erosion adjacent to the groin develop, as expected. With elapsed time, the LSW passes the groin, initially deformed by the presence of the groin, but later resumes its shape as it moves away from the structure.

Application to Southampton Beach

To validate qualitatively the preliminary approach described here to modeling LSW migration, the advective form calculation was applied to the situation at Southampton Beach, Long Island, New York. Here, eleven LSWs present in the early 1990s were identified from aerial photos (Thevenot and Kraus 1995). The LSWs had an average length of 0.75 km, an average amplitude of about 40 m, and an estimated average annual migration speed of 0.35 km/yr. Their speed was found to vary seasonally with the strength of the inferred longshore discharge parameter calculated from the wave hindcast.

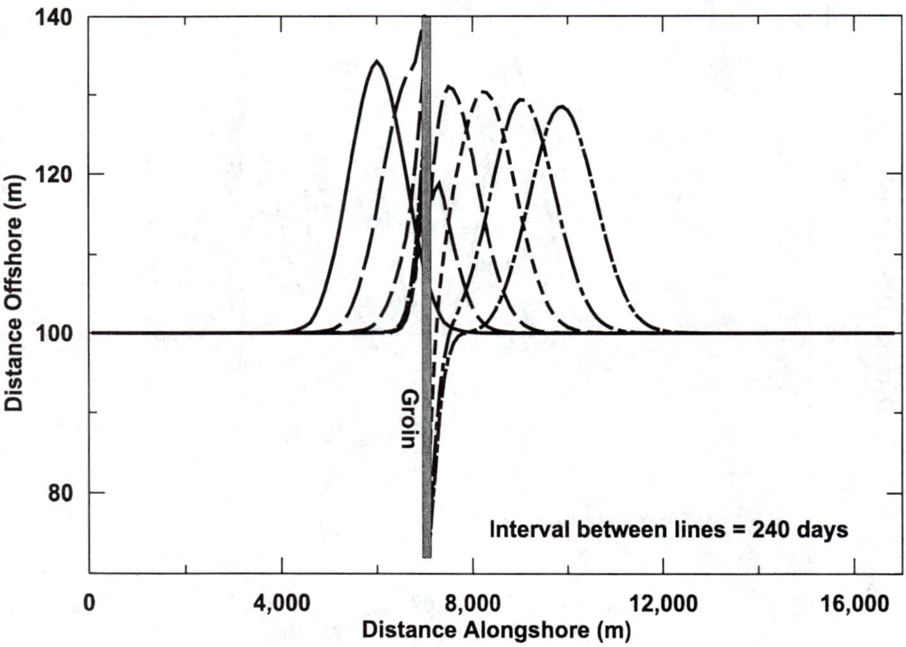

Fig 6. Calculated LSW interaction with a groin using decoupled advection and diffusion.

Available aerial photos dated Sep. 4, 1991, Dec. 20, 1991, and Jan. 2, 1993, served as references for the simulations. The simulated shoreline covers 16.9 km, with a spatial resolution of 100 m and a time step of 3 hr. Results of the simulation are shown in Fig. 7. The same wave hindcast time series as in the above was specified, but now starting on Sep. 4, 1956, and ending Jan. 2, 1958, representing the time period Sep. 4, 1991, to Jan. 2, 1993, because the actual time period of the LSW observation is not covered by the hindcast. The hindcast time period was selected to represent typical conditions at the site.

As expected, the LSWs moved to west (to the right in the simulations). However, it was noted that during the last two weeks of the simulation period, i.e., the end of 1957 (representing the end of 1992), the LSWs moved significantly in the opposite direction, ending on the incorrect side of the intermediate LSW positions. This reversal is an artifact of the wave time series in the hindcast for 1957. The calculations were therefore halted 2 weeks before the end of the simulation period.

The locations of the calculated LSW are in reasonable agreement with most of the LSWs in the photographs. However, representation of the amplitudes is poor. The reason for this is, as indicated previously, that the model cannot produce growth in LSW amplitude,

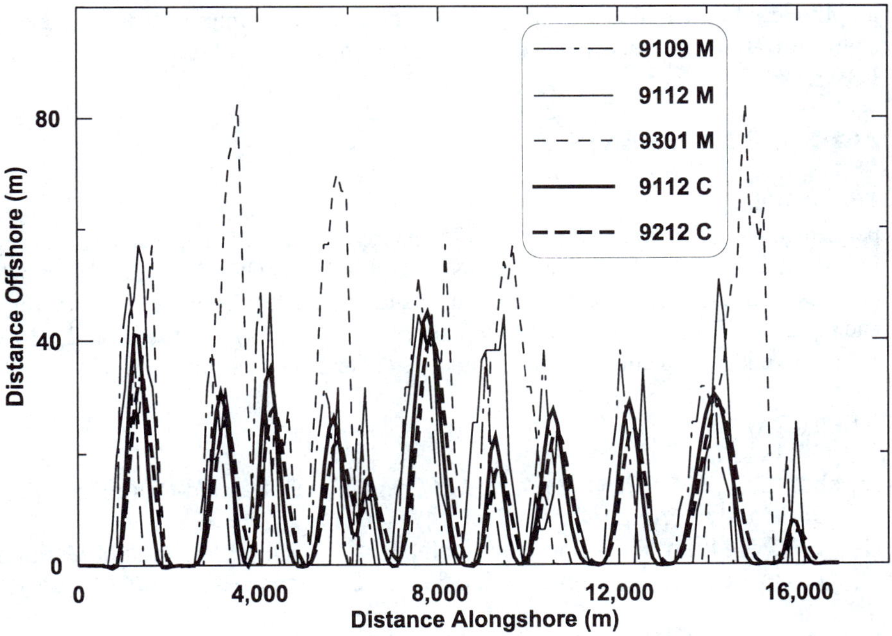

Fig 7. Measured and calculated LSW movement at Southampton Beach, LI, New York.

which appears to be a significant process in the field. It is not known at this point what phenomena are acting to produce a substantial increase in amplitude of the LSWs. It is possible that varying tide level may have contaminated the dimensions of the LSWs measured in the aerial photographs.

CONCLUDING DISCUSSION

Longshore sand waves (LSWs) are a large-scale phenomenon of major significance in coastal engineering and science. The physical processes governing preservation and migration of LSWs have not been firmly identified, no less understood. Predictive models describing LSWs are only at the formative stage, and the present paper has explored three approaches for describing the evolution of LSWs. The interaction between LSWs with groins was also explored.

Comparisons of calculations with measurements of LSWs at Southampton Beach, Long Island, New York, shows that the migration of LSWs may be described by shoreline-change model modified to include a form advection term. However, the temporal and spatial variation in permanence of form and amplitude of the LSWs were not satisfactorily reproduced in the model. Presented results indicate that the advective form calculation approach reproduces the LSW migration better than other methods, whereas the contracted surf zone calculation approach seem. to be superior in preserving LSW amplitude. Further model developments will depend on improved understanding, characterization, and conceptualization of the mechanisms controlling the behavior of LSWs.

ACKNOWLEDGEMENTS

The contribution to this study by H Hanson is based on work in the SAFE-project, performed within the framework of the EU-sponsored Marine Science and Technology Programme (MAST-III), under contract No. MAS3-CT-95-004. N.C. Kraus appreciates permission granted by the Chief of Engineers to publish this information. Our young and talented colleague, Michelle Marie Thevenot passed away on December 21, 1996, after a several-year battle with cancer, and we dedicate this paper to her.

REFERENCES

Berkhoff, J.C.W. 1972. Computation of Combined Refraction-Diffraction. *Proc. 13th Coastal Engrg. Conf.*, ASCE, 471-490.

Dally, W.R., Dean, R.G. and Dalrymple, R.A. 1985. Wave Height Variation Across Beaches of Arbitrary Profile. *J. Geophys. Res.*, 90(C6): 11917-11927.

Hanson, H. and Kraus, N.C., 1989. GENESIS - Generalized Model for Simulating Shoreline Change, Vol. 1: Reference Manual and Users Guide.

Tech. Rep. CERC-89-19, US Army Engineer Waterways Experiment Station, Coastal Engineering Research Center, 247 pp.

Hanson, H. and Larson, M. 1987. Comparison of Analytic and Numerical Solutions of the One-Line Model of Shoreline Change. *Proc. Coastal Sediments '87*, ASCE, 500-514.

Inman, D.L. 1987. Accretion and Erosion Waves on Beaches. *Shore & Beach,* 55(3 & 4): 61-66.

Komar, P.D. and Inman, D.L. 1970. Longshore Sand Transport on Beaches. *J. Geophys. Res.* 75(30): 5914-5927.

Kraus, N.C. and Dean, J.L. 1987. Distribution of the Longshore Sediment Transport Rate Measured by Trap. *Proc. Coastal Sediments '87*, ASCE, 881-896.

Larson, M and Hanson, H. 1996. Schematized Numerical Model of Three-Dimensional Beach Change. *Proc. 10th Congress of the IAHR Asia and Pacific Division*, Lankawi Island, Malaysia.

Larson, M. and Kraus, N.C. 1991. Mathematical Modeling of the Fate of Beach Fill. In: H.D. Niemayer, J. van Overeem, and J. van de Graaf, (Eds.), Artificial Beach Nourishment, *Coastal Engrg.*, 16,: 83-114.

Larson, M, Kraus, N C, and Hanson, H. 1997. Analytical Solutions of the One-Line Model of Shoreline Change Near Coastal Structures. *J. Waterway, Port, Coastal and Ocean Engrg.*, ASCE, in press.

Sonu, C.J. 1968. Collective Movement of Sediment in Littoral Environment. *Proc. 11th Coastal Eng. Conf.*, ASCE, 373-398.

SPM. 1984. *Shore Protection Manual*, 2nd ed., 2 Vols., US Army Corps of Engineers, Coastal Engrg. Res. Cntr., US Govt. Printing Off., Washington, DC.

Thevenot, M. M., and Kraus, N. C. 1995. Longshore Sand Waves at Southampton Beach, New York, Observations and Numerical Simulation of Their Movement. *Mar. Geol.*, 126: 249-269.

Verhagen, H.J. 1989. Sand waves along the Dutch coast. *Coastal Engrg.*, 13: 129-147.

CHAPTER 312

EROSIVE WAVES IN SHORELINE CHANGE DUE TO THE REDUCTION OF A RIVER DELTA

Yoshito Tsuchiya, [1] M. ASCE

ABSTRACT: Using observations of long-term data of shoreline change during the reduction process of a river-mouth delta, space-time shoreline changes and their variations are clarified by the moving-average method. There clearly exist two modes of shoreline changes due to distinct types of erosive wave. The first mode is defined to be an erosive wave propagating down-coast as a diffusion phenomenon. The second was estimated from the variation from the erosive wave and defined to be an erosive wave propagating down-coast as a wave phenomenon. We suggest new terminology for these two kinds of erosive wave. The erosive waves of the first mode are subject to change as positive and negative erosive waves due to the initial and boundary conditions at the river mouth in relation to the sediment input from the river and local change in the submerged river delta. The erosive wave of the second mode propagates down coast faster than the first mode erosive wave. We therefore conclude that shoreline change due to beach erosion can be described as space-time change propagating down-coast as a diffusion phenomenon, upon which shoreline variation propagates as a wave phenomenon faster than the shoreline change. Additionally, a theory for the diffusion-wave phenomenon of shoreline change is attempted using a set of equations of longshore sediment transport rate and continuity of shoreline change to find the dispersion relation of erosive waves.

INTRODUCTION

In the past, it has been considered that shoreline change is a diffusion process, governed by a diffusion-type equation in a one-line theory. It has been observed that, in the reduction process of a river delta, severe shoreline change takes place in the vicinity of the cause of erosion, sometimes with beach collapse, and that shoreline change propagates down-coast like a wave phenomenon. In 1987, Inman defined the propagation velocity of an accretion or erosion wave in terms of the longshore sediment transport rate. By use of this method, Uda and Yamamoto (1994) recently estimated the velocity of a sandy body on the Mihono-matsubara sand spit. In understanding

---

[1]Professor Emeritus, Kyoto University and Professor, Faculty of Urban Science, Meijo University, 4-3-3 Nijigaoka, Kani 509-02, Japan

the shoreline change and establishing methodologies for its control, it is of course very important to recognize whether the shoreline change can be described as a diffusion or a wave phenomenon, or as a diffusion-wave phenomenon.

Recently, Tsuchiya (1995, 1996) studied erosive waves in shoreline change by analyzing long-term shoreline change data from the Shizuoka and Shimoniikawa coasts, using the method of moving average.

In this paper, we analyze data of long-term shoreline change along the Shizuoka Coast facing Suruga Bay by the method of moving average, to find the space-time change of the shoreline and its propagation speed. The narrow, sandy beach was formed on the collision coast due to there being a lot of sediment input from the Abe river. Due to a rapid decrease in sediment input, severe beach erosion and collapse began to take place near the river mouth in 1965. This extends in the direction of longshore sediment transport. The beach erosion can therefore be considered as a reduction process of the river delta. There clearly exist two modes of shoreline change. The first mode is space-time shoreline changes such as positive and negative erosive waves as diffusion phenomena. The second mode is an erosive wave propagating faster than the first mode shoreline change. We demonstrate that shoreline change due to beach erosion must therefore be considered as a diffusion-wave phenomenon.

Additionally, a theory of shoreline change is proposed from a set of equations for longshore sediment transport and continuity of shoreline change. We conclude that the shoreline change can be described by a diffusion-wave phenomenon. The dispersion relation is obtained, and the resulting propagation speed is discussed.

THE DATA

The Shizuoka Coast and Its General Features: Although depth sounding was first performed in 1969, regular surveys of shoreline change and soundings on the

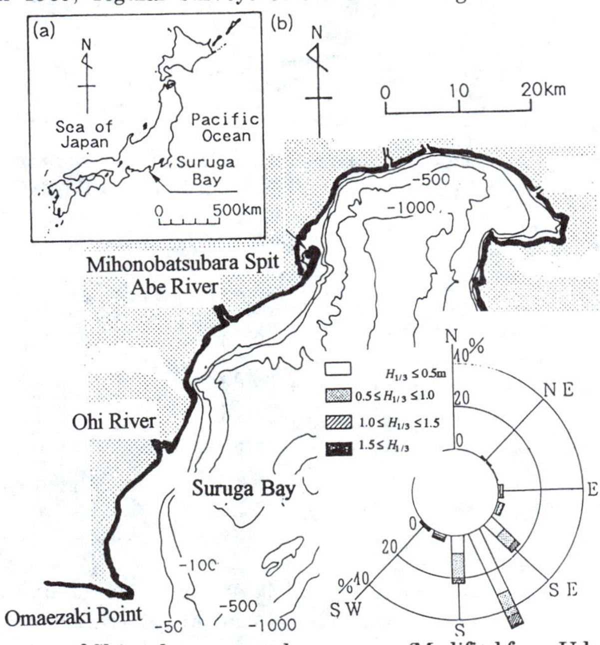

Figure 1. Location of Shizuoka coast and wave rose (Modified from Uda, 1994)

Shizuoka coast have been carried out annually since 1974 by the Shizuoka Prefectural Government.

In this paper, the measured data were used. The Shizuoka coast is one of the Pacific Ocean coasts of the Japanese islands, facing Suruga bay as shown in Figure 1. This coast is a typical collision and mountainous coast according to the coastal classification of Inman and Nordstorm (1971). The width of the coast is very narrow. The beach profiles are very steep and connect with the steep offshore slopes in Suruga bay. Because of the high rate of sediment input from the Abe river, a narrow sandy beach has been formed on the coast by typhoon waves from the south to southeast directions. At the northeast end of the coast, a sand spit called Mihono-batsubara sand spit is beautifully formed by the longshore sediment transport in the northeast direction. Near the sand spit, a sea canyon is located into which part of the longshore sediment is deposited.

Coastal Disasters and the Main Causes of Beach Erosion: Coastal disasters occurred due to the Isewan typhoon in 1959 and Typhoon 6626 in 1966, resulting in

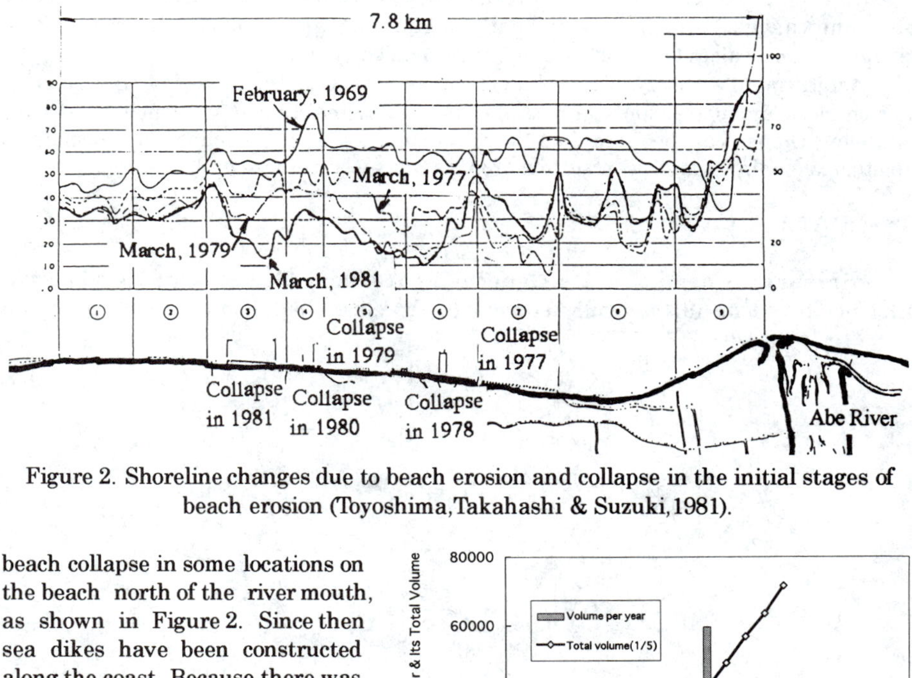

Figure 2. Shoreline changes due to beach erosion and collapse in the initial stages of beach erosion (Toyoshima,Takahashi & Suzuki,1981).

beach collapse in some locations on the beach north of the river mouth, as shown in Figure 2. Since then sea dikes have been constructed along the coast. Because there was remarkable beach erosion at that time, soundings and shoreline surveys have been carried out along the coast since 1969. The sea dikes were sometimes destroyed and the eroded area extended further along the north-east coast. Therefore offshore breakwaters were begun to be

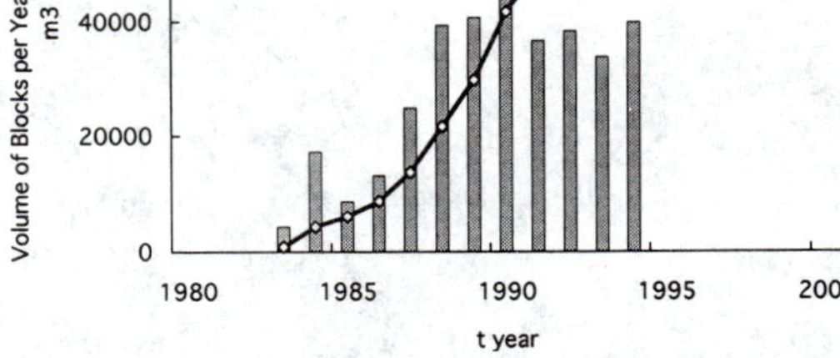

Figure 3. Annual change of total volume of concrete units used for offshore breakwaters.

constructed, and these are now completed along the whole coast. Figure 3 shows the total volume of concrete units used for the offshore breakwaters. In the Abe River sediment mining had been done widely and severely before 1968 for construction works, resulting in severe beach erosion. We therefore conclude that the main cause for beach erosion is due to the decrease in sediment input from the Abe River, and that the beach erosion can be explained as a reduction process of the river delta.

THE METHOD OF DATA ANALYSIS

Purpose of Data Analysis: Shoreline changes are usually measured at intervals of 100m or 200m. As shown in Figure 2, beach collapse takes place locally over several hundred meters, and extends down-coast as it propagates. The spatial magnitude of shoreline change due to beach erosion is wider and reaches sometimes a few kilometers. Therefore, we have to consider shoreline changes extending from several hundred meters up to several kilometers. The shoreline change data include local shoreline changes and some errors in the measurements. In order to clarify the shoreline changes more naturally than we have understood them, it may be effective to analyze the data using spatial smoothing methods.

Methods of Data Analysis: We have two traditional methods for smoothing fluctuating data; 1) the triangle-weighting or uniform-weighting method, and 2) the Fourier transform method. Shoreline changes take place not only due to beach erosion, but also by beach collapse under the action of wave concentration, and extend widely along the coast. In this paper, the uniform-weighting method was used. The number of points for weighting was determined by considering whether the phenomena in question can be clearly observed in the smoothed data. 11, 13 and 15 points were used for large-scale shoreline change as a diffusion phenomenon, and 3 points were used for variations from the large-scale shoreline change as a wave phenomenon. The applicability of the smoothing method was established by comparing the results from using different numbers of smoothing points. We conclude that, in the data measured at intervals of 100 m, 13 and 15 smoothing points are suitable for large-scale shoreline change, but 3 points for variations from the shoreline change.

THE CLASSIFICATION OF EROSIVE WAVES

Spatial and Temporal Shoreline Changes: Using the 15-point smoothing method, smoothed shoreline changes and their variations are shown in Figure 4 at intervals of two every years. In the figure there clearly exist two modes of shoreline change. The first mode, indicated by the thick solid curves, is the large-scale shoreline change. The second mode, indicated by the thin curves with marks, is the variation from the first-mode shoreline change. The large-scale shoreline changes show clearly the annual change of shoreline in spatial form from the river mouth of the Abe river. The shoreline change expands widely down-coast in the north-east direction. This shoreline change clearly demonstrates the total shoreline change under the condition of beach erosion, which we understand as a diffusion phenomenon. This shoreline change propagation may correspond to the accretion or erosion wave proposed by Inman (1987). Referring to the cause of beach erosion, which is a decrease in sediment input from the Abe river, this total shoreline change can be understood to be a reduction process of the river delta. However, the boundary condition for the shoreline change at the river

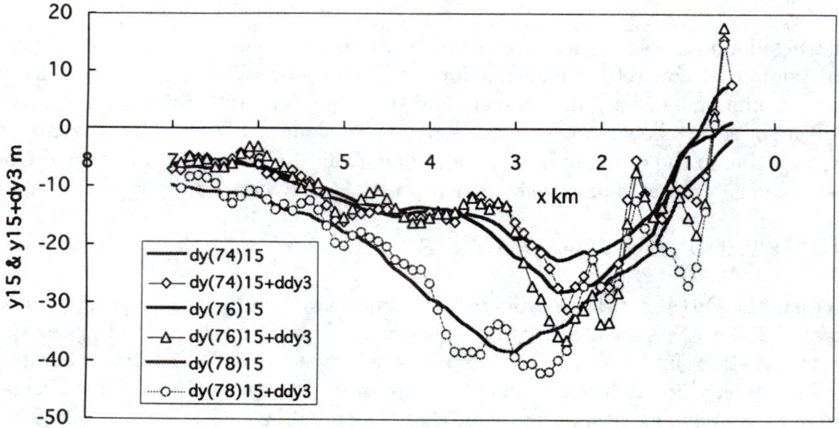

(a) 1974 to 1978.

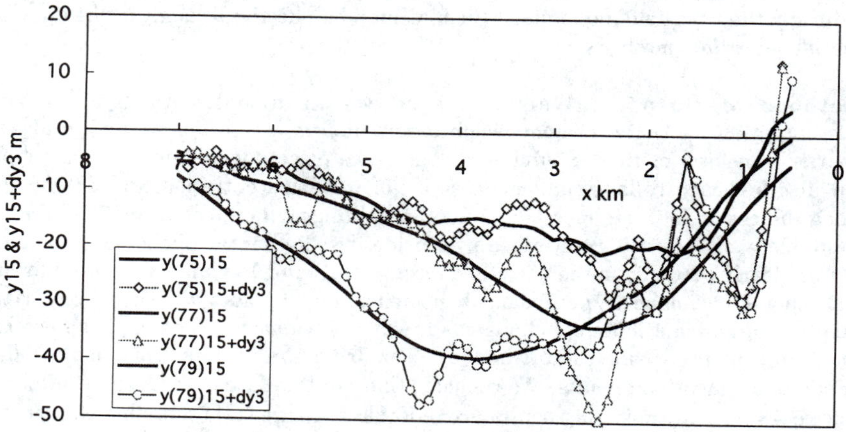

(b) 1975 to 1979.

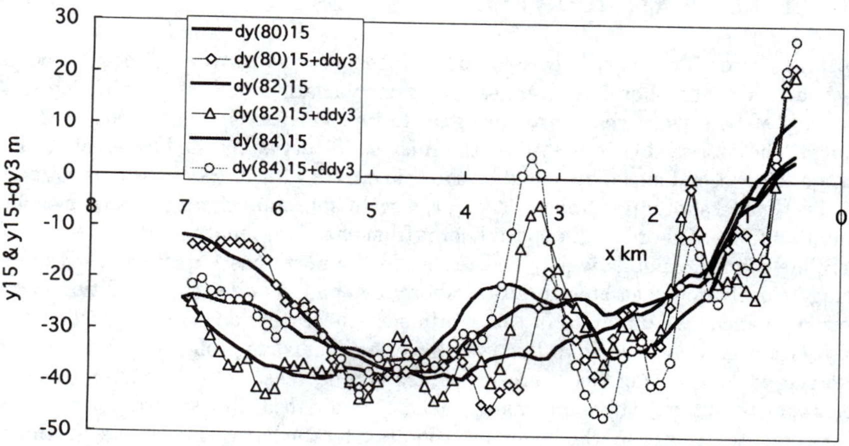

(c) 1980 to 1984.

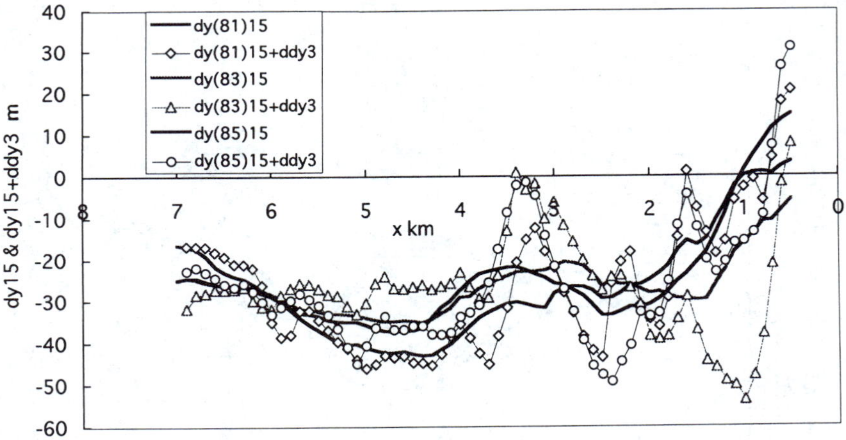

(d) 1981 to 1985.

Figure 4. Total shoreline changes with erosive waves (W) down-coast from the river mouth.

mouth will be discussed in relation to the generation of erosive waves, as well as the decrease in sediment input from the river.

Erosive Waves: The thick curves shown in Figure 5 illustrate annual changes in total shoreline change, for which I define an *erosive wave as a diffusion phenomenon*. The erosive wave propagates down-coast, reducing in amplitude gradually. In this case, it has a wave crest defined by a minimum in the spatial shoreline change, and a trough defined by a maximum. Note the fact that the erosive wave is composed of two elementary waves such as positive and negative erosive waves. The positive erosive wave propagates accompanying the negative erosive wave. Further, note that these two waves were generated by different events at the river mouth, probably rapid decrease and increase in sediment input from the river, or their resulting local changes

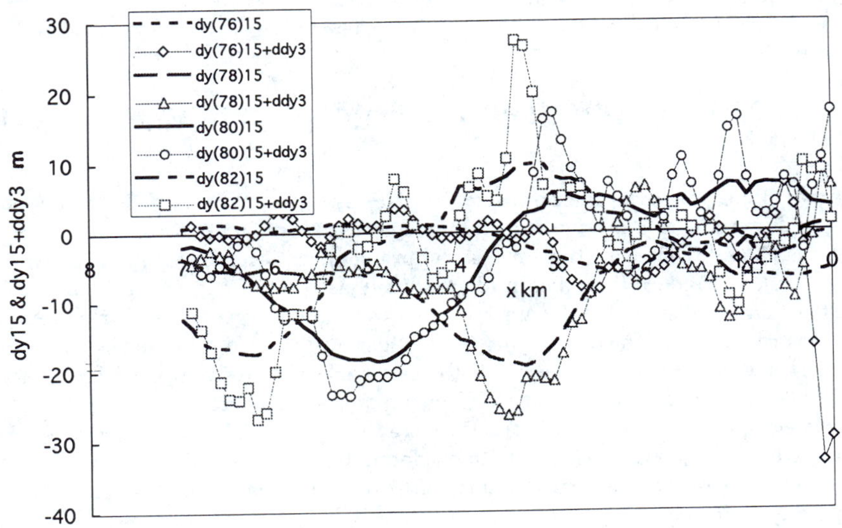

(a) 1976 to 1982.

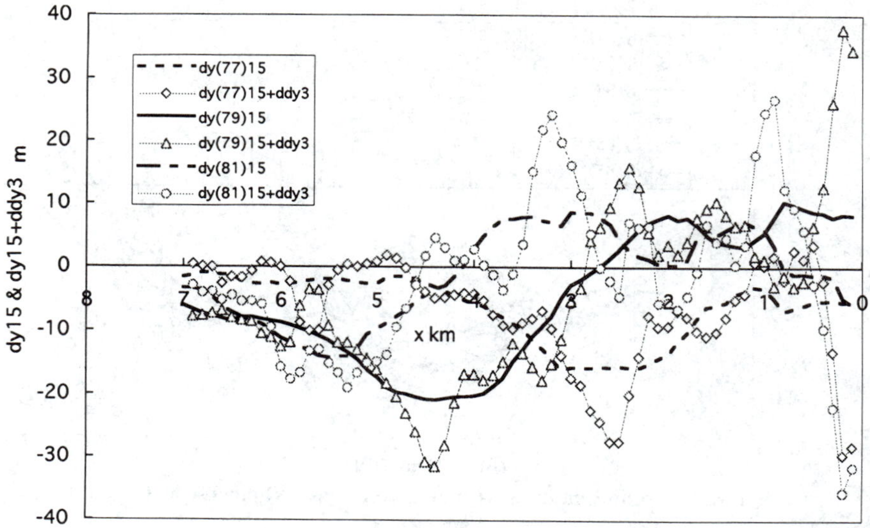

(b) 1977 to 1981.

Figure 5. Spatial profiles and propagation of erosive waves (DP) and (DN) with erosive wave (W) before the construction of offshore breakwaters.

in the submerged river delta. Therefore, I define the erosive wave propagating as a diffusion phenomenon as *erosive wave (D)*, and the positive and negative erosive waves as *erosive wave (DP)* and *erosive wave (DN)*, respectively.

When the erosive waves are specified at the river mouth as the initial and boundary conditions for shoreline change, we may predict the total shoreline change using a common shoreline change model. Actually, beach profiles change in relation to shoreline changes around the river delta, so that the application of the model may need further consideration. The thin curves shown in Figure 6 indicate the variations from the erosive wave (D), for which I define *an erosive wave as a wave phenomenon*. This erosive wave propagates down-coast on the erosive wave (D). Thus, I define it as *erosive wave (W)*. The crest and trough of the erosive wave (W) are denoted as erosive wave (Wc) and (Wt), respectively.

THE EROSIVE WAVE(D) AND ITS PROPAGATION AS A DIFFUSION PHENOMENON

Spatial Profiles and Propagation of the Erosive Waves (D): As previously seen in Figure 5, this erosive wave (D) propagates down-coast accompanying erosive waves (DP) and (DN). Note that the spatial profiles of erosive waves (D) have nearly the same form, but reduce their amplitude gradually as a diffusion phenomenon, until the construction of offshore breakwaters. The erosive waves generated during and after the construction are shown in Figure 6, the thick curves with solid marks indicate the erosive wave (D), and the thin curves with open marks the erosive wave (W) propagating on the erosive wave (D). Note that during this period erosive waves (DN) were generated, and then erosive waves (DP) were then generated. Both the erosive waves propagate slowly down-coast in the same form, but their amplitudes clearly tend to decay down-coast. This fact may be due to the offshore breakwater effect of trapping of longshore sediment.

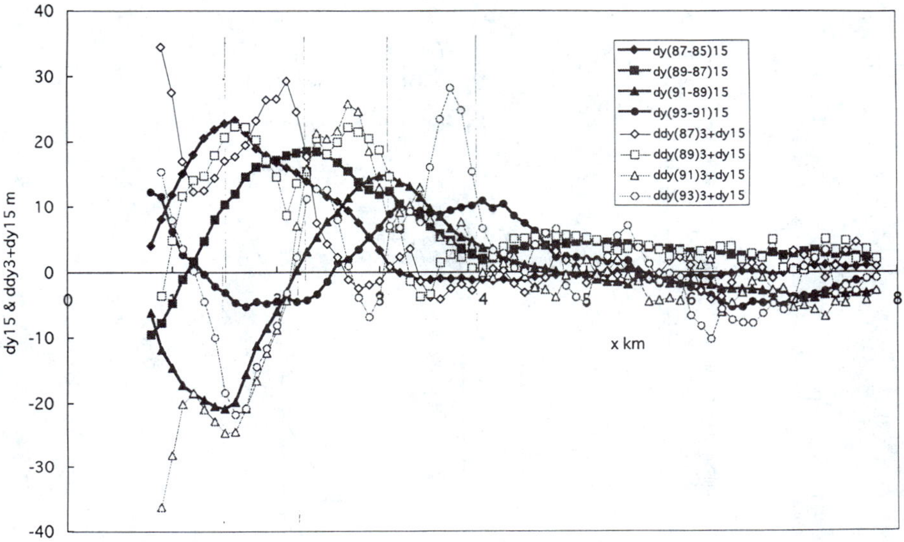

(a) 1987 to 1993.

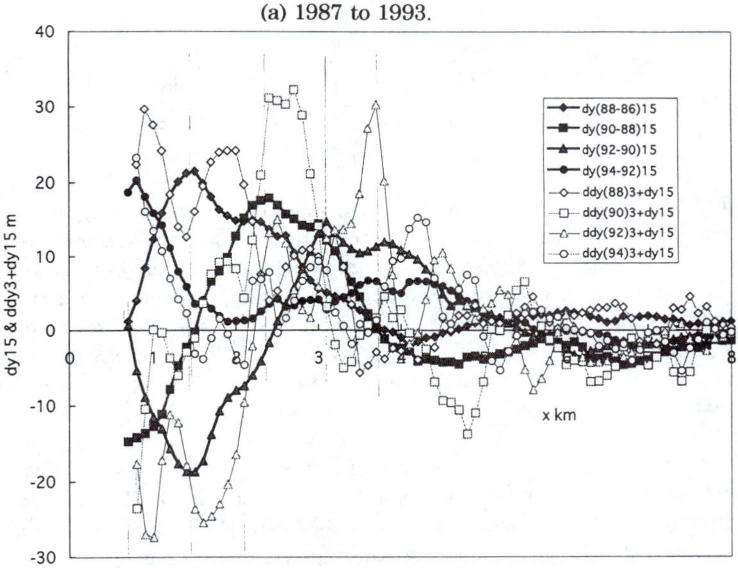

(b) 1988 to 1994.

Figure 6. Spatial profiles and propagation of erosive waves (DP) and (DN) with erosive wave (W) during and after the construction of offshore breakwaters.

Propagation Speeds of Erosive Waves (DP) and (DN): The amount of data for the erosive waves is limited. Therefore, by use of the successive spatial profiles corresponding to the crests of erosive waves (DP) and (DN), the positions of the crests can be determined. Figure 7 shows the time-change of the crest positions where the straight line with solid diamond and square marks indicate respectively erosive waves (DP) generated before and during/after the construction of offshore breakwaters. The propagation speeds of erosive waves are obtained from the gradient of the straight lines.

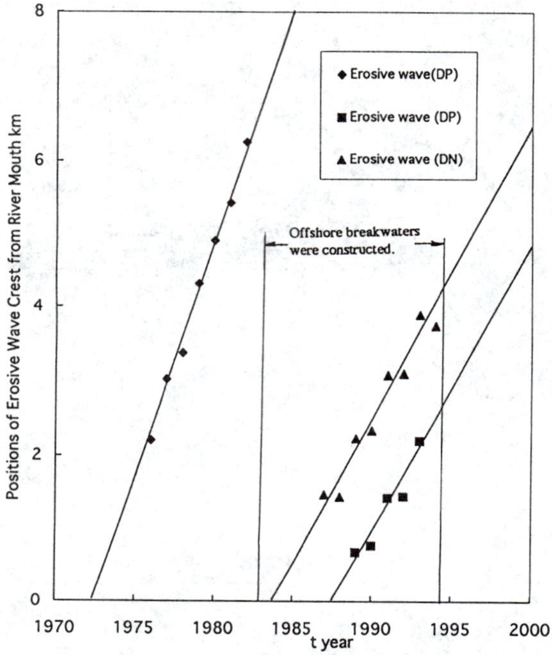

Figure 7. Propagation speed of erosive waves (DP) and (DN) before and during/after the construction of offshore breakwaters.

Note that the propagation speed of erosive waves generated during and after is less than that before the construction. Further, note that the propagation speed of erosive waves (DP) and (DN) are the same. This may explain the fact that the spatial profiles shown in Figure 6 are decaying gradually down-coast, but maintain almost the same form. If the straight line of erosive wave (DP) is extended, the time when the erosive wave was generated is found to be around the year 1972. Alternatively, the times when erosive waves (DP) and (DN) were generated during and after the construction are found to be around the years 1983 and 1987, respectively.

Possible reasons for generation of erosive waves (DP) and (DN): As previously discussed the main cause of beach erosion is due to the decrease in sediment input from the Abe river. Uda, Misawa and Matsui (1996) recently qualified this fact considering river bed changes due to mining of sand and gravel in the river. They conclude that the river bed had been lowered within the alluvial reach near the river mouth during mining until 1968, but changed to deposition of sediment during the period between 1970 and 1973. It therefore seems that within the period, there was no sediment input into the beach, resulting in either the direct generation of the erosive wave(DP) shown in Figure 7, or changing the submerged river delta which may have generated erosive waves. They also described that after the generation of the erosive wave no remarkable floods have been experienced, but a huge flood occurred in 1982, of which the maximum discharge was 3,860 m^3/sec. It also seems that the flood generated the erosive wave(DN), and that variations in the submerged river-mouth delta in relation to floods might generate erosive waves .

THE EROSIVE WAVE(W) AND ITS PROPAGATION AS A WAVE PHENOMENON

Erosive Wave (W) and Its Propagation: Figures 8 and 9 show erosive waves (W) generated before and during/after the construction of offshore breakwaters respectively. A careful inspection shows that the erosive waves (W) propagate down-coast, with some exceptions. Note that the wave amplitudes show their maximum values near the river mouth, and reach over 30 m. Further, note that they decay in their propagation down-coast. It seems that this tendency is the same both before and during/after the construction of offshore breakwaters.

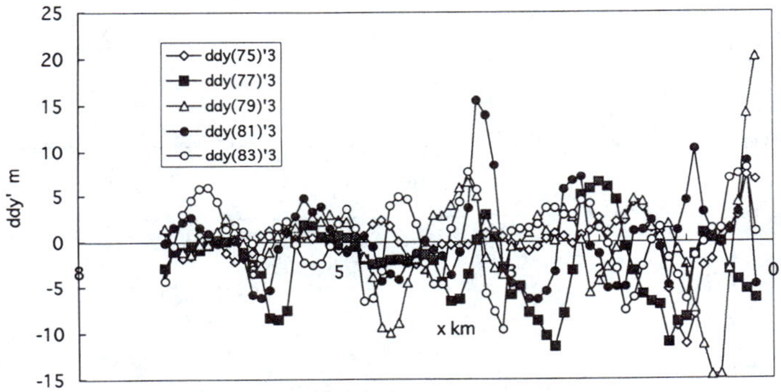

(a) 1975 to 1983.

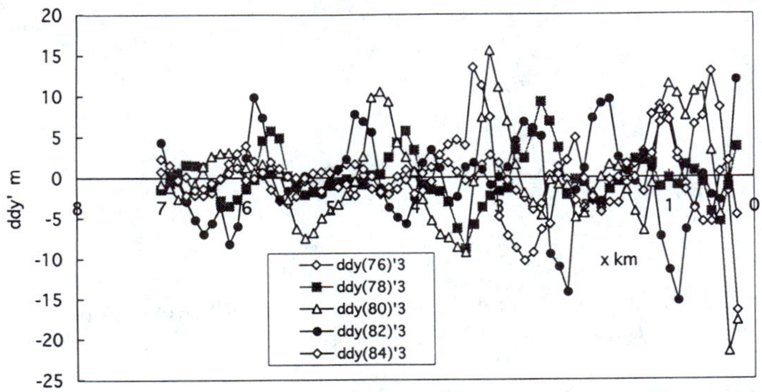

(b) 1976 to 1984.

Figure 8. Erosive waves (W) and their propagation, with wave amplitudes damping down-coast, before the construction of offshore breakwaters.

Wavenumber Spectrum of Erosive Waves (W): The amount of data is not sufficient for calculating wave number spectra of erosive waves (W) accurately. There is also a decay of their wave amplitudes. These may prevent the spectrum from being calculated accurately. However, we tried to estimate them using Mathematica to examine whether a peak wavenumber exists. Figure 10 shows some examples of the spectra. The spectra are not well formed, but there exists a peak wavenumber. The wavelength corresponding to the peak wave number was calculated. Figure 11 shows change in the wave length. It is nearly constant, being about 500 m until 1985, and thereafter varies and increases. This may be due to the trapping effect of longshore

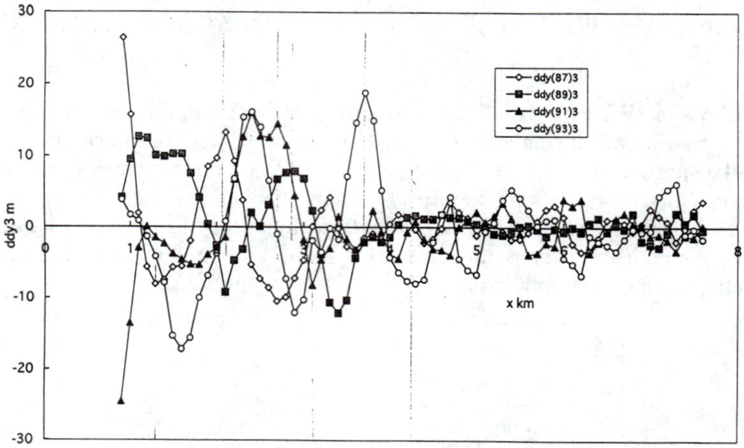

(a) 1987 to 1993.

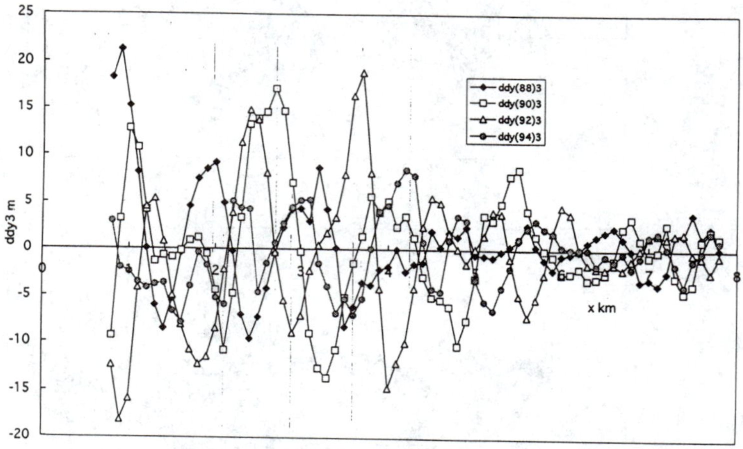

(b) 1988 to 1994.

Figure 9. Erosive waves (W) and their propagation, with wave amplitudes damping down-coast, during and after the construction of offshore breakwaters.

sediment transport by offshore breakwaters.

Propagation Speed of Erosive Waves (W): From successive crests of erosive waves (W) which were selected from the spatial profiles shown in Figures 8 and 9, the positions of the crests were determined. Figure 12 illustrates time changes of the crest positions of erosive waves (W) where the straight line with solid circles indicates the crest position of erosive wave (DP) and those with open squares and triangles mark the crest and trough positions of erosive waves (Wc) and (Wt), respectively. The propagation speeds of the crest and trough of erosive wave (W) are nearly same, and faster than that of erosive wave (DP). We therefore conclude that the erosive wave (W) propagates faster than the erosive wave (DP). This fact may explain why beach collapse or severe erosion takes place locally at some locations down-coast with the sandy beach being eroded.

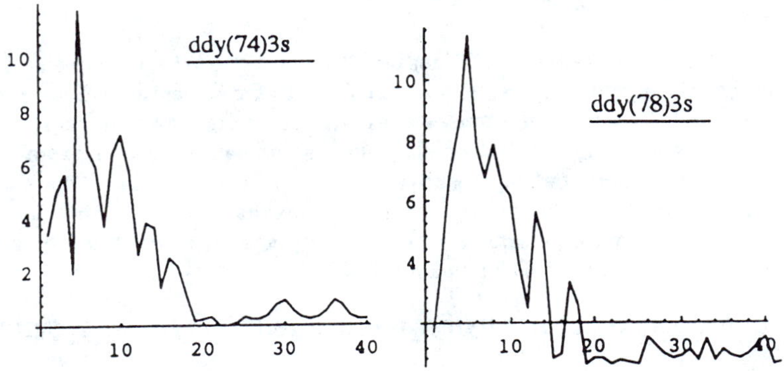

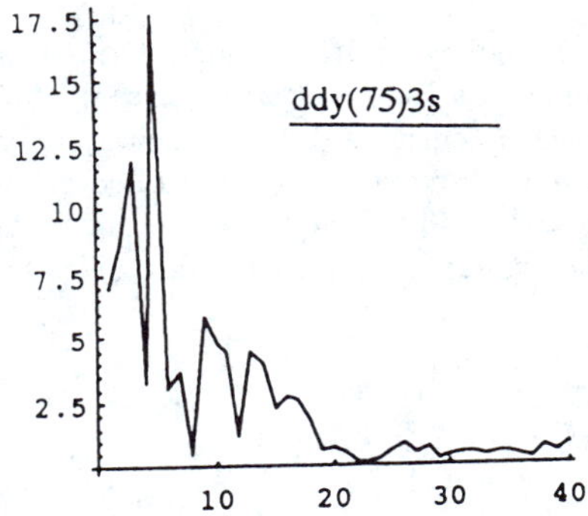

Figure 10. Some examples of wave-number spectra of erosive waves (W).

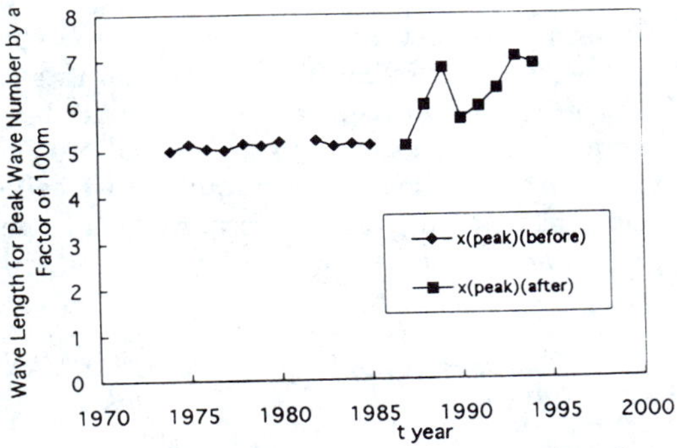

Figure 11. Changes in wavelength corresponding to the peak wavenumber.

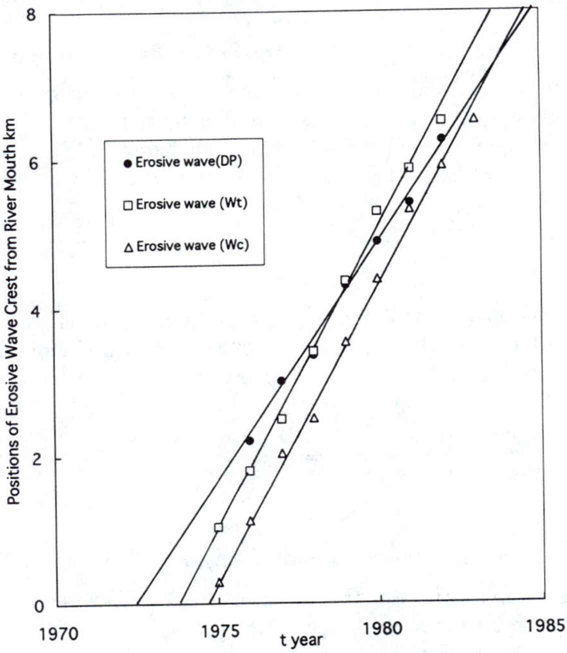

Figure 12. Propagation speed of erosive wave (W) compared with those of erosive waves (DP) and (DN).

Possible Reasons for Generation of Erosive Waves (W): Where there exists a wave phenomenon, there must be a generation force to produce the wave. Such generation forces do exist, such as incoming waves with some frequency of occurrence, but what the restoring forces are should be clarified. As will tentatively be explained later, the local change of shoreline may result in changing the total rate of longshore sediment transport. This may generally result in restoring the shoreline change. Therefore, once a local shoreline change has taken place, the shoreline change propagates down-coast, probably decaying in amplitude.

THE EROSIVE WAVES AS A DIFFUSION-WAVE PHENOMENON AND THEIR PROPAGATION

So far a set of governing equations for shoreline change, composed of the equation of continuity for shoreline change, the total rate of longshore sediment transport and the geometrical relation of shoreline change has been used to derive a diffusion equation for shoreline change. In order to study a natural phenomenon dynamically, a set of both the equations of motion and continuity must be established. The equations of longshore sediment transport rate and continuity of shoreline change are approximately given by (Refaat and Tsuchiya, 1991)

$$\frac{\partial Q}{\partial t}+a\frac{\partial}{\partial x}\left(\frac{Q}{h_b}\right)^2=bgh_b^2\left(\sin 2\alpha_{b0}-2\frac{\partial y_0}{\partial x}\right)-e\sqrt{\frac{g}{h_b}}Q \tag{1}$$

$$\frac{\partial y_0}{\partial t}+\frac{1}{(1-\lambda)h_k}\frac{\partial Q}{\partial x}=0 \tag{2}$$

where y_0 is the shoreline position from the datum line, h_b the breaker depth, h_k the threshold depth of sediment movement, g the acceleration of gravity, λ the porosity of the bottom sediment, and a, b and e are practically constants. The second term in the first term on the right side of (1) demonstrates the restoring force of shoreline change in the wave phenomenon. The linearization of (1) and (2) results in

$$\frac{1}{e}\sqrt{\frac{h_b}{g}}\frac{\partial^2 y_0}{\partial t^2}+\frac{\partial y_0}{\partial t}-\frac{\sqrt{2b}}{e(1-\lambda)}\frac{h_b^2\sqrt{gh_b}}{h_k}\frac{\partial^2 y_0}{\partial x^2}=0 \tag{3}$$

which is the linearized wave-diffusion equation for shoreline change. Neglecting the first term the second and third terms reduces to the usual diffusion equation of shoreline change. The dispersion relation of (3) is given by

$$\frac{(\sigma/k)^2}{gh_b}=2b\left\{\frac{h_b}{(1-\lambda)h_k}\right\}\frac{1}{1-e\sqrt{g/\sigma^2 h_b}} \tag{4}$$

The term $g/\sigma^2 h_b$ in (4) describes the diffusion term effect on the dispersion. When the term vanishes (4) demonstrates the wave celerity in the wave equation which can be obtained by neglecting the second term. The wave celerity shows that the wave celerity of erosive waves is proportional to the long-wave velocity evaluated from the breaker depth and the square root of the ratio of breaker depth to threshold depth of sediment movement.

CONCLUSION

By the use of data on long-term shoreline change in the reduction process of the river delta on the Shizuoka coast, we found that there exist two modes of shoreline change, namely shoreline change as a diffusion phenomenon and variation from shoreline change as a wave phenomenon. We name these erosive waves, which are defined respectively as erosive wave (D) and erosive wave (W) for the two modes of shoreline change. We conclude that the erosive wave (D), the first mode of shoreline change, expands down-coast as a diffusion phenomenon; but the erosive wave (W), the second mode of shoreline change, propagates down-coast as a wave phenomenon faster than the erosive wave (D). The shoreline change in the river delta coast can then be described as a diffusion-wave phenomenon.

The generation of the erosive waves was considered in terms of the sediment input from the river and local change in the submerged river-mouth delta. Finally, a theory for the diffusion-wave phenomenon of shoreline change was attempted using a set of equations of longshore sediment transport rate and continuity of shoreline change. From these the dispersion relation of the phenomenon was found. Further theoretical investigations are now being made to establish a predictive method for shoreline change.

Acknowledgments

I thank Mr. K. Ishida, Chief, Planning Section, Technical Division, Suruga Konsaru Ltd. for his cooperation in the collection of long-term shoreline change data at the Shizuoka coast. Part of this investigation was supported both by Grants-in-Aid for Scientific Research (No. 07555164) from the Ministry of Education, Science, Sports & Culture of Japan, and Research Fund (No. 7-1-4-25) of the Foundation of River and Watershed Environment Management.

References

Inman, D. L. (1987): Accretion and erosion waves on beaches, Shore and Beach, July-October, 61-66.

Refaat, H.E.A.A. and Y. Tsuchiya (1991): Formation and reduction processes of river deltas; Theory and experiments, Bul. DPRI, Kyoto Univ., Vol. 41, 177-224.

Toyoshima, O., S. Takahashi and I. Suzuki (1981): On beach erosion of the Shizuoka coast, Proc. Coastal Eng., JSCE, Vol. 28, 261-265 (in Japanese).

Tsuchiya, Y. (1995): Shoreline change due to beach erosion as a wave phenomenon (1) -The Shizuoka coast-, Proc. Coastal Eng., JSCE, Vol. 42, 551-555 (in Japanese).

Tsuchiya, Y. (1996): Shoreline change due to beach erosion as a wave phenomenon (2) -The Shimoniikawa coast-, Proc. Coastal Eng., JSCE, Vol. 42, 586-590 (in Japanese).

Uda, T., T. Suzuki, M. Oishi, Y. Yamamoto and N. Itahashi (1994): Beach change around a sand spit -in special reference to Mihonomatsubara sand spit-, Proc. Coastal Eng., JSCE, Vol. 41, 536-540 (in Japanese).

Uda, T., H. Misawa and H. Matsui (1996): On the initiation of sand body movement in Shizuoka coast, Proc. Coastal Eng., JSCE, Vol. 43, 581-585 (in Japanese).

CHAPTER 313

DELILAH, DUCK94 & SandyDuck: Three Nearshore Field Experiments

William A. Birkemeier, M. ASCE, Charles E. Long and Kent K. Hathaway[1]

Abstract

Two major field experiments have recently been conducted, and a third is being planned, at Duck, NC, along the mid-Atlantic coast of the USA, to investigate nearshore dynamic processes. Named DELILAH, DUCK94, and SandyDuck, these experiments take advantage of logistical efficiency and relatively uncomplicated, open-coast field conditions provided by the Field Research Facility of the US Army Engineer Waterways Experiment Station. DELILAH occurred in 1990, and emphasized hydrodynamics measurements. DUCK94, in 1994, added measurements of sediment transport and morphologic evolution, and was planned as a comprehensive pilot study for SandyDuck, the most ambitious experiment in the series, scheduled for 1997. Scientific motivation, instrumentation plans, participants, and representative climatic conditions of DELILAH and DUCK94 are described, as are sources of further information and data.

Introduction

Since 1979, the Coastal and Hydraulic Laboratory (formerly the Coastal Engineering Research Center) of the US Army Engineer Waterways Experiment Station has hosted a series of increasingly complex, multi-investigator, multi-agency nearshore field experiments at its Field Research Facility (FRF) located in Duck, North Carolina, USA. Two recent experiments, *DELILAH* and *DUCK94*, and a planned third experiment, *SandyDuck*, all evolved from scientific and pragmatic successes of prior work at this site, and have the basic objectives of improving fundamental understanding and modeling of surf zone physics. The emphasis in DELILAH was surf zone hydrodynamics in the presence of a changing barred bathymetry. DUCK94 and SandyDuck have added components to resolve sediment transport and morphologic evolution at bedform scales from ripples to nearshore bars. DUCK94 was designed as a pilot effort to test instruments and procedures required for the more comprehensive *SandyDuck* experiment. The purpose of this paper is to summarize the two completed experiments, including participants, environmental

---

1) Authors from US Army Engineer Waterways Experiment Station, Coastal and Hydraulic Laboratory, Field Research Facility, 1261 Duck Road, Kitty Hawk, NC 27949

conditions, data collected, and data availability, and thereby suggest the plan for the future experiment.

DELILAH

The DELILAH (***D**uck **E**xperiment on **L**ow-frequency and **I**ncident-band **L**ongshore and **A**cross-shore **H**ydrodynamics*) design (Figure 1) was to make surf zone hydrodynamics measurements to augment directional wave measurements being made in 8-m water depths by the FRF and in 13-m depths during the concurrent SAMSON (***S**ources of **A**mbient **M**icro-**S**eismic **O**cean **N**oise*) experiment (Herbers & Guza, 1994; Nye & Yamamoto, 1994). By agreement among investigators, DELILAH objectives were:

> *To measure the wave- and wind-forced three-dimensional nearshore dynamics with specific emphasis on infragravity waves, shear waves, mean circulation, set-up, runup, and wave transformation, and to monitor bathymetric response to these processes.*

Table 1 lists DELILAH investigators, their organizations at the time of the experiment, and general areas of scientific interest.

The DELILAH array of in situ instruments (Figure 1) consisted of a primary

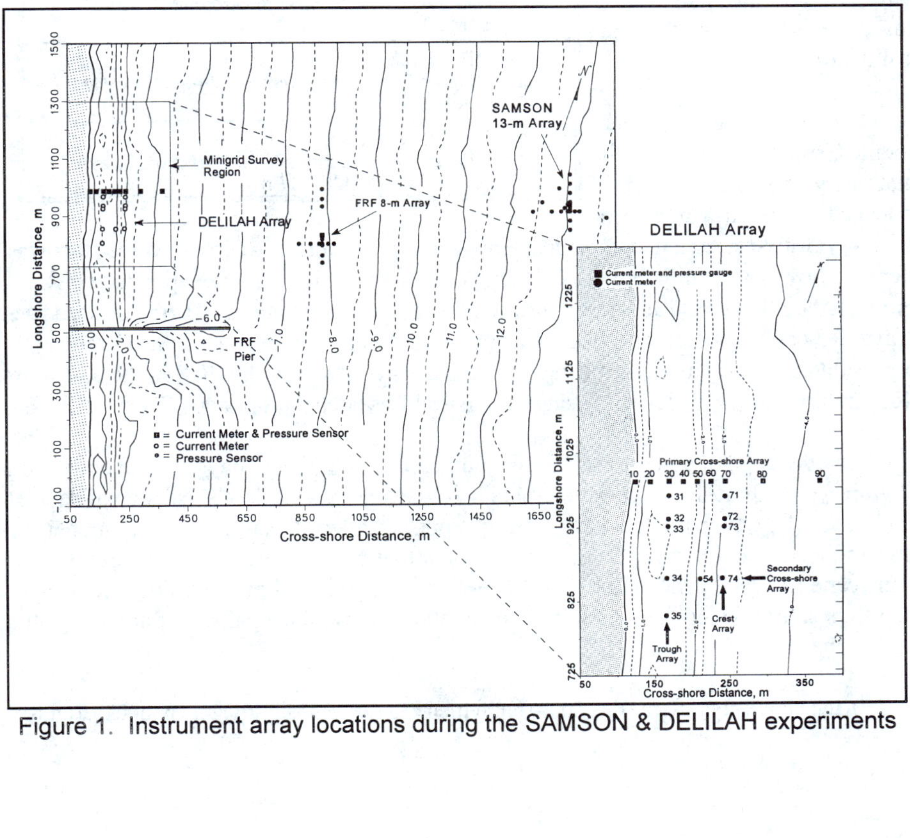

Figure 1. Instrument array locations during the SAMSON & DELILAH experiments

cross-shore array of nine current meters to obtain mean and fluctuating currents, with collocated pressure gauges to enable directional wave measurements. A secondary cross-shore array of three current meters provided long-shore redundancy. Longshore coverage was provided by an array of six current meters in the nearshore trough, and a second array of five current meters located just seaward of the inner bar crest. Instruments were provided by the Naval Postgraduate School (NPS), Scripps Institution of Oceanography (SIO), and the FRF. The DELILAH array was designed by E. B. Thornton, J. M. Oltman-Shay, and P. A. Howd, with Dr. Thornton having primary responsibility for data collection.

Table 1. DELILAH Participants
US Army Engineer Waterways Experiment Station William Birkemeier - morphology Kent Hathaway - runup, infragravity waves, morphology, remote sensing Charles Long - incident and reflected directional wave spectra Nicholas Kraus - longshore currents Jane Smith - vertical current profile, 2-D circulation Todd Walton - runup Edward Thompson - surf beat modeling
Naval Postgraduate School Edward Thornton, Katie Scott - mean circulation, shear waves, rip currents
Naval Research Laboratory Dennis Trizna - radar measurements of waves and currents
Oregon State University Rob Holman - morphology, runup, shear waves, infragravity waves, video remote sensing Peter Howd - infragravity waves, morphology Tom Lippmann - morphology, infragravity waves, video remote sensing Todd Holland (also with the US Geological Survey) - runup, cusp formation
Scripps Institution of Oceanography Robert Guza - cross-shore and longshore currents, infragravity waves, wave transformation, (also SAMSON PI)
Northwest Research Associates Joan Oltman-Shay - shear waves, infragravity waves (also SAMSON PI)
Washington State University Steve Elgar - incident and reflected directional wave spectra (also SAMSON PI)

Additional instruments augmented information from the stationary, in situ array. A mobile sled equipped with a vertical array of five current meters, a pressure gauge, and a surface-piercing wave staff was deployed by J. M. Smith and K. K. Hathaway. Seven video cameras operated by R. A. Holman recorded swash and ocean surface images in the surf zone. D. B. Trizna used five Naval Research Laboratory (NRL) radar systems for remote sensing of waves, currents, and bathymetry. Bathymetric changes were measured daily by the FRF in the 550-m by 400-m minigrid area (Figure 1) using the Coastal Research Amphibious Buggy (CRAB).

A wide range of conditions were encountered during the 21 days of data collection (Figure 2). A short-duration wave event on 1-2 October was followed by several days of low waves ($H_{mo} < 1$ m). On 9-12 October, a "southeaster" built waves of about 2-m height, and induced north-flowing longshore currents that peaked near 1.5 m/s. Passage of Hurricane Lili well offshore on 12-14 October resulted in 2.5-m swell under low-wind conditions. Following Lili, moderately energetic waves and currents continued until the end of the experiment.

In the above conditions, nearshore bathymetry underwent significant changes. Figure 3

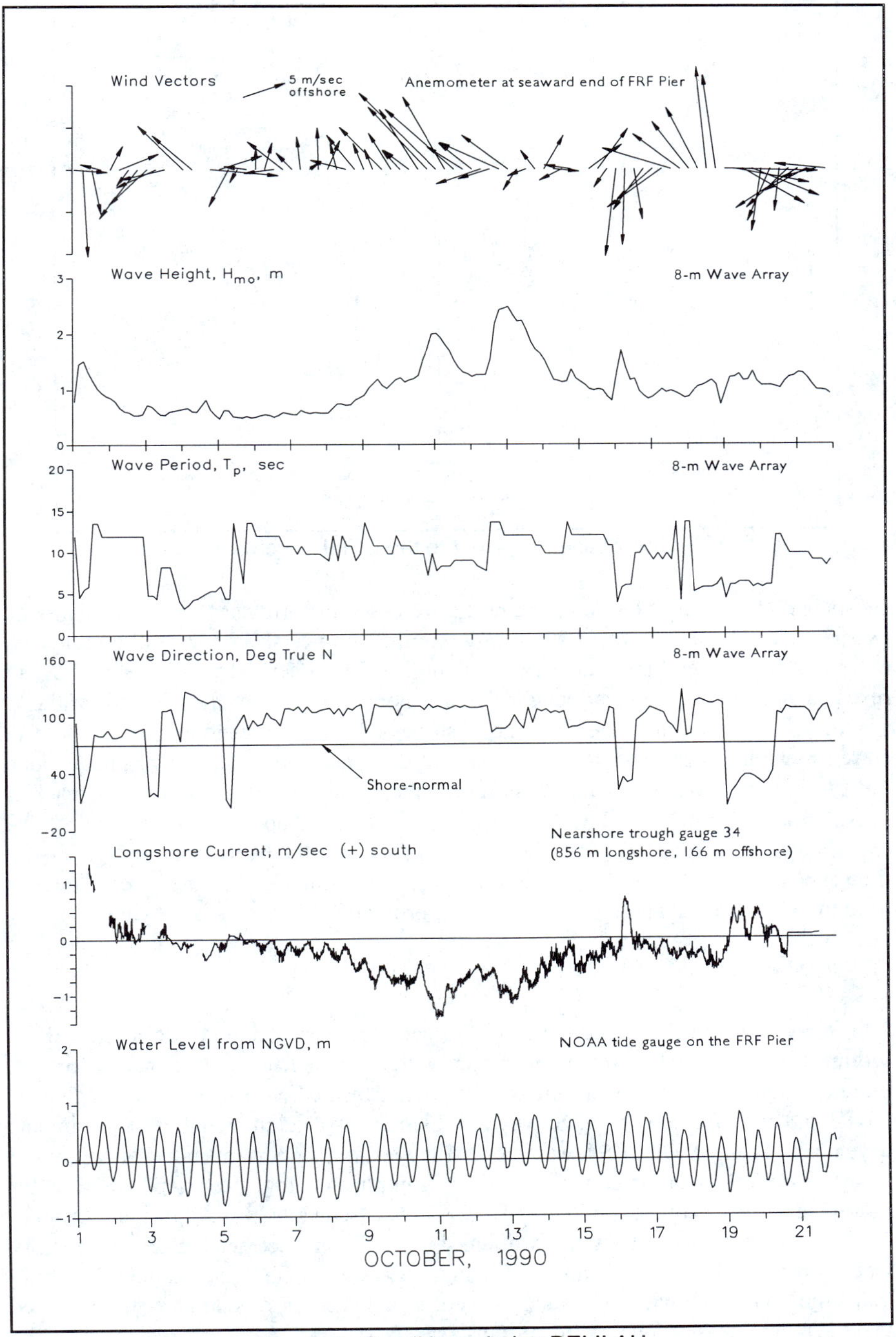

Figure 2. Conditions during DELILAH

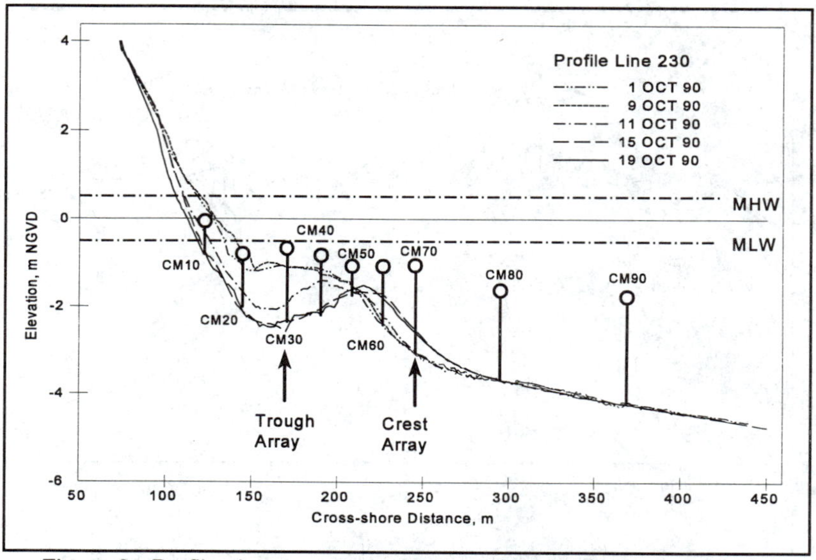

Figure 3. Profile change adjacent to the DELILAH cross-shore array

shows bathymetric profiles along the primary cross-shore array at five times during the experiment, illustrating the offshore migration of the nearshore bar, and notably the extensive deepening of the nearshore trough on 10-13 October. The CRAB surveys also revealed the presence of megaripples in the nearshore trough and on the seaward bar flank. Figure 4 illustrates four of the 20 minigrid surveys. Vertical lines in this figure indicate current meter locations. Bar topography undulated rhythmically in the alongshore direction until 9 October, when it began to assume a more uniform, linear shape. It reached maximum uniformity by 11 October, and retained this shape through the end of the experiment. Well correlated with shoal parts of the nearshore bathymetry in Figure 4 are time exposures of breaking waves patterns shown in Figure 5. Such remote images are used to extend spatial and temporal coverage provided by measured bathymetry.

DUCK94

Success of DELILAH, and the evident need for more detailed information about sediment transport and morphologic evolution that results from hydrodynamic forcing, initiated interest in further field work to be supported by the US Army Corps of Engineers, ONR, and the US Geological Survey. A plan for two additional field experiments developed. The first, DUCK94, was intended as a test run for new instrumentation, a more formal experiment organization, and more complicated logistics in preparation for SandyDuck, the second experiment. DUCK94 was scheduled for August and October 1994 to take advantage of the synergy offered by the National Science Foundation's Coastal Ocean Processes (CoOP) experiment (Butman, 1994), being conducted at the FRF during that time. The following focus topics were established as fundamental to improved understanding of surf zone sediment transport:

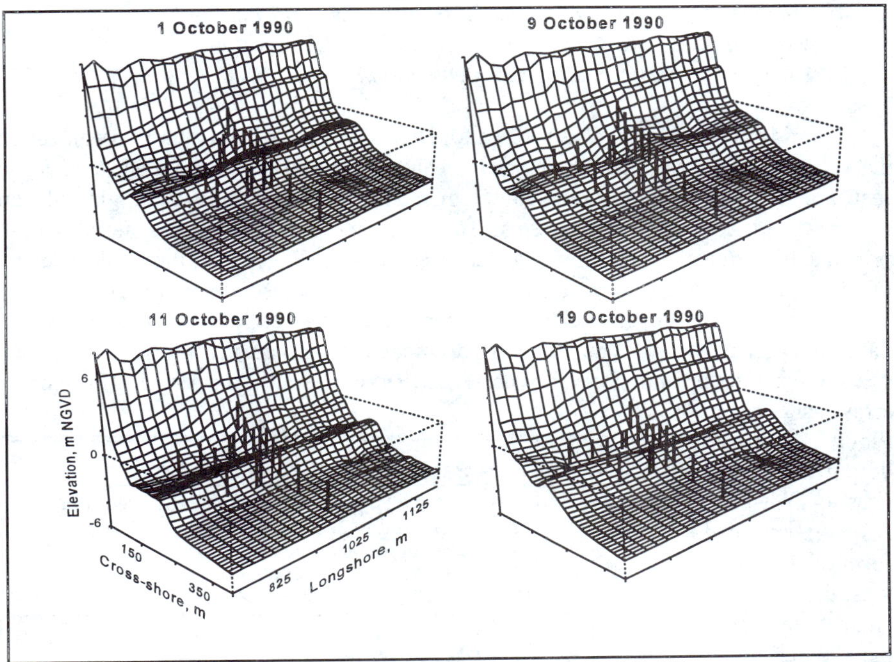

Figure 4. *Minigrid* evolution during DELILAH

Figure 5. Time-exposure video images showing DELILAH morphology.

a. small and medium scale sediment transport and morphology;
b. wave shoaling, wave breaking, and nearshore circulation;
c. swash processes including sediment motion.

Considerable interest was expressed for DUCK94. Table 2 lists the 19 organizations that conducted 31 experiments involving more than 100 scientists, students, and technicians. Instrument measurements were complemented by observations from ground- and aircraft-based radar and video systems. Table 3 lists the 31 basic studies, along with the principal investigators, their primary focus areas, and experiment durations. The extensive instrumentation resulted from consideration of relevant measurement scales required to address SandyDuck science objectives. Guidance was provided by using measured velocity data from DELILAH and sediment transport modeling. Based on this analysis, a general nearshore instrumentation array was designed (Birkemeier & Thornton, 1994). The full array, shown in Figure 6, was used during the October phase of DUCK94. An abbreviated form of this array was used in the August segment of the experiment. Formal dates for DUCK94 were 8-24 August and 1-24 October, though some investigations (Table 3) of various durations were underway between June and November.

Table 2. DUCK94 Participating Organizations

Agencies	1	US Army Engineer Waterways Experiment Station
	2	United States Geological Survey
	3	Office of Naval Research
	4	Naval Research Laboratory
	5	Naval Postgraduate School
Universities	6	Dalhousie University (Canadian)
	7	Duke University
	8	Oregon State University
	9	North Carolina State University
	10	Memorial University of Newfoundland (Canadian)
	11	Scripps Institution of Oceanography
	12	University of Delaware
	13	University of East-Anglia (United Kingdom)
	14	University of Florida
	15	University of Miami
	16	University of Washington
	17	Washington State University
Companies	18	Areté Associates
	19	Neptune Sciences, Inc.

A wide variety of instrumentation was used in DUCK94. Conventional total-station surveying techniques were used in subaerial morphology studies (29, referring to investigations by experiment number in Table 3), minigrid surveys (15), and positioning of all stationary instruments. Central to the main layout were cross-shore arrays of instrument clusters (11), each containing an electromagnetic current meter, a pressure gauge, an acoustic altimeter, and a thermometer (Fedderson, et al.,1997). The altimeters permitted the first comprehensive real-time measurements of bottom changes (8) (Gallagher, Elgar & Guza, 1997). A large number of suspended sediment concentration gauges were deployed, including optical backscattering sensors (16, 22, 26, 30), and less intrusive fiber-optic backscattering sensors (1). The Coherent Acoustic Sediment Probe (Stanton & Thornton, 1997) was mounted on a mobile sled along with current meters, pressure gauges, scanning sonars, and void fraction sensors (20, 25, 26). The Sensor Insertion System,

Table 3. DUCK94 Experiments
(number in parens refers to organization from Table 2)

No.	Investigators	Experiment Title	Participating Months	Wave Shoaling	Nearshore Circulation	Boundary Layers	Swash Processes	Small Scale Sediments	Meso/Macro Morphology	Water Properties
1	Beach(8), Holman, Sternberg	Sediment dynamics in the nearshore environment	Aug,Oct		X	X		X		
3	Church(4), Elgar, Guza	Mine scour, burial, and migration as a function of wave and current forcing	Sep				X			
4	Drake(9), Smith	Nearshore sedimentary structures	Aug,Oct					X		
5	Dugan(17)	Airborne remote sensing of the environment in the littoral zone	Oct	X	X					
6	Earle(18)	Real-time buoy directional wave measurements for driving surf zone numerical models	Aug,Oct	X						
7	Earle(18), Walsh, Boyd	Scanning radar altimeter sea surface topography & high resolution directional wave measurements	Oct	X						
8	Elgar(16)	Temporal and spatial variability of the bathymetry of a natural beach	Aug,Oct						X	
10	Graber(14), Shay, Haus	An investigation of surface currents and internal waves over the inner and mid-shelf	Oct	X	X					
11	Herbers(5), Elgar, Guza, O'Reilly	Surface gravity waves and nearshore circulation	Aug,Oct	X	X					
12	Haines(2), Gelfenbaum	Vertical structure of mean currents & turbulent stresses in the nearshore boundary layer	Aug,Oct		X	X				
13	Hanes(13), Vincent	Near bed intermittent suspension	Aug,Oct		X			X		
14	Hanes(13)	Remote video measurement of mesoscale nearshore processes	Aug,Oct				X		X	
15	Hathaway(1), Leffler	Rip current mapping and minigrid surveys	Aug,Oct		X				X	
16	Hay(6), Bowen	Sediment suspension, local morphology, and bubbles	Oct		X			X	X	X
17	Holman(8), Holland, Plant	Foreshore dynamics	Aug,Oct				X			
18	Howd(7), Hathaway	Processes of shoreface profile adjustment	Aug,Oct		X				X	
19	Jensen(1)	Evolution of wave spectra in shallow water	Aug,Oct	X						
20	Lippmann(11), Thornton, Stanton, Su	Spatial distribution of wave breaking and turbulence	Aug,Oct	X		X				X
21	Long(1)	Wind wave frequency-direction spectral measurements	Aug,Oct	X						
22	Miller(1)	Longshore sediment transport during storms	Aug,Oct					X		
23	Fabre(19), Wilson, Earle	Wave and surf generated ambient noise measurements	Aug,Oct							X
24	Stauble(1), Smith, Birkemeier	Sediment dynamics and profile interactions sampling experiment	Aug,Oct					X		
25	Thornton(5), Dingler	Small-scale morphology in the nearshore	Aug,Oct					X	X	
26	Thornton(5), Stanton	Suspended and bedload sediment transport	Aug,Oct		X			X		
27	Trizna(4)	Radar remote sensing of nearshore processes: bar morphology, directional wave spectra, infragravity waves, wave breaking	Aug,Oct	X	X				X	
28	Walker(4)	Hyperspectral optical characterization of surf zone bottom/resuspended sediment	Aug					X		X
29	Werner(10), Elgar	Swash zone morphology: field manipulation and simulation	Jun,Sep				X		X	
30	White(1)	Field tests of sediment transport theories	Aug,Oct		X			X		
31	Livingston(3), Wolf, Pasewark	Wave and surf noise measurements: supplementation	Oct							X

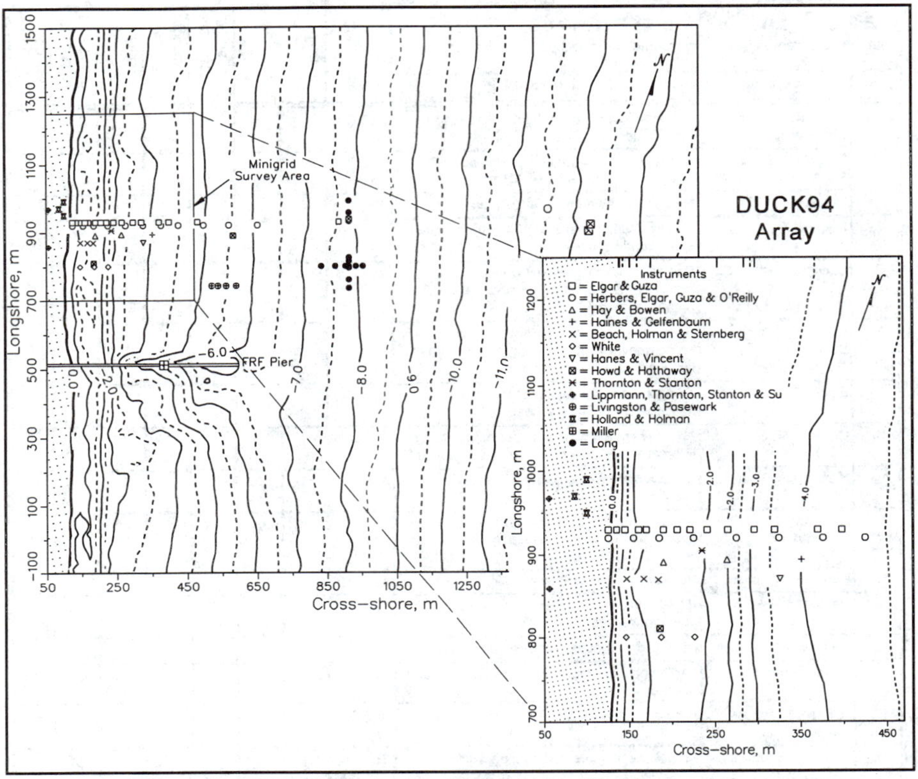

Figure 6. DUCK94 Instrument layout identified by investigator

located on the FRF pier, provided a stable, mobile platform for sediment transport measurements during high-energy conditions (1, 22). In situ (16) and CRAB-mounted (25) side-scan sonars provided observations of bottom bedforms, including megaripples. Most array positions included one or more current meters (1, 3, 11, 12, 13, 15, 16, 18, 22, 26, 30). Incident wave conditions were monitored with directional wave buoys (6, 19), and a direction-sensing array of pressure gauges (21).

Dynamics measurements were complemented by a series of geologic studies that included surface sediment samples (24) (Stauble & Cialone, 1997), short cores, box cores, and vibracores (4). Several remote sensing systems were used. Surf zone and swash processes were observed with tower-mounted video systems (14, 17, 20). Observations were also made with land-based marine radar systems (27), coherent radar systems (10), airborne synthetic aperture radar, topographic lidar, visible and hyperspectral light imaging, and scanning radar altimetry (5, 7, 28). Three studies examined fundamental nearshore acoustic behavior (16, 23, 31).

Environmental conditions during the October phase of DUCK94 are illustrated in Figure 7. Two high-wave events occurred. The first was on 2-4 October, wherein wave

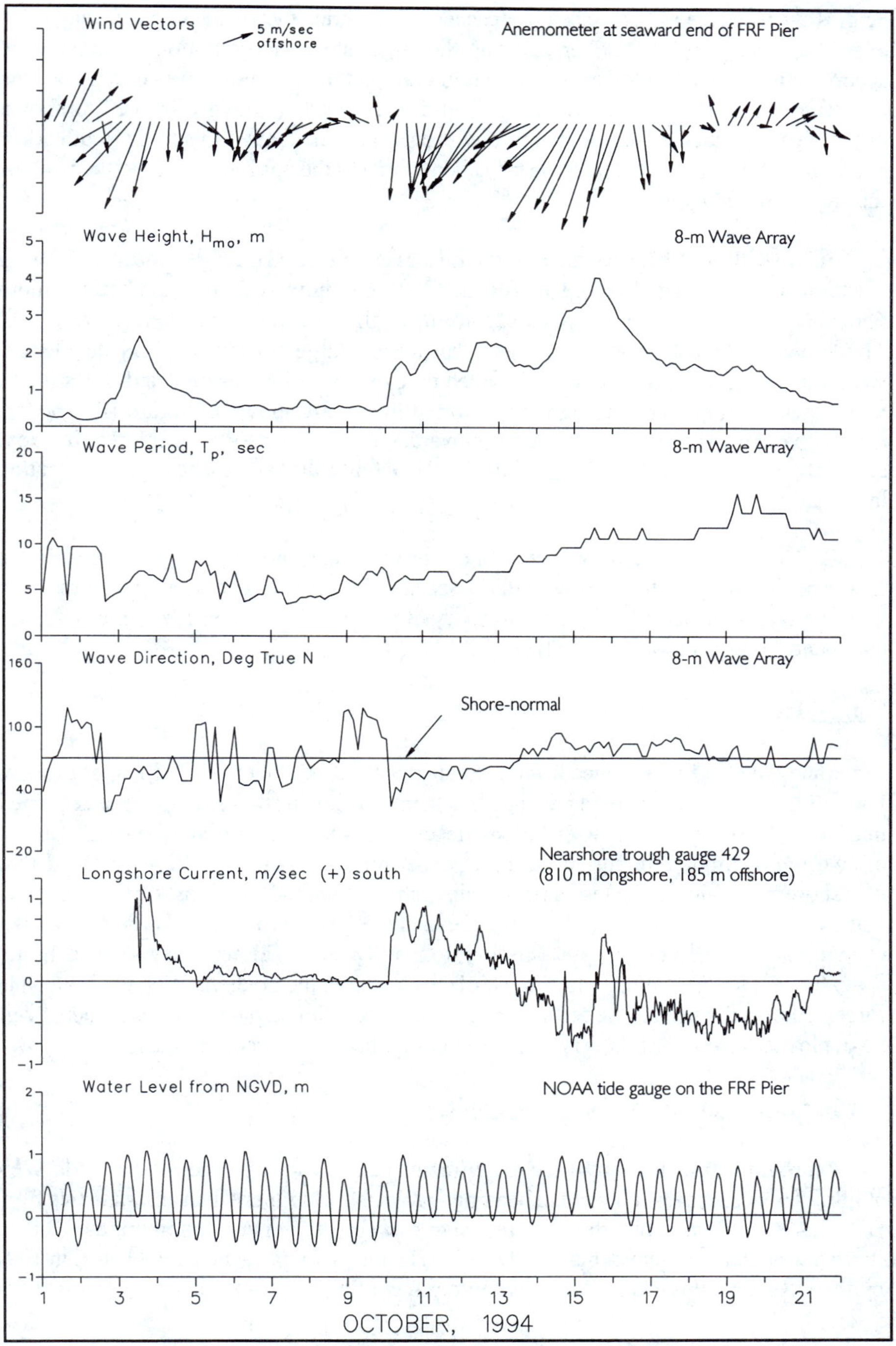

Figure 7. Conditions during the October phase of DUCK94

heights exceeded 2.5 m. Wave heights reached 4.5 m during the second storm, an eight-day event beginning on 10 October. During the larger storm, large bottom changes were accompanied by a complex nearshore circulation pattern wherein wave-driven currents in the surf zone were opposed by strong wind-driven longshore flows offshore. As shown in Figure 7, currents in the nearshore trough changed from about 1 m/s to the south at the beginning of the storm on 10 October to about 1 m/s to the north just prior to the peak of the storm on 15 October.

Figure 8 illustrates four of the 12 minigrid surveys collected during October. Following a pattern similar to that observed in DELILAH, the bar moved offshore and became more linear in the initial part of the 10 October storm. High waves prevented daily surveys until 21 October, when the survey revealed that a very large rip channel had developed. Evolution of this channel is evident in video time exposure images depicted in Figure 9. Sequences of profile data through the region of the rip are shown in Figure 10, where it is seen that the bar crest moved 100 m seaward, causing 1.2 m of deposition at its most seaward observed location on 18 October. By 21 October, the bar crest had begun migrating landward.

DUCK94 data are being analyzed, and research results are beginning to appear in the literature. Preliminary findings were discussed at a post-experiment meeting (summarized by Long & Sallenger, 1995), where adequacy of the DUCK94 experiment plan was also evaluated in preparation for SandyDuck.

SandyDuck

SandyDuck will take place from 22 September to 31 October 1997. Most of the core DUCK94 experiments are being repeated, with improvements based on experience gained in DUCK94, both in keeping with the basic tenets of physics research, and to take advantage of two major improvements in the basic experiment design. DUCK94 revealed that nearshore dynamics is far less uniform alongshore than had previously been assumed. Consequently, instruments will be added to expand longshore coverage of currents, bottom changes, and sediment transport. Missing from all Duck experiments has been accurate, spatially detailed measurements of sea surface elevation, the gradient of which is an O(1) force in the surf zone. As the second change in the experiment plan, new instruments will be deployed to resolve this very important component of nearshore dynamics.

Further Information and Data Availability

More information about these experiments can be found on the World Wide Web at http://frf.wes.army.mil under the heading “projects.” Summary data and statistics from DELILAH are available through the above web site, or via anonymous FTP at ftp://frf.wes.army.mil/pub/delilah. A DELILAH summary report will be published in 1997 by the US Army Engineer Waterways Experiment Station, Vicksburg, MS.

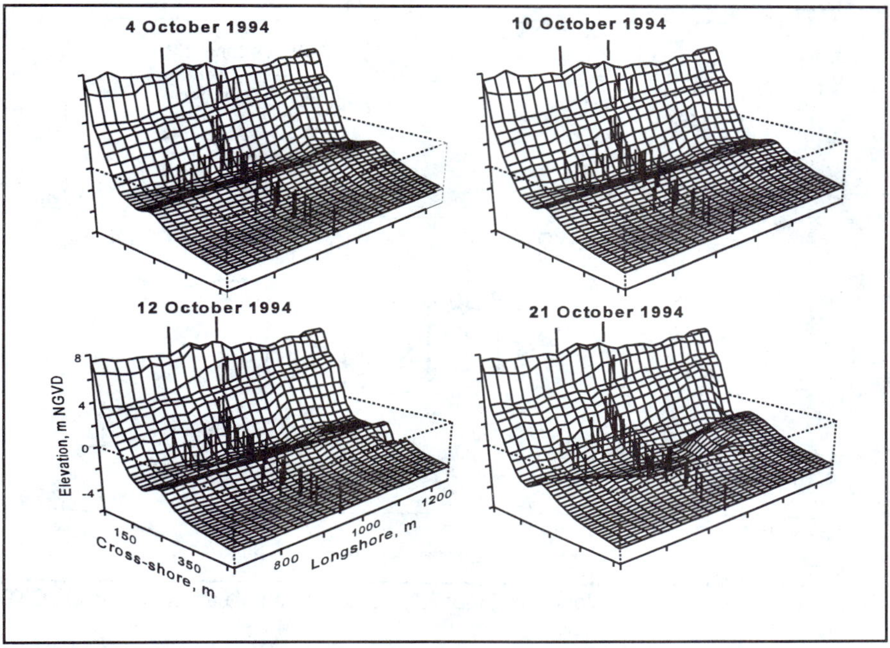

Figure 8. Minigrid evolution during DUCK94

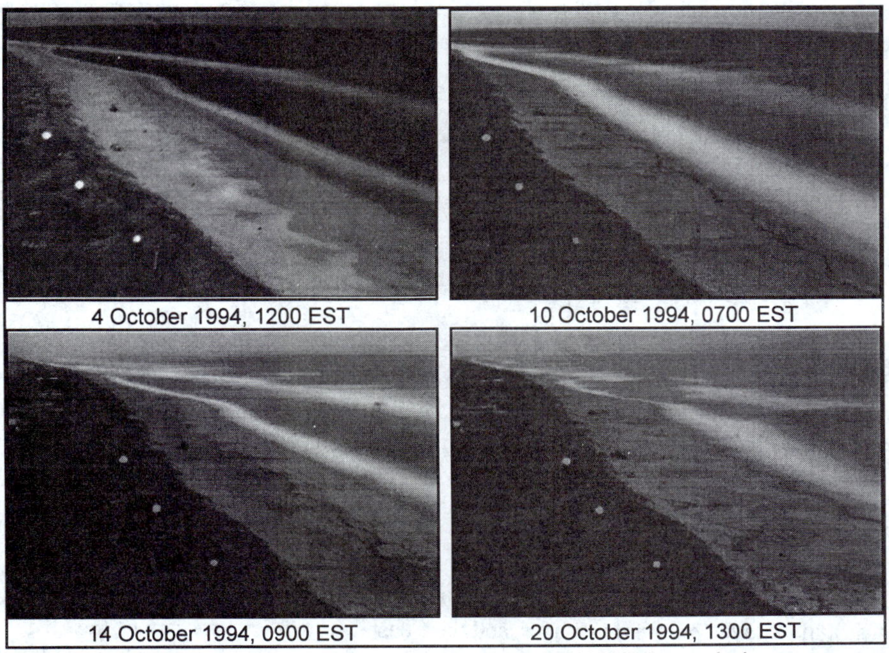

Figure 9. Time-exposure images showing DUCK94 morphology

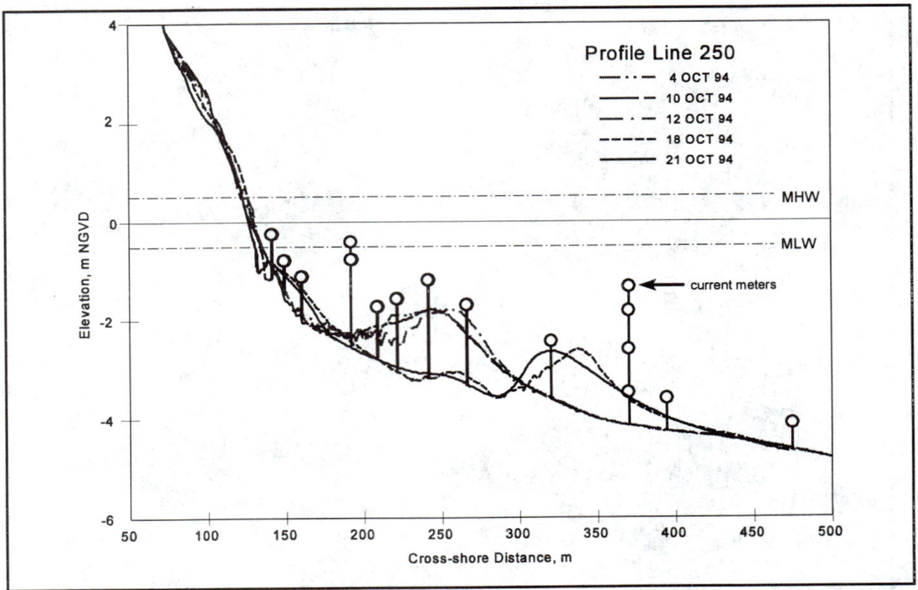

Figure 10. Profile change along the Elgar & Guza cross-shore array during DUCK94

Data from DUCK94 are not yet generally available. The investigators have agreed on a data sharing policy that offers protection of data by collecting investigators, encourages collaboration, and provides for eventual public release. This policy is:

a. global release of all data three years after the experiment;
b. responsible investigators will be identified when data sets are used by others;
c. prior to three-years, data shared by agreement between individual investigators;
d. any manuscript based on shared data must be approved by all responsible investigators prior to submission;
e. no third-party data dissemination;
f. principal investigators control use of their data.

An extensive discussion of the DUCK94 experiments, including tables listing sensors, data sets, and a summary of results, findings and publications, is available through the above web site. It is anticipated that DUCK94 data will become generally available late in 1998, and that SandyDuck data will be released near the turn of the century.

Acknowledgments

This paper was funded through the Field Research Facility Analysis Work Unit of the Coastal Research and Development Program of the US Army Corps of Engineers, Mrs. Carolyn Holmes, program manager, and the Coastal Dynamics Program of the Office of Naval Research, Dr. Thomas Kinder, program manager. We wish to acknowledge the large group of investigators, students, staff, committee members and others who dedicated incredible amounts of time and effort in pursuit of these experiments. Permission to publish was granted by the office Chief of Engineers.

References

Birkemeier, W. A. and Thornton, E. B., 1994, "The DUCK94 Nearshore Field Experiment," Proceedings of the Conference on Coastal Dynamics '94, 815-821.

Butman, C. A., 1994, "CoOP: Coastal Ocean Processes Study," Sea Technology, 35:1, 44-49.

Fedderson, F., Guza, R. T., Elgar, S., and Herbers, T. H. C., 1997, " Cross-shore Structure of Longshore Currents during DUCK94", Proceedings of the 25th International Conference on Coastal Engineering, Orlando, FL, ASCE.

Gallagher, E. L., Elgar, S., and Guza, R. T., 1997, "Observations and Predictions of Sand Bar Motion," Proceedings of the 25th International Conference on Coastal Engineering, Orlando, FL, ASCE.

Herbers, T. H. C., and Guza, R. T., 1994, "Nonlinear Wave Interactions and High-frequency Seafloor Pressure," Journal of Geophysical Research, Vol. 99, No. C5, 10,035-10,048.

Long, C. E. and Sallenger, A., 1995, "Experiment at Duck, N.C. Explores Nearshore Processes," EOS Transactions of the American Geophysical Union, 76:49.

Nye, T. and Yamamoto, T., 1994, "Concurrent measurements of the directional spectra of microseismic energy and surface gravity waves," Journal of Geophysical Research, 99:C7, 14321-14338.

Stanton, T. P. and Thornton, E. B., 1997, "Reynolds Stress and Small-Scale Morphology Measurements during DUCK94," Proceedings of the 25th International Conference on Coastal Engineering, Orlando, FL, ASCE.

Stauble, D. K., and Cialone, M. A., 1997, "Sediment Dynamics and Profile Interactions: DUCK94," Proceedings of the 25th International Conference on Coastal Engineering, Orlando, FL, ASCE.

CHAPTER 314

EULERIAN MEAN VELOCITIES UNDER NON-BREAKING WAVES ON HORIZONTAL BOTTOMS

Peter Nielsen[1] & Zai-Jin You[2]

Abstract

A model is presented for the Eulerian time-mean velocities in combined wave-current flows. The typical, measured profiles of weak currents are well modelled. This includes the "toe" of forward drift in the wave boundary layer and the very flat (as opposed to the parabolic shape of laminar models) current profile between the bottom boundary layer and the wave trough. The model relies on a local force balance like Longuet-Higgins' (1953) diffusion solution rather than on advective influence from the end conditions. This seems justified by experiments, probably because a sufficient amount of turbulent diffusivity is present even for very weak currents. In its present form the model can only handle currents which are so weak that their influence on the wave motion is negligible. However, agreement with measurements of stronger currents can be obtained by modification of the wave Reynolds stress to account the influence of the current on the wave motion. Application to surf zone conditions also require modifications, but the essential structure of the model is globally applicable.

Introduction

Several authors eg. van Doorn & Godefroy (1978), van Doorn (1981), Kemp & Simons (1982), Kaaij & Nieuwjaar (1987), Kampen & Nap (1988), Heiboer (1988), Villaret & Perrier (1992) and Klopman (1994) have found that Eulerian mean velocities under progressive, non breaking waves show different distributions from what has been expected. Following currents reach a maximum value and subsequently decrease towards the surface. See Figure 1. The trend is found for all current strengths as long as the time averaged bed shear stress is in the direction of wave propagation $\left(d\bar{u}/dz > 0 \text{ for } z \to 0^+\right)$ but is not found for currents in the opposite direction. Opposing currents show the oppostive anomaly. That is, compared with a logarithmic profile, the opposing current grows more rapidly towards the surface.

---

[1] Reader, Department of Civil Engineering, University of Queensland AUSTRALIA 4072, Fax +61 7 3365 4599, [2] Victorian Institute of Marine and Freshwater Resources.

Thus, observations of both following and opposing currents contrast with the existing simple wave current boundary layer models, e g, Grant and Madsen (1979), Fredsoe (1984), Coffey & Nielsen (1986) or You (1994) which all predict logarithmic, i e, monotonically increasing current velocities above the wave boundary layer.

It is obvious from Figure 1 that the depth averaged current velocity $<\overline{u}>$ may be significantly overpredicted by the old type of models for following currents. The elevation at which the current maximum occurs depends on the relative strength of the current. For very weak currents, eg. flume experiments with zero net flow, it occurs very close to the bed, i e, below or near the top of the wave boundary layer. For stronger currents, it occurs closer to the mean water surface. This trend was clearly shown by the measurements of Kemp and Simons (1982). The data in Figure 1 were obtained over a fixed bed but numerous profiles over sand beds, measured by Heijboer (1988) and Villaret & Perrier (1992) show the same trends.

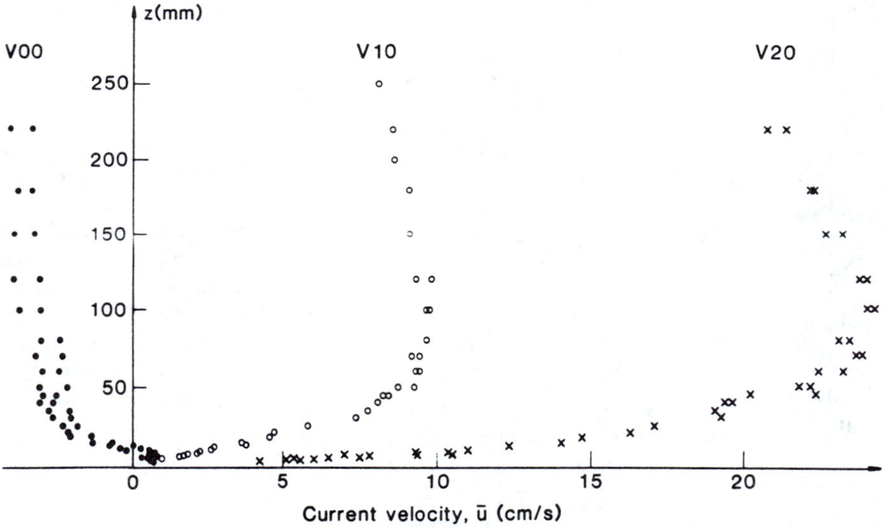

Figure 1: Eulerian mean velocities measured under non breaking waves. van Doorn (1981) test V00, V10, V20. In all tests the wave parametres were (*D,T,H,*)=(*0.3m, 2.0s, 0.12m*), and the bed roughness was artificial ripples with Nikuradse roughness $r = 2.1cm$.

The current maximum is not seen in velocity profiles from U-tube experiments like those of van Doorn (1983). It was therefore suggested by Nielsen (1992) that the profile type in Figure 1 can be understood in terms of the wave Reynolds stress $-\rho\overline{\tilde{u}\tilde{w}}$ which exist under a progressive wave with a bottom boundary layer but not in the U-tubes. The tilde indicates periodic components with zero mean, corresponding to the definition $u(t) = \overline{u} + \tilde{u}(t) + u'(t)$.

Figure 2 shows corresponding examples of following and opposing currents superimposed on the same waves (same period and same wave height) measured by Klopman (1994). Klopman's data show very clearly the difference in profile shape caused by the wave Reynolds stress $-\rho\overline{\tilde{u}\tilde{w}}$. Wave basin measurements by Havinga (1992) with angles of 60°, 90° and 120° between current and wave propagation show intermediate stages of this phenomenon.

The following outlines a model of current profiles in the presence of waves which takes $-\rho\overline{\tilde{u}\tilde{w}}$ into account explicitly in order to explain the shape of the velocity profiles in Figures 1 and 2. The model can successfully predict the distributions of weak currents as in a flume with no recirculation. For strong currents adjustments are needed to the distribution of $-\rho\overline{\tilde{u}\tilde{w}}$ which are as yet not understood. They are most likely due to non linear wave current interaction.

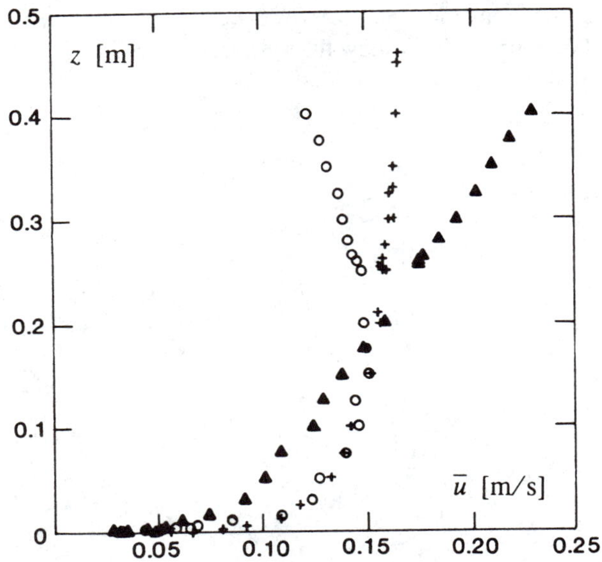

Figure 2: Current profiles in the presence of no waves (+), following waves (o) and opposing waves (triangle) of the same size. Depth D = 0.5m, wave period T = 1.4s and wave height H = 0.12m. After Klopman (1994).

Shear Stresses and Shear

In the following the notation of Nielsen (1992) will be followed. Thus, the symbol τ represents the total shearing force per unit area of a cross section or the total transfer of x-momentum in the z direction. Hence, in a two dimensional flow with steady, periodic and random velocity components $(u,w)=(\overline{u}+\tilde{u}+u',\overline{w}+\tilde{w}+w')$ in the x and z directions we have

$$\overline{\tau}=\rho\nu\frac{d\overline{u}}{dz}-\rho\overline{uw}=\rho\nu\frac{d\overline{u}}{dz}-\rho\,\overline{u}\,\overline{w}-\rho\overline{\tilde{u}\tilde{w}}-\rho\overline{u'w'} \qquad (1)$$

where, as usual, ν is the kinematic viscosity and ρ is the fluid density. The total normal stress (positive as tension) or horizontal transfer of x-momentum is corresponding given by

$$\overline{\sigma}=-\overline{p}-\rho\overline{u^2}+\rho\nu\frac{\partial\overline{u}}{\partial x}=-\overline{p}-\rho\overline{u}^2-\rho\overline{\tilde{u}^2}-\rho\overline{u'^2}+\rho\nu\frac{\partial\overline{u}}{\partial x} \qquad (2)$$

and these two satisfy

$$\frac{\partial \overline{\tau}}{\partial z}+\frac{\partial \overline{\sigma}}{\partial x}=0 \tag{3}$$

in a steady, two dimensional flow, i e,

$$\frac{\partial}{\partial z}\left[\nu\frac{d\overline{u}}{dz}-\overline{uw}-\overline{\tilde{u}\tilde{w}}-\overline{u'w'}\right]=\frac{\partial}{\partial x}\left[\frac{\overline{p}}{\rho}+\overline{u}^2+\overline{\tilde{u}^2}+\overline{u'^2}-\nu\frac{\partial\overline{u}}{\partial x}\right] \tag{4}$$

Here we assume, for the situation of waves and a weak current, that the second term on the left can be neglected and that the first and the third term are dominant on the right hand side so that we have

$$\frac{\partial}{\partial z}\left[\nu\frac{d\overline{u}}{dz}-\overline{\tilde{u}\tilde{w}}-\overline{u'w'}\right]=\frac{\partial}{\partial x}\left[\frac{\overline{p}}{\rho}+\overline{\tilde{u}^2}\right] \tag{5}$$

which is integrated with the bottom boundary condition $\overline{\tau}(0)=\overline{\tau}_o$ to give

$$\nu\frac{\partial\overline{u}}{\partial z}-\overline{\tilde{u}\tilde{w}}-\overline{u'w'}=\frac{\overline{\tau}_o}{\rho}+\int_o^z\frac{\partial}{\partial x}\left[\frac{\overline{p}}{\rho}+\overline{\tilde{u}^2}\right]dz \tag{6}$$

Here we may write the first and third terms on the left in terms of a current eddy viscosity ν_c and solve for the current gradient

$$\frac{\partial\overline{u}}{\partial z}=\frac{1}{\nu_c}\left(\frac{\overline{\tau}_o}{\rho}+\int_o^z\frac{\partial}{\partial x}\left[\frac{\overline{p}}{\rho}+\overline{\tilde{u}^2}\right]dz+\overline{\tilde{u}\tilde{w}}\right) \tag{7}$$

This equation will then directly give the current distribution by integrating with the boundary condition $\overline{u}(0)=0$.

If the main emphasis is on the flow near the bed, it is justified to assume that the mean pressure is hydrostatic. However, in order to get a good representation of the flow closer to the surface it is necessary to use $\overline{p}=\rho g(\overline{\eta}-z)-\rho\overline{\tilde{w}^2}$ in which case Equation (7) can be written.

$$\frac{\partial\overline{u}}{\partial z}=\frac{1}{\nu_c}\left(\frac{\overline{\tau}_o}{\rho}+g\frac{\partial\overline{\eta}}{\partial x}z+\int_o^z\left[\frac{\partial}{\partial x}\left(\overline{\tilde{u}^2}-\overline{\tilde{w}^2}\right)\right]dz+\overline{\tilde{u}\tilde{w}}\right)=\frac{1}{\rho\nu_c}\left(\overline{\tau}_o-\int_o^z\frac{\partial\overline{\sigma}}{\partial x}dz+\overline{\tilde{u}\tilde{w}}\right) \tag{8}$$

This general expression explains qualitatively the shape of the current profiles in Figure 1. For all three cases, $\overline{\tau_o}$ is positive and hence the current gradient is positive at the bed where the other terms are zero. Those terms will however grow in magnitude with increasing z and they are all initially negative for non-breaking waves over a horizontal bed. Thus, the right hand side and hence the current gradient will change sign at a certain level and the current velocity shows a maximum, see Figure 3.

<u>The mean bed shear stress</u>

The bed shear stress $\tau_o=\overline{\tau}(0)$ can be determined from

$$\tau_o=\tau_w-\rho g D\frac{d\overline{\eta}}{dx}-\frac{dS_{xx}}{dx} \tag{9}$$

where $\overline{\tau_w}$ is the time averaged wind stress, D is the local mean depth, $\overline{\eta}$ is the mean surface elevation and S_{xx} is the wave radiation stress in the direction of wave propagation calculated from a suitable wave theory.

The radiation stress term

Assuming linear wave theory, the radiation stress term of Equation (8) is evaluated as

$$\int_o^z \frac{\partial}{\partial x}\left[\overline{\tilde{u}^2} - \overline{\tilde{w}^2}\right]dz = \int_o^z \left[\frac{1}{2}\frac{\partial}{\partial x}\left(\frac{\pi H}{T\sinh kd}\right)^2 \left[\cosh^2 kz - \sinh^2 kz\right]\right]dz$$

$$= H\frac{dH}{dx}\frac{\pi^2}{T^2\sinh^2 kd}\, z \qquad (10)$$

In general the wave height may also vary due to breaking, refraction, diffraction and shoaling, but if only bottom friction is important, the wave height gradient is given by

$$H\frac{dH}{dx} = -\frac{8}{3\pi g}\frac{f_e(A\omega)^3}{c_{g,r} + <\overline{u}>} \qquad (11)$$

where f_e is the energy dissipation factor ($\approx f_w$), $c_{g,r}$ is the wave group velocity relative to the current and $<\overline{u}>$ is the depth averaged current velocity. See e g Nielsen (1992) p 77. By inserting into (10) we obtain

$$\int_o^z \left[\frac{\partial}{\partial x}\left(\overline{\tilde{u}^2} - \overline{\tilde{w}^2}\right)\right]dz = \frac{-8\pi}{3gT^2\sinh^2 kd}\frac{f_e(A\omega)^3}{c_{g,r} + <\overline{u}>}\, z \qquad (12)$$

Hence, the radiation stress contribution to the current shear in Equation (8) has the same form as the surface slope term, i e, both grow linearly in magnitude with distance from the bed. Consequently, the general picture which determines the level of the current maximum ($\frac{\partial \overline{u}}{\partial z} = 0$) is as shown in Figure 3. The current maximum occurs where the terms due to $\frac{\partial \overline{\sigma}}{\partial x}$ and $\overline{\tau}_w$ are balanced by the wave Reynolds stress.

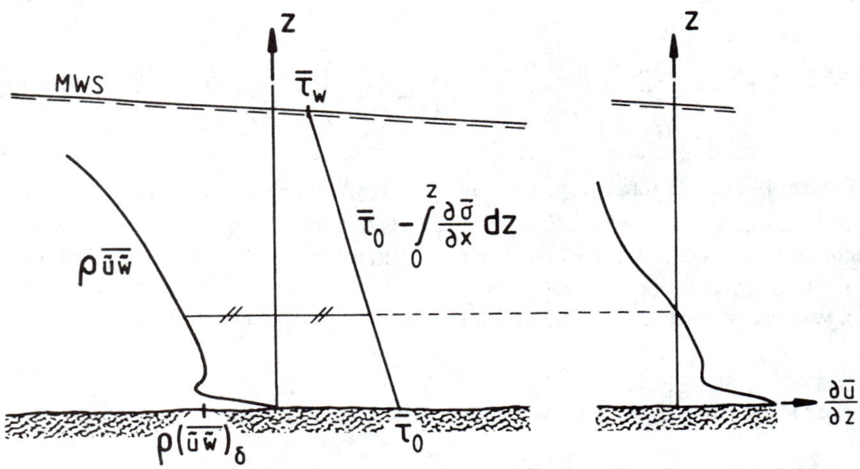

Figure 3: The current shear and the terms that generate it, cf Equation (8), for a typical situation of following current ($\overline{\tau}_o > 0$). The (lowest) current maximum will occur close to the bed for small $\overline{\tau}_o$ and closer to the MWS for larger $\overline{\tau}_o$, cf the measurements in Figure 1.

The wave Reynolds stress

The wave Reynolds stress $-\rho\overline{\tilde{u}\tilde{w}}$ may be seen as consisting of three components as shown by Bijker et al (1974). One due to the phase shifting of velocities in the boundary layer, one due to wave height decay and one due to bed slope.

Longuet-Higgins (1956) derived an expression for the first one for boundary layers with constant eddy viscosity. It varies through the wave boundary layer as

$$-\rho\overline{\tilde{u}\tilde{w}} = \rho\frac{1}{4}(A\omega)^2 k\delta\left[1 - e^{-\zeta}\left(2\cos\zeta - e^{-\zeta} + 2\zeta\sin\zeta\right)\right] \tag{13}$$

where

$$\zeta = \frac{z}{\delta} = \frac{z}{\sqrt{2\nu_t/\omega}} \tag{14}$$

Its asymptotic value above the boundary layer ($z >> \delta$) is

$$-\rho\left(\overline{\tilde{u}\tilde{w}}\right)_\delta = -\rho\frac{1}{4}k\delta\ (A\omega)^2 \tag{15}$$

which can be written in terms of the friction factor as

$$-\rho(\overline{\tilde{u}\tilde{w}})_\delta = \rho\frac{1}{4\sqrt{2}}kA^3\omega^2 f_w \tag{16}$$

cf Nielsen (1992). The wave friction factor may be determined from

$$f_w = \exp\left[5.5\left(\frac{r}{A}\right)^{0.2} - 6.3\right] \quad \text{for } r/A < 1 \tag{17}$$

as suggested by Nielsen (1992).

Estimation of $-\rho\overline{\tilde{u}\tilde{w}}$ in and above a turbulent boundary layer

The expression (13) was derived for wave boundary layers that are laminar or which are turbulent with constant eddy viscosity. For turbulent wave boundary layers with moderate relative roughness: $r/A < 0.06$ it is however necessary to adopt a different description, cf Nielsen 1992, p 47. At present we shall adopt the simple, boundary layer model of Nielsen (1985) by which the horizontal velocity under a decaying sine wave can be written as

$$\tilde{u}(z,t) = A\omega\left[\cosh kz - e^{-[1+i]\left(\frac{z}{z_1}\right)^p}\right]e^{i(\omega t - kx)} \tag{18}$$

Provided the boundary layer is thin: $|kz_1| << 1$. The wave decay is described by the imaginary part of the wave number $k = k_r + ik_i$. We note that the wave decay is not exactly exponential for a turbulent boundary layer, but we have approximately

$$k_i = \frac{1}{H}\frac{dH}{dx} \approx -H\frac{8\pi^2}{3gT_a^3(c_{gr} + <\bar{u}>)\sinh^3 kd}f_e \tag{19}$$

cf Nielsen (1992), p 78. T_a is the period seen by a fixed observer.

The boundary layer parameters z_1 and p may, for hydraulically rough (roughness r) conditions, be found from the following formulae from Nielsen (1985)

$$z_1 = 0.09\sqrt{rA} \tag{20}$$

$$p = 0.59 \exp\left(0.59 \frac{1-(A/38r)^{1.8}}{1+(A/38r)^{1.8}}\right) \tag{21}$$

This model is likely to be satisfactory for all wave boundary layers in the relative roughness range which has so far been observed with beds of loose sand ($0.015 < r/A < 1$) as discussed by Nielsen (1992). For very rough beds, $p \rightarrow 1$, this model corresponds to constant eddy viscosity.

In order to find the wave reynolds stress $-\rho\overline{\tilde{u}\tilde{w}}$ we first find the vertical velocity that corresponds to (18) from the continuity principle:

$$\tilde{w} = \int_o^z -\frac{\partial \tilde{u}}{\partial x} dz = A\omega\, e^{i(\omega t - kx)} [i \sinh k\, z - ik \int_o^z e^{-(1+i)(\frac{z}{z_1})^p} dz] \tag{22}$$

where $\sinh kz \approx \sinh k_r z + i\, k_i z \cosh k_r z$ for $k_i\, z << 1$ so that

$$\tilde{w} \approx A\omega\, e^{i(\omega t - kx)} [-k_i z \cosh k_r z + i \sinh k_r z - ik \int_o^z e^{-(1+i)(\frac{z}{z_1})^p} dz] \tag{23}$$

Similarly the expression (18) for the horizontal velocity can be simplified to

$$\tilde{u} \approx A\omega\, e^{i(\omega t - kx)} [\cosh k_r z + ik_i z \sinh k_r z - e^{-(1+i)(\frac{z}{z_1})^p}] \tag{24}$$

Inside the boundary layer where $k_i z << k_r z << 1$ these expressions lead to an analogous (but very complicated) expression corresponding to (13). For the sake of brevity, we shall use (13) combined with (16) and with ζ replaced by $\xi = (z/z_1)^p$, i e,

$$(\overline{\tilde{u}\tilde{w}})_{BL} \approx -\frac{1}{4\sqrt{2}} k_r A^3 \omega^2 f_w \cosh k_r z \left[1 - e^{-\xi}\left(2\cos\xi - e^{-\xi} + 2\xi \sin\xi\right)\right] \tag{25}$$

where the subscript BL refers to this term being analogous to the boundary layer term considered by Longuet-Higgins (1956). The cosh factor only makes a difference far above the boundary layer where $\xi >> 1$.

If we neglect the boundary layer terms (the last terms) in (22) and (23) we get the contribution due to wave decay:

$$(\overline{\tilde{u}\tilde{w}})_{DECAY} \approx -\frac{1}{2}(A\omega)^2 k_i z(\cosh^2 k_r z + \sinh^2 k_r z) = -\frac{1}{2}(A\omega)^2 k_i z \cosh 2k_r z \tag{26}$$

Figure 4 shows the two $\overline{\tilde{u}\tilde{w}}$-contributions, evaluated for Klopman's (1994) tests.

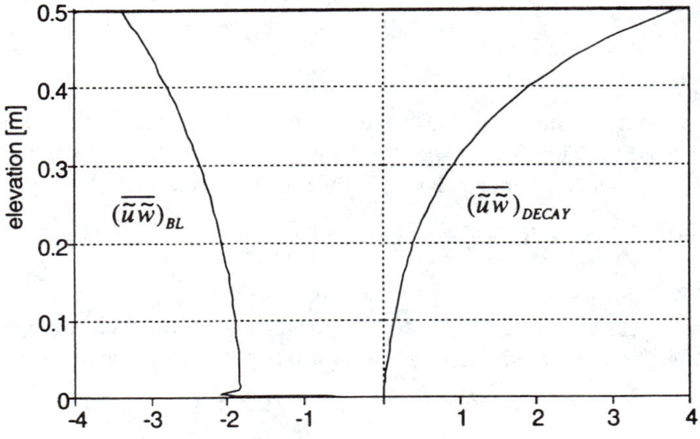

Figure 4: Wave Reynolds stress contributions for (h,T,H,r) = (0.5m, 1.4s, 0.12m, 0.0012m). The units are $m^2/s^2 \times 10^{-5}$.

The eddy viscosity distribution

In order to find the $\bar{u}$-distribution from Equation (8) we must specify the eddy viscosity $\nu_c(z)$ felt by the current. We use the distribution suggeted by Nielsen (1992) which consists of a parabolic/constant part based on the current friction velocity $\overline{u_*} = \sqrt{\tau_o / \rho}$ with a square toe added due to wave boundary layer mixing:

$$\nu_c(z) = Max\left\{\begin{matrix} 2\omega z_1^2 \\ 0.4\overline{u_*}z\left(1-\frac{z}{D}\right) for\, z < D/2,\;\; 0.1\overline{u_*}D\; for\; z > D/2 \end{matrix}\right\} \tag{27}$$

Discharge considerations

Calculation of the total discharge presents a problem because it must include the discharge which occurs above the wave trough where the velocities are not obtained by integration of Equation (8) as they are for the lower part of the flow. One estimate of the net flow between the wave trough and the crest can be obtained from linear wave theory namely

$$Q_{tc} = \frac{1}{8c} g H^2 \tag{28}$$

where c is the wave speed, see eg. Dean & Dalrymple (1991). It is however not obvious how this irrotational result can be combined with an integral over the lower flow domain ($z<D-H/2$) of the velocities obtained from Equation (8).

Identification	Q_{tc} [m²/s] based on Eq (28)	Measured $\int_o^{D-H/2} \bar{u} dz$ [m²/s]	Measured total Q [m²/s]	Total Q [m²/s] based on Eq (29)
Klopman 1994	0.010	-0.012	0	-0.06
" - ": monocrom, following	0.010	0.057	0.080	0.080
" - ": monocrom, opposing	0.010	-0.071	-0.080	-0.091
van Doorn & Godefroy, smooth	0.011	-0.0042	0	0.005
van Doorn & Godefroy, rough	0.011	-0.0062	0	0.001
van Doorn V00	0.011	-0.0073	0	0.0004
van Doorn V10	0.011	0.020	0.030	0.041
van Doorn V20	0.011	0.051	0.060	0.087

Table 1

We shall adopt the approach which is illustrated in Figure 5. That is, velocities below the wave trough are found from integration of Equation (8). Above the trough a parabolic velocity distribution corresponding to the flow rate given by Equation (28) is added to the velocity $\overline{u_{tr}}$ at the trough level. The total flow rate is then

$$Q = \int_o^{D-H/2} \bar{u} dz + \frac{gH^2}{8c} + \overline{u_{tr}} H \tag{29}$$

Some comparison between this approach and measurements is afforded by Table 1. There is obviously some scope for improvement but the message from this limited data set is not very clear.

The surface slope

The surface slope $d\overline{\eta}/dx$, albeit very small influences the current distribution and hence the net flow rate per unit width Q. Hence, it is not possible to achieve agreement with experiments if the surface slope is assumed to be zero. Not even for a "no current situation".

In practical applications it is some times Q which is specified while $d\overline{\eta}/dx$ is not. In such cases the model is easiest applied by trial and error. That is,

1. A reasonable value of $d\overline{\eta}/dx$ is guessed
2. Q is found by integrating Equation (8) and applying (29) to find Q.
3. An improved value of $d\overline{\eta}/dx$ is adopted.

The "necessary" surface slopes to match the usual wave flume conditions with horizontal beds are usually inside the range $\pm 5\times 10^{-5}$. They are thus hardly measureable.

Comparison with experiments

A comparison of the model above with the "monocromatic waves, no current" experiment of Klopman (1994) is shown in Figure 5.

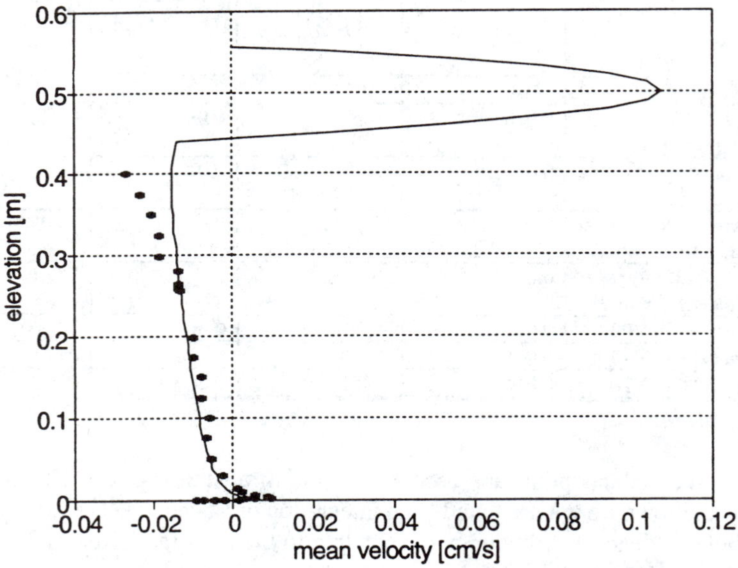

Figure 5: Model comparison with Klopmans data, Monocromatic waves, D=0.5m, T=1.4s, H=0.12m, Q=0m²/s. The flume bed was covered with a single layer of 2mm sand giving a Nikuradse roughness of r=1.2mm. In order to achieve zero nett flow with the model, a surface slope of 0.0000017 must be assumed.

Figure 5 shows some discrepancy between the model and the measurements below $z = 6$mm. Part of this might be corrected by making adjustments to the eddy viscosity distribution in this area. However, the existing detailed data sets: van Doorn & Godefroy (1978), van Doorn (1981), Swan (1990) and Klopman (1994) show quite some variation. In particular, the negative velocities measured by Klopman at $z <$ 1mm are not parallelled by the other experiments. At any rate, comparison is very great detail is not warranted at levels smaller than the roughness height between this model, which is horizontally uniform, and measurements which, at this level, will depend upon the horizontal position of the probe. Measurements may show horizontal variation both on the scale of individual roughness elements and on the scale of the wave length if some reflection and/or circulation cells occur in the flume.

While the agreement between the model and experiments with weak currents is good as shown by Figure 5, adjustments are necessary in order to model stronger currents. In general the data show a stronger deviation from the logarithmic profile than the model predicts. See Figure 6.

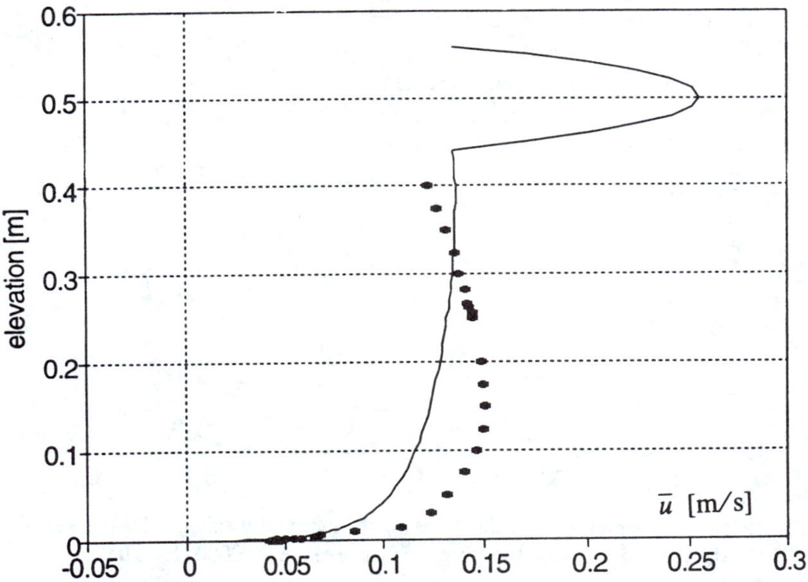

Figure 6: Comparison between the model and the "monocromatic waves, following current" experiment of Klopman (1994). *D*=0.5m, *T*=1.4s, *H*=0.12m, *Q*=0.08m^2/s, *r*=1.2mm. Surface slope -0.00000145.

Figure 6 shows a clear discrepancy between the model described above and Klopman's "following current" experiment. The reason is believed to be that this rather strong current ($<\bar{u}> = 0.92 A\omega$) changes the wave motion and hence $\overline{\tilde{u}\tilde{w}}$ in the upper part of the flow. An indication of this change can also be found the direct $\overline{\tilde{u}\tilde{w}}$-measurements of Supharatid et al (1992).

Figure 7 shows an improved model performance obtained by enhancing the wave Reynolds stress component ($(\overline{\tilde{u}\tilde{w}})_{BL}$ (Equation 25) by the empirical factor

$$C_{WR} = 1 + 100\frac{\overline{u_*}}{A\omega}\frac{z}{D} \tag{30}$$

while $(\overline{\tilde{u}\tilde{w}})_{DECAY}$ is kept unchanged.

This is a purely empirical adjustment but it is thought to mimic changes to $\overline{\tilde{u}\tilde{w}}$ in the upper part of the flow due to extra vorticity of the wave motion caused by interaction with the strong current. There is little reason to believe that $\overline{\tilde{u}\tilde{w}}$ is changed appreciably inside the wave boundary layer by the current, cf the data Supharatid et al (1992) and Nielsen (1985, 1992). It is noted, however, that the same empirical adjustment would also improve the agreement with the zero-net-flow-data in Figure 5.

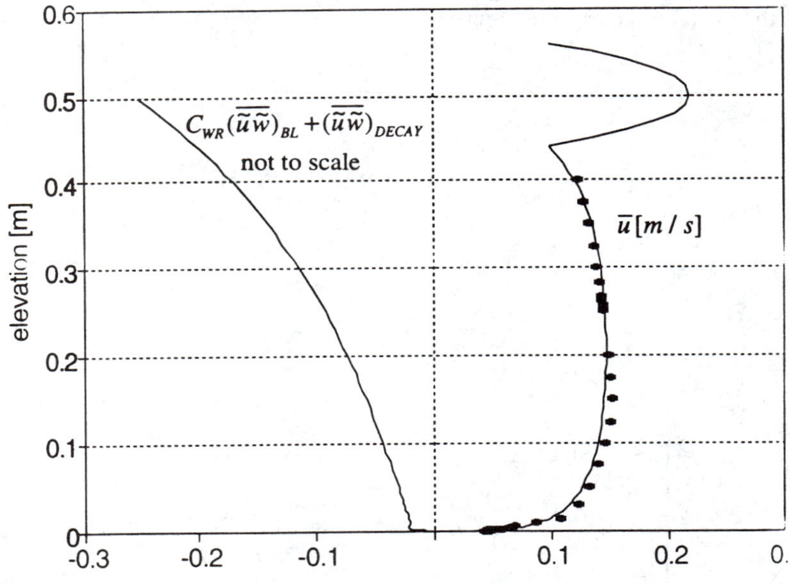

Figure 7: Comparison of the model with empirically enhanced wave Reynolds stress to the same data as in Figure 6. Surface slope -0.000021. The curve on the left shows the enhance distribution of $\overline{\tilde{u}\tilde{w}}$ given by Equations (25), (26) and (30).

Discussion

The magnitude of the eddy viscosity and its capacity to transfer vorticity upwards from the bed influences the nature of the solution as discussed by Longuet-Higgins (1953). If the eddy viscosity is small, the $\bar{u}$ -profile would be strongly influenced by the end conditions through horizontal advection. In that case it would not be determined from the local force balance which leads to Equation (8). However Euation (35) leads to eddy viscosities which are very much larger than the laminar viscosity even in the "no current situations" corresponding to Figure 1a and Figure 4. Thus, in Klopman's "no current" experiment Equation (9) gives a bed shear stress of

0.015N/m^2 corresponding to an eddy viscosity at mid depth of 0.00039m^2/s, about four hundred times the laminar viscosity. This is probably the reason why "no current" experiments quite consistently show $\bar{u}$ -profiles like to ones in Figure 1a and Figure 5 even thogh they may have different end conditions.

In other words, it appears that under conditions like those of van Doorn (1981) and Klopman (1994) and even the smooth bed experiments of van Doorn & Godefroy (1978) (D,T,H) = (0.3m, 2s, 0.12m) the eddy viscosity is strong enough that the net flow is governed by the local conditions as expressed by Equation (8). The experimental conditions of Swan (1990), smooth bed and (D,T,H) = (0.369m, 0.893s, 0.06m) are probably close to the limit although he did show (his Figure 5a) a $\bar{u}$ -profile which is qualitatively similar to those observed under more vigorous flow conditions by can Doorn & Godefroy (1978), can Doorn (1981) and by Klopman (1994).

These observations seem to justify the use of Equation (8) as the basis of modelling currents in the presence of waves under field conditions and in most laboratory situations.

The present version of the model is open to a number of refinements. First of all with respect to the treatment of the flow above the wave trough. Secondly, the enhancement of the wave Reynolds stress in the presence of a stronger current should be investigated theoretically.

Application of the model above model to surf zone conditions may require adjustments in three areas:

1. An additional term due to bottom slope may need to be added to $\overline{\tilde{u}\tilde{w}}$,
2. A different expression for the flow above the wave trough, e g, a roller model may be needed The eddy viscosity must be enhanced in the upper part of the flow. Guidance in these respects may be found in the measuremetns of Nadaoka & Kondoh (1982) and Cox et al (1995).

The strength of the present model compared with previous ones is that a detailed description is achieved at the bottom at the same time as a reasonable overall water balance. This is of course necessary in order to model natural situations with movable beds.

References

Bijker, EW, JP Th Kalkwijk & T Pieters (1974): Mass transport in gravity waves on a sloping bottom. *Proc 14th Int Conf Coastal Eng, ASCE*, pp 447-465.

Coffey, FC & P Nielsen (1986): The influence of waves on current profiles, *Proc 20th Int Conf Coastal Eng, Taipei*, pp 82-96.

Cox, D T, N Kobayashi, & A Okayasu (1995): Experimental and numerical modelling of surf zone hydrodynamics. Research report CACR-95-07, Ocean Engineering Laboratory, University of delaware.

Fredsoe, J (1984): Turbulent boundary layers in wave-current motion. *J Hydraulic Eng, ASCE Vol 110, No 8*, pp 1103-1120.

Grant, WD & OS Madsen (1979): Combined wave current interaction with a rough bottom. *J Geophys Res Vol 84 No C4*, pp 1797-1808.

Havinga, FJ (1992): *Sediment concentrations and sediment transport in case of irregular non-breaking waves with a current.* Draft Thesis, Faculty of Civil Engineering, Delft University of Technology.

Heijboer, D (1988): *Zandconcentratie- en stroomsnelheidsverdelingen onder golven en stroom, parts I, II and III,* Delft University of Technology.

Kaaij, Th van der & MWC Nieuwjaar (1987): *Sediment concentrations and sediment transport in case of irregular non-breaking waves with a current, Parts A: Text* 104pp and *Part B: Tables and Figures.* Delft University of Technology, Faculty of Civil Engineering.

Kemp, PH and RR Simons (1982): The interaction between waves and a current: waves propagating with the current. *J Fluid Mech Vol 116,* pp 227-250.

Klopman, G (1994): Vertical structure of the flow due to waves and currents. *Delf Hydraulics Progress Report H 840.30, Part II,* 38pp.

Longuet-Higgins, MS (1953): Mass transport in Water Waves. *Phil Trans Roy Soc Lond, Vol 245 A,* pp 535-581.

Longuet-Higgins, MS (1956): The mechanics of the boundary layer near the bottom in a progressive wave. *Proc 6th Int Conf Coastal Eng, ASCE, Miami,* pp 184-193.

Nadaoka, K & T Kondoh (1982): Laboratory measurements of velocity field structure in the surf zone by LDV. *Coastal Engineering in Japan, Vol 25,* pp 125-146.

Nielsen, P (1985): On the structure of oscillatory boundary layers. *Coastal Engineering, Vol 9,* pp 261-276.

Nielsen, P (1992): *Coastal bottom boundary layers and sediment transport.* World scientific, Singapore, 324pp.

Supharatid, S, H Tanaka & Nshuto (1992): Interactions of waves and current. *Coastal Eng in Japan, Vol 35, No 2,* pp 167-186.

Swan, C (1990): Convection within an experimental wave flume. *J Hydraulic Res, Vol 28, No 3,* pp 273-282.

van Doorn, Th (1981): *Experimental investigation of near-bottom velocities in water waves without and with a current. Report no M 1423 part 1,* Delft Hydraulics, 66pp.

van Doorn, Th & HWHE Godefroy (1978): *Experimental investigation of the bottom boundary layer under periodic progressive water waves. Report no M 1362 part 1,* Delft Hydraulics, 46pp.

Villaret, C and G Perrier (1992): *Transport of fine sand by combined waves and current: An experimental study. Report no HE-42/92.68,* Electricite de France, 81pp.

You, Z-J (1994): A simple model for current velocity profiles in combined wave current flows. *Coastal Engineering, Vol 23,* pp 289-304.

CHAPTER 315

REYNOLDS STRESS AND SMALL-SCALE MORPHOLOGY MEASUREMENTS DURING DUCK94

Timothy. P. Stanton and Edward B. Thornton
Department of Oceanography, Naval Postgraduate School, Monterey, CA 93943
Tel: (408) 656 3144, FAX 408 656 2712: email stanton@oc.nps.navy.mil

Abstract

A recently developed Coherent Acoustic Sediment Probe (CASP) was deployed from an instrumented sled during the Duck94 nearshore experiment at Duck, North Carolina, in October 1994. The CASP measured acoustic backscatter profiles and three component velocity vectors in an O(1 cm^3) volume 25cm in front of the downward-looking instrument package. The CASP was remotely positioned to selected heights to form profiles of velocity and sediment concentration through the bottom boundary layer. A vertical stack of 8, two-axis electromagnetic current meters extended the horizontal measurements to the surface. A co-located scanning pencil-beam sonar and tilt sensors were used to map the small-scale morphology across a 1m by 4m area centered on the CASP measurements, quantitifying the local roughness and slope elements down to 4cm scales under low acoustic clutter conditions. It is shown that the stress measured near the bed in a gravity oriented vertical reference system is associated with both the turbulent Reynolds' stresses and a wave stress due the vertical velocity induced by the horizontal wave velocity acting on the sloping bottom being correlated with the horizontal velocity. The wave stress is largest at the bed and decreases towards the surface and can be larger than the turbulent Reynolds' stresses.

DUCK94 Experiment

The data presented herein were obtained during the comprehensive nearshore DUCK94 experiment conducted in October 1994 at the U. S. Army Corps of Engineers Field Research Facility (FRF), Duck, North Carolina. The FRF beach usually is a two-bar system composed of a dynamic inner bar (30-120 m offshore) and a secondary bar of lower amplitude (300-400 m offshore). The mean foreshore slope of the beach is approximately 0.08 (1:12) and the mean slope offshore of the bars is approximately 0.006 (1:170) (Lippmann et al., 1993). Sediments have a wide range of coarse sediment size on the beach and foreshore and fine, well sorted sand within the trough and seaward (mean grain diameter ~ 0.2 mm) (Stauble, 1992). The mean tidal range is approximately 1 m.

Boundary layer stress and velocity profile measurements from 11 October have been chosen to show the structure of the bottom boundary layer under strong wave forcing conditions. Storm waves with significant wave heights of 2m approaching at 16° (in 8m depth) produced mean longshore currents of 1 ms^{-1} in the surf zone. The resulting bottom boundary layer was composed of both the steady longshore flow and O(1 ms^{-1}) magnitude cross shore oscillatory flow due to the incoming waves.

Wave, velocity and sediment flux data were acquired using an instrumented sled. A Coherent Acoustic Sediment Probe (CASP) was mounted on a moveable boom (with elevation controlled from shore) deployed from the sled. The CASP measured acoustic backscatter profiles and three-component velocity vectors in an O(1 cm^3) volume 25cm in front of the downward-looking instrument package, as it was remotely positioned to selected heights to form profiles of velocity and sediment concentration through the bottom boundary layer. A co-located scanning pencil-beam sonar and tilt sensors were used to map the small-scale morphology across a 1m by 4m area centered on the CASP measurements, quantitifying the local roughness and slope elements down to 4cm scales under low acoustic clutter conditions

Additional instrumentation on the sled included a vertical stack of eight Marsh-McBirney two-component electromagnetic current meters mounted on a 2.5 m mast to measure vertical profiles of longshore and cross-shore currents. The sled was oriented with the vertical stack of current meters placed on the up-current side to prevent the sled structure from contaminating flow measurements. Surface elevations were measured using an array of five pressure sensors configured in a 3 m square with sensors at the corners and one at the center. For data collection, the sled was towed by the CRAB to its furthest position offshore, dependent upon wave conditions, for the first data run. The sled was tethered to the shore with a chain. A forklift on the beach pulled the sled shoreward approximately 10 to 30 m for each subsequent run. Data were acquired at each location for approximately one hour at four to eight locations. Data were digitized on the sled and transmitted to shore via an armored fibre optic cable married to the chain tether.

Additionally, directional wave spectra were acquired using a linear array of 10 pressure sensors in 8 m depth offshore of the survey area. A fixed current meter in the trough at line 820 m was used to measure longshore currents continuously.

Vertical profiles of the mean longshore currents at various cross-shore location superposed on the bottom profile measured on 11 October are shown in Figure 1. The longshore currents were relatively strong on this day with mean currents exceeding 1 m/s. The bottom profile is barred. The small-scale morphology is near planar on the seaward side and atop the bar and on the foreshore, with well-developed mega-ripples within the trough.

Theory and Analysis

Linear wave theory over a horizontal bed predicts that the horizontal and vertical wave induced velocities are in quadrature, and therefore do not contribute to the Reynolds' stresses. However, for a sloping bottom in a gravity oriented vertical reference system, a vertical velocity is induced at the bed by the horizontal velocity acting on the sloping bottom

$$w = u \tan\beta \quad (1)$$

where u,w are the horizontal and vertical velocities and tan B is the bottom slope. The vertical

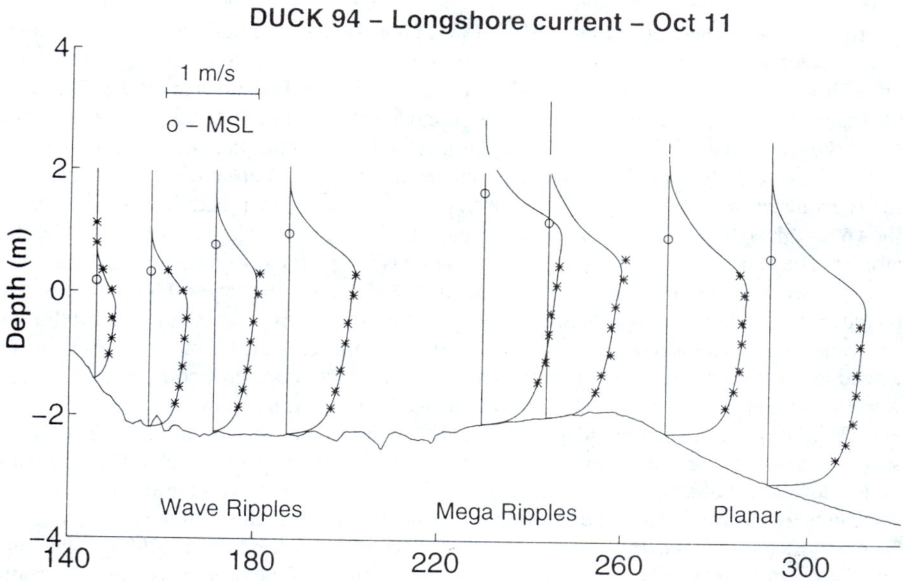

Figure 1. Measured longshore current vertical profiles (*) at various cross-shore locations compared with logarithmic profile (solid line) superposed on the bottom profile.

velocity near the bed will be in-phase with the horizontal velocity, resulting in a contribution to the co-spectrum, <uw>, which is proportional to the Reynolds' stress. The total stress is composed of an induced wave stress plus the stress due to the friction velocity defined as

$$\frac{\tau}{\rho} = u_{*T}^2 = <\vec{u}w> = u_{*Friction}^2 + u_{*Wave}^2 \qquad (2)$$

It is hypothesized that the appropriate coordinates to measure the Reynold's stresses near the bed are oriented normal to the bottom. However, the bed has a broad range of scales in the near shore (Thornton, et al 1997) ranging from the general bottom profile with horizontal scales O(10m) to mega-ripples and ripples with horizontal scales O(1m). A question then arises as to what is the corresponding scale over which to measure the bottom slope? Instead of trying to measure the appropriate bottom slope, we will use hydrodynamic measures to establish the coordinate system. Hydrodynamically, the velocity normal to the bed goes to zero at the bed. Therefore, the bottom slope in (1) is determined three different ways by minimizing the measured near-bed quantities of covariance between horizontal and vertical velocity, <uw>, mean vertical velocity, W, and vertical velocity variance, $<w^2>$.

Results and Conclusions

Direct estimates of the Reynolds stress components <uw,vw> and mean velocity <U,V> were formed at each measurement height with averaging times between 4 and 10 minutes. As a vertical profile of these quantities is formed over approximately 1 hour, measurement stations were selected where the mean currents and wave conditions were relatively constant during this time. Profiles of (U,V) vectors, orbital displacement standard deviations in the x,y directions, the friction velocity vector (u*,v*), and the friction velocity magnitude for station 1 (see Figure 1) are shown in Figure 2. The instrument coordinate velocity vector time series was rotated into a vertically orientated coordinate system to a precision of $\pm 0.2°$ before the mean velocity and stress estimates were made. Therefore the true vertical (rather than bed normal) stresses are shown here. Height above the bed was determined to ±1.6 cm by using the acoustic backscatter profiles. The velocity coordinate system in these plots is v+ offshore and u+ toward the south, and w+ upwards. At both sites, the U/V profiles, which span the lower 0.5m of the bottom boundary layer, show strong offshore and long-shore mean currents decaying toward the bed. The upper CASP-measured mean velocities matched the em-current meter profiles, which extend to the surface (not shown). In the absence of an oscillatory flow component, the stress vectors would be expected to be aligned with the mean current direction within the inner "wall" boundary layer established by the mean flow. However, the mean vertical stress vector has a strong cross-shore wave stress component, increasing toward the bed, which rotates the stress vectors toward the offshore direction (Figure 2). This cross-shore stress is attributed to the wave stress component arising from the shoaling waves interacting with the bottom slope, producing a linearly decreasing wave stress toward the surface (Rivero and Arcilla, 1995). The slight increase in stress in the first 20cm above the bed seen in most of the profiles is likely due to the rotational stress components of the wave boundary layer arising from the strong oscillatory flow and the local bed roughness elements.

An example of the co- and quadrature spectra measured in gravity oriented coordinates at an elevation of 27 cm above the bed is shown in Figure 3. In the gravity oriented vertical reference coordinates, there is a large contribution to both the co- and quadrature spectra (vw, uw) at the wind-wave band of frequencies (0.1 - 0.5 Hz). The co-spectrum is minimized by rotating the coordinates about the x,y axis in 0.1° increments (Figure 4). A definite minimum is found, indicating a mean cross-shore slope of 4.3° and an alongshore slope of 1.6°. The minimized cross-spectra are shown in Figure 5. The co-spectra are greatly reduced, whereas the quadrature spectra are virtually the same, indicating the co-spectral contributions are bottom slope induced, and should not be attributed to turbulent Reynold's stress contributions.

The mean and vertical velocities variance were similarly minimized and the corresponding calculated bottom slopes were essentially the same as calculated from the minimization of the co-spectra. Therefore, the coordinate rotations are based on the minimization of the co-spectra.

The friction velocities calculated at various elevations over the vertical before and after rotations of the coordinates and the cross-shore tilt calculated from the minimization of the co-spectra indicative of the bottom slope are shown for locations offshore the bar and within the trough (Figures 6 and 7). The measured cross-shore bottom slope (averaged over 5m) is also indicated. On the near planar, smooth slope on the seaward side of the bar the calculated slope from the co-spectral minimization corresponds to the bottom slope (averaged over 5m) as the bottom is approached; this indicates that the additional stress is associated with the sloping

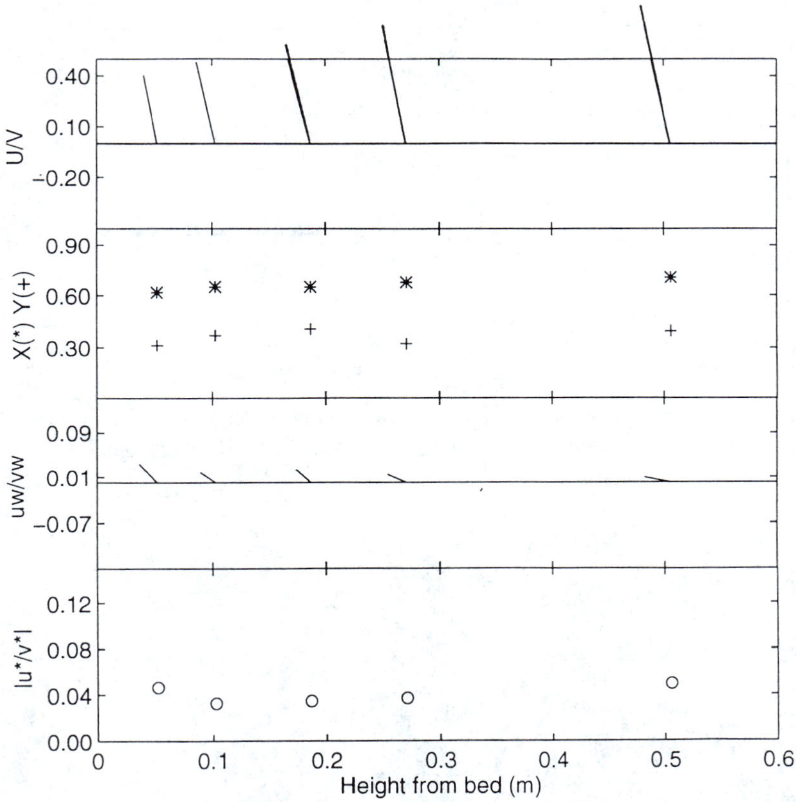

Figure 2. (U,V) mean velocity, orbital displacement magnitudes for x,y components, u*,v* friction velocity vector and friction velocity magnitude profiles from 0 to 0.5m above the bed for the offshore station 1 on October 11.

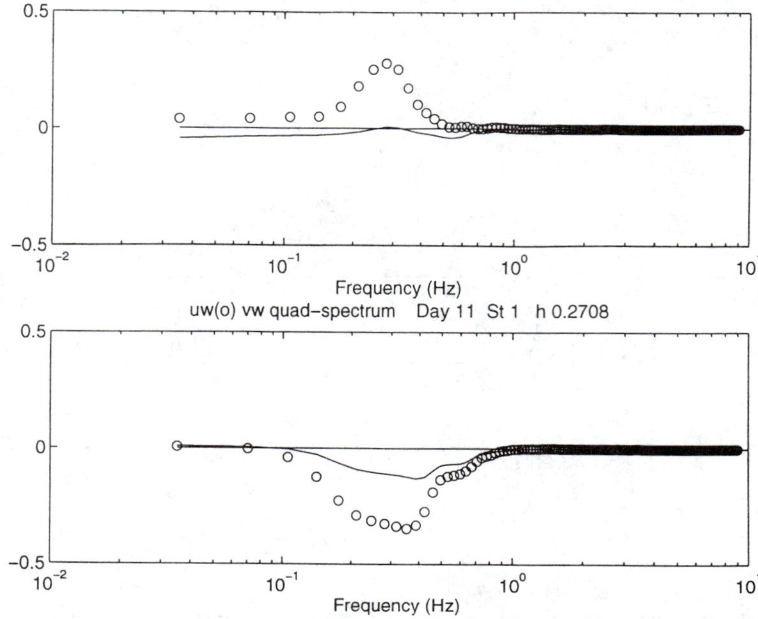

Figure 3. Co-spectra (upper panel) and Quadrature-spectra (lower panel) of uw (o) and vw (-) (m^2-sec) in gravity vertical coordinates, Station 1 at elevation 27cm off the bottom.

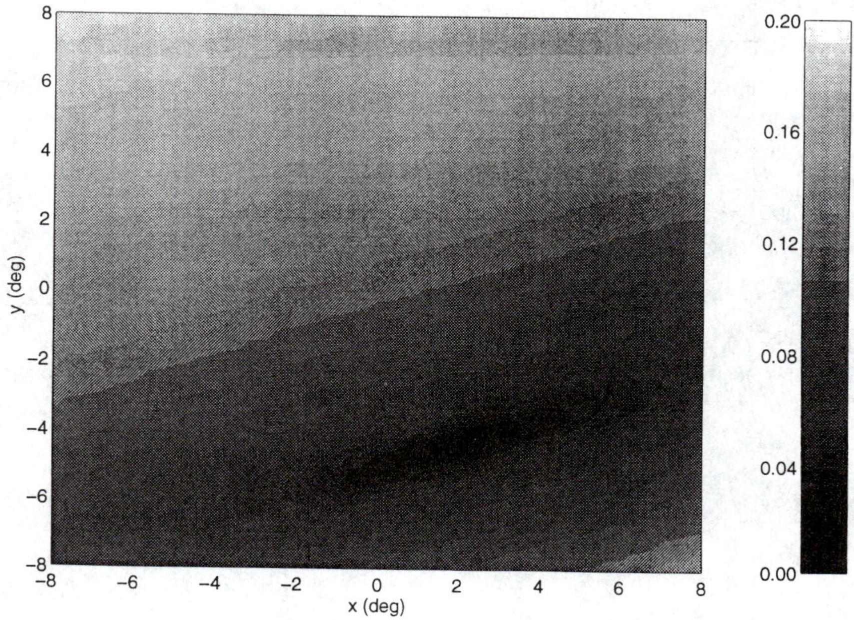

Figure 4. Co-spectra calculated at 0.1 degree rotations in x,y showing minimum, Station 1.

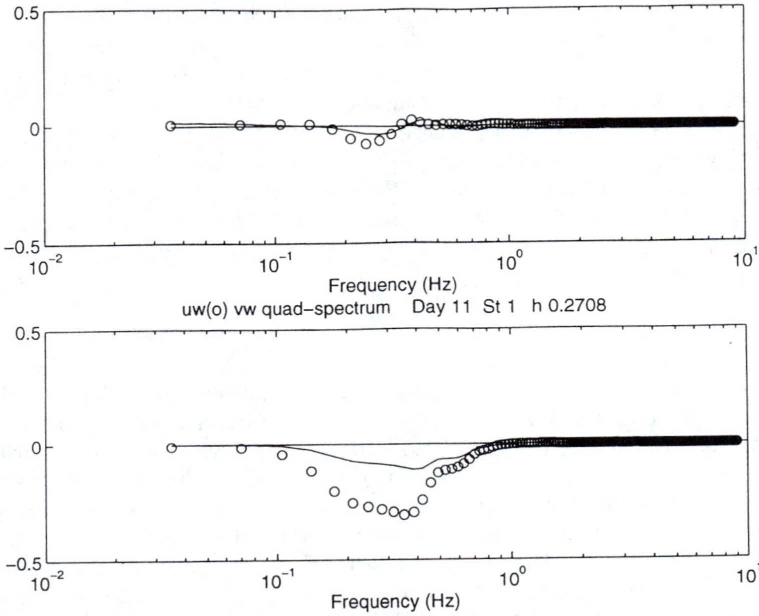

Figure 5. Co-spectra (upper panel) and Quadrature-spectra (lower panel) of uw (o) and vw (-) in rotated coordinates minimizing the co-spectra, Station 1.

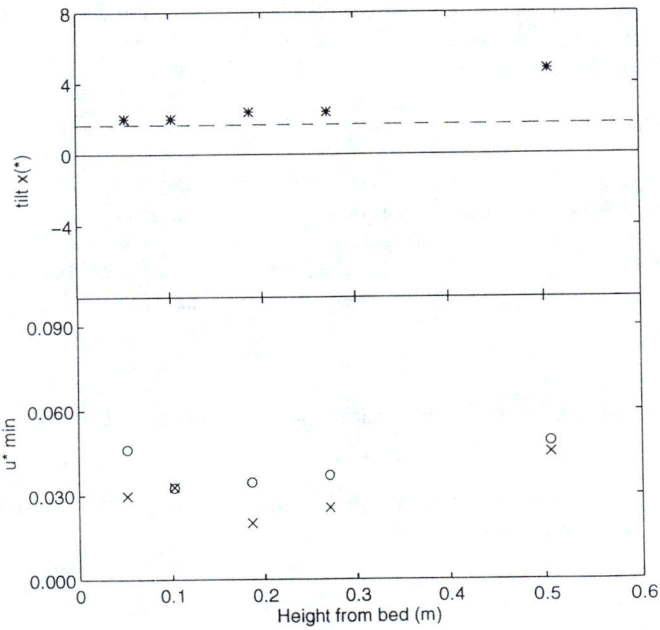

Figure 6. Vertical profiles of cross-shore (x) rotation angles to minimize co-spectra (*) compared with average slope over 5m (-) (upper panel), and total friction velocity (o) and friction velocity in bed normal coordinates (lower panel), Station 1.

bottom. However, within the trough where mega-ripples are prevalent, just the opposite effect is seen where the calculated slope from the minimization of the co-spectra and the bottom slope measure over 5m do not converge until almost 40cm off the bed. The total near bed shear velocities are large. After subtracting u^2_{*Wave} from the total shear velocity, the residual friction velocities are essential constant over the vertical. The detailed bottom profile at station 5 shows the bottom is composed of mega-ripples with heights of approximately 15cm and wave lengths of about 1m (Figure 7). The location of the CASP is at the vertical dotted line and unfortunately information right beneath the CASP is not available. The mean slope is indicated and it can be seen that is can be quite different that the local slope near the bed as indicated by the tilt values calculated in Figure 6.

Conclusions

The total near-bed friction velocity is viewed as the sum of the friction velocity associated with the turbulent Reynolds' stress and a wave friction velocity induced by the sloping bed. The friction velocity is calculated by minimizing the co-spectra to eliminate the contribution by the horizontal wave velocity acting on the bed to induce an in-phase vertical velocity which contributes to the co-spectrum. The horizontal scale of the bed slope which induces the near-bed vertical velocity due to the waves appears to be related to the distance from the bed. Where a multitude of scales exist very close to the bed, such as when mega-ripples are present, the local bed slope (horizontal scale < 1m) appears to dominate, whereas away from the bed the corresponding bed slope can be much larger. For a smooth, near-planar bed, all horizontal scales have essentially the same slope.

The wave-induced friction velocity is largest near the bed and decreases toward the surface, which agrees with the theoretical work of Rivero and Arcilla (1995). The wave-induced velocity near the bed can be much larger than the turbulent Reynolds' stresses.

Acknowledgments

This work was funded by the Office of Naval Research, Coastal Sciences Branch under contract N00114-95-AF-002. The authors wish to express their appreciation to all those who participated in DUCK94 experiment, particularly the staff of the U.S. Arm Field Research Facility under the direction of B. Birkemeier. In addition, special appreciation is expressed to R. Wyland and J. Stockel for their roles in the acquisition and analysis of the data.

References

Rivero, F.J. and A.S. Arcilla, 1995, On the vertical distribution of <uw>, Coastal Engineering, 25, 137-152.

Thornton, E.B., J.L. Swayne and J.R. Dingler, 1997, Small-scale morphology related to waves and currents across the surf zone, (accepted Marine Geology).

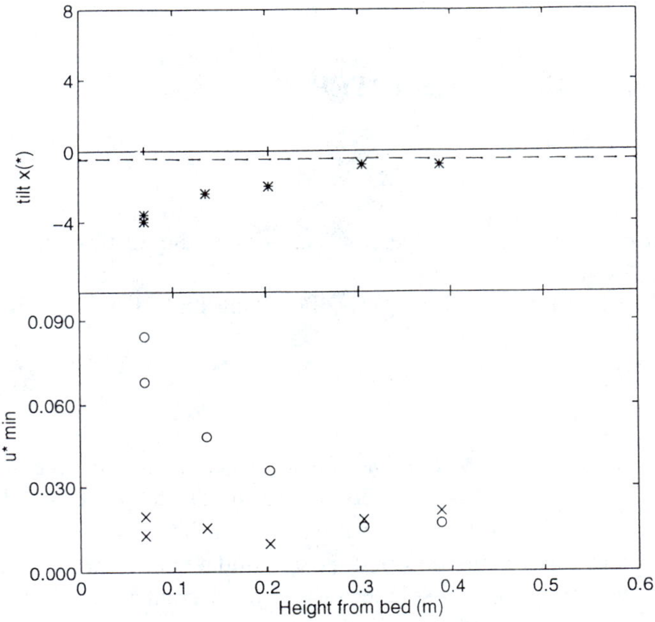

Figure 7. Vertical profiles of cross-shore (x) rotation angles to minimize co-spectra (*) compared with average slope over 5m (-) (upper panel), and total friction velocity (o) and friction velocity in bed normal coordinates (lower panel), Station 5.

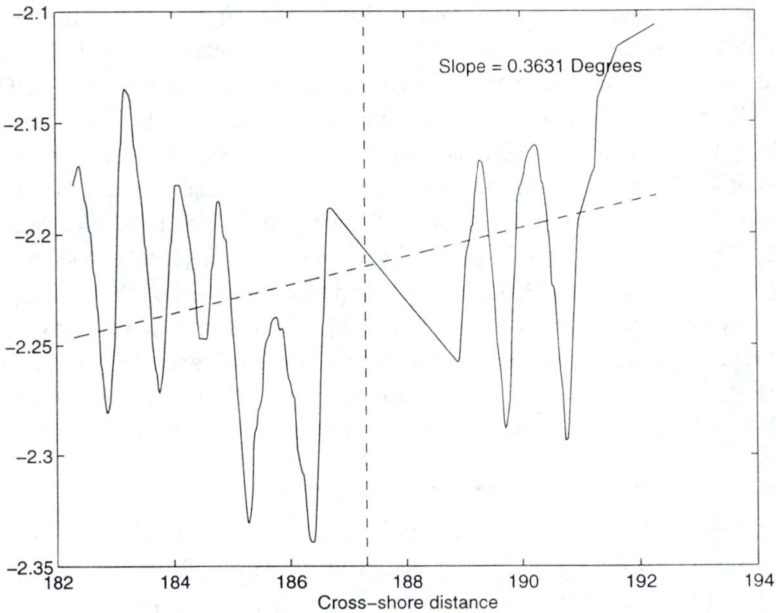

Figure 8. Bottom profile showing mega-ripples superposed on a mean slope (averaged over 5m) of 0.36 degrees.

CHAPTER 316

Studies on the Suspended Concentration in the Surf Zone

Ching-Her Hwang[1] Li-Hung Tsai[2] Po-Ching Lin[2] Chin-Chi Tsai[3]

Abstract

This study is to carry out the field measurements of the suspended concentration by using MTB-16K turbidity records in the surf zone of mild sloping beach (1/144). Then the vertical concentration distributions of the suspended sediment at different water depth d/L_0 and the longitudal concentration distributions of the suspended sediment across the surf zone are analyzed and discussed respectively in this paper.

Indroduction

In order to understand the phenomena of sediment transport in the near shore, it is necessary to investigate the problems of the sediment suspension mechanism induced by waves. Earlier there are many scholars and experts, such as Farichild (1959), Homa and Horikawa (1962), Noda (1967) etc., who had been studied this problems by using siphon, pumped sampler and suspend sampler respectively. In fact, it is difficult to obtain the exact records of suspended sediment variations. Lately there are other researchers such as Homa (1965), Sleath (1974, 1982), Wrigh (1982), Deigaard (1986), Makoto Ifuku (1988), Gotoh (1994), Nielson (1994), Ono (1994) and authors (1993-1995) etc. had also been studied this problems in the laboratory and in the field respectively. And this study is to carry out field measurement of the suspended concentration in the surf zone of mild sloping beach (S=1/144) located at the mid-west coast of Taiwan, R.O.C. under the sponsor of the National Science Council for the integrated research program "Coastal Space Utilities".

---

[1] Chief, the Coastal Engineering Division, Institute of Harbor & Marine Technology, No.83, Lin-Hai Rd., Wuchi, Taichung District, Taiwan, R.O.C.

[2] Assistant researcher, the Coastal Engineering Division, IHMT.

[3] Assist, the Coastal Engineering Division, IHMT.

Field Measurement

Four sets of memory MTB-16K turbidity recorder which set up above the sea-bed 6.5cm, 22cm, 35cm and 76cm etc. near the low water level at the tidal zone are used to measure the vertical concentration distribution of the suspended sediment. Also there are three sets of memory MTB-16K turbidity recorder which set up at a distance of 50m each other and above the sea-bed 6.5cm are used to measure the longitudinal concentration distribution of the suspended sediment across the surf zone. The maximum tidal range is about 5m in this area, hence their concentration distributions of the suspended sediment at different water depth can be measured by the tidal variations. Concurrently a DNW-5M pressure sensor and a UCM 40 Type MKII were also set up respectively near the sea-bed around the low water level and a waverider was set up outside the surf zone to record the oceanography conditions. The schematic diagrams are shown as Fig. 1.

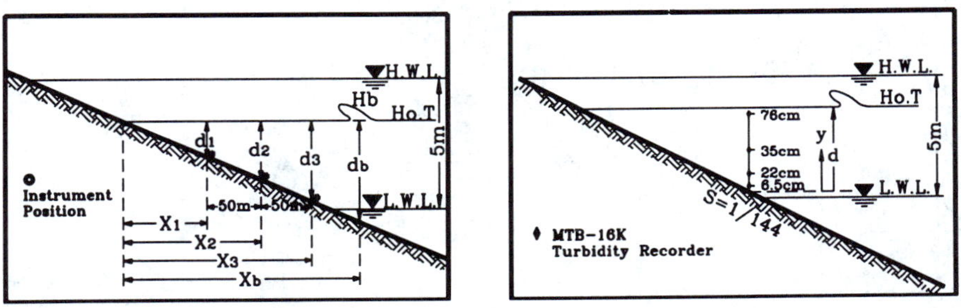

Figure 1 The schematic diagrams of parameters and memory MTB-16K recorders mounted on sea-bed

Field data of wave, water level and suspended concentration are recorded respectively every one hour interval by using the DNW-5M pressure sensor and the MTB-16K turbidity recorder from PM 7, 14 March, 1995 to AM 7, 16 March, 1995. The recorded periods of field data are continuous 10 minutes every times and data is adopted by one second interval.

Data Analysis and Discussions

3.1 The distributions of wave height across the surf zone

Figure 2 shows the parts of time histogram of the wave height and water level field measured data taken from the DNW-5M pressure sensor in the near shore on 14 March, 1995.

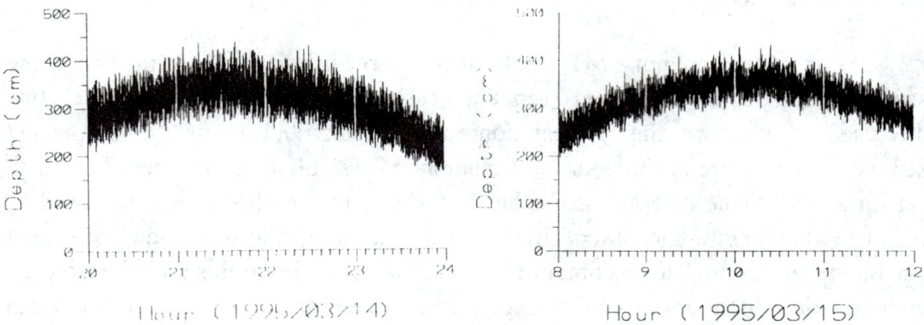

Figure 2 Time series of wave height and water level in the near shore

And the coresponding time series of wave height and wave period taken from the waverider at -20m water depth outside the surf zone are shown as Figure 3.

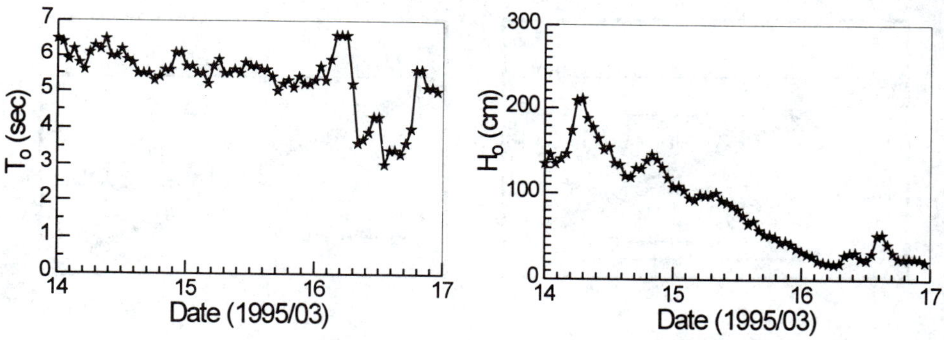

Figure 3 Time series of wave period and wave height at -20m water depth

Results indicate that the distributions of wave height are gradually decreasing from the largest H_0=2.2m occurred at AM 6, 14 March to the smallest H_0=0.20m occurred at AM 7, 16 March and its corresponding distributions of wave period are decreasing from 6~7 secs to 3~4 secs during wholly observing periods.

Figure 4 shows the relationship diagram between the local wave height H and its corresponding water depth d by analyzing the field measured data in the surf zone. Various legends such as "□" and "+" etc. represent the results in the flood or in the ebb. Results also indicate that the local wave height is linearly decreasing with water depth when the H/d value is equal to 0.4. If we convert this relationship with the form $H=r(d+H)$, then the empirical constant is $r \fallingdotseq 0.3$. This means that the distribution of wave height can be expressed as the following empirical form in the surf zone.

$$H = r(d + H) \qquad S = \frac{1}{144} \qquad (1)$$

$$H_0 < 2.2m$$

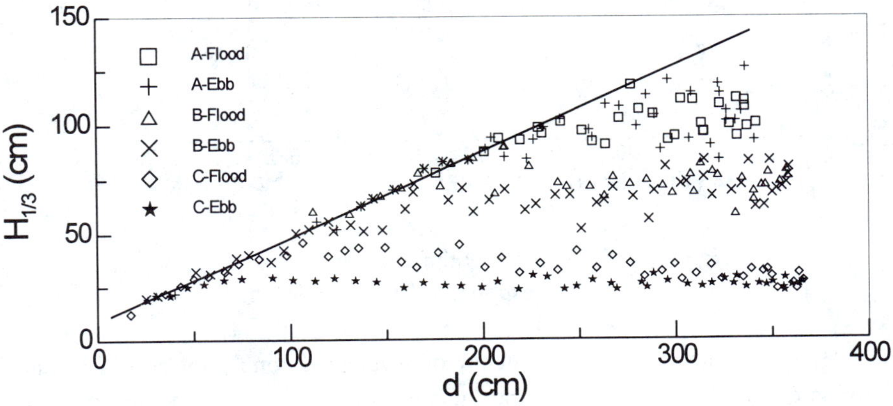

Figure 4 The relationship diagram between wave height and water depth in the surf zone

3.2 The distributions of suspended concentration in the surf zone

3.2.1 The vertical distributions of suspended concentration

Parts of the time series of suspended concentration measured at various positions and its corresponding water level are shown as Figure 5 and Figure 6 respectively.

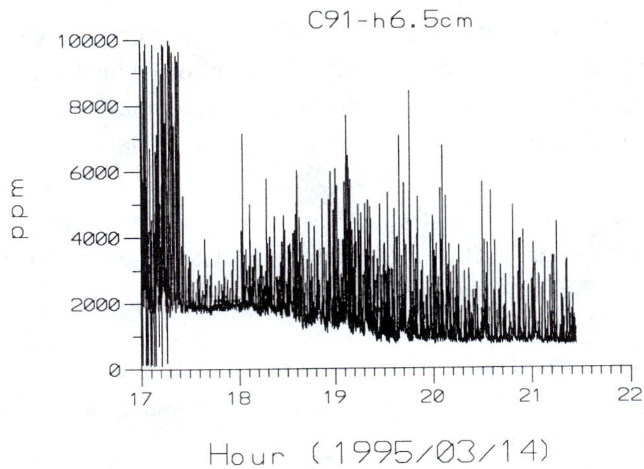

Figure 5 Time series of suspended concentration at various position

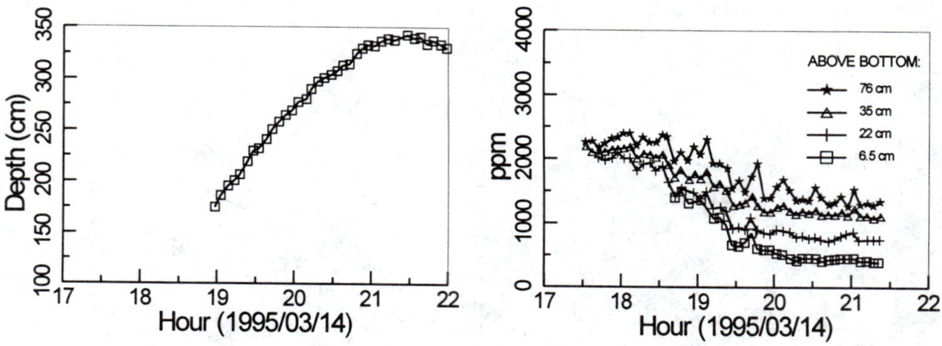

Figure 6 Time series of the water level and its corresponding suspended concentration

Results indicate that the tendency of suspended concentration is decreasing with increasing the water depth. And the higher water depth is, the discrepancy of suspended concentration measured at different height is smaller.

By using the concept of Figure 1 schematic diagram, the concentration distributions of the suspended sediment at different relative water depth d/L_0 can be measured by the tidal variations. Figure 7(a)~(d) shows the vertical distributions of suspended concentration at different relative water depth d/L_0. The dots and solid line in the Figure 7 represent respectively the field measured data and a regressive curve in this study. Results indicate that the vertical distribution of suspended concentration is more uniform for the smaller relative water depth d/L_0.

And results also show that the maximum suspended concentration mostly occurs at the bottom for different relative water depth under the same wave conditions. Figure 8(a)~(d) shows the dimensionless relationship between measured relative suspended concentration (C/C_{max}) and different relative water depth d/L_0.

Then, we can summarize the analyzed results in different way to express the vertical distributions of suspended concentration under various relative water depth d/L_0 as shown in Figure 9.

Moreover, the relationship diagram between the vertical suspended concentration at any relative depth y/d and the relative water depth d/L_0 can be also shown as Figure 10(a)~(c).

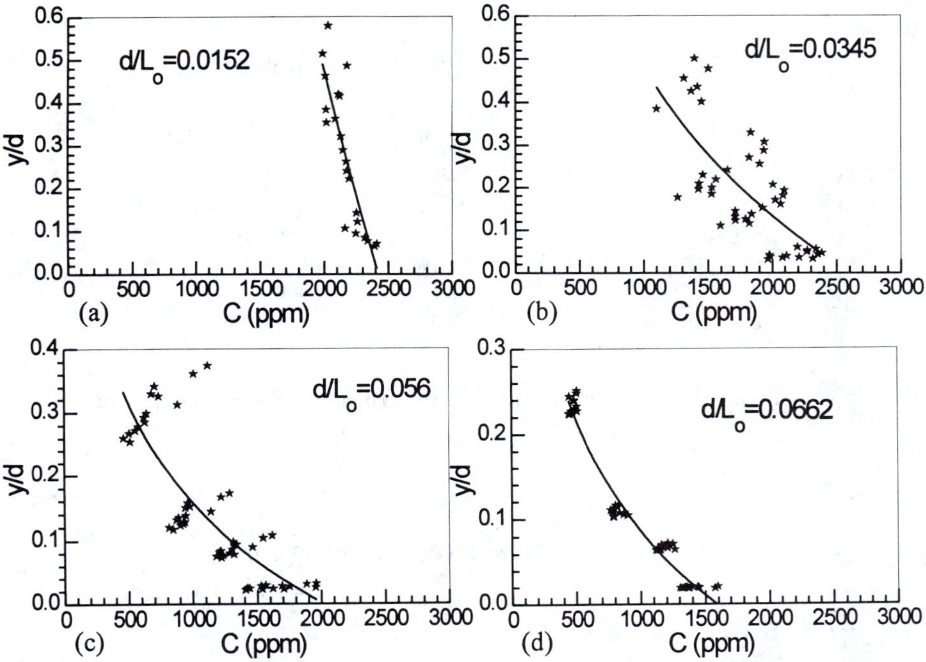

Figure 7 The vertical concentration distribution for the different relative water depth

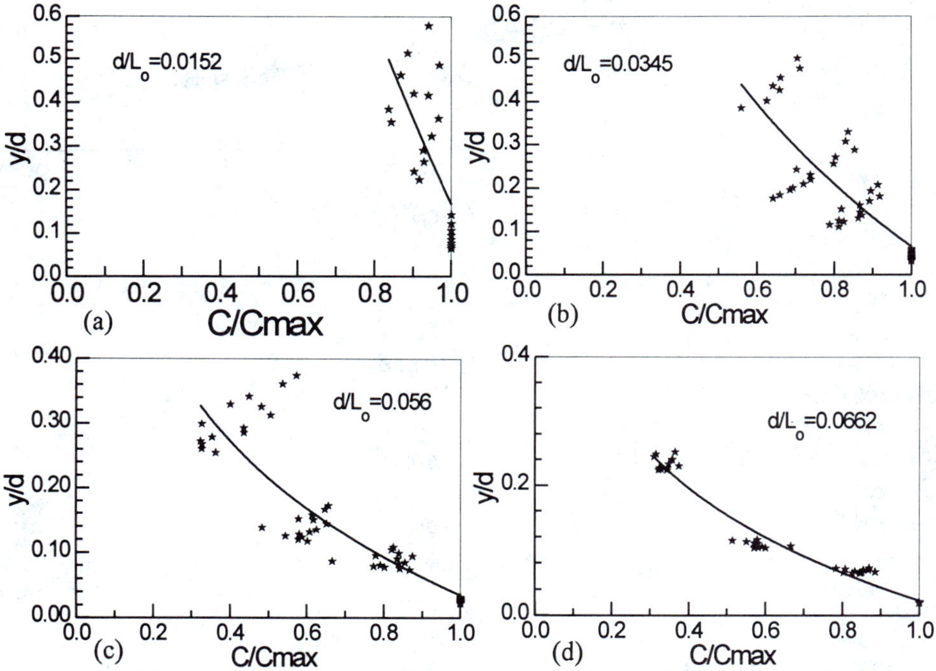

Figure 8 The vertical distribution of dimensionless suspended concentration

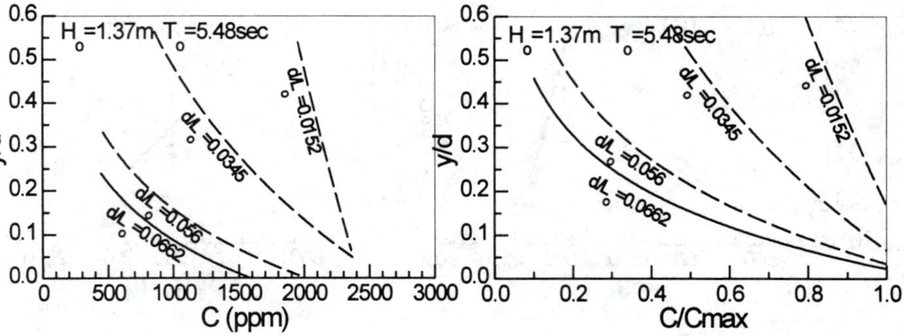

Figure 9 The vertical distributions of suspended concentration under the same wave condition

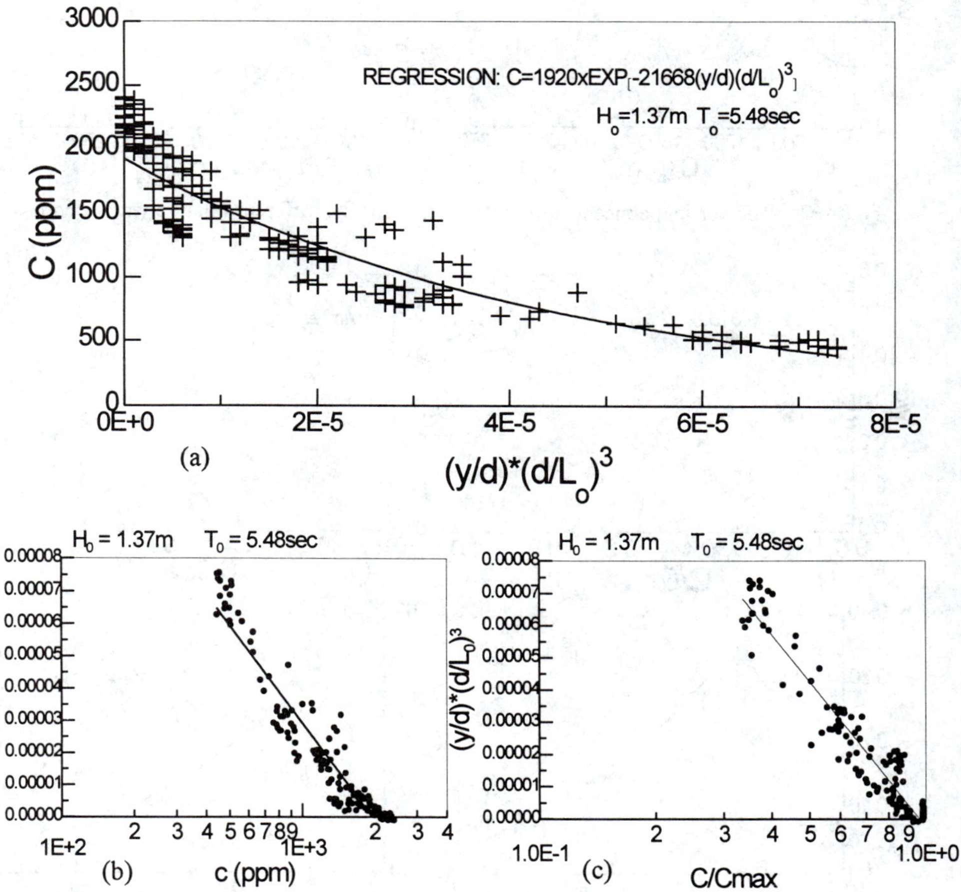

Figure 10 The relationship between suspended concentration C and $(y/d)\cdot(d/L_0)^3$

It means if the water depth d is given, then the suspended concentration at any depth y can be obtained by following regressive formulus.

$$C = 1920 \cdot e^{-21668\left(\frac{y}{d}\right)\cdot\left(\frac{d}{L_o}\right)^3} \tag{2}$$

where d : water depth

y : height from the seabed

C : suspended concentration, its unit is ppm

3.2.2 Suspended concentration near the seabed across the surf zone

Figure 11(a)~(d) shows the concentration distributions measured near the seabed across the surf zone under various wave steepness H_o/L_0. The definitions of parameter X and X_b are shown as Figure 1. The width of the surf zone, X_b, is computed by wave theory here. In general, results indicate the maximum values of suspended concentration usually occur around the breaking point for the same wave condition.

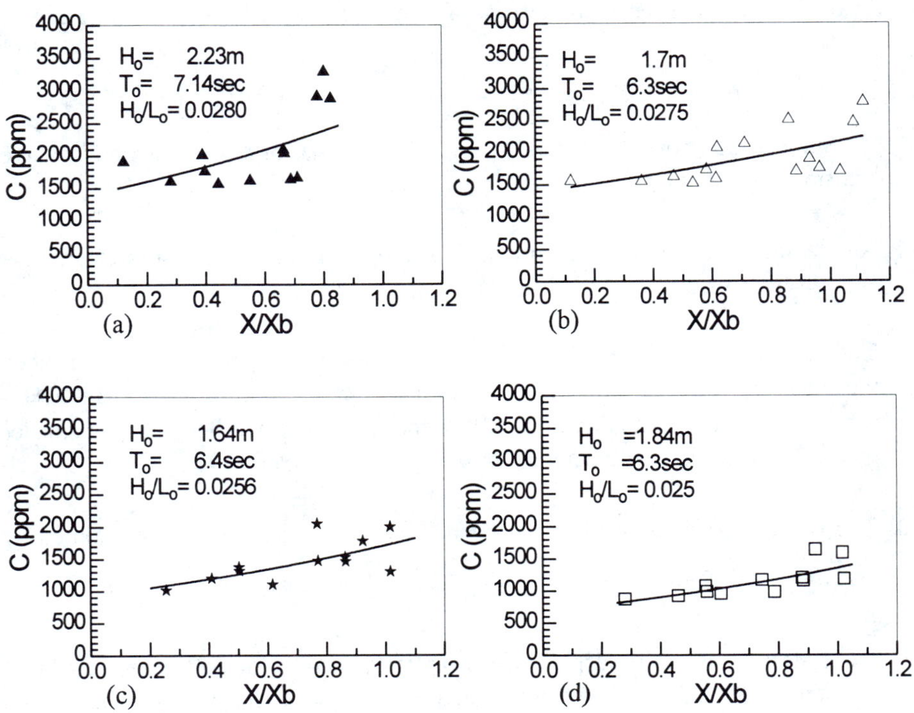

Figure 11 The distributions of concentration near the seabed across the surf zone

We can also summarize above results as Figure 12 under various wave steepness H_o/L_0. It means that the suspended concentration near the seabed will be decreasing with decreasing the depth. But it will be increasing when the incident wave steepness H_0/L_0 is increasing for the same depth.

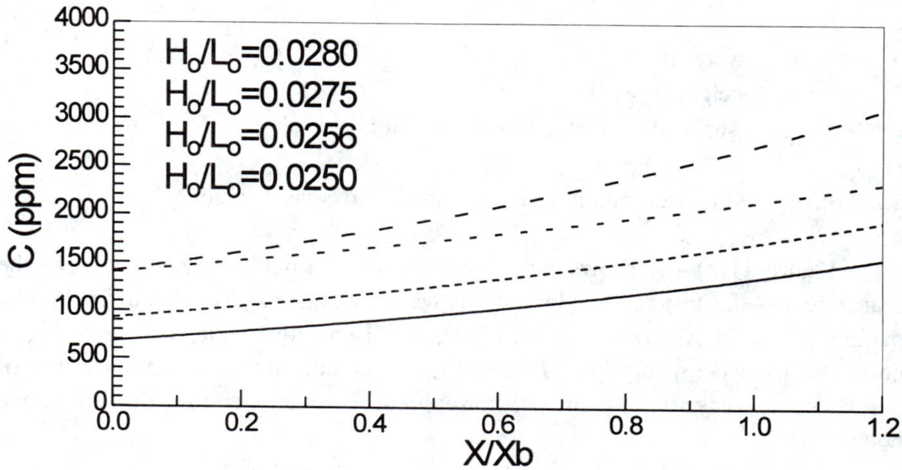

Figure 12 The concentration distributions near the seabed across the surf zone

<u>Conclusions</u>

Summarize above field data analysis and discussions, we can obtain some conclusions as following ：

1. All measured field data can be provided as the quantitative basis of the numerical analysis in this study.

2. The local wave height H will be linearly decreasing onshore with depth d when the incoming wave is broken. Its relationship can be expressed as H=r(d+H) when the beach is a mild slop S=1/144 and incident wave height H_0 is smaller than 2.2m, then the coefficient is r ≒ 0.3.

3. The vertical distribution of the suspended concentration C can be computed by equation $C = 1920 \cdot e^{-21668\left(\frac{y}{d}\right)\cdot\left(\frac{d}{L_o}\right)^3}$ if depth d is known.

4. The distributions of the suspended concentration across the surf zone are related with the incoming wave steepness H_0/L_0. In general, its maximum value usually occurs around the breaking point.

Acknowledgement

This paper is part of the research projects, which is funded by the National Science Council for the integrated research program "Coastal Space Utilities", under contact NSC 84-2611-E-124-002. Authors also deeply appreciate to Miss Chien, E. J. for typing this manuscript.

References

1. Fairchild, J. C. (1959)： Suspended sediment sampling in oscillatory wave action, Beach Erosion Board, Tech. Memo 115 P.
2. Homma, M. and K. Horikawa (1962)： Suspended sediment due to wave action, proc. 8th coastal Eng. Conf. PP.168-193.
3. Sleath, J. F. A. (1982)： The suspension of sand by waves, J. Hydraul Res., Vol.19, PP.439-452.
4. Makoto Ifuku and Tadao Kakinuma (1988)： Suspended sediment concentration in the surf zone, 21th ICCE, Chap.23, PP.1661-1675, ASCE.
5. P. Nielsen (1994)： Suspended sediment particle motion in coastal flows, Proceedings of 24 ICCE, Chap.174, ASCE.
6. Onoetl (1994)： Suspended sediment concentration profiles under Non-breaking and breaking waves, Proceedings of 24 ICCE, Chap.203, ASCE.
7. C. K. Chang and C. H. Hwang (1995)： Studies on Wave, Current and Suspended Sediment Characteristics at the Surf Zone, Coastal Dynamics'95, ASCE, PP.728-738.

CHAPTER 317

Suspended sediment mixing in the surf zone

José A. Jiménez, Francisco J. Rivero, Agustín Sánchez-Arcilla,
Vicente Gracia and Andrés Rodríguez[1]

Abstract

The vertical distribution of suspended sediment in the surf zone is analysed by using field measurements. The analysis focuses in the description of the shape of the profile, *i.e.* mixing coefficient, and to do this, several approaches have been followed: pure diffusion, pure convection and combined diffusion-convection. The obtained results show a strong dependence of the sediment diffusion coefficient with the sediment grain size whereas the convective length scale presented a smaller grain-size influence.

Introduction

The vertical distribution of suspended sediment inside the surf zone may be generally considered as a combination of convective and diffusive processes. The distinction between both can be done in terms of mixing lengths, which in turn depend on the hydrodynamics acting on the sediment. According to Nielsen (1991, 1992), this can be mathematically described in a time averaged form by the convection-diffusion equation

$$w_f c + \varepsilon \frac{dc}{dz} - pF(z) = 0 \qquad (1)$$

where w_f is the fall velocity of the sediment, c is the time averaged suspended sediment concentration, z is the elevation above the bottom, ε is the sediment diffusivity, p is the pick-up function and $F(z)$ is the convective entrainment function.

---

[1] Laboratori d'Enginyeria Marítima (LIM), Catalonia University of Technology (UPC), c/Gran Capitá s/n, 08034 Barcelona, Spain (jimenez@etseccpb.upc.es).

Equation (1) is generally truncated being the pure diffusive description the most common approach (see *e.g.* van de Graff and Roelvink, 1988 among others). In spite of this, the approach may be accurate enough to describe the suspended sediment distribution if sediment is characterised by a mean grain size (without considering the grain size distribution). However, when this distribution is taken into account, a dependence of diffusivity or mixed length with grain size can be often observed (see *e.g.* Nielsen, 1983, van de Graaf and Roelvink, 1988, van Rijn, 1993). Strictly speaking, although a slight variation of the diffusivity of a solid particle (the sediment) with respect to the one of the fluid should be expected (see *e.g.* van Rijn, 1984), large deviations between the sediment diffusivity and the fluid eddy viscosity should indicate that other mechanisms are affecting the sediment mixing.

In this work, the vertical suspended sediment distribution in the surf zone is analysed using field measurements. The analysis focuses on the shape of the distribution from the pure diffusive and/or convective approaches and, an attempt to use the combined diffusion-convection solution is also presented. This last point was introduced due to the observed dependence of the mixing coefficients with sediment characteristics.

Experimental data

The used data were acquired in a field campaign (DELTA' 93) carried out in the Trabucador Bar (Ebro delta, Spanish Mediterranean coast). The measured field data included bathymetry, waves outside and inside the surf zone, mean water levels across the surf zone, velocity field (both horizontal and vertical structure) by means of an instrumented sledge equipped with 6 electromagnetic currentmeters and a step wave gauge (see Rodriguez *et al.* 1995a for a full description of hydrodynamic measurements) and measurements of longshore suspended sediment transport across the surf zone.

Longshore suspended sediment transport were measured by means of portable sediment traps (hereinafter PST) similar to that described in Rosati and Kraus (1989). Each PST was composed by a vertical array of 5 to 6 traps, each one of the cube-type with a nozzle of 5.5x5.5 cm and with a streamer (collection bag) made with a mesh of 100 μm (Ortiz, 1995).

Measurements were taken in the inner part (inside the surf zone in all the cases) of a barred profile under wave conditions characterised by *Hrms* ranging between 0.4 and 0.6 m at 7.5 m depth and with peak periods, *Tp,* from 5 to 6 sec. Spilling breaking waves were representative of surf zone conditions during all the experiments, with induced longshore currents up to 0.7 m/s.

Figure 1 shows some measured vertical profiles of the longshore current, where it can be seen that for relative depths, $z/d \geq 0.1$, current velocity shows small changes in vertical, *i.e.* is nearly constant in vertical.

Longshore suspended sediment transport measurements were taken in three profiles across the surf zone, with 4 to 5 positions each. In each position, PST were operating during a time period of 2 to 5 min. After that time, PST are carried to shore and collected sand is stored in bags to be weighted and sieved at the laboratory.

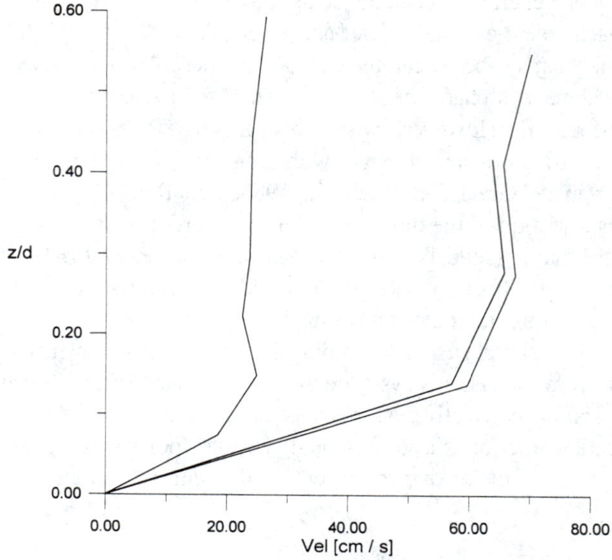

Figure 1. Typical vertical profiles of longshore currents measured during the experiments.

Vertical distribution of suspended sediment concentration

In order to obtain the vertical distribution, the first working hypothesis was that because measurements were acquired inside the surf zone, where a well mixed water column can be assumed, a *constant in vertical* mixing coefficient was considered.

Figure 2 shows the vertical distribution of the vertical eddy viscosity obtained from time series of velocity measurements inside the surf zone (see Rodriguez *et al.* 1995 for further details in the calculation of diffusivity). It can be observed that for relative depths ≤ 0.6 the calculated diffusivity can be considered as constant in vertical, whereas in the upper part of the water column values increase, due to the generation of horizontal turbulent-momentum flux during breaking (Rodríguez *et al.* 1995).

This assumption of a constant in vertical diffusivity leads to an *exponential concentration distribution* (when a pure diffusive process is considered) given by

$$C(z) = Co\ e^{-w_f z/\varepsilon}$$

where Co is the reference concentration at the bottom, w_f is the fall velocity of the sediment, z is the elevation above the bottom and ε diffusion coefficient.

Or in another form (assuming a pure convective process with an exponential function to describe convection, Nielsen, 992) as

$$C(z) = Co\ e^{-z/L}$$

where L is the mixing or length scale.

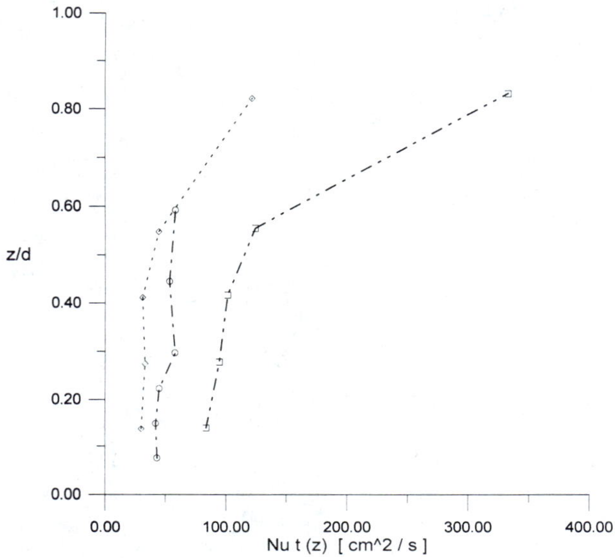

Figure 2. Vertical distribution of eddy viscosity obtained during the experiments.

When experimental profiles were fitted to a vertical distribution given by expressions (2) or (3), a good agreement was obtained. From 15 cases, covering three lines across the surf zone, the mean r^2 value of all the fits was 0.952, with a minimum r^2 value of 0.90 and a maximum of 0.995.

Figure 3 shows an example of the obtained normalised time averaged concentration profiles at different locations across the surf zone with the corresponding exponential fit for one of the monitored cross-shore profiles.

These results indicate that at least during the conditions prevailing during the field experiments, and for measurements taken inside the surf zone, the classical exponential profile describes well the time averaged concentration profiles. In this case, the averaging was done over a relatively long time period (from 20 to 50 waves) and because of this, these results cannot be considered in contradiction to those obtained by Mocke and Smith (1992) since the time scale of their analysis differs.

Across-shore distribution of mixing coefficient

When the cross-shore distribution of the mixing coefficient was analysed, an increasing diffusivity from the shoreline towards the breaking point was observed for all the analysed cases (as expected). Fig. 4 shows the cross-shore distribution of the mixing length, *L*, for one of the control lines, where it can be clearly this increase in mixing towards the breakers where reaches the largest value.

Nielsen (1984) analysing the mixing length -*L*- under non-breaking waves found that it was closely related to the bed ripple height. However, when the surf zone is considered a different "physical entity" of L should be expected, since the mechanism inducing convection is clearly different.

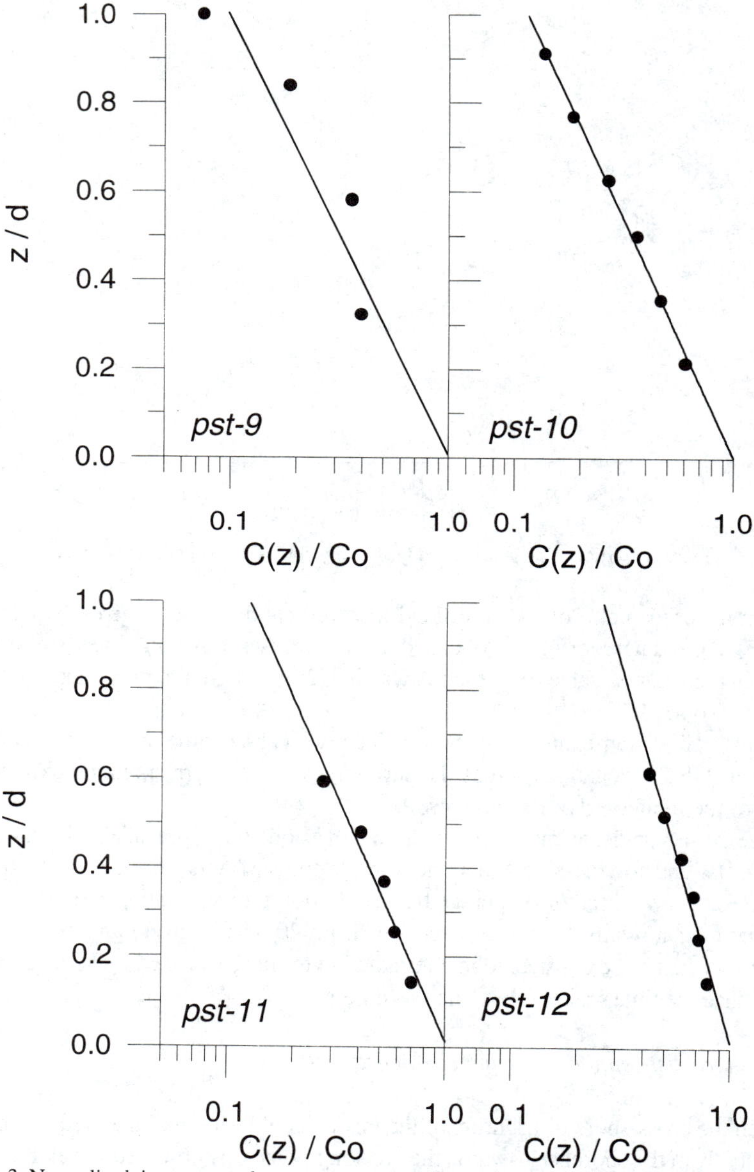

Figure 3. Normalised time averaged concentration profiles at different location across the surf zone.

In this case, it was found that this length scale for convection was closely related to the local H_{rms} (see Figure 4). This fact can be used to validate the assumption of that, inside the surf zone the sediment is picked-up from the bottom and put in suspension at an elevation above the bottom proportional to the wave height (Dean, 1973).

Thus, for the range of measured conditions and under spilling breakers, the local value of H_{rms} was found to be a good predictor of the magnitude of the convective length scale, L.

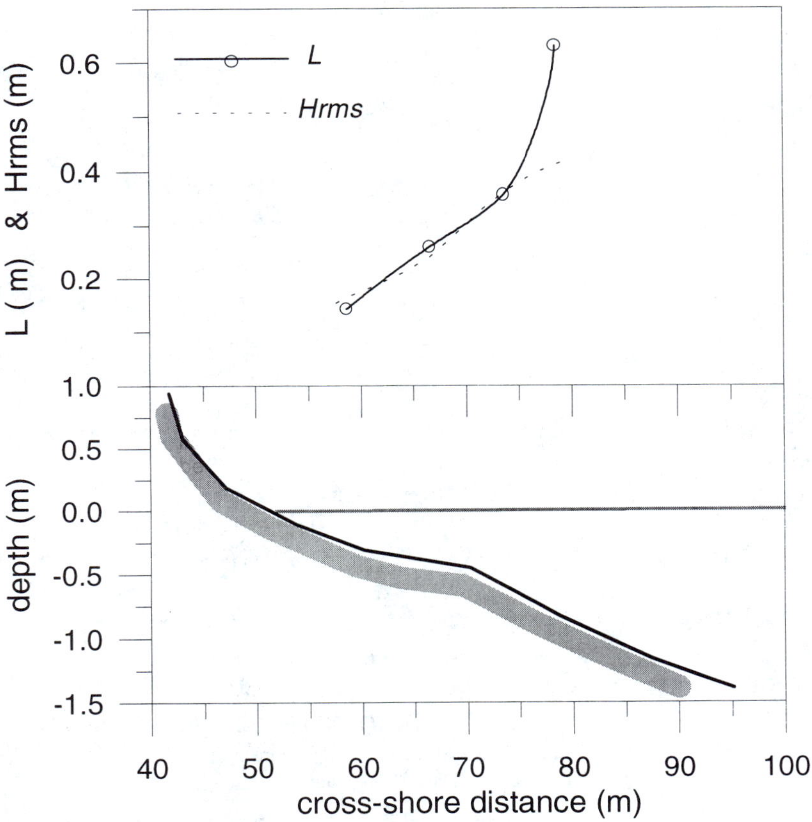

Figure 4. Cross-shore distribution of the mixing length -*L*- and *Hrms*.

When the cross-shore distribution of the mixing coefficient is put in terms of sediment diffusion coefficient, ε, the same pattern than before is observed (Fig. 5). This was due to that from equations (2) and (3) $L = \varepsilon / w_f$, and no significant across-shore variations in sediment grain size was found.

Assuming that this sediment diffusion coefficient must be closely related to the fluid diffusion coefficient, their values should be equivalent to the vertical eddy viscosity for the fluid.

Fig. 5 shows the cross-shore distribution of the eddy viscosity for horizontal mixing using a k-model with production of turbulent energy according to Battjes (1975). It can be seen that both distributions (horizontal and vertical mixing) are qualitatively similar, although the horizontal mixing coefficient is about two order of magnitude larger than the vertical one (see *e.g.* deVriend and Kitou, 1990). A detailed analysis of this discrepancy may be found in Svendsen and Putrevu (1994).

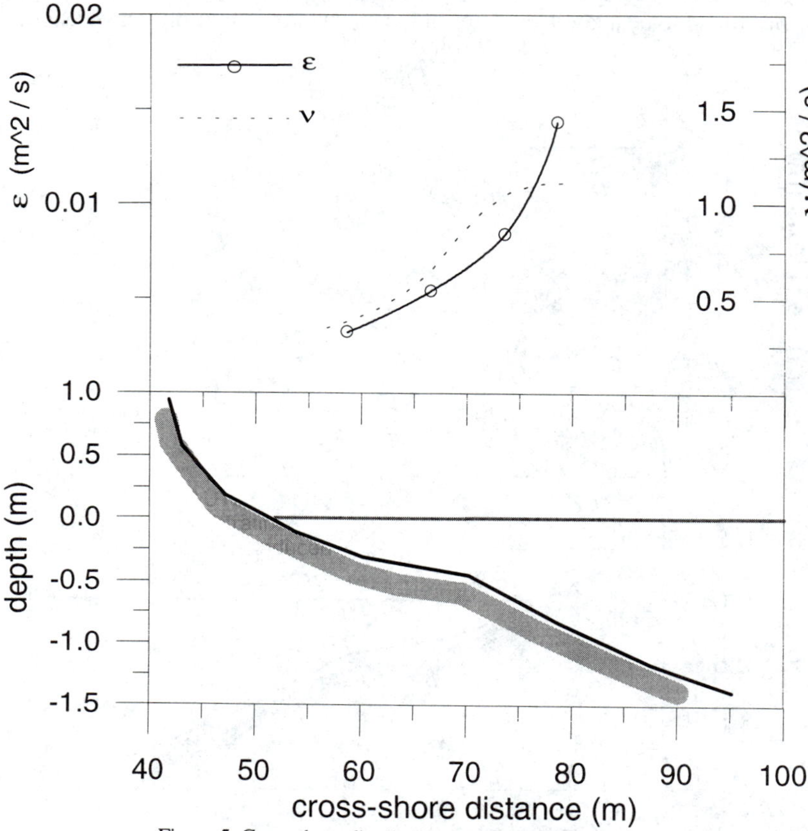

Figure 5. Cross-shore distribution of the sediment diffusion coefficient, ε, and calculated horizontal eddy viscosity ν.

Effects of the grain size

Once the vertical suspended sediment distributions (assuming pure diffusion and convection) were analysed, the effects of the grain size on the sediment mixing was investigated. To do this, each sample was sieved into a series of fractions and, vertical distributions for each fraction were built. Once these classified vertical profiles were obtained, the same analysis than before was applied, *i.e.* fit to equations (2) and (3) to obtain the corresponding mixing coefficients for each fraction..

Figure 6 shows the obtained sediment diffusion coefficients, ε, for the same cases than the ones presented in Fig. 3 but considering different fractions. It can be clearly seen that diffusion coefficient is far from constant for the analysed range.

Focusing on the range of fractions characterised by a sediment fall velocity between 0.02 m/s and 0.06 m/s (out of this range some of the fractions were not statistically representative to build a "classified profile"), a linear relationship between the diffusion coefficient and the fall velocity was found: the larger the fall velocity is, the larger the coefficient will be. Similar results have been obtained for non-breaking waves by van de Graaf and Roelvink (1984) among others (see also Nielsen, 1992).

This result should indicate that for constant fluid diffusion coefficient, the sediment mixing will increase as coarser the sediment is. This discrepancy has been usually explained in terms of a greater influence of the centrifugal forces on the sediment particles with respect to the fluid in a turbulent flow. This would lead to an increase in the effective mixing length and diffusion rate for the sediment (see *e.g.* van Rijn, 1984).

However, although a small change in diffusivity between solid particles and fluid can be expected, the variations shown in figure 6 seems too high to assume that they are due to the above mentioned effect.

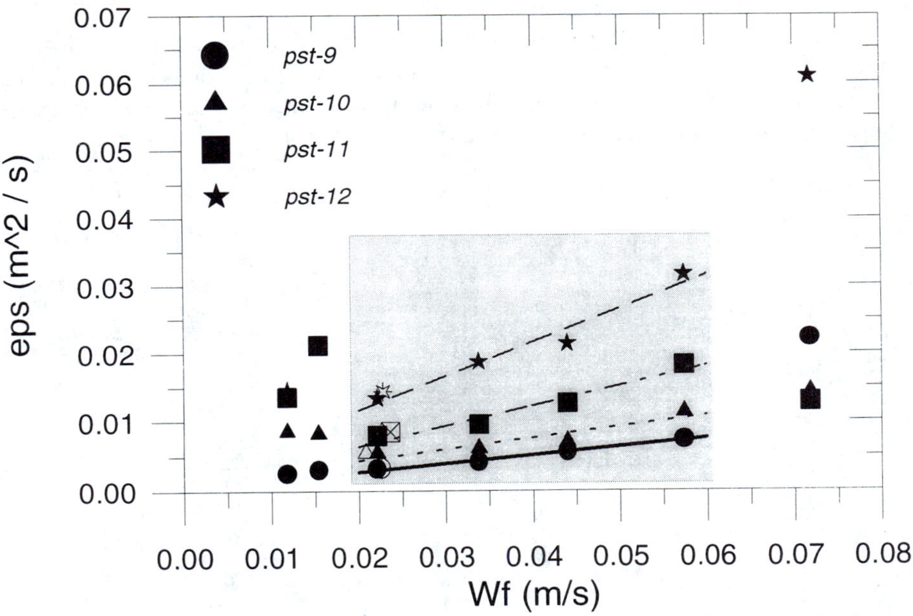

Figure 6. Variation of the diffusion coefficient ε_s *vs* fall velocity of the sediment for different locations across the surf zone. Grey rectangle indicates the range of fall velocities considered in the fit.

In fact, if the variations in diffusivity are measured as

$$\Delta\varepsilon = \frac{\varepsilon_{(w=0.06)} - \varepsilon_{(w=0.02)}}{\varepsilon_{(\text{total sample})}} 100 \quad (4)$$

a mean value of 138% is obtained (considering the four locations across the surf zone).

From figure 6, it seems that the diffusivity dependence with grain size increases from shoreline (pst-9) towards the breaker point (pst-12). Although this is true in absolute terms, when this variation is measured in relative terms (*e.g.* according to eq. 4), they are almost uniform, *i.e.* it seems to be constant across the surf zone.

Thus, the observed variations in the diffusion coefficient for the sediment mixing due to variations in grain size, introduce an uncertainty about the validity of the pure diffusion model to describe the vertical distribution of suspended sediment inside the surf zone.

Applying the same analysis to the case of pure convection model, *i.e.* looking to the variations in L when the different fractions are considered, the results depicted in Figure 7 are obtained.

In this case, although a variation in L is also observed, the direction of such variation is opposite to the above described one. Thus, assuming a pure convective process, as coarser the sediment is, smaller the mixing length will be. In other words, for coarser sediments, the mixing will be smaller.

This can be explained assuming that sediment response to convective process induced by wave breaking will be influenced by the sediment "weight". For heavier sediments (coarser), the theoretical distance from the bottom where they will be put in suspension (entrainment level) will be shortened as coarser the sediment is.

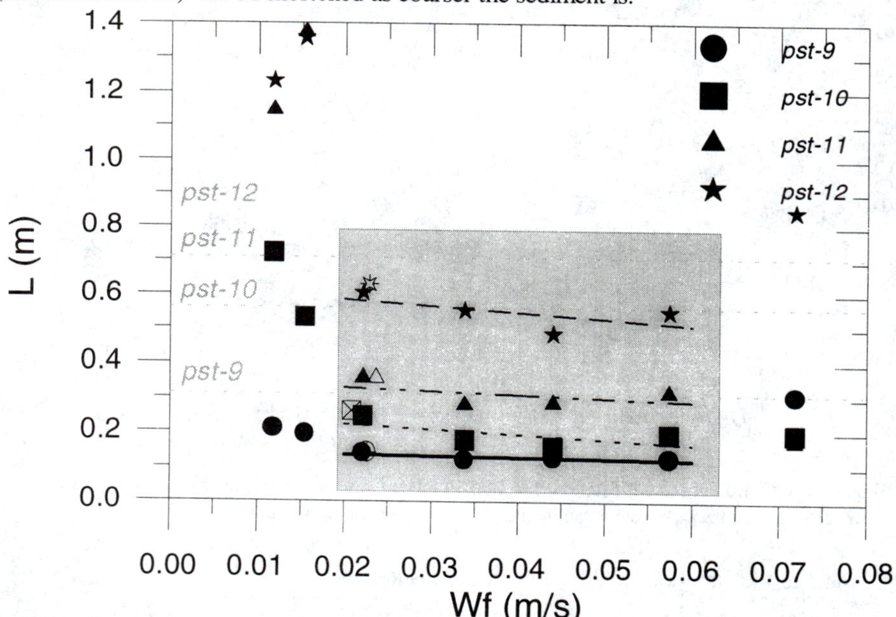

Figure 7. Variation of the length scale, L, *vs* fall velocity of the sediment for different locations across the surf zone. Grey rectangle indicates the range of fall velocities considered in the fit and grid lines represent largest value of L for each location (equal to the local depth).

Measuring the variations in the convective length scale, L, according to the equation 4, a mean value of 10% is obtained. The relative variations across the surf zone are also constant as it was for the previous case.

The main difference between the variations for diffusive and convective mixing lengths is that the former presented a larger variation than the convective one. In other words, the use of equation (3) to describe the vertical distribution of suspended sediment in the surf zone (assuming a pure convective process) is more "stable" in function of the sediment grain size than the use of a diffusive process. In spite of this, a small variation in the length scale was also found.

If the dominance of the process controlling the vertical distribution is measured in terms of its "stability", the convection process should be dominant under the monitored conditions.

Combined diffusion-convection process

Additionally to the above mentioned effect (variation in mixing coefficients for varying grain sizes), when the profiles for the different fractions were analysed, a change in the shape of the profile was also found. Thus, slight convex upward profiles were found to describe the fine fractions, whereas the coarser ones showed a slight concave upward shape. This change in shape was also explained by Nielsen (1992) as an indication of the presence of a combined diffusion-convection process controlling the sediment suspension.

In order to make a first attempt to describe the obtained suspended sediment profiles by using the combined diffusion-convection approach, a full solution of equation (1) was used.

In this attempt, a constant in vertical diffusion coefficient and a convective function given by an exponential law was selected (see *e.g.* Nielsen, 1992). Although this solution has not to be the best one, it was selected due to its simplicity, although presently other combined solutions are being analysed.

The integration of equation (1) with the selected description of diffusion-convection process leads to (Nielsen, 1992)

$$C = Co\left\{\frac{1}{1-\dfrac{\varepsilon}{w_f L}}\,e^{-z/L} + \left(1-\frac{1}{1-\dfrac{\varepsilon}{wL}}\right)e^{-wz/\varepsilon}\right\} \qquad (5)$$

By using the equation (5), the profiles for the different fractions presented in figure 8 were obtained. The main problem to apply this approach is that both diffusion coefficient as well as convective length scale has to be estimated. In this case, it was firstly selected the value of the diffusion coefficient by using the one corresponding to the finest fraction, and afterwards, the convective length was selected by fitting the predicted distribution to the measured profiles. In all the cases, the fitted values of convective length were in a range of 30%-40% of the obtained using a pure convective model.

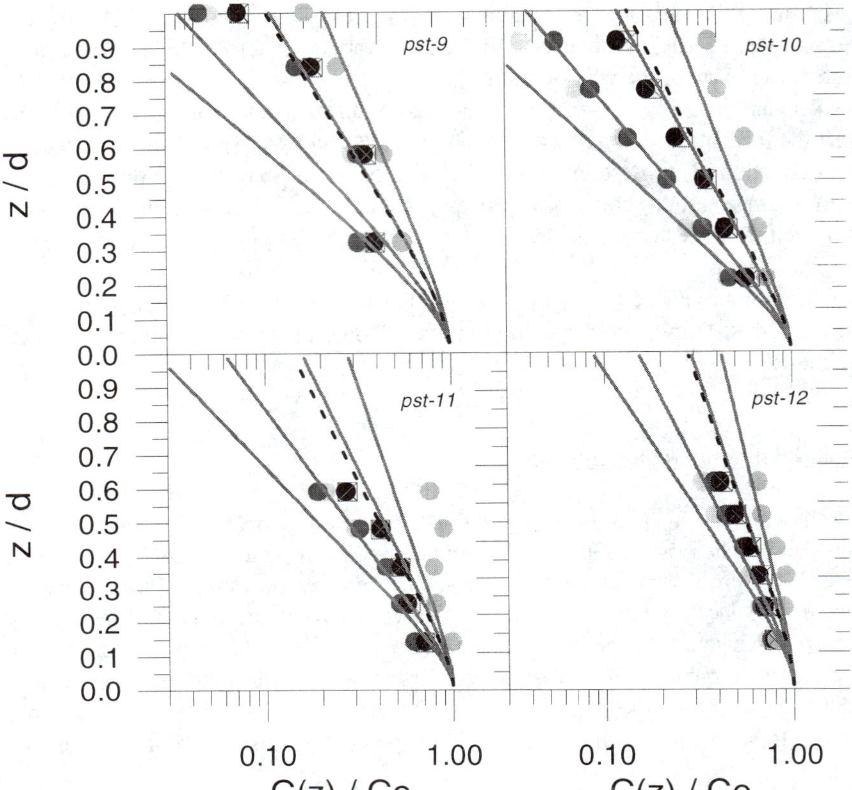

Figure 8. Use of the combined diffusion-convection approach (according to equation 5) to describe the vertical distribution of the different sediment fractions across the surf zone.

Conclusions

For the analysed conditions, several conclusions can be obtained.

- During the measured conditions, the simple *exponential profile* describes well the time-averaged vertical distribution of suspended sediment in the surf zone.
- The *length* scale, L, of the distribution can be well represented by the local value of H_{rms} inside the surf zone.
- The eddy viscosity ν_t for horizontal mixing (obtained using a K-model with production of turbulent energy according to Battjes (1975)) shows a *cross-shore distribution similar* to the vertical mixing coefficient obtained from sediment data, ε_s. Obtained values of ν_t are two orders of magnitude higher than ε_s.

- For the analysed field data, the mixing coefficient ε_s *depends on* sediment characteristics, i.e. w_f. In the range of w_f = 0.02-0.06 m/s, a *linear relationship* between ε_s and w_f. was found. Larger the fall velocity is, larger the diffusion will be.
- Assuming that the vertical distribution can be described by the length scale, L, *a linear dependency* with w_f was also found.
- The magnitude of the measured variation of ε_s with w_f was one order of magnitude larger than the obtained for L.
- These last two conclusions lead to argue the validity of the pure diffusion model to be used in the surf zone.

A first attempt to improve the description of the vertical distribution of suspended sediment considering the different grain has been done. To do this, the approach of Nielsen (1991, 1992) was followed by using a combined solution including diffusion and convection. First results are promising although further works have to be done, specially in the description of the "convective" function.

The generalisation of the obtained conclusions should require additional field data under different energetic conditions. This additional "field effort" is being carried out in the framework of an EU funded Research Project.

Acknowledgements

This work was carried out as part of the *FANS* project. It was supported by the EU *Marine Science and Technology (MAST-III) Programme*, under Contract No. MAS3-CT95-0037. This work was also co-sponsored by the *Spanish Ministry of Education and Research* through *CICYT*. Field data were acquired in the framework of a project funded by *Programa de Clima Marítimo, Spanish Ministry of Public Works.* JAJ was partly supported by a travel grant given by the *DPTOP (Generalitat de Catalunya)* through the *ETSECCPB (UPC).*

References

Battjes, J. 1975. Modeling of turbulence in the surf zone. *Proc. Symp. on Modeling Techniques,* 1050-1061.

Dean, R.G. 1973. Heuristic models of sand transport in the surf zone. *Proc. Conf. on Engineering Dynamics in the Surf Zone,* 208-214.

De Vriend, H.J. and Kitou, N. 1990. Incorporation of wave effects in a 3D hydrostatic mean current model. Delft hydraulics Report H-1295.

Mocke, G.P. and Smith, G.G. 1992. Wave breaker turbulence as a mechanism for sediment suspension. *Proc. 23rd ICCE,* ASCE, 2279-2292.

Nielsen, P. 1983. Entrainment and distribution of different sand sizes under water waves. *Journal of Sedimentary Petrology*, 53, 2, 423-428.

Nielsen, P. 1984. Field measurements of time-averaged suspended sediment concentrations under waves. *Coastal Engineering*, 8, 51-72.

Nielsen, P. 1991. Combined convection and diffusion: A new framework for suspended sediment modelling. *Coastal Sediments' 91*, ASCE, 418-431.

Nielsen, P. 1992. *Coastal Bottom Boundary Layers and Sediment Transport*, World Scientific, Singapore, 324 pp.

Ortiz, V. 1995. *Análisis experimental del transporte longitudinal de sedimento en suspensión en zona de rompientes.* Ms.Sc. Thesis, catalonia University of Technology, Barcelona.

Rodríguez, A., Sánchez-Arcilla, A., Bahía, E. and Mosso, C. 1995a. DELTA' 93: Experimental study of surf-zone hydroynamics at the Ebro delta. *MAST G8M Workshop,* Gdansk.

Rodríguez, A., Sánchez-Arcilla, A., Gómez, J. and Bahía, E. 1995b. Study of surf-zone macroturbulence and mixing using Delta'93 field data. *Coastal Dynamics' 95,* ASCE, 305-316.

Rosati, J.D. and Kraus, N.C. 1989. *Development of a portable sand trap for use in the nerashore.* Tech. Rep. CERC-89-11.

Svendsen, I.A. and Putrevu, U. 1994. Nearshore mixing and dispersion. *Proc. Royal Soc., London,* Series A, 445, 561-576.

van de Graaf, J. and Roelvink, J.A. 1984. Grading effects in concentration measurements. *Proc. 20th ICCE*, ASCE, 1618-1634.

van Rijn, L.C. 1984. Sediment transport, part II: suspended load transport. *Journal of Hydraulic Engineering,* 110, 11, 1613-1641.

CHAPTER 318

Sand Suspension Events and Intermittence of Turbulence in the Surf Zone.

R.D. Kos'yan[1], H. Kunz[2], S.Yu. Kuznetsov[3], N.V. Pykhov[3], M.V. Krylenko[1]

Abstract

The results of the joint field experiment Norderney-94 demonstrates the excellent coincide of fluctuations of suspended sand concentration and kinetic turbulent energy in the nearbottom region of the surf zone. Turbulent fluctuations were separated from the wave induced fluctuations as a part of the motions that are incoherent to the free surface elevation. The time and spatial scales of macroturbulent vortexes had been estimated and discussed.

Introduction

For the prediction of sediment transport in the breaking zone it is necessary to know the distribution of the suspended sediment concentration, of the water velocity and of the cross-correlation between them in the whole frequency spectrum of irregular waves. The analysis of modern approaches to the solutions of this problem published in literature and discussed at the international conferences during the last years (Coastal Dynamics 94, Barcelona/Spain, 1994; 24th Coastal Engineering Conference, Kobe/Japan, 1994; Coastal Dynamics 95, Gdynia/Poland, 1995; MAST 2 - Overall Meeting, Gdynia/Poland, 1995) shows that the most perspective ones are the models based on Boussinesq and Serra equations that afford to adapt these parameters to the time scales of approaching waves and k-e models. But the latter models today allow only to assess mean-time profiles of the suspended sediment concentration in the breaking zone. The possibility to use k-e model for the calculation of the suspended sand fluctuations in time is restricted by the absence of precise descriptions of the physical mechanisms of sediment

[1] *The Southern Branch of the P.P.Shirshov Institute of Oceanology, Russian Academy of Sciences. Gelendzhik-7, 353470 Russia. E-mail: kosyan@sdios.sea.ru;*

[2] *Coastal Research Station. NLOE-Forschungsstelle Kueste. An der Muehle 5, D-26548, Norderney, Germany. E-mail: kunz.crs@t-online.de*

[3] *P.P.Shirshov Institute of Oceanology, Russian Academy of Science, 23, Krasikov str., Moscow, 117851, Russia. E-mail: sergey@shelf.msk.ru; pykhov@coast.msk.ru*

suspension above the bottom, it's distribution in the water column within the breaking zone and by the absence of quantitative correlation between the suspended sediment concentration and parameters of the turbulence for field conditions.

From the physical point of view the presence of instantaneous water flows from the bottom with a velocity that exceeds the sand particle settling velocity is a necessary condition for the occurrence of sediment suspension. In the wave breaking zone such water motions are forced, most probably, by macroturbulent vortexes. Two basic mechanisms of such vortexes formation are possible:

1. Vortexes formation near the water surface under the crests of the breaking waves and its penetration to the bottom. Such a mechanism was observed in laboratory experiments with monochromatic waves above a hard and a movable bottom (Nadaoka, Kondoh, 1989; Zhang et al., 1994). By these experiments it was determined that vortexes were formed as well with horizontal axis as with inclined vertical axis depending on the wave parameters, on the type of their breaking and on the bottom slope. When reaching the bottom, such vortexes, like a tornado, can capture sand from the bottom and suspend it in the flow. The dominant action of this turbulence for sediment suspension is verified by Sato et al. (1990) in a wave flume.

2. Vortexes formation in the bottom boundary layer due to the shear instability of water flow above a plane rough bottom. In this case an explosive character of the generation of turbulent energy and formation of coherent vortexes takes place. Those vortexes are forced from the bottom to the water column. The possibility of realization of such a mechanism has been shown in laboratory experiments (Hino et al., 1983), and its responsibility for the sand suspension is indirectly confirmed in the papers of Foster et al. (1994) and of Pykhov et al. (1995).

The reality of the existence of the first mechanism under field conditions is confirmed by some (not numerous for the present) measurements of the macroturbulence in the surf zone (George et al., 1990; Rodriguez et al., 1995). The results of this research only give a possibility to assess a time-mean value of the turbulent energy, the integral length scale, the turbulent viscosity. But they do not afford to trace the temporal variability of these parameters, which is necessary for the analysis of time fluctuations of the suspended sediment concentration. In spite of the fact that the hypothesis about sediment suspension by such vortexes is shared by the majority of investigators, till now there is no confirmation by direct field measurements.

The aim of our paper based on the data of field measurements is to confirm the existence of turbulent mechanisms of sand suspending by the determination of the scales of fluctuations of turbulent kinetic energy and suspended sand concentration and by ascertaining dependencies between these fluctuations.

Experiment

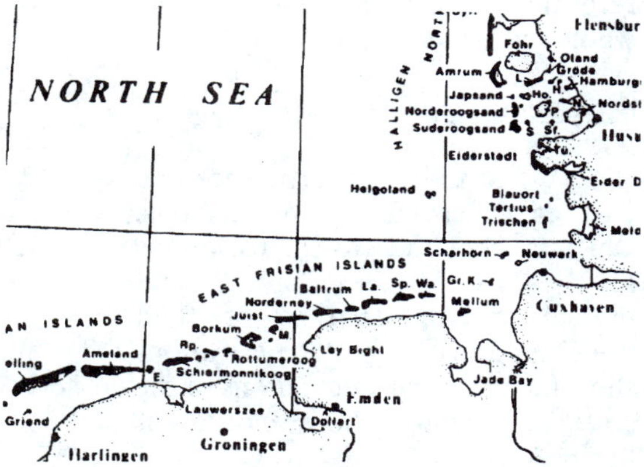

Figure 1. Site of the field experiment.

Field experiment was carried out on a beach section of Norderney Island, North Sea in October 1994 (Fig. 1).

A joint German-Russian team of scientists carried out multipurpose survey at an especially equipped testing site, which is located on the north-western side of the island. In this paper the results of measurements in one point (10 centimeters and more above the bottom) are presented and discussed (Fig. 2). The suspended sand concentration, the free surface elevation and the two horizontal components of water velocity were measured synchronously by optical turbidimeter, pressure gauge and two-component electromagnetic current meter at sampling rate of 18.2 Hz over 20 minutes.

The turbidimeter measures the light beam attenuation at a distance of 6 centimeters that then is recalculated into the suspended sand concentration (Kos'yan et al., 1994). Two-component electromagnetic current meter and pressure sensors (Type NSW) had been used for measuring. Their linear calibration characteristics are in the band of frequency from 0 to 20 Hz. The diameter of the measuring head of the currentmeter is 6 centimeters.

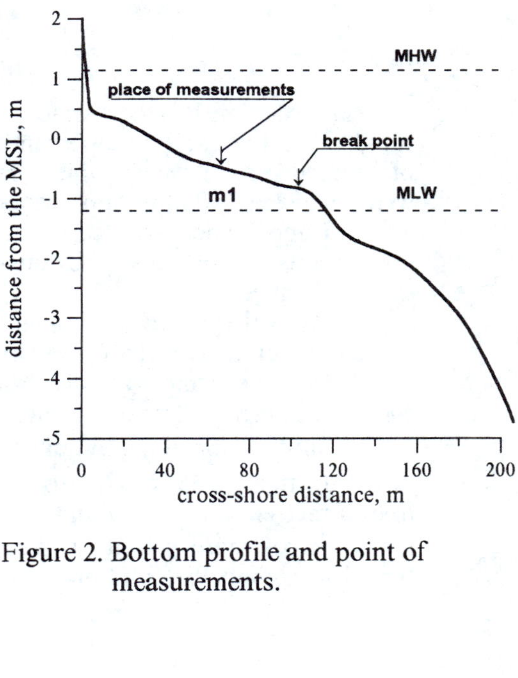

Figure 2. Bottom profile and point of measurements.

The bed consists of sand with a mean diameter of approximately 0.2 mm. The tidal range of about 2.4 m afforded to produce the measurements at a depth range from 0.5 to 2.5 m.. The measurements were carried out in the inner and middle parts of the surf zone under the wind wave and swell conditions. The 25 series of recording were obtained during the whole experiment.

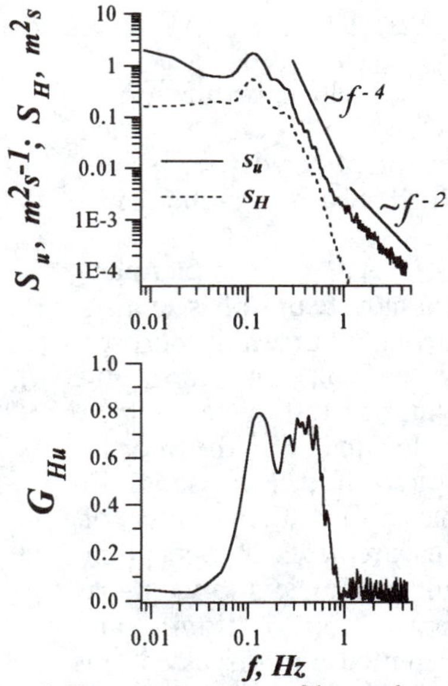

Figure 3. Determination of boundary frequency between waves and turbulence. f_b=0.8 Hz. Record 5a. H_S=1.03 m, T_P=8.7 s, $\bar{h}$=1.53 m, Sp.

Separation of turbulence

We consider the turbulence as a part of water particle fluctuations, which is incoherent to the elevations of the free surface. Spectral and mutual spectral analysis of fluid velocities and free surface elevations show the turbulent fluctuations of the cross- and the long-shore velocities at the frequencies $f > 0.8$ Hz. This is conformed by the absence of the coherence between cross-shore velocity ($u(t)$) and free surface elevations ($H(t)$) at the frequencies $f > 0.8$ Hz and by the change the velocity spectra ($S_u(f)$) gradient from $\sim f^{-4}$ to $\sim f^{-2}$ at $f = 0.8$ Hz (Fig. 3). Turbulent components of the cross-shore $u_t(t)$ and long-shore $v_t(t)$ velocities were obtained by digital filtering.

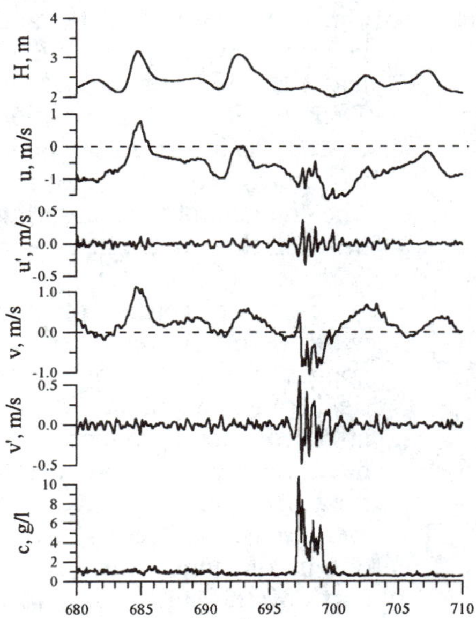

Figure 4. Example of single suspending event. Record 6a. H_S=1.07 m, T_P=8.7 s, $\bar{h}$=1.53 m, Sp.

Examples of events

Analyses of the records show a rapid appearance and disappearance of suspended sand events, which are well coincided with the appearance and disappearance of turbulent fluctuations of the cross- and long-shore velocities.

A typical example for a single suspending event is given by Figure 4. Recorded are wind waves with a significant height H_s=1.07 m, mean period T_P=8.7 s, breaking by spilling at the depth $\bar{h}$=2.36 m. $H(t)$ is the free surface elevation, $u(t)$ and $u'(t)$ are the cross-shore velocity and its turbulent component, $v(t)$ and $v'(t)$

are the long-shore velocity and its turbulent component, C(t) is the suspended sand concentration. The splash of turbulent velocity fluctuations and the splash of concentration between 697 to 700 s may be explained by horizontal advection of turbulent vortexes with trapped sand at the sensor array.

Figure 5 shows the two consecutive events of sand suspending for a more intensive wave regime (H_s=0.90 m, T_P=8.7 s, $\bar{h}$=1.62 m, breaking type is between spilling and plunging). The change of the amplitude of the turbulent fluctuation coincides well in time with the splash of concentration.

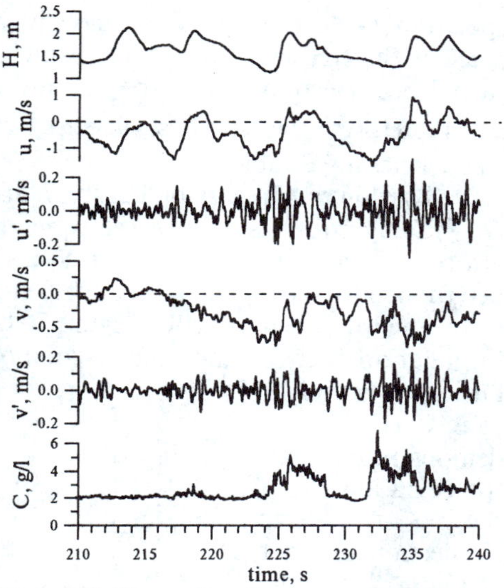

Figure 5. Example of two consecutive events of sand suspending. Record 7f. H_S=0.90 m, T_P=8.7 s, $\bar{h}$=1.62 m. Sp-Pl.

Figure 6 demonstrates the intermittence of turbulence and the corresponded events of sand suspending events during the 500 s for the same record as shown on Figure 5. The fluctuation of turbulent kinetic energy ($E(t)$) coincides well with the fluctuation of suspended sand concentration. Turbulent kinetic energy was calculated as the sum of the squared turbulent velocity components: $E(t) = u'^2(t) + v'^2(t)$

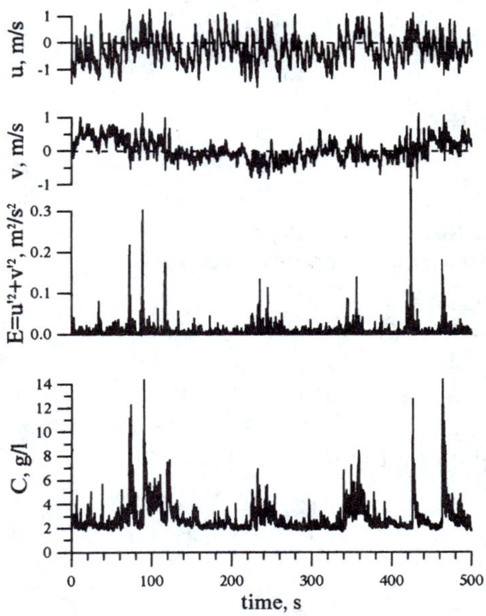

Figure 6. 8-minutes fragment of suspending events. Record 7f. H_S=0.90 m, T_P=8.7 s, $\bar{h}$=1.62 m. Sp-Pl.

Time scales of the turbulence

The upper parts of Figure 7 show the typical spectra describing the concentration fluctuations $S_C(f)$ and the kinetic turbulent energy variations $S_E(f)$ as function of frequency. Both spectra coincide with increasing frequency. The concentration spectrum has a feebly marked peak at the main wave frequency. Turbulent energy spectrum is decreasing monotonously with increasing frequency except the peak at the frequency of 1.6 Hz, which is called forth by the method of the turbulence selection, i.e. by the filtering with a boundary frequency of 0.8 Hz, and therefore it has no physical sense (when squaring ($E(t) = u'^2(t) + v'^2(t)$) the frequencies are doubled).

The lower left part of Figure 7 displays the squared coherence function $G_{CE}(f)$. Significant values of the coherence G_{CE} at the frequencies $f < 0.05$ Hz afford to say, that the low frequency suspended sand fluctuations are determined by the bursts of the turbulent velocity fluctuations.

The lower right part of Figure 7 shows for low-frequencies ($f < 0.05$ Hz) the relation between the related components of concentration (C_{low}) and of kinetic turbulent energy (E_{low}). It demonstrates that the deviations from the linear connection are not caused by the nonlinearity, but by random reasons.

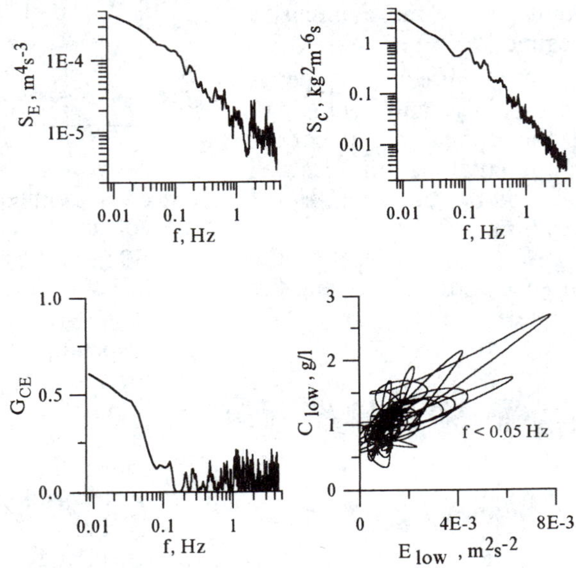

Figure 7. Connection between suspended sand concentration and turbulent energy. Record 5a. H_S=1.07 m, T_P=8.7 s, $\bar{h}$=2.33 m. Sp.

Let us consider the regularities of turbulent fluctuations within one suspension event.

Figure 8 presents the chronograms of turbulent components of velocities and their hodograf for a single suspending event, which was demonstrated at Fig. 4. The hodograf shows, how that the end of the vector of turbulent velocity component runs two full circles during one second. This may correspond to the passing of a series of four vortexes, as principally shown in the lower right part of Figure 8. Neighboring vortexes must rotate in opposite directions.

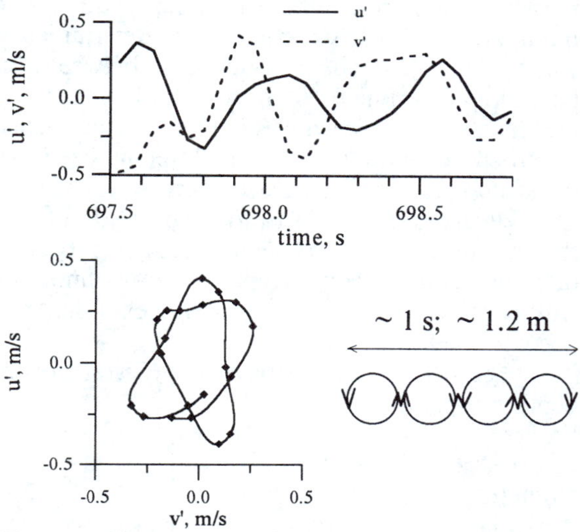

Figure 8. Time scales of turbulent vortexes.

An assumption about the passing of a series of several consecutive vortexes is confirmed by results of laboratory investigations presented at the conference Coastal Dynamics'94 by Zang et al. who presented the nomogram of the type and number of turbulent vortexes that are formed in the moments of wave breaking depending on Reynolds number and breaker-type index given by Galvin. The nomogram shows that all our recorded wave regimes occurred in the region of formation of triple oblique vortexes.

Spatial scales of turbulence

To assess spatial scales of turbulence the chronograms of the turbulent velocity fluctuations were recalculated into series of fluctuations along trajectories of the vortex movement according to Tailor's hypothesis "about frozen turbulence". Wave components (f < 0.8 Hz) were used as carrier velocities. The obtained series of spatial distribution of turbulent velocity fluctuations are presented as stick diagrams, an example is presented on Figure 9. The zero point of the spatial coordinate has been fixed arbitrarily.

The obtained stick diagrams show that turbulent vortexes are grouped in large structures. The passing of these vortex-structures across the point of

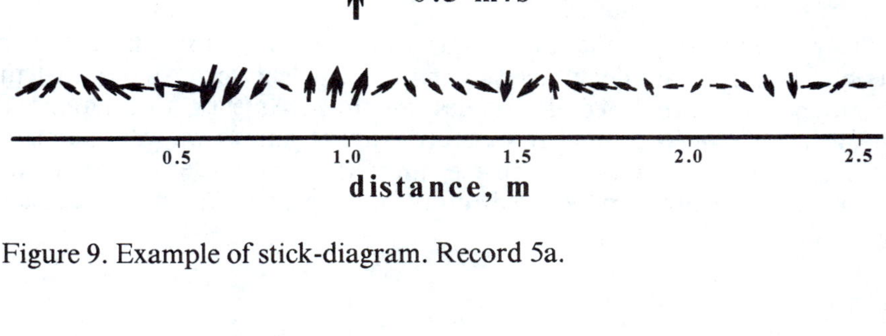

Figure 9. Example of stick-diagram. Record 5a.

measurements coincides with the sand suspending events. The spatial boundaries of such structures were determined visually by using the velocity value of 0.1 m/s as a boundary between the vortex structure and the background turbulence. The length of such structures was varied from 1 to 10 m. Linear scales of vortexes in the recorded structures varied from 0.3 to 1.5 m and there were the vortexes of different size in one structure. There were also the structures of different size in one series of measurements. Since we had only one velocity gauge in operation, we could not determine whether the central or the outline part of turbulent vortex had been recorded. It was a reason of definite incoherence between wave flow dimensions and vortex size. But if we will consider only the largest vortexes in the structures, such correlation becomes obvious.

The dependency between the largest vortex size in a series and a water depth is presented in Figure 10. It demonstrates how the diameter of vortex increases with depth. This dependency confirms the classic idea about a proportionality between sizes of vortexes and flow.

To improve the accuracy of the assessment for the turbulent vortex size the next series of experiments will operate several gauges simultaneously, which were be placed at a distance about one meter.

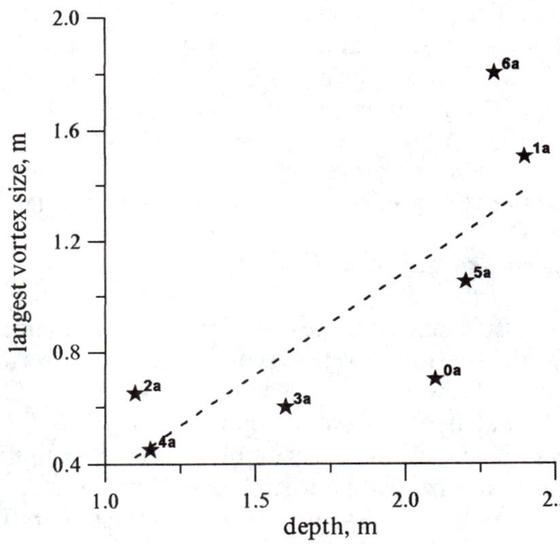

Figure 10. Spatial scales of vortexes.

Conclusions

Our results demonstrate that suspension events in the middle and inner parts of a surf zone are determined by macro-turbulent vortexes and that the intermittence of turbulence determines the low-frequency fluctuations of suspended sand concentration. In our future investigations we will investigate the connections between intermittence of turbulence and wave parameters and the correlation between parameters of turbulence and the suspended sand concentration.

Acknowledgment

This research was supported by the grant 95-05-14343 from the Russian Foundation for Basic Research and by a grant of the German Federal Ministry for Science, Education, Research and Technology (BMBF).

References

Foster D.L., R.A. Holman and R.A. Beach (1994). Sediment suspension events and shear instabilities in to bottom boundary layer. *Proc. Int. Conf. "Coastal Dynamics'94", Barcelona*, 712-716.

George K., R.E. Flick and R.T. Guza (1994). Observation of turbulence in surf zone. *J. of Geoph. Res., Vol. 99, n. C1,* 801-810.

Hino M., M. Kashivayanagi, A. Nakayama and T Hara (1983). Experiments of the turbulence statistics and the structure of a reciprocating oscillatory flow. *J. of Fluid Mech., Vol. 131*, 363-400.

Kos'yan R.D., S.Yu. Kuznetsov, I.S. Podymov, N.V. Pykhov and O.V. Pushkarev (1994). Nearshore suspended sediment concentration measuring during storm. *In: "Littoral'94", Proc. of the second International Symposium. V.1. Lisbon*, 445-452.

Nadaoka K. and T. Kondoh (1989). Turbulent flow field structure of breaking waves in the surf zone. *J. Fluid Mech., Vol. 204*, 359- 387.

Pykhov N., R. Kos'yan and S. Kuznetsov (1995). Time scales of sand suspending by irregular waves. *Proc. of 2nd Int. Conf. "Medcoast-95", Tarragona, Spain,* 1073-1091.

Rodrigues A., A. Sanchez-Arcilla, J. Gomez and E. Bahia (1995). Study of surf-zone macroturbulence and mixing using Delta'93 field data. *Proc. Int. Conf. on Coastal Research in Terms of Large Scale Experiments "Coastal Dynamics'95", Gdansk, Poland,* 305-316.

Sato S., K. Homa and T. Shibayama (1990). Laboratory study on sand suspending due to breaking waves. *Coastal Engineering in Japan, Vol. 33, n. 2.*

Zhang P., T. Sunamura, Sh. Tanaka and K. Yamamoto (1994). Laboratory experiment of longshore bars produced by breaker-induced vortex action. *Proc. Int. Conf. "Coastal Dynamics'94", Barcelona,* 29-43.

CHAPTER 319

MECHANISMS OF BEACH GROUND WATER AND SWASH INTERACTION

Andrew J. Baird[1], Travis E. Mason[2], and Diane P. Horn[3]

ABSTRACT

Previous studies of swash-ground water interactions have given prominence to infiltration of swash and backwash into unsaturated sand as the principal mechanism explaining sedimentation patterns in the inter-tidal zone. Here we consider the role of fluidisation in explaining such sedimentation patterns. It appears that fluidisation due to general ground water outflow from a beach is insufficient to induce fluidisation of surface sand. We propose a different mechanism for explaining fluidisation whereby very small amounts of swash infiltration into the seepage face cause rapid increases in pore water pressure below the beach surface. During the backwash, pressure unloading on the surface causes rapid ground water outflow from the surface sediment. The rate of this outflow is shown, using a simple model, to be sufficient to induce fluidisation. We also consider a model of general beach ground water behaviour and conclude that it can be satisfactorily used to provide boundary conditions for smaller scale models of fluidisation under swash.

1) INTRODUCTION

Grant (1946, 1948) was among the first to suggest a link between beach ground water behaviour and swash zone sediment transport. He proposed that a "dry" beach (one with a low water table, which he equated with unsaturated conditions below the beach surface) allows swash to infiltrate. The reduction in swash depth due to infiltration reduces swash velocity allowing sediment deposition. Therefore, a dry beach promotes accretion. Conversely, on a "wet" beach (one in which the water table is at or near the beach surface) the swash and backwash retain their depth because infiltration is limited. Backwash flows may be augmented by ground water outflow from the beach. Grant (1948) further suggested that the seepage force due to ground water outflow may

[1] Lecturer, Dept. of Geography, University of Sheffield, Sheffield S10 2TN, UK.

[2] Postgraduate research student, Dept. of Oceanography, Southampton Oceanography Centre, Southampton SO14 3ZH, UK.

[3] Lecturer, Dept. of Geography, Birkbeck College, University of London, London W1P 2LL, UK.

cause fluidisation of the sediment at the beach surface and enhance entrainment and sediment transport by backwash.

Infiltration

The logic of Grant's conceptual model has led many researchers to concentrate on the effects of infiltration losses on beach accretion and erosion (e.g. Emery and Foster, 1948; Emery and Gale, 1951; Isaacs and Bascom, 1949; Longuet-Higgins and Parkin, 1962; Duncan, 1964; Strahler, 1966; Harrison 1969,1972; Waddell, 1976; Chappell et al., 1979; Heathershaw, et al. 1981; Lanyon et al., 1982; Carter and Orford, 1993; and Turner 1993). Most of these authors have suggested that infiltration losses during swash provide the main mechanism by which beach accretion occurs above the still water level. However, despite the early suggestion of the importance of infiltration in the swash zone, few attempts have been made to model swash and backwash infiltration. Packwood (1983) developed a numerical model predicting runup due to a single bore incident on an initially dry beach. His justification for the development of this model was based on the assumption that infiltration is the mechanism that controls sediment transport on the beach face. Packwood (1983) argued that, as an incoming tide rises above the water table, the runup advances over dry sand and a proportion of the water mass is lost into the beach. Packwood's results showed that for fine sand (with a porosity, n, of 0.3, and a hydraulic conductivity, K, of 0.01 cm s^{-1}), very little water percolates into the beach. Packwood (1983) found that runup on fine sand and on an impermeable roughened slope are almost the same. In contrast, for medium sand (n = 0.3, K = 0.1 cm s^{-1}), although the maximum runup is reduced very little, the backwash is signficantly reduced in depth. This thin backwash layer drains rapidly into the sand, which will have a significant effect on the bed shear stress and the ability of the flow to transport sediment. Packwood (1983) concluded that the effect of a porous bed is seen much more in the backwash than in the swash. However, he did not consider the effect of a water table nearer to the beach surface on the infiltration process.

Fluidisation

In his 1948 paper Grant appears to suggest that the mechanism for fluidisation of sand grains on the beach surface is due to general ground water outflow from the beach associated with a high water table. A number of researchers have considered this effect but none have found convincing evidence of its importance in sand transport. For example, Oh and Dean (1994) used laboratory experiments and numerical modelling to investigate the influence of the beach water table on profile change and concluded that the effect of the upward flow on sediment transport appeared to be small compared with the effects of a steep slope. It seems likely that ground water outflow from a beach in response to a falling tide alone is insufficient to induce fluidisation, since hydraulic gradients under the beach surface will tend to be relatively small. It is proposed here that fluidisation is only generally possible in the presence of swash and that it occurs over relatively short timescales. As a swash flow advances over the saturated beach surface, there will be a relatively rapid increase in pore water pressures below the beach surface. When under swash flow, the beach sediment behaves like a confined aquifer. The sediment is saturated and movement of water into the beach is extremely limited, since changes in porosity due to expansion and contraction of the mineral 'skeleton' will be minimal. However, because of this limited potential for infiltration, water pressures will propagate rapidly through the sediment. The characteristic time of propagation of any diffusion process is given by

$$t = \frac{x^2}{D} \qquad (1)$$

where x is distance (L) and D is the diffusivity ($L^2\ T^{-1}$), given by the ratio of hydraulic conductivity, K (LT^{-1}), to specific storage, s (L^{-1}). In a sand in the unconfined case, the specific storage multiplied by depth is called the specific yield and typically has values of between 0.1 and 0.3 per unit depth. In the confined case, which is effectively the case for sand under a swash lens, the specific storage is much lower and a value of the order of 0.002 is more realistic. The hydraulic conductivity of beach sands can be expected to be in the range 0.01 to 0.5 cm s^{-1} (Packwood and Peregrine, 1980). For all of the values in this range, the small value of specific storage will give relatively large values of D and small times of propagation, so that the pressure acting on the sand surface under advancing swash will propagate at least into the surface layers of the beach sediment. As the swash retreats there will be a release of pressure on the beach, giving large hydraulic gradients acting vertically upwards immediately below the surface. The resultant seepage force associated with rapid ground water outflow could be sufficient to induce fluidisation of the sand grains at the beach surface. This may occur during the latter stages of backwash and will provide readily entrainable material that can be carried seaward by the backwash flow. To our knowledge this mechanism has not been previously identified and no attempt has been made at modelling it. Later in this paper, we consider in detail how the propagation of pressure through the beach sediment can be modelled. However, whether infiltration into unsaturated or saturated sand is being considered, a first stage in any modelling exercise is the accurate prediction of the position of the seepage face together with predictions of more general ground water flows and pore water pressures within the beach, since these will act as boundary and initial conditions for infiltration and pressure propagation models. Below we consider how the general ground water behaviour of a beach can be modelled.

2) MODELLING BEACH GROUND WATER BEHAVIOUR

Theory

Recently, we have developed a model capable of predicting water table elevations in, and the extent of the seepage face on, sandy beaches. The model, called **GRIST** (**GR**ound water **I**nteraction with **S**wash and **T**ides), has been developed as a modular package and is described in Baird and Horn (1996) and Baird et al. (in review). Two approaches, both based on Darcy's Law and the continuity equation, have been used for modelling water table behaviour in response to tidal forcing in sandy beaches in the

GRIST package. The first approach uses the one-dimensional form of the Boussinesq equation given by

$$\frac{\partial h}{\partial t} = \frac{K}{s}\frac{\partial}{\partial x}\left(h\frac{\partial h}{\partial x}\right) \tag{2}$$

where h is the elevation of the water table (L), t is time (T), K is the hydraulic conductivity of the beach sediment (LT^{-1}), s is the specific yield or drainable porosity (dimensionless), and x is horizontal distance (L). The main assumption in using this equation is that ground water flow in a shallow aquifer can be described using the Dupuit-Forchheimer (D-F) approximation. D-F theory states that in a system of shallow gravity flow to a shallow sink when the flow is approximately horizontal, the lines of equal hydraulic head or potential are vertical, and the gradient of hydraulic head is given by the slope of the water table. Using D-F theory, two-dimensional flow to a sink can be approximated as one-dimensional flow, and the resulting differential equation (equation (2)) is more readily solved. In beaches which are underlain by relatively impermeable solid rock or clay deposits it is likely that D-F theory will provide adequate descriptions of ground water flow. The second approach used in the **GRIST** model considers the beach as a two-dimensional flow system in which the water table is a free surface or flow line. The advantage of the second approach is that it can be used to predict radial flow towards the seepage face if such an approach is necessary; for example, in cases where the flow field in the beach suddenly narrows forcing water to the surface.

Both equations have been solved in the **GRIST** package using standard finite-difference methods and are described in detail in Baird et al. (in review). In the 1D model, a block-centred finite-difference scheme is used and tide-water table decoupling is assumed to occur if

$$h_{i,t} + \left(\frac{(Q_i - Q_{i+1})\Delta t}{s_i}\right) > e_i \tag{3}$$

where Q_i is the rate of ground water discharge ($L^3\ T^{-1}$) (in a seaward direction) into cell i during Δt (from cell i-1), Q_{i+1} is the rate of discharge out of cell i during Δt, and e is the cell height (L). Further details of the model can be foûnd in Baird and Horn (1996).

Field testing of the 1D model

The predictions of the 1D version of the model have been tested against field measurements which are described in detail in Baird et al. (in review). The field measurements were made on a microtidal beach at Canford Cliffs, Poole, Dorset, England. Canford Cliffs beach is a groyned beach and is backed by a sea wall. It has a mean slope of 3.2° ($\tan\beta = 0.056$) on the upper foreshore and a mean slope of 1.5° ($\tan\beta = 0.026$) on the lower foreshore. The tidal range at Canford Cliffs varies between 0.2 and 1.8 m on neap and spring tides respectively. The field measurements which are reported here were taken between the 21st and 28th October 1995 in a period leading from a neap to a spring tide, with a tidal range varying from 0.9 to 1.7

m. The significant wave height (H_s) varied from 0.35 to 0.76 m, while H_{rms} (root mean square wave height) varied between 0.25 and 0.53 m. Zero-crossing period varied between 5.2 and 6 s, and the peak spectral period varied between 5.6 and 9.8 s.

Water table elevations were measured with pressure transducers in ten screened wells placed at distances of 1.9, 3.8, 5.8, 7.8, 9.3, 10.8, 13.1, 15.6, 18.7, and 21.7 m seawards from the sea wall. Wave and tidal frequencies and amplitudes were measured with a seabed-mounted pressure transducer located at the low water mark. All instruments were logged using a PC at the top of the beach and were calibrated before and after the deployment. Beach profiles and instrument locations were measured using a total station. Sediment samples were collected at 0.5 m intervals along a cross-shore transect from the surface of the beach and at a depth of 20 cm for estimation of particle size parameters. Twelve sediment cores were also collected along the cross-shore transect for laboratory determinations of *K* using a constant head permeameter.

Results

The mean grain diameter of the surface sediments was approximately 0.24 mm (2.06 ϕ), the mean sorting was 0.44 ϕ and the mean skewness was -0.71. Results from 20 cm depth were very similar with a mean grain diameter of 0.22 mm (2.17 ϕ), a mean sorting of 0.36 ϕ and a mean skewness of -0.64. The mean *K* of the sediment cores was 0.225 cm s^{-1}, with a range from 0.036 to 1.179 cm s^{-1}, and a coefficient of variation of 143 per cent which is typical for this parameter.

The 1D model was run for Canford Cliffs using 40 computational cells with a width of 1 m. Measured values of tidal elevation were used in the simulation. The inland boundary was a no-flow condition at the sea wall. The bottom boundary was a no-flow condition given by the position of an impermeable barrier located 50 cm below Ordnance Datum. The seaward cell was assumed to be permanently saturated; that is, the water table was assumed to be always at the cell surface. A value of K/s of 0.75 cm s^{-1} was used in the model simulation. This value was calculated using the arithmetic mean of the permeameter measurements of *K* (0.225 cm s^{-1}) and an assumed value of *s* of 0.3.

The predictions of the 1D model are shown for each of the wells in Baird et al (in review). Here we show only two examples of the model's predictions, for wells 6 and 10 (9.3 and 1.9 m seaward of the seawall, respectively). In both cases the remarkably good agreement between modelled and observed water table elevations can be seen. The agreement was similarly good for all ten wells. Thus the model was able to predict with some accuracy water table rise and fall at different positions in the beach, giving a good representation of changes in the shore-normal water table profile over time. For most of the wells the model very accurately describes the lag and asymmetry of water table response. The predictions of lag and asymmetry are less good for wells 8 to 10, with the model predicting a later and steeper rise than was observed. The earlier rise of the observed water table may be due to setup and runup above the shoreline causing a greater flow of water into the beach than would occur due to tidal rise alone (see below). This point is discussed in more detail in Baird et al. (in review). It is also possible that the *K* of the sediment at the back of the beach was lower than that more seaward, giving greater lags than predicted by the model which uses a single value of

K. However, such a pattern of variation in K was not evident from the hydraulic conductivity data although it is recognised that surface sand samples may not be representative of sand at depth.

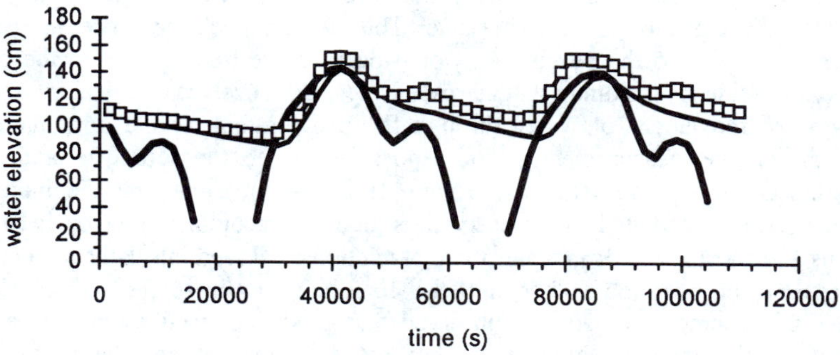

Figure 1. Measured and modelled water table elevations at well 6 (9.3 m seaward of sea wall). Open squares - observed; medium solid line - simulated; thick solid line - observed tide. Elevations given above a datum which represents the position of the impermeable barrier below the beach surface. Time is from 20.00 GMT on 22nd October 1995.

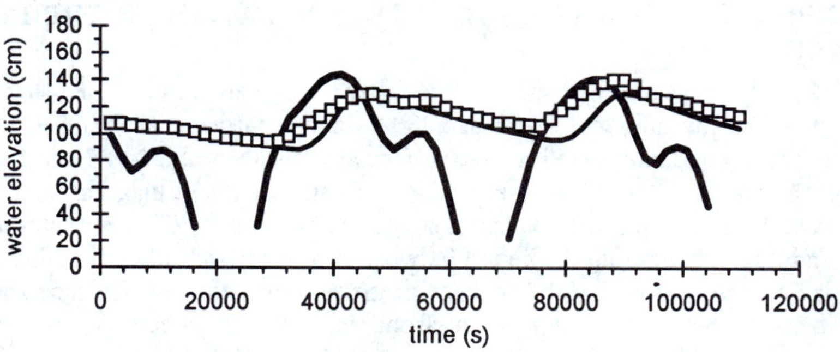

Figure 2. Measured and modelled water table elevations at well 10 (1.9 m seaward of sea wall). Open squares - observed; medium solid line - simulated; thick solid line - observed tide. Elevations given above a datum which represents the position of the impermeable barrier below the beach surface. Time is from 20.00 GMT on 22nd October 1995.

The agreement between observed and simulated water tables is better before high water at 40000 s than after. This is probably due to the fact that wave heights and swash excursions were greater after this time. The wave record shows that through the period of measurement H_s increased from 0.35 to 0.76 m. Setup and swash excursion can be expected to give higher water tables than would be expected due to the tide alone and this is indeed the case with the observed water tables, with levels in well 6 in particular rising above high tide levels (Figure 1). A water table elevation above the high tide elevation is only possible if an additional source of water, such as rainfall, is entering the beach, or if setup and runup raise the mean water surface causing greater rates of flow into the beach. Greater setup due to the larger wave heights at high tide will increase the elevation of the shoreline. This, in turn, will give rise to greater amounts of seepage into the beach than would occur if the free surface on the beach face were at the tide level. Runup will increase water table levels via vertical infiltration and downward movement of water through the unsaturated zone. No rainfall was recorded during the measurement period reported here, so the source of additional water is most likely to be setup and runup. It is also possible that the increased departure between modelled and observed is due to accumulated error in model predictions since a small error in water table predictions will feed into later predictions. However, error propagation will tend to be cumulative whereas the figures clearly show that the discrepancy between observed and predicted occurs more as a step change during the second high water. Thus setup and runup are the more likely explanation for most of the discrepancy between observed and predicted.

The model based on equation (2) provides a good description of water table dynamics in response to tidal forcing although an allowance should be made in future versions of the model for the effects of setup. We suggest, therefore, that the model can be used to provide satisfactory boundary conditions for swash infiltration and pressure propagation models.

3) THE POSITION OF THE WATER TABLE RELATIVE TO THE BEACH SURFACE

Figure 3 shows the measured position of the water table relative to the sand surface for two positions of the tide, low water at 22800 s and high water at 40000 s. At low water the water table intersected the beach face at a distance of about 32 m from the sea wall (the tide was below the level of the beach shown on the diagram) and tended to approach the beach face at a tangent as suggested by Bear (1972). The water table in wells 1 and 2 fell little during the period between low water and subsequent immersion by the rising tide. Thus during the advance of the tide over the seepage face and later across the zone between 18 and 32 m from the sea wall, where the water table remained close to the beach surface, there would have been little opportunity for significant swash infiltration. Any infiltration that did occur is unlikely to have had an effect on swash and backwash dynamics during these conditions because the beach would have behaved like the impermeable case analysed by Packwood (1983). The water table at high water shows a pronounced landward slope and a superelevation of the water table in wells 6 and 7. This superelevation can be attributed to setup and runup as discussed above, and indicates that infiltration into this part of the beach may have the potential to affect swash and backwash hydraulics in the way described by Packwood (1983) for a permeable beach.

It would appear from the above reasoning that swash and backwash infiltration are only likely to be important in affecting swash/backwash hydraulics on a beach such as

Canford Cliffs during the latter part of the rising tide when some swash and backwash move over an unsaturated beach. However, it is important to stress that even small amounts of swash infiltration into the seepage face followed by rapid ground water outflow could induce fluidisation on middle and lower parts of the beach, and that the interaction between beach ground water behaviour and sediment dynamics may be more subtle than suggested by existing swash/backwash infiltration models such as that of Packwood (1983).

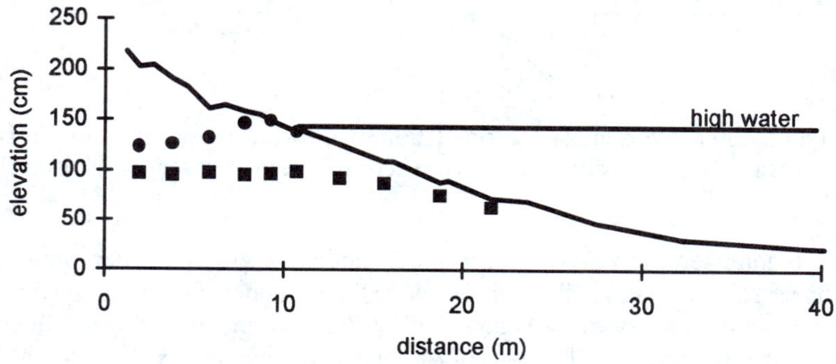

Figure 3. Measured water table elevations along the shore-normal profile at low water after an elapsed time of 22800 s (closed squares), and high water at 40000 s (closed circles).

4) MODELLING FLUIDISATION UNDER SWASH

It was noted earlier that when a swash runs over a saturated beach there will be a relatively rapid increase in pore pressure below the beach surface, and that the pressure release associated with backwash could induce fluidisation of the beach surface. The equation governing water movement through an isotropic and homogeneous saturated beach sediment is

$$\frac{\partial^2 h}{\partial x^2}+\frac{\partial^2 h}{\partial y^2}+\frac{\partial^2 h}{\partial z^2}=\frac{s}{K}\frac{\partial h}{\partial t} \tag{4}$$

where here h is now the hydraulic head (L), and s is the specific storage (L^{-1}). For most purposes this equation can be simplified to its two-dimensional form.

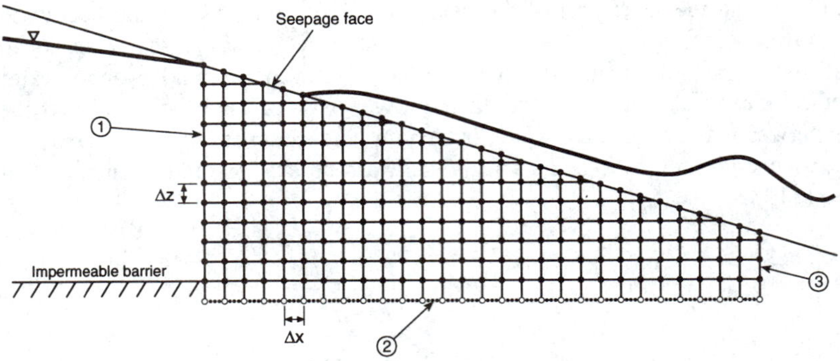

Figure 4. Finite difference grid for the simulation of pressure propagation into a sand bed under swash. ① and ③ are boundaries where the head is given. ② is a no-flow boundary.

The requirements of a two-dimensional model of ground water flow, and therefore pressure propagation, through beach sediment under advancing swash and retreating backwash are shown in Figure 4. The diagram shows a finite-difference, mesh-centred grid representing the body of sand beneath and landward of swash. For the upper boundary condition the model requires a time series of swash depths for that part of the beach face which becomes inundated. It is assumed that the pressure acting on the beach surface at the base of the swash is equivalent to the swash depth; that is, the pressure is hydrostatic. It is this hydrostatic pressure that provides an upper boundary condition to the model. The lower part of the model is assumed to be a no-flow boundary (von Neuman condition) which can be simulated using a set of fictitious nodes shown by open circles. The boundary condition $\partial h/\partial z = 0$ translates in finite-difference form to $h_{i,\ j-1} = h_{i,\ j+1}$ assuming an indexing notation whereby j represents row number (z axis - positive upwards) and i column number (x axis - positive to the right of the diagram). The impermeable barrier is given by j so that j-1 represents the fictitious nodes. The inland boundary condition is a prescribed head condition (Dirichlet condition) which can be provided either from field measurements or, as noted above, from a larger scale beach ground water model such as **GRIST**. In Figure 4 it is shown as the column of nodes below the exit point. The position of the exit point can also be provided from measurements or modelling. The seaward boundary condition is given by prescribed heads at the nodes. This distribution of heads can be assumed to be hydrostatic and equal to wave height at this point. The initial condition for the internal nodes can be estimated from field measurements or can be assumed to be a hydrostatic distribution of hydraulic heads.

Figure 4 represents an ideal solution to the problem of fluidisation under swash and backwash. In the absence of detailed information on the two-dimensional structure of swash, which we are currently measuring using capacitance probes and video techniques, we consider below the effect of water loading and unloading on a one-dimensional column of saturated sand. A simple explicit finite-difference method was used to solve the one-dimensional form of equation (4). The solution is given by

$$\frac{h_{j+1}^n - 2h_j^n + h_{j-1}^n}{(\Delta z)^2} = \frac{s}{K}\left(\frac{h_j^{n+1} - h_j^n}{\Delta t}\right) \tag{5}$$

where the superscript n refers to time.

Five cases of inundation were considered in the model simulations. In all model runs the initial distribution of hydraulic heads below the surface was assumed to be hydrostatic. Nodes were spaced 2 cm apart and the total column length was 60 cm. The base of the column was a no-flow boundary. The cases were:

Case 1. Step increase in pressure on the upper boundary equivalent to a water depth of 10 cm followed by a step return to a condition of no excess pressure on the surface. The pressure loading lasts for 3 seconds and the simulation lasts for 6 s. $K = 0.1$ cm s^{-1}; $s = 0.002$ cm^{-1}.

Case 2. As for Case 1 except that $s = 0.004$ cm^{-1}.

Case 3. Step increase in pressure on the upper boundary equivalent to a water depth of 5 cm followed by a step return to a condition of no excess pressure. The pressure loading lasts for 3 seconds and the simulation lasts for 6 s. $K = 0.025$ cm s^{-1}; $s = 0.002$ cm^{-1}.

Case 4. For a more realistic loading and unloading of water on the upper boundary we used Hughes' (1992) measurements of swash on beaches in Sydney, Australia (case 'a' from Figure 7a of Hughes, 1992). The actual swash depths were approximated using a ramp addition lasting for 2 s and a ramp loss of water lasting for 6.5 s, with a maximum water depth of 20 cm and a total period of inundation of 8.5 s. $K = 0.05$ cm s^{-1}, $s = 0.002$ cm^{-1}.

Case 5. As for Case 4 but with K=0.02 cm s^{-1}.

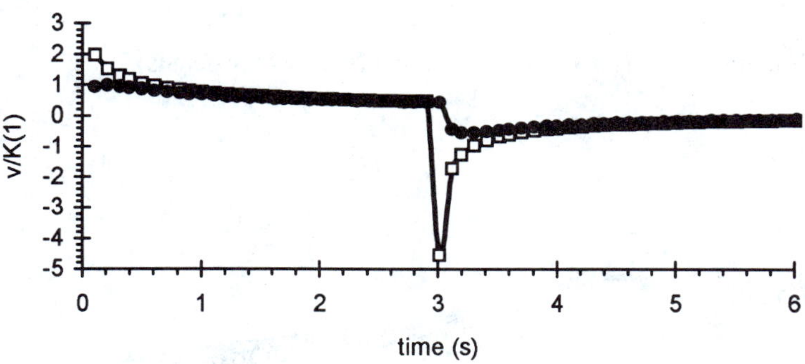

Figure 5. Variations in v/K over time for Case 1 (see text for details).

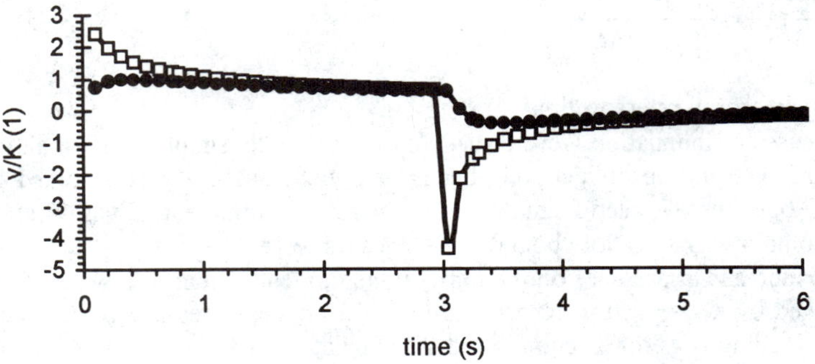

Figure 6. Variations in *v/K* over time for Case 2 (see text for details).

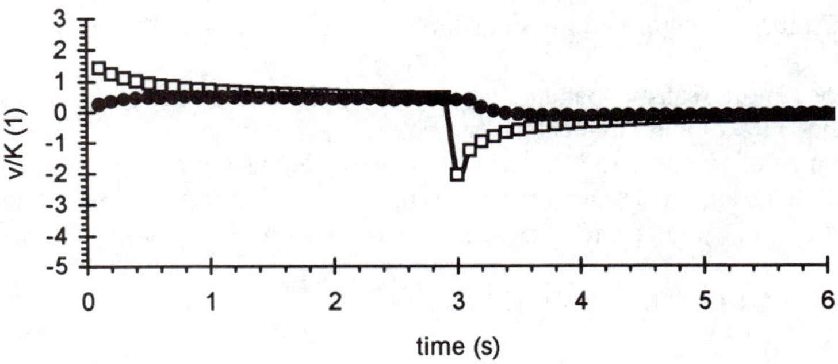

Figure 7. Variations in *v/K* over time for Case 3 (see text for details).

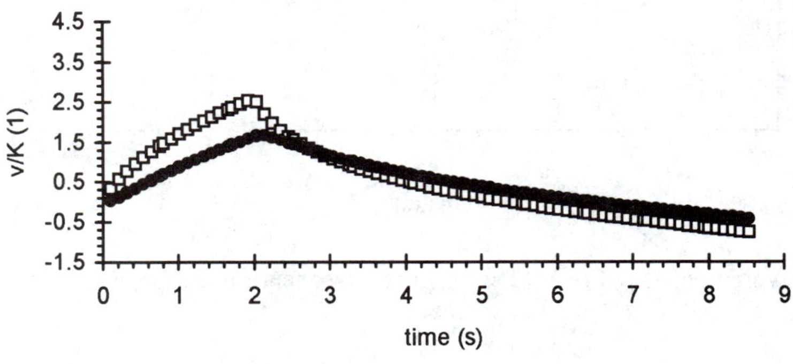

Figure 8. Variations in *v/K* over time for Case 4 (see text for details). Note the change in axes scales.

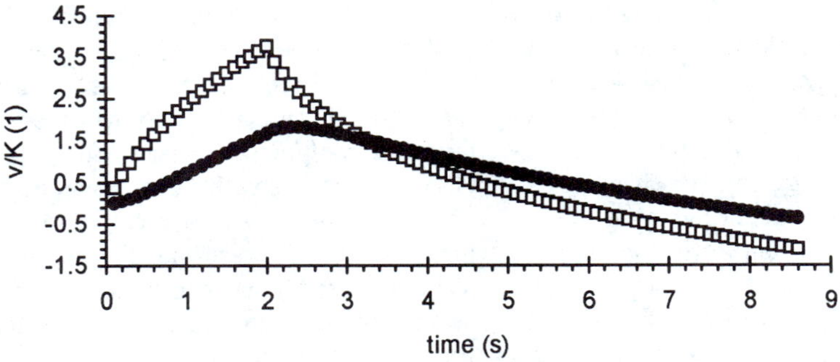

Figure 9. Variations in v/K over time for Case 5 (see text for details).

Cases 1 to 5 are illustrated in Figures 5 to 9, respectively. In each figure, v/K is plotted against time, where v is the specific discharge (cm s^{-1}) through the sand. v/K is equivalent to the hydraulic gradient. Two time series are shown. That with the open squares represents the hydraulic gradient acting between the node 2 cm below the surface and the node representing the beach surface. The closed circles represent the hydraulic gradient between the nodes 4 and 2 cm below the surface. A positive gradient indicates flow downwards into the sand, a negative gradient upward flow. The values of K and s used in Cases 1 to 5 are all within the range of what might be expected for fine to medium sand (see Packwood and Peregrine 1980).

Packwood and Peregrine (1980) noted that for most sands and fine gravels, fluidisation occurs when $v/K < -0.7$ (using the sign convention above). (See also Bear, 1972, pp. 184-189). Fluidisation occurs when the effective stress becomes zero. It should also be noted that downward seepage has the effect of increasing the effective stress and increasing the stability of the sediment. Cases 1 to 3 show that there is an initially large hydraulic gradient associated with the addition of water to the sand surface which declines in an approximately exponential fashion with time. After the removal of the ponded water there is a sudden reversal in the direction of the hydraulic gradient. In all three cases fluidisation can be expected to occur since v/K is less than -0.7. The period of potential fluidisation lasts for approximately 0.4 to 0.6 s. It is interesting to note that large values of v/K are only present in the uppermost layer of sand.

Under real swash, step loading and unloading is unlikely to occur, with the addition and removal of water being more gradual. As noted above, we used the measurements made by Hughes (1992) as a first approximation to loading and unloading under real swash. In Case 4 fluidisation is not predicted by the model. However, a reduction in the ratio of K/s, as shown in Figure 9 (Case 5), does give predicted fluidisation for approximately the last second of the backwash. Much more information is needed on swash and backwash velocities and shear stresses on the bed to understand the implications of this for sediment movement in the swash zone. However, the simple model is revealing in that it suggests that fluidisation of the bed occurs before the beach surface is exposed so that the entrainment and seaward transport of sediment in the latter stages of backwash is enhanced. This is consistent with qualitative observations of backwash by Hughes (1992) who, with other workers, notes that as the swash lens thins during backwash it develops into a slurry of fluid and sediment. The one-dimensional model is obviously an over-simplification of pressure propagation through a sand bed under swash, as is the assumption of a single swash passing over the bed in which the initial condition is a hydrostatic distribution of hydraulic head. However, the model does suggest a mechanism for fluidisation in the swash zone apparently not previously considered by other workers, and one that may exert a greater control on sediment movement in the swash zone than infiltration of swash into the unsaturated part of the beach. It is important to stress that, although the mechanism described here involves infiltration into the sand bed, the amount of water lost from swash and backwash is negligible since the specific storage of saturated sand is so low. Thus it appears that infiltration *sensu* Grant (that is, infiltration into unsaturated sand above the water table) is not needed for ground water to have a significant impact on swash sediment dynamics. The mechanism of fluidisation given here is certainly worthy of further investigation, and to elucidate the processes involved we are engaged in measurements of swash/backwash flows and pressure propagation in both the laboratory and the field site at Canford Cliffs. Results from these studies will be reported in due course.

ACKNOWLEDGEMENTS

Most of the funding for this research came from the United Kingdom Natural Environment Research Council (Grant number GR9/01726). The remainder of the funding was from grants from Birkbeck College, University of London, and the University of Sheffield awarded to Horn and Baird respectively. We would like to thank Mike Wilkin, Paul Bentley, Peter Riley, John Cross, Steve Pearce, Matthew Foote, Simon Gaffney, Richard Kelly, Mark Lee, Chris Turner, Jill Ulmanis, and Steve Wallbridge for their help and advice. Finally, we would like to thank Poole Borough Council who kindly allowed us to use the Surf Lifesaving Club building at Canford Cliffs for housing logging equipment.

REFERENCES

Baird, A.J. and Horn, D.P. 1996. Monitoring and modelling ground water behaviour in sandy beaches. Journal of Coastal Research 12(3): 630-640.

Baird, A.J., Horn, D.P. and Mason, T.E. (in review). Validation of a beach ground water model. Submitted to Marine Geology.

Bear, J. 1972. Dynamics of Fluids in Porous Media, Elsevier, New York, 764 pp.

Carter, R.W.G. and Orford, J.D. 1993. The morphodynamics of coarse clastic beaches and barriers: a short- and long-term perspective. Journal of Coastal Research 15: 158-179.

Chappell, J., Eliot, I.G., Bradshaw, M.P. and Lonsdale, E. 1979. Experimental control of beach face dynamics by water-table pumping. Engineering Geology 14: 29-41.

Duncan, J.R. 1964. The effects of water table and tidal cycle on swash-backwash sediment distribution and beach profile development. Marine Geology 2: 186-197.

Emery, K.O. and Foster, J.F. 1948. Water tables in marine beaches. Journal of Marine Research 7: 644-654.
Emery, K.O. and Gale, J.F. 1951. Swash and swash mark. Transactions of the American Geophysical Union 32(1): 31-36.
Grant, U.S. 1946. Effects of ground water table on beach erosion. Bulletin of the Geological Society of America 57: 1252 (abstract).
Grant, U.S. 1948. Influence of the water table on beach aggradation and degradation. Journal of Marine Research 7: 655-660.
Harrison, W. 1969. Empirical equations for foreshore changes over a tidal cycle. Marine Geology 7, 529-551.
Harrison, W. 1972. Changes in foreshore sand volume on a tidal beach: role of fluctuations in water table and ocean still-water level. Proceedings of the 24th International Geological Conference, Montreal: 159-166.
Heathershaw, A.D., Carr, A.P., Blackley, M.W.L. and Wooldridge, C.F. 1981. Tidal variations in the compaction of beach sediments. Marine Geology 41: 223-238.
Hughes, M.G. 1992. Application of a non-linear shallow water theory to swash following bore collapse on a sandy beach. Journal of Coastal Research 8: 562-578.
Isaaks, J.D. and Bascom, W.N. 1949. Water table elevations in some Pacific coast beaches. Transactions of the American Geophysical Union 30: 293-294.
Lanyon, J.A., Eliot, I.G., and Clarke, D.J. 1982. Observations of shelf waves and bay seiches from tidal and beach ground water-level records. Marine Geology 49, 23-42.
Longuet-Higgins, M.S. and Parkin, D.W. 1962. Sea waves and beach cusps. Geographical Journal 128: 194-201.
Oh, T-M. and Dean, R.G. 1994. Effects of controlled water table on beach profile dynamics. Proceedings of the 24th International Conference on Coastal Engineering: 2449-2460.
Packwood, A.R. 1983. The influence of beach porosity on wave uprush and backwash. Coastal Engineering 7: 29-40.
Packwood, A.R., and Peregrine, D.H. 1980. The propagation of solitary waves and bores over a porous bed. Coastal Engineering 3: 221-242.
Strahler, A.N. 1966. Tidal cycle of changes in an equilibrium beach, Sandy Hook, New Jersey. Journal of Geology 74: 247-268.
Turner, I.L. 1993. The total water content of sandy beaches. Journal of Coastal Research Special Issue 15: 11-26.
Waddell, E. 1976. Swash-ground water-beach profile interactions. In Davis, R.A. and Etherington, R.L. (eds) Beach and nearshore sedimentation. Society of Economic and Paleontological Mineralogists Special Publication 24: 115-125.

CHAPTER 320

The equations for integral and mean flow properties in the swash zone

M. Brocchini † and D.H. Peregrine ‡

Abstract

The swash zone is the area of a beach where the waves move the instantaneous shoreline back and forth. This zone is modelled using the nonlinear shallow-water equations (NLSWE) as equations of motion, as is appropriate for gently sloping beaches. Integrated swash zone boundary conditions are derived for wave resolving models: their use gives considerable numerical advantages. Wave-averaged and swash-zone integrated boundary conditions have also been derived. They include numerous terms for which closures are to be found.

Euations and an analytic swash zone solution

In this section we introduce the basic equations used to describe the flow dynamics near the shore. We choose the still water level to be $z = 0$, and define

$$d(x,t) = h(x) + \eta(x,t) \tag{1}$$

where d is the total water depth, $z = -h(x)$ is the seabed, $z = \eta$ is the position of the free surface (see figure 1).

For a plane beach with $h(x) = \alpha x$ the inviscid equations are

$$\begin{aligned} d^*_{t^*} + (d^*u^*)_{x^*} + (d^*v^*)_{y^*} &= 0 \\ u^*_{t^*} + u^*u^*_{x^*} + v^*u^*_{y^*} + gd^*_{x^*} &= -g\alpha \\ v^*_{t^*} + u^*v^*_{x^*} + v^*v^*_{y^*} + gd^*_{y^*} &= 0. \end{aligned} \tag{2}$$

They can be put in a simple dimensionless form with no explicit dependence on the beach slope α (Meyer & Taylor 1972).

The dimensionless variables which eliminate the beach slope α from the equations are:

$$d = \frac{d^*}{\alpha l_0}, \quad u = \frac{u^*}{u_0}, \quad v = \frac{v^*}{u_0}, \quad u_0 = (gl_0\alpha)^{1/2} \tag{3}$$

$$x = \frac{x^*}{l_0}, \quad y = \frac{y^*}{l_0}, \quad t = \frac{t^*}{t_0}, \quad t_0 = \left(\frac{l_0}{g\alpha}\right)^{1/2} \tag{4}$$

†Research Assistant, School of Mathematics, University of Bristol, University Walk, Bristol, BS8 1TW, UK, (*M.Brocchini@bristol.ac.uk*)

‡Professor of Applied Mathematics, School of Mathematics, University of Bristol, (*D.H.Peregrine@bristol.ac.uk*)

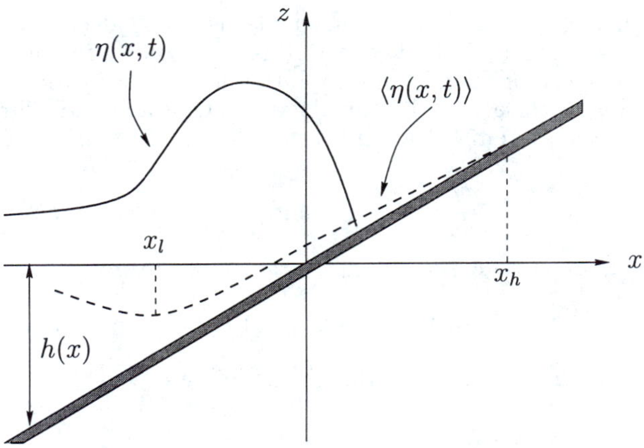

Figure 1: Flow definition in the vicinity of the swash zone.

where l_0 is a reference length which can be specified according to the particular problem under investigation. The equations become:

$$\begin{aligned} d_t + (du)_x + (dv)_y &= 0 \\ u_t + uu_x + vu_y + d_x &= -1 \\ v_t + uv_x + vv_y + d_y &= 0. \end{aligned} \tag{5}$$

For simplicity, friction terms are not included in this analysis and in most of the numerical modelling. Even at laboratory scale they are of secondary importance if beach slope is not too small. Watson et al. (1994) give a quantitative estimate of their importance. On the other hand, the energy dissipation of bores is included in the modelling as discontinuities in the solution.

Equations (5) can be further simplified by approximating for regular waves incident at small angle θ to the beach normal (Ryrie, 1983). We introduce a pseudotime t' and a small parameter ϵ such that:

$$t' = t - \epsilon y \tag{6}$$

gives the only y dependence. It implies that the wave pattern is moving along the shore with phase velocity of $1/\epsilon$. For example, this can occur for a regular train of waves incident towards the shore at a small angle θ to the shore normal, with an alongshore velocity $c/\sin\theta$.

For a weakly three-dimensional flow the following scaling:

$$y' = \epsilon y\ , \quad v' = v/\epsilon, \tag{7}$$

on substitution into the NLSWE (5), neglecting $\mathcal{O}(\epsilon^2)$ terms and dropping primes, gives the equations:

$$\begin{aligned} d_t + (du)_x &= 0 \\ u_t + uu_x + d_x &= -1 \\ v_t + uv_x - d_t &= 0. \end{aligned} \tag{8}$$

The last of the above set of hyperbolic equations is now decoupled from the first two. The v component of the velocity does not appear in the first two equations of (8) ('onshore problem') and a solution for the third equation ('longshore problem') is found once d and u are known. Characteristic directions in the (x,t) plane for the system (8) are given for the onshore and longshore problem respectively by:

$$\frac{dx}{dt} = u \pm c \quad \text{and} \quad \frac{dx}{dt} = u. \tag{9}$$

Equations (8) can be expressed in characteristic form:

$$\alpha_t + (u+c)\alpha_x = 0\ , \quad \beta_t + (u-c)\beta_x = 0\ , \quad \gamma_t + u\gamma_x = 0 \tag{10}$$

where the Riemann invariants are defined as follows:

$$\alpha = 2c + u + t\ , \quad \beta = 2c - u - t\ , \quad \gamma = v - \tfrac{1}{2}u^2 - d - x. \tag{11}$$

Carrier & Greenspan (1958) used a hodograph transformation (velocities are used as coordinates) in solving the onshore problem:

$$\lambda = \alpha - \beta = 2(u+t)\ , \quad \sigma = \alpha + \beta = 4c\ , \quad u(\sigma,\lambda) = \phi_\sigma/\sigma. \tag{12}$$

The characteristic coordinates (σ, λ) are particularly effective for the moving shoreline since the $\sigma = 0$ contour maps the moving shoreline: $d = c = 0$. We note that σ is a space-like coordinate and λ is a time-like coordinate (e.g. see figure 2(d)). Combining the equations for the onshore problem gives a linear equation in ϕ:

$$(\sigma\phi_\sigma)_\sigma - \sigma\phi_{\lambda\lambda} = 0. \tag{13}$$

The simplest periodic solution is that of Carrier and Greenspan (1958):

$$\eta(\sigma,\lambda) = \tfrac{1}{4}\phi_\lambda - \tfrac{1}{2}u^2\ , \quad x(\sigma,\lambda) = \tfrac{1}{4}\phi_\lambda - \tfrac{1}{16}\sigma^2 - \tfrac{1}{2}u^2\ , \quad t(\sigma,\lambda) = \tfrac{1}{2}\lambda - u \tag{14}$$

where

$$\phi(\sigma,\lambda) = AJ_0(\sigma)\sin(\lambda)\ , \tag{15}$$

J_0 is the Bessel function of the first kind. Not all the solutions can be transferred back to the (x,t) plane. This occurs for $A > 1$, i.e. when the Jacobian of the transformation vanishes.

Since the equation for the longshore problem is decoupled we can formally integrate this equation (10c):

$$\gamma_t + u\gamma_x = v_\lambda x_\sigma - v_\sigma x_\lambda + u(v_\sigma t_\lambda - v_\lambda t_\sigma) - (\eta_\lambda x_\sigma - \eta_\sigma x_\lambda) = 0. \tag{16}$$

After some algebra the equation can be rearranged to give a solution valid for any ϕ:

$$v = \tfrac{1}{4}\phi_\lambda + const. \tag{17}$$

and setting $const. = 0$ we get

$$v = \tfrac{1}{4}AJ_0(\sigma)\cos\lambda. \tag{18}$$

Details on some complicated multi-modal solutions are given in Brocchini & Peregrine (1996).

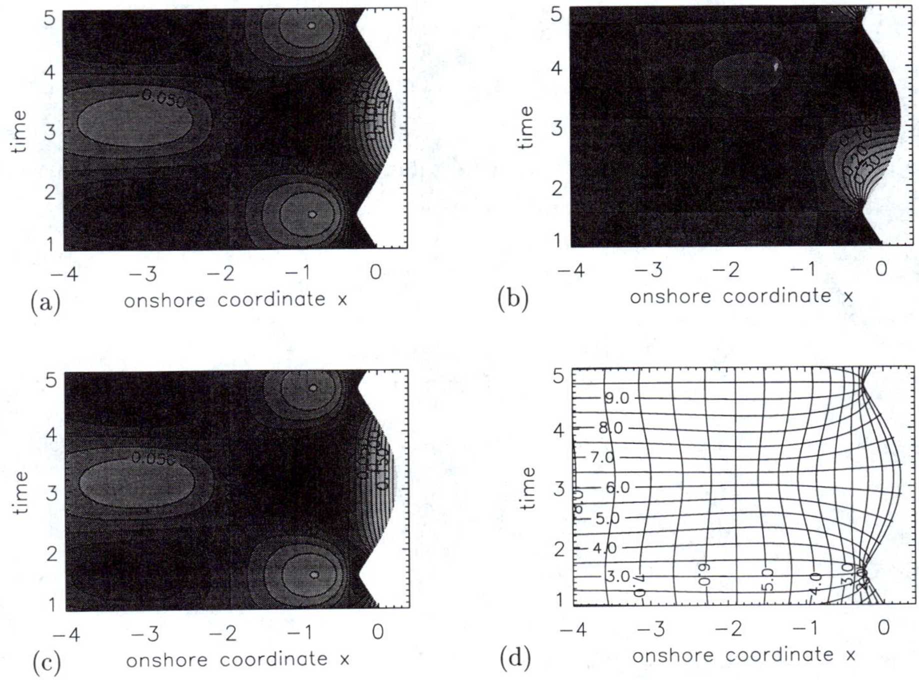

Figure 2: Contour lines for the dimensionless Carrier & Greenspan (1958) solution: (a) the free surface elevation η, (b) the onshore velocity u and (c) the longshore velocity, v, of the weakly three dimensional solution, for the case $A = 1$. Note that the dimensional longshore velocity is a factor ϵ smaller than the dimensionless quantity. (d) shows the (σ, λ) coordinates in the (x, t) plane. Lines of constant λ run from left to right whilst lines of constant σ run from top to bottom.

Contour plots of the free surface elevation and velocity components help visualise the behaviour of the solution, see figure 2. For all the flow properties shown in figure 2 there is a structured pattern of cells. Cells are confined by isolines of zero surface displacement. They also have different sizes and shapes. For all variables the largest values are reached within the most onshore cell. The particular characteristic of the contour structure for u is that it has an antisymmetric cellular pattern within the period. We can therefore state *a priori* that its mean value over a wave period is zero at each $x = const.$ position. This is not true for both the free surface elevation and the longshore velocity. They have similar cell patterns and for both of them, time averaging over a wave period results in a non-zero contribution.

Time averaging of flow properties is rather involved for the Carrier & Greenspan solution because of difficulties in relating (x, t) and (σ, λ). Let

$$\langle G(x,t) \rangle = \frac{1}{T} \int_T G(x,t)\, dt = \frac{1}{T} \lim_{t \to T} H(x,t) \tag{19}$$

be the time average of $G(x,t) = g(\sigma,\lambda)$ where

$$H(x,t) = \int_0^t G(x,s)\, ds. \tag{20}$$

To evaluate this type of average which is based on time integration on curves of $x = const$ we solve the following partial differential equation along the curves $x = const$ in the (σ, λ) plane:

$$\frac{\partial H}{\partial t} = G(x,t) = g(\sigma,\lambda). \tag{21}$$

But for constant x we have

$$\frac{dH}{d\lambda} = \frac{\partial H}{\partial t}\left(\frac{\partial t}{\partial \sigma}\frac{d\sigma}{d\lambda} + \frac{\partial t}{\partial \lambda}\right). \tag{22}$$

Finally a set of two o.d.e.s for $\langle G(x,t)\rangle$ and σ is obtained:

$$\begin{aligned} \frac{d\langle G(x,t)\rangle}{d\lambda} &= \frac{1}{T}\lim_{t\to T} g(\sigma,\lambda)\left[\frac{\partial t}{\partial \lambda}\left(-\frac{\partial x}{\partial \lambda}\Big/\frac{\partial x}{\partial \sigma}\right) + \frac{\partial t}{\partial \sigma}\right] \\ \frac{d\sigma}{d\lambda} &= -\frac{\partial x}{\partial \lambda}\Big/\frac{\partial x}{\partial \sigma}. \end{aligned} \tag{23}$$

The first equation is the total derivative of $\langle G(x,t)\rangle$ in terms of $G(x,t)$ while the second states that we are integrating along curves where $x = const$. This method is used to compute the time average of velocities and mass fluxes.

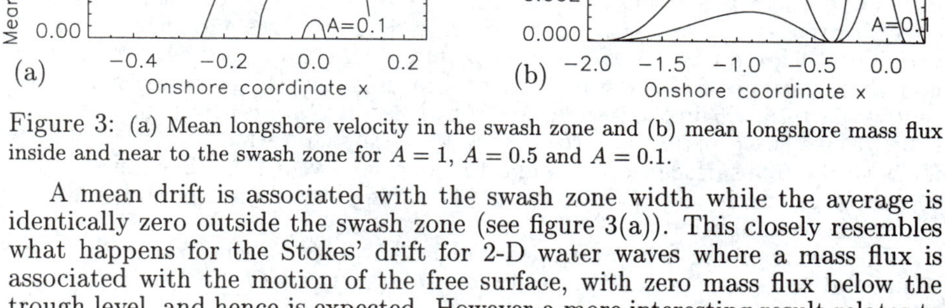

Figure 3: (a) Mean longshore velocity in the swash zone and (b) mean longshore mass flux inside and near to the swash zone for $A = 1$, $A = 0.5$ and $A = 0.1$.

A mean drift is associated with the swash zone width while the average is identically zero outside the swash zone (see figure 3(a)). This closely resembles what happens for the Stokes' drift for 2-D water waves where a mass flux is associated with the motion of the free surface, with zero mass flux below the trough level, and hence is expected. However a more interesting result relates to the mean longshore mass flux (see figure 3(b)). Correlation between solution for the total water depth and the longshore velocity is such that the mean longshore mass flux is non-zero for a wide range of the solution domain.

A more detailed analysis on swash zone mean flow properties is given in Brocchini & Peregrine (1995; 1996).

The model equations for integrated flow properties

In order to average flow properties in the swash zone and to deal with the moving free surface we investigate a model with a boundary chosen at the lower limit of the swash zone.

The swash zone limits are x_l , x_h, the seaward (lower) and shoreward (higher) limits of the swash zone respectively. The seaward boundary may for example be, the lowest limit x_l of the moving shoreline in a group of waves. Integral properties of the water shoreward of that point are considered.

We integrate the basic flow equations across the swash zone. In the following we adopt Einstein's summation convention and use Greek suffices for the two dimensional horizontal flow properties. The general set of equations (5) and the related energy equation are put in conservation form:

$$\begin{aligned} d_t + (ud)_x + (vd)_y &= 0 \\ (ud)_t + \left(u^2 d + \tfrac{1}{2}d^2\right)_x + (uvd)_y + d + \tau_1 &= 0 \\ (vd)_t + (uvd)_x + \left(v^2 d + \tfrac{1}{2}d^2\right)_y + \tau_2 &= 0 \\ \tfrac{1}{2}\left[d(u^2+v^2) + d^2\right]_t + \left(\tfrac{1}{2}u^3 d + \tfrac{1}{2}v^2 ud + ud^2\right)_x & \\ + \left(\tfrac{1}{2}v^3 d + \tfrac{1}{2}u^2 vd + vd^2\right)_y + u(d+\tau_1) + v\tau_2 &= 0. \end{aligned} \tag{24}$$

where a dimensionless bed friction $\boldsymbol{\tau} = (\tau_1, \tau_2)$ is included.

We also assume the wave motion as occurring at a 'fast time scale' t while the swash zone boundaries x_h (run-up) and x_l (run-down) are supposed to vary (if varying in time) on a 'slow time scale' T such as $t = \epsilon T$. We do not use this notation since the separation of the two scales is generally quite clear. Equations (24) are integrated over the swash zone width, with constant x_l, to give:

$$\begin{aligned} \frac{\partial V}{\partial t} &= \int_{x_l}^{x_h} d_t \, dx = Q_1(t)|_{x_l} - \frac{\partial P_2}{\partial y} \\ \frac{\partial P_1}{\partial t} &= \int_{x_l}^{x_h} (ud)_t \, dx = S_{11}(t)|_{x_l} - V(t) - \frac{\partial M_{12}}{\partial y} - \Upsilon_1(t) \\ \frac{\partial P_2}{\partial t} &= \int_{x_l}^{x_h} (vd)_t \, dx = S_{12}(t)|_{x_l} - \frac{\partial M_{22}}{\partial y} - \Upsilon_2(t) \\ \frac{\partial E}{\partial t} &= \int_{x_l}^{x_h} [\tfrac{1}{2}(u^2+v^2)d + d^2]_t \, dx \\ &= F(t)Q_1(t)|_{x_l} - P_1(t) - \frac{\partial}{\partial y}\int_{x_l}^{x_h} FQ_2 \, dx - \Gamma(t). \end{aligned} \tag{25}$$

This introduces a number of new flow properties and flow properties integrated over the swash zone. These are listed in table 1 where $\mathbf{u} = (u_1, u_2)$ is the horizontal velocity vector.

The set of equations (25) for the fully three-dimensional motion can be reduced by Ryrie's approximation to a simpler set of equations valid for weakly three-dimensional motion. Second order terms are neglected after the formal substitution (7) is made. Then, at the leading order the set of equations (25)

Name	Explicit expression	Flow property
Q_μ	$u_\mu d$	Local mass flow
$S_{\mu\nu}$	$u_\mu u_\nu d + \delta_{\mu\nu}\frac{1}{2}d^2$	Local momentum flux tensor
F	$\frac{1}{2}\mathbf{u}^2 + d$	Local energy density
V	$\int_{x_l}^{x_h} d\ dx$	Volume of water in swash zone
P_μ	$\int_{x_l}^{x_h} Q_\mu\ dx$	Momentum of water in swash zone
$M_{\mu\nu}$	$\int_{x_l}^{x_h} S_{\mu\nu}\ dx$	Integrated momentum flux tensor
E	$\int_{x_l}^{x_h} (\frac{1}{2}\mathbf{u}^2 d + d^2)\ dx$	Energy of water in swash zone
Υ_μ	$\int_{x_l}^{x_h} \tau_\mu\ dx$	Friction force in swash zone
Γ	$\int_{x_l}^{x_h} u_\mu \tau_\mu\ dx$	Work done by friction in swash zone

Table 1: Definition of the flow properties adopted in equations (25).

becomes:

$$\begin{aligned}
\frac{\partial V}{\partial t} = \int_{x_l}^{x_h} d_t\ dx &= Q_1(t)|_{x_l} \\
\frac{\partial P_1}{\partial t} = \int_{x_l}^{x_h} (ud)_t\ dx &= S_{11}(t)|_{x_l} - V(t) - \Upsilon_1(t) \\
\frac{\partial P_2}{\partial t} = \int_{x_l}^{x_h} (vd)_t\ dx &= S_{12}(t)|_{x_l} + \frac{\partial M_{22}}{\partial t} - \Upsilon_2(t) \\
\frac{\partial \bar{E}}{\partial t} = \int_{x_l}^{x_h} (\tfrac{1}{2}u^2 d + d^2)_t\ dx &= \bar{F}(t)Q_1(t)|_{x_l} - P_1(t) - \bar{\Gamma}(t).
\end{aligned} \tag{26}$$

Table 2 lists definitions which, because of the approximation, differ from those of table 1. We distinguish newly defined flow properties by using overbars in equations (26) and table 2.

Name	Explicit expression	Flow property
$\bar{F}$	$\frac{1}{2}u_1^2 + d$	Local energy density
$\bar{E}$	$\int_{x_l}^{x_h} \frac{1}{2}(u_1^2 d + d^2)\ dx$	Energy of water in swash zone
$\bar{\Gamma}$	$\int_{x_l}^{x_h} u_1 \tau_1\ dx$	Work done by friction in swash zone

Table 2: Definition of some flow properties adopted in equations (26).

In the idealized case of no dissipation ($\Upsilon_1 = \Upsilon_2 = \bar{\Gamma} = 0$) the set of partial differential equations (26) is such that it can be recursively solved for the integral flow properties in the swash zone (i.e. V, P_1, P_2 and $\bar{E}$) once the local flow properties Q_1, S_{11}, S_{12} and $\bar{F}$ are known at the seaward boundary of the swash zone. The result of the equation for the water volume only depends on the known variable Q_1. Once this equation is solved for V the result can be substituted into the second equation and so on.

We believe that equations (25) can be used as swash zone boundary conditions for wave resolving models. Use of integrated swash zone properties gives considerable numerical advantage, since adequate modelling of the details of the swash zone flows requires small time steps.

The swash zone boundary conditions for wave-averaged models

A natural extension to averaging the swash zone is to seek appropriate boundary conditions at the seaward swash zone limit for wave-averaged models of the incident motions (both long and short period motions). This requires definition of the contributions to the motion of the boundary $x = x_l$ from both long-period motions (e.g. low frequency waves, currents, etc..) and averaged short-period motions. The problem here is that there is no definitive model for the waves approaching the swash zone, especially for the most usual case where waves are already breaking. The following discussion develops swash-zone boundary conditions by assuming that appropriate short-wave properties are known.

To obtain dynamic equations for wave-averaged flow properites we start with the equations of motion and divide the basic flow properties u, v and d into long-period motions and the short-period wave contributions:

$$u = \langle u \rangle + \tilde{u}\ , \quad v = \langle v \rangle + \tilde{v}\ , \quad d = \langle d \rangle + \tilde{d}\ , \qquad \langle \tilde{u} \rangle = \langle \tilde{v} \rangle = \langle \tilde{d} \rangle = 0. \tag{27}$$

Contributions to $\langle u \rangle$, $\langle v \rangle$ and $\langle d \rangle$ come from all motions whose typical time scale is significantly longer than the typical short-wave period. These can be for example either bound long waves associated with the set-down occurring under a group of short-period waves or free long waves caused by a time-varying break-point or any sort of current. On the other hand pure short-period contributions appear as correlations of wave-type terms (e.g. $\langle \tilde{u}\tilde{d} \rangle$, $\langle \tilde{u}\tilde{v} \rangle$, etc..).

By substituting in the shallow water equations (24) in this way for each flow variable and by phase averaging we obtain a set of equations for the long-period flow properties $\langle u \rangle$, $\langle v \rangle$ and $\langle d \rangle$:

$$\frac{\partial \langle d \rangle}{\partial t} + \frac{\partial}{\partial y}\left[\langle u \rangle\langle d \rangle + \langle \tilde{Q}_1 \rangle\right] + \frac{\partial}{\partial y}\left[\langle v \rangle\langle d \rangle + \langle \tilde{Q}_2 \rangle\right] = 0 \tag{28}$$

$$\frac{\partial}{\partial t}\left[\langle u \rangle\langle d \rangle + \langle \tilde{Q}_1 \rangle\right] + \frac{\partial}{\partial y}\left[\langle d \rangle\left(\langle u \rangle^2 + \langle \tilde{u}^2 \rangle + \tfrac{1}{2}\langle d \rangle\right) + 2\langle u \rangle\langle \tilde{Q}_1 \rangle + \langle \tilde{S}_{11} \rangle\right] \tag{29}$$
$$+\frac{\partial}{\partial y}\left[\langle d \rangle\left(\langle u \rangle\langle v \rangle + \langle \tilde{u}\tilde{v} \rangle\right) + \langle u \rangle\langle \tilde{Q}_2 \rangle + \langle v \rangle\langle \tilde{Q}_1 \rangle + \langle \tilde{S}_{12} \rangle\right] + \langle d \rangle = 0$$

$$\frac{\partial}{\partial t}\left[\langle v \rangle\langle d \rangle + \langle \tilde{Q}_2 \rangle\right] + \frac{\partial}{\partial y}\left[\langle d \rangle\left(\langle v \rangle^2 + \langle \tilde{v}^2 \rangle + \tfrac{1}{2}\langle d \rangle\right) + 2\langle v \rangle\langle \tilde{Q}_2 \rangle + \langle \tilde{S}_{22} \rangle\right] \tag{30}$$
$$+\frac{\partial}{\partial y}\left[\langle d \rangle\left(\langle u \rangle\langle v \rangle + \langle \tilde{u}\tilde{v} \rangle\right) + \langle v \rangle\langle \tilde{Q}_1 \rangle + \langle u \rangle\langle \tilde{Q}_2 \rangle + \langle \tilde{S}_{12} \rangle\right] = 0$$

where $\langle \tilde{Q}_\mu \rangle = \langle \tilde{u}_\mu \tilde{d} \rangle$ is the mass flow due to wave motion while the contribution of the momentum flux due to wave motion is given by the radiation stress term $\langle \tilde{S}_{\mu\nu} \rangle$.

The above set can be also written in matrix form in order to clarify the forcing action of the short-period contributions. The set of equations can, thus, be rewritten as:

$$\mathcal{A}\mathcal{U}_t + \mathcal{B}\mathcal{U}_x + \mathcal{C}\mathcal{U}_y + \mathcal{D}\mathcal{U} = \mathcal{E} \tag{31}$$

where

$$\mathcal{U} = \begin{bmatrix} \langle d \rangle \\ \langle u \rangle \\ \langle v \rangle \end{bmatrix} \quad \text{and} \quad \mathcal{E} = -\begin{bmatrix} \langle \tilde{Q}_1 \rangle_x + \langle \tilde{Q}_2 \rangle_y \\ \langle \tilde{Q}_1 \rangle_t + \langle \tilde{S}_{11} \rangle_x + \langle \tilde{S}_{12} \rangle_y + \langle d \rangle \\ \langle \tilde{Q}_2 \rangle_t + \langle \tilde{S}_{12} \rangle_x + \langle \tilde{S}_{22} \rangle_y \end{bmatrix} \tag{32}$$

are respectively the column vector of the unknown variables and the purely short-wave forcing.

This set of equations is not 'closed' unless a particular wave theory for the short-period wave field is adopted to compute terms like the mass flow and the radiation stress. For instance, it is common practice to use linear wave theory while dissipation effects induced by wave-breaking are parameterized in the solution.

However linear wave theory is not a good representation for the wave field near the shore. The Carrier & Greenspan (1958) solutions and their extension described above can be more appropriate. To help crystallise ideas we note that one approach to this problem would be to extend the Carrier & Greenspan solution to $A > 1$ by computing the surf and swash zone behaviour for waves which have incoming Riemann invariants as for the Carrier & Greenspan solution with $A > 1$. The reflection, mass flow, radiation stress and the averaged swash zone properties, V, P and E could be tabulated as functions of the short-wave parameters. We assume that if a suitable model is found for the short-wave motion this can be described in terms of only an amplitude A a frequency ω and a wave number $\mathbf{k}$ once the mean depth and mean flow velocity are given.

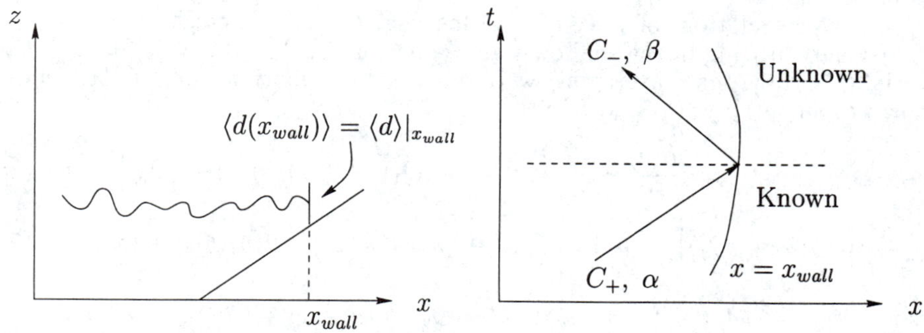

Figure 4: Schematic of: (a) the 'physical' (x, z) plane and (b) the 'characteristic' (x, t) plane for the rigid wall problem.

In order to obtain suitable boundary conditions for wave-averaged models we need to clarify what sort of information is available and what are the unknowns. To achieve this we first use a simple model of a moving boundary that is a rigid wall moving at velocity $U(t)\mathbf{i}$ which varies on a sufficiently slow time scale that its acceleration is unimportant (see figure 4). The flow is one dimensional and the medium is assumed uniform so that A, ω and $\mathbf{k}$ do not depend on (x, t).

A simple exercise of bookkeeping shows what are known and unknown parameters for both long and short wave motions (see table 3 and figure 4).

Natural boundary conditions for this flow are:

$$\langle d \rangle|_{x_{wall}} = const \ , \qquad \frac{dx_{wall}}{dt} = \langle u \rangle = U(t) \tag{33}$$

Consequently it is possible to explicitly compute the long-period motion variable β (i.e. the outgoing Riemann invariant) from the known incoming variable α

$$\alpha = 2\langle c \rangle + \langle u \rangle \ , \quad \beta = 2\langle c \rangle - \langle u \rangle \ , \quad \Longrightarrow \quad \beta = \alpha - 2U. \tag{34}$$

The incoming waves of amplitude A_{in}, wavenumber $\mathbf{k}_{in}$ frequency ω_{in} may be perfectly reflected at the wall, but to determine the outgoing values of A, $\mathbf{k}$ and ω the local behaviour at the moving wall must be considered.

In a frame of reference moving with the wall, the waves have an unchanging intrinsic frequency σ which is readily found from the Doppler shift:

$$\omega_{in} = \sigma + U\mathbf{i}\cdot\mathbf{k}_{in}. \tag{35}$$

If the waves are perfectly reflected then the outward propagating wave has $A_{out} = A_{in}$ and $\mathbf{k}_{out} = -(\mathbf{k}_{in}\cdot\mathbf{i})\mathbf{i} + (\mathbf{k}_{in}\cdot\mathbf{j})\mathbf{j}$ giving:

$$\omega_{out} = \sigma + U\mathbf{i}\cdot\mathbf{k}_{out} = \sigma - U\mathbf{i}\cdot\mathbf{k}_{in} = \omega_{in} - 2U\mathbf{i}\cdot\mathbf{k}_{in}. \tag{36}$$

	Long-period	**Short-period**
Known	$\alpha = \alpha(\langle u \rangle, \langle d \rangle)$, $\langle d \rangle\|_{x_l}$	A_{in} , ω_{in} , $\mathbf{k}_{in}$
Unknown	$\beta = \beta(\langle u \rangle, \langle d \rangle)$, $x_l(t)$	A_{out} , ω_{out} , $\mathbf{k}_{out}$

Table 3: Known and unknown flow variables for the rigid wall problem.

In comparing the moving swash zone boundary with the above simple boundary, we note that it is necessary to model the swash zone boundary as a 'leaky wall' such that fluxes of mass and momentum are allowed across $x = x_{wall}$ which for the 'leaky' case may be identified with $x = x_l$. We also note that:

1. sufficient information must be available to determine the motion of the swash zone boundary, $x_l(t)$, as well as to determine β;

2. as above, a local wave model is needed to determine the properties of the outgoing wave, if any.

We concentrate on the first of these, and look at integration of swash-zone equations similar to (25), or (26), to provide the necessary information. It is clear that in addition to the kinematic type of boundary conditions that occur on the rigid wall (equations 33) a third condition is necessary for the longshore flow in the swash zone.

There is now an important choice to be made: how much of the swash motion, if any, should be assigned to the long time scales? Consideration of the moving rigid wall example leads us to consider all motion relative to the point $x = x_l(y,t)$ to be short wave motion and $x_l(y,t)$ to be 'driven' by the long-period motions. As a result, we consider all the quantities defined in tables 1 and 2 to be defined with velocities relative to x_l and also to be considered as known once incident short-wave parameters A_{in}, $\mathbf{k}_{in}$ and σ are known.

Inside the swash zone most the short-wave quantities averaged over the short wave motion, such as $\langle V \rangle$, $\langle P \rangle$ etc.., are to be determined from a short-wave model of the swash. There are exceptions for the longshore current quantities P_2, $M_{\mu 2}$ which are not entirely dependent on the local waves. When a bore, or flow which was near a bore, enters the swash zone it makes a large contribution to the longshore velocity, which often may only respond to bed friction on the longer time scale, e.g. see Ryrie (1983). This means that a decomposition of long-wave and short-wave contributions is also necessary within the swash zone.

We achieve this decomposition by assuming that swash motion is almost entirely assigned to short-wave contributions and that the only long-wave contribution, other than the motion of x_l, comes from parameterising the longshore drift due to wave-breaking by a longshore current velocity $W = W(y,t)$. In similar fashion to equation (27), which applies outside the swash zone, we separate short-wave from long-wave contribution for the flow properties inside the swash zone:

$$d = \hat{d} \ , \quad u = \frac{\partial x_l}{\partial t} + \hat{u} \ , \quad v = W + \hat{v} \ , \tag{37}$$

where a short-wave property inside the swash zone is defined as $\hat{G}$ rather than $\tilde{G}$ which pertains to short-wave contributions outside the swash zone. This decomposition permits the integrated terms depending on the longshore velocity to be divided between long and short waves:

$$\langle P_2 \rangle = \langle W\hat{V} \rangle + \langle \hat{P}_2 \rangle \ , \tag{38}$$

$$\langle M_{12} \rangle = \langle W\hat{P}_1 \rangle + \langle \hat{M}_{12} \rangle + \frac{\partial x_l}{\partial t}\left(\langle W\hat{V} \rangle + \langle \hat{P}_2 \rangle\right) \ , \tag{39}$$

$$\langle M_{22} \rangle = \langle W^2\hat{V} \rangle + 2\langle W\hat{P}_2 \rangle + \langle \hat{M}_{22} \rangle . \tag{40}$$

All the integral properties which are to be considered as short-wave terms have a hat symbol. In principle it is possible to choose a different short-wave solution for the two regions separated by $x = x_l$ also on many sandy beaches there is a strong difference in the character of the bed in the swash zone compared with the bed just outside the swash zone.

In order to avoid greater complexity, the longshore variation of x_l has been assumed to be negligible in the above derivation. This is essentially a geometrical matter, and the above boundary conditions can be interpretated as being applicable when the x direction is normal to the mean lowest boundary of the swash.

To obtain explicit expression of the first boundary condition we consider the flow of mass into the swash zone relative to $x = x_l$. Average of equation (25) is:

$$\frac{\partial \langle V \rangle}{\partial t} + \frac{\partial \langle P_2 \rangle}{\partial y} = \langle Q_1 \rangle|_{x_l} \tag{41}$$

where now the right hand side contains the relative flow velocity $(u - \partial x_l/\partial t)$. The left hand side is also rewritten in terms of the local variables inside the swash zone and then after using (38) for $\langle P_2\rangle$, after integration across the swash zone width and averaging over the short waves we obtain

$$\frac{\partial\langle\hat{V}\rangle}{\partial t}+\frac{\partial\langle W\hat{V}\rangle}{\partial y}+\frac{\partial\langle\hat{P}_2\rangle}{\partial y}=\left\langle\left(u-\frac{\partial x_l}{\partial t}\right)d\right\rangle=\langle u\rangle\langle d\rangle+\langle\tilde{u}\tilde{d}\rangle-\frac{\partial x_l}{\partial t}\langle d\rangle\,. \tag{42}$$

The first term on the left hand side of this equation is the rate of change of the total volume of water in the swash zone; the second and third terms are the change in volume due to the lateral variation of longshore currents respectively associated with the longshore velocity inside the swash zone W and with the wave contribution $\langle\hat{P}_2\rangle$. The right hand side of (42) is the increase of water in the swash zone through its lower boundary. The right hand side of (42) is evaluated at $x=x_l$ using outer variables.

A similar derivation for the average balance of onshore momentum in the swash zone gives

$$\frac{\partial}{\partial t}\left[\langle\hat{P}_1\rangle+\frac{\partial x_l}{\partial t}\langle\hat{V}\rangle\right]+\frac{\partial\langle W\hat{P}_1\rangle}{\partial y}+\frac{\partial\langle\hat{M}_{12}\rangle}{\partial y}$$

$$+\frac{\partial x_l}{\partial t}\left[\frac{\partial\langle W\hat{V}\rangle}{\partial y}+\frac{\partial\langle\hat{P}_2\rangle}{\partial y}\right]+g\alpha\langle\hat{V}\rangle+\langle\Upsilon_1\rangle=\left\langle\left(u-\frac{\partial x_l}{\partial t}\right)^2d+\tfrac{1}{2}gd^2\right\rangle= \tag{43}$$

$$\left(\langle u\rangle-\frac{\partial x_l}{\partial t}\right)^2\langle d\rangle+\tfrac{1}{2}g\langle d\rangle^2+2\langle\tilde{u}\tilde{d}\rangle\left(\langle u\rangle-\frac{\partial x_l}{\partial t}\right)+\langle\tilde{u}^2\rangle\langle d\rangle+\langle\tilde{u}^2\tilde{d}\rangle+\tfrac{1}{2}g\langle\tilde{d}^2\rangle.$$

In this equation dimensional expressions have been inserted to clarify the origin of terms. The first two terms on the left hand side are the rate of change of the mean momentum in the swash zone. The group of terms with y derivatives are the contribution from longshore velocity gradients which include both long period contributions (terms with W) and short period contributions (terms with $\langle\hat{P}_2\rangle$ and $\langle\hat{M}_{12}\rangle$). The following terms are the action of gravity and friction on the water in the swash zone. The right hand side is the momentum transfer into the swash zone at $x=x_l$ where the mean flow velocity relative to the swash zone limit appears in both long and short period terms. Finally pure short period contributions appear. Conservation of longshore momentum similarly yields

$$\frac{\partial\langle W\hat{V}\rangle}{\partial t}+\frac{\partial\langle\hat{P}_2\rangle}{\partial t}+\frac{\partial\langle W^2\hat{V}\rangle}{\partial y}+2\frac{\partial\langle W\hat{P}_2\rangle}{\partial y}+\frac{\partial\langle\hat{M}_{22}\rangle}{\partial y}+\langle\Upsilon_2\rangle=\left\langle\left(u-\frac{\partial x_l}{\partial t}\right)vd\right\rangle= \tag{44}$$

$$\left(\langle u\rangle-\frac{\partial x_l}{\partial t}\right)\langle v\rangle\langle d\rangle+\langle\tilde{u}\tilde{v}\rangle\langle d\rangle+\langle\tilde{u}\tilde{d}\rangle\langle v\rangle+\langle\tilde{u}\tilde{v}\rangle\langle d\rangle+\langle\tilde{u}\tilde{d}\rangle\langle v\rangle+\langle\tilde{v}\tilde{d}\rangle\left(\langle u\rangle-\frac{\partial x_l}{\partial t}\right)+\langle\tilde{u}\tilde{v}\tilde{d}\rangle.$$

The above three equations are boundary conditions for the long wave motion and x_l if the short wave properties on each side of $x=x_l$ have 'known' models. Whereas in the rigid wall model only the two boundary conditions are required, now we also need to evaluate the changes in W. In addition, the short-wave models may be chosen to permit the reflection of short waves. Note, the short-wave swash model must be used relative to $x=x_l$, hence the motion of x_l is seen as the long period contribution to the swash zone. A delicate matter concerning the different wave models needed on each side of x_l also needs attention.

The short-wave contribution

In this section we briefly analyse the relevance of the short-wave contributions with particular emphasis on those appearing in the equations (42), (43) and (44) which are used to define the boundary conditions at the lower limit of the swash zone. Computation of the contributions has been performed by integrating the analytical solution for $A \leq 1$ and by integrating the fully numerical solution when wave-breaking occurs, i.e. for $A > 1$.

For waves that break momentum and energy are transferred from the short-waves to the longshore currents hence terms depending on the longshore drift velocity W are to be taken into account as major contributions to the swash zone motions in equations (42), (43) and (44) (see figure 5). Note that the curves have been obtained by joining discrete points which represent the results of single computations. This explains the lack of smoothness in some of the curves.

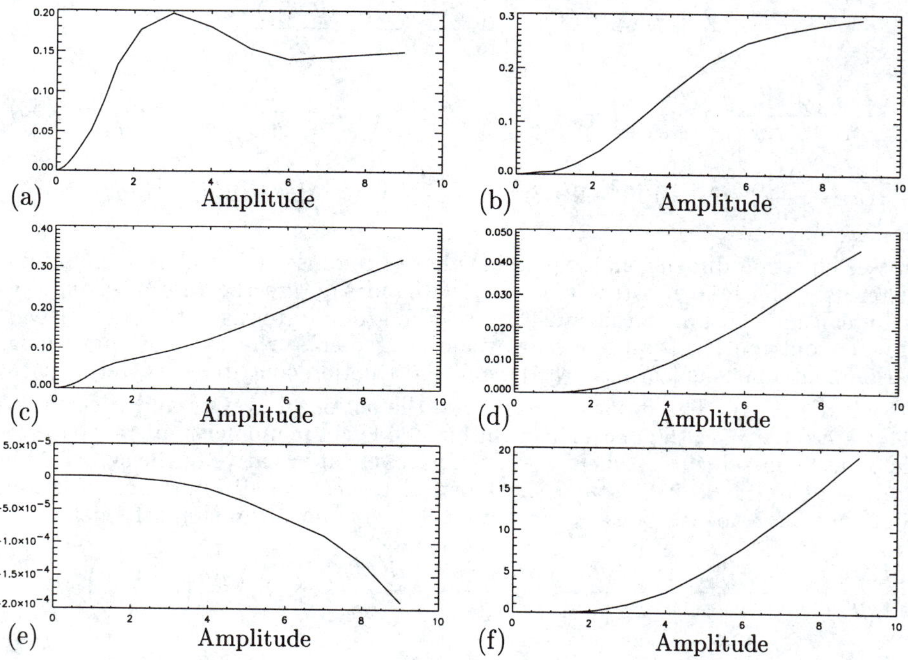

Figure 5: Short-wave contribution at the boundary x_l and inside the swash zone. From top left to bottom right they are the short-wave correlations (a) $\langle \hat{u}^2 \rangle$, (b)$\langle \hat{d^2} \rangle$ the integral properties (c) $\langle \hat{V} \rangle$, and the rate of change of the three global integral properties (d) $\langle P_2 \rangle$, (e) $\langle M_{12} \rangle$ and (f) $\langle M_{22} \rangle$. Note that for $A \leq 1$ numerical integration of the analytical solution is shown while for $A > 1$ fully numerical solution of the NLSWE is represented.

The input of longshore momentum from the bores is balanced by bed friction but this is generally on the longer time scale. Although bores die out when they meet the shoreline and cause run-up, the water that forms the swash has acquired longshore momentum from the bore before it enters the swash zone. A fully detailed consideration of this case also awaits further study; here we present the rate of change in time of the quantities $\langle\hat{V}\rangle$, $\langle\hat{P}_2\rangle$ and $\langle\hat{M}_{22}\rangle$.

The figure presented here indicates that not all the terms that are introduced into the boundary conditions (42, 43 and 44) are necessary. Inspection of figure (5) reveals that the largest term to be retained in case of wave breaking is the rate of change of $\langle\hat{M}_{22}\rangle$. This is not surprising as $\langle\hat{M}_{22}\rangle$ is the only integral term proportional to W^2 (see equations 38) and hence it is more heavily influenced by the longshore drift induced by wave breaking. We can further comment that the non-diagonal contribution $\partial\langle\hat{M}_{12}\rangle/\partial t$ seems to be negligible in this particular case in which we have used $x_l = const.$ However, in view of equation (39), here reported for ease of inspection,

$$\langle M_{12}\rangle = \langle W\hat{P}_1\rangle + \langle\hat{M}_{12}\rangle + \frac{\partial x_l}{\partial t}\left(\langle W\hat{V}\rangle + \langle\hat{P}_2\rangle\right) \tag{45}$$

we believe that this term may give larger contribution for a time-varying x_l as the rate of change of the boundary position explicitely contributes to this term.

Further investigation should help to clarify both which are the most significant terms when considered in the context of the equations and what are the effects of friction terms.

Acknowledgements

Support from E.U., D.G. XII, contracts : ERBCHBICT930678 and MAS2-CT92-0047 is gratefully acknowledged. We wish to thank Dr Gary Watson for many useful conversations.

References

BROCCHINI, M. & PEREGRINE, D.H. (1995). Flow properties of the swash zone. *Proc. Coastal Dynamics '95*, ASCE, 221-232.

BROCCHINI, M. & PEREGRINE, D.H. (1996). Integral flow properties of the swash zone and averaging. *J. of Fluid Mech.* **317**, 241-273.

CARRIER, G.F. & GREENSPAN, H.P. (1958). Water waves of finite amplitude on a sloping beach. *J. of Fluid Mech.* **4**, 97-109.

MEYER, R. E. & TAYLOR, A. D. (1972) Run-up on beaches. In *Waves on beaches.* (ed. R. E. Meyer). Academic Press, New York.

RYRIE, S.C. (1983). Longshore motion generated on beaches by obliquely incident bores. *J. Fluid Mech.* **129**, 193-212.

WATSON, G., BARNES, T.C.D. & PEREGRINE, D.H. (1994). The generation of low frequency waves by a single wave group incident on a beach. *Proc. 24th Internat. Conf. Coast. Engng., ASCE* **1**, 776-790.

CHAPTER 321

Wave Overwash of Subaerial Dunes

Yukiko Tega [1] and Nobuhisa Kobayashi, M. ASCE [2]

Abstract

Small-scale experimental results consisting of 72 runs for minor to major dune overwash are analyzed to improve existing dune erosion models that do not account for wave overwash processes explicitly. The measured cross-shore sediment transport rate is found to vary approximately linearly from the measured rate somewhat below the still water level, which is affected by the overwash intensity, to the overwash rate over the dune crest. The measured overwash sand concentration for the sand with its median diameter of 0.38 mm used in this experiment was in the range of 0.02 to 0.06. The time-dependent one-dimensional numerical model for the overtopping flow combined with two sediment transport formulas indicates that the overwash sand concentraion may be on the order of 0.04 except for very fine or coarse sands.

Introduction

The quantitative understanding of wave overwash of dunes is essential for predicting dune erosion as well as coastal flooding hazards during storms. Existing dune erosion models generally assume that the net sediment transport rate may be extrapolated linearly form the value slightly below the still water level (SWL) to zero at the upper limit of wave runup estimated empirically. This linear extrapolation may be necessary for lack of a simple swash model and appears to be a reasonable first approximation for the case of no overwash.

A simple empirical method for predicting the onshore sand transport rate associated with wave overwash of subaerial dunes is presented herein by expanding the experimental study by Hancock and Kobayashi (1994). The experiment was limited to subaerial dunes, while experiments on submerged dunes were conducted by Steetzel and Visser (1992). Kobayashi et al. (1996) showed that the measured reflection coefficient and overtopping rates could be predicted using the empirical formulas developed for coastal structures and adjusted for dunes. This paper examines the cross-shore sand transport on dunes.

Experiment

Fig.1 shows the setup of the dune overwash experiment conducted in a wave tank that was 30 m long, 2.44 m wide, and 1.5 m high. The experiment was

[1]Technical Research Institute, Nishimatsu Construction Co.Ltd., 2570-4 Shimotsuruma, Yamato, Kanagawa, 242, Japan

[2]Professor and Associate Director, Center for Applied Coastal Research, University of Delaware, Newark, DE 19716.

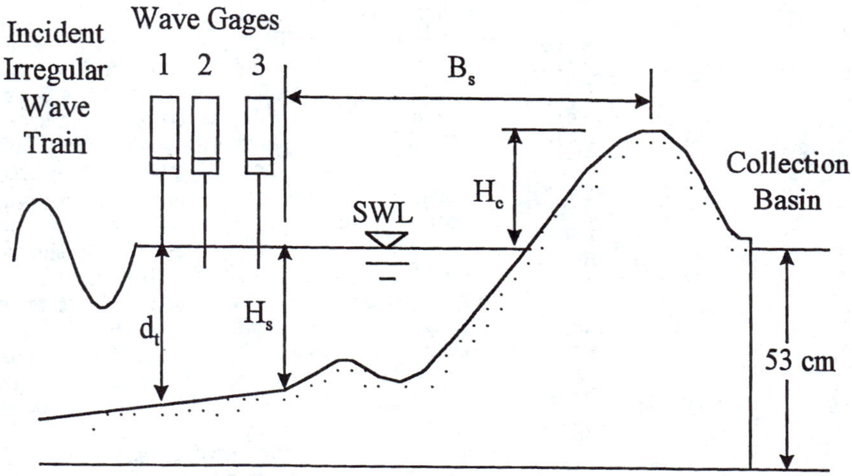

Figure 1: Experimental Setup.

conducted in a 61 cm wide channel by constructing a divider wall to reduce the effects of re-reflection from the wave paddle used to generate irregular waves. The well-sorted sand with a median diameter of 0.38 mm was placed in front of a 53 cm high basin used to collect both overtopped water and overwashed sand. The measured specific gravity, porosity and fall velocity of the sand were 2.66, 0.41 and 5.29 cm/s, respectively. Three wave gages placed immediately seaward of the breaker zone were used to separate the incident and reflected waves at wave gage 1 for each run (Kobayashi and Raichle 1994). The duration of each run was 325 s and the sampling rate for the three gages was 20 Hz.

Seven tests, referred to as tests A to G hereafter, were conducted for minor to major overwash. The dune crest was initially located at approximately 12 cm above SWL and at a horizontal distance of 40cm from the basin wall. Each test consisted of a series of runs to examine the changes of the overtopping rate per unit width, Q , and the overwash rate per unit width, Q_s , as the beach and dune profile evolved. The overtopping and overwash rates per unit width, Q and Q_s, for each run were obtained by dividing the volumes of water and sand, respectively, collected in the basin by the run duration 325s and the channel width 61 cm. The value of Q_s is based on the sediment volume only without any void. The test was terminated just before the collection basin wall would become exposed to direct wave action. Beach and dune profiles were measured along three lines using an ultrasound profiler at the beginning and end of each test and after runs of noticeable profile changes. The average profile of the measured three profiles was used to find the dune crest height H_c above SWL, the location where the depth below SWL equals to the significant wave height H_s measured at wave gage 1, and the horizontal distance B_s from the dune crest to this location as shown in Fig. 1. The equivalent uniform slope for wave overtopping, m_o, is assumed to be given by $m_o = (H_c + H_s)/B_s$. The measured values of m_o were interpolated to obtain the value of m_o in the middle of each run.

Table 1 lists the number of runs in each test and the ranges of the incident wave characteristics at wave gage 1, the dune crest height H_c, the equivalent slope m_o, the overtopping rate Q and the overwash rate Q_s for the seven tests. The water depth d_t below SWL at wave gage 1 for each test was constant where the bottom change at wave gage 1 was negligible. The spectral peak period T_p was

Table 1: Ranges of Incident Wave Characteristics, Dune Crest Height, Overtopping, Overwash Rates and Equivalent Slope for 7 Tests

Test	Number of runs	d_t (cm)	T_p (s)	H_s (cm)	H_c (cm)	Q (cm^2/s)	Q_s (cm^2/s)	m_o
A	10	44.3	1.2	11.8–12.2	7.5–12.2	0–0.40	0–0.0148	0.11–0.26
B	9	43.4	1.4	12.2–12.5	7.1–11.1	0.52–1.08	0.024–0.040	0.11–0.25
C	8	44.8	1.6	12.3–12.8	7.0–10.9	1.51–2.61	0.059–0.089	0.11–0.25
D	7	43.7	1.8	12.1–12.4	7.6–11.0	3.33–4.68	0.12–0.16	0.14–0.24
E	5	43.2	2.0	12.4–12.7	6.7–10.7	6.33–8.72	0.30–0.37	0.16–0.25
F	24	43.2	1.4	12.0–12.3	7.9–11.3	0.31–0.47	0.007–0.026	0.09–0.16
G	9	43.2	2.0	12.0–12.4	7.8–11.1	2.89–4.38	0.10–0.16	0.11–0.16

also constant for each test and $T_p = 1.2 - 2.0$ s for the seven tests. The significant wave height, H_s, based on a zero upcrossing method varied little during each test and among the seven tests because the TMA spectra with the spectral significant wave height $H_{mo} = 12.5$ cm and $T_p = 1.2 - 2.0$ s at the wave paddle were used to generate the incident irregular waves. On the other hand, the dune crest height H_c and the equivalent slopes m_o decreased with the increase of run number as the beach and the dune profile evolved. The data from tests A to E consisting of 39 runs were presented by Hancock and Kobayashi (1994). These tests corresponded to high dunes with large storm surge. Two additional tests, F and G, consisting of 33 runs were conducted by placing more sand in front of the initial dune profile used in tests A to E to reduce the water depth at the toe of the dune.

Data Analyses

First, the measured overwash rate Q_s is related to the measured overtopping rate Q in the form

$$Q_s = C_s Q \qquad (1)$$

where C_s is the avarage volumetric sediment concentration in the overwash. The dimensional values of Q_s and Q are plotted in Fig.2 to show that these values were from the small scale experiment. Fig. 2 shows that the 72 data points fall within the narrow range of $C_s = 0.02 - 0.06$ and that the empirical formula (1) with $C_s \simeq 0.04$ may be used to estimate Q_s for Q predicted separately. The subsequent sections will examine whether the assumption of constant C_s holds for field conditions and other sands.

To predict the overtopping rate Q for normally incident breaking waves on dunes, Kobayashi et al. (1996) adjusted the empirical formula for wave overtopping on coastal structures proposed by Van der Meer and Janssen (1995). The toe of a coastal structure was assumed to corresponds to the location of wave gage 1 immediately seaward of the breaker zone. The equivalent uniform slope m_o was assumed to represent the overall slope effect on wave overtopping of dunes. The adjusted formula was shown to yield the agreement between the measured and empirical values of Q almost within the 95 % confidence limits of the original formula based on about 500 data points. Fig. 3 shows the measured overwash rate Q_s as a function of the empirical overtopping rate Q_e based on the adjusted formula by Kobayashi et al. (1996). The formula (1) with Q replaced by Q_e is still reasonable. However, most of the values of C_s are in the range of $C_s = 0.01 - 0.04$ with a typical value of $C_s = 0.02$. These values of C_s are smaller than those of C_s based on Q by a factor of about two. This is because the adjust formula for Q_e tends to overpredict the overtopping rates for the tests with larger

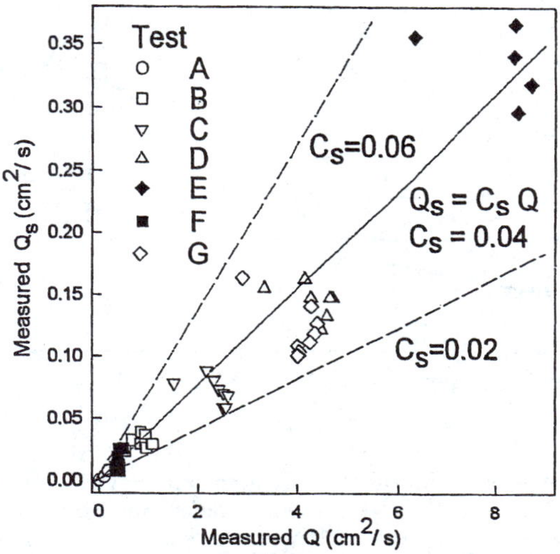

Figure 2: Measured Overwash Rate Q_s vs. Measured Overtopping Rate Q.

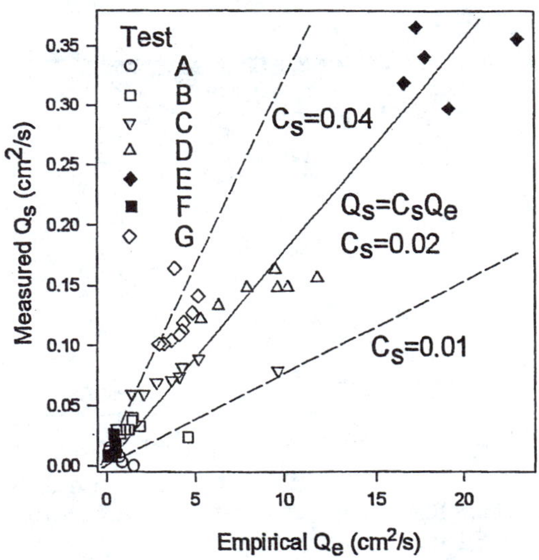

Figure 3: Measured Overwash Rate Q_s vs. Empirical Overtopping Rate Q_e

overtopping rates. The accurate prediction of Q_s using (1) requires the accurate predictions of C_s and Q. Unfortunately, only the order of magnitude of Q, even for coastal structures, can be predicted at present. Consequently, the present study aims at the prediction of C_s within a factor of about two.

In order to predict the beach and dune profile evolution, the cross-shore variation of the net sediment transport rate is required. The net cross-shore sediment transport rate q_s was calculated using the continuity equation of sediment and the measured profiles starting from the seaward location of no profile change as described by Hancock and Kobayashi (1994). The value of q_s at the landward

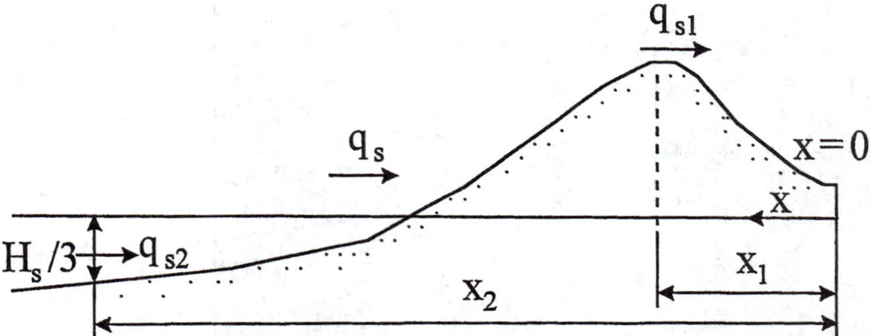

Figure 4: Schematic Representation of Cross-Shore Sediment Transport Rate q_s over Dune

boundary should correspond to the net sediment transport rate into the collection basin but was not exactly the same as the average overwash rate for the same duration calculated from the collected sand because of small profile measurement errors. As a result, the normalized results are presented as explained in the following.

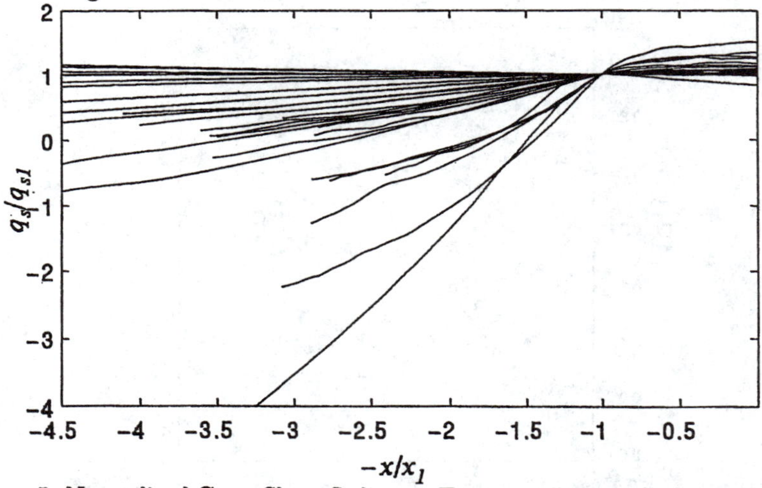

Figure 5: Normalized Cross-Shore Sediment Transport Rate, q_s/q_{s1}, as a Function of $-x/x_1$ with $x = 0$ at Edge of Collection Basin.

The measured cross-shore sediment transport rate q_s is taken to be positive onshore as shown in Fig. 4 where x = cross-shore coordinate taken to be positive seaward with $x = 0$ at the edge of the collection basin. The dune crest is located at $x = x_1$. The seaward limit of this analysis may be taken at the location x_2 where the water depth below SWL equals $H_s/3$ because this depth of $H_s/3$ is larger than the typical smallest depth adopted by existing cross-shore sediment transport models in the surf zone. The values of q_s at $x = x_1$ and x_2 are denoted by q_{s1} and q_{s2}, respectively. The locations of x_1 and x_2 varied somewhat between the two profiles measured consecutively. The first of the two profiles was used to find x_1 and x_2. The average value of H_s during the runs between the two profile measurements was used for the depth $H_s/3$ for x_2. The results presented in the following are not sensitive to the details of the data analysis.

The normalized cross-shore sediment transport rate, q_s/q_{s1}, is plotted in Fig. 5 as a function of $-x/x_1$ in the range $0 < x < x_2$ for all the cross-shore variations of the onshore sand transport rate based on the consecutive profile measurements. Fig. 5 shows that the value of q_s landward of the dune crest is almost constant and may be approximated by the overwash rate Q_s, which is based on the sand volume collected in the basin for each run and should be more accurate than q_s. Furthermore, q_s increases approximately linearly form q_{s2} at $x = x_2$ somewhat below SWL to q_{s1} at the dune crest. Consequently, the cross-shore variation of q_s required for the prediction of the dune profile evolution using the continuity equation for sediment may be represented by $Q_s \simeq q_{s1}$ and q_{s2} where Q_s may be estimated using (1).

Existing dune erosion models estimate the cross-shore sediment transport rate slightly below SWL without regard to wave overwash. To examine the effect of overwash on q_s below SWL, the values of q_{s2}/q_{s1} are plotted as a function of the overtopping rate Q and the overwash rate Q_s in Fig. 6. Since $q_{s1} > 0$ in this experiment, the positive and negative values of q_{s2}/q_{s1} imply the onshore and offshore sediment transport at $x = x_2$ below SWL, respectively. Fig. 6 indicates that the net sediment transport somewhat below SWL in this experiment was offshore for small Q and Q_s but tended to become onshore with the increase of Q and Q_s. This implies that the estimation of q_{s2} will need to account for wave overtopping and overwash.

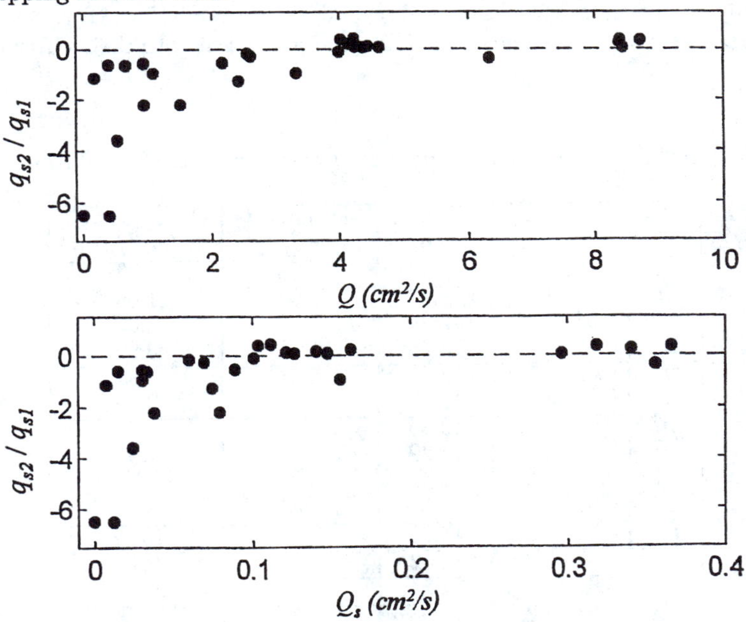

Figure 6: Normalized Sediment Transport Rate ,q_{s2}/q_{s1}, Somewhat below SWL as a Function of Overtopping Rate Q (top) and Overwash Rate Q_s (bottom)

Numerical Modeling

To elucidate the overwash processes and extraporate the small-scale experimental results presented above, use is made of the time-dependent one-dimensional numerical model RBREAK2 (Kobayashi and Poff 1994). RBREAK2 was shown to predict the overtopping rates for revetments in surf zones within about 40 % errors (Kobayashi and Raichle 1994). Two sediment transport formulas are combined with RBREAK2 to predict the net sediment transport rate q_s.

The empirical formula proposed by Trowbridge and Young (1989) for unbroken water waves under sheet flow conditions may be used to express q_s in the form

$$\frac{q_s}{wd} = \frac{A\overline{|u|u}}{g(s-1)d} \tag{2}$$

in which w = fall velocity of the sand particle; g = gravitational acceleration; d = median diameter d_{50} where d is used for brevity; s = specific gravity of the sand; u = instantaneous depth-averaged velocity computed by RBREAK2 ; and A = empirical parameter. The overbar in (2) denotes the time-averaging for the duration of each run. The formula (2) with $A \simeq 0.25$ was employed successfully to predict the onshore sand transport outside the surf zone at the CERC Field Research Facility.

On the other hand, Ribberink and Al-Salem (1994) developed a sediment transport formula using large oscillating water tunnel data. This formula expresses q_s as

$$\frac{q_s}{wd} = af^{1.5}\frac{\overline{u^3}}{w^3} \tag{3}$$

where a = empirical parameter; and f = Jonsson's wave friction factor. The formula with $a = \sqrt{2}$ was shown to yield good agreement with their water tunnel data.

Table 2: Measured and Computed Overtopping Rates for 10 Runs

Test	f	T_p (s)	H_s (cm) (cm)	Measured Q (cm^2/s)	Computed Q_c (cm^2/s)
B6	0.014	1.4	12.36	0.93	0.38
B9	0.014	1.4	12.28	1.08	1.09
B9	0.012	1.4	12.28	1.08	1.06
B9	0.010	1.4	12.28	1.08	0.90
D5	0.012	1.8	12.30	4.63	1.98
D5	0.010	1.8	12.30	4.63	2.19
D7	0.012	1.8	12.29	4.46	1.78
D7	0.010	1.8	12.29	4.46	1.97
E1	0.010	2.0	12.49	6.33	2.16
E2	0.010	2.0	12.67	8.40	3.12
E3	0.010	2.0	12.71	8.38	13.07
G1	0.010	2.0	12.42	2.89	0.12
G2	0.010	2.0	12.34	4.26	0.23
G3	0.010	2.0	12.05	4.38	1.48

The formulas (2) and (3) developed for the sand transport due to prototype large-scale unbroken waves may be applied to the sand transport due to broken waves overtopping on dunes if the empirical parameters A and a are adjusted.

Computations were made for 10 representative runs using the measured profile and incident wave train at wave gagel as input to RBREAK2 . Table 2 summarizes the computed runs where f = friction factor specified in the computation;

T_p = spectral peak period of the incident waves; H_s = incident significant wave height; Q = measured overtopping rate; and Q_c = computed overtopping rate which is the time-averaged volume flux per unit width at the edge of the collection basin during each run. The friction factor f for the overtopping flow is expected to be different from Jonsson's friction factor. The value of $f \simeq 0.01$ is a reasonable estimate from the previous calibration for wave overtopping of revetments performed by Kobayashi and Raichle (1994). RBREAK2 with $f = 0.01 - 0.014$ predicts the order of magnitude of the overtopping rates. For tests E and G, where the overall beach and dune slope was gentler for test G as may be inferred from the range of m_o listed in Table 1, the computed overtopping rates were varied much more than the measured rates for the different profiles of tests E and G and as the profile evolved in each test. This apparent sensitivity of RBREAK2 to the beach and dune profile is not certain for lack of measurements on the depth and velocity of the overtopping flow in this experiment.

The cross-shore variation of the net sediment transport rate q_s is also computed using (2) and (3) where RBREAK2 predicts the temporal and cross-shore variations of the depth-averaged velocity u during each run. The computed cross-shore variation of q_s in the region between the dune crest and the edge of the collection basin is essentially constant and consistent with the measured variation shown in Fig. 5. However, the computed values of q_s seaward of the dune crest vary too rapidly in comparison to these shown in Fig. 5. The assumption of the instantaneous response of sand particles to the depth-averaged velocity u made in (2) and (3) may not be valid seaward of the dune crest. The following analysis is limited to the computed results at the edge of the collection basin which are representative of the computed results landward of the dune crest.

The computed time series of the normalized water depth, $h_* = h/H_s$, where H_s = incident significant wave height for each run, and the normalized depth-averaged velocity, $u_* = u/\sqrt{gH_s}$, are stored at the sampling rate of 20 Hz during each run lasting 325 s. In order to find a relationship between h_* and u_*, the stored values of h_* and u_* for the 10 runs listed in Table 2 are plotted in the form

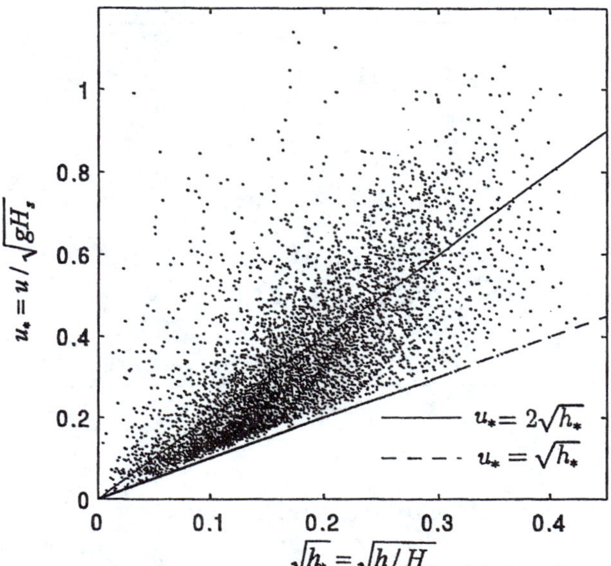

Figure 7: Computed Normalized Velocity, u_*, and Square Root Depth, h_*, of Overwash Flow for 10 Runs

of u_* v.s. $\sqrt{h_*}$ in Fig. 7. RBREAK2 assumes that the overtopping flow at the landward boundary of the computation domain is critical ($u = \sqrt{gh}$ and hence $u_* = \sqrt{h_*}$) or supercritical ($u > \sqrt{gh}$ and hence $u_* > \sqrt{h_*}$). The scattered points in Fig. 7 may be represented crudely by the simple relationship, $u_* \simeq 2\sqrt{h_*}$. This relationship corresponds to that between the bore speed and flow depth on a barrier island measuring using video techniques by Holland et al. (1991) if the celerity and fluid velocity of the bore are assumed to be the same.

The simple relationship of $u_* = 2\sqrt{h_*}$ may describe the overall trend of the overtopping flow but is not accurate enough to predict the instantaneous relationship between u_* and h_*. As an example, Fig. 8 shows the computed time series of u_* and $2\sqrt{h_*}$ for run B9 where $t_* = t/T_p$ is the normalized time with $t_* = 0$ at the beginning of each run. The relationship of $u_* = 2\sqrt{h_*}$ tends to underpredict the peaks of u_* and overpredict the receding segments of u_*.

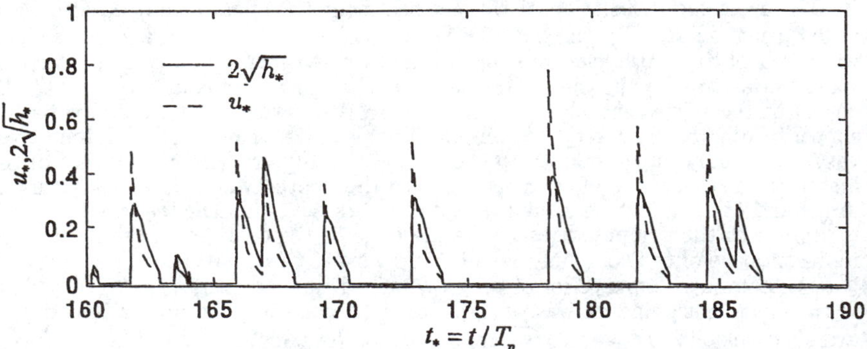

Figure 8: Computed Time Series of u_* and $2\sqrt{h_*}$ for Run B9

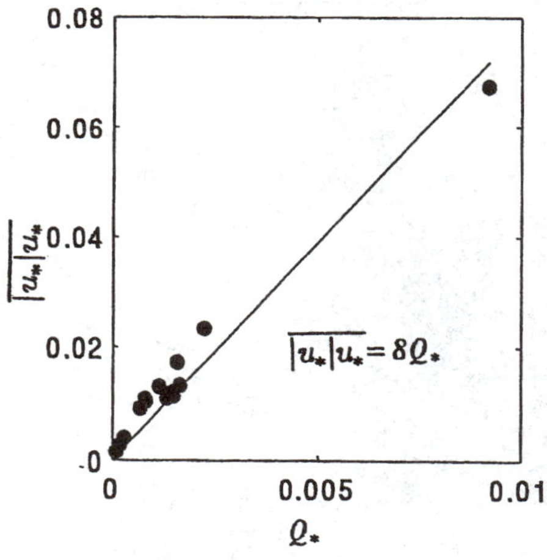

Figure 9: Computed Normalized Velocity Moment $\overline{|u_*|u_*}$ as a Function of Computed Normalized Overtopping Rate Q_*

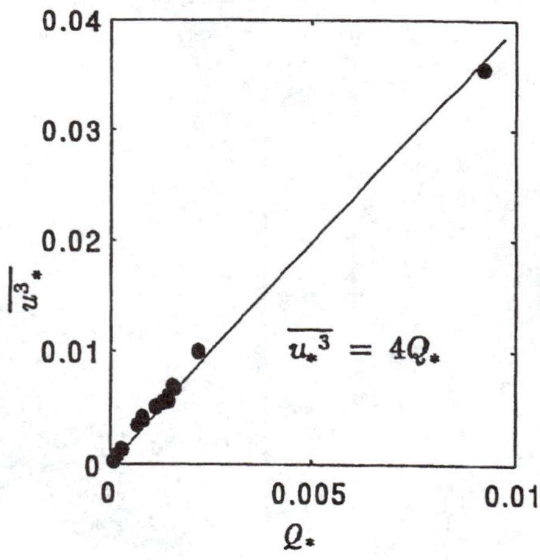

Figure 10: Computed Normalized Velocity Moment $\overline{u^3_*}$ as a Function of Computed Normalized Overtopping Rate Q_*

In order to explain the approximate linear relationship between Q_s and Q expressed by (1) using the formulas (2) and (3), the computed time-averaged values of $\overline{|u_*|u_*}$ and $\overline{u_*^3}$ for each of the runs listed in Table 2 are plotted as a function of the computed normalized overtopping rate, $Q_* = \overline{h_* u_*}$, in Figs. 9 and 10, respectivery.The dimensional overtopping rate Q is given by $Q = H_s\sqrt{gH_s}Q_*$. In Fig.9, $\overline{|u_*|u_*}/Q_* = 7.3 - 18.8$ and the linear regression coefficient is about 8.

In Fig. 10, $\overline{u_*^3}/Q_* = 3.8 - 5.3$ and the relationship of $\overline{u_*^3} = 4Q_*$ based on the assumption of $u_* = 2\sqrt{h_*}$ yields very good agreement. These relationships together with (2) and (3) are used to express the overwash sand concentration C_s in (1) in terms of the sand and wave characteristics.

Overwash Sand Concentration

The overwash rate Q_s is estimated as the values of q_s in (2) and (3) at the edge of the collection basin. Substitution of (1) into (2) and (3) yields the following expressions of C_s:

$$C_s = Aw_* \frac{\overline{|u_*|u_*}}{Q_*} \quad ; \quad w_* = \frac{w}{(s-1)\sqrt{gH_s}} \tag{4}$$

and

$$C_s = af^{1.5} d_* \frac{\overline{u_*^3}}{Q_*} \quad ; \quad d_* = \frac{gd}{w^2} \tag{5}$$

where w_* and d_* are the normalized fall velocity and diameter respectively. To calibrate the empirical coefficients A and a in (4) and (5), the measured concentration C_s is plotted as a function of $w_*\overline{|u_*|u_*}/Q_*$ and $f^{1.5}d_*\overline{u_*^3}/Q_*$ in Figs. 11 and 12, respectively, for the computed runs listed in Table 2. Fig. 11 indicates $A \simeq 0.1$ in (4) instead of $A \simeq 0.25$ calibrated by Trowbridge and Young (1989).

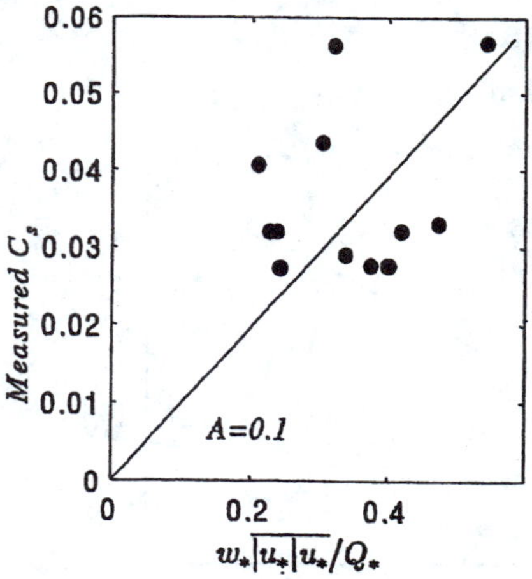

Figure 11: Measured Sand Concentration C_s Compared with Formula (4) with $A = 0.1$

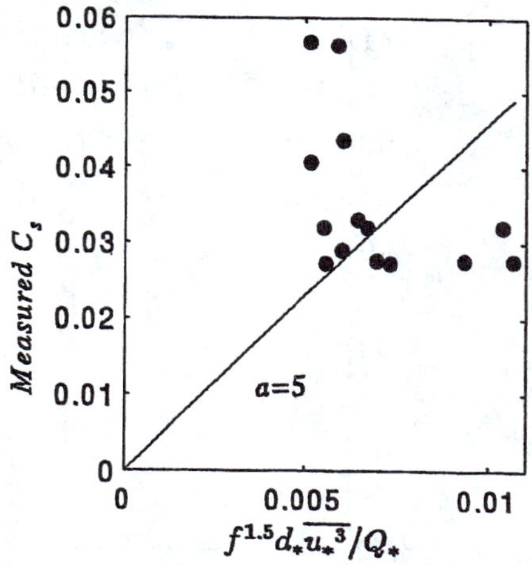

Figure 12: Measured Sand Concentration C_s Compared with Formula (5) with $a = 5$

Fig. 12 suggests $a \simeq 5$ instead of $a \simeq \sqrt{2}$ obtained by Ribberink and Al-Salem (1994). These order-of-magnitude agreements are encouraging in light of the scale differences (small-scale experiment v.s. field or large-scale experiment) and the different wave conditions (breaking wave overwash of dunes v.s. unbroken waves on essentially horizontal bottoms.). Figs. 11 and 12 also reveal the difficulty in predicting the relatively small variation of C_s.

The formula (4) with $A \simeq 0.1$ together with $\overline{|u_*|u_*}/Q_* \simeq 8$ in Fig.9 may be simplified as

$$C_s = Bw_* \tag{6}$$

where B = empirical parameter on the order of 0.8. Furthermore, the formula (5) with $a \simeq 5$ and $f \simeq 0.01$ along with $\overline{u_*^3}/Q_* \simeq 4$ in Fig. 10 may simply be rewritten as

$$C_s = bd_* \tag{7}$$

where b = empirical parameter on the order of 0.02.

For the seven tests consisting of 72 runs listed in Table 1, $d = 0.38$ mm, $w =$ 5.29 cm/s, $s = 2.66$ and $H_s \simeq 12$ cm. As a result, $w_* \simeq 0.03$ and $d_* \simeq 1.33$ for this experiment. For the narrow range of the measured values of C_s for the 72 runs, $B = 0.63-2.22$ with its mean being 0.8 and $b = 0.014-0.049$ with its mean being 0.025. The formulas (6) and (7) are no better than the linear regression between Q_s and Q shown in Fig.2 since w_* is essentially constant and d_* is constant in this experiment. However, (6) and (7) may be applied to estimate C_s for other sands.

The variation of the overwash sand concentration C_s with respect to its median diameter d estimated using (6) with $B = 0.8$ as well as (7) with $b = 0.025$ are shown in Figs. 13 and 14, respectively, where the specific gravity of the sand is assumed to be $s = 2.6$ and its fall velocity w is taken to be equal to that of the spherical particle falling in quiet water whose kinematic viscosity is $\nu = 0.01 cm^2/s$. In Fig.13, use is made of $H_s = 0.1$, 1 and 5 m because w_*

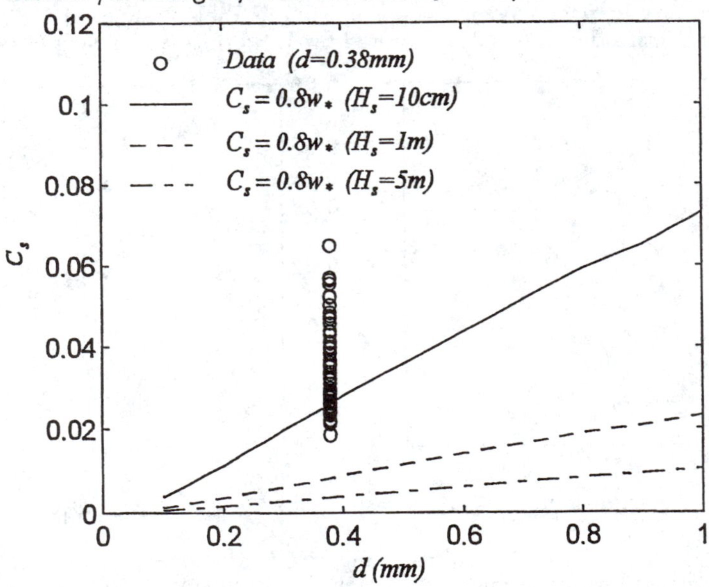

Figure 13: Variation of Sand Concentration C_s with Median Diameter d Estimated by $C_s = 0.8w_*$

defined in (4) depends on the incident significant wave height H_s. Contrary to our expectation, Fig. 13 suggests the increase of C_s as d is increased and H_s is decreased. This indicates the limitation of the formula (2) proposed by Trowbridge and Young (1989) who stated that their formula might be limited to quartz sand whose diameter should not be too different from 0.2 mm. On the other hand, Fig. 14 indicates the increase of C_s with the decrease of d as expected. The formula (3) proposed by Ribberink and Al-Salem (1994) was compared with the data with $d = 0.2 - 1.8$ mm. The rapid increase of C_s with the decrease of d in the range $d < 0.2$ mm in Fig. 14 appears to be unrealistic in view of the physical upper limit of $C_s \simeq 0.6$. The range of $C_s = 0.018 - 0.065$ for the 72 runs in comparison to the variation of $C_s = 0.025d_*$ for $d = 0.2 - 1.0$ mm in Fig. 14 implies that it will be difficult to detect the effect of d on C_s unless very fine or coarse sands are employed in experiments.

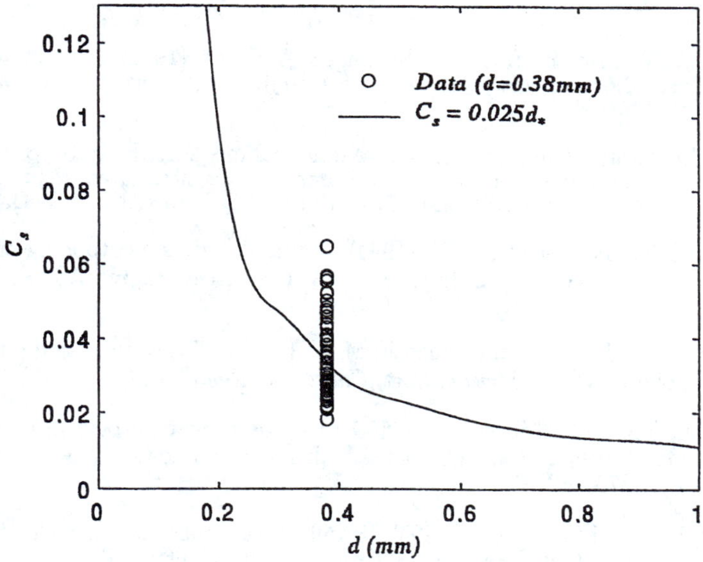

Figure 14: Variation of Sand Concentration C_s with Median Diameter d Estimated by $C_s = 0.025d_*$

Conclusion

This small-scale experiment on wave overwash of subaerial sand dunes indicates that the net cross-shore sediment transport rate q_s varies approximately linearly from $q_s = q_{s2}$ somewhat below SWL to $q_s = q_{s1}$ at the dune crest which is approximately the same as the overwash rate Q_s measured at a short distance landward of the dune crest. The transport rate q_{s2} is affected by the overtopping rate Q and may not be assumed to be the same as that for the case of no overtopping. The overwash rate Q_s may be estimated using the simple relationship of $Q_s = C_s Q$ with $C_s \simeq 0.04$ for the median sand diameter $d = 0.38$ mm but the accuracy of this relationship is limited mostly by the accuracy of the empirical formula for Q which predicts only the order of magnitude of Q. In order to extrapolated these small-scale experimental results, the numerical model RBREAK2 for the overtopping flow is combined with two empirical formulas for C_s. The computed results suggest that C_s is on the order of 0.04 except for very fine or coarse sands. This finding is useful for future experiments.

Acknowledgements

This study is funded by the U.S. Army Research Office, University Research Initiative under Contract No. DAAL03-92-G-0116.

References

Hancock, M.W. and Kobayashi, N. (1994). "Wave overtopping and sediment transport over dunes." *Proc. 24th Coast. Engrg. Conf.*, ASCE, 2, 2028–2042.

Holland, K.T.,Holman, R.A. and Sallenger, A.H.,Jr. (1991). "Estimation of overwash bore velocities using video techniques." *Coastal Sediments '91*, ASCE, 489–497.

Kobayashi, N. and Poff, M.T. (1994). "Numerical model RBREAK2 for random waves on imparmeable coastal structures and beaches." *Res. Rept. CACR-94-12*, Ctr. for Applied Coast. Res., Univ. of Delaware, Newark, Del.

Kobayashi, N. and Raichle, A.W. (1994). "Irregular wave overtopping of revetments in surf zones." *J. Wtrway. Port, Coast. Oc. Engrg.*, ASCE, 120(1), 56–73.

Kobayashi, N., Tega, Y. and Hancock, M.W. (1996). "Wave reflection and overwash of dunes." *J. Wtrway. Port, Coast. Oc. Engrg.*, ASCE, 122(3),150–153.

Ribberink, J.S. and Al-Salem, A.A. (1994). "Sediment transport in oscillatory boundary layers in cases of rippled beds and sheet flow." *J. Geophys. Res.*, 99(C6), 12707–12727.

Steetzel, H.J. and Visser, P.J. (1992) "Profile development of dunes due to overflow." *Proc. 23rd Coast. Engrg. Conf.*, ASCE, 3, 2669–2769.

Trowbridge, J. and Young, D. (1989). "Sand transport by unbroken water waves under sheet flow conditions." *J. Geophys. Res.*, 94(C8), 10971–10991.

Van der Meer, J.W. and Janssen, J.P.F.M. (1995). "Wave run-up and wave overtopping at dikes." *Wave forces on inclined and vertical wall structures*, ASCE, New York, NY, 1- 27.

CHAPTER 322

PRESSURE GRADIENTS WITHIN SEDIMENT BEDS.

Thomas. E. Baldock[1] and Patrick Holmes[2]

ABSTRACT

A new technique for measuring dynamic pressures and pressure gradients within sediments is developed. The pressure is measured with a series of small diameter steel probes connected to pressure transducers via semi-rigid tubing. The dynamic response of the system was evaluated and found to be satisfactory for typical laboratory wave frequencies. Comparisons with measurements of the pressure gradients obtained from conventional transducers show good agreement. The present technique is demonstrated with pressure measurements at the toe of breakwaters and across the surf zone of model beaches.

1) INTRODUCTION

Pressure gradients induced within sediment beds by surface gravity waves may have a significant influence on the transport of sediment and the stability of coastal structures. The pressure gradients may either be steady or transient. For example, the wave induced set-up of the water level at the shoreline will produce a steady flow within a beach (Longuet-Higgins, 1983). This may be increased by the percolation of swash through the upper part of the beach. Alternatively, transient pressure gradients may locally reduce the effective strength of the sediment, possibly temporarily fluidising the bed material (e.g Madsen, 1974; Yamamoto, 1978). This may lead to higher rates of sediment transport and reduce the stability of structures within the surf zone.

However, considerable uncertainty remains as to the effects of pressure gradients on sediment transport rates and previous work has provided conflicting

----

1) Research Associate.

2) Professor.

Dept. of Civil Engineering, Imperial College, London, SW7 2BU, UK.

results (Martin, 1970, Oldenziel and Brink, 1974). Loveless (1994) suggested that vertical wave induced pressure gradients were particularly important during toe scour and Conley and Inman (1992) suggested that bed ventilation could be an important aspect in scale modelling. Seepage within the beach may also be an important feature in both long term beach morphology (Turner, 1995) and short term (storm duration) beach evolution. This effect is highlighted by the performance of coastal drainage systems (e.g Vesterby, 1996; Sato et al, 1994). Nevertheless, the mechanism by which the watertable influences the beach dynamics is not immediately apparent. Sato et al (1994) suggest that the drain reduces offshore flows such as the undertow, in addition to increasing percolation into the upper beach, while Vesterby (1996) suggests that an increase in the beach stability and a reduction in wave induced liquefaction are the dominant factors. However, numerical calculations by Oh and Dean (1994) indicated that the pressure gradients induced by changes in the watertable within a beach were generally far to small to influence sediment stability.

Therefore, although the pressures within sediments have been measured by a number of authors (e.g. Sleath, 1970; Yamamoto et al, 1978), measurements of the pressure gradients will enhance the understanding of both sediment transport mechanisms and beach dynamics. The present study addresses this point and outlines the development of a flexible non-intrusive measurement technique. The method allows the pressure gradients at any position within the fluid or sediment bed to be determined quickly and easily, with minimum disturbance of the local flow field. The technique may also be used close to or under structures such as breakwaters, where it may be difficult to obtain results from conventional pressure transducers.

2) MEASUREMENT TECHNIQUE

Conventional pressure transducers have a number of disadvantages when attempting to measure pressure gradients. Firstly, they are relatively large and therefore when mounted close together may significantly disturb the local flow field. Secondly, a large number of transducers are needed to measure the pressure gradients at a number of points without repeatedly disturbing a sediment bed. Finally, the measurement position must be chosen *a priori* and cannot be changed readily. This is a significant drawback when measuring the pressure gradients within model beaches which evolve over time. For example, in order to measure the pressure gradients over either a bar, or at the toe of a beach, it would first be necessary to run a model test to predetermine the measurement positions.

The aim of the present study is to overcome these limitations and develop a measurement method that is as flexible as possible. The technique has been developed from a method used previously to investigate static pressures within model sand embankments. However, the method has not previously been used to measure either dynamic pressures or pressure gradients. The underlying principle is similar to that proposed by Nielsen et al (1994) and has also been used in medicine to measure blood and lumbar pressures (Kumar et al, 1993).

The pressures within the fluid or sediment are measured by means of a series of 2mm ID steel probes. Each probe is connected to a 0.1 Bar pressure transducer via a three way ball valve and short length of 3mm ID semi-rigid nylon tubing (figure 1). Air may be flushed from the tubing and probe by opening the valve to a small header reservoir mounted above the transducer/valve system. The pressure variation is transmitted into each probe through 12 0.4mm diameter holes drilled into the tip. In order to measure the pressure gradients, five probes are arranged in an array such that both the vertical and horizontal pressure gradients are obtained at a single point. The probes may be moved vertically within the bed by means of a traverse mechanism, with minimal disturbance of the sediment. The probe holder, pressure transducers and header tank are mounted on a carriage running on rails above the flume, which allows the probes to be positioned at any position within the sediment bed.

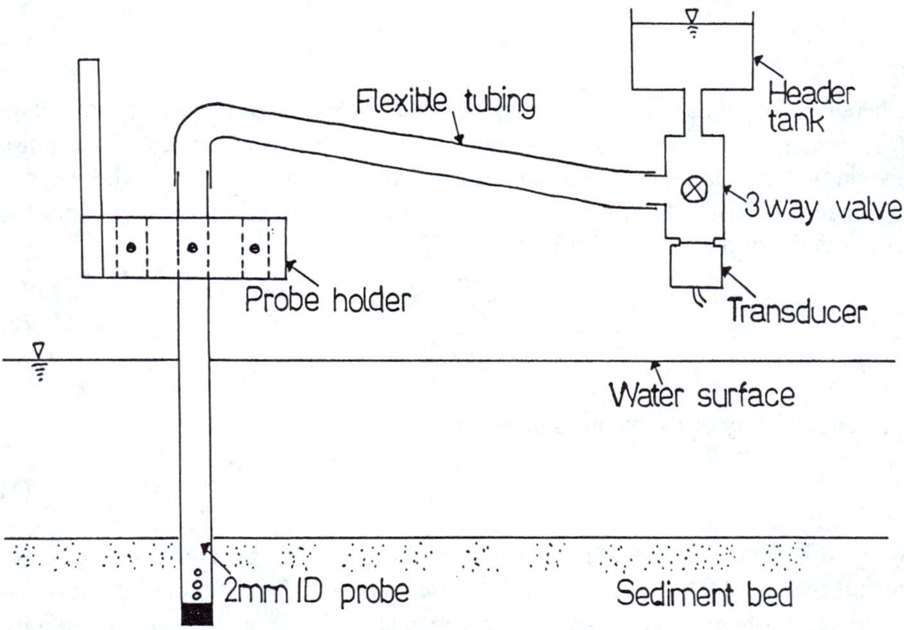

Figure 1. Pressure measurement system.

The static pressure recorded by the probes is the elevation head between the pressure transducers and the mean or still water level. This is generally negative and a variable offset was therefore added to the pressure transducer amplifiers. The "mean pressure" may therefore be set to zero, allowing amplification to be used when sampling the dynamic pressure. This increases the accuracy of the measurements, since the amplitude of the dynamic pressure may be considerably smaller than the static pressure in many practical circumstances. Note that moving the probes vertically does not change the recorded static pressure and that the probes do not measure a velocity head.

3) EVALUATION / CALIBRATION

For the purpose of measuring mean or steady pressure gradients the probe technique requires no calibration. However, for the purpose of measuring dynamic pressure gradients the dynamic response of the measurement system requires evaluation. The dynamic response depends on the length of the probe and tubing, the geometry of the valve/transducer connection and the speed of the pressure pulse. In order to achieve a satisfactory dynamic response the natural frequency of the system must be as high as possible. If it is assumed that the system behaves as a quarter length resonator (Nielsen et al, 1993) then the natural frequency, f_o, is given by

$$f_o = c / 4L \tag{1}$$

where c is the speed of the pressure pulse and L the length of the probe and tubing. L is consequently kept as small as possible ($\langle$1m), while the speed of the pulse should be maximised. The pressure pulse is a one-dimensional wave, whose speed is dependent on the fluid density, ρ, the compressibility of the fluid, C, and the distensibility of the tubing, D (Lighthill, 1978).

$$c = \frac{1}{\sqrt{(\rho(C + D))}} \tag{2}$$

The distensibility of the tubing is given by

$$D = 2a / tE \tag{3}$$

where a is the tube radius, t is the wall thickness and E is the Young's modulus of the tube material (O~10^9 N/m^2). In the present instance D is of order 10^{-9}m^2/N and the compressibility of pure water is approximately 5.10^{-10}m^2/N, giving $c \approx$800m/s. However, the presence of small air bubbles within the tube will increase the compressibility of the water significantly. For example, an air content of 0.4% will give a compressibility of 5.10^{-8}m^2/N. Hence, from (2), a few small air bubbles (O~1mm) are sufficient to reduce c to about 150m/s. This results in a natural frequency of about 35Hz, far higher than the wave frequencies used in typical laboratory studies (0.5-2Hz). In order to confirm the estimates above, the speed of a pressure pulse along a 10m length of tubing was calculated by correlation of this signal with that obtained from a conventional transducer. In this instance the lag along the tube was found to be of order 0.06s, giving a value for *c* of 150m/s (figure 2). The lag along a 1m length of probe and tubing is therefore less than 0.01s. The difference in lag between probes is inconsequential for the purposes of calculating the pressure gradients (figure 3), although in this figure the ±0.02s lag on two of the probes indicates the wave speed.

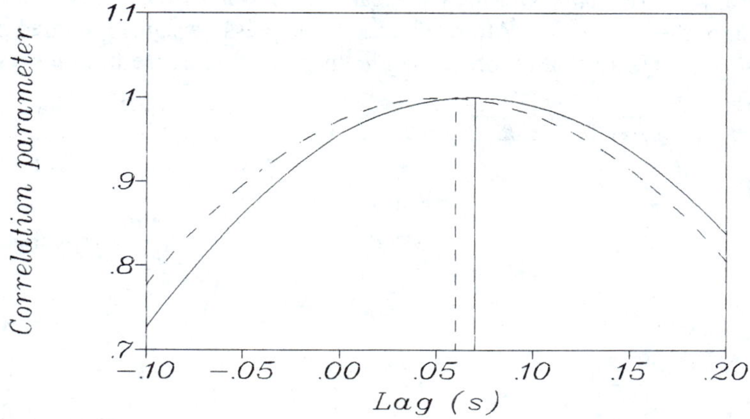

Figure 2. Correlation of the pressure signal from two probes with that from a conventional transducer. L=10m.

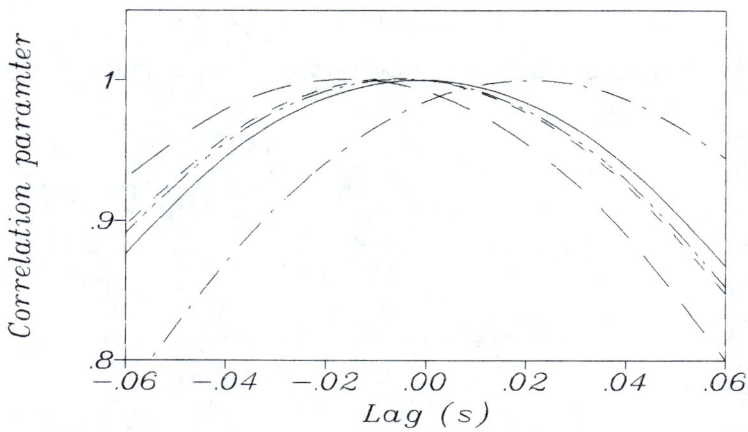

Figure 3. Correlation between the five probes arranged in the measuring array and a conventionally mounted pressure transducer. L=1m.

Regular waves

With a natural frequency of 35Hz, probe/transducer system was expected to have a linear frequency response function and a gain close to 1 over the required range of wave frequencies. However, some damping is to be expected due to energy losses. In regular wave tests the gain function was evaluated by comparing the amplitude of the signal recorded through the probe/transducer system with that obtained from a conventional transducer. Figure 4 shows the measured response functions for a range of frequencies. In each case the response function is very similar, although some difference is observed in the absolute gain values. This is probably due to slight variations in the concentration of air bubbles in each probe. Therefore, when regular waves are used, each probe is individually calibrated at the beginning of the test session. After calibration, the pressure signal recorded by each probe is very close to that recorded by a conventional pressure transducer (figure 5).

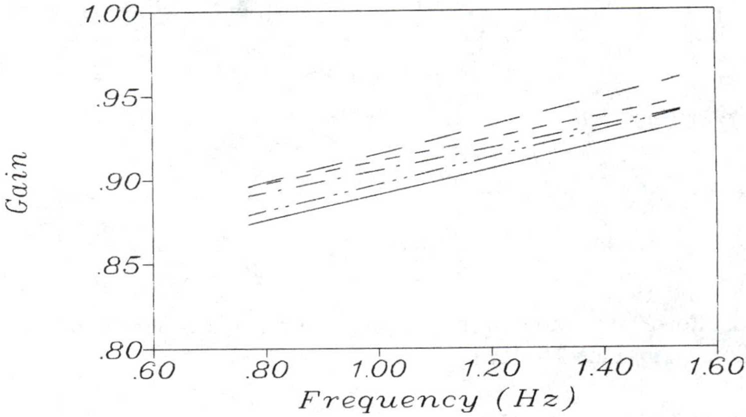

Figure 4. Gain functions measured for each probe/transducer system. L=1m.

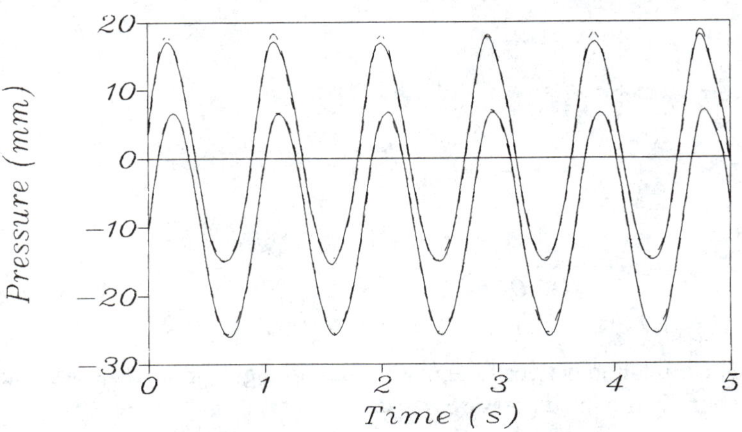

Figure 5. Comparison of the dynamic pressure recorded by two probes with that recorded by conventionally mounted pressure transducers. —— Transducers, – – – probes.

In order to obtain the pressure gradients, the pressure signals from each probe were first low pass filtered at 5Hz and then the gradients calculated with a simple finite difference scheme. The best results were obtained with a probe spacing of 0.06m in the horizontal direction and 0.02m in the vertical direction. The reduced vertical spacing was necessary to obtain the pressure gradients close to the surface of the sediment bed. A comparison of the dynamic horizontal pressure gradient calculated from probe measurements with that from equally spaced conventional transducers clearly shows good agreement (figure 6). Finally, it was necessary to check the performance of the system with all five probes operating and, in particular, to ensure that the phase relationship between the pressure and pressure gradient was accurately reproduced. Figure 7 shows that this was indeed the case.

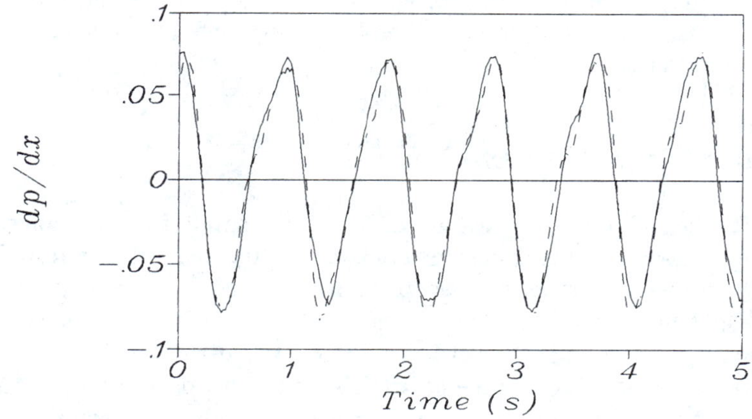

Figure 6. Comparison of the dynamic horizontal pressure gradient.
—— Transducers, – – – Probes.

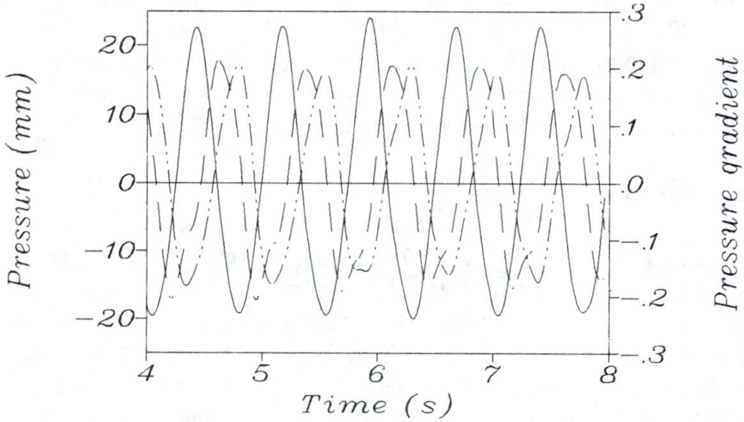

Figure 7. Pressure and pressure gradients within a sand bed.
—— Pressure, – – – Horizontal, — ·· — Vertical.

Random waves

Two approaches are possible when using the present method to measure the pressure variation induced by random waves. The first is simply to calibrate each probe over the appropriate frequency range using regular waves. However, best results are obtained by calibrating the probes at the outset of a model test by using a random wave series and a spectral method. In this instance the pressure signal recorded by each probe is Fourier transformed and compared to the Fourier transform of a pressure signal recorded by a conventional transducer. An impulse response function may then be calculated for each probe and applied to all subsequent measurements. This approach gives gain functions very similar to those for regular waves shown above. Applying the appropriate impulse response function during subsequent tests then results in close agreement between the dynamic pressure obtained from the probe system and that obtained from a conventional transducer.

4) RESULTS / APPLICATIONS

The flexibility of the present technique has considerable advantages for a wide range of pressure measurements. For example, the minimal flow resistance of the probes makes it possible to measure the pressure and pressure gradients at any point within a fluid flow. This is not possible with conventional transducers, which must generally be mounted flush with a surface in order to minimise turbulence effects. A second advantage of the present method is that data may be obtained at a large number of positions within a sediment bed with a minimum number of pressure transducers. This is because the small diameter of each probe allows the probes to be moved vertically within the sediment with virtually no disturbance of the bed material. The variation in the dynamic pressure over the vertical is therefore easily obtained. For example, figure 8 shows measurements of the maximum dynamic pressure within a sand bed and an anthracite bed under regular waves. Data could be collected at a large number of points with only one pressure transducer and gave good agreement with the solution of Sleath (1970).

Pressure gradients

Figures 9 and 10 show the horizontal and vertical pressure gradients at the face of a vertical breakwater and at the toe of a sloping breakwater respectively. The face of the vertical breakwater is an antinode, and therefore the horizontal pressure gradient is small, while the vertical pressure gradient increases rapidly close to the surface of the sediment bed. The sharp discontinuity in the vertical pressure gradient expected at the fluid/sediment interface is also clearly visible, although the finite spacing of the probe tips is apparent. Nevertheless, data of this sort would be difficult to obtain with conventional transducers. The toe of the sloping breakwater lies between a node and antinode and therefore, in this instance, the horizontal and vertical pressure gradients are of similar magnitude at the sediment surface. However, the present measurement technique is sufficiently accurate to detect the change in the rate of decay of the horizontal pressure gradient at the fluid/sediment interface. This change occurs because the variation in the horizontal pressure gradient within the fluid approaches zero at the interface, while the rate of decay within the bed is proportional to the thickness of the sediment bed (e.g. Sleath, 1970).

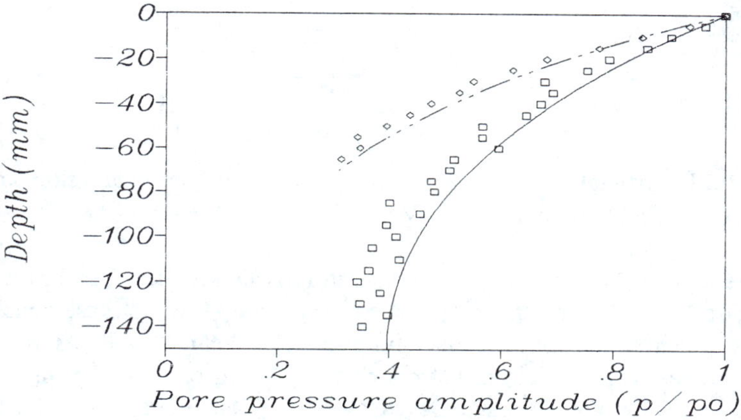

Figure 8. Dynamic pressure within a sand bed and an anthracite bed. □□□□ Sand, T=0.7s, ——— Sleath (1970), ◊◊◊◊ Anthracite, T=0.5s, —··— Sleath (1970).

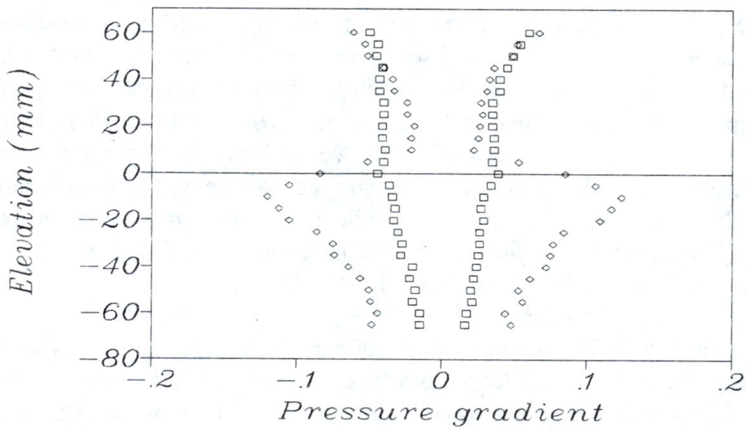

Figure 9. Horizontal and vertical pressure gradients at the face of a vertical porous breakwater. □□□□ Horizontal, ◊◊◊◊ Vertical.

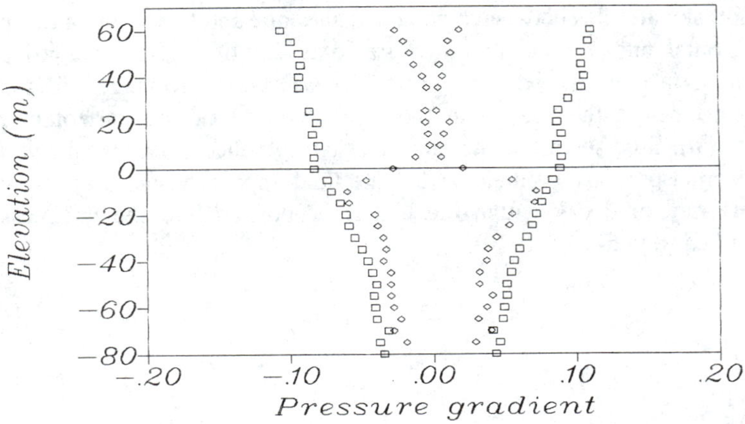

Figure 10. Horizontal and vertical pressure gradients at the toe of a sloping porous breakwater. □□□□ Horizontal, ◊◊◊◊ Vertical.

The pressure gradients within laboratory scale sediment beaches were also investigated with the present measurement technique. The first case considered a coarse sediment beach with a median grain size of 1.5mm and a thickness of 0.1m. Regular waves with a height of 0.1m and wave period of 1.5s were used. Waves were run until the initially plane beach reached a stable shape and the pressure gradients just below the bed surface were then measured at a number of locations across the surf zone. Measurements of this kind would be very difficult to obtain using conventional transducers, which would have to be positioned before the beach attained an equilibrium profile.

Figure 11 shows the mean horizontal pressure gradient increasing across the surf zone due to wave induced set-up and swash zone percolation, reaching a maximum of about 0.1 at the still water level. The vertical pressure gradient has a similar magnitude at this point and is again positive. The flow within the beach is therefore downward and seaward. The vertical pressure gradient reaches a minimum in the middle of the surf zone and, consequently, the flow is out of the beach. However, the mean pressure gradients appear too small to significantly influence the stability of the beach sediment, in general agreement with the numerical calculations of Oh and Dean (1994).

In a second experiment the maximum dynamic pressure gradients were measured across a barred laboratory scale beach. In this instance the beach consisted of sediment with a median grain diameter of 0.5mm and was again 0.1m thick. Regular waves were again run until a stable beach formed and the pressure gradients were measured 20mm below the bed surface in the region of the bar. The data show that both the horizontal and vertical pressure gradients reach a maximum just seaward of the bar and then decrease rapidly after wave breaking (figure 12).

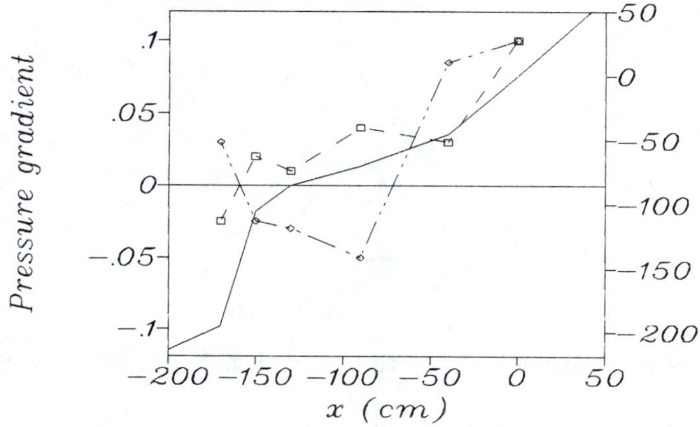

Figure 11. Mean horizontal and vertical pressure gradients within a laboratory scale sand beach. —— Water depth, —□—, horizontal, — ··◇·· — vertical.

The maximum horizontal gradient does not reach the value required for liquefaction of a sand bed (Madsen, 1974), but is nearly twice that required to initiate the failure of an anthracite bed. The negative horizontal pressure gradient is much larger than the positive gradient, indicating steep wave fronts, while the reverse is the case for the vertical pressure gradients, indicating steep crests and flatter wave troughs (figure 13).

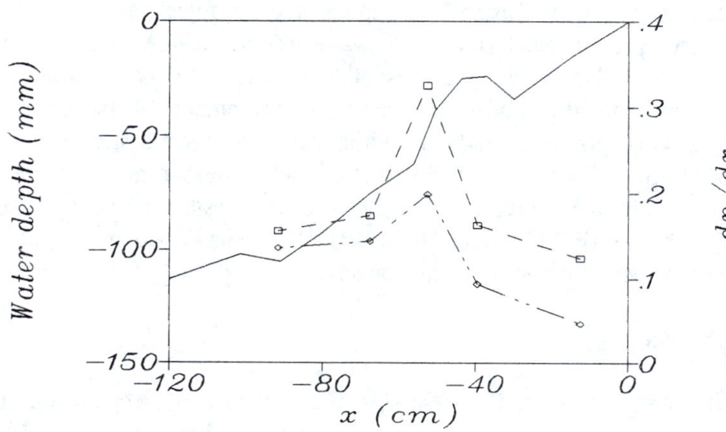

Figure 12. Maximum horizontal dynamic pressure gradient across a laboratory scale sand beach. —— Water depth, —□—, negative, — ··◇·· — positive.

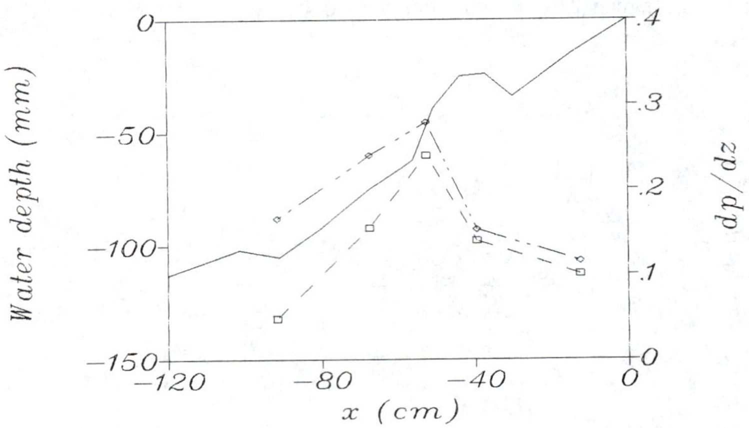

Figure 13. Maximum vertical dynamic pressure gradient across a laboratory scale sand beach.
—— Water depth, —□—, negative, — ..◊.. — positive.

CONCLUSIONS

The development of a new technique for measuring the pressure gradients within sediments is presented. The pressure gradients are measured using a series of small diameter steel probes connected to pressure transducers with semi-rigid flexible tubing. This allows the pressure gradients to be determined quickly and easily with minimal disturbance of the sediment. For the purpose of measuring mean pressure gradients the method requires no calibration. The dynamic response of the system was evaluated using both regular and irregular waves and was found to be satisfactory for typical laboratory wave frequencies. A comparison of the pressure gradients obtained by the present technique with those measured using conventional pressure transducers showed good agreement. Measurements of the mean and dynamic pressure gradients within laboratory scale sand beaches have demonstrated the advantages of the system. The measurement technique has already proved useful for investigating the role of pressure gradients in sediment transport mechanics (Baldock and Holmes, 1996) and will aid investigations into the flow pattern within laboratory scale beaches.

REFERENCES

Baldock, T.E. and Holmes, P., 1996. On the influence of seepage on sediment transport under steady flow and wave motion. Submitted to *J. Hydraulic Research.*

Conley, D.C. & Inman, D.L., 1992. Field observations of the fluid granular boundary layer under near breaking waves. *J. Geo. Res.*, **97** (C6), 9631-9643.

Kumar, M., Werner, W. and Murray, M.J., 1993. The fidelity and dynamic response of fluid-filled catheter systems for direct measurement of lumbar cerebrospinal fluid pressure. *J. Clin. Monit.*, **9**, 314-320.

Lighthill, J., 1978. *Waves in fluids.* Cambridge University Press, Cambridge, UK.

Longuet-Higgins, M.S., 1983. Wave set-up, percolation and undertow in the surf zone. *Proc. R. Soc. Lond.* **A 390**, 283-291.

Loveless, J.H., 1994. Modelling toe scour at coastal structures. *Proc. 2nd Int. Conf. on Hydraulic Modelling*, Stratford, UK. BHR Group, 479-489.

Madsen, O.S., 1974. Stability of a sand bed under breaking waves. *Proc. 14th ICCE*, Copenhagen. ASCE, **2**, 776-794.

Martin, C.S., 1970. Effect of a porous sand bed on incipient sediment motion. *Water Resources Research*, **6**, (4), 1162-1174.

Nielsen, P., Hanslow, D. J. and Apelt, C. J., 1993. A new type of nearshore wave gauge. *Proc 11th Australasian Conf. on Coastal and Ocean Engng.*, Townsville, Australia, pp 247-251.

Oh, T. & Dean, R.G., 1994. Effects of controlled water table on beach profile dynamics. *Proc. 24th ICCE*, Kobe. ASCE, **3**, 2449-2459.

Oldenziel, D. M. & Brink, W.E., 1974. Influence of suction and blowing on entrainment of sand particles. *J. Hydraulics Div.*, ASCE, **100**, 935-949.

Sato, M., Hata, S. and Fukushima, M., 1994. An experimental study on beach transformation due to waves under the operation of a coastal drain system. *Proc. 24th ICCE*, Kobe. ASCE, **3**, 2571-2582.

Sleath, J.F.A., 1970. Wave induced pressures in beds of sand. *J. Hydraulics. Div.*, ASCE, **96**, 367-378.

Turner, I.L., 1995. Simulating the influence of groundwater seepage on sediment transported by the sweep of the swash zone across micro-tidal beaches. *Marine Geology*, 125, 153-174.

Vesterby, H., 1996. Beach drainage - state of the art-. Seminar on Shoreline Management techniques, HR Wallingford, Wallingford, UK.

Yamamoto, T., 1978. Sea bed instability from waves. *Proc. 10th OTC Conf.*, Houston. ASCE, Paper 3262.

Yamamoto, T., Koning, H.L., Sellmeijer, H. & van Hijum, E., 1978. On the response of a poro-elastic bed to water waves. *J. Fluid Mech.*, **87**, 193-206.

ACKNOWLEDGEMENTS

The authors gratefully acknowledge the support of the UK Engineering and Physical Sciences Research Council. The advice of Mr Steve Ackerly (Soil Mechanics) on pressure measurement techniques is much appreciated.

CHAPTER 323

Analysis of Mud Mass Transport under Waves Using an Empirical Rheological Model

Qin Jiang[1] and Akira Watanabe[2]

Abstract

This study is focused on deepening our insight into rheological properties of soft mud under oscillatory forces, on quantifying their effects on the interaction between water waves and mud bed, and especially on the resultant mud mass transport. On the basis of a large number of experiments, an empirical rheological model and formulas for model parameters are proposed. In addition, a simple analytical model for evaluating the temporal change in mud water content ratio under waves is derived. Using these two models, a vertical 2-D numerical model is developed to predict the wave-induced bed mud motion and the resultant mud mass transport velocity. Comparisons between the calculations and measurements are presented.

1. Introduction

Mass transport in a soft mud layer is one of the noteworthy phenomena in a mud bed. Studies on it are of practical importance for both siltation and environmental problems in muddy coasts and esturine areas.

Variety of theories and models have been proposed to describe the wave-mud interaction and relevant phenomena. The theoretical analysis was initiated by Gade (1958), who assumed that the soft mud would behave as a highly viscous fluid and developed an analytical model for the surface wave attenuation in shallow waters. On the same assumption, Dalrymple and Liu (1978) presented a linear theory for water waves propagating in a two-layer viscous fluid system. Instead of viscous fluid assumption, elastic models were proposed by Mallard and Dalrymple (1977) and Dawson (1978), poro-elastic models by Yamamoto

[1]Research Engineer, Penta-Ocean Institute of Technology, 1534-1, Yonku-cho, Nishinasuno-machi, Nasu-gun, Tochigi, 329–27, Japan

[2]Professor, Department of Civil Engineering, University of Tokyo, Hongo 7–3–1, Bunkyo-ku, Tokyo, 113, Japan

et al. (1983, 1987) and Yamamoto and Takahashi (1984), visco-elastic models by Hsiao and Shemdin (1980), Macpherson (1980) and Maa and Mehta (1988, 1990), as well as a Bingham fluid model by Mei and Liu (1987). Recent studies include visco-elasto-plastic models by An *et al.* (1991) and Shibayama and Nguyen (1993).

A primary difference among the above models is in the assumptions of rheological properties of mud. Due to the complexity of the mud properties, it is difficult to judge which of them is most applicable to the wave-mud interaction problems under a certain condition. In addition, many of these models have been proposed either without enough experimental proofs or on the basis of experiments with unidirectional rotary viscometers in which the mud motion is considerably different from the real situation in coastal waters.

On the other hand, Hyunh *et al.* (1990) developed a dynamic rotary shear meter and studied the rheological behaviors of mud under various patterns of loading. Their experiments showed that there existed complicated relationships between the shear stress and shear strain or shear rate when the soft mud was subjected to cyclic loading. Following them, Shen *et al.* (1993) conducted similar experiments and proposed empirical models for the rheological properties and for the mud-wave interaction. However, their experiments only under limited conditions are not sufficient to understand and model the general features of mud behaviors under dynamic wave action.

The present study aims at gaining further understanding on the rheological properties of soft mud under the action of oscillatory forces and the interaction between water waves and a mud bed. An empirical rheological model of soft mud under oscillatory loading will be proposed on the basis of large numbers of measurements. Moreover, a simple analytical model for change in the water content ratio under wave action is derived in order to evaluate its effects on the rheological behaviors and on the motion of bed mud. By incorporating the proposed rheological model as well as the water content ratio model with the linearized Navier-Stokes equations, a vertical 2-D numerical model is constituted for simulating wave-mud interaction and the bed mud motion. Finally, calculated mud mass transport velocities and wave height changes are compared with those measured.

2. Rheological Properties of mud

2.1 Experiments and results

In order to clarify the rheological properties of soft mud under waves, laboratory experiments were conducted for more than 800 runs by using a dynamic rotary shear viscometer that can generate oscillatory forcing like waves (See Hyunh, 1991, for details of the setup). Instead of real seabed mud, pure kaolinite mixed with tap water was used for the test. This is because it has rheological properties similar to and less complicated than those of natural mud in actual coastal areas.

A wide range of conditions was covered in the test, in which the water content ratio was changed from 120% to 300%, the period of cyclic loading was from 1s to 10s, and its angular amplitude was from 0.5° to 28°. With a series of combinations of the motion of the viscometer and the water content ratio, the measurements were made to examine influences of the angular amplitude and period of the oscillatory shear forces as well as the water content ratio on the rheological behaviors of the mud.

Figures 1 (a) and 1 (b) are the experimental results showing the effects of the angular amplitude A of cyclic forcing, which were obtained under the fixed period T and water content ratio W. Examples indicating the effects of the cyclic loading period and water content ratio are given in Figs. 2 (a) through 3 (b).

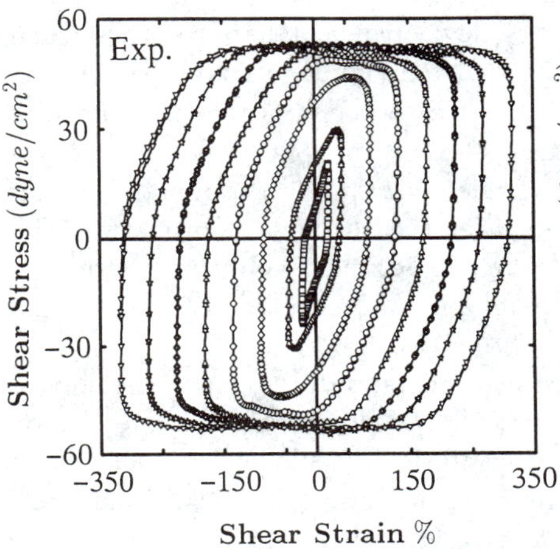

Fig. 1 (a) Shear stress vs. shear strain (effects of the angular amplitude).

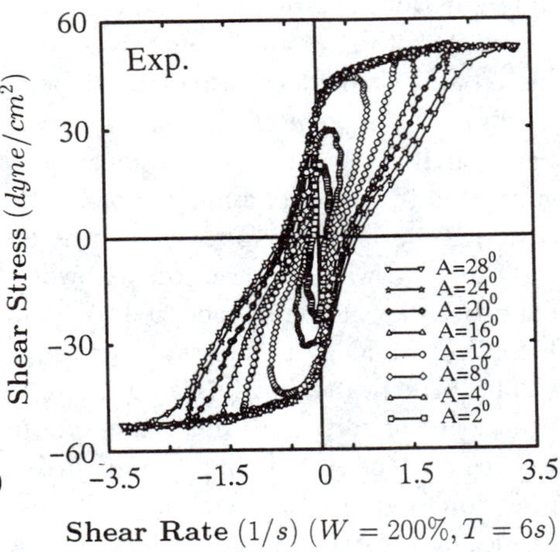

Fig. 1 (b) Shear stress vs. shear rate (effects of the angular amplitude).

From these figures, it can be concluded that under the oscillatory action of external forces, the soft mud is characterized by very complicated visco-elastio-plastic properties. The shear stress of the mud is a function of both the shear strain and shear rate with strong nonlinearity featured by remarkable hysteresisis in their relationships. In addition, the mud properties are significantly influenced by the water content ratio as well as by the period and amplitude of cyclic loading.

2.2 Constitution equation

On the basis of the above experimental results, a constitution equation of the soft mud is constructed for describing its rheological properties under cyclic loading.

Figures 4 (a) through 4 (g) illustrate a basic concept for constructing the constitution equation. Among these figures, Fig. 4 (a) stands for the hysteresis loop of the shear stress τ versus the shear rate γ. It is obvious that this loop can be separated into two parts, namely, a backbone curve shown in Fig. 4 (b) and a

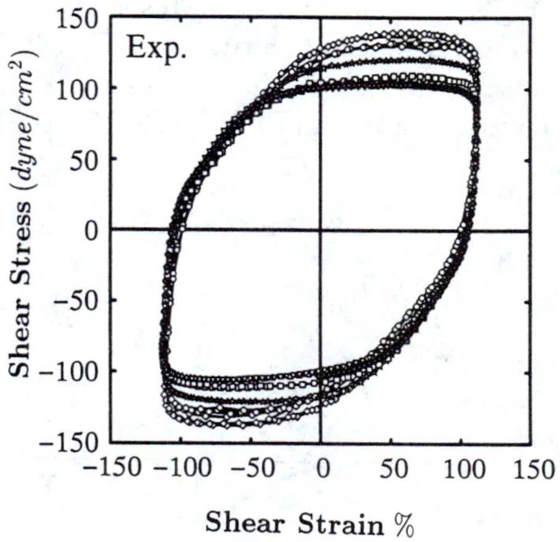

Fig. 2 (a) Shear stress vs. shear strain (effects of the cyclic load period).

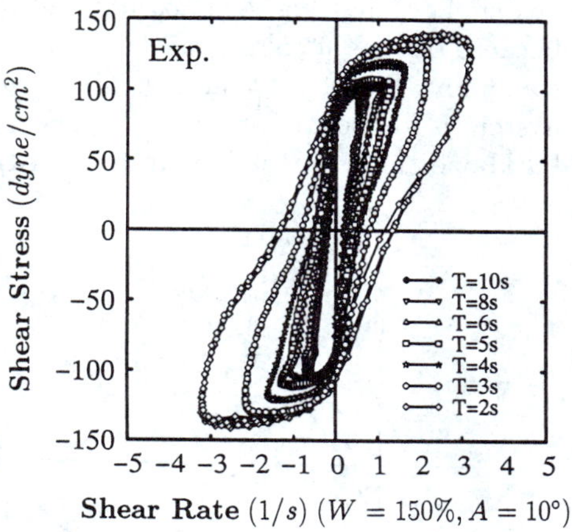

Fig. 2 (b) Shear stress vs. shear rate (effects of the cyclic load period).

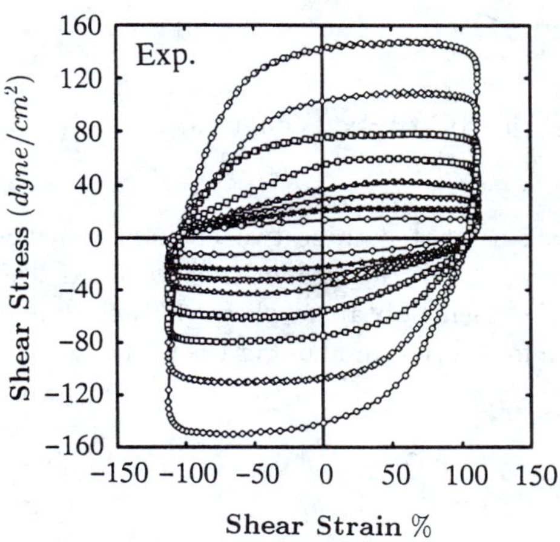

Fig. 3 (a) Shear stress vs. shear strain (effects of the water content).

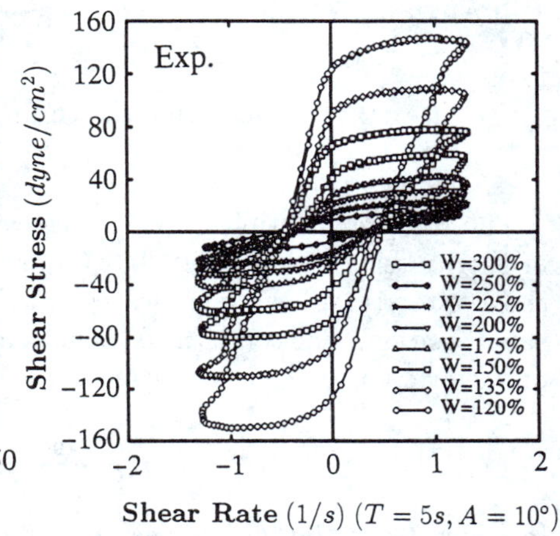

Fig. 3 (b) Shear stress vs. shear rate (effects of the water content).

compensatory curve in Fig. 4 (c). On the other hand, regarding the relationship between the shear stress τ and the shear strain ε shown in Fig. 4 (d), the curves in Fig. 4 (e) and Fig. 4 (f) correspond to Fig. 4 (b) and Fig. 4 (c), respectively, because of the 90° phase lag between the shear strain and shear rate. Namely, combining Fig. 4 (e) and Fig. 4 (f), we obtain Fig. 4 (d) that represents the hysteresis loop of the shear stress versus the shear strain, and Fig. 4 (f) gives the backbone curve for the loop in Fig. 4 (d).

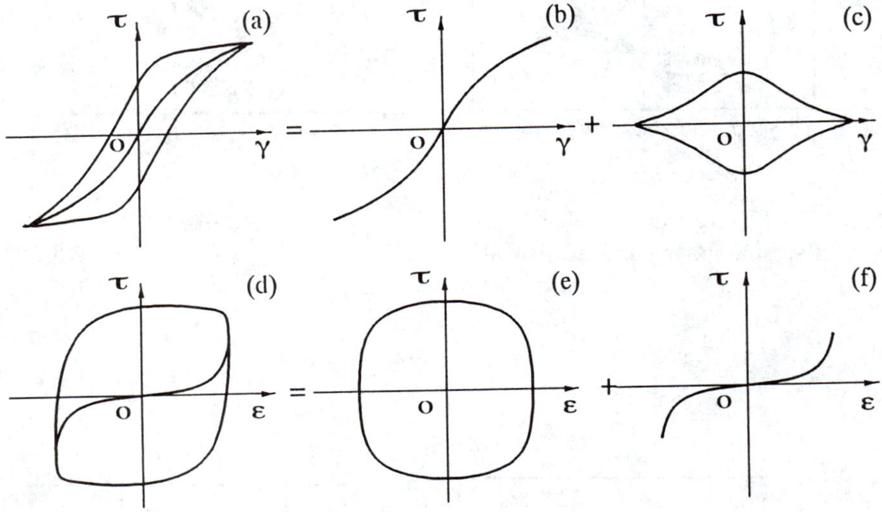

Fig. 4 Definition sketch for the constitution equation.

The concept mentioned above suggests us that if these two backbone curves of the hysteresis loops, Fig. 4 (b) and Fig. 4 (f), are properly determined, the shear stress in mud under waves can be easily obtained through their linear superposition. The constitution equation for the soft mud under the cyclic loading or wave action is thereby written as

$$\tau = \frac{G_0 \varepsilon}{1 - \alpha \mid \varepsilon \mid} + \frac{\mu_0 \gamma}{1 + \beta \mid \gamma \mid} \tag{1}$$

in which G_0 is the initial shear modulus at $\varepsilon = 0$, μ_0 is the initial viscosity at $\gamma = 0$, and α and β are coefficients determining the shapes of the backbone curves. The applicability of the above proposed model is strongly dependent on these four model parameters. With regression analyses, empirical formulas for them are obtained as follows:

$$\begin{aligned}
\alpha &= \{1.0 - (0.8 - 5.0 \times 10^{-3} \cdot T) \\
&\quad \cdot \exp\left(-\left[0.75 \cdot \tanh\left(T/4\right)\right] \cdot \varepsilon_{\max}\right)\}/\varepsilon_{\max} \\
\beta &= (0.45 + 0.11 \cdot T + 0.015 \cdot T^2) \\
&\quad \cdot \varepsilon_{max}{}^{\left(1.05 - 1.42 \cdot T^{0.2}\right)} \\
G_0 &= (5.6 \times 10^5 \cdot W^{-2.8}) \\
&\quad \cdot \{[-7.26 - 560 \cdot (1.0 + \tanh(-0.58 \cdot T))] \\
&\quad + [81.70 + 707 \cdot (1.0 + \tanh(-0.47 \cdot T))] \\
&\quad \cdot \alpha^{1.71/[1.+9.87 \exp(-0.87 \cdot T)]}\} \\
\mu_0 &= (5.6 \times 10^5 \cdot W^{-2.8}) \cdot \left\{37.5 \cdot \left(1.31^T\right)\right\} \\
&\quad \cdot \ln[24.7 \exp(-0.31 \cdot T \cdot \beta + 1.0\}
\end{aligned} \tag{2}$$

where T(s) is the period of the cyclic loading, $\varepsilon_{\max}$ ($0.5 \le \varepsilon_{\max} \le 4.0$) is the shear strain amplitude, and W(%) is the water content ratio of the mud. (Here, for convenience, the angular amplitude of cyclic shear load was replaced by the amplitude of shear strain).

Equations (1) and (2) form a thorough rheological model for the soft mud. Their validity was well confirmed both qualitatively and quantitatively through trial hindcasting of experimental data. Examples of the calculated results are shown in Figs. 5, which corresponds to the measurements shown in Fig. 1 in the previous section.

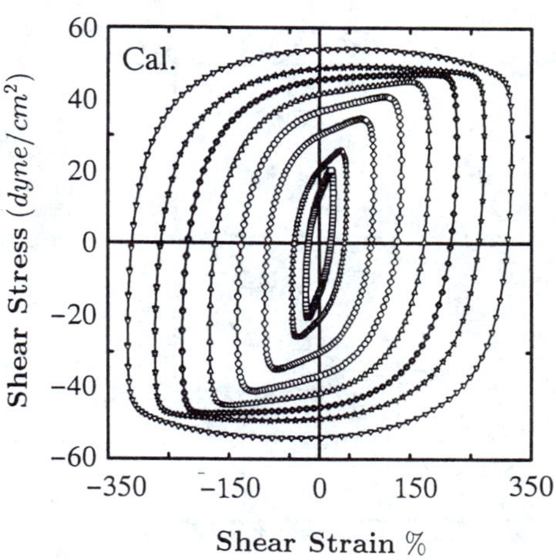

Fig. 5 (a) Shear stress vs. shear strain (effects of the angular amplitude).

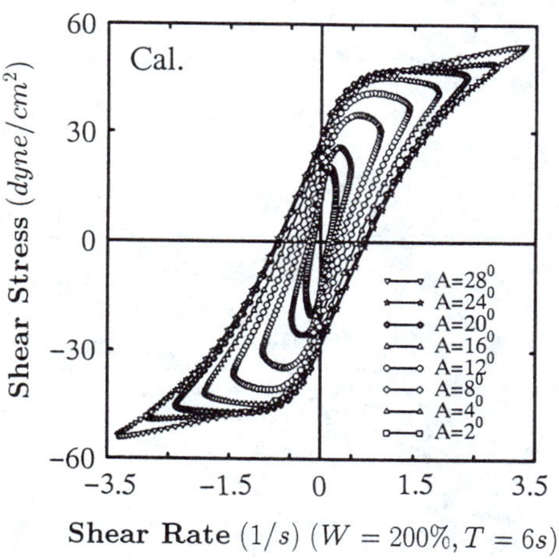

Fig. 5 (b) Shear stress vs. shear rate (effects of the angular amplitude).

3. Water Content Changes in Bed Mud under Waves

The water content ratio is an important factor affecting the rheological properties of the mud. Recent studies revealed that the dynamic change in the water content ratio takes place in a mud bed as a result of wave action. The results of experiments by Shen *et al.* (1993) showed that both the mud self-weight and water waves play significant roles in the variations of water content ratio in mud layers. As compared with the solo consolidation under the self-weight, the wave action causes further complicated change in the water content ratio, including the densification in deeper lower layers of the mud bed and liquefaction or swelling in the near-surface layers. On the basis of these experimental results, a simple analytical model is developed for evaluating the temporal change in water content ratio in bottom mud layers under waves.

3.1 Mechanism of Mud Densification and Liquefaction

Figure 6 illustrates an idealized state of stresses in a mud bed under waves. With consideration for the different effects of wave-induced stresses in mud layers, wave action on a mud bed can be classified into two components which are called "pumping" and "shaking". The pumping means the process of compression and expansion in a mud bed due to the dynamic wave pressure, while shaking refers to the tangential action by the oscillatory bottom shear stresses and the gradient of dynamic pressure in the wave propagation direction.

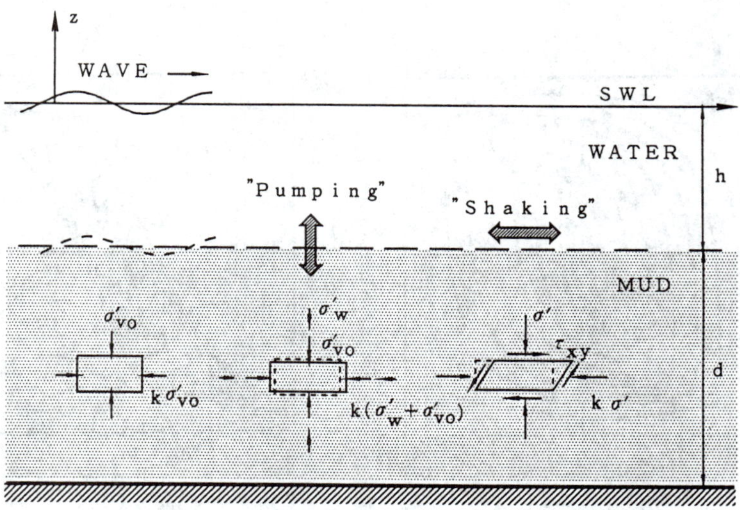

Fig. 6 Idealized stress state in mud bed under waves.

The downward transmission of the pumping and shaking action through the overlying water results in complicated processes in mud layers such as the oscillatory motion of the bed mud, the erosion or settlement of mud particles, and the change of the mud bed structure. Therefore, the water content changes in a mud bed is regarded as the combined results of the pumping and shaking due to waves.

(a) Densification by the Pumping Action of Waves

It is said and often assumed that the effective stress in the underwater deposits remains unchanged even if the water level varies. However, it is not always true under wave action. According to Zen *et al.* (1993), the excess pore pressure and effective stress will be induced whenever the water surface elevation changes so quickly that the associated water pressure on the interface cannot be entirely transmitted into the deposits as the pore pressure. From this fact, the densification of the mud layers can be regarded as the result of the pumping action due to waves.

In this way the pumping action of waves may generally cause oscillatory change of the normal stress in a mud bed. As a result, the excess pore pressure and effective stresses may vary oscillatorily, and the bed mud is thus forced to condense and expand alternatively responding to the up-and-down motion of the water surface. During every wave period, a process of dewatering or densification may occur since the volume compression rate of the mud is bigger than its expansion rate. The cyclic pumping action of waves will thus induce the densification of the bed mud.

(b) Liquefaction Due to the Shaking Action of Waves

A well-known fact as for noncohesive sediment is that shear loading causes the increase in gross sediment volume, which is called dilatancy. In fact, dilatancy is a general soil property both for noncohesive and for cohesive sediments. Therefore, like the liquefaction of noncohesive sediment, the increase of water content ratio in near-surface mud layers can be reasonably interpreted as the dilatancy effect under the cyclic wave action.

This phenomenon can also be regarded as a result of the exchange of water and mud particles. Parchure and Mehta, *et al.* (1985) has found that the erosion of cohesive sediment occurs if the shear force exerted on mud beds is large enough and it takes place much more easily under the cyclic action of waves. This is because the contacts between mud particles are weakened and the shear strength of the mud decreases under the action of external forces in particular under that of oscillatory forces like waves. Thus the mud particles are entrained into the water part and the water particles enter into the new-born pores where the mud particles lodged, resulting in the increase of the water content ratio in near-surface mud layers with the growth of erosion in the mud bed.

3.2 Water Content Change Model

On the basis of the above discussions, the water content change in a mud bed under wave action e_{s+w} can be divided into three parts, which are the consolidation by the self-weight e_{cs} as well as the densification e_{dw} and liquefaction e_{lw} due to waves. Accordingly, an analytical one-dimensional model is developed for evaluating the vertical distribution of water content ratio in mud layers by the linear superposition of the three parts.

Since the mud densification is similar to the soil consolidation and the liquefaction can be treated as its reversal process, all of these processes can be expressed by the conventional consolidation equation.

A general equation for the consolidation of soft mud was given by Gibson *et al.* (1967) as

$$\frac{\partial e}{\partial t} - \frac{\partial e}{\partial z}\left[\frac{k}{\rho_f(1+e)}\frac{d\sigma'}{de}\frac{\partial e}{\partial z}\right] + (\rho_s - \rho_f)\frac{d}{de}\left[\frac{k}{\rho_f(1+e)}\right]\frac{\partial e}{\partial z} = 0 \tag{3}$$

in which e is the void ratio of mud, k is the permeability coefficient, σ' is the effective stress, ρ_f and ρ_s are the densities of the water and mud, and z is the vertical coordinate measured from the rigid bottom elevation. Its linearized form can be written as

$$\frac{\partial e}{\partial t} = C_v \frac{\partial^2 e}{\partial z^2} \quad \text{for} \quad 0 \le z \le z_0 \quad \text{and} \quad t \ge 0 \tag{4}$$

where C_v $(= k/\left[\rho_f(1+e)\right](d\sigma'/de))$ is the coefficient of consolidation, t denotes the time, z_0 is the thickness of mud layer at $t = 0$.

For the consolidation due to the self-weight and densification caused by the pumping action of waves, the initial and boundary conditions are given as follows.

1) Assuming that the distribution of void ratio e_0 at $t = 0$ is uniform, the initial condition is

$$e(z, 0) = e_0 = \text{const.} \tag{5}$$

2) On the mud bed surface, the void ratio remains constant because the effective stress is zero, i.e.,

$$e(z_0, t) = e_0 \tag{6}$$

3) At the rigid bottom underlying the mud, the impervious boundary condition is

$$\left(\frac{\partial e}{\partial z}\right)_{z=0} = \beta \tag{7}$$

where β is a constant depending on the void ratio distribution at the ultimate state of consolidation by the self-weight and densification due to waves, and is expressed as

$$\beta = \frac{(\rho_s - \rho_f)}{\lambda} + \frac{p_0}{z_0 \lambda} \tag{8}$$

After simple manipulation, the solution of Eq. (4) for the void ratio is obtained as

$$e(z,t) = e_0 - \frac{\beta}{z_0}\left[1 - \frac{z}{z_0} - 2\sum_{n=0}^{\infty}\frac{\cos\frac{m\pi z}{z_0}}{m^2\pi^2}\exp\left(-m^2\pi^2\frac{C_v t}{z_0^2}\right)\right] \tag{9}$$

where $m = 2n + 1/2$, $\lambda = d\sigma'/de$, and p_0 is the amplitude of wave induced bottom pressure.

Regarding the fact that the volume change of pore water in a mud bed compensates that of the mud eroded from the bed, the swelling of mud by the shaking action of waves can be derived as

$$\Delta e\,(z,t) = \Delta e_s \left\{ -\frac{z}{z_0} - \frac{2}{\pi} \sum_{n=1}^{\infty} \left[\frac{\sin \frac{n\pi z}{z_0}}{z_0} \exp\left(\left(\frac{n\pi z_0}{z_0} \right)^2 C_l t \right) \right] \right\} \tag{10}$$

where Δe_s is the change of void ratio on the mud surface expressed as in the above equations, $C_l\ (= k_l/[\rho_f\,(1+e)]\,(d\sigma'/de))$ is the coefficient of expansion, N is the number of discretized points in the mud bed layers, and Δz is the grid length in the z direction. The volume of mud particles entrained from the mud bed, V_E, can be estimated by using an empirical formulas for the erosion rate E of mud particles (Parture and Mehta, 1985) as follows:

$$V_E = \sum_{k=1}^{K} \left(\frac{E\Delta t}{\gamma_s} \right) = \sum_{k=1}^{K} \left\{ \frac{E_0 \Delta t}{\gamma_s} \exp\left[\alpha\,(\tau_b - \tau_c)^{0.5} \right] \right\} \tag{11}$$

where Δt is time interval, γ_s is the unit weight of the mud, E_0 is the floc erosion rate (g/cm^2/s), α is a model constant, τ_b is the wave-induced shear stress on the mud bed, and τ_c is the shear strength of the mud.

From Eqs. (9) and (10), the change in water content ratio W in the mud bed can be calculated from,

$$W\,(z,t) = e\,(z,t)\,\frac{\gamma_s}{S} \tag{12}$$

in which S is the saturation degree of the mud.

The validity of the above model is examined through comparisons of the computations with the measurements by Shen *et al.* (1993). Figure 7 shows comparisons between the calculations and measurements for the temporal change in water content ratio distributions under the conditions listed in Table 1. A good agreement is seen through the overall depth for each time step except for little discrepancy near the rigid bottom, where accurate measurement is generally difficult. Hence it is concluded that if values of the model parameters are properly determined for a mud material of interest, this model can be used to evaluate the temporal change in water content ratio distributions under waves with an acceptable accuracy.

Table 1 Computation conditions.

W_0 (%)	$k/[\rho_f(1+e)]$ (m^4/kgs)	$k_l/[\rho_f(1+e)]$ (m^4/kgs)	$\lambda = d\sigma'/de$ (kg/m^2)	T (s)	H_0 (cm)
174	3.6×10^{-11}	2.4×10^{-11}	35.0	1.02	4.5

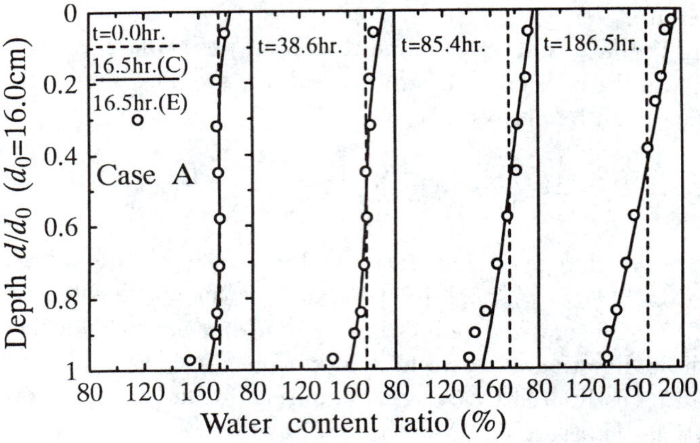

Fig. 7 Comparisons of the change in water content ratio between the calculations and the measurement by Shen *et al.* (1993)

4. Numerical Model of Mud Motion

4.1 Model Equations

A vertical two-dimensional numerical model for the interaction between waves and a mud bed is developed by incorporating the proposed rheological model and the water content ratio model with the Navier-Stokes equations.

On the assumption that the convective acceleration is not significant, the linearized momentum equations and the continuity equation for the incompressible soft mud layer under waves are

$$\rho_m \frac{\partial u}{\partial t} = -\frac{\partial p}{\partial x} + \frac{\partial \tau_{zx}}{\partial z} \tag{13}$$

$$\rho_m \frac{\partial w}{\partial t} = -\frac{\partial p}{\partial z} + \frac{\partial \tau_{xz}}{\partial x} \tag{14}$$

$$\frac{\partial u}{\partial x} + \frac{\partial w}{\partial z} = 0 \tag{15}$$

in which p is the dynamic pressure, u and w are the velocity components in the horizontal x and vertical z directions, ρ_m the mud density, and τ_{xz} and τ_{zx} are the shear stresses which can be obtained by the proposed rheological model Eqs. (1) and (2).

Equations (13) through (15) give a closed set of governing equations for u, w and p. With the non-slip condition ($u_n = w_n = 0$) at the fixed bottom under the mud layers, the zero shear stresses ($\tau_{xz0} = \tau_{zx0} = 0$) and pressure continuity conditions on the mud surface, and the periodic lateral boundary conditions, this equation system can be solved numerically, for instance, by using the SIMPLEC algorithm.

The coefficient of wave damping along the wave propagating direction, D_H, can be estimated as

$$D_H = -\frac{1}{H}\frac{dH}{dx} = \frac{1}{2EC_gL}\overline{\int_x^{x+L}\int_0^d \tau\left(\frac{\partial u}{\partial z}+\frac{\partial w}{\partial x}\right)dzdx} \tag{16}$$

The mass transport velocity of mud, U, is given as the summation of the Stokes drift U_S and the mean Eulerian velocity U_E, which are

$$U_S = \overline{\frac{\partial u}{\partial x}\int_0^t udt'} + \overline{\frac{\partial u}{\partial z}\int_0^t wdt'} \tag{17}$$

$$\mu'\frac{\partial^2 U_E}{\partial z^2} = \overline{\frac{\partial \rho_m u^2}{\partial x}} + \overline{\frac{\partial \rho_m uw}{\partial z}} \tag{18}$$

where μ' is the mean viscosity averaged over one wave period,

$$\mu' = \frac{1}{T}\int_t^{t+T}\frac{\partial \tau}{\partial z}dt = \frac{1}{T}\int_t^{t+T}\frac{\mu_0}{(1+\beta\mid\gamma\mid)^2}dt \tag{19}$$

4.2 Comparisons with Results of Experiment

In order to verify the above-described numerical model, results of calculations are compared with those of experiments by Sakakiyama and Bijker (1989). Figures 8 and 9 give examples of the comparisons with respect to the wave height distribution and the vertical profile of mud mass transport velocity. Agreement is very good for the mass transport velocity distribution, while the wave damping is slightly underestimated by the present model.

To examine the effect of water content ratio on the motion of bed mud, the change of the mud mass transport velocity in response to the temporal change in water content ratio is evaluated as an example. Figure 10 is the predicted mass transport velocity that corresponds to the evolution of water content ratio shown in Fig. 11. It can be seen in these figures that the temporal change in the water content ratio naturally has effect on the mud mass transport but it is not very significant as far as under the conditions in this example.

5. Conclusions

The soft mud is characterized by the very complicated visco-elasto-plastic properties in response to the cyclic action of external forces. In addition, the mud properties are significantly influenced by the water content ratio as well as the period and amplitude of cyclic loading. The proposed empirical rheological model can be regarded more general than most of previous ones, although the formulas for the model parameters may not be directly applicable to actual coastal mud. The temporal change in the mud water content ratio under waves has also been studied and formulated as a simple analytical model. By incorporating these

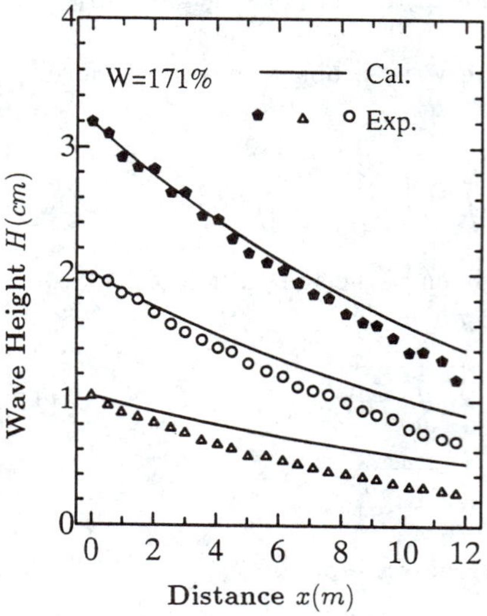

Fig. 8 Comparison between calculated and measured wave damping.

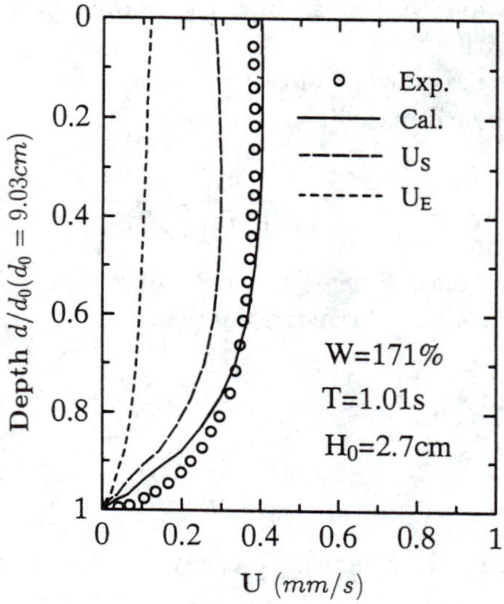

Fig. 9 Calculated and measured mud mass transport velocity.

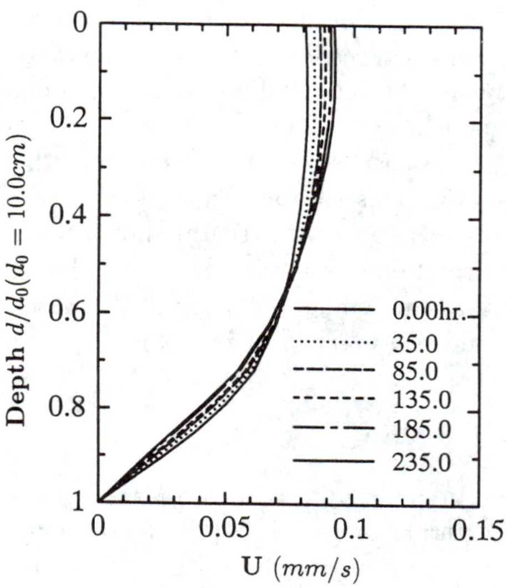

Fig. 10 Temporal changes of mud water content ratio under waves.

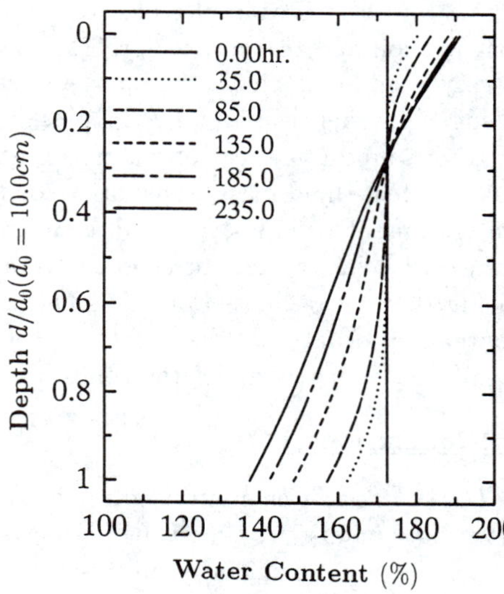

Fig. 11 Evolution of mud mass transport velocity under waves.

models with the Navier-Stokes equations, a vertical two-dimensional numerical model has been constituted to simulate the interaction of waves and mud. It has been verified that the proposed models can reproduce satisfactorily the rheological properties of mud, the temporal change in the water content ratio as well as the mud mass transport velocity under waves.

Acknowledgment

The authors would like to thank Prof. Masahiko Isobe, Department of Civil Engineering, University of Tokyo, for his valuable discussions throughout this study.

References

Been, K. and G. C. Sills, (1981): Self-weight consolidation of soft soils: an experimental and theoretical study, Geotechnique, Vol. 31, No, 4, pp. 519–535.

Dalrymple, R. A. and P. L.-F. Liu, (1978): Wave over soft muds: a two layer fluid model, Jour. Phys. Oceanogr., Vol. 8, pp. 1121–1131.

Gibson, R. K., G. L. England, and M. J. L. Hussey, (1967), The theory of one-dimensional consolidation of saturated clays, (i) Finite non-linear consolidation of thick homogeneous layers, Geotechnique, Vol. 17, No.26, pp. 251–273.

Huynh, N. T., M. Isobe, T. Kobayashi, and A. Watanabe, (1990): An experimental study on rheological properties of mud in the coastal waters, Proc. Coastal Eng., JSCE, Vol., 37, pp225–229. (in Japanese)

Jiang, Q. and A. Watanabe, (1995): Rheological properties of soft mud and a numerical model for its motion under waves, Coastal Engineering in Japan, JSCE, Vol. 38, No. 2, 195–214.

Jiang, Q., (1995): Study on the rheological properties and mass transport of soft mud under water waves, Doctoral dissertation submitted to the University of Tokyo.

Maa, J. P.-Y. and A. J. Mehta, (1990): Soft mud response to water waves, Jour. Waterway, Port, Coastal and Ocean Eng., ASCE, Vol. 116, No.5, pp.634–650.

Macpherson, H., (1980): The attenuation of water waves over a non-rigid bed, Jour. Fluid Mech., Vol. 97, Part 4, pp. 721–742.

Mei, C. C. and K.-F. Liu, (1987): A Bingham-plastic model for a muddy seabed under long waves, Jour. Geophys. Res., Vol. 92, No. C12, pp. 14,581–14,594.

Parture, T. M. and A. J. Mehta, (1985): Erosion of soft cohesive sediment deposits. J. hydraulic Engineering, ASCE, Vol. 111, pp.1308–1326.

Sakakiyama, T. and E. W. Bijker, (1989): Mass transport velocity in mud layer due to progressive waves, J. of Waterways, Ports., Coastal and Ocean Eng., ASCE, Vol. 115, No. 5, pp.614–633.

Shibayama. T., and N. A. Nguyen, (1993): A visco-elastic-plastic model for wave-mud interaction, Coastal Engineering in Japan, JSCE, Vol. 36, No. 1, pp.67–89.

Shen, D. S., H. Yokoki, M. Isobe and A. Watanabe, (1993): Mud transport and muddy bottom deformation due to waves, Proc. Coastal Eng., JSCE, Vol., 40, pp356–360. (in Japanese)

Zen, K. K., (1993): Study on the wave-induced liquefaction in seabed, Tech. Note Port and Harbor Res. Inst., Min. of Transport, Japan, No. 755. (in Japanese)

CHAPTER 324

COUNTERMEASURES AGAINST WIND-BLOWN SAND ON BEACHES

--- A short summary of the present state-of-the-arts and introduction of new methods ---

Shintaro Hotta*
and
Kiyoshi Horikawa**

Abstract

Characteristics and functionings of prevention works for wind-blown sand are briefly reviewed and discussed. Then, guidelines for their use are given. In addition, new prevention works, including trench, movable porous sand fence, and large-scale sand ripples, are proposed according to the result of recent research.

I. Introduction

One of the important problems in beach stabilization is how to control wind-blown sand. Typical problems concerning blown sand are preservation and promotion of the growth of dunes, prevention of channels, inlet and river mouth closure, protection of residential and cultivated land from contamination by sand, and so on. Many kinds of works for preventing blown sand have been employed all over the world. However, no general criteria or technical standards for the use of prevention works exist at present. Therefore, the main objectives of this study are to summarize the operational characteristics of prevention works for wind-blown sand and to provide specific guidelines for their use.

Attempts are made (1) to systematize and classify prevention works for wind-blown sand according to their functioning, and (2) to evaluate conventional prevention works. In addition, (3) new prevention works are proposed according to studies recently carried out.

II. Functioning and Classification of Prevention Works for Wind-Blown Sand

Prevention works for wind-blown sand might have the following two functions: one is to restrain and stop the generation of blown sand, and the other is to trap the blown sand at the upwind side and to store the blown sand at a given location so that it is not transported further downwind. Hereafter, in short for later convenience, the former function will be known as a restrain function and the latter

----

* Shintaro Hotta, Dr., Prof. of Dept. of Construction Engrg. Nihon University, Narashino-dai 7-24-1, Funabashi, Chiba Pref., 274, Japan.

** Kiyoshi Horikawa, Dr., Prof. Emeritus of the University of Tokyo and President of Saitama University, Shimo-Okubo, Urawa, Saitama Pref., 255, Japan function as a trap and fixation function. If a prevention work has a perfect restrain function and can fully stop the generation of blown sand, the work does not need the trap and fixation function. However, it is usually difficult to perfectly stop the generation of blown sand by using a certain prevention work which has the restrain function. It may be expensive or it may prevent an effective use of the beach if we intend to completely stop the generation of blown sand. In such cases, it may be recommendable from an engineering point of view to employ a prevention work which has the trap and fixation function at the downwind site while it may permit the generation of blown sand by decreasing the restrain function, which could be done through changing the kind of prevention work used or downsizing the scale of the prevention work.

The restrain function can be provided by the following methods. One method is to increase the resistance force to the wind which acts on the sand surface and sets sand grains on the surface into motion as blown sand. Another method is to decrease the wind speed that acts near the sand surface to a level so that the sand grains can not begin to move. Examples of methods for increasing the resistance force of the sand surface are spraying water (fresh or salt) or coagulant on the surface and replacement of surface sand. To decrease the wind speed (or the shear stress) that acts on the sand surface, methods to be considered are placing fences, planting shrubs or trees, and covering by grass, nets or mats in order to shut off the wind from direct action on the sand surface.

Two cases are considered for the trap and fixation function. The first case is the forced trap and fixation function that traps coercively blown sand in the air without changing of the basic motion of the sand grains and brings them to rest. The second case involves trapping or bringing to rest the blown sand grains by means of decreasing the wind speed. Cases that catch blown sand grains with streamer-type traps or adhere blown sand grains to a plate coated by an adhesive are considered as cases where the forced trap and fixation function acts. However, it is not possible to trap and store a large amounts of sand by these methods. Thus, these methods can not be employed for engineering use. Another case is where blown sand grains strike a fence stretched by a net with a small mesh and with a high porosity, where the porosity is defined as the ratio of open space area to total projected area, drop and come to rest at the foot of the fence. In this case, we may think of the fence as having the forced trap and fixation function because the fence does not disturb the wind flow and the motion of the sand grains in the air significantly. Such a fence can be applicable for engineering use since the fence can be constructed easily. A trench (which will be described in Section IV) several meters wide and deeper than 1 m can trap almost all blown sand (nearly 100 % of the total amount) without disturbing the wind field (Horikawa *et al.*, 1984). The trench stops coercively the motion of sand grains in the surface creep and saltation at the upwind tip of the trench. Therefore, we may regard the trench as having the forced trap and fixation function. However, inside the trench the wind speed becomes low and sand grains dropped into the trench come to rest. Thus, the trench also has the function of trapping sand by reducing the wind speed.

Prevention works which have the trapping function by reducing the wind speed are fences, foredunes (artificial embankments), planting shrubs, conservation forests for wind-blown sand, artificial large-scale sand ripples, and movable porous fences.

An attempt to systematize and classify the prevention works according to their functioning is given in Fig. 1. In Fig. 1, the prevention works involving trenches, artificial large-scale sand ripples, and movable porous fences are those proposed by the present authors for future use, and they are described in more detail in Section IV.

The prevention works normally employed have both the functions of restrain and of trap and fixation, although some prevention works only have either the function of restrain or trap and fixation. Therefore, the fundamental rule for employment of prevention works is to use more than one type of works, considering the conditions for the generation of blown sand and the general purpose of the works, aiming for the works to perform both functions as effectively as possible.

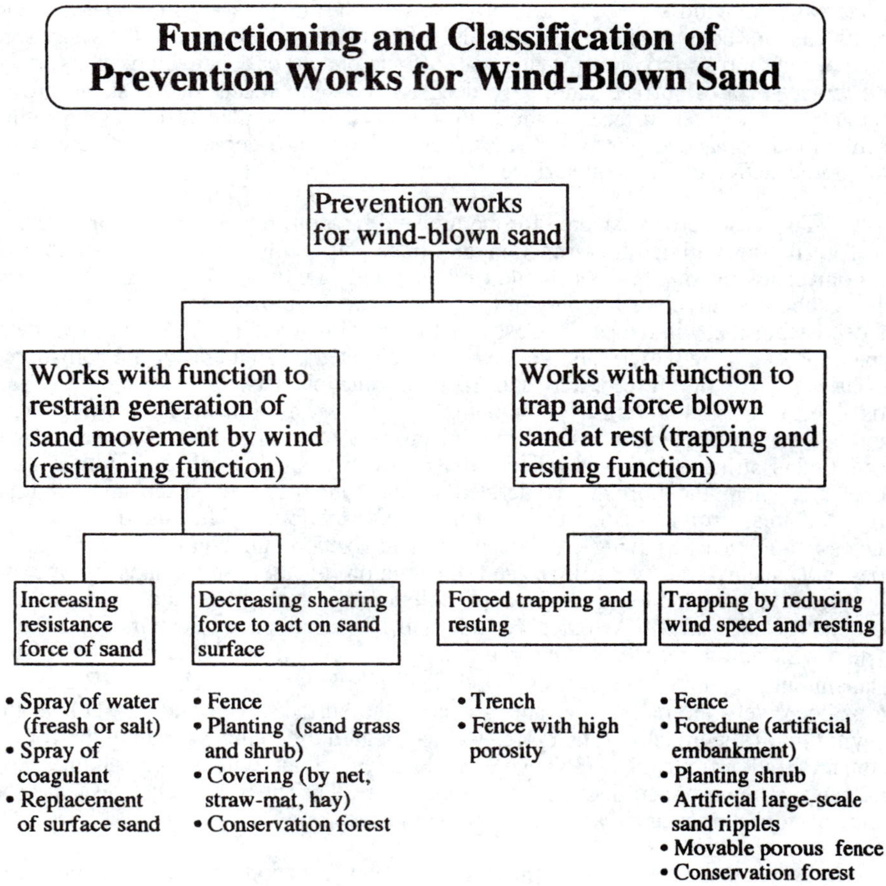

Fig. 1 Functioning and classification of prevention works for wind-blown sand.

III. A Brief Summary of Operational Characteristics and Guidelines for Use of Prevention Works

The prevention works commonly employed worldwide are sand fences, foredunes (including artificial embankment quickly constructed by earthmoving machines), vegetation (planting sand grass or shrubs), and coastal conservation forests. The characteristics and guidelines for practical use of these prevention works are briefly summarized in the following sections.

III.1 Sand Fence

The sand fence is the most commonly employed prevention work throughout the world. Many types of sand fences of varying shape and materials have been designed and employed adapting to the local conditions. The characteristics and guidelines for practical use of sand fences are relatively well understood. Hotta, Kraus and Horikawa (1987) made a critical review of the literature about single row sand fences. They summarized its characteristics and gave guidelines for practical use.

In the guidelines they gave it is recommended to use fence with material of wooden plate, fern, reed, brush, and bamboo which rapidly decompose, expecting that the sand fence will be left in the accumulated sand after the fence is buried. However, if the sand fence is used as a collecting device to interrupt the blown sand intruding into residential area or cultivated land, and if it is planned to remove the accumulated sand for long-term use, it is recommended to use long-life fabrics or metal as the fence material.

For multi-row sand fences, Hotta, Kraus, and Horikawa (1991) also gave a review that concerned the forming of foredunes. Studies regarding multi-row sand fences are limited and many of these studies are not explaining the functioning well. Hotta *et al.* (1992) found important facts about two-row sand fences by means of a wind-tunnel study. The experiment showed that the fence placed downwind would not function to collect blown sand until the fence placed upwind was buried and lost its collecting ability when two-row fences were placed at intervals shorter than $10H$, where H is the height of the fence. This means that two-row fences have no advantage when it comes to collecting blown sand. However, experiments carried out by Manohar and Braun (1970) showed that multi-row sand fences were effective when the wind was strong. Further studies are needed to understand the characteristics of multi-row sand fences and this should be done in the near future.

III.2 Foredune (artificial embankment)

In order to control blown sand, formation of a dune by means of repeated employment of sand fences has been carried out for a long time. The dune formed is usually called a foredune. However, at present formation of a dune is often carried out quickly by using earthmoving machines, and vegetation work is used on the dune surface. A dune formed artificially by construction machines is normally called an artificial embankment.

It is often assumed that the function of a foredune or artificial embankment is almost the same as a non-permeable solid fence. But, the dune has a certain crown width and slopes on both the upwind and downwind sides. Thus, the foredune probably has a somewhat different function from the non-permeable fence. Figure 2 shows a schematic illustration of the wind flow field and sand accumulation characteristics around an artificial embankment. The wind speed profile in the vertical direction begins to distort at a location A affected by the existing embankment. A domain where the wind speed is reduced is formed at the foot of the embankment (around a location B). However, at the upper part of the front slope, the wind will accelerate and the wind speed at the front shoulder of embankment C becomes around 1.2 - 1.8 times of that if no embankment exists. A large circulation cell is formed downwind of the embankment. The wind profile will recover at a location F because of energy supplied from the upper layer. A foredune can catch and store a large amount of blown sand in the wind-reduced domain at the foot, if the foredune is sufficiently high. However, sand is transported beyond the foredune when the wind-reduced domain is filled up by the trapped sand and the storage capacity is lost.

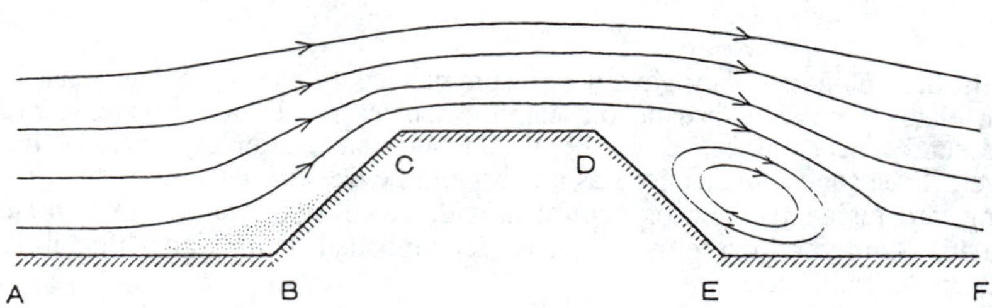

Fig. 2 Schematic illustration of the flow field around an embankment.

There is also a possibility that sand is transported beyond the foredune when the surface of the front slope is bare or paved with some hard material. This can be explained by considering the sand grain motion. A large portion of the sand is transported in a mode of saltation. The sand grains which fall on the hard slope surface can rebound easily and continue the saltation motion. During their stay in the air, sand grains in saltation are accelerated by the wind, and the possibility for the motion to continue increases. Then, sand grains may eventually reach to the top of the foredune. When the surface of the front slope is vegetated by grass or shrubs, the vegetated surface can stop the motion of sand grains and store the grains in the space of the vegetation body. The foredune (or artificial embankment) can control the blown sand as long as the foredune does not lose its storage capacity at the foot and the vegetation on the front slope surface is not buried. Studies concerned with the function of the foredune or embankment for controlling blown sand are limited. Figure 3 shows an example of equi-ratio curves of wind speed reduction in front of an embankment from an experiment by Hotta and Horikawa (1990a). Figure 4 shows the sand accumulation process in front of an embankment under the wind condition corresponding to Fig. 3 (Hotta and Horikawa, 1990b). Note that the scales in the vertical and horizontal directions are not the same in Figs. 3 and 4 and that the figures are distorted.

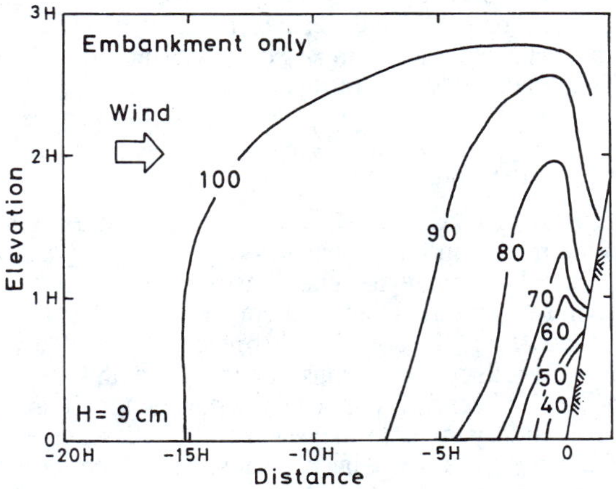

Fig. 3 Equi-ratios curves of wind speed reduction in front of an embankment. (After Hotta and Horikawa, 1990a)

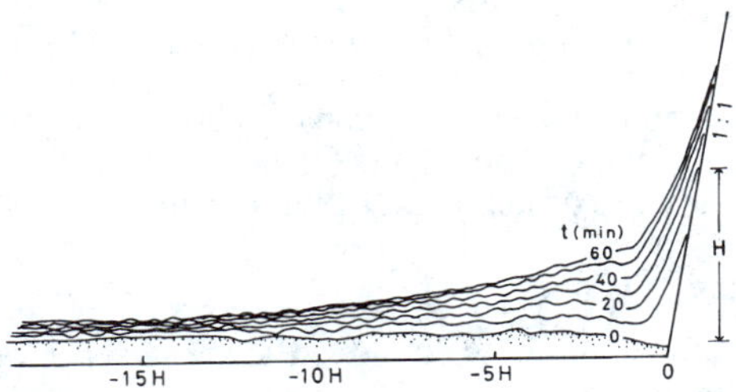

Fig. 4 The sand accumulation process in front of an embankment. (After Hotta and Horikawa, 1990b)

III.3 Vegetation Work

Vegetation is one of the most popular prevention works for controlling blown sand and it has been employed all over the world from early times. Vegetation work has both restrain and trap and fixation functions. Plants (sand grass and shrubs) that cover the sand surface can restrain the generation of blown sand by preventing the wind from acting directly on the surface, and can also trap and fix the blown sand with decreasing wind speed. We can expect a considerable storage capacity for shrubs but not so much for sand grass. The fundamental property required for plants employed in vegetation work is that the rate of growth must be bigger than that of sand accumulation. We can no longer expect any ability for controlling blown sand when plants are buried. There are many previous examples of vegetation works used in the field. The works sometimes succeeded but they occasionally failed. The circumstances surrounding plant growth involve many factors such as weather condition, bed material, and character of plants. No guidelines for engineering use exist at present. The most suitable plants for selection may be the ones that grow at the site under consideration.

III.4 Coastal Conservation Forest

On many beaches a coastal conservation forest is often planted for several purposes. One of the most important purposes of a conservation forest is to control the wind-blown sand. We can regard the conservation forest as an aggregate of porous fences. Therefore, the function of a conservation forest is almost the same as a porous fence. However, the domain where the wind is reduced is much larger for a conservation forest than for porous fence, since the trees composing the conservation forest are considerably higher than a porous fence, and the forest has a certain width. The boundary towards the beach-end at the seaside edge of the forest is a battle field where the trees fight back against the penetration of blown sand. Therefore, to keep the primary function that a conservation forest should control blown sand, the trees planted at the seaside boundary must grew up above the thickness of the accumulating sand layer occurring during strong winds. Growth of trees is usually slow and trees at the seaside boundary are always exposed to the risk of withering due to the attack of the strong wind, blown sand, and salt transported from the sea. Therefore, it is not wise to only use conservation forests for controlling blown sand. In general, sand fences or a foredune will be placed to prevent the withering of the trees. Then, we may as well consider that a conservation forest provides protection in emergency situations.

Trap and fixation function of a conservation forest depends on the height of the trees, planting distance, kind of tree, and width of forest. To control blown sand satisfactorily, a width of the forest of more than 100 m is required.

III.5 Other Prevention Works

It is a technique commonly employed to spray water on the ground surface to reduce the dust generation when the surface becomes dry and dusty during windy days. Spraying water increases the resistance force against the wind, and suppresses the generation of blown sand, or decreases the amount of sand transported. However, in our review, we could not find any examples where spraying water was employed for controlling wind-blown sand, although we found an example where fresh water was sprayed or showered by sprinkler or portable shower vehicles for washing down salt adhered to leaves of pine trees growing in a conservation forest. Taking care of conservation forests by showering fresh water will be reported in detail by the same authors in the near future. We could not find any examples where coagulant was sprayed for controlling wind-blown sand, although we found several examples of spraying coagulant containing fertilizer for promoting vegetation work. We suppose that usage of coagulant will not be recommended because it may contaminate beach sand and obstruct a comfortable use of the beach.

Covering work by net, hay, straw-mat, and others have often been employed to restrain wind erosion of the ground surface. However, it is difficult to cover a large area of the ground surface, such as the beach, and it is also expected to be costly. We can not ignore that there is an engineering possibility to cover a whole beach surface by a kind of plastic net for effectively controlling blown sand. But, the problem has not been studied fully yet. Employment of covering work should be limited to small areas.

IV. New Prevention Works

Based on previous studies carried out by the present authors, we would like to propose the following three new prevention works: trench, movable porous fence, and artificial large-scale sand ripples. These three type of works will be discussed in the following sections.

IV.1 Trench

The idea of a trench in prevention work for wind-blown sand is based on long-term experience and results from our previous studies. Iwagaki(1950) studied the phenomenon of blown sand by using diffusion theory. He predicted that (i) the flying distance of a sand grain in saltation motion was rather short, and that (ii) over 97 % of the sand grains in saltation fell within a range of 5 m from the end of the sand bed when the bed abruptly disappeared like at the shoreline. As a verification of his predictions he pointed out that cultivated land in a sandy beach of Tottori Prefecture where it is facing Japan Sea, was protected from the blown sand by excavating a stream upwind, letting sand grains fall into the stream, and returning the sand to the sea. Considering these results, Horikawa *et al.* (1994) tried to measure the sand transport rate by using a trench which was 1 m deep and several meters wide in the field. The measurements showed that the trench trapped almost 100 % of blown sand transported. Hotta and Horikawa (1993) measured the flying distance of sand grains in saltation using a large wind tunnel. They found that the flying distance for an amount of over 95 % of the blown sand was smaller than 1 m. Using the same technique as Hotta and Horikawa, Shiozawa *et al.* (1993) measured the flying distance of blown sand at a real beach. They found the same result as Hotta and Horikawa (1993). Therefore, the present authors concluded that the trench could be employed as a prevention work for wind-blown sand and propose to use a trench practically in the field. However, it is not clear how many meters of trench width that are suitable. Horikawa *et al.* (1996) carried out a wind tunnel experiment and a numerical experiment to determine the flow field in an open cavity. They found that a width of about 5 or 6 meters would be acceptable when the trench is dug deeper than 1 m. An example of how to use a trench for control of blown sand can be seen in Fig. 7.

IV.2 Movable Porous Fence

This method is based on a simple idea. It is an attempt to erect or place a temporary fence, which is not large and can be movable, on the beach sand surface where the blown sand is generated during the windy season. After the windy season has passed, the fence should be removed to a storage for usage the next windy season. When a strong wind blows, blown sand will be generated even though the sand surface is wet (Hotta *et al.*, 1984). The problem with blown sand is usually most serious when the wind blows from the offshore to the inland. Therefore, the fence should be placed near the inland side of the shoreline where the blown sand is generated, although we have to consider the risk that the fence may be attacked by the runup waves during stormy sea conditions. Figure 5 shows a sketch of a test manufacture of a fence.

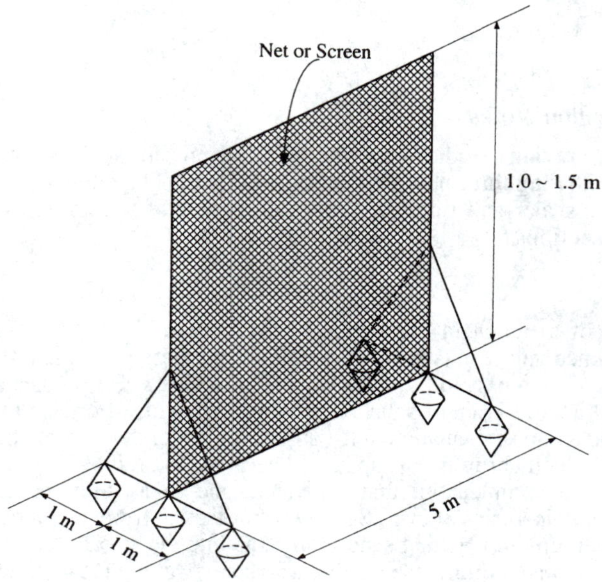

Fig. 5 A test manufacture of a movable porous fence.

IV.3 Artificial Large Scale Sand Ripples

This type of works is based on the same idea as the trench; that is , the flying distance of a sand grain in motion is not so long as described in Section IV. 1. The work is simple, and the purpose is only to form giant ripples 2 or 3 m high and several meters long, as shown in Fig. 6, on the beach surface by using earthmoving machines such as power shovels or backhoes. The blown sand generated from the crest part of the giant ripple falls into the following trough part downwind. The sand is not transported beyond the next crest until the trough fills up. The blown sand will be generated on the whole beach surface and the sand will be transported into the inland when the trough is buried. We must form the ripples again as soon as the trough is buried. This type of works means that the blown sand generated from a small area is caught in a short distance downwind and brought back to the original place.

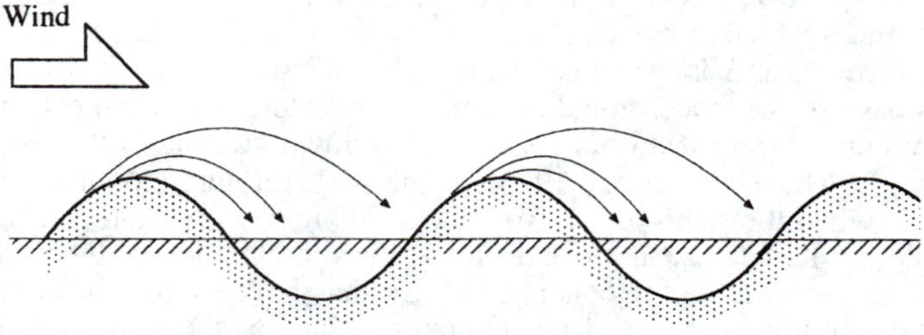

Fig. 6 A schematic illustration of artificial large scale sand ripples.

IV Systematical Use of Prevention Works

Figure 7 shows an example of the systematical use of conventional and new prevention works for wind-blown sand. Movable porous fences are placed on the part of the beach neighboring the shoreline where the blown sand is generated. When fences are buried to more than about two-third of their height, the fences must be pulled up. Then, the fences will recover their restrain and trap and fixation functions. A trench traps the blown sand which pass through the fence field. A fence erected at the downwind tip of the trench with a low fence porosity has a tendency to accumulate sand in front of the fence (Hotta and Horikawa, 1990b). Therefore, we can expect that over 90 % of the amount of the total blown sand can be trapped as far as where the fence is erected at the downwind tip of the trench.

We may remove the sand stored in the trench by using earthmoving machines as a power shovels or backhoes. But, we can also remove sand by using sea water, as described in Fig. 7. The sea water is withdrawn from outside the breaker point, flushing the water in the trench, and draining away the sand at the shoreline. Instead of using seawater, we can also use water from the landside, if enough water is available.

A small amount of blown sand in suspension can be transported beyond the fence. Prevention works arranged behind the trench must control the sand that pass. In the wind field around an embankment, a wind-reduced domain is formed at the foot of the embankment. The domain has a possibility to store the blown sand. Shrubs and grass on the slope of an embankment weaken the wind, and catch and store the blown sand.

The crown of the embankment may be used as a bicycle road or walking road. At the seaside shoulder of the embankment the wind accelerates. To moderate the wind, placement of a fence with a median porosity is recommended. The fence height should be lower than 1 m to maintain a good view.

At the landside shoulder of the crown, placement of a relatively high fence is recommended to reduce the wind and to protect the inland. A low porosity fence is recommended for the upper part of this fence.

V. Concluding Remarks

The characteristics of prevention works for wind-blown sand have mainly been studied in the field of agricultural engineering, forestry, geomorphology and coastal engineering. Studies have been carried out both in the wind tunnel and in the field. The results obtained were qualitative and quantitative evaluations of the effectiveness of the prevention works are scarce. The phenomenon is so complicated that theoretical analysis will be difficult. To evaluate the effectiveness of prevention works quantitatively, systematical and well controlled wind tunnel studies or field studies are needed, although numerical experiment based on the detailed description of the fluid dynamics in computers can assist our understanding.

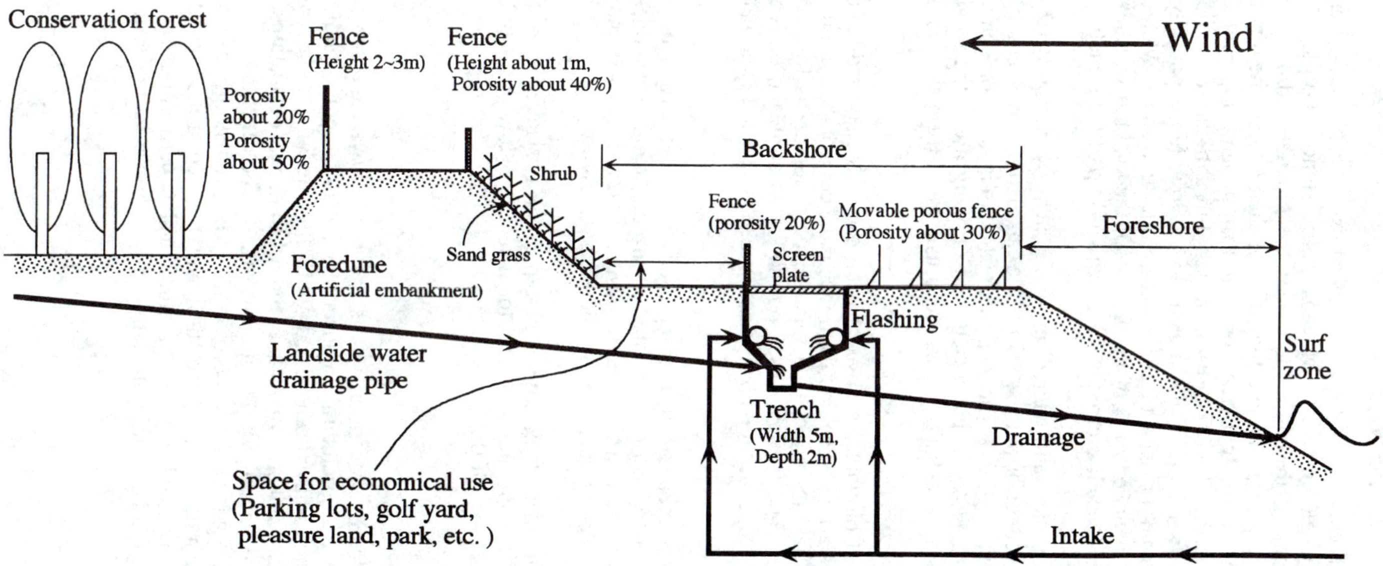

Fig. 7 An example of systematical use of conventional and new prevention works.

References

Iwagaki, Y. (1950): On the effect of the sand-drift on the coast by wind for sand deposition in Ajiro Harbor, Journal of JSCE, Vol. 36 (6), pp. 19-25. (in Japanese)

Hotta, S. and K. Horikawa (1993): Vertical distribution of sand transport rate by wind, Coastal Eng. in Japan, Vol. 36, No. 1, JSCE, pp. 91-110.

Hotta, S., T. Murata, H. Mori and K. Horikawa (1992): Experimental study on function of two-rows sand fences, Proc. of Coastal Eng., Vol. 39, JSCE, pp. 331-335. (in Japanese)

Hotta, S., N.C. Kraus and K. Horikawa (1991): Functioning of Multi-row sand fences in forming foredunes, Proc. Coastal Sediments '91, ASCE, pp. 261-273.

Hotta, S. and K. Horikawa (1990a): Function of sand fence placed in front of embankment, Proc. of Coastal Eng. Vol. 37, JSCE, pp. 444-448. (in Japanese)

Hotta, S. and K. Horikawa (1990b): Function of sand fence placed in front of embankment, Proc. 22nd Coastal Eng. Conf., ASCE, pp. 2754-2767.

Hotta, S., N.C. Kraus and K. Horikawa (1987): Function of sand fences in controlling wind-blown sand, Proc. Coastal Sediments '87, ASCE, pp. 772-787.

Hotta, S., S. Kubota, S. Katori and K. Horikawa (1984): Sand transport by wind on a wet sand surface, Proc. 19th Coastal Eng. Conf., ASCE, pp. 1265-1281.

Horikawa, K., H. Tanaka, K. Hatanaka, N. Yamaya and S. Hotta (1996): Study on trench trap as a prevention work for wind-blown sand, Proc. of Coastal Eng., Vol. 43, JSCE, pp. 671-675. (in Japanese)

Horikawa, K., S. Hotta, S. Kubota and S. Katori (1984): Field measurement of blown sand transport rate by trench trap, Coastal Eng. in Japan, Vol. 27, JSCE, pp. 213-232.

Manohar, M. and P. Bruun (1970): Mechanics of dune growth by sand fences, Dock and Harbour Authority, Vol. LI, No. 600, October, pp. 243-252.

Shiozawa, T., S. Nakayauchi, M. Akazawa, S. Tamashiro and K. Kuroki (1993): Observation of the sand transport rate by wind at Niigata West Beach, Proc. of Coastal Eng., Vol. 40, JSCE, pp. 281-285.

CHAPTER 325

Sea breeze Effects on Nearshore Coastal Processes

Charitha Pattiaratchi[1] and Gerhard Masselink[2]

Abstract

The sea breeze, created by the diurnal solar heating and cooling cycle, is a well known meteorological phenomenon and occurs globally on a regular basis with varying intensity. The impact of the sea breeze system on nearshore coastal processes and sediment budget has received very little attention. In this paper, field data collected from two micro-tidal coastal regions: south-western Australia and Sri Lanka, are presented to illustrate the importance of the sea breeze system in these regions. It is shown that the rapidly changing wave climate, generated by the sea breeze, increases the cross-shore and longshore currents and sediment suspension on the beach. This results in an increase of the longshore sediment flux by up to a factor of 100. The effects of the sea breeze may be present up to 10 hours after the cessation of the sea breeze. The sea breeze system plays a major role in the coastal sediment budget in these regions.

----

[1] Senior Lecturer, Department of Environmental Engineering, Centre for Water Research, The University of Western Australia, Nedlands 6907, Australia.

[2] Research Fellow, Department of Environmental Engineering, Centre for Water Research, The University of Western Australia, Nedlands 6907, Australia.

Introduction

On micro-tidal wave dominated beaches, morphological change is primarily induced by variations in the incident wave climate. These variations are usually ascribed to the passage of storms. A neglected source of change to the wave field is the diurnal sea breeze, generated by the differential solar heating of the land and ocean, which could be obscured by the presence of high-wave energy levels or tidal effects. The sea breeze is a dominant feature along tropical and sub-tropical coastlines and occur along two thirds of the earth's coastlines (Abbs and Physick, 1992; Simpson, 1994). In these areas wind waves generated by the local sea breeze can contribute significantly to the temporal variation in the incident wave climate. In this paper, field studies undertaken in south-western Australia and Sri Lanka are presented to demonstrate the importance of the sea breeze system on the nearshore region and, in particular, the coastal sediment budget.

Data presented in this paper were collected during three field experiments. Two of these experiments were undertaken at City Beach, Western Australia (Figure 1) on 23 January 1992 and 3-9 March 1994. The third experiment was conducted in Ambakandawila, located to the south of Chilaw along the west coast of Sri Lanka (Figure 2). Both sites are located on a relatively straight, north-south oriented shoreline far from any engineering structures and are micro-tidal (tidal range < 0.7 m at spring tides). However, in Western Australia, the tide is diurnal whilst in Sri Lanka the tide is semi-diurnal. The swell waves approach the beach from the south-west at both sites but the direction of the wind (sea breeze) waves differ due to the different direction of the sea breeze. In Western Australia, the sea breeze blows parallel to coastline from the south (an explanation for this effect is given in Pattiaratchi et al., 1997) and thus the sea breeze generated waves approach the beach with a crest of angle 70° (from south-south-west). In Sri Lanka, the sea breeze (in January) blows from north-west and the sea breeze generated waves also approach from the same quadrant. The occurrence of sea breezes in both of these regions is well known to local residents and fisherman in these regions. In Western Australia, the sea breeze is called the 'Fremantle Doctor' due to cooling effect of the breeze during the summer. In Sri Lanka (in the Chilaw region), the sea breeze is termed Kode by the fisherman.

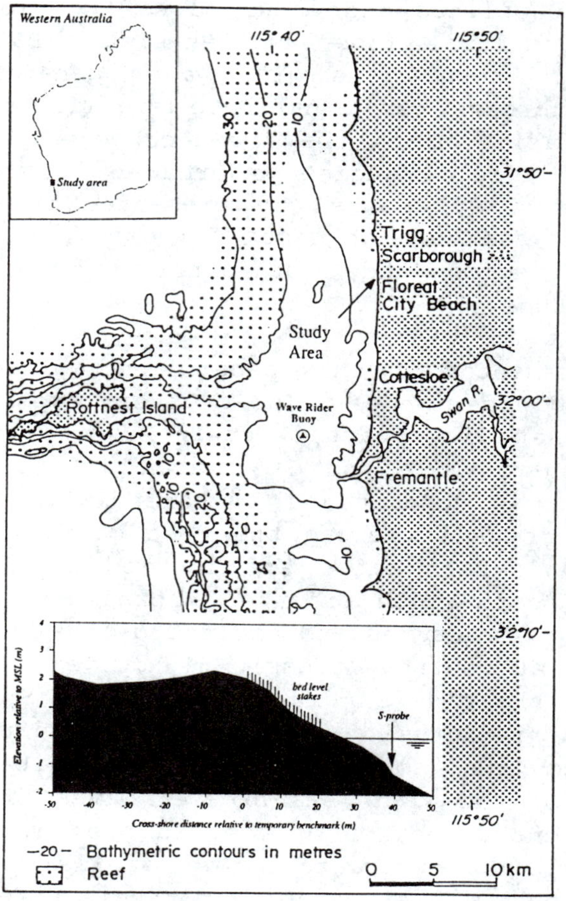

Figure 1 - Location of City Beach, Western Australia

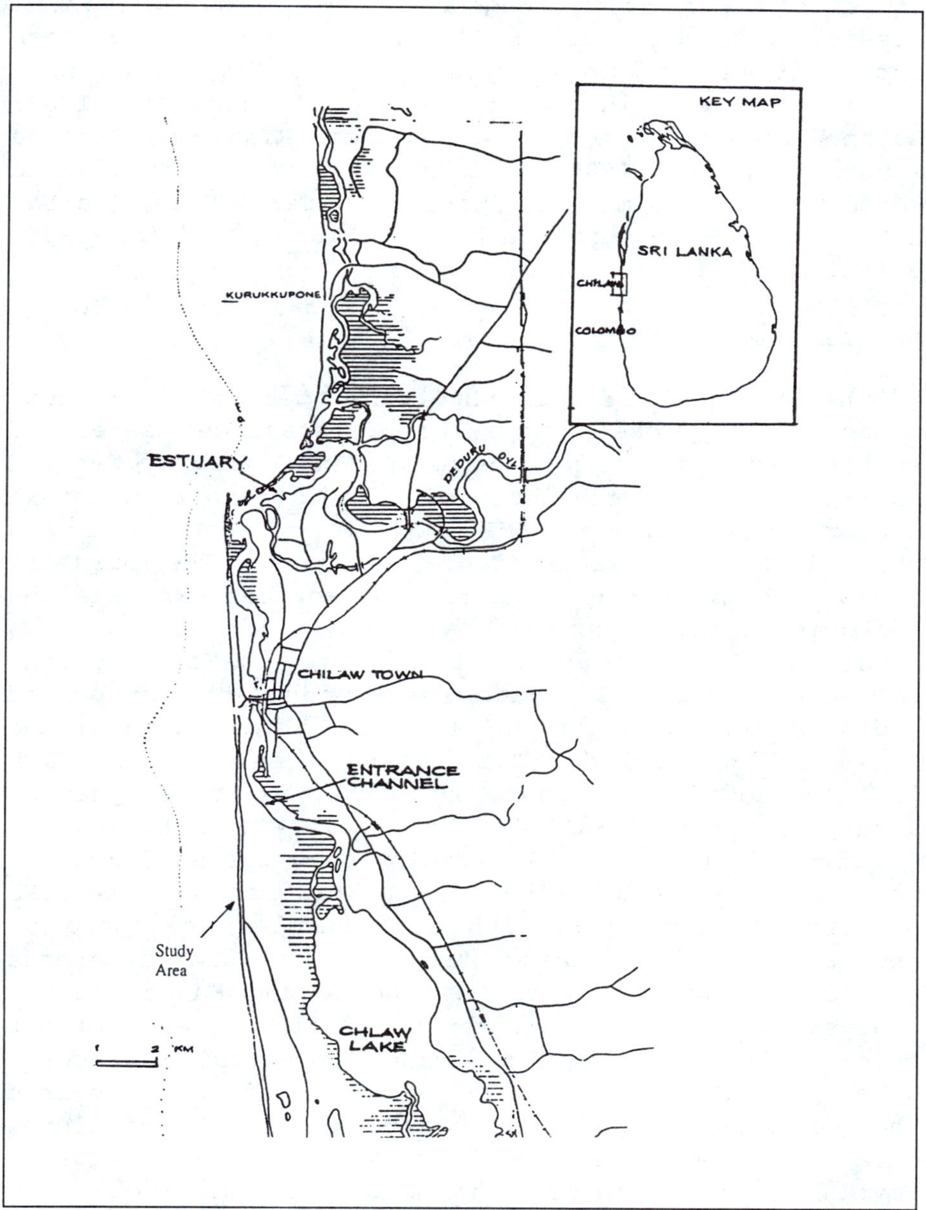

Figure 2 - Location of Ambakandawila, Sri Lanka

Although the sea breeze is a global phenomenon, it's intensity, which depends primarily on the land-ocean temperature gradients, varies geographically. The field data presented in this paper were collected under sea breeze speeds between 7 and 12 ms^{-1}. To allow comparison with other regions, the following maximum speeds, recorded during the sea breeze, have been reported: 14 ms^{-1} in Barcelona, Spain (Redano et al., 1991); 10 ms^{-1} in Greek waters (Prezerakos, 1986; Balopoulos et al., 1986); 9 ms^{-1} near Kingston, Jamaica (Huntley et al., 1988); 8 ms^{-1} in Tokyo (Yoshikado and Kondo, 1989); 7 ms^{-1} in Monterey, California (Banta, 1995);.and, 5 ms^{-1} in Santa Rosa Island, Florida (Sonu et al., 1973).

Methodology

The data presented in this paper were all collected using the 'S-probe', an instrument package, developed at the Centre for Water Research, University of Western Australia. It consists of a Digiquartz pressure sensor, a Neil Brown ACM2 acoustic current meter and three optical backscatterance sensors. The pressure sensor was mounted 0.35 m from the sea bed, whereas the two-dimensional, horizontal current velocity was recorded 0.2 m above the bed. The optical backscatterance (OBS) sensors measured the suspended sediment concentration (SSC) at 0.025 m, 0.125 m and 0.275 m from the bed. The data were transferred via a cable connected to a shore-based computer and logged at a sampling rate of either 5 Hz (exp. 1) or 2 Hz (expts. 2 and 3). The S-probe was deployed in the surf zone at each of the study sites. Calibration of the OBS sensors was conducted with sand samples collected at the survey site, using the methods and apparatus similar to that described by Ludwig and Hanes (1990). Analysis of the water surface elevation, current velocity and suspended sediment concentration records were carried out on time series of 2048 points (approximately 7 or 17 minutes). Although additional data (e.g. offshore wave climate, beach ground water levels, instantaneous shoreline, beach morphology changes etc), have been collected at each of these sites only data collected by the S-probe is presented here. More comprehensive account of the data sets from City Beach may be found in Pattiaratchi et al. (1997),

Masselink and Pattiaratchi, (1997), Masselink et al., (1997).

Data obtained from City Beach on 23 January 1992 (expt 1) is limited to four hours after the onset of the sea breeze. After this time, the S-probe could no longer be maintained in an upright position due to the extremely energetic surf zone conditions. However, continuous data over 3 sea breeze cycles are available from the subsequent survey from City Beach, 3-9 March 1995 (expt 2) and over two days from Ambakandawila, Sri Lanka, 18-20 January 1996 (expt 3).

Results

Experiment 1: City Beach, WA - 23 January 1992

Time series of wind speed and direction obtained on this day indicate a typical sea breeze cycle characterized by weak offshore winds in the morning and early afternoon, and a strong sea breeze starting at 1445 hours which continued into the evening (Figure 3). Wind speeds associated with the sea breeze exceeded 10 ms^{-1} and occur frequently in the summer months in this region. The direction of the sea breeze was consistently from the south and hence the sea breeze was blowing parallel to the shoreline, a feature of the sea breeze system in this region (Pattiaratchi et al., 1997).

Changes in wind speed and direction are reflected in the incident wave field, nearshore currents and suspended sediment concentrations (Figure 3). Prior to the sea breeze, small-amplitude swell (H_S=0.4 m) with zero-upcrossing periods (T_Z) of 7-8 s were present. Mean cross-shore (offshore) and longshore (northerly) currents were less than 0.05 ms^{-1} and 0.1 ms^{-1}, respectively whilst the mean SCC at a distance of 0.275 m above the bed was about 1 gl^{-1}. Sediment was only re-suspended during the passage of large waves in wave groups. The onset of the sea breeze induced an almost instantaneous change in the nearshore hydrodynamics. The wave height increased (H_S=0.7 m), the wave period decreased and assumed a constant value (T_Z=4 s) within one hour of the start of the sea breeze. Offshore-directed cross-shore currents in the surf zone rapidly increased in strength to a maximum of 0.16 ms^{-1} and the northerly longshore current progressively increased and exceeded 1 ms^{-1}. Sand was continuously re-suspended by the waves and the amount of SSC in the water column (at 0.275 m) increased seven-fold to 7 gl^{-1}. The surf zone averaged northward longshore suspended sediment transport rate increased from approximately 1 kg s^{-1} before the start of the sea breeze to 100-200 kg s^{-1} during the sea breeze, an increase by a factor of 100 (Figure 3).

Experiment 2: City Beach, WA - 4-6 March 1995

The results from this experiment are essentially the same as that for experiment 1. However, due to weaker sea breezes (max 7 ms^{-1}), the hydrodynamic changes induced by the sea breeze were less extreme. Continuous measurements of three successive sea breeze cycles were made and these enabled the investigation of the recovery period of the wave conditions after the cessation of the sea breeze. Prior to the sea breezes, offshore winds prevailed with speeds of 2-3 ms^{-1}. All three sea breezes persisted for approximately 7 hours and a relatively constant wind speed of 5-7 ms^{-1} was maintained. After the onset of the sea breeze, immediate changes occurred to the incoming wave field with a decrease in T_z from 10 to 5 s and an increase in the longshore current velocity to 0.15-0.2 ms^{-1}. Around 2045 hrs each day, the sea breezes stopped. The wind direction gradually shifted back to the east and the wind speed decreased to < 4 ms^{-1}. T_z increased and the longshore current velocity decreased immediately after the sea breeze had subsided. Around 0600 hrs on the following day, nine hours after the end of the sea breeze, the wave period and the longshore current velocity had reached pre-breeze levels.

Energy spectra of the cross-shore current were computed to construct a three-dimensional time-frequency plot to illustrate the change in spectral signature over the second sea breeze cycle (Figure 4). The cross-shore current data were used, rather than the

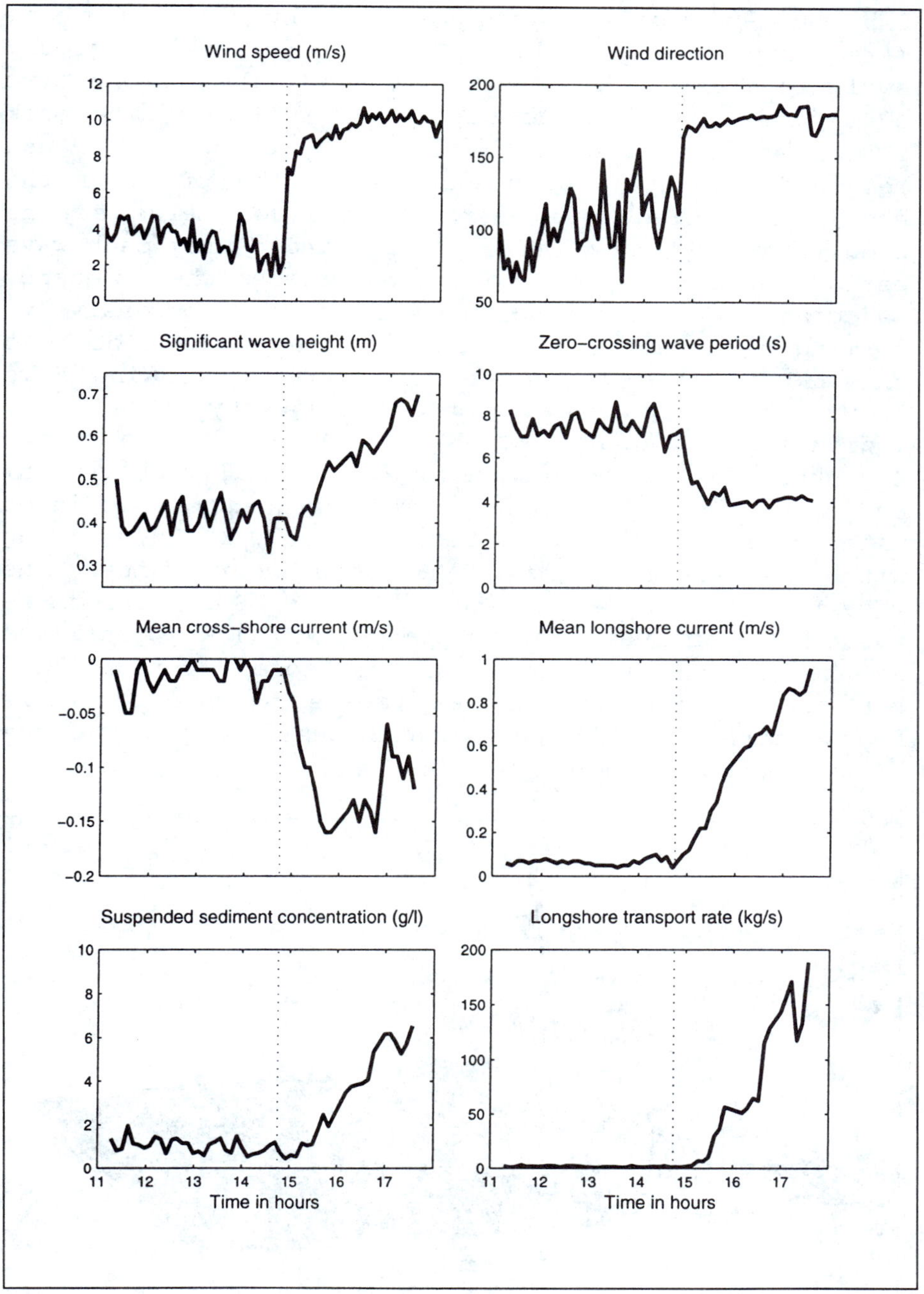

Figure 3 - Time series results for Experiment 1. The vertical dashed line indicates the start of sea breeze.

water surface elevation because they show more clearly the short-period wave energy caused by the sea breeze (frequency > 0.15 Hz). The long-period background swell is present in the time-frequency plot in the form of a linear ridge at frequency 0.07-0.1 Hz. The peak period associated with the swell energy was 12 s and remained relatively constant. After the onset of the sea breeze, wind wave energy begins to emerge at the high-frequency end of the spectra, indicating peak wave periods of 2.5 s. During the sea breeze, the frequency associated with the wind waves decreased progressively, forming a curved ridge in the frequency-time plot. At the end of the sea breeze (2100 hrs), the wind wave energy was concentrated around a frequency of 0.25 Hz, indicating a peak period of 4 s. After the sea breeze had subsided, the wind wave energy gradually decreased, but remained significant. The frequency associated with the wind wave energy decreased to 0.15 Hz, merging with the swell energy. Fifteen hours after the cessation of the sea breeze (1200 hrs 08/03/93), there was still some wave energy present which was generated by the sea breeze in the nearshore zone. This implies that the effect of the sea breeze on the nearshore hydrodynamics may be felt continuously during the sea breeze season (summer).

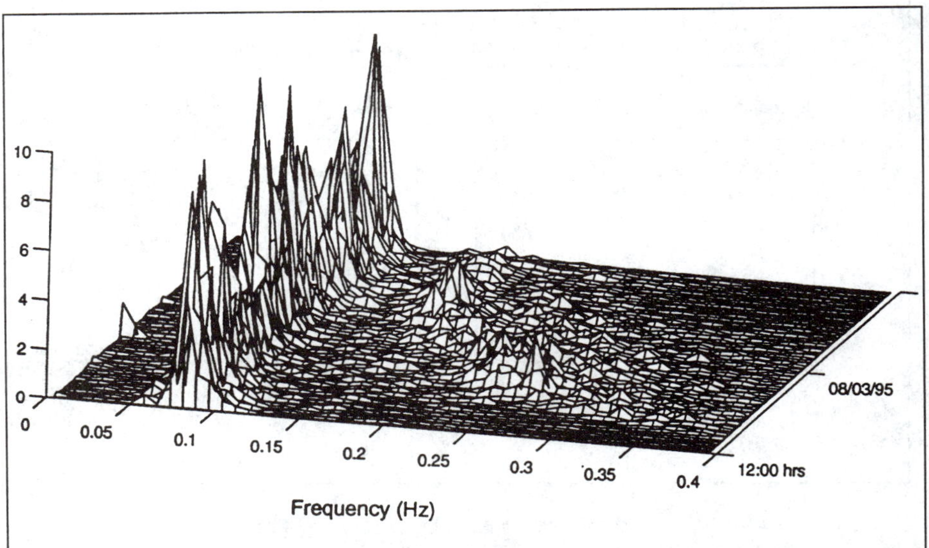

Figure 4- Time frequency plot of cross-shore currents

Experiment 3: Ambakandawila, Sri Lanka - 18-20 January 1996

Data collected from this experiment has not been thoroughly analysed and only a preliminary description is given here. On 18th January 1996, the sea breeze started at 1415 hours and 30 minutes later reached a maximum speed of 10-12 ms^{-1}. Continuous wind data is unavailable at present and the wind speeds quoted here were obtained using a hand-held anemometer on the beach. Coincident with the increase in wind speed, the longshore current increased from 0.05 to 0.75 ms^{-1} and then decreased just before midnight (Figure 5). The cross-shore currents (max = 0.23 ms^{-1}) and SSC (max = 3.8 gl^{-1}) indicated a similar trend (Figure 5). Pre-sea breeze values were reached prior to midnight. The exact time of the sea breeze cessation is unknown at present.

On the following day (19th) the sea breeze started earlier, at 1215 and continued blowing at 6-7 ms^{-1} throughout the afternoon. Here, although the sea breeze was weaker than the previous day, the recorded maximum values are higher (Figure 5): the longshore current max. is 0.80 ms^{-1}; max cross shore current 0.35 ms^{-1}; and, max. SSC 4 gl^{-1}. As per previous day, all values had returned to pre-sea breeze levels before midnight. On both days, the longshore suspended sediment flux (southward) increased by a factor of 60 during the sea breeze.

Discussion

This paper summarizes the results of three field experiments aimed at investigating the impact of sea breeze activity on nearshore processes. The findings are similar to those of Sonu et al. (1973), to date the only other study into sea breeze effects. In summary, the sea breeze results in: (1) an increase of the wave height; (2) a decrease in the wave period; (3) an intensification of the nearshore currents; and, (4) an increase in suspended sediment levels and suspended sediment transport. In Western Australia, due to the predominantly longshore component of the sea breeze,

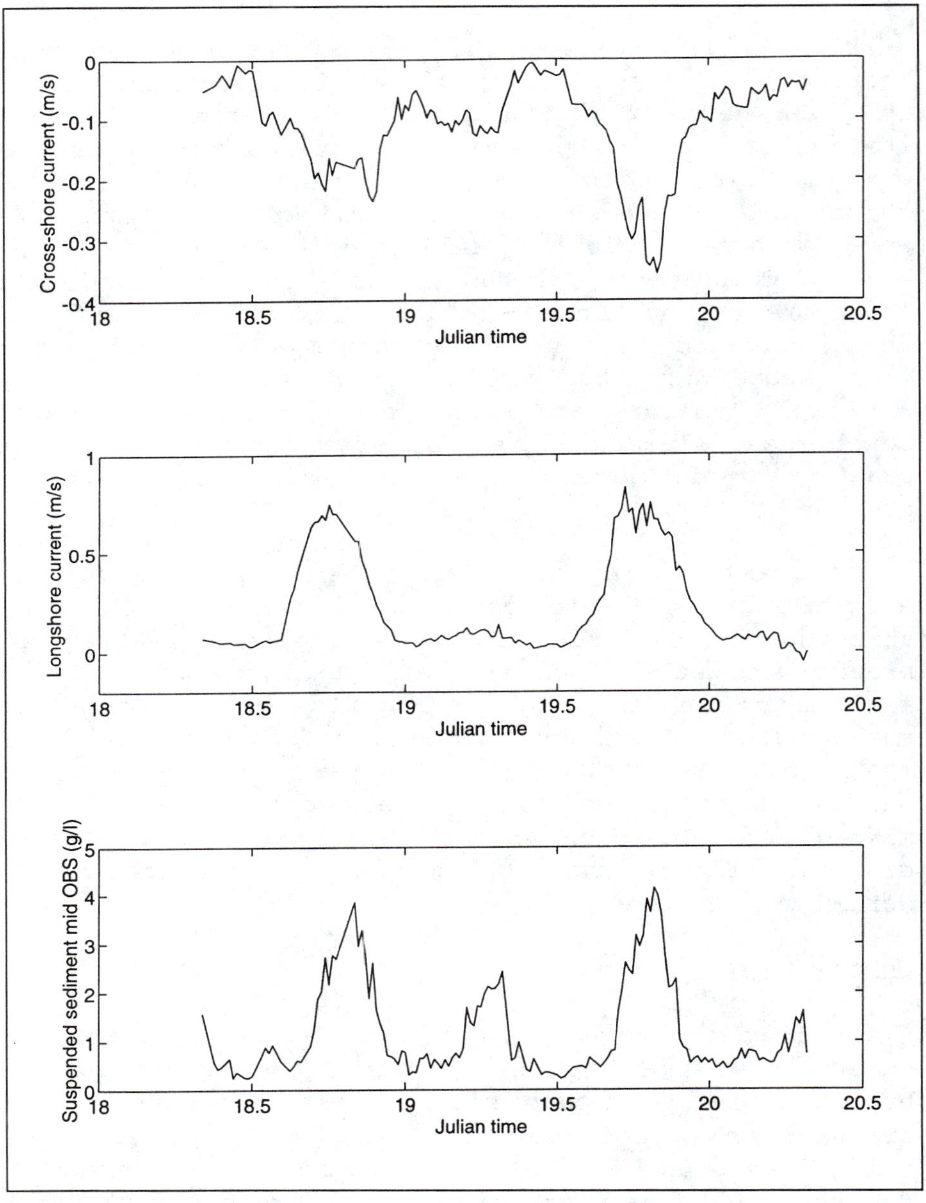

Figure 5 - Time series of cross-shore and longshore currents and SSC from Ambakandawila, Sri Lanka.

the nearshore hydrodynamics are affected long after the sea breeze has ceased to blow. Sonu et al. (1973)

refer to this as the "afterglow effect". Preliminary data from Sri Lanka, shown here, indicate that the impact of the sea breeze did not persist after midnight although visual observations of wave direction on the following morning indicated waves approaching from the direction of the sea breeze. Both the strength and consistency of the sea breeze, and the afterglow effect contribute to the important role that the sea breeze plays on the nearshore processes in both regions.

The sea breeze induces a diurnal cycle of beach change by causing erosion of the upper part of beach and/or planation of beach cusp morphology (Masselink et al., 1997). These changes are reversible in that the beach is usually restored to its pre-breeze state after the cessation of the sea breeze Pattiaratchi et al. (1997). On the larger temporal and spatial scale, the dramatic increase in the longshore sediment transport caused by the sea breeze is important. Masselink and Pattiaratchi (1997) and Pattiaratchi et al. (1997) estimate that along the stable Perth Metropolitan coastline, the annual longshore transport driven by the sea breeze is approximately 100,000 m^3. This estimate of longshore transport due to the sea breeze compared favourably with estimates of sediment accumulation at the southern side of Trigg Island (Perth) at the end of the summer. During the winter, the predominant north-westerly storms transport this sand southwards thus completing the seasonal sediment budget. Hence, it is clear that the sea breeze plays a dominant role in the sediment budget of the coastline around Perth.

Similarly, the sea breeze plays an important role in the annual longshore sediment budget along the west coast of Sri Lanka. During the south-west monsoon (May to September), the longshore sediment transport is predominantly northward. The occurrence of the sea breezes during the north-east monsoon period (October to February) would transport some of this sediment southward as shown by the results obtained from this study. Hence, also in this region, the sea breeze plays a dominant role in the annual sediment budget.

Conclusions

Field data presented from two different micro-tidal areas have shown the importance of the sea breeze system on nearshore processes. The sea breeze system induces a diurnal change in the incident wind wave climate (wave height and period) resulting in large increases to longshore and cross velocities and suspended sediment concentrations. The longshore suspended sediment flux can increase by a factor of 100 or more during the sea breeze which is very important in terms of annual sediment budgets.

Acknowledgments

A large number of individuals assisted in the data collection at the West Australian (WA) site and we would particularly like thank Dr Bruce Hegge for his efforts. Funding for the WA experiments was provided by an Individual Research grant awarded by the Division of Engineering and Computer Science at the University of Western Australia (UWA) to CP. Funding for the development of the S-probe was provided by UWA and the Centre for Environmental Fluid Dynamics. The Perth City Council is acknowledged for granting access to the field site. Funding for the field experiments at Ambakandawila (Sri Lanka) was provided by the Ministry of Fisheries and Aquatic Resources (Sri Lanka) as part of the ADB Chilaw Anchorage project to Lanka Hydraulic Institute and the Centre for Water Research. This paper has Centre for Water Research reference ED1243 CP.

References

Abbs D.J. and Physick W.L. 1992. Sea breeze observations and modelling: a review. Australian Meteorological Magazine, 41, 7-19.

Balopoulos E.Th., Collins M.B. and James A.E. 1986. Satellite images and their use in the numerical modelling of coastal processes. International J. Rem. Sens., 7, 905-919.

Banta R. 1995. Sea breezes shallow and deep on the California coast. Monthly Weather Rev., 123, 3614-3622.

Huntley D.A., Hendry M.D., Haines J. and Greenidge B. 1988. Waves and rip currents on a Caribbean pocket beach, Jamaica. J. Coast. Res., 4, 69-79.

Ludwig K. and Hanes D. 1990. A laboratory evaluation of optical backscatterance suspended sediment solid sensors exposed to sand-mud mixtures. Marine Geology, 94, 173-179.

Masselink G. and Pattiaratchi C.B. 1997. Morphodynamic impact of sea breeze activity on a beach with beach cusp morphology. J. of Coastal Research, (in press).

Masselink G., Hegge B. and Pattiaratchi C.B. 1997. Beach Cusp Morphodynamics. Earth surface processes and landforms, (in press).

Pattiaratchi C., Hegge B., Gould J and Eliot I. 1997. Impact of sea breeze activity on nearshore and foreshore coastal processes in southwestern Australia. Continental Shelf Res. (in press).

Prezerakos N.G. 1986. Characteristics of the sea breeze in Attica, Greece. Boundary Layer Meteorology, 36, 245-266.

Redano A., Cruz J. and Lorente J. 1991. Main features of the sea breeze in Barcelona. Meteorology and Atmospheric Physics, 46, 175-179.

Simpson J. E. 1994. Sea Breeze and Local Wind. Cambridge University Press, 234 pp.

Sonu C. J., Murray S. P., Hsu S. A., Suhayda J. N. and Waddell E. 1973. Sea breeze and coastal processes. EOS, Transactions American Geophysical Union 54:820-833.

Yoshikado H. and Kondo H. 1989. Inland penetration of the sea breeze over the suburban area of Tokyo. Boundary Layer Meteorology, 48, 389-407.

CHAPTER 326

WIND BLOWN SAND AT CASTROVILLE, CALIFORNIA

Douglas J. Sherman[1], Bernard O. Bauer[2], Paul A. Gares[3], Derek W.T. Jackson[4]

ABSTRACT

This paper reports the results of a study designed to evaluate the performance of a set of aeolian transport models, including moisture content corrections, as tested against field measurements. Empirical data were obtained during experiments conducted near Castroville, California, in January, 1993. The comparisons show that the Lettau and Lettau (1977) transport model, coupled with the Belly (1964) moisture correction model, produces estimates most closely approximating measured rates. Nevertheless, the advantages of this approach relative to using an "uncorrected" Bagnold model are small.

INTRODUCTION

Aeolian sand transport is an important component of the sediment budget of many coastal environments. Blowing sand may represent a hazard (Sherman and Nordstrom, 1994) or a resource, especially with regard to dune building and habitat creation (Carter, 1988). For these reasons, it is important to model the aeolian system in a manner consistent with the needs of coastal engineers and resource managers. From a geomorphological perspective, an understanding of the nature of aeolian systems yields insight into fundamental aspects of sediment transport and resulting

---

[1]Professor, Department of Geography, University of Southern California, Los Angeles, CA, 90089-0255, USA.

[2]Associate Professor, Department of Geography, University of Southern California, Los Angeles, CA, 90089-0255, USA.

[3]Associate Professor, Department of Geography, East Carolina University, Greenville, NC, 27858-4353, USA

[4]Science Officer, Department of Environmental Studies, University of Ulster, Coleraine, BT52 1SA Northern Ireland

landform change. Determining which model constitutes the most accurate and physically-meaningful approach to aeolian flux prediction is, therefore, a critical challenge.

A number of models have been developed for the purpose of predicting aeolian sand transport rates (see summary by Namikas and Sherman, in press). Those of most practical interest are based on a common set of assumptions: the wind is steady, uniform, and blowing across an unobstructed, horizontal surface of clean, dry, homogeneous sand. These conditions describe the most basic transport systems that one can posit. Models developed from these assumptions have been discussed extensively elsewhere (e.g., Anderson et al. 1991; McEwan and Willetts 1994; Lancaster 1995), and their derivations are not reviewed here. We note, however, that the idealized conditions they describe are seldom, if ever, appropriate to coastal environments. Complicating effects of sediment moisture content and surface slope, in particular, may exert considerable control on the behavior of the transport system.

Previous field experiments aimed at matching observed and predicted transport rates have usually found poor correspondence. Some successes have been reported in cases where empirical constants have been adjusted to improve the model's performance. Because such adjustments are usually site specific, they are of relatively diminished utility when little is known about the transport environment or where field work is impractical. Indeed, if we are always forced to perform field experiments to calibrate the equations, the practical value of modeling is severely constrained. The purpose of this paper is to report the predictive performances of several transport models, in light of field data and in consideration of slope effects and adjustments for moisture content. The objective is to test the models as they appear in the literature and as they might be employed in an engineering or management context.

MODEL DESCRIPTION

Several previous efforts at comparing observed and predicted transport rates have been reviewed recently by Sherman et al. (in press). Typically, these have omitted slope and/or moisture effects, have used liberally the empirical constants in the models, or have been predicated on potentially flawed experimental designs,. The Sherman et al. study suggests that, of the five models tested, the best statistical explanation (of the order of 20%) was afforded by the Bagnold (1936) and Zingg (1953) models. These two equations are essentially the same, varying only in the selection of the constant. Their conclusions point to the necessity for similar comparative studies using high quality data sets from other environments.

In the present study, we test the transport models of Bagnold (1936), Kawamura (1951), and Lettau and Lettau (1977). They were selected in conformity with the findings of Sherman et al. (in press) and because they are commonly used in coastal applications. We also considered the moisture-content correction equations of Belly (1964), Hotta et al. (1984), and Kawata and Tsuchiya (1976). For all tests, we assumed a sediment density, ρ_s, of 2650 kg m^{-3}, an air density, ρ, of 1.23 kg m^{-3}, and a gravity constant, g, of is 9.8 ms^{-2}. Sediment flux, q, is in kg m^{-1} s^{-1}. A brief description of the models follows.

One of the most frequently used models to predict transport rates is that of **Bagnold (1936)** who suggested that:

$$q = C\sqrt{\frac{d}{D}}\,\frac{\rho}{g}\,u_*^3 \quad \textbf{(1)}$$

where C is a surface-sediment-dependent constant ranging in value from 1.5 to 3.5 (C = 1.8 for typical dune sands and this value is used here), d is mean grain diameter, D is a reference grain diameter of 0.25 mm, and u_* is shear velocity.

Kawamura (1951) proposed a model that includes an explicit threshold shear-velocity term, u_{*t}.

$$q = C\frac{\rho}{g}\,(u_* - u_{*t})\,(u_* + u_{*t})^2 \quad \textbf{(2)}$$

In this expression, C = 2.78, and we adopt this value even though Horikawa et al. (1984) suggest that C = 1.0 might be better.

Lettau and Lettau (1977) proposed an expression with some of the attributes of both previous models:

$$q = C\sqrt{\frac{d}{D}}\,\frac{\rho}{g}\,(u_* - u_{*t})\,u_*^2 \quad \textbf{(3)}$$

where C in this case has a value of 4.2.

The Kawamura (1951) and Lettau and Lettau (1977) models require the threshold shear velocity to be estimated according to local sediment size characteristics. Bagnold (1936) suggested a Shields-type relation for u_{*t} of the following form:

$$u_{*t} = A\sqrt{gd\left(\frac{\rho_s - \rho}{\rho}\right)} \quad \textbf{(4)}$$

where A is a constant (assumed to be 0.085 during saltation).

When moisture is present in the interstices of surface sands, the resulting surface tension acts as a cohesive force. The shear velocity required to entrain sediments must therefore increase (e.g., review by Namikas and Sherman 1995). We considered three models developed to account for increased threshold shear velocity as a result of moisture content, u_{*tw}.

Belly (1964) proposed an empirical model based upon his wind tunnel experiments:

$$u_{*tw} = u_{*t}(1.8 + 0.6 \log_{10} w) \quad \textbf{(5)}$$

where w is percent moisture content by weight. This model applies to conditions where w does not exceed 4%.

Kawata and Tsuchiya (1976, described in Hotta at al., 1984) developed a model based upon direct consideration of the surface-tension-force distribution between particles:

$$U_{*tw} = U_{*t}\sqrt{1+B} \quad \textbf{(6)}$$

where

$$B = \frac{2\sqrt{6}}{5}\sqrt{\alpha_1\alpha_2}\sqrt{n_o}\sqrt{\frac{\rho_s}{\rho_w}\,\frac{T\sqrt{w}\cos\xi}{(\rho_s - \rho)gd}} \quad \textbf{(7)}$$

and α_1, α_2 are constants, n_o is the number of grain contact points, ρ_w is density of water, ξ is the water/grain contact angle, and T is the surface tension of water.

Hotta et al., (1984) proposed a wind-tunnel-derived empirical relationship for moisture effects:

$$U_{*tw} = U_{*t} + 7.5w \quad \textbf{(8)}$$

The prescribed applicability range for this model is 2% to 8% moisture content and sediment sizes between 0.2 mm and 0.8 mm.

In order to obtain the information necessary to evaluate the above models against prototype conditions, measurements or derivations of the following parameters are required: shear velocity, surface slope, mean grain size, transport rate, and moisture content. In order to attribute differences between observed and predicted transport rates to model performance, substantial care must be taken in experimental design. A field study was designed with the aim of obtaining suitable data.

STUDY SITE

Field experiments were conducted at a site along the coast of Monterey Bay, approximately mid-way between Santa Cruz and Monterey, near the town of Castroville. The site is exposed to open-ocean winds and waves from the southwest to northwest, and aeolian activity is therefore common. Instruments were deployed on the upper, flat portion of a well-developed, high-tide berm (Figure 1) fronting a small foredune system that is essentially continuous from the mouth of the Salinas River (to the south of the site) to Moss Landing (to the north). The upper foreshore is relatively steep and often cuspate, and it descends to a much flatter low-tide terrace merging, seaward, with bar topography that is periodically punctuated with rip channels. Morphodynamically, the system falls into the reflective regime during high tide and the dissipative regime during low tide stages. Spring high tides are often associated with foredune scarping and the deposition of flotsam (mostly kelp and woody debris) on the berm flat. Additional details of this site are described in Bauer et al. (1990) and Bauer (1991).

Figure 1: Castroville study site with instrument array.

EXPERIMENTAL DESIGN

Field experiments were conducted over a 30-day period in January, 1993. Only the January 13 data are presented here because this day had the longest duration of continuous sand transport during the study period. Further, winds were blowing nearly parallel to the shore so that it is reasonable to assume that local equilibria between the velocity profiles and the saltation system existed--that is, conditions were as close to ideal as one might expect for a beach environment. This eliminates complications arising from potential fetch effects (e.g. Nordstrom and Jackson 1992), or from internal-boundary-layer development (e.g. Bauer et al. 1996). Measurements began at 05:45 (PST), and continued for about the next five hours. A total of five data runs were completed, each with three to five u_*-q pairings.

Five sampling locations were established across the berm at 5 m intervals (Figure 2). Gill-type, three cup anemometers were hardwired to a PC-based data acquisition system. All sample runs were of 15 minute duration, with sampling at 1 Hz. Tower 1, near the toe of the foredune, had anemometers at elevations of 0.25, 0.5, 1.0, 2.0, 3.0, and 4.0 m above the sand surface. Towers 2, 3, and 4 had anemometers at 0.2, 0.6, 1.2, and 1.8 m. Tower 5 had anemometers at 0.15, 0.3, 0.6, 1.2, 1.8, and 2.4 m. The lowest Tower 5 anemometer was disabled during this experiment. Wind vanes were installed at the tops of Towers 1 and 5. A set of cylindrical sand traps was deployed adjacent to each tower. Moisture samples were taken at Towers 1 and 3 throughout the experiment, and at Tower 5 only near the end of the data collection, by taking scrapings from the upper 3-5 mm of the sand surface. Topographical slope was measured by rod and level survey.

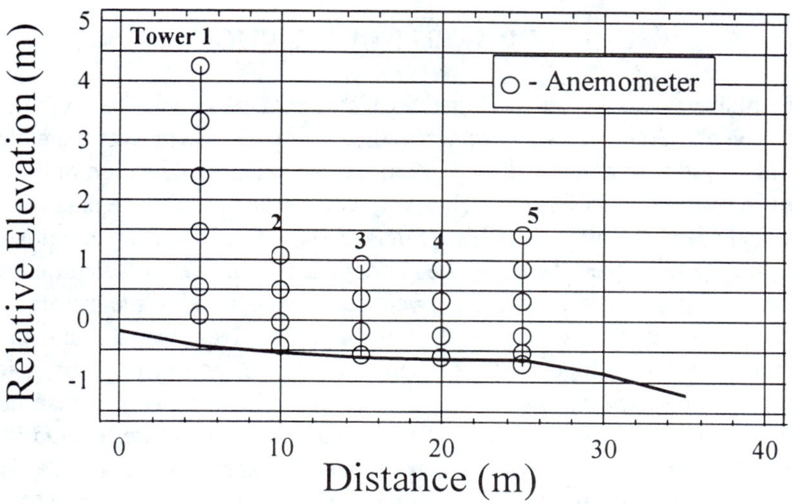

Figure 2: Instrument configuration on January 13, 1993.

Shear Velocity estimates were obtained through analysis of vertical wind-speed profiles. Linear regression was used to obtain "best fit" profile slopes, s, following the methodology described in Bauer et al. (1992). Assuming that the law of the wall applies, estimates of u_* were made using the relationship $u_* = \kappa s$, where κ is von

Karman's constant. To increase the statistical confidence of the u_* estimates, we restricted our analysis to instances where the R^2 associated with the best fit line was greater than 0.98.

Grain Size statistics were estimated using sand samples taken from the sediment traps. Samples were washed, dried, and sieved at 0.25ϕ intervals, and the data were processed according to the method of moments.

Moisture Content was estimated on the basis of pore-water weight relative to dry weight of the sediment sample. Samples were weighed, oven dried, and reweighed in order to solve:

$$w = 100\ \left(\frac{W_t - W_d}{W_d}\right) \tag{9}$$

where W_t is total sample weight and W_d is the dry weight of the sample.

Sediment Transport Rates were measured with cylindrical traps (Leatherman 1978; Rosen 1978). Traps were opened for 15-minute periods to correspond with wind sampling. Trapped sands were oven dried and weighed to obtain the transport rate data.

RESULTS AND MODEL EVALUATION

Grain size estimates were obtained from a total of 19 samples taken during the transport event. Grain-size statistics were surprisingly consistent over the entire measurement period with a mean 'd' of 0.12 mm and a standard deviation of 0.007 mm. **Moisture content** was also almost constant, with a mean 'w' for the data set of 6.9 % and a standard deviation of 0.7 %. **Surface slope** upwind of the instrument array was negligible, largely because the predominant wind fetch was oriented parallel to shore. Three shore-parallel lines were surveyed upwind of the instrument array, and they indicated a maximum relief of the berm flat of less than 0.3 m over a 50 m transect. Common slope-correction models indicate that such minor relief affects predicted transport rates by less than 1%. This value is much less than our confidence level for other measurements and estimates, and therefore, we did not investigate slope effects further. **Sediment transport rate** measurements were also obtained for 19 samples. In a recent field experiment, Greeley et al. (1996) found that the efficiency of the cylindrical traps is approximately 30%. Based on that finding, we adjusted our measured transport rates uniformly by 300%.

Our first evaluation was to test the three transport models without moisture (or slope) corrections. Comparison of observed and predicted rates are presented in Figure 3. It can be seen that the Kawamura model over predicts substantially, while the Bagnold and Lettau and Lettau models fit the observations relatively well. There is very little difference in the degree of statistical explanation associated with each model. With the Bagnold model, $R^2 = 0.50$; for Kawamura, $R^2 = 0.48$; and for Lettau and Lettau, $R^2 = 0.50$. The main difference between the model performances are associated with the slopes and offsets of best-fit lines.

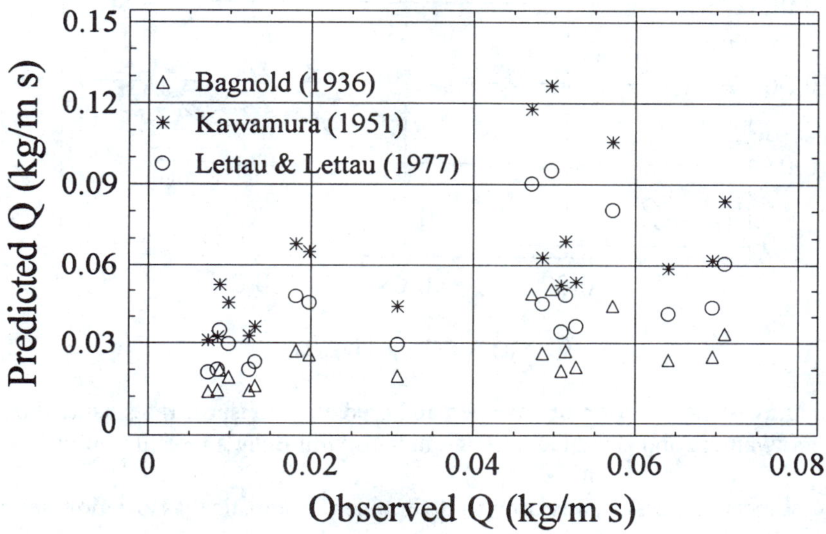

Figure 3: Comparison of observed and predicted transport rates without moisture content adjustments.

A second comparison incorporating moisture-content corrections was made. We evaluated the three moisture models described above (Equations 5 - 8) for their viability with the field data. In several instances, both the Hotta et al. and Kawata and Tsuchiya models yielded predictions of 'no transport' when our measurements clearly showed sediment movement. This peculiar situation arises when the threshold shear velocity predicted by these models on the basis of moisture content is greater than the observed shear velocities. We decided that this was one rational basis upon which the viability of the moisture models, vis-a-vis field data, could be distinguished. Since only the Belly (1964) model was immune to this problem, we used it with the transport equations. Note, however, that because there is no explicit threshold term in the Bagnold model (it is implicit in the parameterization of shear velocity), it was not considered appropriate to apply a moisture correction factor to his equation. Therefore, only the Kawamura and Lettau and Lettau models, with the Belly modification, were compared. Figure 4 shows that there is still considerable scatter in the relationship, with the main effect of accounting for moisture content being an overall reduction in predicted transport rate.

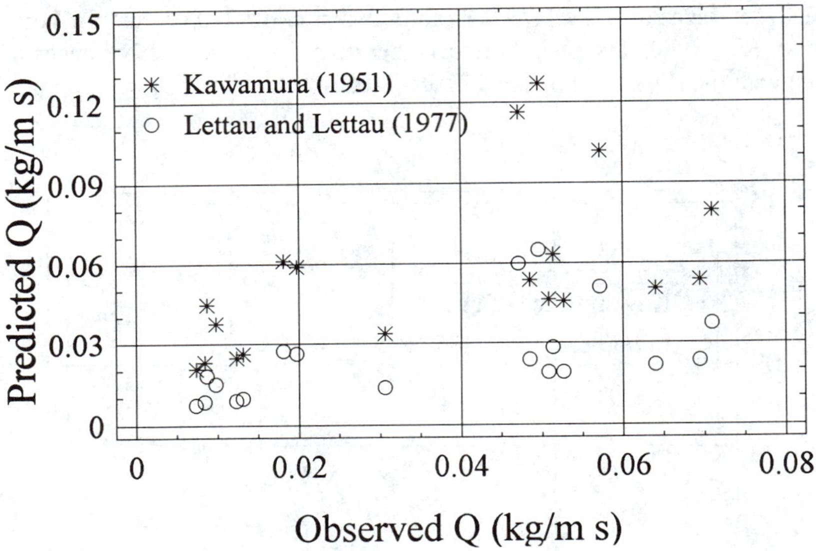

Figure 4: Comparison of observed and predicted transport rates using the Kawamura and Lettau and Lettau models with Belly's moisture model.

Linear regression was used to determine the statistical association between observed and predicted transport rates. On a log-log plot, R^2 for both models was 0.50. However, the Lettau and Lettau/Belly combination yielded predictions that were closer to a one-to-one correspondence with observations. This is best illustrated in Figure 5 using probability density plots (normalized frequency of occurrence) that compare smoothed distributions of ratios of predicted to observed values for both

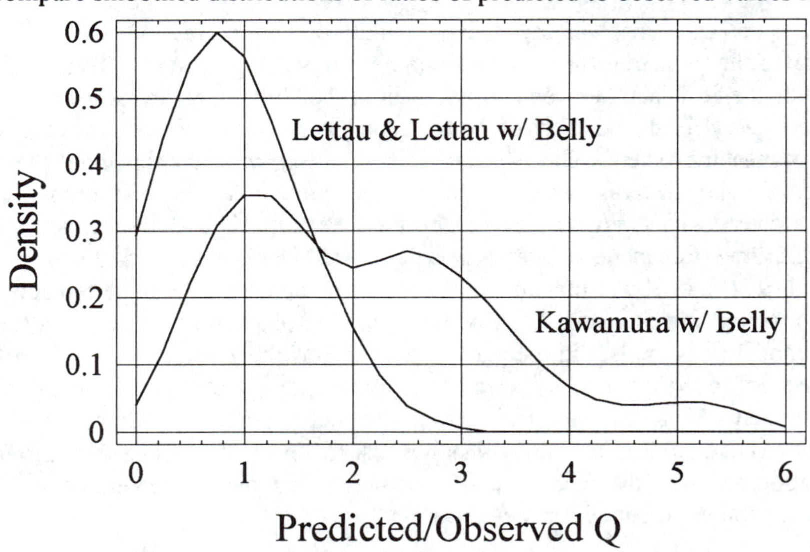

Figure 5: Comparison of ratios of predicted to observed transport rates.

models. An ideal model would have a narrow, peaked distribution centered around one--indicative of a near-one-to-one match between model and prototype conditions. Figure 5 shows that the Lettau and Lettau model, including the moisture content correction, approaches this ideal distribution, whereas the Kawamura and Belly combination over predicts substantially. These distributions of pairwise ratios of observed to predicted transport rates, coupled with the regression analysis, may represent one of the least ambiguous tests of comparative model performances.

DISCUSSION

All statistical comparisons between predicted and measured transport rates show that the scatter around a linear regression fit is approximately the same, regardless of transport model and moisture-content correction (R^2 values range from 0.48 to 0.50). Presumably, this reflects the similarity of the underlying physical bases for the models, and the fact that the variability in sediment moisture content during our experiments was small enough that the Belly model essentially acted as a near-uniform offset in the predictions--that is, all predicted transport rates were approximately uniformly reduced in magnitude.

It is encouraging that, under the conditions reported here, approximately half of the statistical variability in the transport rates monitored can be explained by the models evaluated. The uncertainty regarding the other half of the variability can be ascribed to several sources. First, our estimates of shear velocity were based upon line fitting procedures that, by their very nature, incorporate some degree of statistical uncertainty. Although this uncertainty has been minimized by excluding all wind profiles where R^2 was less than 0.99, it is not totally eliminated (e.g. Bauer et al. 1992). Because errors in estimates of u_* are compounded in the transport models (because third powers of shear velocity are used), 10 % errors translate to about 30% errors in predicted transport rates.

Second, there is substantial uncertainty associated with the measurement of sediment flux using cylindrical-type traps. We implemented a first-order correction to our measurements based on the findings of Greeley et al. (1996), but this is a crude adjustment that forces us to presume that trap efficiency is constant for the range of transport rates. This remains a point of contention.

Third, we feel some degree of unease with the universality of the models themselves and their underlying assumptions concerning grain-size distributions (only mean values are used, and sorting effects are ignored) and the steadiness of the wind field (steady-state conditions are assumed). We are also unable to make substantive distinctions among the physical--theoretical foundations of the different approaches which might provide guidelines as to their applicability. Thus, we are left with statistical performance assessments only.

It is noteworthy that whereas the use of the Lettau and Lettau model with

Belly's moisture correction provides the best predictions of transport for the Castroville system, the results are only marginally better than those obtained using the simple Bagnold model. Figure 6 compares the probability density distributions for the two approaches--the models perform almost identically for this range of conditions. The over-predictions obtained using the Kawamura model may also reflect the use of C = 2.78, instead of the more conservative estimate of 1.0 recommended by Horikawa et al. (1984). The latter value brings the Kawamura results into much closer correspondence with the other models and with our observations.

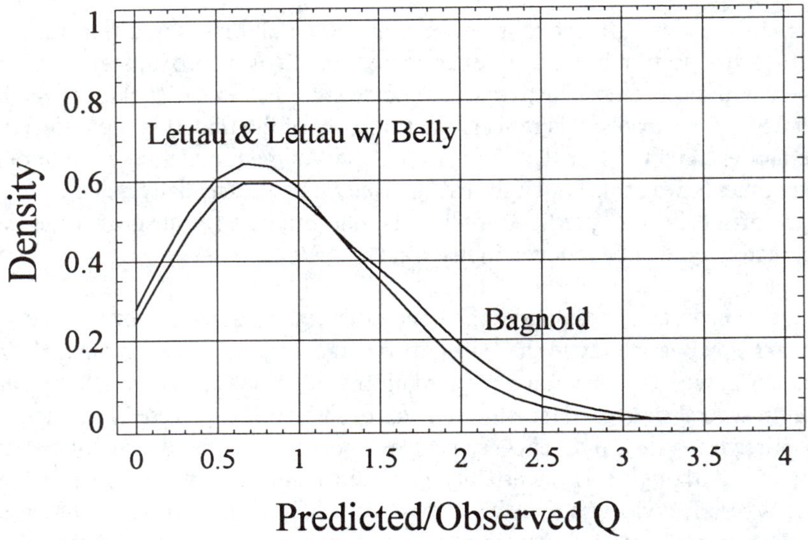

Figure 6: Comparison of ratios of predicted to observed transport rates.

Our results also compare favorably with those of Sherman et al. (in press). In that study of across shore transport conditions, the best models (Bagnold and Zingg) explained only about 20% of the statistical variability found in the observations. The Kawamura and Lettau and Lettau models each accounted for about 15% of the variability in prototype transport rates.

CONCLUSIONS

Conclusions that arise from this study include:

1. Under conditions of relatively uniform slope, grain size, and sediment moisture content, either the Bagnold, Kawamura (with C = 1.0), or Lettau and Lettau models provide reasonable estimates of aeolian sand transport. The performance of the Lettau and Lettau model can be enhanced through its coupling with the Belly moisture

model. Any of these model combinations is able to explain about 50% of the variability in transport rates observed during this part of the Castroville experiments.

2. We believe that it is possible, with carefully designed field experiments, to obtain process data of a quality comparable to that obtained in the laboratory. Nevertheless, continued uncertainty regarding the performance of field instrumentation in general and sediment traps in particular confounds our ability to assess accurately aeolian processes across beaches. The inherent complexity of natural systems may never be captured completely.

ACKNOWLEDGMENTS

This project received financial support through a contract from the California Department of Boating and Waterways and a grant from the North Atlantic Treaty Organization. We are also grateful to Andy Smith and the Monterey Dunes Colony Association for pleasant accommodation, and to the California Department of Parks and Recreation for beach access. Able field assistance was provided by Kevin Parnell, John Wolcott, Deming He, Jianchun Yi, Jinkang Wang, James Tyner, and Steven Namikas.

REFERENCES

Anderson, R.S., M. Sorenson, and B.B. Willetts 1991. A review of recent progress in our understanding of aeolian sediment transport. *Acta Mechanica*, Sup. 1: 1-19.

Bagnold, R.A. 1936. The Movement of Desert Sand. *Proc. Roy. Soc. London.* A157: 594-620.

Bauer, B.O. 1991. Aeolian Decoupling of Beach Sediments. *Annals, Association of American Geographers*, 81(2): 290-303.

Bauer, B.O., R.G.D. Davidson-Arnott, K.F. Nordstrom, J. Ollerhead, and N.L. Jackson. Indeterminancy in aeolian sediment transport across beaches. *J. Coastal Research*, 12: 641-653.

Bauer, B.O., D.J. Sherman, K.F. Nordstrom, and P.A. Gares. 1990. Aeolian transport and measurement across a beach and dune at Castroville, California. In Nordstrom, K.F., N.P. Psuty, and R.W.G. Carter (eds.), *Coastal Dunes: Form and Process*. New York, Wiley, pp. 39-55.

Bauer, B.O., D.J. Sherman, and J.F. Wolcott 1992. Sources of uncertainty in shear stress and roughness length estimates derived from velocity profiles. *Professional Geographer* 44: 453-464.

Belly, P.-Y. 1964. *Sand Movement by Wind.* U.S. Army Corps Eng. CERC. Tech. Mem. 1. Washington D.C. 38 pp.

Carter, R.W.G. 1988. *Coastal Environments: An Introduction to the Physical, Ecological, and Cultural Systems of Coastlines*. Academic Press, London. 617 pp.

Gares, P.A. 1988. Factors affecting eolian sediment transport in beach and dune environments. *J. Coastal Research*, SI3: 121-126.

Greeley, R., D.G. Blumberg, and S.H. Williams 1996. Field measurements of the flux and speed of wind-blown sand. *Sedimentology* 43: 41-52.

Hesp, P.A. 1988. Surfzone, beach and foredune interactions on the Australian SE coast. *J. Coastal Research* SI3: 15-25.

Horikawa, K., S. Hotta, S. Kubota, and S. Katori. 1984. Field Measurement of Blown Sand Transport Rate by Trench Trap. *Coastal Eng. Jpn.* 27: 214-232.

Hotta, S., S. Kubota, S. Katori, and K. Horikawa 1984. Sand transport by wind on a wet sand surface. *Proc., 19th Conf., Coastal Engineering*, ASCE., New York, pp. 1263-1281.

Kawamura, R. 1951. *Study of Sand Movement by Wind.* Translated (1965) as University of California Hydraulics Engineering Laboratory Report HEL 2-8, Berkeley.

Kawata, Y., and Y. Tsuchiya 1976. Influence of water content on the threshold of sand movement and the rate of sand transport in blown sand. *Proc., Japanese Soc. Civil Engineers* 249: 95-100 (in japanese).

Lancaster, 1995. *Geomorphology of Desert Dunes*, Routledge, London, UK. 290 pp.

Leatherman, S.P. 1978. A New Aeolian Sand Trap Design. *Sedimentology* 25: 303-6

Lettau, K., and H. Lettau. 1977. Experimental and Micrometeorological Field Studies of Dune Migration. In K. Lettau and H. Lettau (eds.): *Exploring the World's Driest Climate*. University of Wisconsin-Madison, IES Report 101. 110-147.

McEwan, I.K., and B.B. Willetts 1994. On the prediction of bed-load sand transport rate in air. *Sedimentology* 41: 1241-1251.

Namikas, S.L. and D.J. Sherman, in press. AEOLUS II: An interactive aeolian sedimentation simulation program. *Geomorphology*

Namikas, S.L. and D.J. Sherman 1995. A review of the effects of surface moisture content on aeolian sand transport. In V.P. Tchakerian (ed.), *Desert Aeolian Processes*. New York, Chapman and Hall pp. 269-293.

Nordstrom, K.F., and N.L. Jackson 1992. Effect of source width and tidal elevation changes on eolian transport on an estuarine beach. *Sedimentology*, 39: 769-778.

Rosen, P.S. 1978. An efficient, low-cost aeolian sampling system. *Current Res. A*. Geol. Surv. Can. Pap. 78-1A. 531-2.

Sherman, D.J., D.W.T. Jackson, S.L. Namikas, and J. Wang, in press. Aeolian sediment transport to the coastal dune complex at Inch, Co. Kerry, Ireland. *Geomorphology.*

Sherman, D.J., and K.F. Nordstrom 1995. Wind-blown sand hazard in the coastal zone: a review. *J. Coastal Research* SI12: 263-275.

Zingg, A.W. 1953. Wind tunnel studies of the movement of sedimentary material. *Proc. 5th Hydraulics Conf.* Bull. 34. Iowa City: Inst. of Hydraulics. 111-135.

Heavy surf at Golden Gate Park, San Francisco, California. Photo courtesy of Robert L. Wiegel.

PART V

Coastal, Estuarine and Environmental Problems

Fisher Island adjacent to Miami Beach, Florida. Photo courtesy of Olsen Associates, Inc.

CHAPTER 327

PIPELINE PROTECTION IN THE SURF ZONE

Gerrit J. Schiereck, Henri L. Fontijn[1]

ABSTRACT

Stability relations for rock on a mild slope in breaking waves were investigated, both experimentally and theoretically. Assumptions were made for the influence of turbulence in breaking waves on the load exerted by the wave motion. It appears that with these asssumptions, the mechanism is reasonably described in a qualitative way. For design purposes the method is not accurate enough. This is possibly due to the neglection of the (vertical) velocity field near the bottom in a breaking wave, giving an underestimation of the difference in stability in spilling or plunging breakers. The experimental results seem consistent and can be used provisionally for design purposes. An interesting point is that they also are in line with existing relations for stability on steep slopes.

ACKNOWLEDGEMENT

The authors thank prof.dr. Marcel Stive of Delft Hydraulics for his permission to use the wave model ENDEC. ACZ Marine Contractors is thanked for their initiative on the study on pipeline protection and the continuous interest in the progress.

1. INTRODUCTION

Pipelines on the sea bottom are usually protected in order to prevent damage by anchors or erosion. Where a pipeline crosses a beach, it often lays in a dredged trench, see Figure 1, and is covered with stones. For the design of such a protection, which can be seen as an armour layer on a mild slope, a provisional design rule was established, see Schiereck et al.,1994, based on theoretical considerations and experiments. For non-breaking waves on a mild slope, the

---

[1]*Delft University of Technology, Faculty of Civil Engineering, P.O. Box 5046, 2600 GA Delft, The Netherlands*

Figure 1 Outfall protection

result was reasonably satisfying. For the stability in breaking waves no theoretical concept was available and the number of experiments was not sufficient to give reliable results. In addition, new experiments were carried out, in which the slope, the stone density and the stone size were varied. Also, an attempt was made to derive a theoretical relationship for the stability of stones in breaking waves. The purpose of this attempt was twofold. First, experimental results that can be explained from theory are better understood, decreasing the danger of misusing empirical relations, while, vice versa, theories that can be verified by experiments get more practical value in hydraulic engineering. The second reason comes from didactics. Hydraulic engineering is still heavily based on empirical relations. Presenting all these relations without a link to the theory on fluid motion is considered a weak point in academical engineering education.

2. APPROACH

Experiments are indispensable to establish design rules in hydraulic engineering, so, laboratory tests are the basis of the research in this paper. But, as already stated in the introduction, the experimental results should be connected with the physical background of forces due to moving water. The creation of a link between the fluid motion and experimental results is tried with a simple, but complete description of the phenomena involved. The stability of stones on a slope is governed by the relation between load and strength. The strength is usually satisfactorily described with the effective weight and the friction of the stones. The load is much more complex. The moving water in a breaking wave will exert forces on a stone. Even in a breaking wave, the orbital velocities will play a role in the velocity field. Also, the breaking will cause turbulent eddies, with their own velocity field. The whole of orbital and turbulent velocity field is responsible for the loading on a stone. Another complicating factor is that waves in nature are irregular. Therefore, some statistal description of the waves is necessary.

In the computations the load will be composed of forces caused by the orbital motion, in combination with turbulent velocities due to breaking. For the stability, existing relations between the load and strength of a stone layer in an oscillating water motion will be used.

3. BASIC EQUATIONS

Orbital motion
The oscillatory flow near the bottom is approached with the linear wave theory:

$$\hat{u}_b = \omega\, a_b = \frac{\omega H}{2} \frac{1}{\sinh kh} \tag{1}$$

Turbulent velocities
For the turbulent velocities, an approach as given by Battjes,1975 and Battjes,1987 is used. Battjes coupled the rate of production of turbulence energy to the rate of dissipation of wave energy due to breaking:

$$q \approx (D_B / \rho_w)^{1/3} \tag{2}$$

in which q is the turbulent velocity scale ($q^2 = u_i u_j$). Figure 2 shows a comparison between measured and computed turbulent velocity scale. In this paper the expression for q, equation (2), will be used as a measure for the turbulent velocity.

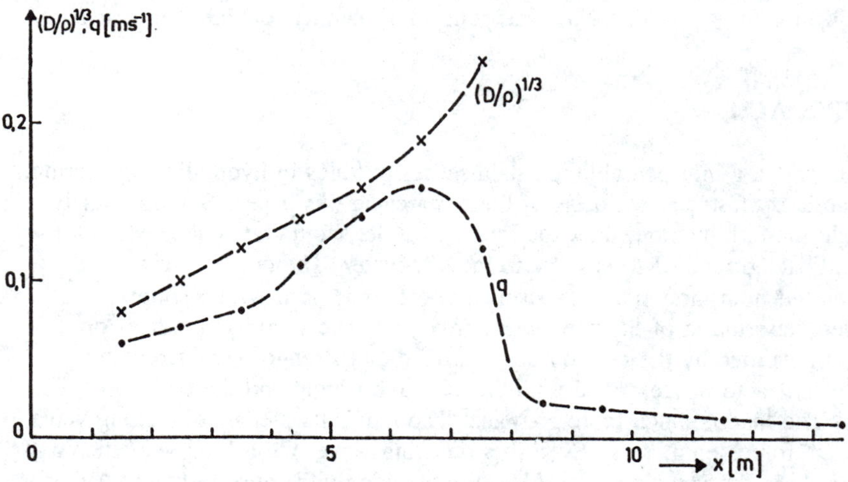

Figure 2 Comparison of computed and measured turbulent velocity scale

The dissipation of wave energy is derived from the analogy between a bore and a breaking wave, see Battjes and Jansen,1978:

$$D_B = \frac{1}{4} \frac{Q_B}{T_p} \rho_w g H_M^2 \tag{3}$$

in which Q_B is the fraction of the (irregular) waves that are broken, derived from:

$$\frac{1 - Q_B}{\ln Q_B} = -\left(\frac{H_{RMS}}{H_M}\right)^2 \tag{4}$$

and where H_M is the maximum wave height, given by:

$$H_M = 0.88 * k^{-1} * \tanh(k\,h) \tag{5}$$

For more details, see Battjes & Janssen,1978. Several models are available in which this concept is implemented, e.g. the 2-dimensional DUT model HISWA, see Holthuijsen et al, 1989. In this study the 1-dimensional model ENDEC (Delft Hydraulics, see Battjes & Stive,1985) was used to compute the various wave parameters along the different slopes, since in this model the wave set-up is explicitely computed, which possibly could be important.

Wave height distribution

As a basis for the wave height distribution the Rayleigh distribution is taken. This distribution is also used in Battjes and Janssen,1978. In shallow water, the wave height distribution deviates from the Rayleigh distribution. $H_{1\%}$ in shallow water, which plays an important role in the stability calculations, is given by, according to Stive, see CUR/CIRIA, 1991:

$$H_{1\%-shallow} = \frac{H_{1\%-Rayleigh}}{\left(1 + H_S/h\right)^{1/3}} \approx \frac{1.5 * H_S}{\left(1 + H_S/h\right)^{1/3}} \tag{6}$$

Load - strength relations

A simple relation to express the stability of stones in oscillating flow is based on experiments in an oscillating water tunnel by Rance and Warren,1968 (see also Schiereck et al.,1994):

$$\frac{a_b}{T^2 \Delta g} = 0.025 \left[\frac{a_b}{d_{50}}\right]^{-\frac{2}{3}} \tag{7}$$

Another approach is given by Sleath,1978. Analogous to the Shields approach in uniform flow, Sleath gives a relation between the shear stress (which is not necessarily the dominant load) and the stone weight, partly based on the experimental data by Rance & Warren (1968). The relation for stones reads:

$$\hat{\tau}_b = 0.056 \,.\, (\rho_s - \rho_w) \cdot g \cdot d \tag{8}$$

d is the equivalent spherical diameter, in this paper approximated by d_{50}, the median sieve diameter, which is easily available and differs only a few percent from the spherical diameter. The shear stress due to orbital velocities can be expressed by:

$$\hat{\tau}_b = \frac{1}{2} \cdot \rho_w \cdot f_w \cdot \hat{u}_b^2 \qquad (9)$$

with f_w and $\hat{u}_b$ depending on the wave height, H, and the period, T. The circumflex over a parameter denotes "amplitude of". Given a certain wave height, the longer the period is, the larger the orbital velocity at the bottom, $\hat{u}_b$. For f_w, the friction coefficient, the opposite holds: the shorter the period, the larger the friction coefficient. In CUR/CIRIA (1991) an expression by Swart is given, based on Jonsson (1966), where f_w is given as a function of the orbital stroke at the bottom, related to the bottom roughness:

$$f_w = \exp[-6 + 5.2 (\frac{a_b}{k_s})^{-0.194}] \qquad (f_{w\,max} = 0.3) \qquad (10)$$

Computations
The combination of orbital velocities, turbulent velocities, wave height distribution and load-strength relations into a design procedure can be done in various ways. In section 5, this is further elaborated.

4. EXPERIMENTS

Experiments were done in a wave tank (length 40 m, width 0.8 m, depth 0.9 m) at the Laboratory of Fluid Mechanics, Delft University of Technology (DUT), see Ye,1996. The slope angles were 1:10 and 1:25. The mass densities of the stones were 2550 and 2850 kg/m^3 while the nominal diameters, dn_{50}, varied between 1 and 1.5 cm. The width of the sieve curves of the stones (d_{85}/d_{15}) used in the experiments was about 1.5. 3 to 4 layers of stone were used, in order to ascertain a proper roughness between stones and slope. The difference with the geometry of a real pipeline cover, which has a filter layer under the top layer, is assumed to be negligible with respect to the stability of the top layer.

The stones were laid in coloured strips of 0.25 m (measured in the wave direction) over the full width of the flume. The total number of stones displaced after every test, n, divided by the number of stones in the width of the flume, was used as the total damage S:

$$S = n\, d_{n50} / w \qquad (11)$$

in which w is the width of the flume. An arbitrary damage level of 2 was chosen for incipient motion.

Irregular waves were generated according to a JONSWAP-spectrum; the number of waves tested per spectrum was 2000. The wave heights and spectra were determined at the toe of the slope. The water depth at that location was 0.6 m.

Figure 3 shows the results of the experiments. The stability is expressed as $H_s/\Delta d$, in which H_s is the significant wave height at the toe of the slope. The stability is plotted against the breaker parameter, ξ, the slope angle relative to the wave steepness.

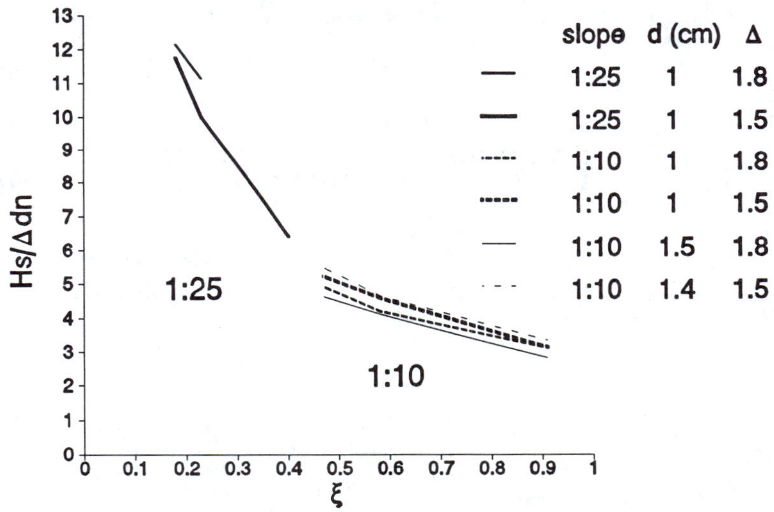

Figure 3 Experimental results

The general tendency is an increase in stability with decreasing ξ. This can partly be explained from the slope angle: the left-hand side of the figure represents the results for the slope 1:25 and the right-hand side those for 1:10. Within the results for each slope angle, there is also clearly an influence of the wave steepness: the larger the wave steepness, the larger the stability. This tendency has to do with the breaker characteristics and was previously found in other experiments, see e.g. Schiereck et al., 1994 and van der Meer,1988. Within this general tendency, there is some influence of stone density and stone dimension. Whether this is a matter of accuracy or comes from physical reasons, remains to be seen.

5. COMPARISON OF COMPUTATIONS AND EXPERIMENTS

In order to compare the computational results, obtained with the approach described in section 2, with the experimental results, at first the wave parameters along the slope were computed with ENDEC, using the measured wave characteristics at the toe of the slope at the threshold of motion. From previous investigations (see Schiereck et al.,1994) it appeared that in irregular waves, the higher waves are responsible for the incipient motion, in particular the wave height that is exceeded by 1% of the waves. This wave height is computed with equation (6) for various locations along the slope. The orbital velocities at the bottom in these waves were computed with equation (1).

Rance & Warren

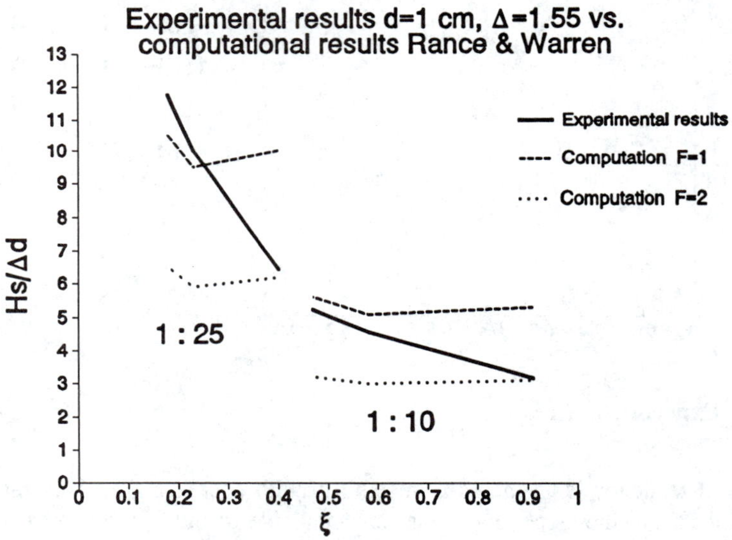

Figure 4 Comparison of experiments with computation according to Rance & Warren

The necessary stone diameter is first computed with the relation of Rance & Warren. With $\hat{u}_b = \omega * a_b$, equation (7) is rewritten as:

$$d_{50} = \frac{2.56 * \hat{u}_b^{2.5}}{\sqrt{T_p} * (\Delta\, g)^{1.5}} \tag{12}$$

This equation is valid for a horizontal bottom. The diameter in this formula is the median sieve diameter, d_{50}. In order to compare the computational results with the experiments, the nominal diameter (dn_{50}) is required, which is approximately 0.84*d_{50}. The turbulent velocity, from equation (2), is simply added to the orbital velocity from equation (1). Together with a correction for the influence of the slope angle on the stability, the equation finally becomes:

$$dn_{50} = \frac{0.84 * 2.56 * (\hat{u}_b + F * q)^{2.5}}{\sqrt{T_P} * (\Delta\, g)^{1.5}} * \frac{\sin\phi}{\sin(\phi - \alpha)} \tag{13}$$

in which F is a calibration factor and ϕ is the angle of repose of the stones, here taken as 45°. q, as defined in equation (2), is taken as the turbulent velocity.

With this equation the necessary diameter along the slope is computed. The maximum computed diameter is used, which is equivalent to the use of a total-damage concept in the experiments, regardless of the location of the damage. For comparison, the results of the experiments with ρ_S = 2550 kg/m³ and dn_{50} = 1 cm are being used. Figure 4 shows the results for F = 1 and F = 2. The influence of the slope angle, as seen in the experiments, is also found in the computations, but the influence of the wave steepness is not reproduced correctly.

Jonsson/Sleath

The relationship as found from the results of Rance & Warren (equation (7), expresses the relation between the stroke of the orbital motion and the necessary diameter to prevent the incipient movement of the stones. This implicitely indicates the influence of the inertia of the orbital motion on the stone stability. The simple addition of a turbulent velocity to the orbital velocity, as done in equation (13), attributes the same influence to the turbulent velocity, which is not logic. Another approach is the following. Consider the forces on a grain in a flow field, see Figure 5. The shear force is exerted by the orbital movement, described with equations (9), (1) and (10). Equation (8) gives the relation between shear stress and stone size for incipient motion. This equation is now rewritten for the equilibrium of forces, where the turbulent velocity is assumed to generate a lift force, reducing the effective weight of the stone. In this way, the influence of turbulence is treated separately from the orbital motion with its specific relation for the friction factor, equation (10).

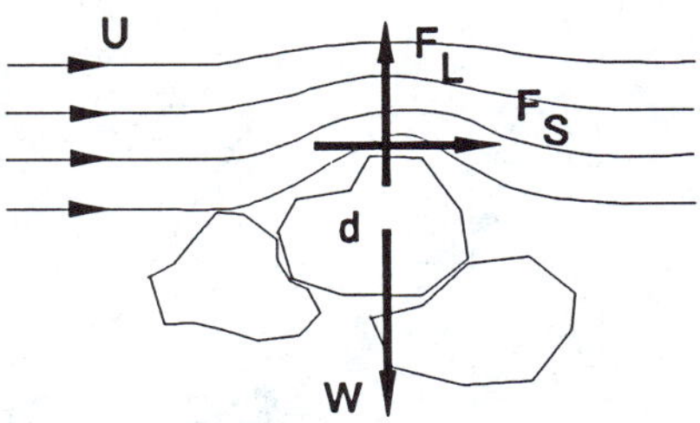

Figure 5 Flow forces on a grain

The equilibrium of forces then leads to the following proportionality:

$$\frac{1}{2}\rho_w f_w u_b^2 A_h \propto (\rho_s - \rho_w) g V - \frac{1}{2}\rho_w C_L A_h q^2 \qquad (14)$$

in which A_h is a representative horizontal area, both for shear and lift. V is the volume of a stone and C_L the lift coefficient. Using equation (8), this leads to:

$$\frac{1}{2}\rho_w f_w u_b^2 = 0.056\,(\rho_s - \rho_w) g d - \frac{1}{2}\rho_w C_L q^2 \qquad (15)$$

giving finally:

$$dn_{50} = \frac{0.84\,\frac{1}{2}(f_w \hat{u}_b^2 + F q^2)}{0.056\,\Delta\, g} * \frac{\sin\phi}{\sin(\phi - \alpha)} \qquad (16)$$

in which C_L is replaced by F, a calibration factor in which both the lift coefficient and the transfer from wave energy dissipation into turbulence is included.

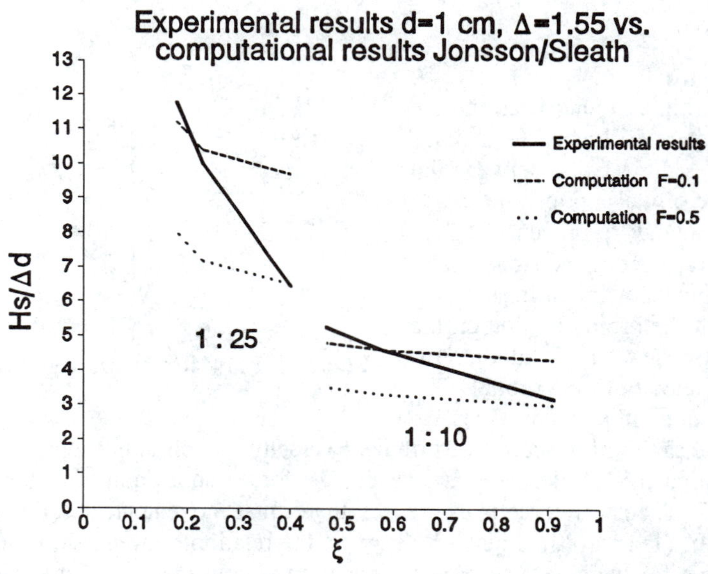

Figure 6 Experimental results compared with computations according to Jonsson/Sleath

Figure 6 shows the results for F = 0.1 and F = 0.5. The agreement is somewhat better than with the relationship of Rance & Warren. The influence of the slope angle is represented just as good as for the Rance & Warren relation, while the tendency of the influence of the wave steepness is qualitatively correct. The difference in stability between low and high values of ξ, however, for one value of the slope angle, is too small in the computations.

Variable friction factor in ENDEC

Until now, in ENDEC only one friction factor has been used, viz. 0.05, which is a relatively high value due to the rough bottom. Equation (10) leads to higher friction factors for steeper waves. Using different friction factors for different wave steepness, (f_w = 0.05, 0.04, 0.03 for s = 0.05, 0.03, 0.01, respectively) is justified, because of equation (10). For the calibration factor F in the computations 0.25 is used, being a value in between the two values of Figure 6.

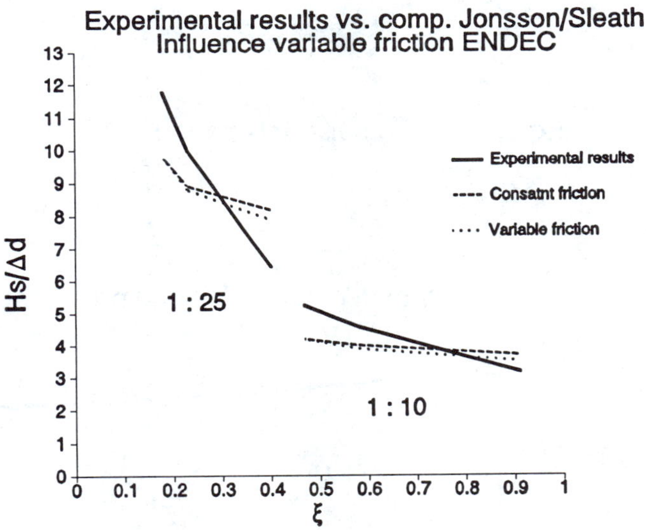

Figure 7 Variable friction factor in ENDEC

The results are now a little bit better, see Figure 7. The models being used are very simple. Considering the fact that the values of the calibration factor F are realistic with respect to the lift coefficient, the description is quite reasonable in a qualitative way. From the point of view of design practice, however, the results are still far from perfect.

The same underestimation of the favourable effect of high wave steepness was found in non-breaking waves, see Schiereck et al., 1994. The explanation has probably to be found in the negelection of the vertical flow pattern in breaking waves. Figure 8 shows a plunging breaker in which a jet obliquely to the bottom occurs, while the jet is absent in a spilling wave. This jet may have an unfavourable influence on the stone stability, which is not described in the models used here. This difference will only appear when using a 2-dimensional wave model for the wave motion on a slope. The agreement so far is encouraging enough to try to couple the experimental results to the water motion, using a better wave-velocity model.

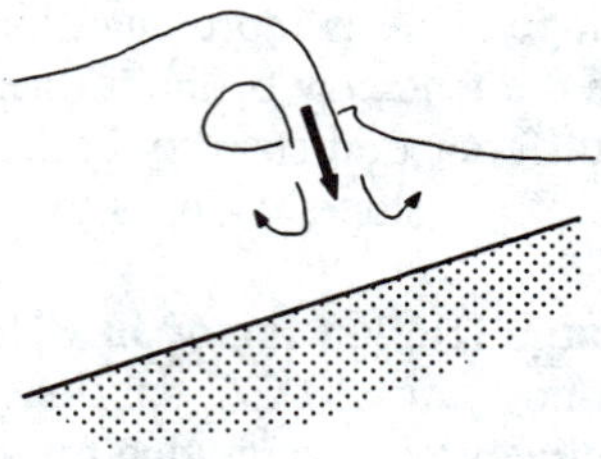

Figure 8 Plunging jet

6. EVALUATION

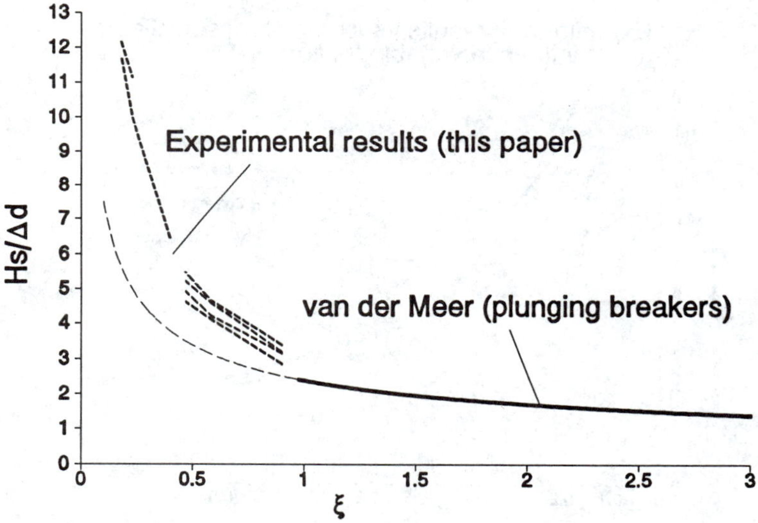

Figure 9 Experimental results compared with van der Meer formula for plunging breakers

Figure 9 shows the experimental results discussed before. The investigations were done in the range $\xi = 0.1 - 1.0$. The figure also shows the relation for stone stability on relatively steep slopes as proposed by van der Meer,1988. This relation was validated for values $\xi > 1$. The trend for both situations is the same, which encourages further research in this field. Together with a 2-dimensional model, describing the wave motion on a slope in more detail than the models used in this paper, it should be possible in the future to give a satisfactorily accurate description of stone stability in (breaking) waves on slopes. For the time being, the experimental results as presented in this paper can be used as a first approximation for stone stability on mild slopes.

7. CONCLUSIONS

1- The computational results give a description that follow the trends of experimental results reasonably well in a qualitative way, when including the following elements from the physical process of the stability of stones in breaking waves on a mild slope:
- orbital movement (from linear wave theory, equation 1)
- wave shear stress (according to Jonssson, equation 10 and using equation 9)
- Rayleigh distribution (with shallow-water correction by Stive, equation 6)
- wave breaking and energy dissipation (according to Battjes/Janssen, equations 3, 4 and 5)
- turbulent velocities (according to Battjes, equation 2)
- stone stability (according to Sleath, equation 8)

.

2- The computed relation between stone stability ($H_S/\Delta d$) and breaker parameter ($\xi = \tan\alpha/\sqrt{(H_S/L_0)}$, is quantitatively insufficient. Probably the fact that the vertical velocities near the bottom in a plunging breaker were not taken into account, is the main reason for this. Other weak points are possibly the turbulence model used and the influence of turbulence on the stability of a single stone.

3- The experimental results seem consistent and are also in line with the (empirical) van der Meer relation for stability of stones on steep slopes. This ressemblance can be used as a basis for future research on stone stability in breaking waves on slopes in general.

4- The experimental results in this paper can be used for the design of a protective rock layer on a mild slope.

SYMBOLS

a_b	orbital stroke at bottom		m
d_{n50}	median nominal diameter of material	$(d_{n50}=(M_{50}/\rho_s)^{0.33})$	m
d_{50}	median sieve diameter of material		m
D_B	energy dissipation due to breaking of waves		$Nm/s/m^2$
f_w	friction coefficient in waves		-
F	tuning factor		-
g	acceleration due to gravity		m/s^2
H	wave height		m
H_M	maximum wave height		m
H_S	significant wave height		m
h	water depth		m
k	wave number	$(k = 2\pi/L)$	1/m
k_s	equivalent sand roughness	$(k_s = d_{50})$	m
L_0	deep-water wave length	$(L_0 = gT_P^2/2\pi)$	m
M	mass		kg
n	number of displaced stones		-
q	turbulent velocity scale in breaking waves		m/s
Q_B	percentage of broken waves		-
S	damage		%
s	wave steepness	$(s = H/L_0)$	-
T_P	peak wave period of spectrum		s
$\hat{u}_b$	amplitude of orbital velocity at bottom		m/s
w	width of flume		m
α	slope angle of structure		-
Δ	relative mass density of material	$(\Delta = (\rho_s-\rho_w)/\rho_w)$	-
ϕ	angle of repose of stones		-
ρ_s	mass density of material		kg/m^3
ρ_w	mass density of water		kg/m^3
ξ	breaker parameter	$(\xi = \tan\alpha/\sqrt{(H/L_0)})$	-
$\hat{\tau}_b$	amplitude of bottom shear stress		N/m^2
ω	angular frequency	$(\omega = 2\pi/T)$	1/s

REFERENCES

Battjes, J.A., 1975 *Modelling of turbulence in the surf zone* Symposium on modeling techniques, ASCE, San Francisco

Battjes, J.A., 1987 *Surfzone turbulence* Seminar on Hydrodynamics of waves in coastal areas, IAHR, Moscow

Battjes, J.A. and Janssen, J.P.F.M., 1978 *Energy loss and set-up due to breaking in random waves*, Proc. 16th Int. Conf. on Coastal Eng., ASCE, New York

Battjes, J.A. and Stive, M.J.F., 1985 *Calibration and verification of a dissipation model for random breaking waves*, Jrnl. of Geophys. Research, 90

CUR/CIRIA, 1991 *Manual on the use of rock in coastal and shoreline engineering,* CUR Report 154/CIRIA Special Publication 83, Balkema, Rotterdam, pp. 197, 216, 217, 298

Holthuijsen, L.A., Booij, N. and Herbers, T.H.C., 1989 *A prediction model for stationary, short-crested waves in shallow water with ambient currents,* Coastal Engineering, 13, pp 23-54

Jonsson, I.G., 1966 *Wave boundary layers and friction factors* Proc. 10th Conf. on Coastal Engineering, Tokyo, pp. 127-148

Meer, J.W. van der, 1988 *Rock Slopes and Gravel Beaches under Wave Attack* Delft Hydraulics, Publication no. 396, The Netherlands

Rance, P.J., and Warren, N.F., 1968, *The threshold of movement of coarse material in oscillatory flow* Proc. 11th Conf. on Coastal Engineering, London, pp. 487-491

Schiereck, G.J., Fontijn, H.L., Grote, W. and Sistermans, P.G.J., 1994 *Stability of rock on beaches*, Proceedings 21th Int. Conf. on Coastal Eng. ASCE, Kobe

Sleath, J.F.A., 1978 *Measurements of bed load in oscillatory flow* Journal of the Waterway, Port, Coastal and Ocean Division, ASCE, Vol.104 no. WW4, pp. 291-307

Ye, Lei, 1996 *Stability of rock on beaches* MSc-thesis, Delft University of Technology

CHAPTER 328

Modeling Tidal Circulation in Florida Bay

Y. Peter Sheng[1], Member ASCE, and Justin Davis[2]

Abstract

A preliminary modeling study on tidal circulation in Florida Bay has been conducted, based on a set of high resolution (20 meters x 20 meters) bathymetric data and a comprehensive set of water level data for Florida Bay obtained by the Everglades National Park in 1993 and 1994. Using the data and a 3-D curvilinear-grid model developed by Sheng (1989, 1994), we conducted preliminary model simulations of tidal, wind-driven and density-driven circulations in Florida Bay (Sheng et al. 1995). This paper presents some results of tidal simulation and comparison with tidal data. The results indicate that the 3-D model is capable of simulating many of the observed circulation features in Florida Bay, including tidal amplitude, tidal phase and residual circulation. Additional work is being conducted with the use of a very fine 50m grid and a robust wetting-and-drying scheme.

Introduction

Florida Bay is a shallow coastal water body located at the southern tip of Florida. It is adjacent to the western Florida Shelf and is separated from the deep Florida Strait by the Florida Keys. In the past few years, Florida Bay has undergone substantial changes in the form of major seagrass die-off and massive algal bloom. Due to the lack of understanding of the extremely complex ecosystem, it is unclear presently as to how Florida Bay can be restored (Johnson and Fennema 1989). One restoration plan, which is based on the assumption that reduced freshwater supply and increase in hypersalinity (salinity reaching up to 35-50 ppt) has caused the seagrass die-off and algal bloom (Figure 1), is to increase the inflow of freshwater into Florida Bay from its northeastern shore. In order to determine the soundness of this

[1]Professor, Coastal & Oceanographic Eng. Dept., Univ. of Florida, 32611-6590, USA
[2]Graduate Assistant, Coastal & Oceanographic Eng. Dept., Univ. of Florida, USA

restoration plan, it is essential to have a quantitative understanding on the circulation and salinity transport in Florida Bay. This paper presents a preliminary study on tidal circulation in Florida Bay. In addition to a comparison between simulated and measured water levels in Florida Bay, we also examine the exchanges between Florida Bay and Florida Strait through the tidal channels along the Florida Keys.

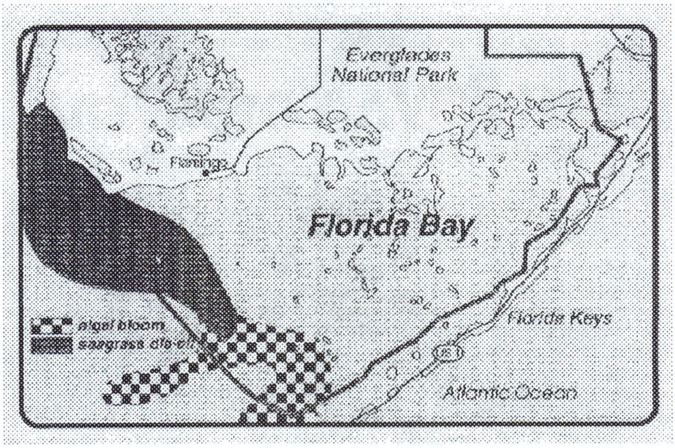

Figure 1. Florida Bay Map with Areas of Massive Algal Bloom and Seagrass Die-off.

Field Data

As a first step, we reviewed a comprehensive set of data which includes the high resolution bathymetric data for Florida Bay and hydrodynamic and hydrographical data at numerous stations in Florida Bay during 1993 and 1994, both compiled by the National Park Service (Sheng et al. 1995). As shown in Figure 2, much of the eastern Florida Bay is less than 1 m deep. Figure 3 shows that mudbanks in Florida Bay actually divide the Bay into numerous lakes during low water. Field data collected in Florida Bay include water level, water temperature, conductivity, rainfall, and evaporation. Water level data were taken at the 27 stations shown in Figure 4. Using these water level data, Smith (1995a) obtained the co-amplitude chart and co-phase chart for the M_2 tide as shown in Figure 5. As can be seen from Figure 5, M_2 tide enters the Florida Bay from both West Florida Shelf and Florida Strait. Diurnal tides such as O_1 and K_1 constituents primarily enter into Florida Bay from West Florida Shelf. Some tide data are from the offshore waters to the south of Florida Keys (Smith 1995b). Wind data are available from 1-2 offshore CMAN stations only. Freshwater discharge data and current meter data are lacking. High evaporation rate often leads to hypersalinity of more than 50 ppt inside Florida Bay even during the wet season.

Using the data indicated above and a 3-D curvilinear-grid model developed by Sheng (1987, 1989, 1994), we conducted model simulations of tidal, wind-driven and density-driven circulations in Florida Bay (Sheng et al. 1995). In this paper, we present some results of preliminary simulation of tidal circulation. model results.

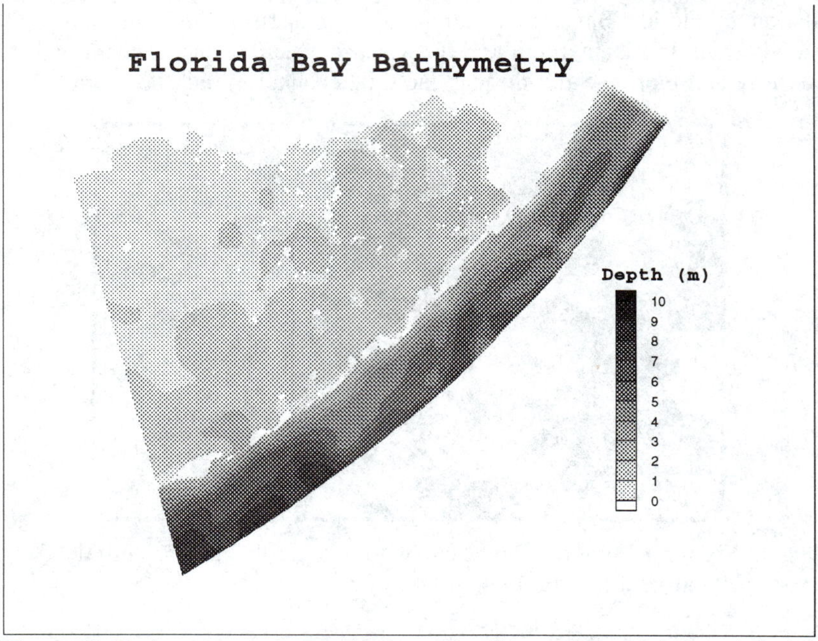

Figure 2. Bathymetry contours in the entire finer-grid model domain. All depths inside the Bay are less than 1-2 meters.

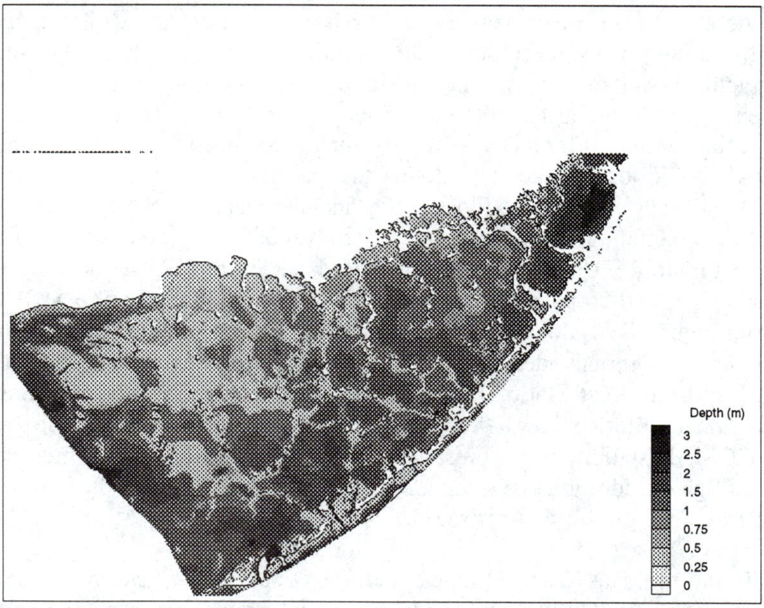

Figure 3. Florida Bay bathymetry provided by Everglades National Park.

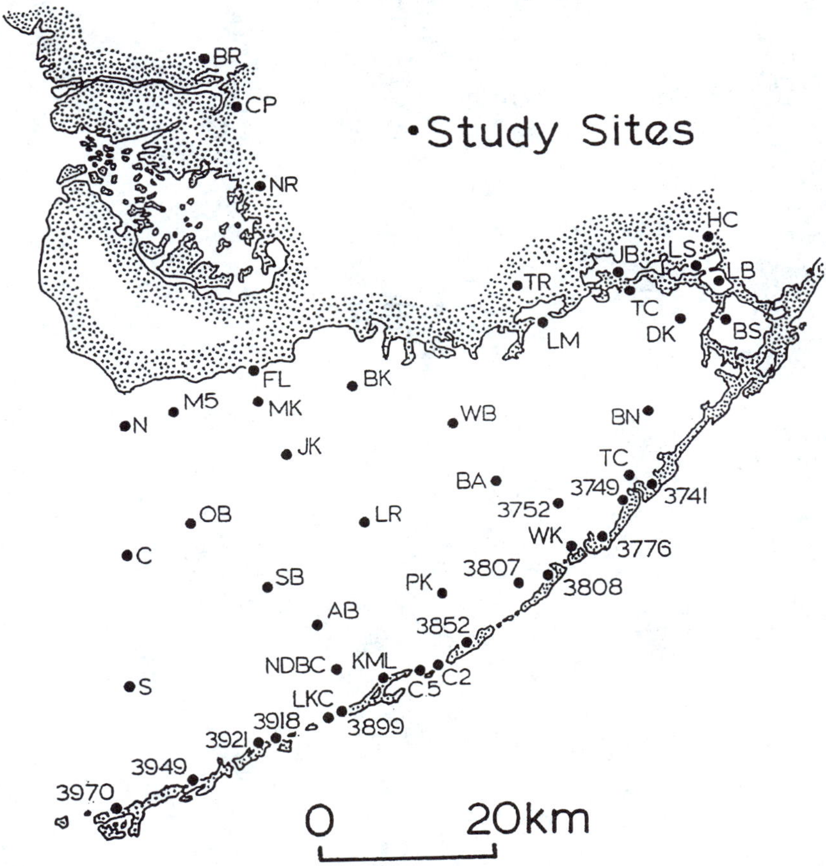

Figure 4. A Map of Monitoring Stations in Florida Bay.

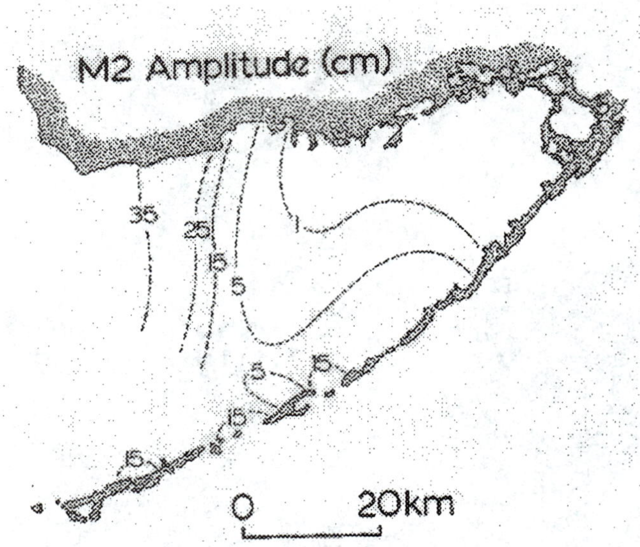

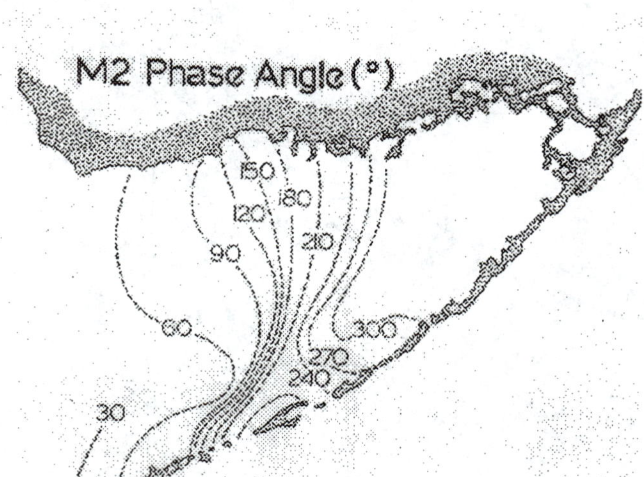

Figure 5. Co-Amplitude and Co-Phase Lines of M_2 Tide in Florida Bay.

Model and Grid

CH3D, the 3-D curvilinear-grid circulation model developed by Sheng (1989, 1994) is modified for application to Florida Bay. In the past few years, through numerous research projects (Chesapeake Bay, James River, Sarasota Bay, Lake Okeechobee, Indian River Lagoon, and Tampa Bay), we have significantly improved the model's features/capabilities for simulating stratified flow, effect of vegetation on circulation, ocean-estuary coupling, and wind-induced flow. The model solves for the three-dimensional equations of motion in terms of the water level and contravariant velocity components. The model solves the vertically-integrated equations of motions as well as the vertical structures of the flow field. Since the same time step is used for the solution of the external mode and the internal mode, the model solutions are consistent with each other and converge. A robust vertical turbulence model (Sheng and Villaret 1989) is used to represent the vertical turbulent mixing. The model allows the use of boundary-fitted grid which can accurately represent the complex shoreline in Florida Bay. In the vertical direction, the sigma grid is used. Sheng et al. (1995) used several boundary-fitted grids to resolve Florida Bay.

The grid shown in Figure 6 includes all the area covered in the detailed bathymetry map in Figure 3 plus the offshore area between the Florida Keys and the reef tract. A total of 98 by 75 grid points are contained in this grid. For 3-D simulations of tidal circulation, a total of 4 vertical cells are generally used.

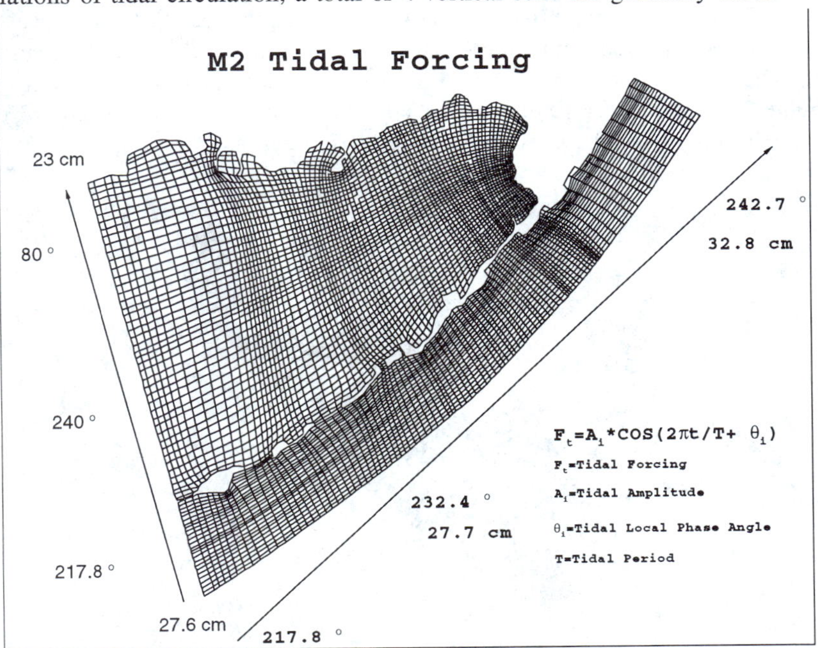

Figure 6. A model grid with 98x75 cells and open boundary conditions for M_2 tide in Florida Bay.

Simulation of Tidal Circulation

Using the horizontal numerical grid shown in Figure 6 and a vertical sigma-grid with 4 cells, we conducted simulations of tidal circulation in Florida Bay with CH3D (Sheng et al. 1995). The major objectives of tidal simulation is to examine the model's ability to simulate tide propagation and dissipation inside Florida Bay, and to examine the simulated residual flow inside the Florida Keys. For the first objective, detailed comparison between model results and data have been made (Sheng et al. 1995). For the second objective, model results can be compared qualitatively with preliminary flow measurement by Smith (1995b) which showed persistent southerly residual flow inside the Florida Keys.

To simulate the tidal dynamics in Florida Bay, it is essential to include part of the offshore water in the numerical grid as shown in Figure 6, since tides in Florida Bay are influenced by tides from the Gulf of Mexico and the Florida Strait. Tidal data along the reef tract were taken at Carysfort Reef, Alligator Reef, Tennessee Reef, Sombrero Key and Sand Key Light. These data were analyzed to produce the amplitudes and phases for various tidal harmonic constituents. Using these data and the tidal data inside Florida Bay, we produced the boundary conditions for the M_2 simulation as shown in Figure 6.

Using the open boundary conditions for M_2 tide described above, the 3-D model CH3D was run for 5 days. The result during the last tidal cycle was then analyzed to produce the maximum water level chart and a corresponding co-phase chart. The simulated maximum water level chart for M_2 tide in Florida Bay is shown in Figure 7. Amplitude in the extremely shallow eastern Florida Bay, where the depth is generally less than 50 cm, is on the order of 1-5 cm only.

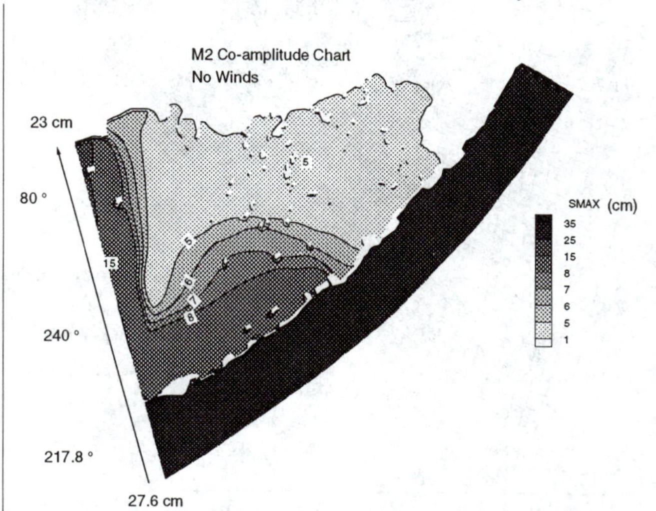

Figure 7. Co-amplitude chart of M_2 tide simulated by the 3-D model.

The results shown in Figure 7 compare favorably with the co-amplitude chart produced from data as shown in Figure 5. The simulated peak water level in the eastern Florida Bay is larger than the observed amplitude, suggesting insufficient tidal dissipation in the model. The simulated co-phase chart, which is presented in Sheng et al. (1995), showed that tides propagated further eastward. This is partly due to the fact that the complicated mudbanks, extremely shallow depths, and narrow Florida Keys in the eastern Florida Bay are not accurately represented in the 3-D model grid and bathymetry. The minimum horizontal grid spacing is approximately 100 meters while a spacing of 20 m may be necessary to represent the detailed bathymetry.

Other reasons for discrepancy between model results and data are that wind and wetting-and-drying were not included in the preliminary simulations. In the absence of wind, the residual (tidally-averaged) water level in the eastern Florida Bay is about 2 cm plus. By including a mild summer wind from the southeast, simulated maximum water level chart as shown in Figure 8 agrees better with the co-tidal chart based on data. In the presence of a summer wind, the residual water level in eastern Florida Bay became approximately -0.5 cm. A robust wetting-and-drying scheme has recently been implemented in CH3D. With the wetting-and-drying scheme and a fine grid (50m minimum grid spacing), we are able to produce much better results with 1 cm tidal amplitude in the eastern Florida Bay.

The tidally-residual vertically-averaged flow field as shown in Figure 9 exhibits significant southerly flow in the Florida Keys, in agreement with recent flow measurement by Smith (1995b). The southerly residual flow results from the higher residual water level in eastern Florida Bay.

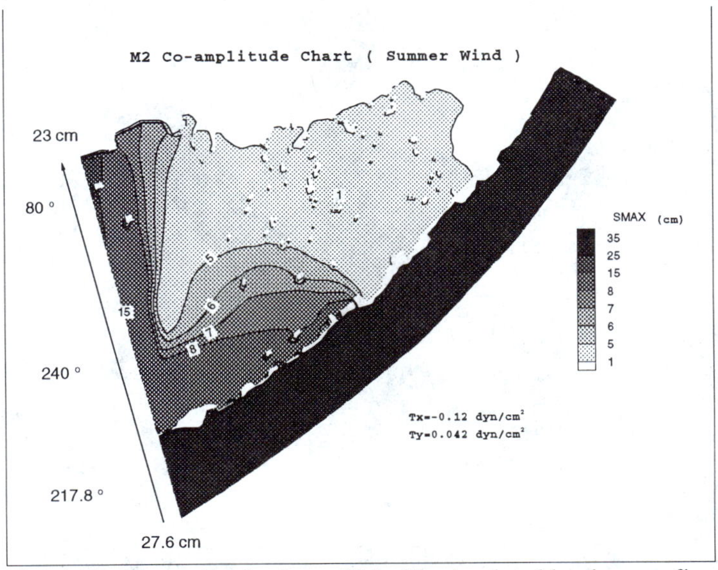

Figure 8. Co-amplitude chart of M_2 tide simulated by the new fine grid model with average summer wind condition.

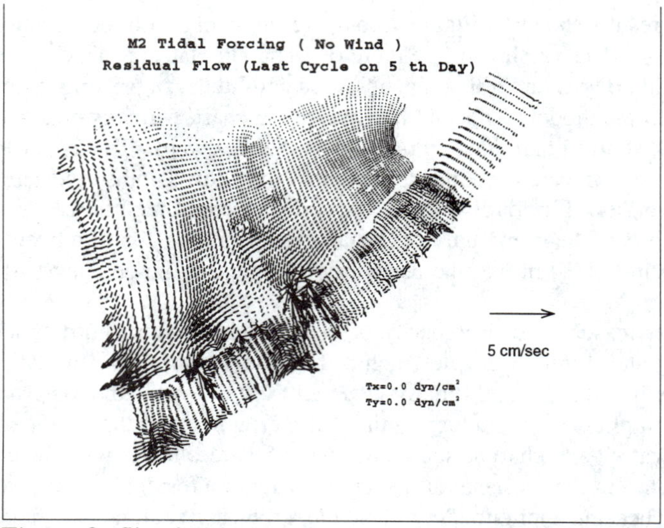

Figure 9. Simulated tidal residual currents in Florida Bay with forcing by pure M_2 tide.

To examine the influence of tides from the southern open boundary, we made a model run with closed Florida Keys. As shown in Figure 10, the maximum water level compares poorly with that in the co-amplitude chart based on the 1-year data. These results also suggest that simulated results in Figure 8 exhibit stronger influence from the southern open boundary than that shown in the data. This could be improved by more accurate representation of the Florida Keys and more accurate open boundary conditions along the southern open boundary.

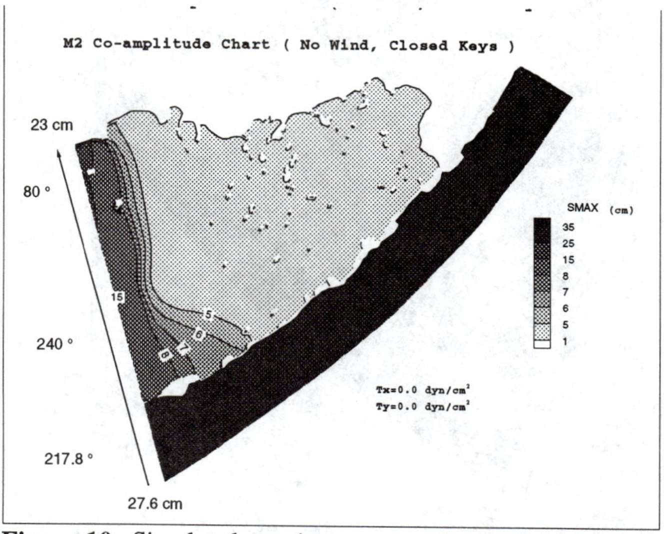

Figure 10. Simulated maximum water level of M_2 tide in Florida Bay with the Florida Keys closed.

Discussion

A curvilinear-grid 3-D circulation model developed by Sheng (1987, 1989, 1994) has been modified to simulate the tidal circulation in Florida Bay. Tide data collected at 27 stations inside Florida Bay during 1993 and 1994 plus offshore tide data during earlier time periods were analyzed in terms various diurnal (O_1 and K_1) and semi-diurnal (M_2) constituents. Using the tidal constituent data to construct open boundary conditions along the southern and western open boundaries, we conducted numerous model simulations with a number of numerical grids and various forcing conditions: first with tide only, then with tide and wind, and finally with tide, wind and salinity field (Sheng et al. 1995). Due to the dominance of M_2 tide from both the Gulf of Mexico and the Florida Strait, this paper focuses on the simulation of M_2 tide. The simulated maximum water level and co-phase chart compare reasonably well with the co-amplitude and co-phase charts produced from the 1-year tide data. Significant dissipation due to the mudbanks and extremely shallow water depths result in very small tidal amplitude on the order of 1-5 cm in the eastern Florida Bay. The residual (tidally-averaged) model results indicate that there is a setup in the eastern Florida Bay which causes southerly residual flow inside the channels along the Florida Keys, which has been found by recent flow measurement.

Additional model simulations of K_1 and O_1 tides and baroclinic circulation have been conducted by Sheng et al. (1995). Model results suggest that data gap in freshwater inflow and evaporation data prevents longterm (from 1 month to 1 year) simulation of Florida Bay circulation at the present time.

We have continued with the modeling effort. Recently, a robust wetting-and-drying scheme has been implemented in CH3D. A very fine grid, with a minimum grid spacing of 50m, has also been used in recent simulations. This grid gives much better representation of the mudbanks and the complex bathymetry. Using the very fine grid and the wetting-and-drying scheme, we are able to significantly improve the model results in the eastern Florida Bay. Maximum M_2 tidal amplitude in the eastern bay was reduced to 1 cm. Details of the wetting-and-drying scheme and the improved model simulations will be reported in two upcoming papers.

Acknowledgement

This study has been supported by the Everglades National Park of the National Park Service, US Department of Interior, and the University of Florida. Data used for this study have been provided by Dewitt Smith and Ned Smith.

References

Johnson, R.A. and R.J. Fennema, 1989: "Conflicts Over Flood Control and Wetland

Preservation in the Taylor Slough and Eastern Panhandle Basins of Everglades National Park,'' in *Wetlands: Concerns and Successes*, American Water Resources Association.

Sheng, Y.P., 1987: ''On Modeling Three-Dimensional Estuarine and Marine Hydrodynamics,'' *Three-Dimensional Models of Marine and Estuarine Dynamics* (J.C.J. Nihoul and B.M. Jamart, Eds.), Elsevier Oceanography Series, Elsevier, pp. 35-54.

Sheng, Y.P., 1989: ''Evolution of A 3-D Curvilinear-Grid Hydrodynamic Model: CH3D,'' *Estuarine and Coastal Modeling,* (Ed. M. Spaulding), ASCE, 40-49.

Sheng, Y.P., 1994: ''Modeling Hydrodynamics and Water Quality Dynamics in Shallow Waters,'' *Keynote Address, International Symposium on Ecology and Engineering (ISEE 1994)*, Taman Negara Resort, Malaysia, 29 Oct. - 3 Nov., 1994.

Sheng, Y.P., 1995: ''On Modeling of Florida Bay Circulation,'' *Abstract Book, AGU Chapman Conference on Circulation in Intra-Americas Sea*, La Parguera, Lajos, Puerto Rico, January 22-26, 1995.

Sheng, Y.P. and C. Villaret, 1989: ''Modeling the Effect of Suspended Sediment Stratification on Bottom Exchange Processes,'' *Journal of Geophysical Research*, **94:C10**, pp. 14429-14444.

Sheng, Y.P. and S.J. Peene, 1993: ''A Field and Modeling Study of Residual Circulation in Sarasota Bay and Tampa Bay, Florida,'' *Coastal & Estuarine Modeling, III*, (Ed. M. Spaulding), ASCE.

Sheng, Y.P. J. Davis, and Y. Liu, 1995: ''Towards the Development of A Comprehensive Model of Circulation and Transport in Florida Bay,'' *Proceedings of the 1995 Scientific Symposium on Florida Bay*, Florida Sea Grant.

Sheng, Y.P. J. Davis, and Y. Liu, 1995: ''Development of A Preliminary Numerical Model of Circulation and Transport in Florida Bay,'' *Final Report to National Park Service*, Coastal & Oceanographic Engineering Department, University of Florida, Gainesville, FL 32611.

Smith, N.P., 1995a: ''Analysis of Water Level and Current Data in Florida Bay,'' *Interim Report to National Park Service*, Harbor Branch Oceanographic Institution.

Smith, N.P., 1995b: Personal Communication.

CHAPTER 329

RECENT DEVELOPMENTS IN COASTAL DEFENCE POLICY AND GUIDANCE IN ENGLAND

Reg Purnell[1]

SUMMARY

In recent years we have seen fundamental changes in the way that coastal defences are planned and implemented. Engineers have recognised the overwhelming public desire to see the environment taken more into account when designing major infrastructure projects and our understanding of coastal processes has improved, allowing the design of coastal defences to become more scientifically based. As a result we have seen attention turn to adapting and supplementing natural coastal processes with the aim of adopting a coastal policy which is both more environmentally acceptable and sustainable in engineering terms.

In England, to reflect these engineering changes, to secure better value for money for our investment in coastal defence and offer a more sustainable approach in the longer term, the Ministry of Agriculture, Fisheries and Food, which has policy responsibility for flood and coastal defences, has invested significant sums of money into research. This has resulted in a better understanding of coastal processes and significant changes to national policy.

INTRODUCTION

As one would expect of a historical maritime nation, England has significant centres of population and economic wealth all around its coastline. The result is that over 5% of the population live in areas which are at risk of coastal flooding, perhaps not significant proportion in international terms. In addition there is an associated risk of coastal erosion which has seen the loss of a significant number of villages and large areas of land over the last few hundred years. The potential economic loss resulting from flooding or erosion is therefore significant in terms of its impact on the GDP. Recognising these risks the UK government spends some $100m per annum on grant aiding capital improvement works in England to achieve its published overall policy aim which is "to reduce risks to people and the developed and natural environment from flooding and coastal erosion by encouraging the provision of technically, environmentally and economically sound and sustainable defence measures".

INSTITUTIONAL ARRANGEMENTS

Whilst the Ministry of Agriculture, Fisheries and Food has overall policy responsibility for flood and coastal defences decisions on where and when defences will be provided are left to the local community. Responsibility for the provision of defences

---

[1]Chief Engineer, Flood and Coastal Defence with Emergencies Division, Ministry of Agriculture, Fisheries and Food, Eastbury House, 30/34 Albert Embankment, London SE1 7TL, England.

against flooding are generally with the Environment Agency, which although a national body, makes such decisions on a regional basis with strong local input. Defences against coastal erosion are provided by Local Councils, some of whom also have additional powers with respect to provision of defences against coastal flooding. The boundaries of these bodies generally take no account of coastal processes being based on historical precedent and geography.

Historically these authorities have not generally cooperated in the provision of coastal defences tending to make their decisions on the basis of local needs and precedence.

STRATEGIC PLANNING

In the past defence measures were planned and implemented on a one-off site specific basis, not the most scientifically based approach. Indeed such site by site appraisal not only restricted engineering options but also ensured that not all costs and impacts were taken into account with the result that options were often limited to merely replacing on a like for like basis. This often perpetuated, previous sometimes, inappropriate solutions such as the replacement of hard reflective walls with more of the same. Improvements in the understanding of coastal processes clearly showed that by considering options over a wider area and a longer timescale more lateral thinking was possible, enabling the preparation of schemes that complemented natural processes rather than fighting against them. It was also felt that this more strategic approach would allow the Ministry to more easily meet its own policy objective.

As noted previously the boundaries of the authorities with primary responsibility for the delivery of coastal defences took no account of coastal processes. In addition these authorities had historically made decisions reflecting the needs and wishes of their own electorate and with little consultation with adjoining authorities. Therefore the first policy objective was to encourage these authorities to work together towards more strategic planning such that decisions taken on the basis of engineering knowledge could cross political boundaries. This was achieved by the establishment of voluntary Coastal Defence Groups, encouraged by MAFF, which took account of coastal process cells. To facilitate this a research study was commissioned from HR Wallingford to define logical boundaries of coastal process cells. Cells were defined as lengths of coastline within which the movement of coarse sediment is largely self-contained. Sub-cells within these were also defined on the basis of limited cross boundary influence due to coarse sediment transport.

Once formed these groups allowed authorities on adjoining lengths of coastline to meet and discuss similar problems and solutions and consider joint approaches to the gathering of data etc. Central government also encouraged collaboration between these groups by setting up the Coastal Defence Forum on which all groups in England and Wales are represented.

Central government then provided guidance and encouragement on the production of Shoreline Management Plans which aim to set out a strategy for coastal defences for a specified length of the coast taking account of the natural coastal processes and human and other environmental influences and needs.

The objective in developing these plans was to:

- improve understanding of the coastal processes operating within the sediment cells;
- predict the likely future evolution of the coast;
- identify all the assets within the area covered by the plan which are likely to be effected by coastal change;
- identify the need for regional or sites specific research investigations;

- facilitate consultation between those bodies with an interest in the coastline.

Once the plan was completed the authorities involved in its preparation would then have the means by which they could:

- assess a range of strategic coastal defence options and agree a preferred approach;
- outline future requirements for monitoring, management of data and research related to the shoreline;
- inform the strategic planning process and related coastal zone planning;
- identify opportunities for maintaining and enhancing the natural coastal environment taking account of any specific targets set by legislation or any locally set targets;
- set out arrangements for continued consultation within the group and other interested parties.

BEACH RECHARGE

The practice of recharging beaches as a form of coastal defence has been recognised worldwide for many years, although its use in England has been more limited than in many other countries in Europe. This is partly because of the greater difficulty of predicting beach behaviour due to the variability of beach material and partly because of the difficulty in gaining sufficient suitable material resulting in greatly increased costs.

Improved knowledge of coastal processes and an increasing public desire to take account of environmental effects when designing major infrastructure projects, increased pressure to use beach recharge as a form of coastal defence in England. Such forms of defence were also considered more sustainable in engineering terms. Initial research allowed improvements to modelling techniques which reduced the perceived financial risks and assisted in educating decision makers on the benefits of beach recharge. However, it was soon recognised that recharged beaches needed management if they were to provide a long term sustainable solution. Unfortunately such management works had not previously attracted Central government funding with the costs falling on the local community. Hence the possibility existed that the chosen solution might have more to do with the availability of funding than finding the right engineering solution.

The resulting policy change was to encourage authorities to produce long term beach management plans with the cost of implementing and monitoring for the first time attracting Central government grant. Thus the choice of solution could now more fully depend on science rather than the source of funds. As an additional benefit this change in policy allowed beach management on its own, without the benefit of an initial recharge, to be seen as an effective means of continuing coastal defences along parts of the coastline. With the possibility of Central government finance, operating authorities are now more willing to seek to understand the periodic movements of beach material and manage it for the benefit of improved coastal defences, thereby achieving better value for money and possibly reducing long term capital expenditure.

MANAGED SET-BACK

It was King Canute who first demonstrated to the English population the strength and power of the sea and the need for man to adapt to the wishes of Neptune rather than impose his will on nature. In the intervening years the lesson has been reinforced by the significant loss of land to the sea, recognising that it is not scientifically or economically sensible to protect all 4000 kilometres of our coastline. However, with the change to more naturally sustainable types of defence the option of doing nothing, or more importantly, allowing the coastline to retreat had to be looked at in a different light. Whilst the retreating of defences did not require any major policy changes, since Mother Nature was doing that already, strategic planning now allowed more informed decisions and the opportunity for environmental enhancement. With improved long term planning it becomes possible to foresee the extent of likely retreat in future years thereby allowing landowners and developers to make more informed decisions but just as importantly the potential beneficial impact of such retreat on other areas of the coastline can be examined.

To compensate for the loss of valuable inter-tidal habitat resulting from coastal squeeze, and man's interference, government also took the opportunity of providing funds to recreate lost tidal wetlands in these setback areas. In areas where setback of the coastline is seen as a viable response mechanism and an opportunity for habitat creation exists, government can now offer payments to landowners for long term management to achieve environmental goals. Obviously merely setting back the coastline may not lead to environmental gain hence a research programme has commenced to determine suitable management regimes.

CARE OF THE ENVIRONMENT

There can be no doubt that flood and coastal defence measures can have a significant impact on the environment with the potential loss of important natural habitats. This is in addition to natural losses which may be exacerbated as a result of the coastline being unable to flexibly respond to natural changes due to man's intervention. These recent changes in coastal defence policy have made it possible to more easily establish the impact of our works on the natural environment and hence consider the possibility of mitigating measures. In addition it may also allow us in the future to understand and quantify some of the natural changes which may result in the loss of important wetland habitat thereby allowing the possibility of adjusting approaches in order to take mitigating measures to retain biodiversity.

FUTURE

As outlined in this paper cooperation between engineers and policymakers has in recent years allowed the scientific advances achieved by coastal engineers to result in consequential policy changes, however, these changes will not necessarily make life any easier for engineers. The move towards long term strategic planning of defences which encourages the maintenance of coastal processes will, without doubt, bring engineers' decisions under closer public scrutiny. Some of their recommendations may not be in accordance with the views of the local population, especially if these include realignment or even the abandonment of some existing defences. With longer term planning an increasing number of people will be involved in decisions affecting the coastline, hence engineers may have to seek consensus views from a wider range of people and interest groups.

The increasing move towards more environmentally friendly solutions to coastal problems will continue to result in change. Earlier in this paper I noted that beach recharge was not as widely used in England as elsewhere in Europe due largely to material supply problems. This is because the material suitable for recharging beaches is also sought by the construction industry. Many prefer to see the use of marine dredged material because of the perceived adverse environmental impact of extracting land based materials whilst the environmental impact of extraction from the sea bed is much less obvious to the general public. Unfortunately in many areas of the North Sea there is now insufficient coarse aggregate to satisfy the long term needs of both the construction industry and beach recharge. Therefore the next policy area that requires consideration is whether the use of marine aggregate sources should be restricted to allow increased use for beach recharge or to leave it to market forces, as at present.

No matter what the decision on any future restrictions in the use of marine aggregate by the construction industry, demand will, at some time, exceed supply. Engineers' response to this possibility is to seek improvements in the predictive tools for long term sediment movement and developments in coastal morphology. An integrated research project is at present underway which should show some interesting results within a few years. In addition estuaries which are an important but little understood part of the sediment chain are a matter for current review.

The continued move towards more natural forms of defence brings forward questions of risk assessment since decisions are now dependent on a greater number of variables. This has been demonstrated in recent years by the unexpected failure of some soft forms of defence as a result of storm series that were not taken into account in the design process. Our response will be improved techniques in risk assessment leading perhaps to greater use of probabilistic design in coastal defences. An interesting perspective for policymakers.

The move towards strategic planning of coastal defences has resulted in a change of public awareness and perception of coastal engineering problems and as a result the public now takes a fuller part in the decision making process, even questioning the advice of engineers, especially when the "do nothing" option is the preferred choice. This move will, without doubt, bring other areas of coastal use into the decision making process leading to more integrated decisions, although implementation of coastal zone planning in its truest sense is, in my opinion, still some way off.

CONCLUSIONS

It is too early to claim complete success for the recent policy changes in this country, but I have no doubt that we are moving in the right direction and with a speed that has surprised many. Clearly the impact and improvements of recent years has been a lesson to us all and the need for engineers and policymakers to work together remains with us. However, if we are to make the best use of our improved scientific knowledge then it should not always be the policymakers responding to new engineering concepts, but engineers should themselves recognise the pressure on policymakers and be ready to respond with help and advice. Not only should changes in engineering techniques help drive policy but changes in policy need should also feed into engineering research.

REFERENCES

Strategy for Flood and Coastal Defence in England and Wales. MAFF/Welsh Office (1993): MAFF Publications PB 1471.

Shoreline Management Plans; a guide for operating authorities. MAFF/Welsh Office (1995): MAFF publication PB 2197.

Coastal Management: Mapping of Littoral Cells. Report SR 328 Hydraulics Research, Wallingford, January 1993.

Purnell RG, 1996 - Shoreline Management Plans: National Objectives and Implementation - proceedings of "Coastal Management: putting policy into practice".

Flood and Coastal Defence Research and Development Implementation Strategy 1994-1998, MAFF publications (1995) ref. PB 2249.

Beach Recharge Materials - Demand and Resources, 1996, Construction Industry Research and Information Association (CIRIA).

DoE Circular 30/92, MAFF Circular FD1/92, WO Circular 68/92: Development and Flood Risk, ISBN 0-11-752737-8.

Planning Policy Guidance Note (PPG) 20: Coastal Planning, DoE/Welsh Office (1992). ISBN 0-11-752711-4.

Coastal Defence and the Environment: a guide to good practice. MAFF (1993). MAFF Publications PB 1191.

The EC Habitats Directive: Implications for flood and coastal defence. MAFF (1995).

MAFF Project Appraisal Guidance Notes for flood and coastal defence. MAFF (1993). MAFF Publications PB 1214.

Policy Guidelines for the Coast - Discussion draft - DoE (1995).

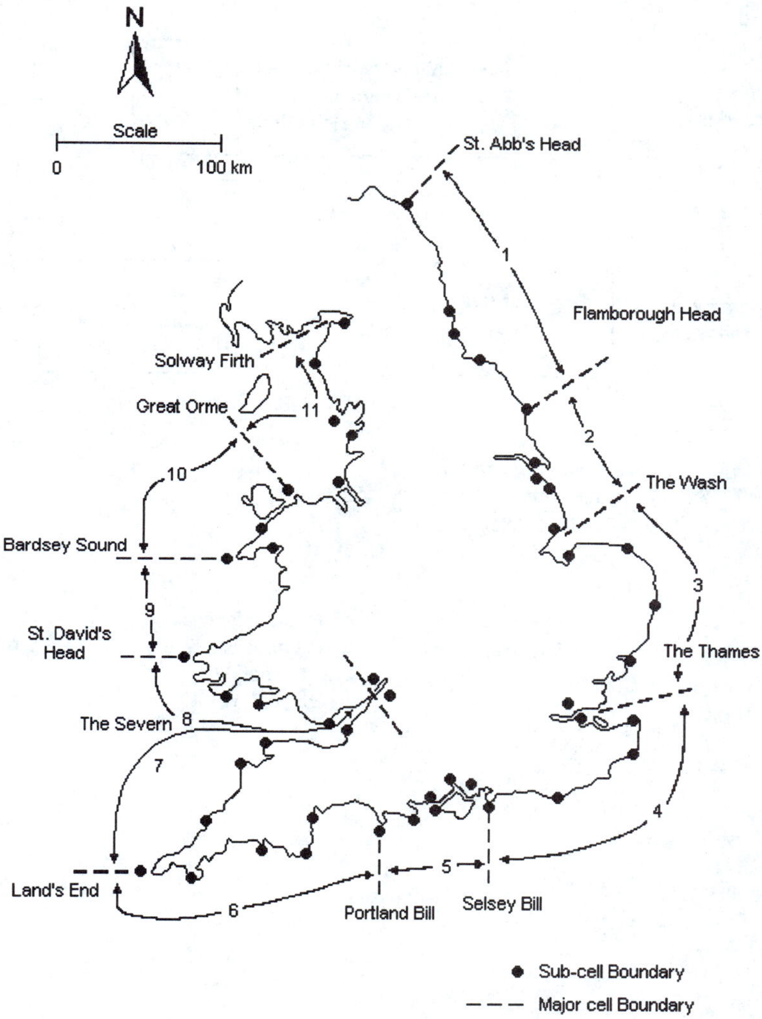

Figure 1: *Boundaries of major sediment cells*

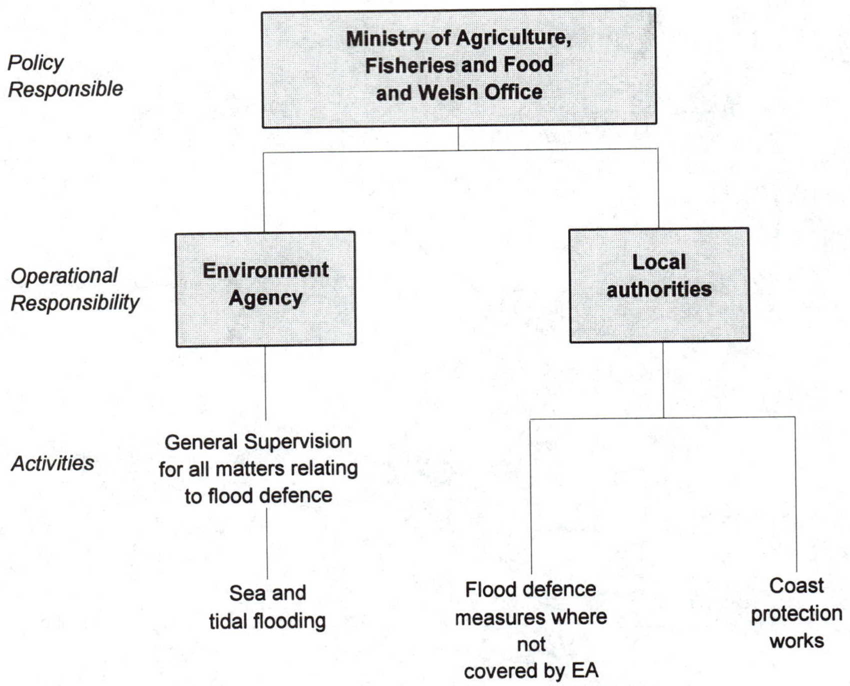

Figure 2 Flood and Coastal Defence: Organisation

CHAPTER 330

STRUCTURAL RESTORATION OF CORAL REEFS DAMAGED BY VESSEL GROUNDINGS

Kevin R. Bodge, Ph.D., P.E.[1]

ABSTRACT

In 1995, a unique project was completed in which two coral reef sites in the Florida Keys were structurally restored after having sustained severe damage from the groundings of large vessels. The project, believed to be the world's first major structural restoration of a damaged reef (vs. in-kind mitigation using artificial reefs), demonstrated numerous innovative materials and marine construction techniques. Restoration focused upon the stabilization of coral rubble and large craters which resulted from the vessel groundings. The project's intent was to re-create a stable foundation which closely emulates the adjacent natural seabed and which would foster future recruitment of local biota. Work at one site included the mechanical transfer of coral rubble back into the craters, placement of limerock boulders atop the rubble, and back-filling the boulders' voids with carbonate sand. Work at the other site included excavation of coral rubble and the precision placement of 40 pre-cast reef-replicating armor units into the crater. The gaps between the units and along the crater's perimeter were filled with a specially-designed, non-separable underwater concrete -- into which coral rubble and soft corals were impressed. Design was complicated by the sites' proximity to environmentally sensitive coral beds and shallow depths (2.5 to 11 m). During construction, semi- real-time video images of the underwater work were relayed to the Engineer's office via the Internet to augment construction review. Construction was successfully completed per the engineering plans with no consequent environmental damage amidst a very active tropical storm season.

BACKGROUND

In two separate incidents in 1989, the 40-m *M/V Alec Owen Maitland* and the 142-m *M/V Elpis* went aground upon living coral reefs in the Key Largo National Marine Sanctuary in the Florida Keys, U.S.A. The two sites were within 6.8-km of one another, about 10.4 km offshore of northern Key Largo; and, about 74 km south-southwest of Government Cut at Miami Beach. The Maitland and Elpis sites, respectively, are located in the vicinity of N 25°11'58.776", W 80°13'34.421"; and N 25°8'54.088", W 80°15'9.20" (see **Figure 1**).

The impact and weight of the vessels upon the reef fractured the underlying coral substrate; and the ships' screws created deep craters in the coralline seabed. Large amounts of coral rubble were created from the fractures and ejected from the craters. Monetary damages were paid by the vessels' operators to the U.S. federal government under the auspices of the National Marine Sanctuaries Act

[1] Sr. Engr, Olsen Assoc., Inc., 4438 Herschel St., Jacksonville, FL 32210 USA; (904) 387-6114.

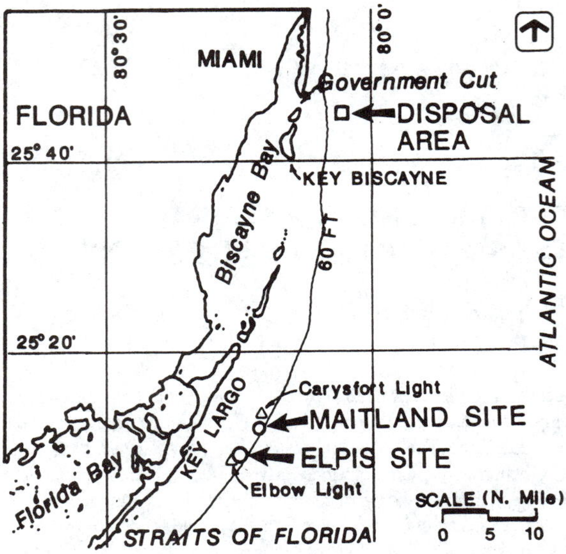

Figure 1: Location Map

for the purposes of site rehabilitation. The author was retained by the U.S. National Oceanic and Atmospheric Administration (NOAA) to quantitatively assess the sites' structural damage, and to design and supervise engineering works to structurally restore the damaged reefs.

ENGINEERING ASSESSMENT

Site Survey and Conditions. A precision survey of the two damage sites was undertaken in July, 1994 using tightly-gridded fathometer transects and diver measurements involving triangulation and taut-wire cross-sections of the craters' geometry and rubble piles. A multi-beam fathometer (such as SEABAT) would have been ideally suited for the remote survey work, but was not yet commercially available at the time of the study.

Damage at the shallower Maitland site (-1.8 to -3.0 m, MLW) consisted primarily of a 370-m^2 crater, varying in relief from 0.1 to 1.0 m, relative to the ambient seabed. The crater was formed by the ships' screw when an attempt was made to free the vessel under her own power. Since its creation, the crater had partially infilled with rubble, but had also doubled in size from what was originally two separate holes. The expansion of the crater was due to wave- and current-induced erosion of its coralline perimeter -- a thin (< 1.5-m) semi-continuous crust overlaying 15-m of densely-packed gravelly-sand. The volume of the crater, measured below the ambient seabed grade, was about 115 m^3. The seabed of the crater featured a matrix of sand and 7- to 20-cm diameter, loose coral rubble with scattered larger pieces measuring up to 50-cm.

Damage at the deeper Elpis site (about -11 m, MLW) primarily included two adjacent craters, measuring 69 m^2 and 163 m^2, likewise formed by the ships' screws as she tried to free herself. The depth of the craters, below ambient grade, varied from 0.5 to 2.3 m. The volumes measured 127 m^3 and 200 m^3. The craters' seabed was composed of sand and coral rubble similar to that of the Maitland site, but featured numerous large coralline boulders of up to 1.8-m diameter. The sand fraction was carbonate, with d_{50} = 0.6 mm. Adjacent to the craters were three berms consisting entirely of relic *Acropora cervicornes* (elkhorn) coral rubble that had been initially ejected from the craters. This naturally branched rubble, primarily 10- to 25-cm long by 3-cm diameter, was interlocked but easily succumbed to dismantling by hand or jet probe. The berms' relief varied from 0.3 to 1.2 m with crest widths of 0.6 to 1.5 m, and totalled about 30 m^3 in volume.

Scattered across a 0.5 by 1.5 km^2 area centered about the craters were over a dozen patches of fractured coral rubble. These patches, varying in size from 3-m diameter to 230 m^2, traced the inbound and outbound paths of the Elpis as her hull clipped and crushed higher-relief coral spurs. The weight of the vessel upon these spurs caused the coral to fracture deep below its surface. As a result, the surface rubble (about 10 to 20 cm) was loose and fairly easily dislodged by hand or jet probe. However, once exposed, the underlying coral could *also* be dislodged -- to an apparently limitless depth -- because of the dense cracks which permeated the substrate. By volume, the rubble measured about 15% at 18 cm, 30% at 13 cm, 20% at 9 cm, 25% at 5 cm, with the remaining 10% being sand. The total area of the rubble fields was estimated to be over 450 m^2.

Both sites, being at the seaward edge of the Florida Keys reef tract, were exposed to ocean waves. Annual average heights and periods were hindcast as about 1.0 m and 5.0 seconds. The largest waves, associated with tropical storms, were assumed to be depth-limited conditions, with nominal 2-m storm surge levels. Both sites featured daily tidal currents on the order of about 0.3 m/s. Hourly wind observations for an 8-year period were analyzed to discern those weeks for which the average hourly-sustained wind speeds were less than 5-, 10- and 15- knots. These data were later utilized to determine the statistically optimum window(s) during which on-site construction activities would be best undertaken. (In brief, the optimum window was found to be July 22 - August 18, with the most expansive window being June 17 - September 22.)

Conceptual Design Alternatives. A range of conceptual engineering alternatives was developed for the sites' structural restoration (Bodge and Creed, 1993). These were evaluated by the Sanctuary's Trustees (the National Oceanic and Atmospheric Administration (NOAA)) in terms of stability, construction feasibility, aesthetics, potential for biotic colonization, and probable cost (Sheehy, Bodge, and Finch, 1995).

The no-action alternative was ultimately not favored primarily because (1) there was little indication that the two sites' injuries were rapidly, naturally healing; (2) there was no indication that the shallower Maitland site's crater would cease its expansion; and (3) there was potential for additional injury by mechanical damage associated with the sites' mobile rubble. (That is, in storm conditions, rubble colonized by juvenile coral larvae would overturn and destroy the colonies and, perhaps, other established corals.) Other alternatives considered for the sites, but ultimately declined in favor of the preferred alternative (described below), included a continuous concrete cap, concrete-filled pillow mats, gabions, cable- and non-cable-stayed revetment mats, conventional boulder fills, and soil stabilization. Particularly at the shallower Maitland site, it was essential that the restoration structurally secure the crater's friable perimeter, and that the work physically fit into the shallow and irregular relief of the crater.

The selected alternative for the deeper *Elpis* site involved temporary relocation of the existing coralline boulders (i.e., those with existing biota), from the craters to the surrounding seabed. The coral rubble berms would then be mechanical relocated back into the craters from which the rubble was ejected. An armor layer of 1.2-m marine limerock boulders (totalling 400 tons) would be then placed atop the rubble. The voids were then to be partially filled with aragonite sand (approx. 60 m^3). The temporarily relocated coralline boulders, originally salvaged from the crater, would be then replaced atop the imported boulders and sand fill. (See **Figure 2**.)

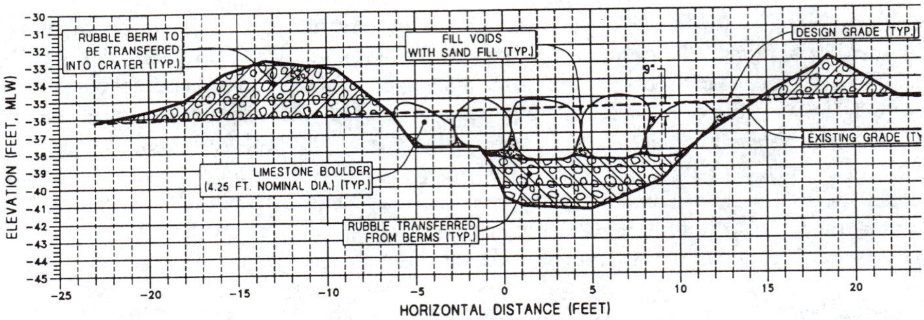

Figure 2: Typical section of Elpis site restoration.

The selected alternative for the shallow *Maitland* site (-2.5 m MLW) involved precision placement of 40 pre-cast "Reef Replicating Armor Units/Living Foundations" (RRAULF's) into the crater, and an underwater pour of specially-designed non-separable concrete (approx. 45 m^3) intended to bind the fragile coral perimeter to the armor units and to fill the gaps between the units (**Figure 3**). The finished elevation of the work was designed to match that of the natural seabed (**Figure 4**). Steel bar, driven into the seabed, and limerock boulders were to be placed between the units and the crater's perimeter to help secure and dress the concrete, respectively (**Figure 5**). Limerock gravel and coral berms were to be used as termination forms for the concrete where necessary (**Figure 6**). As for the Elpis design, the overall stability of the work was designed for a 50-year storm event using stream function wave theory and considerations of drag, inertia, lift and frictional resistance. For stability, the minimum coefficient of friction between an independent armor unit and the seabed was computed as $\mu_f \geq 0.5$ (conservative $C_{Lift} = 0.73$) or $\mu_f \geq 0.38$ ($C_{Lift} = 0.4$).

Each of the 40 armor units featured a highly irregular surface of limerock boulders and exposed aggregate in order to emulate the ambient seabed (**Figure 7**). The units, each 8300 kg (dry weight), were formed in six different sizes to accommodate the irregular shape and depth of the crater. Steel reinforcement for each unit was designed as a perimeter box beam such that each could be cantilevered 50% in any direction -- in anticipation of the irregular seabed. A standard Portland cement mix was specified for the armor units.

Mix design and specifications for the underwater non-separable concrete were developed by Ben C. Gerwick, Inc. (BCG) of San Francisco, CA. The design mix is summarized in **Table 1**. Details of the mix design are available from the author, Bodge (1995), or BCG ((415) 398-8972).

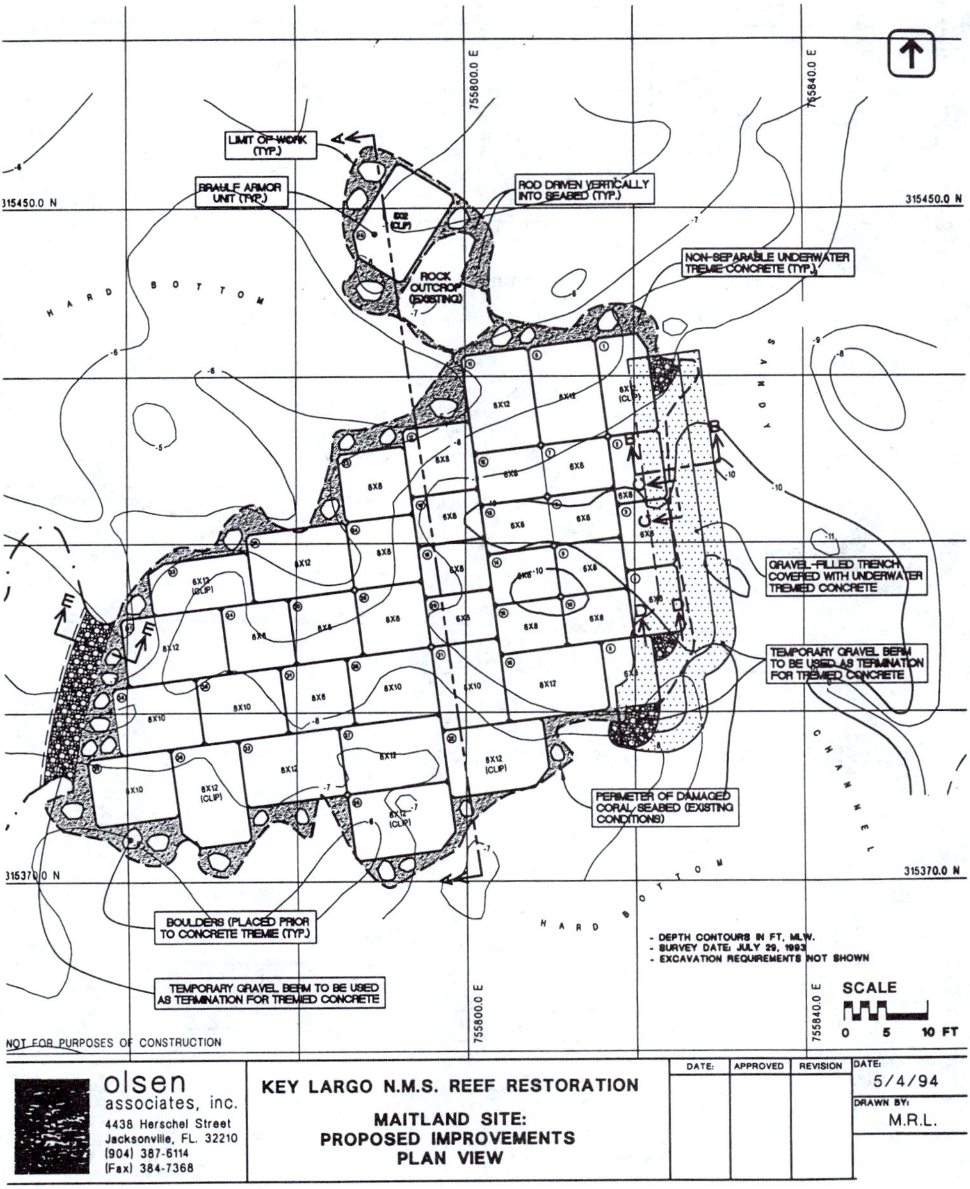

Figure 3: Plan view of Maitland site restoration.

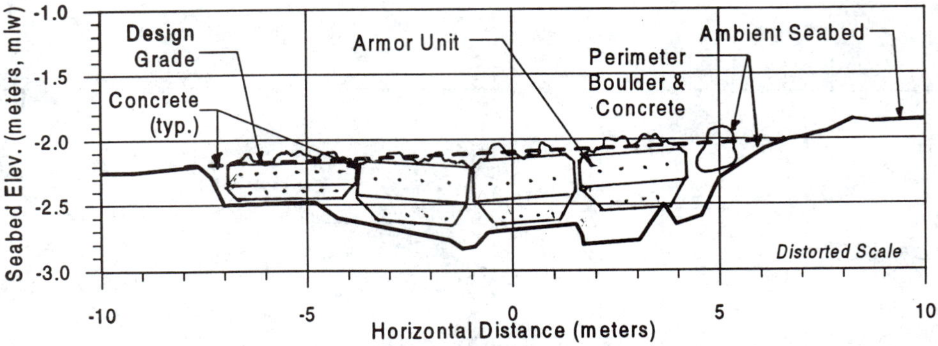

Figure 4: Maitland site restoration (typical section).

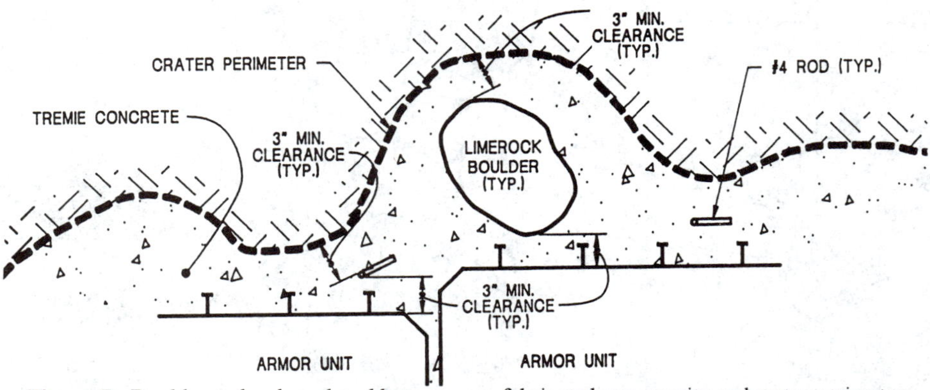

Figure 5: Boulder and re-bar placed between pre-fabricated armor units and crater perimeter.

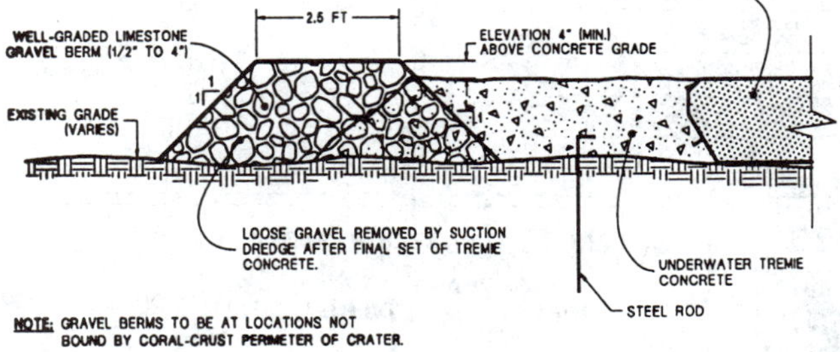

Figure 6: Limerock gravel and coral rubble used as a termination berm for the underwater concrete pour.

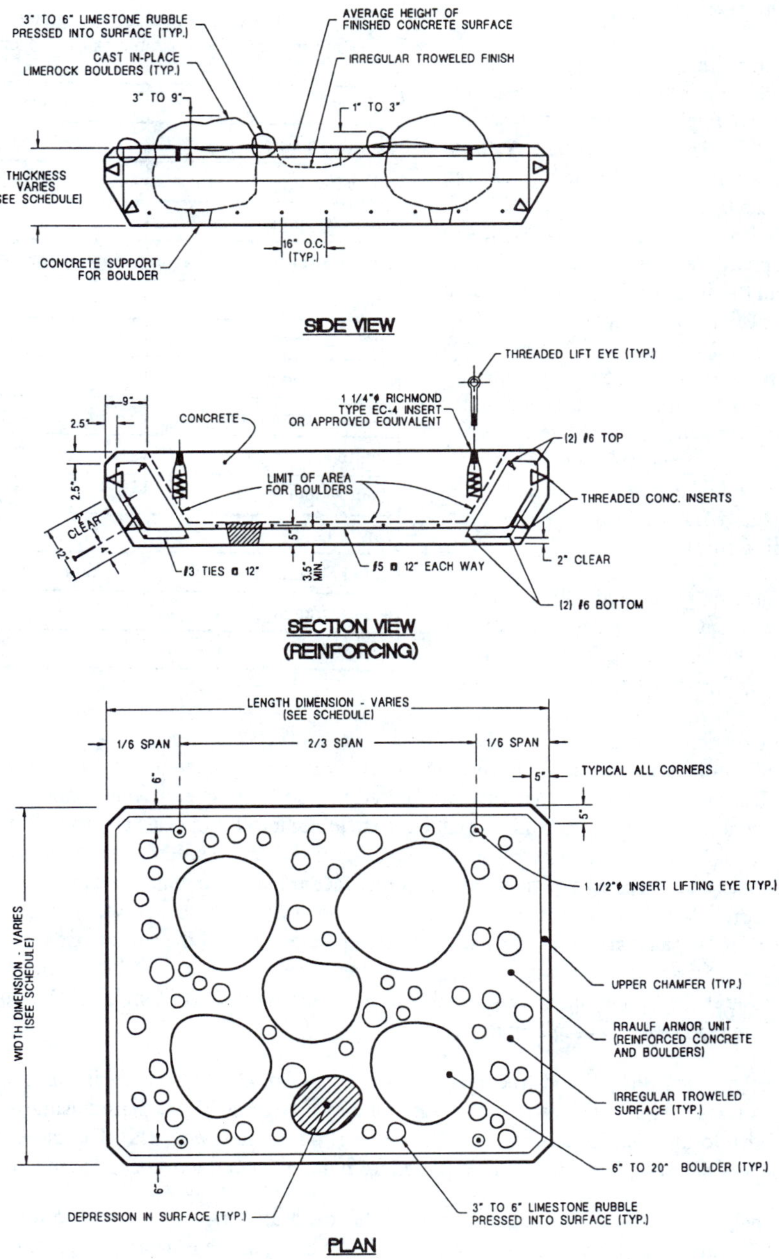

Figure 7: Pre-fabricated (RRAULF) armor unit design for the Maitland site, typical.

IMPLEMENTATION

Table 1: Nominal mix design for underwater, non-separable marine concrete.

Cement	600 lbs./cu.yd.
Fly Ash (class F)	90 lbs./cu.yd.
Silica Fume	43 lbs./cu.yd.
Coarse Aggregate (3/8")	1105 c.y. (40%)
Fine Aggregate	1685 c.y. (60%)
Water	250 lbs/cu.yd.
HRWRA (Sikament 300)	1.3 gals./cu.yd.
AWA (Sikamix 100 SC)	0.6 to 0.9 gals./cu.yd.
Water/Cement Ratio	0.4 ± 0.01
Slump	10" (spec); 9" - 11" (typ.)
Unit Weight (typ.)	142 lb/cu.ft.
% Washout Loss	<1.5%
Initial Setting Time	60 minutes (approx.)
4-day Strength (typ.)	5425 lbs/in^2
28-day Strength (typ.)	6110 lbs/in^2

Permits. Environmental permits for the work were issued by the U. S. Army Corps of Engineers, pursuant to Section 10 of the Rivers and Harbors Act, and Section 103 of the Marine Protection, Research and Sanctuaries Act of 1972. (The latter was required because it was proposed that surplus rubble removed from the sites would be disposed of at an existing ocean disposal site offshore of Miami.) The NOAA Sanctuaries and Reserves Division (SRD) likewise issued a permit for operation within the marine sanctuary. The State of Florida (DEP) was contacted, but did not issue a permit because the activity was outside of State waters (i.e., 4.8-km limit from shore).

Contracting. In mid 1994, NOAA entered into a cooperative agreement with the U. S. Army Corps of Engineers, Jacksonville District, by which the Corps would solicit and manage the construction contract for the work. Final construction plans and specificatons for the work were prepared by the author for the Corps by December, 1994. Procurement was based upon a "best overall value" approach, whereby potential contractors concurrently submitted separate Technical Proposals and Cost Proposals. The former included the Contractor's response, by prescribed form, to specific questions regarding their corporate and personnel experience (among other factors), and their narrative proposals describing their probable technical approach to the work. The offerors' responses were ranked according to prescribed evaluation criteria, previously prepared for the Corps by the author. Cost was to be used only as a "tie-breaker" among the most technically-qualified offerors (which, ultimately, was not necessary). Five proposals were received. The work was awarded to Team Land Development, Inc. of Pompano Beach, FL, in March, 1995.

Construction. Project construction was initiated in April, 1995. The Maitland site's 40 armor units were cast in the Contractor's south Florida yard over a 30-day period using about a dozen wooden forms. The limerock boulders and gravel were locally quarried. The carbonate sand for the Elpis site was oolitic aragonite, imported from the Bahamas and stockpiled in Fort Pierce, FL.

On site, the Contractor utilized conventional shallow draft barge (12 m by 43 m with 1.1-m draft at 150-ton nominal payload), and 27-m tug (1.4 m draft) as the primary work platform. A 10-m pusher tug (0.8 m draft) was used to maneuver the work barge over the shallow Maitland site. At nightfall, the 8- to 12-man work crew bivouaced ashore, at Key Largo, via small workboats.

Virtually all aspects of the work were primarily accomplished or directed by one or two surface-air-supply divers. These divers' communications were audible to the crane operator on the barge deck.

A 3-point mooring was used at the Elpis site, with 120-degree separation at 120-m distance, each, from the work site. Each mooring consisted of a 61-cm, 1.3-cm steel plate anchored to the seabed via four 3-cm, 1.2-m long threaded rods located at the corners of the plate. The rods were anchored to the seabed by pre-drilling deep holes into the coral seabed, placing the rods therein, and grouting with pre-mixed hydraulic cement carried to the seafloor in a 10-cm PVC tube. The plate was secured to the rods by a lockwasher and nut. A padeye was welded to the center of each mooring plate. To each padeye was attached a 3-cm shackle and a buoyant (9-cm polypropylene) mooring line for the barge. Each of the four threaded rods at the corner of each mooring plate was load-tested to vertical pull-out resistance in excess of 66.8 N (15 kip). The barge moorings for the Maitland site consisted of three 61-cm dia. by 1.3-cm wall steel pipes driven approximately 6-m into the seabed by vibro-hammer. Here, the barge was moved offsite at night, during particularly low tides, or in advance of approaching squalls.

At the Elpis site, the rubble berm material was transferred into the cranters using both a crane-operated grapple and diver-operated water jet. The latter was more succesful, by far, than the former. The limerock armor boulders were placed individually by grapple. The fill sand was placed by a crane-supported, diver-operated 3.5-m^3 hopper. Tidal currents carried some of the sand as much as 150-m downcurrent, and resulted in the upcurrent crater being filled with sand to a notably lesser degree than the downcurrent crater (see **Figures 8 and 9**, respectively). As seen in Figure 9, local fishes were attracted to the work within hours of its completion.

Attempts were made at the Elpis site to remove unconsolidated coral rubble using both a 15-cm suction dreged with deck-mounted pump and a 10-cm water lift discharging into metal baskets lowered to the seabed. Neither technique proved satisfactory for the work because the rubble was scattered and productivity was very poor. Larger rubble (>8 cm) frequently jammed the intake. [Subsequent rubble removal activities undertaken with suction dredging and water lifts at other grounding sites -- where the rubble was more spatially concentrated and completely unconsolidated (i.e., where the vessel grounding had occurred within the previous year) demonstrated far greater success than at the Elpis site.]

At the Maitland site, the Contractor preferred to use gravel fill to level the crater's seabed (in preparation for the units' placement) rather than excavation. This choice was primarily dictated by the cumbersome difficulty in excavating the dense rubble/sand matrix of the crater's coralline seabed. Excavation around the crater's perimeter, however, was nonetheless necessary to ensure that the upper surface of the perimeter armor units (thickness < 36 cm) would not extend above the adjacent, ambient seabed grade. Coral rubble excavated from within the crater was used to construct a berm around the crater to serve as an additional termination form for the underwater concrete, and to help blend the units' perimeter with the ambient seabed.

After precision placement of the 40 armor units (placed with 5-cm maximum allowable gap between units, and tighter tolerances for vertical differential), and placement of the perimeter boulders and coral/gravel berms, the underwater concrete was placed using a swing-tube pump with 5-cm hose, PVC nozzle, and diver-operable ball-valve. The concrete was batched aboard the barge using a portable ("junior") mixer.

Figure 8:
Photograph of upcurrent Elpis site crater immediately after restoration. Voids between the boulders provided extensive fish habitat and vertical facing for coral larvae recruitment, but resulted in an aesthetic that contrasted with the ambient seabed.

Figure 9:
Photograph of downcurrent Elpis site crater immediately after restoration. Voids between the boulders were more effectively filled with carbonate sand, resulting in an excellent aesthetic match with the ambient seabed.

Figure 10 (next page) is a photograph of the *M/V Maitland* hull aground upon the reef in 1989. **Figure 11** depicts one of the pre-cast armor units being lowered from the barge deck to the Maitland site seabed. **Figures 12 and 13** are typical photographs of the Maitland site restoration, taken about a week after the work was completed.

Site work was conducted between June 19 and July 12 at the Elpis site, and between July 13 and September 1, 1995, at the Maitland site. For about half of this time, however, no site work was undertaken because of relentlessly bad seas. In all, the 1995 season was the second worst on record in terms of Atlantic tropical storm activity. At one point in August, there were five tropical storms simultaneously lined-up across the Atlantic, moving toward the Florida peninsula.

Construction Review. Real-time construction review was conducted and coordinated by the Corps, with direct participation of NOAA/SRD personnel and the Design Engineer (Olsen Associates, Inc.). The unique and complex relationship between the project's many parties was facilitated, from the outset, by a 2-day *partnering* conference organized by the Corps prior to construction.

Figure 10:
Photograph of Maitland aground on the reef, 1989. The ship screw is seen resting amidst the coral rubble caused by the grounding.

Figure 11:
Photograph of one of the 40 pre-cast reef-replicating armor units being lowered to the seabed. The lifting eyes were unscrewed from the units' threaded inserts after installation.

Figure 12:
Photograph of the top surface of the Maitland site repair, taken several days after project completion. The ambient seabed is in the far background. A grouted joint between adjacent armor units is discernible below the fish, in the foreground.

Between site visits, the Engineer (author) was kept abreast of the work by viewing underwater images that had been video-taped by his on-site representative (Mr. Mark Schroeder, Continental Shelf Associates, Inc.). Selected videotape images were captured as a single frame by Mr. Schroeder, then transmitted from his boat via laptop computer, cellular phone, and E-mail to the author's Jacksonville office -- usually within 20- to 30-minutes of the original photography. On several occassions, the Engineer's access to virtually real-time construction photos permitted timely, cost-saving corrections to the work; or, alerted the Engineer to situations that required an immediate trip to the site and/or instructions to the site-representative.

Environmental Monitoring. Turbidity was monitored using concepts of "exposure" by which "cumulative NTU-hours" were measured and tabulated. There was no observed environmental degradation resulting from the work -- despite the fact that the work was conducted in the midst of sensitive coral resources. Fishes, soft and hard corals have begun re-colonizing the sites (Harold Hudson, NOAA/SRD - personal communication).

Project Cost and Completion. The project's construction cost (bid) was U.S. $1,047,000, with subsequent change orders resulting in a net additional cost of $19,600. The total construction cost was about 10% less than the Engineer's estimate. Overall, the project was completed on-time, within budget, in accordance with the design intent, and with no apparent net adverse impact to the environment.

Figure 13:
Photograph of exposed side-face of Maitland restoration, where the repair was terminated adjacent to an existing sandy channel. The restoration is on the left, the ambient seabed is on the right.

CONCLUSION

To the author's knowledge, this experimental project represents the first large-scale, in-situ structural restoration of a coral reef damaged by mechanical impact. Innovative methods introduced, or resurrected, during this project included

- development of structurally sound seabed restoration with emphasis on aesthetic blending with the ambient environment;
- further application of new non-separable underwater marine concretes,

- contract award by the Corps using "best-overall-value" procurement,
- unique cooperation between multiple federal agencies and the private sector, and
- semi-real-time construction review using underwater-taped video images via the Internet

among other items. The unfortunate, increasing frequency with which vessels ground upon coral resources -- and other environmentally sensitive seabed resources -- suggest that the lessons learned as part of this project will be of increasing utility to future restoration projects.

ACKNOWLEDGEMENT

The engineering assessment, design, and construction review work described herein were conducted in principal part by Olsen Associates, Inc. under subcontract to Industrial Economics Inc. (IEc), Cambridge, Massachussets, as authorized by the U. S. Dept. of Commerce, National Oceanic and Atmospheric Administration (NOAA), Contract No. 52-DGNC-1-00007, under Task Order Nos. 50062 and 50089. The leadership of Dr. Dan Sheehy, of IEc, in this project is specifically noted. Initial site survey work was coordinated by Mr. Keith Spring of Continental Shelf Associates, Inc. (CSA), of Jupiter, Florida. The Engineer's on-site technical representative was Mr. Mark Schroeder, also of CSA. Hydrographic survey work was conducted by ARC Surveying & Mapping, Inc., of Jacksonville, FL. Mr. Chris Creed, of Olsen Associates, Inc., greatly assisted in the site survey, data analysis and preparation of the engineering assessment. The innovative internal structural design and specifications for the Maitland site armor units was principally created by Mr. Steve Klecka of SK Engineering, Inc., Jacksonville, FL. The highly succesful concrete mix design and specifications were provided by Dr. Dale Berner of Ben C. Gerwick, Inc. (BCG), of San Francisco, California; and then field-reviewed by Mr. Patrick Durnal of BCG. Construction management was executed by the U. S. Army Corps of Engineers, Jacksonville District. Construction review for environmental protection was directed by Messrs. Harold Hudson, Bill Goodwin, and John Halas of the NOAA Sanctuaries and Reserves Division, Key Largo. The construction contractor was Team Land Development, Inc., Pompano Beach, Florida; Mr. Ron Coddington, President. For the contractor, Mr. Charles Calloway was the lead engineer. On behalf of the project's sponsor, NOAA, Mr.Tim Osborn (National Marine Fisheries/Restoration Center), Dr. Charles Wahle (Sanctuaries and Reserves Divsion), Ms. Darlene Finch (Damage Assessment Center), and Lt. John Tokar (contracting officer) each managed principal aspects of the project, each with great patience and faith.

REFERENCES

Bodge, K. R., 1995. "Engineering Summary Report: Structural Restoration of the *M/V Alec Owen Maitland* and *M/V Elpis* Vessel Grounding Sites", Prepared for U. S. Dept. of Commerce, National Oceanic and Atmospheric Administration, Silver Spring, MD; by Olsen Associates, Inc., 4439 Herschel St., Jacksonville, FL 32210; December 10, 1995.

Bodge, K. R. and Creed, C. G., 1994. "Conceptual Engineering Alternatives for Structural Restoration of the Matiland and Elpis Grounding Sites", Prepared for U. S. Dept. of Commerce, National Oceanic and Atmospheric Administration, Silver Spring, MD; by Olsen Associates, Inc., 4439 Herschel St., Jacksonville, FL 32210; 112 pp; Oct. 15, 1993.

Sheehy, D. J., Bodge, K. R., and Finch, D., 1995. "Coral Reef Restoration Planning: A Decision Analysis Approach", Proc., 1995 Int'l Conf. of the Society for Ecological Restoration; Seattle, WA.

CHAPTER 331

Dredging and Disposal within the Limits of a National Park

H. A. Manzenrieder and J. M. de Vries[1]

Abstract

In coastal areas under tidal influence, recurrent dredging is an important part of the maintenance work to secure the nautical situation and the functionality of the harbors. In the past, the orientation of dredging activities was purely economic but since the early 80's, environmental concerns have rapidly increased and these concerns often define the limiting or determining factors that cover all parts of such operations in the marine region. At present, the sometimes quite different interests of engineering and ecological groups are becoming more gentle. Besides the natural learning effect from the changes in general, economic conditions are also an important factor.

This situation especially applies when maintenance dredging takes place within the limits of a national park where environmental concerns have the highest priority.

The scope of the engineering investigations was the estimation of the environmental capacity as a function of hydrology, sedimentation and biology leading to a specific management for dredging and disposal areas.

Introduction

Public awareness concerning the environment was highlighted at the United Nations Conference on the Environment in 1970. This triggered the development of legislation concerning the control of all kinds of waste. In Europe, a need for legal action ensued from the following international agreements, some of which were ratified by Germany:

- Oslo Convention of Feb. 15, 1972, with a supplement of June 12, 1991
- Helsinki Convention of April 9, 1972, with a supplement of June 6, 1992
- London Dumping Convention of Dec. 29, 1972
- Paris Convention of June 4, 1974

[1] Consulting Engineering Office Dr. Manzenrieder and Partner, Brookweg 29, 26127 Oldenburg, Germany

The guidelines for the disposal of dredged sediments are designed to assiscontracting parties in the management of dredged sediments in a way that will prevent pollution of the marine environment. The relative importance of the dumping of dredged material within the containment arriving in the North Sea is estimated at approx. 1,600,000 t/y for nitrogen and approx. 10,000 t/y for lead (Neville-Jones 1994).
The guidelines and restrictions apply to the removal as well as the disposal of dredged sediments. A special question is hydraulic injection dredging and transport by the local current capacity.
When making an assessment in general, it will be necessary to consider the central factors as legal requirements, technical aspects, economic considerations, environmental aspects and more complicated subjective local knowledge and emotional aspects or estimations.

In fact, a very important stimulus for this investigation came from the observations of people who have been personally involved in local activities for a long time (even generations), e.g. the tourist guide.

In practice, environmental requirements are reflected in the individual permits which are only valid for a limited time (2-3 years). In Germany, such official permits actually specify the amount of dredging material (sand, mud), the dredging and dumping areas and the duration of dredging and disposal related to the season in the year (preference: winter) and the phase during the astronomic tidal cycle.

The need for investigation which is laid down by the valid guidelines is determined by the pollution load of the dredged material. Guide values r (mg/kg) are respectively defined for the individual heavy metals.
The guide value r results from concentration c related to the 20 μm fraction multiplied with an uncertainty factor α ($r = c \bullet \alpha$), at present $\alpha = 1.5$.
Within the maximum period of three years, the need for investigation is classified by three ranges of values ($c \leq r$, $r < c \leq 5 \bullet r$, $c > 5 \bullet r$).
In addition, permits can also define the need for and determine the scope of investigations on environmental effects as in this case.

Dredging in the National Park

The islands along the southern part of the German North Sea coast, which are predominately oriented to tourism, are located within the eco-system of the National Park's tidal mud flats and have been an important part of this young park, deserving special protection, since 1985.
Based on a mathematical model that gives a rough description of the current field (Dick 1992), the transport system under tidal effects is predominantly from West to East:

The protection status is to safeguard natural development and covers all activities in an area of approx. 2,400 km^2 with some exceptions within the framework of coastal protection (dikes) and maintenance activities (dredging). In 1993, the National Park became part of the UNESCO-Program ''Man and Biosphere'' (MAB). This status was the reason for some interesting economical and engineering requirements e.g. when an expensive tunnel was build for a standard pipeline with a diameter of approx. 1 m (Europipe-Project) or the work to raise a dike was suspended during the operation.

Within this protectorate, the navigable channels to the harbors along the coast and the islands lie mainly in areas with continuous sedimentation, the transport processes of which are defined by a regular average tidal range of approx. 2.6 m, superimposed by random storm surges. The consequence of these conditions is the substantial necessity for quasi permanent dredging activities inside the National Park and subsequent dumping. Alternatives, such as disposal outside the protected area or use of the material for beach nourishment in the vicinity are discussed and carried out in part.

Within the National Park, annual dredging quantities of between 300,000 and 500,000 m^3 arise from the maintenance of navigation channels. Small suction hopper dredges with a cargo hold capacity of 150 t (mud) to 300 t (sand) are used for dredging and disposal, the limiting factor of which is the allowable draught with a maximum of 3 m determined by the shallow depth of the flat tidal area. As a rule, disposal is carried out gradually over the whole disposal area to achieve an even as possible distribution at the bottom and in the space of the water.

In connection with intensive monitoring of all dredging activities within the National Park, the measures are based on complete modernization of the position finding system based on the highest GPS technics in combination with the continuous recording of basic data to objectively document dredging activity.

With these measures, the efficiency of maintenance dredging is to be increased on the one hand and on the other, complete evidence on the executed dredging and disposal processes are to be ensured.

Effects of Dredging and Disposal

Due to the fact that the main problems concentrate more and more on the disposal of dredged material, the following description will focus on this question.

Dependent on material and area specific parameters, the introduction of dredged material into the water space may, as a result, have a variety of effects on the ecological boundary conditions (changes in the environment). This especially applies to the tidal mudflats.

In keeping with the knowledge we have today, an underlying, lasting and negative effect cannot be assumed a priori.

Area specific parameters such as e.g. current conditions, bottom material or sea conditions, are important boundary conditions for the transport and the dispersion behavior of the solid matter introduced into the natural tidal mud flat area by disposal.

The content of suspended solid matter is influenced by a number of parameters:

- Tide (current velocity, water level, turbulence, salinity)
- Meteorology (sea conditions, temperature,) and Climate (duration of ice covering)
- Topography (surrounding tidal mud flats, bottom material, artificial buildings)

In addition, there are chemical parameters such as e.g. oxygen or nutrient content which influence solid matter content, sedimentation behavior and turbidity conditions.

An urgent, independent problem is defining and selecting suitable parameters and characteristic quantities for describing or supplying evidence on the effects against a background of high, natural variance in tidal waters.

In the following, possible effects from the introduction of dredged material into the space of water and therefore the questions associated with it are presented for assessment:

- Local increase of solid matter content
  - Is there a correlation between solid matter content and turbidity?
  - Does a sandblasting effect occur at the bottom in the case of sandy material?
  - Can a natural turbidity or solid matter background be defined?
  - Are changes in light conditions to be evaluated vertically?
  - Is the influence on photosynthesis quantifiable?
  - Will the migration behavior of fish be influenced?
  - Will this reduce the recognition of danger (e.g. fishing nets)?
  - Are there accumulation effects?
- Reduction of oxygen content by introducing oxygen-depleting matter
  - Which material is considered oxygen-depleting?
  - Which depletion potential leads to a reduction in oxygen content?
  - How does the same depletion potential act under dynamic variability?

According to Essink (1993), the disposal of dredged material can, among other things, have the following ecological effects:

- Change of the nutrient balance

  Example: Phosphate redissolution can have the effect of increasing the growth of phytoplankton (Sea grass can be negatively influenced when algae cover the surface).

- Inhibition of growth

  Example: An inhibition of the growth of phytoplankton can occur through increased turbidity resulting in a reduced supply of food for zooplankton and filtering bottom organisms. Turbidity interferes with the search for food of visual predators (flat fish, ocean swallows)

- Inhibition of respiration
  Example: Increased content of suspended solids inhibit gill respiration of fish and invertebrates as well as food intake for zooplankton and filtering organisms (small crabs, mussels)

- Overtaxed adjustment
  Increased sedimentation rates have the effect of smothering bottom animals. Sedentary Benthos species have less chance of survival as opposed to mobile species. Fundamental changes in the composition of sediment can increase the damaging effect on bottom fauna.

The processes that take place in the water space and at the bottom are interdependent and have superpositioning effects. Colonization at the bottom can, for example, lead to increased erosion resistance and therefore influence the suspension content (Manzenrieder 1983).

Relevant Publications

In the past years, several papers have been published on the problem of environmental effects caused by dredging and disposal of sedimentary material and International Conferences in particular have provided an opportunity for exchanging ideas and establishing relationships on a wide range, e.g.. DREDGING ´94 by the ASCE in Orlando.
Some of the projects include the results of field studies and corresponding theoretical models. In Table 1, a partial list gives short information on selected studies based on the results of field programs covering typical engineering tools (sounding, sampling, sensor measurements, etc.) and an indication of extended results from natural science (Chemistry, Biology). The depth given indicates the depth at the dumping area.

The targets, size and executing institutions and - last not least - financial frame for such programs vary. In general, it can be stated first of all that every dredging and disposal area has its own behavior so that a direct transfer of results is not likely. Experience in engineering practice has often indicated that the respective research must be delegated to basic research and research with a direct reference, the latter of which often has narrow limitations on time and budget. Nevertheless, this is often the appropriate solution.

To answer basic questions concerning mud flats, a very large research program named ´´Ecosystem-Research´´ has been underway in Germany since 1989, the main phase expected to end in 1996. Beyond general statements, the results presented could not support the concrete questions formulated in the permits for maintenance dredging.

Table 1: Selected publications on dredging effects with field studies

Literature	Year	Country	Location	Chemistry	Biology
Hübner, H.-J. et al.	1996	Germany	Mudflats	yes	yes
Pejrup, M.	1995	Denmark	Mudflats	no	no
Netzband A. et al.	1995	Germany	Elb River	yes	yes
Borst, W.G. et al.	1994	Netherlands	Haringsvliet	yes	yes
Halka, J. et al.	1994	USA	Chesapeake Bay	no	no
Wiley, M. et al.	1994	USA	Connecticut	no	no
Marsh, J.	1994	U.K.	Cornwall	yes	yes
Stuber,L.M. et al.	1994	USA	Savannah Harbor	yes	yes
Tubman, M. et al.	1994	USA	San Francisco	no	no
Courtney, C.A. et al.	1994	USA	San Francisco	no	no
v. Oostrum, R. et al.	1994	Netherlands		yes	no
Bundesanstalt (BfG)	1993	Germany	Ems River	yes	no
Essink, K. et al.	1993	Netherlands	Ems, Mudflats	no	yes
Thevenot, M. et al.	1992	USA	Tylers Beach	no	no
Paul, J.H. (BfG)	1992	Germany	Weser River	no	no
Dammschneider	1992	Germany	Elb River	yes	no
Gallenne, B.	1989	France	Loire River	yes	yes
Vale, C. et al.	1989	Portugal	Tagus River	yes	yes
Bossinade, J.H. et al.	1988	Netherlands	Eemshaven	no	no
Truitt, C.L.	1986		Duwamish WW	no	no
van der Veer, et al.	1985	Netherlands	Mudflats	no	yes
Tavolaro, J.	1982	USA	NewYork Bight	no	no
Malherbe, B.	1980	Belgium		no	no
Bokuniewicz, H.J. et al.	1978	USA	Lake Erie/Ontario	no	no
			New York Bay		
			New York Bay		
			Saybrook		
			Elliott Bay		
Sustar, J.;Wakeman,T.	1977		Carquinez	no	no
McCauley, J.E. et al.	1976	USA	Coos Bay	yes	yes
Gordon, R.B.	1974	USA	Long Island	no	no
May, E.B.	1973	USA	Mobile Bay	yes	yes

The various public interests in the results of such (expensive) investigations which include social and political components are often much greater than the knowledge that has been established on the processes. The contrary explanations for the appearance of large areas with a oxygen deficit in the upper mud flat surface, so called ′′black spots′′ after a strong winter, was the last remarkable example.

Procedure

The main target of our investigation was and is compliance with the requirements that are a part of the permit, currently valid for a period of 3 years. To achieve this, an investigation program was started in 1994 which is still being executed, the areas of operation being:

- sediments in the subtidal and intertidal flats
- water column
- macrozoobenthos

For a pilot project, disposal sites are selected, which differ in respect to their location within the hydrological units, the quantity of material being disposed as well as the type of disposed material.
In compliance with the concrete time and economic terms of reference and under the responsible coordination of engineering sciences, a pilot study was developed for the interdisciplinary questions, the structure with long-term measurements and detailed surveys was selected as follows:

- Meteorological and hydrological surveys
- Static and dynamic sedimentological surveys
- Biological surveys

To describe biological processes, macrozoobenthos were selected as an indicator group.

Long-term measurements gradually broaden our knowledge and are therefore a part of extended basic research. In view of the investigations listed for the disposal site area, these are primarily oriented to stabilizing the respective causal analysis:

- Analysis of the dredged material (grain distribution, pollutants)
- High resolution topographical surveys using state-of-the-art techniques
- A rough sedimentological and biological survey of the disposal sites
- Taking samples from monitoring stations in the subtidal and intertidal flats
- Sedimentological characterization of surface samples from the mudflats
- Assessment of macrobenthic fauna on the mudflats

The detailed investigations include a great number of individual measurements and are continuously coordinated to developments. Previous investigations have concentrated on the natural variation of turbidity and the change and extension during disposals as well as the changes in oxygen content during disposals.

Previous Results

Part 1: Dredged Material

All objective reflections on the effects of disposal arise from the potential load potential and therefore from the properties of the dredged material with following results:

- The results of laboratory measurements confirm pronounced material-dependent differences.
- As opposed to fine sand, silt (coastal marsh sediments) shows continuously higher depletion rates.
- Silt samples show an up to 45 times higher average depletion activity related to mass than fine sands.
- Depletion curves of the silt samples show a high initial depletion at the beginning which can exceed mean oxygen depletion by 10 times.
- Quantitative results of stabilized laboratory results cannot be directly confirmed by the natural measurements during disposal.
- In individual measurements during the disposal of coastal marsh sediments and sand, no changes in oxygen content beyond the fluctuation margin of measured values before disposal were determined in the space of water over a period of approx. one hour after disposal.
- Heavy metal concentrations determined in the grain size fraction $d < 20$ µm all range below the defined limiting values in the guidelines used (Figure 1).
- On the basis of this assessment, the examined dredged material can be rated as non-polluted.
- At present, there is no uniform assessment concept for the evaluation of heavy metal content and organic pollutants in dredged material.
- In regard to the area-specific loads from sediments, there is a need for uniform recording of data as a basis for documentation and the evaluation of trends while isolating anthropogenetic intervention

Part 2: Disposal Sites

Previously determined results of the engineering and natural science work fields will be presented in abbreviated form.

- Sediment composition and seasonal changes are different at every disposal site.
- Fine and medium grain sands dominate in the sublitoral, the highest variation being shown by the fine grain size fractions.

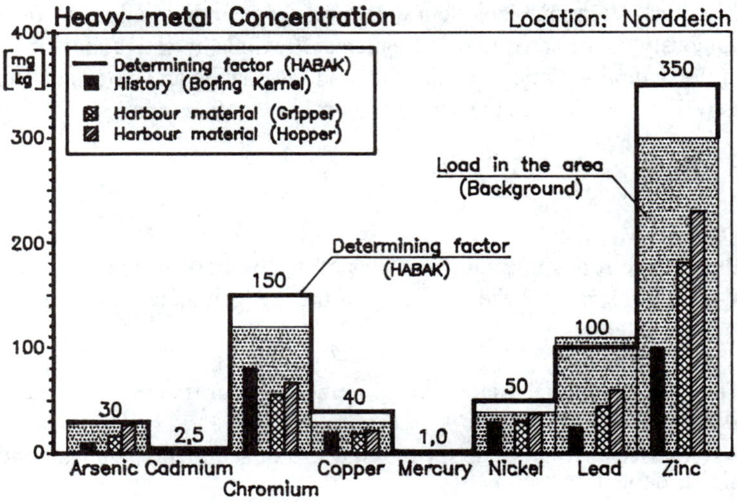

Figure 1. Heavy Metal Concentration and Determining Factors

- With mostly constant main sediment types, changes also occurred on the examined mudflat surfaces during the low-energy summer months in the fine grain areas.
- A sorting of grain size takes place within the immediate disposal site.
- The fine sediment part rises to the edges of a hydraulic/morphological unit (tidal mudflat watershed).

Part 3: Turbidity and Suspension

In order to determine the value range and variation of the natural suspended matter content in connection with current behavior, selective measurements were made over one tide respectively. These results form the basis for an evaluation of concentrated measuring results that were made during the disposal of sand and coastal marsh sediment. The results are characterized as follows:

- The suspension content measured fluctuated during a natural tide cycle (without disposal) in a wide range between 120 mg/l and 660 mg/l. The result over the whole tide was a mean suspension content of c=225 mg/l with the maximum values near the bottom.
- Drogues and continuous current measurements in the area of the examined disposal site support a current potential with the highest velocities close to the surface of up to 1.5 m/s at ebb current which, as a whole, dominates over the flood current.
- The measurements made immediately before disposal at lower current velocities comprised recorded values of 2.5 to 30 TE/F (turbidity unit related to Formazin) with respective suspension samples with solid matter contents of 50 to 200 mg/l.

- During disposals, stationary and dynamic measurements were made at a mean water depth of 15 m and in depth steps of 5 m and 10 m in the disposal site and adjacent lee area to survey the spreading cloud of turbidity.
- In Figure 2 a passage of the turbidity cloud during a disposal of sand (d_{50} ca. 0,15 mm) giving a current velocity and water level above the probe is presented. Distance to the input point was approx. 200 m.
  It was observed that the disposal "trail" clearly stands out against the background values of turbidity. Just a few minutes after input, turbidity values at the selected measuring position rise to approx. 100 TE/F.
  The mean basic turbidity of approx. 15 TE/F is significantly exceeded over a period of approx. 11 minutes with up to 150 TE/F. After that, the level returns to the level before disposal. At a mean current velocity of 40 cm/s, an extension in the length of the turbidity cloud to approx. 270 m results which corresponds to doubling.
- After the disposal of coastal marsh sediment, the maximum level in the turbidity cloud over a range of 3,000 m was reduced to a third of the measured maximum value.
- The oxygen content within the detected turbidity cloud consistently ranged within the measured natural fluctuation margin.

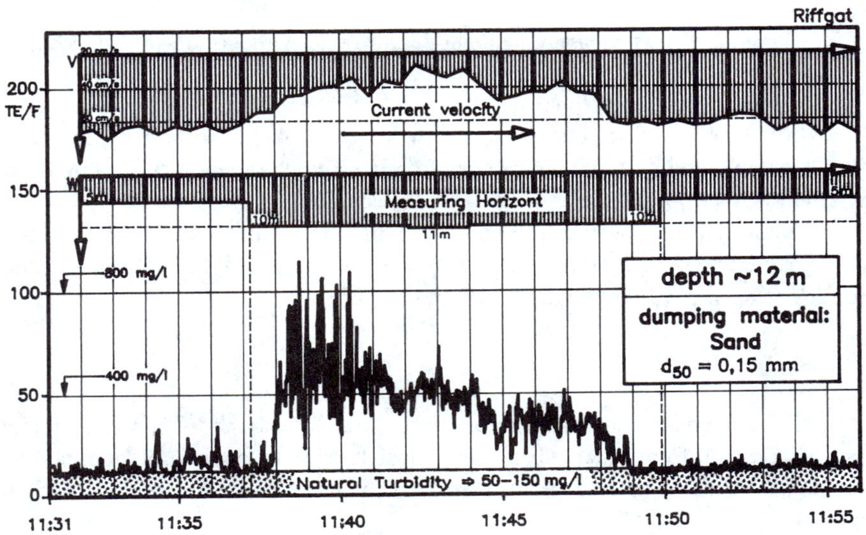

Figure 2. Turbidity Cloud during Dumping passing in the central Lee Side

The newest version of a modern self-sufficient instrument, a so called sand-surface-meter (Manzenrieder 1995) was placed during the disposal of sand and mud direct under the hopper dredger. The real-time history of the stationary measurements is concentrated on Figure 3.

During the observation the disposal of sand is predominant indicated by a stepwise change in the bottom position. The dominant effect of mud disposal is the transient formation of a cloud with very low particle movement. The stability of the mud clouds was detected between 2 and 6 hours during the measurements.

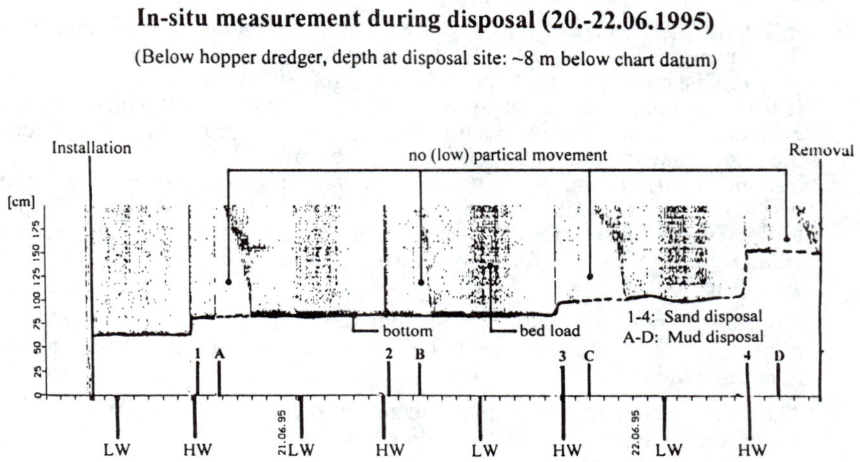

Figure 3. Direct Measurements under a Hopper Dredger during Disposal

Part 4: Benthos Investigations

In the subtidal flats, quantitative samples were taken using van Veen grab sampler supplemented by trawl net catches from a research dredge which were to give a qualitative overview of the epibenthic animal species.

In the intertidal flats, surface samples up to 30 cm in depth were taken respectively. A presentation of the complicated types of analysis and the different extensive groups of results will not be dealt with in this paper.

Assessment (Example)

Since the natural measurements carried out and the model approaches that are available do not form a stable basis for a closed solid matter balance, continuous current and suspension measurements were carried out for a rough model of solid matter loads.

For a defined cross-section in the middle of the disposal site, corresponding suspension loads of 5,800 to 7,700 t per tide phase result for suspension contents of 150 to 200 mg/l. Independent analyses on local bed loads carried by currents show that with an order of magnitude of approx. 10 % of the suspension load, this is a clearly smaller part of the total solid matter load. In Figure 4 the increase of the above presented (natural) suspension transport caused by the input of two dredged material disposals with a quantity of 300 t each is presented for the cross-section of the disposal site. In this case, the first disposal is assumed at approx. 2 hours after the turn of the tide and the second disposal 2 hours before. Under the assumptions made here, both disposals lead to an increase in the quantity of solid matter transported in suspension during the course of a tide of approx. 10%. This value lies within the documented natural fluctuation margin for the location.

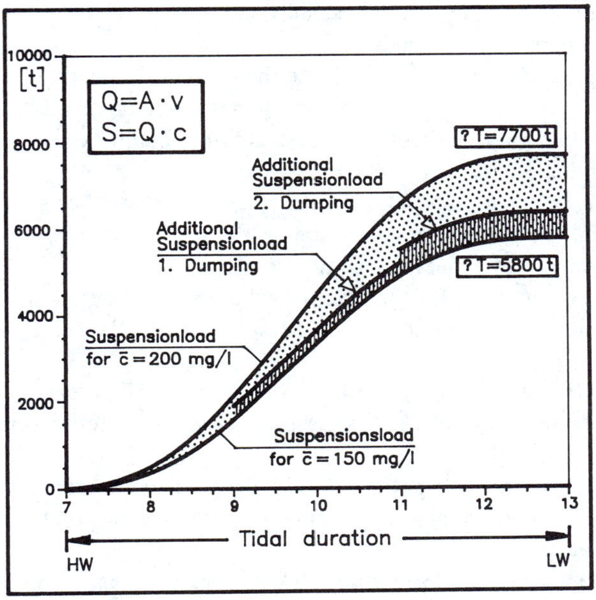

Figure 4. Increase of the natural suspension transport due to 2 disposals (example)

The conditions presented are based as an estimate on assumptions stabilized by measured data and describe the order of magnitude of the solid matter transported.
Greater knowledge about natural variation as a result of seasonal fluctuations or meteorological influences as well as secured data concerning the transport of sediment and suspended solids and their distribution for individual tide phases separately are needed for a secured, quantitative assessment of the effect of dredged material disposal. For this, a corresponding picture of current and transport must be described for the area.

Disposal Site Management

Because of the great number of influencing factors that are to be taken into consideration when managing dredged material, especially in shallow water areas, individual disposal site management is to be connected with corresponding ecological requirements. Here, the qualitative and quantitative assessment of specific boundary conditions provides an opportunity to combine the relevant economic and ecological interests into solutions capable of reaching a consensus.
Sufficient knowledge on the load potential in connection with the tolerance potential forms the basis for such disposal site management.

The load potential (stress factors) comprises all effective area and material specific effects concentrated in disposal.

The tolerance potential (ecological elasticity) describes the available reaction options of the water ecology and especially the bottom of the body of water under natural and/or anthropogenic loads.

A sensitivity classification as a basis for making decisions which defines limit criteria by stabilized guide parameters and therefore contains e.g. statements on allowable sedimentation dynamics (height, duration) should be the goal for selecting main areas for disposal or disposal sites.

When defining possible disposal sites, the question of whether selective or gradual disposal presents the more economical and ecological solution in the course of a year and for different energy input must be answered.

All of the known or the yet to be determined area and material specific parameters supply a contribution to the effect processes which define the eco-system.

Tolerance potential >> Load potential → Mostly uninfluenced environment
No prolonged change or damage to water ecology and little utilization of loading capacity

Tolerance potential > Load potential → Stable environment
No long-term change or damage to water ecology to be expected for high utilization of loading capacity

Tolerance potential = Load potential → Unstable environment
Unstable limit state of water ecology is reached by full utilization of loading capacity

Tolerance potential < Load potential → Destabilized environment
Prolonged changes or damage to water ecology occur when loading capacity is exceeded

Tolerance potential << Load potential → Changed environment
Comprehensive changes or damage to water ecology by clearly exceeding loading capacity

In the following, an evaluation matrix for the individual monitoring stations has been compiled for a disposal site within the National Park (example).

Based on a measuring program carried out over a period of several years in the area of the disposal site, short-termed, local changes could be quantitatively determined.

Long-term changes or damage to the environment which can be directly attributed to the disposals could not be documented in the dynamic ocean area so far.

A valid assessment of the project dream - "stable environment" - was undertaken according to the presented assessment scheme for the evaluation of intervention in the eco-system which forms the basis for the pending decision of the responsible permit authorities.

Parameter	Retake stations in/around the disposal area in relation to LW												
	above (mudflats)											below	
Bottom Sediments	-	-	/	-	-	/	/	-	/	/	+	/	n
Sedimentary Material	n	n	-	-	-	-	-	+	+	n	-	n	n
Oxygen Content	n	n	/	/	/	/	/	+	/	n	+	n	n
Topography	n	n	+	-	/	-	-	-	-	+	+	n	n
Benthos fauna	n	n	+	-	/	/	n	-	-	/	-	/	n
Turbidity	n	n	n	n	n	n	n	n	n	n	n	+	+
Suspension	n	n	n	n	n	n	n	n	n	n	n	+	+
Values in relation to the natural variation: +:above,/:boundary,-:below,n:not determined													

When objectively viewing the (required) state of knowledge on the manifold influencing factors (cluster), it can be determined that there are considerable deficits. Therefore, qualitative natural observations and quantitative natural measurements will still supply the central contributions.

Conclusion

The quantity of the actual dredged material in the National Park is relatively small in relation to the average annual amount in Germany of 35 million m^3.
By complying with the defined requirements, the unit price for dredging in the National Park - up to US $ 13 - has nearly doubled in the past few years. When compared internationally, for example with the United States which has an annual average of 230 million m^3, the relation of costs for 1995 is approx. 10:1 (Hales 1995).

The type and nature of the sediment is often different from that in the harbors (mud) and the connecting navigable channel (sand). This is important for dredging and the area influenced by disposal. The strategy and collected results were used as input for characterizing the environmental behavior of each region as a part of a morphological and hydrological unit. To do this, the natural background needed to be described in its seasonal variation, also with regard to the effect of single, extremely energetic storm periods. The assessment will resume on the central question concerning the following relation between the natural tolerance potential and the additional artificial load potential due to dredging and disposal with the simple effects for the concerned environment:

Tolerance potential > Load potential = Stable environment
Tolerance potential ~ Load potential = Unstable environment
Tolerance potential < Load potential = Destabilized environment

Experience from such investigations with an ecological orientation show that not only the formulated requirements and limit values but also the relevant investigations are to be objectively adjusted to the respective questions. In this connection, the results from natural measurements are of central importance with increasing duration of observation.

In this process, the engineering sciences with their optimizing work methodology must actively shape these projects in a major way, also searching for overlapping areas in other sciences as they proceed.

References

NN (1992). Guiding Rules for the handling of dredged material following the Oslo- and Helsinki-Commission (in German), German Authority for Waterways

NN (1996). Dredging and disposal in the coastal zone, influence on the macrozoobenthos (in German), Mitteilungen der Bundesanstalt für Gewässerkunde (BfG), Heft 11

ESSINK, K. (1993). Ecological effects of dredging and disposal in the Ems-Dollard estuary and mudflats (in Dutch), Dienst Getijdenwateren, Rijksinstitut voor Kust en Zee Haren

Hales, L. (1995). Accomplishments of the Corps of Engineers dredging research program, Journal of Coastal Research, No. 11, pp. 68-88

HÜBNER, H.-J., DE VRIES, J.M., GROTJAHN, M. (1996). Presentation and first results of a pilot study on the effects of disposal in the coastal zone (in German), Mitteilungen der Bundesanstalt für Gewässerkunde (BfG), Heft 11, pp. 30-44

Lozan, L.J., et al. (1990). Warning signals in the North Sea (in German), Paul Parey

Manzenrieder, H. (1983). The biological stabilization of mudflats from an engineering viewpoint (in German), Mitteilungen des Leichtweiß-Instituts der T.U. Braunschweig, No. 79, pp. 135-193

Manzenrieder, H. (1995). Half-life period of sedimentation - model test on a scale 1:1, Proceedings 24th ICCE 94, pp. 3139 -3153

Mc Nair, E.C. -Editor- (1994). Proceedings of the 2nd Conference on dredging and dredged material placement, Florida, ASCE

Neville-Jones, P.; Thompson, G.E. (1994). Coastal pollution, water quality and oil spills, in Abbott, M.B.; Price, W.A.: Coastal estuarial and harbour reference book, pp. 691-703

CHAPTER 332

Storm-Derived Bar/Sill Dynamics in a Dredged Channel

Sean O'Neil, Student Member ASCE,[1] Keith W. Bedford, Member ASCE,[2] and David P. Podber, Student Member ASCE[3]

Abstract

The development of sediment bars and sills due to storm-induced flow events in a dredged Great Lakes tributary is studied. The hydrodynamics associated with long-waves from the lake, create flow reversals at the river mouth, and storm runoff produces large sediment loads delivered to the lake. A laterally-averaged numerical model, including a turbulence closure sub-model, is used to simulate the hydrodynamics. A simple sediment settling, resuspension and transport model is coupled to the hydrodynamic model. Model runs are made for flow and temperature conditions which would be typical of the region during the spring season. Runs are made with and without the sediment settling velocity term, which effectively represents the modeling of two grain sizes; clay particles which have extremely small settling velocities and tend to floc together producing neutrally buoyant particles, and silt sizes which have finite settling velocities.

Introduction

Harbor dredging is the necessary result of long term, persistent deposition of watershed-derived sediments. Periodic redredging is required to ameliorate the occurrence of sills and bars which form as a result of the interaction of wave climate, channel geometry, tributary flow and littoral drift. As opposed to the persistent and predictable tidal forcing on coastal harbors, the harbors on the Great Lakes are moderated by random, long-wave effects derived from storms. The storm surges and resulting seiches, coupled with a high sediment influx from watershed runoff, often

[1]Graduate Research Associate, [2]Professor and Chair, [3]Graduate Research Associate. Department of Civil and Environmental Engineering and Geodetic Science, The Ohio State University, 470 Hitchcock Hall, 2070 Neil Avenue, Columbus, OH 43210-1275, USA. email: sean@superior.eng.ohio-state.edu

conspire to yield a two-sill bottom configuration which motivates maintenance dredging.

As part of a study designed to investigate the impacts of dredging operations, the physical processes responsible for the formation of sills and/or bars is explored by the use of detailed numerical models. The models developed as part of this study will also be used to enhance the tributaries portion of Great Lakes Forecasting System (e.g., Bedford and Schwab, 1994) which currently produces nowcasts four times daily and 24-hour daily forecasts of the state (water levels, wave heights, temperature, currents, etc.) of Lake Erie. This contribution details some of the results of the sill dynamics modeling and analysis for conditions marked by flow reversals and stratified flow.

Study Site

Toledo Harbor, on the Maumee River, is the third busiest port in the Great Lakes shipping arena. The site of the model investigation, shown in Figure 1, extends along

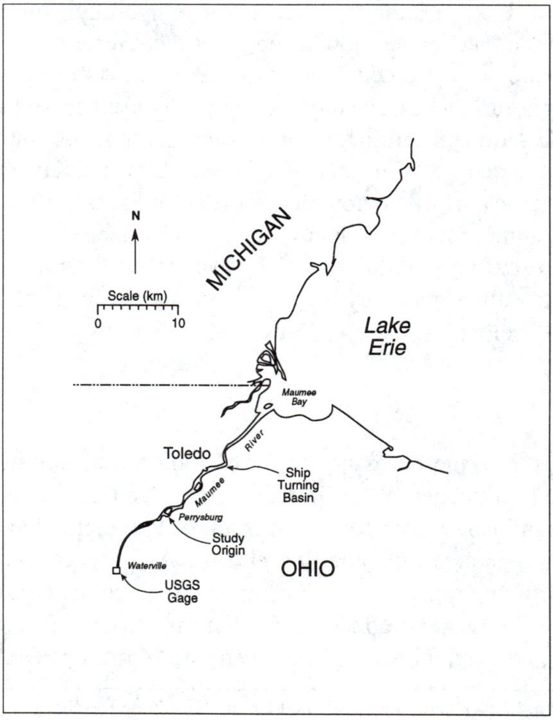

Figure 1: The Maumee River region.

the river, through the Maumee Bay, along a dredged navigation channel. The Maumee River delivers the single largest tributary-derived sediment load to Lake Erie, contributing 44% of the total annual load (Kemp *et al.*, 1976). This high load occurs because 85% of the watershed land use is for agriculture. The extreme shallowness of Maumee Bay, with an average depth 1.5 m (5 ft), and the Western Basin of Lake Erie, average depth 7.6 m (25 ft), necessitates the maintainance of the 152 m (500 ft) wide, 8.5 m (28 ft) deep navigation channel. The dredged portion of the model domain extends from a ship turning basin 9 km upstream in the river, to approximately 6 km into the bay. Figure 2 shows a profile of the modeled portion of the river and navigation channel including the double-silled bathymetric configuration near the mouth of the river.

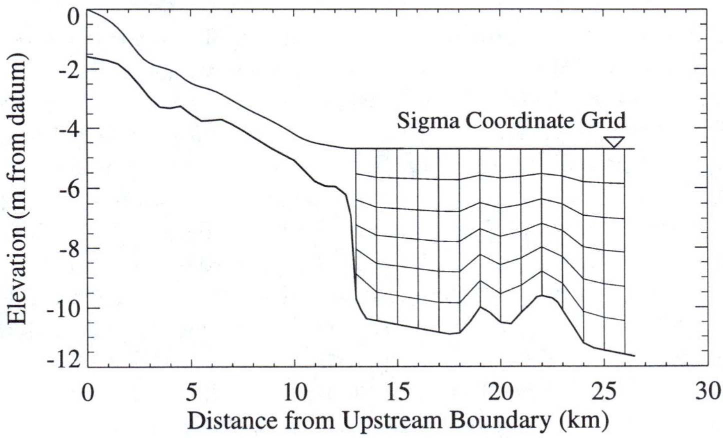

Figure 2: Schematic of model domain profile.

Physical Setting

The physical setting within the Maumee River, Bay and Western Basin of Lake Erie displays all of the features which might be found in a typical marine estuarine system, excluding of course tidal regularity. Storm events in the Lake Erie region occur with a frequency of 5-7 days during the spring and fall, with durations of 1-2 days. The corresponding increase of river flow rate and stage attain values that are significant fractions of the ambient levels.

A typical storm track during these seasons will follow the major axis, west to east, of Lake Erie, producing a significant storm surge at the eastern end of the lake. The storm surge will decay into a lakewide seiche with frequently observed 14.4, 9.1, 5.9 and 4.2 hour longitudinal modes. The seiche typically produces water elevation changes of more than 1 m, which is a significant fraction of the average water depth at Toledo Harbor. After the storm event, the decaying seiche may take 3-4 days to completely disappear. The narrow and deep dredged channel results in a "pipelining" of the excess flow due to runoff into the lake. The oscillations combined with the seasonal variations in water density gradients, here due to temperature differences between river and lake water, result in a system which behaves like a typical estuary during storms, and like a river during calm conditions. Flow reversals are frequently noted as are internal waves and oscillations.

Numerical Models

The numerical methods and schemes employed for this set of model runs have all been well documented and used by several researchers under a variety of conditions, therefore only a brief outline will be given as to how the model couplings occur. The one-dimensional, hyperbolic de St. Venant equations are solved to determine the river stage and discharge from Waterville, OH to the river mouth at Lake Erie, in this case using the upstream discharge at Waterville and the downstream stage at the lake, as boundary conditions. In turn, the values of discharge and stage are used as the boundary conditions for the two-dimensional hydrodynamic model, which is based on the laterally-averaged, hydrostatic, Navier-Stokes equations. The use of this type of model allows the vertical structure of the flow field to be captured without the expense of a fully three-dimensional model. The use of the model is justified by the fact of the narrow, deep dredged channel. The two-dimensional model is used only from the ship turning basin (see Figure 2) to the lake and uses specified temperature and velocity boundary conditions at the upstream, downstream and bottom. A turbulent closure submodel, the Mellor-Yamada level 2.5 scheme (e.g., Blumberg and Mellor, 1987), is employed to calculate the time-varying vertical eddy viscosities and diffusivities. At the water surface, the flux of velocity and heat are nil, and at the bottom a drag-law based on the square of the horizontal velocity just above the bottom, allows for a shear stress or no-slip condition.

The sediment transport component of the model solves the advection-diffusion equation for suspended sediment concentration using an upwind advection scheme, a size specific constant settling velocity term, and source/sink terms for the erosion and deposition of sediment. The erosion and deposition terms are parameterized using the model of Sheng and Lick (1979), where the deposition of sediment is proportional to the sediment concentration and erosion is proportional to an excess shear stress as compared to the shear stress for sediment resuspension. Boundary conditions are specified to be no sediment flux through the water surface, a well mixed upstream condition giving a constant input sediment concentration, and lake-like concentration and temperature profiles imposed downstream. Empirical parameters in the erosion and deposition terms are based on previous studies performed in the Western Basin of Lake Erie and reported in Sheng and Lick (1979).

Conditions and Assumptions

The specific imposed conditions and assumptions which are applied for the spring season case study are outlined. The boundary conditions at the two ends of the model would normally come from data, or possibly, the output from other models. For the spring storm conditions of interest, a typical steady river inflow was applied upstream, the magnitude of 141.6 m^3/s (5000 cfs) was determined from flow hydrographs obtained during the spring months (Pinsak and Meyer, 1976). To approximate the effect of the lake seiche after a storm, a downstream sinusoidal water elevation with an amplitude of 0.61 m (2 ft) and a period of 14 hours, was applied for four complete cycles. Given the short term nature of the simulation, the bottom was assumed to be fixed, so that the sills were not moving or changing shape. This is justified by the fact that (with the exception of the sping snow melt discharge) the bottom probably doesn't evolve much under the influence of a single storm event, but does over the course of a season or longer.

The river temperature was assumed to be a constant upstream value of 12°C and the lake was assumed to have a temperature of 7°C, typical for spring (Shindel *et al.*, 1993). The upstream sediment concentration was assumed to be 1 kg/m^3, and downstream a constant concentration profile, with maximum concentration near the bottom of 1 kg/m^3, imposed if the flow reversed and lake water traveled upstream. However, the form of the profile was found not to affect results in trials with different profile shapes. Model runs were performed for single sediment grain sizes, and two cases were examined. The Stokes' settling velocities were determined to be 1×10^{-5} m/s and 0 for the two cases of silt (median diameter 0.004 to 0.06 mm) and clay particles (median diameter < 0.004 mm).

Results

Analyses examine the impacts of flow reversals introduced by seiches versus regular downchannel river flow during inter-event periods. The following plots depict the water velocity field in the dredged channel as well as the suspended sediment concentration contours. These results were determined at several points within a single seiche cycle, where the cycle fractions are defined as in Figure 3. For brevity, not all of the cycle points designated in Figure 3 will be shown. Depicted in the figures is the third of four complete cycles run due to the fact that the fourth cycle of a run is generally not very different from the third.

The velocity fields shown in Figures 4-7 show that moderate seiche amplitudes are sufficiently strong enough to cause flow reversals in the river, even for riverine flow rates as high as 141 m^3/sec (see also Podber and Bedford, 1993). The upstream sill proximity is consistent with its origination from simple channel deposition of sediments introduced upstream, and the downstream sill is located at the furthest upchannel extent of the 14-hr seiche mode excursion distance. During spring conditions a stable gyre is present above the downstream sill and positioned such that horizontal velocities are zero directly above the peak of the downstream sill, as shown in Figures 5 through 7. As the cycle progresses, the gyre appears to move up and then back down in the water column, but maintaining its position above the downstream sill. This may be an extremely important agent in sill formation and migration. Note also that at all times in the cycle the stratification of the water is maintained, where

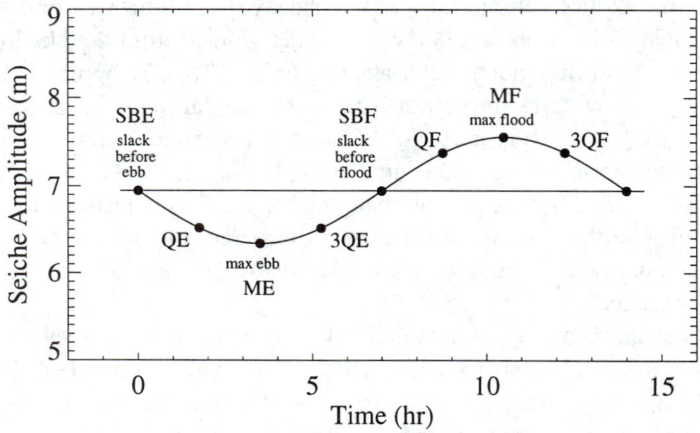

Figure 3: Definitions within a seiche cycle.

the warmer river water flows over the colder lake water. During the fall the inverse situation occurs.

In comparing the suspended sediment contours, it can be seen in all cases that there is a greater mass of the silt sized sediment within the water column than the neutrally buoyant particles. This is seen by fact that the larger valued contours are closer to the bottom and the lake side of the modeled domain. This is most likely due to the fact that neutrally buoyant particles enter the modeled region from the upstream direction, but they can only pass out through the downstream end of the model domain. For the case of the settling silt particles, some portion of the total mass of particles will be settling towards the bottom, and though the flow reversal will bring less sediment-dense water into the river, the settling sediment will tend to push the contours down and closer together. Also note that the extent of the upstream excursion of the intruding lake water has the effect of pushing the neutrally buoyant particles back to the upstream sill, whereas the maximum upstream extent of the silt particles is the downstream sill, which occurs at maximum flood.

Acknowledgements

This research was supported in part by the Coastal Ocean Program of NOAA. The Ohio Supercomputer Center provided computing resources under OSC Grant No. PAS891-1. The support is very much appreciated. This is a contribution from the Great Lakes Forecasting System.

References

Bedford, K. W. and Schwab, D. J. (1994). The Great Lakes Forecasting System: An Overview. In G. V. Cotroneo and R. R. Rumer, editors, *Hydraulic Engineering*

'94, Proceedings of the Conference, pages 197–201, Buffalo, NY. ASCE. See URL http://superior.eng.ohio-state.edu/.

Blumberg, A. F. and Mellor, G. L. (1987). A description of a three-dimensional coastal ocean circulation model. In N. S. Heaps, editor, *Three-dimensional Coastal Ocean Models*, Coastal and Estuarine Sciences 4, pages 1–16. American Geophysical Union.

Kemp, A. L. W., Thomas, R. L., Dell, C. I., and Jaquet, J.-M. (1976). Cultural impact on the geochemistry of sediments in Lake Erie. *J. Fish. Res. Board Can.*, **33**, 440–462.

Pinsak, A. P. and Meyer, T. L. (1976). Environmental baseline for Maumee Bay. Maumee River Basin, Level B Study, MRB Series No. 9, Great Lakes Environmental Research Laboratory.

Podber, D. P. and Bedford, K. W. (1993). Tributary loading with a terrain following coordinate system. In M. L. Spaulding, K. W. Bedford, A. F. Blumberg, R. T. Cheng, and J. C. Swanson, editors, *Estuarine and Coastal Modeling III, Proceedings of the 3rd International Conference*, pages 475–488. ASCE.

Sheng, Y. P. and Lick, W. (1979). The transport and resuspension of sediments in a shallow lake. *J. Geophys. Res.*, **84**(C4), 1809–1926.

Shindel, H., Klingler, J., Mangus, J., and Trimble, L. (1993). Water Resources Data – Ohio, 1992. Volume 2, St. Lawrence River Basin. Technical report, U.S. Geological Survey, Water Resources Division.

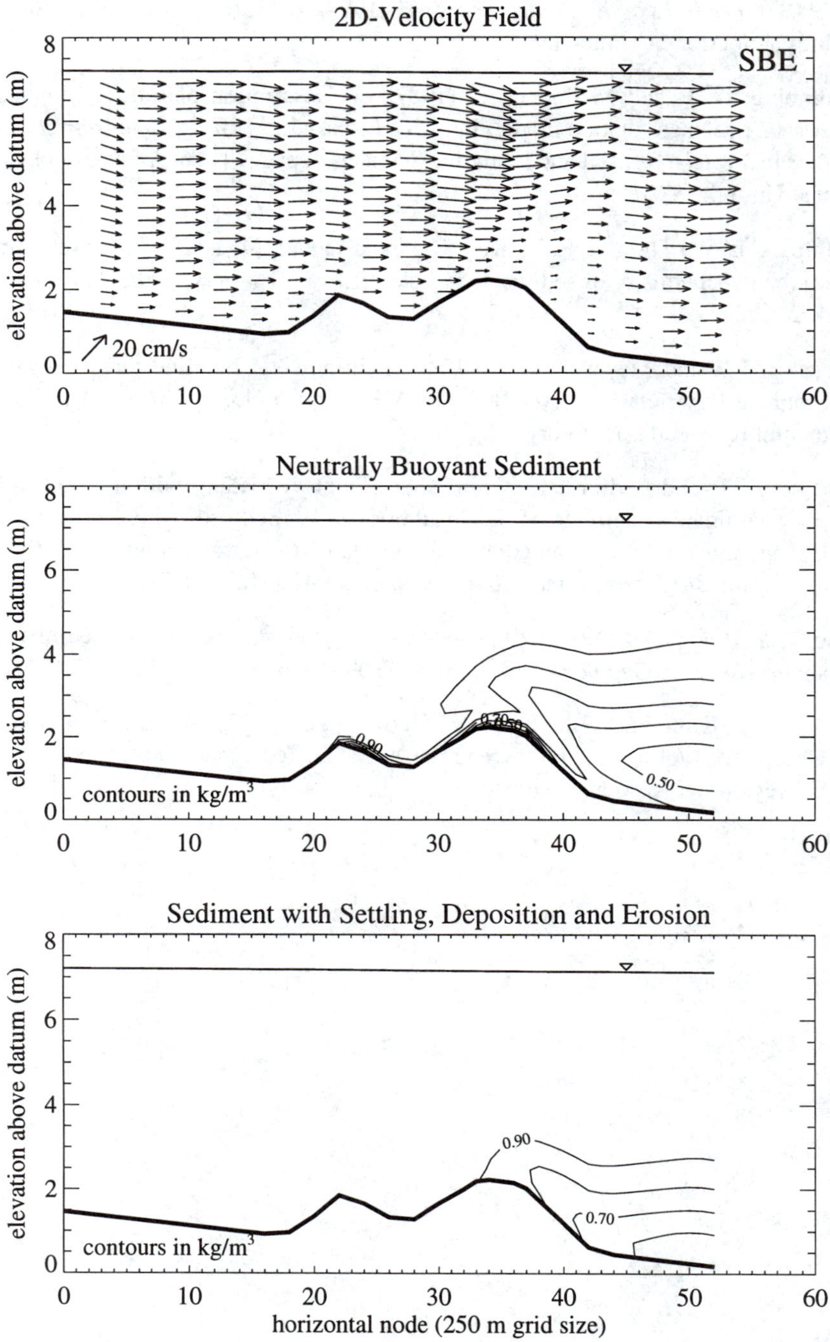

Figure 4: Velocity structure and suspended sediment contours for SBE.

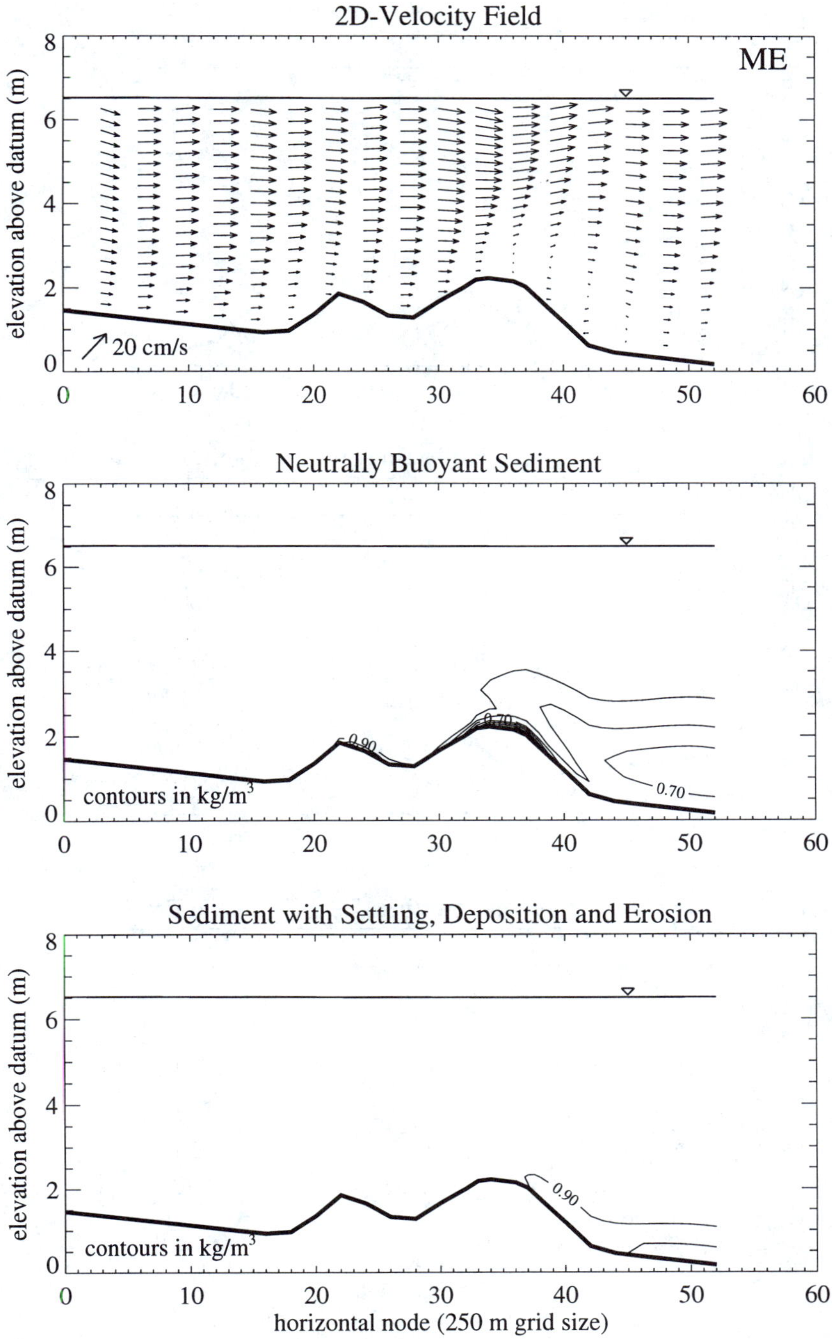

Figure 5: Velocity structure and suspended sediment contours for ME.

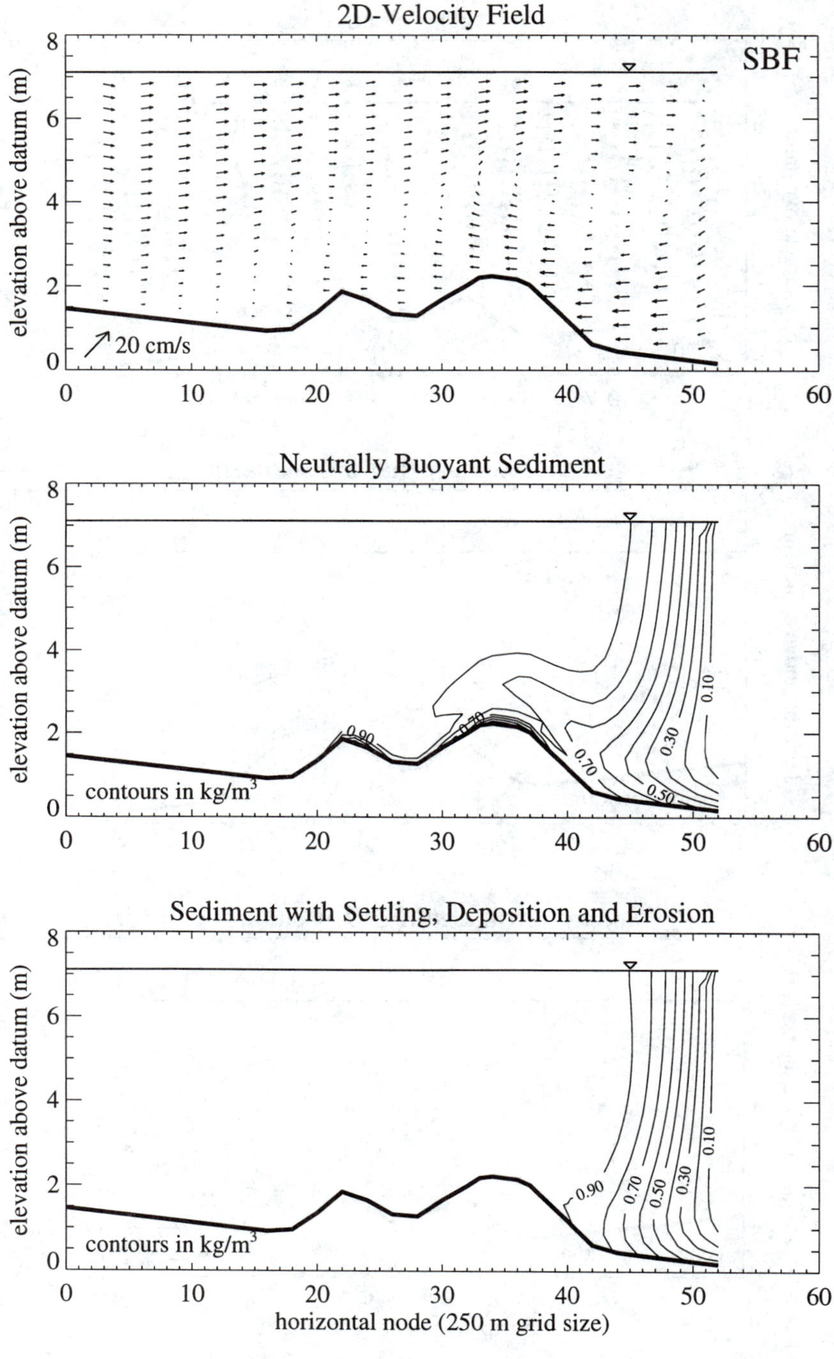

Figure 6: Velocity structure and suspended sediment contours for SBF.

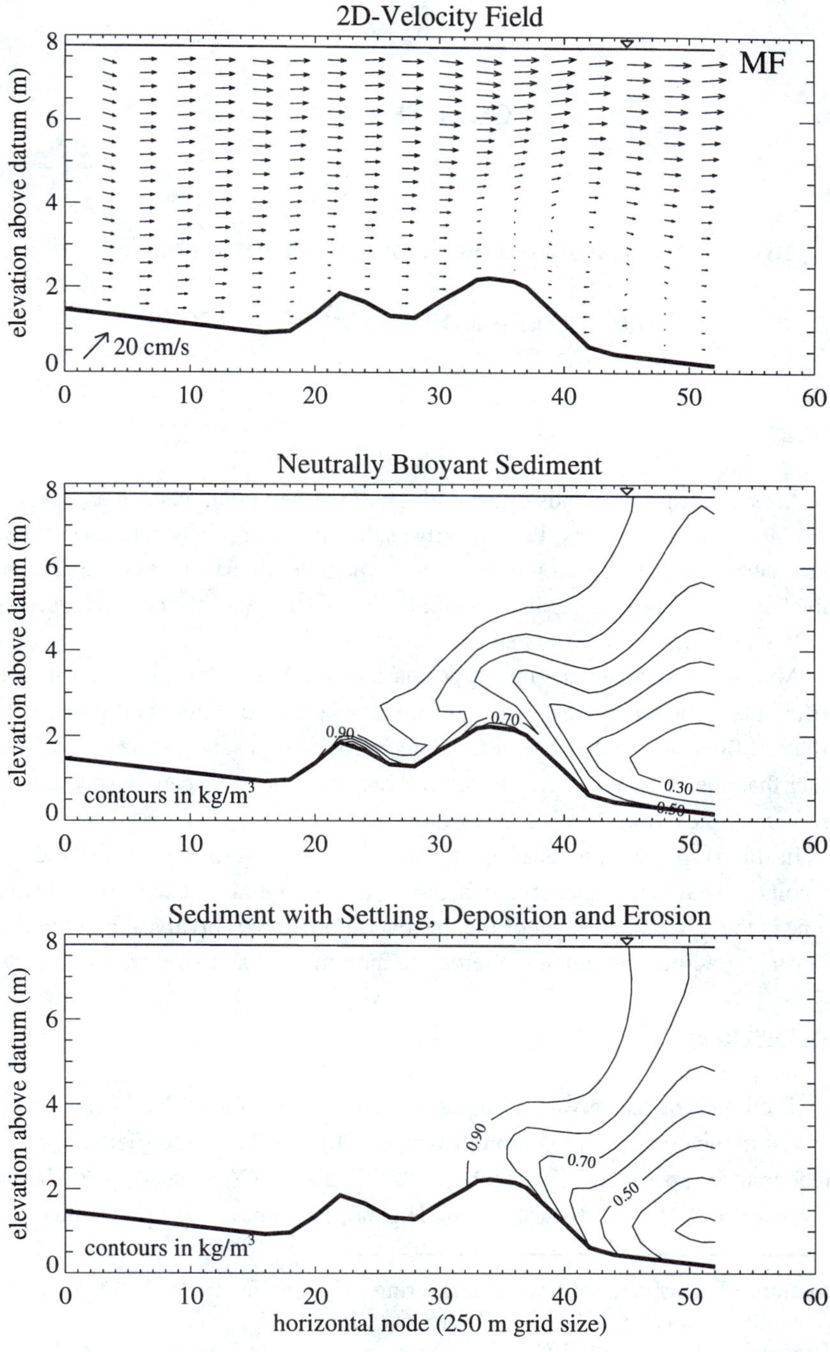

Figure 7: Velocity structure and suspended sediment contours for MF.

CHAPTER 333

Bivalve Habitat Based on Sediment-Transport Mechanics

Hitoshi Gotoh[1] and Tetsuo Sakai[2], M. ASCE

Abstract

After a storm, thousands of dead bivalves washed upon a beach are sometimes observed. Bivalves are important not only as the marine products but also as the index of the quality of coastal environment. Hence, the field observations of the distribution of bivalves in a coastal zone; and the laboratory experiments on the characteristics of the behavior of bivalves have been performed.

An interesting result is found by Yamashita and Matsuoka (1994) in a laboratory experiment on the burrowing process of bivalves under eroding condition due to the oscillatory flow: even when the descending velocity of a sand surface is sufficiently smaller than the burrowing velocity of bivalves, some bivalves cannot stay in the sand layer and are picked-up by oscillatory flow.

In this study, the physical background of their results are considered from a viewpoint of sediment transport mechanics. Two important aspects are investigated: the one is the stochastic aspect of the burrowing process of bivalves; and the other is the reverse grading phenomena observed in motion of mixed-size grains.

Introduction

Thousands of dead bivalves washed upon beach after a storm have been sometimes observed in the coastal region around Japan. Most of the former field observations and laboratory experiments (Watanabe, 1982; Higano and Yasunaga, 1988; Yamashita and Matsuoka, 1994; and Kuwahara and Higano, 1994) have been performed because

---

1 Lecturer, Department of Civil Engineering., Kyoto University
Yoshida Honmachi, Sakyo-ku, Kyoto, 606, Japan
2 Professor, Department of Civil Engineering., Kyoto University

of the importance of bivalves as fishery products. Bivalves, or creatures living in sandy beach, are also important as the index of coastal environment.

There are many irregular factors in a moving process of bivalves such as fluctuation of bottom velocity, descending velocity of sand surface due to scoring, size of bivalves, activity, or moving ability, of bivalves and so on. To take these irregularities into consideration, the moving process of bivalves should be treated as a stochastic process.

The other important aspect in the moving process of bivalves is the mode of sediment motion around bivalves. In general, sediment is transported in various modes; bed load, suspended load over sand ripples and sheetflow. Especially, during storm, or under the action of high-shear force, sheetflow should be a dominant transport mode. In the sediment transport in sheetflow regime, the particle/particle interaction, or the momentum transport due to interparticle collision is a mechanism governing the flow structure. To understand the behavior of bivalves in a sediment layer moving in sheetflow regime is an important subject to clarify the mechanism of bivalves-upward motion in a sand layer.

Two important aspects mentioned above are treated in this study based on the numerical model such as the stochastic model for simulating the probabilistic characteristics of bivalve's behavior and the granular material model for the simulation of sediment/sediment and sediment/bivalves interactions.

Laboratory experiment by Yamashita & Matsuoka

Yamashita and Matsuoka (1994) found out very interesting characteristics of bivalves in their laboratory experiment. They performed the experiment on the burrowing process of infant bivalves, or *Spisula sachalinensis*, maximum length of which is more than 5 mm and less than 20 mm, under the oscillatory flow generated by U-tube type oscillating water tunnel. They investigated the probability of the occurrence of the bivalves picked-out-of sand layer under the various values of the ratio of the descending velocity of sand surface, or v_e, to the burrowing velocity of bivalves, or v_S. Figure 1 shows the one of the results of their study on the relation between the probability of the occurrence of the bivalves picked-out-of sand layer and the velocity ratio v_e/v_S.

When the descending velocity of sand surface due to the erosion is larger than the one-thirds of the burrowing velocity of bivalves, namely in the region where the ratio $v_e/v_S \geq 0.3$, whole of the bivalves in a sand layer are picked-out-of sand. In their experiment, the burrowing velocity of bivalves are measured in a still sand layer. To understand the physics of this phenomena, the behavior of bivalves in a streamwise-moving sand layer should be investigated.

In the region $0.1 \leq v_e/v_S \leq 0.3$, some bivalves are picked-out and others are staying

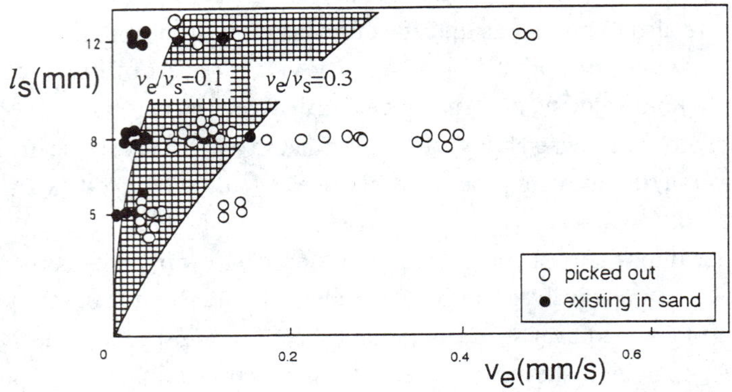

Figure 1. Occurrence of the bivalves picked-out
(Data from Yamashita and Matsuoka, 1994)

in a sand. This means the importance of stochastic consideration of the motion of bivalves to express the probabilistic aspects of the moving process of bivalves.

Stochastic consideration of burrowing process of bivalves

procedures of stochastic calculation

The following assumptions, some of which are based on the previous experiments conducted by Higano, Kimoto and Yasunaga (1993) and Yamashita and Matsuoka (1994), are introduced in this calculation.

(1) The shell length of bivalve l_s=8 mm. (2) The bivalve burrows in a sand layer keeping shell-length axis parallel to the vertical direction. (3) Initially, the top edge of bivalve coincide with the surface of sand layer, as schematically shown in Fig. 2.

(4) Bivalve begins to burrow in a sand layer when its top is exposed due to the scouring. The threshold of exposed length at the beginning of bivalve's burrowing is treated as the probabilistic variables based on the experiment by Higano et al (1993). (5) Bivalve rests in sand layer when the clearance between its top edge and the surface of a sand layer is equal to the half of shell length as shown in Fig. 2. The burrowing and the resting are iterated alternately. (6) Bivalve is defined to be picked out of sand layer when the 80% of the shell length are exposed in water as shown in Fig. 2. (7) The cycle of calculation begins at the moment when the condition (5) is satisfied. If the condition (6) has not been satisfied for 2200 s from the beginning, the calculation is terminated. This situation is defined as the survival of bivalves. (8) As it is mentioned above, bivalve is repeating the burrowing process and resting process, the average of the repeating period of which is equal to 2.2 s, according to Yamashita and

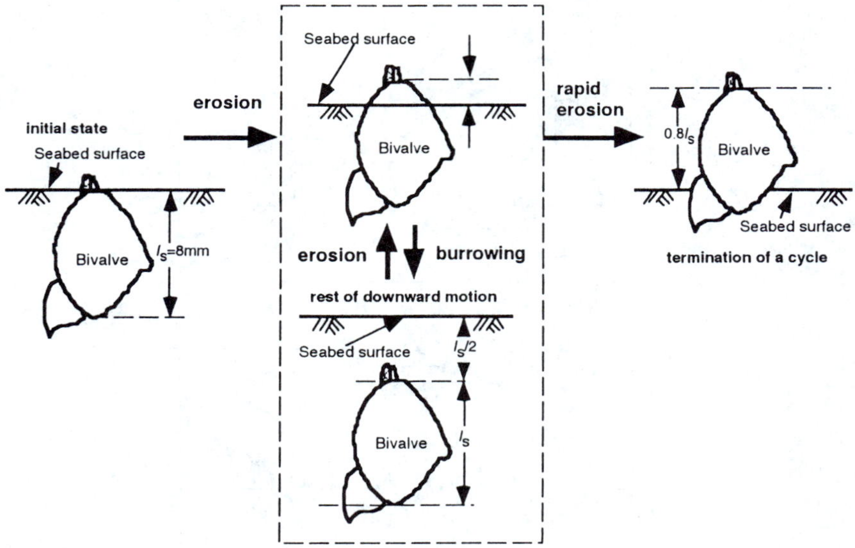

Figure 2. Definition of parameters in bivalves

Matsuoka (1994). They also investigated the distribution of burrowing velocity of bivalves. In this calculation, the burrowing velocity of bivalves changes at every 2.2 s based on the Monte Calro Method, namely by generating the random numbers following the distribution of the burrowing velocity experimentally investigated. (9) The descending velocity of the surface of sand layer is treated as a probabilistic variables following the normal distribution, the standard deviation of which is equal to one-third of averaged descending velocity, or $\sigma=v_e/3$.

results of the stochastic calculation

Figure 3 shows one of the calculated results of the one-cycle of the burrowing process of bivalves under the condition of $v_e/v_S=1.0$. In this figure, the time series of the existing height of bivalves, elevation of sand-layer surface and the thickness of sand layer above the top of bivalves, or δ, are shown. In this case, the thickness δ is decreasing rapidly during $0.0\leq tv_S/l_S\leq 1.0$, and on the verge of the picking-out of bivalves. After $tv_S/l_S=1.0$, the thickness δ increases again. In this case, the bivalve survives through one cycle of calculation. If the decreasing velocity of the thickness δ during $0.0\leq tv_S/l_S\leq 1.0$ is a little more rapidly, this bivalve is picked-out-of sand layer. The survival of bivalve is strongly depends on the accidental-drastic change of the surface elevation of sand layer.

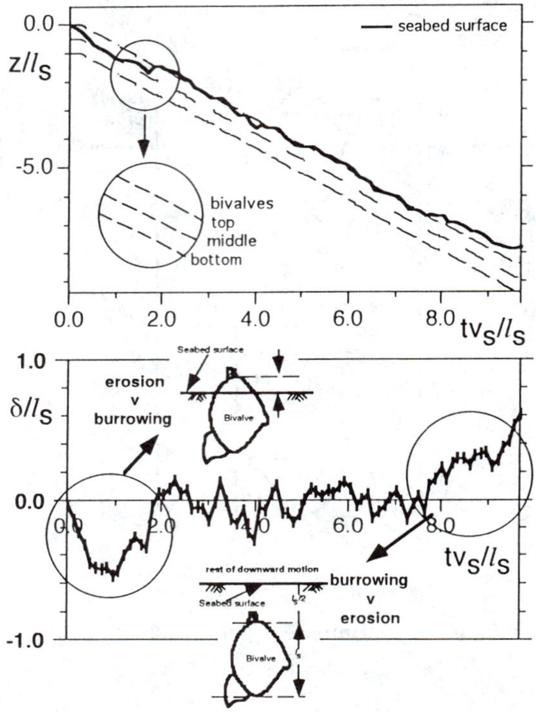

Figure 3. Burrowing process of bivalve

The probability of the survival of bivalves, or P_m, is calculated by continuing the cycle of calculation for m-times with assessing the occurrence of the bivalves picked-out-of a sand layer. Fig. 4 shows the existing probability of bivalves after 100 cycle of calculation, or P_{100}, against the ratio of the descending velocity of sand-bed surface to the burrowing velocity of bivalve, $v_e/v_s(=\alpha)$. Two cases of the calculation are shown in this figure. The one is the calculation in which the probabilistic characteristics of bivalves are only considered; and the other is the calculation in which both of the probabilistic characteristics of bivalves and that of sand-bed surface are considered. The transition range of the probability of the survival of bivalves experimentally investigated by Yamashita and Matsuoka is also shown in this figure.

The transition range of the experiment is $0.1 \leq \alpha \leq 0.3$, while the calculation, in which the probabilistic characteristics of bivalves is only considered, shows the drastic transition around $\alpha = 0.9$. The calculation, in which both of the behavior of bivalves and the motion of sand-bed surface are treated as the probabilistic variables, predicts the transition range in $0.6 \leq \alpha \leq 0.75$. But even if the probabilistic characteristics of

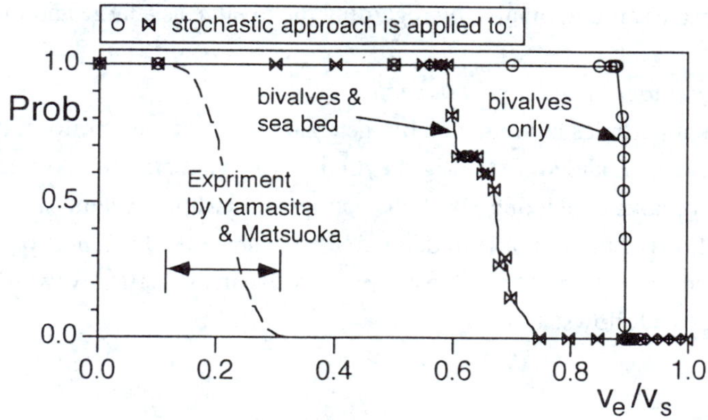

Figure 4. Survival probability of bivalve

sand-bed-surface elevation is considered, the threshold of the emergence of bivalves picked-out-of sand layer are overestimated.

Although the stochastic calculation reproduces the existence of the transition region in the probability of the bivalve's survival against the change of velocity ratio v_e/v_s, the threshold of the bivalve's survival is not reproduced well at least quantitatively. This fact suggests the existence of other mechanism, which promotes the picking-out-of bivalves.

Behavior of bivalves in sheetflow layer

reverse grading as the mechanism of upward vertical motion of bivalves

Boulders are frequently observed to be concentrated at the near-surface region of the front part of debris flow. This phenomenon, namely the existence of the boulders above other grains and gravel, is called reverse grading.

Interparticle collision is the governing mechanism both of debris flow and sediment transport in sheetflow regime. Therefore the same kind of phenomena as the reverse grading of debris flow can be thought to be a mechanism to promote the bivalves to be picked-out-of sand layer. In this study, the behavior of bivalves is traced numerically in the moving sediment particles in sheetflow regime by the distinct element method (=DEM).

distinct element method

Sakai and Gotoh (1995) performed the numerical simulation of the motion of sheared sediment-particle layer based on the distinct element method. In this study, their simulation is applied to simulate the interactive behavior between bivalve and sediment particles around bivalve by regarding the bivalve as a large and light particle.

governing equations of sediomeni particles

Sediment particles are modeled by the rigid cylinders with uniform diameter d; and the bivalve is modeled by the rigid cylinder with diameter D. At inter-cylinder and cylinder-bivalve contacting point, the spring and dashpot systems are introduced, and the equations of motion of cylinder and bivalve are solved by an explicit method.

Equations of motion of the i-th particle or bivalve in the vertically two-dimensional coordinate are as follows:

$$\frac{\pi\sigma_i d_i^2}{4}\frac{d^2x_i}{dt^2}=\sum_j\left\{-f_n(t)\cos\alpha_{ij}+f_s(t)\sin\alpha_{ij}\right\}_j+F_{0i} \tag{1}$$

$$\frac{\pi\sigma_i d_i^2}{4}\frac{d^2y_i}{dt^2}=\sum_j\left\{-f_n(t)\sin\alpha_{ij}+f_s(t)\cos\alpha_{ij}\right\}_j-\frac{\pi(\sigma_i-\rho)d_i^2 g}{4} \tag{2}$$

$$\frac{\pi\sigma_i d_i^3}{16}\frac{d^2\phi_i}{dt^2}=\sum_j f_s(t)_j \tag{3}$$

in which f_n, f_s=the forces acting between the i-th and j-th particles; α_{ij}=contacting angle between the i-th and j-th particles; F_{0i}=shear force acting on the i-th particle; σ_i=density of particle; and d_i=diameter of particle; and g=graviational acceleration. The subscript "n","s" mean normal and tangential components, respectively.

calculation of interparticle-acting force

Figure 5 shows the schematics of the interaction between two contacting particles. Between two contacting particles, springs and dashpots are introduced to describe the dynamic interparticle relation. The acting force between the i-th and j-th particles in normal and tangential direction, f_n and f_s, can be written as follows:

$$f_n(t)=e_n(t)+d_n(t) \quad ; \quad f_s(t)=e_s(t)+d_s(t) \tag{4}$$

$$e_n(t)=\min\left\{e_n(t-\Delta t)+k_n\cdot\Delta\xi_n,\ e_{nmax}\right\} \quad ; \quad d_n(t)=\eta_n\cdot\frac{d\Delta\xi_n}{dt} \tag{5}$$

$$e_s(t)=\min\left\{e_s(t-\Delta t)+k_s\cdot\Delta\xi_s,\ e_{smax}\right\} \quad ; \quad d_s(t)=\eta_s\cdot\frac{d\Delta\xi_s}{dt} \tag{6}$$

in which e_n, e_s=forces working on springs; d_n, d_s=forces working on dashpots; $\Delta\xi_n$, $\Delta\xi_s$=displacement of particle during the time Δt (Δt=time step of the calculation); k_n, k_s=spring constants; and η_n, η_s=damping coefficients. In this simulation, particles

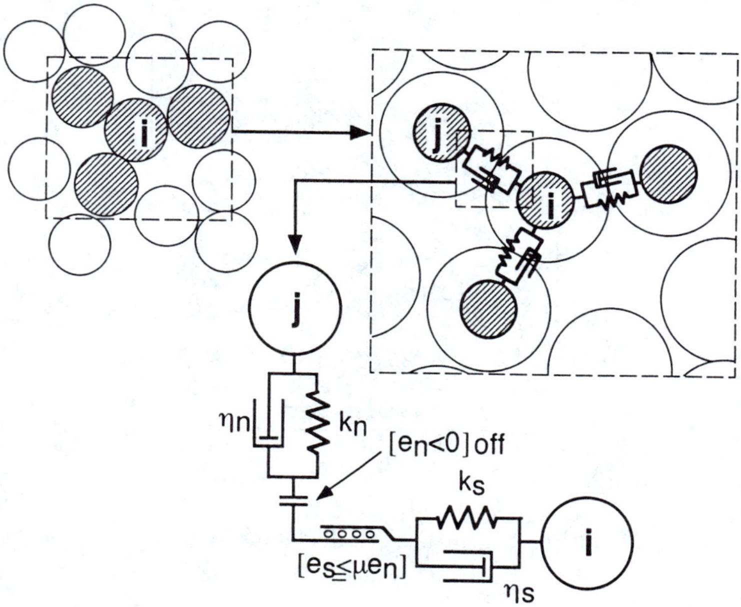

Figure 5. Schematics of the interaction between two contacting particles

are non-cohesive, hence the tensile force does not act between two contacting particles. To describe this characteristics, the joint, which no resistance to the tensile force, is assumed to exist in the normal direction. While, in the tangential direction, the friction force works. To describe this characteristics, the joint, which slips at the limit of the shear stress, is assumed to exist in the tangential direction.

initial conditions and boundary conditions

Figure 6 shows the schematic expression of the calculating domain. In this simulation, streamwise uniform condition is treated, therefore, the both sides of the calculating domain are the periodic boundaries. The bottom boundary is the fixed rough bed constituted by the particles with the same diameter as the moving particles. Before the beginning of the main calculation, the packing to determine the stable initial location of the particles is executed.

The shear stress is distributed to the particles in the neighborhood of the surface of sand layer in a following procedure. Procedure are, firstly, to set the threshold y_{th} (see Fig. 7); secondarily to calculate the area of the particles above the threshold, or

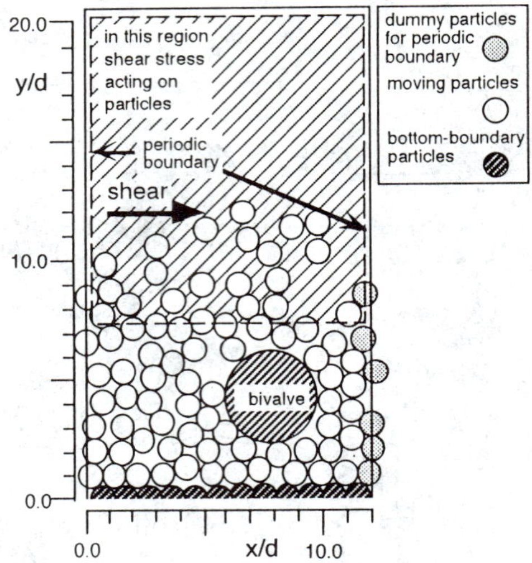

Figure 6. Schematics of calculating domain

the shaded area in Fig. 7, for the particles existing on and above the threshold; and finally to distribute the shear force to the particles existing on and above the threshold with calculating the area of particles above the threshold $S(\theta_i)$ as follows:

$$S(\theta_i)=\begin{cases}\dfrac{r^2}{4}\left(\pi-\theta_i+\dfrac{1}{2}\sin 2\theta_i\right) & ;\ y_i \geq y_{th} \\ \dfrac{r^2}{4}\left(\theta_i-\dfrac{1}{2}\sin 2\theta_i\right) & ;\ y_{th}-\dfrac{r}{2}<y_i<y_{th}\end{cases} \quad ;\ \theta_i=\cos^{-1}\left(\frac{y_i-y_{th}}{r}\right) \tag{7}$$

Distributed shear force to the i-th particle is written as

$$F_{0i}=Lw_i\tau_0 \quad ;\ w_i=S(\theta_i)\Big/\sum_{j=1}^{N}S(\theta_i) \tag{8}$$

in which L=length of the calculating domain in horizontal direction; τ_0=bottom shear stress par unit area. The test particle is 0.5 cm in diameter and 2.65 in specific gravity; and the model of bivalve is 2.0 cm in diameter and 1.30 in specific gravity (see Higano et al., 1993). In the calculating domain, 91 particles and 1 bivalve are traced. The model constants are shown in Table 1.

Initially the bivalve is contacting to the bottom constituting particles. The time step of the calculation is 2.0×10^{-5} s and the totally 80,000 cycles of calculation is executed, hence the motion of sediment particles and bivalve for 1.6 s is traced.

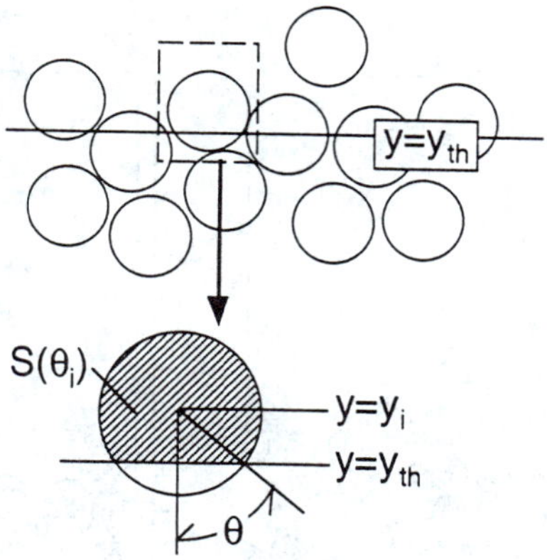

Figure 7. Shear stress distributed to sediment particles

Table 1. Model constants

k_n	9.45×10^6 N/m
k_s	2.36×10^6 N/m
η_n	40.0 Ns/m
η_s	20.0 Ns/m
μ	0.577
e_{nmax}	2.5 N
e_{smax}	0.025 N
Δt	2.0×10^{-5} s

<u>results of simulation</u>

Figure 8 shows the snapshots of the sediment-and-bivalve motion with 0.2 s intervals under the action of the shear stress $\tau_*(\equiv u_*^2/\sqrt{(\sigma/\rho-1)gd})=1.5$ on the surface of sediment layer. Bivalve, which is initially contacted to the bottom constituting particles, moves upward gradually, and it is picked-out-of a sand layer 0.8 s after the initiation of the shear action. Most of the particles in lower layer has laminar motion, hence the momentum exchange in lower layer is inactive. While, in the upper layer the concentration of sediment particles is less than that in the lower layer, the vertical motion of the sediment particles is less frequently obstructed by the particles in the upper layer than in the lower layer.

To understand the physics of the upward motion of bivalve, following hypothesis is proposed. The bivalve exists in the shear layer, in which the sediment particles

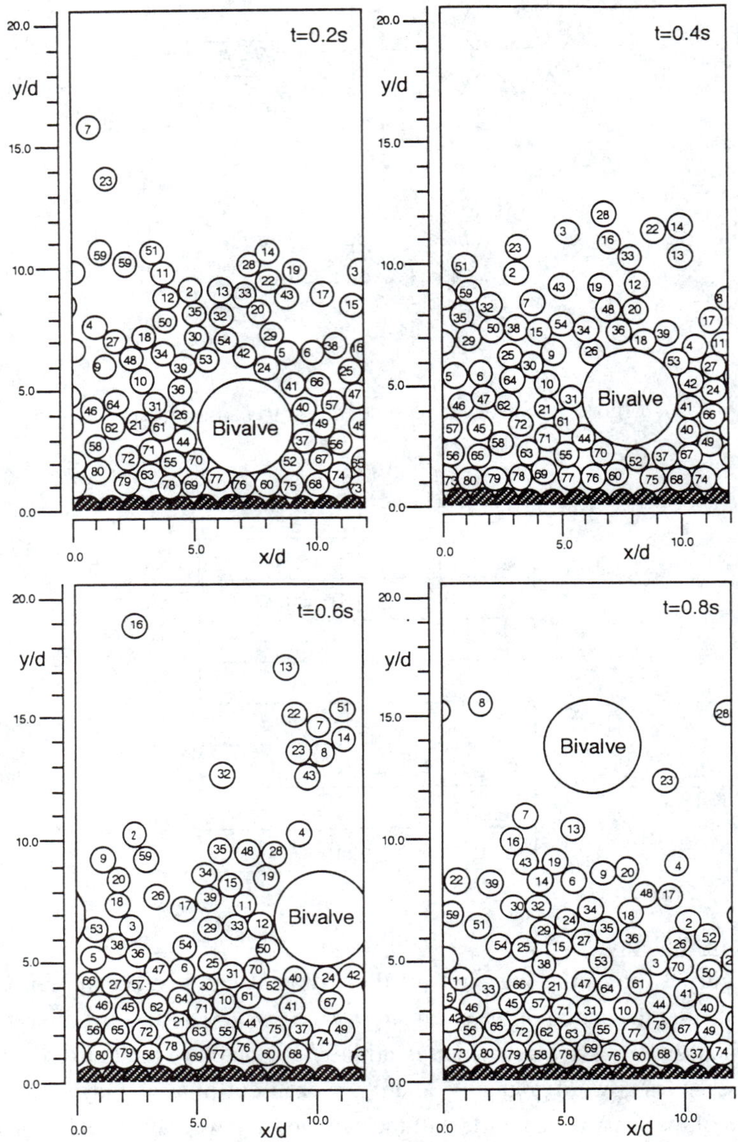

Figure 8. Snapshots of the sediment-and-bivalve motion

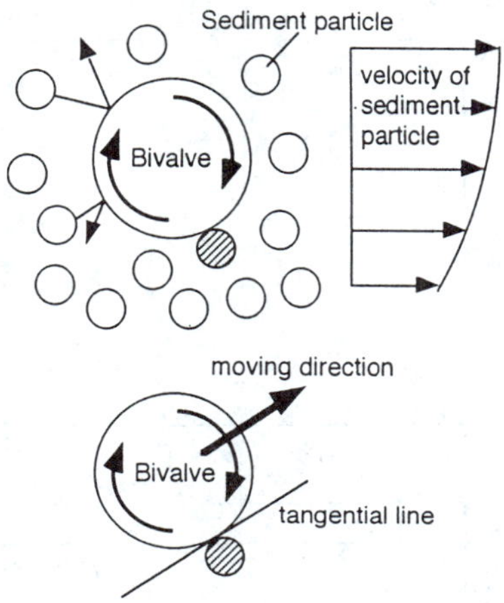

Figure 9. Upward motion of bivalve

have a vertical velocity distribution as schematically shown in Fig.9. Therefore the particle colliding at the upper part of bivalve is faster than that colliding at the lower part of bivalve. This velocity difference of the colliding sediment particles makes the bivalve to rotate in clockwise. Supposing the contacting sediment particle with just the lower part of the bivalve in the downstream section (hatched particle in Fig.9) and the hypothetical plane at the contacting point between bivalve and the particle, the behavior of the bivalve can be simplified as the clockwise rotating cylinder on the slope. And the clockwise rotation moves the cylinder to upward.

Figure 10 shows the time series of the rotating angle, the sign of which is positive in counterclockwise, and the elevation of the centroid of bivalve. The bivalve begins to move upward gradually accelerating the rotational motion. Around 0.6 s, bivalve moves in upward direction drastically, and at the same time the rotational motion is accelerated also drastically. From this fact, the rotational motion of bivalve can be regarded as the driving mechanism of the upward motion of bivalve.

Conclusion

The physical background of the behavior of bivalve in sand layer during storm are considered from a viewpoint of sediment transport mechanics. Two important aspects are investigated: the stochastic aspect of the burrowing process of bivalves and the reverse grading phenomena observed in a motion of mixed-size grains.

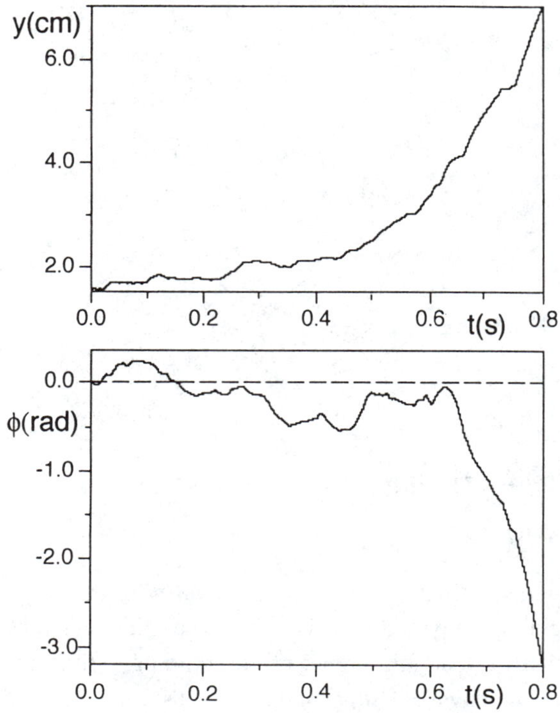

Figure 10. Rotating angle of bivalve

The stochastic model simulates the probabilistic characteristics of environment around bivalves. And the granular material model, or distinct element method, simulates the motion of bivalve in upward direction based on the sediment/sediment and sediment/ bivalves interactions.

References

Higano, J. and Yasunaga, Y. (1988). The change in beach profile and bivalve distribution on a sandy beach facing to open sea. Proc. 35th Japanese Conf. on Coastal Eng. pp.767-771 (in Japanese).

Higano, J., Kimoto K. and Yasunaga, Y. (1993). Influence on the relation between burrowing behavior and physical environment regarding the distribution of sandy beach bivalves *Meretrix lamarckii* and *Gomphina melanaegis*. Bull. Nat. Res. Inst. Fisheries. Eng. 14, pp.65-87 (in Japanese).

Kuwahara, H. and Higano, J. (1994). Model of bivalves on/offshore movement by waves. Proc. 24th ICCE, ASCE, pp.3086-3098.
Sakai, T. and Gotoh, H. (1995). Numerical simulation of sediment transport in sheetflow regime. Proc. XXVIth Cong. IAHR, London, pp.299-304.
Yamashita, T. and Matsuoka, M. (1994). Experimental study on movement and decline of bivalve by water waves. Proc. Civil Eng. in Ocean, vol. 10, pp. 119-122 (in Japanese).
Watanabe, E. (1982). Experimental study on the decrease of surf clam under wave action. Monthly report of Inst. Civil Eng., No.351,pp. 3-15 (in Japanese).

CHAPTER 334

Hydraulic Controls on Tidal Wetlands

R. Eric Katmarian, Philip A. McKee, Timothy W. Kana[1]

Abstract

Tidal wetlands have long been identified as critical habitats for many species of fish, birds, and other wildlife and are protected areas in many states. For these reasons, coastal projects which impact or destroy wetland areas are difficult to permit and often require substantial mitigation. An hydraulic design procedure is presented for the creation or improvement of wetlands based on flooding duration -- the percentage of time an area is inundated by tidal waters. This is followed by a sample application of the procedure to a recent mitigation project in Savannah, Georgia.

Introduction

Tidal wetlands, sometimes called the oceans' nurseries, contain many habitats for both plants and animals. Among the many habitats that exist, the basic ones include open water, tidal flats, low marsh, high marsh, transitional wetlands, and highland (Fig. 1). These habitats are sensitively balanced for existing tidal conditions, wave climate, daily flooding duration, sedimentation rates, and climate (Kana et al., 1986). Perturbations of these factors over an extended period can result in the transformation of one habitat into another.

Because wetlands tend to exist in areas of limited topographic relief, transitions between habitats can be indistinct and often occur over some distance, making precise location of boundaries difficult. Nevertheless, it is generally recognized that the character of flooding controls species distribution, with some wetland species preferring infrequent and irregular inundation, while others have adapted to frequent flooding. As a result, many studies have been undertaken to define wetland boundaries (Teal, 1958; Redfield, 1972; Nixon, 1982).

---

[1] Coastal engineer, research assistant, and senior scientist, respectively; CSE-Baird, PO Box 8056, Columbia, SC 29202

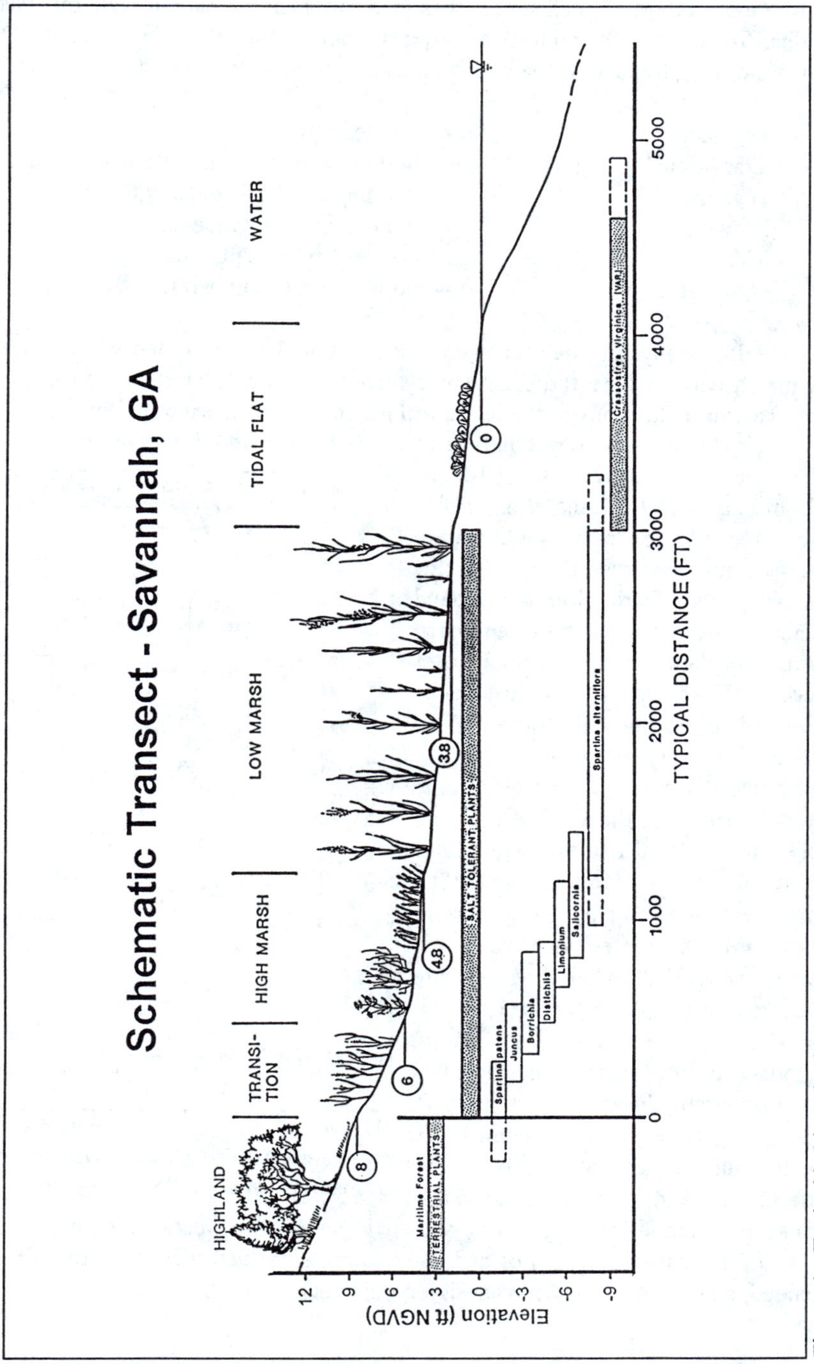

Figure 1. Typical habitats in tidal wetlands near Savannah, Georgia (From CSE 1995).

One method of delineation has been to identify habitats on the basis of flooding frequency – the number of times per day that an area is inundated. On the basis of flooding frequency, the basic habitats occur as follows (Kana et al., 1986):

Highland	-- flooded rarely
Transitional wetlands	-- flooding ranges from biweekly to annually
High marsh	-- flooding ranges from daily to biweekly
Low marsh	-- flooded once or twice daily
Tidal flats	-- flooded up to half the day
Open water	-- flooded more than half the day

Additionally, an often repeated rule of thumb has developed which states that low marsh exists where the substrate elevation is roughly at the mean high water level. However, the above criteria are too general to be incorporated into procedures for quantifying impacts on wetlands or to be used in wetland design criteria.

Many investigators (Adams, 1963 in Lagna, 1975; Kana et al., 1986) have extended the "mean high water" rule, saying that the primary factor controlling the distribution of wetland species is the relationship between marsh substrate elevation and local tidal water levels. Kana et al. (1986) even developed a normalized curve to be applied to arbitrary sites (Fig. 2). To use the normalized curve, a species-specific factor is read from the vertical axis and applied to the local extreme high water level referenced to mean sea level (MSL) datum. For example, for *Spartina alterniflora,* the water level factor is approximately 0.48. Therefore, at a site where extreme high water is 3 meters (m) MSL, *Spartina alterniflora* would be found in low energy environments with a substrate elevation of about 1.44 m MSL. Despite some findings that elevation alone does not control species distribution (Lagna, 1975), the appeal of such simple methods for engineering and environmental applications is clear.

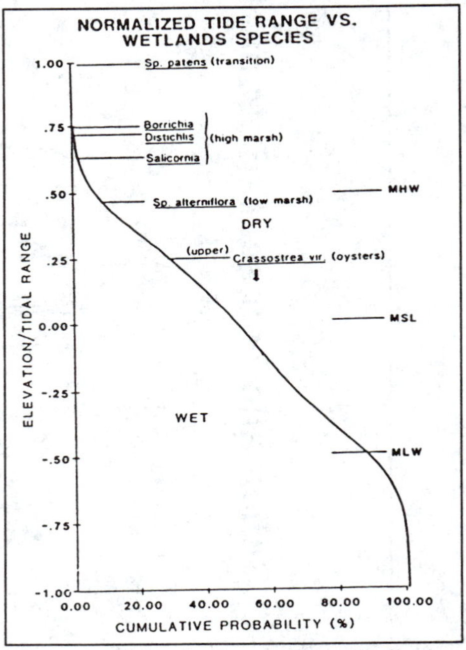

Figure 2. Normalized probability distribution of elevations for various wetland species (From Kana et al., 1986).

In this paper, the idea of hydraulic controls on tidal wetlands are refined and extended, and a general wetlands design method is proposed.

Flooding Duration as a Control on Wetland Habitat Distribution

It is generally accepted that character of inundation is the primary factor governing the distribution of wetland species. As described above, previous investigators have chosen to describe inundation in terms of substrate elevation relative to tidal water levels. However, relative elevation does not necessarily represent identical flooding characteristics at sites where tidal signature differs. For example, if two sites have equal tide ranges but one has an attenuated ebb tide, they will be flooded for differing periods for a given elevation (Fig. 3). When the effects of daily inequality, fortnightly cycles, and atmospherical affects are also introduced, relative elevation becomes an even less reliable measure of flooding character.

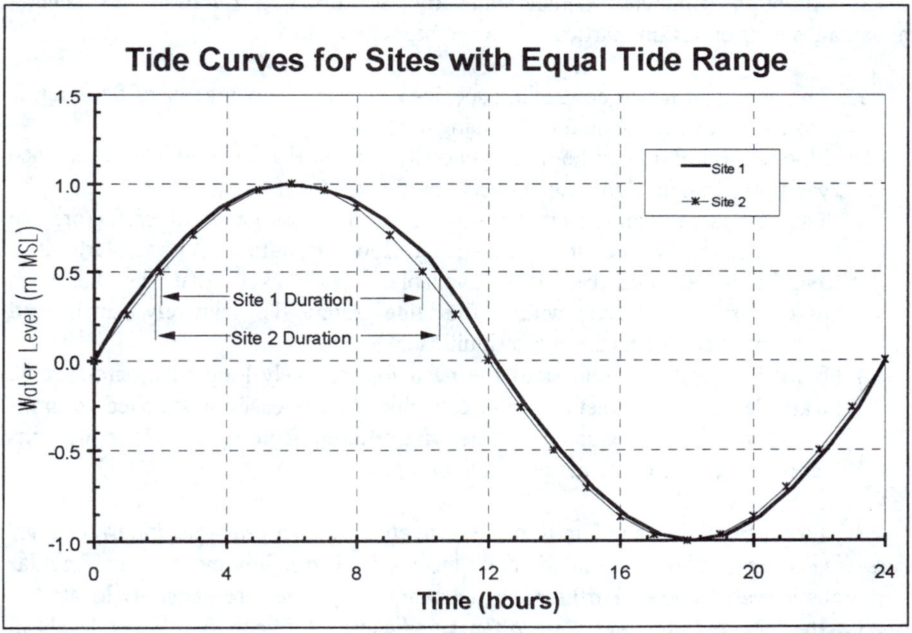

Figure 3. Flooding duration at equal elevations differ at sites where tidal signatures differ.

Flooding duration, the percentage of time that an area is inundated, is proposed as a more accurate measure of flooding character than relative elevation. Flooding duration is applied to species zonation in a manner similar to elevation.

It is assumed that wetland plant species are found in specific ranges of flooding duration. These ranges often overlap, and ranges for some species may be contained in the ranges of others. Indicator species may be chosen to identify specific wetland habitats. Wetland habitats are categorized by flooding duration in Table 1.

The durations listed in Table 1 are based on several wetlands surveys, the project described below, and many years of field observation. Nevertheless, minor variations should be expected on a site-specific basis. Until this method of design has gained wider acceptance, surveys of existing wetlands to confirm flooding durations for specific habitats should be conducted prior to design of new sites.

Table 1. Approximate flooding duration ranges for wetland habitats

Habitat	Flooding Duration
Upland	0 %
Transitional Wetlands	0-2%
High Marsh	2-5%
Low Marsh	5-25%
Tidal Flats	25-50%
Open Water	>50%

Although somewhat cumbersome to use, flooding duration has several advantages over elevation, particularly when broadly applied:

1. The dominant factor controlling species zonation, the character of flooding, is more directly represented by flooding duration.
2. Flooding duration can be more generally applied than relative elevation since variations in tidal signature are accounted for.
3. The effects of frequent inundation due to storm surge or other factors are more accurately accounted for using flooding duration, particularly if a lengthy water level record is available. It is likely that this factor is particularly important where the tide range is relatively small and atmospherical events are more significant.
4. Flooding duration emphasizes the need for relatively long-term, site-specific water level measurements, while elevation data is easily misapplied to areas where the tide is attenuated or otherwise different from open-water conditions (commonly where tide gauges are located).

To calculate flooding duration, *a relatively long-term, site-specific water level record must be used* to account for daily inequality, fortnightly cycles, and irregular variations in water levels. Furthermore, existing tide gauges are generally located in open water where water levels are often significantly different than water levels in wetlands. To obtain an acceptable water level record, a tide gauge may need to be deployed.

The length of the record required will depend on local conditions affecting the variability of the water levels. Judgment must be used in determining the minimum length of record required at a particular site though, in most locations, a 25-day record will suffice.

Once an acceptable water level record is obtained, flooding duration for a given elevation is calculated as:

$$\textbf{Flooding Duration} = \frac{\textbf{Cumulative time with water level above elevation}}{\textbf{Total length of record}} \times \mathbf{100}$$

In some cases, the creation or alteration of a wetland has the potential to significantly change local water levels. This is undesirable since it has the potential to cause habitat transformations in existing wetlands. Additionally, specification of a substrate elevation corresponding to a desired flooding frequency becomes very difficult when postproject water levels are uncertain. Steps must be taken to minimize the impact of projects on local water levels. This can be accomplished by specifying an appropriate substrate elevation for the new wetland or by increasing the hydraulic capacity of connections to open water to accommodate increased flooding areas.

Numerical models can be extremely helpful in predicting postproject water levels when local hydraulics are altered by wetland creation.

Sample Application: Mitigation Site Design

A recent CSE (1995) project involved hydraulic design for a 10-acre wetland at a site previously used for dredge spoil disposal near Savannah, Georgia. The site is approximately 900 m from the Savannah River and has hydraulic connection to the river via a complex network of ditches, culverts and existing wetlands (Fig. 4). The design objective was to create a tidal wetland at the site while minimizing both excavation requirements and impacts to existing wetlands.

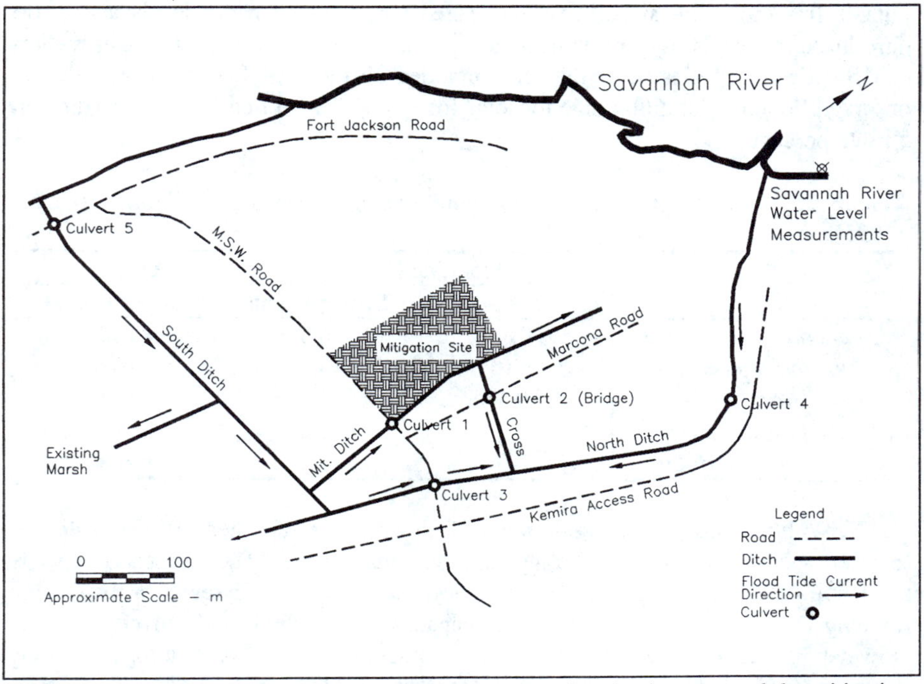

Figure 4. A complicated system of ditches, culverts, and wetlands connected the mitigation site to the Savannah River (From CSE, 1995).

The first step in applying the flooding duration method to the design was to determine the range of flooding duration for the type of wetland desired. A number of transects were surveyed through a nearby "natural" wetland. Elevation and distance were recorded at each point where species changed or where there were noticeable breaks in slope or elevation. Species types were identified with the assistance of a biologist. Elevation for each species was weighted by distance of occurrence to produce an elevation distribution for frequency of occurrence. The elevation data for all observed species were approximately normally distributed.

Local water level data were required to calculate flooding duration for the range of elevations observed for each species. Therefore, water level measurements in the surveyed wetland were taken over several tidal cycles and compared against data from a nearby permanent tide gauge. No significant water levels differences were observed between the surveyed marsh and the permanent gauge, so a 30-day record from the permanent gauge was used for flooding duration calculations. If significant differences had been noted between the marsh and gauge, water levels would have had to be measured at the site for a minimum of 25 days to provide adequate water level data for flooding duration calculations.

Flooding durations were calculated for the average elevation, and average elevation ± one standard deviation of each species observed (Table 2). Elevations outside this range are attributed to localized variations in water levels and errors introduced by weighting approximations. *Spartina alterniflora* was chosen as the low marsh indicator species, as is the usual practice on the East Coast. Therefore, the observed flooding duration range for local low marsh was found to be approximately 5 to 21 percent.

Table 2. Wetland species flooding durations near Savannah, Georgia (CSE, 1995).

	Flooding Duration (% of tidal cycle)		
Species	Mean	Lower Bound	Upper Bound
Spartina cynocuroides	13.7	8.3	18.5
Spartina alterniflora	12.0	4.5	20.5
Carex sp.	9.3	2.7	19.9
Juncus sp.	6.3	3.2	11.3
Typha augustifolia	3.5	1.3	6.8

With the design flooding duration for low marsh established, the next step was to determine how to achieve this range at the mitigation site. The goal was to specify a substrate elevation range for the mitigation site which would result in the desired flooding frequency, given the hydraulic capacity of the ditch and culvert network. However, it was recognized that the addition of the 40,000 m^2 wetland to the existing system would change its tidal characteristics, making postproject water levels different than preproject levels. Without good estimates of postproject water levels, accurate determination of flooding duration was impossible. Furthermore, measurements of flows through existing channels and observations of flow restrictions at culverts indicated that the existing system could not supply adequate tidal exchange to the mitigation site to sustain a low marsh. This meant that improvements would have to be made to the drainage system.

To gain some confidence in postproject conditions, DHI's Mike11 river model was used to simulate the ditch and wetland system. Figure 5 shows a schematic of the model setup. After calibration to existing conditions, the model proved instrumental in arriving at the final design by allowing the effects of ditch and culvert improvements, variation in mitigation site elevation, and impacts to existing wetland areas to be investigated.

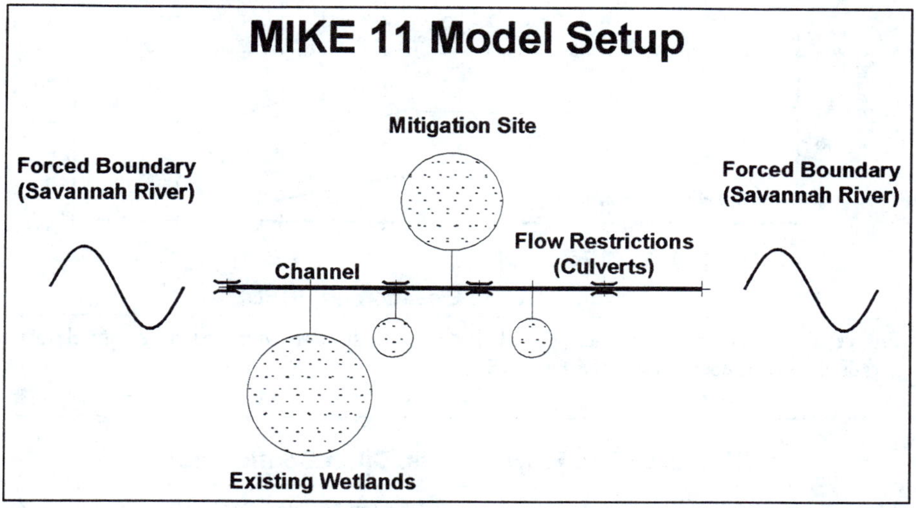

Figure 5. Schematic Mike11 model setup showing the channel/wetlands system modeled (From CSE, 1995).

Model results supported by field observations showed that both the existing wetlands and the site to be mitigated were flooded primarily by the south ditch (Fig. 4). This was primarily due to a topographic high point in the north ditch. With the high point and culvert 4 removed and the north ditch widened, the model showed that flooding of the mitigation site would occur from both the north and south. The model also showed significant increased flooding of the existing wetlands and the potential to convert them to "wetter" habitats. The final design recommendation was to isolate the mitigation site from the existing wetlands by closing culvert 1. Since the north ditch initially provided negligible flooding, the final design did not significantly impact existing wetlands.

With the hydraulic connection set, Mike11 was used to predict postproject water levels at the mitigation site. Using a simulated postproject water level record of 30 days, the variation in flooding duration at the mitigation with mitigation site elevation was established. Mike11 results for water levels at the mitigation site are shown in Figure 6, and corresponding volumetric flow rates in the improved north ditch are given in Figure 7.

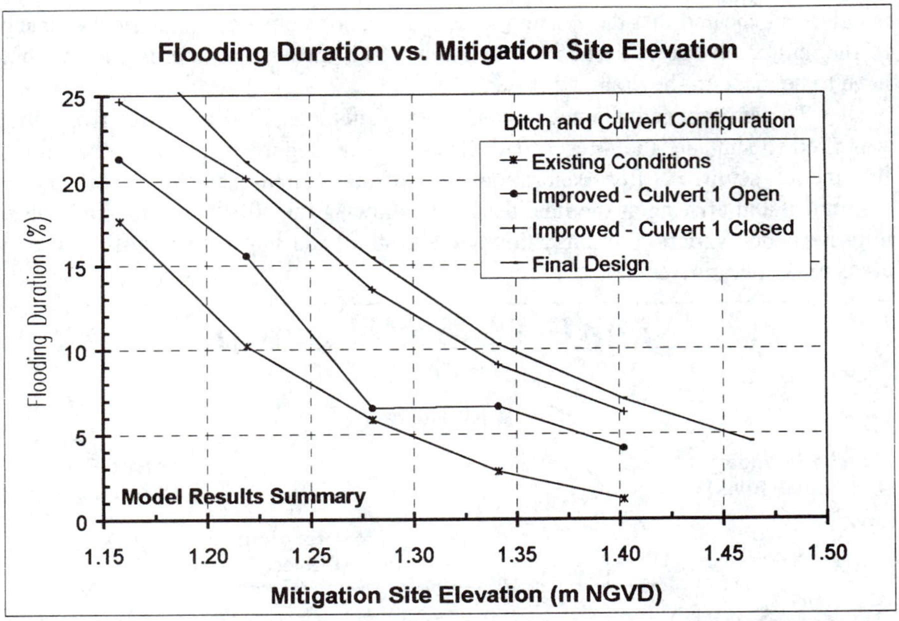

Figure 6. Mike11 model results. Flooding duration variation with mitigation site elevation and ditch/culvert configuration (From CSE, 1995).

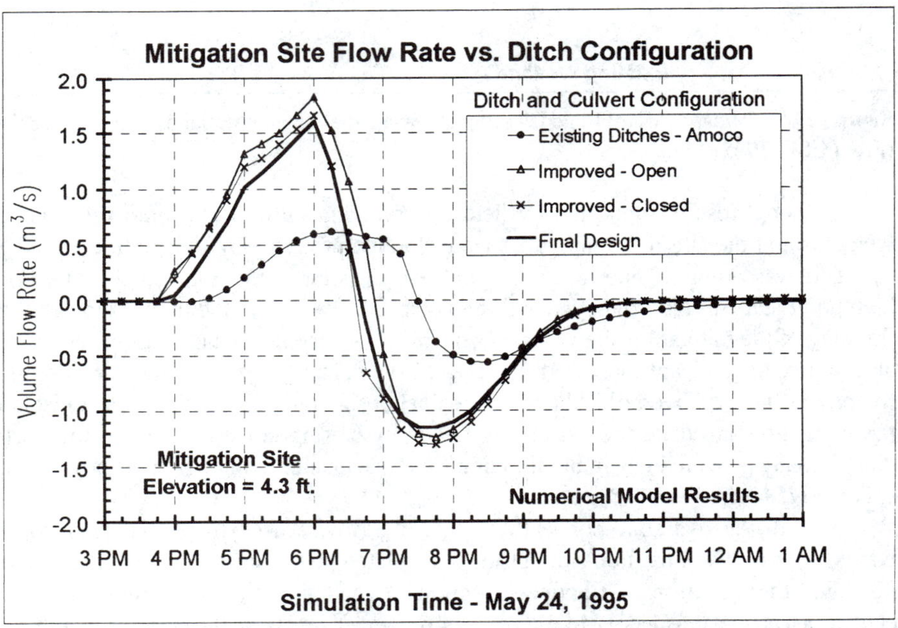

Figure 7. Mike11 model results. Volume flow rates to mitigation site for various ditch and culvert configurations (From CSE, 1995).

Despite the tidal attenuation, the mitigation site elevations were higher than those of the natural marsh. This was a result of greater attenuation of the ebb tide than the flood, which increased flooding duration at the site.

The final design for the mitigation site had the following dimensions:

Wetland area to be created:	40,000 m^2
Wetland substrate elevation:	1.22 m to 1.46 m NGVD
Wetland flooding duration:	5 to 20 %
Length of ditch to Savannah R.:	900 meters
Width of ditch:	3 meters
River to wetland tide attenuation:	0.03 to 0.15 meters
River to wetland phase lag:	30-50 minutes

Conclusions

A procedure for the hydraulic design or improvement of tidal wetlands was proposed based on the principle that wetland plant species inhabit specific ranges of flooding duration. By creating areas with the appropriate flooding duration, desired habitats can be successfully created. When calculating flooding duration, relatively long-term site-specific water levels are required. A 30-day water level record is suggested for typical conditions. Where the creation of wetlands significantly alters local hydraulics, numerical models may be required to predict postproject water levels needed to determine wetland substrate elevations.

References

CSE. 1995. Kemira wetlands mitigation project: tidal marsh hydraulics & analysis. Report for Kemron Environmental Services. Coastal Science & Engineering, Columbia, S. C., 36pp.

Kana, T.W., B.J. Baca, and M.L. Williams. 1986. Potential Impacts of Sea Level Rise on Wetlands Around Charleston, South Carolina. EPA Report 230-10-85-014, Washington, D.C., 65pp.

Lagna, L. 1975. The Relationship of *Spartina Alterniflora* to Mean High Water. New York Sea Grant Institute. NYSSGP-RS-75-002. 43pp.

Nixon, S.W., 1982. The Ecology of New England High Salt Marshes: A Community Profile. U.S. Fish and Wildlife Serv., Washington, D.C., FWS/OBS-81/55, 70pp.

Redfield, A.C., 1972. "Development of a New England salt marsh." Ecol. Monogr., Vol. 42, pp. 201-237.

Teal, J.M., 1958. "Energy flow in the salt marsh ecosystem." In Proc. Salt Marsh Conf., Mar. Inst., Univ. Georgia, pp. 101-107.

CHAPTER 335

Enhanced Mixing Through Perforated Discs on Round Buoyant Jet

Lilun Wu[1], Jiin-Jen Lee[2]

Abstract

Jets have been widely used in submarine outfall diffuser systems for discharging the sewage effluents into large bodies of sea water to facilitate rapid mixing of the effluent with the ambient water. The primary goal of an outfall diffuser system is to accomplish rapid mixing of the effluent with the ambient water. In an effort to enhance the near field mixing, various obstruction devices may be placed exterior to the diffuser nozzle. This experimental study focuses on the enhanced mixing mechanisms of the jets obstructed with perforated discs.

Experiments are conducted in a deep water tank with glass walls on four sides of the tank. The tank has the dimension of 3.35 m in depth with a square cross-section of 1.15 m. The water particle velocities of the resulting flow field are measured by using a portable four-beam, two-component, fiber-optic Laser-Doppler Velocimeter system (LDV). The concentration of the entrained fluid is measured by using a Laser-Induced Fluorescence system (LIF). The data acquisition system used for obtaining data on the concentration profiles is processed by Labview's programmable virtual instruments software.

The experimental data consist of the axial (vertical) and radial (horizontal) velocities, turbulent intensities, Reynolds stresses and concentrations. It is found that velocity fluctuations in both the axial and radial velocities help to generate vorticity and to induce mixings over a larger area in the neighboring flow region. Moreover, the velocity gradient is significantly increased over the region based on the experimental data involving the perforated discs. The results also show that a large reduction in the concentration of the entrained fluid can be achieved due to the

[1] Graduate Research Assistant, Department of Civil Engineering, University of Southern California, Los Angeles, CA 90089.

[2] Professor of Civil Engineering, University of Southern California, Los Angeles, CA 90089-2531, Member of ASCE.

obstructing disc. The results demonstrate quite convincingly the ability of the perforated disc to enhance the mixing and entrainment of the discharging fluid with the ambient fluid.

1.0 Introduction

For the removal of municipal sewage, treated or untreated waste water must be disposed of in such a way that will generate minimal effects to the environment. An ocean discharging system is commonly used to discharge sewage effluents into coastal water and to induce mixing of the effluent with the sea water. The systems are generally comprised of embedded pipes, the main outfall, which delivers the waste water to a diffuser system. The length of the main outfall pipes normally range from 1 km to 9 km, depending upon the sea bed topography, ocean currents and environmental requirements.

There are three major factors that affect the mixing of effluent (Fischer et al. 1979). They are: jet parameters, environmental parameters and geometrical factors of the discharging jet. Both the jet parameters and the environmental parameters have been studied extensively in the last four decades. Geometrical factor is an interesting field for the scientists and engineers to study because better geometrical arrangements may result in greater dilution. Increasing dilution in an ocean discharging system would produce significant saving in the construction costs by reducing needed mixing height of the rising plume.

The present experimental investigation includes measurements of axial and radial velocity, turbulent intensity, turbulent shear stress and concentration profiles. The purpose of the measurements is to accurately measure various flow parameters in the flow field of the jets obstructed with perforated disc to provide an answer to the important question posed for the study.

2.0 Experiments

Experiments were performed in a deep water tank system with vertical glass face. The water tank has the dimensions of 335 cm high, 115 cm wide and 115 cm deep. The facilities also included a mixing tank, a constant-head tank, air bubbling hose and a manometer (flow meter). The mixing tank provides a space for adding dye or salt into effluent fluid. The constant-head tank is used for maintaining a steady flow for the effluent fluid, while the air bubbles provide enhance mixing in both the mixing tank and the glass water tank before conducting experiments. Fluid velocity and Rhodamine 6 G dye concentration can be measured by a Laser Doppler Velocimeter (LDV) and a Laser-Induced Fluorescence system (LIF), respectively. For measuring purposes, this glass walled tank is equipped with an instrument carriage supporting a Laser Doppler Velocimeter or a Laser-Induced Fluorescence device. The carriage is able to move both vertically and horizontally with respect to the tank, thus velocities and concentrations could be measured at various axial and radial locations. The jet nozzle is located on the top of the tank and pointing down. The density of the exit fluid from the jet nozzle is greater (heavier) than that of ambient fluid. Sodium Chloride (salt) is added into the exit fluid to obtain the desired initial density difference between exit fluid and ambient fluid.

A portable four-beam, two component, fiber optic Laser Doppler Velocimeter (LDV) manufactured by TSI, Inc. was used for measuring the water particle velocities. The software of data acquisition for controlling the LDV system, *Flow Information Display (FIND) Data Analysis Package* from TSI, Inc. was used. The concentration is measured by a laser-induced fluorescence system (LIF) at various cross sections downstream from the jet nozzle. This system includes a 2W Argon-Ion laser (Spectra Physics Model 265), a CCD camera (EG&G Reticon LC300A), an analog input/output data acquisition board (National Instrument, AT-MIO-16X) and data acquisition program (Labview). The laser beam is oriented in the direction perpendicular to the centerline of the flow, and it is shot into the water tank across the entire flow field from one side of the tank. A line fluorescence light source is then detected by an array of light-sensitive photo diodes inside the EG&G Reticon LC300A camera. The data acquisition system receives signals from the camera and converts the analog signals to digital data. The data for the concentration profile is processed by Labview's virtual instruments software.

3.0 Results and Discussion

3.1 Velocity Profile and Width of Simple Jets

The results of the velocity measurement conducted for the present work are in agreement with those of Fischer et al (1979) and Papanicolaou and List (1988). This agreement ensures that the instrumentation system used for the present study can yield reliable results. Thus, comparisons can be made between the results conducted for simple jets and for jets obstructed with perforated disc.

Figure 1 shows the normalized diagram by combining all of the velocity profiles of the simple jets into one figure. Data distribution in Figure 1 is a Gaussian distribution. The jet has U_{mean}/U_c value in the range of $|r/z| < 0.2$, beyond this range the U_{mean}/U_c values tend to zero. The normalized diagrams for representing the results of jet obstructed with each perforated disc are presented in Figures 2. Most values of U_{mean}/U_c of each obstructed case are located in between $|r/z| < 0.4$. After comparing the results of velocity profiles for the case of simple jets and the cases of jets obstructed with perforated discs, conclusions can be drawn as follows: (1) For the jets obstructed with different perforated discs compared with the simple jet (Figures 3), the axial (longitudinal) velocity distributions of the jets with perforated disc have more fluctuations exhibited in each profile. (2) The cases of the jets obstructed with perforated disc have larger plume width on each cross section. (3) For jets obstructed with various perforated disc, the axial velocity distributions are more uniform and broader when compared with the case of simple jets. (4) The centerline velocities of the jets obstructed with perforated disc have relatively smaller value with compared with simple jets.

3.2 *Turbulence Properties*

The axial (vertical) and radial (horizontal) turbulent intensities have been measured simultaneously by LDV for both simple jets and jets obstructed with perforated discs. After comparing the results on the turbulent intensity (both axial and radial) for both simple jets and jets obstructed with perforated discs, general conclusions can be made as follows: (1) Simple jets have larger fluctuations of the axial velocity around the boundary of jet. In the outer region of the cross section (beyond the boundary of the plume width), the percentages of axial turbulent fluctuation of jets with each perforated disc are much greater than those of the simple jets. This shows that jets obstructed with perforated discs can produce more turbulent fluctuation of axial velocity than the simple jets in the outer region of the cross section. (2) The distributions of turbulent fluctuations in the axial and radial component for the jets obstructed with perforated discs are much more uniform and much broader than those of simple jets. (3) The jet obstructed with the 2x8 perforated disc has an overall higher U'_{rms}/U_c value than those for other disc arrangements. (4) The percentages of radial turbulent fluctuation of jet obstructed with perforated disc have higher values than those of simple jets at all of the cross sections within the region of the jet width. Moreover, the overall percentages of radial turbulent fluctuation of jets obstructed with perforated discs are higher than those of simple jets in the region beyond the boundary of the jet for each cross section. This shows that the jets obstructed with perforated discs have higher radial (horizontal) turbulent fluctuation than simple jets beyond the boundary region.

The results from the measurement of turbulent intensity indicate that the jets obstructed with perforated discs will enlarge the region of turbulent fluctuation and resulting in more interaction between ambient and effluent fluid. This also supports the experimental hypothesis that using the perforated discs to obstruct the jets will result in larger areas of axial and radial fluctuations in the flow field.

3.3 *Reynolds Stress*

The shear layer between the effluent fluid and the ambient fluid was proved by Fischer et al (1979) as an important factor to entrain the ambient fluid into effluent fluid and to induce mixing of both fluids. The Reynolds stress ($\overline{U'V'}$ or turbulent shear stress) measurements were conducted for both simple jets and jets obstructed with perforated discs. Experimental results have been presented in Figure 4 for the simple jets and in Figure 5 for the jets obstructed with perforated discs. Comparisons between the results of simple jets and jets obstructed, can lead to the following conclusions: (1) Fluctuations can be observed on the $\overline{U'V'}$ profiles of the jets obstructed with perforated discs. This is because the effluent fluid flows through the perforated disc (or holes on the disc) and generates many small vortices inside the boundary of the jets. Thus, it creates a larger area of shear layer and induces more entrainments for the entire flow field. (2) Results show that the jets obstructed with perforated discs have higher value of $\overline{U'V'}$ at the region outside of the jets' boundary, and this is the advantage of using the obstructed perforated discs to broaden the extents of the velocity field and the velocity fluctuation. A broader mixing area is created by the obstructing perforated discs, Thus, mixing is enhanced by a broader region possessing certain value of turbulent shear stress.

3.4 Concentrations and Dilutions

The concentration measurements were conducted and included in the present study. The simple jets' maximum concentrations obtained for the present work are in agreement with the previous studies such as Fischer et al. (1979) and Papanicolaou (1984). The widths of concentration profiles of the simple jets measured in the present study are also in agreement with those of Papanicolaou and List (1988).

After comparing the results of concentration measurements for both simple jets and jets obstructed with perforated discs, conclusions can be made as follows: (1) the results of the simple jets are similar to that reported by Fischer et al. (1979) and Papanicolaou and List (1988), (2) the maximum concentration for the cases of the jets obstructed with perforated discs are quite smaller than those of the simple jets, and (3) the concentration widths (b_c) for the jets obstructed with perforated discs are broader than those of the simple jets.

The calculations for the mean dilution achieved for the simple jets and for jets with perforated discs also were performed. The results indicate that (1) the C_{max}/C_{ave} ratios of the simple jets are in the range of 1.315~1.592, this result is in agreement with that summarized by Fischer et al. (1979) which is 1.4 ± 0.1, (2) for the jets obstructed with perforated discs, the mean dilution is higher than that of the simple jets, (3) the values of mean dilution reflect that both $\frac{\mu}{Q_O}$ (from velocity measurements) and C_{max} (observed from concentration measurement) are in consistant trends, (4) the C_{max}/C_{ave} ratios of the jets obstructed with perforated discs are in the range of 1.052 to 1.369, (5) the μ/Q (mean dilution) versus z/l_Q (normalized distance) of the simple jets and the jets obstructed with perforated discs have been plotted in Figures 6. They show that the jets obstructed with perforated discs have a higher mean dilution than that of the simple jets for a given z/l_Q. Another measure of the effectiveness of enhanced mixing is to compare the maximum dilution for both the case of simple jet and that for jet obstructed with perforated disc. This is shown in Figures 7. It is found that there is a significant increase of the maximum dilution with perforated disc.

4.0 Conclusions

The results of the present work proved that the jets obstructed with perforated discs have wider profiles of velocity, turbulent intensity, turbulent shear stress and concentration than the simple jets. The normalized diagrams from the experimental results also indicate that the data distributions of the jets obstructed with perforated discs are flatter and broader than those of simple jets in the fields of velocity, turbulent intensity, turbulent shear stress and concentration. All of this provides evidence that jets obstructed by perforated discs have larger areas for mixing than those of the simple jets.

It is clear from the concentration measurements that for the jets obstructed with perforated discs, the concentration of the entrained fluid is greatly reduced and that the region for mixing is significantly enlarged. The results from the calculation of mean dilution and maximum dilution also indicate that the jets with perforated discs can produce significantly increased dilutions when compared with simple jets. The overall experimental results demonstrate quite convincingly the ability of the perforated disc to enhance the mixing and entrainment of the discharging fluid with the ambient fluid.

Acknowledgment

This research study has been supported by University of Southern California Foundation for Cross-Connection Control & Hydraulic Research and the San Gabriel Valley Protective Association. The authors extend special thanks to Professor Fredric Raichlen and Professor John List of the California Institute of Technology for their generosity in granting use of the deep water tank and its associated facilities at W. M. Keck Laboratory of Hydraulics and Water Resources of CalTech for the experiments.

References

1. Chen, C. J. and Rodi, W. (1978), Vertical Turbulent Buoyant Jets - A review of Experimental Data. *HMT Vol. 4, Pergamon Press*, 83 pp.

2. Fan, L. N. (1967), Turbulent Buoyant Jets into Stratified or Flowing Ambient Fluids, *Report No. KH-R-15, W.M. Keck Laboratory of Hydraulics and Water Resources*, California Institute of Technology.

3. Fan, L. N.; Brooks, N. H. (1969), Numerical Solution of Turbulent Buoyant Jet Problems, *Report No. KH-R-18, W.M. Keck Laboratory of Hydraulics and Water Resources*, California Institute of Technology.

4. Fischer, H. B.; List, E. J.; Koh, R.C.Y.; Imberger, J.; and Brooks, N. H. (1979), Mixing in Inland and Coastal Waters, *Academic Press*, New York, N. Y., 483 pp.

5. Hannoum, I. A. and List, E. J. (1988) Turbulent mixing at a shear free density interface. *J. Fluid Mechanics*, 189, 211-34.

6. Koh, R. C. Y. and Brooks, N. H. (1975), Fluid mechanics of wastewater disposal in the ocean. *Ann. Rev. Fluid Mech.*, 7, 187-22.

7. Noutsopoulos, G. C. and Demetriou, J. (1980), The round vertical turbulent buoyant jet impinging on a concentric disc, *Progress Water Technology*, 13, 305-314.

8. Noutsopoulos, G. C. and Yannopoulos, P. C. (1989), Axial dilution in obstructed round buoyant jet. *J. of Hydraulic Engineering*, Vol. 115, No. 1, Jan., 1989. 71-81.

9. Papanicolaou, P. N. and List, E. J. (1987), Statistical and spectral properties of tracer concentration in round buoyant jets. *Int. J. Heat Mass Transfer*, Vol. 30, No. 10, 2059-2071.

10. Papanicolaou, P. N. and List, E. J. (1988), Investigations of round vertical turbulent buoyant jets. *J. Fluid Mech.*, Vol. 195, 341-391.

11. Wallace, R. B. and Wright, S. J. (1984), Spreading layer of two-dimensional buoyant jet, *J. Hydraulic Engineering*, Vol. 110, No. 6, June 1984. 813-828.

12. Wu, Lilun (1996), Enhanced mixing through perforated disc on round bouyant jet, Ph.D Thesis University of Southern California, Los Angeles, CA

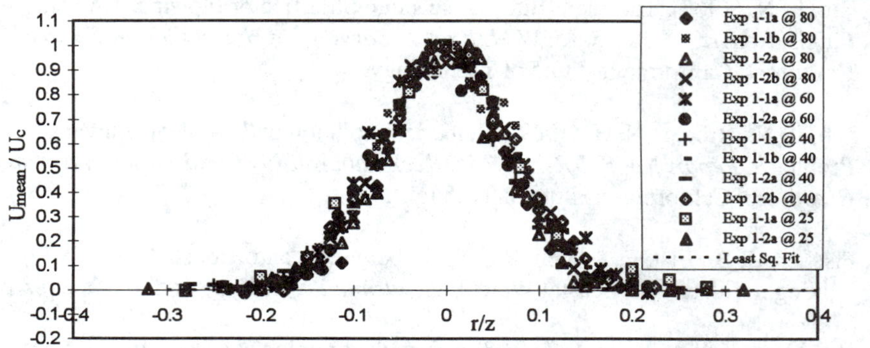

Figure 1 Non-dimensional Mean Axial Velocity Profile for Jets Plotted Against Non-dimensional Distance from Jet Axis, z/D ≥ 50

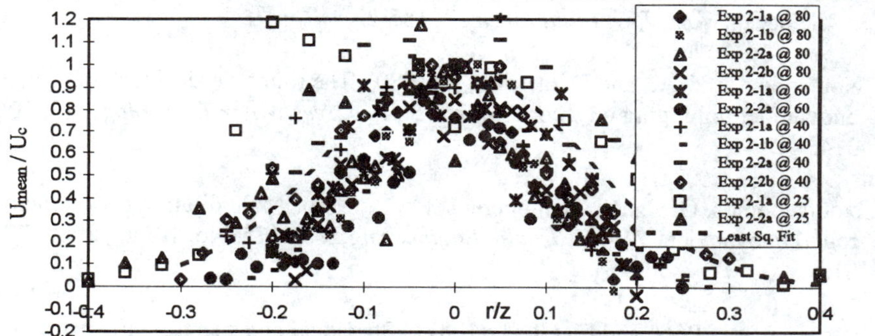

Figure 2a Non-dimensional Mean Axial Velocity Profile for Jet Obstructed with Perforated Disc (2x2) Plotted Against Non-dimensional Distance from Jet Axis, z/D > 50

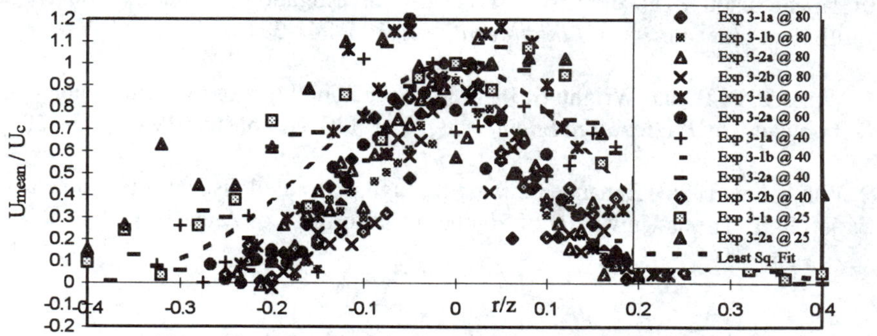

Figure 2b Non-dimensional Mean Axial Velocity Profile for Jet Obstructed with Perforated Disc (2x8) Plotted Against Non-dimensional Distance from Jet Axis, z/D > 50

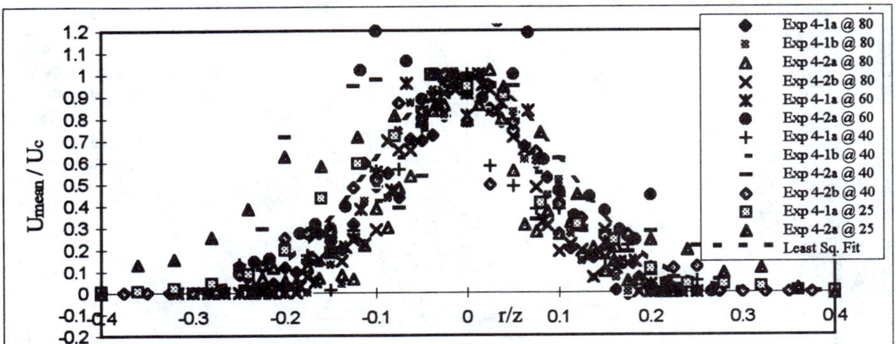

Figure 2c Non-dimensional Mean Axial Velocity Profile for Jet Obstructed with Perforated Disc (B2x8) Plotted Against Non-dimensional Distance from Jet Axis, z/D > 50

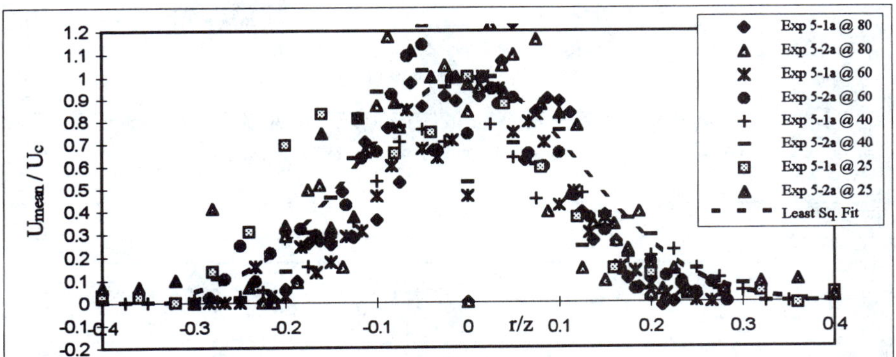

Figure 2d Non-dimensional Mean Axial Velocity Profile for Jet Obstructed with Perforated Disc (323) Plotted Against Non-dimensional Distance from Jet Axis, z/D > 50

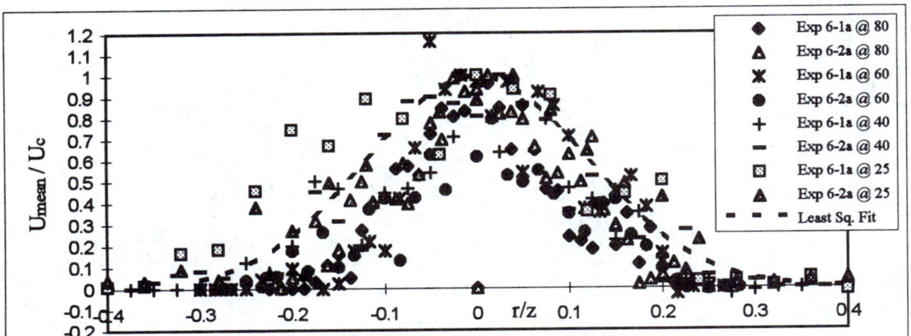

Figure 2e Non-dimensional Mean Axial Velocity Profile for Jet Obstructed with Perforated Disc (4+1) Plotted Against Non-dimensional Distance from Jet Axis, z/D > 50

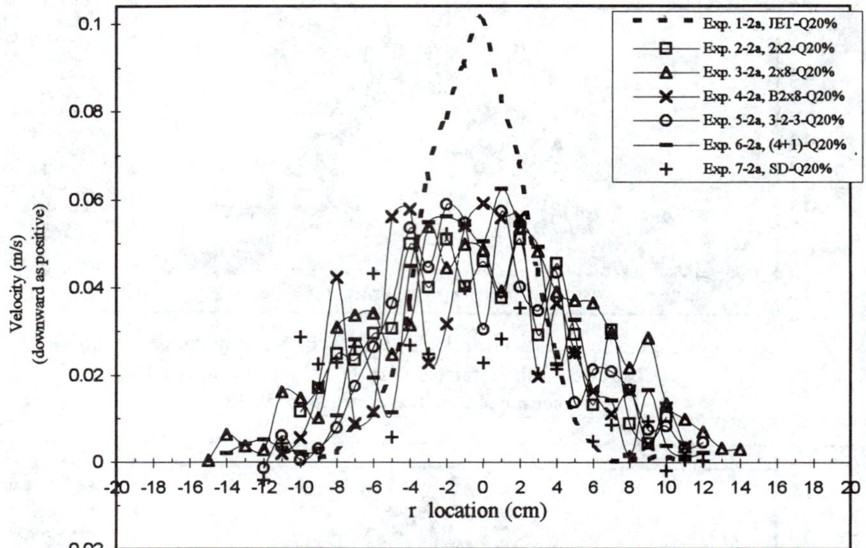

Figure 3a Profiles of Axial Velocity of Jet and Jet Obstructed with Each Perforated Disc at z=40 cm for Q=24.72 cm^3/s

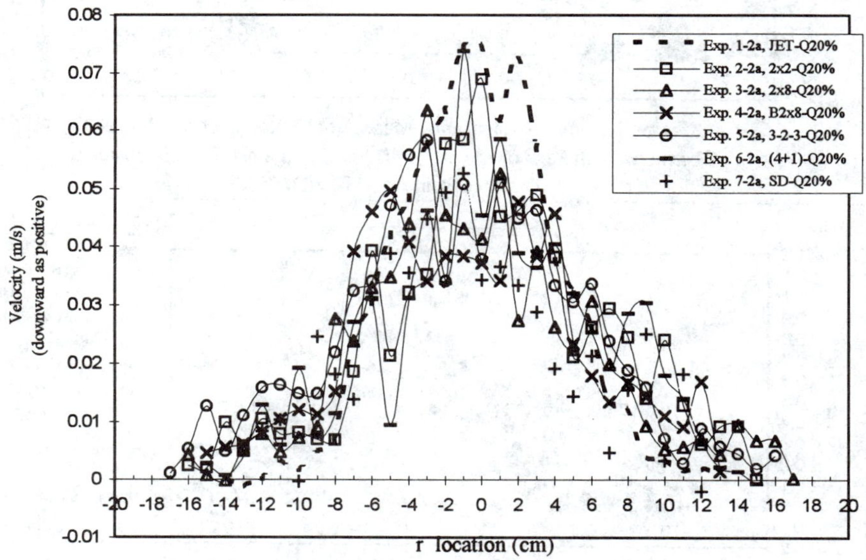

Figure 3b Profiles of Axial Velocity of Jet and Jet Obstructed with Each Perforated Disc at z=60 cm for Q=24.72 cm^3/s

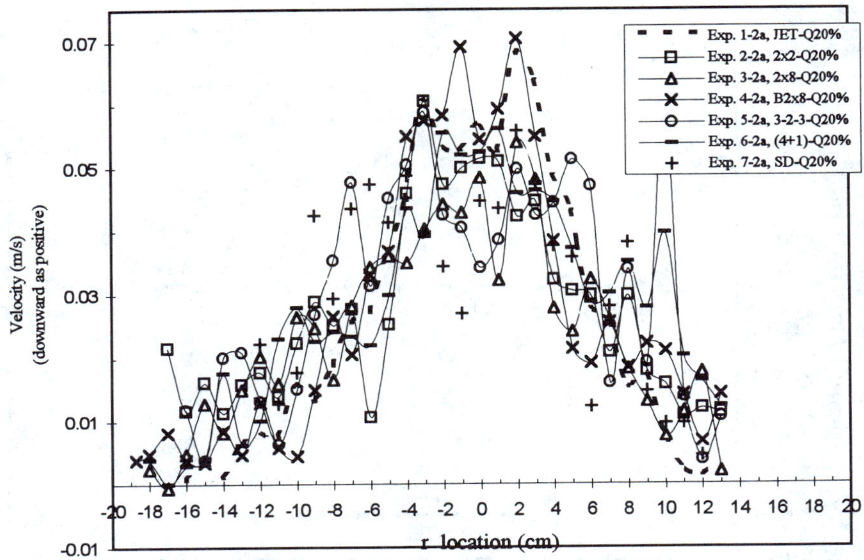

Figure 3c Profiles of Axial Velocity of Jet and Jet Obstructed with Each Perforated Disc at z=80 cm for Q=24.72 cm^3/s

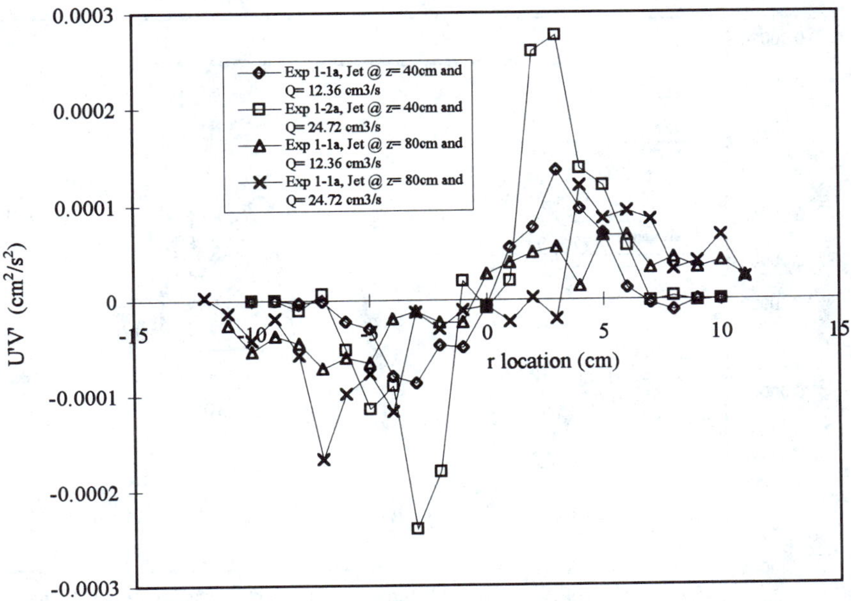

Figure 4 Profiles of Reynolds Stresses (Turbulent Shear Stress) of the Simple Jet

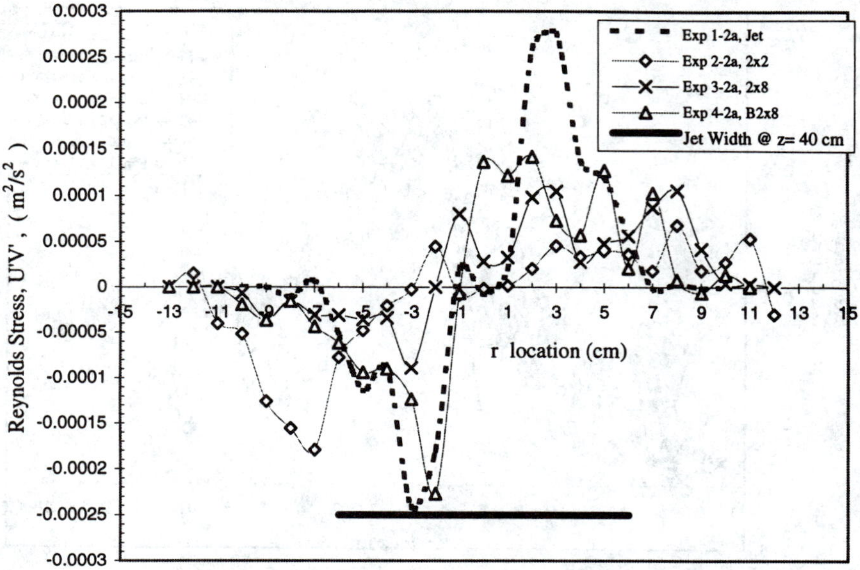

Figure 5a Profiles of Reynolds Stresses (Turbulent Shear Stress) at z=40 cm for Q=24.72 cm^3/s

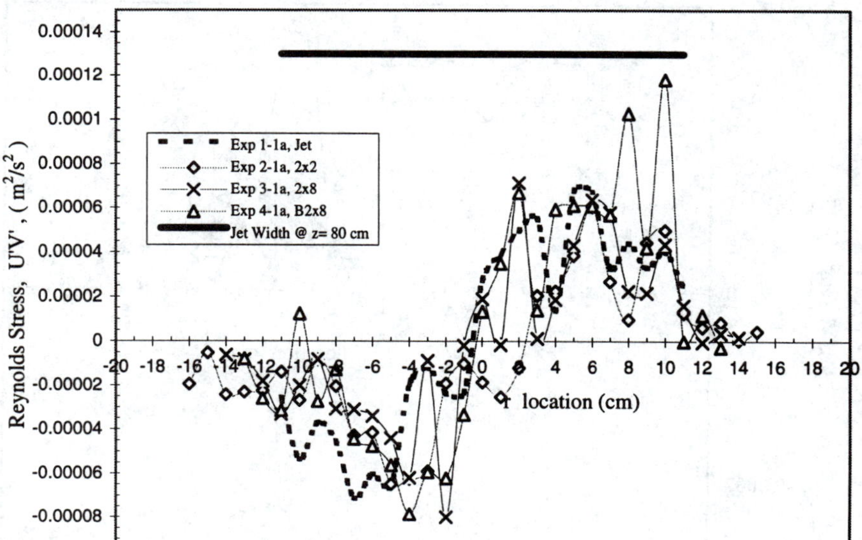

Figure 5b Profiles of Reynolds Stresses (Turbulent Shear Stress) at z=80 cm for Q=24.72 cm^3/s

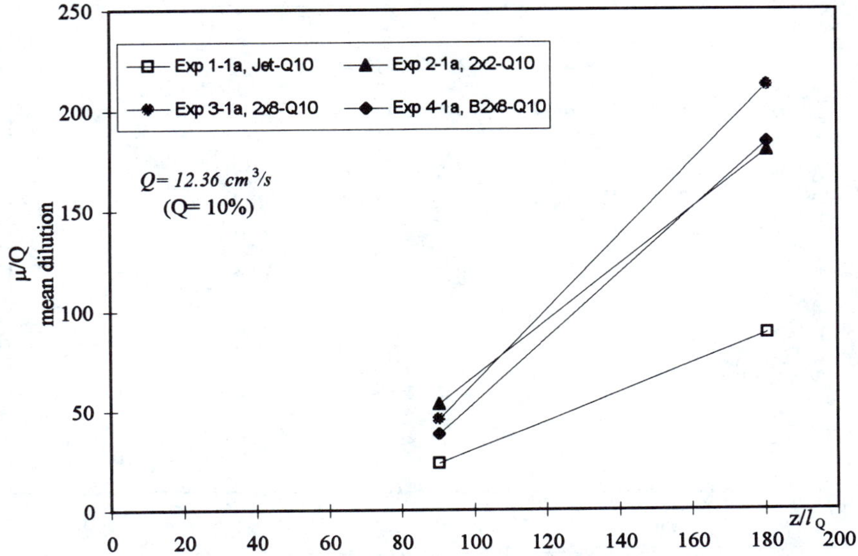

Figure 6a Mean Dilution as a Function of Distance along the axis for Q=12.36 cm^3/s

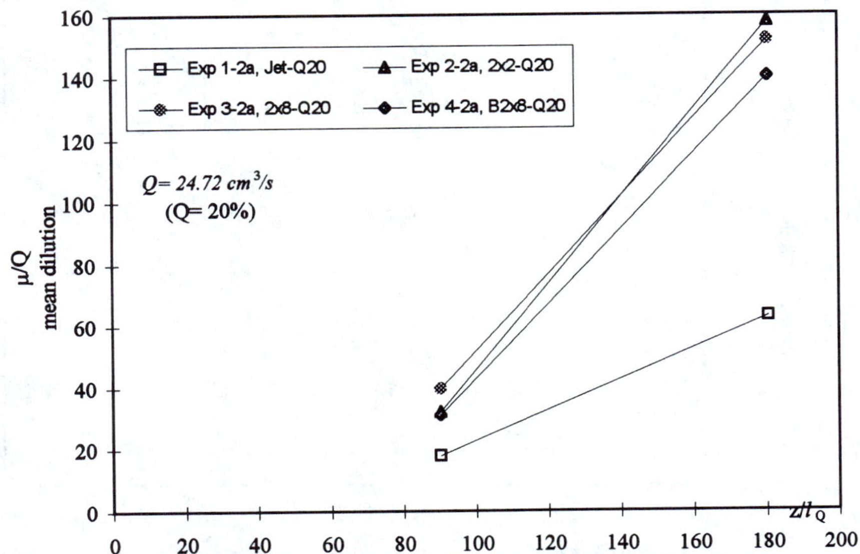

Figure 6b Mean Dilution as a Function of Distance along the axis for Q=24.72 cm^3/s

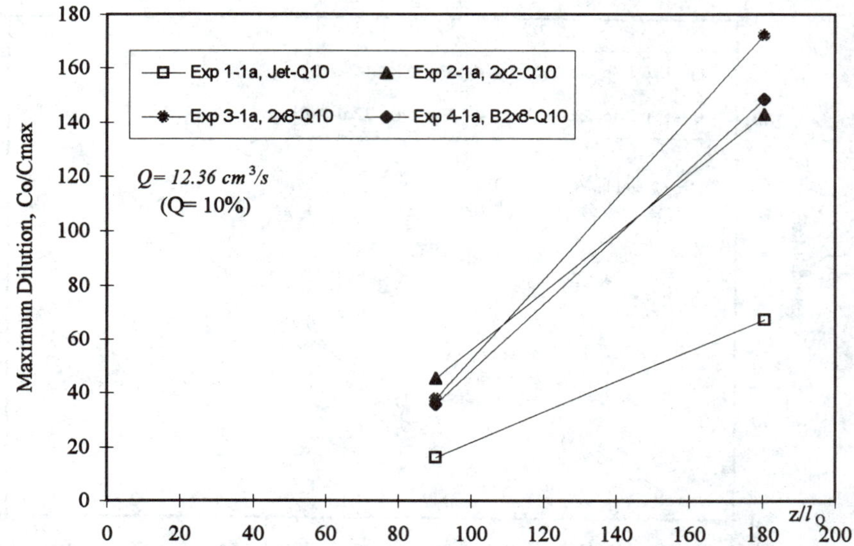

Figure 7a Maximum Dilution as a Function of Distance along the axis for Q=12.36 cm^3/s

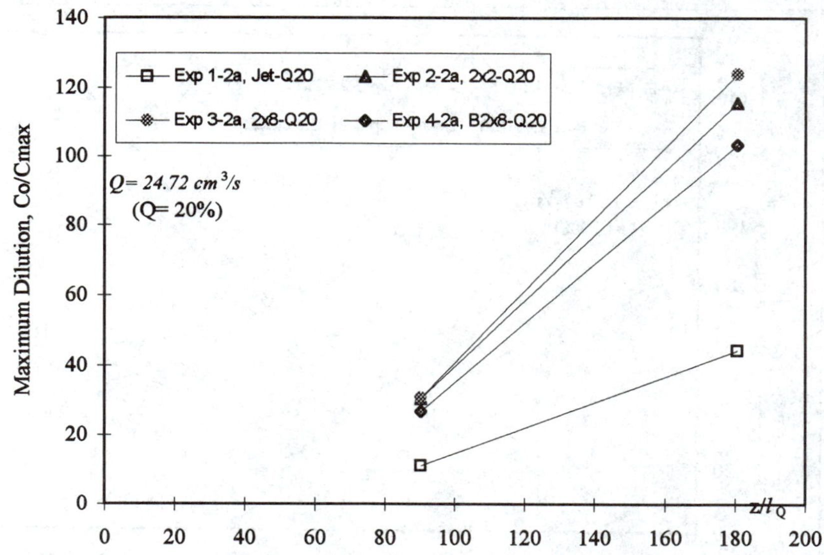

Figure 7b Maximum Dilution as a Function of Distance along the axis for Q=24.72 cm^3/s

CHAPTER 336

ASSESSING COASTAL FLOOD RISKS

by Roger Maddrell[1], Chris Fleming[1] and Chris Mounsey[2]

Abstract

The English and Welsh coasts of the United Kingdom are protected from coastal flooding (as opposed to coastal erosion) by some 1300km of defences, containing about 2000 defence structures. Using data from the sea defence survey of these structures and over 20 years of joint probability data on sea levels and waves, Halcrow assessed the failure risks and coastal flood risks for all these defences. The risks were examined for three Bands or return periods, namely 50 years, 100 years and over 200 years. This analysis included 50 years of relative rise in sea level (isostatic and eustatic changes). The modes of defence failure examined were from overflow, overtopping and toe failure. The results were supplied to the insurance industry, as maps and on disc, as 1km map squares giving the risk band, flood depth and the postal codes in the areas.

Flood risks were shown to be significant and even though reduced when more recent flood defence projects were included, they remain high. The impact of individual storms was later examined and obviously, while the risks were limited to specific areas and coasts, they were still significant. The insurance industry have used these results in order to assess their financial exposure to individual events and to their reinsurers.

Introduction

Potential coastal flood risks areas represent only about 3% of the area of England and Wales, but the value of their insured assets is much higher. Two major storm events in October 1987 and January 1990 sharpened the insurance companies and the "reinsurers" (who insure the insurance companies) concern about their exposure to

---

1 Sir William Halcrow & Partners Limited, Burderop Park, Swindon SN4 0QD, UK

2 Association of British Insurers, 51 Gresham Street, London, EC2V 7HQ, UK

weather and coastal flooding risks. No one wished to see another 1953 type disaster. Consequently, in 1992 the Association of British Insurers (ABI) asked Halcrow to categorise the flood risk from coastal storm events for the benefit of their 450 members

The ABI membership between them account for over 90% of the UK business. ABI's key objectives are to help insurance companies by representing their interests to Government, particularly on public policy issues; providing them with technical services, in particular in respect of industry statistics, market and regulatory information and reducing the incidence of claims; and promoting insurance and insurance companies generally. It also helps its members at the pre-competitive stage, to improve their chance of success in achieving a balanced underwriting account.

Naturally, ABI are the focus for presenting the industry's concerns on major issues to the Government, ministers and civil servants, MPs, European bodies, the media, consumer bodies and other opinion formers. This responsibility is taken very seriously and it is recognised that their value is diminished if based on inaccurate and illfounded information. Thus, ABI concentrates on resolving real live issues which will affect the industry, of which coastal flooding is particularly relevant.

The areas covered in the Halcrow study were those protected by 1300km of coastal flood defence, the defences of only four main estuaries, namely, Tees, Humber, Thames (Upper and Lower) and Severn, but none of the other estuaries and channels. The pilot study in 1993 examined a detailed approach and one using the sea defence survey (SDS) data from the UK's Environmental Agency (EA). As the results in terms of flood risk were similar, the latter was chosen as it had the shortest completion time and was the most cost effective.

The first main study report was presented to the ABI membership in 1994 and was subsequently updated in 1995 to include for the new and improved defences built since the SDS was published in 1990. Additional studies include the prediction of flood risk from individual storms, using data supplied by the UK's Meteorological Office, examining and reporting on those defence lengths shown to be not performing or at greatest risk and the flood risks to London from combined tidal and high fresh water flows. The latter study is nearing completion.

Study Methodology

The assessment of flood risk needs to examine the integrity of the existing sea defences and assess their likelihood of failure and the extent of the subsequent flood area. There are nearly 1300km of coastal flood defences in England and Wales, containing some 2000

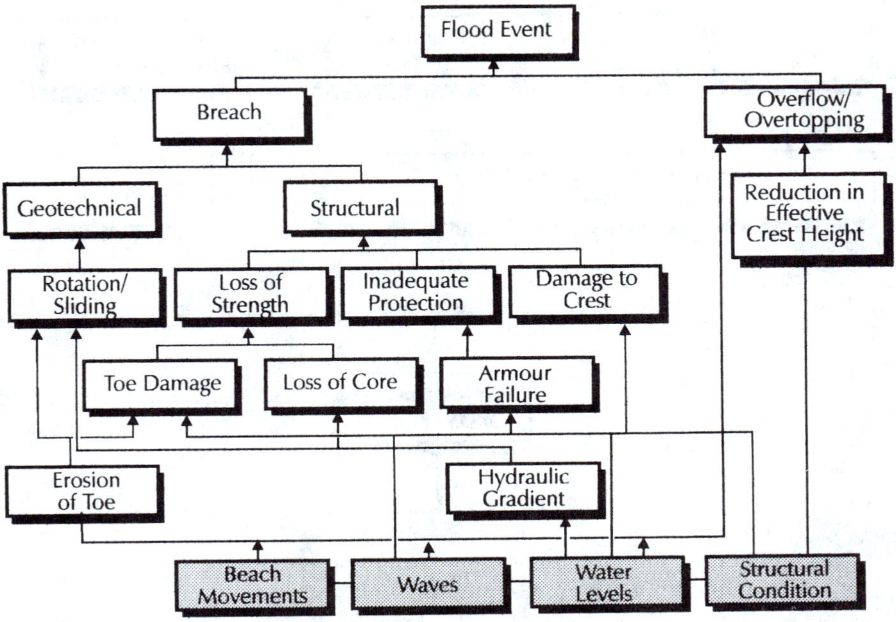

Figure 1. Key failure modes

structures and the integrity of each defence had to be examined. Flood risk can be approached in a number of ways, from highly subjective visual assessments, through to detailed probabilistic analyses based on comprehensive structural data. Whilst detailed probabilistic analyses, including all failure modes, can be applied in specific cases, the pilot study showed it was not economically viable for application on a nationwide basis.

Two methods were examined in the pilot study, which explored the relative costs and the quality of results from each approach. The first method offered a detailed approach, analysing a number of key failure modes (see Figure 1) to provide a quantified assessment of failure risks. This method also incorporated a set of screening tests to reduce the number of defences to be analysed, with the remainder requiring detailed structural information. The major disadvantage of this approach was not the amount of calculation required, which could be automated, but the high costs of acquiring detailed data which could only be acquired through structural inspections (Burgess and Reeve, 1994).

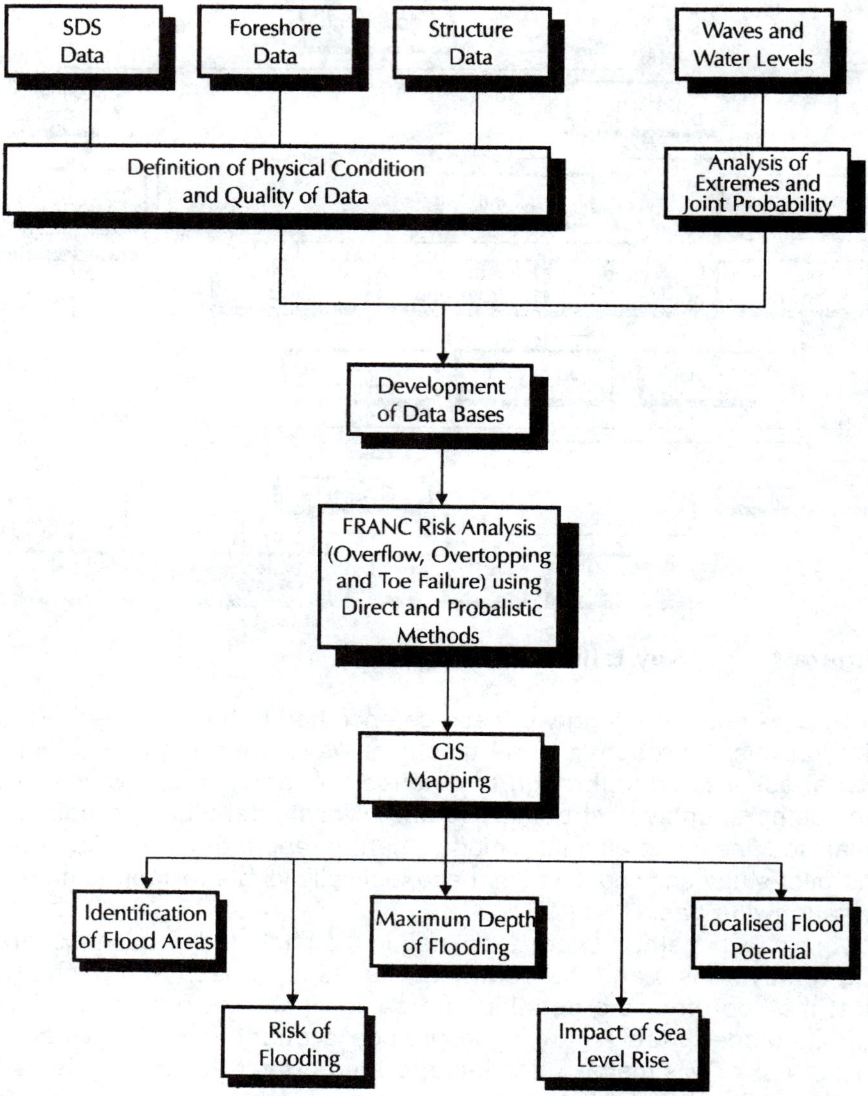

Figure 2. Methodology

The alternative second method made use of information in the UK's Environmental Agency's (EA) sea defence survey (SDS) to estimate risk of failure, supplemented with environmental data. This methodology (see Figure 2) concentrated three primary failure modes namely:

- Overflow - when the still water level exceeds the crest of the defence;

- Overtopping - when the combined effect of waves and water levels results in waves running up and breaking over the defence;
- Toe Failure - when the toe of the structure fails due to low beach levels at the base of the defence (erosion being one of the most common causes of damage to sea defences).

For open coasts, normally overtopping and toe failure and to a certain extent, overflow were critical, but for the estuaries the modes were overflow and toe failure with overtopping being less critical.

The starting point of the study was the 1990 SDS which was a major undertaking and a considerable step forward in flood defence management by updating the original 1980 Herlihy study. An analysis for each defence required a site specific knowledge of waves and water levels. Therefore, inshore study points were defined on the open coast and in the estuaries. Future relative sea level changes were also established. Water level time series covering a period of 30 years were combined with 20 years of offshore wave data, transformed inshore through modelling, to produce marginal and joint probability extremes together with a statistical description of the conditions.

The quality of available structure and beach data was, unfortunately, extremely variable and thus two types of risk assessment were developed for each defence; “direct” and “probabilistic” approaches. In general terms the “direct” approach involved determining the specific conditions and calculating to see whether a particular structure could withstand or would be likely to be at risk of failure under these conditions. The risk of failure and hence flood risk was described by risk bands, with Band 1 defined as a defence potentially vulnerable to a 1 in 50 year event ie with an annual probability of failure greater than 0.02, Band 3 as a defence withstanding conditions with a return period of 200 years, and Band 2 falling between these two limits. The “probabilistic” approach was a first order risk method to calculate the annual probability of failure and defined the risk as Bands 1, 2 or 3, depending on the annual probabilities of failure comparable with the return periods above.

The software package “SANDS” already existed, which could be used for data storage, retrieval and analysis. However, to perform the risk analysis computational software had to be developed for the project, acronym “FRANC”, which enabled all three failure modes to be analysed by both the “direct” and “probabilistic” approaches. This was essential given the volume of data and the need to analyse 3 modes of failure for nearly 2000 defences. “FRANC” read the relevant data direct from the "SANDS" database, performed the calculations, combined the results and assigned the defence classification. The software was designed to establish either an overall probability of failure risk based upon the results of all three mechanisms or to establish the worst case, ie the most likely mode of failure. The results were generated in a format that could be directly linked into the flood area mapping.

To perform the analysis of overflow and overtopping, both wave and water levels were needed. A 30 year water level time series for nearshore locations was obtained from the Proudman Oceanographic Laboratory's (POL) tide and surge model for the United Kingdom. Additional data, including annual maximum levels, were obtained for the Severn, Thames, Tees and Humber estuaries, supplemented by information gathered from the relevant Port Authorities.

Wave data at offshore locations were obtained in the form of a time series output for the North European Storms Study (NESS), which gave wave height and direction for some 32 offshore locations and covered a period of just over 20 years. For the outer areas of the estuaries, wave time series were hindcast from wind measurements covering a period in excess of 14 years for each estuary. Extreme wave heights in the upper estuaries were calculated from monthly maxima wind speeds using shallow water wind-wave hindcasting techniques.

The site specific information on waves and water levels was derived from inshore study points using wave refraction models. This required the digital reproduction of the nearshore bathymetry around the entire English and Welsh coastlines and resulted in the generation of sets of time series wave conditions at each of these locations. Shallow water effects were accounted for by spectral saturation, Bouws et al (1985), whilst wave breaking at each structure related to the beach condition and water level.

In interpolating the water level data to the study points, extreme values were adjusted to reflect the natural variation in Mean High Water Spring (MHWS). Future relative sea level changes in 50 years time for each location required the IPCC's (1990) prediction of global sea level increase due to global warming and isostatic changes described by Shennan (1989).

Extreme values of water levels, wave heights and periods were defined in terms of a return period with the marginal distribution functions determined by fitting a distribution to the data. In this case, the Generalised Extreme Value (GEV) was used to estimate extreme values from the annual maxima of each variable (Carter et al, 1986). Joint extreme values were established empirically from frequencies of occurrence (see Owen, 1980 and Hawkes, 1990), for assessing the risk of overtopping, where the combination of water levels and waves is important. The results of these data processing exercises were incorporated into the database, for use in the risk assessment.

The results of the risk analysis were transferred to a Geographic Information System (GIS) to relate the risk classification of each defence to its associated flood area. The flood areas were defined by establishing the intersect between predicted extreme water levels and the terrain. Flood depths were established from the interrogation of the Ordnance Surveys (OS) base data and any available drawings to derive ground level estimates. Maps were produced on OS backgrounds using the GIS for 10km by 10km coastal flood areas, showing the risk bands and predicted maximum depth of flooding within each 1km grid square. This size grid was chosen mainly because of the large area that had to be covered in the short time available and the lack of detailed survey data below the 5m contour.

Two characteristics of the sea defences were of relevance to the flood risk mapping; the location of a defence and its risk classification. The location of each defence length was defined by start and end co-ordinates in the SDS database. An automated routine was used to interrogate this information and plot the defences within the GIS. The risk classification output from the defence analysis software was loaded into the GIS database and related to the plotted graphic elements using a unique defence length code identifier. When more than one defence structure existed within a defence length, the highest risk probability amongst those structures was assigned to that length.

The risk assigned to a defence length was then transferred to the area being defended. Again, the rationale of assigning the highest risk was applied so that in cases where a single flood area was bordered by a number of defence lengths of differing risk values, the highest risk value amongst those was transferred to the defended area. The seaward extent of the defended area was taken as being the line of defence and the landward extent that line at which the water level, associated with a given event, intersected the land surface ie the same methodology as that used in the SDS. The accuracy with this line could be determined was largely reliant on the availability of ground level information, some of which was poor.

The variation found in the water levels associated with the 50 and 200 year event typically ranged between 0.1m and 0.5m. This made the differentiation of discrete flood boundaries associated with the 50, 100 and 200 year events problematical, given the quality and quantity of ground level data available from OS sheets.

The protected areas were overlaid with 1km grid squares. Each 1km grid square took on the highest risk classification of any defended areas which it wholly or partially covered. Post code sector data was provided by the UK's Post Office and the risk was related to the post code sectors.

In addition to the risk classification assigned to each grid square, an attempt was made to assess the depth of flooding using the very limited ground level data available. The approach was generally simple, making a direct interpolation of the depth of flood water calculated at available data points behind the defence and at the inland boundary of the flood water. Where possible, the influence of features such as inland embankments was included.

This method of interpolation obviously created discrepancies when assigning a single flood value across 1km grid squares, which might contain areas of higher relief. Similarly, where the level of the ground was below that defined by drawings or spot heights, underestimation of the depth of flooding occurred.

Results

Each analysis was rigorously checked for its approach and the correctness of numerical results and a further check was carried out to verify the overall output. Direct verification was restricted due to the lack of recorded instances of defence failures, despite obtaining newspaper reports. The "Effective Level of Service" recorded in the SDS provided an alternative measure against which the analysis results could be compared, details of which are given in Table 1.

Classification	SDS (%)	Halcrows' Study (%)
Band 1, 50 year	33	28
Band 2, 50 to 200 years	15	5
Band 3, > 200 years	52	67

Table 1. Overflow classification results, percentages of defences at risk

The results from the original study showed a good agreement for overflow with those in the SDS (only overflow was examined in the SDS) with 64% of all structures analysed having the same risk classification. Of the remaining 36%, the majority had a lower classification than that given by the SDS. However, notwithstanding the differences in analysis and classification techniques, a slightly lower result is expected as the Halcrow study used probabilistic calculation where possible, in preference to a direct method of determination.

Figure 3. Coastline of England and Wales, outlined by Band 1 risk areas

Consequently, the overall result was considered satisfactory and was confirmed by major coastal flood events. Consequently, the overall results was considered satisfactory and was confirmed by recent major coastal flood events such as those at Towyn in North Wales, which were shown to be in Band 1.

Toe failure results were spread between the risk bands although 22% of them fell within Band 1. Given the widespread erosion of many of the UK's beaches, this was perhaps not too surprising. Of possibly greater significance was the high percentage of defences apparently at risk from overtopping. Whilst the risk criteria were based upon well established and universally accepted overtopping thresholds (see Owen, 1980 and Hawkes, 1990), the question of whether critical damage would only result from sustained overtopping over a reasonable duration, was examined. To take account of this, the risk thresholds were therefore set an order of magnitude higher, ie assuming the defence could withstand ten times the established criteria before failure could occur. All the areas falling into Band 1 are shown on Figure 3.

It should also be noted that, with the Band 3 risk threshold as a 1 in 200 year event, the total number of potentially defences at risk falling within Band 2, was only 12%. This percentage could only be increased if the upper and lower thresholds for Band 2 were an order of magnitude apart ie taking Band 1 as a 20 year event.

The initial study was based upon the structural data in the SDS and did not include new defences or remedial works carried out since 1990. However, minor updating of the SDS information was undertaken where possible, and Halcrows recently completed an updated study for the ABI following liaison and co-operation from the EA Regions. The results, which included the most recent defence improvements, showed a reduction of 7% in the structures vulnerable to Band 1 events.

The maps produced for ABI only indicate the flood areas associated with the defined sea defences and the named estuaries. Other potential risk areas, eg the Fens inland of the Wash and other estuaries were not within the defined scope of this study and were therefore not included on the maps. In addition, the available OS maps were not up to date and recent urban developments may have been missed.

Whilst many defences and thus coastal areas were given a Band 1 classification, this did not mean that all defences protecting such areas had an equivalent risk as, in reality, the failure of only a relatively short length of defence could lead to widespread flooding. As described earlier, the weakest link in any flood area frontage was taken to represent the risk associated with that area, although the majority of the defences might have had a much greater integrity and a relatively minor amount of remedial works could be sufficient to move the areas into Bands 2 or 3.

The results provide a relative measure of flood area risks and the risk of failure does not necessarily imply that a total functional failure of a defence and subsequent flooding would definitely occur. It simply indicates potential risks and vulnerability under particular conditions. No account could be taken of remedial or emergency action, which might occur prior to or after an event. Because of constraints in time, the impact of successive tides could not taken into account, nor the likely time for breach repair or flood duration.

The 1:10,000 maps and Figure 3 show very large areas of the coast and specified estuaries as Risk Band 1, ie being at risk to inundation from a 50 year storm event, it does not mean that all these areas are susceptible to any one single event, simply that they have a greater probability of being affected. Thus, while a major storm event could affect large areas of the East, South and West coastal areas of England and Wales, it will not affect all coastal areas as its impact will be limited to relatively short lengths of coastline, particularly on the South and West coasts. On the East coast a major event could affect larger areas, as experienced during the disastrous 1953 event.

The impact of individual storms was therefore examined following on after the two earlier studies. As described above, the coasts of England and Wales were divided into three coastal areas: West, South and East. Meteorological records going back to the last century were analysed by the UK's Meteorological Office in order to determine generic storm types. In all, the two storms likely to have the most significant impact on each of the coasts and their associated wave and surge conditions were defined for each section of coast. Significant storms affecting more than one section of coast were also defined. While it is possible to define wave and surge levels in terms of their probability of occurrence, it is not possible with storms. This is because they are made up of many different elements eg air pressure, density and spacing of isobars, storm track direction etc, each of which can influence the other. The impact of the conditions produced (wave and surge) also varies along the coast, say being 1 in 50 year near its centre decreasing to 1 in 1 year towards its edge.

The storms produced varying conditions along the coastline and thus assigning a particular return period to an individual storm is problematical. A prime example is the 1953 East coast storm, where shoreline conditions experienced in Yorkshire were estimated to be in the order of 1 in 50 years, whilst by the time the storm surge had been driven south to Lincolnshire and East Anglia, it represented a 1 in 500 year condition.

The analysis defined the impact of potentially different storm types for each coastal area rather than specific storms. Detailed comparison of the analysis against past events was not possible, but a qualitative comparison of results and newspaper reports of flooding was made.

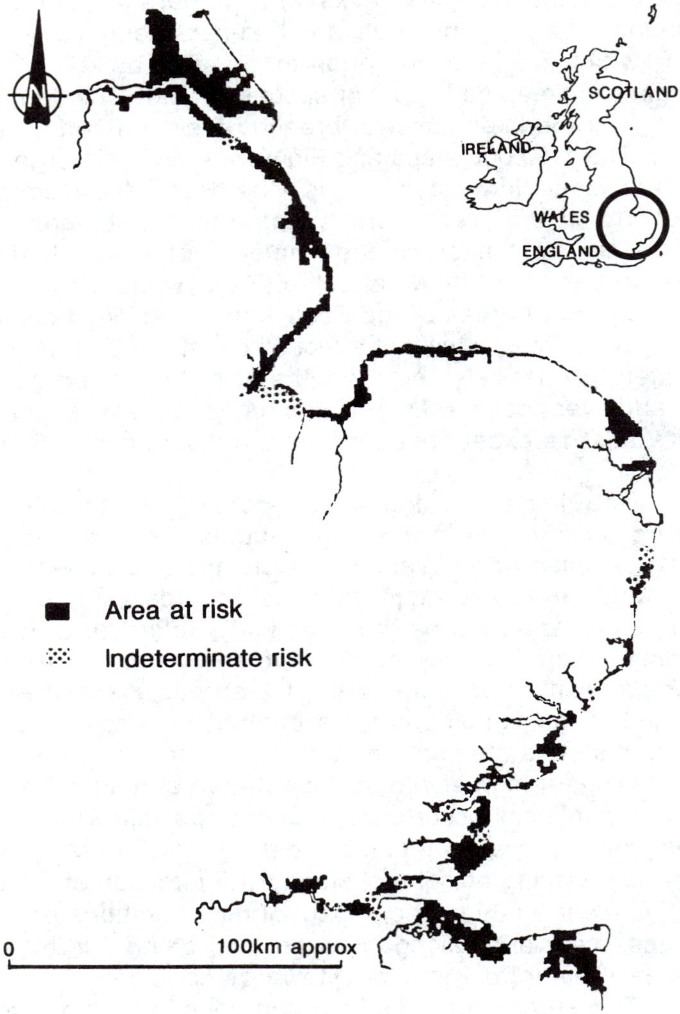

Figure 4. Flood risk areas resulting from a major East coast storm

Two storms affecting individual coasts were examined, together with one affecting both the East and South coasts, one affecting both the West and South coasts and one hypothetical storm for each coast. The hypothetical storms represent small changes of previous storms to modify the timing of the storm surge to coincide with high water. These were then compared as a way of establishing the storm scenarios with the greatest impacts.

Typical variations in the wave and water level conditions adopted for the single storms and the two earlier studies were examined, including the probabilistic wave heights and water levels for return periods of 50, 100 and 200 years. The comparison showed variation in wave height for the East coast locations were markedly different and other differences along the remaining coasts were also apparent. The variations were smaller for water levels.

The results produced by "FRANC" for these storms showed that one of the East coast storm produced the highest number of defences at risk ie 36.5% (see Figure 4) The percentages at risk from the other two East coast storms were very similar, although their distribution was different.

The South coast storms indicate that a high percentage of defences were at risk (approximately 70%). Both have an eastbound surge associated with waves of between 9m and 13m, generated by gale force winds sweeping in from the Atlantic. On the West coast, the hypothetical storm produced the worst conditions with 383 defences at risk (48.1%). Had it been coincident with high tide, the results would have been more severe.

The Insurance Industry's Application of the Results

The information from the studies was applied by the Insurance Industry in two basic ways, namely;

- the practical application of the project output data in overall portfolio exposure and in underwriting individual property risks on a geographic basis according to flood risk; and
- strategic use of the valuable data (which supplements and refines the existing knowledge base) to assist Government, through its agencies (MAFF and the EA), to target resources to exposed locations, always recognising the need for cost/benefit analyses.

The individual insurers try to obtain sufficient premium to cover their potential exposure and to do this they need to understand the underlying risk. The output of the project was therefore applied to this end by analysing the areas at risk against the locations and post codes of the insured properties to establish their aggregate portfolio exposure.

Insurers usually also have to operate with the support of reinsurers (who are also subject to the same regulatory approach) to give them added capacity or to reduce the financial consequences of a catastrophic loss eg 1987/90 storms or coastal flood. These reinsurers need to be satisfied that the insurers they are backing are not themselves overexposed and also require some technical evidence these studies provided in order to take on these risks.

The data analyses from the study were therefore able to assist the insurers to:

- assess their overall exposure;
- determine whether correct rates are being charged; and
- provide information to reinsurers to gain their "capacity" and "catastrophe" support.

It also allows them to put pressure on the relevant Government Departments and to consider their exposure globally.

References

Bouws E, Gunther H, Rosenthal W and Vincent C, 1985. Similarity of the wind wave spectrum in finite depth water. J Geophys Res 90 (CI).

Carter, D J T et al, 1986. Estimating wave climate parameters for engineering applications, Department of Energy Offshore Technology Report OTH 86 228, HMSO.

Diment R P and Deakin R, 1994. The role of GIS in risk assessment, European Geographic Information Systems Conference '94, Paris.

Hawkes P J, 1990. Joint probability of waves and water levels on the north Wales coast, Hydraulics Research Limited, Report EX 2133.

IPCC (Intergovernmental Panel for Climatic Change), 1990. Climate Change, Cambridge University Press.

Herbert D M, 1993. Wave overtopping of vertical walls, Hydraulics Research Wallingford, Report SR316.

Maddrell R J, Burgess K A and Mounsey C, 1995. Coastal flooding: assessing the risks for the insurance industry, ICE Conference, Bournemouth, p167 - 184.

Owen M W, 1980. Design of seawalls allowing for wave overtopping, Hydraulics Research Station Wallingford, Report EX 924.

Shennan I, 1989. Holocene crustal movements and sea level change in Great Britain, Journal of Quaternary Science.

CHAPTER 337

EXPERIMENTAL STUDY ON THE BEHAVIOR OF BIVALVES BY OSCILLATORY FLOW

Toshihiko Yamashita[1], Akira Wada[2], Gaku Matsuoka[3], Kenji Yano[4] and Sadamitsu Akeda[5]

ABSTRUCT

To improve propagation of bivalves living in exposed sandy beach, we need to clarify the behavior of bivalves affected by water waves. Using a U-shape tube, the behavior of bivalves (surf clam and sunray surf clam) in an oscillatory flow was experimentally studied. It was found that the critical condition at which bivalves are forced out of the sand into the water is determined by the ratio between the velocity of bed erosion and the burrowing rate of bivalves, and until the bivalves are forced out into the water their own movement is significant. Also the experimental values measuring their behavior on a smooth surface fixed bed was found to be theoretically explained by setting C_D(drag coefficient), C_M(added mass coefficient) and μ '(coefficient of dynamic friction) at 1.0,0.5 and 0.1, respectively. After the bivalves are forced out into the water from the sand, therefore, their behavior is hardly affected by their own movement and we can regard them as inanimate objects.

INTRODUCTION

Japanese surf clam (*Pseudocardium sybillae)* and sunray surf clam (*Mactra cinensis)*, which live in open sandy beaches along the coast of northern Japan are important fishery resources. However, there is a high mortality rate among young

[1]Associate Professor, Department of Civil Engineering, Hokkaido University, N-13 ,W-8 , Sapporo 060 , JAPAN

[2,3]Graduate Student, Department of Civil Engineering, Hokkaido University

[4,5]Head of section,Civil Engineering Research Institute, Hokkaido Development Bureau, 1-3, Hiragishi, Sapporo 062, JAPAN

bivalves, which must be minimized to ensure their propagation.

There have been reports showing an increase in young bivalves of surf clams in tranquil areas around newly constructed ports (for example, Hayase and Miyamoto , 1985). A large number of young bivalves are often cast onto the shore by high waves in winter (Photo. 1). From these examples and other on-site investigations of the clam's population (Watanabe , 1980), waves, especially high waves during winter, have been thought to be one of the major factors responsible for the mortality of young bivalves. Therefore, the process of mortality and the behavior of young live bivalves due to waves must be clarified. As the first example above shows, bivalves are deeply related to structures built on coastlines. Therefore, for the designe of coastal structures with consideration to the surrounding environment, the influence on the bivalves by waves must be clarified.

The effects of the bivalves' own active movements (burrowing) are important for investigating the behavior of bivalves due to waves. Although some experimental studies have been carried out on the behavior of dead bivalves (Watanabe ,1982; Kuwahara ,1993,1994), except for our previous studies (Yamashita et al. 1994a,1994b,1995), very few experimental studies have been carried out on the behavior and mortality of live bivalves due to waves. The effects of the bivalves' own active movements (burrowing) under water waves have still not been estimated quantitatively.

Bivalves usually dig themselves into the sand. According to our previous studies, when the sea bed erosion and accumulation occurs due to the action of waves, the bivalves keep digging into the sand by reacting to the deformation of the sea bed. However they are sometimes forced out of the sand into the water by a sudden deformation due to high waves. This movement is thought to be the first phase in the mortality process.

In this study, the following five points were investigated in order to understand the first phase of the mortality process: (1) the critical condition of release, (2) the critical condition of burrowing, (3) the burrowing rate of bivalves, (4) the bed erosion velocity and (5) the moving velocity of bivalves on a fixed bed.

Photo.1 Young bivalves cast onto the shore in winter

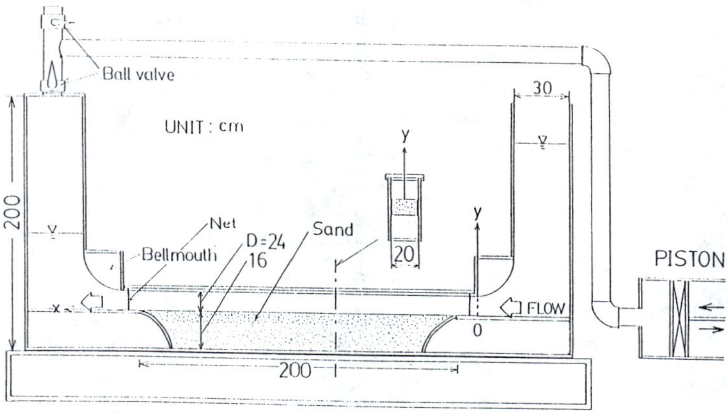

Fig.1 Schematical view of the U-shape tube

EXPERIMENTS

For the experiments, a U-shape oscillatory flow tube made of acrylic resin was used (Fig.1). Sea water and sand were put into it. In the movable bed experiments, a 16cm-deep layer of sand, with a grain size of 0.3mm, was made at the bottom of the tube. In the fixed bed experiment, a plate made of acrylic resin with a smooth surface was used. Young live bivalves (surf clam and sunray surf clam), 5mm～40mm in size, were used in the experiments. The wave period was set at either 3 sec. or 3.5 sec. The amplitude of the flow velocity (Um) ranged from 20 cm/s to 90 cm/s. The water temperature was set at 16℃～22℃.

The following were investigated:

(1)the critical condition at which bivalves are forced out of the sand into the water
(2)the critical condition at which bivalves forced out into the water can burrow back into the sand again
(3)the decrease in the energy (burrowing rate) of bivalves due to a drop in water temperature
(4)the bed erosion velocity at which sand ripples are formed
(5)the behavior of bivalves forced out into the water on a fixed bed with a smooth surface.

Experiments were conducted according to the following methods. Experiment (1): After the bivalves had dug themselves into the sand, we generated an oscillatory flow, and the deformation of the bed and behavior of bivalves were recorded by video camera. Experiment (2): To measure the critical flow velocity at which bivalves can burrow into the sand, bivalves were put into water where stable sand ripples had formed by varying the flow velocity. Experiment (3): We measured the burrowing rate of bivalves by changing the water temperature from 20℃ to 5℃ every 5℃ in a controlled temperature room, and investigated the relationship

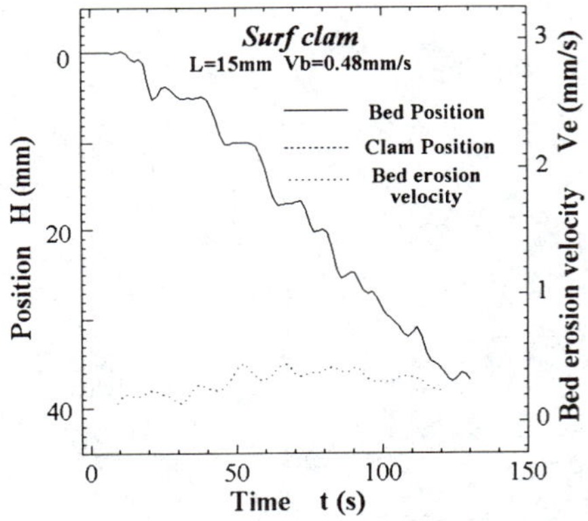

(a)

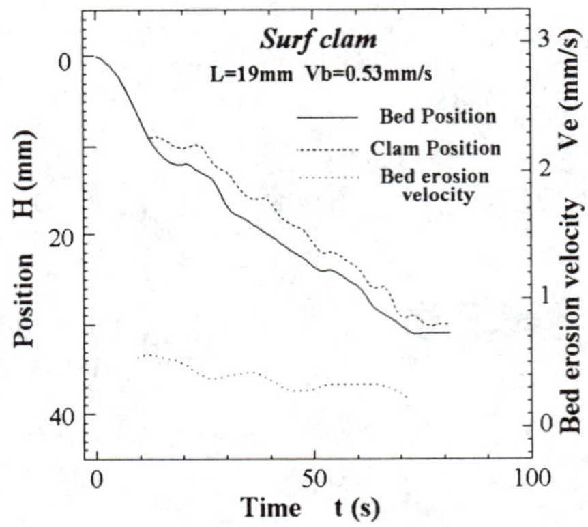

(b)

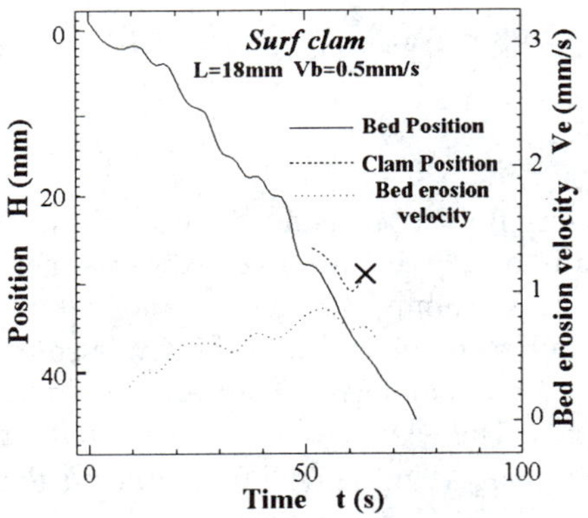

(c)

Fig.2 Relation of bed erosion to burrowing behavior of a bivalve

(a) (b) The case of a bivalve remained in the sand

(c) The case of a bivalve forced out into the water

between water temperature and burrowing rate of bivalves after adapting bivalves to each water temperature for three days. Experiment (4): We measured the bed erosion velocity at which sand ripples were formed from a flat condition using a video camera. Experiment (5): We measured the moving velocities of bivalves, and compared the theoretical and experimental values.

RESULTS

(1) **Critical Condition of Release**

Fig.2 shows typical examples of sea bed deformation and the behavior of the bivalves. The velocity of the sea bed erosion was calculated hourly from the variation in bed form. Fig.2(a) shows an example where the bivalve kept digging into the sand and remained there despite erosion of the sea bed, because the velocity of the

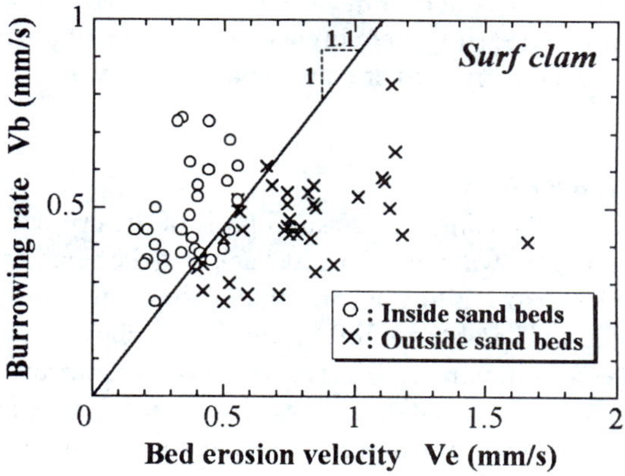

Fig.3 Critical condition at which bivalves are forced out of the sand into the water by the flow

sea bed erosion (Ve) was lower than the burrowing rate of the bivalve (Vb). Fig.2(b) shows another result. At the beginning of the experiment, the velocity of the deformation of the bed exceeded the burrowing rate of the bivalve, and only one-fifth of the shell length of the bivalve was washed out of the sand. However, later as the velocity of the bed erosion became lower than the burrowing rate of the bivalve, the bivalve remained in the sand. Fig.2(c) shows the following example. At the beginning of the experiment, the deformation of the sea bed occurred at a velocity almost equal to the burrowing rate of the bivalve, and the bivalve kept digging into the sand and remained there for 43 sec. after the experiment started. Later, as the velocity of the deformation of the bed exceeded the burrowing rate of the bivalve, the bivalve was at first partly washed out, and then completely forced out into the water. Fig.2 clearly shows that the critical condition at which the bivalves were forced out into the water is related to the bivalves' burrowing rate (Vb) and the velocity of deformation of the sea bed (Ve).

Fig.3 shows the relation between Ve and Vb based on data from 80 samples, using surf clams as Fig.2. Fig.3 shows the conditions of the bivalves with a certain burrowing rate at a certain velocity of deformation of the sea bed, which are categorized into the following: ○ indicates bivalves that remained in the sand completely, and × indicates bivalves that were completely forced out into the water. When the bivalves were washed out of the sand by about half the length of their shell, they tended to be completely forced out of the sand. Thus, the velocity of the deformation of the bed can be defined as the maximum value (Vemax) in the experiment of the mean erosion velocity at which erosion of the sea bed occurred at a depth of half of the shell length. The burrowing rate of bivalves (Vb) can be calculated by dividing the shell length (L) by the time taken for bivalves to burrow into the sand.

Fig.3 shows that the critical condition at which bivalves are forced out of the sand into the water by the flow is determined by the ratio between Ve and Vb, i.e. 1.1. The present experimental results were higher than those using dead bivalves placed on the sand. Thus, until the bivalves are forced out into the water, their own movement is important.

(2) **Critical Condition of Burrowing**

Fig.4 shows the results using sunray surf clams of the critical condition at which bivalves forced out into the water can burrow back into the sand. In Fig.4, ○ indicates bivalves that could burrow into the sand under the water flow, and × indicates bivalves that couldn't. The oblique line in the figure shows the critical condition. The critical flow velocity at which a bivalve with a certain burrowing rate can burrow into the sand is quantitatively obtained from this figure. It was found that when the burrowing rates of the bivalves increased, they can burrow back into the sand by themselves against higher flow velocities.

The critical flow velocity obtained in this experiment is the value that usually occurs at the sea bottom. Therefore, it seems to be difficult for bivalves which are forced out of the sand to burrow into the sand again under a rough sea for several days.

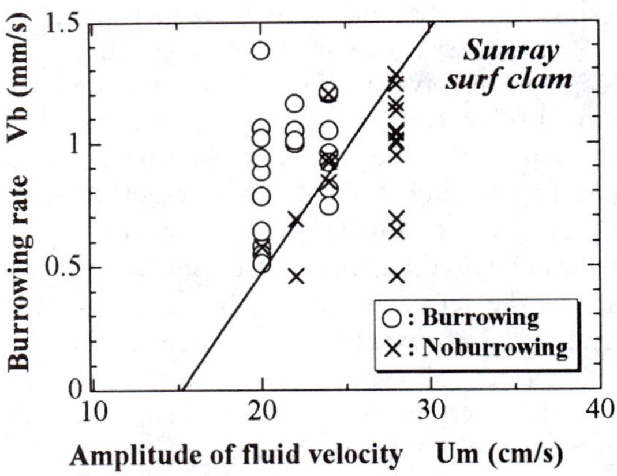

Fig.4 Critical condition at which bivalves forced out into the water can burrow back into sand again

(3) **Borrowing Rate of Bivalves**

The ability of the bivalves to dig into the sand by themselves greatly affects both the critical conditions at which the bivalves are forced out of the sand into the water and at which the bivalves can burrow back into the sand. Fig.5 shows one example (surf clam) of the experimental results concerning the burrowing rates of bivalves with varying water temperatures from 20℃ to 5℃. The straight lines in the figure indicate regression lines that coincide with the origin under each temperature. The result of the experiment was that the burrowing rate of bivalves below a certain water temperature is roughly proportional to the length of the shell, and decreases with a fall in water temperature.

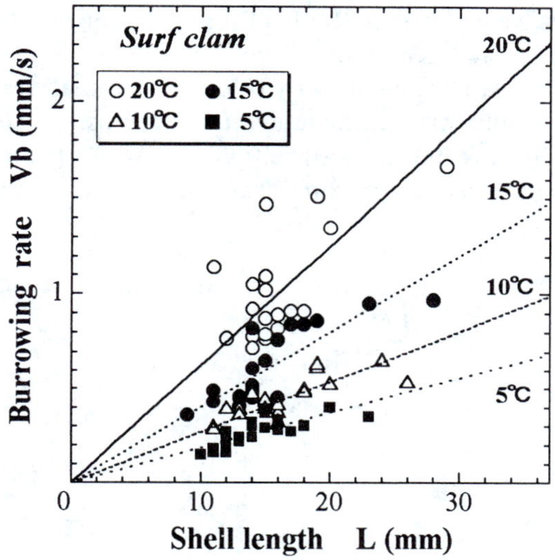

Fig.5 Relation of shel length to burrowing rate

(4) **Bed Erosion Velocity**

It was observed in Experiment (1) that the critical condition at which bivalves are forced out of the sand into the water was related to the velocity of the sea bed erosion. Bivalves tend to be forced out of the sand when sand bed are eroded about shell length in depth for a short time. Major examples of local deformation at a high velocity of topography by waves are the erosion and accumulation of sand resulting from the formation of sand ripples, the movement of sand ripples, and the large scale vortex caused by wave breaking. In this experiment, we first measured the bed erosion velocity under which sand ripples were formed from a flat condition to clarify the bed erosion velocity under which sand ripples were formed by an oscillatory flow, and then the amplitude of the flow velocity Um = 20,40,60,80 cm/s.

Under the condition where sand ripples were formed, the erosion and accumulation of the sea bed occurred complexly. The pattern of deformation of the topography showed a wide variation from 50 sec. to 2000 sec.. The release of bivalves out of the sand into the water depended on whether the velocity of the deformation of the bed exceeded the burrowing rate of the bivalves. Therefore, the probability of the bivalves being released is related locally to the probability of maximum bed erosion velocity occurring. Fig.6 shows the local probability of the maximum bed erosion velocity at which it exceeded a certain erosion velocity of the bed. In the calculation, as in Experiment (1), the maximum bed erosion velocity is defined as the maximum value in the experiment of the mean bed erosion velocity at which the erosion of the bed occurred at a depth of half of the shell length.

Fig.6(a) shows an example of the experimental results of the maximum bed erosion velocity for the bivalve with a shell length of 10mm. Fig.6(b) shows another example of the bivalve with a shell length of 30mm. These figures indicate that under the condition where the same flow velocity or deformation of the bed occurred, the shorter the length of the bivalve, the higher the bed erosion velocity related to the release of the bivalve became. Fig.6 quantitatively shows that as the amplitude of the flow velocity increased, the maximum erosion velocity of the bed increased.

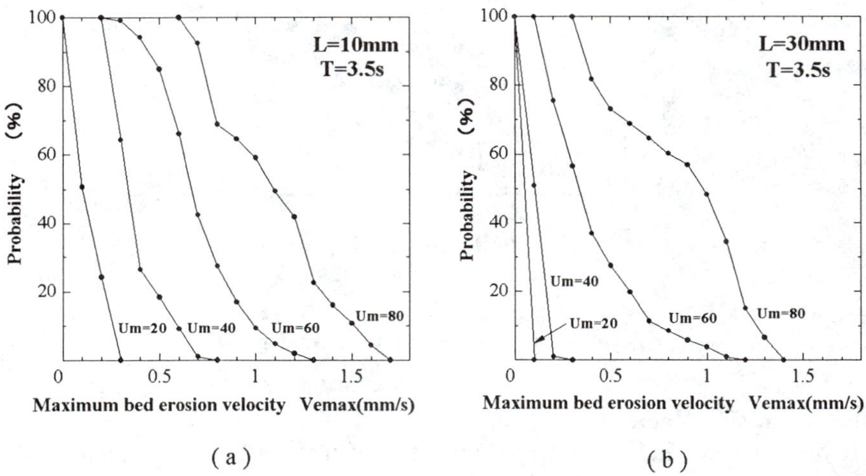

Fig.6 Probability of maximum bed erosion velocity occurring

(5) Moving Velocity of Bivalves on a Fixed Bed

We measured the velocity of moving bivalves which had been forced out into the water from the sand due to the flow, using a fixed bed with a smooth surface. Fig.7 shows a comparison between the experimental and theoretical results. The theoretical values were calculated by the following equation, and we regarded the behavior of bivalves as that of inanimate objects, which can not move by themselves.

$$M\frac{dUs}{dt}+C_m m\frac{d(Us-u)}{dt}=m\frac{du}{dt}+\frac{1}{2}C_D\rho_W A|u-Us|(u-Us)-\mu'(M-m)g\frac{Us}{|Us|}$$

Where, M is the mass of a bivalve, m is the mass of water equal to the volume of the bivalve, Us is the velocity of the moving bivalves and u is the water particle

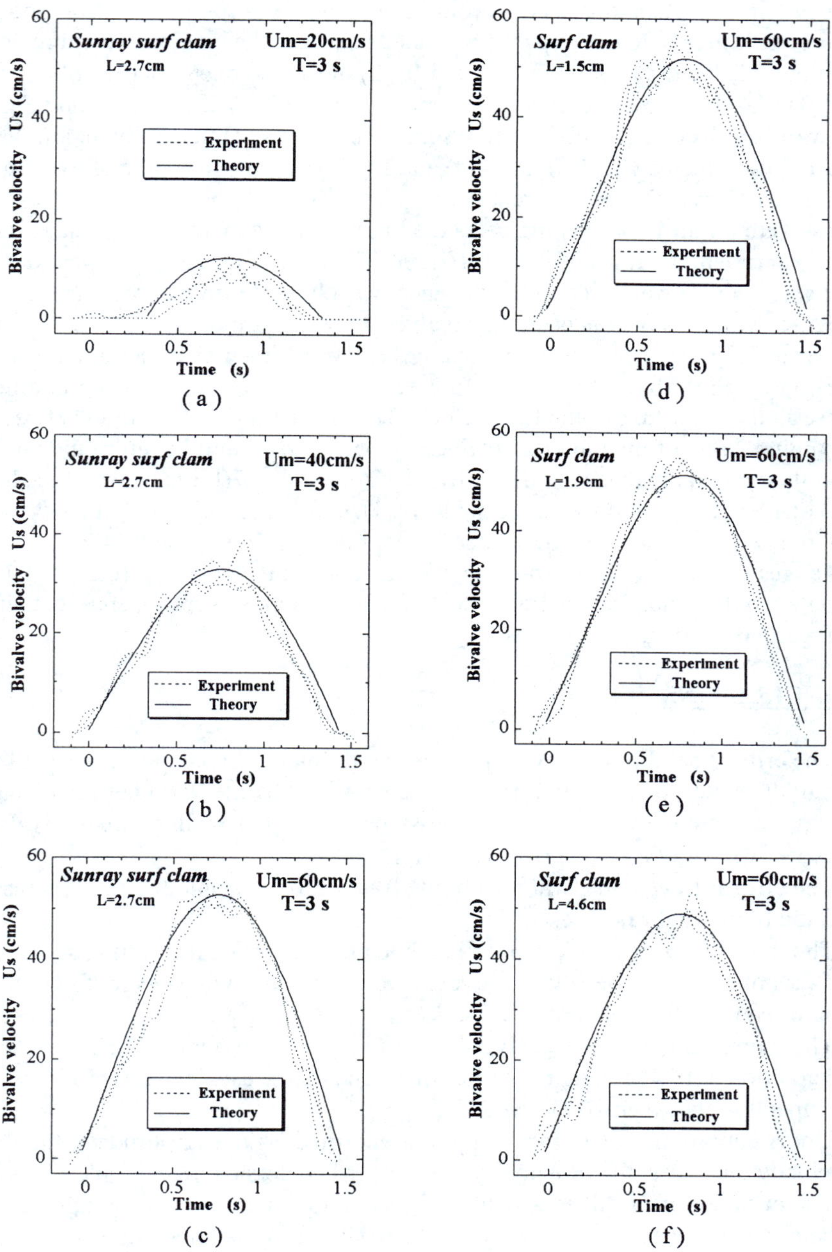

Fig.7 Moving velocity of the bivalves on a smooth surface fixed bed
(a)(b)(c) The case of using sunray surf clams at three different amplitude of the flow velocity
(d)(e)(f) The case of using surf clams of three different sizes

velocity. In the calculation, (1) the volume of a bivalve is converted into a globe and its projected area is regarded as the projected area A of the bivalve in the direction of the flow; (2) the drag coefficient $C_D = 1.0$; (3) the added mass coefficient $C_M = 0.5$; and (4) the coefficient of dynamic friction $\mu' = 0.1$. Fig.7(a), Fig.7(b) and Fig.7(c) show the results of experiments using sunray surf clams at three different amplitudes of the flow velocity. Fig.7(d), Fig.7(e) and Fig.7(f) show the results of experiments using surf clams of three different sizes.

The zero on the hour axis indicates the time when the main stream was reversed. The experimental values of three different velocities of moving bivalves in a one-half wave period were plotted. Turbulence was observed from experimental values because, while the flow acts on the bivalves, they turn around in various directions. However, in general, the theoretical values and experimental values agreed well.

Fig.7(a), Fig.7(b) and Fig.7(c) indicate that bivalves began moving at an early phase as the amplitude of the flow velocity became higher, and the ratio between the maximum value of the moving bivalves' velocity (Usm) and the amplitude of flow velocity (Um) approached 1.0. Fig.7(d), Fig.7(e) and Fig.7(f) indicate that the longer the shell length of the bivalves is, the lower Usm becomes, although there were few differences between shell lengths because of the high flow velocity.

The results of the present study showed that once bivalves are forced out into the water from the sand, there is little effect of their own movement. Therefore, they can be regarded as inanimate objects.

CONCLUSIONS

[1]The critical condition for which bivalves (surf clam) forced out of the sand and into the water by the water flow was determined by the ratio between Ve and Vb, which are the velocity of bed erosion and the burrowing rate of bivalves, respectively. Ve/Vb is approximately 1.1.

[2]The critical flow velocity at which bivalves can borrow into the sand increases as the burrowing rate rises.

[3]The burrowing rate of bivalves is related to both critical conditions, and the effects of the bivalves' own active movements (burrowing) were found to be significant for these conditions.

[4]The burrowing rate of bivalves below a certain water temperature is roughly proportional to the length of the shell, and decre ases with a fall in water temperature between 20℃ and 5℃.

[5]Under conditions where sand ripples are formed, as the amplitude of the flow velocity increases, the probability of a higher bed erosion velocity increases.

[6]The behavior of bivalves which are forced out of sand are hardly affected by their active movements, and they can be regarded as inanimate objects. Supposing that the volume of a bivalve is converted into a globe, the behavior of bivalves could be theoretically explained by setting C_D (drag coefficient) , C_M (added mass coefficient) and μ' (coefficient of dynamic friction) to 1.0, 0.5 and 0.1, respectively.

[7]The shorter the length of the bivalves, the lower the water temperature, and the

higher the wave height, the more easily the bivalves are forced out of the sand into the water by the water flow. These are considered to be one of the major factors affecting the mortality of young bivalves due to waves in winter.

ACKNOWLEDGEMENTS

We wish to thank Hokkaido Central Fisheries Experimental Station, Muroran Branch-Hakodate Fisheries Experimental Station, and National Research Institute of Fisheries Engineering for advice on the ecology of bivalves and cooperation of the collection of bivalves.

REFERENCES

Hayase, Y. and Miyamoto, Y.(1985): Formative Mechanism of Spisula Sachalinensis Fishery by Coastal Structure, Monthly Report of the Civil Engineering Research Institute, Hokkaido Development Bereau, No.386, pp.1-11

Kuwahara, H. and Higano, J.(1993): Analysis Method on On-Off Shore Movement of Bivalves by Waves, Proc. Coastal Eng., JSCE, Vol.40, pp.311-315

Kuwahara, H., Higano, J., Nakamura, Y. and Mimura, N.(1994): Study on mechanium of bivalve movement by wave and validity of numerical method, Proc. Coastal Eng., JSCE, Vol.41, pp.376-380

Watanabe, E.(1980): The relationship between the decrease of surf clam and environmental factors, Monthly Report of the Civil Engineering Research Institute, Hokkaido Development Bereau, No.325, pp.1-12

Watanabe, E(1982): Experimental Study on the Decrease of Surf Clam under Wave Action, Monthly Report of the Civil Engineering Research Institute, Hokkaido Development Bereau, No.351, pp.3-15

Yamashita, T., Matsuoka, G., Yano, K. and Akeda, S.(1994): Experimental Study on Movement and Decline of Bivalve under Oscillatory Flow, Proc. The Japanese Society of Fisheries Engineering, pp.57-58

Yamashita, T. and Matsuoka, G.(1994): Experimental Study on Movement and Decline of Bivalve by Water Waves, Proc. Civil Engineering in the Ocean, JSCE, Vol.10, pp.119-122

Yamashita, T., Wada, A. Matsuoka, G., Yano, K. and Akeda, S.(1995): Experimental Study on Behavior of Bivalves under Oscillatory Flow, Proc. Coastal Eng., JSCE, Vol.42(1), pp.506-510

CHAPTER 338

Wind- and Sea Level-Induced Shore Evolution in Poland

Ryszard B. Zeidler, Marek Skaja, Grzegorz Różyński & Jarka Kaczmarek [1]

Abstract

The effect of climate change factors on Polish shoreline is investigated in mesoscales of decades and tens of kilometres along the entire Baltic coast, with emphasis on the selected segment from Ustka to Łeba. Intensification of westerly wind circulation and sea level rise (SLR) are quantified as an input in computations of shoreline change by a one-line model. Joint probability distributions of wind and sea level derived under an extensive programme employing field data are used to produce input for computations of shoreline change due to rare events. The computations prove that the effects of both wind change and SLR can be perceptible in mesoscales, although it is difficult to clearly single out one from the other. In general, the westerly intensification of wind climate along the Polish coast seems to be slightly less important than SLR. A separate qualitative analysis of the acceleration of Polish shore retreat noted in recent years points to wind change as one of the possible causes.

1. Introduction: Coastal Climate Change (CCC)

The present coastal climate in the Baltic area, including Poland, is believed to undergo perceptible change due to global warming. The sea level rise, an increase in storminess and annual sea level maxima, and the changing patterns of wind circulation are the most pronounced change features affecting the coastal processes. The potential changes in precipitation, evaporation, transpiration, and their various outcomes over the Baltic coast and its drainage area, are also investigated although they are not easy to predict reliably and accurately.

An analysis of the variation of the zonal and meridional components of the geostrophic wind vector allows one to draw conclusions on the change of both components in the 40-year period from 1951 to 1990. For the more pronounced zonal component, this appears to be a trend of 0.0125 m/yr having a 95-% confidence level. By and large, one may note an intensification of the westerly air flow over the Polish coast, reflected in a shift of predominant wind directions towards W and NW, at the expense of the SW-W sector. A schematic change in the wind direction rose postulated for 50 years from 1995 to 2045 is

[1] Professor, Snr Eng, Snr Res Assoc, Snr Asst, respectively; Institute of Hydro-Engineering IBW PAN; 7 Koscierska, 80-953 Gdansk, Poland

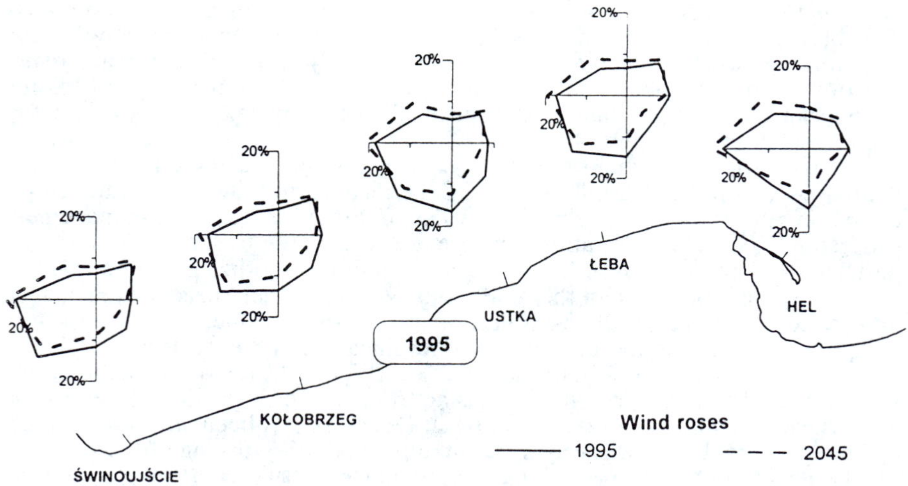

Figure 1. Frequencies of wind along the Polish coastline in 1995 (solid line) and postulated for the year 2045 (dash line).

depicted in Figure 1. In seasonal terms, the intensification of the westerly sector is particularly conspicuous in the winter season from October to March.

The mean sea level (SL) rise in Poland is presently estimated below 30 cm per century but its acceleration due to the greenhouse effect is taken into consideration (Zeidler 1995).

Another factor of climate change, i.e. storm intensity or 'storminess' can be analysed in terms of the annual sea level maximum (occurring once per year). One can prove that such storminess clealy increases over recent decades (Zeidler 1995). The growth trend of annual sea level maxima (ASLM) along the Polish coast is estimated about 10 cm per century.

It is also worthwhile to recollect that sea level increments Δh (four- or six-hourly at the Polish stations) display a certain long-term trend (Zeidler 1995). Moreover, they have been found to be correlated with high sea levels.

The present practice identifies as storm surges the events at which high sea levels alone occur, which brings about ambiguities. In order to determine the linkage between high sea levels and significant sea level increments, an analysis of the two quantities has been carried out (Wróblewski et al. 1996), with the primary objective as the derivation of statistical and probabilistic characteristics for both datasets and subsequent identification of their common features. Taken as a criterion of similarity was the condition that the numbers of occurrences of both quantities (SL and SL increments) are nearly equal. Such similarity has been concluded for 4-hourly SL increments greater than 21 cm at Ustka and Gdańsk (Nowy Port) or 26 cm at Swinoujscie, versus $h \geq 563$ cm at Swinoujscie and Gdańsk and $h \geq 560$ cm at Ustka (mean sea level being 500 cm). Within the above approach, regressional relationships between the time series for SL and their increments have been established.

Further on, the trend of storm winds could have been determined on the basis of the sea level increments measured at Swinoujscie, Ustka and Gdańsk (Wróblewski et al. 1996). This was done by relying on relationships between Δh and storm winds in selected areas; a detailed analysis of statistical characteristics of storm winds and identification of their trend would have required processing of thousands of synoptic maps (e.g. above 100,000 maps taken every 4 hours in the investigated time span of 1955-1990).

The two major factors of CCC, i.e. wind and sea level in Poland are dealt with in this study while other CCC effects (primarily temperature, precipitation and groundwater changes) are incorporated in a coastal management programme led by the first author, where they contribute to changes in land-use patterns, flooding of coastal lowlands, cliff stability and other phenomena.

The basic objective of this study is to determine the effects of wind change and sea level change on the evolution of the Polish coastline.

It is interesting to note that the intensification of the westerly circulation has also been noted along the Danish coast. Figure 2 provides evidence for a time span of 100 years, showing increasing frequency of high wind speeds of the westerlies. The atmospheric circulation in Denmark has been shown to change by enhancement of the westerly winds and the clockwise turning of the predominant winds (from W to NW along the Jutland Peninsula coast investigated by Christiansen and Bowman, 1990). The wind change is accompanied by a more severe wave attack and retreat of the more exposed coastline segments.

In this paper, description of the climate change precedes shore evolution computations, from which conclusions can be drawn as to the evolution of the Polish coast in decadal scales, and these findings can be compared with the prototype data available.

The entire Polish coast measures five hundred kilometers between the mouth of the Oder (Odra) River and Poland's eastern border with Russia across the Vistula Spit. All over the Polish coast one encounters coastal lakes/lagoons of glacial origin. The same origin is assigned to the Hel Peninsula, a spectacular yet vulnerable to storms and climate change 30-km barrier separating the Gulf of Gdańsk from the open sea. Although the large-scale evolution of the Hel Peninsula is also investigated, this paper describes only some results obtained for the coastal stretch from Ustka to Łeba, belonging in part to the former category (lagoonal barrier).

The primary objectives of the study discussed in this paper consist in assessing the effects of climate change on the evolution of Polish coastline. The two basic effects tested are sea level change and wind change, treated both separately and jointly.

2. Joint Characteristics of Coastal Phenomena

In an investigation of shore evolution due to climate change one faces the necessity of focusing on joint characteristics of coastal phenomena, such as waves, sea level and wind. Also, extreme cases of coast evolution should be looked at. The selection and definition of extreme cases brings one to the problem of joint probability distributions for coastal quantities (such as sea level, wave parameter(s), wind etc.). By using the software produced for this programme one is able to compute shoreline changes or coastal evolution for every situation (sea level, wind speed and direction and at the same time the respective wave climate and other derivatives) but the open question remains how wave parameters are correlated with wind and sea level, and which combinations thereof are representative for extreme cases.

Hence four-dimensional probability distributions of wind (speed and di-

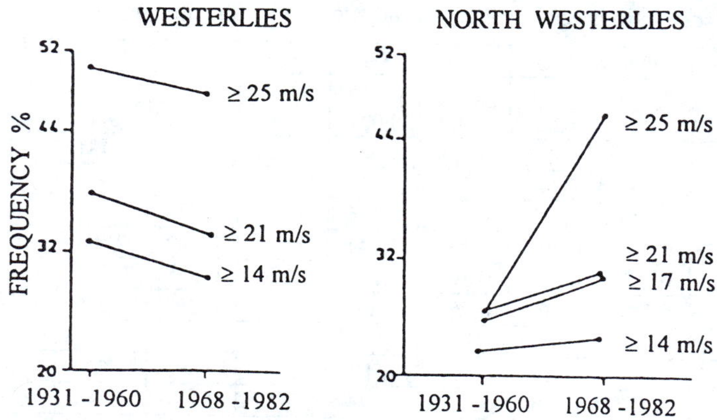

Figure 2. Change in wind direction along a latitudinal segment of NW Jutland (Klitmoller to Blokhus) shown by Christiansen and Bowman (1990).

rection), wave height and sea level have been investigated by the use of a large database for the Polish coastal weather stations, covering the years 1955–1990. At the beginning of the investigations it was realised that the probability of higher sea levels (exceeding 60 cm above the multiyearly mean value) might depend on wind direction. This would mean *inter alia* that the local growth of the mean sea level along the Polish coast due to the expected changes in wind circulation patterns could be quite pronounced by the year 2050.

As the study proceeded, wind-wave parameters were found to be correlated with wind direction as well, which is linked to the effect of fetch. At the same time, the correlation of wave height and sea level is low (about 0.2 for the stations of Hel and Gdynia) for all wind directions lumped, but a clear dependence on particular wind directions is displayed in some classes of water levels.

Figure 3 illustrates the cumulative joint probability of wind speed and sea level for N and NW winds at Hel, while Figure 4 shows the counterparts for Gdańsk (North Harbour). The two graphs stem from different datasets, not only for their locations but also because of the concept by which they were selected. The data for Hel in Figure 3 were taken for all events, i.e. at regular time steps, every three hours over nearly 5 years from 1976 to 1981 (11,500 events). The data for Gdańsk were purposefully biased to represent the impact of storm events (defined as those with significant wave height above 1 m); they also cover a shorter time lapse (1991 to 1992 and some months of 1995; about 4,500 events).

The correlations between sea level and wind speed were tested in eight principal classes of wind directions. Sea levels were arranged in sets exceeding the numbers indicated in the top lines of the tables in Figures 3 and 4. In order to examine the dependence on wind strength, the datasets were truncated in the sense that all events with wind speed below 6 or 8 m/s were rejected.

The correlations at Hel have been found very low in general; they seldom exceed 0.6 (the rightmost figures in the tables are not conclusive because of the low counts of data in those classes). The findings are certainly affected by the

Hel. Correlations sea level (L) - wind speed (v), truncated at v=6m/s

	500	**510**	**520**	**530**	**540**	**550**	**560**	**570**	**580**
N	0,253	0,177	0,146	0,016	-0,03	-0,23	-0,03	0,408	x
NE	0,271	0,326	0,123	0,021	0,017	0,288	0,444	x	x
E	0,131	0,12	0,248	0,071	0,228	0,375	0,431	x	x
SE	-0,02	0,04	0,249	0,041	x	x	x	x	x
S.	0,017	0,087	0,093	0,156	0,478	0,424	x	x	x
SW	0,073	0,104	0,155	0,141	0,232	0,262	-0,06	x	x
W	0,277	0,293	0,306	0,266	0,321	0,335	0,326	0,408	0,116
NW	0,301	0,301	0,281	0,269	0,268	0,267	0,259	0,146	0,319
	N	**NE**	**E**	**SE**	**S**	**SW**	**W**	**NW**	
L - V	0,287	0,199	-0,13	-0,24	-0,11	0,051	0,244	0,246	

Hel. Correlations sea level (L) - wind speed (v), truncated at v=8m/s.

	500	**510**	**520**	**530**	**540**	**550**	**560**	**570**	**580**
N	0,222	0,137	0,097	0,057	0,13	-0,11	0,422	x	x
NE	0,095	0,146	0,062	-0,08	0,163	-0,53	-0,4	x	x
E	0,13	0,06	0,237	0,034	0,251	0,424	0,41	x	x
SE	0,308	0,187	0,622	0,614	x	x	x	x	x
S.	0,069	0,097	0,266	0,264	0,702	x	x	x	x
SW	0,088	0,143	0,195	0,308	0,329	0,29	0,918	x	x
W	0,298	0,316	0,367	0,18	0,359	0,453	0,417	0,225	0,116
NW	0,35	0,333	0,276	0,278	0,301	0,322	0,253	0,041	0,687
	N	**NE**	**E**	**SE**	**S**	**SW**	**W**	**NW**	
L - V	0,238	-0,05	-0,15	-0,18	-0,05	0,049	0,254	0,218	

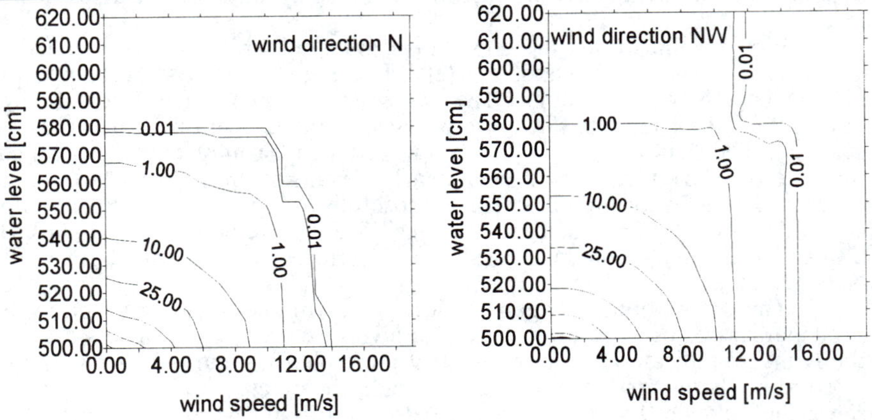

Figure 3. Correlation tables and cumulative exceedance graph for sea level and wind speed at the station of Hel, 1976-1981.

North Harbour. Correlations sea level (L) - wind speed (v), truncated at v=6m/s

	500	510	520	530	540	550	560	570	580
N	x	0,27	0,13	0,17	0,20	-0,20	-0,31	-0,35	x
NE	0,35	0,35	0,35	0,17	0,64	0,41	0,44	-0,24	x
E	0,47	0,39	0,34	0,54	0,54	0,50	-0,57	x	x
SE	-0,05	x	x	x	x	x	x	x	x
S.	-0,05	-0,23	0,45	x	x	x	x	x	x
SW	0,03	-0,27	-0,76	x	x	x	x	x	x
W	0,21	0,47	0,41	0,34	0,06	-0,37	x	x	x
NW	0,29	0,26	0,20	0,07	-0,15	-0,05	x	x	x
	N	**NE**	**E**	**SE**	**S**	**SW**	**W**	**NW**	
L - V	0,66	0,52	0,31	0,01	0,13	0,15	0,23	0,31	

North Harbour. Correlations sea level (L) - wind speed (v), truncated at v=8m/s.

	500	510	520	530	540	550	560	570	580
N	x	0,18	0,07	0,10	0,04	-0,20	-0,31	-0,35	x
NE	-0,09	-0,09	-0,09	0,03	0,03	0,61	0,55	-0,24	x
E	0,39	0,19	0,17	0,54	0,54	0,50	-0,57	x	x
SE	x	x	x	x	x	x	x	x	x
S.	-0,17	-0,05	x	x	x	x	x	x	x
SW	0,22	0,12	-0,52	x	x	x	x	x	x
W	-0,07	0,21	0,19	-0,07	-0,24	-0,37	x	x	x
NW	0,21	0,20	0,17	0,05	-0,11	-0,05	x	x	x
	N	**NE**	**E**	**SE**	**S**	**SW**	**W**	**NW**	
L - V	0,66	0,54	0,33	-0,07	-0,13	0,21	0,13	0,25	

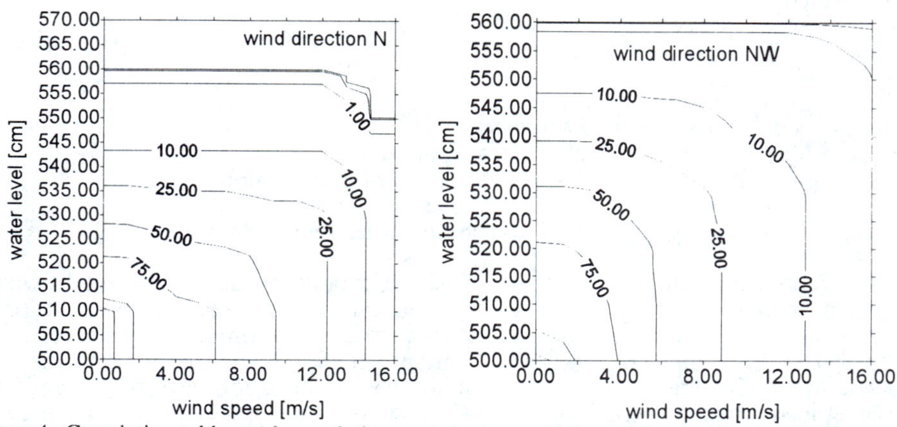

Figure 4. Correlation tables and cumulative exceedance graph for sea level and wind speed at the station of Gdańsk (North Harbour, 1991-1992..1995).

very specific location of the station and the complex nature of the controlling hydrodynamic phenomena.

For the North Harbour, the correlations become significant for NE winds. The relatively high marginal correlations with both N and NE winds (given at the bottom of table) are linked to the fact that all sea levels were taken into account there (also those below 500, i.e. below MSL).

In representation of some characteristic rare events in the subsequent shoreline evolution computations, reference should be made to the exceedance graphs of the type shown in Figures 3 and 4. The following couples of wind and sea level were chosen for the selected isolines of the exceedance probability F:

(a1) wind N 9 m/s + SL 545 (F=2%);
(a2) wind N 16 m/s + SL 500 (F=0.01%);
(b1) wind NW 9 m/s + SL 555 (F=2%);
(b2) wind NW 17 m/s + SL 500 (F=0.01%);
(c1) wind E 7 m/s + SL 535 (F=2%);
(c2) wind E 12 m/s + SL 500 (F=0.01%).

3. Shore Evolution Computations

The wind change trends alone have been taken for granted by Zeidler (1995), who computed the respective sediment transport along the Polish coast in the 50-year time span, and drew conclusions on shoreline evolution. Despite a sharp change in wind direction (but not speed) the alteration of sediment transport rate and the subsequent shoreline evolution was not found dramatic, due to the wind change alone. Yet it was concluded that shoreline change might become quite conspicuous if the wind change is added to the accompanying storm surge or a sea level rise.

The shoreline computations in this study are founded on the one-line theory, which ends up with the diffusion-type equation

$$\frac{dy}{dt} = A\frac{d^2y}{dx^2} - B \tag{1}$$

in which

$A = \frac{2}{(1-p)h_c}\frac{\partial Q_l}{\partial \varphi}$
$B = \frac{2}{(1-p)h_c}(\frac{\partial Q_l}{\partial x} + q_c)$
h_c = depth of closure,
p = porosity of bed sediment,
Q_l = longshore sediment transport rate,
q_c = cross-shore transport rate per unit length along shore,
t = time,
x, y = longshore and cross-shore coordinate of shoreline.

In order to determine the long-term shore evolution due to changes in wind, waves and sea level, by the above one-line scheme, an extensive computational procedure had to be employed. It encompasses five major subroutines: (1) preparation of input data with CCC quantities (wind, wind-induced waves and sea level); (2) arrangement of probability distributions for wave height and sea level in the format accepting CCC; (3) computations of sediment transport for the whole domain of wave height and sea level occurrence; (4) integration of sediment transport rate over representative time spans, with their probability distributions; (5) one-line type assessment of shore evolution over the time span of CCC.

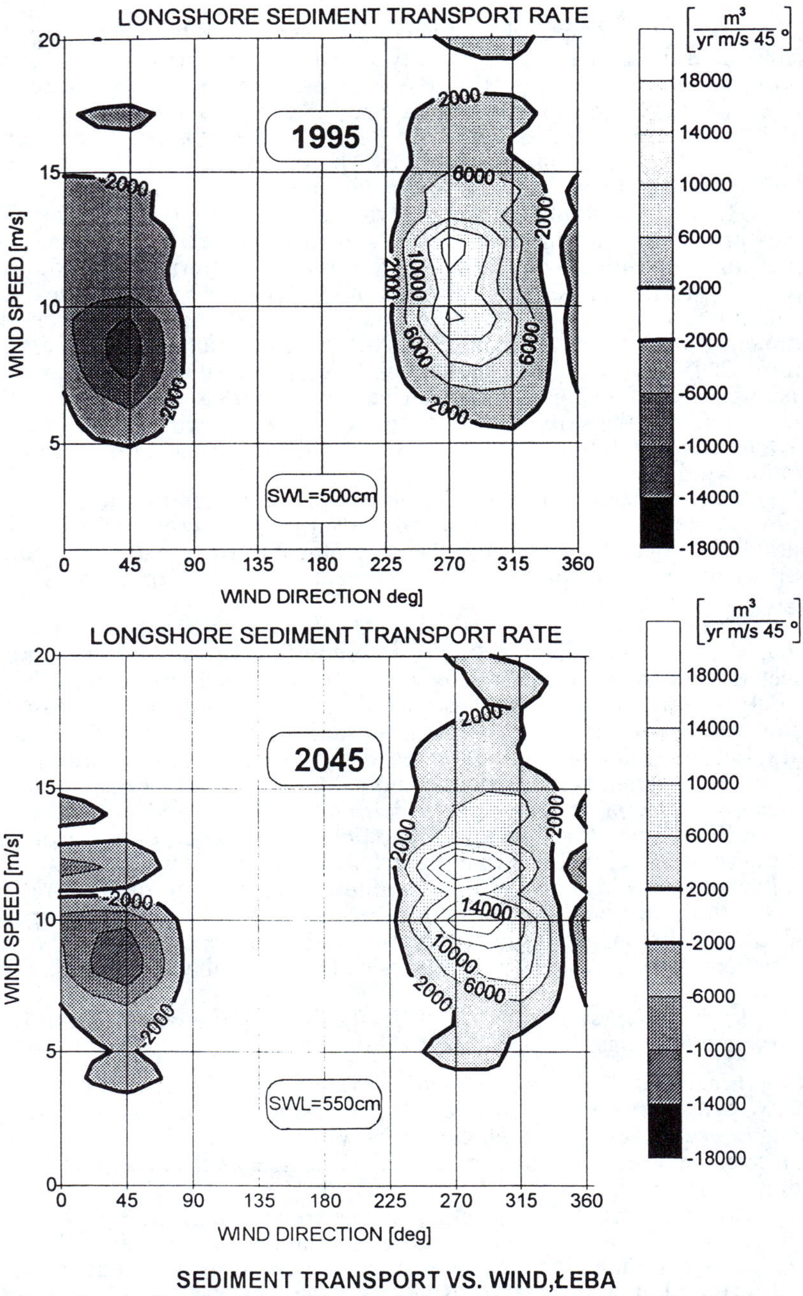

Figure 5. Longshore sediment transport rate as a function of wind speed and direction at Łeba, for the years 1995 (SWL datum 500, top) and 2045 (SL rising by 50 cm to SWL datum 550, bottom).

Waves are computed routinely by Krylov's 'spectral' method calibrated against field data. A Battjes/Jansen-type algorithm for irregular wave transformation is used as an input of the IBW PAN program for wave-induced currents, and the sediment transport formulae are Bijker-type for longshore movement and IBW PAN modified Bailard-type for cross-shore transport. Sea level controls the shore evolution at the places where shoaling and the closure depth intervene but is not included in nonlinear interactions with waves.

Hence in our routine, the input wind change leads to wave climate change, waves producing currents and then sediment transport. On the basis of that input, together with sea level rise, shore evolution patterns are produced.

The results of our computations of the longshore transport are depicted in Figure 5. For every cell of particular wind speed and direction one has the sediment transport rate computed for the present situation (1995) and for the future (2045). It should be noted that the future situation can arise in two versions — wind change only and wind change plus sea level rise. Hence at present one has the status shown at the top of the drawing, while in the future one may face two different situations with sediment transport for sea level change and for wind climate change.

The coefficients A and B in the one-line equation are functions of longshore sediment transport rate (partial derivatives) and cross-shore rates as well (per unit length along the shore). Those coefficients were assessed separately for every location of the study area, with respective closure depths, wave incidence angles, breaking parameters etc.

The study area was confined to the coastal stretch between Ustka and Łeba. Various climate input situations were tested. In addition to the rare cases selected from the joint probability graphs as described in Section 2 (situations a1–c2), the average year was also simulated. The latter encompassed aggregated westerly winds and aggregated easterly winds, for which one obviously had different sediment transport rates, different angles of wave attack etc.

The program USTLEB was compiled for computations of shoreline change in accordance with the one-line model, and was validated against the known analytical solutions (i.a. Larson et al. 1996). It is based on an explicit finite-difference scheme, where instablity problems do not arise in practice. Some peculiarities emerge at points of discontinuity (of sediment transport rate, shoreline position) but the difficulties are overcome by locally decreasing space steps and iterative finding of shoreline derivatives. Estimates of the partial derivatives of the longshore transport rate must also be reasonably derived in a practical way.

In the computations by USTLEB for the 45-km segment Ustka-Łeba, the following conditions were adopted:

(i) the time span 1995–2045 divided in 25 time steps 2 years each;
(ii) two basic versions of climate change: (a) wind climate change only, (b) wind change + sea level change, 50 cm per 50 years;
(iii) 'average year' and 'rare cases' a1–c2 as specified above for the wind ... sediment transport input.

Some results of the shoreline change computations are illustrated in Figure 6. Only the aggregated annual effects are shown to highlight the basic differences resulting from the impacts of the two major directional sectors of wind action along the Polish coast — from west to east (top) and east to west (bottom). The two annual components of shoreline change are strikingly different. The wind change effect is negligible vs. the sea level rise effect for the westerlies, and becomes of the same magnitude for the easterlies. The absolute magnitude of shoreline change is greater for the westerlies than for the winds

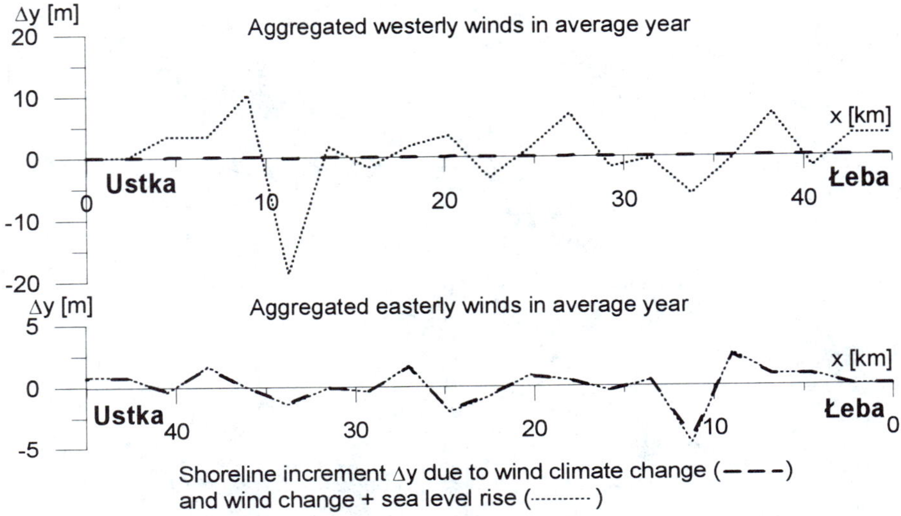

Figure 6. Results of one-line computations for the Polish coast segment from Ustka to Łeba, showing the 50-yr climate change effects — wind change only and wind change + SL change.

blowing from the eastern sector. If one takes into account that the former are more frequent than the latter, then one may venture the conclusion that the effect of wind change on the shoreline change is generally weaker than the impact of sea level rise.

Yet it is appropriate to note that the above conclusion is not so straightforward for the individual 'rare events' (a1–c2) tested in this study. Moreover, the contribution of the cross-shore transport to results of the computations by USTLEB also remains unclear so far. Attempts with various predictors of the cross-shore transport integrated along the shore transects (26 from Ustka to Łeba) have been inconlusive, and therefore are not reflected in Figure 6. More complementary research is required to shed light on the cross-shore effects. Needless to say, the latter may turn out fairly diversified along the Polish coast, if the research is extended far beyond the segment from Ustka to Łeba.

4. Prototype Data and Comparison with Computations

In preparation for the extended research on the entire Polish coast, as signalled at the end of Section 3, one may turn to the prototype evidence collected to date. On the basis of this data one can then venture some general observations and hypotheses relating to the evolution of the Polish coast in large scales of decades and hundreds of kilometres. Similarly, postulates can be formulated for the effects on shoreline of sea level rise and wind climate change.

For a period of slightly above one hundred years, Zawadzka-Kahlau (1994) retrieved the data on the position of Polish beach and shore features. The basic core of that collection consists of a vast database for shoreline ('waterline')and cliff/dune toe stemming from cartographic mapping dating back to

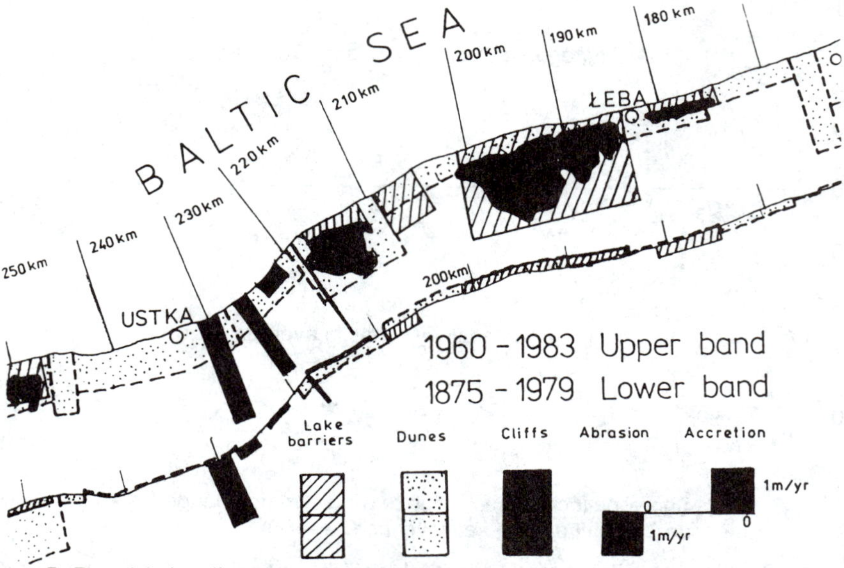

Figure 7. Decadal shoreline change along the coast segment from Ustka to Łeba (Zawadzka-Kahlau 1994).

the 19th century and extending over most of the Polish coastline. The statistical estimates obtained by processing of that database provide an opportunity of verifying the quantities resulting from our computations.

Under this study the database has been re-arranged graphically in the form illustrated in Figure 7 for the area Ustka-Łeba (although the same format was applied to the entire coast). Hence the shoreline change undergone in the years from 1875 till 1979 is shown in the lower band, while that in the more recent 20 years (1960-1983) is presented in the upper band. The purpose of such rearrangement is to provide insight into the general trends and to determine whether one faces some more conspicuous, dramatic shoreline changes in the recent years and if these changes are somehow linked to the change in wind circulation patterns and sea level change.

Consider the intensification of westerly winds, i.e. decadal increase in the frequency of W and NW winds depicted in Figure 1. More pronounced abrasion of concave shore segments could be expected, although to a different degree, depending on the coast type i.e. lake/lagoon barriers, dunes and cliffs. Indeed, such a trend is visible on many Polish coast segments. For majority of concave shore segments it can be concluded that more erosion is encountered in the recent years. There are a number of locations on concave segments where this can be claimed and where dramatic shift from accretion to erosion is noted. As a sound hypothesis, such behaviour can be assigned to the intensification of the westerly circulation. Unfortunately, this qualitative finding suffers i.a. from being embedded in the effect of sea level change. Unless differentiated more clearly, the two effects should be considered equally probable.

Figure 7 has been drawn for the stretch from Ustka to Łeba only. Accretion and erosion have been marked for three different shore types, accretion lines progressing towards the sea and erosion being marked as lines of landward retreat. One very conspicuous feature visible in the drawing is the clear growth of the erosion rate in the years 1960–1983. Identification of the nature and cause of that accelerated growth remains a task of future investigations — the combined effects of the westerly intensification and SL change are slightly more likely than decadal cyclicity of shore evolution change.

5. Summary and Conclusions

The effect of climate change factors (wind and sea level) on Polish shoreline has been investigated in mesoscales of decades and tens of kilometres along the entire Baltic coast. Emphasis has been placed on the selected segment from Ustka to Łeba. Intensification of westerly wind circulation and sea level rise (SLR) have been quantified as an input in computations of shoreline change by a one-line model. Joint probability distributions of wind and sea level derived under an extensive programme employing field data have been used to produce input for computations of shoreline change due to rare events.

The computations carried out in this study prove that the effects of both wind change and SLR can be perceptible in mesoscales, although it is difficult to clearly single out one from the other, and/or from other natural and man-induced effects. In general, the westerly intensification of wind climate along the Polish coast seems to be less important than SLR alone, in terms of decadal and centurial coast evolution patterns. Simple one-line computations of Polish coast evolution confirm the potential effects of wind change and SLR in both interannual coast evolution and extreme events (SLR + Wind Change). Both effects are also equally probable in the light of field data for Polish shore covering the period of more than one century.

The software routine worked out in this study provides a useful tool of long-term coastal planning facing coastal climate change. The data input (wind, wave, sea level, bathymetry) and the software itself pave way for more complex investigations of coast evolution, including the modelling of joint wind, wave and sea level input and chronology effects.

ACKNOWLEDGEMENTS

This paper is based on the work done in the PACE project, in the framework of the EU-sponsored Marine Science and Technology Programme (MAST-III), under contract MAS3-CT95-0002. The work was cosponsored by the Polish Academy of Sciences (PAN) under the programme IBWPAN-Gr-tem-2-96.

REFERENCES

Christiansen Ch. & D. Bowman 1990. Long-term beach and shoreface changes in NW Jutland, Denmark: effects of change in wind direction. *Expected effects of climatic change on marine coastal ecosystems* (J.J. Beukema, ed), 113-122. Kluwer Acad. Publ.

Larson M., H. Hanson & N.C. Kraus 1996. Analytical solutions of the one-line model for shoreline change near coastal structures. *Jour. Waterway, Port, Coastal and Ocean Eng.* (in press).

Wróblewski A., R.B. Zeidler & J. Kaczmarek 1996. Analysis of storm surges, sea level and atmospheric pressure for the Polish Baltic coast. *Arch. Hydr. Envir. Engng* Gdańsk (in press).

Zawadzka-Kahlau E. 1994. *Coast evolution trends on the Polish Baltic* (in Polish). Ph.D. Thesis. Gdańsk University.

Zeidler R.B. 1995. Sea level rise and coast evolution in Poland. *Proceedings 24th Intern. Conf. Coastal Eng.* 3462–3475. ASCE, New York.

CHAPTER 339

ASHDOD PORT'S EFFECT ON THE SHORELINE, SEABED AND SEDIMENT

Abraham Golik[1], Dov S. Rosen[1]-MASCE, Arik Golan[1],
Maxim Shoshany[2], Dan DiCastro[3] and Pinkhas Harari[3]

Abstract

During its 35 years of existence, Ashdod Port, Israel, caused changes to its physical environment. Analysis of aerial photographs and bathymetric surveys show that the port served as a sediment trap, blocking the natural northward sediment transport. Between 1958, prior to the port construction, and 1992 the beach to the south of the port underwent accretion which increased in magnitude from zero, at a distance of 2.5 km south of the port, to more than 100 m near the main breakwater. On the northern side of the port the shoreline was stable during that period. It was found that the beach north of the port did not suffer erosion because the sand of that beach was mined for building purposes prior to the port's construction. When the port was built, it was already a rocky beach.

Comparison between bathymetric surveys, which were carried out in the vicinity of the port, at various periods since prior to its construction until 1995, show that the port has trapped some 4.5 million m^3 of sediments on its southern side. Of these, about 2.2 million m^3 were deposited during the period of 1985-1995. It is estimated that more than half of this volume was deposited in 1992 during three very severe storms. On the basis of the depositional pattern in the vicinity of the port, and assumptions related to the net longshore sediment transport, it is estimated that more than half of the sediment volume bypasses the port northward.

---

[1]National Institute of Oceanography, Israel Oceanographic & Limnological Research, POB 8030, Tel Shikmona, HAIFA 31080, ISRAEL, Fax: +972 48511911
[2]Department of Geography, Bar-Ilan University, Ramat-Gan, Israel
[3]Ports and Railways Authority-Israel, 74 Dereh Petah-Tikva Ave., Tel Aviv, Israel

Introduction

Ashdod Port is located on the Mediterranean coast of Israel some 30 km south of Tel-Aviv. It was built between 1961 and 1964 on a straight sandy beach backed by sand dunes. About 200 m south of the main breakwater of the port Lakhish River discharges into the sea. The length of the existing main breakwater is 2,200 m and that of the lee breakwater is 900 m. The head of the main breakwater is at a water depth of 15 m, and the entrance of the port was 13 m deep when it was built. The port penetrates seaward from the shore to a distance of about 1,000 m.

Since the beginning of its operation in 1964, the volume of traffic in this port has continuously increased. Presently (1996), it handles some 13.7 million tons of cargo a year, but projections are for 15.4 million in the year 2000 and 16.6 million tons a year in 2010. It is therefore planned to expand the port by extending its main breakwater by 1,150 m to a water depth of about 21 m (Figure 1).

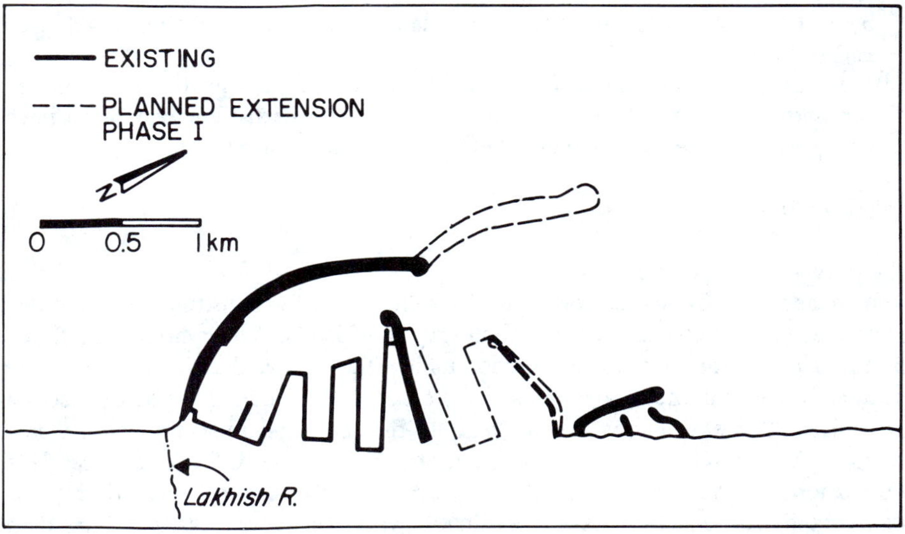

Figure 1. General plan for the planned expansion of Ashdod Port, phase I.

The increasing concern in Israel, as in the rest of the world, regarding the effect of coastal construction on the marine environment, led the authorities in Israel to impose preparation of environmental impact statements (EIS) for each major coastal structure. As part of the requests of the EIS for the expansion of Ashdod port, a numerical sedimentological model had to be carried out. This model should predict the effect of the port's expansion on the environment, and in particular on the nearby bathymetry, coastline and sediments. The present study was carried out in order to gather available information on the sedimentological development resulting from the construction of the existing Ashdod Port, and to evaluate the effect of its breakwaters on the sedimentological balance in the area and the stability, in terms of erosion or accretion, of the seabed and of the shoreline. The results of this study should be used to calibrate and verify the above mentioned model.

Previous studies on the effect of Ashdod Port on the sedimentological processes in its vicinity were carried out by Kran (1980) who analyzed bathymetric profiles which were surveyed between 1964 and 1971 in the vicinity of the breakwaters. Computations of changes in the volume of sediment in the vicinity of the port between 1959 and 1975 were made by Finkelstein (1980) and between 1959 and 1985 by Vajda et al. (1988). Rosen (1985) assessed the longshore sediment transport rate at Ashdod on the basis of wave energy flux calculation.

Three data bases were used for this study:

(a) Aerial photographs, taken since prior to the construction of the port until the present. They enabled to detect shoreline erosion and accretion caused by the port,

(b) Bottom charts and profiles which were prepared before and after the port was built. These were used in order to evaluate changes in the seabed which were caused by the port.

(c) Wave climate resulting from wave observations and measurements carried out near the port since 1957. These were used to evaluate the natural rate of sediment transport in the area, and in particular the longshore transport.

Methodology

Analysis of Aerial Photographs

The changes in shoreline position which resulted from the construction of the port were measured by comparing aerial photographs, which have been taken on different dates since prior to the port construction until 1992. To avoid errors resulting from seasonal changes in the shoreline, only photographs taken during the autumn season were used. This season was selected because the sea is calm and the beach face is steep at that period. Therefore, horizontal change in the waterline due to sea level fluctuations, is rather limited. Also, wave records showed that during the photography sortie and the days preceding it, wave height was less than 1 m. Eight aerial photographic sorties from the years 1958, 1964, 1971, 1976, 1980, 1983, 1988 and 1992 were selected for this study.

The analysis of aerial photographs was that adopted by Shoshany and Degani (1992). The analog format of the photographs was transformed into a digital one using a digital scanner. Prominent objects in the photographs were used to relate photographs from different dates into a uniform geographic system.

The line which separates the wet part of the beach from the dry one was selected to represent the shoreline because it is clear and sharp on the photograph. Also, this line does not fluctuate momentarily as the water line does. This line was further accentuated by enhancing the brightness contrast. The 1988 sortie was selected as a reference sortie, because it contained a large number of common reference points which were seen on photographs from preceding and succeeding sorties. These reference points were used to rectify the aerial photographs and remove distortions from them. Once this was done, the shoreline was digitized for each sortie. It is estimated that the error involved in determination of the shoreline position is less than 10 m.

Bathymetric Analysis

Bathymetric surveys in the vicinity of the port, which were conducted in 1957, 1959, 1970, 1975, 1980, 1983, 1985 and 1995, were used for this study. The surveying methods which were employed in these surveys changed during this period. Until the early 1970's, navigation was carried out by sextant readings from the boat to reference points on the beach, and depth was measured using a Kelvin Hughes surveying echo sounder. After 1972 navigation was carried out by an electronic system, Decca Trisponder, and later by Miniranger. Data processing, interpolation of boat position and depth reading, and collation of these data, were carried out manually until the beginning of the 1980's and gradually changed to computer processing in the early 1980's. The survey of 1995 was carried out using a differential GPS system for navigation and a digital Odom echosounder. The collation of depth and position was carried out on board the surveying boat, using the OCEANOGRAPHER navigation and mapping computer software system, developed and written for such purposes by the third author. The charts were digitized and an interpolated grid of depth points for each survey was prepared. Each grid was subtracted from its predecessor, and depth differential maps which show the magnitude and spatial distribution of deposition or erosion on the seabed were prepared.

Wave Data and Computation of Longshore Sediment Transport

Wave data have been collected from Ashdod Port region since 1957. Some of the data (1957-1975) were based on visual observations, some (1978-1992) on a combination of instrumental measurement (wave height and frequency) and visual observation (wave direction), and for three years, 1992-1995, wave height, frequency and direction were instrumentally measured. In view of the low reliability of the wave directions of the old wave data, the final computation of the longshore sediment transport rate, was based on the April 1992 - March 1995 wave data, gathered off Ashdod with a Datawell Wavec buoy at 3 hour intervals. These data were used as an input in computer programs using the formulas of longshore sediment transport known as the CERC (USArmy CERC-1984), Komar (Komar-1977), LCHF (Migniot et Manoujan-1983) and Bijker (Bijker-1972) formulae. The longshore sediment transport was computed for each sedimentological year, namely years starting in April on one year and ending in March of the following year, to comply with the sedimentological seasonal wave regime in this region.

To account for the longshore sediment transport taking place beyond the surf zone, the joint contribution of wave stirring and geostrophic current transport, current data statistics gathered by Israel Oceanographic & Limnological Reearch off Ashkelon, in 27 m water depth, were used in combination with the Bijker formula.

Results

Shoreline Position

Examination of the position of the shoreline south and north of the port, as derived from the aerial photographs analysis, reveals two phenomena. First, the distance between the positions of the shorelines of 1958, prior to the construction of the port, and 1964, when it was almost completed, is very small and falls within the resolution magnitude of the analysis. The second, is the change with time in the shoreline position south of the port versus those north of it. South of the port, a distinct accretion of the shore with time is noticed, whereas north of the port, the shoreline position is relatively stable with the exception of the sector in the immediate vicinity of the lee breakwater.

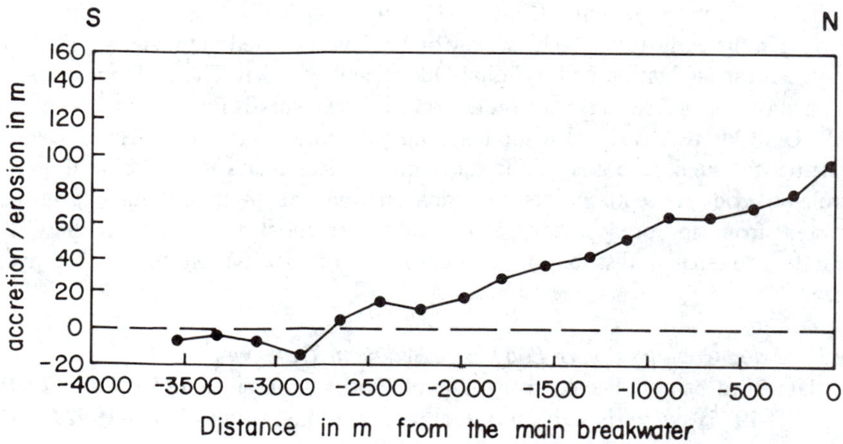

Figure 2. Difference in m between the mean position of the shoreline in 1958-1964 and that of 1983-1992 south of Ashdod Port.

Figures 2 and 3 were prepared with the purpose of showing the general trend in the development of the coastline, south and north of the port. It shows the distance between the mean position of the shoreline in the years 1958 and 1964 and that of the years 1983-1992. Figure 2 shows that during the study period the shoreline has advanced westward gradually from zero at 2,500 m south of the main (southern) breakwater to about 100 m near it. North of the port, Figure 3 shows that accretion has occurred very close to the lee (northern) breakwater, to a distance of about 300 m north of it, but from there on northward, the position of the shoreline was rather stable during the study period.

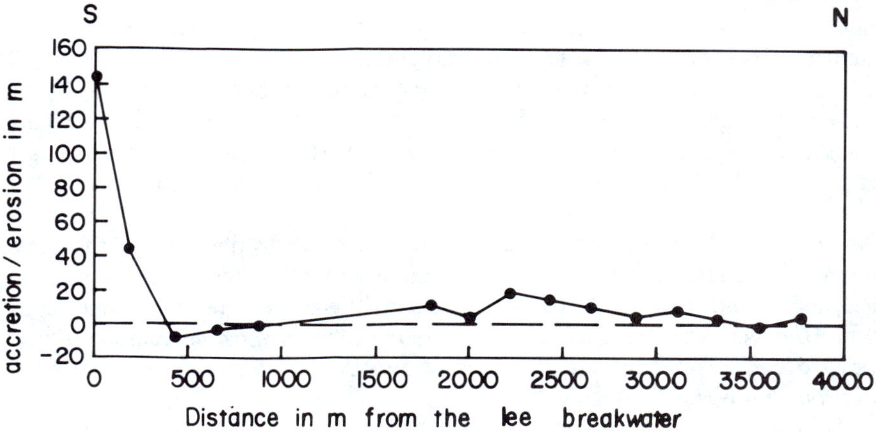

Figure 3. Difference in m between the mean position of the shoreline in 1958-1964 and that of 1983-1992 north of Ashdod Port.

Bathymetric Changes

Examination of the depth differential maps which resulted from the analysis of the bathymetric charts shows that the first impact of the port on the seabed was a severe erosion, up to 2.5 m, north of the port and deposition of up to 3.0 m next to the lee breakwater. With some fluctuations in magnitude, these effects remained throughout the history of the port.

Another phenomenon, which also started following the port's construction, is the increase of deposition near the beach south of the port as well as near the head of the main breakwater. Deposition increased both in thickness as well as in space with time. In 1975, sediment was "creeping" along the main breakwater reaching about a third of its length. In 1980, the depositional area south of the main breakwater increased, and the accumulation at the head of the breakwater increased in thickness. In 1983, deposition followed the same pattern but increased in thickness. Between 1983 and 1985, minor changes occurred in the sea bottom, but between 1985 and 1995, an impressive deposition took place along the southern part of the main breakwater and south of it, parallel to the beach, at a distance of between 700 and 1,000 m. The thickness of this deposition is mostly up to 2.0 m but in restricted areas up to 3.0 m. 4.5 million m^3 of sediment have accumulated in the studied area south of the main breakwater between 1957 and 1995. Of these, 2.3 million m^3 between 1957 and 1985, and 2.2 million m^3 between 1985 and 1995.

Assessment of Wave Climate and Longshore Sediment Transport

Figures 4-6 show the average yearly marginal distributions of the deep water wave characteristics off Ashdod during sedimentological years 1993-1995 (04/1992 -03/95) using data bases of 3 hours data, noon daily data and maximum daily data. Figure 4 shows the marginal distribution of deep water characteristic wave heights, Figure 5 shows the marginal distribution of peak wave periods, and Figure 6 shows the marginal distribution of deep water wave directions for the mentioned period.

Figures 7 and 8 provide the average yearly marginal distributions of the general current characteristics off Ashkelon (15 Km south of Ashdod Port) which were measured in 1992-1993 (Rosen, 1993). The current was measured at a water depth of 27 m, 10 m below the sea surface. Figure 7 shows the marginal distribution of hourly averaged current speeds, and the marginal distribution of hourly averaged current directions is presented in Figure 8.

On the basis of these data and the various formulas mentioned before estimates of the longshore transport were obtained. As can be seen in Figure 9, there are differences of up to four times among the various formulas. The evaluation using the CERC formulae of the net transport is about 3 times larger (720,000 m^3/year) than that of Bijker formula (230,000 m^3/year) without accounting for the presence of currents

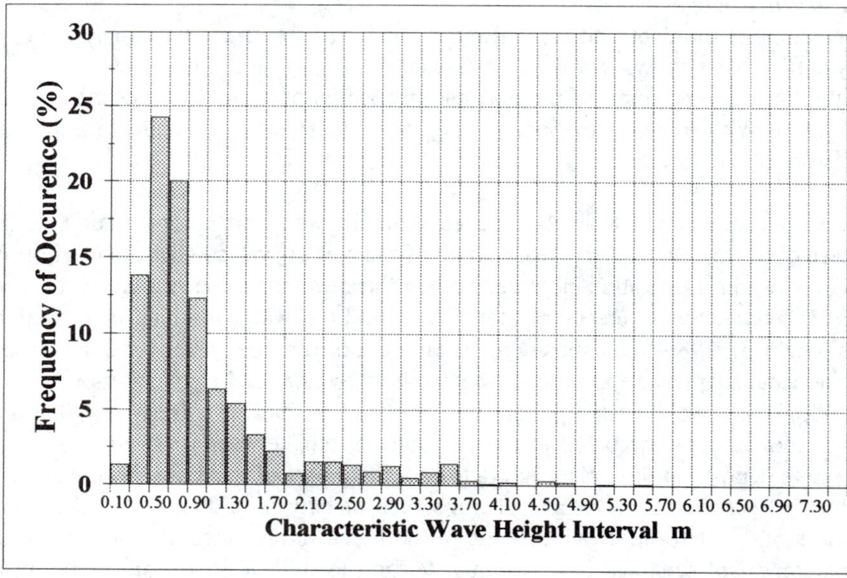

Figure 4. Yearly deep water distribution of characteristic wave height (04/92-03/95).

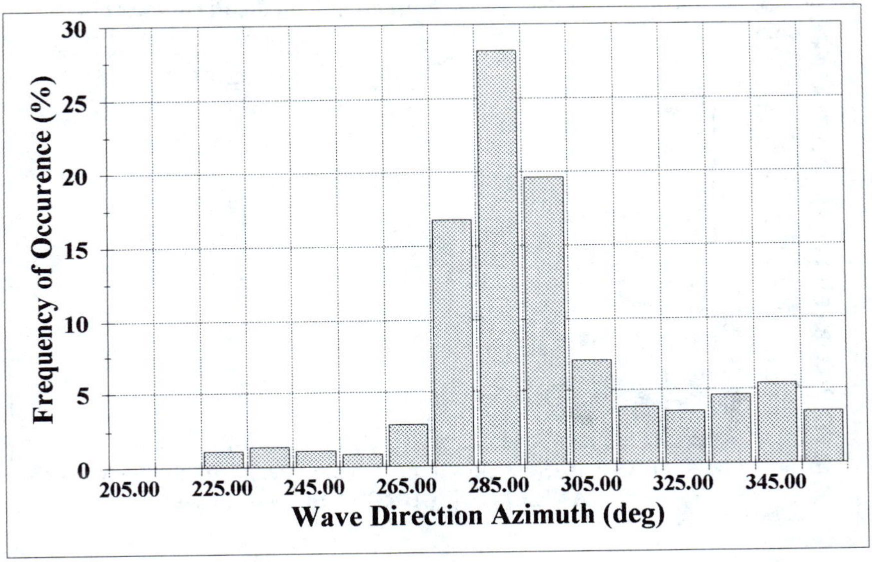

Figure 5. Yearly deep water distribution of peak wave periods (04/92-03/95).

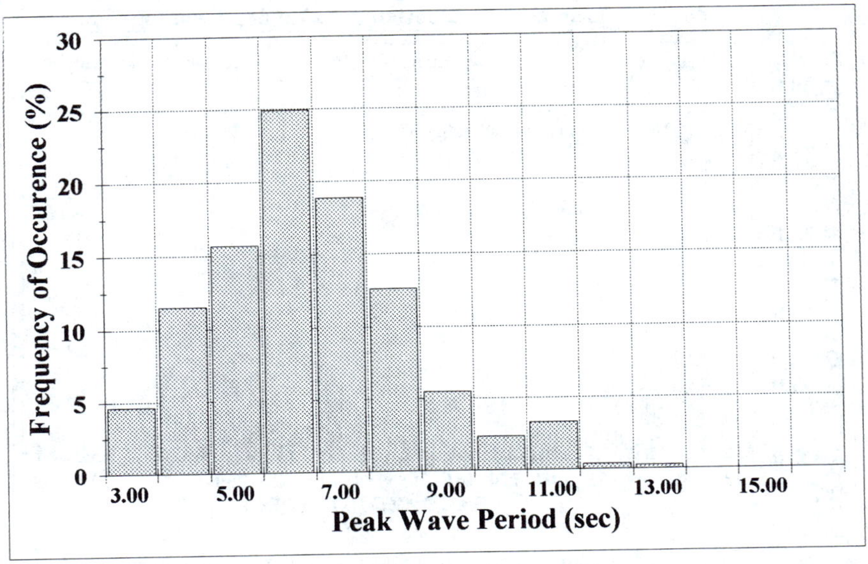

Figure 6. Yearly deep water distribution of wave directions (04/92-03/95).

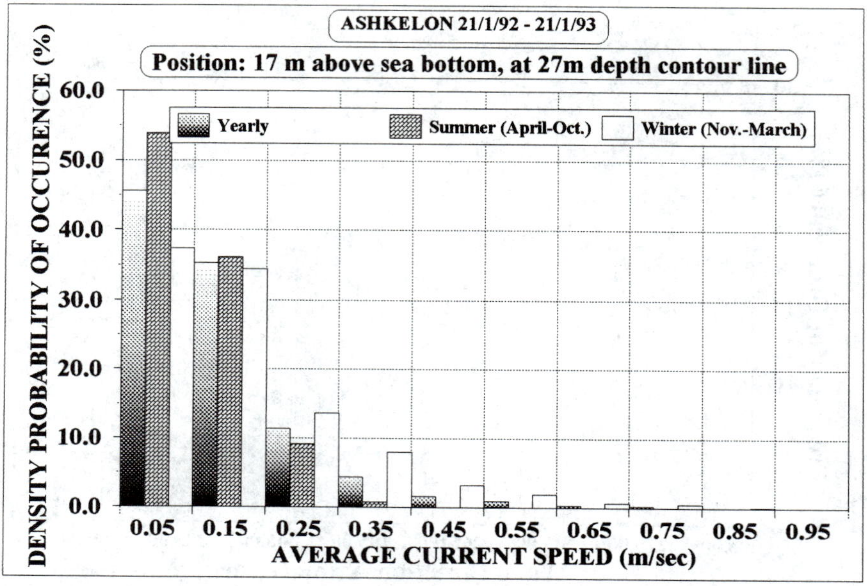

Figure 7. Current speed distribution offshore Ashkelon.

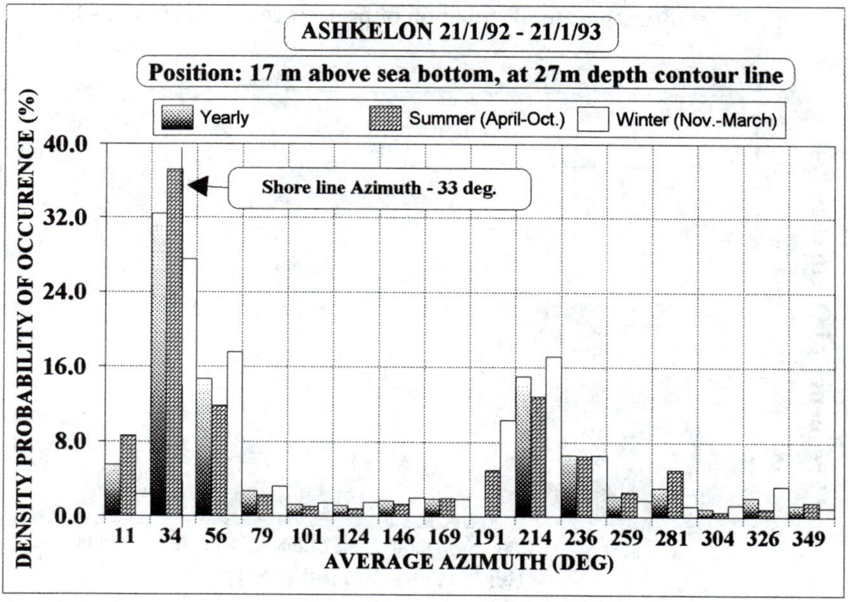

Figure 8. Current direction distribution offshore Ashkelon.

beyond the surf zone. The LCHF provides a value which is about 0.75 of the Bijker formula without currents beyond the surf zone. The comparison between the estimate of the transport with and without accounting for the currents beyond the surf zone is also presented in the same Figure. As one may see, the transport assessment including currents beyond the surf zone on the basis of the current statistics given in Figures 6 and 7 leads to volumes comparable to those of the CERC formula. In Figure 10 the yearly average longshore sediment transport distribution across the shore at Ashdod, based on the period 04/92-03/95, is presented.

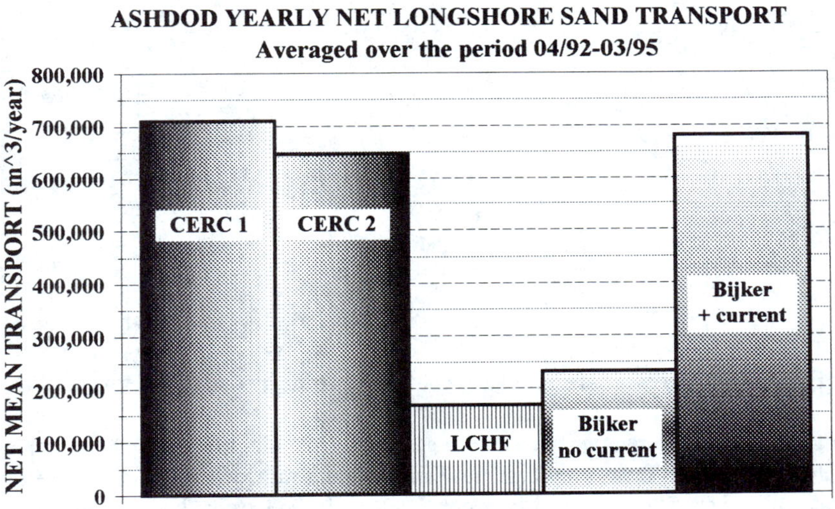

Figure 9. Comparison among methods of evaluation using the 3 hours data set.

Considering the fact that the assessment using currents was based by one year of currents data, during a harsher weather than in the following two years, the average longshore net transport was assessed on a weighted average between assessment with currents and assessment without currents. Thus, an yearly net longshore transport of 350,000 m^3 to the North was assessed as the most probable rate for a normal year.

Discussion

Shoreline Changes

The results of the shoreline analysis clearly indicate that the net sediment transport in the area of Ashdod Port is northward. Under these conditions, one would expect erosion to occur north of Ashdod Port, but no such erosion of the shoreline is noticed from the analysis of the aerial photographs. The reason for this is the sand mining activity which was very intensive on this beach until 1964. This can be seen in Figure 11, which is an aerial photograph taken in 1958, prior to the port construction. The

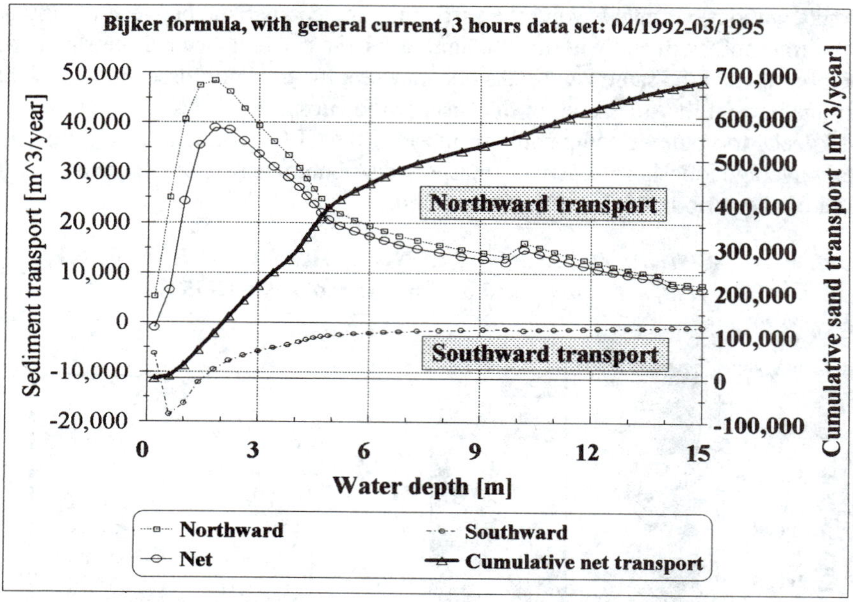

Figure 10. Yearly cross-shore sand transport profile at Ashdod (with currents).

scars of the sand mining are clearly seen on this picture, as well as on aerial photographs taken from the coast both north and south of this point. More than 5 million m^3 of sand were mined between 1949 and 1963 from the beaches stretching along some 30-40 km of the Ashdod coastline (Zifzif Committee, 1964).

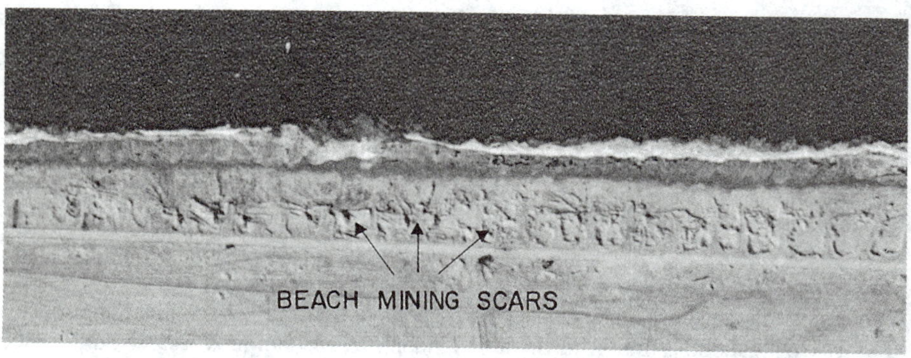

Figure 11. Aerial photograph taken in 1958 showing the scars of beach mining along the Ashdod shoreline.

This mining must have left the coast either devoid of loose sand, or close to it. The sand trapped south of the port after its construction "healed" these scars and widened the beach. However, north of the port, the beach was already rocky before the port construction, lacking sand to be removed by erosion, and this is why the position of the shoreline remained stable. The rocky beach extends today to some 3.5 km north of the port, and from there on northward, sand becomes gradually more ubiquitous on the beach.

Bathymetric Changes

The most striking finding of the analysis of the sea bottom changes is the massive accumulation of sediment, more than 2 million m^3, which took place between 1985 and 1995. This raises a few questions: Was this accumulation a gradual one or an episodic event? What caused this rapid accumulation? Where did the sediment come from?

It was found that a bathymetric survey which was carried out in 1991 offshore Ashdod city, overlapped a small part of our study area in the south. Depth differential maps for the period 1985-1991 and 1991-1995 clearly showed that sediment accumulation which occurred in that area was greater in the period of 1991-1995 by many times than in 1985-1991. This indicates that the massive accumulation of sediment seen in the depth differential map for the period 1985-1995 occurred sometime between 1991 and 1995. The most prominent event that occurred in the period of 1991-1995 is a series of very severe storms which occurred in December 1991, February 1992 and December 1992. In one of them, February 1992, the deep water characteristic wave height was 7.2 m and in the others more than 5 m. Computation of longshore sediment transport rate using a simulated storm with waves of this magnitude resulted in a net transport of up to 400,000 m^3 per storm. The storms of 1991-1992 are therefore responsible for the large deposition of sediment which was detected in the 1985-1995 depth differential map.

There was, however, another source of sediment input into the area. The Lakhish River, which discharges into the sea just south of Ashdod Port, has flooded during the storms, particulary during that of February 1992. According to Hydrological Survey of Israel, the water flow of this river in 1992 was the largest ever to be recorded. Figure 12, which is an aerial photograph taken 10 days after the flood, shows an extensive shoal in front of the river mouth which was formed, at least partly, by the sediment brought by the river to this area.

Sediment Bypass of Ashdod Port

The issue of sediment bypass of Ashdod Port is an important one because it implies to what degree the port acts as an obstacle to the natural transport of sand to the northern beaches of Israel. However, the evidence for such a bypass is only an indirect one. The depth differential map for the period of the port's existence clearly

Figure 12. Aerial photograph taken on 14 February 1992, 10 days after the flooding of Lakhish River, showing the shoal which was formed in front of the river's mouth.

shows that the pattern of accumulation of sediment south of the port, follows the contour of the main breakwater, surrounds it, and there are areas north of the port in which sediment accumulation is already noticed. According to reports of the Ports and Railways Authority the entrance channel to ashdod Port is undergoing siltation. As it is unlikely that this siltation results from an on-offshore transport at this depth, it is another confirmation to the sediment bypass of the port.

It is difficult to provide the rate of sediment bypass because there is no direct way to measure it and we do not have quantitative information for all the parameters which control it. Nevertheless, an attempt was made to come up with an estimate which is based on assumptions and estimates. For the 10 year period between 1985 and 1995 some 3.5 million m^3 entered into the area as a result of the normal net yearly longshore sediment transport (350,000 m^3/yr x 10 years). In addition, the storms of 1992 yielded some 0.9 million m^3 (400,000 m^3/Feb.92 storm + 2 x 250,000 m^3 for the other 2 storms). It is estimated that Lakhish River contributed 0.25 million m^3 during the flood of 1992. All these add up to about 4.6 million m^3 that were input into the area south of Ashdod Port during that period of time. At that time 2.2 million m^3 of sediment were trapped in the area south of the port implying that 2.45 million m^3 , i.e. more than 50% of the sediment managed to bypass the port during that decade.

Acknowledgments

This study was supported by the Ports and Railways Authority, Israel. Wave data from Ashdod were provided by the Coastal & Marine Engineering Research Institute, Haifa, Israel. The authors wish to express their gratitude to E. Kit and J. Zwamborn for their constructive criticism.

References

Bijker E.W., 1972. Lecture notes for Topics in Coastal Engineering, Delft Technical Univ., pp. 20-78.

Finkelstein, A., 1980. Development of Ashdod Port, research in a sedimentological model; model calibration. CAMERI - Coastal and Marine Engineering Research Institute, Report No. P.M. 74/80: 1-49 (in Hebrew).

Kran, N. 1980. Analysis of bathymetric data and sedimentation near Ashdod Harbor (1964-1971). Geol. Surv. Israel Report MG/6/80: 1-9.

Komar P.D., 1977. Beach Sand Transport: Distribution and Total Drift *J.Waterway, Port Coastal and Ocean Div.*, ASCE, **103,** WW2, pp. 225-240 (May 1977).

Manoujan S., and Mignot C., 1983, Ashdod Port Extension, Final Report, Laboratoire Centrale de'Hydraulique de France, Paris, 79p.

Rosen, D.S., 1985. Assessment and Consultation on the Operability of the Coal Unloading Terminal at Ashdod Port North - Stage A, Final Report, CAMERI, PN. 160/85, pp. 13-15. (in Hebrew).

Rosen D.S., 1993, Offshore Coal Unloading Terminal for Rutenberg Power Station Ashkelon Wave and Current Monitoring Program, Reporting Period 21/01/92 - 21/01/93, IOLR, P.N. H5/93.

Shoshany, M. and Degani, A., 1992. Shoreline detection by digital image processing of aerial photography. *J. Coastal Res.*, **8**, 29-34.

U.S. Army Corps of Engineers, Coastal Engineering Research Center, 1984, Shore Protection Manual, vol I,II, 2nd ed.

Vajda, M., Mechrez, E. and Etrog, U., 1988. Disposal of sand from deepening dredging in Ashdod Port in connection with the construction of quay no. 9; examination of alternative solution and analysis of the port impact on the environment. CAMERI - Coastal and Marine Engineering Research Institute, Report No. P.M. 201/88: 1-46 (in Hebrew).

Zifzif Committee, 1964. Zifzif Committee Report. Ministry of Development and Housing (in Hebrew).

CHAPTER 340

Tweed River Sand Bypass: Concepts and Progress

Russell J. Murray,[1] R.P. (Jock) Brodie,[2] Mark Porter,[3] David A. Robinson[4]

ABSTRACT: The objectives of the Tweed River Entrance Sand Bypassing Project on Australia's east coast are to establish and maintain a navigable entrance to the Tweed River and to enhance and maintain the southern Gold Coast beaches, with the objectives to be achieved in perpetuity. The joint project of the New South Wales and Queensland State Governments with the support of the Gold Coast City Council offers the opportunity to achieve this co-operatively in partnership. Agreements have been established. The environmental impact assessment of the first stage (initial dredging and nourishment) has been completed, and the first component of these works involving over 2.2Mm^3 of sand was successfully completed in August 1996. Environmental impact assessment and design studies for the second stage (the permanent bypassing system) are in progress. The paper describes the overall project and its key features, the investigations and analyses undertaken to date, the identified impacts, the initial dredging and nourishment works, and the issues being investigated for the permanent system.

INTRODUCTION

The Gold Coast-Tweed Heads region (28°S, 153.5°E) on Australia's east coast is a major international and national tourism destination, and a significant growing recreational and residential area, with a unique coastal environment. The beaches are subject to cyclonic(hurricane) and storm waves, and are characterised by a

---

1) Project Director, Queensland Dept. of Environment, PO Box 155 Brisbane Albert Street, Qld 4002, Australia.
2) Project Director , New South Wales Dept. of Land and Water Conservation, McKell Building, Level 12, 2-24 Rawson Place, Sydney, NSW 2000, Australia.
3) Manager Coastal, New South Wales Dept. of Land and Water Conservation (as above).
4) Manager Coastal Resource Assessment, Qld Dept. of Environment (as above).

predominantly northwards longshore sand transport averaging 0.5 million cubic metres (Mm^3) per year net, which is driven mainly by the predominant easterly and south-easterly waves (median significant waveheight: 1.4m, median peak period 9s). The beaches are composed of clean fine-grained quartz sand with median diameter of 0.22mm. Mean spring tidal range is 1.3m.

The Tweed River entrance which influences the continuity of the natural littoral system has been plagued by mobile sandbars which have endangered navigation since the early 19th century. Entrance training walls (rock jetties) started in 1890 offered only short-term relief. From 1962 to 1965, the training walls were extended seaward 380m to aid navigation of the entrance. These walls reduced sand transport past the entrance, Letitia Spit to the south of the entrance accreted (total accretion to 1995 of over $7Mm^3$ has filled the walls), and the downdrift southern-most beaches of the Gold Coast experienced erosion (Macdonald and Patterson, 1984). Major beach nourishments totalling over $6Mm^3$ have been successful in substantially offsetting the erosion of these beaches (Murray et al, 1993 and 1994).

Following separate and joint studies by the responsible authorities, and complex negotiations with four jurisdictions, agreement was concluded between the State of New South Wales and the State of Queensland to undertake the Tweed River Entrance Sand Bypassing Project which impacts on both States. It is being supported by the local city, the City of the Gold Coast (Murray, Brodie et al, 1995).

The project agreements are:
(a) the Heads of Agreement signed on 31 March 1994 by the Premiers of New South Wales and Queensland; and
(b) the formal legally-binding Deed of Agreement signed by the responsible Ministers on 2 March 1995.

The New South Wales Parliament has enacted ratifying legislation (the *Tweed River Entrance Sand Bypassing Act 1995* (NSW)), and similar legislation is proposed in Queensland.

OBJECTIVES, PRINCIPLES, AND BENEFITS

The objectives of this joint project (to be achieved in perpetuity) are to:
(a) improve and maintain the navigability of the Tweed River entrance (in New South Wales); and
(b) improve and maintain the southern Gold Coast beaches (in Queensland).

The project comprises two inter-related components:
(a) an initial dredging of the bar and entrance area and beach nourishment; and
(b) an artificial sand bypassing system, to operate in perpetuity.

The benefits are anticipated to include:

(a) the improvement in the safety of navigation of the river entrance with the consequent benefits to recreational boating, tourism, property values and the fishing industry;

(b) improved tidal flushing of the river estuary, improved water quality, and mitigation of flooding; and

(c) the restoration, widening and long-term maintenance of the beaches, with associated benefits to tourism, recreation, property values and the reduction of erosion threats.

To achieve early benefit, initial dredging and nourishment was scheduled to be undertaken while permanent bypassing system options and impacts were being investigated and assessed.

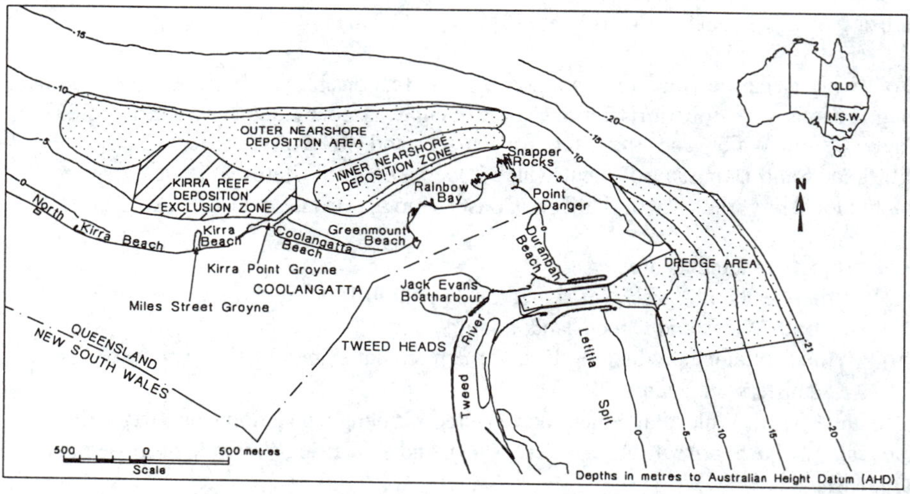

Figure 1: Location diagram, showing approximate dredging and nourishment areas

INITIAL DREDGING AND NOURISHMENT (STAGE 1)

The initial dredging and nourishment involves about 2.5Mm^3 of sand to be dredged from the entrance and placed on the beaches. The environmental impact statement (EIS) was prepared to comply with the requirements of both States. Following exhibition of the document, public meetings and the receipt of twenty-six submissions, an EIS Submission Review Report addressed the submissions received and made recommendations for actions, monitoring and an Environmental Management Plan.

Impacts and issues identified included:

(a) increased operational window for fishing trawlers traversing the bar;
(b) improved ability of beaches of Rainbow Bay, Greenmount, Coolangatta, Kirra and North Kirra to withstand storm erosion;
(c) improved flushing of the lower estuary, to result in improvement in water clarity, nutrients and bacterial loads;
(d) restoration of nearshore bathymetry to a condition similar to pre-1960s (pre-wall extensions) condition enhancing conditions for beach users;
(e) slightly increased tidal range in the estuary (high waters increased by up to 0.04m with possible impact on saltmarsh habitat and roosting and nesting sites for birds, and low waters lowered by up to 0.12m with possible impact on seagrass areas);
(f) erosion of up to 75m from the previously accreted shoreline at the northern end of Letitia Spit;
(g) changed surfing conditions;
(h) concern regarding possible effects on the rocky reef community of Kirra reef if the nourishment produced conditions outside the natural range of seabed fluctuations;
(i) potential disturbance by noise from booster pumps and mechanical equipment during upper beach nourishment; and
(j) possible wrecks in the dredge area.

Each of these issues has been considered in detail and, where appropriate, modifications have been made to the scope and extent of the work (Brodie et al, 1995).

The lowest of six tenders(bids) from Australian and international companies for the first part of the initial dredging and nourishment (Stage 1A) was accepted. Dredeco undertook the work using three dredgers in successive campaigns from 26 April 1995 to 20 August 1996, transferring over 2.2Mm^3 of sand.

The greater proportion of the work was carried out by the (144m long) *Pearl River*, the world's then largest trailing suction hopper dredger. In a six week campaign, it dredged up to approximately 9,000m^3 of sand each load in typically four sweeps. The original design slope was modified to allow bench dredging to a similar volume and shape. The *Pearl River* dredged the bar and entrance area to the limits set by its draft. It deposited 0.93Mm^3 through the bottom-opening doors to a predetermined profile in the outer nearshore deposition area with an upper level of minus 7m Australian Height Datum (AHD, approximately mean sea level). A further 0.59Mm^3 was pumped onto the upper beaches through a single point mooring connected to a 750m long, one metre diameter submerged pipe, which came ashore at Kirra Point, and linked to alongshore pipelines which extended from Kirra Point to the limits of beach nourishment at Rainbow Bay and North Kirra.

The shallower bar areas were dredged by the smaller trailing suction hopper dredger, the *Ngamotu* which has a length of 61m, draft of 3.5m fully laden and a hopper capacity of 490m^3. The *Ngamotu* established a channel through the bar, increasing the ruling depth over the bar from less than 4m AHD to more than 6m AHD, thus allowing access by larger dredgers. During suitable conditions from May to November 1995, it dredged 0.17Mm^3 and placed this sand in the inner nearshore deposition area to an upper level of minus 4.5m AHD.

Subsequently, in two campaigns in January/February and July/August 1996, the 2,300m^3 capacity split-hull trailing suction hopper dredger, the *Krankeloon,* removed a further 0.56Mm^3 and placed this in the inner nearshore deposition zone (0.12Mm^3), the outer nearshore deposition zone (0.39Mm^3), and an eastwards extension of the nearshore deposition zones around Snapper Rocks (0.05Mm^3).

Further dredging and nourishment (Stage 1B) is expected to be carried out in late 1997 following monitoring of the results of this work and assessment of maintenance requirements.

PERMANENT BYPASSING SYSTEM (STAGE 2)

The permanent bypassing component consists of the design, manufacture, supply, delivery and commissioning of a sand bypassing system and the continuing operation of that system including replacement of the capital equipment. Its purpose is to facilitate the natural littoral sand movement processes by conveying past the river entrance the net littoral transport which occurs at the northern portion of Letitia Spit, so as to maintain a clear navigation channel of at least 4.5m AHD and maintain a continuing supply of sand to the beaches.

Artificial sand bypassing alternatives include jetty(pier)-based systems, mobile (jack-up) systems, floating plant, and traditional dredgers. A mobile system could overdredge a buffer during low transport periods to assist in intercepting some of the sand which would otherwise be missed. A mobile system should in principle be capable of bypassing 500,000m^3 per year, even though some is missed from time to time. Fixed jet-pump bypassing system options (such as used at the Gold Coast Seaway 28km to the north where 4.5Mm^3 has been bypassed in the first 10 years) would require fundamental modification or augmentation because there is no buffer sand trap at the project site as, for example, provided by the Seaway training walls (Coughlan and Robinson, 1990), and fixed systems rely on, and are limited by, natural processes in bringing sand to the intakes. No system can be totally effective in intercepting all of the sand transport, particularly during storm/cyclone events when operations may have to be curtailed and sand transport may be very high. Some allowance will have to be made for sand not captured by the system before it reaches the Tweed entrance or which escapes from the operational area of the system. (Robinson, 1993).

The delivery locations are expected to be Snapper Rocks (major quantity), Kirra Point (possible on demand capability), and Duranbah Beach (possible minor quantity capability). As natural sand transport around Snapper Rocks and across the Rainbow Bay and Greenmount/Coolangatta beaches is characterised by some slug-like spurts overlying a more regular transport, there are likely to be some beach management advantages if sand can be discharged on demand at Kirra, where local shortages of sand occur from time to time. However Snapper Rocks could be a sufficient discharge site, provided a suitable discharge mechanism was designed, and operation of the bypassing rate was managed in accordance with changing transport rates. The long term average net littoral transport rate is currently understood to be 500,000m^3 per year, but analysis suggests that it can vary between 270,000m^3 and 900,000m^3 in any single year. The delivery of sand will match the long term average net littoral transport, but will take account of the annual variations in quantities of sand delivered by natural processes to Letitia Spit and the specific objectives of the project, including coastal process issues of the beaches and the entrance, beach usage and navigation requirements. If the long term average net littoral transport rate changes, the rate of delivery will be changed accordingly.

Other issues to be considered for the system include:

(a) system reliability and safety in the high-energy, corrosive, abrasive environment;
(b) outlet design;
(c) water discolouration (though the sand is clean);
(d) public safety (for beach and waterway users);
(e) effects on beach usage (including swimmers, surfers, and anglers);
(f) visual impact;
(g) noise;
(h) operational scheduling;
(i) performance criteria;
(j) commercial and contractual arrangements; and
(k) operational management.

Actions already implemented include the assembly and evaluation of the performance of the Gold Coast Seaway and other existing and planned bypassing systems, a value management study, and the incorporation of data requirements for the permanent system in the current data collection and monitoring program. A comprehensive environmental impact assessment study is in progress, and is scheduled for completion in mid-1997.

CONCLUSIONS

The objectives of the Tweed River Entrance Sand Bypassing Project are to establish and maintain a navigable entrance to the Tweed River and to enhance and maintain the southern Gold Coast beaches, with the objectives to be achieved in perpetuity. The joint project offers the opportunity to achieve this co-operatively in partnership.

The project is located on an open high-energy coastline subject to variable natural forces, in a highly-valued environment, subject to intensive usage. Accordingly, it is recognised that the project must be designed, evaluated and implemented prudently using best practice and in an environmentally sensitive way, if the long-term benefits are to be effectively achieved.

The complex bar dredging and beach nourishment have provided initial benefits, have satisfied strict environmental criteria, and have received widespread public support. Extensive environmental monitoring and public consultation has been incorporated.

The significant engineering, environmental and contractual challenges facing the permanent bypassing system are being progressively addressed.

REFERENCES

Brodie, R.P., Murray, R.J., Porter, M., Robinson, D.A., and Lawson, S. (1995) 'Implementing the Tweed River Entrance Sand Bypassing Project', Proc. NSW Coastal Conf., Tweed Heads, Oct.

Coughlan P. M. and Robinson D.A. (1990) 'The Gold Coast Seaway Queensland, Australia', Shore and Beach, Vol 58, No.1, Jan.

Macdonald, H.V. and Patterson, D.C. (1984) 'Beach Response to Coastal Works, Gold Coast, Australia', Proc. 19th Coastal Eng. Conf., ASCE, Houston. pp 1522-1538.

Murray, R.J., Brodie, R.P., Jackson, L.A., Porter, M., Robinson, D.A., Lawson, S. and Perry, M.P., (1995) 'Tweed River Entrance Sand Bypassing Project: Principles and Progress', 12th Australasian Conf. on Coastal and Ocean Engineering, The Institution of Engineers Australia, Melbourne, 29 May - 2 June, pp7-12.

Murray, R.J., Robinson, D.A, and Soward, C.L. (1993) 'Monitoring of the Southern Gold Coast Beach Nourishment Project', 11th Australasian Conf. on Coastal and Ocean Engineering, The Institution of Engineers Australia, Townsville, 23-27 August, 1993, pp119-125.

Murray, R.J., Robinson, D.A, and Soward, C.L. (1994) 'Southern Gold Coast Beach Nourishment Project: Implementation, Results, Effectiveness', *Terra et Aqua*, 56, Sept. 1994, pp12-23.

Robinson, D.A. (1993) 'Tweed Heads Sand Bypassing System', Bruun Workshop: Beach Nourishment & Sand Bypassing, Dept of Civil Eng., Uni. of Qld, 28-29 Jan.

CHAPTER 341

DESIGN CONSIDERATIONS FOR COASTAL PROJECTS IN COLD REGIONS

Craig B. Leidersdorf, M. ASCE[1]
Peter E. Gadd, M. ASCE[1]
Kennon D. Vaudrey, M. ASCE[2]

ABSTRACT

Design considerations for coastal projects in cold regions include not only the oceanographic factors that tend to assume primary importance in temperate climates, but also ice and thermal factors. This paper reviews the current understanding of each of these three types of design considerations, with emphasis on the lessons learned from recent project experience in the Alaskan Beaufort Sea. It is concluded that the environmental influences unique to cold regions introduce both increased complexities and increased uncertainties into the design process for coastal structures. In consequence, a conservative approach that incorporates damage-tolerant components and non-catastrophic failure modes is recommended for the design of sensitive facilities.

INTRODUCTION

During the past two decades, the number of coastal and nearshore projects constructed in cold regions has increased significantly. The primary motivation for these projects has been petroleum-related activities, which have resulted in the installation of more than 50 artificial islands, causeways, and coastal pads off the Arctic coasts of Alaska and Canada. Public works projects also have been undertaken by communities such as Barrow, Alaska, where bluff stabilization and beach nourishment programs have been carried out in recent years (Wiegel, 1995).

For purposes of this paper, a "cold region" is defined as an area in which the sea surface remains ice-covered for a significant portion of each year. Important

---

[1] Principal, Coastal Frontiers Corp., 9420 Topanga Canyon Blvd., Ste. 101, Chatsworth, CA 91311
[2] President, Vaudrey & Associates, Inc., 1540 Marsh Street, Ste. 105, San Luis Obispo, CA 93401

design considerations for coastal projects in cold regions include not only the oceanographic factors that tend to assume primary importance in temperate climates, but also ice and thermal factors.

The objective of this paper is to review the current understanding of ice, oceanographic, and thermal design considerations for coastal projects in cold regions. Particular emphasis will be placed on recounting the lessons learned from recent project experience in the Alaskan Beaufort Sea.

ICE DESIGN CONSIDERATIONS

Ice-related factors that may warrant consideration in the design of coastal and nearshore facilities include horizontal loads, encroachment onto the structure work surface, gouging of the sea floor by ice floe keels, and "strudel scouring" of the sea bottom by the drainage of river outflow through holes in the ice. The first two factors, horizontal loads and encroachment, can exert a major influence on structures that extend above the sea surface, such as islands, causeways, and coastal pads. The last two factors, ice gouging and strudel scouring, are particularly relevant to engineering works constructed on the sea floor, such as pipelines and toe protection.

Horizontal Ice Loads

The horizontal loads that result from ice movement can be segregated into two categories: global ice loads, which act on large structures such as gravel islands, and local ice loads, which act on small structural elements such as sheet pile walls and caissons. Global ice loads vary with the ice thickness, strength, and mode of failure. In turn, the failure mode depends on the type of ice feature impacting the structure. For exposed locations in the Beaufort Sea in water depths of 10 to 15 m, the governing global ice load is likely to be produced by encounters with multiyear floes (floes that have survived at least one open-water season). For protected locations between the shoreline and the 10-m isobath, the design global ice load results from first-year sheet ice. Such loads can be calculated probabilistically or deterministically, with the methodology for each presented in the American Petroleum Institute Recommended Practice 2N (API RP2N; 1995), and in Sanderson (1988).

In the case of exposed structures such as man-made islands, it is assumed that the kinetic energy of the moving multiyear floe is totally dissipated in ice crushing as the floe either comes to rest against the island or continues moving past while developing a maximum contact width equal to the maximum waterline dimension of the island. If a floe of mass m is traveling at velocity v, it will come to rest after a total penetration X, when:

$$\tfrac{1}{2}mv^2 = \int_0^X P(x)dx \qquad (1)$$

where $P(x)$ is the total force exerted when the floe has moved a distance x after initial contact. The load $P(x)$ may be expressed as:

$$P(x) = \sigma(A)\ A(x) \qquad (2)$$

where the contact area $A(x)$ varies with the penetration distance and $\sigma(A)$ is the average ice failure pressure associated with this contact area. It is assumed that the largest global ice load for any multiyear floe impact will occur at the point of maximum penetration. In order to solve Equations (1) and (2), values are required for the ice floe mass m, the initial velocity v, and the failure pressure $\sigma(A)$. The contact area $A(x)$ is determined from the interaction geometry of the floe and the structure.

Using the above ice loading scenario and Equations (1) and (2), global ice loads can be computed probabilistically by considering the input parameters as random variables. Independent distribution functions are required for the multiyear floe thickness, the floe diameter, the floe velocity, and the ice failure pressure to determine the annual probability distribution of the total ice load P acting against the structure. The maximum annual load developed from an appropriate random sampling of the four input parameters is assumed to approximate an extreme value (e.g., Type I or Gumbel) distribution, from which a design value can be determined. In the Alaskan Beaufort Sea, representative design global ice loads associated with multiyear floe impacts range from 5 to 6 MN per meter of island exposure width for a return period of 100 years.

In the case of protected sites such as coastal pads, the maximum global load P exerted by sheet ice can be estimated on the basis of an ice crushing failure using a modified version of the Korzhavin equation (API, 1995):

$$P/D = p\,t \qquad (3)$$

where P represents the total horizontal ice force acting against the maximum exposure width D of the structure, t represents the ice thickness, and p represents the effective ice crushing or failure pressure (which varies with the ice type, temperature, movement rate, and contact area).

The maximum annual load usually will occur in late winter, when the ice temperature and thickness produce the strongest ice. Values ranging from 1 to 1.5 MPa for the effective ice failure pressure p and 0.5 to 2 m for the ice thickness t are representative of the nearshore first-year ice conditions utilized for design purposes in the Alaskan Beaufort Sea. The resulting global ice loads range from 0.5 to 3 MN per meter of exposure width for a return period of 100 years.

Small structural elements such as sheet pile walls and caissons must be designed to withstand ice pressures exerted over areas of several square meters. These local ice failure pressures can be much larger that the uniaxial compressive ice strength because of the confinement provided by the surrounding ice. In consequence, the design local ice loads on small structural elements always exceed the global loads exerted by the same ice conditions on large structures.

As suggested by Figure 1, local ice pressures increase with decreasing contact areas. The two curves in Figure 1 represent recommended upper and lower bounds derived from an empirical local ice pressure-area relationship set forth in the API RP2N (1995). The lower bound is based upon the mean plus two standard deviations ("M+2SD"), while the upper bound is based upon the mean plus three standard deviations ("M+3SD"). Because these curves were developed from measurements acquired throughout the Arctic, site-specific experience should be taken into consideration when they are applied to specific projects. In addition, the curves should be utilized only for contact areas less than 10 m^2; the product of the local ice pressure and the contact area always should exceed the global ice load calculated according to the appropriate method described above; and the local ice pressure should be regarded as acting normal to the exposed face of the structure.

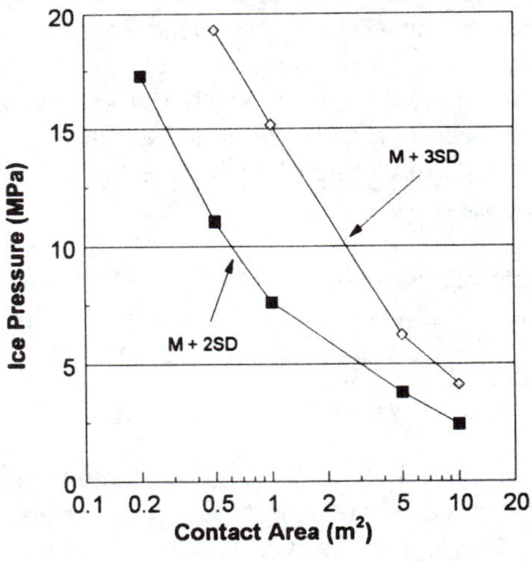

Figure 1. Pressure-Area Curves for Small Structural Elements

Ice Encroachment

In the context of this paper, encroachment refers to the movement of an ice sheet or individual ice blocks onto the work surface of a man-made island or coastal pad (Plate 1). If the sheet ice is driven up the side slope intact by a storm wind, the phenomenon is considered ice "ride-up". If the ice fails in buckling or bending and breaks up into individual blocks, an ice "pile-up" is formed on the side slope. In the Beaufort Sea, ride-up and pile-up events can occur during both freeze-up (early October through mid-December) and break-up (early June through mid-July).

Plate 1. Ice Encroachment on Endeavor Island, Alaska (1982)

A number of factors influence the susceptibility of a given location to ice ride-up, pile-up, and possible encroachment. Although ice movement results from wind stress, the single most important factor in initiating a ride-up or pile-up event is the loss of confinement of the ice sheet. Reversal of the wind direction is the usual cause of confinement loss, due to the formation of cracks or small leads in the nearshore ice. An easterly wind may produce the cracks or leads in the ice, for example, after which a westerly wind can drive the relatively unconfined ice up the slope of a pad or island.

While loss of confinement represents the dominant influence on ice ride-up and pile-up events, the ice thickness and the intensity and duration of storms constitute important secondary influences. The greater the ice thickness, the larger the driving force that will be required to initiate motion. Once the ice begins to move, severe storms of long-duration will maintain sufficient driving force to keep the ice in motion. The ice typically moves at a rate of 100 to 200 m/hr during the formation of a shoreline ice pile-up. Motion ceases when the driving force is balanced by the resistance of an "infinitely" wide obstruction, such as a shoreline.

The most common encroachment event, a combination of ride-up and pile-up, occurs when the sheet ice rides up the structure slope until increasing frictional resistance or a change in slope causes the ice to rubble and form a pile-up. If the pile-up attains sufficient height, ice blocks at the top of the pile can tumble down onto the work surface. Such an event occurred in October 1982 on Endeavor Island, which is located in a water depth of 3.7 m in the Alaskan Beaufort Sea. As is evident in Plate 1, a southwesterly storm with wind speeds of 15-20 m/s created a 7.6 m high pile-up and caused 20-cm thick ice blocks to encroach up to 5 m onto the island work surface (Vaudrey, 1983).

Because our present understanding of pile-up mechanics is limited, encroachment estimates must be based upon statistical extrapolations of pile-up characteristics observed in the general project area. The method involves four primary steps: (1) compiling a database of historical pile-up events; (2) performing an extremal analysis to estimate the pile-up height associated with the desired return period; (3) estimating the pile-up geometry; and (4) computing the encroachment distance based on the structure geometry (Figure 2). A conservative lower bound of 30° is recommended for the slope angle β of the landward side of the pile-up. This value is based on shoreline pile-ups observed on the Alaskan Beaufort Sea coast and on the west coast of Banks Island, Canada, by Kovacs and Sodhi (1980). In the Alaskan Beaufort Sea, the predicted 100-year return period ice encroachment distances typically range from 8 to 16 m for work surface elevations of 3 to 6 m above mean sea level. A buffer zone exceeding the predicted encroachment distance can be maintained around the perimeter of the work surface to insure that sensitive facilities are not impacted.

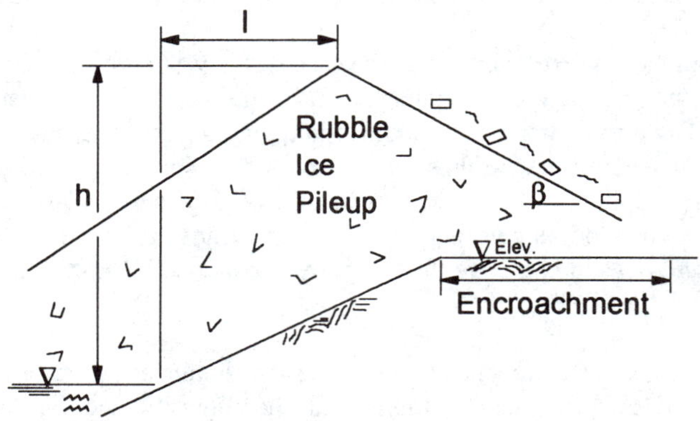

Figure 2. Ice Encroachment Geometry

Ice Gouging

The term "ice gouging" refers to the distinctive furrowing of the sea floor that results from the plowing action of ice keels. A representative example is provided in Figure 3, which displays a multi-beam sonar image of several gouges detected off the Alaskan Beaufort Sea coast in a water depth of 11 m. The single gouges were formed by individual ice keels, while the "gouge multiplet" is believed to have been caused by a multi-keeled ice feature.

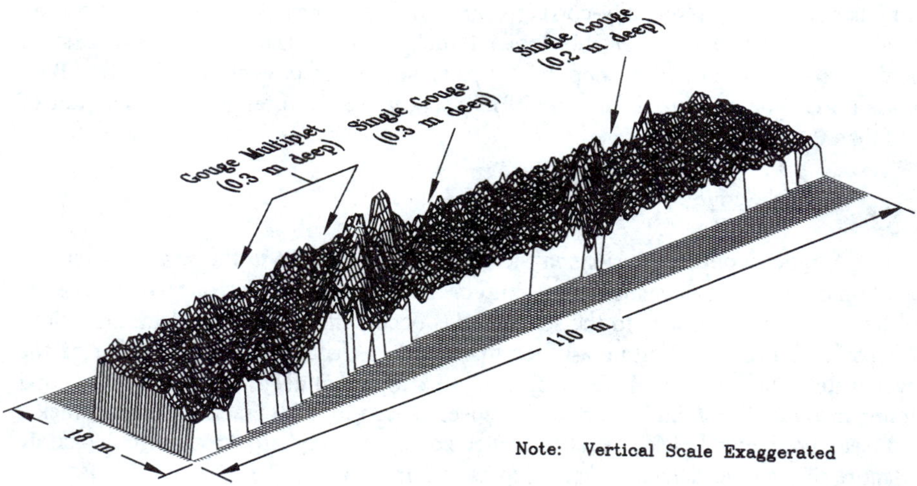

Figure 3. Multi-Beam Sonar Image of Ice Gouges in the Alaskan Beaufort Sea

The frequency and severity of gouging depend upon complex interactions between the dynamic ice environment and the sea floor sediments. Although our knowledge of the process is far from complete, key parameters appear to include the ice movement characteristics, the soil characteristics, the sea floor slope, and the water depth. From a study of ice keel structure and available driving forces, Kovacs and Mellor (1974) concluded that virtually all ice keels possess sufficient strength for gouging.

Extensive ice gouge measurement programs were conducted in the Alaskan Beaufort Sea from 1970 to the mid-1980's by the U.S. Geological Survey (USGS). Based on an analysis of more than 100,000 gouges, Barnes, *et al.* (1984), reported an average gouge density of $70/km^2$ and a maximum density of $490/km^2$. The mean incision depth for individual gouges was 0.56 m (measured below the original sea floor), while the maximum measured value was 4.0 m. The dominant gouge orientation was found to be slightly onshore of contour-parallel. Additional findings from the USGS programs of potential relevance to engineering projects such as subsea pipelines include the following: (1) gouge characteristics (density, depth, and width) attain their maximum values in the zone of active ice rubble formation known as the "stamukhi zone"; (2) sea floor morphology influences the intensity of gouging, with shallower gouges typically occurring in areas protected by offshore shoals; (3) gouge frequency and size are extremely variable from year to year in any one area, and from area to area in any one year (Rearic, 1986; Rearic and Ticken, 1988).

In view of the spatial and temporal variability inherent the gouging process, repeated annual surveys are advisable to obtain an accurate statistical characterization of the gouge population at a particular project site. Multi-beam sonar, when used in conjunction with differentially-corrected Global Positioning System (GPS) position data, has proven to be a valuable tool for such surveys. Based on two successive field programs performed in support of a planned pipeline project near Prudhoe Bay, Alaska, multi-beam sonar is capable of mapping gouge features with a resolution of less than 0.1 m.

Strudel Scour

Strudel scour depressions in the sea floor are formed in the spring, when the breakup of river ice precedes the breakup of ice in the nearshore zone. If the sea ice in the river deltas is frozen to the sea floor (as occurs in the Alaskan Beaufort Sea to a depth of 2 m), this "bottomfast" ice forms a dam that causes overflooding of the river water. The overflood water, typically 0.5 to 1.5 m deep, spreads offshore and drains through discontinuities in the ice sheet caused by tidal cracks, thermal cracks, and seal breathing holes. Such "strudel zones" usually are located immediately offshore of the river deltas, in water depths of 2 to 9 m.

In those instances where the overflood rate is high, the drainage holes become enlarged and powerful strudel jets develop. Sizable scour depressions in the sea floor can result, particularly in the area immediately seaward of the bottom fast ice where the attenuating effect of the water column on the strudel jet is minimal. Although the depressions tend to be circular in plan form, clustered scours and extended linear scours also have been detected. The probable cause of these multiple and extended scours is drainage through linear fractures in the ice sheet, such as tidal cracks.

Measurements of strudel scour characteristics, unlike those of ice gouges, have been extremely sparse. Studies conducted by the USGS in 1972 (Reimnitz, *et al.*, 1974), and 1978-80 (Reimnitz and Kempema, 1982) documented three scour depressions in the vicinity of Prudhoe Bay in water depths of 3.0 to 3.5 m. The scour depths ranged from 1.2 to 4.3 m below the original sea floor, while the maximum horizontal dimensions ranged from 12 to 25 m. Of interest for the design of nearshore structures is the data set acquired by BP Exploration (Alaska) Inc. in the vicinity of Resolution Island, which is located in 2.3 m of water off Alaska's Sagavanirktok River. Annual bathymetric surveys conducted from 1985 through 1995 revealed the formation of seven strudel scours in proximity to the island toe (Coastal Frontiers, 1996). The deepest, 2.8 m below the sea floor, re-filled with sediment to the point that it was no longer detectable five years after its discovery. A probable explanation for the high frequency of strudel scouring at this location is the tendency for the island to induce cracks in the surrounding ice sheet. The primary implication for coastal projects is that the lower slope and toe of a nearshore structure may require additional protection to withstand the strudel jets caused by the presence of that structure.

OCEANOGRAPHIC DESIGN CONSIDERATIONS

Oceanographic design considerations in cold regions are analogous to those in temperate climates, with water levels, waves, and currents often constituting the key parameters. Complications can arise, however, from factors that include a paucity of measured data and exceptionally large storm surges.

Due to low population densities, low volumes of vessel traffic, and the difficulties associated with obtaining measurements in ice-infested waters, oceanographic data tend to be sparse or non-existent in many cold regions. Of particular concern is the situation that arises in areas like the Beaufort Sea, where the most severe storm events of the open-water season often occur just prior to freeze-up. If bottom-mounted instruments such as wave gauges or current meters are removed in advance of this period, the data may be skewed toward milder conditions. Conversely, if the instruments are allowed to remain in place until after freeze-up and then recovered through the ice, the probability of damage or loss from ice keel impacts increases substantially.

In the absence of reliable measurements, the database needed to estimate extreme events at coastal project sites must be derived from hindcast analyses. Difficulties may be encountered in a number of areas, beginning with a lack of meteorological data from which to specify the wind fields. Also problematic is the influence of sea ice, whose damping effect on wave generation and propagation is not fully understood. Specification of the fetch for wave generation is further complicated by the mobility of the ice edge during storm events.

Despite these difficulties, the hindcast approach has proven to be both reliable and cost-effective in the Alaskan Beaufort Sea, where storm surge and wave models were used to hindcast the 16 most severe storms occurring between 1949 and 1980 (Oceanweather, 1982). Initially, the models were implemented for the entire region, with the surge model utilizing a grid spacing of 4.6 km (2.5 nautical miles) and the wave model a grid spacing of 37 km (20 nautical miles). Subsequently, nested grids have been used to predict extreme conditions at numerous individual project sites.

Because Coriolis force increases with the sine of the latitude, coast-parallel winds at high latitude produce storm surges and setdowns of greater magnitude than would occur under similar conditions at low latitude. One obvious implication is that flooding constitutes a major design consideration for many coastal projects in cold regions. In the Beaufort Sea, for example, the predicted 100-year surge heights are typically on the order of 2 m, and driftwood strand lines have been observed as high as 3.4 m above mean sea level (Reimnitz and Maurer, 1979).

Surge-altered water levels are important not only for their direct impact on coastal flooding, but also for their indirect influence on wave and current conditions. In those areas where the wave heights are depth-limited, the elevated water levels that result from storm surge can cause substantial increases in design wave heights. To properly account for this phenomenon in hindcast analyses, a time-series history of water levels should be developed for each historical storm and then utilized when modeling the wave conditions for that event.

The pronounced surge and setdown events that occur at high latitude can induce accelerated coastal boundary currents. As water either builds up or draws down at the coastline, a sloping sea surface is created in the shore-perpendicular direction. The resulting pressure gradient couples geostrophically with Coriolis force to produce a strong shore-parallel flow known as a barotropic coastal jet (Csanady, 1982). Indirect evidence supporting the existence of such currents is provided by ice floe velocities recorded in the Alaskan Beaufort Sea during the 1984 open-water season (Tekmarine, *et al.*, 1985). The floe velocities measured landward of the 30-m isobath during periods of accelerating winds averaged approximately 5% of the wind speed, which is substantially higher than the value of 2 to 3% that might be expected in deep water. The implications of these accelerated currents for the design of nearshore structures include higher velocities for ice floe impacts during periods of open water, and an increased potential for current-induced scour.

THERMAL DESIGN CONSIDERATIONS

Of the many thermal design considerations unique to cold regions, those of particular importance to coastal projects include thermal erosion of ice-bonded bluff sediments, and low-temperature effects on construction materials. In the case of thermally-induced erosion, bluff retreat can result not only from sea water thawing the base of the bluff (Kobayashi and Reimnitz, 1988), but also from elevated air temperatures thawing ice lenses embedded in the bluff face.

Insight into the process of thermal erosion has been provided by an on-going monitoring program at Heald Point Peninsula, which constitutes the northeast boundary of Prudhoe Bay, Alaska. From an analysis of aerial photographs obtained periodically between 1949 and 1988, and survey data acquired annually between 1989 and 1994, it was determined that the central portion of the 5-m high western bluff was retreating at an average rate of 1.6 m/yr. The adjacent bluff areas, by comparison, were retreating at less than 0.5 m/yr (Figure 4). Annual inspections revealed the cause of the accelerated erosion to be the melting of a 2-m thick ice lens embedded in the bluff face near the top of the bluff (Plate 2). Notwithstanding the importance of factors such as wave-induced erosion of the bluff toe and the degree of soil cover on the ice lens, the dominant influence over the annual retreat rate was found to be the summer air temperature. In 1989, for example, when exceptionally high mean monthly temperatures were recorded during the summer season, the bluff edge in the area occupied by the ice lens retreated by up to 7.2 m.

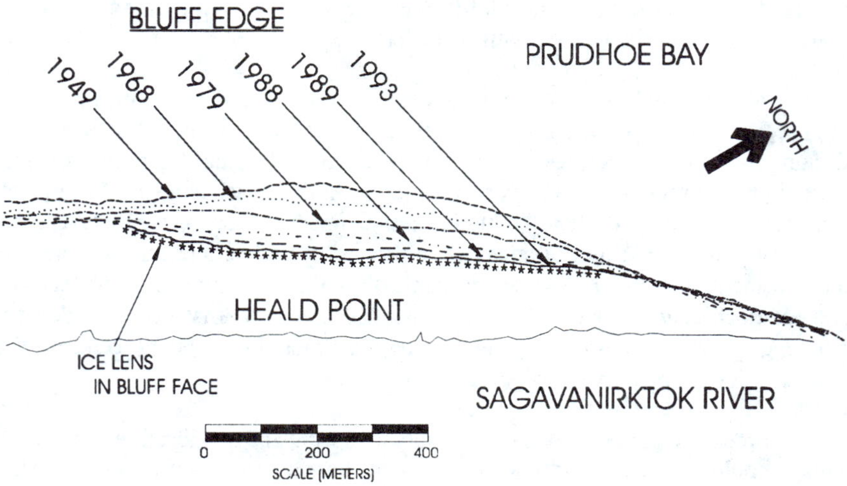

Figure 4. Historical Bluff Edge Retreat at Heald Point, Alaska

Plate 2. Thermal Erosion of Ice Lens in Heald Point Bluff Face (1993)

A particular concern with subaerial melting is that erosion can occur even in the absence of severe oceanographic events. Fortunately, however, the experience acquired with a gravel pad constructed on Heald Point suggests that the problem can be addressed by insulating the bluff face with granular fill material. Since 1993, when gravel was installed from the blufftop to the waterline at a slope of 7(H):1(V), thermal erosion of the western bluff edge has been negligible.

Among the low temperature effects on construction materials encountered in cold regions, freeze-thaw stressing and brittle failure often assume importance for the design of coastal structures. Prior to freeze-up, when air temperatures are sub-freezing but the sea remains unfrozen, materials in the splash zone can be subjected to repeated freeze-thaw cycles. To minimize the impact on concrete structures or armor units, the concrete should be made as impermeable as possible through the use of a high cement factor, a low water-cement ratio, and pozzolan or silica fume (Leidersdorf, *et al.*, 1982). Air entrainment of 6% to 9% is recommended to limit cracking. In addition, sufficient reinforcing steel must be provided to insure that the cracks which do occur will not progressively widen.

Steel components that will be subject to impact or vibration, such as lifting eyes, should remain ductile at the lowest temperatures anticipated. Although corrosion losses over the design life must be taken into consideration, data obtained on Resolution Island suggest that corrosion rates in the Alaskan Beaufort Sea are modest. During a 4-year period in which exposed carbon steel rebar was monitored in the splash zone, corrosion rates were found to be less than 0.05 mm/yr.

CONCLUSIONS

The design of coastal projects in cold regions often is complicated by the need to consider ice and thermal factors as well as oceanographic factors. In some cases, the design process involves identifying a single governing influence; in others, the potential interactions between factors must be taken into account. A representative example of the former situation arises in determining an appropriate work surface elevation for a coastal facility. Two distinctly different phenomena, wave overtopping and ice encroachment, must be evaluated independently before selecting an appropriate minimum value. A situation where interactive environmental factors assume primary importance is in designing a slope protection system that must withstand wave and ice attack. Prototype observations have indicated that ice impacts can predispose the armor to damage under wave conditions that otherwise would be considered benign (Gadd and Leidersdorf, 1990; Leidersdorf, *et al.*, 1990).

In addition to increased complexities, the environmental influences unique to cold regions often introduce increased uncertainties into the design process. At the present level of understanding, quantitative predictions of phenomena that include ice gouging, strudel scouring, and thermal erosion can be derived only from statistical treatments of limited historical databases. In consequence, a conservative approach that incorporates damage-tolerant components and non-catastrophic failure modes is recommended for the design of sensitive facilities.

REFERENCES

American Petroleum Institute, (1995), *Recommended Practice for Planning, Designing, and Constructing Structures and Pipelines for Arctic Conditions* (API RP2N), Dallas, TX.

Barnes, P.W., D.M. Rearic, and E. Reimnitz, 1984, "Ice Gouging Characteristics and Processes", in Barnes, P.W., D.M. Schell and E. Reimnitz, *The Alaskan Beaufort Sea: Ecosystems and Environments*, Academic Press, Orlando, FL, p. 185-214.

Coastal Frontiers Corporation, 1996, "1995 Resolution Island Inspection Program", Chatsworth, CA, 51 pp. + appen.

Csanady, G.P., 1982: *Circulation in the Coastal Ocean*, D. Riedel Publishing Co., Boston, MA, 279 pp.

Gadd, P.E., and C.B. Leidersdorf, 1990, "Recent Performance of Linked Concrete Mat Armor Under Wave and Ice Impact", in Edge, B.L., ed., *Proc. 22nd Coastal Engineering Conference*, ASCE, New York, NY, p. 2768-2781.

Kobayashi, N, and E. Reimnitz, 1988, "Thermal and Mechanical Erosion of Slopes and Beaches", in Chen, A.T. and C.B. Leidersdorf, eds., *Arctic Coastal Processes and Slope Protection Design*, ASCE, New York, NY, p. 46-62.

Kovacs, A. and M. Mellor, 1974, "Sea Ice Morphology and Ice as a Geologic Agent in the Southern Beaufort Sea", in J. Reed and J. Sater, *The Coast and Shelf of the Beaufort Sea*, Arctic Institute of North America, Arlington, VA, p. 113-161.

Kovacs, A. and D. Sodhi, 1980, "Shore Ice Pileup and Rideup", *Cold Regions Science and Technology*, v. 2, p. 209-288.

Leidersdorf, C.B., P.E. Gadd, and W.G. McDougal, 1990, "Arctic Slope Protection Methods, in Edge, B.L., ed., *Proc. 22nd Coastal Engineering Conference*, ASCE, New York, NY, p. 1687-1701.

Leidersdorf, C.B., R.E. Potter, B.C. Gerwick, Jr., and Y.Y. Hsu, 1982, "Modular Slope Protection for the Arctic Environment", in *Proc 14th Offshore Technology Conference*, Houston, TX, p. 689-704.

Oceanweather, Inc., 1992, "Beaufort Sea Wave Hindcast Study: Prudhoe Bay/Sag Delta and Harrison Bay", White Plains, NY, 103 pp. + appen.

Rearic, D.M., 1986, "Temporal and Spatial Character of Newly Formed Ice Gouges in Eastern Harrison Bay, Alaska, 1977-1982", Open-File Report 86-391, U.S. Geological Survey, Menlo Park, CA, 52 pp.

Rearic, D.M., and E.J. Ticken, 1988, "Ice Gouge Processes in the Alaskan Beaufort Sea", in Chen, A.C.T., and C.B. Leidersdorf, eds., *Arctic Coastal Processes and Slope Protection Design*, ASCE, New York, NY, p. 85-107.

Reimnitz, E., and E. Kempema, 1982, "High Rates of Bedload Transport Measured from Infilling Rate of Large Strudel Scour Craters in the Beaufort Sea", *Continental Shelf Research*, v. 1, n. 4, p. 237-252.

Reimnitz, E., and D.K. Maurer, 1979, "Effects of Storm Surges on the Beaufort Sea Coast, Northern Alaska", *Arctic*, v. 32, n. 4, p. 329-344.

Reimnitz, E., C. Roderick, and S. Wolf, 1974, "Strudel Scour: A Unique Arctic Marine Geologic Phenomenon", *Journal of Sedimentary Petrology*, v. 44, n. 2, p. 409-420.

Tekmarine, Inc., Polar Alpine, Inc., and Offshore and Coastal Technologies, Inc., 1985, "1984 Beaufort Sea Multiyear Ice Floe Tracking Study", AOGA Project No. 280, Sierra Madre, CA 233 pp. + appen.

Sanderson, T. 1988, *Ice Mechanics: Risks to Offshore Structures*, Graham & Trotman, London, UK.

Vaudrey, K., 1983, "1982 Freezeup Study of the Barrier Island Chain and Harrison Bay", AOGA Project No. 200, Vaudrey & Associates, Inc., San Luis Obispo, CA..

Wiegel, R.L., 1995, "Shore & Beach Observations, Alaska", *Shore & Beach*, v. 63, n. 2, p. 2, 37.

CHAPTER 342

EXPERIMENTAL STUDY ON DEFORMATION AND FRACTURE OF ICE SHEET BY PROPAGATING WATER WAVE

Shigeki Sakai[1], Xiaodong Liu[2], Makoto Sasamoto[3]
Shigeo Kanada[4], Koh Izumiyama[5]

ABSTRACT

A criteria of ice sheet fracture by propagating waves is proposed based on an elastic bending theory. Experiments with a model ice are carried out to examine the validity of the criteria. Comparisons with experimental results show that the proposed criteria is available for a wide range of ice properties. The crack span induced by waves is between 1/4 and 1/2 of wave length in ice.

INTRODUCTION

In the sub-Arctic sea, wave-induced motion of ice is a dominant factor causing many coastal problems; e.g. damages to structures by ice impact forces, or beach erosion by ice drift. To estimate the wave energy inside the ice zone, it is necessary to understand how waves transform under the ice cover. Previous studies show that the presence of ice causes wave energy attenuation, and the attenuation rate depends on the ice floe size and concentration. If some floes are fractured by waves, the distribution of floe sizes will be changed, and then the effects of the ice zone on waves will also vary. Therefore, for estimation of wave energy inside an ice zone, it is important to consider the ice deformation and fracture caused by waves.

---

1) Associate Professor, Department of Civil and Environmental Eng., Iwate University, 4-3-5 Ueda, Morioka, 020 Japan
2) Graduated Student, Ditto
3) Research Associate, Ditto
4) Researcher, Ship Research Institute, Ministry of Transport 6-38-1 Shinkawa, Mitaka, Tokyo, 181 Japan
5) Senior Researcher, Ditto

Ice-wave interaction has been mainly investigated by field observations, however it is very difficult to observe the deformation and fracture of ice since these occur in stormy sea conditions. A few experiments have been carried out, while in most of them plastic sheets were used as a model ice, and the effects of the difference in material properties between ice and plastic have not been examined. Furthermore, ice fracture cannot be simulated in the experiment with plastic sheets. In the present study, experiments with a model ice were carried out to examine wave-induced deformation of ice sheet and to verify the availability of the ice fracture criterion described hereafter.

EXPERIMENTS

Experimental Equipment and Procedure

Experiments were carried out in an ice tank which is 35m long, 6m wide and 1.8m deep, as illustrated in Figure 1. Water in the tank was doped with a solution of propylene glycol to weaken the ice which would be formed on the water surface. It took 10 hours or more to form an ice sheet of 25mm thickness and another 10 hours to set up experimental equipments, to test material properties and to observe the ice sheet deformation and fracture. Totally, 2 days were needed to perform one experiment, and one more day was necessary for the experiment with the ice sheet of 50mm thickness. The bending strength and the elastic modulus of ice were measured by a cantilever beam failure test and Plate Deflection Method, respectively.

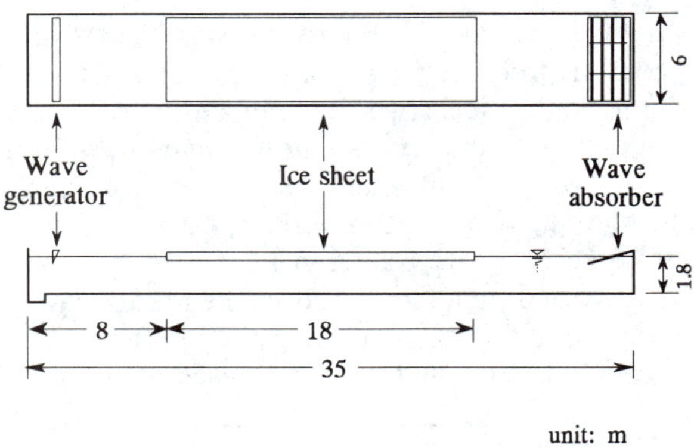

Figure 1 Experimental equipment

The wave was generated by a plunger type wave generator. At the far end of tank, a wave absorber was set up to decrease wave reflection. The vertical displacement of the ice sheet was measured at several points along the center of the tank, by using laser sensors and ultra-sonic sensors. Each experiments started with a very small incident wave height. If the ice sheet fracture was not observed in this condition, another experiment with a little larger wave height was carried out. This procedure was repeated till the ice sheet fracture occurred. After the occurrence of the first fracture, the above procedure was also continued to observe proceeding fracture of ice sheet. The conditions of the incident wave and the material properties of ice are listed in Table 1.

Table 1 Experimental condition

Case No.	Wave Period T(s)	Wave Height in Open Water H_0 (mm)	Ice Thickness h_i (mm)	Bending Strength of Ice σ_t (kPa)	Elastic Modulus of Ice E (MPa)
1	1.19	20.5～30.2	49.2	35.1	21.8
2	0.80	20.6～26.1	25.2	22.6	5.3
3	0.79	12.3～32.0	28.4	36.0	31.4
4	0.80	18.9～32.2	53.3	39.7	29.3
5	0.99	29.7	53.3	39.7	29.3
6	1.40	20.6～27.7	53.3	39.7	29.3
7	1.20	27.7～29.7	54.2	38.1	34.3
8	1.40	14.3～26.7	23.2	119.3	125.4
9	1.20	17.9～23.8	24.3	137.4	202.5
10	1.00	18.4～26.0	23.2	87.9	201.5

Experimental Results

As the incident wave penetrates into the ice-covered water, the wave height changes significantly near the ice edge and attenuates gradually under the ice. Figure 2 shows the effects of incident wave height on the change of

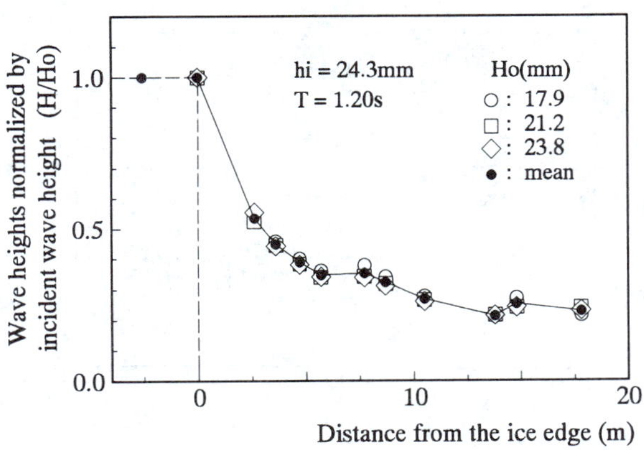

Figure 2 Wave height under ice sheet

wave height under the ice. The wave heights in this figure were normalized by the wave height in the open water. This ratio does not contain the effects of reflected wave from the ice edge since the wave height in the open water were estimated by subtracting reflected wave energy from incident wave energy. Generally, non-linear effects appear clearly with increasing of wave steepness, however there cannot exist steep waves under ice sheet since such waves cause ice sheet fracture. Figure 2 indicates that there is no effect of wave steepness on the wave height change under ice sheet in the condition where the ice sheet fracture does not occur. So, it can be concluded that the non-linearity of waves under ice sheet is to be negligible. Figure 3 gives an example of wave profile under ice sheet. The measured profile is described by a trigonometric function.

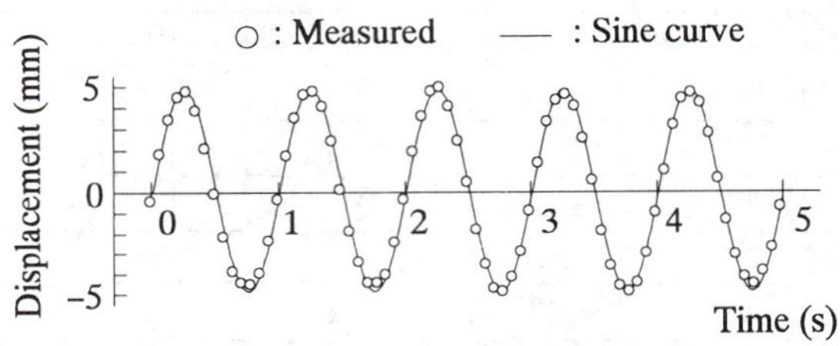

Figure 3 Example of wave profile under ice sheet

The wave celerity under ice sheet differs from that in open water and is almost constant over ice sheet, as shown in Figure 4. The non-linear effects of wave steepness on wave celerity is negligible, as well as on wave height change. A linear theory for waves under elastic plate gives a good explanation to experimental results of wave celerity under ice sheet(see Figure 5).

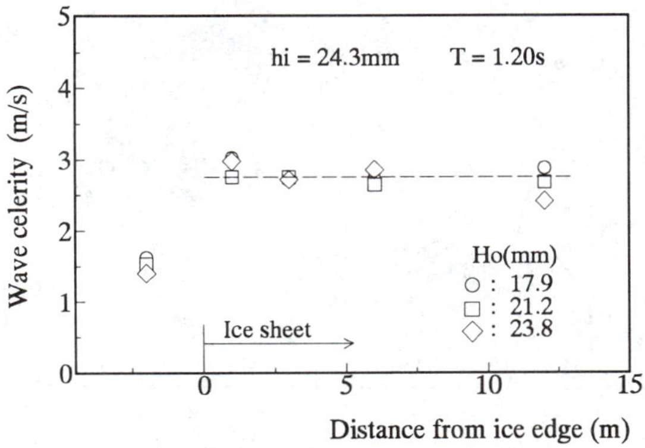

Figure 4 Wave celerity under ice sheet

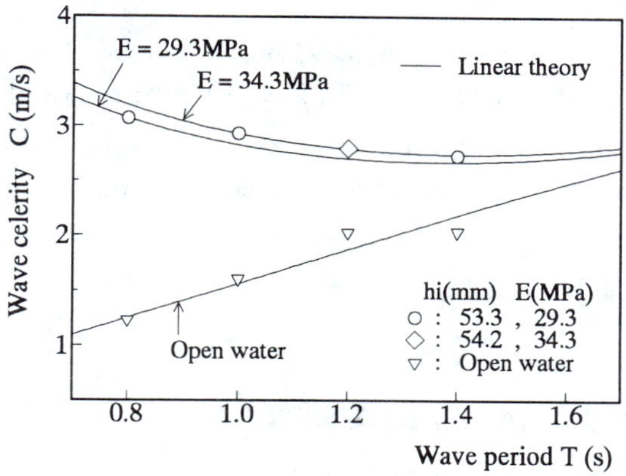

Figure 5 Comparison between measured and calculated wave celerity

THEORY

Ice sheet fracture is to be analyzed based on an elasto-plasticity fracture mechanics. However, it will be helpful for simple theoretical analysis on ice sheet fracture to consider how the elasticity or the plasticity of ice influences on ice-wave interactions. Liu and Sakai(1996) compared the numerical calculation of wave under elastic plate with experimental results of waves under elasto-plastic plate. Their comparisons showed that the calculated wave height near the edge agrees with experimental data fairly well and that the calculated values become somewhat larger than measured one as waves penetrate deeper in the plate. This discrepancy might be brought by a lack of plasticity effects in the numerical calculation. Wadhams(1973) explained the attenuation of wave height under ice sheet by energy loss due to a time dependent plastic strain (creep), and showed that this theory gives a good explanation for field observations of wave height attenuation in sea ice. These studies indicate that the deformation of ice sheet by waves is essentially elastic and the plasticity of ice cannot be negligible with increasing of penetration. Generally, ice sheet fracture by waves occurs near the ice edge. Therefore, it is reasonable to regard ice sheet fracture as a results of elastic deformation induced by waves.

An elastic bending theory gives a stress on the ice sheet surface as;

$$\sigma = \frac{Eh_i}{2(1-\nu^2)}\frac{\partial^2\zeta}{\partial^2 x} \tag{1}$$

where E is elastic modulus of ice, h_i is ice thickness, ν is Poisson's ratio(=0.3), ζ is wave profile of ice sheet and x is direction of wave propagation.

As shown in Figure 3, a wave profile is expressed by a trigonometric function as follows;

$$\zeta = \frac{H}{2}\sin(kx - \omega t) \qquad (2)$$

where H is wave height in ice, k is wave number($= 2\pi/L_i$, L_i is wave length in ice) , ω is angular frequency($= 2\pi/T$, T is wave period) and t is time.

When the bending stress on the surface reaches the bending strength σ_f, ice sheet fracture occurs. Therefore, a fracture criterion is expressed as follows;

$$\frac{Hh_i}{L_i^2} = \frac{1-\nu^2}{\pi^2}\frac{1}{E/\sigma_f} \qquad (3)$$

VALIDITY OF ICE FRACTURE CRITERIA

The measured wave height is not accurate since the ice separated from the main part of ice sheet moves as an ice floe, not as a continuous ice sheet, and this floe dissipates wave energy penetrating into the ice sheet. The wave height can be estimated from the wave height in open water and the distance of fracture point from the ice edge, since the ratio of wave height in ice to that in open water does not depend on the incident wave height for a given wave period, as illustrated in Figure 2. Another quantities in Equation (3) were directly measured in the experiment. The wave length in ice can be also calculated from a linear theory for wave under elastic plate, which gives a good estimation for wave celerity, as shown in Figure 5.

Figure 6 shows that the ice fracture criteria proposed herein is available over a wide range of the ratio of elastic modulus to bending strength of ice. This agreement of the criteria based on an elastic bending theory with experimental results also ensures the applicability of the elastic bending theory for analysis on ice fracture by waves.

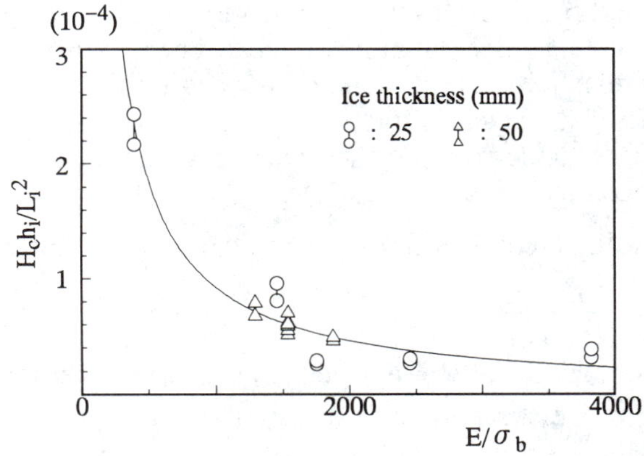

Figure 6 Ice fracture criteria compared with experimental results

CRACK SPAN IN ICE

The cracks in ice are almost perpendicular to the direction of the wave propagation, and the crack span is within 1/4 and 1/2 of the wave length under the ice sheet, as shown in Figure 7. The splits of ice were fractured laterally, resulting in rectangular pieces of almost same size. This regular pattern of ice fracture is also observed in the actual sea ice.

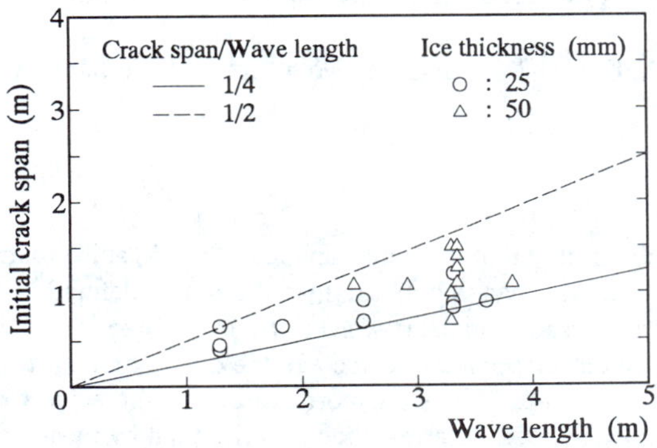

Figure 7 Relation between crack span and wave length in ice

REFERENCES

Liu, X.D. and Sakai, S. (1996): Analysis on the Interaction of Waves with Flexible Floating Structure by BE-FE Combined Method, Proc. of 25th I.C.C.E., (in print)

Wadhams, P. (1973): Attenuation of Swell by Sea Ice, J.G.R., Vol. 78, No. 18, pp. 3552-3563

CHAPTER 343

An Attempt to Determine the Spanish Public Domain Border

R. Medina[1], I.J. Losada[1], F. Méndez[2], A.J. Fernández[3]

Abstract

In this paper the problem of the determination of the Spanish maritime-land public domain border is analyzed. The Spanish public domain border is defined by the Spanish Law as the limit where the waves reach during the worst storm in the area. The law also establishes that this limit must be calculated taking into account the effects of the astronomical tide, the storm surge and the wave run-up. Since the storm surge and the wave run-up are random variables, the extreme flooding level determination is carried out by means of an extension of the joint probability method, first proposed by (Pugh and Vassie, 1978), which includes the effect of the wave run-up. The proposed theoretical approach is applied to a case study of a beach located in the North Coast of Spain.

Introduction

In Spain not only the sea, but also the shoreline is considered to be public property. This consideration is linked to our cultural and legal tradition and can be found in the ancient Roman and medieval Spanish Law.

In the 6th century, for instance, Justiniano established that: *"According to the natural law, the air, the running water, the sea and its shore are common to everybody. These shores have no owner..."*. Later, in the 13th century King Alfonso X, also known as Alfonso "The Wise", ruled that: *"The things that communally belong to all creatures living in this world are: the air, the rainwater, the sea and its shore. No building can be undertaken on the seashore that could restrain the communal use of the people"*.

The present Spanish Legislation on Coasts, passed in 1988, also protects the shore-line and defines what is called the Maritime-Land Public Domain: "The public domain stretches as far as the waves reach (maximum flooding level) during the worst storm in the area, and it includes beaches, dunes, cliffs, swamps and other low wetlands".

No type of trade activities can be allowed in this property. The Legislation on Coasts extends its influence to the private lands adjacent to the public property, Fig. 1,

[1]Associate Professor, [2] Graduate Student, Ocean & Coastal Research Group. Universidad de Cantabria. Dpto. Ciencias y Técnicas del Agua y del Medio Ambiente. Avda. de los Castros s/n. 39005 Santander

[3] Formerly, State Coast Office. Ministry of the Environment. Paseo de la Castellana, 67. 28071 Madrid.

to avoid any use carried out on those areas which could be harmful to this space of high environmental value, (M.O.P.T.M.A., 1994). Therefore, in order to implement this policy, the Ministry of Public Works has undertaken a series of programs of actions with the first being the need to complete a precise indentification of the public domain maritime-land assets in accordance to the definition set down in the Legislation on Coasts. This process has given rise to the current Property Survey Plan, whose purpose it is to fix public domain boundaries throughout the whole of the Spanish coast. This approach to the management of the coast in an integral manner has to embrace all territorial and political levels of government such as central government, states and city councils, as well as any individual with private interest on the coast.

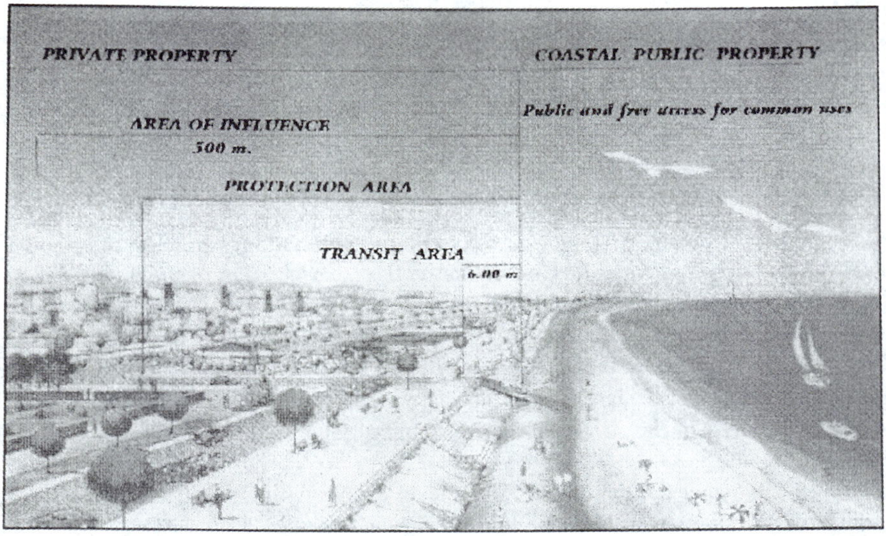

Figure 1. Public and private uses of shoreline

For this reason, the determination of the Spanish maritime-land public domain border is an extremely important element in the Program of Coastal Actions initiated by the Spanish Government, and in the last years it has become a key reason for many law suits between the different agents with interest in the coastal area.

Public Domain Border Determination

In order to determine the limit where "the waves reach during the worst storm in the area", the 1988 Law established the maritime and meteorological dynamics that must be taken into account. In particular, the law established that the calculation of the maximum flooding level must account for the effects of the: (1) astronomical tide, (2) storm surge and (3) wave run-up, Fig. 2. Episodic events such as tsunamis must be omitted from the calculation, and the influence of long term effects such as sea level rise is not explicitly considered in the law.

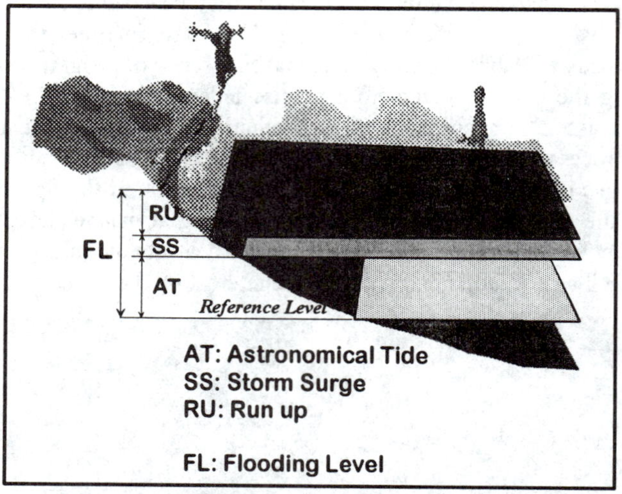

Figure 2. Dynamics to be considered in the determination of the public domain border

Since the storm surge and the wave run-up are random variables, the flooding level is also a random variable. Consequently, the determination of the public domain border is a stochastic extreme problem.

One of the consequences of being a stochastic problem is that there is not a deterministic "limit where the waves reach during the worst storm" but each level will have "a probability of being exceeded during a given storm". This probabilistic aspect is not considered in the law in which the public domain border is thought of as a deterministic line.

In this paper the extreme flooding level determination problem is analyzed by means of the joint probability method, (Pugh and Vassie, 1978), which has to be modified to include the effect of the wave run-up. The theoretical approach proposed to determine the public domain border is then applied to a case study in a beach located in the North Coast of Spain. In this case study, the probabilistic problem is also analyzed and a maritime-land public domain border is proposed.

Theoretical Aproach

The existing methods for estimating the distribution of extreme sea-level at a site, using field data, can be divided into: (1) Direct Methods - in which extremes of the observed water level are analysed-, and (2) Indirect Methods - in which the constituent processes (i.e. tides, surges and waves) are modeled separately and the extreme water level is inferred.

Direct Methods

The direct methods are, mainly, the classical annual maxima approach and the r-largest annual events method.

Annual Maxima Method: The best known, simplest, and most widely used method of analysis of extreme distributions is the annual maxima method, (Gumbel, 1958). This approach has been previously used by several investigators to determine extreme sea-levels, eg. (Lennon, 1963), (Suthons, 1963) and (Graff, 1981).

According to (Tawn and Vassie, 1991), this method is highly inefficient in its use of data and provides inaccurate results. These authors remarked that the assumptions made in using the annual maxima method are namely that hourly sea level heights are (1) independent, (2) identically distributed, and (3) that the number of hours in a year is large enough for the asymptotic approximations to hold. However, because of the nature of the tide and the dependence on the surge sequence, it is clear that (1) and (2) do not hold.

R-largest Annual Events Method: Several proposals have been made to extend the annual maxima method to incorporate all independent extreme sea-level observations into the estimation of the annual maxima distribution: (1) the peaks over thresholds method, (Davison and Smith, 1990); (2) the r-largest events method, (Tawn, 1988); and (3) the point process method, (Smith, 1989).

These methods are based on extreme value limit theory for stationary random sequences and, consequently, it is assumed the r-largest annual extreme levels are from stationary data. This assumption is critical since, due to the tide, the water level is highly non-stationary. According to this argument, the r-largest annual events method is not appropriate to determine extreme sea-levels, (Dixon and Tawn, 1994).

Indirect Methods

The existing indirect approaches, the joint probability method (Pugh and Vassie, 1978) and the revised joint probability method, (Tawn and Vassie, 1991) are the only viable options for estimating extreme levels when short data sets are available.

The main advantage of these methods is to exploit our knowledge of the tide in short data sets to which the annual maxima method could not be applied. Basically, these methods seek the probability distribution function (p.d.f.) of the different variables involved and calculate the sea level p.d.f. by combining those p.d.f.s The nature of the combination of the p.d.f.s depends on whether there is dependence between the variable sequences.

The joint probability method assumes that extreme hourly water levels are independent. Clearly hourly water levels are not independent, but extreme levels may be only weakly dependent. (Tawn and Vassie, 1989) studied this feature and found that the assumption was false but led only to small overestimation of return levels.

In the revised joint probability method the temporal dependence of the variables is considered to be a parameter, called extremal index, θ, which is determined empirically from the data.

Although the revised joint probability method results hold for temporal dependence between variables, the existing records are not long enough to have data in the regions of both extreme tides and surges and, consequently, the external index determination is, many times, impossible.

From the previous analysis it can be concluded that direct methods cannot be used to determine extreme sea levels due to the lack of long records. Furthermore, the revised joint probability method also needs a long enough series in order to estimate θ. Consequently, the only methodology capable of giving reasonable results to the problem under study is the joint probability method.

The Joint Probability Method (J.P.M.)

This method was first introduced by (Pugh and Vassie, 1978) to calculate extreme sea-levels due to the combined effect of tides and storm surges. At any time the observed sea level, after averaging out surface waves, has three components: mean sea level, tidal and storm surge level. Using standard methods, the first two of these components can be removed from the sea level sequence leaving the surge sequence.

Because the tidal sequence is deterministic, the p.d.f. for all tidal levels can be generated from tidal predictions, while the p.d.f. for the surge level is determined from the nontidal residual.

The p.d.f. of the sea level is finally calculated using the tidal and surge p.d.f.s'. Assuming that the tide and the surge can be considered independent, then

$$p(z) = \int_{-\infty}^{\infty} p_T\,(z - y)\, p_S(y)\, dy$$

where $p_T(z - y)$, $p_S(y)$ and $p(z)$ are the probability density functions for tide, surge and sea levels respectively.

Wave-Surge Joint Modelling

As stated previously, wave run-up is an important factor that must be included when determining the maximum flooding level on a beach. The inclusion of wave run-up leads to a two stochastic variables problem, surge and wave run-up, extending the J.P.M. developed by (Pugh and Vassie, 1978), which considers the surge only.

This new problem is one of bivariate extremes and the extrapolation of the distribution tail is required for both the surge and the wave run-up. Furthermore, the dependence between the extremes of theses variables must be accounted for.

If surge and wave run-up are related, the probability density function for the sea level (flooding level) can be written as:

$$p(z) = \int_{-\infty}^{\infty}\int_{-\infty}^{\infty} p_T\,(z - y - x)\, p_{S,Ru}(y, x)\, dx\, dy$$

where $p_{S,\,Ru}(y, x)$ is the combined p.d.f. of surge and wave run-up.

If surge and wave run-up are independent, then

$$p(z) = \int_{-\infty}^{\infty}\int_{-\infty}^{\infty} p_T(z - y - x)\, p_S\,(y)\, p_{Ru}\,(x)\, dx\, dy$$

It should be noted that the availability of wave run-up data for a particular beach is usually scarce and, consequently, wave run-up p.d.f. must be inferred from wave data and wave run-up theory.

The study of wave run-up on artificial and natural beaches has received considerable attention by researchers over the last several years; e.g. (Guza and Thornton, 1981), (Holman, 1986), (Nielsen and Hanslow, 1991), (Holland and Holman, 1993).

In this study, (Nielsen and Hanslow , 1991) results for wave run-up distribution on a natural beach are used. Using run-up distributions measured on a wide spectrum of sandy beaches, these authors showed that, for a sea state, the Rayleigh distribution is a reasonable statistical model for the maximum level reached by individual waves. This result was also obtained theoretically by (Battjes, 1971) for the special case of perfect correlation between wave height, H and wave period, T. That is,

$$F(R_{wm} \geq R) = \exp\left[- \left(\frac{R - R_{100}}{L_{Rwm}} \right)^2 \right]$$

where R_{100} is the highest level transgressed by 100% of the waves and L_{Rwm} is the vertical scale of the distribution.

For extreme wave conditions, the beach can be assumed to be dissipative and, $L_{Rwm} \approx 0{,}05\ (H_{rms}\ L_o)^{0,5}$ where H_{rms} is the mean squared root wave height and L_o is the deep water wave length. Notice that for a typical relationship $T_S = \alpha_1 H_S^{0,5}$, $L_{Rwm} \approx \alpha_2 H_S$, where α_1 and α_2 are two constants, depending on the wave climate of a given region.

Notice that the wave run-up p.d.f. is a Rayleigh distribution that depends on a random variable H_S, which has its own p.d.f. (e.g. Weibull), consequently the flooding level p.d.f. can be written as:

$$p(z) = \int_{-\infty}^{\infty} \int_{-\infty}^{\infty} p_T(z - y - x)\ p_S(y)\ p_{Ru}(x)\ dx\ dy$$

with

$$p_{Ru}(x) = \int_{-\infty}^{\infty} p^*_{Ru}\ (x, H_s)\ p_{H_S}(H_S)\ dH_S$$

p^*_{Ru} ( ) where is the Rayleigh p_{H_S} ( ) p.d.f., and is the significant wave height p.d.f.

Application

In this section, the wave-surge joint modelling is applied to a beach located on the North Coast of Spain.

Study Site

The study site is Oyambre Beach, located close to the city of Santander, on the Cantabrian coast of Spain, Gulf of Biscay, Fig. 3.

The northern coast of Spain is divided into a series of pocket beaches and small inlets isolated between rocky headlands. Most of the headlands extend into deep water and appear to be effective in confining littoral sand to the embayments. Therefore, the coast can be analyzed as a series of littoral cells.

One of these littoral cells is the beach of Oyambre. The beach is bound westward by Cape Oyambre and eastward by Cape El Moro. Several reefs can be found in intermediate water as part of Cape Oyambre along the northwest part of the beach (e.g. La Molar Shoal, San Francisco Shoal, Fig. 3).

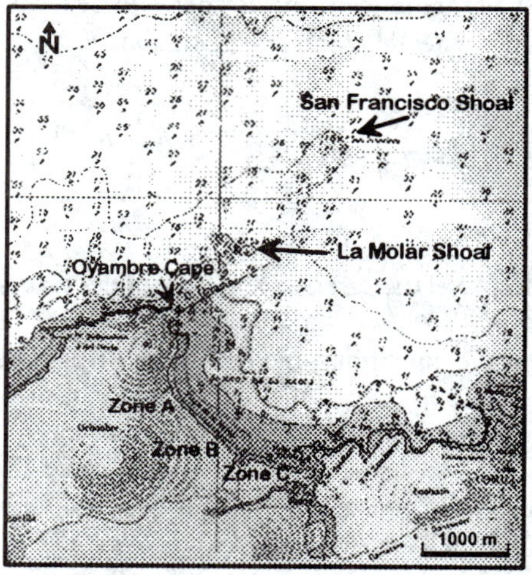

Figure 3. Location Map

More than three-quarters of the deep-water waves approach Oyambre Beach from the northern - northwestern sector. The annual average significant wave height is about 1 m with typical winter storms waves of $H_S \sim 4 \div 6$ m. Tides at Oyambre Beach are semidiurnal with a mean tidal range of 3 m and spring tidal range of 5 m.

Maritime and Meteorological Data

In order to apply the methodology presented in section "Wave-Surge Joint Modelling", different data sets of waves and sea-level were analyzed. These data sets are continuously recorded by the Spanish Network for Maritime Recording and are presented in several publications from the Ministry of Public Works, Transportation and Environment, called R.O.M., (Losada et al, 1996).

Sea-Level Data

Sea-level data used in this study were recorded at Santander. This sea-level data set consists of hourly sea-level measured during the last five years and include three additive components: mean sea-level, tidal, and surge level.

Standard tidal analysis, (Godin 1978), is used to separate the astronomical tide and the surge from the sea-level data. The astronomical tide distribution shown by a probability density function in Fig. 4, is bi-modal. This kind of distribution is typical of semi-diurnal dominated tides, with the modes ocurring at levels corresponding to mean high and low water tides.

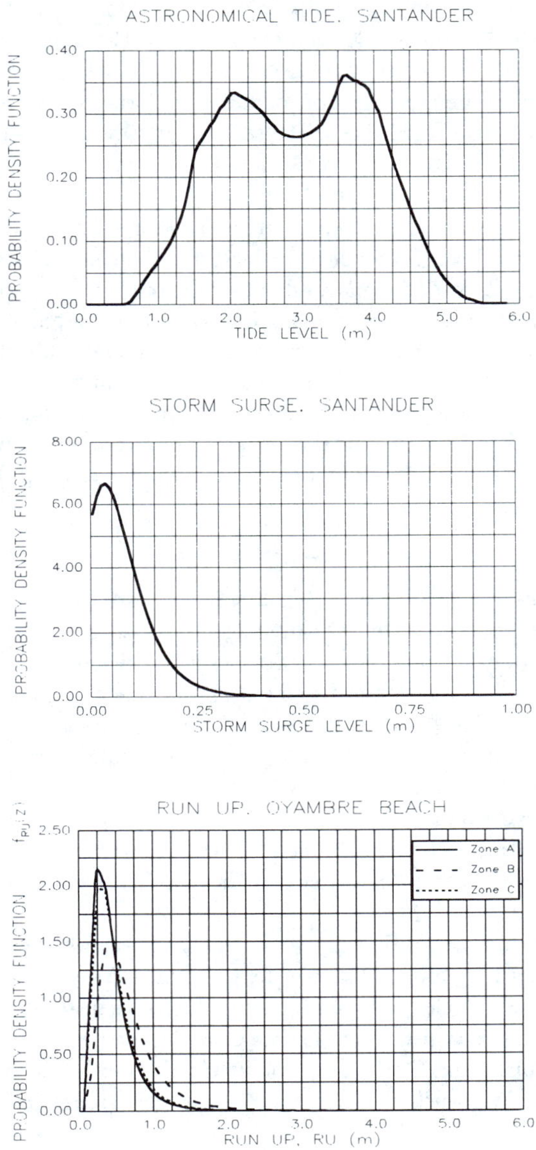

Figure 4. Tide, surge and wave run-up p.d.f.'s

The surge series is defined as the residual of the tidal series; i.e. the difference between the observed sea-level and the predicted tide. The difference between observed levels and tidal predictions is due to meteorological forcing in the form of changes in air pressure and winds which constitutes the surges forcing.

Due to the location of the tidal gauge, inside Santander Bay and in water depth of about 10 m, most of the surge is caused by air pressure. This hypothesis was further corroborated by correlating the 5 year surge data to air pressure data. Using this correlation, a probability density function of the surge was estimated by means of pressure data from the last 50 years, Fig. 4. The surge distribution is uni-modal, with the mode close to zero, and is positively skewed, leading to large positive surge levels being more likely than large negative surges (not shown).

Wind surge cannot be determined from Santander sea-level data. Furthermore, since the variability of wind-surge from coast to coast due to local bathymetry is so high, theoretical wind-surge estimation from wind data is the only feasible way to determine wind-surge at Oyambre Beach. Different simulations were run to estimate the combined surge due to air pressure and wind at Oyambre Beach. The main conclusion was that wind surge can be considered negligible when compared to air pressure surge, especially in extreme events when a low pressure is close to Oyambre. This characteristic of the surge at Oyambre Beach is due to the kind of low pressure reaching the northern Spanish coast and the bathymetry of Oyambre Beach and cannot be generalized for other locations.

Wave Data

Wave data used in this study was recorded at Gijón (about 100 km west of Oyambre). These data are deep water measurements of wave characteristics monitored by the Spanish Network mentioned above. Wave characteristics were propagated to the study area using a numerical wave propagation model. The model used, called REF/DIF, (Kirby and Dalrymple, 1985), solves the parabolic approximation of the mild slope equation and is adequate to estimate the effect of the different shoals located northwestward the beach, Fig. 3.

Since the wave characteristics varies from point to point along the beach, a zoning was made in order to set zones in which the wave characteristics can be assumed to be homogeneous. At Oyambre Beach this classification led to three areas called A, B and C, shown in Fig. 3.

For these three areas, different combinations of wave height, H_S, and wave period, T, at deep water were propagated in order to obtain the wave height distribution at shallow water in front of zones A, B and C. Using this wave distribution and (Nielsen and Hanslow, 1991) empirical distribution for wave run-up, the wave run-up p.d.f. at the different zones is calculated Fig. 4. The wave run-up distributions are also uni-modal and positively skewed. Zone C, which is exposed, has higher run-up probabilities than zones A and B.

Maximum Flooding Level

Using the p.d.f.s shown in Fig. 4 and the Joint Probability method described above, the flooding level p.d.f. is estimated, Fig. 5. The probability that the flooding level is less than x is:

$$F(z) = \int_{-\infty}^{z} p(x)\,dx$$

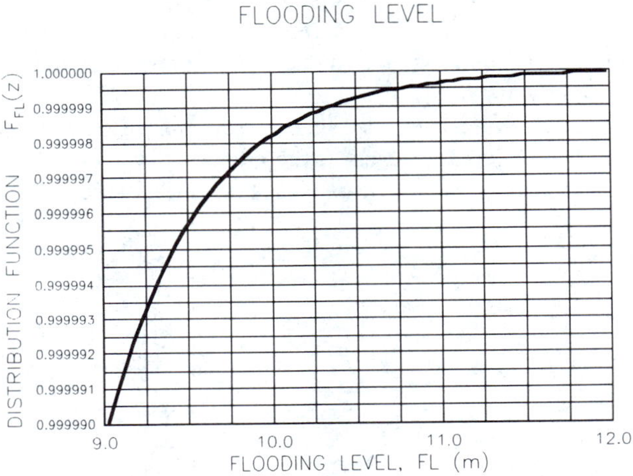

Figure 5. Flooding level p.d.f.

This distribution function *F(z)* is shown, for the maxima, in Fig. 6.

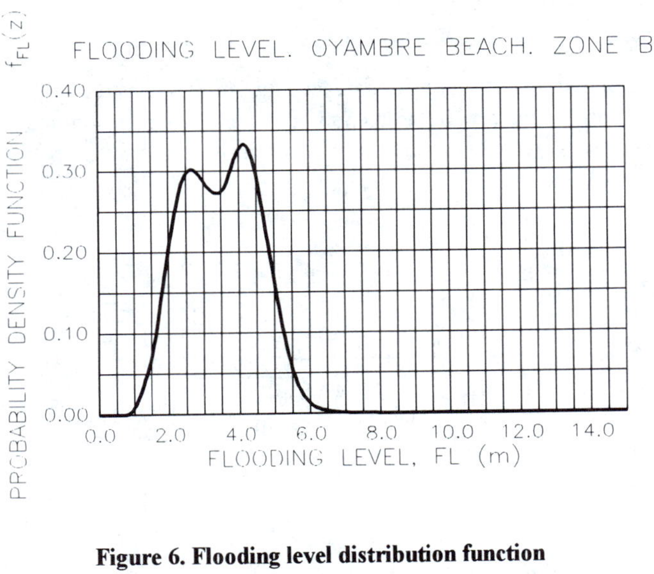

Figure 6. Flooding level distribution function

Although the probability of a level to be exceeded by the combination of tide, surge and wave run-up can be determined from Fig. 6, the application of the results to the determination of the public domain border is not straightforward since the law is not expressed in terms of probability but in terms of number of waves. ("The limit as far as the wave reach during the worst storm").

Changing flooding probability into number of waves is a straightforward technique but it adds a new random variable to the problem; i.e. the number of waves in a given period of time. This new variable can also be determined by means of the data from the wave buoy network.

In Fig. 7, the flooding level distribution, in terms of number of waves, is presented for zone B of Oyambre Beach. If the number of waves is limited to "one wave" we will obtain the limit reached by one wave in a mean year.

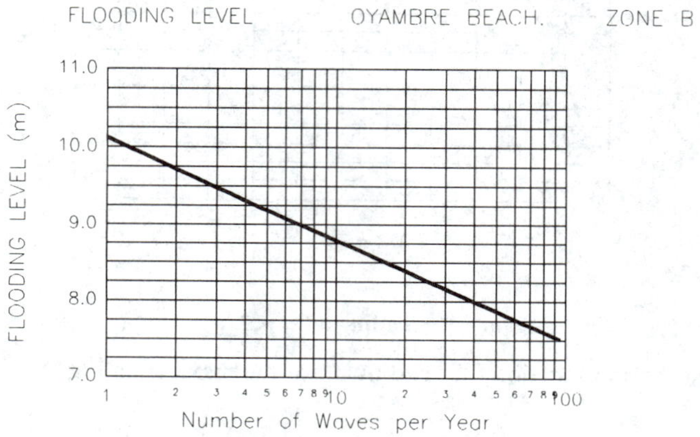

Figure 7. Flooding level distribution in terms of number of waves

Upper flooding levels will have an exceeding probability average less than "one wave per year" but they can be exceeded if we increase the number of years of observation. In this way we can determine the return period (years) of a level to be exceeded by one wave, Fig. 8.

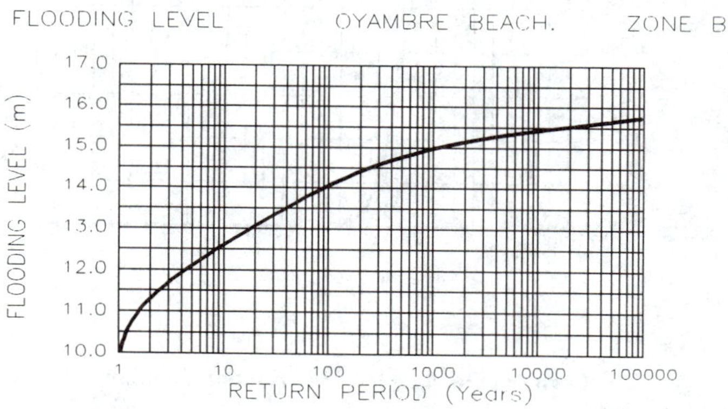

Figure 8. Flooding level exceeder by one wave versus return period (years)

Public Domain Border

From Fig. 8 it is clear that, a given flooding level will be exceeded if we have enough years of observations. At this point a criteria must be selected in order to establish the return period to be adopted in the calculation of the public domain border.

Since the Law of 1988 does not fix that criteria, the Spanish Recommendations for Coastal Works R.O.M. 02-90, (Puertos del Estado, 1990) is used. These recommendations give the designer an estimate of the minimum life and the risk to be adopted as a function of the type of maritime work and the security level demanded. With these two parameters, the return period can be calculated. For a public domain border determination, a mean return period of 500 years has been estimated.

Conclusions

This paper presents a methodology to determine the Spanish public domain border, which is the theoretical line that separates the public and the private properties in the Spanish coastal zone.

The methodology is based on the joint probability method (J.P.M.) first introduced by (Pugh and Vassie, 1978) to estimate extreme sea-levels due to the combined effect of tides and storm surge. In this paper an extension of the J.P.M. has been developed in order to include the wave run-up on a beach.

To account for the probabilistic character of the problem a methodology is presented consistent with other Spanish coastal works recommendations. The given methodology provides an engineering tool to determine the public domain border based on available field data and on hand numerical models.

Acknowledgements

The authors would like to acknowledge the funding provided by the Dirección General de Costas (Ministry of the Environment) and the Comisión Interministerial de Ciencia y Tecnología (CICYT) under grant number MAR96-1135.

References

Battjes, J.A., 1971. Run-up distributions of waves breaking on slopes. Journal of the Waterways, Harbours and Coastal Eng. Div., ASCE, Vol. 97, ww1, pp. 91-114.

Davison, A.C. and R.L. Smith, 1990. Models for exceedances over high thresholds. J. Roy Statist. Soc. B., Vol. 52, pp. 339-442.

Dixon, M.J. and J.A. Tawn, 1994. Extreme sea-levels at the UK A-class sites: Site-by-site analyses. Proudman Oceanographic Laboratory internal document N° 65.

Godin, G., 1991. The analysis of tides and currents. In tidal Hydrodynamics. Bruce B. Parker, ed. John Wiley & Sons, Inc. pp. 675-709.

Graff, J., 1981. An investigation of the frequency distributions of annual sea level maxima at ports around Great Britain. Estuarine, Coastal and Shelf Science, Vol. 18, pp. 389-449.

Gumbel, E.J., 1958. Statistics of extremes. New York: Columbia University Press.

Guza, R.T., and E.B. Thornton, 1981. Wave set-up on a natural beach. Journal of Geophysical. Research, Vol. 86 (C5), pp. 4133-4137.

Holman, R.A., 1986. Extreme value statistics for wave run-up on a natural beach. Coastal Eng. Vol. 9, pp. 527-544.

Holland and R.A. Holman, 1993. The statistical distribution of swash maxima on natural beaches. Journal of Geophysical Research, Vol. 98, N° C6, pp. 10271-10278.

Kirby, J.T. and Dalrymple, R.A., 1993. Combined refraction/diffraction model REF/DIF1. Documentation and user's manual. Report N.° CACR-92-04. Center for Applied Coastal Research. University of Delaware, 106 pp.

Lennon, G.W. , 1963. A frequency investigation of abnormally high tidal levels at certain west coast ports. Proc. Instn. Civ. Engrs., Vol. 25, pp. 451-484.

Losada, M.A., R. Medina, C. Vidal and I.J. Losada, 1996.History of coastal eng. in Spain. In History and Heritage of Coastal Eng. ed. N. Kraus. ASCE, pp. 465-449.

M.O.P.T.M.A., 1994. Recovering the coast. Spanish Ministry of Works, Transportation and Environment report. NIPU: 161-93-019-2.

Nielsen, P., and D.J. Hanslow, 1991. Wave run-up distributions on natural beaches. Journal of Coastal Research. Vol. 7, N° 4, pp. 1139-1152.

Puertos del Estado, 1990. R.O.M. 02-90. Actions in the design of maritime and harbor works. M.O.P.T.M.A. Madrid. 264 pp

Pugh, D.T., and J.M. Vassie, 1978. Extreme sea level from tide and surge probability. Proc. 16th Coastal Eng. Conference, Hamburg. ASCE, Vol. 1, pp. 911-930.

Smith, R.L., 1989. Extreme value analysis of environmental time series: an application to trend detection in ground level ozone. Statist. Sci. Vol. 4, pp. 367-393.

Suthons, C.T., 1963. Frequency of occurrence of abnormally high sea levels on the east and south coast of England. Proc. Instn. Civ. Engrs., Vol. 25, pp. 443-449.

Tawn, J.A., 1988. An extreme value theory model for dependent observations, J. Hyd. Vol. 101, pp. 227-250.

Tawn, J.A. and J.M. Vassie, 1989. Extreme sea-levels: the joint probabilities method revisited and revised. Proc. Instn. Civ. Engr. Part. 2, Vol. 87, pp. 429-442.

Tawn, S.A., and J.M. Vassie, 1991. Recent Improvements in the joint probability method for estimating extreme sea levels. In tidal Hydrodynamics. Ed. Bruce B. Parker John Wiley & Sons, Inc. pp. 813-828.

CHAPTER 344

EDUCATION OF COASTAL ENGINEERS FOR THE 50th ICCE

Henk Jan Verhagen[1]

INTRODUCTION

In general the problem in education is that we teach nowadays students the knowledge from five years ago, to be applied in the next 5 to 15 years (I admit, the student of today will have reached his retirement when he will participate in the 50th ICCE, the title is exaggerating reality). When one analyses the developments in the profession, it is clear that teaching present day facts is not very useful. We have to teach approaches, design-philosophy, concepts, etc. What the meaning is for education of coastal engineers will be worked out in this paper.

CHANGES

In the past engineers only had to look at the technical aspects. Nowadays, the impact on society and environment, the costs and the sustainability are at least as important. The consequence of this is that the role of coastal engineers has changed considerably in the past few years. Recently a joint IAHR-UNESCO panel addressed the current developments in hydraulic engineering and their implications for the education of hydraulic engineers [KOBUS *ET AL.*, 1994]. Being specialized hydraulic engineers, many of the conclusions of this panel are also valid for Coastal Engineers. With reference to the UNCED '92, the panel concluded that "hydraulic engineering should be taught in a broad context of the natural and social environment and must enhance the *engineer's sense of responsibility* for sustainable water development within environmental and societal limitations."

Fortunately, also the tools for the engineers have changed considerably. In the past much was done based on experience, engineering judgement and empirical formula. Verification was usually only possible by means of measuring in prototype or by physical modelling. Worldwide research in coastal sciences and coastal engineering has shown the following two developments:

- a considerable increase in the domain knowledge (knowledge of coastal processes);

---

[1] Associate professor in coastal engineering, Int. Inst. for Infrastructural, Hydraulic and Environmental Engineering (IHE-Delft), P.O. Box 3015, 2601 DA Delft, The Netherlands

- advanced tools have been developed like mathematical models (in various dimensions) and decision support systems. For design and project development the role of the physical models is almost completely taken over by the mathematical models.

A general trend can be observed from applying rules to more conceptual thinking. Rules will change fast, so it is more important for an engineer to know the design philosophy. Also engineers have to learn where to find the most up-to-date knowledge regarding the design which he is making at a certain moment. Because the increased growth in science, rules are outdated faster than in the past; this means that in most cases engineers should not apply the rules they have learned in university. So they should be trained always first to verify if the design method they learned in university is still valid.
Engineers should be trained in a flexible application of the design methodologies they have learned.

For design purposes the use of complicated computer software is increasing. The packages become more and more user friendly, but the insight in the computational process becomes less. This means that the direct link between output and input is less obvious. Engineers have to become more and more aware of the need of checking the output of these programs in inconsistencies (rather than on numerical accuracy). Because of the high quality of presentation methods of modern software, input inconsistencies are often not recognised in time. Engineers have to be trained to become more and more keen on this problem.

TYPES OF ENGINEERS

Because of this development one may distinguish nowadays three types of coastal engineers:

- the research-engineers, involved in expanding the understanding of the coastal processes; using advanced analysis of field data and sophisticated physical models;
- the "toolmakers", engineers involved in translating the newly gained knowledge into mathematical models and encapsulate this knowledge in various knowledge based systems (Hydroinformatics, see ABBOTT [1994]);
- the design engineers (the tool-users), contributing in solving the coastal problems in a multi-disciplinary team.

The first two types of engineers produce products which can be used globally. Results from research done in Australia can easily be used in Iceland. Also (the better) tools can be used anywhere. Tools, developed in the Netherlands can easily be used in Sri Lanka. However, applying the tools for solving local problems requires the specific local knowledge of the various parameters of the system. In his keynote speech to Copedec IV, professor Hildebrando [HILDEBRANDO, 1995] gave an overview of the labour market for coastal engineers (based on Brazilian experiences), and he came to the conclusion that the demand for coastal engineers will grow in the following main fields of activities: Design, Construction, Operation and Research. He came also to the conclusion that the demand for researchers is relatively limited, compared to the other fields.

WHAT IS A (COASTAL) ENGINEER ?

An engineer is a trained person, able to solve practical problems in society (and in our case of course related to the coast), using technical means. In this paper I focus on the engineer with an academic education. From such an engineer it is expected that he is able to solve problems which cannot be solved by applying standard handbook solutions. Therefore he has to be able to understand the problem (analyze the cause of the problem), find in a creative way a path towards a solution of the problem, and implement that solution (make a "construction", however in this sense a "construction" is not necessarily concrete and steel, it may also be a beach nourishment or the implementation of a set-back system). Usually this process is called "design".

An important point is that the academically trained engineer has to bear "design responsibility". This means that he is responsible for the effect of the construction on the human and non-human environment. This includes of course the aimed effects (an erosion protection should protect against erosion), but also not non-aimed effects (an erosion protection should not increase downstream erosion without consultation with the downstream coastal manager).

So the main objective of the engineer is to solve the problem, only understanding is not sufficient. Understanding is only an intermediate step.

At this moment the process description in our profession is in many cases weaker than in other engineering professions. Many formulas used in coastal engineering are empirical or have important empirical coefficients (examples are the sediment transport formulas and the formulas for the stability of armour units). Apart from this, we usually have to work with considerably unreliable input data (e.g. wave data, waterlevels). And last but not least, the materials used in coastal engineering (rock) have a highly variable strength. The result of this all is that the main art of coastal engineering is the handling of uncertainties.

Because uncertainties are so important (and even will become more important in the future), a coastal engineer should have a good knowledge of statistics and probabilistic methods. In many engineering professions uncertainties can be handled by using a "safety-coefficient" or the use of a characteristic value (like the 95% value used in concrete engineering). However, because of the wide standard deviations in the distributions of the parameters common in coastal engineering, often only a full probabilistic approach leads towards acceptable solutions.

In coastal engineering the "client" is often the government. Especially when considering coastal protection works (notably beach nourishments), the link between the paying organisation and the beneficiaries are sometimes unclear. This makes that financing coastal protection works is sometimes quite difficult and prone to an other form of uncertainties.

SCIENCE AND ENGINEERING

This also shows the main difference between a scientist and an engineer. The objective of a scientist is to understand the world. For a coastal scientist it means understanding the coastal processes. Sometimes this understanding is expressed in models. Some scientists think they understand a system when they are able to model the system (usually with mathematical descriptions). Personally I think models can help to understand the process, but the fact that you are able to model a process does not necessarily means that you understand the system. The aim of a scientis is to make this understanding as complete as possible. Also an engineer needs understanding, otherwise he is not able to solve any problem. But his research will be focused towards the needed solution of the problem, not towards a full understanding of the process.

An important point to mention is the worldwide focus on the "pure science" in coastal engineering. Much research in coastal engineering leads to longer formulas, not to better applicable formulas, as is proclaimed by Smith [1994]. Reviewing most journals on coastal engineering, he came to the conclusion that only very few papers were focusing on engineering, while most papers were focused on coastal science.

TRAINING

At most universities emphasis is put on the study of various coastal sciences for increasing our knowledge of the coastal processes. This is very important work, that has to be continued; however the consequence is that most students are only trained for this kind of research, and much less for practical engineering. So, in fact, we are mainly training the first 2 types of engineers, and neglecting the type 3 (the design engineer). Therefore it is necessary that apart from the education of coastal researchers attention is also paid to the education of engineers, i.e. people who are able to solve practical problems and design solutions.

When one analyzes "problem coasts", in most cases the coastal engineer is confronted with several types of problems. Among others, the following types can be distinguished:

- the "obvious" mistake;
- lack of data (no accurate boundary conditions);
- no accurate design method available.

The most usual problem is the "obvious" mistake. For example, buildings constructed too near to the waterline at an eroding coastline, undesirable effects of downstream erosion due to coastal works, etc. In these cases the origin of the problem is not really an engineering problem, but absence of awareness of coastal behaviour at (political) decision making level. It is more a political problem than a technical problem. A better technical training of engineers will never solve this kind of problems, one has to create coastal awareness at the group of clients. This requires communicative skills and the ability to compose clear and understandable reports. Usually this is a weak point of technical people, therefore special attention has to be paid to this point in the education of engineers.

Also very often no detailed input data are available for making an appropriate design. Accurate long-term wave data are very scarce. This means for an engineer:

- he has to learn how to "generate" data (hindcasting, combining several sources of data, etc.)
- he has to learn to make the design in such a way that the influence of unreliable input data is as minimal as possible (for example using probabilistic methods).

In order to solve the third kind of problem (no design method available to the designer), he first has to find out, if such a design method exists somewhere in the world. This means that he should be trained in searching this type of information. And finally in those cases where no design method is available, our engineer has to perform his own research (or write an research assignment to be executed by a specialized institute). In some cases this is necessary. But in 90 % (or more) of the design cases, this is not relevant.

Having made a (far from complete) overview of academic coastal education in the world, the conclusion is that in those universities where coastal engineering is presented, it is usually a specialisation of civil (or hydraulic) engineering, offering only a few graduate courses really in coastal engineering. The most common situation is that in the final year, the hydraulic engineer can follow a few lectures in specialised coastal engineering topics, followed by a thesis on the subject. In the very few more specialized universities, in the last year of the programme more coastal topics are presented and the thesis is usually linked to the current faculty research programme of the institute. Hardly anywhere an M.Sc.-

programme is offered, focusing on coastal engineering only, and give the student the possibility to gather real engineering (design) qualifications.

For a practising engineer, working in a multi-disciplinary team, it is vital to know the physical processes instead of only knowing the mathematical approximation of these processes[2]. In education this point is often underexposed. Especially in the education of engineers from developing countries attention has to be paid to the education of the design (type 3) engineers, see VERHAGEN [1995].

GENERAL REQUIREMENTS FOR COASTAL ENGINEERS

In our institute we have 39 years of experience in giving coastal engineering courses to participants from countries which do not have a tradition in training of coastal engineers. The conclusion is that there is a great need for engineers who are able to plan, design, construct, maintain and manage coastal structures. The need for engineers with abilities to design coastal structures of a very complicated technical nature is much less. Although the interest of both universities and politicians is often different, maintenance and management of coastal structures are important and costly matters. It is important to include maintenance and management aspects already in the design phase, see VERHAGEN AND YAP [1992]. Based on our experience we came to the conclusion that for the design engineers the following requirements are the most important ones:

- good understanding of the basics of the physical processes;
- insight in the effect of reliability of data, used in the design processes (also including the ability to design a construction as less as sensitive to uncertain input data);
- select the appropriate tools for analyzing the problem and designing a solution;
- develop capabilities to organise the design work in a straightforward way, given the limited resources available;
- develop to ability to cooperate with specialists from other disciplines in a inter-disciplinary coastal zone management project;
- develop capabilities to come to practical solutions given the local constraints;
- the ability to formulate research assignments.

Because, according to our experience, the engineering qualities and the design and management skills are the most important requirements for coastal (design) engineers, we started to define the required qualities and skills, instead of a list of subjects to be taught. For the engineering qualities and the design and management skills, we have worked this out into the following general objectives:

<u>Engineering qualities:</u>
The coastal engineer:

- has a good quality engineering attitude, i.e. he is able to analyze practical problems in his profession, is able to generate solutions and is able to implement the engineering

---

[2] A clear example can be found in wave-mechanics. Often teaching starts with the dispersion equation for gravity surface waves (in fact a mathematical model description). And then, for example, refraction is explained by combining the dispersion equation with the continuity of wave crests in order to get a formula for the ray curvature. Usually this is not very clarifying for students with an engineering attitude.

solutions in a sound way; doing this the engineer is knowledgable with respect to his profession, but also aware of the surroundings and the environments of the project, and is able to organize the solving the problem and implementing the solution in a interdisciplinary environment.

- has some knowledge of technical and scientific developments in fields adjacent to his own field and is able to communicate with experts in those fields; examples for coastal engineers are: river engineering aspects, inland navigation, merchant shipping, port management, information technology, environmental sciences.
- has knowledge of modern design methodologies, like the probabilistic approach and the policy analysis approach to select alternatives, and is able to apply them.
- is aware of the fact that hydraulic engineering structures always have a large impact on the environment in its widest meaning; the engineer has to be able to make an inventory the impacts of his work with the aid of specialists in environmental sciences and to execute an environmental impact assessment.
- is able to define the required input data for his problem; he is able to judge the required accuracy of the data; he is able to compose an adequate dataset from an ambiguous dataset.
- is able to search in international literature and databases for relevant design methodologies.

design skills

The coastal engineer is able to:

- analyze coastal (erosion) problems, to generate alternative coastal protection methods (including "soft" methods), to select a protection method, to design the protection and to supervise the construction of the protection.
- protect low-lying coastal areas from flooding by means of the construction of dikes and revetments; he is able to determine the functional requirements of the protection, to generate alternative flood protection methods, to select a protection method, to design the protection and to supervise the construction of the protection.
- evaluate the necessity and to design closure dams, causeways, etc. in a coastal, tidal environment; he is able to determine the functional requirements of the dam, to generate alternative lay-outs and construction methods, to select an alternative, to design it and to supervise the construction of the dam.
- design the entrance to a harbour, i.e. the design of the access-channel and the design of the breakwater; he is able to determine the design boundary conditions, to generate alternative solutions, make a selection, make the detailed design and supervise the construction.
- design mooring facilities, both at offshore jetties as along quay-walls, he is able to develop and select alternatives and is able to design the structure; he is able to specify the boundary conditions for quay-wall and jetty construction; he is able to design a jetty and a quay-wall.
- design earth structures and foundations in the coastal zone (dikes, sheetpiles, soil improvement techniques), the coastal engineer has a workable knowledge of pile foundations.

managerial skills

The coastal engineer:

- is able to organize and cooperate in a multi-disciplinary group to set-up and implement an integrated coastal zone management plan.

- has specific knowledge of construction methods specific for coastal engineering, i.e. wet and dry earth moving; related to dredging he has to be able to define the working method, to know the possibilities of the industry, to prepare terms-of-reference and contract and price estimate; he is able to select the contractor and to supervise the work.
- is able in case of environmental emergencies in the coastal zone (oil-spills, hurricanes, etc) to organize the appropriate actions to mitigate the impact.
- has some knowledge regarding the organization of the profession, the international institutions, the organization of large contractors and consultants.

Having defined these more general qualities and skills, the next step is to define the required knowledge for individual subjects, like "design of breakwaters" or "short waves". Only those aspects of the several subjects should be taught, when they contribute to the general qualities and skills, as defined above. If one neglects this point, the result is usually a programme which consists of a set of hobbies of the teachers. Because of the limited scope of this paper, this step will not be worked out.

ATTAINMENT TARGETS

Of course these objectives, but also a list of topics within a series of lectures gives no information regarding the level, depth or orientation of the subject. Often one sees in curricula of universities descriptions like:

Short waves

linear and non-linear regular gravity waves, wave properties and transformation in shallow water, forecasting and hindcasting of waves.

In fact, such a description gives not very much information. For example, after having followed this course, is the student able to derive refraction equations, is he able to make a refraction calculation by hand, is he able to interpret the results of a refraction calculation by a computer programme, is he able to define whether a refraction calculation is necessary or not, and if yes, what are the input variables ???

Therefore it is more appropriate to describe as detailed as possible the attainment targets a student has to meet in order to get his diploma. For "refraction", such targets could be described as follows:

- *Know the phenomenon of refraction, be able to calculate refraction in case of parallel depth contours, using Snell's law, be able to determine refraction in a qualitative way in case of non-parallel depth contours, know the advantages and problems of ray-refraction programs, be able to interpret two-dimensional wave penetration programs.*

Attainment targets can be worked out for all course objectives, mentioned above. As an example the attainment targets for the design capabilities are given below:

Attainment targets for design capabilities

- be able to translate a vague (political) question into exact requirements and formulate them in terms-of-reference for a study to be performed
- be able to analyze the problem on main aspects and distinguish between essentials and matter of secondary importance
- to be able to analyze the functional requirement of the construction needed, in order to determine design criteria

- be able to collect data for a (real world) project from the general available data sources (like libraries, information retrieval systems, already existing consultancy reports)
- be able to analyze the above mentioned data and to be able to use them in order to come to a complete, relevant and unambiguous set of boundary conditions (including the reliability of the given figures)
- be able to determine scenarios to describe unknown future effects
- be able to determine alternative designs
- be able to make global designs of the alternatives, in order to make a selection; be able to make a choice between the various available computational and design methods; be able to apply these methods
- evaluate the designs on at least the following aspects
  - costs
  - environmental acceptance
  - social acceptance
  - maintenance cost
  - robustness of the design
  - executionability
- be able to read technical drawings, to be aware of the effects of distorted scales (especially in coastal profiles)
- be able to make a bill of quantities given the design drawings
- be able to make a set-up for an Environmental Impact Assessment for the design
- be able to write a systematic report on the design process followed
- be able to write readable well-structured reports, with clear set-up, and with a good quality lay-out.
- be able to organize the design work, define various subtasks and have these subtasks executed by various group members
- be able to make a time planning for a design work, and be able to produce the design report; i.e. to deliver the report (1) in time, (2) with the subjects as determined in the original plan, (3) with the required quality; the sequence of these points is important
- be able to present the results of a study or design work orally in a short period (10-30 minutes) for various types of audiences (such as technical experts, decision makers, financial experts); be able to make good quality overhead sheets (both from a viewpoint of contents as lay-out)
- be able to gather additional skills and knowledge on a rather specialistic field (not treated in detail in the teaching programme) in order to solve the design problem.

THE MASTER DEGREE

The engineering study is usually completed with a Master degree. Unfortunately there is not really a world-wide standard for a master degree. Especially for students from developing countries this is a real problem. For most of them their career depends completely on the fact if they have paper indicating a Master degree, and not on their knowledge or skills. So in fact it is for them very unattractive to follow a high quality course. Usually these courses are long (and therefore costly), with a high risk of failure for the candidate; so why not go to an easy course ? This strong focus on "paper" degrees is a clear problem in the developing world. But on the other hand, for an outsider it is indeed difficult to assess the quality of a diploma. Therefore the detailed described attainment targets can be very useful. In fact, by issuing the diploma, we guarantee that the candidate has reached all the described attainment targets, so a future employer can easily see what is the knowledge and skill of the degree-holder.

There is a variety of Master titles, a taught masters, a master by thesis, a master of engineering, a master of science, etc. In our institute we provide two types, a Taught

Masters and a Master of Science. The first is given to a student who has followed the complete (2000 study hours) program, has done all exercises sufficiently and who has passed all examinations. Students with sufficient high marks for the Taught Masters programme are academically admitted to the Master of Science programme, which consists of an additional individual assignment of approx. 1000 hours. This individual assignment can be a research work, but can also be a design work. Of course not all design work automatically has sufficient level for M.Sc. We use the following list of requirements for a subject for design-work at M.Sc.-level

- It has to be a complex problem, i.e.:
  - The boundary conditions are ill-defined
    The criteria are not clear on beforehand
  - There are several, complex scenarios.
- The technical solution is complicated, this means that it is not possible to apply straightforward existing design formulas.
- It is not possible to determine the best alternative by a simple (one-dimensional) computation.
- The problem has to be realistic and relevant, preferably for the home-country of the participant.

CONCLUSIONS

Because of the focus on research in most universities of the world, the training and education of design engineers is somewhat problematic. Many design engineers have been trained as researchers. In future there is a need for trained designers with understanding of coastal processes, the ability to design in a flexible way coastal "constructions" and able to communicate with others. Training programs can be improved considerably by defining the attainment targets which have to met by the students at the end of their education, and not only by defining the subjects to be dealt with in the curriculum.

REFERENCES

ABBOTT, M.B. [1994] Hydroinformatics: A Copernican revolution in hydraulics; *J. of Hydraulic Research, vol 32, extra Issue, pp 3-13*

HILDEBRANDO DE A. GÓES FILHO [1995] Profile of the new coastal and port engineering technician at the end of the 20th century; *Invited Paper, Copedec IV,* Rio de Janeiro, Brazil

KOBUS, H., PLATE, E., HSIEH, WEN-SHEN, SZÖLLÖSI-NAGI, A. [1994] Education of Hydraulic Engineers; *J. of Hydraulic Research, vol32, nr2, pp 163-181*

SMITH, A.W.S. [1994] The coastal engineering literature and the field engineer; *J. of Coastal Research, vol 10, no2, pp iii-viii*

VERHAGEN, H.J. [1995] The requirements for Coastal Engineers in Integrated Coastal Zone Management; *Copedec IV,* Rio de Janeiro, Brazil

VERHAGEN, H.J. & YAP, J. [1992] Coastal Zone Management with relation to Low Investment Solutions; *proceedings Second seminar on Ports and Inland Waterways in Developing Countries*, PIANC-PCDC; Surabaya 1992

CHAPTER 345

Risk-Based Analysis of Coastal Projects

Edward F. Thompson[1], Member, ASCE, Michael Wutkowski[2], Member, ASCE, and Norman W. Scheffner[1], Member, ASCE

Abstract

Coastal engineering usually embodies large uncertainties about attacking forces and coastal/structure response. The coastal engineer must address these uncertainties with judgement and experience, supplemented with some level of direct probability analysis. The nature and scope of risks related to coastal projects are reviewed. Analysis approaches developed for wider engineering application, where risks are usually better defined (e.g. Ang and Tang 1975, 1984; Harr 1987), may be less effective in coastal engineering. Two accepted general approaches for risk-based analysis of coastal projects are discussed. The approaches are illustrated with an example shore protection project.

Introduction

The approach for analyzing coastal projects is undergoing some fundamental changes, shifting from the traditional deterministic emphasis to a more comprehensive probabilistic, risk-based methodology. The changes strongly impact both planning and engineering phases of project formulation and design.

The changes, which can be expected to be distilled into a new standard for coastal practice by the end of the decade, are driven by several progressive developments. First, our understanding of probabilistic coastal processes continues to advance, particularly due to advances in field measurement, physical modeling, and numerical modeling.

---

[1] Research Hydraulic Engineer, US Army Engineer Waterways Experiment Station, CEWES-CR, 3909 Halls Ferry Road, Vicksburg, Mississippi 39180-6133, USA.

[2] Civil Engineer/Programmer, US Army Engineer District, Wilmington, CESAW-EN-C, P.O. Box 1890, Wilmington, NC 28402-1890, USA.

Second, standard computing capabilities are increasing rapidly, facilitating lengthy probabilistic calculations which would have been impractical in the past. Third, engineers, cognizant of limitations in the traditional approach, are often eager to implement better procedures, provided that they are well-founded and clearly improve the analysis. Finally, the public is becoming more aware and concerned about coastal project performance, and they expect realistic project analyses. In the U.S., public involvement in coastal projects is further intensified by legislation which increases the proportion of costs borne by the client (typically state or local government) in Federal projects.

Traditional vs. Risk-Based Analysis

Traditional analysis treats a coastal project in deterministic terms. The forces of nature are often represented as a design significant wave height, period, and direction, a design water level, etc. Coastal response is described as *the* response if no project is implemented, *the* response if one plan is implemented, *the* response if another plan is implemented, etc., without much formal recognition of the wide variation in possible responses.

In contrast to traditional analysis, some significant developments in probabilistic treatment of coastal projects have appeared during the 1990's. Most relate to coastal structure design (CIRIA/CUR 1991, ICCE 1992). Within the U.S. Army Corps of Engineers (CE), water resource planning guidance has moved from a deterministic to a risk-based approach, which incorporates considerations of risk and uncertainty. Similar concepts are now being adapted to CE coastal engineering studies (U.S. Army Corps of Engineers 1996).

Reasons for Risk-Based Analysis

There are a number of reasons why coastal projects in the broader sense, not just structure design, may be effectively analyzed from a risk-based point of view, as follows (Table 1):

Table 1. Reasons for Risk-Based Analysis of Coastal Projects
1) Forcing is probabilistic
2) Major uncertainties in behavior
3) Damage & functional performance change incrementally
4) Benefits & risks not fully represented in deterministic terms
5) Uncertain effects on adjacent areas

1) *Coastal forcing is probabilistic*. Wave characteristics vary greatly both over short term (individual waves) and long term (from one sea state to another). Similar considerations arise with winds, water levels, infragravity waves, and currents.

2) *Coastal engineering embodies major uncertainties*. Knowledge of both the forcing processes and coastal response usually involves major uncertainties.

Deterministic representations mask the uncertainties and can be misleading.

3) *Damage and functional performance change incrementally.* Coastal projects rarely progress from the design condition to total failure during a single storm event. Damage usually occurs incrementally. For example, damage to a rubble mound breakwater (when it occurs) typically begins during an unusually severe storm and progresses during subsequent severe storms until repairs are done. Similarly, beach fills erode incrementally in response to storms over a period of years. Coastal projects often continue to provide some measure of functional benefit even in a damaged state. A damaged breakwater continues to provide some protection from incident waves; a partially eroded beach fill continues to reduce coastal flooding risks.

4) *Benefits and risks not fully represented in deterministic terms.* Because of the above factors, positive impacts and risks of coastal projects cannot be fully represented in deterministic terms. Some projects provide benefits beyond the design configuration, which are generally ignored in traditional practice. For example, a nearshore berm which is over-built to allow for progressive deterioration provides increased coastal protection during its early life. Another example is an over-dredged entrance channel giving increased vessel access depths until it shoals to the design depth.

5) *Uncertain effects on adjacent areas.* In addition to the uncertainties associated directly with coastal projects, the projects can introduce significant possibilities for changing adjacent areas. While projects are designed with the intent of minimizing adverse impacts on adjacent areas, it is important to recognize that uncertainties and risks can increase beyond the without-project condition. In effect, a project can transfer risk from one area or party to another. When the risks of *all* major aspects of a project are represented as best they can be determined, better-informed final decisions can be made.

Experienced coastal engineers are well-aware of the concerns in Table 1. Even with deterministic methods, they can be expected to produce project plans which include a large measure of professional judgement to insure a technically successful project. However, the ultimate fate of a project can depend upon higher level decision-makers who must weigh technical concerns against economic, environmental, aesthetic, social, and political concerns. By quantifying risks, the coastal engineer can better pass his or her experience and judgement on to other decision-makers, who may not have coastal expertise.

Considerations for Including Risk-Based Analysis in Project Design

Objectives. The main objectives of adopting a risk-based analysis approach rather than a traditional approach are to explicitly identify uncertainties, provide improved information for assessing tradeoffs between risks and cost, and improve decision-making for project optimization (Table 2).

Table 2. Objectives of Risk-Based Analysis
● Explicitly identify uncertainties
● Provide improved information for assessing risk vs. cost tradeoffs
● Improve decision-making for project optimization

Key Variables. Although a large number of variables affect any coastal project, a small subset can usually be identified as *key variables*, that is variables which strongly relate to project performance. The key variables will embody the main forcing mechanisms, project sizing, and project response. For example, some of the key variables for a beach nourishment project might be significant wave height (forcing), beach fill width (project size), and erosion width (response).

Professional Judgement. Coastal engineering requires an unusually large measure of professional judgement because of the number and complexity of processes and responses involved. Analytical and modeling tools help to represent the variability affecting coastal projects, but the judgement of an experienced engineer is a vital ingredient in risk assessment and project optimization.

Resistance and Functional Performance Vary with Time. Both the resistance to damage and functional performance often vary significantly over a coastal project's design life. For example, the resistance (or structural strength) of a rubble mound breakwater may decrease in time due to deterioration of stone such as loss of angular corners, cracking, and breaking. Resistance may also decrease due to displacement of stone and exposure of underlayers to wave attack, which would also decrease protection provided by the breakwater (functional performance). For a beach nourishment project, loss of material to storms decreases the resistance of the beach to future storms. The effectiveness of the beach as a deterrent to coastal flooding is also decreased (functional performance). In some cases, resistance *increases* with time, as in the progressive growth of protective vegetation on coastal dunes and natural cementation of beach sediments rich in calcium carbonate.

Construction Season and Mobilization Concerns. Often maintenance of coastal projects requires major mobilization efforts and is confined to a *construction season* dictated by climate and environmental factors. Therefore the risk during the *interval between construction seasons* rather than during a single storm becomes a key concern. During an unusually stormy winter (such as the winter of 1987-88 in southern California), this risk can be significantly greater than that for individual storms.

Environmental, Aesthetic, Social, and Political Concerns. The role of environmental, aesthetic, social, and political factors in the ultimate planning and design of a coastal project is often at least as important as the technical engineering factors. An optimized final design includes appropriate consideration of these factors and their associated risks and uncertainty.

Frequency-Based vs. Life Cycle Approach

Risk-based analysis of coastal projects can be done by either of two fundamentally different approaches. The *frequency-based approach* deals with frequency-of-occurrence relationships among the key variables. By combining key forcing variables with various occurrence frequencies, information about the frequency-of-occurrence of key project responses can be developed. For example, a traditional stage (water level) vs. frequency curve and a stage-damage curve can be combined to generate a damage vs. frequency curve. This approach can be applied as an add-on to traditional planning and design procedures.

The *life cycle approach* deals with multiple realizations of possible evolution of the project with *time* during the span of its design life. The suite of life cycle realizations is constructed with consideration of the probabilities of key variables. For example, the realistic time variation of key forcing and response variables during a 50-yr life cycle can be generated for 1000 different possible life cycles. Uncertainty in the data and models relating natural forcing to coastal response can be represented as another source of variability. Probabilities and risks associated with the project are then compiled by analyzing project performance over the 1000 life cycles.

An example shore protection project helps to illustrate the life cycle approach in comparison to a more traditional frequency-based approach. The project area is a relatively uniform stretch of beach with several rows of houses along the shore (Fig. 1). The example project is a beach nourishment to widen the existing beach and dune (Fig. 2). The frequency-based approach used for this example is based on a set of six storms representing 5-, 10-, 25-, 50-, 100-, and 500-year events. Expected erosion of the existing or project beach profile and property damages are calculated for each storm. An average annual damage is calculated by integrating over the range of storm probabilities. For each year of the project life span, the shoreline is retreated according to a long term erosion rate and calculations of annual damage are repeated. Total damage over the project life span is the sum of the annual damages. This more traditional approach produces a single result based on a single set of storms and response parameters. There is no indication of confidence level of the answer. The example project shows net benefits of $980,000 over a 50-year life span and a benefit/cost ratio of 2.4.

The life cycle approach in this example embodies sequences of storms (including provisions for multiple storms of varying intensity during each year of the life cycle), erosion and post-storm recovery during each event, partial and complete property damage during each event (depending on water level, waves, extent of storm erosion, and type of building construction), cumulative property damage due to a succession of storms, optional repair or rebuilding after a suitable time lag (with conformance to any stricter building codes in effect), and periodic renourishment of the beach when needed

FIG. 1. Sample Section of Coastline

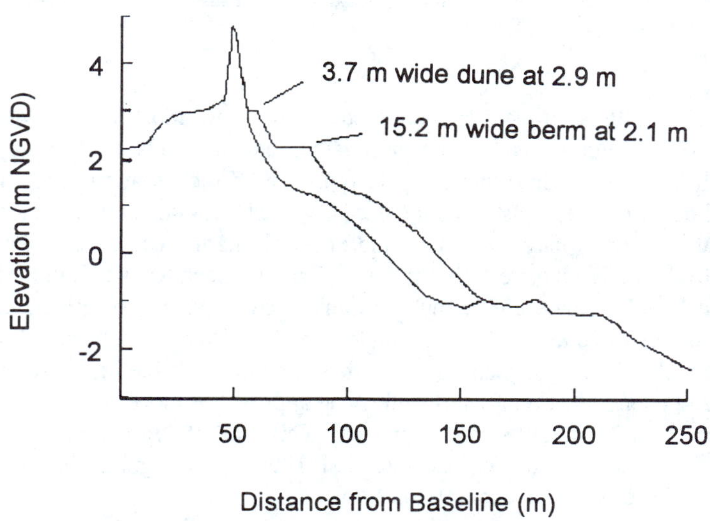

FIG. 2. Example Project Profile

and feasible during the life cycle. A key result from this analysis is the renourishment required during each life cycle, which can be converted to an economic present worth dollar value. The costs and net benefits in this example vary over a wide range, depending on the occurrence of major storms during the life cycle simulation (Fig. 3). The expected cost and economic risks associated with maintaining the beach can be realistically assessed by condensing information from many different life cycle simulations (Figs. 4 and 5). The mode and mean net benefits are $275,000 and $814,000, respectively (Table 3). There is a 65% probability that the net benefits will be less than the $980,000 amount calculated by the more traditional frequency-based approach. There is a 7% probability that the net benefits will be negative (cost of the project exceeds the damages prevented). Negative benefits come from simulations with mild storm climates. While net benefits are used to optimize a project, the benefit/cost ratio is also of interest (Table 3 and Fig. 6).

Table 3. Summary of Benefits, Example Project

Statistic	Net Benefits ($1,000)	Benefit Cost Ratio
Mode	275	1.50
Mean	814	3.09
Standard deviation	649	1.59
Maximum value	3557	8.92
Minimum value	-119	0.67
Number of cases	7000	7000

The life cycle approach appears better suited to most coastal engineering applications. Variation with time is an essential ingredient in most coastal projects, and it is directly incorporated into the life cycle approach. Time variation of resistance and functional performance, constraints imposed by construction season and mobilization, even some economic, environmental, and political factors, can be conveniently and flexibly introduced into the life cycle approach. This approach leads to a unified analysis of technical performance and many economic factors which are critical to project success. As illustrated in the above example, the life cycle approach provides valuable information relative to the objectives of risk-based analysis (Table 2). In addition to its technical and economic strengths, the life cycle approach is more easily understood by nontechnical parties involved with a project. This type of approach is evident in the Empirical Simulation Technique (Scheffner et al. 1996) and CE guidance in preparation (U.S. Army Corps of Engineers 1996).

Typical Project Elements

Risk-based analysis can be integrated into the major planning steps of a coastal engineering project. Typical project elements which are especially well-suited to

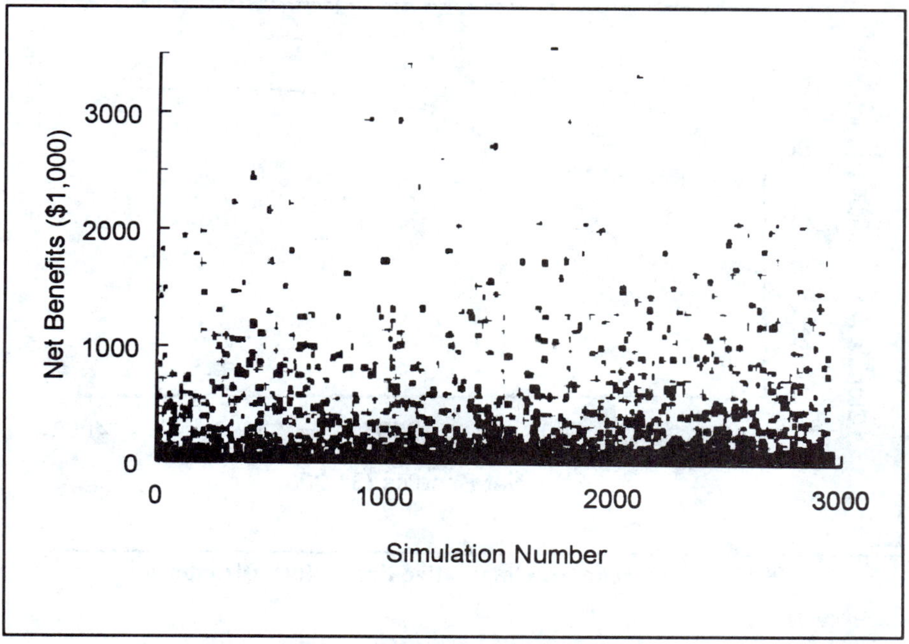

FIG. 3. Unsorted Risk Results

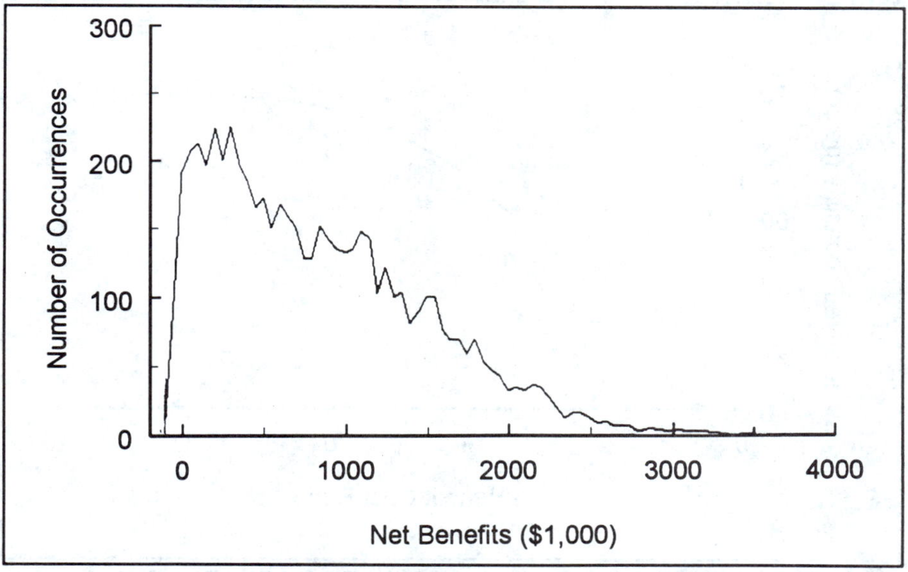

FIG. 4. Net Benefits, Probability Density Distribution

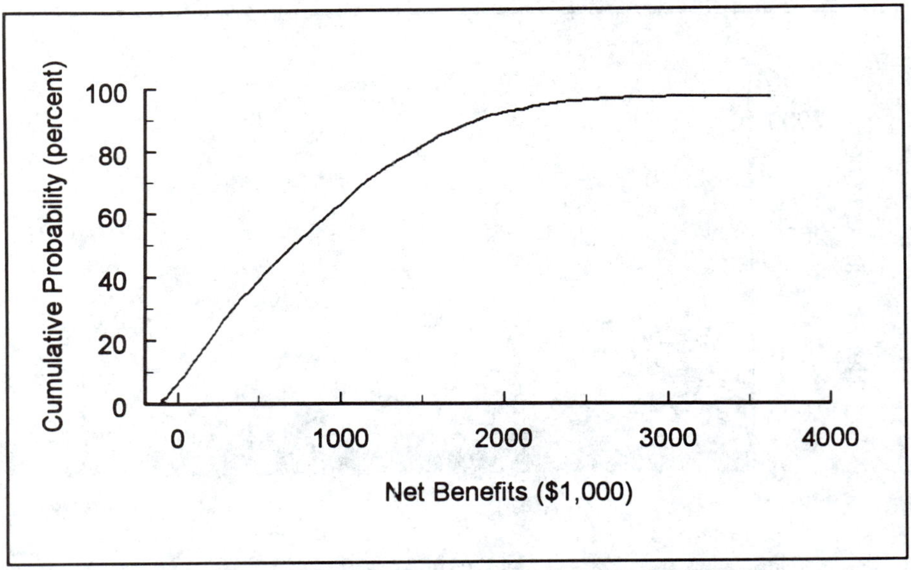

FIG. 5. Net Benefits, Cumulative Probability Distribution

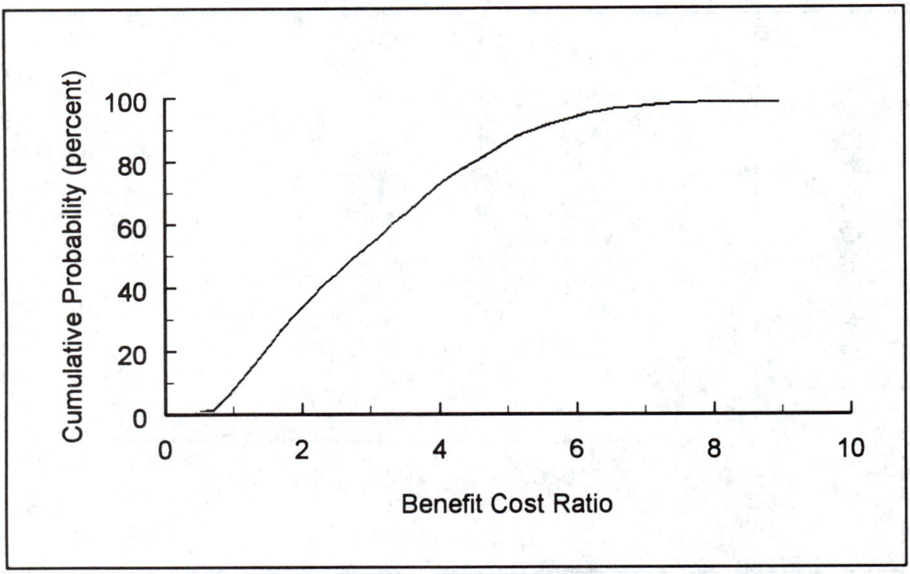

FIG. 6. Benefit Cost Ratio, Cumulative Probability Distribution

risk-based analysis include the following:

Site Characterization. Significant uncertainty can arise in documenting past and present behavior at a site. The uncertainty can be estimated based on data quality and quantity, methodologies used, observed variability, etc.

Without Plan Alternative. Evaluation of what would happen in the future if no project were built involves speculation about the natural processes and human interventions which would affect the site during the proposed project life. The impact of the *without plan* alternative is conveniently described in probabilistic terms.

Formulate, Evaluate, and Compare Alternative Plans. Risk-based analysis can be a powerful tool for formulating and comparing alternative plans. It enables decision-makers to intercompare not only the expected level of performance, but also the probabilities of enhanced or reduced performance levels, which can differ greatly among alternatives. Typically, alternatives involve hard structures (such as walls, revetments, breakwaters, and jetties) and/or soft structures (such as beach nourishment projects, coastal dunes, and nearshore berms). Risk-based analysis of hard structures is increasingly being considered in planning and design (CIRIA/CUR 1991, ICCE 1992). Soft structures involve calculated risks about the movement of sediment through time and the need for future maintenance. Uncertainties arise in forcing processes, sequencing of storms, initial state of nearshore profile when storms occur, and evolution and recovery of storm profiles (especially three-dimensional aspects). The life cycle approach to risk analysis has been shown to be a powerful tool in this type of application.

Conclusions

The following conclusions are reached regarding risk-based analysis of coastal projects:

- Risk-based approaches provide a powerful tool for analyzing coastal projects.
- Risk-based analyses can lead to improved decision-making for project optimization.
- The life cycle approach is especially well-suited to coastal engineering.

Acknowledgements

The work presented herein, unless otherwise noted, was supported by the Coastal Navigation Research Program of the U.S. Army Engineer Waterways Experiment Station, Coastal and Hydraulics Laboratory, under the Coastal Engineering Manual work unit of Coastal Structure Evaluation and Design. The example project was developed by the U.S. Army Engineer District, Wilmington (SAW). Mr. J. T. Jarrett, SAW, contributed significantly to the example project. The work of Dr. David A. Moser, U.S. Army Corps of Engineers (USACE), Institute for Water Resources, and Mr. Steve R. Cone, USACE Headquarters, in developing CE guidance on risk-based analysis is acknowleged. Permission was granted by the Chief of Engineers to publish this information.

Appendix. References

Ang, A. H.-S., and Tang, W. H. (1975). *Probability concepts in engineering planning and design; Volume I: Basic principles.* Wiley, New York, NY.

Ang, A. H.-S., and Tang, W. H. (1984). *Probability concepts in engineering planning and design; Volume II: Desision, risk, and reliability.* Wiley, New York, NY.

CIRIA/CUR. (1991). *Manual on the use of rock in coastal and shoreline engineering.* CIRIA Special Publication 83, Construction Industry Research and Information Association, London, UK; and CUR Report 154, Centre for Civil Engineering Research and Codes, Gouda, The Netherlands.

Harr, M. E. (1987). *Reliability based design in civil engineering.* McGraw-Hill, New York, NY.

ICCE. (1992). *Design and reliability of coastal structures.* Proceedings of a short course attached to the 23rd International Conference on Coastal Engineering, Venice, Italy.

Scheffner, N. W., Borgman, L. E., and Mark, D. J. (1996). "Empirical simulation technique based storm surge frequency analyses." *J. Wtrwy., Port, Coast. and Oc. Engrg.*, 122(2), 93-101.

U.S. Army Corps of Engineers. (1996). "Risk-based analysis for evaluation of hydrology/hydraulics and economics in shore protection studies." *Circular No. 1105-2-xxx* (in preparation), Washington, DC.

CHAPTER 346

DESIGNING FOR PROPELLER ACTION IN HARBOURS

G.A. HAMILL [1], J.A. McGARVEY [2]

Abstract

The scour produced by the action of a ship's propeller is an increasing problem. There is a need for an accurate method of velocity perdition so that adequate steps can be taken to design suitable bed protection systems. The velocity characteristics of the jet produced by the rotating propeller are presented as an aid to the engineer.

Introduction

One consequence of the increases in ship size, installed engine power, and volume of marine transport is an increase attack by propeller slipstreams on the bed and banks of harbour basins and navigation channels. In restricted waterways the situation has been exacerbated since the introduction of bow thrusters in the 1960's which has increased the navigability of large ships to such an extent that they can now operate without tug assistance

A ship's propeller produces thrust by drawing in water, accelerating it and discharging it in the form of a wash. This wash is a highly turbulent submerged jet, the velocities within which depend upon the operating characteristics of the propeller and the speed of advance of the ship. The wash entrains the surrounding water and so the velocities within it decay with distance from the propeller. In this way the jet expands and dissipates its energy by diffusion. If the ship is moving in a confined area, for example in the shallow waters of a harbour basin, the diffusion process within the wash will be restricted and the energy remaining within the jet may cause damage to the bed and nearby quay structures.

Related Research

A survey of harbours in Sweden (Bergh 1981) has shown that 16 out of the 18 ports surveyed have suffered propeller induced damage in recent years. Similar problems have been encountered in several United Kingdom ports and the British Ports Association commissioned a review (Prosser 1986) of existing knowledge. In a survey of all British ports in 1994, Qurrain (1994) found that 42% of these encountered scouring as a result of propeller action, of which 29% regarded this scouring to be a serious problem causing damage to quays, with expensive remedial repairs and costly facility downtime. The diffusion characteristics of a propeller jet have been examined by numerous authors. Investigation which include the influence of the rudder on the diffusion process however, are not extensive and the majority of all predictive equations do not allow for it's inclusion.

---

1) Lecturer 2) Research Student
The Queen's University of Belfast, Department of Civil Engineering, N. Ireland

Robakiewicz (1966), Verhey (1983) and Fuehrer (1985) have studied the velocities within the propeller jet in the presence of a rudder, and presented equations for estimating bottom velocities. These equations are limited in that a full velocity distribution cannot be calculated and they are related to an efflux velocity equation, i.e. the maximum velocity on the propeller face, which has been found to be up to 20% in error.

Experimental Set-up

Detailed velocity measurements were taken within the diffusing jets, produced by two propellers with differing characteristics, up to a distance of 10 diameters from the propeller plane. Four jets, corresponding to four speeds of rotation, were investigated for the first propeller which was 0.076m in diameter, while three jets were investigated for a second propeller 0.131m in diameter. Axial and radial measurements of velocity were taken using a twin component Laser Doppler Anemometry system.

Rudder Effect

The magnitudes of the axial and radial velocity components within the wash were measured within the formation stages of the propeller jet, i.e. up to 2.0 D_p. To provide a better understanding of the jet diffusion, velocity vector diagrams were plotted for all experimental results with and without the rudder present. These provided a 2D schematic representation of the flow within the propeller jet. Figures 1(a) to 1(c) are typical of those obtained during the current investigation.

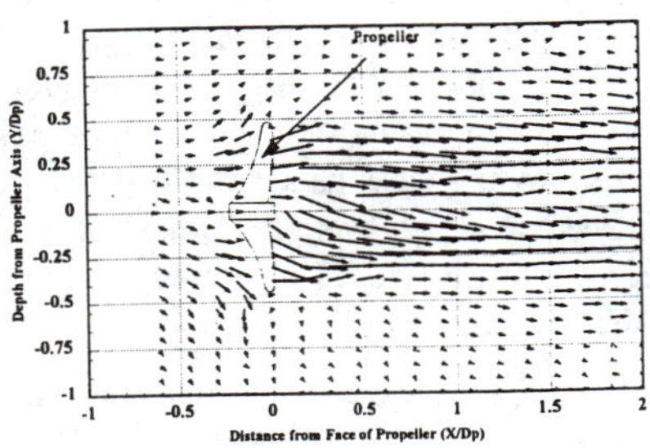

FIGURE 1(a) Resultant velocity vectors (without rudder)

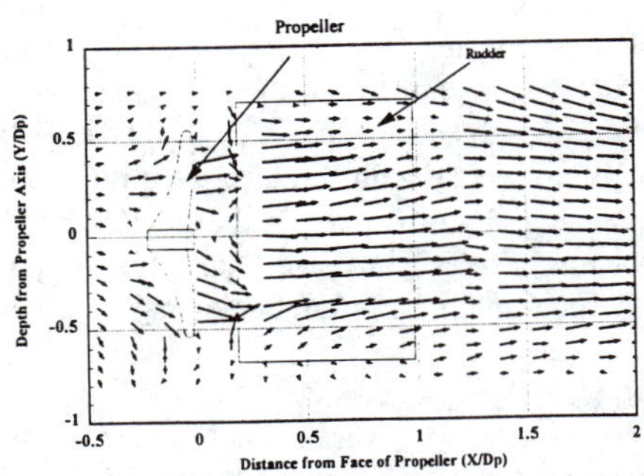

FIGURE 1(b) Resultant Velocity Vectors (Rudder at Zero Degrees)

Figure 1(a) shows the velocity distributions obtained without the rudder. The magnitude of the velocities at the hub is lower than across the blades, and the jet is diffusing away from the propeller centreline due to the radial velocity component. It can be seen that the vectors revert to the horizontal with distance from the propeller, indicating the reduction in the magnitude of the radial velocity component. The velocity distribution at each vertical section from the propeller is approximately symmetrical through the propeller axis which confirms the results of previous work. The results obtained with the rudder situated at zero rudder angle or midships position is shown in figure 1(b).

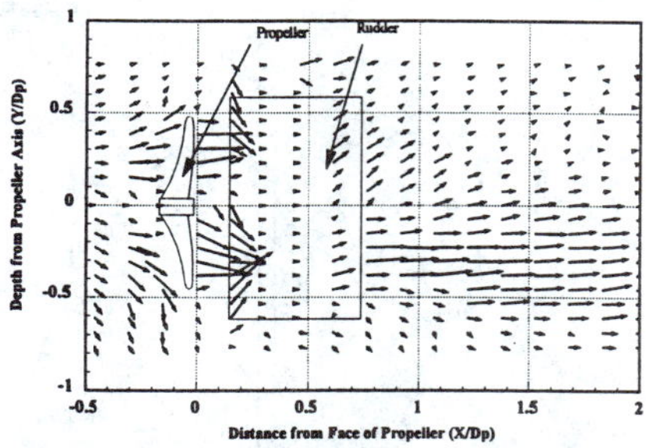

FIGURE 1 (c) Resultant Velocity Vectors (Rudder at +35 Degrees)

It can be seen that the resultant velocity distribution at the face of the propeller is similar to the distribution obtained without the rudder present. There is a significant change in the flow with distance from the propeller, as the jet is circulated around the rudder. There is a rapid reduction in the magnitudes of the velocities at the first section behind the propeller. In subsequent sections the velocities are greater as the flow moves along the sides of the rudder. It can also be seen that the vectors in subsequent section along the rudder are almost to the horizontal. This indicates that the magnitudes of the radial velocities are insignificant, and that the rudder has a straightening effect on the jet. The velocity distribution appears to vary at each section and it can be seen in some sections that there are higher velocities below the propeller centreline than at the corresponding position above the centreline.

Figure 1(c) shows the results obtained when the rudder is placed at an angle of 35 degrees to the propeller axis. The velocities at the face of the propeller are approximately symmetrical through the propeller axis and therefore appear to be unaffected by the rudder. It is the distributions in the subsequent sections that are interesting as the magnitude of the velocity below the centreline can be seen to be greater than above the centreline. This is due to the effects of splitting of the flow by the rudder. The vectors shown represent the velocities in a stream directed towards the bottom. The distribution shows that there is only one position of maximum velocity, unlike the axi-symmetrical profiles found in the absence of the rudder.

In order to provide a better understanding of the propeller diffusion characteristics with the rudder present, velocities were measured throughout the jet using a system of pitot tubes. These results were then plotted in the form of an isovel plot for each speed of rotation, which provided contours of the velocity distribution within the propeller jet. Figures 2(a) to 2(d) show the sequence of isovel plots with the rudder positioned at zero degrees, and is typical of the velocity distributions obtained for all the jets tested at this angle. Figure 2(a) shows the velocities obtained at a distance of 0.46 D_p behind the propeller, at the location of the rudder. The velocities obtained at this section are unlike those obtained for the jet without a rudder as the velocities are not symmetrical through the propeller axis (Stewart 1992). The rudder splits the main wash into two separate streams of jets. one jet is directed towards the water surface, while the other is deflected towards the sea bed. The jets are just forming at this position and the bottom jet below the centreline is clearly defined, with the

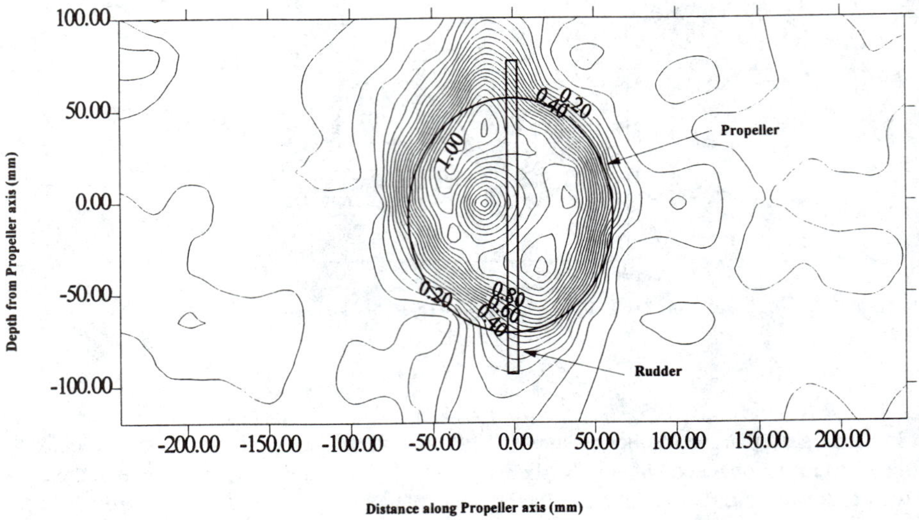

FIGURE 2(a) Isovels at 0.46 Dp for zero rudder angle

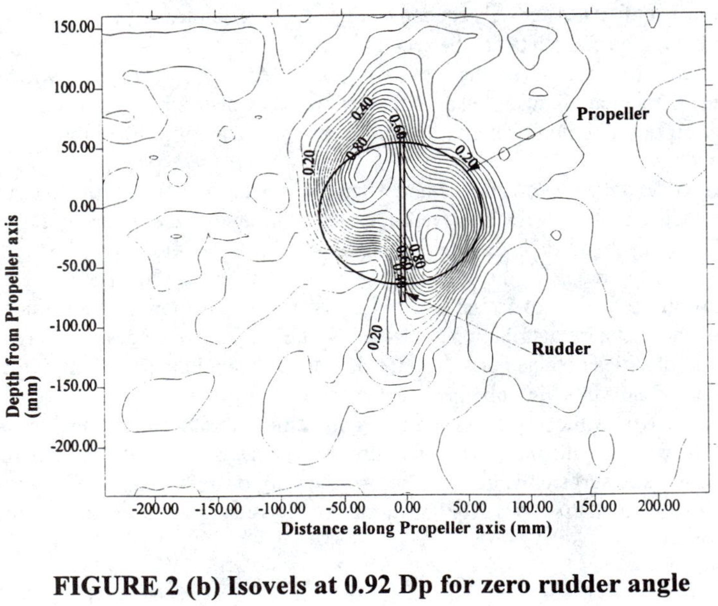

FIGURE 2 (b) Isovels at 0.92 Dp for zero rudder angle

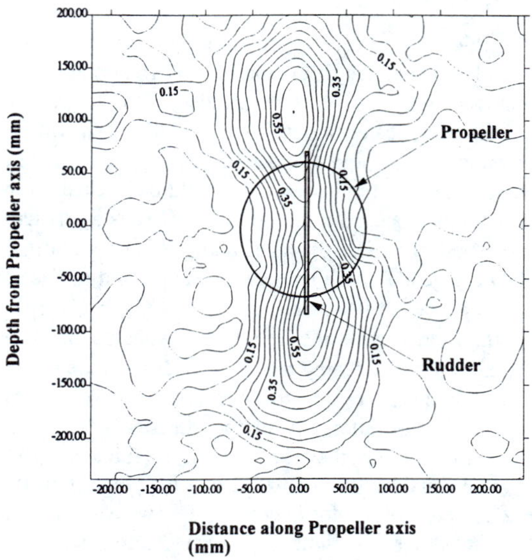

FIGURE 2(c) Isovels at 3.05 Dp for zero rudder angle

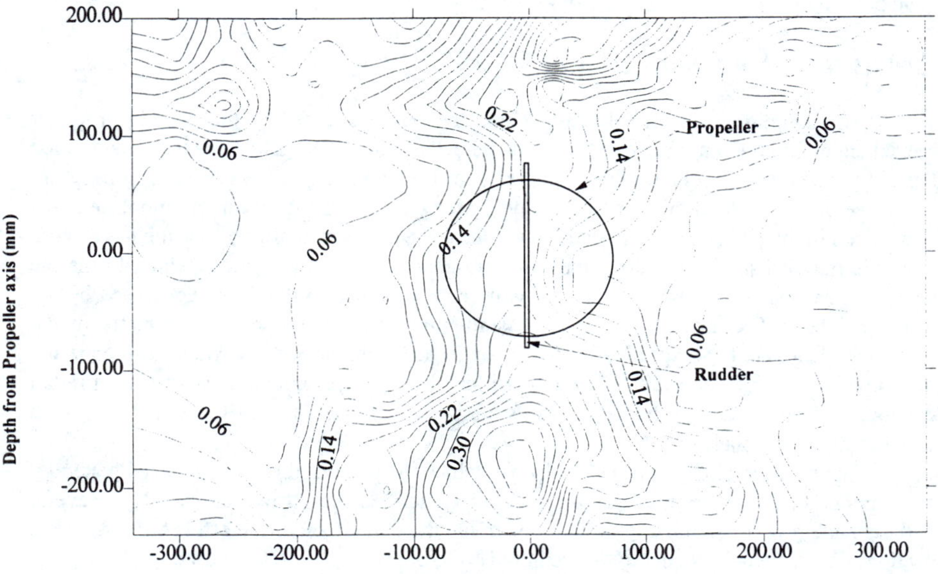

FIGURE 2(d) Isovels at 6.95 Dp for zero rudder angle

surface stream just becoming visible. Figure 2(b) which is located at a section subsequent to the location of the end of the rudder, shows the formation of the two streams, represented by the two high velocity peaks shown on each side of the rudder. This trend continues with distance from the propeller and the two streams are visible at 3.05 D_p as shown in figure 2(c). This position has previously been termed the end of the zone of flow establishment for the jet without a rudder, at which location the maximum velocity would be located on the propeller centreline. It can be seen that with distance from the propeller there is a gradual increase in the distance between the streams, and therefore the maximum velocity is not likely to revert to the propeller centreline. The divergence of the streams continues and this is confirmed in figure 2(d), at 7 D_p, where the two jets are still separating.

It has been shown that the rudder splits the jet into two streams and their deviation continues throughout the remainder of the diffusing jet. The rudder has the effect of altering the diffusion characteristics of a propeller jet and this has resulted in the formation of a different type of jet. There is no further alteration of the streams, other than by natural diffusion with the surrounding fluid, once they have been established. The zone of established flow could therefore be considered to commence at the position of the formation of the two streams. Figure 2(a) showed the formation of the two streams, although not clearly defined. However, it can be seen in figure 2(b) that the two streams are clearly defined and diverging away from each other. This position represents approximately the end of the rudder. It can therefore be concluded that the jet is fully established from the location of the trailing edge of the rudder.

It is evident from the findings of this investigation that the processes of formation and diffusion of a propeller wash, under the influence of a rudder, can in no way be described by the predictive equations that are currently available as the two streams formed by the rudder have been shown to behave as independent jets.

Characteristics of a propeller jet

Efflux Velocity Distribution

The magnitudes of all velocities within the wash produced by the rotating propeller are dependent on the magnitude of the initial velocity within the wash. This velocity, termed the efflux velocity, is found on the cutting edge of the propeller. As can be seen from figure 1 the velocity distribution along the cutting edge, or the efflux distribution, remains unchanged in the presence of the rudder, with an axi-symmetric distribution being developed. It is this initial distribution, along with the subsequent rudder interaction, which forms the wash. The velocity distribution at efflux will be proportional to the thrust developed by the propeller blade. The thrust produced by a ships propeller depends on the geometry of the propeller blades and the speed of rotation. The main characteristics which influence the developed thrust include the pitch and chord length. These characteristics vary with radial distance from the hub, and hence the thrust and therefore the velocity must also vary with distance along the blade.

The position along a propeller blade at which the greatest thrust is produced generally corresponds to the position of maximum velocity, termed the efflux velocity, V_0. Several authors have proposed equations which determine the magnitude of the efflux velocity. The original efflux velocity equation, as reported by Blaauw et al (1978) and adopted by most authors, was based on the Froude momentum theory,

$$V_o = 1.59 n D_p \sqrt{C_t}$$

The assumptions made in the momentum theory have been shown to be invalid. These short-falls in the theory have led to modifications of the theoretical efflux equation by Stewart (1992). These modifications take account of the propeller characteristics as follows,

$$V_o = E_o n D_p \sqrt{C_t} \qquad 1$$

where

$$E_o = \left[\frac{D_p}{D_h}\right]^{-0.323} [C_t]^{-0.1459} [P]^{0.44} [\beta]^{0.513}$$

The magnitude of the axial velocities at the face of the propeller have to date been estimated by solving equations developed by Albertson (1950). He proposed that the velocity distribution within the jet, close to the propeller, follows the Gaussian normal probability function.

Stewart modified the Gaussian equation and presented the following equation, to describe the axial velocity distributions within the propeller jet, based on his own investigations:

$$\frac{V_{xr}}{V_{max}} = EXP\left(-\frac{1}{2}\frac{(r - R_{mo})^2}{\sigma^2}\right) \qquad 2$$

where the standard deviation of velocity, σ, was equal to 0.5 R_{mo} and the maximum velocity, V_{max} at the face of the propeller is equal to the efflux velocity, V_o. Berger et al. (1981) proposed that the distance to the position of the maximum velocity at the face of the propeller, R_{mo}, could be determined as follows,

$$R_{mo} = 0.67 R_p - R_h \qquad 3$$

where R_h is the radius of the hub and R_p is the propeller radius.

The velocity distribution along the face of the propeller can be determined by solving equations 1,2 and 3. Velocity profiles were plotted comparing the results obtained from existing equations and the measured data. The profiles, as shown in figures 3 and 4, indicate that the measured velocity increases from the hub to a distance of approximately 0.7 R_p along the blade, and then decreases rapidly to the blade tips. This is in contrast to the profile predicted by existing equations, where the velocity increases to approximately 0.5 R_p and then gradually decreases to the blade tips. The position of the maximum velocity can also be seen to vary greatly from that determined from existing equations, when compared to the measured velocities. The distance to the maximum velocity, in all of the profiles that were plotted using existing equations, was underestimated when compared to the measured results. The existing equations, which are based on the momentum theory, have been revised and updated on several occasions by replacing the experimental constant. The magnitude of the constant varied in several investigations using different propellers, more recently Stewart updated the equation to relate the constant to the propeller characteristics. This was a calibration of a theory, which is inherently flawed, to suit experimental results. It was decided to investigate a more realistic method by which the magnitude of the axial velocities along the face of the propeller could be obtained.

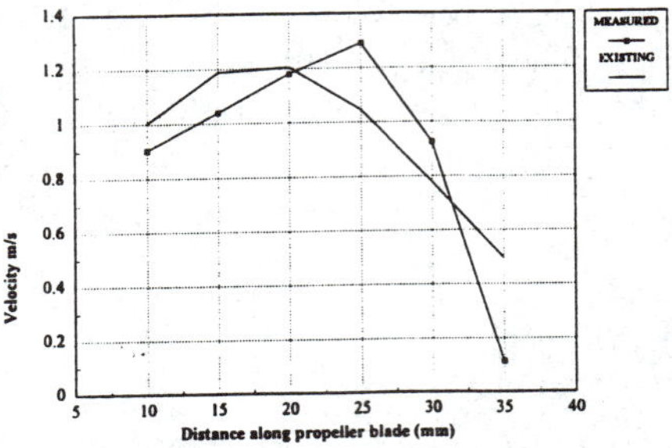

FIGURE 3 Velocity Distribution Comparison for 76mm propeller

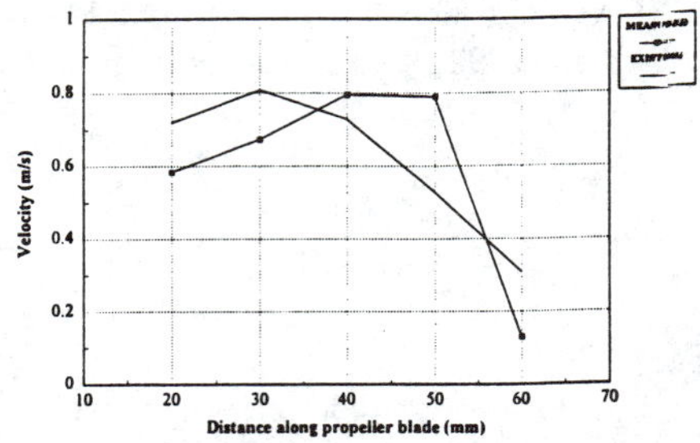

FIGURE 4 Velocity Distribution Comparison for 131mm propeller

A literature survey revealed that Koumbis (1981) investigated the thrust produced by a ship's propeller and developed a computer program which relates the thrust to the characteristics of the propeller blades. The program is based on the Vortice Lattice theory.. In designing a propeller the naval architect determines the most suitable characteristics to provide the optimum thrust for each blade. It therefore seemed reasonable to relate the distribution of velocity on the propeller face to the characteristics which produce this thrust.

The blade cross sections, as shown in figure 5, were obtained for positions along the propeller blade where the velocity measurements had been recorded. It was decided to attempt the derivation of a relationship in which the magnitude of the axial velocity could be expressed in terms of the characteristics of the propeller blades.

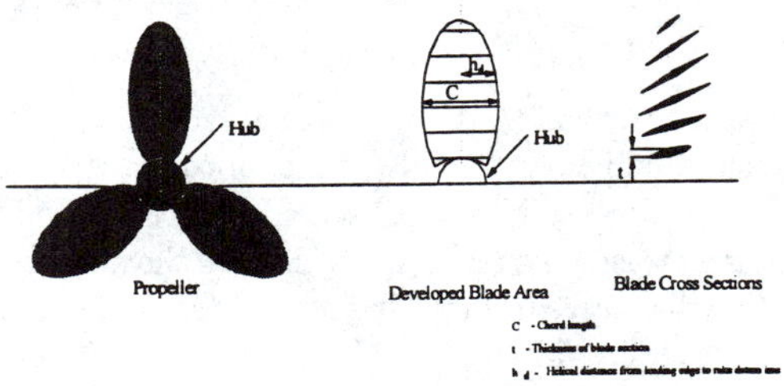

Figure 5 Propeller Blade Characteristics

A literature survey provided detailed information on the design of a propeller, and revealed the properties considered by the naval architect in designing the geometry of the propeller blade. Considering the axial velocity V_{ax}, it can be written as:

$$V_{ax} = f(n, N, r, \mu, \rho, p, t, c, h_D, h_T)$$

where n = number of rev/sec, N = number of blades

r = radius of propeller, μ = kinematic viscosity

ρ = density of fluid, p = pitch of blade

t = thickness of blade, c = chord length

h_D = helical distance from blade section leading edge to rake datum line

h_T = helical distance from blade section leading edge to position of maximum thickness

Using the Pi theorem the equation can be written as:

$$\frac{V_{ax}}{Nnr} = f\left(\frac{\rho n r^2}{\mu}, \frac{p}{r}, \frac{c}{r}, \frac{t}{r}, \frac{h_D}{r}, \frac{h_T}{c}\right)$$

It was found from the literature survey that the effect of viscosity could be neglected if the Reynolds number of the jet was greater than 10^4. In the present investigation the Reynolds numbers were greater than 10^4, and could therefore be neglected from further analysis This analysis resulted in an equation with a correlation coefficient R^2 of 0.985:

$$\frac{V_{ax}}{Nnr} = 1.261 - 0.974\left(\frac{p}{r}\right) + 0.733\left(\frac{c}{r}\right) + 18.527\left(\frac{t}{r}\right) + 5.028\left(\frac{h_d}{r}\right)$$

$$+0.106\left(\frac{p}{r}\right)^2 - 7.277\left(\frac{h_d}{r}\right)^2 - 4.093\left(\frac{h_t}{c}\right)^2 \qquad 4$$

The high correlation obtained for the experimental data shows the assumption that the velocity distribution along the blade was a function of the varying blade characteristics was valid. The velocity distribution obtained from equation 4 and the results obtained from existing equations, were plotted with the experimental results. Figure 6 is typical of that found for the present investigation. It can be seen that there is a significant improvement in the correlation with the measured data, when compared to the results of existing equations. This was found to be the trend of the results obtained for all experiments.

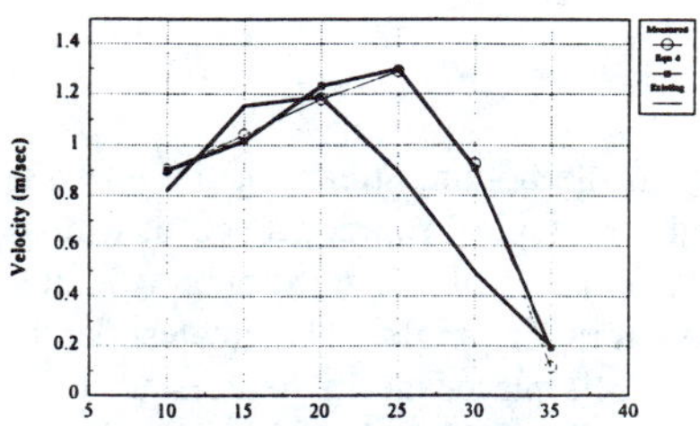

FIGURE 6 COMPARISON OF EXPERIMENTAL DATA WITH THE RESULTS FROM EXISTING AND PROPOSED EQUATION

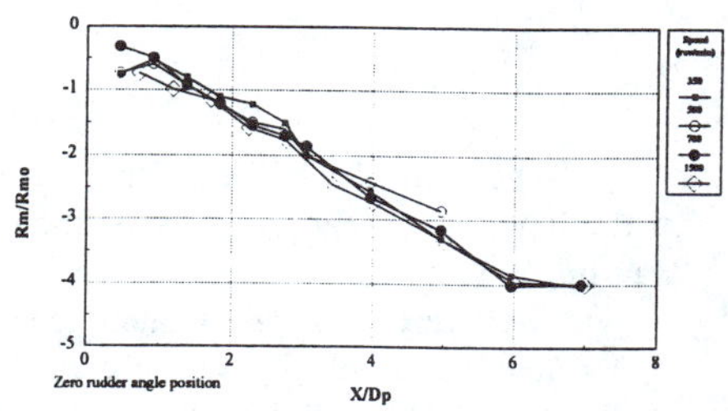

FIGURE 7 LOCATION OF THE MAXIMUM AXIAL VELOCITY WITHIN THE BOTTOM STREAM OF THE DIFFUSING JET

The magnitude of the axial velocity distribution at the face of the propeller can now be determined by solving a single equation, equation 4, instead of the three equations as proposed by existing work. The derivation of this equation is based on the physical properties of the propeller blades rather than the momentum theory which is inherently flawed. Equation 4 will accurately determine the magnitude of the maximum velocity, V_0, and the position at which it occurs along the blade. Its use beyond the scope of the current experimental programme will require further verification. However, when compared to the equations which are currently used, which themselves are based on a limited test programme, it shows great potential.

Location of the maximum velocity

The location of the maximum velocity within the bottom stream was plotted with distance from the propeller. Figure 7 shows the location of the maximum velocity in the bottom stream for each speed of rotation for a zero rudder angle. It can be seen that there is no significant variation in the location of the maximum velocity, with an inclination of approximately 12 degrees observed towards the bottom for each speed of rotation. This is in sharp contrast to that obtained when there was no rudder fitted where the maximum velocity was located along the horizontal propeller axis. It would therefore appear that for a zero rudder angle the location of the maximum velocity is independent of propeller speed of rotation.

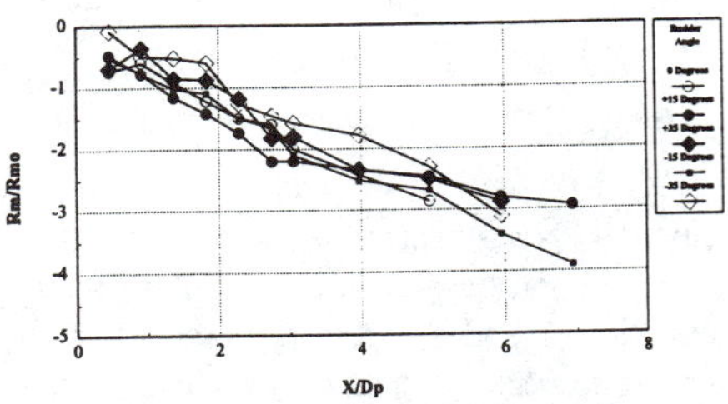

FIGURE 8. LOCATION OF THE MAXIMUM AXIAL VELOCITY WITHIN THE BOTTOM STREAM OF THE DIFFUSING PROPELLER JET FOR CHANGES IN RUDDER CONFIGURATION

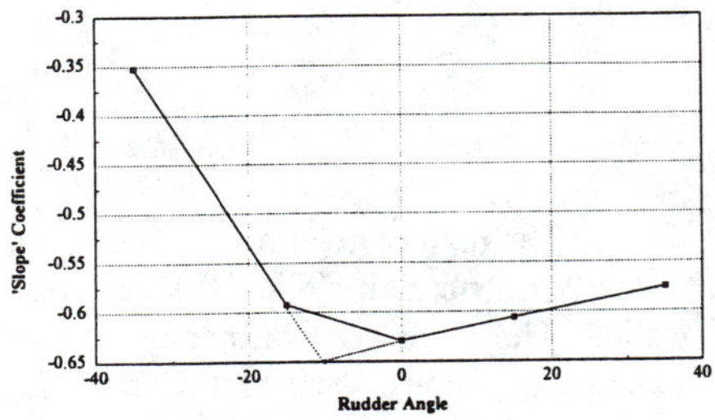

FIGURE 9 EXPERIMANTAL CONSTANT V's RUDDER ANGLE

The vertical location of the maximum velocity in the bottom stream of the jet was then plotted for all experiments and the results obtained for a typical rotational speed, with a varying rudder angle, are as shown in figure 8. The location of the maximum velocity appears to diverge at approximately the same rate for each rudder angle. The greatest depth to the maximum velocity was observed when the rudder is placed at 15 degrees up to a distance of 3 D_p and generally thereafter is greater when the rudder is positioned at zero degrees. The divergence angle was found to be approximately 12 degrees and this was found to be the general trend for this propeller at different rudder angles. The general trend showed that the greatest depths to the maximum velocity were observed when the rudder was placed at positive rudder angles. It can therefore be concluded that the depth to the maximum velocity is dependent of the rudder angle.

Equation for location of maximum velocity

Analysis of the change in position of the maximum velocity showed that it location could best be described by a linear decay equation of the type

$$\frac{R_{mx}}{R_{mo}} = Const.\left(\frac{X}{D_p}\right)$$

where R_{mx} is the maximum velocity location at any distance X from the propeller.

It was obvious that the constant term was a function of the rudder angle, as the location was seen to shift as the rudder angle changed. Figure 9 shows a plot of these constant values against the rudder angle, and it is clear that a significant variation can be observed between the positive and negative rudder angles. This is to be expected as the negative angles cause the rudder to be turned into the bottom stream therefore hindering the flow, while with positive angles the stream is free to expand. Based on this physical interference two separate equations have been developed by which the position of the maximum velocity can be obtained, and these depend on the rudder angle range as follows;

Rudder angle range $-10<\theta<+35$ degrees

$$\frac{R_{mx}}{R_{mo}} = (-0.659 + 0.002\theta)\left(\frac{X}{D_p}\right)$$

Rudder angle range $-35<\theta<-10$ degrees

$$\frac{R_{mx}}{R_{mo}} = (-0.772 - 0.012\theta)\left(\frac{X}{D_p}\right)$$

These equations are a first approximation based on the current experimental investigation and further work would be recommended to provide a clearer understanding of the influence of the rudder on the location of the maximum velocity between the zero and -15 degree rudder position.

Decay of Maximum Axial Velocity

It has been shown that for a jet without a rudder, the decay of the maximum axial velocity is not constant throughout the jet, Qurrain. The present investigation has shown that the jet is fully established from the location of the end of the rudder and therefore a single relationship must be considered to describe the decay of the maximum velocity within the propeller jet with a rudder present.

The rate of decay of the maximum velocity was found to be independent of propeller rotational speed. The rate of decay of the maximum velocity however, varies significantly when compared to previous work due to the presence of the rudder. In order to establish a relationship for the rate of decay, the values of the maximum velocity, V_{max}, within the jet were non-dimensionalised with the maximum velocity at the face of the propeller, V_o, and the distance from the propeller, X, was divided by the propeller diameter, D_p. Analysis of the results was carried out on all the data and it was found that the following type of equation provided the best correlation for the results, as can be seen in figure 10.

$$\frac{V_{\max}}{V_o} = A + B\left(LN\frac{X}{D_p}\right)$$

It was found for all tests that the B coefficient was constant and equal to 0.293. This shows that the actual rate of decay is independent of the rudder angle. There was however a significant variation observed in the values found for the A coefficients. This indicated that although the rudder had no influence on the rate of decay of velocity it clearly was influential in the initial magnitude of the maximum velocity. As was the case in the location of the maximum velocity the A coefficients were compared to the respective rudder angles and again positive and negative angles were seen to influence the magnitude of velocity. Again this is not unreasonable as has already been discussed. As before two equations by which this coefficient should be determined have been produced, again depending on the rudder angle considered.

Rudder angle range $-10<\theta<+35$ degrees

$$A = 0.915 - 0.004(\theta)$$

Rudder angle range -10<θ<-35 degrees

$$A = 1.0 - 0.007(\theta)$$

Concluding Comments

The present investigation has studied the influence of the rudder within a ship's propeller jet, and has confirmed the formation of two high velocity streams, one directed upwards to the surface and the other directed downwards towards the bottom.

The axial velocity distribution at the face of the propeller could only be determined by solving a number of equations based on the results of previous work. The present investigation found that the velocity along the face of the propeller was related to the characteristics of the propeller blade. An equation was developed which related the magnitude of the axial velocity at any point along the blade to the properties of the blade at that point.

It was found that there was an increase in magnitude of the axial velocities by as much as 30% with the rudder present when compared to the jet without a rudder.

The location of the maximum velocity within the propeller jet was on the axis of each stream which were established from the position of the rudder. This led to a redefinition of the zone of established flow. Equations are presented which locate the maximum velocity in the vertical direction allowing for changes in rudder position. The maximum deflection of the bottom stream was observed for the zero ruder angle position and was found to be approximately 14 degrees.

Results have shown that the decay of the maximum velocity within each of the streams was the same at each of the angles tested and was independent of the propeller rotational speed. It was found that there was a logarithmic decay of the maximum velocity at each of the rudder angles tested and the magnitude of the velocity encountered was found to be dependent on rudder position.

By using the equations presented it is now possible for the engineer to establish the maximum velocity at any depth below a rotating propeller. Given the bed clearance conditions it is therefore possible to predict the magnitude of the velocity which will impact on the sea bed. Knowing this velocity, and an understanding of sediment movement prediction techniques, it is possible to design either suitable bed armour units or to predict eroded depths.

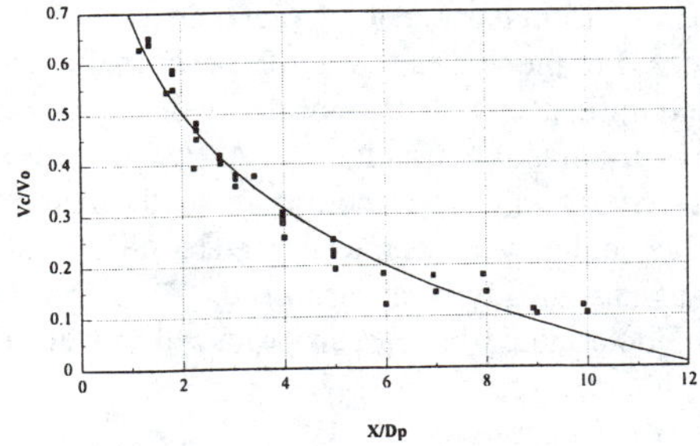

FIGURE 10 DECAY OF THE MAXIMUM AXIAL VELOCITY ALONG THE PROPELLER AXIS FOR ALL EXPERIMENTS AT ZERO RUDDER ANGLE

APPENDIX

REFERENCES

ALBERTSON, M.L., (1950) "Diffusion of submerged jets", Transcript of the A.S.C.E., Paper No. 2409, Vol. 115, pp 639-697.

BERGER, W., FELKEL K., HAGER M., OEBIUS., SCHALE E.,(1981) "Courant rovoque par les bateaux. Protection des berges et solution pour eviter l'erosion du lit du Haut Rhin" PIANC 25th Congress, Edinburgh, 1981, Section I-1

BERGH H., CEDERWALL K., (1981), "Propeller erosion in harbours", Bulletin No. TRITA-VBI-107, Hydraulics Laboratory, Royal Institute of Technology Stockholm, Sweden.

BLAAUW, H.G., KAA, E.J. van de, (1978)," Erosion of bottom and sloping banks caused by the screw race of manoeuvring ships ", Publication No.202, July 1978, Delft Hydraulics Laboratory, Netherlands.

FUEHRER, M., ROMISCH, K., (1985), Dimensionerung von Sohlen- und oschungs -befestigungen an Schiffahrtkanalen. Mitteilungen der Forschungsanstalt fur Schiffhart, Wasser und Grundbau, Berlin. Schriftenreihe Wasser und Grundbau, Heft 45

KOUMBIS A., (1981) "An improved mathemnatical modek of the action of open and ducted thrusters", PhD. thesis, University of Newcastle upon Tyne, 1981.

PROSSER, M.J., (1986), "Propeller induced scour", BHRA report for the British Ports Authority, February 1986.

QURRAIN, R.M.M. (1994), "Influence of the sea bed and berth geometry on the hydrodynamics of the wash from a ship's propeller", Thesis submitted to the Queen's University of Belfast for the degree of Ph.D. 1994.

ROBAKIEWICZ, W., (1966), Wplyw dzialania strumienia zasrubowego na dno przy nabrzezu stanowiska prob na uwiezi na przykladzie baden model owychiernowych ztrawlerem B 20. Rosprawy Hydrotechniczne 19.

STEWART, D.P.J.,(1992), " Characteristics of a ships screw wash and the influence of quay wall proximity ", Thesis submitted to the Queens University of Belfast for theDegree of Doctor of Philosophy.

VERHEY, H.J., (1983), The stability of bottom and banks subjected to the velocities in the propeller jet behind ships. Delft Hydraulics laboratory. Publication No 303

CHAPTER 347

MODELING INLET SAND BYPASSING

Christopher G. Creed[1]

Abstract

Consideration of the magnitude and phasing of sand bypassing events with respect to incident wave conditions is necessary when modeling shoreline change adjacent to tidal inlets where sand bypassing occurs. Likewise, the effects of wave transformation over nearshore bathymetric features (i.e., ebb tidal shoals, bypassing bars, etc.) and the resulting variations in alongshore sand transport rates relative to bypass discharge locations should also be considered. This paper presents a method for developing time-dependent inlet sand bypassing data for use in predicting future shoreline change. The method includes developing a relationship between the local offshore wave climate, longshore sand transport regime and inlet sand bypassing utilizing coincident time series of (1) local wave climate and (2) sand bypassing. The proposed model is compared to two other typical sand bypassing models to demonstrate the importance of relating the wave climate and bypassing. The proposed method was used successfully to create a representative time series of both mechanical and natural sand bypassing data for a shoreline change study conducted for the downdrift shoreline of South Lake Worth Inlet, Florida.

Introduction

Shoreline change adjacent to inlets where sand bypassing occurs is directly related to the characteristics of bypassing conditions. Specifically, sand bypassing is responsible, in part, for the manner in which both the updrift and downdrift shorelines of an inlet respond to the offshore wave climate and longshore sand transport regime. That is, the rate at which sand is removed from an updrift shoreline or added to a downdrift shoreline relative to the magnitude of the longshore transport potential affects the rate of shoreline change. For example, if the bypassing rate is below average during

---

[1] Coastal Engineer, Olsen Associates, Inc., 4438 Herschel Street, Jacksonville, FL 32210.

above average transport conditions, the net effect would be an impoundment of sand along the updrift shoreline and erosion along the downdrift shoreline. Therefore, when modeling shoreline change adjacent to inlets where sand bypassing occurs, it is important to represent both longshore transport and bypassing conditions accurately. This requires that the magnitude and phasing of sand bypassing events with respect to incident wave conditions be considered.

Background

Shoreline change models are typically used to predict future response of shorelines where a beach fill or coastal structure may be constructed. The characteristics of the mechanisms which contribute to future shoreline change (i.e., the wave climate, longshore transport regime, updrift sand supply, bypassing, etc.), however, are generally not known *apriori*. Therefore, accurate prediction of future shoreline change requires assumptions regarding the future characteristics of mechanisms which contribute to shoreline change. Typically, the assumption is made that future conditions can be represented by average historical conditions.

In the instance where a time-dependent model is used, annual and seasonal variations in the shoreline change mechanisms are also represented. Time-dependent shoreline change models often require a time series of input conditions to represent the annual and seasonal variations in shoreline change. For example, a shoreline change model may require a wave time series to represent wave conditions. In predicting future shoreline change, wave conditions may be representative of historical average annual and seasonal conditions.

When modeling shoreline change for a shoreline adjacent to a tidal inlet where sand bypassing occurs, the time dependent nature of bypassing events and the relationship of sand bypassing to incident wave climate and transport regime must also be considered. It is known that inlet sand bypassing events, both mechanical and natural, are related to the incident wave climate and the resulting longshore sand transport regime. Considering this relationship, a *sand bypassing time series* which is correlated to the incident wave climate and resulting longshore sand transport regime is required to represent bypassing conditions accurately.

In instances where it is known that shoreline change adjacent to an inlet is directly affected by bypassing activities, bypassing data must be developed which are representative of the magnitude of bypassing quantities and a function of the incident wave climate. Developing representative wave and bypassing data, without consideration of their interdependence, may not produce an accurate representation of the mechanisms which contribute to shoreline change. A method for the development of this data is presented in the following.

Method

The purpose of the proposed method is to determine the relationship between wave conditions, longshore transport conditions, and sand bypassing. The method is therefore based upon the assumption that sand bypassing around an inlet is a function of both the incident wave climate and the resulting longshore sand transport regime. That is, it is assumed that sand bypassing occurs only when the wave and longshore transport conditions are such that a sufficient volume of material is transported toward an inlet which is mechanical or natural bypassing.

The method requires that coincident time series of historical wave and bypassing data are available. The wave data times series can be either from measured or hindcast sources. Bypassing data may be available from bypass plant records or other measurement efforts. Application of the method results in a correlation expression which describes the relationship between the magnitude and occurrence of bypassing events relative to that of wave and longshore sand transport events. This relationship can then be used along with an assumed representative wave data to formulate bypassing data for use in predicting future shoreline change.

Application of the method includes computing longshore transport conditions from offshore wave time series which correspond to the period for which sand bypassing data area available. Simplistically, the transport conditions may be computed with a longshore sand transport equation such as the CERC formula (Shore Protection Manual, 1984). The general CERC formula can be expresses as

$$Q = K\, H_b^{\frac{5}{2}}\, \sin 2\theta_b \tag{1}$$

where Q is the longshore sand transport rate, H_b is the incipient breaking wave height, θ_b is the local breaking wave angle relative to shore normal and K is a calibration coefficient. Incipient breaking wave heights can be computed from the offshore wave data assuming the wave climate is linear and monochromatic and the offshore bathymetric contours are straight and parallel. A time series of longshore transport data can then be compiled to correspond to the period and frequency of the bypassing data. The resulting longshore sand transport time series can be calibrated with the coefficient, K, to produce average annual transport conditions equivalent to known quantities for the region.

To determine the relationship between the transport and bypassing, the data are plotted and a best-fit expression is derived. This resulting expression implicitly describes the phasing and magnitude of bypassing relative to the offshore wave data.

Alternate Method. In instances where historical bypassing records are not available, an inlet sediment budget may be utilized along with the assumption that bypassing is proportionately related to the local longshore sand transport potential. Although this method does not strictly include the relationship between actual wave and bypassing data, it generally describes annual and seasonal variations in the wave climate and bypassing for modeling purposes. The technique simply describes bypassing as a fraction of longshore sand transport.

Once the relationship between transport and bypassing is determined, the wave time series may be applied to the expression to develop a corresponding bypassing time series which represents future bypassing conditions. This technique was applied for a shoreline change investigation described below.

Application

The proposed method was applied during a shoreline change model study conducted for South Lake Worth Inlet in South Florida. The intent of model study was to simulate shoreline change along approximately 12,000 feet of shoreline immediately downdrift of the inlet for purposes of formulating a Federal Shore Protection Project for the Town of Ocean Ridge, Florida ("Palm" 1996). Analysis of historical shoreline change data, as well as the general morphology and littoral transport patterns, suggested that sand bypassing across the inlet, both mechanical and natural, strongly influenced the downdrift shoreline. The shoreline change model GENESIS, developed by Hanson (1987) and Hanson and Kraus (1989), was used for the investigation. GENESIS is a one-line shoreline change model which computes time-dependent shoreline change using a time series of input wave and sand bypassing conditions. Realizing the importance of sand bypassing to the study shoreline, both mechanical and natural sand bypassing which occur at the inlet were modeled as input to the updrift boundary condition.

Background. South Lake Worth Inlet is located in southern Palm Beach County on the southeastern coast of Florida (Figure 1). The inlet was mechanically cut through the sandy barrier which separates the Atlantic Ocean and Lake Worth in 1927. The inlet is stabilized by two jetties; the north jetty extends about 122 meters farther seaward than the south jetty and is curved towards the southeast. In 1967, the inlet was modified to its current configuration in an attempt to improve its bypassing capacity and reduce interior shoaling. The regional net sand transport in the vicinity of the inlet is from north to south, with approximately eighty-five percent of the gross transport being southerly directed.

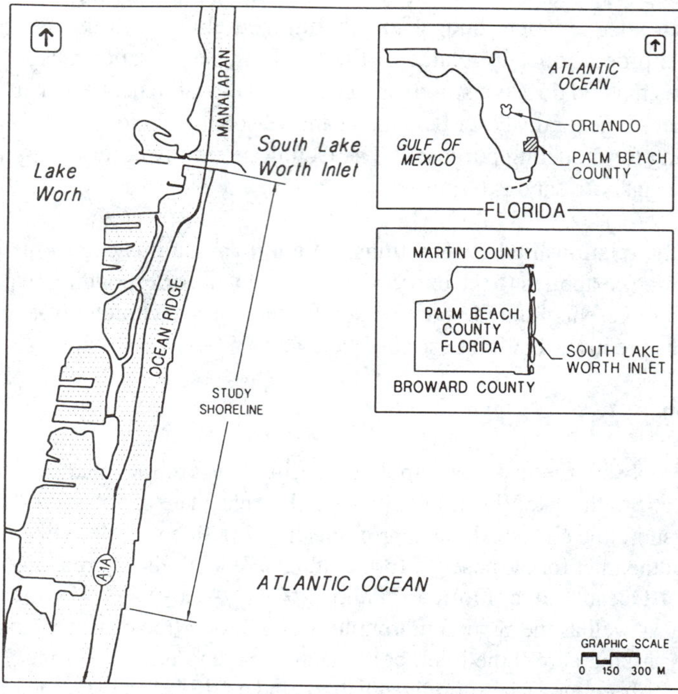

Figure 1: Location of South Lake Worth Inlet, Florida.

Despite natural and mechanical sand bypassing, the inlet acts as a partial littoral barrier to the net southerly drift of sand along Florida's southeast coast. Sand which drifts towards the inlet from the north is intercepted by the inlet's north jetty or diverted to the inlet's ebb and flood tidal shoals. Sand intercepted by the inlet's north jetty is impounded either in the fillet along the northern shoreline or mechanically bypassed to the southern shoreline. Sand diverted to the ebb shoal is stored either in the shoal platform, transported to the flood shoal and sand trap, or naturally bypassed to the southern shoreline. It is estimated that approximately 18 percent of the net southerly drift is intercepted and lost to the inlet system, resulting in a net sediment deficit along the downdrift shoreline. The net southerly sand transport in the vicinity of the inlet may vary between 134,000 and 172,000 cubic meters per year.

Sediment Budget. To quantify sand transport in the vicinity of South Lake Worth Inlet, Olsen Associates, Inc. (1990) developed a detailed nine-component sediment budget which quantitatively described sediment transport paths at South Lake Worth Inlet. The sediment transport components of the sediment budget are highlighted in Figure 2. Six of the nine sediment transport components of the sediment budget were developed from available data. Using these values and the governing equations of the sediment budget, estimates of the natural bar bypassing and the net drift rate south of the inlet were calculated for a range of net littoral drift rates north of the inlet.

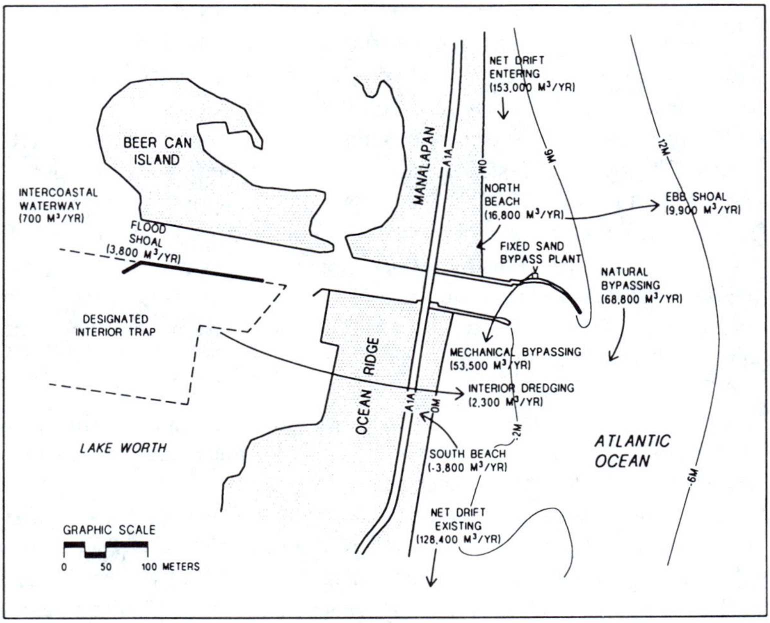

Figure 2: Sediment transport components for South Lake Worth Inlet, Florida.

Data collected since the 1967 modifications suggest that approximately 45 percent of the southerly net drift is bypassed across the ebb tidal shoal and bypassing bar. An additional 35 percent of the net southerly drift is bypassed by the fixed sand transfer plant.

Mechanical bypassing at the inlet is achieved with the inlet's sand transfer plant, which transports sand from the northern shoreline to a location between 60 and 150 meters south of the inlet. The plant is located atop the north jetty and bypasses sand from the northern to southern shoreline at a typical rate of about 53,500 cubic meters per year. Natural sand bypassing occurs along the inlet's ebb tidal shoal and bypassing bar platform. The bypassing bar attaches to the southern shoreline about 600 to 900 meters south of the inlet.

Because approximately 80 percent of the net transport is bypassed across the inlet, it was essential that this input to the study shoreline be included in model simulations.

Therefore, both mechanical and natural sand bypassing data (which were assumed to represent future bypassing conditions) were developed with the method described herein.

Mechanical Sand Bypassing Data. The relationship used to develop mechanical sand bypassing data representative of future conditions was determined from comparison of longshore sand transport data -- computed from offshore wave hindcast data -- and historical sand bypassing records. Mechanical bypassing records are available for the South Lake Worth Inlet sand transfer plant for the period from 1967 to the present. The bypassing rates are compiled from operation records and measured bypassing rates at the plant. WIS hindcast data are available for the area offshore of South Lake Worth Inlet for the period from 1956 to 1995. A time series of longshore sand transport was computed using these WIS hindcast wave data. The period for which the two databases overlap and are most reliable is 1970 to 1990. Therefore, it is this twenty year period for which longshore sand transport and mechanical sand bypassing data were compared.

The sand bypassing data were compared to both the net and southerly longshore transport components. This comparison revealed that mechanical bypassing is more closely correlated to the southerly transport component than the local net transport rate. (In fact, the mechanical sand bypassing plant operates only during periods of southerly directed transport.) The computed southerly transport data are plotted against the measured bypassing data in Figure 3. These data represent the twenty year period from 1970 to 1990. Inspection of the figure suggests a general trend in the data, where the magnitude of bypassing is related to the magnitude of southerly directed sand transport. The best-fit curve is of the form:

$$MB = aQ^{b} \tag{2}$$

where Q is the computed southerly transport rate from Eq. 1 (with K=1.0), MB is the reported mechanical bypassing rate, each in units of cubic meters per month, and a and b are coefficients. In this instance, the coefficients a and b were 29.0 and 0.54, respectively.

To demonstrate the effectiveness of this expression to represent actual mechanical sand bypassing data, a 9-year time series of southerly sand transport was applied to the expression to create a predicted 9-year time series of mechanical bypassing. The actual and predicted bypassing data for this period are presented in Figure 4. It is noted that areas of disagreement between actual and predicted bypassing records (such as the predicted spike in the latter part of 1974) may result from operational limitations of the sand bypass plant. Nonetheless, the comparison suggests that the mechanical sand bypassing at the inlet is correlated reasonably well with the incident wave climate.

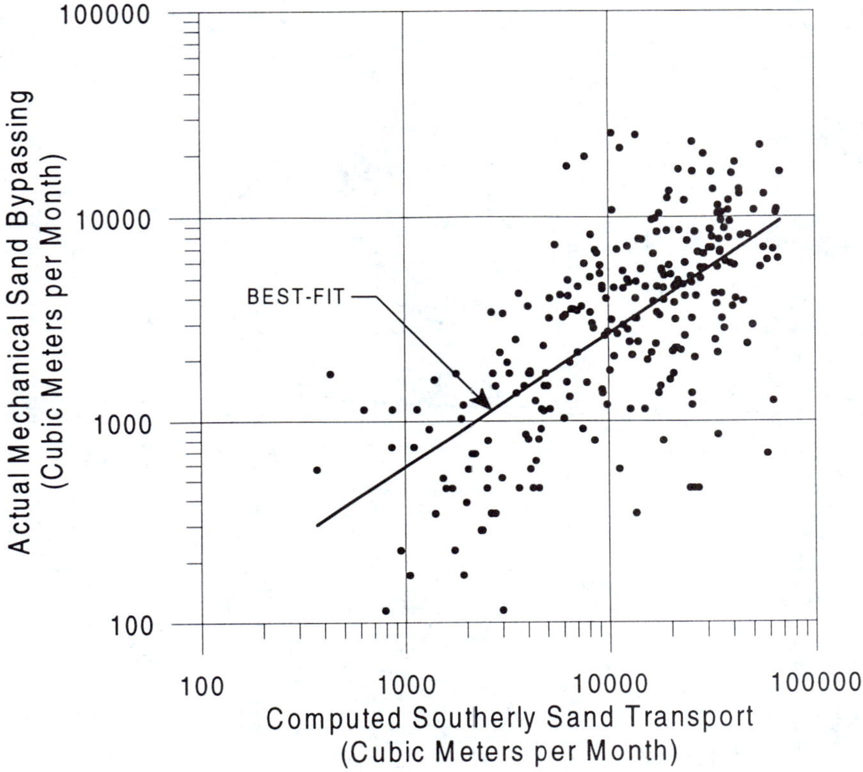

Figure 3: Correlation between actual mechanical sand bypassing (1967-1990) and computed southerly sand transport at South Lake Worth Inlet, Florida.

Natural Sand Bypassing Data. A time series of natural sand bypassing related to the longshore sand transport time series was also required as input to the shoreline change model. Actual measurements of the quantity and occurrence of natural bypassing, however, are not available at most inlets, including South Lake Worth Inlet. Therefore, the inlet's sediment budget and the assumption that natural bypassing occurs only during periods of southerly directed transport were used to formulate a natural bypassing time series. That is, the "natural bypassing" time series was developed by assuming that north to south sand transport across the bypassing bar is proportionately correlated to the southerly longshore sand transport component. This assumption was based, in part, upon the comparison of computed southerly sand transport and recorded mechanical sand bypassing. The proportionality constant for natural bypassing (0.45) was determined from the inlet's sediment budget (Figure 2); i.e.,

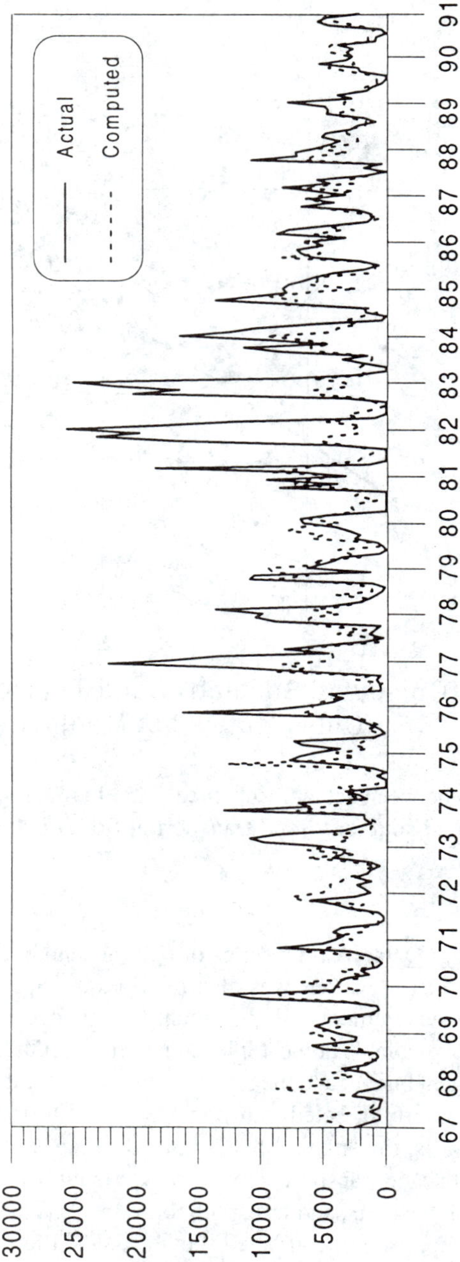

Figure 4: Time Series of actual and computed mechanical sand bypassing at South Lake Worth Inlet, Florida.

$$NB = 0.45Q \quad (3)$$

where *NB* is the rate of natural bypassing and Q is computed southerly transport rate from Eq. 1 (with K=1.0).

Shoreline Change Simulations. For purposes of predicting future shoreline change, a representative four year wave time series was developed for input to the GENESIS model. This time series was selected from the entire 38 year WIS hindcast time database as being most representative of average annual and seasonal variations in wave conditions offshore of the study site. Using this representative wave time series, corresponding mechanical and natural bypassing time series were computed using Eqs. 2 and 3, respectively. Both the wave and bypassing time series were used as input to the GENESIS for model simulations.

Bypassing Locations. The locations along which bypassed sand is supplied to the shoreline were also considered in the shoreline change investigation. As noted above, the study shoreline is supplied with bypassed material along two general areas. The sand which is mechanically bypassed is supplied to the shoreline within 60 to 150 meters south of the inlet. The sand which is bypassed naturally is supplied to the shoreline at the point where the natural bypassing bar attaches to the shoreline. This attachment point is some 600 to 900 meters south of the inlet.

Consideration of the locations along which the sand is supplied to the study shoreline was necessary because the alongshore distribution of sand transport is characterized by extreme accelerations and decelerations due to the effects of the inlet's ebb tidal shoal and bypassing bar on the incident wave climate. Because of these variations, shoreline change is strongly influenced by the locations along which bypassed material is supplied to the shoreline. To accommodate this sensitivity, sand was added to the model at the locations representative of the mechanical sand bypass plant discharge points and the average location at which the natural bypassing bar attaches to the shoreline.

As configured, the GENESIS model does not consider changes to the wave climate caused by such shallow features at South Lake Worth Inlet's ebb tidal shoal and bypassing bar. Because these changes to the wave climate are important to shoreline change and the movement of bypassed sand, a method was developed by Bodge, et.al. (1996) to consider the effects of nearshore bathymetric features on the wave climate. Briefly, the method computes the breaking waves conditions through a grid-based refraction analysis for wave conditions which are representative of the offshore time series. The breaking wave information is then "back-refracted" to a nearshore wave reference line which the GENESIS model is designed to accept. This technique, therefore, includes the variations in breaking wave conditions created by nearshore bathymetric features (such as ebb tidal shoals and bypassing bars) in the wave data input to the GENESIS model. This provides for more accurate representation of the local transport conditions in the areas were sand is supplied to the shoreline through bypassing.

Comparison Bypass Model Comparison. To demonstrate the necessity of accurately representing the relationship between longshore sand transport and sand bypassing, two other representations of sand bypassing were developed. These include assuming that an annual equivalent volume of bypassing is supplied to the shoreline at a constant rate (*uniform*). The other assumes that an annual equivalent volume of bypass sand is supplied at the beginning of each year (*plug*). A graphical representation of these concepts is provided in Figure 5.

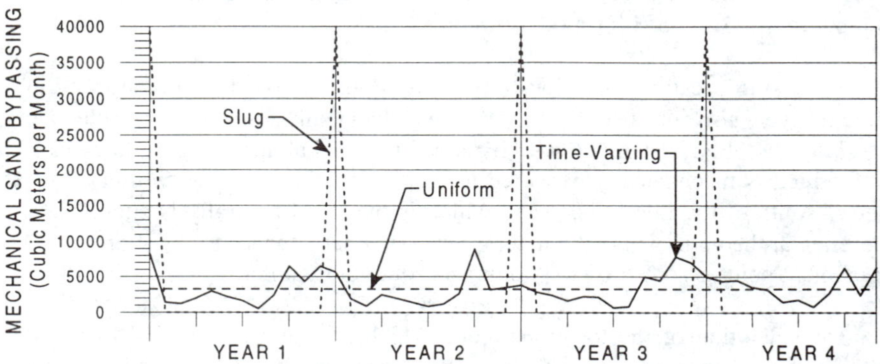

Figure 5: Concept of sand bypassing models.

These two bypassing models along with the bypassing model proposed herein (*time-varying*)were used to develop bypassing data for comparative model simulations. The representative four year wave time series described above was used to develop the bypassing data. Each representation of bypassing supplies an equivalent volume of sand to the model domain over the simulation period. For verification purposes, where both initial and final measured shorelines are available, model simulations were conducted for the 18 year period between 1975 and 1993. The results of the three comparative simulations are shown in Figure 6. In the figure, it is clear that the *time-varying* bypassing model provides a more accurate representation of measured shoreline change. The *uniform* and *slug* bypassing models fall short of the time-varying model by producing a significant build of sand along the northernmost section of the study shoreline. This is due to sand being supplied to the shoreline during periods of inactive longshore transport; thus allowing a build up of material which is not transported away when transport becomes active. To more clearly demonstrate these results, the differences between measured and computed shorelines are provided in Figure 7.

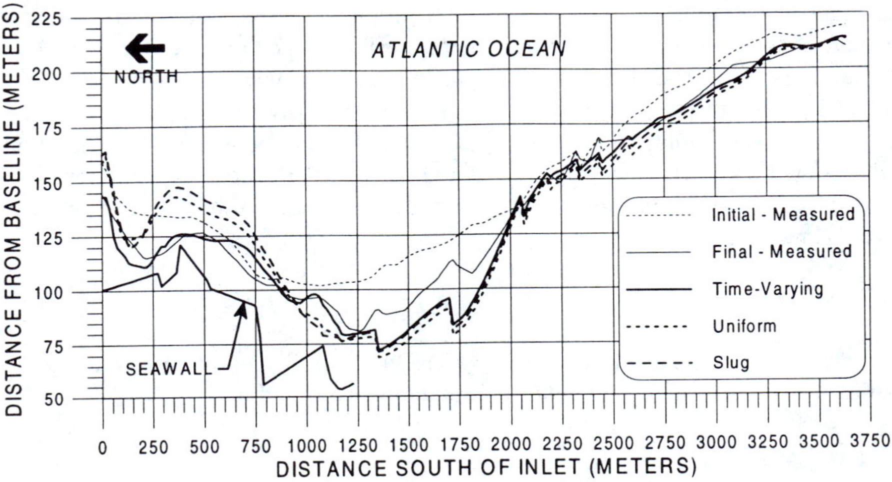

Figure 6: GENESIS shoreline change results using three sand bypassing models.

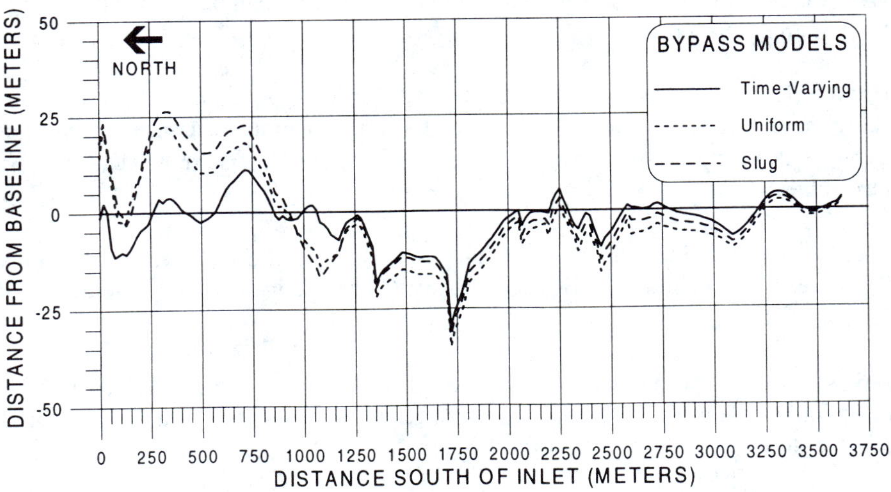

Figure 7: Difference between measured and computed shorelines.

Conclusions

A method is proposed to relate mechanical and natural inlet sand bypassing to the incident wave climate and develop a time series of bypassing for use as input to shoreline change models. Data requirements for the method include coincident times series of (1) local wave climate and (2) sand bypassing are required. Application of the model indicates that the magnitude and phasing of sand bypassing events with respect to incident wave conditions should be considered when modeling shoreline change adjacent to tidal inlets where sand bypassing occurs. The proposed method was used to successfully simulate shoreline change for the downdrift shoreline of South Lake Worth Inlet, Florida. U.S.A.

Acknowledgment

The author is sincerely grateful to Dr. Kevin R. Bodge of Olsen Associates, Inc. for his technical insight and guidance during this investigation.

References

Bodge, K.R., C.G. Creed, and A.W. Raichle, (1996). "Improving input wave data for use with shoreline change models," Technical Note, *J. Wtrwy., Port, Coast., and Oc. Engrg*, ASCE, 122 (5), 259-263.

Hanson, H., (1987). "GENESIS, a generalized shoreline change model for engineering use," Report No. 1007, Department of Water Resources Engineering, University of Lund, Lund, Sweden, 1987.

Hanson, H. and N.C. Kraus, (1989). "GENESIS: generalized model for simulating shoreline change," Technical Reference, Report 1, Coastal Engineering Research Center, U.S. Army Corps of Engineers Waterways Experiment Station, Vicksburg, MS, December, 1989.

Olsen Associates, Inc. (1990)., "South Lake Worth Inlet Sand Management Plan," Submitted to the Palm Beach Board of County Commissioners, Olsen Associates, Inc., Jacksonville, FL.

U.S. Army Corps of Engineers, (1984). "Shore Protection Manual," U.S. Army Corps of Engineers, Coastal Engineering Research Center, U.S. Government Printing Office, Washington, D.C.

"Palm Beach County, Florida, Shore Protection Project, General Design Memorandum for Ocean Ridge Segment," *Rep. Prepared for U.S. Army Corps of Engineers,* Jacksonville , Fla.

CHAPTER 348

AN APPROACH TO MODELING INLET AND BEACH EVOLUTION

Mark B. Gravens[1], A. M., ASCE

Abstract

Improved understanding of the physical processes controlling inlet and adjacent beach morphology is required to manage sand resources in the vicinity of tidal inlets and to reduce costs associated with structural rehabilitation, maintenance dredging of navigation channels, and mitigation of adverse impacts on adjacent shores. A quantitative modeling capability for estimating sediment fluxes, morphology evolution, and the interaction between the inlet and adjacent beaches is a goal of engineering research. This paper presents the initial phase of work aimed at developing empirical relationships between the physical processes and resulting inlet morphologic response, which are needed to formulate a modeling capability. First, the technical and computational requirements of the model are outlined. Then, a conceptual model of inlet sediment bypassing and a brief overview of a recently developed model (De Vriend et al. 1994) which appears to satisfy most of the model requirements is given. Finally, preliminary results of an empirical analysis of inlet ebb shoal geometric characteristics are presented.

Introduction

The presence of tidal inlets along a mainland or barrier island sandy shoreline represents a major morphological perturbation in the otherwise generally linear features (dune, berm, shoreline, longshore bar) that make up the coastal zone. The shoreface morphology at tidal inlets is a product of the changes in the physical forces that form the beachface. In the vicinity of tidal inlets, longshore currents generated by waves breaking at oblique angles to the shoreline interact with concentrated cross-shore currents that pass through the inlet gorge. Littoral sediments carried in the longshore current are swept into the inlet interior or are jetted offshore by the tidal currents. Over time, if the

---

[1] Research Hydraulic Engineer, U.S. Army Engineer Waterways Experiment Station, Coastal and Hydraulics Laboratory, 3909 Halls Ferry Rd., Vicksburg, MS, 39180-6199, USA.

inlet cross-section is stable, depositional shoal features will develop. Walton and Adams (1976), Marino and Mehta (1987), and others have shown that the ebb shoal tends toward an equilibrium volume which can be related to the tidal prism. If flood shoal volumes are also assumed to approach an equilibrium condition or a near constant growth rate, then it follows that under natural conditions, an inlet system could form and in time (10's to 100 years), reach a near-equilibrium condition, in which a large percentage of the littoral drift bypasses the inlet to downdrift shorelines.

However, because navigable tidal inlets represent an important economic resource, many inlets have been stabilized with jetties or improved for navigation by dredging. Inlet stabilization by jetties has been shown to produce a seaward displacement of the ebb shoal (Kraus et al. 1994). Seaward movement of the ebb-tidal shoal can result in an increase in the volume of material comprising the shoal. Dean (1993) presented a conceptual model together with field examples that indicate that maintaining a channel through a tidal inlet that is deeper than the natural channel depth causes sediment to flow into the maintained channel, which comes at the expense of volumetric erosion of the beaches adjacent to the inlet. Consequently, engineering practices involving inlet stabilization by jetties and navigation channel dredging are increasingly cited for their role in the observed persistent erosion of adjacent shorelines.

In the United States, federal responsibility for the design, operation, and maintenance of inlet stabilization structures and navigation channels lies with the U.S. Army Corps of Engineers. Improved sand management in the vicinity of tidal inlets will reduce costs associated with structural rehabilitation, maintenance dredging, and mitigation of adverse shoreline impacts. However, a better understanding of the physical processes controlling inlet morphology is required, including estimating sediment fluxes, morphology evolution, and the interaction between the inlet and adjacent beaches. A quantitative modeling capability can help improve our understanding of the interaction between an inlet system and its adjacent beaches. The model to be developed is intended to aid in the analysis and prediction of inlet adjacent shoreline evolution over mid- to long-term time scales (1 to 10's of years), and spatial scales up to 10's of kilometers.

Model Requirements

Goals and Development Philosophy. The primary goal of the research discussed in this paper is to develop a predictive methodology to estimate quantitatively the potential impacts of engineering activities on shorelines adjacent to modified or engineered inlet systems. The desired use of the numerical model is to predict shoreline response within the reach of shoreline influenced by the inlet in response to different engineering activities. An inlet-shoreline response model is recognized for its value in providing a quantitative basis for the comparison of alternatives. Procedures for application also will include guidance for producing an ensemble of solutions based on Monte Carlo-type simulations produced from a combination of plausible input conditions. The ensemble of solutions then may be examined to determine the range and the frequency of the outcomes within that solution set. Because this model is being designed for use by practicing coastal engineers, model development is being undertaken with the expressed intent of capturing the evolutionary trends of the shoreline and maximizing range of applicability. This development philosophy stems from the observation that models the engineer is most likely to use are those which he or she can understand and defend, and those which require minimum adjustment of calibration parameters. Traditional approaches to coastal engineering design are expected to become more reliable with the addition of objective and quantitative estimates of an inlet's total littoral impact, which will be provided by the numerical model.

Prediction Requirements. Inlets impact adjacent shores primarily by their action as a sediment sink; therefore, a fundamental requirement of the model is the capability to predict whether and to what degree an action will change the net volume of sand lost (trapped against stabilization structures or otherwise stored in inlet associated shoal features) from the adjacent shorelines. A related requirement is the need to predict the distribution of that sand deficit along the adjacent beaches. Furthermore, regulatory agencies asking for a prediction of project performance resulting from both natural and engineered site changes leads to the requirement for model prediction of, for example, sea-level rise, extended mechanical failure of bypassing equipment, and/or planned or unexpected delays in renourishing a project or performing routine maintenance dredging of a navigation channel. The parameter of primary interest in many situations and therefore required from the model predictions is the shoreline location. For other situations the profile geometry is also of keen interest and less frequently sediment volume in various locations around the inlet-adjacent beach system are required. Table 1 provides a list of specific model requirements. Predictions are required for engineering time scales (1 to 10's of years) and spatial scales corresponding with the total longshore extent of the inlet's impact (up to 10's of kilometers).

Physical Processes. Conceptually, inlets impact adjacent shorelines in two fundamental ways. One impact is the inlet channel as a sediment sink which results in a net loss of sediment from the adjacent beach systems. The second impact is the altering of wave and current patterns, which results in a change in sediment transport patterns. Therefore, the physical processes responsible for these fundamental functions of an inlet system must be captured in the numerical model at some appropriate level. The physical processes important to determining an inlet's overall impact to the littoral system are those which result in net sediment impoundment, jetty leakage, channel shoaling, and shoal formation. Sediment conservation requires that the net volumetric impact to the adjacent shorelines equal the net volume of sediment lost to the inlet. To be useful (to estimate the inlet's impact on adjacent shores), the model must compute the rate at which sediment is delivered to the inlet, the rate at which sediment is moved through the inlet (along its axis), and the rate at which sediment is transported across, around and away from the inlet. These quantities will enable estimation of that fraction of the littoral drift which bypasses the inlet from that which is trapped within

Table 1. Model Requirements

Model Requirement	Comments
Foreshore slope changes	Beach profile slopes at stabilized tidal inlets are typically steeper updrift of the inlet and milder downdrift of the inlet as compared to beach profiles not influenced by inlet system.
Sediment point/line source and sink capabilities	To effectively evaluate many typical engineering activities (e.g., sand bypassing, beach nourishment), the capability to model point and/or line sediment sources and sinks is required.
Structure specification	To evaluate the influence of modifying inlet stabilization structures, the model must be sensitive to user-specified characteristics of the structures (e.g., length, orientation and configuration, permeability, weir sections, elevation, etc.).
Wave transformation and the possibility of multiple wave breaker lines	Wave transformation, including shoaling, refraction, diffraction and the possibility of multiple wave breaker lines with differing breaker characteristics, is required to capture the influence of the ebb shoal feature on the incident wave climate.
Hindcast capability	To demonstrate reasonability in predictions, the model must be able to simulate pre-project adjacent beach evolution.
Forecast capability	To be useful in evaluating the relative merits of competing design alternatives, the model must be able to provide mid- to long-term (1 to 10's of years) existing- condition forecasts as well as with-project forecasts.
Scoping mode / design mode applications	To quickly and efficiently provide estimates of the impact of preliminary design considerations, as well as to provide detailed estimates of refined design alternatives, the model is required to have the capability of being operated with differing levels of sophistication and data requirements.
Input data requirements	Data required to operate the model must be consistent with typically available data and/or data which can be reasonably collected.
Compatible with sediment budget methodology	Model input and output should be compatible with classical sediment budget methodologies (e.g., sediment budget information can be input to the model and model outputs can be integrated back into the sediment budget).

the inlet or is otherwise lost to the littoral system. How the net volumetric impact of the inlet is distributed along the adjacent shores is determined by hydrodynamic and sediment transport processes which exist away from the inlet. In short, modeling sediment movement in the vicinity of inlets ultimately involves essentially all of the most complicated processes of coastal sediment transport.

Target Inlet Type and Behavior. In nature, inlet systems vary widely in their characteristics and behaviors, and it is doubtful that one model or even one modeling approach would be equally applicable to all inlet systems. It is expected that the developed model will not be able to adequately predict all inlet behaviors and, therefore, will be more applicable to certain types of inlets than to others. Consequently, it is essential that the target inlet type and behavior are clearly defined. The inlet systems for which the model is being developed may be characterized as being relatively stable, often the result of stabilization structures for the purpose of providing safe navigation. The system is viewed as being in a state of dynamic equilibrium, meaning that the inlet system responds to seasonal variations in the wave and water level climate as well as to storm events. However, it is assumed that these responses are short-term or seasonal variations imposed upon an underlying state of equilibrium and do not have a destabilizing effect on the inlet system nor do they significantly alter the underlying equilibrium state. Engineering activities, on the other hand, can both temporarily displace the system from its equilibrium condition (e.g., sand mining from the ebb shoal) and in some cases change the inlet's equilibrium state (e.g., by changing the tidal prism). In either case, the inlet system will evolve toward either its old or new equilibrium state with impacts ultimately occurring on the adjacent beaches. Natural sediment bypassing is assumed to be a semi-continuous process, and the possibility of random or periodic bar welding on downdrift shorelines is excluded. The model will not allow the inlet to migrate as many natural inlet systems do. Furthermore, because sediment transport along the outer ebb delta is computed as a function of the incident wave energy the model will be more applicable to wave dominated and mixed-energy inlets than it will to tide dominated inlets.

Conceptual Model of Inlet Sediment Bypassing

Inlet sediment bypassing is the process by which littoral material delivered to the inlet from the updrift beach makes its way across the inlet system and is integrated into the littoral material on the downdrift beach. In a series of papers (Bruun and Gerritsen 1959, Bruun 1967, and Bruun 1978), a model has been presented that describes two possible pathways for inlet sediment bypassing by natural action: 1) bypassing on an offshore bar, and 2) bypassing by tidal flow action. In bar bypassing, littoral material that is delivered to the inlet from the updrift shorelines is transported along a submerged bar in front of the inlet system and delivered to the downdrift shoreline. In tidal flow bypassing littoral material is pulled through the inlet by flood currents and then discharged out of the inlet by ebb currents in the downdrift direction. Though sediment may cycle several times through the inlet, it is ultimately bypassed to the downdrift shoreline. Bruun indicated that in most cases a combination of both of these mechanisms for natural bypassing is active but that the dominant process could be identified by examining the ratio of the tidal prism over the total littoral drift.

FitzGerald (1988) presented three conceptual models to explain inlet sediment bypassing along mixed energy coasts. These models are categorized according to the stability of the inlet throat and movement of the main ebb channel. These are: 1) inlet migration and spit breaching, 2) stable inlet processes, and 3) ebb-tidal delta breaching. In the first model, the entire inlet migrates downcoast leaving a trailing spit extending from the updrift barrier. The resulting elongation of the inlet channel becomes hydraulically inefficient for tidal flow between the ocean and the bay. Ultimately the spit is breached during a storm and the new channel which provides a shorter route for tidal exchange will normally remain open while the less efficient old inlet channel closes. In the second model, the inlet throat position and the main ebb channel do not migrate. Sand bypassing at these inlets occurs through the formation, landward migration and welding of large bar complexes to the downdrift shoreline. In the third model, the location of the inlet throat is stable, but the main ebb channel migrates. The dominant direction of longshore sediment transport causes a preferential accumulation of sand on the updrift side of the ebb-tidal delta which results in a deflection of the main ebb channel. Similar to the first model, the migration of the ebb channel ultimately results in a hydraulically inefficient channel and a new channel breaches through an updrift spillover lobe channel. Ebb-tidal delta breaching results in the bypassing of a large portion of the ebb delta sand bodies. Some of this sand fills the abandoned channel while the rest forms a large bar complex that ultimately migrates onshore and attaches to the downdrift shoreline. FitzGerald summarized by indicating that during the history of a particular inlet, all three bypassing processes may dominate at one time or another. He also states that Bruun and Gerritsen's (1959) bar-bypassing mechanism (the movement of sand around an inlet by wave action along the terminal lobe of the ebb-tidal shoal) is an active but secondary process at most mixed-energy tidal inlet systems.

For the target inlet type and behavior described previously we believe that the bypassing models of Bruun and Gerritsen (1959) are more applicable than those of FitzGerald (1988). From a long-term (10's of years) perspective the net effect of FitzGerald's bypassing models could be idealized as a sediment drift along the outer edge or crest of the ebb delta from the updrift to the downdrift island. However, for shorter time periods (up to 10 years) which are of interest to us, FitzGerald's models are not compatible with the approach we have adopted. For the development of a computational model for inlet adjacent beach evolution the mid- to long-term (1 to 10's of years), net transport patterns near the inlet are idealized as shown in Figure 1. This idealization is a substantial simplification of the real situation, but it is consistent with typically available data, the level of schematization envisioned for the computational model and our capability to estimate details of sand transport at tidal inlets.

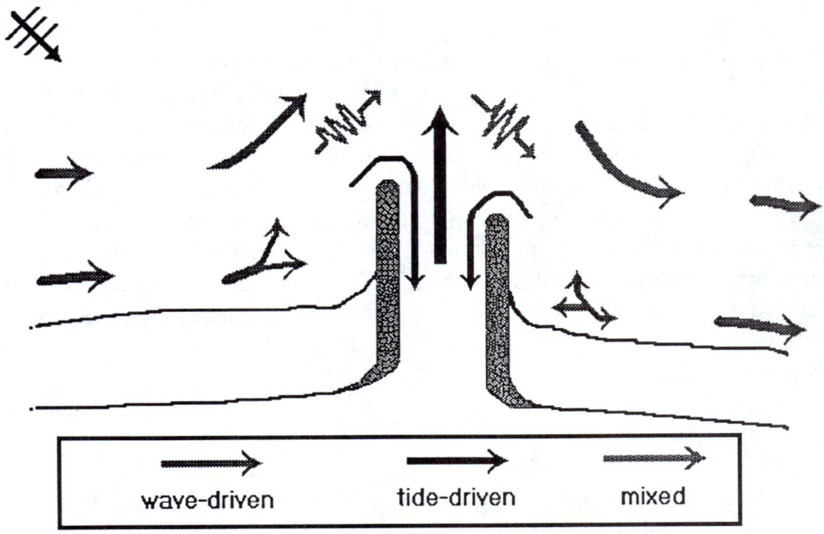

Figure 1. Conceptual model of inlet sediment bypassing.

Computational Approach

The computational approach selected for development of the inlet and adjacent beach evolution model falls into the class of models described by Kraus (1989) as multi-contour line models. This class of model was selected because the capabilities and characteristics of the model best satisfied the needs identified in a requirements analysis for inlet adjacent beach evolution modeling (Truitt and Bodge 1995). The initial development of a morphological behavior model for the outer delta of mixed-energy tidal inlets (De Vriend et al. 1994) tends to confirm that the selected computational approach has some merit and the interest of researchers. In the outer delta model, the two-line model concept of Bakker (1968) is extended to a situation where the coast is interrupted by a tidal inlet. The behavior of adjacent island coasts are modeled directly utilizing a two-line modeling approach (evolution of a beach line and a foreshore line); at the inlet, the beach line is discontinued because there is no subaerial beach profile at this location. The foreshore or inshore line, however, protrudes seaward at the location of the inlet, thereby representing the inlet delta or ebb shoal. Based on empirical relationships for the equilibrium state of the ebb shoal, the protrusion distance and geometric configuration of the ebb shoal are determined. The inlet-adjacent beach system and its evolution are described by: 1) computed wave-driven sediment fluxes along the adjacent island coasts and along the seaward edge of the ebb shoal, 2) specified net sediment fluxes through the inlet channel, and 3) the requirement for sediment continuity at six locations.

The outer delta model includes consideration of sediment fluxes through the main ebb channel as well as sediment fluxes through flood channels located near the beaches adjacent to the inlet. Sediment flux through the flood channels is fed by both the beach line and the inshore line. The equilibrium state of the delta (volume, seaward protrusion) is described using empirical relationships. A sediment balance is maintained throughout the evolution, and the sediment demand or supply by the basin is taken into account as an input parameter. The model output consists of the cross-shore positions of the beach line and the inshore line as a function of time and the longshore coordinate. It is noted, however, that the model developers (De Vriend et al. 1994) caution that the model results should be interpreted at a higher aggregation level (e.g., the trend of predicted change should be given more weight than the predicted magnitude of change). For more details on the outer delta model and its validation the reader is referred to the Master's thesis of Bilse (1993).

Empirical Analysis

Goals. To refine and expand on the initial model developed by De Vriend et al. (1994), a focused empirical analysis and parameterization of inlet geometry and processes has been undertaken. The goal of this research is to characterize the volume, geometric shape, and general location of the ebb shoal based on estimable quantities such as tidal prism, magnitude of net and gross longshore sand transport rates, and tidal range. U.S. inlets with rich temporal and spatial historical data sets are being analyzed. Specific analysis to be conducted include: 1) confirmation of the ebb shoal volume tidal prism relationship proposed by Walton and Adams (1976), 2) investigate the shoal volume to shoal protrusion relationship assumed by De Vriend et al. (1994), 3) the correlation, between the ebb shoal aspect ratio (longshore extent/cross-shore extent) and the tidal prism and/or net or gross longshore sand transport rates, 4) the relative stability of updrift and downdrift shoreline offsets, and 5) the correlation, between ebb shoal skewness (offset between ebb shoal centroid and inlet throat at jetty tips) and net and gross longshore sand transport rates.

Status. At present this analysis has included the geometric analysis of four tidal inlets along the eastern US Atlantic coast. The four inlets are: 1) Ocean City Inlet located in Ocean City, Maryland, 2) Barnegat Inlet located in central New Jersey, 3) Moriches Inlet located between Fire Island and Westhampton Beach on Long Island, New York, and 4) Shinnecock Inlet located between Tiana Beach and Southhampton Beach on Long Island, New York. The analysis has not progressed to the point where new or refined empirical relationships can be proposed, but procedures for the geometric analyses of ebb tidal shoals have been developed, and their application to the four inlet systems will be discussed in the remainder of this paper. (We plan to extend this geometric analysis to a number of other inlet systems in the future at which time a database of the physical driving forces will be integrated for the development of new or updated empirical relationships.)

Methods. The bathymetry data were obtained by three methods, standard hydrographic surveys using a vessel mounted fathometer, sea-sled beach profile surveys, and airborne LIDAR (LIght Detection And Ranging) surveys obtained using the helicopter based SHOALS (Scanning Hydrographic Operational Airborne Lidar Survey) system. In all cases some combination of these survey methods were used to obtain the complete bathymetric coverage data sets. The procedure begins by developing a digital terrain model (DTM) using the original bathymetric data as input. Contours were then generated with reference to the National Geodedic Vertical Datum (NGVD). A polygon was then drawn around the ebb shoal. Typically, the offshore boundary of the polygon followed along the 10- to 11-m NGVD contour (the landward most contour that does not protrude seaward at the ebb shoal). The landward boundary followed along the 0-m NGVD contour. The lateral boundaries were selected at an alongshore location corresponding to where the contours began to protrude seaward around the shoal. A least-squares best-fit line was computed for contours at 1-m intervals outside the ebb shoal polygon. These computed best-fit contours on either side of the ebb shoal polygon were joined and then smoothed using the cubic B-spline method. The "no ebb shoal" bathymetry was estimated by digitizing the smoothed best-fit contours and then generating a DTM of the result. The next step involved taking the difference between the idealized "no ebb shoal" DTM and the original bathymetry DTM within the ebb shoal polygon. The result of this differencing operation is a DTM that we will refer to as the "residual" ebb shoal topography. This terminology describes the morphological residual of the formation of the ebb-tidal shoal; topography is used because the residual is described in terms of positive elevations and no longer represents a bathymetry. The residual ebb shoal topography was used to estimate the volume and geometric configuration of the ebb shoal. Figure 2 provides plots of the ebb shoal bathymetry, the idealized no ebb shoal bathymetry, and the residual ebb shoal topography for each of the four inlet systems examined. This procedure is similar to that used by Dean and Walton (1973) but employs modern terrain modeling software which, to some extent, reduces the subjectivity of the analysis.

Results. Figures 3-6 are contour plots of the residual ebb shoal topography with the extracted geometric characteristics indicated for Barnegat, Ocean City, Shinnecock, and Moriches inlets, respectively. Shown in the figures are the 1-, 3-, and 5-m contours of the residual ebb shoal topography with their alongshore and offshore dimensions. Also indicated is the location of the centroid of the residual ebb shoal topography with respect to the inlet throat at the tips of the inlet jetties. Without examination of the site-specific physical driving forces (e.g., tidal prism, tide range, wave energy, net and gross longshore sand transport rates etc.), we are unable at this time to draw conclusions or explain the differences in the geometric configuration of these four inlet systems.

Conclusion

This paper documents the initial phase of work that has been undertaken to develop a needed modeling capability in the vicinity of tidal inlets. The requirements

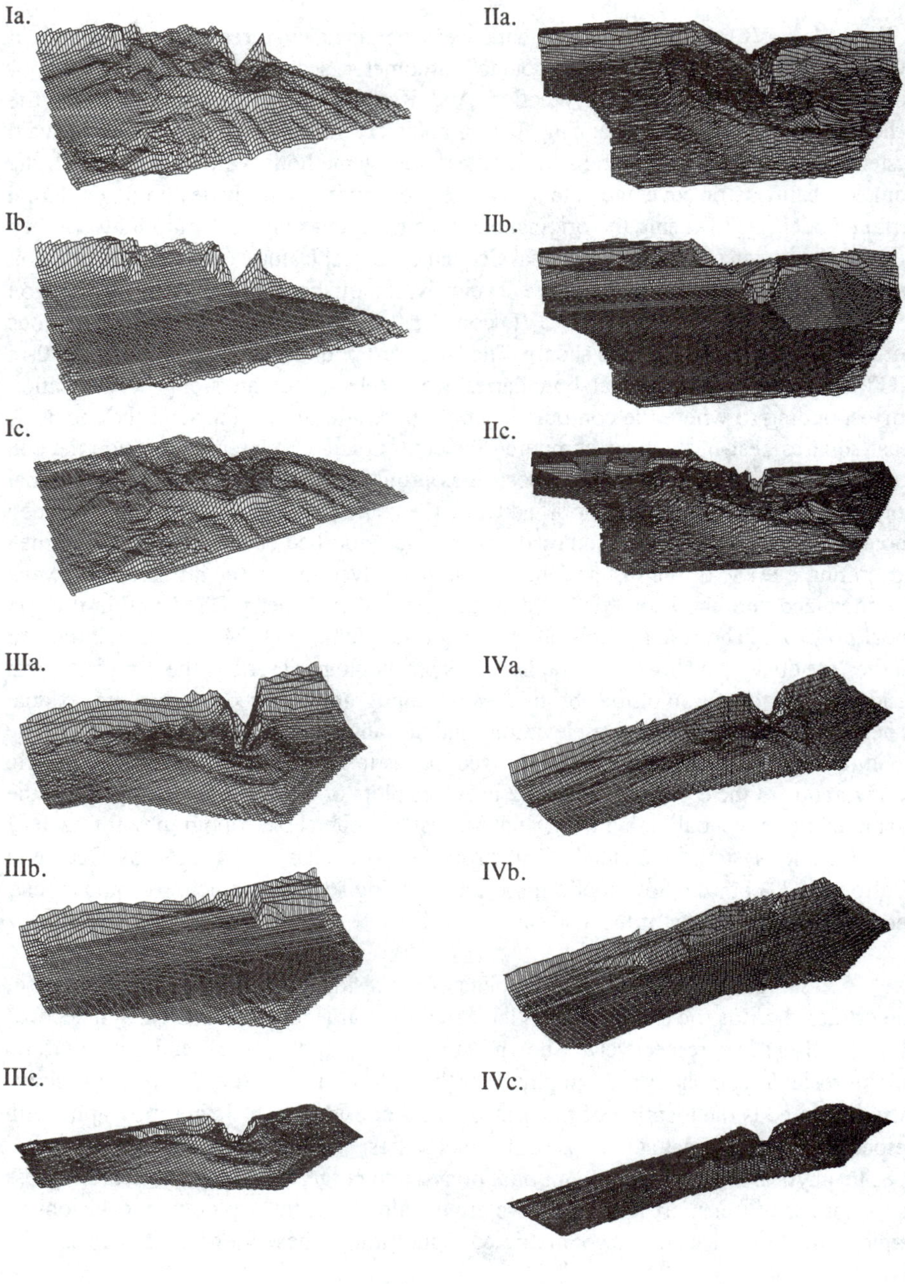

Figure 2. Plots of ebb shoal at Barnegat (I), Ocean City (II), Shinnecock (III), and Moriches (IV) inlets. (a) ebb shoal bathymetry, (b) idealized no ebb shoal bathymetry, and (c) residual ebb shoal topography.

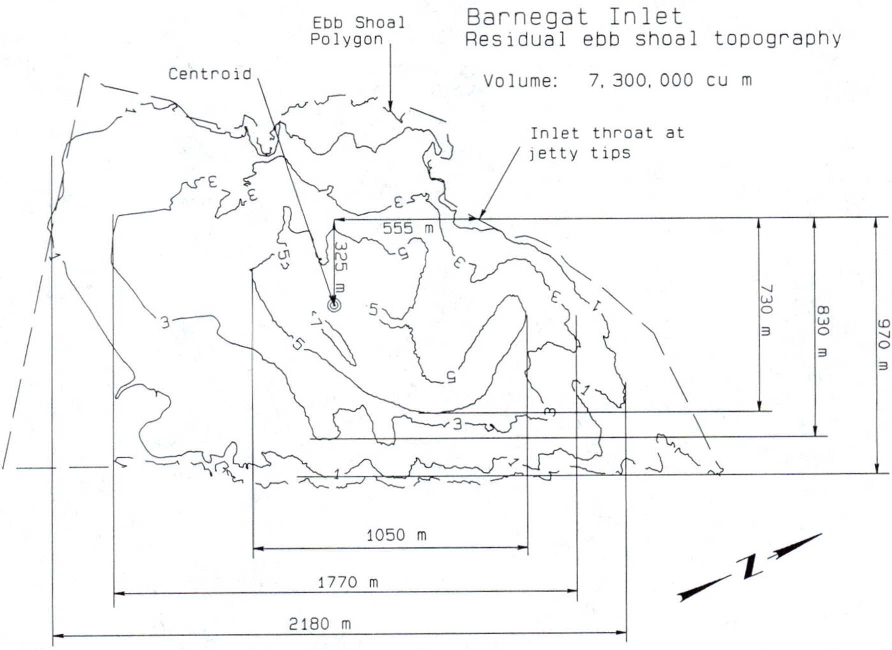

Figure 3. Barnegat Inlet residual ebb shoal topography and geometric attributes.

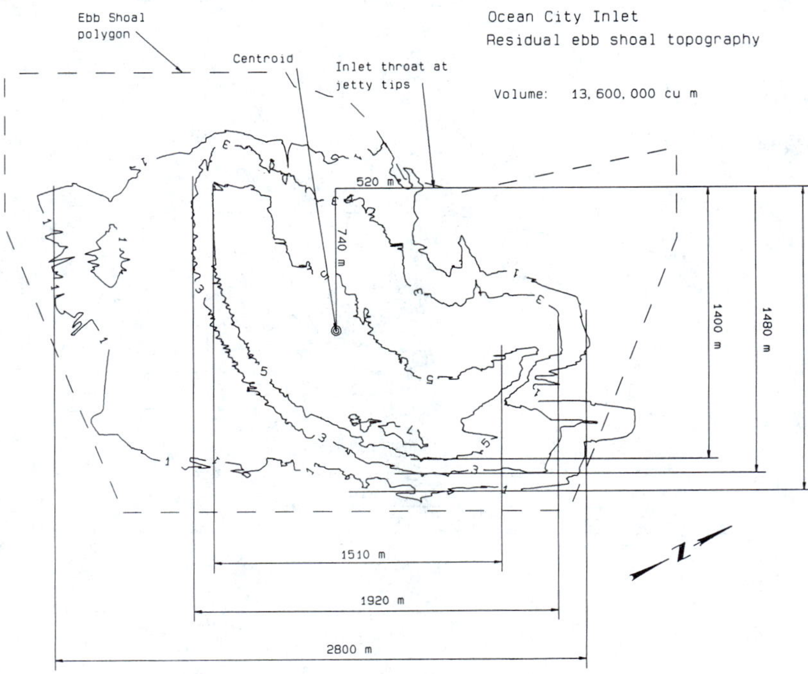

Figure 4. Ocean City Inlet residual ebb shoal topography and geometric attributes.

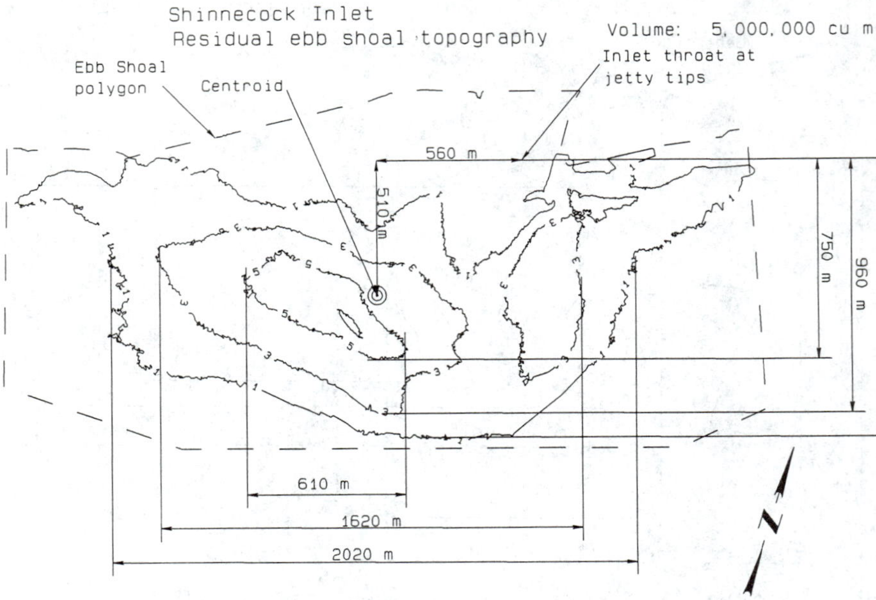

Figure 5. Shinnecock Inlet residual ebb shoal topography and geometric attributes.

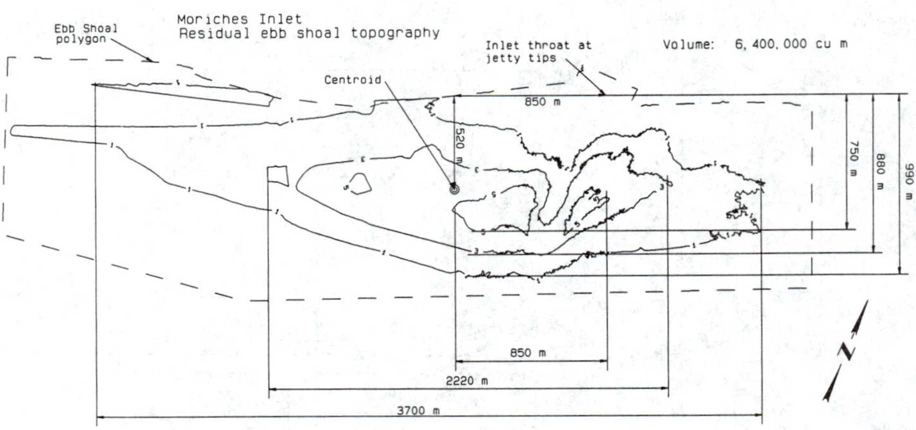

Figure 6. Moriches Inlet residual ebb shoal topography and geometric attributes.

of the numerical model from a philosophical and practical standpoint were documented. The characteristics of the inlet system for which the tool is expected to be applicable were identified. A conceptual model of inlet sediment bypassing consistent with the envisioned computational model and sediment budget methodology was presented. A recently proposed computational approach to modeling the behavior of inlet and adjacent beach systems that appears to satisfy many of the model requirements defined herein and, which we plan to extend through this research, was briefly reviewed. Procedures for characterizing the geometry of an inlet system were presented together with the application of those procedures to four inlet systems located along the central U.S. Atlantic coast.

Acknowledgments

This study was performed under the work unit *Shoreline Change Adjacent to Inlets* of the Coastal Inlets Research Program of the U.S. Army Corps of Engineers, Waterways Experiment Station. Permission was granted by the Chief of Engineers to publish this information.

References

Bakker, W. T. 1968. "The Dynamics of a Coast with a Groyne System," *Proceedings*, 11th International Conference on Coastal Engineering, ASCE, New York, NY. pp. 492-517.

Bilse, D. P. 1993. "A Large Scale Model of the Morphological Behaviour of the Outer Delta of Tidal Inlets," Master Thesis, The International Institute for Infrastructural, Hydraulic and Environmental Engineering, Delft, The Netherlands.

Bruun, P. 1967. "Tidal Inlet Housekeeping," *Journal Hydraulics Division*, ASCE, Vol 93, No. HY5, pp. 167-184.

Bruun, P. 1978. "Stability of Tidal Inlets: Theory and Engineering," Developments in Geotechnical Engineering 23, Elsevier Scientific Pub. Co., New York, 510 p.

Bruun, P., and Gerritsen, F. 1959. "Natural By-passing of Sand at Coastal Inlets," *Journal Waterways and Harbors Division*, ASCE, Vol 85, WW4, pp75-107.

Dean, R. G. 1993. "Terminal Structures at Ends of Littoral Systems", *Journal of Coastal Research*, SI 18, pp.195-210.

Dean, R. G., and Walton, T. L. 1973. "Sediment Transport Processes in the Vicinity of Inlets with Special Reference to Sand Trapping," *Proceedings*, Second International Estuarine Research Conference, Academic Press, 129-149.

De Vriend, H. J., Bakker, W. T., and Bilse, D. P. 1994. "A Morphological Behaviour Model for the Outer Delta of Mixed-Energy Tidal Inlets," *Coastal Engineering*, 23: 305-327.

FitzGerald, D. M. 1988. "Shoreline Erosional-Depositional Processes Associated with Tidal Inlets," In: Aubrey, D. G. and Wisher, L, (eds.), 1988. Hydrodynamics and Sediment Dynamics of Tidal Inlets. Lecture Notes on Coastal and Estuarine Studies, Vol 29, Springer Verlag, New York, 454 pp.

Kraus, N. C. 1989. "Beach Change Modeling and the Coastal Planning Process," *Proceedings*, Coastal Zone '89, ASCE, New York, NY. pp.553-567.

Kraus, N. C., Gorman, L. T., and Pope, J. 1994. "Kings Bay Coastal and Estuarine Physical Monitoring and Evaluation Program: Coastal Studies," Technical Report CERC-94-9, Volumes I and II, US Army Engineer Waterways Experiment Station, Vicksburg, MS.

Marino, J. N., and Mehta, A. J. 1987. "Inlet Ebb Shoals Related to Coastal Parameters," *Proceedings*, Coastal Sediments '87, ASCE, New York, NY. pp. 1608-1623.

Truitt, C., and Bodge, K. R. 1995. "Requirements Analysis for Inlet Adjacent Beach Evolution Research," Mote Marine Laboratory Technical Rep. No. 421, Sarasota, FL.

Walton, T. L., and Adams, W. D. 1976. "Capacity of Inlet Outer Bars to Store Sand," *Proceedings*, 15th International Conference on Coastal Engineering, ASCE, New York, NY. pp. 1919-1937.

CHAPTER 349

Hydrodynamics of a bar in a flood channel - the Westerschelde estuary

Claire Jeuken[1]

Abstract

Several long-term current meter observations (30 days) over a complex flood shield, i.e. an estuarine bar, display major temporal and spatial variations in current asymmetry. The variations in current asymmetry indicate the presence of small-scale circulations of sediment that are induced by smaller channels that penetrate the bar. The current meter observations and depth-averaged current patterns obtained with an ADCP revealed processes of advective flow acceleration and deceleration, as well as flow convergence and divergence, over the bars in the small channels.

Introduction

Estuaries are important coastal systems, providing natural navigation channels as well as habitats for marine flora and fauna, and recreational space. The Westerschelde forms the seaward, marine part (length 60 km) of the Schelde estuary (total length 160 km) and has a well-developed system of channels and intertidal shoals (Fig. 1a). The meandering ebb channels form the main navigation channel for ocean shipping to the harbours of Antwerpen and Gent. The shorter, straight flood channels are only suitable for small ships as the landward channel-margin of the flood channels is marked by a shallow extensive bar. Moreover, the bars in the flood channels display a complex topography due the presence of migrating connecting channels, the smaller channels that penetrate the bar and connect the main ebb and flood channel. To maintain the shipping lane dredging is carried out at the deeper bars in the main ebb channels. At present approximately $8*10^6$ m^3 of sediment is dredged annually. A further deepening of the bars, to enable the passage of larger ships, is planned for the near future. A better understanding of the morphodynamics of channels and shoals and the processes over bars in tidal channels in particular, is important for the management of the estuarine system and for determining the optimal dredging strategies. Recent developments in understanding the morphodynamics of channels and shoals have resulted in the formulation of different types of models (De Vriend (1996), De Vriend and Ribberink (1996)). Bars in tidal channels are however still understudied (Dalrymple and Rhodes, 1995 for a review).

----

[1] Institute for Marine and Atmospheric Research, Utrecht University, Department of Physical Geography, PO box 80.115, 3508 TC, Utrecht, The Netherlands.
c/o Ministry of Transport and Public Works, Directorate Zealand PO box 5014, 4330 KA Middelburg, The Netherlands.

This paper discusses the hydrodynamics of a complex bar at the end of a main flood channel, based on field observations. Flow computations with a 1D-network model indicate that this bar largely determines the distribution of tidal flow between the main ebb and flood channels during the periods of maximum flow (Jeuken, submitted). Objective of this study is to identify spatial and temporal variations of the tidal flow that are important for bar morphology and sediment transport patterns over the bar. Herein two aspects of tidal flow are considered: 1) Spatial velocity gradients and 2) differences between the velocity during ebb and flood, often referred to as current asymmetry. Spatial velocity gradients largely control processes of erosion and sedimentation. In literature tidal current asymmetry is often used as a first indicator for the direction of net sediment transport (e.g. Aubrey, 1986; Dronkers, 1986; Friederichs and Aubrey, 1988; Van de Kreeke and Robaczewska, 1993; Lessa and Masselink, 1995). In most studies current asymmetry is based on the amplitude of velocity components, often derived from a harmonic analysis of the velocity record. In this study a different approach is used to quantify variations in current asymmetry.

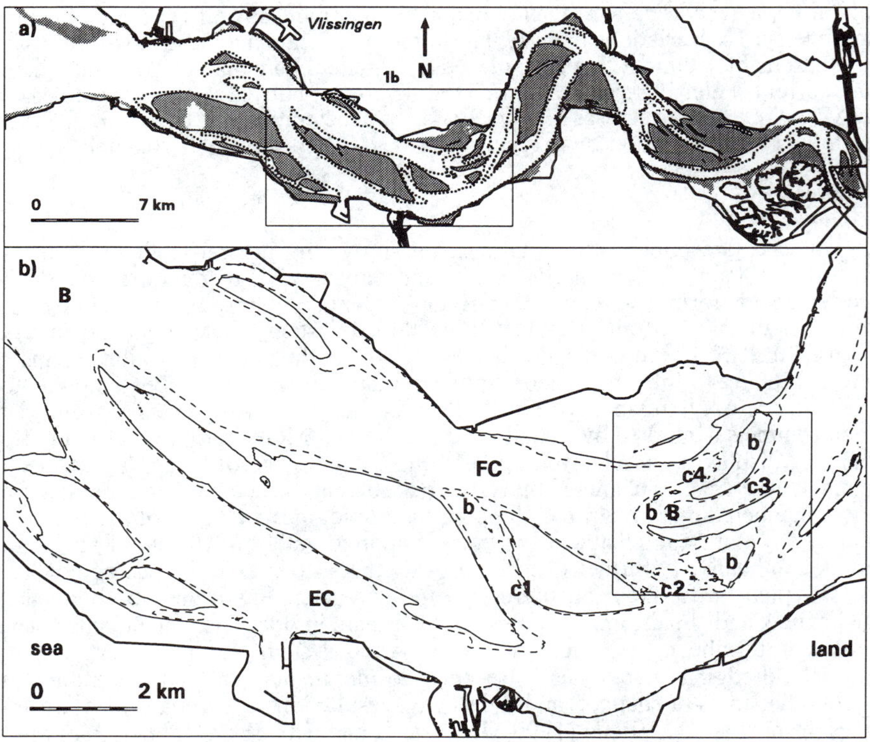

Figure 1 Location and channel configuration in the study area. a) the Westerschelde, b) the main study area. FC= main flood channel, EC=main ebb channel, c#=connecting channel (#=number), B= bar in main channel, b= bar in connecting channel.

The study area

An overview of the general hydraulic and morphologic evolution of the Schelde estuary is given by Van den Berg et.al. (in press). The main characteristics are described in this section. The Schelde is a tide-dominated meso-tidal estuary. The vertical tide displays a general asymmetry characterized by a faster rise than fall. This asymmetry increases in landward direction. In addition the vertical tide shows a pronounced neap-spring tidal variation as do the surface gradients and tidal currents. Maximum depth-averaged current velocities are in the order of 1-1.5m/s.

The larger channels in the Westerschelde display a regular returning pattern consisting of a meander-shaped ebb channel and a straight flood channel, separated by shoals (Fig. 1b). The shoals are bisected by smaller channels. These so called connecting channels form connections between the main ebb and flood channel and owe their existence to water level differences between the channels. Most connecting channels occur in the area of the bar in the main flood channel and tend to display a cyclic behaviour on the timescale of one to several decades. The shallow bar in the main flood channel is marked by various connecting channels, that originated and developed from 1986/1987. The small bars in the connecting channels are superimposed on the large bar in the main flood channel. The connecting channels C3 and C4 together reflect the morphologic characteristics of the main ebb and flood channel with respect to channel alignment and location and depth of the bars. At present (september 1996) channel C3 is rapidly degenerating. A new connecting channel has formed north of connecting channel C4.

Field observations

'Flachsee' impeller-type current meters were deployed at several locations during five measurement campaigns between April 1994 and February 1996 (Fig. 2). The current meters registered current velocity and direction for periods of thirty days with a sampling interval of ten minutes and a sampling period of one minute. At each location two current meters were deployed at the same measurement height, for validation purposes. A comparison of the double current meter deployments revealed an average difference in speed of about five percent, whereas the average difference in current direction approximated three degrees.

Detailed observations of the flow response over the small bars in two connecting channels were obtained in the summer of 1995 with a ship-borne, broad-banded two pulse Acoustic Doppler Current Profiler (ADCP). Measurements were carried out along a straight transect, oriented parallel to the general alignment of each channel, for a period of thirteen hours (Fig. 2b). The measurements along transect 1 were carried out during mean tide in June 1995. The measurements along transect 2 were obtained during spring tide in August 1995. The settings and instrumental accuracy of the ADCP during the two surveys are summarized in Table 1.

Table 1 System configuration of the ADCP during the two surveys

Transect	1	2
Acoustic frequency (kHz)	600	600
Pings per ensemble	5	20
Horizontal sampling interval (m)	10-15	40-50
Vertical bin size (m)	0.5	0.5
Depth range (m)	2.74-20	2.85-20
Velocity precision (m/s)	0.06	0.03
Compass precision (°)	1	1

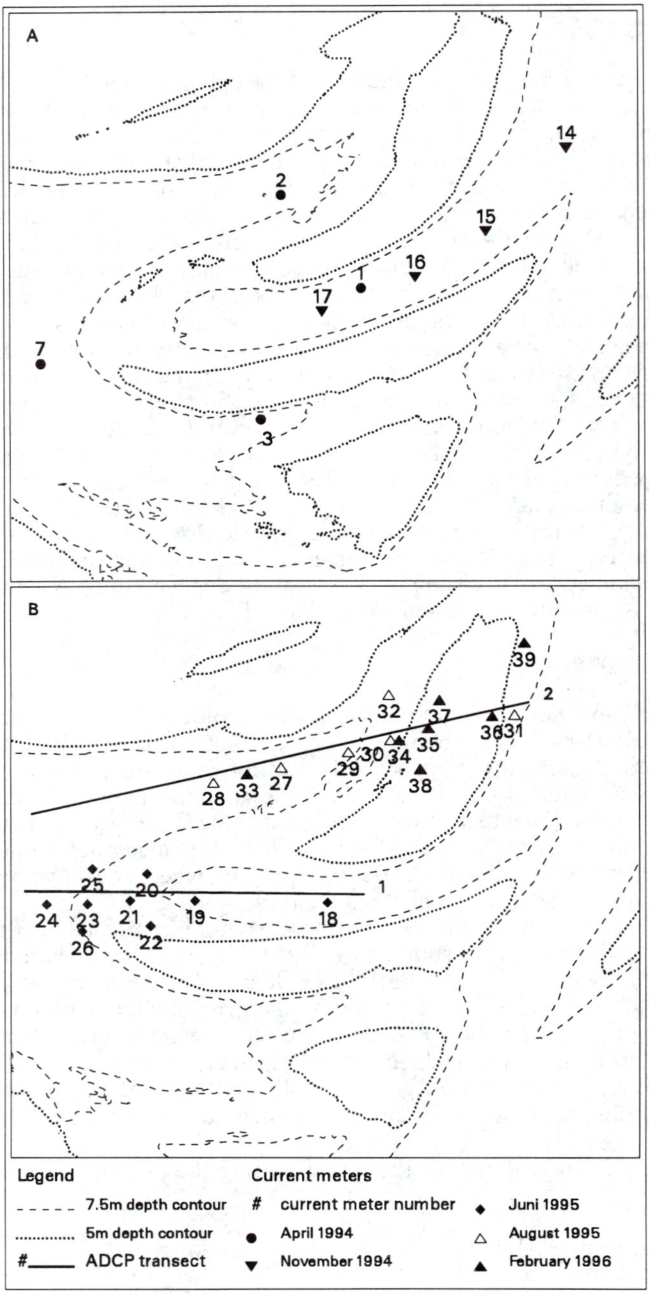

Figure 2 Locations of measurements. a) measurements in 1994 b) measurements in 1995 and 1996. For location of the bar see inset in Figure 1b.

Data-analysis

The current meter data were used to quantify temporal and spatial variations in current asymmetry and to characterize the current patterns over the small bars in the connecting channels. Prior to further analysis, the noise in the velocity data was reduced by applying a low-pass smoothing filter to the North-South and East-West velocity components.

The relative height of the current meters changed with time as a result of the large tidal range (3.3-4.75m) and the small water depths (5-13m). This variation in measurement height inhibit a comparison of measured ebb and flood velocities. Therefore measured current velocities (not direction) were converted to depth-averaged values, by assuming the logarithmic velocity profile of steady and uniform flow and an averaged roughness length based on bedform dimensions (z_o=0.033k_s, k_s=0.4m). In areas of major advective flow acceleration and deceleration this results in under-estimated (flow deceleration) and over-estimated (accelerating flow) depth-averaged velocities. These effects were not taken into account. The obtained time-series of the depth-averaged velocity vectors were then used to compute significant current vectors for each ebb and flood period. The significant velocity is defined as the mean of the 1/3 highest current velocities and represents the conditions near maximum flow over a period of about two hours. For each current meter location relationships between significant velocity (ebb and flood) and observed tidal range were determined by applying a linear regression analysis. The linear relationships were then used to compute significant ebb and flood velocities during mean, neap and spring tide. This was done by substituting the tidal range for neap, mean and spring tide in the linear relationships. Current asymmetry was then defined as the natural logarithm of the ratio of significant ebb velocity over significant flood velocity. A positive current asymmetry indicates ebb-dominated flow. A negative asymmetry indicates flood-dominated flow. The advantage of the significant velocity is, that it gives a better weighting of the tidal variation of current velocities than the often used maximum velocity.

The ADCP observations were used to determine variations in the depth-averaged flow patterns over the small bars in the connecting channels. Prior to further analysis the noise in the ADCP data was reduced by applying a low-pass, infinite impulse response filter to the North-South and East-West velocity components (Stanley et.al. 1984).

The computation of depth-averaged velocities implies extrapolation of the velocity profiles towards bottom and water surface. Extrapolation of the velocity profile was done by fitting a series of three shapefunctions through the data using a least squares method. This method has been derived from the shapefunction approach described by Zitman (1992) and has been previously applied by Van de Meene (1994). As winds were low during the ADCP measurements, the contribution of wind stress to the current velocity was neglected. Then the vertical velocity profile in North-South and East-West direction can be expressed as (Zitman, 1992):

$$u(\zeta) = \sum_{k=1}^{m} F_k \cdot f_k(\zeta) \qquad (1)$$

where f_k is the series of shapefunctions, ζ is the dimensionless vertical coordinate and m is the number of shapefunctions (m=3 in the present analysis). The weights F_k are the unknowns to be determined using the measured velocity profile $u(\zeta)$. Figure 3 shows the applied shapefunctions together with an example of a curve fitted through observations. The shapefunction approach was used only as a statistical, curve-fitting tool, essentially to compute the depth-averaged current vectors in an objective way. The advantage of this approach is that it is possible to describe velocity profiles that deviate from e.g. a parabolic profile.

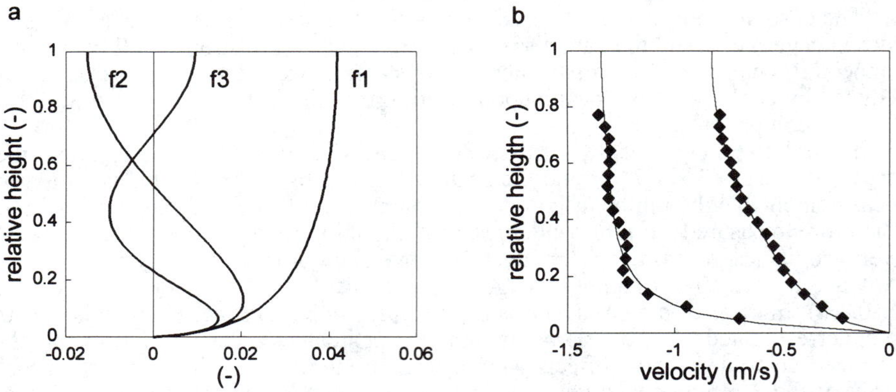

Figure 3 Shapefunction approach. a) the three shapefunctions used in the analysis, b) comparison of measured velocity components (◆) and fit (-).

Variations of current asymmetry over the bar in the main flood channel

Figure 4 shows spatial patterns of current asymmetry over the bar in the main flood channel for mean, neap and spring tide. The spatial variation of current asymmetry in Figure 4a is related to the presence and morphology of the connecting channels. In the connecting ebb channel and at the landward side of the bar in the connecting flood channel current asymmetry ranges between 0 and 0.11, indicating that ebb velocities exceed flood velocities by about 0 to 12 percent. In the connecting flood channels and at the seaward side of the bar in the connecting ebb channel, negative current asymmetries of -0.12 to -0.31 occur, indicating flood-dominated flow. In these areas flood velocities are 12 to 40 percent stronger than the ebb velocities. The bars in the connecting channels form the transition zones between ebb-dominated and flood-dominated flow. Current asymmetry changes with tidal range (Fig.4b). In the connecting ebb channel and at the landward side of the bar in the connecting flood channel, current asymmetry decreases with 6 to 25 percent when tidal range increases from 3.3 to 4.75m. At some locations in the connecting ebb channel tidal flow becomes even slightly flood-dominated during spring tide (locations 15, 20, 21 and 22 in Fig. 2). In the connecting flood channels and at the seaward side of the bar in the connecting ebb channel current asymmetry increases with 10 to 34 percent with increasing tidal range.

The spatial variation in current asymmetry indicates small-scale circulations of net sediment transport over the bar in the main flood channel, that are induced by connecting ebb and flood channels (Fig. 5a). This means net ebb transports in the ebb channel and net flood transports in the flood channel. The circulations confirm the concept of mutually evasive ebb and flood channels of Van Veen (1950), who identified the presence of ebb-dominated and flood-dominated channels on the basis of net water transports and the location of the bars in tidal channels. The changes in current asymmetry with tidal range, has two important implications for the small-scale circulations (Fig. 5b): 1) The intensity of the circulation is not constant in time and 2) the circulation is not closed. The changes in current asymmetry over the neap-spring tidal cycle indicate that the magnitude of net sediment transport, the intensity,

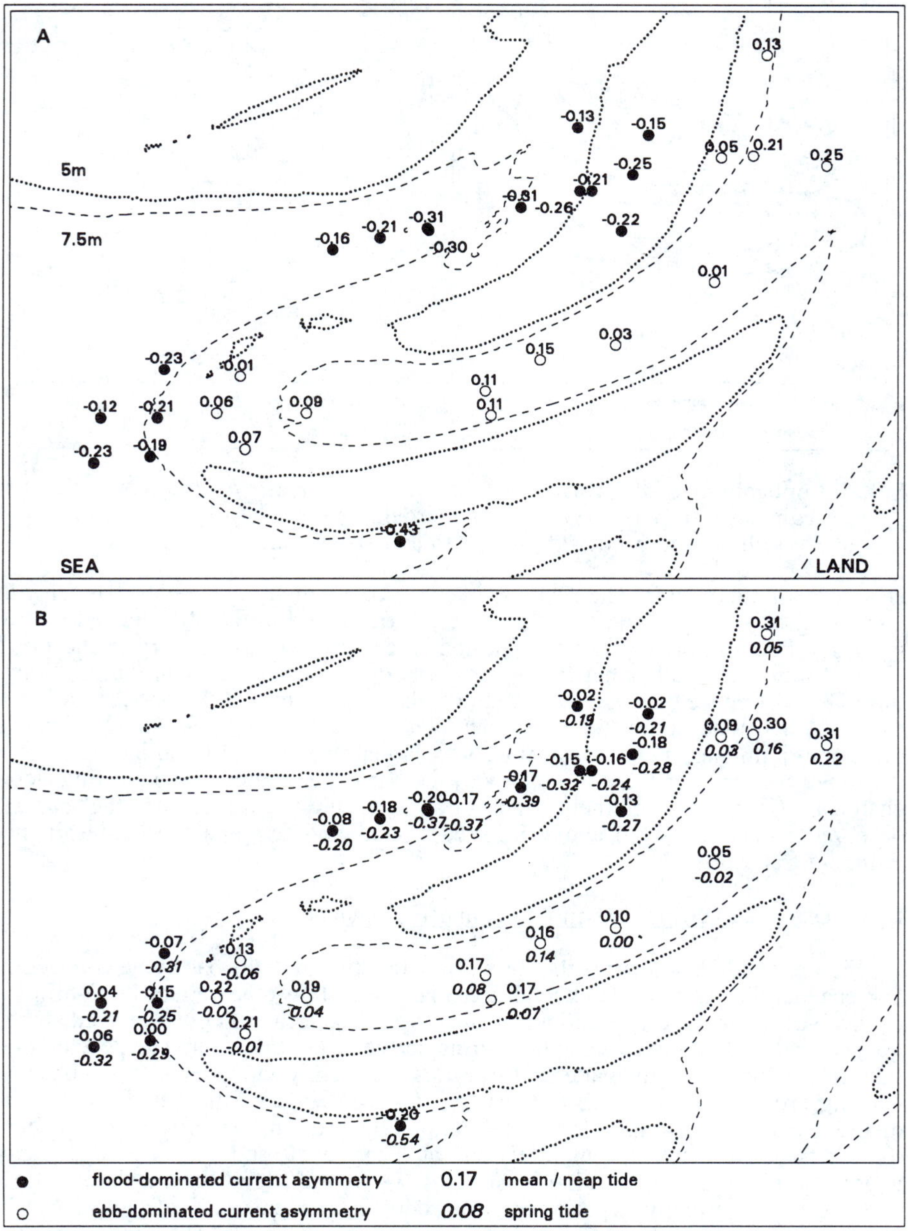

Figure 4 Patterns of current asymmetry. a) mean tide, b) neap and spring tide

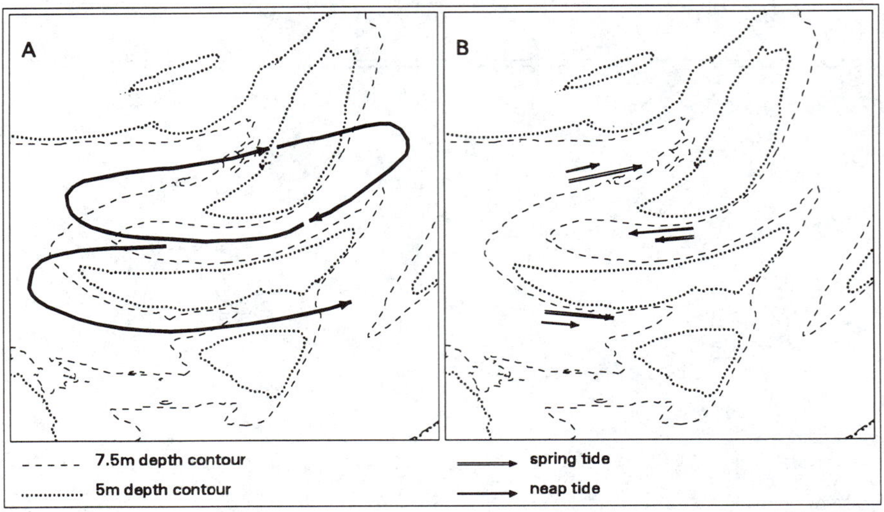

Fig. 5 Implications of the variations in current asymmetry for the patterns of net sediment transport. a) Small-scale circulations, b) the relative magnitude of net sediment transports during neap and spring tide

changes with tidal range. The decrease in current asymmetry in the ebb channel with increasing tidal range indicates larger net ebb transports in the ebb channel during neap tide than during spring tide. The increase of current asymmetry in the flood channel indicates smaller net flood transports during neap tide than during spring tide. This difference between ebb and flood channel means that the circulation is not closed. A net flood-dominated component that increases with tidal range is likely as the flood-dominated current asymmetry is larger than the ebb-dominated current asymmetry. Moreover the morphology of the bar is dominated by connecting flood channels. Computed sediment transports (not shown), based on the current observations, confirm the above inferred spatial and temporal variations in net sediment transport.

Flow response over the bars in two connecting channels

Figure 6 and 7 summarize the flow characteristics over the bar in the connecting ebb channel during mean tidal conditions. Figure 6 shows the pattern of significant current vectors in June 1995. Figure 7 displays the pattern of depth-averaged along-transect and cross-transect velocities during accelerating, maximum and decelerating tidal flow. During ebb the pattern of significant current vectors displays an ebb flow flowing around the bar (Fig. 6a). At the upstream side of the bar the ebb flow diverges and shows minor flow decelerations towards the bar of about 7 percent (0.08m/s). Significant current directions at location 20 and 22 differ by twenty degrees (see Fig. 2b for locations). At the downstream side of the bar the ebb flow converges and strongly decelerates with about 25 percent (0.25m/s). The ADCP observations confirm the patterns of flow deceleration (Fig. 7). The depth-averaged velocity pattern shows an instantaneous drop in current velocity as soon as the ebb flow passes the top of the bar. The depth-averaged velocity reduces with 28 percent (0.3m/s) on average. The flow reduction tends to increase with time and decreasing water level. In addition the ADCP observations show a tendency towards small flow acceleration just before the top of the bar during maximum and decelerating ebb flow. The small cross-channel velocity component, both during ebb and flood, indicates tidal flow approximately parallel to the ADCP-transect (with 5°). Thus the ebb flow diverges and decelerates towards the bar, slightly accelerates near the top of the bar and converges and strongly decelerates after passing the top of the bar.

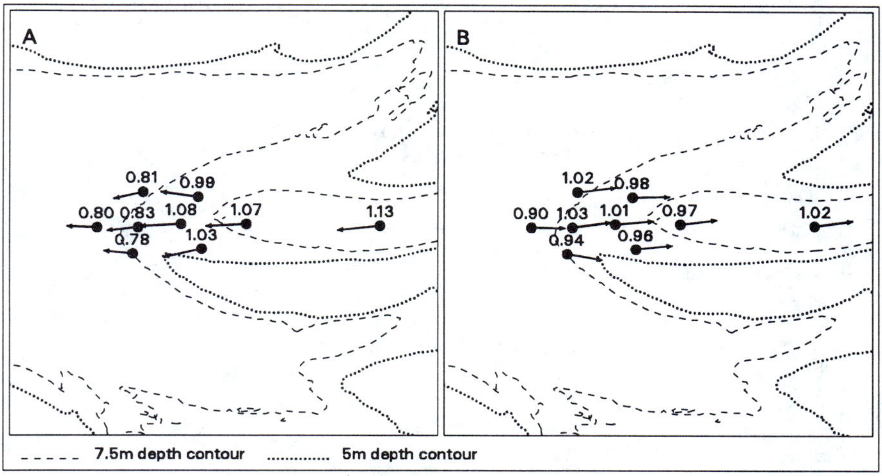

Fig. 6 Mean significant current vectors over the bar in the connecting ebb channel during mean tide in June 1995. a) ebb, b) flood.

The flood displays a small tendency towards flow around the bar (Fig. 6b). The pattern of significant current vectors shows flow divergence at the upstream (seaward) side of the bar that is accompanied by an increase in current velocities of about 14 percent (0.13m/s). At the downstream side of the bar the current vectors are aligned more or less parallel and only minor velocity gradients are observed. The ADCP observations also display flow acceleration at the upstream side of the bar (Fig.7). Depth-averaged velocities increase with about 10 percent (0.1m/s). At the downstream side, between 500 and 1200 meters, reductions in current velocity of about 10 percent (0.1m/s) are observed. Thus the flood flow slightly diverges and accelerates and decelerates over the bar.

The characteristics of the flow pattern over the bar in the connecting flood channel during spring tide, measured in August 1995 and February 1996, are summarized in Figures 8 and 9. During ebb the significant current pattern displays flow deceleration over the bar (Fig 8a). At the downstream side of the bar current velocities are 15 to 30 percent smaller than near the top of the bar (location 36, Fig. 2b). The gradual change in current direction between locations 37 and 38 of fourteen degrees, indicates a tendency towards flow convergence. The ADCP observations (Fig.9) reveal deflection of the ebb flow at the upstream side of the bar, where the current vectors are inclined towards the ADCP-transect by 15-25 degrees (not shown). Despite this current deflection a major velocity gradient over the bar can be identified. On average current velocities at the top of the bar exceed currents at the upstream and downstream side of the bar by 30 percent (0.4-0.5m/s). These large velocity differences indicate flow acceleration at the upstream side of the bar and

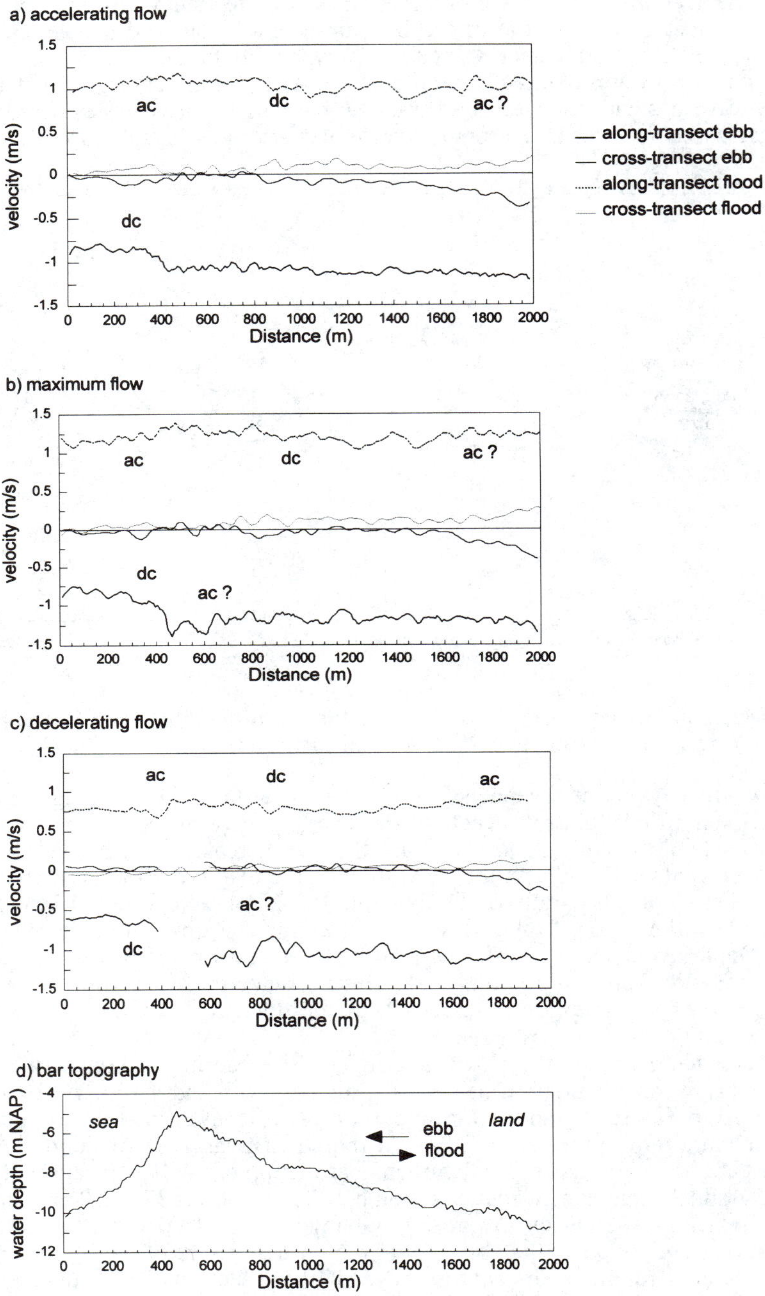

Figure 7 Depth-averaged along-transect and cross-transect velocity over the bar in the connecting ebb channel during ebb and flood on 29 June 1995. Ebb is negative, flood is positive, ac=acceleration, dc=deceleration.

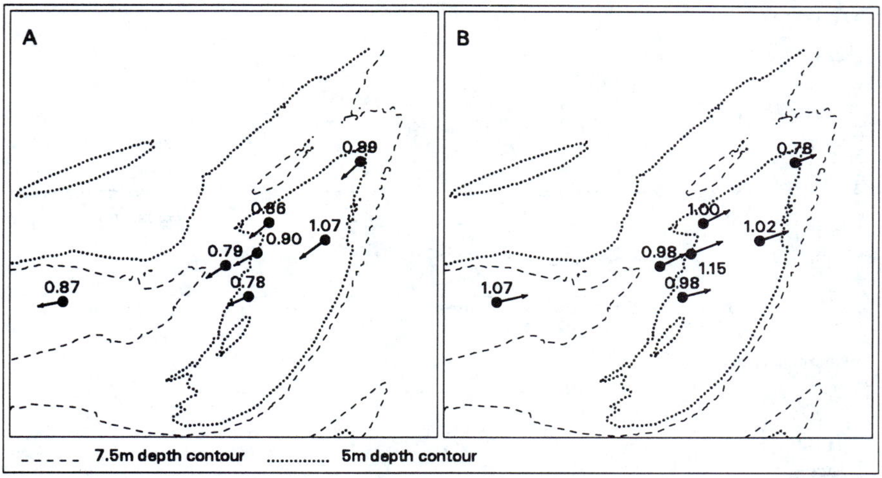

Figure 8 Mean significant current vectors over the bar in the connecting flood channel during spring tide in February 1996. a) ebb, b) flood.

flow deceleration at the downstream side of the bar. Thus the ebb flow accelerates towards the top of the bar and decelerates and converges at the downstream side of the bar. During flood the significant current vectors show a flow acceleration of about 17 percent (0.17m/s) at the upstream side of the bar (Fig. 8b). Significant current directions gradually change with twelve degrees between locations 37 and 38 (see Fig. 2b for locations), indicating minor flow divergence. The current velocity at the upstream side of the bar is 12 to 22 percent (0.13-0.22m/s) stronger than at the downstream side of the bar. This means that the flood flow decelerates over the bar. The ADCP observations show a flow deceleration of about 20 percent just before the bar (between 1700 and 2200m, Fig.9), where the flood flow diverges (not shown). At the upstream side of the bar the flood flow accelerates with about 30 percent (0.3-0.4m/s). At the downstream side flow decelerations of 40 to 50 percent (0.5-0.7m/s) are observed. The cross-transect velocity components indicate that the flood flow is less deflected by the topography of the bar than the ebb flow (inclination of 6-12° with respect to the ADCP-transect, not shown). Thus the flood flow diverges and decelerates towards the bar, accelerates near the top of the bar and strongly decelerates after passing the top of the bar.

The ADCP observations are marked by second-order fluctuations. The amplitude and length scale of these fluctuations increase with time and decreasing water level. The unfiltered velocity data display similar fluctuations. The cause of the fluctuations is not clear. Both turbulence and depth variations may cause such fluctuations.

The pronounced topography of the small bars in the connecting channels is reflected in the hydrodynamics: 1) The bars in these channels demarcate the transition zones between ebb-dominated and flood-dominated flow, 2) Neap-spring tidal variations in current asymmetry change over the bar and 3) The ebb and flood flow tend to flow around the bars of the channels and display non-uniform velocity patterns. In the connecting ebb channel largest velocity gradients are observed during ebb. In the connecting flood channel largest velocities and velocity gradients tend to occur during flood. Fluid continuity across the bar may explain the observed flow

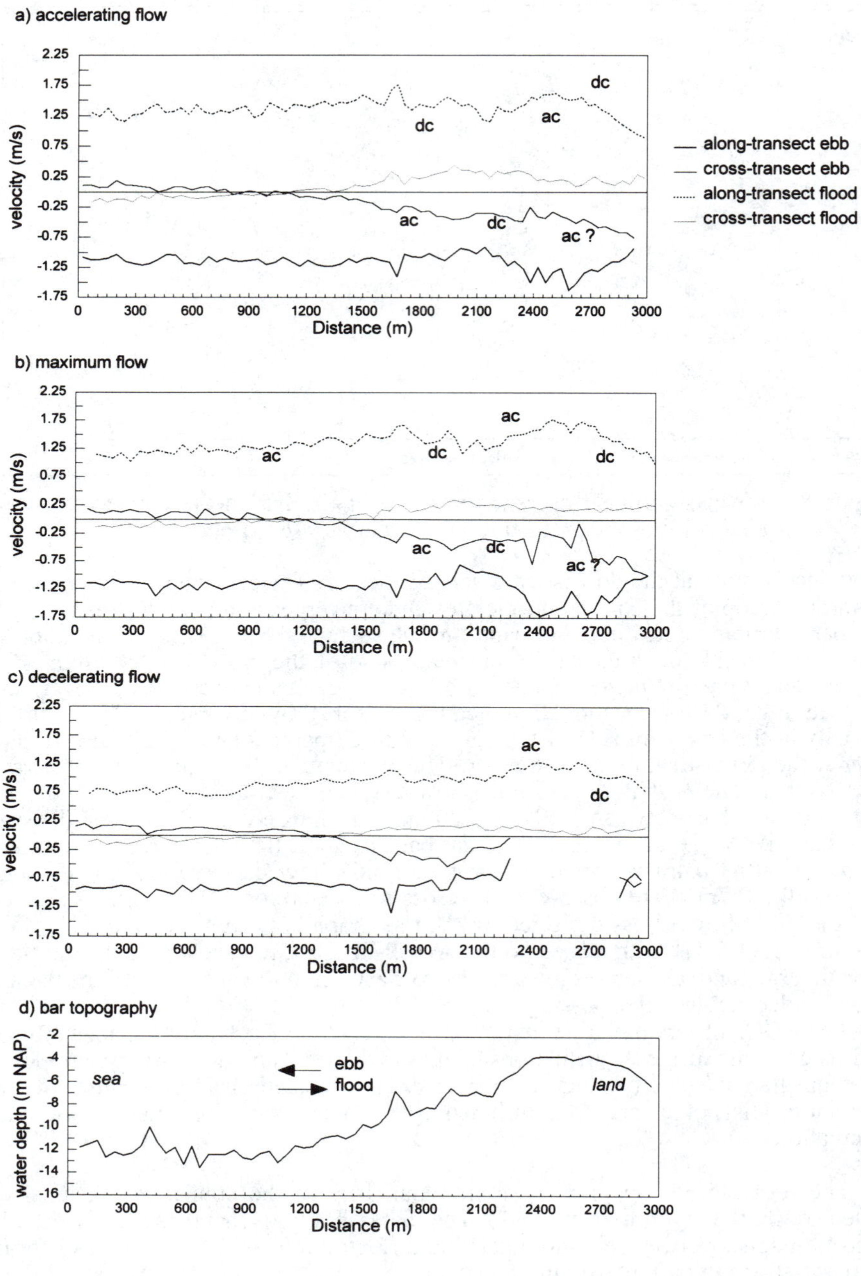

Figure 9 Depth-averaged along-transect and cross-transect velocity over the bar in the connecting flood channel during ebb and flood on 30 August 1995. Ebb is negative, flood is positive, ac=acceleration, dc=deceleration.

around the bars. Fluid continuity implies that the discharge Q (m^3/s) does not change much across the bar. The overall discharge Q is defined as $Q=b.\bar{u}.h$, where b is the width of the imaginary stream tube (m), $\bar{u}$ is the depth-averaged current velocity and h is the water depth. The changes in channel depth over the bar may result in changes in the width of the stream tube and the depth-averaged velocity. At the upstream side of the bars the flow diverges. The width of the stream tube increases. At the downstream side the flow converges. The width of the stream tube decreases. In addition to these changes in width the depth-averaged velocity increases and decreases across the bars, whereas the water depth increases and decreases. The flow over the bars shows a tendency towards flow deceleration and accelerations at the upstream side of the bar and deceleration at the downstream side. The pattern reverses every tidal phase. The implications of these flow phenomena for sediment transports, the patterns of erosion and sedimentation and the net effect over e.g. one semi-diurnal period are not evident. The implications depend on the magnitude and duration of flow accelerations and decelerations, as well as the effect of variations in stream tube width and water depth, differences in current asymmetry and the dominant sediment transport mode. The small bars are stable morphological features despite the observed flow accelerations and decelerations. This indicates that relaxation of the suspended transport may be important for the maintenance of the estuarine bars.

Conclusions

Variations of the tidal flow over the bar in the main flood channel are strongly determined by connecting channels:

1) Spatial variations in current asymmetry indicate the presence of small-scale net circulations of tidal flow and sediment transports over the bar, that are induced by connecting ebb and flood channels.
2) The changes in current asymmetry with tidal range indicate that the intensity of the circulations is not constant and that the circulations are not closed.

The flow over the small bars in the connecting channels is non-uniform and displays tendencies towards divergence, decelerations and accelerations of the flow at the upstream side of the bar and convergence and deceleration of the flow at the downstream side of the bar. The implications of the flow patterns for sediment transports and the maintenance of the bars are not evident and will be further elaborated on the basis of computations with a mathematical model and morphological analysis.

Acknowledgements

This study was funded by Directorate Zealand of the Ministry of Transport Public Works and Water Management. The measurements were carried out by the survey department of Directorate Zealand. Ir A. Langerak, drs A. Van Kleef and dr J.R. Van de Berg are greatly acknowledged for their useful comments.

Appendix

Aubrey, D.G. (1986). Hydrodynamic Controls on Sediment Transport in Well-mixed Bays and Estuaries. In: J. Van de Kreeke (ed.). Physics of shallow estuaries and Bays, p. 245-258.

Dalrymple, R.W. and R.N. Rhodes (1995). In: G.M.E. Perillo (ed.), Estuarine dunes and bars. Geomorphology and Sedimentology of Estuaries. Developments in Sedimentology 53, p. 359-422.

De Vriend, H.J. (1996). Mathematical Modelling of meso-tidal barrier island coasts. Part I: empirical and semi-empirical models. In: P.L.F. Liu (ed.), Advances in coastal and ocean engineering (2), p.115-149.

De Vriend, J. and J.S. Ribberink (1996) Mathematical Modelling of meso-tidal barrier island coasts. Part II: process-based simulation models. In: P.L.F. Liu (ed.), Advances in coastal and ocean engineering (2), p.151-197

Dronkers, J. (1986). Tidal asymmetry and estuarine morphology. Neth. Jour. of Sea Res. 20 (2/3), p. 117-131.

Friederichs C.T. and D. Aubrey (1988). Non-linear Tidal Distortion in Shallow Well-mixed Estuaries: a Synthesis. Est. Coast. and Shelf Sci.(27), p. 521-545.

Jeuken, M.C.J.L. (submitted for review). Morphodynamics of connecting channels in the Westerschelde estuary (the Netherlands). Proc. Conf. Physics of Estuaries and Coastal Seas, The Hague, 1996.

Lessa, G. and G. Masselink (1995). Morphodynamic evolution of a macro-tidal barrier island. Marine Geology 129: 25-46.

Stanley, D.W., G.R. Dougherty and R. Dougherty (1984). Digital Signal Processing, Prentice Hall, 514 p.

Van de Kreeke, J. and K. Robaczewska (1993). Tide-induced residual transport on coarse sediment: application to the Ems estuary. Neth. Jour. of Sea Res. 31 (3), p.209-220.

Van de Meene, J.W.H. (1994). The shoreface-connected ridges along the central Dutch coast, Ph.D. thesis, IMAU, Utrecht Univ., Neth. Geogr. Studies 174, KNAG, Utrecht, 256 p.

Van den Berg, J.H., M.C.J.L. Jeuken and A. F. van der Spek (in press). Hydraulic processes affecting the morphology and evolution of the Westerschelde estuary. In: K.F. Nordstrom and C.J. Roman (eds.). Estuarine Shores: Evolution, Environments and Human Alterations.

Van Veen, J. (1950). Eb- en vloedscharen in de Nederlandse getijwateren. Tijdschr. Kon. Ned. Aardr. Gen. 67, p. 303-335 (in Dutch).

Zitman, T.J. (1992). Quasi three dimensional current modelling based on a modified version of Davies' shapefunction approach. Cont. Shelf Res., 12(1), p. 143-158.

CHAPTER 350

Interactions between a Sand Barrier and Flood Terrace at the Abukuma River Mouth

Akira Mano [1]and Masaki Sawamoto [1]

Abstract

The sand barrier at the Abukuma River mouth flushed by floods in autumn firmly recovered within a half year by stretching its tip to the cross river direction. However, the direction suddenly changed upstream since 1993. This paper analysis the mechanism of the change.

The river has a large terrace formed by the flood in 1986, the maximum in the observed floods over fifty years. Bathymetric data sounded once a year reveal gradual development of a channel on the terrace off the river axis. From wave ray analyses, it is shown the development accompanies change of the refraction characteristics of waves. The channel becomes to function as a half mirror; reflecting waves incoming from one direction and transmitting from the other direction. This is shown to modify direction of dominant waves near the barrier and to turn the stretch direction of the barrier.

The channel evolution is also shown to correlate with the flood outflow.

1 Introduction

Development of sand barriers at river mouths has close influences on the river management such as flood passage and voyage. The barriers also control water salinity inside the river, which relates with water resources and ecological environments. Although the barrier has important roles in the various aspects near the mouth, the mechanism of the development or deformation is not well known.

The difficulties to find the mechanism may mainly come from complexity of the dual external forces of river flows and waves. The complicated topographies inherent to the mouth like flood terraces and sand barriers also tangle the affair, because they cause flow fields and wave fields to considerably deform. Furthermore broadness of the time scale band on the topography change, which would

[1]Department of Civil Engineering, Tohoku University, Sendai 980-77, Japan

relate with the characteristics of the external forces, requires vast set of observed data in the analyses.

Over twelve years, we have been surveying the sand barrier at the Abukuma River mouth in Japan. The details of the observations on the Abukuma River and the characteristics of the barrier deformation in various time scales have been reported (Mano et al. 1995). This river has a catchment of 5,400 km^2 and a main channel of 239 km. The maximum flood over fifty years of observation by Ministry of Construction occurred on August 1986 with the peak discharge of Q=7,900m^3/s near the mouth. This flood flushed out the sand barrier so far and made a large flood terrace off the mouth.

After the flush by the flood, the barrier started to recover. The tip of the barrier advanced in the cross river direction. Then within a half year, it reached equilibrium such that the minimum opening width at the mouth is maintained by the usual river flows and tides. This elongation is the earliest process in the recovery then thickening barrier followed until the next year's flush caused mainly by typhoons in autumn. After the 1986 event, this cycle of flush and recover had continue for seven years always accompanying transverse elongation of the barrier.

However in 1993, the barrier changed its development from cross-river direction to upstream direction. The typical shapes are shown in Fig.1 (a) and (b).

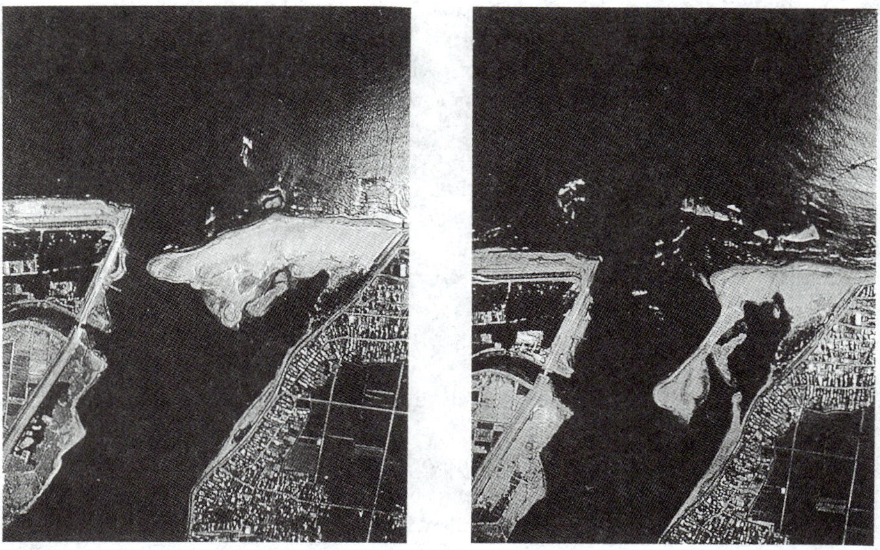

(a) March, 1991. (b) April 1994.

Fig. 1: Aerial photographs for the typical sand barrier shapes.

After a lapse of more than a half year, both photographs show the fully developed barriers in length. However the shapes are quite different. Figure 1 (a) shows the barrier stretching transversely from the right side bank and very narrow opening width, about 70 m. However Fig.1 (b) exhibits the barrier stretching upstream and the widely opened mouth. Waves are intruding inside the river and two typical arced beaches are formed in the river.

The barrier as Fig.1 (b) has less resistance for the flood passage and provides safer routes for the voyage. Therefore it looks like as if the nature is helping river management since then. The purpose of this paper is to find the mechanism that the sand barrier that had firmly stretched in the cross river direction for seven years changed the direction.

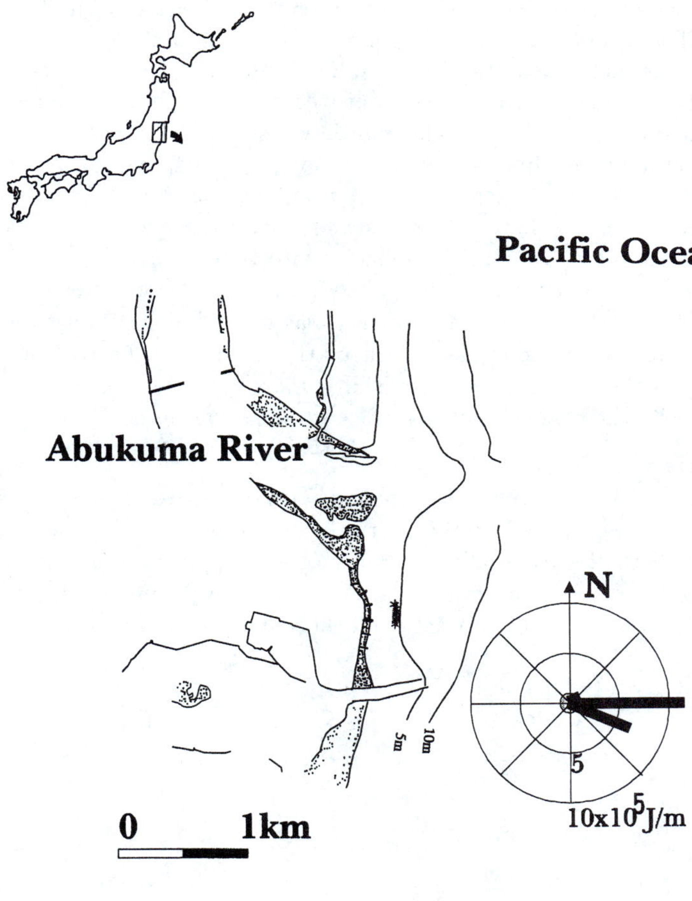

Fig. 2: Topography near the Abukuma River mouth.

2 Topography near the mouth

The Abukuma River flows into the Pacific Ocean, crossing the coast line that has the azimuth, N5°E as shown in Fig.2. No jetties are constructed about the mouth. Waves are observed by the Ministry of Transport off the Sendai Harbor that locates about 25 km north of the mouth. Annual deep water wave energies that are the integration of the energy flux in a year and the averaged values in the period, 1984 to 1992 are also plotted in this figure. The most dominant waves are those of E and have the energy of 11×10^5 J/m that is about 60 % of the total energy. The next dominant waves are those of ESE sharing 35 %. Depending on the directional distribution of the wave energy, the longshore drift around the coast orients north.

The bathymetric charts near the mouth are shown in Fig. 3 (a), (b), (c), and (d) for various stages. Before the 1986 flood, there existed a small terrace in front of the sand barrier and a small channel off the opening, while the barrier was thick. These characteristics correspond to the history of flood activity. Fig. 3 (b), sounded one month after the 1986 flood, showed formation of a large terrace. The top surface of the terrace was very flat with the depth, 4 m. The foot of the terrace reached as deep as 13 m. The mouth was still widely opened.

In 1988, although outline of the terrace was preserved, several deformations were advancing. The shoulder part of the terrace was rounded and the front part of the barrier became shallower. The most remarkable change is the development of a channel on the terrace in the offshore direction from the opening. The opening was usually fixed next to the left side bank by the transverse stretch of the barrier. On the other hand, the terrace was formed along the center line of the river since it was formed by the huge flood that flushed the barrier thoroughly. The position of the channel, therefore, shifted northward from the center of the terrace. In 1993, the channel on the terrace extended offshore.

3 Refraction analysis

To know the influences of the topographic deformations, we have executed wave refraction analysis. Among several numerical methods dealing with the refraction, the wave ray method is especially superior in grasping physical images moreover has the advantage in computational feasibility. Thus we employed the wave ray equation (Mei, 1983) and the refraction coefficient equation (Munk et al., 1952) as follows;

$$\frac{dy}{dx} = \pm\frac{K}{\sqrt{k^2 - K^2}} \tag{1}$$

$$\frac{d^2\beta}{ds^2} + p\frac{d\beta}{ds} + q\beta = 0 \tag{2}$$

$$K = k_o \sin\alpha_o, p = p(c,\alpha), q = q(c,\alpha), K_r = 1/\sqrt{\beta}. \tag{3}$$

Where, x is the local coordinate taken to the steepest gradient of the bottom, y is perpendicular to x, s is the coordinate taken along a wave ray, α is direction

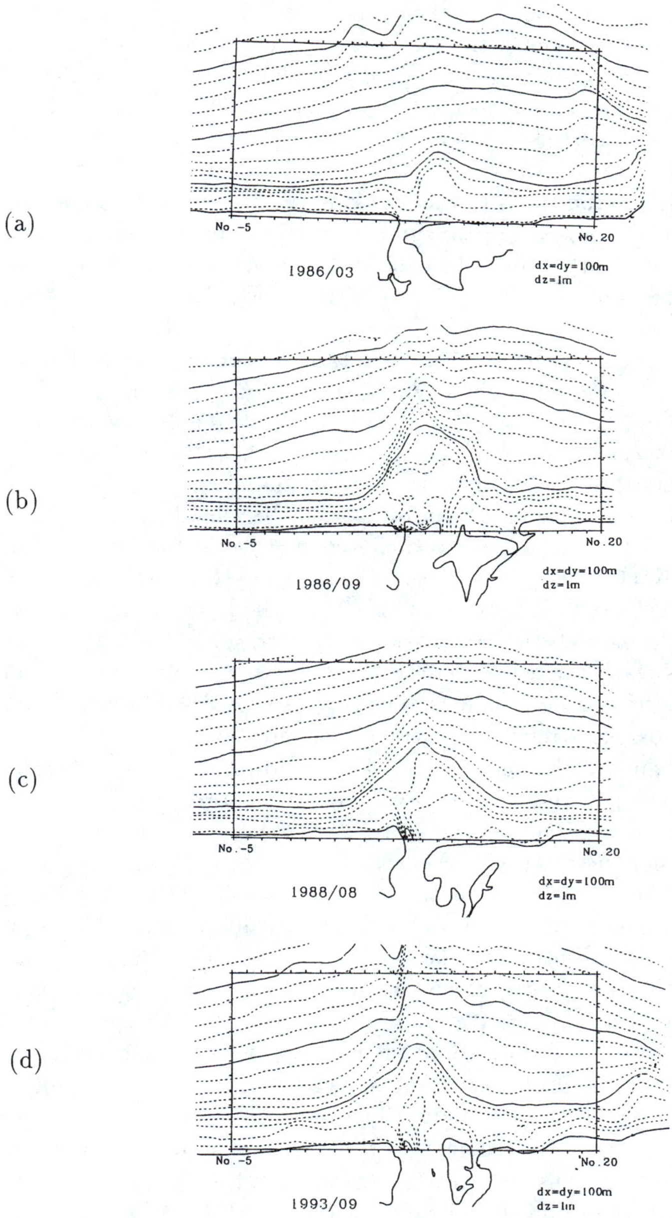

Fig. 3: Bathymetric charts at various stages.(a): March, 1986,(b): September 1986,(c): August 1988,(d): September 1993.

of wave ray, k is the wave number, c is the phase velocity, K_r is the refraction coefficient, and subscript o denotes the boundary value. These equations are solved together with the dispersion relation and boundary conditions by the finite difference method. The employment of the local coordinate x has such an advantage that it is required to solve only once the dispersion relation for each space step marching.

After digitizing the contour lines of the bathymetric charts, we made bathymetric mesh data and expressed the topography by the sets of plane triangle elements. The wave rays started from the offshore boundary were advanced step by step through the integration of the equations with the local coordinate system.

For the topographies mentioned in Figs. 3 and the corresponding annual mean waves, wave rays were computed as shown in Figs. 4 (a), (b), (c), and (d). The incident wave height and period are 0.77 to 0.88 m and 7.4 to 8.4 s respectively. The breaking point for each wave ray was evaluated by using the breaking index (Goda, 1970) and plotted by the filled circle near the shoreline.

In Fig. 4 (a) indicating wave refraction before the 1986 flood, the wave rays approaches uniformly in the longshore direction except very near the river mouth. The small terrace in front of the barrier gathers the wave rays and focuses at the barrier. While for the topography just after the flood, the large terrace gathers more wave rays and focuses about the middle of the terrace. Most of the wave rays passed the focus reach the north coast of the river. Local focuses are again formed. The sand barrier is outside such wave rays and reached rather direct waves that make the barrier stretch in the transverse direction.

In Fig.4 (c), according to the deformation of the terrace, the wave refraction changes. The sand barrier becomes to get two wave groups; direct waves that are almost straight incident and reflected wave at the channel on the terrace. The intensity of the reflection increases with the development of the channel as shown in Fig.4 (d). Along the long channel from the mouth to the contour of 10 m, caustic, that is, the envelop of the wave rays, is formed. Almost all the wave rays reflected at the channel concentrate at the tip of the barrier in such a way that these waves prevent the crosswise stretch or promote the upstream elongation.

Figures 5 (a), (b), (c), (d), and (e) show the effect of the off shore wave direction for the topography of 1993. Here the direction is taken anti-clockwise from the normal to the coastline. For the positive directions, (a) and (b), although the terrace refracts wave rays complicatedly, it transmits the wave rays and makes them concentrate at the tip of the barrier. On the other hand, for the normal to the negative directions, the caustic is firmly formed along the channel moreover makes concentration at the tip of the barrier. Therefor for the waves of positive direction, the terrace is transparent and acts like a convex lens. However for the waves of negative directions it functions as concave reflector. As a whole, the channel functions like a half mirror. Thus the deformed terrace gathers waves of all direction and focuses at the tip. This an-isotropic characteristic of the refraction comes mainly from the position of the channel that is shifted north

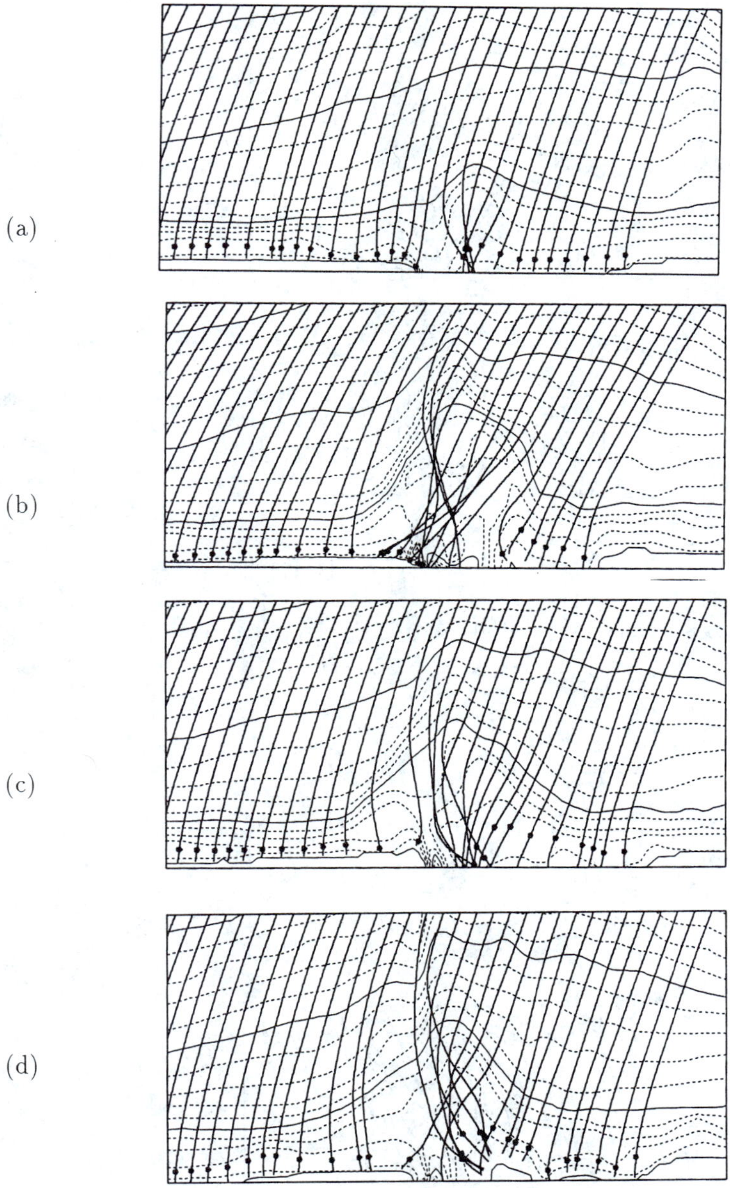

Fig. 4: Wave rays for the topographies at various stages.(a): March, 1986,(b): September 1986,(c): August 1988,(d): September 1993.

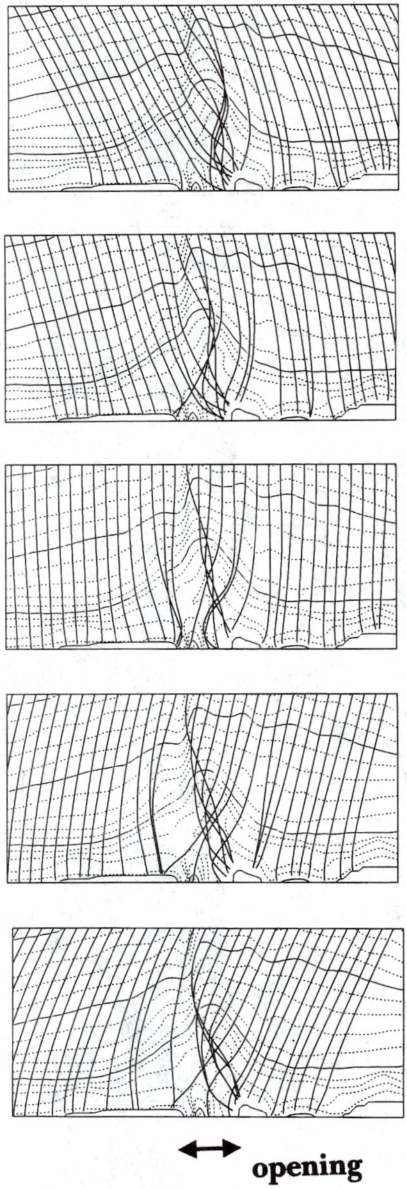

(a) $\theta = +45.0^o$.

(b) $\theta = +22.5^o$.

(c) $\theta = \pm 0.00^o$

(d) $\theta = -22.5^o$.

(e) $\theta = -45.0^o$.

Fig. 5: Effect of wave directions.

from the center of the terrace as stated before.

4 Channel evolution

The next step is to examine how the channel has been developed. The cross sections of the terrace along shore are shown in Figs .6 (a) and (b) for each off shore distance y. The initial shape of the terrace formed by the 1986 flood was almost trapezoidal. Near the mouth there was a broad channel with banks in both sides. Then the banks were disappeared at y = 400m and a very flat and horizontal surface appeared. The depth was about 3 m here.

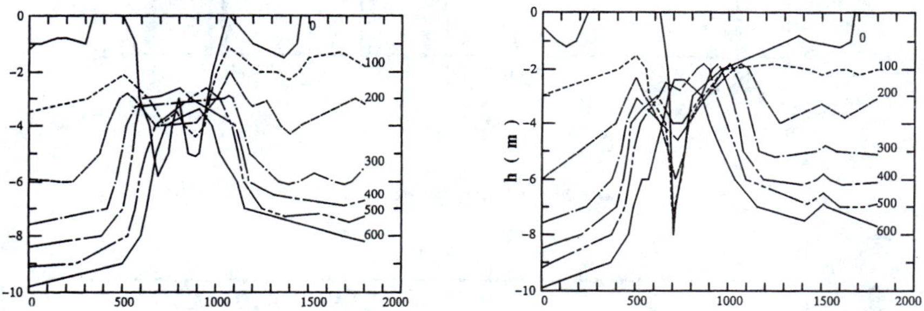

(a) September, 1986. (b) January 1992.

Fig. 6: Profiles of the terrace.

As time passing, the shoulder parts and the banks near the mouth were eroded to become round and the channel on the terrace was developed. The cross sections on January 1992 (Fig. 6 (b)) indicate a wide and triangular channel on the terrace. By using these profiles, we calculated the channel area that is bordered by the line connecting apexes at both sides. Then the volume of the channel was obtained by integrating the area to the offshore direction till y = 600 m, beyond which the evaluation of the area is not so precise.

Time histories of the volume and the outflow that is the integration of the discharge during the flood are shown in Fig. 7. This figure reveals correlation between the flood activity and the channel evolution. In 1991, we had many floods including large one and the channel volume remarkably increased after the floods. While for the inactive year in the floods, the change of the volume is small. Here the case for the 1986 flood is exceptional. After the largest flood of 1986, the increase of the channel was not so large. This is because the 1986 flood mainly deposited considerable amount of sediment off the mouth. The counted volume was due to the local score very near the mouth and bank formation on the terrace near the mouth.

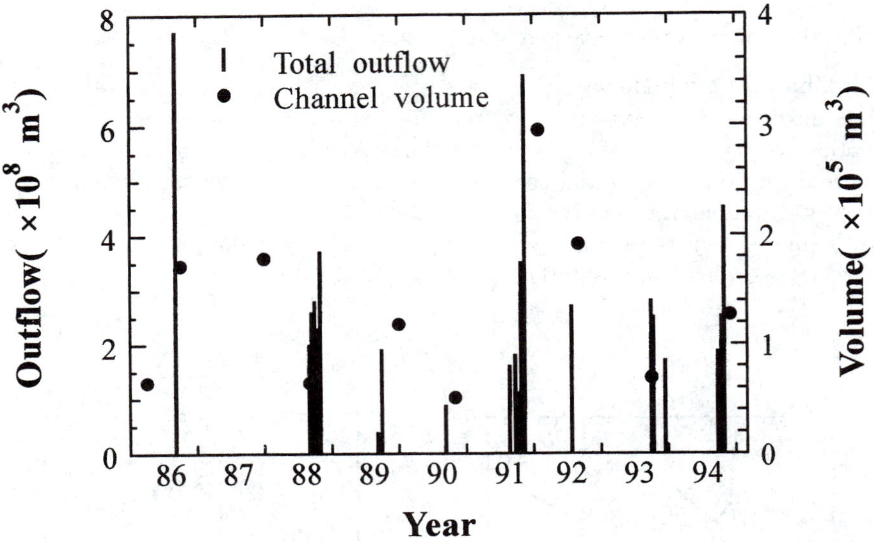

Fig. 7: History of floods and channel evolution.

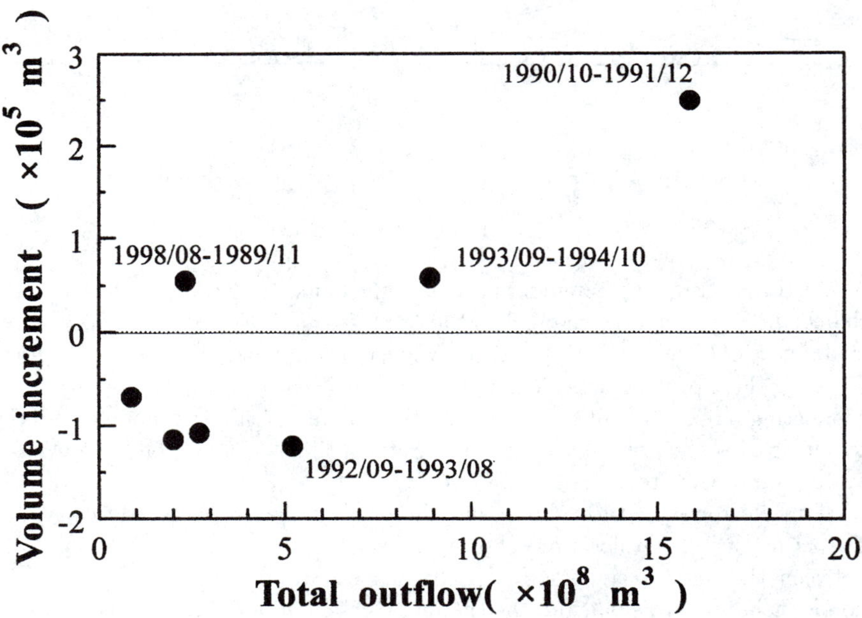

Fig. 8: Correlation between the outflow of the floods and increment of the channel volume.

The correlation between the flood activity and the increase of the volume is examined in Fig. 8. The ordinate indicates the increment of the channel volume obtained from the successive bathymetric charts. The abscissa is the sum of the outflow in the corresponding period. For the large value of the outflow, these two quantities are in linear relationship. While for the small value, the decrement of the channel volume appears, which would correspond to the deposition by waves. Thus the evolution of the channel on the terrace is related with passage of the small to middle scale of floods, except such a huge flood as to make a terrace.

5 Considerations

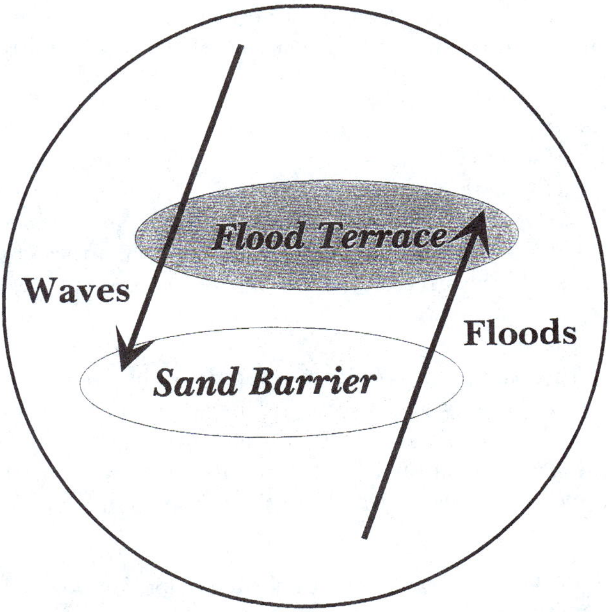

Fig. 9: The river mouth system at the Abukuma River

Starting from the terrace formation by the 1986 flood as an initial stage, here we summarize the flow of the cause and effect in the morphological change in the river mouth system. In the early stage after the terrace formation, the large terrace refracts waves to stretch the sand barrier in the cross river direction. The stretch takes half a year at most. Thus till spring, the opening of the mouth reaches equilibrium maintained by tides and usual river flows with narrow width. Then it is forced to lie adjoining the left bank of the river. Then thickening of the barrier as the following step of the development continues until next flood season. In autumn we have several floods mainly caused by the rainfall due to typhoon. For the small to middle size of floods, they may flush the barrier partly, however the major part of the flow runs out through the old opening. Then it erodes the terrace. The channel on the terrace gradually evolves along the flow axes.

The evolution of the channel also changes the refraction characteristics on the terrace. The reflected waves at the channel are gradually intensified and gathers at the tip of the barrier such a way as to suppress the cross ward stretch. When the reflected waves become dominant to the direct waves, the barrier stops transverse stretch and turns to the upstream direction. The transverse stretch lasted about seven years, while the fixation of the opening to the left bank contributes the channel evolution.

After the turning point, since the opening remained widely open, the reflected waves intrude inside the river and transport the sediment upstream. The waves also make the circular shore lines as shown Fig. 1 (b).

Figure 9 summarizes the interactions between the typical topographies; the sand barrier and flood terrace under the action of the external forces; floods and waves.

6 Conclusions

Based on the bathymetric chart, sand barrier survey, hydro-graphs of floods and wave data over ten years as well as the refraction analysis, we have shown the interactions between the sand barrier and flood terrace under the action of floods and waves. The different time scales on the deformation of these topography also have roles on turning the stretch direction of the barrier.

Acknowledgements

We are grateful for the permission to use the data observed by the Ministry of Transport and the Ministry of Construction in Japan. Our field work was funded by Grant-in-Aid by the Ministry of Education, Science, Sports and Culture in Japan and was partly sponsored by Tohoku Regional Bureau, the Ministry of Construction. We would like to thank for their awards and sponsorships.

References

[1] Goda, Y. (1974): New wave pressure formulae for composite breakwater, 14 th Int. Conf. Coastal Engineering, pp.1702-1720.

[2] Mano A. and M. Sawamoto, and M. Nagao (1995): Response characteristics of river mouth topography in wide time scale range, Proc. 24th Inter. Conf. Coastal Engineering, pp.3126-3138.

[3] Mei, C.C.(1983):The Applied Dynamics of Ocean Waves, John Wiley & Sons.

[4] Munk, W.H. and Arthur, R.S.(1952):Wave intensity along a refracted ray, Natl. Bur. Stand., U.S. , Circ., 521, Gravity Waves.

CHAPTER 351

A Unique Look at Oregon Inlet, NC USA

Herman C. Miller[1], M ASCE, William A. Dennis[2], M ASCE, and Michael J. Wutkowski[2]

Abstract

Oregon Inlet (OI) in North Carolina (NC), the only inlet along a 170 km stretch of coast, supports an active commercial fishing and recreational boating industry. Severe erosion, because of the ongoing migration of OI, resulted in NC constructing a terminal groin to prevent the highway from being severed from the south side of the OI bridge. Construction of this structure provided a unique opportunity to monitor and assess project impacts which could be directly related to the twin jetties which are proposed for this site. The monitoring program included a directional wave gauge, aerial photography, and semi-annual sled-surveys extending 6 km north and south of the inlet. The terminal groin returned the shoreline to its pre-1986 position and has successfully protected the highway abutment to the bridge through many severe storms. This paper presents the results of 6 years of monitoring the morphologic changes. The results document how the coast has adjusted to the construction, a multi-year wave climate reversal, and placement of 1.5 million m^3 of dredged material on the beach. The surveyed area generally lost material both on the up and downdrift sides, much of which apparently has been deposited in the inlet. The effect of these changes on the coast and the inlet's stability are discussed.

Introduction

Oregon Inlet (OI) is the only inlet along a 170 km stretch of coast from Cape Hatteras, North Carolina (NC) to the south, to Rudee Inlet in Virginia to the north, as

---

[1]US Army Engineer Waterways Experiment Station Coastal & Hydraulics Laboratory Field Research Facility, 1261 Duck Road, Kitty Hawk, NC 27949, USA

[2]US Army Engineer Wilmington District, Coastal, Hydrology, & Hydraulics Section, PO Box 1890, Wilmington, NC 28402, USA

shown in Fig 1. The inlet provides an important link between the Atlantic Ocean water and the expansive estuarine Albemarle/Pamlico/Currituck Sound system. The inlet supports an active commercial fishing and recreational boating industry.

During the peak summer months, over 12,000 vehicles a day use the bridge across OI to work and enjoy the beaches in the Cape Hatteras National Seashore. OI and Roanoake Inlet before it, have been migrating south at a rate exceeding 2 km per century. By 1989 this ongoing southerly migration threatened to sever the southern abutment of the bridge to the highway that provides the only land route to the southern beaches. In 1990, NC Department of Transportation (NCDOT) constructed a 953 m rubble mound terminal groin to stabilize the south shoulder of OI. The terminal groin was intended to create a fillet in its lee and return the shoreline to the pre-1986 position. Documenting how the adjacent shoreline and nearshore waters adjusted to construction of the terminal groin was important to the local residents, NC, and the US Army Corps of Engineers (USACE).

This construction provided a unique opportunity to monitor and assess project impacts which could be directly related to the twin jetties which are proposed for this site. Although the function of the terminal groin and jetties (which provide for safe navigation through the inlet) is quite different, it was believed that information gathered concerning the project induced changes resulting from stopping the southerly migration of the inlet would be extremely beneficial in the planning and design of the jetties.

With local and state support for funding, the USACE Wilmington District and the USACE Waterways Experiment Station Coastal and Hydraulics Laboratory', Field Research Facility (FRF) jointly developed a monitoring program to assess the impacts of the terminal groin. The close proximity of OI to the FRF, located in Duck, NC, made it possible to take advantage of the equipment and experienced staff to conduct a long-term monitoring program.

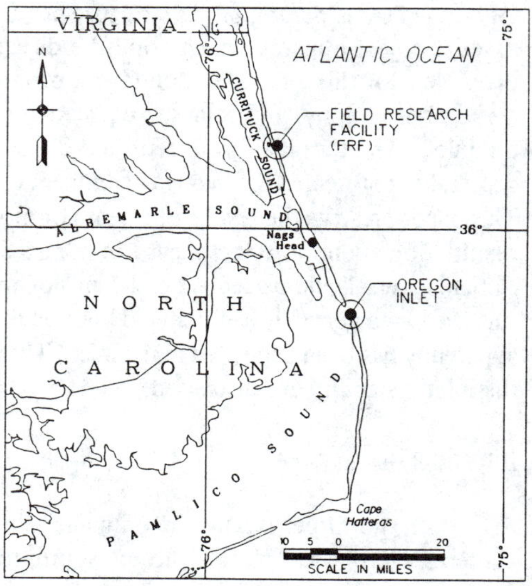

Fig. 1. Site Location. Note, location of USACE Field Research Facility 48 km north of OI. (Kilometers = Miles * 1.61)

This paper presents the results of 6 years of monitoring. In addition to the presence of the terminal groin during that time, there was an unusual long-term wave climate reversal, and material dredged from the inlet was disposed of on the beach on four occasions. The paper organization is generally as follows: monitoring program, monitoring results which document the response of the coast, a discussion of the effect on the inlet stability, and conclusions.

Monitoring Program

The OI area is subject to both hurricanes and northeasters and has one of the highest wave climates on the US east coast. The average annual significant wave height and period for the area are 1 m and 9 sec, respectively (Leffler, et al, 1996). The mean tide range on the open coast is 1.2 m and 0.7 m for the inlet gorge (NOAA 1996). OI separates Bodie Island to the north from Pea Island to the south. The monitoring program consisted of semi-annual nearshore sled-surveys along a portion of the coast extending approximately 6 km on both Bodie and Pea Islands. These surveys were conducted along 38 profile lines spaced at 300 m intervals, Fig 2. The profiles begin behind the dune and extend to the 9 m depth contour. All of the profiles are surveyed in January during the winter storm season and July during the summer recovery season. In addition to these surveys, a directional wave gauge has been operated at the inlet since 1990. Also, aerial photographs were collected approximately every other month.

The sled-survey system (Miller, 1991) consists of an amphibious vehicle and a sled with an 11 m tall mast, Fig 3. The sled, which slides across the ocean bottom virtually unaffected by waves and currents, carries a ring of reflective target-prisms. HYPACK software on a PC is used to collect position and depth measurements every 1 m along a profile using a Geodimeter 140T tracking total-station that is aimed at the prisms. Processing the profile data was facilitated by the Interactive Survey Reduction Program (Birkemeier, 1991). Summaries were generated using Intergraph Inroads/Insite software.

Results

By Halloween in October, 1991, the terminal groin was completed, the fillet had formed, and the shoreline had been returned to the desired location, just in time for a major storm, Fig 4. The road and bridge abutment was protected, accomplishing the primary function of the construction (Dennis and Miller, 1993). As can be seen, the breaking waves define a well formed ebb-shoal which is the main pathway for sand to bypass the inlet. Seeing that this natural pathway was well established, it was not expected that the terminal groin would have a major impact on sediment transport past the inlet. What was expected was that the inlet would narrow, since only the downdrift shoulder was stabilized. The effect of this realignment on the hydraulics of the inlet and on the adjacent coast was of primary interest.

The results are summarized by showing the changes that occurred from the winter of 1991 through the winter of 1994, then from the winter of 1994 through the winter of 1996. This separation shows the different responses of the coast to apparently different processes resulting from an infrequent long-term annual wave climate reversal during 1991 through 1993.

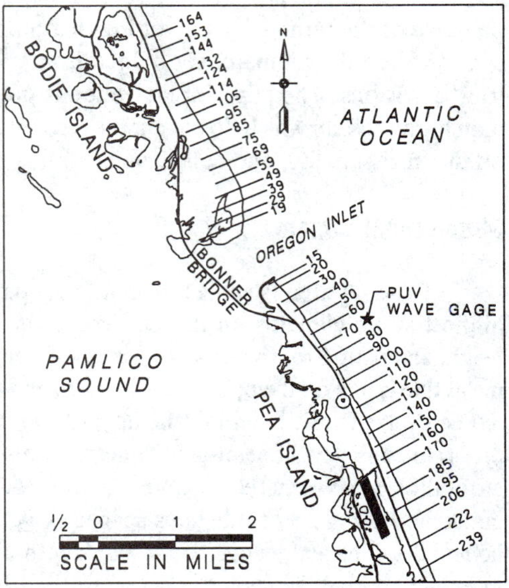

Fig. 2. Sled-survey profile lines. Note, directional wave gauge location at end of Profile #70 off of Pea Island. (Kilometers=Miles* 1.61)

Figure 5 shows the elevation changes over 3 years on the Pea Island (south) side of OI. Every other profile is numbered starting with "15" closest to the inlet and ending with "239," 6.8 km to the south. The February, 1994 shoreline position is included to distinguish the sub-aerial from the

Fig. 3. Survey sled and amphibious vehicle. The sled mast carries the prism array at an 11m elevation.

Fig. 4. OI during "Halloween Storm" on 31 October, 1991. Surge from 4 m, 22 sec. waves have flooded fillet behind the terminal groin at end of the old ferry landing road.

sub-aqueous portions. The terminal groin extends from the "Feb 94" shoreline position at "15" approximately 1 km to the west. The inlet would be to the north (right side) of the terminal groin. Positive elevation changes indicate accretion and are shown as light shaded contour areas. Negative contours in the dark shaded portion of the figure are areas of erosion.

From March, 1991 to February, 1994, near the inlet, there was inshore accretion and offshore erosion along the ebb-shoal. South of "140" there was nearshore erosion that resulted in shoreline retreat, overwash, and destruction of some dunes that were 60 years old. However, the overwhelming feature in the area during this time was the 1.5 million m^3 of dredged material disposed of on the beach between profile lines 50 and 130. As can be seen, much of the disposal material moved offshore. Material also moved north along the inshore bar and trough into the fillet and around the terminal groin into the inlet.

More recently, from February, 1994 to April, 1996, Fig 6, the pattern of changes is quite different in comparison to the earlier time shown in Fig 5. Now the inshore erosion is along the entire length of the survey area. However, there is accretion offshore along a pathway consistent with transport along the ebb shoal toward the south.

Corresponding changes on the Bodie Island (north) side of OI from March, 1991 to February, 1994 are shown in Fig 7. For orientation, the inlet is 0.6 km south (to the left) of "39." Clearly, changes during this time are quite different than those for the Pea Island side shown above. Erosion pervades almost the entire area with the exception of an area of sub-aerial accretion near "59."

An almost opposite response is seen from February, 1994 to April, 1996 as shown in Fig 8. During those 2 years there is sub-aerial and sub-aqueous accretion over most of the area. One exception is the shoreline retreat south of "85" which corresponds to a swing of the Bodie Island spit toward the west. To document what was responsible for these different adjustments before and after February, 1994, it is informative to look at the wave data.

Historic estimates of the annual rate of sediment transport along this portion of coast, (Jarrett, 1978, Birkemeier, et al, 1985, Inman & Dolan, 1989, USAEDW, 1995),

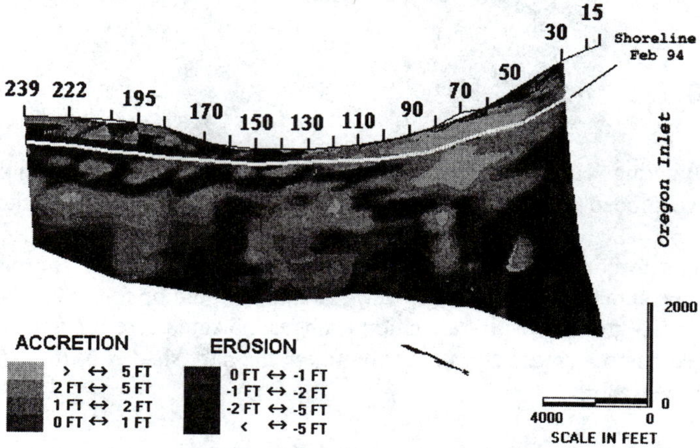

Fig. 5. Pea Island elevation changes from March, 1991 to February, 1994. Terminal groin is at profile 15; inlet is to the north (to right). (Meters = Feet*.3048)

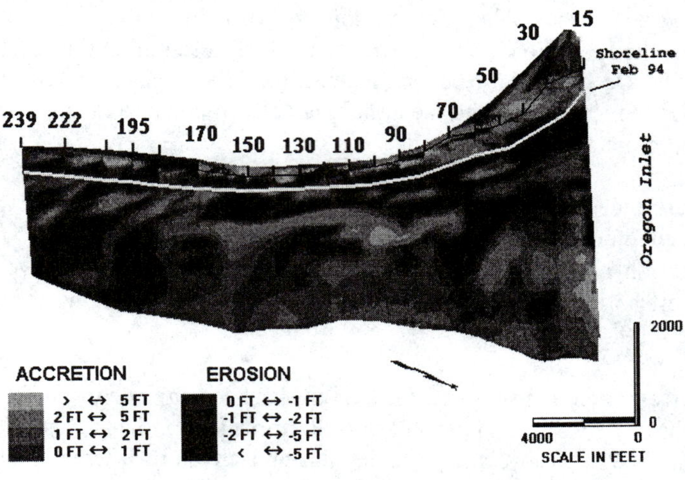

Fig. 6. Pea Island elevation changes from February, 1994 to April, 1996. (M=Ft*.3048)

show that annually approximately 1.5 million m^3 of sediment moves south primarily during the Fall and Winter storm seasons; 800,000 m^3 moves north during the Spring and Summer, for a net southerly transport rate of 700,000 m^3 per year. Potential longshore transport volumes using the energy flux method in the Shore Protection Manual (SPM, 1984) based on hindcasted wave data since 1956 (WIS, 1993), historic wave climatic summaries (Thompson, 1971), and FRF measurements (see FRF WWW page at HTTP://FRF.WES.ARMY.MIL) show that this trend of net annual southerly longshore transport has been consistent over the past 4 decades with one exception in the early 1980's when the southward and northward transport approximately balanced.

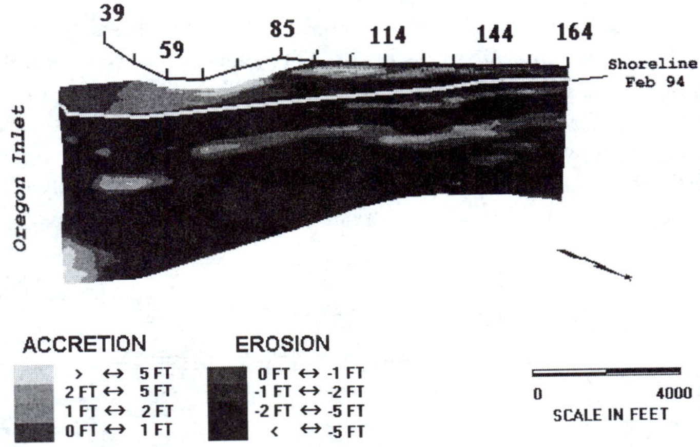

Fig. 7. Bodie Island elevation changes from March, 1991 to February, 1994. Inlet is 2 km south (to left) of Profile #39. (Meters = Feet * .3048)

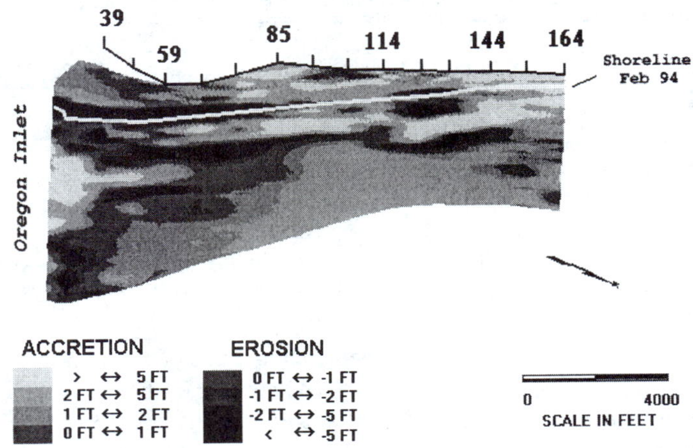

Fig. 8. Bodie Island elevation changes from February, 1994 to April, 1996. Shoreline retreat near inlet accompanied growth of spit toward south. (Meters = Feet * .3048)

Computations based on the FRF's linear directional wave array, (Long and Oltman-Shay, 1991), are shown in Fig 9. The complete data record from the FRF was used, instead of the directional measurement from the self-recording gauge at OI, because of intermittent gaps in the OI data. However, comparison of computations from both gauges, (Miller and Dennis, in prep.), show the wave climate summaries at the FRF are representative of the wave climate at OI. From January, 1990 through February, 1991 and from 1994 through 1996 there is southward transport during the storm seasons, consistent with historic trends. However, during 1991 through 1993, with only a few exceptions, each month the net transport is toward the north. It can also be seen that beginning with the Halloween 1991 storm, frequently, the northward transport rates were quite high. The annual net transport rates during these years approached 2 million m^3 toward the north. One explanation we have considered for this is that there was a long El Nino event in the Pacific Ocean during that time which may have diverted the "jet stream" across the United States causing mid-Atlantic extra-tropical storms to move inland south of OI instead of typically moving up the coast to the north. Whatever the cause, it provided a unique opportunity to study how the coast adjusted to the changing coastal processes.

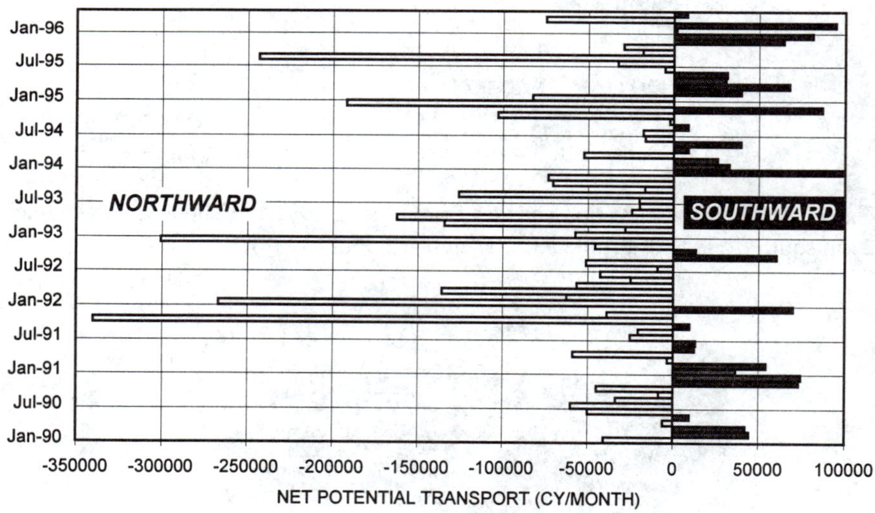

Fig. 9. Monthly net potential transport. Wave climate during 1991-1993 resulted in predominately northward net transport close to 2 million m^3 per year.(M^3=Yd^3*.028)

The effect of an inlet is often categorized in terms of the up drift or downdrift sides of the inlet. Figure 10 summarizes the adjustments of the coast at OI in light of the wave climate reversal that caused the up/downdrift categories to change during the monitoring period. Cumulative volume changes for each of the semi-annual surveys are presented for both Pea and Bodie Islands. During 1991 through 1993, Bodie Island is in the lee because of the wave climate reversal and shows a consistent loss of volume.

By the summer of 1994 the wave climate is more climatologically consistent and the volume shows signs of recovery. On the other hand, the Pea Island side remains approximately neutral through 1994 with some indication of a trend toward loss of material more recently while in the lee of the inlet. However, recall that 1.5 million m^3 of material was deposited on the Pea Island side prior to 1995, and so it too lost volume during the monitoring period. So both sides lost volume, particularly during the times when in the lee of the inlet. Where did the sand go? We believe it went into the inlet. Unfortunately, we do not have volume changes in the inlet. However, using the aerial photography it is possible to gain some insight into the inlet's stability during this time.

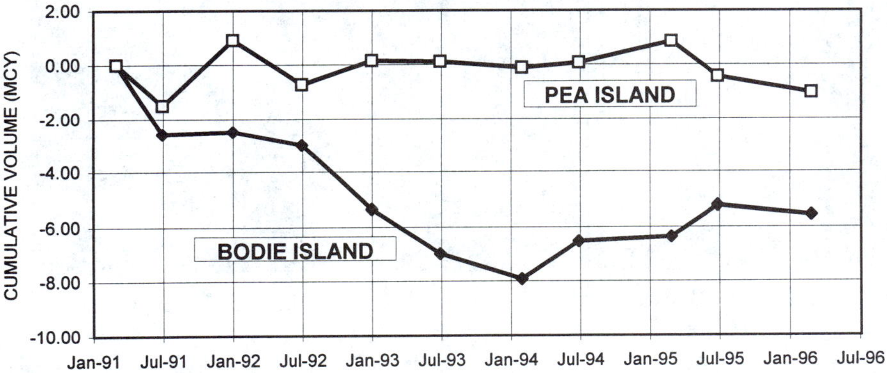

Fig. 10. Pea and Bodie Islands volume changes. Note, while Bodie Island shows loss of volume through July, 1994, Pea Island remains unchanged because of 1.5 million m^3 of dredged material placed on the beach. ($M^3 = Yd^3 * .028$)

Figure 11 was constructed from photography taken on 13 April, 1992. Notice the width of the channel under the bridge. For reference note the "dark spot" on the bridge. This is a repaired section caused by a dredge that washed through the bridge during a storm in October, 1990. Also, note the shape of the ebb shoal. It has a parabolic "flattened bell" shape with a wide base that is asymmetric on the Bodie Island side.

Figure 12 is 1.5 years later on 11 November, 1993. The spit has grown to well south of the repaired bridge section. The Bodie Island side continues to move south as has been the tendency for the past 150 years since OI opened. The narrowing and realignment of the inlet can be seen in the shape of the ebb shoal, which now has an "arrow head" form, less wide at the beach and asymmetric on the Pea Island side. However, the shoal does not extend any further offshore. Higher currents associated with a decreasing inlet cross-section would tend to wash the shoal further offshore which is not seen at this time. The pond in the middle of the fillet behind the terminal groin is the result of mining 250,000 m^3 of sand which was placed on the beach south of the monitoring area.

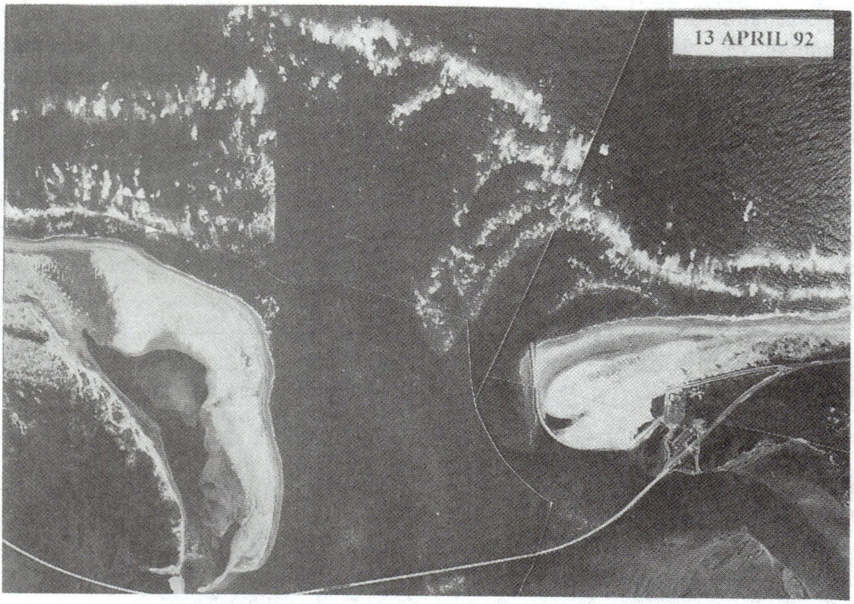

Fig. 11. Mosaic of aerial photography of OI taken on 13 April, 1992.

More recently, as seen in Fig 13 taken on 23 May, 1996, the spit is very near the navigation span under the bridge and an island has formed between the south "Davis" slough and the main OI channel. The volume of material in the inlet, although unknown, appears to have increased dramatically in comparison to Fig 11. We believe this accounts for a large part of the volume losses off of Pea and Bodie Islands.

Inlet Stability

The growth of the Bodie Island spit and development of the shoals have had an effect on the inlet stability. The effect of the shoal under the bridge is that instead of having to dredge to keep the authorized 4.3 m depth in the navigation channel at the bridge, as during the 1991 disposal projects, the depth has now approached 20 m. This equates to approximately 15 m of scour primarily over the past 3 years. Although scour in the navigation channel is unusual, scour has been a problem at OI in the past and has required that remedial measures be taken to reinforce the bridge at "Davis" slough on the south end. The deep scour has not, as yet, been measured in the inlet on the ocean side east of the bridge. In May, 1996 the inlet width, using the traditional method of measuring the minimum distance from the Bodie Island spit to the south side of the inlet (now to the terminal groin), is 820 m. This is the narrowest it has been since 1983, yet the inlet has been more narrow, on a few occasions, such as in 1975 when it was just 640 m wide(USAEDW, 1977). Since the inlet in recent years appears to have maintained a consistent cross-sectional area, (McCafferty, 1996), and is expected to continue to narrow, additional scour can be expected.

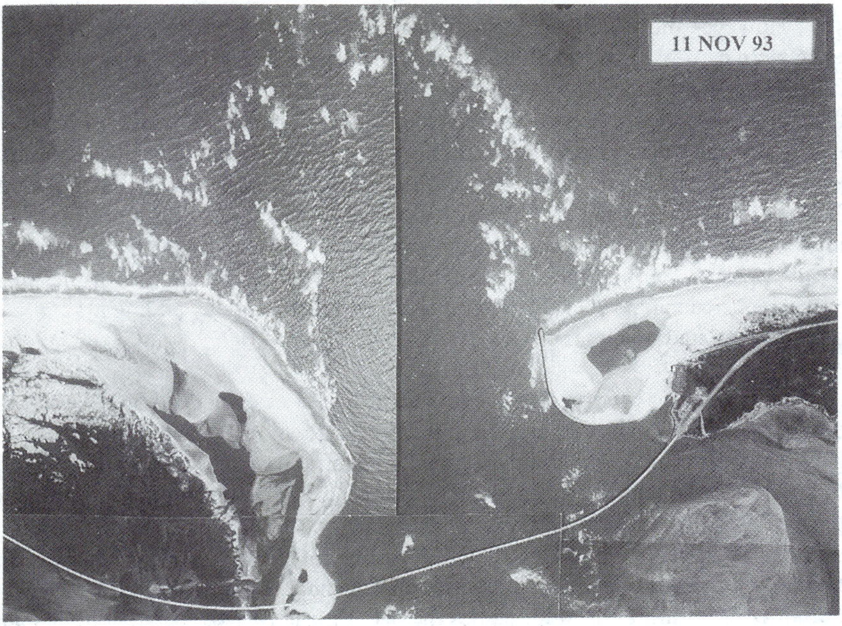

Fig. 12. Aerial photography of OI taken on 11 November, 1993. Note spit now well south of "dark spot" on bridge.

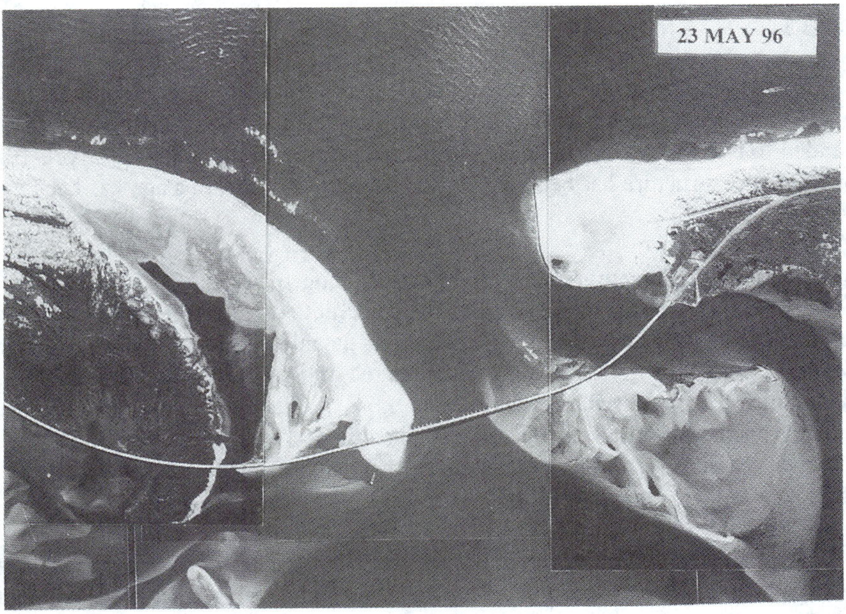

Fig. 13. Aerial photography of OI taken on 23 May, 1996. Development of shoal under bridge has resulted in 15 m of scour in navigation channel.

However, the scour in the navigation channel under the bridge suggests the inlet may be more narrow there. The "effective width" of the inlet might be considered the width measured at the bridge. Considering the shoal, that width is 720 m at this time. As the material continues to enter the inlet and the spit continues to migrate toward the south, it is possible that the main flow channel will shift to "Davis" slough. Should that occur, the spit and island under the bridge would tend to coalesce making it very difficult to maintain the navigation channel at the narrow navigation span of the existing bridge. Also, as the channel moves up against the terminal groin, the scour could increase, possibly, putting the structure in danger. The terminal groin was designed, however, anticipating the channel moving toward the structure by adding a 12 m wide scour apron along the inlet toe.

OI has always been an effective sink for sediment as evidenced by the extensive shoals west of the bridge. The inlet's stability and downdrift beach erosion rates are highly dependant on the natural bypassing of material past the inlet. Unfortunately, with or without the terminal groin, natural bypassing is not efficient at OI. Dredging, in the past used primarily for maintaining the navigation channel, may become an almost mandatory bypassing supplement. This may not be satisfactory since dredging is also not efficient. It requires dredging quantities approaching the annual gross transport of material, which is more than three times the net. An alternative would be to stabilize the inlet with jetties that would prevent the material from entering the inlet and mechanically bypassing the net transport as needed on the downdrift beaches.

Conclusions

A terminal groin was constructed at OI to prevent the important only highway route to popular beaches from being severed from the bridge. The construction was intended to establish a fillet in its lee that would return the shoreline to the pre-1986 position. The structure has been well tested by many large storms and has been very successful.

A long term measurement program, including semi-annual sled surveys, has documented the response of the coast to the construction. Measurements made over the past 6 years captured the unique response of the inlet to an unusual wave climate reversal. During the reversal, erosion was measured on both the up drift and downdrift sides. Since the processes have returned to more normal conditions, there has been recovery on the up drift side and the ebb-shoal is accreting on the downdrift side. However, erosion on both sides indicates that natural bypassing continues to be insufficient at OI.

Apparently, the eroded material is ending up in the inlet. The inlet has adjusted by the growth of the spit toward the south and the rapid development of a shoal under the bridge. The "effective width" of the inlet is now 720 m, near the minimum in half a century, which has caused 15 m of scour in the main channel under the bridge.

The monitoring program at OI has been successful documenting the adjustment of the inlet and adjacent coastal region to construction of the terminal groin. Future efforts will include quantifying the volume changes within the inlet particularly on the shoals. Continued monitoring, as the Bodie Island Spit moves further south, will provide valuable data for future engineering activities that seem inevitable at OI.

References

Birkemeier, W. (1991). "A User's Guide to ISRP: The Interactive Survey Reduction Program." Tech. Rpt. CERC-IR-84-1, Coast. Engrg. Res. Ctr., US Army Eng. Waterways Exp. Stn., Vicksburg, Miss.

Birkemeier, W., Miller, H., Wilhelm, S., DeWall, A., and Gorbics, C. (1985). "A User's Guide to the Coast. Engrg Res. Ctr's (CERC's) Field Research Facility." Tech. Rep. CERC-85-1, US Army Eng. Waterways Exp. Stn., Vicksburg, Miss.

Dennis, W. and Miller, H. (1993). "Shoreline Response Oregon Inlet Terminal Groin Construction." Proc. The Hilton Head Is. S. Carolina USA Int. Coast. Symp. Jour. or Coast. Res., Royal Palm Beach, FL

Inman, D. and Dolan, R. (1989). "The Outer Banks of North Carolina: Budget of Sediment and Inlet Dynamics Along a Migrating Barrier System." Journal of Coastal Research. Vol. 5, No. 2, pp193-237.

Jarrett, J. (1978). "Coastal Processes at Oregon Inlet, North Carolina." Proc. of the 16th Conf. on Coast. Eng., American Society of Civil Engineers.

Leffler, M., Baron, C., Scarborough, B., Hathaway, K., Hodges, P., and Townsend, C. (1996). "Annual Data Sum. for 1994 CERC Field Research Facility." 2 Vols, Tech. Rep. CERC-96-6, Coast. Engrg. Res. Ctr., US Army Eng. Waterways Exp. Stn., Vicksburg, Miss.

Long, C. and Oltman-Shay, J. (1991). "Directional Characteristics of Waves In Shallow Water." Tech. Rep. CERC-91-1, Coast. Engrg. Res. Ctr., US Army Eng. Waterways Exp. Stn., Vicksburg, Miss.

McCafferty, H. (1996) "A Comparative Analysis of The Effects of Storms Upon A Tidal Inlet: Pre- and Post Inlet Stabilization Effects Oregon Inlet, North Carolina." Masters Thesis, Dept. of Env. Sci., Univ. of Virginia.

Miller, H. (1991). "A Topographic Survey System For The Surf and Nearshore." Proc. US Army Corps of Eng. Survey Conf. US Army Eng. Topo. Lab., Ft. Belvoir, VA

Miller, H. and Dennis, W. (in prep). "Predicting Longshore Trans. From Near and Far." National Oceanic and Atmospheric Administration. (1996). "Tide Tables 1992 East Coast of North and South America." US Dept. of Commerce, National Ocean Service, Silver Spring, Maryland

Shore Protection Manual. (1984). 4th Edition, 2 Vols, US Army Eng. Waterways Exp. Station, Coast. Engrg Res. Ctr., US Government Printing Office, Wash., DC.

Thompson, E. (1971). "Wave Climate at Selected Locations Along U.S. Coasts." Tech. Rep. CERC-77-1, Coast. Engrg. Res. Ctr., US Army Eng. Waterways Exp. Stn., Vicksburg, Miss.

US Army Eng. Dist. Wilmington. (1977). "Manteo (Shallowbag) Bay North Carolina General Design Memorandum Phase 1 Plan Formation."

US Army Eng. Dist. Wilmington. (1995). "Manteo (Shallowbag) Bay North Carolina Feature Design Memorandum Sand Bypassing (Management)."

Wave Information Studies of US Coastlines. (1993). "Hindcast Wave Information For the US Atlantic Coast." WIS Rep. 30. Coast. Engrg. Res. Ctr., US Army Eng. Waterways Exp. Stn., Vicksburg, Miss.

Approvals

The authors would like to acknowledge the late, Robert Williams, Chairman of the Oregon Inlet and Waterways Commission, for his support for the monitoring program, efforts to fund the study, and open mind concerning the findings...he will be missed. Permission to publish this paper was granted by the Chief of Engineers.

CHAPTER 352

IMPACTS OF INLET STRUCTURES ON CHANNEL LOCATION

William C. Seabergh[1], Mary A. Cialone[1], Donald K. Stauble[2]

Abstract

Barnegat Inlet, New Jersey, has undergone a variety of structural changes in an attempt to provide a navigable channel from bay to ocean. These structures have included shoreline revetments, arrowhead jetties with their crest elevation at mean tide level, a sand dike to better align interior channel flow, a raised impermeable jetty, and now parallel jetties. Each of these structures has had significant influence on inlet hydraulics and sedimentation, which in turn has impacted channel location.

Introduction and Historical Overview

Barnegat Inlet, New Jersey (Figure 1), is an inlet worth detailed examination as it is rich in history of man's struggle to control nature's scheme. A paper written in the 1970's about the inlet entitled "Barnegat Inlet, Nature Prevails" (Caccese and Spies, 1977) expressed the frustration an inlet can cause coastal engineers and scientists. The natural inlet migrated to the south 1600 m from 1839 to 1932. An inlet shoulder revetment and groin to protect Barnegat lighthouse anchored the south side of the inlet. A pair of "arrowhead" jetties was constructed in the late 1930's, followed in 1943 by a sand dike in the adjacent bay that caused a redirection of flow. Important in understanding the response of the inlet channel in this time frame were the jetties' low crest elevations at mean tide level. These jetties were functioning as "weir jetties," which allowed tidal flow, wave-generated currents, and sand to be transported over these inlet structures. This resulted in creation of sand spits at the inlet gorge and became a new control for channel location, withstanding many dredging attempts to control channel position. During a twenty year period the sinuous channel was completely inverted as this new regime interacted with structural controls.

----

[1]Research Hydraulic Engineer,[2] Research Physical Scientist, U.S. Army Engineer Waterways Experiment Station, Coastal Engineering Research Center, 3909 Halls Ferry Road, Vicksburg, Mississippi 39180-6199

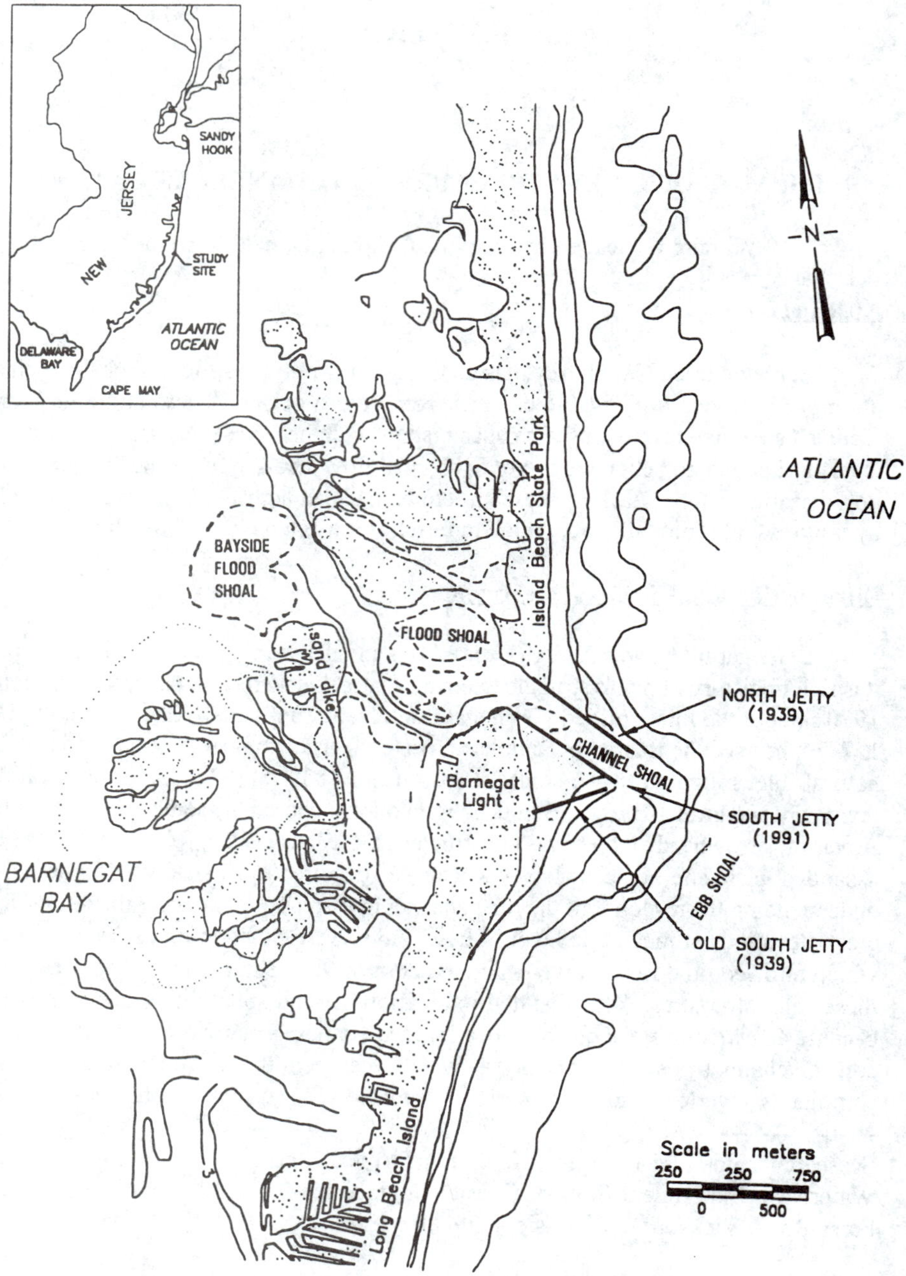

Figure 1. Barnegat Inlet, New Jersey location map and project elements.

In the early 1970's, a sequence of steps to improve channel dynamics was developed from a physical model study (Sager and Hollyfield, 1974) which included raising the north jetty and making it impermeable plus the addition of a new south jetty paralleling the existing north jetty. In 1972-74, the north jetty was raised 1.8 m. Sand was diverted offshore along the impermeable jetty and growth of the ebb shoal resulted. The channels adjusted to a reduction of sediment from the north beach.

A significant amount of dredging was done in the late 1970's to maintain the channel at the inlet throat. However, the increase in ebb shoal volume created additional maintenance dredging in that location. Another phase suggested by the model study was implemented beginning in 1989, with the construction of a new south jetty located within the arrowhead system. This jetty paralleled the original north jetty and served as a replacement to the original low south jetty. Upon completion of the project, a monitoring study was initiated to understand the response of the navigation channel to the most recent inlet structures, but, in doing so, an understanding of the historic interaction of the channel and structures was also necessary.

Physical Factors

The inlet separates Island Beach, a spit to the north, from Long Beach Island, a barrier island to the south. These barriers are characterized by medium to fine sands. Within the inlet region, medium to fine sands (0.25-0.5 mm) are on the ebb and flood shoals and coarser sands (0.50-1.0 mm) are in the deeper channel areas (Stauble and Cialone, 1995). The inlet provides access for commercial fisherman, day fishing excursion boats and small craft. The inlet's design channel is 91.5 m wide by 3 m deep (relative to mean low water), extending through the ebb shoal. The mean ocean tide range is 1.28 m and mean wave height is 1.20 m. Littoral transport estimates at the inlet are 840,000 cubic meters gross transport, with a net of 110,000 cubic meters to the south. These estimates are based on wave heights hindcasted at the 20-m contour near the inlet.

Effect of Inlet Hydraulics on Channel Dynamics

Throughout the recent history of Barnegat Inlet there has been the interaction of structures, changing sedimentation patterns and inlet hydrodynamics. The inlet system now contains four consecutive, fully-developed shoal features (compared to typical one ebb/one flood shoal), with an ebb shoal seaward of the jetties, a shoal in the intra-jetty region (particularly evident for the arrowhead configuration), a large flood shoal contained by the sand dike and finally a bayside flood tidal delta where flow exits into Barnegat Bay (Figure 1). Development of these shoals initially created a higher friction environment, which with the initial structural configuration, created increased sedimentation and a gradual decrease in tidal prism. Raising the north jetty reduced sediment input from the north beach, and coupled with dredging, and the construction of the new south jetty, some flow efficiency was regained as evidenced by increased tidal prism.

Important in relating the channel response to inlet structures is an understanding of the inlet's hydrodynamics. This inlet has maximum flood currents near high water elevations and maximum ebb currents near low water and is typical for an inlet lagoon which has very large surface area relative to channel cross-sectional area. Essentially the lagoon level fluctuates very little and the ocean tide range oscillates about that level, resulting in maximum head differences across the inlet near high and low waters. This phasing of flow relative to structure crest elevation and flow over shallow shoal areas is important to channel location. Figure 2 shows flow patterns for maximum ebb and flood flows as determined for 1968 conditions from a physical model study (Sager and Hollyfield, 1974). For the mean tide level elevation jetties, maximum flood currents (strongest near high water) had a great potential for introducing sediment to the inlet system and thus the development of a large flood shoal complex. Low water ebb currents are more channelized. This permits shoals to be more effective ebb shields; that is, ebb flow will tend to be deflected around the shoal area if the shoal elevation is higher than low water. Also, maximum ebb flows at low water elevation can lead to incising of channels.

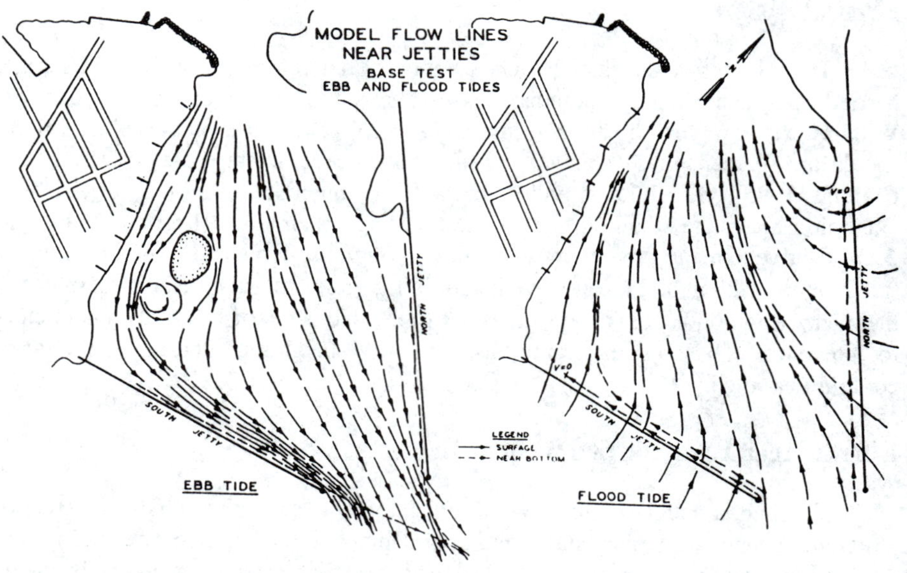

Figure 2. Model study flowlines, ebb and flood currents.(Sager and Hollyfield, 1974)

Historic Bathymetric Analysis

A review of historical bathymetry was performed to provide a basis for interpreting recent bathymetric changes at Barnegat Inlet and in providing guidance to anticipate future changes. Some historical information was derived from US Army Engineer District, Philadelphia, (1981) and Fields and Ashley (1987).

Pre-jetties In 1937 (Figure 3a) the pre-jetty inlet bathymetry shows that the interior channel swept south then turned east to northeast to exit the inlet, then turned southeast, on the ebb shoal, in response to predominant waves from the northeast. The 1937 interior channel was more than 460 m southeast of its current position (1996). The channel probably owed its large curvature to the infilling of the natural inlet on its north side as the inlet and channel both migrated south. The predominant portion of the tidal prism exited the bay from the north, channelizing around the large flood shoals due to a strong ebb shield factor resulting from this inlet's hydraulic phasing.

Arrowhead Jetties In 1939 arrowhead jetties were constructed and a channel, about parallel to the north jetty, was dredged into the bay in an attempt to provide a more direct route to the bay. Figure 3b shows the new interior channel in 1941. Also note the deflection of the ocean channel resulting from having the south jetty placed directly in its path.

Sand Dike By 1943 (Figure 3c), the sand dike was constructed in order to cut off the strong ebb flow from the dominant interior channel which was causing excessive scour on the inside shoulder of the inlet behind the lighthouse. It was anticipated that flow would be diverted to the straight interior channel, providing a deeper direct channel connecting ocean and bay. In addition, groins were constructed along the ocean shoreline inside the south jetty to mitigate shoreline erosion.

The 1946 bathymetry (Figure 4a) indicated a slight deflection of the navigation channel at the inlet's intersection with the shoreline as sediment moved over the low jetties at this location. On the south side of this region there was a shoal extending seaward from the lighthouse area, probably derived from sediment moving toward the inlet gorge along the shoreline inside the south jetty, then deflecting seaward on ebb. A buildup of sand at the shoreline is noted just inside the south jetty indicating an influx from the south beaches.

By 1953 (Figure 4b) the main navigation channel had shifted slightly south and rotated somewhat to the southeast. Sediment was moving over the low north jetty into the inlet gorge region. The interior region between the inlet gorge and the north tip of the sand dike contained flood shoals and three smaller channels. On the ocean side of the inlet, the navigation channel was close to the north jetty, as it had been for the previous ten years.

As of 1959 (Figure 4c) the navigation channel through the inlet gorge rotated away from the north jetty to the south and a scour area reappeared adjacent to the outer portion of the south jetty. The influx of sediment over the north jetty contributed to this rotation. Interesting to note was the shifting of the deepest area at the bayward end of the sand dike, which moved to the southeast side (compared with earlier conditions).

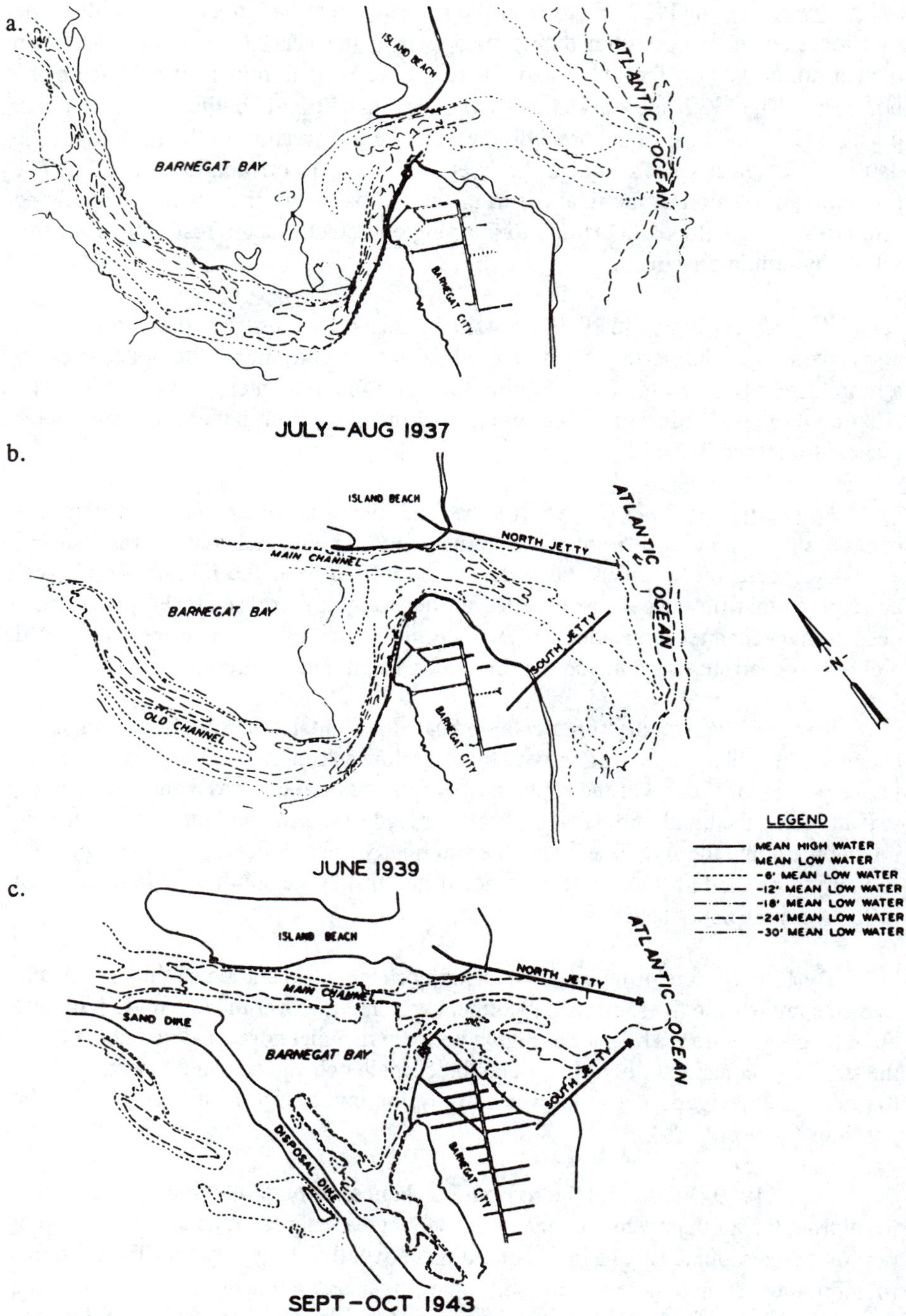

Figure 3. Barnegat Inlet bathymetry, July-August 1937, June 1939 and Sept-October 1943.

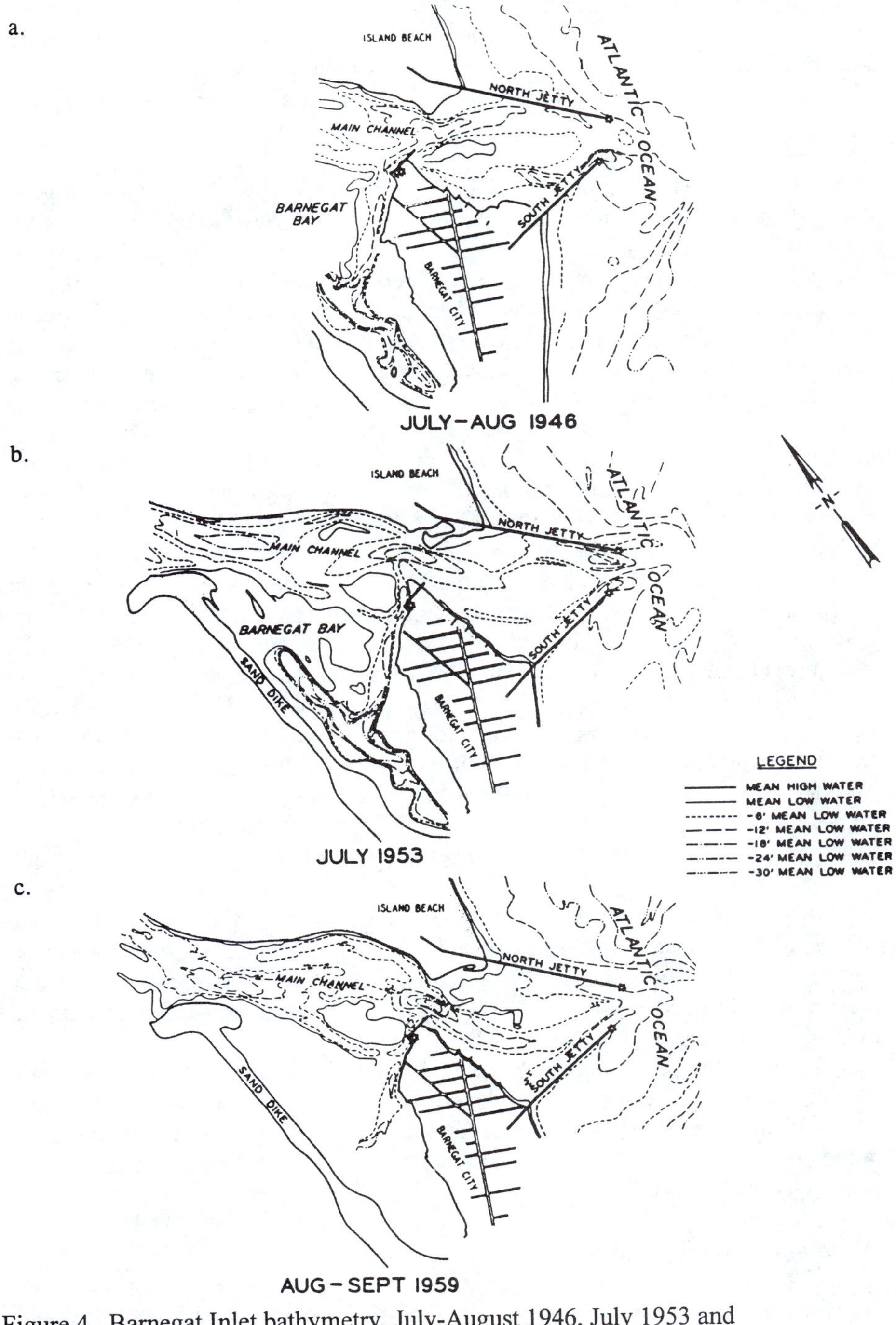

Figure 4. Barnegat Inlet bathymetry, July-August 1946, July 1953 and August-September 1959.

Nineteen-sixty-two (Figure 5a) showed a new part of the flood shoal developing from sediment stripped from a spit which extended further from the north beachline into the inlet gorge. The minimum width of the inlet was reduced considerably due to sediment movement over the north jetty. The seaward portion of the channel migrated against the south jetty.

By 1968 (Figure 5b), the flood shoal occupied the location of the original dredged interior navigation channel. The interior channel was still bifurcated, but the region south of the flood shoal was widening and becoming the main interior ebb channel. It was forming in a similar configuration as was seen for the pre-jetties condition, except not as far southeast due to the presence of the sand dike. Sediment movement over the north jetty almost closed off the inlet gorge. The region between the arrowhead jetties was shoaling considerably except for the channel which had migrated against the south jetty. This sinuous channel was eroding the ocean-facing shoreline inside the south jetty and creating toe erosion which endangered the integrity of the oceanward portion of the south jetty. The trend of flood shoal growth and interior channel shifting south continued until the north jetty was raised 1.8 m from its original mean tide level crest elevation in the 1972-74 period.

Raised North Jetty The 1975 bathymetry (Figure 5c) indicated a major reorientation of the navigation channel through the jetty region. Dredging at the inlet gorge combined with cutting off sediment input by raising the north jetty permitted a straighter channel which was more in alignment with and closer to the north jetty. This channel orientation is reinforced by a concept presented by Keislich (1981) where a channel at a single-jettied inlet migrates toward the structure independent of whether or not the jetty structure is situated on the side of stronger longshore sediment drift. The Barnegat system probably can be considered a single jetty system in this respect due to the free flow of sediment and currents over the mean tide level south jetty, which helps move the channel toward the "single" north jetty.

Raising the north jetty caused a significant change in sediment pathways. The ebb shoal began to increase in magnitude (Figure 6a). This most likely can be attributed to the movement of sediment along the outside of the north jetty which previously had passed over the landward end of the low north jetty and contributed to flood shoal building and the movement of the inlet gorge towards the south. The same trend of channel alignment seen in the 1970's continued through the 1980's (Figure 6b). The interior navigation channel moved more towards its pre-project (1930's) location and the channel between the jetties was concentrated on the north side adjacent to the raised north jetty. This configuration was maintained until the construction of a new south jetty which occurred between late 1987 and 1991.

Hydraulic Response to New South Jetty

In order to build the new south jetty (Figure 7) from the revetted region on the south shoulder of the inlet beneath the lighthouse, shallow shoals were removed from

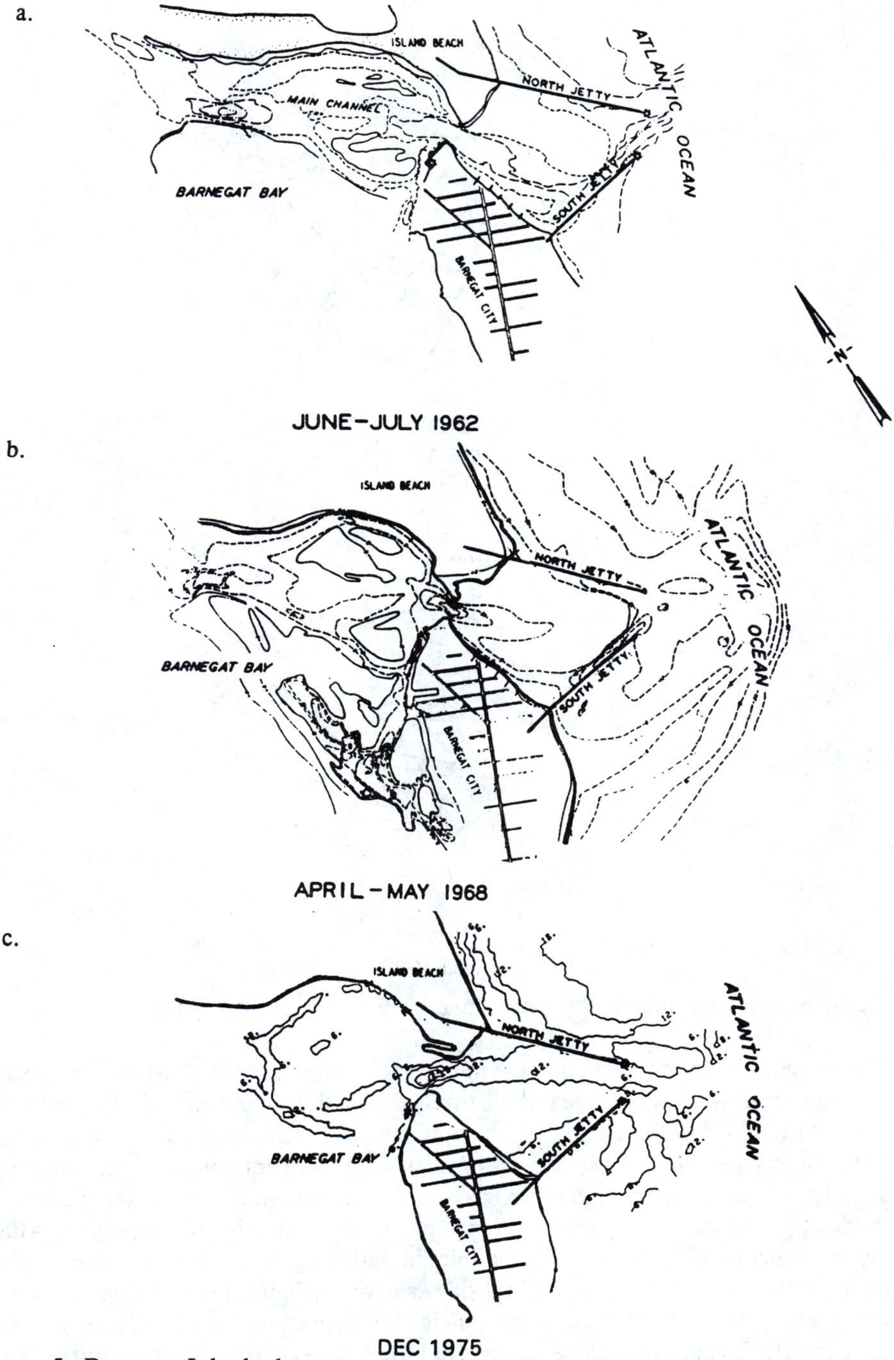

Figure 5. Barnegat Inlet bathymetry, June-July 1962, April-May 1968 and December 1975.

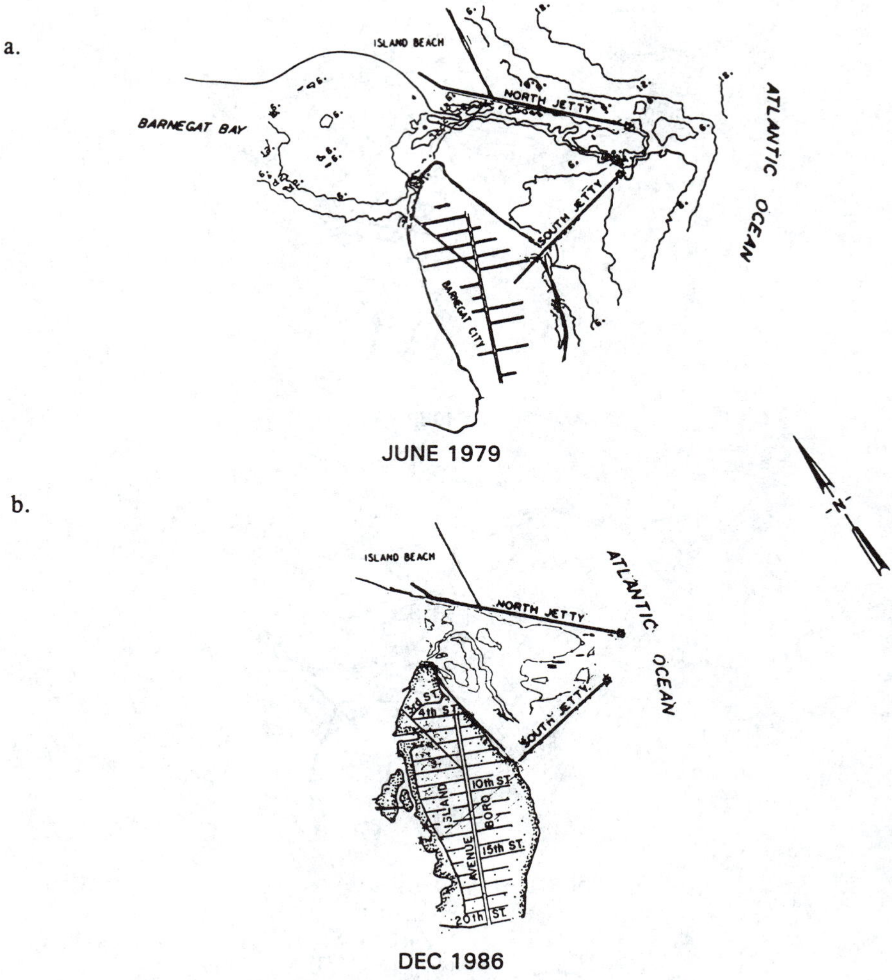

Figure 6. Barnegat Inlet bathymetry, June 1979 and December 1986.

this region. The inlet system became more efficient hydraulically due to an increase in minimum cross-sectional area which resulted from this shoal removal. This follows from O'Brien's (1969) relation between minimum inlet area, A_C (m^2) and tidal prism, P (m^3) at dual jettied systems: $A_C = 7.489 \times 10^{-4} P^{0.86}$. With minimal sediment entering the inlet system the increase in prism has been maintained since completion of the new south jetty. Figure 8 shows the variation in prism for the duration of the project. After the construction of jetties (1941), the inlet initially had the same tidal prism as the natural inlet, but the addition of the inside sand dike (Figure 1) lengthened the inlet channel and the newly dredged interior navigation channel had a reduced channel area compared to the old sinuous one. Sediment influx reduced areas and thus prism. The new south jetty prevented the influx of sediments that had previously occurred for the low south jetty and so the new, larger, minimum area has been maintained.

a.

b.

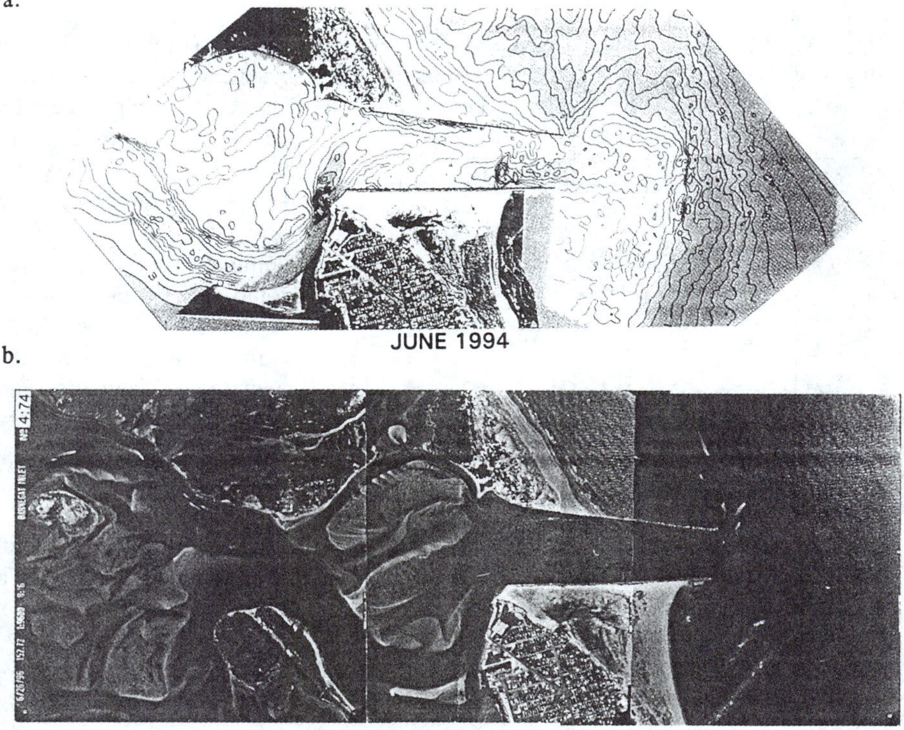

JUNE 1996

Figure 7. Barnegat Inlet 1994 bathymetry and 1996 air photograph.

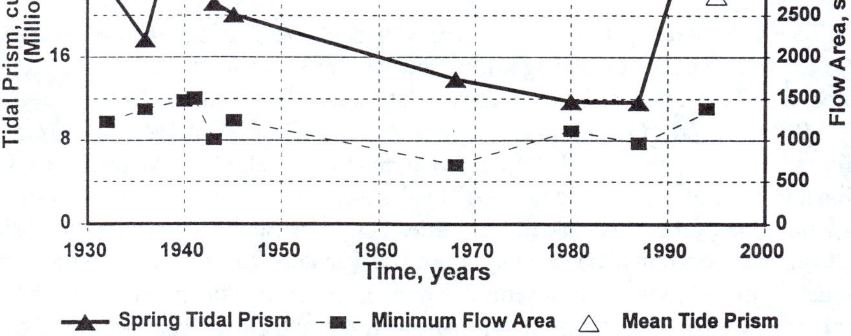

Figure 8. Historical measured tidal prism variation.

Channel Response To New South Jetty

Analysis of the channel response to the most recent addition of the new 1250-m long south jetty reveals what appears to be an evolutionary change in channel depth in the region between the north jetty and new south jetty (Figure 9). The net sediment mass is translating bayward in the region of the parallel jetties. An important aspect of this is the flood dominance of currents on the south side of the intra-jetty region of the channel. Flood currents tend to enter the inlet region from the north jetty side (due to greater ebb shoal depth on the north and shallower ebb shoal depth on the south side, Figure 7a) and sweep toward the south jetty. Also the outer 305 m of the north jetty remains at the mean tide level elevation, permitting maximum flood currents to flow over this low section into the channel perpendicular to flow coming through the oceanward jetty tips. This also helps guide flood currents to the south side of the intra-jetty region. The occurrence of a shoal region between stations 200 and 400 is essentially a nodal point between ebb and flood flow dominance. Flood flow pushes sediment through the channel mostly on the south side, (based on velocity distribution measurements taken in 1994-96 by the authors). Ebb currents from the curved interior channel shear this shoal and sediment moves along the north side of the shoal (where ebb flows are concentrated) towards the ocean. Dredging of the channel shoal and, evidently, a net oceanward sediment circulation out of the intra-jetty region have caused a progressively deeper channel.

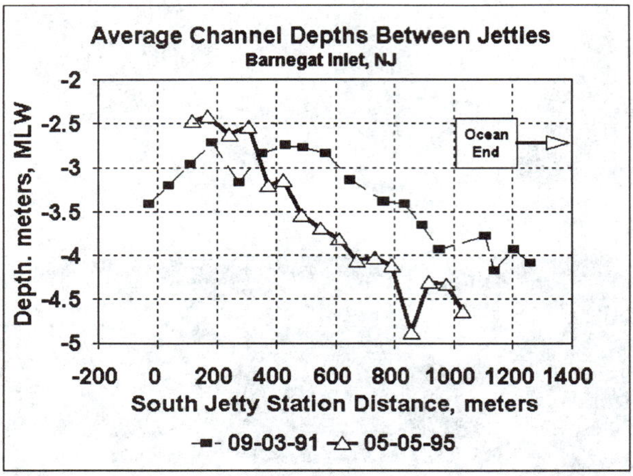

Figure 9. Average channel depths between jetties, 1991 and 1995.

The interior navigation channel is moving southward, as the flood shoal flattens out since sediment is not reaching it from the north or south adjacent shorelines or the ebb shoal. This channel has migrated south about 90 m in the last three years (1994-96). Flood currents plus ocean waves traveling with the currents move sediment towards the bayside of the flood shoal. On ebb, the current shears off sediments from the back edge of the flood shoal and moves it counter-clockwise along the edge of the shoal. Sediment settles out on the south edge of the flood shoal as strong ebb currents move away from the shoal to the outside of the curved channel. This spreading out of sediment may decrease the effectiveness of the flood shoal as an ebb shield and gradually permit more ebb flow over the shoal. Some incised cutting of the center of the flood shoal can be noted in Figure 7b.

Lessons Learned

The concept of arrowhead jetties to provide concentration of ebb flows at the oceanward terminus (in order to cut through the ebb shoal) and for wave attenuation due to diffraction as waves propagated into the expansion area, were positive design attributes. The hydraulics of such a system with regard to the velocity phasing at Barnegat Inlet (maximum flood currents near high water and max ebb near low water) combined with mean tide level jetties would permit broad, less concentrated flood currents to approach the inlet, presumably having less potential to carry sediments into the inlet. Strong ebb flow concentration in the navigation channel would flush sediments out of the channel as water surface elevation dropped. However, the sediment influx over these low jetties overshadowed the positive elements of the plan. Apparently most of the sediment movement was at the shoreline intersection with the jetties.

It was learned from the historical analysis of bathymetries that the interior navigation channel moved back to its pre-structure alignment, probably due to the influx of sand coming over the low north jetty, which enlarged the flood shoal significantly and helped deflect ebb currents coming from the bay towards the southeast. There was a similar situation for the natural inlet, which had been migrating south, thus infilling on the north side and accumulating sediments to help deflect ebb currents to the south.

Raising of the north jetty crest elevation cut off direct sediment influx from the north but sediments from the south maintained the same minimum area at the inlet gorge. Channel migration to the now dominant north jetty, plus dredging, cut off input to the flood shoal and redirected beach sediments to the ebb shoal.

The effects of sediment input into an inlet system in equilibrium usually is balanced by sediment moving out. At Barnegat Inlet, initially structural changes effectively lengthened the channel which led to increased friction, reduced currents, followed by sedimentation and reduction in tidal prism. The addition of a new higher south jetty paralleling the north jetty along with an increase of minimum channel area due to dredging and the prevention of sediments entering from the south into the inlet gorge permitted a larger tidal prism.

It took over 20 years (1941 to 1965) for the straight interior navigation channel to move to the south back to its historic curved configuration. However, recent incising of the flood shoal and the apparent reduction in sediment supply to the flood shoal indicates a potential for ebb currents to eventually "short-cut" across the flood shoal and deepen a channel there.

Conclusions

Inlet channel location is a complex function of inlet hydraulics, littoral influx to the channel system, and inlet structures. Historic analysis of structural effects has provided a clear picture of inlet response and impacts on channel location. Initially low arrowhead jetties followed by an interior sand dike, then sand tightening of one jetty with increased jetty elevation, and finally conversion to a parallel jetty system, affected inlet hydraulics and sediment input, which in turn changed shoaling patterns and thus channel location. With the low arrowhead jetty system, sedimentation reduced channel cross-sectional area with a corresponding reduction in tidal prism. Today's inlet, which is still adjusting to the new parallel south jetty, appears to allow a more stable channel to exist due to the restriction of sediment input into the navigation channel. These factors along with an increase in minimum channel area due to dredging have changed the tidal prism back to pre-structure conditions.

Acknowledgments

The work was funded through the Monitoring Completed Navigation Projects Program and the Coastal Inlets Research program. Assistance of Mr. Keith Watson and others at the US Army Engineer District, Philadelphia, is appreciated. Work performed by previous co-investigators, Ms. Monica Chasten and Mr. John McCormick, is also recognized. Permission to publish was granted by the Office, Chief of Engineers.

References

Caccese, L.A. and Spies, H.R. 1977. "Barnegat Inlet, Nature Prevails" *Proceedings, Coastal Sediments '77,* American Society of Civil Engineers, Charleston, SC, pp 305-310.

Fields, M.L. and Ashley, G.M. 1987. "Barnegat Inlet, New Jersey - A Stabilized Inlet," Proceedings, Coastal Sediments '87,. American Society of Civil Engineers, New Orleans, LA, pp 2006-2021.

Kieslich, J.M. 1981. "Tidal Inlet Response to Jetty Construction," GITI Report 19, U.S. Army, Corps of Engineers, Coastal Engineering Research Center, Fort Belvoir, VA, and U.S. Army Engineer Waterways Experiment Station, Vicksburg, MS.

O'Brien, M.P. **1969**. "Equilibrium Flow Areas of Inlets on Sandy Coasts, *Journal of the Waterways and Harbors* Division, ASCE, Vol 95, No. WW1, pp 43-52.

Sager, R.A. and Hollyfield, N.W. 1974."Navigation Channel Improvements, Barnegat Inlet, New Jersey; Hydraulic Model Investigation,"Technical Report H-74-1, US Army Engineer Waterways Experiment Station, Vicksburg, MS.

Stauble, D.K. and Cialone, M.A. 1995. "Analysis of Sediment Grain Size Trends For Defining Inlet Sediment Processes," *Proceedings of the 8th Annual National Conference on Beach Preservation Technology,* FSBPA, Tallahassee, FL, pp 211-226.

US Army Engineer District, Philadelphia. 1981. Barnegat Inlet, New Jersey, Phase- I General Design Memorandum, Philadelphia, PA.

CHAPTER 353

Complete Closure of The Nanakita River Mouth in 1994

Hitoshi Tanaka[1], Fumihiko Takahashi[2] and Atsushi Takahashi[3]

ABSTRACT

Very severe draught occurred in Japan during the dry season in 1994. Due to considerably reduced river discharge, complete closure was induced five times at the Nanakita River mouth in Japan during this period. Detailed field observations were carried out immediately before and after the closures. By analyzing hydraulic characteristics measured at the mouth, date of the closure occurrence can be estimated. An analytical solution is derived for predicting time-variation of river mouth width under the combined influence of incoming waves and effluent river discharge.

1. INTRODUCTION

There have been a lot of field measurements of topography change at a river mouth in connection with practical problems such as flood control and maintenance of a navigation channel. Most of them are, however, made at relatively large rivers due to their practical importance, and, consequently, very few attention has been paid to small ones.

The authors have been conducting field measurement of topography change at the mouth of the Nanakita River, one of typical small rivers in Japan, since 1988 (Tanaka and Shuto, 1989, 1991, 1992: Tanaka, Kabutoyama and Shuto, 1995: Tanaka and Ito, 1996). The Nanakita River originates at the northern part of Sendai and pours into the Pacific Ocean as shown in Fig.1. The catchment area and the length of the river are 229km^2 and 45km, respectively.

From 1988 to 1993, the complete closure at the Nanakita River mouth has been observed only twice (Tanaka and Shuto, 1991), whereas in 1994, the closure

[1] Professor, Department of Civil Engineering, Tohoku University, Aoba, Sendai 980-77, JAPAN. e-mail: tanaka@tsunami2.civil.tohoku.ac.jp

[2] Civil Engineer, Penta-Ocean Construction Co. Ltd., Koraku, Bunkyoku, Tokyo 125, JAPAN.

[3] Civil Engineer, Niigata Prefecture, 4-1, Sinkocho, Niigata 950, JAPAN.

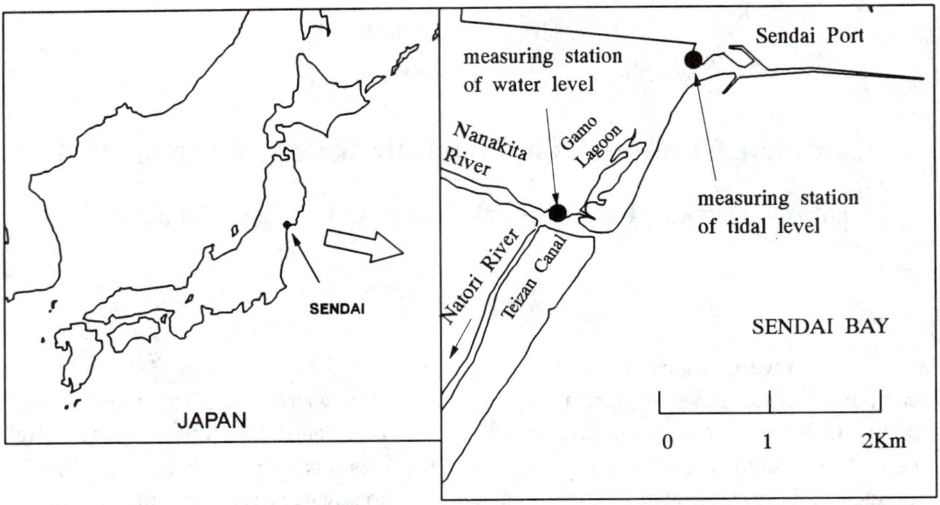

Fig. 1 Location map of the Nanakita River mouth

occurred five times in contrast, probably due to remarkable reduction of the fresh water discharge during the dry season. In this paper, results of detailed field surveying of the topography change will be firstly shown along with corresponding external forces acting at the river mouth. Furthermore, a mathematical model, considering littoral drift by waves and sediment flushing due to river discharge, will be applied to the Nanakita River in order to reproduce the change in the river mouth width.

2. RESULTS OF FIELD OBSERVATION

Figure 2 shows the significant wave height H, the wave period T, the wave incident direction θ measured clockwise from the north, and the river discharge Q_r. The wave characteristics are measured 4 km offshore from the river mouth at the mean depth of 20m, while the fresh water discharge is measured at a gauge station 9km upstream. In Fig.2, the dates when the river mouth closure was firstly observed are also drawn by thick arrows. It should be noted that these are not the dates when each closure occurred, but the dates when the authors firstly noticed the closure. It is seen that the river discharge is considerably small due to draught as compared with other years (Tanaka and Shuto, 1989, 1991) and that very high waves immediately before each closure are common to all events.

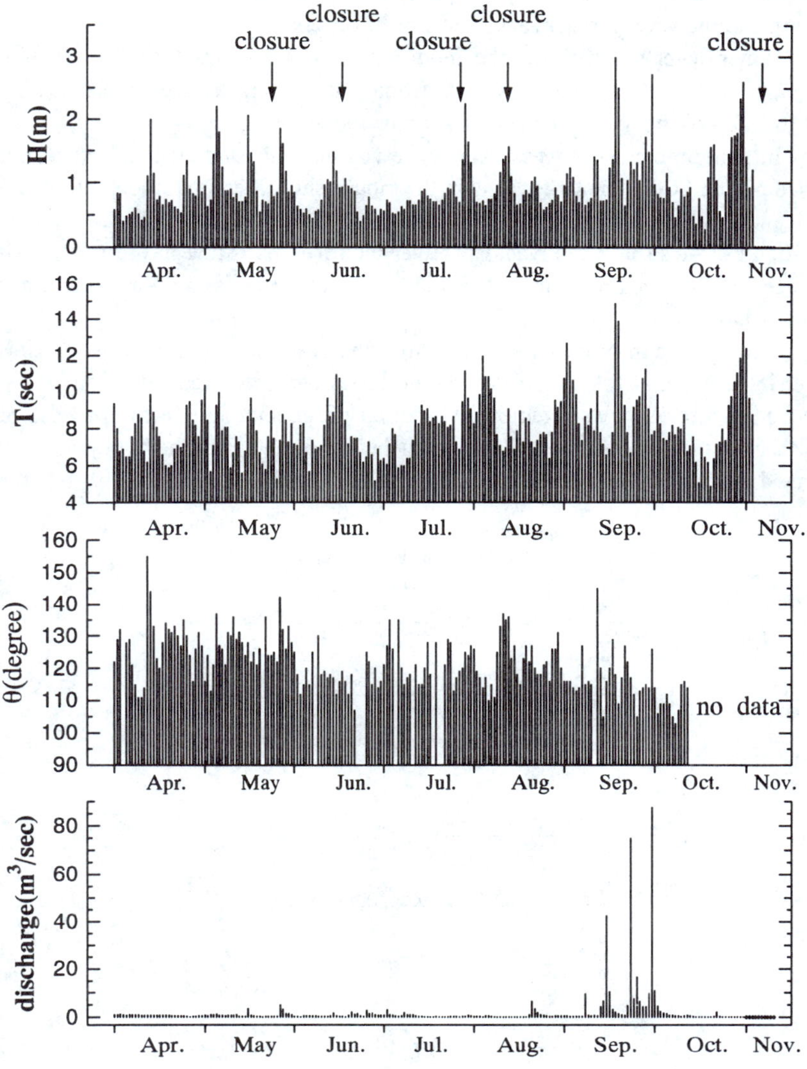

Fig. 2 Time-variation of waves and river discharge

According to Tanaka and Shuto's (1991) study on river mouth closure at the Nanakita River, the incoming waves with incident angle of about 100° are the most unfavorable for keeping the mouth open, suggesting that onshore sediment transport is more effective for the river mouth closure than the longshore sediment movement. In 1994 data set shown in Fig.2, though, such a relationship between the occurrence of closure and the wave incident angle can not be observed.

Water level variation in the mouth is shown in Figs.3 and 4 for June and November respectively, along with the tidal variation measured in the Sendai Port. The location of the measuring stations can be found in Fig.1.

In June, the closure was firstly observed on 17th, and artificial excavation was carried out on 20th. The water level in the mouth shows less variation before and after 17th and is always higher than the tidal level. Even after the closure, slight time-variation can be seen in the Nanakita River mouth. This is due to the tide propagated through the Teizan Canal shown in Fig.1, which connects the Nanakita and Natori River mouths.

The closure in November was firstly observed on 6th. However, judging from Fig.4, it seems that the closure of the mouth progressed gradually from the end of October. Figure 5 shows a temporal variation of the shoreline in the process of river mouth closure in November. Since it can be confirmed that the topography change affected much the water level inside the mouth (Tanaka and Shuto, 1991), the ratio

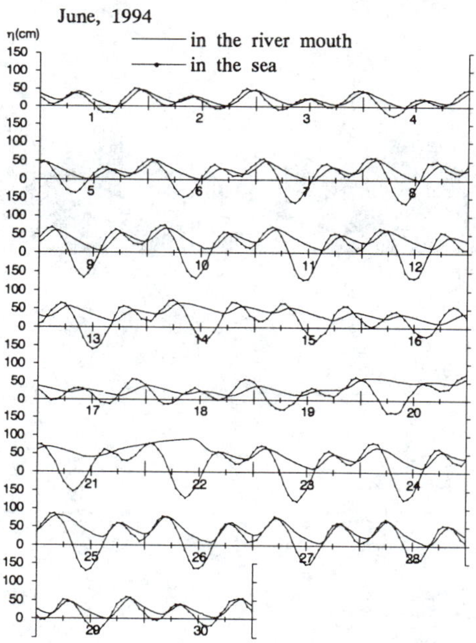

Fig. 3 Water level variation in June

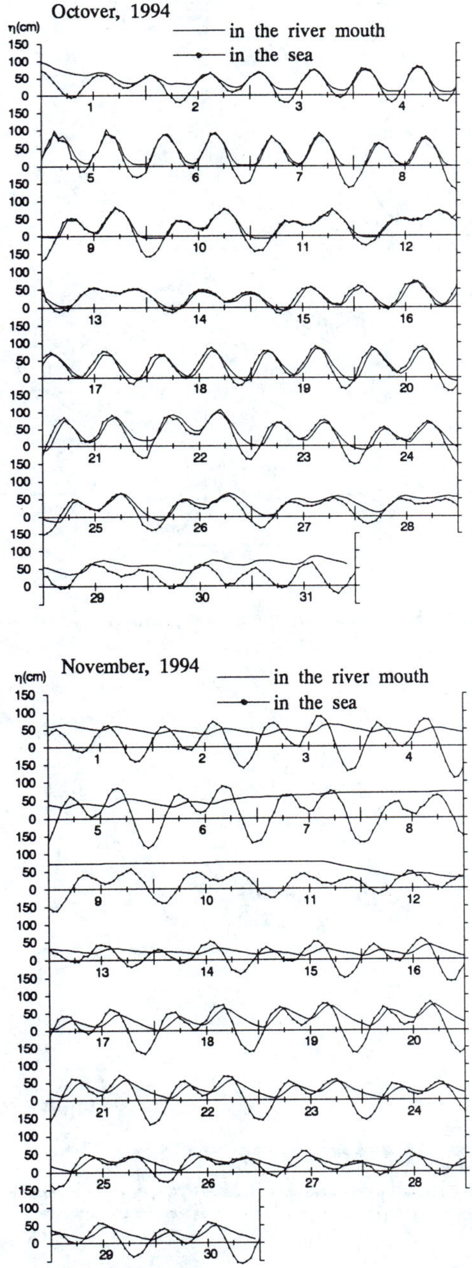

Fig. 4 Water level variation in October and November

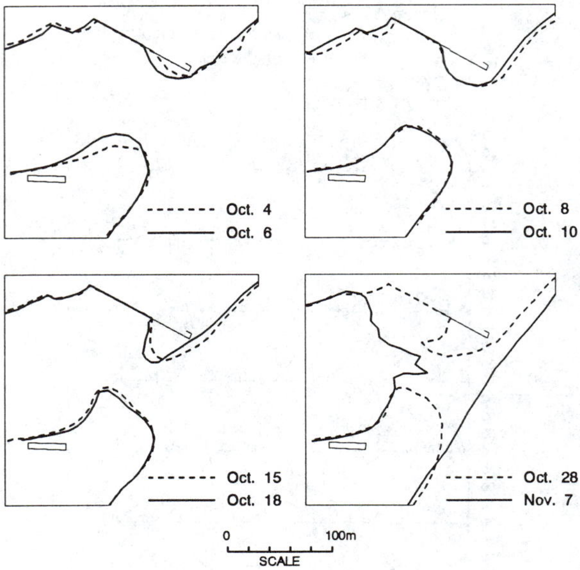

Fig. 5 Shoreline change at the Nanakita River mouth

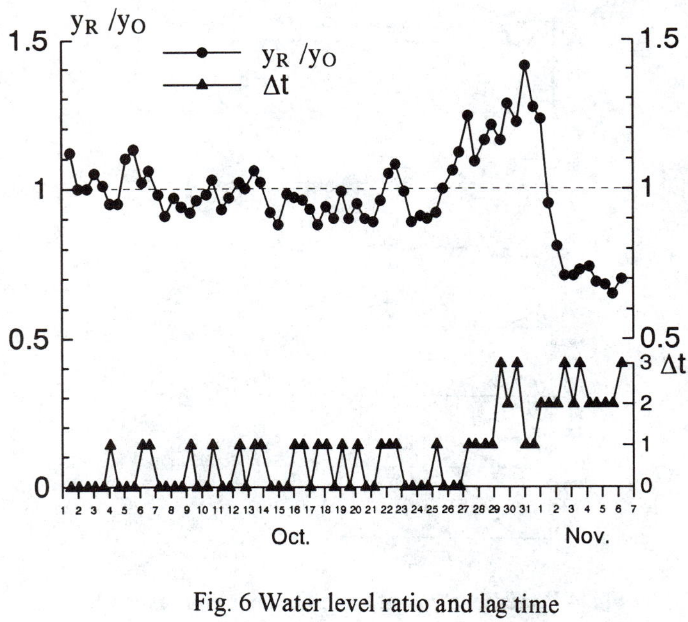

Fig. 6 Water level ratio and lag time

y_R/y_O and the lag time Δt are plotted in Fig.6, where y_R and y_O the daily highest water levels in the river mouth and the sea, respectively, and Δt the time difference between y_R and y_O. At first, the water level ratio shows gradual increase from 1.0 on October 26th to 1.4 on November 1st, and then distinctly decreased to 0.6 on November 3rd, while Δt shows an increase from 0 hour up to 3 hours. Thus, this figure suggests that the river mouth closure was completed on November 3rd. The water level rise in the mouth immediately before the closure in Fig.6 was induced by wave set-up due to wave breaking in front of the river mouth (Tanaka and Shuto, 1992), and the wave set-up height measured in the mouth shows close relationship with the wave height.

3. MATHEMATICAL MODELING

3.1 Governing Equations

A mathematical model proposed by Ogawa et al. (1984) will be employed for predicting closing process at a river mouth owing to predominant wave motion. It is assumed that the wave motion is responsible for the intrusion of sediment into the mouth, while the river discharge is effective for flushing sediment out of the mouth as illustrated in Fig.7. Thus, the corresponding governing equation is expressed as follows (Ogawa et al., 1984).

$$(1-\lambda)Lh\frac{dB}{dt} = e_r q_r B - e_w (1-\lambda) Q_w \qquad (1)$$

where λ the porosity of sand, L the width of sand spit, h the water depth at a mouth, B the width of a river mouth, and e_r and e_w the efficiency of sediment inflow by waves and that of sediment outflow by current, respectively. The sediment transport rate due to current, q_r, and that induced by wave motion, Q_w, can be evaluated by means of conventional formulae, Eq.(2) (Brown, 1950) and Eq.(3) (CERC, 1984), respectively.

$$q_r = K\left(\frac{u_*}{sgd}\right)^m u_* d \qquad (2)$$

$$Q_W = \alpha (Ec_g)_b \sin\theta_b \cos\theta_b \qquad (3)$$

where K is the constant (=10.0), u_* the shear velocity, s the immersed specific weight

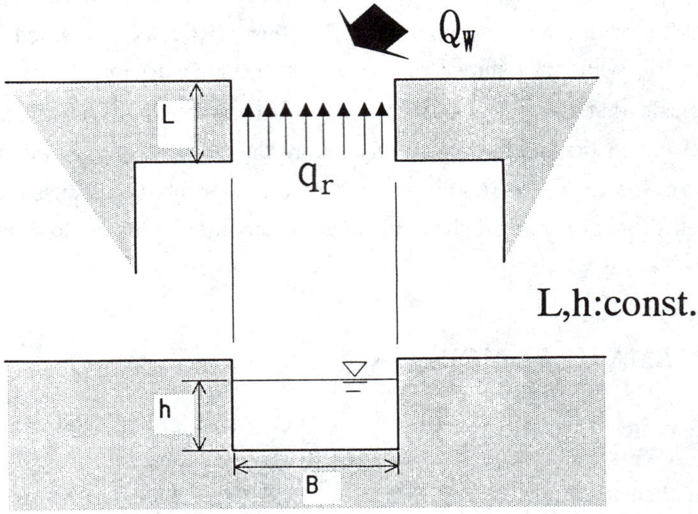

Fig. 7 Schematic explanation for the model

of sediment, g the gravitational acceleration, d the grain diameter, m the power (=2.0), α the empirical coefficient, $(Ec_g)_b$ the incident wave energy flux evaluated at the breaker line, and θ_b the breaking wave angle to the shoreline. The coefficient α in Eq.(3) has already been evaluated to be 0.05 by Tanaka and Shuto (1991) by applying the one-line model to shoreline change adjacent to the Nanakita River mouth.

3.2 Analytical Solution

The following exact solution can be derived from Eq.(1), assuming the wave condition and the river discharge are constant.

$$(B^* - 1) + \frac{A_1^*}{4}\log\left|\frac{(B^* - A_1^*)(1 + A_1^*)}{(B^* + A_1^*)(1 - A_1^*)}\right| - \frac{A_1^*}{2}\left\{\tan^{-1}\left(\frac{B^*}{A_1^*}\right) - \tan^{-1}\left(\frac{1}{A_1^*}\right)\right\} = -t^* \quad (4)$$

where the dimensionless quantities in Eq.(4) are defined as

$$B^* = \frac{B}{B_0}, \quad A_1^* = \sqrt[4]{\frac{A_1/A_2}{B_0}}, \quad t^* = \frac{A_2 t}{B_0} \quad (5)$$

$$A_1 = \frac{e_r K d (\sqrt{g} n Q)^{2m+1}}{(1-\lambda) L h (sgd)^m h^{7(2m+1)/6}} \tag{6}$$

$$A_2 = \frac{e_w Q_w}{Lh} \tag{7}$$

in which B_0 denotes the initial river width, n the Manning's friction coefficient and Q the effluent discharge consists of fresh water and tidal prism.

Some interesting asymptotic behaviors of a river width can be derived from Eq.(4). First of all, if the river discharge is zero, substitution of $A_1^*=0$ into Eq.(4) yields the following simplification.

$$B^* = 1 - t_* \tag{8}$$

Namely, the dimensionless river width shows linear reduction with the passage of time until it closes completely at $t_*=1$. Secondly, the width of equilibrium stage can be easily obtained by substituting $dB/dt=0$ in Eq.(1). The dimensionless form normalized by B_0 is

$$B_\infty^* = \frac{B_\infty}{B_0} = A_1^* \tag{9}$$

According to Eq.(4), it takes infinite duration to reach equilibrium width, though, t_∞ will be herewith defined as the duration when B becomes $0.99B_\infty$. A dimensionless form for t_∞ can be derived from Eq.(4).

$$t_\infty^* = A_1^* \left\{ a^* - \frac{1}{4} \log \left| \frac{1+A_1^*}{1-A_1^*} \right| - \frac{1}{2} \tan^{-1}\left(\frac{1}{A_1^*} \right) \right\} + 1 \tag{10}$$

where $a^*=0.722$ for $A_1^*>1$ and $a^*=0.711$ for $0<A_1^*<1$. As A_1^* approaches infinity, Eq.(10) can be approximated by the simple expression,

$$t_\infty^* = 0.723 A_1^* \tag{11}$$

Time-variation of dimensionless width, B^*, computed from Eq.(4) is depicted in Fig.8. The parameter A_1^* denotes the ratio of sediment transport rate out of a river mouth due to river discharge to sediment intrusion into a river mouth caused by wave

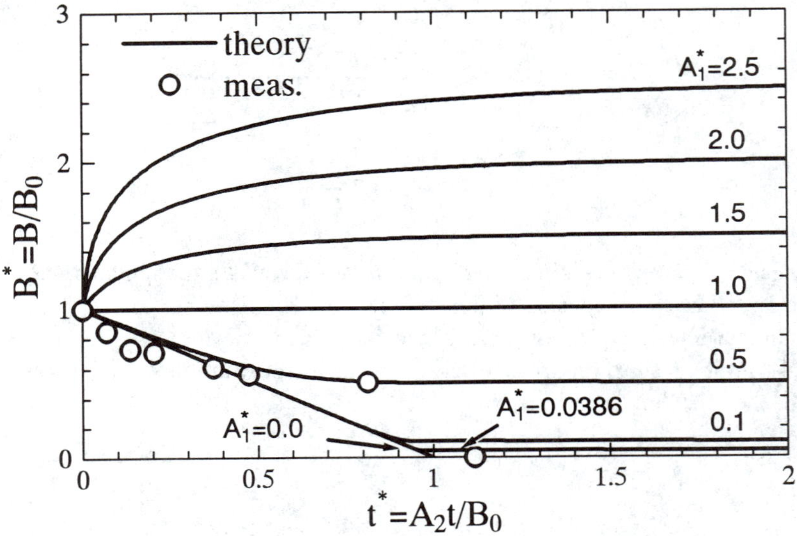

Fig. 8 Time-variation of river mouth width

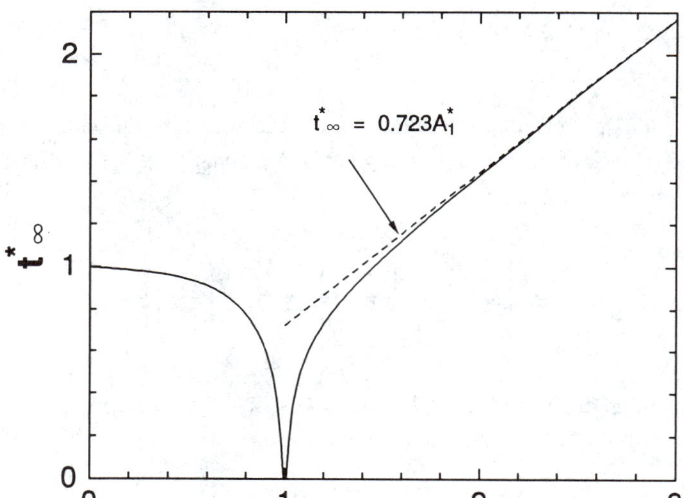

Fig. 9 Relationship between t^*_∞ and $A_1{}^*$

motion, as defined by Eq.(5). Thus, the horizontal line for A_I^*=1.0 in Fig.8 corresponds to dynamic equilibrium state, in which sediment transport rates out of and into a river mouth are balanced, whereas the family of curves for A_I^*>1.0 and A_I^*<1.0 represents widening and reduction processes of river mouth. In case of A_I^*=0.0, the river mouth closes completely at t^*=1.0 as mentioned earlier because of no sediment flushing out of a river mouth. According to the existing theory for river mouth closure without sediment outflow from a river mouth, the solution is given in terms of exponential function, denoting that infinite duration is needed for completion of closure (Tsujimoto et al, 1989). It seems that the present solution, in which closure can be completed within finite duration, is more realistic.

Figure 9 shows the relationship between t^*_∞ and A_I^* obtained from Eqs.(10) and (11). At A_I^*=0, sediment movement at the mouth is in dynamic equilibrium state as described before. Therefore, t^*_∞ is exactly zero at A_I^*=0. The accuracy of Eq.(10) is sufficient as long as A_I^* is higher than 2.0.

3.3 Comparison with Field Observation

The measured shoreline change at the river mouth shown in Fig.5 can be now compared with the present theory. The open circles in Fig.8 denote measured data at the Nanakita River mouth in 1994. Since the wave characteristics and the fresh water discharge are assumed to be constant in the theory, the quantities shown in Fig.2 except wave direction are averaged from October 4th through November 7th, whereas the averaging of wave direction is done from October 4th through November 12th due to lack of the data after November 13th as depicted in Fig.2. As for the coefficients, e_r and e_w in Eq.(1), the values determined by one of the authors (Tanaka, Kabutoyama and Shuto, 1995) are used for the present computation. It is observed that the measured change in the river mouth width shows reasonable agreement with the present theory.

4. CONCLUSIONS

River mouth closure was observed five times at the Nanakita River mouth in Japan in 1994. Conclusions drawn from this study can be summarized as follows:

(1) One of the authors has already reported that the closures of the Nanakita River mouth in 1988 and 1989 were induced by high waves having a incident angle 100° . In 1994, however, the incident angle of high waves which caused the closure did not show this tendency.

(2) By analyzing the ratio and lag time between the water level measured in the mouth and that in the ocean, the date when the closure had completed can be determined. Distinct rise of water level in the mouth caused by wave set-up can be observed immediately before the completion of the closure.

(3) A mathematical model proposed by Ogawa et al. (1984) is used for predicting

time-variation of width at a river mouth under the combined influence of incoming waves and effluent river discharge. Analytical solution can be obtained, assuming constant wave characteristics and fresh water discharge. The measurements and the theory showed reasonable agreement.

5. ACKNOWLEDGMENT

The authors would like to express their grateful thanks to the members of River Engineering Laboratory and Tsunami Engineering Laboratory, Tohoku University, for their cooperation during the course of the field measurements. Their appreciation should be extended to the Shiogama Construction Office, Ministry of Transport, and the Higashi Sendai Construction Office, Miyagi Prefectural Government, for their offer of the field data. A part of this study was supported by a Grant-in-Aid for Scientific Research from the Ministry of Education, Science and Culture of Japan.

6. REFERENCES

Brown, C.B., Sediment Transportation, In H. Rouse (ed.), *Engineering Hydraulics*, pp.769-857, John Wiley & Sons, Inc., New York, 1950.

Coastal Engineering Research Center, *Shore Protection Manual*, U.S. Army Corps of Engineers, 1984.

Ogawa, Y., Fujita, Y. and Shuto, N.: Change in the cross-sectional area and topography at river mouth, *Coastal Eng. in Japan,* Vol.27, pp.233-247, 1984.

Tanaka, H. and Ito, T.: An estimation method of gorge section at a river mouth, *Coast. Eng. in Japan*, Vol.39, No.1, pp.27-38, 1996.

Tanaka, H., Kabutoyama, H. and Shuto, N.: Numerical model for predicting seasonal migration of a river mouth, *Computer Modelling of Seas and Coastal Regions II*, pp.345-352, 1995.

Tanaka, H. and Shuto, N.: Field measurement of topography change at a river mouth, *Proc. Int. Symp. Sediment Transport Modeling*, pp.480-485, 1989.

Tanaka, H. and Shuto, N.: Field measurement of the complete closure at the Nanakita River mouth in Japan, *Proc. Int. Symp. Natural Disaster Reduction and Civil Eng.*, pp.67-76, 1991.

Tanaka, H. and Shuto, N.: Field investigation at a mouth of small river, *Proc. 23rd Int. Conf. Coastal Eng.*, pp.2486-2499, 1992.

Tsujimoto, T., Mori, A., Okabe, T. and Ohmoto, T.: Non-equilibrium sediment transport, *Proc. 33rd Japanese Conf. on Hydr.*, pp.445-461, 1989. (in Japanese)

CHAPTER 354

THE EROSION OF A SALT WEDGE TRAPPED BEHIND A BARRAGE ACROSS AN ESTUARY

Walker[1], S.A., Hamill[2], G.A. , Johnston[3],H.T.

Abstract

The problems associated with the stratification of fluids, particularly in estuaries, are of growing concern. Based on water quality problems which occurred at the River Lagan, N. Ireland, an experimental investigation was undertaken into the fluid mixing process within a stratified flow. High quality temporal measurements have been used to examine the processes of erosion of a saline stratification trapped behind a barrage. Two distinct mechanisms of wedge erosion have been identified, both of which involve the formation of a mixed layer.

Introduction

Density stratification in fluids occur due to variations in momentum, temperature and salinity. Mixing between such strata plays a critical role in various phenomena of meteorology, hydraulics and oceanography. Barrages are now frequently used to improve the visual appearance of estuaries by controlling water levels in a semi-tidal impoundment. In such impoundments fresh river water and salt water from the sea normally mix. However, under certain circumstances a stable stratification occurs between the two fluids. When the flow in a river is low, there is little mixing and therefore limited transfer of oxygen between the fresh and salt water layers. This can lead to oxygen depletion and possibly anaerobic conditions at the bed, with subsequent damage to the aquatic environment.

---

1 Research student, 2 Lecturer, 3 Senior Lecturer
Department of Civil Engineering, The Queen's University of Belfast, David Keir Building, Stranmillis Road, Belfast. BT7 1NN

Background

Belfast developed along a tidal reach of the River Lagan and many industries were established along the banks of the river. At low tide extensive areas of unsightly mudflats were exposed and so a weir was constructed across the river in 1934 to create, on part of the river, a semi-tidal pond which would cover the muds at all tide states. However, the construction of this weir led to periodic water quality problems. During periods of low river flows, the fresh and saline waters entering the pond from the river and the tidal flow stratified, and there was little flushing of the denser saline water trapped behind the weir. The oxygen content of the trapped water was rapidly depleted by the decomposition of organics in the water and the bed muds. This damaged the fish and plant life in the river and led to the production of hydrogen sulphide.

The demise of riverside industries and the poor river water quality resulted in the city turning away from the river front which rapidly became polluted. The result was that extensive inner city areas were blighted and had little development potential.

In 1969 the construction of a new weir downstream of the Queen's Bridge was proposed to regulate river levels and create an attractive water environment. This proposal was renewed by the River Lagan Working Group in the 1970s and specialist studies were instigated. In 1988, the Civil Engineering Department of The Queen's University of Belfast was commissioned to undertake a study seeking to establish design parameters for a structure which would control the water levels, protect the city against tidal flooding and improve the water quality in the river. The proposed impoundment was just under 5 km. in length with a surface area of 40 ha. This pond would have river flow and urban drainage inputs, some of which would have high biochemical oxygen demand levels.

Research Programme

A research programme was undertaken in which a physical model of some 8 km. of the tidal channel and harbour area, including the proposed impoundment, was constructed to a horizontal scale of 1:150 to establish design and operation parameters for the weir. A scale distortion factor of three was used to ensure adequate depths for the study of salinity profiles in the model.

In conjunction with the construction of the physical model a monitoring programme was established on the River Lagan. Velocity and salinity measurements were taken at a number of stations on the river for use in calibrating the model. The measurements were taken for river conditions ranging from low to flood flows, and at a variety of tidal conditions. The velocity measurements were used to establish boundary roughness values in the model at different water levels. The model was then validated by reproducing salinity profiles measured in the river at different tide states.

Salinity profiles at high tide, established after a number of tide cycles in the model, behind the existing weir showed that for a river flow of 2 cumecs the stratification was well defined and extended over the full length of the impounded reach. When the river flow was increased to 5 cumecs, the entrapped wedge retreated showing a significant reduction in both volume and concentration. A new weir placed at the proposed downstream site resulted in a pond with characteristics similar to that behind the existing weir. This showed that the construction of a fixed crest weir at the new site would simply replicate the problems of water quality being experienced behind the existing weir.

An active weir design was considered whereby the impoundment would be controlled by a gated weir structure incorporating a system for withdrawal of water at a low level. A gated weir would have the added advantage of being able to exclude high tides as well as controlling water levels in the impoundment. It was decided to use a gate system which would be able to exclude surge tide levels above that which would be reasonable expected at present but which might occur if the was a rise in tide levels due to global warming.

A low level withdrawal system was included to evacuate saline water from behind the weir on an ebb tide, when stratification problems were present in the impoundment. On a flood tide, the weir would allow saline water into the impoundment over the lowered gates. At high tide, the gates would be raised and the low level withdrawal system would be opened. This would allow drawdown of the impoundment, during the ebb tide period, through the low level withdrawal system which would have a capacity of 25 cumecs.

It became clear from the model study that the mixing process involved in this highly stratified situation was unclear. The lack of information for the design engineers to predict the quantity of saline material which would be removed under normal flow conditions indicated the need for a fundamental investigation of the process. This programme of work commenced in 1993, and the initial findings are reported in this paper.

Related research

It has already been established that the entrainment of the salt water is a function of the Richardson number, Ri, of the flow (Ellison 1959). The Richardson number is a measure of the ratio of the gravity forces to the inertial forces in a shear flow having a vertical density gradient. It is widely used a measure of the stability of the flow with regards to vertical mixing. It generally takes one of two forms, the bulk Richardson number

$$Ri_b = g \frac{\Delta\rho H}{\rho V^2}$$

where $\Delta\rho$ is the density difference, H is the depth of fresh water and V is the average velocity, and the gradient Richardson number

$$Ri_g = -g / \rho \frac{\delta\rho / \delta z}{[\delta\rho / \delta z]^2}$$

where $\delta\rho/\delta z$ is the density gradient and $\delta v/\delta z$ is the velocity gradient. Turner (1973) and Christodolou (1985) further classified the type of mixing to be dependent on the magnitude of the Richardson number. A critical gradient Richardson number of 0.25 was found to be the limiting criteria between stable and unstable flow.

Experimental facility

To allow a wide test range, upto prototype conditions, a new facility was constructed within the Civil Engineering Department. This new facility consisted of a testing area 20m long, 0.75m wide and 0.75m deep a schematic of which is shown in Figure 1. The salt water was retained between two baffles which were 17.5 m apart. The salt water depth was kept constant at 0.25m whilst the fresh water depths were varied by means of an adjustable weir. The ability to produce conditions in which the saline water was either stationary or flowing, either with or against the prevailing river flow, was included thus allowing the simulation of tidal movements. The results

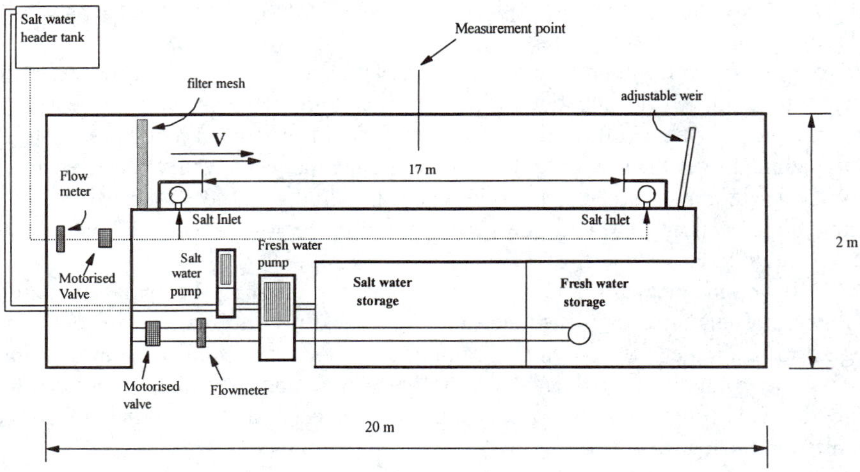

Figure 1 Experimental Facility

presented here represent the condition of a fully trapped saline wedge. Fresh water was mixed with dried salt and dye in the storage tanks and then pumped into the salt water impoundment to a depth of 0.25m. Fresh water was then filled slowly on top, over the course of 12 to 15 hours. Flow was initiated slowly in the upper layer by starting the fresh water pump. Once the flow had reached the required rate the water in the storage tanks was changed slowly by opening the waste valve and filling from the supply. This was necessary due to the contamination of the fresh water which occurred as a result of mixing. The degree of contamination was relatively small in most cases, but large enough to change the salinity of the fresh water over a substantial time period.

The scanning process was commenced after flow was initiated. Velocity and density profiles were obtained at the centre of the channel. Velocity measurements were obtained using LDA with a fibre optic attachment to limit the effect of differing refractive idices. Density measurements were obtained using a conductivity probe with small measuring volume. The probe consisted of two wires 5 mm thick which were completely covered except for 2 rings, 2mm thick, near the bottom. Both probes were mounted on a linear actuator which was computer automated. The probes were placed in the salt water, usually about 20 to 30mm below the interface in order to obtain a reading of maximum salinity. An average was taken over a period of 30 seconds and the probes were then moved in a 5 mm step so that the vertical density profile could be obtained. Due to the temporal nature of the experimental investigation it was necessary to obtain as many profiles per hour as possible. The profile obtained was therefore not a full vertical traverse but was limited to 120mm around the interface. With time the interface moved downwards, as salt water was removed by entrainment from the impoundment, and it was necessary to incorporate movement of the base profile position into the scanning program.

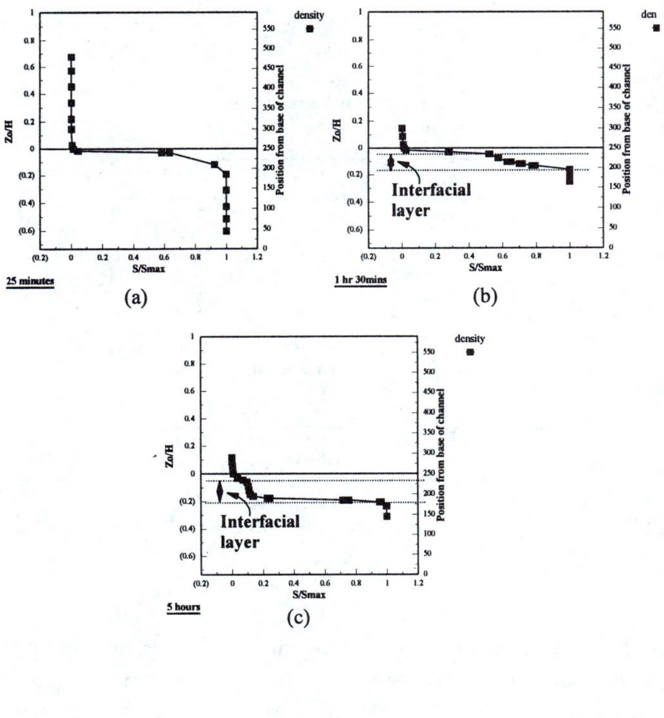

Figure 2 Density profiles (a) After 25 minutes (b) After 1 hour 30 minutes (c) After 5 hours

Development of Interfacial Layer.

The development of the mixing interfacial layer, as fresh water was allowed to flow across the stationary salt wedge, was studied with time. Figure 2 (a) shows the initial condition used in all tests. It clearly displays the fresh and salt water zones. Figures 2 (b) and 2 (c) are for the same situation, but with time having advanced to 1 hour 30 minutes and 5 hours respectively. The presence of a flow in the fresh water can be seen to have developed an interfacial layer, in that there is no longer a clear distinction between the fresh and salt water. It is from within this interfacial layer that the majority of entrainment of the saline liquid takes place.

This is clearly visible by the changing density gradient within the interface. The density within the interfacial layer was taken to be the average value within the density gradient of the layer. Figure 3 shows the interfacial layer in detail. It can be seen that the layer is defined by that position of the mixing region over which the variation of density occurred, this variation being taken as linear for purposes of analysis. The thickness of the interfacial layer was also obtained on this basis.

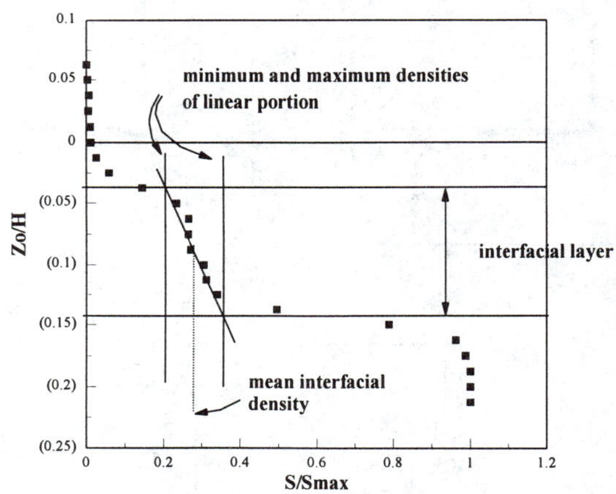

Figure 3 Definition of interfacial layer

It was found that the behaviour of this layer varied with certain combinations of flow, depth and density difference, and two mechanisms of erosion existed. In the first of these the wedge was eroded by sequential stripping of the interfacial layer, this process was labelled as oscillating due to the cyclic nature of the event. The second mechanism showed no such cyclic behaviour, with the density of the mixing layer decreasing to a steady state value. This mechanism was labelled as being non-oscillating. Significant differences in the temporal development and entrainment process was found for each mechanism.

Oscillating mechanism

Using the density profiles obtained from the experimental investigation, the layer position was plotted with time. Figure 4 represents a typical test which displayed the cyclic behaviour. On it can be seen the change, with time, in thickness of the interfacial layer and the density within that layer. The decrease in layer density corresponds with an increase in the layer thickness, this is as would be expected and closely mirrors the initial stage of the steady state condition. However, a critical point is reached where upon the density within the layer starts to increase again, with a corresponding decrease in the layer thickness. This corresponded to a period of rapid upper boundary movement when part of the layer was washed out. When the layer was at its thinnest there appeared to be an increase in the lower boundary movement,

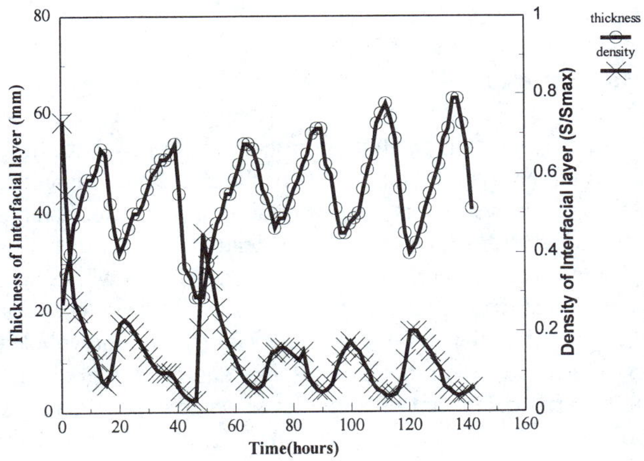

Figure 4 Relative interfacial layer density and thickness (oscillating mechanism)

as the free stream was now much closer to this boundary. This resulted in entrainment from the salt pool within the wedge. It would appear that for the oscillating process as the density of the layer decreases, the interfacial slope increases in order to resist the interfacial shear stresses. Figure 5 shows the variation of typical velocity and density gradients. These also show a definite cyclic behaviour with the maximum velocity gradients occurring when the layer was most dense.

If the change in the gradient Richardson number is examined, as shown on Figure 6, it can be seen that when the upper part of the layer was removed, the value of Ri dropped to approximately 0.25. As this represents the critical Richardson number it implies that the flow becomes unstable at these times, and this was confirmed by the presence of large amplitude breaking waves.

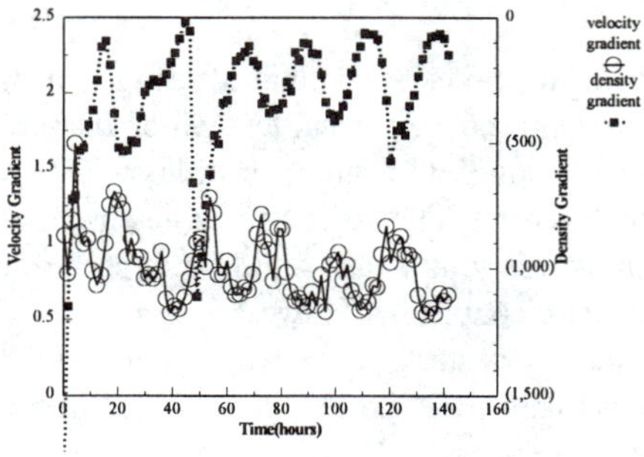

Figure 5 Interfacial velocity and density gradients (oscillating mechanism)

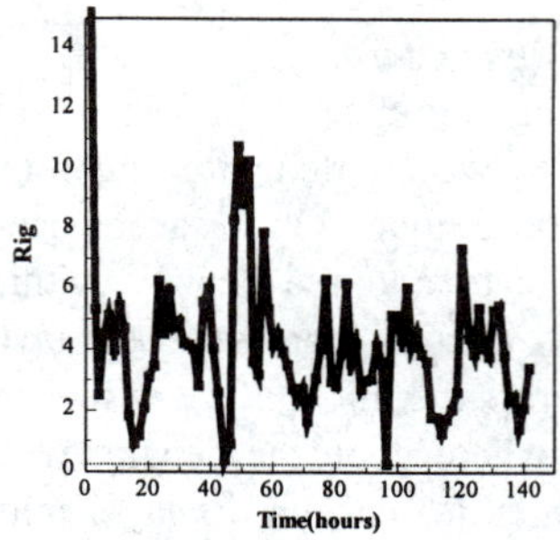

Figure 6 Variation of gradient Richardson number (oscillating mechanism)

Non Oscillating mechanism

Whenever no oscillation of the mixing layer charatceristics occurred the removal of the trapped saline solution progressed steadily with time. The time taken to achieve this varied with the magnitude of the circulation velocity within the mixing layer. Figure 7 shows the typical variation of the interfacial density and layer thickness for those tests which eroded at a steady rate. An initial high growth rate of the mixing layer occurred both in thickness and density, where upon almost constant conditions prevailed, and the saline pool was steadily depleted. Comparison with the equivalent figure for the oscillating test (figure 4) shows clearly the different mechanism of erosion in action. The velocity and density gradients also can be seen, in figure 8, to remain constant once the layer became established, as did the gradient Richardson number. Figure 9 gives an indication of typical variations in the Richardson number for non oscillating tests. The accepted threshold of instability in the flow occurs when the Richardson number fall below a value of 0.25.

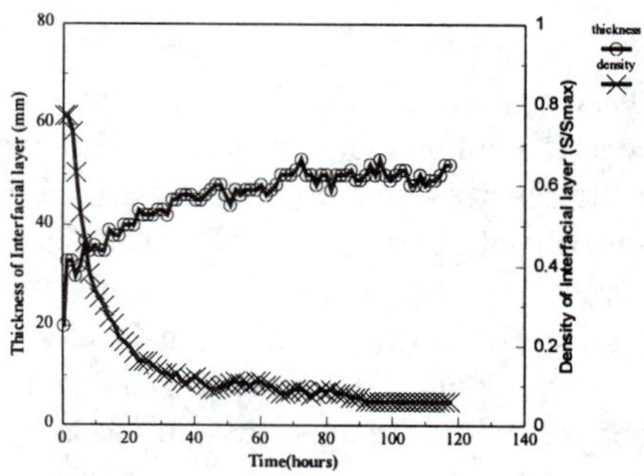

Figure 7 Relative interfacial density and thickness (non oscillating mechanism)

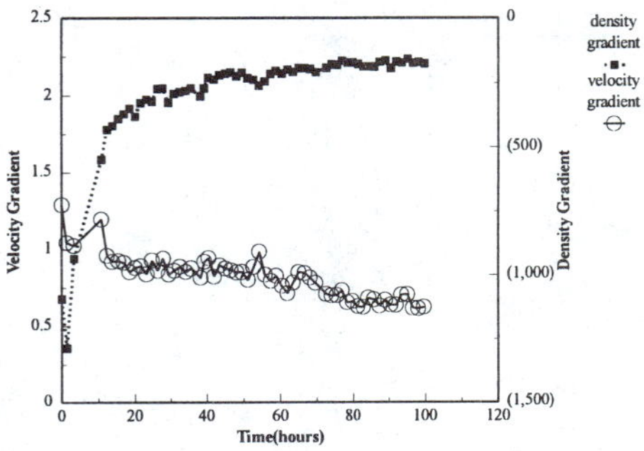

Figure 8 Interfacial velocity and density gradients (non oscillating)

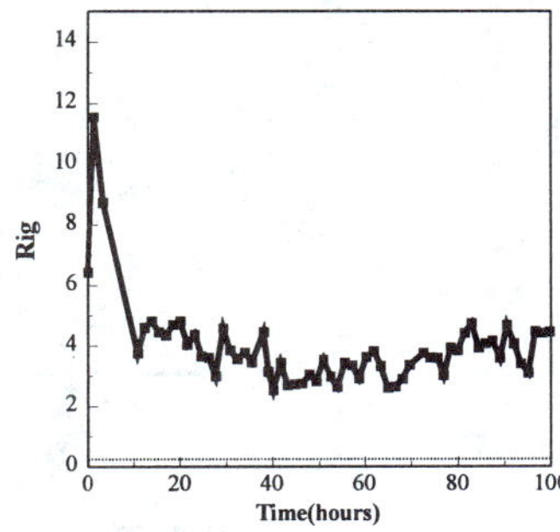

Figure 9 Variation of gradient Richardson number (non oscillating)

At no point during any of the tests which showed no oscillation did the Ri_g number approach this threshold. The key to the two mechanisms is therefore clearly a function of the flow instability.

Transport Rates

The changes in the density profiles obtained at different times during each test were used to determine the rate of erosion of the salt water impoundment. Transport across the upper boundary gave the overall rate of salt water removal during the test. This was calculated by approximating the profile, as can be seen in Figure 10, based on the position of the boundaries and the density of the layer. This allowed the volume of salt to be calculated very easily, with the differences between profiles at different times being taken as the as the volume of salt water transported. Any change in density of the interfacial layer, or drop in position of the upper boundary, was deemed to be due to transport across the boundary. Transport rates across the lower boundary were determined in a similar manner.

Based on a dimensional analysis, a dimensionless form of transport was used. This dimensionless transport. T, was defined as

$$T = \frac{t}{\Delta\rho V H}$$

where t is the average transport (kg/s), Δρ the initial density difference, V the initial average fresh water velocity and H the initial fresh water depth.

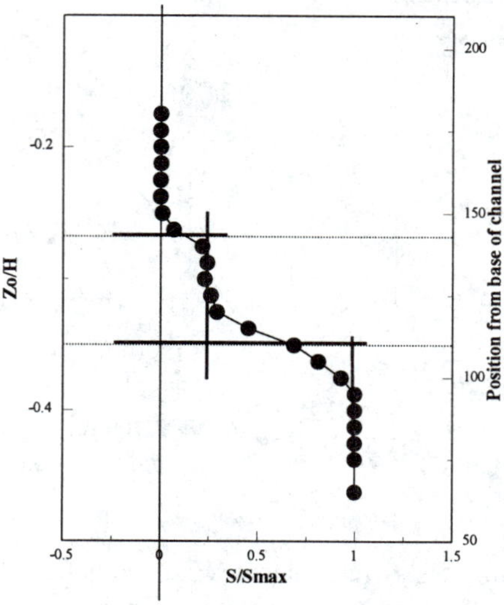

Figure 10 Stepped approximation for density profile

Figures 11 and 12 show the average transport rates obtained for the oscillating and non oscillating mechanisms respectively, plotted against the bulk Richardson number. It can be seen that in each case there exists a strong linear relationship between the rate T, and the Ri_b value, and that the rate obtained for each mechanism is similar, with the major difference occurring in the rates across the upper and lower boundaries.

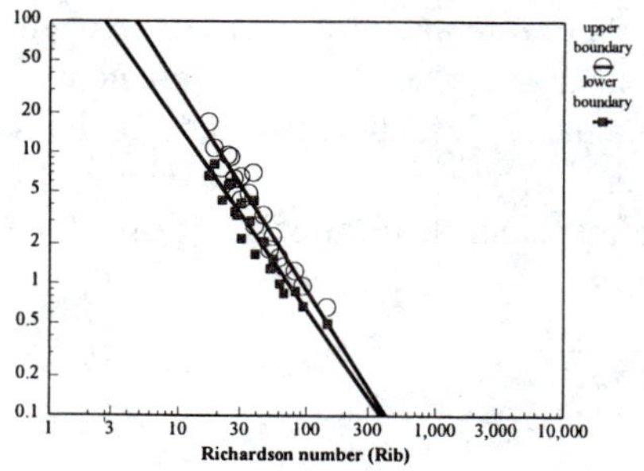

Figure 11 Dimensionless transport rates (oscillating mechanism)

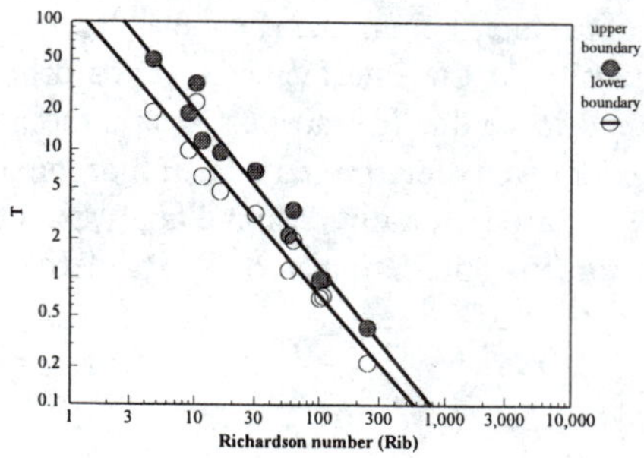

Figure 12 Dimensionless transport rates (non oscillating mechanism)

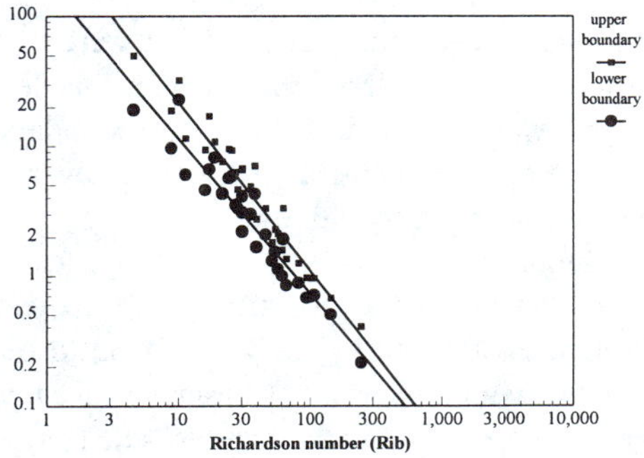

Figure 13 Dimensionless tranport rates for both mechanisms

Figure 13 is a combined plot for the transport rates obtained for both mechanisms of erosion. It can be seen that within each mechanism the transport rates were similar and could be best described by equations of the form

$$Tu = 447.2\ Ri_b^{\ -1.303}$$

for the transport across the upper boundary and

$$Tl = 184.2\ Ri_b^{\ -1.2}$$

for the transport across the lower boundary.

Conclusion

In all cases interfacial mixing resulted in the formation of an interfacial or mixed layer which was around 30 to 40 mm thick. Only those tests conducted with low values of Reynolds number(i.e. <2000) showed no mixing.

Two mechanisms of erosion have been observed. These were classified as oscillating and non-oscillating, depending on the behaviour of the layer in terms of its relative density and thickness. In some tests the density of the layer decreased rapidly, and then remained constant and these were classified as non-oscillating. In a number of cases the density within the layer decreased initially and then increased again after the upper part of the layer was removed. These tests were classified as oscillating as a cyclic behaviour was observed in the layer density.

The gradient Richardson number remained approximately constant during those tests which did not oscillate, indicating a stable flow. For those tests which showed the oscillating behaviour, the value of Rig was found to show a cyclic behaviour. When the layer was washed out, the value of Rig dropped towards 0.25, indicating unstable flow. Mixing occurred as a result of large amplitude breaking interfacial waves.

After the start of a test greater rates of transport were observed at the upper boundary as the layer formed. This resulted in a decrease in density until the layer washed out and a peak rate was observed at the upper boundary. Greater rates of transport were observed at the lower boundary when the layer was thinnest. This caused the density of the layer to increase with time.

Average transport rates were calculated during each test from the change in density profiles. Rates were averaged over the duration of the test and calculated for both boundaries. Dimensionless transport rates were found to be best related to the bulk Richardson number (Rib) with higher rates of transport for low values of Rib.

It is hope that the ability to estimate the removal rate of trapped salt water will aid the process of barrage design, and act as a tool for river management schemes by which the inducement of natural mixing can be accounted for.

References

Christodolou, C.G. (1985) "Interfacial mixing in stratified flows" J. Hyd. Research, Vol. 24, No. 2, pp 77-92

Ellison, T.H. and Turner, J.S., (1959) "Turbulent entrainment in stratified flows" J. Fluid Mech., Vol. 6, pp 423-448

Turner, J.S., (1973) "Buoyancy effects in fluids" Cambridge University Press.

CHAPTER 355

Near-Field Measurements of a Buoyant Jet in Waves and Currents.

K.H. Kwan[1] & C. Swan[2].

Abstract.

The present paper concerns the near-field characteristics of a jet discharged into a wave environment, and contrasts the results of two new experimental studies with a modified integral solution based upon a quasi-Lagrangian description of the flow field. This solution differs from previous attempts to model near-field wave mixing (namely Koole and Swan, 1994) in that it is formulated within an absolute frame of reference, and therefore adequately incorporates the momentum flux associated with the entrained fluid. Having implemented this important modification, the Lagrangian model is shown to provide a good quantitative description of a buoyant hot-water jet discharged into either a steady current or an unsteady wave field. In particular, there is no uncertainty regarding the magnitude of the entrainment coefficients, and as such the model is suitable for design calculations. In addition, detailed comparisons between the temperature profiles ensemble-averaged with respect to the phase of the wave cycle hint at a new mixing mechanism. Experimental data is presented which suggests that the wave-induced oscillatory motion leads to the division of the jet at certain phases of the wave cycle. If this effect is indeed important, then it implies that the average dilution will be enhanced when a jet is discharged into the plane of the wave motion.

Introduction.

Under most practical conditions outfalls discharge their pollutants, whether it be unwanted heat or partially treated effluent, into coastal waters which are dominated by the effects of wave action. However, present design practice largely ignores the wave motion and applies a simplistic approach in which both the

---

[1] Research student & [2] lecturer, Department of Civil Engineering, Imperial College, London. SW7 2BU. United Kingdom.

concentrations and the velocities arising within the near-field are predicted using some form of integral solution (Morton et al, 1956) based upon the conservation of mass, momentum, and where appropriate species concentration. This approach is often justified on the basis that the wave motion will, at worst, produce additional mixing and thus the design calculations are conservative. Unfortunately, recent studies (Koole and Swan, 1995) have shown that this does not necessarily follow, and that under some circumstances the wave motion may produce local increases in the concentration levels. Furthermore, previous studies have also clearly demonstrated that the occurrence of any additional wave-induced mixing is critically dependent upon the flow conditions at the exit nozzle. In particular, the ratio of the exit velocity to the local wave-induced velocity is a significant parameter. This in itself has important design implications given the increasing tendency to build long-outfalls. These are constructed so that firstly, the source of the pollutant is far removed from the adjacent coastline; and secondly, it is discharged in deeper water so as to reduce the pollutant concentrations arising at the water surface. Although, at first sight, these objectives appear desirable, the construction of a long-outfall may inadvertently move the pollutant source to a location at which the effective wave-induced mixing is negligible. As a result, the increased construction costs may actually produce higher concentrations at the water surface, and therefore potentially increase the risk of pollutant contamination arising on an adjacent coastline.

However, in order to incorporate these effects within the design process, and thereby optimise the efficiency of a given outfall design, one first requires a quantitative model capable of predicting the additional wave effects within the near-field. Although previous studies (notably Chin, 1988 and Koole and Swan, 1994) have considered this problem, quantitative comparisons with laboratory data are at best inconclusive. The present paper addresses this point, and provides a brief description of a corrected model which is subsequently shown to be in good quantitative agreement with new laboratory data describing a buoyant hot-water jet discharged into both a steady current and an unsteady (or time-dependent) wave field.

Background.

Although previous studies have clearly demonstrated the potential importance of wave-induced mixing (Sharp (1986), Chyan et al. (1991) and Hwung et al. (1995)); the only papers to have considered an appropriate mixing model are Chin (1988), Koole and Swan (1994) and more recently Chu and Lee (1996). In the first two cases the models were written with the specific intention of incorporating wave effects, while the latter model has to date only been applied to the description of pollutants discharged in a steady current. However, each of these models are similar in that they apply a quasi-Lagrangian representation of the flow field in which elements of the pollutant jet are tracked within the ambient flow.

Detailed comparisons with laboratory data involving both two-dimensional plane jets and three-dimensional round jets suggest that the model outlined by Koole and Swan (1994 and 1995) provides a good qualitative description of a jet discharged in waves. In particular, this model provides a plausible explanation for the non-Gaussian characteristics of the time-averaged radial distributions (both in terms of velocities u(r) and concentrations C(r)), and suggests that the occurrence of a bi-peaked or flat-topped distribution is simply dependent upon the local ratio of the jet diameter to the wave-induced displacement of the jet axis. Furthermore, the multi-stage centre line decay (i.e. u(z) or C(z) at r=0) first noted by Chyan et al. (1991) may also be explained (at least qualitatively) by the wave-induced displacements, and is a clear consequence of applying a time-averaged representation of a cyclic process.

Unfortunately, quantitative comparisons between the laboratory data and the model predictions proved more difficult. In particular, Koole and Swan (1995) demonstrated that good agreement along the time-averaged jet axis could only be obtained if the coefficient of forced entrainment (Winiarski and Frick, 1976) was set at $\beta=1.0$; and, in addition, the coefficient of radial entrainment was increased three-fold with respect to the value typically adopted in a stagnant ambient (i.e. $\alpha=3\times0.055$). The first of these points is consistent with previous studies, and stresses the importance of forced entrainment in non-stagnant ambients. However, the three-fold increase in the coefficient of radial entrainment is perhaps more difficult to explain. Koole and Swan (1994) argued that this change was justified by the large increase in the turbulent shear stresses ($\rho u'v'$) measured along the boundary of the jet discharged in a wave environment. However, no attempt was made to correlate the Reynold's stresses with the increased entrainment, and consequently it remains unclear whether a large increase in the coefficient of radial entrainment represents a genuine increase in the rate of mixing, or merely arises due to some inadequacy within the model formulation.

To help clarify this point the Lagrangian model proposed by Koole and Swan (1994) was applied to the case of a vertical jet discharged in a steady cross-flow (i.e., the ambient fluid flow is perpendicular to the initial jet axis). Although a Lagrangian model is equally applicable to this case, or indeed any other, there is a clear distinction between the cross-flow and waves cases in that previous researchers (Lee and Cheung, 1990) have already confirmed that for the case of a jet discharged into a steady cross-flow the appropriate entrainment coefficients are given by $\beta=1.0$ and $\alpha=0.055$. In other words, in this case it is well known that the ambient flow has no effect upon the coefficient of radial entrainment.

If (x,y,z) represent the usual Cartesian coordinates in which (x,y) defines a horizontal plane and z is measured vertically upwards; the trajectory of a vertical jet subject to a cross-flow orientated along the x axis may be represented on the (x,z) plane. Figure 1 concerns four such cases corresponding to $u_j/u_a=45.1$, 20.0, 16.2 and 11.7; where u_j is the exit jet velocity and u_a is the ambient cross-flow velocity. In each of these cases laboratory data describing the jet trajectory (Patrick, 1967) is compared with the results of the Lagrangian model proposed by Koole and Swan (1994) in which the entrainment coefficients are defined as α=0.055 and β=1.0. The agreement between the measured data and the model predictions is clearly very poor, and suggests (at least in this case) that this Lagrangian description of the flow field is fundamentally incorrect.

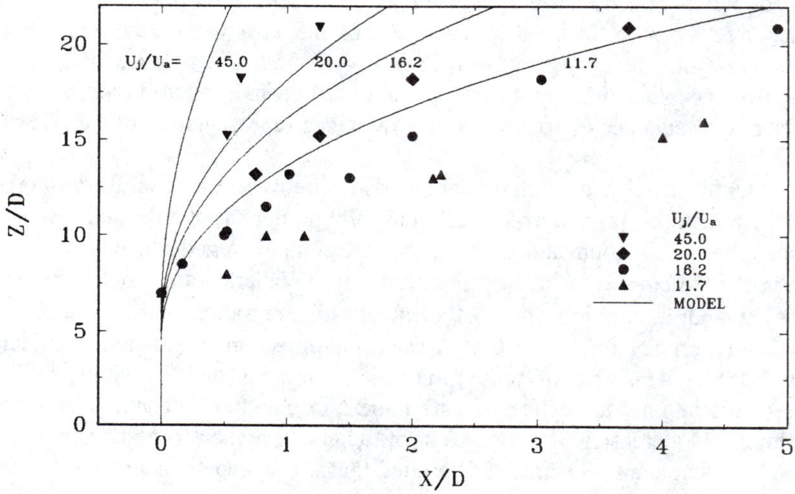

Figure 1. Jet trajectory: comparisons with a Lagrangian model formulated in a relative frame of reference.

Improved Integral Formulation.

In attempting to explain the apparent inadequacy of the previous Lagrangian models (namely Chin 1988 and Koole and Swan 1994) the recent formulation proposed by Chu and Lee (1996) was noted. Although there are many similarities between these three models, the latter solution is quite distinct in that the integral equations are formulated in an absolute frame of reference. The distinction between this and the relative frame of reference, used in the former models, is shown to be very important and in particular accounts for discrepancies within the momentum conservation equation. Figure 2 concerns a simple jet element of area A_j which has a velocity u_j measured relative to the ambient fluid, which itself has a velocity u_a. The vector sum of u_j and u_a defines the instantaneous jet axis such that u_j and u_a are respectively at angles of γ and δ to the ξ-axis. In a relative frame of reference, or one moving with the ambient flow, the conservation equations relating to mass and momentum are given by:

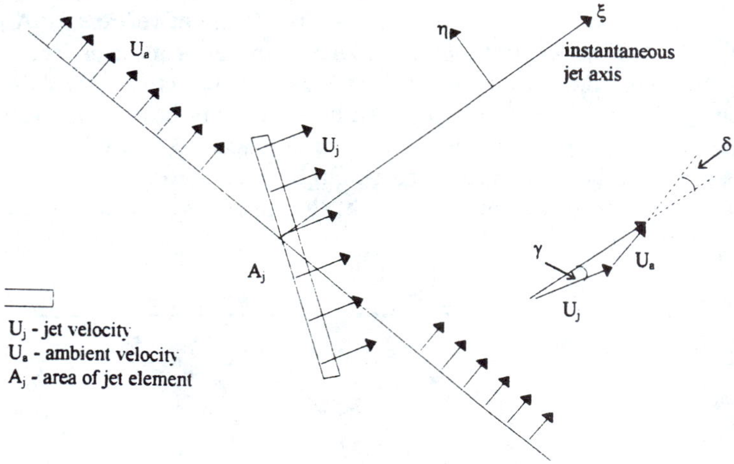

Figure 2. Definition sketch.

$$\frac{\partial}{\partial \xi}\int_{A_j} u_j \cos\gamma d\eta = \alpha u_j 2\pi b_j \cos\gamma + 2b_j\beta\left|u_j \sin\gamma\right| \qquad \text{1a.}$$

$$\frac{\partial}{\partial \xi}\int_{A_j} \left(u_j \cos\gamma\right)^2 d\eta = 0 \qquad \text{1b.}$$

where again α is the coefficient of radial entrainment, β is the coefficient of forced entrainment and b_j is related to the radius of the jet element. In contrast, the same conservation equations written within an absolute frame of reference are given by:

$$\frac{\partial}{\partial \xi}\int_{A_j} u_j \cos\gamma d\eta = \alpha u_j 2\pi b_j \cos\gamma + 2b_j\beta\left|u_j \sin\gamma\right| \qquad \text{2a.}$$

$$\frac{\partial}{\partial \xi}\int_{A_j} \left(u_j \cos\gamma + u_a \cos\delta\right)^2 d\eta = u_a E \qquad \text{2b.}$$

where E corresponds to the total mass of fluid entrained into the jet element, and is given by the right hand side of equation 2a (or 1a). Comparisons between these equations immediately confirm that while the mass conservation equation is unaltered (since the entrainment of fluid is dependant upon a relative velocity effect), the equations defining the conservation of momentum (1b and 2b) differ significantly. In particular, equation 2b incorporates the momentum of the entrained fluid.

In light of these differences a new Lagrangian model was written in which the integral equations were formulated in an absolute frame of reference. A complete description of this model is beyond the scope of the present paper (see Kwan and Swan, 1997), but comparisons with the previously noted cross-flow data (Patrick, 1967) suggest that this modification is sufficient to explain the inadequacy of the previous model highlighted in figure 1. Indeed, figure 3 shows a similar set of comparisons with the new model, and confirms that a good description of the jet trajectory can indeed be achieved with the previously recorded entrainment coefficients (α=0.055 and β=1.0).

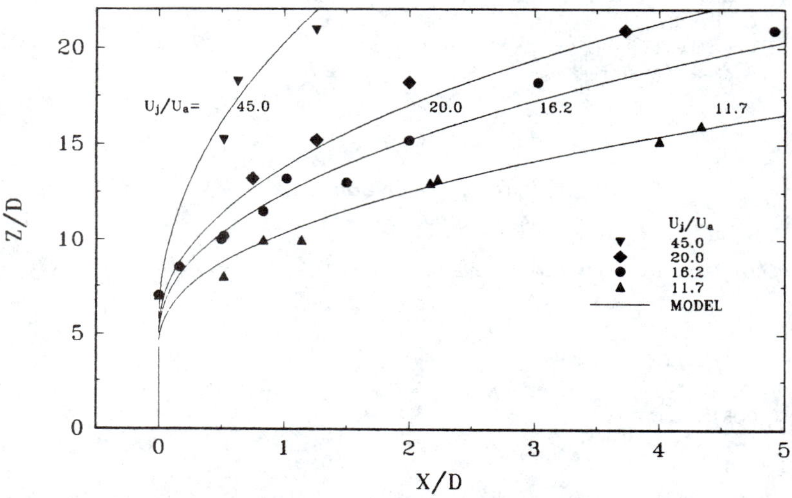

Figure 3. Jet trajectory: comparisons with a Lagrangian model formulated in an absolute frame of reference.

Experimental Work.

To further clarify the near-field behaviour of a jet discharged into a wave environment a new experimental programme was undertaken. In the first stage a buoyant hot-water jet was discharged horizontally beneath a series of regular waves having a period of T=1s and a wave amplitude of a=0.032m. These measurements were undertaken within a small wave basin located in the hydraulics laboratory within the Civil Engineering Department at Imperial College. This facility has a plan area of 6m x 10m, and has an effective working depth of 0.46m. At the upstream end of the basin the waves were generated by 10 individually controlled random wave paddles each 0.6m wide; while at the downstream end the wave energy was dissipated on a one in ten sloping beach, with further passive absorption provided by several large blocks of poly-ether foam. The hot-water jet was introduced in the centre of the wave basin through a horizontal nozzle of diameter D=6mm, with its centre line located 150mm above the bottom boundary. Although several orientations of the jet relative to the direction of wave propagation were considered, the present paper only addresses one case in which the jet was discharged perpendicular to the plane of the wave motion. In this case the exit characteristics of the jet were such that its temperature was 17°C above the ambient water temperature, and its velocity was 0.76m/s. These conditions correspond to a densimetric Froude number of F_{ro}=45 and a velocity ratio of u_j/u_a=10.9, where u_a now represents the maximum wave-induced ambient velocity at the elevation of the discharge nozzle. A sketch describing the layout of this test case is given on figure 4a.

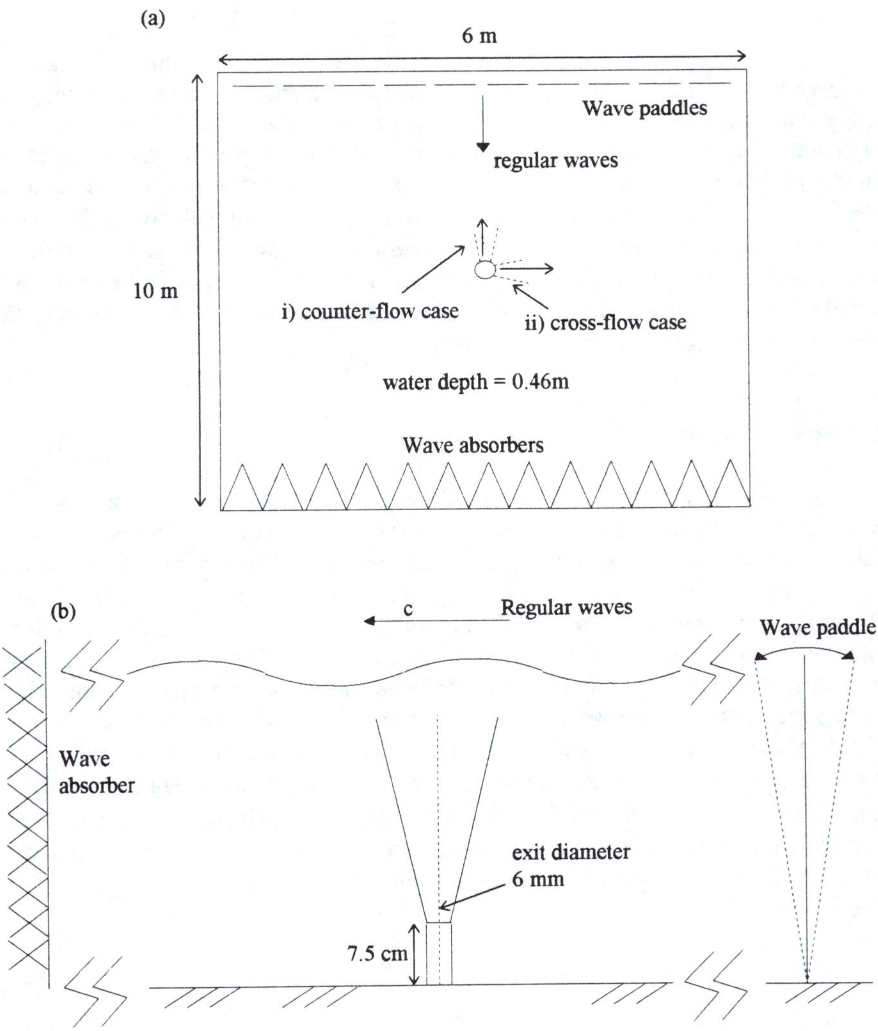

Figures 4a-4b. Experimental apparatus: (a) horizontal jet, and (b) vertical jet.

In a second series of tests a vertical hot-water jet was discharged from a similar nozzle located 75mm above the bed of a two-dimensional wave flume. This facility, which is again located in the Civil Engineering Department at Imperial College, is 25m long, 0.3m wide, and (for the duration of the present tests) had a working depth of 0.62m. The wave conditions again consisted of a regular wave train having a period of T=1.0s and wave amplitude of a=0.3m. Likewise, the exit temperature of the jet was again 17°C above the ambient water temperature and the exit velocity was 0.76m/s. These conditions are similar to those occurring in the previous test case, and correspond to a densimetric Froude number of F_{ro}=37.8 and velocity ratio of u_j/u_a=18. A sketch showing the layout of this case is given on figure 4b.

In both test cases the near-field mixing was quantified using an array of thermocouples. These were supported in a two-dimensional traverse having a positional accuracy of ±0.5mm in both the horizontal and the vertical directions. Each thermocouple was constructed from k-type thermocouple wires having an external diameter of 3 microns, and fused at one end to produce a measuring volume with a diameter of $\phi \leq 0.2$mm. Although probes of this type were difficult to handle, their size was essential to ensure the required accuracy and response (0.1°C in 0.01s). After careful calibration this measuring apparatus was able to resolve the temperature fluctuations within an individual wave cycle, and was thus deemed appropriate to determine the nature of the mixing processes.

Discussion of Results.

Figures 5a and 5b respectively concern the horizontal and vertical jets discussed above, and describe the decay in the time-averaged temperatures measured on the central axis of the jet. In both cases the measured data is compared with two models. The first, indicated by a dashed line, represents a standard integral solution based upon the assumption that the jets were discharged into a stagnant ambient; while the second, indicated by a solid line, represents the results of the new Lagrangian model outlined above. In this latter case it is important to note that although the Lagrangian model is not in perfect agreement with the laboratory data, it provides a significant improvement over the stagnant ambient model. Furthermore, unlike the previous formulations proposed by Koole and Swan (1994), there is no uncertainty with regard the entrainment coefficients (α, β). Indeed, the calculations presented on figures 5a and 5b are based upon the widely accepted entrainment coefficients commonly applied in both stagnant and steady flows (i.e. α=0.055 and β=1.0).

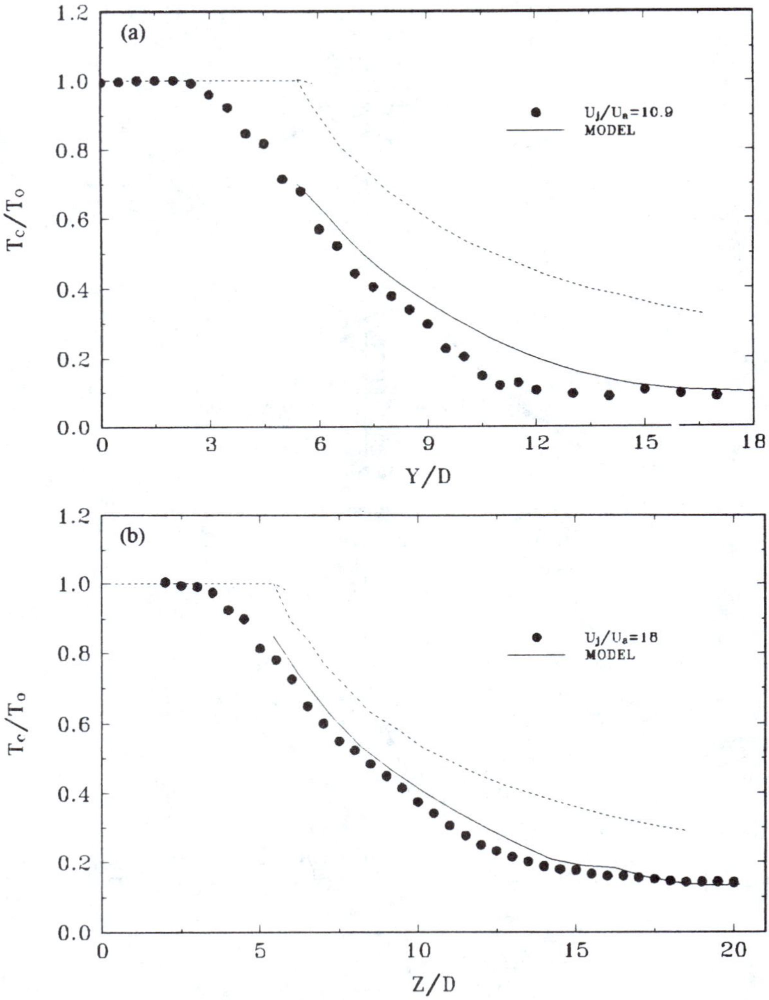

Figures 5a-5b. Centre-line temperature decay: (a) horizontal jet and (b) vertical jet.

To further examine the success of the Lagrangian solution figure 6 concerns the time-averaged temperature profiles at five elevations above the vertical jet, and contrasts the measured data with the model predications. These comparisons clearly suggest that the modified Lagrangian solution is, in general, in good quantitative agreement with the laboratory data. However, on closer inspection it is apparent that the temperature profiles become increasingly asymmetric at higher elevations above the discharge orifice. This pattern has been observed in several other test cases and is, at first sight, very difficult to explain since one would assume that the cyclic nature of the wave motion effects both sides of the jet in an identical manner. However, if one considers that the central core of the jet, at least in the near-field region, provides

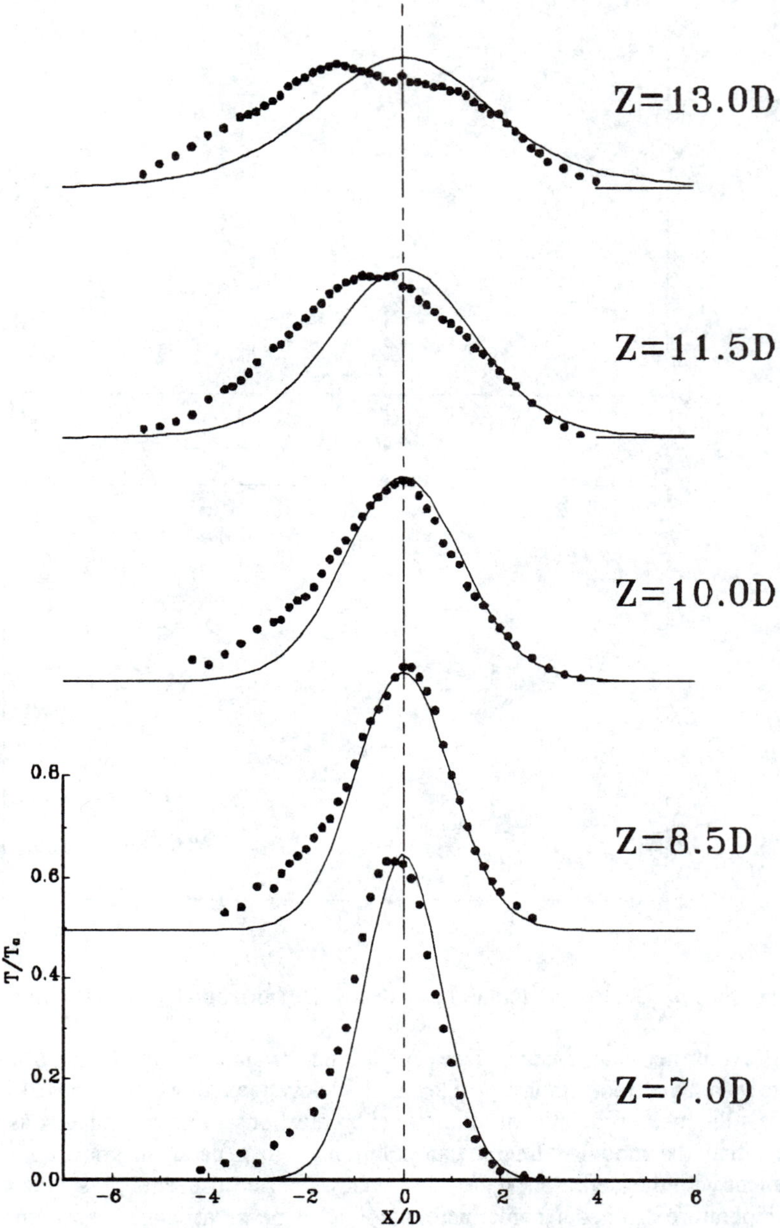

Figure 6. Time-averaged temperature profiles measured above the vertical jet.
• laboratory data, ——— Lagrangian model.

an effective barrier to the wave motion in the vicinity of the jet, it becomes apparent that the upstream and downstream sides of the jet (measured relative to the direction of wave propagation) experience very different flow regimes. Adopting the usual notation whereby a co-flow corresponds to the movement of ambient fluid in the same direction as the jet velocity, and a counter-flow opposes the jet velocity; it can easily be shown that the upstream side of the jet is subject to a transition from a co-flow to a cross-flow, while the downstream side of the jet experiences a transition from a counter-flow to a cross-flow. These changes occur between phase angles $0 \leq \theta \leq 0.25$ and $0.5 \leq \theta \leq 0.75$ respectively, where $\theta=0$ corresponds to a zero up-crossing. The difference between these regimes provides a plausible explanation for the jet asymmetry observed in figure 6.

Indeed, if one further considers the temperature profiles in a phase-averaged sense (or ensemble-averaged at specific phases of the wave cycle) there is evidence of a new wave-induced mixing mechanism which specifically effects the downstream side of the evolving jet. Figures 7a-7e describe the ensemble-averaged temperature profiles at $\theta=0.5$, 0.6, 0.7, 0.8 and 0.9 respectively, where $\theta=0.5$ corresponds to a zero down-crossing and $\theta=0.75$ is in phase with the wave trough. During this portion of the wave cycle, the vertical jet is first subject to a counter-flow ($\theta=0.5$), followed by an increasing cross-flow which is orientated towards the downstream side of the jet. In each of these five figures the large arrow positioned above the temperature data indicates the approximate direction of the ambient flow. In this case the effect of the ambient motion appears to divide the jet so that although a large proportion of the jet is swept upstream (in the positive x direction), a significant proportion also appears to remain almost static. This apparent division of the jet only arises during this stage of the wave cycle, and in particular does not occur 180° later, when the jet is subject to a co-flow ($\theta=0$) followed by an increasing cross-flow orientated towards the upstream face. In this latter case the entire jet is swept downstream (in the negative x direction), and thus the pattern looks very different to that presented on figures 7a-7e.

Concluding Remarks.

The present study has addressed the case of a buoyant hot-water jet discharged in a wave environment, and has sought a quantitative description of the near-field mixing based upon a quasi-Lagrangian formulation of the integral equations. Previous attempts at a solution of this type have provided a good qualitative representation of the flow field, but detailed quantitative comparisons have been limited due to the apparent uncertainty in the entrainment coefficients. In particular, a three-fold increase in the coefficient of radial entrainment was necessary to model the centre-line decay. To investigate this point an existing Lagrangian model, formulated in a relative frame of reference, was used to describe a vertical jet discharged in a steady cross-flow. Although in this case there is no ambiguity regarding the entrainment coefficients, the agreement between the model predictions

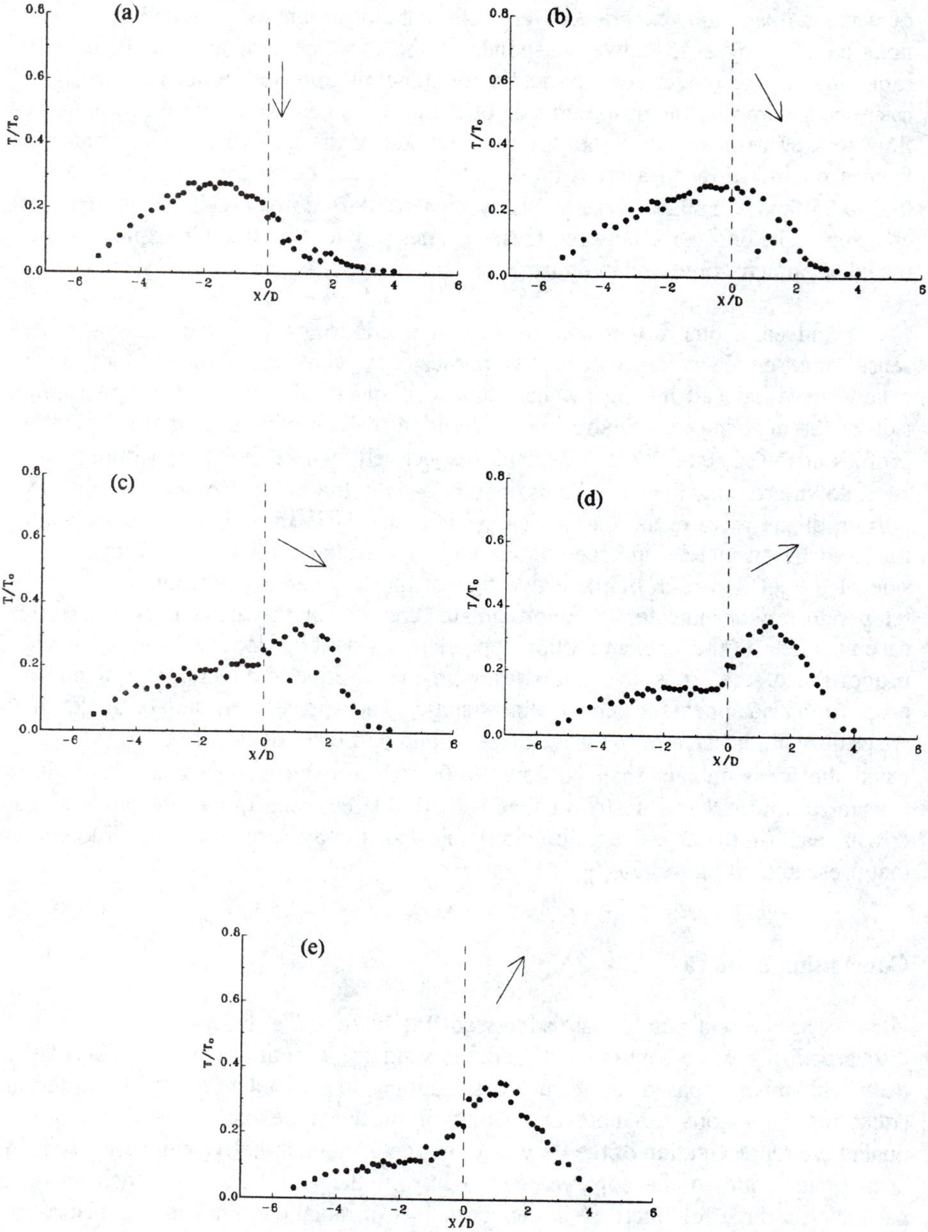

Figures 7a-7e. Phase-averaged temperature profiles at z=13D.
(a) θ=0.5, (b) θ=0.6, (c) θ=0.7, (d) θ=0.8, (e) θ=0.9.

and laboratory data were poor. On closer inspection, it has been shown that a Lagrangian model formulated in a relative frame of reference neglects the momentum flux associated with the entrained fluid. In contrast, an alternative model, formulated in an absolute frame of reference, is shown to be in good agreement with all cases involving buoyant jets discharged in both steady currents and unsteady wave fields.

To further validate this latter model, new experimental studies were undertaken in which both horizontal and vertical jets were discharged within a wave environment. Laboratory data describing the near-field mixing is shown to be in good agreement with the modified Lagrangian solution, and perhaps most importantly there is no ambiguity with regard the entrainment coefficients. As a result, the proposed solution is appropriate to design applications. Furthermore, detailed consideration of the phase-averaged temperature data measured within the near-field provides evidence of a new wave-induced mixing mechanism. This effect provides a plausible explanation for the development of non-symmetric time-averaged temperature profiles. More importantly, it provides the first quantitative evidence which suggest that the average dilution will be enhanced if a pollutant is discharged in the plane of the wave motion.

Acknowledgements.

The authors gratefully acknowledge the financial support provided by the Croucher Foundation, Hong Kong.

References.

Chin, D.A. 1988. Model of buoyant-jet-surface-wave interaction. J. Waterway, Port, Coastal, and Ocean Eng., ASCE, **114**(3), 331-345.

Chu, V.H. & Lee, J.H.W. 1996. General Integral Formulation of the turbulent buoyant jets in cross-flow. J. Hyd. Eng., ASCE. **122**, No. 1, 27-34.

Chyan, J.M., Hwung, H.H. & Chang, Y.H. 1991. Wave effects on the mean flow characteristics of turbulent round jets. In: Environmental Hydraulics, Eds. Lee, J.H.W. & Cheung, Y.K. **1**, 109-114.

Hwung, H.H., Chyan, J.M., Chang, C.Y. & Chen, Y.F. 1995. The dilution processes of alternative horizontal buoyant jets in wave motions. Proc. 24th Int. Con. Coastal Engineering, **3**, 3045-3059. Kobe, Japan 1994.

Kwan, S.H. & Swan, C. 1997. Near-field modelling of a buoyant jet in a wave environment. Submitted to Coastal Engng.

Koole, R. & Swan, C. 1994. Measurements of a 2-D non-buoyant jet in a wave environment. Coastal Eng., **24**, 151-169.

Koole, R. & Swan, C. 1995. Dispersion of pollution in a wave environment. Proc. 24th Int. Con. Coastal Engineering, **3**, 3071-3085. Kobe, Japan 1994.

Lee, J.H.W. & Cheung, Y.K. 1990. Generalised Lagrangian model for buoyant jets in current. J. Envir. Eng., ASCE, **116**(6), 1085-1106.

Morton, B.R., Taylor, G.I. & Turner, J.S. 1956. Turbulent gravitation convection from maintained and instantaneous sources. Proc. Roy. Soc. Lon., SerA: **234**, 1-23.

Patrick, M.A. 1967. Experimental investigation of the mixing and penetration of a round turbulent jet injected perpendicularly into a transverse stream. Trans. Inst. Chem. Eng., **45**, 16-31.

Sharp, J.J. 1986. The effects of waves on buoyant jets. Proc. Inst. Civ. Eng., **81** (2), 471-475.

Winiarski, L.D. & Frick, W.E. 1976. Cooling tower plume model. US EPA, Ecological Res. Series, EPA-600/3-76-100, Corvallis, OR.

CHAPTER 356

COST EFFECTIVENESS OF WAVE POWER EXTRACTION AT EROSIVE COASTS

HIDEO KONDO*

ABSTRACT

Cost of wave power extraction is investigated for the coasts where erosion has been continuing without suitable countermeasures. By a wave power extraction, a shoreline accretion is expected due to the mean sea level drop. Cost effectiveness is discussed with the concept of total cost instead of the generation cost only.

INTRODUCTION

Wave power utilization has not essentially been commercialized yet except small size navigation buoys though remarkable progress has been achieved in the last two decades. The reason is simply that the cost of electricity from waves is still much higher compared with those from fossil fuels and of nuclear. However, it is required to remove unfavorable gases like CO_2 etc. exhausted in the process of burning fuels from the standpoint of global environmental protection. This process needs an additional cost which we may call the environmental cost. The cost is known presently to be no less than that of power production itself.

Meanwhile continuous retreat of coasts by wave and meteorological tidal actions has become serious at most coastal countries in the world. Coastal engineers have worked extensively to solve the problems by constructing coastal structures or by feeding sand in eroding beaches. Extraction of wave power decreases apparently the transmitted wave power shoreward and will weaken beach erosion with similar actions as the above two methods do.

In the present study the author proposes an approach to evaluate the total cost as well as the production cost of wave power extraction at erosive coasts which is subsequently examined for case studies.

* Department of Civil Engineering & Architecture, Muroran Institute of Technology, 27-1 Mizumoto-Cho, Muroran, Hokkaido, 050 Japan

TOTAL COST ESTIMATION

The total cost Pt (¥/ kWh) consists of the market or generation cost Pm, the environmental cost Pe and the social cost Ps as (Kondo et al,1991 and 1993),

$$Pt = Pm + Pe + Ps \qquad (1)$$

Pm is estimated with conventional procedures. Defining $E_{p,j}$ as the environmental value of jth factor, Pe, the cost to repair the environmental damages occurred in power production process is estimated as,

$$Pe = \Sigma \Delta P_{i,j} = \Sigma \frac{\partial E_{p,j}}{\partial x_i} \Delta x_i \qquad (2)$$

In the above equation $\Delta P_{i,j}$ is the repair cost brought by the variation of jth environmental factor by the variation of the environmental factor i in the course of production. Pe of fossil-fired electricity includes expense of removing and treatment of harmful waste gases such as CO_2.

The environmental cost is positive for the cases of fossil-fired and nuclear powers. For the cases of renewable energies it can be minus when they bring better environment than that without extraction. The social cost Ps may be the expenses of moving houses and structures, compensation for damaging fishing industry, and the expenses dispatching troops to keep the sea-lane safe to tankers for imported oil, etc.

WAVE POWER BUDGET AND DEFORMATION

The incident wave power through the extracting device and structure is divided to the reflected, the power extracted, the loss by extracting process and structure, and the transmitted shoreward. The ratios of these depends on types of device, structure and the operating conditions. Approximate figures for the extracted and the transmitted are estimated for later calculation as in the following.

TABLE-1 WAVE POWER BUDGET (Extracted / Transmitted)

TYPE OF SYSTEM	IN OPERATION	STOPPED
Floating (OWC)	30 / 50	0 / 100
Bottom Fixed(Pendulor)	50 / 20	0 / 50

SHORELINE ACCRETION BY MSL DROP

Continuous retreat of shoreline by wave and meteorological tidal action has become one of the most serious problems in all coastal countries. It has been accelerated by the remarkable trend of mean sea level (MSL) rise. The power extraction from waves in offshore or inshore zone decreases the incident power and the height of transmitted wave and the size of MSL.

We know that a MSL rise brings a retreat of shoreline more than that anticipated from the static geometry in sandy coasts. Thus we may apply the relationship between the MSL rise versus the shoreline retreat to MSL drop at beach versus shoreline accretion by the wave power extraction. Several approaches have been proposed for the relationship between MSL rise and beach retreat we will use the one by Mimura et al (1994) for later calculation.

The environmental cost for the wave power extraction is minus since it makes natural beaches wider or saves constructing revetments on shores.

PENDULOR WAVE POWER EXTRACTOR

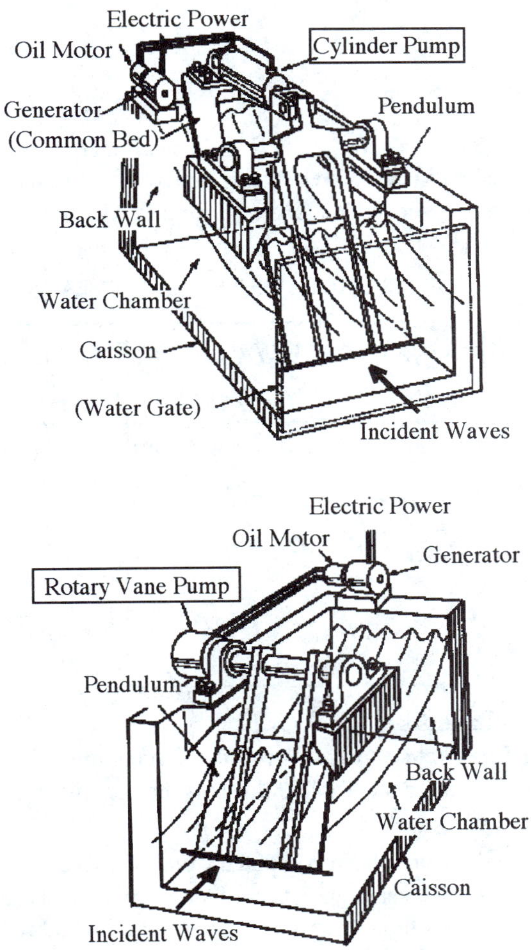

Fig.1 Pendulor Wave Power Extractor in Caisson

We have studied Pendulor power extractor of bottom standing type which can be installed in upright caisson breakwaters since 1980 (Kondo et al, 1984, Yano et al,1985, and Watabe, 1993). Two basic types of it are sketched in Fig.1. It consists of pendulum plate in a concrete caisson with seaside open and a hydraulic power transmission system. A back wall, two side walls and the bottom plate of caisson together with the pendulum plate forms a water chamber where a standing wave system is generated by oscillation of the plate.
Originally it had been developed as the cylinder pump type (the upper one in Fig.1). Laboratory and field experiments has proved it to have high efficiency and to lower the cost compared with the other kinds of extractors (Watabe, 1993).

The rotary pump type, the lower one in Fig. 1, was proposed later and has been being studied (Watabe et al, 1996).

STUDIED LOCATIONS

Three locations of Japanese coast which are typical ones at Hokkaido, Touhoku and Kantou area, respectively of different wave climate have been chosen to estimate the total cost for electric power production employing Pendulor system.

TABLE -2 WAVE CLIMATE OF THREE TYPICAL COASTS

COASTS (Oceans)	AVERAGE WAVE			DESIGN WAVE	
	H1/3	T1/3	W	H1/3	T1/3
Hokkaidou (Japan Sea)	1.1(m)	4.9(sec.)	5.3(kWh/m)	7.5 (m)	12.0(sec.)
Touhoku (Pacific)	1.0	7.4	6.3	7.5	14.0
Kantou (Pacific)	1.5	7.4	11.5	11.0	15.0

COMPUTATION PROCEDURE

The generation or market cost Pm was computed according to a conventional way for the case of electricity in Japan. Table-3 presents the computational process for the case of KANTOU coast.

The environmental cost has been computed from the two approaches:

(a) From the difference of expenses to construct a revetment at shoreline.

The revetments are designed for the two cases of with and without power extraction to the design waves shown in Table-2. The expense to construct a revetment is estimated from which Pe is computed.

(b) From the expense of filling the eroded shores every year.

The volume of beach eroded by the average annual wave shown in Table-2 is estimated for each location. Then the cost of sand filling to eroded shore is estimated from which Pe is computed.

TABLE - 3 Estimation of Market Cost, *Pm* (KANTOU COAST)

(1) DESIGN CONDITION

Water Depth: 10 m below LWL

Tidal Range: 1 m

Mean Wave Power: 11.5 kW/m

Design Wave; $H_{1/3}$ = 11 m, $T_{1/3}$= 15 sec.

Extrator: Pendulor in Bottom Standing Caisson

Overall Efficiency: 0.5

Caisson: Unit Length of 20 m

Rating Output per Unit: 11.5* 20*0.5 = 115

(2) EXPENSE AND COST (Debt Financing Rate 6 % per Year)

A: Cost of Caisson = 12 (M¥ /m)* 100= 1200 M¥

B: Cost of Generating System = 6 0M¥ * 5 = 300 M¥

C: Total Construction Cost = 1500 M¥

D: Specific Cost = C/ Output = 1500/575 = 2.6 M¥ /kW

E: Fixed Cost = Total Cost * Equivalent Coefficient
= 1500 * 0.07581 = 114 M¥ /Year

F: Operation Cost = B *0.05 = 15 M¥ /Yea

G: Expenditure = E + F = 129 M¥ /Year

H: Amount of Generated Electricity
= Output 24hr * 365day * Workig Rate
= 575* 24 * 365* 0.58 = 2.72 * 10^6 kWh/Year

(3) MARKET COST

I: Cost of Electricity = G/H = 44 ¥ /kWh

The result of estimation is shown in Table 4 which reveals that Kantou area shows the lowest market cost, Pm of 44(¥/kWh). It is still about three times than that of oil-fired as of 1994. It is sensitive to the debt financing rate which is 6 % in present estimation. For the case of 2% rate, Pm decreases to 30(¥/kWh).

The total costs, Pt, however, becomes close to it at Kantou area (Table - 5). For the case of the debt ratio of 2 % it becomes less than that of oil-fired.

TABLE - 4 COST OF ELECRICITY FROM WAVE in ¥/kWh

LOCATION	Pm	Pe	Ps	Pt
HOKKAIDOU	62	(a) -18	1	45
	40	(b) - 5	1	58
TOUHOKU	59	(a) -16	1	44
	38	(b) -11	1	49
KANTOU	44	(a) -11	1	34
	30	(b) -13	1	32

In addition to the wave power, the other two ocean energies, namely current and tide, and wind are also investigated for the cost analysis at promising locations in Japan, the costs of which are considerably higher than those of wave power as shown in Table-5, except wind. -5.

TABLE - 5 COST OF ELECTRICITY BY RENEWABLE ENEGIES IN ¥ /kWh (Ratio to Oil-Fired)

ENERGY	LOCATION	Pm	Pe	Ps	Pt
WAVE	KANTOU	44	-13	1	32 (1.3)
CURRENT	TSUGARU STRAIGHT	33	3	1	37 (1.5)
TIDE	ARIAKE BAY	73	3	2	78 (3.1)
WIND	MURORAN	48	-14	1	35 (1.4)
OIL - FIRED	KANTOU	12	12	1	25 (1)

CONCLUSIONS

The present study has revealed that wave power utilization at erosive coasts is found to be a promising way to get inexpensive clean energy in Japan and may be so for most coastal countries though the cost depends on heavily domestic natural and economical condition. The utilization will restore invaluable fossil fuels, and precious beaches for the future generation.

ACKNOWLEDGMENTS

The author would like to thank Dr. Tomiji WATABE, T-Wave Consulting Volunteer and Mr. Senji OSANAI, formerly Manager of Cold Region Port & Harbor Engineering Research Center for their constant support to the wave power study at Muroran Institute of Technology.

Thanks must go to Miss Hiromi HAMATSU, former student who had done a part of this study for her B. E. Thesis and Mr.Noriyuki OOTA, the technical staff who assisted for preparing the manuscript at Department of Civil Eng. & Architecture of the institute.

REFERENCES

KONDO, H. T.WATABE and K.YANO (1984): Wave power extraction at coastal structure by means of *moving body in the chamber, Proc. of 19th International Conf. on Coastal Eng., Vol.4,* ASCE, pp. 2875-2891.

KONDO, H., I. SUGIOKA, S.OSANAI and S.OZAWA (1992): Application of the new cost analysis to wave power extraction at breakwaters, *Renewable Energy, Technology and the Environment, Proc. 2nd World Renewable Energy Congress, 5,* Pergamon Press, pp.2829-2833.

KONDO, H., I. SUGIOKA, S. and OSANAI (1993): The concept of true cost of energy and its application to ocean energies, *Proc. of International Symposium on Ocean Energy Development,* Muroran Inst. Tech. and Cold Region Port & Harbor Eng. Research Center, pp.101-106.

KONDO, Hideo (1997): A hybrid system of wave power extraction and shore protection, *Proc. of 27th IAHR Congress* (submitted).

Mimura, N., H.IKUYO and K.INOUE: Evaluation of effect on sandy shores of the mean sea level rise, *Proc. of 40th Japanese Conf. on Coastal Engineering,* JSCE, pp.1046 -1050 (in Japanese).

OSANAI,S., H.KONDO, Y.MIZUNO and T.WATABE(1997): Feasibility Tests of new Pendular-type wave energy conversion appratus ,*Proc. of 25th International Conf. on Coastal Eng.,*(in press).

Seymour, Richard. J.[Edit.] (1992): *Ocean Energy Recovery - The state of the art-*, ASCE, 309p.

WATABE, Tomiji (1993): Pendulor wave power convertor -15 years study and its future prospect -, *Proc. of International Symposium on Ocean Energy Development*, Muroran Inst. Tech. and Cold Region Port & Harbor Eng. Research Center, pp.101-106.

Watabe, T., H.KONDO, M.NARITA, K.SEINO and E.AKAMA (1996): Studies on a large displacement pump for ocean wave power converter, *JHPS International Symposium on Fluid Power, Yokohama,* JHPS(in press).

YANO, K., H.KONDO and T.WATABE (1985): A device for wave power extraction in coastal structures - Field test of a pendular system - , *Coastal Engineering in Japan*, 28. JSCE, 243- 254.

CHAPTER 357

FAESIBILITY TESTS OF NEW PENDULAR-TYPE WAVE ENERGY CONVERSION APPARATUS

Senji Osanai[1],Hideo Kondo[2],Yuzo Mizuno[3]and Tomiji Watabe[4]

Abstract

In order to utilize natural energy,it is essential to develop a fully compatible system, instead of conventional power- generating systems.

The authors proposed a new pendular-type wave power generation device. Design criteria for a new rotary vane pump systems (NRVP) was studied for several years and a 5-kW-model test plant with a large displacement of 45.9 l/rev pump was established. This simplified pendular system was considered to limit the investment costs. This paper describes the new pendular-type converter and its components, and presents observation data obtained in field tests conducted at sea.

1. Introduction

As Japan is largely dependent on fossil fuels such as coal and petroleum for energy, transition to nuclear power and others non-fossil fuels is being undertaken. Efforts should be made to find energy-saving techniques or new energy sources in order to ensure future development while protecting the earth's environment. It is essential to switch to clean and natural energy sources such as solar radiation, wind power and wave power.

....

1) Adviser, North Japan Port and Harbor Consultants Co.,Ltd.(Formerly, Cold Region Port and Harbor Engineering Research Center) 18-N11 Heiwadori-2 Shiroishi-ku. Sapporo. 003 , Japan

2)Professor, Muroran Institute of Technology , 27-1 Mizumototyo ,Muroran. 050. Japan

3) Dr., Civil Engineering Research Institute, Hokkaido Development Bureau, 1-3 Hiragishi, Sapporo 062, Japan

4)Dr., T-Wave Consulting Volunteer, 5-23-3 Misono, Noboribestu, 059, Japan

However, these sources have problems regarding cost and generation capacity limitations. The biggest obstacle to the practical use of clean and natural energy sources is their generating costs compared with those of fossil fuels.

To address this problem, a new, pendular-type wave energy conversion device, named Pendulor by the authors, was developed.

In 1981, the first device[1), 3)] (20 kW) was installed on a breakwater at the Port of Mashike in a collaborative development project involving the Town of Mashike, the Muroran Institute of Technology and Hitatchi Zosen corp. Full-scale feasibility tests were conducted .

Following these tests, in 1983 a research group from the Muroran Institute of Technology made the first experimental test device by improving the existing one, and conducted test plant experiments. The tests achieved the expected results and the device was proved to be efficient in wave energy conversion. But it was absolutely necessary to further reduce the power generation costs. Therefore, based on the past research and development results, a new, remarkably improved wave energy conversion device was built under a new research plan. Field tests were conducted at sea.

2. Outline of new pendulor device

Wave energy conversion devices can be largely classified into OWC(Oscillating-wave-column) and pendular-types. The principal of the device is shown Fig.1.

Horizontal standing wave motion is converted to motion of the pendular plate, which is then converted to reciprocating rotary motion of a vane -type rotary pump directly connected to the pendulum. This rotation feeds hydraulic oil to the hydraulic motor and drives the generator.

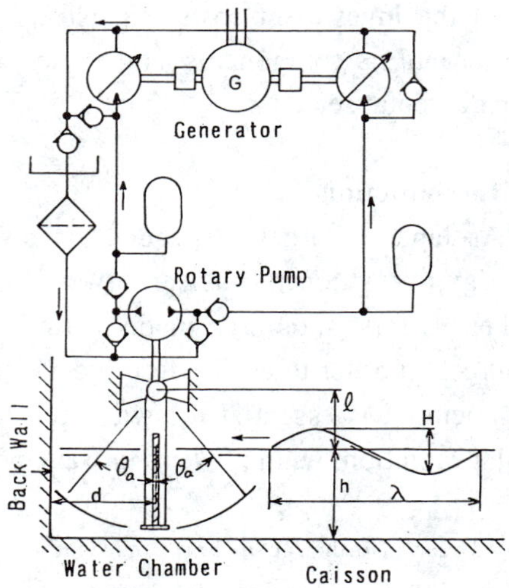

Fig. 1 Principle of the device

Figure 2 shows the original pendular device [4),8)] which had a cylinder pump. It was used to drive an oil motor-generated set.

Since the pendulum actuated the pump , both were fixed on a common bed lest the caisson suffer from the huge reaction of the pump. When the pump and pendulum were coupled together at sea, the system became dangerous because the pendulum moved whenever incident waves exited it. A water gate as illustrated as in the figure was required.

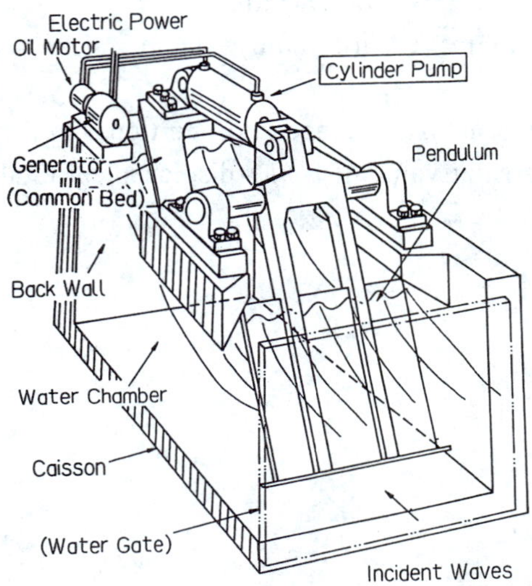

Fig.2 Original pendular device

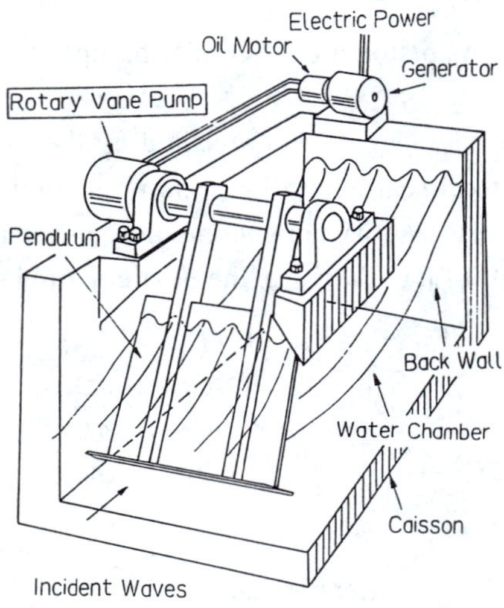

Fig.3 New pendular device

In order to improve cost efficiency, simple combined feature was invented in which a rotary vane pump was coupled with the pendulum.

A long shaft was used to support both the pump rotor and the pendulum, having no coupling component. There were 3 bearings on the shaft, 2 of which were for the pendulum. This combination makes for easier installation work than with the previous model. This system was called a new pendular device[2),5),6)] as seen in Fig.3.

This large-capacity vane -type pump offers the following advantages:

1) The features of the pendular system can be simplified through an integrated pendular plate and rotary vane-type pump.
2) Energy generation costs can be reduced through simplification of installation work at sea.
3) High-wave resistance of the pendular system is also improved.

3. Design of new rotary pump and pendulum

3.1. Rotary pump

Displacement D_P of the pump, which consists of 2 vanes, is described in the following equation, assuming it is the same as the previous pump.

where D_P displacement of the pump

$$D_P = \frac{\pi}{2}\left(d_1^{\ 2} - d_1^{\ 2}\right)B = 2\pi A_P r$$

d_1 vane tip diameter
d_2 rotor cylinder diameter
B vane width
A_P piston area (cylinder pump)=$98.17 \times 10^{-4} m^2$
r arm radius =0.6 m

The leakage is prevented by the metal seals. As leakage is proportional to the total seal length Lp, a pump having greater value of Dp/Lp(displacement/total seal length) can operate more efficiently.
Dp/Lp is shown in the next equation[7]:

$$\frac{D_p}{L_s} = \frac{\pi\left[(d_1/d_2)^2 - 1\right]d_1^2}{4\left(2\frac{d_1}{d_2} + \frac{d_1}{W}\left(\pi + \frac{d_1}{d_2} - 1\right)\right)}$$

where W vane width

Through the results of the research, the following equation was determined to provide the optimal profile of the pump[7]:

$$\frac{d_1}{d_2} = \sqrt{\frac{(0.36 \sim 0.46)(d_2/W)\tau_s}{P_{max}} + 1}$$

where $d_2/W = 1.0 \sim 1.2$, τ_s is shearing stress of the shaft, P_{max} is maximum delivery pressure of the pump.

Applying the research results, the test pump shown in Fig.4 was designed.

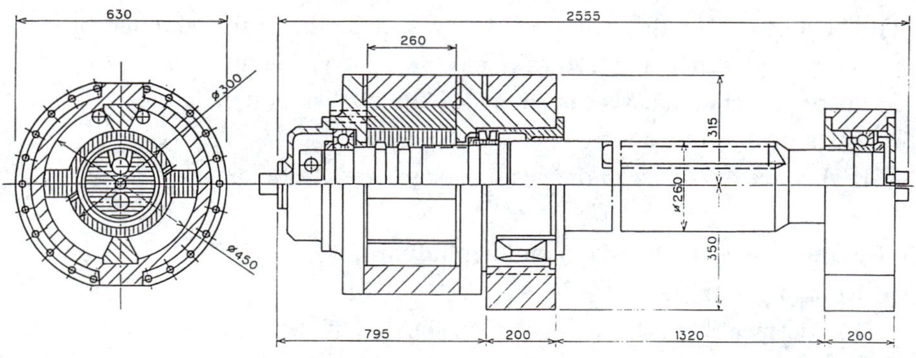

Fig.4 New rotary pump

Its specifications are arranged in Table 1.

Table 1 Specification of the NRVP

Input: $\eta_1 E_1$=16.6kw	Displacement:	Dp=0.0459m^3/rev
Vane tip diameter: d_1=0.45m	Rotor cylinder diameter:	d_2=0.30m
Vane Width: W=0.26m	Shaft diameter:	ds=0.232m
Movable angle: θ max=65°(both sides)		
Pressure(mean/max.): P_{mean}/P_{max}=12.8/25MPa		

3.2. Pendulum

The power E_1 produced by the pendulum can be determined with the following equation[7]:

$$E_1 = 1/8 \times \rho g H^2 C_g B \eta_1$$

where H is wave height = 1.5m , ρ is density of water =1,030kg/m^3, g is acceleration of gravity =9.81 m/sec^2, C_g is group speed of the incident waves=3.46m/s(wave period 4s, water depth:h=2.75m), B is width of the pendulum in water=2m, η_1 is conversion ratio from wave power to pendular power=80%.

In this case, these parameters were adopted in the design, and then it was decided that E_1=16.6kW.

In design, amplitude of pendular motion θ_a was established using the following values.

$\theta_a = M/(\omega N) = 0.285$rad(16.3°) at no load operation

$\theta_a = 16.30° \times 0.625° = 10.20$ at optimal load operation

where M is amplitude of exiting moment =370KNm, N is load factor by wave generation =828KNms, ω is circular frequency of wave =1.571rad/s (T=4s).

As seen in the pendular device shown in Fig.5, the basic style was not changed significantly, though the pendulum is more durable in storms with its rubber tires of heavy track's, it can swing wider

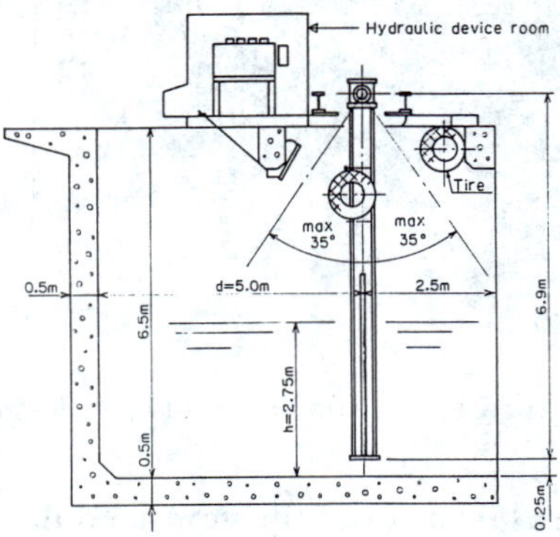

Fig.5 Test pendular device

and more reliable to absorbing shock loads.It is expecting to be relatively maintenance free.

In order to prevent electro-chemical corrosion, the shaft was connected to the earth through brushes and zinc plate were fixed to the pendulum surface and incorporated into the vane- type rotary pump.

3.3.Installation

The device was installed using a large crane in August 1994,on the caisson of the Muroran Institute of Technology around the exterior of the south breakwater in the Etomo area of Muroran. The installation was completed within 2 hours, not included preparation of a gate.

Muroran is centrally located on the middle of Hokkaido and faces the Pacific Ocean. The water in front of the breakwater is approximately 4.0 m in depth. Because of the port's location in a bay, incoming waves are small. The largest significant wave was 2.7 m high and its period was 5.4 second. The mean wave direction was west northwest. Photo 1 shows the installation of the device. Fig.6 shows the location of the port of Muroran.

Photo 1 Installation work

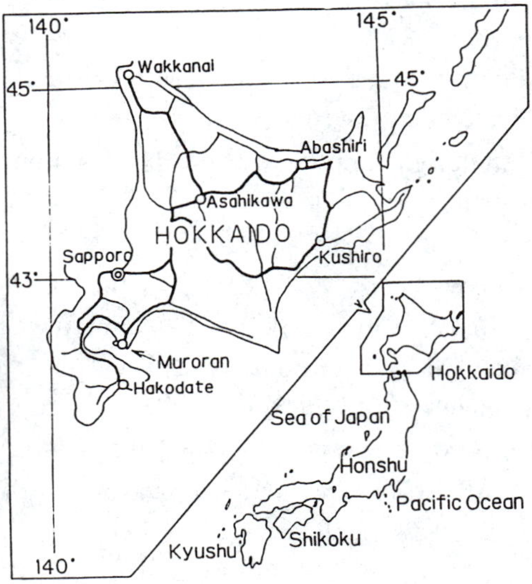

Fig. 6 Location of Muroran port

4. Measurement of the new pendular-type wave conversion device

Measurement were made at 8 points: ① Pump pressure (4 points). ② Motor axis torque (1 point), ③ Motor axis rotation speed (1 point), ④ Pendulum gradient (1 point), ⑤ Wave recorder (1 point).

Eight patterns were successively measured, with load condition changed every 20 minutes .The data was stored on a CVS-type disk. Figure 7 shows the interior structure of the observation system.

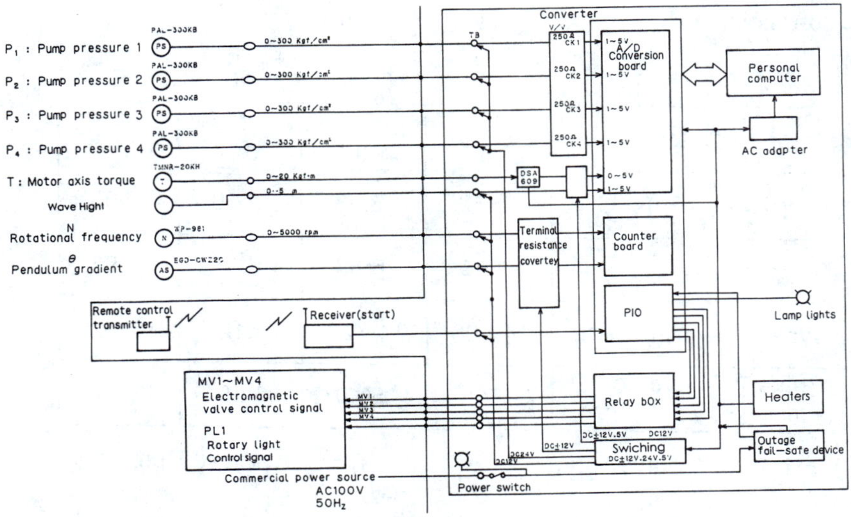

Fig.7 Observation instruments and wiring diagram

5. Observation of the new pendulor-type wave energy conversion device

The new pendulor-type wave energy conversion device was produced and tested in accordance with the development schedule shown in Fig.8.

The device was produced in 1993 and 1994. It was installed at the site, along with observation instruments, and ancillary electrical work was carried out before conducting observations.

	1993	1994	1995	1996	97~98
Development of device	—				
Testing		—	—	—	—
Data analysis					—
Other					

Fig.8 Research and development schedule

The device was installed in August 1994 in front of the port of Muroran shown fig-6. Since then, tests have been conducted for 20 months. In order to analyze efficiency , measurements were conducted at 8 points during operation. The data is presented in Table1. Figure 9 shows spectrum response of the system . Two spectrums, wave and angular displacement of the pendulum, showed a close correlation.

Year / Date	95 11.8	95 11.8	96 1.04	96 1.07
$H_{1/3}$ (m)	2.52	2.53	2.36	2.63
$T_{1/3}$ (sec)	6.45	6.57	6.04	5.93
Wave Power (kw)	14.426	14.727	14.125	18.401
Primary Conversion Efficiency (%)	61.71	56.0	53.55	42.16

$H_{1/3}$ Wave Height, $T_{1/3}$ Wave Perid

Table 1 Observation data

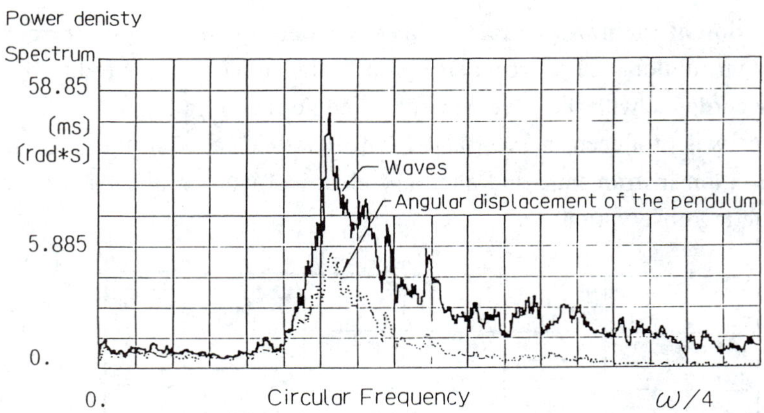

Fig.9 Responsiveness of the system

Figure 10 shows the efficiency of the new pendulor. The efficiency η_1 is shown in relation to wave period,T_{13}. It can be seen that η_1 =0.5~0.9, and it reaches peak value at approximately T_{13} =4.5~5.2second.

During the field test ,the pump operated without a breakdown, but pump efficiency was only 27-37%.

No contamination of the oil has been observed, so it is presumed the cause of low efficiency may be malfunction of the seal .

Besides the power conversion test, remarkable durability of the pump has been proved .There were no troubles around the pendulum or the pump such as loosening of bolts , leaking of oil or wear of sliding parts.

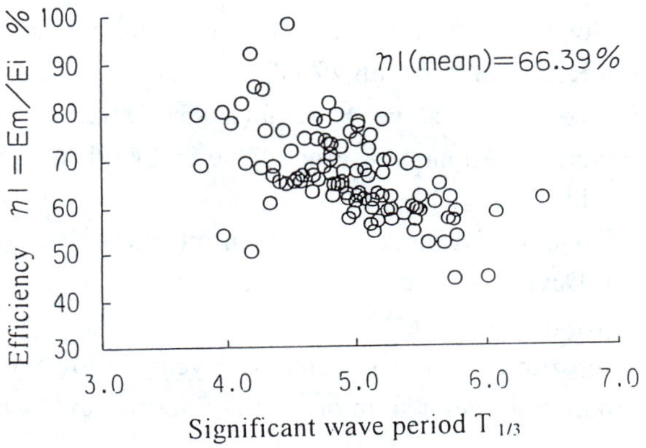

Em= mech. energy converted.
Ei=incident wave energy.

Fig.10 Efficiency of the new pendulor

6.Conclusions

Through testing of the new pendular device , the following conclusions were obtained.

1) The new rotary-vane pump system proposed for the pendulor makes installation easier, without the need for a water gate. It can lower energy costs.
2) The new pendular system improves durability not only for the pump, but also for the pendulor.
3) Efficiency of the new pendular was approximately 50～70 %(It was the same as the old pendulor coupled with a cylinder pump.)
4) Pump efficiency did not achieve the predicted value, as derived in the indoor test.

Therefore, there is still the need to solve the problems regarding pump efficiency and the seals on the pump unit. Further research to this end is expected .

After a great deal of research and development, the new pendular system was realized. It employed unique technique developed in Hokkaido, and which must be further developed. To put the device to practical use, we must continue field tests. Observation data, particularly that for durability and efficiency versus load, will be collected for application in design.

Acknowledgments

The authors would like to express their sincere gratitude to Mr. Seino, Muroran works of Narasaki Co. Ltd., and the Port Authority of Muroran for their contributions in the installation of the measuring instruments and to the 1st Regional Marine Safety Headquarters for their cooperation in the use of a power source.

References

1) M.Kuroi (1982), Wave Power Generation at the Port of Mashike. Sea and Port, Hokkaido Port and Coastal Research Assn.,pp.79~80.

2)S.Osanai, H.Kondo, Y.Mizuno, T.Watabe, K.Seino(1995), Development of a New Model of Wave Conversion Apparatus, Proc.of the Hokkaido Civil Engineering Lectures. pp.308~311

3) T.watabe, H.Kondo, K.Yano, H.Takeda and M.Kuroi(1982), Research on Wave Energy Absorption Devices Attached to Breakwater. Proc.of the 29th Coastal Engineering lectures, pp486~490.

4)T.Watabe(1993), Pendulor Wave Power Conversion:15 years Study and Future Prospect, Pro.of International Symposium on Ocean Energy Development, Muroran,pp41~52.

5)T.Watabe, H.Kondo, H.Shirai, K.Seino(1994), Remolding of Muroran Wave Test Plant, ISOPE 94. pp.353~358

6)T.Watabe, H.Kondo, M.Narita, E.Matumoto, K.Seino, M.Inoya and F.Akama(1995),New Pendulor and its Field Test, Pro.of 2nd European Wave Power Conference, Lisbon. (in printing)

7)T.Watabe, H.kondo, M.Narita, K.Seino(1996), Studies on a large Displacement Pump for a Ocean Wave Power Converter, JHPS International Symposium on Fluid.

8) K.Yano(1992), Study of Methods for Pendulum-type Wave energy conversion Device, Report of the Civil Engineering Research Institute, Hokka ido Development Bureau, No.97. pp1~48.

CHAPTER 358

Watertable Dynamics in Coastal Areas

Hong-Yoon Kang[1] and Peter Nielsen[2]

Abstract

Fresh water lens measured under the coastal barrier is much thinner than that predicted by the classical Ghyben-Herzberg theory . Landward downsloping watertable which is usually seen at coastal areas results in a landward flow of salty water or waste water released into aquifer. Comparison between the laboratory experiments with regular waves and no tides and the field data has revealed significant difference in magnitudes of the infiltration velocity. It is obvious that the tidal phase is very important. Infiltration velocities are large when the shoreline moves landward on a partially saturated beach on a rising tide. General magnitudes of U_I are $0.08K$ for steady laboratory conditions to $0.45K$ for field conditions during the rising tides. A mathematical model of the watertable which includes the effects of runup infiltration is obtained in a steady state. A finite-difference numerical model of the watertable is also presented with the glassy/dry boundary as a boundary condition.

Introduction

Watertable dynamics in coastal areas are of obvious interest in relation to problems such as salt water intrusion to the aquifer and wastewater disposal from coastal developments, but research into these processes has been scarce. In addition, the modeling of swash zone sediment transport requires a better knowledge of the beach groundwater dynamics.

Extensive field studies along the east coast of Australia have revealed that the overheight in the coastal watertable due to wave runup and tidal action on the ocean side are sufficient to create a steady drift of salty ground water under narrow coastal islands

---

[1]Center for Applied Coastal Research, University of Delaware, Delaware 19716, USA
Fax: +1 (302) 831 1228, Email: Kang@coastal.udel.edu

[2]Department of Civil Engineering, University of Queensland, Brisbane 4072, Australia
Fax: +61 (7) 3365 4599, Email: nielsen@uq_civil.civil.uq.oz.au

and barriers. Figure 1 details a typical situation showing the landward downsloping watertable and the very thin fresh water lens monitored at the narrow northern end of Bribie Island near Brisbane, Australia.

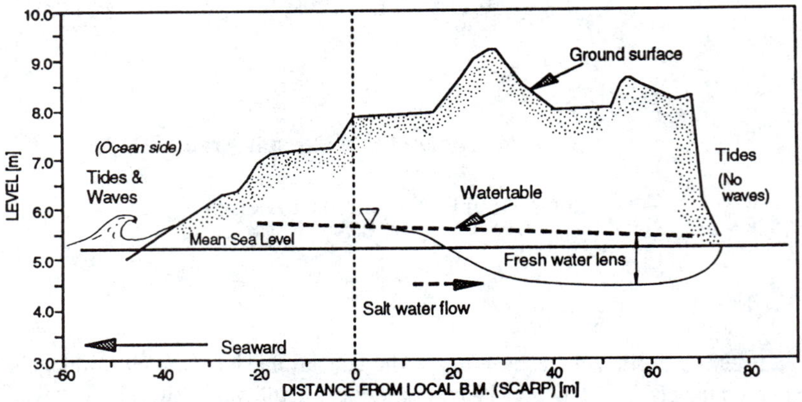

Figure 1. Field data showing the landward downsloping watertable and the thin fresh water lens (Bribie Island North, 14 July 1994).

In this case, the difference in groundwater level between the ocean side which is exposed to waves and the landward side which is relatively sheltered from waves is 0.4 meters. The magnitude of difference varies depending on the wave conditions, the slopes on the two sides and the landmass width. These features are important for understanding ecosystems in these areas and for pollution control. The difference in groundwater levels shown in Figure 1 is due to the combined effect of tides and waves.

The fresh water lens observed at the barrier island is seen to be much thinner than that predicted by the classical Ghyben-Herzberg (see Domenico and Schwartz, 1990) theory. The Ghyben-Herberg result was obtained from simple hydrostatics with no consideration of the details of the wave runup infiltration near the high water mark. The measured fresh water lens under the barrier also shows an asymmetry, i.e. it opens up gradually on the ocean side, but closes abruptly on the landward side.

This paper outlines the recently developed theory for wave effects significantly affecting coastal watertable dynamics and presents comparison with field and laboratory measurements. Watertable modeling is also given mathematically and numerically.

Wave Induced Watertable Overheights

Ocean waves raise the coastal watertable through their lifting of the free mean water surface seaward of the shoreline, i.e. wave setup, and due to the infiltration from wave runup landward of the shoreline. See Figure 2.

Wave Setup

The phenomenon of wave setup has been treated extensively in the water wave

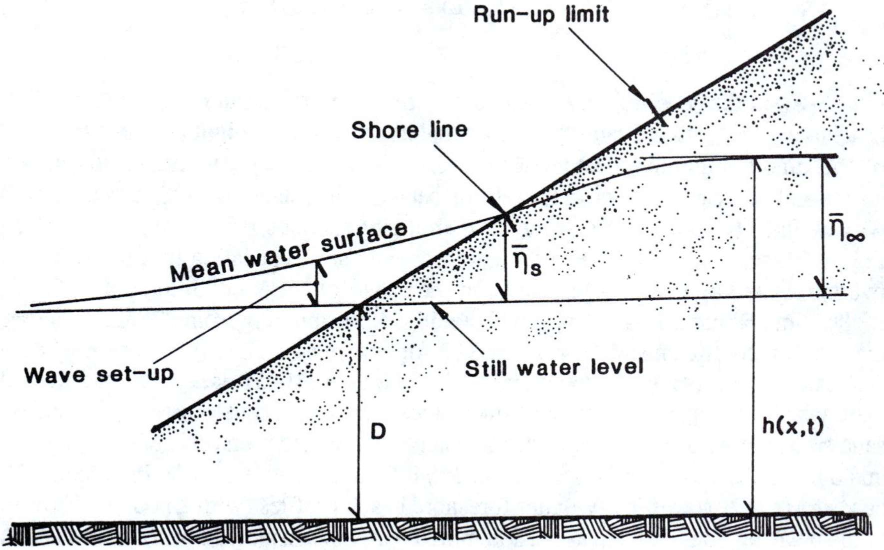

Figure 2. Definition sketch

literature. Extensive field data have been reported by Nielsen [1988] and Hanslow and Nielsen [1993]. They indicate that the shoreline setup $\overline{\eta_s}$

$$\overline{\eta}_S = 0.38 H_{orms} \tag{1}$$

or

$$\overline{\eta}_S = 0.048\sqrt{H_{orms} L_o} \tag{2}$$

where H_{orms} is the root mean square deep water wave height and L_o the deep water wave length ($L_o = gT^2/2\pi$, where T is the wave period).

<u>Regular Waves, No Tide</u>

Experimental studies were undertaken in a wave flume with regular waves and no tide to gain insight into process of infiltration from wave runup and the resulting watertable overheight as reported by Kang et al. (1994). It was found from the flume experiments that the asymptotic inland overheight $\overline{\eta}_\infty$ depends on the beach profile, the wave height H_i and the wave period T, but not directly on the hydraulic conductivity K of the beach material. Kang et al. [1994] recommended the following formula in relation to Hunt's [1959] formula for the runup of regular waves:

$$\overline{\eta}_\infty = 0.62(Z_R - SWL) = 0.62\tan\beta_F\sqrt{H_i L_O} \tag{3}$$

where Z_R is the level of maximum runup, SWL the still water level and tanβ the beachface slope. This indicates that the asymptotic inland overheight amounts to be 62% of the runup height irrespective of sand size. However, separate regression lines for coarse and fine sands, as seen in the relation between the inland overheight scaled on the wave height (h_∞-SWL/H_i) and the surf similarity parameter ξ (=$\tan\beta_F(L_o/H_i)^{0.5}$) (see Figure 3), show a weak effect/correlation of sand size, indicating that the watertable overheight is larger for larger sands, in the range of wave conditions of 1.2sec≤ T ≤2.8sec and 58mm ≤ H_i ≤ 200mm with aquifer depths ranging from 372mm to 436mm (regression coefficients: 0.64 with r^2=0.73 for coarse sands and 0.48 with r^2=0.86 for fine sands). This trend is consistent with Gourlay's [1985] data. The fact that the watertable overheight is larger for coarser sands may not be caused directly by different sand sizes, but by different beach profile shapes associated with different breaker types and different beach materials (see Rector, 1954; Gourlay, 1980; 1985). In general, finer sands form characteristically flatter foreshore beach profiles (with an outer bar formed at the breaking point of the waves), and more energy is dissipated through the surf zone than for beaches with coarse sands for the same wave conditions. The former results in less runup on the beach face than the latter. Smaller wave runup heights hence produce less watertable overheight.

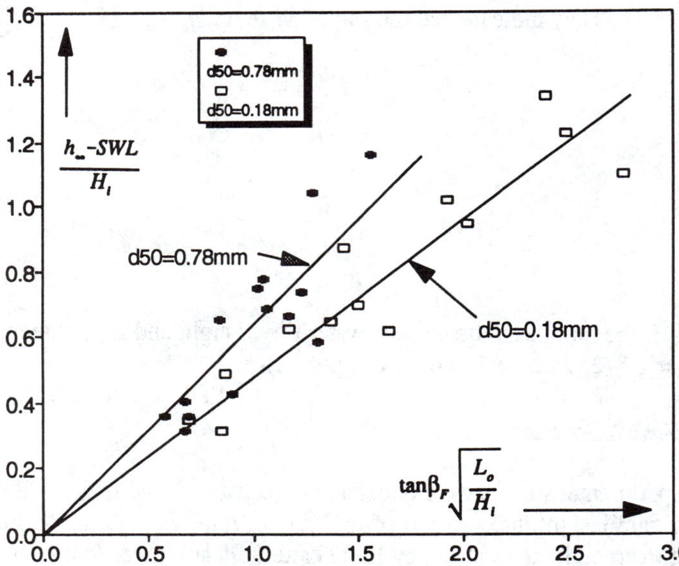

Figure 3. The normalized inland overheight as a function of the surf similarity parameter $\xi(=\tan\beta_F(L_o/H_i)^{0.5})$.

On the other hand, the present result is not directly consistent with Gourlay's [1992] example for a flat wave (H_i/L_o=0.007) which shows that the elevation of the watertable increased as the hydraulic conductivity of the beach material decreased. Equation (3) may apply when the breaking waves are not surging (for H_i/L_o>0.01: see Figure 4 of Gourlay [1985]) and the beach material is sand, but it may not hold for very flat waves (H_i/L_o<0.01). The wave steepness is an important factor governing onshore or offshore movement of beach material which determines the beach profile and breaking point/characteristic. It thus influences the beach watertable elevation/profile.

Wave Runup Infiltration

Recently, the wave runup infiltration contributing to significant lifting of the coastal watertable was stressed by several researchers (e.g. Nielsen et al., 1988; Hegge and Masselink, 1991 and Kang et al., 1994). Figure 4 outlines the processes of wave runup/infiltration and the coastal watertable.

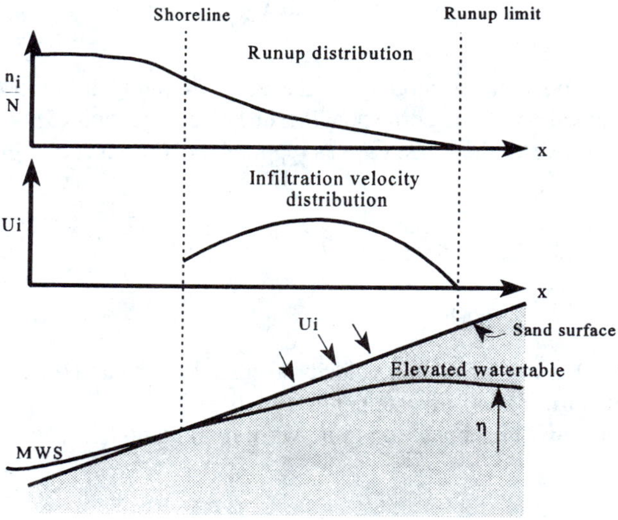

Figure 4. Relationships between wave runup, infiltration and the coastal watertable.

The top part of the figure shows the runup distribution (transgression statistics) of irregular waves, i.e. the fraction of the waves which transgressed certain points during the recording interval. The middle part shows the infiltration velocity distribution from wave runup, and the bottom part shows the corresponding elevated coastal watertable. With no sinks or sources landward of the runup limit the watertable will be horizontal landward of the runup limit.

Runup infiltration velocities indirectly determined from the watertable measurements and the parameters n and K of the porous medium, using the Finite

Difference approach, have maximum values roughly midway between the shoreline and the runup limit from both the field and laboratory data (see Kang et al., 1994 and Kang, 1996). General magnitudes of the infiltration velocity U_I are $0.08K$ for the steady laboratory conditions to $0.45K$ for field conditions during the rising tides. This difference is probably due to the fact that the field data were taken during rising tides where the beach is less saturated than in the steady state of the laboratory experiments. These magnitudes may vary according to characteristics of the porous medium.

Mathematical Modeling of the Coastal Watertable

Consider homogeneous sand body, with the specific yield n and the hydraulic conductivity K overlying a horizontal impermeable stratum at depth D. Under the Dupuit-Forchhermer assumption, the local height h of the watertable above the impermeable layer obeys the Boussinesq equation:

$$n\frac{\partial h}{\partial t} = K\frac{\partial}{\partial x}\left(h\frac{\partial h}{\partial x}\right) \tag{4}$$

In order to model the watertable in the zone of runup infiltration, the modified Boussinesq equation or the corresponding linearized equation (5) which includes the runup infiltration effect should be used as reported by Nielsen et al. (1988) and Kang et al. (1994):

$$n\frac{\partial h}{\partial t} = KD\frac{\partial^2 h}{\partial x^2} + U_I(x,t) \tag{5}$$

where $U_I\ (x,t)$ is the infiltration velocity averaged over numerous uprush-backwash cycles or the infiltration flow rate per unit area.

For steady flume condition (pure wave forcing, no tide), equation (2) leads to:

$$\frac{\partial^2 h}{\partial x^2} = -\frac{1}{KD}U_I(x) \tag{6}$$

The nature of $U_I(x)$ may reasonably be expected to be:

$$U_I(x) = \begin{cases} C_I K f(x) & \text{for} \quad x_s \le x \le x_R \\ 0 & \text{for} \quad x > x_R \end{cases} \tag{7}$$

where C is a dimensionless infiltration coefficient and $f(x)$ is a dimensionless function of x. x_s and x_R are the horizontal shoreline and the runup limit coordinates, respectively.

In order to define the solution for equation (6) under the steady laboratory conditions, consider a simple situation where the infiltration velocity $U_I(x)$ is distributed evenly between the shoreline x_S and the runup limit x_R. That is

$$U_I(x) \equiv U_I \tag{8}$$

Then equation (8) is written as:

$$\frac{\partial^2 h}{\partial x^2} = -\frac{U_I}{KD} \tag{9}$$

The boundary condition at the runup limit ($x = x_R$) may be written as:

$$\frac{dh}{dx}\bigg|_{x=x_R} = 0 \tag{10}$$

and the boundary condition at the shoreline can be stated as:

$$h_{x=x_S} = D + \overline{\eta}_S \tag{11}$$

The solution to equation (9) with the boundary conditions (10) and (11) is :

$$h(x) = D + \overline{\eta}_S + \frac{U_I(x_R - x_S)^2}{KD}\left[\frac{x - x_S}{x_R - x_S} - \frac{1}{2}\left(\frac{x - x_S}{x_R - x_S}\right)^2\right] \quad , \qquad x_S \le x \le x_R \tag{12}$$

and for $x = x_R$, the asymptotic inland overheight is obtained as:

$$h_\infty = h_{x=x_R} = D + \overline{\eta}_S + \frac{1}{2}\frac{U_I(x_R - x_S)^2}{KD} \tag{13}$$

Hence we see that the asymptotic inland overheight is proportional to $(x_R - x_S)^2/D$. If $U_I(x)$ is always distributed in the same way between x_S and x_R, i.e.

$$U_I = KF\left(\frac{x - x_S}{x_R - x_S}\right) \tag{14}$$

where F is a universal function, equation (13) can be written as:

$$h_{\infty} = D + \overline{\eta}_S + C\frac{(x_R - x_S)^2}{D} \tag{15}$$

where C is a constant. A universal function F would give a universal constant C.

Numerical Modeling of the Watertable

Model Formulation

The numerical technique used here is the finite-difference approach. Consider the typical four-node grid with node spacings of δ_x and δ_t in each coordinate direction. The dependent variable $h(x, t+\delta_t)$ will be obtained from the following equation, which is a Taylor series expansion about node $h(x,t)$:

$$h(x,t+\delta_t) = h(x,t) + \delta_t\left(\frac{\partial h}{\partial t}\right) \tag{16}$$

By substituting the watertable equation (4) into (16), we get the governing equation:

$$h(x,t) = h(x,t) + \delta_t\frac{K}{n}\frac{\partial}{\partial x}\left(h\frac{\partial h}{\partial x}\right) \tag{17}$$

for the groundwater surface level h. In this way one equation is obtained for each node, which can be used to find $h(x, t+\delta_t)$.

In order to get the solution for the most seaward and landward wells, we need two boundary conditions. The first of these can be obtained using the glassy/dry boundary (G/D) which corresponds to the watertable exit point, as the seaward watertable boundary condition (see Figure 5). Hence, since the G/D is moving horizontally as well as vertically, the spacing δ_x from the G/D to the first well is not constant, i.e.

$$\delta_{x1} = \delta_{x2}\,, \qquad \textit{for the first well landward of MSS} \tag{18}$$

where δ_{x1} is the distance between the glassy/dry boundary and the first well landward of MSS, δ_{x2} is the distance between the first and second well landward of MSS and MSS defines the intersection between the mean sea level (MSL) and the beachface (cf. Figure 5).

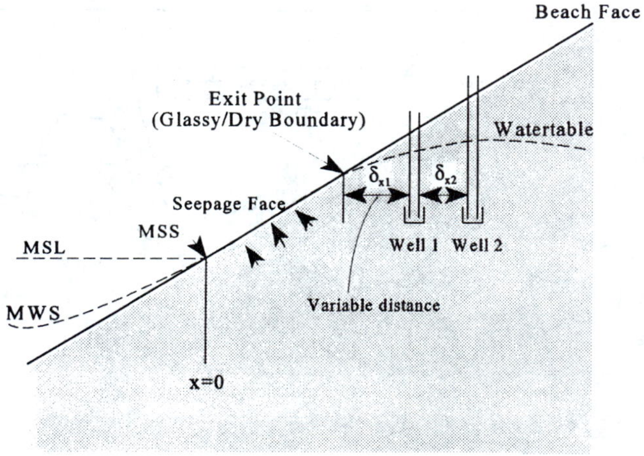

Figure 5. The moving glassy/dry boundary as the seaward watertable boundary condition

Secondly, at the landward boundary, a reflecting, no flux boundary condition is imposed. This is reasonably realistic since the watertable waves decay exponentially with distance from the shore. That is

$$h(x-\delta_x,t) = h(x+\delta_x,t)\ , \qquad for \quad x \to \infty \tag{19}$$

Numerical solution of equation (17) subject to the periodic boundary condition at the ocean and to (19) is then readily obtained by iteration. That is, in order to overcome the influence of the initial conditions, the model is run for several tidal cycles.

Comparison with Experiments

As an example, the watertable data measured for 25 hours from Kings Beach, Queensland are used because tides were approximately sinusoidal at the time of measurement.

Figures 6 and 7 show the time variations of simulated and measured watertables at 15m, 20m and 30m landward from the mean sea level shoreline with the variation of glassy/dry boundaries during two semidiurnal tidal cycles (25 hours period), respectively. The amplitudes of the watertable variation from the model give quite good agreement with those of the measured watertable. It is however seen that the simulated watertable levels always sit a bit higher than the measured watertable levels. This difference may be reduced by using the shoreline as a boundary condition because the shoreline always sits lower than the glassy/dry boundary level. Some improvement may also be obtained by varying hydraulic conductivity K and aquifer depth D at each well. The compactness of the beach sand body may vary in the shorenormal direction and hence hydraulic conductivity may vary shorenormally.

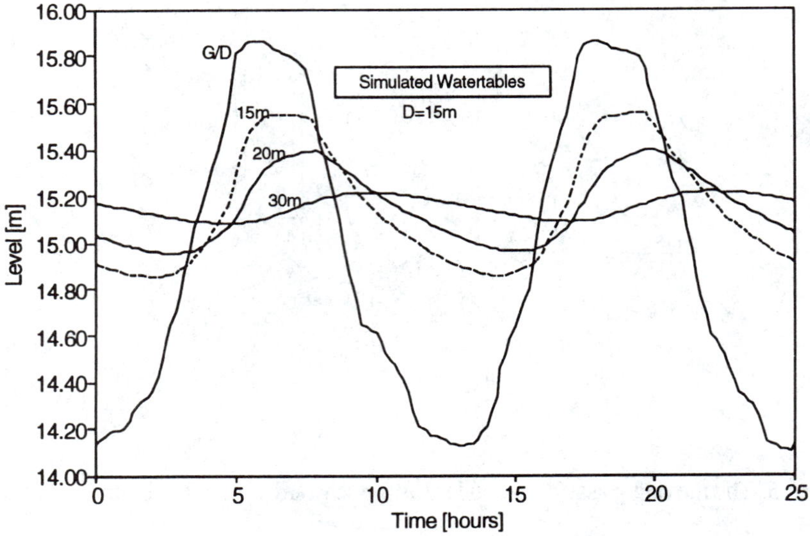

Figure 6. Simulated watertable variations during two semidiurnal tidal cycles at wells landward of the mean sea level shoreline. The numbers on the curves indicate distances landward from MSS.

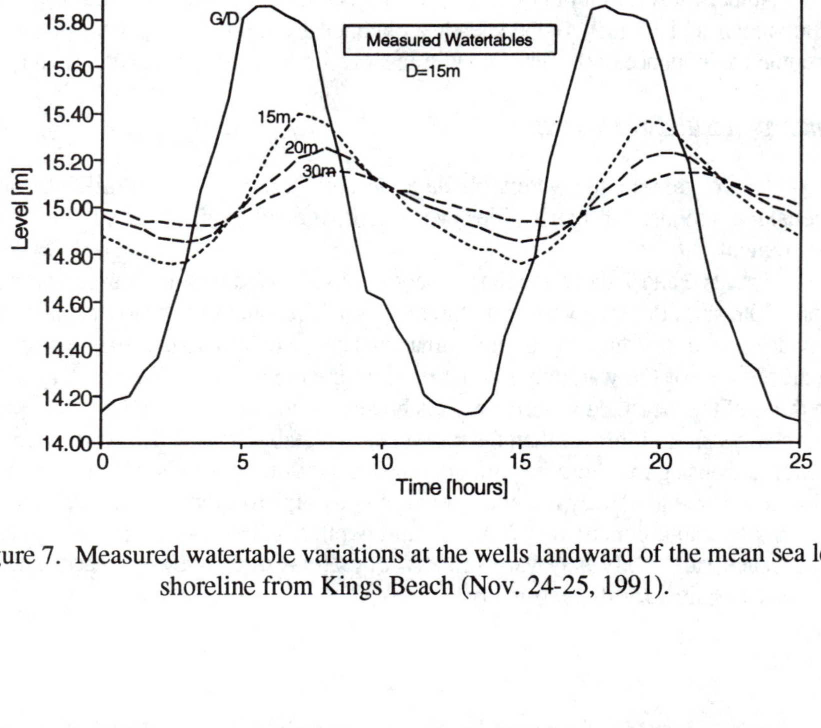

Figure 7. Measured watertable variations at the wells landward of the mean sea level shoreline from Kings Beach (Nov. 24-25, 1991).

Conclusions

Field measurements of groundwater dynamics in coastal barriers show landward downsloping watertable which results in a landward flow of salt water or waste water released into aquifer. Fresh water lens under the barrier tends to be much thinner than that predicted by the classical Ghyben-Herzberg principle.

Comparison between the laboratory experiments with regular waves and no tides and the field data (see Kang, 1996) has revealed significant difference in magnitudes of the infiltration velocity. It is obvious that the tidal phase is very important. Infiltration velocities are large when the shoreline moves landward on a partially saturated beach on a rising tide. General magnitudes of U_I are $0.08K$ for the steady laboratory conditions and $0.45K$ for field conditions during the rising tides.

A mathematical model of the watertable which includes the effects of wave runup infiltration was obtained in a steady state. The development of the numerical watertable model was also attempted with the glassy/dry boundary as a boundary condition. However, the numerical model of the watertable in this study is based on tidal fluctuations with the use of the wave-influenced G/D boundary condition.

References

Domenico P.A. and Schwartz F.W. (1990): Physical and chemical hydrogeology, *John Wiley & Sons, Inc.*, 824pp.

Gourlay M.R.(1980): Beaches: profiles, processes, permeability, *Proc. 17th International Conf. On Coastal Engrg,* ASCE, Sydney, pp 1320-1339.

Gourlay M.R. (1985): Beaches: states, sediments and set-up, *Preprints 1985 (7th) Australasian Conf. Coastal and Ocean Engrg,* Vol.1, Christchurch, pp 347-356.

Gourlay M.R. (1992): Wave set-up, wave runup and beach water table: Interaction between surf zone hydraulics and groundwater hydraulics, *J. Coastal Engineering*, 17, pp 93-144.

Hanslow D.J. and Nielsen P.(1993): Shoreline setup on natural beaches. *J. Coastal Research*, Special Issue 15, pp 1-10.

Hegge B.J. and Masselink G.(1991): Groundwater-table responses to wave runup: an experimental study from Western Australia, *J. Coastal Res.*, 7(3), pp 623-634.

Hunt I.A. (1959): Design of sea walls and breakwaters, *Proc. Paper 2172*, ASCE, Vol.85 No.WW3, pp 123-152.

Kang H.Y., Nielsen P. and Hanslow D.J.(1994): Watertable overheight due to wave runup on a sandy beach, *Proc. 24th International Conf. On Coastal Engineering*, ASCE, Kobe, pp 2115-2124.

Kang H.Y. (1996): Watertable dynamics forced by waves, *PhD thesis*, Department of Civil Engineering, University of Queensland, Australia, 200pp.

Nielsen P., Davis G.A., Winterbourne J. And Elias G. (1988): Wave setup and the watertable in sandy beaches, *Tech. Memo.*, 88/1, NSW Public Works Department, Australia, 132pp.

Nielsen P. (1988): Wave setup: a field study. *J. Geophys Res*, Vol 93, No C12, pp 15643-15652.

Rector R.L.(1954): Laboratory study of equilibrium profiles of beaches, *Technical Memo.* No.41, Beach Erosion Board, U.S. Army Corps of Engineers, pp 1-38.

CHAPTER 359

LONG RANGE POSITIVE EFFECTS OF THE DELRAY BEACH NOURISHMENT PROGRAM

Kim E. Beachler, P.E.[1]
and
Douglas W. Mann, P.E.[2]

INTRODUCTION

The success of beach nourishment projects is often questioned due to the observed losses of sand from the fill area and what appears to be a loss of benefits. Observations made by the authors indicate that there have been noticeable advances in the shoreline both to the north and the south of the Delray Beach, Florida beach nourishment project. The purpose of this evaluation is to show that much of the sand which moves out of the project can actually be accounted for and benefits the neighboring beaches.

The City of Delray Beach is located on the southeast coast of Florida approximately 80 kilometers north of Miami (Figure 1). A portion of the City encompasses about 4.8 km of the barrier island. The nourishment project is maintained over a distance of 4.3 km.

BACKGROUND HISTORY

In the 1960's, the Delray Beach shoreline was experiencing severe erosion. The public beach was almost non-existent and State Road A1A was damaged annually in winter storms (Figure 2). To combat this problem, the City constructed concrete and rock revetments along the public beach. In 1971, the concrete revetment failed and the City initiated a beach nourishment program.

The initial beach nourishment was completed in 1973. The project totaled 1,249,739 cubic meters of sand which was dredged from an offshore borrow area and placed in the project area. Subsequent to the initial nourishment the project area was renourished in 1978, 1984 and 1992 using sand volumes of 536,000, 994,000 and 770,000 cubic meters, respectively. The location of the various fill placements is shown in Figure 2.

The annualized cost of the beach nourishment program is approximately US$980,000 (Coastal Planning & Engineering, Inc.,1991) . The program is a Federally authorized shore protection project. As a result, the nourishments are designed consistent with Federal (U.S. Army Corps of Engineers) standards and are funded at both the federal, and local level.

The variation of fill volumes in this program (1973 to the present) is indicative of the expanded knowledge gained from both the monitoring of the beach fill and the advancement in beach nourishment design over this time period. A review of the initial 1973 fill placement has indicated that only enough material was placed to construct the design (storm protection) section of the beach. Virtually no sand was placed to account for end losses or background erosion losses (advanced nourishment). Subsequent nourishments have placed sufficient sand in an advanced nourishment section to maintain the design section through the nourishment interval. This design procedure is consistent with the National Research Council (NRC) recommendations (NRC, 1995).

---

[1] Formerly, Vice President, Coastal Planning & Engineering, Inc., 2481 NW Boca Raton Blvd., Boca Raton, FL 33431, USA.

[2] Coastal Engineer, Coastal Planning & Engineering, Inc., 2481 NW Boca Raton Blvd., Boca Raton, FL 33431, USA.

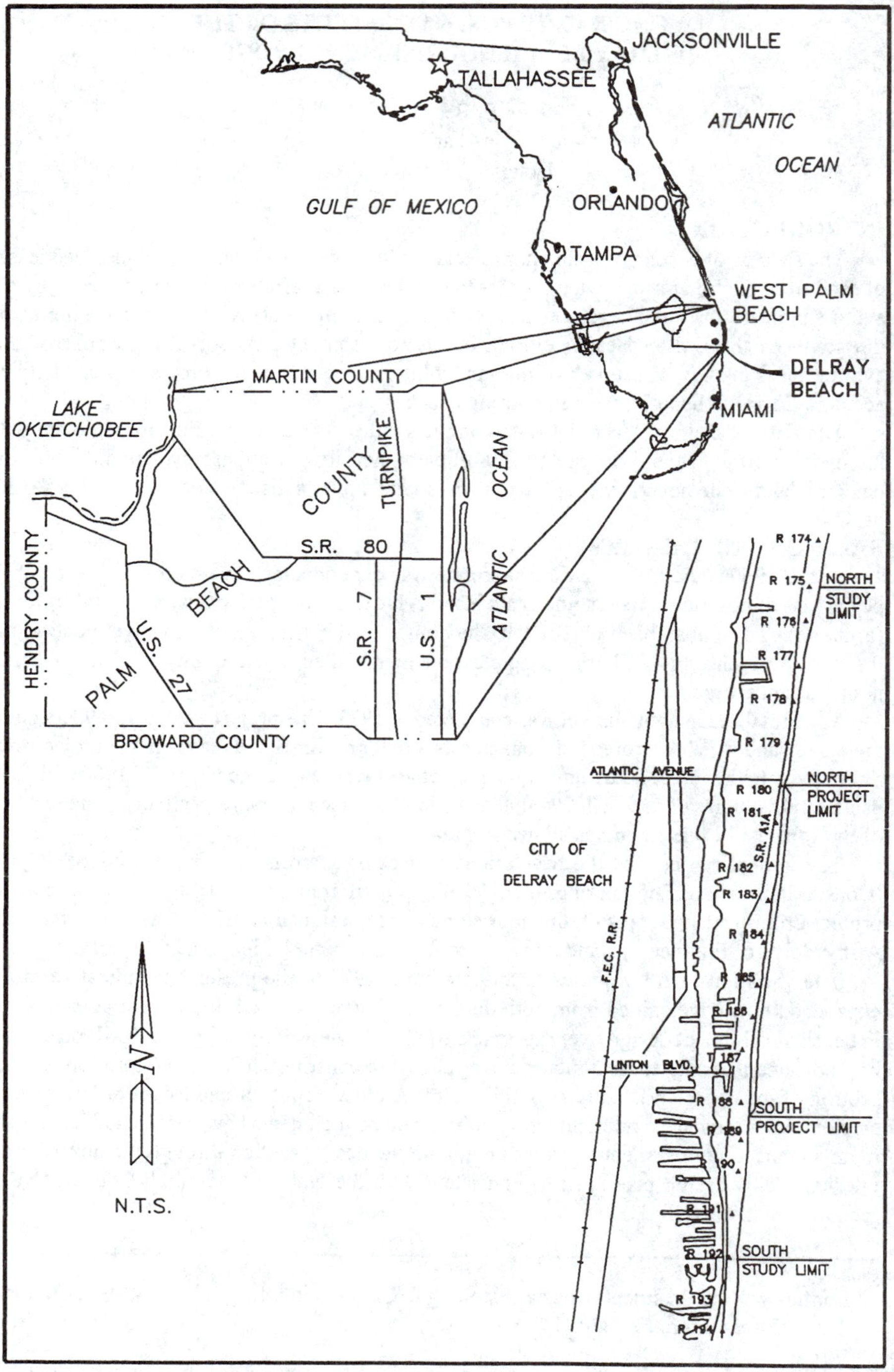
JACKSONVILLE
TALLAHASSEE
ATLANTIC
OCEAN
GULF OF MEXICO
ORLANDO
TAMPA
WEST PALM BEACH
DELRAY BEACH
MIAMI
MARTIN COUNTY
LAKE OKEECHOBEE
TURNPIKE
COUNTY
OCEAN
S.R. 80
HENDRY COUNTY
BEACH
ATLANTIC
U.S. 27
S.R. 7
U.S. 1
PALM
BROWARD COUNTY
NORTH STUDY LIMIT
ATLANTIC AVENUE
NORTH PROJECT LIMIT
CITY OF DELRAY BEACH
S.R. A1A
F.E.C. R.R.
LINTON BLVD.
SOUTH PROJECT LIMIT
SOUTH STUDY LIMIT
N
N.T.S.
R 174
R 175
R 176
R 177
R 178
R 179
R 180
R 181
R 182
R 183
R 184
R 185
R 186
T 187
R 188
R 189
R 190
R 191
R 192
R 193
R 194

FIGURE 1

BEACH MONITORING, ANALYSIS, AND RESULTS

Annual monitoring studies have been conducted by the City of Delray Beach since the initial restoration, providing data for the performance assessment. Survey data from the State of Florida Department of Environmental Protection collected in 1974, 1990, and 1995 was used to evaluate changes outside of the project area. The State of Florida maintains a series of beach profiles which are spaced approximately every 330 meters along the sandy shoreline of the State. The profiles in the fill area are shown in Figure 2. All beach profiles are surveyed from the upland limit to the depth of closure (approximately -10 meters).

Figure 3 shows the mean high water shoreline changes from 1974 to 1995 both within the fill area and on the adjacent shorelines. While the initial fill project occurred in 1973, no beach profiles were measured on the adjacent shorelines until 1974. As a result, the shoreline response in Figure 3 is not complete; nevertheless, the survey comparison represents one of the few extended (spatially and temporally) monitorings of a beach nourishment program.

The beach fill plan form demonstrates end loss or diffusion characteristics which are similar to those predicted by Pelnard-Considere'(1956) though a direct comparison has not been made. Accretion outside of the fill area is highest immediately adjacent to the fill and tapers to lesser amounts further from the fill (Figure 3). The length of the taper to the south is longer than the taper to the north which is consistent with the net littoral transport to the south. There are distinct points to the north (R165) and south (R201) where the data showed no change or erosion over the 21-year comparison. These end points are used as the lateral limits of the impact of the beach fill on adjacent beaches. It is probable that the beach fill impacts beaches further to the south, but that the accretion due to the fill is less than the existing erosion. The north limit is in the vicinity of a large natural limestone rock outcrop which tends to act as a groin and divides the beach into small littoral cells.

The monitoring data indicates that some of the sand from the nourishment project has moved about 3 kilometers north and south of the fill limits. This movement results in an advancement of the shoreline as well as increased volumes of sand on the beaches within the neighboring towns of Gulfstream and Highland Beach. The nourishment program has placed about 3.5 million cubic meters of sand on Delray Beach between 1973 and 1992. The beach profiles indicate that a total of 43% of that sand, about 1.5 million cubic meters, has been eroded from the fill area; however, the analysis found that 1.3 million cubic meters (85%) of that sand could be located between 2.7 km north of the project to 3.5 km south of the project. The total volume accreted to the north is 629,000 cubic meters over 2.7 km and to the south the accretion is 673,000 cubic meters over 3.5 km. Therefore, the majority of the sand that has eroded from the project has not been "lost" but rather has been deposited on adjacent beaches.

Nourishment Interval

The nourishment interval for the Delray Beach nourishment project has increased from 5 years to the present 8 years (1984 to 1992). The design of the fill places an economically optimum amount of advanced nourishment on the beach. Currently, the advanced nourishment is designed to last 9 years (U. S. Army Corps of Engineers, 1992). The increase in the actual nourishment interval is a result of better estimation of end losses and the spreading of the beach fill. The diffusion theory indicates that a longer fill project will last longer than a shorter fill project. The accretion of sand on adjacent beaches has extended the "effective" length of the beach and has resulted in less diffusion loss and a longer nourishment interval.

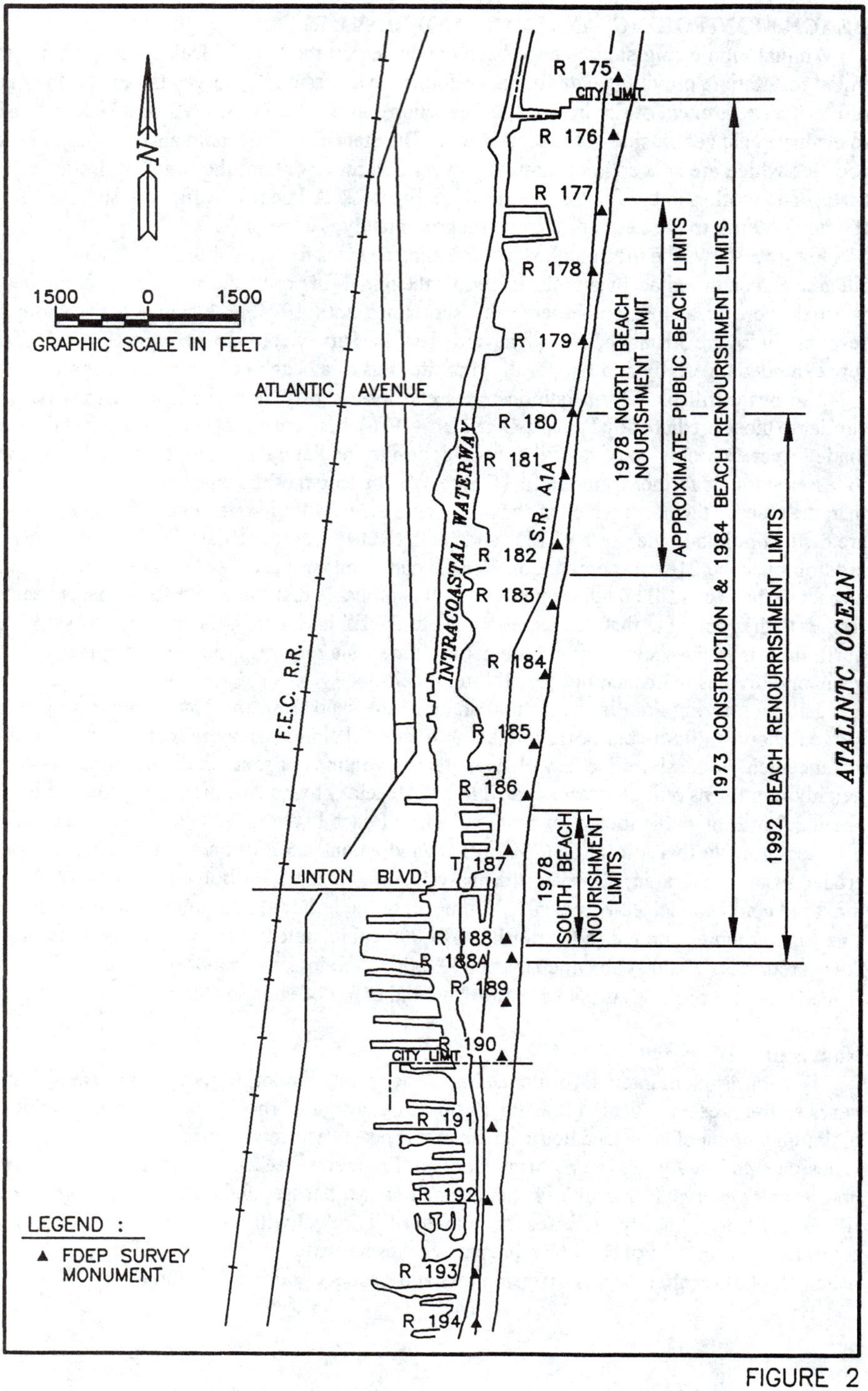
1500 0 1500
GRAPHIC SCALE IN FEET
N
ATLANTIC AVENUE
LINTON BLVD.
F.E.C. R.R.
INTRACOASTAL WATERWAY
S.R. A1A
CITY LIMIT
R 175
R 176
R 177
R 178
R 179
R 180
R 181
R 182
R 183
R 184
R 185
R 186
T 187
R 188
R 188A
R 189
R 190
R 191
R 192
R 193
R 194
1978 NORTH BEACH NOURISHMENT LIMIT
APPROIXIMATE PUBLIC BEACH LIMITS
1973 CONSTRUCTION & 1984 BEACH RENOURISHMENT LIMITS
1992 BEACH RENOURRISHMENT LIMITS
1978 SOUTH BEACH NOURISHMENT LIMITS
ATALINTC OCEAN
LEGEND :
FDEP SURVEY MONUMENT

FIGURE 2

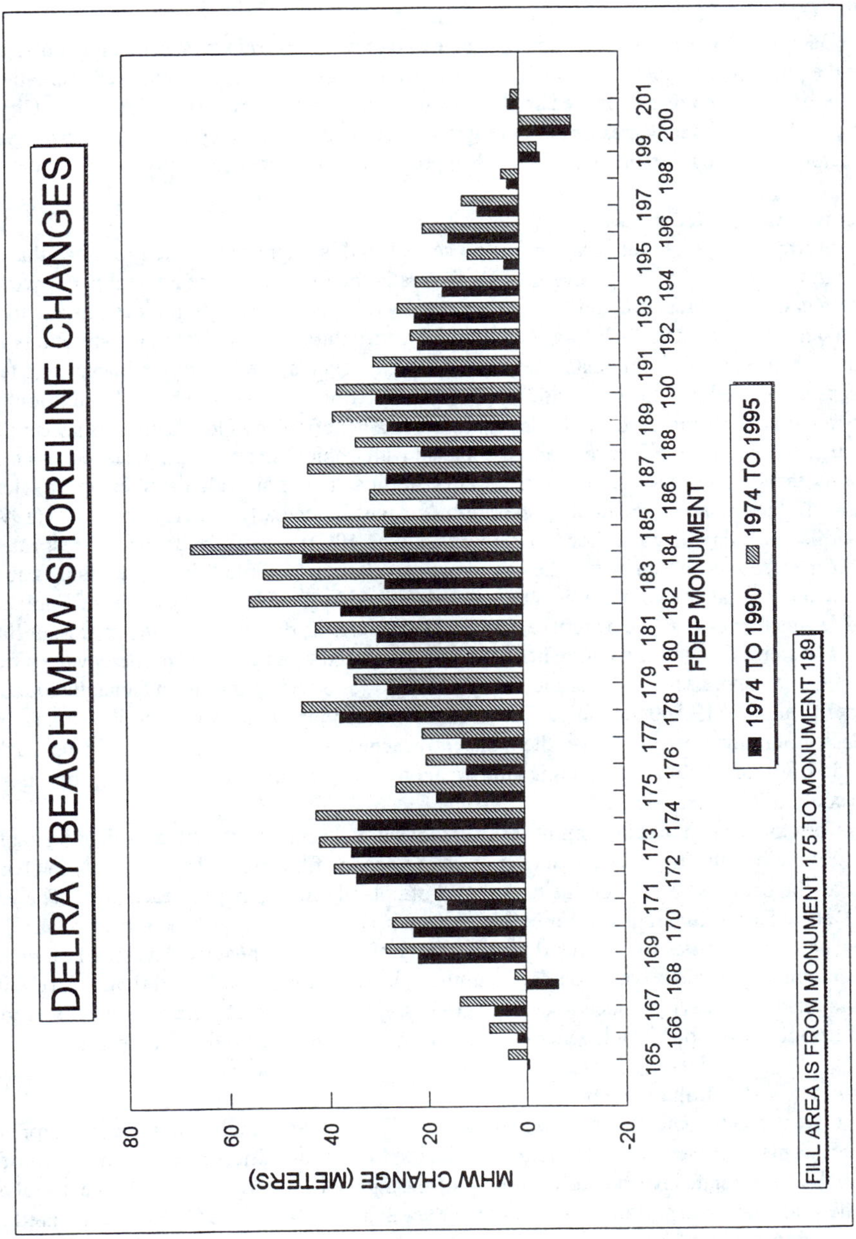
DELRAY BEACH MHW SHORELINE CHANGES
MHW CHANGE (METERS)
80
60
40
20
0
-20
165 166 167 168 169 170 171 172 173 174 175 176 177 178 179 180 181 182 183 184 185 186 187 188 189 190 191 192 193 194 195 196 197 198 199 200 201
FDEP MONUMENT
1974 TO 1990
1974 TO 1995
FILL AREA IS FROM MONUMENT 175 TO MONUMENT 189

Figure 3

PROJECT BENEFITS

Benefits of a beach nourishment project are attained in several ways. The most important are the protection of upland properties from coastal storm damage, the improved recreational value of the maintained beaches and the resulting increase in property values. The City of Delray Beach, who manages its beach program, requested that these benefits be determined to understand the positive impact of the beach program on the local economy.

Storm Damage Reduction

Storm damage prevention benefits presented in this paper are limited to those that are recognized by the U.S. Army Corps of Engineers in their evaluation of Federal participation. Storm damage reduction benefits are estimated by comparing the with project damage to the with out project damage. Damages are limited to structural damage to upland structures and roads, the loss of fill from upland areas from extreme events, the loss of land from long term recession of the beach profile, and the damage to existing coastal armoring. In addition, the prevention of coastal armoring is also included as a benefit. The Florida coast is susceptible to large storm surge and wave events associated with tropical storms and hurricanes. As these storms are acknowledged to be random events, storm damages are calculated in a probablistic method. The method of computing the storm damages is generally described by Bodge(1991). For this analysis, the computer program of Thomas(1990) was utilized. An interest rate of 7.625 percent was assumed for the economic calculations. While this interest rate is about twice the U.S. rate of inflation, it is the rate prescribed by the U.S. Army Corps of Engineers and results in conservative benefits estimates. All economic data is calculated in U.S. dollars.

The storm damage prevention benefits were calculated using land and property values from the 1995 tax appraisers roll. For the fill area only(R175 to R189), the storm damage reduction benefits are $3,819,800 annually. The annual storm damage reduction benefit of the Delray Beach nourishment projects on the adjacent beaches is $5,863,000. This can be further divided; $2,552,500 for properties in Gulfstream to the north and $3,310,500 for properties to the south in Highland Beach.

The storm damage reduction benefits are thus greater outside the fill area than inside the fill area. While this may appear counterintuitive since the fill material is placed only in Delray Beach, the benefits are higher due to the level of upland development in each area. The City of Delray Beach maintains a public beach which is accessed by a beach front road. The adjacent communities of Highland Beach and Gulfstream are heavily developed along the ocean with expensive private homes and multi-unit condominiums. In addition, the Town of Gulfstream was heavily eroded in the early 1970's. The accretion of sand on a heavily eroded and heavily developed beach generates significant storm damage reduction benefits.

Property Value Enhancement

Recent beach nourishments within the state of Florida have resulted in increased property values. This increase in property values results in beach front owners receiving higher profits from their real estate investments as well as increasing the local property tax. A recent analysis of the study area was performed by Regional Research Associates(1996). They compared the unit property values($/square meter) for similar properties located on the barrier island and on the mainland(Figure 1). The Delray Beach nourishment project has resulted in beachfront properties costing 15-20% more than mainland properties due to the maintained beach. This increase is $109.6 million within the fill area, $31.0 million and $58.4 million in Gulfstream and Highland Beach, respectively. The property value increase results in annual local taxes of $2.8 million, $1.1 million, and $1.9 million for Delray Beach, Gulfstream and Highland Beach, respectively.

Recreation Benefits

Recreational benefits attributed to a beach nourishment project are due to the increase in available beach space. The Corps of Engineers limits the recreational benefits by the demand to go to the beach and by the availability of public parking. The maintenance of the beach nourishment results in the annual benefits of $496,000 in Delray Beach (Coastal Planning & Engineering, Inc., 1991). Because there is no public parking or access within Gulfstream or Highland Beach, no recreational benefits accrue due to the project.

Benefit Summary

The Delray Beach beach maintenance program generates approximately $10.2 million annually in storm damage reduction benefits and recreational benefits to the City of Delray Beach and the adjacent towns of Highland Beach and Gulfstream. A common economic decision making principle is the ratio of annual benefits to annual costs. For this project, the benefit to cost ratio is 10.4. This excludes the increase in annual tax revenues due to the effect of the beach on property values.

CONCLUSIONS

The Delray Beach nourishment program has had a positive effect on a total of 11.3 km of shoreline as a result of filling about 4.3 km on a periodic basis. The analysis showed average shoreline advances of 28.3 meters to the north of the fill and 23.1 monitoring to the south. This was confirmed by an analysis of total volume change over the 23-year life of this project. The study accounted for 85% of the sand lost from the nourishment project.

As a result of this accretion outside of the fill area in neighboring cities, it was found that storm damage prevention benefits amounted to $3,819,800 within the fill area and $5,863,000 outside of the fill area but as a direct result of sand moving from Delray Beach to neighboring shorelines. The fact that there is a beach in place within the fill area provides an increase of 15 to 20 percent in property values.

ACKNOWLEDGMENT

The property value analysis data from Regional Research Associates is appreciated.

REFERENCES

Pelnard-Considere', R., "Essai de Theorie de l'Evolution des forms de Rivage en Plage de Sable et de Galets," *4th Journees de l'Hydraulique*, Les Energies de la Mer, Question III, No. 1, pp.289-298, 1956.

Bodge, K., "Damage Benefits and Cost Sharing for Shore Protection Projects", *Shore and Beach*, Vol 59, No. 2, pp. 11-18, April 1991.

Coastal Planning & Engineering, Inc., "Palm Beach County, Florida From Martin County Line to Lake Worth Inlet and from South Lake Worth Inlet to Broward County Line, General Design Memorandum Addendum for Third Periodic Nourishment at Delray Beach," report to the City of Delray Beach, 1991.

National Research Council, *Beach Nourishment and Protection*, National Academy Press, Washington D.C., 1995.

Regional Research Associates, "The Economic and Fiscal Impact of the Delray Beach Maintenance Program on Delray Beach, Gulfstream and Highland Beach Barrier Island Properties", unpublished report to the City of Delray Beach , 1996.

Thomas, C., "STORM02, Shore Protection Project Economic Analysis," copyrighted software, 1990.

CHAPTER 360

EFFECTIVENESS OF A COMBINED BEACH AND SHOREFACE NOURISHMENT ON THE ISLAND OF NORDERNEY/EAST FRISIA, GERMANY

Hanz D. Niemeyer[1], Ralf Kaiser[1] & Heiko Knaack[1]

Abstract

In order to improve the effectiveness of artificial nourishments a replenishment outside of the beach itself was taken into consideration for a prototype test. Contradictory to the earlier six conventional beach nourishments the material was deposited both on the shoreface and on the beach. The fate of the combined shoreface and beach nourishment was as well as that of the preceding conventional beach replenishment intensively monitored including field investigations of bathymetry, hydro- and sediment dynamics. On the basis of these data and additional application of mathematical models the effectiveness of this new type of nourishment has been evaluated in comparison with a conventional beach replenishment.

Introduction

The East Frisian island of Norderney experiences structural erosion in its western parts since the beginning of last century which is caused primarily by large scale tidal inlet and ebb delta migration [LUCK 1977; NIEMEYER 1995]and secondary by a gradient in long shore transport in the generally west-eastern directed drift [KRAMER 1960; NIEMEYER 1991]. In addition to earlier erected solid structures eight beach nourishments have been executed here since 1951/52 [KRAMER 1960; KUNZ 1991; NIEMEYER et al. 1995a]. Hydro- and morphodynamical boundary conditions have already been described in earlier publications [NIEMEYER 1991].

In 1992 a combined beach and shoreface nourishment was chosen as an alternative to conventional beach nourishments (fig. 1) [NIEMEYER et. al 1995a]. In order to evaluate its effectiveness an intensive monitoring programme including sedimentological, morphological and hydrodynamical site

----

[1]Coastal Research Station of the Lower Saxonian Central State Board for Ecology, P.O. Box 1221, 26534 Norderney/East Frisia, Germany

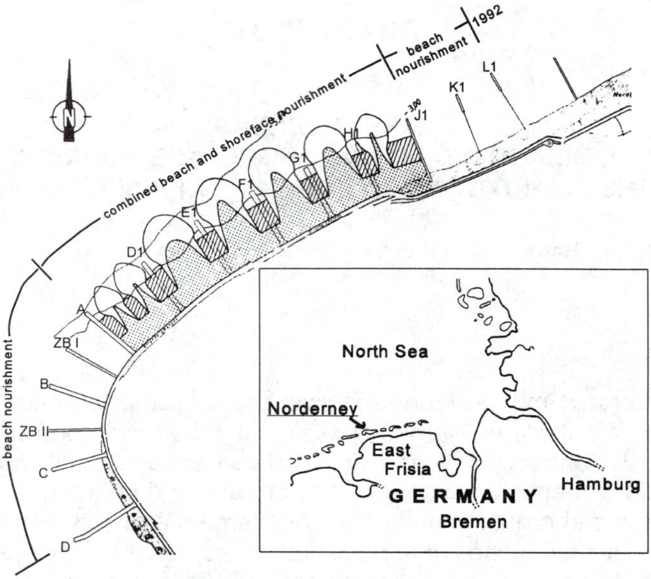

Figure 1. Island of Norderney/East Frisia in the southern North Sea; placements of the combined beach and shoreface nourishment in 1992 and of the preceding conventional beach replenishment in 1989

measurements has been carried out [KNAACK et al. 1996] as well as for the preceding conventional beach nourishment [KUNZ 1991; NIEMEYER 1991]. The investigations were carried out in cooperation with Danish and Dutch partners in the project "Innovative nourishment techniques evaluation (NOURTEC)" in the framework of the MAST II-programme of the European Communities and are additionally funded on a national basis. Results of the investigations carried out in respect to the combined shoreface and beach nourishment on the island of Norderney are subject of this paper focussing on its effectiveness by comparison with the preceding conventional one.

Design and initial behavior of the combined beach and shoreface nourishment

On the western and northwestern beaches of Norderney structural erosion is counterbalanced by artificial beach nourishments since 1951 [KRAMER 1960; KUNZ 1991; NIEMEYER 1991] though already between 1899 and 1909 dredging material has been dumped on the shoreface of the western beaches in order to improve beach stability [NIEMEYER et al. 1996]. In 1992 both nourishment techniques were combined for the replenishment of the northwestern beaches whereas on the western beaches only the beach was nourished (fig. 1). Basic idea of the deposition of material on the shoreface was to benefit from observed transport mechanisms explaining the pending of material between beach and shoreface [NIEMEYER 1991]. On the western beaches a deposition on the steep shoreface was considered as inconvenient; high initial irreversible losses into the close deep inlet channel were anticipated. A pure shoreface nourishment on the northwestern beaches was excluded because its execution by dumping or the rainbow-method on the shoreface between the groynes was regarded as dangerous for the executing vessels. Therefore the nourishment of the shoreface was carried out similar to that of the beach after finishing there the replenishment: the pipes were lengthened in the middle of the groyne fields and large humps were pumped into that area (fig. 2). It was anticipated that the local hydodynamical forces would create a dispersion of the sand stored in the artificial humps across the whole shoreface within a few weeks leading to its more or less uniform distribution.

Explanatory the reshaping of the hump in the groyne field E_1-F_1 is documented here by depth lines with reference to German datum = NN which is approximately equivelant to mean sea-level whereas mean high tide is about NN + 1,2 m and mean low tide about NN - 1,3 m (fig. 3): It is evident that the hump is lesser pronounced on the shoreface than on the beach. The highest initial losses occur therefore also on the intertidal beach and not on the shoreface which lower parts even benefit, but to a much lesser extent than the reduction of the hump's volume on the beach. A suitable explanation might be

Figure 2. Execution of the shoreface nourishment by pumping humps of sand form the shore

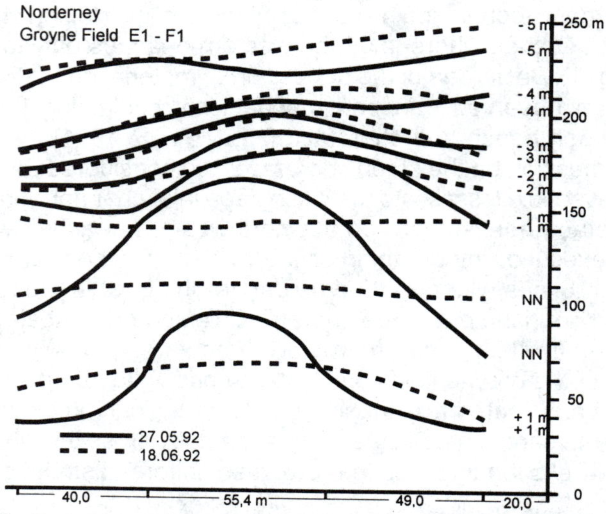

Figure 3. Reshaping of a sand hump after nourishment execution

that the geometry of the humps enforced the initilization of strong ripp currents evacuationg large amounts of the nourished material from the beach across the shoreface directly offshore being there unavailable for the anticpated pending between beach and shoreface.

Behavior of the combined beach and shoreface nourishment

Behavior of the total nourishment

The analysis of the behavior of the nourished beaches is focussed on the areas where the replenish-ment has been carried out both on the beach and on the shoreface. For this area also the corresponding data of the preceding conven-tional nourishment are taken into consideration in order to compare the effectiveness of both replenishment techniques. A comparison for the relative volumes per m beach length of the two replenishments above certain levels with reference to NN across the whole stretch between the groynes A and J_1 (fig. 1) highlights again that in the case of the combined nourishment the volumes on the shoreface below NN -1,5 m are not significantly large in relation to that ones on the beach, particularly not if compared with volumes of the preceding conventional replenishment (fig. 4a). This fact is evidently confirmed by a comparison of the volumes between the chosen levels (fig. 4b): the relative volumes on the beach of the combined nourishment are higher than those of the conventional one. But nevertheless the data highlight also that a comparison of the two distinct nourishment on the basis of their lifetime will remain insufficient, because the remaining relative volumes of the combined beach and shoreface nourishments were at the end of its lifetime in 1994 higher than those of the preceding conventional one in its final stage.

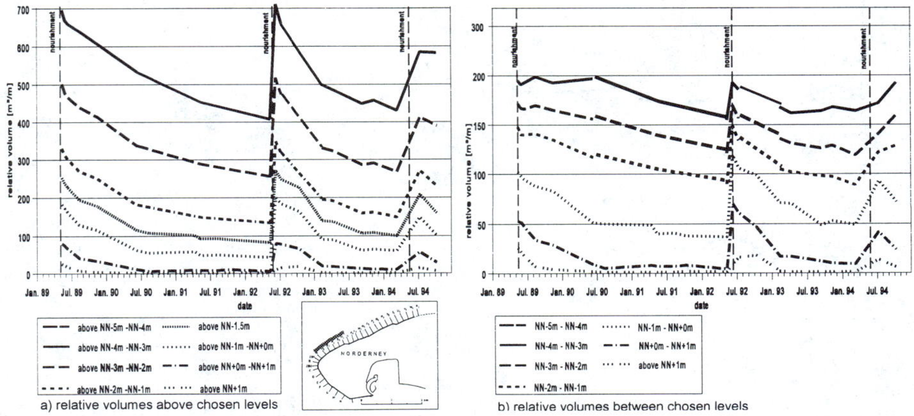

Fig. 4: Relative volumes per m beach length between the groynes A and J_1 after the conventional beach replenishment in 1989 and the combined beach and shoreface nourishment in 1992

Behavior of the nourishment in specific groyne fields

In order to get a deeper insight into the behavior of the two nourishments the development of their volumes in specific groyne fields has been evaluated with reference to the differences of shape and relative volumes. In the groyne field A-D_1 (fig. 1) at the downdrift edge of the combined beach and shoreface nourishment the relative volumes become between nearly all levels continuously smaller after the conventional beach nourishment of 1989. Only a slight increase occurred in the lower part of the shoreface in the very beginning and on the intertidal beach after a lifetime of more than two years (fig. 5). Contradictory the relative volumes of the lower shoreface increase during the whole lifetime of the combined nourishment and there is after initial erosion accretion on the upper part of the beach above mean sea-level during the first autumn and winter after implementation of the nourishment.

In the updrift succeeding groyne field D_1-E_1 (fig. 1) the intertidal beach looses relative volume immediately after the conventional beach replenishment in 1989 (fig. 6). After the first winter following the implementation of the nourishment here the relative volumes have again the same order of magnitude as before. They remain on the lower part of the intertidal beach then rather stable and decrease slightly in the final stage of the lifetime. The relative volumes on the shoreface increased during the first stage of the lifetime or experienced even a transient increase. After a period of about one year in coincidence with reduction of the relative volumes on the beach to the pre-nourishment level a continuos erosion of the shoreface started. It is itself suggesting that probable losses on the shoreface might have been compensated by eroded material from the beach as long as this source was available.

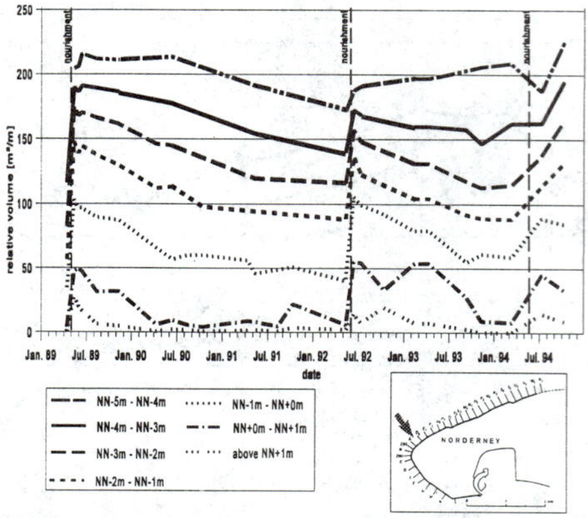

Fig. 5: Relative volumes per m beach length between the groynes A and D_1 between distinct levels after the conventional beach replenishment in 1989 and the combined beach and shoreface nourishment in 1992

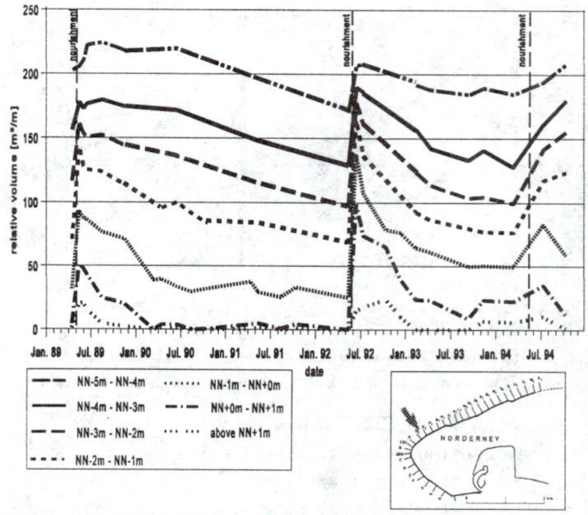

Fig. 6: Relative volumes per m beach length between the groynes D_1 and E_1 between distinct levels after the conventional beach replenishment in 1989 and the combined beach and shoreface nourishment in 1992

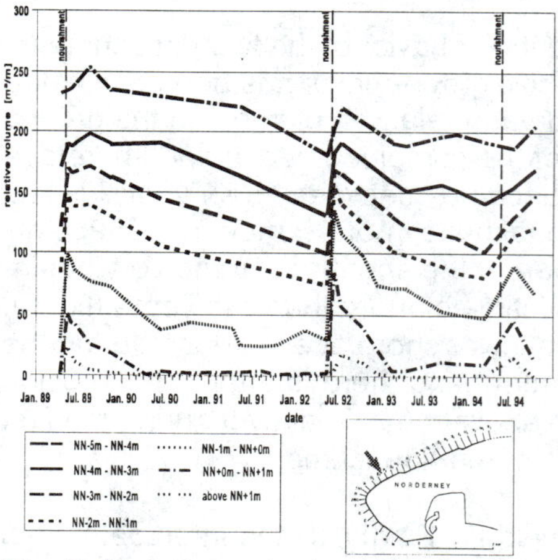

Fig. 7: Relative volumes per m beach length between the groynes E_1 and F_1 between distinct levels after the conventional beach replenishment in 1989 and the combined beach and shoreface nourishment in 1992

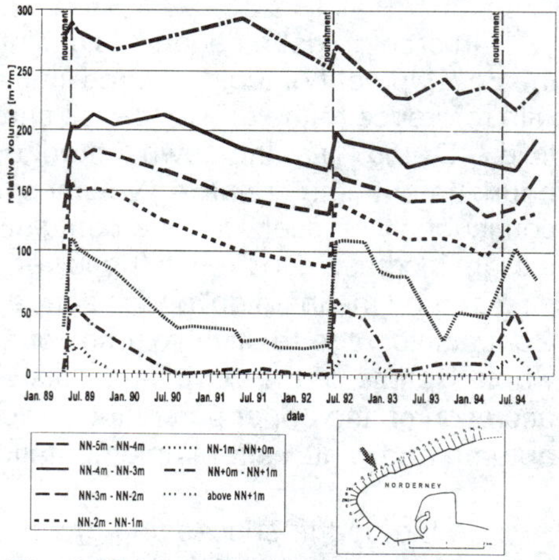

Fig. 8: Relative volumes per m beach length between the groynes F_1 and G_1 between distinct levels after the conventional beach replenishment in 1989 and the combined beach and shoreface nourishment in 1992

The nourished volumes in this groyne field have been larger for the subsequent combined nourishment leading to remarkable initial losses beside a slight increase on the highest part of the beach during the following five months. But after more than one year of lifetime a stabilization occurred on a

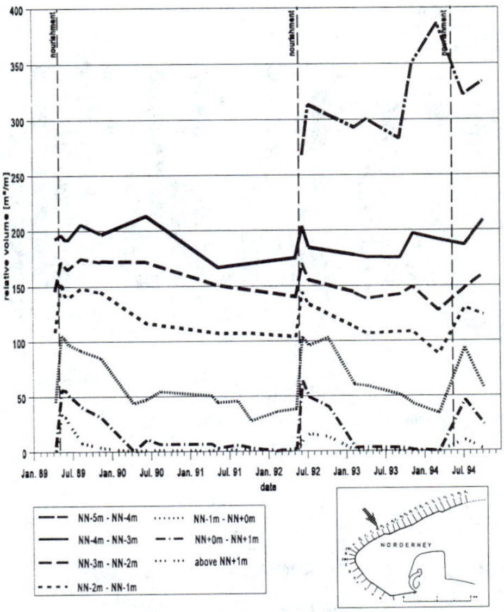

Fig. 9: Relative volumes per m beach length between the groynes G_1 and H_1 between distinct levels after the conventional beach replenishment in 1989 and the combined beach and shoreface nourishment in 1992

level significantly higher than the one before the implementation of the nourishment. Moreover the higher parts of the beaches recovered where the material is in respect of the primary aim to reduce wave loads more effective than anywhere else in the profile. This highlights once more that the morphological situation did not demand for the subsequent nourishment in 1994.

Similar results are gained for the next two groyne fields in updrift direction E_1 - F_1 and F_1 - G_1 (fig. 1) though less pronounced than for D_1-E_1: There are both on the upper part of the beach as well as on the lower shoreface transient phases with accretion after initial erosion (fig. 7 + 8). Particularly on the lower part of the beach the relative volumes are significant larger at the final stage of the combined beach and shoreface nourishment than before.

Contradictory the relative volumes after the combined nourishment in the groyne field G_1-H_1 follow another tendency. Whereas their values remain rather stable or increase even on the lower shoreface those on the beach reflect continuous erosion. Particularly there the preceding conventional beach replenishment has not experienced a higher amount of erosion though its lifetime lasted longer (fig. 9). This effect requires further explanation, because the behavior of the nourished beach and foreshore in this area has changed in comparison with preceding nourishments. Erosion has become relatively higher with respect to the rates in the groyne fields being located updrift.

Total balance

For all groyne fields the relative volumes between certain levels have been determined as well for the combined beach and shoreface nourishment in 1992 as for the preceding conventional beach replenishment in 1989 on the following basis: total amount before and after execution (fig. 10.1 + 10.2), losses and nourished volume (fig. 11.1 + 10.2). The relative volumes after the execution of the combined nourishment are higher than after the implementation of the preceding conventional one in the groyne field E_1-F_1, H_1-J_1 and particularly in D_1-E_1. According to the relative volumes before

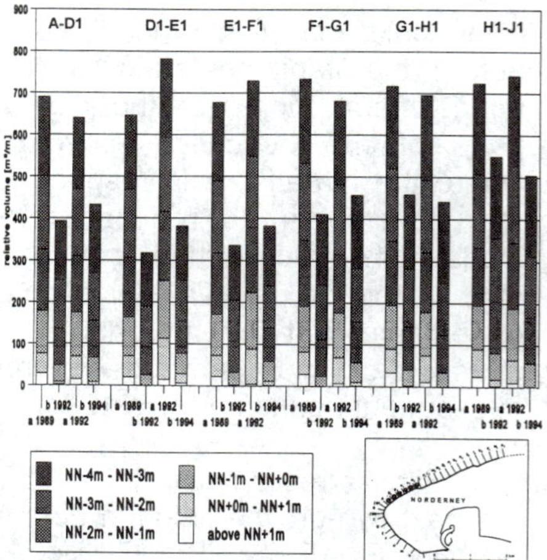

10.1 Groyne fields A-D1 to H1-J1 above NN -4m

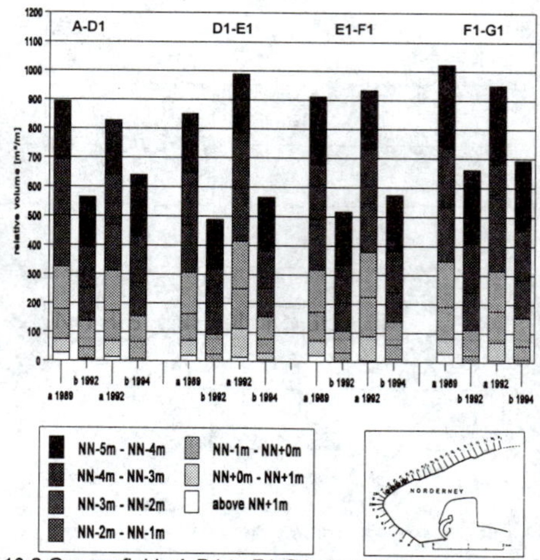

10.2 Groyne fields A-D1 to F1-G1 above NN -5m

Fig. 10: Relative volumes per m beach length between distinct levels after implementation (a) and at the end of the lifetime (b) for the nourishments in 1989 and 1992

implementation this was aimed at counterbalancing an anticipated higher rate of erosion in these areas (fig. 10.1 + 10.2). The same design procedure has been applied for the replenishment in 1989 (fig. 11.1) for the groyne field A-D_1 which has been rather successful if the remaining relative volumes in total are taken into consideration. In the groyne fields A-D_1, F_1-G_1 and G_1-H_1 the total relative volumes have been smaller for the combined nourishment in comparison to that one in 1989. The remaining relative volumes at the end of the lifetime of the combined nourishment are higher at that of the conventional replenishment between the groynes A and G_1 and smaller in the two following groyne fields G_1-H_1 and H_1-J_1 (fig. 10). Of high importance with respect to the primary aim to reduce wave attack at the seawall are the significantly higher remaining relative volumes on the beach and there particularly in the upper part above mean sea-level in the area between the groynes A and F_1 and to a lesser extent in the groyne field F_1-G_1. With respect to the necessity of a subsequent nourishment a comparison of the remaining relative volumes at the end of the lifetime of both nourishments makes evident that the minimum values existing before the combined beach and shoreface nourishment were still by about 25% and more exceeded in any place at the end of its lifetime (fig. 10)

Beside the groyne field A-D_1 the losses after the conventional beach nourishment exceed the nourished relative volumes (fig. 11). Though the losses in A-D_1 were not compensated by the combined nourishment it delivered on overish ot which is particularly evident for the lower shoreface where instead of losses a net win has taken place (fig. 11.2). The comparison

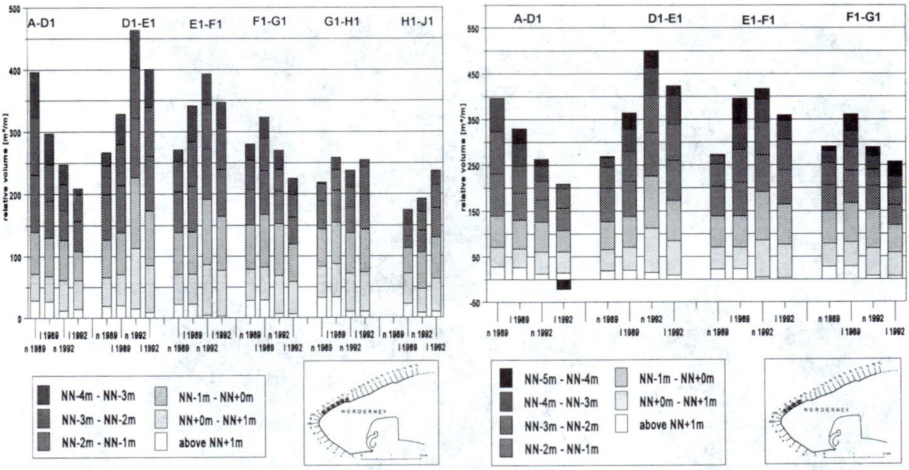

11.1 Groyne fields A-D1 to H1-J1 above NN -4m

11.2 Groyne fields A-D1 to F1-G1 above NN -5m

Fig. 11: Relative nourished volumes (n) and afterward losses (l) per m beach length between distinct levels for the nourishments in 1989 and 1992

between nourished relative volumes and subsequent losses makes evident that particular for the stretch between the groynes A and G_1 where for both nourishments the highest losses occur the combined beach and nourishment left over an overshot of material at the end of its lifetime. In the groyne fields A-D_1, F_1-G_1, G_1-H_1 (fig. 11) and if the lower part of the shoreface is included even in the groyne field E_1-F_1 the total losses after the implementation of the combined nourishment have been remarkably smaller than for the preceding conventional beach replenishment. Only in the groyne fields D_1-E_1 and H_1-J_1 the total losses after the execution of the combined nourishment have been larger than those of the preceding one (fig. 11.1). The difference decreases by about 50 % for the groyne field D_1-E_1 if the lower shoreface is taken into consideration (fig. 11.2). With respect to of the major purpose of artificial nourishments in this place to reduce wave attack on the seawall it is of high importance to check the losses on the upper part of the beach. The behavior of combined beach and shoreface nourishment in the area with the largest losses makes evident that particularly there the relation between nourished and lost relative volumes above mean sea-level is favourable in comparison to that of the precending conventional beach replenishment (fig. 11).

Impacts of changing local wave climate

measurements in the offshore area and on the shoreface of the island of Norderney highlighted enormous changes in wave damping on the ebb delta of the tidal inlet. In comparison to results from earlier investigations [NIEMEYER 1987] waves are lesser attenuated when propagating across the ebb delta, even for conditions with no remarkable set-up on mean high tide the

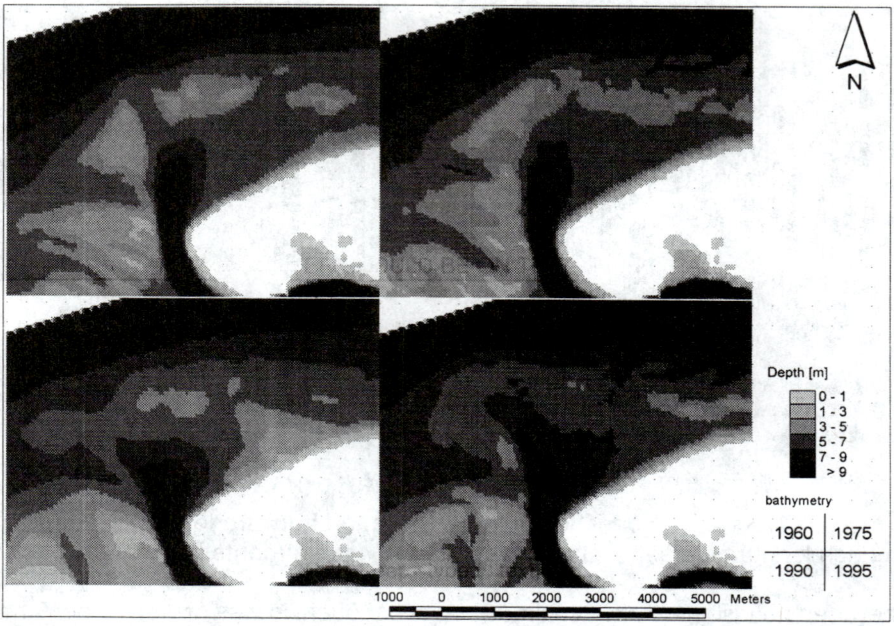

Fig. 12: Topography for the area seaward of the tidal inlet and the northwestern shore of the island of Norderney according to surveys in 1960, 1975, 1990 and 1995

shelter effect of the ebb delta for the northwestern beaches of the island of Norderney was remarkably reduced. In order to get a deeper and more systematic insight into background on these processes runs with the mathematical wave model HISWA [HOLTHUIJSEN & BOOIJ 1987; BOOIJ & HOLTHUIJSEN 1992] have been carried out for four distinct morphological situations of the area seaward of the tidal inlet and the northwestern shore of the island of Norderney (fig. 12): the bathymetries of 1960, 1975, 1990 and 1995 were used as a basis for model topography. The model HISWA has already been successfully applied to this specific area and similar ones and is regarded as a suitable tool for the anticipated purpose [DEN ADEL et al. 1991; NIEMEYER et al. 1995b].

Already the simple comparison of the topography for the disinct four situations makes signifkant morphological changes evident (fig. 12): Whereas the ebb delta is becoming even more pronounced after 1960 with a climax in 1975 the later surveys in 1990 and 1995 make a reduction of the shallows with heights above NN - m evident. Accompanied is this process by a migration and seaward directed deepening of the main channel of the inlet between 1990 and 1995: Both morphological changes enhance onshore wave penetration in direction of the island's northwestern shores.

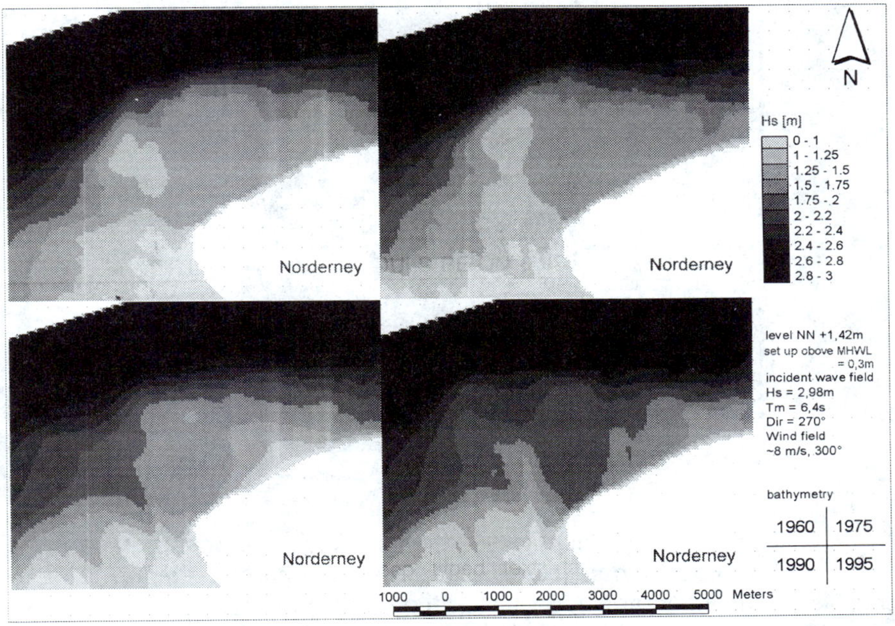

Fig. 13: Significant wave heights for the area seaward of the tidal inlet and the northwestern shore of the island of Norderney according to the bathymetry in 1960, 1975, 1990 and 1995 (computed with HISWA)

The model runs confirm this first estimate with respect to penetration of higher waves into the northwestern shoreface of the island of Norderney (fig. 13): For the situations of 1960 and 1975 no significant changes occur. But already for the topography of 1990 higher waves with 1,75 m $\leq H_s \leq$ 2,00 m propagate across the ebb delta though not appearing nearshore. For the topography of 1995 waves with heights with this order of magnitude occur on a large part of the shoreface. This increase in nearshore wave energy is not only confirmed by prototype data; the measured wave heights on the shoreface are even higher than those being computed with the model for the same offshore conditions. Additional investigations have shown that this effect does not increase for storm surges with higher set-up and higher waves. The higher waves enforce also a more intensive wave energy dissipation due to breaking in the ebb delta area.

In order to make the effect of this change in local wave climate for the dynamics on the northwestern shoreface and beaches of the island of Norderney more evident the computed significant wave heights for all four morphological situations are compared at 26 selected points at the edge of shoreface and beach parallel to the island's northwestern shore (fig. 14). For the situations of 1960 and 1975 there are no remarkable changes in wave climate. The significant wave heights are of the same order of magnitude on

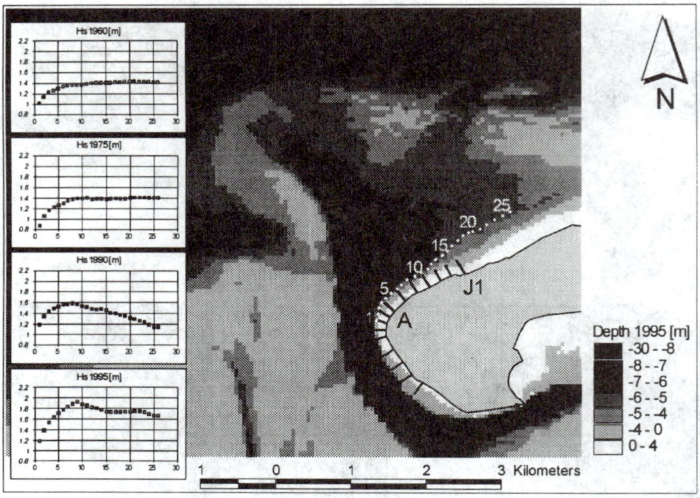

Fig. 14: Significant wave heights at the edge of shoreface and beach parallel to the northwestern shore of the island of Norderney according to the bathymetry in 1960, 1975, 1990 and 1995 (computed with HISWA)

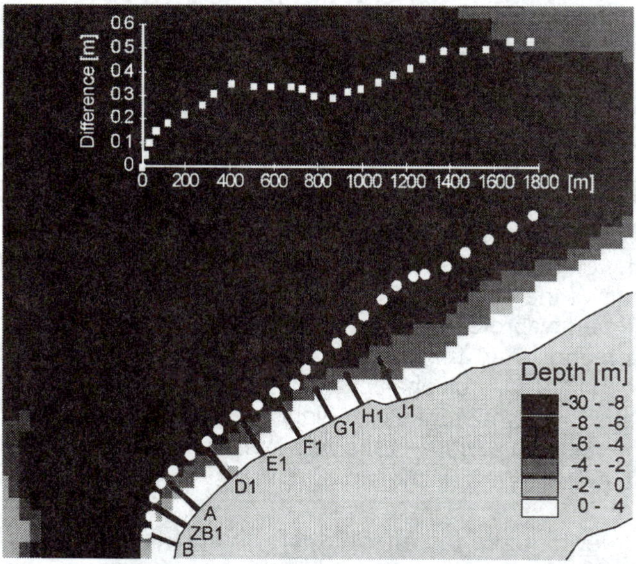

Fig. 15: Differences of significant wave heights at the edge of shoreface and beach parallel to the north-western shore of the island of Norderney according to the bathymetry in 1990 and 1995 (computed with HISWA)

the whole stretch of the northwestern shoreface. In 1990 there is an increase of wave heights particularly in the stretch between the groynes A and F_1 but they decrease further downdrift. In 1995 wave height have increased in all places downdrift of groyne A (fig. 15). This comparison highlights that the combined nourishment experience for the same offshore conditions higher wave loads with stronger driving forces than the preceding conventional one. This unfavorable boundary conditions are another indication for the superior effectiveness of the combined nourishment. The change in wave climate explains also the higher losses occurring after the implementation of the combined beach and shoreface nourishment in the area downdrift of groyne H_1.

Conclusions

The comparison between the combined beach and shoreface nourishment and the preceding conventional replenishment with respect to their effectiveness has lead to the following conclusions:

1. Lifetime is in this case no suitable measure for comparing the effectiveness because the remaining relative volumes are for the combined nourishment in any place larger than the accepted minimum values for the preceding conventional for which no damages of the seawall occurred.

2. High initial losses of the combined nourishment might be credited to the chosen execution method with the deposition of humps which might have provoked strong ripp currents in the groyne fields. A less concentrated dumping of supply material on the shoreface could improve the measure.

3. Relatively higher losses in the downdrift part of the combined nourishment in comparison to the preceding conventional one do not occur due to the changes in the execution method. They are enforced by changes in local wave climate due to migration of the ebb delta leading to an increase of the driving forces for longshore currents and littoral drift.

4. The application of reliable mathematical wave models in combination with available actual bathymetries is suitable to make changing wave loads evident allowing an optimized adaption of the spatial distribution of relative nourished volumes in distinct groyne fields.

5. Particularly in the area with high rates of erosion the behavior of the combined nourishment delivered indications for the anticipated pending of the sand between shoreface and beach leading there to a recovery of the upper part of the beach which is beneficial with respect to the major aim of these nourishments to enforce wave energy dissipation in front of the landward placed seawall.

6. An further optimization of artificial nourishments on the northwestern shore of the island of Norderney requires a definition of acceptable wave loads for the existing seawall in order to evaluate the ultimately necessary minimum of relative volumes on the beach for sufficient wave energy dissipation.

Acknowledgement

This investigatitions have been carried out in the framework of the NOURTEC project of THE COMMISSION OF THE EUROPEAN COMMUNITIES (Contract MAS-CT93-0049) and additionally benefitted from national funds of the GERMAN FEDERAL MINISTRY FOR EDUCATION AND RESEARCH and the LOWER SAXONIAN MINISTRY FOR THE ENVIRONMENT. The authors have substantially benefitted from the joint cooperation in this European project with their Danish and Dutch partners. They are also very grateful for the assistance and support by Mr. Günther Brandt, Detlef Glaser, Thomas Hartkens, Peter Heddinga, Holger Karow, and Georg Münkewarf.

References

den Adel, J.D., H.D. Niemeyer, A.F. Franken, N. Booij, J. Dekker, J.A. Vogel [1991]: Wave model application in a Wadden Sea area. Proc. 22nd Int. Conf. o. Coast. Eng. Delft, ASCE, New York

Booij, N. & L.H. Holthuijsen [1992]: HISWA user manual. Prediction of stationary, short-crested waves with ambient currents. TU Delft, Faculteit d. Civ. Techn.

Holthuijsen, L.H. & N. Booij [1992]: A grid model for shallow water waves. Proc. 20th Intern. Conf. o. Coast. Eng. Taipei, ASCE, New York

Knaack, H., R. Kaiser, H.D. Niemeyer [1996]: Behaviour and effectiveness of the NOURTEC experimental test sites - Norderney case. Rep. MAS II-CT93-0049 (unpubl.)

Kramer, J. (1960): Beach Rehabilitation by Use of Beach Fills and Further Plans for the Protection of the Island of Norderney. Proc. 7th Conf. o. Coast. Eng. Richmond, ASCE

Kunz, H. [1991]: Artificial Beach Nourishment on Norderney, a Case Study. Proc. 22nd Int. Conf. o. Coast. Eng. Delft, ASCE, New York

Laustrup, C., H. Toxvig Madsen, P. Sörensen, I. Bröker [1997]: Comparison of beach and shoreface nourishment-Torsminde Tange, Denmark. Proc. 25th Intern. Conf. o. Coast. Eng. Orlando/Fl., USA, ASCE, New York

Luck, G. [1977]: Long-termed development of tidal inlets between Eastfrisian islands. Proc. 15th Intern. Conf. o. Coast. Eng. Honolulu, ASCE, New York

Mulder, J.P.M., J. van de Kreeke, P. v. Vessem [1995]: Experimental shoreface nourishment, Terschelling (NL). Proc. 24th Intern. Conf. o. Coast. Eng. Kobe/Japan, ASCE, New York

Niemeyer, H. D. [1987]: Changing of wave climate due to breaking on a tidal inlet bar. Proc. 20th Intern. Conf. o. Coast. Eng. Taipei, ASCE, New York

Niemeyer, H.D. [1991]: Field measurements and analysis of wave-induced nearshore currents. Proc. 22nd Intern. Conf. o. Coast. Eng. Delft/The Netherlands, ASCE, New York

Niemeyer, H.D. [1995]: Long-term morphodynamical development of East Frisian Islands and Coast. Proc. 24th Intern. Conf. o. Coast. Eng. Kobe/Japan, ASCE, New York

Niemeyer, H.D. & R. Kaiser [1994]: NOURTEC-design report: Combined shoreface and beach nourishment, island of Norderney, East Frisia. Com. Europ. Comm. - MAST II-Progr., MAS 2-CT93-0049

Niemeyer, H.D., R. Kaiser, J.D. den Adel [1995b]: Anwendung des mathematischen Seegangsmodells HISWA auf Wattenmeerbereiche. Die Küste, H. 57

Niemeyer, H.D., E. Biegel, R. Kaiser, H. Knaack, C. Laustrup, J.P.M. Mulder, R. Spannhof, H. Toxvig [1995a]: General aims of the NOURTEC project -effectiveness and execution of beach and shoreface nourishments-. Proc. 4th Conf. o. Coast. & Port Eng. i. Develop. Countr., Rio de Janairo/Brazil

Niemeyer, H.D., H. Rohde & H. Eiben [1996]: History and heritage of German coastal engineering. In: N. Kraus (ed.): History and heritage of coastal engineering. ASCE, New York

Lake Worth sand bypass plant, Palm Beach, Florida.
Photo courtesy of Billy L. Edge.

PART VI

Case Studies

Hillsboro Inlet, Hillsboro, Florida. Photo courtesy of
Coastal Systems International, Inc.

CHAPTER 361

Nearshore Berm Performance at Newport Beach, California, USA

Chuck Mesa[1]

Abstract

In 1992, approximately 1,276,000 cubic yards of littoral material was hydraulically placed into a berm configuration in nearshore water depths at Newport Beach, California, USA. A monitoring program conducted included directional wave measurements, beach and bathymetric profiles, surficial sediment sampling, and controlled aerial photography. Analysis of the bathymetric profiles collectively indicate the berm is experiencing a shoreward-directed dispersal; there is no evidence suggesting offshore or alongshore directed berm movement. Sediment physical characteristics and grain size distribution indicating the berm effects on the seabed are discussed. Recently proposed models for berm stable/active categorization and migration rate are examined. Surfing at the site was significantly enhanced; the berm created breaking wave conditions never before experienced at this location.

Introduction

This paper describes a nearshore disposal monitoring project at Newport Beach, California, USA. The nearshore placement activity was conducted during the period Jan-Nov 1992. Approximately 1,276,000 cubic

---

[1] U.S. Army Corps of Engineers, Los Angeles District, P.O. Box 2711, Los Angeles, CA, 90053, USA

yards of suitable littoral material was hydraulically placed into a berm configuration in water depths ranging from -5 ft to -30 ft MLLW (Note: All elevations and/or depths cited herein are in feet referred to Mean Lower Low Water (MLLW)). These depths are both within and outside the day-to-day surf zone. The purpose of the monitoring plan was to determine the fate of the disposal berm by providing a quantitative and qualitative description of the movement of disposal material in the cross-shore and alongshore directions. The material was obtained as the result of the Lower Santa Ana River flood control channel expansion project.

The project site is located in the City of Newport Beach, California, USA. The shoreline immediately landward of the disposal area is a mixed use area of recreation and residential development. Coastal structures in the project area include three shore perpendicular jetties at the mouth of the Santa Ana River, and a field of eight groins. The disposal site was selected to confine the material between the river mouth system and the groin field.

Monitoring Plan

The monitoring program conducted during the period Dec 91 - May 95 included directional wave measurements, beach and bathymetric profiles, surficial sediment sampling, and controlled aerial photography of the project area (Fig 1).

The nearshore directional wave climate was monitored by installation of a S_{xy} slope array wave recording instrument during the period 3/92 - present. The slope array was placed offshore of Huntington Beach approximately 1.5 miles northwest of the project site at a water depth of 33 ft.

The 24 beach and bathymetric profile lines were established as follows: 8 lines placed throughout the groin field system; 11 higher spatial resolution lines placed at 300 ft intervals over the disposal area; 1 line placed over the Santa Ana River delta; 3 lines spaced at even intervals; 1 line (control) placed approximately 1.5 miles northwest of the project area. The control line is not expected to be influenced by movements of the nearshore disposal material, and acts as an indicator for seasonal or gross shoreline movements. Profiles were measured from a fixed point on the backshore to depths of greater than -40 ft. Profile survey dates were: 12/91 pre-construction; 5/92 1st interim; 7/92 2nd interim; 11/92 1st post-construction; 3/93 2nd post-construction; 5/93 3rd post-construction; 1/94 4th post-construction; 11/94 5th post-construction; and 5/95 6th post-construction.

Sediment samples were obtained by surficial grab sampling techniques along 6 profile lines (Line 1, Line 5, Line 9, Line 14, Line 19, Line 24) at 8 elevations corresponding to +6 ft, 0 ft, -6 ft, -12 ft, -18 ft, -24 ft, -30 ft, and -36 ft.

Controlled aerial photography flown over the project area included: 12/91, 6/92, 11/92, 12/92, 3/93, and 5/93.

Wave Climate Description

Directional wave measurements indicate a multi-directional, seasonally dependent, sea/swell climate is experienced at the project site. The nearshore slope array recorded 6905 observations during the measurement period. The sample histogram of wave heights included a maximum measured wave of 10.6 ft; the mean, mode, and standard deviation are given as 2.7 ft, 2.5 ft and 0.9 ft respectively. The sample histogram of spectral peak period indicates double peaks at T=7 sec and T=15 sec. In southern California this is typically associated with sea and swell respectively (not withstanding the respective classical definitions dependent on the point of origin). The combined sea/swell climate was assumed a priori; a double peaked distribution of wave climate was expected.

Profile Comparison

Repeated bathymetric mapping of the cross-shore profiles indicates the nearshore disposal material formed a significant clearly distinguishable feature on the local seabed relative to the pre-project profile.

Fig 2 illustrates a time series of the cross-shore distribution of berm material at Line 19 (located at the disposal area northwestern boundary). Fig 2 indicates the berm is distributed along the profile approximately between depths of -2 ft and -32 ft. The berm toe appears located at the -32 ft contour; the berm indicates a relatively sharp relief on the seaward side and is diffused gradually towards shallower water. The berm maximum vertical relief (e.g. "centroid") appears located at a depth of -29 ft.

Fig 3 is a time series comparison of the profile *differences* developed by subtraction of the pre-project profile from the post-construction profiles. Inspection of Fig 3 clearly illustrates the berm cross-shore behavior. The berm centroid is located 1900 ft from the survey baseline, or about 1200 ft from the beach. The berm had a maximum vertical relief of 14.5 ft from the pre-project seabed, diminishing successively to 11.5 ft, 9.0 ft, 8.5 ft, and finally diminishing to 8.0 ft by the sixth survey episode (31 months). Thus, the berm vertical relief appears to diminish rapidly initially, with the deflation rate decreasing over time. Close inspection of Fig 3 indicates as the berm crest erodes, the crest material is sheared off in the *landward* direction. The survey data indicates there is significant accretion of material throughout the region between the berm and the foreshore. This landward directed movement is particularly indicated by the accretion of material in a bar formation as indicated in the 5/93 survey episode. Further inspection of Fig 3 indicates the berm centroid location is stationary, and not migratory as the berm erodes. There is no indication the berm migrates as a solitary feature; migration is measured as a function of the crest material dispersal. This observation is relevant to the following discussion of berm migration rate.

There is no evidence in the data to suggest offshore or alongshore directed movement of the berm material. Analysis of all measured profiles indicates no signal within the resolution of the survey data which substantiates seaward berm movement. There is little or no indication of significant alongshore movement of the berm. Due to the prevailing wave climate, net sediment transport is a priori expected to be in the southeasterly direction. Analysis of the profile closest to the disposal area southeastern boundary (Line 8 within the groin field) indicates no signal to substantiate southeasterly alongshore movement. Likewise, there is no significant indication of berm migration in the northwesterly direction. This is unambiguously no movement across the northwesterly control line. The profile closet to the northwestern boundary exhibits some accretionary signal, however, that was determined to be remnant storm flow material from the Santa Ana River.

Sediment Physical Characteristics and Distribution

The sediment characteristics of the nearshore disposal material are known from pre-project geotechnical analyses. The material was approximately 83% sand and 17% fines (defined as passing the #200 U.S. Std sieve), with a d_{50} median diameter of approximately 0.27 mm which is classified as fine sand under the Unified Soils Classification.

Fig 4 illustrates a time series comparison of sediment grain size distribution along a selected profile (Line 19) and depth (-24 ft). The pre-project (12/91) sediment distribution is a poorly graded fine sand material, approximately 17% fines, with a d_{50} median diameter of 0.09 mm. Fig 4 indicates a well behaved coarsening of the grain size distribution between the 1st, 2nd, and 5th survey episodes. At the 5th survey episode (11/94) the sediment achieved its coarsest distribution, a poorly graded fine sand, 1% fines, with a d_{50} median diameter of 0.22 mm. There appears to be a rebound of the grain size distribution at the 6th survey episode.

Fig 5 illustrates a cross-shore time series comparison of d_{50} median diameters at Line 19. The data is fairly well behaved and indicates the coarsest fraction is located on the upper portions of the profile while the fine grained fraction is dispersed towards deeper water. The berm material is coarser than the pre-project seabed and appears to be maintaining its position on the seabed.

Fig 6 illustrates a cross-shore time series comparison of percent fines at Line 19. The data clearly shows the percent fines is significantly higher in water depths greater than -24 ft. Over time the percent fines remain less than or approximately equal to the pre-project seabed condition.

Berm Categorization

Hands & Allison (1991) present a empirically based method to categorize stable or active nearshore berms. The method calculates the long-term distribution of wave-induced, near-bed velocities as an appropriate criterion to discriminate between stable and active behavior. Used conjunctively with the Hallermeier proposed inner and outer limits of profile zonation, the method successfully discriminates between stable and active berms in the "buffer" zone bounded by the inner and outer limits.

The results of the present analysis, shown in Fig 7, delineates the regions of berm stability or activity based on the wave-induced near-bed velocities. The berm in the present analysis is classified as *stable*, particularly at the 75-95 percentile which Hands & Allison describe as possessing the best correlation with berm behavior. It is worth noting that the present berm velocity distribution crosses the boundary delineating the stable-active regions, and also tends to converge towards the active classification near the 99-percentile (indicating a tendency for berm movement during extreme events).

The results using the Hallermeier profile zonation limits, shown in Fig 8, indicate the berm falls in the "buffer" zone. Hands & Allison further showed that berms stable in the buffer zone were less than 50% below the outer limit, while active berms were more than 50%

above the outer limit. Thus, with this methodology the Newport Beach berm gains an *active* classification. This directly contradicts the previously determined *stable* classification based on near-bed velocities. However, it should be noted that the berm could be considered weakly active based on it's relative position as slightly greater than the 50% outer limit within the buffer zone.

Berm Migration Rate

Douglass (1995) proposed a model for landward migration rate estimation of nearshore sand berms. The model is formulated on the assumption that net shoreward sediment transport is due primarily to the velocity asymmetry characteristics of finite amplitude waves. The model parameters can be estimated using a joint probability density function (height-period, direction is assumed onshore) of site-specific wave climate data. The "expected value" of berm migration rate can be estimated as a function of depth.

The results estimate the berm migration rate to be approximately 100 ft/yr (Fig 9). Douglass provides no guidelines for choosing a depth from the functional relationship to use for berm migration estimation. An observation made during this analysis indicates that the results are critically dependent on the depth chosen. The depth used in the current study is the depth at the berm center of mass. However, this depth no longer exists after berm placement. Since the method is based on the velocity asymmetry of finite amplitude waves, the depth used should be the depth at the berm crest. This appears to be an inherent inconsistency, yet still the model appears to produce results that are in reasonable agreement with observed values and are the correct order of magnitude. The decaying semi-logarithmic function of depth appears intuitively correct. However, with decreasing depths at the shore, the model expected value approaches unrealistically high values. The model also assumes landward migration only and does not consider offshore movement.

Surfing

The nearshore berm disposal project temporarily significantly enhanced conditions for the recreational sport of surfing. This study photographically recorded breaking waves never before experienced at this location. The pre-project wave-breaking condition typically consisted of long crested swell waves propagating over straight and parallel contours with wave breaking occurring relatively close to shore. Typical pre-project surfing can be described as beach-break resulting in short, fast spilling/plunging "lefts" and "rights".

The hydraulic placement techniques employed by the construction contractor resulted in a series of small mounds constructed throughout the disposal area footprint. These disposal mounds were dispersed cross-shore throughout the nearshore zone, from near the typical breaker line to upwards of 1,200 feet offshore. Refraction effects focused the wave energy directly onto the individual mounds, which were effectively a severe perturbation to the straight and parallel bathymetric contours, tending to scatter the local wave field (Photo 1). The result was many wave "peaks" throughout the nearshore zone. Due to the extremely sharp relief of the mounds, waves tended to shoal extremely quick, increasing the local wave steepness, and creating a very "fast", "hollow" surfing wave (Photo 2). The surf lasted throughout the period of nearshore disposal, and several months thereafter. The surf was best at lower tidal levels with longer period swell.

Conclusions

In 1992, approximately 1,276,000 cubic yards of littoral material was hydraulically placed into a nearshore berm configuration at Newport Beach, California, USA. The monitoring program established included directional wave measurements, beach and bathymetric profiles, surficial sediment sampling, and controlled aerial photography. Bathymetry measurements indicate the berm is eroding by undergoing dispersal of the crest material shoreward. The berm base appears stable. There is no evidence suggesting either offshore or alongshore movement of the berm material. Sediment physical characteristics indicate the coarser sediments are remaining within the active littoral system, while the berm is not adversely affecting the percentage of fines. Results using recently proposed models for berm stable/active categorization and migration rate are mixed. The berm gained both a stable and active classification. A calculated migration rate of 100 ft/yr appears correct in order of magnitude, but some question remains on functional application based on selection of appropriate water depth. Surfing at the site was significantly improved. Berm influence on the local wave field created shoaling and breaking conditions never before experienced at this location.

Acknowledgments

The author wishes to acknowledge the Commander, U.S. Army Corps of Engineers Los Angeles District, for authorizing publication of this paper. It was prepared as a result of the Lower Santa Ana River Flood Control Improvement study. I would like to thank the following individuals for their assistance and participation in this project: Dr. Scott L. Douglass at the University of South Alabama and Mr. Edward B. Hands at the Coastal Engineering Research Center. A special thanks is expressed to the boys at The Frog House Surf Shop.

References

Douglass, Scott L. (1995). "Estimating Landward Migration of Nearshore Constructed Sand Mounds." *Journal of Waterway, Port, Coastal, and Ocean Engineering.*, ASCE, New York, N.Y., Vol 121, No. 5.

Hands, E.B., and Allison, M.C. (1991). "Mound Migration in Deeper Water and Methods of Categorizing Active and Stable Depths." *Proceedings, Coastal Sediments 91 Conference.*, ASCE, New York, N.Y., Vol. 2, 1985-1999.

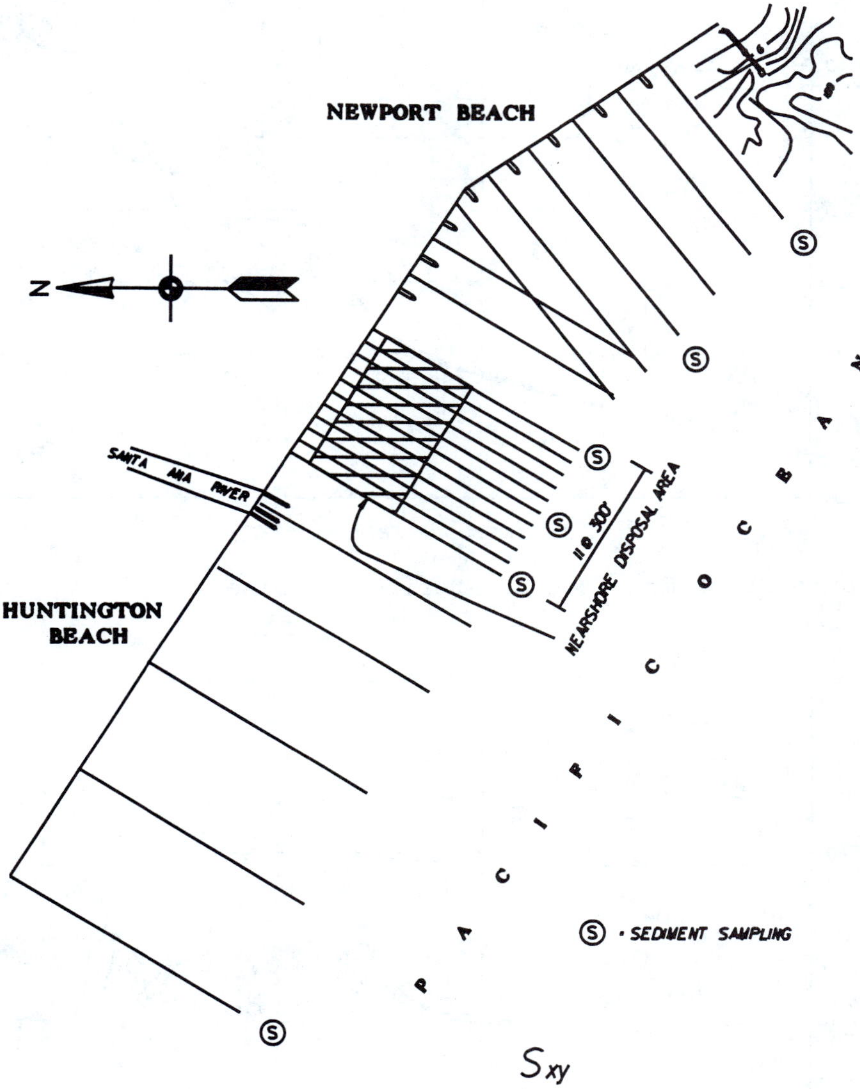

Figure 1. Monitoring Program Schematic.

Note: Survey lines numbered sequentially 1-24 from southeast to northwest.

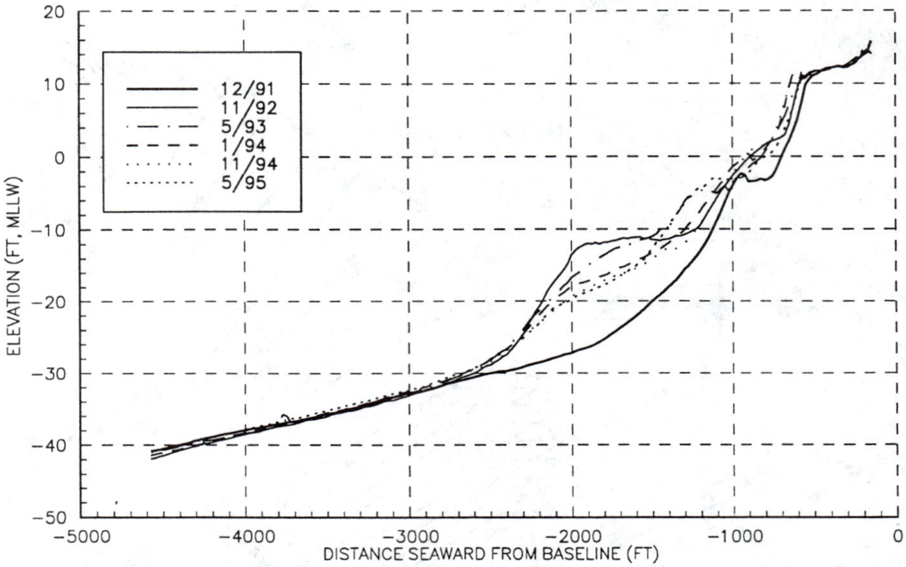

Figure 2. Comparison of Profiles at Line 19.

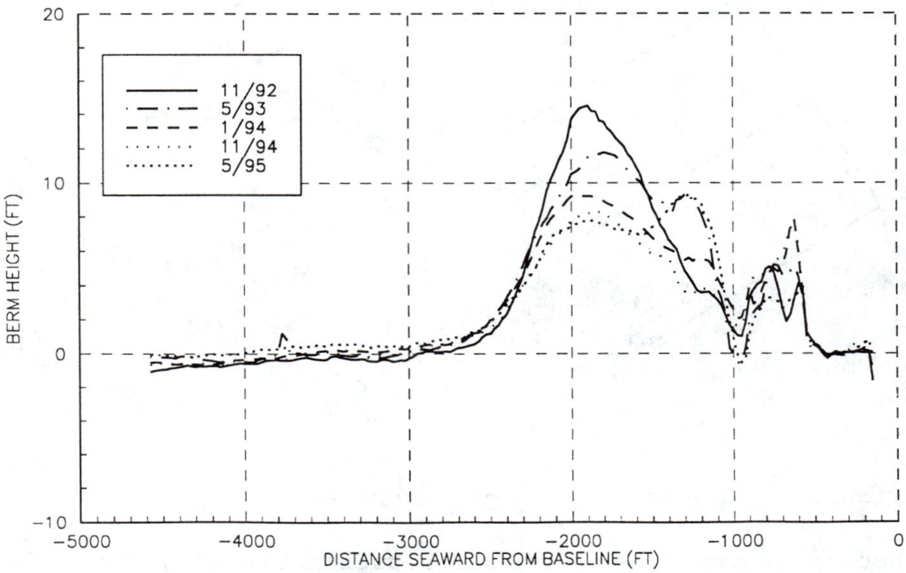

Figure 3. Comparison of Profile Differences at Line 19 Relative to Pre-project Condition (12/91).

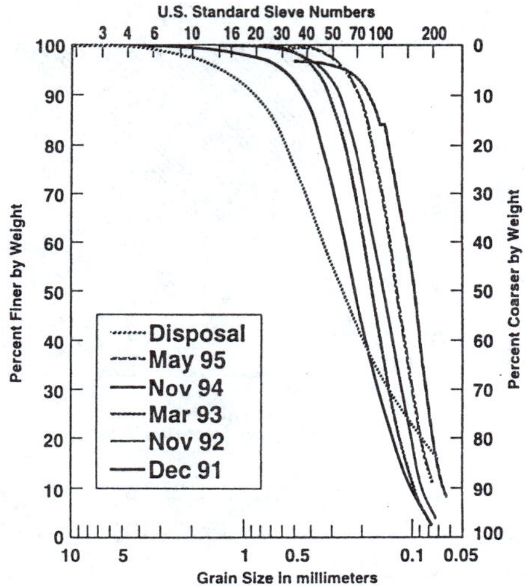

Figure 4. Grain Size Distribution Comparison, Line 19, d = -24 ft

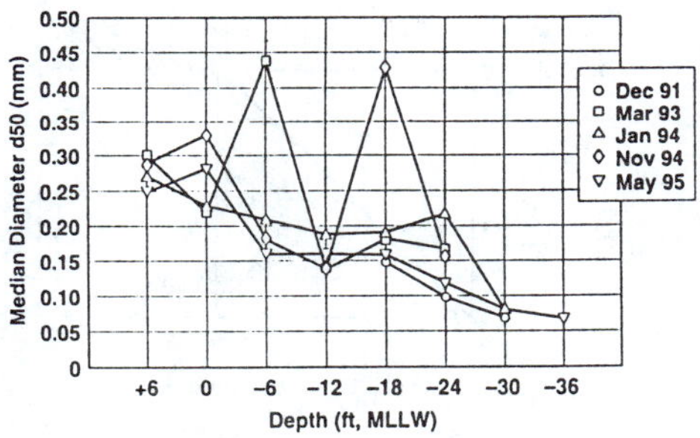

Figure 5. Cross-shore Comparison of d_{50} Median Diameter, Line 19

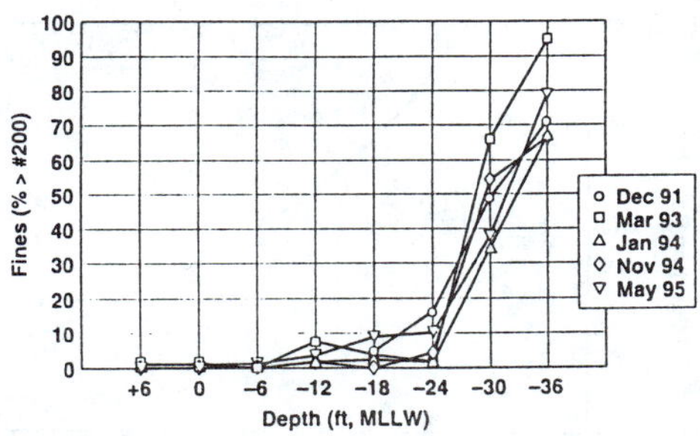

Figure 6. Cross-shore Comparison of Percent Fines, Line 19

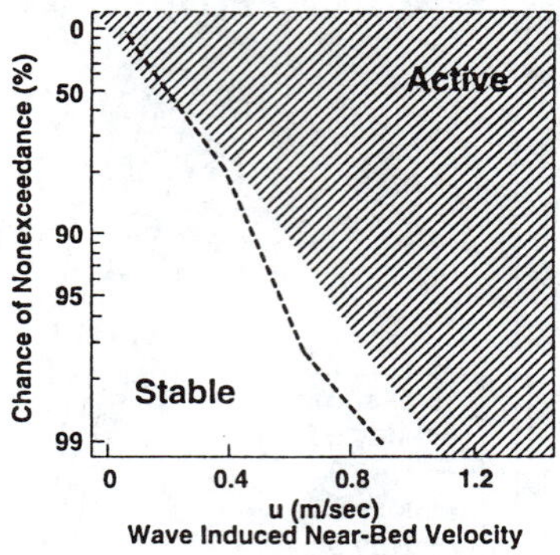

Figure 7. Berm Categorization Based on Wave Induced Near-bed Velocity. Adapted from Hands & Allison (1991).

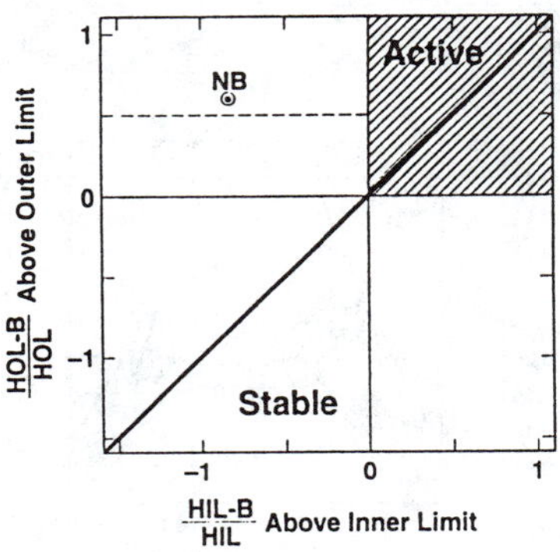

Figure 8. Berm Categorization Based on Hallermeier Limits. Adapted from Hands & Allison (1991).

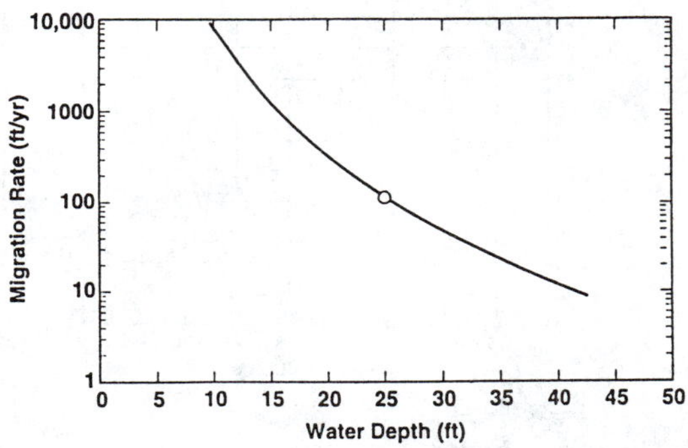

Figure 9. Berm Migration Rate.

Photo 1. Wave Refraction due to Nearshore Berm. 12/92.

Photo 2. Wave Breaking due to Nearshore Berm. Summer 1992.

CHAPTER 362

San Gabriel River to Newport Bay Erosion Control Project
Orange County, California
30 Years of Periodic Beach Replenishment

Arthur T. Shak, P.E.[1] ,M. ASCE and Joseph A. Ryan, P.E.[2]

Abstract: Monitoring of a Beach Erosion Control project consisting of a protective and feeder beach with periodic beach nourishment in Orange County, California is assisted with Computer Aided Design and Drafting technology. Time series of depth changes and profile volumes are analyzed, and predicted renourishment requirements compared to 31 years of project history.

Introduction

The U.S. Congress authorized an erosion control project in 1962, recognizing the impacts of flood works, coastal harbors and other factors in causing beach erosion along the northern Orange County, California shoreline. With legislatively established cost-sharing between the State and federal government, an initial beachfill was constructed in 1964, with periodic nourishment in 1971, 1979, 1984, 1989-90 and 1996-97. Future beachfills are projected to be needed on a 5 year cycle, indefinitely.

The construction of the project was modified from the originally formulated plan with the use of sand sources of opportunity, the addition of a groin field, and deferring a detached breakwater/sand trap feature. A comprehensive analysis was performed of available profile data to compute the shoreline and volumetric history from pre- and post project to the present. This analysis was compiled by meticulously reconstructing historic data from the 1960's to the present to create controlled digital terrain maps which can easily be analyzed with CADD software.

---

1) Civil Engineer, U.S. Army Corps of Engineers, Los Angeles District, Los Angeles, California 90053.
2) Civil Engineer, U.S. Army Corps of Engineers, Los Angeles District, Los Angeles, California 90053.

Project History

The erosion control project formulated in the 1960's consisted of a protective and feeder beach to be located at the updrift end of the littoral cell at Surfside and Sunset Beach, with periodic beach nourishment by back-passing from the downdrift end of the cell near the Santa Ana River. The project area of approximately 20 km alongshore had a planned fill volume of three million cubic yards (mcy) for the initial protective beach, and a projected periodic renourishment rate of 1.75 mcy every five years. A detached offshore breakwater was also planned as a sand trap to be located at the downdrift location. The overall project plan is shown on Figure 1.

The initial protective and feeder beach was built in 1964 with four mcy of material from the adjacent Anaheim Bay/Seal Beach Naval Weapons Station (NWS). Between the initial fill and 1995, four renourishment cycles and three dredging projects of the NWS placed 8.6 mcy of dredge material - measured at the dredge site - at the feeder beach, as listed in Table 1. Of dredge material, 5.0 mcy were borrowed from offshore dredge pits. Photos 1 and 2 show typical pre and post beachfill conditions.

Construction of the detached breakwater/sand trap is deferred pending a demonstration of need, and no backpassing from the downdrift beach area has been performed. A groin field and beachfill at west Newport Beach was added to the project and constructed between 1969 and 1973. Eight groins were constructed and their cells filled with about 1.5 mcy of beachfill from the Santa Ana River or the adjacent Balboa peninsula.

Table 1 Surfside-Sunset Beach Fills (1963-90)

Completion Date	Dredge Volume (cy)	Cumulative Volume (cy)	Description /Borrow Sit
June 1964	4,000,000	4,000,000	Stage 1/Seal Beach NWS
May 1971	2,260,000	6,260,000	Stage 4A/Seal Beach NWS
June 1979	1,644,000	7,904,000	Stage 7/Offshore Borrow Pit
May 1983	400,000	8,304,000	Deepening NWS Channel
April 1984	1,500,000	9,804,000	Stage 8/Offshore Borrow Pit
April 1984	783,000	10,587,000	Seal Beach NWS
March 1989	180,000	10,767,000	Deepening NWS Channel
June 1990	1,300,000	12,067,000	Stage 9/Offshore Borrow Pit
Sep 1990	522,000	12,589,000	Stage 9/Offshore Borrow Pit

Analysis Methodology

Commercial Computer Aided Drafting and Design (CADD) software, MicroStation and InXpress from Bentley Systems, Inc., were utilized in creating digital terrain models (DTM) of the beach and nearshore hydrography. The CADD design files (.dgn) and digital terrain models (.dtm) are full scale models referenced to the California State Plane Coordinate System (SPCS27, Zone 6) as described in Stem(1989). The coordinate system was chosen for compatibility with an existing Geographic Information System (GIS) database made

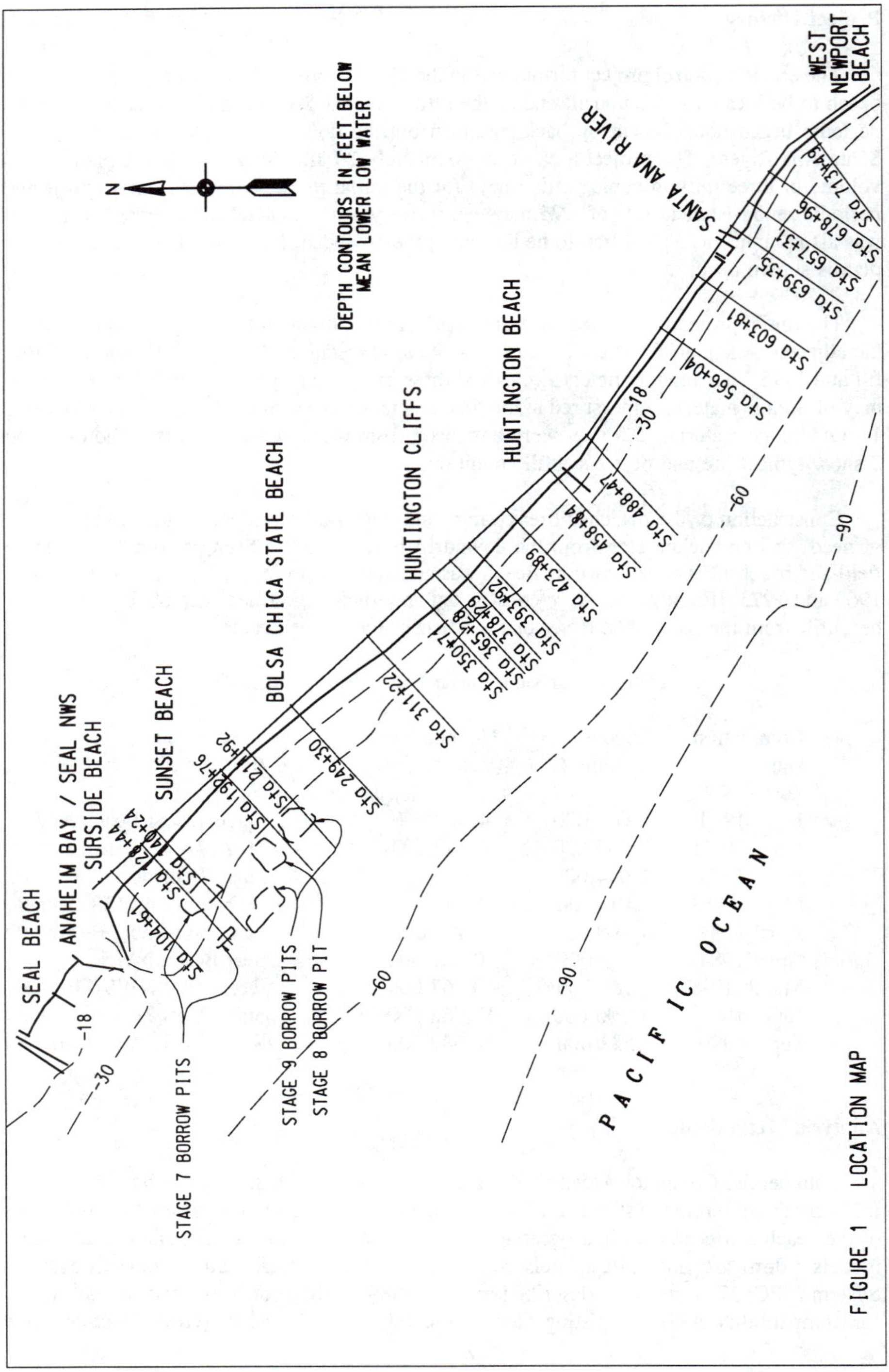

FIGURE 1 LOCATION MAP

PHOTO 1 Feeder Beach Area - Post Stage 7 Beachfill (Oct 31, 1979)

PHOTO 2 Feeder Beach Area - Pre Stage 8 Beachfill (April 16, 1982)

available by Orange County, California, allowing overlays of the beach profile data to map data having property line resolution and orthogonally rectified aerial photography. Programs, such as CORPSCON developed by the U.S. Army Corps of Engineers Topographic Engineering Center, are available for conversion to any of the common horizontal datums with an accuracy as good as the original positioning data. The vertical datum used for all of the analysis is mean lower low water.

Available surveys used to construct DTM models, also referred to as Triangulated Irregular Network (TIN) models, include condition surveys, and pre- and post construction surveys of the San Gabriel River to Newport Bay Beach Erosion Control project; dredging of the Seal Beach NWS; the Santa Ana River Mainstem (SAR) project; the Coast of California Storm and Tidal Wave Study (CCSTWS); and surveys of the National Ocean Survey (NOS) conducted along the coasts in 1934 and 1975-77. Surveys were performed with varying degrees of spatial coverage and resolution, as well as duration. The recent profile surveys (1990's) were collected in time periods as short as a week while the 1960's surveys were collected over several months and the NOS surveys collected over multiple years. A tabulation of surveys utilized are listed in Tables 2 and 3.

After the tedium of reconstructing survey notes to the SPCS27 datum and data input into the DTM models, analysis alignments or baselines and control volumes were established. Volume estimates can be made with three different computational methods: 1) surface to surface comparisons between each facet of TIN models, 2) comparison of gridded surfaces fit to the TIN models, and 3) the traditional average-end area method applied to cross-sections of an alongshore alignment. The surface to surface approach provides the most numerically precise computation of the available data, however, the most useable volumetric comparisons were provided by the average-end area method with carefully selected sections located near surveyed profile lines. The reason that the surface to surface or gridded model computed erroneous results were due to differences in alongshore spacing of profiles between the different surveys resulting in the comparison of measured to interpolated surfaces. Interpolation of the beach profile across even slight embayments can result in large errors. The location of the selected sections used are shown on Figure 1. Sample cross-sections generated from the TIN model are shown on Figure 2.

Closure Depth and Profile Control Volume

The depth of closure, "pinch-out" depth, or location where measurable depth changes do not occur could not be readily found in the profile comparisons. Some reasons why the profiles have significant depth changes in deeper than typical depth of closure water depths are the dredge borrow pits, nearshore dredge material disposal, and subsidence associated with mineral extraction, besides natural sediment transport occurring on the nearshore shelf. The depth of closure concept defined as the seaward limit of significant sediment transport is a misnomer since near-bottom wave induced currents do mobilize large volumes of sediment, and small depth changes on the nearshore shelf result in large volume changes in comparison to the volumes contained on the beach and within the limits of the surf zone. Utilizing a depth of closure concept as a control volume boundary in plan, with its location determined from an abrupt change in the standard deviation of a time series of elevation is a more useful definition along this coastline. This boundary approximates the seaward limit of "significant" wave generated longshore currents and is numerically more consistent with

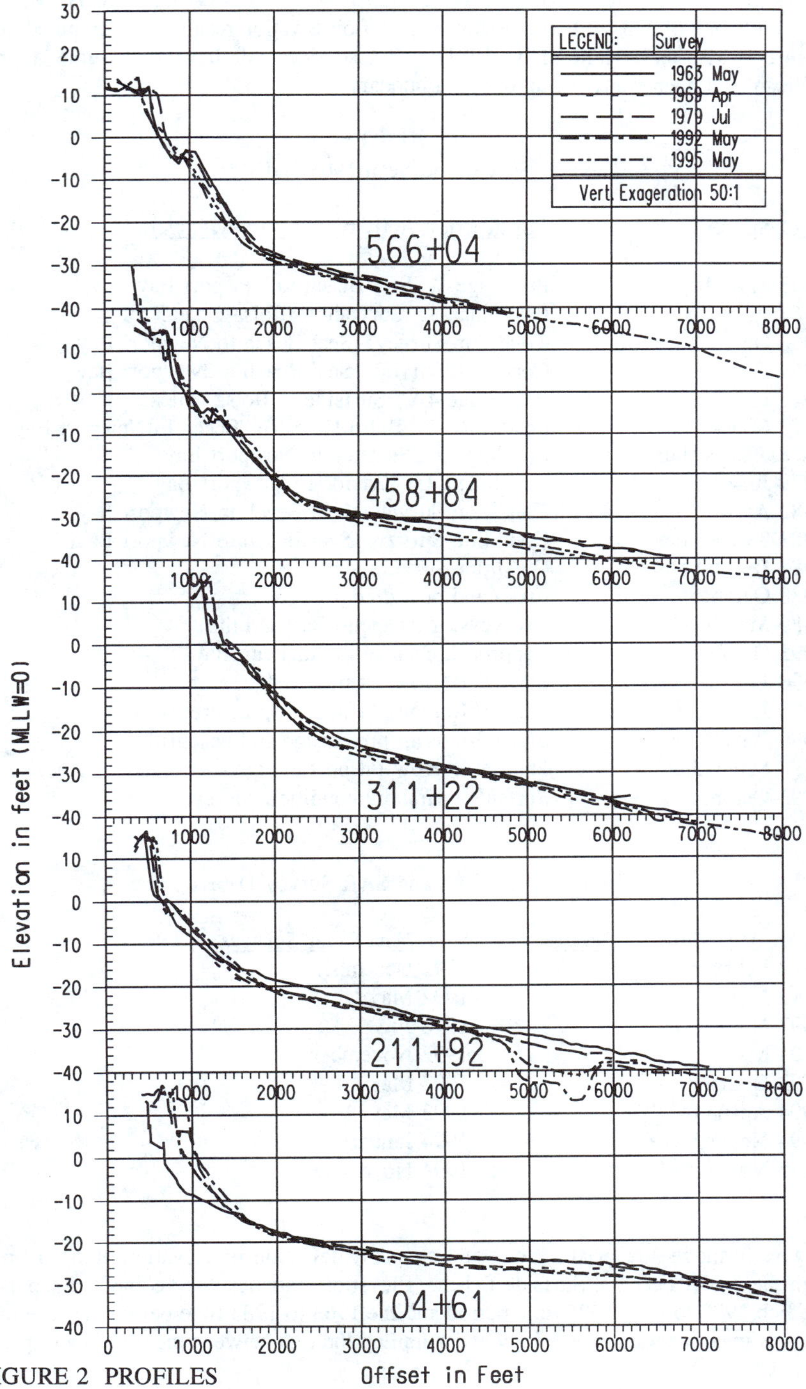

FIGURE 2 PROFILES

values estimated from formulae and wave statistics, ie. the 12-hour annual wave at Huntington Beach of 233 cm based on four years of observation results in a computed depth of closure of 16 feet (Hallermeir, 1981). It also does not have to assume a priori insignificant sediment transport across the boundary.

Table 2 Surveys between 1963 and 1993

Date of Survey	Construction Activity and Survey Coverage
1963 May-August	Pre-Stage 1. Seal Beach to Newport Bay.
1964 July	Post Stage 1. Surfside to Bolsa Chica.
1966 October	Condition Survey. Seal Beach to Newport Bay.
1969 April	Condition Survey. Seal Beach to Newport Bay.
1971 June	Post Stage 4A. Surfside to Bolsa Chica.
1973 May	Post Stage 5. Bolsa Chica St. Beach to Newport Bay.
1978-79 Dec-Jan	Pre-Stage 7. Surfside to Newport Bay.
1979 June	Post-Stage 7. Surfside to Newport Bay.
1982 April	Condition Survey. Seal Beach to Newport Bay.
1982-83 Dec-Jan	Condition Survey. Seal Beach to Newport Bay.
1987 December	Condition surveys and
1988 October	pre- and post-dredge
1989 March	surveys of Anaheim Bay and the
1993 October	approach channel to Anaheim Bay
1990 January	Stage 9 Borrow Pit pre-dredge.
1990 June	Stage 9 Borrow Pit interim post-dredge.
1990 July	Stage 9 interim pre-dredge and beachfill.
1990 September	Stage 9 Borrow Pit post-dredge.
1993 October	Surfside nearshore condition survey.

Table 3 CCSTWS and SAR Survey Dates

Coast of California Surveys	Santa Ana River Surveys
1992 February	1991 December
1992 May	1992 May
1992 November	1992 July
1993 May	1992 November
1993 October	1993 Mar
1994 April	1993 May
1994 November	1994 January
1995 May	1994 November

Figure 3 shows the mean, range and standard deviation of elevation along a profile through the feeder beach at Surfside Beach. Elevation statistics were computed separately for the Feb 1992 to May 1995 time period and the 1963 to 1983 time period because of the differences in collection methodology and sampling interval between the two data sequences

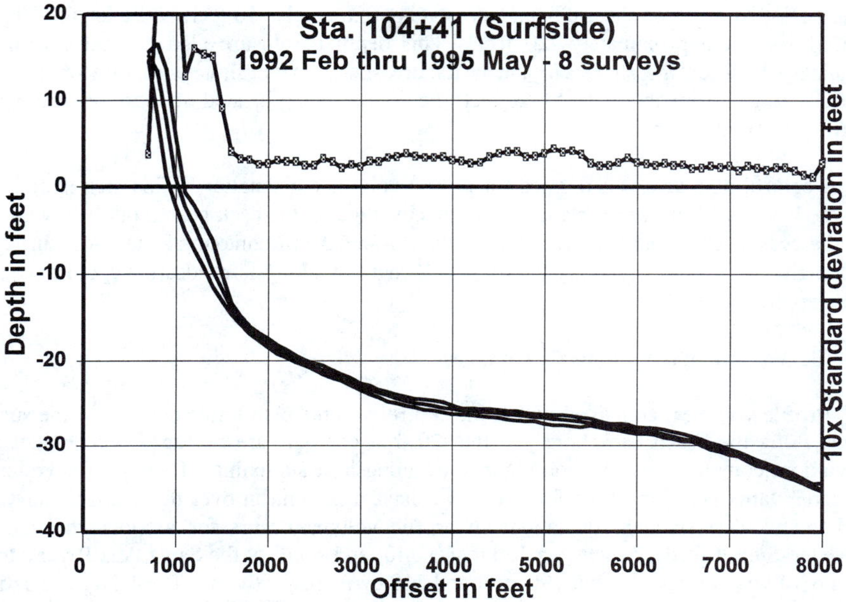

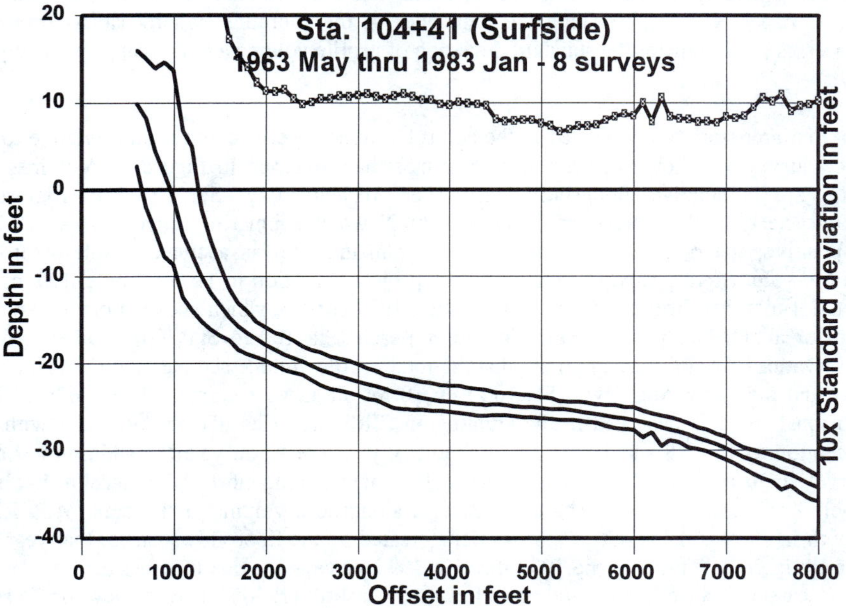

FIGURE 3 MEAN, RANGE and STANDARD DEVIATION of PROFILE DEPTH

-- ie. multiple surveys per year compared to multiple years per survey. For the recent 3-year period, a depth of closure boundary is at approximately -16 feet, while for the 20-year period it is at approximately -20 feet. This depth of closure also coincides with the boundary between the shoreface and nearshore shelf. Alongshore variation of depth of closure ranges from about -16 to -28 feet, the deepest being located offshore of Huntington Cliffs and Huntington Beach.

A control plan area was defined for profile volume computation. This area is bounded by the landward limit of overlapping survey coverage on the back beach, but below the toe of the coastal cliffs, and the location of the 20-foot depth contour of the 1963 survey -- although a seaward boundary that varies with depth by alongshore location could have also been applied.

Profile Volume and Volumetric Changes

Profile volumes, contained within the control volume plan limits defined by the survey limit along the beach backshore and the -20 foot contour, are averaged with ranges and deviations computed as shown on Figure 4 by alongshore stationing. These profile volumes, in cubic yards per alongshore foot of beach, have a zero mean over the available surveys. Relatively large standard deviations in profile volume occurs for sections through the Surfside-Sunset feeder beach (Sta 100 to Sta 140) and south of the Santa Ana River Mouth (Sta 630) with values of about 190 cy/ft and 110 cy/ft, respectively. These large variations are due primarily to beach nourishment, groin field and nearshore berm construction (see Mesa, Paper No. 330). Along the remainder of the shoreline, where natural transport processes predominate, the standard deviation of profile volume averages approximately 60 cy/ft.

End-area volumes contained in the control volume described above and relative to the 1963 survey are shown in the format of a mass-haul diagram in Figure 5. Volumes, by date, are cumulative alongshore starting at Anaheim Bay (Sta 97+71) progressing southeasterly to West Newport Beach and the Newport submarine canyon (Sta 757+74). A positive, flat or negative slope of the diagram indicates an accreted, stable or eroded profile volume, respectively, relative to pre-project condition in 1963. The effects of the initial 4 mcy beachfill can be seen in the July 1964 survey, which shows a profile volume gain of about 3 mcy along Surfside-Sunset Beach (Sta 100 to Sta 200). Other general observations from this diagram are that the total profile volume accreted between Anaheim Bay and the Santa Ana River (Sta 100 to 630) for all surveys compared to 1963; a slight erosional trend is observed in the Huntington Cliffs area (Sta 350 to 410); and with the exception of Dec 78 and Jan 83, both abnormally severe storm years in addition to being surveys before renourishment stages, the profile volume along most of the shoreline has been stable or accretional. The sharp increase in alongshore volume at the Santa Ana River mouth in 1969 is the result of a broad delta created by the flood flows earlier that year. A similar increase shown in the 1992 through 1995 surveys are due to construction of a 1.3 mcy nearshore berm (1992), and groin field and beachfill (1970's) at West Newport Beach.

Other control volumes were defined such as the area bounded by the back beach and the MLLW datum plane (dry beach volume), and the nearshore shelf area bound by the -20 and -30 foot contour in 1963. The dry beach volume follows similar trends to that of the profile

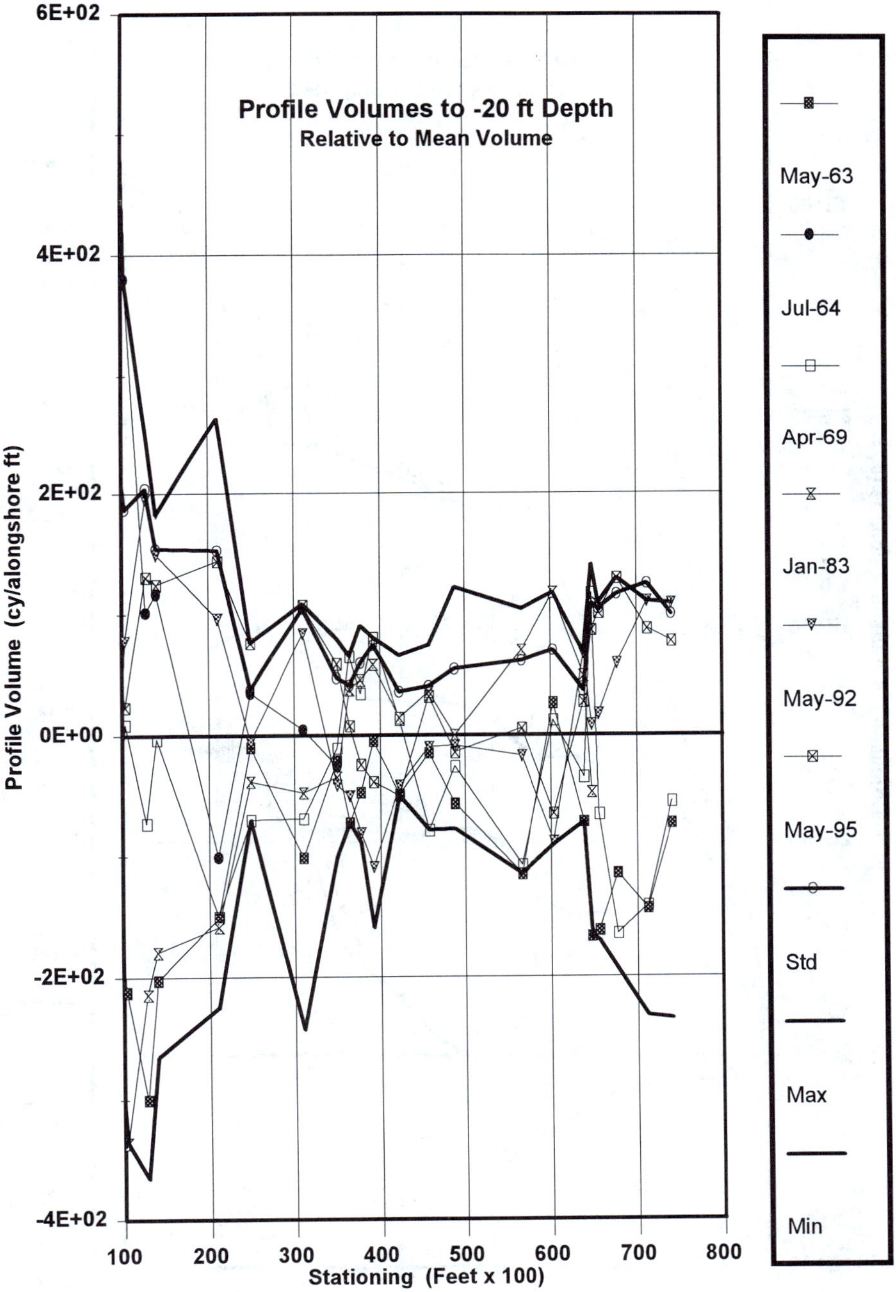

FIGURE 4 PROFILE VOLUMES

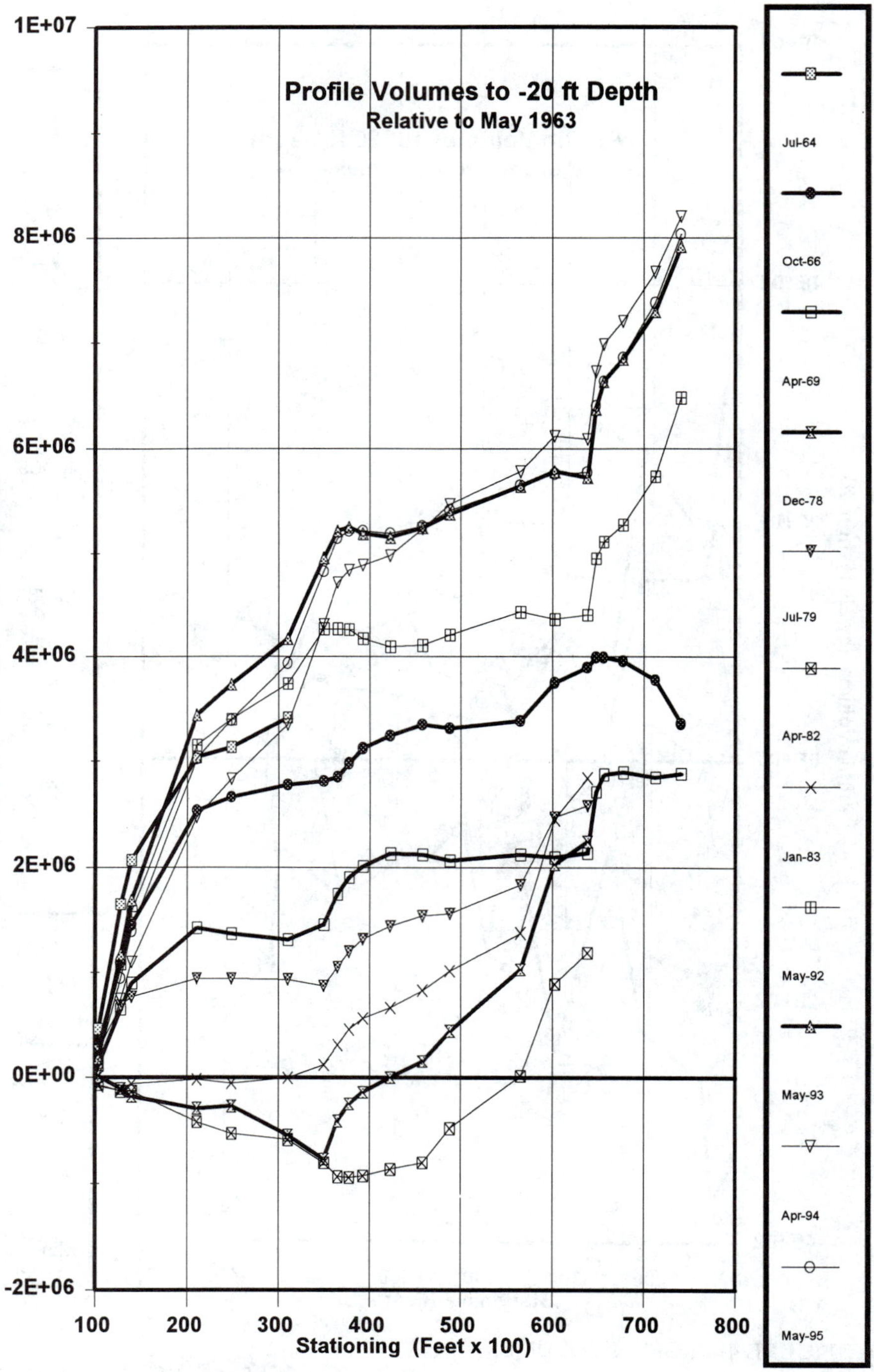

FIGURE 5 CUMULATIVE PROFILE VOLUMES

volume of the -20 foot depth, whereas, the nearshore shelf (-20 to -30 ft) has experienced deepening over a large portion of this coastal segment. Some of this deepening is the direct result of sand mining for beach replenishment while other deepened areas appear to be directly related to subsidence associated with mineral extraction. The latter has been documented through differential leveling along the coast highway and is most evident in comparison of the 1934 to 1975 NOS nearshore hydrography.

Conclusions

Available surveys for the northern Orange County coastal segment have been analyzed with the use of CADD and triangulated irregular network models. Historic surveys with U.S. Coast and Geodetic Survey control can be reduced and used to build a shoreline and profile history database accessible to GIS systems. State Plane Coordinate Systems can be easily related to geodetic coordinates allowing utilization of data from multiple historic sources. Once the TIN models are created, time series of depths, volumes and shoreline positions can be queried for analysis.

As for the San Gabriel River to Newport Bay Erosion Control Project, monitoring over 30 years of beach nourishment has shown a transport rate from the feeder beach to be remarkably close to the originally projected beachfill requirements published by the U.S. Army Engineer District, Los Angeles(1962). Comparison of profiles generally show wider beaches and a deepening of the nearshore. By comparison to the pre-project condition, artificial beach fills have created a protective and feeder beach such that the profile volume has always been above that which existed in 1963, except during the severe winter seasons of 1978 and 1982-83. At the feeder beach location, 772,000 more cubic yards existed on the beach above MLLW datum in May 1995 than in 1963, and 1.39 million cubic yards more existed within the profile volume bound by the beach and historic -20 foot contour. The initial beach construction, NWS channel deepening and four nourishment cycles placed dredge material totaling 12.6 mcy on the feeder beach. Assuming 15 percent of the dredge prism to be composed of fine grain sediments which would quickly wash away, a beach nourishment volume of 10.8 mcy can be estimated. Only 1.4 mcy remain within the profile volume of the feeder beach after 31 years, yielding an average loss rate of 303,000 cy/year. The predicted periodic renourishment volume of 1.75 mcy about every five years in the 1962 study was based on an annual loss of 350,000 cy/year, consisting of 300,000 cy/yr loss from the littoral cell and 50,000 cy/yr loss to the offshore, wind and other causes.

While periodic beach nourishment has maintained profile volumes above pre-project levels, protective beach widths are not always provided for all locations. This is particularly true for about 3,000 feet adjacent to the Anaheim Bay East Jetty (sta 100+00 to 130+00) which appears to have a component of southeasterly directed longshore transport from westerly swells and reflected southerly swell as can be seen in the approaching wave fronts of Photos 3 and 4.

PHOTO 3 Surfside Beach with a South Swell Wave Pattern (June 7 , 1980)

PHOTO 4 Surfside Beach with a West Swell Wave Pattern (April 17, 1981) Sunset Gage: 0904, Hs=110.3 cm; Tp=13s.

Acknowledgements

The authors are grateful to their predecessors in the Los Angeles District and the Beach Erosion Board who collected the historic profile data and had the foresight of using national datums with documented survey control. We also acknowledge the National Ocean Surveys for providing their hydrographic database, the Coastal Data Information Program for making wave data available online, and Orange County Beaches and Harbors Department and Survey Department for providing the GIS database and performing the recent surveys.

References

Hallermeir, R. J. (1981). "Seaward Limit of Significant Sand Transport by Waves: an Annual Zonation for Seasonal Profiles." Coast. Engr. Tech. Aid No. 81-2, Coast. Engrg. Res. Ctr., U.S. Army Engr. Wtrwys. Exp. Sta., Vicksburg, Miss.

Stem, James E. (1989). "State Plane Coordinate System of 1983." NOAA Manual NOS NGS 5, U.S. Dept. of Commerce, NOAA, NOS, Charting and Geodetic Services, (Reprinted Jan. 1993), Rockville, MD.

U.S. Army Engineer District, Los Angeles (1962). "Beach Erosion Control Report on Cooperative Study of Orange County, California." Appendix V, Phase 2, (Mar. 1), Los Angeles, California.

CHAPTER 363

A Ten-Year History of Dolos Monitoring at Crescent City

William S. Appleton[1], Thomas Kendall[1], M. ASCE, and Jeffrey A. Melby[2], M. ASCE

ABSTRACT

Ten years of monitoring data are now available from the Crescent City dolos monitoring program. This extensive field data collection program was initiated in 1986, following the rehabilitation of the dolos section of the main stem of the Crescent City Outer Breakwater. The rehabilitation involved the placement of 680, fiber-reinforced, 38-tonne dolosse. Unique field monitoring techniques have been employed to collect data at Crescent City: movement data have been collected with ground-truthed, low-altitude (helicopter) photogrammetric survey techniques and stress data have been derived from strain gages mounted internally in select dolosse grouped together near the center of the dolos field. Results presented in this paper include 10 year histories on: wave climate, dolos movement/breakage, stress build up within dolosse, and concrete strength and fatigue. In addition, internal strain gage instrumentation verification test results are presented.

INTRODUCTION

The hydraulic stability of dolos armor units has been a large factor in their success as an armor unit. For this reason, the use of dolos armor units has helped to allow for the construction and long-term stability of shore and navigation protection structures sited in extremely energetic wave environments. The use of dolos armor units has been extensive and in most cases these armor units have performed well; however, due to the slender nature of the dolos design, these units are susceptible to breakage especially as the size of the units increases. To gain further insight into the mechanisms which govern the design life of these

---

[1]USACE, San Francisco District, 333 Market Street, San Francisco, CA 94105-2197, USA

[2]USACE, Waterways Experiment Station, 3909 Halls Ferry Road ,Vicksburg, MS 39180-199,USA

breakwater armor units, the Corps has collected an extensive data set from the dolos armor layer located on the Crescent City, California Outer Breakwater.

There are now 10 years worth of monitoring data available addressing the role of dolos movement, breakage, fatigue, and static stress build-up which provide insight into mechanisms affecting the life of these armor units.

BACKGROUND

The Crescent City Outer Breakwater has had a history of being badly damaged since it was lengthened in 1930 to include the section which now contains dolos armoring. In particular, the last 100 meters of the structure's main stem have been especially difficult to maintain and, consequently, dolosse have been used in this section for the past 22 years. The most recent round of extensive maintenance work came in 1986 when, following years of wave damage from winter storms, the dolos section of the breakwater was rehabilitated. The rehabilitation work involved the placement of 680, 38-tonne fiber-reinforced dolosse on the seaward slope of the structure (Figure 1) and 80 on the lee side slope for future use. Placement of the armor units was aimed at recreating, to the extent possible, the design tested in physical model studies which included the use of trenching and buttressing (with stone) along the dolosse southern perimeter for increased stability at the dolosse-stone transition. A thorough background description of the project site can be found in Kendall (1988) and Kendall and Melby (1989).

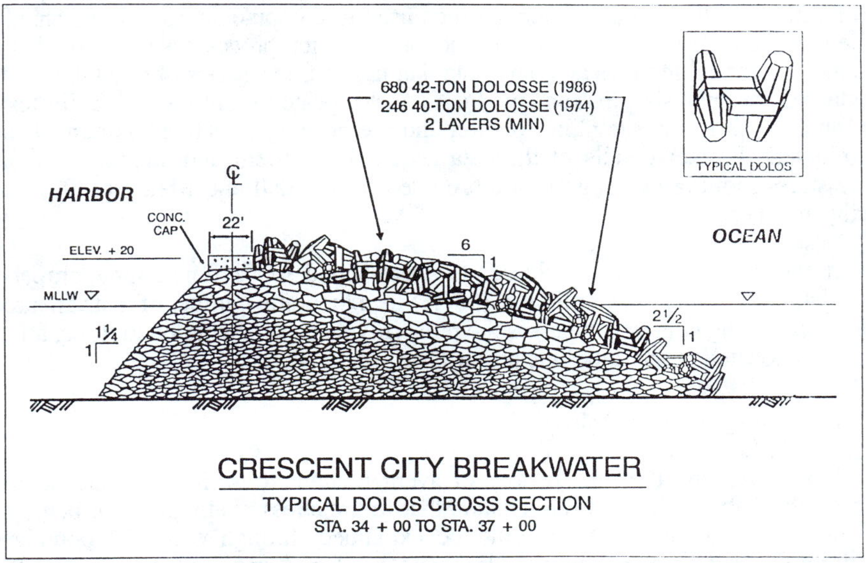

Figure 1: Typical cross-section through the dolos section of the Crescent City Outer Breakwater.

MONITORING TECHNIQUES

Displacement and Breakage Monitoring

Dolos displacements and breakage have been monitored since the completion of the rehabilitation project. Targeted and non-targeted dolos displacements have been detected and assessed using photogrammetric techniques. A helicopter continues to be used as the platform for mapping photography work; this technique allows for the measurement of targeted dolos displacements to be reported with an accuracy of ±2 cm and non-targeted dolos displacements with an accuracy of ±15 cm. This work has been previously reported by Kendall (1988) and Kendall and Melby (1992). Changes in the subsurface slope and toe of the structure are monitored using both side scan sonar and diving inspections. However, given the conditions under which this work is conducted, only gross estimates of subsurface displacements can be made.

Dolos breakage counts are monitored through walk-over inspections of the dolos field as well as by helicopter reconnaissance. To date, no techniques have been identified which would allow for determining subsurface breakage.

Static Stress Monitoring

As part of the 1986 rehabilitation work, 20 of the dolosse were instrumented with internal strain gauges mounted on rebar rosettes. An internal microprocessor allowed for data to be digitized within the dolos. The instrumentation measures two moments and a torque, which are then used to compute stresses at the shank-fluke interface. The data stream from the internal microprocessors is sampled at a rate of 1 Hz for six minutes and only data that has passed a series of reliability tests is retained for analysis. Instrumented units that repeatedly fail the reliability tests are removed from the sampling rotation and are no longer retained as part of the sample set. Further details of the data acquisition system and summary of the static stress monitoring program are provided by Kendall and Melby (1992) and Melby and Turk (1995).

Of the 20 instrumented dolosse, 17 were found to be functioning properly during the first round of sampling in 1987. Since then, the number of instrumented dolosse returning data which passes the reliability tests has dwindled to three, all of which are located in the top of the armor layer.

Artificial Loading of Dolos

Upon reviewing the time history of average static stress in the instrumented dolosse through 1994, it was anticipated that future stress readings might begin to fall outside of a range which could be explained through observed boundary conditions. Therefore, a testing method was developed and implemented in 1995 on one dolos (dolos C), to determine if the instrumentation was giving false readings. Dolos C was selected because it had a low RMS error on the data channels, the magnitude of the recorded displacements were not very reasonable and the dolos was resting on the top layer with the upper surface of the shank nearly horizontal.

A steel load frame was bolted onto the shank of the dolos and designed to allow for the placement of a load on the extreme end of the dolos fluke equivalent to the full weight of another dolos. A 90.6-tonne (100-ton) hydraulic jack was used to apply the load, which was done in increments of 6.9 MPa (1000psi) corresponding to a force of 8,858 kg (19,511 lbs).

Fatigue and Concrete Strength

Given the constant cyclical loadings experienced by concrete armor units placed in the near-shore wave environment, the question of what role, if any, fatigue plays in the life of concrete armor units, was asked. In the most recent round of monitoring activities, this question was addressed.

Although fatigue testing in its formal sense, strength versus number of blows and destructive testing, was not carried out on the armor units, 7.6 cm (3 in) diameter core samples were taken from dolosse placed in both 1974 and 1986 to determine changes in the concrete strength over time. If the armor units were experiencing significant fatigue, it was expected that the compressive and flexural-tensile strength of the samples would reflect it. Both of the armor units were located near the cap since heavy sampling equipment presented a significant mobility limitation.

WAVE CLIMATE

Presented in Figure 2 is a time history of wave power off of Crescent City. Buoy data, taken primarily from the Point St. George buoy #46026 (41.9° latitude, 124.4° longitude, 60 m of water), was used to generate the plot shown. Nearby buoys were used to fill in gaps in the Point St. George Buoy data set. The extreme local wave energetics that the structure is subjected to annually are clearly evidenced in Figure 2.

Hales (1985) provides a thorough description of depth-limited waves at the outer breakwater at Crescent City. Hales applied RCPWAVE, a numerical model which calculates the combined wave refraction/diffraction/shoaling effects in shallow water, using the bathymetry off Crescent City and tidal stages of -0.3 m (-1 ft) mean lower low water (MLLW) and +3.04 m (+10 ft) to obtain the design wave criteria for the 1986 rehabilitation. Based on Hales work, a depth-limited breaking wave of approximately 10.5 m was determined to be the design wave for the structure. Hale's work as well as comparison with a small data set from a buoy placed just offshore of the structure in 12 m of water, indicated that wave transformation coefficients are on the order of unity or greater for typical storm events out of the west to west-southwest directions.

Approximately 10 years of buoy data (buoy # 46014) were analyzed to determine whether the structure has in fact seen its design wave. Presented in Table 1 are the probabilities of exceedence for wave heights (H) greater than 5 m for measured significant wave heights (H_S) equal to or exceeding 5m. It was assumed that the wave heights were Rayleigh distributed. The buoy data set was filtered for events generating significant wave heights greater than 5m, of which 29 were found for a total of 80 hours. Given the near unity wave transformation coefficients, deep water wave heights were likely to have existed just seaward of the structure as well. The maximum tidal stage was then found for each of the events using standard tide tables. Presented in Figure 3 is a plot of the results of this analysis.

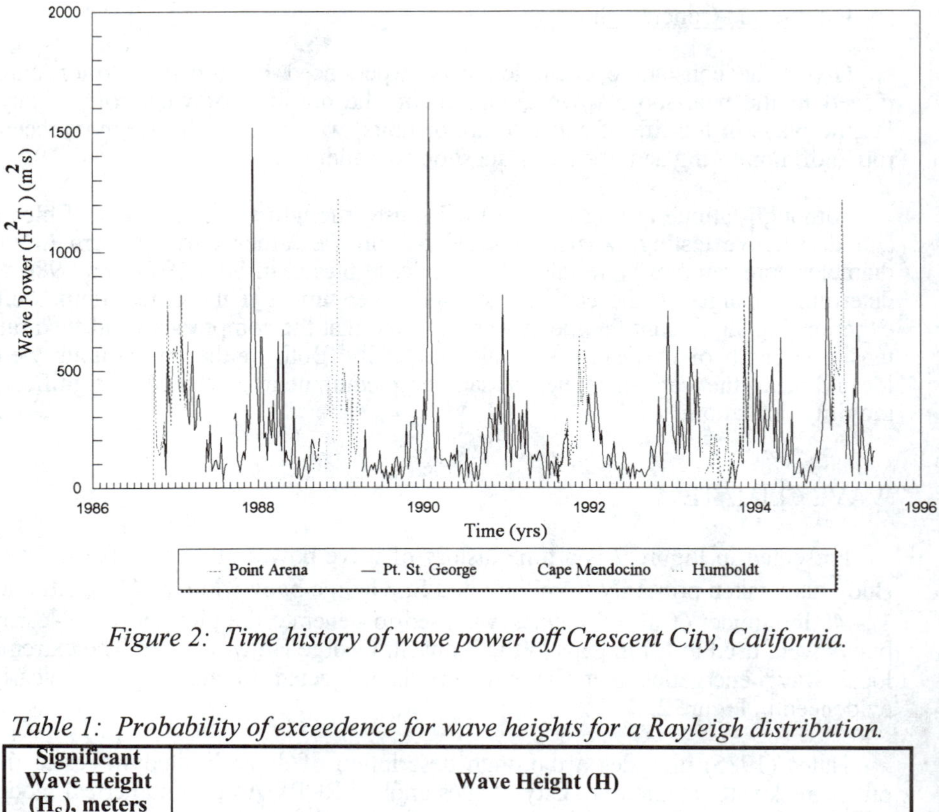

Figure 2: Time history of wave power off Crescent City, California.

Table 1: Probability of exceedence for wave heights for a Rayleigh distribution.

Significant Wave Height (H_S), meters	**Wave Height (H)**					
	5 m	**6 m**	**7 m**	**8 m**	**9 m**	**10 m**
5 m	13.5 %	5.6 %	1.9 %	0.6 %	0.15 %	0.03 %
6 m	24.9 %	13.5 %	6.6 %	2.9 %	1.1 %	0.39 %
7 m	36.0 %	23.0 %	13.5 %	7.3 %	3.7 %	1.7 %
8 m	45.8 %	32.5 %	21.6 %	13.5 %	8.0 %	4.4 %
9 m	53.9 %	41.1 %	29.8 %	20.6 %	13.5 %	8.5 %
10 m	60.7 %	48.7 %	37.5 %	27.8 %	19.8 %	13.5 %

Of the 29 extreme events identified, 9 of them, or approximately 30%, occurred during peak tidal stages that equaled or exceeded +1.8 m (6 ft), giving a minimum water depth at the toe of the repair section of 10.6 m (35 ft). The storm setup contribution to the water depth at the structure during the events is not known since measured tidal records at Crescent City have not yet been analyzed, however, it is reasonable to assume that the storm setup contribution would be on the order of at least a half meter. Therefore, there have been at least 9 events to-date which may have generated waves equaling and possibly exceeding the design wave in water depths that would have allowed these waves to reach the structure prior to breaking. Generally speaking, the above events correspond with wave power readings in excess of 800 m^2s in Figure 2.

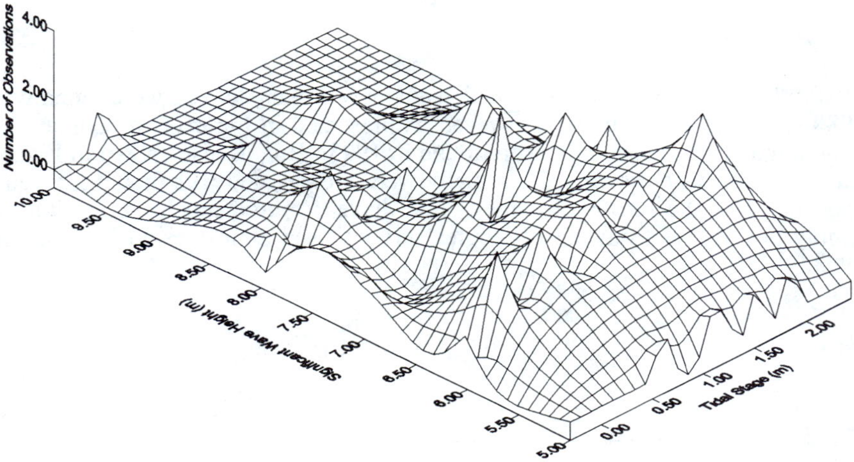

Figure 3: Number of significant wave events and corresponding tidal stage.

MONITORING RESULTS THROUGH THE TENTH YEAR

Evidence of Long-term Nesting

The history of average cumulative dolos displacement is shown in Figure 4. Kendall and Melby (1992) reported that for those units having undergone translation, the dominant direction of movement was up-slope. For those units having experienced rotation, the dominant axis about which rotation occurred was the z axis (yaw). Since 1992, significant displacements have occurred for four units, one in 1994 and three in 1996. In all four cases, translation up-slope was observed, however there did not appear to be an indication that rotation about the z axis was more prevalent than rotation about any of the other axes. It is interesting to note that while the curve describing average cumulative translations continues to behave as one might expect for a fully nested structure, the jump in the cumulative rotation curve is not as intuitive. It would appear that as equilibrium nesting conditions are approached, significant rotations are more likely to occur than significant translations. In other words, a nested quasi-equilibrium state is more apt to allow for rotations than translations of the armor units. A similar jump in the cumulative rotation curve corresponding with a relatively flat translation curve was seen in the 1992-1993 data as well. Considering that translation requires that the centroid of the unit be displaced, while rotations can occur via a pivot point, it is understandable that rotations should be more likely to occur under nested conditions. Results appear to indicate that despite near equilibrium nesting conditions, movements, especially rotations, will continue to occur although they seem to be small enough not to be considered significant.

Side scan sonar records as well as dive inspections indicate that there were no significant changes in the position of the toe or slope of the dolos rehabilitation section.

Breakage

Prior to 1993, the last year in which any new broken dolosse were found was in 1988 (one unit was found). However, as presented in Figure 5, a comparatively large number of broken units were found in 1993. But, as also shown in Figure 5, it is uncertain when these units actually were broken, and it is thought, given the appearance of the breaks, that they could have occurred much earlier, but were simply not visible from the helicopter surveys used almost exclusively up until 1993.

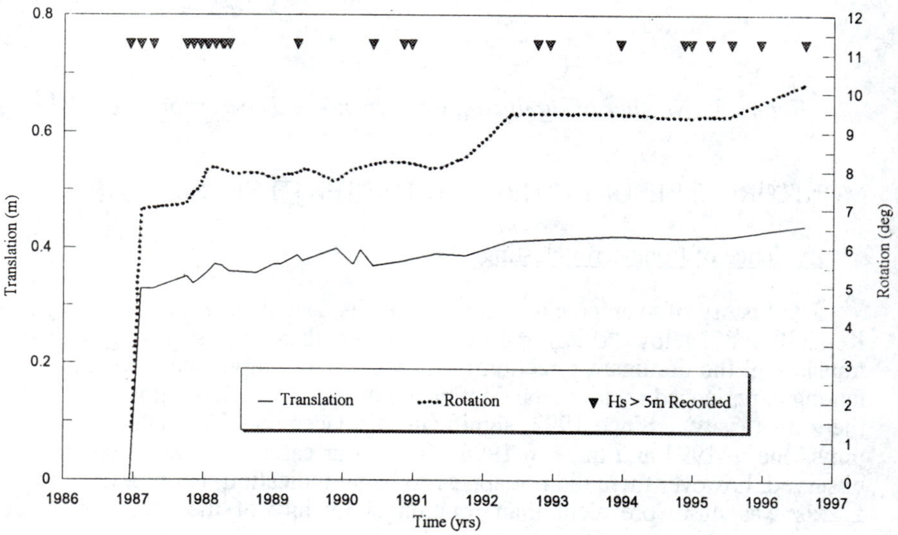

Figure 4: Time history of cumulative dolos displacement.

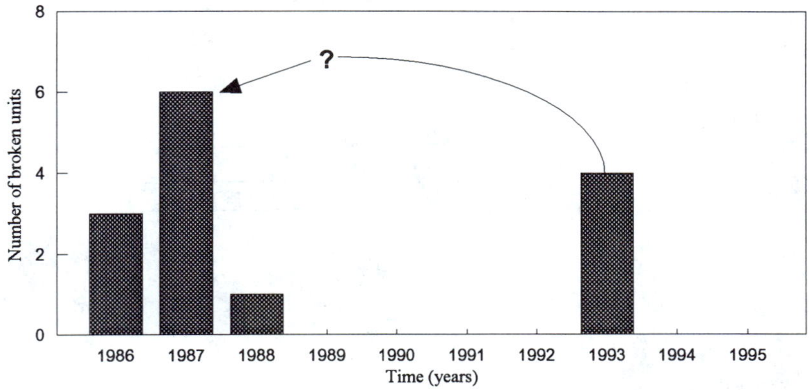

Figure 5: Dolos breakage results.

Static Stress History

Burcharth (1988) defined the various loads on armor units, the three primary loads being static loads, impact loads and pulsating loads. The history of average static stress measured within the dolosse is presented in Figure 6. Each data point represents an average of all of the reporting units on that date. Also shown in Figure 6 is the dwindling sample set from which these readings were taken. Initial measurements, made in 1987, were taken from 17 units, 3 in the lower armor layer and 14 in the top layer. 1995 measurements were taken from 3 units, all of which were located in the top layer. The average static stress for the three (3) reporting units was found to be approximately 3 MPa (434 psi). The trend in this curve is not an artifact of the dwindling sample size as the history of the three remaining units reflects the same trend exhibited.

The newest data point appears to be consistent with the trend apparent in the data from prior years, showing a linear increase in static stress. However, the continuing linear increase in static stress levels within the dolosse is cause for concern, since boundary conditions leading to such stress levels are not at all apparent. Therefore a check was made to determine if the static stress levels being reported by the instrumentation were reasonable. An extreme boundary condition was considered, one where a dolos was assumed to be loaded at the extreme end of a fluke by the weight of another 38-tonne unit. Simple beam theory was applied to obtain an approximation of the normal stress at the section A - A shown in Figure 7. This simple calculation showed that the maximum normal stress (located at the extreme fibers of the cross section) for the boundary condition shown is approximately 2.5 MPa (358 psi). Hence, the latest stress reading has exceeded the estimated "full dolos loading" level.

If the post nesting growth of static stress within the dolosse is to be believed, then the flexural-tensile strength of the concrete is rapidly being approached. The slope of the plot in Figure 6 indicates that the reserve concrete strength is being consumed at a rate of approximately 3% per year. However, if this was in fact true, given the wave events that these units are being subjected to, one would

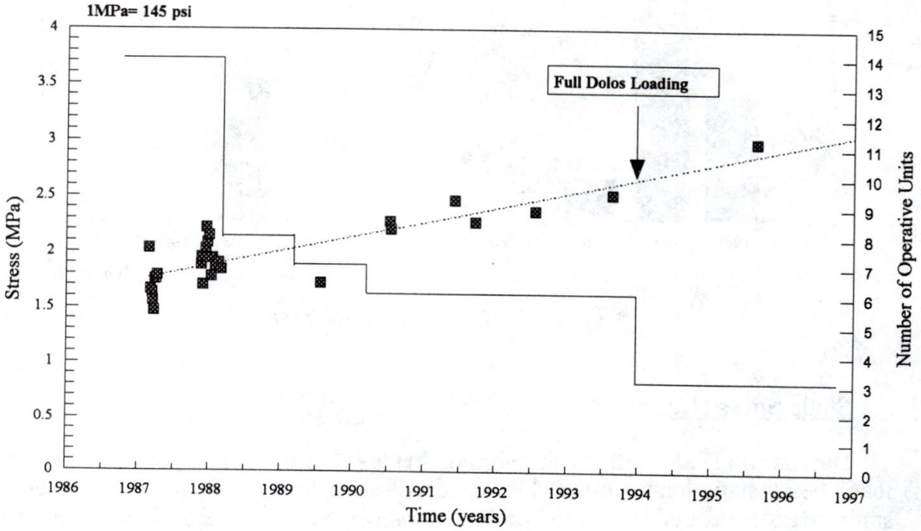

Figure 6: Time history of average static stress levels

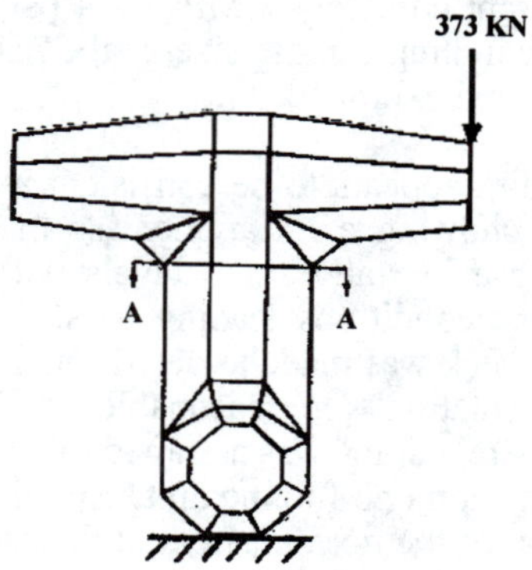

Figure 7: Caricature of dolos loading

expect a much higher rate of breakage in recent years than what has been observed. In order for the instrumentation to be functioning properly while still returning data that indicates a continuing rise in static stress, the concrete must be relaxing while the strain gauges continue to deform. Shrinkage, thermal contraction/expansion and creep were considered as mechanisms that may be responsible for creating permanent strains on the instrumentation, while at the same time allowing the concrete to recover via plastic deformations. Creep or some form of plastic deformation within the concrete is the only plausible explanation. However, as pointed out by Neville and Brooks (1987), the effectiveness of stress relaxation by creep is reduced with time. Most stress that is relieved via creep is done so in the very early stages of concrete curing. Shrinkage, swelling and thermal expansions/contractions may have influenced the static stress trend during the first year, but after that they would play a very minimal role in any static stress buildup. Therefore, material properties of concrete and boundary conditions are unable to fully explain the recorded high measurements of static stress.

Artificial Load Testing

Results from the artificial load testing showed that the recorded strain was of the correct order of magnitude given the load, but half what one would expect from an analytical check using simple beam theory. Since the actual moment arm length is difficult to quantify due to unknown boundary conditions and since the dolos is not a slender beam, the difference between the analytical check and the measured value is considered to be within reason (Melby and Turk, 1995).

Fatigue and Concrete Strength

Core samples that were returned to the lab for testing were subjected to traditional compressive and tensile testing methods. The compressive tests were done in accordance with American Standard for Testing Materials (ASTM) C 39, the splitting in accordance with ASTM 496 and the flexure in accordance with ASTM C 293. Shown in Table 2 are the concrete strength test results.

Table 2: Concrete strength test results from core samples taken in 1995.

Dolos/Core Number	Compressive Strength (MPa)	Flexural Tensile Strength (MPa)	Splitting Tensile Strength (MPa)
559 / N1	-	9.79	4.31
559 / N2	53.24	8.55	2.14
559 / N3	56.41	-	4.59
752 / N1	70.69	-	4.48
752 / N2a	-	11.93	3.48
752 / N2b	-	-	3.03
752 / N4	60.00	10.03	3.48
752 / N6	73.45	-	-
236 / N1	52.34	-	4.17
236 / N2	-	8.76	-
236 / N3	62.83	7.55	4.14
236 / N4	-	-	2.93

Non-shaded area contains data obtained from 1986 dolosse and the shaded area is data obtained from 1974 dolosse.

Lab results indicated that on average, the flexural-tensile strength of the concrete samples from the 1986 dolosse had increased 48% and the compressive strength had increased 70%. These increases in concrete strength are based upon a 28-day compressive strength of 36.4 MPa (5,280 psi) reported by Bevins (1989) and a 28-day flexural-tensile strength of 6.8 MPa (984 psi) reported by Gutschow (1989). For standard type II cement, a 50% increase in 28-day compressive strength after five years can be expected. Therefore, although the concrete used in the dolos is steel-fiber reinforced, the 70% increase over 10 years is significant considering that most of the changes within concrete occur in the first five years.

Conclusions regarding the compressive and flexural-tensile strength of the concrete used in the 1974 dolos could not be made due to a lack of data prior to 1995. However, the compressive strength of the cores tested was within 2% of the average compressive strength obtained from the 1986 dolos concrete samples while the average flexural-tensile strength was within 20% of the 1986 values.

CONCLUSIONS

- It is highly likely that the dolos section of the Crescent City Outer Breakwater has been subjected to its design wave and possible that it has experienced design wave conditions as many as nine time.

- While the time history of dolos displacements does continue to show some rotational displacements, the dolos field is believed to have reached a relatively nested state following the first storm season.

- Strain gages in the tested dolos were shown to be functional within acceptable parameters. Moments measured were on the same order as those predicted by simple linear beam theory.

- Although the internal instrumentation within dolos C was found to be functional and recorded strain readings brought about by artificial loading were within reason, static stress readings still need to be questioned. The material properties of concrete, i.e. creep, shrinkage/swelling, thermal expansion/contraction, do not appear to explain the observed measurements.

- Both compressive and flexural-tensile concrete strengths measured in 1995 on the 1986 dolosse, were found to be significantly higher than design strengths (28-day); 70% and 48%, respectively.

- Fatigue weakening within dolosse does not appear to be significant.

RECOMMENDATIONS

- Conduct aerial photogrammetric work every 2 years. Exceptions would be made following extremely energetic winters having repeated events during which there is a greater than 1% chance of the design wave occurring (approximately when $H_s^2T > 800\ m^2s$).

- Investigate the possibility of picking up a dolos and measuring static stress rebound as an additional step in verifying internal instrumentation measurements.

ACKNOWLEGEMENTS

The work described in this paper has been conducted as part of the Operations and Maintenance program of the U.S. Army Corps of Engineers (USACE) San Francisco District.

REFERENCES

Bevins, T.L., 1989, U.S. Army Engineer Waterways Experiment Station, Vicksburg, MS, personal communication.

Burcharth, H. F. and Howell, G.L., 1988. "On Methods of Establishing Design Diagrams for Structural Integrity of Slender Complex Types of Breakwater Armor Units," Seminaire International Entreien des Infrastructure Maritimes, Casablanca, Morocco.

Gutschow, R., 1989, U.S. Army Engineer District, Los Angeles, CA, personal communication.

Hales, L.Z., 1985. Water Wave Refraction/Diffraction/Shoaling Investigation, Crescent City, California, Miscellaneous Paper CERC-85-3, U.S. Army Engineer Waterways Experiment Station, Vicksburg, MS.

Kendall, T.R., 1988. "Analysis of 42-Ton Dolos Motions at Crescent City," Coastal Engineering, 1988 Proceedings of the International Conference, ASCE, New York, NY.

Kendall, T.R. and Melby, J.A., 1989. "Continued Monitoring of 42-Ton Dolos Movements and Static Stresses at Crescent City," Proceeding of Stresses in Concrete Armor Units, Vicksburg, MS, ASCE, New York, NY.

Kendall, T.R. and Melby, J.A., 1992. "Movements and Static Stresses in Dolosse: Six Years of Field Monitoring at Crescent City," Coastal Engineering, 1992 Proceeding of the International Conference, ASCE, New York, NY.

Melby, J.A. and Turk, G., 1995. "Crescent City, California, Breakwater, Dolos Structural Response Study," Letter Report, U.S. Army Corps of Engineers Waterways Experiment Station, Vicksburg, MS.

Neville, A.M. and Brooks, J.J., 1987. "Concrete Technology", Longman Scientific & Technical, New York, NY.

CHAPTER 364

BREAKWATER DAMAGE IN OKUSHIRI PORT DUE TO THE HOKKAIDO NANSEI-OKI EARTHQUAKE TSUNAMI

Katsutoshi KIMURA[1]
Yuzo MIZUNO[1]
Hiroichi TSURUYA[2]
Yasuyuki NAKAGAWA[2]

Abstract

A tsunami caused by the Hokkaido-Nansei-oki earthquake struck Okushiri Island and the southwest coast of Hokkaido damaging many port facilities. This paper describes typical breakwater damage at Okushiri Port and the applicability of Tanimoto's formula for tsunami wave forces on vertical walls based on prototype failure analyses and the result of numerical simulations. The armor stability and the rubble mound scourings were examined from the results of 3-D model tests at a scale of 1/15. The damage patterns at the narrow opening of breakwaters were reproduced by the tests. The necessary weight of armor blocks and foot-protection blocks for breakwater heads were calculated by the C.E.R.C. formula using averaged flow speed at the peak of tsunami.

Introduction

The Hokkaido Nansei-oki earthquake of July 12, 1993, had a Richter scale magnitude of 7.8, and generated a huge tsunami. Okushiri Island, 70 km from the seismic center, was struck by the tsunami. In addition to a death toll exceeding 200, the tsunami damaged many public facilities in the coastal areas and port facilities, mainly breakwaters.

In general, breakwater damage from tsunamis are classified into the following two patterns;

- Sliding of caissons by tsunami wave forces
- Scouring of rubble mound foundations at breakwater heads by the strong tsunami flows

---

1) Civil Engineering Research Institute, Hokkaido Development Bureau, 1-3 Hiragishi, Sapporo, 062, JAPAN
2) Port and Harbour Research Institute, Ministry of Transport, 3-1-1 Nagase, Yokosuka, 239, JAPAN

The caissons of the Kawaragi breakwaters in Hachinohe Port were damaged along a total length of 328m in the 1968 Tokachi-oki Earthquake. Ito et al. (1968) analyzed the tsunami height by numerical simulations and confirmed that the difference in the water level between the inside and outside of the breakwater led to the damage. In Noshiro Port, the offshore seawalls of a power plant were heavily damaged by the 1983 Nihonkai-Chubu earthquake tsunami. Tanimoto et al. (1983) analyzed the tsunami wave forces based on the field survey and model tests. In Okushiri Port at the eastern side of Okushiri Island, similar damage patterns appeared in the north breakwater. In this paper, the relationship between the sliding distances and tsunami height is examined from the results of numerical simulations

The armor stability at the breakwater head is a very important factor for the design of tsunami protection breakwaters. Tanimoto et al. (1988) made the three-dimensional (3-D) model tests for the offshore breakwaters in Kamaishi Port, and proposed armor stability coefficients for practical design. A remarkable aspect of tsunami damage appeared at the narrow opening between the outer east breakwater and the east breakwater in Okushiri Port. The head caissons of both breakwaters

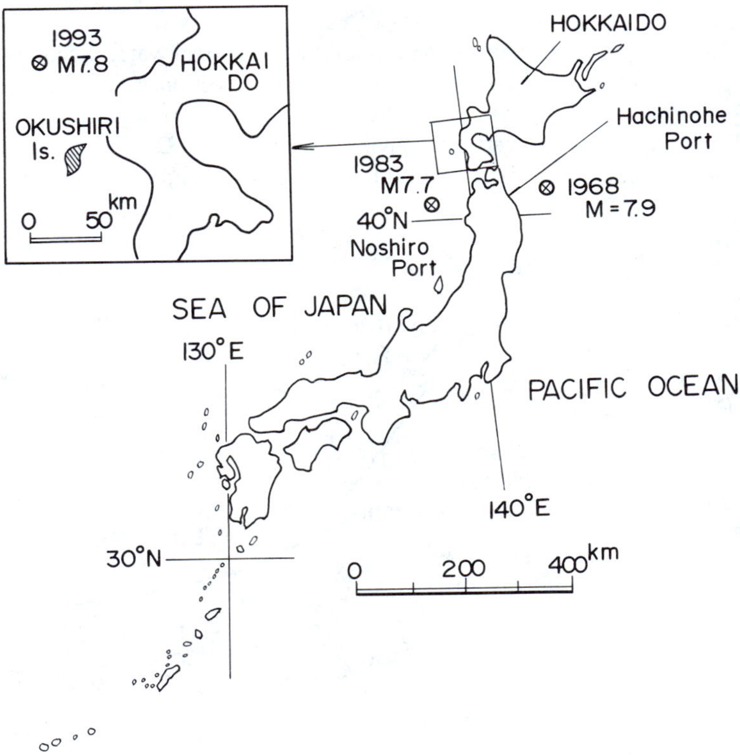

Figure 1. Location of Epicenter

collapsed due to scourings of rubble mound covered with armor blocks by the strong tsunami flow. To reproduce the damage pattern, large-scale 3-D hydraulic model tests were made at a model scale of 1/15. In this paper, the relationship between the flow speed and the damage ratio is examined from test results, and the applicability of armor stability coefficients is confirmed for the practical design of breakwaters against tsunamis.

Outline of breakwater damages

On 12 July 1993, a tsunami, caused by seismic disturbance at latitude 42° 47' and longitude 139° 12' struck Okushiri Island and along the southwest coast of Hokkaido. Figure 1 shows the location of the epicenter and Okushiri Island.

Okushiri Port on the eastern side of Okushiri Island is the only port of the island, and includes the ferry terminal to mainland Hokkaido. Figure 2 shows the breakwaters and port facilities in Okushiri port. The outer east breakwater and the east breakwater are the main barriers against wind waves. The opening between the outer east breakwater and the north breakwater is used for the main waterway, and a narrow opening between the outer east breakwater and the east breakwater is also used for a waterway for small craft.

Figure 3 shows the damage pattern of the north breakwater. Sections A, B, C, and D are caisson-type composite breakwaters, and section E is a concrete block type. According to the eyewitnesses, the tsunami waves propagated parallel to the shore line and attacked vertically against sections, C, D, and E of the north breakwater. In contrast, caissons slid little in sections A and B near the head of the breakwater.

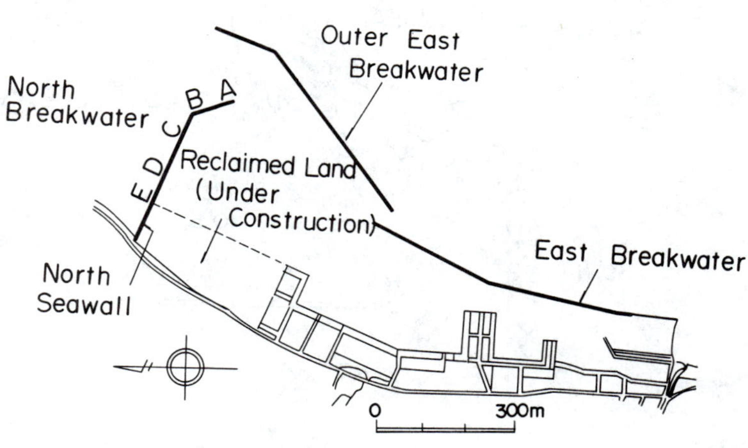

Figure 2. Breakwaters in Okushiri Port

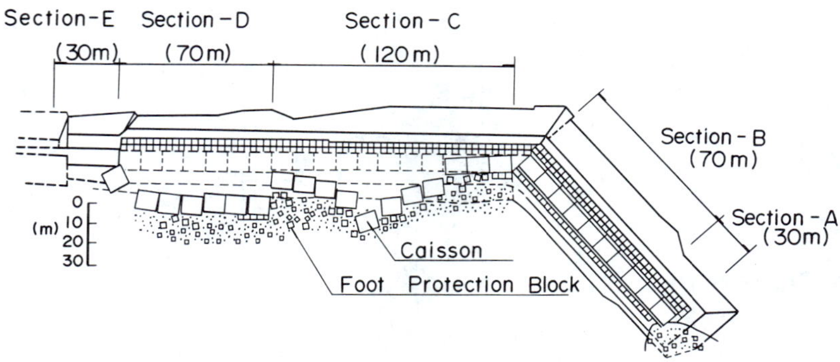

Figure 3. Caisson Sliding at the North Breakwater

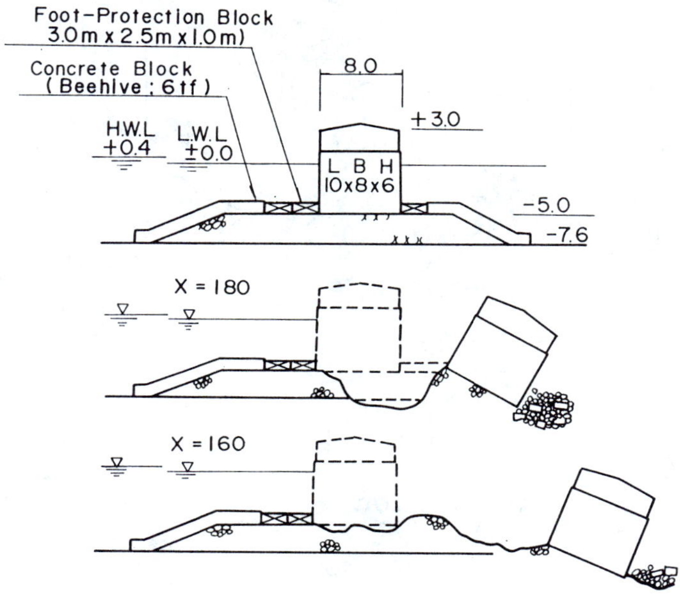

Figure 4. Damage Patterns of Section C of the North Breakwater

Figure 4 (upper diagram) shows a standard cross section of section C. The depth was 7.6m and the caisson was settled at a depth of 5.0m. Figure 4 (lower diagrams) shows typical damage patterns. At the position of X=180 m, the caissons slid 20m and pushed foot-protection blocks (3.0m long, 2.5m wide and 1.0m thick) and concrete blocks (6tf beehive). The scouring of rubble stones (30~300kg)

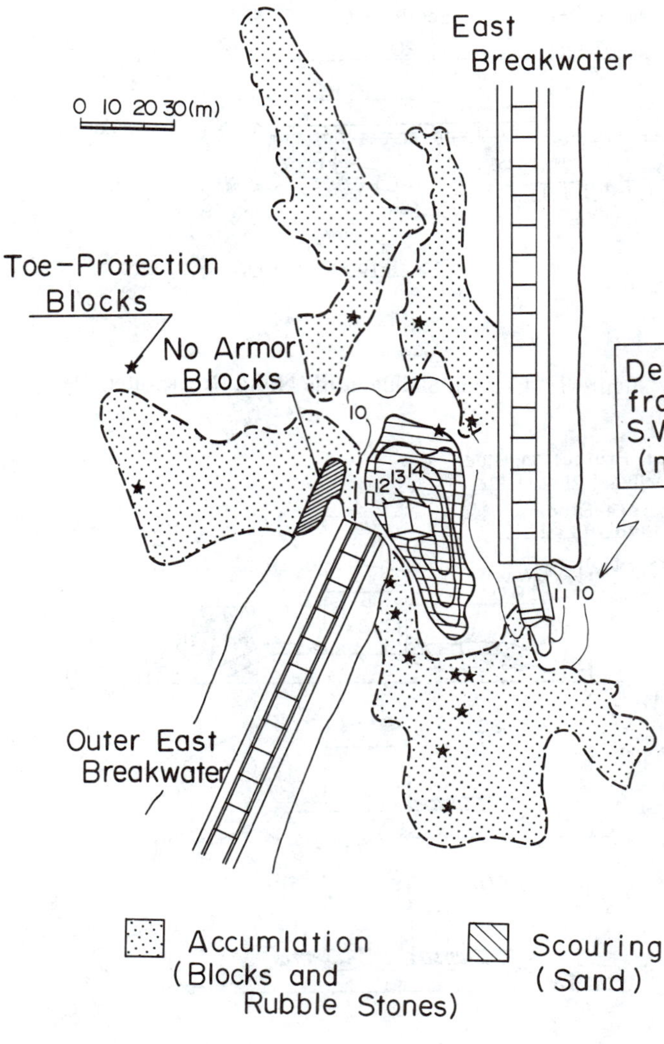

Figure 5. Scouring at the Narrow Opening

progressed along with the action of tsunami waves even after the caissons had slid, and mound rubble collected at X = 160 m.

Figure 5 shows the scouring at the narrow opening between the east and the outer east breakwaters. The width of the opening is 35m and the depth was 10m. Rubble mounds of both breakwaters were covered by concrete blocks (1 tf beehive) and foot-protection blocks (1.5m wide, 2.5m long and 1.5m thick). These armor blocks were moved and scattered as shown in Fig. 5. The maximum scouring depth was 4 m from the original sea bed, and the head caissons at both breakwaters collapsed.

Tsunami forces on vertical walls

The lower part of Fig.6 shows the sliding distance S of each section in the north breakwater. The upright structures of Section E (seawalls for reclaimed land) were heavily damaged, because the sand fillings were not completed when the tsunami struck. Other sections were caisson breakwaters with depths of 5~11 m, and widths of 7.5~9.0 m.

Tanimoto (1983) proposed tsunami wave pressure distributions (Fig.7). The pressure intensity under the still water level p_1 and the crest tsunami height η^* can be calculated as follows:

$$p_1 = 2.2 w_0 a_I \quad \text{-----------(1)}$$

$$\eta^* = 3.0 a_I \quad \text{-----------(2)}$$

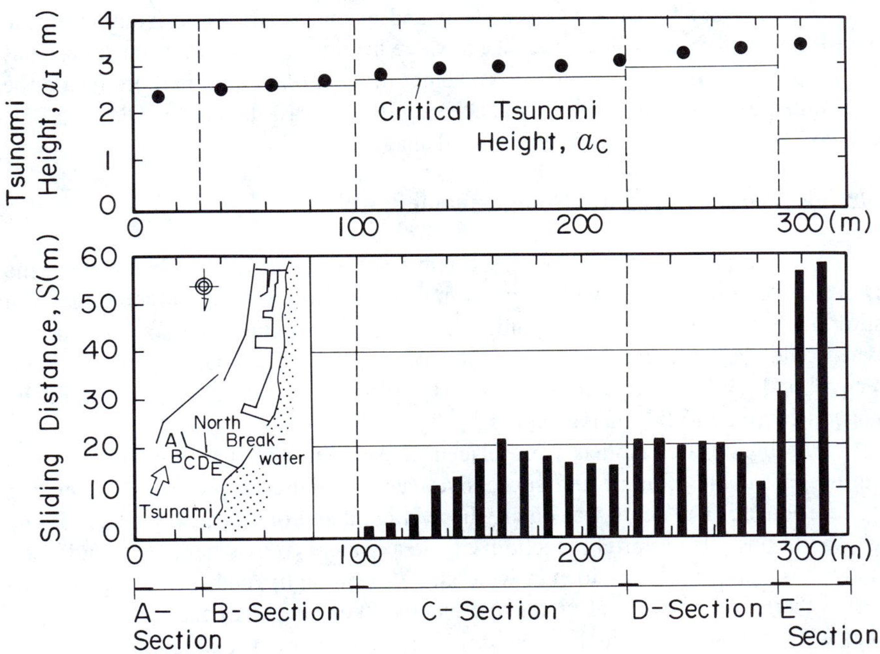

Figure 6. Tsunami Heights and Sliding Distances of the North Breakwater

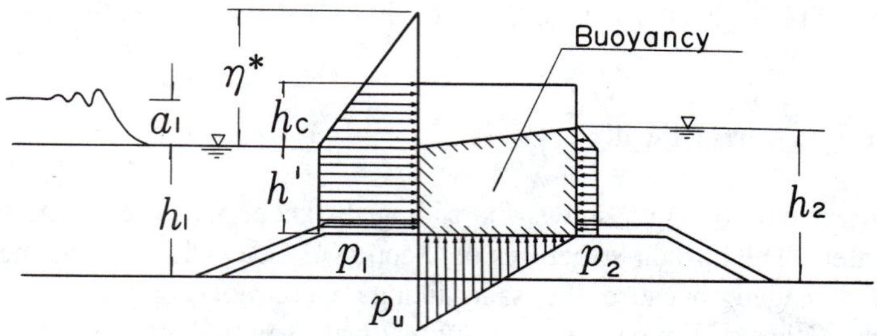

Figure 7. Tsunami Wave Pressure Distribution

where a_I is the incident tsunami height and w_0 is the unit weight of water. h_1 and h_2 are the depth at the offshore side and the harbor side, respectively.

The upper part of Fig.6 shows the critical tsunami heights a_C (horizontal lines) calculated by Tanimoto's formula, and the incident tsunami heights a_I (solid circles) estimated by 3-D numerical simulations. The sliding distances S increase when $a_I \geq a_C$, which confirms the applicability of Tanimoto's formula for prototype failures.

The distance between Okushiri Port and mainland Hokkaido is about 30km. Because the wind fetch is short, the design wave height for the breakwater H_{max} was 3.5 ~ 5.1 m. However, the design wave height in most ports and harbors facing the Sea of Japan exceeds H_{max} of 10 m. The small design wave height for Okushiri Port seems to have contributed to the extensive damages.

Experiments on scouring by strong tsunami flows

Large-scale hydraulic model tests with a scale of 1/15 were also made in the 3-D basin (50 m long, 20 m wide and 3.5 m deep) at the Port and Harbour Research Institute (Fig.8). Four axial flow pumps were installed in the basin so that tsunami waves could be reproduced by the action of the flow. By changing the rotation direction and cycles of the axial pumps, the flow direction and speed could be changed according to the analog signals.

The breakwater models were placed in the center of the basin (Fig.9). The scourings of rubble mounds and successive caisson failures at the narrow opening were reproduced. The depth was 66cm, the average depth of this area considering the tidal conditions. The upright section of breakwaters were made of reinforced concrete box (66.7cm long, 56.7cm wide and 73.3cm high) and their weights were adjusted by the filling materials. Concrete blocks (296gf beehive) and foot protection blocks (10cm wide, 16.7cm long and 6.7cm thick) were produced to scale. To facilitate the observations, concrete blocks were classified into four areas (Fig.9) and painted different colors.

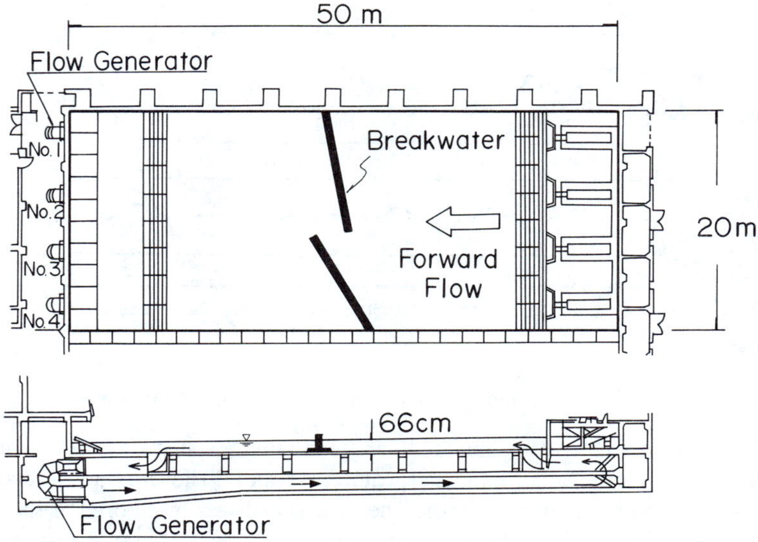

Figure 8. 3-D Basin with Flow Generators

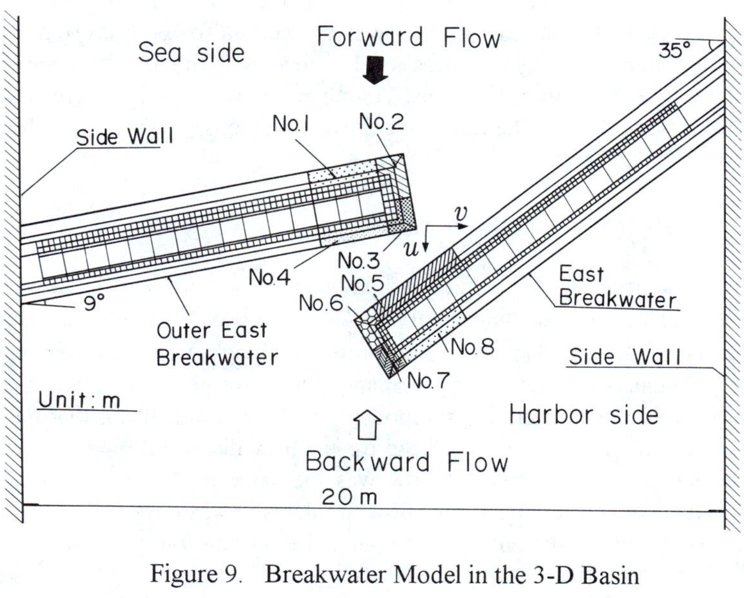

Figure 9. Breakwater Model in the 3-D Basin

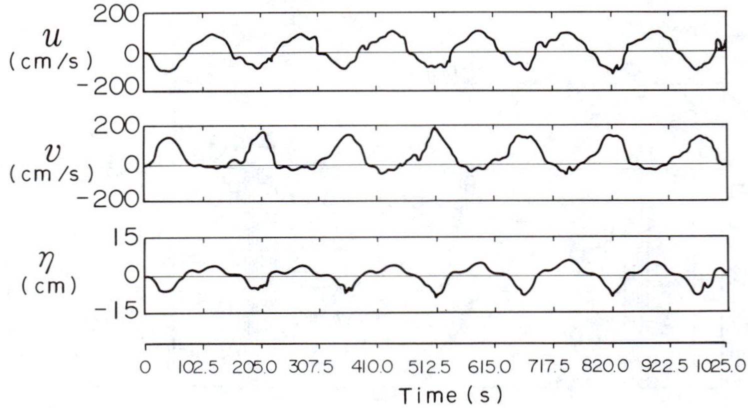

Figure 10. Flow and Water Level at the Center of Opening

Tsunami waves were simulated by generating oscillatory flows with the observed tsunami period of 10 minutes (prototype scale). The forward direction is defined as flow coming into the harbor. The flow speed was monitored at the center of the opening (Fig. 8) with a 2-D electromagnetic flow meter at the depth of 10 cm above the still water level. The peak flow speed of the tsunami (U_0) was defined as the square root of each flow components. The water level was also measured at the same point by a wave gauge.

Five kinds of tsunami flow with different peak speeds (U_0= 112, 151, 169, 180 and 196 cm/s) were used for stability tests. How many tsunami waves attacked this area is not clear, but at least three waves are certain to have covered this area from information given by witnesses. In this experiment, tsunami waves corresponding to a duration of 6.5 times as long as the wave period were created to observe the displacement of the armor blocks and caissons. Figure 10 shows the flow and water level changes at U_0= 196 cm/s.

Damage patterns around breakwater heads

Armor damage was found first at area No.3 in the outer east breakwater at a flow rate U_0=151 cm/s. The number of blocks moved was 18, and the damage ratio in this area was 20 %. The damage of foot-protection blocks was not observed at this flow rate. Tanimoto et al. (1988) examined the flow patterns at the mouth of tsunami protection breakwaters in 3-D model tests. They pointed out that the armor damage was heavier at the projecting head for the flow direction. In our experiment, the damage pattern for the forward flow was the same as for the experiments of Tanimoto. However, for the backward flow, no damage was observed at the area of No.7 of the east breakwater, probably because the opening ratio of our experiment was much lower than usual for tsunami protection breakwaters

Photo 1. Damaged Caisson and Mound of the Outer East Breakwater

At the flow rate U_0=169 cm/s (6.5 m/s in prototype), all armor blocks were damaged at the outer east breakwater and rubble stones were also scattered. At the east breakwater, 80% of the armor blocks were damaged. The damage of foot protection blocks was observed as 80% at the head of the outer east breakwater.

At the flow rate U_0=196 cm/s (7.6 m/s in prototype), the first tsunami wave scoured most of rubble stones around the head caissons, and the head caissons of both breakwaters tilted due to the scouring of the rubble mound. In the outer east breakwater, the damage progressed further to the head caisson failure by the second tsunami wave. The maximum flow rate at the failure was recorded to be 2.3 m/s (8.9 m/s in prototype). Photo 1 shows the damaged caisson and mound of the outer east breakwater after the experiment. The prototype damage around the breakwater heads as shown in Fig. 5 was successfully reproduced by these model tests.

Necessary weight of Armors

The weight required to withstand the flow can be calculated by the following Coastal Engineering Research Center (C.E.R.C.:1984) formula;

$$W = \frac{\pi w U_o^6}{48 y^6 g^3 (w/w_o - 1)^3 (\cos\theta - \sin\theta)^3} \quad (3)$$

where W is the weight of armor unit, U_0 is the flow speed, w_0 is the unit weight of fluid, w is the unit weight of armor units, g is the acceleration of gravity and θ is the angle of mound slope. This equation (3) is commonly accepted as giving the

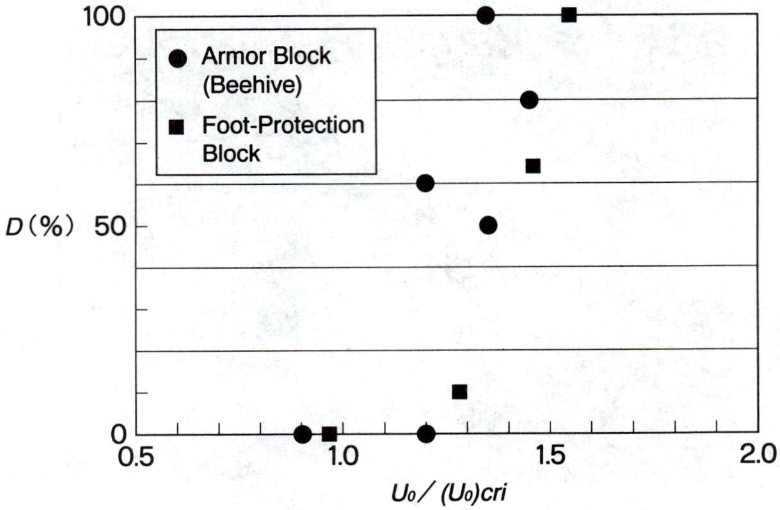

Figure 11. Damage Ratios of Armor Blocks and Foot-Protection Blocks

necessary weight of stones and armor blocks to withstand steady flows. Tanimoto et al.(1988) carried out the armor stability tests of tsunami protection breakwaters, and showed the applicability of this formula for tsunami flows.

Tsuruya et. al. (1994) recommended that the Isbash parameter y for beehive blocks should be 1.2 in the actual settled condition, and 0.967 for individual blocks separately settled on the sea bottom. Kimura(1995) proposed that y=0.6~0.7 can be used based on the previous tsunami damage for foot-protection blocks.

In the 3-D model test for breakwater heads, damage ratios were also observed for each tsunami flows. Figure 11 shows the damage ratio of beehive and foot-protection blocks for U_0=151, 169, 180 and 196 cm/s. The critical flow speed $(U_0)_{cri}$ is calculated from Eq.(3) for beehive (y = 1.2) and foot-protection block (y = 0.65). Damage starts from$U_0/(U_0)_{cri}$ =1.0, and therefore the applicability of these coefficients is proved.

Conclusions

The damage analyses of breakwaters in Okushiri Port due to the 1993 Hokkaido Nansei-oki earthquake tsunami gave the following results and conclusions;

- The caisson breakwaters, whose design wave height due to storm waves were small, were damaged by the tsunami attack.
- The applicability of Tanimoto's formulae for tsunami forces on vertical walls was confirmed by the prototype failure analyses of the north breakwaters in Okushiri Port.

- The failure at breakwaters heads was reproduced by the 3-D model tests. The maximum flow speed of the tsunami was estimated as 8.9 m/s in the prototype.
- The necessary weight of armors around the head of breakwaters can be estimated by the C.E.R.C. formula using the Isbash parameter of previous model tests.

The 3-D experiment in this report did not reproduce the scouring of sandy sea beds. The sea bed was possibly scoured at a smaller flow velocity than indicated in this report, which resulted in the destruction of mounds and in turn led to the collapse of caissons at the breakwater head. Future studies should consider the influence of sediment movement.

Acknowledgments

The field survey on breakwater damages was made by the Port and Harbor Division of Hokkaido Development Bureau and the Cold Region Port and Harbor Engineering Research Center. The authors would like to express their gratitude to Professor N. Shuto at Tohoku University for advice on the numerical calculations on tsunami height.

References

Coastal Engineering Research Center(1984): Shore Protection Manual, Volume II, pp.7-253.

Ito, Y., Tanimoto, K., and Kihara, T.(1969): Desital computation on the effect of breakwaters against long-period waves(5th Report), Repot of Port and Harbour Res. Inst., Vol.8, No.3, pp.16~46.

Kimura, K. (1995): Stability of rubble mound foundation for composite breakwaters against storm waves and tsunamis, Report of Civil Engineering Res. Inst., No. 110, pp. 82.

Tanimoto, K.(1983): On the Hydraulic Aspects of Tsunami Breakwaters in Japan, Proc. of the International Tsunami Symposium by IUGG Tsunami Commission, pp.423~435.

Tanimoto, K., et al. (1983): Field and laboratory investigations of tsunami caused by 1983 Nihonkai Chubu Earthquake, Technical Note of the Port and Harbour Res. Inst., No. 470, 299p.

Tanimoto, K., Kimura, K., Ikegami, M. and Miyazaki, K. (1988): Study on stability of submerged dike at the opening section of tsunami protection breakwaters, Report of Port and Harbor Res. Inst., Vol.27, No.4, pp.93~121.

Tsuruya, H. and Nakagawa, Y. (1994): Model experiment for reproduction of disaster at Okushiri-Higashi breakwater by Hokkaido-Nanseioki earthquake Tsunami, Technical Note of the Port and Harbour Res. Inst., No.789, 19p.

CHAPTER 365

RICHARDS BAY NORTH BREAKWATER - REPAIR OF A ROUNDHEAD: MONITORING, MODEL TESTING, DESIGN AND CONSTRUCTION

D Phelp[1] , A McClarty[2] and A Bartels[3]

ABSTRACT

The head section of the north breakwater, at the entrance to the Port of Richards Bay, has suffered some damage since its construction in 1973. This paper briefly discusses the changes in the design conditions, the first unsuccessful repair in 1986/87, using 15t dolosse, the annual photographic monitoring results showing the rates of damage to the roundhead since then, and the model testing of various repair options, using available spare 20t and 30t dolosse. Finally the construction of the optimal repair, carried out in 1996, using a heavy duty mobile crane (with 48m boom reach) and DGPS positioning, is described. Construction was still in progress at the time of publishing this paper.

INTRODUCTION

The Port of Richards Bay, on the east coast of South Africa, has two dolos breakwaters, a shorter straight breakwater on the northern side of the harbour entrance channel and a longer curved breakwater on the southern side. The north breakwater consists of a straight rubble mound structure, constructed in 1973, which stretches for approximately 600m perpendicular to the coastline. The original armouring on this breakwater consists mainly of 5t dolosse on both sides of the trunk, but includes 15t dolosse on the roundhead (Figure 1). Like many similar breakwaters, the neck of the roundhead was found to have experienced the worst damage to the armour layer.

[1] Environmentek, CSIR, P O Box 320, Stellenbosch 7600 South Africa
[2] Portnet, Port of Richards Bay, P O Box 464, Richards Bay 3600 (SA)
[3] Entech Consultants (Pty) Ltd, P O Box 752, Stellenbosch 7600 (SA)

Annual photographic monitoring of the north breakwater showed an increase in damage in localised areas on the head, particularly on the southern side (Figures 1 and 2), despite temporary repairs using 15t dolosse, carried out by Portnet in 1986/87. A previous model study (CSIR, 1992a) indicated that the spare 20t and 30t dolosse, available from a stockpile of spare dolosse at the root of the south breakwater, would be suitable for the repairs. It was therefor decided to test various repair options using 20t and 30t dolosse.

In June 1994 the CSIR was commissioned by Portnet to carry out a physical model study to determine the most cost effective design for the repair work. These included repair solutions in which a toe of anchored 30t dolosse was constructed around the head and 20t dolosse were placed in the cusps and over the anchor chains. After an initial calibration run, a total of four different repair options were tested. Due to delays in the commissioning of a suitable crane, the construction of the final repair was delayed to the end of 1996. This paper does however include construction details of a temporary repair using 5t dolosse (placed using an available smaller crane) and the start of the final repair, using the 20t and 30t dolosse.

FIGURE 1: View of North Breakwater and Cross-Section through head.

MONITORING RESULTS

The crane/helicopter photographic survey method (Phelp *et al,* 1994) was used to annually monitor the breakwater. The photographic survey stations are spaced at 25m intervals, and both sides of the breakwater are monitored; the breakwater was however already 13 years old before annual photographic monitoring was commenced. The areas with the highest damage (over 25% breakage) continued to deteriorate the fastest. At stations P2 and P3 on the round-head, the damage was mostly localized in cusps, which are visible in Figure 1. A sharp increase in the rate of damage in the worst areas, since 1992 (72 months), can be seen in Figure 2. This damage has mainly been attributed to bottom scour resulting in increased wave heights reaching the breakwater, ie. increased depth limited wave heights. Station P1 also lies at the transition between the 5t and 15t dolosse, which is at an inherent weak point in the armouring.

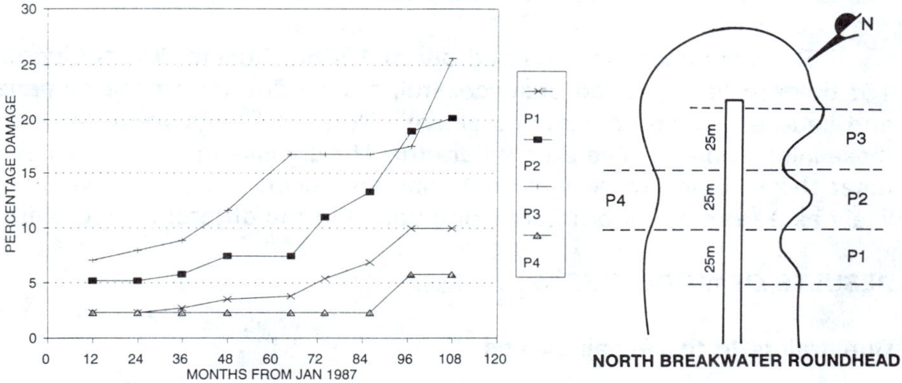

FIGURE 2: Rates of Damage to the Dolos Armouring Since 1987.

FACTORS CONTRIBUTING TO THE HIGH DAMAGE

The occurrence of low pressure cyclonic storms (cyclones Demoina and Emboa in 1984) subjected the breakwater to wave heights exceeding the 7,9m 1:50 year design H_{mo}. Storm wave and low atmospheric pressure set-up associated with these storms also had the effect of raising the water level, thereby raising the depth limited wave heights reaching the breakwater. A previous model study on wave penetration in the entrance channel (CSIR 1974) also showed, based on wave refraction that wave conditions between SE and S (deep sea) result in the worst conditions opposite the north breakwater head.

Bathometric surveys carried out regularly by Portnet revealed a general erosion of sand, of up to 2m, around the head and seawards of the breakwater, especially between 1989 and 1994. This would have resulted in further increasing the depth limited heights of the waves reaching the breakwater. Besides the scour at the toe (seismic surveys carried out by CSIR in August 1993 confirmed the toe erosion) the breakwater was also subjected to severe storms in October 1990 (max H_{mo} of 6.1m) and moderate storms since then, resulting in accelerated damage to the dolos armouring behind the weakened toe.

Like many similar breakwaters, both sides of the neck of the roundhead were found to have experienced the worst damage to the armour layer. Jensen (Jensen, 1984) found that for tetrapods, relative to the stability coefficients for the breakwater trunk, the roundhead area at 135° from the wave direction showed the lowest stability. This indicates that some of the damage found on the southern side of the north breakwater could also have been caused by more northerly waves. Model tests also showed evidence of eddies on the lee-side of the roundhead where toe dolosse were pulled off the slope.

Repairs which were carried out in 1986/87 using 200 additional 15t dolosse have proved unsuccessful, mainly due to the toe damage and sea-bed erosion mentioned above. No model tests were done for these initial repairs. The area where the 15t dolosse on the roundhead meet the 5t dolos trunk (Figure 1) has also been a focus of damage, likely as a result of poor interlocking between the different sized units.

RESULTS OF MODEL TESTS

Constraints to the Repair Design

Model tests were carried out in an existing 3D model of Richards Bay (CSIR, 1995), to save both time and costs. This available model was built at a scale of 1:110 and covered the harbour entrance and part of the inner channel. The model test options were restricted to using available 200 20t and 100 30t dolosse, which have been brought across the harbour from a stockpile near the southern breakwater. The proposed repair consists of a double row of 30t dolosse at the toe, with a double layer of 20t dolosse behind. 5t dolosse were also available to wedge into the gaps behind the 20t repair and between the existing 5t slope where the repair overlapped the breakwater trunk section.

The removal of rubble and pre-repair slope preparation was limited to the removal of only large broken dolos pieces (assisted by divers), and the filling of holes at the toe of the armour slope. The crane reach was limited to 45m for a 30t dolos, which was achieved by using a specially designed crane which could fit onto the 6m wide mass-concrete capping of the north breakwater.

Choice of Model Scale

The scale of 1:110 used for the tests gives a Reynolds number of approximately 1×10^4, which is just within the minimum range recommended by Van der Meer (Van der Meer, 1988) but less than the value recommended by CERC. Some scale effects should therefore be expected, especially on the 5t dolosse, but these scale effects have been found (Oumeraci, 1984) to make the model results conservative. The scale effects, being similar for all the test runs, still allowed the comparisons made between the options tested. The results were however interpreted as giving an indication rather than an exact prediction of the possible damage. The calibration test showed that the hindcast of the damage which occurred in the 1990 storm is in qualitative agreement with the observed prototype damage confirming the validity of the physical model. A plan of the model is shown in Figure 3 and the "as-built" cross-section through the head in Figure 1.

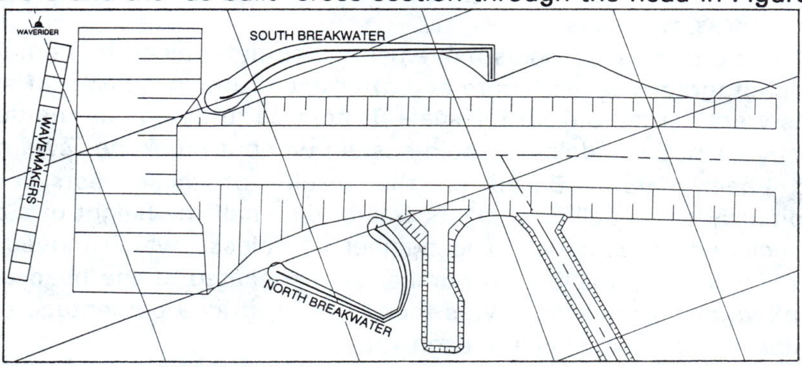

FIGURE 3: Plan showing Model Layout

Wave Generation and Measurement

Eight standard wire resistance wave probes were used which were coupled so that measurements could be carried out over prescribed areas. The wave generator bank was 9,75 m long situated at the 1 km mark (Figure 3). The direction of the generators was 30° N, ie: waves coming perpendicularly out of the wave boxes have a direction of **120°**. Based on a review of existing data, the following **main test conditions:**

- Wave direction, entrance area (12s) **145°**
- Storm input, **Richards Bay Spectrum,** γ = 2,74 with the following approximate steps H_{mo} = **4, 5, 6, 8 m** with following wave per respectively: T_p = **10, 12, 13, 15,5 s**
- Water level for main test MHWS = **2,0 m CD**

DESCRIPTION OF THE TESTS

Test procedure

In order to calibrate the physical model, a calibration test was carried out in which the prototype damage resulting form the 1990 storm was reproduced in the model. A number of repair options were then investigated, starting with the simplest option and then extending the repair until a stable solution was found. Before each repair option was constructed, the original damage (after the 1990 storm) was replicated in the model. The repaired breakwater was than exposed to the five different wave conditions described above. After each test, the repair dolosse were removed and the 1990 damage reconstructed, after which the next repair option was implemented.

Measurement of Damage

Movements of Dolosse

The number of dolos movements was determined by comparing the photographs taken before and after each run. A number of swing tests were carried out on full-scale 9t dolosse to determine the degree of movement these dolosse could sustain without breakage (Zwamborn and Phelp, 1989). Based on the results of these tests it was recommended that all movements greater than half the height of a dolos be included as damage. The number of dolosse which moved was determined separately for a number zones located at the head of the breakwater. The damage was then expressed as a percentage of the number of dolosse placed in each zone.

Rocking Dolosse

In addition to determining the movements, the number of dolosse which were rocking was estimated visually during each test. However due to the difficulty in observing these movements over the whole test area, it was decided to use the small movements (less than h/2), which were measured from the photographs, as an estimation of rocking dolosse. In (CSIR, 1992b) this alternative approach was also adopted, where it was found that dividing the movements less than h/2 by nine, would give a reliable estimation of the number of rocking dolosse.

DETAILS OF THE REPAIR OPTIONS TESTED

Discussion on the Repair Strategy Followed

The strategy of placing 20t or 30t dolosse on top of the damaged 5t and 15t dolosse was originally tested and found to be feasible in previous model tests (CSIR, 1992a and 1992b). Static tests on dolosse have shown that a dolos can carry 4 to 6 times its own weight without breaking; this implies that a number of layers can be constructed without breakage under static load. Thus it was considered feasible to place a 1 to 2 layer thick 20t dolos strengthening layer, safely on top of the existing damaged 15t dolosse. Although the quality of the underlying 15t dolosse is questionable, the dynamic loading over the past 8 years has caused the weaker dolosse to break, and careful placing of new dolosse should not result in significant further breakages. However, since most parts of the repair sections will consist of already broken units, the repair itself was designed as well interlocked armour, finished to a uniform slope, which should be able to stand on its own.

Although stresses cannot be modelled, extensive prototype observations and structural tests on full size dolosse support the above conclusions. For example the main breakwater at Gansbaai, on the South African southern coast, consists of an original layer of 4.5t dolosse covered by one layer of 12.4t units, covered by one layer of 17.1t dolosse overlapped with a double layer of 25t dolosse. This repair has thus far proved to be successful.

Repair Option 1

The first repair option tested involved the filling of the cusps with the minimum number of dolosse, ie. starting with the simplest option and then extending the repair until a stable solution was found. This was a temporary (intermediate) repair solution to protect the breakwater core from further damage, but would also eventually form part of a more permanent repair. The number of repair dolosse per hole was approximately 10 30t and 20 to 30 20t dolosse. The total number of dolosse needed to repair all the cusps (northern and southern sides) was approximately 40 30t and 150 20t. The 30t dolosse were used to construct a toe at the bottom in front of the cusps to anchor the 20t dolosse filling up the cusps behind.

Repair Option 2

After repair option 1 had shown extensive damage to the edges of the repaired areas, it was decided to extend the repaired area to provide a more uniform profile. The second repair option was similar to the first repair option but with a more continuous row of 30t dolosse at the toe around the whole breakwater head. The construction of a toe with a double layer of 30t dolosse was an attempt to increase the stability of the repaired slope. The 20t repair layer was then placed on top of the 15t dolosse. The shape of the head was modified slightly to have a slightly flatter slope. Approximately 80 30t and 200 20t dolosse were used.

Repair Option 3

In this repair option the cusps were also filled with 20t dolosse, but the 30t dolosse on the southern-side of the breakwater were fixed using anchor chains (scrap, ship's anchor chains). First, a double row of 30t dolosse were placed to construct the toe, after which the 20t dolosse were used to repair the cusps and to cover the chains. This method proved to be very successful in the construction of the Port of Cape Town breakwater where similar conditions exist at the toe (CSIR, 1985 and Zwamborn *et al*, 1990). As was done in Cape Town, the end of the chains should be connected to a piece of scrap rail to provide improved anchorage. Approximately the same number of dolosse were used as for the temporary repair option (option 1). The chains used to fix the 30t dolosse can be clearly seen. Figure 4 shows the anchor chains in the model and as used in prototype.

FIGURE 4: Anchor Chains used to Secure 30t Toe Dolosse

Repair Option 4

In this repair option, a double row of anchored 30t dolosse was used to construct the toe around a slightly wider head. The 20t dolosse were then placed in the cusps and over the chains. Also, before the 20 t dolosse were placed on top of the 15t dolosse, the damaged slope beneath the repair was reprofiled to allow for a double layer of 20t dolosse, to provide a more uniform repair slope. In this last repair option all the 30t dolosse were fixed by anchor chains.

INTERPRETATION OF RESULTS

Comparison of Damage to the Repair Options Tested

A comparison of the damage at the end of each run is shown graphically in Figure 5. The x-axis presents the section of the breakwater which starts at the northern-side of the breakwater, around the head to the southern side of the breakwater. These location numbers are given in Figure 5. Repair Option 4 is clearly the best option.

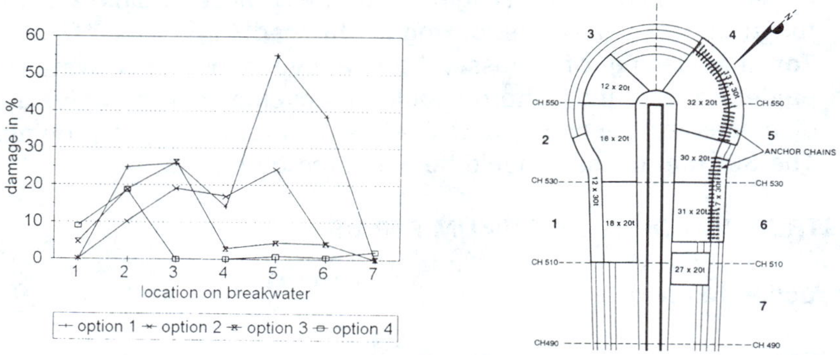

FIGURE 5: Comparison of Damage to the Repair Options Tested

For all the repair options, the existing 15t dolosse were well protected by the additional dolosse. For repair options 1 and 2 it was clear that again, similar to the prototype, the highest damage could be observed just behind the head at the southern side of the breakwater. The eddies caused by the waves passing around the head were responsible for rolling the toe dolosse along the surface of the rubble. Once a dolos had been extracted from the slope it was easily rolled along by big waves.

The 30t dolosse, which were fixed by anchor chains, reduced the damage in that specific area considerably. Repair Option 4 showed acceptable damage levels for all the repair dolosse, apart from the area on the northern side of the head where the 20t dolosse showed higher damage - poor interlocking of this section may have been the cause. The damaged 15 t dolosse on this side were not reprofiled as was done on the southern side, which meant that the 20t repair dolosse were not interlocked properly in a double layer. Although the original prototype damage (CSIR, 1994) on the northern side of the head is less than one third of that on the southern side, it is felt that the repairs would have been more effective if the whole northern side had also been properly prepared (reprofiled) before the repair.

Based on the results of the model study and from previous experience, the following recommendations for the repair works were made:

a) Before any repairs, as many as possible of the broken pieces of dolosse should be removed without disturbing adjacent dolosse.
b) Crane and ball surveys should be used to identify all underwater depressions, both before and during construction of the repair.
c) Particular care should be taken when placing the dolosse to ensure that dolosse interlock as well as possible with the

underlying dolosse. Re-slinging of a newly placed dolos is worth the effort if a better interlocking is obtained.

d) For the placing of dolosse, both above and below water, a predetermined grid pattern should be developed with angles and distances for the crane, or x,y references for DGPS positioning.

e) The 30t toe dolosse should be anchored by chains.

CONSTRUCTION OF THE OPTIMUM REPAIR

Construction Methods

Based on the results of the model tests, 20t and chained 30t dolosse were to be used for the repair. The 20t and 30t dolosse were brought across from the south side of the entrance channel to a stockpile at the root of the north breakwater, while the 5t dolosse were also already available from a part of the root of the original breakwater which was now covered with sand. Three double direction trailers were then used to transport the dolosse onto the breakwater. These trailers could only pass when unloaded, which meant that only one 20t or 30t dolos could be brought onto the breakwater at any one time (Figure 6).

Initial crane and ball surveys were done with 5m profile intervals over the damaged areas. A certain amount of reprofiling was then carried out to remove irregularities in the overall profile. Broken pieces of dolosse were removed and placed in erosion holes. Another crane and ball survey was then carried out to get the underlayer profile, from which the repair dolos placing grid could be calculated. The smoother the underlayer profile, the easier it was to set a placing grid with uniform packing density.

The repair dolosse were then placed, starting with the double row of chained 30t dolosse at the toe. The chains, which were lifted on a separate hook were laid at an angle of 45° seawards along the toe of the slope, later to be covered with 20t dolosse. The crane hook was fitted with a 15m sling (to ensure the hook and pulley remained out of the seawater), a quick release hook (Figure 6) and a double cable sling. The double slings which support the dolosse were hand spliced (instead of swage joined) to allow easy removal of the slings once the dolos was in position. The quick release hook was hung from a swivel and fitted with two torque bars, which allowed easy rotation of the dolosse to ensure good interlocking. The torque bars were attached to 10mm nylon (light and water resistant) ropes, which were pulled perpendicularly from the mass capping to orientate the dolosse.

It was found that to ensure correct packing density, the dolos placing must be kept as close as possible to the grid coordinates. The final orientation and positioning of the dolos is then done by eye to ensure good interlocking. Dolosse are placed with a minimum of three contact points to reduce the change of rocking under wave action. After all the grid positions were full, it was found that up to 10% additional dolosse had to be placed "in holes" to ensure a well interlocked uniform profile. To identify these "holes" it is advisable to get an aerial view of the slope form a helicopter, or from a basket hung from the crane.

DGPS for Crane Positioning

For both the crane and ball surveys of the slope profiles, and the correct placing of the dolosse, there was a need to accurately position the hook of the crane. The original method used was triangulation from two theodolites, but this proved labourious and time consuming. Another method was to fit the crane with a pendulum boom angle indicator and a horizontal slew protractor whereby the horizontal and vertical angles of the boom could be measured and converted from polar to x,y grid coordinates.

Recently a differential GPS system has been introduced using satellite positioning linked to a portable computer onboard the crane (Figure 7). The satellite receiver is positioned on top of the crane boom, directly above the position of the hook. The pre-determined positions are entered into AutoCAD software on the computer, and standard survey software enters the real time navigation parameters which indicate the position of the boom. By entering the standing position of the crane along the breakwater, the boom reach and safety circle can also be indicated on the screen. The crane operator can then immediately see which dolosse can be placed from the present position of the crane. The AutoCAD dolos placing grid is shown in Figure 8.

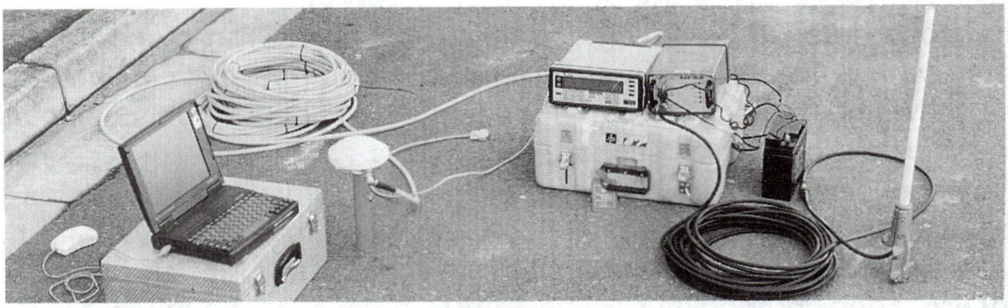

The positioning software includes the following useful features:

- Zoom in and out, and centring the cranes position on the screen.
- The entry of up to 20 predetermined crane standing positions on the breakwater, including facility to orientate and offset.
- The entry of up to 500 top and 500 bottom layer dolosse, including an indication of size and numbering (colour options)
- The facility to import and editing of an AutoCAD or other CAD drawing of the breakwater eg: the "as-built" layout.
- Indication and editing of the safe radius of the cranes reach.
- The input and storage of the placed positions of the dolosse.
- A backup system where the polar coordinates can be entered to position the crane, should the DGPS signal fail.

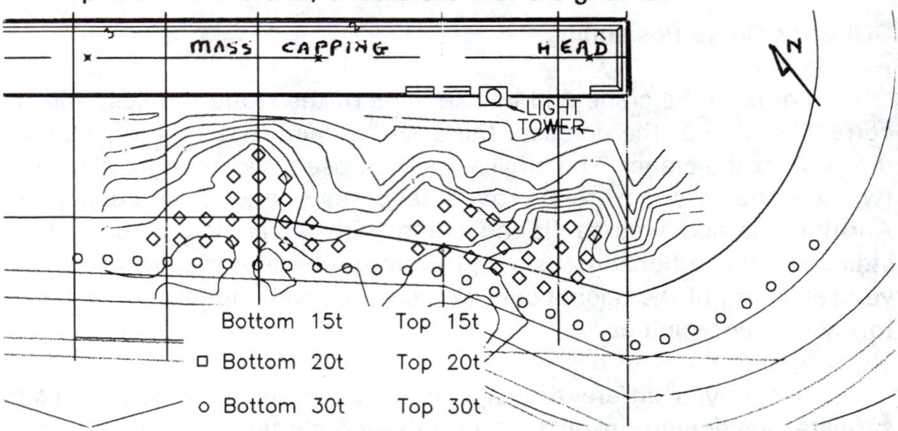

FIGURE 8: Example of DGPS AutoCAD Dolos Placing Grid

Emergency Temporary Repair and Final Repair

Planning for the repair was well underway by mid 1996, when a breakdown of the special crane (the only crane which could fit onto the breakwater and have sufficient reach for placing the 30t dolos toe) resulted in having to delay the final repair to the end of 1996. As the localised damage on the southern side of the roundhead was starting to expose the underlayer rock, it was decided to carry out emergency temporary repairs. These took the form of filling the cusps with 5t dolosse (which could be placed with a smaller mobile crane). This repair was not extended further than the original slope profile so that it would form part of the underlayer of the final repair. The same construction method was used as described above. Figure 9 shows this section before and after the emergency repair. Figure 10 shows the chained 30t dolosse at the start of the final repair in December 1996.

ACKNOWLEDGEMENTS

The first author would like to thank Mr McClarty of Portnet and Mr Bartels of Entech for their contributions to this paper and for the team spirit which prevailed between Portnet, the Client/Port Authority, Entech, the Coastal Consulting Engineers and the CSIR Research Laboratory.

REFERENCES

CSIR (1974). Design of Richards Bay Harbour Entrance Wave Model Tests on Entrance Layout. CSIR report C/SEA 74/2.

CERC (1984). shore protection manual. US Army Coastal Engineering Research Centre, Vol. 2. US Government Printing Office, WashingtonDC

CSIR (1992a). Port of Richards Bay: Model Study to Optimise Breakwater Repairs. CSIR report EMAS-C 93049.

CSIR (1992b). Mossel Bay Breakwater: Model Study of Repair Options. CSIR report EMAS-C 92051.

CSIR (1994). Port of Richards Bay: Photographic crane and helicopter surveys of the north and south breakwaters, March 1994. CSIR report EMA/C 94115.

CSIR (1995). Port of Richards Bay - North Breakwater Repair: Results of Model Tests. CSIR Report EMAS-C 95033, Stellenbosch.

Jensen (1984). A Monograph on Rubble Mound Breakwaters, O. Juul Jensen, DHI, Denmark. November 1984.

Oumeraci H (1984). Scale effects in coastal hydraulic models. Symposium on scale effects in modelling hydraulic structures. International Association for Hydraulic Research.

Phelp *et al* (1994). Results of Extensive Field Monitoring of Dolos Breakwaters. 24 ICCE 1994, Kobe, Japan.

Van der Meer J (1988). Rock slopes and gravel beaches under wave attack. Doctoral thesis. Delft University of Technology.

Zwamborn J A and Phelp D (1989). Structural tests on dolosse. Seminar on stresses in concrete armour units. Vicksburg, USA.

Zwamborn *et al* (1990). Redesign, Repair, and Monitoring of the Table Bay Breakwater. 26 PIANC 1990, Osaka, Japan.

CHAPTER 366

THYBORØN COASTAL INVESTIGATIONS 1995: NEW LESSONS FROM AN OLD COASTAL PROBLEM

by

I. Brøker[1], J.A. Zyserman[2] and Per Roed Jakobsen[3]

INTRODUCTION AND BACKGROUND

Thyborøn channel is located on the West coast of Jutland, Denmark, and connects the Limfjord with the North Sea, see figure 1. The channel was naturally opened following a major storm event in 1862. In the years after the opening of the channel, the neighbouring coasts experienced substantial erosion, which put the town of Thyborøn at risk. For historical background and previous investigations and analysis, please see References.

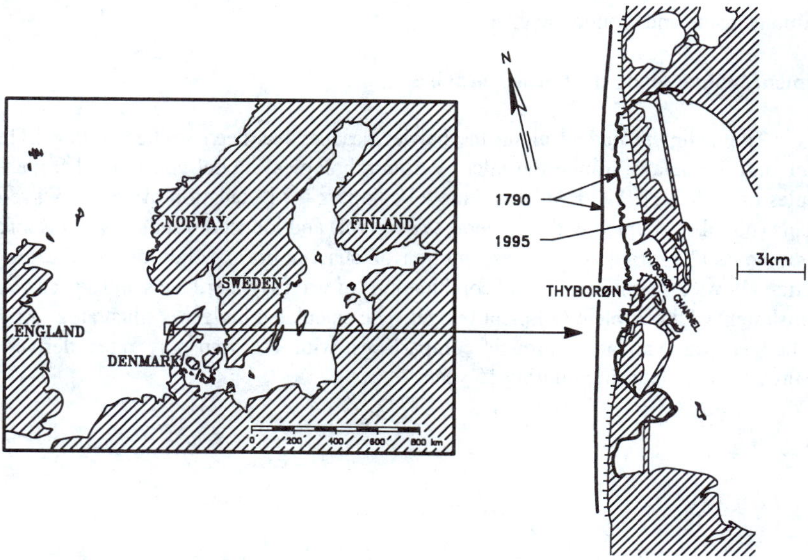

Figure 1 Location map, including indication of development over 205 years

[1] Chief Engineer, Danish Hydraulic Institute (DHI), Agern Allé 5, DK-2970 Hørsholm, Denmark
[2] Coastal Engineer, Danish Hydraulic Institute (DHI), Agern Allé 5, DK-2970 Hørsholm, Denmark
[3] Director, Danish Coastal Authority, Højbovej 1, DK-7620 Lemvig, Denmark

The geology in the area can be briefly described as postglacial sediments, mostly fine to medium sand, overlaying hard clay in depths of 10 to 12 m below seabed level. The area is exposed to severe storms from directions between southwest and northwest. Along the coast the conditions are influenced by a northgoing sea current. In the inlet area the currents are determined by the differences in water level across the peninsula of Jutland. The tidal variations are small, with a maximum tidal range of ~40 cm. The yearly net sediment transport along the narrow barrier beaches is directed towards the inlet from both sides. The undisturbed net transport rates are of the order of magnitude of 1 million m^3/year. A dense groyne system reduces the longshore sediment transport rates to about 2-300,000 m^3/year from each side; the sediment transported into the channel ends up on large shoals inside the Limfjord, giving a yearly deposition of approximately 0.5 mill m^3.

The groynes along the coast have decreased the rate of shoreline retreat to about 50% of the natural erosion. This means that the land ends of the groynes have to be extended inland in pace with the erosion. The tips of the groynes experience continuous deterioration due to severe pressure arising from the wave action. The submerged groyne elements are left on the seabed as they are still considered as active parts of the coastal system.

NUMERICAL MODELLING STUDIES

As a part of the study of the processes around this inlet numerical models were established, both for the quantification of the overall sediment budget along the barrier beaches, and for the study of the details of the morphological evolution under storm conditions inside and around the inlet.

Sediment budget along the barrier beaches

The sediment budget along the barrier beaches has been studied by the LITPACK model, and the details around the inlet by a morphological model constructed from various modules of MIKE 21. **LITPACK** is a model complex for calculation of waves, wave-driven currents (possibly coupled with a general sea current) and longshore sediment transport on an arbitrary coastal profile. A basic assumption of quasi-uniform conditions in the longshore direction allows for simulation of all combinations of waves, water levels and sea currents and establishment of a complete sediment budget. The calculated budget is illustrated in figure 2. The budget was validated through comparisons with the transport rates derived from measured erosion/deposition during 15 years.

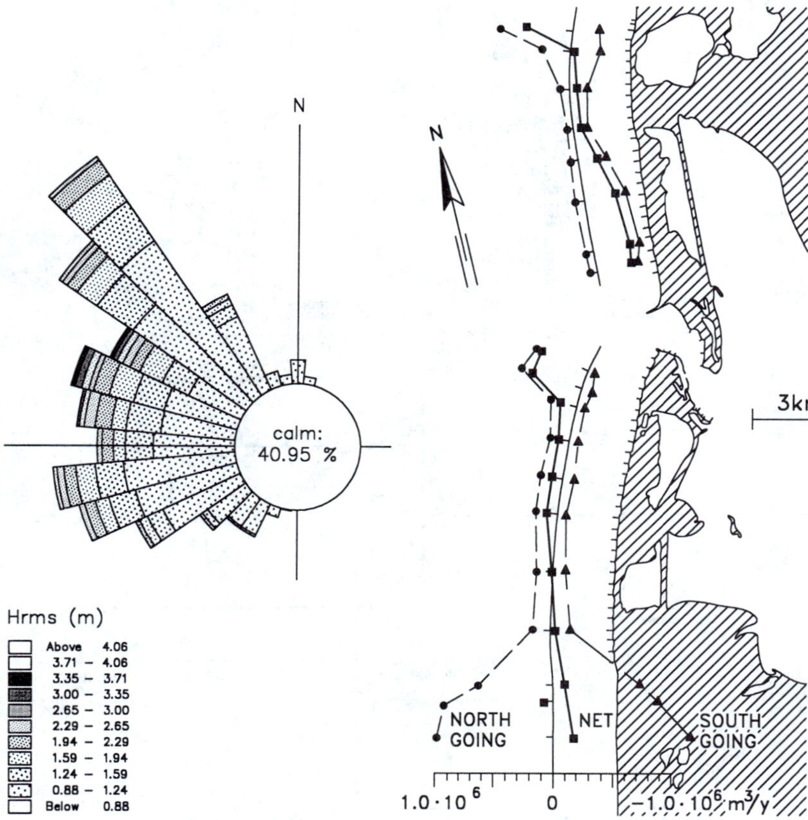

Figure 2 Left: Illustration of wave statistics
Right: Sediment budget along the barrier beaches

Wave level, wind and directional wave data were available for a three-year period. Two characteristic storms (Storm 1 and Storm 3) were selected for detailed studies of sediment transport and morphological evolution in the area around the inlet. Furthermore, the littoral transport along the barrier islands during each storm was calculated using LITPACK to give an indication of the representativeness of the two storms. The distributions of the littoral drift in the longshore direction and along a selected profile are shown in figure 3. The comparison showed that the storms represented approx. 20 % and 15 % of the southgoing and northgoing yearly transport, respectively. The distribution along the coastal profile is, however, more concentrated over the bars under storm conditions than in average over the year.

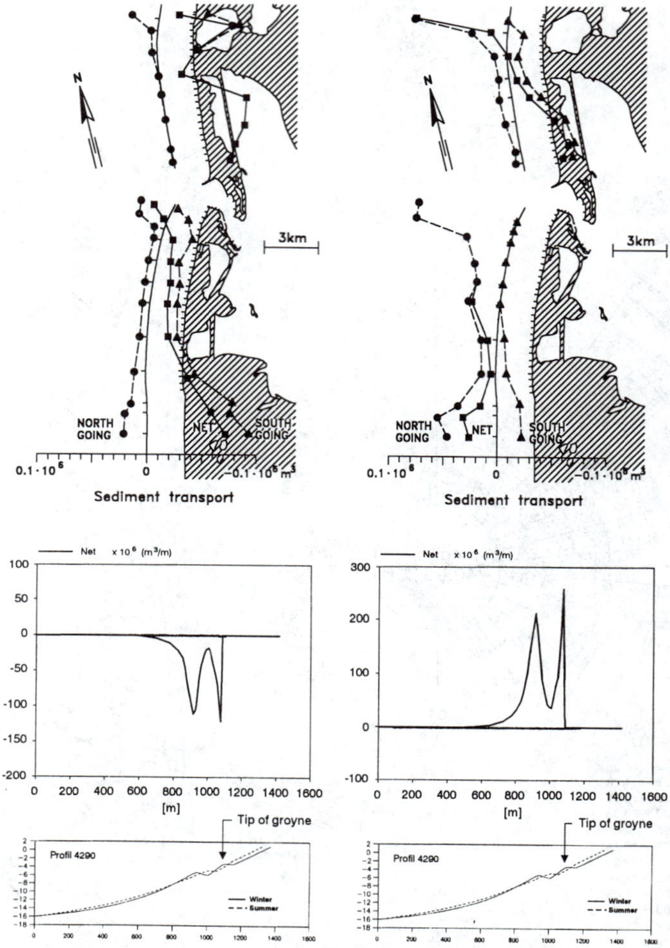

Figure 3 Upper, left: Storm 1, littoral drift along the coasts
Upper, right: Storm 3, littoral drift along the coasts
Lower, left: Storm 1, littoral transport along a profile
Lower, right: Storm 3, littoral transport along a profile

Detailed studies around the inlet

Model setup and boundary conditions for the morphological model

A comprehensive set of models was established in order to transfer measured data into boundary conditions for a morphological model covering the area of interest. The measured data comprised: directional wave measurements at Fjaltring, water levels at Thorsminde, Thyborøn, Aalborg and Hanstholm, wind at Thyborøn and Aalborg, see figure 4. Current speed and direction were measured in the channel during one month for calibration purposes. The model areas and the positions of measurements are indicated in figure 4. It must be noted that the fluxes across the eastern boundary of the general hydrodynamic model are calculated from a 1D river model, MIKE 11, which covers the entire Limfjord from Thyborøn to Aalborg, see Figure 4.

The general and the regional hydrodynamic models have a resolution of 400 m and 100 m respectively. The model is MIKE 21 HD, which is depth-integrated. In the general model the driving forces are wind and variations in tidal levels and fluxes along the boundaries. The effect of Coriolis force is included.

The regional wave model is a parameterised, spectral wind wave model, MIKE 21 NSW.

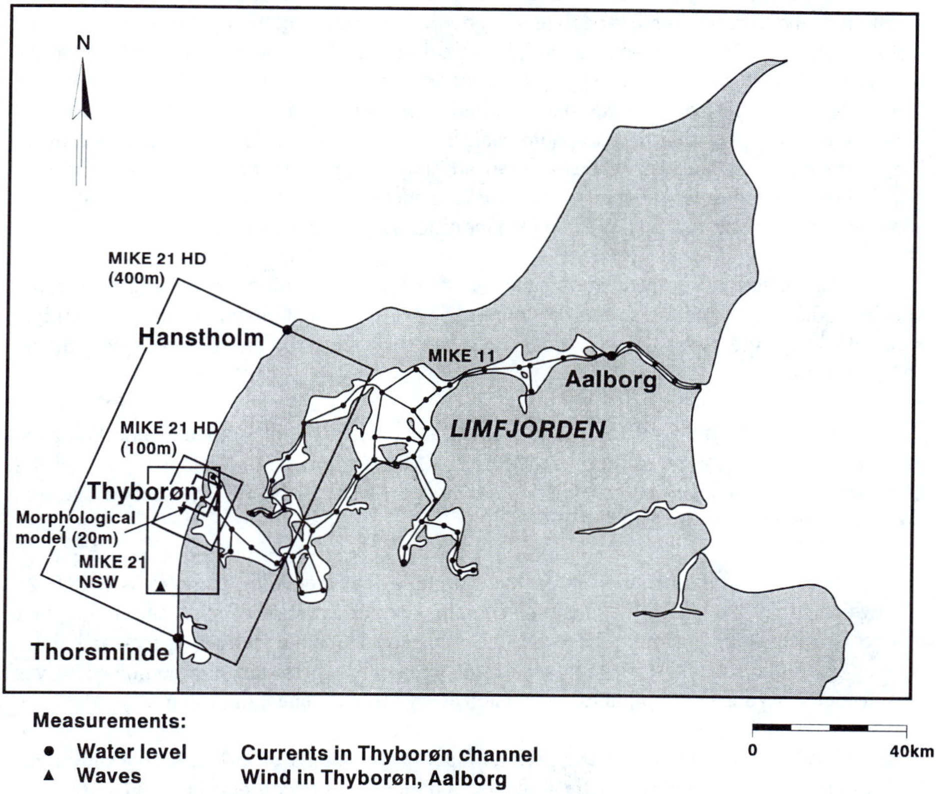

Figure 4 Areas covered by the general and regional hydrodynamic model
Area covered by the regional wave model
Area covered by the morphological model
Locations of measured water levels, wind, waves, currents

The regional models give boundary conditions to the morphological model, see figure 4 for the extent of this model. In the morphological model, wave-, current- and sediment transport modules are coupled as described below. A mild-slope wave model, based on the parabolic approach, was applied in the morphological model complex. Inside the morphological model area, the time- and space-varying radiation stresses were included as an extra forcing term for the hydrodynamic model. The resolution within the morphological model area is 20 m.

Brief description of the morphological modelling complex

DHI's morphological modelling complex is centred around the hydrodynamic model. A morphological simulation is 'warmed up' by calculation of initial wave and current fields on a fixed bed. After the 'warm-up period' the sediment transport field and the corresponding rates of bed level change rates, $\partial z/\partial t$, are calculated. $\partial z/\partial t$ is incorporated in the continuity equation of the hydrodynamic model, and the hydrodynamic simulation proceeds for a period of time corresponding to a so-called 'morphological timestep', which is determined so that the maximum Courant number for the migration of bed forms is less than 1. At the end of this (typically short) period of time the sediment transport field is recalculated using the bathymetry evolved within the morphological timestep. The wave field is recalculated every k'th morphological timestep, where k is an arbitrary integer. (In the present simulation k equals 3). A second-order accurate finite-difference method is used for the calculation of $\partial z/\partial t$ and the truncation errors are eliminated by a modified Lax-Wendroff scheme.

This setup meets the requirements of offering the capability of running with time-varying boundary conditions, and of minimizing the computational effort by avoiding recalculation in the hydrodynamic model. A block flow chart of the morphological model complex is shown in Figure 5.

Three different types of **wave models** are presently built into the morphological complex: an elliptic mild slope model, MIKE 21 EMS, a parabolic mild slope model, MIKE 21 PMS, and a spectral nearshore wind wave model, MIKE 21 NSW. The second one was used in the simulations described here.

The **hydrodynamic module** in the morphological modelling complex is MIKE 21 HD, which solves the vertically integrated equations of conservation of mass and momentum in two horizontal dimensions. The hydrodynamic model operates with a space- and time-varying hydraulic roughness, accounting for the apparent bed resistance in combined waves and current. This apparent roughness is updated every morphological time step.

The deterministic intra-wave period model for **transport of non-cohesive sediment**, STP, is applied. The model covers the range from pure current to combined current and waves (breaking or non-breaking), see Deigaard et al (1986). It is a basic assumption in the present model complex that the transport capacity is a function of the local conditions.

The hard-clay layer described in the Introduction section was included in the model as a non-erodible surface over which transport was allowed to take place.

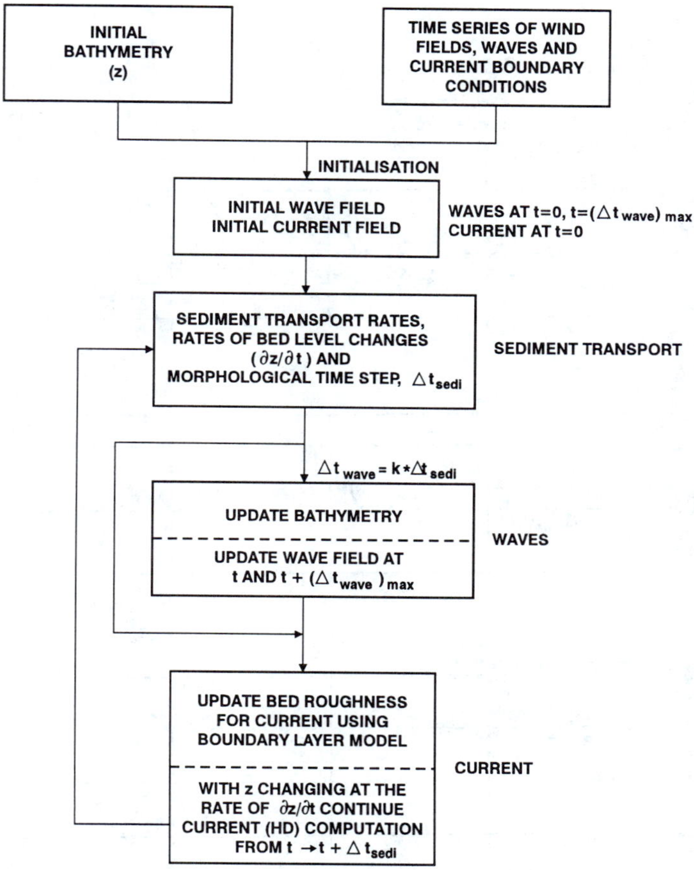

Figure 5 Block flow chart for the morphological modelling complex

Examples of results from the morphological modelling

As mentioned above, two historical storms, each of a duration of 4 days, were modelled in detail. The two storms represented typical storms from the two dominant directions, namely southwest and northwest.

Results from one of the storms, covering the period 20/1/1993 0:00 - 24/1/1993 0:00, are presented in figures 7 and 8. Figure 6 gives a brief overview of some of the important parameters, i.e. water levels at 3 stations, wind speed and direction and wave height and direction. Figure 7 shows wave-, current and sediment transport patterns at the peak of the storm, 22/1/1993 12:00. Figure 8 shows the bed level changes after 1 day, 2 days, 3 days and 4 days. The results show that sediment is 'sucked' out of the cells between the groynes, some of it is deposited off the groyne system in this storm, but most of it, as well as littoral drift which enters the model area, is trapped in the inlet area.

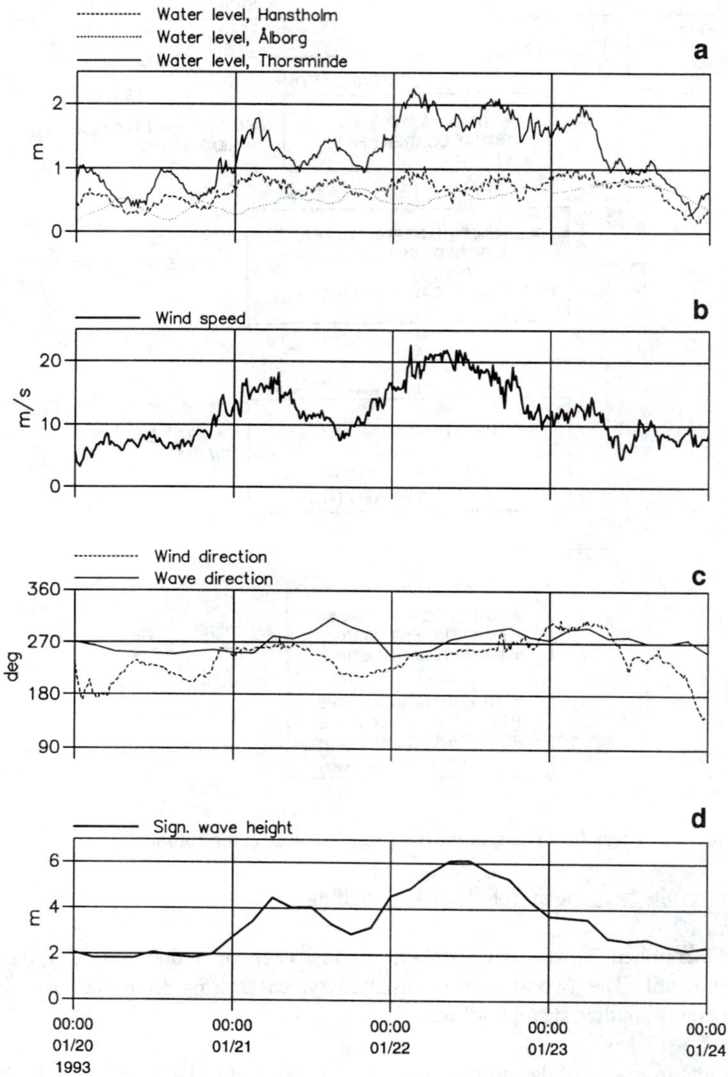

Figure 6 Time series of: a) water levels at Thorsminde, Thyborøn, Hanstholm and Aalborg
b) wind speed at Thyborøn
c) wind direction at Thyborøn, wave direction at Fjaltring
d) wave height at Fjaltring

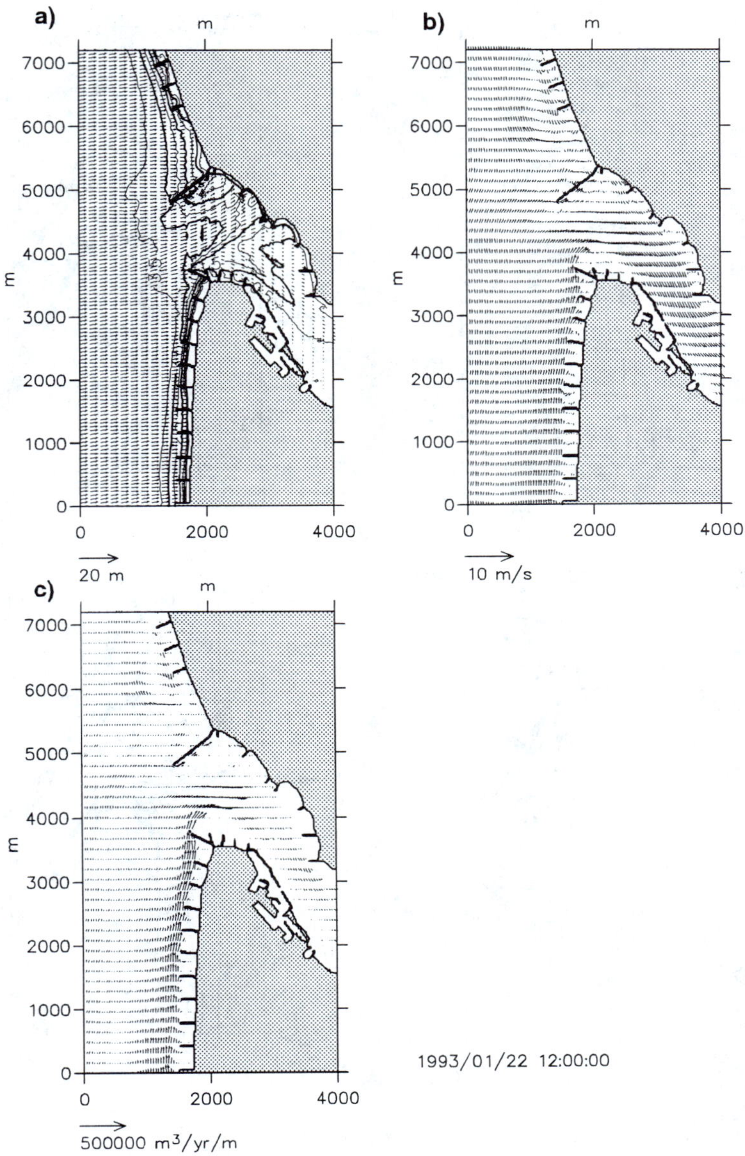

Figure 7 a) Calculated wave field, 22/1 1993 12:00
b) Calculated flow field, 22/1 1993 12:00
c) Calculated sediment transport field, 22/1 1993 12:00

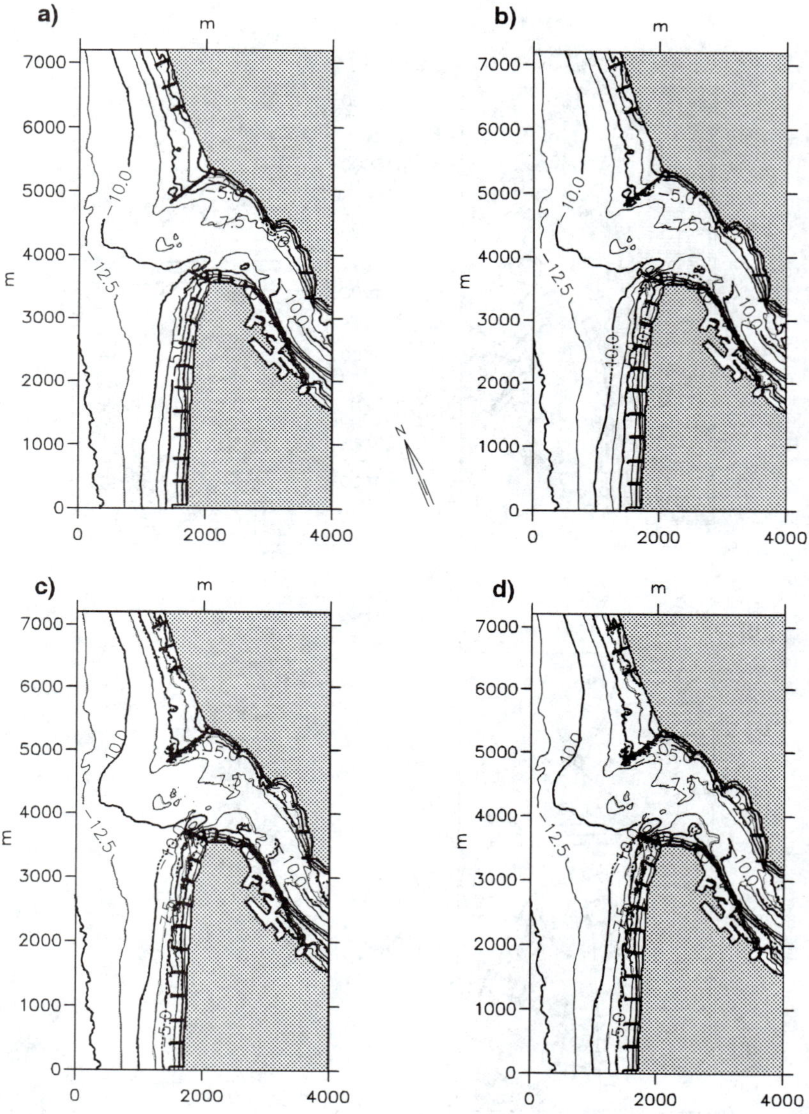

Figure 8 Calculated morphological evolution

a) From 20/1 1993 0:00 to 21/1 1993 0:00

b) From 20/1 1993 0:00 to 22/1 1993 0:00

c) From 20/1 1993 0:00 to 23/1 1993 0:00

d) From 20/1 1993 0:00 to 24/1 1993 0:00

New understanding of the complicated processes in the area was gained from the modelling work. Examples of details which were learned from the simulations are given below.

One example is the fact that the inlet over the last years has started to erode more severely on the opposite side, (the east side) to that where erosion previously occurred. It was clearly seen from the modelling that storms from southwest were responsible for this development, and over the last years the intensity of these storms has actually increased.

This effect becomes more clear if the model is run without any non-erodible clay layer. Figure 9 shows the comparison of the accumulated bed level changes over 4 days with and without the clay layer. It appears that a potential for erosion at the east side exists. (It should be mentioned that erosion does take place in the clay, just at a much slower pace than for sand. Erosion in the clay is therefore not included in the present model.) One effect of the clay layer on the sediment transport is seen from figure 10, where the accumulated transport across the western limit of Thyborøn channel, defined by the tips of the breakwaters, is compared for Storm 3 with and without the clay layer. It appears that the transport capacity is larger when the clay layer is included than when this is not the case. This is connected to the fact that the transport capacity decreases for increasing depth for constant flux.

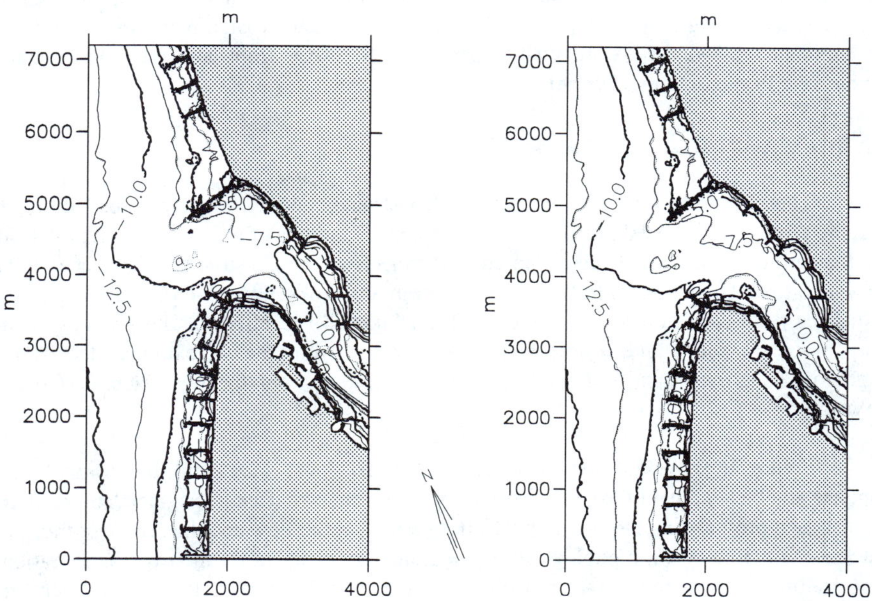

Figure 9 Comparison of accumulated bed level changes over 4 days.
Left: without clay layer
Right: with clay layer

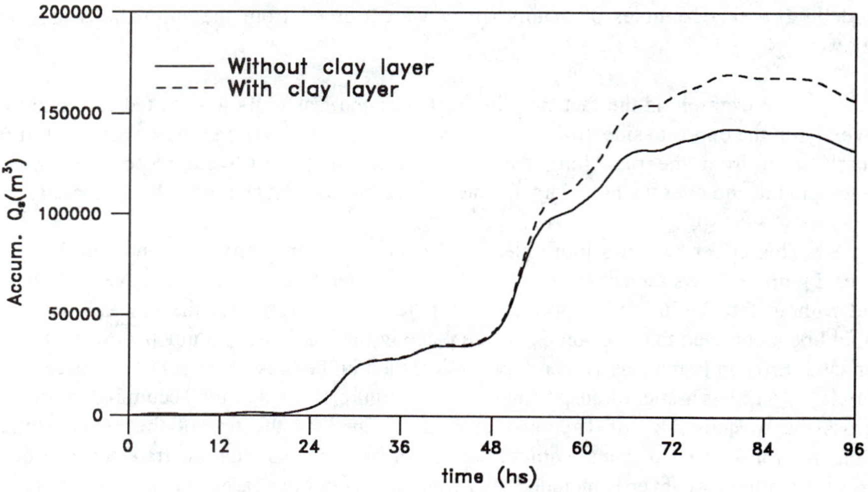

Figure 10 Comparison of sediment transport through the inlet with and without the clay layer

Another example was the observed tendency for more severe wave overtopping along the harbour protection works, which has moved further inside the channel than earlier observed. This could now be explained by the deposition and erosion patterns related to the predominant southwesterly storms which permitted the waves from northwesterly directions to penetrate further into the channel.

Sensitivity to the length of the groynes

The groyne system at Thyborøn was constructed about 100 years ago. Due to the ongoing erosion, the coastal profiles have steepened off the tips of the groynes with the result that the ends of the groynes have collapsed, and rocks from the old structures have been left at the sea bottom. The length of today's groyne system and the way in which its submerged part influences the littoral transport are therefore not known. The effect of the groynes on the morphological evolution has consequently been investigated for the two extremes: the original length of the groynes and today's length, as determined from the overwater part of the groynes.

Figure 11 shows the sediment transport on 22/1/1993 15:00 and the morphological response in the vicinity of some of the groynes in the two cases: original length of the groynes, and present length of the groynes. It appears that the flow and the sediment transport patterns and thereby the morphological response are very different for the two configurations. The example demonstrates the capability of the modelling complex to represent the interaction between the longshore current that is generated within the cells and forms a rip current along the updrift side of each groyne and the eddies generated in the lee of each groyne.

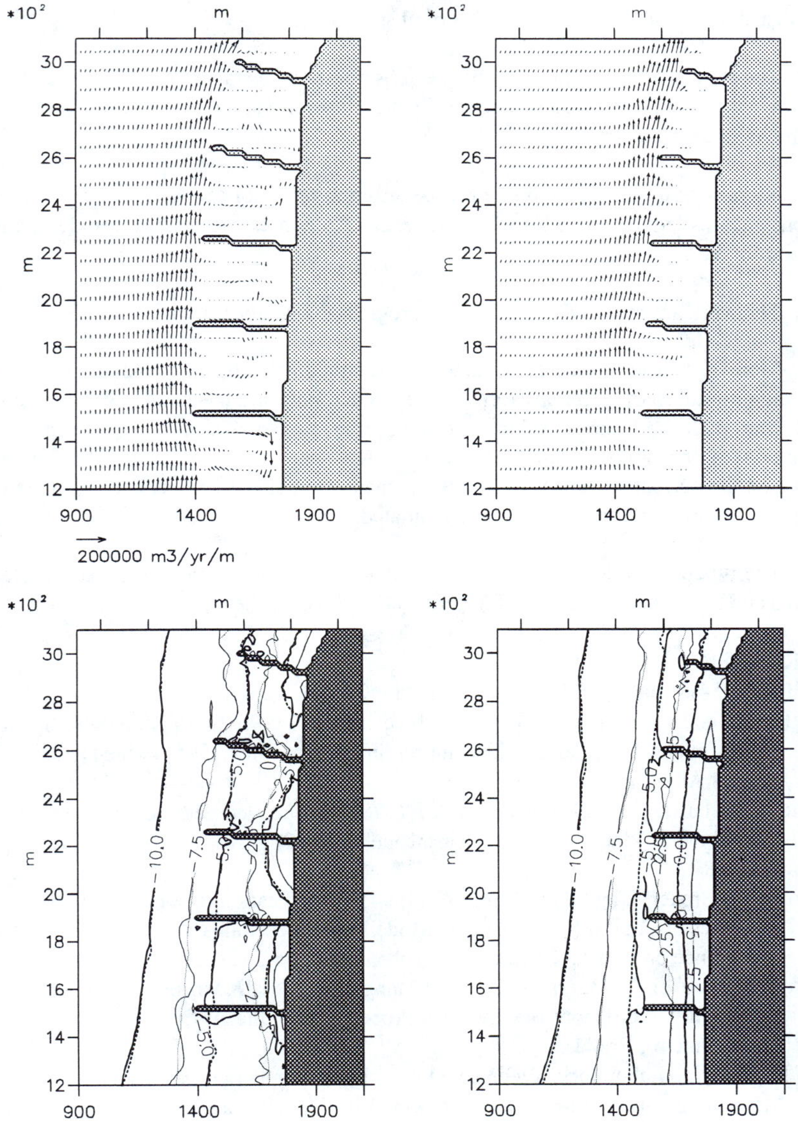

Figure 11 Sediment transport (instantaneous) and morphological response (20/1/1993 0:00 - 22/1/1993 15:00) in a groyne field during a storm. Two different lengths of the groynes are tested

CONCLUSIONS AND FUTURE PROSPECTS OF THE STUDY

Use of mathematical coastal models has cast new light on well-known coastal problems in the area and has allowed to quantify important parameters such as transport rates, spatial distributions of wave heights and current speeds, etc.

The results from the models have been useful in understanding some of the observed processes, cf. changes in erosion patterns in the channel and increased wave overtopping along the harbour structures.

Numerical models have proved to be an important tool for the analysis of different situations such as length of the groynes, existence of a non-erodible surface underlaying the sandy bottom, etc.

The work around Thyborøn is still in progress, November 1996.

More historical storms are planned to be modelled. The intention is to simulate a large enough number of different storms to reproduce both the observed sediment transport through the inlet and the development along the barrier beaches through a suitable combination of the storm events. At the present stage, only very scarce information is available on the current statistics in the inlet. A measuring campaign has been planned based on the model results, and this has now been initiated.

The model complex will be used to guide the design of possible coastal protection structures and to evaluate the effect of different nourishment schemes.

REFERENCES

Bruun, Per (1954): Coast Stability. Atelier Elektra, Copenhagen.

Danish Hydraulic Institute (1995): Thyborøn Investigations 1995. Simulation of Sediment Transport and Morphological Evolution (in Danish), May 1995 for the Danish Coastal Authority.

Deigaard, R., Fredsøe, J., Brøker Hedegaard, J. (1986). Suspended Sediment in the Surf Zone, Journal of the Waterway, Port, Coastal and Ocean Eng. ASCE, Vol 112, No 1, pp 115-128.

Johnson, H.K., Brøker, I. and Zyserman, J.A. (1994): Identification of some Relevant Processes in Coastal Morphological Modelling. Proceedings of the 24th International Conference on Coastal Engineering, Kobe, Japan, pp. 2871-2885.

Johnson, H.K., Brøker, I., Zyserman, J.A. and Mangor, K. (1995): Morphological Response in the Vicinity of Offshore Breakwaters. Proceedings of the IV COPEDEC Conference, Rio de Janeiro, Brazil.

Coastal Engineering Laboratory, Technical University of Denmark (1959): Coastal Morphology of the Barrier Beaches of the Limfjord (in Danish). Report No 3 to the Thyborøn Commission of 1957.

Mikkelsen, Søren, C. (1997): The Effect of Groynes on Beach Erosion and Channel Stability at the Limfjord Barriers, Denmark. To appear in Coastal Sediments 1997.

Sørensen, Torben (1961): The Development of Coastal Profiles on a Receding Coast Protected by Groynes. Proceedings 7th International Conference on Coastal Engineering (ICCE), The Hague, Netherlands, pp. 836-846.

CHAPTER 367

Probabilistic Risk Assessment of Beach Erosion at Pevensey Bay in England

Ping Dong[1] and Keith J Riddell[2]

Abstract

This paper presents a comprehensive case study in which probabilistic methods were applied to assess the risk of damage to properties protected by a shingle beach on the south coast of England. Both the short term storm response and longer-term longshore drift effects were considered. The mathematical model was calibrated against data from an extensive series of physical model tests and was used to predict the shoreline evolution for a number of beach management design options. A full cost-benefit analysis was carried out based on the predicted damage probabilities. The general methodology developed is believed to be applicable to other situations with different beach material types and various design options.

Introduction

In recent years, the limitations of a conventional semi-probabilistic approach in coastline studies and scheme development have become more widely recognised. In studying beach processes, hydraulic parameters such as extreme waves, tides and surges are usually specified in terms of single parameter return periods or joint return periods. The safety of the structure, or the level of protection it provides, is ensured by selecting design conditions (generally waves and water levels) which, individually and/or in combination, have a sufficiently remote chance of occurring. No explicit reliability calculations are undertaken and the precise level of damages over the design life, due to overtopping, breaching and erosion, is unknown. Consequently, the standard of protection that coastal defence schemes may be expected to achieve

---

[1] Formerly Principal Engineer, Babtie Group Ltd, now Lecturer, Department of Civil Engineering, University of Dundee, Dundee DD1 4HN, United Kingdom

[2] Divisional Director, Babtie Group Ltd, Simpson House, 6 Cherry Orchard Road, Croydon, Surrey, CR9 6BE, United Kingdom

can only be given in terms of the return periods of these input parameters rather than the damages which may be expected to occur. The success of such an approach would require that the return period of any damages is not lower than the return periods of the corresponding input parameters, a condition that can not be guaranteed *a priori* in a non-linear multi-parameter dynamic system such as a beach. By comparison with traditional methods, a direct technique which aims at evaluating the probabilities of occurrence of the beach processes that are actually responsible for the flood damages gives more meaningful design conditions. It also enables the evaluation of the variabilities and uncertainties in the damage estimates and provides a rational basis for cost-benefit assessment to be used as an integral part of scheme development.

The probabilistic design concept in coastal engineering is not new. It has been used in the design of breakwaters, seawalls and sand dunes as shown in a recent review on probabilistic design of flood defences by Vrijling (1990). Nearly all of the previous applications were restricted to dealing with the short term response of coastal defence structures under extreme forcing conditions, and the methodology used was largely based upon standard theories and techniques developed for the reliability of structures. Only very recently has some attention been paid to longshore processes. Vrijling and Swart (1992) presented a probability Level II method for berm breakwater design to evaluate the cumulative damage to the breakwater as the result of the longshore transport of rocks under extreme angular wave attack. Both Level II and Level III methods were employed by Vrijling and Meijer (1992) to assess long term shoreline evolution using a one-line shoreline model as the transfer function. Both the practical value of such approaches in engineering assessment and the uncertainties involved were highlighted by these workers.

The advantages of adopting a probabilistic approach and some important practical issues concerning its application in scheme development were presented by Riddell (1993). Probably, due to the lack of quality long term data and the perceived complexity of probability methods, the coastal engineering community are still showing signs of reluctance in adopting such methods in the design of beach management schemes. However, this situation is about to change in the UK with the introduction of new design guidelines from the Government which emphasise the need for rational cost-benefit analysis, the long term sustainability of engineering works and the need for all coastal works to be carefully evaluated in the light of global considerations.

The paper presents a comprehensive case study in which a probabilistic methodology was applied to assess the risk of damage to properties protected by a shingle beach on the south coast of England. Both the short term storm response and longer-term longshore drift effects were considered. The mathematical model was calibrated against data from an extensive series of physical model tests and was used to predict the shoreline evolution for a number of beach management design options. A full cost-benefit analysis was carried out based on the predicted damage probabilities. The general methodology developed is believed to be applicable to other situations with different beach material types and various design options.

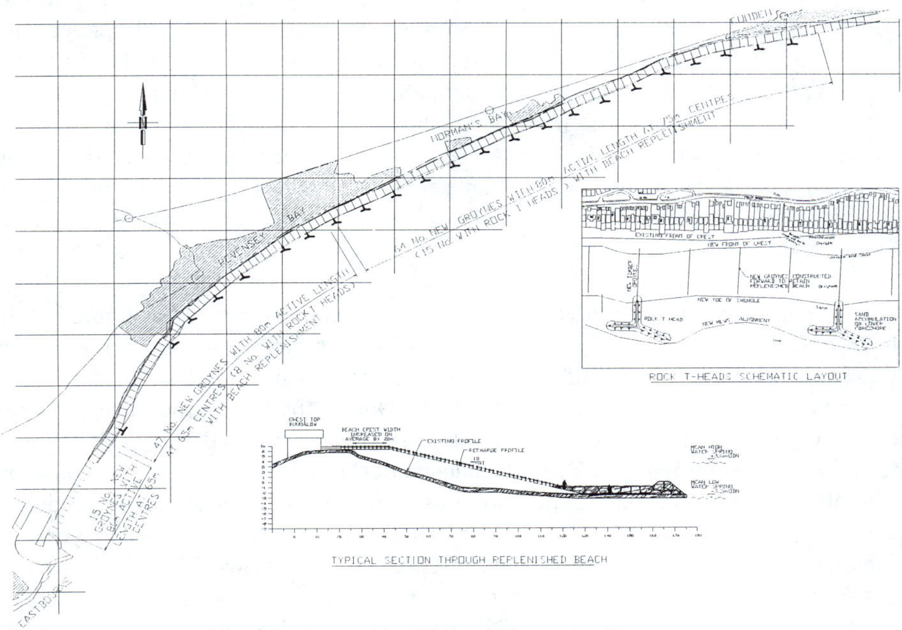

Figure 1 - Site Plan of Pevensey Bay

Site characteristics and design objectives

Pevensey Bay (Figure 1) has a shoreline 9km in length curving from a north/south alignment to east/west in a fairly classical bay configuration. Its beach is formed by shingle material above the mean water level grading rapidly to sand below this point. The shingle ridges, which are generally low and narrow, stand as a fragile

Figure 2

line of defence to the moderately well-developed low lying areas immediately behind the beach crest. Along several sections of the frontage, the beach has suffered from severe depletion of the mobile material and a large part of the existing timber groyne system has also been badly damaged, being, effectively, near the end of its useful life. Over the last twenty years extreme storms, combined with high tides, have, on a number of occasions, resulted in overtopping, beach crest recession and even small-scale breaching. These events have brought about considerable damages, not least of which were the costs of mobilising emergency services and carrying out rehabilitation measures to ensure some continued standard of protection. This situation can be best illustrated by an aerial view and a land-based photograph (Figure 2) of one of many critical locations showing the close proximity of housing to the crest of the shingle beach.

The threat of much larger breaching of the defences in the near future led the former National Rivers Authority (now part of the Environment Agency) to engage Babtie Group to carry out a comprehensive study of the Pevensey frontage lasting for a period of well over four years (1991-1995). During the first phase of this project, extensive data (both contemporary and historic) were collected and analysed with regard to waves, currents, sediment characteristics, beach profiles and past damages. Deterministic mathematical models were used to predict the beach response during extreme storm attack and the long term shoreline evolution over many years of average wave climate. These field data and predictions were used in an approximate cost-benefit analysis and for the preliminary design of a range of management options. During the second phase of the project, a detailed assessment of a limited number of potential scheme options was carried out, involving an extensive physical modelling programme and numerical model predictions.

Having realised the deficiencies of the deterministic methods that were used in the first phase, the decision was made to apply probabilistic techniques for both beach recession predictions and cost-benefit analysis. In this paper some of the work carried out in the second phase of the project is presented, with special emphasis on the use and validation of the probabilistic methods for beach recession predictions and cost-benefit analysis.

Methodology

General approach

The study methodology identified storm beach response as the primary process to be assessed in order to determine the type and severity of erosion, breach and levels of flooding. The probabilistic approach used for crest recession and overtopping was a simplified version of a full Level III approach with the erosion due to long-term longshore drift gradient being treated as an independent addition to the short-term crest recession.

Extensive physical model tests were carried out in the random wave flume and wave basin at HR Wallingford. The purpose of these tests was to assess the beach response under selected design storm waves and average morphological conditions over a period of 5 years. For the flume tests, 9 conditions were tested having estimated joint wave and water level return periods ranging from 1 in 1 year to 1 in 250 years. These tests were carried out on three profiles and two different gradings of beach material. Beach profile development for each test condition was recorded and used to calibrate a mathematical model. In the basin tests, one morphological condition and three (two for some cases) storm conditions were used. The full wave climate was represented by four equivalent wave spectra with increasing energy but decreasing probability of occurrence. The water levels were controlled to follow a predetermined tidal cycle. The data from the physical model tests were used to adjust the predictions made by the mathematical models. The return periods of overtopping and crest recession were calculated using a twenty year hindcast joint wave and water level time series, both including and excluding an allowance for long term sea level rise.

From wave refraction analysis and inshore monitoring results, it was found that the inshore wave climate exhibits considerable change from one end of the frontage to the other. In order to account for this variation, the use of different design wave conditions at different sections of frontage, for the purpose of cross-shore process analysis, was necessary. Due to these longshore changes of wave climate, as well as the change in shoreline orientation, large gradients in littoral drift exist at a number of sections of the shoreline. The effect of this was confirmed by observations that these beaches were known to be prone to erosion and required continuous maintenance. Therefore, longshore structural erosion was included in the overall risk analysis.

In developing design options, the probability of various levels of damage was determined for each beach management scenario, and these probabilities were then used to carry out a comprehensive cost-benefit analysis. This methodology is summarised in Box 1.

Transfer functions

A key element in the design methodology is the transfer function which determines the response of the system for a given set of input parameters. In coastal engineering, the transfer functions are usually in the form of empirical equations or models or process-based mathematical models. Deterministic beach profile responses during storms have been parameterised by Dean (1972), Van der Meer (1991) and Powell (1991). Among these models the **B**each **P**rofile **P**rediction **M**odel by Powell was developed specifically for shingle beaches based on extensive physical model tests and validated against a certain amount of (mainly UK) field data. This was therefore selected for the present work.

In order to limit the number of random variables that have to be considered, the model parameters were treated as deterministic and adjustable to give the best fit to the site specific data from the flume tests. In the case of Pevensey Bay such adjustments were required for only one of the three profiles considered. This was largely due to the influence of the limited thickness of the shingle layer at the location of this profile. It should be pointed out that the relationship as described by the BPPM model, between the equilibrium profile shapes and the input wave and water-level parameters, is highly non-linear and dependent upon the characteristics of the initial profile. Therefore, the predicted probability distribution of crest recession for different profiles could be quite different not just in terms of numeric values but also in the shape of the overall distribution.

APPROACH

Response of beach to specific storm events

↓

Generalisation to cover complete wave climate

↓

Application to joint wave and water level series

↓

Probability of crest recession, overtopping and breach

↓

Probability of damages

↓

Preferred Design

Box 1

Input Data

The input data can be classified into two main types: hydraulic data and morphological data. The hydraulic data consisted of a twenty year inshore joint wave and high water level time series, obtained by transferring a hindcast offshore time series to a number of inshore locations using a spectral back-tracking model (OUTRAY by HR Wallingford). The hindcasting model was calibrated against one year of offshore directional wave data obtained at the site during the study period. The refraction model was calibrated against simultaneous inshore records obtained at two locations.

The morphological data consists of surveyed beach profiles, sediment sizes and shoreline positions. Cross-sections at 150m intervals of this and adjacent frontages had been taken annually for a period of twenty years and were used for the detailed assessment of recent shoreline evolution and sediment budget considerations. For simplicity of modelling, the morphological inputs were treated as deterministic. The appropriate mean values and variations for these morphological input data were derived using historical charts and the long term beach survey data.

Prediction procedures

When considering cross-shore processes, it is important to preserve the essential correlation in the wave and water time series. This rules out a probability Level II method which assumes the parameters to be independent and to follow Gaussian distributions. All the wave and water level records in the time series, apart from some small waves deemed to be insignificant, were directly input into the calibrated BPPM model to obtain a single time series for the beach crest movement. The predicted beach movements were then ranked and fitted with appropriate logarithmic distributions. The same calculations were performed for each of the three profiles, taken as being representative of sections of frontage, using the appropriate time series for different inshore points.

Longshore processes are more difficult to deal with, because they essentially have two time scales; one for long-term average movements of sediment and the other for short-term storm effects. Although the techniques of Vrijling and Swart (1992) could, at least in principle, be adopted for predicting the probability of long term average shoreline changes, difficulties arise when trying to combine the two sets of recession probabilities from cross-shore and alongshore calculations. The short-term effects due to oblique storm waves on the coastline development would also need to be determined separately.

Short of using a 3D or quasi-3D model as a transfer function, further assumptions on the potential correlation between short- and long-term processes and between cross-shore and longshore effects have to be made. Due to the lack of research concerning the above problems, longshore processes were treated as deterministic. As a result, the crest recession probabilities would be invariant although the reference shoreline position would change with time as predicted by a one-line model. The corrections to the reference shoreline position were introduced in the design assessment based on the basin test results.

Predictions of crest recession

Model calibration

In order to ensure that the transfer function used was reliable the numerical model (BPPM) was firstly calibrated against the crest recession data from the flume tests. It was found that the model predictions agreed well with the measurements from Profiles 2 and 3 but some adjustment was necessary for Profile 1 in order to achieve best overall fit. Similarly the one-line model predictions were also compared with the actual shoreline positions determined from the annual surveys for the same data period (20 years). To achieve best agreement, it was found that the transport parameter in the CERC formula 'K' should be about 0.04 assuming the existing groynes are virtually ineffective.

Cross-shore beach crest recession

Assuming that an equilibrium beach profile can be established over the period of one tidal cycle, under given wave and water level conditions, the time series of crest recession and overtopping rates can be calculated for a given beach configuration. The exceedance probability and the return periods of crest recession can then be determined using ranking statistics. Some of the results are shown in Figures 3 and 4. It can be seen that the active crest tends to establish seaward of the initial crest for most wave conditions during the twenty year record period whilst crest retreat takes place during only a small number of severe events. In order to determine the extreme crest recession, a logarithmic curve was fitted to the computational data. The crest recessions for the more extreme events such as 1 in 250 year, can be obtained by extrapolation of the fitted recession curves.

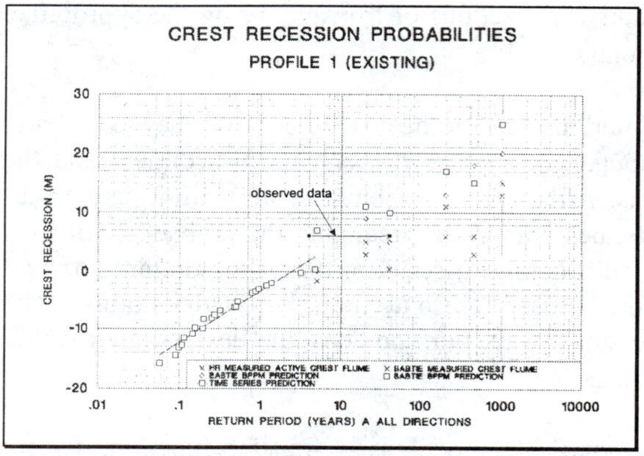

Figure 3

The above method is more easy to apply than a more conventional approach which involves extrapolating the joint probability density of wave and water level and then calculating the overtopping and crest recession for the extended records. This is due to the fact that the joint probability density function is very irregular at small values, and it is difficult to find a sensible surface to fit to these values. Whatever probability distributions is assumed, large errors are unavoidable in extrapolation.

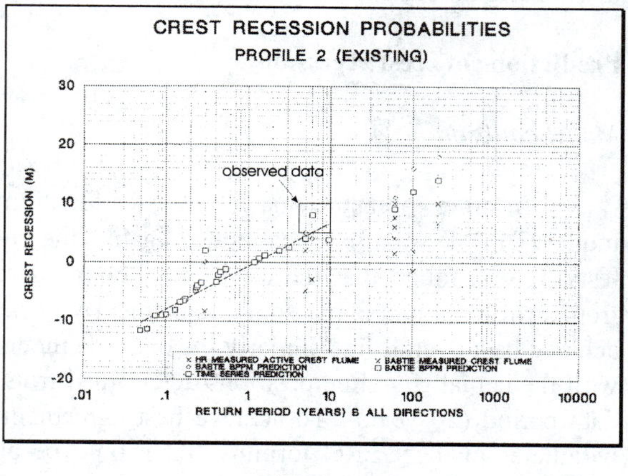

Figure 4

In general, the accuracy of the simpler method has been shown to depend on the data length, regularity of crest movement and the transfer functions adopted. From the model tests it was found that the beach can be 'washed off' by overtopping when its crest is very thin. This means that extrapolation of the predicted recession probability can not be extended indefinitely because the crest movement can change from progressive erosion to wash-out with only a very small change of input hydraulic conditions. A minimum crest width has been assumed on the basis of the model tests which represents this change of mode of beach damage. A 250 year return period was determined to be the upper cut-off limit for the extrapolation of beach processes at the study frontage.

Figures 3 and 4 also show the observed recession ranges during previous observed storms. Although the data are rather limited, and the quality of these data is far from ideal, it was quite satisfying to note that the predictions were consistent with the field data on all three profiles. Since the numerical model was only calibrated against data from the physical model, and for only a limited number of extreme storm conditions, the predictions obtained are considered to be remarkably good.

Effects Of Long Term Shoreline Evolution

Although the storm response of the beach is dominant in the determination of the level of risk at any particular point in time, the long-term shoreline evolution trend must be taken into account if the true risk level of a particular section of the

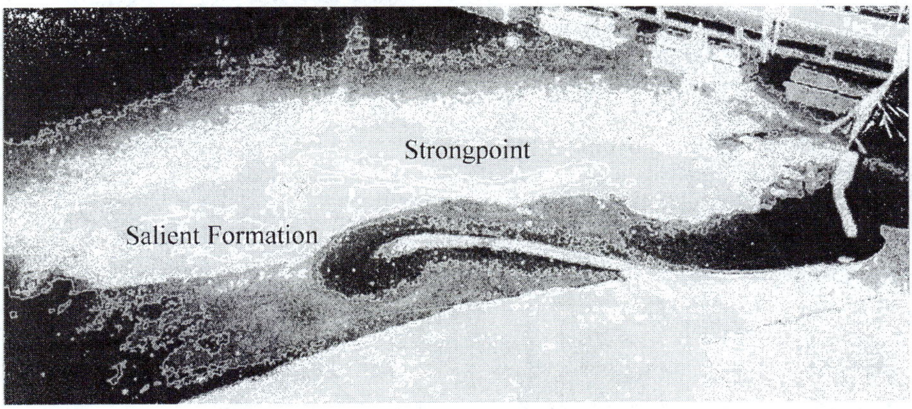

Figure 5, Strongpoint Basin Test in Progress

coastline, over a significant length of time, is to be established. For an open beach, the long term erosion due to the longshore transport gradient is usually fairly regular at a given site, depending, primarily, on the average wave climate, the supply of material and local shoreline orientation. The mean erosion rates for the Pevensey Bay frontage were calculated using a standard one-line model. In order to account for variability in the predictions of wave climate and sediment supply, the average shoreline position was determined for 5, 10, 20 and 50 year periods. The storm wave response probability was assumed to be invariant at any time within the design life of the defence. Although it is possible to introduce a higher level of probabilistic representation for the shoreline position than that used, such an approach was rejected on the grounds that the transfer function would have become too complicated to be practicable and no suitable method was found which was capable of dealing with the correlation between longshore and cross-shore processes in a consistent and robust way.

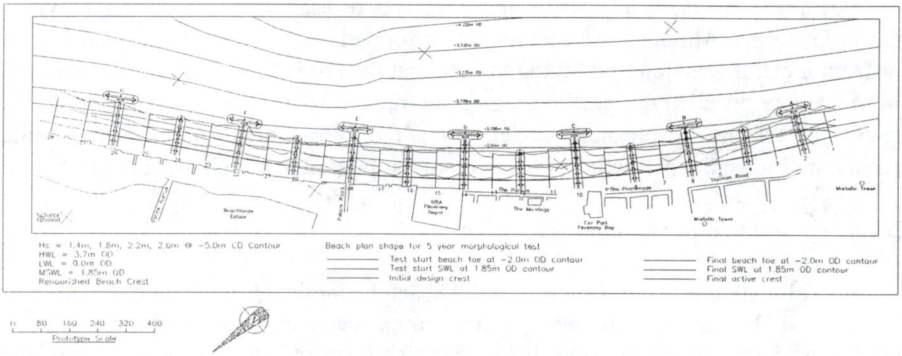

Figure 6 - Basin Test Results for mixed Rock T-head and Groyne Scheme

In developing the scheme options a number of beach management scenarios, ranging from timber groynes to large shore-attached rock structures, were tested in the wave basin. The crest movements under morphological and extreme storm conditions were obtained. Using similar techniques as described for the open beach case, the crest recession probabilities for each of the beach management options were estimated, taking into account the effects of the longshore gradient of drift rates during storm events. An example of a test in progress is shown in Figure 5 and an example of scheme test results in Figure 6. A comparison of the effect of different

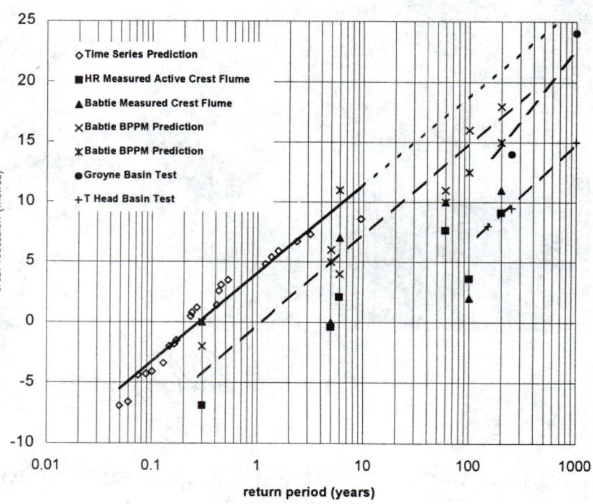

Figure 7 - Crest Recessions for various schemes

management options on the probability distribution of crest recession is shown in Figure 7.

Cost Benefit Analysis

The ultimate aim of all the preceding analyses is to generate damage probability graphs for all kinds of damage/loss considered. A typical graph for an open beach 'do-nothing' option is shown in Figure 8. In developing various design options, losses due to erosion, overtopping and breaching were all included wherever appropriate. The damage cost was assessed for each year during a nominal 50 year design life using appropriate discount factors. Since the total loss for any area is not always the simple sum of all probable losses, depending on the physical characteristics and economic values of the shoreline and hinterlands, and the interaction between them, care must be taken in deriving the total losses/damages.

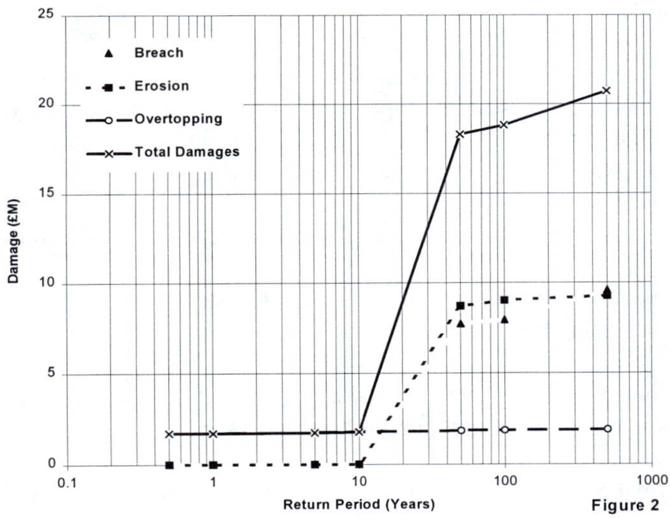

Figure 8 - Damage / Probability Profile

Discussion And Conclusions

A probabilistic design methodology has been developed which can be applied to shoreline studies, scheme design and the selection of management options. It provides a rational basis for the assessment of flooding risk and the standard of protection.

As long as the joint time series of wave and water level can be made available, it is straightforward to calculate the probability of run-up, overtopping and breaching using either empirical formulae or mathematical models. Physical modelling significant confidence to the results of the above and is essential for the consideration of non-standard situations.

Beach response is not just dependent on waves and water levels. Many other parameters are important such as the initial beach profile, material variability and storm duration and sequencing. A true probabilistic design methodology needs to take into account all of these parameters on a physical basis including systematic sensitivity assessments.

The transfer functions which provide the link between hydraulic parameters and beach processes are dependent on the configuration of the beach and detailed material properties. Great care must be taken when applying a transfer function developed for one type of beach condition to another.

Based on this study the following conclusions can be drawn

1. Predicted erosion probabilities are consistent with experimental data and some limited field observations.

2. The probabilistic method is no more difficult to apply than a conventional method, although it does require more long-term data and is therefore more time consuming.

3. Predictions are site dependent and sensitive to changes in the parameters characterising the beach and within the numerical model transfer function.

4. The damage modes treated in this study were limited and many other damage modes such as scouring, abrasion and offshore loss of beach material could be important at other beaches. Hence, design decisions made at one site should not be transferred to another site despite apparently similar features.

5. It has been found to be virtually impossible to deal with both long term and short term processes using the same high level of probabilistic techniques and within the prediction framework adopted. Research in this area is urgently required.

References

CIRIA, 1995 Beach Management Manual.

HR Wallingford, 1995, Langney Point to Cooden Sea Defences-Physical Model Studies,

HR Wallingford, 1995, Report No. EX3241.

Riddell, K.J., 1993, Probabilistic Methods in Scheme Development, Proceedings of MAFF River and Coastal Engineering Conference.

Vrijling, J.K. and G.J. Meijer, 1992, Probabilistic coastline position computations, Coastal Engineering, 17, pp. 1-23.

Vrijling, J.K. and P.F. de Swart, 1992, Berm breakwater design - the longshore transport case: a probabilistic approach, Coastal Structures and Breakwaters, Thomas Telford, London, pp. 513-525.

Vrijling, J.K., 1990, Probabilistic Design of Flood Defences, in Coastal Protection (ed. K.W Pilarczyk), Publisher: AA Balkema 1990.

CHAPTER 368

BEACH NOURISHMENT IN ALTAFULLA, SPAIN: VERIFICATION OF THEORETICAL MODELS

J. Galofré [1], F. J. Montoya [2] and R. Medina [3]

ABSTRACT

This paper describes the comparison between results from theoretical methods and field data measurements on a beach nourishment project, in Altafulla beach, Spain. A brief review of the processes involved in the behaviour of beach nourishment projects are made in order to analyse beach evolution after filling it. Theoretical results can be obtained applying theoretical methods. A monitoring program was carried out after nourishment works and field data measurements were taken in order to verify theoretical models. Capability, application and validity of models to the prediction of a project performance are discussed comparing with field data results.

The conclusion of the comparison is made for the Altafulla case study and a discussion of the results is included in order to understand beach evolution method by field data results.

1.- INTRODUCTION

Beach nourishment works, as an integrated coastal zone management plan, are used as a coastal engineering tool in order to guarantee the functions of a beach. The most important functions of a beach are: - Energy dissipation mechanism an Useful free space for every one.

Theoretical, technical-empirical and numerical methods are needed in order to design the solution and optimize nourishment efficiency. They must be calibrated in order to know the right beach parameters and analyze their behaviour.

---

1) Coastal Civil Engineer. Tarragona Coastal Service. Environment Ministry. Pl.Imperial Tarraco 4-4ª. 43005 Tarragona, Spain.

2) Coastal Civil Engineer. Head of Tarragona Coastal Service. Environment Ministry. Pl. Imperial Tarraco 4-4ª. 43005 Tarragona, Spain.

3) Associated professor. Ocean & Coastal Research Group. Universidad de Cantabria. Avda. de los Castros s/n. Santander 39005. Spain.

Analysis of beach behaviour has received considerable attention during the last decades. Although the sediment processes involved in the changes of a beach are non linear and have great variability both in space and time, the comparative analysis of beach evolution with theoretical models and field data is necessary in order to know how powerful the predictive models are. These models can be used to assist in the determination of project design and/or beach evolution. Furthermore, models describing the response of beaches to different coastal forcings have become increasingly numerous and sophisticated in recent years (see Work and Dean 1995, as a general reference). At present, however, there is no model that can be used to solve all the spatial and temporal scales of variability involved in beach nourishment evolution and, consequently, different models must be used. Monitoring field data are needed in order to calibrate and verify the models. It is necessary to know how well every model works and when it can be used.

The goal of this paper is to analyse the theoretical and real behaviour of Altafulla beach, comparing monitoring field data to technical-empirical and numerical model results. Various types of models are used, they are classified by their spatial and temporal domains of applicability.

2.- ANALYSIS METHOD

The analysis of processes, scales and tools is necessary in order to advance the knowledge of beach behaviour.

In order to understand beach performance it is necessary to analyse the beach forcings that are acting on the beach, these are the cause of all the processes.

The beach responses are the consequence of the beach forcings, they are the phenomena that appear along the coast. Beach behaviour analysis implies many methods working with the forcings and responses give a comprehension of beach performance.

Erosion, accretion and beach change in offshore bottom topography are controlled by beach forcings described in table 1. They are divided into three time scales of variability: Short term (less than 15 days), middle term (15 days to 6 months) and long term (years). The space scale considered is the meso scale that has a range from 100 m to 10 km shoreline. There are others spatial scales like microscale, less than 100 m shoreline, and macroscale, more than 10 km shoreline. This case study involves the mesoscale. The responses of beaches to these perturbations and variable forcings can be found in a wide range of time scales, see table 2.

Besides the wide range of temporal and spatial scales of variability, coastal evolution processes are often three-dimensional. In spite of this, important aspects of the coastal behaviour can be understood and prediction on the bases of lower-

PROCESS – FORCINGS		
SHORT TERM	**MIDDLE TERM**	**LONG TERM**
• Waves • Tides • Wave currents • Wind • Atmospheric pressure • Sediment budget	• Platform–Currents • Fortnight–Tides • Storms • Sediment budget	• Winter–Summer waves • Platform–Currents • M.S.L. Variation • Sediment budget

TABLE 1 FORCINGS INVOLVED IN BEACHES

RESPONSES		
SMALL SCALE	**MESO SCALE**	**LARGE SCALE**
• Bed forms: • Beach cusps • Bars • Morphological changes • Sediment transportation and distribution	• Profile changes • Planform changes • Shoreline changes • Crescentic bars • Morphodynamic states variation • Sediment distribution	• Coastline acretion–erosion • Beach equilibrium planform • Beach equilibrium profile • Sediment distribution • Eustatic reponse

TABLE 2 RESPONSES INVOLVED BEACHES

	MODELS		
	SHORT TERM SMALL SCALE	MIDDLE TERM MESO SCALE	LONG TERM LARGE SCALE
FORCINGS	• Wave propagation • Tide propagation • Wave – currents	• Wave propagation • Tide propagation • Wave – currents	• Wave propagation • Tide propagation • Waves – currents
RESPONSES	• Profile models • Local sediment transport	• Profile models • N–Lines models • Sediment transport	• Parametric Models (plan, profile and granulometric) • N–Lines models • Sediment transport • Statistic models

TABLE 3 MODELS USEFULS IN BEACH BEHAVIOUR ANALYSIS

dimensional models. These models take advantatge of the circumstance that the response of a beach often exhibits a different behaviour with essentially different length scales in three mutually orthogonal space directions (vertical, cross-shore and longshore), De Vriend, (1992).

The next point is to define the models that it is possible to use for the analysis of the forcings and responses. Table 3 shows a wide range of different models.

To improve predictive models for beach responses, an accurate description of the forcings is necessary. From table 3, it can be seen that existing hydrodynamic models (wave, tide propagation and wave induced currents) can be used for solving the forcings at almost all spatial scales of interest. The choice of a particular model should be made in relation to the response model to be used.

When selecting a response model, several physical facts must be taken into account. Cross-shore transport is very important just after the fill. A nourishment beach reaches its equilibrium profile within the first year after the fill (Kamphuis and Moir 1977). Several existing models can describe the post-fill evolution of the profile (usually neglecting longshore transport). However, cross-shore transport at greater time scales (months to years) remains a challenging problem that has not received a great deal of attention (Work and Dean 1995).

Longshore sediment transport is found to be important near the shoulders of the fill in the beginning. The effects of the longshore gradients propagate afterwards into the nourished region. N-line models can represent these coastline changes in the mid-long term. The one-line approach imposes limitations by neglecting the influence of cross-shore transport. However, this can be overcome if the model is calibrated adequately (Hanson and Kraus 1977). Technical empirical models can be useful in order to know beach equilibrium planform (Hsu et al, 1989) and profile equilibrium form (Dean, 1995). For the long term evolution prediction, parametric models (e.g. equilibrium profile-coast line models) and statistical models (e.g. P.C.A. models) can help N-line models.

The next step is to analyse this wide range of models and select the most interesting in order to know the beach behavior. In beach analysis from input data with the analysis method, must be obtained the output data. In table 4 an analysis model is proposed for it.

The case study will compare the models propounded in table 4 with the field data obtained from the monitoring program.

INPUT DATA

Historical analysis
Maritime climate data
Topobatimetry
Granulometric data

ANALYSIS			
	SHORT TERM	MIDDLE TERM	LONG TERM
TECNICAL–EMPIRICAL METHODS		Morphodynamics state distribution	Shoreline equilibrium model Profile equilibrium model Grain size distribution study Empirical evolution methods
NUMERICAL METHODS	Waves propagation Breaking current system Sediment transport		Numerical evolution shoreline methods (One line) Stadistical model (3PCA)

OUTPUT DATA

Beach Evolution Behaviour:

- Beach planform
- Beach profile form
- Grain size

TABLE 4 ANALYSIS METHOD IN BEACH BEHAVIOUR

3.- FIELD SITE AND DATA COLLECTION

The site of the field study is Altafulla (Fig.1), a sandy beach located 10 km north of Tarragona and 80 km south of Barcelona, in Catalonia, on the Mediterranean Coast of Spain. Altafulla is a half-opened beach 2.3 km long located between two capes, “Els Munts” to the east and “Tamarit” to the west. A small river flows during storms in the middle of the beach.

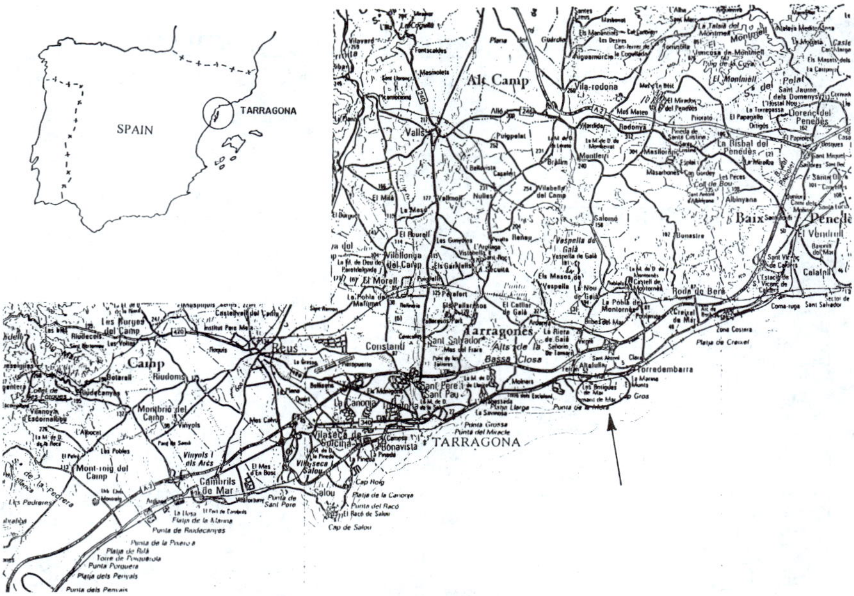

FIGURE 1 SITE LOCATION MAP

There are two predominant directions of wave aproach: SW and E. More than three quarters of deep water waves approach Altafulla from those sectors. The annual average significant wave height is about 0.5 m with typical winter storm waves of H_s of about 3.0 meters. Tides at Altafulla are negligible. In figure 3 a visual wave distribution is made and the affected area is shown, sea and swell limits are defined.

The native beach sand had a mean diameter between D_{50} = 0.12 to 0.2 mm and the beach profile slope changed from 1.2 % to 2.0 % from shoreline to bathymetric -5 meters. This value contour is considered the profile closure depth at Altafulla. The bottom of the sea is sandy up to the 10 m bathymetric contour line.

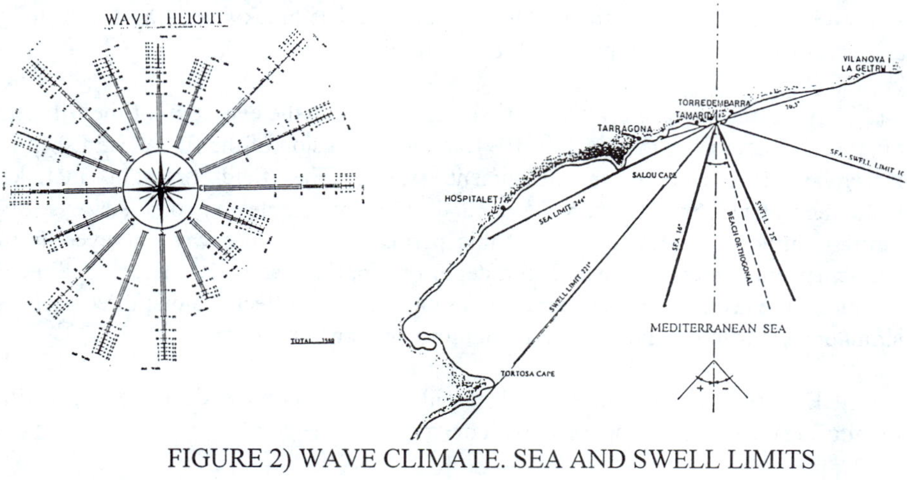

FIGURE 2) WAVE CLIMATE. SEA AND SWELL LIMITS

4.- BEACH NOURISHMENT AND MONITORING PROGRAM

In figure 3, a bathymetric map is shown in order to present beach problems and their solution.

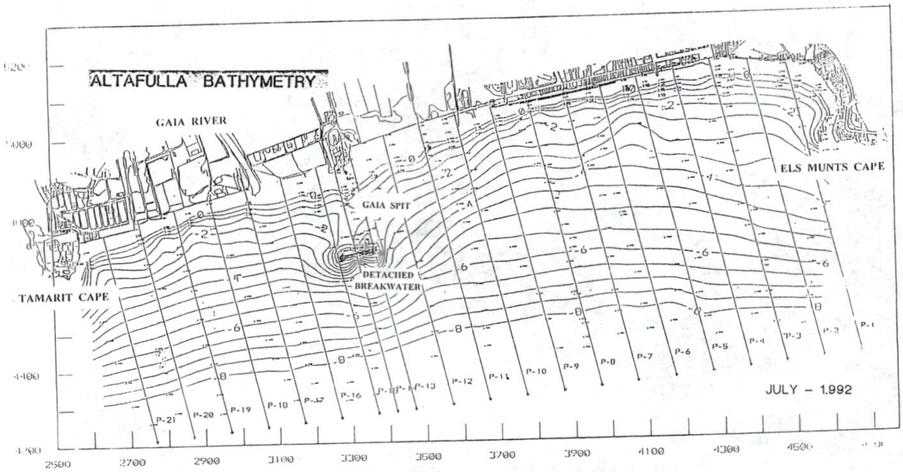

FIGURE 3 MORPHOLOGICAL AND BATHYMETRIC MAP

Several erosion problems occured in the northern part of the beach. A seawall was built to prevent backshore building damage. Southward littoral drift reduction of sediment appeared because of the construction of a dam 8 km up to the Gaià river and sand extraction during the 50's and 60's for construction works and filling the surrounding marsh. These have been theorized as major factors in the erosion that has been witnessed in Altafulla. A beach nourishment project was undertaken in 1989. Beach nourishment started in late 1990 and was completed in 1991. The beach nourishment works consisted of 160.000 m^3 of borrowed sand volume. The borrowed sand had a mediam diameter averaging D_{50} = 0.6 mm. A detached breakwater was also built in the middle of the beach, see figure 3. The breakwater was 110 m long and was placed at the -5 m bathymetric contour line.

A monitoring project was carried out to evaluate the evolution of the fill. The monitoring program started in July 1991, at the conclusion of the fill, and finished in December 1993, before renourishment works. The field program includes bathymetric beach profile every survey and sediment samples on some surveys. Six profile surveys were taken in this period. Each profile was surveyed from permanent monuments landward to a depth of approximately 10 meters. Sediment samples were taken in the last survey and samples were collected along three profiles simultaneous with beach equilibrium planform and profile survey.

Historical aerial pictures and visual maritime climate data are used with monitoring data in the comparison with the models.

5.- MODEL VERIFICATION

In table 3 a wide range of models which affect forcings and responses have been described. In table 4, a selection of these models have been chosen in order to define the analysis method to study beach nourishment evolution. Beach planform, beach profile form and grain size distribution can be obtained from historical analysis, maritime climate data, topobatymetric and granulometric data using the analysis methods.

The field data are needed in order to validate the models, some data are input, some are used to calibrate the models and others are output data. A planform and profile form are analysed in order to know their equilibrium. Grain size distribution is evaluated.

Qualitative and quantitative verification have been made. Qualitative results have been obtained studying beach dynamics and comparing with aerial pictures and topobatymetrics. Beach dynamics were computed by means of the numerical models.

Computations were carried out for conditions before and after the fill and construction of the detached breakwater. Different wave heights, wave periods and wave approach directions were used applying the REFDIF program, from Kirby and Dalrymple (1983-1986) computed by the parabolic wave propagation model.

Different wave heights were composed, the refraction and difraction phenomena were combined. By analysing aerial pictures, the protected areas behaviour, predicted by the models, can be observed. Figure 4 shows the wave induced currents determined from the wave field, as shown in the last figure. It can be seen that the breaking waves currents direction, induced by different wave height, change near the capes and the breakwater by difracting phenomena. Aerial pictures

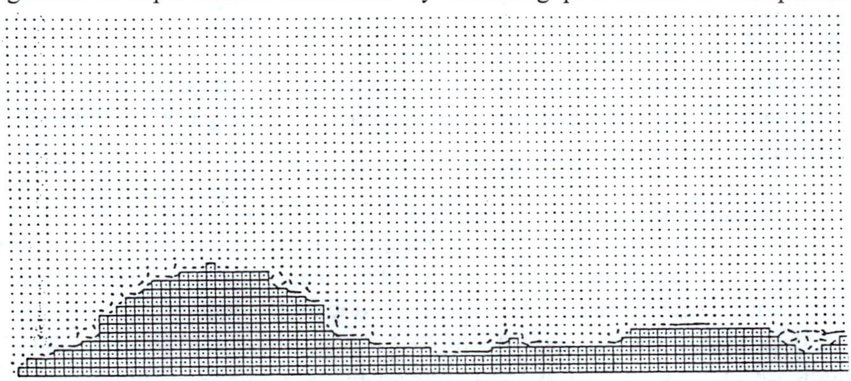

FIGURE 4.a) WAVE INDICED CURRENTS H= 0.5 m; $\alpha = +45°$; T = 10s.

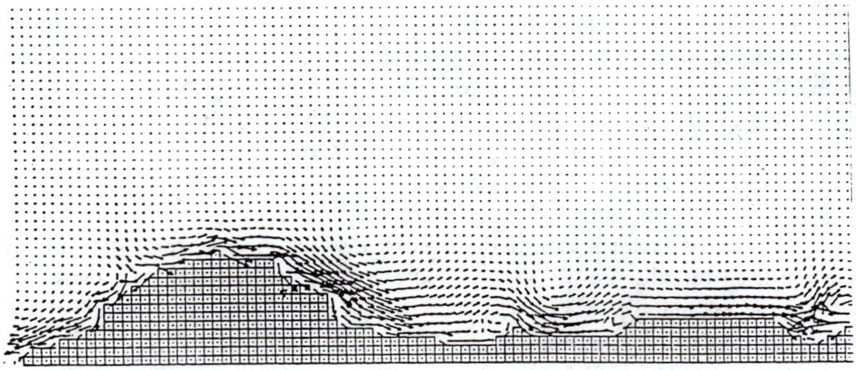

FIGURE 4.b) WAVE INDUCED CURRENTS H= 3.0 m; α = + 45°; T = 10s.

confirm these results showing sand accumulation in the calm areas. The spit formation landward of the breakwater can be explained with these results as is shown in the monitoring batymetry.

Quantitative verification can be made in order to analyse beach behaviour. Beach profile, planform and prediction methods are verified.

Beach field data profiles were compared with theoretical profiles before and after nourishment works. The Dean profile was used and compared with field data. The beach profile before nourishment can be approached by $y = A.\ X^{2/3}$, with the corresponding A according with D_{50} = 0.16 mm., the average of D_{50} in the native beach. After nourishment the beach profile can be approached by the same formula with A corresponding with D_{50} = 0.35 mm., the average of D_{50} in the nourished beach in the last topobatymetry field data.

Closure depth profile is calculated by the Birkermeier formula and compared with monitoring field data. Bathymetric -5 m. contour is considered the closure depth at Altafulla. This value was determined from the monitoring profiles and by technical-empirical model.

Planform was analysed by technical-empirical and numerical models and the results have been compared with monitoring field data results. The technical-empirical Hsu and Silvester method was used for analysing planform on the east part of the beach. Figure 5 shows the results with a great similarity with theoretical and field data results. The spit formed landward of the detached breakwater built in the middle of the beach is analysed using the method proposed by Gonzalez (1995), to calculate spits and tombolos. Figure 6 shows the application of this method and the conclusion is compared with the results of the monitoring field data results.

Numerical models can be used in order to understand beach performance and beach evolution. The basic procedures followed in this case study to analyse the beach nourishment evolution were the one-line model and statistical model. Field data are needed in order to calibrate and verify the one-line model and as input data for statistical models that can show theoretical tendencies from field data.

The one-line model used was GENESIS, (Hanson and Krauss, 1991). The model was calibrated with the profile evolution data following the procedure by Hanson and Krauss (1989). The climate waves were taken from visual data and calibrated from Tarragona harbour buoy data . Figure 7 shows the results of the shoreline position obtained from the model. The field data used for the calibration was 92 - 02 - 28 to 93 - 10 - 17.

The one-line model is a powerful tool that must be used carefully. To model the reality means to simplify the problem in order to reach some logical results that represent because two capes exist at the ends of the beach. These capes are not very long but it is not know if sand bypassed them.

In the model the capes are represented by two diffracting groings 550 m long. This hypothesis works well in a small and mesotime scale. It was calibrated and verified using two years monitoring field data and the results were quite reasonable. But if we apply long-term scales, more than 10 years, the results must be discussed more accurately. The boundary conditions are critical, especially for the evaluation of the sand beach losing outside the study stretch. Extra data are needed in order to calibrate the program better. Length, sand bypassing and sand transmission must be discussed in order to aplly long term.

Wave climate is another data that must be chosen carefully. In this case visual data, calibrated with Tarragona harbour buoy data, were used. During the monitoring period not many storms occured. This means that the medium wave climate that visual data represent was not absolutley according to real wave climate.

Sand granulometry is another aspect to consider. In the GENESIS model and program homogeneous sand is used, in this case it was supposed that all the sand had D_{50} = 0,35 mm. Before restoration D_{50} = 0,2 mm and in the mouth os the Gaià river a lot of rocks and gravel existed. This phenomena was not considered in the GENESIS program. In the places where gravel exists erosion problems are less than in sandy areas. In long term analysis these aspects have importance and the results must be taken carefully.

Combined phenomena, boundary conditions, wave climate and sand granulometry have incidences in the prediction results in long term scale and it is difficult to plan long term future by this kind of program in this case.

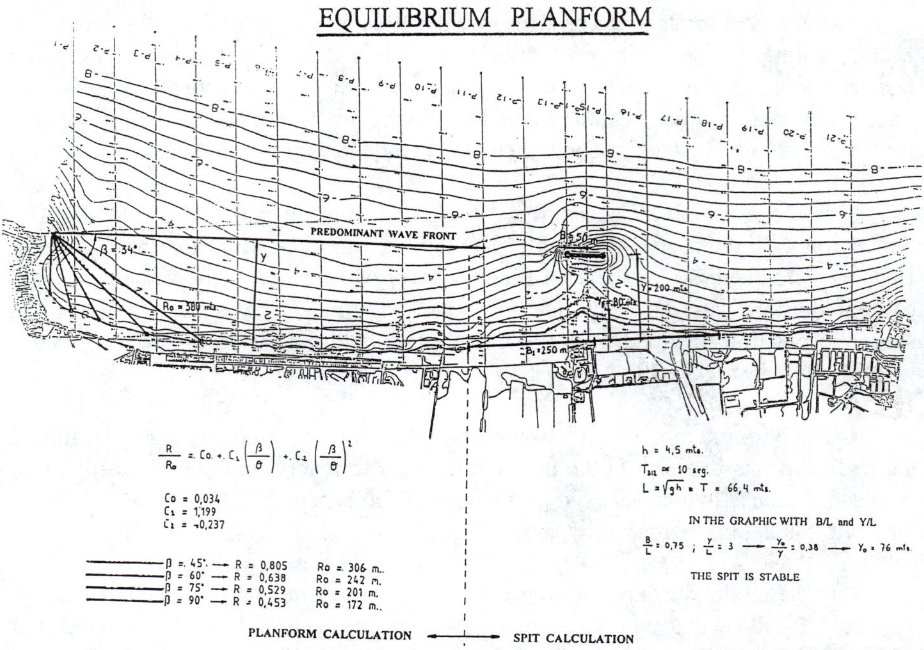

FIGURE 5 THECNICAL-EMPIRICAL METHODS INPLANFORM EVALUATION

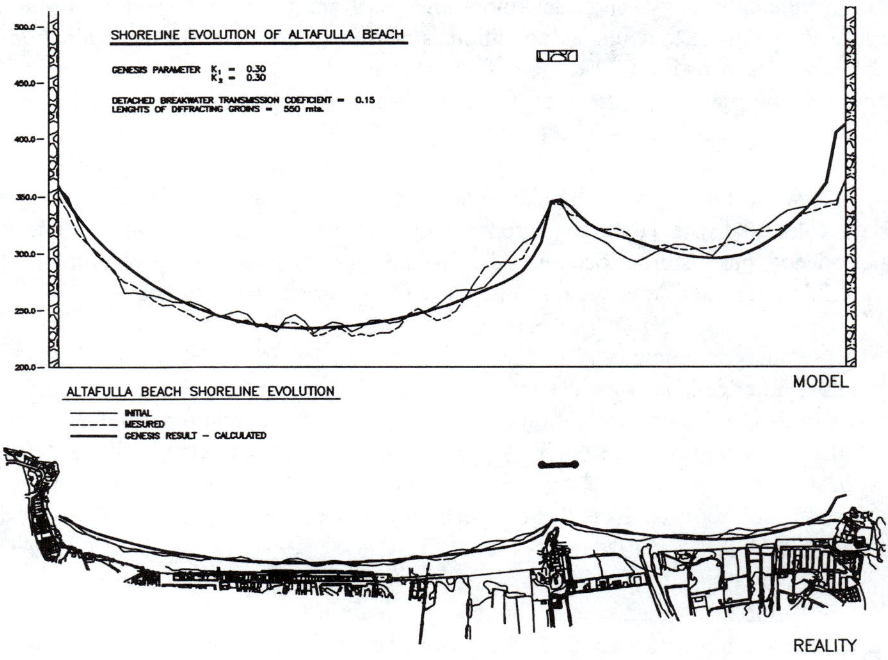

FIGURE 6 NUMERICAL METHOD, GENESIS, FOR SHORELINE EVOLUTION ANALYSIS

In order to understand beach behaviour, after the analysis of beach profiles and planform, a brief study of longshore transport was made. Theoretical formulas can be used, some of them only considere oblique wave incidence: CERC, KOMAR, etc. In this case heigth wave gradient, because of difraction phenomena, are also important. That is why it is necessary to use models that consider both phenomena. The GENESIS program also calculates longshore transport ratios along the shoreline and the average transport ratio considering obliquity wave incidence and height wave gradient. In this calculation initial and final shoreline data were used to calibrate the model. The average of long shore transport is about 5.000-9.000 m^3/year, this result agrees with some theoretical studies made near this area.

A statistical model was used computing the monitoring topobathymetry field data in order to find the tendencies of beach behaviour. The statistical model used was Principal Component Analysis (three way PCA). The method was used to objectively separate the spatial and temporal variability of the beach profile, planform and the sediment grain size data, as described by Medina et al 1994. The results show a seasonal variability in the beach profile data and a long-term trend. The bases of these statistical methods are the concentration of all topobathymetric information and to find some intrinsical information that can explain the trend of the beach evolution.

Three-way PCA is a statistical tool that can be used in the knowledge of beach behaviour. In Figure 7 the first, second and third cross-shore components, eigenvector are shown. The firts one defines the mean of the variable, in this case representing the medium average profile. The second cross-shore eigenvector is very important in the upper part of the profile and it will play an important role in determining the upper profile slope. The third cross-shore eigenvector shows a berm-bar variability. In our case the second and third eigenvectors have influence on the detached breakwater comparing COMPL ∅ and COMPL 1. A similar analysis can be made for long-shore eigenvectors.

The shows temporal eigenvectors, have been calculated the first one represents medium average temporal variation. The second one represents a seasonal dependence. Both of these are constant, and in the case of a two-year monitoring program, the temporal variability is not important. This means that the beach is in a quite equilibrium position with this wave climate. The combination of these eigenvectors explains most of the variability of the evolution of the beach. In order to combine them we will use the corresponding Bartussek core matrix values and the percentatge of variation explained. This matrix shows how to combine eigenvectors. In order to better interpret the variability they account for, it is useful to examine the profile or the bathymetry that is obtained by the product or one alongshore eigenvector with one cross-shore eigenvector, and use the corresponding temporal function to determine how the obtained profile or bathymetry evolves in time.

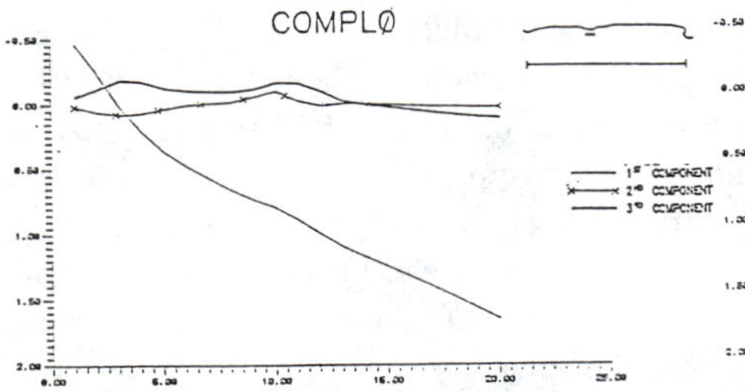

FIGURE 7 a) CROSS-SHORE EIGENVECTOR AFTER NOURISHMENT WORKS

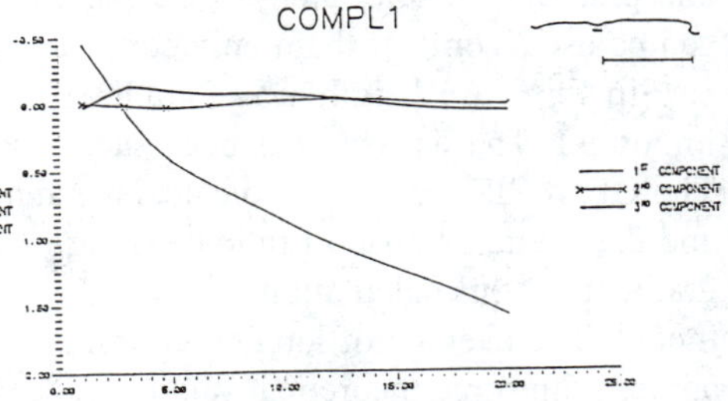

FIGURE 7 b) CROSS-SHORE EIGENVECTORS BEFORE NOURISHMENT WORKS

6.- CONCLUSION AND DISCUSSION

1.- Good data is needed in order to know the beach behaviour. It is the most important aspect for Coastal Engineers.

1.1 Wave climate must be obtained from:, Directional buoy, Non directional bouy with the direction data, Non directional bouy, Visual climate data.
This is the classification from the most valuable data to the least. Also, it depends on the range of information that it is possible to use. It is necessary to check different data information, if possible, and analyse it carefully before choosing the calculation data.

1.2 Topobathymetry data are the physical suport of Coastal Engineers. Accurate data is necessary for all king of model. It is necessary to control: Study are limits, Profile distance,Profile length, Points per profile, Position system and errors, Bathymetric and althymetric measurement system and errors,Working weather conditions, Technical and worker personnel,Data analysis and computation,Topobathymetry frequency.

1.3 Sand sampling is needed as a basic component of beach performance. It is necessary to control: Study area limits, Sampling points, Sampling system, Sampling analysis and computation, Sampling frequency.

2.- Technical empirical models are easy to use and give important qualitative and some quantitative results. Data errors can be detected.

3.- Numerical models are more complicated to use than technical-empirical and sometimes can take data errors obtaining results with more errors.

4.- No model, technical-empirical or numerical, gives good results with bad data.

5.- Numerical models need a lot of data, a calibration step and a verification step. The whole process requires a lot of time and money. For this reason, it is necessary to be rigorous choosing them on quantity and quality.
6.- Statistical models must be studied in depth in order to find the beach trends.
7.- Altafulla beach has been studied comparing monitoring fiel data and theoretical models. It is a quite stable beach with a sand drift of 5.000 m^3/year, from east to west. Sand is accumulated in calm areas and eroded in open areas.

REFERENCES

* Dean R.G., 1995. "Equilibrium Beach Profiles: Characteristics and Application" J. Coastal Research, vol. 7 nº 1, pp. 53-84.
* González E.M., 1995. "Morfología de playas en equilibrio: Planta y Perfil". Universidad de Cantabria, p.p. 183-198.
* Hanson, H. and N.C. Kraus, 1991. "Comparison of Shoreline Change obtained with Physical and Numerical Models". Proc. Coastal Sediments'91, ASCE, pp. 1785-1799.
* Hanson, H. and N.C. Kraus, 1989. "GENESIS-Generalized Model for Simulating Shoreline Change", vol. 1. Reference Manual and Users Guide, Tech. Report CERC-89-19, U.S. Army of Engineers, Waterway Experiment Station, 247 pp.
* Hsu, J.R.C., Silvester R., Yi-Min Xia 1989. "Static Equilibrium Bays: New Relationship". Journal of Waterway, Port, Coastal, and Ocean Engineering, vol. 115, nº 3, pp. 285-298.
* Kamphuis, J.W. and J.R. Moir, 1977. "Mean Diameter Distribution of Sediment Sizes before and after Artificial Beach Nourishment". Proc. Coastal Sediments'77. ASCE, New York, pp. 115-125.
* Medina, R., M.A. Losada, I.J. Losada and C. Vidal, 1994. "Temporal and Spatial Relationship between Sediment Grain Size and Beach Profile". Marine Geology, 118, pp. 195-206.
* de Vriend, H.J., 1992. "Mathematic Modelling of 3D Coastal Morphology". Proc. Short Course on Design and Reliability of Coastal Structure. 23rd I.C.C.E. Venecia, Chapter 10,25 pp.
* Work, P.A. and R.G. Dean 1995. "Assessment and Prediction of Beach-Nourishment Evolution". Journal of Waterway, Port, Coastal and Ocean Engineering, vol. 121, nº 3, May/June, pp. 182-189.

CHAPTER 369

ALTERNATIVES TO BEACH STABILIZATION: CAMBRILS COAST CASE STUDY (SPAIN)

Nuria Lupón(1), Alfonso Vidaor(2), Jordi Galofré(3) and F. Javier Escartín(4)

ABSTRACT

A study of alternatives including a shoreline evolution numerical modelization has been carried out in order to both diagnose the erosion problem at the beaches located between Cambrils Harbour and Pixerota delta (Tarragona, Spain) and select nourishment alternatives.

INTRODUCTION

Cambrils and Montroig are two villages located in a zone of big tourism growth in the Tarragona coast (northeast of Spain) 35 km from Tarragona and 125 km from Barcelona, very well communicated by road, motorway and railway (see figure 1).

Due to the importance of this tourist zone the Spanish Ministery planned a beach nourishment project for the area, including six beaches (6,000 meter long): Verge del Camí, Llosa, Torrent d'en Gené, Ardiaca, Riudecanyes and Pixerota.

The Cambrils Harbour construction in the thirties with a breakwater reaching the active depth, acting as a sediment transport barrier, strongly modified the littoral dynamics on beaches located southwest, suffering a severe regression only partially controlled by stabilization works carried out, which consisted basically in the construction of detached breakwaters and sand contribution. The efficiency of those works wasn't completely satisfactory, and the coastal zone between Cambrils Harbour and Pixerota delta presents nowadays a clear lack of stability with a general trend to regression.

---

(1) Senior Engineer; Europrincipia, sl., Centre Tecnològic Europroject, Parc Tecnològic del Vallès, 08290 Cerdanyola, Barcelona, SPAIN; Tel: 34-3-5820187; Fax: 34-3-5824394; e-mail: nuria@europroject.es

(2) Director; Europrincipia, sl., Centre Tecnològic Europroject, Parc Tecnològic del Vallès, 08290 Cerdanyola, Barcelona, SPAIN; Tel: 34-3-5820187; Fax: 34-3-5824394; e-mail: alfonso@europroject.es

(3) Projects and Works Service Chief; Tarragona Coasts Demarcation, General Directorate of Coasts, Environtment Ministery, Plaza Imperial Tarraco 4, 43005 Tarragona, SPAIN.

(4) Senior Engineer; Europrincipia, sl., Centre Tecnològic Europroject, Parc Tecnològic del Vallès, 08290 Cerdanyola, Barcelona, SPAIN; Tel: 34-3-5820187; Fax: 34-3-5824394; e-mail: fjeg@europroject.es

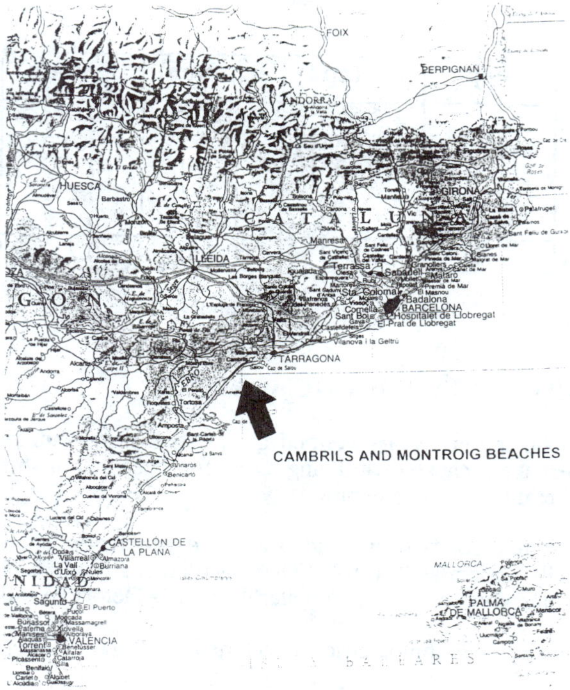

Figure 1: Location plan

MARITIME CLIMATE

The wave data used were:

- Cap Tortosa directional buoy, belonging to the General Directorate of Ports and Coasts of the Catalonia Autonomous Government during the interval 90/6 to 93/6, located at 0º58'6'' E, 40º44'6'' N coordinates in a depth of 45 m.
- Visual data registered at Casablanca oil platform during the interval 1983-1988, located at 1º21'30'' E, 40º43'3'' N coordinates in a depth of 170 m.
- Visual data registered by ships in the zone 0º-2º20'E, 40º-41º30'N and treated by the National Oceanic and Atmospheric Administration in Asheville.

The three data are valid though each of them presents some inconvenients:

- The Cap Tortosa buoy data have great reliability for being instrumental registers, but encompass a period of only three years.
- The ship visual data present a longer temporal interval, enough to represent the average zonal climate, but they have lesser reliability because of their origin.
- The Casablanca visual data present an intermediate reliability compared to other data.

NATIVE SAND GRANULOMETRIC ANALYSIS

A complete topographic and bathymetric survey of the studied zone was made including from dry beach to depth -13, deeper than active depth. At the same time a native seabed sand sampling campaign was performed: sixty sand samples in twelve profiles at five depths (+2.5, ±0.0, -3.0, -5.0 and -8.0) were analyzed obtaining the results shown in table 1.

Sample	D_{16} (mm)	D_{50} (mm)	D_{84} (mm)
+2.5	1.568	0.549	0.290
±0.0	1.684	0.553	0.296
-3.0	0.453	0.299	0.180
-5.0	0.439	0.312	0.221
-8.0	0.439	0.311	0.212
Total	0.859	0.390	0.232

Table 1: Native sand granulometry

BORROW SAND GRANULOMETRIC ANALYSIS

Studies performed in the last years verified the existence of submarine sand banks close to the project area, capable of being used for beach nourishment (as in the L'Hospitalet de l'Infant beach regeneration).

Another possible source for borrow sand was the material originated by the granite meteorization, locally known as "sauló". There existed four quarries near the project area that had been already used by the Ministery with satisfactory results.

The representative mixture granulometric characteristics for the two borrow sand types are shown in the table 2

	D_{16} (mm)	D_{50} (mm)	D_{84} (mm)
"Sauló"	1.568	0.549	0.290
Seabed sand	1.684	0.553	0.296

Table 2: Borrow sand granulometry

REGENERATION CHARACTERISTICS

There isn't a general rule that relates submerged beach slope with sand granulometry. However it's known that the slope proportionaly increases with the average sand grain size.

R. Silvester in his classical publication "Coastal Engineering" showed relationships between the beach slope and the sand size for three types of beaches: protected, semi-exposed and exposed.

For the different native sand samples the relation between beach slope and grain size was analyzed, obtaining figure 2, in which the three theoretical curves before mentioned were drawn, too. The values measured in Cambrils and Montroig beaches were located on the curves, representing exposed beaches. The borrow sand average size (0.42 or 0.92 mm) was located in the upper zone of the figure, where an additional dotted line representing Cambrils and Montroig beaches was drawn.

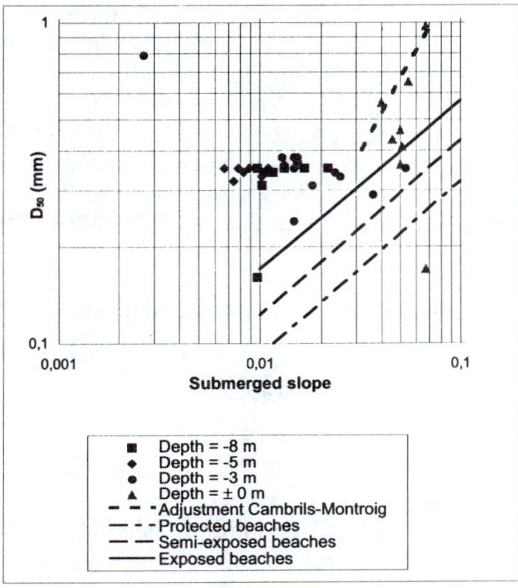

Figure 2: Submerged beach slope vs. Average grain size

In 1984 Garau and Friedman suggested a beach profile model for the case of sand classified by littoral processes for open beaches in the Mediterranian. The beach slope could be calculated using the average diameter D_{50} by means of the expression

$$m = 0.075 + 0.035 \bullet \ln D_{50}$$

Finally, from eroded profiles analysis Vellinga (1984) got a relationship between beach slope (with a certain granulometry) and slope (with another grain size):

$$m_1 = m_2 \left(\frac{w_1}{w_2} \right)^{0,56}$$

where *w* is the grain fall velocity. Once known the Cambrils and Montroig beaches grain sizes (D_{50} = 0.39 mm) and slopes (between 1V:30H and 1V:57H) and borrow sand sizes (D_{50} = 0.42 or 0.92 mm) the design beach slope could be calculated. The Ministery finally chose the seabed sand option with 1V:50H as submerged slope.

The estimation of longshore sediment transport rate induced by breaking wave actionwas used to obtain the zero transport direction, that is the shoreline orientation for which the average sand movement is zero.

A first estimation using the Komar's formulation was made in the assumption of straight uniform beach. The calculations were made for different shoreline orientations between 10ºN and 115ºN and for the three wave data (see figure 3). It can be noticed that the differences between the ship and Casablanca visual data curves are relatively little whereas the Tortosa buoy data curve differs much more (this is probably due to that the differences between the ship and Casablanca visual data curves are relatively little whereas the Tortosa buoy data curve differs much more (this is probably due to the shorter time interval which could be not completely representative of average local climate). It can be observed too, that for the real shoreline orientation (around 60° N) the transport direction is NE-SW, as it could be deduced from sand accumulation north of Cambrils Harbour (that acts as total barrier to sediment transport). A further analysis indicated that sediment transport values estimated with the ship visual data seemed to be overestimated, specially gross transport, so the most representative data appeared to be those obtained from the Casablanca data. In this case the zero transport orientation was approximately 38°N.

The average incident wave direction was also estimated according to

$$D_{av} = \frac{\sum_i D_i \cdot p_i}{\sum_i p_i}$$

where p_i is the presentation probability, obtaining a value D_{av} = 135,1°N, very close to the zero net transport direction. Therefore this was considered as the average wave direction, value very similar to the perpendicular to the shoreline orientation north of Cambrils Harbour, were there is no sand movement.

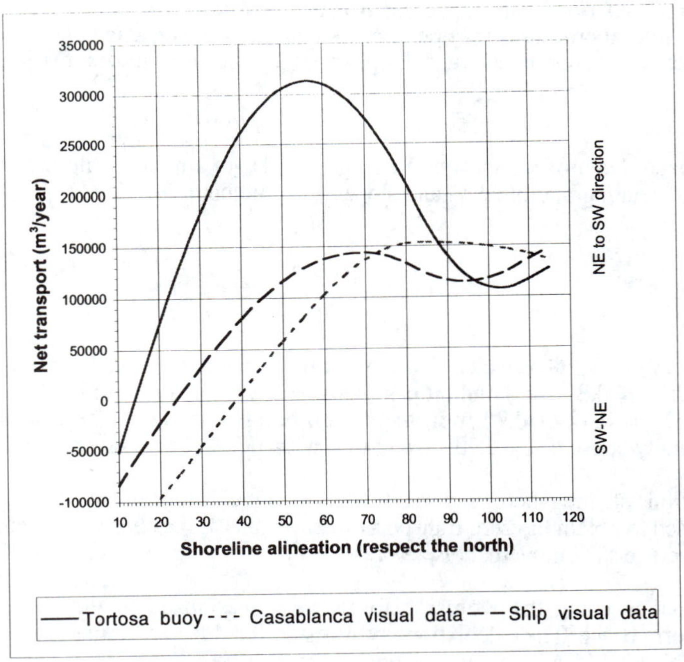

Figure 3: Net sediment transport

HISTORICAL EVOLUTION

A previous step to alternatives study was the shoreline historical evolution analysis in order to know the situation in the area between Cambrils Harbour and Pixerota delta, and how the existing beaches were formed. Five aerial photographs restitutions corresponding to 1947 (February), 1957 (June), 1965 (August), 1973 (June) and 1977 (March), two aerial photopraghs (not restituted, for qualitative estimation, June of 1983 and May of 1990) and May 1994 bathymetry were used for the analysis.

Historically the Cambrils and Montroig coast presented a continuous beach nourished by the sand transported by littoral drift in NE-SW direction. The Cambrils Harbour, built in the late thirties, became the main artificial barrier to sediment transport.

The main local sedimentary sources are the existing streams (Alforja, Verge del Camí, Torrent d'en Gené, Riudecanyes and Pixerota) characterized by a discontinuous flow, only important during storms periods. Some of them have created small deltaic formations with coarser material (gravels).

The correspondence between the beaches located south of the harbour and its distance or kilometric point (KP) is

Verge del Camí (**V**)	100 to 900	L'Ardiaca (**A**)	2300 to 2800
Llosa (**L**)	900 to 1800	Riudecanyes (**R**)	2800 to 3400
Torrent d'en Gené (**T**)	1800 to 2300	Pixerota (**P**)	3400 to 5400

The 1947 shoreline included the Cambrils Harbour presence and became the comparison base for the study. In the period 1947-1957 the **V**, **L**, **T** and **A** beaches eroded in average 2.7, 2.8, 1.8 and 2.1 meters/year respectively, although the first 200 meters of **V** beach advanced 0.8 m/year. The first 400 meters of **R** beach eroded 1 m/year whereas its last 200 meters advanced 0.4 m/year. **P** beach eroded an average of 0.9 m/year. The cause of this behaviour was the interruption of sediment transport by Cambrils Harbour that in the zone had the NE-SW direction: therefore beaches located SW of the port became the main nourish source for the littoral current, so its erosion process started.

Between 1957 and 1965 the generalized erosion process continued with an absolute lower value in the northern half. The average regressions were 0.7, 1.4, 1.1, 1.1 and 1.1 m/year respectively, though the first 350 meters of **V** beach advanced 1.5 m/year. The first 200 meters of **P** beach advanced 0.4 m/year whereas the rest eroded in average 1.1 m/year.

In the period 1965-1973 the **V**, **L**, **T**, **A** and **R** beaches eroded an average of 1.4, 1.9, 1.9, 1.3 and 0.2 m/year respectively, although the first 300 meters of **R** beach advanced 0.3 m/year. The first 200 meters of **P** beach advanced 0.5 m/year whereas the rest eroded 1.0 m/year.

This means that in 1973 the average shoreline erosion for the sector was between 20 and 55 meters in comparision to 1947 (with extreme values of 70 meters). So in 1976 the first stabilization project was performed, consisting in the construction of four detached breakwaters, the enlargement of an existing breakwater and the nourishment of sand to generate tombolos behind the detached breakwaters that stabiliced the **V**, **L**, and **T** beaches. Nevertheless the lack of sediment proceeding from the three first beaches which in the past had nourished partially the littoral current accelerated the erosion process in the rest of beaches: the **A**, **R** and **P** beaches eroded in average 1.7, 3.6 and 1.9 m/year.

Thus, in 1985 a second stabilization project was carried out, with the construction of two additional detached breakwaters to protect the **A** and **R** beaches and six artificial islands between the detached breakwaters and the coastline to optimize its behaviour that had caused important erosions at the center of the protected beaches. Since then the northern beaches shoreline position hasn't varied substantially, indicating its stability, while the Riudecanyes delta and **P** beach have continued to erode with an average of 0.6 and 1.6 meters/year.

All this values are summarised in table 3 whereas all the shoreline evolution is represented in figure 4.

Beach	Verge del Camí		Llosa	Torrent d'en Gené	L'Ardiaca	Riudecanyes		Pixerota	
1947-57	+0.8	-2.7	-2.8	-1.8	-2.1	-1.0	+0.4	-0.9	
1957-65	1.5	-0.7	-1.4	-1.1	-1.1	-1.1		+0.4	-1.1
1965-73	-1.4		-1.9	-1.9	-1.3	+0.3	-0.2	+0.5	-1.0
1973-77	Tombolos generation		Tombolos generation	Tombolos generation	-1.7	-3.6		-1.9	
1977-94	Tombolos generation		Tombolos generation	Tombolos generation	Tombolos generation	Tomb. gen.	-0.6	-1.6	

Table 3: Average annual accretion and erosion summary

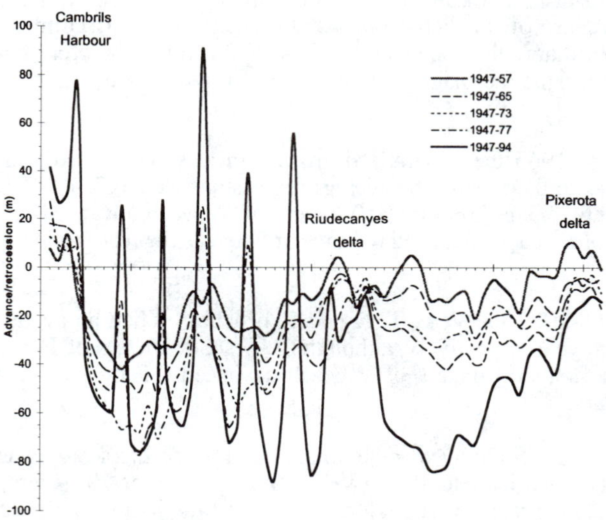

Figure 4: Shoreline evolution between Cambrils Harbour and Pixerota delta

From shoreline variations the volume of sand mobilized by wave action between Cambrils Harbour and Pixerota delta was estimated in 35,000 m^3/year with a direction NE-SW.

ALTERNATIVES STUDY

In the study carried out nine stabilization alternatives were analysed, taking into account flexible solutions (just sand nourishment) and rigid ones (new breakwaters). All of them included the demolition of the existing detached groins until the depth -2.0 and considered a minimum dry beach width of 60 meters (see figure 5). The quarry and sand volumes required and budgets for each alternative are presented in table 4.

- *Alternative 0:* It consisted exclusively in sand contribution between Cambrils Harbour and Pixerota delta. From the historical shoreline evolution study it was estimated that 8 years after nourishment the erosion in the northern beach would be important enough (about 20 m) to consider a periodical sand contribution south of the harbour at a rate of 35,000 m^3/year.
- *Alternative 1:* Compared to alternative 0 it included a 142 m long groin at the Pixerota delta to ensure a beach support. The groin was designed short enough to let the sediment pass (and nourish southern beaches as Rifà and Porquerola), so the same periodical contributions that in alternative 0 were necessary.
- *Alternatives 1a, 2, 2a:* Compared to option 1, they incorporated one, two and three groins respectively with the same purpose, being also permeables to sand movement.
- *Alternative 2b:* In this case the groins position was very similar to alternative 2a, but they were much longer and included submerged toes in order to stabilize completely the new three northern beaches (untill Riudecanyes delta), as littoral transport was completely interrupted. The southern groin (at Pixerota delta) was the same than in alternatives 0 to 2a. From historical shoreline evolution study it was estimated that 17,500 sand m^3/year would be mobilized in direction NE-SW between Riudecanyes and Pixerota deltas and would cross the last groin to nourish partially the southern beaches (Rifà, Porquerola...). Therefore there would exist in those beaches an insufficiency of 17,500 m^3/year that should be added periodically south of Pixerota delta to avoid its erosion. Moreover, the sand mobilized along the new Pixerota beach would produce important erosions in the northern part, so it was estimated that 8 years after the regeneration periodical sand contributions (17,500 m^3/year) south of Riudecanyes delta would be necessary.

Alternative	0	1	1a	2	2a	2b	3	4	5
KTn of demolished quarry	183	183	183	183	183	183	183	183	183
KTn of quarry for new groins	0	15	37	53	76	352	379	350	291
Km^3 of sand for initial nourishm.	1,695	1,930	2,613	2,684	2,742	3,954	2,841	2,691	2,699
Km^3 of sand for periodical contributions	770	770	770	770	770	787.5	787.5	787.5	787.5
Budget (M$)	13.3	14.3	17.0	17.3	17.6	24.3	20.4	19.5	19.0

Table 4: Quarry and sand volumes required and budgets

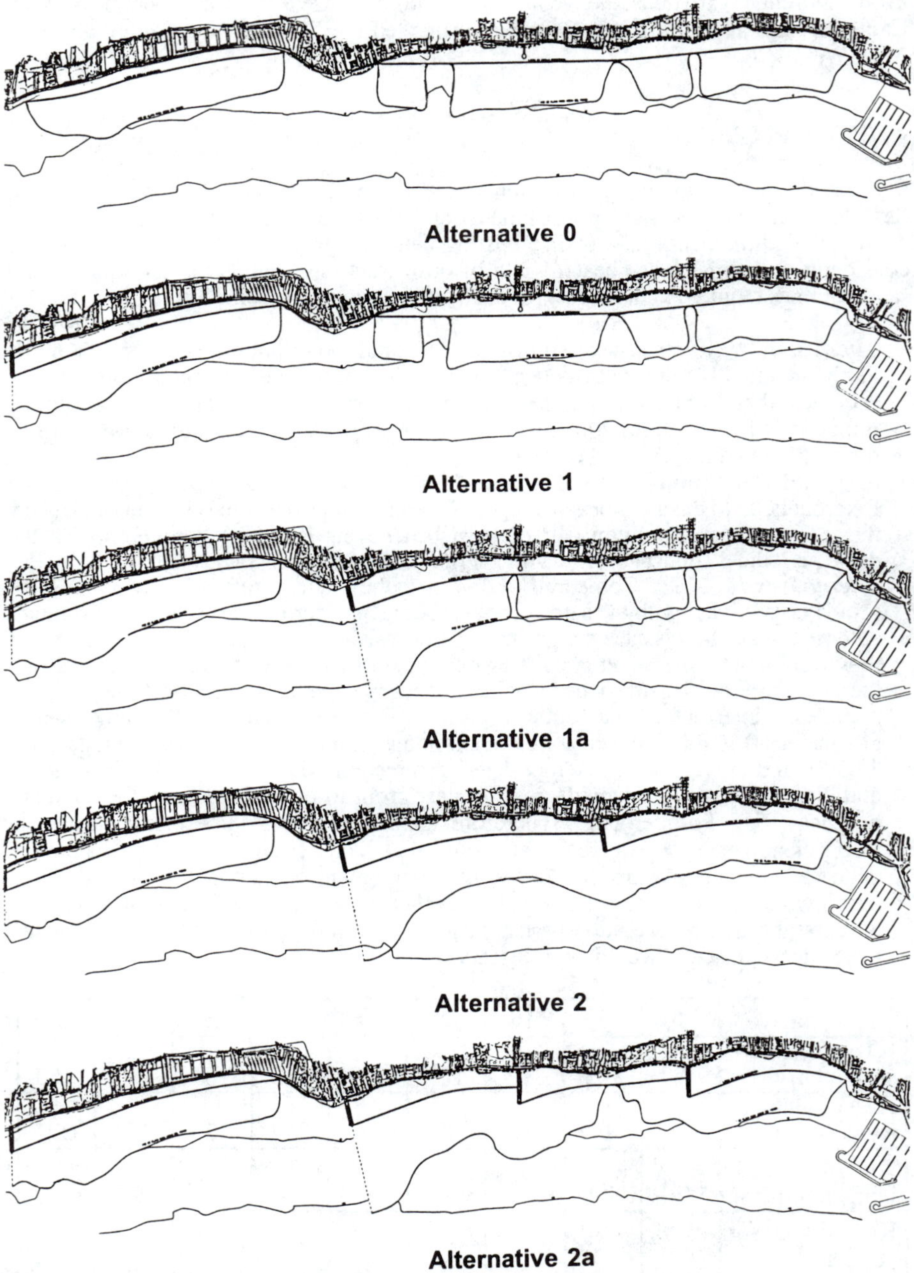

Figure 5a: Suggested regeneration alternatives

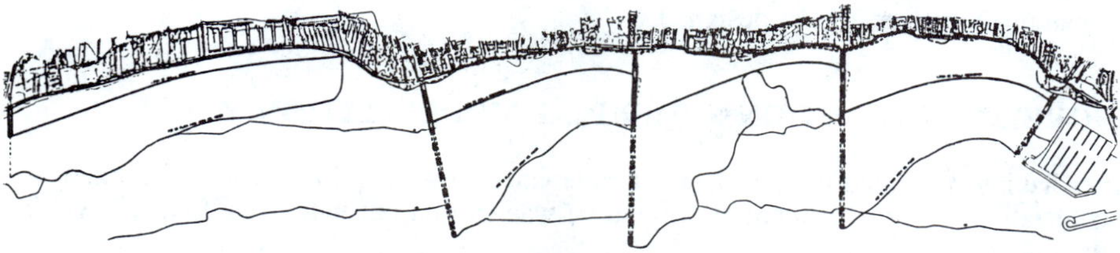

Alternative 2b

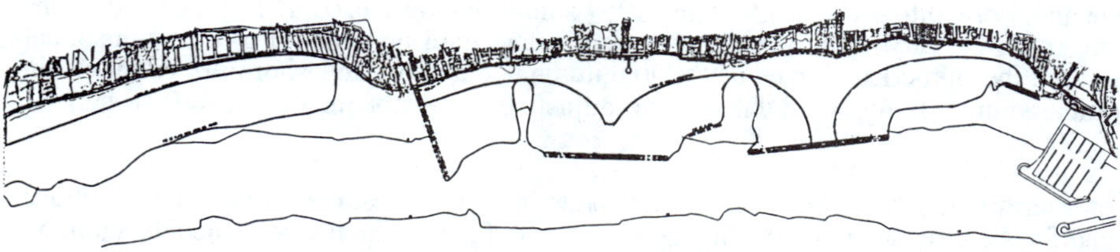

Alternative 3

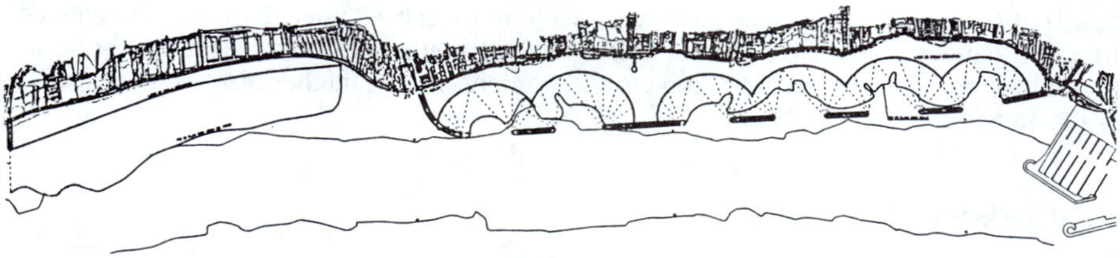

Alternative 4

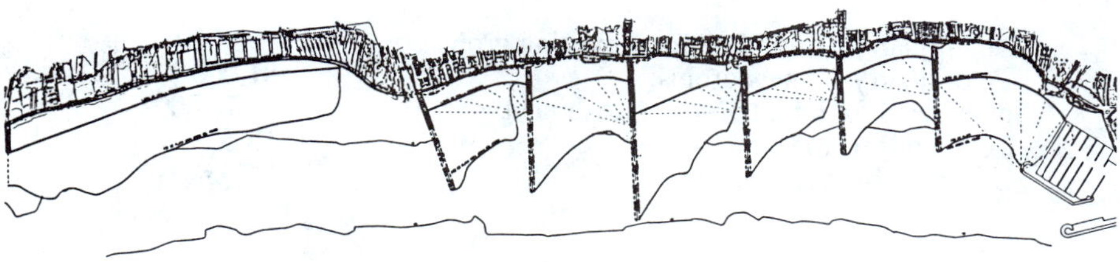

Alternative 5

Figure 5b: Suggested regeneration alternatives

- *Alternatives 3, 4 and 5:* They had the same purpose than the previous alternative but the complete stabilization of the northern beaches was obtained respectively by means of: *a)* two long detached groins and one transversal groin with a submerged toe, *b)* seven detached groins and *c)* six transversal groins between 157 and 214 meters long with submerged toes respectively. In the Pixerota delta the same groin than in alternative 2b was designed.

SHORELINE EVOLUTION NUMERICAL MODELIZATION

After having presented the nourishment alternatives, the shoreline evolution numerical modelization for three of them (1, 1a and 5) was carried out with the GENESIS v 3.0 program.

A thorough calibration process was done in order to obtain transport coefficients for the area, necessary for long term coastal evolution predictions of the alternatives selected. For calibration the 1947 and 1957 shorelines were used. Different simulations were made by varying the formulation parameters until obtaining a proper adjustment between the numerical values of the longitudinal transport and shoreline evolution in comparision with measured values. This adjustment was obtained for $K_1 = 0.40$ and $K_2 = 0.20$

Long term simulations (10 years) were made in order to estimate the beach evolutive trend for the three alternatives. In figure 6 results for the alternative 1 modelization are presented. It can be noticed the main problem is that the shoreline erodes half of the beach amplitude (won with the nourishment) five years after the regeneration at certain points. Therefore a second prediction including periodical sand contributions was made (figure 7). After some tests it was estimated that a 38,000 m^3/year contribution stabilized the beach. This value was very similar to that estimated in the alternatives study. At Riudecanyes and Pixerota deltas the maximum erosion was limitated due to its granulometric characteristics (gravels and boulders). This methodology was repeated for the 1a and 5 alternatives.

CONCLUSIONS

The alternative 1a was selected by the Ministery and a construction detailed design was developed in order to prepare the tendency of the works. Main quantities and budget for the project are:

Tn of quarry demolished from existing groins:	193,000
Tn of quarry for new groins:	81,000
m^3 of sand for initial beach nourishment:	2,616,000
Total budget (M$):	15.6
Length of beach stabilized:	5.2 km

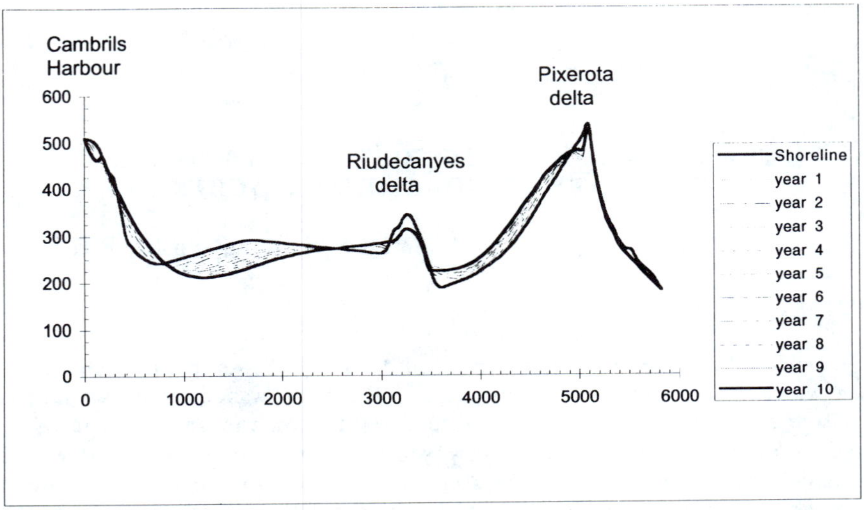

Figure 6: Alternative 1 shoreline evolution prediction without periodical sand contributions

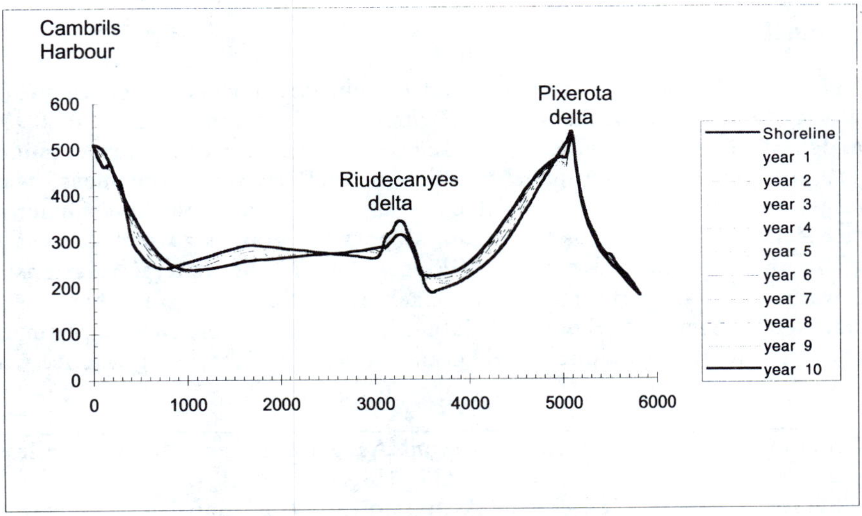

Figure 7: Alternative 1 shoreline evolution prediction with periodical sand contributions

CHAPTER 370

THE EFFECT OF GRAIN SORTING ON PROFILE STABILITY OF NOURISHED BEACHES

Ram K. Mohan, Ph.D., P.E.[1] and Timothy W. Kana, Ph.D., P.G.[2]

Abstract

Dean's (1983) model is often used to predict the profile shape and berm width after nourishment. Ideally, if the borrow sediment is identical to native sediment, the adjusted profile will take on the same shape as the existing profile. But, if the borrow sediment is coarser than native, the nourished profile will adjust to a steeper configuration, producing a wider dry beach. If borrow sediments are finer, the fill will adjust to a gentler slope, producing a narrow dry beach. A conceptual model of the effect of grain sorting on nourished profile stability is presented in this paper and verified by comparison with actual beach measurements. The results indicate that the post-nourishment performance can be improved measurably by minor variations in grain size distribution, specifically where a coarse fraction is present (i.e., negative skewness). Fill placement technique and other environmental factors, of course, also impact performance.

Introduction

Numerous researchers have developed empirical models for beach profile stability and equilibrium slopes (Brunn, 1954; Dean, 1983 and 1991; Hands, 1981; and Krumbein and James, 1965). However, these models relate primarily to environmental factors such as wave steepness, wave energy, tide levels, and sea-level changes. Historical data from numerous beaches indicate a trend of increasing slope with grain size and decreasing slope with increasing wave energy (Wiegel, 1964). The general relationship of wave energy, grain size, and beach face slope on sand beaches is illustrated in Figure 1. Empirical data from numerous beaches confirm the trend of increasing slope with grain size and decreasing slope with

---

[1] Senior Engineer, Gahagan & Bryant Associates, Inc., 9008-0 Yellow Brick Road, Baltimore, MD 21237, USA.

[2] Senior Scientist, CSE-Baird, A division of W.F. Baird & Associates, Ltd., P.O. Box 8056, Columbia, SC 29210, USA.

increasing wave energy (Wiegel, 1964). Larger grain sizes allow increased percolation of wave uprush and a corresponding decrease in backrush volumes (Waddell, 1973). This results in steeper beach face slopes with increasing grain size.

Intense mixing under breaking waves will further affect the distribution of grain sizes over the shore face. Coarsest material is common at the breakpoint ("step" at the toe of the beach face), and fining will occur in either direction: toward the foredunes and back shore (Mason and Folk, 1958; Fox el al., 1966) and toward the lower foreshore (Swift et al., 1971). Multimodal size distributions across the beach occur in some areas [e.g., Duck, North Carolina (Kana et al., 1980)], but more commonly, a single mode distribution occurs. Data from Swift et al. (1971) revised by Swift (1976) show variations in the modal grain size from the berm to the lower shoreface (Figure 2). Sediment coarsening near the breaker line is reflected in the size distribution in Swift's transect B, as well as by coarse-skewed distributions for this zone of profile in transects A and C. As Swift's data show, sorting is generally good across the shoreface. Dean (1983; 1991) performed extensive work in modeling equilibrium profiles, assuming a uniform grain size distribution across the active profile.

Work and Dean (1991) suggest that varying the grain size across the profile within limited size ranges typical for beaches does not significantly improve the simpler, and often used, profile model (Brunn, 1954; Dean, 1983):

$$h(y) = Ay^{0.667}$$

where h is the water depth at a distance y offshore of the mean water line and A is a scale parameter dependent on sediment characteristics.

Dean's model (Dean, 1983; 1991) has been used to predict the resulting profile shape and berm width (dry beach width) after nourishment. In simple terms, if borrow sediment is identical to the native size distribution, the adjusted profile will take on the same shape as the existing profile. But if the borrow sediment is coarser than native, the nourishment profile will adjust to a steeper configuration, leaving more sand along the back shore (and producing a wider dry beach). If borrow sediments are finer, the fill will adjust to a gentler slope with more material shifting offshore (producing a narrow dry beach) as shown in Figure 3. As a result of these important relationships, the Shore Protection Manual (USACE, 1984) and common practice recommends use of the same or slightly coarser size distribution for nourishment purposes. The authors followed these guidelines for selecting the borrow area for the 1991 Hunting Island nourishment project (CSE, 1991). The borrow area sediments for this project contained slightly coarser sands and the grain size distributions confirmed the negative skewness of the samples produced by the presence of broken shells. Overfill ratios for this project were favorable in the range 1.10 to 1.20 and the percentage mud was relatively low at 5 percent. Based on these results and the findings of other investigators just outlined, some changes in the profile shape and slope after nourishment would be expected. This change would be due to a general coarsening of the beach sediment, or due to a change in the grain size distributions across the shoreface after nourishment.

A conceptual model of the grain size distribution shift for hydraulic nourishment projects will be presented in this paper and verified by comparison with actual beach measurements taken after the fifth Hunting Island beach nourishment project.

Effect of Method of Placement on Profile Stability

Given a different size distribution of borrow sediments compared to the native sediments, the method of placement would affect the size distribution across the profile. Fill by trucks along intertidal and dry beach is likely to produce uniform admixtures which are gradually eroded along the seaward margin. Back shore fill in this case may initially contain the same distribution of fines as the borrow area. In fact, a legitimate concern with such projects is the possibility of creating hard pan because of the binding effect of muddy sediment in the mixture (Siah et al., 1985; Kana and Jones, 1988). These fines may not erode until large storms occur or more seaward material has shifted downslope through profile adjustment.

For projects built by pipeline dredge, the response will be different. Studies on dredged material disposal (Montgomery, 1978; Palermo et al., 1978) show that coarse material settles closest to the discharge point, whereas the fine material moves further away in the slurry. If the pipeline is placed along the back shore, this leads to sorting down-profile, with coarse material concentrating along the berm and finer material settling downslope. The result will be a systematic shift in the grain size distributions from the back shore to the lower foreshore.

Conceptual Model of Nourished Beach Profile Stability

A conceptual model of the grain size distribution shift for hydraulic nourishment projects is presented in Figure 4. As the material is discharged onto the beach, the coarser fraction settles in the upper beach face. This modifies the prenourished grain size frequency curve by shifting it to the left, thereby indicating that the coarser fraction has increased (Figure 4a). The mid-sized fraction moves further down the beach profile and settles along the beach face. Since the grain size distribution of this section of the beach is very similar to the modal size fraction, the prenourished grain size frequency curve retains its shape more or less (Figure 4b). The finer fraction moves further down the profile and eventually settles at the lower foreshore. This modifies the prenourished grain size frequency curve by shifting it right, thereby indicating that the finer fraction has increased (Figure 4c).

An important consequence of this is that the upper beach face and berm becomes more stable and wider than would occur if the borrow area sediments matched the native beach. This means that the recreational life may be longer as well because dry-beach width often defines recreational benefits. A negative consequence of such a shift in the grain size distributions is the possibility of accelerated erosion of the lower shoreface.

Comparison with Actual Beach Observations

Comparison of prenourished (1990) and postnourished (1993) grain size distribution curves for the fifth Hunting Island beach nourishment project (Figure 5) show a clear trend of the sediments at the berm becoming coarser after nourishment while those along the beach face becoming finer after nourishment. The 1993 beach samples retained a coarse tail but the proportion of coarse material decreases with distance offshore. The bulk of the sample population is seen to shift toward finer sized down profile and has a smaller modal size than native in the lower shoreface.

Two schemes of statistical analysis were used to compare the native beach (1990) and nourished beach (1993) sediments (Krishna Mohan et al., 1993). In the first method (Method A), coarse sand, very coarse sand and gravel were grouped together to form percent coarse material (ie., <1.0 Phi). Similarly, very fine sand, silt, and clay were grouped together to form percent fine material (ie., >3.0 Phi). According to this scheme, the total sediment volume of the borrow area (757,644 cy) consisted of 15.35 percent (116,298 cy) coarse material and 19.17 percent (145,240 cy) fines. While the results of this analysis (see Table 1) show the expected trend of increased coarsening in the berm and increased fines in the lower beach, they do not include a large portion of the fill represented by size classes 1.0 Phi to 3.0 Phi. Therefore, the authors prepared a second analysis whereby the entire sediment grain size range was divided into two broad subdivisions comprising of coarse and fine groups.

In the second method of analysis (Method B), materials with grain sizes less than 2.5 Phi were classified as coarse and those with grain sizes greater than 2.5 Phi were classified as fines. Accordingly, most fine sands, medium sands, coarse sands, very coarse sands and gravel fall into the "coarse" category. Similarly, some fine sands, very fine sands, silts and clays fall into the "fine" category. According to this scheme, the total sediment volume of the borrow area (757,644 cy) is comprised of about 51.44 percent (389,732 cy) coarse material and about 48.56 percent (367,912 cy) fines. Table 2 summarizes the results obtained by this method which clearly supports the propositions of the conceptual model. A close examination of Table 2 indicates that coarse sediments in the berm increased from 36 percent to about 78 percent of the sample population after nourishment, whereas those at the beach face decreased from 56 percent to about 42 percent of the sample population. These trends correspond to the leftward shift of the upper beach sediments and the rightward shift of the lower beach sediments as predicted by the conceptual model. The grain size statistics in Tables 1 and 2 confirm these trends but also show that two years after nourishment the fill retains a coarse tail and is more graded. Sorting decreased significantly between 1990 prefill and 1993 postfill conditions (about 0.4 Phi to 1.0 Phi, respectively); and skewness became more negative in the 1993 samples. Skewness after nourishment was higher on the beach face than on the berm or lower foreshore.

Table 1. Comparison of native beach (1990) versus nourished beach (1993) composite samples. [Coarse <1.0ϕ; fine >3.0ϕ]

Location	Mean Grain Size		%Coarse		%Fine	
	[1990]	[1993]	[1990]	[1993]	[1990]	[1993]
Berm	2.73	1.57	0.20	32.52	26.70	4.35
UBF	--	1.93	--	23.05	--	11.39
LBF	2.49	2.41	0.80	8.66	9.15	17.01
LSF	--	2.56	--	0.31	--	12.83

Note:UBF=Upper Beach Face; LBF=Lower Beach Face; LSF=Lower Shore Face.

Kana and Andrassy (1993) conducted a volume change analysis for the postnourished beach profile which showed that there has been a rapid loss of sand in the project area. By April 1993, north beach had lost 50 percent of the fill, and south beach had lost almost 80 percent of its fill. The central project area compartment lost about 45 percent of the fill by April 1993. Figure 6 (lower) illustrates the average unit volume beach change since nourishment (February 1991) as a function of contour interval within the project area. The dry beach to MHW (+10 ft to +3.2 ft NGVD) within the project area retained 70 percent of the fill through April 1993. The intertidal beach (MHW to MLW; ie., +3.2 ft to -2.2 ft NGVD) retained about 45 percent of the fill two years later. In contrast, the underwater lenses (-2.2 ft to -12.0 ft NGVD) retained only 27 percent of the fill by April 1993. A close review of these results reveals an interesting factor: the loss rate for the upper beach lenses was much lower. These results support the predictions of the conceptual model that more fill is retained at the berm than on the lower beach.

Table 2. Comparison of native beach (1990) versus nourished beach (1993) composite samples. [Coarse <2.5ϕ; fine >2.5ϕ]

Location	Mean Grain Size		%Coarse		%Fine	
	[1990]	[1993]	[1990]	[1993]	[1990]	[1993]
Berm	2.73	1.57	35.98	78.04	64.02	21.96
UBF	--	1.93	--	60.71	--	39.29
LBF	2.49	2.41	56.02	43.25	43.98	56.75
LSF	--	2.56	--	46.28	--	53.72
	Sorting (ϕ)					
Berm	0.42	1.13				
UBF	--	1.14				
LBF	0.41	0.82				
LSF	--	0.41				
	Skewness					
Berm	0.02	-0.29				
UBF	--	-0.48				
LBF	-0.58	-0.91				
LSF	--	-0.13				

Note:UBF=Upper Beach Face; LBF=Lower Beach Face; LSF=Lower Shore Face.

Comparison with Previous Projects

Unpublished profile data following the December 1968 and December 1971 federal projects at Hunting Island were used to develop an estimate of the average fill volumes remaining (Figure 6 top and middle). The reported mean grain sizes prior to nourishment are given in Tabel 3.

Table 3. Mean Grain Size of Native Beach and Borrow Area Sediments.

Project Year	Native Beach Grain Size (mm)	Borrow Area Grain Size (mm)
1968	0.16	0.18
1971	0.16-0.21	0.18
1975	--	--
1980	0.14-0.20	--
1991	0.20	0.20-0.23

A review of Figure 6 indicates that fill losses have been exceedingly rapid after all projects. However, the proportion of fill remaining above low water has been many times higher following the 1991 nourishment. In fact, upwards of 75% of the fill remaining following the 1991 project has been situated above the low water contour. In contrast, most of the fill above low water was lost soon after the 1968 and 1971 projects.

Conclusions

A conceptual model of profile stability for beach nourishment projects was presented in this paper. The model predicts a trend of sediments in the upper beach becoming coarser by means of selective sorting at the discharge point (hydraulic nourishment) along the back shore, while those at the lower beach become finer after nourishment. While the conceptual model of nourished profile stability is not quantitative, it can be used as a basis for predicting trends and formulating alternatives. Comparison with actual beach observations support this prediction since the actual (observed) data showed the berm sediments becoming measurably coarser in both mean grain size and degree of grading (ie., poor sorting). Comparative profiles and a detailed volume change analysis revealed that the dry beach retained 70 percent of its fill whereas the underwater lenses retained only 27 percent of their fill after two years. This further supports the idea that use of coarser borrow material will improve longevity of the upper beach. Profiles, in this case, became steeper after nourishment. In conclusion, three factors become important in controlling the postnourishment grain size distribution and profile stability (in order): *(i) Grain size distribution:* This fundamentally controls the overall slope and distribution as predicted by equilibrium profile theory. Basic engineering logic suggests increasing coarse material tends to improve beach profile stability. The finer fraction washes out faster, thereby decreasing longevity and, hence, should be kept to a bare minimum whenever possible; *(ii) Placement of fill material:* The most common method of placement of fill material in a nourishment project is by the use of pipelines. If discharge is along the back shore, coarse material will settle near the berm and fines will shift downslope with the slurry. This type of placement improves berm longevity. If the discharge point is along the lower foreshore (ie., profile nourishment, Brunn, 1988), coarse material will have less chance of concentrating on the berm; and, *(iii) Environmental factors:* These include the magnitude and interrelationship of the following variables: background erosion rates, shoreline morphology, waves, currents, tides, and storm frequency. These factors produce site-specific responses and are independent variables in nourishment design.

References

Brunn, P. 1954. Coastal Erosion and the Development of Beach Profiles. Tech. Memorandum No. 44, Beach Erosion Board, Washington, D.C.

CSE. 1991. Geotechnical Studies for the Hunting Island Beach Nourishment

Project. Coastal Science & Engineering, Inc., Columbia, SC.
Dean, R.G. 1983. Principles of Beach Nourishment. In Handbk. of Coastal Processes and Erosion (Komar, ed.), CRC Press, Boca Raton, FL.
Dean, R.G. 1991. Equilibrium Beach Profiles: Characteristics and Applications. Journal of Coastal Research, Vol. 7(1).
Fox, W.T., J.W. Ladd, and M.K. Martin. 1966. A Profile of the Four Moment Measures Perpendicular to a Shoreline. Journal of Sediment Petrology, Vol. 36(4).
Hands, E.B. 1981. Predicting Adjustments in Shore and Offshore Sand Profiles on the Great Lakes. CETA81-4, U.S. Army Corps of Engineers, Coastal Engineering Research Center, Ft. Belvoir, VA.
Kana, T.W., and C.J. Andrassy. 1993. Hunting Island State Park 1991 Beach Nourishment Report. Coastal Science & Engg., Columbia, SC.
Kana, T.W., and C.P. Jones. 1988. Myrtle Beach Nourishment Project. Coastal Science & Engineering, Inc., Columbia, SC.
Kana, T.W., L.G. Ward., and H.M. Johnson. 1980. Suspended Sediment Transport at Duck, NC. Coastal Science & Engg., Inc., Columbia, SC.
Komar, P.D. 1976. Beach Processes and Sedimentation. Prentice Hall, Englewood Cliffs, N.J.
Krishna Mohan, R., Kana, T.W., and C.J. Andrassy. 1993. Comparison of Geotechnical Properties of Beach Fill Versus Borrow Area Sediments for the Fifth Hunting Island (SC) Beach Nourishment Project. Coastal Science & Engineering, Inc., Columbia, SC.
Krumbein, W.D., and W.R. James. 1965. A Lognormal Size Distribution Model for Estimating Stability of Beach Fill Material. Tech. Memo. No. 16, U.S. Army Corps of Engineers, Vicksburg, MS.
Mason, C.C., and R.K. Folk. 1958. Differentiation of Beach, Dune and Aeolian Flats by Size Analysis. J. of Sediment Petrology, Vol. 28(2).
Montgomery, R.L. 1978. Methodology for Design of Fine-Grained Dredged Material Containment Areas for Solids Retention. TR D-78-56, U.S. Army Corps of Engg., Waterways Experiment Station, Vicksburg, MS.
Palermo, M.R., R.L. Montgomery, and M.E. Poindexter. 1978. Guidelines for Designing, Operating and Managing Dredged Material Containment Areas. TR DS-78-10, U.S. Army Corps of Engineers, Vicksburg, MS.
Siah, S.J., E.J. Olsen, and T.W. Kana. 1985. Myrtle Beach Nourishment Project. Coastal Science & Engg, Inc., Columbia, SC.
Swift, D.J.P. 1976. Coastal Sedimentation. In Marine Sediment Transport and Environmental Mgmt. (Stanley & Swift, ed.), Wiley & Sons, N.Y.
Swift, D.J.P., R.B. Sanford, C.E. Dill, Jr., and N.F. Avignone. 1971. Textural Differentiation on the Shoreface During Erosional Retreat of an Unconsolidated Coast. Sedimentology, Vol. 16.
USACE. 1984. Shore Protection Manual. U.S. Army Corps of Engineers, Washington, DC.
Waddell, E. 1973. Dynamics of Swash and its Implications to Beach Response. Louisiana State University, Baton Rouge, LA.
Wiegel, R.L. 1964. Oceanographical Engg. Prentice Hall, Engle Cliffs, NJ.
Work, P.A., and R.G. Dean. 1991. Effect of Varying Sediment Size on Equilibrium Beach Profiles. ASCE Coastal Sediments'91, New York, NY.

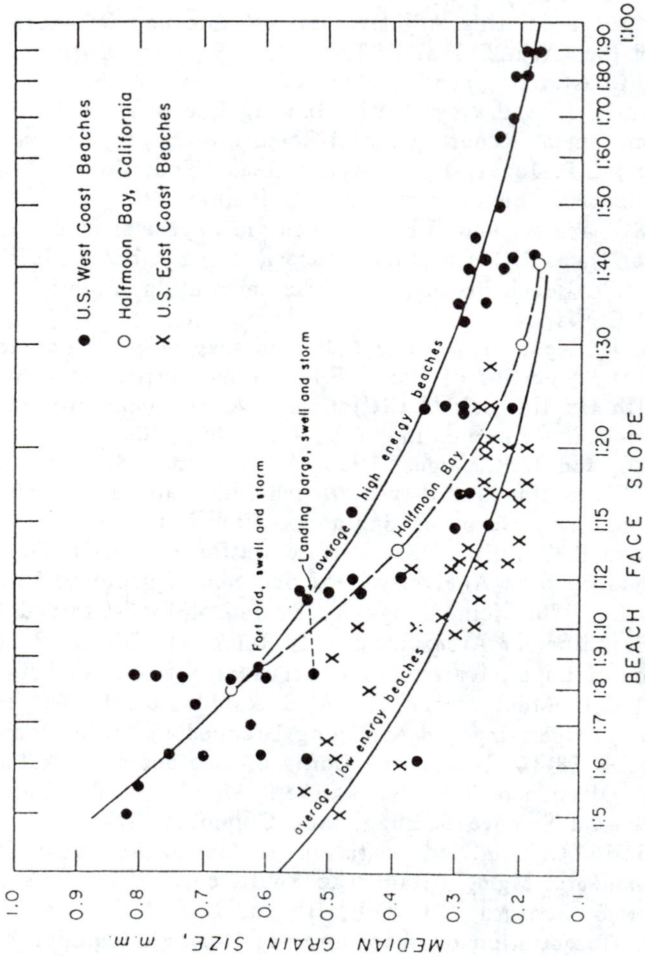

Figure 1. Relationship of grain size, wave energy, and beach face slope [from Komar, 1976]

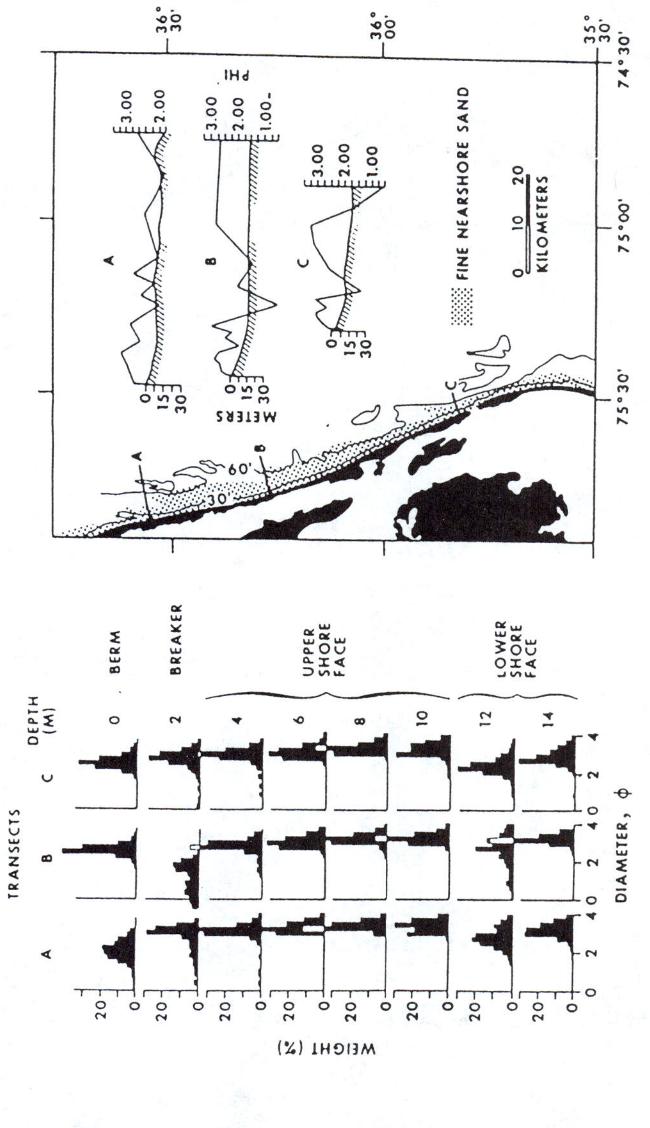

Figure 2. Distribution of grain sizes (in Phi units) along three shoreface transects of Virginia-North Carolina coast [from USACE, 1984, after Swift et al, 1971]

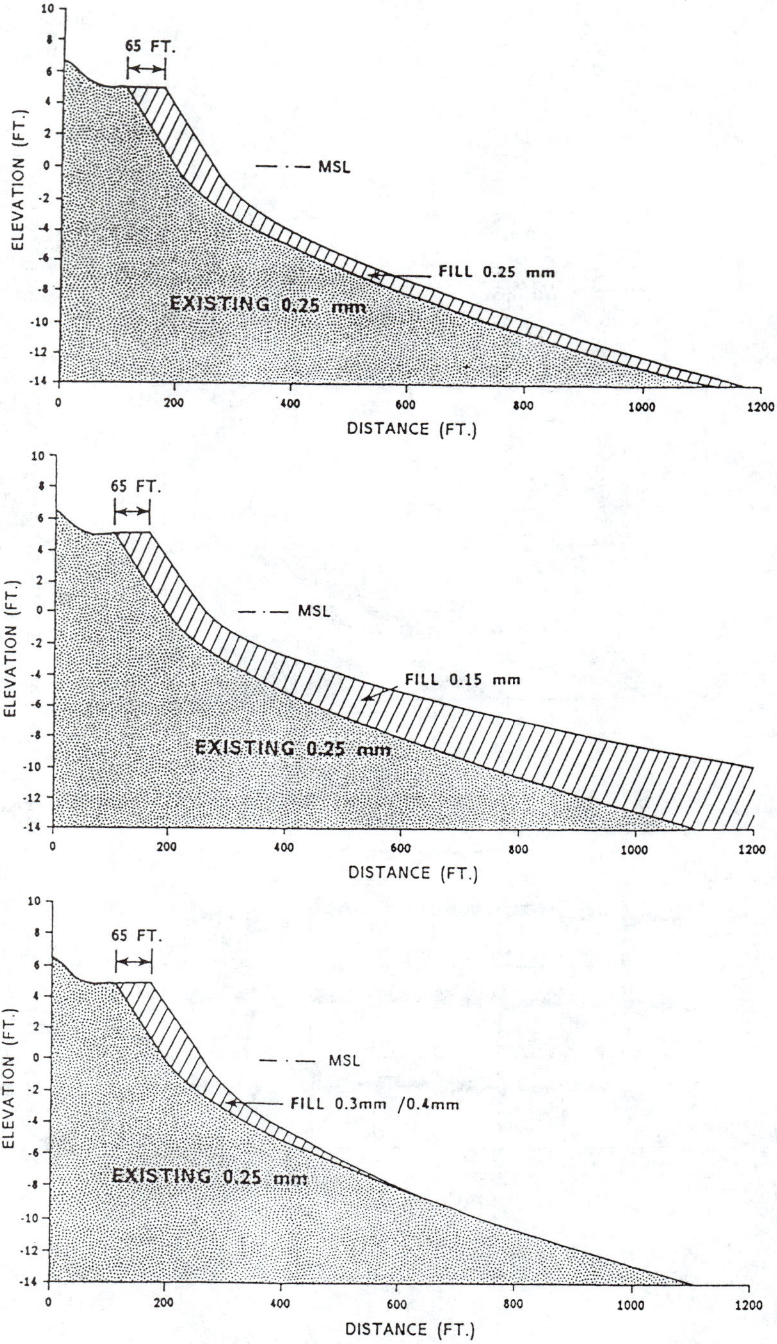

Figure 3. Theoretical equilibrium profiles to produce a uniform berm width using borrow sediments that: matches native sediment (top), is finer than native sediment (middle), and, is coarser than native sediment (bottom) [from Siah et al., 1985, after Dean, 1983]

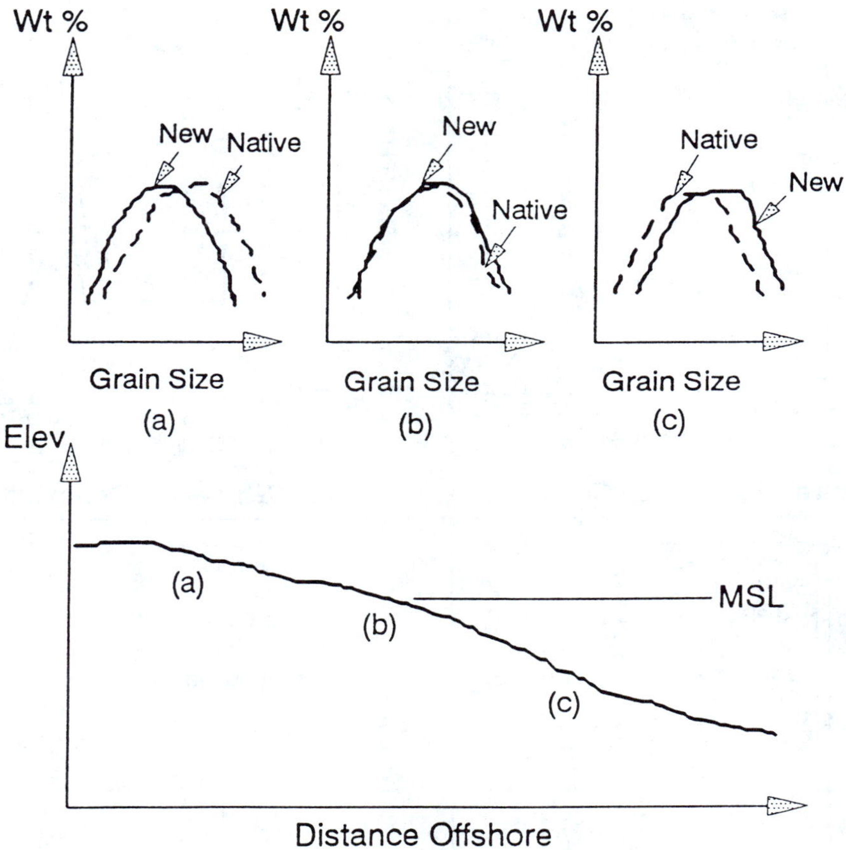

Figure 4. Schematic of the conceptual model relating to a shift in grain size distribution under a hydraulic fill with discharge along the back shore. [This assumes that the borrow source contains some fraction of sediment which is coarser than the native sediment]

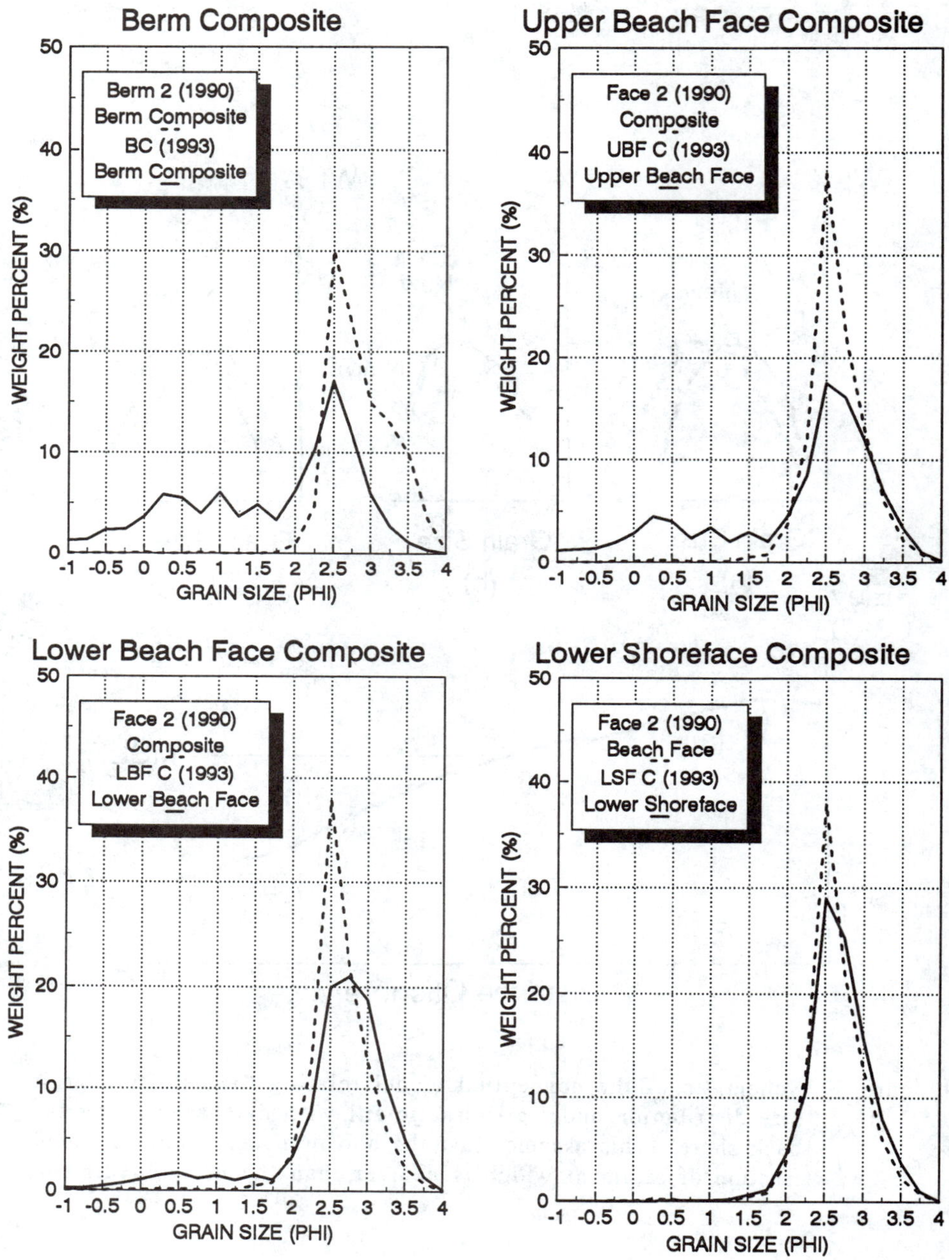

Figure 5. Grain-size comparison plots of 1990 versus 1993 beach composites for the Hunting Island beach nourishment project.

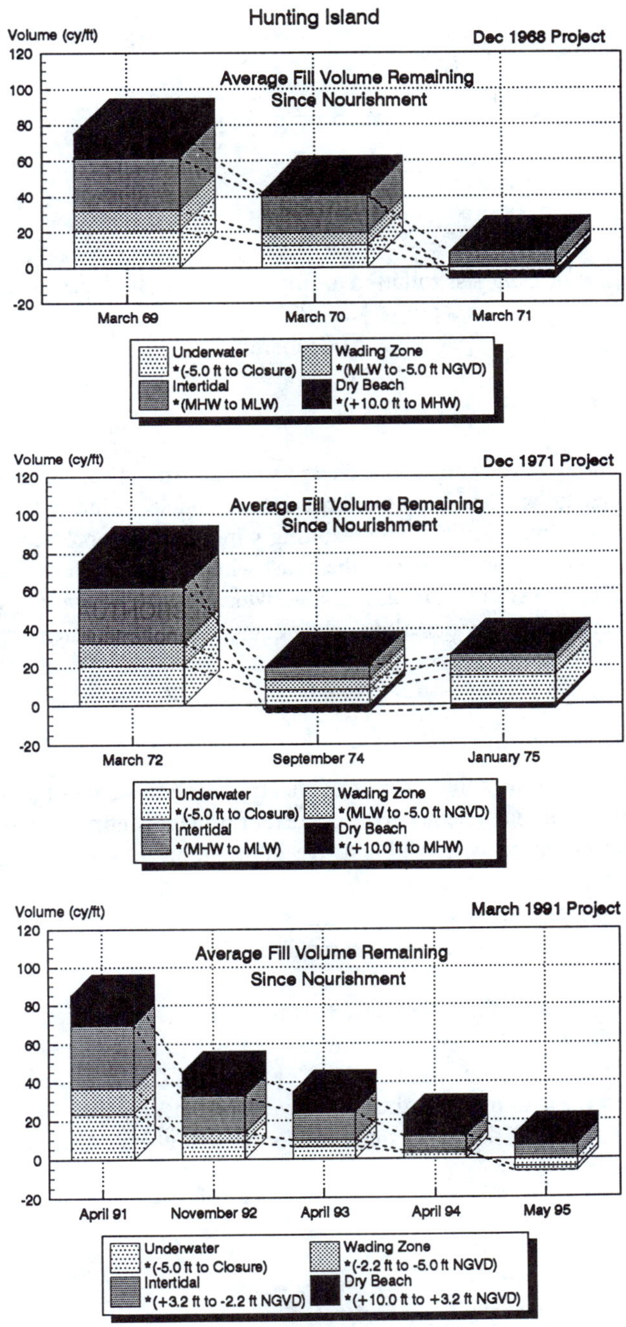

Figure 6. Estimated average unit volume fill remaining after the December 1968, December 1971, and March 1991 nourishment projects as a function of contour interval* within the Hunting Island project limits (based on surveys by USACE and CSE).

CHAPTER 371

The Use of Data Assimilation to Improve Wave Hindcast Results

Jon M. Hubertz[1]

Abstract

A sequential data assimilation scheme was used to assimilate wind speed and direction measured at land and buoy platforms into a modeled wind field over the Gulf of Mexico during January 1996. The resulting wind field is more accurate than the modeled field alone for locations near the coast where winds are affected locally by transition from land to water. The assimilated winds result in improved hindcasts of wave height over modeled winds alone when compared to wave measurements from buoys.

Introduction

Data assimilation is the process of incorporating observations of a dynamic system into a model of the system. In the context of wave modeling, this is the process of incorporating observed wind and/or wave information into a wave model or the results from a wave model. At this stage of wave model development, the most significant and timely improvement in the accuracy of model results lies in the combination of measurements and model results.

The Wave Information Study (WIS) at the Coastal and Hydraulics Laboratory (CHL) has developed and is applying a WIS Data Assimilation System (WDAS) to make optimal use of data collected under the Coastal Field Data Collection Program at CHL. The objective is to improve the accuracy of hindcast results, especially in high energy events where available wind fields and/or present wave models are sometimes not sufficiently accurate.

----

[1]Coastal and Hydraulics Laboratory, U.S. Army Engineer Waterways Experiment Station, 3909 Halls Ferry Road, Vicksburg, MS 39180-6199

Accuracy of Present Wave Models

Cardone et al. (1996) conclude that first, second, and third generation wave models presently in use give equivalent results when used to hindcast extreme wave heights in two severe storms off the U.S. East Coast. Their conclusion is based on comparison of model results to measurements from nine buoys. They also note that all models underestimate the highest waves over 12 m. Tracy and Cialone (1996) provide statistics from comparison of results from the same second generation model used by Cardone et al. to 16 buoys off the U.S. East Coast for one year, 1994. They conclude there is little bias in wave height and peak period results from the model compared to measurements, and root mean square differences are about 0.5 m and 2 sec, respectively. Until a significant improvement in wave modeling emerges, the most expedient approach to improved accuracy is the use of data assimilation in wave hindcasting.

Assimilation Techniques

All data assimilation techniques attempt to improve the results of a model by using some method to minimize the difference between model results and a set of observations of model variables, spaced closely in location and time to model values and considered more accurate than model values. Data assimilation approaches can be grouped into two general categories. One is referred to as sequential and the other variational.

Sequential techniques use observations to improve model results only at the time the observations occur; that is, current observations are discarded as soon as assimilated. Their success increases with the number of observations in space each time data are assimilated. A number of different approaches are included in the sequential method. These are, in ascending order of complexity, direct insertion, blending, nudging, optimal interpolation, successive corrections, and Kalman filtering.

The adjoint method is a commonly used variational approach. It considers a set of observations in space over a certain time interval which by ergodicity is equivalent to a distribution in probability space. Again, the objective is to minimize the difference between model results and observations or a cost function. This is done by solving a set of equations consisting of the model equations and adjoint equations. The adjoint equations are formed by differentiating the cost functions with respect to a control variable and setting the result equal to zero. This differentiation can be very difficult due to the dynamical coupling between state variables in the cost function and control variables, for example, wave height and wind speed, respectively.

Hubertz, Thompson, and Wang (1996) provide a summary of recent studies which employed these techniques to improve wind, wave, and water level/current model results. The WDAS employs a sequential technique similar to optimal interpolation which allows assimilation of vector or scalar measurements from random points on a latitude, longitude grid with model data and interpolation to a target grid using minimization of a quadratic form to provide fit and smoothness.

WIS Data Assimilation System

Application of the system consists of three steps. First, available measurements pass through a quality control filter to ensure only valid data are used and they are consistent in characteristics, e.g., units, conventions, elevation, etc. Second, a target grid is specified upon which the results of the assimilation process will be available for use. Randomly spaced measurements and/or data from a model are interpolated to this target grid. Finally, a set of equations (one for each target grid point) is solved which minimizes the difference between input data and final values on the target grid. The equations allow a weight for data type if one type of data is considered more important, and a smoothing weight on a derivative constraint to control smoothness of the final field. The equations to be solved have the form

$$w_d(F_d - F)^2 + \beta^2[(\delta F/\delta x)^2 + (\delta F/\delta y)^2] = Minimum$$

where F refers to a vector or scalar variable, d to input data, w a weight, β a smoothing parameter, δ the derivative operator, and x, y the longitudinal and latitudinal directions, respectively. The equations are solved with a Gauss-Seidel iteration scheme. Details of the system are provided by Oceanweather Inc. (1996).

Results - Time Series at a Point

Wind speed and direction from buoys and coastal land stations in the Gulf of Mexico were assimilated into an approximate 1 degree latitude, longitude background wind field from the National Meteorological Center (NMC) during January 1996. Figure 1 shows the locations of measured and modeled wind data. Measured data locations are indicated by either the alphabetic Coastal-Marine Automated Network (CMAN) designation or the National Data Buoy Center (NDBC) numeric identifier. Modeled National Center for Environmental Prediction (NCEP) wind results are indicated by the meteorological wind barb symbols.

Measured data are available every hour while NCEP data are available at a 6-hour interval. The NCEP data are interpolated in time to every hour. NCEP values are removed within a specified radius of measured values to give added influence to the measured data. Figure 2 provides an example of the difference between modeled NCEP data, measured data from the buoy and the assimilation product WDAS using both NCEP modeled and buoy data at the grid point closest to the buoy. The WDAS and buoy data are almost identical reflecting the influence of the assimilation process at the grid point closest to the buoy. The NCEP model data depart from the other two most notably at times of higher wind speeds, which in these cases are due to fronts moving offshore. These are at approximately hours 175, 275, 450, and 650 in Figure 2.

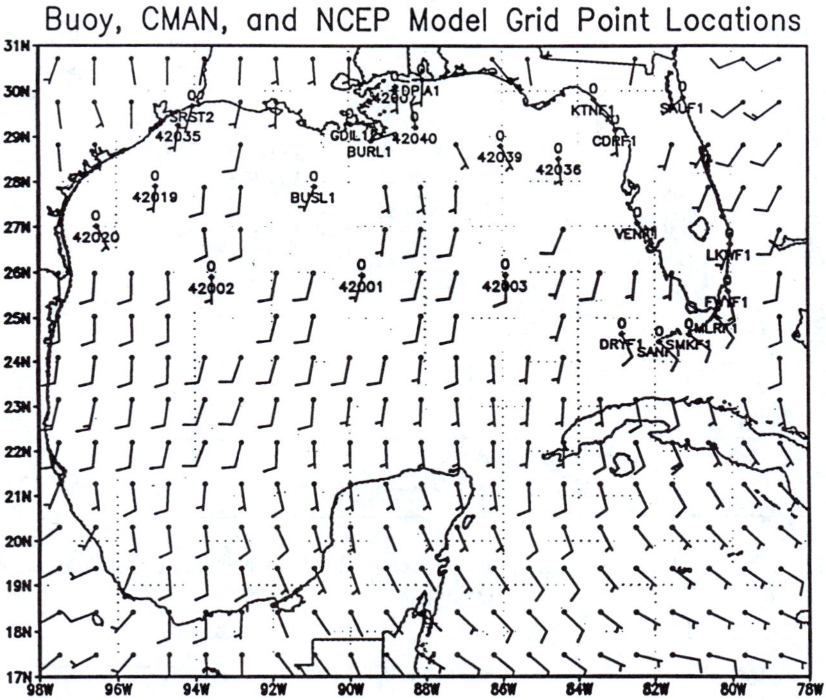

Figure 1. Locations of Measured and Modeled Wind Data

Wind fields from NCEP and from the assimilation process were input to the WISWAVE model to hindcast wave conditions over the Gulf of Mexico during January 1996. The hindcast results were then compared to wave measurements made at the buoys shown in Figure 1. Hindcast wave height results at the grid point closest to the buoy and measured values at the buoy are compared in Figure 3. Wave heights are overestimated at the same times wind speeds are overestimated, the NCEP winds being higher with respect to hindcast values using assimilation with measurements at the buoy. The first 50-75 hours in Figure 3 represent the "spin-up" of the model and should not be used to compare differences in wind input. At other times there is little difference between using NCEP winds and the assimilated wind product.

In general there was little improvement in hindcast wave height and peak period using the assimilated winds over using the NCEP winds alone, as measured by monthly values of bias and root mean square differences at the eight buoys recording wave information during January 1996. However, these monthly statistics mask cases where

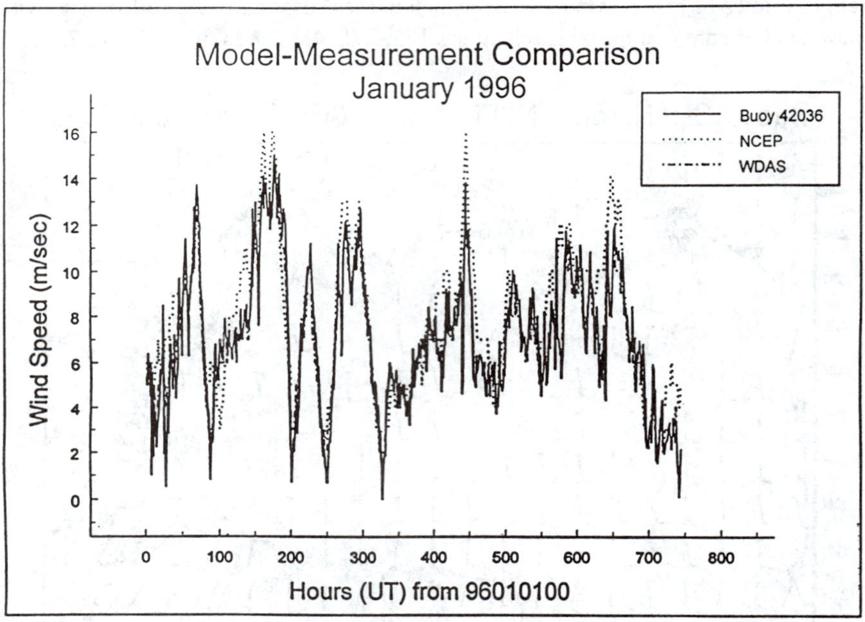

Figure 2. Wind Speed Versus Time at Buoy 42036 for Measured, Modeled, and Assimilated Data

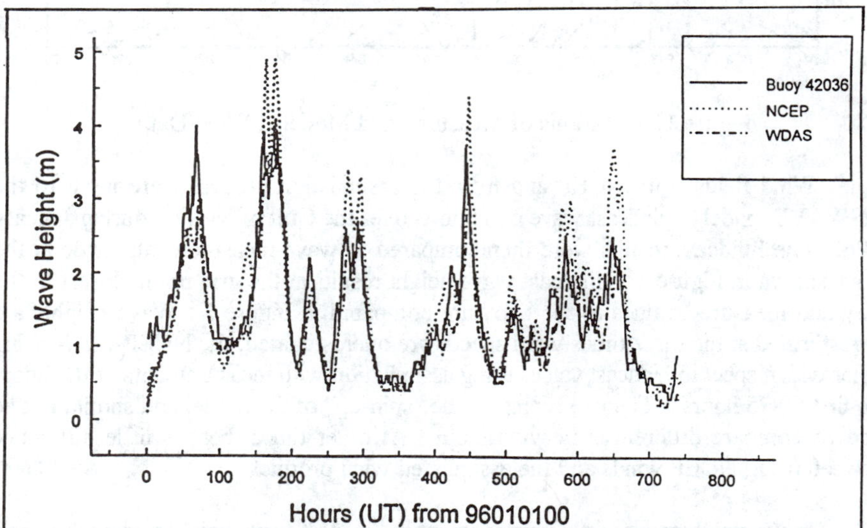

Figure 3. Wave Height Versus Time from WISWAVE Model Using Wind Input from NCEP Model and Winds from WDAS Compared to Measurements from Buoy

the NCEP winds did not accurately represent local wind speeds and directions. In these cases, improved hindcast values of wave height and peak period were obtained.

Results - Wave Height on Grid at One Time

Next, differences in hindcast wave heights over the entire grid at one time are shown for the case of using NCEP winds alone and using the product of the assimilation process. Figure 4 shows the winds over the Gulf of Mexico at 0100 UT on January 28, 1996. This corresponds to hour 650 on the previous time series plots. The contour interval is 1 m/s. Wind speed is relatively high (10 m/s) and offshore along the northwest coast of Florida. It was in this area assimilated winds were lower than NCEP winds (Figure 2) and resulted in lower wave heights of about 1 m (Figure 3).

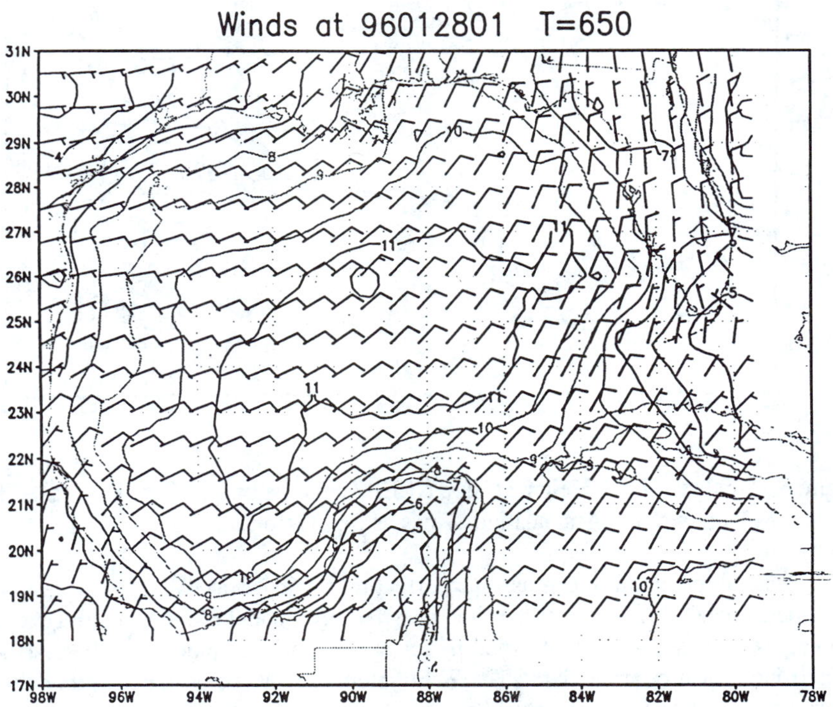

Figure 4. NCEP Wind Field Over Gulf of Mexico at 0100 UT, January 28, 1996

The difference in hindcast wave height using NCEP winds as input and using the assimilated winds as input over the whole Gulf at hour 650 is shown in Figure 5. The contour interval is 0.4 m. Most of the difference is in the northeast Gulf where the land/water boundary has an influence on offshore winds. These differences illustrate the spatial effect of assimilating data when the NCEP winds do not accurately model local conditions. Note, that in other areas differences are small indicating there is little effect of using assimilation. In these areas at this time, the NCEP winds provide accurate wave results.

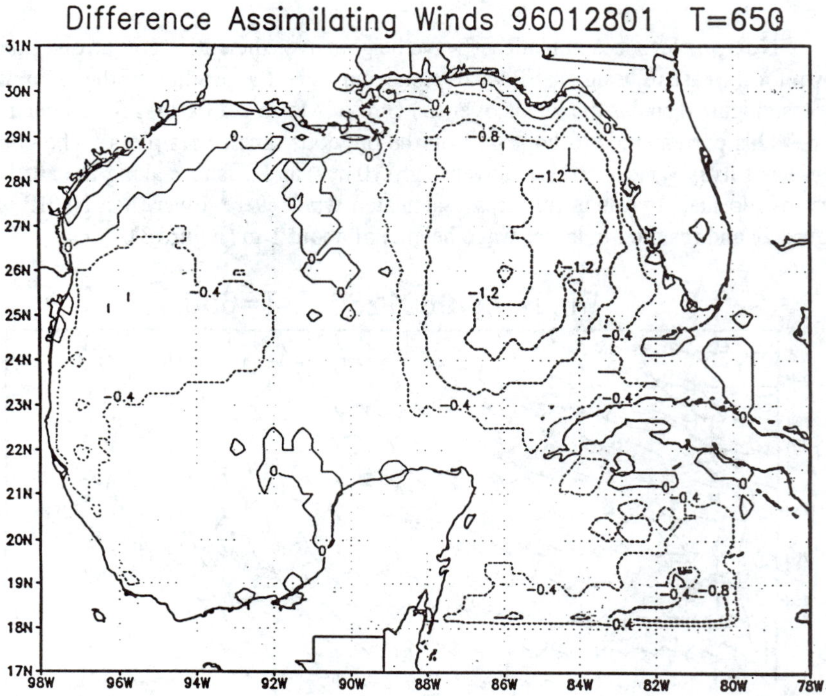

Figure 5. Hindcast Wave Heights Using NCEP Winds as Input Minus Hindcast Wave Heights Using Assimilation Winds as Input (Contour Interval 0.4)

The above applications assume correcting a wind field will lead to acceptable wave model results. If this is not the case, then the wave model is deficient, and accurate hindcast results can only be obtained by fixing the model or modifying the results by assimilating measured wave data. The WIS system can also be used in this mode using either scalar input such as wave height and peak period or vector input from wave height and direction values.

This approach can also be extended to modifying wave spectra. Typically, a WIS hindcast spectra has 20 frequency bands. For a given wave model grid over a geographic region at a given time, one has 20 stacked grids of energy densities representing the energy spectrum at each grid point. Measured energy density corresponding to model bands can be assimilated over the grid at each frequency level. This amounts to applying the system 20 times in a scalar mode using energy density. The resulting volume of data accurately reproduces the measured spectrum near the measurement point and spreads to surrounding points based on the value of a smoothing parameter.

Conclusions

At this time, first, second, and third generation wave models in use give equally good or bad results. Until advances in the theories of wave models are made, the most expedient improvement in the accuracy of wave hindcast results lies in the use of available measurements to either improve wind input or modeled wave results through data assimilation. Since accurate wave hindcast results are critically dependent on accurate winds, a system has been developed to improve hindcast wind input. The system was tested in the Gulf of Mexico using one month of modeled hindcast winds and available wind measurements from land and buoy platforms.

The system was successful in improving hindcast wave heights for certain times and locations dependent on the land/water boundary and weather systems. Improvements of up to 1 m in wave height were realized. The system will be implemented in the WIS nowcast procedure in a test mode in 1997.

Acknowledgment

This work was carried out under the Coastal Field Data Collection Program, Coastal Engineering Functional Area of Civil Works Research and Development. The Office of the Chief of Engineers, U.S. Army Corps of Engineers is acknowledged for authorizing publication of this paper.

References

Cardone, V. J., Jensen, R. E., Resio, D. T., Swail, V. R., Cox, A. T. (1996). "Evaluation of Contemporary Ocean Wave Models in Rare Extreme Events: The Halloween Storm of October 1991 and the Storm of the Century of March 1993," Journal of Atmospheric and Oceanic Technology, Vol. 13, No. 1 pp.198-230.

Hubertz, J. M., Thompson, E. T., Wang, H. V. (1996). "Annotated Bibliography on Coastal and Ocean Data Assimilation," WIS Report 36, U.S. Army Corps of Engineers Waterways Experiment Station, 3909 Halls Ferry Road, Vicksburg, MS 39180-6199.

Oceanweather Inc. (1996). "Interactive Objective Kinematic Analysis Program for Marine Surface Wind Field Analysis, Documentation and User's Guide," Contractor's Report, Wave Information Study, U.S. Army Corps of Engineers Waterways Experiment Station, Vicksburg, MS 39180-6199.

Tracy, B. T., Cialone, A. (1996). "Wave Information Study Annual Summary Report, Atlantic 1994," WIS Report 34, U.S. Army Corps of Engineers Waterways Experiment Station, 3909 Halls Ferry Road, Vicksburg, MS 39180-6199.

CHAPTER 372

Grain-size influence on sand transport in oscillatory sheet flow

C. Marjolein Janssen[1], Jan S. Ribberink[2]

Abstract

Experimental investigations (Ribberink & Chen, 1993; Ribberink & Al-Salem, 1994; Katopodi et al., 1994) indicated that the grain diameter is an important parameter for the prediction of the *net* (time-averaged) transport rates in oscillatory sheet flow. As only little is known about the influence of the grain diameter on the net transport rates and the transport mechanisms, new experiments with fine sand (D_{50} = 0.13 mm) were carried out in the Large Oscillating Water Tunnel (LOWT) of Delft Hydraulics. Since 1992 the LOWT is equipped with a recirculation system, such that experiments with combined wave-current flow can be performed at full scale (1:1).

Several existing transport models are based on the assumption that the transport can be described in a quasi-steady way (see Janssen, 1995). In order to investigate whether or not this assumption is valid, both a 'quasi-steady' (Bailard, 1981) and a 'semi-unsteady' (Dibajnia & Watanabe, 1992) transport model are verified with existing and new experimental data.

Description of two existing sand transport models

a) Bailard (1981)

The first model described here, is the one developed by Bailard (1981), which is one of the 'quasi-steady' models. These models are based on the assumption that the transport rates under oscillatory flow can be described in a quasi-steady

---

1) PhD-student, Delft University of Technology, Dep. of Civil Engineering, p/a Delft Hydraulics, PO Box 152, 8300 AD Emmeloord, The Netherlands. e-mail: marjolein.janssen@wldelft.nl

2) Formerly Delft Hydraulics; University of Twente, Section Civil Engineering and Management, PO Box 217, 7500 AE Enschede, The Netherlands. e-mail: j.s.ribberink@sms.utwente.nl

way, which means that it is assumed that the sand transport reacts immediately to changes in flow condition. Therefore a direct relation exists between the instantaneous near-bed free stream velocity or instantaneous bed shear stress and the instantaneous transport rate. The assumption of quasi-steadiness is often applied in sheet flow conditions, where the majority of the sand transport takes place in a thin layer close to the bed and a quick sediment response to the oscillatory flow can be expected (see Ribberink et al., 1994).

The model of Bailard consists of a bed-load and a suspended-load part. Both parts contain a term which depends on the bed slope. Because the model is compared with measurements of sand transport above a horizontal bed, these terms become zero. In order to calculate the *net* sand transport rate, the equation is averaged over the wave-period, resulting in:

$$\langle q_s \rangle = \frac{\rho_w \, c_f}{(\rho_s - \rho_w)\, g} \, \frac{\varepsilon_b}{\tan\varphi} \langle u^3 \rangle + \frac{\rho_w \, c_f}{(\rho_s - \rho_w)\, g} \, \frac{\varepsilon_s}{w_s} \langle |u|^3 u \rangle \qquad \textbf{(1)}$$

Where $\langle .. \rangle$ means time-averaging over the wave period, q_s is the sand transport rate per unit width (m^2/s), ρ_w and ρ_s are the density of the water and the sediment (kg/m^3), c_f is a friction coefficient (-), g is the gravity acceleration (m/s^2), ε_b and ε_s are efficiency factors for the bed-load and the suspended-load transport (-), $\tan\varphi$ is the tangent of the dynamic friction angle (-), w_s is the settling velocity of the sediment (m/s) and u is the horizontal near-bed velocity, directly above the wave boundary layer (m/s).

Because of the assumption of quasi-steadiness *and* the fact that q_s is related to the third- and fourth-power velocity moment, the *net* sand transport rate, predicted by this model, will always be in the direction of the largest velocity, i.e. under real asymmetric waves, generally this will be in the onshore direction.

b) Dibajnia & Watanabe (1992)

Dibajnia & Watanabe developed a 'semi-unsteady' model, which is able to take into account unsteady effects, like the time-lag between velocity and concentration, although it does not contain fully unsteady equations. Considering the amount of sand which is entrained during a positive half wave cycle, the model determines the part which will be transported directly by the positive half cycle and the part that will still be in suspension as the flow reverses and will therefore be transported by the negative velocity, during the successive half cycle. The relative magnitude of these parameters depends on the ratio of the fall time of the particles (determined by their settling velocity and the height to which the particles are suspended) and the period of the positive half wave cycle. Of course the same is valid for the negative half cycle. The net sediment transport rate is the difference between the two contributions in positive and the two contributions in negative direction. For the exact formulations, the reader is referred to the original paper.

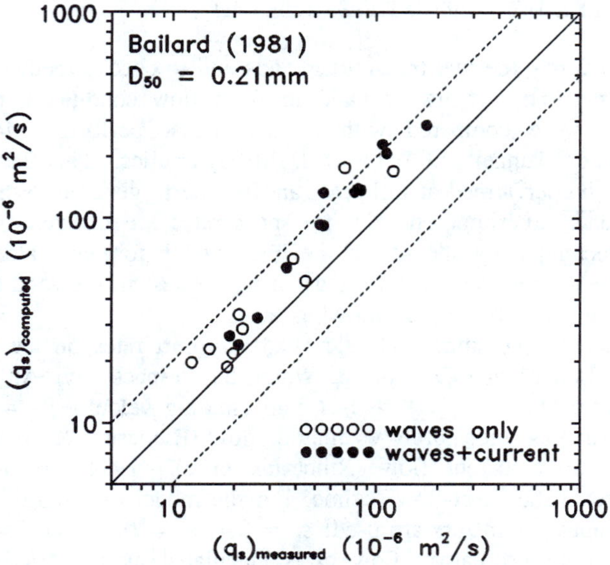

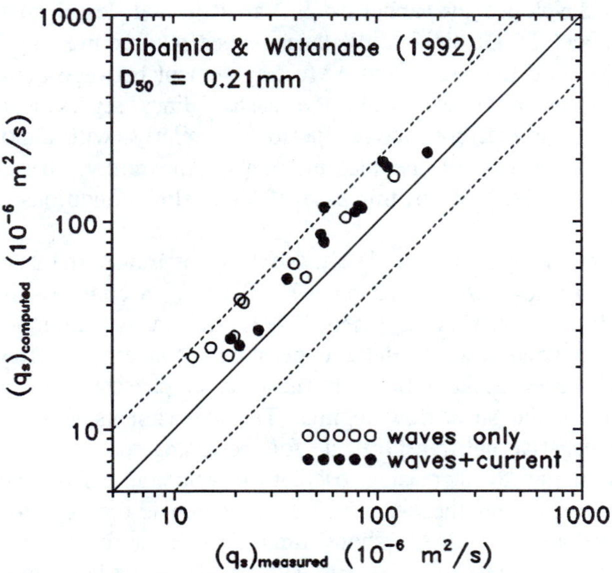

Figure 1: Comparison between measured and calculated net transport rates for sand with D_{50} = 0.21 mm. Model results of Bailard (1981) and Dibajnia & Watanabe (1992).

Comparison between two models and existing data

In order to verify the two transport models and to check whether the assumption of quasi-steadiness is always valid in sheet flow conditions, predicted net transport rates were compared with measurements, performed in the Large Oscillating Water Tunnel (LOWT) of Delft Hydraulics. This facility enables experiments to be performed at full scale and is able to generate oscillatory flows with net currents superimposed. Net transport rates are derived from a mass-conservation technique by measuring the bed level before and after a test and determining the weight of sand collected during the test in the sand traps. (For a more detailed description see the following pages).

Fig.1 shows the measured and calculated transport rates on log-log scale for the models of Bailard and Dibajnia & Watanabe, respectively. All data are for sand with a mean diameter D_{50} = 0.21 mm (settling velocity w_s = 0.026 m/s) and both experiments with purely oscillatory flow (Ribberink & Al-Salem, 1994) and combined wave-current flow (Ribberink et al., 1994) are included. All conditions were in the sheet-flow regime. For the model of Bailard the following values of the input parameters are used: $\varepsilon_b = 0.1$, $\varepsilon_s = 0.02$, $\tan\varphi = 0.625$. For the case of purely oscillatory flow c_f is calculated as the friction factor of Jonsson, slightly modified by Swart (1974) with $k_s = D_{50}$. For the case of combined wave-current flow, c_f is calculated as a combined wave-current flow friction factor, developed by Ribberink & Van Rijn and described by Koelewijn (1994). The model of Dibajnia & Watanabe does not require any further input parameters. The solid lines represent perfect agreement between the predicted and the measured transport rates, while the dashed lines represent a factor two difference. It is clear that both models perform similarly, with almost all predictions within a factor two of the measurements. Apparently, the assumption of quasi-steadiness is reasonable in this case of sheet flow conditions and relatively coarse sand.

However, the situation is completely different for transport rates of sand with a mean grain diameter D_{50} of 0.13 mm. Fig.2 shows a comparison between the experimental data (Ribberink & Chen, 1993) and the predictions by the two models. The test conditions of these experiments concerned purely oscillatory, asymmetric (2nd-order Stokes) flow. It must be emphasized that again all test conditions were in the sheet-flow regime. The experiments show the surprising behaviour of a negative net transport rate for increasing root-mean-square velocity (U_{rms}), corresponding to increasing oscillatory velocities. This means that for large oscillatory velocities the *net* sand transport rate is opposite to the direction of the largest velocity (i.e. in 'offshore' direction). Because of the assumption of quasi-steadiness, the model of Bailard is not capable to predict this behaviour and apparently, the assumption of quasi-steadiness is not valid for this fine sand.

In contrast to the model of Bailard, the model of Dibajnia & Watanabe predicts decreasing and even negative net transport rates for increasing oscillatory velocities. This is caused by the part of the entrained sand which remains in suspension and is transported in opposite direction. Due to the asymmetry of the flow this results in negative net transport rates. Although the trend shown in the data is followed much better by the model of Dibajnia & Watanabe, the magnitudes are still different.

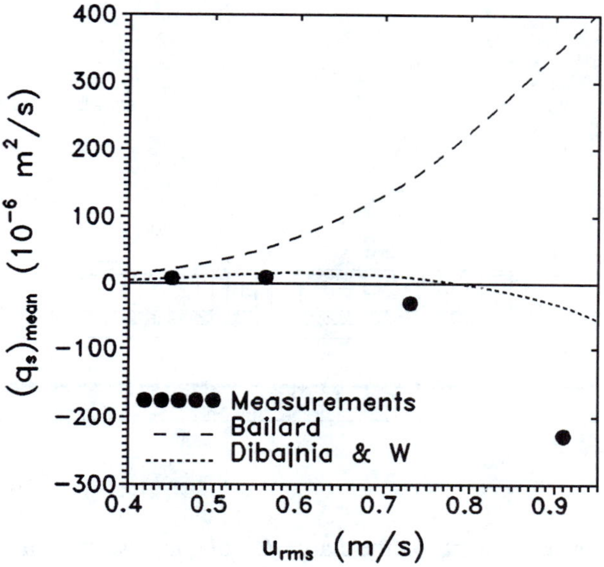

Figure 2: Net transport rates as a function of the root-mean-square velocity (U_{rms}). Measurements by Ribberink & Chen (1993), model results of Bailard (1981) and Dibajnia & Watanabe (1992)

<u>New experiments</u>

From the previous section it is clear that the grain diameter is an important parameter in the prediction of the net transport rate. Apparently, the assumption of quasi-steadiness is not valid in the case of sand transport of very fine sand in purely oscillatory sheet flow conditions. In order to investigate the influence of the grain diameter on the net transport rates and the transport mechanisms in *combined wave-current flow*, a new set of experiments with fine sand (D_{50} = 0.13 mm) was performed in the LOWT of Delft Hydraulics. The experimental research was carried out between October 1995 and January 1996 and consisted of two parts. In the first part, net sand transport rates were measured and in the second part, detailed time-dependent measurements of velocity and concentration were performed. This paper focuses on the first part of the experiments, i.e. the net transport measurements.

i) The Large Oscillating Water Tunnel

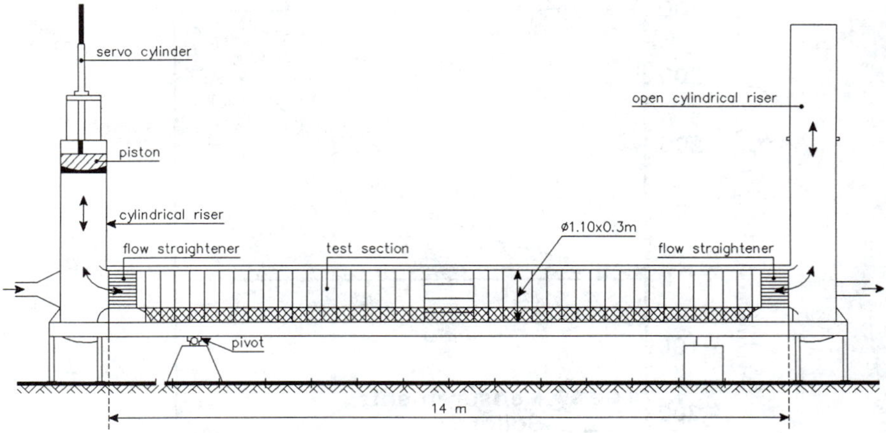

Figure 3: General outline of the Large Oscillating Water Tunnel of Delft Hydraulics.

Fig.3 shows a general outline of the LOWT. The tunnel has the shape of a vertical U-tube with a rectangular horizontal test section and two cylindrical risers on either end. The desired oscillatory water motion inside the test section is imposed by a steel piston in one of the risers. The other riser is open to the atmosphere. The test section is 14 m long, 1.1 m high and 0.3 m wide and is provided with flow straighteners on either end. A 0.3 m thick sand bed can be brought into the test section, leaving 0.8 m height for the flow above the bed. Two sand traps are located underneath the cylindrical risers.

The maximum piston amplitude is 0.75 m, which means a maximum semi-excursion length of the water particles in the test section of 2.45 m. The range of velocity amplitudes is 0.2 - 1.8 m/s and the range of oscillation periods is 4 - 15 seconds. It is possible to generate purely sinusoidal, regular asymmetric and irregular oscillatory motions with the piston.

In 1992 the tunnel was extended with a recirculating flow system connected to the cylindrical risers, so that a steady current can be superimposed onto the oscillatory motion. The maximum superimposed current velocity in the test section of the tunnel is 0.5 m/s. The recirculating flow system is also provided with a sand trap, consisting of a 12 m long pipe with a diameter of 1.2 m. The trap was designed for trapping 90% of the suspended sediments (minimum grain size 100 microns) at maximum flow discharge.

ii) Measurement procedure

During these experiments net transport rates as well as near-bed velocities were measured. The velocities were measured at 10 cm above the bed, using a 2D forward scatter Laser-Doppler system, developed by Delft Hydraulics.

Net transport rates were derived from a mass-conservation technique. Therefore the bed level along the test section is measured before and after the test and the weight of sand, collected in the traps during the test, is determined.

The bed levels in earlier experiments were measured by hand. However, for the present experiments a bed level profiling system (BLPS) was built. The BLPS consists of three bed profilers, mounted on a measurement carriage, which can be moved along the test section. The bed level measurement technique is based on a conductivity measurement. The profilers measure the bed level every cm along the test section, accurate to about 0.4 mm in the vertical. From the difference between the bed level before and after a test, the total volume of sand, eroded from the test section (including pores), can be determined. This sand is collected in the three sandtraps and weighed under water, giving the total volume (without pores).

The transport rate in the middle undisturbed part of the test section can now be calculated by solving the continuity equation for the sediment over the test section, starting either from the left- or from the right-hand side, resulting in two estimates of the net transport rates.

iii) Test conditions

Table 1 Test conditions present experiments

Condition	T (s)	U_m (m/s)	U_a (m/s)
H2	7.2	0.23	0.70
H3	7.2	0.23	0.95
H4	7.2	0.23	1.10
H5	7.2	0.23	1.30
H6	*7.2*	*0.23*	*1.47*
H7	7.2	0.44	0.50
H8	7.2	0.44	0.70
H9	*7.2*	*0.44*	*0.95*
H24	4.0	0.23	0.70
H4S	4.0	0.23	1.10
H12	12.0	0.23	0.70
H12S	12.0	0.23	1.10

The test conditions of the present experiments consisted of combinations of sinusoidal oscillatory flow, with varying amplitudes of the velocity (U_a) and varying oscillation periods (T), superimposed onto varying net current velocities (U_m). Again all test conditions are in the sheet-flow regime. The test conditions are summarized in Table 1, where the italic typed conditions represent the two test conditions which are equal to the test conditions of earlier experiments with coarser sand (Katopodi et al., 1994).

Results of new experiments in comparison with model predictions

Fig.4 shows the measured net transport rates of the new experiments (D_{50} = 0.13 mm) in comparison with the predictions of the two models. The results for a constant net current velocity (U_m) of 0.23 m/s and a constant wave period (T) of 7.2 s are shown as a function of the amplitude of the oscillatory velocity (U_a).

Because of the assumption of quasi-steadiness in the model of Bailard, this model predicts strongly increasing net transport rates for increasing oscillatory velocities. The measured net transport rates have much smaller values and show a weaker increase for increasing oscillatory velocities. The predictions of the model of Dibajnia & Watanabe are much closer to the measured values. However, a more detailed comparison (Fig.4b) shows that this model predicts *decreasing* transport rates for amplitudes of the oscillatory velocity larger than about 1.1 m/s. This is again caused by the unsteady effects. Although the increase in the measured net transport rates is much smaller than predicted by the quasi-steady model, the measurements do not yet show a decrease in transport rate. Thus apparently the unsteady effects are smaller than predicted by the model of Dibajnia & Watanabe. This is opposite to the situation for purely oscillatory flow, where the measurements showed a larger unsteady effect, than predicted by the model of Dibajnia & Watanabe (see Fig.2).

Fig.5 shows a comparison between the measured and calculated net transport rates as a function of the wave period (T), for a constant amplitude of the oscillatory velocity (U_a) of 1.1 m/s and a constant net current velocity (U_m) of 0.23 m/s. In the quasi-steady model of Bailard the wave period only has an effect on the friction factor. Increasing wave periods lead to decreasing friction factors, resulting in decreasing sand transport rates, as shown in Fig.5. However, the measurements show decreasing net transport rates for *de*creasing oscillation periods. The same trend is predicted by the model of Dibajnia & Watanabe. It is caused by the fact that the unsteady effects become relatively more important for decreasing wave periods.

Grain-size influence on net sand transport rates in sheet flow conditions.

The influence of the grain size can be analyzed by just comparing the experimental data and by analyzing the performance of both models for all data of fine and coarser sand. In Fig.6 the net sand transport rates are plotted as a function of the third power velocity moment for all experiments in combined wave-current flow (Ribberink et al., 1994 and present tests). From this figure it is clear that the net transport rates of fine sand are smaller than those of coarser sand. This is completely opposite to the behaviour in uni-directional steady flow, where finer sediments would result in larger transport rates. The decrease in net transport rate

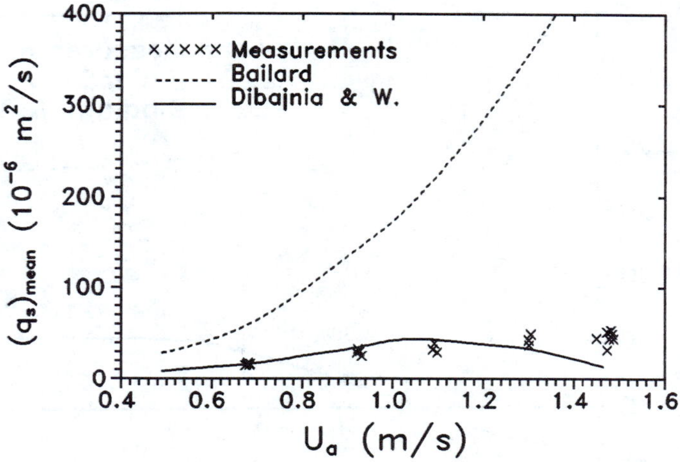

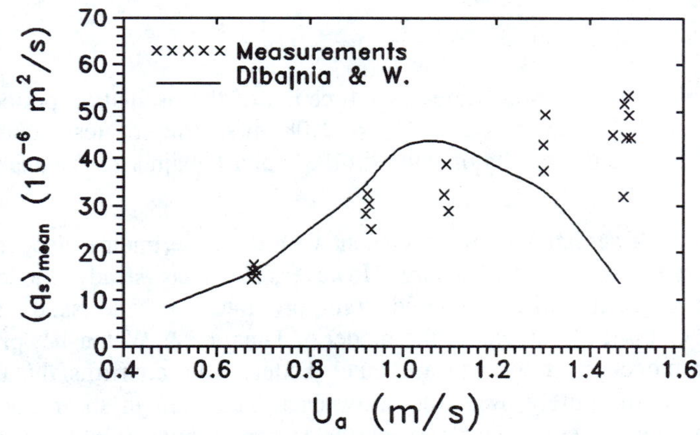

Figure 4: Net transport rates as a function of the amplitude of the oscillatory velocity U_a, for U_m = 0.23 m/s and T = 7.2 s. Present tests compared with model results of Bailard (1981) and Dibajnia & Watanabe (1992).

is caused by the unsteadiness, as the amount of sand in motion (or gross transport rate) will be larger for the finer sand. However, part of the sand remains suspended and is transported in opposite direction, resulting in smaller net transport rates.

Fig.7 is similar to Fig.1, but now also includes the data for fine sand. The two tests with negative transport rates in purely oscillatory flow (see Fig.2) are left out, because they cannot be plotted on a log scale. As observed before, both

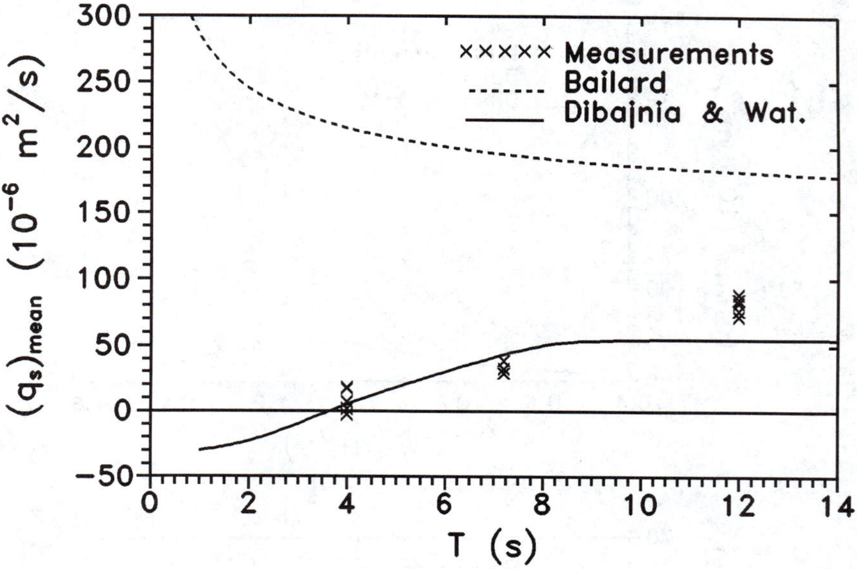

Figure 5: Net transport rates as a function of the oscillation period T, for $U_m = 0.23$ m/s and $U_a = 1.08$ m/s. Present tests compared with model results of Bailard (1981) and Dibajnia & Watanabe (1992).

models show a similar kind of agreement with the experimental data for sand with a mean grain diameter of 0.21 mm. However, the 'quasi-steady' model of Bailard largely overpredicts the measured transport rates of fine sand. Due to the inclusion of unsteady effects in the model of Dibajnia & Watanabe, predictions by the latter model for fine sand are much better. This confirms the fact, already observed for the purely oscillatory flow data, that even in sheet flow conditions the assumption of quasi-steadiness of the transport rates is not always valid for fine sand.

Analysis of unsteady behaviour of net transport rate of fine sand

In order to analyze the unsteady behaviour of the transport rate of fine sand, use is made of the results from the second part of the experiments (time-dependent measurements of velocity and concentration). The net sand transport rate can be calculated from:

$$\langle q_s \rangle = \int_0^h \langle u(z,t) * c(z,t) \rangle dz = \int_0^h \langle \varphi(z,t) \rangle dz \tag{2}$$

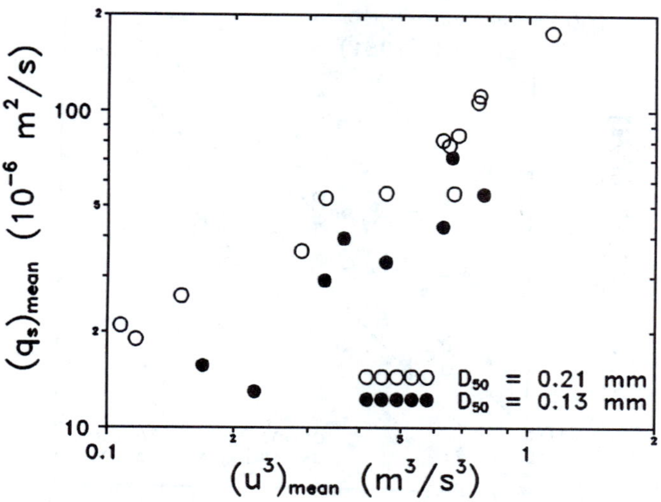

Figure 6: Measured net transport rates for fine (D_{50} = 0.13 mm) and coarse (D_{50} = 0.21 mm) sand as a function of the third-power velocity moment.

Where c is the sediment concentration (m^3/m^3), φ is the sediment flux (m/s), z is the height above the bed (m), t is time (s) and h is the water depth (m).

Because both the velocity and concentration consist of a time-averaged ($\langle..\rangle$) and an oscillatory ($\sim$) component, the time-averaged sediment flux $\langle\varphi(z)\rangle$ can be split up into a current-related $\langle\varphi_c(z)\rangle$ and a wave-related $\langle\varphi_w(z)\rangle$ contribution, as follows:

$$\langle u(z,t)*c(z,t)\rangle = \langle u(z)\rangle*\langle c(z)\rangle + \langle\tilde{u}(z,t)*\tilde{c}(z,t)\rangle = \langle\varphi_c\rangle+\langle\varphi_w\rangle \qquad (3)$$

The vertical distributions of these two contributions, as well as the total (time-averaged) flux profile, are shown in Fig.8 (for more detail about the time-dependent measurements, see Janssen et al., 1996). The results are for test condition H6 (see Table 1). It is clear that the total net transport rate (derived by integrating the total flux profile over the height), will be positive (i.e. in the direction of the net current). Of course the current-related contribution is positive everywhere, because it is just the product of the time-averaged velocity (which is positive by definition, because it is the velocity of the net current) and the time-averaged concentration. However, the wave-related contribution has a value which is of the same order of magnitude, but in opposite direction (against the net current). This must be caused by a time-lag effect between the oscillatory component of the concentration and the oscillatory velocity. Because of this negative wave-related transport rate, the total net transport rate will be much smaller than

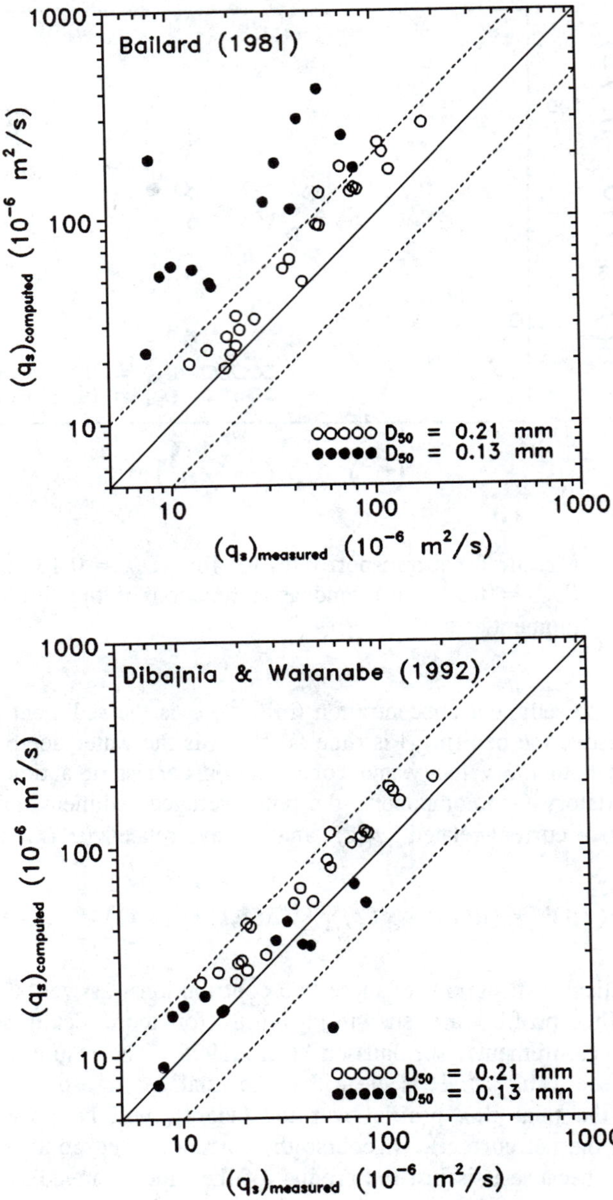

Figure 7: Comparison between measured and calculated net transport rates for fine sand (D_{50} = 0.13 mm) and coarse sand (D_{50} = 0.21 mm). Model results of Bailard (1981) and Dibajnia & Watanabe (1992).

predicted by the 'quasi-steady' model, which does not include this effect. Additionally, the total net transport rate will be smaller than in the case of coarser sand, where the larger particles result in a smaller time-lag between concentration and velocity and therefore less negative wave-related transport rates. So apparently, the behaviour, as present in the concept of the model of Dibajnia & Watanabe is not only occurring in rippled bed conditions, but also occurs in sheet flow conditions with fine sand.

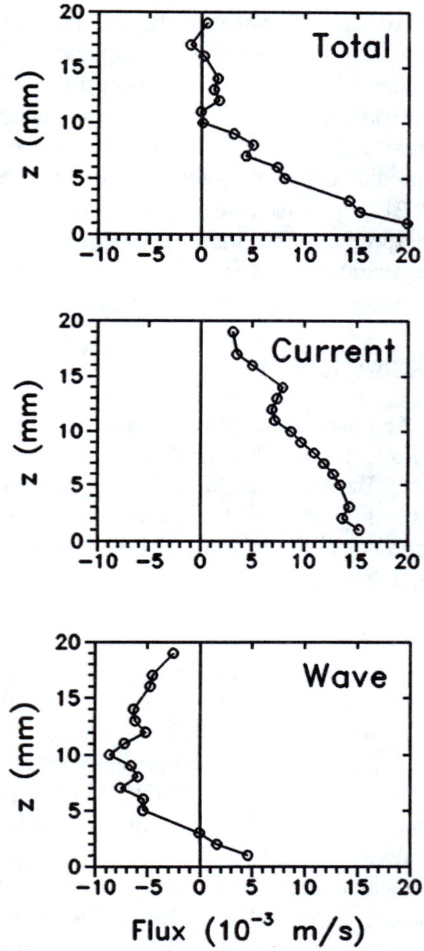

Figure 8: Vertical distribution of time-averaged total, current- and wave-related sediment flux, determined from time-dependent velocity and concentration measurements (experiment H6).

Conclusions

Modelling the sand transport in a quasi-steady way leads to the following predictions for net sand transport rates: i) increasing net transport rates for decreasing grain diameter, ii) largely increasing net transport rates for increasing oscillatory velocities and iii) increasing net transport rates for decreasing wave periods. However, experimental data of net transport rates in oscillatory sheet flow show the following results:

- *Smaller* net transport rates for 0.13 mm sand than for 0.21 mm sand.
- A much *weaker increase* in net transport rates for 0.13 mm sand than for 0.21 mm sand, with increasing oscillatory velocities.
- *Decreasing* net transport rates with decreasing wave periods for sand with D_{50} = 0.13 mm and conditions with sufficiently large oscillatory velocities.

All these 'unsteady effects' are caused by a time-lag between velocity and concentration, resulting in negative wave-related transport rates and therefore smaller total net transport rates than without these effects. Apparently, even in sheet flow conditions, the net sand transport rate cannot always be described in a quasi-steady way. Obviously, the quasi-steady model of Bailard (1981) is not able to predict these phenomena. The 'semi-unsteady' model of Dibajnia & Watanabe (1992) is at least able to predict the behaviour qualitatively. However, this model underpredicts the unsteady effects in purely oscillatory asymmetric flow and overpredicts the unsteady effects in the case of combined wave-current flow (compare Fig.2 and Fig.4b).

Acknowledgements

The authors want to thank the Dutch Ministry of Transport and Public Works (RIKZ), for the financial support to perform the experiments. The help of Mr. W.N. Hassan and Mr. R.J. v.d. Wal during the experiments and the data-processing is greatly appreciated. The first author also wants to thank the Technology Foundation (STW) for their financial support during the complete research project.

References

Bailard, J.A. (1981). An energetics total load sediment transport model for a plane sloping beach. J. Geoph. Res., Vol. 86, No. C11, pp. 10938-10954.

Dibajnia, M. & A. Watanabe (1992). Sheet flow under nonlinear waves and currents. Proc. 23rd ICCE, Venice, pp. 2015-2028.

Janssen, C.M. (1995). Sand transport in oscillatory sheet-flow; a literature review. Delft University of Technology, Communications on Hydraulic and Geotechnical Engineering, Rep. No. 95-6, ISSN 0169-6548.

Janssen, C.M., W.N. Hassan, R. v.d. Wal, J.S. Ribberink (1996). Net sand transport rates and transport mechanisms of fine sand in combined wave-current sheet flow conditions. Delft Hydraulics, Data report H2462, Part IV.

Katopodi, I., J.S. Ribberink, P. Ruol & C. Lodahl (1994). Sediment transport measurements in combined wave-current flows. Proc. Coast. Dyn.'94, Spain.

Ribberink, J.S. & Z. Chen (1993). Sediment transport of fine sand under asymmetric oscillatory flow. Rep. H840, Part VII, Jan 1993, Delft Hydraulics.

Ribberink, J.S. & A.A. Al-Salem (1994). Sediment transport in oscillatory boundary layers in cases of rippled bed and sheet-flow. J.Geoph.Res., Vol.99, C6.

Ribberink, J.S., I. Katopodi, K.A.H. Ramadan, R. Koelewijn & S. Longo (1994) Sediment transport under (non)linear waves and currents. Proc.24th ICCE.

Jonsson, I.G. (1966): Wave boundary layers and friction factors. Proc.10th ICCE.

Koelewijn, H. (1994): Sediment transport under sheet-flow conditions. M.Sc. Thesis, Delft University of Technology, October 1994, 98 pp.

Swart, D.H. (1974): Offshore sediment transport and equilibrium beach profiles. Delft Hydr. Lab. Publ., No. 131, Delft Hydraulics, The Netherlands.

CHAPTER 373

Cross-shore sediment transport mechanisms in the surfzone on a timescale of months to years

Klaas T. Houwman[1] and Gerben Ruessink[1]

Abstract

The relevance of several cross-shore sediment transport mechanisms on a timescale of months to years was studied. Data from a large measuring campaign carried out in the outer surfzone of a barred surfzone was used to calculate sediment fluxes. The transport rates were calculated using the measured near bed velocities and sediment concentrations, and two sediment transport models. Both fluxes and the transport rates are only significant during breaking waves. The high-frequency fluxes/transports are onshore directed, the mean fluxes/transports are offshore directed, and the low frequency waves also give offshore directed fluxes/transports. Taking into account the frequency of occurrence of the conditions, the breaking wave conditions turned out to be the most important for the cross-shore sediment transport on a timescale of a year. The cross-shore sediment transport is a delicate balance between the onshore directed high frequency term and the offshore directed mean term. Therefore the smaller terms, like the low frequency term and the bedload term, may still be of importance to the net sediment transport.

Introduction

In the surfzone mean currents and high- and low frequency waves are responsible for sediment transport. Most field experiments presented in literature have focussed a typical time scale of a single storm-event. It is therefore unknown whether sediment transport on the scale of months to years is dominated by infrequent high energy conditions or by the often ocurring calm to moderate conditions.
In the frame work of the NOURTEC-project (Hoekstra et al., 1994) a shoreface nourishment was carried out in the multiple bar system of the Dutch island of Terschelling. A large-scale field measuring program, started to monitor this nourishment, offered an opportunity to determine the most important sediment

[1] Institute for Marine and Atmospheric Research, Utrecht University, Department of Physical Geography, P.O.Box 80.115, 3508 TC Utrecht, The Netherlands.

transport mechanisms at different locations in a barred surfzone and to evaluate the relevance of these mechanisms on a time scale of months to years. The sediment transport mechanisms were evaluated in the following way. Firstly, sediment fluxes were calculated using the measured velocity and sediment concentrations to determine the importance of mean currents, and high- and low-frequency waves to the *suspended* sediment transport on a time scale of a few weeks. These fluxes were related to the ratio of the local wave height and the local waterdepth (H_{m0}/h), incorporating both changes in wave conditions and tidal water levels. Secondly, the Bailard (1981)-approach was used to evaluate the relative importance of currents, high- and low frequency waves to the suspended sediment transport. The calculated contributions of the currents and waves were combined with a probability density function of H_{m0}/h to obtain insight in the effects on a timescale of years. Finally, a convection-diffusion sediment transport model (Van Rijn, 1993) was used, again in combination with probability density functions to evaluate the influence of waves and currents.

The Terschelling field site and data acquisition

The field experiment site is fully exposed to the North Sea with an annual mean significant offshore wave height of 1.1m (Van Beek, 1995). The tide is semidiurnal with a range of 1.5 and 2.5m during neap and spring tide, respectively. The nearshore area can be described as very dissipative with a very mild overall slope (1:200) and is characterized by the presence of several breaker bars (Hoekstra et al., 1994). In the period May 1994 to November 1995 three measuring campaigns were carried out: T2- (May and June 1994), T3- (October and November 1994), and the T4- (October and November 1995) campaign of five weeks each. The offshore significant wave height during the whole period ranged from 0.2 to 5m. During these campaigns four tripods (F2 to F5) were positioned in a cross-shore array (see Figure 1). Each tripod was equipped with two electro-magnetic flow meters (EMF) at an initial height of 0.25 and 1.2m above the bed and with a pressure sensor at 2.2m above the bed. On the tripods F3, F4 and F5 also two optical back scatter sensors (OBS) at 0.15m and 0.25m above the bed were installed. Timeseries were acquired with a sampling frequency of 2 Hz for 2048 seconds per hour. During the whole period from May 1994 to November 1995 the tripod F3 operated continuously (however without OBSs). At a waterdepth of 15m a wave directional buoy (WAVEC) was operational during the same period. The morphology landward of F4 evolved from a more or less alongshore uniform bar-trough system during the T2 and T3 campaigns to a three dimensional bar system during the T4 campaign. Seaward of F4 no large morphological changes were observed.

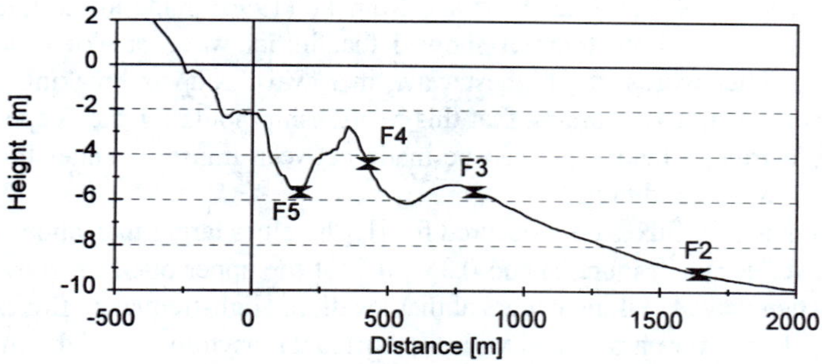

Figure 1. Instrumented positions and cross-shore profile of June 1994

Field observations of suspended sediment fluxes

Method

Fluxes were calculated from the measured timeseries of the cross-shore velocity u and concentration c. These timeseries were divided into a mean- and a high and low frequency oscillating term.

$$<u.c> = <\bar{u}.\bar{c}> + <\tilde{u}_h.\tilde{c}_h> + <\tilde{u}_l.\tilde{c}_l> \quad (1)$$

The left-hand side of (1) is the net suspended sediment flux; the first term on the right-hand side is the mean flux and the second and the third one are the high frequency and low frequency oscillating flux, respectively. The brackets denote time-averaged values. A positive flux is directed onshore. The separation frequency between high- and low-frequency waves was set to 0.04 Hz. Because of the different height at which u and c were measured, a correction for $\bar{u}$ had to be applied. Here, a simple engineering rule was used (Van Rijn, 1991), which reads as

$$\bar{u}_1 = \bar{u}_2 \left(\frac{z_1}{z_2} \right)^{0.25} \quad (2)$$

As a result, $\bar{u}$ at $z = 0.15$ m was about 88% of that at $z = 0.25$ m. Very small offsets in $\bar{c}$ (< 1 kgm^{-3}) were encountered and corrected for. For this reason, the background concentration equaled 0 kgm^{-3}.

Results

Burst-averaged values of the high- and low-frequency oscillating fluxes at z = 0.15 m at F3 (T2 and T4-), F4 (T2 +T3-) and F5 (T2 campaign) are plotted in Figure 2 as a function of the local relative wave height (spectral wave height based on short wave band) H_{m0}/h. A study of Van Enckevort and Reincke (1996) made in the intertidal zone of the same cross-shore transect showed that initial wave breaking starts at $H_{m0}/h = 0.33$. In other words, the highest wave in a wavegroup is breaking at this local relative wave height. Assuming that this relationship holds for the deeper parts of the nearshore zone, a distinction can be made between transport under breaking and nonbreaking wave conditions.

F3 : At F3 non-zero fluxes only occured for H_{m0}/h values larger than about 0.3 - 0.35 at the lower sensor (Figure 2) and 0.35 - 0.4 at the upper one (not shown). In other words; when wave breaking occurs at that location. High-frequency fluxes were onshore directed and were associated with the horizontal asymmetry of the incident short waves, i.e. the difference in magnitude between onshore and offshore orbital motion. Low-frequency fluxes were rather small and seaward directed and were likely related to bound long waves.

For $H_{m0}/h < 0.3$ sediment was not resuspended on the time scale of individual short waves nor on that of wave groups. The negligible oscillating fluxes for $H_{m0}/h < 0.3$ thus seemed to be mainly due to a lack of coherence between u and c for almost all frequencies. In addition, the variance in c was very low at both sensors compared to more energetic events.

For $H_{m0}/h > 0.3$ sediment was stirred on both the individual short-wave and wave-group time scale. The co-spectrum between u and c showed the 'classic' shape of a large positive (onshore) peak at the peak frequency of the short waves and a broad negative (offshore) peak at infragravity frequencies (cf. Huntley and Hanes, 1987). The phases at the peak frequency were between -45 and 0 degrees, which indicates that c lagged u slightly. Phase lags between both sensors were, in general, negligible at both high and low frequencies.

The oscillating fluxes at $z = 0.25$ m were considerably smaller than at $z = 0.15$ m. The sediment fluxes at the upper OBS were about 29 % and 4 % of that at the lower OBS for the high- and low-frequency sediment flux, respectively.

Time-averaged values of the mean flux at $z = 0.15$ m are also shown in Figure 2 as a function of H_{m0}/h. The largest fluxes, which were in a seaward direction, occured for conditions with H_{m0}/h above 0.3 to 0.35. This indicates that these negative fluxes were associated with wave-driven undertows.

F4 : The same pattern for the oscillating fluxes and the mean fluxes were also found at F4. However, during one wave event offshore directed high frequency fluxes occured as well, caused by phase lags between $\tilde{u}_h$ and $\tilde{c}_h$ larger than 90 degrees. The reason for these large phase lags is not fully understood. In general, phase lags are ascribed to the presence of ripples (e.g., Vincent and Green, 1990), though they have also been observed in the sheetflow regime (e.g., Ribberink and Al-Salem, 1995). Why such large phase lags developed during this particular storm (presumably sheet flow conditions) and not during other ones or at other positions is unknown. The mean fluxes at F4 were not only offshore directed due to undertows, but also onshore directed velocities and fluxes were observed (Figure 2). The origin of these onshore currents is not clear. On average, the mean fluxes were offshore directed and the averaged high-frequency fluxes onshore.

F5 : In the trough at F5, no breaking wave conditions occured ($H_{m0}/h < 0.3$). Although some scatter is present at the high frequency fluxes (Figure 2), no systematic contribution of this term was found. The scatter at the mean fluxes was not the result of enhanced cross-shore currents, but was caused by high sediment concentrations unrelated to the cross-shore velocity signal. These high sediment concentrations were probably the result of the tidal longshore currents in combination with bedforms.

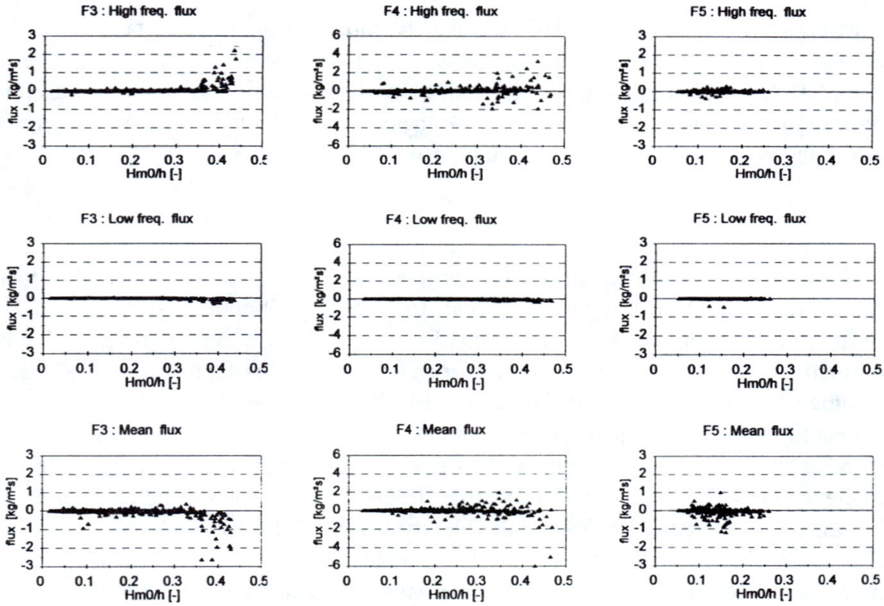

Figure 2. Measured fluxes at 15 cm above the bed at F3, F4 and F5

Long-term suspended sediment transport: an energetics approach

Introduction

The analysis of the suspended sediment fluxes at a height z above the bed has given important insight into the magnitude, direction and relative importance of different processes to the total suspended sediment transport. However, the findings strictly apply to the height z and are, therefore, not necessarily representative for the depth-averaged fluxes. Furthermore, the analysis presented sofar has not taken the long-term distribution of the hydrodynamical conditions into account. The magnitude of the fluxes increased with H_{m0}/h, but this does not necessarily mean that on a time scale of months to years the sediment transport is limited to conditions which occurrence in time is only infrequent. The cumulative effect of small, but often occurring transports may outweigh that of large, but rarely present ones. The cross-shore transport rate under mean and oscillating flows can relatively simple be modelled with a formulation in which the sediment transport rate is assumed to respond instantaneously to fluctuations in velocity above the bed (Bowen, 1980; Bailard, 1981). This rate is then expressed as a linear combination of terms containing products of the three different velocity components $\bar{u}$, $\tilde{u}_h$ and $\tilde{u}_l$. These terms, referred to as 'velocity moments', represent the depth-averaged sediment fluxes.

Roelvink and Stive (1989) showed that the most important contributions to the magnitude and the direction of suspended load transport is given by the term $<|u(t)|^3 u(t)>$. This term can be expanded in several terms after replacing the instantanous velocity *u(t)* by a sum of the mean, high and the low frequency oscillating velocity. The most important terms (Roelvink and Stive, 1989) are:

$$<|u|^3 u> \;=\; 4<|\tilde{u}_h|^3 \bar{u}> \;+\; <|\tilde{u}_h|^3 \tilde{u}_h> \;+\; 4<|\tilde{u}_h|^3 \tilde{u}_l> \qquad (3)$$

In these terms the stirring of sand is proportional to $|\tilde{u}_h|^3$. The sediment is then transported by any mean current, by high-frequency (HF) and low-frequency (LF) oscillations. Under breaking wave conditions in shallow water other velocity moments, such as those related to the stirring of sediment by long waves ($|\tilde{u}_l|^3$), may become of importance as well (Foote et al., 1994; Kroon, 1994), but these terms are neglected in the present analysis. The velocity moments were scaled with the inverse of the fall velocity *w* to investigate their cross-shore variability. The drag - and the efficiency coefficient (Roelvink and Stive, 1989) were assumed to be constant over the profile. The median sediment fall velocity at F2, F3, F4 and F5 were 15.1, 17.7, 27.6 and 20.5 mm/s, respectively (Guillén and Hoekstra, 1996). The velocities of the lower EMF were used to calculate the velocity moments. It should be emphasized that the energetics approach is only applied to investigate trends; quantitative information on the time-averaged transport rates is not a subject of this section.

Results

The w^{-1}-scaled velocity moments are shown in Figure 3 as a function of H_{m0}/h. The lines shown in Figure 3 are based on the average values of a particular H_{m0}/h-class (width =0.02). At each position the w^{-1} -scaled velocity moments were clearly related to H_{m0}/h , and the growth of the velocity moments start at H_{m0}/h=0.3. In general, the contribution of $<|\tilde{u}_h|^3\tilde{u}_h>$ (HF) and $<|\tilde{u}_h|^3 \bar{u}>$ (mean) to the total fourth-order velocity moment were about equal, but of opposite sign. Consequently, the LF-term $<|\tilde{u}_h|^3\tilde{u}_l>$, which had a negative sign and was smaller than both $<|\tilde{u}_h|^3\tilde{u}_h>$ and $<|\tilde{u}_h|^3 \bar{u}>$, often strongly influenced the size and the direction of the net fourth-order velocity moment. The fourth-order velocity moments at F5 were minimal compared to those at the other postions indicating negligible suspended sediment transport in the middle trough.

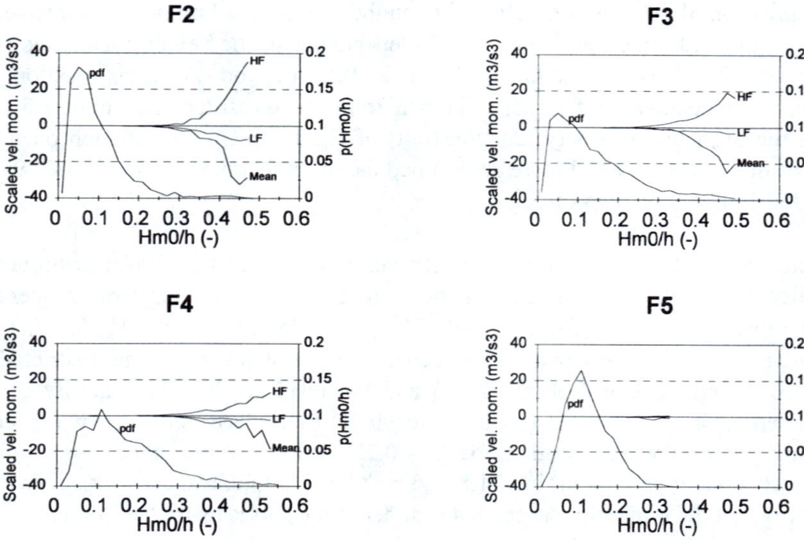

Figure 3. Measured scaled velocity moments and probability density function of (Hm0/h)

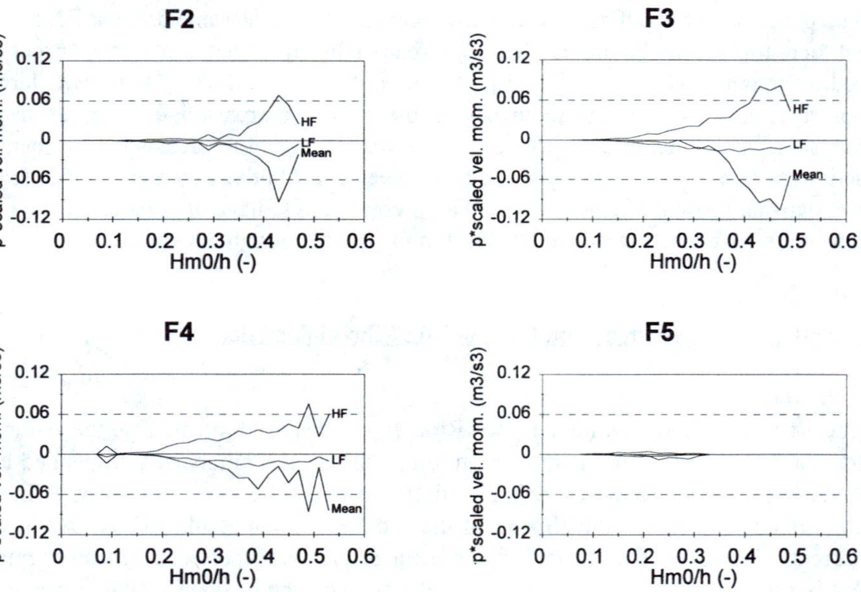

Figure 4. Long term transport rates weighted for the frequency of occurrence

An impression of the long-term effect, i.e. on the time scale of months to a few years, can be obtained by multiplying the H_{m0}/h-dependence of the velocity moments with the probability density function of H_{m0}/h at that location. At each location the frequency of occurence of H_{m0}/h is known from the measurements. Figure 3 also shows the probability density function (pdf) of H_{m0}/h ($p(H_{m0}/h)$) at each position. The pdf of F2, F3, F4 and F5 are determined based on measured data of 2682, 9220, 1551 and 1541 hours, respectively.

The product of the pdf and the w^{-1}-scaled velocity moments is shown in Figure 4. Significant contributions of the separate processes to the long-term suspended sediment transport started from a value of $H_{m0}/h = 0.25$ at F2, and of $H_{m0}/h = 0.15$ at F3 and F4, which means that the suspended sediment transport was restricted to 3.4%, 32% and 51% in time at F2, F3 and F4, respectively. The majority of the suspended sediment transport (about 70 to 90 %) took place, however, when wave breaking occurred at that location ($H_{m0}/h > 0.33$) and as a consequence, was limited to a much smaller amount in time (1.8, 5.6 and 8.5% in time respectively). This does not imply that the long-term suspended sediment transport is restricted to truely extreme events, except perhaps at F2. For decreasing water depths the occurence of the long-term suspended sediment transport shifts towards more common conditions, although it remains largely restricted to conditions with breaking waves.
The products of the pdf and the velocity moments are in the same order at F2, F3 and F4 and much smaller at F5. The approximately equal net long-term suspended sediment transports at F2, F3 and F4 are caused by the fact that the two largest contributions, viz. the high-frequency oscillating and current-related one, are nearly the same and, therefore, cancel out. This implies that the cross-shore suspended sediment transport is a delicate balance between contributions by mean currents, and by high- and low-frequency waves. The present results have also indicated that the long-term suspended sediment transport in the middle trough is very small.

Long-term sediment transport: the Van Rijn/Ribberink model

Method

The Van Rijn/Ribberink model (Van Rijn, 1993) was used to calculate the sediment transport. This model is a combination of two different submodels: a suspended load (Van Rijn, a convection-diffusion model) and a bedload one (Ribberink). In the suspended load model the stirring of the sediment is the result of both waves and currents. However, the suspended sediment is only transported by mean currents; oscillating suspended load transport is not included. The bedload model, however, is an instantaneous model. As a result, the sediment can be transported due to both waves (short and long waves) and currents. The model has been modified in such a way that timeseries of measured velocities and waterlevels can be used as input instead of the vertical mean velocity and wave height (Houwman and Hoekstra, 1994). Timeseries of the T2- and T4-campaigns at position F2, F3, F4 and F5 were used to calculate the sediment transport. The measured timeseries of velocities *u* and *v* (at 0.25 m above the bed) and the measured waterlevels (measured by the pressure sensor at 2.2 m above bed) were used as input into the model. From these velocity signals the near bed velocity distribution was constructed assuming a logarithmic velocity profile for the mean current and a constant orbital velocity distribution in the vertical. Only measured velocity signals were used, therefore the effect of the wave induced near bed velocity ('Longuet-Higgins streaming') was not included in the computations. For the computations the local median grain size was used (d_{50} is 154, 163, 172 and 178 μm for F2, F3, F4 and F5 respectively) and a constant bed roughness height $k_s = 0.3$ cm.

The previous analysis of the fluxes and the velocity moments showed that the high-frequency suspended load transport cannot be neglected. To get insight in contribution of this term to the net transport, the Van Rijn model is modified in the following way. The model produces a time-averaged sediment concentration distribution over the vertical. At a certain height above the bed this mean sediment concentration is thought to be the time averaged value of two sediment concentration peaks per wave cycle, one during the onshore directed wave motion and the other during the offshore directed peak. According to the velocity moments approach the shape of the sediment concentration peaks can be assumed to be equal to the shape of $|u(t)|^3$. If one also assumes that each half wave cycle can be described with linear wave theory with different amplitudes but with equal duration, equation (4) is derived.

$$\bar{c}(z) = k\left[\left|\frac{2}{T}\int_0^{T/2} U_{on}^3 \sin^3\omega t\,dt\right| + \left|\frac{2}{T}\int_{T/2}^{T} U_{off}^3 \sin^3\omega t\,dt\right|\right]$$
$$= k*[U_{on}^3 + U_{off}^3]*\left|\frac{2}{T}\int_0^{T/2}\sin^3\omega t\,dt\right| \tag{4}$$

In equation (4), the left term is the time averaged sediment concentration at height z above the bed, U_{on} and U_{off} refer to the onshore and offshore peak orbital velocity, T and ω are the wave period and angular frequency, respectively and k is a constant. The oscillating suspended flux at a certain height *z* above the bed can be related to the fourth order moment $u^*|u^3|$. The wave averaged oscillating sediment transport rate *S(z)* through a layer *dz* is then described by (5):

$$S(z) = \frac{1}{T}\int_0^T k*u(t)\,|u(t)^3|\,dt\ dz = k[U_{on}^{\ 4} - U_{off}^{\ 4}]\,dz*\frac{1}{T}\int_0^T \sin^4\omega t\,dt \tag{5}$$

$$S(z) = k' * \bar{c}(z) * \frac{U_{on}^4 - U_{off}^4}{U_{on}^3 + U_{off}^3} * dz$$

$$with: \quad k' = \frac{1}{2}\int_0^T \sin^4\omega t\, dt * \left[\int_0^{T/2} \sin^3\omega t\, dt\right]^{-1} \approx 0.884$$

After integration over the vertical the oscillating suspended sediment transport can be obtained. The mean concentration at height *z* above the bed is calculated with the Van Rijn/Ribberink model, and the measured significant onshore and offshore orbital motions can be used.

This approach gives an upper limit for the suspended oscillating sediment transport, because one assumes that there are no phase lags present between u(t) and c(t) and that the sediment always responds instantaneously to the third power of the orbital velocity. However, the measurements have shown that phase lags were present and it is unclear if the sediment always responded to $u(t)^3$. Therefore, the parameter *k'* in (5) is replaced by a new parameter, the "efficiency coefficient" (*eff*) which incorporates these effects. This parameter was determined from the OBS data.

The measured fluxes were divided by the mean concentration for both OBSs at F3 and were related to the significant orbital velocity. The relative (to the mean) response of the sediment is important, because the mean concentration *c(z)* is obtained from the model. The values are averaged for an orbital velocity class (width of 0.1m/s), based on data of the T2, T3 and T4 campaigns. A comparison between the measured values and the values of the fourth-order moment show that no fixed *k'* values can be taken. Based on the two lines from both OBSs the efficiency coefficient was determined as:

$$\begin{aligned} &eff(U_{on}, z) = [(2.5 * z - 0.292) * U_{on} - 3 * z + 0.42] * \tanh(6.5(z_{\max} - z)) \\ &IF \ eff(U_{on}, z) < 0 \rightarrow eff(U_{on}, z) = 0 \end{aligned} \qquad (6)$$

To prevent that large oscillating fluxes are calculated far away from the bottom, a tanh function was used to force the efficiency coefficent down to zero again at a particular height (z_{max}) above the bed. The model outcomes were not very sensitive for the values of z_{max}. In the calculations a value of 0.6 m was chosen for z_{max}. A comparison between the 'theoretical' efficiency coefficient *k'* and the 'observed' efficiency coefficient makes clear that the theoretical one is significantly higher than the observed one. At a height of 0.25 m above the bed the observed efficiency coeficient ranges from 0 to 0.2 (at U_{on} =1.6 m/s) instead of having a fixed value of

0.884. The oscillating suspended load sediment transport was also calculated at F4 using the same formulation for the efficiency coefficient. The observed values were described reasonably by the model. Therefore the approach described above was used for all tripods.

Results

From the calculated cross-shore transport rates an average sediment transport can be calculated for a particular relative wave height class (H_{m0}/h). Figure 5 shows the transport rates calculated at four locations and split up into the contribution of the three transport mechanisms: bed(load), mean- and osc(illating suspended load) transport. The p(H_{m0}/h) for each position are also given. At F2, F3 and F4 the mean and oscillating suspended load transport increased rapidly for situations with a relative wave height larger than 0.3.

The 'yearly-averaged' transport rates shown in Figure 6 were created in a similar way with the pdfs as for the velocity moments. For both F2 and F3 the oscillating and mean suspended load sediment transport were very high during breaking wave conditions ($H_{m0}/h > 0.33$). The yearly net sediment transport is a sensitive balance between the offshore directed mean suspended load transport and the onshore directed oscillating suspended load transport, the contribution of the bedload transport at these positions is only small. However, the difference between the relative large contributions of the suspended load terms may also be very small and, consequently the influence of the bedload transport on the net transport is not clear yet. Again at location F4 the mean suspended load is neutralized by the oscillating suspended load component. The magnitudes of the different yearly averaged transport rates at F2, F3 and F4 were of the same order of magnitude.

In the trough at the position F5 the cross-shore sediment transport rates were much lower and offshore directed. The lack of any onshore transport mechanism in the trough region indicates that no large onshore exchange of sediment across the trough may be expected.

Discussion and conclusions

The field measurements at Terschelling have shown that high- and low-frequency waves, as well as mean currents, contribute to the net suspended sediment transport. In general, the high-frequency oscillating transport is onshore directed and is caused by the positive skewness of the short waves. The low-frequency oscillating transport is in a seaward direction owing to the dominance of bound long waves in the total long wave field and the stirring of sediment on the wave-group time scale. The current-related flux is in an offshore direction in most cases and is caused by wave-driven undertows.

The "measured" fluxes start to increase at a relative wave height $H_{m0}/h > 0.33$, which coincides with breaking wave conditions. The velocity moments are increasing at a lower H_{m0}/h ratio (0.15 - 0.25) but also here the highest transport rates were found during breaking wave conditions. The suspended load transport rates calculated with

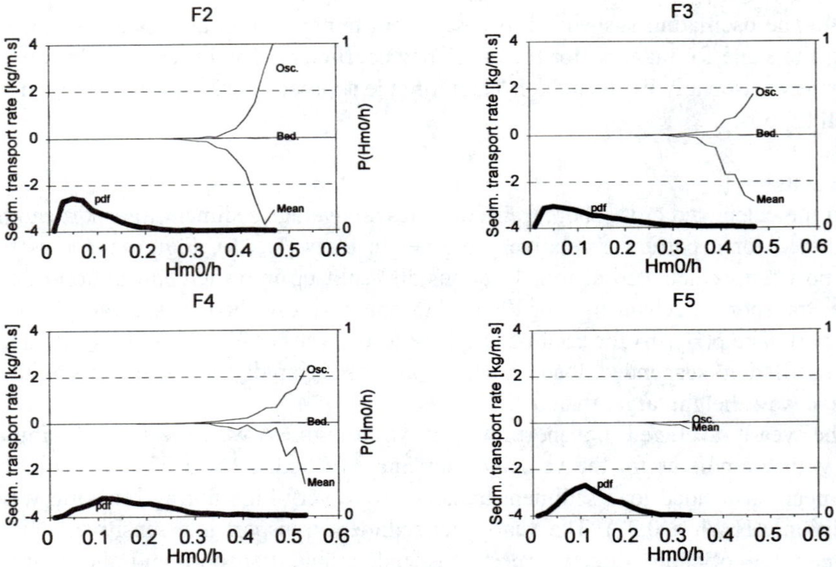

Figure 5. Modelled transport rates with pdf

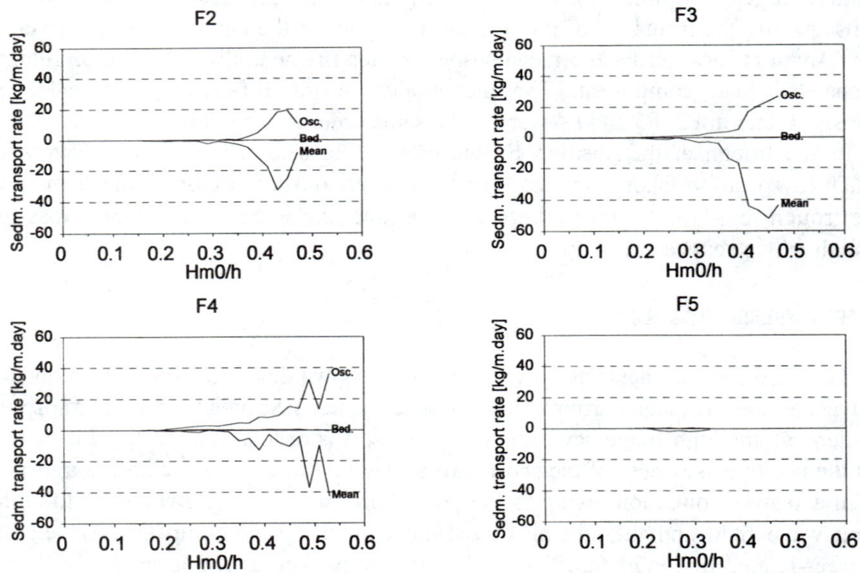

Figure 6. Long term transport rates weighted for the frequency of occurrence

the Van Rijn model were only non-zero during these conditions ($H_{m0}/h > 0.33$) as well. In general, the suspended load transport rates are well related to the relative wave heights H_{m0}/h.

All three applied methods show that the contributions of the onshore directed oscillating suspended load transport or flux is more or less balanced by the offshore directed mean transport/flux.

The combination of a local probability density function (pdf) and the calculated velocity moments and transport rates gives insight in the long term contributions of different transport mechanisms. The yearly averaged suspended load transport rates obtained in this way, start to increase at $H_{m0}/h > 0.33$ for F2, F3 and F4. This indicates that on the long run only breaking wave conditions are important at these positions. But again, at this time scale the suspended sediment transport is a delicate balance between the onshore directed (short wave) oscillating transport and the offshore directed mean transport.

The contribution of the low frequency suspended load transport is rather small compared to the high frequency and mean transport but is systematic and therefore still significant. The calculations of the bedload transport have shown that this contribution is very small compared to the other terms. Although the contributions of the low frequency suspended transport and the bedload transport are small their importance will increase when the mean suspended transport is neutralized by the oscillating high frequency transport.

The results obtained from the calculated long term transport rates are not in agreement with the beach-shoreface exchange mechanism proposed by Niederoda et al. (1984). They found an offshore directed sediment transport from the beach/inner surfzone to the upper shoreface ($h > 4$m) during a storm and a slow return of this sediment during low energetic conditions in response of asymmetric waves propagating to the coast. If this mechanism is also valid for the Terschelling coast one expects a large contribution of the short wave transport during non-breaking wave conditions ($H_{m0}/h < 0.33$) and a clear dominance of the mean transport during breaking wave conditions. However, for the whole range of H_{m0}/h values the terms are in balance and no clear dominance of a term at a particular H_{m0}/h value has been found. In addition, the morphological observations do not support the theory of such a systematic exchange mechanism between beach and deeper water.

The lack of significant sediment transport in the trough at location F5 also prevents a large exchange between the shallow water and deep water region. In the trough no waves break, and the calculated transport rates are very low. It seems that the landward-located trough acts as a barrier against cross-shore sediment transport.

References

Bailard, J.A., 1981. An energetics total load sediment transport model for a plane sloping beach. Journal of Geophysical Research, 86 (C11), p. 10938-10954.

Bowen, A.J., 1980. Simple models of nearshore sedimentation; beach profiles and longshore bars. In: McCann, S.B. (ed.), The coastline of Canada, Geological Survey Canada, Paper 80-10, p.1-11.

Foote, Y.L.M. and Huntley, D.A., 1994. Velocity moments on a macrotidal intermediate beach. Proc. of Coastal Dynamics'94, ASCE, p. 794-808.

Guillén, J. and Hoekstra, P., 1996. The "equilibrium" distribution of grain size fractions and its implications for cross-shore sediment transport: a conceptual model. Marine Geology (in press).

Hoekstra, P., Houwman, K.T., Kroon, A., Van Vessem, P., Ruessink, B.G., 1994. The NOURTEC experiment of Terschelling: process-oriented monitoring of a shoreface nourishment (1993-1996), Proc. of Coastal Dynamics'94, ASCE, p. 402-416.

Houwman, K.T., and Hoekstra, P., 1994. Shoreface sediment dynamics; Field measurements Egmond aan Zee, IMAU R94-3, Utrecht University, 25 pp.

Huntley, D.A. and Hanes, D.M., 1987. Direct measurement of suspended sediment transport. Proc. of Coastal Sediments'87, ASCE, p. 723-737.

Kroon, A., 1994. Sediment transport and morphodynamics of the beach and nearshore zone near Egmond, the Netherlands, PhD Thesis, Utrecht University, 275 pp.

Niedoroda, A.W., Swift, D.J.P., Sawyer Hopkins, T. and Chen-Mean Ma, 1984. Shoreface morphodynamics on wave-dominated coasts, Marine Geology, 60, p. 331-354.

Ribberink, J.S. and Al-Salem, A.A., 1995. Sheet flow and suspension of sand in oscillatory boundary layers. Coastal Engineering, 25, p. 205-225.

Roelvink, J.A. and Stive, M.J.F., 1989. Bar-generating cross-shore flows on a beach. Journal of Geophysical Research, 94 (C4), p. 4785-4800.

Van Beek, L.P.H., 1995. Wave climate of Terschelling. A statistical analysis for the Nourtec project. IMAU R 95-10, Utrecht University, 48 pp.

Van Enckevort, I. and Reincke, E., 1996. Longshore currents in the intertidal zone of Terschelling. IMAU V96.09, Utrecht University, 62 pp.

Van Rijn, L.C., 1991. Database sand concentrations and sand transport in currents and/or waves. Delft Hydraulics, Report H1148-04/05.

Van Rijn, L.C., 1993. Principles of sediment transport in rivers, estuaries and coastal seas, Aqua Publications, Amsterdam, The Netherlands, 614 pp

Van Rijn, L.C. and Havinga, F.J., 1995. Transport of fine sands by currents and waves. II.,Journal of Waterway, Port, Coastal, and Ocean Engineering, p. 123-133.

Vincent, C.E. and Green, M.O., 1990. Field measurements of the suspended sand concentration profiles and fluxes and of the resuspension coefficient $\gamma 0$ over a rippled bed. Journal of Geophysical Research, 95 (C7), p. 11591-11601.

CHAPTER 374

NUMERICAL PREDICTION OF BREAKING WAVES AND CURRENTS WITH A BOUSSINESQ MODEL

Okey George Nwogu[1]

ABSTRACT

This paper describes the extension of a comprehensive numerical model for simulating the propagation and transformation of ocean waves in coastal regions and harbours to include wave breaking, runup and breaking-induced currents. The numerical model is based on a time-domain solution of a fully nonlinear set of Boussinesq-type equations for wave propagation in intermediate and shallow water depths. The equations are able to describe most of the phenomena of interest in the nearshore zone including shoaling, refraction, diffraction, reflection, wave directionality and nonlinear wave-wave interactions. The Boussinesq model is extended to the surf and swash zones by coupling the mass and momentum equations with a one-equation model for the temporal and spatial evolution of the turbulent kinetic energy produced by wave breaking. The waves are assumed to start breaking when the horizontal component of the orbital velocity at the wave crest exceeds the phase velocity of the waves. Numerical and experimental results are presented for the shoaling and runup of regular and irregular waves on a constant slope beach and wave-induced currents behind a detached breakwater.

1. INTRODUCTION

The processes of wave breaking, runup, setdown and setup of the mean water level, turbulent energy production, generation of nearshore currents and generation of infragravity waves are important driving mechanisms for the transport of sediments and pollutants in coastal regions. A full mathematical description of the complex hydrodynamics in the surf and swash zones is difficult due to the highly nonlinear and turbulent nature of flow. By making different simplifying assumptions, however, numerous mathematical models been developed which reproduce with varying degrees of success the different hydrodynamic phenomena that occur in the surf and swash zones.

For waves propagating over simple bathymetries where shoaling, refraction and breaking are the dominant wave transformation processes, models based on the conservation of energy flux with an appropriate energy dissipation term have successfully been used to model the wave height variation in the surf zone (e.g. Battjes and Janssen 1978, Dally et al. 1985). Battjes and Janssen (1978) used a hydraulic jump analogy to derive

[1]Canadian Hydraulics Centre, National Research Council, Ottawa, Canada K1A 0R6

the form of dissipation term while Dally et al. (1985) assumed that the dissipation rate was proportional to the difference between the local energy flux and a stable energy flux. Similar energy dissipation terms have also been incorporated into the mild-slope equation (e.g. Isobe 1987, Özkan and Kirby 1993).

Most surf zone wave transformation models are based on time-averaged integral wave properties and do not follow the breaking process in a time dependent manner. For applications such as irregular wave propagation and sediment transport over barred beaches, time domain modeling provides a more accurate description of the breaking and reformation process. A time-dependent breaking model is also able to simulate the transition region after the onset of wave breaking, where there is a rapid decay in wave energy with almost no change in the setdown of the mean water level. The transition zone plays an important role in the prediction of wave-induced currents and sediment transport in the surf zone (Nairn et al., 1990).

Time-dependent numerical models based Boussinesq-type equations have successfully been used to simulate the nonlinear transformation of multidirectional ocean waves in intermediate and shallow water depths (e.g. Nwogu, 1994). Boussinesq equations can accurately describe most wave transformation processes outside the surf zone including shoaling, refraction, diffraction, reflection and nonlinear wave-wave interactions. Recently, several authors have extended Boussinesq models to simulate wave breaking in the surf zone including Zelt (1991), Karambas and Koutitas (1992), Schäffer et al. (1993), and Sato and Kabiling (1994). The models essentially incorporate a dissipative term due to turbulence stresses or the presence of a surface roller into the momentum equation. The models differ on how they treat the onset of breaking and the rate of wave energy dissipation.

In this paper, we employ a one-equation turbulence model to describe the temporal and spatial evolution of the turbulent kinetic energy produced by wave breaking. The waves are assumed to start breaking when the horizontal component of the orbital velocity at the wave crest exceeds the phase velocity of the waves. The rate of production of turbulent kinetic energy is assumed to be proportional to the vertical gradient of the horizontal water particle velocity at the wave crest. The time-dependent model is applicable to both periodic and non-periodic unidirectional and multidirectional waves.

One advantage of extending Boussinesq-type models to the surf zone is the ability to implicitly model interactions between hydrodynamic processes occurring at different time scales. Wave-induced currents and mean water level fluctuations are implicitly included in the wave propagation model and are derived from a time-average of the predicted velocities and surface elevation respectively, without having to explicitly calculate radiation stresses and separately solve a time-averaged hydrodynamic model.

2. THEORETICAL MODEL

2.1 Fully Nonlinear Boussinesq Equations

Boussinesq equations represent the depth-integrated equations for the conservation of mass and momentum for nonlinear dispersive waves, propagating in water of variable depth. The velocity potential, ϕ, can be expanded as power series in the vertical coordinate, z. An approximation is introduced for frequency dispersion by truncating the series at second-order, resulting in a quadratic variation for the horizontal velocity over depth, and a linear variation for the vertical velocity:

$$\mathbf{u}(z) = \mathbf{u}_\alpha + (z_\alpha - z)\left[\nabla(\mathbf{u}_\alpha \cdot \nabla h) + (\nabla \cdot \mathbf{u}_\alpha)\nabla h\right] + \left[\frac{(z_\alpha + h)^2}{2} - \frac{(z+h)^2}{2}\right]\nabla(\nabla \cdot \mathbf{u}_\alpha) \tag{1}$$

$$w(z) = -\left[\mathbf{u}_\alpha \cdot \nabla h + (z+h)\nabla \cdot \mathbf{u}_\alpha\right] \tag{2}$$

where $\nabla = (\partial/\partial x, \partial/\partial y)$, $h(\mathbf{x})$ is the water depth and $\mathbf{u}_\alpha = \nabla\phi|_{z=z_\alpha}$. Given a vertical profile for the flow field (Eqns. 1 & 2), the governing equations of fluid motion can be integrated over depth, reducing the three-dimensional problem to a two-dimensional one. The depth-integrated mass conservation equation can be written as:

$$\eta_t + \nabla \cdot \int_{-h}^{\eta} \mathbf{u}\, dz = 0, \tag{3}$$

where $\eta(\mathbf{x}, t)$ is the free surface elevation. Although Boussinesq equations were originally derived for weakly nonlinear waves, a fully nonlinear variant of the equations was recently derived by Wei et al. (1995). The momentum equation, fully nonlinear up to the order of truncation of the dispersive terms, is derived from the dynamic boundary condition at the free surface as:

$$\begin{aligned}&\mathbf{u}_{\alpha t} + g\nabla\eta + (z_\alpha - \eta)\left[\nabla(\mathbf{u}_{\alpha t} \cdot \nabla h) + (\nabla \cdot \mathbf{u}_{\alpha t})\nabla h\right] \\ &+ \left[\frac{(z_\alpha + h)^2}{2} - \frac{(h+\eta)^2}{2}\right]\nabla(\nabla \cdot \mathbf{u}_{\alpha t}) + \frac{1}{2}\nabla(\mathbf{u}_s \cdot \mathbf{u}_s + w_s^2) \\ &-\left[(\mathbf{u}_{\alpha t} \cdot \nabla h) + (h+\eta)(\nabla \cdot \mathbf{u}_{\alpha t})\right]\nabla\eta + \frac{f_w}{(h+\eta)}\mathbf{u}_b|\mathbf{u}_b| + \frac{\partial}{\partial z}\left(\overline{\mathbf{u}'w'}\right) = 0\end{aligned} \tag{4}$$

where g is the gravitational acceleration, $\mathbf{u}_b = \mathbf{u}(-h)$, $\mathbf{u}_s = \mathbf{u}(\eta)$ and $w_s = w(\eta)$. Two additional terms have been introduced into the momentum equation to simulate the dissipation of wave energy due to bottom friction and wave breaking. f_w is an empirical bottom friction factor while $\overline{\mathbf{u}'w'}$ represents the shear stress due to breaking-induced turbulent velocity fluctuations (u', v', w'). A semi-empirical turbulence closure model (Section 2.2) is used to relate the turbulence stresses to the wave orbital velocities.

The elevation of the velocity variable z_α is a free parameter and is chosen to minimize the differences between the linear dispersion characteristics of the Boussinesq model and linear theory. An optimum depth for the velocity variable, $z_\alpha = -0.53h$, gives errors of less than 2% in the phase speed from shallow water depths up to the deep water depth limit.

2.2 Turbulence Model

Several turbulence models have been proposed for fluid flow problems (see review by Rodi, 1980). In this paper, the standard one-equation turbulence model is used to describe the spatial and temporal evolution of the turbulent kinetic energy, k, produced by wave breaking. We do not attempt to model details of the turbulent motion, but rather, investigate the effect of turbulence on the wave field.

The key assumptions made in developing the model are:

1. the breaking process is assumed to be "spilling"
2. turbulence is initiated in the wave crest region due to large vertical gradients of the horizontal velocity, $\partial \mathbf{u}/\partial z$
3. turbulence is produced only when the horizontal velocity at the wave crest exceeds the phase velocity of the waves
4. the rate of production of turbulent kinetic energy is proportional to the vertical gradient of the horizontal velocity at the wave crest
5. turbulence is primarily convected in the crest front region with the horizontal component of the orbital velocity at the crest

The Boussinesq eddy viscosity concept can be used to relate the turbulence shear stresses to the velocity gradients:

$$\overline{\mathbf{u}'w'} = -\nu_t \frac{\partial \mathbf{u}}{\partial z} \tag{5}$$

where ν_t is the eddy viscosity. By substituting the expression for $\mathbf{u}(z)$ given in equation (1), the dissipative term in the momentum equation (Eqn. 4) becomes:

$$\frac{\partial}{\partial z}\left(\overline{\mathbf{u}'w'}\right) = -\frac{\partial}{\partial z}\left(\nu_t \frac{\partial \mathbf{u}}{\partial z}\right) = \nu_t \nabla(\nabla \cdot \mathbf{u}_\alpha) \tag{6}$$

The rate of wave energy dissipation is thus governed by the magnitude of the eddy viscosity which is related to the turbulent kinetic energy, k, and a turbulence length scale, ℓ_t, by:

$$\nu_t = \sqrt{k}\,\ell_t \tag{7}$$

The turbulent kinetic energy is determined from a semi-empirical transport equation with a source term for turbulent kinetic energy production by wave breaking:

$$k_t + \mathbf{u}_s \cdot \nabla k = \nu_k \nabla^2 k + B\nu_t \left[\left(\frac{\partial u}{\partial z} \right)^2 + \left(\frac{\partial v}{\partial z} \right)^2 \right]_{z=\eta} - C_D \frac{k^{3/2}}{\ell_t} \quad (8)$$

The first term on the right hand side represents the horizontal diffusion of turbulent kinetic energy. This term is usually much smaller than the convective term, $\mathbf{u}_s \cdot \nabla k$. The second term represents the production of turbulent kinetic energy due to wave breaking. The parameter B is introduced to ensure that turbulence is produced only when horizontal velocity at the wave crest, $\mathbf{u}_s$, exceeds the phase velocity of the waves, C, i.e.

$$B = \begin{cases} 0 & |\mathbf{u}_s| < C \\ 1 & |\mathbf{u}_s| \geq C \end{cases} \quad (9)$$

This criterion is valid for progressive waves and has been found to accurately predict the breaker location observed in experiments with regular waves shoaling on a constant slope beach. For irregular waves, an approximate phase velocity can be calculated using the average zero-crossing period, although an instantaneous phase velocity can also be determined using $C = -\eta_t/|\nabla\eta|$.

The last term on the right hand side of the transport equation (8) represents the dissipation of turbulent kinetic energy into heat. C_D is an empirical constant with a representative value of 0.08 for most turbulent shear flows (Rodi, 1980).

The rate of production of turbulent kinetic energy depends on ν_t which initially is unknown. Using the mixing length hypothesis which assumes a local balance between production and dissipation of turbulent kinetic energy, the value of ν_t for the production term is determined as:

$$\nu_t = \frac{\ell_t^2}{\sqrt{C_D}} \left[\left(\frac{\partial u}{\partial z} \right)^2 + \left(\frac{\partial v}{\partial z} \right)^2 \right]_{z=\eta}^{1/2} \quad (10)$$

By choosing an arbitrary value for C_D, the turbulence length scale ℓ_t becomes the only free parameter in the turbulence model which governs the rate of wave energy dissipation and turbulent kinetic energy production. ℓ_t is expected to be of the order of the wave height and is determined from comparisons of the numerical model results with experimental results.

2.3 Numerical Solution

The governing Boussinesq equations have been solved with an iterative Crank-Nicolson finite difference method, with a predictor-corrector scheme used to provide the initial estimate. The computational domain is discretized using a rectangular grid, with the dependent variables η, u_α and v_α defined at the grid points in a staggered manner. The numerical solution procedure consists of solving an algebraic expression for η at all grid points, tridiagonal matrices for u_α along lines in the x direction and tridiagonal matrices for v_α along lines in the y direction at every time step. An explicit second-order scheme is used for the turbulence transport equation. Details of the numerical solution technique can be found in Nwogu (1996).

The boundaries of the computational domain may be specified as wave input boundaries or solid walls. Along external or internal wave generation boundaries, time histories of $(u_\alpha, u_{\alpha,xx}, v_{\alpha,xy})$ or $(v_\alpha, v_{\alpha,yy}, u_{\alpha,xy})$, corresponding to periodic or non-periodic unidirectional or multidirectional sea states are input. Waves propagating out of the domain are artificially absorbed in damping regions placed next to solid wall boundaries. Artificial damping of wave energy is accomplished by introducing terms out of phase with the surface velocity and fluid acceleration into the continuity and momentum equations respectively.

A simple extrapolation scheme is used to simulate wave runup. Consider the one-dimensional case where the shoreline boundary condition is given by zero mass flux, or equivalently, $\eta_x = 0$. When the water surface elevation η_i at grid point i exceeds the elevation of an adjacent land point, h_{i+1}, we assume that the land point will be flooded at the next time step. The initial value of the water level on the the land point for the next time step is then extrapolated from the surface elevation at adjacent water points:

$$\eta_{i+1} = 4\eta_i - 3\eta_{i-1} \qquad \text{if } \eta_i > -h_{i+1} \tag{11}$$

Outputs of the numerical model are time histories of the surface elevation and two components of the horizontal velocity at desired grid points in the computational domain, corresponding frequency and directional wave spectra at those locations, the instantaneous water surface elevation at specified time steps, and the significant wave height distribution and time-averaged horizontal velocities over the entire computational domain.

3. NUMERICAL AND EXPERIMENTAL RESULTS

The numerical model was initially evaluated using data obtained from experiments carried out by Nwogu (1993) in the three-dimensional wave basin of the Canadian Hydraulics Center. A 1:25 constant slope concrete beach was constructed in the 30 m $\times$ 20 m $\times$ 3 m basin, equipped with a 60-segment directional wave generator. The toe of the slope was located 4.6 m away from the wave boards. The water depth in the constant depth portion of the basin was 0.56 m. Tests were carried out for a wide variety of regular and irregular, unidirectional and multidirectional waves. The water surface elevation along the centerline of the basin was measured with a linear array of 23 water level gauges. The experimental setup is described in greater detail by Nwogu (1993).

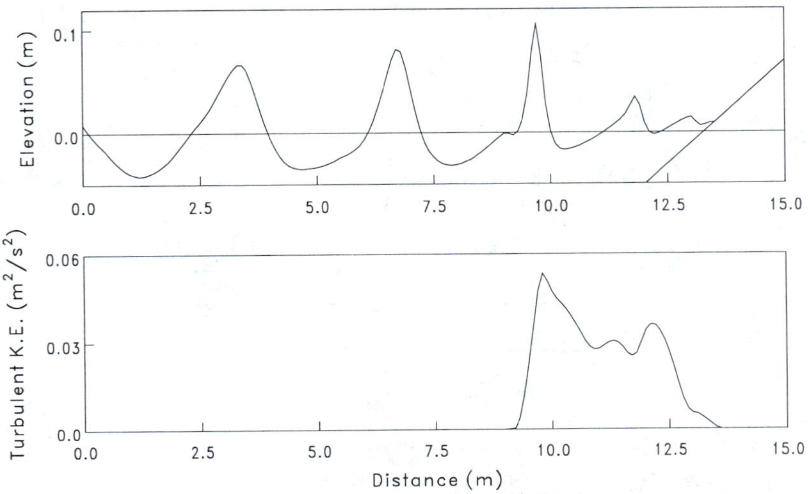

Figure 1. Breaking and runup of a regular wave (T = 2 s) on a 1:25 beach.

3.1 Shoaling of Regular Waves

The numerical model was used to investigate the shoaling and breaking of a regular wave with period, T = 2 s, and height, H = 0.09 m, on a 1:25 beach. The computations were carried out using a grid size Δx = 0.1 m and time step size Δt = 0.04 s. A bottom friction coefficient, f_w = 0.01, was used for all the computations. Figure 1 shows the spatial profile of the water surface elevation at an instant of time and the corresponding turbulent kinetic energy distribution. The turbulent kinetic energy is maximum close to the breaker location and decreases towards the shoreline. The turbulent kinetic energy exhibits an oscillatory behaviour in space due to the periodic nature of the production of turbulence from each breaking wave event.

Figure 2 shows a comparison of the measured and predicted time histories of the surface elevation at four water depths, both outside and inside the breaker zone. The numerical model is able to reproduce the highly asymetric wave profile in the surf zone. The spatial variation of the average zero-crossing wave height and mean water level are plotted in Figure 3. Fairly good agreement is obtained between the numerical and experimental results. A turbulence length scale l_t = 0.35 m with C_D = 0.08 gave the best match of rate of wave energy decay through the surf zone. The solution is not unique, however, as other combinations of C_D, ℓ_t, and the breaking criterion also give reasonable matches of the wave height decay. An extensive investigation is presently being carried out to recommend optimum values of the turbulence model parameters as a function of incident wave parameters.

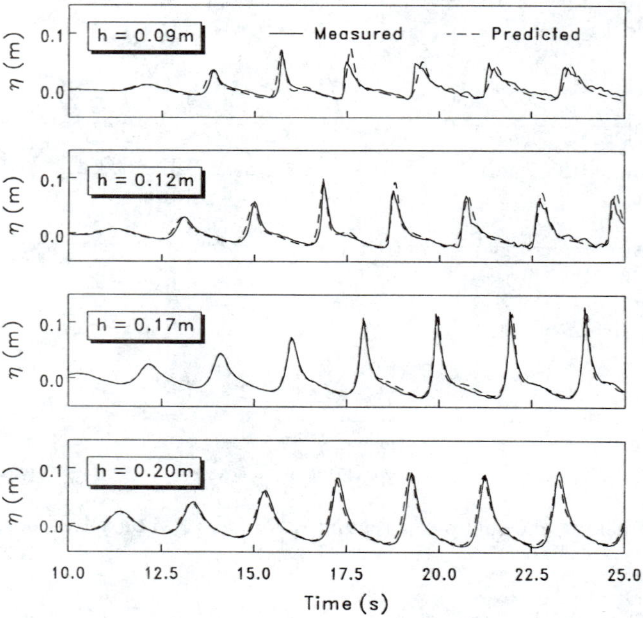

Figure 2. Measured and predicted time histories for a regular wave ($T = 2$ s) shoaling on a 1:25 beach.

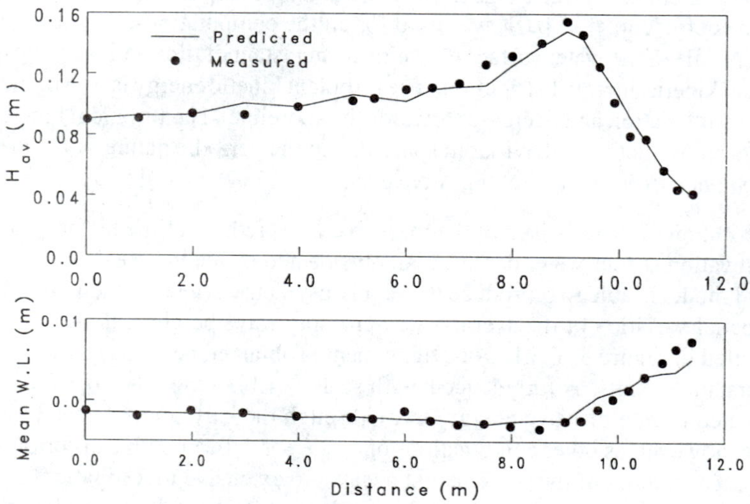

Figure 3. Measured and predicted spatial variation in wave height for a regular wave ($T = 2$ s) shoaling on a 1:25 beach.

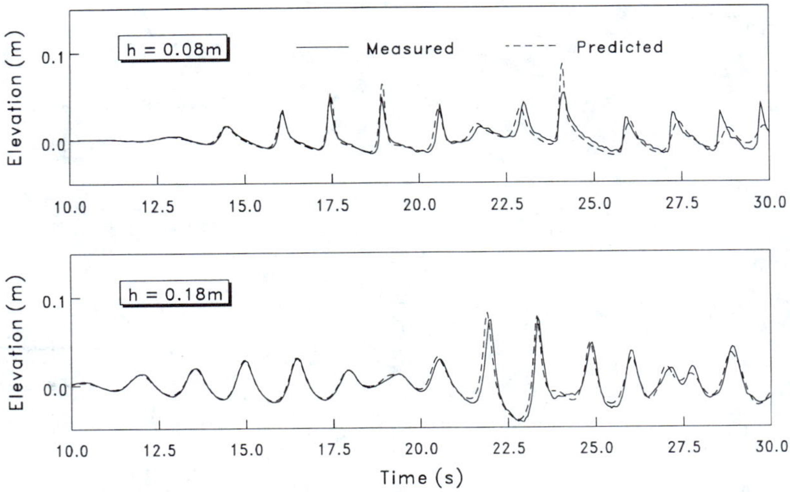

Figure 4. Measured and predicted water surface elevations for an irregular wave with $T_p = 1.5$ s shoaling on a 1:25 beach.

3.2 Shoaling of Irregular Waves

The numerical model was applied to the shoaling of an irregular wave train on the 1:25 beach. A sea state with a duration of 819.2 s was synthesized from a JONSWAP spectrum with significant wave height, $H_{mo} = 0.09$ m, peak period, $T_p = 1.5$ s, and peak enhancement factor, $\gamma = 3.3$, using the random phase method. The computations were carried out using $\Delta x = 0.075$ m, $\Delta t = 0.025$ s and $\ell_t = 0.1$ m. Figure 4 shows a comparison of the measured and predicted water surface elevations at two shallow water depths. The numerical model is able to reproduce reasonably well the time-domain characteristics of individual wave breaking in an irregular wave train.

The spectral densities of the measured and predicted water surface elevation time histories at different water depths are compared in Figure 5. The spectral estimates were averaged over 0.04 Hz frequency bands. Fairly good agreement is obtained between the measured and predicted wave spectra. The numerical model is able to accurately describe the cross-spectral transfer of energy due to nonlinear wave-wave interactions as well as the decrease in wave energy through the surf zone.

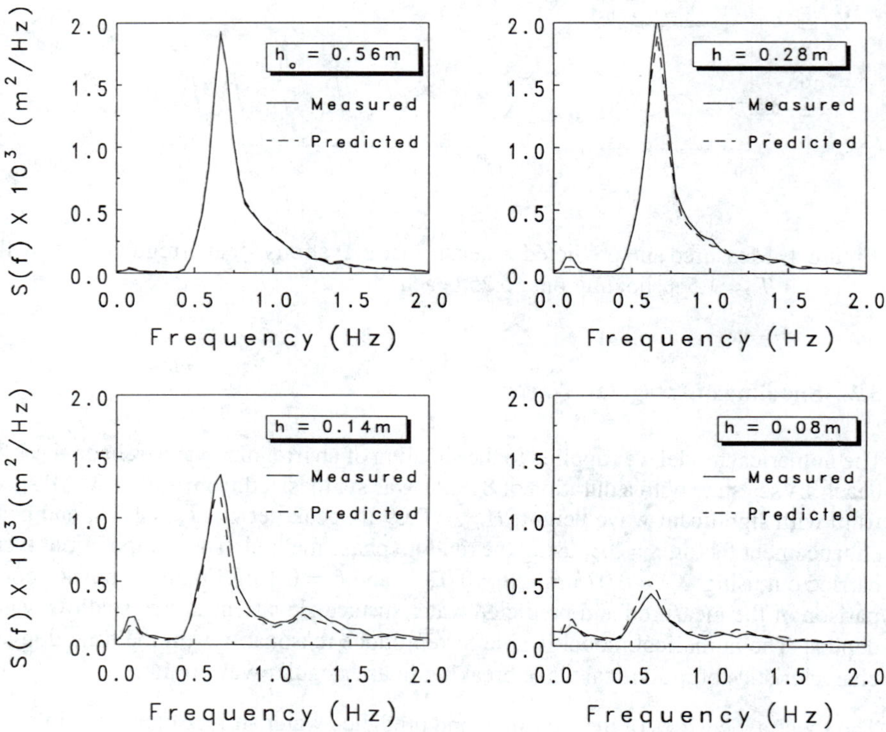

Figure 5. Spectral densities of surface elevation for an irregular wave with $T_p = 1.5$ s shoaling on a 1:25 beach.

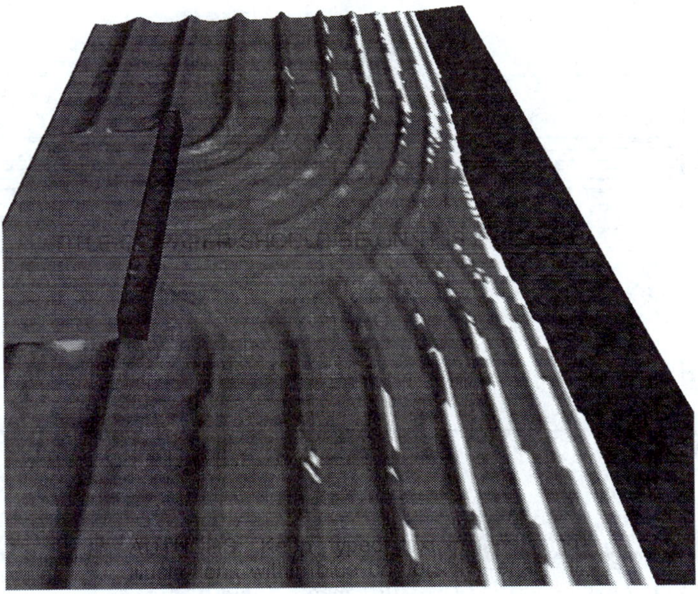

Figure 6. Three-dimensional view of the instantaneous water surface elevation around an offshore breakwater a 1:50 beach.

3.3 Wave-Induced Currents behind a Detached Breakwater

The calculation of currents induced by breaking waves is important in modeling the transport of sediments and pollutants in the surf zone. The two-dimensional Boussinesq model was used to investigate the generation of currents behind an offshore breakwater on a constant slope beach by shoaling and breaking waves. A generic example was used with a 400 m long breakwater placed 350 m from the shoreline on a 1:50 beach. A normally incident regular wave with period T = 10 s and height H = 2 m was generated at the 10 m depth. The computations were carried out with $\Delta x = \Delta y$ = 5 m, Δt = 0.2 s, and f_w = 0.01.

Figure 6 shows a three-dimensional view of the instantaneous water surface elevation around the offshore breakwater. Regions of the crest front with turbulent kinetic energy present have been highlighted in white. The predicted wave height distribution is shown in Figure 7 while the time and depth-averaged velocity pattern is shown in Figure 8. Two circulation cells are observed behind the detached breakwater. This pattern is qualitatively consistent with known observations of wave-induced currents and the formation of tombolos behind offshore breakwaters.

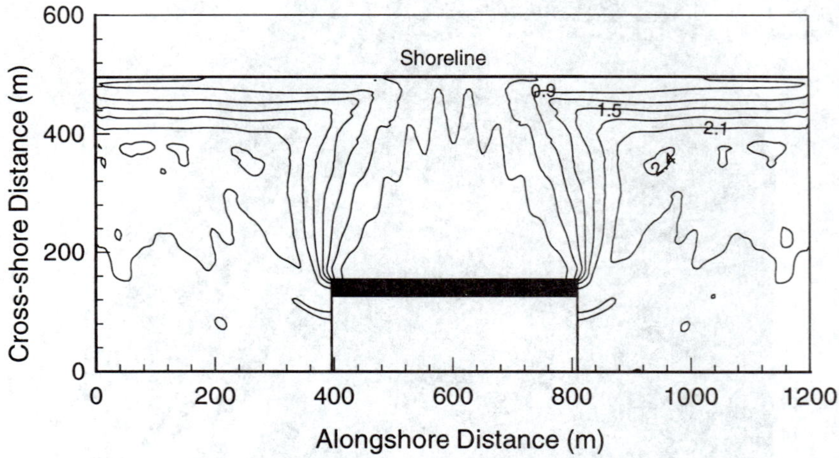

Figure 7. Average wave height distribution behind an offshore breakwater for a regular wave ($T = 10$ s, $H = 2$ m) shoaling on a 1:50 beach.

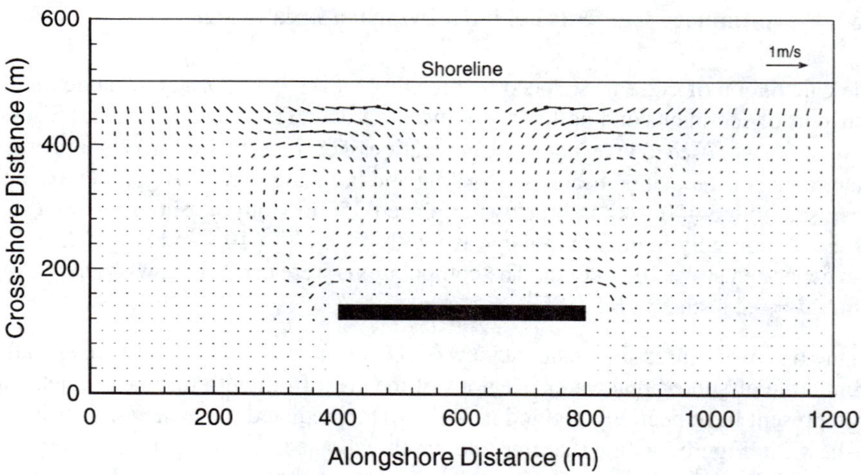

Figure 8. Predicted time and depth-averaged velocity behind an offshore breakwater for a regular wave (T = 10 s, H = 2 m) shoaling on a 1:50 beach.

4. CONCLUSIONS

A fully nonlinear Boussinesq wave propagation model has been extended to the surf and swash zones by coupling the mass and momentum equations with a one-equation turbulent kinetic energy transport model. The waves are assumed to start breaking when the horizontal velocity at the wave crest exceeds the phase velocity of the waves. The rate of production of turbulent kinetic energy is assumed to be proportional to the vertical gradient of the horizontal velocity at the wave crest. The numerical model has been compared to experimental data for the shoaling of regular and irregular waves on 1:25 beach. The model is able to reproduce reasonably well the frequency and time domain characteristics of waves in the surf zone. The results presented in this paper are of preliminary nature and further work is being carried to compare the numerical model with measured data on the spatial and temporal evolution of turbulence induced by wave breaking, the vertical distribution of currents in the surf zone, wave runup and nearshore circulation patterns.

REFERENCES

Battjes, J.A. and Janssen, J.P.F.M. 1978. Energy loss and setup due to breaking of random waves. *Proc. of the 16th Int. Conference on Coastal Engineering*, ASCE, Hamburg, 569-589.

Dally, W.R., Dean, R.G., and Dalrymple, R.A. 1985. Wave height variation across beaches of arbitrary profile. *Journal of Geophysical Research*, **90**(C6), 11917-11927.

Isobe, K. 1987. A parabolic equation model for transformation of irregular waves due to refraction, diffraction and breaking. *Coastal Engineering in Japan*, **30**, 33-47.

Karambas, Th. V. and Koutitas, C. 1992. A breaking wave propagation model based on Boussinesq equations. *Coastal Engineering*, **18**, 1-19.

Nairn, R.B., Roelvink, J.A., and Southgate, H.N. 1990. Transition zone width and implications for modelling surfzone hydrodynamics. *Proc. of the 22nd Int. Conference on Coastal Engineering*, ASCE, Delft, 68-81.

Nwogu, O. 1993. Alternative form of Boussinesq equations for nearshore wave propagation. *Journal of Waterway, Port, Coastal and Ocean Engineering*, ASCE, **119**(6), 618-638.

Nwogu, O. 1994. Nonlinear evolution of directional wave spectra in shallow water. *Proc. of the 24th Int. Conference on Coastal Engineering*, ASCE, Kobe, 467-481.

Nwogu, O. 1996. WaveSim^{TM} - A comprehensive wave simulation model for coastal regions and harbours. *Canadian Hydraulics Centre Technical Report*, National Research Council, Canada (in preparation).

Rodi, W. 1980. Turbulence models and their application in hydraulics - A state of the art review. International Association for Hydraulic Research, Delft.

Sato, S., and Kabiling, M.B. 1994. A numerical simulation of beach evolution based on a nonlinear dispersive wave-current model. *Proc. of the 24th Int. Conference on Coastal Engineering*, ASCE, Kobe, 2557-2570.

Schäffer, H.A., Madsen, P.A., and Deigaard, R. 1993. A Boussinesq model for waves breaking in shallow water. *Coastal Engineering*, **20**, 185-202.

Özkan, H.T. and Kirby, J.T. 1993. Evolution of breaking directional spectral waves in the nearshore zone. *Proc. of the 2nd Int. Symposium on Ocean Wave Measurement and Analyis*, WAVES '93, ASCE, 849-863.

Wei, G., Kirby, J.T., Grilli, S.T., and Subramanya, R. 1995. A fully nonlinear Boussinesq model for surface waves. Part 1. Highly nonlinear unsteady waves. *Journal of Fluid Mechanics*, **294**, 71-92.

Zelt J.A. 1991. The run-up of nonbreaking and breaking solitary waves. *Coastal Engineering*, **15**, 205-246.

CHAPTER 375

THE EFFECT OF THE CL-VORTEX FORCE IN 3D WAVE-CURRENT INTERACTION

M.W. Dingemans[1], J.A.Th.M. van Kester[2], A.C. Radder[3] and R.E. Uittenbogaard[2]

Abstract

Experiments in flumes (i.e., Kemp and Simons, 1982, 1983, and Klopman, 1994) have shown that the mean current profile for following waves is more uniformly distributed than the corresponding current-only case, whereas the case of an opposing current leads to a more straight profile. This paper is concerned with proving that inclusion of the so-called Craik-Leibovich (CL) vortex force in the mean-current equation gives, at least for a part, an explanation of this phenomenon.

Introduction

For coastal dynamics applications it is essential to have an accurate prediction of the vertical structure of the mean current. The mean current, however, is strongly influenced by the free surface wave motion. Experiments in flumes (i.e., Kemp and Simons, 1982, 1983, and Klopman, 1994) have shown that the mean-current profile for following waves is more uniformly distributed than the corresponding current-only case, whereas the case of an opposing current leads to a more straight profile (see Figure 1). The purpose of this paper is to give a possible expanation of this behaviour. Hereto we consider the so-called Craik-Leibovich (CL) equation, derived by Craik and Leibovich (1976) in which a vortex force is present. The effect of this vortex force on the underlying flow is studied.

First we consider the CL-equation. As it is essential to have shear in the velocity profile, the used turbulence-closure model is subsequently described. Because the CL-equation is also used to describe the formation of Langmuir circulation cells, some properties of these are succintly described next. For numerical evaluations the treatment of the vertical momentum equation is possible in a number

[1]Senior Researcher, Delft Hydraulics, P.O. Box 152, 8300 AD Emmeloord, The Netherlands
[2]Senior Researcher, Delft Hydraulics, P.O. Box 177, 2600 MH Delft, The Netherlands
[3]Senior Researcher, Rijkswaterstaat-RIKZ, P.O. Box 20907, 2500 EX The Hague, The Netherlands

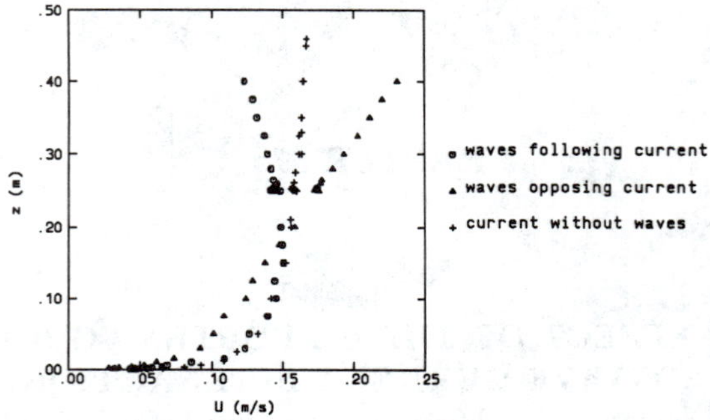

Figure 1: Mean current profiles, after Klopman (1994).

of ways. The most accurate one is described next. Finally the results of some numerical experiments are shown and compared with Klopman's measurements.

The CL equation

When deriving the mean-current equations in so-called Generalised Lagrangian Mean (GLM) coordinates, the following equation is obtained (see Andrews and McIntyre, 1978, Eq. (3.8)):

$$\overline{D}^L\left(\overline{u}_i^L\right)+\frac{1}{\rho_0}\frac{\partial\overline{p}^L}{\partial x_i}-\frac{\overline{\rho}^L}{\rho_0}g_i=\overline{D}^L\left(\overline{P}_i^L\right)+\frac{\partial}{\partial x_i}\left(\frac{1}{2}\left(\overline{u_m^\ell u_m^\ell}\right)\right)+\overline{P}_k^L\frac{\partial\overline{u}_k^L}{\partial x_i}\,, \qquad (1)$$

where $\overline{P}_i^L$ is the pseudo-momentum and $\overline{()}^L$ is the Generalised Lagrangian mean, i.e. the mean over the perturbed position. Summation over repeated indices is used, i.e., the Einstein convention is applied.

Leibovich (1980) has shown that under mild conditions, Eq. (1) reduces to the so-called Craik-Leibovich equation in Eulerian coordinates, which can be written as

$$\frac{\partial\overline{\boldsymbol{u}}}{\partial t}+(\overline{\boldsymbol{u}}\cdot\mathrm{grad})\,\overline{\boldsymbol{u}}+\mathrm{grad}\,\overline{\pi}=\overline{\boldsymbol{u}}^S\wedge\mathrm{curl}\,\overline{\boldsymbol{u}}\,, \qquad (2)$$

where the pressure term $\overline{\pi}$ is given by $\overline{\pi}=\frac{\overline{p}}{\rho}+gz+\left\langle\frac{1}{2}\tilde{\boldsymbol{u}}\cdot\tilde{\boldsymbol{u}}\right\rangle$, $\overline{\boldsymbol{u}}$ is the (Eulerian) mean velocity and $\tilde{\boldsymbol{u}}$ is the wave part of the velocity. The so-called mild conditions for which the approximations are valid, amount to the condition that the waves are primarily dominated by their irrotational part. This implies that either the mean shear or the mean current is relatively weak and we thus have to impose the condition that the current is small with respect to the orbital velocity.

Notice that with $\overline{\boldsymbol{u}}^S=\left(\overline{u}^S,\overline{v}^S,0\right)^T$ and $\overline{\boldsymbol{u}}=(\overline{u},\overline{v},\overline{w})^T$ we have, writing $\overline{\boldsymbol{\omega}}=\mathrm{curl}\,\overline{\boldsymbol{u}}$,

$$\overline{\boldsymbol{T}}=\overline{\boldsymbol{u}}^S\wedge\overline{\boldsymbol{\omega}}=$$

$$= \left(\overline{v}^S\left(\frac{\partial\overline{v}}{\partial x}-\frac{\partial\overline{u}}{\partial y}\right), -\overline{u}^S\left(\frac{\partial\overline{v}}{\partial x}-\frac{\partial\overline{u}}{\partial y}\right), \overline{u}^S\left(\frac{\partial\overline{u}}{\partial z}-\frac{\partial\overline{w}}{\partial x}\right)-\overline{v}^S\left(\frac{\partial\overline{w}}{\partial y}-\frac{\partial\overline{v}}{\partial z}\right)\right)^T$$

$$\cong \left(0, \overline{u}^S\frac{\partial\overline{u}}{\partial y}, \overline{u}^S\frac{\partial\overline{u}}{\partial z}\right)^T \equiv \overline{\boldsymbol{T}}_0\,, \tag{3}$$

where the latter approximations are those of Craik (1982), valid for a flume with x directed along the flume.

Although the inclusion of viscosity is not directly needed, it is needed in an indirect way, otherwise no shear would be present in the mean-current profile. Moreover, once generated vortices should remain bounded in magnitude and this is achieved by taking viscosity into account. Also in view of the application to a wave flume, where the lateral boundary conditions play an important role, we include the viscous stress tensor in the description. Equation (2) then is extended to

$$\frac{\partial\overline{\boldsymbol{u}}}{\partial t}+(\overline{\boldsymbol{u}}\cdot\mathrm{grad})\,\overline{\boldsymbol{u}}+\mathrm{grad}\left(gz+\frac{\overline{p}}{\rho}+\left\langle\frac{1}{2}\tilde{\boldsymbol{u}}\cdot\tilde{\boldsymbol{u}}\right\rangle\right)=\overline{\boldsymbol{u}}^S\wedge\overline{\boldsymbol{\omega}}+\frac{1}{\rho}\,\mathrm{div}\,\overline{\boldsymbol{\sigma}}'\,, \tag{4}$$

where $\mathrm{div}\,\overline{\boldsymbol{\sigma}}' \equiv \partial\overline{\sigma}'_{ki}/\partial x_k$ and $\overline{\sigma}'_{ki}$ is given by $\overline{\sigma}'_{ki} = \left(\eta^{(i)}\partial u_k/\partial x_i + \eta^{(k)}\partial u_i/\partial x_k\right)$, while no summation over the indices on η should be performed and the viscosity is usually non-isentropic. However, we take the (eddy) viscosity coefficient to be isentropic because of the scales on which the flow occurs here and the directions. Applying the Boussinesq-hypothesis, the stresses $\overline{\sigma}'_{ki}$ are approximated as

$$\overline{\sigma}'_{ki} = \nu_T\left(\frac{\partial\overline{u}_i}{\partial x_k}+\frac{\partial\overline{u}_k}{\partial x_i}\right)\,, \tag{5}$$

while the eddy viscosity ν_T has still to be determined. In the present case we consider a cross section of the flume, i.e., we take $\partial_{x_1} \equiv 0$.

The turbulence-closure model

For the specification of the eddy-viscosity coefficient ν_T three possibilities were considered.

1. For a constant eddy-viscosity coefficient the velocity profiles near the bottom and the side walls are parabolic. This is physically not correct as for such situations the profiles should be logarithmic; ν_T should be flow-dependent.

2. Prandtl mixing-length model.

   In the Prandtl mixing-length model, the mixing length ℓ is prescribed as a function of the geometry. Because in our case the lateral boundaries are also of importance, the following variation of Bakhmetev's (1932) choice for shallow-water models is used for the mixing length:

$$\ell = \kappa y\,(b-y)\,(z+h)\left(1-\frac{z+h}{h+\zeta}\right)^{1/2} , \tag{6}$$

where b is the width of the flume, h is the still-water depth, ζ is the free-surface elevation and κ is von Karman's costant. This choice for ℓ yields a parabolic profile near the boundaries. As ℓ only depends on the distance to the boundaries, it is independent of the secondary circulations induced by the CL vortex force $\overline{\boldsymbol{T}}_0$. No results from this model are shown.

3. $\mathsf{k}-\epsilon$ model.

   In our application of the $\mathsf{k}-\epsilon$ model the advection and dissipation terms in the direction of the flume (the $x-$direction) have been neglected. The transport equations for the kinetic energy k and the dissipation ϵ are then given by

$$\frac{D\mathsf{k}}{Dt} = \nabla\cdot\left[\left(\nu+\frac{\nu_T}{\sigma_k}\right)\nabla\mathsf{k}\right] + P_k - \epsilon \tag{7a}$$

$$\frac{D\epsilon}{Dt} = \nabla\cdot\left[\left(\nu+\frac{\nu_T}{\sigma_\epsilon}\right)\nabla\epsilon\right] + P_\epsilon - c_{2\epsilon}\frac{\epsilon^2}{\mathsf{k}} , \tag{7b}$$

where

$$D/Dt \equiv \partial/\partial t + \boldsymbol{v}\cdot\nabla \quad , \quad \nabla = (0,\partial_y,\partial_z)^T \quad , \quad \boldsymbol{v} = (0,v,w)^T \; , \tag{7c}$$

$$P_k = \frac{1}{2}\nu_T\left|\frac{\partial\overline{u}_k}{\partial x_j}+\frac{\partial\overline{u}_i}{\partial x_k}\right|^2 \quad \text{with} \quad \partial_{x_1}\equiv 0 \; , \tag{7d}$$

$$P_\epsilon = c_{1\epsilon}\frac{\epsilon}{\mathsf{k}}P_k \quad \text{and} \quad \nu_T = c_\mu\frac{\mathsf{k}^2}{\epsilon} \tag{7e}$$

and the constants are given by the standard setting:

$$c_\mu = 0.09 \; , \; c_{1\epsilon} = 1.44 \; , \; c_{2\epsilon} = 1.92 \; , \; \sigma_k = 1 \text{ and } \sigma_\epsilon = 1.3 \; . \tag{7f}$$

Boundary conditions for the $\mathsf{k}-\epsilon$ model

The turbulent kinetic energy is imposed both at the side walls and the bed. The pertinent value is obtained by applying local equilibrium between production and dissipation of kinetic energy. This leads to the Dirichlet boundary conditions

$$\mathsf{k}\big|_{y=0} = \mathsf{k}\big|_{y=b} = (u_*^s)^2/\sqrt{c_\mu} \quad \text{and} \quad \mathsf{k}\big|_{z=-h} = \left(u_*^b\right)^2/\sqrt{c_\mu} \tag{8a}$$

with u_*^s and u_*^b the shear-stress velocity at the sidewalls and the bottom respectively. The shear-stress velocity u_* is obtained by applying the logarithmic boundary law for the velocity point nearest to the boundary. With δ half the grid size near the side walls and the bed and z_0 the roughness length, the following relations hold

$$\frac{\overline{u}(\delta)}{u_*} = \frac{1}{\kappa} \log \left(\frac{\delta}{z_0} \right) \quad \text{for a rough wall} \tag{8b}$$

and

$$\frac{\overline{u}(\delta)}{u_*} = \frac{1}{\kappa} \log \left(\frac{\delta u_*}{\nu} \right) + 5.2 \quad \text{for a smooth wall ,} \tag{8c}$$

where log is the natural logarithm. In the latter case the value u_* is found by iteration. Because a staggered grid is used, the velocity $\overline{u}(\delta)$ is indeed known at the previous time step. At the free surface we take $\mathsf{k} = 0$.

For the transport equation for dissipation rate ϵ we take the following Neumann-type boundary condition:

$$\left. \frac{\partial \epsilon}{\partial z} \right|_{\delta} = -\frac{u_*^3}{\kappa \delta^2} , \tag{8d}$$

and at the free surface is taken $\epsilon = 0$.

Langmuir circulations

Equation (4) is basic to the generation of Langmuir circulations by an instability mechanism (Leibovich, 1983). The essential point is that a combination of a 2D shear flow in the vertical xz plane with an irrotational Stokes drift of the wave field in the same direction is unstable to spanwise (i.e., in y) disturbances. Write now the vorticity vector as $\boldsymbol{\omega} = (\omega_x, \omega_y, \omega_z)^T$. Two different forms of CL-instability are recognised:

1. **CL1-type** in which the vorticity-component ω_x can be generated by a Stokes drift which varies in the lateral direction

2. **CL2-type** in which the vorticity component ω_x may be generated by a lateral uniform Stokes drift acting on a pre-existing vertical vorticity ω_z.

The CL2-type instability is described by Leibovich (1977) and amounts to the following. It is supposed that the wind stress has produced a parallel shear flow $\overline{u}(z)$ and a Stokes drift $\overline{u}^S(z)$. A perturbation of the shear flow is now considered such that the total flow is described by $(\overline{u}(z) + \hat{u}, \hat{v}, \hat{w})^T$. A perturbation of the pressure is not needed here. Then, the following behaviour of the perturbations is considered:

$$\hat{w} = \phi(z) \exp\left[\sigma t + i\left(my + kz\right)\right] \quad , \quad \hat{u} = \frac{1}{(m^2 + k^2)} \left(ik\hat{w}_z - m^2 \hat{w} \frac{\tilde{u}}{f} \right)$$

$$\hat{v} = \frac{m}{(m^2 + k^2)} \left(i\hat{w}_z + k\hat{w} \frac{\tilde{u}}{f} \right) \quad , \quad f = \sigma + ik\left(\overline{u} + \overline{u}^S \right) ,$$

with $\tilde{u}$ being the wave part of the current. For $\phi(z)$ then is obtained in the non-viscous approximation the Rayleigh equation

$$\frac{\partial^2\phi}{\partial z^2} - \left(m^2 + k^2\right)\phi + \left\{\frac{k}{if}\frac{\partial^2\overline{u}}{\partial z^2} + \frac{m^2}{f^2}\frac{\partial\overline{u}}{\partial z}\frac{\partial\overline{u}^S}{\partial z}\right\}\phi = 0\,. \tag{9}$$

With the further simplification of rolls aligned with the mean flow (i.e., $k = 0$) we get

$$\sigma^2\left(\frac{\partial^2\phi}{\partial z^2} - m^2\phi\right) + m^2\mathcal{M}(z)\phi = 0 \quad \text{with} \quad \mathcal{M}(z) = \frac{\partial\overline{u}^S}{\partial z}\frac{\partial\overline{u}}{\partial z}\,. \tag{10}$$

The condition for instability now reads $\mathcal{M}(z) > 0$ anywhere in the interval for z. The corresponding maximum growth is $\mathcal{M}^{1/2}(0)$. For the growth of Langmuir circulations it is thus necessary that $\mathcal{M}(z)$ is positive anywhere in the fluid, in other words, a positive shear of the current is required because the Stokes drift diminishes for increasing depth. However, Nepf and Monismith (1991) found experimentally, in a flume, that vortices developped for both $\mathcal{M}(z) > 0$ and $\mathcal{M}(z) < 0$, see also the discussion in Nepf et al. (1995).

Manipulation of the vertical momentum equation

The numerical evaluation will be performed within an existing free-surface flow model in which the treatment of the vertical momentum equation will be worked out in a number of ways.

Hydrostatic approach

For the handling of the vertical momentum equation several possibilities exist. A very simple one consists of ignoring the vertical momentum equation altogether and determining the vertical velocity afterwards by an integration of the continuity equation in which the solutions u and v from the horizontal momentum equations are used. These horizontal momentum equations in most simple form are

$$\frac{du}{dt} + \frac{\partial\overline{\pi}}{\partial x} = \frac{1}{\rho}\frac{\partial\overline{\sigma}'_{k1}}{\partial x_k} \quad \text{and} \quad \frac{dv}{dt} + \frac{\partial\overline{\pi}}{\partial y} = \frac{1}{\rho}\frac{\partial\overline{\sigma}'_{k2}}{\partial x_k} + \overline{u}^S\frac{\partial u}{\partial y}\,. \tag{11}$$

One of the ways of explaining the measured mean current profiles is the following qualitative one. We consider waves and current in a laboratory flume. In the lateral (y) direction, changes in the velocity field should occur. This has to be generated by the wall effects. The basic idea is as follows. Suppose that a circulation cell has developped. Suppose that the cell is such that in the middle of the flume the flow is in downward direction. The situation is as sketched in Figure 2. In the upper part of the fluid, water with a small amount of momentum is transported towards the middle of the flume and consequently decelerates, while in the lower part of the fluid the reverse situation occurs. In the upper half of

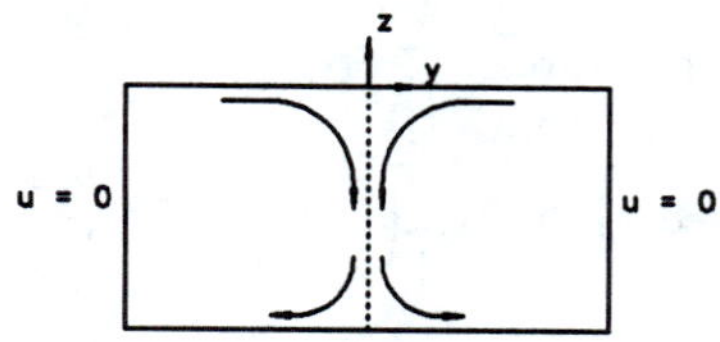

Figure 2: Flow in a lateral vertical section

the fluid the downward velocity increases and in the lower half it decreases. In the middle of the flume $\partial v/\partial y = 0$ because of symmetry. From the continuity equation $u_x + w_z = 0$ in the middle of the flume then follows that $\partial u/\partial x < 0$ in the upper half and $\partial u/\partial x > 0$ in the lower half of the fluid. Consequently, the profile $u(z)$ turns back in the upper part of the fluid and increases in the lower part. The result is a more uniform horizontal velocity profile in the vertical.

A semi-Poisson approach

A more precise formulation consists of splitting the pressure in a hydrostatic and a non-hydrostatic part. The solution procedure then consists of two steps. In the first step only the hydrostatic part of the pressure is used and in the second step a Poisson equation for the non-hydrostatic part has to be solved. Such a method has also been used by Casulli (1995). We write the pressure as the sum of a hydrostatic part p_h and a non-hydrostatic part q: $p = p_h + q$. The computation is split in two parts.

Step 1

Consider the continuity equation and for the momentum equations we only use the hydrostatic part of the pressure. These momentum equations, which are to be solved in the first step then are, in vectorial form

$$\frac{d\overline{\boldsymbol{u}}}{dt} + \frac{1}{\rho}\operatorname{grad} p_h = -\boldsymbol{g} + \frac{1}{\rho}\operatorname{div}\overline{\boldsymbol{\sigma}}' \quad \text{with} \quad \boldsymbol{g} = (0, 0, -g)^T \ . \tag{12}$$

Step 2

In the second step we consider the momentum equations without the convective terms and we only account for the non-hydrostatic part of the pressure. Furthermore the vortex force is accounted for here. We then have the system:

$$\frac{\partial\overline{\boldsymbol{u}}}{\partial t} + \frac{1}{\rho}\operatorname{grad} q = \overline{\boldsymbol{T}}_0 \tag{13}$$

with the vortex force $\overline{\boldsymbol{T}}_0$ given by (3).

We now apply the operation div on (13). For the velocities it is mandatory that $\operatorname{div}\overline{\boldsymbol{u}} = 0$ (i.e., the continuity equation is applied as a side condition because it was already used in the first step). The result is a Poisson equation for the non-hydrostatic part of the pressure:

$$\operatorname{div}\left(\frac{1}{\rho}\operatorname{grad} q\right) = \operatorname{div}\left[\left(0, \overline{u}^S \frac{\partial u}{\partial y}, \overline{u}^S \frac{\partial u}{\partial z}\right)^T\right] \tag{14}$$

After solving the pressure part q of this Poisson equation, the solution q is substituted in Eq. (13) yielding new estimates for the velocities. This completes step 2.

Notice that the splitting is such that in step 1 a usual free-surface flow computation is performed. The second step gives the correction to the hydrostatic pressure, which correction is due to the vortex force. The vortex force, figuring at the right-hand side of the Poisson equation, is computed with the velocities estimated in the first step.

Numerical experiments

We now perform some computations to compare with the results of measurements of Klopman's physical experiments in the so-called Scheldt flume, which has a width of 1 m while the mean water depth in the experiments is 0.5 m. The walls of the flume are made of glass and are therefore hydraulically smooth. The bed was roughened with sand with a grain size of approximately 2 mm. The roughness of the bed was determined experimentally and the Nikuradse roughness was found to be 1 mm. The experimental set-up consists of two wave boards near each end of the flume; both wave boards generated waves and, at the same time, absorbed reflected waves. A discharge was generated with a constant outflow of 80 litres/s. For further details about the lay-out of these experiments we refer to Klopman (1994).

We only consider the regular-wave experiments. The main parameters are

wave period	$T = 1.44\,\mathrm{s}$
wave amplitude	$a = 0.06\,\mathrm{m}$
current velocity	$\overline{u} = 0.16\,\mathrm{m/s}$
mean depth	$h = 0.50\,\mathrm{m}$
bottom roughness height	$z_0 = 0.04\,\mathrm{mm}$

Following waves

In order to obtain enough shear for the CL-vortex force to be effective, the computations are started first without the CL force included and as initial value a uniform velocity of 0.16 m/s is used, whereas the turbulent kinetic energy **k** and the dissipation ϵ are initially taken to be zero. After 200 s the turbulent logarithmic shear flow has developped sufficiently for the CL-vortex force to be effective. From that moment on the CL-vortex force is included in the computations. At t = 300 s a new steady state is obtained with two eddies in the cross-section of the flume. For the situation of waves following the current, the velocity profile in a cross section of the flume has been shown in Fig. 3. The development of the

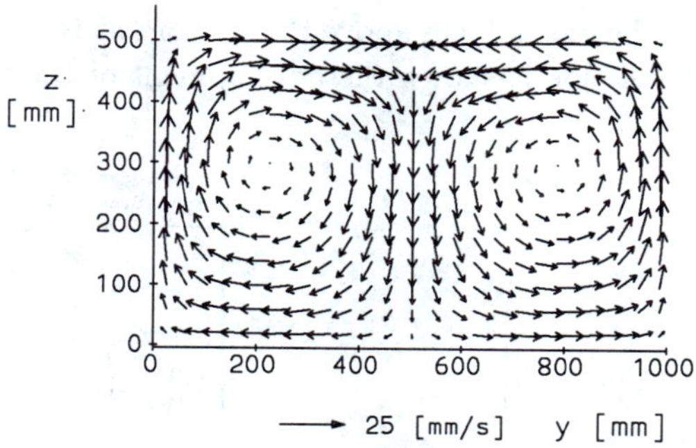

Figure 3: Velocity field in cross section at time $t = 300$ s.

vertical profile of the horizontal velocity in time is shown at the left-hand side of Fig. 4; the time that logarithmic profile is developped is now taken as the origin of the time-axis for the further discussion.

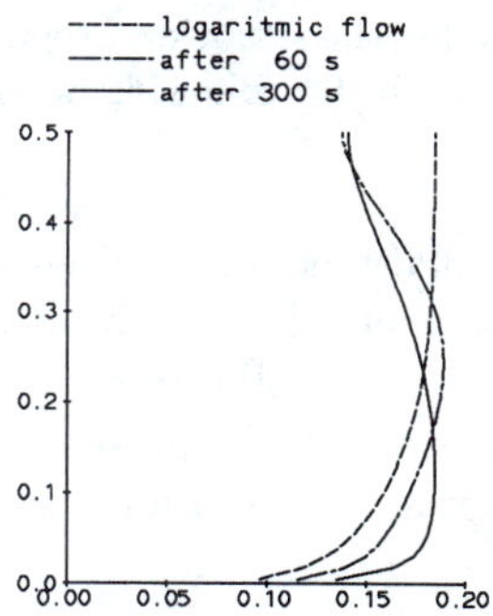

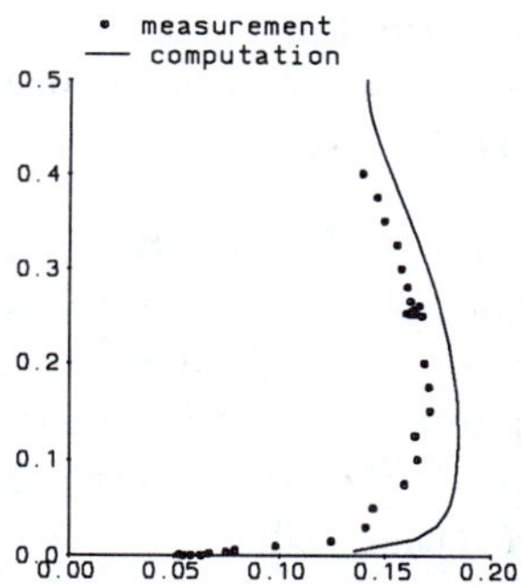

Figure 4: Development in time of the vertical profile of the horizontal velocity for the situation of waves following the current *(left)*; Comparison of the vertical profile for following waves with the measured profile *(right)*.

Because in the measurements a reference velocity, averaged over the depth, was based on the total flow, with the assumption of uniform flow over the width of the flume, a correction procedure is necessary now to account for the side-wall effects. For the situation of current alone (no waves) the computed vertically averaged horizontal velocity, evaluated in the centre of the flume, was taken as reference for the vertically averaged velocity as resulting from the measurements, likewise taken in the centre of the flume. This requires an enlargement of the measured velocity in the centre of the flume by a factor 1.135. This factor has been applied to all measurements, also those with following and opposing waves. The comparison of the vertical profile of the computation and the measurement with this correction is shown at the right-hand side of Fig. 4. The inclusion of the CL-vortex force thus has, for following waves, the effect of increasing the near-bottom horizontal velocity and decreasing the horizontal velocity near the free surface, as was also measured. The profile of the current over the width of the flume is shown for a height of 5 cm above the bottom at the left-hand side of Fig. 5. At the right-hand side is shown the development of the vertical current in the centre of the flume.

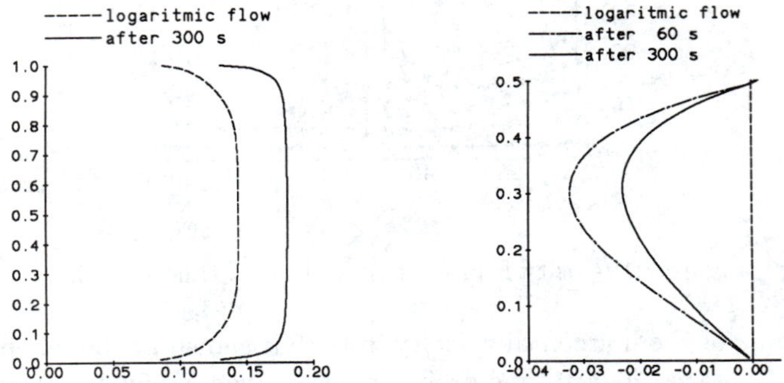

Figure 5: The horizontal current across the width of the flume for following waves, without waves (the logarithmic profile) and 300 s later (with waves); development in time of the vertical profile of the vertical velocity for the situation of waves following the current *(right)*; Comparison of the vertical profile for following waves withe the measured profile *(right)*.

A test on the possibility of Langmuir circulations was devised by taking a very wide flume with otherwise the same parameters. In this situation also only two circulation cells developped and therefore the possibility of obtaining true Langmuir circulations is ruled out for this situation. In the present situation the flow is boundary-driven. The rigid side walls act as an agent for the development of the circulation cells.

Waves propagating against the current

Also in this case two eddies are generated, rotating in opposite directions as those generated for waves propagating in the same direction as the current. This is as was expected. The development of the vertical profile for the horizontal current is shown in Fig. 6. This shows that the flow near the bed is reduced, and that the flow near the free surface is not increased as was to be expected. Notice that Nepf and Monismith (1991) found from measurements in a flume for both waves following and against the current that circulation cells developped. This is also the case in the present computations. Why almost no effect on the current profile is found for opposing waves is not yet clear. This deserves further investigation.

Conclusion

The numerical experiments lead to the following conclusions.

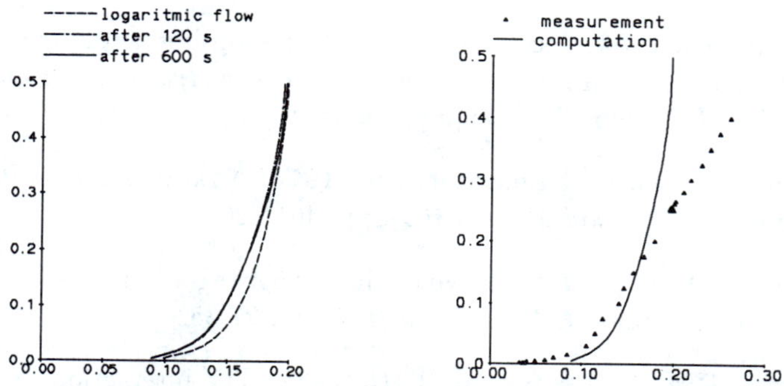

Figure 6: Development in time of the vertical profile of the horizontal velocity for the situation of waves propagating against the current *(left)*; Comparison of the vertical profile of the horizontal velocity for waves propagating against the current with the measured profile *(right)*.

1. Because also for a very wide flume only two circulation cells will develop, we conclude that the possibility of Langmuir circulation is ruled out for this situation. The flow is boundary-driven.

2. The principal mechanism of change of the near-surface velocities is the lateral transport of low-momentum fluid from the rigid side walls to the centre of the flume. Therefore, we conclude that the CL-vortex force is, at least partly, responsible for the observed change in velocities due to waves upon currents.

3. For both following and opposing waves circulation cells are found in the computations, which is in accord with the experiments of Nepf et al. (1995). Why the case for opposing waves has so little effect on the change of the current profile compared to the case without waves is not yet clear and deserves further investigation.

Acknowledgement

This work was commissioned by the National Institute for Coastal and Marine Management/RIKZ of the Dutch Public Works Department.

References

[1] **Andrews, D.G. and McIntyre, M.E. (1978)**. An exact theory of non-linear waves on a Lagrangian-mean flow. *J. Fluid Mech.* **89(4)**, pp. 609-646.

[2] **Bakhmetev, B.A., 1932**. *Hydraulics of Open Channels*, Eng. Soc. Monograph, McGraw-Hill.

[3] **Casulli, V., 1995**. Recent developments in semi-implicit numerical methods for free surface hydrodynamics. In: *Advances in Hydroscience and Engineering*, Vol. II, March 1995, Beijing, China, 8 pp.

[4] **Craik, A.D.D. and Leibovich, S., 1976**. A rational model for Langmuir circulations. *J. Fluid Mech.* **73(3)**, pp. 401-426.

[5] **Craik, A.D.D., 1982**. The generalized Lagrangian-mean equation and hydronamic stability. *J. Fluid Mech.* **125**, pp. 27-35.

[6] **Kemp, P.H. and Simons, R.R., 1982**. The interaction between waves and a turbulent current: waves propagating with the current. *J. Fluid Mech.* **116**, pp. 227-250.

[7] **Kemp, P.H. and Simons, R.R., 1983**. The interaction between waves and a turbulent current: waves propagating against the current. *J. Fluid Mech.* **130**, pp. 73-89.

[8] **Klopman, G. (1994)**. Vertical Structure of the flow due to waves and currents; Laser-Doppler flow measurements for waves following or opposing a current. *Delft Hydraulics*, Report **H840.30, Part II**.

[9] **Leibovich, S., 1977**. Convective instability of stably stratified water in the ocean. *J. Fluid Mech.* **82(3)**, pp. 561-581.

[10] **Leibovich, S., 1980**. On wave-current interaction theories of Langmuir circulations. *J. Fluid Mech.* **99(4)**, pp. 715-724.

[11] **Leibovich, S., 1983**. The form and dynamics of Langmuir circulations. *Annual Review of Fluid Mechanics* **15**, pp. 391-427.

[12] **Nepf, H.M. and Monismith, S.G., 1991**. Experimental study of wave-induced longitudinal vortices. *J. of Hydraulic Engineering* **117(12)**, pp. 1639-1649.

[13] **Nepf, H.M., Cowen, E.A., Kimmel, S.J. and Monismith, S.G., 1991**. Longitudianl vortices beneath breaking waves. *J. Geophys. Res.* **100(C8)**, pp. 16211-16221.

SUBJECT INDEX

Page number refers to the first page of paper

AUTHOR INDEX

Page number refers to the first page of paper